Teacher's Edition

Algebra 2

$r = 6.37 \times 10^6$

Explorations and Applications

Authors

Senior Authors

Miriam A. Leiva   Richard G. Brown

Loring Coes III
Shirley Frazier Cooper
Joan Ferrini-Mundy
Amy T. Herman
Patrick W. Hopfensperger
Celia Lazarski
Stuart J. Murphy
Anthony Scott
Marvin S. Weingarden

McDougal Littell
A HOUGHTON MIFFLIN COMPANY
Evanston, Illinois ◆ Boston ◆ Dallas

Authors

Senior Authors

Richard G. Brown Mathematics Teacher, Phillips Exeter Academy, Exeter, New Hampshire

Miriam A. Leiva Cone Distinguished Professor for Teaching and Professor of Mathematics, University of North Carolina at Charlotte

Loring Coes III Chair of the Mathematics Department, Rocky Hill School, E. Greenwich, Rhode Island

Shirley Frazier Cooper Curriculum Specialist, Secondary Mathematics, Dayton Public Schools, Dayton, Ohio

Joan Ferrini-Mundy Professor of Mathematics and Mathematics Education, University of New Hampshire, Durham, New Hampshire

Amy T. Herman Mathematics Teacher, Atherton High School, Louisville, Kentucky

Patrick W. Hopfensperger Mathematics Teacher, Homestead High School, Mequon, Wisconsin

Celia Lazarski Mathematics Teacher, Glenbard North High School, Carol Stream, Illinois

Stuart J. Murphy Visual Learning Specialist, Evanston, Illinois

Anthony Scott Assistant Principal, Orr Community Academy, Chicago, Illinois

Marvin S. Weingarden Supervisor of Secondary Mathematics, Detroit Public Schools, Detroit, Michigan

The authors wish to thank **Jane Pflughaupt**, Mathematics Teacher, Pioneer High School, San Jose, California, and **Martha E. Wilson**, Preparatory Mathematics Specialist, Mathematical Sciences Teaching and Learning Center, University of Delaware, Newark, Delaware, for their contributions to this Teacher's Edition.

ISBN: 0-395-86299-X

123456789—VH—01 00 99 98 97

Contents of the Teacher's Edition

Philosophy of Algebra 2

Explorations and Applications

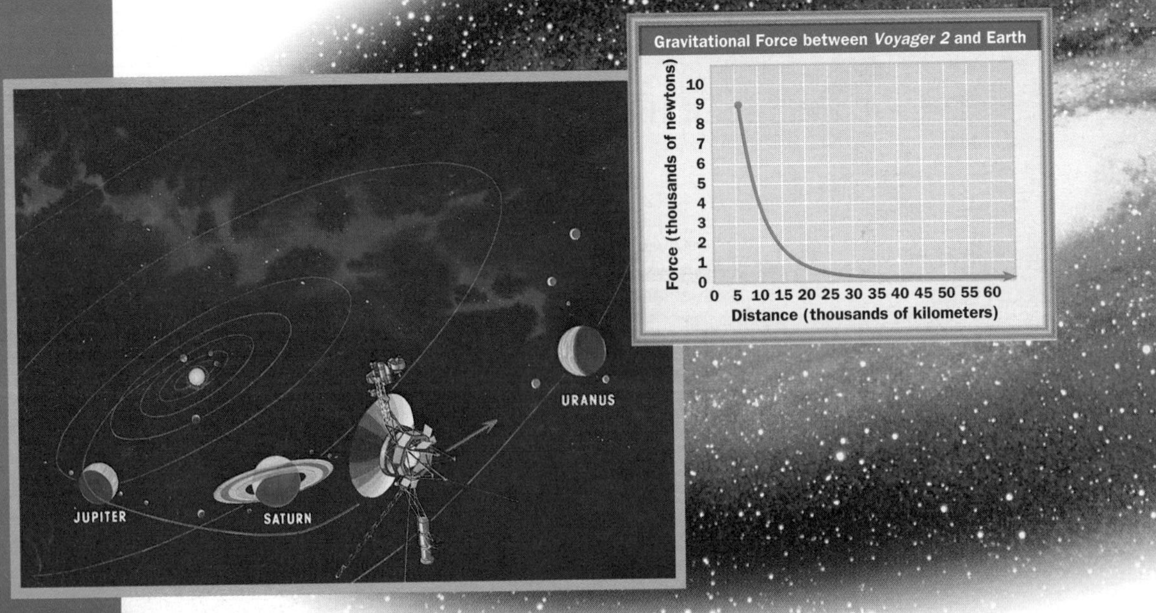

Goals of the Course

This course has been designed to make mathematics accessible and inviting to the wide range of students who are studying algebra today. It helps you prepare today's students for tomorrow's world by:

- Building understanding of the concepts that provide a strong foundation for future courses and careers
- Connecting algebra to the real world and to other subjects and math topics
- Involving students in exploring and discovering math concepts
- Assessing students' progress in ways that support learning

Mathematical Content

The content and the teaching strategies in the textbook reflect the curriculum, teaching, and assessment standards of the National Council of Teachers of Mathematics. The fresh, new course outline:

- Uses functions as a unifying theme
- Emphasizes graphing, and the relationship between graphs and equations
- Integrates technology as a problem-solving tool
- Connects algebra to geometry, data analysis, probability, and discrete mathematics

Teaching Strategies

The flexible course design offers frequent opportunities for you to incorporate:

- Exploratory activities that build conceptual understanding
- Applications that strengthen problem-solving skills
- Discussion and writing questions that develop communication skills
- Strategies for using technology to visualize and solve problems

Assessment

You and your students can measure their mathematical growth throughout the course in a variety of ways, including:

- Cooperative learning activities
- Open-ended problems
- Journal writing
- Portfolio projects

Program Overview

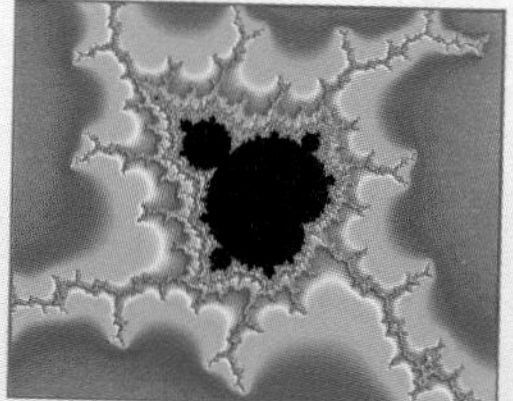

Pages T6-T42 give an overview of *Algebra 2: Explorations and Applications* and the teaching materials that support it. These pages provide information about:

Contents

CHAPTER 1

Modeling Using Algebra

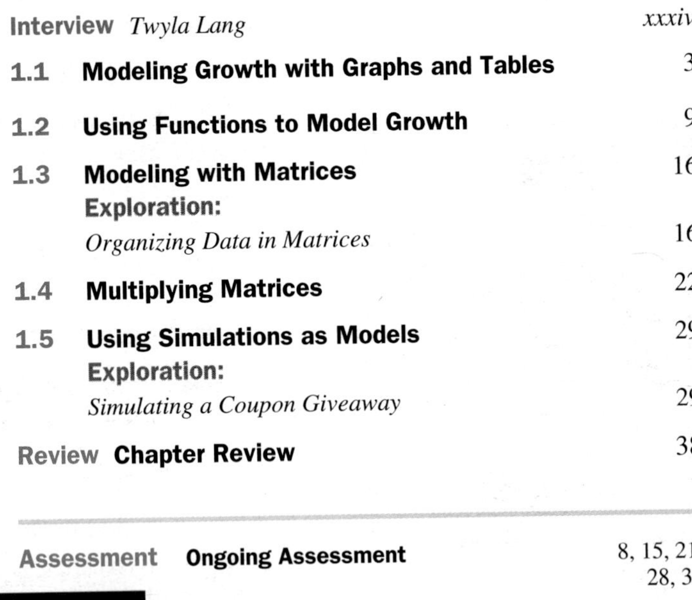

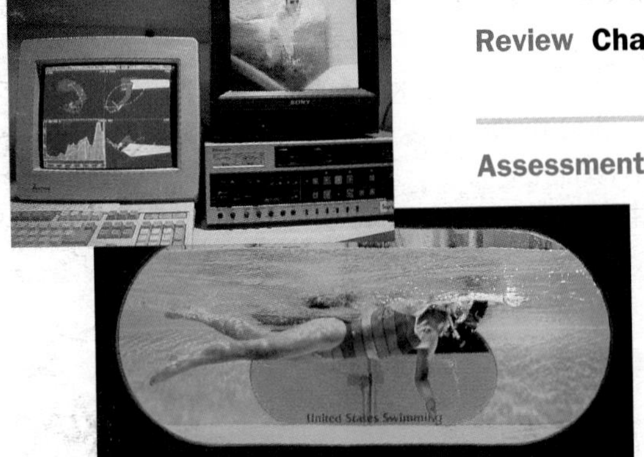

Applications

CHAPTER 1

This chapter establishes some major themes of the course. Students explore mathematical modeling using graphing technology and spreadsheets. Comparison of constant and proportional growth functions sets the stage for later study of linear and exponential functions in Chapters 2 and 3. Matrices are introduced as tools of data analysis.

CHAPTER
2

Linear Functions

Applications

Additional applications include: precision skating, geometry, sports, manufacturing, personal finance, temperature scales, oceanography, economics, health, business, physics

CHAPTER 2

The familiar notion of a **linear function** becomes a platform for the broader study of **functions** and **curve-fitting** throughout the course. Students develop and test linear models. **Parametric equations** are introduced for modeling motion in two dimensions in this and later chapters.

CHAPTER

3 Exponential Functions

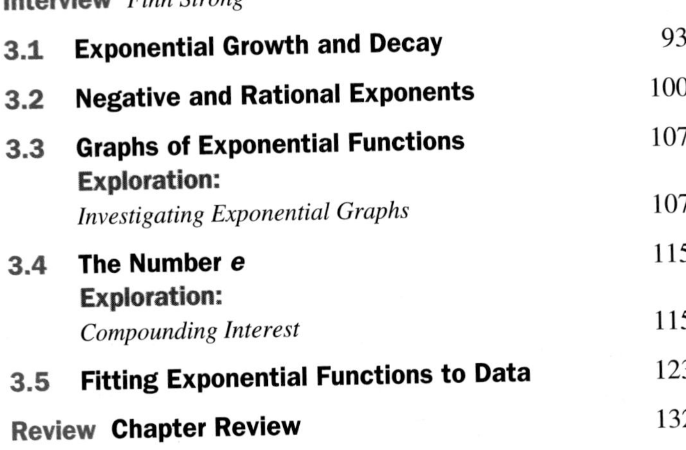

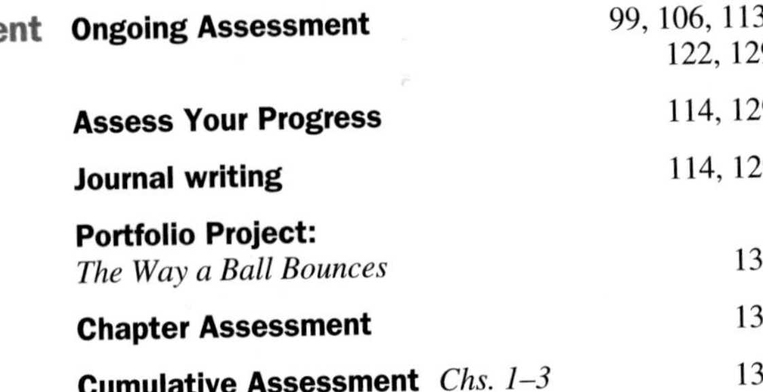

Applications

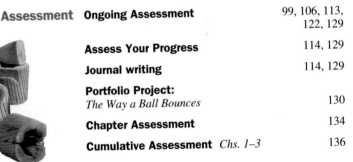

Additional applications include: mythology, geometry, banking, physiology, chemistry, biology, psychology, astronomy

CHAPTER 3

Building on earlier work with exponents, students use graphing technology to **explore** equations and graphs of **exponential and logistic functions**. Based on their knowledge of patterns of **proportional growth and decay**, students develop exponential models for real-world situations.

CHAPTER

4 Logarithmic Functions

Number of Cranes in Izumi, Japan

Population

Years since 1945

4.5 *Using logarithms to model data* 168

JAPAN

Tokyo

Kyoto

Izumi

Applications

Interview:
Ednaly Ortiz
Archaeologist
146, 159

Connection

Transportation	147	**Radiology**	166
Astronomy	153	**Social Studies**	172
Acoustics	160	**Biology**	174

Additional applications include: business, forestry, recreation, mountain climbing, population, earth science, cooking, personal finance, ecology

CHAPTER 4

With linear and exponential functions as a foundation, this chapter introduces **inverse functions**. Students use algebra to solve real-world problems modeled by exponential and logarithmic equations. By graphing **scatter plots** with graphing technology, students **model data** with exponential and power functions.

CHAPTER

5 Quadratic Functions

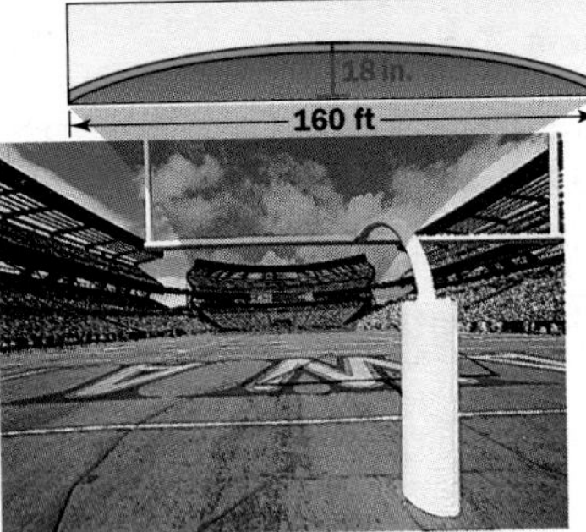

5.3 *Parabolic athletic fields* 204

Applications

Interview: ***Mark Thomas***
Automotive Engineer
190, 220

Connection

Bicycling	189	**Basketball**	212
Personal Finance	197	**Physiology**	219
Engineering	204	**Astronomy**	226
Biology	211		

Additional applications include: geometry, business, physics, sports science, government, driving, electricity

CHAPTER 5

Students first **analyze graphs** of quadratic functions and then relate the visual representations to their equations. Graphing technology is used to find **maximum and minimum values** for real-world situations modeled with quadratic functions. Students begin an ongoing study of the effects of **geometric transformations** on functions.

CHAPTER

6

Investigating Data

6.3 *Archaeological sampling* 254

Applications

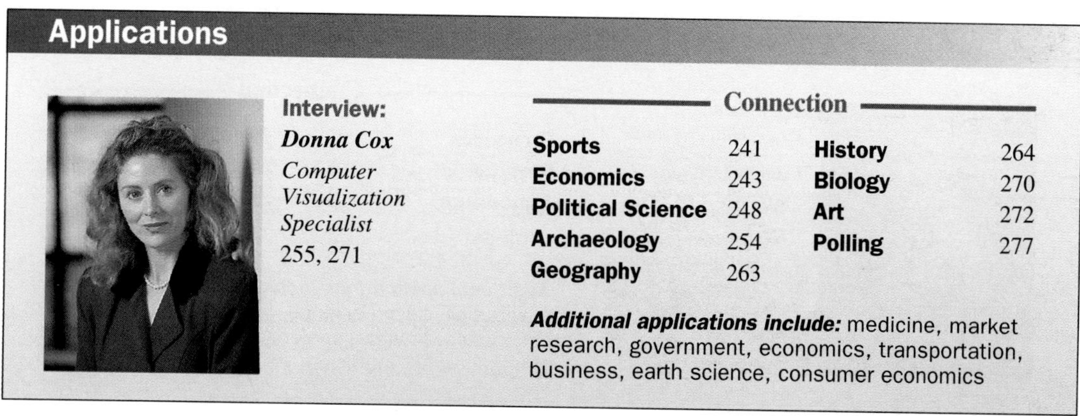

Interview:
Donna Cox
Computer Visualization Specialist
255, 271

Connection

Sports	241	**History**	264
Economics	243	**Biology**	270
Political Science	248	**Art**	272
Archaeology	254	**Polling**	277
Geography	263		

Additional applications include: medicine, market research, government, economics, transportation, business, earth science, consumer economics

CHAPTER 6

This chapter shifts the focus from functions to **collecting and analyzing data**, a process essential for understanding our world. The sequence of topics follows the order in which social and natural scientists carry out the steps of the process. The chapter culminates with a section on **making decisions** from data.

CHAPTER

7 Systems

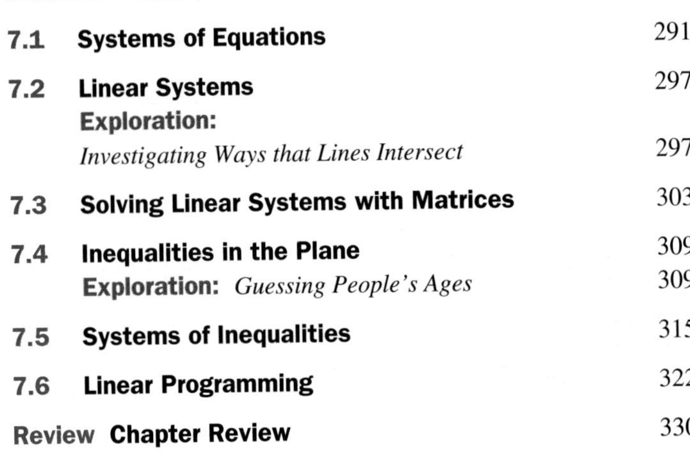

Applications

Interview:
Gina Oliva
Fitness Director
299, 316, 319, 326

Additional applications include: personal finance, advertising, social science, recreation, consumer economics, finance, cooking, art, cartography, manufacturing, medicine, geology, fitness

CHAPTER 7

Earlier work with linear systems is extended by the use of **matrices** as another method of solution. Techniques of solving systems will be applied to nonlinear systems when conic sections are studied later in the course. Earlier work with linear inequalities serves as the basis for solving **optimization problems** by linear programming.

CHAPTER

8 Radical Functions and Number Systems

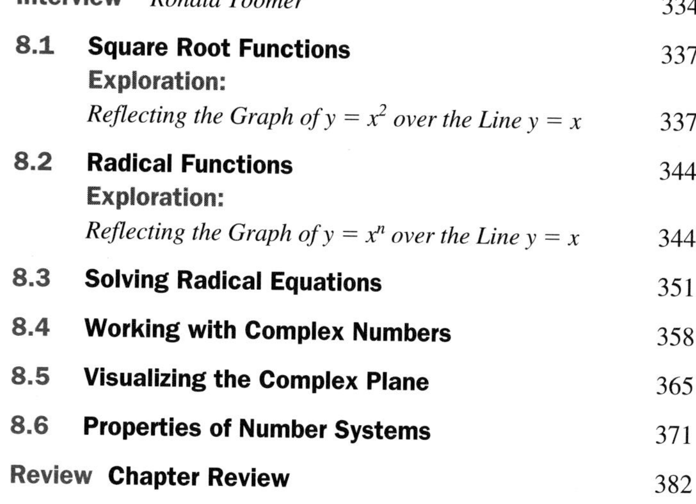

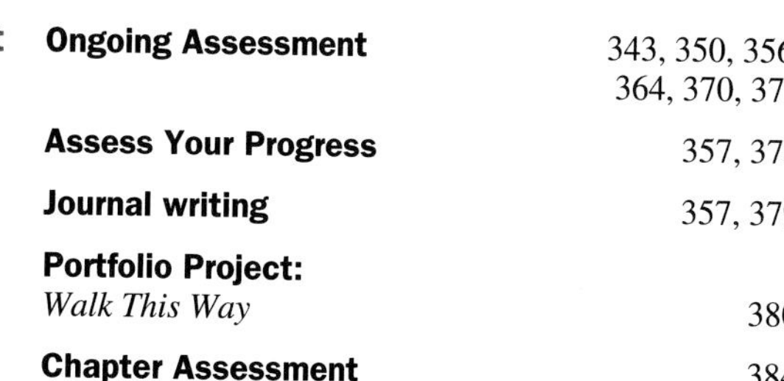

8.3 *America's Cup boats* 356

Applications

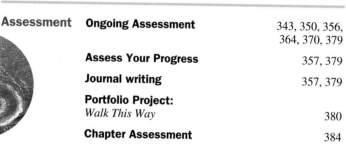

Interview:
Ronald Toomer
Roller Coaster Designer
342, 348

Connection

Oceanography	341	**Electronics**	363
Sports	349	**Mandelbrot Set**	369
Meteorology	354	**Dance**	377
Sports	356	**Games**	378

Additional applications include: biology, geometry, astronomy, wind chill

CHAPTER 8

This chapter resumes the work with functions from the first five chapters. Quadratic functions and inverses are used to develop **radical functions**. Work with radicals leads to the introduction of **imaginary and complex numbers** and an examination of the formal **structure** of number systems.

CHAPTER

9

Polynomial and Rational Functions

Planet	Mean distance from sun (millions of miles)	Apparent size of sun
Mercury	36.0	2.58
Venus	67.2	1.38
Earth	93.0	1.00
Mars	142.0	0.655
Jupiter	484.0	0.192
Saturn	885.0	0.105
Uranus	1780.0	0.0522
Neptune	2790.0	0.0333
Pluto	3660.0	0.0254

Applications

Interview:
Vera Rubin
Astronomer
430, 446

Connection

Additional applications include: biology, psychology, physiology, space travel, education, metallurgy

CHAPTER 9

Earlier work with polynomials is extended by using **graphing technology** to examine the **end behavior** of polynomial and rational functions and to solve problems by locating **maximums and minimums**. To prepare for future courses, the ideas of infinity and limits are explored.

CHAPTER

10

Sequences and Series

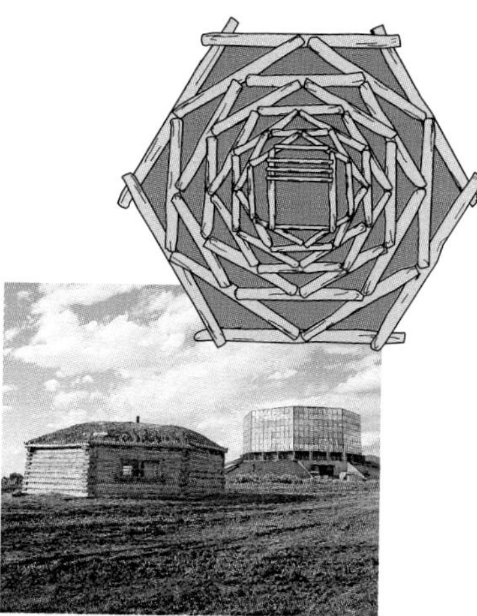

Applications

Interview:
Jhane Barnes
Textile Designer
471, 479, 502

Additional applications include: chemistry, biology, architecture, medicine, language arts, physics, economics, genealogy

CHAPTER 10

Students make the transition from **continuous** functions **to discrete** functions by comparing arithmetic and geometric sequences with linear and exponential functions. Novel **applications** plus **cooperative learning** activities maintain student involvement. Topics include sums of infinite series and fractals defined by **recursion**.

CHAPTER

11 Analytic Geometry

Additional applications include: architecture, geometry, flag making, astronomy, carpentry, language arts

CHAPTER 11

This chapter integrates geometry with algebra by reconsidering parabolas and hyperbolas from a geometric standpoint and connecting them with other conic sections. Students use graphing technology to graph equations of conic sections and solve nonlinear systems. Applications are drawn from fields such as art, medicine, and aviation.

CHAPTER 12

Discrete Mathematics

12.5 *Investigating Pascal's triangle* 593

Applications

Interview:
Maria Rodriguez
Electrical Engineer
574, 582

Connection

Marine Biology	567	**Sports**	589
Sports	575	**Games**	591
Genetics	584	**History**	597

Additional applications include: cartography, scheduling, chemistry, ecology, medicine, business, fashion, forensics, music, literature, hobbies, manufacturing, computers, advertising, fundraising, industry, recreation, government

CHAPTER 12

Familiar discrete mathematics topics like Pascal's triangle and combinatorics are presented along with newer ones like **graph theory**. Continuing a major theme of the course, students **model** real-life situations using **multiple representations** in the form of graphs and matrices.

CHAPTER

13 Probability

Applications

Additional applications include: sports, astronomy, manufacturing, physics, meteorology, movies, electronics, health, market research, architecture, driver's education, biology, agriculture

CHAPTER 13

Interconnections between data analysis, combinatorics, and probability are highlighted. Probability tree diagrams serve as **visual representations** of situations. Students develop **critical thinking skills** for predicting future events, interpreting and judging the credibility of evidence, and making decisions in real-life situations.

CHAPTER

14 Triangle Trigonometry

Applications

Additional applications include: archaeology, exercising, parasailing, aviation, meteorology, sports, navigation, dancing, architecture, cooking

CHAPTER 14

In this chapter and the following one, the focus is on the three basic trigonometric functions. The sequence builds on students' experience with right triangles. Students use technology to evaluate basic trigonometric functions and the inverses of basic trigonometric functions.

CHAPTER

15 Trigonometric Functions

15.2 *Testing stereo speakers* 715

Applications

Interview: *Joe Lopez*
Record Producer
715, 718, 724

Connection

Carousels	711	**Architecture**	732
Solar Power	717	**Sports**	738
Surveying	725		

Additional applications include: amusement park rides, geometry, automotive mechanics, music, hot air ballooning, astronomy, engineering

CHAPTER 15

Students use **graphing technology** to analyze the graphs of the three basic trigonometric functions. The emphasis is on **modeling periodic phenomena**. Continuing the work begun in Chapter 5 and continued in Chapters 8, 9, and 11, students explore the effects of **geometric transformations** on equations of periodic functions.

Student Resources

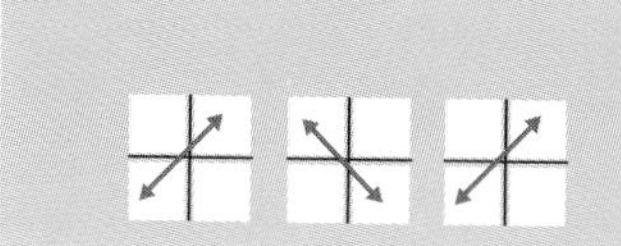

Student Resources

These resources include **practice**, **review**, and **reference** material to help students with the course. The *Toolbox* provides a quick brush-up on **prerequisite skills** that includes worked examples and practice exercises. The *Technology Handbook* introduces students to useful features provided by **graphing calculators**.

Algebra 2

Explorations and Applications

This new program helps you prepare today's students for tomorrow's world.

Complete Teaching Resources

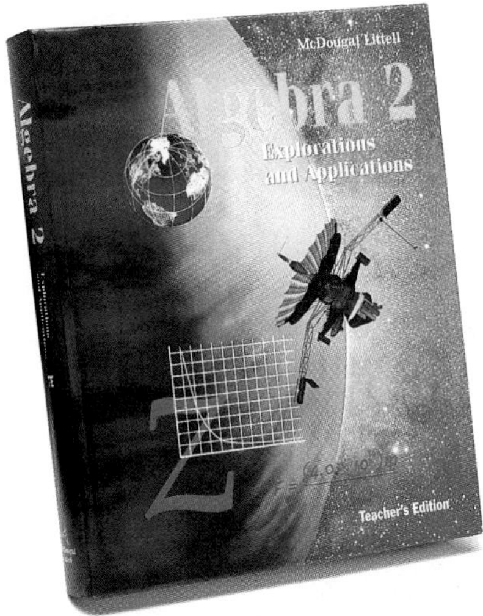

Teacher's Edition includes complete support for planning and teaching your lessons.

- Student pages with complete answers
- Point-of-use teaching and exercise notes
- Planning pages for each chapter, with assignment guides for standard and block scheduling

Teacher's Resources provides a complete selection of support materials, with suggestions for implementing new ideas for teaching and assessment.

- Lesson Plans
- Study Guide
- Practice Bank
- Assessment Book
- Challenge Problems
- Portfolio Project Book
- Explorations Lab Manual
- Professional Handbook
- Warm-Up Transparencies
- Preparing for College Entrance Tests
- Technology Book
- Course Guides
- Solution Key

Also available:

- Overhead Visuals
- Multi-Language Glossary
- Assessment Book, Spanish Edition
- Study Guide, Spanish Edition
- Test and Practice Generator (Macintosh and IBM)
- McDougal Littell Mathpack software

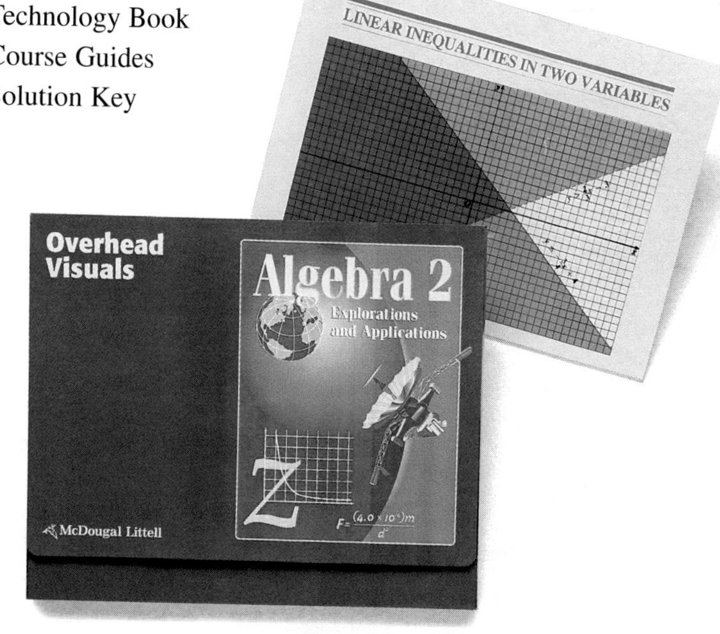

Support Materials
for all teaching and learning needs

Of Special Interest . . .

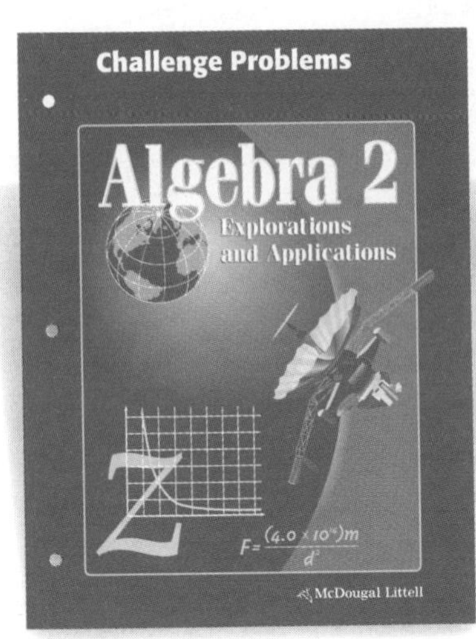

Assessment Book

- Chapter tests — 2 versions
- Performance-based assessment
- Short quizzes and cumulative tests

Also available in Spanish

Challenge Problems

- Additional challenging exercises and problems for each section of the textbook
- Chapter-review challenge sets, including Extension problems

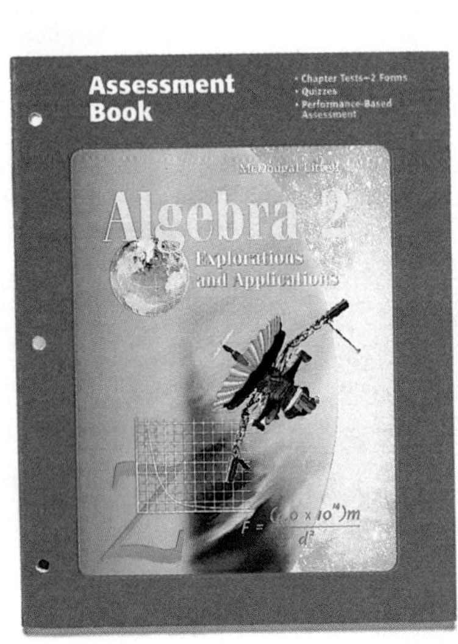

. . . for You and your Students

Preparing for College Entrance Tests

- Student handbook with test-preparation strategies
- Preview tests with questions in standardized-test formats

Study Guide

- Study Guide lesson for each section of the textbook
- Examples and key terms
- Practice and review

Also available in Spanish

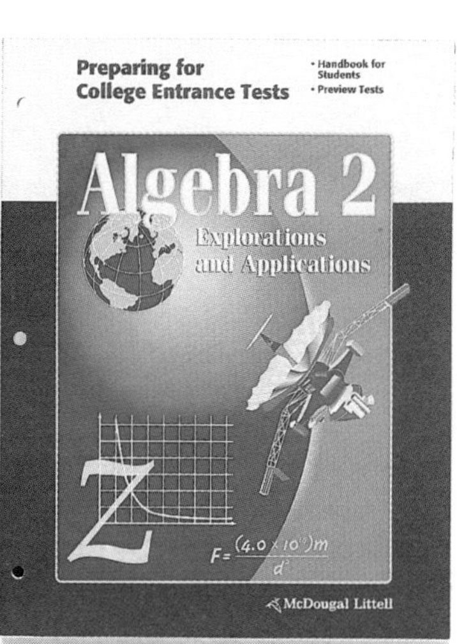

Please turn the page for Technology Support ➤

Technology Support

Integrating technology The technology tools and print support materials that you need to integrate technology successfully into your courses are available with this program.

McDougal Littell Mathpack

Software for Exploring and Applying Mathematics

- Function and statistical graphing
- Activities linked to the textbook

Technology Book

Includes activities for:

- TI-81, T1-82, and TI-92 graphing calculators
- Spreadsheets
- Calculator-Based Laboratory

Calculators

- Texas Instruments TI-34 calculator and TI-81 and TI-82 graphing calculators

Test Generator

Technology-based assessment with:

- Test-generating software for Macintosh and IBM
- Test bank with user's guide

Algebra 2 Explorations and Applications

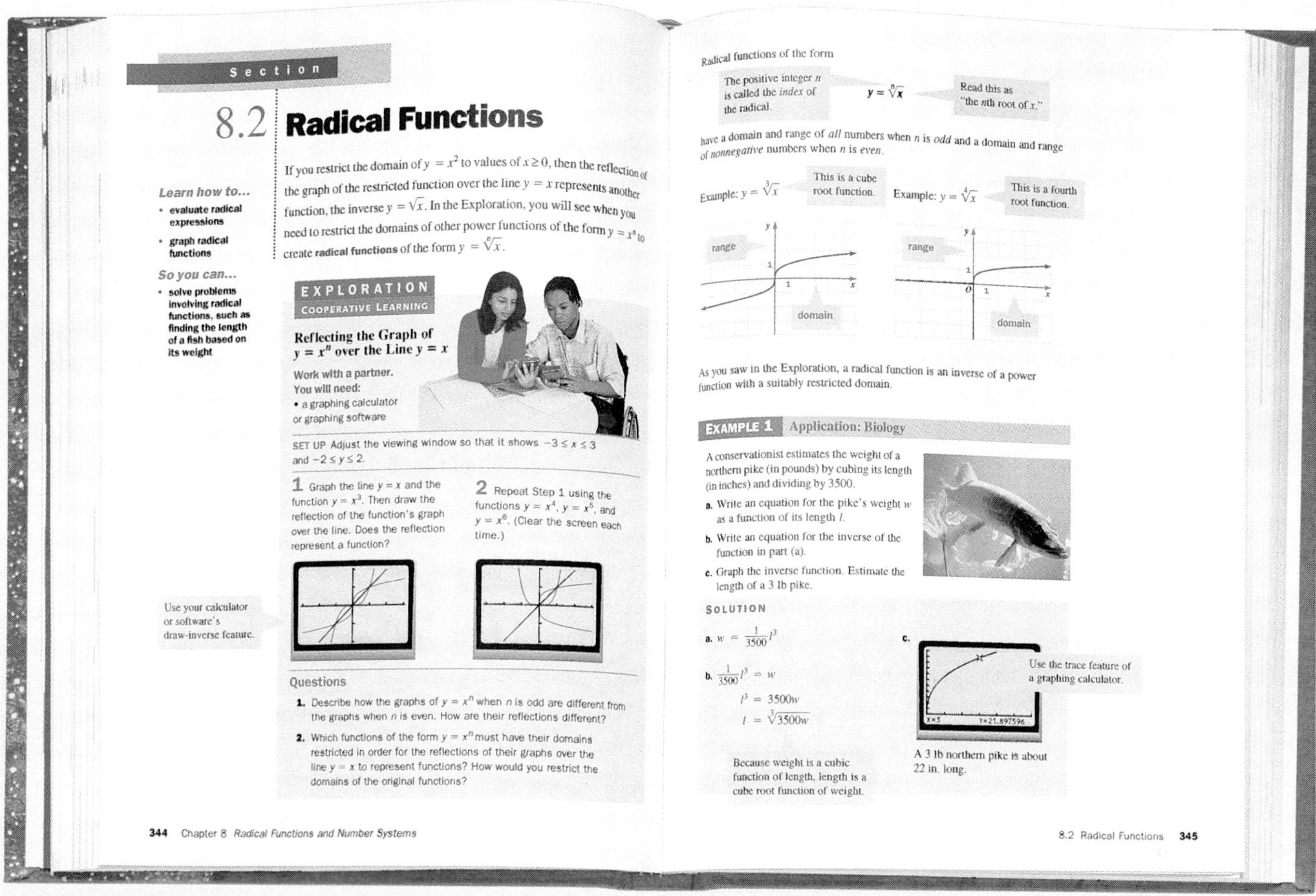

Section

8.2 Radical Functions

Learn how to...

- evaluate radical expressions
- graph radical functions

So you can...

- solve problems involving radical functions, such as finding the length of a fish based on its weight

If you restrict the domain of $y = x^2$ to values of $x \geq 0$, then the reflection of the graph of the restricted function over the line $y = x$ represents another function, the inverse $y = \sqrt{x}$. In the Exploration, you will see when you need to restrict the domains of other power functions of the form $y = x^n$ to create **radical functions** of the form $y = \sqrt[n]{x}$.

EXPLORATION
COOPERATIVE LEARNING

Reflecting the Graph of $y = x^n$ over the Line $y = x$

Work with a partner.
You will need:
- a graphing calculator or graphing software

SET UP Adjust the viewing window so that it shows $-3 \leq x \leq 3$ and $-2 \leq y \leq 2$.

1 Graph the line $y = x$ and the function $y = x^3$. Then draw the reflection of the function's graph over the line. Does the reflection represent a function?

2 Repeat Step 1 using the functions $y = x^4$, $y = x^5$, and $y = x^6$. (Clear the screen each time.)

Use your calculator or software's draw-inverse feature.

Questions

1. Describe how the graphs of $y = x^n$ when n is odd are different from the graphs when n is even. How are their reflections different?
2. Which functions of the form $y = x^n$ must have their domains restricted in order for the reflections of their graphs over the line $y = x$ to represent functions? How would you restrict the domains of the original functions?

344 Chapter 8 *Radical Functions and Number Systems*

Radical functions of the form

$y = \sqrt[n]{x}$

The positive integer n is called the *index* of the radical.

Read this as "the nth root of x."

have a domain and range of *all* numbers when n is *odd* and a domain and range of *nonnegative* numbers when n is *even*.

Example: $y = \sqrt[3]{x}$ — This is a cube root function.

Example: $y = \sqrt[4]{x}$ — This is a fourth root function.

As you saw in the Exploration, a radical function is an inverse of a power function with a suitably restricted domain.

EXAMPLE 1 Application: Biology

A conservationist estimates the weight of a northern pike (in pounds) by cubing its length (in inches) and dividing by 3500.

a. Write an equation for the pike's weight w as a function of its length l.

b. Write an equation for the inverse of the function in part (a).

c. Graph the inverse function. Estimate the length of a 3 lb pike.

SOLUTION

a. $w = \frac{1}{3500}l^3$

b. $\frac{1}{3500}l^3 = w$

$l^3 = 3500w$

$l = \sqrt[3]{3500w}$

Because weight is a cubic function of length, length is a cube root function of weight.

c. Use the trace feature of a graphing calculator.

A 3 lb northern pike is about 22 in. long.

8.2 Radical Functions 345

Turn to pages T28–T35 to see how this book integrates these teaching strategies:

- **Involve students in exploring and discovering math concepts.**
- **Connect algebra to the real world and to other subjects and math topics.**
- **Build understanding of the concepts that provide a strong foundation for future courses and careers.**
- **Assess students' progress in ways that support learning.**

Program Overview

Involve *students in exploring and discovering math concepts.*

Section 3.4 The Number *e*

Learn how to...
- model exponential growth and decay using the base *e*, and model logistic growth

So you can...
- make predictions about real-life growth and decay, such as compound interest and biological populations

In Section 3.3 you used an exponential function to describe compounding interest annually in a bank account. Most banks offer interest that is compounded more than once a year. When interest is compounded *n* times per year for *t* years at an interest rate *r* (expressed as a decimal), a principal of *P* dollars grows to the amount *A* given by this formula:

$$A = P\left(1 + \frac{r}{n}\right)^{nt}$$

EXPLORATION
COOPERATIVE LEARNING

Compounding Interest

Work in a group of four students.
You will need:
- a scientific calculator

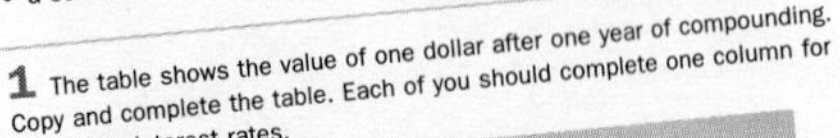

1 The table shows the value of one dollar after one year of compounding. Copy and complete the table. Each of you should complete one column for one of the interest rates.

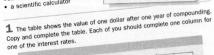

Compounding	n	Formula	r = 0.05	r = 0.10	r = 0.50	r = 1.00
annually	1	$(1+\frac{r}{1})^1$	1.05			
semiannually	2	$(1+\frac{r}{2})^2$		1.1025		
quarterly	4	$(1+\frac{r}{4})^4$			1.6018	
monthly	12	$(1+\frac{r}{12})^{12}$				2.6130
daily	365	$(1+\frac{r}{365})^{365}$				
hourly						
every minute						
every second						

2 Describe how increasing the frequency of compounding interest affects the value of a dollar that is invested for one year.

3 If a bank compounded interest more often than every second, would this make much difference? Explain.

3.4 The Number *e* **115**

Cooperative Learning *A Team Approach*

Mathematics is more meaningful for students when they can explore ideas, discover solutions, and discuss results.

Section 11.4 Ellipses

Learn how to...
- write an equation of an ellipse
- graph an equation of an ellipse

So you can...
- describe an elliptical object, such as the Oval Office in the White House

At one time, astronomers thought that the orbit of a planet around the sun was a circle or a combination of circles. In 1609, the German astronomer Johannes Kepler realized that the orbit of a planet follows an elliptical path, with the sun located at one focus of the elliptical orbit. In the Exploration, you will see how to draw an *ellipse*.

EXPLORATION
COOPERATIVE LEARNING

Drawing an Ellipse

Work with a partner.
You will need:
- graph paper
- a 10-inch piece of string
- two pushpins
- a piece of cardboard

1 Draw axes on a piece of graph paper. Label the points (−4, 0) and (4, 0). Place the graph paper on the cardboard. Use the pins to hold one end of the string at (−4, 0) and the other end at (4, 0).

2 Use a pencil to pull the string until it is taut. Move the pencil above and below the *x*-axis, keeping the string taut, until you have sketched a closed geometric figure called an *ellipse*.

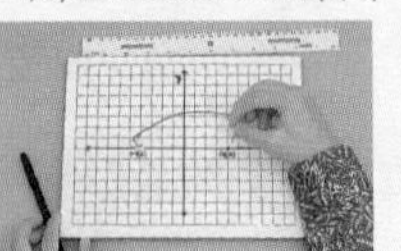

$P(x, y)$, $Q(x_1, y_1)$, $F_1(-4, 0)$, $F_2(4, 0)$

Questions

1. If P and Q are any two points on the figure you drew, how do the lengths F_1P and F_2P compare with the lengths F_1Q and F_2Q?
2. Give a definition for the figure you drew using the relationship between F_1P, F_2P, and the length of the string.
3. Move the pins closer together and repeat Steps 1 and 2. How did the ellipse change? What do you think happens to the shape of the ellipse when the pins are moved even closer together?

11.4 Ellipses **535**

Modeling with Manipulatives *Building Understanding*

Exploring with manipulatives helps students develop an intuitive base for understanding concepts and techniques.

Section

9.3 Exploring Graphs of Polynomial Functions

Learn how to...

- recognize graphs of polynomial functions and describe their important features

So you can...

- understand how the volume of air in a person's lungs changes as the person breathes, for example

You already know what the graphs of some polynomial functions look like. For example, the graph of $y = 2x + 5$ is a line, and the graph of $y = 4x^2 - 8x - 3$ is a parabola. You'll look at graphs of higher-degree polynomial functions in the Exploration.

EXPLORATION
COOPERATIVE LEARNING

Looking for Patterns in Graphs

Work with a partner.
You will need: • a graphing calculator or graphing software

SET UP Adjust the viewing window for your calculator or software so that the intervals $-5 \le x \le 5$ and $-20 \le y \le 20$ are shown on the axes. Complete Steps 1 and 2 for each of these functions:

- $y = 4x^2 - 8x - 3$
- $y = x^3 - x^2 - 3x + 1$
- $y = x^4 + 2x^3 - 5x^2 - 7x + 3$
- $y = 2x^5 + 6x^4 - 2x^3 - 14x^2 + 5$
- $y = 3x^6 - 13x^4 + 15x^2 + x - 17$
- $y = x^7 - 8x^5 + 18x^3 - 6x$

1 Describe what happens to the graph as x takes on large positive and large negative values.

Example: $y = 2x^2 - 2x - 11$

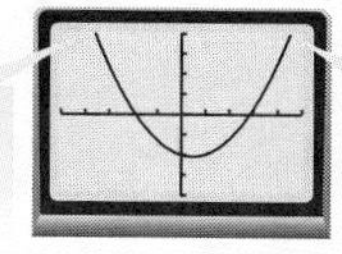

As x takes on large negative values, the graph rises.

As x takes on large positive values, the graph rises.

2 Find the number of *turning points.* Give the approximate coordinates of each turning point.

Example: $y = 2x^2 - 2x - 11$

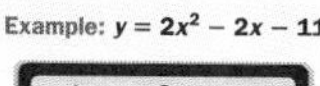

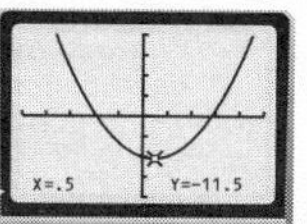

The graph has one turning point. Its coordinates are $(0.5, -11.5)$.

Exploring with Technology *Visualizing Patterns*

With technology, students can explore a wider range of topics than in the past. Graphing calculator activities help students understand the relationship between equations and graphs.

PORTFOLIO PROJECT

Walk This Way

Try walking across a flat, open space at an increasing speed. You'll notice that at some point you'll feel the urge to switch to a run. Your body recognizes when it's more efficient to run than to walk.

Now think about a small child trying to keep up with an adult who's walking briskly. With shorter legs than the adult, the child often has to run. Obviously, how fast someone can walk (without breaking into a run) has something to do with leg length.

PROJECT GOAL **Examine the relationship between leg length and walking speed.**

Doing an Experiment

Work with a partner to design and carry out an experiment in which you measure the leg length, in inches, of ten subjects, and then time the subjects as they walk. Be sure to choose subjects with a variety of leg lengths. You will need a tape measure and a stopwatch. Here are some guidelines for doing your experiment.

1. MEASURE each subject's leg length from hip to heel.
2. CHOOSE a straight, flat, 60 ft course that is free of obstacles.
3. INSTRUCT each subject to walk the 60 ft course as fast as he or she can. With each step, the walker must be sure that the rear foot does not leave the ground before the forward foot lands.
4. TIME each subject as he or she walks the course. Divide the distance traveled (60 ft) by each subject's time to calculate his or her speed in feet per second.

Real-World Experiments *Working with Data*

Group projects give students opportunities to gather and analyze data relating to real-world applications.

Connect *algebra to the real world and to other subjects and math topics.*

Real-World Applications *Mathematics in Context*

Throughout this course new concepts are introduced, practiced, and extended in the context of real-world applications. Seeing the wide variety of settings in which algebra is useful helps students value mathematics.

Section

4.1 Using Inverses of Linear Functions

Learn how to...
- graph and find equations for inverses of linear functions

So you can...
- solve problems about bicycling and home construction, for example

Most bicycles today have multiple gears to help cyclists ride comfortably over different terrain. By choosing the right gear, a cyclist can conserve energy and ride longer distances.

Lower gears are used for going up hills.

Higher gears are used on level ground.

THINK AND COMMUNICATE

The table shows the distance a bicycle travels for different numbers of pedal rotations when the bicycle is in first gear. Use the table and graph for Questions 1–4.

Number of pedal rotations	Distance traveled (meters)
0	0
1	3.1
2	6.2
3	9.3

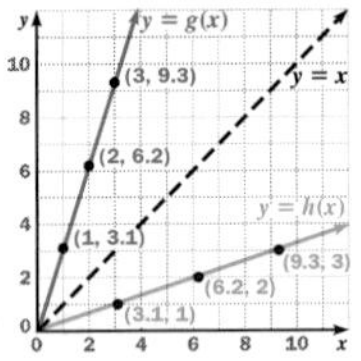

$y = g(x)$
$y = x$
$y = h(x)$
(3, 9.3)
(2, 6.2)
(1, 3.1)
(9.3, 3)
(6.2, 2)
(3.1, 1)

BY THE WAY

In 1884, John Starley invented the Rover "safety" bicycle. His bicycle was the first one to have a geared chain drive. Gears made pedaling easier for riders.

1. Replace each _?_ with *number of pedal rotations* or *distance traveled.*
   a. g gives _?_ as a function of _?_. b. h gives _?_ as a function of _?_.
2. a. Which function would you use to find the distance traveled in 2.5 pedal rotations? Why?
   b. Which function would you use to find the number of pedal rotations needed to travel 8 m? Why?
3. Complete these statements: If the point (a, b) is on the graph of g, then the point _?_ is on the graph of h. So if $g(a) = b$, then $h(\underline{?}) = \underline{?}$.
4. How are the graphs of g and h geometrically related to the line $y = x$?

4.1 Using Inverses of Linear Functions 141

For each graph, tell why the graph may misrepresent the data. Then suggest a way to improve the graph.

11.

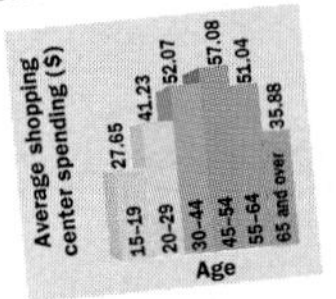

Average shopping center spending ($)
27.65
41.23
52.07
57.08
51.04
35.88
15–19
20–29
30–44
45–54
55–64
65 and over
Age

12.

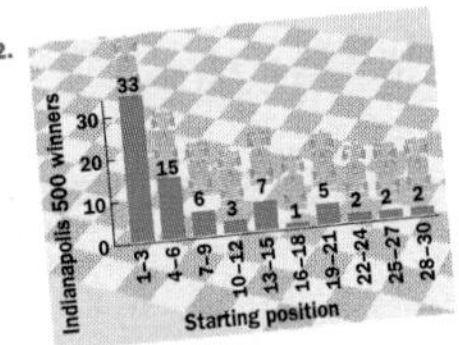

Indianapolis 500 winners
33
15
6
3
7
1
5
2
2
2
1–3
4–6
7–9
10–12
13–15
16–18
19–21
22–24
25–27
28–30
Starting position

13. **Writing** If you are given only an absolute frequency histogram, can you create a relative frequency histogram? Can you create an absolute frequency histogram given only a relative frequency histogram? Explain.

Connection HISTORY

European settlers began arriving in what is now the United States in the sixteenth century. In 1787, Delaware, Pennsylvania, and New Jersey became the first former colonies to achieve statehood. The box plots compare the year in which a present-day state was settled by Europeans and the year in which it achieved statehood.

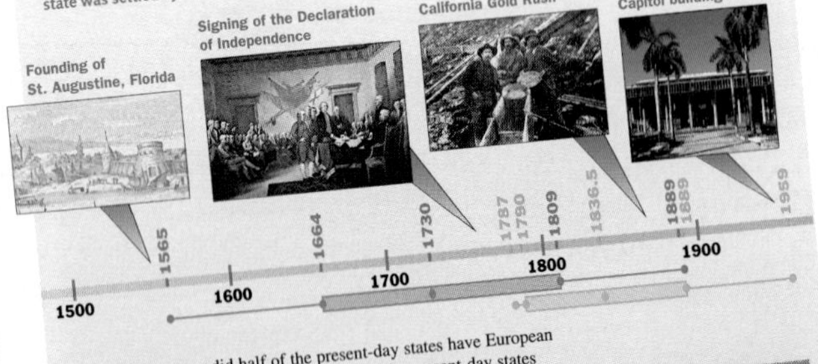

Founding of St. Augustine, Florida
Signing of the Declaration of Independence
California Gold Rush
Opening of Hawaii's State Capitol building
1565
1664
1730
1787
1790
1809
1836.5
1889
1889
1959
1500
1600
1700
1800
1900

14. By what year did half of the present-day states have European settlements? By what year did half of the present-day states achieve statehood?
15. In 1788, eight former colonies became states—the largest number achieving statehood in one year. This number is twice the next closest number to achieve statehood in one year. What effect does this have on the statehood box plot?
16. **Research** Find out the year your state was settled by Europeans and the year it achieved statehood. In what quartile is the year your state was settled? In what quartile is the year your state achieved statehood?

BY THE WAY

The oldest city founded by Europeans in what is now the United States is St. Augustine, Florida. It was founded in 1565 by settlers from Spain.

264 Chapter 6 *Investigating Data*

Interdisciplinary Problems *Connecting Learning*

Each chapter has several clusters of exercises that focus on the connections of algebra to careers and to other subject areas, both within and outside of mathematics.

Careers *Preparing for the Future*

The *Interviews* at the start of each chapter, and the related exercises throughout the chapter, give students a sense of the types of mathematical skills they will need in their future careers.

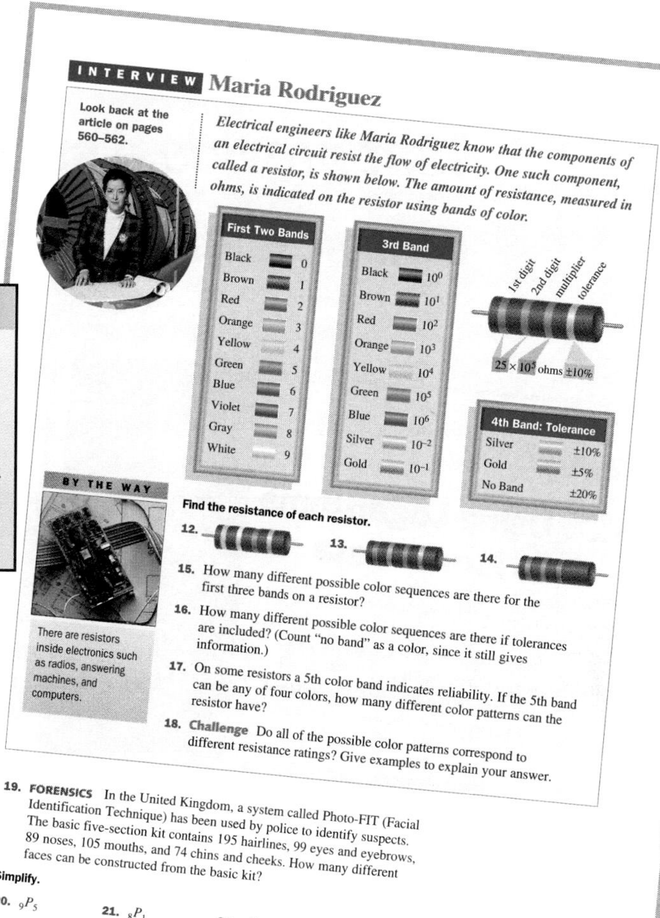

INTERVIEW Maria Rodriguez

Look back at the article on pages 560–562.

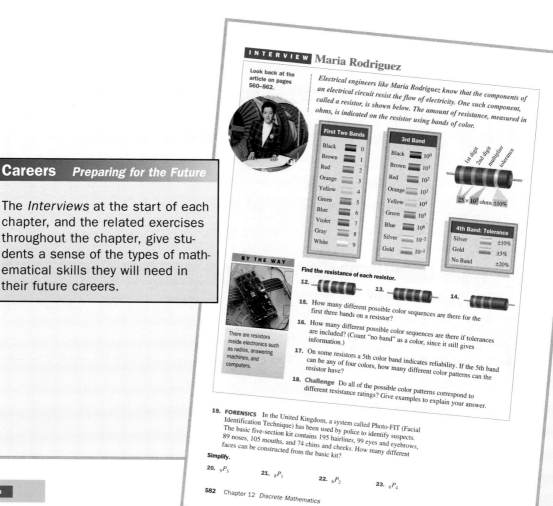

Electrical engineers like Maria Rodriguez know that the components of an electrical circuit resist the flow of electricity. One such component, called a resistor, is shown below. The amount of resistance, measured in ohms, is indicated on the resistor using bands of color.

First Two Bands	
Black	0
Brown	1
Red	2
Orange	3
Yellow	4
Green	5
Blue	6
Violet	7
Gray	8
White	9

3rd Band	
Black	10^0
Brown	10^1
Red	10^2
Orange	10^3
Yellow	10^4
Green	10^5
Blue	10^6
Silver	10^{-2}
Gold	10^{-1}

4th Band: Tolerance	
Silver	±10%
Gold	±5%
No Band	±20%

1st digit, 2nd digit, multiplier, tolerance

25×10^3 ohms ±10%

BY THE WAY

There are resistors inside electronics such as radios, answering machines, and computers.

Find the resistance of each resistor.

12. 13. 14.

15. How many different possible color sequences are there for the first three bands on a resistor?

16. How many different possible color sequences are there if tolerances are included? (Count "no band" as a color, since it still gives information.)

17. On some resistors a 5th color band indicates reliability. If the 5th band can be any of four colors, how many different color patterns can the resistor have?

18. **Challenge** Do all of the possible color patterns correspond to different resistance ratings? Give examples to explain your answer.

19. **FORENSICS** In the United Kingdom, a system called Photo-FIT (Facial Identification Technique) has been used by police to identify suspects. The basic five-section kit contains 195 hairlines, 99 eyes and eyebrows, 89 noses, 105 mouths, and 74 chins and cheeks. How many different faces can be constructed from the basic kit?

Simplify.

20. ${}_9P_5$ 21. ${}_8P_1$ 22. ${}_6P_2$ 23. ${}_6P_4$

582 Chapter 12 *Discrete Mathematics*

Section

12.2 Directed Graphs and Matrices

Learn how to...
- represent situations with directed graphs
- represent directed graphs with matrices

So you can...
- analyze situations involving direction, such as predator-prey relationships and electricity transfers

A *food web* shows how energy, in the form of food, is transferred through an ecosystem. The illustration below shows a food web for polar seas.

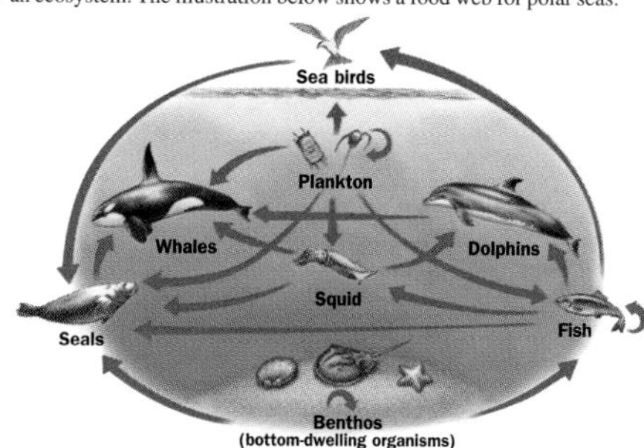

BY THE WAY

There are several kinds of whales in the polar seas. Bottlenose whales and sperm whales eat squid, killer whales eat dolphins and seals, and whalebone whales eat plankton.

THINK AND COMMUNICATE

1. In the food web above, what do the arrows represent?
2. Why is there an arrow from fish back to fish?
3. Which animals eat squid?
4. Seals eat benthos directly. They also eat benthos indirectly by eating fish that have eaten benthos. How else do seals eat benthos indirectly?
5. List the direct and indirect food sources for dolphins.
6. How would squid be affected if all of the plankton died?
7. Which animal is the most important food source for the rest of the animals? Explain.

You can represent the food web by a *directed graph* as shown at the right. A **directed graph** is a graph in which the **edges** are **arrows**.

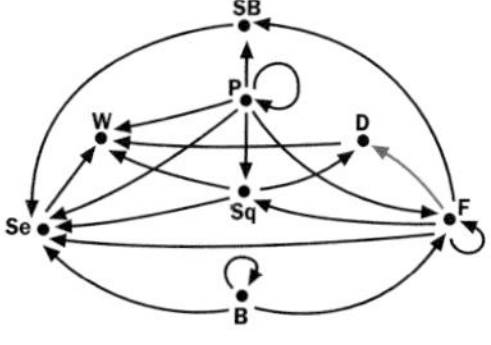

12.2 Directed Graphs and Matrices 569

Mathematical Connections *Integrated Problem Solving*

In this course students see that integrating algebra with concepts from geometry, statistics, probability, and discrete mathematics enables them to solve a greater variety of problems.

Build understanding of the concepts that provide a strong foundation for future courses and careers.

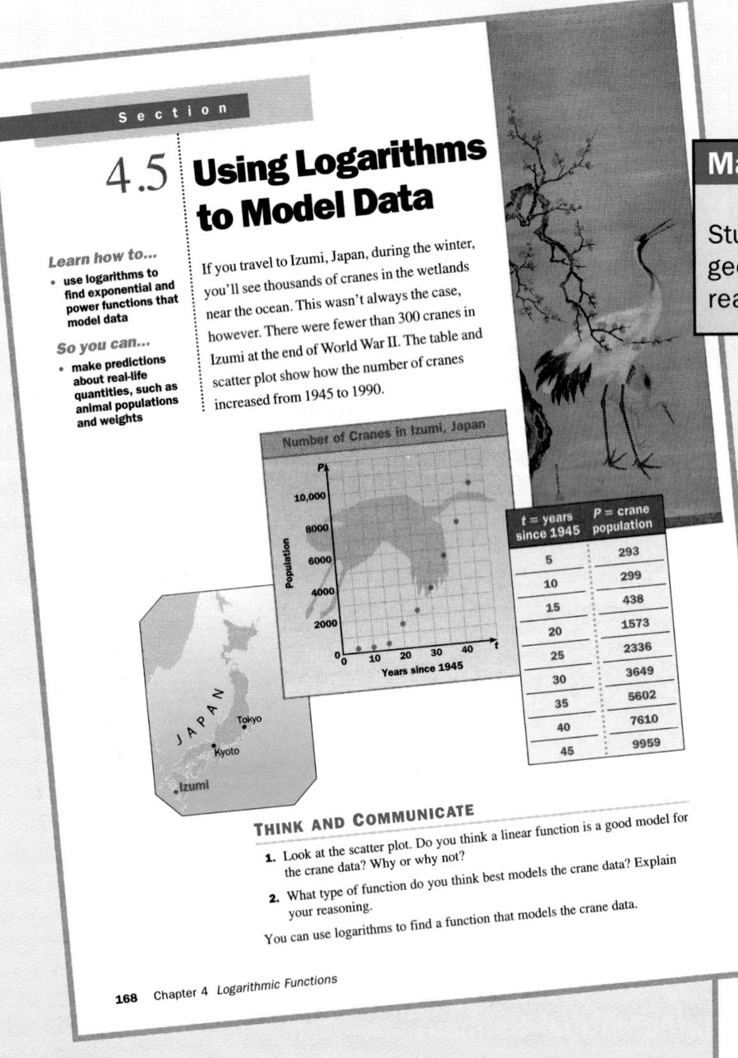

Section

4.5 Using Logarithms to Model Data

Learn how to...

- use logarithms to find exponential and power functions that model data

So you can...

- make predictions about real-life quantities, such as animal populations and weights

If you travel to Izumi, Japan, during the winter, you'll see thousands of cranes in the wetlands near the ocean. This wasn't always the case, however. There were fewer than 300 cranes in Izumi at the end of World War II. The table and scatter plot show how the number of cranes increased from 1945 to 1990.

t = years since 1945	P = crane population
5	293
10	299
15	438
20	1573
25	2336
30	3649
35	5602
40	7610
45	9959

THINK AND COMMUNICATE

1. Look at the scatter plot. Do you think a linear function is a good model for the crane data? Why or why not?
2. What type of function do you think best models the crane data? Explain your reasoning.

You can use logarithms to find a function that models the crane data.

168 Chapter 4 *Logarithmic Functions*

Mathematical Modeling *A Tool for Predicting*

Students use ideas from statistics, algebra, geometry, and probability to model a variety of real-world situations.

EXAMPLE 1 Application: Ecology

Use the table on page 168.

a. Make a scatter plot of the data pairs (t, log P). What relationship exists between t and log P?

b. Find an equation giving P as a function of t. What type of function is this equation?

SOLUTION

a. Make a table of data pairs (t, log P). Plot the pairs in a coordinate plane. The scatter plot suggests that there is a linear relationship between t and log P.

Crane Data

	A	B	C
1	t	P	log P
2	5	293	2.47
3	10	299	2.48
4	15	438	2.64
5	20	1573	3.20
6	25	2336	3.37
7	30	3649	3.56
8	35	5602	3.75
9	40	7610	3.88
10	45	9959	4.00

log P
4
3
2
1
0
0 10 20 30 40 t

An equation of this fitted line is log $P = 0.0431t + 2.18$.

b. Use the equation of the fitted line.

$$\log P = 0.0431t + 2.18$$
$$P = 10^{0.0431t + 2.18}$$
Write the equation in exponential form.
$$P = (10^{2.18})(10^{0.0431t})$$
$$P = (10^{2.18})(10^{0.0431})^t$$
$$P = 151(1.10)^t$$

An equation is $P = 151(1.10)^t$. This equation is an exponential function.

BY THE WAY

The crane is the most popular shape in *origami*, the Japanese art of paper folding. The tradition of folding and stringing together one thousand cranes is thought to guarantee that a wish will come true.

THINK AND COMMUNICATE

3. **Technology** Use a graphing calculator or graphing software to graph the data pairs (t, P) on page 168 and the function $P = 151(1.10)^t$ in the same coordinate plane. Do you think the given function is a good model for the crane data? Explain.
4. Predict the number of cranes in Izumi in 1998.
5. Based on the exponential model in Example 1, by about what percent did the Izumi crane population increase each year from 1945 to 1990?

4.5 Using Logarithms to Model Data 169

Visualizing Mathematics *Multiple Representations*

Tables, graphs, and functions are among the tools students use to visualize and analyze mathematical relationships.

Functions *A Unifying Theme*

Functions are investigated throughout the program as students model and solve problems involving linear and nonlinear equations.

Section

1.2 Using Functions to Model Growth

Learn how to...
- write a function to model a set of data

So you can...
- make predictions about a situation, such as plastics production in the United States

What do nylon stockings, polyester pants, and vinyl siding have in common? Each item contains plastic in both the product and its name! Nylon, polyester, and vinyl are just some of the types of plastics that have been developed since the 1930s. Look at the graph of plastics production in the United States from 1961 to 1991.

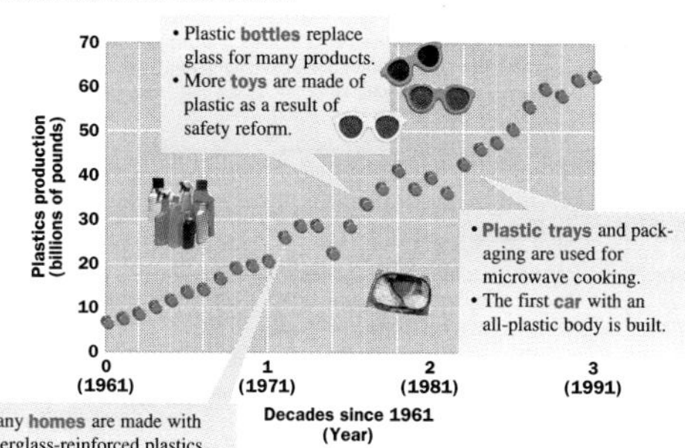

• Plastic bottles replace glass for many products.
• More toys are made of plastic as a result of safety reform.
• Plastic trays and packaging are used for microwave cooking.
• The first car with an all-plastic body is built.
• Many homes are made with fiberglass-reinforced plastics.
Plastics production (billions of pounds)
70
60
50
40
30
20
10
0
0 (1961)
1 (1971)
2 (1981)
3 (1991)
Decades since 1961 (Year)

Think and Communicate

1. Copy and complete the table.

Decades since 1961	Plastics production (billions of pounds)	Change per decade	Percent change per decade
0 (1961)	7	—	—
1 (1971)	21	?	?
2 (1981)	40	?	?
3 (1991)	63	?	?

2. What was the average change?
3. What was the average percent change?
4. How many decades since 1961 is 2001? Estimate the plastics production in 2001.

Section

9.4 Solving Cubic Equations

Learn how to...
- solve cubic equations
- find equations for graphs of cubic functions
- find zeros of cubic functions

So you can...
- analyze the flight of the space shuttle after launch, for example

The space shuttle is powered by three liquid-fuel motors in its tail and two external solid-fuel booster rockets. When the shuttle reaches a speed of about 3000 mi/h, the booster rockets fall off and return to Earth. The liquid-fuel motors continue firing and propel the shuttle into orbit.

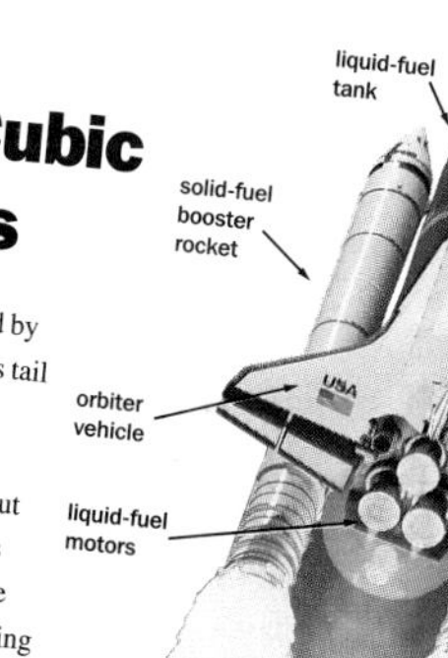

liquid-fuel tank
solid-fuel booster rocket
orbiter vehicle
liquid-fuel motors

Example 1 Application: Space Travel

The speed of the space shuttle t seconds after launch can be approximated using the function

$$s(t) = 0.000559t^3 + 0.0313t^2 + 13.6t$$

where $s(t)$ is measured in miles per hour. How long after launch do the booster rockets fall off?

Solution

The booster rockets fall off when $s(t) = 3000$, so you need to solve this equation:

$$0.000559t^3 + 0.0313t^2 + 13.6t = 3000$$

Method 1 Use a graphing calculator or graphing software.

Graph $y = 0.000559x^3 + 0.0313x^2 + 13.6x$ and $y = 3000$ in the same coordinate plane.

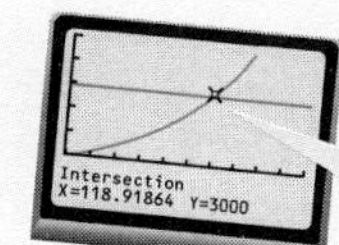

Intersection
X=118.91864 Y=3000

Find the x-coordinate of the point where the graphs intersect. The x-coordinate is about **119**.

The booster rockets fall off about 119 s (or about 2 min) after launch.

Problem Solving Strategies *A Variety of Approaches*

Students use a number of techniques to investigate and understand mathematical concepts and applications.

Assess students' progress in ways that support learning.

Embedded Assessment *A Part of Learning*

Ongoing Assessment questions in each section ask students to apply concepts and explain their thinking. *Journal* writing gives students a chance to reflect on their own learning process.

ONGOING ASSESSMENT

16. **Open-ended Problem** Choose a real-world application of a rational function from Section 9.7 or 9.8. Write a question based on this application that can be answered by solving a rational equation. Exchange questions with a classmate, and answer your classmate's question.

SPIRAL REVIEW

Perform the indicated operation. *(Section 8.4)*

17. $(5 - 7i) + (8 + 3i)$ 18. $(2 + i) - (6 - 11i)$ 19. $(3 + i)^2$

Technology Use a graphing calculator or software with matrix calculation capabilities to find an equation of the parabola passing through each set of points. *(Section 7.3)*

20. (1, 2), (2, 8), and (3, 18) 21. (1, 0), (2, 2), and (3, 6)

ASSESS YOUR PROGRESS

VOCABULARY

rational equation (p. 450)

For each function:

a. Find the vertical asymptotes of the function's graph.

b. Describe the function's end behavior using infinity notation. *(Section 9.8)*

1. $f(x) = \dfrac{2x - 7}{(x - 1)(x - 5)}$ 2. $g(x) = \dfrac{-x^2 - 2x + 8}{x + 3}$

3. $h(x) = \dfrac{x^4}{x^2 + 3x - 10}$ 4. $f(x) = \dfrac{3x^3 + 1}{x^3 - 9x}$

Solve each equation. *(Section 9.9)*

5. $\dfrac{5}{3} + \dfrac{2x + 9}{x + 4} = 4$ 6. $\dfrac{w}{w - 5} = \dfrac{2}{w - 3} + \dfrac{4}{(w - 3)(w - 5)}$

7. **SPORTS** The 1994 major league baseball season ended prematurely because of a players' strike. When the season ended, Tony Gwynn of the San Diego Padres had the league's highest batting average (the ratio of hits to at-bats). Gwynn had 165 hits in 419 at-bats, for an average of .394. If there had been no strike and Gwynn had continued playing, how many consecutive hits would he have needed to raise his average to .400? *(Section 9.9)*

Tony Gwynn

BY THE WAY

If Tony Gwynn had batted at least .400 in 1994, he would have been the first major league player to do so since 1941, when Ted Williams batted .406.

8. **Journal** Explain how the end behavior of a rational function $f(x) = \dfrac{p(x)}{q(x)}$ is related to the degrees of the polynomials $p(x)$ and $q(x)$.

9.9 Solving Rational Equations **453**

CHAPTER 9 **Assessment**

SECTIONS 9.6 and 9.7

15. **HOME REPAIR** On some tubes of caulking, the diameter of the round nozzle opening from which the caulking flows can be adjusted by the user. As the diameter d increases, the length l of caulking obtained from the tube decreases, as shown in the table.

d (in.)	l (in.)
$\frac{1}{8}$	1440
$\frac{1}{4}$	360
$\frac{3}{8}$	160
$\frac{1}{2}$	90

a. Does l vary inversely with d? Explain.

b. Find the area A of each nozzle opening whose diameter is given in the table. Does l vary inversely with A? Explain.

c. Write an equation giving l as a function of A.

16. Let $f(x) = \dfrac{30x + 17}{6x + 1}$. Find the asymptotes of the graph of f, and tell how the graph is related to a hyperbola with equation of the form $y = \dfrac{a}{x}$.

SECTIONS 9.8 and 9.9

17. Let $h(x) = \dfrac{x^4}{4x^2 + 11x - 3}$. Find the vertical asymptotes of the graph of h, and describe the end behavior of h using infinity notation.

18. **SPORTS** Janice is a member of her high school's varsity basketball team. So far this season, she has made 26 of the 80 three-point shots she has attempted, for a shooting percentage of 32.5%. How many consecutive three-point shots must she make to raise her shooting percentage to 40%?

19. Solve the equation $\dfrac{2}{x(x - 2)} - \dfrac{1}{x - 2} = 1$.

PERFORMANCE TASK

20. This problem was posed 700 years ago by the Chinese mathematician Ch'in Chiu-shao:

There is a circular town of unknown diameter having four gates. A tall tree stands 3 li from the north gate as shown. When you exit the south gate and turn east, you must walk 9 li before you can see the tree. Find the diameter of the town. (*Note:* 1 li ≈ 0.33 mi)

a. Show that the radius r of the town must satisfy the polynomial equation $4r^4 + 12r^3 + 9r^2 - 486r - 729 = 0$.

b. Find all real and imaginary solutions of the equation in part (a). Explain why only one of the solutions makes sense in the given situation. What is this solution? What is the diameter of the town?

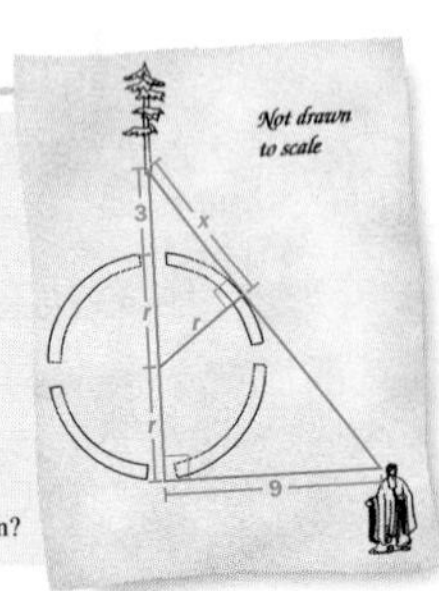

Assessment **459**

Monitoring Progress *Check for Understanding*

A variety of different types of assessment questions are included throughout the course. A *Performance Task* is included in each chapter assessment.

Portfolio Assessment *Demonstrating Growth*

The *Portfolio Projects* provide opportunities for original student work that can be part of a mathematical portfolio. In each project, students are asked to present their results and assess their work.

PORTFOLIO PROJECT

Predicting Basketball Accuracy

In 1979 the National Basketball Association instituted a *three-point rule.* According to this rule, a basket shot from a point outside the three-point "line" is worth three points, while a basket shot from a closer distance is worth two points. Why do you think the rules were changed?

The **three-point line** is two inches wide. A shot counts for only two points if a player's foot touches the three-point line.

PROJECT GOAL Do an experiment to find how your ability to make a basket changes with distance.

Investigating Your Basketball Ability

For this investigation, you will need a piece of wadded-up paper to be your ball, a wastebasket, a yardstick, and graph paper. You may find it helpful to have a graphing calculator or graphing software.

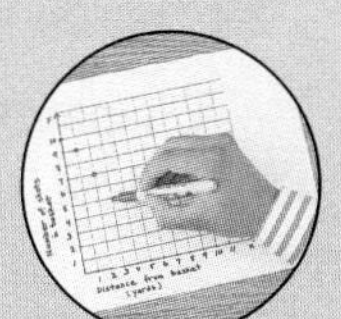

1. **STAND** 0 yards from the wastebasket. Shoot your paper ball into the basket 10 times and record the number of shots that go into the basket.

2. **STEP** back one yard and record your results for 10 shots from this distance. Continue stepping back one yard until you reach a distance where none of your 10 shots go into the basket.

3. **MAKE** a scatter plot of your data. Let the horizontal axis show the distance from the basket, and let the vertical axis show the number of shots that went into the basket.

Making a Model

• **DESCRIBE** your graph. Do you think the data decreases *linearly*? *exponentially*? At what distance did the number of shots that went into the basket decrease the most?

• **ORGANIZE** your data in a table like the one shown. Find the average change and the average percent change.

Distance from basket (yd)	Number of shots that went into basket	Change	Percent change
0	?	—	—
1	?	?	?
2	?	?	?
⋮	⋮	⋮	⋮

• **MODEL** your data with a linear function and with an exponential function.

• **GRAPH** your linear function, exponential function, and data points in the same coordinate plane. Which model do you think most accurately represents your data? Why?

Writing a Report

Write a report about your experiment that includes your data, graphs, models, and explanations. You may also want to investigate and report on these things:

• How do you think your results would change if you repeated the experiment 5 times? 10 times? 50 times? Explain your reasoning. Then repeat the experiment at least 5 times. How do your results compare to your prediction?

• Repeat the experiment using a real basketball and hoop. How could a basketball player use this experiment to decide whether to try for a two-point shot or a three-point shot?

• Why do you think the three-point rule was created? What other sports have rules that take the *likelihood* of scoring points into account? Describe the method of scoring for each sport.

Self-Assessment

Describe any difficulties you had in finding a linear and an exponential model to fit your data. How did you resolve these difficulties?

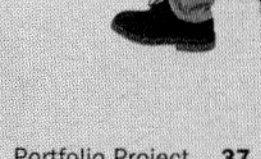

Planning Pages for Every Chapter

OVERVIEW

The **Overview** provides a summary of connections to prior and future learning and highlights the chapter interview, explorations, portfolio project, and use of technology.

OBJECTIVES

Chapter **Objectives** gives objectives and NCTM Standards for each section.

INTEGRATION

The **Integration** chart highlights the mathematical and interdisciplinary connections and applications found throughout the chapter.

TECHNOLOGY

The **Technology** chart highlights opportunities to use technology in both the student book and support materials.

CHAPTER 5

Quadratic Functions

OVERVIEW

Connecting to Prior and Future Learning

⇔ This chapter opens with a review of graphing and solving simple quadratic functions in Section 5.1. This topic is expanded in Section 5.2 to include translations of parabolas. Related topics reviewed in the **Student Resources Toolbox** include translating a figure on page 796, reflecting a figure and line symmetry on page 797, and dilating a figure on page 798.

⇔ Students continue to study graphs of quadratic functions in Section 5.3, where they learn to write quadratic functions in intercept form. A review of factoring variable expressions is available on page 782 of the **Student Resources Toolbox**.

⇔ The last three sections of this chapter present different ways to solve quadratic equations: completing the square, the quadratic formula, and factoring. Students will use these methods again as they continue to study mathematics.

Chapter Highlights

Interview with Mark Thomas: Engineers use mathematics in many aspects of their work. Exercises relating to this interview can be found on pages 190 and 220.

Explorations: In Section 5.2, students use a graphing calculator or graphing software to study the translation of parabolas. In Sections 5.4 and 5.6, students use algebra tiles to explore completing the square and factoring.

The Portfolio Project: Students conduct an experiment to find out how the water level in a can affects the time it takes water to flow out of the can and use a quadratic model to describe their results.

Technology: Throughout this chapter, graphing calculators or graphing software are used to graph functions and spreadsheets are used to create tables.

OBJECTIVES

Section	Objectives	NCTM Standards
5.1	• Recognize and draw graphs of quadratic functions. • Solve simple quadratic equations. • Examine quadratic relationships in real-world situations.	1, 2, 3, 4, 5, 6
5.2	• Graph equations in the form $y = a(x - h)^2 + k$. • Solve quadratic equations. • Solve real-world problems using quadratic equations.	1, 2, 3, 4, 5, 6
5.3	• Write quadratic equations in intercept form. • Maximize or minimize quadratic functions. • Use quadratic functions to find maximums and minimums in real-world situations.	1, 2, 3, 4, 5, 6
5.4	• Complete the square to write quadratic functions in vertex form. • Use completing the square to find maximums or minimums in real-world situations.	1, 2, 3, 4, 5, 6
5.5	• Solve equations using the quadratic formula. • Use the discriminant to determine how many solutions an equation has. • Use the quadratic formula to solve real-world problems.	1, 2, 3, 4, 5, 6
5.6	• Factor quadratic expressions. • Solve quadratic equations using factoring. • Solve real-world problems using factoring.	1, 2, 3, 4, 5, 6

182A

INTEGRATION

Mathematical Connections	5.1	5.2	5.3	5.4	5.5	5.6
algebra	**185–191***	**192–198**	**199–205**	**206–213**	**214–220**	**221–227**
geometry	189		204			225
data analysis, probability, discrete math	185, 187, 189–191	194–198	201–205	208–213	214–216, 218–220	224–227
patterns and functions	**185–191**	**192–198**	**199–205**	**206–213**	**214–220**	**221–227**
logic and reasoning	**185–191**	192, 194, 196–198	201–205	207, 208, 210–213	214–216, 218–220	222–227
Interdisciplinary Connections and Applications						
biology and earth science		198		211		
chemistry and physics		194		208, 210		
business and economics			201, 203			
sports and recreation	185, 189			212	218	
astronomy	190					226
government					219	
personal finance, engineering, physiology, driving		197	204		219	224, 225

* ***Bold page numbers*** *indicate that a topic is used throughout the section.*

TECHNOLOGY

Section	*opportunities for use with* Student Book	Support Material
5.1	graphing calculator McDougal Littell Mathpack *Function Investigator* *Stats!*	
5.2	graphing calculator or graphing software McDougal Littell Mathpack *Stats!*	**Technology Book:** Calculator Activity 5 Spreadsheet Activity 5
5.3	graphing calculator or graphing software McDougal Littell Mathpack *Function Investigator*	
5.4	graphing calculator McDougal Littell Mathpack *Function Investigator*	
5.5	graphing calculator	
5.6	graphing calculator McDougal Littell Mathpack *Function Investigator*	

182B

PLANNING GUIDE

Regular Scheduling (45 min)

Section	Materials Needed	Core Assignment	Extended Assignment	*exercises that feature* Applications	Communication	Technology
5.1	graphing calculator, graph paper	1–4, 6–17, 24–26, 28–32	1–5, 7, 9, 11–32	19–25	18	20, 27
5.2	graphing calculator or software, graph paper	1–4, 9–24, 27–36	1–8, 9–19 odd, 21–36	5–8, 25	4, 5, 7, 27	7
5.3	graphing calculator or software, graph paper	1–10, 15, 17–28, AYP*	1–28, AYP	12–14	7, 11, 18	
5.4	algebra tiles, graph paper, graphing calculator	**Day 1:** 1–14, 17 **Day 2:** 25–47	**Day 1:** 1–24 **Day 2:** 25–47	14–23	15, 22	16
5.5	graph paper	**Day 1:** 1–17 **Day 2:** 18–23, 30, 31, 35–44	**Day 1:** 1–17 **Day 2:** 18–44	11 25–29, 31–34	25, 26	
5.6	algebra tiles, graph paper, graphing calculator	**Day 1:** 1–10, 12–21 **Day 2:** 26–39, AYP	**Day 1:** 1–25 **Day 2:** 26–39, AYP	10, 23–25	21 30	
Review/ Assess		**Day 1:** 1–11 **Day 2:** 12–23 **Day 3:** Ch. 5 Test	**Day 1:** 1–11 **Day 2:** 12–23 **Day 3:** Ch. 5 Test	15, 19	6	
Portfolio Project		Allow 2 days.	Allow 2 days.			

Yearly Pacing (with Portfolio Project)	**Chapter 5 Total**	**Chapters 1–5 Total**	**Remaining**	**Total**
	14 days	66 days	94 days	160 days

Block Scheduling (90 min)

	Day 27	Day 28	Day 29	Day 30	Day 31	Day 32	Day 33	Day 34
Teach/Interact	Ch. 4 Test 5.1	5.2: Exploration, page 192 5.3: Exploration, page 199	5.4: Exploration, page 206	5.5	5.6: Exploration, page 221	Review Port. Proj.	Review Port. Proj.	Ch. 5 Test 6.1
Apply/Assess	**Ch. 4 Test** **5.1:** 1–4, 6–17, 24–26, 28–32	**5.2:** 1–4, 9–24 27–36 **5.3:** 1–10 15, 17–28, AYP*	**5.4:** 1–14, 17, 25–47	**5.5:** 1–23, 30, 31, 35–44	**5.6:** 1–10 12–21, 26–39, AYP	**Review:** 1–11 **Port. Proj.**	**Review:** 12–23 **Port. Proj.**	**Ch. 5 Test** **6.1:** 1–11, 13–17, 23–31

NOTE: A one-day block has been added for the Portfolio Project—timing and placement to be determined by teacher.

Yearly Pacing (with Portfolio Project)	**Chapter 5 Total**	**Chapters 1–5 Total**	**Remaining**	**Total**
	7 days	$33\frac{1}{2}$ days	$48\frac{1}{2}$ days	82 days

*** AYP** *is Assess Your Progress.*

182C

PLANNING GUIDE

The **Planning Guide** gives materials, pacing, and suggested assignments for each section, and block scheduling assignments.

LESSON SUPPORT

Section	Practice Bank	Study Guide*	Assessment Book*	Visuals	Explorations Lab Manual	Lesson Plans	Technology Book
5.1	26	5.1		Warm-Up 5.1	Master 2 Add. Expl. 7	5.1	
5.2	27	5.2		Warm-Up 5.2 Folder 5	Masters 1, 2, 9	5.2	Calculator Act. 5 Spreadsheet Act. 5
5.3	28	5.3	Test 17	Warm-Up 5.3	Masters 2, 10	5.3	
5.4	29	5.4		Warm-Up 5.4 Folder B	Masters 2, 3, 4, 11	5.4	
5.5	30	5.5		Warm-Up 5.5	Master 2 Add. Expl. 8	5.5	
5.6	31	5.6	Test 18	Warm-Up 5.6 Folder B	Masters 2, 3, 4	5.6	
Review Test	32	Chapter Review	Tests 19, 20 Alternative Assessment			Review Test	Calculator Based Lab 5

***Spanish versions** of *Study Guide* and *Assessment Book* are available.

Chapter Support

- Course Guide
- Lesson Plans
- Portfolio Project Book
- Preparing for College Entrance Tests
- Multi-Language Glossary
- *Test Generator* Software
- Professional Handbook

Software Support

McDougal Littell Mathpack
Stats!
Function Investigator

Internet Support

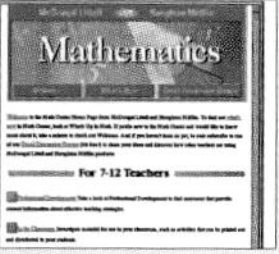

http://www.hmco.com
Next go to McDougal Littell; then the Education Center; then Secondary Math.

OUTSIDE RESOURCES

Books, Periodicals

Edwards, Thomas G. "Exploring Quadratic Functions: From *a* to *c*." *Mathematics Teacher* (February 1996): pp. 144–146.

Metz, James. "Sharing Teaching Ideas: Seeing *b* in $y = ax^2 + bx + c$." *Mathematics Teacher* (January 1994): pp. 23–25.

Jones, Graham A., Carol A. Thornton, Carol A. McGehe, and David Cole. "Rich Problems—Big Payoff." *Mathematics Teaching in the Middle School* (November–December 1995): pp. 520–525.

Activities, Manipulatives

Breuningsen, Chris, Bill Bower, Linda Antinone, and Elisa Breuningsen. "That's the Way the Ball Bounces." *Real-World Math with the CBL System.* Activity 9: pp. 49–54. Texas Instruments, 1995.

Leiva, Miriam, Joan Ferrini-Mundy, and Loren P. Johnson. "Playing With Blocks: Visualizing Functions." *Mathematics Teacher* (November 1992): pp. 641–646.

Breuningsen, Chris, Bill Bower, Linda Antinone, and Elisa Breuningsen. "What Goes Up...." *Real-World Math with the CBL System.* Activity 8: pp. 45–48. Texas Instruments, 1995.

Software

Meridian Creative Group. "Exploration: Modeling Quadratic Functions I" and "Exploration: Modeling Quadratic Functions II." Worksheets and software included for use with the CBL. *Explorations in Precalculus for the TI-82:* pp. 13–24.

Videos

Videodisk: "Projectile Motion." *Science: Forces and Energy:* (Side A). Macmillan/McGraw-Hill, 1993.

Internet

Join the Math Forum at Swarthmore by connecting to:
http://forum.swarthmore.edu

182D

LESSON SUPPORT

The **Lesson Support** chart lists all support materials for each section.

OUTSIDE RESOURCES

Outside Resources lists books, periodicals, manipulatives and activities, software, videos, and Internet addresses.

A Teaching Plan for Every Section

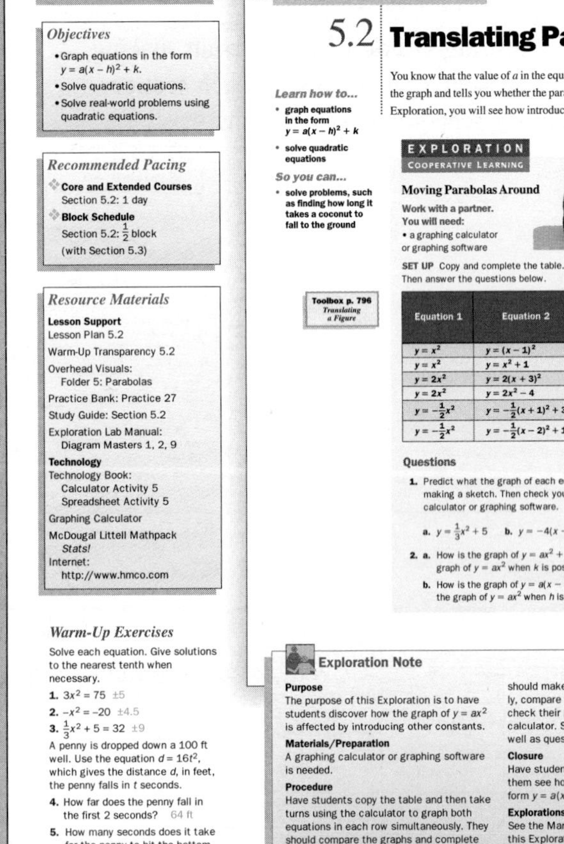

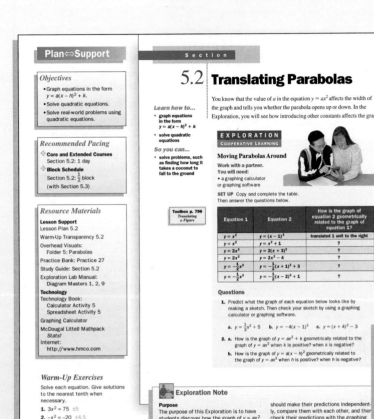

Plan⇔Support

Objectives

- Graph equations in the form $y = a(x - h)^2 + k$.
- Solve quadratic equations.
- Solve real-world problems using quadratic equations.

Recommended Pacing

Core and Extended Courses
Section 5.2: 1 day
Block Schedule
Section 5.2: $\frac{1}{2}$ block (with Section 5.3)

Resource Materials

Lesson Support
Lesson Plan 5.2
Warm-Up Transparency 5.2
Overhead Visuals: Folder 5: Parabolas
Practice Bank: Practice 27
Study Guide: Section 5.2
Exploration Lab Manual: Diagram Masters 1, 2, 9
Technology
Technology Book: Calculator Activity 5, Spreadsheet Activity 5
Graphing Calculator
McDougal Littell Mathpack *Stats!*
Internet: http://www.hmco.com

Warm-Up Exercises

Solve each equation. Give solutions to the nearest tenth when necessary.

1. $3x^2 = 75$ ±5
2. $-x^2 = -20$ ±4.5
3. $\frac{1}{3}x^2 + 5 = 32$ ±9

A penny is dropped down a 100 ft well. Use the equation $d = 16t^2$, which gives the distance d, in feet, the penny falls in t seconds.

4. How far does the penny fall in the first 2 seconds? 64 ft
5. How many seconds does it take for the penny to hit the bottom of the well? 2.5 s

Section

5.2 Translating Parabolas

Learn how to...
- graph equations in the form $y = a(x - h)^2 + k$
- solve quadratic equations

So you can...
- solve problems, such as finding how long it takes a coconut to fall to the ground

You know that the value of a in the equation $y = ax^2$ affects the width of the graph and tells you whether the parabola opens up or down. In the Exploration, you will see how introducing other constants affects the graph.

EXPLORATION
COOPERATIVE LEARNING

Moving Parabolas Around

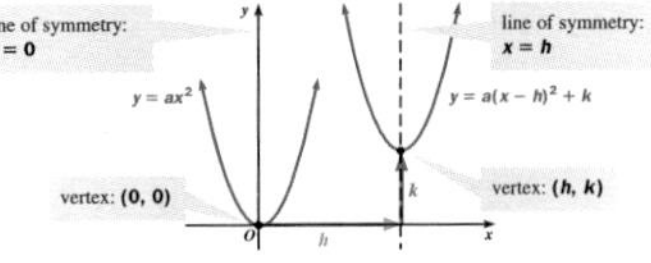

Work with a partner.
You will need:
- a graphing calculator or graphing software

Toolbox p. 796 *Translating a Figure*

SET UP Copy and complete the table. Then answer the questions below.

Equation 1	Equation 2	How is the graph of equation 2 geometrically related to the graph of equation 1?
$y = x^2$	$y = (x - 1)^2$	translated 1 unit to the right
$y = x^2$	$y = x^2 + 1$	?
$y = 2x^2$	$y = 2(x + 3)^2$	?
$y = 2x^2$	$y = 2x^2 - 4$	?
$y = -\frac{1}{2}x^2$	$y = -\frac{1}{2}(x + 1)^2 + 3$	?
$y = -\frac{1}{2}x^2$	$y = -\frac{1}{2}(x - 2)^2 + 1$	?

Questions

1. Predict what the graph of each equation below looks like by making a sketch. Then check your sketch by using a graphing calculator or graphing software.
   a. $y = \frac{1}{3}x^2 + 5$ b. $y = -4(x - 1)^2$ c. $y = (x + 4)^2 - 3$
2. a. How is the graph of $y = ax^2 + k$ geometrically related to the graph of $y = ax^2$ when k is positive? when k is negative?
   b. How is the graph of $y = a(x - h)^2$ geometrically related to the graph of $y = ax^2$ when h is positive? when h is negative?

Exploration Note

Purpose
The purpose of this Exploration is to have students discover how the graph of $y = ax^2$ is affected by introducing other constants.
Materials/Preparation
A graphing calculator or graphing software is needed.
Procedure
Have students copy the table and then take turns using the calculator to graph both equations in each row simultaneously. They should compare the graphs and complete the third column. For question 1, students should make their predictions independently, compare them with each other, and then check their predictions with the graphing calculator. Students can use the table as well as question 1 to answer question 2.
Closure
Have students discuss question 2 to help them see how to graph equations in the form $y = a(x - h)^2 + k$.
Explorations Lab Manual
See the Manual for more commentary on this Exploration.
Diagram Master 9

192

Plan⇔Support

- Section Objectives
- Recommended Pacing
- Resource Materials
- Warm-Up Exercises

The equations that you graphed in the Exploration have the general form $y = a(x - h)^2 + k$. To graph an equation in this form, you can start with the graph of the simpler equation $y = ax^2$ and translate it h units horizontally and k units vertically.

Because you can read the vertex (as well as other information about the graph) directly from the equation $y = a(x - h)^2 + k$, the equation is called the **vertex form** of a quadratic equation.

Vertex Form of a Quadratic Equation

The graph of $y = a(x - h)^2 + k$ is a parabola that:
- has its vertex at (h, k),
- has the line $x = h$ as its line of symmetry,
- opens up if $a > 0$ and opens down if $a < 0$.

EXAMPLE 1

Describe the graph of $y = -2(x + 3)^2 + 4$ and then sketch it.

SOLUTION

Rewrite the equation as $y = -2(x - (-3))^2 + 4$. The vertex is $(-3, 4)$. The line of symmetry is $x = -3$. The parabola opens down.

For x-values 1 unit to the left or right of the line of symmetry, the corresponding y-value is $y = -2(\pm 1)^2 + 4 = 2$.

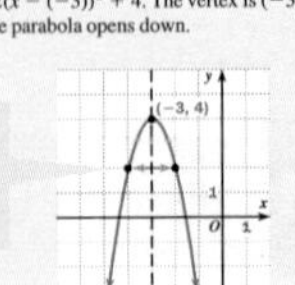

5.2 Translating Parabolas 193

Teach⇔Interact

Section Notes

Topic Spiraling: Review
Some students may need a brief reminder concerning the equation of a vertical line. They need to realize that the expression $x = h$ for the line of symmetry represents a line and not merely the x-coordinate of a single point.

Reasoning
Given a point (x_1, y_1) on the graph of $y = x^2$, ask students to find the coordinates of its image point after a translation of h units horizontally and k units vertically. ($(x_1 + h, y_1 + k)$) Then have them verify that this point lies on the graph of $y = a(x - h)^2 + k$. ((x_1, y_1) is on $y = ax^2$, so $y_1 = ax_1$ and $a((x_1 + h) - h)^2 + k = ax_1^2 + k = y_1 + k$. Thus, $(x_1 + h, y_1 + k)$ lies on $y = a(x - h)^2 + k$.)

About Example 1

Mathematical Procedures
The method used to find points on the graph of a parabola takes advantage of the vertical reflection symmetry these graphs have over their axis of symmetry. Students could extend this method to find other points as well. For example, to find points 3 units to the right or left of the vertex, replace the $x + 3$ in the expression of the function with ±3: $y = -2(\pm 3)^2 + 4 = -2(9) + 4 = -14$, yielding points (−6, 14) and (0, 14).

Additional Example 1

Describe the graph of $y = \frac{1}{2}(x - 1)^2 - 3$ and then sketch it.
Rewrite the equation as $y = \frac{1}{2}(x - 1)^2 + (-3)$. The vertex is $(1, -3)$. The line of symmetry is $x = 1$. The parabola opens up. For x-values 2 units to the left or right of the line of symmetry, the corresponding y-value is $y = \frac{1}{2}(\pm 2)^2 - 3 = -1$.

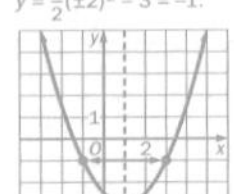

ANSWERS Section 5.2

Exploration

Equation 1	Equation 2	How is the graph of equation 2 geometrically related to the graph of equation 1?
$y = x^2$	$y = (x - 1)^2$	translated 1 unit to the right
$y = x^2$	$y = x^2 + 1$	translated up 1 unit
$y = 2x^2$	$y = 2(x + 3)^2$	translated 3 units to the left
$y = 2x^2$	$y = 2x^2 - 4$	translated down 4 units
$y = -\frac{1}{2}x^2$	$y = -\frac{1}{2}(x + 1)^2 + 3$	translated 1 unit to the left, up 3 units
$y = -\frac{1}{2}x^2$	$y = -\frac{1}{2}(x - 2)^2 + 1$	translated 2 units to the right, up 1 unit

Questions

1. a. The graph is the graph of $y = x^2$ shrunk vertically by a factor of 3 and translated 5 units up.
   b. The graph is the graph of $y = x^2$ translated 1 unit to the right, stretched vertically by a factor of 4, and reflected over the x-axis.
   c. The graph is the graph of $y = x^2$ translated 4 units to the left and down 3 units.
2. See answers in back of book.

193

Teach⇔Interact

- Additional Examples
- Closure Questions
- Notes on the student lesson, including:

Technology
Explorations
Learning Styles
Communication
Multicultural Information
Second-Language Learners

Apply⇔Assess

- Suggested Assignment
- *Practice Bank* facsimile
- Notes on the exercises, including:

Applications
Problem Solving
Technology
Cooperative Learning
Integrating the Strands
Assessment

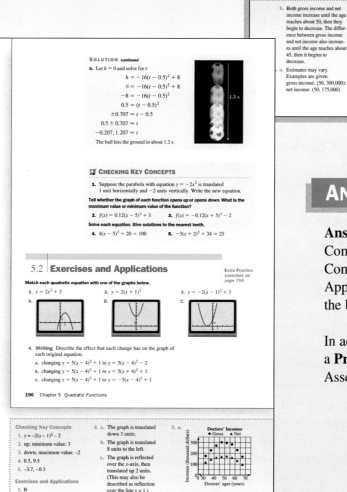

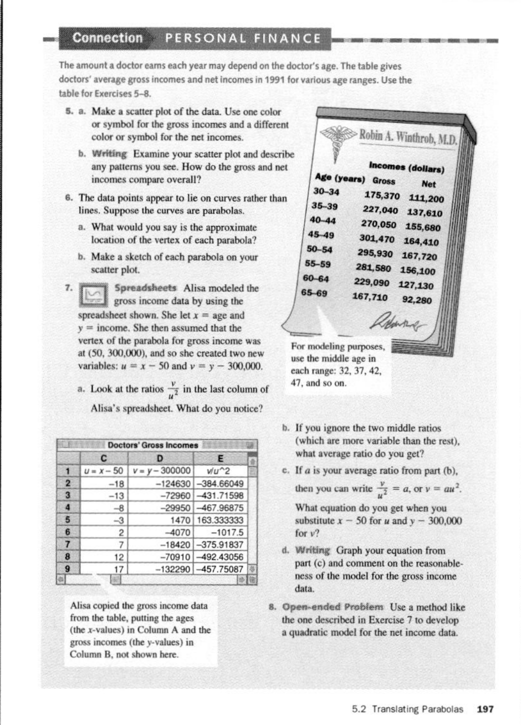

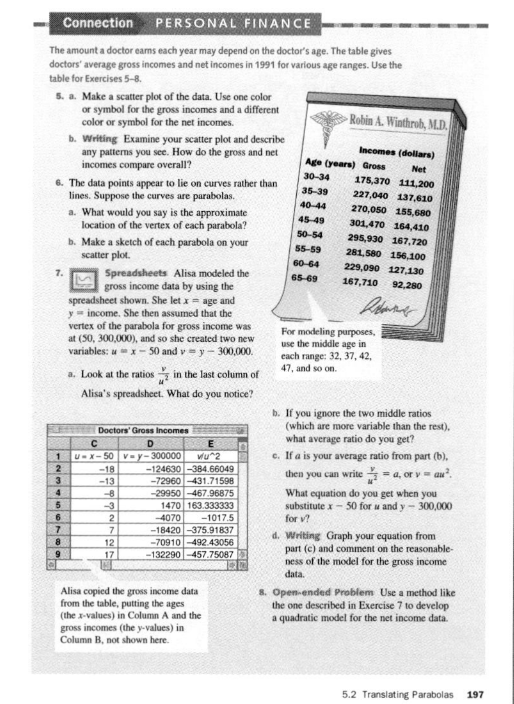

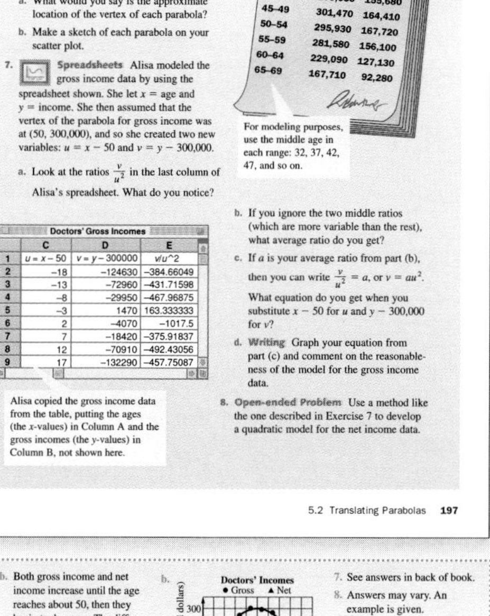

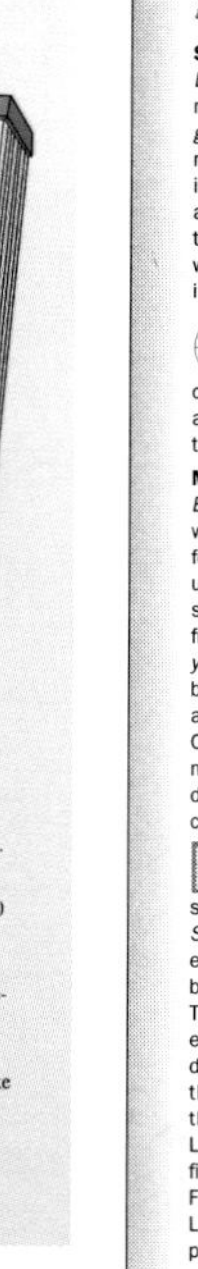

Connection PERSONAL FINANCE

The amount a doctor earns each year may depend on the doctor's age. The table gives doctors' average gross incomes and net incomes in 1991 for various age ranges. Use the table for Exercises 5–8.

5. a. Make a scatter plot of the data. Use one color or symbol for the gross incomes and a different color or symbol for the net incomes.
b. **Writing** Examine your scatter plot and describe any patterns you see. How do the gross and net incomes compare overall?

6. The data points appear to lie on curves rather than lines. Suppose the curves are parabolas.
a. What would you say is the approximate location of the vertex of each parabola?
b. Make a sketch of each parabola on your scatter plot.

7. **Spreadsheets** Alisa modeled the gross income data by using the spreadsheet shown. She let x = age and y = income. She then assumed that the vertex of the parabola for gross income was at (50, 300,000), and so she created two new variables: $u = x - 50$ and $v = y - 300{,}000$.
a. Look at the ratios $\frac{v}{u^2}$ in the last column of Alisa's spreadsheet. What do you notice?
b. If you ignore the two middle ratios (which are more variable than the rest), what average ratio do you get?
c. If a is your average ratio from part (b), then you can write $\frac{v}{u^2} = a$, or $v = au^2$. What equation do you get when you substitute $x - 50$ for u and $y - 300{,}000$ for v?
d. **Writing** Graph your equation from part (c) and comment on the reasonableness of the model for the gross income data.

8. **Open-ended Problem** Use a method like the one described in Exercise 7 to develop a quadratic model for the net income data.

Robin A. Winthrob, M.D.

Age (years)	Incomes (dollars) Gross	Net
30–34	175,370	111,200
35–39	227,040	137,610
40–44	270,050	155,680
45–49	301,470	164,410
50–54	295,930	167,720
55–59	281,580	156,100
60–64	229,090	127,130
65–69	167,710	92,280

For modeling purposes, use the middle age in each range: 32, 37, 42, 47, and so on.

Doctors' Gross Incomes

	C	D	E
1	u = x − 50	v = y − 300000	v/u^2
2	−18	−124630	−384.66049
3	−13	−72960	−431.71598
4	−8	−29950	−467.96875
5	−3	1470	163.333333
6	2	−4070	−1017.5
7	7	−18420	−375.91837
8	12	−70910	−492.43056
9	17	−132290	−457.75087

Alisa copied the gross income data from the table, putting the ages (the x-values) in Column A and the gross incomes (the y-values) in Column B, not shown here.

5.2 Translating Parabolas 197

5. b. Both gross income and net income increase until the age reaches about 50, then they begin to decrease. The difference between gross income and net income also increases until the age reaches about 45, then it begins to decrease.

6. a. Estimates may vary. Examples are given. gross income: (50, 300,000); net income: (50, 175,000)

b. Doctors' Incomes ● Gross ▲ Net

7. See answers in back of book.

8. Answers may vary. An example is given. $y = -287(x - 50)^2 + 175{,}000$

Apply⇔Assess

Exercise Notes

Second-Language Learners
Ex. 5 Students learning English may not be familiar with the terms *gross income* and *net income.* If necessary, explain that net income is the total income left after taxes and other deductions have been taken out of the original income, which is also known as the gross income.

Application
Exs. 5–8, 25 These exercises show students some of the applications of quadratic functions to finance and biology.

Mathematical Procedures
Exs. 7, 8 There are many different ways to develop quadratic models for a set of data. These exercises use an approximate vertex and a spreadsheet to find the constant a from the vertex form $y = a(x - h)^2 + k$. Students should be able to see how this method is an extension of the method used in Chapter 2 to find direct variation models. In this case, $y - k$ varies directly with $(x - h)^2$ and a is the constant of variation.

Using Technology
Exs. 7, 8 You may want to suggest that students use the *Stats!* software for these exercises. The problem in Ex. 7 can also be done using the list features of a TI-82 graphing calculator. After entering the age and gross income data as lists L1 and L2, position the cursor over L3 at the top of the third column and enter L1−50. List L3 should now be identical to the first column of the spreadsheet. For L4, enter L2−300000, and for L5, enter L4/L3^2. A similar procedure can be used for Ex. 8.

Topic Spiraling: Preview
Exs. 7, 8 This is the first of several methods students will see for finding quadratic models. Ask them what they think are some advantages and disadvantages of this method.

Communication: Listening
Exs. 9–14 These exercises can be done orally in class. As one student answers, have the other students raise their hands if they disagree with the solution. Any disagreements can then be discussed.

197

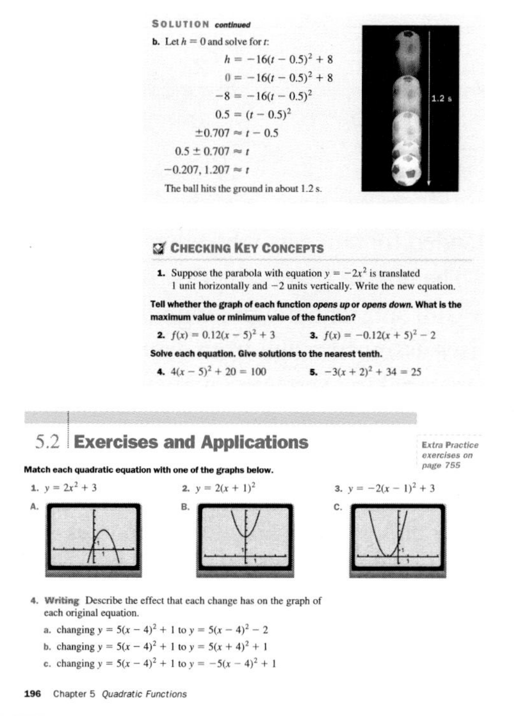

Teach⇔Interact

Section Note

Common Error
Students often think that when a is positive, the function has a maximum value. Encourage them to visualize the graph. When a is positive the graph curves up and thus has a minimum value.

Checking Key Concepts

Teaching Tip
Students having difficulty with questions 4 and 5 should be encouraged to write a journal entry listing the steps required to solve such equations. They can also use examples to illustrate the steps.

Closure Question
Describe the characteristics of the function $y = a(x - h)^2 + k$ and its graph. If a is positive, the function has a minimum value of k at $x = h$ and the graph opens up with vertex (h, k). If a is negative, the function has a maximum value of k at $x = h$ and the graph opens down with a vertex (h, k). In either case, the graph has an axis of symmetry $x = h$.

Apply⇔Assess

Suggested Assignment
Core Course
Exs. 1–4, 9–24, 27–36
Extended Course
Exs. 1–8, 9–19 odd, 21–36
Block Schedule
Day 28 Exs. 1–4, 9–24, 27–36

Exercise Notes

Communication: Writing
Ex. 4 For part (c), students should be able to explain why $y = -5(x - 4)^2 + 1$ is not merely a reflection of $y = 5(x - 4)^2 + 1$ across the x-axis but is instead a reflection followed by a translation. ($y = -f(x)$ is a reflection of $y = f(x)$ across the x-axis. The reflection of $y = 5(x - 4)^2 + 1$ is $y = -(5(x - 4)^2 + 1) = -5(x - 4)^2 - 1$. This curve is then translated up 2 units to get the desired curve.)

196

SOLUTION *continued*

b. Let $h = 0$ and solve for t:

$$h = -16(t - 0.5)^2 + 8$$
$$0 = -16(t - 0.5)^2 + 8$$
$$-8 = -16(t - 0.5)^2$$
$$0.5 = (t - 0.5)^2$$
$$\pm 0.707 \approx t - 0.5$$
$$0.5 \pm 0.707 \approx t$$
$$-0.207, 1.207 \approx t$$

The ball hits the ground in about 1.2 s.

1.2 s

CHECKING KEY CONCEPTS

1. Suppose the parabola with equation $y = -2x^2$ is translated 1 unit horizontally and −2 units vertically. Write the new equation.

Tell whether the graph of each function *opens up* or *opens down*. What is the maximum value or minimum value of the function?

2. $f(x) = 0.12(x - 5)^2 + 3$
3. $f(x) = -0.12(x + 5)^2 - 2$

Solve each equation. Give solutions to the nearest tenth.

4. $4(x - 5)^2 + 20 = 100$
5. $-3(x + 2)^2 + 34 = 25$

5.2 Exercises and Applications

Extra Practice exercises on page 755

Match each quadratic equation with one of the graphs below.

1. $y = 2x^2 + 3$
2. $y = 2(x + 1)^2$
3. $y = -2(x - 1)^2 + 3$

A. B. C.

4. **Writing** Describe the effect that each change has on the graph of each original equation.
a. changing $y = 5(x - 4)^2 + 1$ to $y = 5(x - 4)^2 - 2$
b. changing $y = 5(x - 4)^2 + 1$ to $y = 5(x + 4)^2 + 1$
c. changing $y = 5(x - 4)^2 + 1$ to $y = -5(x - 4)^2 + 1$

196 Chapter 5 *Quadratic Functions*

Checking Key Concepts
1. $y = -2(x - 1)^2 - 2$
2. up; minimum value: 3
3. down; maximum value: −2
4. 0.5, 9.5
5. −3.7, −0.3

Exercises and Applications
1. B
2. C
3. A

4. a. The graph is translated down 3 units.
b. The graph is translated 8 units to the left.
c. The graph is reflected over the x-axis, then translated up 2 units. (This may also be described as reflection over the line $y = 1$.)

5. a. Doctors' Incomes ● Gross ▲ Net

ANSWERS

Answers to Explorations, Think and Communicate questions, Checking Key Concept exercises, and Exercises and Applications are conveniently located at the bottom of each page.

In addition to the section side-column notes, a **Progress Check** is provided for each Assess Your Progress in the student book.

Pacing and Making Assignments

PACING CHART

A yearly Pacing Chart and daily assignments are provided for three courses—a core course, an extended course, and a block-scheduled course. The core and extended courses require 160 days, and the block-scheduled course requires 82 days. These time frames include days for using the Portfolio Projects and time for review and testing. The Pacing Chart below shows the number of days allotted for each of the three courses. Semester and trimester divisions are indicated by red and blue rules, respectively. For teachers who wish to include the chapters on probability and trigonometry, pacing charts utilizing Chapters 13-15 are provided on page T42 and in the *Course Guides* publication.

Chapter	1	2	3	4	5	6	7	8	9	10	11	12
Core Course	13	13	13	13	14	14	12	15	17	12	13	11
Extended Course	13	13	13	13	14	14	12	15	17	12	13	11
Block Schedule	7	$6\frac{1}{2}$	$6\frac{1}{2}$	$6\frac{1}{2}$	7	7	6	$7\frac{1}{2}$	$8\frac{1}{2}$	6	$7\frac{1}{2}$	6

trimester semester trimester

Core Course

The Core Course is intended for students who enter with typical mathematical and problem-solving skills. The course covers the first twelve chapters. The daily assignments provide students with substantial work with the skills and concepts presented in each lesson. The exercises assigned range from exercises that involve straightforward application of the new material to exercises involving higher-order thinking skills.

Extended Course

The Extended Course is intended for students who enter with strong mathematical and problem-solving skills and who are able to understand new concepts quickly. The course covers the first twelve chapters. The daily assignments include all material in the core course plus additional exercises that focus on higher-order thinking skills.

Block-Scheduled Course

The Block-Scheduled Course is intended for schools that use longer periods, typically 90-minute blocks, for instruction. The course covers the first twelve chapters. The exercises assigned range from exercises that involve straightforward application of the new material to exercises involving higher-order thinking skills. All material in the core course is included, plus some additional exercises requiring higher-order thinking skills.

Part of the Block-Scheduled Course for Chapter 5 is shown on the facing page. The entire chart for each chapter is located on the interleaved pages preceding the chapter.

PLANNING GUIDE

The Planning Guide for each chapter is located on the interleaved pages preceding the chapter. Part of the Planning Guide for Chapter 5 is shown here.

Regular Scheduling (45 min)

Section	Materials Needed	Core Assignment	Extended Assignment	*exercises that feature* Applications	Communication	Technology
5.1	graphing calculator, graph paper	1–4, 6–17, 24–26, 28–32	1–5, 7, 9, 11–32	19–25	18	20, 27
5.2	graphing calculator or software, graph paper	1–4, 9–24, 27–36	1–8, 9–19 odd, 21–36	5–8, 25	4, 5, 7, 27	7
5.3	graphing calculator or software, graph paper	1–10, 15, 17–28, AYP*	1–28, AYP	12–14	7, 11, 18	
5.4	algebra tiles, graph paper, graphing calculator	**Day 1:** 1–14, 17 **Day 2:** 25–47	**Day 1:** 1–24 **Day 2:** 25–47	14–23	15, 22	16

Applications

Each section contains exercises that relate the mathematics of that section to real-world applications. These exercises are usually assigned in the daily assignments and are listed in the Planning Guide under the *Applications* head.

Communication

Each section contains exercises that require students to communicate mathematically. These exercises have students discuss or write about the mathematical concepts presented in the section and are usually assigned in the daily assignments. These exercises are denoted by in-line heads in the Exercises and Applications sets and are listed in the Planning Guide under the *Communication* head.

Technology

Each chapter contains exercises that involve technology, usually graphing calculators, spreadsheets, or computer software. Technology-based exercises are usually assigned in the daily assignments. Exercises that require technology or are especially appropriate for using technology have a logo (shown in the chart above) beside them in the textbook. These exercises are listed in the Planning Guide under the *Technology* head.

Block Scheduling (90 min)

	Day 27	Day 28	Day 29	Day 30	Day 31	Day 32	Day 33	Day 34
Teach/Interact	Ch. 4 Test 5.1	5.2: Exploration, page 192 5.3: Exploration, page 199	5.4: Exploration, page 206	5.5	5.6: Exploration, page 221	Review Port. Proj.	Review Port. Proj.	Ch. 5 Test 6.1
Apply/Assess	**Ch. 4 Test** **5.1:** 1–4, 6–17, 24–26, 28–32	**5.2:** 1–4, 9–24 27–36 **5.3:** 1–10 15, 17–28, AYP*	**5.4:** 1–14, 17, 25–47	**5.5:** 1–23, 30, 31, 35–44	**5.6:** 1–10 12–21, 26–39, AYP	**Review:** 1–11 **Port. Proj.**	**Review:** 12–23 **Port. Proj.**	**Ch. 5 Test** **6.1:** 1–11, 13–17, 23–31

Pacing and Making Assignments continued ➤

Alternative Pacing Charts

The following alternative pacing charts are intended for use in programs that include trigonometry or have a greater emphasis on discrete math and data analysis. The first chart shows the pacing for a course that includes trigonometry and de-emphasizes discrete math. The second chart shows an alternative pacing of the Core curriculum which allows a chapter on probability to be included in the course.

Pacing for Algebra 2 with Trigonometry

Chapter	1	2	3	4	5	7	8	9	10	11	14	15	12 chapters
Days	**13**	**13**	**13**	**13**	**14**	**12**	**15**	**17**	**12**	**13**	**12**	**13**	**160 days**
Days in Core Course	**13**	**13**	**13**	**13**	**14**	**12**	**15**	**17**	**12**	**13**	**11**	**11**	**157 days**

In contrast to the Core course, the pacing guide for Algebra 2 with Trigonometry includes additional days to cover sections 14.5, 15.1, and 15.3.

Pacing for Algebra 2 with Data Analysis/Discrete Math Emphasis

Chapter	1	2	3	4	5	6	7	8	9	10	11	12	13	13 chapters
Days	**12**	**12**	**12**	**12**	**13**	**13**	**11**	**15**	**16**	**12**	**12**	**11**	**9**	**160 days**
Days in Core Course	**13**	**13**	**13**	**13**	**14**	**14**	**12**	**15**	**17**	**12**	**13**	**11**	**9**	**169 days**

In contrast to the Core course, the added empahsis on data analysis and discrete math may require that each of the following lessons be completed in one day: 1.4, 2.4, Chapter 3 review, Chapter 4 review, 5.5, 6.6, 7.5, Chapter 9 review, and 11.1.

Data Analysis and Discrete Math Alternatives

Teachers who wish to emphasize data analysis and discrete math in the second half of the course may decide to position Chapter 6 (*Investigating Data*) after Chapter 9 (*Polynomial and Rational Functions*).

If a teacher chooses to spend more time on data analysis and discrete math than shown in the pacing chart above, it is suggested that some parts of Chapter 9 (*Polynomial and Rational Functions*) or Chapter 11 (*Analytic Geometry*) be omitted.

Explorations and Applications

Authors

Senior Authors

Miriam A. Leiva Richard G. Brown

Loring Coes III
Shirley Frazier Cooper
Joan Ferrini-Mundy
Amy T. Herman
Patrick W. Hopfensperger
Celia Lazarski
Stuart J. Murphy
Anthony Scott
Marvin S. Weingarden

McDougal Littell
A HOUGHTON MIFFLIN COMPANY
Evanston, Illinois ◆ Boston ◆ Dallas

Authors

Senior Authors

Richard G. Brown Mathematics Teacher, Phillips Exeter Academy, Exeter, New Hampshire

Miriam A. Leiva Cone Distinguished Professor for Teaching and Professor of Mathematics, University of North Carolina at Charlotte

Loring Coes III Chair of the Mathematics Department, Rocky Hill School, E. Greenwich, Rhode Island

Shirley Frazier Cooper Curriculum Specialist, Secondary Mathematics, Dayton Public Schools, Dayton, Ohio

Joan Ferrini-Mundy Professor of Mathematics and Mathematics Education, University of New Hampshire, Durham, New Hampshire

Amy T. Herman Mathematics Teacher, Atherton High School, Louisville, Kentucky

Patrick W. Hopfensperger Mathematics Teacher, Homestead High School, Mequon, Wisconsin

Celia Lazarski Mathematics Teacher, Glenbard North High School, Carol Stream, Illinois

Stuart J. Murphy Visual Learning Specialist, Evanston, Illinois

Anthony Scott Assistant Principal, Orr Community Academy, Chicago, Illinois

Marvin S. Weingarden Supervisor of Secondary Mathematics, Detroit Public Schools, Detroit, Michigan

Editorial Advisors

Martha A. Brown Mathematics Supervisor, Prince George's County Public Schools, Capitol Heights, Maryland

Diana Garcia Mathematics Supervisor, Laredo Independent School District, Laredo, Texas

Sue Ann McGraw Mathematics Teacher, Lake Oswego High School, Lake Oswego, Oregon

Editorial Advisors helped plan the concept, teaching approach, and format of the book. They also reviewed draft manuscripts.

ISBN: 0-395-86298-1   123456789—VH—01 00 99 98 97

Review Panel

Judy B. Basara Curriculum Chair, St. Hubert's High School, Philadelphia, Pennsylviania

Dane Camp Mathematics Teacher, New Trier High School, Winnetka, Illinois

Pamela W. Coffield Mathematics Teacher, Brookstone School, Columbus, Georgia

Kathleen Curran Mathematics Teacher, Ball High School, Galveston, Texas

Randy Harter Mathematics Specialist, Buncombe County Schools, Asheville, North Carolina

William Leonard Assistant Director of Mathematics, Fresno Unified School District, Fresno, California

Betty McDaniel Coordinator of Mathematics, Florence School District, Florence, South Carolina

Roger O'Brien Mathematics Supervisor, Polk County Schools, Bartow, Florida

Leo Ramirez Mathematics Teacher, McAllen High School, McAllen, Texas

Michelle Rohr Director of Mathematics, Houston Independent School District, Houston, Texas

May Samuels Mathematics Chairperson, Weequahic High School, Newark, New Jersey

Betty Takesuye Mathematics Teacher, Chaparral High School, Scottsdale, Arizona

Members of the review panel read and commented upon outlines, sample lessons, and research versions of the chapters.

Manuscript Reviewers

Dane Camp Mathematics Teacher, New Trier High School, Winnetka, Illinois

Pamela W. Coffield Mathematics Teacher, Brookstone School, Columbus, Georgia

Ravi Kamat Mathematics Teacher, Roosevelt High School, Dallas, Texas

William Leonard Assistant Director of Mathematics, Fresno Unified School District, Fresno California

Joseph E. Montaño Mathematics Teacher, Martin High School, Laredo, Texas

Betty Takesuye Mathematics Teacher, Chaparral High School, Scottsdale, Arizona

Straight Line Editorial Development, Inc. Editorial Consultants, San Francisco, California

Manuscript Reviewers read and reacted to draft manuscript, focusing on its effectiveness from a teaching/learning viewpoint.

Student Advisors

Iruma Bello, Henry H. Filer Middle School, Hialeah, FL; Leslie Blaha, Walnut Springs Middle School, Westerville, OH; Tai-Ling Bloomfield, Alyeska Central School, Anchorage, AK; Gabriel Bonilla, G.W. Carver Middle School, Miami, FL; Anne Burke, Nichols School, Buffalo, NY; Eric de Armas, G.W. Carver Middle School, Miami, FL; Beth Donaldson, Bettendorf Middle School, Bettendorf, IA; Megan Foreman, Allen High School, Allen, TX; Clifton Gray, Riverside University High School, Milwaukee, WI; Colleen Kelly, Chaparral High School, Scottsdale, AZ; Tony Liberati, Hampton Middle School, Gibsonia, PA; Christian Maiden, Hathaway Brown School, Shaker Heights, OH; J.P. Marshall, Lincoln High School, Tallahassee, FL; Jeff Phi, Irving Junior High School, Kansas City, MO; Jaret Radford, Fondren Middle School, Houston, TX; Marisa A. Sharer, Chaparral High School, Scottsdale, AZ; Scott Terrill, Swartz Creek School, Swartz Creek, MI; Gabriela Zúñiga, San Marcos High School, San Marcos, TX.

Acknowledgment
The authors wish to thank Pamela W. Coffield and her students at Brookstone School in Columbus, Georgia, for using and providing comments on the Portfolio Projects in this book. Ms. Coffield's students are Kate Baker, Bo Bickerstaff, Lizzie Bowles, Bradford Carmack, Lucy Cartledge, Henry Dunn, Patrick Graffagnino, Charles Haines, Daniel McFall, Sravanthi Meka, Blake Melton, Jason Pease, Ann Phillips, Dorsey Staples, and Jeffrey Usman.

Contents

CHAPTER 1

Modeling Using Algebra

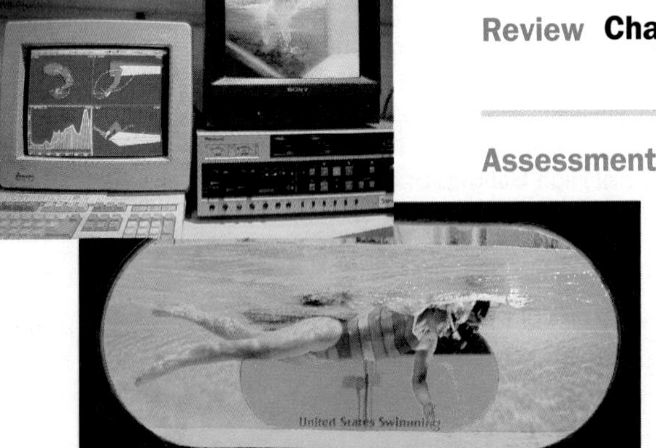

Applications

Additional applications include: business, manufacturing, industry, sports, entertainment, personal finance

CHAPTER

2 Linear Functions

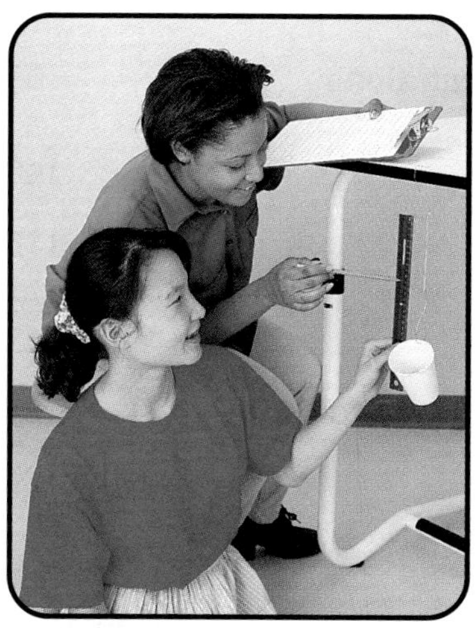

Applications

CHAPTER 3

Exponential Functions

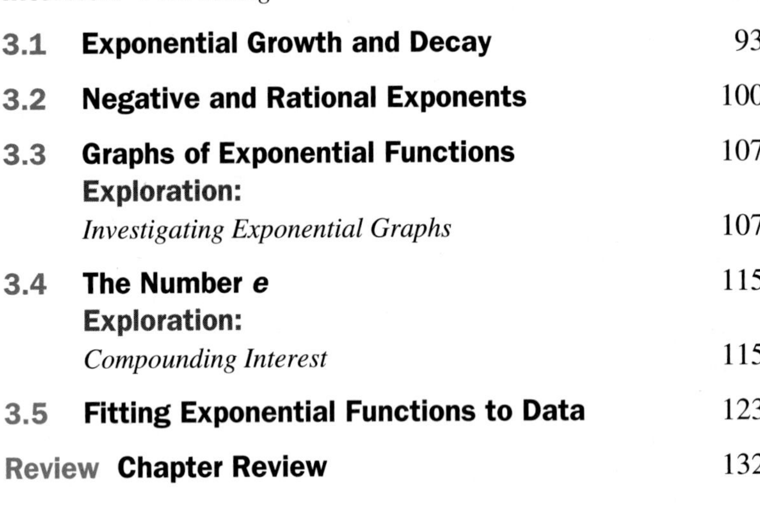

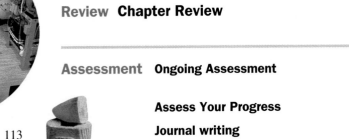

Applications

Additional applications include: mythology, geometry, banking, physiology, chemistry, biology, psychology, astronomy

CHAPTER 4

Logarithmic Functions

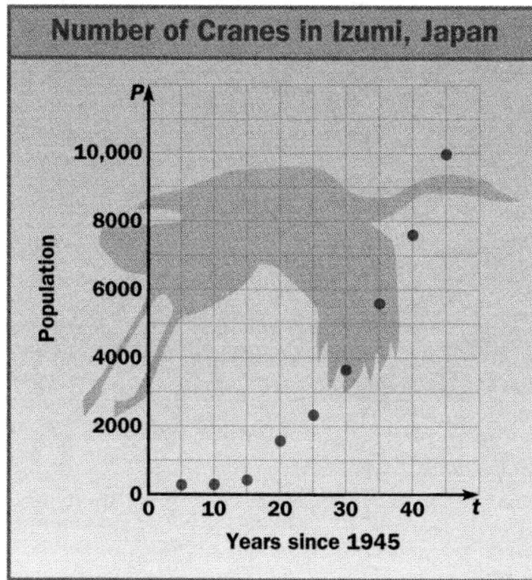

Applications

Additional applications include: business, forestry, recreation, mountain climbing, population, earth science, cooking, personal finance, ecology

CHAPTER

5 Quadratic Functions

Applications

CHAPTER

6

Investigating Data

6.3 *Archaeological sampling* 254

Applications

Interview:
Donna Cox
Computer Visualization Specialist
255, 271

Additional applications include: medicine, market research, government, economics, transportation, business, earth science, consumer economics

CHAPTER

7 Systems

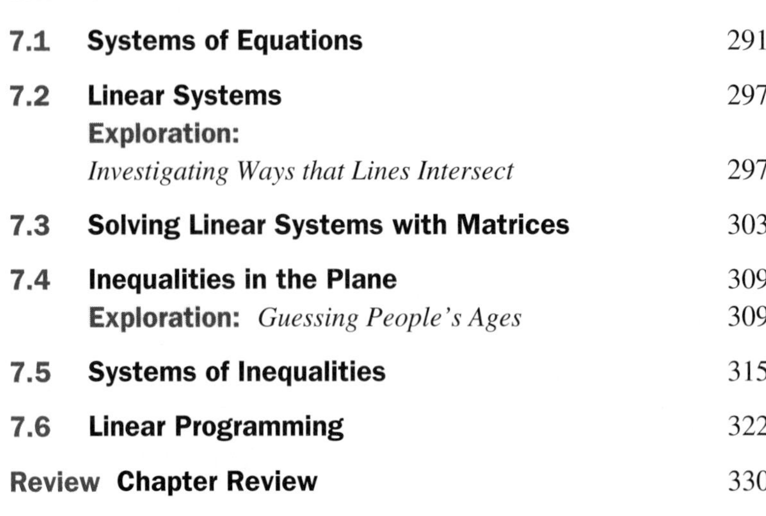

7.5 *Comparing desert climates* 315

Applications

Interview:
Gina Oliva
Fitness Director
299, 316, 319, 326

Connection

Additional applications include: personal finance, advertising, social science, recreation, consumer economics, finance, cooking, art, cartography, manufacturing, medicine, geology, fitness

CHAPTER

8 Radical Functions and Number Systems

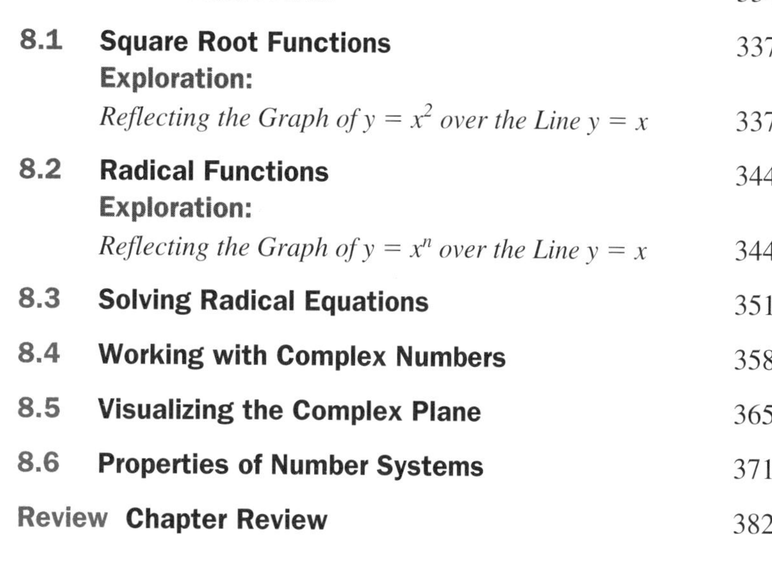

Applications

Additional applications include: biology, geometry, astronomy, wind chill

CHAPTER

9 Polynomial and Rational Functions

Planet	Mean distance from sun (millions of miles)	Apparent size of sun
Mercury	36.0	2.58
Venus	67.2	1.38
Earth	93.0	1.00
Mars	142.0	0.655
Jupiter	484.0	0.192
Saturn	885.0	0.105
Uranus	1780.0	0.0522
Neptune	2790.0	0.0333
Pluto	3660.0	0.0254

Applications

Interview:
Vera Rubin
Astronomer
430, 446

Additional applications include: biology, psychology, physiology, space travel, education, metallurgy

CHAPTER

10

Sequences and Series

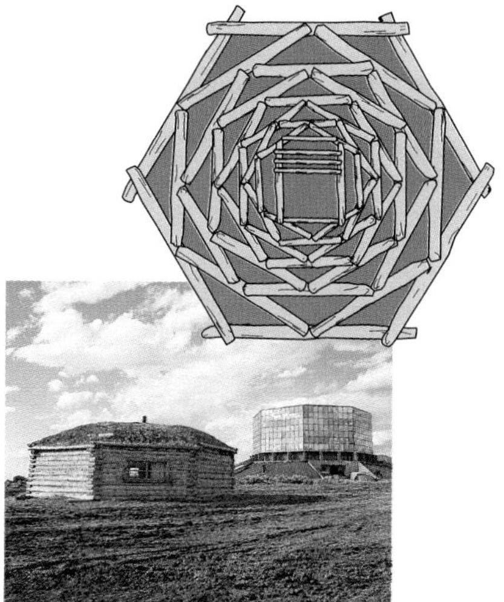

10.2 *Building hogans* 478

Applications

Interview:
Jhane Barnes
Textile Designer
471, 479, 502

Additional applications include: chemistry, biology, architecture, medicine, language arts, physics, economics, genealogy

CHAPTER

11

Analytic Geometry

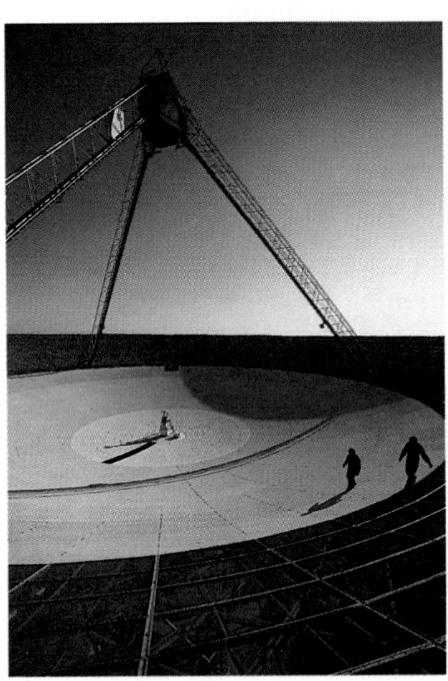

11.2 *Parabolic radio telescope* 525

Applications

Interview:
Kija Kim
Cartographer
517, 532

Additional applications include: architecture, geometry, flag making, astronomy, carpentry, language arts

CHAPTER 12

Discrete Mathematics

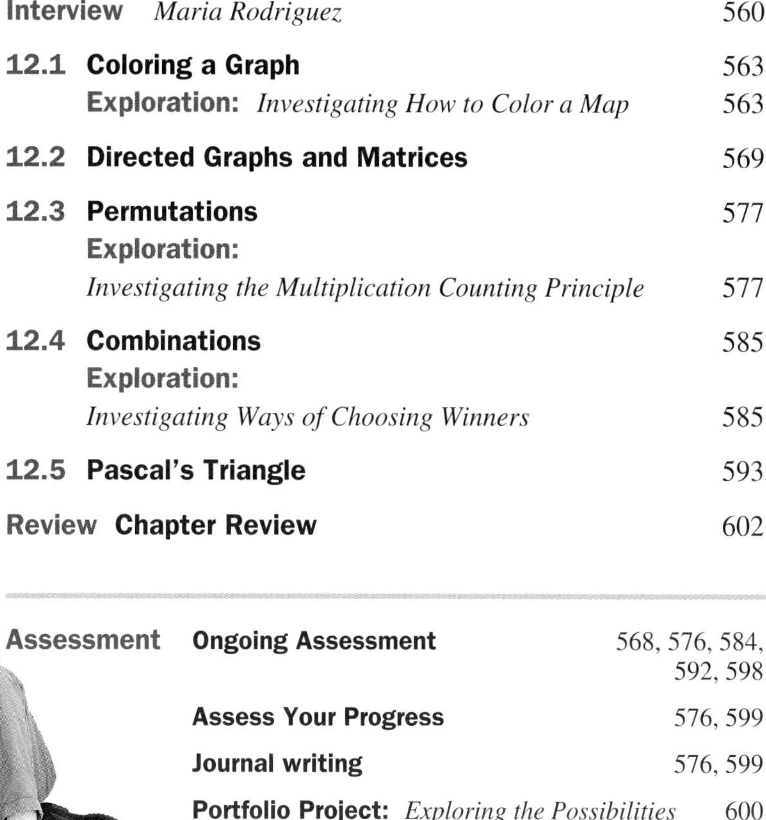

12.5 *Investigating Pascal's triangle* 593

Applications

Interview:
Maria Rodriguez
Electrical Engineer
574, 582

Connection

Marine Biology	567	**Sports**	589
Sports	575	**Games**	591
Genetics	584	**History**	597

Additional applications include: cartography, scheduling, chemistry, ecology, medicine, business, fashion, forensics, music, literature, hobbies, manufacturing, computers, advertising, fundraising, industry, recreation, government

CHAPTER

13 Probability

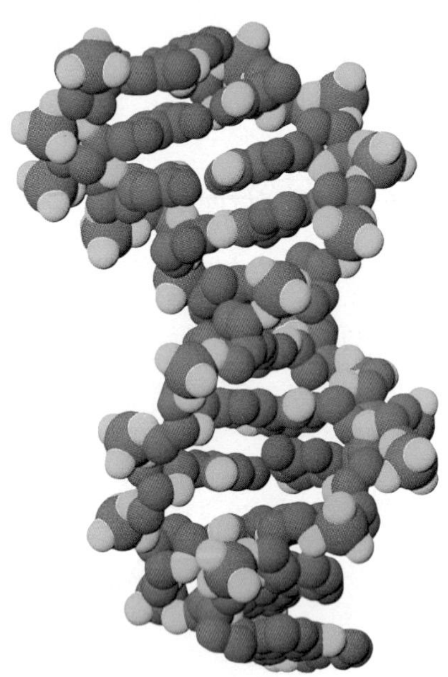

Applications

Additional applications include: sports, astronomy, manufacturing, physics, meteorology, movies, electronics, health, market research, architecture, driver's education, biology, agriculture

CHAPTER

14

Triangle Trigonometry

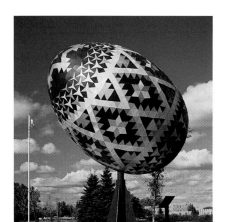

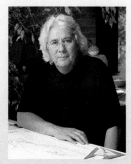

CHAPTER

15

Trigonometric Functions

Applications

Additional applications include: amusement park rides, geometry, automotive mechanics, music, hot air ballooning, astronomy, engineering

Student Resources

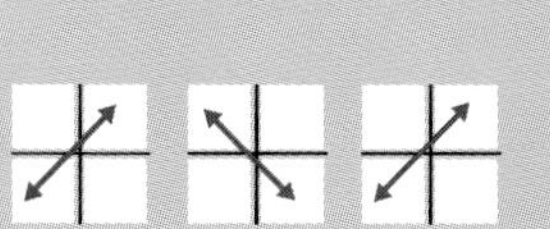
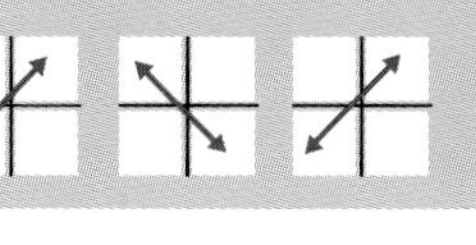

About the Interviews

Using Mathematics in Careers

Each chapter of this book starts with a personal interview with someone who uses mathematics in his or her career. You may be surprised by the wide range of careers that are included. These are the people you will be reading about:

- **Business Owner** ***Twlya Lang***
- **Underwater Photographer** ***Norbert Wu***
- **Aquarium Designer** ***Finn Strong***
- **Archaeologist** ***Ednaly Ortiz***
- **Automotive Engineer** ***Mark Thomas***
- **Computer Visualization Specialist** ***Donna Cox***
- **Fitness Director** ***Gina Oliva***
- **Roller Coaster Designer** ***Ronald Toomer***
- **Astronomer** ***Vera Rubin***
- **Textile Designer** ***Jhane Barnes***
- **Cartographer** ***Kija Kim***
- **Electrical Engineer** ***Maria Rodriguez***
- **Lawyer** ***Robert Ward***
- **Zoo Designer** ***Johnpaul Jones***
- **Record Producer** ***Joe Lopez***

Ednaly Ortiz

Mark Thomas

Jhane Barnes

xx

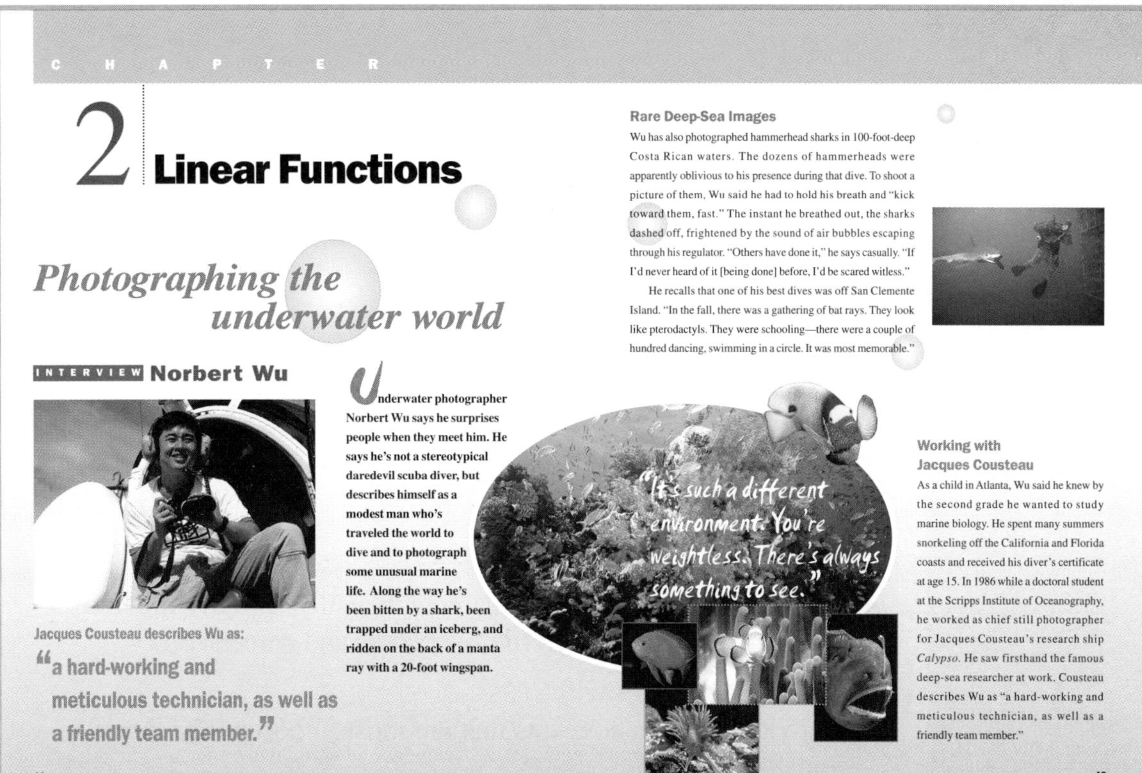

CHAPTER

2 Linear Functions

Photographing the underwater world

INTERVIEW **Norbert Wu**

Jacques Cousteau describes Wu as:

"a hard-working and meticulous technician, as well as a friendly team member."

Underwater photographer Norbert Wu says he surprises people when they meet him. He says he's not a stereotypical daredevil scuba diver, but describes himself as a modest man who's traveled the world to dive and to photograph some unusual marine life. Along the way he's been bitten by a shark, been trapped under an iceberg, and ridden on the back of a manta ray with a 20-foot wingspan.

"It's such a different environment. You're weightless. There's always something to see."

Rare Deep-Sea Images

Wu has also photographed hammerhead sharks in 100-foot-deep Costa Rican waters. The dozens of hammerheads were apparently oblivious to his presence during that dive. To shoot a picture of them, Wu said he had to hold his breath and "kick toward them, fast." The instant he breathed out, the sharks dashed off, frightened by the sound of air bubbles escaping through his regulator. "Others have done it," he says casually. "If I'd never heard of it [being done] before, I'd be scared witless."

He recalls that one of his best dives was off San Clemente Island. "In the fall, there was a gathering of bat rays. They look like pterodactyls. They were schooling—there were a couple of hundred dancing, swimming in a circle. It was most memorable."

Working with Jacques Cousteau

As a child in Atlanta, Wu said he knew by the second grade he wanted to study marine biology. He spent many summers snorkeling off the California and Florida coasts and received his diver's certificate at age 15. In 1986 while a doctoral student at the Scripps Institute of Oceanography, he worked as chief still photographer for Jacques Cousteau's research ship *Calypso*. He saw firsthand the famous deep-sea researcher at work. Cousteau describes Wu as "a hard-working and meticulous technician, as well as a friendly team member."

42 43

Applying What You've Learned

At the end of each interview, there are ***Explore and Connect*** questions that guide you in learning more about the career being discussed. Some of these questions involve research that is done outside of class. In addition, in each chapter there are ***Related Examples and Exercises*** that show how the mathematics you are learning is directly related to the career highlighted in the interview.

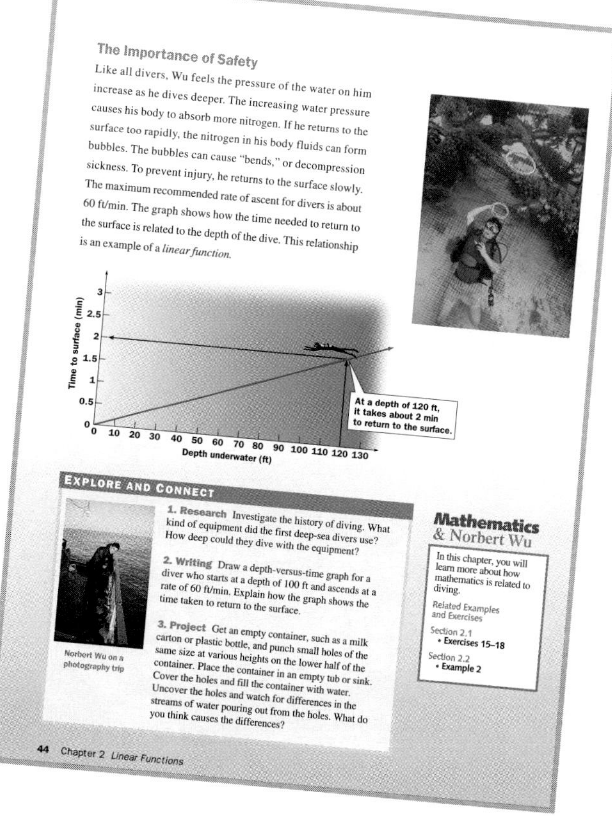

The Importance of Safety

Like all divers, Wu feels the pressure of the water on him increase as he dives deeper. The increasing water pressure causes his body to absorb more nitrogen. If he returns to the surface too rapidly, the nitrogen in his body fluids can form bubbles. The bubbles can cause "bends," or decompression sickness. To prevent injury, he returns to the surface slowly. The maximum recommended rate of ascent for divers is about 60 ft/min. The graph shows how the time needed to return to the surface is related to the depth of the dive. This relationship is an example of a *linear function*.

EXPLORE AND CONNECT

Norbert Wu on a photography trip

1. **Research** Investigate the history of diving. What kind of equipment did the first deep-sea divers use? How deep could they dive with the equipment?

2. **Writing** Draw a depth-versus-time graph for a diver who starts at a depth of 100 ft and ascends at a rate of 60 ft/min. Explain how the graph shows the time taken to return to the surface.

3. **Project** Get an empty container, such as a milk carton or plastic bottle, and punch small holes of the same size at various heights on the lower half of the container. Place the container in an empty tub or sink. Cover the holes and fill the container with water. Uncover the holes and watch for differences in the streams of water pouring out from the holes. What do you think causes the differences?

Mathematics & Norbert Wu

In this chapter, you will learn more about how mathematics is related to diving.

Related Examples and Exercises

Section 2.1
- Exercises 15–18

Section 2.2
- Example 2

44 Chapter 2 Linear Functions

xxi

Welcome to Algebra 2

Explorations and Applications

Goals of The Course

This book will help you use mathematics in your daily life and prepare you for success in future courses and careers.

In this course you will:

- Study the algebra concepts that are most important for today's students
- Apply these concepts to solve many different types of problems
- Learn how calculators and computers can help you find solutions

You will have a chance to develop your skills in:

- Reasoning and problem solving
- Communicating orally and in writing
- Studying and learning independently and as a team member

xxii

MATHEMATICAL CONTENT

This contemporary algebra course gives you a strong background in the types of mathematical reasoning and problem solving that will be important in your future.

The book emphasizes:

- Using functions, equations, and graphs to model problem situations
- Investigating connections of algebra to geometry, statistics, probability, and discrete mathematics

ACTIVE LEARNING

To learn algebra successfully, you need to get involved!

There will be many opportunities in this course for you to participate in:

- Explorations of mathematical concepts
- Cooperative learning activities
- Small-group and whole-class discussions

So don't sit back and be a spectator. If you join in and share your ideas, everyone will learn more.

Course Overview

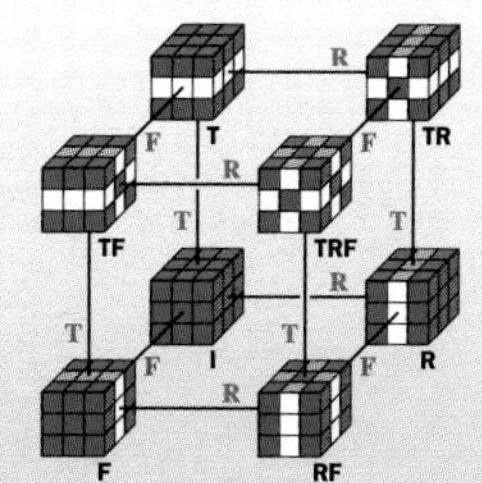

To get an overview of your course, turn to pages xxiv–xxxiii to see some of the types of problems you will solve and topics you will explore.

xxiii

COURSE OVERVIEW

"What does algebra have to do with me?"

Applications and Connections

Algebra is about you and the world around you.

In this course you'll learn how algebra can help answer many different types of questions in daily life and in careers.

Section

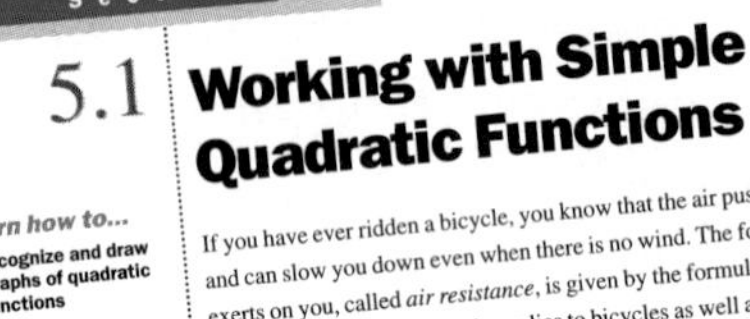

5.1 Working with Simple Quadratic Functions

Learn how to...

- **recognize and draw graphs of quadratic functions**
- **solve simple quadratic equations**

So you can...

- **examine the relationship between air resistance and speed when cycling, for example**

If you have ever ridden a bicycle, you know that the air pushes against you and can slow you down even when there is no wind. The force that the air exerts on you, called *air resistance*, is given by the formula on the preceding page. (The formula applies to bicycles as well as cars.)

EXAMPLE 1 Application: Bicycling

Shirley Scott is cycling on a day with no wind. She is in a casual, upright position on her bicycle. In this situation, the drag coefficient is 1.1 and the frontal area is 5.5 ft^2. Write an equation that gives the air resistance R (in pounds) as a function of the bicycle's speed s (in miles per hour). Graph the function.

SOLUTION

Use the formula shown on the preceding page.

$$R = 0.00256(1.1)(5.5)s^2$$
$$= 0.0155s^2$$

Substitute 1.1 for C_D and 5.5 for frontal area.

Make a table of values, plot the points, and connect them with a smooth curve.

$R = 0.0155s^2$	
s	R
0	0
4	0.248
8	0.992
12	2.232
16	3.968
20	6.200

THINK AND COMMUNICATE

1. What happens to the air resistance when Shirley Scott doubles her speed?
2. What would you say is a reasonable domain for the air resistance function? What is the corresponding range?

5.1 Working with Simple Quadratic Functions 185

Recreation

How does increasing your speed on a bicycle affect air resistance?

(Chapter 5, page 185)

Business Management

How can past sales growth be used to estimate future sales of Kente cloth scarves?

(Chapter 1, page 6)

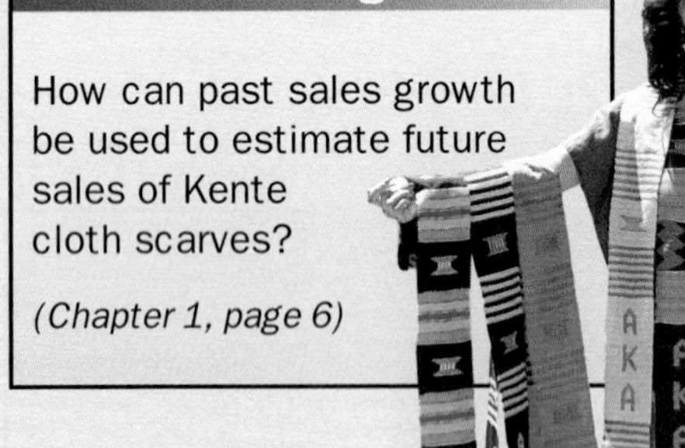

Year	Scarf sales
1992	100
1993	350
1994	650

xxiv

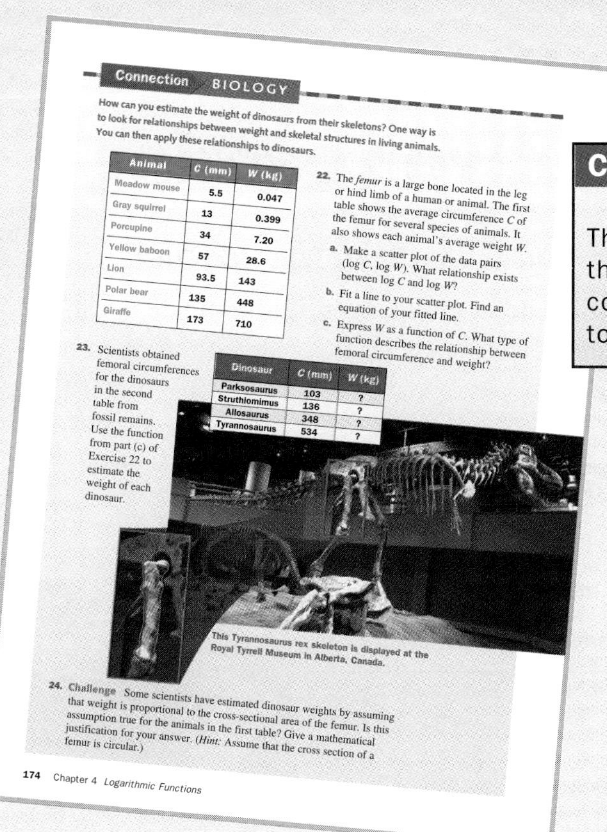

Connection BIOLOGY

How can you estimate the weight of dinosaurs from their skeletons? One way is to look for relationships between weight and skeletal structures in living animals. You can then apply these relationships to dinosaurs.

Animal	C (mm)	W (kg)
Meadow mouse	5.5	0.047
Gray squirrel	13	0.399
Porcupine	34	7.20
Yellow baboon	57	28.6
Lion	93.5	143
Polar bear	135	448
Giraffe	173	710

22. The *femur* is a large bone located in the leg or hind limb of a human or animal. The first table shows the average circumference C of the femur for several species of animals. It also shows each animal's average weight W.
    a. Make a scatter plot of the data pairs $(\log C, \log W)$. What relationship exists between $\log C$ and $\log W$?
    b. Fit a line to your scatter plot. Find an equation of your fitted line.
    c. Express W as a function of C. What type of function describes the relationship between femoral circumference and weight?

23. Scientists obtained femoral circumferences for the dinosaurs in the second table from fossil remains. Use the function from part (c) of Exercise 22 to estimate the weight of each dinosaur.

Dinosaur	C (mm)	W (kg)
Parksosaurus	103	?
Struthiomimus	136	?
Allosaurus	348	?
Tyrannosaurus	534	?

This Tyrannosaurus rex skeleton is displayed at the Royal Tyrrell Museum in Alberta, Canada.

24. **Challenge** Some scientists have estimated dinosaur weights by assuming that weight is proportional to the cross-sectional area of the femur. Is this assumption true for the animals in the first table? Give a mathematical justification for your answer. (*Hint:* Assume that the cross section of a femur is circular.)

174 Chapter 4 *Logarithmic Functions*

Connections Exercises

These clusters of exercises, which appear throughout each chapter, focus on the connections of algebra to a particular topic, career, or branch of mathematics.

Oceanography

What is the water pressure 35,800 ft below the surface of the ocean?

(Chapter 2, page 49)

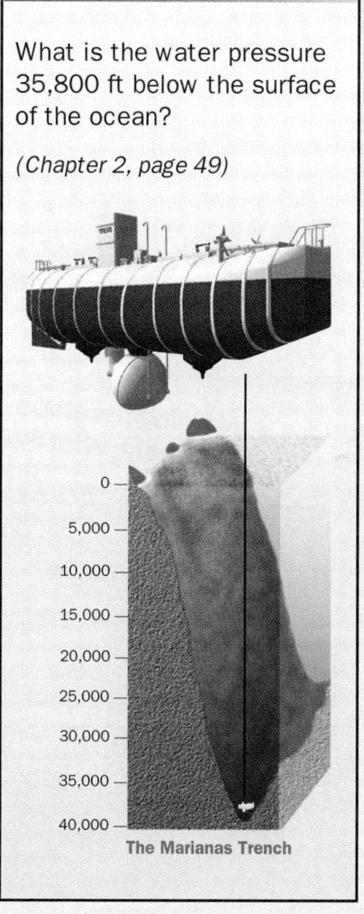

Earth Science

How did the height of the Mississippi River change during a 1993 flood?

(Chapter 2, page 69)

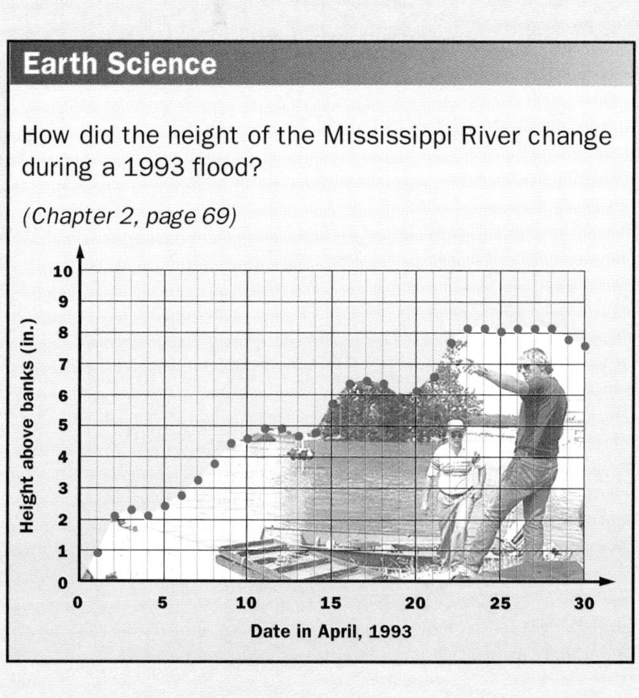

xxv

COURSE OVERVIEW

“Do we just sit back and listen?”

Explorations and Cooperative Learning

In this course you’ll be an active learner.

Working individually and in groups, you’ll investigate questions and then present and discuss your results. Here are some of the topics you’ll explore.

Drawing a Parabola

How is a point on a parabola related to the focus and the directrix?

(Chapter 11, page 520)

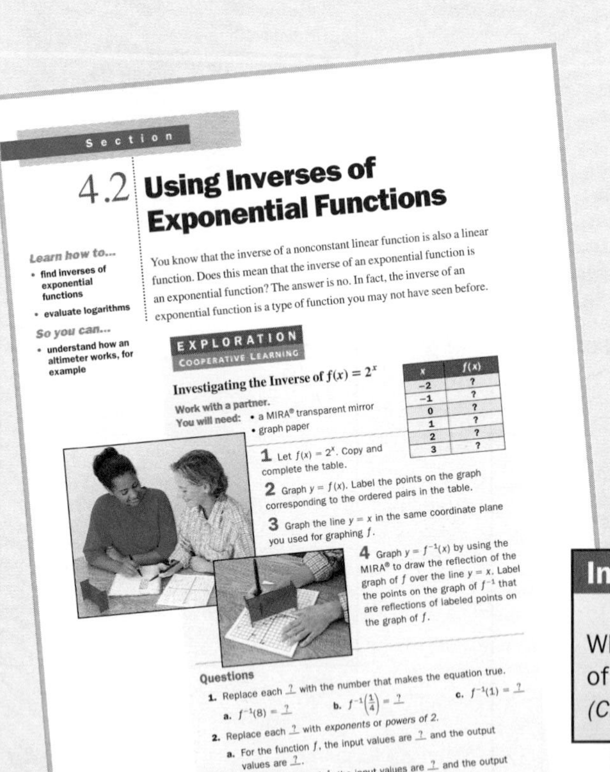

Section

4.2 **Using Inverses of Exponential Functions**

Learn how to...
- find inverses of exponential functions
- evaluate logarithms

So you can...
- understand how an altimeter works, for example

You know that the inverse of a nonconstant linear function is also a linear function. Does this mean that the inverse of an exponential function is an exponential function? The answer is no. In fact, the inverse of an exponential function is a type of function you may not have seen before.

EXPLORATION
COOPERATIVE LEARNING

Investigating the Inverse of $f(x) = 2^x$

Work with a partner.
You will need: • a MIRA® transparent mirror • graph paper

x	$f(x)$
−2	?
−1	?
0	?
1	?
2	?
3	?

1 Let $f(x) = 2^x$. Copy and complete the table.

2 Graph $y = f(x)$. Label the points on the graph corresponding to the ordered pairs in the table.

3 Graph the line $y = x$ in the same coordinate plane you used for graphing f.

4 Graph $y = f^{-1}(x)$ by using the MIRA® to draw the reflection of the graph of f over the line $y = x$. Label the points on the graph of f^{-1} that are reflections of labeled points on the graph of f.

Questions

1. Replace each ? with the number that makes the equation true.
   a. $f^{-1}(8) = ?$ b. $f^{-1}\left(\frac{1}{4}\right) = ?$ c. $f^{-1}(1) = ?$
2. Replace each ? with exponents or powers of 2.
   a. For the function f, the input values are ? and the output values are ?.
   b. For the function f^{-1}, the input values are ? and the output values are ?.

148 Chapter 4 *Logarithmic Functions*

Investigating Inverse Functions

What type of function is the inverse of an exponential function?
(Chapter 4, page 148)

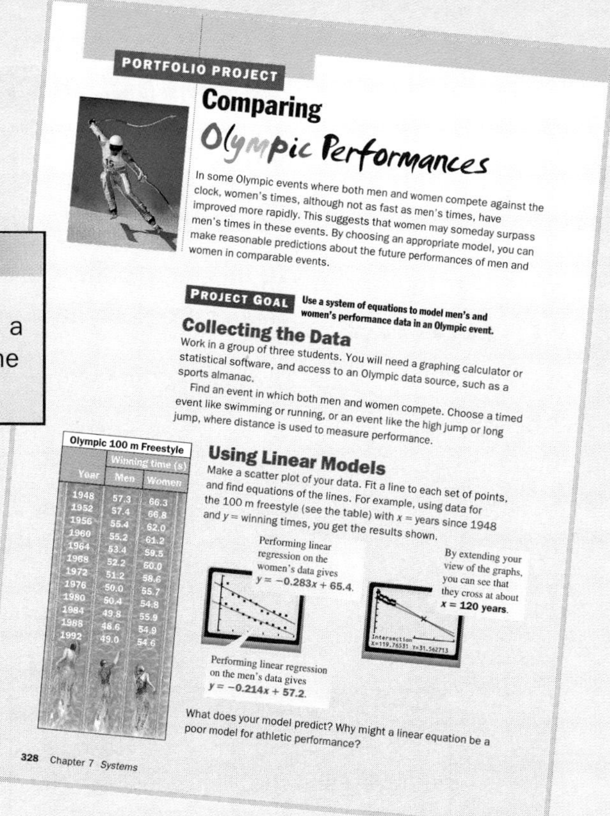

PORTFOLIO PROJECT

Comparing Olympic Performances

In some Olympic events where both men and women compete against the clock, women's times, although not as fast as men's times, have improved more rapidly. This suggests that women may someday surpass men's times in these events. By choosing an appropriate model, you can make reasonable predictions about the future performances of men and women in comparable events.

PROJECT GOAL Use a system of equations to model men's and women's performance data in an Olympic event.

Collecting the Data

Work in a group of three students. You will need a graphing calculator or statistical software, and access to an Olympic data source, such as a sports almanac.

Find an event in which both men and women compete. Choose a timed event like swimming or running, or an event like the high jump or long jump, where distance is used to measure performance.

Olympic 100 m Freestyle

Year	Winning time (s)	
	Men	Women
1948	57.3	66.3
1952	57.4	66.8
1956	55.4	62.0
1960	55.2	61.2
1964	53.4	59.5
1968	52.2	60.0
1972	51.2	58.6
1976	50.0	55.7
1980	50.4	54.8
1984	49.8	55.9
1988	48.6	54.9
1992	49.0	54.6

Using Linear Models

Make a scatter plot of your data. Fit a line to each set of points, and find equations of the lines. For example, using data for the 100 m freestyle (see the table) with x = years since 1948 and y = winning times, you get the results shown.

Performing linear regression on the women's data gives $y = -0.283x + 65.4$.

Performing linear regression on the men's data gives $y = -0.214x + 57.2$.

By extending your view of the graphs, you can see that they cross at about $x = 120$ years.

What does your model predict? Why might a linear equation be a poor model for athletic performance?

328 Chapter 7 Systems

Portfolio Projects

These open-ended projects give you a chance to explore applications of the topics you have studied.

Finding an Exponential Model

Can you use an equation to model the temperature of cooling water?

(Chapter 4, page 167)

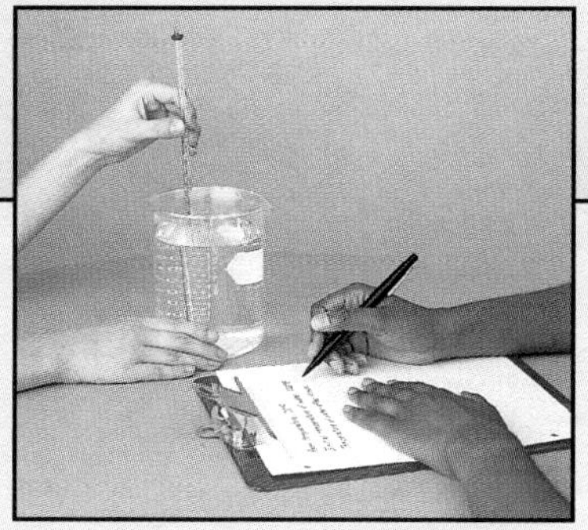

Modeling an Infinite Geometric Series

Is it possible for an infinite geometric series to have a sum?

(Chapter 10, page 497)

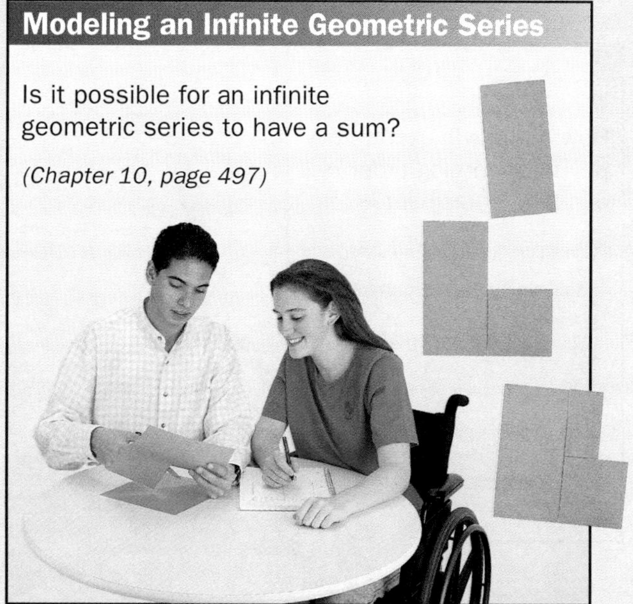

xxvii

"How can I visualize the problem?"

Using Technology

Calculators and computers can help you see mathematical relationships.

In this course there are many opportunities to use technology to model problem situations, identify patterns, and find solutions.

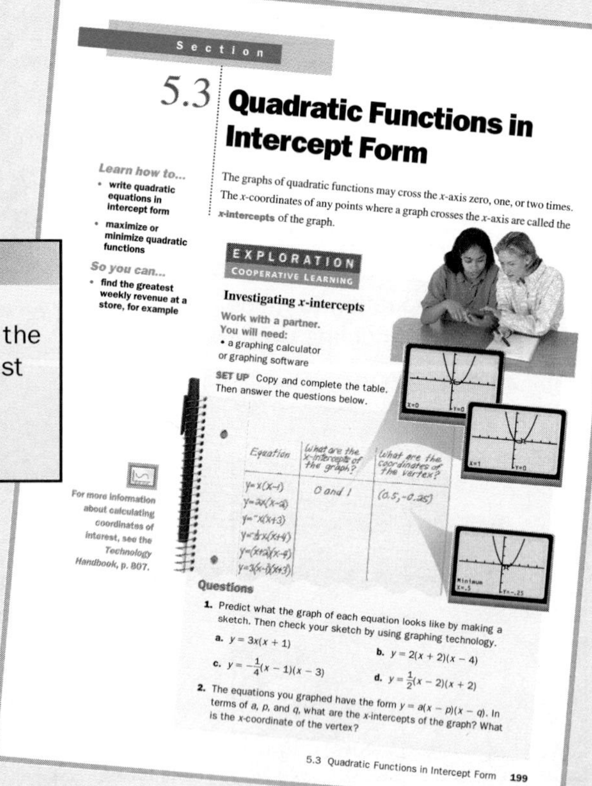

Graphing Calculator

Can you find the x-intercepts of the graph of a quadratic equation just by looking at the equation?

(Chapter 5, page 199)

Matrix Operations

How much will each family pay for a catalog order of camping equipment?

(Chapter 1, pages 24-25)

	Tents	Sleeping bags	Back-packs	Cook sets
Graham	1	4	2	0
Piscitelli	0	2	3	1
Brewer	2	5	4	2

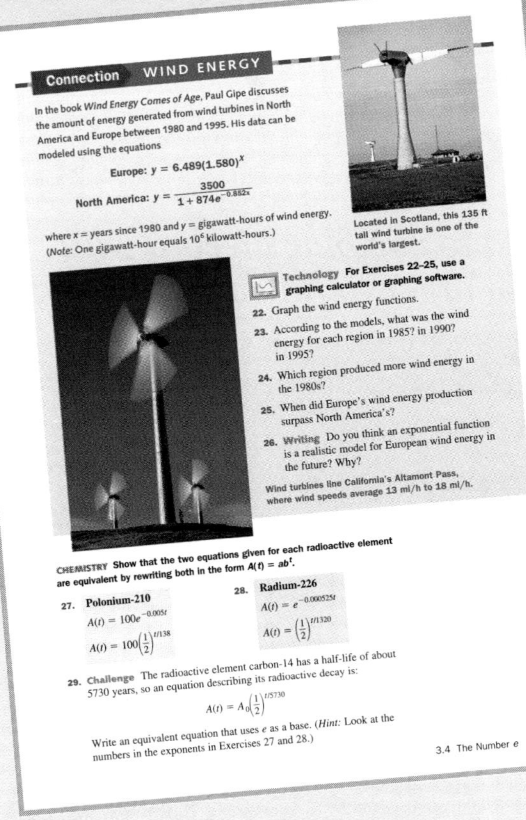

Connection WIND ENERGY

In the book *Wind Energy Comes of Age*, Paul Gipe discusses the amount of energy generated from wind turbines in North America and Europe between 1980 and 1995. His data can be modeled using the equations

$$\text{Europe: } y = 6.489(1.580)^x$$

$$\text{North America: } y = \frac{3500}{1 + 874e^{-0.852x}}$$

where x = years since 1980 and y = gigawatt-hours of wind energy. (*Note:* One gigawatt-hour equals 10^6 kilowatt-hours.)

Located in Scotland, this 135 ft tall wind turbine is one of the world's largest.

Technology For Exercises 22–25, use a graphing calculator or graphing software.

22. Graph the wind energy functions.
23. According to the models, what was the wind energy for each region in 1985? in 1990? in 1995?
24. Which region produced more wind energy in the 1980s?
25. When did Europe's wind energy production surpass North America's?
26. **Writing** Do you think an exponential function is a realistic model for European wind energy in the future? Why?

Wind turbines line California's Altamont Pass, where wind speeds average 13 mi/h to 18 mi/h.

CHEMISTRY Show that the two equations given for each radioactive element are equivalent by rewriting both in the form $A(t) = ab^t$.

27. **Polonium-210**
$A(t) = 100e^{-0.005t}$
$A(t) = 100\left(\frac{1}{2}\right)^{t/138}$

28. **Radium-226**
$A(t) = e^{-0.000525t}$
$A(t) = \left(\frac{1}{2}\right)^{t/1320}$

29. **Challenge** The radioactive element carbon-14 has a half-life of about 5730 years, so an equation describing its radioactive decay is:

$$A(t) = A_0\left(\frac{1}{2}\right)^{t/5730}$$

Write an equivalent equation that uses e as a base. (*Hint:* Look at the numbers in the exponents in Exercises 27 and 28.)

3.4 The Number e 121

Technology Exercises

In these exercises you will be using graphing technology or spreadsheets to practice, apply, and extend what you have learned.

Spreadsheet

How much should you save each year for college, assuming that the cost of college continues to grow?

(Chapter 10, page 505)

College Planning

	A	B	C
1	Year	College cost	Savings
2	1	24876	1200
3	2	B2*1.059	C2*1.1+120

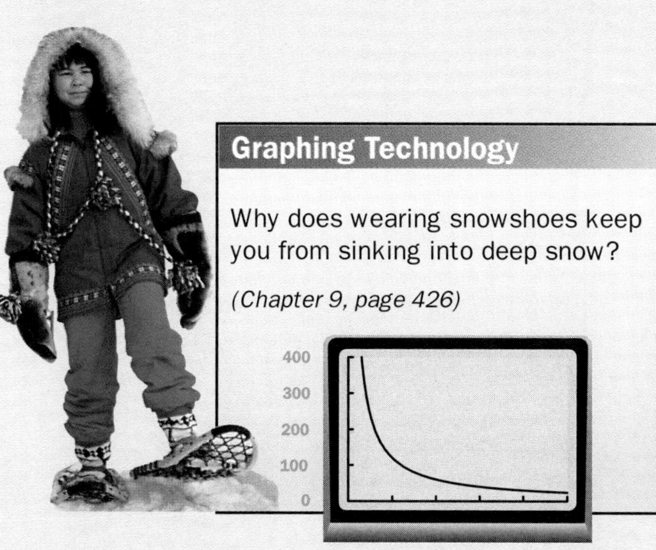

Graphing Technology

Why does wearing snowshoes keep you from sinking into deep snow?

(Chapter 9, page 426)

xxix

"Can I solve this problem with algebra?"

Integrating Math Topics

Sometimes you need to combine algebra with other math topics in order to find a solution.

In this course you'll see how you can solve problems by integrating algebra with geometry, statistics, probability, and discrete mathematics.

Discrete Mathematics

How is food energy transferred through an ecosystem?

(Chapter 12, page 569)

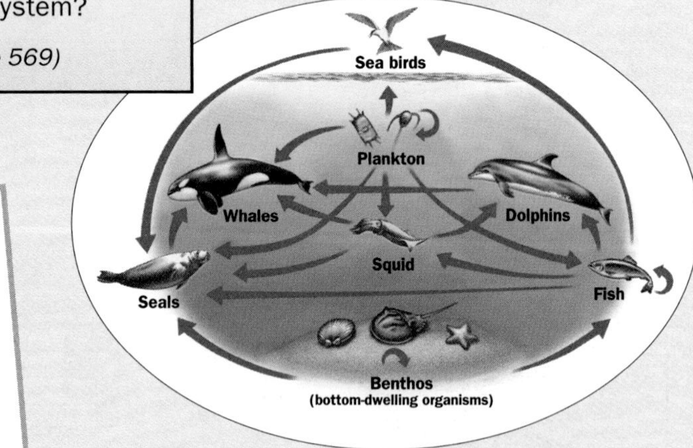

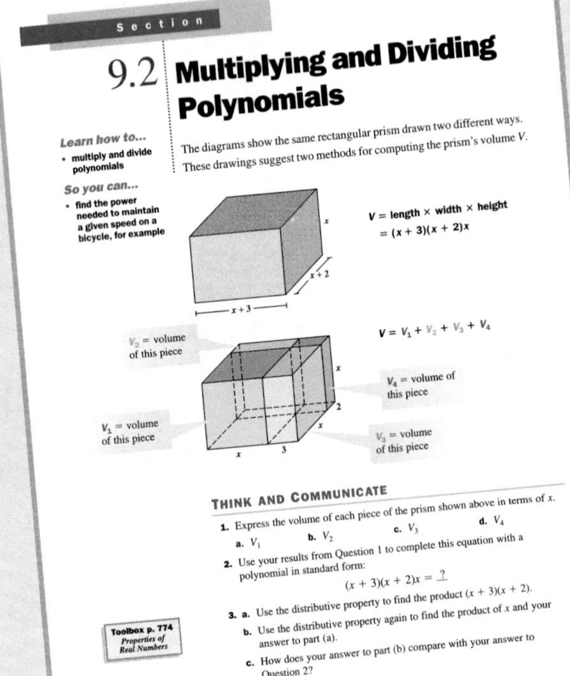

Section

9.2 Multiplying and Dividing Polynomials

Learn how to...

- multiply and divide polynomials

So you can...

- find the power needed to maintain a given speed on a bicycle, for example

The diagrams show the same rectangular prism drawn two different ways. These drawings suggest two methods for computing the prism's volume V.

THINK AND COMMUNICATE

1. Express the volume of each piece of the prism shown above in terms of x.
   a. V_1 **b.** V_2 **c.** V_3 **d.** V_4
2. Use your results from Question 1 to complete this equation with a polynomial in standard form:
   $$(x + 3)(x + 2)x = \underline{?}$$
3. **a.** Use the distributive property to find the product $(x + 3)(x + 2)$.
   b. Use the distributive property again to find the product of x and your answer to part (a).
   c. How does your answer to part (b) compare with your answer to Question 2?

Toolbox p. 774 *Properties of Real Numbers*

9.2 Multiplying and Dividing Polynomials **397**

Geometry

How can thinking about the volume of a rectangular prism help you understand multiplying polynomials?

(Chapter 9, page 397)

xxx

Probability

What is the probability of a complete match between fingerprints from two randomly chosen people?

(Chapter 13, page 630)

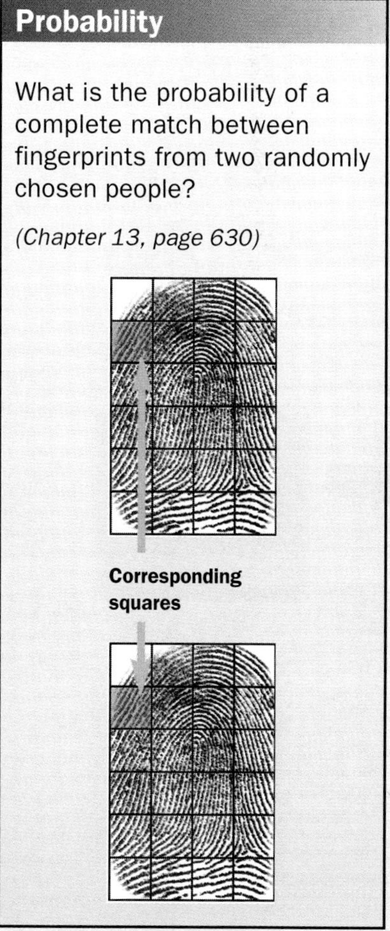

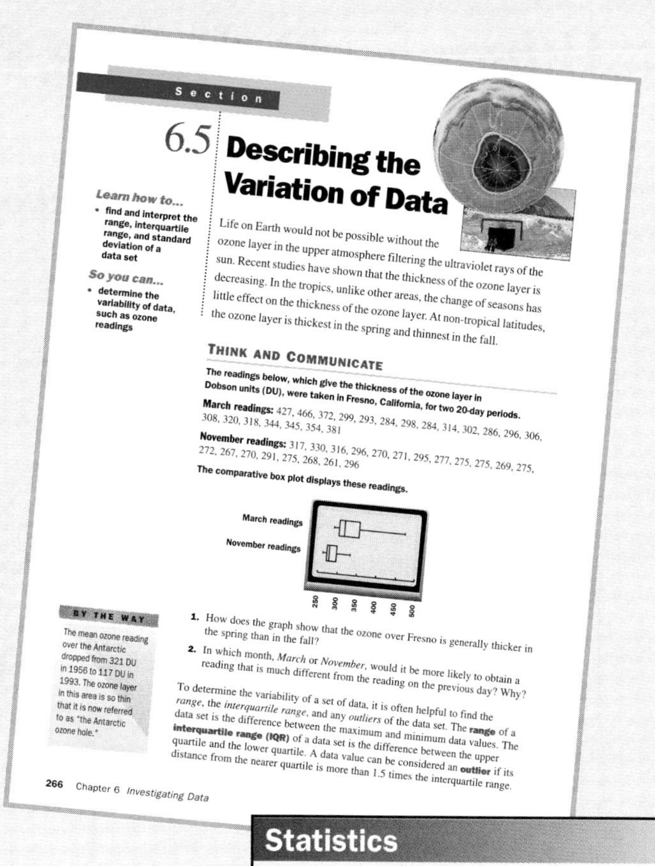

Section

6.5 **Describing the Variation of Data**

Learn how to...

- **find and interpret the range, interquartile range, and standard deviation of a data set**

So you can...

- **determine the variability of data, such as ozone readings**

Life on Earth would not be possible without the ozone layer in the upper atmosphere filtering the ultraviolet rays of the sun. Recent studies have shown that the thickness of the ozone layer is decreasing. In the tropics, unlike other areas, the change of seasons has little effect on the thickness of the ozone layer. At non-tropical latitudes, the ozone layer is thickest in the spring and thinnest in the fall.

THINK AND COMMUNICATE

The readings below, which give the thickness of the ozone layer in Dobson units (DU), were taken in Fresno, California, for two 20-day periods.

March readings: 427, 466, 372, 299, 293, 284, 298, 284, 314, 302, 286, 296, 306, 308, 320, 318, 344, 345, 354, 381

November readings: 317, 330, 316, 296, 270, 271, 295, 277, 275, 275, 269, 275, 272, 267, 270, 291, 275, 268, 261, 296

The comparative box plot displays these readings.

BY THE WAY

The mean ozone reading over the Antarctic dropped from 321 DU in 1956 to 117 DU in 1993. The ozone layer in this area is so thin that it is now referred to as "the Antarctic ozone hole."

1. How does the graph show that the ozone over Fresno is generally thicker in the spring than in the fall?
2. In which month, *March* or *November*, would it be more likely to obtain a reading that is much different from the reading on the previous day? Why?

To determine the variability of a set of data, it is often helpful to find the *range*, the *interquartile range*, and any *outliers* of the data set. The **range** of a data set is the difference between the maximum and minimum data values. The **interquartile range (IQR)** of a data set is the difference between the upper quartile and the lower quartile. A data value can be considered an **outlier** if its distance from the nearer quartile is more than 1.5 times the interquartile range.

266 Chapter 6 *Investigating Data*

Statistics

How do the changing seasons affect the thickness of the ozone layer?

(Chapter 6, page 266)

Analytic Geometry

What curves are produced by slicing through a double cone?

(Chapter 11, page 550)

xxxi

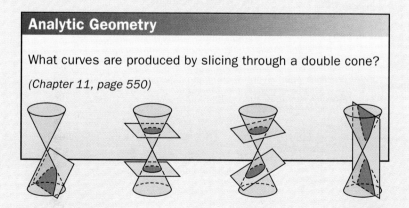

"When will I ever use this?"

Building for the Future

The skills you'll learn in this course will form a strong foundation for the future.

They'll prepare you for more advanced courses and increase your career opportunities.

Problem Solving and Communication

The exercise sets help you develop your problem solving and communication skills.

3.5 Exercises and Applications

Extra Practice exercises on page 753

Write an exponential function whose graph passes through each pair of points.

1. (0, 5), (1, 8) **2.** (1, 8), (2, 10) **3.** (0, 10), (1, 8) **4.** (5, 7), (6, 6)

5. (3, 3), (8, 7) **6.** (0, 6), (7, 9) **7.** (20, 4), (25, 3) **8.** (0, 17), (7, 15)

9. ASTRONOMY The *apparent magnitude* of a star is a measure of its apparent brightness. A star of magnitude 1 is 100 times brighter than a star of magnitude 6.

a. Find an equation of the exponential decay function whose graph passes through (1, 100) and (6, 1).

b. Graph the function from part (a).

c. Copy and complete the table showing comparative brightness values for various sky objects.

Sky object	Apparent magnitude	Brightness value
Uranus	6	1
Aldebaran	1	100
Vega	0	?
Sirius	−1.5	?
Full moon	−12.5	?
Sun	−26.7	?

d. **Research** Use a book about astronomy to find the names and magnitudes of some stars. Plot them on your graph from part (b). Calculate their brightness values and add them to your table from part (c).

10. SAT/ACT Preview Which of the equations represents an exponential decay function whose graph passes through (2, 1)?

A. $y = (0.5)^x$ **B.** $y = (0.5x)^2$ **C.** $y = \frac{1}{4}(0.5)^{-x}$ **D.** $y = 4(0.5)^x$ **E.** $y = \frac{1}{4} \cdot 2^x$

11. Open-ended Problem The table shows the increase in the number of host sites on the Internet computer network.

Year	1988	1989	1990	1991	1992	1993	1994
Years since 1988	0	1	2	3	4	5	6
Internet hosts (1000s)	56	159	313	617	1136	2056	3864

a. Choose any two years and write an exponential function to represent the growth.

b. Check your function by seeing how its predictions compare to the other data values in the table.

c. **Technology** Enter all the data into a graphing calculator or statistical software and find an exponential equation that gives a good fit. Compare the results to your results from part (a).

126 Chapter 3 *Exponential Functions*

Open-ended Problems test your problem solving skills.

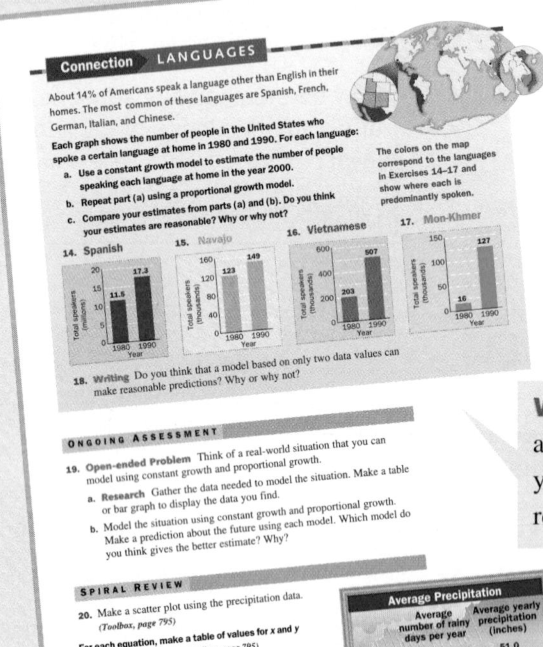

Connection LANGUAGES

About 14% of Americans speak a language other than English in their homes. The most common of these languages are Spanish, French, German, Italian, and Chinese.

Each graph shows the number of people in the United States who spoke a certain language at home in 1980 and 1990. For each language:

a. **Use a constant growth model to estimate the number of people speaking each language at home in the year 2000.**

b. **Repeat part (a) using a proportional growth model.**

c. **Compare your estimates from parts (a) and (b). Do you think your estimates are reasonable? Why or why not?**

The colors on the map correspond to the languages in Exercises 14–17 and show where each is predominantly spoken.

14. Spanish **15. Navajo** **16. Vietnamese** **17. Mon-Khmer**

18. Writing Do you think that a model based on only two data values can make reasonable predictions? Why or why not?

Writing questions ask you to express your mathematical reasoning in words.

ONGOING ASSESSMENT

19. Open-ended Problem Think of a real-world situation that you can model using constant growth and proportional growth.

a. **Research** Gather the data needed to model the situation. Make a table or bar graph to display the data you find.

b. Model the situation using constant growth and proportional growth. Make a prediction about the future using each model. Which model do you think gives the better estimate? Why?

SPIRAL REVIEW

20. Make a scatter plot using the precipitation data. *(Toolbox, page 795)*

For each equation, make a table of values for x and y when $x = 0, 1, 2,$ and 3. *(Toolbox, page 795)*

21. $y = -3x + 6$ **22.** $y = 4.5(2^x)$

Simplify. Give answers with the appropriate number of significant digits. *(Toolbox, page 789)*

23. $3.2 + 0.15$ **24.** 42.1×3.6

Average Precipitation	Average number of rainy days per year	Average yearly precipitation (inches)
Olympia, WA	164	51.0
Honolulu, HI	100	23.5
Charleston, WV	151	42.4
Madison, WI	118	30.8

8 Chapter 1 *Modeling Using Algebra*

xxxii

Career Preview

Throughout the exercise sets there are problems that show how math is used in a variety of different careers.

Acoustical Engineering, page 160

Cartography, page 512

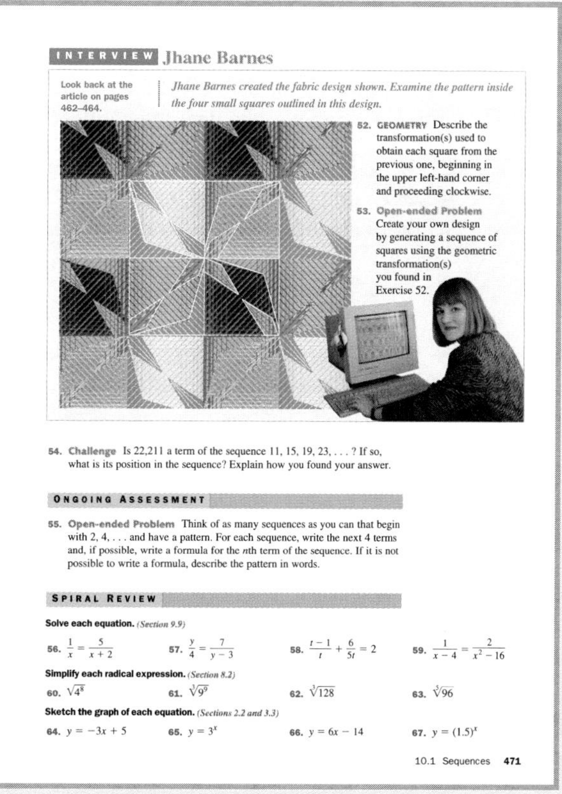

INTERVIEW Jhane Barnes

Look back at the article on pages 462–464.

Jhane Barnes created the fabric design shown. Examine the pattern inside the four small squares outlined in this design.

52. **GEOMETRY** Describe the transformation(s) used to obtain each square from the previous one, beginning in the upper left-hand corner and proceeding clockwise.

53. **Open-ended Problem** Create your own design by generating a sequence of squares using the geometric transformation(s) you found in Exercise 52.

54. **Challenge** Is 22,211 a term of the sequence 11, 15, 19, 23, . . . ? If so, what is its position in the sequence? Explain how you found your answer.

ONGOING ASSESSMENT

55. **Open-ended Problem** Think of as many sequences as you can that begin with 2, 4, . . . and have a pattern. For each sequence, write the next 4 terms and, if possible, write a formula for the nth term of the sequence. If it is not possible to write a formula, describe the pattern in words.

SPIRAL REVIEW

Solve each equation. *(Section 9.9)*

56. $\frac{1}{x} = \frac{5}{x+2}$ 57. $\frac{y}{4} = \frac{7}{y-3}$ 58. $\frac{t-1}{t} + \frac{6}{5t} = 2$ 59. $\frac{1}{x-4} = \frac{2}{x^2-16}$

Simplify each radical expression. *(Section 8.2)*

60. $\sqrt{4^8}$ 61. $\sqrt[3]{9^9}$ 62. $\sqrt[3]{128}$ 63. $\sqrt[5]{96}$

Sketch the graph of each equation. *(Sections 2.2 and 3.3)*

64. $y = -3x + 5$ 65. $y = 3^x$ 66. $y = 6x - 14$ 67. $y = (1.5)^x$

10.1 Sequences 471

Textile Design, page 471

xxxiii

CHAPTER

1 Modeling Using Algebra

OVERVIEW

Connecting to Prior and Future Learning

⇔ In Chapter 1, students review their knowledge of graphs, tables, functions, and matrices as they use these concepts to model growth. Students will find the review of percent change on page 775 and mean, median, and mode on page 790 of the **Student Resources Toolbox** helpful as they complete their work.

⇔ Scalar multiplication and multiplying matrices are introduced in this chapter. Students will use these skills again when they study systems of equations in Chapter 7. Students may find it helpful to review ratios and proportions on page 785 of the **Students Resources Toolbox.**

⇔ This chapter also includes an introduction to using a simulation to model a situation. Students learn how they can use the results of a simulation to make predictions. The concept of making predictions is used again in Chapter 6.

Chapter Highlights

Interview with Twyla Lang: The use of mathematics in making business decisions is highlighted in this interview, with related exercises on pages 6 and 26.

Explorations in Chapter 1 involve organizing data in matrices in Section 1.3 and simulating a coupon giveaway in Section 1.5.

The Portfolio Project: Students do an experiment to collect data about how their ability to make baskets in basketball changes with distance. They use both linear and exponential functions to model the data they collect.

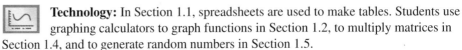

Technology: In Section 1.1, spreadsheets are used to make tables. Students use graphing calculators to graph functions in Section 1.2, to multiply matrices in Section 1.4, and to generate random numbers in Section 1.5.

OBJECTIVES

Section	Objectives	NCTM Standards
1.1	• Model real-world situations with tables and graphs. • Compare different models describing the same situation. • Make predictions about future trends.	1, 2, 3, 4
1.2	• Write a function to model a set of data. • Make predictions about situations using functions.	1, 2, 3, 4, 5, 6
1.3	• Organize information in a matrix. • Add matrices. • Multiply a matrix by a scalar. • Use matrices to model data.	1, 2, 3, 4, 5, 12
1.4	• Multiply matrices. • Solve problems involving matrix multiplication.	1, 2, 3, 4, 5, 12
1.5	• Model a situation with a simulation. • Make predictions using simulations.	1, 2, 3, 4

INTEGRATION

Mathematical Connections	1.1	1.2	1.3	1.4	1.5
algebra	**3–8***	**9–15**	**16–21**	**22–28**	**29–35**
geometry			20		
data analysis, probability, discrete math	**3–8**	**9–15**	**16–21**	**22–28**	**29–35**
patterns and functions		**9–15**	21		
logic and reasoning	5, 6, 8	9, 11, 13–15	17–19, 21	26–28	29, 31, 33–35
Interdisciplinary Connections and Applications					
history and geography		14			
reading and language arts	8				
business and economics	3, 4				
arts and entertainment		15	21		33
sports and recreation		12, 13		22, 24, 27	
education, manufacturing, industry, agriculture, personal finance	7	10–12	19, 20		

* ***Bold page numbers*** *indicate that a topic is used throughout the section.*

TECHNOLOGY

Section	opportunities for use with Student Book	Support Material
1.1	scientific calculator spreadsheet software	**Stats! Activity Book:** Activity 11
1.2	scientific calculator graphing calculator McDougal Littell Mathpack *Function Investigator*	**Technology Book:** Calculator Activity 1 Spreadsheet Activity 1
1.3	graphing calculator McDougal Littell Mathpack *Matrix Analyzer*	**Stats! Activity Book:** Activity 3
1.4	scientific calculator graphing calculator McDougal Littell Mathpack *Matrix Analyzer*	**Technology Book:** Spreadsheet Activity 1 **Function Investigator with Matrix Analyzer Activity Book:** Matrix Analyzer Activity 1
1.5	scientific calculator graphing calculator	

Planning Guide

Regular Scheduling (45 min)

Section	Materials Needed	Core Assignment	Extended Assignment	*exercises that feature* Applications	Communication	Technology
1.1	scientific calculator, spreadsheet software, graph paper	1–5, 7–13, 19–24	1–24	2, 3, 7–18	4, 5, 18, 19	
1.2	scientific calculator, graphing calculator or graphing software	1–8, 10–12, 16–21, AYP*	1–21, AYP	1–12, 14, 15	9, 11–13, 16	10, 13, 14
1.3	graph paper, graphing calculator	**Day 1:** 1–18 **Day 2:** 19–25, 28–41	**Day 1:** 1–18 **Day 2:** 19–41	12–15 26, 30–32	3 29	
1.4	scientific calculator, graphing calculator or matrix software	**Day 1:** 1–16 **Day 2:** 23–27, 29–41	**Day 1:** 1–16 **Day 2:** 17–41	14, 15 17–22	16 18, 22, 29	23
1.5	dice, coins, scientific calculator, graphing calculator with random number generator, graph paper	**Day 1:** 1–9 **Day 2:** 12, 16–23, AYP	**Day 1:** 1–11 **Day 2:** 12–23, AYP	9–11	10, 11 16	12–15
Review/ Assess		**Day 1:** 1–7 **Day 2:** 8–15 **Day 3:** Ch. 1 Test	**Day 1:** 1–7 **Day 2:** 8–15 **Day 3:** Ch. 1 Test	4 14	3	
Portfolio Project		Allow 2 days.	Allow 2 days.			

Yearly Pacing (with Portfolio Project)	**Chapter 1 Total** 13 days	**Remaining** 147 days	**Total** 160 days

Block Scheduling (90 min)

	Day 1	Day 2	Day 3	Day 4	Day 5	Day 6	Day 7
Teach/Interact	1.1	1.2 1.3: Exploration, page 16	Continue with 1.3 1.4	Continue with 1.4 1.5: Exploration, page 29	Continue with 1.5 Review	Review Port. Proj.	Port. Proj. Ch. 1 Test
Apply/Assess	**1.1:** 1–5, 7–13, 19–24	**1.2:** 1–8, 10–12, 16–21, AYP* **1.3:** 1–18	**1.3:** 19–25, 28–41 **1.4:** 1–16	**1.4:** 23–27 29–41 **1.5:** 1–9	**1.5:** 12, 16–23, AYP **Review:** 1–7	**Review:** 8–15 **Port. Proj.**	**Port. Proj.** **Ch. 1 Test**

NOTE: A one-day block has been added for the Portfolio Project—timing and placement to be determined by teacher.

Yearly Pacing (with Portfolio Project)	**Chapter 1 Total** 7 days	**Remaining** 75 days	**Total** 82 days

* ***AYP*** *is Assess Your Progress.*

Lesson Support

Section	Practice Bank	Study Guide*	Assessment Book*	Visuals	Explorations Lab Manual	Lesson Plans	Technology Book
1.1	1	1.1		Warm-Up 1.1	Master 1	1.1	
1.2	2	1.2	Test 1	Warm-Up 1.2		1.2	Calculator Act. 1 Spreadsheet Act. 1
1.3	3	1.3		Warm-Up 1.3	Master 2	1.3	
1.4	4	1.4		Warm-Up 1.4 Folder 1	Add. Expl. 1	1.4	Spreadsheet Act. 1
1.5	5	1.5	Test 2	Warm-Up 1.5	Master 1	1.5	
Review Test	6	Chapter Review	Tests 3, 4 Alternative Assessment			Review Test	

***Spanish versions** of *Study Guide* and *Assessment Book* are available.

Chapter Support

- Course Guide
- Lesson Plans
- Portfolio Project Book:
  Additional Project 1:
  Exploring Algebraic Concepts
- Preparing for College Entrance Tests
- Multi-Language Glossary
- *Test Generator* Software
- Professional Handbook

Software Support

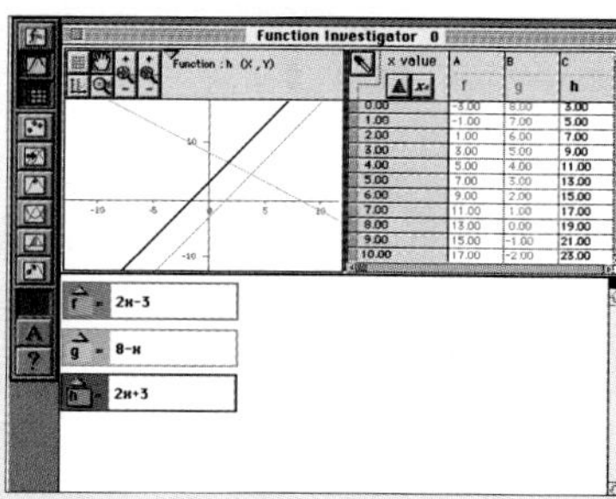

McDougal Littell Mathpack
Function Investigator
Matrix Analyzer

Internet Support

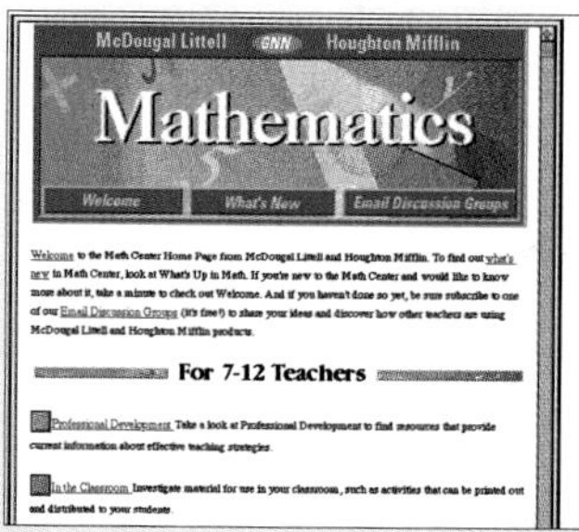

http://www.hmco.com
Next go to McDougal Littell; then the Education Center; then Secondary Math.

Outside Resources

Books, Periodicals

Schielack, Vincent P., Jr. "The Football Coach's Dilemma: Should We Go for 1 or 2 Points First?" *Mathematics Teacher* (December 1995): pp. 731–733.

"Exploring Data," from the *Quantitative Literacy Series*; written by members of the Joint Committee on the Curriculum on Statistics and Probability of the American Statistical Association and the National Council of Teachers of Mathematics. Dale Seymour Publications.

Ballew, Hunter. "Sherlock Holmes, Master Problem Solver." *Mathematics Teacher* (November 1994): pp. 596–601.

"The Art and Techniques of Simulation," from the *Quantitative Literacy Series.* Dale Seymour Publications.

Activities, Manipulatives

Hollowell, Kathleen A. "The Case of the Blue Wooden Flower." *Mathematics Teacher* (May 1995): pp. 366–370.

Nord, Gail D. and John Nord. "An Example of Algebra in Lake Roosevelt." *Mathematics Teacher* (February 1995): pp. 116–120.

Software

Department of Mathematics and Computer Science of the North Carolina School of Science and Mathematics. *Matrices.* Materials and software. IBM. Reston, VA: NCTM.

Department of Mathematics and Computer Science of the North Carolina School of Science and Mathematics. *Data Analysis.* Materials and software. IBM. Reston, VA: NCTM.

Videos/Internet

The American Institute of Small Business (AISB) has released a video: *The Internet—What to Know and How to Get On.* Minneapolis, MN.

Interview Notes

Background

Kente Cloth

The Ashanti and Ewe people of West Africa are the weavers of Kente cloth. It has been suggested that the word *kente* is derived from an African word "kenten," meaning "basket," possibly because the cloths were carried in baskets or perhaps because of the patterns on the baskets themselves. However, a direct connection is still unsure. The African people do not use this more modern expression, but instead define the cloths by their patterns and colors.

The Ashanti and Ewe people also use kente cloth for more than just ceremonial purposes. The weavers may create many long strips of cloth and then sew them together to make a larger piece of cloth that has a more uniform length and width. Individually, the strips may not appear to have any pattern or design, but when sewn together, a pattern or design is created by the positioning of the strips. The cloth is then worn as a toga-like wrap. Each piece of cloth is created specifically for a woman or a man, but both men and women of both people wear the cloth.

The meaning of the cloth, however, is determined by its color and by the motifs woven into it. For example, a light-colored cloth may appeal for a happy occasion, and the motifs may symbolize a person's prestige or rank, or perhaps simply key elements of everyday life. Airplanes, anteaters, combs, hands, letters, and other abstract forms provide meaning about the owner and wearer of the cloth.

CHAPTER

1 Modeling Using Algebra

Starting a new tradition

INTERVIEW **Twyla Lang**

"Travel is so broadening," author Sinclair Lewis wrote. Twyla Lang agrees wholeheartedly. "My parents always told me that travel was educational," Lang says. "Since I was a little girl, they encouraged me to visit countries all over the world so that I could learn about other cultures and broaden my horizons. Well, travel used to be part of my education, but now it's part of my business."

"No matter where I travel, everybody understands when it comes to numbers. Math truly is the universal language."

Showing Cultural Pride

In 1989, while still a college student, Lang founded a company that imports and sells jewelry, purses, and clothing from Africa. Her main business revolves around the sale of Kente cloth scarves made in Ghana and Togo. Kente cloth is a handwoven fabric made from cotton, rayon, or silk. "In Africa, it's a symbol of royalty, unity, and achievement," Lang says.

Some Ghanaians and Togolese wear wraps of Kente cloth on special occasions.

Lang sells the scarves to students graduating from high school or college. "Students wear Kente cloth as a way of showing pride in their achievements," she explains. "The idea is to add a touch of cultural diversity to the graduation ceremony."

Getting Started

Lang's company sprung from her habit of collecting gifts and souvenirs from her far-ranging trips. "People always seemed interested in what I brought back, so I thought maybe I could make a living selling these things," she says. She traveled throughout Africa, gathering samples and testing out their appeal, before settling on the Kente products from Ghana and Togo.

The business started small, initially with just a suitcase full of merchandise. But now Lang receives two shipments a month and sells hundreds of items each year.

1

Interview Notes

Background

Twyla Lang

Twyla Lang has traveled to a different part of the world every year. In addition to visiting Ghana and Togo, Ms. Lang has visited Morocco and Nigeria, as well as Côte d'Ivoire. The idea for her company, Motherland Imports, came from observing both Korean and Mexican American people who would travel to their native countries, purchase goods there, and then sell them in the United States. To date, her main clientele include graduating seniors, fraternities and sororities, professional groups, and customers at trade shows, but she would like to expand to other areas as well. Since she began her company, Ms. Lang has graduated from college, taught 5th and 6th graders for two years, and earned her MBA. Today, she continues to manage her growing company.

The Ashanti and Ewe People

The Ashanti people live in central Ghana as well as in Togo and Côte d'Ivoire in West Africa. Primarily farmers, the Ashanti number more then 1,000,000. As a people, the Ashanti trace a child's ancestry through the lineage of their mother but believe that a child's spirit comes from the father. Some of the Kente cloth woven by the Ashanti contain symbolism depicting The Golden Stool, which is the royal throne that is believed to contain all the souls of the Ashanti people.

The Ewe people also live in West Africa, primarily in Togo and Ghana. They are the largest group in Togo and their numbers have been estimated at about 1.5 million. The Ewe people are farmers, craftsmen, traders, and fishermen. The lineage of a child is traced through the father's line. Most Ewe people believe in animism, or the philosophy that all living things—including people, trees, and animals—possess souls and that their souls live on after their physical bodies die.

Interview Notes

Second-Language Learners

Students learning English may be unfamiliar with the author Sinclair Lewis. If necessary, inform them that he was a Nobel Prize-winning American author who lived from 1885 to 1951. Also, the idiomatic expressions *Travel is so broadening* and *broaden my horizons* may need to be explained. The first expression means "to expand one's knowledge through travel" and the second expression means "to add further to my knowledge."

Mathematical Connection

Predicting the amount of growth and finding an appropriate way to display the information are key ideas discussed in Section 1.1. In this section, students use past sales figures of Kente cloth items to predict future sales. Then, the concept of modeling growth is further explored in Section 1.4 with the use of matrices. Students need to use matrices to analyze the sale of Kente scarves and other items using the figures from sales at a trade show.

Explore and Connect

Research
It may help students' understanding to select a business that sells products they are familiar with, such as a CD or music store, a bookstore, or an ice cream store. Students may also select a business where they are working or would like to work, or a business where a friend or family member is employed. Also, discuss what information should be included in a profile of a business before the students write their profiles.

Writing
Ask students to make a general and specific prediction. Perhaps have them give both a descriptive (or qualitative) estimate as well as a numerical (or quantitative) estimate.

Project
Other students in your school may be able to provide some traditions or ideas in this area. The library may also be a good source of information.

Planning for the Future

Lang's company has experienced dramatic growth in recent years. Continued success will depend on careful planning, among other things. She uses information about her sales to create a *model* that can help her predict what the future may hold for her business.

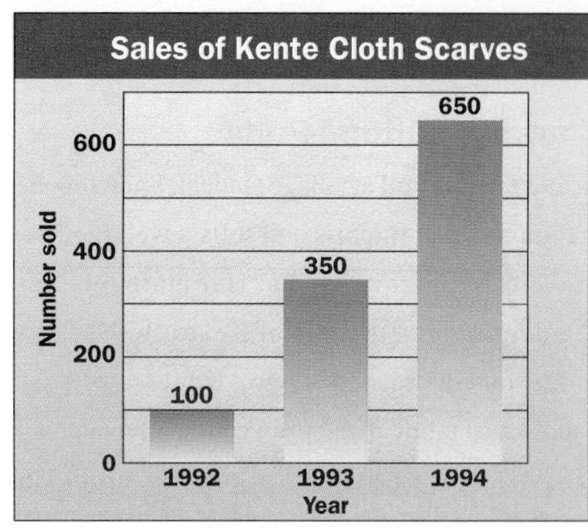

Note: Lang's actual sales figures are confidential.

"When I draw up a business plan, I chart past sales in order to see what I might expect in the future," she says. "An easy way to do that is with a graph. I can look at what I sold in the past and try to make realistic goals for the future."

EXPLORE AND CONNECT

Twyla Lang displays some of her Kente cloth products.

1. Research Investigate how a company in your community started. How long has it been in business? Did the owner start the company? Did the owner hold other positions in the company? Write a profile of the company based on your research.

2. Writing Based on the graph shown above, what prediction would you make for 1995 sales? Explain.

3. Project Lang was able to adapt the traditional use of Kente cloth in a way that made it meaningful for American students. Learn more about a tradition from another culture. Think of a way that this tradition can be adapted to be meaningful in American life.

Mathematics & Twyla Lang

In this chapter, you will learn more about how mathematics is used to make decisions in business.

Related Exercises

Section 1.1
- **Exercises 2 and 3**

Section 1.4
- **Exercises 14 and 15**

1.1 Modeling Growth with Graphs and Tables

Learn how to...

- **model real-world situations with tables and graphs**
- **compare different models describing the same situation**

So you can...

- **make predictions about future trends, such as growth in sales**

Three years ago the Carter High School Volunteer Club began selling T-shirts. The graph shows the amount of money the club raised each year. The club members want to know what amount they might expect to raise in the school year ahead. How much do you think the club will raise?

EXAMPLE 1 Application: Business

Toolbox p. 790
Mean, Median, and Mode

a. Find the average (mean) yearly increase in the amount raised.

b. Use your answer to part (a) and the initial amount raised to estimate the amount the club raises each year after the initial year. Then estimate the amount the club will raise in the school year ahead.

SOLUTION

a. Find the increase for each year.

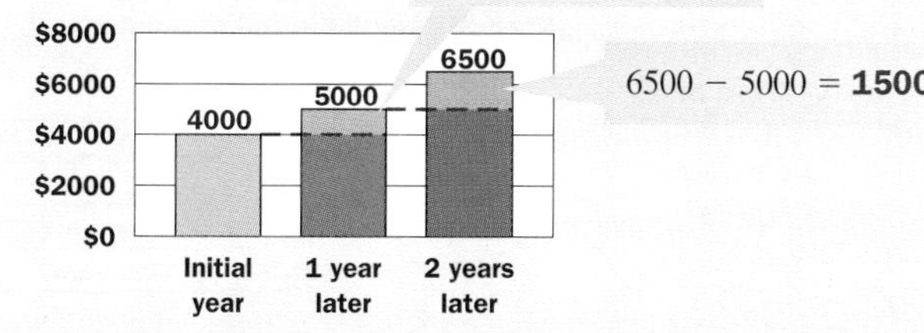

The average yearly increase is $\frac{1000 + 1500}{2}$, or $1250.

Solution continued on next page.

Plan⇔Support

Objectives

- Model real-world situations with tables and graphs.
- Compare different models describing the same situation.
- Make predictions about future trends.

Recommended Pacing

Core and Extended Courses
Section 1.1: 1 day

Block Schedule
Section 1.1: 1 block

Resource Materials

Lesson Support
Lesson Plan 1.1
Warm-Up Transparency 1.1
Practice Bank: Practice 1
Study Guide: Section 1.1
Exploration Lab Manual: Diagram Master 1

Technology
Scientific Calculator
Spreadsheet Software
McDougal Littell Mathpack *Stats! Activity Book:* Activity 11
Internet: http://www.hmco.com

Warm-Up Exercises

Find the average of each group of numbers.

1. 23, 37, 30, 18 27

2. 400, 460 430

3. 76, 79, 83, 94, 88 84

Find the percent increase or decrease from the first year to the second.

4. 1995 sales: $500,000
1996 sales: $700,000
40% increase

5. 1995 dental expenses: $800
1996 dental expenses: $100
87.5% decrease

Teach⇔Interact

Additional Example 1

For three years, the Carter High School Glee Club has given a benefit concert to raise money for local charities. The graph shows the amount raised each year.

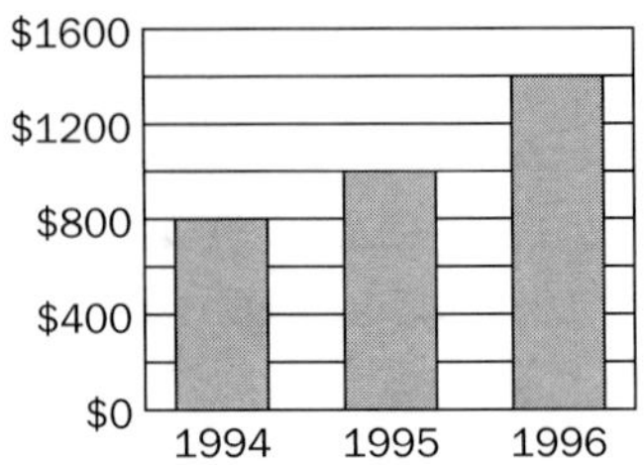

a. Find the average (mean) yearly increase in the amount raised. Find the increase for each year.

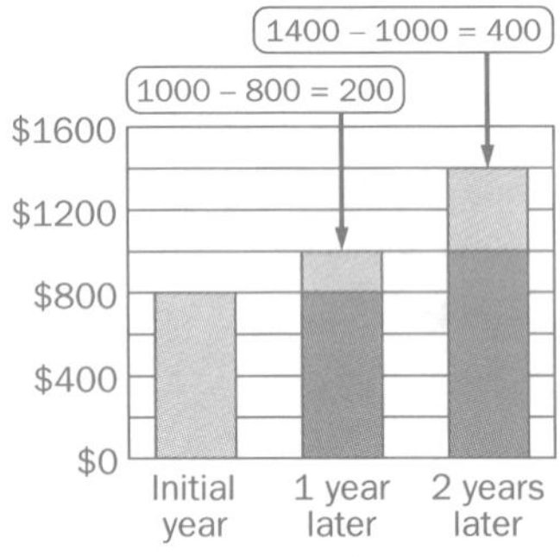

The average yearly increase is $\frac{200 + 400}{2}$, or \$300.

b. Use your answer to part (a) and the initial amount raised to estimate the amount the Glee Club raises each year after the initial year. Then estimate the amount the club will raise in the school year ahead. Make a table or use a spreadsheet.

Glee Club

	A	B	
1	Year	Amount raised (\$)	
2	Initial year	800	+ 300
3	1 year later	1100	+ 300
4	2 years later	1400	+ 300
5	3 years later	1700	

The club can expect to raise about \$1700 in the school year ahead.

Additional Example 2

Refer to Additional Example 1. Suppose the amount of money raised by the Glee Club's benefit concerts grows by about the same percent from year to year. Create a model for this situation. Then use the model to estimate the amount the club will raise in the school year ahead.

Step 1 Find the average yearly percent change.

For more information about using a spreadsheet, see the *Technology Handbook*, p. 816.

SOLUTION *continued*

b. Make a table. You may want to use a spreadsheet.

Volunteer Club

	A	B	
1	Year	Amount raised (\$)	
2	Initial year	4000	+1250
3	1 year later	5250	+1250
4	2 years later	6500	+1250
5	3 years later	7750	

The club can expect to raise about \$7750 in the school year ahead.

When you describe a situation using graphs, tables, or equations, you are making a **mathematical model**. Different assumptions can lead to different models.

EXAMPLE 2 Application: Business

Toolbox p. 775 *Percent Change*

Suppose that the amount of money raised from T-shirt sales grows by about the same *percent* from year to year. Create a model for this situation. Then use the model to estimate the amount the club will raise in the year ahead.

SOLUTION

Step 1 Find the average yearly percent change.

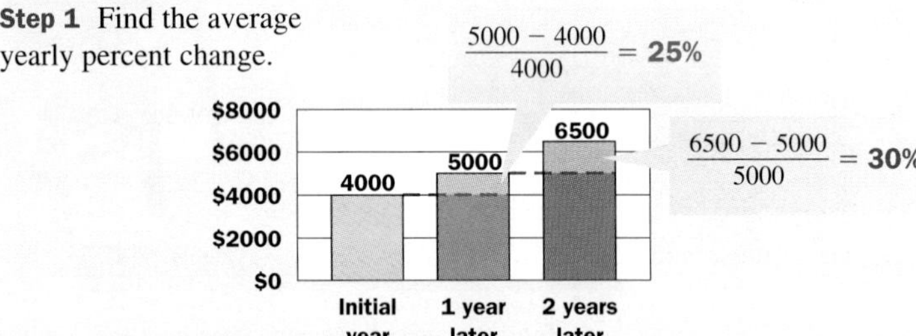

The average yearly percent change is $\frac{25 + 30}{2}$, or 27.5%.

Step 2 Find the *growth factor*. The growth factor is 100% of the previous year plus the average yearly percent change:

$$100\% + 27.5\% = 127.5\%, \text{ or } 1.275.$$

Step 3 Use the initial amount and the growth factor to estimate the amount the club will raise in the school year ahead.

Volunteer Club

	A	B	
1	Year	Amount raised (\$)	
2	Initial year	4000	×1.275
3	1 year later	5100	×1.275
4	2 years later	6502.5	×1.275
5	3 years later	8290.6875	

The club will raise about \$8300 in the school year ahead.

BY THE WAY

About 100 students from the Chicago-area Niles West High School belong to a club called West Helping Others. The students volunteer at schools, hospitals, and shelters. They frequently organize bake sales and car washes to finance activities.

ANSWERS Section 1.1

Think and Communicate

1. Yes; both models give values that are quite close to the actual values.
2. Answers may vary. An example is given. I think both estimates are reasonable, since in both models, the values are close to the actual data.

Checking Key Concepts

1. about 1,178,000 people
2. about 1,208,000 people
3. Yes; the number of degrees (in thousands) predicted by the constant growth model for the second and third years are 921 and 1049 and by the proportional growth model are about 912 and 1049. Both sets of estimates are reasonable, since they are close to the actual values for those years.

The model in Example 1 is based on an assumption that sales will increase by the same *amount* each year. This is an example of a *constant growth* model. The model in Example 2 is based on an assumption that sales will increase by the same *percent* each year. This is an example of a *proportional growth* model.

One way to see which model is more accurate is to compare the values from each model to the actual data.

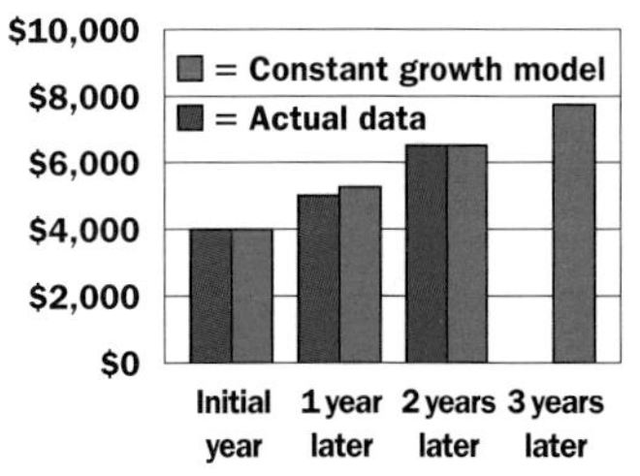

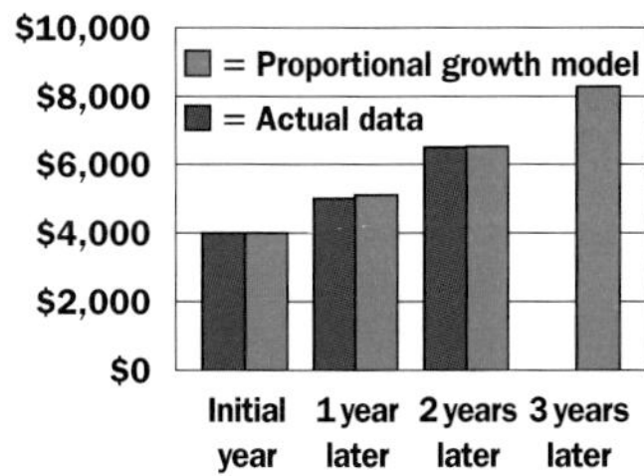

THINK AND COMMUNICATE

1. Does each model above accurately estimate the actual data? Explain.
2. Compare the predictions for the year ahead. What do you think is a reasonable estimate for the amount the club might raise in the school year ahead? Why?

CHECKING KEY CONCEPTS

1. Use a constant growth model to estimate the number of people earning a bachelor's degree in 2000.
2. Repeat Question 1 using a proportional growth model.
3. Are your estimates in Questions 1 and 2 reasonable? Why?

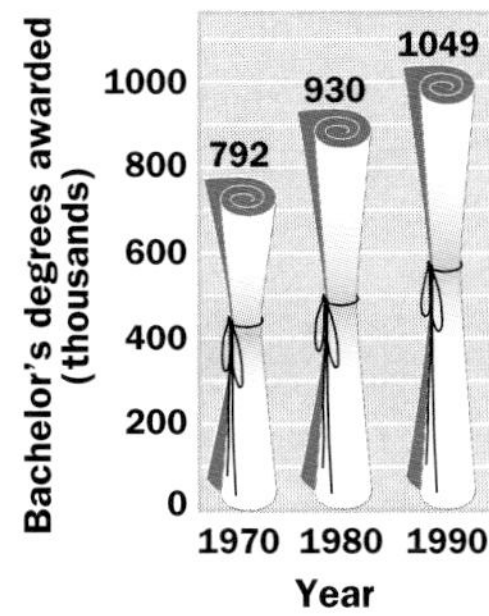

1.1 Exercises and Applications

Extra Practice exercises on page 749

1. The table shows the average daily hospital cost in the United States.

Average Daily Hospital Cost			
Year	**Daily charge ($)**	**$ Increase**	**% Increase**
1975	134	—	—
1980	245	245 − 134 = 111	111/134 = 83%
1985	460	?	?
1990	687	?	?

a. Copy and complete the table.

b. Use a constant growth model to estimate the daily cost in 1995.

c. Use a proportional growth model to estimate the daily cost in 1995. How does this estimate compare to the one you found in part (b)?

Teach⇔Interact

Additional Example 2 (continued)

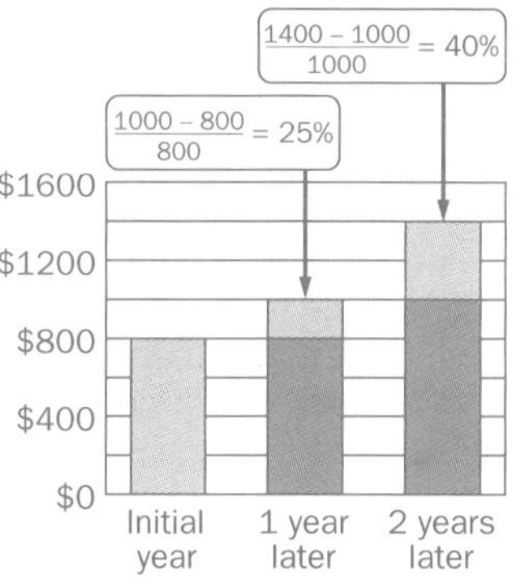

The average yearly percent change is $\frac{25 + 40}{2}$, or 32.5%.

Step 2 Find the growth factor. The growth factor is 100% of the previous year plus the average yearly percent change:

100% + 32.5% = 132.5%, or 1.325.

Step 3 Use the initial amount and the growth factor to estimate the amount the club will raise in the school year ahead.

	Glee Club		
	A	B	
1	Year	Amount raised ($)	
2	Initial year	800	× 1.325
3	1 year later	1060	× 1.325
4	2 years later	1404.5	× 1.325
5	3 years later	1860.9625	

The club will raise about $1860 in the school year ahead.

Closure Question

How does the procedure for predicting future trends with a constant growth model differ from the procedure for predicting with a proportional growth model? For a constant growth model, you start with the initial data value and add the same number repeatedly. For a proportional growth model, you start with the initial data value and multiply by the same number repeatedly.

Exercises and Applications

1. Prediction estimates may vary due to rounding. Increases are rounded to the nearest dollar or the nearest percent.

a.

Average Daily Hospital Cost			
Year	Daily charge ($)	$ Increase	% Increase
1975	134	—	—
1980	245	111	83%
1985	460	215	88%
1990	687	227	49%

b. about $870

c. about $1200; It is much higher.

Apply⇔Assess

Suggested Assignment

Core Course
Exs. 1–5, 7–13, 19–24

Extended Course
Exs. 1–24

Block Schedule
Day 1 Exs. 1–5, 7–13, 19–24

Apply⇔Assess

Exercise Notes

Common Error

Exs. 1, 2, 5, 7–10, 13 When students compute the average increases and the average percent increases for these exercises, they may divide by the wrong number. They often divide by the number of data items and not by that number minus 1. This can be corrected by pointing out that students should divide by the number of items they are adding to compute the average increase or percent increase, not by the number of items in the raw data column.

Application

Ex. 5 This exercise affords an excellent opportunity to stress a key assumption often overlooked when mathematical models are used to predict trends. The assumption is that things will behave in the future as they have in the past. It is not valid to assume that increases in the minimum hourly wage are automatic. On the contrary, they are greatly affected by economic and political conditions.

Challenge

Ex. 5 You may wish to challenge some students to discover shortcuts for computing the average increase and the growth factor for the two models used in this exercise. For the average increase, subtract the 1974 hourly wage from the 1981 hourly wage and divide by 1 less than the number of years. For the growth factor, divide the hourly wage for each year by the hourly wage for the preceding year. (This ratio cannot be computed for 1974, since no data are given for 1973.) Then find the average of the resulting quotients and subtract 1. Suggest that students look at why these procedures work. They can consider the matter algebraically, using four hypothetical data items a, b, c, and d. For the average increase, they must compute the average of the differences $b - a$, $c - b$, and $d - c$. Then, by adding these differences and dividing by 3, the result $\frac{d-a}{3}$ is found. The algebra is more complicated for the growth factor, but some students may be able to handle it.

INTERVIEW Twyla Lang

Look back at the article on pages xxxiv–2.

Twyla Lang estimates her company's future sales so that she can stock her inventory and decide how much she can afford to spend on advertising. Her estimates depend on the assumptions she makes about the company's growth.

2. The table shows the approximate sales of Kente cloth scarves from 1992 to 1994.

Year	Scarf sales
1992	100
1993	350
1994	650

a. Use a constant growth model to estimate the sales of Kente cloth scarves in 1995.

b. Repeat part (a) using a proportional growth model.

c. What do you think is a reasonable estimate of the sales of Kente cloth scarves for 1995? Why?

3. Estimate the sales of Kente cloth scarves for the year 2000. Do you think your estimate is accurate? Why or why not?

4. Writing What assumptions do you make about a situation when you use a constant growth model? a proportional growth model? What other factors may influence how well a model predicts future trends?

5. Cooperative Learning Work with a partner. Use the graph showing the minimum wage in the United States since 1974.

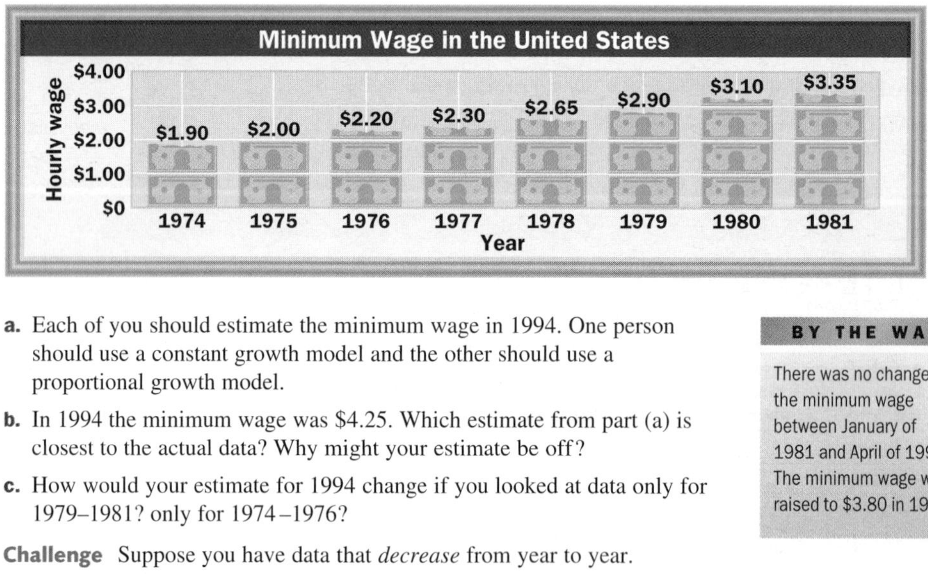

Minimum Wage in the United States

Year	1974	1975	1976	1977	1978	1979	1980	1981
Hourly wage	$1.90	$2.00	$2.20	$2.30	$2.65	$2.90	$3.10	$3.35

(Vertical axis: Hourly wage, $0 to $4.00; horizontal axis: Year)

a. Each of you should estimate the minimum wage in 1994. One person should use a constant growth model and the other should use a proportional growth model.

b. In 1994 the minimum wage was $4.25. Which estimate from part (a) is closest to the actual data? Why might your estimate be off?

c. How would your estimate for 1994 change if you looked at data only for 1979–1981? only for 1974–1976?

BY THE WAY

There was no change in the minimum wage between January of 1981 and April of 1990. The minimum wage was raised to $3.80 in 1990.

6. Challenge Suppose you have data that *decrease* from year to year.

a. How would this change be reflected in a constant growth model?

b. How would this change be reflected in a proportional growth model?

2, 3. Prediction estimates may vary due to rounding.

2. a. about 925 scarves

b. about 1740 scarves

c. Answers may vary. An example is given. I think the lower figure is more reasonable. I do not think the high percentage of annual growth in sales will continue.

3. Answers may vary. An example is given. I think a reasonable estimate is about 2300 scarves. This is based on the constant growth model, which I think is a reasonable model.

4. that the amount of change will be constant; that the percent change will be constant; Answers may vary. An example is given. The amount and accuracy of the available data could affect the predictions.

5. Prediction estimates may vary due to rounding.

a. constant growth model: about $6.10; proportional growth model: about $8.86

b. the estimate obtained from the constant growth model; Both estimates will be off because of the fact that the minimum wage did not increase at all for 9 years.

c. 1979–1981: The estimate with the constant growth model would be slightly higher; the estimate with the

Connection EDUCATION

A college education may be costly, but it may be more expensive *not* to go to college. In the early 1990s the average salary for a graduate of a four-year college was about 85% more than the average salary for someone without a college degree.

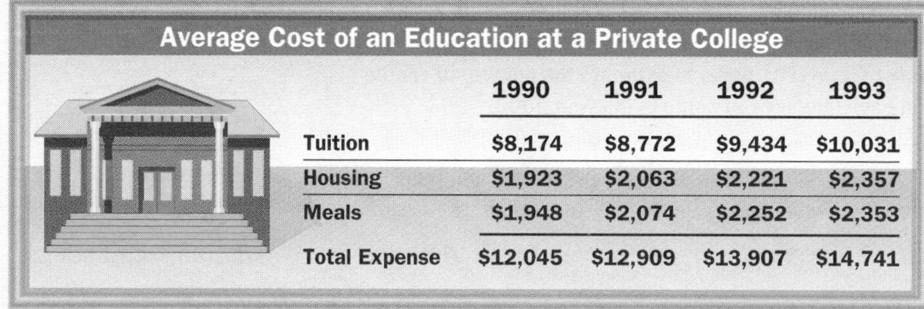

Average Cost of an Education at a Private College

	1990	1991	1992	1993
Tuition	\$8,174	\$8,772	\$9,434	\$10,031
Housing	\$1,923	\$2,063	\$2,221	\$2,357
Meals	\$1,948	\$2,074	\$2,252	\$2,353
Total Expense	\$12,045	\$12,909	\$13,907	\$14,741

For each expense, estimate the cost in 1998. Describe the model you use.

7. tuition **8.** housing **9.** meals **10.** total expenses

11. Which 1990–1993 expense increased by the most dollars? by the largest percentage?

12. a. Suppose you begin college right after you graduate from high school. Estimate the total cost for your first year at a private college.

b. What do you think an education at a private college would cost you, assuming you complete college in four years? Do you think your predictions are reasonable? Why or why not?

c. What are some reasons you should use caution when you use a model to predict far into the future? Explain your reasoning.

BY THE WAY

Sixty-two percent of the 1992 high school graduates in the United States enrolled in either a 2-year or 4-year college in the fall after graduation.

The graph shows the average annual income for people in the United States in 1990 based on the highest degree earned.

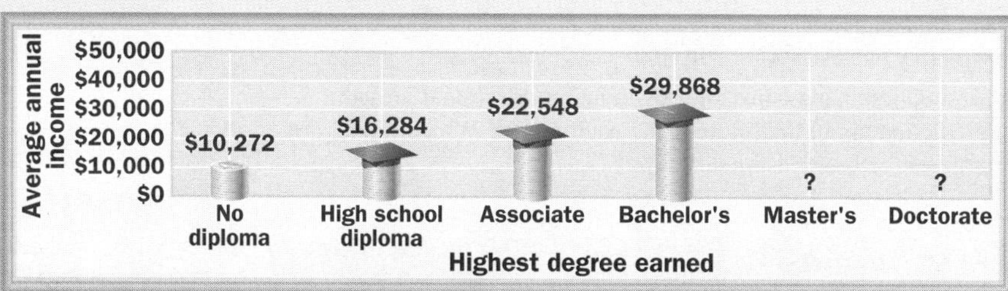

13. a. What is the average percent increase in annual income for each increase in the level of degree earned?

b. Estimate the annual incomes for a person with a master's degree and a person with a doctorate.

c. The average annual salary is about \$38,532 for a person with a master's degree and about \$54,540 for a person with a doctorate. Are these figures close to your estimates from part (b)? Why might the actual figures and your estimates be different?

Apply⇔Assess

Exercise Notes

Multicultural Note

Exs. 7–12 In 1881, Booker T. Washington opened Tuskegee Institute, a school for African American teachers, on an abandoned plantation purchased with borrowed money. Washington worked hard to raise funds for the school, while his students constructed many of the school buildings. By 1888, there were over 400 students in attendance, and the school had become a leader in educating African Americans.

Research

Exs. 7–12 Some students may wish to research general cost-of-living data for the years from 1990 through 1993. They can then compare the trends in the data they obtain to the trends they observed in the college cost data.

Career Connection

Exs. 7–13 The economy of the United States has been changing significantly during the past 20 to 25 years. Many jobs involved in the manufacture of consumer products, for example, have been lost to other countries where the cost of labor is less than in the U.S. At the same time, the demand for people with knowledge skills has increased. This shift in the availability of good paying jobs is still continuing, and the graph for Ex. 13 shows vividly the financial benefits of a career requiring higher education.

Communication: Reading

Ex. 13 Ask students how the graph in this exercise differs from other graphs used in this section. They should notice that the horizontal axis does not use numerical measures.

proportional growth model would be slightly lower. 1974–1976: The estimate with the constant growth model would be significantly lower; the estimate with the proportional growth model would be about the same.

6. a. A constant amount would be subtracted each year.

b. The average yearly percent change would be negative, so the growth factor would be less than 1.

7–10. Estimates may vary due to rounding. Answers may vary. Two examples are given for each item. The first example is based on a constant growth model, the second on a proportional growth model.

7. about \$13,126; about \$14,100

8. about \$3082; about \$3300

9. about \$3028; about \$3250

10. about \$19,236; about \$20,650

11. total expense; tuition

12. a, b. Answers may vary.

c. Answers may vary. Examples are given. A model is based on the assumption that the conditions under which it was created will remain in effect over the course of the prediction. The farther into the future you try to predict, the less likely it is that those conditions will remain unchanged.

13. a. about 43%

b. about \$42,954; about \$61,424

c. No; the actual statistics are significantly lower than the estimates. Answers may vary. An example is given. The difference between a master's degree and a doctorate may not have as significant an effect on income as the difference between lower education levels.

Apply⇔Assess

Exercise Notes

Communication: Discussion
Ex. 18 This exercise should be discussed in class to make sure all students understand why two data values cannot be relied upon to make reasonable predictions.

Assessment Note
Ex. 19 Students can work on this exercise in groups of 3 or 4. They can share ideas about good sources of data and then work independently to gather the data. Each group member can participate in making the table or graph and in developing the two growth models.

Topic Spiraling: Preview
Exs. 21, 22 These exercises prepare students for Section 1.2.

Practice 1 for Section 1.1

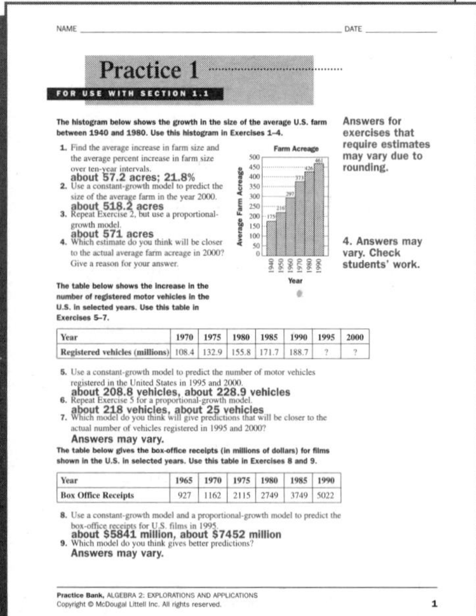

NAME ______________________ DATE ____________

Practice 1

FOR USE WITH SECTION 1.1

The histogram below shows the growth in the size of the average U.S. farm between 1940 and 1980. Use this histogram in Exercises 1–4.

Answers for exercises that require estimates may vary due to rounding.

1. Find the average increase in farm size and the average percent increase in farm size over ten-year intervals.
   about 57.2 acres; 21.8%
2. Use a constant-growth model to predict the size of the average farm in the year 2000.
   about 518.2 acres
3. Repeat Exercise 2, but use a proportional-growth model.
   about 571 acres
4. Which estimate do you think will be closer to the actual average farm acreage in 2000? Give a reason for your answer.

4. Answers may vary. Check students' work.

The table below shows the increase in the number of registered motor vehicles in the U.S. in selected years. Use this table in Exercises 5–7.

Year	1970	1975	1980	1985	1990	1995	2000
Registered vehicles (millions)	108.4	132.9	155.8	171.7	188.7	?	?

5. Use a constant-growth model to predict the number of motor vehicles registered in the United States in 1995 and 2000.
   about 208.8 vehicles, about 228.9 vehicles
6. Repeat Exercise 5 for a proportional-growth model.
   about 218 vehicles, about 25 vehicles
7. Which model do you think will give predictions that will be closer to the actual number of vehicles registered in 1995 and 2000?
   Answers may vary.

The table below gives the box-office receipts (in millions of dollars) for films shown in the U.S. in selected years. Use this table in Exercises 8 and 9.

Year	1965	1970	1975	1980	1985	1990
Box Office Receipts	927	1162	2115	2749	3749	5022

8. Use a constant-growth model and a proportional-growth model to predict the box-office receipts for U.S. films in 1995.
   about $5841 million, about $7452 million
9. Which model do you think gives better predictions?
   Answers may vary.

Practice Bank, ALGEBRA 2: EXPLORATIONS AND APPLICATIONS

1

Connection LANGUAGES

About 14% of Americans speak a language other than English in their homes. The most common of these languages are Spanish, French, German, Italian, and Chinese.

Each graph shows the number of people in the United States who spoke a certain language at home in 1980 and 1990. For each language:

a. Use a constant growth model to estimate the number of people speaking each language at home in the year 2000.

b. Repeat part (a) using a proportional growth model.

c. Compare your estimates from parts (a) and (b). Do you think your estimates are reasonable? Why or why not?

The colors on the map correspond to the languages in Exercises 14–17 and show where each is predominantly spoken.

14. Spanish

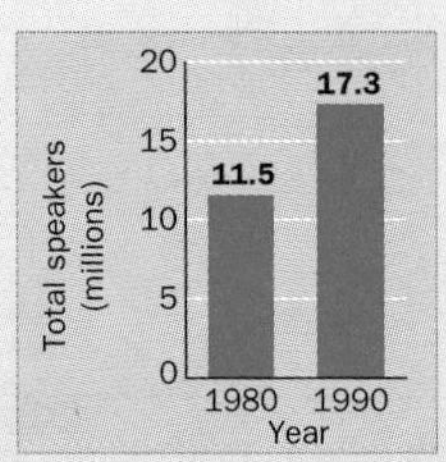

15. Navajo

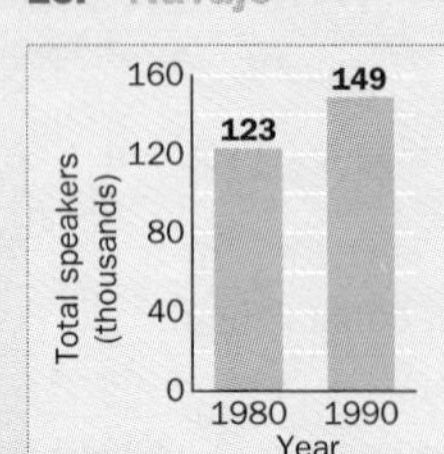

16. Vietnamese

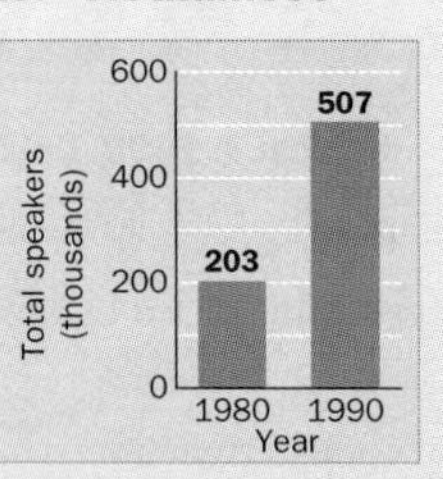

17. Mon-Khmer

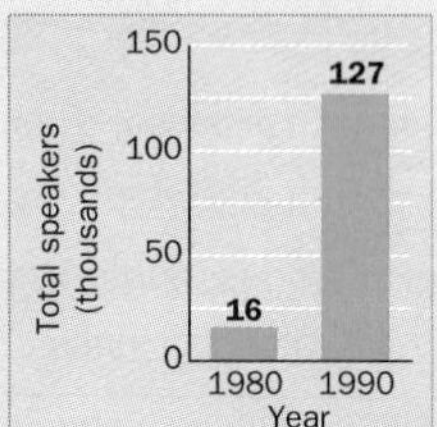

18. Writing Do you think that a model based on only two data values can make reasonable predictions? Why or why not?

ONGOING ASSESSMENT

19. Open-ended Problem Think of a real-world situation that you can model using constant growth and proportional growth.

a. **Research** Gather the data needed to model the situation. Make a table or bar graph to display the data you find.

b. Model the situation using constant growth and proportional growth. Make a prediction about the future using each model. Which model do you think gives the better estimate? Why?

SPIRAL REVIEW

20. Make a scatter plot using the precipitation data. *(Toolbox, page 795)*

Average Precipitation

	Average number of rainy days per year	Average yearly precipitation (inches)
Olympia, WA	164	51.0
Honolulu, HI	100	23.5
Charleston, WV	151	42.4
Madison, WI	118	30.8

For each equation, make a table of values for *x* and *y* when *x* = 0, 1, 2, and 3. *(Toolbox, page 795)*

21. $y = -3x + 6$

22. $y = 4.5(2^x)$

Simplify. Give answers with the appropriate number of significant digits. *(Toolbox, page 789)*

23. $3.2 + 0.15$

24. 42.1×3.6

14. a. about 23.1 million people
    b. about 26 million people
    c. Answers may vary. See the answer for Ex. 18.
15. a. about 175,000 people
    b. about 180,500 people
    c. Answers may vary. See the answer for Ex. 18.
16. a. about 811,000 people
    b. about 1,266,000 people
    c. Answers may vary. See the answer for Ex. 18.
17. a. about 238,000 people
    b. about 1,008,000 people
    c. Answers may vary. See the answer for Ex. 18.
18. Answers may vary. An example is given. I think you cannot make reasonable predictions based on two data values. You have no way of knowing if the data are part of a trend or if one or both of the values reflect some unusual condition. In Exs. 14–17, for example, the statistics might be significantly affected by a one-time relaxing of immigration laws due to political or economic issues.
19. Answers may vary. Possible topics include the population of one's school, town, or state.
20.

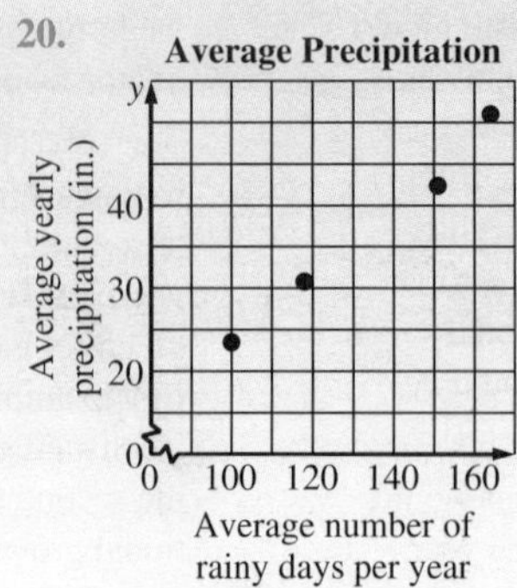

21.

x	y
0	6
1	3
2	0
3	−3

22.

x	y
0	4.5
1	9
2	18
3	36

23. 3.4 24. 151,6

Section

1.2 Using Functions to Model Growth

Learn how to...

- write a function to model a set of data

So you can...

- make predictions about a situation, such as plastics production in the United States

What do nylon stockings, polyester pants, and vinyl siding have in common? Each item contains plastic in both the product and its name! Nylon, polyester, and vinyl are just some of the types of plastics that have been developed since the 1930s. Look at the graph of plastics production in the United States from 1961 to 1991.

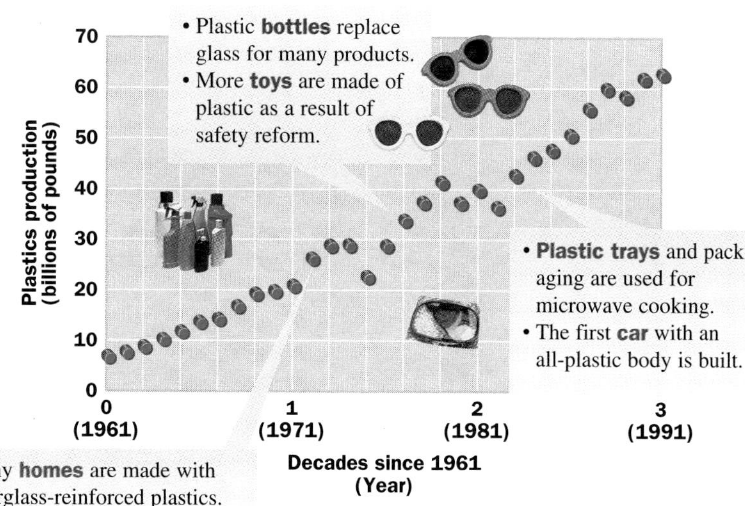

• Plastic **bottles** replace glass for many products.
• More **toys** are made of plastic as a result of safety reform.
• **Plastic trays** and packaging are used for microwave cooking.
• The first **car** with an all-plastic body is built.
• Many **homes** are made with fiberglass-reinforced plastics.
Plastics production (billions of pounds)
0
10
20
30
40
50
60
70
0 (1961)
1 (1971)
2 (1981)
3 (1991)
Decades since 1961 (Year)

Think and Communicate

1. Copy and complete the table.

Decades since 1961	Plastics production (billions of pounds)	Change per decade	Percent change per decade
0 (1961)	7	—	—
1 (1971)	21	?	?
2 (1981)	40	?	?
3 (1991)	63	?	?

2. What was the average change?

3. What was the average percent change?

4. How many decades since 1961 is 2001? Estimate the plastics production in 2001.

Plan⇔Support

Objectives

- Write a function to model a set of data.
- Make predictions about situations using functions.

Recommended Pacing

Core and Extended Courses
Section 1.2: 1 day

Block Schedule
Section 1.2: $\frac{1}{2}$ block
(with Section 1.3)

Resource Materials

Lesson Support
Lesson Plan 1.2
Warm-Up Transparency 1.2
Practice Bank: Practice 2
Study Guide: Section 1.2
Assessment Book: Test 1

Technology
Technology Book:
Calculator Activity 1
Spreadsheet Activity 1
Scientific Calculator
Graphing Calculator
McDougal Littell Mathpack
Function Investigator
Internet:
http://www.hmco.com

Warm-Up Exercises

Find the value of y when x has the given value.

1. $y = 7x + 19, x = 0$ 19

2. $y = -23x + 14, x = 8$ −170

3. $y = 12.8x - 3.6, x = 5$ 60.4

4. $y = 3(5.2)^x, x = 2$ 81.12

5. $y = 8.4(0.5)^x, x = 3$ 1.05

ANSWERS Section 1.2

Think and Communicate

1. Percents are rounded to the nearest tenth.

Decades since 1961	Plastics production (billions of pounds)	Change per decade	Percent change per decade
0 (1961)	7	—	—
1 (1971)	21	14	200%
2 (1981)	40	19	90.5%
3 (1991)	63	23	57.5%

2. $18\frac{2}{3}$, or about 18.7 billion pounds

3. 116%

4. 4 decades; about 82 billion pounds (constant growth model); about 152 billion pounds (proportional growth model)

Teach⇔Interact

Additional Example 1

The table shows data on shopping bags made by Brenda Marino's company from 1983 to 1989.

Two-year periods since 1983	Shopping bags manufactured (hundreds of thousands)
0 (1983)	2.8
1 (1985)	4.7
2 (1987)	13.3
3 (1989)	28.0

Brenda used a constant growth model and a proportional growth model to study the data. She found that the average change per two-year period was 8.4 (in hundreds of thousands) and that the average percent change was about 120%.

a. Write an equation to model the number of shopping bags manufactured using the average change that Brenda computed.
Let x = the number of two-year periods since 1983. Let y = the number of shopping bags manufactured (in hundreds of thousands). Find y by starting with the number of bags manufactured in 1983 and adding the constant growth rate x times.
y = (bags manufactured in 1983) + (constant growth rate) • x
$y = 2.8 + 8.4x$

b. Write an equation to model the number of bags manufactured based on the average percent change that Brenda computed.
Find y by starting with the number of bags manufactured in 1983 and multiplying by the proportional growth factor x times.
y = (bags made in 1983) • (proportional growth factor)^x
$y = 2.8(2.2)^x$

Section Note

Topic Spiraling: Review
Students have probably encountered the terms *function, independent variable*, and *dependent variable* in earlier courses. Since it is important for students to have a firm understanding of their meanings, you may wish to engage them in a brief discussion of these concepts. Ask for volunteers to suggest some examples of functions and have others identify the independent and dependent variables.

BY THE WAY

By the thirteenth century B.C., people in China were using the first industrial plastic: lacquer. This extremely durable natural plastic was tapped from lacquer trees, and was used for items such as kitchen utensils, screens, and shoes.

EXAMPLE 1 Application: Manufacturing

Write an equation to model the plastics production since 1961 using:

a. the average change from *Think and Communicate* Question 2.

b. the average percent change from *Think and Communicate* Question 3.

SOLUTION

Let x = the number of decades since 1961.
Let y = the plastics production (in billions of pounds).

a. You can find y by starting with the plastics production in 1961 and adding the constant growth rate x times.

$$y = \begin{pmatrix}\text{amount of plastic}\\ \text{produced in 1961}\end{pmatrix} + \begin{pmatrix}\text{constant}\\ \text{growth rate}\end{pmatrix} \cdot x$$

In 1961 there were 7 billion pounds of plastics produced.

The average constant growth rate is 18.7.

$$y = 7 + 18.7x$$

b. You can find y by starting with the plastics production in 1961 and multiplying by the proportional growth factor x times.

$$y = \begin{pmatrix}\text{amount of plastic}\\ \text{produced in 1961}\end{pmatrix}\begin{pmatrix}\text{proportional}\\ \text{growth factor}\end{pmatrix}^x$$

In 1961 there were 7 billion pounds of plastics produced.

Each decade about 116% more plastics are produced. The proportional growth factor is 216%.

$$y = 7(2.16)^x$$

A **function** pairs each input value, x, with exactly one output value, y. The two equations from Example 1 are functions.

A model based on constant growth is called a **linear** function.

$y = 7 + 18.7x$	
x (input)	y (output)
0	7
1	25.7
2	44.4
3	63.1

A model based on proportional growth is called an **exponential** function.

$y = 7(2.16)^x$	
x (input)	y (output)
0	7
1	15.1
2	32.6
3	70.5

Since you input an x-value to represent a decade of your choice, x is called the **independent variable.** Since the plastics production *depends* on the decade you choose, y is called the **dependent variable**.

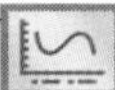

Technology Note

Some students may know they can use graphing calculators or graphing software to find linear or exponential regression equations that model sets of data. These equations will generally not be the same as those obtained by using the methods of this section. The algorithms that are used by calculators and graphing software typically use a least-squares method to find the key constants for the equations.

EXAMPLE 2 Application: Manufacturing

a. Use a graphing calculator or graphing software to graph the equations from Example 1 and the plastics production data for 1961, 1971, 1981, and 1991 in the same coordinate plane.

b. Does each equation reasonably model the data? Why or why not?

c. Use each model to estimate the plastics production in 2001.

For more information about graphing equations and graphing data points, see the *Technology Handbook*, p. 804 and p. 814.

SOLUTION

a. Graph $y = 7 + 18.7x$, $y = 7(2.16)^x$, and the data points for 1961, 1971, 1981, and 1991.

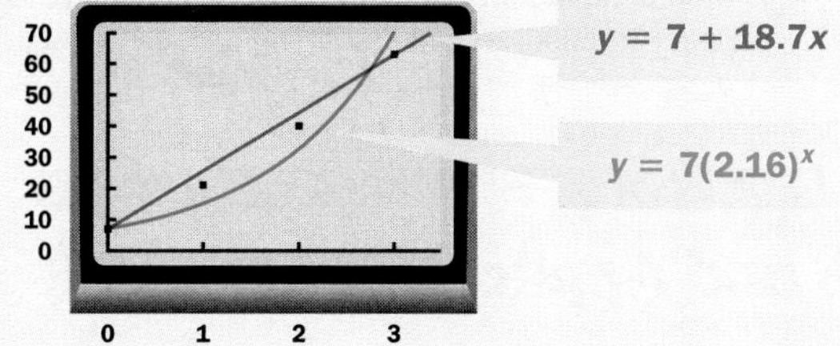

b. Both equations reasonably model the data, although the linear equation appears to lie a little closer to the data points.

c. The year 2001 is four decades after 1961. Trace along each graph until $x = 4$.

Linear Model

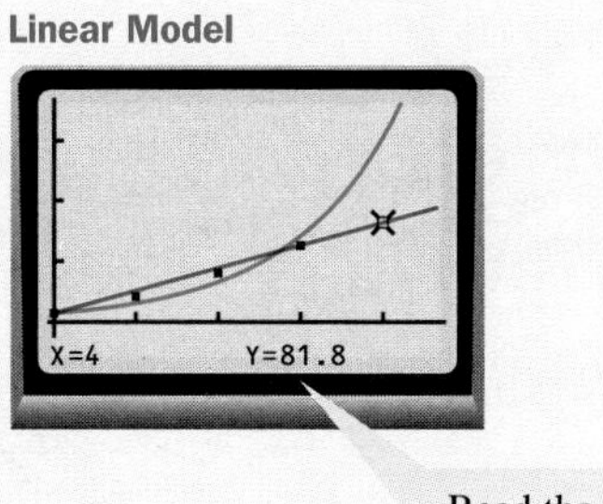

Exponential Model

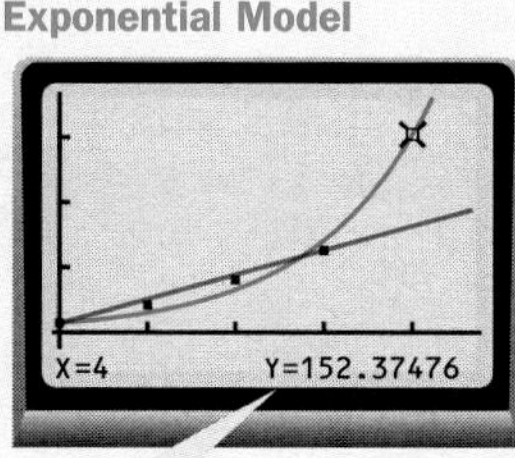

Read the value of y when $x = 4$.

A low estimate of the plastics production in 2001, based on the linear model, is about 81 billion pounds. A high estimate, based on the exponential model, is about 152 billion pounds.

THINK AND COMMUNICATE

5. What would be a mid-range estimate for part (c) of Example 2?

6. Which function grows faster, the *exponential function* or the *linear function*? What does this mean when you use these functions to make long-term predictions?

7. Which function do you think more accurately predicts the growth of plastics production in the future? Why?

Technology Note

On the TI-82, if the graphing window does not permit tracing to a graph point with a specific *x*-coordinate, select 1:value from the CALCULATE menu. If part of a graph or scatter plot is obscured by coordinate read-outs at the bottom of the screen, change the window setting for Ymin and display a slightly modified graph. (This was done for the graphs in the answer for part (c) of Example 2.)

Teach⇔Interact

Additional Example 2

Refer to Additional Example 1.

a. Use a graphing calculator or graphing software to graph the equations from Additional Example 1 and the data on the number of shopping bags made in 1983, 1985, 1987, and 1989 in the same coordinate plane.

Graph $y = 2.8 + 8.4x$, $y = 2.8(2.2)^x$, and the data points for 1983, 1985, 1987, and 1989.

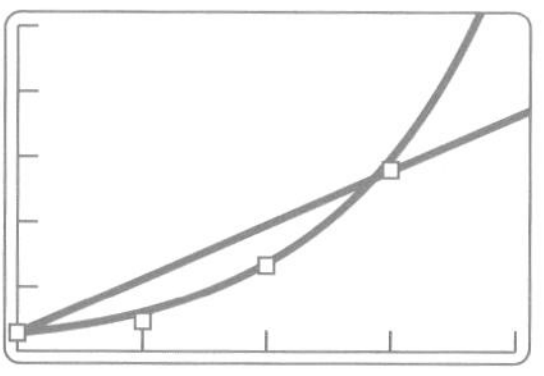

b. Does each equation model the data? Why or why not?

Both equations reasonably model the data, although the exponential equation's graph is closer to the data points.

c. Use each model to estimate the bag production in 1991.

The year 1991 is four 2-year periods after 1983. Trace along each graph until $x = 4$.

Linear Model

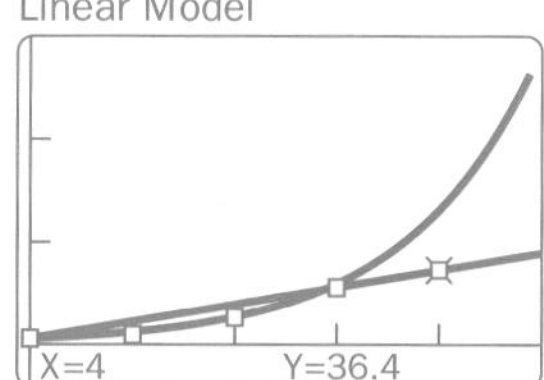

Exponential Model

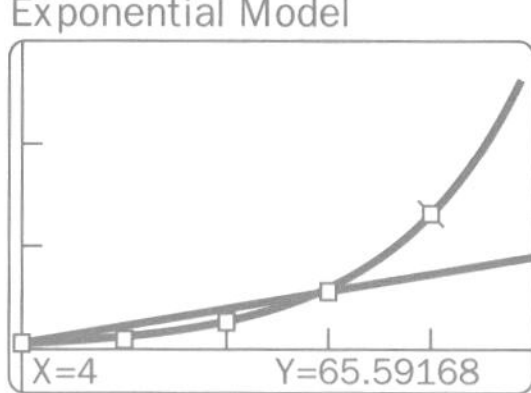

A low estimate of the bag production in 1991, based on the linear model, is about 3,640,000. A high estimate, based on the exponential model, is about 6,560,000.

Think and Communicate

Students can consider whether there is a way to find *equations* that would give mid-range estimates for question 5. One possibility is to experiment with the value of the growth factor in the exponential equation.

For answers to Think and Communicate, see answers in back of book.

Teach⇔Interact

Checking Key Concepts

Communication: Discussion When discussing question 4, be sure to examine students' estimates based on both types of models, linear and exponential.

Closure Question

In most everyday situations, which kind of model is more likely to be unsatisfactory for making long-range predictions? Explain your thinking.

Answers may vary. An example is given. If the data values used for the models increase over time, the exponential model will be less satisfactory because it will eventually increase too rapidly to be realistic.

Apply⇔Assess

Suggested Assignment

- **Core Course**
  Exs. 1–8, 10–12, 16–21, AYP
- **Extended Course**
  Exs. 1–21, AYP
- **Block Schedule**
  Day 2 Exs. 1–8, 10–12, 16–21, AYP

Exercise Notes

Reasoning

Ex. 4 You may wish to ask students how Jon-Paul's equation can be used to get an estimate for 1980. (Use –5 as the value of x in $y = 3x + 31$.)

Year	Number of workers
1987	7,147
1988	11,400
1989	15,927
1990	21,382
1991	26,327
1992	34,348

CHECKING KEY CONCEPTS

INDUSTRY The table shows the number of people working in the cellular telephone industry from 1987 to 1992. The average change was 5440 workers per year and the average percent change was 37.4% per year.

1. Define the dependent and independent variables in this situation.
2. Write a linear function to model the number of people working in the cellular telephone industry since 1987.
3. Write an exponential function to model the situation.
4. Estimate the number of people working in the cellular telephone industry in 2000. Explain how you arrived at your estimate.

1.2 Exercises and Applications

Extra Practice exercises on page 749

SPORTS Jon-Paul and Marya each used an equation to model the number of youth softball teams, in thousands, in the United States. Refer to each model to answer Exercises 1–8.

Marya
Suppose x is number of years since 1980 and y is the number of youth softball teams in the United States, in thousands. An equation that models the situation is $y = 18(1.10)^x$.

Jon-Paul
Let x represent the number of years since 1985 and y represent the number of youth softball teams in the United States, in thousands. An equation that models the situation is $y = 3x + 31$.

1. Which student created an exponential model? a linear model?
2. Which variable is the independent variable? the dependent variable?
3. What year is represented by an x-value of 0 in Jon-Paul's model? in Marya's model? Why are these values different?
4. How many youth softball teams were there in 1980? in 1985? How did you determine each value?
5. What is the average change in the number of youth softball teams? Which model did you use to find this information?
6. What is the average percent change in the number of youth softball teams each year? Which model did you use to find this information?
7. Use Marya's model to find the number of youth softball teams in 1988.
8. Use Jon-Paul's model to find the number of youth softball teams in 1995.

BY THE WAY

George W. Hancock developed softball as an indoor game in 1887. In 1895, Lewis Robe adapted the game for outdoor play.

Checking Key Concepts

1–4. See answers in back of book.

Exercises and Applications

1. Marya; Jon-Paul
2. x; y
3. 1985; 1980; Jon-Paul's data begins in 1985, while Marya's begins in 1980.
4. 18,000 teams; 31,000 teams; Substitute 0 for x in Marya's equation to get the figure for 1980, and substitute 0 for x in Jon-Paul's equation to get the figure for 1985.
5. 3000; Jon-Paul's (the linear model)
6. 10%; Marya's (the exponential model)
7. about 39,000 teams
8. about 61,000 teams
9. Answers may vary. An example is given. I think since there are only statistics for years in which Olympic Summer games were held, it would be better to define the independent variable as the number of four-year periods since 1960.

10, 11. Answers may vary slightly due to rounding. Let x = the number of four-year periods since 1960 and y = time in seconds.

10. a. $y = 290.6 - 5.43x$; The winning times are decreasing.
    b. $y = 290.6(0.98)^x$
    c. linear

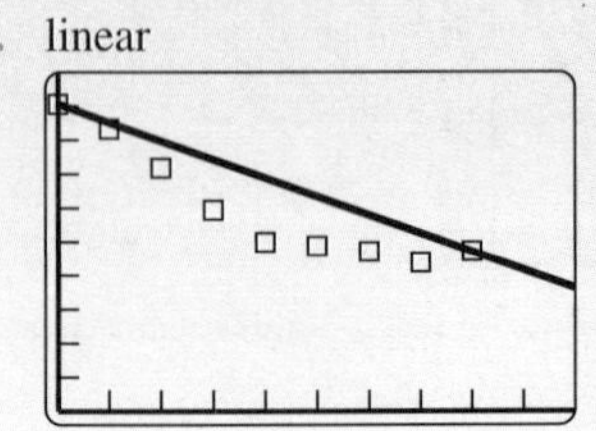

exponential

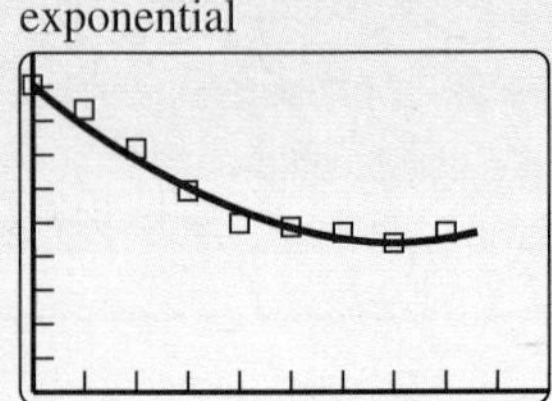

Answers may vary. An example is given. Yes; the data points are reasonably near both graphs.

Connection OLYMPIC EVENTS

The table shows the winning times, in seconds, in the 400 m Olympic freestyle swimming event from 1960 to 1992. Modern training methods are one reason why the winning times have improved so greatly.

Year	Men	Women
1960	258.3	290.6
1964	252.2	283.3
1968	249.0	271.8
1972	240.27	259.44
1976	231.93	249.89
1980	231.31	248.76
1984	231.23	247.10
1988	226.95	243.85
1992	225.00	247.18

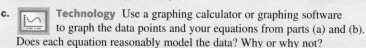

(Top) Janet Evans of the United States swims in the women's 400 m freestyle at the 1988 Olympics in Seoul. (Above) An Olympic swimmer trains for a competition by swimming against a pump-generated current. While she swims, a computer analyzes her stroke and oxygen intake.

9. **Writing** To analyze the swimming data, Fredric wants to let x be the number of years since 1960. Do you think this is a good definition for the independent variable? Why or why not?

10. a. Use a linear function to model the women's winning swim times. Why is the growth rate negative?

    b. Use an exponential function to model the women's times. (*Hint:* If the average percent change is -5%, the growth factor is $1.00 + (-0.05)$, or 0.95.)

    c. **Technology** Use a graphing calculator or graphing software to graph the data points and your equations from parts (a) and (b). Does each equation reasonably model the data? Why or why not?

    d. Estimate the women's winning swim times in 1996 and 2000.

11. a. Repeat Exercise 10 using the data for the men's winning times in the 400 m Olympic freestyle swimming event.

    b. **Writing** Use your models from Exercise 10 and part (a) of Exercise 11 to estimate the year in which the women's winning time will be better than the men's winning time. Do you think your estimate is a reasonable one? Explain.

    c. **Challenge** What x-value would you use in your models from part (a) to estimate the men's winning time in the 1956 Olympics? Why? Use this value to estimate the men's winning time in 1956.

12. **Writing** Do you think that the winning swim times will continue to decrease? Why or why not?

d. linear: about 241.73 s, about 236.30 s; exponential: about 242.29 s, about 237.44 s

11. a. (a) $y = 258.3 - 4.16x$ (b) $y = 258.3(0.983)^x$

(c) linear exponential

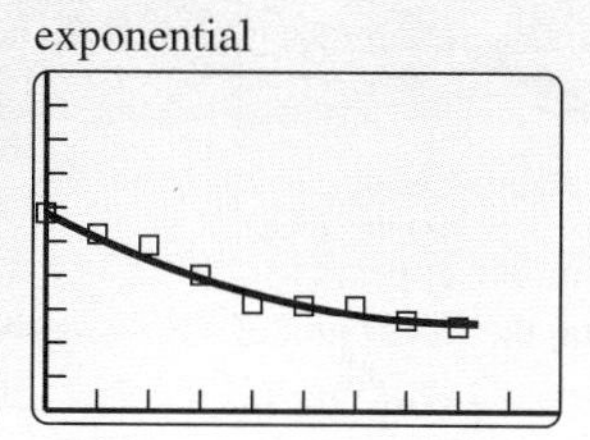

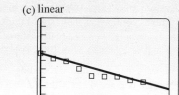

Answers may vary. An example is given. Yes; the data points are reasonably near both graphs. (d) linear: about 220.86 s, about 216.70 s; exponential: about 221.36 s, about 217.60 s

b. linear: 2064; exponential: 2116; Answers may vary.

c. –1; If 0 represents the year 1960, it makes sense that –1 would represent the previous Olympic year. linear: about 262.46 s; exponential: about 262.77 s

12. Answers may vary. An example is given. I think that winning swim times will continue to decrease. However, I think there must be some lower limit determined by physical limitations.

Apply⇔Assess

Exercise Notes

Multicultural Note

Exs. 9–12 Students interested in the feats of swimming champions may want to research the career of U.S. swimmer Pablo Morales. After a narrow loss at the 1984 Olympics, Morales set the world record for the men's 100 m butterfly in 1986, a record that stood for nine years. In 1988, Morales failed to make the Olympic team and decided to retire. He attended Cornell University Law School for three years, and then decided to try out for the 1992 Olympics. After training for only seven months, Morales made the team. He won the gold medal at the age of 27, making him one of the oldest Americans to win a gold medal in swimming.

Cooperative Learning

Exs. 10, 11 Students can benefit from working with a partner on these exercises, which require that a number of calculations be made. Students may choose to do some parts of the exercises together and others independently, for which they can compare answers.

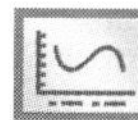

Using Technology

Exs. 10, 11, 13–15 You may want to suggest that students use the *Function Investigator* software for this group of exercises.

For Ex. 11, part (b), students can use equations, tables, or graphs to arrive at an estimate. Students who are proficient with the TI-82 graphing calculator should have no difficulty with any of these methods, although graphs are probably the fastest. To find where two graphs intersect, students using the TI-82 should select 5:intersect from the CALCULATE menu.

Communication: Drawing

Ex. 12 Students may find it possible to present a more convincing case for their thinking if they use graphs to accompany their written responses.

Apply⇔Assess

Exercise Notes

Mathematical Procedures

Ex. 13 Encourage students to specify what the variables in the equation models represent and the scales they are using for their graphs.

Communication: Discussion

Ex. 13 Students can have an interesting discussion of how political developments might affect the reliability of the estimates in part (c). For the writing exercise in part (d), students might also wish to consider efforts to reduce the national debt.

Reasoning

Ex. 14 Discuss possible reasons for the fact that neither model gives a very good estimate for the year 1992. (Possible answers: There are very few data points. It is difficult to anticipate technological advances that might nullify the validity of predictions based on past data. Also, the data for 1492, 1692, and 1892 involve vessels that were constructed to carry many passengers or a significant amount of cargo.)

Interdisciplinary Problems

Exs. 14, 15 Many students probably think that mathematics could not be used to understand historical events, but these exercises illustrate a nice connection to history, using linear and exponential functions.

Assessment Note

Ex. 16 This exercise should make it possible to determine whether students understand the basic differences between the graphs of linear and exponential functions. Students may wish to use graphs to clarify their ideas and should be encouraged to do so.

13. Cooperative Learning Work with a partner. You will need graphing calculators or graphing software.

a. Create a linear model and an exponential model of the growth of the United States national debt for your data set. One of you should use the 1950–1990 data and the other should use the 1990–1994 data.

b. **Technology** Each of you should graph your equations and the data points in the same coordinate plane. With your partner, decide which equation is a better model for each data set.

c. Estimate the national debt in 2000. Compare your estimates.

d. **Writing** Which model do you think gives a better estimate of the national debt in 1996? in 2010? Explain your reasoning.

National Debt

Year	Trillions of $
1950	0.26
1960	0.29
1970	0.38
1980	0.91
1990	3.27
1991	3.68
1992	4.08
1993	4.44
1994	4.70

Connection HISTORY

Christopher Columbus sailed across the Atlantic Ocean in 71 days. In 1927 it took Charles Lindbergh 33.5 hours to fly across the Atlantic Ocean. As transportation has improved, the amount of time needed to cross the Atlantic Ocean has changed.

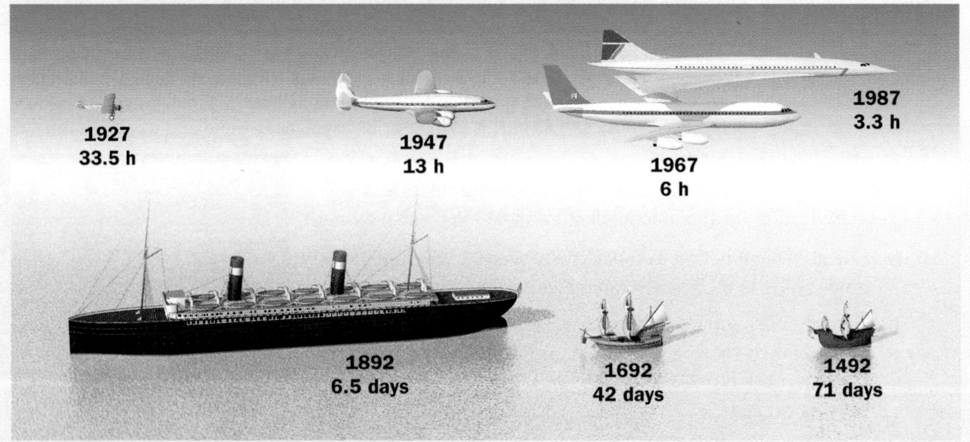

Use the information in the diagram above for Exercises 14 and 15.

14. a. Write a linear function and an exponential function to model the amount of time it takes to cross the Atlantic Ocean by sea. What interval of time does each x-value represent?

b. **Technology** Use a graphing calculator or graphing software to graph the data points and your equations from part (a) in the same coordinate plane. Does each equation reasonably model the data?

c. Use your model to estimate the amount of time it would take to cross the Atlantic Ocean by sea in 1992. A powerboat crossed the Atlantic Ocean in about 2.5 days in 1992. How close is your estimate?

15. Repeat Exercise 14 using the data for the amount of time it takes to cross the Atlantic Ocean by air. Use your model to estimate the amount of time it will take to cross the Atlantic Ocean by air in 2007.

13. a. Answers may vary due to rounding. Examples are given. For the 1950–1990 data, let x = the number of decades since 1950 and y = the debt in trillions of dollars; linear model: $y = 0.26 + 0.75x$; exponential model: $y = 0.26(2.1)^x$. For the 1990–1994 data, let x = the number of years since 1990 and y = the debt in trillions of dollars; linear model: $y = 3.27 + 0.36x$; exponential model: $y = 3.27(1.095)^x$

b. 1950–1990

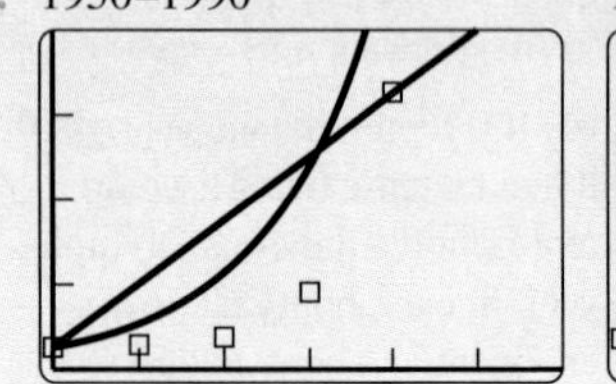

1990–1994

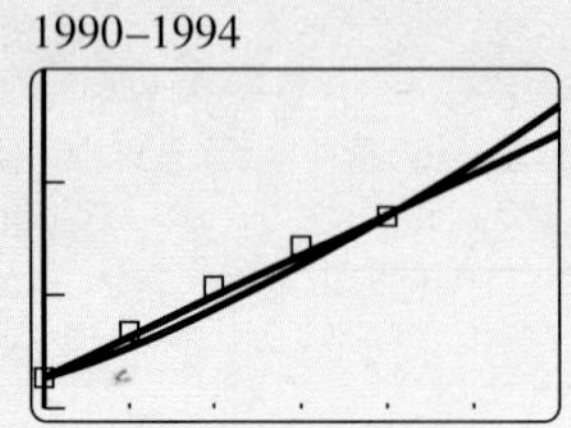

For the 1950–1990 data, the exponential model appears to be a better model. For the 1990–1994 data, the linear model appears to be better.

c. Answers may vary. Examples are given. about \$4.01 trillion (1950–1990 data, linear model); about \$10.62 trillion (1950–1990 data, exponential model); about \$6.87 trillion (1990–1994 data, linear model); about \$8.10 trillion (1990–1994 data, exponential model)

d. Answers may vary. An example is given. If trends continue as they have, I think the linear model gives a better estimate of the national debt for both 1996 and 2010.

14. a. Answers may vary due to rounding. An example is given. Let x = the number of 200-year periods since 1492 and y = the number of days to cross the Atlantic by sea. linear model: $y = 71 - 32.25x$; exponential model: $y = 71(0.37)^x$; 200 years

ONGOING ASSESSMENT

16. **Writing** When do you think it is best to model a situation using a linear model? an exponential model?

SPIRAL REVIEW

17. The table shows the revenue from book sales in the United States. *(Section 1.1)*
    a. Use a constant growth model to estimate the revenue in 1993.
    b. Repeat part (a) using a proportional growth model.

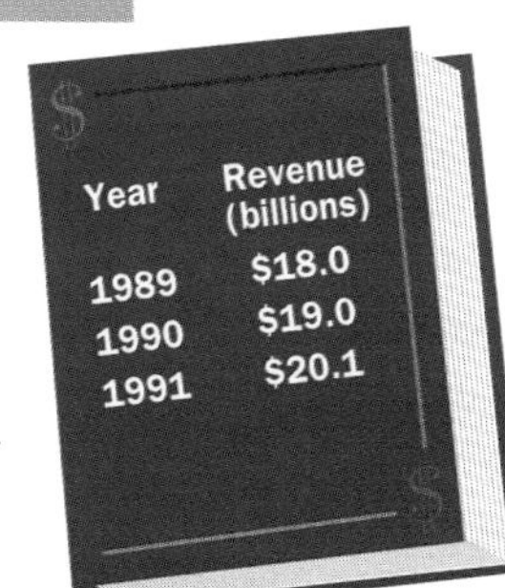

Year	Revenue (billions)
1989	\$18.0
1990	\$19.0
1991	\$20.1

Evaluate each expression when $a = 5$ and $b = -4$. *(Toolbox, page 780)*

18. $-(a + b)^2$

19. $-a - b$

20. $a - 2b + 14$

21. The table shows the number of major hurricanes from 1900 to 1980. Find the mean, median, and mode of the data. *(Toolbox, page 790)*

Period	1901–1920	1921–1940	1941–1960	1961–1980
Major hurricanes	13	13	18	10

ASSESS YOUR PROGRESS

VOCABULARY

mathematical model (p. 4)
function (p. 10)
independent variable (p. 10)
dependent variable (p. 10)

1. **Writing** What calculation do you perform when you model a situation using constant growth? proportional growth? *(Section 1.1)*

2. **ENTERTAINMENT** The table shows the average cost of a movie ticket from 1951 to 1991. *(Section 1.2)*

   a. Write a linear function to model the situation.
   b. Write an exponential function to model the situation.
   c. **Technology** Use a graphing calculator or graphing software to graph the data points and your equations from parts (a) and (b) in the same coordinate plane. Does one model fit the data better? If so, which one? Explain your choice.
   d. Estimate the price of a movie ticket in 2001.

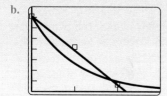

3. **Journal** When do you think it is best to model a situation using a table? an equation? a graph? Which method do you prefer? Why?

Apply⇔Assess

Assess Your Progress

Review the vocabulary terms, and use Exs. 9–12 from this section to illustrate their meanings. Ex. 2 in Assess Your Progress will show whether students understand the definitions and calculations used to model data with linear and exponential equations. It will also give additional insight into students' ability to use appropriate technology to study constant growth and proportional growth models of real-world situations.

Journal Entry
Students can refer to their responses to Ex. 16 as they work on this journal entry.

Progress Check 1.1–1.2

See page 38.

Practice 2 for Section 1.2

NAME ______________________ DATE __________

Practice 2
FOR USE WITH SECTION 1.2

The table below shows the number of children enrolled in the U.S. government's Head Start program in selected years. Use this table in Exercises 1–4.

Year	1990	1991	1992	1993	1994
Enrollment (thousands)	548.5	583.5	621.1	713.9	?

1. Find the average annual growth in the number of children enrolled in the Head Start program. Also, find the average annual percent growth. **about 55,100 children; about 9%**
2. Let x = the number of years since 1990. Write an equation that is a linear model for the growth of the program. $y = 55{,}100x + 548{,}500$
3. Using the same variable x as in Exercise 2, write an equation that is an exponential model of the growth of the program. $y = 548{,}500(1.09)^x$
4. The House Ways and Means Committee, which compiled this table, estimated that the number of children enrolled in the Head Start program in 1994 was 750,000. Which model comes closest to giving the estimate that the committee used? **the linear model**

The histogram at the right shows the growth of the total prize money (in thousands of dollars) for the annual Indianapolis 500 Auto Race in selected years. Use this table in Exercises 5–7.

Indy 500 Purse
Thousands of Dollars: 0, 200, 400, 600, 800, 1000, 1200, 1400, 1600
Year: 1940, 1950, 1960, 1970, 1980
86, 201, 369, 1000, 1503

5. Let x = number of decades since 1940, and let y = the total prize money in thousands of dollars. Write an equation that is a linear model for the growth of the prize money. $y = 354.25x + 86$
6. Using the same variables as in Exercise 5, write an equation that is an exponential model for the growth. $y = 86(2.1)^x$
7. Use both the linear and exponential models to estimate the total Indy 500 prize money for the year 1990. The actual total prize money awarded in 1990 was about \$6,326,000. Which model gives a prediction closest to this figure? **\$1,857,000; \$3,541,000; the exponential model**
8. **Open-ended Problem** Estimate the average daily amount of time you spent talking to friends on the telephone during each of the last several years. Create a linear model and an exponential model for this situation. Which model more accurately predicts the time you will spend talking to friends on the telephone next year? **Answers may vary. Check students' work.**

2 **Practice Bank,** ALGEBRA 2: EXPLORATIONS AND APPLICATIONS

b. Both models are reasonable; the linear model is somewhat better.

c. linear model: about –9.6 days, which is not reasonable; exponential model: about 5.9 days, which is not a very close estimate

15. a. Answers may vary due to rounding. An example is given. Let x = the number of 20-year periods since 1927 and y = the number of hours to cross the Atlantic by air. linear model: $y = 33.5 - 10.06x$; exponential model: $y = 33.5(0.47)^x$; 20 years

b.

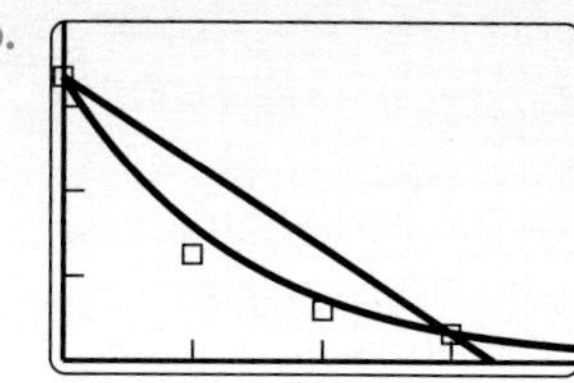

The exponential model fits the data more closely.

c. linear model: about –6.8 h, which is not reasonable; exponential model: about 1.6 h

16. Answers may vary. Examples are given. If you graph the data points and they appear to lie on or near a line, a linear model might be best. If you graph the data points and they appear to lie on or near a curve, an exponential model might be best. Or, if the growth appears to be constant, use a linear model. If growth appears to be proportional, use an exponential model.

17–21. See answers in back of book.

Assess Your Progress

1–3. See answers in back of book.

Plan⇔Support

Objectives

- Organize information in a matrix.
- Add matrices.
- Multiply a matrix by a scalar.
- Use matrices to model data.

Recommended Pacing

Core and Extended Courses
Section 1.3: 2 days

Block Schedule
Section 1.3: 2 half-blocks (with Sections 1.2 and 1.4)

Resource Materials

Lesson Support
Lesson Plan 1.3
Warm-Up Transparency 1.3
Practice Bank: Practice 3
Study Guide: Section 1.3
Exploration Lab Manual: Diagram Master 2

Technology
Graphing Calculator
McDougal Littell Mathpack
Matrix Analyzer
Stats! Activity Book: Activity 3

Internet:
http://www.hmco.com

Warm-Up Exercises

The table shows color preferences of people surveyed at a shopping mall. Use the data in the table to answer the questions.

	Blue	Red	Yellow
Children	329	460	128
Teens	553	288	31
Adults	307	195	44

1. How many teens preferred red? 288 teens
2. How many adults preferred yellow? 44 adults
3. Which two colors were most preferred? blue and red
4. Which color was most preferred by children? red
5. How many people were surveyed? 2335 people

Section

1.3 Modeling with Matrices

Learn how to...

- organize information in a matrix
- add matrices
- multiply a matrix by a scalar

So you can...

- use matrices to model data, such as characteristics of people

Did you know that about ten percent of the population is left-handed? Scientists who studied subjects in works of art dating from more than 5000 years ago found that the percentage of left-handed people in the world has stayed fairly constant over time.

EXPLORATION
COOPERATIVE LEARNING

Organizing Data in Matrices

Work with half of your class.

1 Find out which hand each person in your group uses to write. Record the results in a *matrix* like the one shown.

	Left	Right	Both
Males	?	?	?
Females	?	?	?

Use the "Both" column if a person writes well with both hands.

2 Give your matrix to the other group. Create a new matrix by adding the number in each position of your matrix to the number in the same position in the other group's matrix. What does the new matrix represent?

3 How many females in your class are right-handed? How many males in your class are left-handed?

4 How do you think you could use your data to predict what a similar matrix for your whole school might be?

A **matrix** is a group of numbers arranged in rows and columns. Each number in a matrix is called an **element**. The **dimensions of a matrix** have the form $r \times c$ where r is the number of rows and c is the number of columns.

$$A = \begin{bmatrix} 3 & 1 & -4 \\ 0 & 7 & 15 \end{bmatrix}$$

Matrix A has 2 rows and 3 columns. Its dimensions are $\mathbf{2 \times 3}$ (read "two by three").

You write $a_{2,1}$ to represent an element in the second row and first column of matrix A.

Exploration Note

Purpose
The purpose of this Exploration is to show students how to organize data in a matrix, to have students understand how to add two matrices, and to have students discover how to multiply a matrix by a number and use the data to make predictions.

Materials/Preparation
No special materials are required.

Procedure
For Step 1, have each group draw a matrix like the one shown. One student can tally the data for the group, or the page can be passed around for students to place tally marks in the appropriate boxes. When all the tally marks have been recorded, have the group make a new matrix using numbers instead of tally marks. The two groups can use their final matrices for Steps 2–4.

Closure
Discuss Steps 2–4. Be sure all students understand how to find the sum in Step 2 and the product in Step 4.

Explorations Lab Manual
See the Manual for more commentary on this Exploration.

EXAMPLE 1

The groups in one of Kim Timpf's algebra classes got the matrices shown when they did the Exploration. Find the class total.

Group A: (columns: Left, Right, Both)

$$\begin{matrix}\text{Males}\\\text{Females}\end{matrix}\begin{bmatrix}2 & 8 & 0\\1 & 5 & 0\end{bmatrix} = A$$

Group B: (columns: Left, Right, Both)

$$\begin{matrix}\text{Males}\\\text{Females}\end{matrix}\begin{bmatrix}0 & 5 & 0\\2 & 9 & 0\end{bmatrix} = B$$

SOLUTION

The class total is the sum $A + B$. You can add the two matrices because the corresponding entries in each matrix represent the same characteristic. For example, $a_{1,1}$ and $b_{1,1}$ both represent left-handed males.

$$A + B = \begin{bmatrix}2 & 8 & 0\\1 & 5 & 0\end{bmatrix} + \begin{bmatrix}0 & 5 & 0\\2 & 9 & 0\end{bmatrix}$$

$$= \begin{bmatrix}2+0 & 8+5 & 0+0\\1+2 & 5+9 & 0+0\end{bmatrix}$$

$$= \begin{bmatrix}2 & 13 & 0\\3 & 14 & 0\end{bmatrix}$$

Add each element in matrix A to the element in the same position in matrix B.

THINK AND COMMUNICATE

1. What is the position of the 5 in matrix B of Example 1? What is the value of $a_{1,3}$ in matrix A of Example 1?
2. Based on the solution to Example 1, state how many students in Kim Timpf's class are:
   a. right-handed females b. left-handed males
3. Can you add matrices G and H? Explain why or why not.

$$G = \begin{bmatrix}3 & 9 & -8\\-4 & 1 & 1\end{bmatrix} \qquad H = \begin{bmatrix}6 & 4\\-2 & 0\\3 & 5\end{bmatrix}$$

EXAMPLE 2

Find $C - D$ when $C = \begin{bmatrix}4 & -2\\7 & 0\end{bmatrix}$ and $D = \begin{bmatrix}1 & 1\\0 & 2\end{bmatrix}$.

SOLUTION

$$C - D = \begin{bmatrix}4 & -2\\7 & 0\end{bmatrix} - \begin{bmatrix}1 & 1\\0 & 2\end{bmatrix}$$

$$= \begin{bmatrix}4-1 & -2-1\\7-0 & 0-2\end{bmatrix}$$

$$= \begin{bmatrix}3 & -3\\7 & -2\end{bmatrix}$$

Subtract corresponding elements.

ANSWERS Section 1.3

Exploration

1–3. Answers may vary.

4. Multiply every number in the matrix by $\frac{\text{number of students in the school}}{\text{number of students in your class}}$.

Think and Communicate

1. first row, second column; 0
2. a. 14 students
   b. 2 students
3. No; the matrices do not have the same dimensions, so they do not have corresponding elements to add.

Teach⇔Interact

Additional Example 1

Luis Ramos surveyed the students in his two physics classes to see how many lived in single-family houses, multifamily houses, or condominiums. He recorded the results in two matrices.

$$\begin{matrix}\text{Males}\\\text{Females}\end{matrix}\overset{\begin{matrix}S & M & C\end{matrix}}{\begin{bmatrix}3 & 9 & 1\\1 & 4 & 0\end{bmatrix}} = P$$

$$\begin{matrix}\text{Males}\\\text{Females}\end{matrix}\overset{\begin{matrix}S & M & C\end{matrix}}{\begin{bmatrix}1 & 10 & 0\\2 & 8 & 1\end{bmatrix}} = Q$$

Find the total for the two classes.

The total is the sum $P + Q$. The matrices can be added since corresponding entries represent the same characteristics.

$$P + Q = \begin{bmatrix}3 & 9 & 1\\1 & 4 & 0\end{bmatrix} + \begin{bmatrix}1 & 10 & 0\\2 & 8 & 1\end{bmatrix}$$

$$= \begin{bmatrix}3+1 & 9+10 & 1+0\\1+2 & 4+8 & 0+1\end{bmatrix}$$

$$= \begin{bmatrix}4 & 19 & 1\\3 & 12 & 1\end{bmatrix}$$

Think and Communicate

When discussing question 3, ask students if they can state a general rule about when it is possible to add two matrices. If necessary, suggest that they use the terminology introduced on page 16. Students should recognize that two matrices can be added if and only if they have the same dimensions.

About Example 2

Mathematical Procedures

Students should see that the conditions under which matrices can be subtracted are the same as for matrix addition. Matrices can be subtracted if and only if they have the same dimensions.

Additional Example 2

Find $M - N$ when $M = \begin{bmatrix}8 & -1 & 4\\7 & 5 & -3\end{bmatrix}$ and $N = \begin{bmatrix}4 & 0 & 1\\-2 & 12 & -5\end{bmatrix}$.

$$M - N = \begin{bmatrix}8 & -1 & 4\\7 & 5 & -3\end{bmatrix} - \begin{bmatrix}4 & 0 & 1\\-2 & 12 & -5\end{bmatrix}$$

$$= \begin{bmatrix}8-4 & -1-0 & 4-1\\7-(-2) & 5-12 & -3-(-5)\end{bmatrix}$$

$$= \begin{bmatrix}4 & -1 & 3\\9 & -7 & 2\end{bmatrix}$$

Teach⇔Interact

About Example 3

Alternate Approach

Another approach for this Example is to find the ratio of 160 to 32, which is 5 to 1. Thus, each element of $\begin{bmatrix} 2 & 13 & 0 \\ 3 & 14 & 0 \end{bmatrix}$ can be multiplied by 5 to 1, or 5.

$$5\begin{bmatrix} 2 & 13 & 0 \\ 3 & 14 & 0 \end{bmatrix} = \begin{bmatrix} 5\cdot 2 & 5\cdot 13 & 5\cdot 0 \\ 5\cdot 3 & 5\cdot 14 & 5\cdot 0 \end{bmatrix} = \begin{bmatrix} 10 & 65 & 0 \\ 15 & 70 & 0 \end{bmatrix}$$

Additional Example 3

Refer to Additional Example 1. Luis Ramos has 120 students in all the classes that he teaches. Use the solution from Additional Example 1 to estimate the totals for all 120 of his students.

The elements of $P + Q$ are counts of students. There are 40 students in all.

Step 1 Convert the counts in $P + Q$ to ratios by multiplying each element of $P + Q$ by the scalar $\frac{1}{40}$.

$$\frac{1}{40}\begin{bmatrix} 4 & 19 & 1 \\ 3 & 12 & 1 \end{bmatrix} = \begin{bmatrix} \frac{4}{40} & \frac{19}{40} & \frac{1}{40} \\ \frac{3}{40} & \frac{12}{40} & \frac{1}{40} \end{bmatrix}$$

Step 2 Estimate the counts for all 120 of Luis Ramos' students by multiplying each element of the matrix from Step 1 by the scalar 120.

$$120\begin{bmatrix} \frac{4}{40} & \frac{19}{40} & \frac{1}{40} \\ \frac{3}{40} & \frac{12}{40} & \frac{1}{40} \end{bmatrix} = \begin{bmatrix} 120\left(\frac{4}{40}\right) & 120\left(\frac{19}{40}\right) & 120\left(\frac{1}{40}\right) \\ 120\left(\frac{3}{40}\right) & 120\left(\frac{12}{40}\right) & 120\left(\frac{1}{40}\right) \end{bmatrix} = \begin{bmatrix} 12 & 57 & 3 \\ 9 & 36 & 3 \end{bmatrix}$$

Closure Question

If you want to add or subtract two matrices in a situation where the matrices model real-world data, what must be true?

The matrices must have the same dimensions, and corresponding elements must represent the same characteristics.

Scalar Multiplication

You can multiply a matrix by a number, or **scalar**. This process, known as **scalar multiplication**, is shown in the following example.

EXAMPLE 3

Use the solution from Example 1 to estimate the totals for all 160 of Kim Timpf's algebra students.

SOLUTION

The elements of $A + B$ are *counts* of students. There are 32 students in all.

Toolbox p. 785 *Ratios and Proportions*

Step 1 Convert the counts to *ratios* by multiplying each element of $A + B$ by the scalar $\frac{1}{32}$.

$$\frac{1}{32}\begin{bmatrix} 2 & 13 & 0 \\ 3 & 14 & 0 \end{bmatrix} = \begin{bmatrix} \frac{2}{32} & \frac{13}{32} & \frac{0}{32} \\ \frac{3}{32} & \frac{14}{32} & \frac{0}{32} \end{bmatrix}$$

Step 2 Estimate the counts for all 160 of Kim Timpf's students by multiplying each element of the matrix from Step 1 by the scalar 160.

$$160\begin{bmatrix} \frac{2}{32} & \frac{13}{32} & \frac{0}{32} \\ \frac{3}{32} & \frac{14}{32} & \frac{0}{32} \end{bmatrix} = \begin{bmatrix} 160\left(\frac{2}{32}\right) & 160\left(\frac{13}{32}\right) & 160\left(\frac{0}{32}\right) \\ 160\left(\frac{3}{32}\right) & 160\left(\frac{14}{32}\right) & 160\left(\frac{0}{32}\right) \end{bmatrix} = \begin{bmatrix} 10 & 65 & 0 \\ 15 & 70 & 0 \end{bmatrix}$$

THINK AND COMMUNICATE

4. What is an estimate for the number of right-handed males in Kim Timpf's classes? for the number of left-handed females?
5. What assumption do you make when "scaling up" from the 32 students in Example 1 to the 160 students in Example 3?

BY THE WAY

People have a preferred foot, as well as a preferred hand. Most people use a particular foot to kick a ball or to pick up an object with their toes.

CHECKING KEY CONCEPTS

For Questions 1–6, use the matrices below to evaluate each matrix expression.

$$A = \begin{bmatrix} 1 & 3 \\ 6 & 2 \end{bmatrix} \quad B = \begin{bmatrix} 4 & 1 \\ 0 & 3 \end{bmatrix} \quad C = \begin{bmatrix} 5 & -2 & 1 \\ 0 & -8 & 6 \\ -3 & 4 & -7 \end{bmatrix} \quad D = \begin{bmatrix} -1 & 2 & 1 \\ 6 & -6 & 4 \\ 2 & -8 & 3 \end{bmatrix}$$

1. $A + B$
2. $C + D$
3. $A - B$
4. $2B$
5. $C + 3D$
6. $3C - 2D$

Think and Communicate

4. about 65 right-handed males; about 15 left-handed females
5. that the 32 students in one class are representative of Kim Timpf's classes as a whole

Checking Key Concepts

1. $\begin{bmatrix} 5 & 4 \\ 6 & 5 \end{bmatrix}$
2. $\begin{bmatrix} 4 & 0 & 2 \\ 6 & -14 & 10 \\ -1 & -4 & -4 \end{bmatrix}$
3. $\begin{bmatrix} -3 & 2 \\ 6 & -1 \end{bmatrix}$
4. $\begin{bmatrix} 8 & 2 \\ 0 & 6 \end{bmatrix}$
5. $\begin{bmatrix} 2 & 4 & 4 \\ 18 & -26 & 18 \\ 3 & -20 & 2 \end{bmatrix}$
6. $\begin{bmatrix} 17 & -10 & 1 \\ -12 & -12 & 10 \\ -13 & 28 & -27 \end{bmatrix}$

Exercises and Applications

1. 2×4
2. 2; 3
3. No; the matrices do not have the same dimensions, so they do not have corresponding elements.
4. $\begin{bmatrix} 8 & 11 \\ -11 & -\frac{5}{2} \end{bmatrix}$
5. $\begin{bmatrix} 14 & 6 \\ 19 & 18 \\ 11 & -17 \end{bmatrix}$
6. undefined
7. $\begin{bmatrix} -2 & 10 \\ -19 & 6 \\ 9 & 9 \end{bmatrix}$

1.3 Exercises and Applications

Extra Practice exercises on page 749

For Exercises 1–3, use the matrix $A = \begin{bmatrix} -1 & 2 & 0 & 11 \\ 3 & 6 & -9 & 4 \end{bmatrix}$.

1. What are the dimensions of A?
2. What is the value of $a_{1,2}$? the value of $a_{2,1}$?
3. **Writing** Let $B = \begin{bmatrix} 1 & 8 & -3 \\ 4 & 2 & 0 \end{bmatrix}$. Can you add matrix A to matrix B? If so, describe the method you use. If not, explain why not.

For Exercises 4–11, use matrices *P, Q, R,* and *S* to evaluate each matrix expression. If an operation cannot be performed, write *undefined*.

$$P = \begin{bmatrix} 7 & 11 \\ -15 & 3 \end{bmatrix} \quad Q = \begin{bmatrix} 1 & 0 \\ 4 & -\frac{11}{2} \end{bmatrix} \quad R = \begin{bmatrix} 8 & -2 \\ 19 & 6 \\ 1 & -13 \end{bmatrix} \quad S = \begin{bmatrix} 6 & 8 \\ 0 & 12 \\ 10 & -4 \end{bmatrix}$$

4. $P + Q$
5. $R + S$
6. $Q + R$
7. $S - R$
8. $2P$
9. $-4P + 2Q$
10. $3R - \frac{1}{2}S$
11. $-Q + 5S$

Connection AGRICULTURE

Farmers in the United States harvest crops from more than 300 million acres of land. Almost 25% of the crops are grown in Illinois, Iowa, and Kansas. The table shows the number of bushels, in thousands, harvested for four leading crops.

	Corn		Wheat		Soybeans		Oats	
	1990	1991	1990	1991	1990	1991	1990	1991
Illinois	1,320,800	1,177,000	91,200	44,800	354,900	341,250	11,560	6,600
Iowa	1,562,400	1,427,400	3,375	1,700	323,900	350,325	40,800	21,250
Kansas	188,000	206,250	472,000	363,000	46,800	43,700	6,600	5,830

12. a. Create a 3 × 4 matrix to represent the data for 1990. Label this matrix A.
    b. Create a 3 × 4 matrix using the data for 1991. Label this matrix B.
    c. For matrices A and B, what does each row represent? What does each column represent?
13. Write an expression in terms of A and B to represent the total amount of each crop produced in each state during 1990 and 1991. Then evaluate your expression.
14. Calculate the matrix $B - A$. What does this matrix represent? What does a negative number in matrix $B - A$ represent?
15. How would the dimensions of each matrix change if Nebraska and Missouri were added to the list of states? if barley were added to the list of crops?

8. $\begin{bmatrix} 14 & 22 \\ -30 & 6 \end{bmatrix}$
9. $\begin{bmatrix} -26 & -44 \\ 68 & -23 \end{bmatrix}$
10. $\begin{bmatrix} 21 & -10 \\ 57 & 12 \\ -2 & -37 \end{bmatrix}$
11. undefined

12. a. $A = \begin{bmatrix} 1{,}320{,}800 & 91{,}200 & 354{,}900 & 11{,}560 \\ 1{,}562{,}400 & 3{,}375 & 323{,}900 & 40{,}800 \\ 188{,}000 & 472{,}000 & 46{,}800 & 6{,}600 \end{bmatrix}$

    b. $B = \begin{bmatrix} 1{,}177{,}000 & 44{,}800 & 341{,}250 & 6{,}600 \\ 1{,}427{,}400 & 1{,}700 & 350{,}325 & 21{,}250 \\ 206{,}250 & 363{,}000 & 43{,}700 & 5{,}830 \end{bmatrix}$

    c. one state's production of the four major crops; production of a particular crop

13. $A + B$; $\begin{bmatrix} 2{,}497{,}800 & 136{,}000 & 696{,}150 & 18{,}160 \\ 2{,}989{,}800 & 5{,}075 & 674{,}225 & 62{,}050 \\ 394{,}250 & 835{,}000 & 90{,}500 & 12{,}430 \end{bmatrix}$

14. $\begin{bmatrix} -143{,}800 & -46{,}400 & -13{,}650 & -4{,}960 \\ -135{,}000 & -1{,}675 & 26{,}425 & -19{,}550 \\ 18{,}250 & -109{,}000 & -3{,}100 & -770 \end{bmatrix}$; the change in production from 1990 to 1991; a decrease in production from 1990 to 1991

15. The dimensions would be 5 × 4; the dimensions would be 3 × 5 (or 5 × 5 if Nebraska and Missouri were added as well).

Apply⇔Assess

Suggested Assignment

Core Course
Day 1 Exs. 1–18
Day 2 Exs. 19–25, 28–41

Extended Course
Day 1 Exs. 1–18
Day 2 Exs. 19–41

Block Schedule
Day 2 Exs. 1–18
Day 3 Exs. 19–25, 28–41

Exercise Notes

Using Technology *Exs. 4–11, 23–25* You may want to suggest that students use the *Matrix Analyzer* software for this group of exercises.

Exs. 4–11 can also be done on graphing calculators that have matrix capabilities. The TI-82 can handle matrices with up to 99 rows and 25 columns. Entering a matrix is simple. Matrices on the TI-82 are called [A], [B], [C], [D], or [E]. To enter the matrix P, press MATRX ▶ ▶ 1 to select [A]. The calculator will display a screen where one can enter the matrix dimensions and elements. To enter the dimensions, press 2 ENTER 2 ENTER. The dimensions 2 × 2 will be displayed on the first line of the screen and the cursor will move to row 1, column 1. Press the keys for the number that goes in that position and then press ENTER. Type the other number for row 1 and press ENTER. The element is entered in the matrix and the cursor goes to row 2, column 1. Enter the second row in the same way as row 1. (The keystrokes for entering a matrix on the TI-81 are different from those for the TI-82.) To enter matrix Q, use the name [B] and follow the same procedure. Note that the element $-\frac{11}{2}$ can be entered in the row 2, column 2 position by pressing (–) 11 ÷ 2 ENTER. To add P and Q (Ex. 4), on the homescreen, press MATRX 1 + MATRX 2 ENTER. To multiply P by 2, press 2 MATRX 1 ENTER. Note that the scalar must be placed *before* the matrix. (On the TI-81, the scalar may be placed either before or after the matrix.)

Apply⇔Assess

Exercise Notes

Visual Thinking

Exs. 16–21 Ideas for helping students develop their visual thinking skills are provided at strategic points throughout the side-column notes. Visual thinking skills include:

1. Observation
2. Identification
3. Recognition
4. Recall
5. Interpretation
6. Exploration
7. Correlation
8. Generalization
9. Inference
10. Perception
11. Communication
12. Self-Expression

Integrating the Strands

Exs. 16–21 Matrices have many important applications in the study of geometry. These exercises give students an idea of how matrices can be used to translate and dilate geometric figures. Matrices can also be used to describe rotations of figures.

Reasoning

Exs. 23–25 Ask students to state in mathematical terms any generalizations that these exercises suggest. Then have them explain why the generalizations are true. (If A, B, and C are any matrices that have the same dimensions and r is any scalar, then $A + B = B + A$, $A + (B + C) = (A + B) + C$, and $r(A + B) = rA + rB$. The matrix operations have these properties because matrix addition and scalar multiplication are performed on an element-by-element basis and hence inherit the corresponding properties of real number operations.)

Application

Ex. 26 This exercise shows students how the operation of adding matrices can be applied to a situation involving personal finance.

Connection GEOMETRY

Have you ever used a computer drawing program to enlarge a figure or change its position? The computer can use scalar multiplication and matrix addition to find the coordinates of the changed figure.

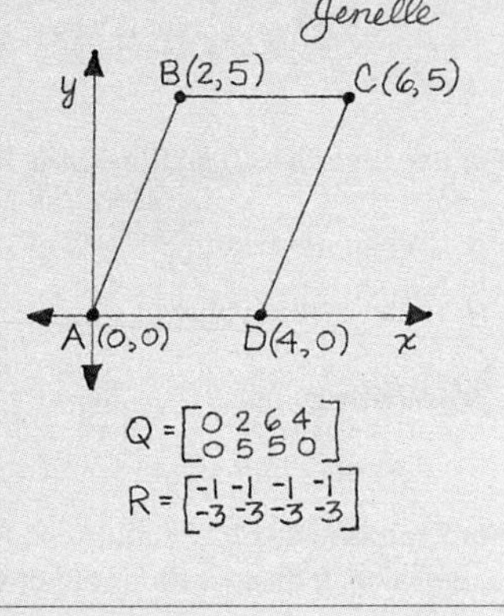

Jenelle wrote the coordinates of quadrilateral *ABCD* in matrix *Q*.

16. What does each column of matrix Q represent?

17. a. Find $Q + R$.

b. Matrix $Q + R$ defines a new quadrilateral $EFGH$. Sketch $EFGH$. How is $EFGH$ related to $ABCD$?

18. a. Find $2Q$.

b. Matrix $2Q$ defines a new quadrilateral $IJKL$. Sketch $IJKL$. How is $IJKL$ related to $ABCD$?

For Exercises 19–21, the coordinates of a geometric figure are given in the black matrix, and a transformation is indicated in red. Describe each transformation. Then sketch the original figure and its image to check your answer.

19. $3\begin{bmatrix} 0 & 1 & 4 \\ 0 & 2 & 0 \end{bmatrix}$ **20.** $\frac{1}{2}\begin{bmatrix} 0 & 8 & 2 & -2 \\ 2 & 4 & -8 & -2 \end{bmatrix}$ **21.** $\begin{bmatrix} 1 & 3 & 5 & 3 \\ 3 & 6 & 3 & -3 \end{bmatrix} + \begin{bmatrix} -1 & -1 & -1 & -1 \\ 3 & 3 & 3 & 3 \end{bmatrix}$

22. Suppose A is any matrix.

a. Describe the matrix $A - A$.

b. How are the matrices $A + A$ and $2A$ related?

Use matrices *A*, *B*, and *C* for Exercises 23–25.

$$A = \begin{bmatrix} 3 & 1 & -2 \\ -1 & 5 & 6 \\ 4 & 13 & 0 \end{bmatrix} \quad B = \begin{bmatrix} -9 & 10 & -3 \\ 0 & 6 & 1 \\ 14 & 7 & -8 \end{bmatrix} \quad C = \begin{bmatrix} 7 & -2 & 0 \\ 15 & 1 & 19 \\ -4 & 12 & 2 \end{bmatrix}$$

23. Show that $A + B = B + A$.

24. Show that $A + (B + C) = (A + B) + C$.

25. Show that $r(A + B) = rA + rB$ where r is any scalar.

26. PERSONAL FINANCE Roger and Meredith Stone each have a savings account and a checking account. The money they had in each account at the end of May is represented by matrix A. Matrix D represents the total deposits Roger and Meredith made in June, and matrix W represents the total withdrawals each made in June.

	Savings	Checking
Roger	3000	540
Meredith	9500	1200

$= A$

Savings	Checking
300	100
0	250

$= D$

Savings	Checking
0	−120
−1500	−1000

$= W$

a. Write an expression to show how to calculate the amount of money in each account at the end of June.

b. Simplify the expression you wrote in part (a) to write a matrix showing the amount of money in each account at the end of June.

16. the coordinates of a vertex of $ABCD$

17. a. $\begin{bmatrix} -1 & 1 & 5 & 3 \\ -3 & 2 & 2 & -3 \end{bmatrix}$

b.

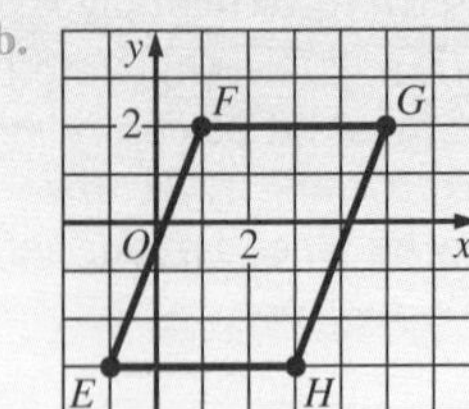

$EFGH$ is $ABCD$ moved 1 unit in the negative x-direction and 3 units in the negative y-direction.

18. See answers in back of book.

19. enlarged by a factor of 3

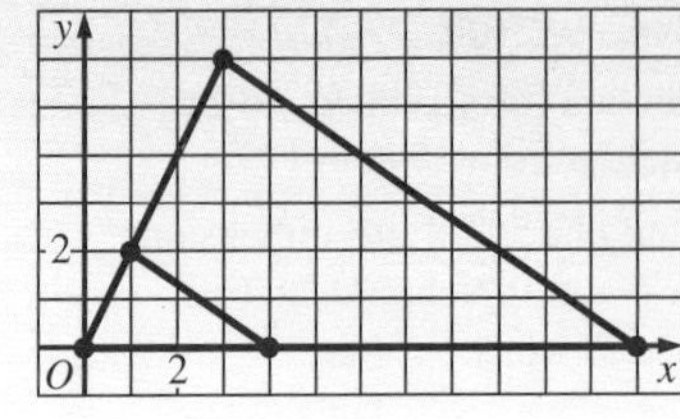

20. reduced by $\frac{1}{2}$

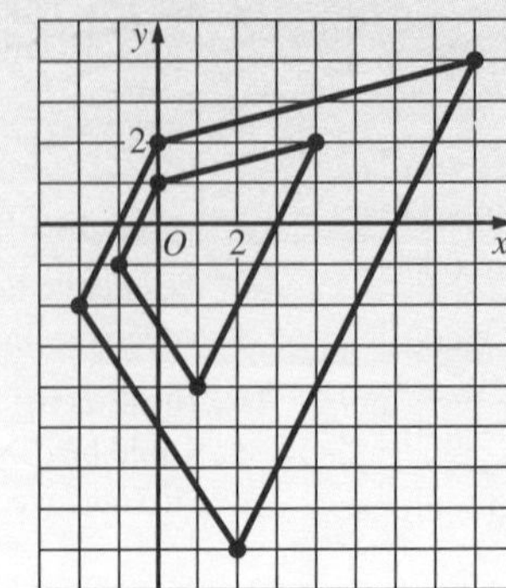

21–24. See answers in back of book.

25. The element in row m, column n of $r(A + B)$ is $r(a_{m,n} + b_{m,n})$. The element in row m, column n of $rA + rB$ is $ra_{m,n} + rb_{m,n}$, which is equal to $r(a_{m,n} + b_{m,n})$.

$r(A + B) = $

$$r\begin{bmatrix} -6 & 11 & -5 \\ -1 & 11 & 7 \\ 18 & 20 & -8 \end{bmatrix} = \begin{bmatrix} -6r & 11r & -5r \\ -r & 11r & 7r \\ 18r & 20r & -8r \end{bmatrix};$$

27. Challenge Find X if X is a matrix and $2X + \begin{bmatrix} 9 & -2 \\ 4 & 7 \end{bmatrix} = \begin{bmatrix} 3 & 10 \\ -6 & 12 \end{bmatrix}$.

28. SAT/ACT Preview What value of a makes the equation true?

$$\begin{bmatrix} 2 & -5 \\ -1 & 4 \end{bmatrix} + \begin{bmatrix} 1 & 0 \\ 2a & -5 \end{bmatrix} = \begin{bmatrix} 3 & -5 \\ -9 & -1 \end{bmatrix}$$

A. 5 **B.** -5 **C.** -10 **D.** -4 **E.** 2

ONGOING ASSESSMENT

29. Cooperative Learning Work in a group of four. Each of you should choose a class (freshman, sophomore, junior, or senior) and ask ten to twenty people from the class how they most often get to school. Complete a matrix like the one below for each class.

	Walk	Ride bike	Ride bus	Ride in car	Drive
Females	?	?	?	?	?
Males	?	?	?	?	?

a. Find out the total number of students in each class in your school. Scale each matrix up to estimate how many males and females in each class use each form of transportation.

b. Add the four matrices you found in part (a) to create a matrix estimating how many students in your school use each form of transportation.

c. Why wouldn't you add the matrices first and then scale up?

SPIRAL REVIEW

ENTERTAINMENT The table shows the number of people, in millions, subscribing to cable television systems from 1960 to 1990. *(Section 1.2)*

Decades since 1960	Cable users (millions)
0	0.65
1	4.5
2	16
3	50

30. Write a linear function to model the situation.

31. Write an exponential function to model the situation.

32. Estimate the number of cable television subscribers in 2000. Explain how you arrived at your estimate.

Simplify. *(Toolbox, page 778)*

33. $4(-5) + 7(3)$ **34.** $16 \div 8 + 12 \div 3$ **35.** $15 \cdot 2 + 3 \cdot 6$

Tell whether each property is true for real numbers *a*, *b*, and *c*. Give an example to support your answer. *(Toolbox, page 774)*

36. $a + b = b + a$ **37.** $a - b = b - a$ **38.** $ab = ba$

39. $a \cdot \frac{1}{a} = 1$ **40.** $b \cdot 1 = b$ **41.** $|a| + |b| = |a + b|$

Apply⇔Assess

Exercise Notes

Challenge

Ex. 27 Students can find X by using procedures similar to the ones they used when solving equations.

Problem Solving

Ex. 28 Encourage students to solve this problem mentally and to check the answer.

Assessment Note

Ex. 29 Decide how to provide time for each group to gather data. This exercise presents an opportunity for students to demonstrate their ability to use matrices and matrix operations to model and study a real-world situation. Different groups will find it beneficial to compare their results for parts (b) and (c).

Practice 3 for Section 1.3

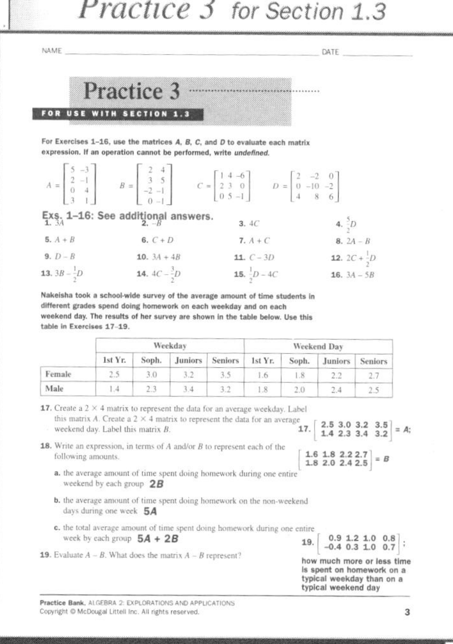

NAME ______ DATE ______

Practice 3

FOR USE WITH SECTION 1.3

For Exercises 1–16, use the matrices *A*, *B*, *C*, and *D* to evaluate each matrix expression. If an operation cannot be performed, write *undefined*.

$A = \begin{bmatrix} 5 & -3 \\ 2 & -1 \\ 0 & 4 \\ 3 & 1 \end{bmatrix}$ $B = \begin{bmatrix} 2 & 4 \\ 3 & 5 \\ -2 & -1 \\ 0 & -1 \end{bmatrix}$ $C = \begin{bmatrix} 1 & 4 & -6 \\ 2 & 3 & 0 \\ 0 & 5 & -1 \end{bmatrix}$ $D = \begin{bmatrix} 2 & -2 & 0 \\ 0 & -10 & -2 \\ 4 & 8 & 6 \end{bmatrix}$

Exs. 1–16: See additional answers.

1. $3A$ 2. $-B$ 3. $4C$ 4. $\frac{5}{2}D$
5. $A + B$ 6. $C + D$ 7. $A + C$ 8. $2A - B$
9. $D - B$ 10. $3A + 4B$ 11. $C - 3D$ 12. $2C + \frac{1}{2}D$
13. $3B - \frac{1}{2}D$ 14. $4C - \frac{3}{2}D$ 15. $\frac{1}{2}D - 4C$ 16. $3A - 5B$

Nakeisha took a school-wide survey of the average amount of time students in different grades spend doing homework on each weekday and on each weekend day. The results of her survey are shown in the table below. Use this table in Exercises 17–19.

	Weekday				Weekend Day			
	1st Yr.	Soph.	Juniors	Seniors	1st Yr.	Soph.	Juniors	Seniors
Female	2.5	3.0	3.2	3.5	1.6	1.8	2.2	2.7
Male	1.4	2.3	3.4	3.2	1.8	2.0	2.4	2.5

17. Create a 2×4 matrix to represent the data for an average weekday. Label this matrix A. Create a 2×4 matrix to represent the data for an average weekend day. Label this matrix B.

17. $\begin{bmatrix} 2.5 & 3.0 & 3.2 & 3.5 \\ 1.4 & 2.3 & 3.4 & 3.2 \end{bmatrix} = A$; $\begin{bmatrix} 1.6 & 1.8 & 2.2 & 2.7 \\ 1.8 & 2.0 & 2.4 & 2.5 \end{bmatrix} = B$

18. Write an expression, in terms of A and/or B to represent each of the following amounts.

a. the average amount of time spent doing homework during one entire weekend by each group **2B**

b. the average amount of time spent doing homework on the non-weekend days during one week **5A**

c. the total average amount of time spent doing homework during one entire week by each group **5A + 2B**

19. Evaluate $A - B$. What does the matrix $A - B$ represent?

19. $\begin{bmatrix} 0.9 & 1.2 & 1.0 & 0.8 \\ -0.4 & 0.3 & 1.0 & 0.7 \end{bmatrix}$; **how much more or less time is spent on homework on a typical weekday than on a typical weekend day**

Practice Bank, ALGEBRA 2: EXPLORATIONS AND APPLICATIONS

3

$rA + rB =$

$$\begin{bmatrix} 3r & r & -2r \\ -r & 5r & 6r \\ 4r & 13r & 0 \end{bmatrix} + \begin{bmatrix} -9r & 10r & -3r \\ 0 & 6r & r \\ 14r & 7r & -8r \end{bmatrix} =$$

$$\begin{bmatrix} -6r & 11r & -5r \\ -r & 11r & 7r \\ 18r & 20r & -8r \end{bmatrix}$$

26. a. $A + D + W$ b. $\begin{bmatrix} 3300 & 520 \\ 8000 & 450 \end{bmatrix}$

27. $\begin{bmatrix} -3 & 6 \\ -5 & 2\frac{1}{2} \end{bmatrix}$

28. D

29. a, b. Answers may vary.
c. The scale factors for the four classes are probably not the same.

30, 31. Answers may vary due to rounding. Let x = the number of decades since 1960 and y = the number (in millions) of cable users.

30. $y = 0.65 + 16.45x$

31. $y = 0.65(4.53)^x$

32. about 66.45 million (linear model); about 273.72 million (exponential model)

33. 1

34. 6

35. 48

36. True; for example, $2 + 3 = 5 = 3 + 2$.

37. False; for example, $5 - 1 = 4$, but $1 - 5 = -4$.

38. True; for example, $2 \cdot 3 = 6 = 3 \cdot 2$.

39. True for every real number a except 0; for example, $7 \cdot \frac{1}{7} = 1$. ($\frac{1}{0}$ is not defined.)

40. True; for example, $9 \cdot 1 = 9$.

41. False; for example, $|-5| + |5| = 5 + 5 = 10$, while $|-5 + 5| = |0| = 0$.

Plan⇔Support

Objectives

- Multiply matrices.
- Solve problems involving matrix multiplication.

Recommended Pacing

Core and Extended Courses
Section 1.4: 2 days

Block Schedule
Section 1.4: 2 half-blocks (with Sections 1.3 and 1.5)

Resource Materials

Lesson Support
Lesson Plan 1.4
Warm-Up Transparency 1.4
Overhead Visuals:
  Folder 1: Multiplying Matrices
Practice Bank: Practice 4
Study Guide: Section 1.4
Exploration Lab Manual:
  Additional Exploration 1

Technology
Technology Book:
  Spreadsheet Activity 1
Scientific Calculator
Graphing Calculator
McDougal Littell Mathpack
  Function Investigator with Matrix Analyzer Activity Book: Matrix Analyzer Activity 1
Internet:
  http://www.hmco.com

Warm-Up Exercises

Give the dimensions of each matrix.

1. $\begin{bmatrix} 5 & 8 & -1 \\ -7 & 0 & 0 \\ 3 & 6 & 0 \end{bmatrix}$ 3×3

2. $\begin{bmatrix} 4 & -9 & 3 & 6 \\ -4 & 8 & 7 & -2 \end{bmatrix}$ 2×4

3. $\begin{bmatrix} 0 \\ 3 \\ 1 \end{bmatrix}$ 3×1

Evaluate each expression.

4. $3(-15) + 8(20) + (-9)(-8)$ 187

5. $2(100) + (-3)(400) + 8(200)$ 600

Section

1.4 Multiplying Matrices

Learn how to...

- multiply matrices

So you can...

- solve problems involving matrix multiplication, such as planning a canoe trip

Have you ever gone canoeing? Depending on the route you choose, you may need to carry the canoes and supplies over land to get to another waterway or to avoid an obstacle. This process is called *portage*.

EXAMPLE 1 Application: Canoeing

A group of canoers arrives at the Canoe Centre at Opeongo Campgrounds in Algonquin Provincial Park. There are about seven hours of daylight left, six of which they can use to travel to a campsite. They can canoe 1 km in 10 min, but they need 20 min to portage 1 km. They would like to camp at one of the four sites shown on the map. Which route should they take?

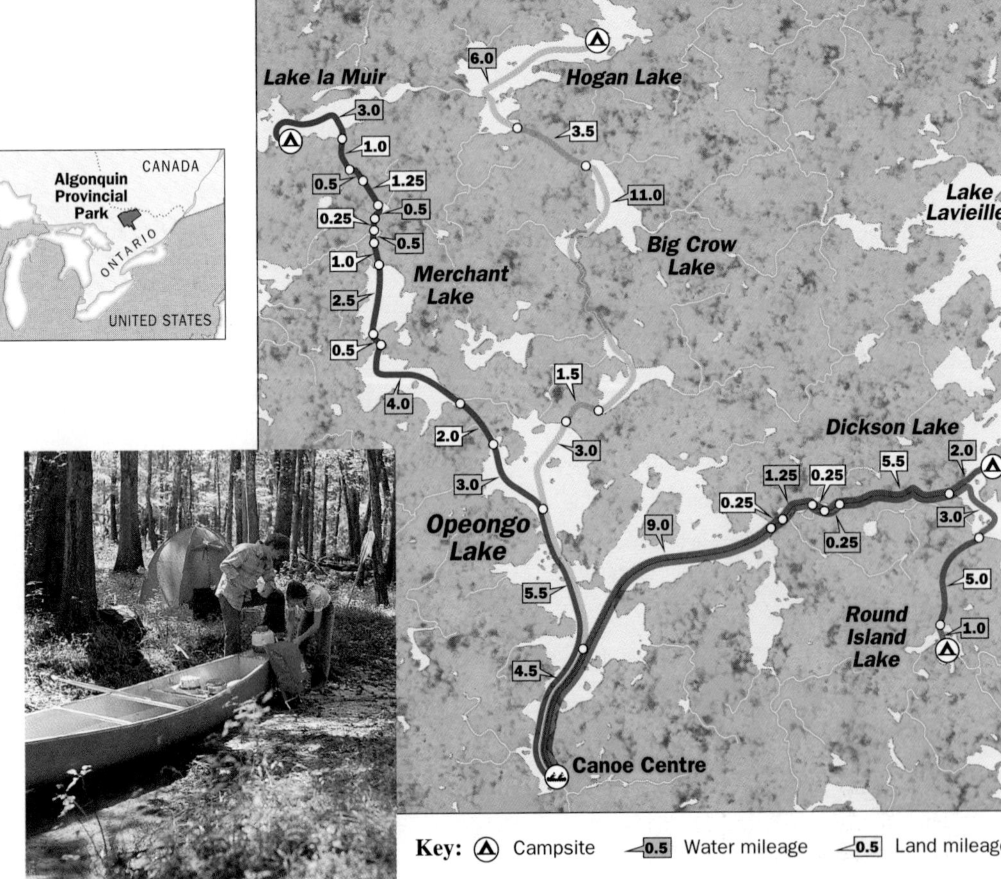

SOLUTION

Let matrix A show the canoeing and portaging distances, in kilometers, along each route. Let matrix B show the number of minutes per kilometer that the group can travel using each method of travel.

	Canoe (km)	Portage (km)
Route to Lake la Muir	24	6
Route to Hogan Lake	30	5
Route to Dickson Lake	17	6
Route to Round Island Lake	19	11

$= A$

	Time (min/km)
Canoe	10
Portage	20

$= B$

Use matrices A and B to create a matrix of total travel times to each lake.

	Time (min)		
Route to Lake la Muir	24(10) + 6(20)	=	360
Route to Hogan Lake	30(10) + 5(20)		400
Route to Dickson Lake	17(10) + 6(20)		290
Route to Round Island Lake	19(10) + 11(20)		410

Since the group has 6 h, or 360 min, they have time to get to either Lake la Muir or Dickson Lake.

Matrix Multiplication

Example 1 illustrates a process called **matrix multiplication**.

$$\begin{bmatrix} 24 & 6 \\ \mathbf{30} & \mathbf{5} \\ 17 & 6 \\ 19 & 11 \end{bmatrix} \begin{bmatrix} \mathbf{10} \\ \mathbf{20} \end{bmatrix} = \begin{bmatrix} 24(10) + 6(20) \\ \mathbf{30(10) + 5(20)} \\ 17(10) + 6(20) \\ 19(10) + 11(20) \end{bmatrix} = \begin{bmatrix} 360 \\ \mathbf{400} \\ 290 \\ 410 \end{bmatrix}$$

Dimensions: $4 \times 2 \quad 2 \times 1 \qquad 4 \times 1$

This element, which is in the second row and first column of the product matrix, is the sum of the products of the corresponding elements in the second row of A and the first column of B: **400 = 30(10) + 5(20).**

For the product to be defined, the number of elements in a **row of A** must equal the number of elements in a **column of B**.

The product matrix has the same number of rows as A and the same number of columns as B.

When you multiply matrices with labels, the column labels of the first matrix must have the same attributes as the row labels of the second matrix.

These labels are the same.

	Canoe	Portage
Route to Lake la Muir	24	6
Route to Hogan Lake	30	5
Route to Dickson Lake	17	6
Route to Round Island Lake	19	11

$\times$

	Time
Canoe	10
Portage	20

$=$

	Time
Route to Lake la Muir	360
Route to Hogan Lake	400
Route to Dickson Lake	290
Route to Round Island Lake	410

Teach⇔Interact

Additional Example 1

Refer to Example 1. Another group of campers camped for 3 days at Lake la Muir. They wanted to return to the Canoe Center, renew their supplies, and then go on to another campsite. They assumed that they could canoe 1 km in 15 min and portage 1 km in 25 min. Would it be possible for them to start at daybreak, allow 2.5 hours at the Canoe Center, and move on to reach another campsite in a 15 hour day?

Let matrix C show the canoeing and portaging distances, in kilometers, along each route. Let matrix D show the number of minutes per kilometer that the group can travel using each method of travel. Remember that each distance must include the distance from Lake la Muir to the Canoe Center.

	Canoe (km)	Portage (km)
To H. Lake	54	11
To D. Lake	41	12
To R.I. Lake	43	17

$= C$

	Time (min/km)
Canoe	15
Portage	25

$= D$

Use matrices C and D to create a matrix of total travel times to each of the other campsites.

	Time (min)
To H. Lake	54(15) + 11(25)
To D. Lake	41(15) + 12(25)
To R.I. Lake	43(15) + 17(25)

$$= \begin{bmatrix} 1085 \\ 915 \\ 1070 \end{bmatrix}$$

There are 900 min in 15 hours. The shortest time in the final matrix is greater than 900 min. The group cannot make it to another campsite in a 15 hour day, especially if they need another 2.5 hours at the Canoe Center.

Section Note

Using Technology

Matrix Analyzer Activity 1 in the *Function Investigator with Matrix Analyzer Activity Book* provides an opportunity to discover properties of matrix addition, subtraction, multiplication, scalar multiplication, and powers. Students also explore whether matrix operations are commutative or associative, and compare matrix operations with operations on real numbers.

Teach⇔Interact

Additional Example 2

The Linn, Schwartz, Mellon, and Leon families ordered camping supplies from a catalog. They also ordered binoculars for those who were interested in bird watching. The tables show the items ordered and the weight and cost for each item.

	Tents	Sleep bags	Backpacks	Cook sets	Binoculars
Linn	1	3	3	1	1
Schwartz	1	4	2	1	0
Mellon	2	2	3	2	1
Leon	2	4	3	1	1

	Weight (lb)	Cost ($)
Tents	8.5	375
Sleeping bags	4	210
Backpacks	3	185
Cook sets	4	30
Binoculars	0.5	150

Find the shipping weight and total cost for each family's order.

Let matrix C show each family's camping supplies order, and let matrix D show the shipping weight and cost of each item.

Columns: Tents, Sleep bags, Backpacks, Cook sets, Binoculars; rows: Linn, Schwartz, Mellon, Leon

$$\begin{bmatrix} 1 & 3 & 3 & 1 & 1 \\ 1 & 4 & 2 & 1 & 0 \\ 2 & 2 & 3 & 2 & 1 \\ 2 & 4 & 3 & 1 & 1 \end{bmatrix} = C$$

Columns: Weight, Cost; rows: Tents, Sleeping bags, Backpacks, Cook sets, Binoculars

$$\begin{bmatrix} 8.5 & 375 \\ 4 & 210 \\ 3 & 185 \\ 4 & 30 \\ 0.5 & 150 \end{bmatrix} = D$$

CD will be a 4×2 matrix showing the shipping weight and total cost for each family. Find CD.

Method 1 Use paper and pencil.

$$CD = \begin{bmatrix} 1 & 3 & 3 & 1 & 1 \\ 1 & 4 & 2 & 1 & 0 \\ 2 & 2 & 3 & 2 & 1 \\ 2 & 4 & 3 & 1 & 1 \end{bmatrix}\begin{bmatrix} 8.5 & 375 \\ 4 & 210 \\ 3 & 185 \\ 4 & 30 \\ 0.5 & 150 \end{bmatrix}$$

$$= \begin{bmatrix} 34 & 1740 \\ 34.5 & 1615 \\ 42.5 & 1935 \\ 46.5 & 2325 \end{bmatrix}$$

EXAMPLE 2 Application: Camping

The Graham, Piscitelli, and Brewer families order supplies for an upcoming camping trip from the catalog shown. The number of items needed by each family is given in the table. Find the shipping weight and total cost for each family's order.

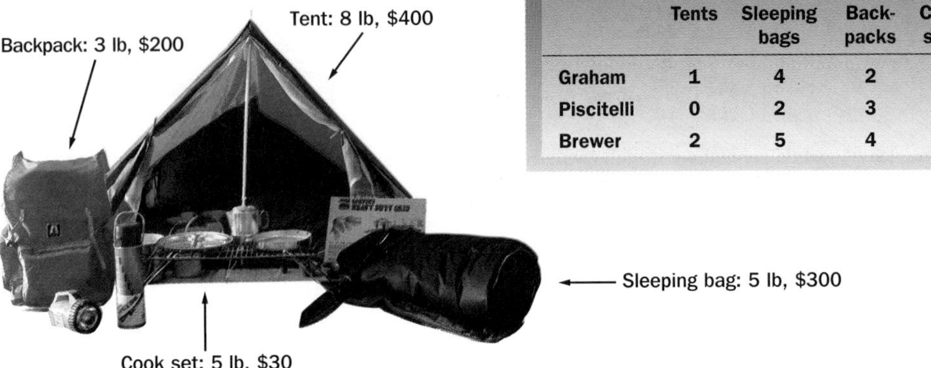

	Tents	Sleeping bags	Back-packs	Cook sets
Graham	1	4	2	0
Piscitelli	0	2	3	1
Brewer	2	5	4	2

SOLUTION

Let matrix A show each family's camping supplies order, and let matrix B show the shipping weight and cost of each item.

Columns: Tents, Sleeping bags, Backpacks, Cook sets; rows: Graham, Piscitelli, Brewer

$$\begin{bmatrix} 1 & 4 & 2 & 0 \\ 0 & 2 & 3 & 1 \\ 2 & 5 & 4 & 2 \end{bmatrix} = A$$

Columns: Weight (lb), Cost ($); rows: Tents, Sleeping bags, Backpacks, Cook sets

$$\begin{bmatrix} 8 & 400 \\ 5 & 300 \\ 3 & 200 \\ 5 & 30 \end{bmatrix} = B$$

The product AB will be a 3×2 matrix showing the shipping weight and total cost for each family. Find AB.

Method 1 Use paper and pencil.

$$AB = \begin{bmatrix} 1 & 4 & 2 & 0 \\ 0 & 2 & 3 & 1 \\ 2 & 5 & 4 & 2 \end{bmatrix}\begin{bmatrix} 8 & 400 \\ 5 & 300 \\ 3 & 200 \\ 5 & 30 \end{bmatrix}$$

$$= \begin{bmatrix} 1(8) + 4(5) + 2(3) + 0(5) & 1(400) + 4(300) + 2(200) + 0(30) \\ 0(8) + 2(5) + 3(3) + 1(5) & 0(400) + 2(300) + 3(200) + 1(30) \\ 2(8) + 5(5) + 4(3) + 2(5) & 2(400) + 5(300) + 4(200) + 2(30) \end{bmatrix}$$

$$= \begin{bmatrix} 34 & 2000 \\ 24 & 1230 \\ 63 & 3160 \end{bmatrix}$$

The order placed by the Graham family weighs 34 lb and costs $2000. The order placed by the Piscitelli family weighs 24 lb and costs $1230. The order placed by the Brewer family weighs 63 lb and costs $3160.

Method 2 Use a graphing calculator or software with matrix calculation capabilities.

Use the matrix feature to enter matrices A and B. Then find AB.

The order placed by the Graham family weighs 34 lb and costs \$2000. The order placed by the Piscitelli family weighs 24 lb and costs \$1230. The order placed by the Brewer family weighs 63 lb and costs \$3160.

For more information about matrix calculations, see the *Technology Handbook*, pp. 811–812.

CHECKING KEY CONCEPTS

Find each product. If the matrices cannot be multiplied, state that the product is *undefined*.

$P = \begin{bmatrix} 2 & 0 \\ 5 & 1 \end{bmatrix}$ $Q = \begin{bmatrix} 4 & 3 & -1 \\ -2 & 5 & 1 \end{bmatrix}$ $R = \begin{bmatrix} 1 & 2 & -1 \\ 4 & 3 & -3 \\ 5 & 1 & -1 \end{bmatrix}$ $S = \begin{bmatrix} 8 & -10 & 0 \\ 3 & -6 & 4 \\ -2 & 1 & 5 \end{bmatrix}$

1. a. PQ **b.** QP

2. a. QR **b.** RQ

3. a. RS **b.** SR

4. a. QS **b.** SQ

5. Which pair of products in Questions 1–4 were both defined? Were any of the products commutative?

1.4 Exercises and Applications

Extra Practice exercises on page 750

1. In Example 2, suppose each family also orders some compasses. How would the dimensions of matrices A, B, and AB change?

Tell whether *LM* and *ML* are defined. If so, give the dimensions of the product matrices. If not, explain why not.

2. L is any 4×2 matrix. M is any 1×4 matrix.

3. L is any 8×6 matrix. M is any 6×6 matrix.

4. L is any 3×3 matrix. M is any 3×1 matrix.

5. L is any 4×5 matrix. M is any 3×4 matrix.

For Exercises 6–13, use matrices *A, B, C, D,* and *E* to find each product. If the matrices cannot be multiplied, state that the product is *undefined*.

$A = \begin{bmatrix} 2 \\ -4 \\ 6 \end{bmatrix}$ $B = \begin{bmatrix} 3 & 1 \\ 5 & 0 \end{bmatrix}$ $C = \begin{bmatrix} 2 & 0 & 7 \\ -1 & -2 & 0 \\ 3 & 7 & 4 \end{bmatrix}$ $D = \begin{bmatrix} 0 & 3 & -2 \\ 2 & 5 & 1 \end{bmatrix}$ $E = \begin{bmatrix} 1 & -2 \\ 5 & 3 \end{bmatrix}$

6. CA **7.** BE **8.** CD **9.** DA

10. AB **11.** ED **12.** DC **13.** BA

Teach⇔Interact

Additional Example 2 (continued)

Method 2 Use a graphing calculator or software with matrix calculation capabilities.

```
[C][D]
  [[34    1740]
   [34.5  1615]
   [42.5  1935]
   [46.5  2325]]
```

The orders placed by the families have the following weights and costs:

Linn family:	34 lb, \$1740
Schwartz family:	34.5 lb, \$1615
Mellon family:	42.5 lb, \$1935
Leon family:	46.5 lb, \$2325

Closure Question

How can you decide, for a given pair of matrices A and B, whether AB exists? If it does exist, how can you find the element for a given row and column of the product?

AB exists if the number of columns in matrix A is the same as the number of rows in matrix B. If $k_{i,j}$ is the element for row i column j of the product, the value of $k_{i,j}$ can be found by going across row i of A and down column j of B, multiplying the numbers element by element. Add all the products to find $k_{i,j}$.

Apply⇔Assess

Suggested Assignment

Core Course
Day 1 Exs. 1–16
Day 2 Exs. 23–27, 29–41

Extended Course
Day 1 Exs. 1–16
Day 2 Exs. 17–41

Block Schedule
Day 3 Exs. 1–16
Day 4 Exs. 23–27, 29–41

ANSWERS Section 1.4

Checking Key Concepts

1. a. $\begin{bmatrix} 8 & 6 & -2 \\ 18 & 20 & -4 \end{bmatrix}$ b. undefined

2. a. $\begin{bmatrix} 11 & 16 & -12 \\ 23 & 12 & -14 \end{bmatrix}$ b. undefined

3. a. $\begin{bmatrix} 16 & -23 & 3 \\ 47 & -61 & -3 \\ 45 & -57 & -1 \end{bmatrix}$ b. $\begin{bmatrix} -32 & -14 & 22 \\ -1 & -8 & 11 \\ 27 & 4 & -6 \end{bmatrix}$

4. a. $\begin{bmatrix} 43 & -59 & 7 \\ -3 & -9 & 25 \end{bmatrix}$ b. undefined

5. RS and SR; No.

Exercises and Applications

1. A: 3×5; B: 5×2; AB: 3×2

2. LM: No; the number of columns in L does not equal the number of rows in M. ML: Yes; 1×2.

3. LM: Yes; 8×6. ML: No, the number of columns in M does not equal the number of rows in L.

4. LM: Yes; 3×1. ML: No; the number of columns in M does not equal the number of rows in L.

5. LM: No; the number of columns in L does not equal the number of rows in M. ML: Yes; 3×5.

6. $\begin{bmatrix} 46 \\ 6 \\ 2 \end{bmatrix}$

7. $\begin{bmatrix} 8 & -3 \\ 5 & -10 \end{bmatrix}$

8–13. See answers in back of book.

Apply⇔Assess

Exercise Notes

Common Error

Exs. 6–13 Many students have difficulty multiplying matrices and tend to make a variety of conceptual and computational errors. When multiplying matrices, it may help students to think of lifting a column of the second matrix and laying it on its side above a row of the first matrix so that the elements match. Then multiply matching elements and add the products. The sum is the element for the corresponding row and column of the product matrix.

Using Technology

Exs. 6–17, 19, 20, 23, 30–33 You may want to suggest that students use the *Matrix Analyzer* software for this group of exercises.

Exs. 14, 15, 17, 19, and 20 can also be solved with the aid of a graphing calculator. Note that if a matrix has not been properly constructed (for example, if elements are missing that should have been entered), the TI-82 has no way of knowing this. However, if you try to find a product that does not exist (is not defined), the calculator will show an error message that indicates a dimension mismatch.

INTERVIEW Twyla Lang

Look back at the article on pages xxxiv–2.

In addition to receiving orders through the mail, Twyla Lang sells her products at conventions and trade shows. Along with the Kente scarves, Twyla's company offers other items, such as earrings and shirts.

14. The Kente scarves that Lang sells come in four different styles. Suppose matrix A shows the number of each style of scarf sold for 1992–1994.

	Class	Plain	Sorority	2-color sorority
1992	0	0	100	0
1993	250	100	15	20
1994	450	85	65	75

$= A$

a. The costs of the different scarves are shown in the photo below. Create a matrix C to represent the cost of each style of Kente scarf.

b. Find AC. What does this matrix represent?

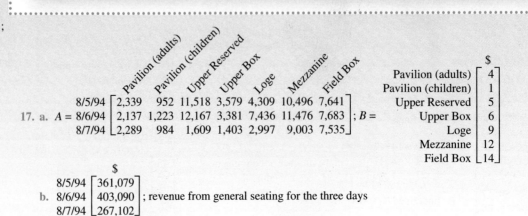

Class year scarf: **$25**
Day 1 sales: **40**
Day 2 sales: **32**

Plain Kente scarf: **$20**
Day 1 sales: **45**
Day 2 sales: **52**

Shirts: **$12**
Day 1 sales: **65**
Day 2 sales: **45**

Earrings: **$15**
Day 1 sales: **25**
Day 2 sales: **18**

2-color sorority scarf: **$25**
Day 1 sales: **12**
Day 2 sales: **24**

Sorority Kente scarf: **$30**
Day 1 sales: **28**
Day 2 sales: **16**

15. Suppose the sales for each item from one show are those shown above.

a. Create matrix S to show the sales for each item and matrix T to show the cost of each item. What are the dimensions of each matrix?

b. Find ST. What does this matrix represent?

16. Cooperative Learning Work with a partner. Use matrices A, B, C, and D.

$$A = \begin{bmatrix} 4 & 0 \\ -2 & 1 \end{bmatrix} \quad B = \begin{bmatrix} 1 & 0 \\ 0 & 1 \end{bmatrix} \quad C = \begin{bmatrix} -2 & -3 \\ 3 & 5 \end{bmatrix} \quad D = \begin{bmatrix} -5 & -3 \\ 3 & 2 \end{bmatrix}$$

a. Multiply all possible pairs of matrices. (You do not need to multiply a matrix by itself.) Is the product always commutative?

b. For real numbers, the number 1 is called the *multiplicative identity* because $1 \cdot n = n$ and $n \cdot 1 = n$. Which matrix acts as a multiplicative identity for matrices A, B, C, and D? Give an example.

c. For real numbers not equal to zero, n and $\frac{1}{n}$ are *multiplicative inverses* because $n\left(\frac{1}{n}\right) = 1$ (the multiplicative identity). Which two matrices are multiplicative inverses of each other? Explain your reasoning.

14, 15. See answers in back of book.

16. a. $AB = \begin{bmatrix} 4 & 0 \\ -2 & 1 \end{bmatrix}$; $BA = \begin{bmatrix} 4 & 0 \\ -2 & 1 \end{bmatrix}$;

$AC = \begin{bmatrix} -8 & -12 \\ 7 & 11 \end{bmatrix}$; $CA = \begin{bmatrix} -2 & -3 \\ 2 & 5 \end{bmatrix}$;

$AD = \begin{bmatrix} -20 & -12 \\ 13 & 8 \end{bmatrix}$;

$DA = \begin{bmatrix} -14 & -3 \\ 8 & 2 \end{bmatrix}$;

$BC = \begin{bmatrix} -2 & -3 \\ 3 & 5 \end{bmatrix}$; $CB = \begin{bmatrix} -2 & -3 \\ 3 & 5 \end{bmatrix}$;

$BD = \begin{bmatrix} -5 & -3 \\ 3 & 2 \end{bmatrix}$; $DB = \begin{bmatrix} -5 & -3 \\ 3 & 2 \end{bmatrix}$;

$CD = \begin{bmatrix} 1 & 0 \\ 0 & 1 \end{bmatrix}$; $DC = \begin{bmatrix} 1 & 0 \\ 0 & 1 \end{bmatrix}$;

The product is not always commutative.

b. B; $AB = A$ and $BA = A$

c. C and D; $\begin{bmatrix} 1 & 0 \\ 0 & 1 \end{bmatrix}$ is the multiplicative identity and $CD = DC = \begin{bmatrix} 1 & 0 \\ 0 & 1 \end{bmatrix}$.

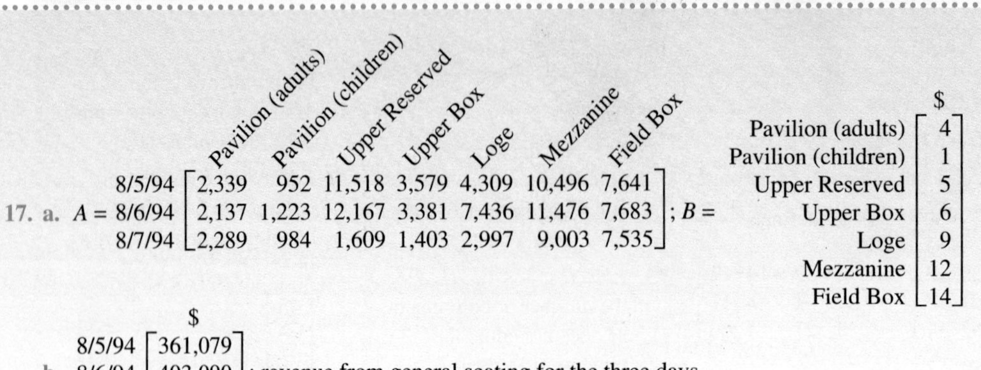

17. a. $A =$

	Pavilion (adults)	Pavilion (children)	Upper Reserved	Upper Box	Loge	Mezzanine	Field Box
8/5/94	2,339	952	11,518	3,579	4,309	10,496	7,641
8/6/94	2,137	1,223	12,167	3,381	7,436	11,476	7,683
8/7/94	2,289	984	1,609	1,403	2,997	9,003	7,535

; $B =$

	$
Pavilion (adults)	4
Pavilion (children)	1
Upper Reserved	5
Upper Box	6
Loge	9
Mezzanine	12
Field Box	14

b.

	$
8/5/94	361,079
8/6/94	403,090
8/7/94	267,102

; revenue from general seating for the three days

Connection BASEBALL

On August 5, 6, and 7, 1994, the San Francisco Giants played the Houston Astros at the Houston Astrodome. The Astrodome has many different types of seating available, as shown in the diagram. The attendance for the general seating at each game is given in the table.

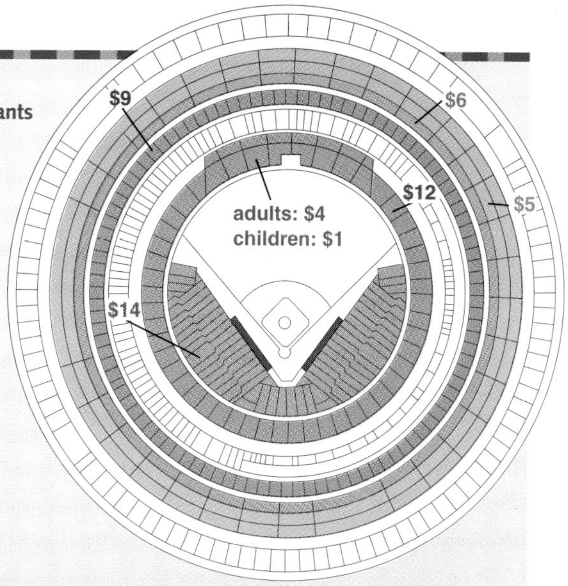

	Pavilion (adults)	Pavilion (children)	Upper Reserved	Upper Box	Loge	Mezzanine	Field Box
8/5/94	2,339	952	11,518	3,579	4,309	10,496	7,641
8/6/94	2,137	1,223	12,167	3,381	7,436	11,476	7,683
8/7/94	2,289	984	1,609	1,403	2,997	9,003	7,535

17. **a.** Create matrix A showing the attendance for each type of seating by date. Create matrix B showing the cost for each type of seating.

b. Find AB. What does this matrix represent?

18. **Writing** Which type of seating has the most consistent revenue? Which has the most varied revenue? Explain your reasoning.

19. In addition to the general seating, the Astrodome has executive seats. The attendance for these seats is given in the table below. Create matrix C showing the attendance for each type of seating by date. Create matrix D showing the cost for each type of seating. Find CD.

	Skybox ($10)	Club Level ($15)	Bullpen Room ($15)	Star Deck ($17)	Star Suite ($30)	Owner's Club ($50)	Diamond Level ($100)
8/5/94	1325	47	65	2715	947	101	94
8/6/94	1768	52	50	2696	1140	80	94
8/7/94	1225	44	50	2718	721	51	93

20. Find the sum of AB and CD. What does this matrix represent?

21. Which game had the largest revenue?

22. **Writing** Which type of seat do you think is most popular? least popular? Explain your reasoning.

Apply⇔Assess

Exercise Notes

Integrating the Strands

Ex. 16 This exercise relates properties of real numbers to matrices. For part (b), students should see that multiplication by matrix B, the multiplicative identity matrix, is commutative, even though matrix multiplication, in general, is not commutative. For part (c), you may wish to ask students to see if they can construct another pair of matrices that are multiplicative inverses.

Second-Language Learners

Exs. 17–22 Some students learning English may not be familiar with the sport of baseball and baseball stadiums, particularly the many different types of seating available. You may wish to assign these students to cooperative learning groups having English-speaking students who can explain the meanings of the many terms used in these exercises.

20. $\begin{array}{c} \\ 8/5/94 \\ 8/6/94 \\ 8/7/94 \end{array} \begin{array}{c} \$ \\ \begin{bmatrix} 465{,}024 \\ 515{,}732 \\ 360{,}448 \end{bmatrix} \end{array}$; total revenue for the three days

21. the game played on 8/6/94

22. Answers may vary. An example is given. I think it is difficult to decide which seats are most popular and which are least popular. If the most popular seats are the ones for which the greatest number of tickets are sold, then mezzanine seating is most popular and, similarly, club seats are least popular. However, that is not necessarily true. If the most popular seats are the ones people would most like to choose if they could afford to do so, the diamond level seats might be the most popular. The least popular might be the upper reserved, which appear to be farthest from the field.

18. The pavilion seating for adults and the field boxes have the most consistent revenue because they have the most consistent attendance. The upper reserved seating has the most varied revenue because it has the most varied attendance.

19. $C = \begin{array}{c} \\ 8/5/94 \\ 8/6/94 \\ 8/7/94 \end{array} \begin{array}{c} \begin{array}{ccccccc} \text{Skybox} & \text{Club Level} & \text{Bullpen Room} & \text{Star Deck} & \text{Star Suite} & \text{Owner's Club} & \text{Diamond Level} \end{array} \\ \begin{bmatrix} 1325 & 47 & 65 & 2715 & 947 & 101 & 94 \\ 1768 & 52 & 50 & 2696 & 1140 & 80 & 94 \\ 1225 & 44 & 50 & 2718 & 721 & 51 & 93 \end{bmatrix} \end{array}$; $D = \begin{array}{r} \\ \text{Skybox} \\ \text{Club Level} \\ \text{Bullpen Room} \\ \text{Star Deck} \\ \text{Star Suite} \\ \text{Owner's Club} \\ \text{Diamond Level} \end{array} \begin{array}{c} \$ \\ \begin{bmatrix} 10 \\ 15 \\ 15 \\ 17 \\ 30 \\ 50 \\ 100 \end{bmatrix} \end{array}$;

$CD = \begin{array}{c} \\ 8/5/94 \\ 8/6/94 \\ 8/7/94 \end{array} \begin{array}{c} \$ \\ \begin{bmatrix} 103{,}945 \\ 112{,}642 \\ 93{,}346 \end{bmatrix} \end{array}$

Apply⇔Assess

Exercise Notes

Communication: Writing

Ex. 23 This exercise provides a good opportunity for students to check whether they understand matrix multiplication. A good, clearly worded response would make a good journal entry.

Reasoning

Exs. 25, 27 Ex. 25 leads to the conclusion that matrix multiplication, in general, is not commutative. Ask what this implies about the distributive property for matrix multiplication and addition. (If $A(B + C)$ exists, then $A(B + C) = AB + AC$, but it does not follow that $(B + C)A = BA + CA$, since $(B + C)A$ may not exist.)

Topic Spiraling: Preview

Ex. 28 This exercise previews ideas that students will examine in greater depth in Chapter 7.

Practice 4 for Section 1.4

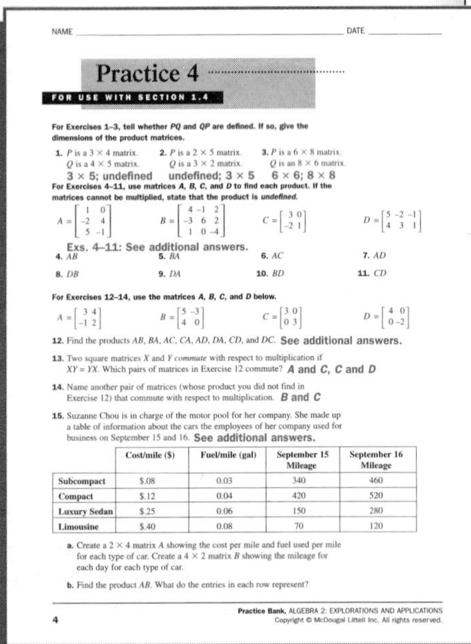

NAME ______________________ DATE ________

Practice 4

FOR USE WITH SECTION 1.4

For Exercises 1–3, tell whether *PQ* and *QP* are defined. If so, give the dimensions of the product matrices.

1. P is a 3×4 matrix. Q is a 4×5 matrix. **3×5; undefined**
2. P is a 2×5 matrix. Q is a 3×2 matrix. **undefined; 3×5**
3. P is a 6×8 matrix. Q is an 8×6 matrix. **6×6; 8×8**

For Exercises 4–11, use matrices *A*, *B*, *C*, and *D* to find each product. If the matrices cannot be multiplied, state that the product is *undefined*.

$$A = \begin{bmatrix} 1 & 0 \\ -2 & 4 \\ 5 & -1 \end{bmatrix} \quad B = \begin{bmatrix} 4 & -1 & 2 \\ -3 & 6 & 2 \\ 1 & 0 & -4 \end{bmatrix} \quad C = \begin{bmatrix} 3 & 0 \\ -2 & 1 \end{bmatrix} \quad D = \begin{bmatrix} 5 & -2 & -1 \\ 4 & 3 & 1 \end{bmatrix}$$

Exs. 4–11: See additional answers.

4. AB 5. BA 6. AC 7. AD
8. DB 9. DA 10. BD 11. CD

For Exercises 12–14, use the matrices *A*, *B*, *C*, and *D* below.

$$A = \begin{bmatrix} 3 & 4 \\ -1 & 2 \end{bmatrix} \quad B = \begin{bmatrix} 5 & -3 \\ 4 & 0 \end{bmatrix} \quad C = \begin{bmatrix} 3 & 0 \\ 0 & 3 \end{bmatrix} \quad D = \begin{bmatrix} 4 & 0 \\ 0 & -2 \end{bmatrix}$$

12. Find the products AB, BA, AC, CA, AD, DA, CD, and DC. **See additional answers.**
13. Two square matrices X and Y *commute* with respect to multiplication if $XY = YX$. Which pairs of matrices in Exercise 12 commute? **A and C, C and D**
14. Name another pair of matrices (whose product you did not find in Exercise 12) that commute with respect to multiplication. **B and C**
15. Suzanne Chou is in charge of the motor pool for her company. She made up a table of information about the cars the employees of her company used for business on September 15 and 16. **See additional answers.**

	Cost/mile ($)	Fuel/mile (gal)	September 15 Mileage	September 16 Mileage
Subcompact	$.08	0.03	340	460
Compact	$.12	0.04	420	520
Luxury Sedan	$.25	0.06	150	280
Limousine	$.40	0.08	70	120

a. Create a 2×4 matrix A showing the cost per mile and fuel used per mile for each type of car. Create a 4×2 matrix B showing the mileage for each day for each type of car.

b. Find the product AB. What do the entries in each row represent?

4

Practice Bank, ALGEBRA 2: EXPLORATIONS AND APPLICATIONS

23. **Technology** Use a graphing calculator or computer software with matrix calculation capabilities. Enter each matrix.

$$S = \begin{bmatrix} 10 & 14 \\ -3 & 23 \\ 0 & 18 \end{bmatrix} \qquad T = \begin{bmatrix} 12 & -26 & 6 & 36 \\ -45 & 25 & 0 & 24 \end{bmatrix}$$

a. What result do you get when you compute ST?

b. What result do you get when you compute TS? What does this result mean?

c. **Open-ended Problem** Create additional rows or columns to change matrices S and T so that you avoid the result you got in part (b). Write down your new matrices. Then find TS.

SAT/ACT Preview Suppose that *M* and *N* are matrices and that $M \neq N$. Tell if each statement is *always true, sometimes true,* or *never true.*

24. If MN exists, then NM exists.
25. $MN = NM$
26. $M + N = N + M$
27. $3M + 3N = 3(M + N)$
28. **Challenge** Use the equation to answer parts (a) and (b).

$$\begin{bmatrix} 3 & -1 \\ 2 & 1 \end{bmatrix}\begin{bmatrix} x \\ y \end{bmatrix} = \begin{bmatrix} 1 \\ 9 \end{bmatrix}$$

a. Multiply the matrices on the left. Then equate corresponding elements to write two equations in terms of x and y.

b. Solve the equations from part (a) for x and y. Use matrix multiplication to check your work.

ONGOING ASSESSMENT

29. a. **Writing** Explain how you can use the dimensions of two matrices to determine if their product exists. If the product does exist, how are you able to determine the dimensions of the product matrix before you complete the multiplication?

b. **Open-ended Problem** Create a poster that demonstrates, step by step, the process for multiplying two 2×2 matrices. Use color-coding.

SPIRAL REVIEW

For Exercises 30–32, use matrices *A* and *B*. *(Section 1.3)*

$$A = \begin{bmatrix} 1 & 0 \\ 2 & 4 \end{bmatrix} \qquad B = \begin{bmatrix} 2 & 2 \\ 1 & 3 \end{bmatrix}$$

30. Find $A + B$.
31. Find $-\frac{3}{2}A$.
32. Find $4A + 2B$.
33. Find $B - A$.

Express each of the following as a decimal between 0 and 1. *(Toolbox, page 788)*

34. a 30% chance
35. a 1 in 4 chance
36. a 99% chance
37. a 1 in 100 chance

Solve each inequality. *(Toolbox, page 787)*

38. $2x + 1 < 5$
39. $-(x + 1) \geq 8$
40. $\frac{1}{2}x \geq 3$
41. $-5x + 1 \leq 4$

23. a. $\begin{bmatrix} -510 & 90 & 60 & 696 \\ -1071 & 653 & -18 & 444 \\ -810 & 450 & 0 & 432 \end{bmatrix}$

b. An error statement results. TS cannot be computed because the number of columns of T and the number of rows of S are not equal.

c. Answers may vary. An example is given. Let $S = \begin{bmatrix} 10 & 14 \\ -3 & 23 \\ 0 & 18 \\ 1 & 1 \end{bmatrix}$; then $TS = \begin{bmatrix} 234 & -286 \\ -501 & -31 \end{bmatrix}$.

24. sometimes true
25. sometimes true
26. always true
27. always true
28. a. $3x - y = 1$; $2x + y = 9$

b. $x = 2$; $y = 5$; $\begin{bmatrix} 3 & -1 \\ 2 & 1 \end{bmatrix}\begin{bmatrix} 2 \\ 5 \end{bmatrix} = \begin{bmatrix} 6 - 5 \\ 4 + 5 \end{bmatrix} = \begin{bmatrix} 1 \\ 9 \end{bmatrix}$

29. a. The product AB of two matrices A and B exists only if the number of elements in a row of A is the same as the number of elements in a column of B, that is, A has dimensions $m \times n$ and B has dimensions $n \times k$ for some positive integers m, n, and k. AB has dimensions $m \times k$.

b. Check students' work.

30. $\begin{bmatrix} 3 & 2 \\ 3 & 7 \end{bmatrix}$
31. $\begin{bmatrix} -\frac{3}{2} & 0 \\ -3 & -6 \end{bmatrix}$
32. $\begin{bmatrix} 8 & 4 \\ 10 & 22 \end{bmatrix}$
33. $\begin{bmatrix} 1 & 2 \\ -1 & -1 \end{bmatrix}$
34. 0.3
35. 0.25
36. 0.99
37. 0.01
38. $x < 2$
39. $x \leq -9$
40. $x \geq 6$
41. $x \geq -\frac{3}{5}$

Section

1.5 Using Simulations as Models

Learn how to...

- model a situation with a simulation

So you can...

- make predictions, such as how long it takes to get a certain coupon in a giveaway

Have you ever played a game that imitated life? Some games allow players to experiment with topics such as real estate finance or life decisions. These games are examples of *simulations.* A **simulation** is an experiment that you use to model a situation and make predictions. You can use coins, dice, spinners, and calculators with a random number feature to simulate events.

EXPLORATION
COOPERATIVE LEARNING

Simulating a Coupon Giveaway

Work with a partner.
You will need:
- a die

SET UP Suppose a music store gives each customer a discount coupon. The amount of the discount is known when a sales clerk removes the seal on the coupon to reveal either a 10%, 15%, or 30% discount.

1 Suppose there is an equal chance of getting each discount. Let each face of a die represent the type of discount you receive.

Roll of the die	1 or 2	3 or 4	5 or 6
Discount	10%	15%	30%

Roll the die until you get a 5 or a 6, and then record the number of rolls it took you to get a 5 or a 6. Repeat the simulation 10 times.

2 What was the greatest number of coupons you needed to have to get a 30% discount? What was the least number? the average number?

3 The store manager decides that in the next coupon order, she will request that half of the coupons give a 10% discount, a third give a 15% discount, and a sixth give a 30% discount. Which numbers on the die will you use to represent a discount of 10%? 15%? 30%? Why?

4 Run a simulation for the new coupon values. About how many coupons do you need to collect to get a 30% discount?

Exploration Note

Purpose
The purpose of this Exploration is to have students learn what a simulation is and how one can be used to model a real-world situation.

Materials/Preparation
Each pair of students should have a die or a number cube with sides numbered from 1 to 6.

Procedure
For Step 1, one student can roll the die and the other can record the numbers that come up. For Step 4, students can switch their roles of rolling the die and recording the results.

Closure
Have several pairs of students describe their results for Step 2. Combine the results of all pairs to get one result for the whole class. When discussing Step 3, students should understand that the fractions chosen by the manager ($\frac{1}{2}$, $\frac{1}{3}$, and $\frac{1}{6}$) have a sum of 1.

Explorations Lab Manual
See the Manual for more commentary on this Exploration.

For answers to the Exploration, see following page.

Plan⇔Support

Objectives

- Model a situation with a simulation.
- Make predictions using simulations.

Recommended Pacing

Core and Extended Courses
Section 1.5: 2 days

Block Schedule
Section 1.5: 2 half-blocks (with Section 1.4 and Review)

Resource Materials

Lesson Support
Lesson Plan 1.5
Warm-Up Transparency 1.5
Practice Bank: Practice 5
Study Guide: Section 1.5
Exploration Lab Manual: Diagram Master 1
Assessment Book: Test 2

Technology
Scientific Calculator
Graphing Calculator
Internet: http://www.hmco.com

Warm-Up Exercises

1. When you flip a single coin, what is the probability that you will get heads? $\frac{1}{2}$
2. If you roll a single die, what is the probability that you will get a number less than 6? $\frac{5}{6}$
3. List the possible sums that you could get by rolling two dice at the same time.
   2, 3, 4, 5, 6, 7, 8, 9, 10, 11, 12
4. Suppose you flip three pennies at the same time. List all the ways they can land. Use H for heads and T for tails.
   HHH, HHT, HTH, HTT, THH, THT, TTH, TTT

Teach⇔Interact

About Example 1

Using Technology
On the TI-82 graphing calculator, use the following procedure to generate random numbers between 0 and 1. Press [MATH] [▶][▶] (or [MATH][◀]) to go to the PRB menu. Press 1 to select rand, and then press [ENTER] repeatedly. Each time you press [ENTER], the calculator will respond with a random number between 0 and 1.

Additional Example 1

Mr. and Mrs. Abrams want to know how many children they can expect to have if they plan to have no more children once they have two girls. Create a simulation to estimate the number of children they can expect to have. Run the simulation 5 times. About how many children will be in the Abrams family?

Assume the chances of having a girl or a boy are the same.

Method 1 Use a coin. Let heads be the birth of a girl and tails be the birth of a boy. Flip the coin until it has landed heads up (H) twice.

Simulation	Results	Children
1	HTH (GBG)	3
2	THTH (BGBG)	4
3	THTH (BGBG)	4
4	TTTHH (BBBGG)	5
5	TTTHTTH (BBBGBBG)	7

The average number of children is $\frac{3+4+4+5+7}{5} = 4.6$. Based on these results, the Abrams family will have either 4 or 5 children.

Method 2 Use a calculator. Use the random number feature, and let the birth of a girl be represented by getting a number 0.5 or greater.

EXAMPLE 1

Mr. and Mrs. Skotzke are planning to have children. Create a simulation to estimate the number of children they need to have before their family includes at least one girl and one boy. Repeat the simulation 5 times. About how many children do you think will be in the Skotzke family?

SOLUTION

Assume that the chances of having a girl are the same as the chances of having a boy.

Method 1 Use a coin.

Let "heads up" represent the birth of a girl and "tails up" represent the birth of a boy. Flip the coin until at least 1 head (H) and 1 tail (T) land face up.

Simulation	1	2	3	4	5
Toss results	HHHT (GGGB)	HHT (GGB)	HHT (GGB)	TH (BG)	TTTTH (BBBBG)
Number of children	4	3	3	2	5

The average number of children is $\frac{4+3+3+2+5}{5} = 3.4$. Based on these results, the Skotzke family will have either 3 or 4 children.

Method 2 Use a calculator.

For more information about the random number feature on a calculator, see the *Technology Handbook*, p. 814.

The random number feature on a calculator gives a random decimal between 0 and 1. Let the birth be of a girl if the number is less than 0.5. Let the birth be of a boy if the number is 0.5 or greater.

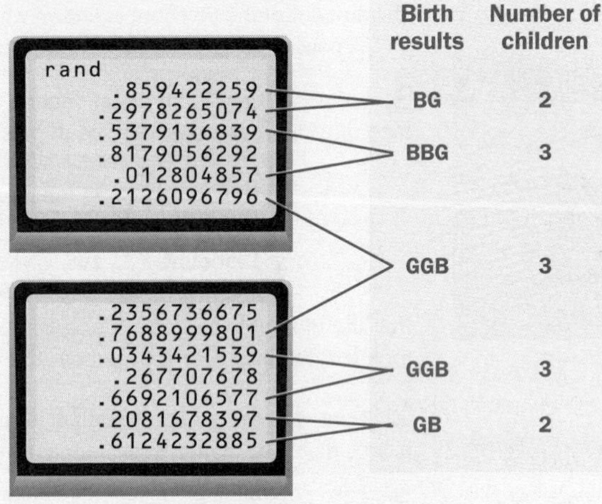

The average number of children is $\frac{2+3+3+3+2}{5} = 2.6$. Based on these results, the Skotzke family will have either 2 or 3 children.

ANSWERS Section 1.5

Exploration

1, 2. Answers may vary. Sample trial results are given.

1. 1, 2, 2, 1, 9, 3, 1, 2, 5, 3
2. for the sample above, 9 coupons; 1 coupon; about 3 coupons
3. Answers may vary. An example is given. I would let 1, 2, or 3 represent a 10% discount because the chance of rolling a 1, 2, or 3 is $\frac{1}{2}$. I would let a 4 or 5 represent a 15% discount because the chance of rolling a 4 or 5 is $\frac{1}{3}$. I would let a 6 represent a 30% discount because the chance of rolling a 6 is $\frac{1}{6}$.
4. Answers may vary. Sample trial results are given. 12, 9, 2, 13, 7, 3, 4, 5, 14, 2; You need about 7 coupons.

EXAMPLE 2

Emily used a program to repeat the simulation from Example 1.

```
Get N, the number of repetitions.              →  :Disp "NO. OF SIMULATIONS":Input N
T totals children for all simulations.         →  :0→T
S, the simulation counter, runs from 1 to N.   →  :For (S,1,N)
B and G are the boy and girl counters.         →  :0→B:0→G
                                                  :Lbl 1
A child is born; update either B or G.         →  :If rand>0.5:Then:B+1→B
                                                  :Else:G+1→G:End
Repeat until at least 1 boy and 1 girl.        →  :If B=0 or G=0:Goto 1
Add the boy and girl counts to the total.      →  :T+B+G→T:End
Report results.                                →  :Disp "AVERAGE NUMBER OF CHILDREN
                                                  IS", T/N
```

a. Find the average number of children the Skotzke's will have if you use Emily's program to run the simulation 250 times.

b. **Writing** Which method do you think is most accurate: Method 1 from Example 1, Method 2 from Example 1, or Emily's method? Explain.

For more information about programming, see the *Technology Handbook*, p. 815.

SOLUTION

a.

```
prgmCHILDREN
HOW MANY TIMES?
?250
AVERAGE NUMBER
OF CHILDREN IS
          2.98
```

According to these results, the Skotzke family will have **3 children**.

b. Emily's method seems most reliable. Emily got her results after repeating the simulation 250 times. In Example 1 the simulations were repeated only 5 times for each method. Because there were so few results for the methods of Example 1, any unusual results had more of an effect on the average.

THINK AND COMMUNICATE

1. Tell what each line does in the program from Example 2.
   a. `If rand>0.5:Then:B+1→B:Else:G+1→G`
   b. `If B=0 or G=0:Goto 1`
2. Do you think your results would be much different if you ran Emily's program 500 times? 1000 times? Explain your reasoning.
3. Research has shown that the chance of a baby being a girl is 48.8% and the chance of it being a boy is 51.2%. How would you simulate this situation? Do you think you would get different results? Explain.

Teach⇔Interact

Additional Example 1 (continued)

The average number of children is $\frac{4+4+3+5+2}{5} = 3.6$. Based on these results, the Abrams family will have either 3 or 4 children.

Additional Example 2

Serge used a program to repeat the simulation from Additional Example 1.

```
:Disp "No. of simulations":Input N
:0→T
:For (S,1,N)
:0→B:0→G
:Lbl 1
:If rand<0.5:Then:B+1→B:Else:G+1→G
:End:If G<2:Goto 1
:T+B+G→T:End
:Disp "AVERAGE NUMBER OF CHILDREN
IS",T/N
```

a. Find the average number of children the Abrams will have if you use Serge's program to run the simulation 250 times.

```
Prgm2GIRLS
HOW MANY TIMES?
?250
AVERAGE NUMBER OF
CHILDREN IS
            4.048
```

According to the results, the Abrams family will have 4 children.

b. Which method do you think is most accurate: Method 1 from Additional Example 1, Method 2 from Additional Example 1, or Serge's method? Explain.
Serge's method seems most reliable. He got his results after running the simulation 250 times. In Additional Example 1, the simulation was run only 5 times. Since there were so few results for the methods of Additional Example 1, any unusual results had more of an effect on the average.

Think and Communicate

When students discuss question 2, you may wish to have them run Emily's program using N = 500 and N = 1000. This will take some time, but they can let the program run while they discuss question 3.

Think and Communicate

1. a. The line assigns "boy" or "girl" to the random number produced and adds 1 to the appropriate counter.
   b. This line ends the program unless B = 0 or G = 0, that is, unless the simulation has produced at least one boy and one girl. Otherwise, the program returns to the beginning and continues the simulation.
2. Answers may vary. An example is given. I think that 250 trials will produce a reasonable result and that running the program 500 or 1000 times would not produce significantly different results.
3. Change "If rand > 0.5" to "If rand > 0.512." Since 0.512 is so close to 0.5, the results would probably not differ significantly.

Teach⇔Interact

Checking Key Concepts

Teaching Tip
When discussing these questions, you may wish to have students actually run the simulations they describe (see Exs. 6–8). This can be especially instructive if students suggest different simulation methods.

Closure Question

What are some advantages and disadvantages of using simulations to make predictions about real-world situations? Possible advantages: Simulations are more convenient and less time-consuming than dealing with a real situation. Computer or calculator simulations often make it possible to use a large number of runs. Possible disadvantages: If there is inadequate information about all of the important factors that influence outcomes in a situation, then the simulation cannot take them into account and may lead to unreliable predictions.

Apply⇔Assess

Suggested Assignment

Core Course
Day 1 Exs. 1–9
Day 2 Exs. 12, 16–23, AYP

Extended Course
Day 1 Exs. 1–11
Day 2 Exs. 12–23, AYP

Block Schedule
Day 4 Exs. 1–9
Day 5 Exs. 12, 16–23, AYP

Exercise Notes

Communication: Discussion
Exs. 1–4 The situations in these and other exercises in this section involve probability concepts and thus are mathematical models of the situations. Nonmathematical simulations are sometimes used in other school subjects, particularly in business courses. Ask students if they have been involved in using simulations in other courses and, if so, how they differ from those of this section.

✓ CHECKING KEY CONCEPTS

Suppose there is one of six bonus stickers in each pack of baseball cards and there is an equal chance of getting each sticker.

1. Describe a simulation to find the number of packs of baseball cards you need to buy in order to get a particular sticker. In your simulation, how do you decide if a pack contains the desired sticker?
2. Describe a simulation to find the number of packs of baseball cards you need to buy in order to collect all six stickers.
3. Suppose that there is an 80% chance of getting a sticker of an individual player and a 20% chance of getting a sticker of a team. Describe a simulation to find the number of packs of baseball cards you need to buy in order to get a sticker of a team.

1.5 Exercises and Applications

Extra Practice exercises on page 750

For Exercises 1–3, describe how you would simulate each situation.

1. guessing the correct answer on at least 7 of 10 true/false questions
2. choosing a bulb for a red tulip from a bin if one in six of the bulbs is for a red tulip
3. getting a hit in a baseball game if the batter averages one base hit every three times at bat
4. **a.** Simulate the situation from Exercise 3 to find the number of hits the player gets in a game if he bats 4 times. Repeat the simulation 30 times. Record the number of hits the player gets in each game.
   b. In how many games did the player get no hits? 1 hit? 2 hits? 3 hits? 4 hits? Do you think your results are reasonable? Why or why not?
5. **SAT/ACT Preview** Which simulation techniques are appropriate for randomly choosing between two equally likely events, A and B?
   I. Toss a coin. Choose A if you get heads, B if you get tails.
   II. Use the spinner. Choose A if you get red, B if you get blue.
   III. Roll a die. Choose A if you get an even number, B if you get an odd number.
   A. I only **B.** I and III only **C.** I, II, and III **D.** none of these
6. **a.** Do the simulation you described in *Checking Key Concepts* Question 1. Repeat your simulation 10 times.
   b. What is the greatest number of packs of baseball cards you need to buy to get the desired sticker? What is the least number? the average number?
7. Repeat Exercise 6 for the simulation you described in *Checking Key Concepts* Question 2.
8. Repeat Exercise 6 for the simulation you described in *Checking Key Concepts* Question 3.

Checking Key Concepts

Answers may vary. Examples are given. In each case, repeat the simulation a number of times, say 10, and compute the average.

1. Roll a die and let rolling a 6 represent getting the particular sticker. Count the number of trials needed to roll a 6.
2. Roll a die and let the numbers on the die represent the six stickers. Count the number of trials needed to roll each number at least once.
3. Use the random number feature on a calculator. Let random numbers greater than or equal to 0.2 represent individual stickers and numbers less than 0.2 represent team stickers. Count the number of trials needed to generate a number less than 0.2.

Exercises and Applications

1–3. Answers may vary. Examples are given. In each case, repeat the simulation a number of times, say 10, and compute the average.

1. Use the random number feature on a calculator. Let the first digit of each random number represent the number of correct answers guessed. Count the number of trials needed to generate a number greater than or equal to 7.

Connection GAMES

In many societies people play games that involve tossing some sort of die or dice. One such game, played by the Paiute people of what is now Nevada, involves tossing twelve sticks painted on one side. A tossed stick will land with the painted side up about half the time.

9. To play, one player tosses all twelve sticks. If *exactly* five sticks land with the painted side up, the player gets a point. The players take turns tossing the sticks until one player gets fifty points.

a. Simulate the Paiute game, but only go to five points. Describe how you decide whether a stick lands with the painted side up or down. Record the results of each player's toss and the number of points each player has.

b. **Cooperative Learning** How many turns did your game take? Compare your results to those found by other students. What do you think is the average number of turns it takes to play the Paiute game to fifty points?

c. Do you think each player has an equal chance of winning? Explain.

10. **Writing** Explain why a player who has exactly seven sticks land with the painted side *down* will get a point.

BY THE WAY

Sarah Winnemucca was an educator, interpreter, and spokeswoman for Native American Rights. In 1883, she wrote *Life Among the Paiutes.*

11. **Cooperative Learning** Work in a group of three students. The table shows the chance of 0, 1, 2, or 3 people arriving at a bus stop in a given minute. Suppose that a bus arrives about once every 6 min.

Number of people arriving in a given minute	Chance of the given number of people arriving
0	20%
1	40%
2	30%
3	10%

a. Describe a way to simulate the number of people who arrive at the bus stop. Describe a way to simulate whether a bus arrives.

b. Simulate the activity at the bus stop for a 60 min period. Use a table like the one below to record your results.

Minute of simulation	1	2	3	4	5	...	60
Number of people who arrive	1	3	0	2	0	...	3
Bus arrives	0	0	0	1	0	...	0

c. How many people arrived at the bus stop in 60 min? How many buses arrived? What was the average number of people who got on each bus? (Assume that any people who arrive in the same minute as a bus get on that bus.)

d. About how many minutes did a person have to wait before a bus arrived? What was the longest waiting time? the shortest?

Apply⇔Assess

Exercise Notes

Multicultural Note

Exs. 9, 10 The Paiute people were once one of the most widespread groups in the Great Basin region. Due to the limited natural resources in this dry, hot region, many of the Paiute people remained largely unaffected by changes brought about by white settlers until the late 1850s, when the discovery of silver deposits drew many settlers to the Nevada area, and the Paiutes' traditional ways of life suffered disruption.

Problem Solving

Ex. 11 Some general suggestions that students may find helpful in designing simulations are as follows. (1) Identify key mathematical aspects of the situation and assign probabilities for the possible outcomes. (2) Select a random device that embodies the important mathematical aspects of the situation. Possible devices are things such as coins, dice, spinners, or calculator/computer programs. (3) Define clearly what constitutes a run of the simulation. (4) Conduct a large number of runs. (5) Find a simulation estimate for the original situation by computing the ratio of the number of successful runs to the total number of runs.

2. Roll a die. Let rolling a 1 represent choosing a bulb for a red tulip. Count the number of trials needed to roll a 1.

3. Roll a die. Let rolling a 1 or a 2 represent a hit. Count the number of trials needed to roll a 1 or a 2.

4. Answers may vary. Sample trial results are given.

a. 1, 3, 2, 1, 0, 2, 2, 1, 0, 0, 1, 1, 2, 2, 2, 2, 1, 0, 0, 2, 3, 1, 0, 0, 2, 1, 1, 1, 1, 0

b. for the preceding sample, 8 games; 11 games; 9 games; 2 games; no games; The results are reasonable, since the batter got 35 hits in 120 times at bat for an average of about 0.292, which is close to his average of $\frac{1}{3}$.

5. B

6–8. Answers may vary. Sample trial results are given.

6. a. 9, 4, 4, 9, 4, 2, 4, 3, 1, 2

b. for the preceding sample, 9 packs; 1 pack; about 4 packs

7. a. 7, 15, 18, 10, 9, 9, 13, 13, 16, 15

b. for the preceding sample, 18 packs; 7 packs; about 13 packs

8. a. 4, 13, 8, 9, 12, 2, 4, 4, 4, 14

b. for the preceding sample, 14 packs; 2 packs; about 7 packs

9. a. Simulation methods may vary. For example, toss a coin. Let heads represent painted side up and tails represent painted side down. To simulate each player's turn, toss the coin 12 times. The player gets 1 point if heads come up exactly five times.

b. Answers may vary.

c. Answers may vary. An example is given. On each turn, each player has an equal chance of getting 1 point. However, I think the player who throws first has a slight advantage. Consider, for example, if both players score 1 point on each turn, Player 1 will win.

10. If exactly seven sticks land with the painted side down, then exactly five sticks landed with the painted side up.

11. See answers in back of book.

Apply⇔Assess

Exercise Notes

Challenge

Ex. 12 Since many simulations use random numbers, it may be useful for students to generate "custom-made" random numbers. Challenge students to find an expression that can be entered on a calculator (such as the TI-82) to generate a random integer from *a* to *a* + *b*. (The expression int((b+1)*rand)+a will generate random integers from *a* to *a* + *b*.)

Research

Exs. 12–15 In connection with these exercises, point out that simulations involving the use of random numbers give probabilistic approximations of mathematical or real-world situations. Some students may wish to research Monte Carlo methods. Possible sources of information are encyclopedias and books on probability and statistics.

Student Study Tip

Exs. 13, 14 Mention to students that they may find it helpful to compare the programs in these exercises with the program that was explained in Example 2 on page 31.

Using Manipulatives

Exs. 13, 14 Suggest that students consider how they might use dice, spinners, or other devices to simulate the situations in these exercises.

12. **Technology** Genna used a formula on her calculator to generate the numbers 1, 2, 3, 4, and 5 randomly.

This function cuts off the decimal part of a positive number.

Put the number you want to start with here.

Put the number of numbers you want here.

a. What formula would you use to generate the numbers 4, 5, 6, 7, and 8 randomly? the numbers 0–9 randomly?

b. Use Genna's method to repeat Exercise 6 if there are 8 stickers available.

Technology For Exercises 13 and 14, use the given program or adapt it for use on your graphing calculator or computer.

a. Enter the program and run it at least 10 times. Enter a different number of simulations each time. Record each set of results.

b. What results, if any, are unusual? Explain.

c. What do you think is a reasonable result? Why?

13. This program simulates a best-of-seven championship series between two evenly matched teams. It tells how many times each team won the series and the average number of games the teams played in each series.

```
:Disp "HOW MANY TIMES?"
:Input N
:0→T:0→C:0→D
:For (S,1,N)
:0→A:0→B:0→G
:Lbl 1
:G+1→G
:If rand>0.5:Then:A+1→A
:Else:B+1→B:End
:If A≠4 and B≠4:Goto 1
:If A=4:C+1→C:If B=4:D+1→D
:T+G→T:End
:Disp "TIMES A WON",C
:Disp "TIMES B WON",D
:Disp "AVG. NO. OF GAMES",T/N
```

14. This program simulates the average number of times a basketball player makes both baskets when the player is given two free throw shots. Enter a player's free throw percentage as a number between 0 and 100.

```
:Disp "WHAT IS PLAYER'S
FREE-THROW PERCENTAGE?"
:Input P
:Disp "HOW MANY TIMES?":Input N
:0→C
:For (S,1,N)
:0→B
:For (X,1,2)
:If int (100*rand)+1≤P
:Then:B+1→B:End
:If B=2:Then:C+1→C:End
:End:End
:Disp "NO. OF 2 PT SHOTS",C
:Disp "PERCENT 2 PT SHOTS"
:Disp int(100*(C/N))
```

15. **Challenge** Change the program in Exercise 13 to simulate the winner of a best-of-seven championship series in which one team is twice as likely to win each game as the other team. Then repeat Exercise 13.

12. a. int(5*rand)+4; int(10*rand)
    b. Use int(8*rand)+1
       (a) sample trial results: 6, 3, 16, 4, 1, 8, 33, 6, 5, 5
       (b) for the preceding sample, 33 packs; 1 pack; about 9 packs

13. For use on a graphing calculator, use int(2*rand)+1 to generate random numbers. Let 1 represent a win for one team and 2 a win for the other. (Assume that ties are not allowed.) For each trial, generate numbers until you get four 1's or four 2's, whichever comes first. The number of numbers needed represents the number of games in the series.
    a, b. Answers may vary.
    c. It is reasonable to expect that each team will win about half the time, because the teams are evenly matched.

14. For use on a graphing calculator, define a random number function based on the player's percentage. For example, if the player's free-throw percentage is 60%, use int(5*rand)+1. Generate pairs of random numbers and let numbers from 1 to 3 represent a shot made and 4 and 5 represent a missed shot.
    a, b. Answers may vary.
    c. The player in the example above should make both shots about $(0.6)^2 = 0.36$, or 36% of the time.

15. Change "If rand>0.5" to "If int (3*rand)+1>1."
    a, b. Answers may vary.
    c. The team more likely to win should win about $\frac{2}{3}$ of the time.

16. Answers may vary.

17. $\begin{bmatrix} 14 \\ 3 \end{bmatrix}$

18. $\begin{bmatrix} 31 & 35 \\ 28 & 68 \\ 30 & 80 \end{bmatrix}$

19. $\begin{bmatrix} 7 & -5 & 10 \\ -3 & 4 & -7 \\ -4 & -12 & 16 \end{bmatrix}$

20.

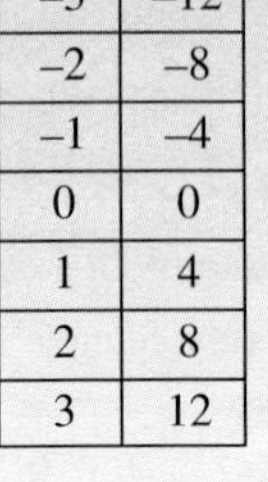

x	y
−3	−12
−2	−8
−1	−4
0	0
1	4
2	8
3	12

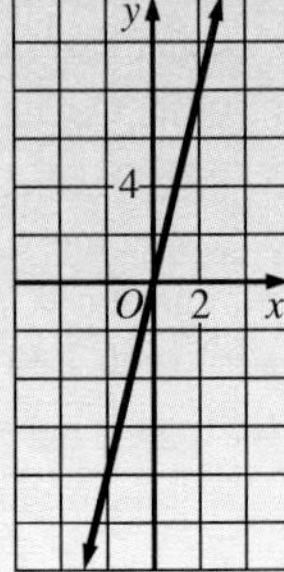

ONGOING ASSESSMENT

16. Open-ended Problem Simulate a situation of your choice. Describe the situation and the simulation. Record your results. Repeat your simulation enough times so that you have confidence in your results.

SPIRAL REVIEW

Multiply. *(Section 1.4)*

17. $\begin{bmatrix} 4 & 2 \\ 1 & 0 \end{bmatrix}\begin{bmatrix} 3 \\ 1 \end{bmatrix}$ **18.** $\begin{bmatrix} 2 & 1 & 7 \\ 0 & 8 & 4 \\ 3 & 6 & 2 \end{bmatrix}\begin{bmatrix} 4 & 10 \\ 2 & 8 \\ 3 & 1 \end{bmatrix}$ **19.** $\begin{bmatrix} 2 & -3 \\ -1 & 2 \\ 0 & -4 \end{bmatrix}\begin{bmatrix} 5 & 2 & -1 \\ 1 & 3 & -4 \end{bmatrix}$

For each equation, make a table of values for *x* and *y*. Use −3, −2, −1, 0, 1, 2, and 3 as values for *x*. Then use your points to graph each equation. *(Toolbox, page 795)*

20. $y = 4x$ **21.** $y = -x$ **22.** $y = 3x - 1$ **23.** $y = -2x + 5$

ASSESS YOUR PROGRESS

VOCABULARY

matrix, element (p. 16)
dimensions of a matrix (p. 16)
scalar (p. 18)
scalar multiplication (p. 18)
matrix multiplication (p. 23)
simulation (p. 29)

For Exercises 1–10 use the matrices below. If the operation cannot be performed, write "undefined." *(Sections 1.3 and 1.4)*

$A = \begin{bmatrix} 2 & 0 & -1 \\ -3 & 1 & 4 \end{bmatrix}$ $B = \begin{bmatrix} -1 & 6 & 2 \\ 0 & 3 & -3 \end{bmatrix}$ $C = \begin{bmatrix} 3 & -2 & 2 \\ 5 & -1 & -3 \\ 0 & 4 & 1 \end{bmatrix}$ $D = \begin{bmatrix} 1 & 0 \\ 4 & -2 \\ 3 & -1 \end{bmatrix}$

1. $A + B$ **2.** $3C$ **3.** $C + 2D$ **4.** $C - 4D$ **5.** $-2A - B$

6. AC **7.** BD **8.** BA **9.** CD **10.** CB

11. Suppose there is a one in six chance that any employee of a 20-person company will leave the company in a given year. *(Section 1.5)*

a. Simulate how long each employee will work for the company. Describe your simulation. Record your results in a table.

Employee	1	2	3	4	…	20
Simulation results	?	?	?	?	…	?
Number of years an employee stays	?	?	?	?	…	?

b. What is the average number of years an employee stays with the company? How many of the current 20 employees are still with the company after 5 years? after 10 years?

12. Journal What do you think are the most difficult steps of a simulation? Why? What methods of simulation do you like best? Why?

Apply⇔Assess

Exercise Notes

Assessment Note
Ex. 16 Students can work on this exercise with a partner.

Assess Your Progress

Review the vocabulary terms. Students can illustrate the terms having to do with matrices by referring to Exs. 1–10.

Journal Entry
In discussing Ex. 12, lead students to identify the simulation steps listed in the Problem Solving note on page 33.

Progress Check 1.3–1.5

See page 39.

Practice 5 for Section 1.5

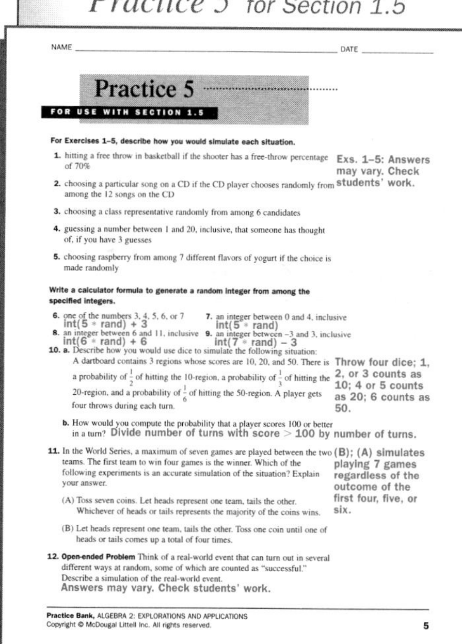

NAME ________ DATE ________

Practice 5
FOR USE WITH SECTION 1.5

For Exercises 1–5, describe how you would simulate each situation.

1. hitting a free throw in basketball if the shooter has a free-throw percentage of 70%

2. choosing a particular song on a CD if the CD player chooses randomly from among the 12 songs on the CD

Exs. 1–5: Answers may vary. Check students' work.

3. choosing a class representative randomly from among 6 candidates

4. guessing a number between 1 and 20, inclusive, that someone has thought of, if you have 3 guesses

5. choosing raspberry from among 7 different flavors of yogurt if the choice is made randomly

Write a calculator formula to generate a random integer from among the specified integers.

6. one of the numbers 3, 4, 5, 6, or 7 int(5 * rand) + 3

7. an integer between 0 and 4, inclusive int(5 * rand)

8. an integer between 6 and 11, inclusive int(6 * rand) + 6

9. an integer between −3 and 3, inclusive int(7 * rand) − 3

10. a. Describe how you would use dice to simulate the following situation: A dartboard contains 3 regions whose scores are 10, 20, and 50. There is a probability of $\frac{1}{2}$ of hitting the 10-region, a probability of $\frac{1}{3}$ of hitting the 20-region, and a probability of $\frac{1}{6}$ of hitting the 50-region. A player gets four throws during each turn. Throw four dice; 1, 2, or 3 counts as 10; 4 or 5 counts as 20; 6 counts as 50.

b. How would you compute the probability that a player scores 100 or better in a turn? Divide number of turns with score > 100 by number of turns.

11. In the World Series, a maximum of seven games are played between the two teams. The first team to win four games is the winner. Which of the following experiments is an accurate simulation of the situation? Explain your answer. (B); (A) simulates playing 7 games regardless of the outcome of the first four, five, or six.

(A) Toss seven coins. Let heads represent one team, tails the other. Whichever of heads or tails represents the majority of the coins wins.

(B) Let heads represent one team, tails the other. Toss one coin until one of heads or tails comes up a total of four times.

12. Open-ended Problem Think of a real-world event that can turn out in several different ways at random, some of which are counted as "successful." Describe a simulation of the real-world event.
Answers may vary. Check students' work.

Practice Bank, ALGEBRA 2: EXPLORATIONS AND APPLICATIONS

5

21.

x	y
−3	3
−2	2
−1	1
0	0
1	−1
2	−2
3	−3

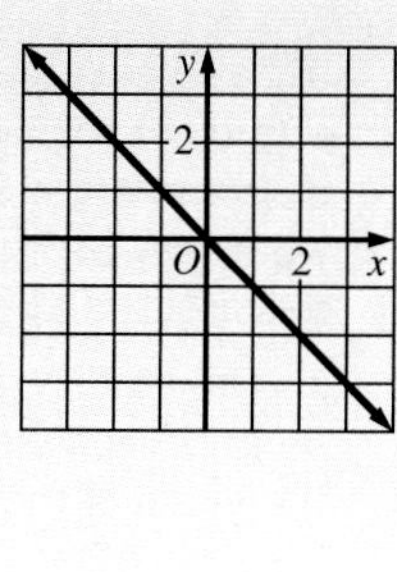

22.

x	y
−3	−10
−2	−7
−1	−4
0	−1
1	2
2	5
3	8

23.

x	y
−3	11
−2	9
−1	7
0	5
1	3
2	1
3	−1

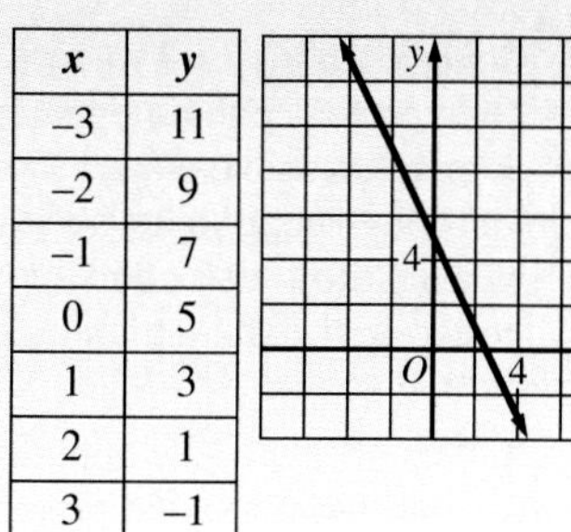

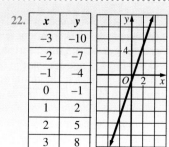

Assess Your Progress

1. 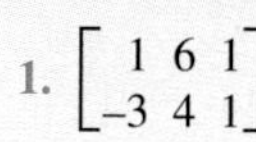 $\begin{bmatrix} 1 & 6 & 1 \\ -3 & 4 & 1 \end{bmatrix}$

2. 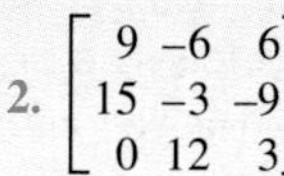 $\begin{bmatrix} 9 & -6 & 6 \\ 15 & -3 & -9 \\ 0 & 12 & 3 \end{bmatrix}$

3. undefined

4. undefined

5. $\begin{bmatrix} -3 & -6 & 0 \\ 6 & -5 & -5 \end{bmatrix}$

6. $\begin{bmatrix} 6 & -8 & 3 \\ -4 & 21 & -5 \end{bmatrix}$

7. $\begin{bmatrix} 29 & -14 \\ 3 & -3 \end{bmatrix}$

8. undefined

9–11. See answers in back of book.

12. Answers may vary.

Project Notes

Mathematical Goals

- Collect and organize data using a scatter plot and table.
- Model data with a linear function and an exponential function.
- Graph linear and exponential functions.
- Select the best model for the data collected.

Planning

Materials
- piece of paper
- wastebasket
- yardstick
- graph paper
- graphing calculator or graphing software

Project Teams
Students may choose to work with a partner and assist each other in completing the experiment. Then they can decide which model best fits each of their own data.

Guiding Students' Work

To get a more accurate set of data, suggest that students repeat the experiment two or three times and use the results of the combined trials in their table and graph. If students are working in pairs, they may want to have each person complete the experiment and then combine the data into one table and graph. Also, if students are using a graphing calculator, encourage them to use the value of r, the correlation coefficient, that the calculator displays for both the linear model and the exponential model in their report. Once students have graphed their two equations and the scatter plot, discuss the indications of the graphs verbally before having them write their reports.

Second-Language Learners
A peer tutor or aide might help students learning English discuss and organize their ideas for the written report.

PORTFOLIO PROJECT

Predicting Basketball Accuracy

In 1979 the National Basketball Association instituted a *three-point rule.* According to this rule, a basket shot from a point outside the three-point "line" is worth three points, while a basket shot from a closer distance is worth two points. Why do you think the rules were changed?

The **three-point line** is two inches wide. A shot counts for only two points if a player's foot touches the three-point line.

PROJECT GOAL **Do an experiment to find how your ability to make a basket changes with distance.**

Investigating Your Basketball Ability

For this investigation, you will need a piece of wadded-up paper to be your ball, a wastebasket, a yardstick, and graph paper. You may find it helpful to have a graphing calculator or graphing software.

1. STAND 0 yards from the wastebasket. Shoot your paper ball into the basket 10 times and record the number of shots that go into the basket.

2. STEP back one yard and record your results for 10 shots from this distance. Continue stepping back one yard until you reach a distance where none of your 10 shots go into the basket.

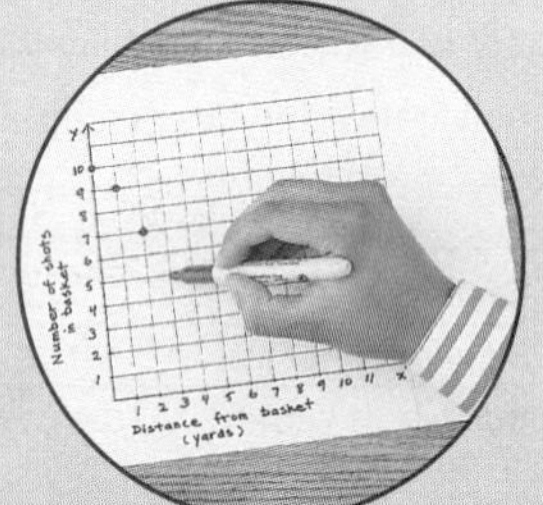

3. MAKE a scatter plot of your data. Let the horizontal axis show the distance from the basket, and let the vertical axis show the number of shots that went into the basket.

General Rubric for Projects
Each project can be evaluated in many possible ways. The following rubric is just one way to evaluate these open-ended projects. It is based on a 4-point scale.

4 The student fully achieves all mathematical and project goals. The presentation demonstrates clear thinking and explanation. All work is complete and correct.

3 The student substantially achieves the mathematical and project goals. The main thrust of the project and the mathematics behind it is understood, but there may be some minor misunderstanding of content, errors in computation, or weakness in presentation.

2 The student partially achieves the mathematical and project goals. A limited grasp of the main mathematical ideas or project requirements is demonstrated. Some of the work may be incomplete, misdirected, or unclear.

1 The student makes little progress toward accomplishing the goals of the project because of lack of understanding or lack of effort.

Making a Model

• **DESCRIBE** your graph. Do you think the data decreases *linearly*? *exponentially*? At what distance did the number of shots that went into the basket decrease the most?

• **ORGANIZE** your data in a table like the one shown. Find the average change and the average percent change.

Distance from basket (yd)	Number of shots that went into basket	Change	Percent change
0	?	—	—
1	?	?	?
2	?	?	?
⋮	⋮	⋮	⋮

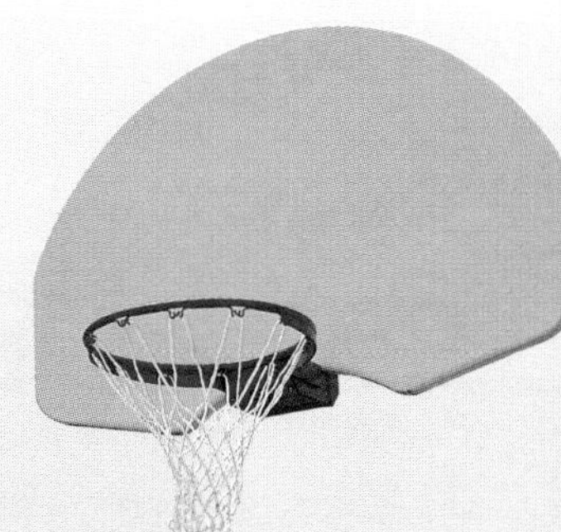

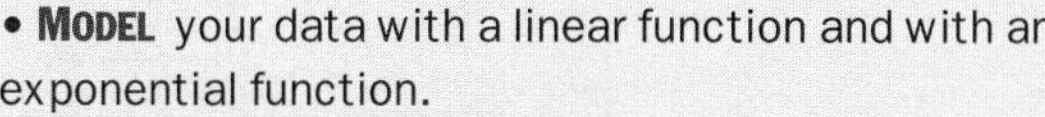

• **MODEL** your data with a linear function and with an exponential function.

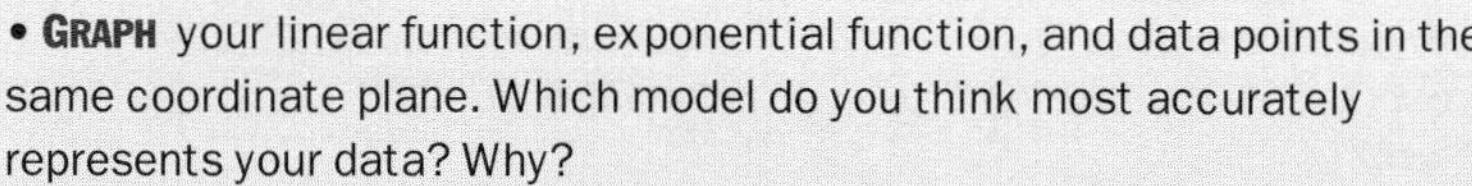

• **GRAPH** your linear function, exponential function, and data points in the same coordinate plane. Which model do you think most accurately represents your data? Why?

Writing a Report

Write a report about your experiment that includes your data, graphs, models, and explanations. You may also want to investigate and report on these things:

• How do you think your results would change if you repeated the experiment 5 times? 10 times? 50 times? Explain your reasoning. Then repeat the experiment at least 5 times. How do your results compare to your prediction?

• Repeat the experiment using a real basketball and hoop. How could a basketball player use this experiment to decide whether to try for a two-point shot or a three-point shot?

• Why do you think the three-point rule was created? What other sports have rules that take the *likelihood* of scoring points into account? Describe the method of scoring for each sport.

Self-Assessment

Describe any difficulties you had in finding a linear and an exponential model to fit your data. How did you resolve these difficulties?

Project Notes

Guiding Students' Work

Rubric for Chapter Project

4 Students complete the experiment and record the results correctly. They make an accurate scatter plot of the data with axes and scale shown. Their descriptions are clear and correctly relate the data collected with the mathematical ideas. Both their linear and exponential models are calculated and written correctly, then graphed accurately. They repeat the experiment a number of times, and also repeat the experiment with a real basketball and hoop. Their final report is well thought out and clearly presented.

3 Students complete the experiment and table and calculate both a linear and an exponential model, but they may not make all of the calculations correctly or do not relate all of the mathematical ideas to the data. Students include all of the information required for the report and show they extended the project by completing the experiment more times and using a real basketball and hoop. The report is clear and shows a considerable effort.

2 Students complete the experiment but may not accurately record, calculate, or describe the data. Only some of the models are drawn on the graph, and the table may be incomplete. They repeat only the wastebasket experiment, but do not do the experiment with a real basketball and hoop. A report is written but may not include all of the ideas necessary to convey the meaning of the activity.

1 Students do not show that they understand the mathematical ideas of the chapter and their work is incomplete. The descriptions of the graph are inaccurate and the calculations in the table are not complete. The linear and exponential functions models are not correctly related to the data and the report does not convey an understanding of the material or a significant effort to complete the activity. Students should be encouraged to speak with the teacher as soon as possible to review their work and to make a new start on the project.

Chapter Support

Course Guide: Chapter 1
Lesson Plans: Chapter 1
Practice Bank: Cumulative Practice 6
Study Guide: Chapter 1 Review
Assessment Book:
Chapter Tests 3 and 4
Chapter 1 Alternative Assessment
Test Generator Software
Portfolio Project Book:
Additional Project 1:
Exploring Algebra Concepts
Preparing for College Entrance Tests
Professional Handbook

Progress Check 1.1–1.2

Months after release	CDs sold (thousands)
0	4.6
1	8.3
2	14.2
3	32.0

Round numerical answers to the nearest hundredth.

1. What is the average increase (in thousands) for the number of CDs sold? *(Sec. 1.1)* about 9.13
2. What is the average percent increase? *(Sec. 1.1)* about 92%
3. Use your results from Exs. 1 and 2 to write a linear function and an exponential function that model the data. *(Sec. 1.2)* linear: $y = 4.6 + 9.13x$; exponential: $y = 4.6(1.92)^x$
4. Use a graphing calculator to graph the data points and the functions from Ex. 3 on the same coordinate plane. *(Sec. 1.2)*

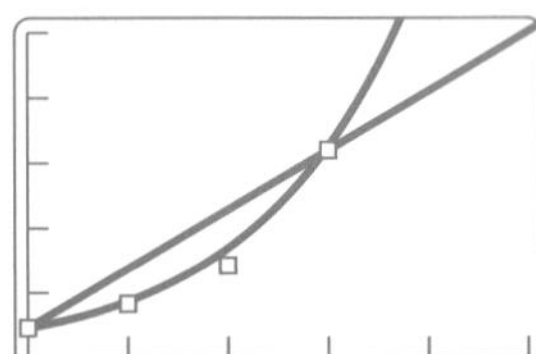

5. Use the linear model and the exponential model to estimate the CD sales (in thousands of CDs) for the 4th month. *(Sec. 1.2)* linear model: about 41.12; exponential model: about 62.51
6. Which of the estimates from Ex. 5 would you expect to be closest to the actual number of CDs sold in month 4? Why? *(Sec. 1.2)* Answers may vary. An example is given. the estimate from the exponential model; The graph fits the given data points more closely.

CHAPTER 1

Review

STUDY TECHNIQUE

What study techniques have you tried before? Write two brief paragraphs starting with these phrases:

- **To study for a mathematics test I usually...**
- **A study technique that has not helped me in the past is...**

VOCABULARY

mathematical model (p. 4)
function (p. 10)
independent variable (p. 10)
dependent variable (p. 10)
matrix (p. 16)
element (p. 16)
dimensions of a matrix (p. 16)
scalar (p. 18)
scalar multiplication (p. 18)
matrix multiplication (p. 23)
simulation (p. 29)

SECTIONS 1.1 *and* 1.2

Mathematical models help you understand and make predictions about real-world situations, such as the federal funding for research and development in transportation.

Years since 1990	Funding (billions of $)	Change	Percent change
0	1.05	—	—
1	1.23	0.18	17%
2	1.52	0.29	24%
3	1.78	0.26	17%
averages		0.24	19%

Linear Model
$y = 1.05 + 0.24x$

x (input)	y (output)
0	1.05
1	1.29
2	1.53
3	1.77

Exponential Model
$y = 1.05(1.19)^x$

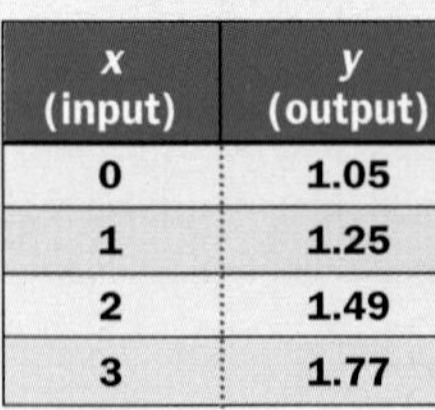

x (input)	y (output)
0	1.05
1	1.25
2	1.49
3	1.77

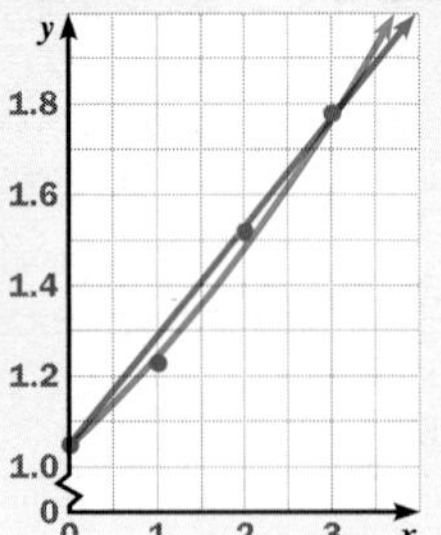

SECTIONS 1.3 *and* 1.4

A **matrix** is a group of numbers arranged in rows and columns. For example, matrix I represents the current inventory of wooden toys and art in a small craft and souvenir store. Matrix S represents sales for one week, and matrix A represents additional inventory that has come in during the week.

$$\begin{array}{r} \\ \text{New} \\ \text{Antique} \end{array}\begin{array}{c}\text{Toys Art}\\ \begin{bmatrix} 15 & 9 \\ 10 & 21 \end{bmatrix}\end{array} = I \qquad \begin{array}{r} \\ \text{New} \\ \text{Antique} \end{array}\begin{array}{c}\text{Toys Art}\\ \begin{bmatrix} 4 & 1 \\ 2 & 7 \end{bmatrix}\end{array} = S \qquad \begin{array}{r} \\ \text{New} \\ \text{Antique} \end{array}\begin{array}{c}\text{Toys Art}\\ \begin{bmatrix} 5 & 2 \\ 3 & 0 \end{bmatrix}\end{array} = A$$

At the end of the week, the stock is $I - S + A$.

$$I - S + A = \begin{bmatrix} 15-4+5 & 9-1+2 \\ 10-2+3 & 21-7+0 \end{bmatrix} = \begin{array}{c}\text{Toys Art}\\ \begin{bmatrix} 16 & 10 \\ 11 & 14 \end{bmatrix}\end{array}\begin{array}{l} \\ \text{New} \\ \text{Antique} \end{array}$$

You can sometimes multiply two matrices to find information. For example, matrix A represents the amount of labor and paint required for a body shop to repair damage to different parts of a car. Matrix B represents the amount the body shop charges for each hour of labor and each pint of paint.

$$\begin{array}{r} \\ \text{Door} \\ \text{Hood} \\ \text{Rear bumper} \end{array}\begin{array}{c}\text{Labor} \quad \text{Paint}\\ \text{(hours)} \; \text{(pints)}\\ \begin{bmatrix} 4.0 & 1.2 \\ 3.0 & 1.0 \\ 2.5 & 1.0 \end{bmatrix}\end{array} = A \qquad \begin{array}{r} \\ \text{Labor} \\ \text{Paint} \end{array}\begin{array}{c}\text{Charge (\$)}\\ \begin{bmatrix} 32 \\ 18 \end{bmatrix}\end{array} = B$$

To determine the total cost to repair each car part, multiply A by B.

$$AB = \begin{bmatrix} 4.0 & 1.2 \\ \mathbf{3.0} & \mathbf{1.0} \\ 2.5 & 1.0 \end{bmatrix}\begin{bmatrix} \mathbf{32} \\ \mathbf{18} \end{bmatrix} = \begin{bmatrix} (4.0)32 + (1.2)18 \\ \mathbf{(3.0)32 + (1.0)18} \\ (2.5)32 + (1.0)18 \end{bmatrix} = \begin{array}{c}\text{Charge (\$)}\\ \begin{bmatrix} 149.6 \\ \mathbf{114} \\ 98 \end{bmatrix}\end{array}\begin{array}{l} \\ \text{Door} \\ \text{Hood} \\ \text{Rear bumper} \end{array}$$

SECTION 1.5

A **simulation** is an experiment that models a situation and can be used to make predictions. For example, suppose Zack is selecting from a tray of spring rolls. One half of the rolls contain only vegetables and one half contain meat. He wants to get at least one vegetable roll. How many rolls should he take? To simulate this situation, toss a coin repeatedly, letting heads (H) represent meat rolls and tails (T) represent vegetable rolls. Stop when you get a T.

Simulation	1	2	3	4
Toss results	HT	T	HHHHT	T
Number of rolls	2	1	5	1

The average is $\frac{2+1+5+1}{4} = 2.25$, so he should take two or three rolls.

Progress Check 1.3–1.5

For Exs. 1–5, use the matrices given below. If the operation cannot be performed, write "undefined." *(Sections 1.3 and 1.4)*

$$M = \begin{bmatrix} 7 & 0 \\ 3 & 8 \\ -1 & 5 \end{bmatrix} \qquad N = \begin{bmatrix} 4 & -1 \\ 1 & 6 \end{bmatrix}$$

$$P = \begin{bmatrix} 2 & 1 \\ 3 & 9 \\ 2 & 0 \end{bmatrix} \qquad Q = \begin{bmatrix} 1 & -1 & 1 \\ 0 & 2 & 3 \\ 2 & 0 & -2 \end{bmatrix}$$

1. $-3M$ $\begin{bmatrix} -21 & 0 \\ -9 & -24 \\ 3 & -15 \end{bmatrix}$

2. MN $\begin{bmatrix} 28 & -7 \\ 20 & 45 \\ 1 & 31 \end{bmatrix}$

3. $M + N$ undefined

4. QP $\begin{bmatrix} 1 & -8 \\ 12 & 18 \\ 0 & 2 \end{bmatrix}$

5. $QM - 2P$ $\begin{bmatrix} -1 & -5 \\ -3 & 13 \\ 12 & -10 \end{bmatrix}$

6. A cereal manufacturer puts a card with a whole number from 1 to 3 in each box of cereal it produces. Anyone who can present cards whose numbers total 20 or more wins a set of 6 cereal bowls. Describe a simulation that uses a spinner to estimate the number of boxes a person would have to buy in order to win a set of bowls. Assume that the chances of getting a 1, 2, or 3 in a given box of cereal are equal. *(Section 1.5)*
Make a circular spinner that has three equal sections. Label one section with the number 1, another section with the number 2, and the remaining section with the number 3. Spin the spinner and record the numbers that come up. Keep a running total. A complete run is achieved when the total equals or exceeds 20. Record the number of times it was necessary to spin the spinner to complete the run. Conduct several runs and calculate the average number of spins per run. Round up to the nearest whole number. Use the result as the estimate for the number of boxes you would have to buy to win a set of bowls.

Chapter 1 Assessment
Form A Chapter Test

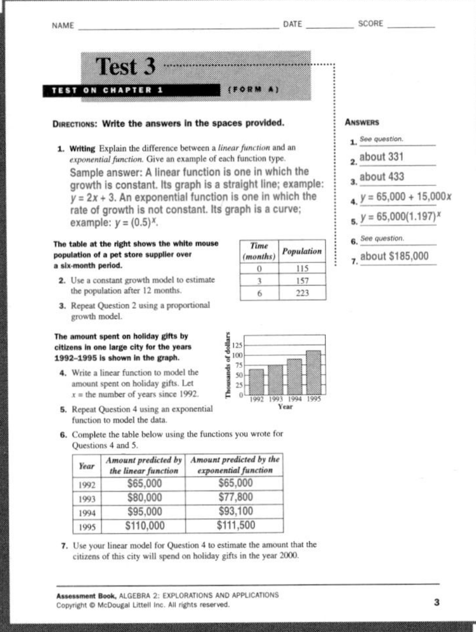
NAME ______ DATE ______ SCORE ______

Test 3

TEST ON CHAPTER 1 (FORM A)

DIRECTIONS: Write the answers in the spaces provided.

1. **Writing** Explain the difference between a *linear function* and an *exponential function*. Give an example of each function type.
   Sample answer: A linear function is one in which the growth is constant. Its graph is a straight line; example: $y = 2x + 3$. An exponential function is one in which the rate of growth is not constant. Its graph is a curve; example: $y = (0.5)^x$.

The table at the right shows the white mouse population of a pet store supplier over a six-month period.

Time (months)	*Population*
0	115
3	157
6	223

2. Use a constant growth model to estimate the population after 12 months.
3. Repeat Question 2 using a proportional growth model.

The amount spent on holiday gifts by citizens in one large city for the years 1992–1995 is shown in the graph.

4. Write a linear function to model the amount spent on holiday gifts. Let x = the number of years since 1992.
5. Repeat Question 4 using an exponential function to model the data.
6. Complete the table below using the functions you wrote for Questions 4 and 5.

Year	*Amount predicted by the linear function*	*Amount predicted by the exponential function*
1992	\$65,000	\$65,000
1993	\$80,000	\$77,800
1994	\$95,000	\$93,100
1995	\$110,000	\$111,500

7. Use your linear model for Question 4 to estimate the amount that the citizens of this city will spend on holiday gifts in the year 2000.

ANSWERS

1. *See question.*
2. about 331
3. about 433
4. $y = 65{,}000 + 15{,}000x$
5. $y = 65{,}000(1.197)^x$
6. *See question.*
7. about \$185,000

Assessment Book, ALGEBRA 2: EXPLORATIONS AND APPLICATIONS

3

Chapter 1 Assessment
Form B Chapter Test

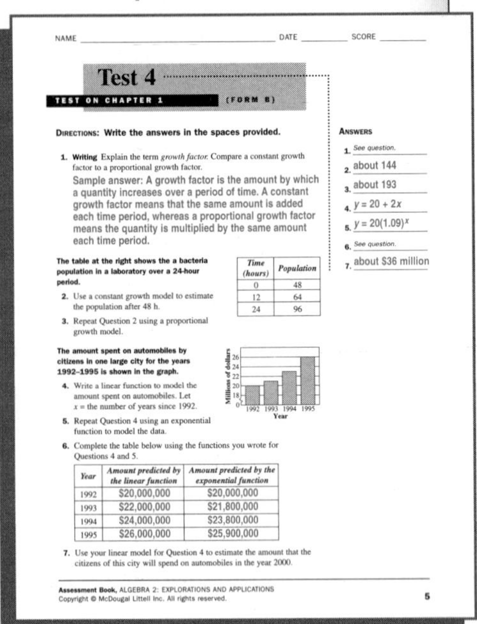
NAME ______ DATE ______ SCORE ______

Test 4

TEST ON CHAPTER 1 (FORM B)

DIRECTIONS: Write the answers in the spaces provided.

1. **Writing** Explain the term *growth factor*. Compare a constant growth factor to a proportional growth factor.
   Sample answer: A growth factor is the amount by which a quantity increases over a period of time. A constant growth factor means that the same amount is added each time period, whereas a proportional growth factor means the quantity is multiplied by the same amount each time period.

The table at the right shows the a bacteria population in a laboratory over a 24-hour period.

Time (hours)	*Population*
0	48
12	64
24	96

2. Use a constant growth model to estimate the population after 48 h.
3. Repeat Question 2 using a proportional growth model.

The amount spent on automobiles by citizens in one large city for the years 1992–1995 is shown in the graph.

4. Write a linear function to model the amount spent on automobiles. Let x = the number of years since 1992.
5. Repeat Question 4 using an exponential function to model the data.
6. Complete the table below using the functions you wrote for Questions 4 and 5.

Year	*Amount predicted by the linear function*	*Amount predicted by the exponential function*
1992	\$20,000,000	\$20,000,000
1993	\$22,000,000	\$21,800,000
1994	\$24,000,000	\$23,800,000
1995	\$26,000,000	\$25,900,000

7. Use your linear model for Question 4 to estimate the amount that the citizens of this city will spend on automobiles in the year 2000.

ANSWERS

1. *See question.*
2. about 144
3. about 193
4. $y = 20 + 2x$
5. $y = 20(1.09)^x$
6. *See question.*
7. about \$36 million

Assessment Book, ALGEBRA 2: EXPLORATIONS AND APPLICATIONS

5

CHAPTER

1 Assessment

VOCABULARY QUESTIONS

For Questions 1 and 2, complete each paragraph.

1. A(n) __?__ pairs every input value with exactly one output value. The input variable, x, is called the __?__. The output variable, y, is called the __?__ because its value depends upon what value you choose for x.
2. A(n) __?__ is a group of numbers arranged in rows and columns. Each number is called a(n) __?__.

SECTIONS 1.1 *and* 1.2

3. The table shows United States Census figures for the population of California.

Year	Population (millions)
1970	20.0
1980	23.7
1990	29.8

   a. What was the average change in population per decade?
   b. What was the average percent change in population per decade?
   c. Use a constant growth (linear) model to estimate California's population in the year 2000.
   d. Repeat part (c) using a proportional growth (exponential) model.
   e. **Writing** Compare your answers to parts (c) and (d). Which do you think is a more reasonable estimate for the population of California in the year 2000? Why?

4. The graph shows the amount Americans spent on dental care for 1986–1989.

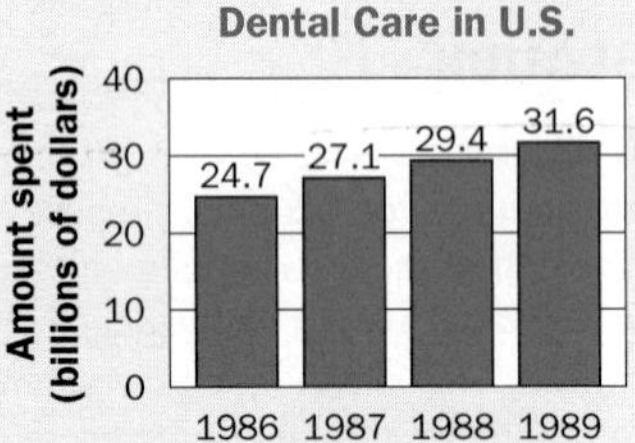

   a. Write a linear function to model the amount spent on dental care. Let x = the number of years since 1986.
   b. Repeat part (a) using an exponential function to model the data.
   c. Make a table showing the actual amount spent, the amount predicted by your linear function, and the amount predicted by your exponential function for each of the four years shown in the graph.
   d. Based on your answer to part (c), which model do you think is more reasonable—the *linear model* or the *exponential model*? Explain.

ANSWERS Chapter 1

Assessment

1. function; independent variable; dependent variable
2. matrix; element
3. a. 4.9 million
   b. 22%
   c, d. Let x = the number of decades since 1970.
   c. $y = 20.0 + 4.9x$; about 34.7 million
   d. $y = 20.0(1.22)^x$; about 36.3 million
   e. Answers may vary. An example is given. I think both results are reasonable.
4. a. $y = 24.7 + 2.3x$
   b. $y = 24.7(1.09)^x$
   c. Amounts are in millions of dollars.

Year	Actual Amount Spent	Amount predicted by linear function	Amount predicted by exponential function
1986	24.7	24.7	24.7
1987	27.1	27.0	26.9
1988	29.4	29.3	29.3
1989	31.6	31.6	32.0

   d. The values predicted by the linear model are slightly closer to the actual values than are the values predicted by the exponential model. However, both models are reasonable.

SECTIONS 1.3 *and* 1.4

For Exercises 5–10, use matrices *S, T, U, and V* to evaluate each matrix expression. If an operation cannot be performed, write *undefined*.

$$S=\begin{bmatrix}1 & 0\\ -2 & 1\end{bmatrix}\quad T=\begin{bmatrix}5 & 2\\ 0 & -1\\ 7 & 11\end{bmatrix}\quad U=\begin{bmatrix}13\\ -6\end{bmatrix}\quad V=\begin{bmatrix}3 & -4\\ -2 & 1\\ 0 & 20\end{bmatrix}$$

5. $T+V$ **6.** $U+S$ **7.** $4V-T$

8. TU **9.** $3S$ **10.** ST

For Exercises 11–13, use the matrix $P=\begin{bmatrix}8 & 6 & 5 & 1\\ -2 & 17 & 0 & -4\end{bmatrix}$.

11. What are the dimensions of P?

12. What is the value of the element $p_{2,1}$?

13. If Q is a 3×2 matrix, what are the dimensions of the product QP?

SECTION 1.5

14. The makers of ABC, a new juice drink, have devised a contest for consumers to win a free bottle of ABC. One of the letters A, B, or C is printed on the liner of each bottle cap. One half of the caps have an A, one third of the caps have a B, and one sixth of the caps have a C. If you collect all 3 letters, you win a free bottle of ABC.

a. Describe how you would simulate this giveaway using a die.

b. The table shows some results from a simulation to determine how many bottle caps a person needs to collect to win a free bottle. Based on these results, how many bottle caps (on average) would a person need in order to win a free bottle?

Simulation	1	2	3	4	5
Results	B, A, A, A, A, A, C	A, A, A, A, A, A, C, C, B	A, A, B, C	B, A, C	A, B, B, A, B, A, A, B, A, A, B, A, C

PERFORMANCE TASK

15. Interview a local businessperson. See if the person is willing to share information about sales, inventory, pricing, special promotions, and so on. Use the information you collect to develop mathematical models involving functions, matrices, or simulations. Prepare a report to present to the class.

Chapter 1 Assessment
Form C Alternative Assessment

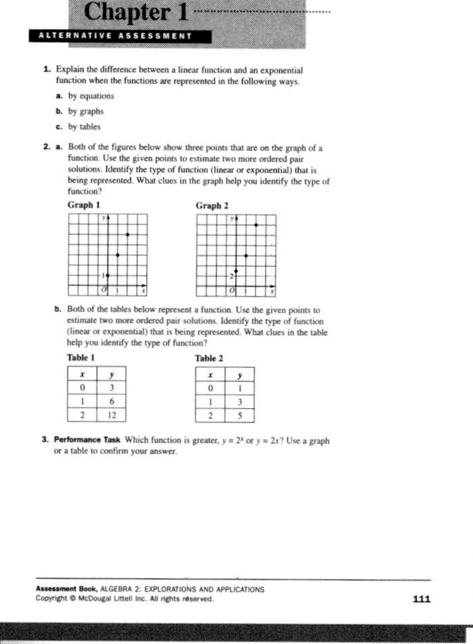
Chapter 1
ALTERNATIVE ASSESSMENT

1. Explain the difference between a linear function and an exponential function when the functions are represented in the following ways.

a. by equations

b. by graphs

c. by tables

2. a. Both of the figures below show three points that are on the graph of a function. Use the given points to estimate two more ordered pair solutions. Identify the type of function (linear or exponential) that is being represented. What clues in the graph help you identify the type of function?

Graph 1 Graph 2

b. Both of the tables below represent a function. Use the given points to estimate two more ordered pair solutions. Identify the type of function (linear or exponential) that is being represented. What clues in the table help you identify the type of function?

Table 1

x	y
0	3
1	6
2	12

Table 2

x	y
0	1
1	3
2	5

3. Performance Task Which function is greater, $y = 2^x$ or $y = 2x$? Use a graph or a table to confirm your answer.

Assessment Book, ALGEBRA 2: EXPLORATIONS AND APPLICATIONS

111

4. Performance Task Which of the following can be best represented by a linear function? by an exponential function? Explain how you arrived at your decision.

a. The independent variable is the number of hours you worked. The dependent variable is the salary you earned. You are earning minimum wage.

b. The independent variable is the number of minutes that have lapsed. The dependent variable is the number of bacteria in a petri dish.

c. The independent variable is number of years that an amount of money has been in a savings account. The dependent variable is the amount of money in the account.

d. The independent variable is the number of people in the room. The dependent variable is the time it takes to evacuate the room during a fire drill.

e. The independent variable is the number of years that have passed. The dependent variable is the value of a particular car. The car depreciates approximately 20% each year.

f. The independent variable is the number of days that have passed. The dependent variable is the amount of radon gas that remains. Radon has a half-life of 3.5 days.

5. In some mathematical situations the independent and dependent variables are similar to a cause and an effect. Explain how the independent variable is similar to the cause and the dependent variable is similar to the effect.

6. Project Find examples of matrix operations in your everyday life. Demonstrate your understanding of matrices by describing one real-world situation in detail. Include real examples of matrix addition, scalar multiplication, and matrix multiplication. Perform the actual operations and interpret your answers.

7. Research Project Sometimes problems cannot be solved by simple mathematical computation. Simulations can be used to model the situation and give answers that are fairly accurate. Describe some situations in which modeling would be the best way to solve a problem or make a prediction. Look up the Monte Carlo Method in a statistics book and write about the five-step process. Discuss some of the tools that are used in the Monte Carlo Method to simulate complicated problems.

8. Open-ended Problem A high school basketball player shoots free throws with 75% accuracy. Design an experiment that models this player's accuracy.

9. Research Project Ask a computer programmer if the computer random number generator is truly random. Find out what programmers do to improve the randomness.

112
Assessment Book, ALGEBRA 2: EXPLORATIONS AND APPLICATIONS

5. $\begin{bmatrix}8 & -2\\ -2 & 0\\ 7 & 31\end{bmatrix}$

6. undefined

7. 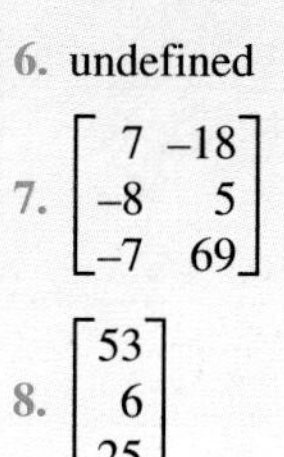 $\begin{bmatrix}7 & -18\\ -8 & 5\\ -7 & 69\end{bmatrix}$

8. $\begin{bmatrix}53\\ 6\\ 25\end{bmatrix}$

9. $\begin{bmatrix}3 & 0\\ -6 & 3\end{bmatrix}$

10. undefined

11. 2×4

12. –2

13. 3×4

14. a. Let three numbers, say 1, 2, and 3, represent the letter A. Let two numbers, say 4 and 5, represent the letter B, and let the remaining number represent the letter C. Roll the die until a number representing each of the three letters comes up. Record the number of rolls it took. Repeat a number of times, say 20, and calculate the average number of rolls.

b. 7 or 8 bottle caps

15. Answers may vary. Check students' work.

CHAPTER

2 Linear Functions

OVERVIEW

Connecting to Prior and Future Learning

⇔ The study of linear functions, begun in Algebra 1, is continued in this chapter. Linear equations are used to describe direct variations, to model events, and to make predictions. Function notation is reviewed and expanded. Students use ratios and proportions and linear inequalities in their work. A review of these concepts can be found on pages 785 and 787 in the **Student Resources Toolbox**.

⇔ The line of fit and the correlation coefficient are revisited in this chapter. Students use scatter plots, which are reviewed on page 795 of the **Student Resources Toolbox**.

⇔ The concept of using equations to model situations is continued with the introduction of parametric equations. Students will use this concept in future mathematics courses.

Chapter Highlights

Interview with Norbert Wu: The interview and related exercises and example on pages 49 and 54, respectively, emphasize the use of mathematics in underwater photography.

Explorations in Chapter 2 involve writing an equation to model water levels in Section 2.2, finding a line of fit by plotting data about the thickness of books in Section 2.4, and seeing how parametric equations can be used to analyze motion in two dimensions in Section 2.6.

The Portfolio Project: Students perform an experiment to study the relationship between pulling force and rubber band length.

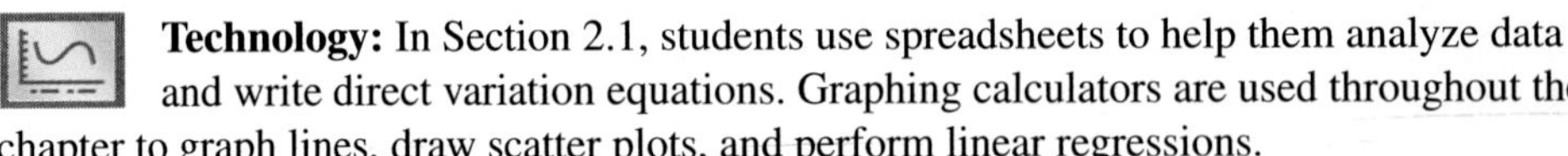

Technology: In Section 2.1, students use spreadsheets to help them analyze data and write direct variation equations. Graphing calculators are used throughout the chapter to graph lines, draw scatter plots, and perform linear regressions.

OBJECTIVES

Section	Objectives	NCTM Standards
2.1	• Recognize direct variation. • Write and use direct variation equations. • Analyze data and make predictions.	1, 2, 3, 4, 5, 6, 8
2.2	• Write an equation of a line in slope-intercept form. • Graph a linear equation in slope-intercept form. • Model events.	1, 2, 3, 4, 5, 6, 8
2.3	• Write the point-slope equation of a line. • Use function notation. • Find the domain and range of a linear function. • Make predictions about linear data.	1, 2, 3, 4, 5, 6, 8
2.4	• Draw a line of fit and find its equation. • Make predictions from linear data.	1, 2, 3, 4, 5, 6, 10, 12
2.5	• Interpret a correlation coefficient. • Recognize positive and negative correlation. • Determine the strength of a linear relationship.	1, 2, 3, 4, 5, 6, 10, 12
2.6	• Write and graph a pair of linear parametric equations. • Rewrite a pair of parametric equations as a single equation. • Model situations with parametric equations.	1, 2, 3, 4, 5, 6

INTEGRATION

Mathematical Connections	2.1	2.2	2.3	2.4	2.5	2.6
algebra	**45–51***	**52–57**	**58–64**	**65–71**	**72–77**	**78–83**
geometry	48	57				
data analysis, probability, discrete math	**45–51**	**52–57**	58, 61–64	**65–71**	**72–77**	**78–83**
patterns and functions	**45–51**	**52–57**	**58–64**	**65–71**	**72–77**	**78–83**
logic and reasoning	46, 48, 50, 51	54–57	59, 61–64	67–71	72, 74–77	80–83
Interdisciplinary Connections and Applications						
biology and earth science		56	63	69, 71	76	
chemistry and physics						79
business and economics				68	75	
arts and entertainment						82
sports and recreation	45, 48, 50		63			81
government				70	77	
manufacturing, personal finance, temperature scales, meteorology, oceanography, health	48, 51	55	61, 62	66, 67, 71	73	

****Bold page numbers** *indicate that a topic is used throughout the section.*

TECHNOLOGY

Section	*opportunities for use with* Student Book	Support Material
2.1	graphing calculator spreadsheet software McDougal Littell Mathpack *Stats!*	**Function Investigator with Matrix Analyzer Activity Book:** Function Investigator Activity 3
2.2	graphing calculator McDougal Littell Mathpack *Function Investigator*	**Function Investigator with Matrix Analyzer Activity Book:** Function Investigator Activities 2 and 4–6
2.3	graphing calculator McDougal Littell Mathpack *Function Investigator*	**Technology Book:** Spreadsheet Activity 2 TI-92 Activity 1
2.4	graphing calculator McDougal Littell Mathpack *Stats!*	**Technology Book:** Spreadsheet Activity 2
2.5	graphing calculator McDougal Littell Mathpack *Stats!* *Function Investigator*	**Technology Book:** Calculator Activity 2 Spreadsheet Activity 2
2.6	graphing calculator McDougal Littell Mathpack *Function Investigator*	**Technology Book:** Spreadsheet Activity 2

PLANNING GUIDE

Regular Scheduling (45 min)

Section	Materials Needed	Core Assignment	Extended Assignment	*exercises that feature* Applications	Communication	Technology
2.1	spreadsheet software, graphing calculator, graph paper	1–14, 19–25, 29–32	1–5, 7–9, 11–32	13–18, 22–25, 28	26, 29	28
2.2	glass container, water, marbles, metric ruler, graph paper, paper cups	1–13, 15–22, 27–34	1–34	22–24	21, 26, 29	26
2.3	graph paper	**Day 1:** 1–5, 11–19 **Day 2:** 21–33, AYP*	**Day 1:** 1–19 **Day 2:** 20–33, AYP	5–10, 14, 15 20	 25	
2.4	set of encyclopedias, metric ruler, graph paper, graphing calculator or statistical software	**Day 1:** 1–6, 9 **Day 2:** 10–19	**Day 1:** 1–9 **Day 2:** 10–19	1, 3–9 10	1, 7, 8 10	2, 5
2.5	graphing calculator or statistical software, graph paper	1–8, 10, 12–14, 16–23	1–23	12–15	4, 9, 10, 16	12, 13, 15
2.6	graphing calculator with parametric mode, graph paper	1–8, 10–12, 15–20, AYP	1–20, AYP	10–14	12	13, 14
Review/ Assess		**Day 1:** 1–12 **Day 2:** 13–23 **Day 3:** Ch. 2 Test	**Day 1:** 1–12 **Day 2:** 13–23 **Day 3:** Ch. 2 Test	18	19	18
Portfolio Project		Allow 2 days.	Allow 2 days.			

Yearly Pacing (with Portfolio Project)	**Chapter 2 Total** 13 days	**Chapters 1–2 Total** 26 days	**Remaining** 134 days	**Total** 160 days

Block Scheduling (90 min)

	Day 8	Day 9	Day 10	Day 11	Day 12	Day 13	Day 14
Teach/Interact	2.1 2.2: Exploration, page 52	2.3	2.4: Exploration, page 65	2.5 2.6: Exploration, page 78	Review Port. Proj.	Review Port. Proj.	Ch. 2 Test 3.1
Apply/Assess	**2.1:** 1–14, 19–25, 29–32 **2.2:** 1–13, 15–22, 27–34	**2.3:** 1–5, 11–19, 21–33, AYP*	**2.4:** 1–6, 9–19	**2.5:** 1–8, 10, 12–14, 16–23 **2.6:** 1–8, 10–12, 15–20, AYP	**Review:** 1–12 **Port. Proj.**	**Review:** 13–23 **Port. Proj.**	**Ch. 2 Test** **3.1:** 1–16, 22, 27–34

NOTE: A one-day block has been added for the Portfolio Project—timing and placement to be determined by teacher.

Yearly Pacing (with Portfolio Project)	**Chapter 2 Total** $6\frac{1}{2}$ days	**Chapters 1–2 Total** $13\frac{1}{2}$ days	**Remaining** $68\frac{1}{2}$ days	**Total** 82 days

* ***AYP*** *is Assess Your Progress.*

Lesson Support

Section	Practice Bank	Study Guide*	Assessment Book*	Visuals	Explorations Lab Manual	Lesson Plans	Technology Book
2.1	7	2.1		Warm-Up 2.1 Folder A	Master 2	2.1	
2.2	8	2.2		Warm-Up 2.2 Folder A	Masters 2, 5	2.2	
2.3	9	2.3	Test 5	Warm-Up 2.3 Folder A	Masters 1, 2	2.3	Spreadsheet Act. 2 TI-92 Act. 1
2.4	10	2.4		Warm-Up 2.4	Master 1 Add. Expl. 2	2.4	Spreadsheet Act. 2
2.5	11	2.5		Warm-Up 2.5 Folder 2	Master 1	2.5	Calculator Act. 2 Spreadsheet Act. 2
2.6	12	2.6	Test 6	Warm-Up 2.6	Masters 1, 2	2.6	Spreadsheet Act. 2
Review Test	13	Chapter Review	Tests 7, 8 Alternative Assessment			Review Test	Calculator Based Lab 1

***Spanish versions** of *Study Guide* and *Assessment Book* are available.

Chapter Support

- Course Guide
- Lesson Plans
- Portfolio Project Book
- Preparing for College Entrance Tests
- Multi-Language Glossary
- *Test Generator* Software
- Professional Handbook

Software Support

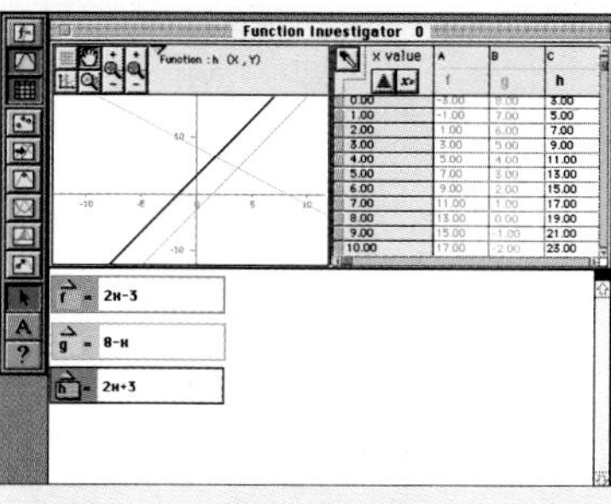

McDougal Littell Mathpack
Stats!
Function Investigator

Internet Support

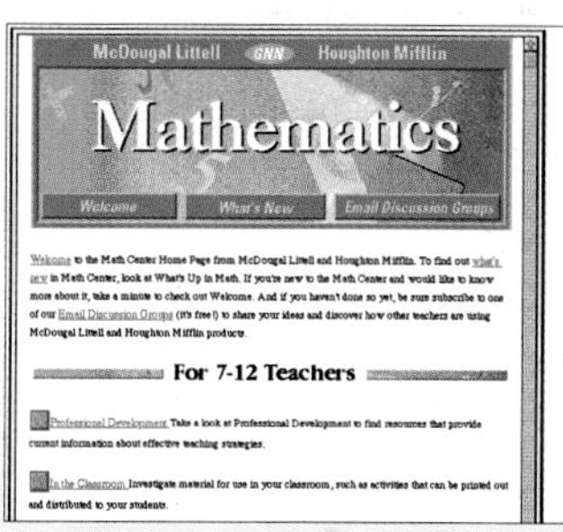

http://www.hmco.com
Next go to McDougal Littell; then the Education Center; then Secondary Math.

Outside Resources

Books, Periodicals

Wallace, Edward C. "Exploring Regression with a Graphing Calculator." *Mathematics Teacher* (December 1993): pp. 741–743.

Alper, Lynne, Dan Fendel, Sherry Fraser, and Diane Reseck. "Is This a Mathematics Class?" *Mathematics Teacher* (November 1995): pp. 632–638.

Sandefur, James T. "Technology, Linear Equations, and Buying a Car." *Mathematics Teacher* (October 1992): pp. 562–567.

Activities, Manipulatives

Breuningsen, Chris, Bill Bower, Linda Antinone, and Elisa Breuningsen. "Making Cents of Math." *Real-World Math with the CBL System.* Activity 2: pp. 17–22. Texas Instruments, 1995.

Coes, Loring III. "What Is the *r* For?" *Mathematics Teacher* (December 1995): pp. 758–760.

Software

Spinnaker Software. *Better Working Spreadsheet.* Apple. Reston, VA: NCTM.

Meridian Creative Group. "Exploration: Modeling Direct Proportions." Worksheets and software included for use with the CBL. *Explorations in Precalculus for the TI-82:* pp. 31–36.

Internet

The U.S. Dept. of Education's Online Library includes a Technology Resource Guide at:
http://www.ed.gov/
and also at:
gopher.ed.gov

Join the discussions on Kidsphere (excellent mailing list for K–12 education) by sending a message to:
kidsphere-request@vms.cis.pitt.edu
In the body of the message, type
subscribe kidsphere [first name, last name]

Interview Notes

Background

Scuba Diving

The word *scuba* stands for self-contained underwater breathing apparatus, and scuba diving has been occurring from as early as 4500 B.C., when breath-holding divers dove for shells in the Mediterranean Sea. Hollow reeds were used for snorkels about the year 100 A.D., and goggles made from polished tortoise shells were used in the Persian Gulf as early as 1300 A.D. Today, scuba diving can involve the use of many different types of equipment. The breathing gear is composed of one or more tanks and a demand regulator, which consists of the air hose and the mouth piece. The demand regulator controls the flow of air to the diver so that the pressure within the diver's lungs is equal to the pressure of the surrounding water. Divers also use computers, worn as a wristwatch, to allow them to figure maximum bottom times and minimum surface intervals. Divers may also use such equipment as a mask, wet suit (or dry suit in frigid waters), weighted belt, fins, snorkel, and buoyancy compensator.

To take pictures under water, the camera must either be waterproof or enclosed in a special housing. Many housings are made of plastic or aluminum and surround the camera to protect it from the water. A camera is also subject to the pressure of the water and should not be used below the recommended depth.

Norbert Wu

Norbert Wu has done many things involving wildlife photography and writing. His photographs and articles have appeared in numerous books and magazines including *National Geographic, Audubon, Harper's, International Wildlife, Omni, Outside,* and *Smithsonian.* He is the writer and photographer for six books on marine life, including one children's book on the kelp forest. He serves as associate and contributing editor for five magazines and leads diving expeditions all around the world. The locations of his diving expeditions have ranged from the freezing waters of the Arctic to the warm waters of the Pacific, including such places as the Galapagos, East Africa, the Red Sea, Australia, and the Bahamas.

CHAPTER

2 Linear Functions

Photographing the underwater world

INTERVIEW **Norbert Wu**

Jacques Cousteau describes Wu as:

"a hard-working and meticulous technician, as well as a friendly team member."

Underwater photographer Norbert Wu says he surprises people when they meet him. He says he's not a stereotypical daredevil scuba diver, but describes himself as a modest man who's traveled the world to dive and to photograph some unusual marine life. Along the way he's been bitten by a shark, been trapped under an iceberg, and ridden on the back of a manta ray with a 20-foot wingspan.

Rare Deep-Sea Images

Wu has also photographed hammerhead sharks in 100-foot-deep Costa Rican waters. The dozens of hammerheads were apparently oblivious to his presence during that dive. To shoot a picture of them, Wu said he had to hold his breath and "kick toward them, fast." The instant he breathed out, the sharks dashed off, frightened by the sound of air bubbles escaping through his regulator. "Others have done it," he says casually. "If I'd never heard of it [being done] before, I'd be scared witless."

He recalls that one of his best dives was off San Clemente Island. "In the fall, there was a gathering of bat rays. They look like pterodactyls. They were schooling—there were a couple of hundred dancing, swimming in a circle. It was most memorable."

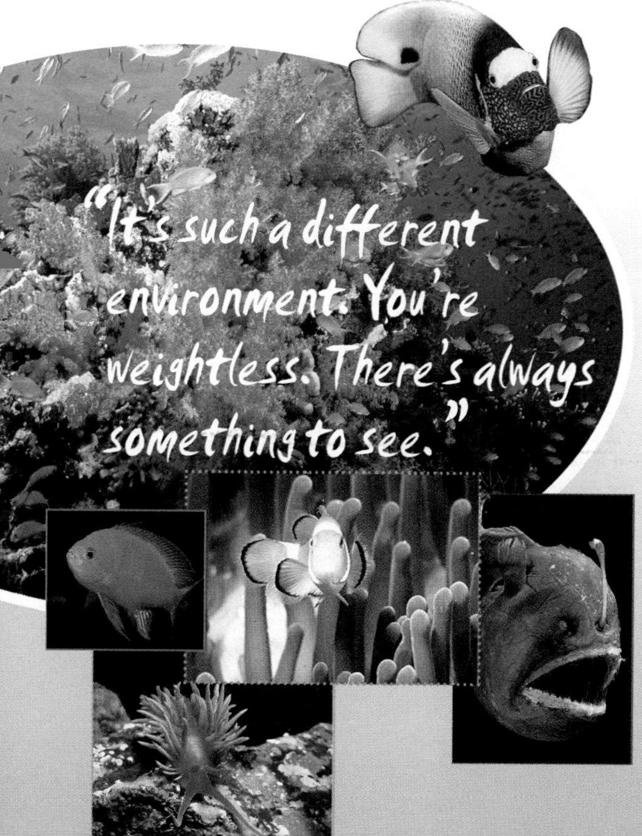

"It's such a different environment. You're weightless. There's always something to see."

Working with Jacques Cousteau

As a child in Atlanta, Wu said he knew by the second grade he wanted to study marine biology. He spent many summers snorkeling off the California and Florida coasts and received his diver's certificate at age 15. In 1986 while a doctoral student at the Scripps Institute of Oceanography, he worked as chief still photographer for Jacques Cousteau's research ship *Calypso.* He saw firsthand the famous deep-sea researcher at work. Cousteau describes Wu as "a hard-working and meticulous technician, as well as a friendly team member."

Interview Notes

Background

The Ama

The ama, fisherwomen who live along the coast of Japan, are among the world's most accomplished divers. They harvest over four million pounds of abalone a year and may dive as deep as seventy-five feet. Following a 2000-year-old tradition, the ama use no modern diving equipment, except for goggles and wet suits. Attached by a safety line to a boat, the ama may dive as many as one hundred times in a day, with just their fishing tool to hunt among the ocean's corals.

Second-Language Learners

Students learning English may not understand what is meant by the expressions *stereotypical daredevil scuba diver* and *meticulous technician.* If necessary, explain that the first phrase means "a scuba diver who is quite brave, which most scuba divers are" and the second phrase means "a worker who pays a great amount of attention to detail." Some students may also be unfamiliar with *pterodactyls,* which are prehistoric flying reptiles. Also, the sport of snorkeling may be unfamiliar to some students. If necessary, explain that snorkeling enables a swimmer to view plant and fish life from just beneath the surface of the water, using a special breathing apparatus (a snorkel) that allows in air from above the water.

Interview Notes

Mathematical Connection

The pressure that is exerted on an object is often measured in pounds per square foot, or lb/ft^2. There is a relationship between pressure that the surrounding water puts on a person and the depth that a diver dives. This relationship has been found to be linear through experimentation and discovery, and can be modeled by a direct variation equation. In Section 2.1, modeling equations and calculating pressure and depth are two of the ideas explored in four exercises of the section. In addition to the pressure being exerted on a diver by the surrounding water, pressure is also exerted by the surrounding air. Example 2 of Section 2.2 looks at the total pressure exerted on a diver using modeling by linear equations and graphing.

Explore and Connect

Research
Students may also find scuba diving listed under skin diving. The school or local library is a good source to research the history of diving. Students may also include pictures in their report to illustrate some old diving equipment.

Writing
Give students a chance to draw their own graph and make their own scale. After they have finished drawing their graphs, compare their choices of scale and discuss techniques for choosing an appropriate scale.

Project
Students can watch for differences in the intensity with which the water streams out from the carton. As the water level lowers, the streams do not flow as hard due to the decreasing pressure.

The Importance of Safety

Like all divers, Wu feels the pressure of the water on him increase as he dives deeper. The increasing water pressure causes his body to absorb more nitrogen. If he returns to the surface too rapidly, the nitrogen in his body fluids can form bubbles. The bubbles can cause "bends," or decompression sickness. To prevent injury, he returns to the surface slowly. The maximum recommended rate of ascent for divers is about 60 ft/min. The graph shows how the time needed to return to the surface is related to the depth of the dive. This relationship is an example of a *linear function.*

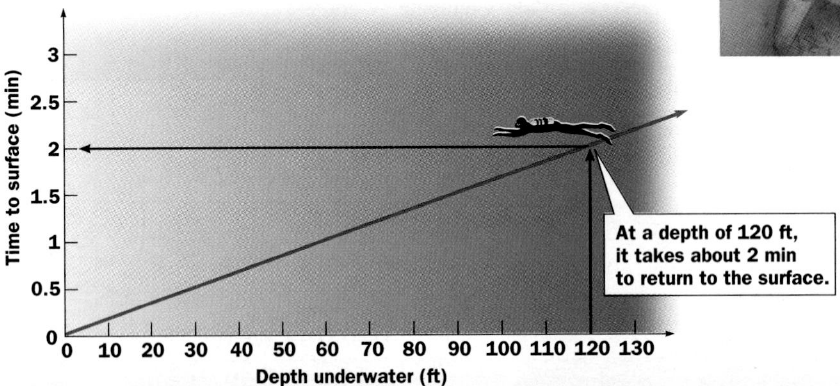

EXPLORE AND CONNECT

Norbert Wu on a photography trip

1. Research Investigate the history of diving. What kind of equipment did the first deep-sea divers use? How deep could they dive with the equipment?

2. Writing Draw a depth-versus-time graph for a diver who starts at a depth of 100 ft and ascends at a rate of 60 ft/min. Explain how the graph shows the time taken to return to the surface.

3. Project Get an empty container, such as a milk carton or plastic bottle, and punch small holes of the same size at various heights on the lower half of the container. Place the container in an empty tub or sink. Cover the holes and fill the container with water. Uncover the holes and watch for differences in the streams of water pouring out from the holes. What do you think causes the differences?

Mathematics & Norbert Wu

In this chapter, you will learn more about how mathematics is related to diving.

Related Examples and Exercises

Section 2.1
- **Exercises 15–18**

Section 2.2
- **Example 2**

Section

2.1 Direct Variation

Learn how to...

- recognize direct variation
- write and use direct variation equations

So you can...

- analyze data and make predictions about skating speeds, for example

When precision skaters perform a maneuver called *the wheel,* they form a revolving line. A skater's speed depends upon the skater's distance from the center of rotation. How are speed and distance mathematically related in this situation? Consider the following table of data.

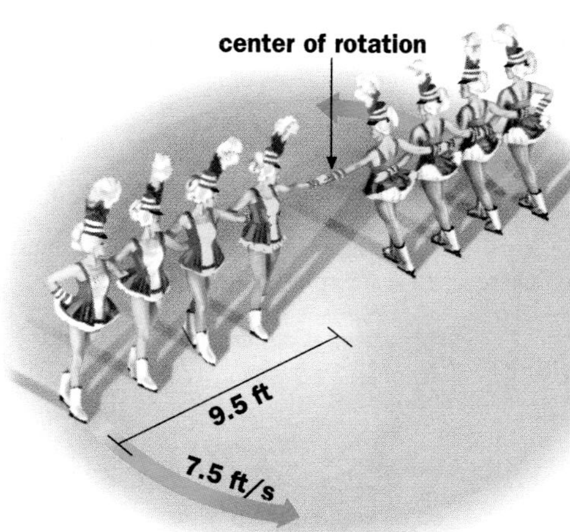

Distance from center of rotation (ft)	Speed (ft/s)
2	1.6
4.5	3.5
7	5.5
9.5	7.5
12	9.4

Toolbox p. 785 *Ratios and Proportions*

EXAMPLE 1 Application: Precision Skating

Use a spreadsheet to see if there is a common speed-to-distance ratio for the data given in the table.

SOLUTION

Calculate the ratios $\frac{\text{speed}}{\text{distance}}$ using a spreadsheet.

For more information about using a spreadsheet, see the *Technology Handbook*, p. 816.

Precision Skating Data

	A	B	C
1	Distance *d*	Speed *s*	*s/d* ratio
2	2	1.6	0.8
3	4.5	3.5	0.77777778
4	7	5.5	0.78571429
5	9.5	7.5	0.78947368
6	12	9.4	0.78333333

The ratios are approximately equal.

There is a common ratio of about 0.79 ft/s of speed per 1 ft of distance from the center of rotation.

Technology Note

As an alternative to doing a spreadsheet, students can use the list features of a TI-82 graphing calculator. Using the STAT EDIT menu, enter distance values into list L1 and speed values into list L2. Using the home screen, enter L2/L1→L3, where → is the STO▶ key. This stores the quotients into list L3. Use the STAT EDIT menu again to study the three lists.

You can extend the ideas of this section by assigning Function Investigator Activity 3 in the *Function Investigator with Matrix Analyzer Activity Book*. This activity leads students to explore the relationship between the value of the constant of variation, *a*, in the direct variation equation $y = ax$, the slope of the line that is the graph of the equation, and the angle of inclination of the line.

Plan⇔Support

Objectives

- Recognize direct variation.
- Write and use direct variation equations.
- Analyze data and make predictions.

Recommended Pacing

Core and Extended Courses
Section 2.1: 1 day

Block Schedule
Section 2.1: $\frac{1}{2}$ block
(with Section 2.2)

Resource Materials

Lesson Support
Lesson Plan 2.1
Warm-Up Transparency 2.1
Overhead Visuals:
Folder A: Multi-Use Graphing Packet, Sheets 1–3
Practice Bank: Practice 7
Study Guide: Section 2.1
Exploration Lab Manual:
Diagram Master 2

Technology
Graphing Calculator
Spreadsheet Software
McDougal Littell Mathpack
Function Investigator with Matrix Analyzer Activity Book: Function Investigator Activity 3
Stats!
Internet:
http://www.hmco.com

Warm-Up Exercises

1. $d = 2.5$ and $s = 1.5$. Find $\frac{s}{d}$. 0.6
2. Find an ordered pair that is a solution of $y = 0.5x$. Answers may vary. (5, 2.5)
3. Find the slope of $y = -3x - 2$. −3
4. A class has 18 boys and 16 girls. Find the ratio of boys to girls. $\frac{9}{8}$
5. If $P = 3t$ is the profit on t tickets, what is the profit on 250? If the number of tickets is doubled, what is the profit? $750; $1500

When the ratio of two variables is constant, the variables are in **direct variation.** This means that if $\frac{y}{x} = a$ for some nonzero constant a, then:

$$y = ax$$

You can say that ***y** varies directly* with **x**, or **y** is *proportional* to **x**.

The number **a** is the **constant of variation.**

EXAMPLE 2

Write a direct variation equation for the data in Example 1, and use a graphing calculator or graphing software to graph the equation. Then predict the speed of a skater 19.5 ft from the center of rotation.

SOLUTION

In Example 1, you found that $\frac{s}{d} = 0.79$, so $s = 0.79d$.

A graph of this equation is a line. To check the reasonableness of the graph, compare it to a scatter plot of the data.

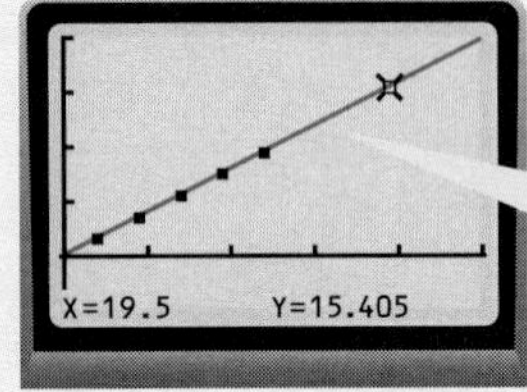

On a graphing calculator or graphing software, graph **y = 0.79x** instead of $s = 0.79d$.

Notice how well the graph fits the plotted data. For a skater 19.5 ft from the center of rotation, $d = 19.5$ and $s = 0.79(19.5) \approx 15.41$. The skater's speed is about 15.4 ft/s.

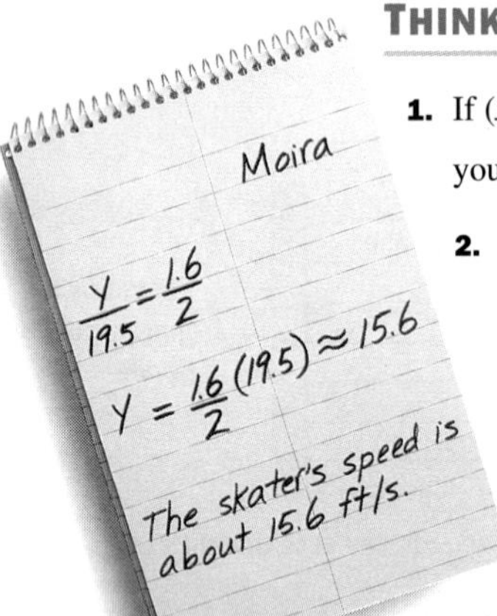

Moira
$\frac{Y}{19.5} = \frac{1.6}{2}$
$Y = \frac{1.6}{2}(19.5) \approx 15.6$
The skater's speed is about 15.6 ft/s.

THINK AND COMMUNICATE

1. If (x_1, y_1) and (x_2, y_2) are two data pairs from a direct variation, what can you say about $\frac{y_1}{x_1}$ and $\frac{y_2}{x_2}$? Explain.
2. Moira found the speed of the skater in Example 2 a different way.
   a. Describe the method Moira used.
   b. What explains the difference between Moira's answer and the answer given in Example 2?
3. Notice that the graph in Example 2 includes the origin. What does this mean in terms of the rotating line of skaters? Why does this make sense?

Teach⇔Interact

Section Note

Communication: Discussion You may wish to ask students to explain why a skater's speed is greater when the skater is farther from the center of rotation.

Additional Example 1

Although thunder and lightning occur together, thunder is heard after the lightning is seen because light travels faster than the speed of sound. Use a spreadsheet to see if there is a common distance-to-time ratio for the data given in the table.

Time between lightning and thunder (s)	Distance from lightning strike (km)
4.7	1.5
9.3	3
13.9	4.5
18.9	6
23.2	7.5

Calculate the ratios $\frac{\text{distance}}{\text{time}}$ using a spreadsheet.

Lightning Thunder Data

	A	B	C
1	Time t	Distance d	d/t ratio
2	4.7	1.5	0.31914894
3	9.3	3	0.32258065
4	13.9	4.5	0.32374101
5	18.9	6	0.31746032
6	23.2	7.5	0.32327586

There is a common ratio of about 0.32 km of distance per 1 second of time between seeing the lightning and hearing the thunder.

Additional Example 2

Write a direct variation equation for the data in Additional Example 1, and use a graphing calculator or graphing software to graph the equation. Then predict the distance from the lightning if the thunder was heard 26 s after the lightning was seen. In Additional Example 1, you found that $\frac{d}{t} = 0.32$, so $d = 0.32t$. A graph of this equation is a line. To check the reasonableness of the graph, compare it to a scatter plot of the data.

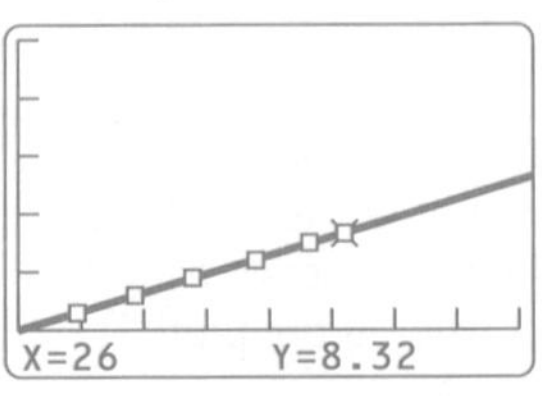

Technology Note

For Example 2, students can use the table features of a TI-82 calculator to find x-values to the nearest tenth. Enter Tblset:ΔTbl = 0.5 and look at the table under the TABLE option to see the value of y that corresponds to $x = 19.5$.

The graph of $y = ax$ is a line that passes through the origin. In a previous course you learned that the ratio $\frac{\text{vertical change}}{\text{horizontal change}}$ gives the **slope** of a line. For a direct variation graph, the slope is just the constant of variation, a.

The graph of $y = 2x$ is a line that passes through the origin.

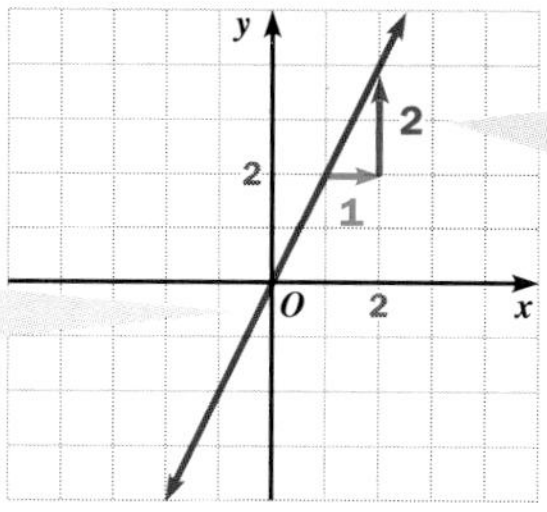

When x increases by 1 unit, y increases by 2 units. The slope of the line is $\frac{2}{1}$, or 2.

EXAMPLE 3

Write an equation for the direct variation graph shown below.

SOLUTION

The graph passes through the point $(-2, 3)$, so the constant of variation is:

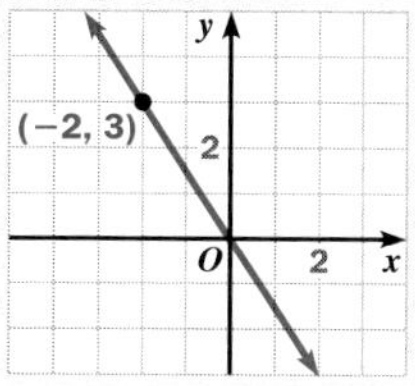

$$a = \frac{3}{-2} = -\frac{3}{2} = -1.5$$

An equation of the graph is $y = -1.5x$.

CHECKING KEY CONCEPTS

1. Tell whether y varies directly with x. Explain your reasoning.

a. $y = 3.2x$ **b.** $y = -3.2x$ **c.** $y = 3.2x + 1$

2. Tell whether r and s are proportional. Explain your reasoning.

a. $\frac{r}{s} = \frac{2}{5}$ **b.** $r = 8s$ **c.** $r + s = 2$

3. For each graph, tell whether y varies directly with x. If so, give an equation for the graph.

a.

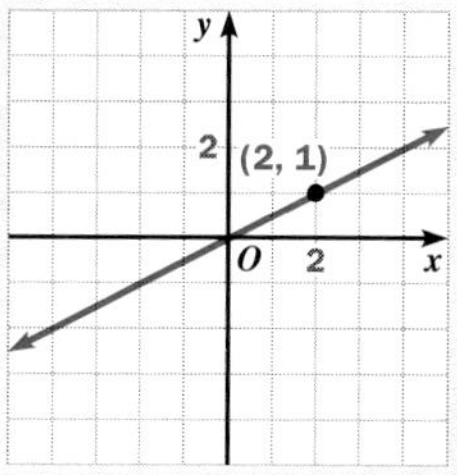

b.

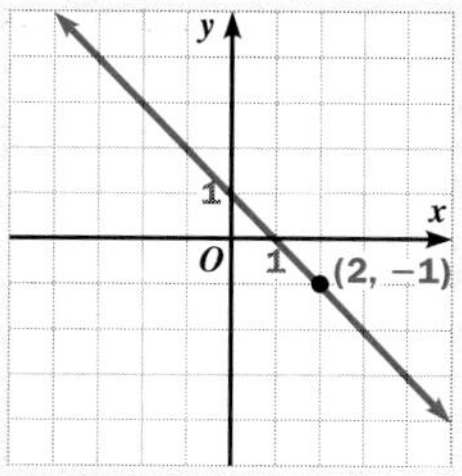

4. Graph $y = 3x$. Describe the change in y when:

a. x increases by 1 unit **b.** x doubles **c.** x is halved

Teach⇔Interact

Additional Example 2 (continued)

Notice how well the graph fits the plotted data. For thunder heard 26 s after the lightning is seen, $t = 26$ and $d = 0.32(26) \approx 8.3$. The lightning is about 8.3 km away.

Learning Styles: Visual

The graph at the top of this page shows that the slope of the line $y = ax$ is the constant of variation a. Encourage students to keep this picture in mind when they think about slope as being vertical change divided by horizontal change.

Section Note

Topic Spiraling: Review

Be certain that students have a thorough understanding of the concept of slope before proceeding further.

Additional Example 3

Write an equation for the direct variation graph shown below.

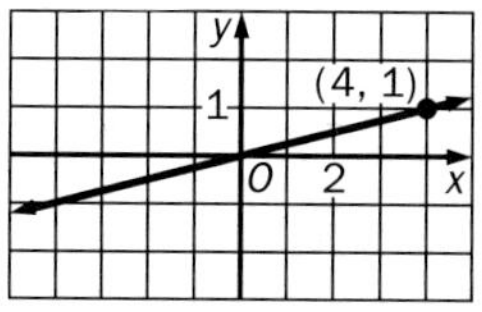

The graph passes through the point (4, 1), so the constant of variation is $a = \frac{1}{4}$. An equation of the graph is $y = \left(\frac{1}{4}\right)x$.

Checking Key Concepts

Common Error

Some students may think that the graph of any line is a direct variation graph. Stress that line graphs are direct variation graphs only when they contain the origin.

Closure Question

Explain how to determine if data vary directly.

Arrange the data in a table in numerical order. Check to see if there is a common ratio for the data. If there is, the data vary directly.

ANSWERS Section 2.1

Think and Communicate

1–3. See answers in back of book.

Checking Key Concepts

1. a. Yes; the ratio of y and x is constant; $\frac{y}{x} = 3.2$.

b. Yes; the ratio of y and x is constant; $\frac{y}{x} = -3.2$.

c. No; the ratio of y and x is not constant. The graph of $y = 3.2x + 1$ does not pass through the origin.

2. a. Yes; the ratio of r and s is constant; $\frac{r}{s} = \frac{2}{5}$.

b. Yes; the ratio of r and s is constant; $\frac{r}{s} = 8$.

c. No; the ratio of r and s is not constant. The graph of $r + s = 2$ does not pass through the origin.

3. a. Yes; $y = \frac{1}{2}x$.

b. No.

4.

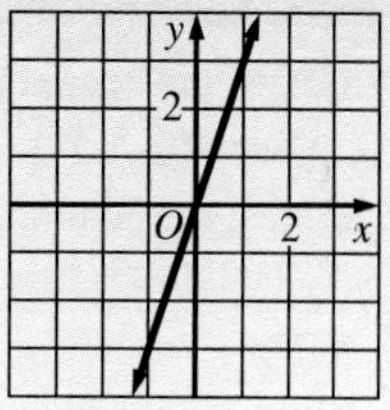

a. y increases by 3 units.

b. y doubles.

c. y is halved.

Apply⇔Assess

Suggested Assignment

Core Course
Exs. 1–14, 19–25, 29–32

Extended Course
Exs. 1–5, 7–9, 11–32

Block Schedule
Day 8 Exs. 1–14, 19–25, 29–32

Exercise Notes

Using Technology
Exs. 1–3 Students can use a TI-82 graphing calculator to calculate the ratios of the data items and display them as they would in a spreadsheet. The procedure is simple and efficient. First, press [STAT] 1 and enter the data from the table as lists L1 and L2. Use the arrow keys to move the cursor to the column for L3 and hold the [▲] key down until the cursor is on L3, at the very top of the column. Then enter L2 ÷ L1 by pressing [2nd] [L2] [÷] [2nd] [L1] [ENTER]. The calculator will calculate the ratio of each pair of data items and display the results in the L3-column. For Ex. 3, note that each fraction for list L1 can be typed by using the [÷] key. When the students press [ENTER], the calculator converts the fraction to a decimal.

Cooperative Learning
Exs. 1–3 Working in small groups will allow students to compare their answers to these exercises. Have each person explain how he or she determined the common ratio for one of the data tables.

Visual Thinking
Exs. 1–3 Ask students to create graphs of the data presented in each of the three tables. Discuss how these graphs demonstrate whether there is a *common ratio* or not. This activity involves the visual skills of *identification* and *interpretation*.

Integrating the Strands
Ex. 4 This exercise relates the concept of direct variation to a geometric situation. In so doing, students can see how an algebraic concept can be used to understand the formulas for the perimeter and the area of a square.

2.1 Exercises and Applications

Extra Practice exercises on page 750

For each table, tell whether there is a common ratio. If there is, write an equation relating the two variables.

1.

Calories in Lean Hamburger	
Serving size (oz)	Calories
2	125
2.5	156
3	188
3.5	219
4	250

2.

Income for a Software Company	
Sales (billion \$)	Profit (billion \$)
1.18	0.28
1.84	0.46
2.76	0.71
3.75	0.95
4.65	1.15

3.

Importer's Prices of Diamonds	
Size (carats)	Price (dollars)
$\frac{1}{3}$	199
$\frac{1}{2}$	399
$\frac{3}{4}$	499
1	599
2	1499

4. GEOMETRY The length of a side of a square is s meters.

a. Find formulas in terms of s for the perimeter P and area A of the square. In what units are P and A measured?

b. Does P vary directly with s? Does A vary directly with s? Explain your answers.

For each equation, tell whether *y* varies directly with *x*. If so, graph the equation.

5. $y = \frac{5}{2x}$ **6.** $y = \frac{5x}{2}$ **7.** $y = 5x + 2$ **8.** $y = \frac{2}{5}x$

9. $y = x + 4$ **10.** $y = \frac{4}{x}$ **11.** $y = 4$ **12.** $y = 4x$

13. SPORTS At the 1994 Winter Olympics in Hamar, Norway, speed skaters were required to complete 1.25 laps around the track for the 500 m race. Find a direct variation equation that relates the number of laps to the length of a race in meters. Then find the number of laps required to complete the 3000 m race.

14. MANUFACTURING Machinists often cut hexagonal shapes from round stock. Many times the machinist knows the distance f between the "flats" or sides of the hexagon. The direct variation equation $d = 1.1547f$ is then used to determine the distance d between opposite corners of the hexagon.

a. Find the diameter of the round stock needed to cut a hex nut measuring 1.3750 in. between the flats.

b. Use what you know about the geometry of a regular hexagon to derive the direct variation equation relating d and f.

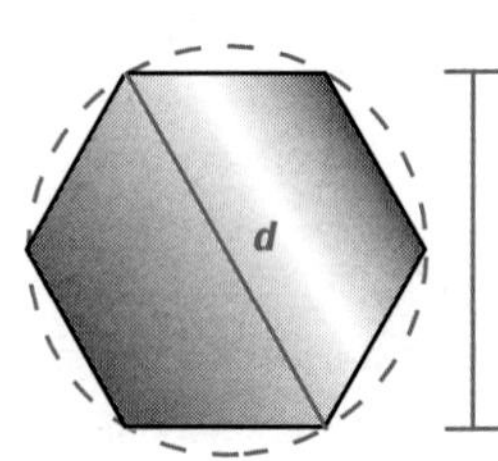

BY THE WAY

The Hamar Olympic Hall, shaped like an inverted Viking ship, is the world's largest indoor ice skating hall.

Exercises and Applications

1. Yes; the common ratio is about 62.5; $y = 62.5x$.
2. Yes; the common ratio is about 0.25; $y = 0.25x$.
3. No.
4. a. $P = 4s$; $A = s^2$; meters; square meters
   b. Yes; $P = 4s$ is an equation of a direct variation. No; $A = s^2$ is not an equation of a direct variation.
5. No.
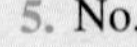
6. Yes.

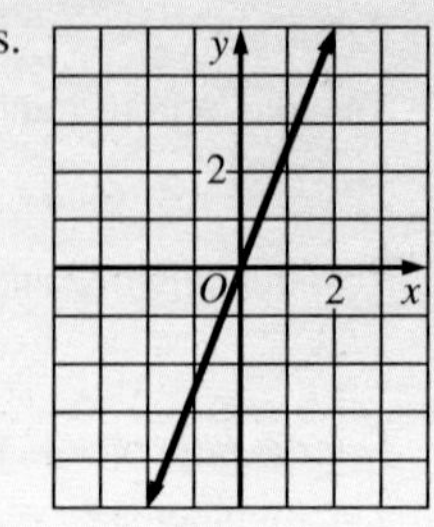

7. No.
8. Yes.

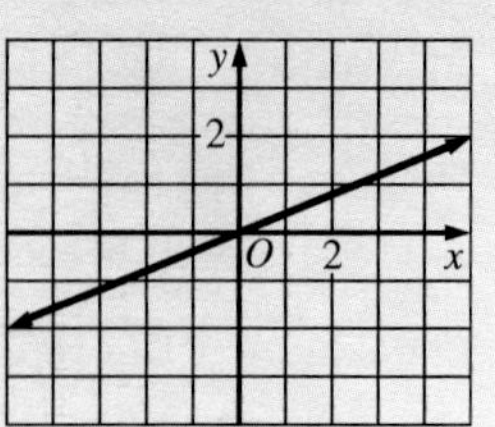

9. No.
10. No.
11. No.

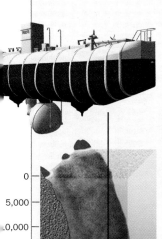

INTERVIEW Norbert Wu

Look back at the article on pages 42–44.

The deeper Norbert Wu dives, the more pressure the water exerts on him. This is true of anything else in the water, including animals and submarines.

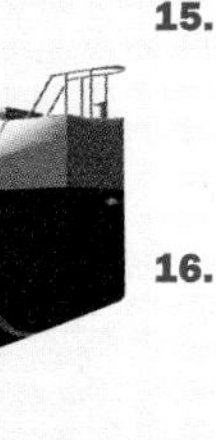

The Marianas Trench

15. The water pressure P on a submerged object is proportional to the object's depth d. At a depth of 50 ft, the water pressure is 3200 lb/ft^2. Use this information to write an equation relating P and d.

16. Using standard gear, a scuba diver can go to a maximum depth of about 130 ft. What is the water pressure on a scuba diver at that depth?

17. Some of the fish that Norbert Wu has photographed live at a depth of 3000 ft. (These fish have to be brought to the surface with a net.) What is the water pressure on the fish living at that depth?

18. The deepest ocean descent on record was achieved in 1960 by Dr. Jacques Piccard of Switzerland and Lt. Donald Walsh of the United States Navy. They used a small research submarine called a bathyscaph to descend 35,800 ft into the Pacific Ocean's Marianas Trench. At that depth, what was the water pressure on the hull of the bathyscaph?

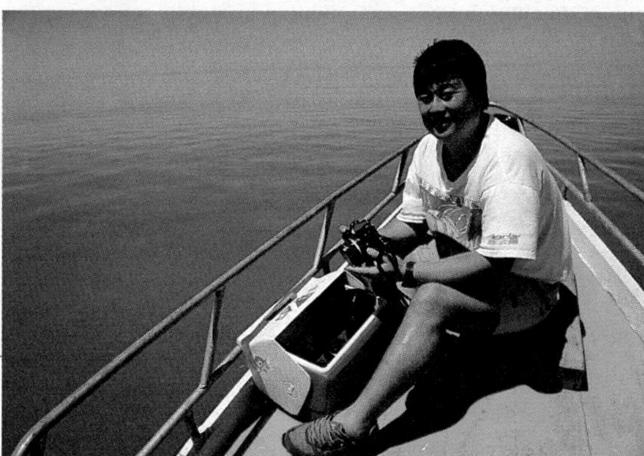

For each graph, tell whether y varies directly with x. If so, give an equation for the graph.

19.

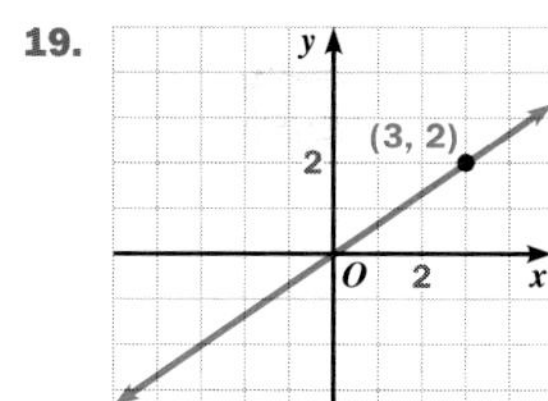

20.

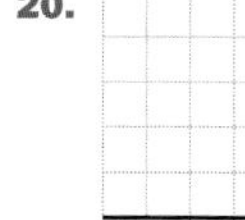

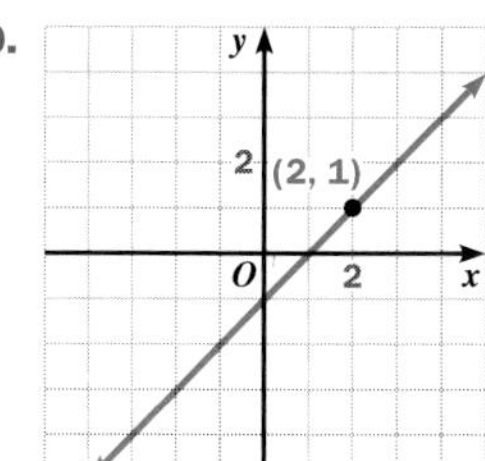

21.

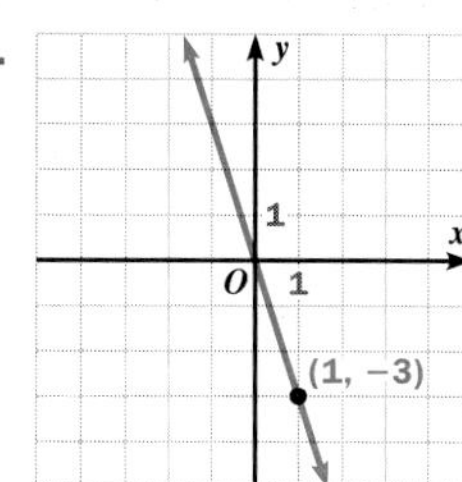

12. Yes.

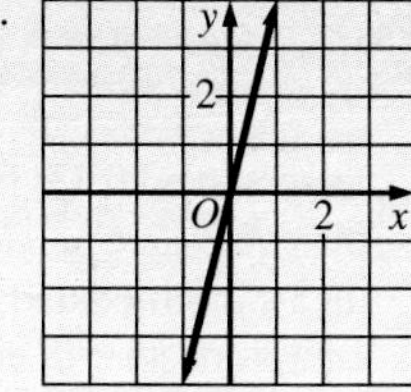

13. Let l = the number of laps and d = the distance of the race in meters; $l = 0.0025d$; 7.5 laps.

14. a. about 1.59 in.

b. Answers may vary. An example is given. The hexagon can be divided into eight congruent equilateral triangles with sides of length $\frac{1}{2}d$ and height $\frac{1}{2}f$. In each of the triangles, f is the length of the longer leg of a 30°-60°-90° triangle with hypotenuse of length d, so $d = \frac{2}{\sqrt{3}}f \approx 1.1547f$.

15. $P = 64d$

16. about 8320 lb/ft^2

17. 192,000 lb/ft^2

18. 2,291,000 lb/ft^2

19. Yes; $y = \frac{2}{3}x$.

20. No.

21. Yes; $y = -3x$.

Apply⇔Assess

Exercise Notes

Application

Ex. 13 After discussing this exercise, some students may be interested in finding a direct variation equation that applies to a sport they play or enjoy watching. Encourage them to use their own data to generate the equation.

Multicultural Note

By the Way The people known as the Vikings had their origins in northwestern Europe. About 4000 years ago, they began migrating to what is now Sweden, Denmark, and Norway. Since much of the land was surrounded by water, the Vikings needed to develop their shipbuilding and sailing skills. Among their innovations in ship design was the use of a keel, a long narrow piece of wood that ran along the entire underside of the ship's center. The keel helped to balance a ship in calm or rough seas.

Reasoning

Ex. 14 This exercise provides an opportunity for students to review a number of geometric ideas about a regular hexagon. When deriving the equation relating d and f, students should support their work with valid geometric reasons for each step.

Interview Note

Exs. 15–18 Students may find it challenging to create a graphing calculator viewing window that can be used for these exercises. A typical window is Xmin = 0, Xmax = 40,000, Ymin = 0, Ymax = 2,500,000. Have students graph the equation in Ex. 15 and trace the graph to answer Exs. 16–18. Also, it may help some students gain a better appreciation of the pressures involved by converting pounds per square feet to pounds per square inch.

Common Error

Exs. 19, 21 The graph for Ex. 19 shows that as the variable x increases, the variable y also increases. Some students intuitively think that this is true for all direct variation situations. The graph for Ex. 21, however, shows that this is not the case; as x increases, y decreases, but y still varies directly with x.

Apply⇔Assess

Exercise Notes

Research
Ex. 27 You may wish to have students find the actual cost of electrical usage for their own homes. Have them express the cost as a function of usage and explain if the function is a direct variation. Then students can calculate a new constant of variation for part (b).

Second-Language Learners
Ex. 28 Point out to second-language learners that the *log* here refers to a record of how much fuel the car uses. Also explain that an *odometer* is an instrument that shows the distance a vehicle has traveled.

Using Technology
Ex. 28 As an alternative approach to the method shown in Example 2 on page 46, students can use the list and linear regression features of a TI-82 calculator. If they can recall how to find the line of fit from Algebra 1 (which is presented in Section 2.4 of this book), then, after entering the data as lists, students can find the line of fit, $y = ax + b$, to determine if the data represent a direct variation. If so, b should be near zero and a will be the constant of variation. To do this, use the STAT CALC menu, and choose LinReg(ax+b).

Visual Thinking
Ex. 28 Ask students to work in groups to collect mileage and fuel information for two different cars. Have them create a graph that compares the data from the two cars and discuss the graph with the class. Encourage them to answer questions such as "How do the two cars compare?" and "Does either data set suggest *direct variation*? If so, how?" This activity involves the visual skills of *interpretation* and *generalization*.

Connection RECREATION

The number of calories (Cal) you use when exercising is proportional to the amount of time you exercise.

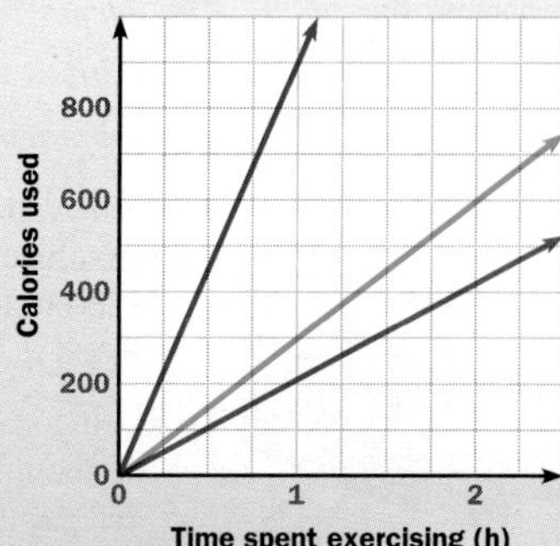

22. Visual Thinking For a 150 lb person, the graphs of calories used when walking at 2.5 mi/h, when walking at 3.25 mi/h, and when running at 10 mi/h are shown. Which graph do you think applies to each type of exercise? Explain your reasoning.

23. Alex and Robert both weigh 150 lb. For 30 min, Alex walks at 3.25 mi/h while Robert walks at 2.5 mi/h. About how many more calories does Alex use than Robert?

24. Suppose Joe, who weighs 150 lb, has just eaten a snack containing 300 calories. He wants to "walk it off." About how long will it take if he walks at 2.5 mi/h? at 3.25 mi/h?

25. a. Find an equation for the running graph. What is the constant of variation? (Include its units.)

b. Do you think the constant would be different for a person who weighs less than 150 lb? If so, do you expect it to be more or less than the constant in part (a)? Why?

26. Cooperative Learning Work in a group of six or more students. Your group will need a stopwatch or a watch with a second hand.

a. Starting with one person and adding another at each stage of the experiment, time how long it takes to pass a "wave" (an up-and-down motion of the arms) along a row of people.

b. Do your data suggest that "wave" time varies directly with the number of people? If so, write an equation and compare its graph to a scatter plot of your data.

27. Challenge The equation $u = 0.33t$ gives the electrical usage u, measured in kilowatt-hours (kW•h), as a function of the time t, in hours, that Betty's television set is on. Also, the equation $C = 0.0918u$ gives the cost of electricity C, in dollars, as a function of Betty's electrical usage u.

a. For Betty's television, can you say that the cost C of using it varies directly with the time t it is used? If so, what equation relates these two variables?

b. Explain how you obtained the constant of variation for your equation in part (a).

22. The slope of the graph is the number of calories used per hour, so the exercise that uses the most calories per hour should have the steepest graph. The blue graph applies to running, the green to walking at 3.25 mi/h, and the red to walking at 2.5 mi/h.

23, 24. Answers may vary.

23. about 50 Cal

24. about 1 h 23 min; about 1 h

25. a. Let x = the number of hours spent exercising and y = the number of calories used; $y = 900x$; 900 Cal/h.

b. Yes; the constant would be less for a person who weighs less than 150 lb, because it requires less energy to move a smaller mass.

26. Answers may vary.

27. a. Yes; $C = 0.03t$.

b. I substituted $0.33t$ for u in the equation $C = 0.0918u$, then simplified and rounded the constant of variation.

28. PERSONAL FINANCE Sonya keeps a log of her car's fuel economy. Each time she stops for gasoline, she records the mileage from the car's odometer and the number of gallons of gas needed to fill the fuel tank. The beginning of Sonya's log is shown.

a. **Spreadsheets** Enter Sonya's data in a spreadsheet. Include two other columns, one labeled "Distance driven since last fuel stop (mi)" and one labeled "Fuel economy (mi/gal)." Have the spreadsheet perform the appropriate calculations for these two columns.

b. Does part (a) suggest a direct variation? If so, write an equation relating *d*, the distance driven, and *g*, the gas used.

c. The fuel tank of Sonya's car holds about 13.5 gal. How far can Sonya travel on a full tank of gas?

Odometer reading (mi)	Gas used since last fuel stop (gal)
18	Car is new with a full tank of gas.
232	8.2
534	11.6
621	3.4
871	9.5
1033	6.2
1308	10.7

ONGOING ASSESSMENT

29. Writing Find an example of direct variation in everyday life. Create a table of data for the two variables and use your data to find the constant of variation. Write a brief explanation of why one variable varies directly with the other.

SPIRAL REVIEW

30. When Leah took her little brother Johnny shopping, Johnny noticed a machine that dispenses small toys in plastic capsules. The machine holds five types of toys, including a high-bounce ball that Johnny would like to have. It costs $.25 to get a toy from the machine. If the five types of toys are equally likely to drop from the machine, how much would it cost, on average, to get the toy that Johnny wants? Use a simulation to answer this question. *(Section 1.5)*

31. The linear function $C = 3 + 0.25n$ gives a checking account's monthly cost *C*, in dollars, in terms of the number of checks written *n*. Tell what the numbers 3 and 0.25 mean in this situation. *(Section 1.2)*

32. Which pair(s) of matrices can be added together? Which pair(s) can be multiplied? *(Sections 1.3, 1.4)*

$A = \begin{bmatrix} 1 & 2 \\ 3 & 4 \end{bmatrix}$ $\quad B = \begin{bmatrix} 2 \\ 3 \end{bmatrix}$ $\quad C = \begin{bmatrix} 4 & 2 \\ 1 & 0 \end{bmatrix}$

Apply⇔Assess

Exercise Notes

Communication: Writing
Ex. 29 Writing about an example of direct variation in everyday life will help students to organize and clarify their thoughts about this fundamental concept. You may wish to call upon two or three students to explain their examples and read their explanations to the class.

Topic Spiraling: Preview
Ex. 31 This exercise previews linear equations in slope-intercept form, which is the topic of the next section.

Practice 7 for Section 2.1

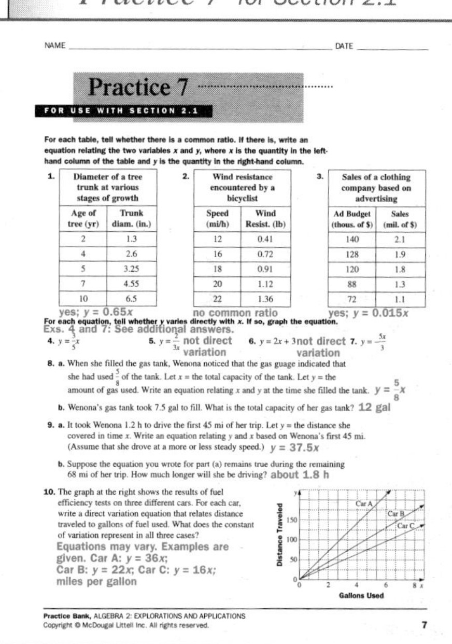

NAME ______________________ DATE __________

Practice 7

FOR USE WITH SECTION 2.1

For each table, tell whether there is a common ratio. If there is, write an equation relating the two variables *x* and *y*, where *x* is the quantity in the left-hand column of the table and *y* is the quantity in the right-hand column.

1.

Diameter of a tree trunk at various stages of growth	
Age of tree (yr)	Trunk diam. (in.)
2	1.3
4	2.6
5	3.25
7	4.55
10	6.5

yes; $y = 0.65x$

2.

Wind resistance encountered by a bicyclist	
Speed (mi/h)	Wind Resist. (lb)
12	0.41
16	0.72
18	0.91
20	1.12
22	1.36

no common ratio

3.

Sales of a clothing company based on advertising	
Ad Budget (thous. of $)	Sales (mil. of $)
140	2.1
128	1.9
120	1.8
88	1.3
72	1.1

yes; $y = 0.015x$

For each equation, tell whether *y* varies directly with *x*. If so, graph the equation.
Exs. 4 and 7: See additional answers.

4. $y = \frac{3}{5}x$ 5. $y = \frac{2}{3x}$ not direct variation 6. $y = 2x + 3$ not direct variation 7. $y = \frac{5x}{3}$

8. a. When she filled the gas tank, Wenona noticed that the gas guage indicated that she had used $\frac{5}{8}$ of the tank. Let x = the total capacity of the tank. Let y = the amount of gas used. Write an equation relating x and y at the time she filled the tank. $y = \frac{5}{8}x$

b. Wenona's gas tank took 7.5 gal to fill. What is the total capacity of her gas tank? 12 gal

9. a. It took Wenona 1.2 h to drive the first 45 mi of her trip. Let y = the distance she covered in time x. Write an equation relating y and x based on Wenona's first 45 mi. (Assume that she drove at a more or less steady speed.) $y = 37.5x$

b. Suppose the equation you wrote for part (a) remains true during the remaining 68 mi of her trip. How much longer will she be driving? about 1.8 h

10. The graph at the right shows the results of fuel efficiency tests on three different cars. For each car, write a direct variation equation that relates distance traveled to gallons of fuel used. What does the constant of variation represent in all three cases?
Equations may vary. Examples are given. Car A: $y = 36x$; Car B: $y = 22x$; Car C: $y = 16x$; miles per gallon

Practice Bank, ALGEBRA 2: EXPLORATIONS AND APPLICATIONS

7

---

28. a.

Odometer reading (mi)	Gas used since last fuel stop (mi)	Distance driven since last fuel stop (mi)	Fuel economy (mi/gal)
18	Car is new with a full tank of gas.	—	—
232	8.2	214	26.09756098
534	11.6	302	26.03448276
621	3.4	87	25.58823529
871	9.5	250	26.31578947
1033	6.2	162	26.12903226
1308	10.7	275	25.70093458

b. Yes. Answers may vary due to rounding. $d = 26g$

c. about 351 mi

29. Answers may vary.

30. Answers may vary. An example is given. Using the function int(5 * rand) + 1 to generate random numbers, I let 1 represent getting the toy Johnny wanted. I generated numbers until I got a 1. I did this 10 times with the following numbers of tries needed: 4, 2, 3, 4, 1, 13, 11, 1, 2, 6. The average number of tries was 4.7 or about 5. Using these results, it would cost about $1.25 to get the toy that Johnny wants.

31. the basic monthly cost of the account; the cost per check

32. *A* and *C*; *A* and *B* can be multiplied in that order, that is, *AB* is defined. *C* and *B* can be multiplied in that order, that is, *CB* is defined. *A* and *C* can be multiplied in either order, that is, *AC* and *CA* are both defined.

Plan⇔Support

Objectives

- Write an equation of a line in slope-intercept form.
- Graph a linear equation in slope-intercept form.
- Model events.

Recommended Pacing

Core and Extended Courses
Section 2.2: 1 day

Block Schedule
Section 2.2: $\frac{1}{2}$ block
(with Section 2.1)

Resource Materials

Lesson Support
Lesson Plan 2.2
Warm-Up Transparency 2.2
Overhead Visuals:
Folder A: Multi-Use Graphing Packet, Sheets 1–3
Practice Bank: Practice 8
Study Guide: Section 2.2
Exploration Lab Manual:
Diagram Masters 2, 5

Technology
Graphing Calculator
McDougal Littell Mathpack
Function Investigator with Matrix Analyzer Activity Book: Function Investigator Activities 2 and 4–6
Internet:
http://www.hmco.com

Warm-Up Exercises

1. Find the ordered pairs that are solutions to $y = \frac{1}{2}x + 3$ for x-values −4, −2, 0, 2, 4. (−4, 1), (−2, 2), (0, 3), (2, 4), (4, 5)

Tell whether each equation is *linear* or *nonlinear.*

2. $y = -3x + 4$ linear

3. $y = 2x^2$ nonlinear

4. $y = 2$ linear

5. What is the slope of the line $y = 0.02x$? 0.02

Section

2.2 Linear Equations and Slope-Intercept Form

Learn how to...

- **write an equation of a line in slope-intercept form**
- **graph a linear equation in slope-intercept form**

So you can...

- **model events, such as rising water levels**

You've probably noticed that when you get into a bathtub, the water level rises. In the Exploration you will investigate a similar situation and develop a linear model.

EXPLORATION
COOPERATIVE LEARNING

Measuring Water Height

Work with a group of three or four students.
You will need:

- a glass container about half filled with water
- twenty marbles of the same size
- a centimeter ruler

A narrow container with straight sides works best.

Measure the water height to the nearest 0.1 cm. Be sure to put your eyes at the level of the water before measuring.

Number of marbles	Water height (cm)
0	?
5	?
10	?
15	?
20	?

1 Copy the table. Measure and record the water height before you add any marbles to the container.

2 Add five marbles at a time to the container. Measure and record the water height each time.

3 Make a scatter plot of your data. Put the number of marbles on the horizontal axis and the water height on the vertical axis. What do you notice about your data points?

4 Calculate the change in water height per marble.

5 Let m = the number of marbles and h = the water height. Use your data to write an equation of the form:

water height = starting height + change in water height per marble × number of marbles

Exploration Note

Purpose
The purpose of this Exploration is to have students discover that the water level in a container is a linear function of the number of marbles added.

Materials/Preparation
Each group needs a glass container, water, 20 marbles, a centimeter ruler, and graph paper.

Procedure
Students record the initial height of water in a container. They then add 5 marbles at a time to the container and measure and record the water height each time, adding a total of 20 marbles. They make a scatter plot of their data and use it to write an equation that represents the water height as a function of the number of marbles added.

Closure
Students should understand that the height of the water in a glass container increases by a fixed amount for each marble added.

Explorations Lab Manual
See the Manual for more commentary on this Exploration.
Diagram Master 5

In doing the Exploration, Maria and Anthony obtained the graph shown.

The water height started at **5.2 cm**. This number is called the **vertical intercept** because it indicates where the graph crosses the vertical axis.

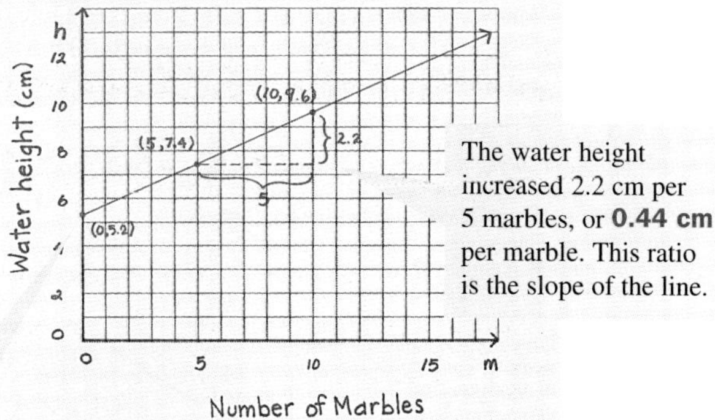

The water height increased 2.2 cm per 5 marbles, or **0.44 cm** per marble. This ratio is the slope of the line.

An equation for Maria and Anthony's graph is $h = 5.2 + 0.44m$, or $h = 0.44m + 5.2$. This equation is in **slope-intercept form** because it fits this pattern:

$$y = ax + b$$

The number a tells you the slope.

The number b tells you the vertical intercept.

Notice how the slope of Maria and Anthony's graph is calculated:

$$\text{slope} = \frac{\text{vertical change}}{\text{horizontal change}} = \frac{9.6 - 7.4}{10 - 5} = \frac{2.2}{5} = 0.44$$

In general, if (x_1, y_1) and (x_2, y_2) are two points on a line, then the slope a is:

$$a = \frac{\text{vertical change}}{\text{horizontal change}} = \frac{y_2 - y_1}{x_2 - x_1}$$

WATCH OUT!

You may remember the slope-intercept form as $y = mx + b$ from a previous course. The letter a is used in place of m here because that is what you will see on some common graphing calculators.

EXAMPLE 1

Find the slope-intercept equation of the line through the points (−4, 1) and (0, 2).

SOLUTION

Step 1 The line crosses the y-axis at (0, 2), so $b = 2$ is the vertical intercept, or y-intercept.

Step 2 To find the slope a, use the points (−4, −1) and (0, 2) on the line:

$$a = \frac{2 - (-1)}{0 - (-4)} = \frac{3}{4}$$

Substituting $a = \frac{3}{4}$ and $b = 2$ in $y = ax + b$ gives the slope-intercept equation:

$$y = \frac{3}{4}x + 2$$

ANSWERS Section 2.2

Exploration

1–5. Answers may vary.

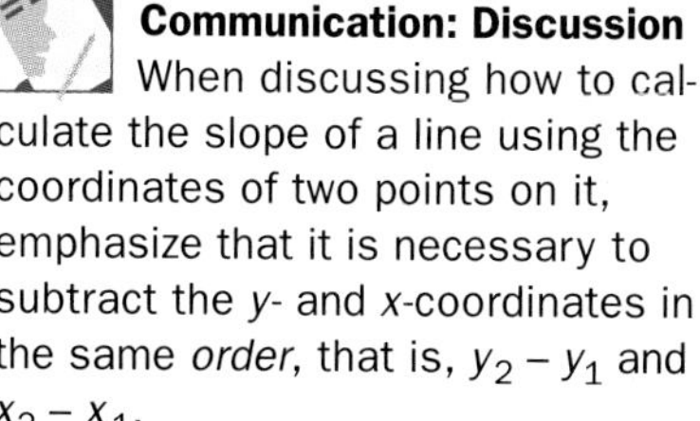

Section Notes

Communication: Discussion
When discussing how to calculate the slope of a line using the coordinates of two points on it, emphasize that it is necessary to subtract the y- and x-coordinates in the same *order*, that is, $y_2 - y_1$ and $x_2 - x_1$.

Common Error
The two most common errors students make when finding slope are not subtracting the y- and x-coordinates in the same order (as discussed above) and inverting the slope formula. To help overcome these errors, students should memorize the formula $a = \frac{y_2 - y_1}{x_2 - x_1}$ and, when using it, actually designate on paper which ordered pair is (x_1, y_1) and which is (x_2, y_2).

Student Progress
The concepts of vertical intercept, slope-intercept form of a line, and the formula for finding a slope should be familiar to most students from previous mathematics courses.

Using Technology
Function Investigator Activities 4 and 5 in the *Function Investigator with Matrix Analyzer Activity Book* can help students discover how changing the parameters a and b in the equation $y = ax + b$ affects the graph of the equation.

Additional Example 1

Find the slope-intercept equation of the line through the points (0, −3) and (5, −2).

Step 1 The line crosses the y-axis at (0, −3), so $b = -3$ is the vertical intercept, or y-intercept.

Step 2 To find the slope a, use the points (0, −3) and (5, −2) on the line:

$$a = \frac{-2 - (-3)}{5 - 0} = \frac{1}{5}$$

Substituting $a = \frac{1}{5}$ and $b = -3$ in $y = ax + b$ gives the slope-intercept equation:

$$y = \frac{1}{5}x - 3$$

Teach⇔Interact

About Example 2

Teaching Tip
Some students may be familiar with the air pressure at sea level as being 14.7 lb/in.^{2}, which is called *one atmosphere*. A pressure of 2100 lb/ft^2 is approximately the same as 14.7 lb/in.^{2} ($14.7 \times 144 = 2116.8$).

Additional Example 2

Geothermal energy is heat produced in Earth's interior. The underground temperature increases as you go deeper. An estimate of the temperature T in degrees Celsius is $T = 35d + 20$, where d is the depth in kilometers and 20°C is the temperature at a particular place on the surface of Earth. Graph the equation for temperature.

Step 1 Plot the point (0, 20) where the line crosses the T-axis.
Step 2 Use the slope to plot a second point on the line by going up 35 units when you move right 1 unit from (0, 20).

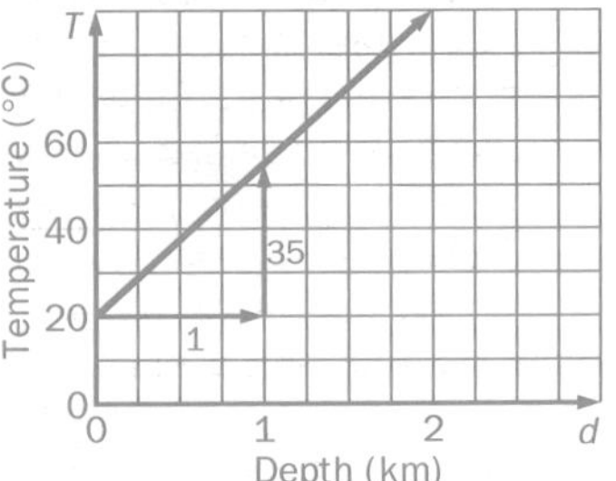

Checking Key Concepts

Reasoning
For questions 1–3, students should understand that finding the slope of the line passing through each pair of points does not depend on which point is (x_1, y_1) or (x_2, y_2) in the equation $a = \frac{y_2 - y_1}{x_2 - x_1}$. Students can verify this fact by calculating the slope both ways.

Closure Question

Explain how to find the equation of a line in slope-intercept form if you are given the vertical intercept and a point on the line.
Use the two given points and the formula for slope to find the slope of the line. Then write the equation of the line as y = slope times x plus the vertical intercept.

EXAMPLE 2 Interview: Norbert Wu

The water pressure on a diver increases at a rate of 64 lb/ft^2 for every 1 ft of depth d. The air pressure on a diver is 2100 lb/ft^2, which is constant for anyone at sea level. The total pressure P on a diver is the sum of the water pressure and the air pressure:

$$P = 64d + 2100$$

Graph the equation for total pressure.

SOLUTION

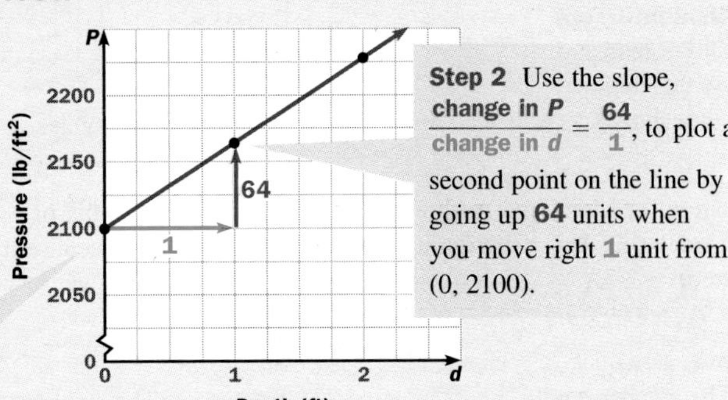

Step 1 Plot the point **(0, 2100)** where the line crosses the P-axis.

Step 2 Use the slope, $\frac{\text{change in } P}{\text{change in } d} = \frac{64}{1}$, to plot a second point on the line by going up **64** units when you move right **1** unit from (0, 2100).

THINK AND COMMUNICATE

1. Does the total pressure on a diver vary directly with depth? Explain.
2. How is the graph in Example 2 geometrically related to the graph of the equation for water pressure only (see Exercise 15 on page 49)?

CHECKING KEY CONCEPTS

Find the slope of the line passing through each pair of points.

1. (1, 3) and (3, 7)
2. (−4, 2) and (5, −3)
3. (−1, 6) and (1, 6)

State the slope and *y*-intercept of each line with the given equation.

4. $y = 8x - 5$
5. $y = \frac{1}{4} - \frac{3}{7}x$
6. $y = 1$

Find the slope-intercept equation of each line.

7. 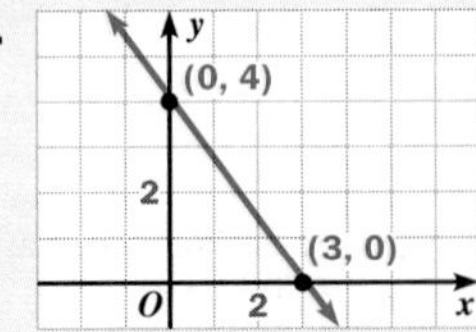

8. 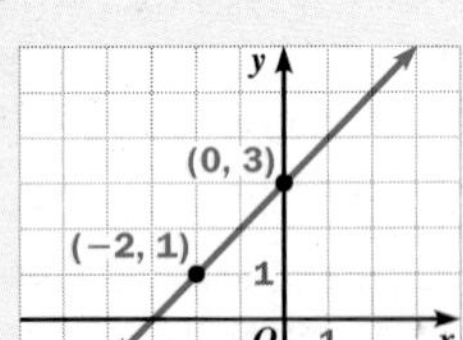

Think and Communicate

1. No. Answers may vary. An example is given. If the pressure varied directly with depth, the graph of the equation would pass through the origin.
2. The graph is the same graph shifted 2100 units in the positive y-direction.

Checking Key Concepts

1. 2
2. $-\frac{5}{9}$
3. 0
4. 8; –5
5. $-\frac{3}{7}; \frac{1}{4}$
6. 0; 1
7. $y = -\frac{4}{3}x + 4$
8. $y = x + 3$

Exercises and Applications

1. 3; 7
2. $-\frac{2}{3}; \frac{1}{3}$
3. 1; 0
4. –5; 2
5. $y = -2$
6. $y = \frac{2}{3}x + 2$
7. $y = \frac{1}{2}x - 1.5$
8. Let x = the number of rides and y = the total cost in dollars; $y = x + 4$.

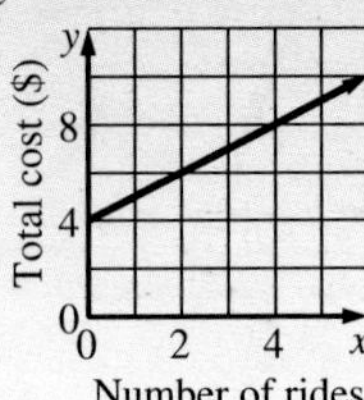

2.2 Exercises and Applications

Extra Practice exercises on pages 750–751

State the slope and y-intercept of each line with the given equation.

1. $y = 3x + 7$ **2.** $y = -\frac{2}{3}x + \frac{1}{3}$ **3.** $y = x$ **4.** $y = 2 - 5x$

Find the slope-intercept equation of each line.

5.

(0, −2) (5, −2)

6.

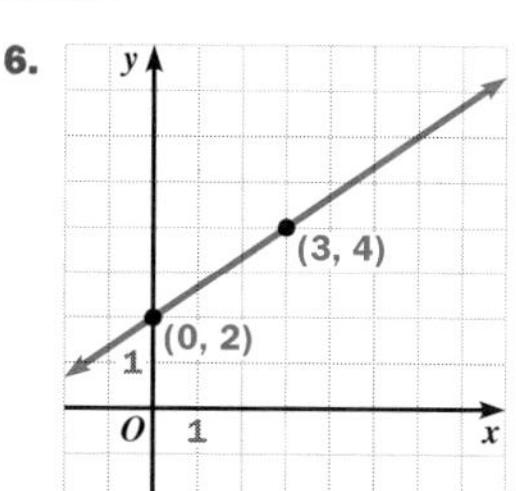

7.

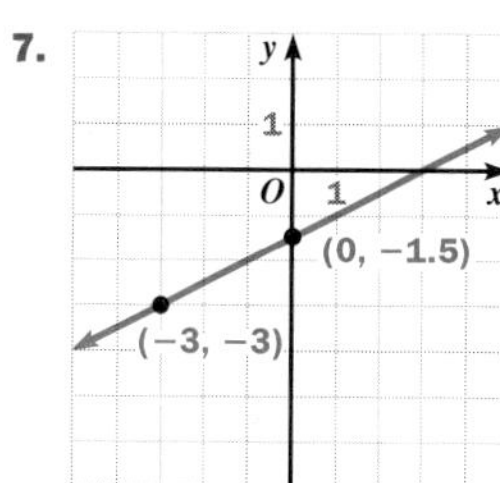

For Exercises 8–13, model each situation with an equation and a graph. Be sure to identify the independent and dependent variables.

8. An amusement park charges \$4.00 admission and \$1.00 per ride.

9. An empty 0.75 gal pitcher weighs 1.2 lb. Water weighs 8.3 lb/gal.

10. A film club membership costs \$10.00 and \$2.00 per movie.

11. A test is worth 100 points with 5 points deducted for each wrong answer.

12. A candle is 8 in. tall and burns at a rate of 1.5 in./h.

13. The gas tank of a van holds 16 gal. The van uses $\frac{1}{15}$ gal/mi.

14. Open-ended Problem Describe a situation that could be modeled by the equation $y = 0.75x + 4.25$.

Graph each equation.

15. $y = 1 - 2x$ **16.** $y = -4 + \frac{2}{3}x$ **17.** $y = 4$

18. $y = -3 - 3x$ **19.** $y = 7x - 12$ **20.** $y = x$

21. Writing Describe how the sign of the slope affects the graph in Exercises 15–20.

22. TEMPERATURE SCALES The drawing shows the temperatures at which water freezes and boils in both degrees Celsius and Fahrenheit. The Fahrenheit temperature F is a linear function of the Celsius temperature C.

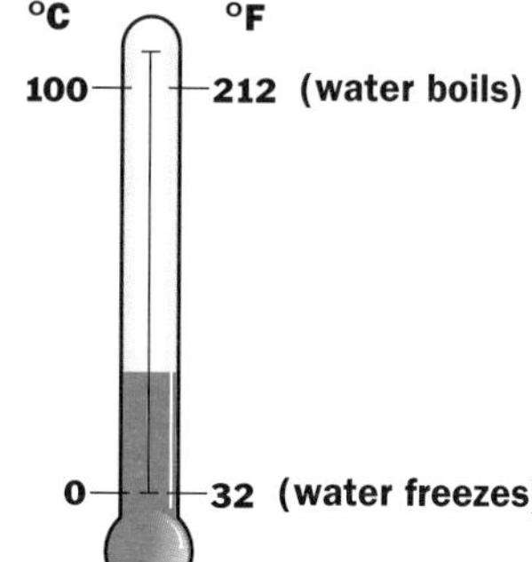

a. Plot the two data pairs (C, F) in a coordinate plane, and draw a line through the plotted points.

b. Find the slope-intercept equation of the line.

c. Find the Fahrenheit temperature equivalent to 56°C.

d. The equation from part (b) gives F as a function of C. Rewrite this equation so that it expresses C as a function of F.

e. Use your equation from part (d) to find the Celsius temperature equivalent to 77°F.

9. Let x = the number of gal of water in the pitcher and y = the total weight in pounds of the pitcher and water; $y = 8.3x + 1.2$.

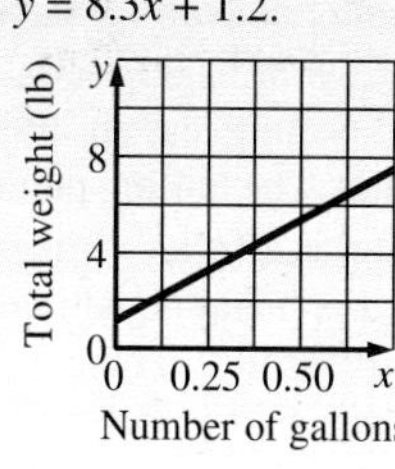

10. Let x = the number of movies attended and y = the total cost in dollars; $y = 2x + 10$.

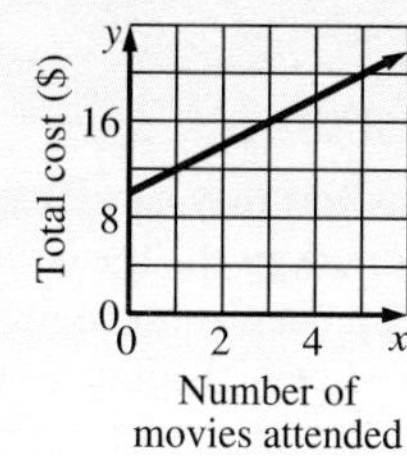

11. Let x = the number of wrong answers and y = the test score; $y = -5x + 100$.

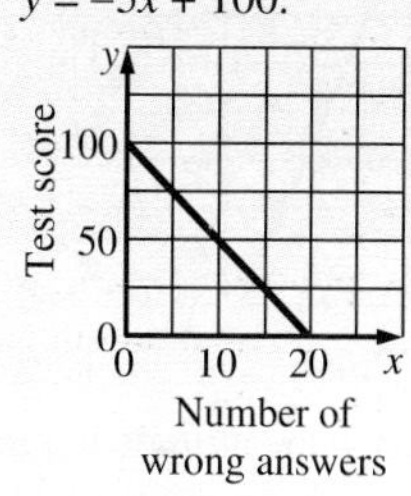

12–22. See answers in back of book.

Apply⇔Assess

Suggested Assignment

Core Course
Exs. 1–13, 15–22, 27–34

Extended Course
Exs. 1–34

Block Schedule
Day 8 Exs. 1–13, 15–22, 27–34

Exercise Notes

Communication: Drawing
Exs. 5–7 These exercises will give students opportunities to connect the equation of a line with its graph. This will be especially useful to visual thinkers.

Using Technology
Exs. 5–7 Students can use a graphing calculator or the *Function Investigator* software to find the slope-intercept equation for a nonvertical line if the coordinates of two points on the line are known. For example, here is the procedure to use with Ex. 6 and a TI-82 graphing calculator. Press STAT 1 and enter the x-coordinates of the points as list L1. Enter the corresponding y-coordinates as list L2. Then press STAT ▶ 5 ENTER. The calculator will display values a and b for an equation of the form $y = ax + b$. Note that a and b will be in decimal form. To convert the decimals to fractions, press VARS 5 ▶ ▶ 1 or VARS 5 ▶ ▶ 2 and then MATH 1 ENTER.

Cooperative Learning
Exs. 8–13 You may wish to have students work on these exercises with a partner. They can do them individually and then compare their equations and graphs.

Alternate Approach
Exs. 15–20 As an alternative approach to the method shown in Example 2 on page 54, students can graph these equations by generating a table of values.

Reasoning
Ex. 22 After students work through all parts of this exercise, ask them what Celsius temperature corresponds to a Fahrenheit temperature of zero degrees (−17.78°C) and what this temperature represents on a graph showing C as a function of F (the vertical intercept).

Apply⇔Assess

Exercise Notes

Second-Language Learners

Exs. 23, 24 In the paragraph preceding these exercises, students learning English may be unfamiliar with the meaning of the word *quartz*. If necessary, explain that it is a type of hard crystal.

Multicultural Note

Ex. 23 The United Republic of Tanzania, located on the east coast of Africa, was formed in 1964 when the countries of Tanganyika and Zanzibar joined together after gaining independence from Britain. Tanzania is composed of more than 120 peoples, the majority of whom speak their own dialect of Swahili. Swahili literally means *coast people* and is the country's official language. Within Tanzania are two world renowned regions, the Serengeti National Park and the Olduvai Gorge. The Serengeti National Park is a 5700 square mile wildlife preserve, the habitat of diverse species such as the wildebeest, African lion, cheetah, and black rhinoceros. In the Olduvai Gorge, archeologist Mary Leakey discovered fossils more than one million years old; she and other scientists believe these indicate that the first human beings may have lived in East Africa.

Connection EARTH SCIENCE

When European explorers first saw Mount Kenya and Mount Kilimanjaro in the 1840s, they sent back reports of snow-covered mountains in central Africa. European geographers of the time thought that the explorers had really seen white quartz, because they thought that snow could not exist near the equator.

Mount Kilimanjaro in Tanzania is the highest mountain in Africa.

23. The scale above gives the standard air temperature at different altitudes.

   a. Plot the data in a coordinate plane. Draw a line through the points.

   b. Calculate the change in temperature per 1 km increase in altitude.

   c. Write an equation to model the temperature at different altitudes. What is the slope? What is the vertical intercept?

   d. What is the predicted temperature on top of Mount Kilimanjaro, which is 5.9 km high? Is snow possible at this temperature?

24. The boiling temperature of water is 100°C at sea level. For every 1 km increase in altitude, the boiling temperature decreases by about 3.5°C. (The boiling temperature of water decreases with altitude because the air pressure decreases.)

   a. Write the slope-intercept equation giving boiling temperature as a function of altitude.

   b. What is the boiling temperature of water in Mexico City, which is at altitude 2.2 km?

   c. What is the boiling temperature of water on top of Mount Kenya, which is 5.2 km high?

   d. Research At the top of Mount Everest (8.85 km high), water boils at about 69°C. This makes cooking food difficult. Use a cookbook to learn about the problems of cooking at high altitude, or use a chemistry book to find out what factors affect the boiling point of water. Write a brief report.

BY THE WAY

Mount Kilimanjaro has several distinct vegetation zones, including semiarid scrub at the base, a dense cloud forest, an alpine desert, and a moss and lichen community at the summit.

23. a.

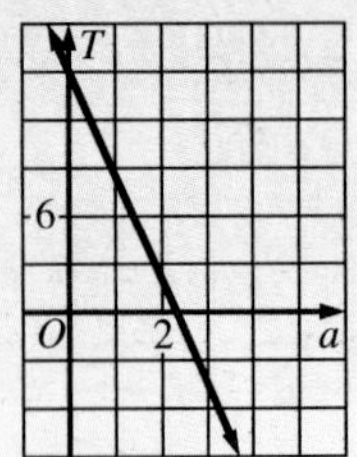

b. −6.5°/km

c. Let a = the altitude in km and T = the temperature in °C; $T = -6.5a + 15$; −6.5; 15.

d. −23.35°C; Yes.

24. a. Let a = the altitude in km and b = the boiling point in °C; $b = -3.5a + 100$.

b. 92.3°C

c. 81.8°C

d. Answers may vary. Examples are given. The cooking process most complicated by high altitudes is baking, because of the change in the boiling point of water, but also because with reduced air pressure, baked products tend to over-rise. This may be counteracted by using smaller amounts of baking powder or baking soda and beating less air into a batter. The boiling point of water is affected by many factors including altitude, air pressure, and the presence of any chemical substance in the water.

25. **Challenge** In this exercise, you will use algebra to show that the slope of the line $y = ax + b$ is a.

a. Let (x_1, y_1) and (x_2, y_2) be any two distinct points on the line $y = ax + b$. Write an expression for the slope of the line in terms of x_1, y_1, x_2, and y_2.

b. Use the fact that $y_1 = ax_1 + b$ and $y_2 = ax_2 + b$ to write your expression from part (a) in terms of only x_1 and x_2.

c. Show that your expression from part (b) equals a.

26. **Technology** Miguel Santos wants to rent a car for a day while on vacation. Carla's Cars charges \$22 for an economy car plus \$.08 per mile driven. Friendly Rentals charges \$38 for the same model of car with no additional charge per mile.

a. For each company, find the slope-intercept equation giving the cost C to rent a car as a function of the distance d driven.

b. Use a graphing calculator or graphing software to graph the two equations from part (a) in the same coordinate plane.

c. **Writing** Discuss the circumstances for which each company offers Miguel the better deal.

27. **SAT/ACT Preview** If A = $-4(x + 1)$ and B = $-4x - 3$, choose the statement that is true for all values of x.

A. A > B B. A < B C. A = B

D. relationship cannot be determined

28. **GEOMETRY** *Without* plotting the points $(-2, -1)$, $(1, 5)$, and $(3, 10)$, determine whether the points lie on the same line. Explain your reasoning.

ONGOING ASSESSMENT

29. **Cooperative Learning** Work with a partner. You will need a set of identical paper cups. Measure the height of a single cup. Then add cups to form a stack, and measure the height of the stack each time a cup is added. Develop a linear model relating the height of the stack to the number of cups added to the first cup in the stack.

SPIRAL REVIEW

30. A typist types at a rate of 50 words per minute. Write a direct variation equation that relates the number of words typed to the amount of time spent typing. *(Section 2.1)*

31. Which of the following expressions is equivalent to $-2x + 6$? *(Toolbox, page 781)*

A. $8 - 2(x + 1)$ B. $2 - 2(x - 3)$ C. $-2(x + 3)$

Solve each proportion. *(Toolbox, page 785)*

32. $\frac{x - 3}{3 + x} = 2$ 33. $\frac{4}{5 - 2x} = \frac{1}{2}$ 34. $12 = \frac{x + 2}{4x - 2}$

Apply⇔Assess

Exercise Notes

Using Technology
Ex. 26 For part (b), you may need to remind students to replace C with y and d with x in order to graph the functions from part (a).

Reasoning
Ex. 27 Ask students to justify mathematically their choice of an answer. Some students may need to use numerical examples as an explanation.

Integrating the Strands
Ex. 28 This exercise illustrates how an algebraic calculation involving the slope of a line can be used to answer an essentially geometric question; that is, are the points collinear?

Practice 8 for Section 2.2

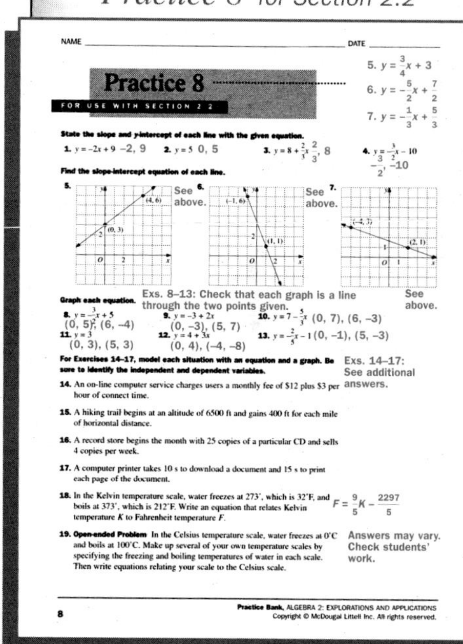

NAME ____ DATE ____

Practice 8

FOR USE WITH SECTION 2.2

5. $y = \frac{3}{4}x + 3$
6. $y = -\frac{5}{2}x + \frac{7}{2}$
7. $y = -\frac{1}{3}x + \frac{5}{3}$

State the slope and y-intercept of each line with the given equation.

1. $y = -2x + 9$ $-2, 9$ 2. $y = 5$ $0, 5$ 3. $y = 8 + \frac{2}{3}x$ $\frac{2}{3}, 8$ 4. $y = -\frac{3}{2}x - 10$ $-\frac{3}{2}, -10$

Find the slope-intercept equation of each line.

5. See above. 6. See above. 7. See above.

Graph each equation. Exs. 8–13: Check that each graph is a line through the two points given. See above.

8. $y = -\frac{3}{2}x + 5$ (0, 5), (6, −4) 9. $y = -3 + 2x$ (0, −3), (5, 7) 10. $y = 7 - \frac{5}{3}x$ (0, 7), (6, −3)

11. $y = 3$ (0, 3), (5, 3) 12. $y = 4 + 3x$ (0, 4), (−4, −8) 13. $y = -\frac{2}{5}x - 1$ (0, −1), (5, −3)

For Exercises 14–17, model each situation with an equation and a graph. Be sure to identify the independent and dependent variables. Exs. 14–17: See additional answers.

14. An on-line computer service charges users a monthly fee of \$12 plus \$3 per hour of connect time.

15. A hiking trail begins at an altitude of 6500 ft and gains 400 ft for each mile of horizontal distance.

16. A record store begins the month with 25 copies of a particular CD and sells 4 copies per week.

17. A computer printer takes 10 s to download a document and 15 s to print each page of the document.

18. In the Kelvin temperature scale, water freezes at 273°, which is 32°F, and boils at 373°, which is 212°F. Write an equation that relates Kelvin temperature K to Fahrenheit temperature F. $F = \frac{9}{5}K - \frac{2297}{5}$

19. **Open-ended Problem** In the Celsius temperature scale, water freezes at 0°C and boils at 100°C. Make up several of your own temperature scales by specifying the freezing and boiling temperatures of water in each scale. Then write equations relating your scale to the Celsius scale. Answers may vary. Check students' work.

8 Practice Bank, ALGEBRA 2: EXPLORATIONS AND APPLICATIONS

25. a. $\frac{y_2 - y_1}{x_2 - x_1}$

b. $\frac{ax_2 + b - (ax_1 + b)}{x_2 - x_1}$

c. $\frac{ax_2 + b - (ax_1 + b)}{x_2 - x_1} = \frac{a(x_2 - x_1)}{x_2 - x_1} = a$

26. a. Carla's Cars: $C = 0.08d + 22$; Friendly Rentals: $C = 38$

b.

X=200 Y=38

For 200 mi of driving, the two companies charge the same amount, \$38.

c. If Miguel plans to drive more than 200 mi, Friendly Rental will charge less. If he plans to drive fewer than 200 mi, Carla's Cars will charge less.

27. B

28. No; the slope of the line containing (−2, −1) and (1, 5) is 2. The slope of the line containing (−2, −1) and (3, 10) is $\frac{11}{5}$.

29. Answers may vary. Check students' work.

30. Let x = the number of minutes spent typing and y = the number of words typed; $y = 50x$, or $\frac{y}{x} = 50$.

31. A

32. −9

33. $-\frac{3}{2}$

34. $\frac{26}{47}$

Plan⇔Support

Objectives

- Write the point-slope equation of a line.
- Use function notation.
- Find the domain and range of a linear function.
- Make predictions about linear data.

Recommended Pacing

Core and Extended Courses
Section 2.3: 2 days

Block Schedule
Section 2.3: 1 block

Resource Materials

Lesson Support
Lesson Plan 2.3
Warm-Up Transparency 2.3
Overhead Visuals:
Folder A: Multi-Use Graphing Packet, Sheets 1–3
Practice Bank: Practice 9
Study Guide: Section 2.3
Exploration Lab Manual:
Diagram Masters 1, 2
Assessment Book: Test 5

Technology
Technology Book:
Spreadsheet Activity 2
TI-92 Activity 1
Graphing Calculator
McDougal Littell Mathpack
Function Investigator
Internet:
http://www.hmco.com

Warm-Up Exercises

1. Find the slope of the line through the points (–4, 6) and (1, –3). $-\frac{9}{5}$
2. Find the equation of a line through the points (0, 4) and (2, –5). $y = -\frac{9}{2}x + 4$
3. Give the slope of the line $y = -1.5x - 4.2$. –1.5
4. If $y = \frac{5}{3}x - 2$, find y when $x = 9$. 13
5. Is the point (–3, 2) a solution of the equation $y = 4 + 2(x + 2)$? Yes.

Section

2.3 Point-Slope Form and Function Notation

Learn how to...

- **write the point-slope equation of a line**
- **use function notation**
- **find the domain and range of a linear function**

So you can...

- **make predictions about a runner's finishing time, for example**

You probably find it harder to run in a headwind (a wind blowing against your direction of motion) than to run in still air or with the wind at your back. In fact, the time required for a runner to finish a 1 mi race increases an average of 1.5 s for every 1 mi/h of headwind speed.

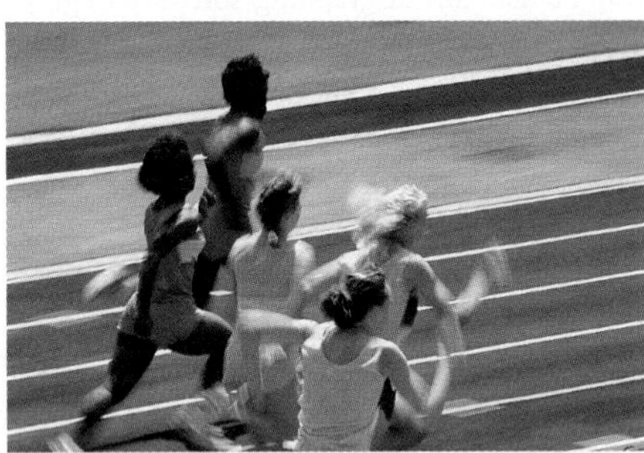

EXAMPLE 1

Lisa runs a 1 mi race in a 12 mi/h headwind. She finishes the race in 5:20 (5 min 20 s). Find an equation giving Lisa's finishing time as a function of headwind speed.

SOLUTION

Since Lisa's finishing time t increases 1.5 s for each 1 mi/h increase in headwind speed h, the graph giving t as a function of h is a line with slope 1.5. Also, $t = 5(60) + 20 = 320$ when $h = 12$, so the line passes through the point (12, 320).

To find an equation of the line, let (h, t) be any point on the line other than (12, 320).

The slope between these points should equal 1.5.

$$\frac{t - 320}{h - 12} = 1.5$$

$$t - 320 = 1.5(h - 12)$$

$$t = 320 + 1.5(h - 12)$$

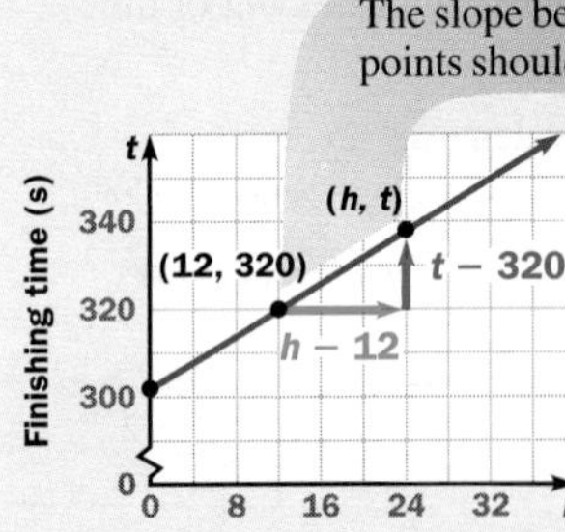

The equation $t = 320 + 1.5(h - 12)$ gives Lisa's finishing time t as a function of headwind speed h.

THINK AND COMMUNICATE

1. In Example 1, predict Lisa's finishing time in still air.
2. Write the equation $t = 320 + 1.5(h - 12)$ in the slope-intercept form $t = ah + b$. What does the vertical intercept tell you?

The Point-Slope Equation of a Line

In Example 1, the equation

$$t = 320 + 1.5(h - 12)$$

is based on knowing the point (12, 320) and the slope 1.5. In general, if you know that a line passes through a point (x_1, y_1) and has slope a, you can write the following **point-slope equation** of the line:

$$y = y_1 + a(x - x_1)$$

You can also use a point-slope equation when you know two points.

EXAMPLE 2

Write a point-slope equation of the line through the points (−4, 7) and (3, 1).

SOLUTION

Step 1 Find the slope a:

$$a = \frac{1 - 7}{3 - (-4)} = \frac{-6}{7} = -\frac{6}{7}$$

Step 2 Write a point-slope equation using either (−4, 7) or (3, 1) as the point (x_1, y_1). If you choose (−4, 7), your equation is:

$$y = 7 + \left(-\frac{6}{7}\right)(x - (-4))$$

$$y = 7 - \frac{6}{7}(x + 4)$$

A point-slope equation of the line is $y = 7 - \frac{6}{7}(x + 4)$.

THINK AND COMMUNICATE

3. a. Write a point-slope equation of the line in Example 2 using (3, 1) for (x_1, y_1). Compare your answer with the equation obtained using (−4, 7). Do the two equations look the same?
   b. Write the two equations in slope-intercept form. What do you notice?
4. A line has equation $y = -2 + \frac{1}{3}(x + 1)$.
   a. What is the line's slope?
   b. Name a point through which the line passes.

ANSWERS Section 2.3

Think and Communicate

1. about 302 s
2. $t = 1.5h + 302$; the finishing time in still air (that is, with a headwind of 0 mi/h)
3. a. $y = 1 - \frac{6}{7}(x - 3)$; The equations do not look the same.
   b. In slope-intercept form, the equations are identical: $y = -\frac{6}{7}x + \frac{25}{7}$.
4. a. $\frac{1}{3}$
   b. Answers may vary. Examples are given. (−1, −2), $\left(0, -\frac{5}{3}\right)$, (2, −1)

Teach⇔Interact

About Example 1

Problem Solving
The solution to the problem in this Example, which is a line in point-slope form, prepares students for the discussion of the general point-slope equation of a line on this page.

Additional Example 1

The finishing time of a racing boat may decrease 0.8 min for every 1 km/h of tailwind speed. Jonathan races his boat in a 100 km race in a 9 km/h tailwind. He finishes the race in 2:45 (2 h 45 min). Find an equation giving Jonathan's finishing time as a function of tailwind speed.

Since Jonathan's finishing time t decreases 0.8 min for each 1 km/h increase in tailwind speed w, the graph giving t as a function of w is a line with slope −0.8. Also, $t = 2(60) + 45 = 165$ when $w = 9$, so the line passes through the point (9, 165). To find an equation of the line, let (w, t) be any point on the line other than (9, 165).

$$\frac{t - 165}{w - 9} = -0.8$$

$$t - 165 = -0.8(w - 9)$$

$$t = 165 - 0.8(w - 9)$$

The equation $t = 165 - 0.8(w - 9)$ gives Jonathan's time t as a function of tailwind speed w.

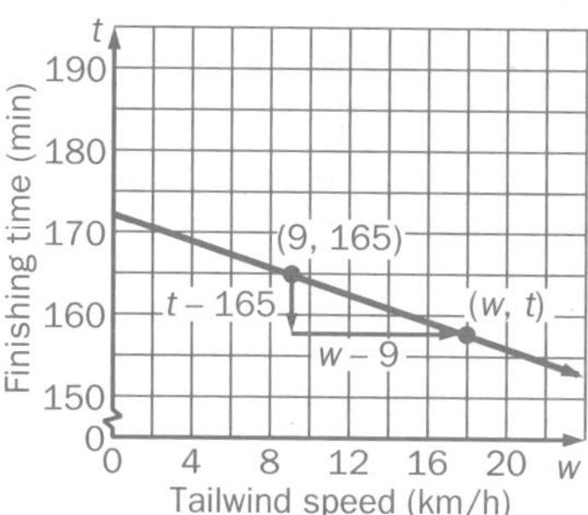

Additional Example 2

Write the point-slope equation of the line through the points (2, 5) and (4, 7).

Step 1 Find the slope a:

$$a = \frac{7 - 5}{4 - 2} = \frac{2}{2} = 1$$

Step 2 Write a point-slope equation using either (2, 5) or (4, 7) as the point (x_1, y_1). If you choose (2, 5), your equation is $y = 5 + (x - 2)$. A point-slope equation of the line is $y = 5 + (x - 2)$.

Teach⇔Interact

Section Notes

Historical Connection
The function notation $f(x)$ can be attributed to the Swiss mathematician Leonhard Euler (1707–1783).

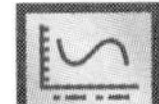 **Using Technology**
On a TI-82, an alternative way to evaluate a function is to enter the function as y_1 on the Y= list. Then go to the home screen and choose Y1 from the Y-VARS 1:Function menu and enter the value in parentheses. For example, $f(6)$ would be Y1(6).

Additional Example 3

Suppose the function $f(x) = -\frac{1}{2}x + 4$ is defined for $x \le 4$.

a. Find $f(-2)$.

$f(-2) = -\frac{1}{2}(-2) + 4 = 5$

b. Find the domain and range of f.

Method 1 Use algebra.
The domain is $x \le 4$. You can obtain the range from the domain.

$$x \le 4$$
$$-\frac{1}{2}x \ge -\frac{1}{2}(4)$$
$$-\frac{1}{2}x + 4 \ge -2 + 4$$
$$f(x) \ge 2$$

Method 2 Look at a graph.

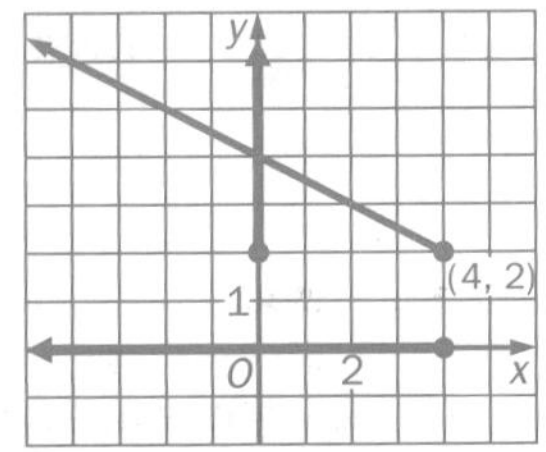

The range consists of the y-coordinates of the points on the graph. The range of f is $y \ge 2$. The domain consists of the x-coordinates of the points on the graph. The domain of f is $x \le 4$.

Function Notation

WATCH OUT!
$f(x)$ means "the value of f at x." $f(x)$ does *not* mean "multiply f by x."

The **function notation** $y = f(x)$ tells you that y is a function of x. If there is a rule relating y to x, such as $y = 2x + 3$, then you can also write:

$$f(x) = 2x + 3$$

The name of the function is $\boldsymbol{f}$. Other letters are also used to name functions, especially $\boldsymbol{g}$ and $\boldsymbol{h}$.

$\boldsymbol{f(x)}$, read "$\boldsymbol{f}$ **of** $\boldsymbol{x}$," represents the value of the function at x.

The **domain** of a function f is the set of values of x for which f is defined. The **range** of a function f is the set of all values of $f(x)$, where x is in the domain of f.

For instance, if the function in Example 1 applies to headwind speeds up to 20 mi/h, then the domain of the function is $0 \le h \le 20$. The corresponding range of the function is $302 \le t \le 332$.

EXAMPLE 3

Suppose the function $f(x) = \frac{3}{2}x - 2$ is defined for $x \ge 2$.

a. Find $f(6)$.

b. Find the domain and range of f.

SOLUTION

a. $f(6) = \frac{3}{2}(6) - 2 = 7$

b. **Method 1**
Use algebra.
The domain is $x \ge 2$. You can obtain the range from the domain.

Toolbox p. 787 *Solving Linear Inequalities*

Transform the inequality so that the left side becomes $\frac{3}{2}x - 2$.

$$x \ge 2$$
$$\frac{3}{2}x \ge \frac{3}{2}(2)$$
$$\frac{3}{2}x - 2 \ge 3 - 2$$
$$f(x) \ge 1 \quad \longleftarrow \text{ This is the range of } f.$$

Method 2
Look at a graph.

The range consists of the y-coordinates of the points on the graph. The **range** of f is $\boldsymbol{y \ge 1}$.

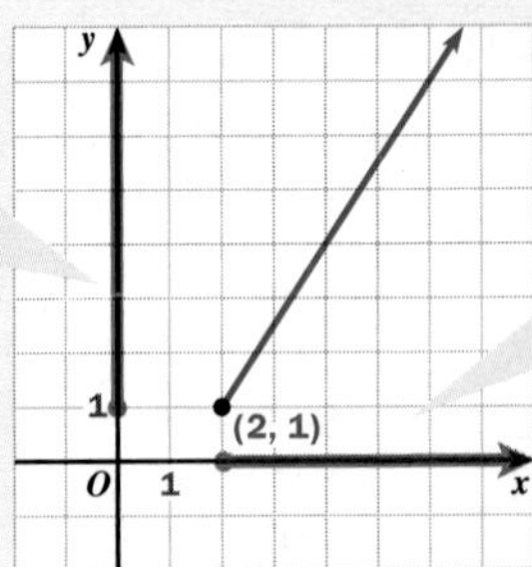

The domain consists of the x-coordinates of the points on the graph. The **domain** of f is $x \ge 2$.

Checking Key Concepts

1. $y = 3 + (x - 2)$
2. $y = 6 - \frac{1}{4}(x + 4)$
3. $y = 3 - \frac{3}{2}x$
4. $y = 3 - 4(x - 1)$ or $y = -5 - 4(x - 3)$
5. $y = 3 + (x + 2)$ or $y = 2 + (x + 3)$
6. $y = 8 - \frac{7}{4}x$ or $y = 1 - \frac{7}{4}(x - 4)$
7. a. -1
   b. $x \ge -2; f(x) \ge -5$

Exercises and Applications

1. a. $y = 2 + (x - 3)$
   b.

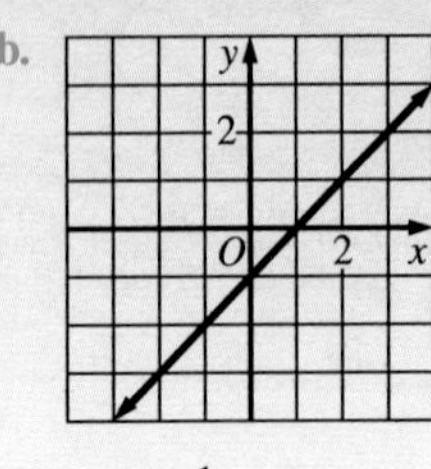

   c. $y = x - 1$

2. a. $y = 5 - \frac{1}{4}(x + 1)$
   b.

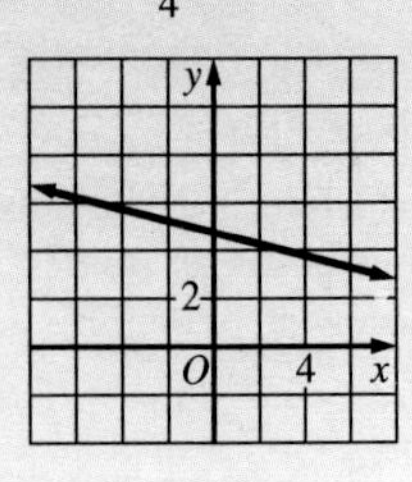

   c. $y = -\frac{1}{4}x + \frac{19}{4}$

CHECKING KEY CONCEPTS

For Questions 1–3, write a point-slope equation of the line passing through the given point and having the given slope.

1. point: (2, 3) slope $= 1$

2. point: $(-4, 6)$ slope $= -\frac{1}{4}$

3. point: (0, 3) slope $= -\frac{3}{2}$

For Questions 4–6, write a point-slope equation of the line passing through the given points.

4. (1, 3) and $(3, -5)$

5. $(-2, 3)$ and $(-3, 2)$

6. (0, 8) and (4, 1)

7. Suppose the function $f(x) = 4x + 3$ is defined for $x \geq -2$.

a. Find $f(-1)$.

b. Find the domain and range of f.

2.3 Exercises and Applications

Extra Practice exercises on page 751

For Exercises 1–4:

a. Write a point-slope equation of the line passing through the given point and having the given slope.

b. Graph the equation.

c. Write the equation in slope-intercept form.

1. point: (3, 2) slope $= 1$

2. point: $(-1, 5)$ slope $= -\frac{1}{4}$

3. point: (3, 1) slope $= -\frac{2}{3}$

4. point: $(-5, -2)$ slope $= \frac{4}{5}$

5. OCEANOGRAPHY The speed of sound in the ocean depends on the temperature, salinity, and depth of the water. However, in the open ocean below approximately 1000 m, temperature and salinity are fairly constant and the speed of sound increases linearly at a rate of 17 m/s for every 1000 m of depth.

a. The speed of sound at a depth of 1000 m is about 1484 m/s. Write a point-slope equation giving the speed of sound s in meters per second as a function of the depth d in meters.

b. Use the equation you found in part (a) to predict the speed of sound at a depth of 2000 m. Then use this value to determine the distance between the red and black submarines if it takes 3.2 s for sound to travel from one to the other.

c. The Marianas Trench, at a depth of 10,924 m, is the deepest point in Earth's oceans. What is the domain and range of the function you found in part (a) in Earth's oceans? Explain.

BY THE WAY

Understanding how the speed of sound varies in water is essential for mapping the sea floor with sonar, and for communicating underwater.

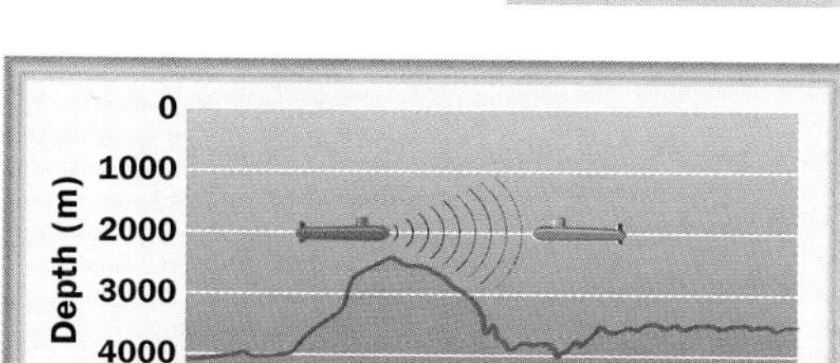

3. a. $y = 1 - \frac{2}{3}(x - 3)$

b.

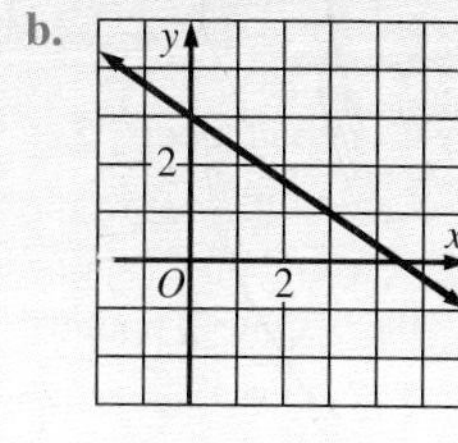

c. $y = -\frac{2}{3}x + 3$

4. a. $y = -2 + \frac{4}{5}(x + 5)$

b.

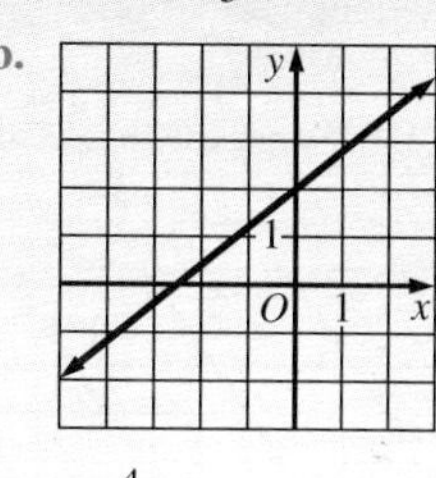

c. $y = \frac{4}{5}x + 2$

5. a. $s = 1484 + 0.017(d - 1000)$

b. about 1501 m/s; about 4803 m

c. The domain is $1000 \leq d \leq 10{,}924$ since the function describes the relation at depths below approximately 1000 m and the deepest point is 10,924 m. Using the domain, you can find that the range is $1484 \leq s \leq 1653$. (The speed at 10,924 m is $1652.708 \approx 1653$ m/s.)

Teach⇔Interact

Checking Key Concepts

Teaching Tip

Encourage students to check their answers to questions 1–6 by verifying that each point is a solution to the equation of the line. Have students describe how to check their answers by graphing.

Closure Question

Describe how to find the equation of a line through two given points. Find the slope a of the line through the two points. Then write a point-slope equation $y = y_1 + a(x - x_1)$ using either point as (x_1, y_1).

Apply⇔Assess

Suggested Assignment

Core Course
Day 1 Exs. 1–5, 11–19
Day 2 Exs. 21–33, AYP

Extended Course
Day 1 Exs. 1–19
Day 2 Exs. 20–33, AYP

Block Schedule
Day 9 Exs. 1–5, 11–19, 21–33, AYP

Exercise Notes

Using Technology

Exs. 1–4 Students can verify that the slope-intercept equations in part (c) are equivalent to the point-slope equations in part (a) by graphing both equations on the same coordinate plane. The graphs should be coincident.

Interdisciplinary Problems

Ex. 5 This exercise illustrates how the mathematical concept of this section can be used to study the speed of sound in an ocean. Oceanography today is concerned with studying Earth's oceans to find new food, minerals, and energy resources; to control pollution; and to conserve the biological resources of the oceans. There is much that is still unknown about Earth's oceans, which cover two-thirds of the planet.

Apply⇔Assess

Exercise Notes

Second-Language Learners

Exs. 6–10 Students learning English may benefit from a review of the concepts of longitude, latitude, and the four different time zones of the contiguous United States.

Career Connection

Exs. 6–10 Students interested in weather may wish to research meteorology as a career. A meteorologist is a person who studies the typical weather patterns we observe. For the meteorologist, there are six elements that make weather: air temperature, barometric pressure, wind velocity, clouds, humidity, and precipitation. Weather forecasting models using computers have been improved in recent years, but because of the dynamic nature of weather, existing models are still only rough approximations to reality.

Challenge

Ex. 10 Ask students to think about approximately how many degrees of longitude are covered by a time zone. (15°)

Connection METEOROLOGY

Sunrise and sunset occur earliest at the eastern edge of a time zone. They occur four minutes later for each degree west you travel within that time zone, if you remain at roughly the same latitude.

6. If the sun rises at 6:05 A.M. in Philadelphia, then for any location at about the same latitude as Philadelphia and at longitude l within the Eastern time zone, sunrise occurs at the time t given by:

$$t = 6{:}05 + ({:}04)(l - 75)$$

Cities near 40°N latitude	Longitude (°W)
Philadelphia, PA	75
Indianapolis, IN	86

a. Use the equation to predict the time the sun will rise in Indianapolis.

b. The Eastern time zone includes longitudes from about 67°W to 90°W. State the domain and range of the given function.

7. a. Suppose the sun sets at 6:17 P.M. in Sweetwater, Texas. Write a point-slope equation giving sunset time as a function of longitude for locations at about 32°N within the Central time zone.

b. Use the equation to predict the time the sun will set in Marshall, Texas.

Cities near 32°N latitude	Longitude (°W)
Sweetwater, TX	100
Marshall, TX	94

8. Open-ended Problem Use an almanac to find the longitude for your town. Check a newspaper to find today's sunset time for your town. Write an equation giving sunset times for locations along the same latitude as your town and in your time zone. Find another town at about the same latitude and predict the sunset time there today.

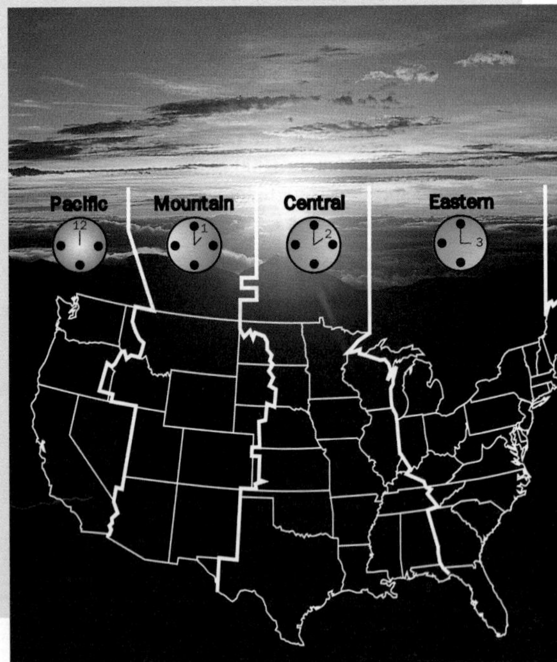

9. Anton receives a New York newspaper in Akron, Ohio, located at 82°W longitude. He notices that the sunrise time given for New York City, located at the same latitude as Akron, is 32 min earlier than the sunrise time printed in his Akron newspaper. Determine the longitude of New York City.

10. Challenge Draw a graph of sunrise times across the United States at 40°N latitude if you know that sunrise occurs in Philadelphia at 6:15 A.M. on a given day. (*Note:* Your graph should include the loss of 1 h each time you pass into a new time zone going west. At 40°N latitude, the United States extends from about 74°W to 124°W longitude.)

6. a. 6:49 A.M.
   b. domain: $67 \le l \le 90$; range: $5{:}33 \le t \le 7{:}05$
7. a. Let l = longitude (°West) and t = time; $t = 6{:}17 + ({:}04)(l - 100)$.
   b. 5:53 P.M.
8. Answers may vary.
9. 74°W
10. Answers will vary depending on estimates of the longitude of time zone changes.

U.S. Sunrise Times at 40°N

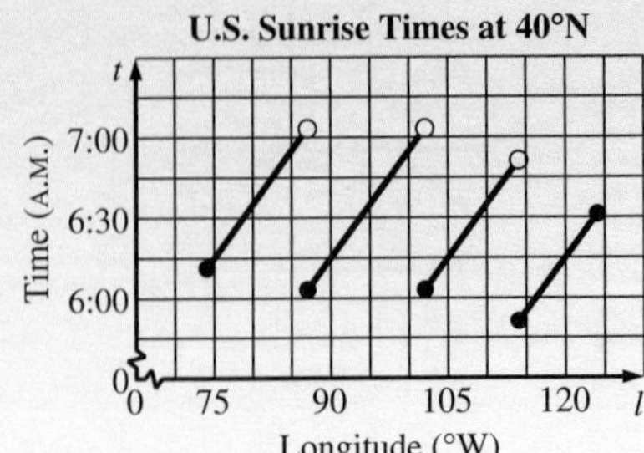

11. all real numbers; all real numbers
12. $x \le 3$; $y \ge 0$
13. all real numbers; 2
14. Let a = the age of the cow in months and w = the weight in pounds; $w = 150 + 32.5(a - 3)$ or $w = 540 + 32.5(a - 15)$; about 702.5 lb.

For each function f whose graph is shown, find the domain and range.

11. 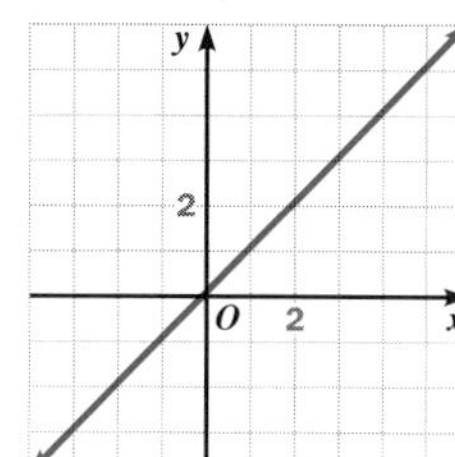

12. 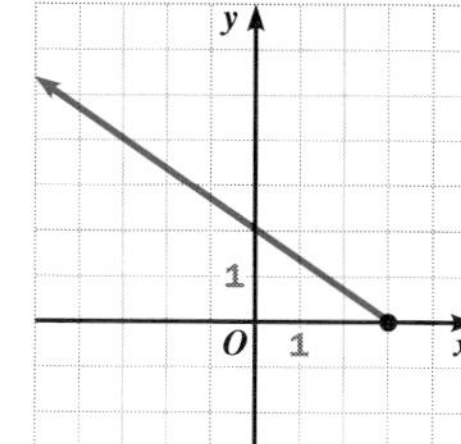

13. 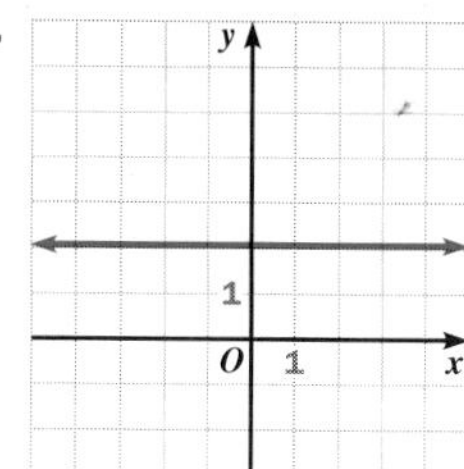

14. **BIOLOGY** The weight of Jersey cattle increases roughly linearly with age. A typical Jersey cow weighs 150 lb at 3 months and 540 lb at 15 months. Write a point-slope equation giving the weight of a Jersey cow as a function of age. Predict the weight of a Jersey cow at 20 months.

15. **SPORTS** Although a headwind hurts a runner's performance, a tailwind helps a runner move more quickly. Studies have shown that the time required for a runner to finish a 1 mi race decreases an average of 0.5 s for every 1 mi/h of tailwind speed.

    a. Running with an 8 mi/h tailwind, Alicia finishes a 1 mi race in 6:30 (6 min 30 s). Find a point-slope equation giving Alicia's finishing time t in seconds as a function of tailwind speed w in miles per hour.

    b. Predict Alicia's finishing time in still air.

16. **SAT/ACT Preview** Which ordered pair is a solution to the equation $y = 15(x - 3) + 7$?

    A. $(-3, 7)$ **B.** $(0, 0)$ **C.** $(3, 15)$ **D.** $(7, 3)$ **E.** $(3, 7)$

Find a point-slope equation of each line.

17. 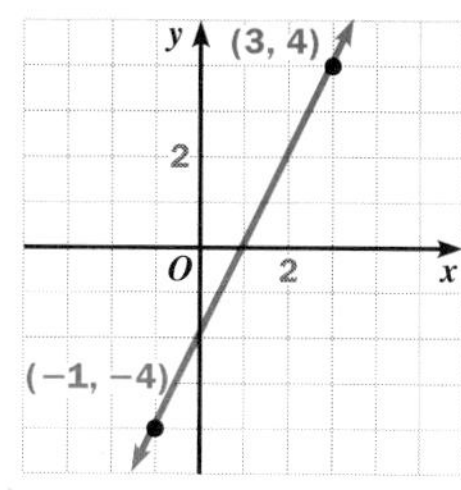

18. 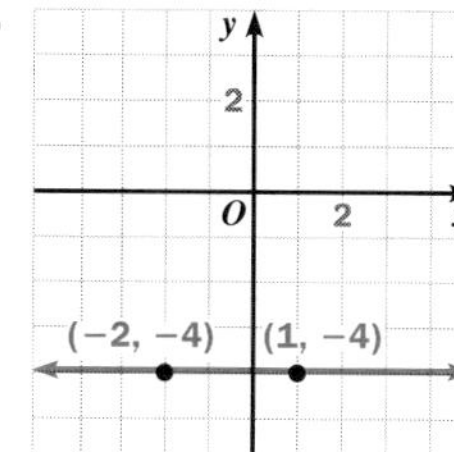

19. 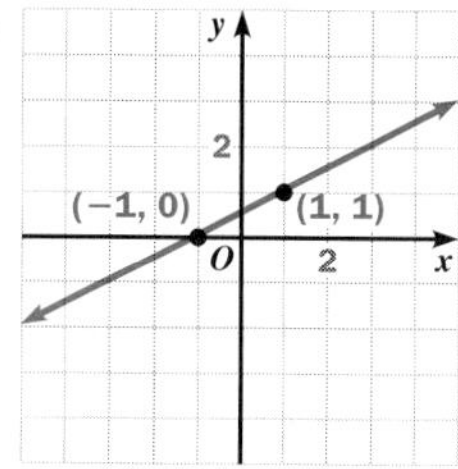

20. **SPORTS** A 120 lb in-line skater burns 7.4 Cal/min skating at 10 mi/h. For every extra 20 lb of weight, a skater skating at 10 mi/h burns an extra 0.9 Cal/min.

    a. Write a point-slope equation giving c, the number of calories burned per minute, as a function of w, the weight of the skater.

    b. **Open-ended Problem** What is a reasonable domain of the function you found in part (a)? What is the range?

For Exercises 21–24, find the domain and range of each function.

21. $f(x) = 5x - 2$ for $x > 0$

22. $f(x) = 2x - 1$ for $x \leq 0.5$

23. $f(x) = 7.2x + 320$ for $x \geq 14$

24. $f(x) = 18$ for $0 \leq x \leq 5$

Apply⇔Assess

Exercise Notes

Student Progress

Exs. 11–24 The exercises on this page cover all the objectives of this section. Thus, they provide students with an opportunity to assess their understanding of both the concepts and procedures developed in the section. A thorough review of the answers to all of these exercises will help students check their work.

Using Technology

Exs. 17–19 Students can check their answers to these exercises by graphing their point-slope equations. You may want to challenge them to find the exact viewing window that produces a graph matching the graph in the book.

Using Technology

Exs. 21–24 You may wish to have students use the graphing capabilities and the logic features of a TI-81 or TI-82 calculator to find the domain and range of each function. For example, for Ex. 21, they would graph Y1=(5X−2)(X>0), with ">" found under the TEST menu.

15. a. $t = 390 - 0.5(w - 8)$

    b. about 394 s or 6 min 34 s

16. E

17. $y = 4 + 2(x - 3)$ or $y = -4 + 2(x + 1)$

18. $y = -4$

19. $y = 1 + \frac{1}{2}(x - 1)$ or $y = \frac{1}{2}(x + 1)$

20. a. $c = 7.4 + \frac{0.9}{20}(w - 120)$ or $c = 7.4 + 0.045(w - 120)$

    b. Answers may vary. Examples are given. $120 \leq c \leq 225$; $7.4 \leq c \leq 12.125$

21. $x > 0$; $f(x) > -2$

22. $x \leq 0.5$; $f(x) \leq 0$

23. $x \geq 14$; $f(x) \geq 420.8$

24. $0 \leq x \leq 5$; 18

Apply⇔Assess

Exercise Notes

Assessment Note

Ex. 25 Having three or four students read their descriptions to the class would provide useful feedback for all students to evaluate what they have written.

Assess Your Progress

Journal Entry

For Ex. 4, a written explanation will help students who are having difficulty understanding that the point-slope form and the slope-intercept form are two different ways to write an equation of a given line.

Progress Check 2.1–2.3

See page 86.

Practice 9 for Section 2.3

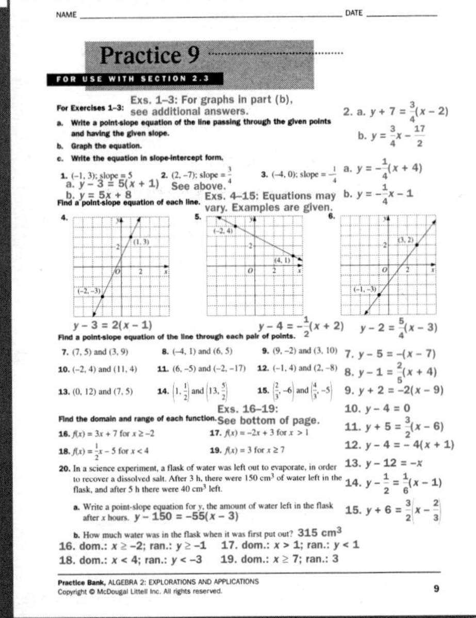

NAME ______________________ DATE __________

Practice 9

FOR USE WITH SECTION 2.3

For Exercises 1–3: Exs. 1–3: For graphs in part (b), see additional answers.

a. **Write a point-slope equation of the line passing through the given points and having the given slope.**

b. **Graph the equation.**

c. **Write the equation in slope-intercept form.**

1. (−1, 3); slope = 5 a. $y - 3 = 5(x + 1)$ b. $y = 5x + 8$

2. (2, −7); slope = $\frac{3}{4}$ See above. 2. a. $y + 7 = \frac{3}{4}(x - 2)$ b. $y = \frac{3}{4}x - \frac{17}{2}$

3. (−4, 0); slope = $-\frac{1}{4}$ a. $y = -\frac{1}{4}(x + 4)$ b. $y = -\frac{1}{4}x - 1$

Find a point-slope equation of each line. Exs. 4–15: Equations may vary. Examples are given.

4. (graph through (1, 3) and (−2, −3)) $y - 3 = 2(x - 1)$

5. (graph through (−2, 4) and (4, 1)) $y - 4 = -\frac{1}{2}(x + 2)$

6. (graph through (3, 2) and (−1, −3)) $y - 2 = \frac{5}{4}(x - 3)$

Find a point-slope equation of the line through each pair of points.

7. (7, 5) and (3, 9) **8.** (−4, 1) and (6, 5) **9.** (9, −2) and (3, 10)

10. (−2, 4) and (11, 4) **11.** (6, −5) and (−2, −17) **12.** (−1, 4) and (2, −8)

13. (0, 12) and (7, 5) **14.** $\left(1, \frac{1}{2}\right)$ and $\left(13, \frac{5}{2}\right)$ **15.** $\left(\frac{2}{3}, -6\right)$ and $\left(\frac{4}{3}, -5\right)$

7. $y - 5 = -(x - 7)$
8. $y - 1 = \frac{2}{5}(x + 4)$
9. $y + 2 = -2(x - 9)$
10. $y - 4 = 0$
11. $y + 5 = \frac{3}{2}(x - 6)$
12. $y - 4 = -4(x + 1)$
13. $y - 12 = -x$
14. $y - \frac{1}{2} = \frac{1}{6}(x - 1)$
15. $y + 6 = \frac{3}{2}\left(x - \frac{2}{3}\right)$

Find the domain and range of each function. Exs. 16–19: See bottom of page.

16. $f(x) = 3x + 7$ for $x \geq -2$ **17.** $f(x) = -2x + 3$ for $x > 1$

18. $f(x) = \frac{1}{2}x - 5$ for $x < 4$ **19.** $f(x) = 3$ for $x \geq 7$

20. In a science experiment, a flask of water was left out to evaporate, in order to recover a dissolved salt. After 3 h, there were 150 cm^3 of water left in the flask, and after 5 h there were 40 cm^3 left.

a. Write a point-slope equation for y, the amount of water left in the flask after x hours. $y - 150 = -55(x - 3)$

b. How much water was in the flask when it was first put out? 315 cm^3

16. dom.: $x \geq -2$; ran.: $y \geq -1$ 17. dom.: $x > 1$; ran.: $y < 1$

18. dom.: $x < 4$; ran.: $y < -3$ 19. dom.: $x \geq 7$; ran.: 3

Practice Bank, ALGEBRA 2: EXPLORATIONS AND APPLICATIONS

9

ONGOING ASSESSMENT

25. Writing Describe how to find the domain and range of a function from its graph.

SPIRAL REVIEW

For Exercises 26–28, find the slope-intercept equation of the line passing through each pair of points. *(Section 2.2)*

26. (0, 2) and (4, 5) **27.** (0, −3) and (3, −5) **28.** (−1, −2) and (0, −1)

29. Michael is studying the number of families with preschoolers in his neighborhood for a social studies project. His data appear in the table. Write a linear equation to model the number of families with preschoolers each year. *(Section 1.2)*

Year	Families
1993	11
1994	10
1995	8
1996	8

Solve each equation. *(Toolbox, page 784)*

30. $x + \frac{3}{2} = \frac{7}{8}$ **31.** $20 - x = 3x + 4$

32. $5n = 125$ **33.** $\frac{s}{8} = 7s + 11$

ASSESS YOUR PROGRESS

VOCABULARY

direct variation (p. 46)
constant of variation (p. 46)
slope (p. 47)
vertical intercept (p. 53)
slope-intercept form (p. 53)
point-slope equation (p. 59)
function notation (p. 60)
domain (p. 60)
range (p. 60)

1. a. The table shows the Loquasto family's driving time and distance traveled for each day of their car trip from Boston to Denver. Show that the distance traveled each day varies directly with driving time.

Time (h)	Distance (mi)
8.5	468
6	328
7	385
6.5	360
8	438

b. From Denver, the Loquastos plan to drive to Salt Lake City, a distance of 493 mi. Estimate how long it will take them. *(Section 2.1)*

2. Graph the line $y = 3 - \frac{2}{3}x$. *(Section 2.2)*

3. Suppose the domain of the function $f(x) = 5 - 2x$ is $x \leq 3$. What is the range? Graph the function. *(Section 2.3)*

4. Journal How are the point-slope and slope-intercept forms of a linear equation related? How are they different?

---

25. The domain consists of the x-coordinates of all the points on the graph of the function. The range consists of the y-coordinates of all the points on the graph.

26. $y = \frac{3}{4}x + 2$ 27. $y = -\frac{2}{3}x - 3$

28. $y = x - 1$

29. Let x = the number of years since 1993 and y = the number of families with preschoolers; $y = 11 - x$.

30. $-\frac{5}{8}$ 31. 4 32. 25 33. $-\frac{8}{5}$

Assess Your Progress

1. a. For every data pair, the ratio $\frac{\text{distance}}{\text{time}} \approx 55$. Since the ratio is constant, distance varies directly with time.

b. Estimates may vary; about 9 h.

2.

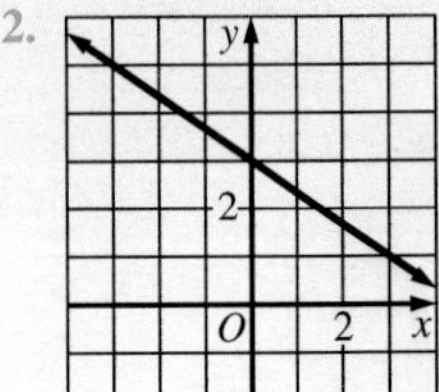

3. $f(x) \geq -1$

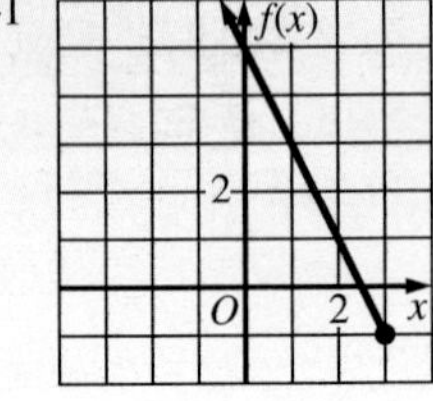

4. Answers may vary. Examples are given. The slope-intercept form of an equation can be obtained by simplifying the point-slope form. The two forms look different and the slope and y-intercept can be read directly from the slope-intercept form, while only the slope can usually be read directly from point-slope form. Also, a line has only one slope-intercept equation but any number of equations in point-slope form.

Section

2.4 Fitting Lines to Data

Learn how to...

- draw a line of fit and find its equation

So you can...

- make predictions about bicycle production, for example

You know that the more pages a book has, the thicker it is. The Exploration will help you decide if the relationship between page count and thickness is linear.

EXPLORATION
COOPERATIVE LEARNING

Measuring the Thickness of Books

Work with a group of three or four students.
You will need:

- a set of encyclopedias
- a metric ruler
- graph paper and a pencil

1 Choose ten of the volumes from a set of encyclopedias. For each, find the number of pages P and measure the volume's thickness T to the nearest 0.1 cm. Record your data in a table.

2 Plot the data pairs (P, T) in a coordinate plane.

Toolbox p. 795
Making a Scatter Plot

3 Use the ruler to draw a *line of fit.*

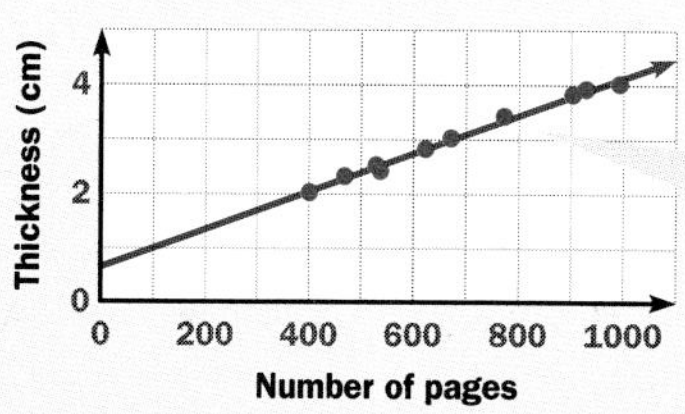

A **line of fit** is a line that lies as close as possible to all the points in a scatter plot. It does not have to pass through any of them.

Use your graph to answer the following questions.

1. Find the slope of your line. What does the slope represent in this situation? (*Hint:* Think about the relationship between the number of pages and the number of pieces of paper.)
2. a. Find an equation of your line of fit. What is the significance of the T-intercept?
   b. Measure the thickness of a volume not included in your data set. Use the equation to predict how many pages it has. Then use the book to check your prediction.

Exploration Note

Purpose
The purpose of this Exploration is to introduce students to a *line of fit.*

Materials/Preparation
A set of encyclopedias, metric rulers, and graph paper are needed.

Procedure
Students record the number of pages and the thickness for each of ten books as data pairs. They then plot the data pairs and draw a line of fit through the points. From their graph, they find the slope and an equation of the line. Students should understand that since the data are generated experimentally, the data pairs and the line of fit may vary.

Closure
Students should conclude that the thickness of a book can be represented by a linear model $y = ax + b$, where x is the number of pages in the book, a is half the thickness of a sheet of paper (two pages per sheet), and b is the thickness of both covers combined.

Explorations Lab Manual
See the Manual for more commentary on this Exploration.

For answers to the Exploration, see following page.

Plan⇔Support

Objectives

- Draw a line of fit and find its equation.
- Make predictions from linear data.

Recommended Pacing

Core and Extended Courses
Section 2.4: 2 days

Block Schedule
Section 2.4: 1 block

Resource Materials

Lesson Support
Lesson Plan 2.4
Warm-Up Transparency 2.4
Practice Bank: Practice 10
Study Guide: Section 2.4
Exploration Lab Manual:
Additional Exploration 2, Diagram Master 1

Technology
Technology Book:
Spreadsheet Activity 2
Graphing Calculator
McDougal Littell Mathpack
Stats! Activity Book: Activities 8, 16, and 21
Internet:
http://www.hmco.com

Warm-Up Exercises

1. Find the slope of the line through (–2, 4) and (5, –1). $-\frac{5}{7}$
2. Find the slope-intercept equation of a line through (0, –4) and (1, 6). $y = 10x - 4$
3. If $C = 25 + 0.10m$ represents the cost in dollars for renting a car, where m is the number of miles driven, predict the rental cost for 500 miles. \$75
4. Find a point-slope equation of the line through (1, –5) and (3, 2). $y = -5 + \frac{7}{2}(x - 1)$ or $y = 2 + \frac{7}{2}(x - 3)$
5. What is the vertical intercept of the graph of $y = -3.6x + 10.2$? 10.2

Teach⇔Interact

Section Note

Multicultural Note

Bicycles are a popular means of transportation all over the world. In China and in the Netherlands, for instance, many people use bicycles as their primary means of transportation. In China, it is estimated that there are more than 210 million bicycles. During rush hour in major Chinese cities, the streets are filled with cyclists. In the Netherlands, there are as many bicycles as there are people and separate bike paths are found in many parts of the country.

About Example 1

Teaching Tip

When discussing how to find the equation of the line of fit, it should be clear to students that choosing a different pair of points may generate a different point-slope equation of the line.

Additional Example 1

The table gives a typical yearly tuition cost at a public university for the years 1987 to 1992.

Year	Tuition
1987	$2450
1988	2525
1989	2600
1990	2680
1991	2760
1992	2850

a. Show that the data have a linear relationship by making a scatter plot and drawing a line of fit.
Let t be the time (in years) since 1987, and c be the tuition cost.

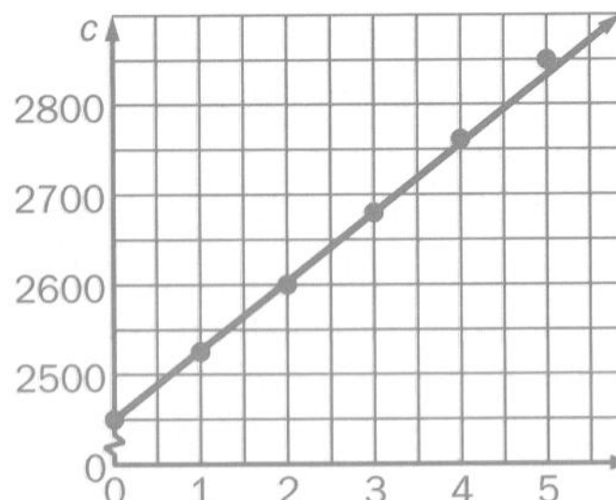

b. Find an equation of the line of fit.
Choose any two points that appear to lie on the fitted line, for example, (1, 2525), and (3, 2680). The slope of the line is: $\frac{2680 - 2525}{3 - 1} = \frac{155}{2} = 77.5$.
Therefore, a point-slope equation of the line is:
$y = 2525 + 77.5(x - 1)$.

Lines of fit are often used to model data, such as data obtained in scientific studies, opinion polls, and sporting events, and to predict what happens between and beyond plotted data points. The "eyeball method" used in the Exploration is the simplest way to determine a line of fit.

BY THE WAY

More people are transported by bicycles in Asia alone than are transported by automobiles worldwide.

EXAMPLE 1 Application: Manufacturing

The table gives the number of bicycles produced worldwide for various years between 1965 and 1990.

Year	1965	1970	1975	1980	1985	1990
Number of bicycles (in millions)	21	36	43	62	79	95

a. Show that the data have a linear relationship by making a scatter plot and drawing a line of fit.

b. Find an equation of the fitted line.

c. Use the equation to predict the number of bicycles produced in the year 2000.

SOLUTION

a. Let t be the time (in years) since 1965, and let b be the number of bicycles (in millions) produced.

t	b
0	21
5	36
10	43
15	62
20	79
25	95

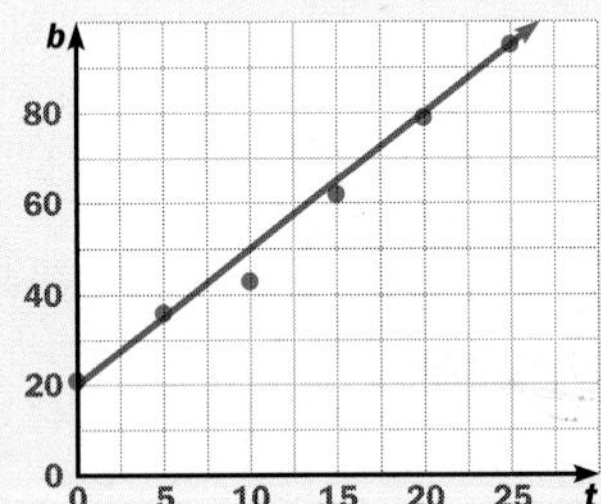

b. Choose any two points that *appear* to lie on the fitted line, for example (5, 35) and (20, 80). The slope of the line is:

$$\frac{80 - 35}{20 - 5} = \frac{45}{15} = 3$$

Therefore a point-slope equation of the line is:

$$y = 35 + 3(x - 5)$$

c. The year 2000 is 35 years after 1965, so substitute $x = 35$ into the equation from part (b):

$$y = 35 + 3(35 - 5)$$
$$= 125$$

You can predict that about 125 million bicycles will be produced in the year 2000.

ANSWERS Section 2.4

Exploration

1. Answers may vary. The slope represents the change in thickness per page of the book. Since there is one piece of paper for every two pages, the thickness of a piece of paper is twice the slope.

2. a. Answers may vary. The T-intercept represents the thickness of the book without any pages, that is the total thickness of the covers.
   b. Answers may vary.

Think and Communicate

1. No.

2. a. probably not; When eyeballing a line of fit, there is no single correct answer.
   b. Answers may vary. An example is given. You could use the equations of both lines to estimate data points and see which does a better job.

THINK AND COMMUNICATE

1. Does a line of fit have to pass through at least one of the data points?
2. a. If two different students eyeballed a line of fit for the same data, would you expect the lines of fit to be the same? Explain.
   b. How would you know which student's line was a better fit for the data?

The Least-Squares Line

Different people tend to "eyeball" different lines of fit for a set of data points. To avoid this problem, mathematicians have developed a standard line of fit called the *least-squares regression line*, or just the **least-squares line.**

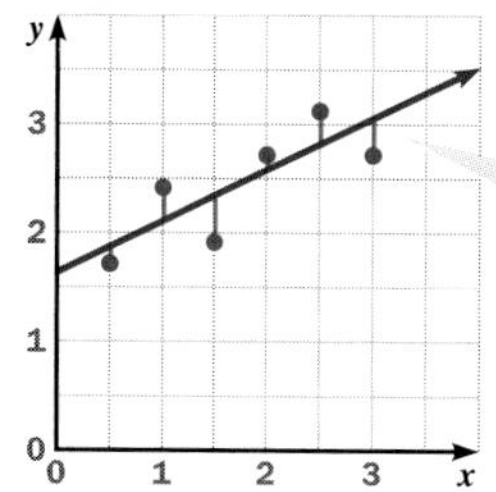

The sum of the squares of the vertical distances shown in green is smallest for the least-squares line.

Most graphing calculators and statistical software will find an equation of the least-squares line for you.

EXAMPLE 2 Application: Manufacturing

Refer to the bicycle data in Example 1.

a. Use a graphing calculator or statistical software to find an equation of the least-squares line for the data.

b. Use the equation from part (a) to predict the number of bicycles produced in the year 2000.

SOLUTION

a. Enter the data and have the calculator or software perform a linear regression. You should get results like those shown. An equation of the least-squares line is $y = 2.96x + 19$.

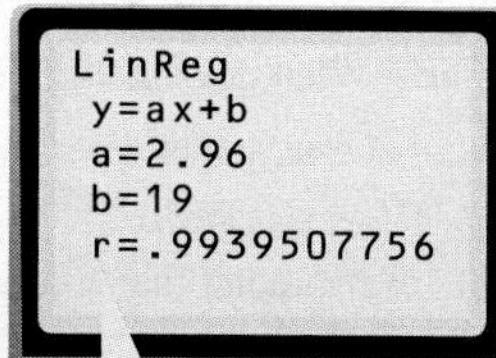

```
LinReg
 y=ax+b
 a=2.96
 b=19
 r=.9939507756
```

For now, ignore the variable r in the display.

b. Substitute $x = 2000 - 1965 = 35$ into the equation from part (a):

$$y = 2.96(35) + 19$$
$$= 122.6$$

You can predict that about 123 million bicycles will be produced in the year 2000.

For more information about linear regression, see the *Technology Handbook*, p. 814.

WATCH OUT!

On some calculators the roles of a and b are reversed: $y = a + bx$. Be sure to check your calculator's instruction manual.

Technology Note

On a graphing calculator, students can show the line of fit on a scatter plot of the data by graphing the regression line over the scatter plot. The line can then be traced to do predictions for any year. The following procedure is appropriate for the TI-82 graphing calculator.

Enter the data and display a scatter plot. Use the procedure described in the text to perform a linear regression and display the coefficients for the least-squares line. Next, press [Y=] and clear the Y= list. Position the curser after Y1= on the first line of the Y= list. Press [VARS] [▶] [▶] 7. The regression equation is automatically placed on the Y= list. Press [GRAPH] to display simultaneously the scatter plot and line of fit.

To predict y-values for given x-values, press [2nd] [CALC]. When Eval X= appears at the bottom of the screen, type an x-value and press [ENTER]. The calculator will display the corresponding y-value. (Note: The value of x must be within the range of x-values for the graphing window. If it is not, change the graphing window settings to include it.)

Teach⇔Interact

Additional Example 1 (continued)

c. Use the equation to predict the typical yearly tuition cost in the year 2000.
The year 2000 is 13 years after 1987, so substitute $x = 13$ into the equation from part (b):
$y = 2525 + 77.5(13 - 1)$
$= 3455$
You can predict that a typical tuition at a public university in the year 2000 will be about $3455.

Think and Communicate

For question 2, you may wish to mention that different lines of fit are expected when technology is not used. In Example 2, students see that when using a graphing calculator only one line of fit is produced for the data points.

About Example 2

Teaching Tip
Some students may be interested in the variable r in the display. Point out that the closer the value r is to 1 (or -1 for a negative slope), the better is the line of fit for the data.

Additional Example 2

Refer to the tuition data in Additional Example 1.

a. Use a graphing calculator or statistical software to find an equation of the least-squares line for the data.
Enter the data and have the calculator or software perform a linear regression. You should get results like those shown. An equation of the least-squares line is $y = 79.57x + 2445$.

```
LinReg
 y=ax+b
 a=79.57142857
 b=2445.238095
 r=.9994696657
```

b. Use the equation from part (a) to predict the typical cost of tuition in the year 2000.
Substitute $x = 2000 - 1987 = 13$ into the equation from part (a):
$y = 79.57(13) + 2445$
$= 3479.41$
You can predict that the typical cost of tuition will be about $3479 in the year 2000.

Teach⇔Interact

Checking Key Concepts

Topic Spiraling: Review
For question 1, students can do a quick check that the data have a linear relationship by finding a similar ratio of calories to fat for each data pair. Ask students why this method works in this situation. (because 0 grams of fat has 0 calories; If these data have a linear relationship, they are a direct variation.)

Closure Question

Explain how to find a line of fit for data having a linear relationship. By hand, make a scatter plot of the data and draw a line of fit. Choose any two points that appear to lie on the fitted line and use them to find a point-slope equation of the line. Using technology, enter the data and have the calculator or software perform a linear regression.

Apply⇔Assess

Suggested Assignment

Core Course
Day 1 Exs. 1–6, 9
Day 2 Exs. 10–19

Extended Course
Day 1 Exs. 1–9
Day 2 Exs. 10–19

Block Schedule
Day 10 Exs. 1–6, 9–19

Exercise Notes

Teaching Tip
Ex. 1 Remind students to consider 1958 as "year 0," the vertical intercept of the scatter plot.

Applications
Exs. 1–10 Many applications of mathematics involve the analysis of data to make predictions. These exercises cover situations in economics, sports, earth science, government, biology, and health. One common use of data that appears frequently in the news media is that of medical experiments involving large groups of people to determine the effects of various medicines or vitamins.

CHECKING KEY CONCEPTS

1. The table gives the grams of fat and the number of calories in 100 g of various salad dressings. The lowest in fat and calories is low-fat French, and the highest in fat and calories is mayonnaise.

Fat (g)	4.3	42.3	50.2	52.3	60.0	79.9
Calories	96	435	502	504	552	718

a. Show that the data have a linear relationship by making a scatter plot and drawing a line of fit.

b. Find an equation of the line.

c. Use the equation to predict the number of calories in a salad dressing that has 30 g of fat.

2. a. **Technology** Use a graphing calculator or statistical software to find an equation of the least-squares line for the salad dressing data in Question 1.

b. Use the equation from part (a) to predict the number of calories in a salad dressing that has 30 g of fat.

2.4 Exercises and Applications

Extra Practice exercises on page 751

1. ECONOMICS If a person works an eight-hour day, how much of that time goes just toward paying taxes? The table shows how this figure has changed from 1958 to 1994.

Year	1958	1964	1970	1976	1982	1988	1994
Hours per day for taxes	2.2	2.3	2.5	2.6	2.7	2.7	2.8

a. Make a scatter plot of the data and draw a line of fit.

b. Find an equation of the line. What does the line's slope represent?

c. Predict the number of hours per day that a person living in 2010 will have to work just to pay taxes.

d. Writing Does the model ever predict that a person's entire workday will go to pay taxes? Discuss the limitations of the model.

2. **Technology** The table shows the winning times for men in the Boston Marathon for various years between 1955 and 1995. Use a graphing calculator or statistical software to find an equation of the least-squares line for winning time as a function of year.

Year	1955	1960	1965	1970	1975	1980	1985	1990	1995
TIME (MIN)	138.37	140.90	136.55	130.50	129.92	132.18	134.08	128.32	129.37

Checking Key Concepts

1. Answers may vary. Examples are given.

a.

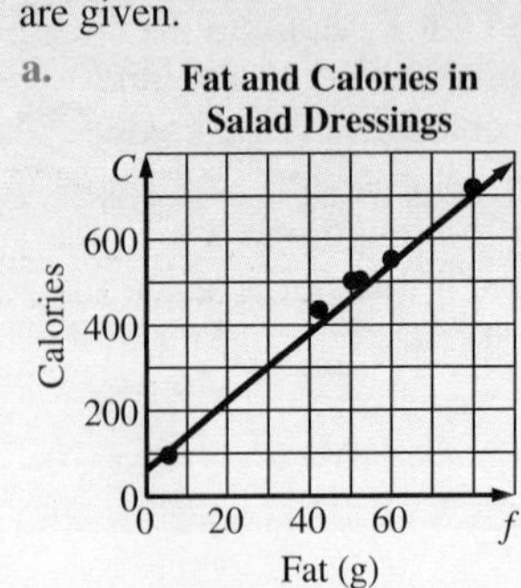

b. $C = 8f + 62$

c. about 302 g

2. Answers may vary due to rounding. Examples are given.

a. $C = 8.21f + 72.62$

b. about 319 g

Exercises and Applications

1. Answers may vary. Examples are given.

a.

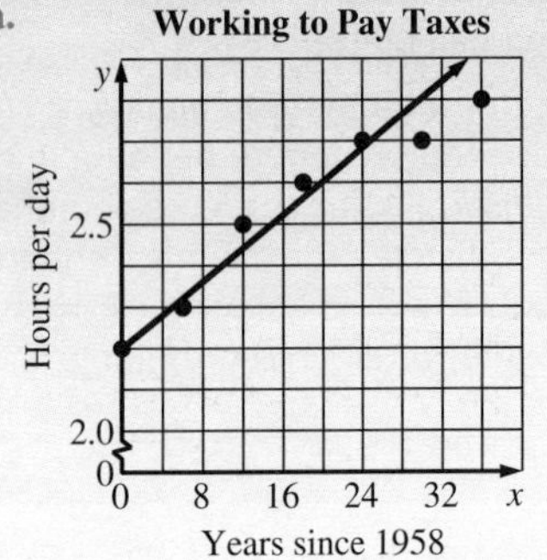

Connection EARTH SCIENCE

In 1993 some of the worst flooding in U.S. history occurred when the Mississippi and Missouri rivers overflowed their banks after reaching record crests. The flooding was caused by downpours in the Midwest.

3. The table gives the height of the Mississippi River above its banks in Hannibal, Missouri, for one week in April, 1993.

Date in April	Height above banks (in.)
4	2.1
5	2.4
6	2.8
7	3.2
8	3.8
9	4.4
10	4.6

   a. Show that the data have a linear relationship by making a scatter plot and drawing a line of fit.

   b. Find an equation of the line of fit. Predict the height of the river on April 1st and on April 30th. Which prediction do you think is more accurate?

4. The scatter plot shows the height of the Mississippi River above its banks in Hannibal, Missouri, for the entire month of April, 1993.

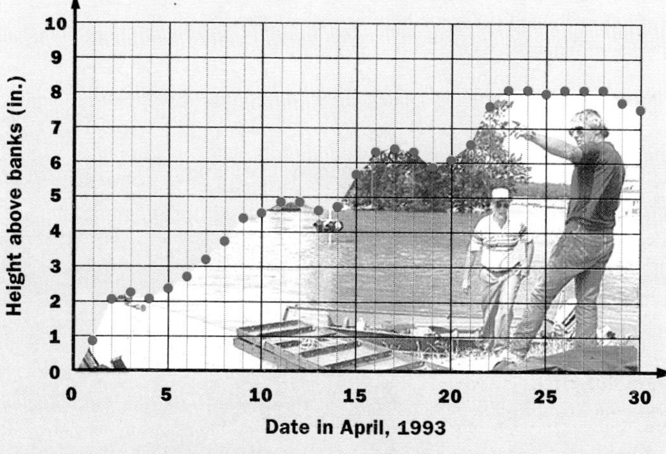

   a. **Visual Thinking** Does the height of the water still appear to increase roughly linearly over time? Use a pencil or ruler to eyeball a line of fit for the entire month of April, and for April 4–10. What do you notice?

   b. Using the graph, find an equation for a new line of fit for the height of the water as a function of time for the month of April.

Hannibal Courier-Post

The Flood of '93

And a river runs through it

The Flood of '93

As disaster hits, we pull together, survive

Exercise Notes

Using Technology
Exs. 1–3, 5, 9–12 Graphing calculators and the *Stats!* software are suitable for all of these exercises.

Interdisciplinary Problems
Exs. 3, 4 There are various sciences that can be considered to be a part of earth science. In general, earth science deals with any aspect of Earth, such as its composition, configuration, or weather. Thus, geology, geography, meteorology, and oceanography are all earth sciences.

Communication: Discussion
Ex. 4 Ask students to discuss their answers to this exercise. While the data have an overall appearance of being linear for the entire month of April, they are not as good a fit as in the beginning of April. Some students may be able to share their knowledge or experience with the phenomena being discussed to help other students understand the situation.

b. Let x = the number of years since 1958 and y = the number of hours worked per day to pay taxes; $y = 0.02x + 2.2$; the increase per year.

c. about $3\frac{1}{4}$ h

d. Yes. According to the model, in about 2248 a person will work a full 8 h workday just to pay taxes. Obviously, this does not make sense. The model is based on current conditions which cannot be expected to be in effect nearly 250 years from now.

2. Answers may vary due to rounding. An example is given. Let x = the number of years since 1955 and y = the winning time in minutes; $y = -0.26x + 138.49$.

3. Answers may vary. Examples are given.

   a.

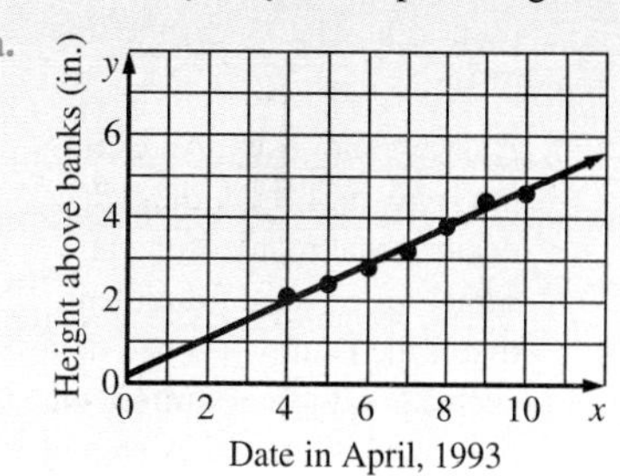

   b. Let x = the date in April, 1993 and y = the height of the river above its banks; $y = 0.45x + 0.2$; 0.65 in.; 1.37 in.; 1st.

4. Answers may vary. Examples are given.

   a. Yes; the second line is steeper than the first.

   b. $y = 0.24x + 1.52$

Apply⇔Assess

Exercise Notes

Second-Language Learners
Exs. 5–7 Some students learning English may benefit from more background information on how members of a state senate are elected, particularly if they worked on these exercises independently. They may also benefit from working with an aide or peer tutor when writing a response to Ex. 7.

Using Technology
Ex. 5 On a TI-82 calculator, enter the last four columns of the table as lists L1, L2, L3, and L4. Then on the home screen, enter L2–L1→L5 and L4–L3→L6. L5 will store the values of variable *m* and L6 will be the values of the variable *a*. Under the STAT CALC menu, choose SetUp and change Xlist to L5 and Ylist to L6. Then when a scatter plot is made, it will be of *m* and *a*. Also, if LinReg(ax+b) is used to calculate the line of fit, the values of *m* and *a* will be used. After the data have been entered in the lists, a complete window to display the scatter plot can be found by pressing ZOOM and selecting 9:ZoomStat.

Communication: Discussion
Exs. 7, 8 Have students discuss their answers to these exercises. They should use the data to justify their statements.

Connection GOVERNMENT

Lines of fit have been used in court to interpret data about election results.

Year	District	Democratic machine vote	Republican machine vote	Democratic absentee vote	Republican absentee vote
1982	2	47,767	21,340	551	205
1982	4	44,437	28,533	594	312
1982	8	55,662	13,214	338	115
1986	2	39,034	23,363	609	316
1986	4	52,817	16,541	666	306
1986	8	48,315	11,605	477	171
1990	2	27,543	26,843	660	509
1990	4	39,193	27,664	482	831
1990	8	34,598	8,551	308	148

5. The table shows the results in three Philadelphia voting districts for state senate elections. "Machine vote" means votes cast at official polling places.

a. **Technology** Let m = the difference between the Democratic and the Republican machine vote. Let a = the difference between the Democratic and the Republican absentee vote. Use a graphing calculator or statistical software to add two columns to the table for m and a.

b. Make a scatter plot of the data, with m on the horizontal axis and a on the vertical axis. Find a line of fit.

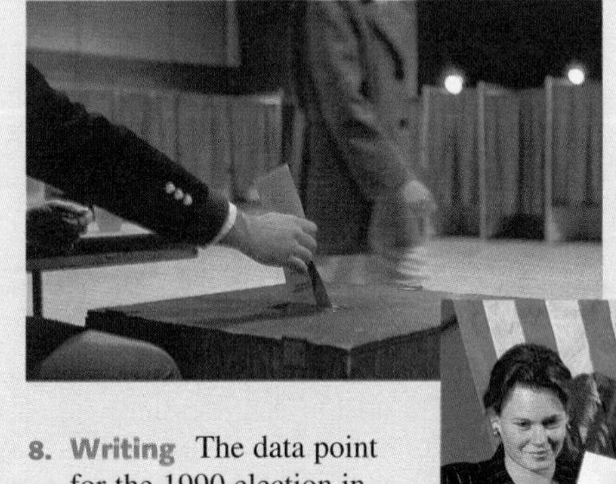

6. Visual Thinking In a special runoff election in District 2 in 1993, the Democratic machine vote was 19,127, and the Republican machine vote was 19,691. The Democratic absentee vote was 1396, and the Republican absentee vote was 371. Find m and a for this election and include the new point on your scatter plot. What do you notice?

7. Writing The special runoff election was challenged in court. The Republicans charged that many of the absentee ballots were fraudulent. The Democrats argued that they had done a good job turning out the absentee vote. The judge awarded the seat to the Republican. Why do you think the judge made this decision?

8. Writing The data point for the 1990 election in District 4 is also far off the line of fit. Why do you think this election was not challenged? Should it have been?

BALLOTS

5. a.

Year	District	m	a
1982	2	26,427	346
1982	4	15,904	282
1982	8	42,448	223
1986	2	15,671	293
1986	4	36,276	360
1986	8	36,710	306
1990	2	700	151
1990	4	11,529	–349
1990	8	26,047	160

b. Line of fit may vary.

Voting Results

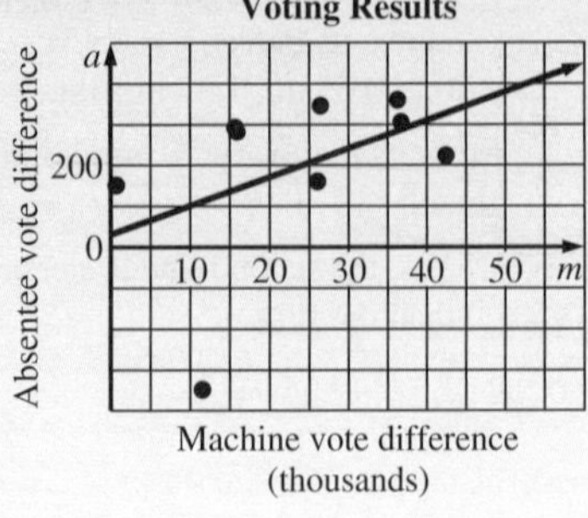

6. $m = -564$; $a = 1025$; The new point is nowhere near the line of fit.

7. Answers may vary. An example is given. I think the judge decided that many of the absentee votes were fraudulent, that if the Democrats had done such a good job of turning out the absentee vote, they should have been able to turn out the machine vote as well.

8. Answers may vary. An example is given. I think the election was not challenged because the party that appeared to have too many absentee votes lost anyway. I think if fraud was suspected, a challenge should have been made anyway.

9. Answers may vary. Examples are given.

a. Weight and Oxygen Consumption—Harbor Seals

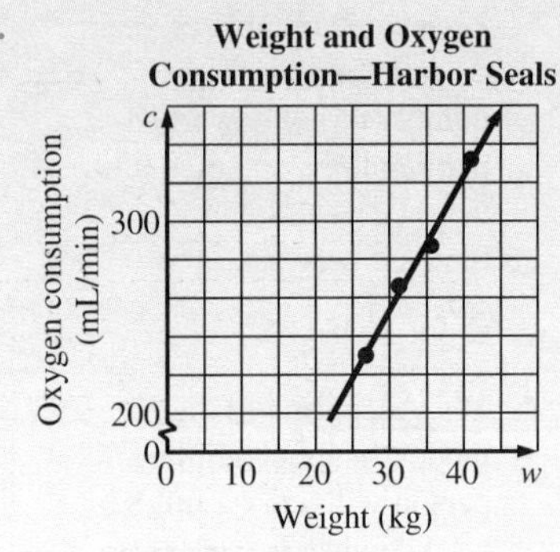

b. Let w = weight in kg and c = oxygen consumption in mL/min; $c = 7w + 43$.

9. **BIOLOGY** The table gives the weight and oxygen consumption of four harbor seals.

Weight (kg)	26.8	31.3	35.6	41.0
Oxygen consumption (mL/min)	230	266	287	332

   a. Show that the data have a linear relationship by drawing a scatter plot and a line of fit.
   b. Find an equation of the line of fit.
   c. Predict the oxygen consumption of a 25 kg harbor seal.

10. **HEALTH** The table gives average weights for men and women of different heights. The data are for people of medium frame, aged 30–39 years.

Women		Men	
Height (in.)	Weight (lb)	Height (in.)	Weight (lb)
60	120	64	145
62	126	66	153
64	132	68	161
66	139	70	170
68	146	72	179
70	154	74	188
72	164	76	199

   a. Make two scatter plots, one for men and one for women, in the same coordinate plane. Put height on the horizontal axis and weight on the vertical axis, and use different symbols or colors to distinguish the two sets of data.
   b. For each scatter plot, draw a line of fit and find its equation.
   c. Predict the average weights of men 80 in. tall and women 76 in. tall.
   d. **Writing** Compare the slopes of the two lines of fit. Did you expect them to be the same? Explain.

ONGOING ASSESSMENT

11. **Open-ended Problem** Look through a newspaper or almanac and find a set of real-world data that appears to be linear. Make a scatter plot of the data, draw a line of fit, and find an equation of the line. Use your equation to make predictions.

SPIRAL REVIEW

12. a. The table gives the number of eighth graders, in millions, enrolled in public schools in the United States. Make a scatter plot of the data. *(Toolbox, page 795)*

Year	Eighth graders (millions)
1960	2.70
1970	3.60
1980	3.09
1990	2.98

   b. Does it make sense to fit a line to the data?

Write a point-slope equation of the line passing through the given point and having the given slope. *(Section 2.3)*

13. point: (7, 3)
    slope = −4.2

14. point: (12, 3)
    slope = 2

15. point: (−2, −3)
    slope = 0.5

For each equation, state whether y varies directly with x. *(Section 2.1)*

16. $y = \frac{1}{2}x$

17. $y = \frac{1}{3}x$

18. $y = 2(x + 1)$

19. $y = -3x$

Apply⇔Assess

Exercise Notes

Assessment Note

Ex. 11 By sharing their results for this exercise, students can broaden their experience with different sets of real-world data that may be linear.

Challenge

Ex. 12 Ask students how they could examine these data points and determine if there is a linear relationship without graphing. (Since the change in x-coordinate is constant, the change in y-coordinate should also be approximately constant.)

Topic Spiraling: Preview

Ex. 12 You may want students to do this exercise using technology, which will give them the correlation coefficient, the topic of the next section.

Practice 10 for Section 2.4

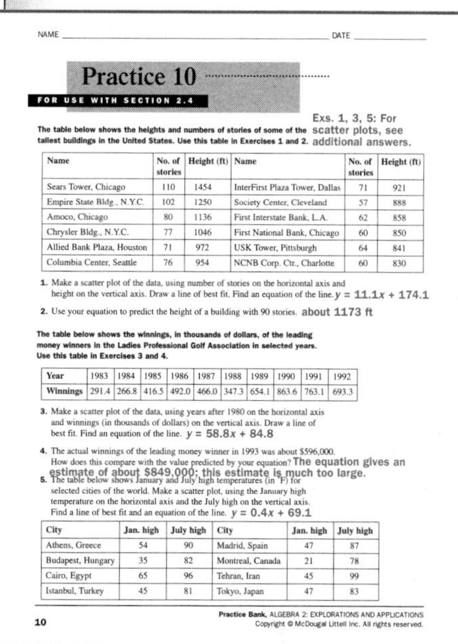

NAME ________________ DATE ________

Practice 10

FOR USE WITH SECTION 2.4

The table below shows the heights and numbers of stories of some of the tallest buildings in the United States. Use this table in Exercises 1 and 2. Exs. 1, 3, 5: For scatter plots, see additional answers.

Name	No. of stories	Height (ft)	Name	No. of stories	Height (ft)
Sears Tower, Chicago	110	1454	InterFirst Plaza Tower, Dallas	71	921
Empire State Bldg., N.Y.C.	102	1250	Society Center, Cleveland	57	888
Amoco, Chicago	80	1136	First Interstate Bank, L.A.	62	858
Chrysler Bldg., N.Y.C.	77	1046	First National Bank, Chicago	60	850
Allied Bank Plaza, Houston	71	972	USK Tower, Pittsburgh	64	841
Columbia Center, Seattle	76	954	NCNB Corp. Ctr., Charlotte	60	830

1. Make a scatter plot of the data, using number of stories on the horizontal axis and height on the vertical axis. Draw a line of best fit. Find an equation of the line. $y = 11.1x + 174.1$
2. Use your equation to predict the height of a building with 90 stories. about 1173 ft

The table below shows the winnings, in thousands of dollars, of the leading money winners in the Ladies Professional Golf Association in selected years. Use this table in Exercises 3 and 4.

Year	1983	1984	1985	1986	1987	1988	1989	1990	1991	1992
Winnings	291.4	266.8	416.5	492.0	466.0	347.3	654.1	863.6	763.1	693.3

3. Make a scatter plot of the data, using years after 1980 on the horizontal axis and winnings (in thousands of dollars) on the vertical axis. Draw a line of best fit. Find an equation of the line. $y = 58.8x + 84.8$
4. The actual winnings of the leading money winner in 1993 was about $596,000. How does this compare with the value predicted by your equation? The equation gives an estimate of about $849,000; this estimate is much too large.
5. The table below shows January and July high temperatures (in °F) for selected cities of the world. Make a scatter plot, using the January high temperature on the horizontal axis and the July high on the vertical axis. Find a line of best fit and an equation of the line. $y = 0.4x + 69.1$

City	Jan. high	July high	City	Jan. high	July high
Athens, Greece	54	90	Madrid, Spain	47	87
Budapest, Hungary	35	82	Montreal, Canada	21	78
Cairo, Egypt	65	96	Tehran, Iran	45	99
Istanbul, Turkey	45	81	Tokyo, Japan	47	83

10 **Practice Bank,** ALGEBRA 2: EXPLORATIONS AND APPLICATIONS

c. about 218 mL/min

10. Answers may vary. Examples are given.

a.

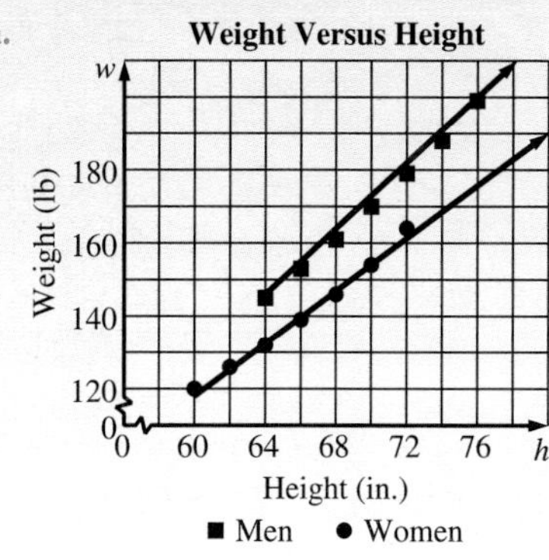

b. men: $w = 4.5h - 142$; women: $w = 3.6h - 98$

c. about 218 lb; about 176 lb

d. The slope of the line of fit for men is greater than that for women. Answers may vary. An example is given. The results are as I expected; I think men's weights tend to increase by a greater amount for the same increase in height than women's do.

11. Answers may vary.

12. a.

Eighth Graders

Number of eighth graders (millions): 2.60, 3.00, 3.40; n

Years since 1960: 0, 10, 20, 30; y

b. No; the data points do not lie near a line.

13. $y = 3 - 4.2(x - 7)$

14. $y = 3 + 2(x - 12)$

15. $y = -3 + 0.5(x + 2)$

16. Yes.

17. Yes.

18. No.

19. Yes.

Plan⇔Support

Objectives

- Interpret a correlation coefficient.
- Recognize positive and negative correlation.
- Determine the strength of a linear relationship.

Recommended Pacing

Core and Extended Courses
Section 2.5: 1 day

Block Schedule
Section 2.5: $\frac{1}{2}$ block
(with Section 2.6)

Resource Materials

Lesson Support
Lesson Plan 2.5
Warm-Up Transparency 2.5
Overhead Visuals:
Folder 2: Correlation Coefficient
Practice Bank: Practice 11
Study Guide: Section 2.5
Exploration Lab Manual:
Diagram Master 1

Technology
Technology Book:
Calculator Activity 2
Spreadsheet Activity 2
Graphing Calculator
McDougal Littell Mathpack
Stats!
Function Investigator
Internet:
http://www.hmco.com

Warm-Up Exercises

1. Find the slope of the line through (4, –3) and (6, 1). 2
2. Does the line through (1, 4) and (–1, 3) have a positive or negative slope? positive
3. Describe a line with positive slope. The line rises from left to right.
4. Describe a line with negative slope. The line falls from left to right.
5. Find some values of x so that $|x| \le 1$. Answers may vary. An example is given. –0.5, 0.5

Section

2.5 The Correlation Coefficient

Learn how to...

- interpret a correlation coefficient
- recognize positive and negative correlation

So you can...

- determine the strength of a linear relationship, such as between longitude and the highest January temperature

Is there a strong linear relationship between a city's location and the highest January temperature in that city? Examine the two scatter plots.

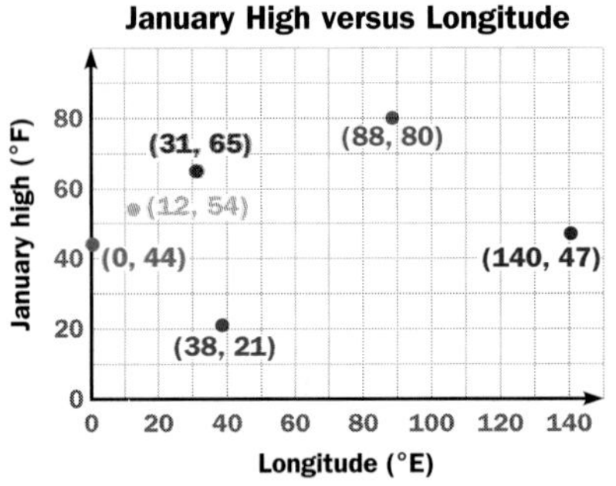

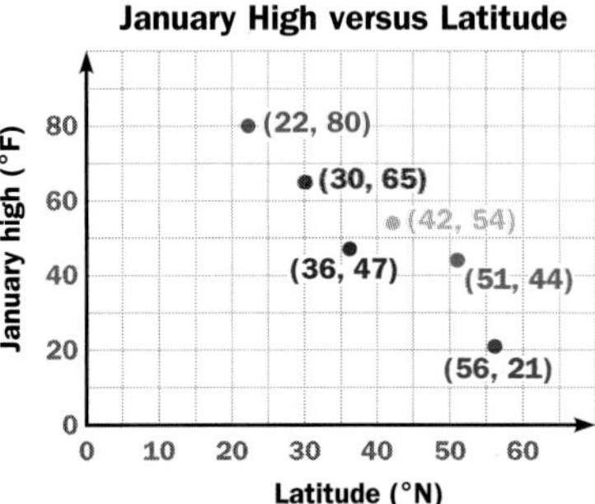

THINK AND COMMUNICATE

1. For which scatter plot do the data points "line up" better?
2. Do you think the longitude of a city or the latitude of a city more accurately predicts the highest January temperature in that city? Explain why your answer makes sense.

A **correlation coefficient**, denoted r, is a number between -1 and $+1$ that measures how well data points line up. There is a **positive correlation** $(0 < r \le 1)$ between the variables x and y when y tends to increase as x increases. There is a **negative correlation** $(-1 \le r < 0)$ between the variables when y tends to decrease as x increases.

ANSWERS Section 2.5

Think and Communicate

1. latitude versus January high
2. latitude; Answers may vary. An example is given. Temperature is more closely related to a location's distance from the equator than from the prime meridian.

Perfect Positive Correlation
$r = +1$

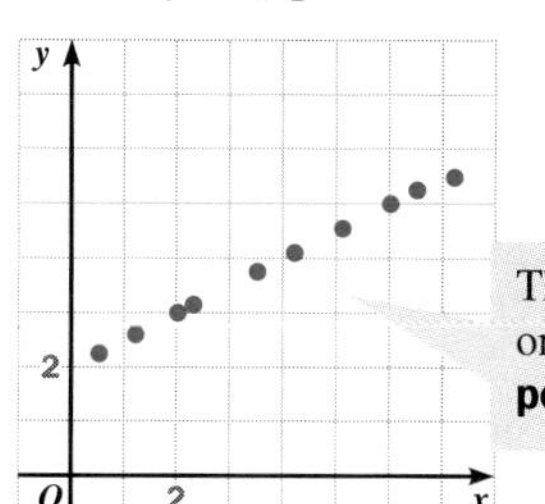

The points lie on a line with **positive slope**.

Perfect Negative Correlation
$r = -1$

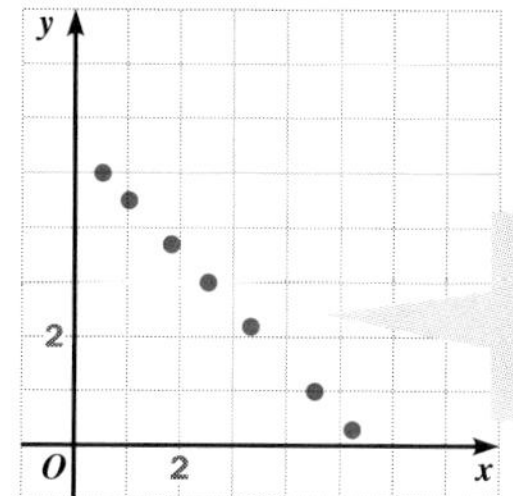

The points lie on a line with **negative slope**.

Below are examples of correlations between -1 and $+1$. If $|r|$ is near 1, the data points almost lie on a line. If $|r|$ is near 0, the data points tend not to lie on any line.

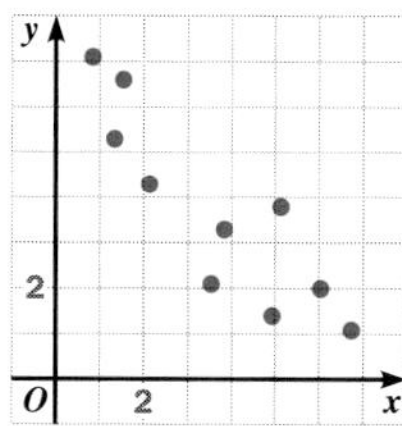

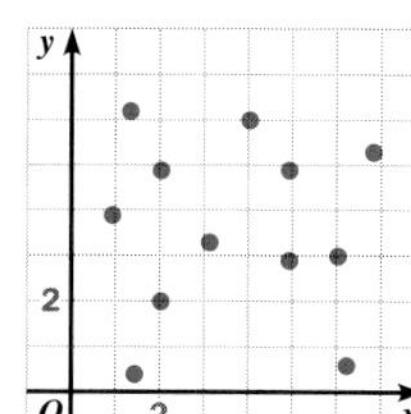

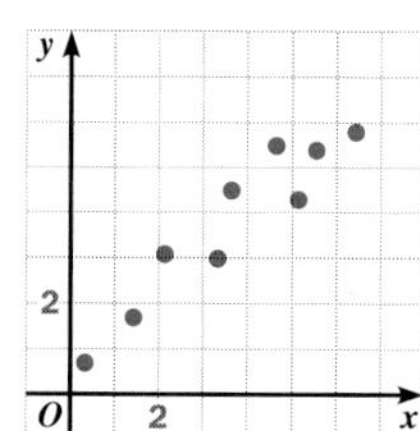

-1	$r \approx -0.88$ strong negative correlation	$r \approx 0$ weak correlation	$r \approx 0.93$ strong positive correlation	$+1$

The formula for the correlation coefficient r is complicated, but graphing calculators and statistical software often find r when they compute the least-squares line.

EXAMPLE 1 Application: Meteorology

Refer to the scatter plots on the previous page.

a. Use a graphing calculator or statistical software to find the correlation coefficient for the scatter plot of highest January temperature versus longitude.

b. Repeat part (a) for the scatter plot of highest January temperature versus latitude.

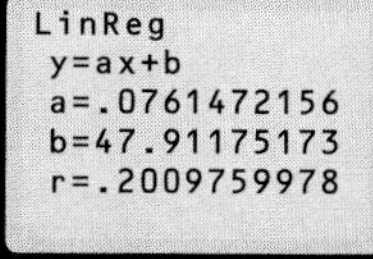
```
LinReg
 y=ax+b
 a=.0761472156
 b=47.91175173
 r=.2009759978
```

```
LinReg
 y=ax+b
 a=-1.457596095
 b=109.4083791
 r=-.93144244
```

SOLUTION

a. Enter the data pairs (longitude, January high) into a graphing calculator or statistical software. The top calculator screen shows that the correlation coefficient r is about 0.20.

b. For the data pairs (latitude, January high), the bottom calculator screen shows that the correlation coefficient r is about -0.93, which is close to -1.

Teach⇔Interact

Section Note

Communication: Drawing
Encourage students to think of drawing an oval shape around the data points for each graph on this page. If the oval includes the points and is long and thin, then there is a strong positive or negative correlation. If the oval looks more like a circle, then there is a weak correlation.

About Example 1

Topic Spiraling: Review
You may need to remind students that the values a and b given by a graphing calculator or statistical software are the slope and the vertical intercept of the least-squares line of fit.

Additional Example 1

Refer to the scatter plots for the data in Exercise 10 on page 71.

a. Use a graphing calculator or statistical software to find the correlation coefficient for the scatter plot of women's heights versus weights.
Enter the women's data pairs (height, weight) into a graphing calculator or statistical software. The calculator screen below shows that the correlation coefficient r is about 0.996, which is close to 1.

```
LinReg
 y=ax+b
 a=3.607142857
 b=-97.92857143
 r=.9960532893
```

b. Repeat part (a) for the scatter plot of men's heights versus weights.
For the men's data pairs (height, weight), the calculator screen below shows that the correlation coefficient r is about 0.999, which is close to 1.

```
LinReg
 y=ax+b
 a=4.464285714
 b=-141.7857143
 r=.9988180986
```

Teach⇔Interact

Think and Communicate

You can use question 5 to emphasize why finding correlations is an important statistical tool. If data have a strong correlation, they can be used to make predictions about similar data.

About Example 2

Second-Language Learners
Some of the text and the tone of the article for this Example might prove challenging to students learning English. You may wish to explain vocabulary such as *dour, prudence, longevity, conscientious,* and *free-wheeling.* You may also wish to explain that the opening words *Score one for the pious voices of prudence* means "those people who have made wise and careful choices in their lives are now reaping the benefits."

Additional Example 2

How Much Broccoli is Needed?

In a recent study of college graduates, researchers found that those who consume broccoli at least 4 times per month were more likely to live longer lives.

The article suggests that eating broccoli causes people to live longer. Explain what might account for this correlation.

There are probably many factors, such as eating a healthy diet, exercising, and leading a healthy lifestyle, that are highly correlated with eating broccoli. These are the factors that actually cause people to live longer.

Checking Key Concepts

Communication: Discussion
When discussing question 4, encourage students who are baseball fans to give their opinions about the correlation between a baseball team's season record and home attendance. This will give them an opportunity to share their knowledge with the class.

THINK AND COMMUNICATE

3. In Example 1, are highest January temperature and longitude positively or negatively correlated? Is this correlation strong or weak?
4. Repeat Question 3 for highest January temperature and latitude.
5. Berlin, Germany, is located at 53°N latitude and 13°E longitude. Make two estimates of the highest January temperature in Berlin, one based on latitude and the other on longitude. Which estimate do you think is more accurate? Why?

Correlation and Causation

You've probably noticed that there is a strong positive correlation between the amount of time you study for a test and the grade you receive. Plenty of study time is one cause of high test scores. Correlation does not always imply causation, however.

The Secret of Long Life? Be Dour and Dependable

Score one for those pious voices of prudence: being cautious and somewhat dour is a key to longevity, according to a 60-year study of more than 1000 men and women.

Those who were conscientious as children were 30 percent less likely to die in any given year of adulthood than their most free-wheeling peers.

"We don't really know why conscientious people live longer—it's not as simple as wearing your sweater when it's cold outside," said Dr. Howard S. Friedman, who did the research.

EXAMPLE 2

The article suggests that being conscientious causes people to live longer. Explain what might account for this correlation.

SOLUTION

There are probably many factors, such as leading a healthy lifestyle, setting long-term life goals, and avoiding risks, that are highly correlated with being conscientious. These are the factors that actually cause people to live longer.

CHECKING KEY CONCEPTS

For Questions 1–3, match each scatter plot to one of the following values of r: 0.6, −0.3, −0.9.

1.
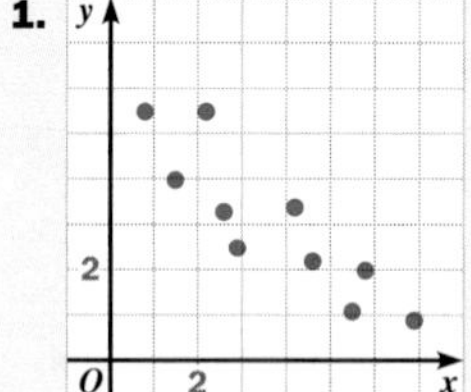

2.
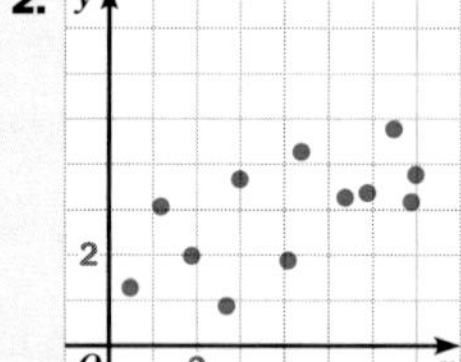

3.
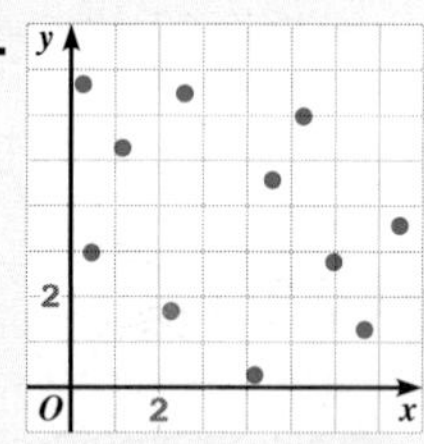

4. Brahim found a strong positive correlation between a baseball team's season record and home attendance. Can you say that a high season record causes a high home attendance? Explain.

3. positively; weak
4. negatively; strong
5. Estimates may vary; about 32° (based on latitude); about 49° (based on longitude); 32°. The correlation for the latitude data is much stronger than that for the longitude data.

Checking Key Concepts

1. −0.9
2. 0.6
3. −0.3
4. No; a strong correlation does not imply causation. While it is possible that a good record causes high home attendance, it is also possible that the high home attendance is affected by good weather, the popularity of one or more of the players, or lack of alternative entertainment possibilities.

Exercises and Applications

1–3. Estimates may vary.
1. about −0.9
2. about 0.2
3. about 1
4. No; since the correlation coefficient is −1, the data points lie on a line with negative slope. The slope of the line containing both (1, 2) and (4, 5) is positive.

2.5 Exercises and Applications

Extra Practice exercises on pages 751–752

For Exercises 1–3, estimate the correlation coefficient for the given scatter plots.

1\.

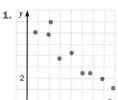

2\. 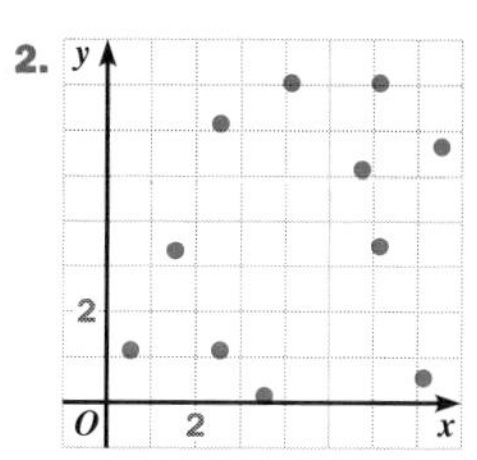

3\. 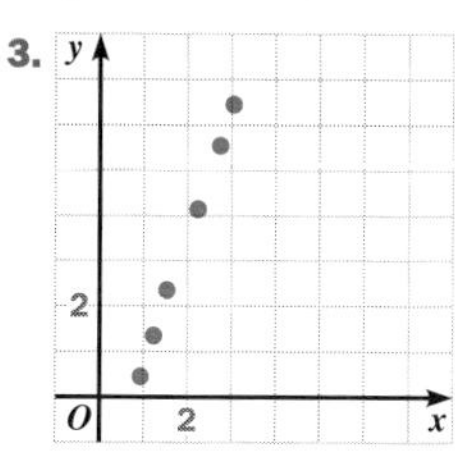

4\. **Writing** Can the points (1, 2) and (4, 5) both lie on a scatter plot with correlation coefficient $r = -1$? Explain your answer.

For Exercises 5–8, tell whether you would expect the correlation between the two quantities to be positive, negative, or about zero.

5\. the height and weight of a person

6\. the age of a car and its value

7\. the shoe size and salary of an adult

8\. the outside temperature and parka sales

9\. **Writing** The correlation coefficient for variables u and v is 0.43, while the correlation coefficient for variables x and y is -0.92. Elvia says that u and v are more strongly correlated than x and y since $0.43 > -0.92$. Write a sentence or two explaining Elvia's mistake.

10\. **Writing** There is a strong positive correlation between the number of mailboxes in a city and the amount of light pollution at night in that city. This certainly does not mean that mailboxes cause light pollution. Explain what might account for the correlation.

11\. **Challenge** If variables x and y are positively correlated and variables y and z are negatively correlated, what can be said of the correlation between x and z? Explain your reasoning.

12\. **BUSINESS** The bar graphs give information about sales figures and profits for a major United States company.

a. **Technology** Use a graphing calculator or statistical software to find the equation of the least-squares line and the correlation coefficient for x = total sales and y = total profits.

b. Use the equation for the least-squares line to predict the total profits when the total sales for the company are 6.0 billion dollars.

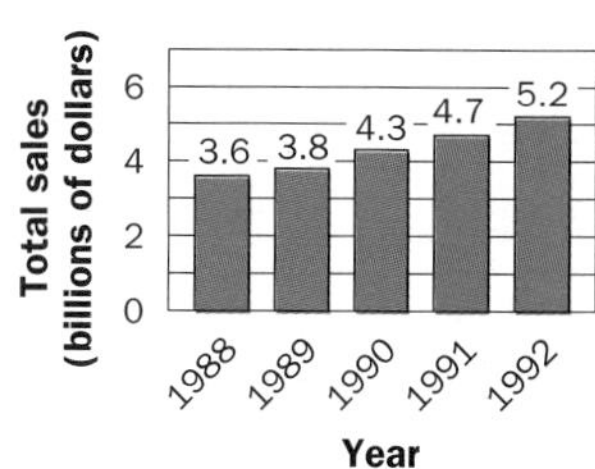

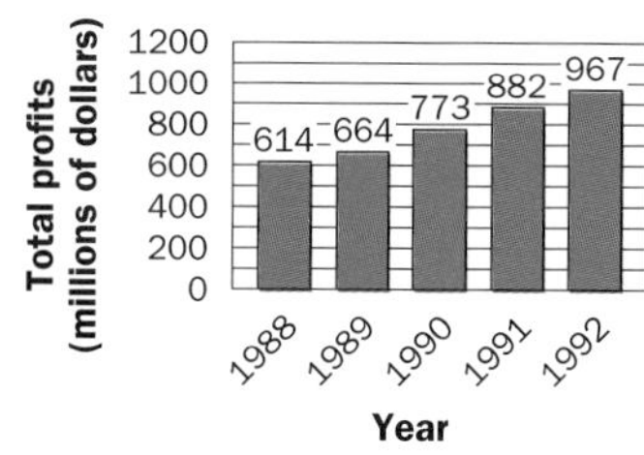

BY THE WAY

Many people consider artificial light a form of pollution because it makes the environment unfit for certain activities, such as astronomy.

5\. positive

6\. negative (unless the car is a collectible antique)

7\. about zero 8. negative

9\. To determine whether a correlation is weak or strong, you consider the absolute value of the correlation coefficient. If $|r| \approx 1$, the correlation is strong. Since $|-0.92| > |0.43|$, x and y are more strongly correlated than u and v.

10\. Answers may vary. An example is given. The more mail boxes a city has, the larger the city is. That is, a city with many mail boxes probably has many homes, offices, factories, shopping malls, parking lights, and other facilities giving off light at night.

11\. x and z are negatively correlated. Answers may vary. An example is given. If x and y are positively correlated, as x increases, y increases. If y and z are negatively correlated, as y increases, z decreases. Then as x increases, z decreases and x and z are negatively correlated.

12\. Answers may vary due to rounding.

a. $y = 224.41x - 189.47$; 0.998

b. about $1157 million

Teach⇔Interact

Closure Question

What steps would you follow to find the correlation coefficient for a data set?

Use a graphing calculator or statistical software to make a scatter plot of the data and do a linear regression. Read the correlation coefficient from the given output.

Apply⇔Assess

Suggested Assignment

Core Course
Exs. 1–8, 10, 12–14, 16–23

Extended Course
Exs. 1–23

Block Schedule
Day 11 Exs. 1–8, 10, 12–14, 16–23

Exercise Notes

Reasoning

Ex. 4 For this exercise, students need to remember that a perfect correlation of –1 means that all the data points lie on the least-squares line and that this line has a negative slope. The unique line containing the two given points has a positive slope.

Second-Language Learners

Exs. 9–11 These writing and challenge activities may prove difficult for some second-language learners. Consider having students work in small groups to brainstorm ideas before beginning each activity.

Common Error

Ex. 9 A common error that students make when thinking about correlation coefficients is to interpret a negative correlation as being weaker than any positive correlation. This exercise will help students understand that the *sign* of the correlation is not related to the strength of the correlation.

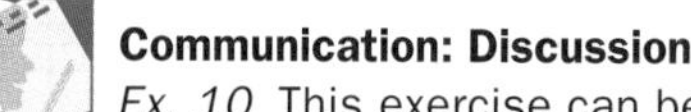

Communication: Discussion

Ex. 10 This exercise can be used to emphasize the fact that a correlation between two sets of data does not always imply causation. Ask students to suggest other examples to help reinforce this idea.

Apply⇔Assess

Exercise Notes

Interdisciplinary Problems

Exs. 13, 14 The science of biology is the science of life in all its forms, including both plant and animal life. Biology is considered to be a life science, whereas geology, for example, is an earth science, and chemistry is a physical science. The first applications of mathematics were to the physical sciences. The use of mathematics in the life sciences, such as biology or medicine, are relatively recent developments.

Reasoning

Ex. 13 Use part (c) of this exercise to help students understand that the value of a correlation coefficient does not depend on which variable is the dependent variable or which variable is the independent variable.

Multicultural Note

Ex. 15 The Mexican presidential elections are held every six years. The voting age is eighteen and women won the right to vote in 1953. In the election held in 1994, Ernesto Zedillo Ponce de León, the candidate of the Institutional Revolutionary Party (PRI), won the presidency. PRI candidates have won every presidential election since the party was formed in 1929. Today, there are six significant political parties in Mexico: Institutional Revolutionary Party, National Action Party (PAN), Democratic Revolutionary Party (PRD), Party of Cárdenas Front for National Reconstruction (PFCRN), Authentic Party of the Mexican Revolution (PARM), and Popular Socialist Party (PPS).

Connection BIOLOGY

Large animals have many advantages over small animals, including being able to move more rapidly.

13. **Technology** The table gives the average body lengths and the highest observed flying speeds of various animals.

a. Use a graphing calculator or statistical software to make a scatter plot of the data. Estimate the correlation coefficient from your scatter plot.

b. Find the correlation coefficient of flying speed versus body length.

c. Find the correlation coefficient of body length versus flying speed. Compare this coefficient with the answer to part (b). Does the value of the correlation coefficient depend on whether body length is the independent or dependent variable?

Species	Length (cm)	Flying speed (m/s)
Horsefly	1.3	6.6
Ruby-throated hummingbird	8.1	11.2
Dragonfly	8.5	10.0
Willow warbler	11	12.0
Flying fish	34	15.6
Whimbrel	41	23.2
Common pintail	56	22.8

14. The two scatter plots show the highest observed running speeds and swimming speeds versus body lengths of various animals.

a. Which scatter plot shows the stronger correlation? Explain what this tells you.

b. What variables besides body length might be correlated with speed?

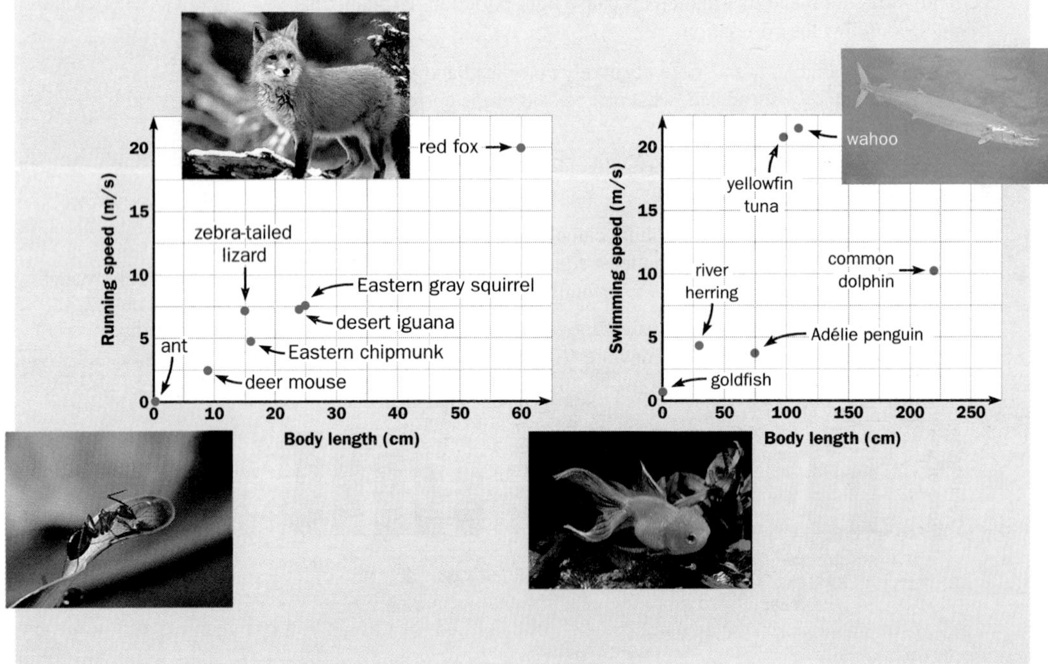

13. a.

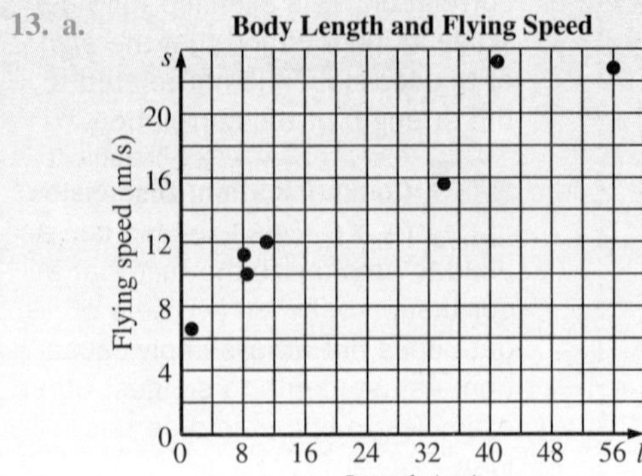

Estimates may vary; about 0.9.

b. 0.95

c. 0.95; They are the same. No.

14. a. running speed; Body length is a better predictor of running speed than of swimming speed.

b. Answers may vary. Examples are given. weight, age, leg or fin length

15. a. 0.9997671913 ≈ 1

b. positive; strong (nearly perfect)

c. There is a no correlation between a state's population and its number of senators; every state has two senators regardless of population. There is a very strong correlation between population and number of representatives; the number of representatives is determined by the population.

16. Answers may vary. The age and the odometer reading should be positively correlated, since, in general, the longer a car is owned, the more miles it is driven.

15. GOVERNMENT The table lists six political parties in Mexico, the percent of votes cast for each party in the 1991 election, and the size of their delegations to the governing body called the *Cámara Federal de Diputados.*

Political party	Percent of votes cast	Total seats
PRI	61.46	320
PAN	17.72	89
PRD	8.26	41
PFCRN	4.36	23
PARM	2.15	15
PPS	1.80	12

a. **Technology** Use a graphing calculator or statistical software to find the correlation coefficient for the data.

b. Is the correlation between percent of votes and delegation size positive or negative? Is it strong or weak?

c. **Research** Read about the United States Senate and House of Representatives in an encyclopedia or some other source. Is there a strong correlation between a state's population and its number of senators? between a state's population and its number of representatives? Explain.

ONGOING ASSESSMENT

16. Cooperative Learning Work in a group to collect data on the ages and odometer readings of cars. Make a scatter plot of the data and estimate the correlation coefficient. Write a brief explanation of why the variables are correlated.

SPIRAL REVIEW

17. The table shows the percent of women marrying who are within various age groups. For example, 37.1% of the women marrying in 1980 were between 20 and 24 years old. *(Section 2.4)*

Year	20–24 years old	25–29 years old	65 years and older
1980	37.1	18.7	1.0
1985	34.4	22.1	1.0
1988	31.5	24.1	1.0

a. For each age group make a scatter plot of percent of women marrying versus year. Then draw a line of fit.

b. Find an equation for each line.

c. Predict the percent of women marrying who will be in each age group in 1990 and 2000.

Write an equation of the line with the given slope and *y*-intercept. *(Section 2.2)*

18. slope = 4, y-intercept = 7

19. slope = 0, y-intercept = −2

20. slope = −3, y-intercept = 1

Solve each linear equation. *(Toolbox, page 784)*

21. $3 + 2t = 5$

22. $\frac{t-7}{3} = 2$

23. $3(t - 2) = 6$

Apply⇔Assess

Exercise Notes

Assessment Note

Ex. 16 This exercise will show how clearly students understand the ideas in this section. Have each group present their findings to the class.

Practice 11 for Section 2.5

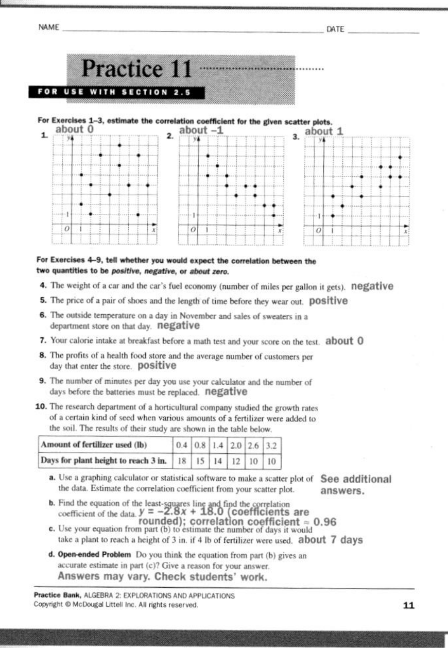
NAME ______ DATE ______

Practice 11

FOR USE WITH SECTION 2.5

For Exercises 1–3, estimate the correlation coefficient for the given scatter plots.

1. about 0
2. about −1
3. about 1

For Exercises 4–9, tell whether you would expect the correlation between the two quantities to be *positive, negative,* or *about zero.*

4. The weight of a car and the car's fuel economy (number of miles per gallon it gets). **negative**
5. The price of a pair of shoes and the length of time before they wear out. **positive**
6. The outside temperature on a day in November and sales of sweaters in a department store on that day. **negative**
7. Your calorie intake at breakfast before a math test and your score on the test. **about 0**
8. The profits of a health food store and the average number of customers per day that enter the store. **positive**
9. The number of minutes per day you use your calculator and the number of days before the batteries must be replaced. **negative**
10. The research department of a horticultural company studied the growth rates of a certain kind of seed when various amounts of a fertilizer were added to the soil. The results of their study are shown in the table below.

Amount of fertilizer used (lb)	0.4	0.8	1.4	2.0	2.6	3.2
Days for plant height to reach 3 in.	18	15	14	12	10	10

a. Use a graphing calculator or statistical software to make a scatter plot of the data. Estimate the correlation coefficient from your scatter plot. **See additional answers.**

b. Find the equation of the least-squares line and find the correlation coefficient of the data. **$y = -2.8x + 18.0$ (coefficients are rounded); correlation coefficient ≈ 0.96**

c. Use your equation from part (b) to estimate the number of days it would take a plant to reach a height of 3 in. if 4 lb of fertilizer were used. **about 7 days**

d. **Open-ended Problem** Do you think the equation from part (b) gives an accurate estimate in part (c)? Give a reason for your answer. **Answers may vary. Check students' work.**

Practice Bank, ALGEBRA 2: EXPLORATIONS AND APPLICATIONS

11

17. a.

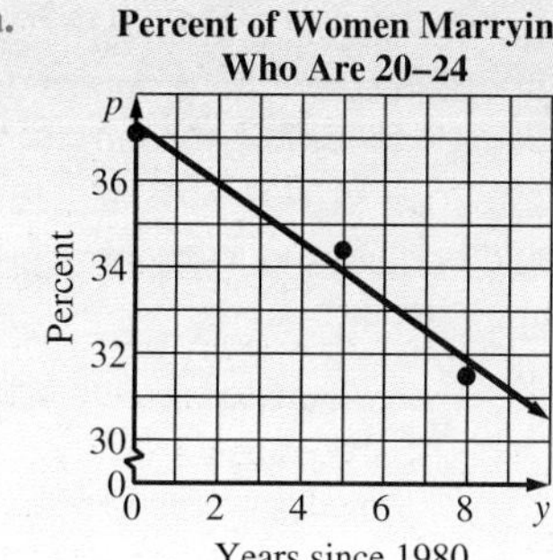

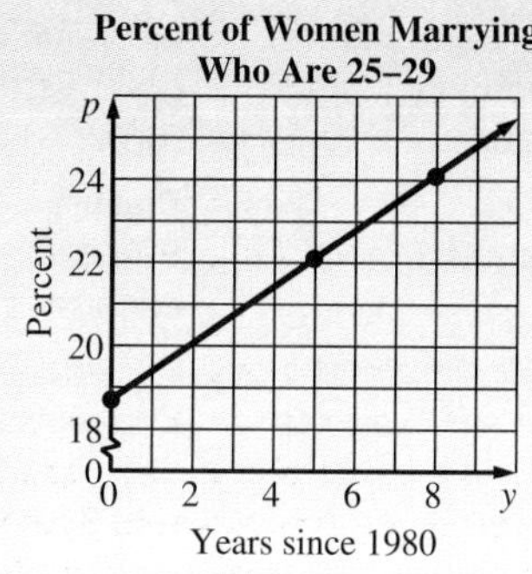

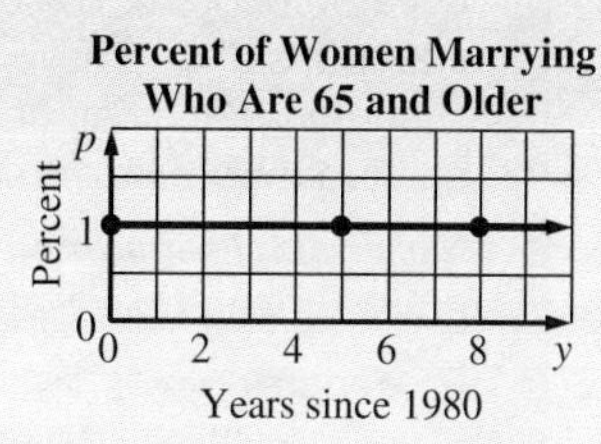

b. Answers may vary. Examples are given. Let y = the number of years since 1980 and p = the percent of women marrying who are the given age.

20–24: $p = -0.68y + 37.30$; 25–29: $p = 0.68y + 18.71$; 65 and older: $p = 1$

c. 20–24: about 30.5% in 1990, about 23.7% in 2000; 25–29: about 25.5% in 1990, about 32.3% in 2000; 65 and older: 1% in 1990, 1% in 2000

18. $y = 4x + 7$ **19.** $y = -2$

20. $y = -3x + 1$ **21.** 1

22. 13 **23.** 4

Plan⇔Support

Objectives

- Write and graph a pair of linear parametric equations.
- Rewrite a pair of parametric equations as a single equation.
- Model situations with parametric equations.

Recommended Pacing

Core and Extended Courses
Section 2.6: 1 day

Block Schedule
Section 2.6: $\frac{1}{2}$ block
(with Section 2.5)

Resource Materials

Lesson Support
Lesson Plan 2.6
Warm-Up Transparency 2.6
Practice Bank: Practice 12
Study Guide: Section 2.6
Exploration Lab Manual:
Diagram Masters 1, 2
Assessment Book: Test 6

Technology
Technology Book:
Spreadsheet Activity 2
Graphing Calculator
McDougal Littell Mathpack
Function Investigator
Internet:
http://www.hmco.com

Warm-Up Exercises

1. If $f(x) = -4x - 6$, find f when $x = -3$. 6
2. If $y = -4x$, find y when $x = 0.5$. -2
3. If $y = 2x + 1$, find y when $x = 3t$. $6t + 1$
4. Solve the equation $x = 3t - 4$ for t. $t = \frac{x+4}{3}$
5. If $f(x) = x - 6$ for $x \geq 0$, find the domain and range of f. domain: $x \geq 0$; range: $y \geq -6$

Section

2.6 Linear Parametric Equations

Learn how to...

- **write and graph a pair of linear parametric equations**
- **rewrite a pair of parametric equations as a single equation**

So you can...

- **model situations, such as the motion of a plane**

When an airplane takes off or lands, it moves both horizontally (forward) and vertically (up or down). To analyze motion in two dimensions, you can use *parametric equations*.

EXPLORATION
COOPERATIVE LEARNING

Landing a Plane

Work with a partner.
You will need:
- a graphing calculator with parametric mode

SET UP Imagine you are piloting a small airplane at **12,000 ft**, preparing to land. Once you begin your descent to the runway, your altitude changes at a rate of **−15 ft/s**. Your horizontal speed is **200 ft/s**.

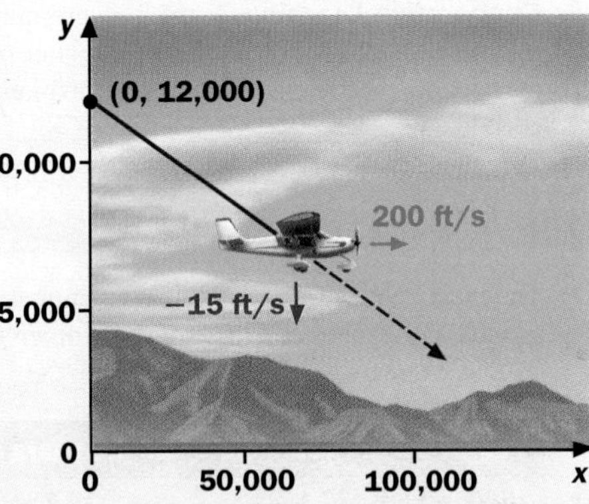

1 The plane is initially at the point (0, 12,000). Complete the equations for x and y.

The plane's horizontal position at time t (in seconds) is given by:

$$x = 0 + \underline{?}\,t$$

The plane's vertical position at time t (in seconds) is given by:

$$y = \underline{?} + \underline{?}\,t$$

For more information about parametric mode, see the *Technology Handbook*, p. 810.

2 Enter your equations from Step 1 into a graphing calculator set in parametric mode. Then graph the equations, adjusting the viewing window and t-values so you can see where the graph crosses the x-axis.

3 Use your calculator's trace feature to determine how long it takes the plane to reach ground level. What is the value of x at this time? At what horizontal distance from the airport should the airplane begin its descent? (*Note:* 1 mi = 5280 ft)

Exploration Note

Purpose
The purpose of this Exploration is to introduce students to the concept of parametric equations and to have them understand that such equations can be used to model and analyze motion in two dimensions.

Materials/Preparation
graphing calculator with parametric mode

Procedure
Students initially determine the parametric equations for the descent of a small airplane. They then graph the equations in parametric mode on a calculator after determining the appropriate values of Tmin and Tmax for the viewing window. Encourage students to estimate Tmax by solving the first parametric equation for t and using 150,000 for a possible value of x. After seeing a complete graph, students then trace the graph to determine how long it takes the plane to land.

Closure
Students should understand that parametric equations can be used to model situations that involve motion in two dimensions.

Explorations Lab Manual
See the Manual for more commentary on this Exploration.

In the Exploration, the airplane's x- and y-coordinates were not related through an equation of the form $y = f(x)$. Instead, you expressed x and y as separate functions of t:

$$x = g(t) \qquad y = h(t)$$

The variable t is called a **parameter**.

The equations $x = g(t)$ and $y = h(t)$ are called **parametric equations**.

EXAMPLE 1 Application: Physics

Tom lives directly west across a river from his grandfather's house. The river flows south at 3 mi/h and is 0.5 mi wide. If Tom tries to row east across the river at 2 mi/h, how far from his grandfather's house will he land?

SOLUTION

Let x = Tom's east-west position and y = Tom's north-south position, in miles.

Step 1 Draw a sketch and show a coordinate system with the origin at Tom's house.

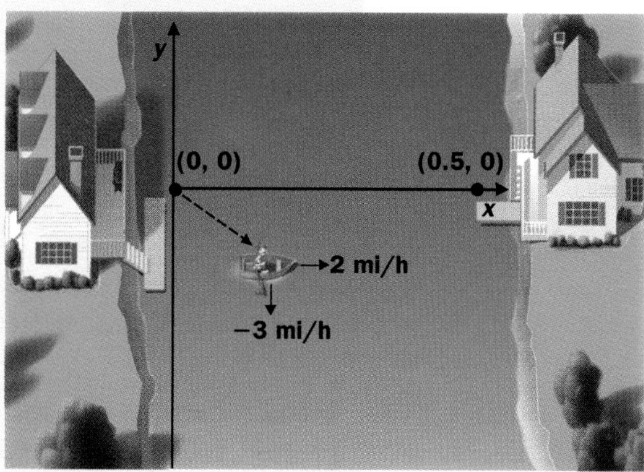

Step 2 Write parametric equations for x and y.

$$x = 2t \qquad y = -3t$$

Step 3 Find t when $x = 0.5$.

$0.5 = 2t$, so $t = 0.25$

Step 4 Find y when $t = 0.25$.

$y = -3(0.25)$, so $y = -0.75$

Tom will land 0.75 mi downstream from his grandfather's house.

EXAMPLE 2

Graph the parametric equations from Example 1 for $0 \le t \le 0.25$.

SOLUTION

Use a table to plot a few points.

t	$x = 2t$	$y = -3t$
0	0	0
0.1	0.2	−0.3
0.2	0.4	−0.6
0.25	0.5	−0.75

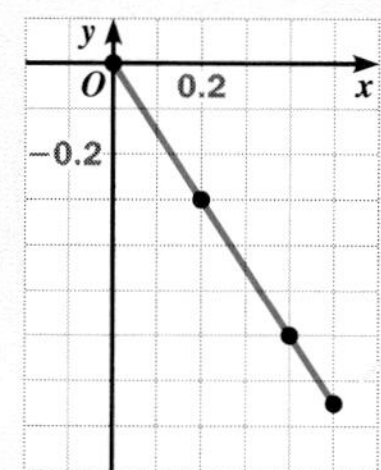

For $0 \le t \le 0.25$, $0 \le x \le 0.5$ and $-0.75 \le y \le 0$.

Teach⇔Interact

Additional Example 1

An airplane at 15,000 ft has a horizontal speed of 250 mi/h. If the plane is descending at a rate of 12 mi/h for 0.06 h, what will be its altitude after it has traveled 10 mi as measured along the ground?

Let x = the plane's east-west position and y = the plane's north-south position, in miles.

Step 1 Show a coordinate system with 15,000 ft ≈ 2.84 mi as the vertical intercept.

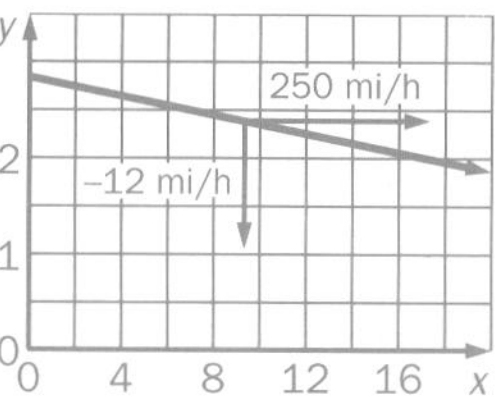

Step 2 Write parametric equations.

$x = 250t \qquad y = 2.84 - 12t$

Step 3 Find t when $x = 10$.

$10 = 250t$, so $t = \frac{1}{25}$.

Step 4 Find y when $t = \frac{1}{25}$.

$y = 2.84 - 12\left(\frac{1}{25}\right)$, so $y = 2.36$.

The altitude of the plane will be about 2.36 mi, or 12,461 ft.

About Example 2

Alternate Approach

An alternative way to graph the equations in Example 2 is to use a graphing calculator in parametric mode. Enter the equations X1T=2T and Y1T=−3T. Set Tmin = 0 and Tmax = 0.25 in the WINDOW and press [GRAPH]. Experiment with different values of Tstep, Xmin, Xmax, Ymin, and Ymax to get the desired graph.

Additional Example 2

Graph the parametric equations from Additional Example 1 for $0 \le t \le 0.06$.

t	$x = 250t$	$y = 2.84 - 12t$
0	0	2.84
0.02	5	2.6
0.04	10	2.36
0.06	15	2.12

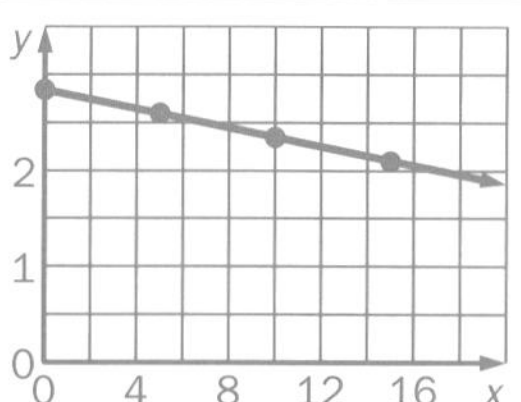

ANSWERS Section 2.6

Exploration

1. 200; 12,000; −15
2. 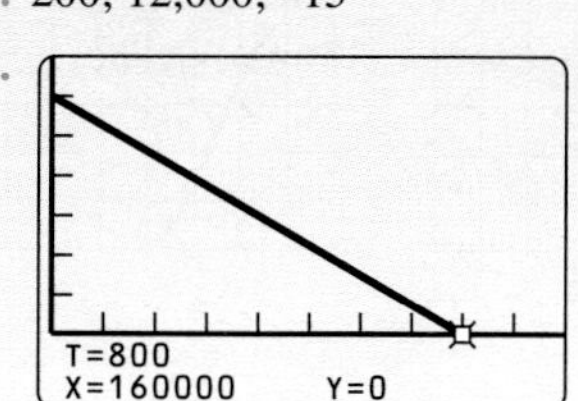

3. 800 s; 160,000; about 30.3 mi

Teach⟺Interact

Think and Communicate

Students' answers to question 3 should be a good indicator as to how well they understand what parametric equations represent.

Additional Example 3

Using the parametric equations from Additional Example 1, express y as a function of x.

Your goal is to eliminate the parameter t. To do this, first solve $x = 250t$ for t.

$$250t = x$$
$$t = \frac{1}{250}x$$

Then substitute $\frac{1}{250}x$ for t in $y = 2.84 - 12t$.

$$y = 2.84 - 12t$$
$$= 2.84 - 12\left(\frac{1}{250}x\right)$$
$$= 2.84 - 0.048x$$

For the equation $y = 2.84 - 0.048x$, the restriction on x is $0 \le x \le 2.12$ because $0 \le t \le 0.06$.

Think and Communicate

For question 5, students may need to make several attempts before finding a pair of parametric equations to represent this function. Because of this, and the fact that there is more than one possible answer, you may wish to have students work with a partner. After the groups have answered the question, the various answers could be presented and discussed in class.

Checking Key Concepts

Using Technology Students may check their answers to questions 1–3 by graphing. See the Alternate Approach note on page 79.

Closure Question

Describe how you can analyze the linear motion of an object in two dimensions.

Parametric equations can be written for x and y in terms of time t. A graphing calculator can be used to graph the parametric equations. The graph would show the motion of the object.

THINK AND COMMUNICATE

1. In Example 1, what would the parametric equations be if the origin were located at Tom's grandfather's house?
2. In Example 2, what are the domain and range of the function $x = 2t$? of the function $y = -3t$?
3. What does the graph in Example 2 represent?

EXAMPLE 3

Using the parametric equations from Example 1, express y as a function of x.

SOLUTION

Your goal is to eliminate the parameter t. To do this, first solve $x = 2t$ for t:

$$2t = x$$
$$t = 0.5x$$

Then substitute $0.5x$ for t in $y = -3t$:

$$y = -3t$$
$$= -3(0.5x)$$
$$= -1.5x$$

Bear in mind that a restriction on t creates a restriction on x. For the equation $y = -1.5x$, the restriction on x is $0 \le x \le 0.5$ because $0 \le t \le 0.25$.

THINK AND COMMUNICATE

4. What information do the parametric equations give you that is missing from the equation you found in Example 3?
5. Consider the graph of the function $y = 4 + x$ for $x \ge -1$. Find a pair of parametric equations having the same graph. Is more than one answer possible?

CHECKING KEY CONCEPTS

For each pair of parametric equations, find x and y when $t = 2$.

1. $x = 2t - 1$, $y = -5t - 2$, $t \ge 0$
2. $x = 0.5t$, $y = t + 1$, $0 \le t \le 12$
3. $x = 12 - t$, $y = 14$, $t \ge 5$

4–6. Express y as a function of x using the equations in Questions 1–3. State any restriction on x.

Think and Communicate

1. $x = 2t - 0.5$; $y = -3t$
2. $0 \le t \le 0.25, 0 \le x \le 0.5$; $0 \le t \le 0.25, -0.75 \le y \le 0$
3. the path of Tom's boat
4. the horizontal and vertical positions of the boat at time t
5. More than one answer is possible. An example is given. $x = t - 1$, $y = t + 3$, for $t \ge 0$

Checking Key Concepts

1. 3; −12
2. 1; 3
3. The equations are not defined for $t < 5$.
4. $y = -\frac{5}{2}x - \frac{9}{2}$; $x \ge -1$
5. $y = 2x + 1$; $0 \le x \le 6$
6. $y = 14$; $x \le 7$

Exercises and Applications

1.

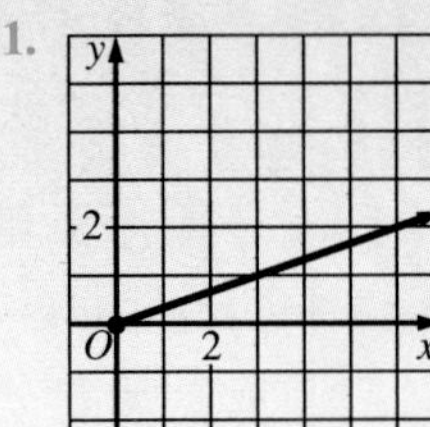

2.

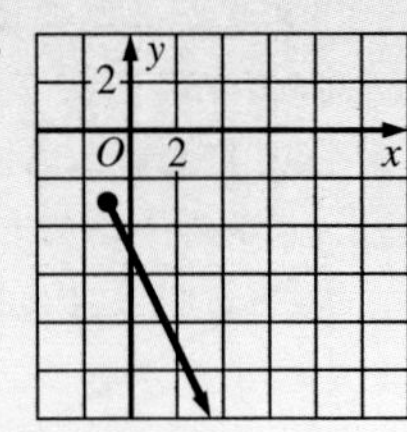

2.6 Exercises and Applications

Extra Practice exercises on page 752

Graph each pair of parametric equations for the given restriction on *t*.

1. $x = 3t$
$y = t$
$t \geq 0$

2. $x = -1 - t$
$y = -3 + 2t$
$t \leq 0$

3. $x = -5 + 5t$
$y = 4t$
$0 \leq t \leq 1$

4. $x = 1 + t$
$y = -2 + t$
no restriction on t

5–8. Express y as a function of x using the equations in Exercises 1–4. State any restriction on x.

9. Challenge Find a pair of parametric equations to describe the graph of $y = 2x - 3$ for $x \geq 2$.

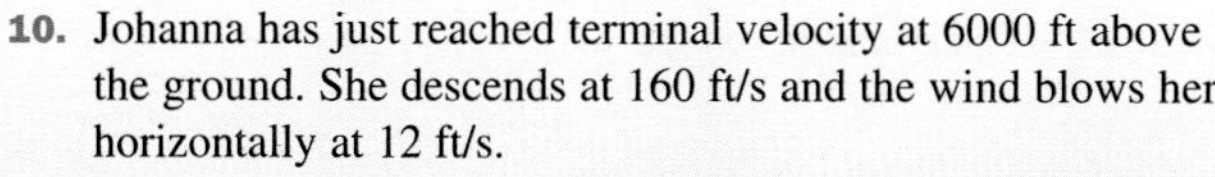

Connection RECREATION

Immediately after jumping from a plane, a parachutist falls with increasing speed. After about 9 s, however, the speed of a parachutist in a flat stable position levels off at about 110 mi/h, or 160 ft/s. This speed is called *terminal velocity*. The parachutist is in *free fall* until the parachute is opened.

10. Johanna has just reached terminal velocity at 6000 ft above the ground. She descends at 160 ft/s and the wind blows her horizontally at 12 ft/s.

a. Let $t = 0$ correspond to the time when she reaches terminal velocity, 6000 ft above the ground. Write a pair of parametric equations to represent her motion for $0 \leq t < 20$.

b. Sketch a graph of her descent in free fall.

11. Johanna deploys her parachute at time $t = 20$ s, when she is 2800 ft above the ground. Almost instantly her rate of descent changes to 10 ft/s.

a. When does she land?

b. She is still traveling horizontally at 12 ft/s. Write a pair of parametric equations to model her descent with the open parachute. What are the restrictions on t?

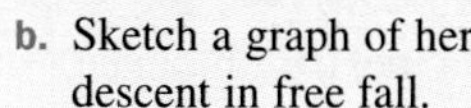
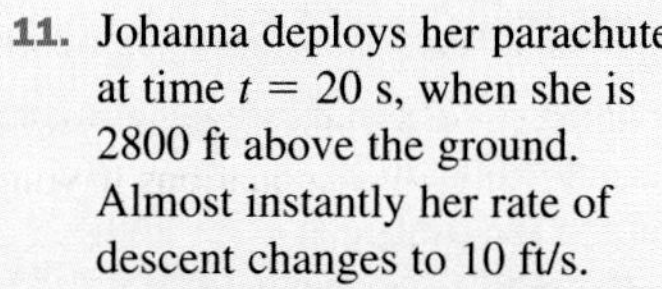

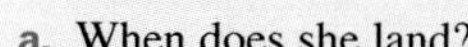

Apply⇔Assess

Suggested Assignment

Core Course
Exs. 1–8, 10–12, 15–20, AYP

Extended Course
Exs. 1–20, AYP

Block Schedule
Day 11 Exs. 1–8, 10–12, 15–20, AYP

Exercise Notes

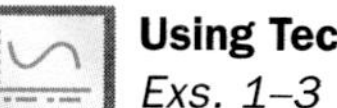

Using Technology
Exs. 1–3 Encourage students to use the restrictions on *t* to determine appropriate values of Tmin and Tmax when graphing in parametric mode. Explain that the Tstep determines how many points will actually be plotted by the graphing calculator.

Using Technology
Ex. 9 Graphs such as the one in this exercise can be displayed on a graphing calculator that is operating in function mode. For example, on the TI-82 or TI-81, enter the function Y1=(2X–3)/(X≥2) on the Y= list. The calculator evaluates (X≥2) as 0 for all values of X less than 2. Since division by 0 is undefined, the calculator will not display points for such values of X. But for values of X greater than or equal to 2, the calculator will evaluate (X≥2) as 1 and display a point on the graph.

Multicultural Note
Exs. 10, 11 The Chinese invented parachutes over two thousand years ago. There are references to the use of parachutes in some Chinese legends, where parachutes take the form of conical straw hats and umbrellas. Chinese and Thai acrobats used umbrellas to land safely after jumping from great heights. In the late 17th century, the French ambassador to Thailand wrote an account of having watched such acrobats perform high-air feats with umbrellas, and the use of parachutes spread to Europe. These accounts resulted in a Frenchman named A.J. Garnerin successfully parachuting from a hot-air balloon at the end of the 18th century.

3.
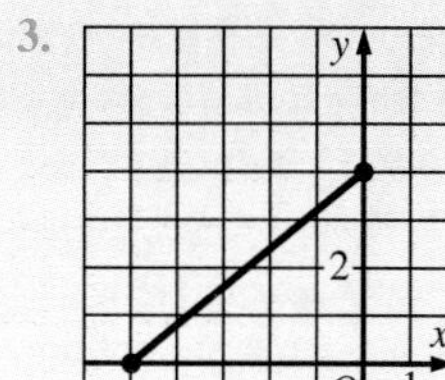

4.

5. $y = \frac{1}{3}x$; $x \geq 0$

6. $y = -2x - 5$; $x \geq -1$

7. $y = \frac{4}{5}x + 4$; $-5 \leq x \leq 0$

8. $y = x - 3$; no restriction on x

9. Answers may vary. An example is given. $x = t + 2$, $y = 2t + 1$, $t \geq 0$

10. a. $x = 12t$, $y = 6000 - 160t$; $0 \leq t < 20$

b.
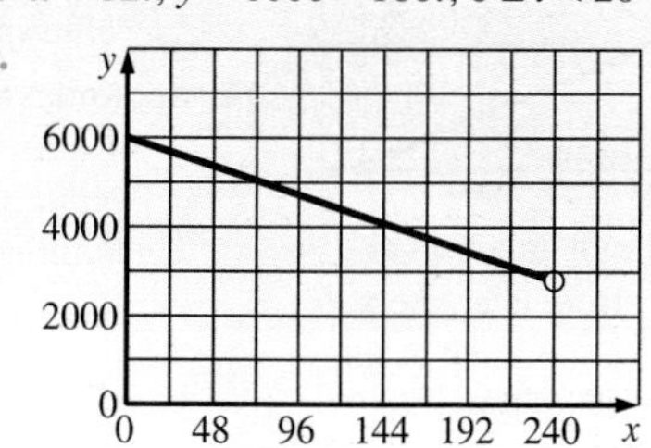

11. a. 280 s after deploying the parachute

b. $x = 12t$, $y = 3000 - 10t$, $20 \leq t \leq 300$

Apply⇔Assess

Exercise Notes

Second-Language Learners
Ex. 12 For part (d), students learning English may benefit from working cooperatively with English-fluent students to write paragraphs about mathematical situations.

Using Technology
Ex. 14 For part (b), it is important that students set Tmax = 3 for the viewing window. A higher value may show Bao intersecting the other two dancers. Students should also set Tmin = 0 to make the graph realistic.

12. An object moves according to the parametric equations $x = 3t$ and $y = 4t$ where t is in seconds and x and y are in meters.

a. Draw a graph of the object's path for $0 \le t \le 1$.

b. What is the object's speed along its path in meters per second? (*Hint:* Find the distance traveled and divide it by the travel time.)

c. Another object moves according to the parametric equations $x = 6t$ and $y = 8t$ for $0 \le t \le 0.5$. What is the relationship between the path of this object and the path of the first object? What is the relationship between their speeds?

d. **Writing** Write a paragraph about a situation that could be modeled by these parametric equations.

Connection PERFORMING ARTS

When you direct a performance, it is important to know where the performers will be at different times.

13. The drama club is putting on a play set in a haunted house. The script calls for two characters to back into each other. Glen starts at (0, 15), and his path is described by the equations $x = 2.5t$ and $y = 15 - t$, with t in seconds and x and y in feet. Neepa starts at (40, 10), and her path is described by the equations $x = 40 - 3t$ and $y = 10$.

a. **Technology** Use a graphing calculator or a table of values to plot Glen and Neepa's paths.

b. Will they bump into each other? How do you know?

c. Could you have answered part (b) if you were given Glen and Neepa's paths with y expressed as a function of x? Explain.

14. Yuki is choreographing a modern dance performance on the same stage as the play in Exercise 13. She wants three dancers to run past each other at the same instant. The dancers' movements for $0 \le t \le 3$ are described by the equations in the table below, with x and y in feet and t in seconds.

Anita	$x = 5t$	$y = 10 - \frac{5}{3}t$
Bao	$x = 27 - 4t$	$y = 8$
Claire	$x = 5t$	$y = 16 - \frac{5}{3}t$

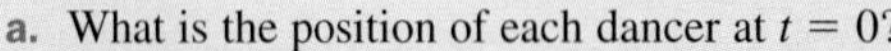

a. What is the position of each dancer at $t = 0$?

b. **Technology** Enter the equations for the three dancers into a graphing calculator set in parametric mode. Do the dancers pass each other at the same time? (*Hint:* To pass, they should be at different points on the same line.)

12. a.

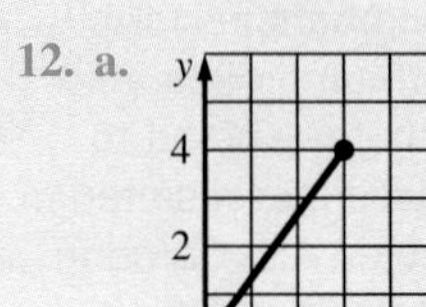

b. 5 m/s

c. The paths are the same; the second object is moving twice as fast as the first object.

d. Answers may vary.

13. a.

b. No; the graphs intersect at (12.5, 10). Both characters cross that spot but not at the same time. Glen is at the spot after 5 s, Neepa after about 9 s.

c. No; with no indication of time, you could not tell if the two performers were ever in the same spot at the same time.

14. a. Anita: (0, 10); Bao: (27, 8); Claire: (0, 16)

b.

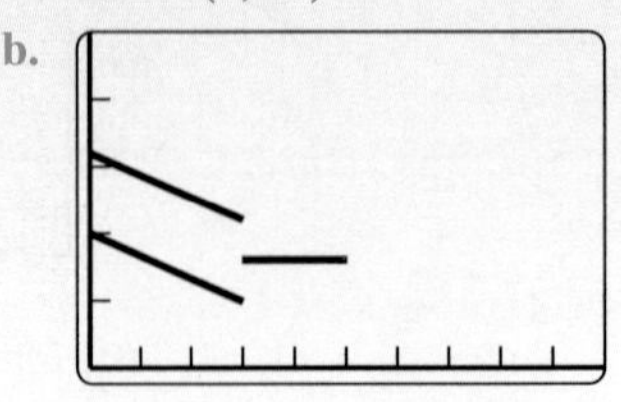

Yes; at $t = 3$.

15. Answers may vary. One possible example involves several people traveling by different methods and meeting at a given time and place.

16. Answers may vary. An example is given. Correlation is a measure of how two variables change in relation to each other. If two variables are strongly positively correlated, it means that one increases as the other does. That does not mean that the change in one causes the change in the other. Causation refers to one thing causing another to

ONGOING ASSESSMENT

15. Open-ended Problem Think of a situation you can model with two or three objects following paths defined by parametric equations.

a. Write up your problem. Some questions to consider are: What are the restrictions on each object's movement? Where are the objects initially? Do you want the objects to meet or to miss each other?

b. Writing Write a paragraph or short story about what happens in the problem you wrote.

SPIRAL REVIEW

16. Explain the difference between correlation and causation. *(Section 2.5)*

17. Sales of computer books at The Science Bookshop are growing at an average of 15% a year. In 1990 the bookshop sold 560 computer books. *(Section 1.2)*

a. Write an equation to model the growth in computer book sales.

b. How many computer books did the bookshop sell in 1994?

Write a point-slope equation of the line through each pair of points. *(Section 2.3)*

18. (13, 8) and (−1, 1) **19.** (−3, 7) and (2, −5) **20.** (−3, 2) and (1, 4)

ASSESS YOUR PROGRESS

VOCABULARY

least-squares line (p. 67)
correlation coefficient (p. 72)
positive correlation (p. 72)
negative correlation (p. 72)
parameter (p. 79)
parametric equations (p. 79)

Exercises 1 and 2 refer to the table, which shows the number of English-language newspapers published in the United States.

Year	Newspapers
1980	1745
1982	1711
1984	1688
1986	1657
1988	1642
1990	1611

1. Show that the data have a linear relationship by making a scatter plot and drawing a line of fit. *(Section 2.4)*

2. Technology Use a graphing calculator or statistical software to find the correlation coefficient for the data. Is the correlation strong or weak? positive or negative? *(Section 2.5)*

3. a. Graph the parametric equations $x = 2 + t$ and $y = 1 - t$ for $t \geq 0$.

b. Express y as a function of x and give the function's domain. *(Section 2.6)*

4. Journal Find a real-life example of correlation and explain whether it involves causation.

Apply⇔Assess

Exercise Notes

Assessment Note
Ex. 15 This exercise gives students an opportunity to display their understanding of parametric equations in an original and creative way. A discussion of some responses to this activity can be both interesting and enlightening to the entire class.

Assess Your Progress

Journal Entry
For Ex. 4, you may wish to have students discuss their answers in small groups. Have them explain to each other how they know if their example involves causation.

Progress Check 2.4–2.6

See page 87.

Practice 12 for Section 2.6

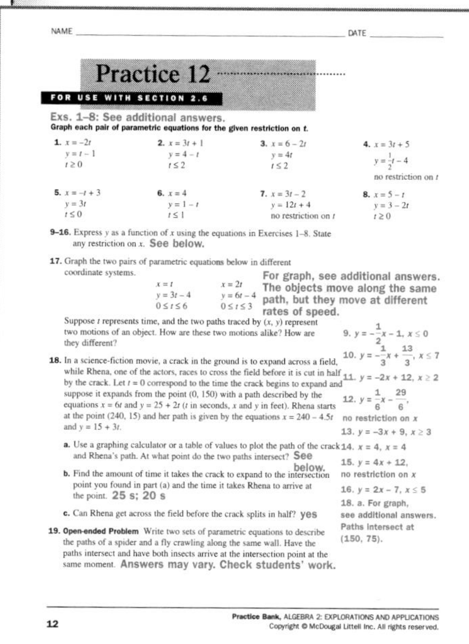

NAME ______________________ DATE __________

Practice 12

FOR USE WITH SECTION 2.6

Exs. 1–8: See additional answers.
Graph each pair of parametric equations for the given restriction on t.

1. $x = -2t$, $y = t - 1$, $t \geq 0$
2. $x = 3t + 1$, $y = 4 - t$, $t \leq 2$
3. $x = 6 - 2t$, $y = 4t$, $t \leq 2$
4. $x = 3t + 5$, $y = \frac{1}{2}t - 4$, no restriction on t
5. $x = -t + 3$, $y = 3t$, $t \leq 0$
6. $x = 4$, $y = 1 - t$, $t \leq 1$
7. $x = 3t - 2$, $y = 12t + 4$, no restriction on t
8. $x = 5 - t$, $y = 3 - 2t$, $t \geq 0$

9–16. Express y as a function of x using the equations in Exercises 1–8. State any restriction on x. See below.

17. Graph the two pairs of parametric equations below in different coordinate systems.

$x = t$, $y = 3t - 4$, $0 \leq t \leq 6$ $x = 2t$, $y = 6t - 4$, $0 \leq t \leq 3$

For graph, see additional answers. The objects move along the same path, but they move at different rates of speed.

Suppose t represents time, and the two paths traced by (x, y) represent two motions of an object. How are these two motions alike? How are they different?

18. In a science-fiction movie, a crack in the ground is to expand across a field, while Rhena, one of the actors, races to cross the field before it is cut in half by the crack. Let $t = 0$ correspond to the time the crack begins to expand and suppose it expands from the point (0, 150) with a path described by the equations $x = 6t$ and $y = 25 + 2t$ (t in seconds, x and y in feet). Rhena starts at the point (240, 15) and her path is given by the equations $x = 240 - 4.5t$ and $y = 15 + 3t$.

a. Use a graphing calculator or a table of values to plot the path of the crack and Rhena's path. At what point do the two paths intersect? See below.

b. Find the amount of time it takes the crack to expand to the intersection point you found in part (a) and the time it takes Rhena to arrive at the point. 25 s; 20 s

c. Can Rhena get across the field before the crack splits in half? yes

19. Open-ended Problem Write two sets of parametric equations to describe the paths of a spider and a fly crawling along the same wall. Have the paths intersect and have both insects arrive at the intersection point at the same moment. Answers may vary. Check students' work.

9. $y = -\frac{1}{2}x - 1$, $x \leq 0$
10. $y = -\frac{1}{3}x + \frac{13}{3}$, $x \leq 7$
11. $y = -2x + 12$, $x \geq 2$
12. $y = \frac{1}{6}x - \frac{29}{6}$, no restriction on x
13. $y = -3x + 9$, $x \geq 3$
14. $x = 4$, $x = 4$
15. $y = 4x + 12$, no restriction on x
16. $y = 2x - 7$, $x \leq 5$
18. a. For graph, see additional answers. Paths intersect at (150, 75).

12 Practice Bank, ALGEBRA 2: EXPLORATIONS AND APPLICATIONS

happen. For example, since people tend to buy snow shovels in winter and winter is also flu season, if you were to compare snow shovel sales and flu cases, the data might show a positive correlation. But snow shovels do not cause flu.

17. a. Let x = the number of years since 1990 and y = the number of computer books sold; $y = 560(1.15)^x$.

b. about 979 books

18. $y = 8 + \frac{1}{2}(x - 13)$ or $y = 1 + \frac{1}{2}(x + 1)$

19. $y = 7 - \frac{12}{5}(x + 3)$ or $y = -5 - \frac{12}{5}(x - 2)$

20. $y = 2 + \frac{1}{2}(x + 3)$ or $y = 4 + \frac{1}{2}(x - 1)$

Assess Your Progress

1. Lines of fit may vary.

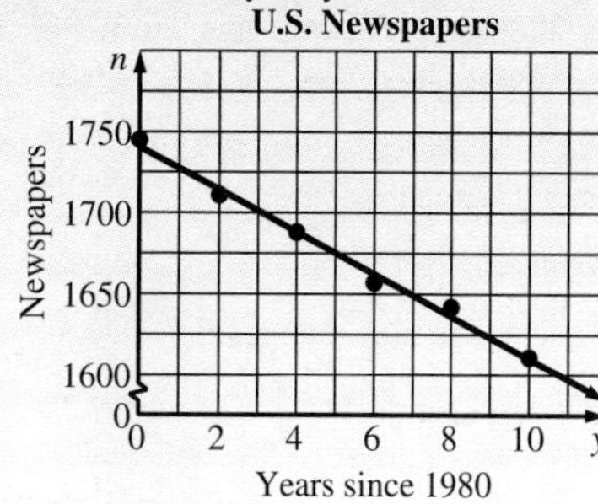

2. −0.996; strong; negative

3. a.

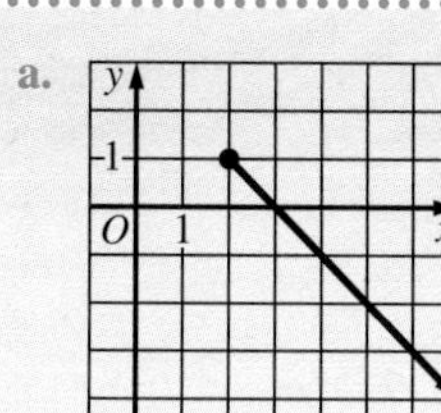

b. $y = -x + 3; x \geq 2$

4. Answers may vary. An example is given. An adult's weight and the number of calories needed daily for proper nutrition are positively correlated. In this case, there is causation.

Project Notes

Mathematical Goals

- Conduct an experiment and record data in a table.
- Draw a scatter plot of the data.
- Write a linear equation based upon the data and use it to predict values beyond the data.

Planning

Materials

- strong rubber band
- 2 large paper clips
- a paper cup
- 20 marbles
- a ruler (preferably transparent)
- graph paper
- pencil

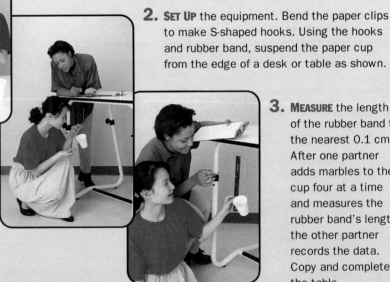

Project Teams

Students can choose a partner and decide how to conduct the experiment together. After collecting their data, they can share the work of analyzing the data and writing the report.

Guiding Students' Work

Before students conduct the experiment, it may be helpful to work with the whole class on a list of goals for the experiment, and then have each pair of students discuss the reasoning for the goals in their report. Students should share responsibilities for the project and contribute to each part of it.

Second-Language Learners

Students learning English are likely to benefit from working cooperatively with English-fluent students when writing their reports.

PORTFOLIO PROJECT

Stretching a Rubber Band

When you put fruit on a grocery store scale, the scale's dial tells you the weight. If you could look inside the scale, you would see a spring that controls the movement of the dial. This spring stretches as you add more fruit.

A rubber band behaves in a similar way: the harder you pull on it, the longer it stretches. These situations both involve a pulling force that increases length. Have you ever wondered how the force used to stretch a spring or a rubber band is related to its length?

PROJECT GOAL **Perform an experiment to study the relationship between pulling force and rubber band length.**

Conducting an Experiment

Work with a partner. Here are a few suggestions for carrying out the experiment.

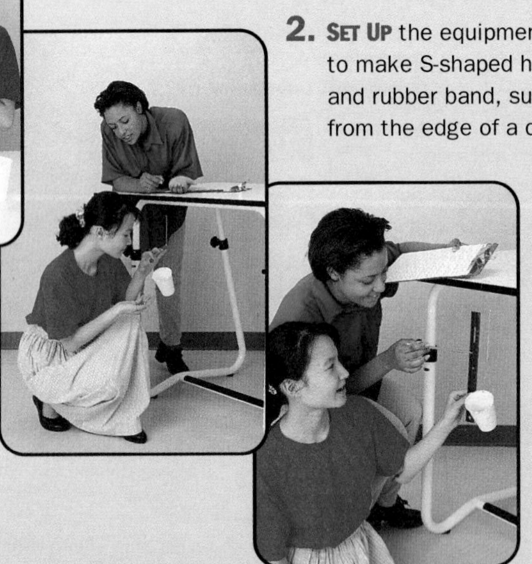

1. **GATHER** a rubber band, 2 large paper clips, a paper cup, 20 marbles, a ruler (preferably transparent), graph paper, and a pencil.
2. **SET UP** the equipment. Bend the paper clips to make S-shaped hooks. Using the hooks and rubber band, suspend the paper cup from the edge of a desk or table as shown.
3. **MEASURE** the length of the rubber band to the nearest 0.1 cm. After one partner adds marbles to the cup four at a time and measures the rubber band's length, the other partner records the data. Copy and complete the table.

m = number of marbles in cup	0	4	8	12	16
l = length of rubber band (cm)	?	?	?	?	?

Analyzing the Data

- **DRAW** a scatter plot of the data pairs (m, l) in a coordinate plane. The plotted points should almost line up. Use the ruler or a graphing calculator to obtain a line of fit.
- **WRITE** an equation for your line. Use the equation to find l when $m = 20$.
- **CHECK** the prediction by putting 20 marbles in the cup and measuring the rubber band length. How did your predicted value for $m = 20$ compare to your experimental value?

Writing a Report

Write a report about your experiment. Include a paragraph on each of these points:

- goals of the experiment
- descriptions of your procedure
- tables and graphs of your data
- conclusions based on your results

You may want to extend your report by investigating and reporting on some other ideas.

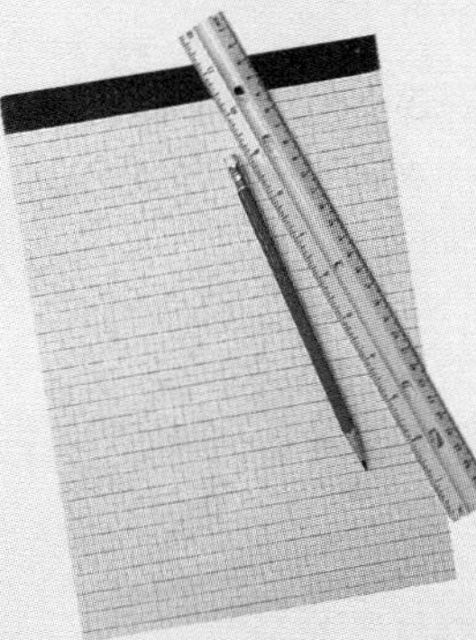

- Use your graph from the project to sketch a predicted graph for this experiment: You start with 20 marbles in the cup and remove four at a time. Then conduct the experiment to check your graph.
- Repeat the project using various types of rubber bands. Do some rubber bands stretch more than others? What factors do you think contribute to the stretchability of the rubber bands?
- Research Hooke's law from a physics book and explain how it applies to this project.

Self-Assessment

Were you satisfied with your prediction of rubber band length based on your line of fit? If so, what factors influenced your success? If not, how can you adjust your methods to improve your results?

Project Notes

Guiding Students' Work

Rubric for Chapter Project

4 Students conduct the experiment and accurately record the data. They draw the scatter plot correctly and all labels are made on the graph. They write an accurate equation to model the data and correctly calculate an estimate for the length of the rubber band with 20 marbles in the cup. The report is well written and contains a clear analysis of the four points listed in the text. The report also indicates an understanding of the mathematical ideas of the project. Students complete the extension ideas and investigate the three possibilities listed.

3 Students complete the experiment and record the data accurately. Their graphs are drawn correctly, and the calculations and equations made are correct for the data collected. The written report touches on the four points listed in the text and reflects an understanding of the mathematical ideas, but students did not extend the project to check their work or look at other factors.

2 Students complete the experiment and record the data in the table. A mistake is made in the calculation of the equation for the line or in the estimate for 20 marbles. The report is written and is complete but does not convey a thorough understanding of the mathematical ideas of the project. Students did not extend the project using the ideas listed in the text.

1 Students did not complete the experiment or complete only part of it and then guess the results for the remaining parts. Students draw the graphs incorrectly or not at all, and miscalculate the equation of the line. A report is written but is incomplete and does not convey any understanding of the concepts of the project. Students should be encouraged to speak with the teacher as soon as possible to review their work and to make a new start on the project.

Chapter Support

Course Guide: Chapter 2
Lesson Plans: Chapter 2
Practice Bank: Cumulative Practice 13
Study Guide: Chapter 2 Review
Assessment Book: Chapter Tests 7 and 8; Chapter 2 Alternative Assessment
Function Investigator with Matrix Analyzer Activity Book: Function Investigator Activity 8
Technology Book: Calculator Based Lab 1
Portfolio Project Book
Preparing for College Entrance Tests
Professional Handbook

Progress Check 2.1–2.3

For each equation, tell whether y varies directly with x. If so, graph the equation. *(Section 2.1)*

1. $y = 3x$ Yes. **2.** $y = \frac{5}{3x}$ No.

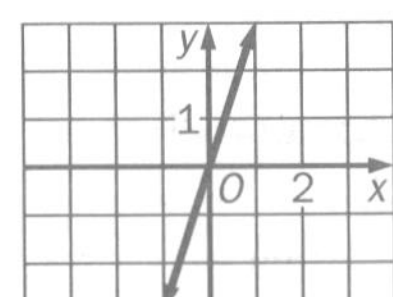

Give the slope and vertical intercept of each line. *(Section 2.2)*

3. $y = 6 - 2x$ -2; 6

4. $y = \frac{2}{3}x + \frac{1}{4}$ $\frac{2}{3}$; $\frac{1}{4}$

5. A health club membership costs \$29.95 and \$3.00 per visit. Model the situation with an equation. *(Section 2.2)*
$M = 29.95 + 3x$

6. Write a point-slope equation of the line passing through the given point and having the given slope. *(Section 2.3)*
point: $(4, -5)$; slope $= -\frac{3}{4}$
$y = -5 - \frac{3}{4}(x - 4)$

7. Graph the equation in Ex. 6. *(Section 2.3)*

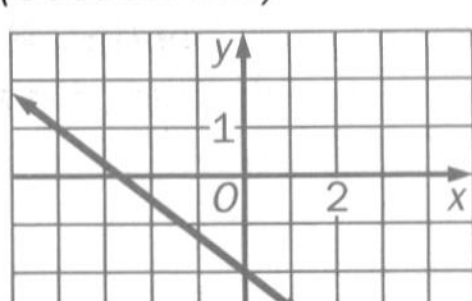

8. Find the domain and range for the function $f(x) = 4x - 3$ for $x \geq 2$. *(Section 2.3)*
domain: $x \geq 2$: range: $y \geq 5$

CHAPTER 2

Review

STUDY TECHNIQUE

Go back through the sections in this chapter. For each section, write two questions that could appear on the chapter test. One should be a short-answer question. One should be a more involved question about general ideas. Exchange your questions with another student and answer the set you receive.

VOCABULARY

direct variation (p. 46)
constant of variation (p. 46)
slope (p. 47)
vertical intercept (p. 53)
slope-intercept form (p. 53)
point-slope equation (p. 59)
function notation (p. 60)
domain (p. 60)
range (p. 60)
least-squares line (p. 67)
correlation coefficient (p. 72)
positive correlation (p. 72)
negative correlation (p. 72)
parameter (p. 79)
parametric equations (p. 79)

SECTIONS 2.1, 2.2, *and* 2.3

The variables x and y are in **direct variation** if $y = ax$ for some nonzero constant a. A direct variation graph is a straight line passing through the origin.

For example, a 39 oz box of laundry detergent can clean 13 loads of clothing. The number of loads l that you can clean varies directly with d, the amount of detergent in ounces, so $l = ad$.

To find a, the **constant of variation**, rewrite $l = ad$ as $a = \frac{l}{d}$.

$$a = \frac{13}{39} = \frac{1}{3}$$

The direct variation equation is $l = \frac{1}{3}d$.

The slope of the graph is the constant of variation, $\frac{1}{3}$.

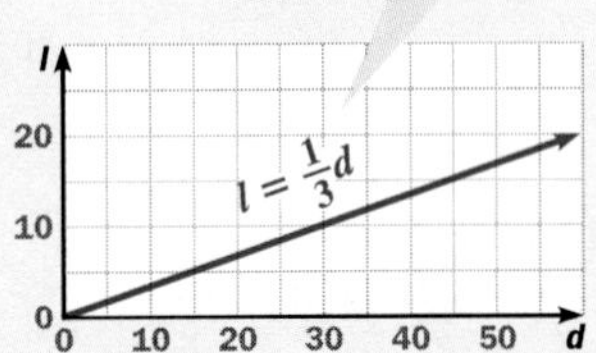

To find the **slope** of the line that passes through two points (x_1, y_1) and (x_2, y_2), use this formula:

$$\text{slope} = \frac{\text{vertical change}}{\text{horizontal change}} = \frac{y_2 - y_1}{x_2 - x_1}$$

For example, the line shown passes through $(-2, -1)$ and $(0, 3)$. The slope a is:

$$a = \frac{-1 - 3}{-2 - 0} = 2$$

The **slope-intercept equation** of a line is $y = ax + b$, and a **point-slope equation** is $y = y_1 + a(x - x_1)$. For the line shown, the equations are:

Slope-intercept $y = 2x + 3$

Point-slope $y = -1 + 2(x + 2)$

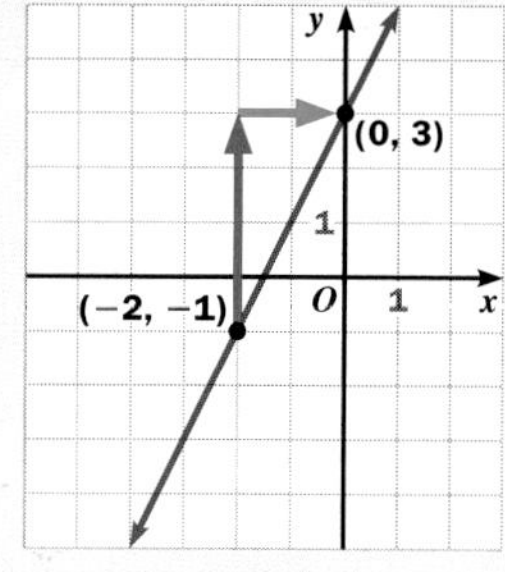

SECTIONS 2.4 *and* 2.5

For a set of paired data, you can draw a line of fit by hand, or use a graphing calculator or statistical software to find the **least-squares line**.

For example, the table gives six data pairs, which are plotted below.

x	y
0	2.5
2	2.6
4	3.6
6	5.4
8	5.5
10	7.4

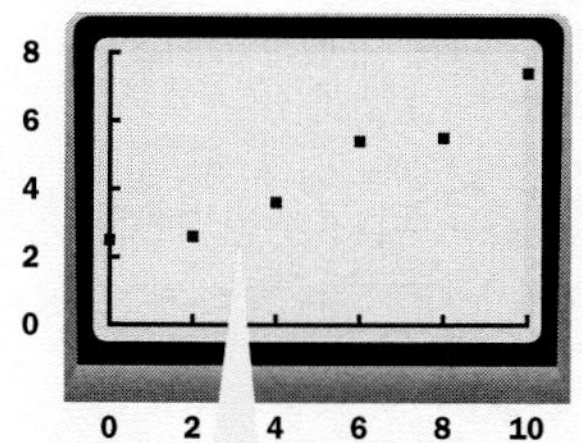

The scatter plot shows that the plotted points "line up," which implies a **strong correlation**.

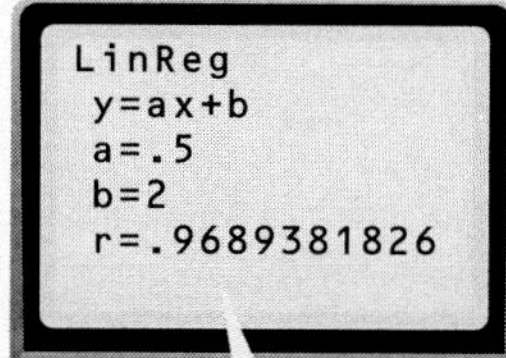

The least-squares line has slope 0.5 and vertical intercept 2, so an equation is $\mathbf{y = 0.5x + 2}$.

SECTION 2.6

Use **parametric equations** to solve problems about two-dimensional motion.

For example, the graph of the parametric equations $x = t + 1$ and $y = 3t - 2$ for $1 \le t \le 3$ is shown.

You can express y as a function of x by eliminating the **parameter** t. For the graph shown, $y = 3x - 5$ for $2 \le x \le 4$.

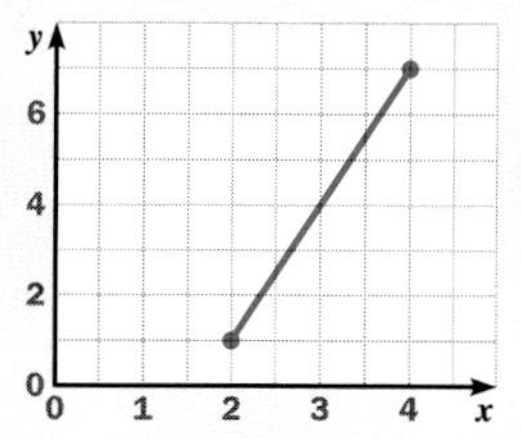

Progress Check 2.4–2.6

1. The estimated population in a rural county in the United States is shown in the table below. *(Section 2.4)*

Year	Population
2010	830
2020	720
2030	560
2040	430
2050	310

a. Make a scatter plot of the data and draw a line of fit.

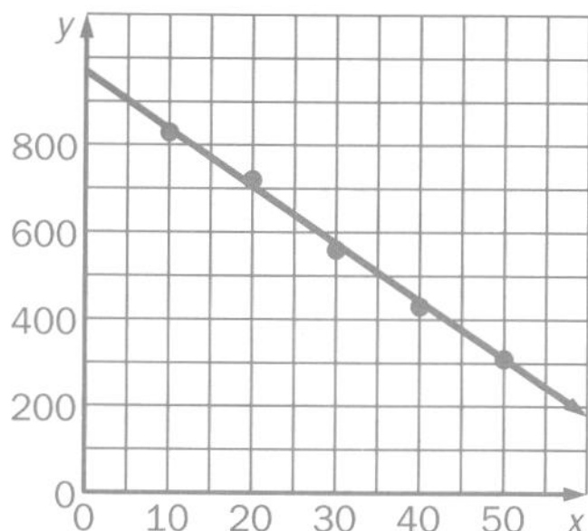

b. Find an equation of your line of fit. Answers may vary. An example is given. If x = the years since 2000, an equation is $y = -13.3x + 969$.

c. Predict the population for the year 2060. 171

2. Refer to Ex. 1 above. *(Section 2.5)*

a. Find the correlation coefficient. −0.999

b. Is the correlation strong or weak? strong

3. Tell whether you would expect the correlation between the wing span of a bird and its weight to be positive, negative, or zero. *(Section 2.5)* positive

4. a. Graph the parametric equations $x = 4t$ and $y = -3 + t$ for $0 \le t \le 1$. *(Section 2.6)*

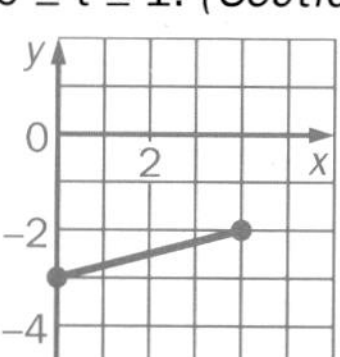

b. Express y as a function of x and give the domain of the function.

$y = -3 + \frac{x}{4}$; domain: $0 \le x \le 4$

Chapter 2 Assessment
Form A Chapter Test

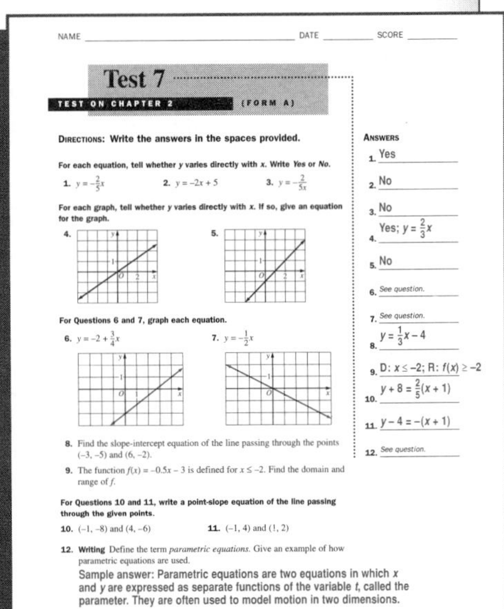
NAME ______ DATE ______ SCORE ______

Test 7

TEST ON CHAPTER 2 (FORM A)

DIRECTIONS: Write the answers in the spaces provided.

For each equation, tell whether *y* varies directly with *x*. Write *Yes* or *No*.

1. $y = -\frac{2}{5}x$ 2. $y = -2x + 5$ 3. $y = -\frac{2}{5x}$

For each graph, tell whether *y* varies directly with *x*. If so, give an equation for the graph.

4. 5.

For Questions 6 and 7, graph each equation.

6. $y = -2 + \frac{3}{4}x$ 7. $y = -\frac{1}{2}x$

8. Find the slope-intercept equation of the line passing through the points (–3, –5) and (6, –2).
9. The function $f(x) = -0.5x - 3$ is defined for $x \le -2$. Find the domain and range of f.

For Questions 10 and 11, write a point-slope equation of the line passing through the given points.

10. (–1, –8) and (4, –6) 11. (–1, 4) and (1, 2)

12. **Writing** Define the term *parametric equations*. Give an example of how parametric equations are used.
Sample answer: Parametric equations are two equations in which x and y are expressed as separate functions of the variable t, called the parameter. They are often used to model motion in two dimensions.

ANSWERS
1. Yes
2. No
3. No
4. Yes; $y = \frac{2}{3}x$
5. No
6. See question.
7. See question.
8. $y = \frac{1}{3}x - 4$
9. D: $x \le -2$; R: $f(x) \ge -2$
10. $y + 8 = \frac{2}{5}(x + 1)$
11. $y - 4 = -(x + 1)$
12. See question.

Assessment Book, ALGEBRA 2: EXPLORATIONS AND APPLICATIONS

9

Chapter 2 Assessment
Form B Chapter Test

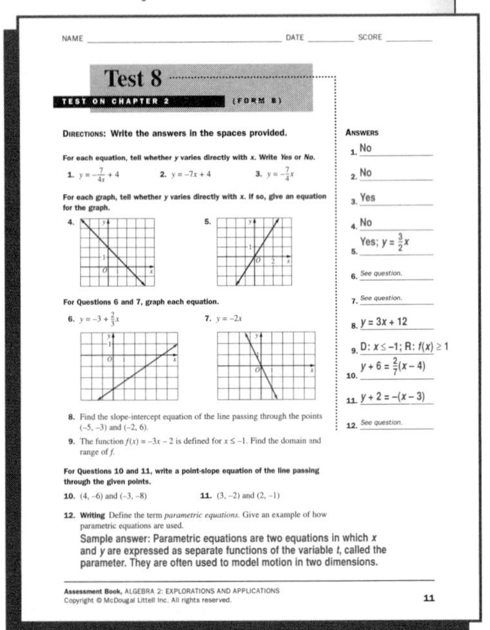
NAME ______ DATE ______ SCORE ______

Test 8

TEST ON CHAPTER 2 (FORM B)

DIRECTIONS: Write the answers in the spaces provided.

For each equation, tell whether *y* varies directly with *x*. Write *Yes* or *No*.

1. $y = -\frac{7}{4x} + 4$ 2. $y = -7x + 4$ 3. $y = -\frac{7}{4}x$

For each graph, tell whether *y* varies directly with *x*. If so, give an equation for the graph.

4. 5.

For Questions 6 and 7, graph each equation.

6. $y = -3 + \frac{2}{3}x$ 7. $y = -2x$

8. Find the slope-intercept equation of the line passing through the points (–5, –3) and (–2, 6).
9. The function $f(x) = -3x - 2$ is defined for $x \le -1$. Find the domain and range of f.

For Questions 10 and 11, write a point-slope equation of the line passing through the given points.

10. (4, –6) and (–3, –8) 11. (3, –2) and (2, –1)

12. **Writing** Define the term *parametric equations*. Give an example of how parametric equations are used.
Sample answer: Parametric equations are two equations in which x and y are expressed as separate functions of the variable t, called the parameter. They are often used to model motion in two dimensions.

ANSWERS
1. No
2. No
3. Yes
4. No
5. Yes; $y = \frac{3}{2}x$
6. See question.
7. See question.
8. $y = 3x + 12$
9. D: $x \le -1$; R: $f(x) \ge 1$
10. $y + 6 = \frac{2}{7}(x - 4)$
11. $y + 2 = -(x - 3)$
12. See question.

Assessment Book, ALGEBRA 2: EXPLORATIONS AND APPLICATIONS

11

CHAPTER 2 Assessment

VOCABULARY QUESTIONS

For Questions 1–3, complete each paragraph.

1. The equation $y = ax$ tells you that x and y are in _?_.
2. The _?_ form for the equation of a line is $y = ax + b$. It has this name because a is the _?_ of the line, and b is the line's _?_.
3. When you enter data pairs into a graphing calculator or statistical software, you can obtain an equation for a line of fit called the _?_. You can also obtain the value of r, called the _?_, which measures how well this line fits the data points.

SECTIONS 2.1, 2.2, and 2.3

For each equation, tell whether *y* varies directly with *x*. If so, graph the equation.

4. $y = \frac{4}{3x}$
5. $y = \frac{4x}{3}$
6. $y = 4x + 3$

7. Suppose p varies directly with s. When $s = 12$, $p = 5$.
   a. Find an equation relating p and s.
   b. Find p when $s = 27$.

Give the slope and *y*-intercept of the line described by each equation.

8. $y = 6x$
9. $y = 2 - 5x$
10. $y = -3$

Graph each equation.

11. $y = \frac{5}{3}x - 4$
12. $y = -x$
13. $y = 2.4$

Find a point-slope equation of each line.

14.
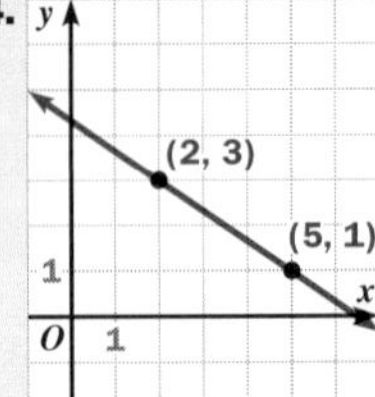

15.
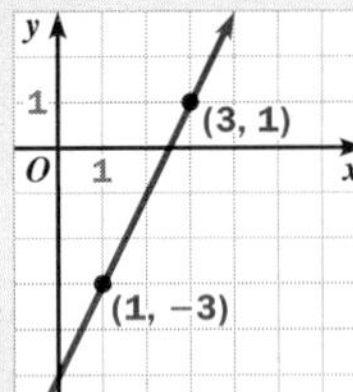

16.
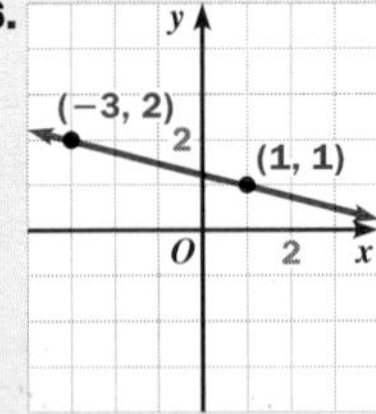

17. Suppose the function $f(x) = -2x + 5$ is defined for $x \le 3$.
   a. Find the domain and range of f.
   b. Graph the function.

ANSWERS Chapter 2

Assessment

1. direct variation
2. slope-intercept; slope; y-intercept
3. least-squares line; correlation coefficient
4. No.
5. Yes.

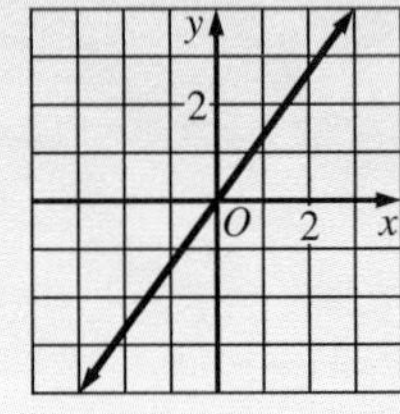

6. No.
7. a. $p = \frac{5}{12}s$
   b. $11\frac{1}{4}$
8. 6; 0
9. –5; 2
10. 0; –3
11.

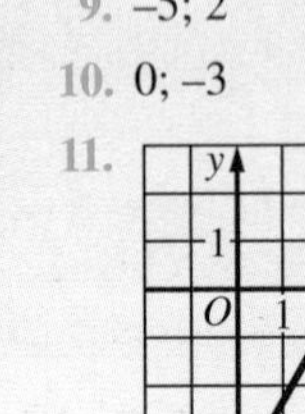

12.

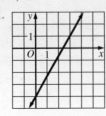

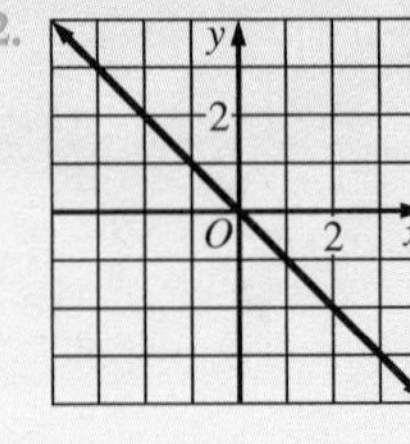

13.

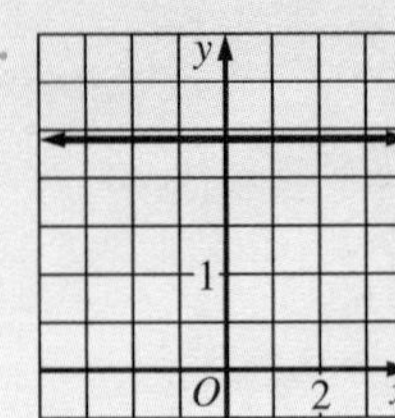

14. $y = 3 - \frac{2}{3}(x - 2)$ or $y = 1 - \frac{2}{3}(x - 5)$
15. $y = -3 + 2(x - 1)$ or $y = 1 + 2(x - 3)$
16. $y = 1 - \frac{1}{4}(x - 1)$ or $y = 2 - \frac{1}{4}(x + 3)$

Chapter 2 Assessment

Form C Alternative Assessment

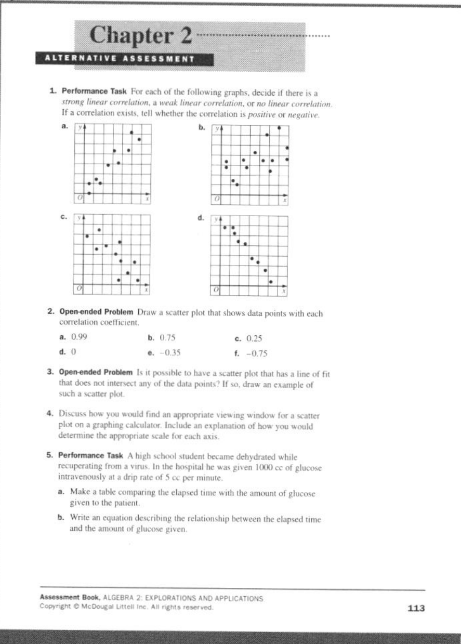

Chapter 2

ALTERNATIVE ASSESSMENT

1. **Performance Task** For each of the following graphs, decide if there is a *strong linear correlation*, a *weak linear correlation*, or *no linear correlation*. If a correlation exists, tell whether the correlation is *positive* or *negative*.

a. b. c. d.

2. **Open-ended Problem** Draw a scatter plot that shows data points with each correlation coefficient.

a. 0.99 b. 0.75 c. 0.25
d. 0 e. −0.35 f. −0.75

3. **Open-ended Problem** Is it possible to have a scatter plot that has a line of fit that does not intersect any of the data points? If so, draw an example of such a scatter plot.

4. Discuss how you would find an appropriate viewing window for a scatter plot on a graphing calculator. Include an explanation of how you would determine the appropriate scale for each axis.

5. **Performance Task** A high school student became dehydrated while recuperating from a virus. In the hospital he was given 1000 cc of glucose intravenously at a drip rate of 5 cc per minute.

a. Make a table comparing the elapsed time with the amount of glucose given to the patient.

b. Write an equation describing the relationship between the elapsed time and the amount of glucose given.

Assessment Book, ALGEBRA 2: EXPLORATIONS AND APPLICATIONS

113

c. Draw a graph. Discuss the meaning of the x-intercept and the y-intercept in this real-world situation.

d. What information from the table in part (a) and the graph in part (b) might be valuable to the doctor?

6. **Open-ended Problem** Describe a situation that can be modeled by the equation $y = 8.25x + 15$.

7. **Performance Task** Describe the domain and range of the function $f(x) = -6x + 2$, where $x > -2$. Graph $f(x)$. Explain how you can verify the domain and range by looking at your graph.

8. **Performance Task** Two parents whose children are all in college generate five baskets of laundry per week. When their three youngest children are home from college, the family generates eleven baskets of laundry per week. Write an equation that shows the relationship between the number of people and the number of baskets of laundry. Use your equation to predict the amount of laundry generated by the family when their oldest daughter comes home with her husband and two children for the holidays. Assume the other children are still home from college. What factors might affect the accuracy of your answer?

9. **Open-ended Problem** Write at least five statements about the graph below. Use correct mathematical terminology in describing its features.

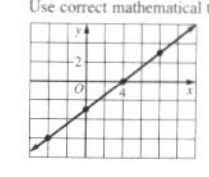

10. **Group Activity** Your group will need an empty plastic drink bottle.

a. Punch a small hole in the side of an empty drink bottle 10 cm from its base. Fill the bottle with water. Make observations about the stream of water that comes out of the hole. Is the water coming out of the bottle at the same rate when the bottle is almost empty as when the bottle was almost full? Ask a physics teacher why this happens.

b. Refill the bottle. Measure the height of the water in the bottle every 10 s. Record the time and the height of the water. Which data would be represented by the independent variable? Which data would be represented by the dependent variable? Graph the data and find an equation of a line of fit. Why is a line not a good fit for the data?

c. Refill the bottle. Measure the height of the water. Record the time in seconds each time the water level drops 3 cm. Which data would be represented by the independent variable? Which data would be represented by the dependent variable? Graph the data and find an equation of a line of fit. Why is a line not a good fit for the data?

d. Why were the dependent and independent variables different in parts (b) and (c)?

114
Assessment Book, ALGEBRA 2: EXPLORATIONS AND APPLICATIONS

SECTIONS 2.4 *and* 2.5

18. SPORTS The table gives the 1993 populations of six states and the total number of professional baseball, basketball, football, and hockey teams in each state at that time.

State	Population (millions)	Number of sports teams
CA	30	16
GA	6	3
IL	11	5
MA	6	4
NY	18	9
PA	12	7

a. Show that the data have a linear relationship by making a scatter plot.

b. Is the correlation between the population and the number of teams *positive* or *negative*? Is it *strong* or *weak*?

c. **Technology** Use a graphing calculator or statistical software to find the correlation coefficient for the data.

d. Fit a line to the data. Find an equation for your line of fit.

e. Use the equation to predict the number of professional sports teams in New Jersey, with population 8 million. How does your prediction compare with New Jersey's actual figure of two professional teams?

19. Open-ended Problem Do you think there is a correlation between the amount of time a person spends reading each week and the amount of time the person spends watching television each week? Explain your reasoning. If there is a correlation, is it *positive* or *negative*? Why? Is there causation? Explain.

SECTION 2.6

For each of Exercises 20–22:

a. Graph the pair of parametric equations for the given restriction on *t*.

b. Express *y* as a function of *x*, and state any restriction on *x*.

20. $x = -t$
$y = 5t - 3$
$t \leq 0$

21. $x = 2 + t$
$y = 3t + 4$
$0 \leq t \leq 3$

22. $x = \frac{1}{2}t + 1$
$y = -3t + 6$
$2 \leq t \leq 6$

PERFORMANCE TASK

23. Design and carry out an experiment that you think will result in linear data. Make a scatter plot of the data, draw a line of fit, and find an equation of the line. Use your equation to make predictions. In addition to the graph, equation, and predictions, your report should include a description of the experiment and an assessment of how well a linear model fits your data.

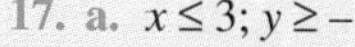

17. a. $x \leq 3$; $y \geq -1$

b.

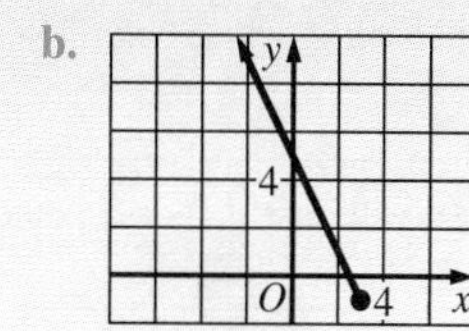

18. a.

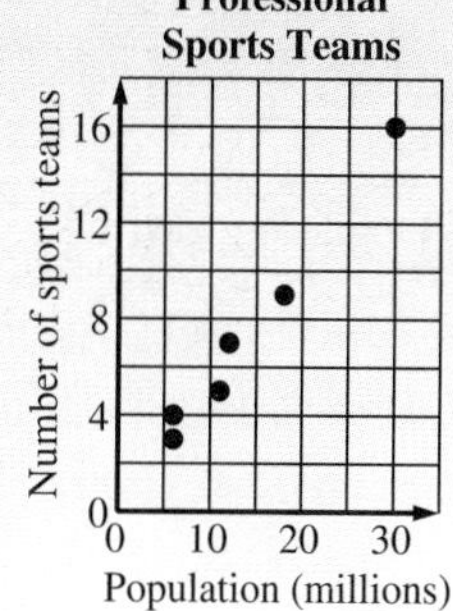

b. positive; strong

c. 0.99

d. Answers may vary. The least-squares line is $y = 0.519x + 0.151$.

e. Answers may vary. The estimate based on the least-squares line is about 4 teams; that is higher than the actual figure.

19. Answers may vary. An example is given. I think there is both a negative correlation and causation. The more time you spend reading, the less time you have to spend watching television.

20. a.

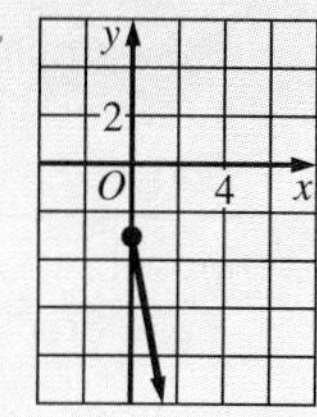

b. $y = -5x - 3$; $x \geq 0$

21–23. See answers in back of book.

CHAPTER

3 Exponential Functions

OVERVIEW

Connecting to Prior and Future Learning

⇔ The development of concepts relating to exponential functions, such as exponential growth and decay and negative exponents, is continued from Algebra 1. In addition, students are introduced to rational exponents. The **Student Resources Toolbox** provides students with a review of exponents and powers on page 776, rational numbers and irrational numbers on page 779, and logical statements on page 799.

⇔ Students learn to graph exponential functions and fit exponential functions to data. Students will find the review of simplifying variable expressions on page 781 and reflecting a figure and line symmetry on page 797 of the **Students Resources Toolbox** helpful.

⇔ The number e is introduced and used to model exponential growth and decay. These models can be used to make predictions about real-life growth and decay. These concepts will be used by students in future mathematics courses.

Chapter Highlights

Interview with Finn Strong: The use of mathematics to describe the relationship between temperature and the amount of water vapor the air can hold is emphasized in this interview. The related exercises on pages 112 and 128 allow students to explore this application further.

Explorations focus on using a graphing calculator to investigate exponential graphs in Section 3.3 and on using a scientific calculator to study compound interest in Section 3.4.

The Portfolio Project: Students collect and analyze data about a bouncing ball. They use this information to write an exponential decay equation which is then used to make predictions.

Technology: Graphing calculators are used throughout the chapter to graph equations and evaluate expressions. The ability of graphing calculators to perform exponential regressions is discussed in Section 3.5. Spreadsheets are used in Sections 3.1 and 3.2 to show tables of data. Scientific calculators can be used throughout the chapter to evaluate expressions.

OBJECTIVES

Section	Objectives	NCTM Standards
3.1	• Describe growth and decay using tables of data and equations. • Use exponential equations to model real-life situations.	1, 2, 3, 4, 5, 6
3.2	• Evaluate expressions that use negative and rational exponents. • Describe situations involving continuous exponential growth or decay.	1, 2, 3, 4, 5, 6
3.3	• Draw graphs of exponential functions. • Interpret how different values of a and b affect the graph of $y = ab^x$. • Determine the doubling time or the half-life in real-world situations.	1, 2, 3, 4, 5, 6
3.4	• Model exponential growth and decay using the base e, and model logistic growth. • Make predictions about real-life growth and decay situations.	1, 2, 3, 4, 5, 6
3.5	• Write exponential functions that fit sets of data. • Make predictions about exponential growth and decay situations.	1, 2, 3, 4, 5, 6

INTEGRATION

Mathematical Connections	3.1	3.2	3.3	3.4	3.5
algebra	**93–99***	**100–106**	**107–114**	**115–122**	**123–129**
geometry	99				128
data analysis, probability, discrete math	96	**100–106**	108–114	**115–122**	**123–129**
patterns and functions	**93–99**	**100–106**	**107–114**	**115–122**	**123–129**
logic and reasoning	94–99	101, 102, 104–106	108, 109, 111–114	115, 116, 120–122	124–129
Interdisciplinary Connections and Applications					
biology and earth science				118, 122	125, 129
chemistry and physics				117, 121	
arts and entertainment		106			123
sports and recreation			114		127
bookmaking and graphic design	93, 95, 97, 99				
physiology and health			109		123, 129
banking			108, 112	117, 119	124
mythology, census, traffic, snowmaking, carbon dating, farming, wind energy, psychology, astronomy	98	100, 105	113	120, 121	125, 126

* ***Bold page numbers*** *indicate that a topic is used throughout the section.*

TECHNOLOGY

Section	Student Book (*opportunities for use with*)	Support Material (*opportunities for use with*)
3.1	graphing calculator spreadsheet software McDougal Littell Mathpack *Stats!* *Function Investigator*	
3.2	graphing calculator McDougal Littell Mathpack *Function Investigator*	
3.3	graphing calculator McDougal Littell Mathpack *Function Investigator*	**Technology Book:** Calculator Activity 3 **Function Investigator with Matrix Analyzer Activity Book:** Function Investigator Activity 29
3.4	scientific calculator graphing calculator McDougal Littell Mathpack *Function Investigator*	**Technology Book:** Spreadsheet Activity 3
3.5	graphing calculator McDougal Littell Mathpack *Function Investigator*	

Planning Guide

Regular Scheduling (45 min)

Section	Materials Needed	Core Assignment	Extended Assignment	*exercises that feature* Applications	Communication	Technology
3.1	graphing calculator or software, spreadsheet software, graph paper, dictionary	1–16, 22, 27–34	1–4, 6–11, 13–34	22–26	19, 21	17, 20
3.2	graphing calculator, graph paper	**Day 1:** 1–25 **Day 2:** 26–30, 34, 35, 41–48	**Day 1:** 1–25 **Day 2:** 26–48	26–32, 35–40	9 41	
3.3	graphing calculator, graph paper	**Day 1:** 1–9, 11–14, 16–22 **Day 2:** 25–41, AYP*	**Day 1:** 1–22 **Day 2:** 23–41, AYP	16–22 23–30	10, 22	2, 3
3.4	scientific calculator, graphing calculator, graph paper	**Day 1:** 1–17 **Day 2:** 18–21, 27, 28, 30–38	**Day 1:** 1–17 **Day 2:** 18–38	18–30	17 26, 31	17 22–25, 30, 35
3.5	graphing calculator, graph paper	1–8, 10–12, 17, 18, 20–23, AYP	5–23, AYP	9, 14, 17–19, 21	12, 20	11, 14
Review/ Assess		**Day 1:** 1–9 **Day 2:** 10–18 **Day 3:** Ch. 3 Test	**Day 1:** 1–9 **Day 2:** 10–18 **Day 3:** Ch. 3 Test	7 17	12	16
Portfolio Project		Allow 2 days.	Allow 2 days.			

Yearly Pacing (with Portfolio Project)	**Chapter 3 Total** 13 days	**Chapters 1–3 Total** 39 days	**Remaining** 121 days	**Total** 160 days

Block Scheduling (90 min)

	Day 14	Day 15	Day 16	Day 17	Day 18	Day 19	Day 20
Teach/Interact	Ch. 2 Test 3.1	3.2	3.3: Exploration, page 107	3.4: Exploration, page 115	3.5 Review	Review Port. Proj.	Ch. 3 Test Port. Proj.
Apply/Assess	**Ch. 2 Test** **3.1:** 1–16, 22, 27–34	**3.2:** 1–30, 34, 34, 35, 41–48	**3.3:** 1–9, 11–14, 16–22, 25–41, AYP*	**3.4:** 1–21, 27, 28, 30–38	**3.5:** 1–8, 10–12, 17, 18, 20–23, AYP **Review:** 1–9	**Review:** 10–18 **Port. Proj.**	**Ch. 3 Test** **Port. Proj.**

NOTE: A one-day block has been added for the Portfolio Project—timing and placement to be determined by teacher.

Yearly Pacing (with Portfolio Project)	**Chapter 3 Total** $6\frac{1}{2}$ days	**Chapters 1–3 Total** 20 days	**Remaining** 62 days	**Total** 82 days

* ***AYP*** *is Assess Your Progress.*

Lesson Support

Section	Practice Bank	Study Guide*	Assessment Book*	Visuals	Explorations Lab Manual	Lesson Plans	Technology Book
3.1	14	3.1		Warm-Up 3.1 Folder 3	Master 2 Add. Expl. 3	3.1	
3.2	15	3.2		Warm-Up 3.2	Master 2	3.2	
3.3	16	3.3	Test 9	Warm-Up 3.3	Master 2	3.3	Calculator Act. 3
3.4	17	3.4		Warm-Up 3.4	Masters 1, 6	3.4	Spreadsheet Act. 3
3.5	18	3.5	Test 10	Warm-Up 3.5	Master 2 Add. Expl. 4	3.5	
Review Test	19	Chapter Review	Tests 11, 12 Alternative Assessment			Review Test	

*__Spanish versions__ of *Study Guide* and *Assessment Book* are available.

Chapter Support

- Course Guide
- Lesson Plans
- Portfolio Project Book: Additional Project 2: Functions
- Preparing for College Entrance Tests
- Multi-Language Glossary
- *Test Generator* Software
- Professional Handbook

Software Support

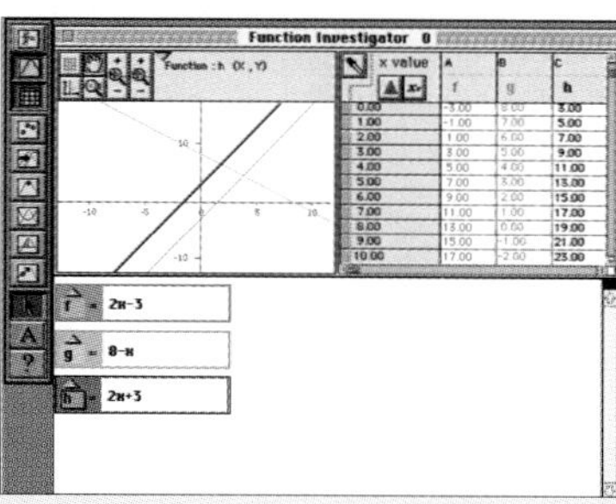

McDougal Littell Mathpack
Stats!
Function Investigator

Internet Support

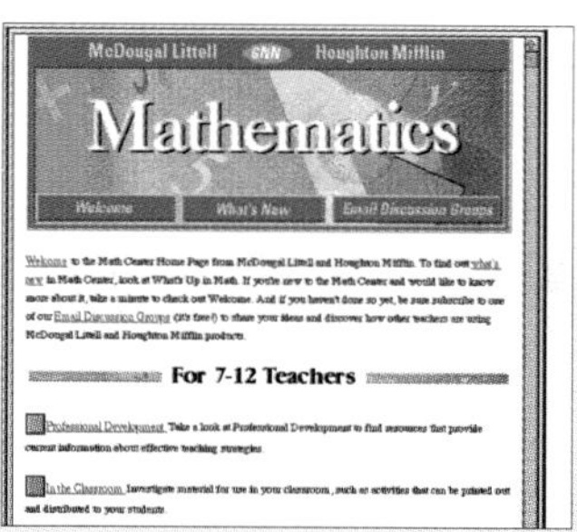

http://www.hmco.com
Next go to McDougal Littell; then the Education Center; then Secondary Math.

Outside Resources

Books, Periodicals

Shell. *The Language of Functions and Graphs*, Unit B3: "Looking at Exponential Functions," pp. 120–125. The Shell Centre for Mathematical Education at the University of Nottingham in the UK.

Jones, Graham A. "Mathematical Modeling in a Feast of Rabbits." *Mathematics Teacher* (December 1993): pp. 770–773.

Flores, Alfinio. "Connections: A Lottery, a Computer, and the Number *e*." *Mathematics Teacher* (November 1993): pp. 652–655.

Masalski, William J. *How to Use the Spreadsheet as a Tool in the Secondary Mathematics Classroom*: "Topic: Compound Interest": pp. 16–19. Reston, VA: NCTM, 1990.

Activities, Manipulatives

Kincaid, Charlene, Guy Mauldin, Deanna Mauldin. "The Marble Sifter: A Half-Life Simulation." *Mathematics Teacher* (December 1993): pp. 748–759.

Breuningsen, Chris, Bill Bower, Linda Antinone, and Elisa Breuningsen. "Charging Up, Charging Down." *Real-World Math with the CBL System.* Activity 12: pp. 63–68. Texas Instruments, 1995.

Videos

Southern Illinois University at Carbondale. *World Population Review.* 1990.

Internet

To search for economical local access to the information super highway, investigate the non-profit National Public Telecommunications Network by sending e-mail to: info@nptn.org

Interview Notes

Background

Aquariums
To set up a home aquarium, the necessary equipment would include a tank, a tank cover, one or more filters, a heater, and a thermometer. An aquarium should also have an air pump as well as some plants and gravel to provide a healthier environment for the fish. A good gauge for the amount of water necessary for an aquarium is one gallon of water for every inch of fish.

There are many public aquariums around the world. Their sizes range from small tanks that hold 5 gallons of water to huge ones that hold more than 5 million gallons of water. The oldest public aquarium is at the London Zoo in England. It opened in 1853. The largest aquarium in terms of volume of water is the Living Seas Aquarium at the Epcot Center in Florida. It has a total capacity of 6.25 million gallons of water and contains over 3000 fish. The largest aquarium in terms of marine life is the Monterey Bay Aquarium in California. It contains over 6500 specimens.

Finn Strong
Finn Strong's river tank has received much attention and many awards, including first place at the New England Inventors Exhibition in September, 1992. But the final product was not completed without perseverance and problem solving. In addition to the water evaporation problem, Strong also had to overcome difficulties with the lighting system in the tank. The fluorescent bulbs either burned the plants or did not allow them enough room to grow. He attempted to solve the problem by moving the light up higher, but then the glass in front became foggy. He then decided to put ventilation holes on top of the tank, which worked. The heated air escaped through the holes, while fresh air was drawn through a vent in the front. Strong resides in Providence, Rhode Island where he runs his business, Finn Strong Designs, Inc. He graduated from The Rhode Island School of Design and uses his degree to design, market, and sell his products.

CHAPTER

3 Exponential Functions

A river running through it

INTERVIEW Finn Strong

When Finn Strong was two years old, he started pouring water over rock piles. At the age of 12, he built a rock garden with plants and fish in the living room of his family's home. The garden included a waterfall that tumbled into a fish tank. At the age of 24, Strong started a company to sell a new kind of fish tank. He invented the "river tank" to show what a river might look like if you put a window along one side. Inside the fish swim through rapids, waterfalls, and eddies. Strong says, "They're much more active than fish in normal tanks."

"If you have a crazy idea that everyone says can't be done, you still ought to try."

The Rainmaker

The water in the tank stays clear because it is constantly flowing, cleansing itself like an actual river. The continual churning of the water helps provide more oxygen for the fish as well. "People put chemicals in our tanks when they're not supposed to," Strong says. "The key is to set it up and leave it alone." Evaporation posed a problem for Strong's river tank because water had to be added frequently. Strong solved this problem by designing a rainmaking device. With this device, water has to be added only once every one to two weeks. Beads of water condense on the surfaces of the rainmaker until they periodically fall as rain.

"Everything runs
according to nature."

A Natural Interdependency

Frogs, toads, lizards, salamanders, snails, and plants live in the water and on the riverbanks of Strong's river tank. "The tank shows the interdependence of plants and animals," Strong says. The fish can't live without the plants, because the plants keep the water clean. The plants, in turn, live off the fish wastes. Even the bacteria have a role, eating nutrients that would otherwise cloud the water and make it toxic. The idea, Strong adds, is "to provide the right ingredients so that everything runs according to nature."

Interview Notes

Background

Precipitation

Precipitation in terms of weather refers to the forms of water particles that form in the atmosphere and fall to Earth. Precipitation can fall in five forms: hail, sleet, snow, rain, and drizzle. Dew and frost are not included because they form on solid surfaces and do not fall to the ground.

Second-Language Learners

Some students learning English may be unfamiliar with *rapids* and *eddies.* If necessary, explain that *rapids* are stretches of river in which the water moves swiftly, and *eddies* are parts of a river where the water swirls in a direction opposite to the flow of the river.

Interview Notes

Mathematical Connection

The idea of saturation of air involves the maximum amount of water vapor that air can hold at a particular temperature and pressure. When air is saturated, it cannot hold any more water vapor. The saturation point, however, varies with the temperature. This relationship, relating the amount of water vapor that the air can hold and the temperature, can be described using exponential equations and is explored in Section 3.3. In this section, students calculate and write the different parts of exponential equations. Another application of exponential equations deals with the amount of oxygen that water can hold. Its relationship to temperature and chlorinity is explored in Section 3.5, where students use data values to write exponential equations and predict other data values.

Explore and Connect

Writing/Research
Exploring the idea of relative humidity in the research section can help students answer the questions in the writing section. Also, have students verbally describe the connections between humidity and relative humidity, hot weather and high humidity, and cold weather and dry air, to clarify their thoughts and solidify their understanding of the concepts.

Project
Students can use the school or local library to find information on the hydrologic cycle. The diagrams should explain how water moves below, on, and above the surface of Earth. The best diagrams can be displayed in class.

Raindrops Keep Falling

How does the rainmaker work? The amount of water vapor that air can hold depends on the temperature. At higher temperatures, air can hold much more water. The graph shows how the maximum amount of water vapor that air can hold depends on the temperature. In Strong's river tank, the warm moist air near the surface of the water rises to the top. Raindrops form when the air touches the cool surfaces of the rainmaker.

The relationship between temperature and the amount of water vapor the air can hold can be modeled by an *exponential function*. The graph of an exponential function rises slowly at first and then takes off dramatically. For example, the graph shows that when the air is cold, a small change in temperature has a small effect on the amount of water vapor the air can hold. At higher temperatures, a small increase in temperature produces a large increase in the amount of water vapor the air can hold.

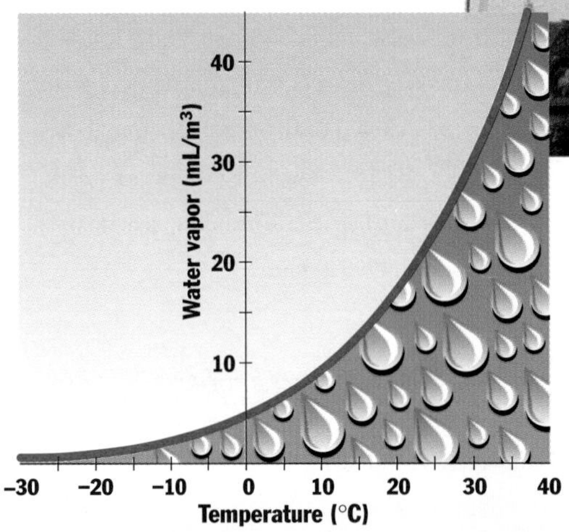

Water vapor (mL/m³)
10
20
30
40
–30
–20
–10
0
10
20
30
40
Temperature (°C)

EXPLORE AND CONNECT

Strong points out a shelf that he designed for lizards. They can keep dry by staying in the upper areas of the tank.

1. Writing In many parts of the world, hot weather is often accompanied by high humidity. Cold weather is often accompanied by dry air. How does the graph help explain these phenomena?

2. Project In some ways, Finn Strong's river tank and rainmaking device are a model of the *hydrologic cycle.* This phrase is used to describe the constant movement of water between the atmosphere and the surface of Earth. Find out more about the hydrologic cycle. Draw a diagram to show the cycle.

3. Research When you listen to a weather report, you may hear the forecaster talk about *relative humidity.* Find out what is meant by relative humidity.

Mathematics & Finn Strong

In this chapter, you will learn more about how mathematics is related to the water content of air and the oxygen content of water.

Related Exercises

Section 3.3
- **Exercises 23 and 24**

Section 3.5
- **Exercises 17–19**

Section

3.1 Exponential Growth and Decay

Learn how to...

- **describe growth and decay using tables of data and equations**

So you can...

- **use exponential equations to model real-life situations, such as making books**

Did you know that this book was made by printing on both sides of large sheets of paper and folding them many times? Each large sheet forms a group of pages called a *signature,* which is trimmed and then bound with other signatures. An exponential function relates the number of times a sheet is folded to the number of pages in the signature.

EXAMPLE 1 Application: Bookmaking

Each time you fold a sheet of paper used in bookmaking, you double the number of sheets of paper in the signature. Each sheet of paper, called a *leaf,* has a front and a back, so there are two pages for every leaf.

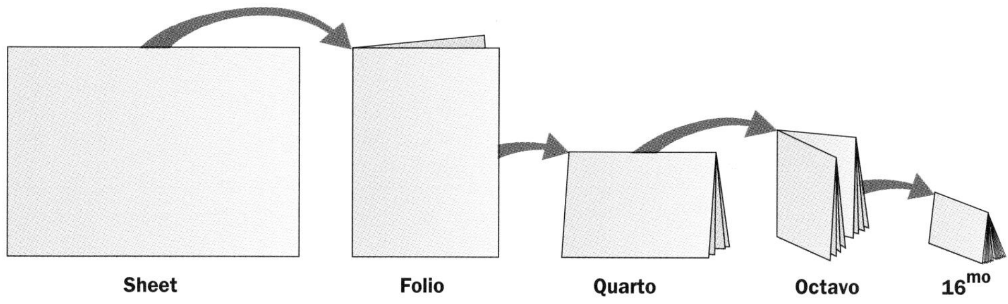

a. Make a table showing the number of leaves and pages when a sheet of paper is folded many times. The names of page sizes are given in the table.

b. Write equations that show how the number of leaves and the number of pages depend on the number of folds.

Name of page size	Number of folds	Number of leaves	Number of pages
—	0		
folio	1		
quarto	2		
octavo	3		
16mo	4		
32mo	5		
64mo	6		

BY THE WAY

The abbreviation "16mo" is short for the Latin *sextodecimo,* meaning *one sixteenth.*

Plan⇔Support

Objectives

- Describe growth and decay using tables of data and equations.
- Use exponential equations to model real-life situations.

Recommended Pacing

Core and Extended Courses
Section 3.1: 1 day

Block Schedule
Section 3.1: $\frac{1}{2}$ block
(with Chapter 2 Test)

Resource Materials

Lesson Support
Lesson Plan 3.1
Warm-Up Transparency 3.1
Overhead Visuals:
Folder 3: Exponential Growth and Decay
Practice Bank: Practice 14
Study Guide: Section 3.1
Exploration Lab Manual:
Additional Exploration 3,
Diagram Master 2

Technology
Graphing Calculator
Spreadsheet Software
McDougal Littell Mathpack
Stats!
Function Investigator
Internet:
http://www.hmco.com

Warm-Up Exercises

Evaluate.

1. 2^5 32
2. 4^3 64
3. $3.85(1.4)^2$ 7.546
4. $8 \cdot 5^4$ 5000
5. $2^3 \cdot 2^3 \cdot 2^2$ 256
6. $\frac{9^4}{9^3}$ 9

Teach⇔Interact

Additional Example 1

A large paper equilateral triangle is folded three times at each of several successive stages. Each fold matches a vertex of the triangle with the midpoint of the opposite side. The result is a smaller equilateral triangle, to which the same process is then applied to obtain the triangle for the next stage.

a. Imagine that at the end of some stage, the paper is pierced near a vertex and a wire ring is put through the hole. Imagine, too, that the paper can be sliced along the edges. The result would be a "book" with pages in the shape of congruent equilateral triangles. Make a table to show the number of leaves and pages at various stages. Let stage 0 be the stage at which no folds have been made in the original triangle.

Stage	Number of leaves	Number of pages
0	1	2
1	4	8
2	16	32
3	64	128
4	256	512

b. Write equations that show how the number of leaves and the number of pages depend on the number of the stage at which the edges are cut. Let n = the stage number. At each new stage, the three folds produce four equilateral triangles for each of the equilateral triangles from the preceding stage. So, the number of leaves L is given by this equation: $L = 4^n$.
Each leaf has two pages, so the number of pages P is given by this equation:

$$\begin{aligned} P &= 2 \cdot L \\ &= 2 \cdot 4^n \\ &= 2 \cdot (2^2)^n \\ &= 2 \cdot 2^{2n} \\ &= 2^{2n+1} \end{aligned}$$

Section Note

Teaching Tip
Ask students what special result they obtain from the quotient property of exponents when the integers m and n are equal.
($\frac{b^m}{b^m} = b^{m-m} = b^0$. Since $\frac{b^m}{b^m}$ is also equal to 1, it follows that $b^0 = 1$.)

SOLUTION

a.

Name of page size	Number of folds	Number of leaves	Number of pages
—	0	1	2
folio	1	2	4
quarto	2	4	8
octavo	3	8	16
16mo	4	16	32
32mo	5	32	64
64mo	6	64	128

b. Let n = the number of folds.

Each fold doubles the number of leaves, so the number of leaves L is given by this equation:

$$L = 2^n$$

Each leaf has two pages, so the number of pages P is given by this equation:

$$\begin{aligned} P &= 2 \cdot L \\ &= 2 \cdot 2^n \\ &= 2^{n+1} \end{aligned}$$

Since $2 = 2^1$ and since 2^1 and 2^n have the same base, you can **add** the exponents.

THINK AND COMMUNICATE

1. Are the names of page sizes related to the number of leaves or the number of pages after folding?
2. Suppose a piece of paper is folded 7 times. How many leaves would there be? What might the page size be named?

Toolbox p. 776
Exponents and Powers

To rewrite $2 \cdot 2^n$ as 2^{n+1} in part (b) of Example 1, you use the product property of exponents. That property, as well as two others you learned in a previous course, are stated below.

Properties of Exponents

For any $b > 0$ and positive integers m and n:

		Examples:
Product Property:	$b^m \cdot b^n = b^{m+n}$	$2^3 \cdot 2^4 = 2^7$
Quotient Property:	$\frac{b^m}{b^n} = b^{m-n}$	$\frac{5^6}{5^4} = 5^2$
Power Property:	$(b^m)^n = b^{mn}$	$(3^6)^2 = 3^{12}$

When paper is folded to make a signature for a book, each fold is perpendicular to the previous fold. You can use this information to see how the page width, height, and area are affected by folding.

ANSWERS Section 3.1

Think and Communicate

1. the number of leaves
2. 128; 128mo

EXAMPLE 2 Application: Bookmaking

a. The "Royal" paper size is 46 cm wide and 60 cm high. Make a table of the width, height, and area of the page sizes made by folding a sheet of Royal-sized paper in half many times.

b. Write an equation for the page area as a function of the number of folds.

SOLUTION

a.

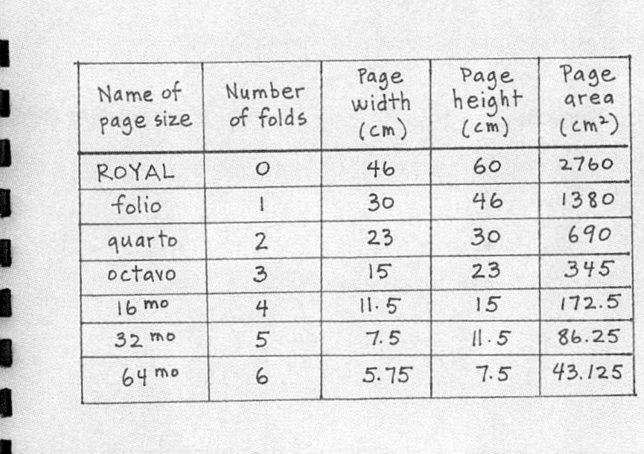

Name of page size	Number of folds	Page width (cm)	Page height (cm)	Page area (cm²)
ROYAL	0	46	60	2760
folio	1	30	46	1380
quarto	2	23	30	690
octavo	3	15	23	345
16 mo	4	11.5	15	172.5
32 mo	5	7.5	11.5	86.25
64 mo	6	5.75	7.5	43.125

b. The original page area is 2760 cm². Each fold halves the page size, so the page area A is given by:

$$A = 2760\left(\frac{1}{2}\right)^n$$

BY THE WAY

The word *paper* comes from *papyrus*, a writing material made by ancient Egyptians from the stem of the papyrus plant. Paper, however, consists of dissolved fibers of wood, cotton, or linen and is made by a process invented in China about the second century B.C.

EXAMPLE 3

Suppose a stack of 400 leaves of a mathematics textbook measures 1 in.

a. What is the thickness of one leaf?

b. Write an equation for the thickness T of the stack of leaves when a sheet is folded n times.

SOLUTION

a. $\frac{1 \text{ in.}}{400 \text{ leaves}} = 0.0025$ in./leaf

b. $T = 0.0025(2^n)$

To find T, multiply the total number of leaves by the thickness per leaf.

THINK AND COMMUNICATE

3. Suppose a stack of 900 leaves of a history textbook measures 2 in. How does the thickness equation given in part (b) of Example 3 change?

4. Compare the equations from Examples 2 and 3. How are they similar?

Teach⟺Interact

Additional Example 2

a. Refer to Additional Example 1. Suppose the original equilateral has sides of length 100 cm. The height of the triangle is $\frac{100\sqrt{3}}{2}$ or $50\sqrt{3}$. Make a table of the page lengths, heights, and areas formed at the various stages.

Stage	Page length (cm)	Page height (cm)	Page area (cm²)
0	100	$50\sqrt{3}$	$2500\sqrt{3}$
1	50	$25\sqrt{3}$	$625\sqrt{3}$
2	25	$12.5\sqrt{3}$	$156.25\sqrt{3}$
3	12.5	$6.25\sqrt{3}$	$39.0625\sqrt{3}$
4	6.25	$3.125\sqrt{3}$	$9.765625\sqrt{3}$

b. Write an equation for the page area as a function of the stage number.

The original page area is $2500\sqrt{3}$. At each new stage, the new equilateral triangles have an area one-fourth the area of the previous triangles. So, the page area A is given by: $A = 2500\sqrt{3}\left(\frac{1}{4}\right)^n$.

Additional Example 3

Refer to the situation in Additional Example 1. Suppose a stack of 500 triangular leaves measures 1.25 in.

a. What is the thickness of one leaf?

$\frac{1.25}{500} = 0.0025$ in./leaf

b. Write an equation for the thickness T of the stack of leaves at the nth stage of the folding process.

To find T, multiply the total number of leaves by the thickness per leaf.
$T = 0.0025(4^n)$

Think and Communicate

3. 0.0025 is replaced by $\frac{2}{900} \approx 0.002$. The equation is $T = 0.002(2^n)$.

4. Answers may vary. An example is given. The equations have the same form, $y = k(a^x)$, for constants k and a. The values of the constants are different.

Teach⇔Interact

Learning Styles: Visual

Students can remember the concepts of exponential growth or decay by visualizing the graphs on this page. It may help students to refer to the tables that were used to plot the points. This will focus attention on how the values of the functions change as *n* increases.

Section Notes

Teaching Tip
Ask students to give the values for *a* and *b* for each of the equations graphed. Discuss the statements below each graph. Point out that for $b = 1$, the graphs and equations model neither exponential growth nor exponential decay. Ask students why this is so. (The graph would be a set of points on a horizontal line, meaning that the *y*-values neither increase nor decrease.)

Integrating the Strands
The presentation of the concepts of exponential growth and decay integrates the strands of algebra, geometry, and discrete mathematics because each fold of the paper is a discrete operation and is represented by a single point.

Closure Question

If $a > 0$ and $b > 0$, how can you tell whether an equation of the form $y = ab^x$ models exponential growth, exponential decay, or neither?
If $b > 1$, the equation models exponential growth. If $b < 1$, the equation models exponential decay. If $b = 1$, the equation models neither growth nor decay.

The equations from Examples 2 and 3, $A = 2760\left(\frac{1}{2}\right)^n$ and $T = 0.0025(2^n)$, both have this general form:

$y = ab^x$

- amount after *x* events
- number of events
- original amount
- proportional growth or decay factor

There is an important difference in their graphs, however.

WATCH OUT!
In both graphs, curves are *not* drawn through the points, because it does not make sense to consider non-integral values for the number of times paper is folded.

Exponential Growth

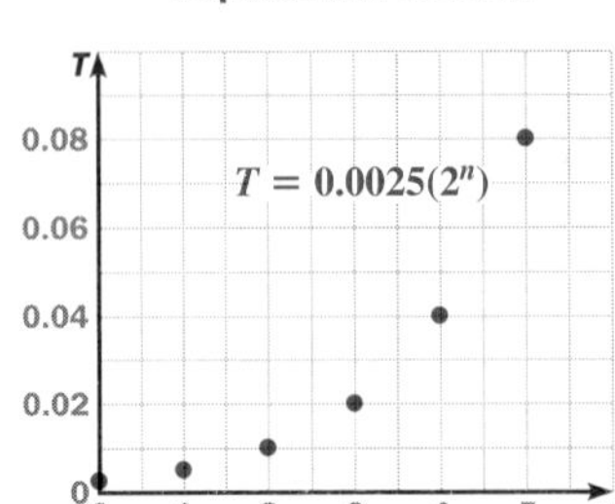

When $a > 0$ and $b > 1$, the graph of $y = ab^x$ *rises* from left to right.

Exponential Decay

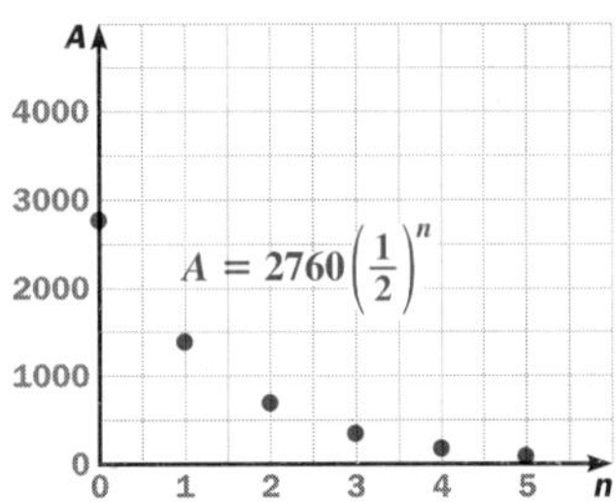

When $a > 0$ and $0 < b < 1$, the graph of $y = ab^x$ *falls* from left to right.

THINK AND COMMUNICATE

Use the equations for stack thickness and page area given with the graphs above.

5. What is the original amount for the stack thickness function? for the page area function?
6. What is the proportional growth factor for the stack thickness function?
7. What is the proportional decay factor for the page area function?
8. What "event" is represented by the variable *n*?

☑ CHECKING KEY CONCEPTS

Tell how many leaves would be formed if a sheet of paper could be folded the given number of times.

1. 7 2. 8 3. 9 4. 10
5. The "Demy" paper size is 38 cm wide and 51 cm high.
   a. Make a table of the width, height, and area of the page sizes made by folding a sheet of Demy-sized paper.
   b. Write an exponential decay equation for the page area as a function of the number of folds.

Apply⇔Assess

Suggested Assignment

Core Course
Exs. 1–16, 22, 27–34

Extended Course
Exs. 1–4, 6–11, 13–34

Block Schedule
Day 14 Exs. 1–16, 22, 27–34

Think and Communicate
5. 0.0025; 2760
6. 2
7. $\frac{1}{2}$
8. folding the paper *n* times

Checking Key Concepts
1. 128 leaves
2. 256 leaves
3. 512 leaves
4. 1024 leaves
5. See answers in back of book.

Exercises and Applications
1. 4 folds
2. 6 folds
3. 3 folds
4. 7 folds
5. 2^4
6. 2^7
7. 2^6
8. 2^{12}
9. 160
10. 6400
11. 200
12. 4
13. linear
14. exponential
15. neither
16. exponential

3.1 Exercises and Applications

Extra Practice exercises on page 752

Tell how many folds are needed in a sheet of paper to create the given number of leaves.

1. 16 **2.** 64 **3.** 8 **4.** 128

Write each expression as a power of 2.

5. $2 \cdot 2 \cdot 2 \cdot 2$ **6.** $4 \cdot 32$ **7.** $2 \cdot 2^5$ **8.** $2^5 \cdot 2^7$

Evaluate each expression when $x = 5$.

9. $5(2^x)$ **10.** $200(2^x)$ **11.** $6400\left(\frac{1}{2}\right)^x$ **12.** $128\left(\frac{1}{2}\right)^x$

Tell whether each equation represents growth that is *linear, exponential,* or *neither.*

13. $y = (3^2)x$ **14.** $y = 3(2^x)$ **15.** $y = 3x^2$ **16.** $y = 2(3^x)$

17. **Technology** Use a graphing calculator or graphing software to graph the equation from part (b) of Example 3. Use the trace feature to find how many folds are needed to make a stack of pages taller than you are.

18. **Visual Thinking** Fold a piece of paper in half four times and unfold it. How can you number the small rectangles on both sides so that when you fold the paper again and trim it along three edges, you get a small book with 32 correctly numbered pages? (*Hint:* One side of one possible answer is shown. How should the other side be numbered?)

24	9	16	17
25	8	1	32
28	5	4	29
21	12	13	20

19. **Investigation** Try folding a sheet of paper in half six times. What do you notice? Why do you think signatures of 64 leaves or more are not commonly used in bookmaking? Look at some hardcover books to see how many folds were used to create the signatures.

20. **Spreadsheets** You can use a spreadsheet to make a table of the width, height, and area of the page sizes made by folding a sheet of paper. An example, using the "Demy" paper size, is shown.

a. The "Pott" paper size is 31 cm wide and 39 cm high. Use a spreadsheet to make a table of the width, height, and area of the page sizes made by folding a sheet of Pott-sized paper.

b. Write an exponential decay equation for the page area as a function of the number of folds.

c. Graph the equation from part (b).

Bookmaking Page Sizes

C9 =0.5*D8

	A	B	C	D	E
1	Name of page size	Number of folds	Page width (cm)	Page height (cm)	Page area (cm^2)
2	DEMY	0	38	51	1938
3	folio	1	25.5	38	969
4	quarto	2	19	25.5	484.5
5	octavo	3	12.75	19	242.25
6	16mo	4	9.5	12.75	121.125
7	32mo	5	6.375	9.5	60.5625
8	64mo	6	4.75	6.375	30.28125
9	128mo	7	3.1875	4.75	15.140625

This is half of the amount in cell D8.

This is the same as the amount in cell C8.

Apply⇔Assess

Exercise Notes

Common Error

Exs. 9–16 When doing exercises such as these, students may make order-of-operations errors. With paper and pencil, the most common error is to multiply before raising the number in parentheses to the appropriate power. Students who get incorrect answers when using a calculator usually have forgotten that the parentheses are important when entering fractions, such as in Exs. 11 and 12. Review the rules for the order of operations to help students correct these kinds of errors.

All of the equations in Exs. 13–16 contain an exponent. Students who classify the equations incorrectly have probably not paid close attention to the definitions of linear and exponential functions. Stress that in order for an equation to describe exponential growth or decay, the independent variable must occur as an exponent.

Using Technology

To solve Ex. 17, it is necessary to enter the equation on the Y= list and display a continuous graph, but this graph will contain points other than those that represent thickness after various folds. To focus students' attention on the points of the continuous graph that are also points of the discrete graph, have them choose 8:ZInteger from the ZOOM menu. After moving the cursor to the center of the window desired, press ENTER and TRACE. By using the ▶ key, the cursor will move only to points with integer-valued x-coordinates. You may also want to suggest that students use the *Function Investigator* software for this exercise.

Problem Solving

Ex. 18 One way for students to begin this exercise is to fold a sheet of paper as directed and then cut the pages only halfway. Students can then open the "book" enough to write page numbers on each page. Then have them unfold the sheet and look for patterns and relationships. Students should note that there are various ways to number the sections to achieve a solution.

Using Technology

Ex. 20 You may want to suggest that students use the *Stats!* software for this exercise.

17.

-20

18

Answers may vary. For example, 15 folds would be required for a person 5 ft 6 in. tall.

18. Answers may vary. The numbering for the other side of the possible answer given in the text is shown. The page is shown flipped horizontally. The section numbered "18" is the other side of the section numbered "17."

18	15	10	23
31	2	7	26
30	3	6	27
19	14	11	22

19. It is difficult to fold a sheet of paper 6 times, no matter how large a sheet of paper you begin with. For 64 or more leaves, at least 6 folds are required. Answers may vary. Signatures for hardcover books are commonly produced using 3 or 4 folds.

20. See answers in back of book.

Apply⇔Assess

Exercise Notes

Second-Language Learners
Ex. 21 Invite students to play this game with dictionaries of languages other than English.

Research
Ex. 21 Students may find it interesting to see what they can find out about the number of words in various dictionaries. This leads naturally to the question of how many words there are in the English language. Students should realize that this raises a number of interesting issues: Are slang terms included, specialized technical terms, and so on? An informative class discussion can consider these and other questions.

Interdisciplinary Problems
Ex. 21 It may surprise students that the concepts of exponential growth or decay can be applied to an analysis of this dictionary word game. You can use this fact to enhance students' understanding of the power of mathematical concepts and reasoning to analyze and solve real-world problems in many disciplines.

Multicultural Note
Ex. 22 The Kalmyks are a Mongolian people who live in the autonomous republic of Kalmykia, part of the Russian Federation. They traditionally have held Tibetan Buddhist beliefs. In the seventeenth century, they migrated from western Mongolia and settled in a region northwest of the Caspian Sea. They were persecuted during World War II and many of them were deported to Siberia. After they returned to their homeland in 1957, they were allowed to establish their own autonomous republic. Today, most individuals make their living through sheep herding or by raising crops.

Teaching Tip
Exs. 23–25 Students should note that the areas in column D of the spreadsheet have been rounded to the nearest thousandth of a square meter.

21. Cooperative Learning Here is a way to discover a word someone is thinking of just by asking 20 questions:

Ask your friend to think of a word. Open a dictionary at the halfway point and ask a question such as, "Is your word before or after *lifesaver* in a dictionary?"

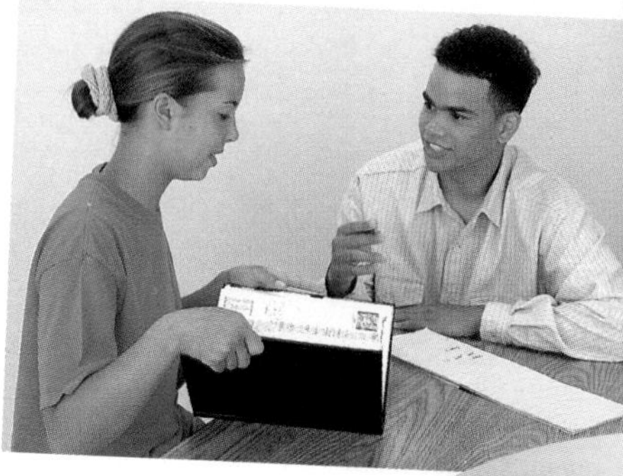

Suppose your friend is thinking of the word *electricity*. Your friend says, "Before."

Look in the middle of the first half and ask another question: "Is your word before or after *diesel*?" Your friend says, "After."

Divide the indicated section in half and continue asking questions this way until you either find the word or use up 20 questions.

a. Try the dictionary game with a friend or family member. Are you able to discover the word in fewer than 20 questions?

b. About how many words are in the dictionary you used in part (a)?

c. How many words would a dictionary have to include to require you to use more than 20 questions?

d. **Writing** Explain how the dictionary game works. Use powers of $\frac{1}{2}$ or powers of 2 in your explanation.

22. MYTHOLOGY The Mongol people known as Kalmyks in Central Asia believed the proportions of the universe were fixed by mathematical rules. In their tales, a central mountain called Sumeru rises 80,000 leagues above the surface of the world-ocean. Around it are seven circular mountain chains, the first and innermost being 40,000 leagues high, the next 20,000 leagues high, and so on.

a. Find the height of the seventh mountain chain.

b. Suppose the mountain chains continue beyond the seventh. Write an equation for the height of the nth mountain chain.

21. a, b. Answers may vary. Examples are given.

a. Yes.

b. about 200,000 words

c. more than 2^{20} or 1,048,576 words

d. Every time you ask a question, you eliminate half the remaining words. Let n = the number of questions and w = the number of words remaining. For a dictionary with 200,000 words, the number of remaining words is given by $w = 200{,}000\left(\frac{1}{2}\right)^n$. For $n = 18$, $w < 1$.

22. a. 625 leagues

b. Let h = the height in leagues of the mountain chain; $h = 80{,}000\left(\frac{1}{2}\right)^n$.

23–25. Let n = the paper-size number; that is, for example, 0 represents paper size A0.

23. Let w = paper width in millimeters; $w = 841(0.707)^n$ or $w = 841\left(\sqrt{\frac{1}{2}}\right)^n$.

24. Let h = paper height in millimeters; $h = 1189(0.707)^n$ or $h = 1189\left(\sqrt{\frac{1}{2}}\right)^n$.

25. Let A = paper area in square millimeters; $A = \left(\frac{1}{2}\right)^n$.

26. $\sqrt{2}$; Let W and L be the original paper width and length.

Connection GRAPHIC DESIGN

In many countries, the International Standards Organization "A series" of paper sizes is used, with metric measurements as shown.

ISO A Series Paper Sizes

	A	B	C	D
1	Name of paper size	Paper width (mm)	Paper height (mm)	Paper area (m^2)
2	A0	841	1189	1.000
3	A1	594	841	0.500
4	A2	420	594	0.249
5	A3	297	420	0.125
6	A4	210	297	0.062
7	A5	148	210	0.031
8	A6	105	148	0.016

A0 paper size

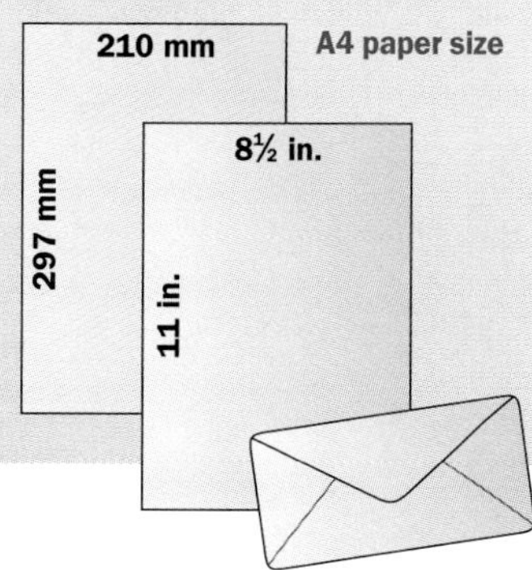

A4 paper size

Write an exponential decay equation for each variable. Use the numbers in the paper size names as the independent variable.

23. paper width **24.** paper height **25.** paper area

26. Challenge What is the ratio of paper height to paper width for any sheet in the "A series"? Explain why this ratio allows you to write exponential decay functions for paper width and paper height. Is there any other ratio that would allow you to write such functions?

The A0 paper size is 16 times larger than the A4 paper size.

ONGOING ASSESSMENT

27. GEOMETRY The edge length of each red triangle in the pattern is twice the edge length of the triangle above it.

a. Make a table showing edge length, perimeter, and area.

b. Write exponential growth equations for each column of the table.

c. Graph the equations from part (b).

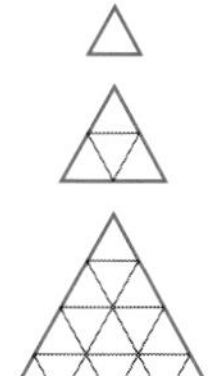
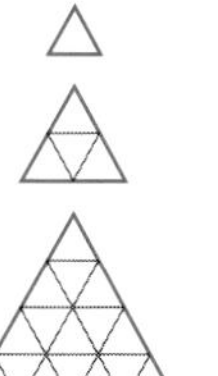

SPIRAL REVIEW

For each pair of parametric equations, express y as a function of x. *(Section 2.6)*

28. $x = 0.5t$, $y = 3(t - 1)$ **29.** $x = t + 1$, $y = t + 2$ **30.** $x = 4t + 1$, $y = t$

Tell whether each equation shows direct variation. *(Section 2.1)*

31. $y = 2.5x$ **32.** $y + 2 = x + 2$ **33.** $y = 2x - 1$

34. Write a linear equation that gives the number of decades, d, in y years. Is this an example of direct variation? *(Section 2.1)*

Apply⇔Assess

Exercise Notes

Visual Thinking

Ex. 27 Ask students to create other visual models that express the concepts of *exponential growth* and *exponential decay* without using numbers. Examples might include drawing a house or a person, then drawing one that is three times as big and one that is three times bigger, and drawing one that is one third the size and one that is one third the size of that. Encourage students to display and explain their models. This activity involves the visual skills of *interpretation* and *communication*.

Practice 14 for Section 3.1

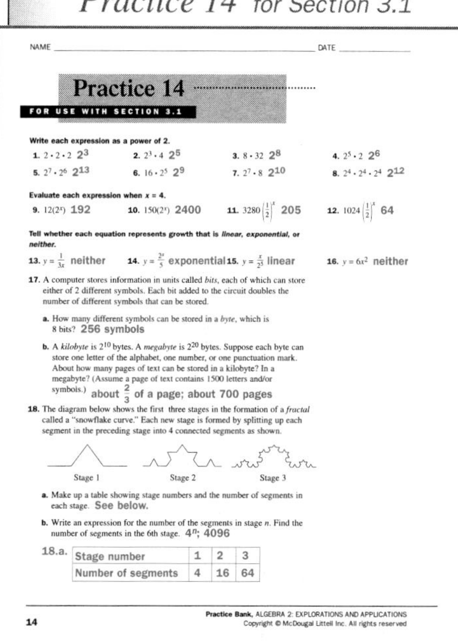
NAME ______ DATE ______

Practice 14

FOR USE WITH SECTION 3.1

Write each expression as a power of 2.

1. $2 \cdot 2 \cdot 2$ 2^3 2. $2^3 \cdot 4$ 2^5 3. $8 \cdot 32$ 2^8 4. $2^5 \cdot 2$ 2^6
5. $2^7 \cdot 2^6$ 2^{13} 6. $16 \cdot 2^5$ 2^9 7. $2^7 \cdot 8$ 2^{10} 8. $2^4 \cdot 2^4 \cdot 2^4$ 2^{12}

Evaluate each expression when $x = 4$.

9. $12(2^x)$ 192 10. $150(2^x)$ 2400 11. $3280\left(\frac{1}{2}\right)^x$ 205 12. $1024\left(\frac{1}{2}\right)^x$ 64

Tell whether each equation represents growth that is *linear, exponential,* or *neither.*

13. $y = \frac{1}{3x}$ neither 14. $y = \frac{2^x}{5}$ exponential 15. $y = \frac{x}{2^5}$ linear 16. $y = 6x^2$ neither

17. A computer stores information in units called *bits*, each of which can store either of 2 different symbols. Each bit added to the circuit doubles the number of different symbols that can be stored.

a. How many different symbols can be stored in a *byte*, which is 8 bits? 256 symbols

b. A *kilobyte* is 2^{10} bytes. A *megabyte* is 2^{20} bytes. Suppose each byte can store one letter of the alphabet, one number, or one punctuation mark. About how many pages of text can be stored in a kilobyte? In a megabyte? (Assume a page of text contains 1500 letters and/or symbols.) about $\frac{2}{3}$ of a page; about 700 pages

18. The diagram below shows the first three stages in the formation of a *fractal* called a "snowflake curve." Each new stage is formed by splitting up each segment in the preceding stage into 4 connected segments as shown.

Stage 1 Stage 2 Stage 3

a. Make up a table showing stage numbers and the number of segments in each stage. See below.

b. Write an expression for the number of the segments in stage n. Find the number of segments in the 6th stage. 4^n; 4096

18.a.

Stage number	1	2	3
Number of segments	4	16	64

14 **Practice Bank,** ALGEBRA 2: EXPLORATIONS AND APPLICATIONS

Because of the way widths and lengths are related after folding, a table looks like this:

Name of paper size	Paper width (cm)	Paper length (cm)
A0	W	L
A1	$L/2$	W
A2	$W/2$	$L/2$
…	…	…

Exponential decay functions can be written only if the ratios are constant, that is, if $\frac{W}{L/2} = \frac{L/2}{W/2}$ or $\frac{2W}{L} = \frac{L}{W}$. This is true if $\frac{L^2}{W^2} = 2$ or $\frac{L}{W} = \sqrt{2}$. No other ratio would allow you to write such functions.

27. See answers in back of book.

28. $y = 6x - 3$

29. $y = x + 1$

30. $y = \frac{1}{4}(x - 1)$

31. Yes.

32. Yes.

33. No.

34. $d = \frac{y}{10}$; Yes.

Plan⇔Support

Objectives

- Evaluate expressions that use negative and rational exponents.
- Describe situations involving continuous exponential growth or decay.

Recommended Pacing

Core and Extended Courses
Section 3.2: 2 days

Block Schedule
Section 3.2: 1 block

Resource Materials

Lesson Support
Lesson Plan 3.2
Warm-Up Transparency 3.2
Practice Bank: Practice 15
Study Guide: Section 3.2
Exploration Lab Manual:
Diagram Master 2

Technology
Graphing Calculator
McDougal Littell Mathpack
Function Investigator
Internet:
http://www.hmco.com

Warm-Up Exercises

Evaluate.

1. $1.48\left(\frac{1}{16}\right)$ 0.0925

2. $3.2\left(\frac{1}{125}\right)$ 0.0256

3. $720\left(\frac{1}{64}\right)$ 11.25

Use the properties of exponents to simplify each expression.

4. $(3^2)^5$ 3^{10}

5. $\frac{2^7}{2^2}$ 2^5

6. Write as a single fraction $3 \cdot 2^{-1}$
$\frac{3}{2}$

Section

3.2 Negative and Rational Exponents

Learn how to...

- evaluate expressions that use negative and rational exponents

So you can...

- describe situations involving continuous exponential growth or decay, such as population increases

Every ten years the federal government does a census to estimate the number of U.S. residents. It is not practical to do a census every year, but by using a model to describe the growth, you can estimate the population between census years.

EXAMPLE 1 Application: Census

Find an exponential model for the U.S. census data given in the spreadsheet.

U.S. Population Data

D3 =C3/C2

	A	B	C	D
1	Census year	Decades since 1800	Population (millions)	Proportional growth rate
2	1800	0	5.31	
3	1810	1	7.24	1.363
4	1820	2	9.64	1.331
5	1830	3	12.87	1.335
6	1840	4	17.07	1.326
7	1850	5	23.19	1.359

SOLUTION

Column D in the spreadsheet calculates the proportional growth factors between consecutive census years. The average of these proportional growth factors is about 1.34.

An exponential model for the data is

$$P = 5.31(1.34)^d$$

where P is the population in millions and d is the number of decades since 1800.

Negative Exponents

You can use the model from Example 1 to predict the population of the United States a decade before the 1800 census. Consider a pictograph of the first few population values predicted by the model.

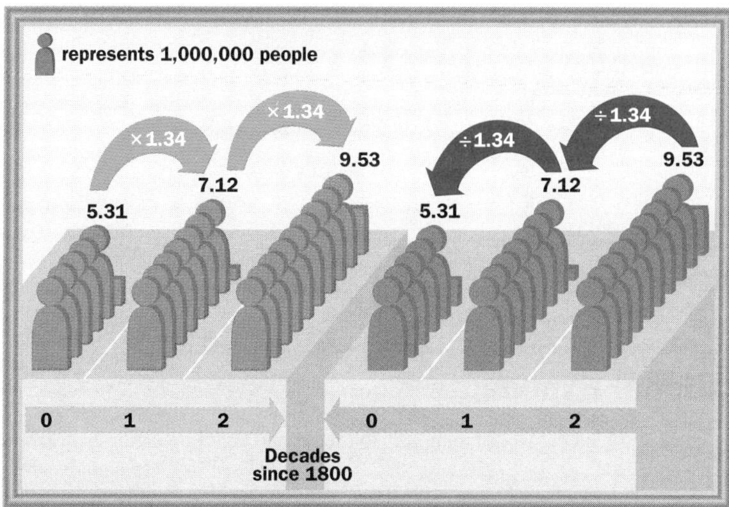

Moving forward in time means multiplying the population values by the proportional growth factor.

Moving backward in time means **dividing** the population values by the proportional growth factor.

THINK AND COMMUNICATE

1. Divide 5.31 by 1.34 to predict the U.S. population in 1790, one decade *before* the 1800 census.
2. a. What value of d would you use in the model $P = 5.31(1.34)^d$ to get the population in 1790?
   b. Use a calculator to evaluate $5.31(1.34)^d$ using the value of d from part (a). Do you get the same result as in Question 1?

You have seen that $5.31(1.34)^{-1} = \frac{5.31}{1.34}$. This implies that $1.34^{-1} = \frac{1}{1.34}$, which is an example of the following property of exponents.

The Meaning of b^{-n}

For any base $b > 0$ and any positive integer n:

$$b^{-n} = \frac{1}{b^n}$$

Example:

$$2^{-3} = \frac{1}{2^3} = \frac{1}{8}$$

Teach⇔Interact

Additional Example 1

Find an exponential model for the U.S. census data given in the table.

Census Year	Decades since 1900	Population (millions)	Prop. growth rate
1900	0	75.99	
1910	1	91.97	1.210
1920	2	105.71	1.149
1930	3	122.78	1.162
1940	4	131.67	1.072
1950	5	150.76	1.145

The last column of the table gives the proportional growth factors between consecutive census years. The average of these proportional growth factors is about 1.148.

An exponential model for the data is $P = 75.99(1.148)^d$, where P is the population in millions and d is the number of decades since 1900.

Section Notes

Reasoning
Students should understand that the definition of b^{-n} extends previously stated rules for multiplying powers to situations involving negative exponents. They may find it helpful to observe that for any base $b > 0$ and any positive integer n, $b^{-n} \cdot b^n = b^{-n+n} = b^0 = 1$. From $b^{-n} \cdot b^n = 1$, it follows that $b^{-n} = \frac{1}{b^n}$.

Mathematical Procedures
Negative exponents may seem strange to some students because of their prior experiences with the meaning of positive exponents. Emphasize that to evaluate expressions with negative exponents, the expression can always be rewritten in terms of positive exponents and then evaluated.

ANSWERS Section 3.2

Think and Communicate

1. about 3.96 million
2. a. $d = -1$
   b. about 3.96; Yes.

Teach⇔Interact

Additional Example 2

Predict the U.S. population in 1870.

Use the model $P = 75.99(1.148)^d$ from Additional Example 1. Since 1870 is three decades before the 1900 census, $d = -3$.

```
75.99(1.148)^-3
      50.22625304
```

You can get the same result by evaluating $\frac{75.99}{(1.148)^3}$. The predicted population in 1870 is about 50.23 million.

Think and Communicate

Students are used to thinking linearly. They may assume that because $\frac{1}{2}$ is halfway between 0 and 1 that $5.31(1.34)^{1/2}$ is halfway between $5.31(1.34)^0$ and $5.31(1.34)^1$. Remind students that any part of an exponential growth curve increases more slowly in the first half than in the second half.

Section Notes

Student Progress

It may take time for some students to fully assimilate the meaning of a rational exponent as defined for $b^{1/n}$. You may wish to use a few more examples by having students complete the following:

$4^{1/2} = ?$

$27^{1/3} = ?$

$25^{1/2} = ?$

$64^{1/3} = ?$

Using Technology

Point out to students that it is important to use parentheses around the exponent when evaluating expressions of the form $b^{1/n}$ on a calculator. To evaluate $(1.34)^{1/2}$, for example, they should use (1.34)^(1/2) or 1.34^(1/2). If parentheses are not used around the exponent, the calculator will calculate 1.34^1 and divide the result by 2.

EXAMPLE 2

Predict the U.S. population in 1780.

SOLUTION

Use the model $P = 5.31(1.34)^d$ from Example 1.

Since 1780 is two decades before the 1800 census, $d = -2$.

You can also get this result by evaluating $\frac{5.31}{1.34^2}$.

The predicted population in 1780 is about 2.96 million.

Rational Exponents

You can use the model from Example 1 to predict the U.S. population *between* census years, too.

Just as 1.34 is the population's 10-year proportional growth factor, $(1.34)^{1/2}$ is the population's 5-year growth factor.

Year	1800	1805	1810
Decades since 1800	0	$\frac{1}{2}$	1
Population (millions)	$5.31(1.34)^0$	$5.31(1.34)^{1/2}$	$5.31(1.34)^1$

THINK AND COMMUNICATE

3. a. You can use a calculator to find $(1.34)^{1/2}$. What do you get? Store this result in memory.
   b. Multiply 5.31 by the number stored in memory. What does the result represent?
   c. Multiply the result from part (b) by the number stored in memory. What does the result represent?
4. a. Assume that the power property, $(b^m)^n = b^{mn}$, applies to rational as well as integral values of m and n. Apply the property to $\left[(1.34)^{1/2}\right]^2$. What do you get?
   b. Describe the mathematical relationship between $(1.34)^{1/2}$ and 1.34.

Toolbox p. 799
Logical Statements

The Meaning of $b^{1/n}$

For any base $b > 0$ and any positive integer n, $b^{1/n}$ is the number c whose nth power is b.

$b^{1/n} = c$ if and only if $b = c^n$

Example:

$8^{1/3} = 2$ because $8 = 2^3$

Think and Communicate

3. a. 1.15758369
   b. 6.146769395 ≈ 6.15; the predicted population in millions in 1805
   c. 7.1154 ≈ 7.12; the predicted population in millions in 1810
4. a. $[(1.34)^{1/2}]^2 = 1.34^{(1/2)2} = 1.34^1 = 1.34$
   b. $(1.34)^{1/2} = \sqrt{1.34}$

By thinking of $b^{m/n}$ as $\left(b^{1/n}\right)^m$, you can see that it is possible to raise a base to any rational power, such as $\frac{2}{3}$ or -2.4. The properties of exponents listed in Section 3.1 also apply to all rational exponents.

Toolbox p. 779
Rational Numbers and Irrational Numbers

EXAMPLE 3

Use the properties of exponents to simplify each expression.

a. $5^{2/3} \cdot 5^{5/3}$ **b.** $\dfrac{2^{7/3}}{2^{2/3}}$ **c.** $\left(6^{1/3}\right)^{3/5}$

SOLUTION

a. $5^{2/3} \cdot 5^{5/3} = 5^{(2/3\ +\ 5/3)}$
$= 5^{7/3}$

b. $\dfrac{2^{7/3}}{2^{2/3}} = 2^{(7/3\ -\ 2/3)}$
$= 2^{5/3}$

c. $\left(6^{1/3}\right)^{3/5} = 6^{(1/3 \cdot 3/5)}$
$= 6^{1/5}$

EXAMPLE 4

Predict the U.S. population in 1776.

SOLUTION

Use the model $P = 5.31(1.34)^d$.

Since 1776 is 24 years, or 2.4 decades, before 1800, $d = -2.4$.

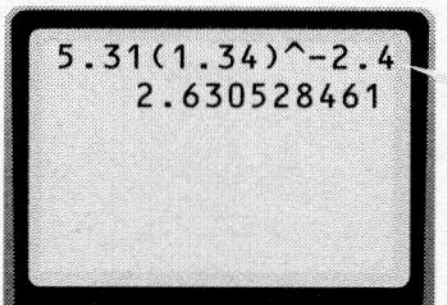

A calculator will accept negative rational exponents as well as positive ones.

The predicted population in 1776 is about 2.63 million.

BY THE WAY

In 1776, Philadelphia had a population of 40,000, which was more than the populations of Boston and New York combined.

EXAMPLE 5

Use the equation $P = 5.31(1.34)^d$ to write an equation that gives the U.S. population in terms of y years after 1800 instead of d decades.

SOLUTION

The number of decades is one tenth the number of years: $d = \frac{1}{10}y$.

$P = 5.31(1.34)^d$

$= 5.31(1.34)^{(1/10)y}$ ← **Substitute $\frac{1}{10}y$ for d.**

$= 5.31\left[(1.34)^{1/10}\right]^y$ ← $b^{mn} = \left(b^m\right)^n$

$= 5.31(1.03)^y$ ← **$(1.34)^{1/10} \approx 1.03$**

Teach⇔Interact

Additional Example 3

Use the properties of exponents to simplify each expression.

a. $6^{1/5} \cdot 6^{7/5}$

$6^{1/5} \cdot 6^{7/5} = 6^{(1/5 + 7/5)}$
$= 6^{8/5}$

b. $\dfrac{7^{9/5}}{7^{2/5}}$

$\dfrac{7^{9/5}}{7^{2/5}} = 7^{(9/5 - 2/5)}$
$= 7^{7/5}$

c. $(3^{1/2})^{2/5}$

$(3^{1/2})^{2/5} = 3^{(1/2 \cdot 2/5)}$
$= 3^{1/5}$

Additional Example 4

Predict the U.S. population in 1887.

Use the model $P = 75.99(1.148)^d$. Since 1887 is 13 years, or 1.3 decades, before 1900, $d = -1.3$.

```
75.99(1.148)^-1.
3
        63.5085197
```

The predicted population in 1887 is about 63.51 million.

Additional Example 5

Use the equation $P = 75.99(1.148)^d$ to write an equation that gives the U.S. population in terms of 5-yr periods after 1900 instead of d decades.

The number of 5-yr periods is one fifth the number of years: $d = \frac{1}{5}y$.

Substitute $\frac{1}{5}y$ for d in the model and use the properties of exponents.

$P = 75.99(1.148)^d$
$= 75.99(1.148)^{(1/5)y}$
$= 75.99[(1.148)^{1/5}]^y$
$= 75.99(1.028)^y$

Teach⇔Interact

Checking Key Concepts

Teaching Tip

To provide additional practice with the properties of exponents, have students evaluate the expressions in questions 2 and 4 by using different methods. For example, the expression in question 4 can be evaluated by entering any one of the following expressions on the home screen of a calculator.

5.31(1.34)^–1.5
5.31/1.34^1.5
5.31(1.34)^(–3/2)
5.31(1.34)$^{-1}$^1.5 (Use $\boxed{x^{-1}}$ for $^{-1}$.)

Closure Question

Explain the meaning of b^{-n} and $b^{1/n}$ if $b > 0$ and n is any positive integer.

b^{-n} is the reciprocal of b^n, that is, $b^{-n} = \frac{1}{b^n}$. $b^{1/n}$ is the number c whose nth power is b, that is, $b^{1/n} = c$ if and only if $b = c^n$.

Apply⇔Assess

Suggested Assignment

Core Course
Day 1 Exs. 1–25
Day 2 Exs. 26–30, 34, 35, 41–48

Extended Course
Day 1 Exs. 1–25
Day 2 Exs. 26–48

Block Schedule
Day 15 Exs. 1–30, 34, 35, 41–48

Exercise Notes

Common Error

Exs. 10–25 Watch for students who try to evaluate expressions involving negative or rational exponents by multiplying the base by the exponent. This error may result from a failure to understand the definitions of b^{-n} and $b^{1/n}$. Review the definitions and illustrate them using some of these exercises. Point out to students that although the minus sign is being used in the expression b^{-n}, its meaning is entirely different in this context. b^{-n} does not indicate a negative number, $-b^n$, but a reciprocal, $\frac{1}{b^n}$.

THINK AND COMMUNICATE

5. What is the significance of the number 1.03 in Example 5?

6. Evaluate the function $P = 5.31(1.03)^y$ when $y = 10$. How does the result help you check the solution to Example 5?

CHECKING KEY CONCEPTS

Evaluate each expression. Explain the significance of the result in terms of the population model in this section.

1. $5.31(1.34)^0$ **2.** $5.31(1.34)^{-2}$

3. $5.31(1.34)^2$ **4.** $5.31(1.34)^{-1.5}$

5. Explain why the values predicted by the model $P = 5.31(1.34)^d$ do not exactly match the population data in the spreadsheet for Example 1.

Evaluate each expression.

6. $64^{1/2}$ **7.** $64^{1/3}$ **8.** $64^{3/2}$

9. $64^{2/3}$ **10.** $100^{1/2}$ **11.** $100^{1/3}$

3.2 Exercises and Applications

Extra Practice exercises on page 752

1. Use the model from Example 1. Explain why the population in 1805 is not halfway between the population in 1800 and the population in 1810.

Use the model $P = 5.31(1.34)^d$ to estimate the population for each year.

2. 1830 **3.** 1850 **4.** 1870

Use the model $P = 5.31(1.03)^y$ to estimate the population for each year.

5. 1812 **6.** 1799 **7.** 1950

8. The actual U.S. population in 1950 was 151.3 million. Does this agree with the prediction in Exercise 7? What may account for any lack of agreement?

9. Writing Explain why it does not make sense to use negative or non-integral exponents in the examples in Section 3.1. Explain why it *does* make sense to use such exponents in the examples in this section.

Simplify.

10. 3^{-2} **11.** 2^{-4} **12.** 5^{-3} **13.** 7^{-1}

Simplify using the properties of exponents.

14. $5^{1/2} \cdot 5^{1/2}$ **15.** $6^{1/2} \cdot 6^{3/2}$ **16.** $3^{3/2} \cdot 3^{-7/2}$ **17.** $6^{2/3} \cdot 6^{-2/3}$

18. $\frac{2^{5/2}}{2^{1/2}}$ **19.** $\frac{4^{13/5}}{4^{3/5}}$ **20.** $\frac{25^{3/4}}{25^{1/4}}$ **21.** $\frac{4^{7/3}}{2^{11/3}}$

22. $(7^4)^{1/2}$ **23.** $(25^3)^{1/6}$ **24.** $(8^{2/3})^{3/2}$ **25.** $(16^0)^{3/7}$

Think and Communicate

5. It is the growth factor per year.

6. about 7.14; The value of the function $P = 5.31(1.03)^y$ when $y = 10$ should be about 7.12, the value of the function $P = 5.31(1.34)^d$ when $d = 1$.

Checking Key Concepts

1–4. Answers are rounded to two decimal places.

1. 5.31; the predicted population in millions in 1800

2. about 2.96; the predicted population in millions in 1780

3. about 9.53; the predicted population in millions in 1820

4. about 3.42; the predicted population in millions in 1785

5. The growth rate is an estimate. Also, the proportional growth rates in Column D of the spreadsheet are not all equal, only approximately equal. This indicates that the exponential equation models the data only approximately.

6. 8 7. 4
8. 512 9. 16
10. 10 11. about 4.64

Exercises and Applications

1. The growth is exponential, not linear.

2–7. Answers are rounded to two decimal places.

Connection TRAFFIC

The Central Artery, a highway through Boston, was opened in the late 1950s. It was designed to accommodate 75,000 vehicles per day. The table shows how the traffic has increased over time. By 1989, this stretch of highway was more crowded than any interstate in America.

Year	Vehicles per day (thousands)
1959	75
1969	100
1979	140
1989	190

26. Find the average growth factor per decade.

27. Write an equation modeling the Central Artery traffic as a function of the number of decades after 1959.

In the 1990s Boston planners began building a larger highway underground to replace the Central Artery. The project was called "The Big Dig."

Use the equation from Exercise 27 to estimate the traffic for each year.

28. 1964 **29.** 1974 **30.** 1984

31. Use the method of Example 5 to rewrite the equation from Exercise 27 in terms of years rather than decades after 1959.

32. Use the new equation from Exercise 31 to predict the traffic in 2001. Why do you think Boston planners are replacing the Central Artery?

33. Open-ended Problem Write several numerical examples that confirm that the properties of exponents in Section 3.1 can be extended to include:

a. negative integer exponents b. fractional exponents

34. SAT/ACT Preview Which of the following is *true*?

A. $2^{1/3} \cdot 2^3 = 2^1$ B. $5^{1/2} + 5^{1/2} = 5^1$

C. $4^{-2} \cdot 4^2 = 8^0$ D. $6^{3/2} - 6^{1/2} = 6^1$

E. none of these

35. SNOWMAKING Artificial snow is made by combining air and water in a ratio that depends on the outdoor temperature. The table shows the rate of air flow needed per gallon of water as a function of temperature.

a. Write an exponential equation for the rate of air flow needed per gallon of water as a function of temperature.

b. Use the equation to predict the rate of air flow needed at 5°F, at 15°F, and at 25°F.

Temperature (°F)	Air flow per gallon of water (ft³/min)
0	3.0
10	4.7
20	9.8

Apply⇔Assess

Exercise Notes

Communication: Reading
Exs. 26–32 In this connection to traffic, students should observe that the year numbers in the table are spaced 10 years apart (in decades). Some students may overlook this fact because the year numbers are not multiples of 10.

Career Connection
Exs. 26–32 The creation of an underground highway involves the cooperation of people working in many different occupations. First, a comprehensive plan has to be developed by experts in urban planning, engineering, traffic control, landscape architecture, and so on. The actual building of such a highway employs people with diverse skills in many occupations. You might wish to have students consider the many different types of jobs that would be involved and list them on the board.

Journal Entry
Ex. 33 Students who have done a thorough job of choosing their examples should include their responses for this exercise in their journals.

28–30. Answers may vary due to rounding.

28. about 87,500 vehicles per day

29. about 119,000 vehicles per day

30. about 162,000 vehicles per day

31. $v = 75(1.03)^y$

32. about 260,000 vehicles per day; The Central Artery is already carrying more than 2.5 times the traffic it was intended to bear. The situation will continue to worsen.

33. Answers may vary.

34. C

35. a. Answers may vary depending on rounding and how the independent variable is defined. An example is given. Let d = the number of degrees above zero and a = the air flow per gallon of water in ft³/min; $a = 3.0(1.06)^d$.

b. Answers may vary due to rounding; about 4.0 ft³/min; about 7.2 ft³/min; about 12.9 ft³/min

2. 12.78 million 3. 22.94 million

4. 41.19 million 5. 7.57 million

6. 5.16 million 7. 447.38 million

8. No; a decrease in the population growth rate.

9. The examples in Section 3.1 deal with the physical process of folding paper. The concept of a negative or non-integral number of folds has no meaning. The examples in this section deal with time, with an arbitrary point selected to represent 0. It makes sense to let positive numbers represent times since the chosen 0 point, and negative numbers represent times before that time. Since time is continuous, it also makes sense to use non-integral exponents.

10. $\frac{1}{9}$ 11. $\frac{1}{16}$

12. $\frac{1}{125}$ 13. $\frac{1}{7}$

14. 5 15. 36

16. $\frac{1}{9}$ 17. 1

18. 4 19. 16

20. 5 21. 2

22. 49 23. 5

24. 8 25. 1

26. about 1.36

27. Let d = the number of decades after 1959 and v = the number of vehicles per day in thousands; $v = 75(1.36)^d$.

Apply⇔Assess

Exercise Notes

Communication: Discussion
Exs. 36–39 Students who are knowledgeable about reading or composing music may wish to help answer their classmates questions when discussing these exercises.

Research
Exs. 36–39 Students who are interested in music may wish to research ideas related to those studied here for non-Western music.

Assessment Note
Ex. 41 This exercise not only helps to check whether students have understood the important ideas of this section, but it also helps prepare them for work in Section 3.3.

Practice 15 for Section 3.2

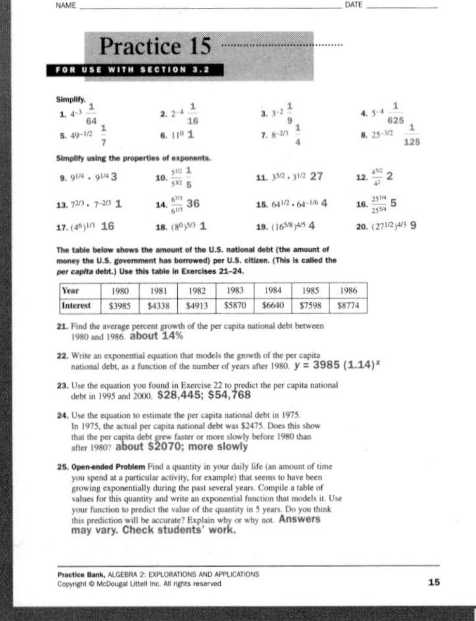

NAME ______ DATE ______

Practice 15

FOR USE WITH SECTION 3.2

Simplify.

1. 4^{-3} $\frac{1}{64}$
2. 2^{-4} $\frac{1}{16}$
3. 3^{-2} $\frac{1}{9}$
4. 5^{-4} $\frac{1}{625}$
5. $49^{-1/2}$ $\frac{1}{7}$
6. 11^0 1
7. $8^{-2/3}$ $\frac{1}{4}$
8. $25^{-3/2}$ $\frac{1}{125}$

Simplify using the properties of exponents.

9. $9^{1/4} \cdot 9^{1/4}$ 3
10. $\frac{5^{1/2}}{5^{3/2}}$ $\frac{1}{5}$
11. $3^{5/2} \cdot 3^{1/2}$ 27
12. $\frac{4^{5/2}}{4^2}$ 2
13. $7^{2/3} \cdot 7^{-2/3}$ 1
14. $\frac{6^{7/3}}{6^{1/3}}$ 36
15. $64^{1/2} \cdot 64^{-1/6}$ 4
16. $\frac{25^{7/4}}{25^{5/4}}$ 5
17. $(4^6)^{1/3}$ 16
18. $(8^0)^{5/3}$ 1
19. $(16^{5/8})^{4/5}$ 4
20. $(27^{1/2})^{4/3}$ 9

The table below shows the amount of the U.S. national debt (the amount of money the U.S. government has borrowed) per U.S. citizen. (This is called the *per capita* debt.) Use this table in Exercises 21–24.

Year	1980	1981	1982	1983	1984	1985	1986
Interest	\$3985	\$4338	\$4913	\$5870	\$6640	\$7598	\$8774

21. Find the average percent growth of the per capita national debt between 1980 and 1986. **about 14%**
22. Write an exponential equation that models the growth of the per capita national debt, as a function of the number of years after 1980. $y = 3985\,(1.14)^x$
23. Use the equation you found in Exercise 22 to predict the per capita national debt in 1995 and 2000. **\$28,445; \$54,768**
24. Use the equation to estimate the per capita national debt in 1975. In 1975, the actual per capita national debt was \$2475. Does this show that the per capita debt grew faster or more slowly before 1980 than after 1980? **about \$2070; more slowly**
25. **Open-ended Problem** Find a quantity in your daily life (an amount of time you spend at a particular activity, for example) that seems to have been growing exponentially during the past several years. Compile a table of values for this quantity and write an exponential function that models it. Use your function to predict the value of the quantity in 5 years. Do you think this prediction will be accurate? Explain why or why not. **Answers may vary. Check students' work.**

Practice Bank, ALGEBRA 2: EXPLORATIONS AND APPLICATIONS

15

Connection MUSIC

The musical note "concert A" has a frequency of 440 vibrations per second. The A note that is one octave higher has a frequency of 880 vibrations per second. The frequencies of the notes between these two A's follow an exponential pattern:

$$F = 440 \cdot 2^{x/12}$$

The bassoon can play an A with a frequency of 110 vibrations per second, which is two octaves lower than concert A.

36. Explain why the growth factor between two adjacent notes is $2^{1/12}$.

37. Copy and complete the table of frequencies shown below. Give answers to the nearest tenth.

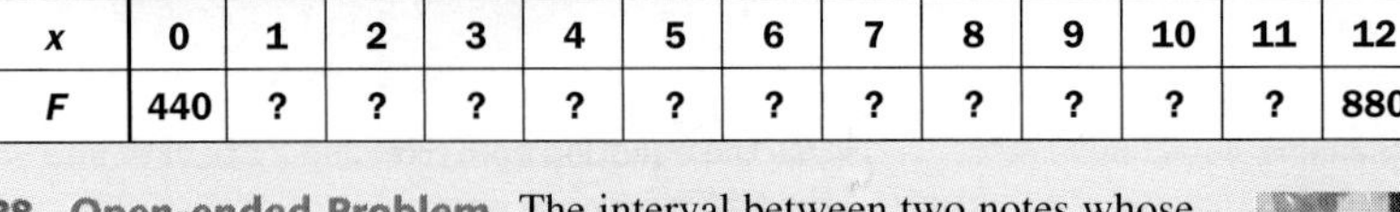

Note	A	A#	B	C	C#	D	D#	E	F	F#	G	G#	A
x	0	1	2	3	4	5	6	7	8	9	10	11	12
F	440	?	?	?	?	?	?	?	?	?	?	?	880

38. Open-ended Problem The interval between two notes whose frequencies have a ratio of 2:3 is called a "perfect fifth." Find two notes whose interval is close to a perfect fifth.

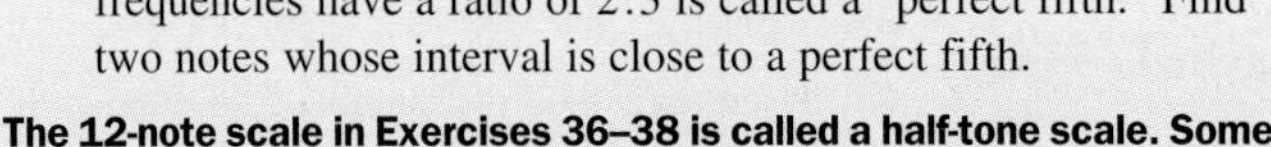

The 12-note scale in Exercises 36–38 is called a half-tone scale. Some composers have used scales that include more than 12 notes.

Julián Carrillo

39. The quarter-tone scale has 24 notes. Its frequency equation is $F = 440 \cdot 2^{x/24}$. Complete the table of frequency equations for sixth-, eighth-, and sixteenth-tone scales.

Name of scale	Notes in scale	Frequency equation
half-tone	12	$F = 440 \cdot 2^{x/12}$
quarter-tone	24	$F = 440 \cdot 2^{x/24}$
sixth-tone	36	$F = 440 \cdot 2^{x/?}$
eighth-tone	48	$F = 440 \cdot 2^{x/?}$
sixteenth-tone	96	$F = 440 \cdot 2^{x/?}$

40. The Mexican composer Julián Carillo used the sixteenth-tone scale. Find the frequencies of the first 10 notes in his 96-note scale. Start with the frequency 440.

ONGOING ASSESSMENT

41. Cooperative Learning Work with another student. One of you should graph $y = 8^{x/3}$ for $x = 0, 1, 2,$ and 3. The other should graph $y = 2^x$ for $x = 0, 1, 2,$ and 3. Compare your graphs. What do you notice?

SPIRAL REVIEW

Evaluate each expression. *(Section 3.1)*

42. $7(2^3)$ **43.** $3.2(0.5)^4$ **44.** $4.8(2^1)$

Compare each graph to the graph of $y = 3 + 2x$. *(Section 2.2)*

45. $y = 6 + 2x$ **46.** $y = 3 + 6x$ **47.** $y = 1 + 2(x + 1)$

48. Carry out a simulation to estimate how many people are needed on average before two have the same birth month. *(Section 1.5)*

36. $440 \cdot 2^{x/12} = 440(2^{1/12 \cdot x}) = 440(2^{1/12})^x$

37–40. See answers in back of book.

41. [graph with axes x and y, marks 2 and 2, origin O, points plotted]

The graphs are identical.

42. 56

43. 0.2

44. 9.6

45. The graphs have the same slope but different y-intercepts. The graph of $y = 6 + 2x$ is the graph of $y = 3 + 2x$ shifted 3 units in the positive y-direction.

46. The graphs have the same y-intercept, but the slope for $y = 3 + 6x$ is greater. The graph of $y = 3 + 6x$ is steeper than the graph of $y = 3 + 2x$.

47. The equation is simply a point-slope equation of the same line. The graphs are identical.

48. Simulation methods and results may vary. An example is given. I used the function int(12 * rand) + 1 to generate random numbers. For each trial, I generated random numbers until I had a repetition. I did this 10 times and each time recorded the number of numbers needed to produce a match. My results were: 6, 5, 5, 7, 5, 8, 5, 2, 4, 6. The average number of tries was 5.3 or about 5. Using these results, I estimate that five people on average are needed before two have the same birth month.

Section

3.3 Graphs of Exponential Functions

Learn how to...

- draw graphs of exponential functions
- interpret how different values of a and b affect the graph of $y = ab^x$

So you can...

- determine the doubling time for an investment or the half-life of caffeine, for example

In this section you will see how different values of a and b affect the graphs of exponential functions in the form $y = ab^x$.

EXPLORATION
COOPERATIVE LEARNING

Investigating Exponential Graphs

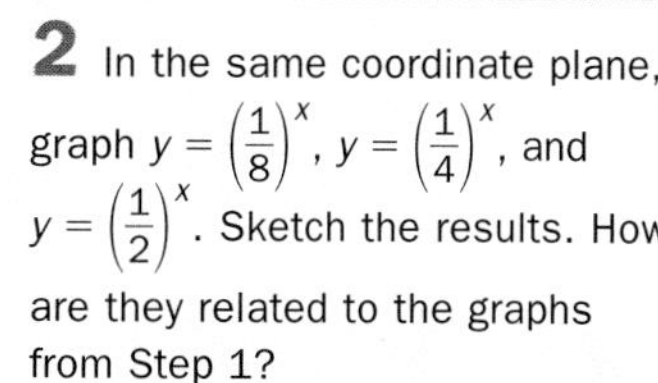

Work in a group of four students.
You will need:
- a graphing calculator

1 In the same coordinate plane, graph $y = 8^x$, $y = 4^x$, $y = 2^x$, and $y = 1^x$. Sketch the results. What do you notice about the y-intercepts?

2 In the same coordinate plane, graph $y = \left(\frac{1}{8}\right)^x$, $y = \left(\frac{1}{4}\right)^x$, and $y = \left(\frac{1}{2}\right)^x$. Sketch the results. How are they related to the graphs from Step 1?

3 Choose a positive value for b and graph $y = b^x$ and $y = \left(\frac{1}{b}\right)^x$. What do you notice about the graphs? Draw the graphs from your group on one piece of paper. Which graphs represent exponential growth? Which represent exponential decay?

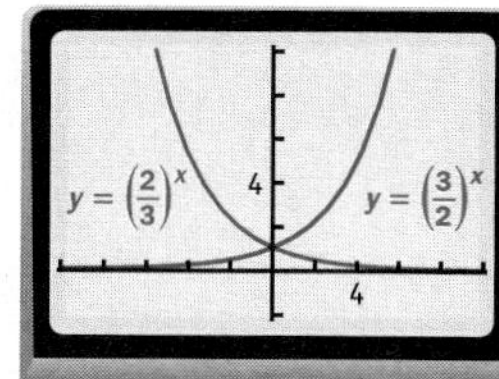

4 In the same coordinate plane, graph $y = 8 \cdot 2^x$, $y = 4 \cdot 2^x$, $y = 2 \cdot 2^x$, and $y = 1 \cdot 2^x$. Sketch the results. What do you notice?

5 Choose a positive value for a and graph $y = a \cdot 3^x$ and $y = -a \cdot 3^x$. What do you notice about the graphs? Draw the graphs from your group on one piece of paper.

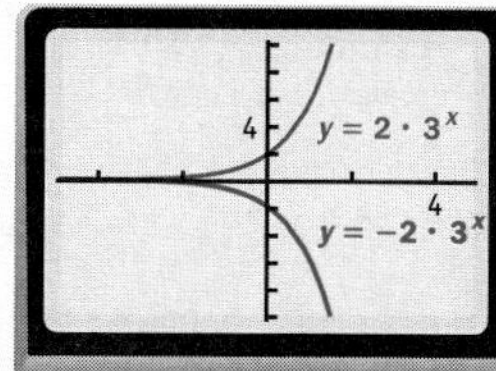

6 Discuss how the values of a and b affect the graph of $y = ab^x$.

Exploration Note

Purpose
The purpose of this Exploration is to have students discover how the values of a and b affect the graph of $y = ab^x$.

Materials/Preparation
Each group needs a graphing calculator.

Procedure
Urge students to experiment with different windows to obtain satisfactory graphs. Students should sketch how the graphs appear on the calculator screen and should indicate the scales used on the axes. Instruct students to record the values of b and a that they use in Steps 3 and 5, respectively.

Closure
Discuss the graphs obtained by different groups. Ask students to use their results to predict the general appearance of graphs of functions such as $y = 5^x$, $y = \left(\frac{1}{5}\right)^x$, $y = 2\left(\frac{4}{3}\right)^x$, and $y = -2\left(\frac{4}{3}\right)^x$. Check the predictions on a graphing calculator. Use Step 6 to summarize how the values of a and b affect the graph of $y = ab^x$.

Explorations Lab Manual
See the Manual for more commentary on this Exploration.

For answers to the Exploration, see following page.

Plan⇔Support

Objectives

- Draw graphs of exponential functions.
- Interpret how different values of a and b affect the graph of $y = ab^x$.
- Determine the doubling time or the half-life in real-world situations.

Recommended Pacing

Core and Extended Courses
Section 3.3: 2 days

Block Schedule
Section 3.3: 1 block

Resource Materials

Lesson Support
Lesson Plan 3.3
Warm-Up Transparency 3.3
Practice Bank: Practice 16
Study Guide: Section 3.3
Exploration Lab Manual: Diagram Master 2
Assessment Book: Test 9

Technology
Technology Book: Calculator Activity 3
Graphing Calculator
McDougal Littell Mathpack *Function Investigator with Matrix Analyzer Activity Book:* Function Investigator Activity 29
Internet: http://www.hmco.com

Warm-Up Exercises

For each function, tell whether the given point is on the graph of the function or not.

1. $y = 5^x$, (3, 125) Yes.
2. $y = 3^x$, (–2, –6) No.
3. $y = \left(\frac{1}{5}\right)^x$, (0, 1) Yes.
4. $y = 6(2)^x$, (0, 6) Yes.
5. $y = -4(0.5)^x$, (–1, 2) No.

Teach⟺Interact

About Example 1

Second-Language Learners
Some students learning English may not be familiar with the phrase *interest compounded annually.* If necessary, explain that the initial sum of money (in this case, \$300) will gain an additional 5% (\$15) if it is left in the bank for a year, and that in the year that follows, 5% interest will be paid on the compounded amount (\$315).

Additional Example 1

Suppose you have \$475 invested in a bank that offers 6% interest compounded annually.

a. Write an equation of the form $y = ab^x$ for the amount in your account after x years.
The interest rate is 6%, so the growth factor is 1.06.
$y = 475(1.06)^x$

b. How many years will it take to double your money?
You want to find the value of x for which $475(1.06)^x = 2 \cdot 475$.
Find the intersection of the graphs of $y = 950$ and $y = 475(1.06)^x$.

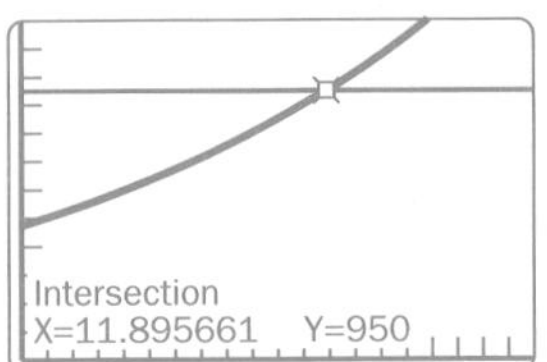

It will take about 11.9 years to double your money if the interest rate is 6%.

Section Note

Reasoning
You may wish to generalize the ideas in Example 1 to other growth factors. Suppose $b > 0$ and $b \neq 1$. Let k be a positive number and define c as the solution to $ab^c = ak$. Then $b^c = k$. Raising each side of this last equation to the power $\frac{1}{c}$ gives

$$(b^c)^{1/c} = k^{1/c}$$
$$b^{c/c} = k^{1/c}$$
$$b = k^{1/c}$$

Thus, it follows that $y = ab^x$ can be rewritten as $y = a(k^{1/c})^x = ak^{x/c}$, where k and c are defined so that $ab^c = ak$. In particular, if $k = \frac{1}{2}$, then c is the half-life and the formula $y = a\left(\frac{1}{2}\right)^{x/c}$ is derived.

Doubling Time

When something is growing exponentially, the time it takes for an initial amount to double, called the **doubling time**, is one way of describing its growth.

EXAMPLE 1 Application: Banking

Suppose you have \$300 invested in a bank that offers 5% interest compounded annually.

a. Write an equation in the form $y = ab^x$ for the amount in your account after x years.

b. How many years will it take to double your money?

For more information about finding intersections, see the *Technology Handbook*, p. 808.

SOLUTION

a. $y = 300(\mathbf{1.05})^x$

The interest rate is 5%, so the **growth factor** is 1.05.

b. You want to find the value of x for which $300(1.05)^x = 2 \cdot 300$.

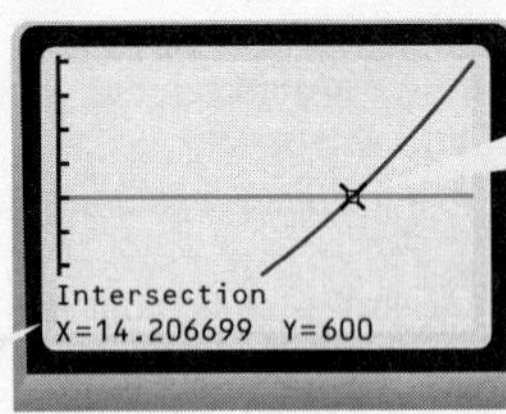

Find the intersection of the graphs of $y = 600$ and $y = 300(1.05)^x$.

It will take about **14.2 years** to double your money if the interest rate is 5%.

You saw in the Exploration that an exponential function in the form $y = ab^x$ represents *growth* when a is positive and b is greater than 1. Such a growth function can always be rewritten in this form:

$$y = a \cdot 2^{x/d}$$ d is the doubling time.

For instance, the equation in Example 1 could be rewritten this way:

$$y = 300 \cdot 2^{x/14.2}$$

THINK AND COMMUNICATE

Toolbox p. 781 *Simplifying Variable Expressions*

1. Show that the equations $y = 300(1.05)^x$ and $y = 300 \cdot 2^{x/14.2}$ are equivalent equations by:
a. comparing their graphs **b.** evaluating $2^{1/14.2}$

2. Joey says, "If each year the bank increases the amount by 5%, then in 14.2 years the total increase is only 71%. That's not double!" Explain what is wrong with his reasoning.

ANSWERS Section 3.3

Exploration

1. 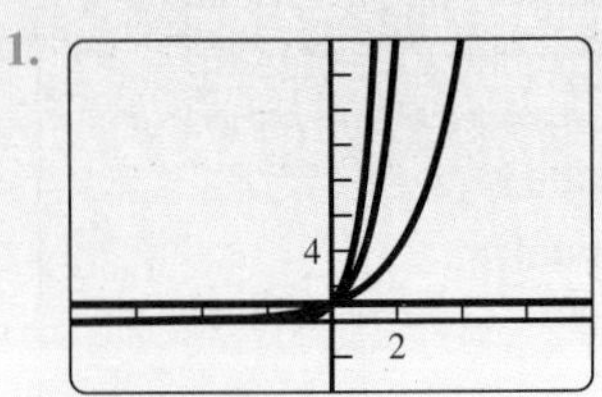

The graphs have the same y-intercept, 1.

2. 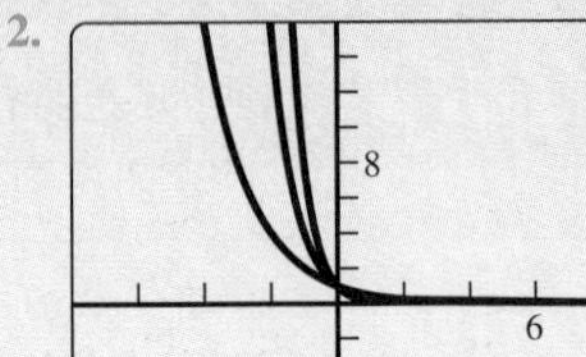

The graph of $y = b^x$ is the image of the graph of $y = \left(\frac{1}{b}\right)^x$ reflected over the y-axis.

3. Graphs may vary. Each graph is the reflection of the other over the y-axis. For $b > 1$, the graph of $y = b^x$ represents exponential growth. Since $0 < \frac{1}{b} < 1$, the graph of $y = \left(\frac{1}{b}\right)^x$ represents exponential decay.

4. 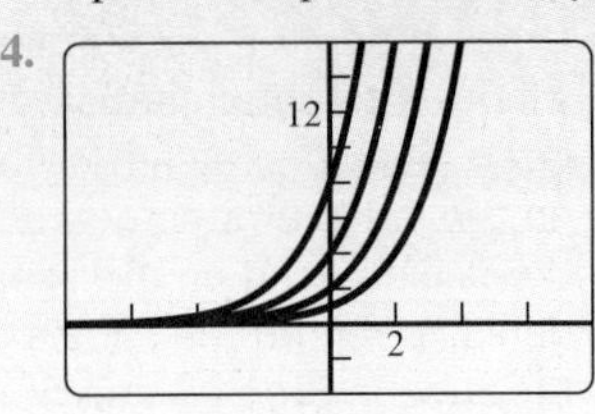

For each graph, the numerical coefficient is the y-intercept.

Half-Life

A function in the form $y = ab^x$ represents *decay* when a is positive and b is between 0 and 1. A decay function can always be rewritten in this form:

$$y = a \cdot \left(\frac{1}{2}\right)^{x/h}$$ h is the *half-life*.

The **half-life** is the amount of time needed for the function to be half its initial value.

EXAMPLE 2 Application: Physiology

A 5 oz cup of coffee has about 120 mg of caffeine. Caffeine is eliminated from the bloodstream at a rate of about 12% per hour in adults.

a. Write an equation of the form $y = ab^x$ for the amount of caffeine in the bloodstream after x hours.

b. How many hours will it take for half the caffeine to be eliminated?

c. Write an equation of the form $y = a \cdot \left(\frac{1}{2}\right)^{x/h}$ for the amount after x hours.

SOLUTION

a. $y = 120(0.88)^x$

The **decay factor** is $1 - 0.12$, or 0.88.

b. You want to find the value of x for which $120(0.88)^x = \frac{120}{2}$.

Find the intersection of the graphs of $y = 60$ and $y = 120(0.88)^x$.

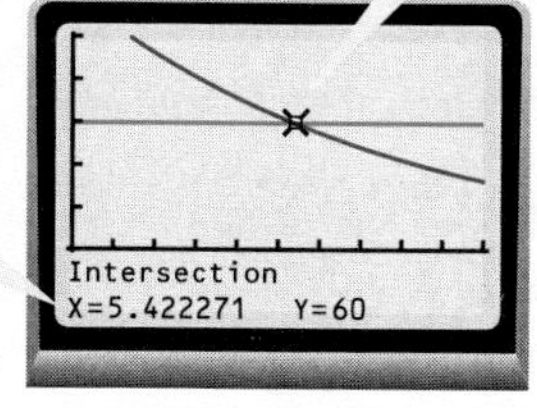

It will take about **5.4 h** for half the caffeine to be eliminated.

c. $y = 120 \cdot \left(\frac{1}{2}\right)^{x/5.4}$

Use **5.4** as the **half-life**.

THINK AND COMMUNICATE

3. Check part (c) of Example 2 by showing that $0.88 \approx 0.5^{1/5.4}$.
4. After how many hours will the amount of caffeine be 30 mg? 15 mg?
5. Suppose the initial amount of money in Example 1 was \$500 instead of \$300. How long would \$500 take to double?
6. Suppose the initial amount of caffeine in Example 2 was 80 mg instead of 120 mg. What would the half-life be?

5. Graphs may vary. Each graph is the reflection of the other over the x-axis.
6. Summaries may vary. The value of a determines the y-intercept of the graph. The table further describes the graphs.

	Above or below x-axis?	Represents exponential:
$a > 0, b > 1$	above	growth
$a > 0, 0 < b < 1$	above	decay
$a < 0, b > 1$	below	decay
$a < 0, 0 < b < 1$	below	growth

Think and Communicate

1. See answers in back of book.
2. Joey would be correct if the bank paid simple interest. Instead, the bank compounds the interest, that is, pays interest each year on the total amount in the account, including the original investment as well as previously earned interest.
3. Using a calculator, $0.5^{1/5.4} \approx 0.8795361709 \approx 0.88$.

4–6. Estimates may vary.

4. about 10.8 h; about 16.2 h
5. about 14.2 years
6. about 5.4 h

Teach⇔Interact

Think and Communicate

When discussing question 1, be sure students understand why evaluating $2^{1/14.2}$ helps show that the two equations are equivalent.

Additional Example 2

A particular medication is eliminated from a patient's bloodstream at the rate of 15% per hour. Suppose a tablet contains 20 mg of the medication.

a. Write an equation of the form $y = ab^x$ for the amount of the medication in the bloodstream after x hours.

The decay factor is $1 - 0.15$, or 0.85.

$y = 20(0.85)^x$

b. How many hours will it take for half the medication from a tablet to be eliminated?

Find the value of x for which $20(0.85)^x = \frac{20}{2}$. To do this, find the intersection of the graphs of $y = 10$ and $y = 20(0.85)^x$.

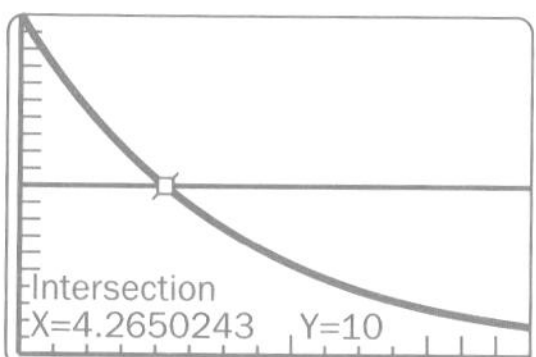

It will take about 4.27 h for half of the medication to be eliminated.

c. Write an equation of the form $y = a \cdot \left(\frac{1}{2}\right)^{x/h}$ for the amount after x hours.

Use 4.27 as the half-life.

$y = 20 \cdot \left(\frac{1}{2}\right)^{x/4.27}$

Think and Communicate

When discussing question 4, ask students if they observe a pattern in the number of hours for 60 mg, 30 mg, and 15 mg to remain in the bloodstream. (The number of hours increases by 5.4.) Ask students to generalize their answers for questions 5 and 6. (The doubling time for an exponential growth situation depends only on the rate of growth. The half-life for an exponential decay situation depends only on the rate of decay. The initial amount is irrelevant in determining doubling time and half-life.)

Teach⇔Interact

Section Note

Reasoning

You may wish to ask students whether a function such as $y = -3\left(\frac{1}{2}\right)^{x/2}$ describes exponential growth or decay. The values of *y* increase as one moves along the graph from left to right, and the function is an exponential function. However, the definitions at the top of this page rule out this function as representing exponential growth or decay since the factor –3 is not positive.

Checking Key Concepts

Communication: Discussion
Ask students which of the functions in questions 4–12 represent exponential growth. (4, 5, 6, 10, and 12). Ask students to explain how to identify the functions that represent exponential growth. (Look for functions in which the growth factor is greater than 1 and for which the initial amount *a* (in $y = ab^x$) is positive.)

Closure Question

How can you tell by looking at the graph of an exponential function of the form $y = ab^x$ whether the function represents exponential growth or exponential decay?

Exponential growth functions and exponential decay functions have curved graphs that lie in quadrants I and II. Exponential growth functions increase from left to right. Exponential decay functions decrease from left to right.

Exponential Growth and Decay

Exponential Growth

The function $y = ab^x$ represents exponential growth when $a > 0$ and $b > 1$.

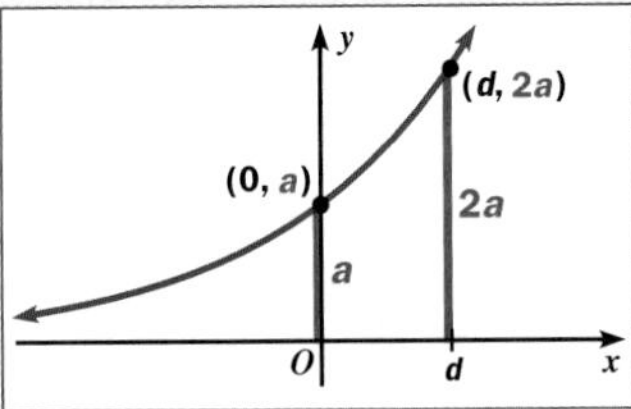

The graph shows that the initial amount *a* is doubled when $x = d$, the doubling time. Replacing *b* with $2^{1/d}$ gives:

$$y = a \cdot 2^{x/d}$$

Exponential Decay

The function $y = ab^x$ represents exponential decay when $a > 0$ and $0 < b < 1$.

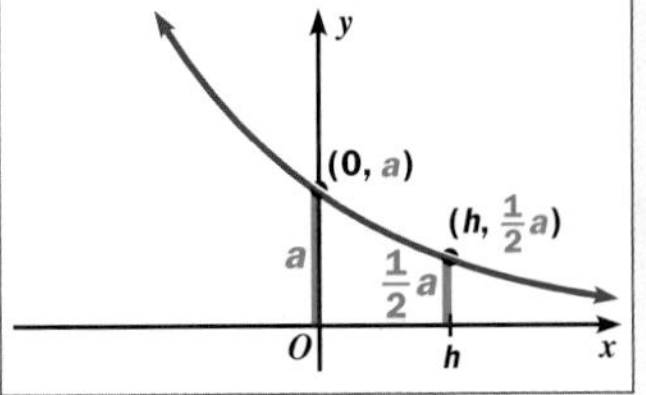

The graph shows that the initial amount *a* is halved when $x = h$, the half-life. Replacing *b* with $\left(\frac{1}{2}\right)^{1/h}$ gives:

$$y = a \cdot \left(\frac{1}{2}\right)^{x/h}$$

CHECKING KEY CONCEPTS

Tell whether each equation is an example of *exponential growth* or *exponential decay*.

1. $y = (3.2)^x$ **2.** $y = (0.6)^x$ **3.** $y = \left(\frac{4}{3}\right)^x$

Find the *y*-intercept of the graph of each equation.

4. $y = 3^x$ **5.** $y = 3 \cdot 2^x$ **6.** $y = 2 \cdot 3^x$

7. $y = -4^x$ **8.** $y = -7 \cdot 6^x$ **9.** $y = -2.5(0.2)^x$

Find the doubling time or half-life of each function.

10. $y = 3.4 \cdot 2^{x/7}$ **11.** $y = 7 \cdot \left(\frac{1}{2}\right)^{x/4}$ **12.** $y = 4(2.7)^x$

3.3 Exercises and Applications

Extra Practice exercises on *page 752*

1. a. In the same coordinate plane, graph $y = 10^x$, $y = 5^x$, $y = 4^x$, and $y = (1.25)^x$.

b. Which of the graphs represents the fastest growth? the slowest growth?

c. What are the *y*-intercepts of the graphs?

Apply⇔Assess

Suggested Assignment

Core Course
Day 1 Exs. 1–9, 11–14, 16–22
Day 2 Exs. 25–41, AYP

Extended Course
Day 1 Exs. 1–22
Day 2 Exs. 23–41, AYP

Block Schedule
Day 16 Exs. 1–9, 11–14, 16–22, 25–41, AYP

Checking Key Concepts

1. exponential growth
2. exponential decay
3. exponential growth
4. 1
5. 3
6. 2
7. –1
8. –7
9. –2.5
10. 7 (doubling time)
11. 4 (half-life)
12. Estimates may vary; about 0.7 (doubling time).

Exercises and Applications

1. a.

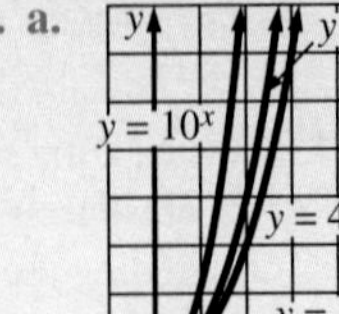
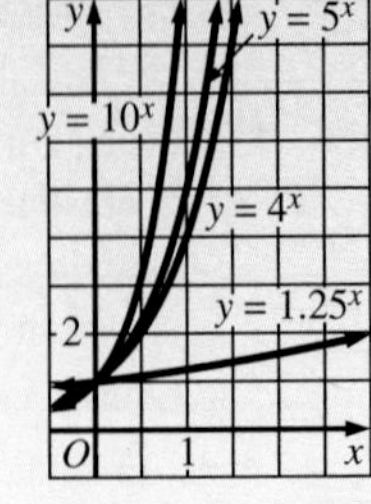

b. $y = 10^x$; $y = (1.25)^x$

c. The *y*-intercept of each of the graphs is 1.

2. a.

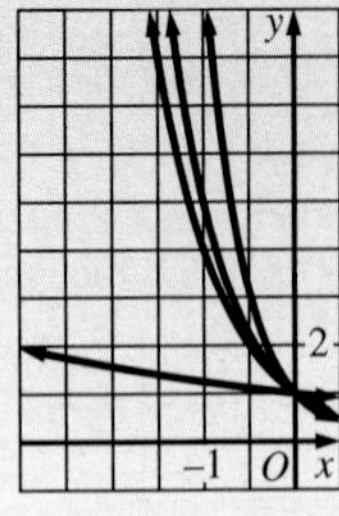

b. $y = \left(\frac{1}{10}\right)^x$; $y = \left(\frac{1}{5}\right)^x$; $y = \left(\frac{1}{4}\right)^x$; $y = \left(\frac{4}{5}\right)^x$

c. Check students' work.

2. a. Sketch the reflections of the graphs in Exercise 1 over the y-axis.

b. Write equations for the graphs you sketched in part (a).

c. **Technology** Check your equations in part (b) by graphing them using a graphing calculator or graphing software.

3. a. Sketch the reflections of the graphs in Exercise 1 over the x-axis.

b. Write equations for the graphs you sketched in part (a).

c. **Technology** Check your equations in part (b) by graphing them using a graphing calculator or graphing software.

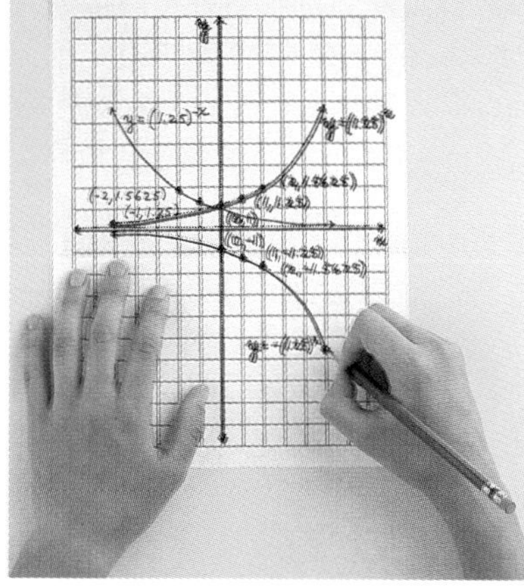

For each function in Exercises 4–7, do the following:

a. Find the y-intercept of the graph.

b. Tell whether the graph represents *exponential growth* or *exponential decay.*

c. Sketch the graph.

Toolbox p. 797 *Reflecting a Figure; Line Symmetry*

4. $y = 4(1.05)^x$ **5.** $y = (4.5)^x$ **6.** $y = 6^x$ **7.** $y = 2.5(0.8)^x$

8. a. Explain why the y-intercept of $y = ab^x$ is a when $b > 0$.

b. **Investigation** What does the graph of $y = ab^x$ look like when $b < 0$ and the domain of the function is all integers? What happens if the domain is real numbers? $\left(\textit{Hint:} \text{ Try evaluating the function when } x = \frac{1}{2}.\right)$

9. a. For what value of b is the graph of $y = ab^x$ a horizontal line? Describe the line.

b. For what value of a is the graph of $y = ab^x$ a horizontal line? Describe the line.

10. **Writing** What does the graph of $y = ab^{-x}$ look like for different values of a and b? How does it compare to the graph of $y = ab^x$? How does it compare to the graph of $y = a\left(\frac{1}{b}\right)^x$?

Match the graphs and the equations.

11. $y = -3 \cdot 2^x$

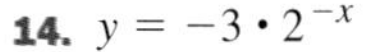

12. $y = 3 \cdot 2^x$

13. $y = 3(0.5)^x$

14. $y = -3 \cdot 2^{-x}$

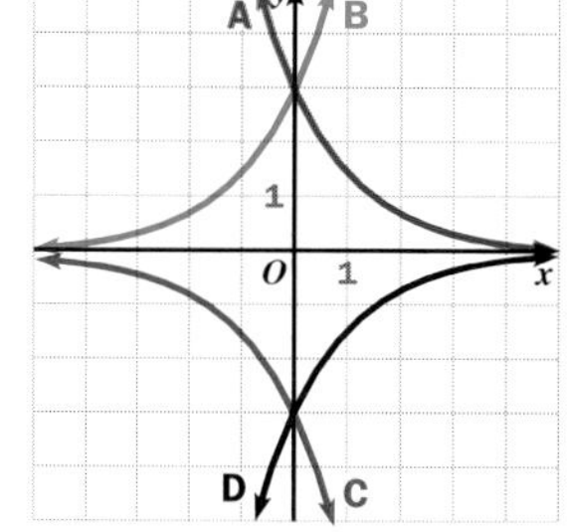

15. **Challenge** You may have noticed in the Exploration that the graph of $y = 8 \cdot 2^x$ looks like the graph of $y = 2^x$ translated to the left by 3 units. Explain why this is so.

Apply⇔Assess

Exercise Notes

Using Technology

Exs. 1–3 Students may use graphing calculators or the *Function Investigator* software to help sketch the graphs in Ex. 1. Ask students to show the graphs for Ex. 1 on the same axes along with the graphs for Exs. 2 and 3.

You may want to assign Function Investigator Activity 29 in the *Function Investigator with Matrix Analyzer Activity Book.*

Problem Solving

Ex. 8 Students may find it helpful to choose a variety of values for b for which $b < 0$ and to use several values of x. Graphing functions for particular values of a and b may also help. If students use a TI-82 graphing calculator, suggest a graphing window that uses Xmin = –9.4, Xmax = 9.4. If they use a TI-81 graphing calculator, suggest Xmin = –9.4, Xmax = 9.6.

Communication: Writing

Ex. 15 Students may find it interesting to see if they can generalize this result. If students are able to generalize the result, urge them to write a summary that explains their reasoning. The resulting work would make a good journal entry.

3. a.

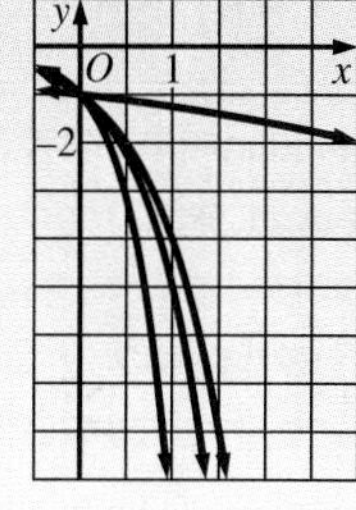

b. $y = -10^x$; $y = -5^x$; $y = -4^x$; $y = -(1.25)^x$

c. Check students' work.

4. a. 4

b. exponential growth

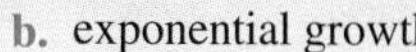

c.

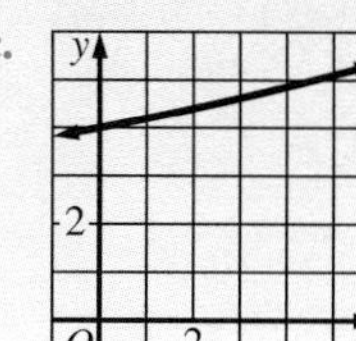

5. a. 1

b. exponential growth

c.

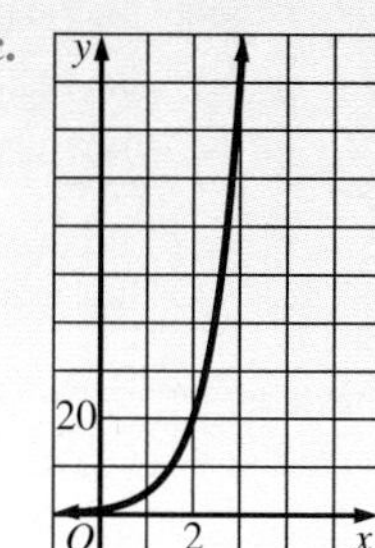

6. a. 1

b. exponential growth

c.

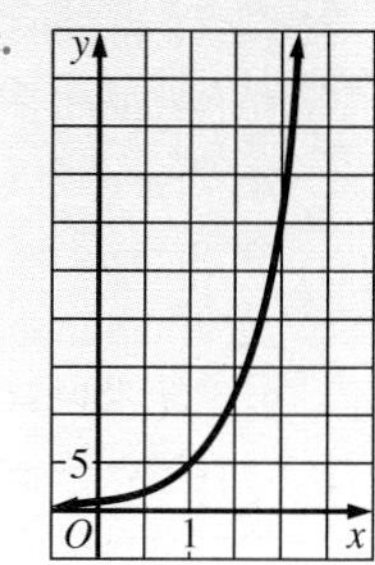

7. a. 2.5

b. exponential decay

c.

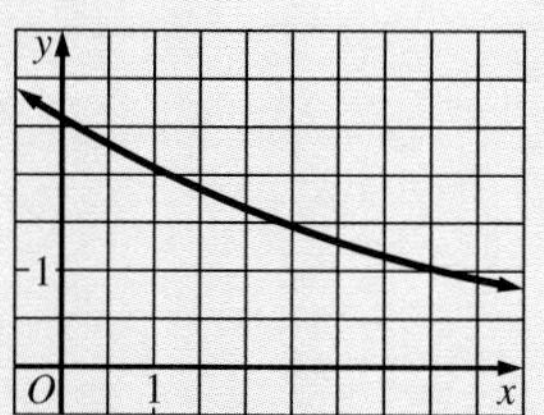

8. a. For $b \neq 0$, $b^0 = 1$, so when $x = 0$, $y = a \cdot 1 = a$. That is, the y-intercept of $y = ab^x$ is a.

b. If the domain is all integers, the graphs consist of points that are alternately above and below the x-axis. The function cannot be defined for all real numbers if $b < 0$. For example, for any integer n, if $x = \frac{1}{2n}$, then y is not defined.

9. a. For $b = 1$, the graph of $y = ab^x$ is a horizontal line through $(0, a)$.

b. For $a = 0$, the graph of $y = ab^x$ is the x-axis.

10–15. See answers in back of book.

Apply⇔Assess

Exercise Notes

Application
Ex. 22 The "Rule of 72" is a good example of a useful rule of thumb that provides convenient estimates in many situations where questions arise about doubling time. Caution students, however, that such rules of thumb should not be used in situations where precise doubling times are required.

Cooperative Learning
Exs. 23, 24 These exercises involve some lengthy computations. If students work with a partner, they can check their calculations and look for more efficient procedures.

Second-Language Learners
Exs. 23, 24 Students learning English may benefit from having a peer tutor or aide read the introductory information and exercises aloud, and then discussing with him or her the meanings of terms that are unfamiliar.

Mathematical Procedures
Ex. 24 Some students may need help with part (a). To get a 1-degree growth factor, raise the average 5-degree growth factor to the 0.2 power; that is, use $(1.35)^{0.2}$.

BANKING Find the doubling time of money in a bank account earning interest compounded annually at each interest rate. Round each doubling time to the nearest whole number of years.

16. 8% **17.** 6% **18.** 4%

19. 9% **20.** 12% **21.** 18%

22. **Cooperative Learning** One way to estimate how long it will take money to double at various interest rates is called the "Rule of 72." Discuss your answers to Exercises 16–21 and what you think the "Rule of 72" is. Write the rule in a way that you can remember.

INTERVIEW Finn Strong

Look back at the article on pages 90–92.

The "rainmaker" on top of Finn Strong's fish tank is cooler than the air near the water's surface. Water vapor in the air that touches the rainmaker condenses into rain droplets, because air that is cool can hold less water vapor than air that is warm. The table shows the amount of water vapor that can exist in a cubic meter of air at various temperatures.

Temperature (°C)	Saturation (mL of water per m^3 of air)
0	4.847
5	6.797
10	9.399
15	12.83
20	17.30
25	23.05
30	30.38
35	39.63

23. a. Find the growth factors for the saturation column of the table. What is the average growth factor associated with a 5-degree difference in temperature?

b. Write an exponential equation for the saturation that uses the average 5-degree growth factor as a base.

24. a. Write an exponential equation for the saturation that uses a 1-degree growth factor as a base.

b. Write an exponential equation for the saturation that uses 2 as a base.

c. What change of temperature allows you to double the amount of water in a cubic meter of air?

16. 9 years
17. 12 years
18. 18 years
19. 8 years
20. 6 years
21. 4 years
22. If the annual interest rate is r%, then the doubling time is about $\frac{72}{r}$ years.

23. a. Answers are given to two decimal places. The growth factors are 1.40, 1.38, 1.37, 1.35, 1.33, 1.32, and 1.30, with an average growth rate of 1.35.

b. Let n = the number of 5° units above zero and s = the saturation in mL of water per m^3 of air; $s = 4.847(1.35)^n$.

24. a. $s = 4.847(1.062)^x$

b. Answers may vary due to rounding; $y = 4.847(2)^{x/11.5}$.

c. about 11.5°C

25–28. Answers are given to two decimal places.

25. 0.74
26. 0.55
27. 0.30
28. 0.09

Connection CARBON DATING

When a plant or animal dies, it stops acquiring carbon-14 and carbon-12 from the atmosphere. Because carbon-14 is an unstable radioactive isotope, it decays over time, with a half-life of about 5730 years. The fraction of the original amount A of carbon-14 that remains in a sample after t years is given by $A = \left(\frac{1}{2}\right)^{t/5730}$.

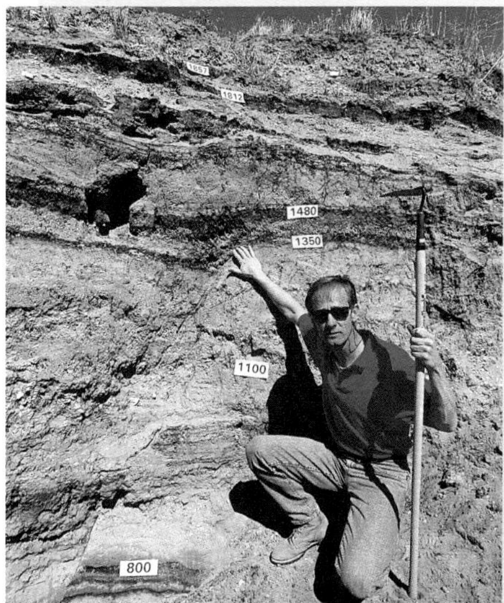

Tell what fraction of the original carbon-14 remains in a sample after each number of years.

25. 2500 **26.** 5000

27. 10,000 **28.** 20,000

29. As a graduate student, Kerry Sieh discovered a way to document earthquakes that happened in the past. Using carbon dating on the peat found in streambeds that had been offset by earthquakes, he found that about 64% of the original carbon-14 remained. When was the earthquake that disturbed the stream?

Kerry Sieh walked a 250 mile stretch of the San Andreas fault to search for streambeds.

30. Chess historians were puzzled when archaeologists originally dated these animal-bone chess pieces to the first century A.D., 500 years before chess was invented in India. More recent carbon dating revealed that about 87.6% of the original carbon-14 remains in the pieces. What is a better estimate of their age?

For Exercises 31 and 32, assume exponential growth or decay.

31. A company wants to double its sales in 7 years. By what percent must sales increase each year to achieve this goal?

32. A neighborhood committee wants to cut the amount of littering in half in 5 years. By what percent must littering be reduced each year to achieve this goal?

ONGOING ASSESSMENT

33. Open-ended Problem Write an equation representing an exponential growth or decay situation with a factor other than 2 or $\frac{1}{2}$. Show how the initial amount and the growth factor appear in your equation. Graph the equation and determine the doubling time or half-life.

29–32. Estimates may vary.

29. about 3700 years ago

30. about 1100 years old

31. about 10.4%

32. about 12.9%

33. Answers may vary. An example is given. Suppose the population of a town in 1900 was 2500 and the population has grown at a rate of 20% every 10 years. Its population, P, d decades after 1900 is modeled by the equation $P = 2500(1.2)^d$. The doubling time is about 4 decades.

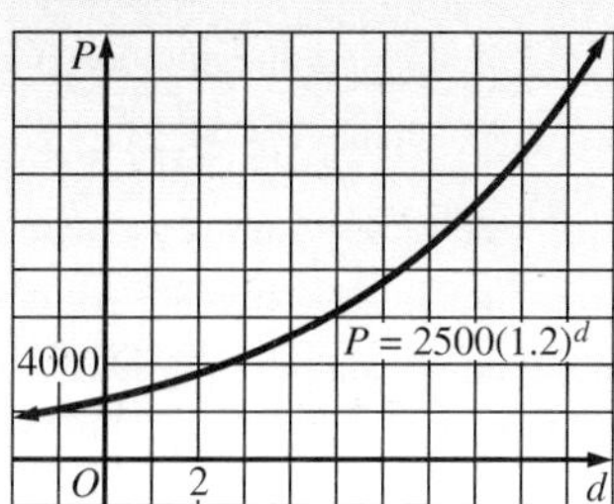

Apply⇔Assess

Exercise Notes

Application

Exs. 25–30 Students may wonder how scientists determine what fraction of the original amount of carbon-14 actually remains in samples used for dating. The key is to determine the ratio of carbon-14 to carbon-12 in the sample. Carbon-14 decays, but carbon-12 is stable. Cosmic rays bombard the atmosphere and create new carbon-14 to replace the naturally decaying carbon-14. This keeps the ratio of carbon-14 to carbon-12 in the atmosphere relatively constant. As long as an organism lives and breathes, the ratio of carbon-14 to carbon-12 in its tissues is the same as that of the atmosphere. Once it dies and respiration ceases, the ratio declines, since the carbon-14 in the tissues decays and is not replaced by atmospheric carbon-14. Thus, by carefully measuring the ratio of carbon-14 to carbon-12, scientists can deduce what fraction of the original carbon-14 remains. This in turn makes it possible to estimate the age of the sample.

Problem Solving

Exs. 31, 32 Students are given the doubling time or half-life and need to determine the growth or decay factor. Remind them that they need to subtract 1 from the growth factor or subtract the decay factor from 1 to get the percent.

Visual Thinking

Exs. 31, 32 Ask students to create graphs of the information presented in either Ex. 31 or 32. Ask them to use their graphs to show how many years it would take to double sales again, or to cut the amount of littering in half again. This activity involves the visual skills of *interpretation* and *exploration.*

Assessment Note

Ex. 33 A discussion of students' responses to this open-ended problem would provide all students with an opportunity to evaluate their understanding of the concepts involved.

Apply⇔Assess

Assess Your Progress

You may wish to review the vocabulary terms in class. Be sure that students can write examples of exponential growth equations and exponential decay equations. They should also be able to sketch the general shape of the graphs and show their locations in the coordinate plane. Ex. 4 will help show whether students can write an equation for a real-world situation that involves exponential decay.

Journal Entry

Students may wish to draw on their answers to the exercises in the Exercises and Applications for ideas that will help them with their journal entries.

Progress Check 3.1–3.3

See page 132.

Practice 16 for Section 3.3

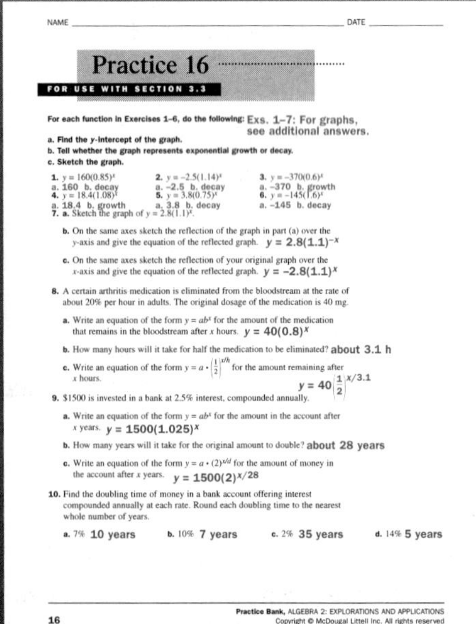

NAME ______________________ DATE __________

Practice 16

FOR USE WITH SECTION 3.3

For each function in Exercises 1–6, do the following: Exs. 1–7: For graphs, see additional answers.

a. Find the y-intercept of the graph.
b. Tell whether the graph represents exponential growth or decay.
c. Sketch the graph.

1. $y = 160(0.85)^x$ a. 160 b. decay
2. $y = -2.5(1.14)^x$ a. –2.5 b. decay
3. $y = -370(0.6)^x$ a. –370 b. growth
4. $y = 18.4(1.08)^x$ a. 18.4 b. growth
5. $y = 3.8(0.75)^x$ a. 3.8 b. decay
6. $y = -145(1.6)^x$ a. –145 b. decay

7. a. Sketch the graph of $y = 2.8(1.1)^x$.

b. On the same axes sketch the reflection of the graph in part (a) over the y-axis and give the equation of the reflected graph. $y = 2.8(1.1)^{-x}$

c. On the same axes sketch the reflection of your original graph over the x-axis and give the equation of the reflected graph. $y = -2.8(1.1)^x$

8. A certain arthritis medication is eliminated from the bloodstream at the rate of about 20% per hour in adults. The original dosage of the medication is 40 mg.

a. Write an equation of the form $y = ab^x$ for the amount of the medication that remains in the bloodstream after x hours. $y = 40(0.8)^x$

b. How many hours will it take for half the medication to be eliminated? about 3.1 h

c. Write an equation of the form $y = a \cdot \left(\frac{1}{2}\right)^{x/h}$ for the amount remaining after x hours. $y = 40\left(\frac{1}{2}\right)^{x/3.1}$

9. \$1500 is invested in a bank at 2.5% interest, compounded annually.

a. Write an equation of the form $y = ab^x$ for the amount in the account after x years. $y = 1500(1.025)^x$

b. How many years will it take for the original amount to double? about 28 years

c. Write an equation of the form $y = a \cdot (2)^{x/d}$ for the amount of money in the account after x years. $y = 1500(2)^{x/28}$

10. Find the doubling time of money in a bank account offering interest compounded annually at each rate. Round each doubling time to the nearest whole number of years.

a. 7% 10 years **b.** 10% 7 years **c.** 2% 35 years **d.** 14% 5 years

16

Practice Bank, ALGEBRA 2: EXPLORATIONS AND APPLICATIONS

SPIRAL REVIEW

The equation $P = 11.7(1.02)^x$ models the population P of Chile in millions, with x = years since 1985. Find P for each year. *(Section 3.2)*

34. 1990 **35.** 1980 **36.** 1984

37. What is the doubling time for the population of Chile? *(Section 3.3)*

Simplify. *(Section 3.2)*

38. $8^{1/3}$ **39.** 3^{-2} **40.** $16^{3/2}$

41. The table shows the water content and calories of several soups. *(Section 2.4)*

a. Graph the data and draw a line of fit.

b. Find an equation for the line of fit.

c. Make a prediction about the number of calories in vegetable beef soup, which is 91.9% water.

Soup	Percent water	Calories
Bean with pork	84.4	67
Beef noodle	93.2	28
Minestrone	89.5	43
Split pea	85.4	59
Tomato	90.5	36

ASSESS YOUR PROGRESS

VOCABULARY

exponential growth (p. 96) **doubling time** (p. 108)
exponential decay (p. 96) **half-life** (p. 109)

Evaluate each expression when $n = 4$. *(Section 3.1)*

1. 2^n **2.** $\left(\frac{1}{2}\right)^n$ **3.** $3 \cdot 2^{n-1}$

4. SPORTS A National Collegiate Athletic Association Basketball Championship starts with 64 teams in the first round. Each subsequent round eliminates half the teams. Write an equation for the number of teams that remain after n rounds. *(Section 3.1)*

Simplify using the properties of exponents. *(Section 3.2)*

5. $\frac{2^{5/4}}{2^{1/2}}$ **6.** $7^{3/4} \cdot 7^{1/4}$ **7.** $\left(4^{1/3}\right)^{3/2}$

A city has been reducing its traffic accident rate by 2% a year. In 1993 there were 278 traffic accidents. Estimate the number of accidents in each year. *(Section 3.2)*

8. 1995 **9.** 2000 **10.** 1985

11. A mail order company wants to reduce complaints by 5% a year. If the company is successful, when will the company be receiving half as many complaints as the company is now? *(Section 3.3)*

12. Journal Explain how the values of a and b affect the doubling time or half-life of an exponential function $y = ab^x$.

---

34. about 12.9 million

35. about 10.6 million

36. about 11.5 million

37. Estimates may vary. An example is given. about 35 years

38. 2

39. $\frac{1}{9}$

40. 64

41. a. Lines of fit may vary.

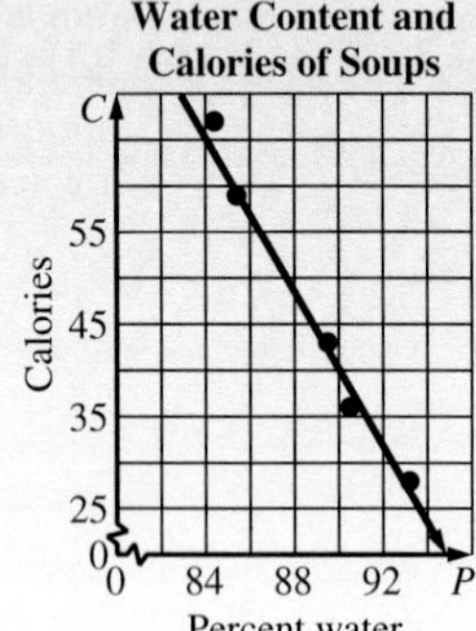

b. Answers may vary. An example is given. Let P = percent of water and C = the number of calories; $C = -4.4P + 435$.

c. Estimates may vary; about 31 calories.

Assess Your Progress

1. 16 **2.** $\frac{1}{16} = 0.0625$

3. 24

4. Let T = the number of teams; $T = 64\left(\frac{1}{2}\right)^n$.

5. $2^{3/4} \approx 1.68$

6. 7 **7.** 2

8. about 267 accidents

9. about 241 accidents

10. about 327 accidents

11. Estimates may vary. An example is given. in about $13\frac{1}{2}$ years

12. Answers may vary. An example is given. The value of a does not affect the doubling time or half-life; a is simply the initial value. In cases of exponential growth, the greater the value of b, the shorter the doubling time. In exponential decay cases, the greater the value of b, the longer the half-life.

Section

3.4 The Number *e*

Learn how to...

- model exponential growth and decay using the base *e*, and model logistic growth

So you can...

- make predictions about real-life growth and decay, such as compound interest and biological populations

In Section 3.3 you used an exponential function to describe compounding interest annually in a bank account. Most banks offer interest that is compounded more than once a year. When interest is compounded n times per year for t years at an interest rate r (expressed as a decimal), a principal of P dollars grows to the amount A given by this formula:

$$A = P\left(1 + \frac{r}{n}\right)^{nt}$$

EXPLORATION
COOPERATIVE LEARNING

Compounding Interest

Work in a group of four students. You will need:

- a scientific calculator

1 The table shows the value of one dollar after one year of compounding. Copy and complete the table. Each of you should complete one column for one of the interest rates.

Compounding	n	Formula	r = 0.05	r = 0.10	r = 0.50	r = 1.00
annually	1	$(1+\frac{r}{1})^1$	1.05			
semiannually	2	$(1+\frac{r}{2})^2$		1.1025		
quarterly	4	$(1+\frac{r}{4})^4$			1.6018	
monthly	12	$(1+\frac{r}{12})^{12}$				2.6130
daily	365	$(1+\frac{r}{365})^{365}$				
hourly						
every minute						
every second						

2 Describe how increasing the frequency of compounding interest affects the value of a dollar that is invested for one year.

3 If a bank compounded interest more often than every second, would this make much difference? Explain.

Exploration Note

Purpose
The purpose of this Exploration is to have students discover that as the interest rate approaches 100% and the frequency of compounding increases that the value of an account with one dollar in it approaches a constant value.

Materials/Preparation
Each group should have a scientific or graphing calculator.

Procedure
Students can work together to complete the second and third columns of the table. Then each student should complete one of the four remaining columns.

Closure
Discuss the result of increasing the frequency of compounding as the interest rate approaches 100%. Students should see that the value of the dollar invested for a one-year period eventually levels off to about \$2.72.

Explorations Lab Manual
See the Manual for more commentary on this Exploration.

Diagram Master 6

For answers to the Exploration, see answers in back of book.

Plan⇔Support

Objectives

- Model exponential growth and decay using the base *e*, and model logistic growth.
- Make predictions about real-life growth and decay situations.

Recommended Pacing

Core and Extended Courses
Section 3.4: 2 days

Block Schedule
Section 3.4: 1 block

Resource Materials

Lesson Support
Lesson Plan 3.4
Warm-Up Transparency 3.4
Practice Bank: Practice 17
Study Guide: Section 3.4
Exploration Lab Manual: Diagram Masters 1, 6

Technology
Technology Book: Spreadsheet Activity 3
Scientific Calculator
Graphing Calculator
McDougal Littell Mathpack *Function Investigator*
Internet: http://www.hmco.com

Warm-Up Exercises

Evaluate the expression $\left(1 + \frac{r}{n}\right)^n$ for the given values of r and n. Use a calculator. Round answers to four decimal places.

1. $r = 0.03$, $n = 100$ 1.0304
2. $r = 0.03$, $n = 200$ 1.0305
3. $r = 0.5$, $n = 400$ 1.6482
4. $r = 0.65$, $n = 500$ 1.9147
5. $r = 0.95$, $n = 1000$ 2.5845

Teach⇔Interact

Section Notes

Communication: Discussion When discussing the definition of e with students, emphasize that no value of n will ever make the value of $\left(1+\frac{1}{n}\right)^n$ equal to e. The number e is a limiting value.

Integrating the Strands
Ask students if they can think of another famous number in mathematics that can be defined as a limiting value. Ask them to explain how this number can be described in terms of limits. (The number π can be defined in terms of limits. For example, if a regular n-sided polygon is inscribed in a circle whose radius is 1 unit, the areas of the polygons approach the number π as the number of sides n increases without limit.)

Historical Connection
Euler was not the first mathematician to encounter the number 2.718..., but he was one of the first to realize that this number was appearing frequently in a wide variety of mathematical situations. Students may find it interesting to consult works on the history of mathematics for more information about this famous number.

Think and Communicate

In connection with question 6, some students may accidentally come upon some unexpected behavior on the part of their calculators. If $\left(1+\frac{1}{n}\right)^n$ is evaluated for $n = 10^{13}$ using a TI-82 graphing calculator, the calculator gives a result of 2.718281828. If $n = 10^{14}$ is then used to see if an even more accurate approximation results, the calculator gives a value of 1. This strange behavior results from the inherent limitations of the calculator itself. Similar behavior occurs with the TI-81, but the unexpected behavior occurs for $n = 10^{13}$ (not, however, for $n = 10^{12}$).

The last column of the table in the Exploration shows what happens to a dollar when 100% interest is compounded more and more frequently. As the frequency of compounding increases, the amount in the account approaches \$2.718. . . . Because the number 2.718. . . is special in mathematics, it is given a name: **e**.

BY THE WAY

The letter e was first used in 1727 to refer to the number 2.718. . . by a Swiss mathematician named Leonhard Euler (pronounced "oiler"). Even after losing his vision at age 65, Euler continued to make mathematical discoveries and write books about mathematics.

The Number e

As n increases, $\left(1+\frac{1}{n}\right)^n$ approaches e.

$$e = 2.718281828...$$

THINK AND COMMUNICATE

Use a calculator to evaluate each power of e.

1. $e^{0.05}$ 2. $e^{0.10}$ 3. $e^{0.50}$ 4. e^1
5. Compare your answers to Questions 1–4 with the last row of the table from the Exploration. What do you notice?
6. What does your answer to Question 5 suggest is the value of $\left(1+\frac{r}{n}\right)^n$ as n gets very large?
7. When compounding takes place without interruption, it is called *continuous compounding*. Explain why it is reasonable that the formula for continuous compounding of interest is $A = Pe^{rt}$.
8. **a.** If $r = 0$, what is the value of e^r?
   b. If $r > 0$, what can you say about the value of e^r?
   c. If $r > 0$, does the graph of $A = Pe^{rt} = P(e^r)^t$ represent *exponential growth* or *exponential decay*? Why does this make sense?

Because $e > 1$, $y = e^x$ is an exponential growth function. Also, because $e^{-1} = \frac{1}{e} < 1$, $y = e^{-x}$ is an exponential decay function.

The Graphs of $y = e^x$ and $y = e^{-x}$

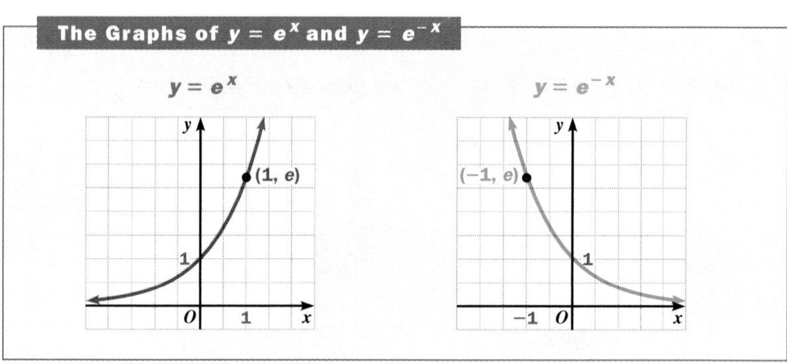

Exponential functions with the base e are often used to describe continuous growth or decay, such as continuously compounded interest or radioactive decay, as shown in Examples 1 and 2 on the next page.

ANSWERS Section 3.4

Exploration

See answers in back of book.

Think and Communicate

1–4. Answers are given to nine decimals places.

1. 1.051271096
2. 1.105170918
3. 1.648721271
4. 2.718281828
5. The table values are the answers to Questions 1–4 rounded to four decimal places.
6. e^r
7. As we have seen, when n is very large, $\left(1+\frac{r}{n}\right)^n$ is very close to e^r. Then for continuous compounding, we can substitute e^r in the interest formula: $A = P\left(1+\frac{r}{n}\right)^{nt} = P\left(\left(1+\frac{r}{n}\right)^n\right)^t = Pe^{rt}$.
8. a. 1 b. $e^r > 1$
   c. exponential growth; In mathematical terms, it makes sense because the function can be written in exponential form with $a > 0$ and $b > 1$, conditions that represent exponential growth. It makes common sense because it represents the amount in an account that earns interest continuously, which should certainly not model decay.

EXAMPLE 1 Application: Banking

Suppose a bank offers 3.5% interest compounded continuously. If an account starts with $1000, what will its value be after 1 year, 5 years , and 10 years? Use the formula $A = Pe^{rt}$.

SOLUTION

Evaluate the formula $A = Pe^{rt}$ when $P = 1000$, $r = 0.035$, and $t = 1, 5$, and 10.

```
1000e^(.035*1)
      1035.619709
1000e^(.035*5)
      1191.246217
1000e^(.035*10)
      1419.067549
```

After 1 year, the account's value is about **$1035.62**.

After 5 years, the account's value is about **$1191.25**.

After 10 years, the account's value is about **$1419.07**.

EXAMPLE 2 Application: Chemistry

Polonium-210 is a radioactive element. An initial amount of 100 micrograms (μg) decays to an amount $A(t)$ in t days according to this formula:

$$A(t) = 100e^{-0.005t}$$

a. Use a graphing calculator or graphing software to graph the decay function.

b. What is the half-life of polonium-210? Write an equation for the amount of polonium-210 using $\frac{1}{2}$ as a base instead of e.

SOLUTION

a.

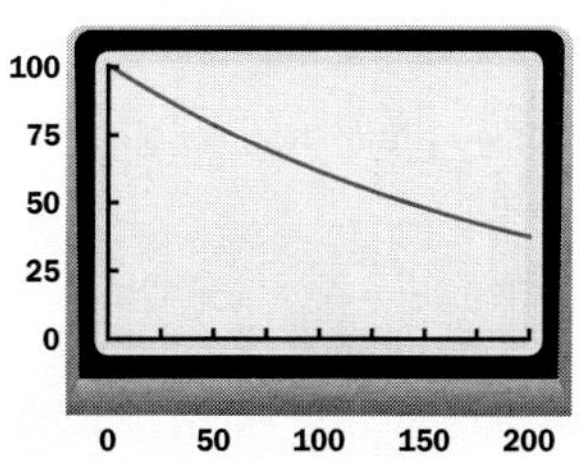

The **half-life** of polonium-210 is about **138 days**.

b. Find the value of t when $A(t) = 50$.

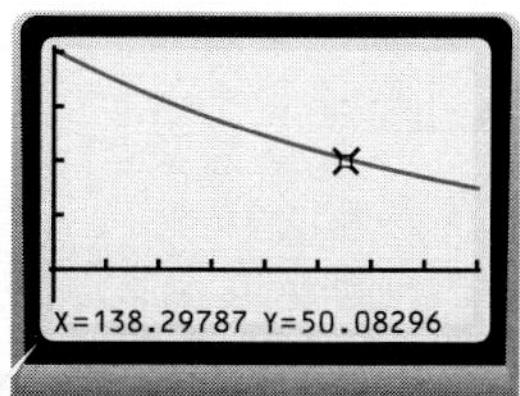

An equation for the decay of 100 μg of polonium-210 is:

$$A(t) = 100\left(\frac{1}{2}\right)^{t/138}$$

BY THE WAY

Polonium was discovered by Marie Curie in 1898. She named it after her native country, Poland.

Teach⇔Interact

Additional Example 1

Suppose a bank offers 4.2% interest compounded continuously. If an account starts with $600, what will its value be after 2 years, 4 years, and 8 years? Use the formula $A = Pe^{rt}$.

Evaluate the formula $A = Pe^{rt}$ when $P = 600$, $r = 0.042$, and $t = 2, 4$, and 8.

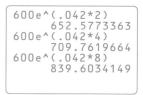

```
600e^(.042*2)
      652.5773363
600e^(.042*4)
      709.7619664
600e^(.042*8)
      839.6034149
```

After 2 years, the account's value is about $652.58.

After 4 years, the account's value is about $709.76.

After 8 years, the account's value is about $839.60.

Additional Example 2

Fluorine-21 is a radioactive element. An initial amount of 14 grams decays to an amount $A(t)$ in t seconds according to the formula $A(t) = 14e^{-0.1386t}$.

a. Use a graphing calculator or graphing software to graph the decay function.

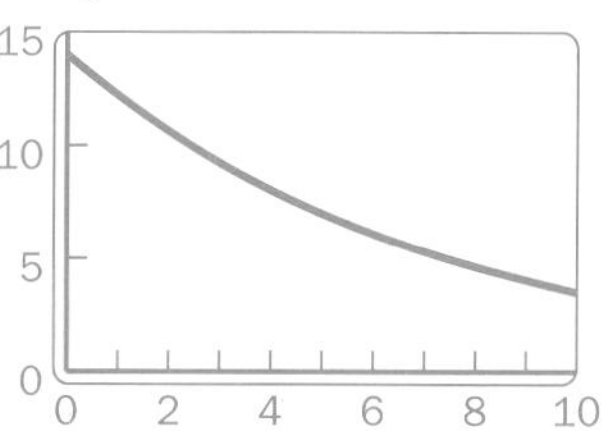

b. What is the half-life of fluorine-21? Write an equation for the amount of fluorine-21 using $\frac{1}{2}$ as a base instead of e.

Find the value of t when $A(t) = 7$.

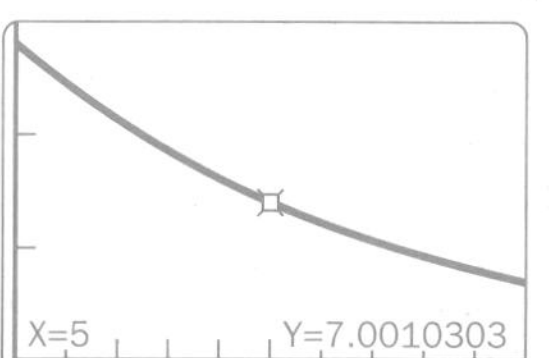

The half-life of fluorine-21 is about 5 seconds. An equation for the decay of 14 g of fluorine-21 is $A(t) = 14\left(\frac{1}{2}\right)^{t/5}$.

Additional Example 3

In 1976, the population of Earth was estimated to be 4 billion people. Suppose that subsequent data give the following equation as a model for the population $P(t)$ (in billions) as a function of t, where t is the number of years since 1976:

$$P(t) = \frac{280}{4 + 65.99e^{-0.0208t}}$$

a. Graph the population function using a graphing calculator or graphing software. Evaluate the function when $t = 0$. Does the answer make sense?

The graphing calculator screen shows the graph of $y = \frac{280}{4 + 65.99e^{-0.0208x}}$.

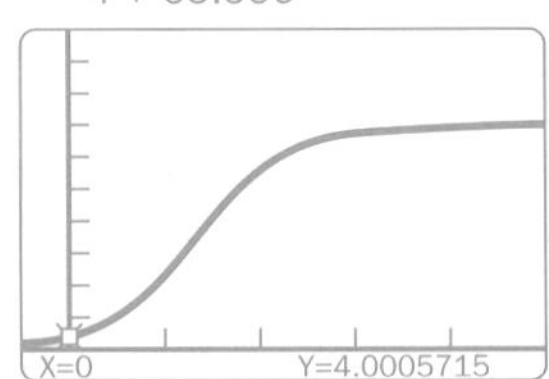

When $t = 0$, $e^{-0.0208t} = e^0 = 1$. Therefore, $P(0) = \frac{280}{4 + 65.99(1)} \approx 4$. This makes sense, because the population of Earth in 1976 was 4 billion.

b. Estimate the year when the population will be 50 billion people.

Find the intersection of the graph from part (a) and the line $y = 50$.

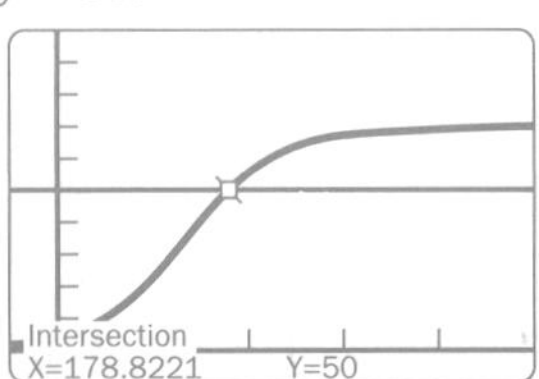

The population will reach 50 billion about 179 years after 1976, around the year 2155.

c. Investigate the value of $P(t)$ for large values of t. According to the model, at what value will the population of Earth eventually stabilize?

The expression $e^{-0.0208t}$ is an example of exponential decay. When t is large, this expression approaches 0. Therefore, for large values of t:

$$P(t) \approx \frac{280}{4 + 65.99(0)} = 70.$$

According to the model, the population of Earth will eventually stabilize at about 70 billion.

Logistic Growth

In many situations, exponential growth does not continue forever, because environmental or other conditions limit the growth. One kind of limited growth, called **logistic growth**, begins with rapid growth but eventually levels off. Equations for logistic growth are often written using e.

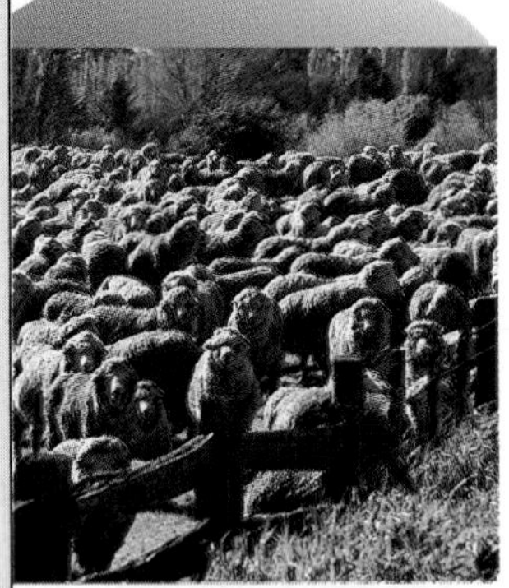

EXAMPLE 3 Application: Biology

About 187,000 sheep inhabited the island of Tasmania in 1819. As European settlers began to develop the land for sheep farming, this population increased until it stabilized about 70 years later.

The sheep population $P(t)$ can be modeled by a logistic function, where t is the number of years since 1819:

$$P(t) = \frac{1{,}670{,}000}{1 + 7.915e^{-0.131t}}$$

a. Graph the sheep population function using a graphing calculator or graphing software. Evaluate the function when $t = 0$. Does your answer make sense?

b. Estimate the date when the sheep population reached 1,500,000.

c. Investigate the value of $P(t)$ for large values of t. According to the model, at what value did the sheep population eventually stabilize?

SOLUTION

a. The graphing calculator screen shows the graph of

$$y = \frac{1{,}670{,}000}{1 + 7.915e^{-0.131x}}.$$

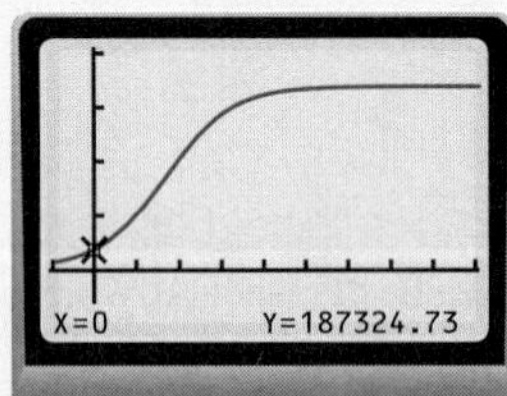

When $t = 0$, $e^{-0.131t} = e^0 = 1$. Therefore,

$$P(0) = \frac{1{,}670{,}000}{1 + 7.915(1)} \approx 187{,}000.$$

This makes sense, because the sheep population in 1819 was 187,000.

b. Find the intersection of the graph from part (a) and the line

$$y = 1{,}500{,}000.$$

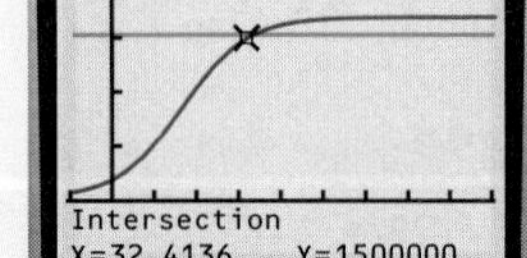

The sheep population reached 1,500,000 about 32 years after 1819, in the year 1851.

c. The expression $e^{-0.131t}$ is an example of exponential decay. When t is large, this expression approaches 0. Therefore, for large values of t:

$$P(t) \approx \frac{1{,}670{,}000}{1 + 7.915(0)} = 1{,}670{,}000.$$

According to the model, the sheep population on Tasmania stabilized at about 1,670,000 sheep.

CHECKING KEY CONCEPTS

1. Suppose you deposit $1000 in the account described in the advertisement. What amount would be in the account after 1 year? What do you think *effective annual yield* means?
2. Suppose Neighborhood Bank offered continuous compounding instead of daily compounding. Find the value of a $1000 deposit after 1 year. How does this result compare with the value from Question 1?
3. Radium-226 is a radioactive element that decays after t years according to this formula:

$$A(t) = A_0e^{-0.000525t}$$

Suppose A_0 is 1 gram. Find the amount of radium-226 left after:

a. 2640 years b. 3960 years c. 5280 years

4. Estimate the half-life of radium-226 and write a decay equation using $\frac{1}{2}$ as a base instead of e.
5. Use the logistic function $L(t) = \frac{2000}{1 + 19e^{-0.1t}}$.
   a. Find $L(0)$.
   b. Investigate the value of $L(t)$ for large values of t.
   c. For what value of t is $L(t) = 1000$?
   d. For what value of t is $L(t) = 4000$?

NEIGHBORHOOD BANK

INTRODUCING THE

— PASSPORT SAVINGS — ACCOUNT

3.20% Interest Compounded Daily*	3.25% Effective Annual Yield

*Based on a Minimum Daily Balance of $500

3.4 Exercises and Applications

Extra Practice exercises on page 753

Suppose a bank compounds interest 360 times a year. For each interest rate, find the *effective annual yield.* (The effective annual yield is the annual interest rate that would yield the same amount after one year.)

1. 2.50% 2. 5.00% 3. 7.50% 4. 10.00%

Suppose a bank offers interest compounded continuously. Use the formula $A = Pe^{rt}$ to find the value of $1000 after 10 years at each interest rate.

5. 3.75% 6. 6.25% 7. 9.50% 8. 11.00%

Find the value of $\left(1 + \frac{1}{n}\right)^n$ for each value of n. Round each answer to six decimal places.

9. 1 10. 10 11. 10^2 12. 10^3
13. 10^4 14. 10^5 15. 10^6 16. 10^7

Teach⇔Interact

Checking Key Concepts

Application
For question 1, try to have available some data about local investment opportunities that illustrate compounding several times per year. Ask what information is provided or can be deduced about the effective annual yield.

Reasoning
In connection with question 5, some students may wonder whether there is a general definition of a logistic function. Point out that one commonly used definition states that a logistic function is modeled by an equation of the form $y = \frac{k}{1 + e^{a+bx}}$, where $b < 0$. Ask students to show that the equation for the function in question 5 can be written in this form. What are the values of k, a, and b when the given equation is written in this form? ($k = 2000$, $a \approx 2.94444$, $b = -0.1$)

Closure Question

Describe the concepts of exponential growth and logistic growth.
When a quantity grows by a proportional factor over time the growth is said to be exponential. If the exponential growth is limited, that is, it eventually levels off, it is called logistic growth.

Apply⇔Assess

Suggested Assignment

Core Course
Day 1 Exs. 1–17
Day 2 Exs. 18–21, 27, 28, 30–38

Extended Course
Day 1 Exs. 1–17
Day 2 Exs. 18–38

Block Schedule
Day 17 Exs. 1–21, 27, 28, 30–38

Checking Key Concepts

1. $1032.52; the annual interest rate that would yield the same amount after one year
2. $1032.52; The results are identical.
3. a. about 0.2501 g
   b. about 0.1251 g
   c. about 0.0625 g
4. about 1320 years; $A(t) = A_0\left(\frac{1}{2}\right)^{t/1320}$
5. a. 100
   b. When t is very large, $e^{-0.1t} \approx 0$, so $L(t) \approx 2000$.
   c. Estimates may vary. An example is given. about 29.4
   d. none

Exercises and Applications

1–4. Answers are given to two decimal places.

1. 2.53% 2. 5.13%
3. 7.79% 4. 10.52%
5. $1454.99 6. $1868.25
7. $2585.71 8. $3004.17
9. 2 10. 2.593742
11. 2.704814 12. 2.716924
13. 2.718146 14. 2.718268
15. 2.718280 16. 2.718282

Apply⇔Assess

Exercise Notes

Common Error

Exs. 5–8 If students enter exponents without parentheses, they will get incorrect results. Point out that for Ex. 5, the calculator will interpret 1000e^0.0375*10 to mean $1000 \cdot e^{0.0375} \cdot 10$, not $1000 \cdot e^{(0.0375 \cdot 10)}$. Students must enter Ex. 5 as 1000e^(0.0375*10).

Using Technology

Exs. 9–16 Most graphing calculators make it possible to do these calculations rapidly. On the TI-82, for example, you can calculate the value for Ex. 11 by using the expression (1+1/10^2)^(10^2) and the [ENTER] key. To calculate the value for Ex. 12, press [2nd] [ENTRY] and use the [◄] key to edit the expression for Ex. 11. Change each 2 to a 3, then press [ENTER].

Communication: Reading

Ex. 17 Group members should be urged to read and discuss all parts of this exercise carefully before they start to work on their answers.

Interdisciplinary Problems

Exs. 18–21 The application of scientific methods and mathematics to the study of farming and agriculture have helped to increase food yields in the United States.

Problem Solving

Ex. 20 Students should show how they obtained the equation in their final answers. Be sure they understand that the value of *b* gives the yearly proportional decay factor.

17. Cooperative Learning Work in a group of four students to investigate the interest rates that double your money in a year if interest is compounded *n* times a year. Each of you should choose a different value of *n* from the table shown below.

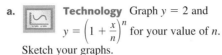

a. **Technology** Graph $y = 2$ and $y = \left(1 + \frac{x}{n}\right)^n$ for your value of *n*. Sketch your graphs.

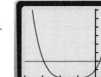

b. For your value of *n*, what interest rate doubles your money in a year?

c. Combine your results to complete the table. Discuss the results.

d. **Writing** Describe how your group's graphs compare to the graph of $y = e^x$.

e. If interest is compounded continuously, what interest rate will double your money in one year?

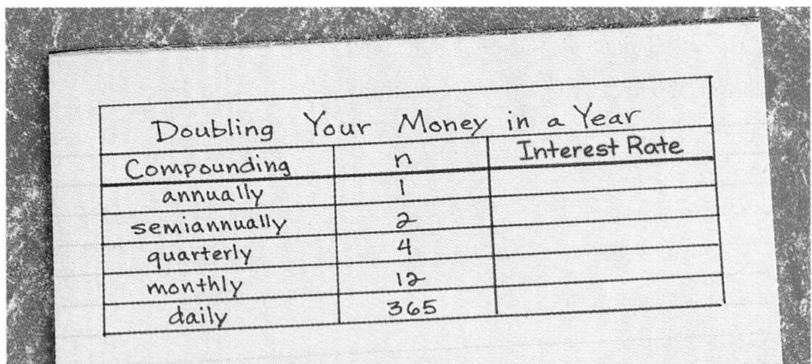

Doubling Your Money in a Year		
Compounding	n	Interest Rate
annually	1	
semiannually	2	
quarterly	4	
monthly	12	
daily	365	

Connection FARMING

Data gathered in the 1920s suggested that the number of eggs *E* a Leghorn chicken produces per year declines exponentially with its age *t* in years, according to this function:

$$E(t) = 179.2e^{-0.12t}$$

18. Graph the egg production function.

19. Estimate how many years it takes for a chicken's egg production to be cut in half. Use your estimate to write an egg production function in the form $E(t) = a \cdot \left(\frac{1}{2}\right)^{t/h}$.

20. Write an egg production function in the form $E(t) = ab^t$. What is the yearly proportional decay factor?

21. By the 1990s certain breeds of 1-year-old chickens were laying about 250 eggs per year. Compare the egg production of one of these chickens to the egg production of a 1-year-old Leghorn from the 1920s. By what percent does a modern chicken out-produce the Leghorn?

17. a. Graphs may vary. Check students' work.

b. Answers may vary. See table in part (c).

c. Estimates may vary.

Doubling Your Money in a Year		
Compounding	*n*	Rate
annually	1	100%
semiannually	2	82.8%
quarterly	4	75.7%
monthly	12	71.4%
daily	365	69.4%

As the frequency of compounding increases, the interest rate needed to double your money drops. However, the rate is very high even for daily compounding. It would be extremely unusual to find an investment that offered an annual interest rate of 69.4%.

d. As *n* increases, the graph of $y = \left(1 + \frac{x}{n}\right)^n$ approaches the graph of $y = e^x$.

e. Estimates may vary; about 69.3%.

18.

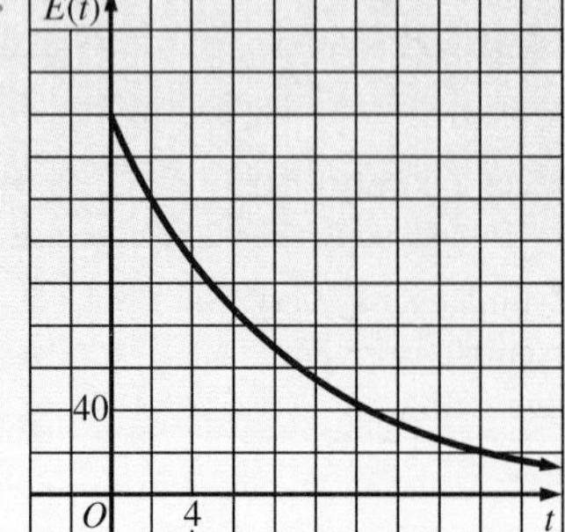

19. Estimates and equations may vary depending on rounding. An example is given. about 5.8 years; $E(t) = 179.2\left(\frac{1}{2}\right)^{t/5.8}$

20. $E(t) = 179.2(0.89)^t$; 0.89 (Egg production decreases by about 11% per year.)

21. A 1-year-old Leghorn chicken of the 1920s produced about 159 eggs per year. The modern chickens out-produce the Leghorns by about 57%.

Connection WIND ENERGY

Located in Scotland, this 135 ft tall wind turbine is one of the world's largest.

In the book *Wind Energy Comes of Age*, Paul Gipe discusses the amount of energy generated from wind turbines in North America and Europe between 1980 and 1995. His data can be modeled using the equations

$$\textbf{Europe: } y = 6.489(1.580)^x$$

$$\textbf{North America: } y = \frac{3500}{1 + 874e^{-0.852x}}$$

where x = years since 1980 and y = gigawatt-hours of wind energy. (*Note:* One gigawatt-hour equals 10^6 kilowatt-hours.)

Technology For Exercises 22–25, use a graphing calculator or graphing software.

22. Graph the wind energy functions.

23. According to the models, what was the wind energy for each region in 1985? in 1990? in 1995?

24. Which region produced more wind energy in the 1980s?

25. When did Europe's wind energy production surpass North America's?

26. **Writing** Do you think an exponential function is a realistic model for European wind energy in the future? Why?

Wind turbines line California's Altamont Pass, where wind speeds average 13 mi/h to 18 mi/h.

CHEMISTRY Show that the two equations given for each radioactive element are equivalent by rewriting both in the form $A(t) = ab^t$.

27. **Polonium-210**

$$A(t) = 100e^{-0.005t}$$

$$A(t) = 100\left(\frac{1}{2}\right)^{t/138}$$

28. **Radium-226**

$$A(t) = e^{-0.000525t}$$

$$A(t) = \left(\frac{1}{2}\right)^{t/1320}$$

29. **Challenge** The radioactive element carbon-14 has a half-life of about 5730 years, so an equation describing its radioactive decay is:

$$A(t) = A_0\left(\frac{1}{2}\right)^{t/5730}$$

Write an equivalent equation that uses e as a base. (*Hint:* Look at the numbers in the exponents in Exercises 27 and 28.)

Apply⇔Assess

Exercise Notes

Multicultural Note

Exs. 22–26 For many centuries, windmills have been used to power irrigation mechanisms. Evidence of such use dates back 3700 years to Babylon and 2400 years to India. Windmills were also widely used in early Islamic civilizations, such as Persia. Today, many Asian and Pacific countries, including Sri Lanka, Thailand, Malaysia, New Zealand, and Australia, use windmills for agricultural tasks, such as crushing sugar cane and grinding grains, as well as for industrial work such as mixing concrete and grinding compost. India and China are among the countries that conduct ongoing research on how best to harness the power of the wind.

Challenge

Exs. 22–26 In connection with these exercises, ask what kind of function is described by the equation for North America's wind energy production. (a logistic function) Challenge students to show that this function can be written in the form given in the Reasoning note on page 119.

Application

Exs. 27–29 Point out that half-life calculations for radioactive materials are important for many reasons. One application of great importance has to do with planning the safe disposal of nuclear waste materials.

Research

Exs. 27–29 Students can use chemistry and physics textbooks, encyclopedias, or other reference works to collect information about other radioactive elements and their half-lives. Have them write an equation to model radioactive decay for a sample of each element they have information about.

22.

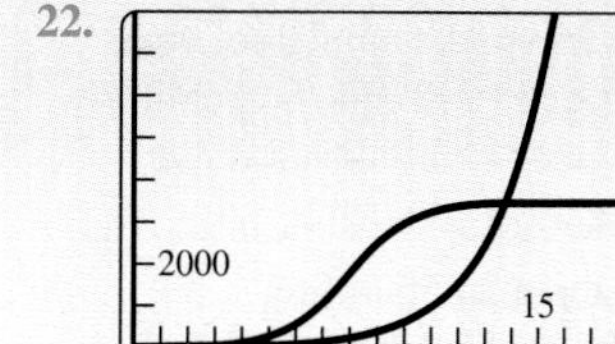

23. Answers are rounded to the nearest gigawatt-hour.

Europe	
Year	Energy (gigawatt-hours)
1985	64
1990	629
1995	6195

North America	
Year	Energy (gigawatt-hours)
1985	262
1990	2980
1995	3491

24. North America (except for 1980 and 1981)

25. 1994

26. Answers may vary. An example is given. I think it is unlikely that Europe will be able to continue to increase production at this rate.

27. $y = 100(0.995)^t$

28. $y = (0.9995)^t$

29. Estimates of exponent may vary due to rounding. $A(t) = A_0e^{-0.00012t}$

Apply⇔Assess

Exercise Notes

Assessment Note

Ex. 31 Students' graphs should indicate the scales used on the axes. Since calculations for the table in part (b) will have been performed on a calculator, ask students to show exactly what expression they entered for one of the table entries. This will help uncover any syntax errors that students may have made.

Spiral Review: Preview

Exs. 34–36 These exercises can prepare students for work with regression equations in Section 3.5.

Practice 17 for Section 3.4

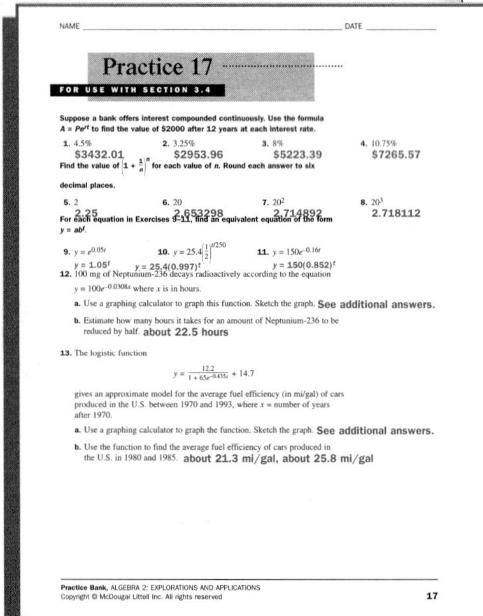

NAME ____________ DATE ____________

Practice 17

FOR USE WITH SECTION 3.4

Suppose a bank offers interest compounded continuously. Use the formula $A = Pe^{rt}$ to find the value of $2000 after 12 years at each interest rate.

1. 4.5% **$3432.01**
2. 3.25% **$2953.96**
3. 8% **$5223.39**
4. 10.75% **$7265.57**

Find the value of $\left(1 + \frac{1}{n}\right)^n$ for each value of n. Round each answer to six decimal places.

5. 2 **2.25**
6. 20 **2.653298**
7. 20^2 **2.714892**
8. 20^3 **2.718112**

For each equation in Exercises 9–11, find an equivalent equation of the form $y = ab^t$.

9. $y = e^{0.05t}$ **$y = 1.05^t$**
10. $y = 25.4\left(\frac{1}{2}\right)^{t/250}$ **$y = 25.4(0.997)^t$**
11. $y = 150e^{-0.16t}$ **$y = 150(0.852)^t$**

12. 100 mg of Neptunium-236 decays radioactively according to the equation $y = 100e^{-0.0308x}$ where x is in hours.
    a. Use a graphing calculator to graph this function. Sketch the graph. **See additional answers.**
    b. Estimate how many hours it takes for an amount of Neptunium-236 to be reduced by half. **about 22.5 hours**

13. The logistic function

$$y = \frac{12.2}{1 + 65e^{-0.435x}} + 14.7$$

gives an approximate model for the average fuel efficiency (in mi/gal) of cars produced in the U.S. between 1970 and 1993, where x = number of years after 1970.
    a. Use a graphing calculator to graph the function. Sketch the graph. **See additional answers.**
    b. Use the function to find the average fuel efficiency of cars produced in the U.S. in 1980 and 1985. **about 21.3 mi/gal, about 25.8 mi/gal**

Practice Bank, ALGEBRA 2: EXPLORATIONS AND APPLICATIONS 17

30. BIOLOGY The temperature at which red-eared slider turtle eggs are incubated affects the percent that are hatched as males. A logistic decay function that models this phenomenon is:

$$M(T) = \frac{100}{1 + e^{4.99T - 11.2}}$$

where $M(T)$ = the percent of males and T = the incubation temperature measured in degrees Celsius above 27°.

a. **Technology** Use a graphing calculator or graphing software to graph the function.

b. At what temperature will about half the turtles be male?

c. At what temperature will about 10% of the turtles be female?

ONGOING ASSESSMENT

31. Cooperative Learning Work with a partner.

Radon-222 gas decays according to the function $A(t) = A_0e^{-0.18t}$, where t is measured in days.

a. Graph the decay function for radon-222.

b. Work with your partner to make a table showing the number of days it takes for $A(t)$ to reach 90%, 80%, 70%, . . . , 10% of its initial value.

SPIRAL REVIEW

Exercises 32 and 33 refer to the graph of $y = ab^x$. *(Section 3.3)*

32. For what values of a and b will the graph show exponential decay?

33. For what values of a and b will the graph show exponential growth?

Exercises 34–36 refer to the table that shows public spending, in billions of dollars, on maternal and child health care programs in the United States over 7 years. *(Section 2.4)*

Years since 1985	Public funds (billions)
0	1.3
1	1.4
2	1.6
3	1.7
4	1.8
5	1.9
6	2.0

34. Show that the data have a linear relationship by making a scatter plot and drawing a line of fit.

35. **Technology** Use a graphing calculator or statistical software to find an equation of the least-squares line that fits the data.

36. Predict public spending on maternal and child health care in the year 2012.

Multiply. *(Section 1.4)*

37. $\begin{bmatrix} 5 & 2 \\ 0 & 8 \end{bmatrix}\begin{bmatrix} 3 & -4 & 0 \\ 12 & 6 & -5 \end{bmatrix}$

38. $\begin{bmatrix} 14 & 2 & 1 \\ 0 & 2 & 3 \\ 11 & 7 & 6 \end{bmatrix}\begin{bmatrix} 1 \\ 3 \\ 1 \end{bmatrix}$

30. a.

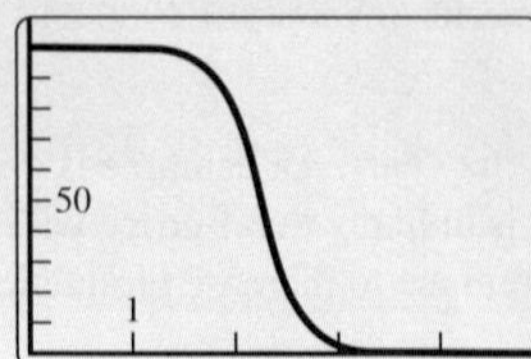

b. Estimates may vary. An example is given. about 29.2°C

c. Estimates may vary. An example is given. about 28.8°C

31. See answers in back of book.

32. $a > 0$ and $0 < b < 1$

33. $a > 0$ and $b > 1$

34. Lines of fit may vary.

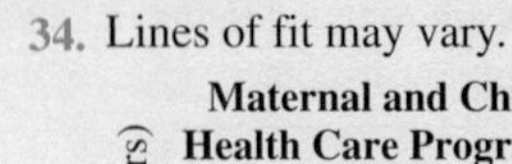

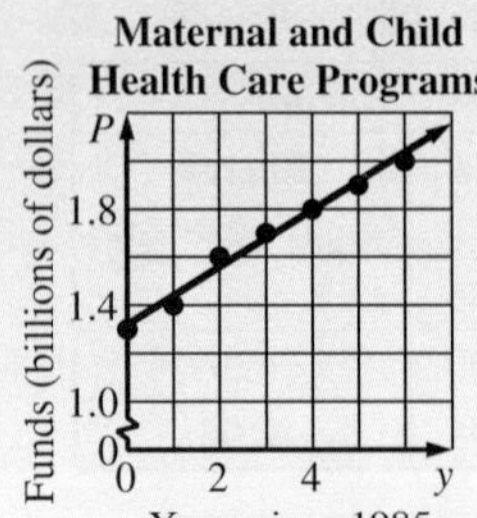

35. Let y = the number of years since 1985 and P = public funds in billions of dollars; $P = 0.12y + 1.32$

36. about $4.56 billion

37. $\begin{bmatrix} 39 & -8 & -10 \\ 96 & 48 & -40 \end{bmatrix}$

38. $\begin{bmatrix} 21 \\ 9 \\ 38 \end{bmatrix}$

Section

3.5 Fitting Exponential Functions to Data

Learn how to...

- write exponential functions that fit sets of data

So you can...

- make predictions about exponential growth and decay situations, such as learning curves and computer passwords

When data appear to grow or decay exponentially, you can try to find an exponential function that is a "good fit." You will need to find appropriate values of a and b for an equation of the form $y = ab^x$.

EXAMPLE 1 Application: Psychology

Psychologists studying how people learn timed a young piano student as she played Chopin's "Minute Waltz." On her second try, she took 2.367 min to perform the piece. On her third try she took 2.067 min.

Suppose the student's learning curve is exponential. Write an equation that relates the duration of her performance to the number of times she has played the piece.

SOLUTION

You want to find an exponential function whose graph passes through the points (2, 2.367) and (3, 2.067).

Step 1 Substitute the coordinates of each point into the general equation $y = ab^x$.

$$2.067 = ab^3$$
$$2.367 = ab^2$$

Step 2 Divide the two equations to solve for b.

$$\frac{2.067}{2.367} = \frac{ab^3}{ab^2}$$
$$0.873 \approx b$$

Step 3 To solve for a, use the value of b from Step 2 in either equation.

$$2.367 = a(0.873)^2$$
$$\frac{2.367}{(0.873)^2} = a$$
$$3.104 \approx a$$

An equation that describes the piano student's performance duration y on the xth attempt is

$$y = 3.104(0.873)^x.$$

Plan⇔Support

Objectives

- Write exponential functions that fit sets of data.
- Make predictions about exponential growth and decay situations.

Recommended Pacing

Core and Extended Courses
Section 3.5: 1 day

Block Schedule
Section 3.5: $\frac{1}{2}$ block (with Review)

Resource Materials

Lesson Support
Lesson Plan 3.5
Warm-Up Transparency 3.5
Practice Bank: Practice 18
Study Guide: Section 3.5
Exploration Lab Manual: Additional Exploration 4, Diagram Master 2
Assessment Book: Test 10

Technology
Graphing Calculator
McDougal Littell Mathpack *Function Investigator*
Internet: http://www.hmco.com

Warm-Up Exercises

Solve each equation for x.

1. $x^3 = 125$ 5
2. $x^5 = 32$ 2
3. $x^4 = 810{,}000$ ±30
4. $x^2 = 0.0576$ ±0.24
5. $x^7 = 0.0002187$ 0.3

Teach⇔Interact

Additional Example 1

Eloy is participating in an experiment to determine how many seconds it takes him to decide which button to push in reaction to a signal that is flashed on a computer screen at random time intervals. On his third try, his time was 1.09 seconds. On his fourth try, his time was 0.97 seconds. Suppose Eloy's time is modeled by an exponential function. Write an equation that relates his decision time to the number of tries he has made.

You want to find an exponential function whose graph passes through the points (3, 1.09) and (4, 0.97).

Step 1 Substitute the coordinates of each point into the general equation $y = ab^x$.

$0.97 = ab^4$

$1.09 = ab^3$

Step 2 Divide the two equations to solve for *b*.

$\frac{0.97}{1.09} = \frac{ab^4}{ab^3}$

$0.89 \approx b$

Step 3 To solve for *a*, use the value of *b* from Step 2 in either equation.

$1.09 = a(0.89)^3$

$\frac{1.09}{(0.89)^3} = a$

$1.546 \approx a$

An equation that describes Eloy's decision time *y* on the *x*th attempt is $y = 1.546(0.89)^x$.

Additional Example 2

Planners in a worldwide communications company are considering various plans that would permit the company to expand its customer base as needed in future years. They are considering, among other things, customer ID numbers that could consist of combinations of the letters A–Z and numbers 1–9. An ID number could consist of all numbers, all letters, or any combination of numbers and letters. An ID number that is 4 characters long allows for 1,500,625 customers. An ID number 6 characters long allows for 1,838,265,625 customers. Write an exponential equation that relates the number of passwords to the password length.

You need to find an exponential function whose graph passes through the points (4, 1,500,625) and (6, 1,838,265,625).

THINK AND COMMUNICATE

1. Predict how many times the student must play the "Minute Waltz" before she will be able to play it in about one minute.
2. Explain why an exponential decay model is not useful after the student has mastered playing the waltz in one minute.
3. How would you find an exponential equation that passes through two points if the *x*-coordinates of the points are *not* exactly 1 unit apart?

EXAMPLE 2 Application: Banking

To use an automated teller machine, you need to enter a password. Longer passwords are better because they are more difficult for other people to guess.

If a password uses the letters A–Z and is four letters long, there are 456,976 possible passwords. If the password is seven letters long, there are 8,031,810,176 possible passwords. Write an exponential equation that relates the number of passwords to the password length.

SOLUTION

You need to find an exponential function whose graph passes through the points (4, 456,976) and (7, 8,031,810,176).

Step 1 Write two equations.

$$8{,}031{,}810{,}176 = ab^7$$
$$456{,}976 = ab^4$$

Use substitution as in Example 1.

Step 2 Solve for *b*.

$$\frac{8{,}031{,}810{,}176}{456{,}976} = b^3$$
$$17{,}576 = b^3$$
$$(17{,}576)^{1/3} = b$$
$$26 = b$$

If $b^n = c$, then $b = c^{1/n}$.

Dividing the equations eliminates *a*.

Step 3 Solve for *a*.

$$456{,}976 = a \cdot 26^4$$
$$\frac{456{,}976}{26^4} = a$$
$$1 = a$$

To solve for *a*, use **26** as the value of ***b*** in either equation.

The number *y* of all possible passwords when *x* letters are used is $y = 1 \cdot 26^x$, or $y = 26^x$.

ANSWERS Section 3.5

Think and Communicate

1. 8 times
2. Mastery implies that the student has reached a goal and does not need to continue trying to play the minute waltz in a shorter time.
3. Use the method of Example 1. However, the process yields $n = b^m$, rather than $n = b$. Solve to get $b = n^{1/m}$.

THINK AND COMMUNICATE

4. Explain why the base 26 makes sense if the password includes only the letters A–Z.
5. What would the exponential function be if a password includes a mix of the letters A–Z, the letters a–z, and the numbers 0–9?

You can also use a graphing calculator or statistical software to find an exponential equation that is a good fit to a set of data.

EXAMPLE 3 Application: Biology

The table shows that the mean weight of Atlantic cod from the Gulf of Maine increases with age. Find an exponential function that models the data.

Atlantic Cod from the Gulf of Maine

Age (y)	Weight (kg)
1	0.751
2	1.079
3	1.702
4	2.198
5	3.438
6	4.347
7	7.071
8	11.518

SOLUTION

First enter the data values.

L1	L2	L3
2	1.079	
3	1.702	
4	2.198	
5	3.438	
6	4.347	
7	7.071	
8	11.518	

L2(8)=11.518

Then use the exponential regression feature of a graphing calculator.

```
ExpReg
 y=a*b^x
 a=.508948569
 b=1.459711295
 r=.99728485
```

If the correlation coefficient ***r* is close to 1** (or −1), then **exponential growth** (or decay) is a good model.

An exponential function that models the Atlantic cod weight data is $y = 0.509(1.460)^x$ where x is age in years and y is weight in kilograms.

CHECKING KEY CONCEPTS

Write an exponential function whose graph passes through each pair of points.

1. (0, 5), (1, 3)
2. (0, 10), (2, 20)
3. (2, 48), (5, 3072)
4. **HEALTH** Tuberculosis cases in the United States fell exponentially from about 84,000 in 1953 to about 22,000 in 1985. Write an exponential function that models this decay.

Think and Communicate

4. There are 26 possible choices for each letter in the password.
5. $y = 1 \cdot 62^x$

Checking Key Concepts

1. $y = 5\left(\frac{3}{5}\right)^x$
2. $y = 10(\sqrt{2})^x$ or $y = 10(1.414)^x$
3. $y = 3 \cdot 4^x$
4. Let x = the number of years after 1953 and y = the number of cases of tuberculosis; $y = 84{,}000(0.959)^x$.

Teach⇔Interact

Additional Example 2 (continued)

Step 1 Write two equations.

$1{,}838{,}265{,}625 = ab^6$

$1{,}500{,}625 = ab^4$

Step 2 Divide the equations to eliminate a. Then solve for b.

$$\frac{1{,}838{,}265{,}625}{1{,}500{,}625} = b^2$$
$$1225 = b^2$$
$$(1225)^{1/2} = b$$
$$35 = b$$

Step 3 To solve for a, use 35 for b in either equation.

$1{,}500{,}625 = a \cdot 35^4$

$\frac{1{,}500{,}625}{35^4} = a$

$1 = a$

The number of possible customer ID numbers when x characters are used is $y = 1 \cdot 35^x$ or $y = 35^x$.

Think and Communicate

After discussing question 5, ask students to generalize to the situation where K characters are available for passwords. If K characters are available, then the function that relates the number of characters x in a password and the number of possible passwords y is $y = K^x$.

Additional Example 3

Suppose that a biologist is raising Atlantic cod under controlled conditions. By modifying their diets, she is able to produce larger fish.

Atlantic Cod raised under controlled conditions

Age (y)	Weight (kg)
1	0.741
1.5	0.950
2	1.105
2.5	1.483
3	1.821
3.5	2.176
4	2.513

Find an exponential function that models the data.

First enter the data values and then use the exponential regression feature of a graphing calculator.

```
ExpReg
 y=a*b^x
 a=.5012334325
 b=1.515539361
 r=.9962239489
```

Since the correlation coefficient is close to 1, the exponential function is a good model. An exponential function that models the Atlantic cod weight data is $y = 0.501(1.516)^x$, where x is age in years and y is weight in kilograms.

Teach⇔Interact

Closure Question

How can you write an exponential function to model data? Use the equation $y = ab^x$ and find appropriate values of a and b by choosing the coordinates of two data points. Substitute the coordinates of the points into $y = ab^x$ to write two equations. Solve the equations for b. Use the value of b to solve for a. Replace a and b in the equation $y = ab^x$ to write the exponential function.

Apply⇔Assess

Suggested Assignment

Core Course
Exs. 1–8, 10–12, 17, 18, 20–23, AYP

Extended Course
Exs. 5–23, AYP

Block Schedule
Day 18 Exs. 1–18, 10–12, 17, 18, 20–23, AYP

Exercise Notes

Alternate Approach
Exs. 1–8 Students may find it helpful to solve these exercises by first using the method of Examples 1 and 2. They can then check their answers by using the method of Example 3. Except for minor variations due to rounding, the equations obtained by the two methods should be the same.

Using Technology
Ex. 11 This exercise can be solved with the aid of a TI-81 or TI-82 graphing calculator or with the *Function Investigator* software. Some students may want to use paper and pencil for some of the work in part (a), though all of the work can be done by using a graphing calculator or statistical software.

Visual Thinking
Ex. 11 Students can draw a graph in part (c) and use it to make future predictions. How many Internet hosts will there be in 2020? In what year will there be double the number of sites that existed in 1994? How might logistic growth impact future predictions? This activity involves the visual skills of *exploration* and *perception.*

3.5 Exercises and Applications

Extra Practice *exercises on page 753*

Write an exponential function whose graph passes through each pair of points.

1. (0, 5), (1, 8) **2.** (1, 8), (2, 10) **3.** (0, 10), (1, 8) **4.** (5, 7), (6, 6)

5. (3, 3), (8, 7) **6.** (0, 6), (7, 9) **7.** (20, 4), (25, 3) **8.** (0, 17), (7, 15)

9. ASTRONOMY The *apparent magnitude* of a star is a measure of its apparent brightness. A star of magnitude 1 is 100 times brighter than a star of magnitude 6.

a. Find an equation of the exponential decay function whose graph passes through (1, 100) and (6, 1).

b. Graph the function from part (a).

c. Copy and complete the table showing comparative brightness values for various sky objects.

Sky object	Apparent magnitude	Brightness value
Uranus	6	1
Aldebaran	1	100
Vega	0	?
Sirius	−1.5	?
Full moon	−12.5	?
Sun	−26.7	?

d. Research Use a book about astronomy to find the names and magnitudes of some stars. Plot them on your graph from part (b). Calculate their brightness values and add them to your table from part (c).

10. SAT/ACT Preview Which of the equations represents an exponential decay function whose graph passes through (2, 1)?

A. $y = (0.5)^x$ **B.** $y = (0.5x)^2$ **C.** $y = \frac{1}{4}(0.5)^{-x}$ **D.** $y = 4(0.5)^x$ **E.** $y = \frac{1}{4} \cdot 2^x$

11. Open-ended Problem The table shows the increase in the number of host sites on the Internet computer network.

Year	1988	1989	1990	1991	1992	1993	1994
Years since 1988	0	1	2	3	4	5	6
Internet hosts (1000s)	56	159	313	617	1136	2056	3864

a. Choose any two years and write an exponential function to represent the growth.

b. Check your function by seeing how its predictions compare to the other data values in the table.

c. **Technology** Enter all the data into a graphing calculator or statistical software and find an exponential equation that gives a good fit. Compare the results to your results from part (a).

Exercises and Applications

1. $y = 5\left(\frac{8}{5}\right)^x$
2. $y = 6.4\left(\frac{5}{4}\right)^x$
3. $y = 10\left(\frac{4}{5}\right)^x$
4. $y = 15.130\left(\frac{6}{7}\right)^x$
5. $y = 1.804(1.185)^x$
6. $y = 6(1.060)^x$
7. $y = 12.642(0.944)^x$
8. $y = 17(0.982)^x$
9. See answers in back of book.
10. D
11. a, b. Answers may vary. Check students' work.
    c. Let x = the number of years since 1988 and y = the number of Internet hosts in thousands; $y = 71.171(1.979)^x$. Comparisons may vary. Check students' work.
12. Answers may vary. Check students' work.

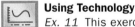

12. Cooperative Learning Work with another student.

a. Each of you should write an exponential function that your partner cannot see.

b. Find two points that are on the graph of your exponential function. Give their coordinates to your partner.

c. Find the exponential function that passes through the two points your partner gives you.

d. Check your answers by revealing the original exponential functions.

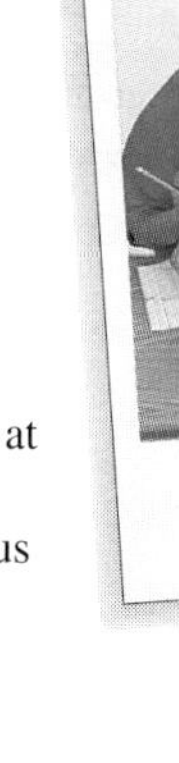

13. Investigation Suppose a computer program tests passwords at a rate of 10,000 characters per second. Make a table showing how long it will take to go through all the passwords of various lengths if the password uses:

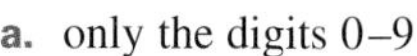

a. only the digits 0–9

b. only the letters A–Z

c. the letters A–Z, a–z, and the digits 0–9

14. SPORTS The table shows the breathing rate (in liters of air per minute) of bicyclists traveling at various speeds on two types of bikes.

Touring bike speed (mi/h)	Breathing rate (L/min)	Racing bike speed (mi/h)
0	7	0
1.8	9	3.2
6	13	7.2
8.3	18	10.5
12	29	14.5
16	50	19
18.5	72	22
21	93	25
22.5	115	27

a. Write an exponential function to show how the breathing rate of a bicyclist on a racing bike depends on speed.

b. Write an exponential function to show how the breathing rate of a bicyclist on a touring bike depends on speed.

c. **Technology** Use a graphing calculator or graphing software to graph the two functions from parts (a) and (b) in the same coordinate plane.

d. For a given speed, who breathes at a faster rate, a bicyclist on a racing bike or a bicyclist on a touring bike? Explain why your answer makes sense.

15. Challenge Explain why the growth or decay factor for an exponential function whose graph passes through (p, q) and (r, s) is $\left(\frac{s}{q}\right)^{1/(r-p)}$.

Apply⇔Assess

Exercise Notes

Integrating the Strands
Ex. 13 The equations that describe how the number of passwords depends on the number of characters available can be obtained by the methods of this section. However, they can also be obtained by using ideas about permutations. The latter approach could be the basis for an interesting class discussion.

Cooperative Learning
Ex. 14 Students can work in pairs on this exercise.

Second-Language Learners
Ex. 14 Some students learning English may benefit from having a peer tutor explain or paraphrase the meaning of phrases in this exercise.

Multicultural Note
Ex. 14 The first African American to participate in integrated national bicycle races was Marshall W. "Major" Taylor. He was known as the "Fastest Bicycle Rider in the World" until his retirement in 1910. He became a cycling champion in 1898, winning his first professional start. By the end of that year, he had won twenty-one races, placed second in thirteen, and placed third in eleven. Taylor is now recognized as the first African American champion in any professional sport.

Problem Solving
Ex. 15 Students who are uncertain about how to proceed with this exercise should review the method used in Examples 1 and 2.

13. Answers may vary. Examples are given.

Password length	Testing Time: Using only 0–9	Using only A–Z	Using A–Z, a–z, 0–9
4	1 s	45.70 s	24.63 min
6	1.67 min	8.58 h	65.74 days
7	16.67 min	9.30 days	11.17 years
10	278 h	447.64 years	2,661,401 years

14. For each function, let x = the speed of the bicycle in mi/h and y = the breathing rate of the rider in L/min.

a. $y = 6.374(1.113)^x$

b. $y = 6.696(1.134)^x$

c.

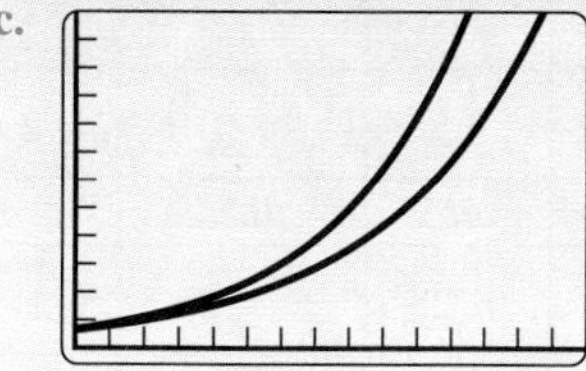

d. a bicyclist on a touring bike; Answers may vary. An example is given. Racing bicycles are extremely lightweight compared to touring bicycles and are designed for high speeds.

15. Let $y = ab^x$ be the function. Then $q = ab^p$ and $s = ab^r$, so $\frac{ab^r}{ab^p} = \frac{s}{q}$, and $b^{r-p} = \frac{s}{q}$. Therefore, the growth factor, b, is $\left(\frac{s}{q}\right)^{1/(r-p)}$.

Apply⇔Assess

Exercise Notes

Mathematical Procedures
Ex. 16 Students may complete this exercise using various methods. Some may complete the whole table using a paper model or sketches to calculate the area, side length, and perimeter after each fold, then determine the growth factor for each column and use the initial values given. Others may complete the first 2 rows and then use either the algebraic or calculator-based methods of this section to find the regression equations and use these equations to get the remaining table values. If different members of the class use different methods, ask them to write their solutions on the board so they can be reviewed and compared.

Using Manipulatives
Ex. 16 Many students would benefit by folding an actual square of paper and measuring the side lengths. Then ask them to justify their results mathematically.

Interview Note
Exs. 17–19 These exercises are appropriate for small groups of 2 or 3 students.

Using Technology
Exs. 17–19 Students may see more clearly the effect of chlorinity by graphing the equations found in Exs. 17 and 18. The procedure to do this on the TI-82 graphing calculator is as follows. After entering the data and finding the first exponential regression equation on the home screen, go to Y=, clear any equations from the list, and position the cursor at Y_1=. Choose 5:Statistics from the VARS menu, move the cursor to EQ and choose 7:RegEQ. This copies the exponential equation to Y_1. Next find the exponential regression equation for 2% chlorinity and repeat the procedure to copy this equation to Y_2 on the Y= list.

Assessment Note
Ex. 20 A thorough discussion of this exercise can provide a good comprehensive review of the major concepts and skills of this chapter. Although this is primarily a writing exercise, students should be encouraged to illustrate their responses with appropriate sketches or graphs.

16. GEOMETRY If you fold the corners of a square piece of paper to the center of the square, you get a smaller square. You can perform the folding process x times to get smaller and smaller squares.

a. Copy and complete the table.

b. Write an exponential function for each column of the table.

x	Area	Side length	Perimeter
0	1	1	4
1	0.5	?	?
2	?	?	?
3	?	?	?
4	?	?	?

BY THE WAY

Folding the four corners of a square to the center is called a "blintz" fold in the art of origami. "Blintz" is a Yiddish word of Ukrainian origin that refers to a thin folded pancake.

To make the candy dish, flip the square over before folding a second time.

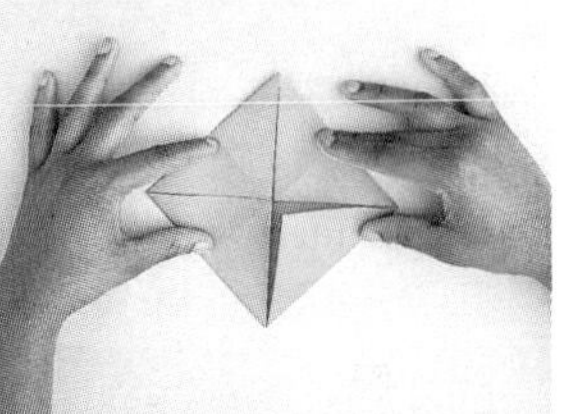
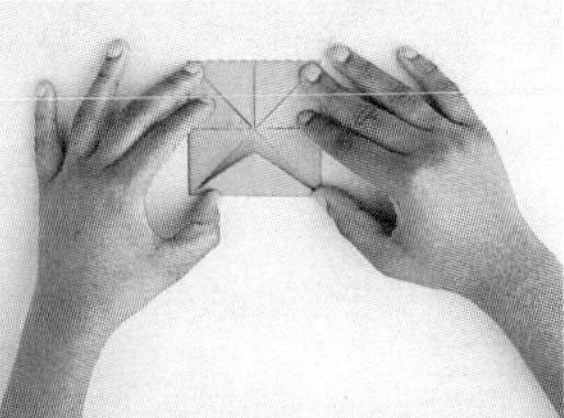

INTERVIEW Finn Strong

Look back at the article on pages 90–92.

In Finn Strong's fish tanks, the fish need oxygen to survive. The amount of oxygen that water can hold varies exponentially with the temperature and depends on the water's chlorinity (a measure related to its saltiness).

17. Use the table to write an exponential function that fits the data for fresh water (with 0% chlorinity), such as the water in Finn Strong's tanks.

18. Use the table to write an exponential function that fits the data for ocean water (with 2% chlorinity).

19. Open-ended Problem For any given temperature, describe how the chlorinity of the water affects the amount of oxygen the water can hold. Predict table values for water of 0.5% and 1.5% chlorinity. Justify your predictions.

Oxygen Absorbed by Water (mL/L)

Temperature (°C)	Chlorinity 0%	1%	2%
0	10.29	9.13	7.97
10	8.02	7.19	6.35
15	7.22	6.50	5.79
20	6.57	5.95	5.31
30	5.57	5.01	4.46

16. a.

x	Area	Side length	Perimeter
0	1	1	4
1	0.5	$\frac{\sqrt{2}}{2} \approx 0.71$	$2\sqrt{2} \approx 2.83$
2	0.25	0.5	2
3	0.125	$\frac{\sqrt{2}}{4} \approx 0.35$	$\sqrt{2} \approx 1.41$
4	0.0625	0.25	1

b. Let x = the number of folds, A = area, s = side length, and P = perimeter; $A = \left(\frac{1}{2}\right)^x$; $s = \left(\frac{\sqrt{2}}{2}\right)^x$ or $s = (0.707)^x$; $P = 4\left(\frac{\sqrt{2}}{2}\right)^x$ or $P = 4(0.707)^x$.

17. Let x = the number of degrees Celsius above zero and y = the oxygen absorbed in mL/L; $y = 10.015(0.98)^x$.

18. Let x = the number of degrees Celsius above zero and y = the oxygen absorbed in mL/L; $y = 7.823(0.98)^x$.

ONGOING ASSESSMENT

20. Writing Suppose you have to explain to a friend who missed school what you have learned about exponential modeling. Write a page reminding your friend what modeling is, explaining how you determine if a relationship is exponential, and suggesting some ways to find an exponential function that describes a set of data.

SPIRAL REVIEW

21. BIOLOGY The growth of the bacteria *Mycobacterium tuberculosis* can be modeled by the equation $P(t) = P_0e^{0.116t}$, where $P(t)$ is the population after t hours and P_0 is the population when $t = 0$. *(Section 3.4)*

a. At 1:00 P.M. there are 30 *Mycobacterium tuberculosis* bacteria in a sample. Write a function for the number of bacteria after 1:00 P.M.

b. What is the population at 5:00 P.M.?

c. What was the population at noon?

d. How would you find the population at 3:45 P.M.?

22. Use the equation $y = 8 - 2x$. *(Section 2.2)*

a. Find y when $x = -3$.   **b.** Find x when $y = 5$.

23. a. Graph the parametric equations $x = 2 - t$ and $y = 3t + 1$ for $t \geq 0$. *(Section 2.6)*

b. Express y as a function of x. State any restriction on x.

ASSESS YOUR PROGRESS

VOCABULARY

e (p. 116)   **logistic growth** (p. 118)

Find the value of an investment of $2000 after four years if the interest rate is 4.5% and interest is compounded as specified. *(Section 3.4)*

1. annually   **2.** weekly   **3.** continuously

4. Graph $y = 2^x$, $y = e^x$, and $y = 3^x$ in the same coordinate plane. What do you notice?

5. PHYSIOLOGY Your visual *near point* is the closest point at which your eyes can see an object distinctly. Your near point moves farther away from you as you grow older, as shown by the data in the table. *(Section 3.5)*

Age (years)	Near point (cm)
10	10
20	12
30	15
40	25
50	40
60	100

a. Choose two data pairs and write an exponential function to represent how near point distance varies with age.

b. **Technology** Enter the data into a graphing calculator or statistical software and perform an exponential regression. What equation do you get?

c. Predict the near point distance in centimeters at age 80.

6. Journal Compare exponential and logistic growth. Give some examples of situations in which one model is better than the other.

Apply⇔Assess

Assess Your Progress

Students should know the formal definition of *e* and a decimal approximation for *e* to at least three decimal places. They should be able to illustrate logistic growth functions by referring to specific examples in the chapter.

Journal Entry
Encourage students to include sketches or graphs of particular functions to illustrate their written explanations.

Progress Check 3.4–3.5

See page 133.

Practice 18 for Section 3.5

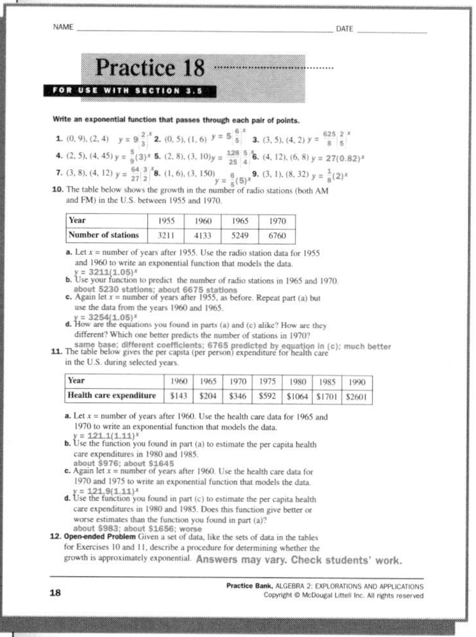
NAME ______ DATE ______

Practice 18

FOR USE WITH SECTION 3.5

Write an exponential function that passes through each pair of points.

1. (0, 9), (2, 4) $y = 9\left(\frac{2}{3}\right)^x$ **2.** (0, 5), (1, 6) $y = 5\left(\frac{6}{5}\right)^x$ **3.** (3, 5), (4, 2) $y = \frac{625}{8}\left(\frac{2}{5}\right)^x$

4. (2, 5), (4, 45) $y = \frac{5}{9}(3)^x$ **5.** (2, 8), (3, 10) $y = \frac{128}{25}\left(\frac{5}{4}\right)^x$ **6.** (4, 12), (6, 8) $y = 27(0.82)^x$

7. (3, 8), (4, 12) $y = \frac{64}{27}\left(\frac{3}{2}\right)^x$ **8.** (1, 6), (3, 150) $y = \frac{6}{5}(5)^x$ **9.** (3, 1), (8, 32) $y = \frac{1}{8}(2)^x$

10. The table below shows the growth in the number of radio stations (both AM and FM) in the U.S. between 1955 and 1970.

Year	1955	1960	1965	1970
Number of stations	3211	4133	5249	6760

a. Let x = number of years after 1955. Use the radio station data for 1955 and 1960 to write an exponential function that models the data.
$y = 3211(1.05)^x$

b. Use your function to predict the number of radio stations in 1965 and 1970.
about 5230 stations; about 6675 stations

c. Again let x = number of years after 1955, as before. Repeat part (a) but use the data from the years 1960 and 1965.
$y = 3254(1.05)^x$

d. How are the equations you found in parts (a) and (c) alike? How are they different? Which one better predicts the number of stations in 1970?
same base; different coefficients; 6765 predicted by equation in (c); much better

11. The table below gives the per capita (per person) expenditure for health care in the U.S. during selected years.

Year	1960	1965	1970	1975	1980	1985	1990
Health care expenditure	$143	$204	$346	$592	$1064	$1701	$2601

a. Let x = number of years after 1960. Use the health care data for 1965 and 1970 to write an exponential function that models the data.
$y = 121.1(1.11)^x$

b. Use the function you found in part (a) to estimate the per capita health care expenditures in 1980 and 1985.
about $976; about $1645

c. Again let x = number of years after 1960. Use the health care data for 1970 and 1975 to write an exponential function that models the data.
$y = 121.9(1.11)^x$

d. Use the function you found in part (c) to estimate the per capita health care expenditures in 1980 and 1985. Does this function give better or worse estimates than the function you found in part (a)?
about $983; about $1656; worse

12. Open-ended Problem Given a set of data, like the sets of data in the tables for Exercises 10 and 11, describe a procedure for determining whether the growth is approximately exponential. **Answers may vary. Check students' work.**

18

Practice Bank, ALGEBRA 2: EXPLORATIONS AND APPLICATIONS

19.

Temperature (°C)	Chlorinity	
	0.5%	1.5%
0	9.71	8.55
10	7.60	6.77
15	6.86	6.15
20	6.26	5.63
30	5.29	4.74

The amount of oxygen absorbed by water at a given temperature is a linear function of chlorinity.

20. Summaries may vary. They should include the following.
- Modeling is the use of functions, tables, graphs, equations, or inequalities to describe situations.
- You can determine if a relationship is exponential by using a table to determine if the data show a constant growth or decay factor, by drawing a scatter plot and observing if the points lie on an exponential curve, by trying to fit the data to an exponential function, or by using technology to perform an exponential regression and seeing whether the correlation coefficient is close to 1 or –1.

21. a. Let t be the number of hours after 1:00 P.M. and $P(t)$ = the number of bacteria; $P(t) = 30e^{0.116t}$.

b. about 48   **c.** about 27

d. Substitute 2.75 for t in the equation.

22. a. 14   **b.** 1.5

23. a.

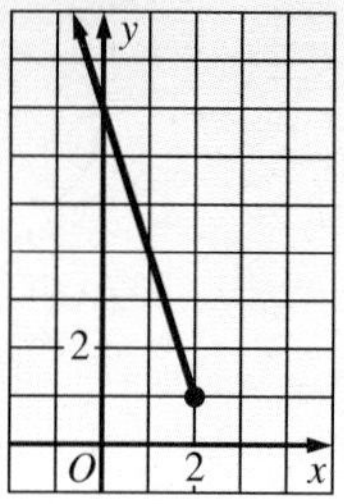

b. $y = -3x + 7;\ x \leq 2$

Assess Your Progress

1–6. See answers in back of book.

Project Notes

Mathematical Goals

- Collect and organize experimental data in a table.
- Calculate a bounce factor from the data.
- Write an exponential equation that describes rebound height as a function of the number of bounces.

Planning

Materials

- a ball
- two metersticks or a tape measure
- paper and pencil

Project Teams
Students can choose their partners and decide on how to divide up their project responsibilities. They should choose a convenient place to do the experiment and record their results.

Guiding Students' Work

To ensure enough trials in the experiment, students should use a specific number of drop heights, such as 10 or 15. This will also ensure a better mean bounce factor. Also, discuss with students that the mean bounce factor is always a number greater than 0 and less than 1 because all of the recorded bounce factors are greater than 0 and less than 1.

Second-Language Learners
Students learning English are likely to benefit from working cooperatively with English-fluent students when writing their reports.

PORTFOLIO PROJECT

The Way a Ball Bounces

Have you ever watched a ball bounce repeatedly? Due to the loss of energy each time a bouncing ball hits the floor, the ball never rebounds to the same height from which it fell.

The bounciness, or "bounce factor," of a ball can be determined by comparing the ball's rebound height to the original height from which the ball was dropped. Then, using the bounce factor of the ball, you can predict the height of the ball after any number of bounces.

PROJECT GOAL **Create a model that relates the bounce height of a ball to the number of times the ball has bounced.**

Conducting an Experiment

Work with a partner to design and carry out an experiment in which you measure the bounce height of a ball after dropping the ball from varying heights. You will need a ball and two metersticks or a tape measure. Here are some guidelines for conducting your experiment.

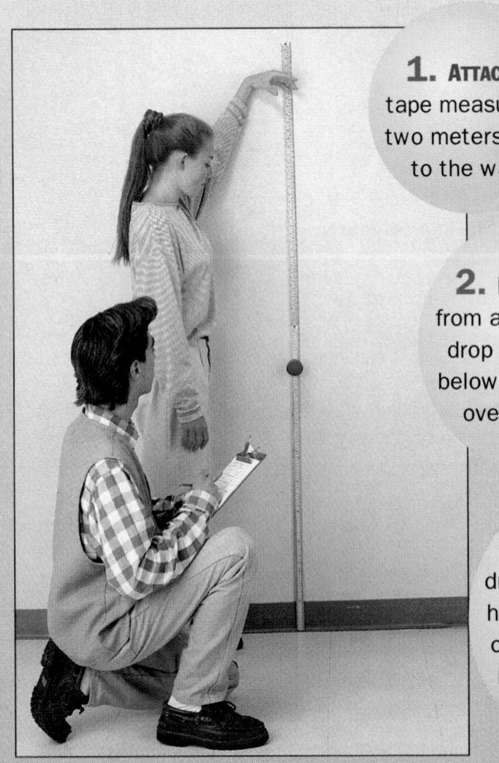

1. **ATTACH** a tape measure or two metersticks to the wall.

2. **DROP** the ball from a wide range of drop heights, from below your knees to over your head.

3. **MEASURE** the drop height and bounce height from the bottom of the ball to the floor. Round to the nearest centimeter.

4. **RECORD** your results in a table.

Drop height (cm)	Bounce height (cm)
210	134
190	124
170	113
150	101
130	87
110	76
90	64
50	34

	Bounce Factor		
C10		=SUM (C2:C9)/8	
	A	B	C
1	Drop height (cm)	Bounce height (cm)	Bounce factor
2	210	134	0.64
3	190	124	0.65
4	170	113	0.66
5	150	101	0.67
6	130	87	0.67
7	110	76	0.69
8	90	64	0.71
9	50	34	0.68
10	Mean bounce factor:		0.67

Analyzing the Data

• **CALCULATE** the bounce factor of your ball for each drop height.The bounce factor of your ball is equal to the ratio $\frac{\text{bounce height}}{\text{drop height}}$.

• **FIND** the mean of the bounce factors. Use this figure as the average bounce factor of your ball.

Predicting Exponential Decay

Suppose you dropped your ball from the height of the top of your head, and watched it fall and rebound, over and over.

• **MAKE** a table of predicted rebound heights. Start with the ball's initial height, and calculate the ball's height after one bounce, after two bounces, and so on. Continue for several bounces.

• **FIND** an equation that describes rebound height as a function of the number of bounces. Determine the proper values for a and b to substitute into the exponential decay model $y = ab^x$. What does a stand for in your model? What does b stand for?

• **PREDICT** how many bounces it will take until the ball's rebound height is less than the height of your knee. Test your prediction.

Writing a Report

Write a report about your experiment. State the project goal, describe your procedures, and present data that support the conclusions you make. Be sure to explain the role that drop height and bounce factor have in your model and evaluate how well your exponential decay model predicts bounce height.

You may wish to extend your report to include data about other types of balls, or do some research about the official rules concerning the bounciness of a ball in a sport like tennis, basketball, or football.

Self-Assessment

Does the model that you developed for this project make sense to you? If so, in what ways? If not, where did you get confused? What would help make things clearer for you?

Project Notes

Guiding Students' Work

Rubric for Chapter Project

4 Students complete the experiment and record the results in a table. The data from the table are analyzed correctly and the bounce factors and mean bounce factor are calculated correctly. Students write a clear and thorough report relating the project goals with the mathematical concepts of the chapter. They include prediction tables as well as a correctly written exponential model for the experiment. Students extend the project using other balls and do research on a ball in a sport.

3 Students complete the experiment and record the results in a table. Students also calculate all of the bounce factors correctly, and make predictions about the number of bounces in their report. The report is written clearly and indicates an understanding of the mathematics, but it also shows a minimum amount of work in the prediction tables and in the analysis of the ideas. Students do not extend the project beyond the specific tasks in the text.

2 Students perform the experiment and make a table for the data but do not calculate some of the bounce factors or the mean bounce factor correctly. Students make predictions from the data, but they do not indicate an understanding of the mathematical ideas about exponential equations and their meanings, and the values of a or b in the equation are not correct. Students turn in a report and make an attempt at understanding the project, but the report is unclear and does not extend the project beyond the bounce experiment or the analysis of the data.

1 Students record only a few bounce heights in a table and do not show a record of the data. Students do not make a prediction table but guess what the equation will be and what the number of bounces will be without experimentation. A report is turned in, but it is incomplete and does not state project goals, describe procedures, or present data. Students should be encouraged to speak with the teacher as soon as possible to review their work and to make a new start on the project.

Chapter Support

Course Guide: Chapter 3
Lesson Plans: Chapter 3
Practice Bank:
Cumulative Practice 19
Study Guide: Chapter 3 Review
Assessment Book:
Chapter Tests 11 and 12
Chapter 3 Alternative Assessment
Test Generator Software
Portfolio Project Book:
Additional Project 2: Functions
Preparing for College Entrance Tests
Professional Handbook

Progress Check 3.1–3.3

Evaluate when $n = 5$. *(Section 3.1)*

1. $8 \cdot 5^n$ 25,000
2. $130\left(\frac{1}{2}\right)^n$ 4.0625
3. Suppose $y = 1316(1.15)^x$ models the population on an island, where x is the number of decades since 1920. *(Section 3.2)*
   a. What was the population of the island in 1920? 1316
   b. What was the population in 1945? about 1866
   c. Predict the population in 1910. about 1144
4. The diagram shows the graphs of three functions. Tell whether each graph represents *exponential growth, exponential decay,* or *neither. (Section 3.3)*

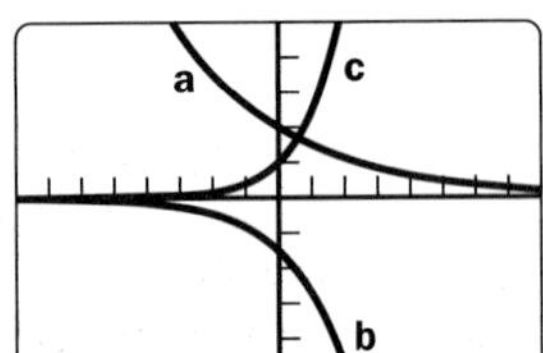

Graph a: exponential decay
Graph b: neither
Graph c: exponential growth

5. Find the doubling time for a colony of bacteria that is increasing at a proportional growth rate of 1.84 per hour. *(Section 3.3)* about 1.14 h
6. Find the half-life for an exponential decay function that is modeled by $y = 417(0.83)^x$. Then rewrite the function in the form $y = 417\left(\frac{1}{2}\right)^{x/k}$. *(Section 3.3)*

   3.72; $y = 417\left(\frac{1}{2}\right)^{x/3.72}$

CHAPTER

3 Review

STUDY TECHNIQUE

Review your homework from this chapter. Find one exercise from each section that gave you the most difficulty. Write and solve new exercises like the ones you chose or choose similar exercises from the text to solve. Exchange your solutions with another student and check each other's work.

VOCABULARY

exponential growth (p. 96)
exponential decay (p. 96)
doubling time (p. 108)
half-life (p. 109)
e (p. 116)
logistic growth (p. 118)

SECTIONS 3.1 *and* 3.2

You can simplify expressions involving exponents by using the properties of exponents and the definitions of negative and rational exponents.

Properties of Exponents

For any $b > 0$:

		Examples:
Product Property:	$b^m \cdot b^n = b^{m+n}$	$4^3 \cdot 4^2 = 4^5$
Quotient Property:	$\frac{b^m}{b^n} = b^{m-n}$	$\frac{7^{1.8}}{7^{0.8}} = 7^1 = 7$
Power Property:	$(b^m)^n = b^{mn}$	$3^{2x} = (3^2)^x = 9^x$

Negative and Rational Exponents

For any base $b > 0$:

Negative Exponent	**Example:**
$b^{-n} = \frac{1}{b^n}$	$2^{-5} = \frac{1}{2^5} = \frac{1}{32}$
Rational Exponent	**Example:**
$b^{1/n} = c$ if and only if $b = c^n$	$81^{1/2} = 9$ because $81 = 9^2$

SECTION 3.3

Exponential growth and **exponential decay** can be modeled by the equation $y = ab^x$.

Suppose you invested \$100 at 4% interest compounded yearly. You can model the growth of the investment using $y = 100(1.04)^x$.

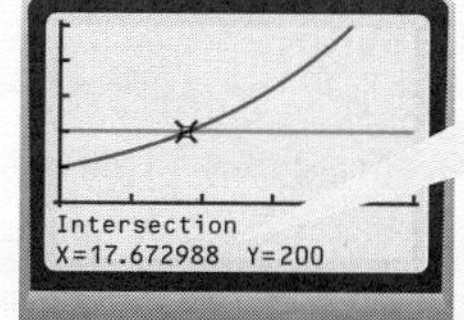

Because the investment has a **doubling time** of about 17.7 years, you can rewrite $y = 100(1.04)^x$ as $y = 100 \cdot 2^{x/17.7}$.

The trade-in value of an \$8000 used car decreases by 14% per year. You can model the decay in the car's value using $y = 8000(0.86)^x$.

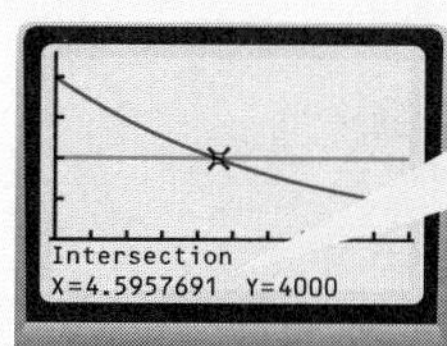

Because the car's value has a **half-life** of about 4.6 years, you can rewrite $y = 8000(0.86)^x$ as $y = 8000\left(\frac{1}{2}\right)^{x/4.6}$.

SECTIONS 3.4 *and* 3.5

The irrational number ***e***, approximately equal to 2.718, is often used as the base of exponential functions.

$A(t)$ gives the amount in t years when \$100 is invested at a 4% annual rate compounded continuously.

$$A(t) = 100e^{0.04t}$$
$$A(2) = 100e^{0.04(2)}$$
$$\approx 108.3$$

Evaluating $A(2)$ gives the amount in 2 years.

The graph shows the **logistic growth** of a population modeled by:

$$P(t) = \frac{1000}{1 + 9e^{-0.05t}}$$

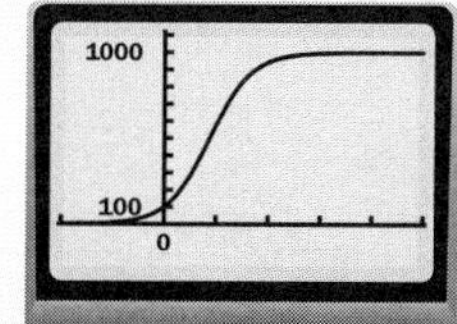

At time $t = 0$, the population is **100**. According to the model, the population stabilizes at **1000**.

You can fit an exponential function to paired data by finding an exponential curve that passes through two representative data points or by performing exponential regression on all the data points. For example, the table shows how the sales of home fax machines increased since 1989.

Using these two data points, you get $y = 306(1.14)^x$ for an exponential model.

x (years)	y (thousands)
1	350
2	400
3	520
4	800

Using all the data points, you can perform exponential regression on a graphing calculator or graphing software to get $y = 247(1.32)^x$.

Progress Check 3.4–3.5

1. How much money will be in an account after 15 years if the initial amount is \$1200, compounding is continuous, the annual interest rate is 4%, and there are no deposits or withdrawals? *(Section 3.4)* \$2186.54

2. Does $y = e^{-x}$ describe an exponential growth function or an exponential decay function? To the nearest thousandth, what is the proportional growth factor? *(Section 3.4)*
exponential decay; 0.368

3. Records show that the number of nondefective items out of every 100 items made by a piece of equipment has increased with each day of its use. An analyst has found that the percent of nondefective items is modeled by $P(t) = \frac{100}{1 + 25e^{-0.015t}}$, where t is the number of days the equipment has been used, and $P(t)$ is the number of nondefective items out of every 100 items. What kind of growth does this represent? What percent of nondefective items will be produced after 365 days of use? *(Section 3.4)*
logistic growth; about 90.5%

4. Find an equation for the exponential function whose graph contains the points (4, 18) and (10, 92). *(Section 3.5)*
A possible equation is $y = 6.066(1.312)^x$.

5. A chain of record stores found that the number of CDs that it sold grew exponentially over the first 10 years after it opened as shown in the table.

Year number	2	4	6	8	10
CDs sold (thousands)	3.8	6.4	11.2	21	36.8

Use a graphing calculator to find an exponential function that models the data. What is the correlation coefficient? *(Section 3.5)*
A possible equation is $y = 2.081(1.332)^x$; about 0.9995.

6. How many exponential functions have graphs that pass through a given pair of points? (Assume that the points are not on the same horizontal or vertical line.) *(Section 3.5)* one

Chapter 3 Assessment
Form A Chapter Test

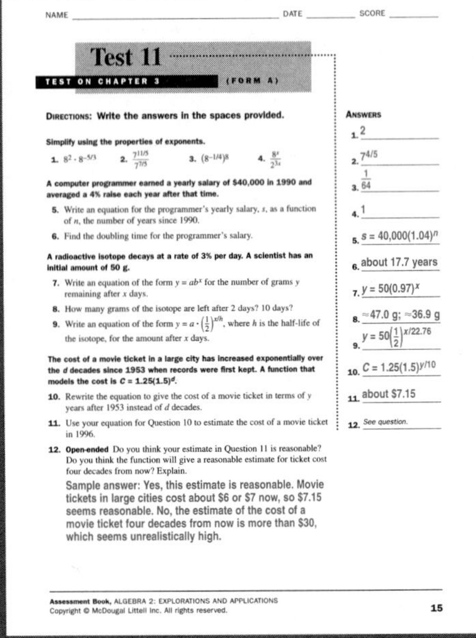

NAME ______ DATE ______ SCORE ______

Test 11

TEST ON CHAPTER 3 (FORM A)

DIRECTIONS: Write the answers in the spaces provided.

Simplify using the properties of exponents.

1. $8^2 \cdot 8^{-5/3}$ 2. $\frac{7^{11/5}}{7^{7/5}}$ 3. $(8^{-1/4})^8$ 4. $\frac{8^x}{2^{3x}}$

A computer programmer earned a yearly salary of $40,000 in 1990 and averaged a 4% raise each year after that time.

5. Write an equation for the programmer's yearly salary, s, as a function of n, the number of years since 1990.
6. Find the doubling time for the programmer's salary.

A radioactive isotope decays at a rate of 3% per day. A scientist has an initial amount of 50 g.

7. Write an equation of the form $y = ab^x$ for the number of grams y remaining after x days.
8. How many grams of the isotope are left after 2 days? 10 days?
9. Write an equation of the form $y = a \cdot \left(\frac{1}{2}\right)^{x/h}$, where h is the half-life of the isotope, for the amount after x days.

The cost of a movie ticket in a large city has increased exponentially over the d decades since 1953 when records were first kept. A function that models the cost is $C = 1.25(1.5)^d$.

10. Rewrite the equation to give the cost of a movie ticket in terms of y years after 1953 instead of d decades.
11. Use your equation for Question 10 to estimate the cost of a movie ticket in 1996.
12. **Open-ended** Do you think your estimate in Question 11 is reasonable? Do you think the function will give a reasonable estimate for ticket cost four decades from now? Explain.

    Sample answer: Yes, this estimate is reasonable. Movie tickets in large cities cost about \$6 or \$7 now, so \$7.15 seems reasonable. No, the estimate of the cost of a movie ticket four decades from now is more than \$30, which seems unrealistically high.

ANSWERS

1. 2
2. $7^{4/5}$
3. $\frac{1}{64}$
4. 1
5. $s = 40{,}000(1.04)^n$
6. about 17.7 years
7. $y = 50(0.97)^x$
8. ≈ 47.0 g; ≈ 36.9 g
9. $y = 50\left(\frac{1}{2}\right)^{x/22.76}$
10. $C = 1.25(1.5)^{y/10}$
11. about $7.15
12. *See question.*

Assessment Book, ALGEBRA 2: EXPLORATIONS AND APPLICATIONS

15

Chapter 3 Assessment
Form B Chapter Test

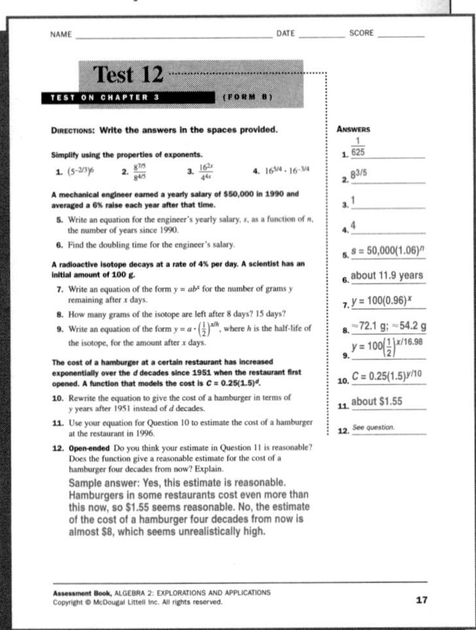

NAME ______ DATE ______ SCORE ______

Test 12

TEST ON CHAPTER 3 (FORM B)

DIRECTIONS: Write the answers in the spaces provided.

Simplify using the properties of exponents.

1. $(5^{-2/3})^6$ 2. $\frac{8^{7/5}}{8^{4/5}}$ 3. $\frac{16^{2x}}{4^{4x}}$ 4. $16^{5/4} \cdot 16^{-3/4}$

A mechanical engineer earned a yearly salary of $50,000 in 1990 and averaged a 6% raise each year after that time.

5. Write an equation for the engineer's yearly salary, s, as a function of n, the number of years since 1990.
6. Find the doubling time for the engineer's salary.

A radioactive isotope decays at a rate of 4% per day. A scientist has an initial amount of 100 g.

7. Write an equation of the form $y = ab^x$ for the number of grams y remaining after x days.
8. How many grams of the isotope are left after 8 days? 15 days?
9. Write an equation of the form $y = a \cdot \left(\frac{1}{2}\right)^{x/h}$, where h is the half-life of the isotope, for the amount after x days.

The cost of a hamburger at a certain restaurant has increased exponentially over the d decades since 1951 when the restaurant first opened. A function that models the cost is $C = 0.25(1.5)^d$.

10. Rewrite the equation to give the cost of a hamburger in terms of y years after 1951 instead of d decades.
11. Use your equation for Question 10 to estimate the cost of a hamburger at the restaurant in 1996.
12. **Open-ended** Do you think your estimate in Question 11 is reasonable? Does the function give a reasonable estimate for the cost of a hamburger four decades from now? Explain.

    Sample answer: Yes, this estimate is reasonable. Hamburgers in some restaurants cost even more than this now, so \$1.55 seems reasonable. No, the estimate of the cost of a hamburger four decades from now is almost \$8, which seems unrealistically high.

ANSWERS

1. $\frac{1}{625}$
2. $8^{3/5}$
3. 1
4. 4
5. $s = 50{,}000(1.06)^n$
6. about 11.9 years
7. $y = 100(0.96)^x$
8. ≈ 72.1 g; ≈ 54.2 g
9. $y = 100\left(\frac{1}{2}\right)^{x/16.98}$
10. $C = 0.25(1.5)^{y/10}$
11. about $1.55
12. *See question.*

Assessment Book, ALGEBRA 2: EXPLORATIONS AND APPLICATIONS

17

CHAPTER

3 Assessment

VOCABULARY QUESTIONS

For Questions 1 and 2, complete each paragraph.

1. The graph of the function $y = ab^x$ describes __?__ when $b > 1$ or __?__ when $0 < b < 1$.

2. As n gets larger and larger, the value of $\left(1 + \frac{1}{n}\right)^n$ approaches __?__.

SECTIONS 3.1 *and* 3.2

Simplify using the properties of exponents.

3. $4^2 \cdot 4^{-3/2}$ **4.** $\frac{6^{3/4}}{6^{9/4}}$ **5.** $\left(7^{-1/2}\right)^4$ **6.** $\frac{9^x}{3^{2x}}$

7. AGRICULTURE The number of eggs E a Leghorn chicken produces per year declines exponentially with its age y in years. A function that models the data is:

$$E = 179.2(0.89)^y$$

a. Rewrite the equation to give the number of eggs as a function of age w in weeks.

b. Use your equation from part (a) to estimate egg production for a 104-week-old hen.

SECTION 3.3

8. A computer systems analyst earned a salary of $30,000 in 1990 and averaged 5% more per year after that time.

a. Write an equation for s, the annual salary, as a function of n, the number of years since 1990.

b. Find the doubling time for the analyst's salary.

9. The radioactive isotope Germanium-71 decays at a rate of about 6% per day. A scientist has an initial amount of 10 g.

a. Write an equation of the form $y = a \cdot \left(\frac{1}{2}\right)^{x/h}$ for the number of grams remaining after x days.

b. How many grams are left after 2 days?

ANSWERS Chapter 3

Assessment

1. exponential growth; exponential decay
2. e
3. 2
4. $\frac{1}{6\sqrt{6}}$ or $\frac{\sqrt{6}}{36}$
5. $\frac{1}{49}$
6. 1
7. a. $E = 179.2(0.89)^{w/52}$
   b. about 142 eggs per year
8. a. $s = 30{,}000(1.05)^n$
   b. 14.2 years
9. a. $y = 10\left(\frac{1}{2}\right)^{x/11.2}$
   b. 8.84 g
10. a. \$3136.62
    b. \$4738.15
    c. \$11,225.04
11. a. $E(t) = 50(0.970)^t$
    b. about 27 errors
12. Answers may vary. Population growth can usually be modeled with a logistic function.
13. $y = 4.43(1.17)^x$
14. $y = 7.40(1.04)^x$
15. $y = 0.943(2.12)^x$
16. a. $y = 0.766(2.17)^x$
    b. \$36.9 trillion

SECTIONS 3.4 *and* 3.5

10. Suppose a bank offers interest compounded continuously. Use the formula $A = Pe^{rt}$ to find the value of \$2000 after 15 years at each rate.

a. 3.00% **b.** 5.75% **c.** 11.50%

11. The function $E(t) = 50e^{-0.03t}$ gives the average number of typing errors a student makes on a typing test after t days of typing instruction.

a. Rewrite the function in the form $E(t) = ab^t$.

b. Find the average number of errors after 20 days of typing instruction.

12. Writing Describe a situation for which a logistic function would provide a good model. Include a rough sketch in your answer.

Write an exponential function whose graph passes through each pair of points.

13. (3, 7), (8, 15) **14.** (2, 8), (5, 9) **15.** (3, 9), (1, 2)

16. Technology The United States Department of Health and Human Services estimates that national health care expenses will grow exponentially according to the data in the table.

a. Enter the data into a graphing calculator or statistical software and perform an exponential regression. What equation do you get?

b. Estimate the expenses in 2040 using the equation from part (a).

National Health Care Expenses

Decades since 1990	Expenses (trillions of dollars)
0	0.7
1	1.8
2	3.8
3	7.8
4	16.0

17. ECONOMICS The Consumer Price Index (CPI) for consumers in the United States compares the cost of goods and services with their cost at another time. Assume that the value of the CPI increases exponentially over time. The CPI was 130.7 in 1990 and 136.2 in 1991.

a. Write an exponential function to model the value of the CPI.

b. Use your model to estimate the value of the CPI in 1988 and in 1992.

PERFORMANCE TASK

18. The table gives the percent of the world population in urban areas for decades since 1800. Use the data to write exponential models based on:

- the average proportional growth rate
- an estimate of the doubling time or half-life
- two representative data pairs
- exponential regression

Compare your models and use one that gives a good fit to the data to make predictions about the percent of the world population living in urban areas in 2030.

You can also use other data that appear to be exponential. A good source for data is an almanac.

World Population

Decades since 1800	Percent in urban areas
0	2
5	4
10	9
15	21

Chapter 3 Assessment
Form C Alternative Assessment

Chapter 3
ALTERNATIVE ASSESSMENT

1. **a.** Explain why the Product Property of Exponents, $b^m \cdot b^n = b^{m+n}$, makes sense by analyzing expanded examples.

   b. Explain why the Quotient Property of Exponents, $\frac{b^m}{b^n} = b^{m-n}$, makes sense by analyzing expanded examples.

   c. Explain why the Power Property of Exponents, $(b^m)^n = b^{mn}$, makes sense by analyzing expanded examples.

2. **Open-ended Problem** Think of an example of a real-world situation that can be modeled by a function that has a domain of integers only. Describe your function. Explain why your function cannot have all real numbers as its domain.

3. **Performance Task** Explain why $2^3 \cdot 2^4$ does not equal 4^7.

4. **Group Activity** Make up values for a and b in the equation $y = ab^x$. (Use a wide variety of values.) Draw graphs for each of your examples. Discuss patterns in the graphs. Is $y = ab^x$ an increasing or decreasing function? Use your graphs to support your answer.

5. **Performance Task** Suppose you are away at college. You have just received a letter from your little brother in high school. He has missed a week of school and needs some help with negative and fractional exponents. Write a letter to your brother explaining about these types of exponents. Use examples to help clarify your explanation.

6. **Project** Put some warm water into a small container. Place a thermometer in the water. Record the temperature of the water. Add some ice cubes and record the temperature every 10 s. Graph your results. Use exponential regression to find an equation of a line of fit.

7. **Project** Investigate the differences in money placed in a savings account that is compounded annually, quarterly, monthly, daily, and continuously. Find out when and why compounding continuously became popular.

Assessment Book, ALGEBRA 2: EXPLORATIONS AND APPLICATIONS

115

17. a. Let x = the number of years since 1990 and y = the CPI; $y = 130.7(1.04)^x$.

b. 120.8; 141.4

18. Answers may vary. Examples are given. Let x = the number of decades since 1800 and y = the percent of the world population in urban areas.

Method	Equation	Prediction for 2030
average proportional growth rate	$y = 2(2.19)^{x/5}$ or $y = 2(1.17)^x$	about 74%
estimate of doubling time	$y = 2 \cdot 2^{x/4.4}$	about 75%
data pairs ((5, 4) and (15, 21))	$y = 1.75(1.18)^x$	about 79%
exponential regression	$y = 1.92(1.17)^x$	about 71%

Cumulative Assessment

CHAPTERS 1–3

CHAPTER 1

SPORTS The table shows the winning times for the men's 400 m dash in the Olympics from 1980 to 1992. Use the table in Questions 1–3.

Year	1980	1984	1988	1992
Time (s)	44.60	44.27	43.87	43.50

1. Let x be the number of 4-year intervals since 1980. Find a linear function and an exponential function to model the winning times.

2. Estimate the winning time in 1996.

3. **Writing** Which model do you think gives a better estimate of the winning time in 1996? in 2096? Explain your reasoning.

For Questions 4–7, use the matrices below to evaluate each matrix expression. If an expression cannot be evaluated, write *undefined*.

$$A = \begin{bmatrix} 2 & -7 & 4 \\ 3 & 0 & -1 \end{bmatrix} \quad B = \begin{bmatrix} -9 & 0 & 4 \\ 3 & 6 & 1 \end{bmatrix} \quad C = \begin{bmatrix} 8 & -4 \\ 4 & 5 \end{bmatrix} \quad D = \begin{bmatrix} 7 & 2 \\ -1 & -3 \end{bmatrix}$$

4. $3A + 2B$ **5.** BD **6.** CD **7.** $A - B$

8. **Open-ended Problem** Suppose that in a family with three children, each child is as likely to be a girl as a boy. Simulate the number of girls and boys in 20 different families, each with three children. Describe the method you used for your simulation. How many of the families have 3 girls? 2 girls? 1 girl? no girls?

CHAPTER 2

For each equation, tell whether *y* varies directly with *x*. If so, graph the equation.

9. $y = 3x - 1$ **10.** $y = \frac{4}{5x}$ **11.** $y = \frac{4}{5}x$

12. Write an equation that relates the number of meters, m, to the number of centimeters, c. Tell whether m varies directly with c. Explain your answer.

Find the slope-intercept equation of the line passing through each pair of points.

13. (1, 3) and (4, 7) **14.** (2, 5) and (6, 9) **15.** (3, 7) and (5, 8)

Graph each equation.

16. $y = 3x - 7$ **17.** $y = -2x + 9$ **18.** $y = \frac{1}{2}x + 8$

ANSWERS Chapters 1–3

Cumulative Assessment

1. Let y = the winning time in seconds; $y = 44.6 - 0.367x$; $y = 44.6(0.992)^x$.
2. 43.1 s (linear model); 43.2 s (exponential model)
3. Answers may vary. An example is given. I think both models give reasonable estimates for 1996. However, I do not think either would give a reasonable estimate for 2096. I think it is unlikely that performance will continue to improve without limit for close to 100 years. Eventually, the decrease in winning times will slow down.
4. $\begin{bmatrix} -12 & -21 & 20 \\ 15 & 12 & -1 \end{bmatrix}$
5. undefined
6. $\begin{bmatrix} 60 & 28 \\ 23 & -7 \end{bmatrix}$
7. $\begin{bmatrix} 11 & -7 & 0 \\ 0 & -6 & -2 \end{bmatrix}$
8. Answers may vary. An example is given. I generated random numbers on a calculator, letting numbers less than 0.5 represent a girl and numbers greater than 0.5 represent a boy. I generated 20 sequences of three numbers representing the children in 20 families. My results represented GGG, BGG, BBB, GGG, GBG, GGB, GBB, GBB, GGB, BGG, BBB, GGB, GGG, GBG, BBG, GGB, BBG, GGB, BBB, BGG. There are 3 families that have 3 girls, 10 families that have 2 girls, 4 families that have 1 girl, and 3 families that have no girls.
9. No.
10. No.
11. Yes.

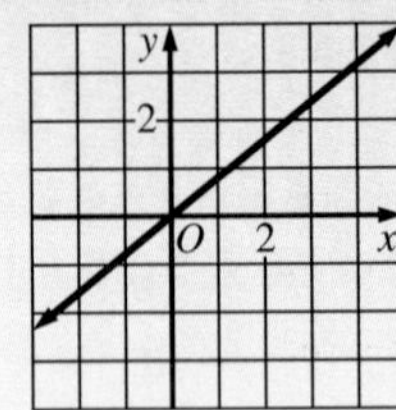

12. $m = 0.01c$; Yes; since $\frac{m}{c} = 0.01$, the ratio of the two variables is constant.

13. 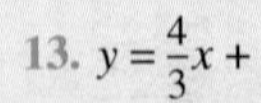 $y = \frac{4}{3}x + \frac{5}{3}$

14. $y = x + 3$

15. $y = \frac{1}{2}x + \frac{11}{2}$

16.

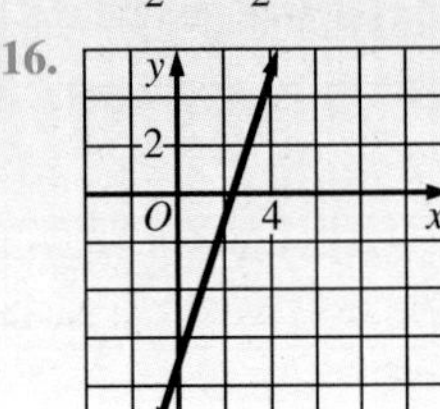

17.

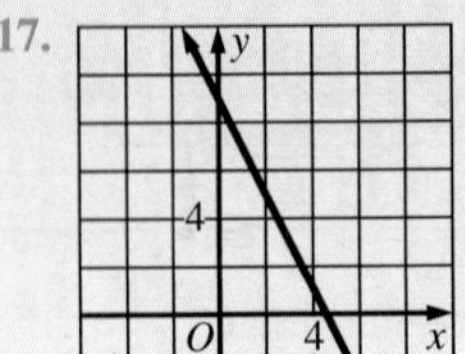

18.

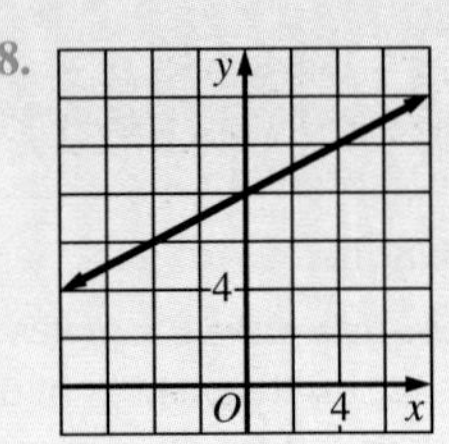

19. a. 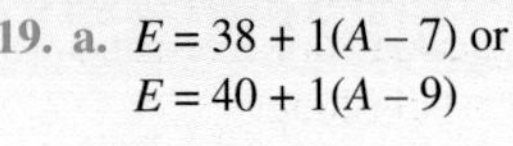 $E = 38 + 1(A - 7)$ or $E = 40 + 1(A - 9)$

b. 41

19. **MANUFACTURING** American women's shoe sizes 7 and 9 correspond to European women's shoe sizes 38 and 40.

a. Write a point-slope equation that gives the European shoe size, E, as a linear function of the American shoe size, A.

b. Find the European size that corresponds to an American size 10.

20. **Technology** Refer to the table used in Questions 1–3. Use a graphing calculator or statistical software to find the equation of the least-squares line and the correlation coefficient for the data. Is the correlation *positive* or *negative*? *strong* or *weak*?

21. **Open-ended Problem** Give an example of two real-world quantities for which the correlation would be positive and strong.

22. a. Graph the parametric equations $x = 10 - t$ and $y = 5 + t$ for $t \leq 0$.

b. Express y as a function of x and give the function's domain.

c. **Writing** Explain why you could find a different pair of parametric equations for the graph in part (a).

CHAPTER 3

Simplify using the properties of exponents.

23. $\frac{9^{5/8}}{9^{1/8}}$ **24.** $\left(6^{-9}\right)^{1/3}$ **25.** $3^5 \cdot 3^{-5}$ **26.** $16^{1/8} \cdot 16^{5/8}$

27. a. Sketch the graphs of $y = 100(2^x)$ and $y = 100\left(\frac{1}{2}\right)^x$ in the same coordinate plane.

b. **Writing** Compare the graphs in part (a).

c. **Open-ended Problem** Describe a real-life situation that could be modeled by each equation in part (a).

28. **BANKING** Find the number of years needed to double the value of a \$500 investment if the 7.2% annual interest is compounded:

a. annually **b.** monthly **c.** continuously

29. **CHEMISTRY** The amount that remains of a 50 g sample of the radioactive element carbon-11 after t seconds is modeled by the formula $A(t) = 50e^{-0.0347t}$.

a. Find the amount remaining after 30 s.

b. Find the half-life h and use it to write a formula in the form $A(t) = a \cdot \left(\frac{1}{2}\right)^{t/h}$.

30. **Technology** Use the table from Questions 1–3. Let $x =$ the number of years since 1980. Enter the data into a graphing calculator or statistical software and find an exponential equation that gives a good fit. Predict the winning time in 1996 and compare it with the answer to Question 3.

20. $y = -0.37x + 44.6$; -0.9993432334; There is a strong negative correlation.

21. Answers may vary. Examples are given. weights of roasting chickens and their cost; capacity of a gas tank in gallons and the cost of filling it

22. a.

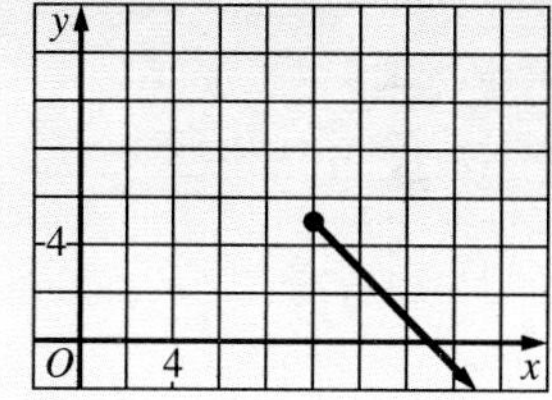

b. $y = -x + 15$; $x \geq 10$

c. You could choose another point to represent $t = 0$.

23. 3

24. $\frac{1}{216}$

25. 1

26. 8

27. a.

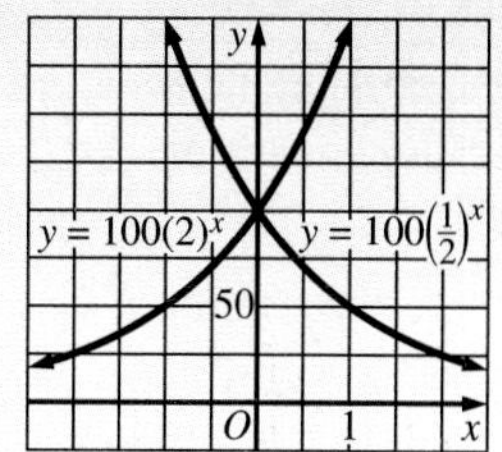

b. Summaries may vary. Both graphs are exponential curves intersecting the y-axis at 100. Both increase without limit in one direction and approach the x-axis in the other direction. (The graph of $y = 100(2)^x$ approaches the x-axis as x decreases, the graph of $y = 100\left(\frac{1}{2}\right)^x$ approaches the x-axis as x increases.) Each is the image of the other reflected over the y-axis.

c. Answers may vary. Examples are given. $y = 100(2)^x$: growth of bacteria; $y = 100\left(\frac{1}{2}\right)^x$: teams remaining in a single-elimination athletic tournament involving 100 teams

28. Estimates may vary.

a. about 10 years

b. about 9.7 years

c. about 9.6 years

29. a. 17.7 g

b. 20.0 s; $A(t) = 50\left(\frac{1}{2}\right)^{t/20.0}$

30. $y = 44.6(0.998)^x$; 43.2 s; The estimates are similar.

CHAPTER

4 Logarithmic Functions

OVERVIEW

Connecting to Prior and Future Learning

- ⇔ This chapter opens with a discussion of inverses of linear and exponential functions. Inverses of exponential functions are used to introduce students to common and natural logarithms. Students learn to solve exponential and logarithmic equations in Section 4.4.
- ⇔ Rules for working with products, quotients, and powers of logarithms are discussed in Section 4.3. These rules will be used throughout this and future mathematics courses.
- ⇔ Chapter 4 concludes with a section on using logarithms to model data. As with the previous models of data presented, these logarithmic models will be used to make predictions about real-life situations.

Chapter Highlights

Interview with Ednaly Ortiz: The relationship between mathematics and archaeology is emphasized in this interview, with related exercises on pages 146 and 159.

Explorations involve investigating the inverse of $f(x) = 2^x$ in Section 4.2 and investigating properties of logarithms in Section 4.3.

The Portfolio Project: Students use Zipf's law to model the relationship between population and rank for cities in a country of their choice.

Technology: Graphing calculators and scientific calculators are used throughout the chapter to evaluate logarithmic expressions. As in previous chapters, graphing calculators are used to graph functions and to perform linear and exponential regressions. Spreadsheets are used to create tables of values.

OBJECTIVES

Section	Objectives	NCTM Standards
4.1	• Graph and find equations for inverses of linear functions. • Solve problems using inverses of linear functions.	1, 2, 3, 4, 5, 6
4.2	• Find inverses of exponential functions. • Evaluate logarithms. • Understand how logarithmic functions are applied.	1, 2, 3, 4, 5, 6
4.3	• Use the properties of logarithms. • Understand logarithmic scales.	1, 2, 3, 4, 5, 6
4.4	• Solve exponential and logarithmic equations. • Solve real-world applications of exponential and logarithmic equations.	1, 2, 3, 4, 5, 6
4.5	• Use logarithms to find exponential and power functions that model data. • Make predictions about real-life quantities.	1, 2, 3, 4, 5, 6

INTEGRATION

Mathematical Connections	4.1	4.2	4.3	4.4	4.5
algebra	**141–147***	**148–154**	**155–161**	**162–167**	**168–175**
data analysis, probability, discrete math	141, 145–147	151–154	157, 159–161	166, 167	**168–175**
patterns and functions	**141–147**	**148–154**	**155–161**	**162–167**	**168–175**
logic and reasoning	143–147	150, 152–154	156, 158–161	163, 165–167	168, 169, 171–175
Interdisciplinary Connections and Applications					
biology and earth science			157, 159		170, 174, 175
business and economics	144				
sports and recreation	146	151			
population and social studies		154	161		172
forestry, transportation, astronomy, acoustics, cooking, radiology, personal finance, ecology	145, 147	153	160	162, 166	169

* ***Bold page numbers*** *indicate that a topic is used throughout the section.*

TECHNOLOGY

Section	*opportunities for use with* Student Book	Support Material
4.1	graphing calculator McDougal Littell Mathpack *Function Investigator*	**Function Investigator with Matrix Analyzer Activity Book:** Function Investigator Activity 7 **Geometry Inventor Activity Book:** Activities 16–18
4.2	graphing calculator McDougal Littell Mathpack *Function Investigator*	**Function Investigator with Matrix Analyzer Activity Book:** Function Investigator Activity 30
4.3	scientific calculator graphing calculator McDougal Littell Mathpack *Function Investigator*	
4.4	graphing calculator McDougal Littell Mathpack *Function Investigator*	**Technology Book:** Calculator Activity 4 Spreadsheet Activity 4 TI-92 Activity 2
4.5	graphing calculator spreadsheet software McDougal Littell Mathpack *Stats!* *Function Investigator*	

Planning Guide

Regular Scheduling (45 min)

Section	Materials Needed	Core Assignment	Extended Assignment	*exercises that feature* Applications	Communication	Technology
4.1	graph paper, graphing calculator	1–9, 11, 12, 16–23	1–23	10, 12–14	18	16, 19
4.2	MIRA® transparent mirror, graph paper, graphing calculator	**Day 1:** 1–26 **Day 2:** 31–40, 43–52	**Day 1:** 1–26 **Day 2:** 27–52	27–30, 39	26 30, 43	
4.3	scientific calculator, graphing calculator, graph paper	**Day 1:** 1–17, 20–25 **Day 2:** 27, 29–31, 34–37, AYP*	**Day 1:** 1–25 **Day 2:** 26–37, AYP	18–20 26–28	9, 19 34	30, 37
4.4	graphing calculator or software	1–8, 12–19, 22–37	1–13, 15, 17, 19–37	20, 21, 30	21, 31, 32	
4.5	graphing calculator, spreadsheet software, graph paper	**Day 1:** 2–7, 10–15, 17 **Day 2:** 21–23, 25–34, AYP	**Day 1:** 1–20 **Day 2:** 21–34, AYP	8, 9 22–24	1, 17–19 30	16, 18
Review/ Assess		**Day 1:** 1–18 **Day 2:** 19–36 **Day 3:** Ch. 4 Test	**Day 1:** 1–18 **Day 2:** 19–36 **Day 3:** Ch. 4 Test	28	7 36	35
Portfolio Project		Allow 2 days.	Allow 2 days.			

Yearly Pacing (with Portfolio Project)	**Chapter 4 Total** 13 days	**Chapters 1–4 Total** 52 days	**Remaining** 108 days	**Total** 160 days

Block Scheduling (90 min)

	Day 21	Day 22	Day 23	Day 24	Day 25	Day 26	Day 27
Teach/Interact	4.1 4.2: Exploration, page 148	Continue with 4.2 4.3: Exploration, page 155	Continue with 4.3 4.4	4.5	Review Port. Proj.	Review Port. Proj.	Ch. 4 Test 5.1
Apply/Assess	**4.1:** 1–9, 11, 12, 16–23 **4.2:** 1–26	**4.2:** 31–40, 43–52 **4.3:** 1–17, 20–25	**4.3:** 27, 29–31, 34–37, AYP* **4.4:** 1–8, 12–19, 22–37	**4.5:** 2–7, 10–15, 17, 21–23, 25–34, AYP	**Review:** 1–18 **Port. Proj.**	**Review:** 19–36 **Port. Proj.**	**Ch. 4 Test** **5.1:** 1–4, 6–17, 24–26, 28–32

NOTE: A one-day block has been added for the Portfolio Project—timing and placement to be determined by teacher.

Yearly Pacing (with Portfolio Project)	**Chapter 4 Total** $6\frac{1}{2}$ days	**Chapters 1–4 Total** $26\frac{1}{2}$ days	**Remaining** $55\frac{1}{2}$ days	**Total** 82 days

****AYP*** *is Assess Your Progress.*

Lesson Support

Section	Practice Bank	Study Guide*	Assessment Book*	Visuals	Explorations Lab Manual	Lesson Plans	Technology Book
4.1	20	4.1		Warm-Up 4.1 Folder 4	Master 2 Add. Expl. 5	4.1	
4.2	21	4.2		Warm-Up 4.2 Folder 4	Masters 1, 2, 7	4.2	
4.3	22	4.3	Test 13	Warm-Up 4.3	Masters 2, 8	4.3	
4.4	23	4.4		Warm-Up 4.4		4.4	Calculator Act. 4 Spreadsheet Act. 4 TI-92 Act. 2
4.5	24	4.5	Test 14	Warm-Up 4.5	Master 1 Add. Expl. 6	4.5	
Review Test	25	Chapter Review	Tests 15, 16 Alternative Assessment			Review Test	Calculator Based Lab 2

*__Spanish versions__ of *Study Guide* and *Assessment Book* are available.

Chapter Support

- Course Guide
- Lesson Plans
- Portfolio Project Book
- Preparing for College Entrance Tests
- Multi-Language Glossary
- *Test Generator* Software
- Professional Handbook

Software Support

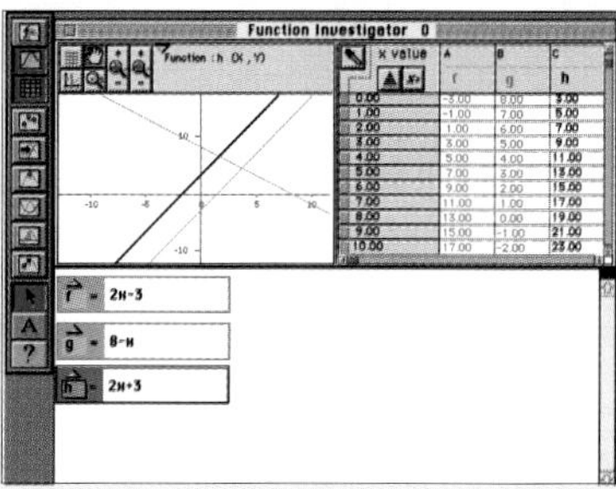

McDougal Littell Mathpack
Stats!
Function Investigator

Internet Support

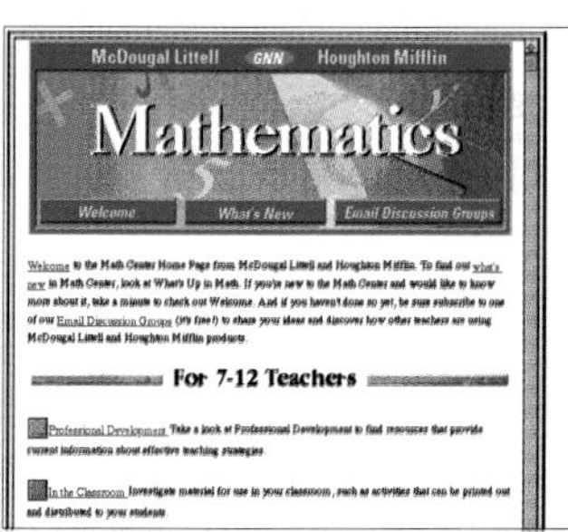

http://www.hmco.com
Next go to McDougal Littell; then the Education Center; then Secondary Math.

Outside Resources

Books, Periodicals

Hammack, Richard and David Lyons. "A Simple Way to Teach Logarithms." *Mathematics Teacher* (May 1995): pp. 374–375.

Rahn, James R. and Barry A. Berndes. "Using Logarithms to Explore Power and Exponential Functions." *Mathematics Teacher* (March 1994): pp. 161–168.

Jacobs, Harold R. *Mathematics: A Human Endeavor.* Chapter 4: "Large Numbers and Logarithms." San Francisco, CA: W. H. Freeman and Company, 1994.

Activities, Manipulatives

Van Dyke, Frances. "The Inverse of a Function." *Mathematics Teacher* (February 1996): pp. 121–137.

Breuningsen, Chris, Bill Bower, Linda Antinone, and Elisa Breuningsen. "Sour Chemistry." *Real-World Math with the CBL System.* Activity 14: pp. 75–78. Texas Instruments, 1995.

Software

$f(g)$ *Scholar.* IBM-comp. Southampton, PA: Future Graph.

Internet

For great discussions among mathematics teachers, subscribe to the NCTM list by sending e-mail to:
majordomo@forum.swarthmore.edu
In the body of the message, type:
subscribe nctm-l [first name, last name]

Interview Notes

Background

Acid Rain

Acid rain is rain in which pH levels have been lowered because of the presence of atmospheric pollutants, such as sulfuric acid and nitrous oxides. Normally, rainwater has a pH level of between 5.6 and 6.0, but in acid rain the pH level can fall to 4.0 or 5.0. This acidic water is then added to the lakes and rivers, lowering its pH level as well. This, in turn, adversely affects the fish and plants that live in the water.

Ednaly Ortiz Camacho

Ednaly Ortiz Camacho attended José S. Alegria High School in Puerto Rico while she was working on her research project. The archaeological sites of the excavations are Ensenada Breñas and Ojo del Buey, both of which are not far from her home. She completed the four parts of the project over the years 1989–1993. During these years, Ms. Camacho participated in the International Science and Engineering Fair in the Earth and Space Sciences division four times, and won awards three out of the four years. She has also attended many other meetings and symposiums where she explained and displayed her research.

CHAPTER

4 Logarithmic Functions

Digging for new ideas

INTERVIEW Ednaly Ortiz

It's not often that a young person impresses a society of professional scientists. By the time Ednaly Ortiz Camacho entered college, she had done that several times at presentations in Spain and throughout the United States. Ortiz earned international fame while still in high school for her soil studies on archaeological sites in her native Puerto Rico.

"I went to the symposium and everyone was surprised I was a high school student!"

As a young girl, Ortiz often accompanied her father, a history teacher, on archaeological digs. Now her work provides archaeologists with new methods for detecting a historical human presence at excavations—even if no artifacts are found.

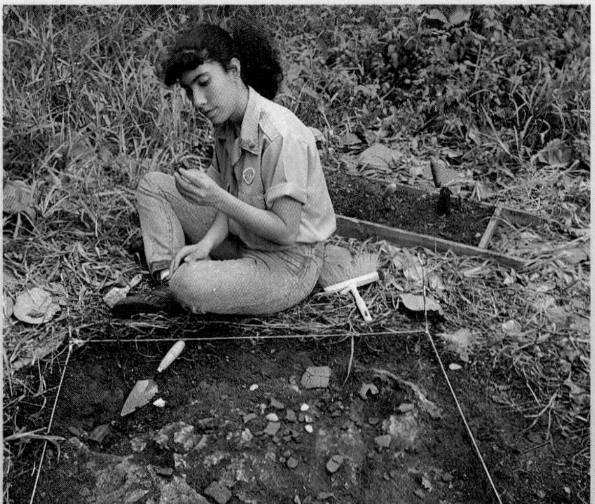

"When people ask me what I do," she says, "I tell them it's not only an excavation for artifacts, it's also a study involving chemistry, geology, and topography."

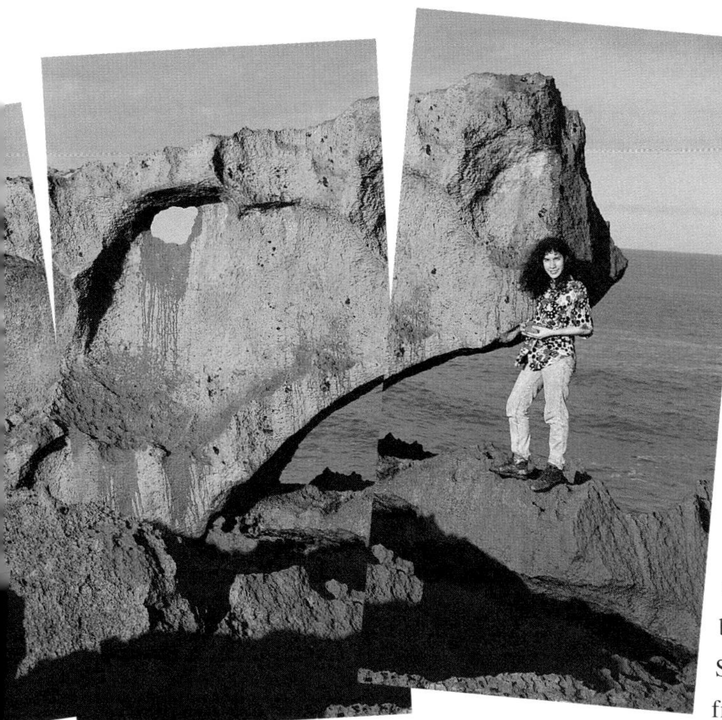

Examining the Soil

Ortiz set out to determine the impact of these cultures on the soil by conducting a chemical analysis of soil samples. It was difficult to find a laboratory to work with her, she says, because she was only 14 years old. Ortiz concluded through chemical testing of the soil that erosion of the artifacts affected the soil's pH level. The high concentrations of calcium, due to the breakdown of the mollusk shells and animal bones, made the soil more alkaline. She compared the soil to other soil from pits where no cultural evidence was found. "In areas where the Elenoid or Chicoid people did not live," she says, "the components were normal."

New Evidence from Old Cultures

Her four-year, four-phase project began in 1989 at a site near the northeastern shore of Puerto Rico. Armed with trowels, sifters, brushes, and other tools, Ortiz and her fellow freshman classmates uncovered pieces of ceramic bowls, stone weights, pestles, shellfish remains, and the bones of an extinct rabbit-like animal called *jutía*. The artifacts belonged to two pre-Columbian cultures, the Elenoid and Chicoid peoples, who lived from around A.D. 900 to 1500. Both groups of people grew crops and fished for food, which explains the abundance of shell remains. It was the first time evidence of these native people had been discovered in this area of Puerto Rico.

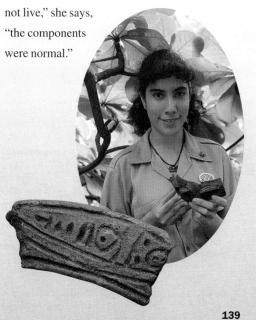

Interview Notes

Background

Puerto Rico

The island of Puerto Rico, which lies 1000 miles off the southeast coast of Florida, is a commonwealth of the United States. Its people are American citizens; however, the island maintains and governs its local affairs under its own constitution. Puerto Rico has both a Senate and a House of Representatives. The governor is the highest official in the commonwealth. Two-thirds of the population live in urban areas. The manufacturing industry provides much of Puerto Rico's gross domestic product and includes machinery, clothing, and chemicals. The rest of the gross domestic product is provided by tourism, agriculture, mining, and fishing.

Second-Language Learners

Students learning English may not be familiar with a number of terms related to archaeology, such as *archaeological digs, sites, excavation, soil samples, artifacts,* and *topography.* You may wish to review these terms with students and then read the interview together. The English names of the tools *trowels* and *sifters* may not be familiar to some students as well. Explain that a trowel is "a small digging hand tool with a scooped blade" and a sifter is "a kind of straining device that separates sand or dirt." You might invite students to tell the class what these tools are called in their native languages. Also, if the term *pre-Columbian cultures* is not familiar, explain that it refers to the many distinct cultures which flourished in the Americas before Columbus arrived.

Interview Notes

Mathematical Connection

Because logarithmic functions and exponential functions are inverses, they have a special relationship. The idea of inverse is introduced in Section 4.1. In this section, students relate excavation techniques with inverse functions. In the pH scale, the pH levels increase as the number of hydrogen ions decreases. Therefore, the pH scale is actually an example of a negative logarithmic function. This relationship is explored in Section 4.3, where students are asked to investigate the pH scale and its meaning.

Explore and Connect

Project
Students can also test the pH level of different sources of water, such as tap water, rain water, and ground water. Then have the class compare the readings.

Writing
Take time to ensure that students understand the concept being explained here. It is a good example of the characteristics of logarithmic functions and exponential functions.

Research
Students should ask professional gardeners or farmers these questions if possible, but libraries may also be a source of information.

The pH Scale

When Ortiz measured the pH level of soil, she really measured the concentration of hydrogen ions in the soil (pH stands for "potential of hydrogen"). The pH scale runs from 0 to 14 where 7 is neutral, alkalinity increases as the numbers increase, and acidity increases as the numbers decrease. Since each whole-number increase in pH represents a ten-fold decrease in the concentration of hydrogen ions, the pH scale is an example of a *logarithmic scale*.

"What I do is not only an excavation for artifacts, it's also a study involving chemistry, geology, and topography."

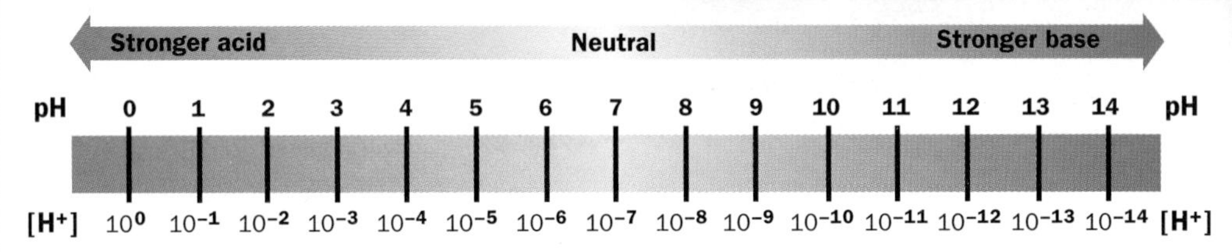

EXPLORE AND CONNECT

Ednaly Ortiz stands near the shore of Ojo del Buey, the site of her investigation into pre-Columbian cultures.

1. **Project** The acidity of many common foods and drinks can be measured using the pH scale. Ask a science teacher for litmus paper and use it to measure the pH of various foods and drinks in your house. Make a chart comparing the acidity of the items you test.

2. **Writing** A soft drink has a pH of 3.0 and black coffee has a pH of 5.0. Which is more acidic? How many times greater is the concentration of hydrogen ions in the soft drink? Explain your reasoning.

3. **Research** Find out what plants are best suited for the pH level of the soil in your area. What do gardeners or farmers do to change the pH level of their soil?

Mathematics & Ednaly Ortiz

In this chapter, you will learn more about how mathematics relates to archaeology.

Related Exercises

Section 4.1
- **Exercises 13 and 14**

Section 4.3
- **Exercises 18 and 19**

Section

4.1 Using Inverses of Linear Functions

Learn how to...

- graph and find equations for inverses of linear functions

So you can...

- solve problems about bicycling and home construction, for example

Most bicycles today have multiple gears to help cyclists ride comfortably over different terrain. By choosing the right gear, a cyclist can conserve energy and ride longer distances.

Lower gears are used for going up hills.

Higher gears are used on level ground.

THINK AND COMMUNICATE

The table shows the distance a bicycle travels for different numbers of pedal rotations when the bicycle is in first gear. Use the table and graph for Questions 1–4.

Number of pedal rotations	Distance traveled (meters)
0	0
1	3.1
2	6.2
3	9.3

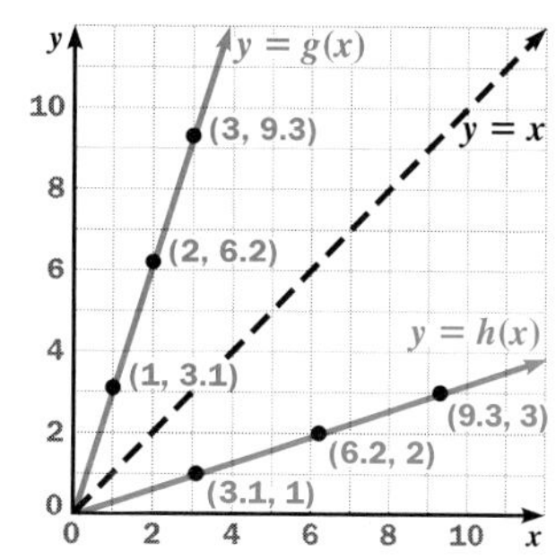

1. Replace each _?_ with *number of pedal rotations* or *distance traveled.*
   a. g gives _?_ as a function of _?_. b. h gives _?_ as a function of _?_.
2. a. Which function would you use to find the distance traveled in 2.5 pedal rotations? Why?
   b. Which function would you use to find the number of pedal rotations needed to travel 8 m? Why?
3. Complete these statements: If the point (a, b) is on the graph of g, then the point _?_ is on the graph of h. So if $g(a) = b$, then $h(\underline{?}) = \underline{?}$.
4. How are the graphs of g and h geometrically related to the line $y = x$?

BY THE WAY

In 1884, John Starley invented the Rover "safety" bicycle. His bicycle was the first one to have a geared chain drive. Gears made pedaling easier for riders.

Plan⇔Support

Objectives

- Graph and find equations for inverses of linear functions.
- Solve problems using inverses of linear functions.

Recommended Pacing

Core and Extended Courses
Section 4.1: 1 day

Block Schedule
Section 4.1: $\frac{1}{2}$ block
(with Section 4.2)

Resource Materials

Lesson Support
Lesson Plan 4.1
Warm-Up Transparency 4.1
Overhead Visuals:
Folder 4: Functions and Inverses
Practice Bank: Practice 20
Study Guide: Section 4.1
Exploration Lab Manual:
Additional Exploration 5, Diagram Master 2

Technology
Graphing Calculator
McDougal Littell Mathpack
Function Investigator with Matrix Analyzer Activity Book: Function Investigator Activity 7
Geometry Inventor Activity Book: Activities 16–18
Internet:
http://www.hmco.com

Warm-Up Exercises

1. $g(x) = 3x + 1$. Find $g(3.2)$. 10.6
2. Verify that $(4, -9)$ is on the graph of $f(x) = -2x - 1$. Since $-9 = -2(4) - 1$, $(4, -9)$ is on the graph.
3. Describe the domain and range of the function relating the distance a bicycle travels to the number of pedal revolutions. D and R: all real numbers ≥ 0
4. Give an example of a constant linear function. $y = 6$
5. Solve for T: $S = 500 + 3T$. $T = \frac{1}{3}(S - 500)$

ANSWERS Section 4.1

Think and Communicate

1. a. distance traveled; number of pedal rotations
   b. number of pedal rotations; distance traveled
2. a. g; g gives the distance traveled for a number of pedal rotations.
   b. h; h gives the number of pedal rotations for a given distance traveled.
3. (b, a); b, a
4. Each is the reflection of the other over the line $y = x$.

Teach⇔Interact

Learning Styles: Visual

Some students may have difficulty understanding the symbolic definition of an inverse. Encourage these students to remember that the graph of the inverse of a function is the reflection of the graph of the function over the line $y = x$.

Section Note

Communication: Reading Some students may question why the term *inverse* is used here instead of *inverse function.* Point out that in Chapter 8, students will find inverses, but that these inverses are not functions unless the domain is restricted.

About Example 1

Reasoning

In Example 1, students should understand that the definition of the inverse of a function requires that when (0, 6) is a point of the graph of f, (6, 0) is a point of the graph of the inverse of f.

Additional Example 1

Let $g(x) = 2x - 4$. Graph $y = g(x)$ and $y = g^{-1}(x)$ in the same coordinate plane.

Step 1 Graph $y = g(x)$. The graph of g is a line passing through the points (0, –4) and (3, 2).

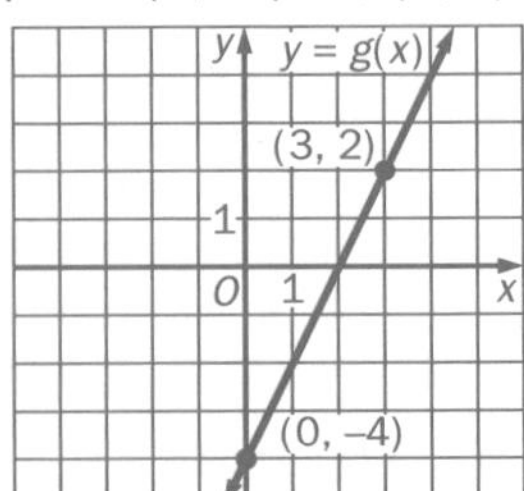

Step 2 Draw the line $y = x$.

Step 3 Graph $y = g^{-1}(x)$ by reflecting the graph of g over the line $y = x$. Since the points (0, –4) and (3, 2) are on the graph of g, the points (–4, 0) and (2, 3) are on the graph of g^{-1}.

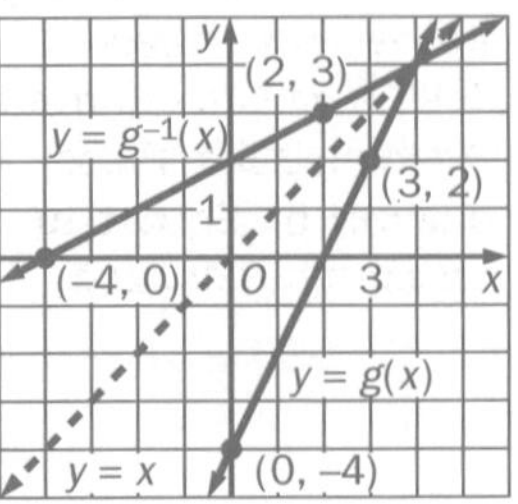

In the *Think and Communicate* questions, the function h is the *inverse* of the function g. In general, the **inverse** of a function f, denoted f^{-1}, is a function that satisfies this property:

WATCH OUT! The symbol -1 in f^{-1} is *not* an exponent. In general, $f^{-1}(x) \neq \frac{1}{f(x)}$.

$$f^{-1}(b) = a \text{ if and only if } f(a) = b.$$

You read f^{-1} as "f inverse."

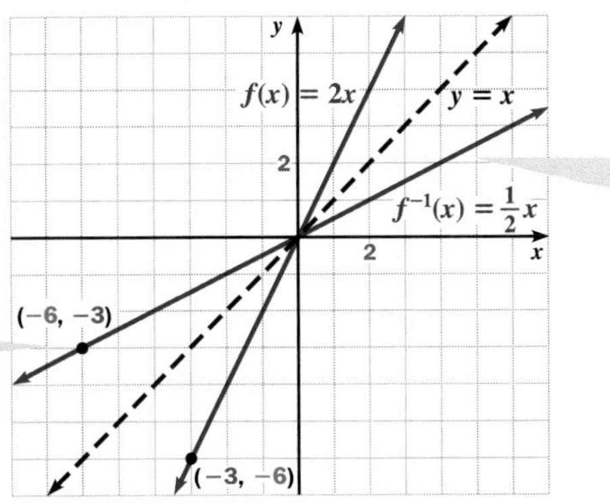

A point (b, a) is on the graph of f^{-1} if and only if (a, b) is on the graph of f.

The graph of f^{-1} is the reflection of the graph of f over the line $y = x$.

Given the graph of a function f, you can find the graph of f^{-1}.

Toolbox p. 797 *Reflecting a Figure; Line Symmetry*

EXAMPLE 1

Let $f(x) = -3x + 6$. Graph $y = f(x)$ and $y = f^{-1}(x)$ in the same coordinate plane.

SOLUTION

Note that f is a linear function with $f(0) = 6$ and $f(1) = 3$.

Step 1 Graph $y = f(x)$. The graph of f is a line passing through the points **(0, 6)** and **(1, 3)**.

Step 2 Draw the line $y = x$.

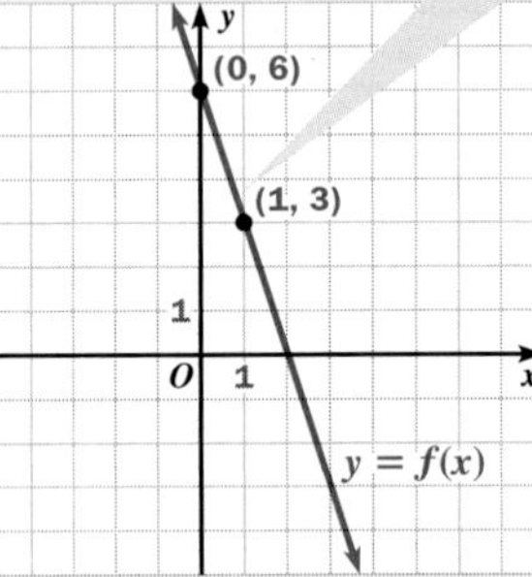

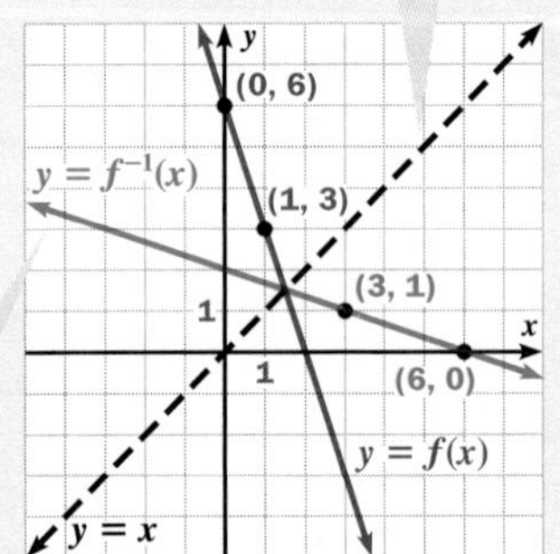

Step 3 Graph $y = f^{-1}(x)$ by reflecting the graph of f over the line $y = x$. Since the points **(0, 6)** and **(1, 3)** are on the graph of f, the points **(6, 0)** and **(3, 1)** are on the graph of f^{-1}.

THINK AND COMMUNICATE

5. Is the inverse of a nonconstant linear function always a nonconstant linear function? Explain.
6. Does a constant function have an inverse? Why or why not? (*Hint:* What is the reflection of the graph of a constant function over the line $y = x$?)

Given an equation for a function f, you can find an equation for f^{-1}.

EXAMPLE 2

Let $g(x) = 7x - 4$. Find an equation for g^{-1}.

SOLUTION

Step 1 Replace $g(x)$ with y in the equation for g.

$$y = 7x - 4$$

Step 2 Solve for x.

Toolbox p. 788 *Rewriting Equations and Formulas*

$$y + 4 = 7x$$
$$\frac{1}{7}(y + 4) = \frac{1}{7}(7x)$$
$$\frac{1}{7}y + \frac{4}{7} = x$$

Step 3 Switch x and y so that x is the independent variable of the inverse function.

$$\frac{1}{7}x + \frac{4}{7} = y$$

Step 4 Replace y with $g^{-1}(x)$.

$$\frac{1}{7}x + \frac{4}{7} = g^{-1}(x)$$

An equation for g^{-1} is $g^{-1}(x) = \frac{1}{7}x + \frac{4}{7}$.

THINK AND COMMUNICATE

7. How can you use graphs to verify that $y = \frac{1}{7}x + \frac{4}{7}$ is the inverse of $y = 7x - 4$?
8. Use the function f in Example 1.
   a. Find an equation for f^{-1} using the fact that its graph is a line passing through the points (6, 0) and (3, 1).
   b. Find an equation for f^{-1} using the method in Example 2. Compare this equation with the equation you found in part (a). Which method do you prefer? Why?

Think and Communicate

5. Yes. Answers may vary. An example is given. The graph of a nonconstant linear function is a line that is not horizontal. The reflection of such a line over the line $y = x$ is a line that is not vertical, and thus is the graph of a linear function. Since the original graph was not a vertical line, the reflection is not a horizontal line, so the inverse function is nonconstant.
6. No. Answers may vary. An example is given. The graph of a constant function is a horizontal line. The reflection of a horizontal line over the line $y = x$ is a vertical line, which is not the graph of a function.
7. Draw one graph and reflect it over the line $y = x$, showing that its image is the other graph.
8. a. The slope is $\frac{1-0}{3-6} = -\frac{1}{3}$. Using the point-slope form, an equation is $y = -\frac{1}{3}(x - 6)$.
   b. $y = -3x + 6$; $y - 6 = -3x$; $x = -\frac{1}{3}(y - 6)$; $y = -\frac{1}{3}(x - 6)$; The equations should be the same or equivalent equations. Preferences may vary.

Teach⇔Interact

Think and Communicate

For question 6, students should see that a constant function does not have an inverse because its reflection over the line $y = x$ is a vertical line, which is not the graph of a function. Ask students to explain why a vertical line cannot be a function.

About Example 2

Topic Spiraling: Review
It is important that students have a thorough understanding of how to solve equations in two variables for one variable in terms of the other variable.

Mathematical Procedures
Students should understand that switching the x and y variables in Step 3 is necessary because of how the inverse function is defined. The roles of the dependent and independent variables are reversed.

Additional Example 2

Let $h(x) = \frac{1}{3}x + 5$. Find an equation for $h^{-1}(x)$.

Step 1 Replace $h(x)$ with y in the equation for h.

$$y = \frac{1}{3}x + 5$$

Step 2 Solve for x.

$$y - 5 = \frac{1}{3}x$$
$$3(y - 5) = 3\left(\frac{1}{3}x\right)$$
$$3y - 15 = x$$

Step 3 Switch x and y so that x is the independent variable of the inverse function.

$$3x - 15 = y$$

Step 4 Replace y with $h^{-1}(x)$.

$$3x - 15 = h^{-1}(x)$$

An equation for h^{-1} is $h^{-1}(x) = 3x - 15$.

Think and Communicate

For question 7, you may want students to graph the functions with a graphing calculator and then use the trace features to verify that the functions are inverses of each other. Use the up and down arrows to move between graphs. Press ZOOM and select 5:ZSquare to see a more accurate reflection.

Teach⇔Interact

About Example 3

Application
This business application shows how the concept of an inverse function can be used to solve a practical real-world problem.

Additional Example 3

Sales associates at The Home Place department store earn a base salary of \$100 per week plus a 5% commission on the sales they had during the week.

a. Find an equation giving the weekly salary E of a sales associate as a function of his or her total weekly sales S.

Note that:
Weekly salary = Base salary + (Commission rate × Total sales)
$E = 100 + 0.05S$
An equation is $E = 100 + 0.05S$.

b. Find the inverse of the function from part (a).

Solve the equation in part (a) for S.

$$E - 100 = 0.05S$$
$$\frac{1}{0.05}(E - 100) = \frac{1}{0.05}(0.05S)$$
$$20E - 2000 = S$$

The inverse is $S = 20E - 2000$.

c. How much does a sales associate have to sell in order to earn \$500?

Use the inverse to find S when $E = 500$.

$$S = 20(500) - 2000$$
$$= 10{,}000 - 2000$$
$$= 8000$$

A sales associate must have sales of \$8000 in order to earn \$500 in a week.

Think and Communicate

For question 10, point out that x is the independent variable of each function and its inverse.

Checking Key Concepts

Common Error
For question 3, some students may think that two functions are inverses if the graph of one can be reflected onto the graph of the other. Point out that reflecting over the x-axis does not generate the points of the inverse function but reflecting over the line $y = x$ does.

EXAMPLE 3 Application: Business

Jaime Ramirez is a builder who is developing a subdivision of a small town. He sells housing lots in the subdivision for \$84,000 each. He also builds houses on the lots for \$60 per square foot of floor space.

The cost of a lot is **\$84,000**.

The cost per square foot is **\$60**.

a. Find an equation giving the cost C of building a house as a function of the house's size S in square feet.

b. Find the inverse of the function from part (a).

c. How large a house can be built for \$225,000?

SOLUTION

a. Note that:

Cost of building a house = **Cost of a lot** + (**Cost per square foot** × **Number of square feet**)

$$C = 84{,}000 + 60S$$

An equation is $C = 84{,}000 + 60S$.

b. Solve the equation in part (a) for S.

$$C - 84{,}000 = 60S$$
$$\frac{1}{60}(C - 84{,}000) = \frac{1}{60}(60S)$$
$$\frac{1}{60}C - 1400 = S$$

The inverse is $S = \frac{1}{60}C - 1400$.

WATCH OUT!
You could write the functions in Example 3 as:
$f(x) = 84{,}000 + 60x$
and
$f^{-1}(x) = \frac{1}{60}x - 1400.$
However, in real-life applications you should use variables like C and S that are easily associated with the quantities they represent.

c. Use the inverse to find S when $C = \mathbf{225{,}000}$.

$$S = \frac{1}{60}(\mathbf{225{,}000}) - 1400$$
$$= 3750 - 1400$$
$$= 2350$$

A house with 2350 ft^2 of floor space can be built.

THINK AND COMMUNICATE

9. Replace each _?_ with *cost* or *square footage*.

a. The equation $C = 84{,}000 + 60S$ gives _?_ as a function of _?_.

b. The equation $S = \frac{1}{60}C - 1400$ gives _?_ as a function of _?_.

10. If you want to graph the two functions in Example 3 on a graphing calculator, you have to enter them as $y = 84{,}000 + 60x$ and $y = \frac{1}{60}x - 1400$. For each function, tell what x and y represent.

Think and Communicate

9. a. cost; square footage
b. square footage; cost

10. In the first function, x represents square footage and y represents cost. In the second function, x represents cost and y represents square footage.

Checking Key Concepts

1. 2
2. c
3. No.
4. Yes.

Exercises and Applications

1. a.

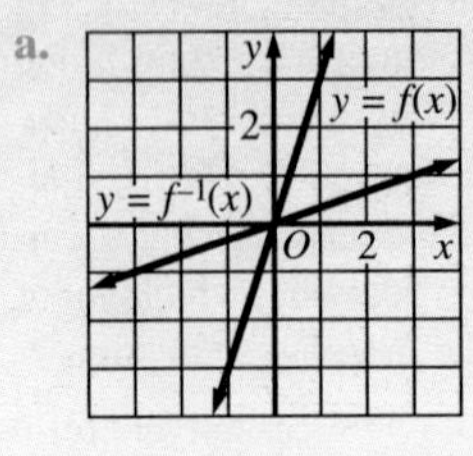

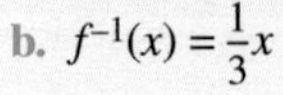

b. $f^{-1}(x) = \frac{1}{3}x$

2. a.

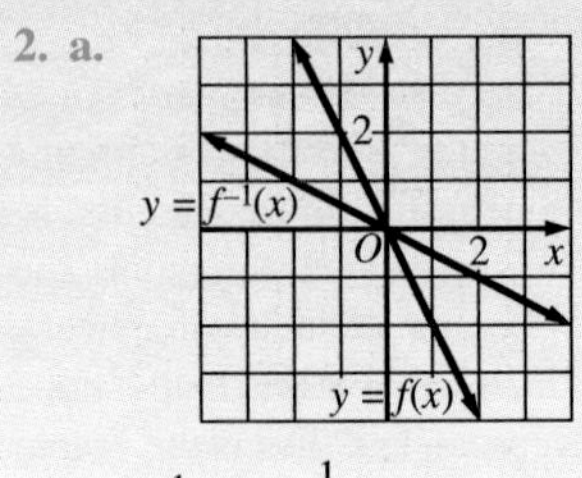

b. $f^{-1}(x) = -\frac{1}{2}x$

☑ CHECKING KEY CONCEPTS

Complete each statement.

1. If $f(2) = 6$, then $f^{-1}(6) = \underline{?}$. **2.** If $h^{-1}(c) = d$, then $h(d) = \underline{?}$.

For each pair of functions whose graphs are shown, tell whether *g* is the inverse of *f*.

3.

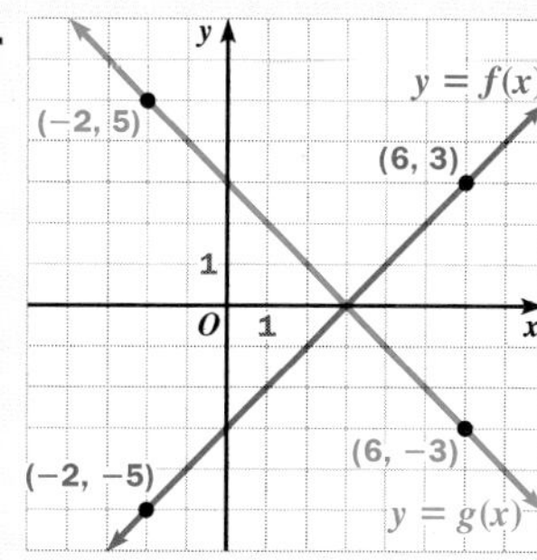

4.

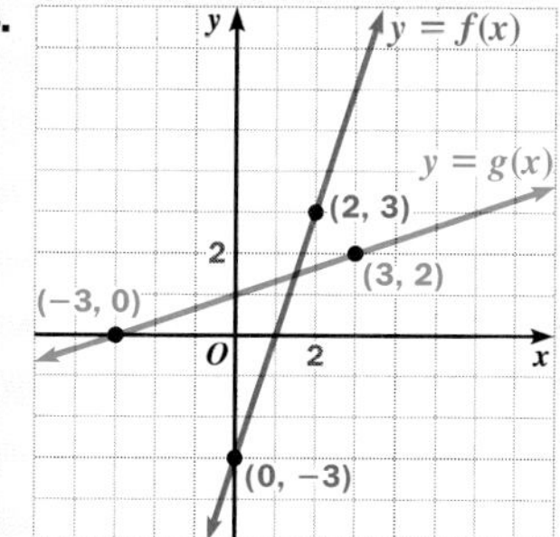

4.1 Exercises and Applications

Extra Practice exercises on page 753

For each function:

a. Graph the function and its inverse in the same coordinate plane.

b. Find an equation for the inverse.

1. $f(x) = 3x$ **2.** $f(x) = -2x$ **3.** $y = -\frac{1}{4}x$ **4.** $y = \frac{1}{5}x$

5. $g(x) = 4x + 7$ **6.** $h(x) = -5x + 1$ **7.** $y = \frac{1}{3}x - 2$ **8.** $y = -\frac{3}{2}x - \frac{1}{2}$

9. SAT/ACT Preview Which function is the inverse of $y = 6x - 11$?

A. $y = -6x + 11$ **B.** $y = \frac{1}{6x - 11}$ **C.** $y = \frac{1}{6}x - \frac{11}{6}$ **D.** $y = \frac{1}{11 - 6x}$ **E.** $y = \frac{1}{6}x + \frac{11}{6}$

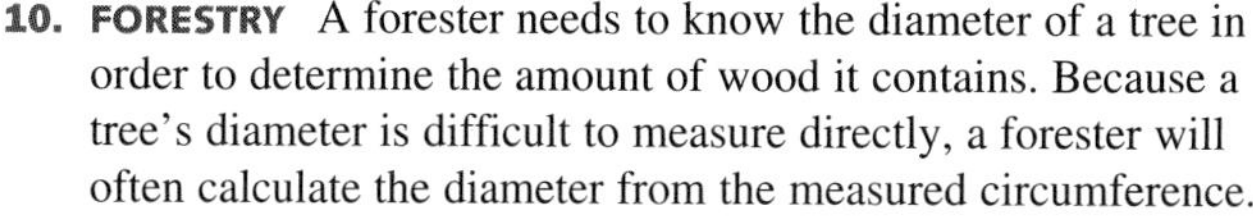

10. FORESTRY A forester needs to know the diameter of a tree in order to determine the amount of wood it contains. Because a tree's diameter is difficult to measure directly, a forester will often calculate the diameter from the measured circumference.

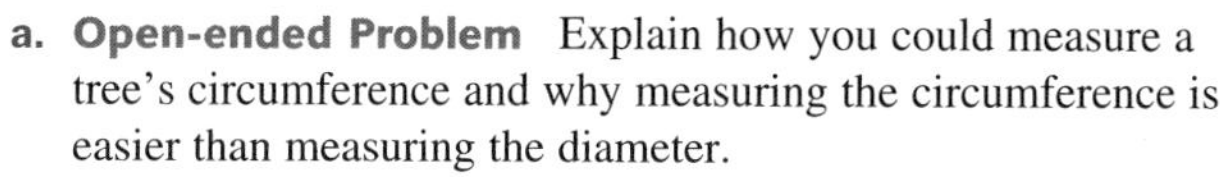

a. Open-ended Problem Explain how you could measure a tree's circumference and why measuring the circumference is easier than measuring the diameter.

b. Write an equation giving a tree's circumference C as a function of its diameter d.

c. Find the inverse of the function from part (b).

d. A forester measures the circumference of a tree to be 75 in. Find the tree's diameter.

3. a.

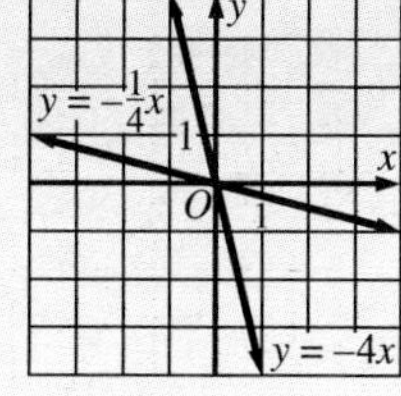

b. $y = -4x$

4. a.

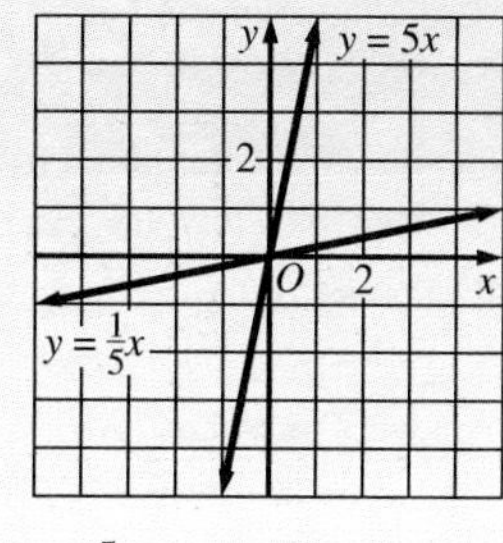

b. $y = 5x$

5. a.

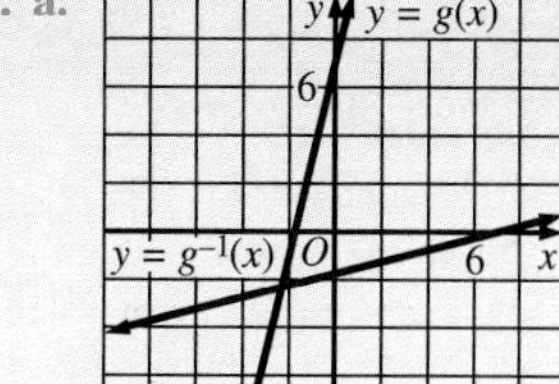

b. $y = \frac{1}{4}x - \frac{7}{4}$

6–10. See answers in back of book.

Teach⇔Interact

Closure Question

Explain how to find the inverse of a function. Either graph the function and then reflect it over the line $y = x$ or find the equation of the inverse function. To find the equation of the inverse of the function $y = f(x)$, replace $f(x)$ with y and solve for x. Then switch x and y so that x is the independent variable. Replace y with $f^{-1}(x)$.

Apply⇔Assess

Suggested Assignment

Core Course
Exs. 1–9, 11, 12, 16–23

Extended Course
Exs. 1–23

Block Schedule
Day 21 Exs. 1–9, 11, 12, 16–23

Exercise Notes

Using Technology
Exs. 1–8 Students can use a TI-82 graphing calculator to check their solutions for these exercises. If a function and its inverse are graphed as Y1 and Y2, the Y-VARS key can be used to see that for a point (x, y) on the function, there is a point (y, x) on the inverse. Enter Y1(2), for example. If Y1(2) = 0, use the Y-VARS key again to check that Y2(0) = 2.

Problem Solving
Ex. 9 Students can eliminate possibilities for this multiple choice question by picking a point of the function, reversing the x- and y-coordinates to give a point of the inverse, and then checking the new point in each of the choices.

Cooperative Learning
Ex. 10 Working in pairs will allow students to compare their answers to this exercise. Have each person explain why measuring the circumference of a tree is a good way to estimate its diameter.

Research
Ex. 10 After discussing this exercise, some students may be interested in researching other real-world measurements that can be made by finding the inverse function.

Apply⇔Assess

Exercise Notes

Topic Spiraling: Review

Ex. 11 This exercise provides an opportunity for students to review the domain and range of a function. They can then compare the domains and ranges of the inverse functions in this exercise.

Second-Language Learners

Ex. 12 Second-language learners may benefit from a review of the terminology used in this exercise. Invite volunteers familiar with bicycling to explain the terms.

Career Connection

Exs. 13, 14 Archaeology is the study of historic or prehistoric peoples and their cultures, usually by an analysis of artifacts. It is a wide-ranging subject that covers a time span of over 3 million years. The field of archaeology is usually organized into the time spans of prerecorded history and recorded history. Archaeologists are trained in colleges and universities and often specialize in a particular time period or culture. Their work enriches our understanding of the past and therefore our understanding of cultures that exist today.

Common Error

Ex. 15 Some students think that every function has an inverse. In part (c) of this exercise, students find that the inverse of $f(x) = ax + b$ is $f^{-1}(x) = \frac{1}{a}x - \frac{b}{a}$. Since $a = 0$ for a constant function, students should be convinced that a constant function does not have an inverse, since division by zero is impossible.

Using Technology

Exs. 16, 17 You may want to suggest that students use the list features of a TI-82 calculator to make a scatter plot of the data for Ex. 16. Enter the data as lists in L1 and L2, use the STAT PLOT menu, and choose Xlist:L1 and Ylist:L2. To plot the inverse function, use the STAT PLOT menu, and choose Xlist:L2 and Ylist:L1.

For Ex. 17, students can plot the inverse function in part (a) by using the parametric mode of a TI-82 or TI-81 calculator. Have them set the calculator to parametric mode and then enter the function from part (b) of Ex. 16 into Y1T. Also enter X1T = T. Enter X2T = Y1T and Y2T = X1T. Graph these functions to see the inverse graphs. Using this method reinforces the idea that the domain and range values get switched for inverses.

11. Look back at Example 3. Find the domains and ranges of the functions $C = 84{,}000 + 60S$ and $S = \frac{1}{60}C - 1400$. What do you notice?

12. RECREATION One kind of bicycle travels 7.7 m per pedal rotation when in tenth gear.

a. Find an equation giving the distance d that the bicycle travels in tenth gear as a function of the number n of pedal rotations.

b. Find the inverse of the function from part (a).

c. Complete this statement: The inverse gives ? as a function of ?.

d. How many pedal rotations are needed to ride the bicycle 200 m?

INTERVIEW Ednaly Ortiz

Look back at the article on pages 138–140.

Archaeologists such as Ednaly Ortiz often make excavation grids like the one below on sites where they are searching for artifacts. An excavation grid consists of squares of earth separated by partitions called balks.

13. Writing What do you think excavation grids are used for?

14. Suppose a group of 10 people is excavating a section of a large field. The group creates an excavation grid such that the area of each grid square and its surrounding balks is 25 m². Each person can excavate a grid square in 20 days.

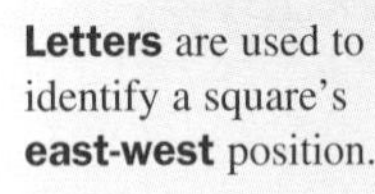

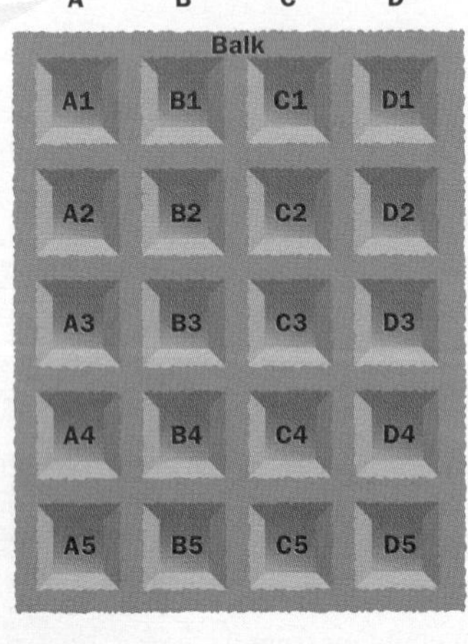

a. Find an equation giving the area A that the group can excavate in n days.

b. Find the inverse of the function from part (a).

c. How many days will it take the group to excavate 125 m² of the field?

This excavation site, located in Maine, has been divided into grid squares.

15. Challenge Consider the general linear function $f(x) = ax + b$.

a. Find an equation for f^{-1}. (Your equation will involve a and b.)

b. Use the equation from part (a) to find the inverse of $f(x) = -8x + 3$.

c. How does the equation from part (a) show that a constant function does *not* have an inverse?

d. What must be true about a and b if $f(x) = f^{-1}(x)$ for all real numbers x?

11. For C, the domain is the nonnegative real numbers and the range is the real numbers greater than or equal to 84,000. For S, the domain is the real numbers greater than or equal to 84,000 and the range is the nonnegative real numbers. The domain of C is the range of S and the range of C is the domain of S.

12. a. $d = 7.7n$ b. $n = \frac{d}{7.7}$
c. the number of pedal rotations; distance traveled
d. about 25.97 or 26 pedal rotations

13. Answers may vary. An example is given. I think the grids allow the archaeologists to identify and record an exact location within the general site for each object found.

14. a. $A = 12.5n$
b. $n = 0.08A$
c. 10 days

15. a. $f^{-1}(x) = \frac{1}{a}x - \frac{b}{a}$
b. $f^{-1}(x) = -\frac{1}{8}x + \frac{3}{8}$
c. For a constant function, $a = 0$, so $\frac{1}{a}x - \frac{b}{a}$ is not defined.
d. $a = 1$ and $b = 0$ or $a = -1$ and b is any real number

Connection TRANSPORTATION

The table shows the average fuel economy of cars in the United States for the years from 1980 to 1990.

Year	Fuel economy (mi/gal)
1980	15.46
1981	15.94
1982	16.65
1983	17.14
1984	17.83
1985	18.20
1986	18.27
1987	19.20
1988	19.87
1989	20.31
1990	21.02

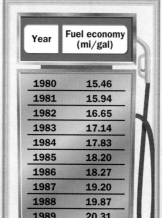

16. **Technology** Let x = the number of years since 1980. Let y = the average fuel economy.

a. Enter the data pairs (x, y) into a graphing calculator or statistical software. Make a scatter plot of the data. What do you notice about the plotted points?

b. Find an equation of the least-squares line for the scatter plot.

c. Predict the average fuel economy in the year 2000.

17. a. Find the inverse of the function from part (b) of Exercise 16. What quantity does the independent variable of the inverse represent? What quantity does the dependent variable represent?

b. Use your answer to part (a) to predict the year in which the average fuel economy will reach 30 mi/gal.

ONGOING ASSESSMENT

18. Writing Write a note to a friend explaining the relationship between the graph of a function f and the graph of f^{-1}. Also list the steps needed to find an equation for f^{-1} given an equation for f.

SPIRAL REVIEW

19. **Technology** The table shows the volume of a landfill for the years from 1990 to 1994. *(Section 3.5)*

x = years since 1990	y = volume (m^3)
0	50,000
1	63,000
2	79,000
3	102,000
4	124,000

a. Enter the data pairs (x, y) into a graphing calculator or statistical software. Find an exponential function that models the data.

b. Predict the volume of the landfill in 1998.

c. Estimate the volume of the landfill in 1987.

Find a point-slope equation of the line passing through the given points. *(Section 2.3)*

20. $(1, 2)$ and $(2, 5)$ **21.** $(-3, 4)$ and $(0, -2)$ **22.** $(-6, -3)$ and $(4, 9)$

23. Suppose Chen deposits \$500 in a savings account earning 4% annual interest compounded continuously. Find the amount of money in Chen's account after 3 years, assuming he makes no more deposits or withdrawals. *(Section 3.4)*

Apply⇔Assess

Exercise Notes

Application

Exs. 16, 17 These exercises help students to strengthen their understanding of functions and their inverses by using a real-world application of the average fuel economy of cars in the United States during a span of 10 years. Ask students why the average fuel economy has improved during the period from 1980 to 1990.

Assessment Note

Ex. 18 Writing about the relationship between a function and its inverse will help students to organize and clarify their thoughts about these concepts. You may wish to call upon two or three students to present their responses to the class.

Practice 20 for Section 4.1

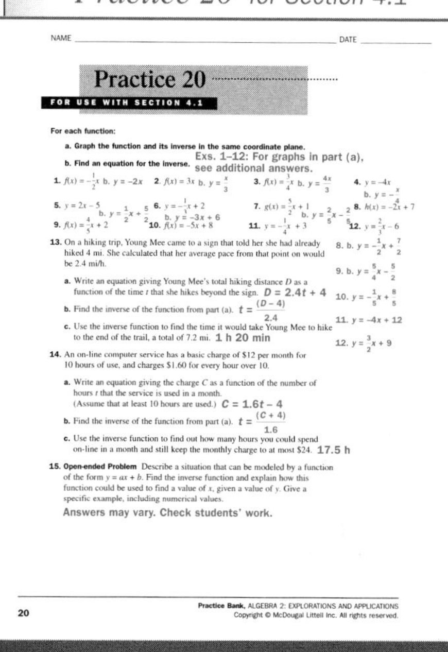

NAME ______________________ DATE __________

Practice 20

FOR USE WITH SECTION 4.1

For each function:

a. Graph the function and its inverse in the same coordinate plane.

b. Find an equation for the inverse. Exs. 1–12: For graphs in part (a), see additional answers.

1. $f(x) = -\frac{1}{2}x$ b. $y = -2x$
2. $f(x) = 3x$ b. $y = \frac{x}{3}$
3. $f(x) = \frac{3}{4}x$ b. $y = \frac{4x}{3}$
4. $y = -4x$ b. $y = -\frac{x}{4}$
5. $y = 2x - 5$ b. $y = \frac{1}{2}x + \frac{5}{2}$
6. $y = -\frac{1}{3}x + 2$ b. $y = -3x + 6$
7. $g(x) = \frac{5}{2}x + 1$ b. $y = \frac{2}{5}x - \frac{2}{5}$
8. $h(x) = -2x + 7$
9. $f(x) = \frac{4}{5}x + 2$
10. $f(x) = -5x + 8$
11. $y = -\frac{1}{4}x + 3$
12. $y = \frac{2}{3}x - 6$

8. b. $y = -\frac{1}{2}x + \frac{7}{2}$
9. b. $y = \frac{5}{4}x - \frac{5}{2}$
10. $y = -\frac{1}{5}x + \frac{8}{5}$
11. $y = -4x + 12$
12. $y = \frac{3}{2}x + 9$

13. On a hiking trip, Young Mee came to a sign that told her she had already hiked 4 mi. She calculated that her average pace from that point on would be 2.4 mi/h.

a. Write an equation giving Young Mee's total hiking distance D as a function of the time t that she hikes beyond the sign. $D = 2.4t + 4$

b. Find the inverse of the function from part (a). $t = \frac{(D - 4)}{2.4}$

c. Use the inverse function to find the time it would take Young Mee to hike to the end of the trail, a total of 7.2 mi. 1 h 20 min

14. An on-line computer service has a basic charge of \$12 per month for 10 hours of use, and charges \$1.60 for every hour over 10.

a. Write an equation giving the charge C as a function of the number of hours t that the service is used in a month. (Assume that at least 10 hours are used.) $C = 1.6t - 4$

b. Find the inverse of the function from part (a). $t = \frac{(C + 4)}{1.6}$

c. Use the inverse function to find out how many hours you could spend on-line in a month and still keep the monthly charge to at most \$24. 17.5 h

15. Open-ended Problem Describe a situation that can be modeled by a function of the form $y = ax + b$. Find the inverse function and explain how this function could be used to find a value of x, given a value of y. Give a specific example, including numerical values.

Answers may vary. Check students' work.

20

Practice Bank, ALGEBRA 2: EXPLORATIONS AND APPLICATIONS

---

16. a. They appear to lie near a line.

b. $y = 0.5409x + 15.47$

c. about 26.29 gal

17. a. $y = 1.849x - 28.60$; average fuel economy; number of years since 1980

b. 2007

18. The graph of the function f and the graph of the function f^{-1} are reflections of each other over the line $y = x$. To find an equation for f^{-1} given an equation for f, replace $f(x)$ with y in the equation, solve for x in terms of y, then switch x and y in the resulting equation. Finally, replace y with $f^{-1}(x)$.

19. a. $y = 50{,}100(1.26)^x$

b. about 318,000 m^3

c. about 25,000 m^3

20. $y = 2 + 3(x - 1)$ or $y = 5 + 3(x - 2)$

21. $y = 4 - 2(x + 3)$ or $y = -2 - 2x$

22. $y = -3 + \frac{6}{5}(x + 6)$ or $y = 9 + \frac{6}{5}(x - 4)$

23. \$563.75

Plan⇔Support

Objectives

- Find inverses of exponential functions.
- Evaluate logarithms.
- Understand how logarithmic functions are applied.

Recommended Pacing

Core and Extended Courses
Section 4.2: 2 days

Block Schedule
Section 4.2: 2 half-blocks (with Sections 4.1 and 4.3)

Resource Materials

Lesson Support
Lesson Plan 4.2
Warm-Up Transparency 4.2
Overhead Visuals:
Folder 4: Functions and Inverses
Practice Bank: Practice 21
Study Guide: Section 4.2
Exploration Lab Manual:
Diagram Masters 1, 2, 7

Technology
Graphing Calculator
McDougal Littell Mathpack
Function Investigator with Matrix Analyzer Activity Book: Function Investigator Activity 30
Internet:
http://www.hmco.com

Warm-Up Exercises

Find the value of x in each of the following equations.

1. $2^x = 8$ 3
2. $x^4 = 81$ 3
3. $10^4 = x$ 10,000
4. $x^{1/2} = 5$ 25
5. Evaluate 4^{-3}. $\frac{1}{64}$

Section

4.2 Using Inverses of Exponential Functions

Learn how to...

- **find inverses of exponential functions**
- **evaluate logarithms**

So you can...

- **understand how an altimeter works, for example**

You know that the inverse of a nonconstant linear function is also a linear function. Does this mean that the inverse of an exponential function is an exponential function? The answer is no. In fact, the inverse of an exponential function is a type of function you may not have seen before.

EXPLORATION
COOPERATIVE LEARNING

Investigating the Inverse of $f(x) = 2^x$

Work with a partner.
You will need: • a MIRA® transparent mirror • graph paper

x	$f(x)$
−2	?
−1	?
0	?
1	?
2	?
3	?

1 Let $f(x) = 2^x$. Copy and complete the table.

2 Graph $y = f(x)$. Label the points on the graph corresponding to the ordered pairs in the table.

3 Graph the line $y = x$ in the same coordinate plane you used for graphing f.

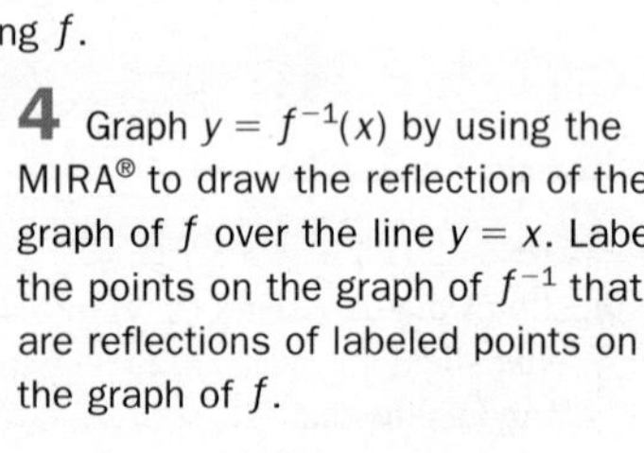

4 Graph $y = f^{-1}(x)$ by using the MIRA® to draw the reflection of the graph of f over the line $y = x$. Label the points on the graph of f^{-1} that are reflections of labeled points on the graph of f.

Questions

1. Replace each ? with the number that makes the equation true.
   a. $f^{-1}(8) = \underline{?}$ b. $f^{-1}\left(\frac{1}{4}\right) = \underline{?}$ c. $f^{-1}(1) = \underline{?}$
2. Replace each ? with *exponents* or *powers of 2*.
   a. For the function f, the input values are ? and the output values are ?.
   b. For the function f^{-1}, the input values are ? and the output values are ?.

Exploration Note

Purpose
The purpose of this Exploration is to have students investigate the inverse of the function $f(x) = 2^x$.

Materials/Preparation
Each group needs a MIRA® transparent mirror and graph paper.

Procedure
Students use a table of values to graph $f(x) = 2^x$ and label the points. They then graph the line $y = x$ on the same coordinate plane and reflect each point of $f(x) = 2^x$ with the MIRA®. They label the reflected points and use them to explain how the input values of a function are related to the output values of the inverse function.

Closure
Students should understand that input and output values for a function are reversed for a function and its inverse. In other words, the domain and range values are reversed.

Explorations Lab Manual
See the Manual for more commentary on this Exploration.
Diagram Master 7

Logarithmic Functions

The inverse of an exponential function is called a **logarithmic function**. You write the inverse of the exponential function $f(x) = b^x$ as $f^{-1}(x) = \log_b x$. The number b, which must be positive and not equal to 1, is called the *base* of the logarithmic function. The expression $\log_b x$ is called the **base-*b* logarithm** of x.

For example, the inverse of $y = 3^x$ is $y = \log_3 x$. The graphs of these functions are shown.

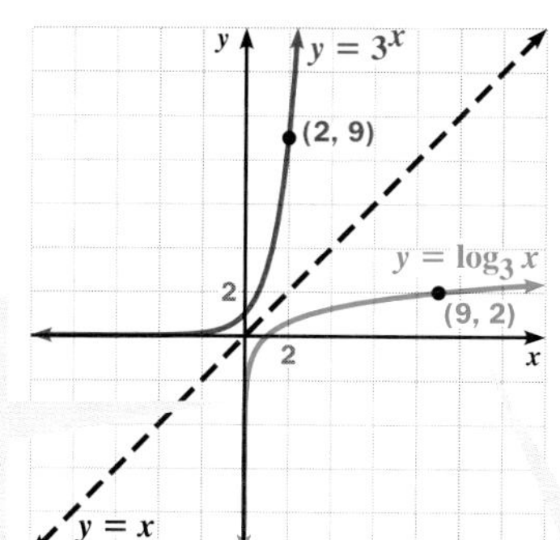

The domain of $y = \log_b x$ is $x > 0$. The range is all real numbers.

The graph of $y = \log_b x$ is the reflection of the graph of $y = b^x$ over the line $y = x$.

Since $3^2 = 9$, it follows that $\log_3 9 = 2$.

Since $f^{-1}(x) = \log_b x$ is the inverse of $f(x) = b^x$, you know that:

$$f^{-1}(N) = k \text{ if and only if } f(k) = N$$

Therefore:

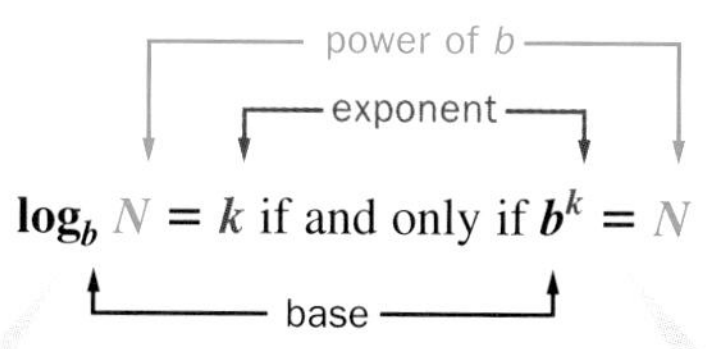

$$\log_b N = k \text{ if and only if } b^k = N$$

This equation is in **logarithmic form**.

This equation is in **exponential form**.

The exponential and logarithmic equations are equivalent. This means that an equation given in one form can be rewritten in the other form.

EXAMPLE 1

a. Write $4^3 = 64$ in logarithmic form.

b. Write $\log_5 \frac{1}{25} = -2$ in exponential form.

SOLUTION

a. $\log_4 64 = 3$

b. $5^{-2} = \frac{1}{25}$

Teach⇔Interact

Section Notes

Historical Connection
Logarithms were invented by John Napier (1550–1617), a Scotsman, in about 1594. Napier was interested in facilitating calculations he was doing in spherical trigonometry for astronomical problems. In the early 1700s, the mathematicians John Wallis, Isaac Newton, Gottfried Leibniz, and John Bernoulli showed that the logarithmic function is the inverse of the exponential function.

Communication: Discussion
When discussing the domain and range of the logarithmic function, emphasize that the base must be positive because the base of the exponential function is positive.

Student Progress
The graph of the exponential function should be familiar to most students from Chapter 3. Thus, you should expect students to be able to reflect the points of the exponential function to get the graph of the logarithmic function.

Using Technology
You can help students explore the inverses of exponential functions by assigning Function Investigator Activity 30 in the *Function Investigator with Matrix Analyzer Activity Book.*

About Example 1

Teaching Tip
Some students may notice that the value of the logarithm in part (b) of Example 1 is negative. Point out that a logarithm is negative when it represents the negative power of the base of the exponential function. Also point out that the positive or negative power of any positive number is always positive.

Additional Example 1

a. Write $\left(\frac{1}{3}\right)^3 = \frac{1}{27}$ in logarithmic form.
$\log_{1/3} \frac{1}{27} = 3$

b. Write $\log_6 36 = 2$ in exponential form. $6^2 = 36$

ANSWERS Section 4.2

Exploration

1.

x	$f(x)$
−2	$\frac{1}{4}$
−1	$\frac{1}{2}$
0	1
1	2
2	4
3	8

2–4.

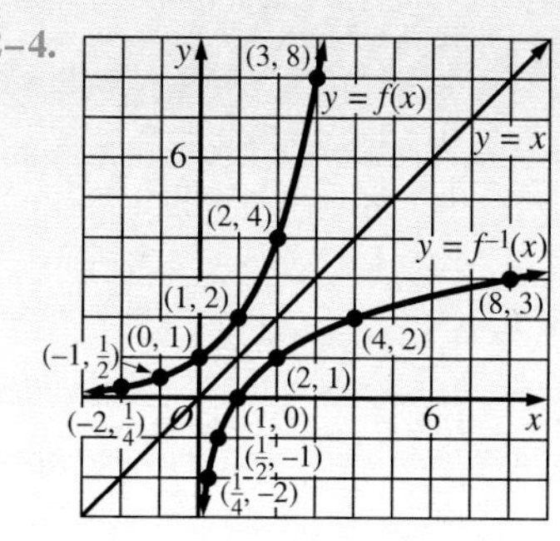

Questions

1. a. 3
b. −2
c. 0

2. a. exponents; powers of 2
b. powers of 2; exponents

Teach⇔Interact

About Example 2

Teaching Tip
Example 2 uses the principle that if the bases of equal powers are the same, then the exponents must be equal. In part (b), the first step in solving the equation is to rewrite the exponential equation so that the bases match. Students will see later that this is not always possible.

Additional Example 2

Evaluate each logarithm.

a. $\log_{1/2} 8$

Let $\log_{1/2} 8 = k$.

Then: $\left(\frac{1}{2}\right)^k = 8$

$2^{(-k)} = 2^3$

$-k = 3$

$k = -3$

$\log_{1/2} 8 = -3$

b. $\log_{27}\left(\frac{1}{9}\right)$

Let $\log_{27}\left(\frac{1}{9}\right) = k$.

Then: $27^k = \frac{1}{9}$

$(3^3)^k = 3^{-2}$

$3^{3k} = 3^{-2}$

$3k = -2$

$k = -\frac{2}{3}$

$\log_{27}\left(\frac{1}{9}\right) = -\frac{2}{3}$

Think and Communicate

For question 3, students should understand that the logarithm of a negative number is not defined because the power of a positive number is always positive. Since a negative power represents the reciprocal of a number, any power, positive or negative, of a positive number is also positive. Students can verify this by doing some examples.

Section Note

Student Study Tip
Some students may enjoy reading about the history of the development of logarithms. In so doing, they will see how many of the ideas of this section were invented and systematized.

EXAMPLE 2

Evaluate each logarithm.

a. $\log_6 36$ **b.** $\log_{16} 8$

SOLUTION

a. Let $\log_6 36 = k$.

Then: $6^k = 36$ ← Write the equation in exponential form.

$6^k = 6^2$ ← Make the bases the same.

$k = 2$ ← Equate exponents.

$\log_6 36 = 2$

b. Let $\log_{16} 8 = k$.

Then: $16^k = 8$

$(2^4)^k = 2^3$

$2^{4k} = 2^3$

$4k = 3$

$k = \frac{3}{4}$

$\log_{16} 8 = \frac{3}{4}$

THINK AND COMMUNICATE

1. To evaluate $\log_6 36$ mentally, you should ask yourself the question, "What power of 6 equals 36?" What question should you ask to evaluate each of these logarithms mentally?

a. $\log_2 8$ **b.** $\log_7 \frac{1}{7}$ **c.** $\log_9 3$

2. Mentally evaluate each logarithm in Question 1.

3. Explain why $\log_6(-36)$ is not defined.

4. a. Evaluate these logarithms: $\log_2 1$, $\log_3 1$, $\log_4 1$.

b. In general, what is the value of $\log_b 1$?

5. a. Evaluate these logarithms: $\log_2 2$, $\log_3 3$, $\log_4 4$.

b. In general, what is the value of $\log_b b$?

6. a. Complete this equation: $\log_b b^k = \underline{\ ?\ }$.

b. Explain why your answer to part (a) is consistent with your results from Questions 4 and 5.

Most calculators have LOG and LN keys for evaluating common and natural logarithms.

Common and Natural Logarithms

Two kinds of logarithms are used so often that they are given special names and symbols. The **common logarithm** of a positive number N is the base-10 logarithm of N and is denoted **log *N*** (rather than $\log_{10} N$). The **natural logarithm** of N is the base-e logarithm of N and is denoted **ln *N*** (rather than $\log_e N$).

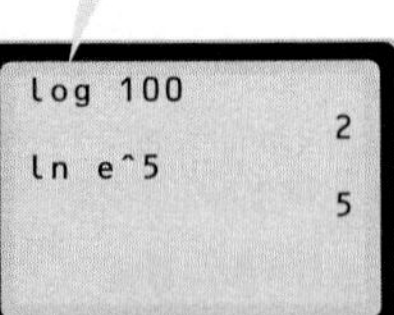

Think and Communicate

1. a. What power of 2 equals 8?
   b. What power of 7 equals $\frac{1}{7}$?
   c. What power of 9 equals 3?
2. a. 3
   b. −1
   c. $\frac{1}{2}$
3. There is no number k for which $6^k = -36$. Every power of 6 is a positive number.
4. a. 0; 0; 0 b. 0
5. a. 1; 1; 1 b. 1
6. a. k
   b. For every number $b \neq 0$, $b^0 = 1$ and $b^1 = b$. Then $\log_b 1 = \log_b b^0 = 0$ and $\log_b b = \log_b b^1 = 1$.

EXAMPLE 3 Application: Mountain Climbing

Mountain climbers use an instrument called an *altimeter* to help them navigate. An altimeter finds a climber's height above sea level by measuring the air pressure. The height h and air pressure P are related by the function

$$P = 101{,}300e^{-h/8005}$$

where h is in meters and P is in pascals.

a. Find the inverse of the given function.

b. In 1975, Junko Tabei of Japan became the first woman to reach the summit of Mount Everest, the highest mountain in the world. What was the reading on Tabei's altimeter when the air pressure was 60,000 pascals?

Junko Tabei stands on "the top of the world." She had to wear an oxygen mask due to the low level of oxygen at that high altitude.

BY THE WAY

Mount Everest is located on the border of Nepal and China, two countries in south-central Asia. The mountain is part of the Himalaya range.

SOLUTION

a. Solve the equation $P = 101{,}300e^{-h/8005}$ for h.

$$\frac{P}{101{,}300} = e^{-h/8005}$$

$$\ln \frac{P}{101{,}300} = -\frac{h}{8005}$$ Write the equation in logarithmic form.

$$-8005 \ln \frac{P}{101{,}300} = h$$

The inverse is $h = -8005 \ln \frac{P}{101{,}300}$.

b. Use the inverse to find h when $P = 60{,}000$.

$$h = -8005 \ln \frac{60{,}000}{101{,}300}$$ Substitute **60,000** for ***P***.

$$\approx (-8005)(-0.5237)$$

$$\approx 4190$$ Use a calculator to evaluate the logarithm.

The reading on Junko Tabei's altimeter was about 4190 m above sea level.

Teach⇔Interact

About Example 3

Teaching Tip

Example 3 has students solve an exponential equation by writing the equation in logarithmic form and then solving for the variable. Point out that this is equivalent to finding the inverse of the exponential equation.

Additional Example 3

Aerospace engineers use instruments to measure the amount of power generated by the power supply aboard a satellite. The time t and the power P are related by the function $P = 50e^{-t/300}$, where t is in days and P is in watts.

a. Find the inverse of the given function.

Solve the equation $P = 50e^{-t/300}$ for t.

$$\frac{P}{50} = e^{-t/300}$$

$$\ln \frac{P}{50} = -\frac{t}{300}$$

$$-300 \ln \frac{P}{50} = t$$

The inverse is $t = -300 \ln \frac{P}{50}$.

b. How long does it take the power supply to generate 10 watts of power?

Use the inverse to find t when $P = 10$.

$$t = -300 \ln \frac{10}{50}$$

$$\approx (-300)(-1.6094)$$

$$\approx 483 \text{ days}$$

It takes about 483 days to generate 10 watts of power.

Teach⇔Interact

Think and Communicate

To help answer question 7, have students view the graph of the pressure function on a graphing calculator. Enter the function using window values of Xmax = 2000 and Ymax = 150,000. The graph shows that the y-values (pressure) decrease as x-values (height) increase.

Checking Key Concepts

Reasoning
For questions 10–12, you may wish to see if students can discover the formula $\log_b b^k = k$.

Closure Question

State four facts about the inverse of an exponential function.
Answers may vary. An example is given. The inverse of an exponential function is called a logarithmic function. For an exponential function of the form $b^k = N$, the inverse is $\log_b N = k$, where b is the base of each function. The base b must be positive and not equal to one. The graph of the logarithmic function is the reflection of the related exponential function over the line $y = x$.

Apply⇔Assess

Suggested Assignment

Core Course
Day 1 Exs. 1–26
Day 2 Exs. 31–40, 43–52

Extended Course
Day 1 Exs. 1–26
Day 2 Exs. 27–52

Block Schedule
Day 21 Exs. 1–26
Day 22 Exs. 31–40, 43–52

Exercise Notes

Common Error
Exs. 10–17 Remind students who make errors in converting equations from logarithmic form to exponential form that the logarithm of a number is the exponent in the exponential form.

THINK AND COMMUNICATE

7. In Example 3, does a decrease in air pressure indicate an *increase* or *decrease* in height above sea level? How do you know?
8. In the solution to part (a) of Example 3, why would it be incorrect to write $\log \frac{P}{101{,}300} = -\frac{h}{8005}$?

CHECKING KEY CONCEPTS

Find the inverse of each function.

1. $f(x) = 2^x$ 2. $g(x) = \left(\frac{1}{3}\right)^x$ 3. $y = e^x$

Write each equation in logarithmic form.

4. $3^4 = 81$ 5. $8^0 = 1$ 6. $10^{-2} = \frac{1}{100}$

Write each equation in exponential form.

7. $\log_6 216 = 3$ 8. $\log \frac{1}{10} = -1$ 9. $\log_{27} 3 = \frac{1}{3}$

Evaluate each logarithm.

10. $\log_2 32$ 11. $\log_{125} 25$ 12. $\ln e^3$

Use a calculator to find the value of each logarithm to the nearest hundredth.

13. $\log 146$ 14. $\ln 15$ 15. $\log 0.32$

4.2 Exercises and Applications

Extra Practice exercises on page 753

1. Look back at the graph of $y = \log_3 x$ on page 149. For what value(s) of x is $\log_3 x$:
 a. positive? b. negative? c. zero? d. undefined?

Write each equation in logarithmic form.

2. $3^2 = 9$ 3. $2^3 = 8$ 4. $10^4 = 10{,}000$ 5. $e^0 = 1$
6. $5^{-1} = \frac{1}{5}$ 7. $64^{-1/3} = \frac{1}{4}$ 8. $(0.3)^2 = 0.09$ 9. $\left(\frac{4}{5}\right)^3 = \frac{64}{125}$

Write each equation in exponential form.

10. $\log_4 16 = 2$ 11. $\log_2 16 = 4$ 12. $\log_3 \frac{1}{9} = -2$ 13. $\log_2 \frac{1}{32} = -5$
14. $\log 1 = 0$ 15. $\log_{1/6} \frac{1}{36} = 2$ 16. $\log_{0.2} 0.008 = 3$ 17. $\log_{64} 16 = \frac{2}{3}$

Evaluate each logarithm.

18. $\log_3 27$ 19. $\log_7 49$ 20. $\ln e^{-6}$ 21. $\log \frac{1}{1000}$
22. $\log_{1/2} \frac{1}{8}$ 23. $\log_9 1$ 24. $\log_4 8$ 25. $\log_{81} 27$

26. **Writing** Explain why the base b of $y = \log_b x$ cannot be 1.

Think and Communicate

7. increase; As P decreases, $\ln \frac{P}{101{,}300}$ decreases, so $-8000 \ln \frac{P}{101{,}300}$ increases.
8. The logarithmic equation is based on an exponential equation with base e, not base 10.

Checking Key Concepts

1. $f^{-1}(x) = \log_2 x$
2. $g^{-1}(x) = \log_{1/3} x$
3. $y = \ln x$
4. $\log_3 81 = 4$
5. $\log_8 1 = 0$
6. $\log \frac{1}{100} = -2$
7. $6^3 = 216$
8. $10^{-1} = \frac{1}{10}$
9. $27^{1/3} = 3$
10. 5
11. $\frac{2}{3}$
12. 3
13. 2.16
14. 2.71
15. −0.49

Exercises and Applications

1. a. $x > 1$ b. $0 < x < 1$ c. 1 d. $x \le 0$
2. $\log_3 9 = 2$
3. $\log_2 8 = 3$
4. $\log 10{,}000 = 4$
5. $\ln 1 = 0$
6. $\log_5 \frac{1}{5} = -1$
7. $\log_{64} \frac{1}{4} = -\frac{1}{3}$
8. $\log_{0.3} 0.09 = 2$
9. $\log_{4/5} \frac{64}{125} = 3$
10. $4^2 = 16$
11. $2^4 = 16$
12. $3^{-2} = \frac{1}{9}$
13. $2^{-5} = \frac{1}{32}$

Connection ASTRONOMY

Star	Apparent magnitude
Sirius	−1.5
Tau Ceti	3.5
Barnard's Star	9.5
Luyten 726-8	12.5
Wolf 359	13.5

Recall from Section 3.5 that the *apparent magnitude* of a star is a measure of how bright the star looks to a person on Earth. The greater a star's apparent magnitude, the *less* bright the star appears. The table shows the apparent magnitudes of several stars.

You can see stars with apparent magnitudes of up to 6.5 without a telescope. To see a star with magnitude $M > 6.5$, you need a telescope having an objective lens or mirror whose diameter is at least D mm, where:

$$D = 10^{(M-2)/5}$$

This time-lapse photograph shows the motion of Barnard's Star over a four-year period.

BY THE WAY

Barnard's Star, located in the constellation of Ophiuchus, moves across the sky faster than any other star.

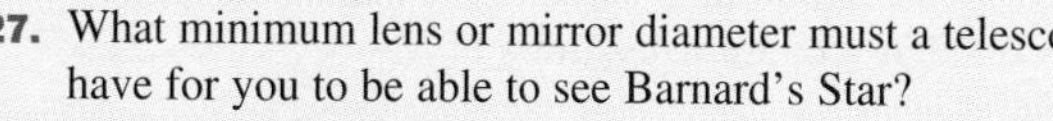

27. What minimum lens or mirror diameter must a telescope have for you to be able to see Barnard's Star?

28. Find the inverse of $D = 10^{(M-2)/5}$.

29. A telescope's *limiting magnitude* is the apparent magnitude of the dimmest star that can be seen with the telescope.

a. A typical amateur telescope might have a lens or mirror diameter of about 75 mm. Use your answer to Exercise 28 to find the limiting magnitude of such a telescope.

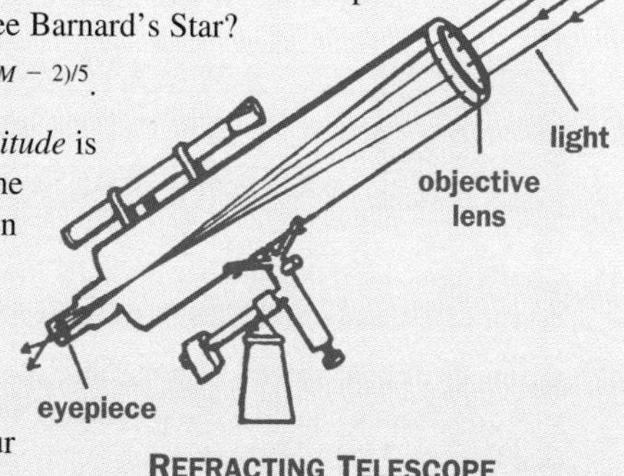

REFRACTING TELESCOPE

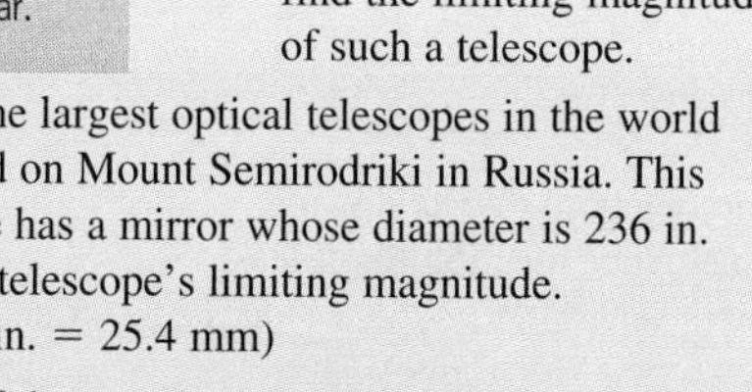

b. One of the largest optical telescopes in the world is located on Mount Semirodriki in Russia. This telescope has a mirror whose diameter is 236 in. Find the telescope's limiting magnitude. (*Hint:* 1 in. = 25.4 mm)

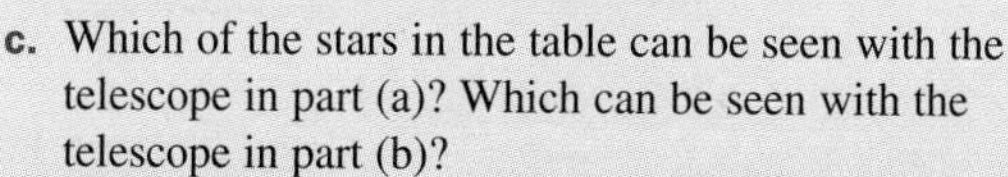

c. Which of the stars in the table can be seen with the telescope in part (a)? Which can be seen with the telescope in part (b)?

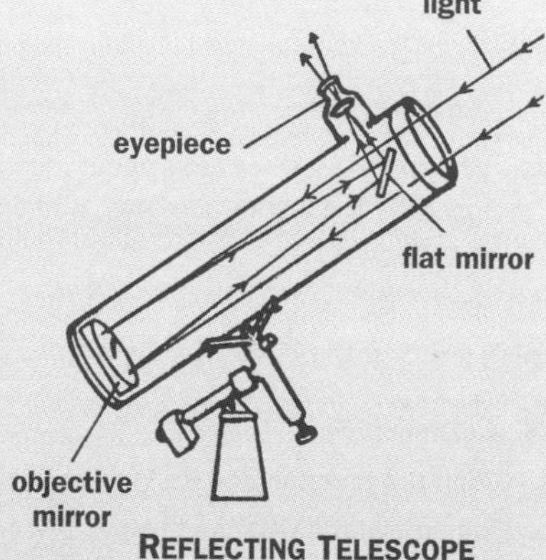

REFLECTING TELESCOPE

30. Writing Suppose the apparent magnitude of star A is less than the apparent magnitude of star B. Does star A necessarily give off more light than star B? Explain.

Use a calculator to find the value of each logarithm to the nearest hundredth.

31. log 12 **32.** log 257 **33.** ln 6.5 **34.** $\ln \frac{3}{8}$

For each function:

a. Graph the function and its inverse in the same coordinate plane.

b. Find an equation for the inverse.

35. $f(x) = 4^x$ **36.** $h(x) = 10^x$ **37.** $y = \left(\frac{1}{2}\right)^x$ **38.** $y = 2 \cdot 3^x$

Apply⇔Assess

Exercise Notes

Cooperative Learning
Exs. 18–25 You may wish to have students work on these exercises with a partner. Have them do mental calculations for Exs. 18–21 and 23, and then discuss how the other exercises can be evaluated.

Reasoning
Ex. 26 Students should understand that the base b of $y = \log_b x$ cannot be 1 because the equivalent exponential equation, $1^y = x$, is not a function.

Multicultural Note
Exs. 27–30 The science of astronomy has been practiced for many thousands of years. The ancient Chinese charted the position of the stars in the sky, and recorded solar and lunar eclipses more than 3000 years ago. The ancient Babylonians also studied these positions 2700 years ago and were able to make predictions about certain planets' courses in relation to the sun. The ancient Egyptians could predict the coming of each season by observing the position of specific stars. All of these ancient civilizations left careful documentation of their astronomical observations; their records of observations are studied with great interest by modern scholars.

Teaching Tip
Ex. 29 You may want to point out that this exercise can be done easily by using the result of Ex. 28.

Research
Ex. 30 Some students may wish to research how astronomers calculate the apparent magnitude of a star. They should discover that to determine the actual brightness of a star, its distance from Earth must be known. Thus, a star that looks dim in a telescope may be very bright but very far away.

Communication: Discussion
Exs. 35–38 Have students discuss how the table of values for the graphs in these exercises are related to the table of values for the graphs of the inverse functions. Reflecting points over the line $y = x$ generates a table of values which switches the x- and y-values.

Teaching Tip
Ex. 38 Students may have difficulty finding the equation for the inverse if they do not divide both sides of the equation by 2 first.

14. $10^0 = 1$
15. $\left(\frac{1}{6}\right)^2 = \frac{1}{36}$
16. $(0.2)^3 = 0.008$
17. $64^{2/3} = 16$
18. 3
19. 2
20. −6
21. −3
22. 3
23. 0
24. $\frac{3}{2}$ 
25. $\frac{3}{4}$
26. It does not make sense to speak of the base-1 logarithm of a number x. For every real number x, $1^x = 1$. No number except 1 can be written as a power of 1.
27. about 31.6 mm
28. $M = 5 \log D + 2$
29. a. about 11.4
b. about 20.9
c. Barnard's Star, Tau Ceti, and Sirius; all the listed stars
30. No; the apparent magnitude is affected not only by the amount of light given off by the star but also by the star's distance from Earth.
31. 1.08
32. 2.41
33. 1.87
34. −0.98
35. a.

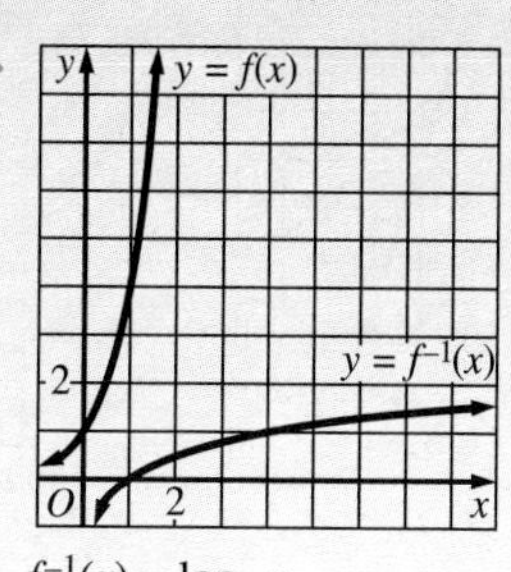

b. $f^{-1}(x) = \log_4 x$

36–38. See answers in back of book.

Apply⇔Assess

Exercise Notes

Cooperative Learning

Ex. 42 You may wish to have students discuss their choices for exponential functions in small groups. Ask students to check each other's answers and discuss how the function and its inverse can be used to solve a real-world problem.

Visual Thinking

Ex. 42 You can extend this exercise by asking students to create diagrams that demonstrate the relationships between an *exponential function* and a *logarithmic function.* Encourage them to use color, arrows, and other devices to make the relationships clear. Have them try out their diagrams on the class. This activity involves the visual skills of *correlation* and *generalization.*

Practice 21 for Section 4.2

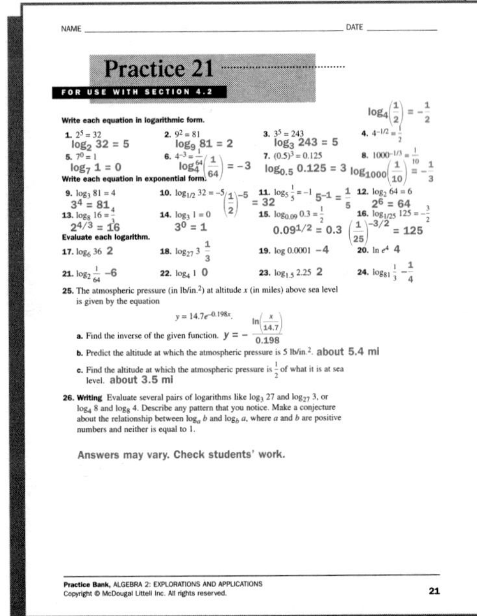
NAME ______________________ DATE __________

Practice 21

FOR USE WITH SECTION 4.2

Write each equation in logarithmic form.

1. $2^5 = 32$ $\log_2 32 = 5$
2. $9^2 = 81$ $\log_9 81 = 2$
3. $3^5 = 243$ $\log_3 243 = 5$
4. $4^{-1/2} = \frac{1}{2}$ $\log_4\left(\frac{1}{2}\right) = -\frac{1}{2}$
5. $7^0 = 1$ $\log_7 1 = 0$
6. $4^{-3} = \frac{1}{64}$ $\log_4\left(\frac{1}{64}\right) = -3$
7. $(0.5)^3 = 0.125$ $\log_{0.5} 0.125 = 3$
8. $1000^{-1/3} = \frac{1}{10}$ $\log_{1000}\left(\frac{1}{10}\right) = -\frac{1}{3}$

Write each equation in exponential form.

9. $\log_3 81 = 4$ $3^4 = 81$
10. $\log_{1/2} 32 = -5$ $\left(\frac{1}{2}\right)^{-5} = 32$
11. $\log_5 \frac{1}{5} = -1$ $5^{-1} = \frac{1}{5}$
12. $\log_2 64 = 6$ $2^6 = 64$
13. $\log_8 16 = \frac{4}{3}$ $2^{4/3} = 16$
14. $\log_3 1 = 0$ $3^0 = 1$
15. $\log_{0.09} 0.3 = \frac{1}{2}$ $0.09^{1/2} = 0.3$
16. $\log_{1/25} 125 = -\frac{3}{2}$ $\left(\frac{1}{25}\right)^{-3/2} = 125$

Evaluate each logarithm.

17. $\log_6 36$ 2
18. $\log_{27} 3$ $\frac{1}{3}$
19. $\log 0.0001$ -4
20. $\ln e^4$ 4
21. $\log_2 \frac{1}{64}$ -6
22. $\log_4 1$ 0
23. $\log_{1.5} 2.25$ 2
24. $\log_{81} \frac{1}{3}$ $-\frac{1}{4}$

25. The atmospheric pressure (in lb/in.^{2}) at altitude x (in miles) above sea level is given by the equation

$$y = 14.7e^{-0.198x}.$$

a. Find the inverse of the given function. $y = -\dfrac{\ln\left(\frac{x}{14.7}\right)}{0.198}$

b. Predict the altitude at which the atmospheric pressure is 5 lb/in.^{2}. about 5.4 mi

c. Find the altitude at which the atmospheric pressure is $\frac{1}{2}$ of what it is at sea level. about 3.5 mi

26. **Writing** Evaluate several pairs of logarithms like $\log_3 27$ and $\log_{27} 3$, or $\log_4 8$ and $\log_8 4$. Describe any pattern that you notice. Make a conjecture about the relationship between $\log_a b$ and $\log_b a$, where a and b are positive numbers and neither is equal to 1.

Answers may vary. Check students' work.

Practice Bank, ALGEBRA 2: EXPLORATIONS AND APPLICATIONS 21

39. POPULATION The population P of Nepal (in millions) can be modeled by the function

$$P = 19.1e^{0.025t}$$

where t is the number of years since 1990.

a. Find the inverse of the given function. What information can you obtain with the inverse?

b. Predict the year in which the population of Nepal will reach 30 million.

c. Estimate the year in which the population of Nepal was 15 million.

Bhadgaon, Nepal

40. Investigation Copy and complete the table. (The first column has been done for you.) What is the relationship between the number of digits in a positive integer N and the common logarithm of N?

N	7	10	52	100	613	1,000	4,849	10,000	91,770
Number of digits in N	1	?	?	?	?	?	?	?	?
log N	0.85	?	?	?	?	?	?	?	?

41. Challenge Find the number of digits in 3^{1000}. (*Hint:* Express 3 as a power of 10, and use the result from Exercise 40.)

42. Open-ended Problem Choose an exponential function of the form $y = ae^{kx}$ from Chapter 3. Your function should model something from real life. Find the inverse of your function. Describe what information the inverse gives, and use the inverse to solve a problem.

ONGOING ASSESSMENT

43. Writing List three facts about logarithmic functions. Write down a logarithmic function $y = \log_b x$, and evaluate your function for at least two values of x.

SPIRAL REVIEW

For each function:

a. Graph the function and its inverse in the same coordinate plane.

b. Find an equation for the inverse. *(Section 4.1)*

44. $f(x) = 5x$ **45.** $g(x) = -\frac{1}{3}x$ **46.** $h(x) = 2x - 10$ **47.** $y = \frac{3}{8}x + \frac{7}{8}$

48. Let $A = \begin{bmatrix} 2 & 5 \\ -8 & 0 \end{bmatrix}$ and $B = \begin{bmatrix} -1 & 3 \\ -5 & 6 \end{bmatrix}$. Find each matrix. *(Sections 1.3 and 1.4)*

a. $A + B$ b. $B - A$ c. $2A + 3B$ d. AB

Tell whether each equation represents growth that is *linear, exponential,* or *neither.* *(Section 3.1)*

49. $y = 7x^2$ **50.** $y = 7(2^x)$ **51.** $y = 2(7^x)$ **52.** $y = (7^2)x$

39–43. See answers in back of book.

44. a.

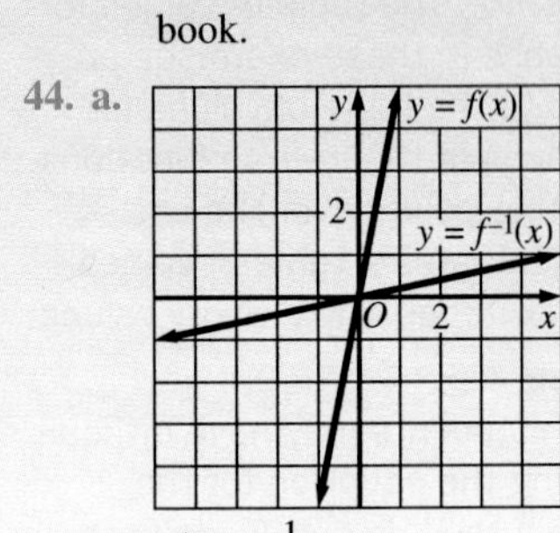

b. $f^{-1}(x) = \frac{1}{5}x$

45. a.

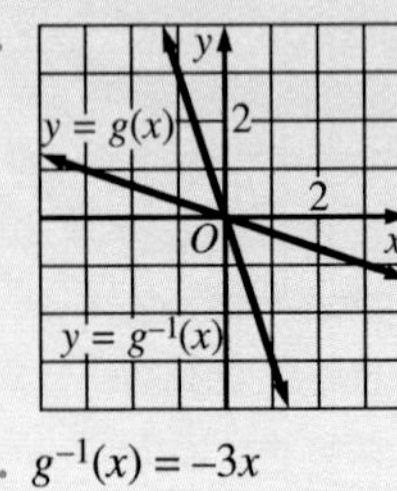

b. $g^{-1}(x) = -3x$

46. a.

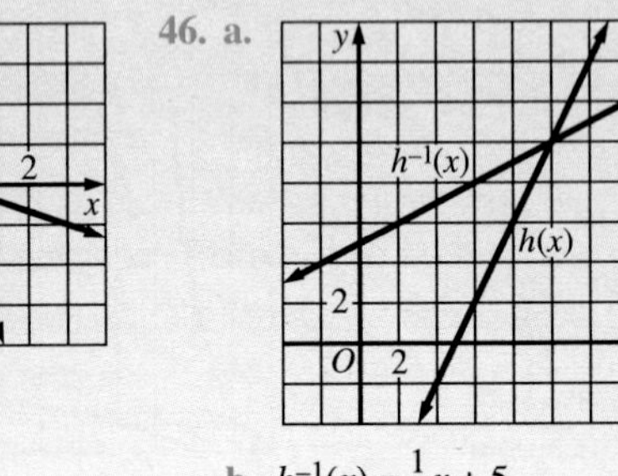

b. $h^{-1}(x) = \frac{1}{2}x + 5$

47. a.

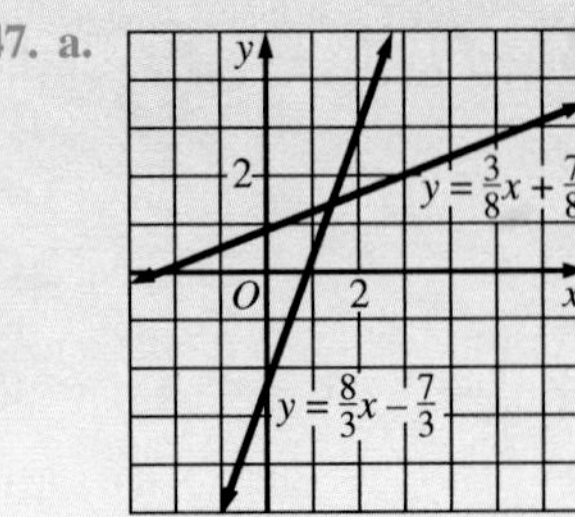

b. $y = \frac{8}{3}x - \frac{7}{3}$

48. a. $\begin{bmatrix} 1 & 8 \\ -13 & 6 \end{bmatrix}$

b. $\begin{bmatrix} -3 & -2 \\ 3 & 6 \end{bmatrix}$

c. $\begin{bmatrix} 1 & 19 \\ -31 & 18 \end{bmatrix}$

d. $\begin{bmatrix} -27 & 36 \\ 8 & -24 \end{bmatrix}$

49. neither

50. exponential

51. exponential

52. linear

Section

4.3 Working with Logarithms

Learn how to...
- use the properties of logarithms

So you can...
- understand logarithmic scales, such as the Richter scale for earthquakes

In Chapter 3, you learned rules for simplifying products, quotients, and powers of powers. As you'll see in the Exploration, there are corresponding rules for logarithms.

EXPLORATION
COOPERATIVE LEARNING

Investigating Properties of Logarithms

Work with a partner.
You will need:
- a scientific calculator

SET UP Copy the table. Use a calculator to complete the table. Round each logarithm to four decimal places.

N	1	2	3	4	5	6	7	8	9	10	20	30	40	50	60	70	80	90	100	1000
$\log N$	?	?	?	?	?	?	?	?	?	?	?	?	?	?	?	?	?	?	?	?

Questions

1. a. Use your table to find $\log 2 + \log 4$.
   b. Look for a value of N in your table for which $\log N$ equals your answer to part (a). Complete this equation: $\log 2 + \log 4 = \log \underline{?}$.
2. Use your table to complete these equations.
   a. $\log 8 + \log 5 = \log \underline{?}$ b. $\log 4 + \log 20 = \log \underline{?}$
   c. $\log 7 + \log 10 = \log \underline{?}$ d. $\log 9 + \log 10 = \log \underline{?}$
3. Generalize your results from Questions 1 and 2: $\log M + \log N = \log \underline{?}$.
4. Use your table to complete these equations.
   a. $\log 8 - \log 2 = \log \underline{?}$ b. $\log 50 - \log 5 = \log \underline{?}$
   c. $\log 60 - \log 10 = \log \underline{?}$ d. $\log 90 - \log 30 = \log \underline{?}$
5. Generalize your results from Question 4: $\log M - \log N = \log \underline{?}$.
6. Use your table to complete these equations.
   a. $2 \log 3 = \log \underline{?}$ b. $2 \log 10 = \log \underline{?}$ c. $3 \log 10 = \log \underline{?}$
7. Generalize your results from Question 6: $k \log M = \log \underline{?}$.

Exploration Note

Purpose
The purpose of this Exploration is to have students discover the product, quotient, and power properties of logarithms.

Materials/Preparation
A scientific calculator is needed.

Procedure
Students use a calculator to find the common logarithm of the numbers 1 to 9 and also the common logarithm of some multiples of 10. Using the table, they complete logarithm equations which demonstrate the product property, the quotient property, and the power property.

Closure
Students should conclude that the logarithm of the product of two numbers is the sum of the logarithms of each number, the logarithm of the quotient of two numbers is the difference of the logarithms of each number, and that the logarithm of the power of a number is the exponent of the number times the logarithm of its base.

Explorations Lab Manual
See the Manual for more commentary on this Exploration.
Diagram Master 8

For answers to the Exploration, see following page.

Plan⇔Support

Objectives
- Use the properties of logarithms.
- Understand logarithmic scales.

Recommended Pacing

Core and Extended Courses
Section 4.3: 2 days

Block Schedule
Section 4.3: 2 half-blocks (with Sections 4.2 and 4.4)

Resource Materials

Lesson Support
Lesson Plan 4.3
Warm-Up Transparency 4.3
Practice Bank: Practice 22
Study Guide: Section 4.3
Exploration Lab Manual: Diagram Masters 2, 8
Assessment Book: Test 13

Technology
Scientific Calculator
Graphing Calculator
McDougal Littell Mathpack *Function Investigator*
Internet: http://www.hmco.com

Warm-Up Exercises

Evaluate each of the following.
1. $\log 10$ 1
2. $\log 20$ 1.3010
3. $\log 2 + \log 10$ 1.3010
4. $\log_2 64$ 6
5. $\log_2 16$ 4
6. $\log_2 \frac{64}{16}$ 2

Teach⇔Interact

Section Notes

Communication: Writing
A worthwhile writing activity would be to have students copy the properties of logarithms in their journals. This should help them to remember each property and be able to use it in solving problems.

Reasoning
Since a logarithm is essentially an exponent, the properties of logarithms can be derived from the laws of exponents. Ask students if they see any relationship of these properties to any other laws they are familiar with. (See Exs. 31–33 on page 160.)

Additional Example 1

Write $\log_5 \frac{\sqrt{x^3y}}{z^4}$ in terms of $\log_5 x$, $\log_5 y$, and $\log_5 z$.

$\log_5 \frac{\sqrt{x^3y}}{z^4} = \log_5 \frac{(x^3y)^{1/2}}{z^4}$
$= \log_5 \frac{x^{3/2}y^{1/2}}{z^4}$
$= \log_5 x^{3/2} + \log_5 y^{1/2} - \log_5 z^4$
$= \frac{3}{2}\log_5 x + \frac{1}{2}\log_5 y - 4\log_5 z$

Additional Example 2

Write $2\log_b 6 + \frac{1}{2}\log_b 3 - \log_b 12$ as a logarithm of a single number.

$2\log_b 6 + \frac{1}{2}\log_b 3 - \log_b 12$
$= \log_b 6^2 + \log_b \sqrt{3} - \log_b 12$
$= \log_b 36 + \log_b \sqrt{3} - \log_b 12$
$= \log_b 36\sqrt{3} - \log_b 12$
$= \log_b \frac{36\sqrt{3}}{12}$
$= \log_b 3\sqrt{3}$

Think and Communicate

For question 1, remind students that since the base is 10, they should use the [log] key rather than the [ln] key. To stress how the computation relates to the properties of logarithms, it may help to use variables. Store the numbers 48, 2, and 5 in variables by using 48→A, 2→B, and 5→C. Then evaluate $\frac{A}{B^3} \cdot C$ and store the answer in D by using (A/B^3)C→D. Finally, have students evaluate log A – 3 log B + log C and log D.

You can use the following properties to rewrite logarithms of products, quotients, and powers.

WATCH OUT!
There are no properties that apply to logarithms of sums and differences. Note especially that, in general:
$\log_b (M + N) \neq \log_b M + \log_b N$
$\log_b (M - N) \neq \log_b M - \log_b N$

Properties of Logarithms

Let M, N, and b be positive numbers with $b \neq 1$. Then:

Product Property: $\log_b MN = \log_b M + \log_b N$

Quotient Property: $\log_b \frac{M}{N} = \log_b M - \log_b N$

Power Property: $\log_b M^k = k \log_b M$

EXAMPLE 1

Write $\log_3 \frac{p^4q^5}{r^{1/2}}$ in terms of $\log_3 p$, $\log_3 q$, and $\log_3 r$.

SOLUTION

$$\log_3 \frac{p^4q^5}{r^{1/2}} = \log_3 p^4q^5 - \log_3 r^{1/2}$$
Use the **quotient property**.

$$= \log_3 p^4 + \log_3 q^5 - \log_3 r^{1/2}$$
Use the **product property**.

$$= 4\log_3 p + 5\log_3 q - \frac{1}{2}\log_3 r$$
Use the **power property**.

EXAMPLE 2

Write $\log_a 48 - 3\log_a 2 + \log_a 5$ as a logarithm of a single number.

SOLUTION

$$\log_a 48 - 3\log_a 2 + \log_a 5 = \log_a 48 - \log_a 2^3 + \log_a 5$$
Use the **power property**.
$$= \log_a 48 - \log_a 8 + \log_a 5$$
$$= \log_a \frac{48}{8} + \log_a 5$$
Use the **quotient property**.
$$= \log_a 6 + \log_a 5$$
$$= \log_a (6 \cdot 5)$$
Use the **product property**.
$$= \log_a 30$$

THINK AND COMMUNICATE

1. Use a calculator to evaluate $\log_a 48 - 3\log_a 2 + \log_a 5$ and $\log_a 30$ when $a = 10$. Are your results consistent with Example 2?

ANSWERS Section 4.3

Exploration

N	1	2	3	4	5	6	7	8	9	10
$\log N$	0	0.3010	0.4771	0.6021	0.6990	0.7782	0.8451	0.9031	0.9542	1.0000
N	20	30	40	50	60	70	80	90	100	1000
$\log N$	1.3010	1.4771	1.6021	1.6990	1.7782	1.8451	1.9031	1.9542	2.0000	3.0000

Questions

1. a. 0.9031 b. 8; 8
2. a. 40 b. 80 c. 70 d. 90
3. MN
4. a. 4 b. 10 c. 6 d. 3
5. $\frac{M}{N}$
6. a. 9 b. 100 c. 1000
7. M^k

EXAMPLE 3 Application: Earth Science

The *Richter scale* is used to rate the severity of earthquakes. An earthquake is assigned a *Richter magnitude* based on the amount of energy it releases. The Richter magnitude M and energy E are related by the equation

$$M = \frac{2}{3}\log\frac{E}{10^{11.8}}$$

where E is measured in units called *ergs*. Suppose the difference in the Richter magnitudes of two earthquakes, A and B, is 1. How many times as much energy does earthquake A release as earthquake B?

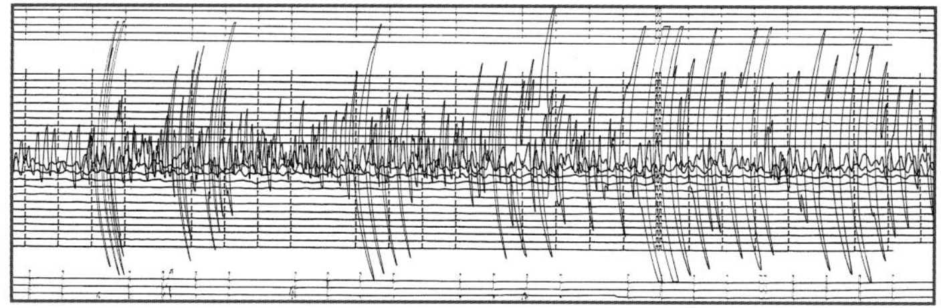

A seismograph draws a graph like this to show the movement of the earth during an earthquake. Scientists use these graphs to determine the amount of energy released.

BY THE WAY

The smallest earthquakes that can be felt have Richter magnitudes of about 2.0. The most powerful earthquake ever recorded had a Richter magnitude of 8.9. It occurred in Japan on March 2, 1933.

SOLUTION

Let E_A and M_A be the energy and Richter magnitude for earthquake A. Let E_B and M_B be the energy and Richter magnitude for earthquake B. Then:

$$M_A - M_B = 1$$

The difference in magnitudes is 1.

$$\frac{2}{3}\log\frac{E_A}{10^{11.8}} - \frac{2}{3}\log\frac{E_B}{10^{11.8}} = 1$$

Use the given equation.

$$\frac{2}{3}\left(\log\frac{E_A}{10^{11.8}} - \log\frac{E_B}{10^{11.8}}\right) = 1$$

$$\log\frac{E_A}{10^{11.8}} - \log\frac{E_B}{10^{11.8}} = \frac{3}{2}$$

Use the quotient property.

$$\log\left(\frac{E_A/10^{11.8}}{E_B/10^{11.8}}\right) = \frac{3}{2}$$

$$\log\left(\frac{E_A}{10^{11.8}} \cdot \frac{10^{11.8}}{E_B}\right) = \frac{3}{2}$$

$$\log\frac{E_A}{E_B} = \frac{3}{2}$$

Write the equation in exponential form.

$$\frac{E_A}{E_B} = 10^{3/2}$$

$$E_A = 10^{3/2} \cdot E_B \approx 32E_B$$

Earthquake A releases about 32 times as much energy as earthquake B.

Toolbox p. 782
Factoring Variable Expressions

Teach⇔Interact

About Example 3

Teaching Tip

Example 3 helps students see that adding 1 to a logarithmic scale is equivalent to multiplying a quantity by some constant based on the scale.

Additional Example 3

The Henderson-Hasselbach formula is used to find the level of acidity, known as the pH, in a person's blood. The ratio R of the concentration of bicarbonate and carbonic acid in the blood is related to the pH of the blood by the equation pH = 6.1 + log R. Most people have a blood pH of 7.4. Suppose the difference in pH in two samples of blood is 0.8. How many times higher is the ratio of bicarbonate to carbonic acid in one sample than in the other?

Let pH_A and R_A be the pH and bicarbonate/carbonic acid ratio for sample A. Let pH_B and R_B be the pH and bicarbonate/carbonic acid ratio for sample B. Then:

$$pH_A - pH_B = 0.8$$

$$(6.1 + \log R_A) - (6.1 + \log R_B) = 0.8$$

$$\log R_A - \log R_B = 0.8$$

$$\log\frac{R_A}{R_B} = 0.8$$

$$\frac{R_A}{R_B} = 10^{0.8}$$

$$R_A = 10^{0.8} \cdot R_B \approx 6.3R_B$$

Blood sample A has about 6.3 times the bicarbonate/carbonic acid ratio as blood sample B.

Think and Communicate

1. Rounded to four decimal places, log 48 – 3 log 2 + log 5 = 1.4771; log 30 = 1.4771; Yes.

Teach⇔Interact

Section Note

Teaching Tip
In terms of energy release for an earthquake, as the Richter unit R increases one unit, the amount of energy released increases by a factor of 32. In terms of seismic wave amplitude, an increase of one unit in the Richter scale increases the intensity of the wave by a factor of 10.

Think and Communicate

For question 3, students should be able to redo Example 3 by starting with the equation $M_A - M_B = 2$.

Checking Key Concepts

Teaching Tip
Encourage students to check their answers to questions 5–7 by substituting 10 for each base and using a scientific calculator to verify the solutions.

Closure Question

State the product, quotient, and power properties of logarithms.
$\log_b MN = \log_b M + \log_b N$;
$\log_b \frac{M}{N} = \log_b M - \log_b N$;
$\log_b M^k = k \log_b M$

Apply⇔Assess

Suggested Assignment

Core Course
Day 1 Exs. 1–17, 20–25
Day 2 Exs. 27, 29–31, 34–37, AYP

Extended Course
Day 1 Exs. 1–25
Day 2 Exs. 26–37, AYP

Block Schedule
Day 22 Exs. 1–17, 20–25
Day 23 Exs. 27, 29–31, 34–37, AYP

The Richter scale is an example of a *logarithmic scale*. In general, *adding* 1 on a logarithmic scale corresponds to *multiplying* a related quantity (such as energy) by some number. Logarithmic scales are also used to measure acidity and loudness. (See Exercises 18, 19, and 26–28.)

THINK AND COMMUNICATE

2. Summarize the result in Example 3 by completing this statement: *Adding* 1 on the Richter scale corresponds to *multiplying* the energy by __?__.

3. In 1982, an earthquake measuring 6.0 on the Richter scale struck Yemen. In 1993, an earthquake measuring 8.0 on the Richter scale struck Guam. Compare the amounts of energy released by the two earthquakes.

CHECKING KEY CONCEPTS

Write each expression in terms of $\log_2 p$, $\log_2 q$, and $\log_2 r$.

1. $\log_2 p^3$ **2.** $\log_2 pq^5$ **3.** $\log_2 \frac{p^2}{r^4}$ **4.** $\log_2 \frac{p^7q^{10}}{r^{3/8}}$

Write as a logarithm of a single number or expression.

5. $5 \log_a 2$ **6.** $\log_b \frac{1}{8} + 3 \log_b 4$

7. $9 \log_a x - 2 \log_a y$ **8.** $2 \ln 3 + \ln 6 - \frac{3}{4} \ln 81$

4.3 Exercises and Applications

Extra Practice *exercises on page 754*

Write each expression in terms of $\log_7 p$, $\log_7 q$, and $\log_7 r$.

1. $\log_7 p^4$ **2.** $\log_7 q^{1/7}$ **3.** $\log_7 q^2r^3$ **4.** $\log_7 \frac{p^{2/5}}{q^8}$

5. $\log_7 \frac{1}{q}$ **6.** $\log_7 p^6q^3r^{1/4}$ **7.** $\log_7 \frac{p^9q^{11}}{r^7}$ **8.** $\log_7 \frac{r^2}{pq^{1/3}}$

9. Writing Tamara expressed the product property of logarithms in words to help her understand what it means. Express the quotient and power properties in words.

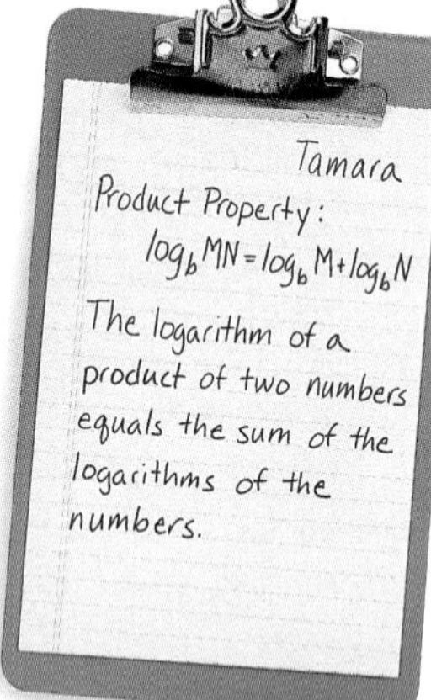
Tamara
Product Property:
$\log_b MN = \log_b M + \log_b N$
The logarithm of a product of two numbers equals the sum of the logarithms of the numbers.

Write as a logarithm of a single number or expression.

10. $4 \log_a 3$ **11.** $-\frac{1}{2} \log_a 49$

12. $\log_3 p + 2 \log_3 q$ **13.** $6 \log_5 u - 10 \log_5 v$

14. $2 \log_b 10 - \frac{2}{3} \log_b 125$ **15.** $12 \log x^2 + \frac{4}{5} \log x^{10}$

16. $3 \ln p - \ln q - 7 \ln r$ **17.** $\frac{1}{3} \log_b 27 + 2 \log_b 6 - \frac{1}{2} \log_b 144$

Think and Communicate

2. about 32
3. The Guam earthquake released 1000 times as much energy as the Yemen earthquake.

Checking Key Concepts

1. $3 \log_2 p$
2. $\log_2 p + 5 \log_2 q$
3. $2 \log_2 p - 4 \log_2 r$
4. $7 \log_2 p + 10 \log_2 q - \frac{3}{8} \log_2 r$
5. $\log_a 32$
6. $\log_b 8$
7. $\log_a \frac{x^9}{y^2}$
8. $\ln 2$

Exercises and Applications

1. $4 \log_7 p$
2. $\frac{1}{7} \log_7 q$
3. $2 \log_7 q + 3 \log_7 r$
4. $\frac{2}{5} \log_7 p - 8 \log_7 q$
5. $-\log_7 q$
6. $6 \log_7 p + 3 \log_7 q + \frac{1}{4} \log_7 r$
7. $9 \log_7 p + 11 \log_7 q - 7 \log_7 r$
8. $2 \log_7 r - \log_7 p - \frac{1}{3} \log_7 q$
9. Answers may vary. An example is given. The logarithm of a quotient of two numbers equals the difference of the logarithms of the two numbers. The logarithm of a power of a number is the product of the exponent and the logarithm of the number.
10. $\log_a 81$
11. $\log_a \frac{1}{7}$

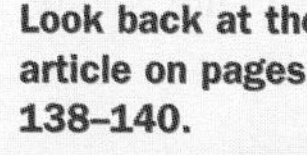

Ednaly Ortiz

Look back at the article on pages 138–140.

Archaeologists like Ednaly Ortiz use the pH scale to rate the acidity of soils. The pH of a soil is given by the equation

$$\text{pH} = -\log\,[\text{H}^+]$$

where $[H^+]$ is the soil's hydrogen ion concentration in moles per liter. The greater a soil's pH, the more alkaline (or less acidic) the soil is.

An *ion* is an electrically charged atom or group of atoms. A *mole* of ions contains about 6.02×10^{23} ions.

18. The diagram shows the different soil layers found in one of Ortiz's excavation pits, as well as the hydrogen ion concentration in each layer.

a. Find the pH of each soil layer.

b. Ortiz found that the presence of human artifacts increases a soil's alkalinity. Which of the soil layers in the diagram do you think contained the most artifacts? Explain.

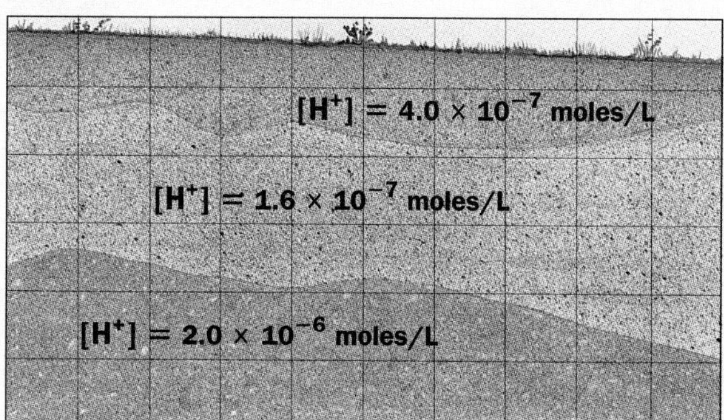

19. a. Suppose the difference in pH of two soils, A and B, is 1. Compare the hydrogen ion concentrations in the soils.

b. Writing Explain why the pH scale is a logarithmic scale.

20. EARTH SCIENCE Look back at Example 3.

a. On June 9, 1994, a powerful earthquake struck La Paz, Bolivia. The earthquake released about $10^{24.1}$ ergs of energy. Find the Richter magnitude of the earthquake to the nearest tenth.

b. On March 27, 1964, an earthquake in Alaska released about twice as much energy as the La Paz earthquake. Find the Richter magnitude of the Alaska earthquake to the nearest tenth. Compare this magnitude with the magnitude of the La Paz earthquake.

21. Open-ended Problem For each statement, find positive numbers M, N, and b (with $b \neq 1$) that show that the statement is false in general.

a. $\log_b (M + N) = \log_b M + \log_b N$

b. $\log_b (M - N) = \log_b M - \log_b N$

Let $x = \log_a 5$ and $y = \log_a 10$. Write each expression in terms of x and y.

22. $\log_a 50$ **23.** $\log_a 2$ **24.** $\log_a 100$ **25.** $\log_a 250$

Apply⇔Assess

Exercise Notes

Common Error
Exs. 4, 5, 7, 8 Some students may apply the quotient property incorrectly as $\log_b \frac{M}{N} = \log_b (M - N)$. Have them study Example 1 on page 156 carefully.

Second-Language Learners
Ex. 9 Students learning English may benefit from working with a peer tutor or aide.

Using Technology
Exs. 10, 11, 14, 15, 17 Students can verify their solutions to these exercises by substituting 10 for the base and using a scientific calculator.

Interview Note
Exs. 18, 19 Students who have not studied chemistry may benefit from working in small groups with students who have.

Application
Ex. 19 Students from families interested in gardening may have used a pH test kit at one time or another to test the pH of their soil. Ask these students if they can show the kit to the class and explain how it is used.

Multicultural Note
Ex. 20 Situated 12,000 feet above sea level, La Paz, the capital of Bolivia, is the highest capital city in the world. La Paz lies on the slopes and at the bottom of a canyon, and is made up of a mixture of traditional adobe houses and modern high-rise buildings. Half of the population is made up of Aymara Indians, while the other half consists of people of either European ancestry or mixed European and Indian ancestry. City life centers on the *Plaza Murillo*, a garden-filled plaza that faces the presidential palace, a national museum, and a major cathedral. Most of the business conducted in La Paz is of the import/export nature and includes canned foods, glass, and textiles.

Problem Solving
Ex. 21 Open-ended problems such as this one provide students with an opportunity to explore mathematics. Have some students describe to the class how they proved the statements false.

12. $\log_3 pq^2$ 13. $\log_5 \frac{u^6}{v^{10}}$

14. $\log_b 4$ 15. $\log x^{32}$

16. $\ln \frac{p^3}{qr^7}$ 17. $\log_b 9$

18. a. layer 1: 6.4; layer 2: 6.8; layer 3: 5.7

b. layer 2: Since the presence of human artifacts increases a soil's alkalinity, the layer with the greatest alkalinity (that is, the highest pH) should have the most artifacts.

19. a. The hydrogen ion concentration in the soil with the higher pH is 0.1 times that of the soil with the lower pH.

b. Adding 1 on the pH scale corresponds to multiplying the hydrogen ion concentration by 0.1.

20. a. 8.2

b. 8.4; The magnitude is 0.2 greater than that of the La Paz earthquake.

21. Answers may vary. Examples are given.

a. Let $b = 10$, $M = 6$, and $N = 4$; $\log_b (M + N) = \log 10 = 1$, but $\log_b M + \log_b N \approx 1.3802$.

b. Let $b = 10$, $M = 11$, and $N = 1$; $\log_b (M - N) = \log 10 = 1$, but $\log_b M - \log_b N \approx 1.0414$.

22. $x + y$ 23. $y - x$

24. $2y$ 25. $2x + y$

Apply⇔Assess

Exercise Notes

Interdisciplinary Problems
Exs. 26–28 The study of acoustics, a branch of physics, involves many ideas that can be modeled by logarithmic functions. These exercises discuss the intensities of some common sounds.

Reasoning
Ex. 28 Ask students to explain why doubling the intensity of sound does not double the number of decibels. Have them also explain their answers to part (b). For example, if a speaker has a dB level of 80, two speakers have an dB level of 83.

Mathematical Procedures
Ex. 30 This exercise illustrates the need to pay attention to the domain of a function. When $x > 0$, $\log_b x^2 = 2 \log_b x$. The logarithm function $f(x) = 2 \log_b x$ is not defined for negative numbers.

Historical Connection
Exs. 31–33 It is an interesting historical fact that although logarithms are now universally regarded as exponents, they were, in fact, discovered before exponents were in use. At first, in the early seventeenth century, logarithms were not defined as exponents because fractional and irrational exponents were not in use. In 1742, William Jones gave the first systematic approach of logarithms in terms of exponents.

Cooperative Learning
Exs. 32, 33 You may wish to have students go over their proofs for these exercises by working together in groups of two or three.

Connection ACOUSTICS

The loudness of a sound is measured on the *decibel scale* and depends on the sound's *intensity*, or power per unit area. The loudness L and intensity I are related by the equation $L = 10 \log \frac{I}{I_0}$ where L is measured in decibels (dB), I is measured in watts per square meter (W/m²), and $I_0 = 10^{-12}$ W/m² is the intensity of a barely audible sound.

Sound	Intensity (W/m²)
Falling pin	10^{-11}
Quiet conversation	10^{-6}
Subway train	10^{-3}
Jet at takeoff	1

An acoustical engineer studies the noise created by an industrial machine.

26. The table shows the intensities of some common sounds. Find the loudness of each sound.

27. a. By what percent must the intensity of a sound increase in order for the loudness to be raised by 1 dB?

b. Writing Explain why the decibel scale is a logarithmic scale.

28. a. Suppose the intensity of sound from a stereo speaker is doubled. By how many decibels does the loudness increase?

b. Challenge Suppose the intensity of sound from a speaker is increased by a factor of n. By how many decibels does the loudness increase? (Your answer will involve n.)

29. SAT/ACT Preview Which of the following does $2 \log_3 6 - \log_3 4$ equal?

A. 2 **B.** 3 **C.** 9 **D.** 48 **E.** none of these

30. **Technology** Graph $y = \log x^2$ and $y = 2 \log x$ using a graphing calculator or graphing software. Are the graphs of the two functions the same? If not, why not? Do the graphs contradict the power property of logarithms? Explain.

31. Antonio proved the product property of logarithms as shown. Justify each step of Antonio's proof.

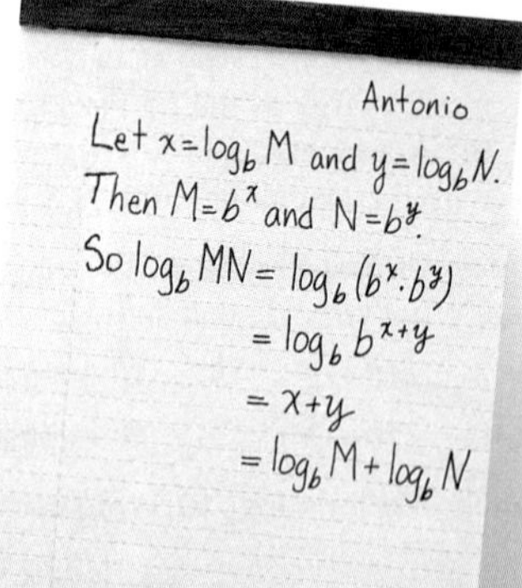
Antonio
Let $x = \log_b M$ and $y = \log_b N$.
Then $M = b^x$ and $N = b^y$.
So $\log_b MN = \log_b (b^x \cdot b^y)$
$= \log_b b^{x+y}$
$= x+y$
$= \log_b M + \log_b N$

32. Prove the power property: $\log_b M^k = k \log_b M$. (*Hint:* Let $x = \log_b M$. Then $M = b^x$, so that $\log_b M^k = \log_b (b^x)^k$.)

33. Use the product and power properties to prove the quotient property:

$$\log_b \frac{M}{N} = \log_b M - \log_b N.$$

$\left(\textit{Hint: } \text{Write } \frac{M}{N} \text{ as } MN^{-1}.\right)$

26. falling pin: 10 dB; quiet conversation: 60 dB; subway train: 90 dB; jet at takeoff: 120 dB

27. a. about 26%

b. Yes; adding 1 on the decibel scale corresponds to multiplying loudness by about 1.26.

28. a. about 3 dB

b. $10 \log n$ dB

29. A

30. $y = \log x^2$

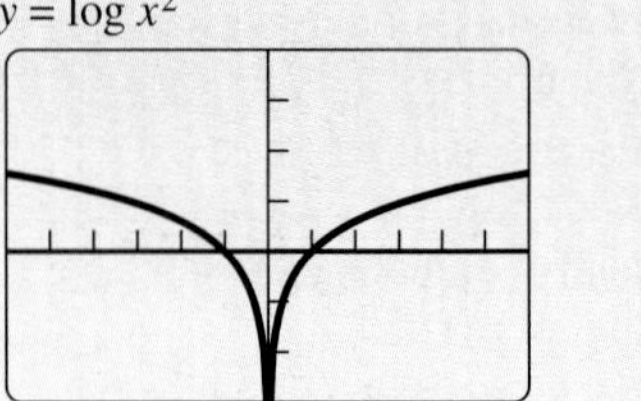

$y = 2 \log x$

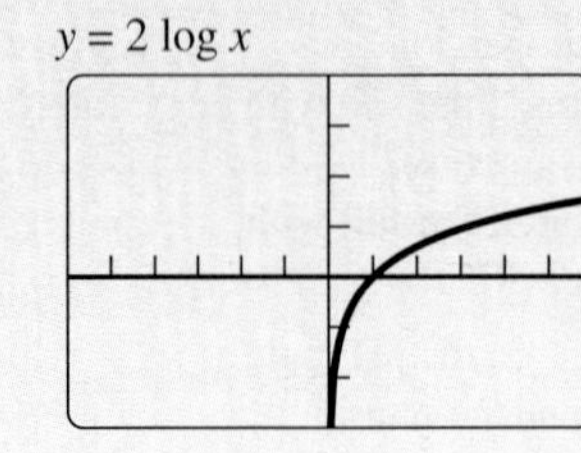

No; the functions do not have the same domain. The domain of $y = \log x^2$ is all real numbers except 0. The domain of $y = 2 \log x$ is all positive real numbers. The graphs are the same over the common part of the domains. This does not contradict the power property of logarithms because the property applies only to positive numbers M.

31. substitution; the product property of exponents; definition of logarithm; substitution

32. Let $x = \log_b M$. Then $M = b^x$ and $\log_b M^k = \log_b (b^x)^k = \log_b b^{xk} = xk = kx = k \log_b M$.

33. $\log_b \frac{M}{N} = \log_b MN^{-1} = \log_b M + \log_b N^{-1} = \log_b M + (-1) \log_b N = \log_b M - \log_b N$

34. Since $\log E = 11.8 + 1.5M$, $\log E = \log 10^{11.8} + \frac{3}{2}M$. Then $\frac{3}{2}M = \log E - \log 10^{11.8} = \log \frac{E}{10^{11.8}}$, so $M = \frac{2}{3} \log \frac{E}{10^{11.8}}$.

ONGOING ASSESSMENT

34. Writing An encyclopedia gives the equation relating the energy and Richter magnitude of an earthquake in this form:

$$\log E = 11.8 + 1.5M$$

Show that this equation is equivalent to the equation in Example 3. (*Hint:* Note that $11.8 = \log 10^{11.8}$.)

SPIRAL REVIEW

35. Write $4^2 = 16$ in logarithmic form. *(Section 4.2)*

36. Write $\log 1000 = 3$ in exponential form. *(Section 4.2)*

37. **Technology** Use a graphing calculator or graphing software to find the solution of $30e^{0.1x} = 75$ to the nearest hundredth. *(Section 3.4)*

ASSESS YOUR PROGRESS

VOCABULARY

inverse of a function (p. 142)
logarithmic function (p. 149)
base-*b* logarithm (p. 149)
common logarithm (p. 150)
natural logarithm (p. 150)

For each function:

a. Graph the function and its inverse in the same coordinate plane.

b. Find an equation for the inverse. *(Section 4.1)*

1. $f(x) = 2x$ **2.** $g(x) = 3x + 2$ **3.** $h(x) = \frac{1}{4}x - 1$ **4.** $y = -6x - 5$

Evaluate each logarithm. *(Section 4.2)*

5. $\log_2 4$ **6.** $\log_4 2$ **7.** $\log_3 \frac{1}{81}$ **8.** $\ln e^{5/11}$

9. POPULATION The population P of Argentina (in millions) can be modeled by the function $P = 33.9e^{0.0109t}$ where t is the number of years since 1994. *(Section 4.2)*

a. Find the inverse of the given function.

b. Predict the year in which Argentina's population will reach 35 million.

10. Write $\log \frac{p^3q^4}{r^5}$ in terms of $\log p$, $\log q$, and $\log r$. *(Section 4.3)*

11. Write $\log_b 18 - 2\log_b 3 + \log_b 4$ as a logarithm of a single number. *(Section 4.3)*

12. Journal Explain what a logarithmic scale is. Give two examples of a logarithmic scale used in real life.

Apply⇔Assess

Exercise Notes

Assessment Note

Ex. 34 Having three or four students show their calculations to the class at the board would provide useful feedback for all students to evaluate what they have written.

Topic Spiraling: Preview

Ex. 37 This exercise foreshadows solving exponential equations in the next section.

Assess Your Progress

Journal Entry

Giving examples of logarithmic scales used in real life will be more helpful if students can explain how the scale is used.

Progress Check 4.1–4.3

See page 178.

Practice 22 for Section 4.3

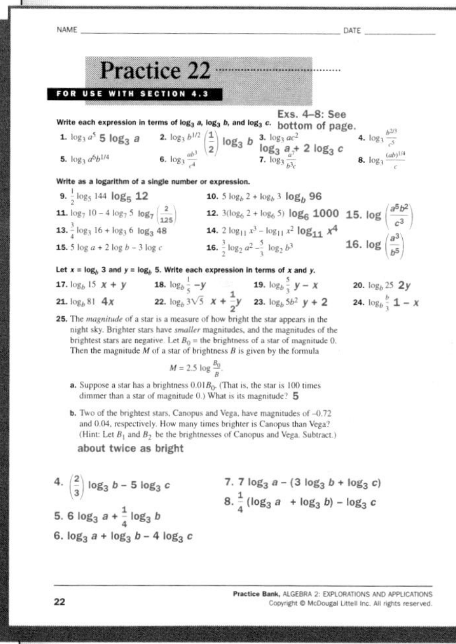
NAME ______ DATE ______

Practice 22

FOR USE WITH SECTION 4.3

Write each expression in terms of $\log_3 a$, $\log_3 b$, and $\log_3 c$. Exs. 4–8: See bottom of page.

1. $\log_3 a^5$ $5\log_3 a$
2. $\log_3 b^{1/2}$ $\left(\frac{1}{2}\right)\log_3 b$
3. $\log_3 ac^2$ $\log_3 a + 2\log_3 c$
4. $\log_3 \frac{b^{2/3}}{c^5}$
5. $\log_3 a^6b^{1/4}$
6. $\log_3 \frac{ab^3}{c^4}$
7. $\log_3 \frac{a^7}{b^3c}$
8. $\log_3 \frac{(ab)^{1/4}}{c}$

Write as a logarithm of a single number or expression.

9. $\frac{1}{2}\log_5 144$ $\log_5 12$
10. $5\log_b 2 + \log_b 3$ $\log_b 96$
11. $\log_7 10 - 4\log_7 5$ $\log_7\left(\frac{2}{125}\right)$
12. $3(\log_6 2 + \log_6 5)$ $\log_6 1000$
13. $\frac{3}{4}\log_3 16 + \log_3 6$ $\log_3 48$
14. $2\log_{11} x^3 - \log_{11} x^2$ $\log_{11} x^4$
15. $5\log a + 2\log b - 3\log c$
16. $\frac{3}{2}\log_2 a^2 - \frac{5}{3}\log_2 b^3$

15. $\log\left(\frac{a^5b^2}{c^3}\right)$
16. $\log\left(\frac{a^3}{b^5}\right)$

Let $x = \log_b 3$ and $y = \log_b 5$. Write each expression in terms of x and y.

17. $\log_b 15$ $x + y$
18. $\log_b \frac{1}{5}$ $-y$
19. $\log_b \frac{5}{3}$ $y - x$
20. $\log_b 25$ $2y$
21. $\log_b 81$ $4x$
22. $\log_b 3\sqrt{5}$ $x + \frac{1}{2}y$
23. $\log_b 5b^2$ $y + 2$
24. $\log_b \frac{b}{3}$ $1 - x$

25. The *magnitude* of a star is a measure of how bright the star appears in the night sky. Brighter stars have *smaller* magnitudes, and the magnitudes of the brightest stars are negative. Let B_0 = the brightness of a star of magnitude 0. Then the magnitude M of a star of brightness B is given by the formula

$$M = 2.5 \log \frac{B_0}{B}.$$

a. Suppose a star has a brightness $0.01B_0$. (That is, the star is 100 times dimmer than a star of magnitude 0.) What is its magnitude? 5

b. Two of the brightest stars, Canopus and Vega, have magnitudes of -0.72 and 0.04, respectively. How many times brighter is Canopus than Vega? (Hint: Let B_1 and B_2 be the brightnesses of Canopus and Vega. Subtract.)
about twice as bright

4. $\left(\frac{2}{3}\right)\log_3 b - 5\log_3 c$
5. $6\log_3 a + \frac{1}{4}\log_3 b$
6. $\log_3 a + \log_3 b - 4\log_3 c$
7. $7\log_3 a - (3\log_3 b + \log_3 c)$
8. $\frac{1}{4}(\log_3 a + \log_3 b) - \log_3 c$

22 Practice Bank, ALGEBRA 2: EXPLORATIONS AND APPLICATIONS

35. $\log_4 16 = 2$ 36. $10^3 = 1000$

37. 9.16

Assess Your Progress

1. a.

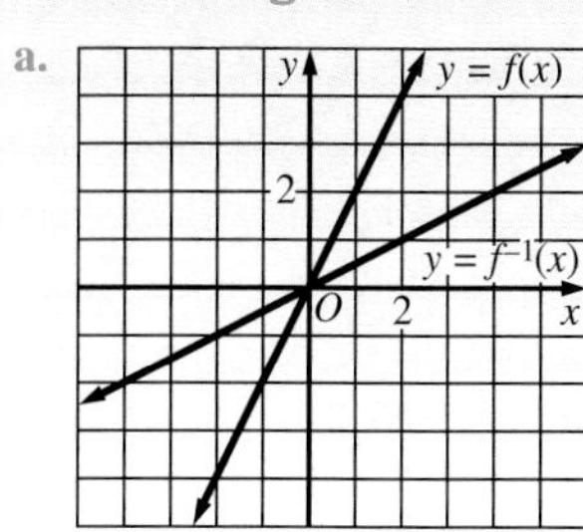

b. $f^{-1}(x) = \frac{1}{2}x$

2. a.

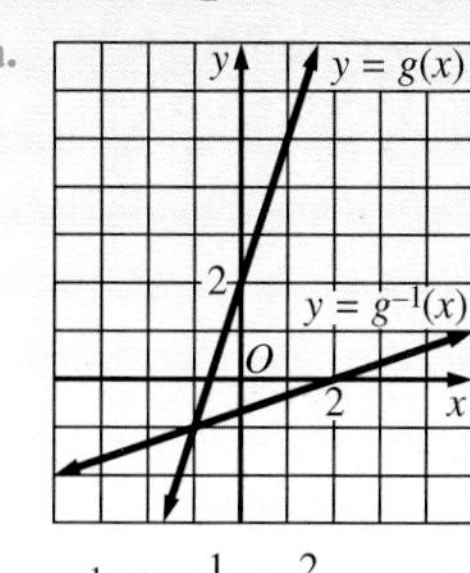

b. $g^{-1}(x) = \frac{1}{3}x - \frac{2}{3}$

3. a.

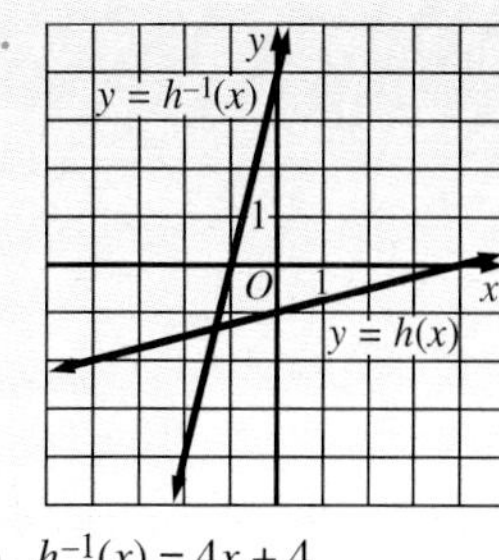

b. $h^{-1}(x) = 4x + 4$

4. a.

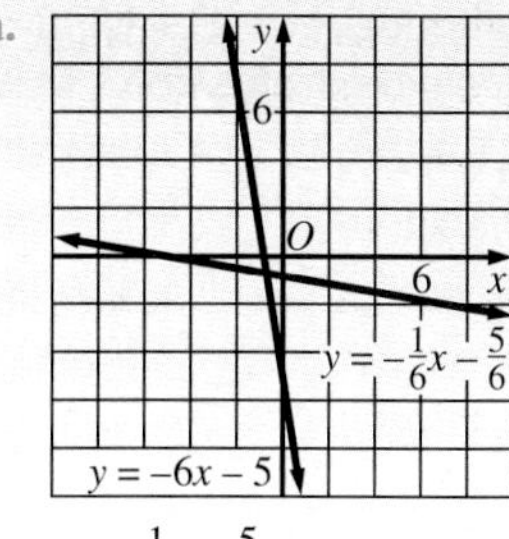

b. $y = -\frac{1}{6}x - \frac{5}{6}$

5–12. See answers in back of book.

Plan⇔Support

Objectives

- Solve exponential and logarithmic equations.
- Solve real-world applications of exponential and logarithmic equations.

Recommended Pacing

Core and Extended Courses
Section 4.4: 1 day

Block Schedule
Section 4.4: $\frac{1}{2}$ block
(with Section 4.3)

Resource Materials

Lesson Support
Lesson Plan 4.4
Warm-Up Transparency 4.4
Practice Bank: Practice 23
Study Guide: Section 4.4

Technology
Technology Book:
Calculator Activity 4
Spreadsheet Activity 4
TI-92 Activity 2
Graphing Calculator
McDougal Littell Mathpack
Function Investigator
Internet:
http://www.hmco.com

Warm-Up Exercises

Solve each of the following equations.

1. $2^x = 32$ 5
2. $\log 1000 = x$ 3
3. $\ln e^3 = y$ 3
4. Solve for h: $P = 48e^{-h/200}$. $h = -200 \ln \frac{P}{48}$
5. Rewrite $\log_b (x + 1) - \log_b (x - 2)$ as a single expression. $\log_b \frac{x+1}{x-2}$

Section

4.4 Exponential and Logarithmic Equations

Learn how to...

- **solve exponential and logarithmic equations**

So you can...

- **find how long it takes fudge to cool, for example**

The notecard below shows a recipe for fudge. Notice that before adding powdered milk and vanilla, you have to let the fudge mixture cool.

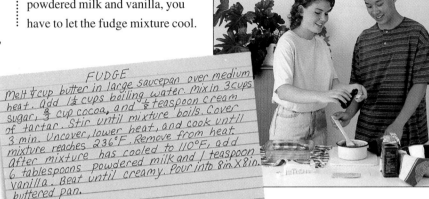

FUDGE
Melt $\frac{1}{4}$ cup butter in large saucepan over medium heat. Add $1\frac{1}{2}$ cups boiling water. Mix in 3 cups sugar, $\frac{2}{3}$ cup cocoa, and $\frac{1}{8}$ teaspoon cream of tartar. Stir until mixture boils. Cover 3 min. Uncover, lower heat, and cook until mixture reaches 236°F. Remove from heat. After mixture has cooled to 110°F, add 6 tablespoons powdered milk and 1 teaspoon vanilla. Beat until creamy. Pour into 8in.X8in. buttered pan.

EXAMPLE 1 Application: Cooking

The equation $T = 164e^{-0.041t} + 72$ gives the temperature T of a fudge mixture t minutes after it begins cooling from 236°F. How long does it take for the mixture to cool to 110°F?

SOLUTION

Find t when $T = 110$.

$$164e^{-0.041t} + 72 = \mathbf{110}$$

Substitute **110** for T in the given equation.

$$164e^{-0.041t} = 38$$

$$e^{-0.041t} \approx 0.232$$

$$-0.041t \approx \ln 0.232$$

Write the equation in logarithmic form.

$$t \approx \frac{\ln 0.232}{-0.041}$$

$$t \approx 35.6$$

It takes about 36 min for the mixture to cool from 236°F to 110°F.

THINK AND COMMUNICATE

1. **Technology** Use a graphing calculator or graphing software to graph $y = 164e^{-0.041x} + 72$.
   a. Check the solution of Example 1 by using the trace or intersection feature to find x when $y = 110$.
   b. To what temperature will the fudge mixture eventually cool? How do you know?

EXAMPLE 2

Solve $3^x = 20$.

SOLUTION

$$3^x = 20$$
$$\log 3^x = \log 20$$
$$x \log 3 = \log 20$$
$$x = \frac{\log 20}{\log 3}$$
$$x \approx 2.727$$

Take the common logarithm of both sides.

Use the power property.

To the nearest hundredth, the solution is 2.73.

THINK AND COMMUNICATE

2. Use natural logarithms to solve the equation in Example 2.
3. Compare the steps you would use to find decimal solutions of $e^x = 5$ and $2^x = 5$. Which equation is easier to solve? Why?

In Example 2, the solution of $3^x = 20$ was expressed as a quotient of common logarithms:

$$x = \frac{\log 20}{\log 3}$$

You can also express the solution as a single base-3 logarithm by writing $3^x = 20$ in logarithmic form:

$$x = \log_3 20$$

So $\log_3 20 = \frac{\log 20}{\log 3}$. This illustrates the *change-of-base formula*.

Change-of-Base Formula

Let M, b, and c be positive numbers with $b \neq 1$ and $c \neq 1$. Then:

$$\log_b M = \frac{\log_c M}{\log_c b}$$

Teach⇔Interact

Additional Example 1

The equation $N = Pe^{0.0462t}$ gives the number N of e. coli bacteria present after P bacteria have reproduced by mitosis for t minutes. How long does it take for 100 bacteria to grow to 1000?

Find t when $P = 100$ and $N = 1000$.

$$1000 = 100\, e^{0.0462t}$$
$$10 = e^{0.0462t}$$
$$\ln 10 = 0.0462t$$
$$t = \frac{\ln 10}{0.0462}$$
$$t \approx 49.8$$

It takes about 50 minutes for 100 e. coli bacteria to grow to 1000.

Think and Communicate

To answer part (b) of question 1, students need to make the connection between the graph of the exponential function and the asymptote at $y = 72$.

About Example 2

Mathematical Procedure
When solving exponential equations, taking either the common logarithm or the natural logarithm of both sides is a useful technique when the base of the exponential equation is a number other than 10 or *e*.

Additional Example 2

Solve $\left(\frac{1}{2}\right)^x = 15$.

$$\left(\frac{1}{2}\right)^x = 15$$
$$\log \left(\frac{1}{2}\right)^x = \log 15$$
$$x \log \left(\frac{1}{2}\right) = \log 15$$
$$x = \frac{\log 15}{\log \left(\frac{1}{2}\right)}$$
$$x \approx -3.907$$

To the nearest hundredth, the solution is –3.91.

Section Note

Teaching Tip
The change-of-base formula is useful in evaluating logarithms and solving exponential equations in other bases.

ANSWERS Section 4.4

Think and Communicate

1. a.

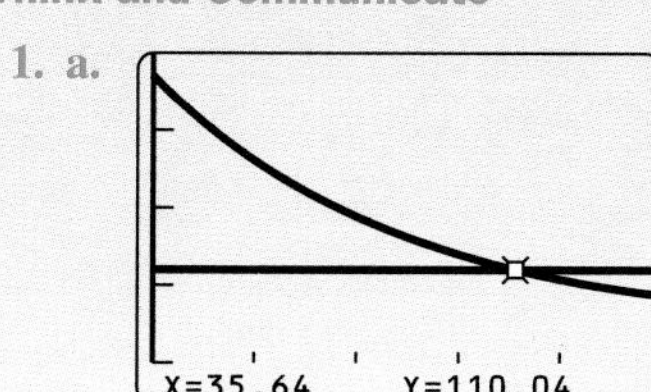

b. 72°; When x is very large, $e^{-0.041x}$ is very close to 0, so $164e^{-0.041x} + 72$ is very close to 72.

2. $\ln 3^x = \ln 20$; $x \ln 3 = \ln 20$; $x = \frac{\ln 20}{\ln 3} \approx 2.73$

3. To find a decimal solution for $e^x = 5$, you would take the natural logarithms of both sides. Since $\ln e^x = x$, the solution is simply $x = \ln 5 \approx 1.61$. To find a decimal solution for $2^x = 5$, you would take either the common logarithms or the natural logarithms of both sides. For example, $\ln 2^x = \ln 5$; $x \ln 2 = \ln 5$; $x = \frac{\ln 5}{\ln 2} \approx 2.32$. The first equation is easier to solve because the arithmetic is simpler.

Teach⇔Interact

About Example 3

Teaching Tip

The technology method of Example 3 has students use the change-of-base formula to express the first equation in terms of common logarithms. Students do not have to use the properties of logarithms to simplify expressions.

Additional Example 3

Solve $\log_3(x+1) + \log_3(x-1) = 2$.

Method 1 Use a graphing calculator or graphing software.

Graph $y = \log_3(x+1) + \log_3(x-1)$ and $y = 2$ in the same coordinate plane. Use the change-of-base formula to express the first equation in terms of common logarithms when entering it into the calculator or software.

$$y = \frac{\log(x+1)}{\log 3} + \frac{\log(x-1)}{\log 3}$$

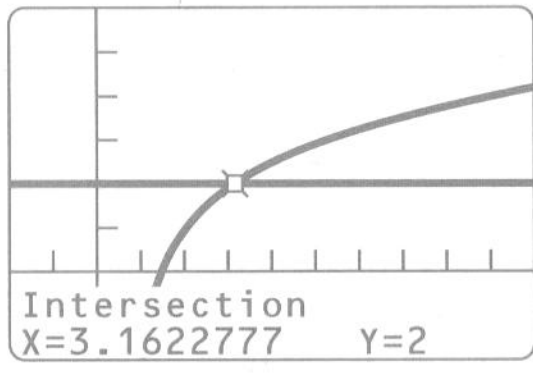

Find the x-coordinate of the point where the graphs intersect.

The solution is about 3.16.

Method 2 Use algebra.

$$\log_3(x+1) + \log_3(x-1) = 2$$
$$\log_3[(x+1)(x-1)] = 2$$
$$(x+1)(x-1) = 3^2$$
$$x^2 - 1 = 9$$
$$x^2 = 10$$
$$x = \pm\sqrt{10}$$

Check $x = \sqrt{10}$:

$$\log_3(\sqrt{10}+1) + \log_3(\sqrt{10}-1) = 2$$
$$\log_3[(\sqrt{10}+1)(\sqrt{10}-1)] = 2$$
$$(\sqrt{10}+1)(\sqrt{10}-1) = 3^2$$
$$10 - 1 = 9$$
$$9 = 9 \checkmark$$

Check $x = -\sqrt{10}$:

$$\log_3(-\sqrt{10}+1) + \log_3(-\sqrt{10}-1) = 2$$

Since negative numbers are not in the domain of the logarithmic function, $-\sqrt{10}$ cannot be a solution. The solution is $\sqrt{10}$, or about 3.16.

Section Note

Teaching Tip

Stress the importance of checking answers when solving a logarithmic equation. The domain restrictions on the logarithmic function mean that the function is not defined for negative values of x.

You can use the change-of-base formula and a graphing calculator to evaluate any logarithm or graph any logarithmic function.

To evaluate $\log_2 6$, calculate the quotient $\frac{\log 6}{\log 2}$ or $\frac{\ln 6}{\ln 2}$.

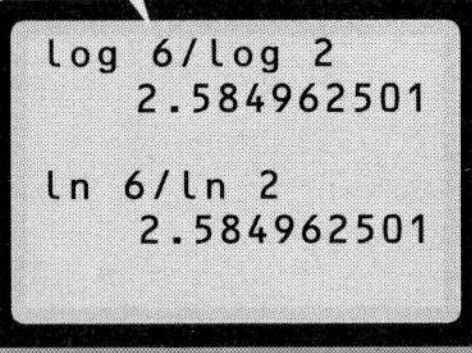

To graph $y = \log_2 x$, enter the equation $y = \frac{\log x}{\log 2}$ or $y = \frac{\ln x}{\ln 2}$.

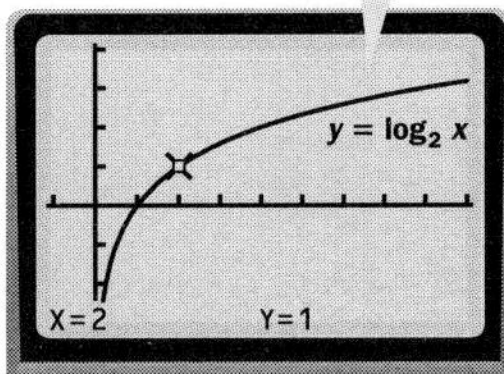

EXAMPLE 3

Solve $\log_2(5x+4) - \log_2(x-1) = 3$.

SOLUTION

Method 1

Use a graphing calculator or graphing software.

Graph $y = \log_2(5x+4) - \log_2(x-1)$ and $y = 3$ in the same coordinate plane. Use the change-of-base formula to express the first equation in terms of common logarithms when entering it into the calculator or software:

$$y = \frac{\log(5x+4)}{\log 2} - \frac{\log(x-1)}{\log 2}$$

Find the x-coordinate of the point where the graphs intersect.

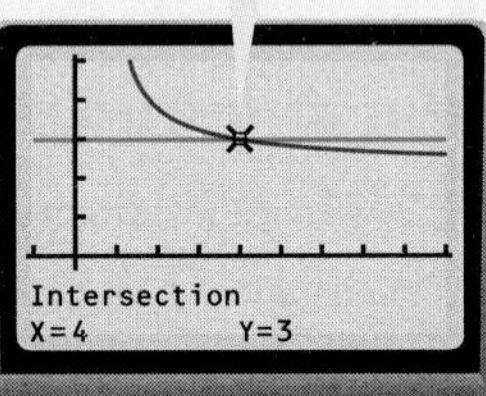

The solution is 4.

Method 2

Use algebra.

$$\log_2(5x+4) - \log_2(x-1) = 3$$

Use the quotient property.

$$\log_2 \frac{5x+4}{x-1} = 3$$

Write the equation in exponential form.

$$\frac{5x+4}{x-1} = 2^3$$

Multiply both sides by $x - 1$.

$$5x + 4 = 8(x-1)$$
$$5x + 4 = 8x - 8$$
$$-3x + 4 = -8$$
$$-3x = -12$$
$$x = 4$$

The solution is 4.

Check

$$\log_2(5 \cdot 4 + 4) - \log_2(4-1) \stackrel{?}{=} 3$$
$$\log_2 24 - \log_2 3 \stackrel{?}{=} 3$$
$$\log_2 \frac{24}{3} \stackrel{?}{=} 3$$
$$\log_2 8 \stackrel{?}{=} 3$$
$$3 = 3 \checkmark$$

Technology Note

Students may find the x-coordinate of the point where two graphs intersect by using the zoom and trace features, by using the table features, or by using the intersect feature of the CALC menu. To use the table features of a TI-82 calculator, have students set TblMin = 0 and ΔTbl = 0.1 under the TblSet menu. Choose TABLE and scroll through the Y1 and Y2 lists until the values both equal 2.

When solving a logarithmic equation algebraically, it is possible to obtain a solution that does *not* satisfy the original equation. Such a solution is called an **extraneous solution**. Be sure to check your solutions to eliminate any that are extraneous.

THINK AND COMMUNICATE

4. Use Kim's work.

a. Substitute Kim's solution into the original equation. Explain why the solution is extraneous.

b. **Technology** Use a graphing calculator or graphing software to graph $y = \log_2 (x - 3) - \log_2 (x - 2)$ and $y = 1$. How do the graphs show that Kim's equation has no solution?

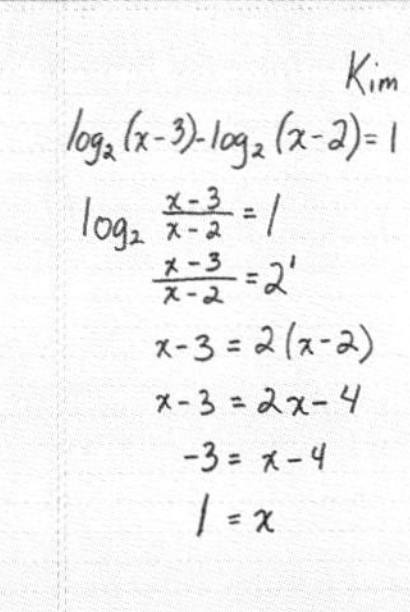
Kim
$\log_2 (x-3) - \log_2 (x-2) = 1$
$\log_2 \frac{x-3}{x-2} = 1$
$\frac{x-3}{x-2} = 2^1$
$x-3 = 2(x-2)$
$x-3 = 2x-4$
$-3 = x-4$
$1 = x$

CHECKING KEY CONCEPTS

Solve each equation. Round your answers to the nearest hundredth.

1. $e^x = 3$ **2.** $2^y + 7 = 60$ **3.** $5^t = 4$ **4.** $9 \cdot 10^{-0.2x} = 2$

Evaluate each logarithm. Round your answers to the nearest hundredth.

5. $\log_2 5$ **6.** $\log_6 4$ **7.** $\log_8 14.2$ **8.** $\log_3 \frac{7}{5}$

Solve each equation. Be sure to check your solutions.

9. $\log_2 x = 5$ **10.** $\log_4 (11x + 20) = 3$

11. $\log_2 u - \log_2 (u - 6) = 2$ **12.** $\log_3 (5y + 2) - \log_3 (y - 4) = 1$

4.4 Exercises and Applications

Extra Practice exercises on page 754

Solve each equation. Round your answers to the nearest hundredth.

1. $e^x = 7$ **2.** $2e^{-0.3t} = 1.6$ **3.** $10^{4y} = 5$ **4.** $3^x = 8$

5. $\left(\frac{1}{2}\right)^x = \frac{3}{4}$ **6.** $6e^k - 13 = 29$ **7.** $1 + 8(0.43)^{2u} = 3$ **8.** $4 \cdot 7^w = 3 \cdot 2^w$

9. Open-ended Problem Write an exponential equation. Exchange equations with a classmate, and solve your classmate's equation.

10. Look back at Example 1. How long does it take for the fudge mixture to cool to 150°F?

11. Challenge Find the solution of $ab^x = cd^x$ in terms of a, b, c, and d.

Evaluate each logarithm. Round your answers to the nearest hundredth.

12. $\log_2 9$ **13.** $\log_9 2$ **14.** $\log_5 7$ **15.** $\log_{16} 75$

16. $\log_7 \frac{1}{8}$ **17.** $\log_{1/3} 6$ **18.** $\log_4 0.93$ **19.** $\log_{0.1} 0.4$

Teach⇔Interact

Think and Communicate

For part (a) of question 4, you may suggest that students explain why the solution is extraneous in terms of the definition of the logarithmic function.

Closure Question

Explain how to solve a logarithmic equation. Using algebra, simplify expressions using the properties of logarithms. Then write the equation in exponential form and solve. Using technology, use the change-of-base formula, if necessary, to graph each side of the equation in the same coordinate plane. Find the x-coordinate of the point where the graphs intersect.

Apply⇔Assess

Suggested Assignment

Core Course
Exs. 1–8, 12–19, 22–37

Extended Course
Exs. 1–13, 15, 17, 19–37

Block Schedule
Day 23 Exs. 1–8, 12–19, 22–37

Exercise Notes

Teaching Tip
Exs. 6–8 Remind students that they should not write the equation in logarithmic form until the equation is in the form $b^x = \frac{y}{a}$.

Challenge
Ex. 11 Some students will answer in terms of common logarithms and others in terms of natural logarithms. Have students discuss whether the two forms of the answers are equivalent.

Mathematical Procedures
Exs. 12–19 These exercises exemplify the important use of logarithms to find any power of a number, including a fractional power.

Think and Communicate

4. a. If you try to substitute Kim's answer into the original equation, you get $\log_2 (1 - 3) - \log_2 (1 - 2)$, which is not defined.

b. 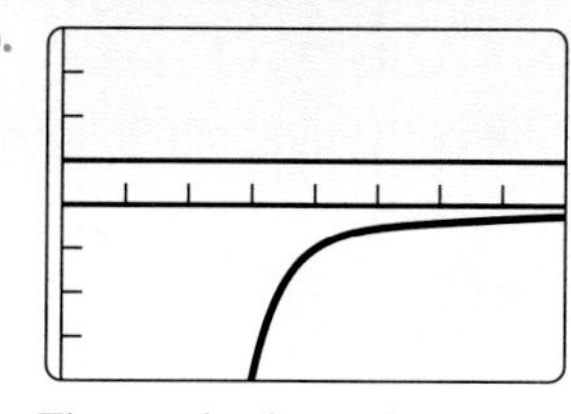

The graphs do not intersect.

Checking Key Concepts

1. 1.10 2. 5.73 3. 0.86 4. 3.27 5. 2.32 6. 0.77 7. 1.28 8. 0.31 9. 32 10. 4 11. 8 12. no solution

Exercises and Applications

1. 1.95 2. 0.74 3. 0.17 4. 1.89 5. 0.42 6. 1.95 7. 0.82 8. −0.23 9. Answers may vary. 10. about 18 min

11. $x = \frac{\log \frac{c}{a}}{\log \frac{b}{d}}$ or $x = \frac{\ln \frac{c}{a}}{\ln \frac{b}{d}}$

12. 3.17 13. 0.32 14. 1.21 15. 1.56 16. −1.07 17. −1.63 18. −0.05 19. 0.40

Apply⇔Assess

Exercise Notes

Application
Exs. 20, 21 This real-world application of mathematics involves solving an exponential equation using logarithms. Since it is likely that some students have had X-rays taken, you may wish to discuss these exercises in class.

Second-Language Learners
Exs. 20, 21 Be sure that students learning English understand that the word *lead*, as used here, refers to the metal. You may also need to explain the meaning of a *lead apron*.

Teaching Tip
Ex. 25 You may want to have students solve this equation in two ways: by applying the power property and by not applying the power property. Have them verify that they get the same answer.

Using Technology
Ex. 30 As an alternative method in part (a), students can use the following TI-82 program to find the monthly payment, where A stands for A_0.

```
:Input "DECIMAL RATE",R
:R/12→R
:Input "NUMBER OF YEARS",N
:Input "AMOUNT OF LOAN",A
:AR/(1–(1+R)^(–12N))→P
:Disp "MONTHLY PAYMENT",P
```

Visual Thinking
Ex. 30 Ask students to use a graphing calculator or graphing software to create a graph of the loan repayment plan described in part (c). Ask them to use the graph to answer questions such as: How much money will be owed in 10 years? How much will be owed in 20 years? How much would the monthly payments be if the loan is paid back in 20 years instead of 30? This activity involves the visual skills of *recognition* and *exploration*.

Connection RADIOLOGY

If X-rays of a fixed wavelength strike a material x cm thick, then the intensity $I(x)$ of the X-rays transmitted through the material is given by $I(x) = I_0e^{-\mu x}$ where I_0 is the initial intensity and μ is a number that depends on the type of material and the wavelength of the X-rays.

Material	Value of μ
Aluminum	0.43
Copper	3.2
Lead	43

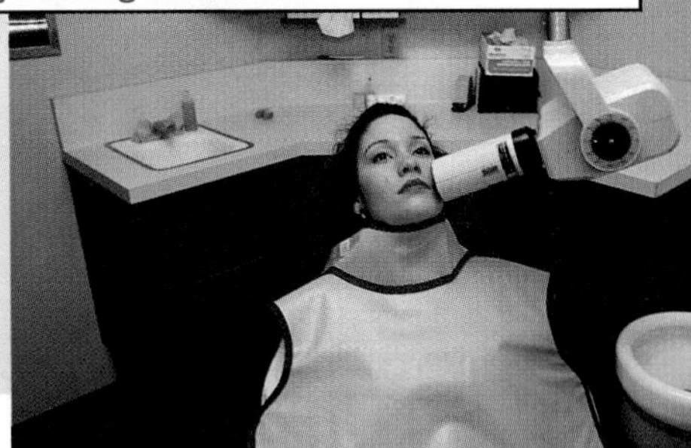

x cm
I(x) = I0e−μx
I0 is the initial intensity of the X-rays striking the material.

20. The table shows values of μ for various materials. These μ-values apply to X-rays of medium wavelength.

a. Find the thickness of aluminum shielding that reduces the intensity of X-rays to 30% of their initial intensity. (*Hint:* Find the value of x for which $I(x) = 0.3I_0$.)

b. Repeat part (a) for copper shielding.

c. Repeat part (a) for lead shielding.

21. Writing Your dentist puts a lead apron on you before taking X-rays of your teeth to protect you from harmful radiation. Based on your results from Exercise 20, explain why lead is a better material to use than aluminum or copper.

Solve each equation. Be sure to check your solutions.

22. $\log_3 x = -2$

23. $\ln \frac{x}{5} = 4$

24. $\log_6 (7t + 13) = 1$

25. $3 \log_8 (u - 1) = 2$

26. $\log_2 (x - 1) - \log_2 (x - 8) = 3$

27. $\log_4 (3y - 4) = -1 + \log_4 (y + 6)$

28. $\log_5 (9x + 2) - \log_5 (3x + 8) = 2$

29. $\log (\log w) = 0$

30. PERSONAL FINANCE When you take out a loan to buy a house, you must repay a portion of the loan with interest each month. The amount $A(n)$ you still owe after making n monthly payments is given by

$$A(n) = \left(A_0 - \frac{P}{r}\right)(1 + r)^n + \frac{P}{r}$$

where A_0 is the original amount of the loan, P is the monthly payment, and r is the monthly interest rate expressed as a decimal. (The monthly rate is the annual rate divided by 12.)

a. Kachina Tome borrows \$140,000 at 12% annual interest to buy a house. She plans to pay back the loan over 30 years, or 360 months. Calculate Kachina's monthly payment. (*Hint:* Use the fact that $A(360) = 0$.)

b. Write an equation giving the amount Kachina still owes after making n monthly payments.

c. After how many months will Kachina have paid back half the money she borrowed? Does your answer surprise you?

20. a. about 2.8 cm
b. about 0.38 cm
c. about 0.03 cm

21. It takes a much thinner piece of lead than of aluminum or copper to provide the same protection.

22. $\frac{1}{9}$ 23. 272.99
24. –1 25. 5
26. 9 27. 2
28. no solution 29. 10

30. a. \$1440.06
b. Let A = the amount still owed after n monthly payments; $A = 144{,}006 - 4006(1.01)^n$.
c. 293 months or about $24\frac{1}{2}$ years; Answers may vary. An example is given. I think it is surprising that it takes about 80% of the loan period to pay back half the loan.

31. a.
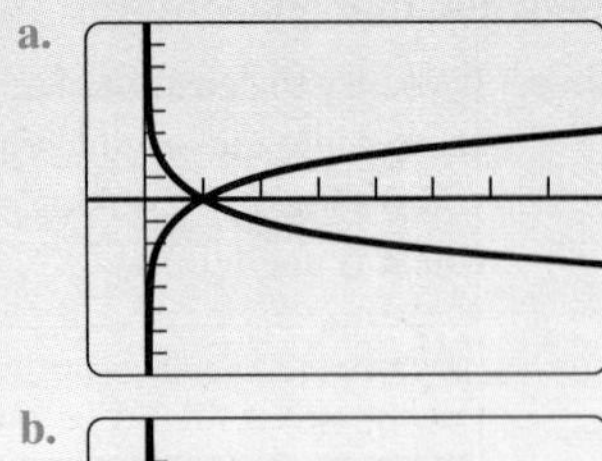

b.
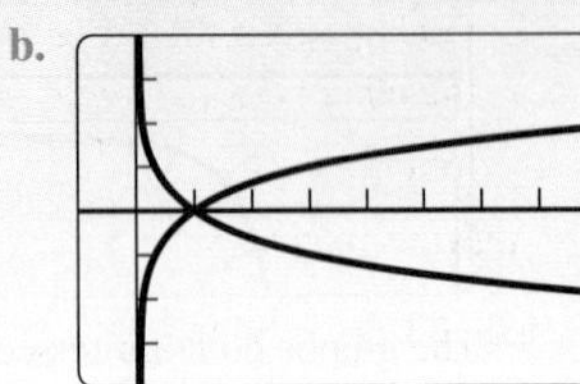

31. Visual Thinking For parts (a)–(c), use a graphing calculator or graphing software to graph the two equations in the same coordinate plane.

a. $y = \log_2 x$
$y = \log_{1/2} x$

b. $y = \log_3 x$
$y = \log_{1/3} x$

c. $y = \log_4 x$
$y = \log_{1/4} x$

d. How are the graphs of $y = \log_b x$ and $y = \log_{1/b} x$ related?

e. Write an equation expressing $\log_{1/b} x$ in terms of $\log_b x$.

ONGOING ASSESSMENT

32. Cooperative Learning The equation $T = 164e^{-0.041t} + 72$ in Example 1 is a special instance of *Newton's law of cooling*. This law states that the temperature T of a heated substance t minutes after it begins cooling is given by this equation:

$$T = (T_0 - T_R)e^{-kt} + T_R$$

T_0 is the substance's initial temperature.
k is a constant.
T_R is room temperature.

In this activity, you will find an equation that models the temperature of cooling water. Work in a group of four students. You need a transparent glass, an outdoor thermometer, and a watch. Follow these steps:

Step 1 Record the room temperature T_R from the thermometer.

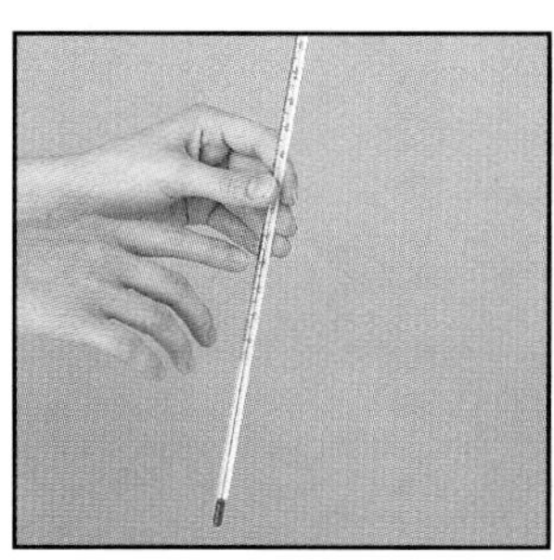

Step 2 Fill the glass with hot tap water, and measure the water's initial temperature T_0.

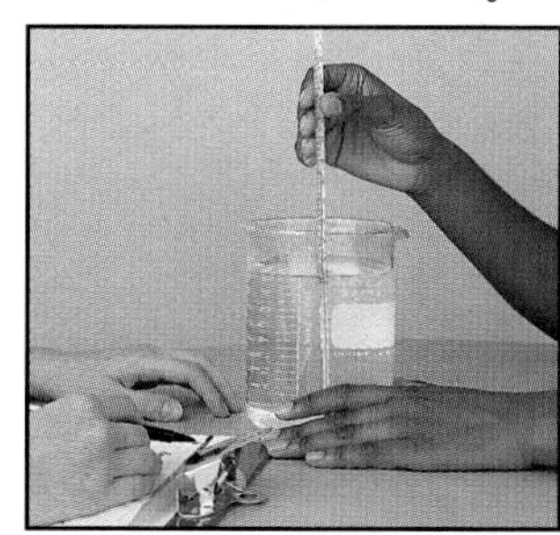

Step 3 Wait 10 min and measure the water's new temperature T_{10}.

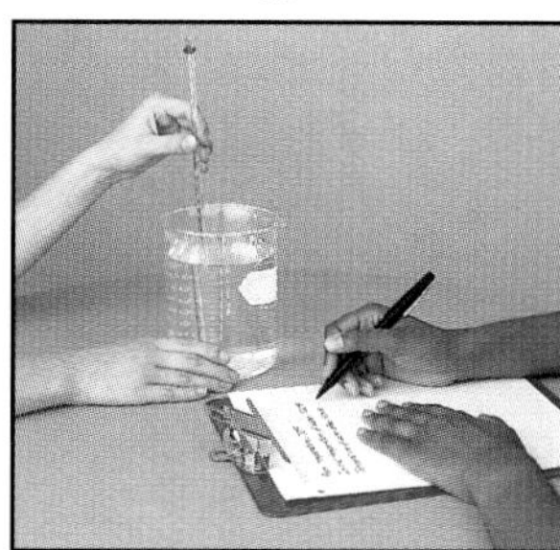

a. Substitute your values of T_0 and T_R into the general cooling equation above. Then use the fact that $T = T_{10}$ when $t = 10$ to find k.

b. Use your values of T_0, T_R, and k to write an equation giving the water's temperature T after t minutes.

c. Use your equation from part (b) to predict the temperature of the water 20 min after you measured the initial temperature T_0. Check your prediction by actually measuring this temperature with the thermometer. Compare the predicted and actual temperatures.

BY THE WAY

Newton's law of cooling was discovered by Sir Isaac Newton, the great English scientist and mathematician. Newton is most famous for his theories of motion and gravitation.

SPIRAL REVIEW

Write as a logarithm of a single number or expression. *(Section 4.3)*

33. $2 \log_a 3 + \frac{1}{2} \log_a 49$

34. $6 \ln u + \ln v - 8 \ln w$

Graph each function. *(Section 3.3)*

35. $y = 3^x$

36. $y = 3^{-x}$

37. $y = -3^x$

Apply⇔Assess

Exercise Notes

Using Technology

Ex. 31 A good viewing window for the graphs is Xmin = –1, Xmax = 10 and Ymin = –5, Ymax = 5. Students should be careful with parentheses when they enter the equation for parts (a)–(c). For example, in part (c) the second equation should be entered on the Y= list as Y2=log X/log (1/4) or as Y2=ln X/ln (1/4). Omitting the parentheses will result in an incorrect graph.

Assessment Note

Ex. 32 Since many variables affect the results of an experiment, students are likely to obtain different results for this activity. Have students discuss what variables may affect the results of the experiment and how they may make their results more reliable.

Practice 23 for Section 4.4

NAME ______________________ DATE __________

Practice 23

FOR USE WITH SECTION 4.4

Solve each equation. Round your answers to the nearest hundredth.

1. $10^{2x} = 57$ 0.88 **2.** $e^{4y/3} = 18$ 2.17 **3.** $2 \cdot 10^{r-1} = 39$ 2.29 **4.** $7e^{x/4} = 56$ 8.32

5. $5^t = 26$ 2.02 **6.** $4e^{n+2} + 3 = 45$ 0.35 **7.** $3(0.8)^{2a} - 5 = 19$ −4.66 **8.** $5 \cdot 3^x = 4 \cdot 2^x$ −0.55

Evaluate each logarithm. Round your answer to the nearest hundredth.

9. $\log_3 14$ 2.40 **10.** $\log_8 93$ 2.18 **11.** $\log_{12} 5$ 0.65 **12.** $\log_{14} 87$ 1.69

13. $\log_{1/2} 9$ −3.17 **14.** $\log_5 \frac{3}{4}$ −0.18 **15.** $\log_6 0.28$ −0.71 **16.** $\log_{1.5} 6.3$ 4.54

Solve each equation. Be sure to check your solutions. Round decimal answers to the nearest hundredth.

17. $\log_2 x = -4$ $\frac{1}{16}$ **18.** $5 \ln \frac{3x}{4} = 2$ 1.99

19. $\log_3 (1 - 8v) = 2$ −1 **20.** $4 \log_7 (3w + 1) = 14$ 302.16

21. $\log_5 (a + 3) - \log_5 (a - 1) = 1$ 2 **22.** $\log_2 (y + 1) = 3 + \log_2 (y - 6)$ 7

23. $\log_9 (4x + 1) - \log_9 (2x - 7) = \frac{1}{2}$ 11 **24.** $\log_3 (\log_5 x) = 1$ 125

25. A *mortgage* is a kind of loan that is usually repaid in equal monthly installments. The following formula gives the monthly payment M in terms of the amount of the loan P, the number of months n before the loan is fully repaid, and the monthly interest rate r:

$$M = \frac{rP}{1 - (1 + r)^{-n}}$$

a. Ariella borrows \$150,000 at 8.5% annual interest for 25 years, or 300 months. Calculate her monthly payment. \$1207.84

b. Suppose Ariella would like to have \$1150 as a monthly payment. With the same interest rate as in part (a), how many months would it be before the loan was repaid? 365 months

Practice Bank, ALGEBRA 2: EXPLORATIONS AND APPLICATIONS

23

31. c.

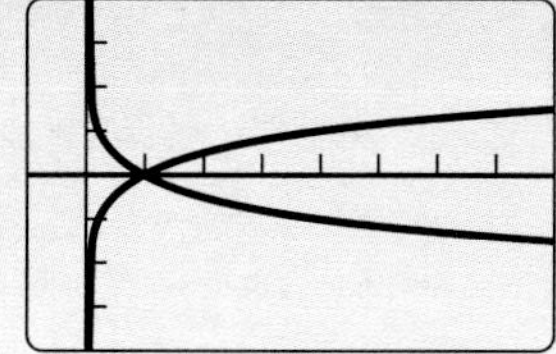

d. Each graph is the reflection of the other over the x-axis.

e. $\log_{1/b} x = -\log_b x$

32. Answers may vary.

33. $\log_a 63$

34. $\ln \frac{u^6 v}{w^8}$

35.

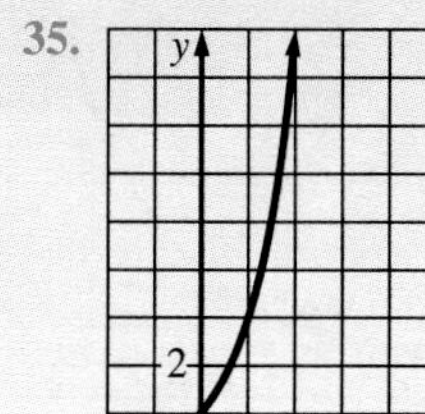

36.

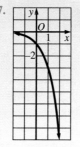

37.

Plan⇔Support

Objectives

- Use logarithms to find exponential and power functions that model data.
- Make predictions about real-life quantities.

Recommended Pacing

Core and Extended Courses
Section 4.5: 2 days

Block Schedule
Section 4.5: 1 block

Resource Materials

Lesson Support
Lesson Plan 4.5
Warm-Up Transparency 4.5
Practice Bank: Practice 24
Study Guide: Section 4.5
Exploration Lab Manual: Additional Exploration 6, Diagram Master 1
Assessment Book: Test 14

Technology
Graphing Calculator
Spreadsheet Software
McDougal Littell Mathpack *Stats!* *Function Investigator*
Internet: http://www.hmco.com

Warm-Up Exercises

1. Which of the following is an example of a linear function? b

   a. $y = 3x^2$ b. $y = 3x + 2$

2. Find an equation of the line through (5, 0.04) and (15, 0.025). $y = -0.0015x + 0.0475$ or $y - \frac{1}{25} = \frac{-3}{2000}(x - 5)$
3. Solve $\log(x + 1) = 1$. 9
4. Solve $\log P = 0.2t$ for P. $P = 10^{0.2t} \approx 1.6^t$
5. Solve $\log M - \log N = 0.05$ for M. $M = N(10^{0.05}) \approx 1.12N$

Section

4.5 Using Logarithms to Model Data

Learn how to...

- use logarithms to find exponential and power functions that model data

So you can...

- make predictions about real-life quantities, such as animal populations and weights

If you travel to Izumi, Japan, during the winter, you'll see thousands of cranes in the wetlands near the ocean. This wasn't always the case, however. There were fewer than 300 cranes in Izumi at the end of World War II. The table and scatter plot show how the number of cranes increased from 1945 to 1990.

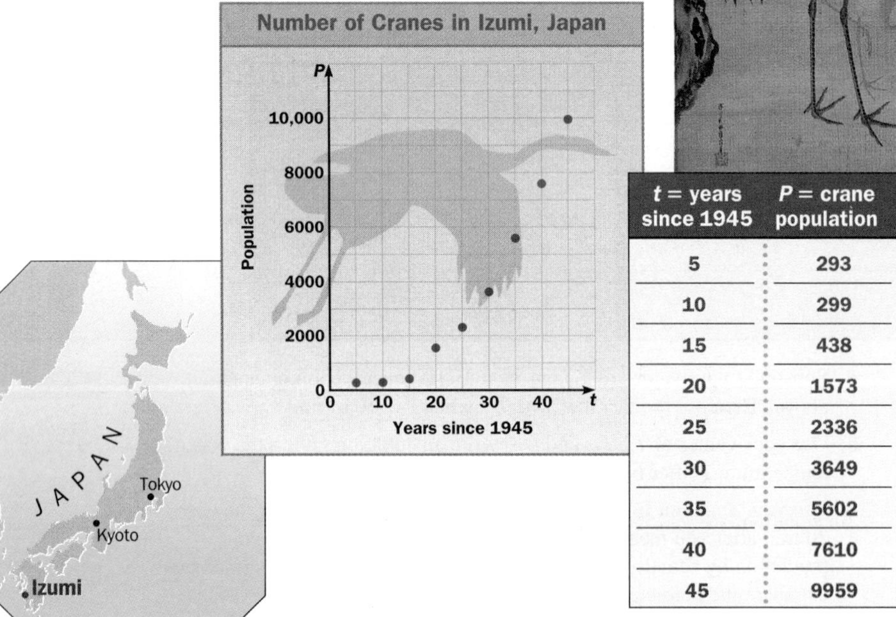

t = years since 1945	P = crane population
5	293
10	299
15	438
20	1573
25	2336
30	3649
35	5602
40	7610
45	9959

THINK AND COMMUNICATE

1. Look at the scatter plot. Do you think a linear function is a good model for the crane data? Why or why not?
2. What type of function do you think best models the crane data? Explain your reasoning.

You can use logarithms to find a function that models the crane data.

ANSWERS Section 4.5

Note: Answers may vary slightly due to differences in technology or to rounding.

Think and Communicate

1. No; the points do not appear to lie near a line.
2. exponential; The points appear to lie near an exponential curve.
3. 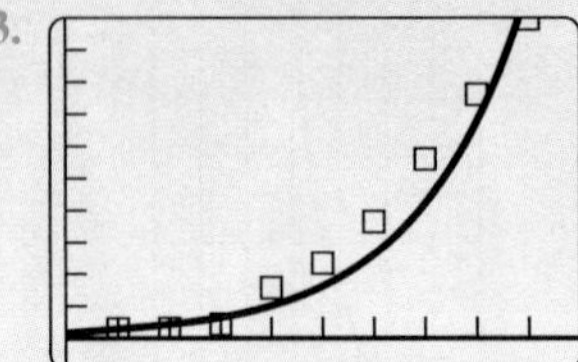

   The function is a reasonable model for the data.
4. about 23,593 cranes
5. about 10%

EXAMPLE 1 Application: Ecology

Use the table on page 168.

a. Make a scatter plot of the data pairs (t, log P). What relationship exists between t and log P?

b. Find an equation giving P as a function of t. What type of function is this equation?

SOLUTION

a. Make a table of data pairs (t, log P). Plot the pairs in a coordinate plane. The scatter plot suggests that there is a linear relationship between t and log P.

Crane Data

	A	B	C
1	t	P	log P
2	5	293	2.47
3	10	299	2.48
4	15	438	2.64
5	20	1573	3.20
6	25	2336	3.37
7	30	3649	3.56
8	35	5602	3.75
9	40	7610	3.88
10	45	9959	4.00

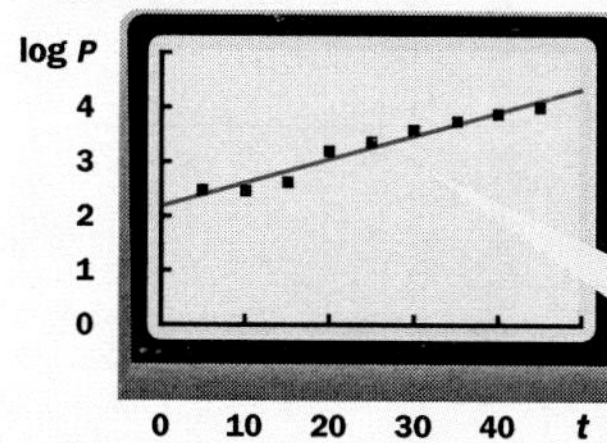

An equation of this fitted line is **log P = 0.0431t + 2.18.**

b. Use the equation of the fitted line.

$$\log P = 0.0431t + 2.18$$

$$P = 10^{0.0431t + 2.18}$$ Write the equation in exponential form.

$$P = (10^{2.18})(10^{0.0431t})$$

$$P = (10^{2.18})(10^{0.0431})^t$$

$$P = 151(1.10)^t$$

An equation is $P = 151(1.10)^t$. This equation is an exponential function.

BY THE WAY

The crane is the most popular shape in *origami*, the Japanese art of paper folding. The tradition of folding and stringing together one thousand cranes is thought to guarantee that a wish will come true.

THINK AND COMMUNICATE

3. **Technology** Use a graphing calculator or graphing software to graph the data pairs (t, P) on page 168 and the function $P = 151(1.10)^t$ in the same coordinate plane. Do you think the given function is a good model for the crane data? Explain.

4. Predict the number of cranes in Izumi in 1998.

5. Based on the exponential model in Example 1, by about what percent did the Izumi crane population increase each year from 1945 to 1990?

Technology Note

Students may enter all data into a graphing calculator by using the list features. Enter t values into list L1, P values into list L2, then go to the top of L3 and enter log L2 to enter log P quickly. Turn STAT PLOT on and set Xlist to L1, and Ylist to L3 to see a scatter plot of the data pairs in L1, L3.

Teach⇔Interact

Think and Communicate

For question 1, the scatter plot offers a good visual estimate of the model which may fit the data. Encourage students to use the statistics functions of a graphing calculator to find different models for the data.

Additional Example 1

The U.S. farm population from 1910 to 1990 is given in the table.

t = years since 1900	P = population (millions)
10	32.077
20	31.974
30	30.529
40	30.547
50	23.048
60	15.635
70	9.712
80	6.051
90	4.801

a. Make a scatter plot of the data pairs (t, log P). What relationship exists between t and log P?

Make a table of data pairs (t, log P). Plot the pairs in a coordinate plane.

Farm Data

	A	B	C
1	t	P	log P
2	10	32.077	1.5062
3	20	31.974	1.5048
4	30	30.529	1.4847
5	40	30.547	1.485
6	50	23.048	1.3626
7	60	15.635	1.1941
8	70	9.712	0.9873
9	80	6.051	0.7818
10	90	4.801	0.6813

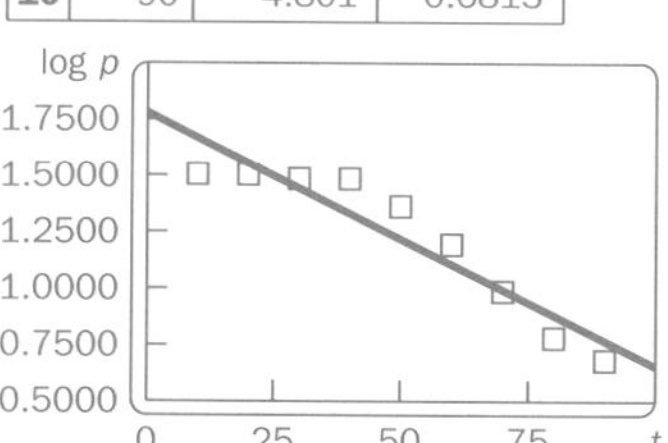

The scatter plot suggests that there is a linear relationship between t and log P.

b. Find an equation giving P as a function of t. What type of function is this equation?

Use the equation of the fitted line.

$$\log P = -0.0113t + 1.78$$

$$P = 10^{-0.0113t + 1.78}$$

$$P = (10^{-0.0113t})(10^{1.78})$$

$$P = 60.256(10^{-0.0113})^t$$

$$P = 60.256(0.97)^t$$

An equation is $P = 60.256(0.97)^t$. This equation is an exponential function.

Teach⇔Interact

About Example 2

Teaching Tip

This example has students find a power model for data. A power model is a good fit when ($\log x$, $\log y$) is linear for data pairs (x, y).

Additional Example 2

The table shows the U.S. nuclear generation of electricity from 1975 to 1991.

Years t since 1970	Kilowatt hours K (billions)
5	172.5
10	251.1
15	383.7
16	414.0
17	455.3
18	527.0
19	529.4
20	576.9
21	612.6

a. Make a scatter plot of the data pairs ($\log t$, $\log K$). What relationship exists between $\log t$ and $\log K$? Make a table of data pairs ($\log t$, $\log K$). Plot the pairs in a coordinate plane.

Electricity Data

	A	B	C	D
1	t	K	log t	log K
2	5	172.5	0.70	2.24
3	10	251.1	1	2.40
4	15	383.7	1.18	2.58
5	16	414	1.20	2.62
6	17	455.3	1.23	2.66
7	18	527	1.26	2.72
8	19	529.4	1.28	2.72
9	20	576.9	1.30	2.76
10	21	612.6	1.32	2.79

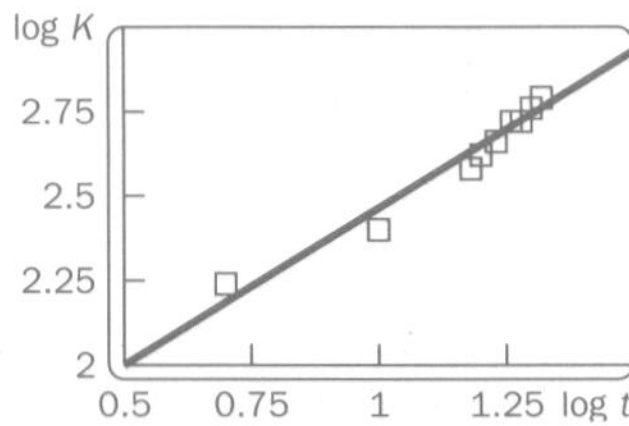

The scatter plot suggests that there is a linear relationship between $\log t$ and $\log K$.

b. Find an equation giving K as a function of t. Use the equation of the fitted line.

$$\log K = 0.9 \log t + 1.57$$
$$\log K = \log t^{0.9} + 1.57$$
$$\log K - \log t^{0.9} = 1.57$$
$$\log \frac{K}{t^{0.9}} = 1.57$$
$$\frac{K}{t^{0.9}} = 10^{1.57}$$
$$K = 37.2t^{0.9}$$

An equation is $K = 37.2t^{0.9}$.

EXAMPLE 2 Application: Biology

The table shows how the weight of a chicken embryo inside an egg changes over time.

a. Make a scatter plot of the data pairs ($\log d$, $\log W$). What relationship exists between $\log d$ and $\log W$?

b. Find an equation giving W as a function of d.

d = days after egg is laid	W = weight of embryo (grams)
1	0.0002
4	0.05
8	1.15
12	5.07
16	15.98
20	30.21

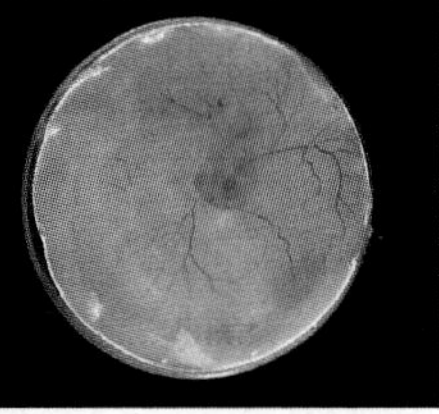

Chicken embryo after 4 days of incubation

SOLUTION

a. Make a table of data pairs ($\log d$, $\log W$). Plot the pairs in a coordinate plane. The scatter plot suggests that there is a linear relationship between $\log d$ and $\log W$.

Embryo Data

	A	B	C	D
1	d	W	log d	log W
2	1	0.0002	0.00	−3.70
3	4	0.05	0.60	−1.30
4	8	1.15	0.90	0.06
5	12	5.07	1.08	0.71
6	16	15.98	1.20	1.20
7	20	30.21	1.30	1.48

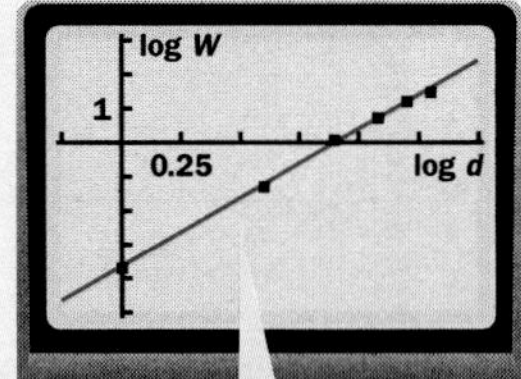

An equation of this fitted line is $\log W = 4.04 \log d - 3.69$.

b. Use the equation of the fitted line.

$$\log W = 4.04 \log d - 3.69$$
$$\log W = \log d^{4.04} - 3.69$$ Use the power property.
$$\log W - \log d^{4.04} = -3.69$$
Use the quotient property. $$\log \frac{W}{d^{4.04}} = -3.69$$
$$\frac{W}{d^{4.04}} = 10^{-3.69}$$ Write the equation in exponential form.
$$W = 0.000204d^{4.04}$$

An equation is $W = 0.000204d^{4.04}$.

The function $W = 0.000204d^{4.04}$ in Example 2 is called a *power function*. A **power function** has the form $y = ax^b$ where a and b are constants.

Technology Note

Students may be using a graphing calculator or software that can find power regression equations. If so, they can obtain their equations by using that feature. The regression equation must be generated by using the original data (not the logarithms of the data).

Regardless of which method is used to find the equation, suggest that students use the original data to display a scatter plot. They should then graph the equation on the same screen as the scatter plot to see just how well the equation models the data. If the calculator or software displays a correlation coefficient r, then students have a measure of how well the equation models the data. For most students, however, this does not take the place of direct visual confirmation.

THINK AND COMMUNICATE

6. A chick hatches 21 days after the egg is laid. Estimate the weight of a chick when it hatches.
7. Tell whether each function is a power function.

   a. $y = 0.3(1.6)^x$ b. $y = 0.3x^{1.6}$ c. $y = \frac{0.3}{x + 1.6}$ d. $y = 5x^{-2}$

Using Exponential and Power Functions to Model Data

1. You can use an **exponential function** to model a set of data pairs (x, y) if there is a **linear relationship** between x and $\log y$.
2. You can use a **power function** to model a set of data pairs (x, y) if there is a **linear relationship** between $\log x$ and $\log y$.

CHECKING KEY CONCEPTS

For each table:

a. Make a scatter plot of the data pairs (x, log y).

b. Make a scatter plot of the data pairs (log x, log y).

c. Tell whether an *exponential function* or a *power function* is a better model for the data. Find an equation giving y as a function of x.

1.

x	y
1	2.4
2	2.8
3	3.3
4	3.9
5	4.7
6	5.5

2.

x	y
5	6.3
10	15.7
15	26.7
20	39.0
25	52.2
30	66.3

3.

x	y
2	1.25
4	1.05
6	0.94
8	0.87
10	0.82
12	0.79

4.5 Exercises and Applications

Extra Practice exercises on page 754

1. **Writing** Describe how you can use logarithms to determine if an exponential function is a good model for a set of data pairs (x, y). Describe how you can determine if a power function is a good model.

Write y as a function of x.

2. $\log y = 0.06x + 1.3$
3. $\log y = 0.2x + 3.8$
4. $\log y = -0.35 + 0.45x$
5. $\log y = -0.04x - 0.091$
6. $\ln y = x + 2$
7. $\ln y = 0.83 - 3.27x$

Think and Communicate

6. about 44.81 g
7. a. No.
   b. Yes.
   c. No.
   d. Yes.

Checking Key Concepts

1–3. See answers in back of book.

Exercises and Applications

1. To determine if an exponential function is a good model for a set of data pairs (x, y), draw a scatter plot of the data pairs $(x, \log y)$. If the scatter plot suggests a linear relationship, then the original data pairs can be modeled by an exponential function. To determine if a power function is a good model for a set of data pairs (x, y), draw a scatter plot of the data pairs $(\log x, \log y)$. If the scatter plot suggests a linear relationship, then the original data pairs can be modeled by a power function.
2. $y = 20.0(1.15)^x$
3. $y = 6310(1.58)^x$
4. $y = 0.447(2.82)^x$
5. $y = 0.811(0.912)^x$
6. $y = 7.39e^x$
7. $y = 2.29(0.0380)^x$

Teach⇔Interact

Think and Communicate

You may wish to have students work on question 7 in groups. Have them list the characteristics of linear, exponential, and power models and apply them to the exercises.

Checking Key Concepts

Communication: Discussion When discussing questions 1–3, encourage students to give their reasons for choosing an exponential model or a power model. These models look quite similar to some students, so a discussion may help clear up any misconceptions.

Closure Question

When can you use an exponential function and when can you use a power function to model a set of data points (x, y)? You can use an exponential function if there is a linear relationship between x and $\log y$ and a power function if there is a linear relationship between $\log x$ and $\log y$.

Apply⇔Assess

Suggested Assignment

Core Course
Day 1 Exs. 2–7, 10–15, 17
Day 2 Exs. 21–23, 25–34, AYP

Extended Course
Day 1 Exs. 1–20
Day 2 Exs. 21–34, AYP

Block Schedule
Day 24 Exs. 2–7, 10–15, 17, 21–23, 25–34, AYP

Exercise Notes

Student Progress
Ex. 1 This exercise can be used to make sure students understand the content of this section.

Cooperative Learning
Exs. 2–7 These exercises are all exponential functions. You may wish to have students check their answers in small groups.

Apply⇔Assess

Exercise Notes

Multicultural Note

Ex. 8 Brazil, the largest country in South America, is a leading industrial and agricultural power. Its chief agricultural exports include coffee, soybeans, cacao beans, oranges, and livestock, while its industrial exports include cars, machinery, and aircraft. Most of the population live in cities, such as São Paulo, Rio de Janeiro, and the capital Brasilia. The official language, Portuguese, is spoken by most Brazilians, though the indigenous people of the Amazon region still speak their traditional languages. In 1843, Brazil became one of the first countries to issue adhesive postage stamps. These oval-shaped stamps were called "Bull's-Eyes."

Teaching Tip

Ex. 8 For part (c), students who use the curve fitting option of a graphing calculator may get a different answer than if they use a software package. Caution them to round as little as possible when finding a model.

Common Error

Ex. 16 Students who use graphing calculators to find a power function regression curve may run into difficulty if one of the data values is 0. Since the logarithm of 0 does not exist, it generates an error in the calculator.

Using Technology

Ex. 16 To give students a deeper insight into the significance of the values of *r* for the two models, have students display the crane data in a scatter plot. Then have them graph the exponential and power functions on the same axes. The procedure is straightforward. After displaying the information for the exponential function, press [Y=]. Clear any equations that may be on the Y= list. Position the blinking cursor after Y1= and press [VARS] 5 [▶] [▶] 7 [ENTER]. This enters the exponential function on the Y= list. In a similar manner, display the information for the power function on the home screen. Press [Y=], position the cursor after Y2=, and enter the equation for the power function using the same procedure as above. Press [GRAPH]. Note that the graph of the exponential function fits the data points better than the graph of the power function.

Connection SOCIAL STUDIES

Use the table. The variables in the table are defined as follows:

t = number of years since 1900

N = total varieties of stamps issued through year 1900 + t

D = varieties of stamps issued in the previous decade

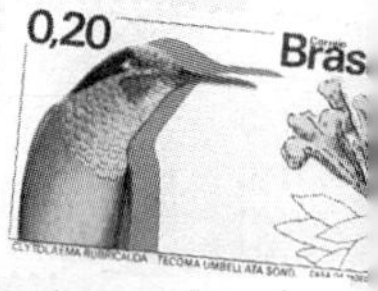

8. a. Make a scatter plot of the data pairs $(t, \log N)$. What relationship exists between t and $\log N$?

b. Fit a line to your scatter plot. Find an equation of your fitted line.

c. Find an equation giving N as an exponential function of t.

d. Predict the number of different stamps Brazil will have issued by the year 2010.

9. a. Copy and complete the third column of the table. The first entry has been done for you.

b. Challenge Show that D is an exponential function of t *without* using logarithms. (*Hint:* Use the equation from part (c) of Exercise 8 and the fact that $D(t) = N(t) - N(t - 10)$.)

c. Make a scatter plot of the data pairs $(t, \log D)$. Use your scatter plot to find an equation giving D as an exponential function of t. Compare this function with the one you found in part (b).

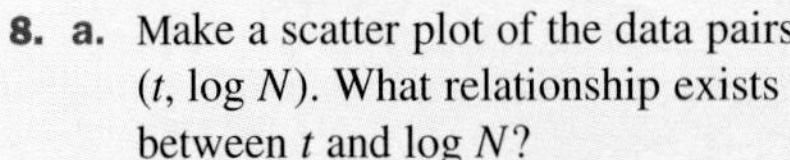

Stamps Issued in Brazil

t	N	D
0	145	—
10	173	28
20	216	?
30	299	?
40	477	?
50	658	?
60	868	?
70	1125	?
80	1668	?
90	2186	?

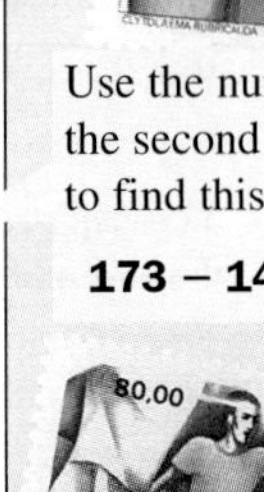

Use the numbers in the second column to find this value:

173 − 145 = 28

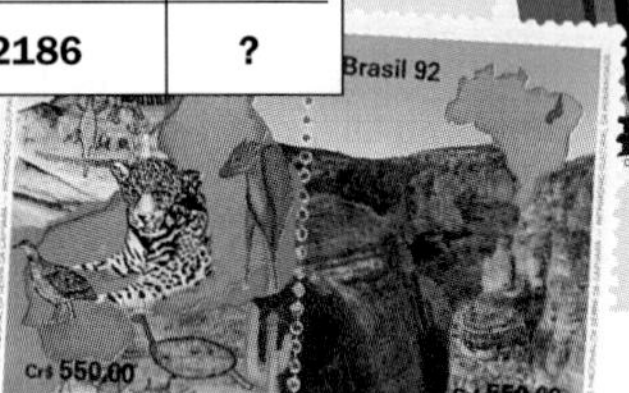

Write *y* as a function of *x*.

10. $\log y = 0.4 \log x + 0.8$
11. $\log y = 0.72 \log x + 2.34$
12. $\log y = -0.3 + 4.85 \log x$
13. $\log y = -0.61 \log x - 0.039$
14. $\ln y = 3.7 \ln x + 3.7$
15. $\ln y = 1.48 - \ln x$

16. **Technology** Some graphing calculators and statistical software will find an exponential or power function for a set of data pairs (x, y). They may also give a correlation coefficient r that tells you how well a line fits the data points $(x, \log y)$ for exponential functions or $(\log x, \log y)$ for power functions. The closer $|r|$ is to 1, the better the fit is. The screen shows the exponential function and correlation coefficient given by one calculator for the crane data on page 168.

a. Use a graphing calculator or statistical software to find a power function that models the crane data. What is the correlation coefficient?

b. Compare the correlation coefficients for the exponential and power functions. Based on these coefficients, which type of function is the better model?

```
ExpReg
y=a*b^x
a=152.4460214
b=1.104262659
r=.9790116222
```


8. a. (scatter plot: axes log N and t; tick labels 3, 2, O, 20)

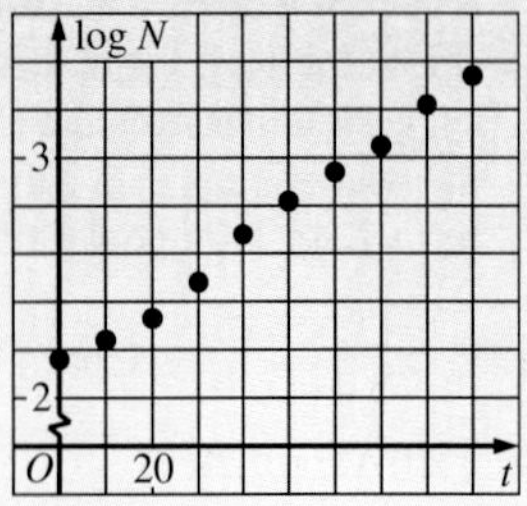

linear

b. $\log N = 0.0137t + 2.11$

c. $N = 129(1.03)^t$

d. about 3330 stamps

9. a.

Stamps Issued in Brazil

t	N	D
0	145	—
10	173	28
20	216	43
30	299	83
40	477	178
50	658	181
60	868	210
70	1125	257
80	1668	543
90	2186	518

b. $D(t) = N(t) - N(t - 10) =$
$129(1.03)^t - 129(1.03)^{t-10} =$
$129(1.03)^t - 129(1.03)^t(1.03)^{-10} =$
$129(1.03)^t - 96.0(1.03)^t =$
$33(1.03)^t$

17. Use the table of chicken embryo data in Example 2.

  a. Make a scatter plot of the data pairs (d, W). Is it clear from the scatter plot that a power function is a better model for the data than an exponential function? Explain.

  b. Make a scatter plot of the data pairs $(d, \log W)$. Fit a line to your scatter plot.

  c. **Writing** Compare the scatter plot from part (b) with the scatter plot of the data pairs $(\log d, \log W)$ in Example 2. Explain why these scatter plots show that a power function is a better model for the chicken embryo data than an exponential function.

18. **Spreadsheets** You can use a spreadsheet to compare the rates at which different types of functions increase.

  a. Consider the linear function $y = 2x$, the exponential function $y = 2^x$, the logarithmic function $y = \log_2 x$, and the power function $y = x^2$. Use a spreadsheet to find the values of these functions for $x = 1, 2, 3, \ldots, 12$. Rank the functions in order from fastest to slowest rate of increase.

Comparing Functions

	A	B	C	D	E
1	x	y = 2*x	y = 2^x	y = log(x)/log(2)	y = x^2
2	1	?	?	?	?
3	2	?	?	?	?
4	3	?	?	?	?
5	4	?	?	?	?
6	5	?	?	?	?
7	6	?	?	?	?
8	7	?	?	?	?
9	8	?	?	?	?
10	9	?	?	?	?
11	10	?	?	?	?
12	11	?	?	?	?
13	12	?	?	?	?

  b. Repeat part (a) several times using functions of the form $y = bx$, $y = b^x$, $y = \log_b x$, and $y = x^b$ where b is a number greater than 2.

  c. **Writing** Write a paragraph summarizing your results from parts (a) and (b). Your paragraph should include a general comparison of the rates of increase of linear, exponential, logarithmic, and power functions.

19. **Cooperative Learning** Work with a partner. One person should choose an exponential function f to generate a table of 10 ordered pairs $(x, f(x))$. The other person should use logarithms to "discover" an equation for f. Reverse roles with your partner, and repeat the activity using a power function.

20. **Research** Choose a two-column table of data from a magazine, almanac, or other source. Find an exponential function and a power function that model the data. Decide which function is the better model.

21. **SAT/ACT Preview** Suppose $\log y = 2 \log x + 1$. Which equation gives y as a function of x?

  A. $y = 2x + 1$ **B.** $y = x^2 + 1$ **C.** $y = 10^{2x+1}$ **D.** $y = 10x^2$ **E.** none of these

Apply⇔Assess

Exercise Notes

Topic Spiraling: Review
Ex. 17 Have students generate each model with a graphing calculator to also get a value of *r*, the correlation coefficient, with the model. This can help review the concept of correlation.

Second-Language Learners
Ex. 17 Students learning English may benefit from discussing the data with a partner before writing their explanations for part (c).

Integrating the Strands
Ex. 18 Using a spreadsheet to compare the rates at which different types of functions increase integrates the strands of discrete mathematics and algebra.

Research
Ex. 20 The population data in an almanac is a good source of exponential data. Some almanacs also give projected growth rates which can be compared to the rates generated by the data.

c.

$D = 25.1(1.04)^t$; The equations are similar.

10. $y = 6.31x^{0.4}$

11. $y = 219x^{0.720}$

12. $y = 0.501x^{4.85}$

13. $y = 0.914x^{-0.61}$

14. $y = 40.4x^{3.70}$

15. $y = 4.39x^{-1}$

16. a. $y = 7.31x^{1.82}$; 0.9434628588

  b. The correlation coefficient for the exponential function is closer to 1. The exponential function is the better model.

17, 18. See answers in back of book.

19. Answers may vary. Check students' work.

20. Answers may vary. Check students' work.

21. D

Apply⇔Assess

Exercise Notes

Integrating the Strands

Exs. 22–24 These exercises integrate algebraic and geometric concepts. As the linear dimension of an object increases, the volume and therefore the weight of the object goes up by the third power of the dimension increase. In this case, the dimension is the circumference of the femur. Students may also use geometry to answer Ex. 24.

Career Connection

Exs. 22–24 The branch of biology that is concerned with the forms of life that existed in previous geologic periods is called *paleontology.* These forms of life, both animal and plant life, are known by studying their fossil remains. Scientists who do this kind of work are called *paleontologists.* They are most often employed in universities, teaching and doing research that attempts to understand the evolution of life on planet Earth. Paleontologists also work for museums, such as the Museum of Natural History in New York City.

Connection BIOLOGY

How can you estimate the weight of dinosaurs from their skeletons? One way is to look for relationships between weight and skeletal structures in living animals. You can then apply these relationships to dinosaurs.

Animal	C (mm)	W (kg)
Meadow mouse	5.5	0.047
Gray squirrel	13	0.399
Porcupine	34	7.20
Yellow baboon	57	28.6
Lion	93.5	143
Polar bear	135	448
Giraffe	173	710

22. The *femur* is a large bone located in the leg or hind limb of a human or animal. The first table shows the average circumference C of the femur for several species of animals. It also shows each animal's average weight W.

a. Make a scatter plot of the data pairs $(\log C, \log W)$. What relationship exists between $\log C$ and $\log W$?

b. Fit a line to your scatter plot. Find an equation of your fitted line.

c. Express W as a function of C. What type of function describes the relationship between femoral circumference and weight?

23. Scientists obtained femoral circumferences for the dinosaurs in the second table from fossil remains. Use the function from part (c) of Exercise 22 to estimate the weight of each dinosaur.

Dinosaur	C (mm)	W (kg)
Parksosaurus	103	?
Struthiomimus	136	?
Allosaurus	348	?
Tyrannosaurus	534	?

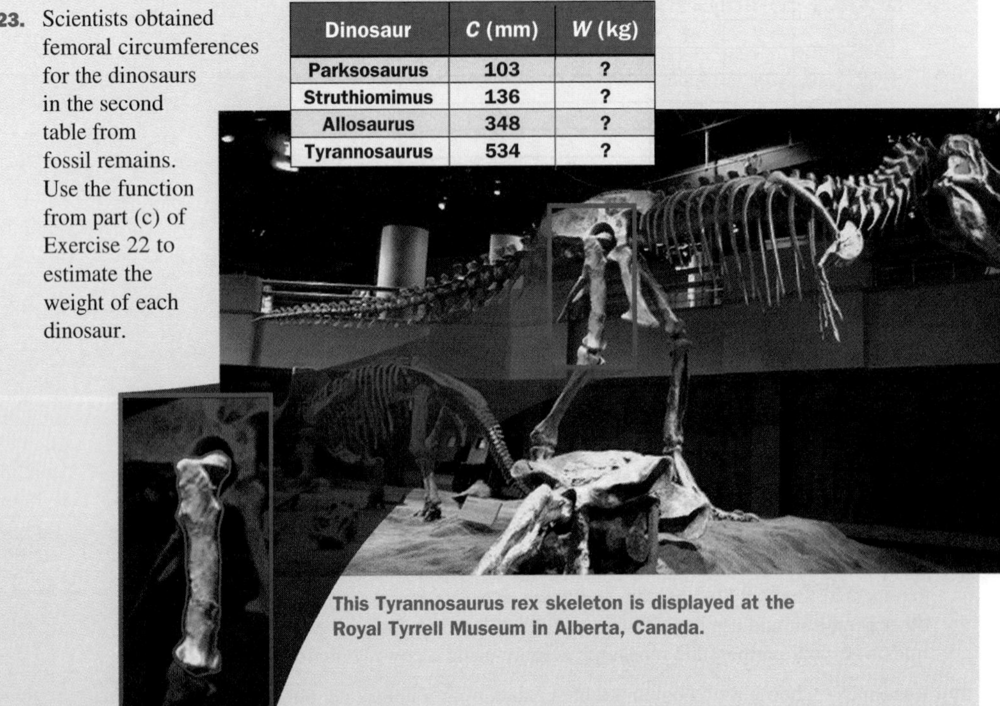

This Tyrannosaurus rex skeleton is displayed at the Royal Tyrrell Museum in Alberta, Canada.

24. Challenge Some scientists have estimated dinosaur weights by assuming that weight is proportional to the cross-sectional area of the femur. Is this assumption true for the animals in the first table? Give a mathematical justification for your answer. (*Hint:* Assume that the cross section of a femur is circular.)

22. a.

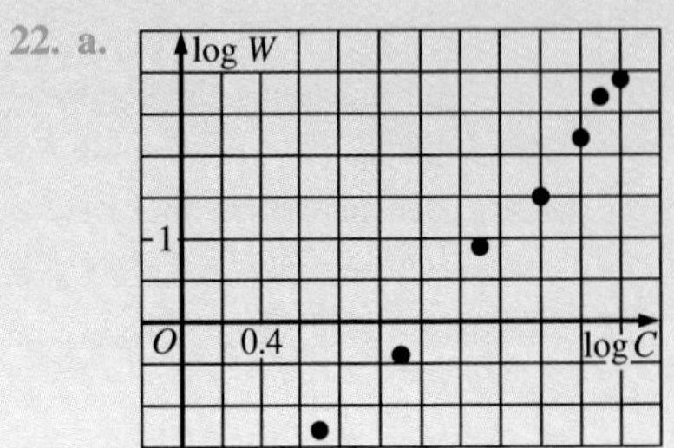

There is a linear relationship between $\log C$ and $\log W$.

b. $\log W = 2.86 \log C - 3.51$

c. $W = 0.000309C^{2.86}$; a power function

23. Parksosaurus: about 176 kg; Struthiomimus: about 391 kg; Allosaurus: about 5740 kg; Tyrannosaurus: about 19,500 kg

24. Answers may vary. An example is given. The area of a circle with circumference C is $\frac{C^2}{4\pi}$, so $C = 2\sqrt{\pi A} \approx 3.54A^{0.5}$. Substituting in the equation in part (c) of Ex. 22, $W = 0.000309(3.5A^{0.5})^{2.86} \approx 0.011A^{1.43}$, which is a power function, not a proportion.

25. See answers in back of book.

26. 2.26

27. 0.94

28. −0.63

29. −0.85

ONGOING ASSESSMENT

25. **Open-ended Problem** The table shows the number of African-American elected officials in the United States for various years. Find at least two models for the data. Discuss how well each of your models fits the data. Use your models to predict the number of African-American elected officials for future years.

Year	1982	1984	1986	1988	1990	1992
Number of officials	5115	5654	6384	6793	7335	7517

SPIRAL REVIEW

Evaluate each logarithm. Round your answers to the nearest hundredth. *(Section 4.4)*

26. $\log_3 12$ 27. $\log_8 7$ 28. $\log_5 0.36$ 29. $\log_2 \frac{5}{9}$

30. **Writing** Explain the relationship between the slope of the least-squares line and the sign of the correlation coefficient for a set of data points (x, y). *(Section 2.5)*

Evaluate each expression when $x = 3$ and $x = -2$. *(Toolbox, page 780)*

31. x^2 32. $5x^2$ 33. $-4x^2$ 34. $\frac{1}{2}x^2$

BY THE WAY

The first African-American United States senator was Hiram Revels of Mississippi, who was elected in 1870.

ASSESS YOUR PROGRESS

VOCABULARY

extraneous solution (p. 165) **power function** (p. 170)

Solve each equation. Round your answers to the nearest hundredth. *(Section 4.4)*

1. $e^x = 6$ 2. $4^t = 23$ 3. $10(1.09)^{2k} + 7 = 51$

Solve each equation. Be sure to check your solutions. *(Section 4.4)*

4. $\log_4 x = 3$ 5. $\log_2 (w - 9) - \log_2 (w - 2) = 3$

6. **BIOLOGY** The table shows the average weight W (in grams) of several rats t days after birth for $t = 1, 2, \ldots, 6$. *(Section 4.5)*

t	W
1	5.64
2	6.21
3	7.07
4	8.37
5	9.74
6	10.7

a. Find an exponential function that models the data.

b. Find a power function that models the data.

c. Which of the functions from parts (a) and (b) is the better model? Explain your reasoning.

7. **Journal** Discuss how logarithms can be used to model data.

Apply⇔Assess

Exercise Notes

Assessment Note

Ex. 25 This exercise will help you to assess how clearly students understand the ideas in this section. Have several students present their models to the class for discussion.

Assess Your Progress

Journal Entry

Ask students to share their ideas about how logarithms can be used to model data for Ex. 7. Have them justify their examples.

Progress Check 4.4–4.5

See page 179.

Practice 24 for Section 4.5

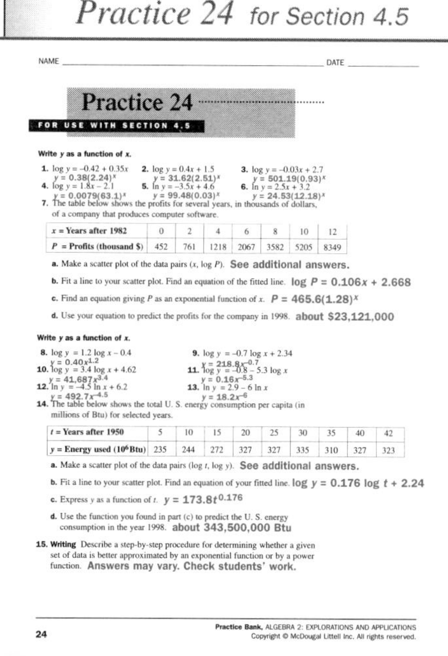

NAME ______ DATE ______

Practice 24

FOR USE WITH SECTION 4.5

Write y as a function of x.

1. $\log y = -0.42 + 0.35x$ — $y = 0.38(2.24)^x$
2. $\log y = 0.4x + 1.5$ — $y = 31.62(2.51)^x$
3. $\log y = -0.03x + 2.7$ — $y = 501.19(0.93)^x$
4. $\log y = 1.8x - 2.1$ — $y = 0.0079(63.1)^x$
5. $\ln y = -3.5x + 4.6$ — $y = 99.48(0.03)^x$
6. $\ln y = 2.5x + 3.2$ — $y = 24.53(12.18)^x$
7. The table below shows the profits for several years, in thousands of dollars, of a company that produces computer software.

x = Years after 1982	0	2	4	6	8	10	12
P = Profits (thousand \$)	452	761	1218	2067	3582	5205	8349

a. Make a scatter plot of the data pairs $(x, \log P)$. **See additional answers.**

b. Fit a line to your scatter plot. Find an equation of the fitted line. **$\log P = 0.106x + 2.668$**

c. Find an equation giving P as an exponential function of x. **$P = 465.6(1.28)^x$**

d. Use your equation to predict the profits for the company in 1998. **about \$23,121,000**

Write y as a function of x.

8. $\log y = 1.2 \log x - 0.4$ — $y = 0.40x^{1.2}$
9. $\log y = -0.7 \log x + 2.34$ — $y = 218.8x^{-0.7}$
10. $\log y = 3.4 \log x + 4.62$ — $y = 41{,}687x^{3.4}$
11. $\log y = -0.8 - 5.3 \log x$ — $y = 0.16x^{-5.3}$
12. $\ln y = -4.5 \ln x + 6.2$ — $y = 492.7x^{-4.5}$
13. $\ln y = 2.9 - 6 \ln x$ — $y = 18.2x^{-6}$
14. The table below shows the total U. S. energy consumption per capita (in millions of Btu) for selected years.

t = Years after 1950	5	10	15	20	25	30	35	40	42
y = Energy used (10^6 Btu)	235	244	272	327	327	335	310	327	323

a. Make a scatter plot of the data pairs $(\log t, \log y)$. **See additional answers.**

b. Fit a line to your scatter plot. Find an equation of your fitted line. **$\log y = 0.176 \log t + 2.24$**

c. Express y as a function of t. **$y = 173.8t^{0.176}$**

d. Use the function you found in part (c) to predict the U. S. energy consumption in the year 1998. **about 343,500,000 Btu**

15. **Writing** Describe a step-by-step procedure for determining whether a given set of data is better approximated by an exponential function or by a power function. **Answers may vary. Check students' work.**

24 **Practice Bank,** ALGEBRA 2: EXPLORATIONS AND APPLICATIONS

---

30. The sign of the correlation coefficient for a set of data points is the same as the sign of the slope of the least-squares line.

31. 9; 4

32. 45; 20

33. −36; −16

34. $\frac{9}{2}$; 2

Assess Your Progress

1. 1.79
2. 2.26
3. 8.60
4. 64
5. no solution
6. a. $W = 4.83(1.14)^t$

   b. $W = 5.19t^{0.365}$

   c. The exponential function is the better model since the correlation coefficient for the exponential function, 0.9963581796, is closer to 1 than the correlation coefficient for the power function, 0.9550944564.

7. Answers may vary. An example is given. Given a set of data points (x, y), you can make scatter plots of the points $(x, \log y)$ and $(\log x, \log y)$. If the scatter plots suggest a linear relationship between x and $\log y$, the data can be modeled by an exponential function of the form $y = ab^x$. If the scatter plots suggest a linear relationship between $\log x$ and $\log y$, the data can be modeled by a power function of the form $y = ax^b$.

Project Notes

Mathematical Goals

- Collect, record, organize, and analyze data.
- Make a scatter plot of data pairs and fit a line to the plotted points.
- Find an equation of the line of fit.
- Express rank as a power function of population using the equation of the fitted line.
- Make predictions about data.

Planning

Materials

- graph paper or graphing calculator
- access to encyclopedias

Project Teams

Students with an interest in a specific country may wish to work together. Students can select group members and decide how to share project duties.

Guiding Students' Work

Students may be able to find the populations in other sources such as *The Statesman's Yearbook.* Have students use data that are as current as possible. Also, explain the relationship between the points (log P, log R) and the line of fit with the equation for Zipf's Law. The equation for the line of fit is $\log R = a \cdot \log P + b$, and the equation for Zipf's Law is $R = 10^b \cdot P^a$. Students can use the value of r, the correlation coefficient, to find how well the data values fit, and apply it to their explanation of how well Zipf's Law models their data values.

Second-Language Learners

Students learning English can benefit from working on the research and writing the report with the help of a peer tutor or aide.

PORTFOLIO PROJECT

Investigating Zipf's Law

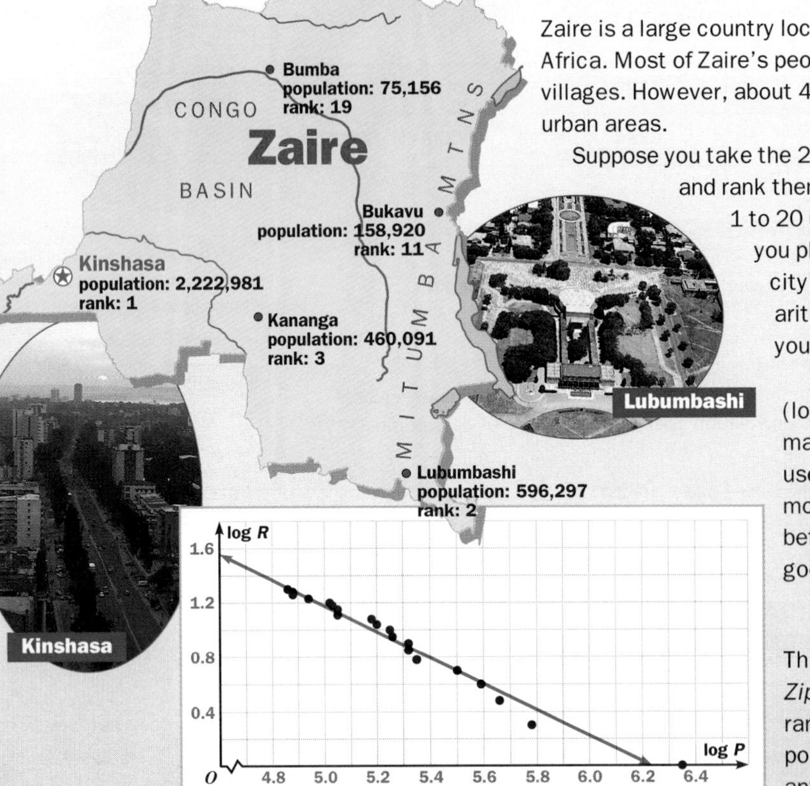

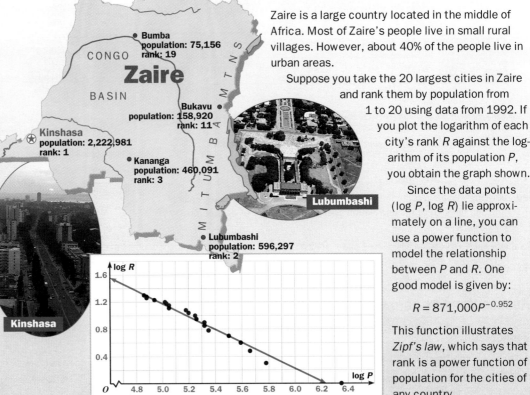

Zaire is a large country located in the middle of Africa. Most of Zaire's people live in small rural villages. However, about 40% of the people live in urban areas.

Suppose you take the 20 largest cities in Zaire and rank them by population from 1 to 20 using data from 1992. If you plot the logarithm of each city's rank R against the logarithm of its population P, you obtain the graph shown.

Since the data points (log P, log R) lie approximately on a line, you can use a power function to model the relationship between P and R. One good model is given by:

$$R = 871{,}000P^{-0.952}$$

This function illustrates *Zipf's law*, which says that rank is a power function of population for the cities of any country.

PROJECT GOAL Your goal is to use Zipf's law to model the relationship between population and rank for cities in a country.

Collecting the Data

Work in a group of three students. Your group needs graph paper or a graphing calculator and access to a set of encyclopedias. Choose a country that interests you. Use an encyclopedia to make a list of the country's largest cities, their populations, and their ranks. Your list should contain at least 10 cities.

Some possibilities:

- "The Twenty Largest Cities in India"
- "Chinese Cities Having at Least One Million People"

The Twenty Largest Cities in Zaire

City	Population	Rank
Kinshasa	2,222,981	1
Lubumbashi	596,297	2
Kananga	460,091	3
Mbuji-Mayi	391,845	4
Kisangani		

Analyzing the Data

1. MAKE a scatter plot of the data pairs ($\log P$, $\log R$) for your cities. Fit a line to the plotted points, and find an equation of the line.

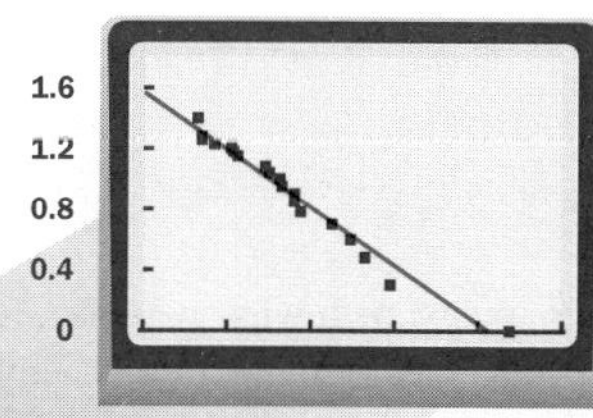


```
LinReg
 y=ax+b
 a=-.9521288837
 b=5.940205715
 r=-.9876633883
```

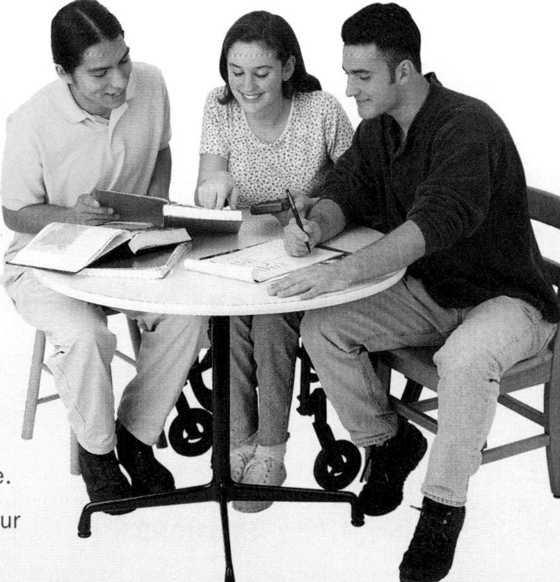

2. EXPRESS rank as a power function of population, using the equation of your fitted line.

3. PREDICT the rank of some city not used in your scatter plot. How well does the predicted rank match the actual rank?

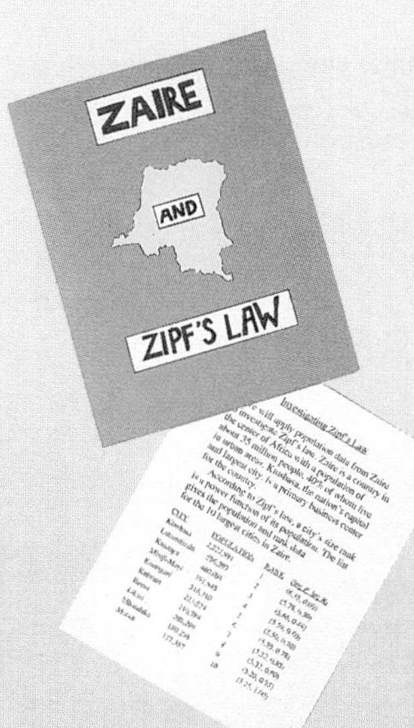

Writing a Report

Write a report summarizing your results. Your report should include:

- a statement of the goal of your project
- a brief description of your chosen country and its people
- a map of your country
- the population and rank data you collected
- a scatter plot and fitted line for the data pairs ($\log P$, $\log R$)
- an equation giving rank as a power function of population
- an assessment of how well Zipf's law describes the relationship between population and rank for cities in your country

Self-Assessment

In your report, explain how this project increased your understanding of logarithms and their role in mathematical modeling. How well did you work with your partners as a group? What could you do to improve your project?

Project Notes

Guiding Students' Work

Rubric for Chapter Project

4 Students select a country and rank the 10 most populated cities. They correctly make a scatter plot of the data and find an equation for the line of fit. Students write a correct power function prediction and use it to predict the rank of another city. An accurate and complete report is written that contains all of the information requested and indicates a thorough understanding of the mathematics of the project.

3 Students choose a country and rank the 10 most populated cities correctly. Students also make a scatter plot, but slight calculation errors are made in the points ($\log P$, $\log R$), in finding the line of fit, or in using the calculated values to find the power function. A report is written that shows an understanding of the mathematics, but some connections in the concepts of the project are not made. The report contains all of the information requested in the text.

2 Students choose a country and rank the 10 most populated cities, but do not correctly find a scatter plot related to the data or a line of fit. Students do not correctly calculate a power function, but they do use it to predict rank. Students complete a written report that contains only some of the requested information, and the mathematics of the project is not shown to be clearly understood.

1 Students select a country but do not rank 10 cities. Students do not draw a scatter plot with a line of fit or calculate a power function, or neither is done. Students do not predict the rank of any other cities. A report is done, but it is incomplete and does not show that students understand the mathematics of the project. Students should be encouraged to speak with the teacher as soon as possible to review their work and to make a new start on the project.

Chapter Support

Course Guide: Chapter 4
Lesson Plans: Chapter 4
Practice Bank: Cumulative Practice 25
Study Guide: Chapter 4 Review
Assessment Book: Chapter Tests 15 and 16; Chapter 4 Alternative Assessment
Test Generator Software
Technology Book: *Calculator Based Lab 2*
Portfolio Project Book
Preparing for College Entrance Tests
Professional Handbook

Progress Check 4.1–4.3

For each function:

a. Graph the function and its inverse in the same coordinate plane.

b. Find an equation for the inverse. *(Section 4.1)*

1. $y = -2x + 4$

a. 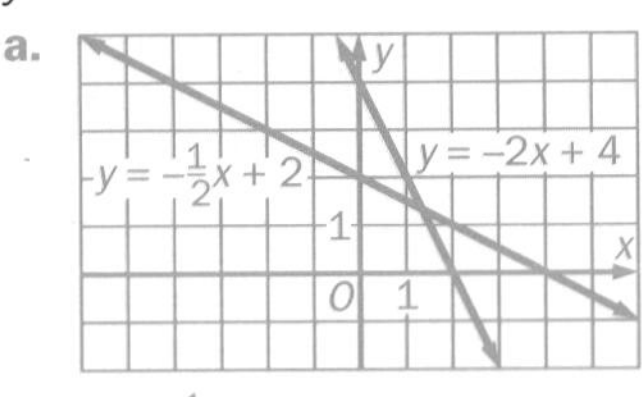

b. $y = -\frac{1}{2}x + 2$

2. $h(x) = \frac{1}{3}x - 2$

a. 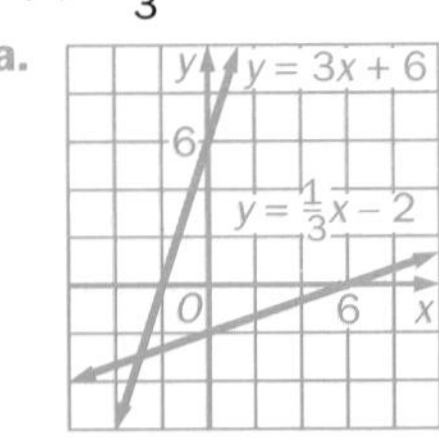

b. $y = 3x + 6$

3. Write $3^4 = 81$ in logarithmic form. *(Section 4.2)* $\log_3 81 = 4$

4. Write $\log_{16} 256 = 2$ in exponential form. *(Section 4.2)* $16^2 = 256$

Evaluate each logarithm. *(Section 4.2)*

5. $\log_8 64$ 2 **6.** $\log_{32} 8$ $\frac{3}{5}$

7. Write $\log_6 \frac{p^8q^5}{r^{1/3}}$ in terms of $\log_6 p$, $\log_6 q$, and $\log_6 r$. *(Section 4.3)*
$8 \log_6 p + 5 \log_6 q - \frac{1}{3} \log_6 r$

8. Write $\frac{2}{3} \log_5 p - 3 \log_5 q + 9 \log_5 r$ as a logarithm of a single expression. *(Section 4.3)*
$\log_5 \frac{p^{2/3}r^9}{q^3}$

CHAPTER

4 Review

STUDY TECHNIQUE

Compare what you learned in Chapter 4 with what you learned in Chapter 3. Make a list of the important ideas you learned in Chapter 3 and then revisited and extended in Chapter 4. Also list any new ideas discussed in Chapter 4.

VOCABULARY

inverse (p. 142)
logarithmic function (p. 149)
base-*b* logarithm (p. 149)
common logarithm (p. 150)
natural logarithm (p. 150)
extraneous solution (p. 165)
power function (p. 170)

SECTIONS 4.1 *and* 4.2

Given the graph of a function f, you can find the graph of f^{-1}.

A point **(b, a)** is on the graph of f^{-1} if and only if **(a, b)** is on the graph of f.

The graph of f^{-1} is the reflection of the graph of f over the line $y = x$.

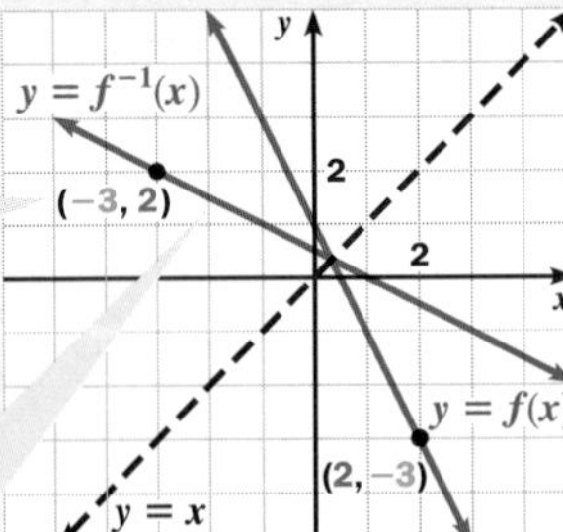

You can find the **inverse** of a function $y = f(x)$ by solving for x and switching x and y.

$$f(x) = -2x + 1$$
$$y = -2x + 1$$
$$-\frac{1}{2}y + \frac{1}{2} = x \quad \text{Solve for } x.$$
$$y = -\frac{1}{2}x + \frac{1}{2} \quad \text{Switch } x \text{ and } y.$$
$$f^{-1}(x) = -\frac{1}{2}x + \frac{1}{2}$$

The inverse of an exponential function is a **logarithmic function**.

	Exponential form	Logarithmic form	
base 10	$y = 10^x$ $100 = 10^2$	$\log y = x$ $\log 100 = 2$	common log
base e	$y = e^x$ $20 \approx e^3$	$\ln y = x$ $\ln 20 \approx 3$	natural log
base b ($b > 0, b \neq 1$)	$y = b^x$ $81 = 3^4$	$\log_b y = x$ $\log_3 81 = 4$	base-b log

SECTIONS 4.3 *and* 4.4

You can use properties of logarithms to simplify logarithmic expressions and to solve exponential and logarithmic equations.

Let M, N, b, and c be positive numbers with $b \neq 1$ and $c \neq 1$. Then:

Product Property

$$\log_b MN = \log_b M + \log_b N$$

$$\log 4 + \log 25 = \log 100 = 2$$

Quotient Property

$$\log_b \frac{M}{N} = \log_b M - \log_b N$$

$$\log_2 x - \log_2 8 = 1$$
$$\log_2 \frac{x}{8} = 1$$
$$\frac{x}{8} = 2^1$$
$$x = 16$$

Power Property

$$\log_b M^k = k \log_b M$$

$$5^x = 8$$
$$\log 5^x = \log 8$$
$$x \log 5 = \log 8$$
$$x = \frac{\log 8}{\log 5} \approx 1.29$$

Change-of-Base Property

$$\log_b M = \frac{\log_c M}{\log_c b}$$

This property is often used to change base-b logarithms to common or natural logarithms.

$$\log_3 5 = \frac{\log 5}{\log 3} \approx 1.46$$

$$\log_4 7 = \frac{\ln 7}{\ln 4} \approx 1.40$$

SECTION 4.5

You can use exponential functions or **power functions** to make predictions. For example, you can predict the United States per capita personal income I for some number of years t in the future. The table at the right shows the data for six years, and the table below gives an exponential and a power model for the data.

	Exponential Model	Power Model
Equation of the fitted line	$\log I = 0.026t + 1.01$	$\log I = 0.431 \log t + 0.834$
Function	$I = 10.2(1.06)^t$	$I = 6.82t^{0.431}$

t (years since 1980)	I (thousands of dollars)
5	13.9
6	14.6
7	15.6
8	16.6
9	17.7
10	18.6

In general, a set of data pairs (x, y) can be modeled by:

- an exponential function $y = ab^x$ if the points $(x, \log y)$ lie on or close to a line.
- a power function $y = ax^b$ if the points $(\log x, \log y)$ lie on or close to a line.

Progress Check 4.4–4.5

Solve each equation. Round answers to the nearest hundredth. *(Section 4.4)*

1. $4^x = 18$ 2.08

2. $3e^{-0.2t} = 2.6$ 0.72

3. $\log_4 x + \log_4 (x - 3) = 1$ 4

4. Evaluate $\log_7 9$. 1.13

Write y as a function of x. *(Section 4.5)*

5. $\log y = 0.08x + 2.4$
$y = 251.19(1.202)^x$

6. $\ln y = 0.76 + 2.8x$
$y = 2.138(16.44)^x$

7. $\log y = 0.84 \log x - 2.1$
$y = 0.00794x^{0.84}$

8. a. Make a scatter plot of the data pairs $(t, \log P)$ for a small city. *(Section 4.5)*

Time t (years after 1960)	Population P (in thousands)
5	128
10	132
15	140
20	148
25	158
30	170

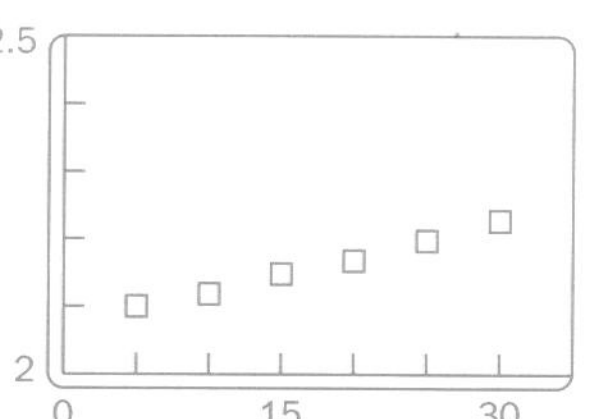

b. Find an equation giving P as an exponential function of t.
$P = 120(1.01)^t$
(by calculator only: $P = 118.78(1.012)^t$)

Chapter 4 Assessment
Form A Chapter Test

NAME ______________ DATE ________ SCORE ________

Test 15

TEST ON CHAPTER 4 (FORM A)

DIRECTIONS: Write the answers in the spaces provided.

Graph each function and its inverse in the same coordinate plane, then find an equation for the inverse.

1. $y = -\frac{2}{3}x$ **2.** $y = -2x + 5$

An hourglass containing sand measuring 12 in. high allows the sand to flow from the top to the bottom at a rate of 0.2 inches per hour.

3. Write an equation giving the height h of the sand remaining in the top of the hourglass as a function of the time t since the hourglass was turned over.

4. Find the inverse of the function you wrote for Question 3.

5. At what time is the height of the sand 6 in.?

Write each equation in logarithmic form.

6. $3^4 = 81$ **7.** $e^{-2} = \frac{1}{e^2}$ **8.** $(0.2)^3 = 0.008$

Write each equation in exponential form.

9. $\log_{81} 3 = 0.25$ **10.** $\ln e = 1$ **11.** $\log_{1/4} 32 = -\frac{5}{2}$

For Questions 12–15, evaluate each logarithm.

12. $\log_{64} 4$ **13.** $\ln e^{-5}$ **14.** $\ln e^4$ **15.** $\log 0.001$

16. Writing Without using a calculator, explain why $3 < \log_2 12 < 4$.
Notice that $3 = \log_2 8$ and $4 = \log_2 16$. Since $8 < 12 < 16$, $\log_2 8 < \log_2 12 < \log_2 16$. Therefore, $3 < \log_2 12 < 4$.

ANSWERS

1. inverse: $y = -\frac{3}{2}x$
2. inverse: $y = -\frac{1}{2}x + \frac{5}{2}$
3. $h = 12 - 0.2t$
4. $h = -5t + 60$
5. 30 h
6. $\log_3 81 = 4$
7. $\ln\left(\frac{1}{e^2}\right) = -2$
8. $\log_{0.2} 0.008 = 3$
9. $81^{0.25} = 3$
10. $e^1 = e$
11. $\left(\frac{1}{4}\right)^{-5/2} = 32$
12. $\frac{1}{3}$
13. -5
14. 4
15. -3
16. *See question.*

Assessment Book, ALGEBRA 2: EXPLORATIONS AND APPLICATIONS

21

Chapter 4 Assessment
Form B Chapter Test

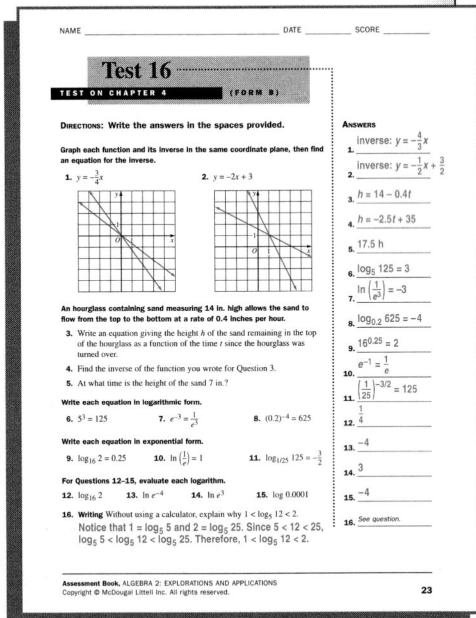

NAME ______________ DATE ________ SCORE ________

Test 16

TEST ON CHAPTER 4 (FORM B)

DIRECTIONS: Write the answers in the spaces provided.

Graph each function and its inverse in the same coordinate plane, then find an equation for the inverse.

1. $y = -\frac{3}{4}x$ **2.** $y = -2x + 3$

An hourglass containing sand measuring 14 in. high allows the sand to flow from the top to the bottom at a rate of 0.4 inches per hour.

3. Write an equation giving the height h of the sand remaining in the top of the hourglass as a function of the time t since the hourglass was turned over.

4. Find the inverse of the function you wrote for Question 3.

5. At what time is the height of the sand 7 in.?

Write each equation in logarithmic form.

6. $5^3 = 125$ **7.** $e^{-3} = \frac{1}{e^3}$ **8.** $(0.2)^{-4} = 625$

Write each equation in exponential form.

9. $\log_{16} 2 = 0.25$ **10.** $\ln\left(\frac{1}{e}\right) = 1$ **11.** $\log_{1/25} 125 = -\frac{3}{2}$

For Questions 12–15, evaluate each logarithm.

12. $\log_{16} 2$ **13.** $\ln e^{-4}$ **14.** $\ln e^3$ **15.** $\log 0.0001$

16. Writing Without using a calculator, explain why $1 < \log_5 12 < 2$.
Notice that $1 = \log_5 5$ and $2 = \log_5 25$. Since $5 < 12 < 25$, $\log_5 5 < \log_5 12 < \log_5 25$. Therefore, $1 < \log_5 12 < 2$.

ANSWERS

1. inverse: $y = -\frac{4}{3}x$
2. inverse: $y = -\frac{1}{2}x + \frac{3}{2}$
3. $h = 14 - 0.4t$
4. $h = -2.5t + 35$
5. 17.5 h
6. $\log_5 125 = 3$
7. $\ln\left(\frac{1}{e^3}\right) = -3$
8. $\log_{0.2} 625 = -4$
9. $16^{0.25} = 2$
10. $e^{-1} = \frac{1}{e}$
11. $\left(\frac{1}{25}\right)^{-3/2} = 125$
12. $\frac{1}{4}$
13. -4
14. 3
15. -4
16. *See question.*

Assessment Book, ALGEBRA 2: EXPLORATIONS AND APPLICATIONS

23

CHAPTER

4 Assessment

VOCABULARY QUESTIONS

For Questions 1 and 2, complete each paragraph.

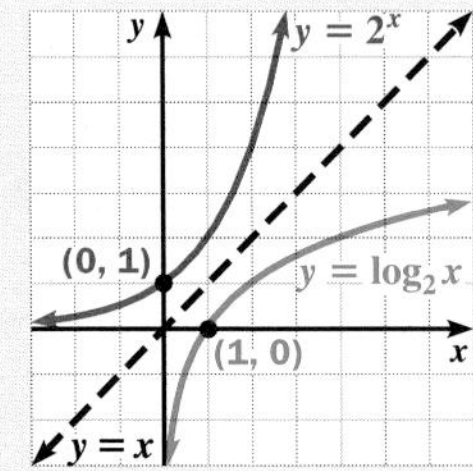

1. The function $y = \log_2 x$ is the _?_ of the function $y = 2^x$. The graph of $y = \log_2 x$ is the _?_ of the graph of $y = 2^x$ over the line $y = x$.

2. The expressions log 12, ln 12, and $\log_5 12$ represent the _?_ logarithm, the _?_ logarithm, and the _?_ logarithm of 12. The _?_ formula can be used to evaluate $\log_5 12$.

SECTIONS 4.1 *and* 4.2

For each function:

a. Graph the function and its inverse in the same coordinate plane.

b. Find an equation for the inverse.

3. $f(x) = -\frac{1}{2}x$ **4.** $y = -x + 3$ **5.** $g(x) = 10^{x/2}$ **6.** $y = e^{-x}$

7. a. Look at your answers to Questions 3–6. What is interesting about the equation in Question 4?

b. Writing Explain why the equation in Question 4 has the characteristic you noted in part (a).

8. A 10 in. tall cylindrical candle burns at a rate of 0.25 in./h.

a. Write an equation giving height as a function of time.

b. Find the inverse of the function in part (a).

c. When is the height of the candle 6 in.?

Write each equation in logarithmic form.

9. $2^6 = 64$ **10.** $e^{-1} = \frac{1}{e}$ **11.** $(0.5)^{-3} = 8$

Write each equation in exponential form.

12. $\log_4 2 = 0.5$ **13.** $\ln 1 = 0$ **14.** $\log_{1/9} 27 = -\frac{3}{2}$

Evaluate each logarithm.

15. $\log_{27} 3$ **16.** $\ln e^{-2}$ **17.** $\ln e^4$ **18.** $\log 0.01$

19. If $0 < x < 1$, what can you say about $\log x$? Explain your answer.

20. Without using a calculator, explain why $2 < \log_3 15 < 3$.

ANSWERS Chapter 4

Assessment

1. inverse; reflection
2. common; natural; base-5; change-of-base

3. a.

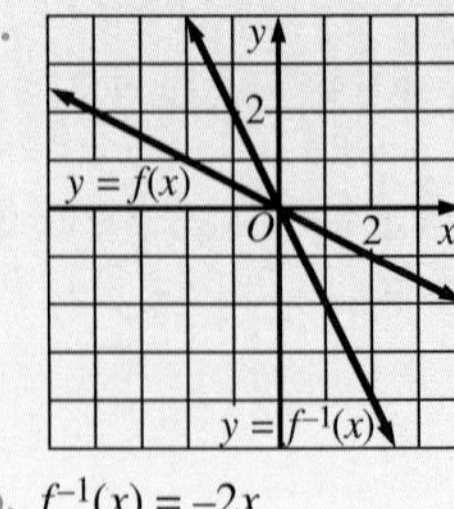

b. $f^{-1}(x) = -2x$

4. a.

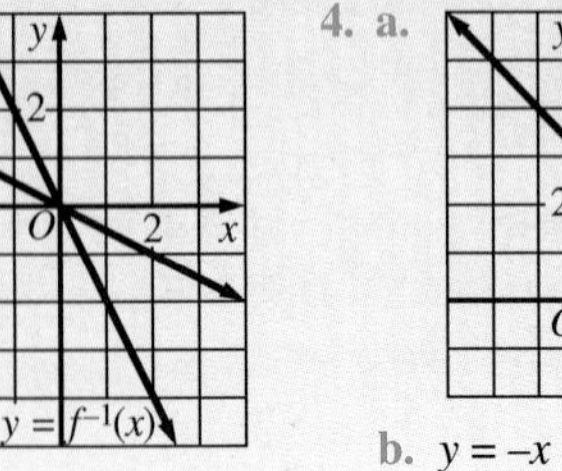

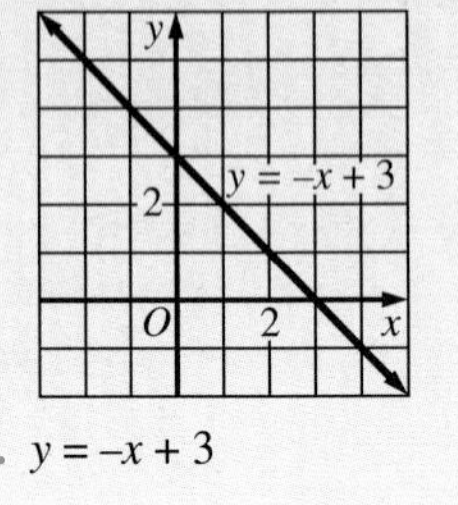

b. $y = -x + 3$

5. a.

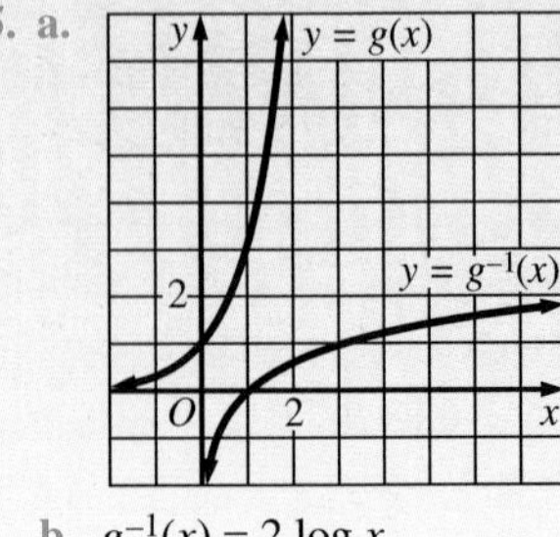

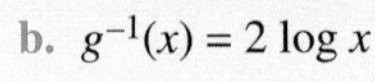

b. $g^{-1}(x) = 2 \log x$

6. a.

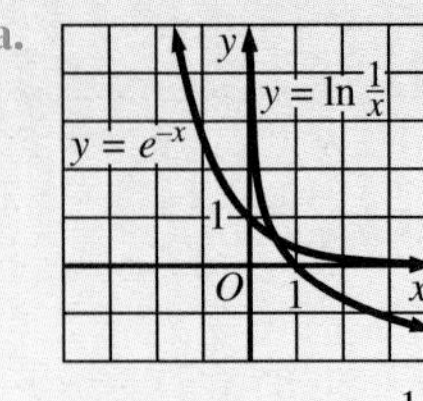

b. $y = -\ln x$ or $y = \ln\frac{1}{x}$

7. a. The equation for the function and the equation for its inverse are the same.

SECTIONS 4.3 *and* 4.4

Write each expression in terms of $\log_5 p$, $\log_5 q$, and $\log_5 r$.

21. $\log_5 pq^2$ **22.** $\log_5 \frac{5}{r}$ **23.** $\log_5 \frac{r^{1/4}}{p^3 q}$

Write as a logarithm of a single number or expression.

24. $\frac{1}{4} \ln 16$ **25.** $3 \log_a u + 2 \log_a v$ **26.** $3 \log_b 5 - \frac{1}{2} \log_b 25$

27. Open-ended Problem Find positive numbers M, N, and b (with $b \neq 1$) to show that, in general, $\log_b MN \neq \log_b M \cdot \log_b N$.

28. CHEMISTRY The formula used to calculate pH is $\text{pH} = -\log [\text{H}^+]$ where $[\text{H}^+]$ is the concentration of hydrogen ions in moles per liter. Find the pH of acid rain, for which $[\text{H}^+] = 3 \times 10^{-5}$ moles/L. Round your answer to the nearest hundredth.

Solve each equation. Round each answer to the nearest hundredth.

29. $4e^x = 7.2$ **30.** $12(0.6)^{2x} = 15$ **31.** $2 \log_4 (t + 3) = 6$

Evaluate each logarithm. Round each answer to the nearest hundredth.

32. $\log_9 42$ **33.** $\log_{1/6} 3$ **34.** $\log_3 0.7$

SECTION 4.5

35. The table shows the median price for single-family homes in the United States.

Years since 1980	Price (thousands)
5	75.5
9	93.1
10	95.5
11	100.3

a. **Technology** Find an exponential function and a power function that model the data.

b. Which of the functions from part (a) is the better model? Explain your choice.

c. Use the model you named in part (b) to predict the median price for single-family homes in the United States in 1992. Compare this with the actual median price of $103,700.

PERFORMANCE TASK

36. Cooperative Learning Work with a partner. One of you should write several functions for exponential growth. The other should write several functions for exponential decay. (See Chapter 3 for ideas.)

Exchange your functions and use logarithms to find the doubling times for the growth functions or the half-lives for the decay functions. Generalize your results using $y = ab^x$ with $b > 1$ (for growth) and with $0 < b < 1$ (for decay). Compare your generalizations and describe any similarities.

Soon Yi

$y = 50(1.02)^x$

$100 = 50(1.02)^x$

$2 = (1.02)^x$

$\ln 2 = x \ln (1.02)$

$\frac{\ln 2}{\ln (1.02)} = x$

$35.0 \approx x$

Chapter 4 Assessment

Form C Alternative Assessment

Chapter 4

ALTERNATIVE ASSESSMENT

1. **Open-ended Problem** A plumber charges \$20 an hour. For each job he will charge an additional \$25 for the service call. Write an equation that represents the plumber's total charge with respect to the number of hours worked. Find the inverse of your equation. What does the inverse equation represent? Describe a real-world situation involving the plumber that could be represented by the inverse equation.

2. **Group Activity** Work in groups of four. Draw the graphs of ten functions. Use a wide variety of functions. Exchange graphs with another group. Members of the group should draw the inverse of each function on the same pair of axes. Return the graphs to the original group. The original group members should check the accuracy of the inverse graphs by folding the graphs along the line $y = x$ and verifying that the inverse graphs are reflections of the original graphs.

3. **Project** Draw examples of various functions and their inverses. Use your drawings to answer the following questions.
   a. Is the inverse of a function always a function? Why or why not?
   b. Describe the cases in which the inverse of a function is also a function.

4. **Performance Task** Show that $\log_b \frac{M}{N}$ is not the same as $\frac{\log_b M}{\log_b N}$ by finding values for b, M, and N that make the statement false.

5. **Performance Task** What is the value of $b^{\log_b N}$? Explain why.

6. **Performance Task** What is the value of $\log_b 1$? Explain why.

7. **Performance Task** Use the Product Property of Logarithms and the Product Property of Exponents to explain why $\log_b M^k = k \log_b M$.

8. **Performance Task** Graph $y = \log_3 x$ and $y = 3^x$ at the same time on a graphing calculator. (*Hint:* Use the Change of Base Property to rewrite the logarithm equation in a graphable form.) Use the table on your calculator to get a table of values for each equation. How can you tell from the tables that these equations are inverses?

9. **Project** Graph the data below to compare the men's and women's running times. Find the equations of a line of fit for each graph. Compare linear, exponential, logarithmic, and power regressions. Will the women's record ever be lower than the men's record?

Men's and Women's World Records in the 800 Meter Run

Year	Men's Record	Women's Record
1905	113.4	—
1915	111.9	—
1925	111.9	144.0
1935	109.7	135.6
1945	106.6	132.0
1955	105.7	125.0
1965	104.3	118.0
1975	104.1	117.5
1985	101.73	113.3

b. The function is its own inverse. A point (b, a) is on the graph of the inverse if (a, b) is on the graph of the function, that is, if $b = -a + 3$. Then for every point (b, a) on the graph of the inverse, $(b, a) = (-a + 3, a) = (-a + 3, -(-a + 3) + 3)$, which is on the graph of the original function.

8. a. Let t = time in hours after the candle is lit and h = the height of the candle in inches; $h = 10 - 0.25t$.

b. $t = 40 - 4h$

c. 16 h after the candle is lit

9. $\log_2 64 = 6$

10. $\ln \frac{1}{e} = -1$ **11.** $\log_{0.5} 8 = -3$

12. $4^{0.5} = 2$ **13.** $e^0 = 1$

14. $\left(\frac{1}{9}\right)^{-3/2} = 27$ **15.** $\frac{1}{3}$

16. −2 **17.** 4

18. −2

19. If $0 < x < 1$, then $\log x < 0$. Raising 10 to any positive power produces a number greater than 1.

20. $3^2 = 9$ and $3^3 = 27$; Since $9 < 15 < 27$, $2 < \log_3 15 < 3$.

21. $\log_5 p + 2 \log_5 q$

22. $1 - \log_5 r$

23. $\frac{1}{4} \log_5 r - 3 \log_5 p - \log_5 q$

24. ln 2 **25.** $\log_a u^3v^2$ **26.** $\log_b 25$

27. Answers may vary. An example is given. Let $b = 10$, $M = 2$, and $N = 5$; $\log_b MN = \log 10 = 1$; $\log_b M \cdot \log_b N = \log 2 \cdot \log 5 \approx 0.21$.

28. 4.52 **29.** 0.59 **30.** −0.22

31. 61 **32.** 1.70 **33.** −0.61

34. −0.32

35. See answers in back of book.

36. Answers may vary. Check students' work.

CHAPTER

5 Quadratic Functions

OVERVIEW

Connecting to Prior and Future Learning

⇔ This chapter opens with a review of graphing and solving simple quadratic functions in Section 5.1. This topic is expanded in Section 5.2 to include translations of parabolas. Related topics reviewed in the **Student Resources Toolbox** include translating a figure on page 796, reflecting a figure and line symmetry on page 797, and dilating a figure on page 798.

⇔ Students continue to study graphs of quadratic functions in Section 5.3, where they learn to write quadratic functions in intercept form. A review of factoring variable expressions is available on page 782 of the **Student Resources Toolbox**.

⇔ The last three sections of this chapter present different ways to solve quadratic equations: completing the square, the quadratic formula, and factoring. Students will use these methods again as they continue to study mathematics.

Chapter Highlights

Interview with Mark Thomas: Engineers use mathematics in many aspects of their work. Exercises relating to this interview can be found on pages 190 and 220.

Explorations: In Section 5.2, students use a graphing calculator or graphing software to study the translation of parabolas. In Sections 5.4 and 5.6, students use algebra tiles to explore completing the square and factoring.

The Portfolio Project: Students conduct an experiment to find out how the water level in a can affects the time it takes water to flow out of the can and use a quadratic model to describe their results.

Technology: Throughout this chapter, graphing calculators or graphing software are used to graph functions and spreadsheets are used to create tables.

OBJECTIVES

Section	Objectives	NCTM Standards
5.1	• Recognize and draw graphs of quadratic functions. • Solve simple quadratic equations. • Examine quadratic relationships in real-world situations.	1, 2, 3, 4, 5, 6
5.2	• Graph equations in the form $y = a(x - h)^2 + k$. • Solve quadratic equations. • Solve real-world problems using quadratic equations.	1, 2, 3, 4, 5, 6
5.3	• Write quadratic equations in intercept form. • Maximize or minimize quadratic functions. • Use quadratic functions to find maximums and minimums in real-world situations.	1, 2, 3, 4, 5, 6
5.4	• Complete the square to write quadratic functions in vertex form. • Use completing the square to find maximums or minimums in real-world situations.	1, 2, 3, 4, 5, 6
5.5	• Solve equations using the quadratic formula. • Use the discriminant to determine how many solutions an equation has. • Use the quadratic formula to solve real-world problems.	1, 2, 3, 4, 5, 6
5.6	• Factor quadratic expressions. • Solve quadratic equations using factoring. • Solve real-world problems using factoring.	1, 2, 3, 4, 5, 6

INTEGRATION

Mathematical Connections	5.1	5.2	5.3	5.4	5.5	5.6
algebra	**185–191***	**192–198**	**199–205**	**206–213**	**214–220**	**221–227**
geometry	189		204			225
data analysis, probability, discrete math	185, 187, 189–191	194–198	201–205	208–213	214–216, 218–220	224–227
patterns and functions	**185–191**	**192–198**	**199–205**	**206–213**	**214–220**	**221–227**
logic and reasoning	**185–191**	192, 194, 196–198	201–205	207, 208, 210–213	214–216, 218–220	222–227
Interdisciplinary Connections and Applications						
biology and earth science		198		211		
chemistry and physics		194		208, 210		
business and economics			201, 203			
sports and recreation	185, 189			212	218	
astronomy	190					226
government					219	
personal finance, engineering, physiology, driving		197	204		219	224, 225

***Bold page numbers** *indicate that a topic is used throughout the section.*

TECHNOLOGY

	opportunities for use with	
Section	**Student Book**	**Support Material**
5.1	graphing calculator McDougal Littell Mathpack *Function Investigator* *Stats!*	**Function Investigator with Matrix Analyzer Activity Book:** Function Investigator Activity 12 **Geometry Inventor Activity Book:** Activities 16, 23, and 29
5.2	graphing calculator or graphing software McDougal Littell Mathpack *Stats!*	**Technology Book:** Calculator Activity 5 Spreadsheet Activity 5 **Function Investigator with Matrix Analyzer Activity Book:** Function Investigator Activity 15 **Geometry Inventor Activity Book:** Activity 22
5.3	graphing calculator or graphing software McDougal Littell Mathpack *Function Investigator*	**Function Investigator with Matrix Analyzer Activity Book:** Function Investigator Activity 16
5.4	graphing calculator McDougal Littell Mathpack *Function Investigator*	
5.5	graphing calculator	
5.6	graphing calculator McDougal Littell Mathpack *Function Investigator*	

PLANNING GUIDE

Regular Scheduling (45 min)

Section	Materials Needed	Core Assignment	Extended Assignment	*exercises that feature* Applications	Communication	Technology
5.1	graphing calculator, graph paper	1–4, 6–17, 24–26, 28–32	1–5, 7, 9, 11–32	19–25	18	20, 27
5.2	graphing calculator or software, graph paper	1–4, 9–24, 27–36	1–8, 9–19 odd, 21–36	5–8, 25	4, 5, 7, 27	7
5.3	graphing calculator or software, graph paper	1–10, 15, 17–28, AYP*	1–28, AYP	12–14	7, 11, 18	
5.4	algebra tiles, graph paper, graphing calculator	**Day 1:** 1–14, 17 **Day 2:** 25–47	**Day 1:** 1–24 **Day 2:** 25–47	14–23	15, 22	16
5.5	graph paper	**Day 1:** 1–17 **Day 2:** 18–23, 30, 31, 35–44	**Day 1:** 1–17 **Day 2:** 18–44	11 25–29, 31–34	25, 26	
5.6	algebra tiles, graph paper, graphing calculator	**Day 1:** 1–10, 12–21 **Day 2:** 26–39, AYP	**Day 1:** 1–25 **Day 2:** 26–39, AYP	10, 23–25	21 30	
Review/ Assess		**Day 1:** 1–11 **Day 2:** 12–23 **Day 3:** Ch. 5 Test	**Day 1:** 1–11 **Day 2:** 12–23 **Day 3:** Ch. 5 Test	15, 19	6	
Portfolio Project		Allow 2 days.	Allow 2 days.			

Yearly Pacing (with Portfolio Project)	**Chapter 5 Total**	**Chapters 1–5 Total**	**Remaining**	**Total**
	14 days	66 days	94 days	160 days

Block Scheduling (90 min)

	Day 27	Day 28	Day 29	Day 30	Day 31	Day 32	Day 33	Day 34
Teach/Interact	Ch. 4 Test 5.1	5.2: Exploration, page 192 5.3: Exploration, page 199	5.4: Exploration, page 206	5.5	5.6: Exploration, page 221	Review Port. Proj.	Review Port. Proj.	Ch. 5 Test 6.1
Apply/Assess	**Ch. 4 Test** **5.1:** 1–4, 6–17, 24–26, 28–32	**5.2:** 1–4, 9–24 27–36 **5.3:** 1–10 15, 17–28, AYP*	**5.4:** 1–14, 17, 25–47	**5.5:** 1–23, 30, 31, 35–44	**5.6:** 1–10 12–21, 26–39, AYP	**Review:** 1–11 **Port. Proj.**	**Review:** 12–23 **Port. Proj.**	**Ch. 5 Test** **6.1:** 1–11, 13–17, 23–31

NOTE: A one-day block has been added for the Portfolio Project—timing and placement to be determined by teacher.

Yearly Pacing (with Portfolio Project)	**Chapter 5 Total**	**Chapters 1–5 Total**	**Remaining**	**Total**
	7 days	$33\frac{1}{2}$ days	$48\frac{1}{2}$ days	82 days

* ***AYP*** *is Assess Your Progress.*

LESSON SUPPORT

Section	Practice Bank	Study Guide*	Assessment Book*	Visuals	Explorations Lab Manual	Lesson Plans	Technology Book
5.1	26	5.1		Warm-Up 5.1	Master 2 Add. Expl. 7	5.1	
5.2	27	5.2		Warm-Up 5.2 Folder 5	Masters 1, 2, 9	5.2	Calculator Act. 5 Spreadsheet Act. 5
5.3	28	5.3	Test 17	Warm-Up 5.3	Masters 2, 10	5.3	
5.4	29	5.4		Warm-Up 5.4 Folder B	Masters 2, 3, 4, 11	5.4	
5.5	30	5.5		Warm-Up 5.5	Master 2 Add. Expl. 8	5.5	
5.6	31	5.6	Test 18	Warm-Up 5.6 Folder B	Masters 2, 3, 4	5.6	
Review Test	32	Chapter Review	Tests 19, 20 Alternative Assessment			Review Test	Calculator Based Lab 3

*__Spanish versions__ of *Study Guide* and *Assessment Book* are available.

Chapter Support

- Course Guide
- Lesson Plans
- Portfolio Project Book
- Preparing for College Entrance Tests
- Multi-Language Glossary
- *Test Generator* Software
- Professional Handbook

Software Support

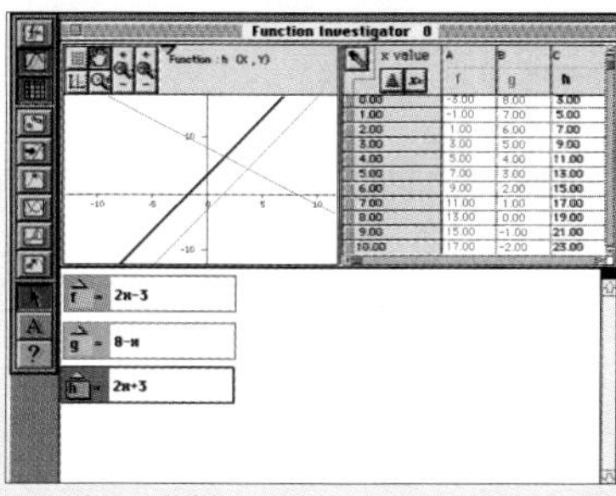

McDougal Littell Mathpack
Stats!
Function Investigator

Internet Support

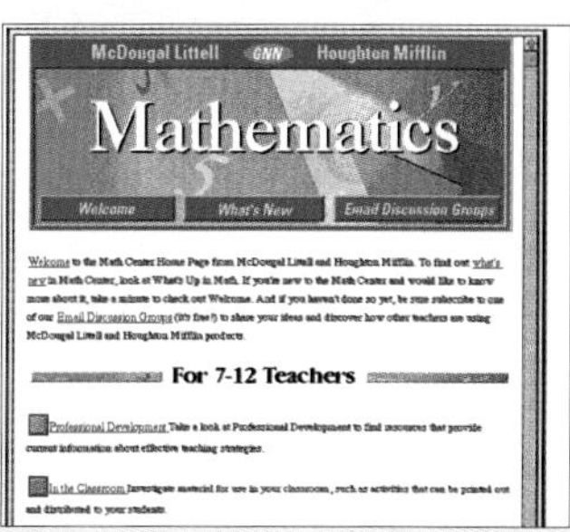

http://www.hmco.com
Next go to McDougal Littell; then the Education Center; then Secondary Math.

OUTSIDE RESOURCES

Books, Periodicals

Edwards, Thomas G. "Exploring Quadratic Functions: From *a* to *c*." *Mathematics Teacher* (February 1996): pp. 144–146.

Metz, James. "Sharing Teaching Ideas: Seeing b in $y = ax^2 + bx + c$." *Mathematics Teacher* (January 1994): pp. 23–25.

Jones, Graham A., Carol A. Thornton, Carol A. McGehe, and David Cole. "Rich Problems—Big Payoff." *Mathematics Teaching in the Middle School* (November–December 1995): pp. 520–525.

Activities, Manipulatives

Breuningsen, Chris, Bill Bower, Linda Antinone, and Elisa Breuningsen. "That's the Way the Ball Bounces." *Real-World Math with the CBL System.* Activity 9: pp. 49–54. Texas Instruments, 1995.

Leiva, Miriam, Joan Ferrini-Mundy, and Loren P. Johnson. "Playing With Blocks: Visualizing Functions." *Mathematics Teacher* (November 1992): pp. 641–646.

Breuningsen, Chris, Bill Bower, Linda Antinone, and Elisa Breuningsen. "What Goes Up...." *Real-World Math with the CBL System.* Activity 8: pp. 45–48. Texas Instruments, 1995.

Software

Meridian Creative Group. "Exploration: Modeling Quadratic Functions I" and "Exploration: Modeling Quadratic Functions II." Worksheets and software included for use with the CBL.

Explorations in Precalculus for the TI-82: pp. 13–24.

Videos

Videodisk: "Projectile Motion." *Science: Forces and Energy:* (Side A). Macmillan/McGraw-Hill, 1993.

Internet

Join the Math Forum at Swarthmore by connecting to:
http://forum.swarthmore.edu

Interview Notes

Background

Engineering

An engineer is a person who takes scientific information and puts it to practical use. The word *engineering* comes from the Latin word *ingenaire*, which means to *design* or to *create*. The work of engineers has very practical uses and a wide variety of applications. Virtually every aspect of our lives has been influenced by engineering, including transportation, housing, energy use, the environment, and consumer products. While many careers in engineering are fairly well known, there are some branches of the profession that are less known, such as acoustical, agricultural, marine, textile, and transportation engineering. The field of engineering is constantly changing, and new areas are continuously developing.

Mark Thomas

Mark Thomas and the Lyons brothers came up with the idea for the Paper Vehicle Project in the fall of 1990, when the three men were working on their master's degrees in mechanical engineering at Stanford University. They were also teaching engineering at a high school in the San Francisco Bay area, and designed the project to develop an awareness of science and engineering in middle and high school students. The project began at Flood Middle School in Menlo Park, California and has since spread to high schools in Detroit, Michigan and Cincinnati, Ohio. The designed and constructed cars must be less than 10 feet in length, shorter than 5 feet in height, and must support at least 300 pounds.

Mark Thomas has received two national awards for the Paper Vehicle Project, including the Black Engineer of the Year for Community Service from *US Black Engineer* magazine in 1993.

CHAPTER

5 Quadratic Functions

Engineering a paper car

INTERVIEW Mark Thomas

Mark Thomas had not thought about engineering until a high school guidance counselor suggested it would be a good career for someone like him who did well in mathematics and science. "At the time, I didn't have the faintest idea as to what engineering was," Thomas admits. "But I decided to check it out anyway." Now, as a senior project engineer for a large auto manufacturer, Thomas is giving students an early introduction to the field he has grown to love. With the help of his coworkers Vincent Lyons and Elliot Lyons, Thomas has started the Paper Vehicle Project, a program that is offering Detroit-area high school students hands-on experience in engineering. The challenge for the students is to design and build a steerable, movable car made entirely out of paper and capable of supporting at least 300 lb.

> **"Never say never, no matter how big the challenge. If you work hard, you can do whatever it takes to make it happen."**

Mark Thomas stands with *Predator*, a paper vehicle constructed by students at Western High School in Detroit, Michigan.

182

Teamwork and Communication

"At first, they don't have any idea how to do it," Thomas explains. "But eventually they realize that a big, complex task can be accomplished through teamwork and by breaking it down into smaller, more manageable chunks."

In addition to gaining technical skills, the students learn how to communicate their ideas and collaborate successfully. "That's essential in any field," Thomas says, "whether they go on to become lawyers or janitors. The popularity of the program is obvious," he adds, "given that 90% of our kids come to school for three hours on Saturday mornings to build their cars. That's pretty amazing!"

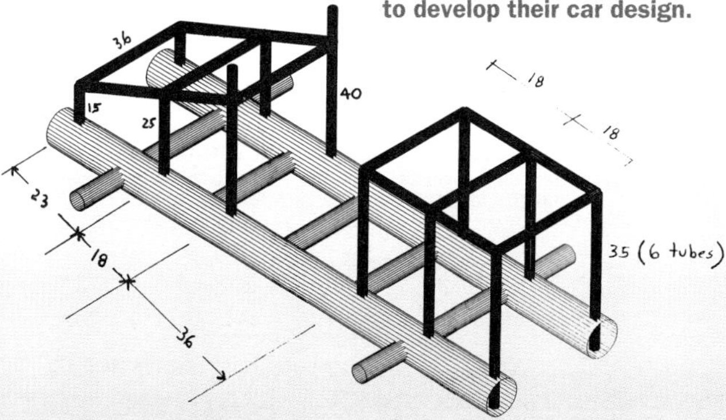

Students must spend many hours preparing to build a car before they can actually begin. They use CAD (Computer-Aided Design) software to develop their car design.

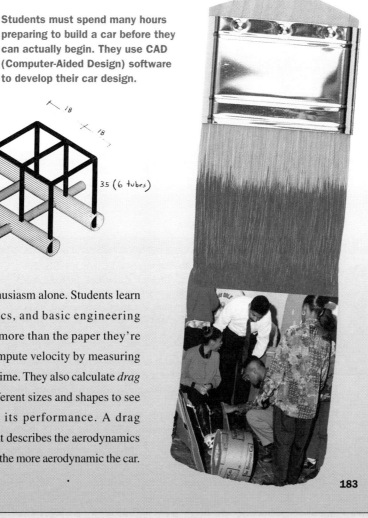

Basic Engineering

Still, the cars won't run on enthusiasm alone. Students learn enough mathematics, physics, and basic engineering to make their vehicles worth more than the paper they're constructed from. Students compute velocity by measuring how far a car travels in a given time. They also calculate *drag coefficients* for vehicles of different sizes and shapes to see how a car's design affects its performance. A drag coefficient, C_D, is a number that describes the aerodynamics of a car—the lower the number, the more aerodynamic the car.

Interview Notes

Background

Women and Automobile Factories
During World War II, significant numbers of women came to automobile factories and other heavy machinery operations to work. By September, 1943, ten million American men had gone to war, and women were called upon to replace them as factory workers, producing war supplies such as trucks, planes, and tanks. In Detroit, where many car factories were converted to war production, over 80% of the new hirees in September and October, 1942 were women. Many women were employed as factory line workers; others enlisted in machine shop courses to become skilled operators. In November, 1942, *Newsweek* reported that "depending on the industry, women today make up from 10 to 88 percent of total personnel in most war plants."

Second-Language Learners

Students learning English may wish to work on the research and writing of this project with the help of a peer tutor or aide.

Interview Notes

Mathematical Connection

The concept of increasing by a power instead of by a multiple is one of the key ideas to understanding quadratic functions. The effect of the squared term on data is investigated in Section 5.1. In this section, students look at the air resistance equation to find what component of the equation has the greatest impact on air resistance. Quadratic equations and their graphs have many applications in nature. In Section 5.5, students explore the meanings of the vertex and solutions of quadratic equations in a real-world application about fuel efficiency.

Explore and Connect

Research
The research may have a greater impact on students if they can talk to actual engineers. If possible, obtain a list of engineers in various fields who would be willing to be interviewed by students to discuss the mathematics involved in their jobs.

Project
Students can rank the pictures, then arrange them in a time line and display them on a poster to be hung in class. The time line should illustrate how aerodynamics has improved over the years.

Writing
Students' explanations should convey an understanding of the formula for air resistance.

Air Resistance

Aerodynamics is concerned with a force called *air resistance* that pushes against a car when it moves. This force depends on several variables, all of which are constant for a given car in a given place, except for the speed of the air relative to the car. The formula shows how air resistance is calculated. Air resistance is measured in pounds, frontal area in square feet, and relative speed in miles per hour.

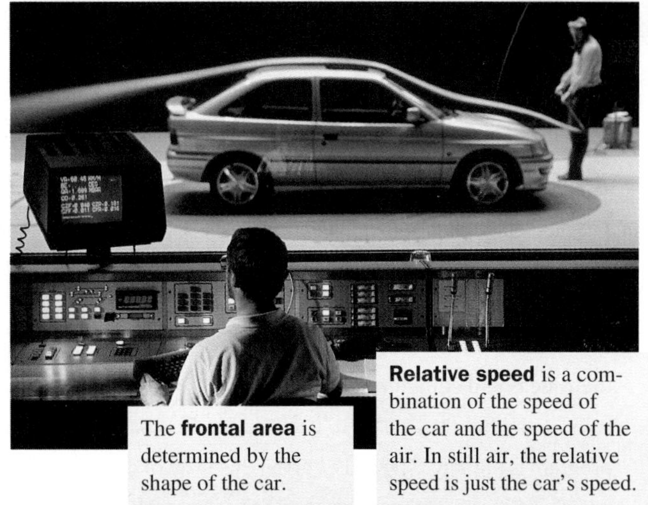

The **frontal area** is determined by the shape of the car.

Relative speed is a combination of the speed of the car and the speed of the air. In still air, the relative speed is just the car's speed.

$$\text{air resistance} = 0.00256 \times C_D \times \text{frontal area} \times (\text{relative speed})^2$$

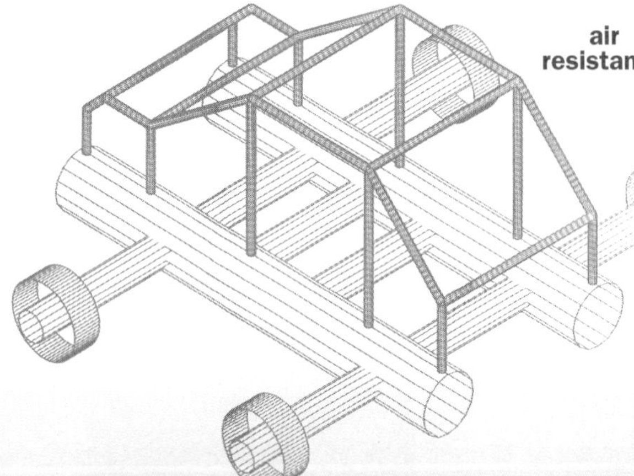

Because air resistance depends on the square of the relative speed, the functional relationship between these two variables is *quadratic*. With quadratic functions, small changes in the independent variable can cause large changes in the dependent variable.

Explore and Connect

The Lyons brothers lend a helping hand to Mark Thomas as he completes an inspection of another paper vehicle.

1. Research Engineers work in many different fields and have different specialties. Find out more about one of these specialties (some examples are civil, environmental, electrical, mechanical, and nuclear engineering) and how mathematics is used in the work.

2. Project Gather pictures of old and new cars and trucks. Rank them according to their aerodynamics. Explain your ranking.

3. Writing What do you think car designers do to minimize air resistance? Which parts of the formula for air resistance can they control?

Mathematics & Mark Thomas

In this chapter, you will learn more about air resistance and how the speed of a car affects fuel economy.

Related Exercises

Section 5.1
- **Exercises 22 and 23**

Section 5.5
- **Exercises 32–34**

Section

5.1 Working with Simple Quadratic Functions

Learn how to...

- **recognize and draw graphs of quadratic functions**
- **solve simple quadratic equations**

So you can...

- **examine the relationship between air resistance and speed when cycling, for example**

If you have ever ridden a bicycle, you know that the air pushes against you and can slow you down even when there is no wind. The force that the air exerts on you, called *air resistance*, is given by the formula on the preceding page. (The formula applies to bicycles as well as cars.)

EXAMPLE 1 Application: Bicycling

Shirley Scott is cycling on a day with no wind. She is in a casual, upright position on her bicycle. In this situation, the drag coefficient is 1.1 and the frontal area is 5.5 ft^2. Write an equation that gives the air resistance R (in pounds) as a function of the bicycle's speed s (in miles per hour). Graph the function.

SOLUTION

Use the formula shown on the preceding page.

$$R = 0.00256(1.1)(5.5)s^2$$
$$= 0.0155s^2$$

Substitute **1.1** for C_D and **5.5** for **frontal area**.

Make a table of values, plot the points, and connect them with a smooth curve.

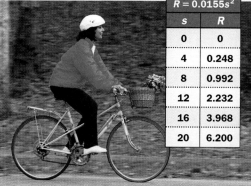

$R = 0.0155s^2$	
s	R
0	0
4	0.248
8	0.992
12	2.232
16	3.968
20	6.200

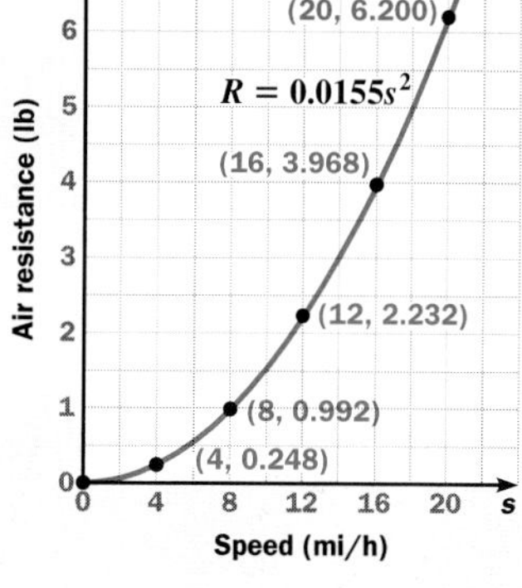

THINK AND COMMUNICATE

1. What happens to the air resistance when Shirley Scott doubles her speed?
2. What would you say is a reasonable domain for the air resistance function? What is the corresponding range?

Plan⇔Support

Objectives

- Recognize and draw graphs of quadratic functions.
- Solve simple quadratic equations.
- Examine quadratic relationships in real-world situations.

Recommended Pacing

Core and Extended Courses
Section 5.1: 1 day

Block Schedule
Section 5.1: $\frac{1}{2}$ block
(with Chapter 4 Test)

Resource Materials

Lesson Support
Lesson Plan 5.1
Warm-Up Transparency 5.1
Practice Bank: Practice 26
Study Guide: Section 5.1
Exploration Lab Manual: Additional Exploration 7, Diagram Master 2

Technology
Graphing Calculator
McDougal Littell Mathpack *Stats!* *Function Investigator with Matrix Analyzer Activity Book:* Function Investigator Activity 12 *Geometry Inventor Activity Book:* Activities 16, 23, and 29

Internet: http://www.hmco.com

Warm-Up Exercises

Evaluate each expression for $x = 3$ and $x = -0.5$.

1. $y = -2x^2$ −18; −0.5
2. $y = 0.0095x^2$ 0.0855; 0.002375

A triangle is formed by (−2, 1), (0, 4), and (2, 9). Find their images after each transformation.

3. a vertical stretch by a factor of 2 (−2, 2), (0, 8), (2, 18)
4. a horizontal shrink by a factor of $\frac{1}{4}$ $\left(-\frac{1}{2}, 1\right)$, (0, 4), $\left(\frac{1}{2}, 9\right)$
5. a reflection over the x-axis (−2, −1), (0, −4), (2, −9)

ANSWERS Section 5.1

Think and Communicate

1. Air resistance is multiplied by 4.
2. Answers may vary. An example is given. I think $0 \le s \le 30$ is a reasonable domain. The corresponding range is $0 \le R \le 13.95$.

Teach⇔Interact

Additional Example 1

Maya Raman built a soapbox racer to compete in the city-wide derby. She estimated that the drag coefficient for the racer is 0.57 and the frontal area is 4.8 ft^2. Write an equation that gives the air resistance R (in pounds) as a function of the racer's speed s (in miles per hour). Graph the function.

Use the formula shown on page 184. Substitute 0.57 for C_D and 4.8 for frontal area.

$R = 0.00256(0.57)(4.8)s^2$
$\approx 0.007s^2$

Make a table, plot the points, and connect them with a curve.

s	R
0	0
4	0.112
8	0.448
12	1.008
16	1.792
20	2.800

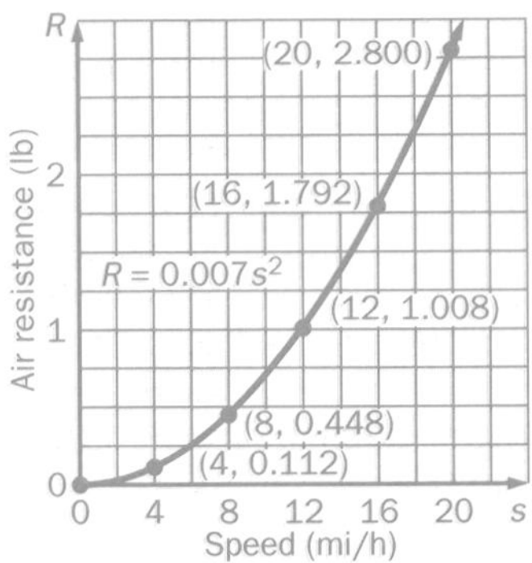

Think and Communicate

Suggest to students having difficulty with question 1 that they try specific examples such as 4 mi/h and 8 mi/h.

Additional Example 2

Use the graph of $y = x^2$ to sketch the graphs of the functions $y = 4x^2$, $y = \frac{1}{4}x^2$, and $y = -4x^2$.

Choose points on the graph of $y = x^2$, and on the graphs of the other functions, plot points that have the same x-coordinate.

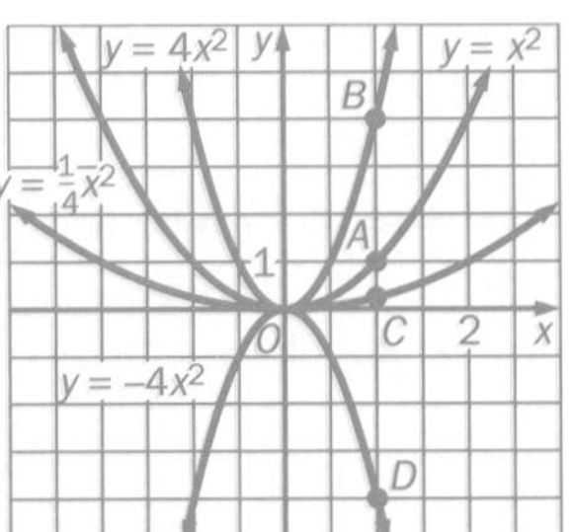

186

The air resistance function in Example 1 is an example of a **quadratic function** having the general form $y = ax^2$. The graph of the quadratic function $y = x^2$ with a domain of all real numbers is shown below.

Toolbox p. 797
Reflecting a Figure; Line Symmetry

The graph of a quadratic function is called a **parabola**.

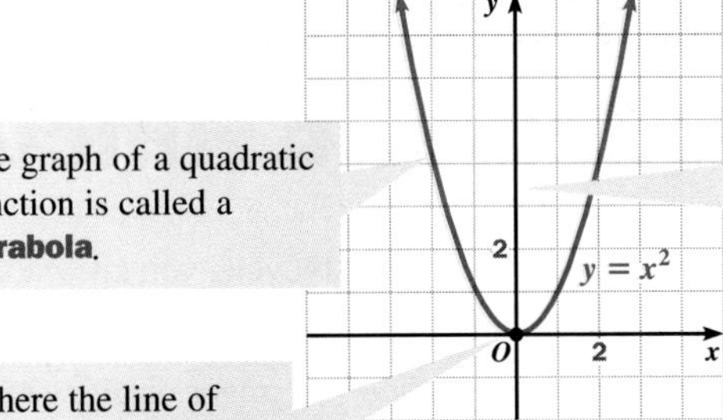

A parabola has a **line of symmetry**. This means that the part of the parabola on one side of the line is a reflection of the part on the other side.

The point where the line of symmetry crosses the parabola is called the **vertex**.

THINK AND COMMUNICATE

3. For the graph of the function $y = x^2$, what is the vertex? What is the line of symmetry?
4. What is the range of the function $y = x^2$?
5. The vertex of a parabola is sometimes called the *turning point*. Explain the significance of this name.
6. Why did the graph in Example 1 lack symmetry?

EXAMPLE 2

Use the graph of $y = x^2$ to sketch the graphs of the related functions $y = 2x^2$, $y = \frac{1}{2}x^2$, and $y = -x^2$. (In this case, $y = x^2$ is called a *parent* function.)

SOLUTION

Choose a point such as (1, 1) on the graph of $y = x^2$. Then plot the points with the same x-coordinate on the graphs of the other functions.

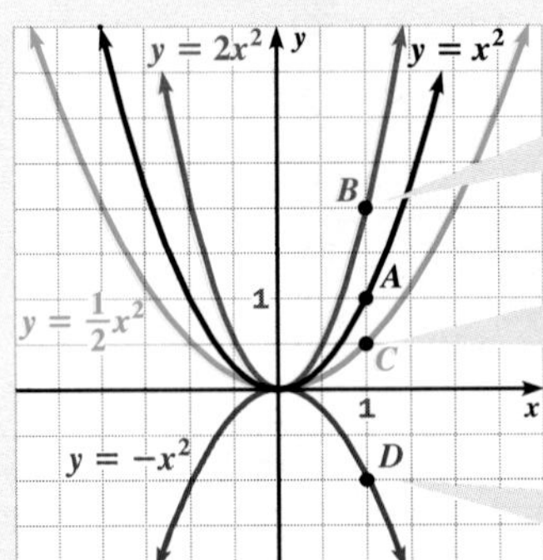

The y-coordinate of **B(1, 2)** is twice the y-coordinate of **A(1, 1)**.

The y-coordinate of $C\left(1, \frac{1}{2}\right)$ is one half of the y-coordinate of **A(1, 1)**.

The y-coordinate of **D(1, −1)** is the opposite of the y-coordinate of **A(1, 1)**.

Think and Communicate

3. (0, 0); the y-axis
4. the nonnegative real numbers
5. It is at this point that the curve of the parabola turns direction. For example, in the parabola at the top of page 186, the left side of the parabola curves downward. At the vertex, the parabola begins to curve upward.
6. The domain of the function included only positive numbers.

THINK AND COMMUNICATE

Toolbox p. 798
Dilating a Figure

7. You can think of the graphs of $y = 2x^2$ and $y = \frac{1}{2}x^2$ as being the result of vertical stretches and vertical shrinks of the graph of $y = x^2$. Explain.
8. Complete each statement using *vertical stretch* or *vertical shrink.*
   a. When $0 < a < 1$, the graph of $y = ax^2$ is the result of a _?_ of the graph of $y = x^2$.
   b. When $a > 1$, the graph of $y = ax^2$ is the result of a _?_ of the graph of $y = x^2$.
9. Complete this statement: The graph of $y = -3x^2$ is the *reflection over the x-axis* of the graph of $y =$ _?_.

You know that for every input x the function $y = x^2$ gives a unique output y. But what happens when you know y and want to find x?

The graph shows that two x-values are paired with each positive y-value. The x-values are called the *square roots* of y.

A number x is a **square root** of a number y if it satisfies the equation $x^2 = y$.

If $y = 16$, then $x = 4$ or $x = -4$.

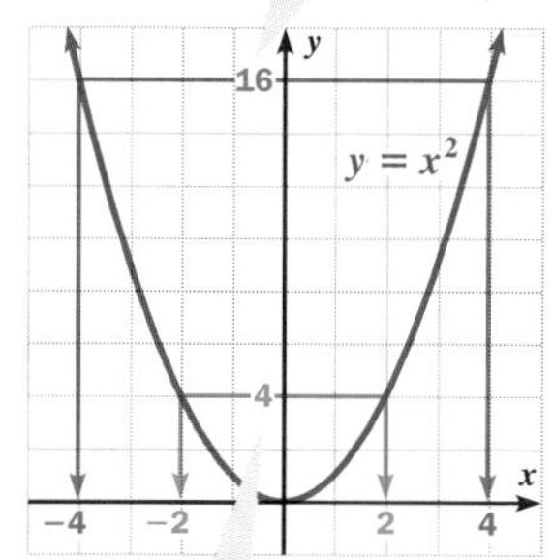

If $x^2 = y$, then $x = \begin{cases} 0, \text{ if } y = 0. \\ \pm\sqrt{y}, \text{ if } y > 0. \end{cases}$

Read "plus-or-minus square root of y."

If $y = 4$, then $x = 2$ or $x = -2$.

BY THE WAY

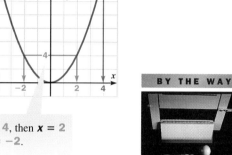

The experiment shows that objects of different weights fall the same distance in the same amount of time.

EXAMPLE 3

According to legend, the Italian scientist Galileo Galilei dropped objects of different weights from the top of the Leaning Tower of Pisa, about 180 ft high. The equation $d = 16t^2$ gives the distance d, in feet, an object falls as a function of time t, in seconds. How many seconds would it take an object to fall 180 ft?

SOLUTION

Use $d = 16t^2$ to find t when $d = 180$.

$$d = 16t^2$$

Substitute **180** for *d*.

$$180 = 16t^2$$

$$\frac{180}{16} = t^2$$

Divide by 16 to get t^2 alone on one side of the equation.

$$11.25 = t^2$$

$$\pm\sqrt{11.25} = t$$

$$\pm 3.35 \approx t$$

A calculator gives this decimal approximation for $\sqrt{11.25}$.

Since t represents time, use the positive solution. An object would fall 180 ft in about 3.4 s.

Teach⇔Interact

Learning Styles: Visual

Using the graph of $y = x^2$ to introduce square roots will help visual learners to understand why there is both a positive and negative square root for each positive number.

About Example 3

Historical Connection
Galileo Galilei (1564–1642) was an Italian astronomer who made important contributions to mathematics in the early part of the seventeenth century. At the age of 25, he performed experiments that showed, contrary to the teachings of Aristotle, that heavy objects do not fall faster than light ones. He discovered the formula $d = 16t^2$, which states that the distance an object falls is proportional to the square of the time of falling.

Common Error
Many applications involving quadratic equations, such as Example 3, use only the positive square root. This can lead students to forget that positive numbers have two square roots. Caution students that they should always find both square roots and then eliminate those answers to real-world problems that do not make sense when writing the final solution.

Additional Example 3

The equation $d = 4.9t^2$ gives the distance, in meters, an object falls as a function of time t, in seconds. How many seconds would it take a pebble to fall 211 m?

Use $d = 4.9t^2$ to find t when $d = 211$. Substitute 211 for d and divide by 4.9 to get t^2 alone on one side of the equation.

$$d = 4.9t^2$$
$$211 = 4.9t^2$$
$$\frac{211}{4.9} = t^2$$
$$43.06 \approx t^2$$
$$\pm\sqrt{43.06} \approx t$$
$$\pm 6.56 \approx t$$

Since t represents time, use the positive solution. The pebble would fall 211 m in about 6.6 s.

Think and Communicate

7. Answers may vary. An example is given. If you could physically stretch the graph of $y = x^2$ so that each point on the new graph were twice as far from the x-axis as the corresponding point on the original graph, the result would be the graph of $y = 2x^2$. If you could physically shrink the graph of $y = x^2$ so that each point on the new graph were half as far from the x-axis as the corresponding point on the original graph, the result would be the graph of $y = \frac{1}{2}x^2$.
8. a. vertical shrink
   b. vertical stretch
9. $3x^2$

Teach⇔Interact

Checking Key Concepts

Cooperative Learning
Students would benefit from having time to compare their work on these questions with a partner. This would assure that all students understand the major concepts of the section before beginning the exercises.

Closure Question

Describe two different methods for graphing an equation of the form $y = ax^2$. Make a table of values and plot points or compare the equation to $y = x^2$ and identify any vertical stretch, shrink, or reflection over the x-axis.

Apply⇔Assess

Suggested Assignment

Core Course
Exs. 1–4, 6–17, 24–26, 28–32

Extended Course
Exs. 1–5, 7, 9, 11–32

Block Schedule
Day 27 Exs. 1–4, 6–17, 24–26, 28–32

Exercise Notes

Visual Thinking
Exs. 5 Encourage students to discuss how they would change the equation to make the parabola wider, to make it less wide, and to have it open down instead of up. Ask them to make sketches of the parabolas described on the same coordinate plane. This activity involves the visual skills of *recognition and* generalization.

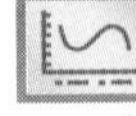

Using Technology
Exs. 6–13, 20 Students can check their solutions to Exs. 6–13 using a graphing calculator or graphing software. For example, to solve the equation $x^2 + 3 = 15$, graph $y = x^2 + 3$ and $y = 15$ and use the trace feature to find the intersection points. The x-coordinate of any intersection will be a solution to $x^2 + 3 = 15$. You may want to suggest that students use the *Function Investigator* software for all of these exercises.

Checking Key Concepts

1. Sketch the graph of $y = 3x^2$. Label the parabola, the line of symmetry, and the vertex.
2. Graph $y = -5x^2$ and $y = \frac{1}{5}x^2$ in the same coordinate plane. Compare these graphs with the graph of $y = x^2$.

Solve each equation. Give solutions to the nearest tenth when necessary.

3. $x^2 = 25$
4. $x^2 = 8$
5. $x^2 + 3 = 15$
6. $55 = 1.1x^2$
7. $200 = 0.8x^2$
8. $-0.15 = -0.05x^2$

5.1 Exercises and Applications

Extra Practice exercises on page 754

Match each graph with its equation.

1. 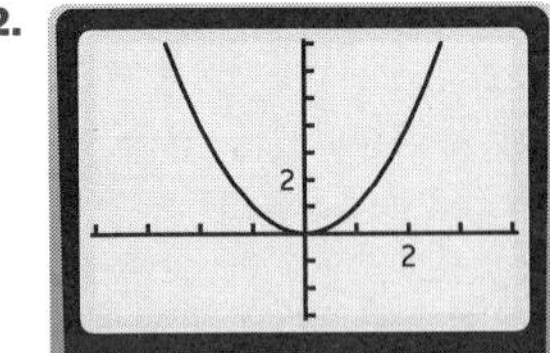

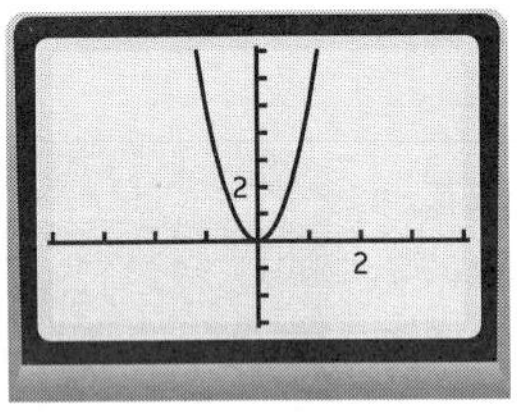

2. 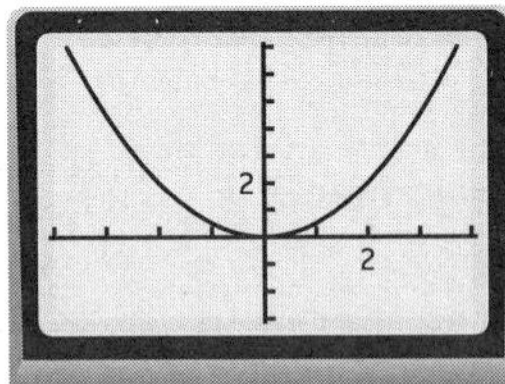

3. 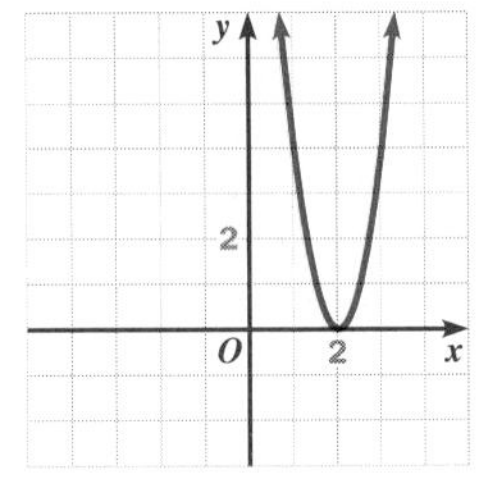

A. $y = x^2$ B. $y = 5x^2$ C. $y = 0.5x^2$

4. **Open-ended Problem** Explain why the graph shown at the right is not symmetric about the y-axis. Then write an equation whose graph *is* symmetric about the y-axis. Tell why your equation is symmetric about the y-axis.

5. **Challenge** Prove that the graph of $y = ax^2$ is symmetric about the line $x = 0$. (*Hint:* If (x, y) is any point on the graph, what are the coordinates of the point's reflection over the y-axis? What must be true about this image point?)

Solve each equation. Give solutions to the nearest tenth when necessary.

6. $x^2 = 16$
7. $x^2 = 12$
8. $8x^2 = 24$
9. $36x^2 = 9$
10. $3x^2 = 27$
11. $\frac{1}{3}x^2 - 27 = 0$
12. $-4x^2 = -64$
13. $-5x^2 = -35$

Each graph has an equation of the form $y = ax^2$. Find a.

14. 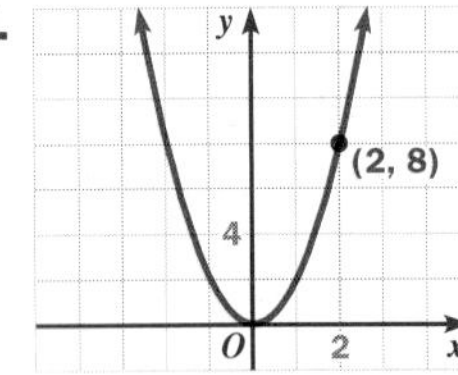

15. 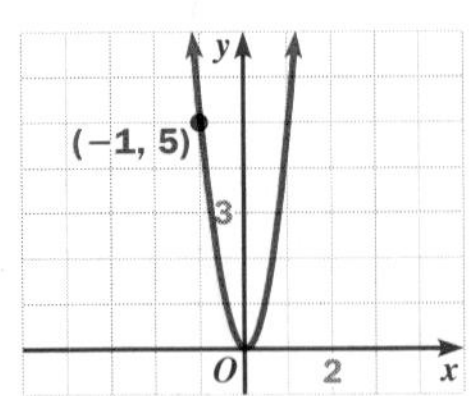

16. 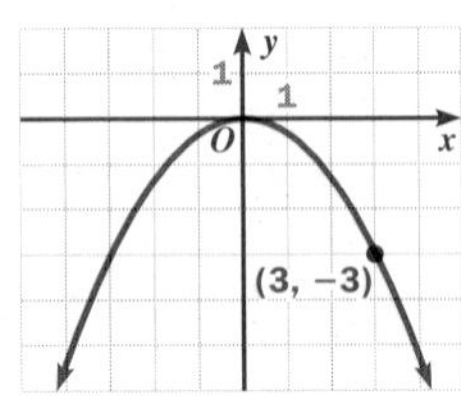

Checking Key Concepts

1. 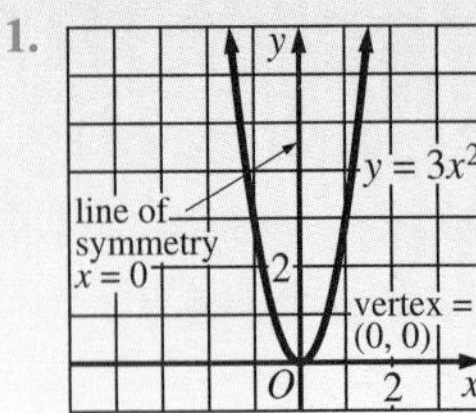

2. 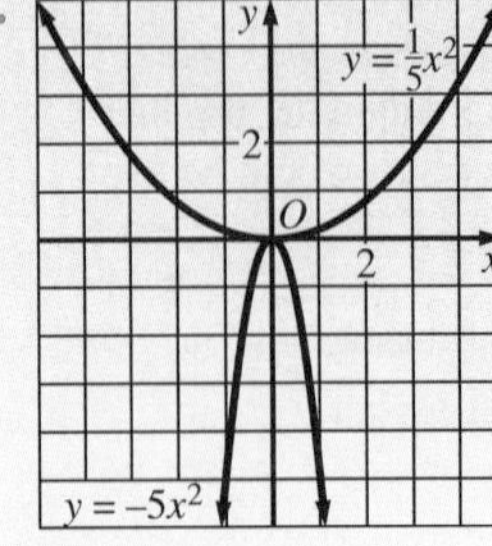

The graph of $y = -5x^2$ is a vertical stretch of the graph of $y = x^2$ reflected over the x-axis. The graph of $y = \frac{1}{5}x^2$ is a vertical shrink of the graph of $y = x^2$.

3. ± 5
4. ± 2.8
5. ± 3.5
6. ± 7.1
7. ± 15.8
8. ± 1.7

Exercises and Applications

1. B
2. A
3. C
4. Answers may vary. An example is given. If the graph is

17. GEOMETRY An equation for the volume of a cylinder is $V = \pi r^2 h$.

a. Suppose $h = 8$ in. and $V = 600$ in.^{3} Write an equation you could use to find r.

b. Find r. Use $\pi \approx 3.14$.

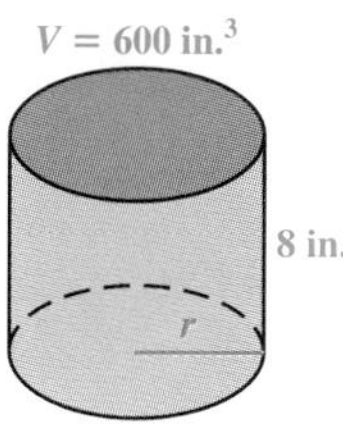

18. Challenge In Section 3.2, you learned about the meaning of rational powers. In this section, you learned about the meaning of square roots.

a. Use a calculator to compare the decimal values of $16^{1/2}$ and $\sqrt{16}$, $5^{1/2}$ and $\sqrt{5}$, and $(0.1)^{1/2}$ and $\sqrt{0.1}$. What do you notice?

b. **Writing** State a general rule about $\frac{1}{2}$ powers and square roots. Then give a convincing argument in support of your rule.

Connection BICYCLING

In Example 1 on page 185, the values of the drag coefficient and the frontal area determined the air resistance on a bicyclist in a particular riding position. These values change for different riding positions and bicycle designs.

For Exercises 19–21, use the information about air resistance shown with the photographs below for various bicycling positions.

19. Write an equation that gives the air resistance R on a bicyclist traveling at speed s in each situation shown (assuming still air).

With complete fairing: $C_D = 0.11$, frontal area = 4.56 ft^2

20. **Technology** Use a graphing calculator or graphing software to graph each equation from Exercise 19 in the same coordinate plane. How do the graphs show which rider experiences the least air resistance at a given speed?

A *fairing* is an external surface that reduces drag.

21. What is the speed of a crouched, racing bicyclist who experiences 4 lb of air resistance?

Upright position: $C_D = 1.1$, frontal area = 5.5 ft^2

Racing crouch position: $C_D = 0.83$, frontal area = 3.9 ft^2

Partial fairing and racing crouch: $C_D = 0.7$, frontal area = 4.1 ft^2

Closely following another bicycle: $C_D = 0.5$, frontal area = 3.9 ft^2

Apply⇔Assess

Exercise Notes

Integrating the Strands

Exs. 17, 24, 25 These exercises illustrate how geometric properties of solids, such as volume and surface area, can be calculated using quadratic equations.

Topic Spiraling: Review

Ex. 18 This exercise helps students see how rational powers studied in Chapter 3 are related to square roots.

Application

Exs. 19–25 Quadratic functions are used extensively in real-world situations. These exercises allow students to work with some applications in bicycling, engineering, and astronomy.

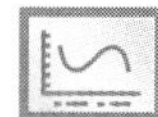

Using Technology

Ex. 20 Ask students to consider carefully their choice of graphing windows before graphing these equations. You might suggest that they start with Ymin = –0.5 and Ymax = 0.5.

folded along the y-axis, the part of the parabola on one side of the axis is not a reflection of the part of the parabola on the other side. The graph of any equation of the form $y = ax^2$ is symmetric about the y-axis. If the graph is folded along the y-axis, the part of the parabola on one side of the axis is a reflection of the part of the parabola on the other side.

5. Suppose (x, ax^2) is any point on the graph. Its reflection over the y-axis has coordinates $(-x, ax^2)$. Since $a(-x)^2 = ax^2$, $(-x, ax^2)$ is also on the graph of $y = ax^2$. Then if the graph is folded along the y-axis, the part of the parabola on one side of the axis is a reflection of the part of the parabola on the other side.

6. ± 4 **7.** ± 3.5

8. ± 1.7 **9.** ± 0.5

10. ± 3 **11.** ± 9

12. ± 4 **13.** ± 2.6

14. 2 **15.** 5

16. $-\frac{1}{3}$

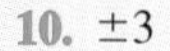

17. a. $r^2 = \frac{V}{\pi h} = \frac{75}{\pi}$

b. about 4.9 in.

18. a. $16^{1/2} = \sqrt{16} = 4$; $5^{1/2} = \sqrt{5} \approx 2.2$; $(0.1)^{1/2} = \sqrt{0.1} \approx 0.3$

b. Summaries may vary. For $x > 0$, $x^{1/2} = \sqrt{x}$. $x^{1/2} \cdot x^{1/2} = x^{1/2 + 1/2} = x^1 = x$. Then by definition, $x^{1/2} = \sqrt{x}$.

19. upright position: $R = 0.0155s^2$; racing crouch position: $R = 0.00829s^2$; partial fairing and racing crouch: $R = 0.00735s^2$; closely following another bicycle: $R = 0.00499s^2$; with complete fairing: $R = 0.00128s^2$

20.

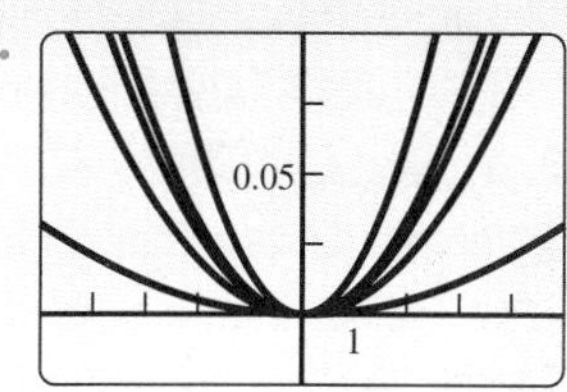

The graph that is the flattest indicates the rider that experiences the least air resistance.

21. 22.0 mi/h

Apply⇔Assess

Exercise Notes

Interview Note

Exs. 22, 23 In air resistance problems, *relative speed* refers to a car's speed relative to the air (or the air's speed relative to the car; the two are identical). The velocities of the air and the car relative to each other are equal with opposite signs. So, driving 60 mi/h into a headwind of 10 mi/h results in a relative speed of 70 mi/h. This may be easier to understand with bicycling, where a headwind clearly makes you work harder against air resistance.

Problem Solving

Ex. 23 Suggest to students that they use a particular example to answer the question in part (b). Ask them how they can use their answer to part (a). (Calculate the speed the bus must travel for it to have 100 lb of air resistance, a 50% reduction from part (a). Then compare this answer with the one to part (a) to determine the reduction factor.)

Reasoning

Ex. 26 Ask students to justify mathematically their choice of an answer. Some students may need to use numerical examples as an explanation.

Using Technology

Ex. 27 You may want to suggest that students use the *Stats!* software to do part (b). Students may also complete this exercise using the list features of a TI-82 graphing calculator. Enter d as L1 and w as L2. On the home screen, enter L2/L1²→L3. This will store the ratios $\frac{w}{d^2}$ in list L3. To find the average ratio, choose SetUp from the STAT CALC menu, and under 1-Var Stats, change the Xlist to L3. Then choose 1-Var Stats from the STAT CALC menu and press [ENTER].

INTERVIEW Mark Thomas

Look back at the article on pages 182–184.

Automotive engineers like Mark Thomas know that the more aerodynamic a vehicle is, the less power the vehicle's engine must provide to overcome air resistance. As the formula on page 184 shows, the air resistance that a vehicle encounters depends on the vehicle's drag coefficient, its frontal area, and the square of the relative speed. The table below gives typical values of the drag coefficient and frontal area for several types of vehicles.

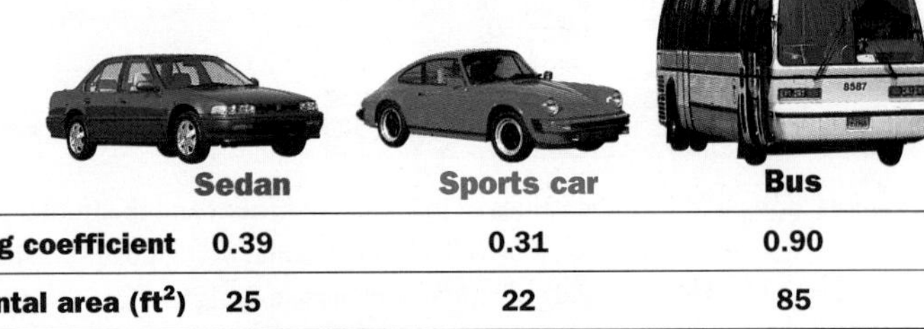

	Sedan	Sports car	Bus
Drag coefficient	0.39	0.31	0.90
Frontal area (ft²)	25	22	85

Use the table above and the formula on page 184 for Exercises 22 and 23.

22. **a.** Write an equation giving the air resistance R on a sedan as a function of the sedan's speed s.

b. Write an equation giving the air resistance on a sports car as a function of its speed.

c. Find the air resistance on both a sedan and a sports car at 30 mi/h and at 60 mi/h. What do you notice?

23. **a.** Suppose the air resistance on a bus is 200 lb. About how fast is the bus traveling?

b. By what factor should the bus reduce its speed in order to reduce the air resistance by 50%?

ASTRONOMY The surface area of a sphere with radius r units is given by the formula S.A. = $4\pi r^2$. Use this formula in Exercises 24 and 25.

24. **a.** Earth's radius is about 4000 mi. What is its surface area?

b. What fraction of Earth's surface is in daylight at any given moment?

c. What is the surface area of the part of Earth that is in daylight at any given moment?

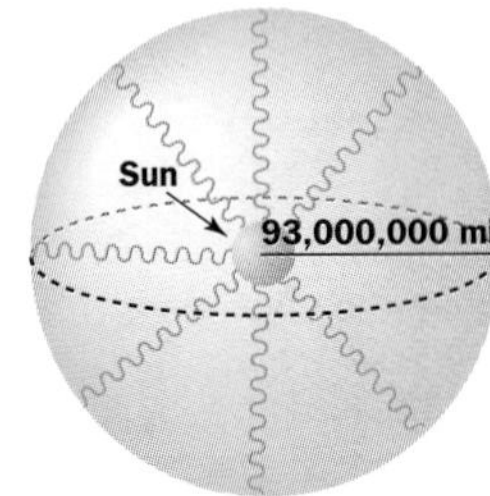

25. Sunlight travels about 93,000,000 mi to reach Earth's surface. Because light from the sun radiates out in all directions, you can think of the available sunlight at 93,000,000 mi as the surface of a sphere with a radius equal to that distance.

a. Find the fraction of available sunlight that hits Earth's surface at any given moment.

b. How large would Earth's radius have to be for Earth to receive just 1% of the available sunlight?

22. a. $R = 0.0250s^2$
 b. $R = 0.0175s^2$
 c. sedan at 30 mi/h: 22.5 lb; sedan at 60 mi/h: 90 lb; sports car at 30 mi/h: 15.75 lb; sports car at 60 mi/h: 63 lb; Answers may vary. Examples are given. At both speeds, the resistance on the sedan is nearly $1\frac{1}{2}$ times that on the sports car. Doubling the speed of either car quadruples the resistance.

23. a. about 31.9 mi/h
 b. $1 - \frac{1}{\sqrt{2}} \approx 29.3\%$

24. a. about 201,000,000 mi²
 b. $\frac{1}{2}$
 c. about 100,500,000 mi²

25. a. 9.25×10^{-10}
 b. about 13,000,000 mi

26. A

27. a. If the rope's cross section is a circle with diameter d and the breaking weight is proportional to the area, then $w = kA = k\left(\pi\left(\frac{d}{2}\right)^2\right) = \frac{k\pi}{4} \cdot d^2$. This means that w is proportional to d^2 with constant of proportionality $\frac{k\pi}{4}$.
 b. Ratios are given to three significant digits.

26. SAT/ACT Preview If A = $\sqrt{x}$ and B = $\sqrt{y}$ for positive integers x and y, and $x > y$, then:

A. A > B **B.** B > A
C. A = B **D.** relationship cannot be determined

27. The spreadsheet shows the breaking weights w, in pounds, for polyester ropes of various maximum diameters d, in inches. In theory, the breaking weight for a rope is proportional to the rope's cross-sectional area A: $w = kA$.

Rope Data		
	A	B
1	d = diameter (in.)	w = breaking weight (lb)
2	0.1875	1200
3	0.25	2000
4	0.3125	3000
5	0.375	4400
6	0.4375	6000
7	0.5	8200

a. Explain why w should be proportional to d^2 if w is proportional to A.

b. **Spreadsheets** Copy the spreadsheet shown and create a third column that gives the ratio $\frac{w}{d^2}$ for each size of rope. Are the ratios about equal? What is the average ratio?

c. **Open-ended Problem** Use the average ratio to write an equation relating w and d. If someone wants to use your equation to find the breaking weight of a rope, would you still use the average ratio? Explain.

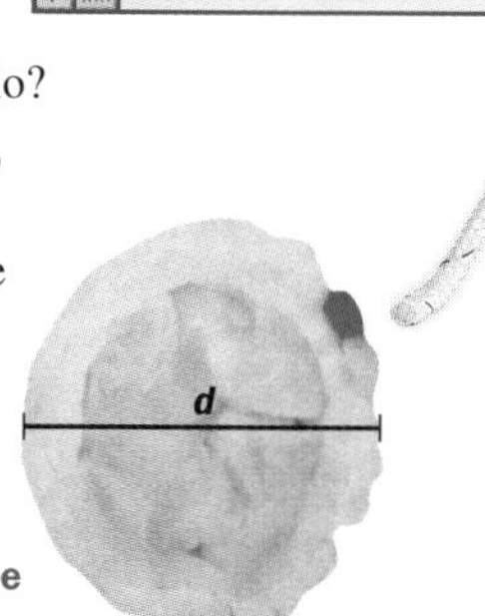

Cross-section of rope

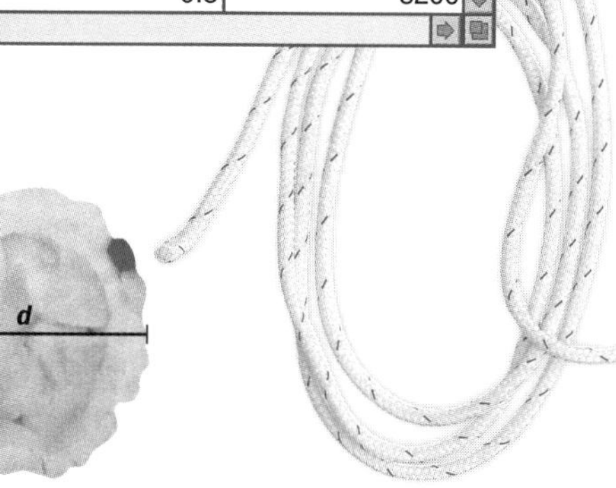

ONGOING ASSESSMENT

28. Open-ended Problem The *period* of a pendulum is the time the pendulum takes to swing back and forth. The period t (in seconds) of a pendulum of length l (in meters) is given by $l = 0.25t^2$. Create a word problem that can be solved by using this quadratic function. Solve the problem.

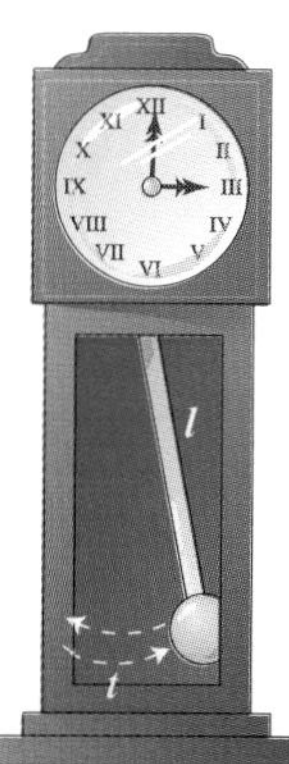

SPIRAL REVIEW

29. Suppose $\log y = 2 \log x + 1.3$. Write y as a function of x. *(Section 4.5)*

30. Graph each equation. *(Section 2.2)*

a. $y = 3x - 1$ **b.** $y = 3x$ **c.** $y = 3x + 2$

31. How are the graphs in Exercise 30 alike? How are they different? *(Section 2.2)*

32. Suppose Maia invests \$1200 in a bank account that offers 7% interest compounded continuously. *(Section 3.4)*

a. How much money does Maia have in the account after five years? after ten years?

b. Find the doubling time for the investment.

Apply⇔Assess

Exercise Notes

Historical Connection
Ex. 28 Galileo was the first to discover that the period of a pendulum is independent both of the size of the arc of oscillation and the weight of the pendulum's bob.

Assessment
Ex. 28 Asking students to write their own problem concerning a pendulum will help them to connect the mathematics of this section to real-world situations with which they are already familiar.

Topic Spiraling: Review
Ex. 29 This exercise should help students realize that a quadratic function of the form $y = ax^2$ is a special kind of power function.

Practice 26 for Section 5.1

NAME ______ DATE ______

Practice 26
FOR USE WITH SECTION 5.1

Solve each equation.

1. $x^2 = 144$ ±12
2. $x^2 = 80$ $\pm 4\sqrt{5}$
3. $3x^2 = 147$ ±7
4. $50x^2 = 18$ $\pm\frac{3}{5}$
5. $\frac{3}{5}x^2 = 60$ ±10
6. $-7x^2 = -175$ ±5
7. $\frac{2}{3}x^2 - 150$ ±15
8. $0.009x^2 = 6.3$ $\pm\sqrt{70}$

Each graph has an equation of the form $y = ax^2$. Find a.

9. (4, –8)
10. (–2, 12)
11. (4, 4)

12. A music tent is to be built in the shape of a cone of height 12 ft atop a cylinder of height 16 ft. The volume V of the tent is given by the formula
$$V = 20\pi r^2,$$
where r is the radius of both the cone and the cylinder.
a. Graph the function. See additional answers.
b. Find the value of r that makes the volume of the tent 100,000 ft^3. about 39.9 ft

13. The *period* T of a pendulum is the time it takes the pendulum to make one complete back-and-forth swing. The length L (in feet) that a pendulum must have in order to have a period T (in seconds) is given by the formula
$$L = \frac{8T^2}{\pi^2}.$$
a. Graph this function. See additional answers.
b. Find the length of a pendulum with a period of 1 s. about 0.81.ft
c. What is the period of a pendulum of length 4 ft? about 2.22 s

26 **Practice Bank,** ALGEBRA 2: EXPLORATIONS AND APPLICATIONS

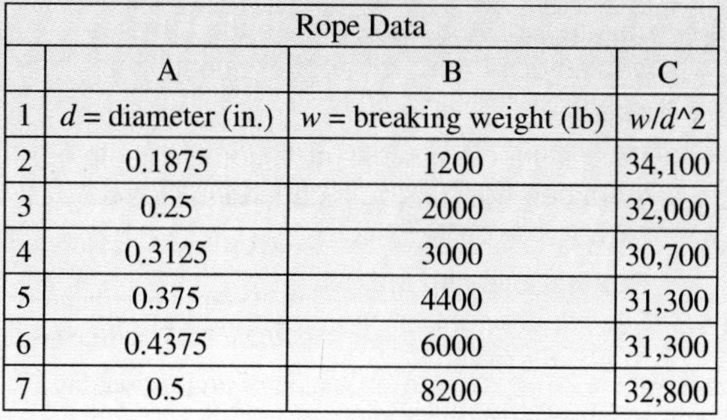

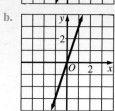

Rope Data			
	A	B	C
1	d = diameter (in.)	w = breaking weight (lb)	w/d^2
2	0.1875	1200	34,100
3	0.25	2000	32,000
4	0.3125	3000	30,700
5	0.375	4400	31,300
6	0.4375	6000	31,300
7	0.5	8200	32,800

Yes; the data vary from the average by at most about 7%. The average ratio is about 32,000.

c. $w = 32{,}000d^2$; Answers may vary. An example is given. I think it would be safer to use the minimum ratio rather than the average ratio.

28. Answers may vary. An example is given. Find the period of a pendulum that is 1.2 m long. ($1.2 = 0.25t^2$; $t^2 = 4.8$; $t \approx 2.19$; about 2.19 s)

29. $y = 20.0x^2$

30. a.

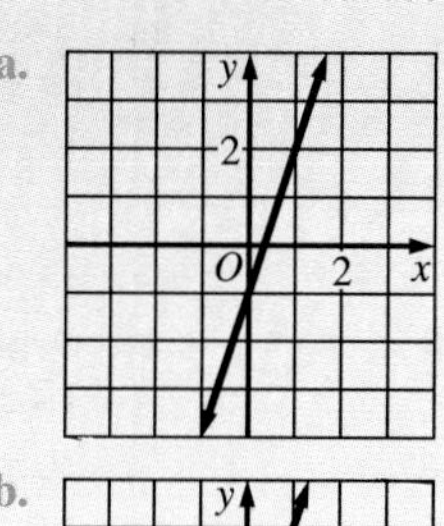

b.

c.

31. They all have the same slope, 3. Each graph has a different vertical intercept.

32. a. \$1702.88; \$2416.50
b. 9.9 years

Plan⇔Support

Objectives

- Graph equations in the form $y = a(x - h)^2 + k$.
- Solve quadratic equations.
- Solve real-world problems using quadratic equations.

Recommended Pacing

Core and Extended Courses
Section 5.2: 1 day

Block Schedule
Section 5.2: $\frac{1}{2}$ block
(with Section 5.3)

Resource Materials

Lesson Support
Lesson Plan 5.2
Warm-Up Transparency 5.2
Overhead Visuals:
Folder 5: Parabolas
Practice Bank: Practice 27
Study Guide: Section 5.2
Exploration Lab Manual:
Diagram Masters 1, 2, 9

Technology
Technology Book:
Calculator Activity 5
Spreadsheet Activity 5
Graphing Calculator
McDougal Littell Mathpack
Function Investigator with Matrix Analyzer Activity Book: Function Investigator Activity 15
Geometry Inventor Activity Book: Activity 22
Internet:
http://www.hmco.com

Warm-Up Exercises

Solve each equation. Give solutions to the nearest tenth if necessary.

1. $3x^2 = 75$ ±5
2. $-x^2 = -20$ ±4.5
3. $\frac{1}{3}x^2 + 5 = 32$ ±9

A penny is dropped down a 100 ft well. Use the equation $d = 16t^2$, which gives the distance d, in feet, the penny falls in t seconds.

4. How far does the penny fall in the first 2 seconds? 64 ft
5. How many seconds does it take for it to hit the bottom? 2.5 s

Section

5.2 Translating Parabolas

Learn how to...

- graph equations in the form $y = a(x - h)^2 + k$
- solve quadratic equations

So you can...

- solve problems, such as finding how long it takes a coconut to fall to the ground

You know that the value of a in the equation $y = ax^2$ affects the width of the graph and tells you whether the parabola opens up or down. In the Exploration, you will see how introducing other constants affects the graph.

EXPLORATION
COOPERATIVE LEARNING

Moving Parabolas Around

Work with a partner.
You will need:
- a graphing calculator or graphing software

SET UP Copy and complete the table. Then answer the questions below.

Toolbox p. 796
Translating a Figure

Equation 1	Equation 2	How is the graph of equation 2 geometrically related to the graph of equation 1?
$y = x^2$	$y = (x - 1)^2$	translated 1 unit to the right
$y = x^2$	$y = x^2 + 1$	?
$y = 2x^2$	$y = 2(x + 3)^2$	?
$y = 2x^2$	$y = 2x^2 - 4$	?
$y = -\frac{1}{2}x^2$	$y = -\frac{1}{2}(x + 1)^2 + 3$	?
$y = -\frac{1}{2}x^2$	$y = -\frac{1}{2}(x - 2)^2 + 1$	?

Questions

1. Predict what the graph of each equation below looks like by making a sketch. Then check your sketch by using a graphing calculator or graphing software.

   a. $y = \frac{1}{3}x^2 + 5$ **b.** $y = -4(x - 1)^2$ **c.** $y = (x + 4)^2 - 3$

2. **a.** How is the graph of $y = ax^2 + k$ geometrically related to the graph of $y = ax^2$ when k is positive? when k is negative?

   b. How is the graph of $y = a(x - h)^2$ geometrically related to the graph of $y = ax^2$ when h is positive? when h is negative?

Exploration Note

Purpose
The purpose of this Exploration is to have students discover how the graph of $y = ax^2$ is affected by introducing other constants.

Materials/Preparation
A graphing calculator or graphing software is needed.

Procedure
Have students copy the table and then take turns using the calculator to graph both equations in each row simultaneously. They should compare the graphs and complete the third column. For question 1, students should make their predictions independently, compare them with each other, and then check their predictions with the graphing calculator. Students can use the table as well as question 1 to answer question 2.

Closure
Have students discuss question 2 to help them see how to graph equations in the form $y = a(x - h)^2 + k$.

Explorations Lab Manual
See the Manual for more commentary on this Exploration.

Diagram Master 9

The equations that you graphed in the Exploration have the general form $y = a(x - h)^2 + k$. To graph an equation in this form, you can start with the graph of the simpler equation $y = ax^2$ and translate it h **units horizontally** and k **units vertically**.

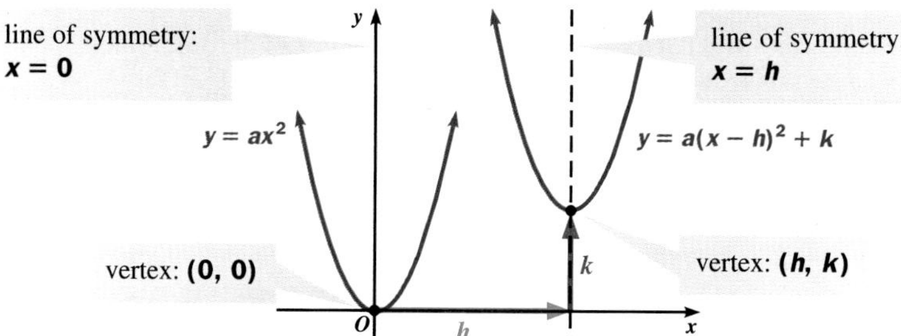

Because you can read the vertex (as well as other information about the graph) directly from the equation $y = a(x - h)^2 + k$, the equation is called the **vertex form** of a quadratic equation.

Vertex Form of a Quadratic Equation

The graph of $y = a(x - h)^2 + k$ is a parabola that:

- **has its vertex at (h, k),**
- **has the line $x = h$ as its line of symmetry,**
- **opens up if $a > 0$ and opens down if $a < 0$.**

EXAMPLE 1

Describe the graph of $y = -2(x + 3)^2 + 4$ and then sketch it.

SOLUTION

Rewrite the equation as $y = -2(x - (-3))^2 + 4$. The vertex is $(-3, 4)$. The line of symmetry is $x = -3$. The parabola opens down.

For x-values 1 unit to the left or right of the line of symmetry, the corresponding y-value is $y = -2(\pm 1)^2 + 4 = 2$.

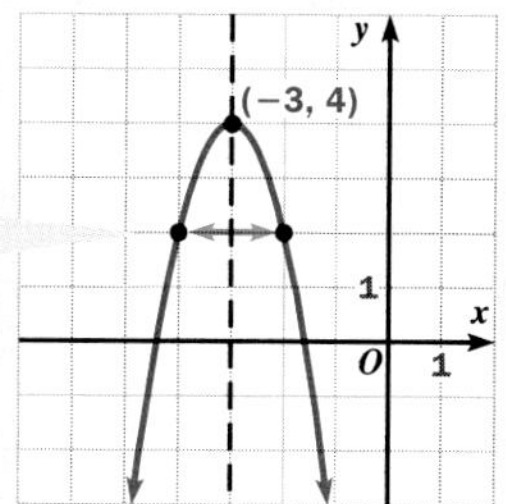

Teach⇔Interact

Section Notes

Topic Spiraling: Review

Some students may need a brief reminder concerning the equation of a vertical line. They need to realize that the expression $x = h$ for the line of symmetry represents a line and not merely the x-coordinate of a single point.

Reasoning

Given a point (x_1, y_1) on the graph of $y = x^2$, ask students to find the coordinates of its image point after a translation of h units horizontally and k units vertically. $((x_1 + h, y_1 + k))$ Then have them verify that this point lies on the graph of $y = a(x - h)^2 + k$. ((x_1, y_1) is on $y = ax^2$, so $y_1 = ax_1$ and $a((x_1 + h) - h)^2 + k = ax_1{}^2 + k = y_1 + k$. Thus, $(x_1 + h, y_1 + k)$ lies on $y = a(x - h)^2 + k$.)

About Example 1

Mathematical Procedures

The method used to find points on the graph of a parabola takes advantage of the vertical reflection symmetry these graphs have over their axis of symmetry. Students could extend this method to find other points as well. For example, to find points 3 units to the right or left of the vertex, replace the $x + 3$ in the expression of the function with ±3: $y = -2(\pm 3)^2 + 4 = -2(9) + 4 = -14$, yielding points $(-6, -14)$ and $(0, -14)$.

Additional Example 1

Describe the graph of $y = \frac{1}{2}(x - 1)^2 - 3$ and then sketch it.

Rewrite the equation as $y = \frac{1}{2}(x - 1)^2 + (-3)$. The vertex is $(1, -3)$. The line of symmetry is $x = 1$. The parabola opens up. For x-values 2 units to the left or right of the line of symmetry, the corresponding y-value is $y = \frac{1}{2}(\pm 2)^2 - 3 = -1$.

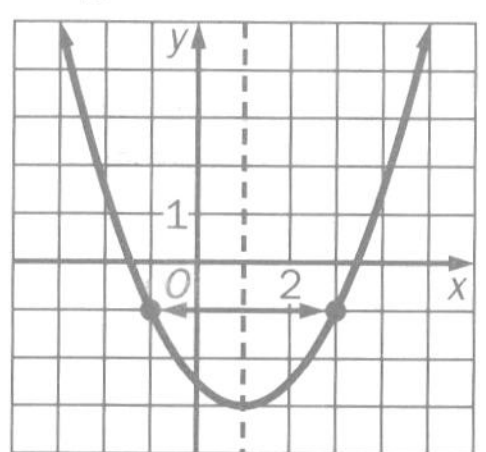

ANSWERS Section 5.2

Exploration

Equation 1	Equation 2	How is the graph of equation 2 geometrically related to the graph of equation 1?
$y = x^2$	$y = (x - 1)^2$	translated 1 unit to the right
$y = x^2$	$y = x^2 + 1$	translated up 1 unit
$y = 2x^2$	$y = 2(x + 3)^2$	translated 3 units to the left
$y = 2x^2$	$y = 2x^2 - 4$	translated down 4 units
$y = -\frac{1}{2}x^2$	$y = -\frac{1}{2}(x + 1)^2 + 3$	translated 1 unit to the left, up 3 units
$y = -\frac{1}{2}x^2$	$y = -\frac{1}{2}(x - 2)^2 + 1$	translated 2 units to the right, up 1 unit

Questions

1. a. The graph is the graph of $y = x^2$ shrunk vertically by a factor of 3 and translated 5 units up.

   b. The graph is the graph of $y = x^2$ translated 1 unit to the right, stretched vertically by a factor of 4, and reflected over the x-axis.

   c. The graph is the graph of $y = x^2$ translated 4 units to the left and down 3 units.

2. See answers in back of book.

Teach⇔Interact

About Example 2

Using Technology
Students can use the table features of a TI-82 graphing calculator to help them sketch the graph of $h = -16t^2 + 80$. Enter $y = -16x^2 + 80$ on the Y= list. Then press [2nd][TblSet] and set TblMin to 0 and ΔTbl to 1. Press [2nd][TABLE]. Scan the table to find when the height becomes negative (between 2 and 3 seconds). Adjust the ΔTbl setting until a sufficient number of points are given in the table before the height becomes negative and use the (x, y) pairs to sketch the graph. By continuing to adjust the TblMin and ΔTbl settings, the table can also be used to get a good approximation of t when $h = 0$. Set TblMin to the last x-value from the previous table before the y-value became negative and continue to decrease the ΔTbl setting until the desired accuracy is reached.

Additional Example 2

A diver dives off a 63 m cliff. A photographer waits on an outcropping 20 m above the water to take the diver's picture.

a. Express the height of the diver as a function of time. Use the fact that the distance d, in meters, an object falls in t seconds is given by $d = 4.9t^2$. The relationship between the distance fallen, d, and the height above the water, h, is given by $d + h = 63$.
Since $d = 4.9t^2$, substitute and solve for h:

$$4.9t^2 + h = 63$$
$$h = 63 - 4.9t^2$$
$$h = -4.9t^2 + 63$$

b. Sketch the graph of the equation. When is the diver 20 m above the water? The graph of the height function has its vertex at (0, 63) and opens down.

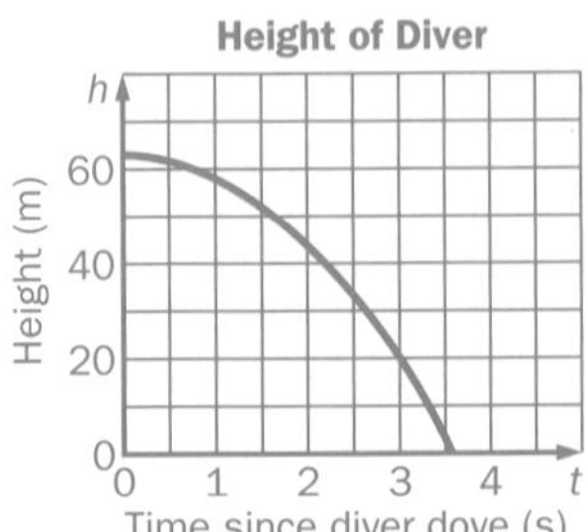

The diver is 20 m above the water at time $t \approx 3.0$ s.

EXAMPLE 2 Application: Physics

The people living on many islands in the South Pacific harvest coconuts by climbing coconut palm trees and cutting the stems of the fruit. The coconuts then fall to the ground. Suppose a coconut falls from a height of 80 ft.

a. Express the height of the falling coconut as a function of the time t since it began falling. Use the fact that the distance d, in feet, an object falls in t seconds is given by $d = 16t^2$. (See Example 3 on page 187.)

b. Sketch the graph of the equation from part (a). When does the coconut hit the ground?

SOLUTION

a. The diagram shows that the relationship between the distance fallen, d, and the height above the ground, h, is given by:

$$d + h = 80$$

Since $d = 16t^2$, substitute and solve for h:

$$16t^2 + h = 80$$
$$h = 80 - 16t^2$$
$$= -16t^2 + 80$$

b. The graph of the height function has its vertex at (0, 80) and opens down.

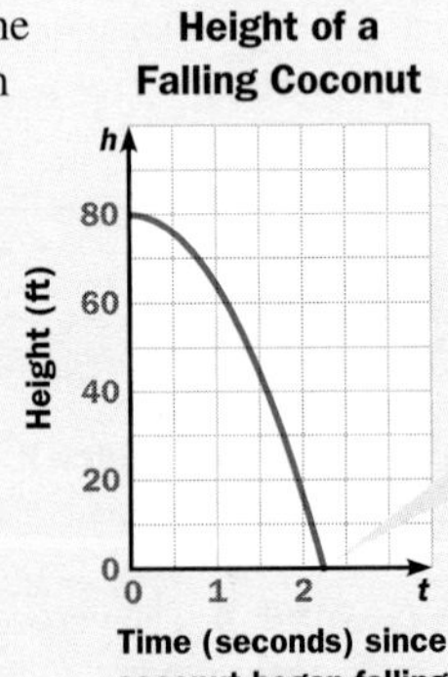

The coconut hits the ground when $h = 0$. The graph suggests that this happens at time $t \approx$ **2.2 s**.

BY THE WAY

The coconut palm is one of over 3000 species of palm tree. In the tropics, people use palm trees to obtain food, shelter, clothing, fuel, paper, starch, oil, sugar, wax, and other products.

THINK AND COMMUNICATE

1. Substitute 0 for h in the equation $h = -16t^2 + 80$. Solve for t. Does your answer agree with the solution of part (b) of Example 2?
2. Does the graph in Example 2 represent the *path* of the falling coconut? Explain.
3. The formula $d = 16t^2$ gives the distance an object falls in feet. Similarly, the formula $d = 4.9t^2$ gives the distance an object falls in meters. If a coconut falls from a height of 25 m, express the height of the falling coconut as a function of the time t since it began falling.

Think and Communicate

1. $0 = -16t^2 + 80$; $t^2 = 5$; $t \approx 2.2$; Yes.
2. No; the path is a vertical line.
3. $h = -4.9t^2 + 25$

Minimum and Maximum Values

Because the vertex (h, k) represents the *lowest* point on the graph of $y = a(x - h)^2 + k$ when $a > 0$, k is called the **minimum value** of the quadratic function.

Likewise, because the vertex (h, k) represents the *highest* point on the graph of $y = a(x - h)^2 + k$ when $a < 0$, k is called the **maximum value** of the quadratic function.

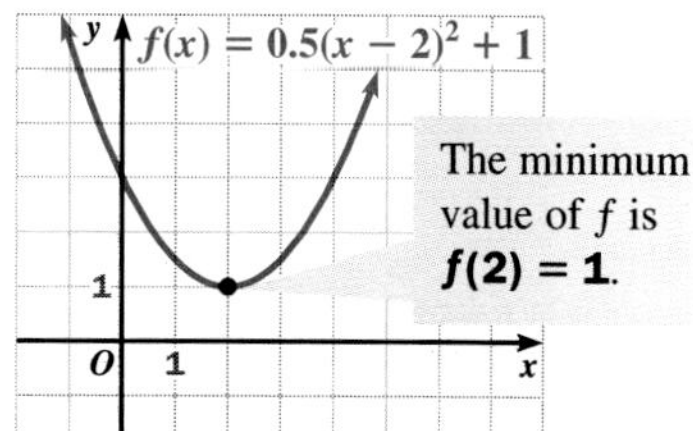

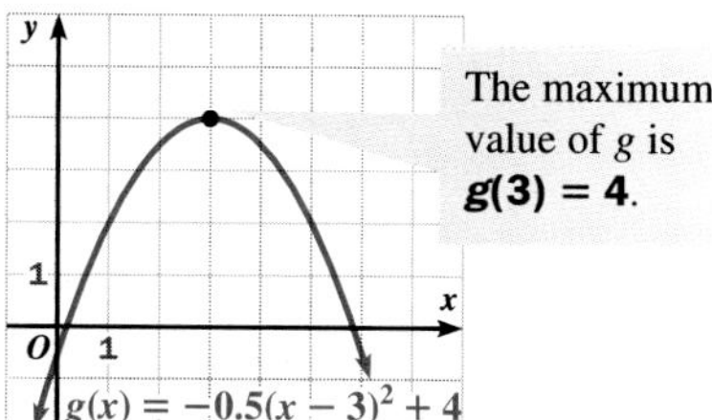

EXAMPLE 3

Nyasha throws a ball straight up into the air. The ball reaches a maximum height of 8 ft above the ground in 0.5 s.

a. Find an equation giving the ball's height h as a function of time t.

b. When does the ball hit the ground?

SOLUTION

a. If you measure time from the moment the ball reaches its maximum height, a graph of the height function looks like this:

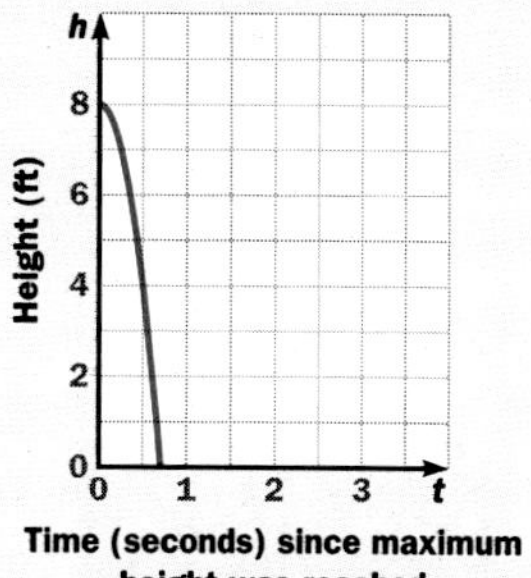

An equation for this graph is $h = -16t^2 + 8$. This is just like the falling coconut's equation in Example 2.

But if you measure time from the moment the ball leaves Nyasha's hands, a graph of the height function looks like this:

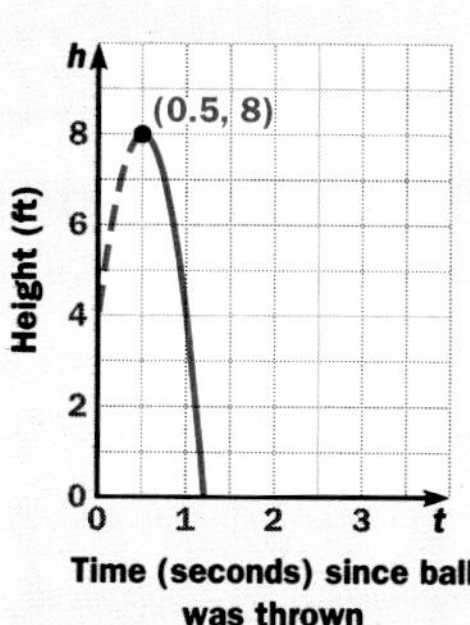

The previous graph has been translated 0.5 units to the right. The equation is now $h = -16(t - 0.5)^2 + 8$.

Solution continued on next page.

Teach⇔Interact

Additional Example 3

Use the information from Additional Example 2. Suppose there is a spring board at the top of the cliff so that the diver jumps up at the beginning of the dive to reach a maximum height of 67 meters 0.9 seconds after jumping.

a. Find an equation of the diver's height h as a function of time t.

If you measure time from the moment the diver reaches the maximum height, a graph of the height function looks like this.

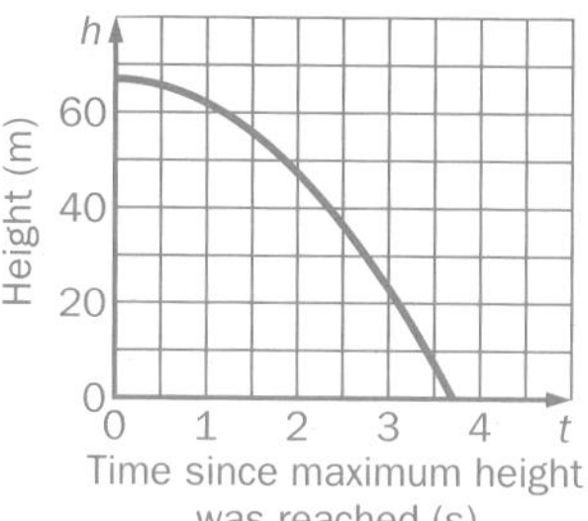

An equation for this graph is $h = -4.9t^2 + 67$. This is just like the diver's equation in Additional Example 2. But if you measure time from the moment the diver leaves the cliff, a graph of the height function looks like this.

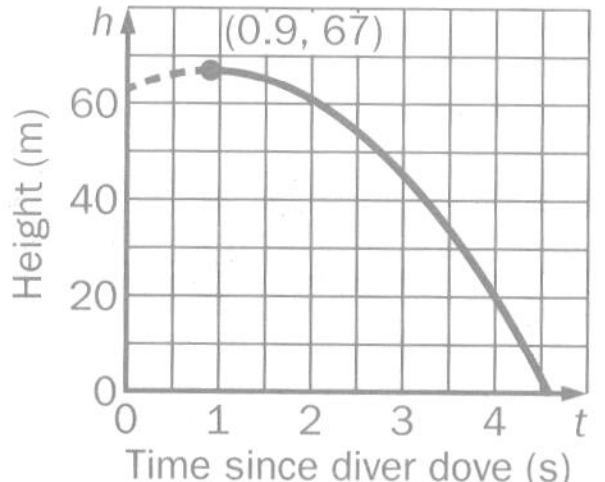

The previous graph has been translated 0.9 units to the right. The equation is now $h = -4.9(t - 0.9)^2 + 67$.

b. When is the diver 20 m above the water?

$$h = -4.9(t - 0.9)^2 + 67$$
$$20 = -4.9(t - 0.9)^2 + 67$$
$$-47 = -4.9(t - 0.9)^2$$
$$9.59 \approx (t - 0.9)^2$$
$$\pm 3.097 \approx t - 0.9$$
$$0.9 \pm 3.097 \approx t$$
$$-2.197, 3.997 \approx t$$

The diver is 20 m above the water after about 4 s.

Teach⇔Interact

Section Note

Common Error
Students often think that when a is positive, the function has a maximum value. Encourage them to visualize the graph. When a is positive the graph curves up and thus has a minimum value.

Checking Key Concepts

Teaching Tip
Students having difficulty with questions 4 and 5 should be encouraged to write a journal entry listing the steps required to solve such equations. They can also use examples to illustrate the steps.

Closure Question

Describe the characteristics of the function $y = a(x - h)^2 + k$ and its graph. If a is positive, the function has a minimum value of k at $x = h$ and the graph opens up with vertex (h, k). If a is negative, the function has a maximum value of k at $x = h$ and the graph opens down with a vertex (h, k). In either case, the graph has an axis of symmetry $x = h$.

Apply⇔Assess

Suggested Assignment

Core Course
Exs. 1–4, 9–24, 27–36

Extended Course
Exs. 1–8, 9–19 odd, 21–36

Block Schedule
Day 28 Exs. 1–4, 9–24, 27–36

Exercise Notes

Communication: Writing
Ex. 4 For part (c), students should be able to explain why $y = -5(x - 4)^2 + 1$ is not merely a reflection of $y = 5(x - 4)^2 + 1$ across the x-axis but is instead a reflection followed by a translation. ($y = -f(x)$ is a reflection of $y = f(x)$ across the x-axis. The reflection of $y = 5(x - 4)^2 + 1$ is $y = -(5(x - 4)^2 + 1) = -5(x - 4)^2 - 1$. This curve is then translated up 2 units to get the desired curve.)

SOLUTION *continued*

b. Let $h = 0$ and solve for t:

$$h = -16(t - 0.5)^2 + 8$$
$$0 = -16(t - 0.5)^2 + 8$$
$$-8 = -16(t - 0.5)^2$$
$$0.5 = (t - 0.5)^2$$
$$\pm 0.707 \approx t - 0.5$$
$$0.5 \pm 0.707 \approx t$$
$$-0.207, 1.207 \approx t$$

The ball hits the ground in about 1.2 s.

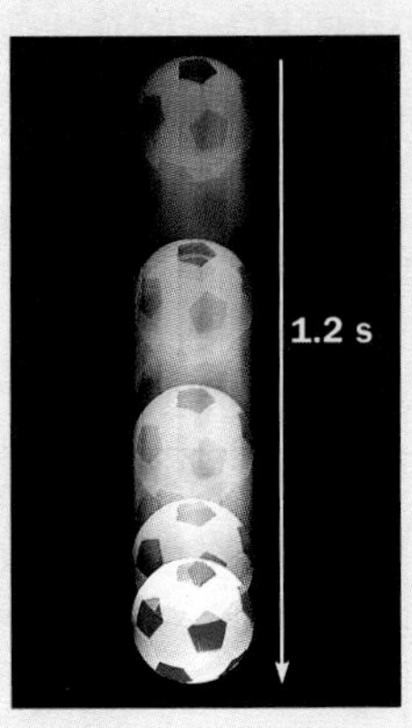

CHECKING KEY CONCEPTS

1. Suppose the parabola with equation $y = -2x^2$ is translated 1 unit horizontally and -2 units vertically. Write the new equation.

Tell whether the graph of each function *opens up* or *opens down*. What is the maximum value or minimum value of the function?

2. $f(x) = 0.12(x - 5)^2 + 3$

3. $f(x) = -0.12(x + 5)^2 - 2$

Solve each equation. Give solutions to the nearest tenth.

4. $4(x - 5)^2 + 20 = 100$

5. $-3(x + 2)^2 + 34 = 25$

5.2 Exercises and Applications

Extra Practice exercises on page 755

Match each quadratic equation with one of the graphs below.

1. $y = 2x^2 + 3$

2. $y = 2(x + 1)^2$

3. $y = -2(x - 1)^2 + 3$

A.

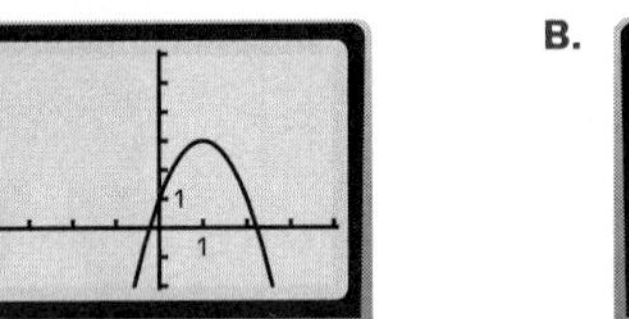

B.

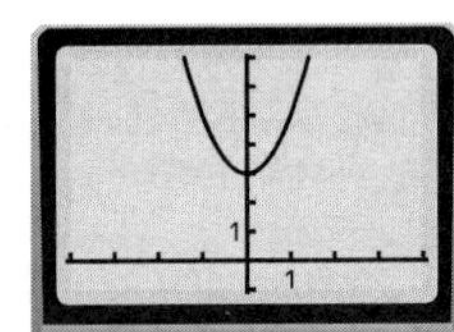

C.

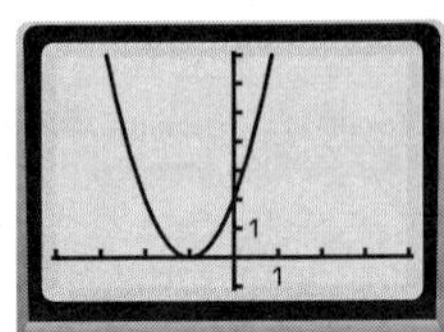

4. Writing Describe the effect that each change has on the graph of each original equation.

a. changing $y = 5(x - 4)^2 + 1$ to $y = 5(x - 4)^2 - 2$

b. changing $y = 5(x - 4)^2 + 1$ to $y = 5(x + 4)^2 + 1$

c. changing $y = 5(x - 4)^2 + 1$ to $y = -5(x - 4)^2 + 1$

Checking Key Concepts

1. $y = -2(x - 1)^2 - 2$
2. up; minimum value: 3
3. down; maximum value: -2
4. 0.5, 9.5
5. $-3.7, -0.3$

Exercises and Applications

1. B
2. C
3. A
4. a. The graph is translated down 3 units.
   b. The graph is translated 8 units to the left.
   c. The graph is reflected over the x-axis, then translated up 2 units. (This may also be described as reflection over the line $y = 1$.)
5. a.

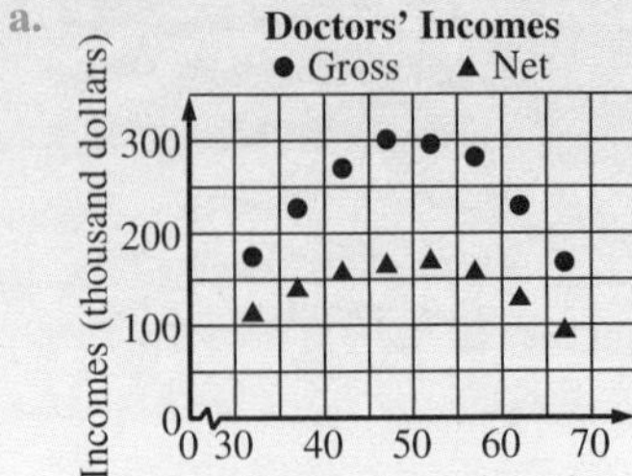

Connection PERSONAL FINANCE

The amount a doctor earns each year may depend on the doctor's age. The table gives doctors' average gross incomes and net incomes in 1991 for various age ranges. Use the table for Exercises 5–8.

Robin A. Winthrob, M.D.

Age (years)	Incomes (dollars) Gross	Net
30–34	175,370	111,200
35–39	227,040	137,610
40–44	270,050	155,680
45–49	301,470	164,410
50–54	295,930	167,720
55–59	281,580	156,100
60–64	229,090	127,130
65–69	167,710	92,280

For modeling purposes, use the middle age in each range: 32, 37, 42, 47, and so on.

5. **a.** Make a scatter plot of the data. Use one color or symbol for the gross incomes and a different color or symbol for the net incomes.

b. **Writing** Examine your scatter plot and describe any patterns you see. How do the gross and net incomes compare overall?

6. The data points appear to lie on curves rather than lines. Suppose the curves are parabolas.

a. What would you say is the approximate location of the vertex of each parabola?

b. Make a sketch of each parabola on your scatter plot.

7. **Spreadsheets** Alisa modeled the gross income data by using the spreadsheet shown. She let x = age and y = income. She then assumed that the vertex of the parabola for gross income was at (50, 300,000), and so she created two new variables: $u = x - 50$ and $v = y - 300{,}000$.

a. Look at the ratios $\frac{v}{u^2}$ in the last column of Alisa's spreadsheet. What do you notice?

Doctors' Gross Incomes

	C	D	E
1	u = x – 50	v = y – 300000	v/u^2
2	–18	–124630	–384.66049
3	–13	–72960	–431.71598
4	–8	–29950	–467.96875
5	–3	1470	163.333333
6	2	–4070	–1017.5
7	7	–18420	–375.91837
8	12	–70910	–492.43056
9	17	–132290	–457.75087

Alisa copied the gross income data from the table, putting the ages (the x-values) in Column A and the gross incomes (the y-values) in Column B, not shown here.

b. If you ignore the two middle ratios (which are more variable than the rest), what average ratio do you get?

c. If a is your average ratio from part (b), then you can write $\frac{v}{u^2} = a$, or $v = au^2$. What equation do you get when you substitute $x - 50$ for u and $y - 300{,}000$ for v?

d. **Writing** Graph your equation from part (c) and comment on the reasonableness of the model for the gross income data.

8. **Open-ended Problem** Use a method like the one described in Exercise 7 to develop a quadratic model for the net income data.

Apply⇔Assess

Exercise Notes

Second-Language Learners

Ex. 5 Students learning English may not be familiar with the terms *gross income* and *net income*. If necessary, explain that net income is the total income left after taxes and other deductions have been taken out of the original income, which is also known as the gross income.

Application

Exs. 5–8, 25 These exercises show students some of the applications of quadratic functions to finance and biology.

Mathematical Procedures

Exs. 7, 8 There are many different ways to develop quadratic models for a set of data. These exercises use an approximate vertex and a spreadsheet to find the constant a from the vertex form $y = a(x - h)^2 + k$. Students should be able to see how this method is an extension of the method used in Chapter 2 to find direct variation models. In this case, $y - k$ varies directly with $(x - h)^2$ and a is the constant of variation.

Using Technology

Exs. 7, 8 You may want to suggest that students use the *Stats!* software for these exercises. The problem in Ex. 7 can also be done using the list features of a TI-82 graphing calculator. After entering the age and gross income data as lists L1 and L2, position the cursor over L3 at the top of the third column and enter L1–50. List L3 should now be identical to the first column of the spreadsheet. For L4, enter L2–300000, and for L5, enter L4/L3^2. A similar procedure can be used for Ex. 8.

Topic Spiraling: Preview

Exs. 7, 8 This is the first of several methods students will see for finding quadratic models. Ask them what they think are some advantages and disadvantages of this method.

Communication: Listening

Exs. 9–14 These exercises can be done orally in class. As one student answers, have the other students raise their hands if they disagree with the solution. Any disagreements can then be discussed.

b. Both gross income and net income increase until the age reaches about 50, then they begin to decrease. The difference between gross income and net income also increases until the age reaches about 45, then it begins to decrease.

6. a. Estimates may vary. Examples are given. gross income: (50, 300,000); net income: (50, 175,000)

b.

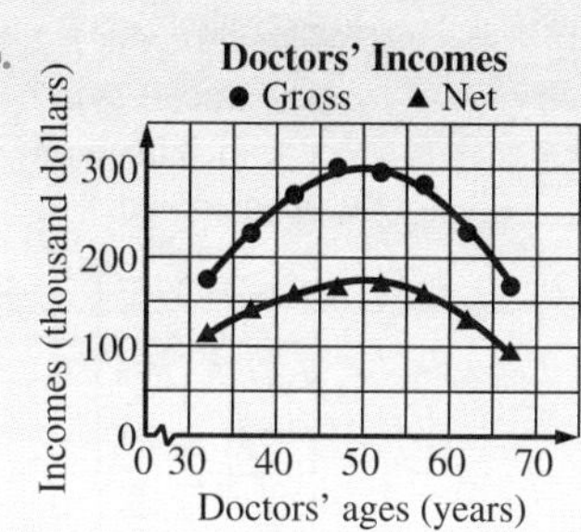

7. See answers in back of book.

8. Answers may vary. An example is given. $y = -287(x - 50)^2 + 175{,}000$

Apply⇔Assess

Exercise Notes

Cooperative Learning
Exs. 15–20 Students who need more practice on these exercises can work with a partner. Each partner can make up an equation which the other partner can graph.

Teaching Tip
Exs. 21–23 Encourage students to check their work by testing both given points in their equations.

Assessment
Ex. 27 Have several students read their descriptions and illustrate them at the board. This will help all students to evaluate their own descriptions and to solidify their understanding.

Topic Spiraling: Preview
Exs. 31–33 Students will use factoring in the next section and in Section 5.6 to find the intercept form of a quadratic equation.

Practice 27 for Section 5.2

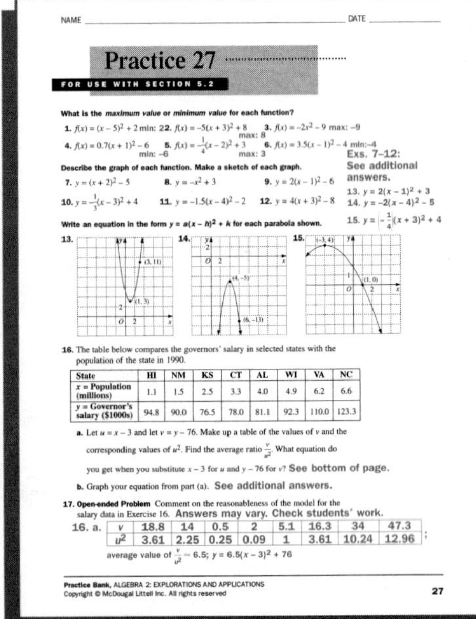

NAME ______ DATE ______

Practice 27

FOR USE WITH SECTION 5.2

What is the *maximum value* or *minimum value* for each function?

1. $f(x) = (x - 5)^2 + 2$ min: 2
2. $f(x) = -5(x + 3)^2 + 8$ max: 8
3. $f(x) = -2x^2 - 9$ max: −9
4. $f(x) = 0.7(x + 1)^2 - 6$ min: −6
5. $f(x) = -\frac{1}{4}(x - 2)^2 + 3$ max: 3
6. $f(x) = 3.5(x - 1)^2 - 4$ min: −4

Describe the graph of each function. Make a sketch of each graph. Exs. 7–12: See additional answers.

7. $y = (x + 2)^2 - 5$
8. $y = -x^2 + 3$
9. $y = 2(x - 1)^2 - 6$
10. $y = -\frac{1}{3}(x - 3)^2 + 4$
11. $y = -1.5(x - 4)^2 - 2$
12. $y = 4(x + 3)^2 - 8$

Write an equation in the form $y = a(x - h)^2 + k$ for each parabola shown.

13. $y = 2(x - 1)^2 + 3$
14. $y = -2(x - 4)^2 - 5$
15. $y = -\frac{1}{4}(x + 3)^2 + 4$

16. The table below compares the governors' salary in selected states with the population of the state in 1990.

State	HI	NM	KS	CT	AL	WI	VA	NC
x = Population (millions)	1.1	1.5	2.5	3.3	4.0	4.9	6.2	6.6
y = Governor's salary ($1000s)	94.8	90.0	76.5	78.0	81.1	92.3	110.0	123.3

a. Let $u = x - 3$ and let $v = y - 76$. Make up a table of the values of v and the corresponding values of u^2. Find the average ratio $\frac{v}{u^2}$. What equation do you get when you substitute $x - 3$ for u and $y - 76$ for v? See bottom of page.

b. Graph your equation from part (a). See additional answers.

17. **Open-ended Problem** Comment on the reasonableness of the model for the salary data in Exercise 16. Answers may vary. Check students' work.

16. a.

v	18.8	14	0.5	2	5.1	16.3	34	47.3
u^2	3.61	2.25	0.25	0.09	1	3.61	10.24	12.96

; average value of $\frac{v}{u^2} \approx 6.5$; $y = 6.5(x - 3)^2 + 76$

What is the maximum value or minimum value of each function?

9. $f(x) = -4(x + 4)^2 + 1$
10. $f(x) = 0.3(x + 1)^2 + 4$
11. $f(x) = 3(x - 1)^2 - 4$
12. $f(x) = -0.4(x + 4)^2 - 1$
13. $f(x) = 5(x - 3)^2 - 2$
14. $f(x) = -2(x + 2)^2 + 6$

Describe the graph of each function. Make a sketch of each graph.

15. $y = -3(x + 3)^2 - 3$
16. $y = 6(x - 1)^2$
17. $y = -(x + 2)^2$
18. $y = \frac{1}{2}(x - 2)^2 - 5$
19. $y = -2(x + 4)^2 + \frac{1}{2}$
20. $y = -\frac{1}{3}(x - 3)^2 - 4$

Write an equation in the form $y = a(x - h)^2 + k$ for each parabola shown.

21.

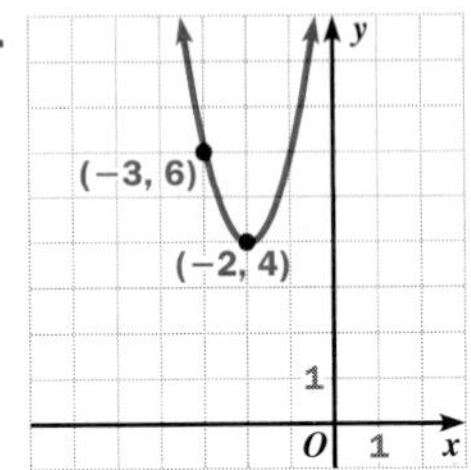

22.

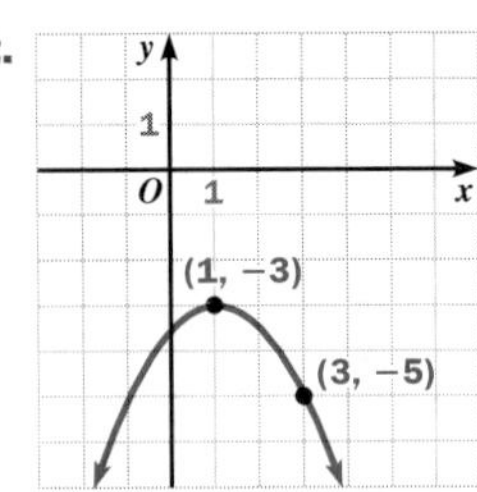

23.

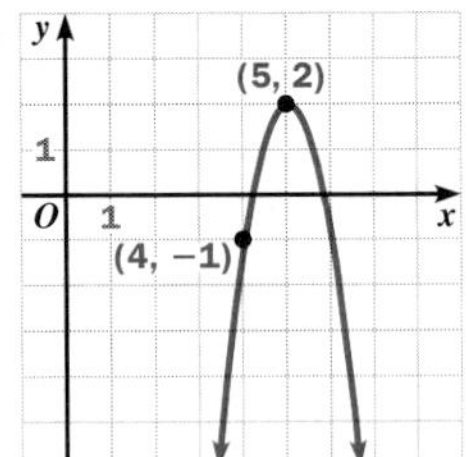

24. Use the equation $h = -16(t - 0.5)^2 + 8$ from part (b) of Example 3 on page 196. Find the possible times the ball is at each height.

 a. 5 ft b. 3 ft c. 1 ft

25. **BIOLOGY** Many birds drop clams or other shellfish in order to break the shell and get to the food inside. Crows along the west coast of Canada use this method to break the shells of whelks. Suppose a crow drops a whelk from a height of 17 ft.

 a. Write a quadratic equation for this situation.

 b. How long will it take the whelk to reach the ground?

26. **Challenge** Show that the line $x = h$ is the line of symmetry for the graph of $y = a(x - h)^2 + k$. (*Hint:* Consider the x-values $x = h \pm b$ where b is an arbitrary number.)

17 ft

ONGOING ASSESSMENT

27. **Writing** Describe how to graph an equation in vertex form. Include sketches illustrating your description.

SPIRAL REVIEW

Solve each equation. (*Section 5.1*)

28. $x^2 = 0.25$
29. $5x^2 = 50$
30. $2x^2 - 7 = 11$

Rewrite each expression in factored form. (*Toolbox, page 782*)

31. $3x + 6$
32. $35x - 21y$
33. $5x^2 + 2x$

Find the inverse of each function. (*Sections 4.1 and 4.2*)

34. $f(x) = \ln x$
35. $g(x) = 3x + 15$
36. $h(x) = 2^x$

9. 1 (maximum)
10. 4 (minimum)
11. −4 (minimum)
12. −1 (maximum)
13. −2 (minimum)
14. 6 (maximum)

15. The vertex is (−3, −3). The line of symmetry is $x = -3$. The parabola opens down.

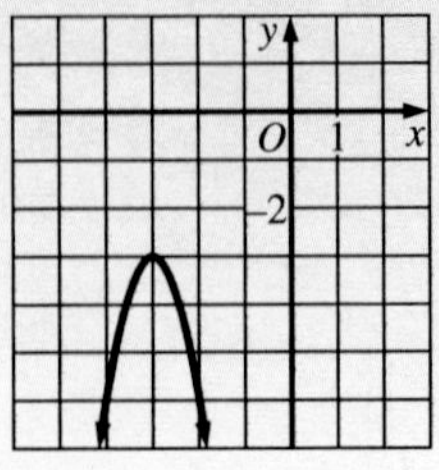

16. The vertex is (1, 0). The line of symmetry is $x = 1$. The parabola opens up.

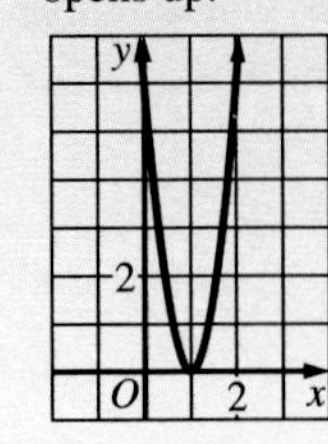

17. The vertex is (−2, 0). The line of symmetry is $x = -2$. The parabola opens down.

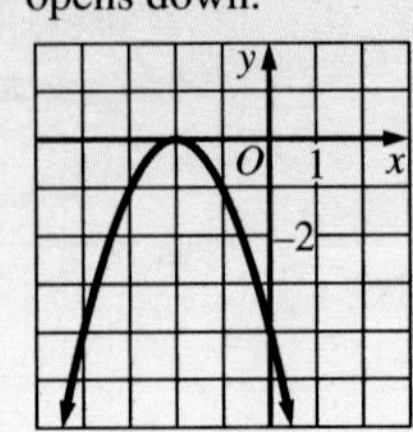

18. The vertex is (2, −5). The line of symmetry is $x = 2$. The parabola opens up.

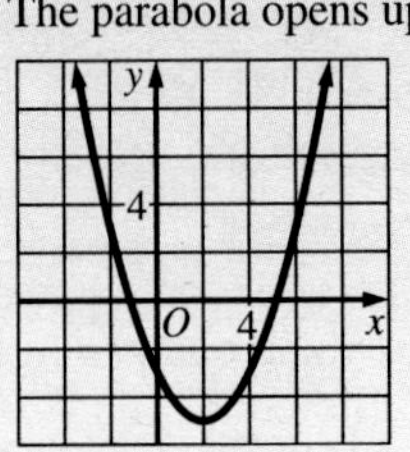

19–36. See answers in back of book.

Section

5.3 Quadratic Functions in Intercept Form

Learn how to...
- write quadratic equations in intercept form
- maximize or minimize quadratic functions

So you can...
- find the greatest weekly revenue at a store, for example

The graphs of quadratic functions may cross the x-axis zero, one, or two times. The x-coordinates of any points where a graph crosses the x-axis are called the **x-intercepts** of the graph.

EXPLORATION
COOPERATIVE LEARNING

Investigating x-intercepts

Work with a partner.
You will need:
- a graphing calculator or graphing software

SET UP Copy and complete the table. Then answer the questions below.

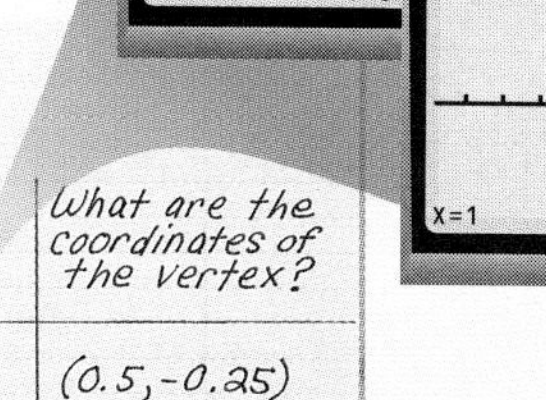

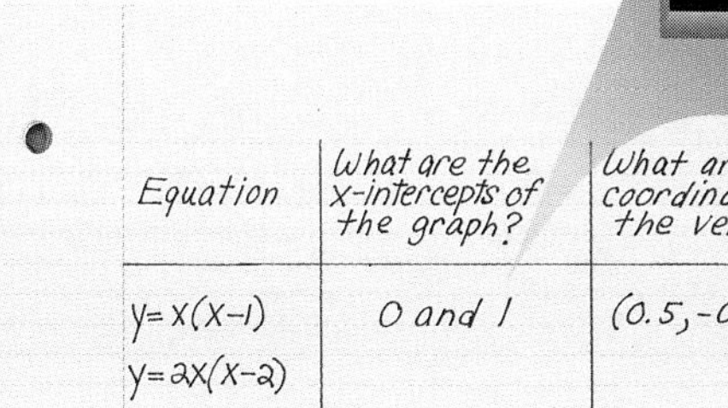

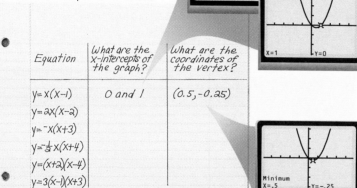

Equation	What are the x-intercepts of the graph?	What are the coordinates of the vertex?
$y = x(x-1)$	0 and 1	(0.5, −0.25)
$y = 2x(x-2)$		
$y = -x(x+3)$		
$y = -\frac{1}{3}x(x+4)$		
$y = (x+2)(x-4)$		
$y = 3(x-1)(x+3)$		

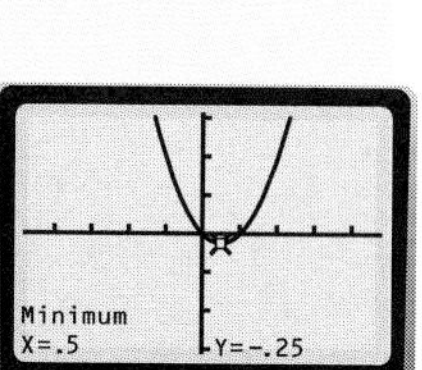

For more information about calculating coordinates of interest, see the *Technology Handbook*, p. 807.

Questions

1. Predict what the graph of each equation looks like by making a sketch. Then check your sketch by using graphing technology.

   a. $y = 3x(x + 1)$
   b. $y = 2(x + 2)(x - 4)$
   c. $y = -\frac{1}{4}(x - 1)(x - 3)$
   d. $y = \frac{1}{2}(x - 2)(x + 2)$

2. The equations you graphed have the form $y = a(x - p)(x - q)$. In terms of a, p, and q, what are the x-intercepts of the graph? What is the x-coordinate of the vertex?

Exploration Note

Purpose
The purpose of this Exploration is to have students predict the x-intercepts and the x-coordinate of the vertex of a quadratic function written in intercept form.

Materials/Preparation
Each pair of students needs a graphing calculator or graphing software.

Procedure
Students should complete the table by graphing each equation and finding the x-intercepts and vertex. Students using a TI-82 can complete the table using the CALCULATE menu. They should choose 2:root to find the x-intercepts and either 3:minimum or 4:maximum to find the vertex. After completing the table, they should look for patterns.

Closure
Review question 2 as a class activity. Ask students to give the equation of a parabola with particular x-intercepts. Is more than one answer possible? (Yes.)

Explorations Lab Manual
See the Manual for more commentary on this Exploration.

Diagram Master 10

For answers to the Exploration, see following page.

Plan⇔Support

Objectives

- Write quadratic equations in intercept form.
- Maximize or minimize quadratic functions.
- Use quadratic functions to find maximums and minimums in real-world situations.

Recommended Pacing

Core and Extended Courses
Section 5.3: 1 day

Block Schedule
Section 5.3: $\frac{1}{2}$ block
(with Section 5.2)

Resource Materials

Lesson Support
Lesson Plan 5.3
Warm-Up Transparency 5.3
Practice Bank: Practice 28
Study Guide: Section 5.3
Exploration Lab Manual: Diagram Masters 2, 10
Assessment Book: Test 17

Technology
Graphing Calculator
McDougal Littell Mathpack *Function Investigator with Matrix Analyzer Activity Book:* Function Investigator Activity 16
Internet: http://www.hmco.com

Warm-Up Exercises

Factor each expression.

1. $7x - 21$ $7(x - 3)$
2. $4 - 2x$ $-2(x - 2)$
3. Evaluate $y = 5(7x + 1)(2 - x)$ when $x = -0.5$. -31.25

Find the average of each pair of numbers.

4. $-7, 3$ -2
5. $3, 10$ 6.5
6. Solve $-3 - 3x = 12$. -5

Teach⇔Interact

Additional Example 1

Graph $y = (3x - 4)(2x + 6)$.

Step 1 Rewrite $y = (3x - 4)(2x + 6)$ in the form $y = a(x - p)(x - q)$.

$$y = (3x - 4)(2x + 6)$$
$$= [3(x - \tfrac{4}{3})][2(x + 3)]$$
$$= 6(x - \tfrac{4}{3})(x + 3)$$
$$= 6(x - \tfrac{4}{3})(x - (-3))$$

Step 2 Identify the x-intercepts and the coordinates of the vertex.

The x-intercepts are $\frac{4}{3}$ and -3.

The x-coordinate of the vertex is $x = \frac{\frac{4}{3} + (-3)}{2} \approx -0.83$.

The y-coordinate of the vertex is

$$y \approx [3(-0.83) - 4][2(-0.83) + 6]$$
$$\approx (-6.49)(4.34)$$
$$\approx -28.2$$

Step 3 Graph the function.

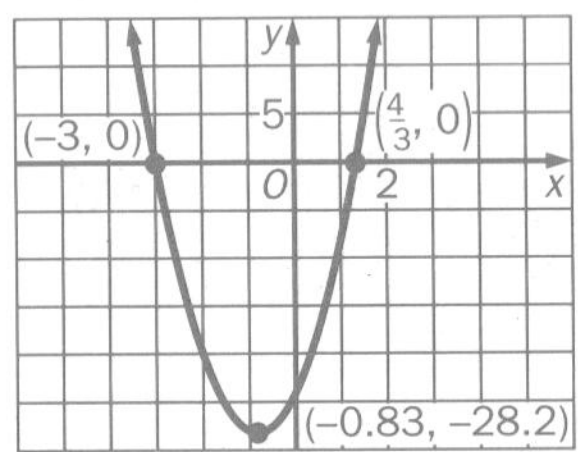

Additional Example 2

Write an equation for the parabola shown below.

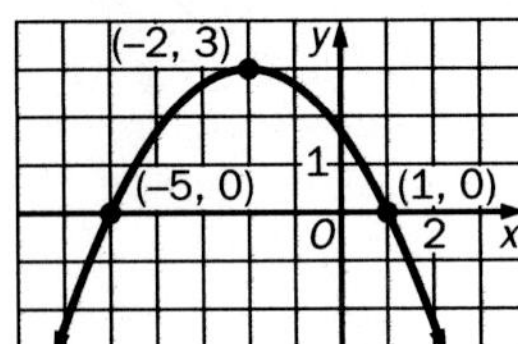

From the graph, you see that the x-intercepts are -5 and 1, and the vertex is $(-2, 3)$. Use the intercept form $y = a(x - p)(x - q)$.

Substitute -5 and 1 for p and q.

$$y = a(x - p)(x - q)$$
$$y = a(x - (-5))(x - 1)$$
$$y = a(x + 5)(x - 1)$$

Substitute -2 for x and 3 for y to find a.

$$3 = a(-2 + 5)(-2 - 1)$$
$$3 = -9a$$
$$-\frac{1}{3} = a$$

Note that a is negative. This agrees with the fact that the graph opens down. An equation in intercept form is $y = -\frac{1}{3}(x + 5)(x - 1)$.

Because you can read x-intercepts directly from an equation in the form $y = a(x - p)(x - q)$, it is called the **intercept form** of a quadratic function. As you saw in the Exploration, there is a relationship between the graph of the function and the three factors a, $x - p$, and $x - q$.

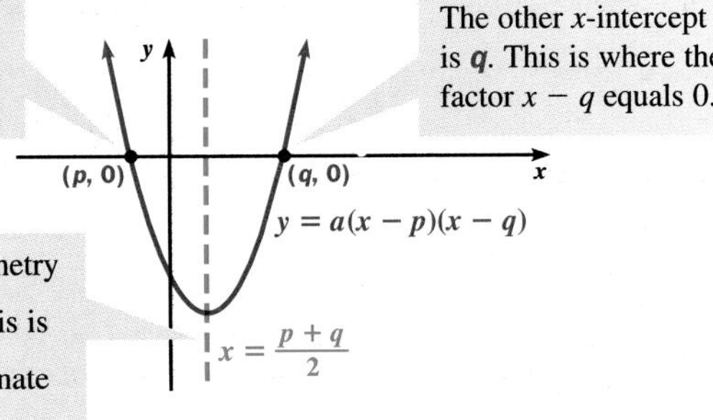

EXAMPLE 1

Graph $y = (3x + 6)(1 - x)$.

Toolbox p. 782
Factoring Variable Expressions

SOLUTION

Step 1 Rewrite $y = (3x + 6)(1 - x)$ in the form $y = a(x - p)(x - q)$.

$$y = (3x + 6)(1 - x)$$
$$= [3(x + 2)][-1(x - 1)]$$
$$= -3(x + 2)(x - 1)$$
$$= -3(x - (-2))(x - 1)$$

Step 2 Identify the x-intercepts and the coordinates of the vertex.

The x-intercepts are -2 and 1.

The x-coordinate of the vertex is:

$$x = \frac{-2 + 1}{2} = -0.5$$

The y-coordinate of the vertex is:

$$y = [3(-0.5) + 6][1 - (-0.5)]$$
$$= 4.5 \cdot 1.5$$
$$= 6.75$$

Step 3 Graph the function.

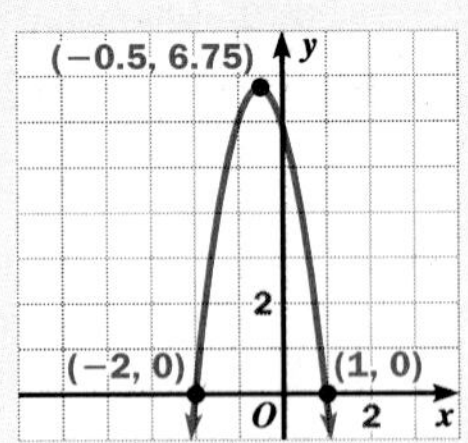

ANSWERS Section 5.3

Exploration

Equation	What are the x-intercepts of the graph?	What are the coordinates of the vertex?
$y = x(x - 1)$	0 and 1	(0.5, −0.25)
$y = 2x(x - 2)$	0 and 2	(1, −2)
$y = -x(x + 3)$	0 and −3	(−1.5, 2.25)
$y = -\frac{1}{2}x(x + 4)$	0 and −4	(−2, 2)
$y = (x + 2)(x - 4)$	−2 and 4	(1, −9)
$y = 3(x - 1)(x + 3)$	1 and −3	(−1, −12)

1. a.

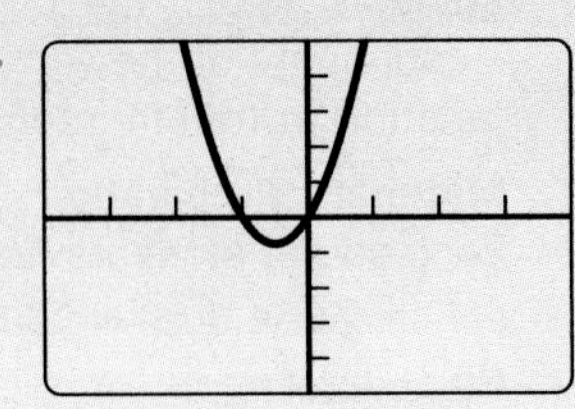

b.

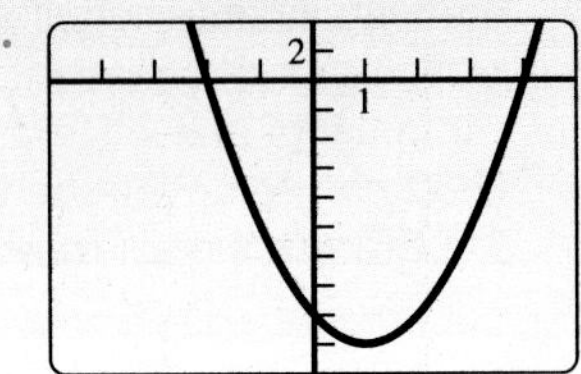

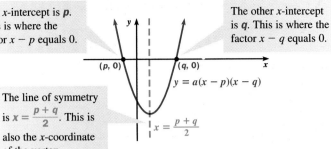

THINK AND COMMUNICATE

1. In Step 1 of Example 1, Joanna found the x-intercepts by setting each factor, $3x + 6$ and $1 - x$, equal to 0. Does this method work? Explain.
2. What is the maximum value of the function in Example 1?

EXAMPLE 2

Write an equation for the parabola shown below.

SOLUTION

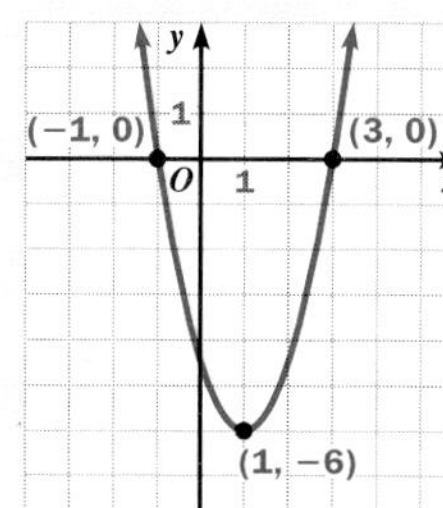

From the graph, you see that the x-intercepts are -1 and 3, and the vertex is $(1, -6)$. Use the intercept form $y = a(x - p)(x - q)$.

$$y = a(x - p)(x - q)$$

Substitute **−1** and **3** for ***p*** and ***q***.

$$y = a(x - (-1))(x - 3)$$

$$y = a(x + 1)(x - 3)$$

Substitute 1 for x and −6 for y to find a.

$$-6 = a(1 + 1)(1 - 3)$$

$$-6 = -4a$$

$$\frac{3}{2} = a$$

Note that ***a* is positive**. This agrees with the fact that the graph opens up.

An equation in intercept form is $y = \frac{3}{2}(x + 1)(x - 3)$.

EXAMPLE 3 Application: Business

A store manager receives 100 dresses with a suggested retail price of \$70.00. The projected sales are 20 dresses per week. Based on past experience, the store manager knows that sales will increase by 2 dresses per week for every \$5.00 decrease in price.

a. Find a function that gives the store's weekly revenue as a function of the number of \$5.00 price decreases.

b. What price maximizes weekly revenue?

SOLUTION

a. Let $x =$ the number of \$5 decreases in price.

Then $70 - 5x =$ the cost of each dress after x decreases in price,

and $20 + 2x =$ the number of dresses sold after x decreases in price.

The weekly revenue R is the product of the cost per dress, $70 - 5x$, and the number sold, $20 + 2x$. So $R = (70 - 5x)(20 + 2x)$.

Without any price decreases, the weekly revenue is **(\$70)(20) = \$1400**.

Solution continued on next page.

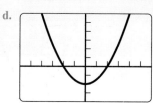

Teach⇔Interact

Additional Example 3

The manager from Example 3 also received 80 sport coats with a suggested retail price of \$120. The projected sales are 18 sport coats per week. Based on experience, the store manager knows that sales will increase by 3 sport coats per week for every \$10 decrease in price.

a. Find a function that gives the store's weekly revenue as a function of the number of \$10 price decreases.

Let $x =$ the number of \$10 decreases in price. Then $120 - 10x =$ the cost of each sport coat after x decreases in price, and $18 + 3x =$ the number of sport coats sold after x decreases in price. The weekly revenue R is the product of the cost per sport coat, $120 - 10x$, and the number sold, $18 + 3x$. So $R = (120 - 10x)(18 + 3x)$.

b. What price maximizes weekly revenue?

The number of price decreases that will maximize weekly revenue is the x-coordinate of the vertex of the graph of $R = (120 - 10x)(18 + 3x)$.

Step 1 Find the x-intercepts. Substitute 0 for R.

$R = (120 - 10x)(18 + 3x)$
$0 = (120 - 10x)(18 + 3x)$

The product of factors equals 0 if and only if at least one of the factors equals 0.

$0 = 120 - 10x$ or $0 = 18 + 3x$
$10x = 120$ $-3x = 18$
$x = 12$ $x = -6$

The x-intercepts are 12 and −6.

Step 2 Find the x-coordinate of the vertex. The x-coordinate of the vertex is $x = \frac{12 + (-6)}{2} = 3$.

The store manager will maximize weekly revenue by decreasing the price of a sport coat by 3(\$10), or \$30, making the price \$120 − \$30, or \$90.

Section Note

Using Technology
Function Investigator Activity 16 in the *Function Investigator with Matrix Analyzer Activity Book* can help students understand how the values of a, p, and q affect the graph of $y = a(x - p)(x - q)$.

c.

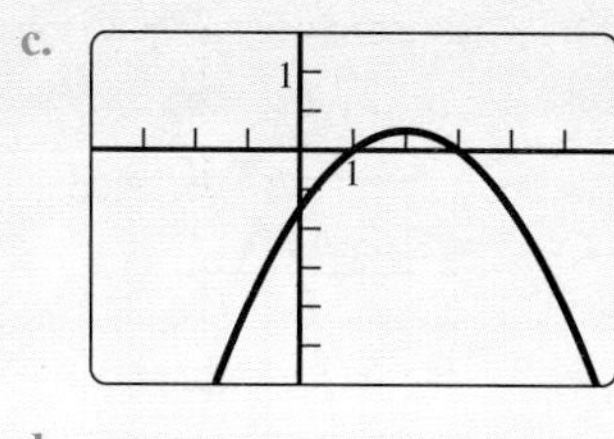

d.

2. $p, q; \frac{p + q}{2}$

Think and Communicate

1. Yes; Joanna determined the values of x for which the factors are equal to 0 rather than rewriting the equation in intercept form and reading the values from the equation.
2. 6.75

Teach⇔Interact

Think and Communicate

Discuss question 4 in class. This would help students realize that many factors contribute to the pricing of a product.

Checking Key Concepts

Communication: Drawing When sketching the graphs of quadratic functions, it is important to consider the scales on the axes before beginning. Encourage students to use the *x*-intercepts to find a good scale for the *x*-axis and to use the vertex to find a good scale for the *y*-axis. In part (c) of question 1, the *x*-intercepts are –7 and 6. A good choice of scale would include both of these points. The *x*-axis could go from –10 to 10 with a scale of 1. The vertex is (–0.5, –253.5). The *y*-axis could go from –275 to 25 with a scale of 25.

Closure Question

Describe how to find the maximum or minimum value of a function whose equation is given in intercept form.

Read the *x*-intercepts from the equation. Average these to get the *x*-coordinate of the vertex. Then substitute that value into the equation to get the maximum value if *a* is negative or the minimum value if *a* is positive.

SOLUTION *continued*

b. The number of price decreases that will maximize weekly revenue is the *x*-coordinate of the vertex of the graph of $R = (70 - 5x)(20 + 2x)$.

Step 1 Find the *x*-intercepts.

$R = (70 - 5x)(20 + 2x)$

Substitute **0** for *R*. $0 = (70 - 5x)(20 + 2x)$

$0 = 70 - 5x$ or $0 = 20 + 2x$

$5x = 70$ $\quad -2x = 20$

$x = 14$ $\quad x = -10$

The product of the factors equals 0 if and only if at least one of the factors equals 0.

The *x*-intercepts are 14 and -10.

Step 2 Find the *x*-coordinate of the vertex.

The *x*-coordinate of the vertex is $x = \frac{14 + (-10)}{2} = 2$.

The store manager will maximize weekly revenue by decreasing the price of a dress by 2($5), or $10, making the price $70 – $10, or $60.

THINK AND COMMUNICATE

3. In Example 3, if the manager sells the dresses at $60, what will the store's weekly revenue be?

4. Does maximizing weekly revenue also maximize *total* revenue from selling the dresses? Why would a store manager discount the price?

CHECKING KEY CONCEPTS

1. Use the function $y = 6(x + 7)(x - 6)$.

a. What are the *x*-intercepts of the function's graph?

b. What are the coordinates of the graph's vertex?

c. Sketch the graph.

2. Write an equation for the parabola shown.

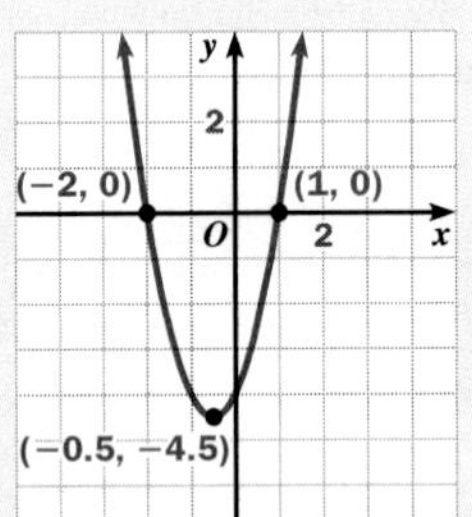

3. What is the maximum value of the function $f(x) = (4x + 12)(6 - 2x)$?

Think and Communicate

3. $1440

4. No; answers may vary. Examples are given. to make space for new inventory, to generate cash

Checking Key Concepts

1. a. –7; 6 b. $\left(-\frac{1}{2}, -253.5\right)$

c.

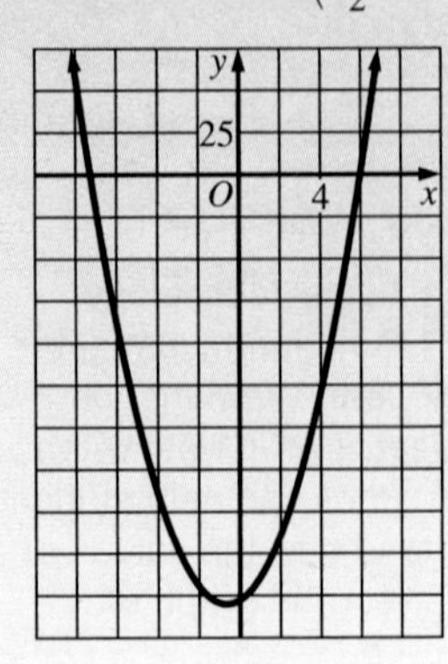

2. $y = 2(x + 2)(x - 1)$ 3. 72

Exercises and Applications

1. a. 1; –3

b. (–1, –16)

c.

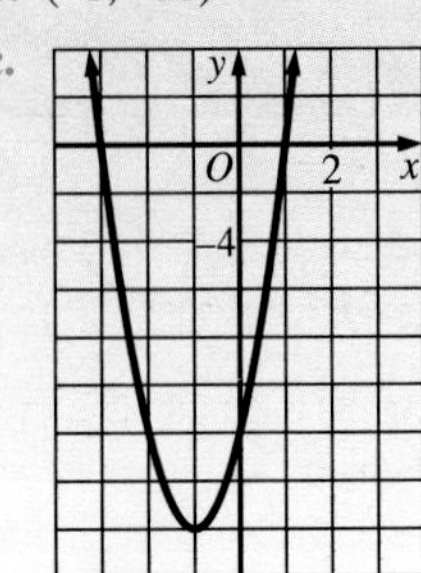

2. a. –1; 5

b. (2, 27)

c.

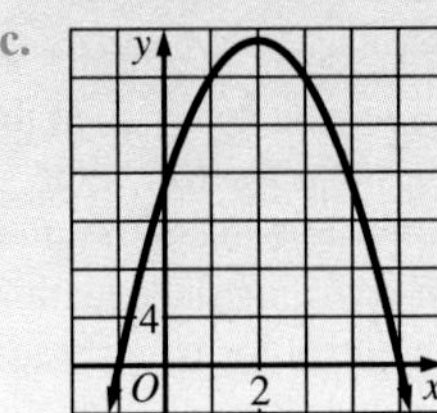

3. a. 4; –4

b. (0, 16)

c.

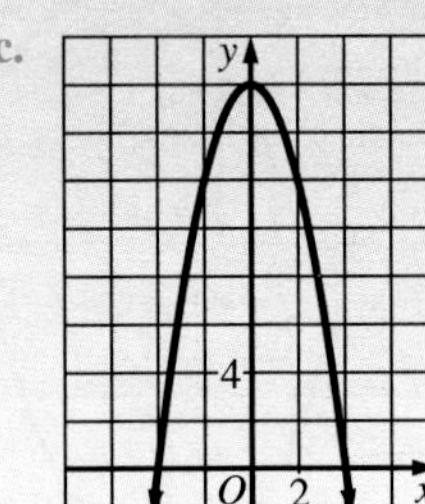

5.3 Exercises and Applications

Extra Practice exercises on page 755

For each equation in Exercises 1–6:

a. Find the x-intercept(s). **b.** Find the vertex. **c.** Sketch the graph.

1. $y = 4(x - 1)(x + 3)$
2. $y = -3(x + 1)(x - 5)$
3. $y = -(x - 4)(x + 4)$
4. $y = (2x + 4)(3x + 3)$
5. $y = (6 - x)(5x + 20)$
6. $y = 0.2(x - 3)(x - 3)$

7. Writing Cheryl and Juan each wrote different equations for the graph shown. Are the students' answers equivalent? Explain why or why not.

Write an equation for each graph.

8.

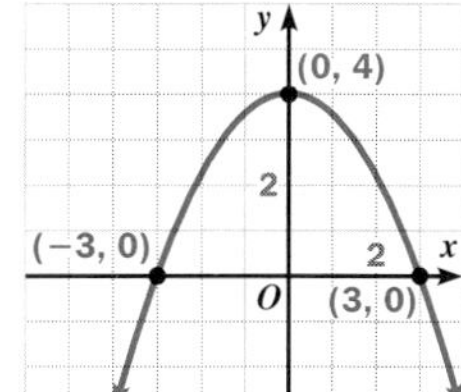

9.

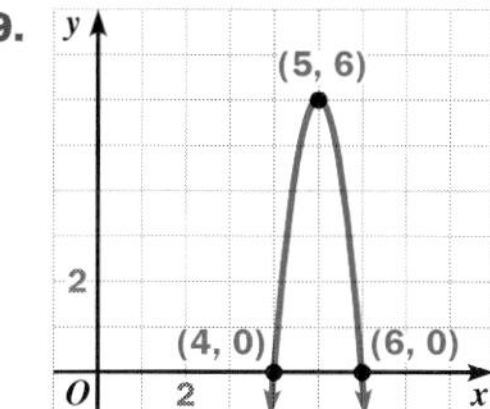

10.

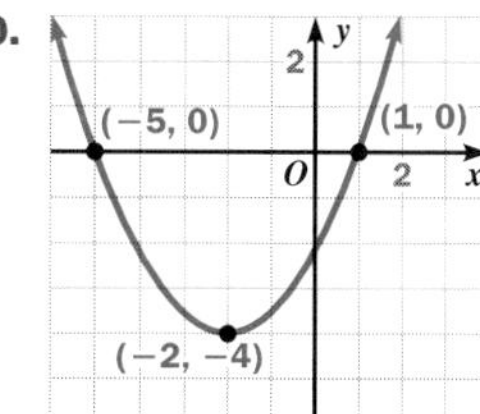

11. Cooperative Learning Work with two other students. You may want to use a graphing calculator or graphing software. Each student should choose one of these three equations:

$y = (x + 2)(3x - 12)$ $y = (0.5x + 1)(x - 4)$ $y = (2x + 4)(4 - x)$

a. Graph your equations in the same coordinate plane. How are the three graphs alike? How are they different?

b. Write your equations in intercept form. How are the three equations alike? How are they different?

12. BUSINESS A magazine publisher polled its subscribers to see how much more they would be willing to pay for a year's subscription to the magazine. The magazine currently has 20,000 subscribers, and the subscription price is \$16. Based on the results of the poll, the publisher knows that, on average, for every \$4 increase in the price of a magazine subscription, 1000 people will decide not to renew their subscriptions.

a. Write a quadratic function that gives the revenue R as a function of x, the number of subscription price increases.

b. What subscription price will give the maximum revenue?

c. What is the maximum revenue for this situation?

Apply⇔Assess

Suggested Assignment

Core Course
Exs. 1–10, 15, 17–28, AYP

Extended Course
Exs. 1–28, AYP

Block Schedule
Day 28 Exs. 1–10, 15, 17–28, AYP

Exercise Notes

Student Progress
Exs. 1–6 Students should be able to complete these exercises quickly and correctly. Any student having difficulty should be encouraged to reread the section.

Second-Language Learners
Ex. 7 Students learning English may benefit from discussing their ideas with a peer tutor or aide before writing their explanations.

Teaching Tip
Ex. 7 In Section 5.4, students will need to rewrite expressions of the form $ax + b$ as $a\left(x + \frac{b}{a}\right)$ even when b is not evenly divisible by a. This exercise is a good opportunity to discuss this type of algebraic manipulation. Ask students how they can rewrite $y = (2x - 1)(3x + 4)$ in the form $y = a(x - p)(x - q)$. They should see that they need to factor out a 2 from the first factor. This can be done by dividing both terms by the coefficient of x. Thus, $2x$ is divided by 2 and -1 is divided by 2 to get $2\left(x - \frac{1}{2}\right)$. Have students multiply the factored form to verify that it is equivalent to $2x - 1$. Then ask them to write $3x + 4$ in factored form.

Cooperative Learning
Ex. 11 This exercise allows students the opportunity to work together to examine three different quadratic functions whose graphs have the same x-intercepts.

Problem Solving
Ex. 12 Students who have difficulty setting up this problem should be encouraged to use Example 3 on page 201 as a guide.

4. a. −2; −1
b. $\left(-\frac{3}{2}, -\frac{3}{2}\right)$
c.

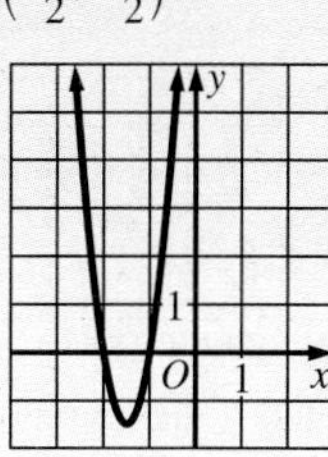

5. a. 6; −4
b. (1, 125)
c.

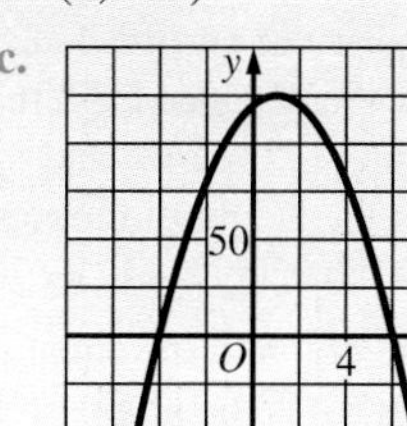

6. a. 3
b. (3, 0)
c.

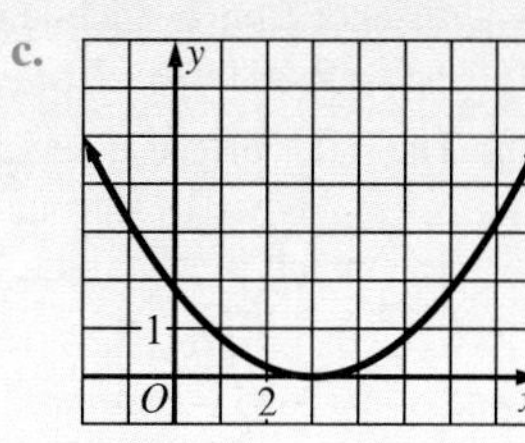

7–12. See answers in back of book.

Apply⇔Assess

Exercise Notes

Application
Exs. 13, 14 In these problems, students can see how several everyday objects, such as the surface of a football field and a road, are shaped like parabolas. Some other examples are the cross section of a satellite dish and the reflector in a flashlight.

Multicultural Note
Exs. 13, 14 Some of the earliest road systems were built by people in China about 5000 years ago to accommodate the growing trade between villages and towns. The expansion of the silk trading routes necessitated the development of road networks between China and its trading partners in Persia and Europe. Approximately 2300 years ago, the ancient Romans began constructing intricate roadway systems to provide efficient transportation for their armies. About 100 years later, they built more than 50,000 miles of highway connecting Rome to places as far away as Spain and Asia Minor.

Integrating the Strands
Ex. 16 This is a classic problem. Students who continue in mathematics will most likely see a variation of this problem in their calculus course. Students might be interested in knowing that they can solve a calculus problem using only algebra.

Common Error
Ex. 17 Most students who choose B think that the square of a sum is equal to the sum of its squares. Remind them that this is only true of products and quotients. If these students need additional convincing, have them substitute values for x and see that the two expressions do not yield the same value. Encourage students to write $(3x + 2)^2$ as $(3x + 2)(3x + 2)$ and then multiply.

Assessment
Ex. 18 Ask several students to read their answers to this question to the class and then write their examples at the board. This would ensure that all students have an accurate summary of the major concepts of this section.

Connection ENGINEERING

Engineers and designers often need to consider the effects of weather on objects and surfaces to be constructed. For example, the surfaces of some roads and athletic fields are shaped like parabolas to allow rain to run off to either side.

13. The diagram shows the parabolically shaped surface of a football field with synthetic turf.

a. Use the fact that football fields are 160 ft wide. Find an equation that gives the height h of a field at a distance d from a sideline.

b. How much higher is the field surface 15 ft from a sideline than at the sideline?

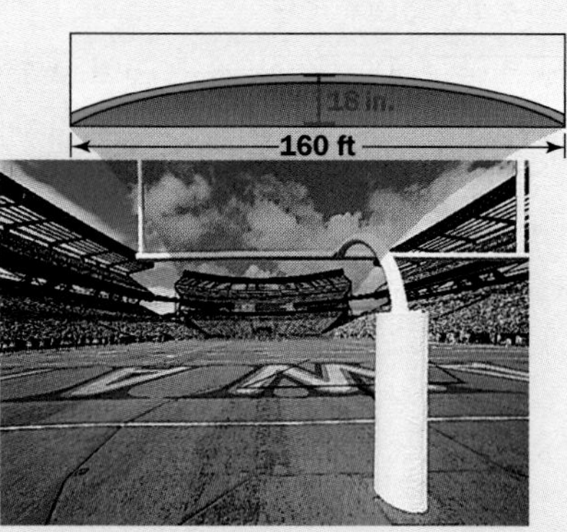

14. The cross-sectional sketch of a two-lane road below shows that each lane is 12 ft wide and the surface at the center line of the road is 3 in. higher than at the edges of the road.

a. Suppose you set up a coordinate plane with the origin at one side of the road shown. Then one of the x-intercepts of the road surface is 0. What is the other x-intercept? What are the coordinates of the vertex? Write an equation for the cross section of the road surface.

b. Suppose you use the same x-axis as in part (a), but you put the y-axis on the line of symmetry for the road surface. What are the coordinates of the vertex now? What are the x-intercepts? Write an equation for the road surface.

c. How is the graph of the equation in part (a) geometrically related to the graph of the equation in part (b)?

d. Use either the equation from part (a) or the equation from part (b) to find how much higher the road surface is at 4 ft from an edge of the road than at the edge of the road.

15. Open-ended Problem Make a sketch of a parabola that has each given number of x-intercepts. Write an equation for each graph.

a. two b. one c. none

16. Challenge A gardener has 50 ft of fencing and plans to create a rectangular garden using a side of a house as one edge of the garden. What dimensions maximize the area of the garden?

17. SAT/ACT Preview If $y = (3x + 2)^2$, then $y =$ __?__.

A. $9x + 4$ **B.** $9x^2 + 4$ **C.** $9x^2 + 6x + 4$
D. $9x^2 + 12x + 4$ **E.** none of these

13. a. If h = the height of the field surface in inches and d = the distance from a sideline in feet, then $h = -0.0028(d)(d - 160)$.
b. The field surface is about 6.1 in. higher 15 ft from a sideline than it is at the sideline.

14. Assume that the distance from the side of the road is measured in feet and the height of the road surface is measured in inches.
a. 24; (12, 3); Let x = the distance from the side of the road and y = the height of the road; $y = -0.0208x(x - 24)$.
b. (0, 3); –12 and 12; Let x = the distance from the center of the road and y = the height of the road; $y = -0.0208(x - 12)(x + 12)$.
c. The graph of the equation in part (b) is the graph of the equation in part (a) translated 12 units to the left.
d. 1.7 in.

15. See answers in back of book.

16. 12.5 ft by 25 ft

17. D

18. I know that the x-intercepts are p and q, the x-coordinate of the vertex is $\frac{p+q}{2}$, and the line of symmetry is $x = \frac{p+q}{2}$. I can tell if the graph of the equation opens up ($a > 0$) or down ($a < 0$). If the graph opens up, the minimum value

ONGOING ASSESSMENT

18. Writing Given a quadratic function in the form $y = a(x - p)(x - q)$, describe all you know about its graph from the values of a, p, and q. Give an example.

SPIRAL REVIEW

Graph each function in the same coordinate plane. Describe the relationships among the four graphs. *(Section 5.2)*

19. $y = x^2$ **20.** $y = (x - 1)^2$

21. $y = (x - 1)^2 - 1$ **22.** $y = -(x - 1)^2 - 1$

Solve each equation. Give solutions to the nearest tenth when necessary. *(Sections 5.1 and 5.2)*

23. $16t^2 = 400$ **24.** $3(x - 1)^2 = 12$ **25.** $5\left(x + \frac{1}{2}\right)^2 - 10 = 0$

Write as a logarithm of a single number or expression. *(Section 4.3)*

26. $-\frac{1}{3}\log_b 8$ **27.** $4 \log_2 x + \frac{1}{2}\log_2 x^4$ **28.** $6 \log_a p - 3 \log_a q$

ASSESS YOUR PROGRESS

VOCABULARY

quadratic function (p. 186)
parabola (p. 186)
line of symmetry (p. 186)
vertex (p. 186)
square root (p. 187)
vertex form (p. 193)
minimum value (p. 195)
maximum value (p. 195)
***x*-intercept** (p. 199)
intercept form (p. 200)

For Exercises 1–3:

a. Graph each function in the same coordinate plane.

b. Solve for x when $y = 12$. Give solutions to the nearest tenth when necessary. *(Section 5.1)*

1. $y = x^2$ **2.** $y = 4x^2$ **3.** $y = \frac{1}{4}x^2$

Write an equation in the form $y = a(x - h)^2 + k$ using the given coordinates of the vertex and of another point on the graph. *(Section 5.2)*

4. vertex: (3, 7); (0, −20) **5.** vertex: (−1, 4); (−5, 12)

6. vertex: (5, −2); (7, 6) **7.** vertex: (−6, −1); (2, 7)

Rewrite each function in intercept form. Then graph each function. *(Section 5.3)*

8. $y = (3x + 27)(x + 4)$ **9.** $y = (6x - 18)(4 - x)$

10. $y = (4x - 12)(x + 3)$ **11.** $y = (10 - 5x)(2x + 4)$

12. Journal Describe how you can find the coordinates of the vertex and the x-intercepts of the graph of a quadratic function given in intercept form. How can you do the same for a function given in vertex form?

Apply⇔Assess

Assess Your Progress

Students should be able to illustrate all but the first of the vocabulary terms with a sketch. Exs. 1–3 may be solved by using technology or by using algebra. You may want to require students to do the problems using both methods to ensure proficiency.

Journal Entry

This entry will help students solidify what they have learned about the intercept form of a quadratic function. Students also need to differentiate between the methods used to find the vertex and x-intercepts of the graph when the function is given in intercept form or vertex form.

Progress Check 5.1–5.3

See page 230.

Practice 28 for Section 5.3

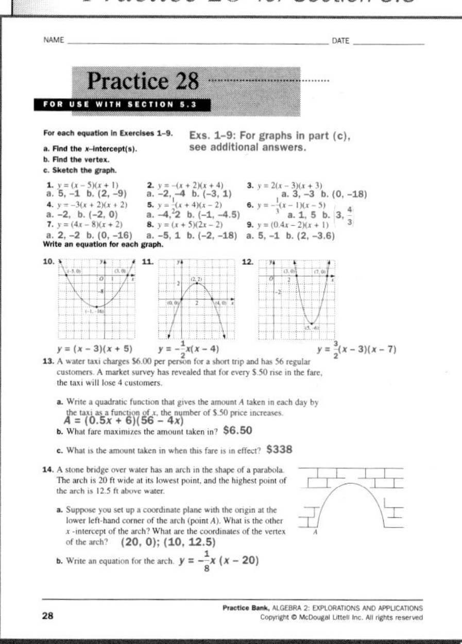

NAME ______ DATE ______

Practice 28

FOR USE WITH SECTION 5.3

For each equation in Exercises 1–9. Exs. 1–9: For graphs in part (c), see additional answers.

a. Find the x-intercept(s).
b. Find the vertex.
c. Sketch the graph.

1. $y = (x - 5)(x + 1)$ a. 5, −1 b. (2, −9)
2. $y = -(x + 2)(x + 4)$ a. −2, −4 b. (−3, 1)
3. $y = 2(x - 3)(x + 3)$ a. 3, −3 b. (0, −18)
4. $y = -3(x + 2)(x + 2)$ a. −2, b. (−2, 0)
5. $y = \frac{1}{2}(x + 4)(x - 2)$ a. −4, 2 b. (−1, −4.5)
6. $y = -\frac{1}{3}(x - 1)(x - 5)$ a. 1, 5 b. $\left(3, \frac{4}{3}\right)$
7. $y = (4x - 8)(x + 2)$ a. 2, −2 b. (0, −16)
8. $y = (x + 5)(2x - 2)$ a. −5, 1 b. (−2, −18)
9. $y = (0.4x - 2)(x + 1)$ a. 5, −1 b. (2, −3.6)

Write an equation for each graph.

10. $y = (x - 3)(x + 5)$ **11.** $y = -\frac{1}{2}x(x - 4)$ **12.** $y = \frac{3}{2}(x - 3)(x - 7)$

13. A water taxi charges \$6.00 per person for a short trip and has 56 regular customers. A market survey has revealed that for every \$.50 rise in the fare, the taxi will lose 4 customers.

a. Write a quadratic function that gives the amount A taken in each day by the taxi as a function of x, the number of \$.50 price increases. $A = (0.5x + 6)(56 - 4x)$

b. What fare maximizes the amount taken in? **\$6.50**

c. What is the amount taken in when this fare is in effect? **\$338**

14. A stone bridge over water has an arch in the shape of a parabola. The arch is 20 ft wide at its lowest point, and the highest point of the arch is 12.5 ft above water.

a. Suppose you set up a coordinate plane with the origin at the lower left-hand corner of the arch (point A). What is the other x-intercept of the arch? What are the coordinates of the vertex of the arch? **(20, 0); (10, 12.5)**

b. Write an equation for the arch. $y = -\frac{1}{8}x(x - 20)$

28 **Practice Bank,** ALGEBRA 2: EXPLORATIONS AND APPLICATIONS

of the function occurs at $x = \frac{p+q}{2}$. If the graph opens down, the maximum value of the function occurs at $x = \frac{p+q}{2}$. For example, the x-intercepts of the graph of $y = 3(x - 2)(x + 10)$ are 2 and −10, the x-coordinate of the vertex is −4, the line of symmetry is $x = -4$, the graph opens up, and the minimum value occurs at $x = -4$.

19–22.

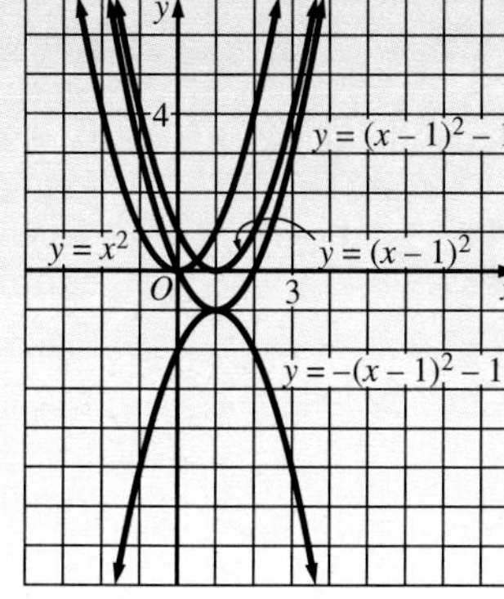

The graphs all have the same basic shape. Each graph is the image of every other graph under translation, reflection, or both. For example, the graph in Ex. 22 ($y = -(x - 1)^2 - 1$) can be obtained by translating the graph in Ex. 19 ($y = x^2$) 1 unit to the right and 1 unit down, then reflecting it over the line $y = -1$.

23. ±5 24. −1; 3

25. −1.9; 0.9 26. $\log_b \frac{1}{2}$

27. $\log_2 x^6$ 28. $\log_a \frac{p^6}{q^3}$

Assess Your Progress

1–12. See answers in back of book.

Plan⇔Support

Objectives

- Complete the square to write quadratic functions in vertex form.
- Use completing the square to find maximums or minimums in real-world situations.

Recommended Pacing

Core and Extended Courses
Section 5.4: 2 days

Block Schedule
Section 5.4: 1 block

Resource Materials

Lesson Support
Lesson Plan 5.4
Warm-Up Transparency 5.4
Overhead Visuals:
Folder B: Algebra Tiles
Practice Bank: Practice 29
Study Guide: Section 5.4
Exploration Lab Manual:
Diagram Masters 2, 3, 4, 11

Technology
Graphing Calculator
McDougal Littell Mathpack
Function Investigator
Internet:
http://www.hmco.com

Warm-Up Exercises

Solve. Round your answers to the nearest tenth when necessary.

1. $3(x + 1)^2 = 4$ −2.2, 0.2
2. $-16(t - 2)^2 = -64$ 0, 4

For Exs. 3–5, use the formula $d = 16t^2$ for the distance, in feet, an object falls in t seconds. Consider an object falling from a 130 ft building.

3. How far does the object fall in 1.5 s? 36 ft
4. How far from the ground will the object be after 2 s? 66 ft
5. How long will it take for the object to hit the ground? about 2.9 s

Section

5.4 Completing the Square

Learn how to...

- complete the square to write quadratic functions in vertex form

So you can...

- find maximums or minimums, such as the maximum height of an object thrown into the air

You can write a quadratic function in many different forms. You have already used the vertex form and the intercept form. The **standard form** of a quadratic function is $y = ax^2 + bx + c$. A process called *completing the square* is sometimes used to change quadratic functions in standard form to vertex form. In the Exploration, you will model this process with algebra tiles.

EXPLORATION
COOPERATIVE LEARNING

Using Algebra Tiles to Complete the Square

Work with a partner.
You will need:
- algebra tiles

1 Use tiles to model the expression $x^2 + 6x$.

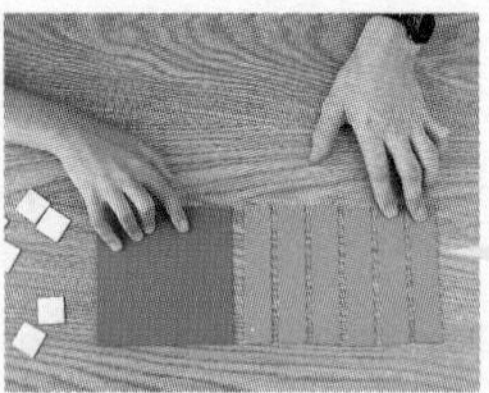

You will need one x^2-tile and six x-tiles.

2 If possible, arrange the tiles in a square. Your arrangement may have a "hole."

You want the length and width of your "square" to be equal.

3 Determine the number of 1-tiles needed to fill the hole.

By adding nine 1-tiles, you see that $\mathbf{x^2 + 6x + 9 = (x + 3)^2}$.

Exploration Note

Purpose
The purpose of this Exploration is to have students discover a pattern that they can use to complete the square.

Materials/Preparation
Each pair of students needs the following algebra tiles: 1 x^2-tile, 12 x-tiles, and 36 1-tiles. Students may need a brief reminder of how to use algebra tiles.

Procedure
Students work with a partner to follow Steps 1–3 for completing the square of $x^2 + 6x$. They then repeat this process for three other expressions to complete the table in question 1. For question 2, students look for patterns that can be used to complete the square without using algebra tiles.

Closure
Discuss questions 1 and 2 as a class activity. Students should understand the answers to question 2 and be able to complete the square for expressions of the form $x^2 + bx$ correctly.

Explorations Lab Manual
See the Manual for more commentary on this Exploration.
Diagram Master 11

Questions

1. Copy and complete the table by following the procedure described on the previous page.

Expression	Number of 1-tiles needed to complete the square	Expression written as a square
$x^2 + 6x +$?	9	$x^2 + 6x + 9 = (x + 3)^2$
$x^2 + 4x +$?	?	?
$x^2 + 8x +$?	?	?
$x^2 + 12x +$?	?	?

2. Look for patterns in the last column of your table. Consider the general statement $x^2 + bx + c = (x + d)^2$.

a. How is d related to b in each case?

b. How is c related to d in each case?

c. How could you obtain the numbers in the second column of your table directly from the coefficients of x in the expressions from the first column?

EXAMPLE 1

Write $y = x^2 + 3x + 2$ in vertex form.

SOLUTION

Use the method of completing the square.

$$\begin{aligned} y &= x^2 + 3x + 2 \\ &= (x^2 + 3x + \underline{?}) + 2 - \underline{?} \\ &= \left(x^2 + 3x + \frac{9}{4}\right) + 2 - \frac{9}{4} \\ &= \left(x + \frac{3}{2}\right)^2 - \frac{1}{4} \end{aligned}$$

Whatever number you **add** to complete the square, you must also **subtract** to avoid changing the equation.

The number you want is the square of half of the coefficient of x: $\left(\frac{3}{2}\right)^2 = \frac{9}{4}$.

In vertex form, the equation is $y = \left(x + \frac{3}{2}\right)^2 - \frac{1}{4}$.

THINK AND COMMUNICATE

1. **Technology** Use a graphing calculator or graphing software to graph $y = x^2 + 3x + 2$ and $y = \left(x + \frac{3}{2}\right)^2 - \frac{1}{4}$ in the same coordinate plane. What do you notice?

2. The graph of the equation in Example 1 has what point as its vertex?

Teach⇔Interact

Learning Styles: Kinesthetic

Students with a kinesthetic learning style will benefit from the Exploration. They will be better able to understand how each step in the process of completing the square corresponds to a manipulation of the tiles. Later, when students complete the square without the aid of tiles, they should be encouraged to visualize the physical process using the tiles.

Section Note

Topic Spiraling: Preview
Students will use completing the square when working with circles, ellipses, and hyperbolas in Chapter 11.

About Example 1

Common Error
Some students forget to subtract the constant that is added to complete the square. Encourage students to write in the parentheses and blanks first and then to divide and square the coefficient of x and put this number in both blanks.

Additional Example 1

Write $y = x^2 + 5x + 3$ in vertex form.

Complete the square.

$$\begin{aligned} y &= x^2 + 5x + 3 \\ &= (x^2 + 5x + \underline{\ ?\ }) + 3 - \underline{\ ?\ } \\ &= \left(x^2 + 5x + \frac{25}{4}\right) + 3 - \frac{25}{4} \\ &= \left(x + \frac{5}{2}\right)^2 + 3 - \frac{25}{4} \\ &= \left(x + \frac{5}{2}\right)^2 - \frac{13}{4} \end{aligned}$$

In vertex form, the equation is $y = \left(x + \frac{5}{2}\right)^2 - \frac{13}{4}$.

Think and Communicate

Students should recognize that question 1 gives them a method for checking their work when completing the square. Question 2 focuses students' attention on a reason for completing the square: to transform a quadratic equation in standard form to vertex form to determine its vertex.

ANSWERS Section 5.4

Exploration

1.

Expression	Number of 1-tiles needed to complete the square	Expression written as a square
$x^2 + 6x +$?	9	$x^2 + 6x + 9 = (x + 3)^2$
$x^2 + 4x +$?	4	$x^2 + 4x + 4 = (x + 2)^2$
$x^2 + 8x +$?	16	$x^2 + 8x + 16 = (x + 4)^2$
$x^2 + 12x +$?	36	$x^2 + 12x + 36 = (x + 6)^2$

2. a. $d = \frac{1}{2}b$ b. $c = d^2$

c. Divide the coefficient by 2 and square the result.

Think and Communicate

1. 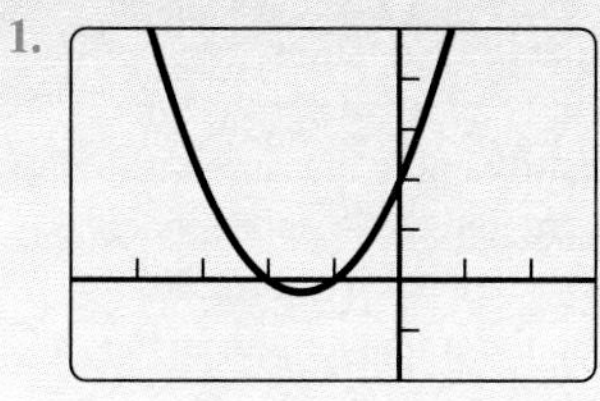

The graphs are the same. The equations are equivalent.

2. $\left(-\frac{3}{2}, -\frac{1}{4}\right)$

Teach⇔Interact

Section Notes

Teaching Tip
Students may wonder what is meant by the term *vertical velocity*. Explain that all velocities can be written as the sum of vertical and horizontal components. The vertical component contributes only to the height of the object, while the horizontal component contributes only to the horizontal distance traveled. The height of the object is independent of whether or not it was given any initial horizontal velocity. Exs. 19–23 give students an example where both horizontal and vertical velocities are involved.

Communication: Reading
When reading any text that contains both a graph and commentary about the graph, it is important that students go back and forth between the commentary and the graph as each new point is made. The commentary and graph at the top of this page provide an excellent opportunity to practice this technique. Point out to students that after reading the sentence to the left of the graph for example, they should find the graph of $h = v_0 t + h_0$ and note that it is linear. Ask them what would happen to the ball if it had this height equation.

Interdisciplinary Problems
The mathematics on this page and Example 2 illustrate how quadratic equations and the method of completing the square can be used to solve projectile problems in physics. Similar problems can be found in Exs. 9, 10, and 14.

About Example 2

Mathematical Procedures
The coefficient of the quadratic term in Example 2 is –16 rather than 1, as it has been for all previous examples of completing the square. There are two additional steps necessary to complete the square of $h = -16t^2 + 32t + 4$ that were not present in Example 1: (1) before completing the square, the coefficient of x^2 must be factored out of the x^2- and x-terms, and (2) when the new constant is added to complete the square, you must subtract that constant multiplied by the number outside the parentheses.

Projectiles

You know that a dropped object falls a distance of $16t^2$ feet in t seconds. (See Example 3 on page 187.) When an object is not simply released but is thrown or launched, it is called a *projectile*. What happens to a projectile that is launched with some initial vertical velocity v_0 (measured in feet per second) at some initial height h_0 (measured in feet)?

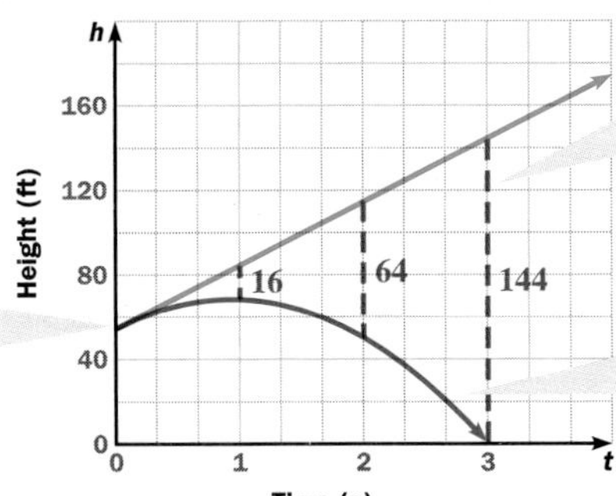

Without gravity to pull the projectile down, its height h would increase according to the equation $h = v_0 t + h_0$.
h
Height (ft)
160
120
80
40
0
16
64
144
0
1
2
3
t
Time (s)
With gravity, the projectile falls $16t^2$ feet in t seconds.
So the projectile's height at any time t is given by $h = -16t^2 + v_0 t + h_0$.

The function $h = -16t^2 + v_0 t + h_0$ is a model for the height of a projectile, in feet, as a function of time, in seconds. The model has its limitations, since it does not take into account factors such as air resistance, which can affect the projectile's vertical movement.

THINK AND COMMUNICATE

3. Why is the coefficient of t^2 negative in $h = -16t^2 + v_0 t + h_0$ but positive in $d = 16t^2$ (from Example 3 on page 187)?
4. Consider a baseball thrown upward and a leaf thrown upward. Which situation is more accurately modeled by $h = -16t^2 + v_0 t + h_0$? Why?
5. Use the projectile graph shown in red above.
   a. Estimate the object's initial height and maximum height.
   b. What is the domain of the object's height function? What is the range?
   c. Does the graph tell you anything about the object's *horizontal* position at time t? Explain.
   d. How would the graph of the object's height function change if the object's initial vertical velocity were increased?

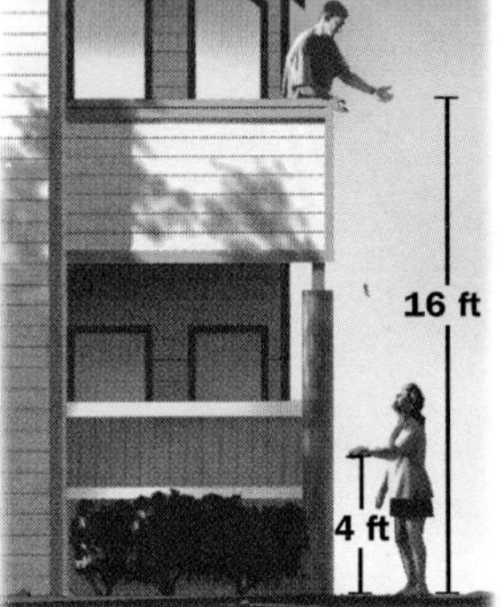

16 ft
4 ft

EXAMPLE 2 Application: Physics

Shelly throws her keys up in the air, releasing them from a height of 4 ft above the ground, with an initial vertical velocity of 32 ft/s.

a. What maximum height do the keys reach?

b. Her brother Mark is standing on a balcony above her. If his outstretched arms are 16 ft above the ground, at what time(s) can he catch the keys?

Think and Communicate

3. In the first equation, the coefficient is negative because the quantity being measured is height and the distance fallen due to gravity is subtracted from the initial height. In the second equation, the coefficient is positive because the quantity being measured is distance traveled, which is positive.

4. the baseball; A leaf that is broad and flat would encounter significant air resistance and its path would be altered by even a minor breeze.

5. a. Estimates may vary; about 55 ft; about 67 ft.

b. $t \geq 0$; $0 \leq h \leq 67$

c. No; the graph only gives information about the vertical position of the projectile at given times after it is launched or thrown.

d. The maximum point of the graph would be higher and the parabola would be wider.

SOLUTION

a. The maximum height of the keys is at the vertex of the parabola described by the equation $h = -16t^2 + 32t + 4$. Use the method of completing the square to rewrite the equation in vertex form.

$$\begin{aligned} h &= -16t^2 + 32t + 4 \\ &= -16(t^2 - 2t) + 4 \\ &= -16(t^2 - 2t + 1) + 4 - (-16) \\ &= -16(t - 1)^2 + 20 \end{aligned}$$

By putting a **1** inside the parentheses to complete the square, you have added **−16 · 1**, or **−16**, to the equation. So you must also subtract **−16**.

The vertex is at (1, 20). So the keys are at a maximum height of 20 ft after 1 s.

b. Use the equation $h = -16(t - 1)^2 + 20$ from part (a). Find t when $h = 16$.

$$\begin{aligned} -16(t-1)^2 + 20 &= 16 \\ -16(t-1)^2 &= -4 \\ (t-1)^2 &= \frac{1}{4} \\ t - 1 &= \pm\frac{1}{2} \\ t &= 1 \pm \frac{1}{2} \\ t &= \frac{1}{2}, \frac{3}{2} \end{aligned}$$

Substitute **16** for h.

The keys will be at a height of 16 ft after 0.5 s and again after 1.5 s.

CHECKING KEY CONCEPTS

1. Find the number that completes the square for each expression.

a. $x^2 + 14x + \underline{?}$ **b.** $x^2 + 5x + \underline{?}$ **c.** $x^2 + 0.20x + \underline{?}$

2. Write each equation in vertex form.

a. $y = 3x^2 + 6x + 1$ **b.** $y = x^2 + \frac{3}{4}x + 1$ **c.** $y = -5x^2 + 25x + 2$

3. What is the maximum value of the function $f(x) = -2x^2 + 8x + 3$?

4. Hamish throws a baseball straight up with a velocity of 24 ft/s from an initial height of 6 ft.

a. What equation describes the height of the ball as a function of time?

b. What is the ball's maximum height? When does the ball reach that height?

c. If Hamish doesn't catch the ball, when does it hit the ground?

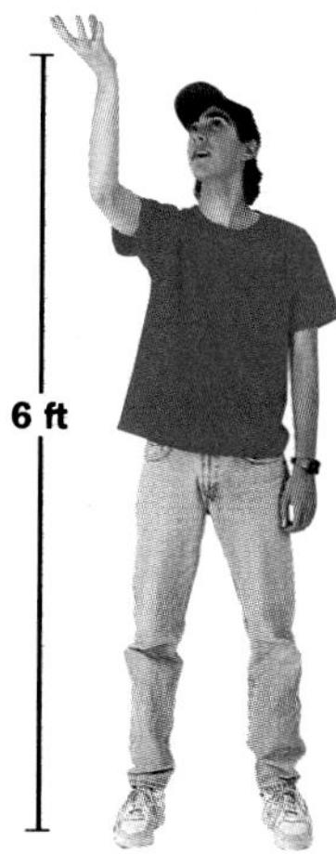

Checking Key Concepts

1. a. 49

b. $\frac{25}{4}$

c. 0.01

2. a. $y = 3(x + 1)^2 - 2$

b. $y = \left(x + \frac{3}{8}\right)^2 + \frac{55}{64}$

c. $y = -5\left(x - \frac{5}{2}\right)^2 + \frac{133}{4}$

3. 11

4. a. $h = -16t^2 + 24t + 6$

b. 15 ft; 0.75 s after it is thrown

c. 1.7 s after it is thrown

Teach⇔Interact

Additional Example 2

Mark's football is stuck in a tree. He hopes to dislodge it by standing on a small hill and throwing a baseball at it. Use the fact that a dropped object falls a distance of $4.9t^2$ meters in t seconds. Mark throws the baseball with an initial vertical velocity of 14.7 m/s from a height of 3 m.

a. What maximum height does the baseball reach?

The maximum height of the baseball is at the vertex of the parabola. Complete the square to rewrite the equation in vertex form.

$$\begin{aligned} y &= -4.9t^2 + 14.7t + 3 \\ &= -4.9(t^2 - 3t) + 3 \\ &= -4.9(t^2 - 3t + 2.25) + 3 - (-11.025) \\ &= -4.9(t - 1.5)^2 + 14.025 \end{aligned}$$

The vertex is at (1.5, 14.025). So, the baseball is at a maximum height of 14.025 m after 1.5 s.

b. If the football is 12 m above the ground, at what time(s) could the baseball hit it?

Use the equation $y = -4.9(t - 1.5)^2 + 14.025$ from part (a). Find t when $h = 12$.

$$\begin{aligned} -4.9(t-1.5)^2 + 14.025 &= 12 \\ -4.9(t-1.5)^2 &= -2.025 \\ (t-1.5)^2 &\approx 0.41 \\ t - 1.5 &\approx \pm 0.64 \\ t &\approx 1.5 \pm 0.64 \\ t &\approx 0.86, 2.14 \end{aligned}$$

The baseball will be at a height of 12 m after about 0.86 s and again after about 2.14 s, assuming that it did not hit the football on the way up.

Checking Key Concepts

Using Manipulatives

Students who have difficulty with question 1 should be encouraged to use algebra tiles.

Closure Question

Consider the equation $x^2 + bx + c = (x + d)^2 + e$. How is d related to b? How is c related to e?

d is half of b. To get e, b was divided by two, squared, and then subtracted from c.

Suggested Assignment

Core Course
Day 1 Exs. 1–14, 17
Day 2 Exs. 25–47

Extended Course
Day 1 Exs. 1–24
Day 2 Exs. 25–47

Block Schedule
Day 29 Exs. 1–14, 17, 25–47

Exercise Notes

Using Manipulatives
Exs. 1–8 Allow those students having difficulty completing the square algebraically to use algebra tiles for these exercises. Encourage them, however, to then go back and try to do each exercise without using the tiles.

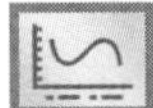
Using Technology
Exs. 1–8 Students can check their answers to these exercises by using the CALC features of a TI-82 graphing calculator. Enter the function on the Y= list. Then press [2nd][CALC] and choose either 3:minimum or 4:maximum. Use the right and left arrow keys to move the cursor to the left of the vertex and press [ENTER]. Then move the cursor to the right of the vertex and press [ENTER] twice. The coordinates of the vertex will appear at the bottom of the screen.

Visual Thinking
Ex. 10 Ask students to sketch graphs of the situations described in parts (a)–(c). Encourage them to discuss their responses to part (d) based on their sketches. This activity involves the visual skills of *correlation* and *interpretation.*

Alternate Approach
Exs. 11–13 One approach to doing these exercises would be to have the class discuss the different methods students used individually and then consider which were the most efficient. Some possible methods include putting the equation in vertex form, looking at the *y*-intercepts, or seeing if the coordinates of the vertex satisfy the given equation.

5.4 Exercises and Applications

Extra Practice exercises on page 756

Write each function in vertex form.

1. $y = x^2 + 4x + 3$
2. $y = -x^2 + 8x - 1$
3. $y = x^2 - 7x - 12$
4. $y = 2x^2 + 7x + 3$
5. $y = x^2 + 6x$
6. $y = 8x^2 + 2x + 9$
7. $y = 2x^2 - 5x - 3$
8. $y = -x^2 + 4x - 7$

9. Carol throws a softball upward from a height of 3 ft above the ground and with an initial velocity of 24 ft/s.
   a. Write an equation to model this situation.
   b. What is the maximum height the ball reaches?
   c. When does the ball reach its maximum height?
   d. When does the ball hit the ground?

10. Andrew tosses a rock into an empty well from a height of 54 ft above the bottom of the well. Write an equation giving the rock's height above the bottom of the well as a function of time in the following situations.
   a. He throws the rock upward with an initial velocity of 32 ft/s.
   b. He drops the rock, with no initial velocity.
   c. He throws the rock downward, with an initial velocity of −32 ft/s.
   d. Find the coordinates of the vertex of each of the parabolas you found in parts (a)–(c). What do you notice?

Match each equation with its graph. Explain your choices.

11. $y = -x^2 - 3x + 5$
12. $y = x^2 - 3x - 2$
13. $y = 2x^2 + 4x + 2$

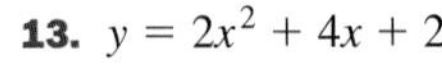

A.
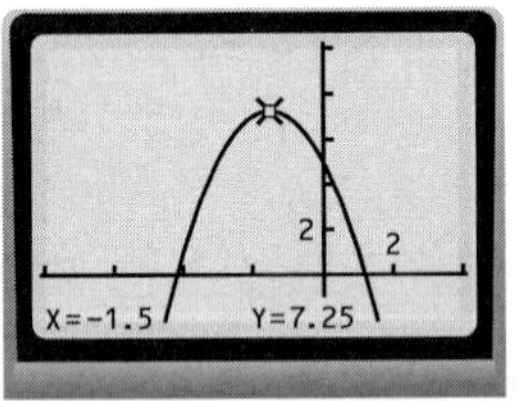

B.
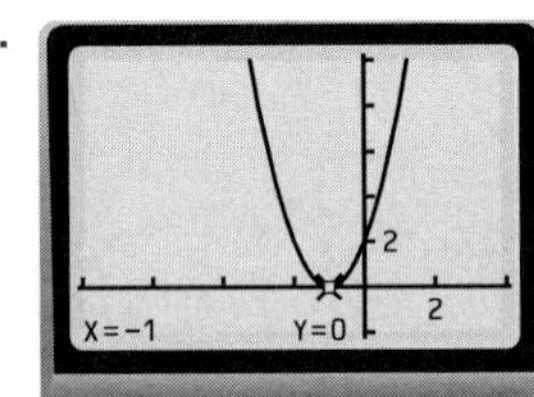

C.
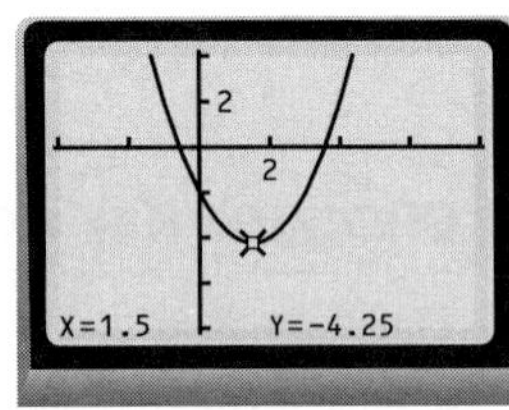

14. **PHYSICS** To jump, a frog pushes itself into the air using its long hind legs. Suppose this jumping frog's body is $\frac{1}{3}$ ft above the ground when the frog's outstretched hind legs leave the ground. At this point, you can treat the frog as a projectile.
   a. Suppose the frog jumps with a vertical velocity of 4 ft/s. Write an equation to model the height of the frog's body above the ground.
   b. What maximum height does the frog's body reach?

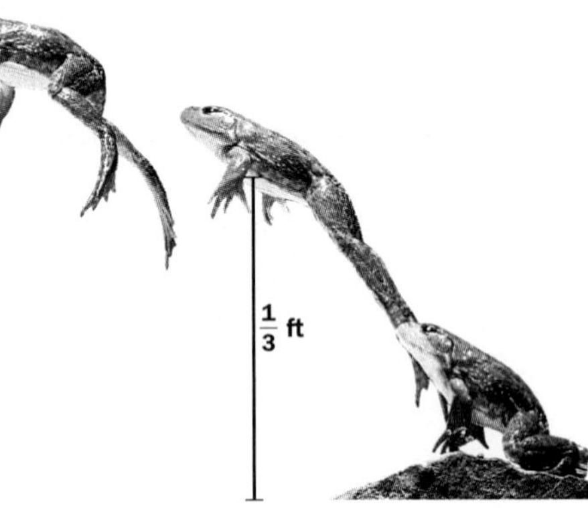

Exercises and Applications

1. $y = (x + 2)^2 - 1$
2. $y = -(x - 4)^2 + 15$
3. $y = \left(x - \frac{7}{2}\right)^2 - \frac{97}{4}$
4. $y = 2\left(x + \frac{7}{4}\right)^2 - \frac{25}{8}$
5. $y = (x + 3)^2 - 9$
6. $y = 8\left(x + \frac{1}{8}\right)^2 + \frac{71}{8}$
7. $y = 2\left(x - \frac{5}{4}\right)^2 - \frac{49}{8}$
8. $y = -(x - 2)^2 - 3$
9. a. $h = -16t^2 + 24t + 3$
   b. 12 ft
   c. 0.75 s after it is thrown
   d. 1.6 s after it is thrown
10. a. $h = -16t^2 + 32t + 54$
   b. $h = -16t^2 + 54$
   c. $h = -16t^2 - 32t + 54$
   d. (a) (1, 70); (b) (0, 54); (c) (−1, 70); Answers may vary. An example is given. In the cases where the initial velocity is nonnegative, the *x*-coordinate of the vertex indicates the time the maximum height is reached and the *y*-coordinate indicates the maximum height. As you would expect, the *y*-coordinate is greater when the rock is thrown up than when it is dropped. When the initial velocity is negative, the vertex has no meaning in relation to the problem, since the *x*-coordinate is negative. Also, since the rock was thrown downward from a height of

Connection BIOLOGY

To determine the energy cost of motion (walking, running, and so on) for a human or an animal, biologists often measure the rate at which oxygen is consumed. As you might expect, oxygen consumption depends on the level and type of exertion.

15. The table shows the rate of oxygen consumed by a test subject running at a speed of 16 km/h using various stride lengths. The graph shows the data from the table and a parabola that fits the data. The parabola has the equation $y = 0.001292x^2 - 0.3784x + 31.56$.

Stride length (cm)	Oxygen consumption (L/min)
135	4.03
149	3.87
169	4.52

stride length

a. Write the equation in vertex form. What are the coordinates of the vertex of the parabola? What does this vertex mean to the runner?

b. **Writing** At 16 km/h, the test subject freely chose the stride length of 149 cm when allowed to run in a way that felt natural. What do you notice about this stride length and the x-coordinate of the vertex from part (a)? What conclusion might you draw from this observation?

BY THE WAY

Olympic long-distance runner Paavo Nurmi of Finland won six gold medals in the 1920s. Nurmi's unusually long stride length was the most economical for *him*, but not for the other runners who tried to copy it.

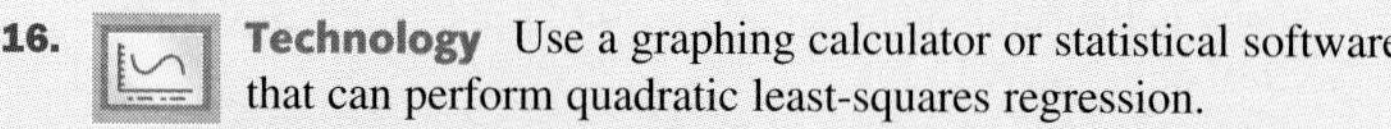

16. **Technology** Use a graphing calculator or statistical software that can perform quadratic least-squares regression.

a. Find an equation of a parabola that is a good fit for the data below. Here, the test subject described in Exercise 15 ran at 14 km/h.

Stride length (cm)	119	135	153
Oxygen consumption (L/min)	3.55	3.35	3.75

b. The subject's "natural" stride length in this case is 135 cm. Does this confirm your conclusion from part (b) of Exercise 15? Explain.

17. The graph shows the oxygen consumption of a parakeet flying level at various speeds. The data can be modeled by the function $y = 0.06x^2 - 4.0x + 87$, where x is the speed in kilometers per hour and y is the oxygen consumption in milliliters of oxygen per gram of body mass per hour.

a. Write the equation in vertex form.

b. What is the minimum value of the function $y = 0.06x^2 - 4.0x + 87$? What does it mean for the parakeet?

18. **Challenge** Estimate the oxygen consumption of a 35 g parakeet flying level at 20 km/h for 5 min.

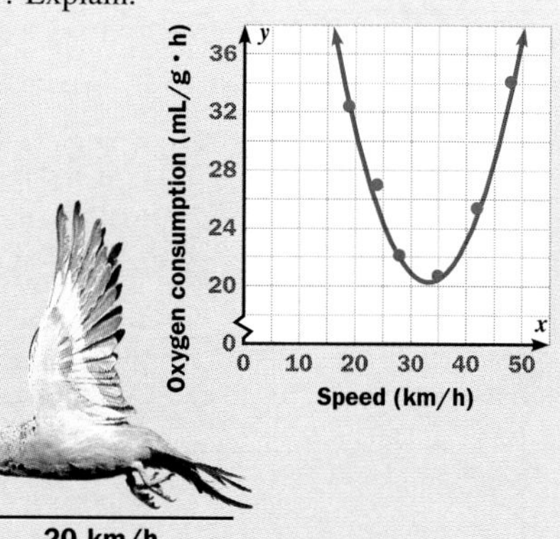

Apply⇔Assess

Exercise Notes

Interdisciplinary Problems
Exs. 15–18 These exercises allow students to see how quadratic functions and the method of completing the square can be used to solve problems in biology.

Second-Language Learners
Ex. 15 Students learning English may benefit from working with a peer tutor or aide to solve the problem and justify their responses.

Mathematical Procedures
Ex. 15–17 Point out to students that they must use division to find the factored form of the first two terms of a quadratic function before they can complete the square. For example, in Ex. 15, since $\frac{-0.3784}{0.001292} \approx -292.88$, then $y = 0.001292(x^2 - 292.88x) + 31.56$. Now the square can be completed to find the vertex form of the equation.

Teaching Tip
Exs. 15, 17 Students should compare the vertex they find with the graph to see if their answer is reasonable.

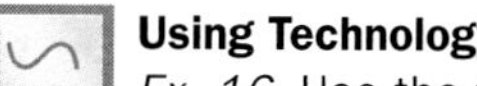

Using Technology
Ex. 16 Use the following procedure to find the quadratic least-squared regression on a TI-82 graphing calculator. Enter the stride length as L1 and the oxygen consumption as L2. Under the STAT CALC menu, choose 6:QuadReg and press ENTER twice.

54 ft, the y-coordinate of the vertex, which is 70, could clearly not represent the maximum height.

11–13. Explanations may vary. Examples are given.

11. A; This graph is the only one in which the parabola opens down.

12. C; The vertex form of this equation is $y = (x - 1.5)^2 - 4.25$, so the vertex of the graph is $(1.5, -4.25)$.

13. B; The vertex form of this equation is $y = 2(x + 1)^2$, so the vertex of the graph is $(-1, 0)$.

14. a. $h = -16t^2 + 4t + \frac{1}{3}$

b. $\frac{7}{12}$ ft or 7 in.

15. a. $y = 0.001292(x - 146.4)^2 + 3.854$; $(146.4, 3.854)$; The vertex indicates the minimum oxygen consumption required (3.854 L/min) and the stride length required to achieve the minimum.

b. The two values are very close. Answers may vary. An example is given. I think this indicates that a runner feels most natural when running with a stride that allows him or her to achieve minimum oxygen consumption.

16. a. $y = 0.001021x^2 - 0.2719x + 21.44$

b. Yes; the vertex of the parabola in part (a) is $(133.2, 3.338)$. Again the runner's "natural" stride allowed for minimum oxygen consumption.

17. a. $y = 0.06(x - 33.33)^2 + 20.33$

b. 20.33; The minimum oxygen consumption of which the parrot is capable is about 20.33 mL/g • h. This is achieved by flying at a speed of about 33.33 km/h.

18. 90.4 mL

Exercise Notes

Topic Spiraling: Review
Exs. 19–23 These exercises demonstrate a natural meshing of the theories of quadratic functions and parametric equations.

Application
Exs. 19–23 In doing these exercises, students will see how quadratic functions and the method of completing the square can be applied to a situation in sports.

Multicultural Note
Exs. 19–23 Basketball is one of the world's most popular indoor sports and an important Olympic event. There are professional women's leagues in countries such as Greece, Israel, and Hungary, and professional men's leagues in countries such as Spain, Israel, Brazil, and Greece. The rules of international competition differ from American rules in several important ways. For example, after the ball is shot, both offensive and defensive players may touch the ball, regardless of its position, once it has hit the rim. (American rules forbid touching the ball when it is within an imaginary cylinder extending upward from the rim.)

Using Technology
Exs. 19–23 The trace features of a graphing calculator or graphing software can be used to check students' solutions to Exs. 19(c), 20(b), and 21. Remind students to change their calculators to parametric mode for Exs. 19–22 and back to function mode for Ex. 23. You may want to suggest that students use the *Function Investigator* software for Ex. 22.

Teaching Tip
Ex. 19 Students would benefit from making a table of values for *t*, *x*, and *y* and graphing these points by hand. This will help them see how the three variables and two parametric equations work together to describe the path of the ball at different times.

Cooperative Learning
Ex. 22 This exercise allows students to see how different combinations of vertical and horizontal velocities can be used to achieve the same result: a basket. Working with a partner will enable students to solidify their understanding of this problem.

Connection BASKETBALL

When distances are measured in feet and time in seconds, you can determine the position (x, y) of a projectile at any time t using this pair of parametric equations:

$$x = v_x t + x_0$$
$$y = -16t^2 + v_y t + y_0$$

In these equations, v_x and v_y are the initial velocities in the horizontal and vertical directions, and x_0 and y_0 are the initial horizontal and vertical positions of the projectile. (*Note:* For help in using parametric equations, look back at Section 2.6.)

19. Suppose a basketball player throws the ball as shown above. The initial velocity in the horizontal direction is 20 ft/s and the initial velocity in the vertical direction is 16 ft/s.

a. **Open-ended Problem** Where would you place the coordinate axes in this situation? Draw a sketch.

b. Write a pair of parametric equations to model the *trajectory*, or path, of the basketball.

c. Find the y-coordinate of the ball when the x-coordinate of the ball equals the x-coordinate of the hoop. Does the player make the shot?

20. Suppose the basketball player in Exercise 19 jumps up 1 ft before throwing the ball.

a. How would your parametric equations change?

b. Would the player make the basket?

21. Complete the square on the parametric equation for the height of the basketball in part (b) of Exercise 19. What maximum height does the ball reach? When does it reach that height?

22. **Cooperative Learning** Work with a partner to find other initial horizontal and vertical velocities that are likely to result in a basket for the player in Exercise 19. You may wish to use a graphing calculator or graphing software to help you examine the possibilities graphically.

23. **Challenge** Using the parametric equations from part (b) of Exercise 19, express y as a function of x. Graph your function in an xy-plane to confirm that the graph is the trajectory of the basketball.

19. a. Answers may vary. An example is given. I would put the origin at the player's feet, with the y-axis along the player's body and the x-axis along the floor. The initial position of the ball would be (0, 6) and that of the basket would be (15, 10).
b. $x = 20t$; $y = -16t^2 + 16t + 6$
c. 9; No.

20. a. The equation for y would become $y = -16t^2 + 16t + 7$.
b. Yes.

21. $y = -16(t - 0.5)^2 + 10$; 10 ft; 0.5 s after the ball is thrown

22. Answers may vary. One situation that will produce a basket is an initial horizontal velocity of 15 ft/s and an initial vertical velocity of 20 ft/s.

23. $y = -0.04x^2 + 0.8x + 6$

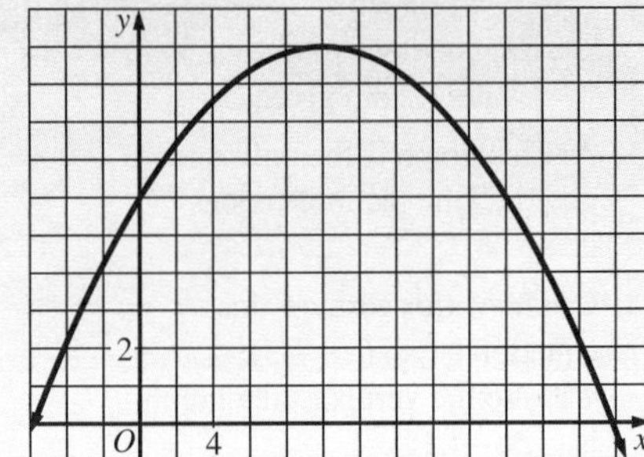

24. Challenge The Buckingham Fountain in Chicago shoots water to a maximum height of 125 ft above the fountain jet. Find the initial vertical velocity of the water. (*Hint:* Remember that the water falling from the maximum height drops $16t^2$ feet in t seconds.)

125 ft

State whether each function has a *maximum value* or a *minimum value*. Then find that value.

25. $y = -16x^2 + 32x + 4$ **26.** $y = 3x^2 - 6x + 5$

27. $y = -x^2 - 6x + 3$ **28.** $y = 3x^2 - 12x + 2$

29. $y = -x^2 - 6x + 2$ **30.** $y = x^2 - x - 1$

31. $y = 2x^2 - x + \frac{1}{3}$ **32.** $y = \frac{2}{3}x^2 - x$

33. $y = -x^2 + 8x + 5$ **34.** $y = -3x^2 + 9x - 4$

35. a. Write the functions $y = x^2 + x$, $y = x^2 + 2x$, and $y = x^2 + 3x$ in vertex form. Examine the graph of each function and note the following characteristics:

- the x-intercepts
- the coordinates of the vertex

b. Investigation Using integer values of b, investigate the graphs of functions of the form $y = x^2 + bx$ and describe the effect of b on the characteristics noted in part (a).

c. Challenge The vertices of the graphs of $y = x^2 + bx$ all lie on a single parabola. Find an equation for that parabola.

ONGOING ASSESSMENT

36. Open-ended Problem Write any quadratic function in standard form. Put the function in vertex form.

SPIRAL REVIEW

Write an equation in intercept form for each parabola shown. *(Section 5.3)*

37.

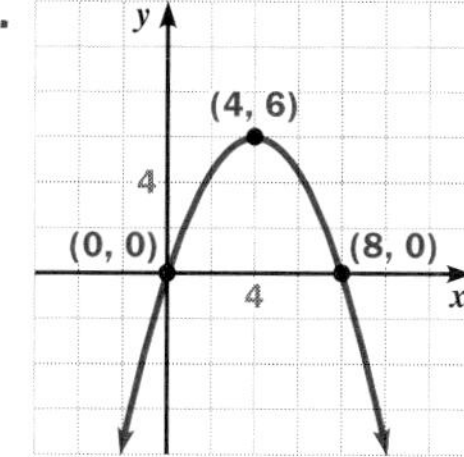

38.

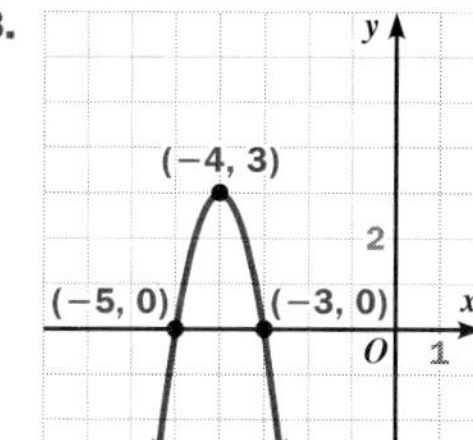

39.

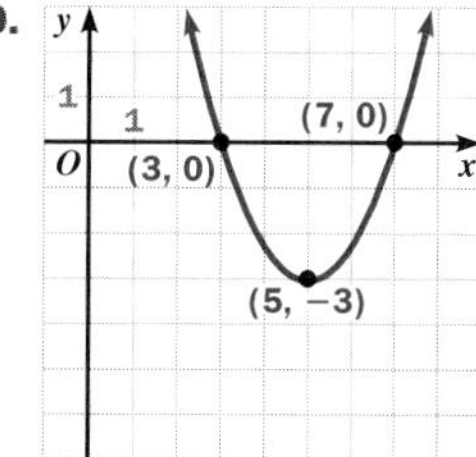

Solve each equation. Give solutions to the nearest tenth when necessary. *(Section 5.1)*

40. $200 = 4x^2$ **41.** $12x^2 = 48$ **42.** $32x^2 = 50$ **43.** $\frac{2}{9}x^2 = \frac{1}{2}$

Find the inverse of each function. *(Section 4.1)*

44. $f(x) = 2x - 5$ **45.** $f(x) = 2.5x - 7.8$ **46.** $g(x) = x + 7$ **47.** $h(x) = 4 - 2x$

Apply⇔Assess

Exercise Notes

Student Progress

Exs. 25–34 Students should be able to complete these problems correctly. You may want to check students' work on some of them to ensure that all students have mastered the content of this section.

Challenge

Ex. 36 After competing this problem, ask students to generalize the result and change $y = ax^2 + bx + c$ to vertex form. This gives them a relatively simple formula to find the x-coordinate of the vertex, $\frac{-b}{2a}$, and it is also the first step used to develop the quadratic formula in the next section.

Topic Spiraling: Review

Exs. 37–39 Students will again use the intercept form in Section 5.6. These exercises help them review this form.

Practice 29 for Section 5.4

NAME ______ DATE ______

Practice 29

FOR USE WITH SECTION 5.4

Write each function in vertex form. **Exs. 4–9: See bottom of page.**

1. $y = x^2 - 10x + 19$ **$y = (x - 5)^2 - 6$**
2. $y = x^2 + 8x - 2$ **$y = (x + 4)^2 - 18$**
3. $y = -x^2 + 6x - 2$ **$y = -(x - 3)^2 + 7$**
4. $y = 3x^2 + 12x - 9$
5. $y = -2x^2 - 6x - 4$
6. $y = 3x^2 + 3x$
7. $y = 4x^2 - 10x + 8$
8. $y = 5x^2 + 30x + 15$
9. $y = \frac{1}{4}x^2 - \frac{1}{2}x + 1$

State whether each function has a *maximum* or *minimum* value. Then find that value.

10. $y = x^2 - 2x + 3$ **min; 2**
11. $y = 5x^2 + 30x + 4$ **min; −41**
12. $y = -3x^2 + 18x - 2$ **max; 25**
13. $y = 2x^2 + 28x - 5$ **min; −103**
14. $y = -\frac{1}{2}x^2 + 9x - 2$ **max; $\frac{77}{2}$**
15. $y = -4x^2 - 6x + 2$ **max: $\frac{17}{4}$**

16. In a springboard dive, Ho Chan's center of gravity starts at a point 7 ft above the water, and she takes off with an upward velocity of 16 ft/s.
   a. Write an equation to model the height of Ho Chan's center of gravity above the water at time t (in seconds) after she takes off. **$y = -16t^2 + 16t + 7$**
   b. How long after her take off does her center of gravity reach its maximum height? **0.5 s**
   c. What maximum height does her center of gravity reach? **11 ft**

17. Playing miniature golf, Marta hit the ball with an initial vertical velocity of 12 ft/s upward from the bottom of an incline. If the ball were placed at any point on the incline, it would roll down, losing $8t^2$ ft of height in t seconds.
   a. Write an equation to model the height of Marta's golf ball above the bottom of the incline after t seconds. **$y = -8t^2 + 12t$**
   b. What maximum height does Marta's ball reach? **4.5 ft**

18. **Open-Ended Problem** Describe a situation in which an object is thrown or propelled upward and then undergoes free fall. Estimate the object's initial upward velocity, and then calculate the maximum height it reaches.

Answers may vary. Check students' work.

4. $y = 3(x + 2)^2 - 21$ **5.** $y = -2\left(x + \frac{3}{2}\right)^2 + \frac{1}{2}$ **6.** $y = 3\left(x + \frac{1}{2}\right)^2 - \frac{3}{4}$

7. $y = 4\left(x - \frac{5}{4}\right)^2 + \frac{7}{4}$ **8.** $y = 5(x + 3)^2 - 30$ **9.** $y = \frac{1}{4}(x - 1)^2 + \frac{3}{4}$

Practice Bank, ALGEBRA 2: EXPLORATIONS AND APPLICATIONS

29

---

24. 89.4 ft/s
25. maximum; 20
26. minimum; 2
27. maximum; 12
28. minimum; –10
29. maximum; 11
30. minimum; $-1\frac{1}{4}$
31. minimum; $\frac{5}{24}$
32. minimum; $-\frac{3}{8}$
33. maximum; 21
34. maximum; $2\frac{3}{4}$
35. a. $y = \left(x + \frac{1}{2}\right)^2 - \frac{1}{4}$, $y = (x + 1)^2 - 1$, $y = \left(x + \frac{3}{2}\right)^2 - \frac{9}{4}$; x-intercepts: 0 and –1, 0 and –2, 0 and –3; vertex: $\left(-\frac{1}{2}, -\frac{1}{4}\right)$, $(-1, -1)$, $\left(-\frac{3}{2}, -\frac{9}{4}\right)$

   b. The x-intercepts are 0 and $-b$. The vertex is $\left(-\frac{b}{2}, -\left(\frac{b}{2}\right)^2\right)$.

   c. $y = -x^2$
36. Answers may vary. An example is given. The vertex form of the equation $y = 3x^2 + 2x + 1$ is $y = 3\left(x + \frac{1}{3}\right)^2 + \frac{2}{3}$.
37. $y = -\frac{3}{8}x(x - 8)$
38. $y = -3(x + 5)(x + 3)$
39. $y = \frac{3}{4}(x - 3)(x - 7)$
40. ±7.1
41. ±2
42. ±1.3
43. ±1.5
44. $f^{-1}(x) = \frac{1}{2}x + \frac{5}{2}$
45. $f^{-1}(x) = 0.4x + 3.12$
46. $g^{-1}(x) = x - 7$
47. $h^{-1}(x) = -\frac{1}{2}x + 2$

Plan⇔Support

Objectives

- Solve equations using the quadratic formula.
- Use the discriminant to determine how many solutions an equation has.
- Use the quadratic formula to solve real-world problems.

Recommended Pacing

Core and Extended Courses
Section 5.5: 2 days

Block Schedule
Section 5.5: 1 block

Resource Materials

Lesson Support
Lesson Plan 5.5
Warm-Up Transparency 5.5
Practice Bank: Practice 30
Study Guide: Section 5.5
Exploration Lab Manual: Additional Exploration 8, Diagram Master 2

Technology
Graphing Calculator
Internet: http://www.hmco.com

Warm-Up Exercises

Transform each equation into vertex form.

1. $y = x^2 - 6x + 2$ $y = (x - 3)^2 - 7$
2. $y = 3x^2 + 12x - 3$ $y = 3(x + 2)^2 - 15$

Use the height model for a projectile, $h = -16t^2 + v_0t + h_0$.

3. Write an equation for a ball thrown with a vertical velocity of 20 ft/s from a height of 3 ft. $h = -16t^2 + 20t + 3$
4. Use the method of completing the square to find how high the ball will go. 9.25 ft
5. Simplify -2 ± 1.5. $-3.5, -0.5$

Section

5.5 Using the Quadratic Formula

Learn how to...

- **solve equations using the quadratic formula**
- **use the discriminant to determine how many solutions an equation has**

So you can...

- **find how long it takes a projectile to reach the ground, for example**

Michael and his algebra class want to use what they know about quadratic functions while playing baseball. Suppose Michael hits a baseball with a vertical velocity of 60 ft/s from a height of 4 ft. He wants to know how high the ball will travel and how long it will take the ball to hit the ground in the outfield.

Michael can use the height model for a projectile, $h = -16t^2 + v_0t + h_0$. For the baseball he hit, the equation is:

$$h = -16t^2 + 60t + 4$$

The questions below are answered using both the equation for Michael's baseball and a general equation, $y = ax^2 + bx + c$.

Question 1: *How high will the baseball go?*

Use the method of completing the square to find the vertex of the graph of each equation.

MICHAEL'S BASEBALL

$$\begin{aligned} h &= -16t^2 + 60t + 4 \\ &= -16\left(t^2 + \left(-\tfrac{15}{4}\right)t\right) + 4 \\ &= -16\left(t^2 - \frac{15}{4}t + \frac{225}{64}\right) + 4 - \left(-16 \cdot \frac{225}{64}\right) \\ &= -16\left(t - \frac{15}{8}\right)^2 + \frac{241}{4} \end{aligned}$$

GENERAL EQUATION

$$\begin{aligned} y &= ax^2 + bx + c \\ &= a\left(x^2 + \frac{b}{a}x\right) + c \\ &= a\left(x^2 + \frac{b}{a}x + \frac{b^2}{4a^2}\right) + c - \left(a \cdot \frac{b^2}{4a^2}\right) \\ &= a\left(x + \frac{b}{2a}\right)^2 + \frac{4ac - b^2}{4a} \end{aligned}$$

So the vertex of the height function's graph is at $\left(\frac{15}{8}, \frac{241}{4}\right)$. Michael's baseball will reach a maximum height of $\frac{241}{4}$ ft, or $60\frac{1}{4}$ ft.

THINK AND COMMUNICATE

1. The vertex of the graph of $y = ax^2 + bx + c$ has what x-coordinate?
2. The function $y = ax^2 + bx + c$ has what maximum or minimum value? How can you tell from the equation whether the value is a maximum or a minimum?

ANSWERS Section 5.5

Think and Communicate

1. $-\frac{b}{2a}$
2. $\frac{4ac - b^2}{4a}$; The value is a maximum if $a < 0$ and a minimum if $a > 0$.

Question 2: *How long will it take the baseball to reach the ground?*

The ground is at height $h = 0$. Find t when $h = 0$. Use the completed-square form from Question 1.

MICHAEL'S BASEBALL	**GENERAL EQUATION**
$h = -16\left(t - \frac{15}{8}\right)^2 + \frac{241}{4}$	$y = a\left(x + \frac{b}{2a}\right)^2 + \frac{4ac - b^2}{4a}$
$0 = -16\left(t - \frac{15}{8}\right)^2 + \frac{241}{4}$	$0 = a\left(x + \frac{b}{2a}\right)^2 + \frac{4ac - b^2}{4a}$
$-\frac{241}{4} = -16\left(t - \frac{15}{8}\right)^2$	$\frac{b^2 - 4ac}{4a} = a\left(x + \frac{b}{2a}\right)^2$
$\frac{241}{64} = \left(t - \frac{15}{8}\right)^2$	$\frac{b^2 - 4ac}{4a^2} = \left(x + \frac{b}{2a}\right)^2$
$\pm\sqrt{\frac{241}{64}} = t - \frac{15}{8}$	$\pm\sqrt{\frac{b^2 - 4ac}{4a^2}} = x + \frac{b}{2a}$
$\frac{15}{8} \pm \sqrt{\frac{241}{64}} = t$	$-\frac{b}{2a} \pm \sqrt{\frac{b^2 - 4ac}{4a^2}} = x$
$\frac{15}{8} \pm \frac{\sqrt{241}}{8} = t$	$-\frac{b}{2a} \pm \frac{\sqrt{b^2 - 4ac}}{2a} = x$
$\frac{15 \pm \sqrt{241}}{8} = t$	$\frac{-b \pm \sqrt{b^2 - 4ac}}{2a} = x$

THINK AND COMMUNICATE

3. Find decimal values for $\frac{15 + \sqrt{241}}{8}$ and $\frac{15 - \sqrt{241}}{8}$. Which value is the answer to Question 2?

4. What are the two solutions for the general equation?

You can use the vertex that you found in Question 1 and the solutions when $h = 0$ or $y = 0$ from Question 2 to graph each equation.

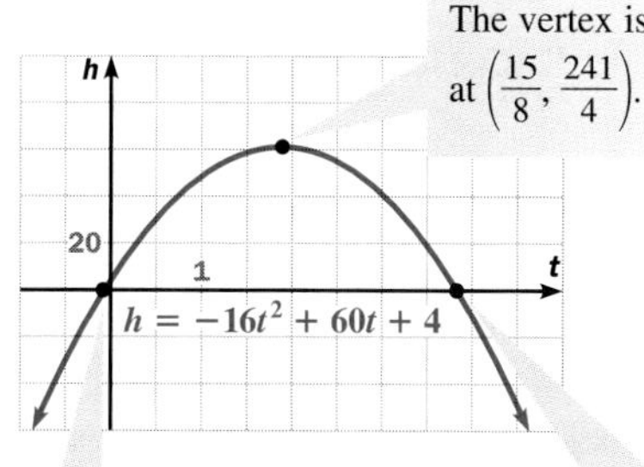

The vertex is at $\left(\frac{15}{8}, \frac{241}{4}\right)$.

One t-intercept is -0.066.

The other t-intercept is 3.82.

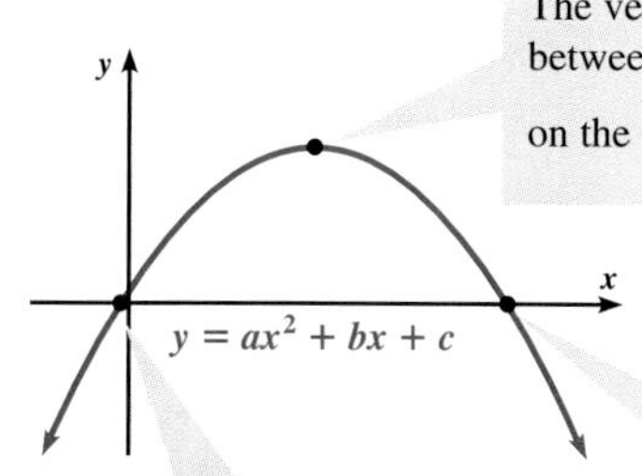
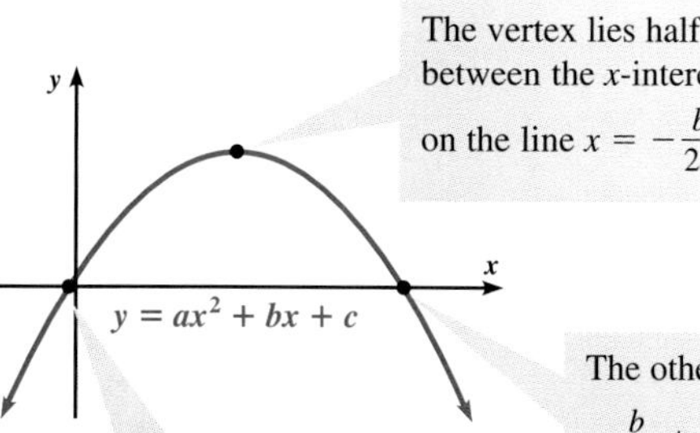

The vertex lies halfway between the x-intercepts, on the line $x = -\frac{b}{2a}$.

One x-intercept is $-\frac{b}{2a} - \frac{\sqrt{b^2 - 4ac}}{2a}$.

The other x-intercept is $-\frac{b}{2a} + \frac{\sqrt{b^2 - 4ac}}{2a}$.

Think and Communicate

3. 3.82 and –0.066; 3.82

4. $\frac{-b + \sqrt{b^2 - 4ac}}{2a}$ and $\frac{-b - \sqrt{b^2 - 4ac}}{2a}$

Teach⇔Interact

Section Note

Communication: Reading

While reading the solutions to Questions 1 and 2 on pages 214 and 215, it is important for students to go back and forth between Michael's baseball and the general equation to see what is being done for every line. Students should identify the step from completing the square for each line. For example, on the second line, a student might identify the step as "Factor out the coefficient of x^2 from the x^2- and x-terms." Students should read the second line of both examples carefully to verify that this step is indeed followed. In the baseball example, they should confirm that $\frac{60}{-16} = -\frac{15}{4}$.

Think and Communicate

Questions 1 and 2 focus students' attention on the information that can be gotten from the general equation in vertex form, namely, a formula for the x-coordinate of the vertex of a parabola and a formula for the minimum or maximum value of the function.

Learning Styles: Visual

The labeled graphs at the bottom of this page will help visual learners see and understand the algebraic results derived for the general equation.

Section Notes

Reasoning

Students should recall that to find the x-coordinate of the vertex of a parabola, they can average the two x-intercepts. Ask students to show that the average of the two general expressions for the x-intercepts given here is $\frac{-b}{2a}$, the formula for the x-coordinate of the vertex.

Teaching Tip

Some students may not fully understand what information the quadratic formula gives you about a quadratic equation and about a quadratic function. Emphasize that the results of the quadratic formula are the *solutions* to the quadratic equation $ax^2 + bx + c = 0$ and the *x-intercepts* of the graph of the quadratic function $ax^2 + bx + c = y$.

Teach⇔Interact

About Example 1

Many students initially feel that the quadratic formula is cumbersome. Students may gain a greater appreciation for the formula, however, when they compare the procedure used for answering Questions 1 and 2 on pages 214 and 215 with that of Example 1. Both methods are used to answer equivalent questions. Clearly, using the quadratic formula is simpler than completing the square and solving the resulting equation.

Additional Example 1

A tennis ball is hit with a vertical velocity of 50 ft/s from an initial height of 2 ft.

a. How long does it take the tennis ball to reach the ground?

The height model for the tennis ball is $h = -16t^2 + 50t + 2$. Solve $-16t^2 + 50t + 2 = 0$ using the quadratic formula. Substitute –16 for a, 50 for b, and 2 for c.

$$t = \frac{-50 \pm \sqrt{50^2 - 4(-16)(2)}}{2(-16)}$$
$$= \frac{-50 \pm \sqrt{2628}}{-32}$$
$$\approx 3.16, -0.04$$

The tennis ball will return to the ground in about 3.16 s.

b. When is the tennis ball at its highest point?

The greatest time is at $t = -\frac{b}{2a}$.

Substitute –16 for a and 50 for b.

$$t = -\frac{50}{2(-16)} \approx 1.56$$

The tennis ball is at its highest point about 1.6 s after it is hit.

Think and Communicate

Question 6 leads students to consider the case in which $b^2 - 4ac$ is a negative number. After answering this question, ask students what happens when $b^2 - 4ac$ is zero. Also ask them what effect either situation will have on the x-intercepts of the graph. See if they can draw an example of a graph for each situation.

The solution for the general equation in Question 2 is called the *quadratic formula*. You can use the **quadratic formula** to find the solutions of any quadratic equation in the form $ax^2 + bx + c = 0$ directly, without having to complete the square each time.

The Quadratic Formula

The solutions of any quadratic equation in the form $ax^2 + bx + c = 0$ are:

$$x = \frac{-b + \sqrt{b^2 - 4ac}}{2a} \quad \text{and} \quad x = \frac{-b - \sqrt{b^2 - 4ac}}{2a}$$

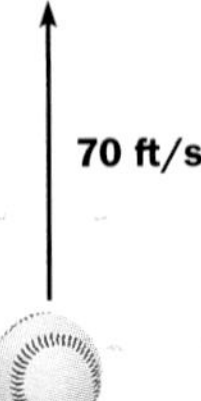

EXAMPLE 1

Suppose Michael hits the baseball with a vertical velocity of 70 ft/s instead of 60 ft/s.

a. How long does it take the baseball to reach the ground?

b. When is the baseball at its highest point?

SOLUTION

The height model for Michael's baseball is $h = -16t^2 + 70t + 4$.

a. Solve $-16t^2 + 70t + 4 = 0$ using the quadratic formula.

$$t = \frac{-70 \pm \sqrt{70^2 - 4(-16)(4)}}{2(-16)}$$
$$= \frac{-70 \pm \sqrt{5156}}{-32}$$
$$\approx 4.43, -0.06$$

Substitute **−16** for ***a***, **70** for b, and **4** for ***c*** in the quadratic formula.

The baseball will return to the ground in about 4.4 s.

b. The greatest height is at time $t = -\frac{b}{2a}$.

$$t = -\frac{70}{2(-16)} \approx 2.19$$

Substitute **−16** for ***a*** and **70** for b.

The baseball is at its highest point about 2.2 s after it is hit.

THINK AND COMMUNICATE

5. Why is the value $t \approx -0.06$ not a solution in part (a) of Example 1?

6. a. Is it possible for the value of $b^2 - 4ac$, the expression under the radical sign in the quadratic formula, to be negative? Explain why or why not.

b. Can you find the square root of a negative number?

c. Discuss whether you can use the quadratic formula to find the solutions of an equation when $b^2 - 4ac$ is negative.

Think and Communicate

5. t is the number of seconds after the ball is hit and must be nonnegative.

6. a. Yes; for example, consider the equation $y = 2x^2 + x + 1$; $b^2 - 4ac = -7$.

b. No; there are no real numbers whose squares are negative.

c. When $b^2 - 4ac$ is negative, the equation $y = ax^2 + bx + c$ has no real solutions.

Using the Discriminant

The value $b^2 - 4ac$ in the quadratic formula is called the **discriminant**. You can use the sign of the discriminant to determine the number of solutions that a quadratic equation has. You can also tell how many x-intercepts the graph of the related function $y = ax^2 + bx + c$ has.

Value of the discriminant	Quadratic equation $ax^2 + bx + c = 0$	Related function $y = ax^2 + bx + c$ (with examples)
$b^2 - 4ac > 0$	Two solutions: $x = \frac{-b + \sqrt{b^2 - 4ac}}{2a}$, $x = \frac{-b - \sqrt{b^2 - 4ac}}{2a}$	Two x-intercepts
$b^2 - 4ac = 0$	One solution: $x = -\frac{b}{2a}$	One x-intercept
$b^2 - 4ac < 0$	No solution	No x-intercept

EXAMPLE 2

Find the number of x-intercepts that the graph of each function has.

a. $y = x^2 - 3x - 28$ **b.** $y = x^2 - 6x + 9$ **c.** $y = 2x^2 + 3x + 6$

SOLUTION

	Equation when $y = 0$	Value of $b^2 - 4ac$	Number of x-intercepts
a.	$x^2 - 3x - 28 = 0$	121	2
b.	$x^2 - 6x + 9 = 0$	0	1
c.	$2x^2 + 3x + 6 = 0$	−39	none

Teach⇔Interact

Section Notes

Reasoning

Point out to students that if they remember that the discriminant is the part of the quadratic formula underneath the square root sign, it will be easier for them to decide what the value of the discriminant indicates about the solutions to the quadratic equation and the x-intercepts of the related function. Since the square root of a negative number has no real number value, a quadratic equation has no real solutions when its discriminant is negative, and the graph of the related function will have no x-intercepts. Similar reasoning works for the other two cases as well.

Challenge

Ask students if they can develop a test similar to that of the discriminant for finding the number of solutions to a quadratic equation in vertex form $0 = a(x - h)^2 + k$. Tell them that they will need to consider when a is positive and negative. It also helps to make sketches of possible parabolas. (When $k = 0$, there is one solution. When a and k have opposite signs, there are two solutions, and when a and k have the same sign, there is no solution.)

Additional Example 2

Find the number of x-intercepts that the graph of each function has.

a. $y = x^2 + 4x + 31$
Equation when $y = 0$:
$x^2 + 4x + 31 = 0$
Value of $b^2 - 4ac$: −108
Number of x-intercepts: none

b. $y = x^2 + 5x - 1$
Equation when $y = 0$:
$x^2 + 5x - 1 = 0$
Value of $b^2 - 4ac$: 29
Number of x-intercepts: 2

c. $y = 3x^2 + 12x + 12$
Equation when $y = 0$:
$3x^2 + 12x + 12 = 0$
Value of $b^2 - 4ac$: 0
Number of x-intercepts: 1

Teach⇔Interact

Closure Question

What do the quadratic formula and the discriminant tell you about a quadratic equation? The quadratic formula gives you the value of the solutions if there are any and also the x-intercepts of the related function's graph. The discriminant tells you how many real solutions the equation has: two solutions if it is positive, one solution if it is zero, and no solutions if it is negative. The discriminant also tells you how many x-intercepts the related functions graph has: two if the discriminant is positive, one if it is zero, and none if it is negative.

Apply⇔Assess

Suggested Assignment

Core Course
Day 1 Exs. 1–17
Day 2 Exs. 18–23, 30, 31, 35–44

Extended Course
Day 1 Exs. 1–17
Day 2 Exs. 18–44

Block Schedule
Day 30 Exs. 1–23, 30, 31, 35–44

Exercise Notes

Using Technology
Exs. 1–9 Once students have had sufficient practice using the quadratic formula to find solutions of quadratic equations, you might challenge some students to write a program for a computer or graphing calculator that can calculate the values given by the formula when given *a*, *b*, and *c* for the standard expression of the equation . One simple example of a program for the TI-82 is given below.

```
:Prompt A
:Prompt B
:Prompt C
:(-B-√(B²-4AC))/2/A→D
:(-B+√(B²-4AC))/2/A→E
:Disp D,E
```

CHECKING KEY CONCEPTS

1. Suppose a soccer player standing on a sideline throws the ball onto the field from an initial height of 5 ft and with an initial vertical velocity of 16 ft/s.

a. How long will it take the ball to reach the ground?

b. How long will it take the ball to reach its maximum height?

c. What is the ball's maximum height?

2. Determine whether the graph of $y = 3x^2 + 6x + 6$ has *two x-intercepts*, *one x-intercept*, or *no x-intercepts*.

5.5 Exercises and Applications

Extra Practice *exercises on page 756*

For Exercises 1–9, find the solution(s) of each quadratic equation. Use the quadratic formula.

1. $2x^2 + 8x - 13 = 0$ **2.** $3x^2 + 18x + 5 = 0$ **3.** $5x^2 - 2x - 3 = 0$

4. $4x^2 - 9x + 1 = 0$ **5.** $x^2 + 2x - 15 = 0$ **6.** $x^2 + 6x + 9 = 0$

7. $2x^2 + 5x + 3 = 9$ **8.** $x^2 = 6x - 2$ **9.** $3x^2 - 8 = 4x$

10. A circus performer is launched up off a springboard. The height in feet that the performer reaches in t seconds is given by $h = -16t^2 + 18t$.

a. After how many seconds will the performer land on the ground?

b. After how many seconds will the performer reach maximum height?

c. What maximum height does the performer reach?

11. SPORTS SCIENCE A researcher testing javelins found that the horizontal distance d (in feet) that a particular javelin traveled when launched at a speed of 100 ft/s and at an angle a (in degrees) could be modeled by $d = -0.251a^2 + 16.5a + 53.4$.

a. Find the angle at which the javelin should be launched to maximize the horizontal distance it travels.

b. What maximum horizontal distance does the javelin travel?

BY THE WAY

In Russia, circus artists must complete years of rigorous training at a professional school in Moscow before they can join one of the many circuses that perform around the country.

Tell whether each equation has *two solutions*, *one solution*, or *no solution*.

12. $x^2 + 12x + 3 = 0$ **13.** $4x^2 - 12x + 9 = 0$ **14.** $7x^2 - x + 2 = 0$

15. $24x^2 - 14x + 5 = 0$ **16.** $16x^2 + 40x + 25 = 0$ **17.** $3x^2 - 21 = 0$

Find the *x*-intercepts, if any, for the graph of each quadratic function.

18. $y = x^2 + 1$ **19.** $y = x^2 + 3x + 4$ **20.** $y = 3x^2 + 5x$

21. $y = -2x^2 + 4x - 2$ **22.** $y = 6x^2 - 7x + 2$ **23.** $y = -0.1x^2 + 0.2x + 0.1$

24. Challenge Suppose the x-intercepts of the graph of the quadratic function $y = x^2 + bx + c$ are $\frac{1 - \sqrt{3}}{2}$ and $\frac{1 + \sqrt{3}}{2}$. Find b and c.

Checking Key Concepts

1. a. 1.25 s b. 0.5 s c. 9 ft
2. no x-intercepts

Exercises and Applications

Answers may vary slightly due to rounding.

1. –5.2; 1.2
2. –5.7; –0.3
3. –0.6; 1
4. 0.1; 2.1
5. –5; 3
6. –3
7. –3.4; 0.9
8. 0.4; 5.6
9. –1.1; 2.4
10. a. 1.1 s b. 0.6 s c. 5.06 ft
11. a. 32.9° b. 324.6 ft
12. two solutions
13. one solution
14. no solution
15. no solution
16. one solution
17. two solutions
18. no x-intercepts
19. no x-intercepts
20. $0; -\frac{5}{3}$
21. 1
22. $\frac{1}{2}; \frac{2}{3}$
23. –0.4; 2.4
24. $b = -1; c = -\frac{1}{2}$
25. Answers may vary. An example is given. Reaction times to both stimuli decrease with age until early adulthood, then begin to increase.

Connection PHYSIOLOGY

For a science fair project, a student used audio and visual stimuli to test the reaction times of people of various ages. His data led to the following models where x is a person's age and y is the reaction time (measured in an arbitrary time unit):

Audio stimulus: $y = 15.0008 - 0.3185x + 0.0051x^2$

Visual stimulus: $y = 22.0036 - 0.2287x + 0.005x^2$

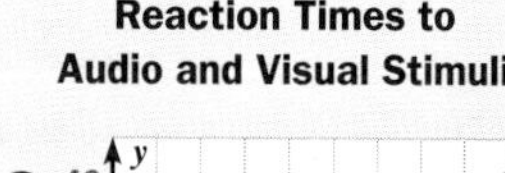

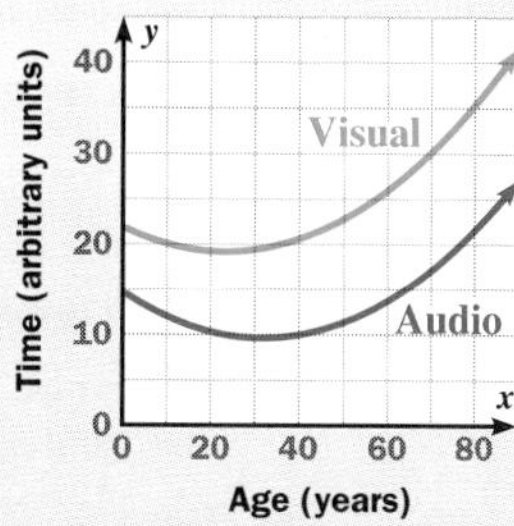

25. Visual Thinking Examine the graphs of the two models. What general conclusions can you draw about reaction times?

26. a. Find the age at which the minimum value of each model occurs.

b. Writing Compare your answers to part (a) and comment on their significance.

27. Suppose a person's reaction time to the visual stimulus is 30 time units. About how old is the person?

28. Suppose a person's reaction time to the audio stimulus is 12 time units. About how old is the person?

29. Open-ended Problem What do the models predict for *your* reaction times to audio and visual stimuli?

30. Suppose a parabola has vertex $(-1, 5)$ and crosses the x-axis at $(-6, 0)$ and $(4, 0)$.

a. Make a sketch of the parabola.

b. Write an equation for the parabola in intercept form, $y = a(x - p)(x - q)$.

c. Write an equation for the parabola in vertex form, $y = a(x - h)^2 + k$.

d. Compare your equations. Are the values for a the same?

e. Put each equation in standard form. Are the equations the same?

f. Let $y = 0$ in your equation from part (e). Solve the resulting equation for x using the quadratic formula. Do the solutions agree with the information you were given about the parabola?

31. GOVERNMENT Each year, money not spent by the federal hospital insurance program is put in a trust fund. Reports issued in 1995 indicated that this trust fund was in danger of going bankrupt. The function $y = -3.59x^2 + 5.09x + 135$ described the projected year-end balance y for each year x since 1994. In what year was the fund projected to go bankrupt?

BY THE WAY

The federal hospital insurance (HI) program pays for inpatient hospital care for those age 65 or over, and for the long-term disabled. The HI program is financed primarily by payroll taxes.

Apply⇔Assess

Exercise Notes

Common Error

Exs. 7–9 Some students may try to apply the quadratic formula to these equations as given. Remind these students that one side of the equation must be zero in order to use the quadratic formula.

Interdisciplinary Problems

Exs. 25–29 These exercises illustrate how quadratic modeling and the quadratic formula can be used in physiology to study the reaction times of individuals of different ages to audio and visual stimuli.

Communication: Writing

Ex. 26 Being able to provide a written interpretation of data is an important skill. Ask several students to share their responses to the question in part (b) so that all students can see the variety of responses that are possible.

Problem Solving

Exs. 27–29 Students sometimes have difficulty deciding when they need to use the quadratic formula and when it is unnecessary. Stress that the quadratic formula can help them find the independent variable when the dependent variable is known. This is the case in Exs. 27 and 28. The quadratic formula is not needed in Ex. 29 because it is the independent variable that is known. In this case, students can substitute the value for the independent variable and simplify.

Student Progress

Ex. 30 This is an excellent summary question about the three forms of quadratic functions presented in this chapter. Ask students to turn in their responses to this exercise so you can check that they all have a solid understanding of this material.

---

26. a. audio: about 31 years; visual: about 23 years

b. At every age, reaction time to audio stimuli is less than to visual stimuli. Minimum reaction time to audio stimuli occurs about 8 years later than minimum reaction time to visual stimuli. Summaries may vary. A person's reaction time to visual stimuli is best when the person is about 23 years old, while his or her reaction time to audio stimuli continues to improve until about the age of 31.

27. about 69 years old

28. about 12 years old or about 51 years old

29. Answers may vary. An example is given. For a 17-year-old, the predicted response time to an audio stimulus is about 11.06 s. The predicted response time to a visual response is about 19.56 s.

30. a.

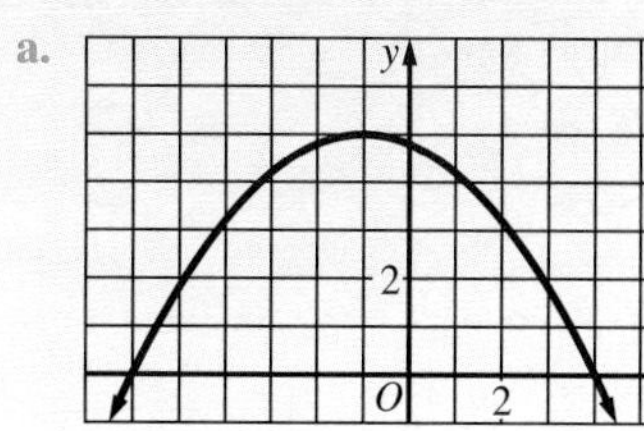

b. $y = -0.2(x + 6)(x - 4)$

c. $y = -0.2(x + 1)^2 + 5$

d. Yes.

e. Yes; $y = -0.2x^2 - 0.4x + 4.8$.

f. Yes; the x-intercepts are $\frac{0.4 \pm \sqrt{(-0.4)^2 - 4(-0.2)(4.8)}}{2(-0.2)}$ or -6 and 4. The vertex is $\left(-\frac{-0.4}{2(-0.2)}, \frac{4(-0.2)(4.8) - (-0.4)^2}{4(-0.2)}\right)$ or $(-1, 5)$.

31. 2001

Apply⇔Assess

Exercise Notes

Interview Note

Exs. 32–34 The quadratic formula is applied to the fuel economy of a car in Exs. 32 and 33. In Ex. 34, students must find the maximum value of the fuel efficiency function to determine the speed at which the best fuel economy is obtained.

Assessment

Ex. 35 Students must pick values of *a*, *b*, or *c* to obtain a discriminant whose sign will indicate the desired number of *x*-intercepts. Students can check their answers quickly by graphing the functions on a graphing calculator or computer.

Practice 30 for Section 5.5

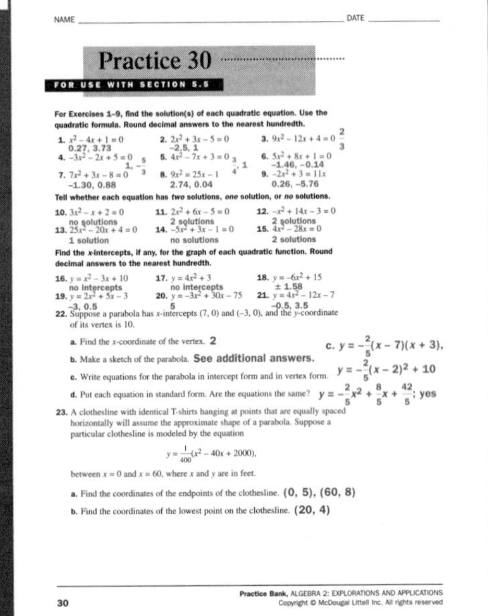

NAME ______________________ DATE __________

Practice 30

FOR USE WITH SECTION 5.5

For Exercises 1–9, find the solution(s) of each quadratic equation. Use the quadratic formula. Round decimal answers to the nearest hundredth.

1. $x^2 - 4x + 1 = 0$ **0.27, 3.73**
2. $2x^2 + 3x - 5 = 0$ **−2.5, 1**
3. $9x^2 - 12x + 4 = 0$ $\frac{2}{3}$
4. $-3x^2 - 2x + 5 = 0$ $1, -\frac{5}{3}$
5. $4x^2 - 7x + 3 = 0$ $\frac{3}{4}, 1$
6. $5x^2 + 8x + 1 = 0$ **−1.46, −0.14**
7. $7x^2 + 3x - 8 = 0$ **−1.30, 0.88**
8. $9x^2 = 25x - 1$ **2.74, 0.04**
9. $-2x^2 + 3 = 11x$ **0.26, −5.76**

Tell whether each equation has *two* solutions, *one* solution, or *no* solutions.

10. $3x^2 - x + 2 = 0$ **no solutions**
11. $2x^2 + 6x - 5 = 0$ **2 solutions**
12. $-x^2 + 14x - 3 = 0$ **2 solutions**
13. $25x^2 - 20x + 4 = 0$ **1 solution**
14. $-5x^2 + 3x - 1 = 0$ **no solutions**
15. $4x^2 - 28x = 0$ **2 solutions**

Find the *x*-intercepts, if any, for the graph of each quadratic function. Round decimal answers to the nearest hundredth.

16. $y = x^2 - 3x + 10$ **no intercepts**
17. $y = 4x^2 + 3$ **no intercepts**
18. $y = -6x^2 + 15$ **± 1.58**
19. $y = 2x^2 + 5x - 3$ **−3, 0.5**
20. $y = -3x^2 + 30x - 75$ **5**
21. $y = 4x^2 - 12x - 7$ **−0.5, 3.5**
22. Suppose a parabola has *x*-intercepts (7, 0) and (−3, 0), and the *y*-coordinate of its vertex is 10.
   a. Find the *x*-coordinate of the vertex. **2**
   b. Make a sketch of the parabola. **See additional answers.**
   c. Write equations for the parabola in intercept form and in vertex form. **c.** $y = -\frac{2}{5}(x - 7)(x + 3)$, $y = -\frac{2}{5}(x - 2)^2 + 10$
   d. Put each equation in standard form. Are the equations the same? $y = -\frac{2}{5}x^2 + \frac{8}{5}x + \frac{42}{5}$; **yes**
23. A clothesline with identical T-shirts hanging at points that are equally spaced horizontally will assume the approximate shape of a parabola. Suppose a particular clothesline is modeled by the equation

$$y = \frac{1}{400}(x^2 - 40x + 2000),$$

between $x = 0$ and $x = 60$, where x and y are in feet.

   a. Find the coordinates of the endpoints of the clothesline. **(0, 5), (60, 8)**
   b. Find the coordinates of the lowest point on the clothesline. **(20, 4)**

30 — **Practice Bank,** ALGEBRA 2: EXPLORATIONS AND APPLICATIONS

INTERVIEW Mark Thomas

Look back at the article on pages 182–184.

The cars built by students in Thomas's Paper Vehicle Project do not need fuel to run, but cars and trucks on the roads and highways do. Fuel consumption can be harmful to the environment and expensive as well, so it makes sense to drive at speeds that use fuel most efficiently.

The fuel efficiency of an average car is given by

$$y = -0.0177x^2 + 1.48x + 3.39$$

where x is the speed in miles per hour and y is the fuel economy in miles per gallon.

32. Find the speed(s) at which an average car will get 20 mi/gal.

33. Challenge Determine the range of speeds that yield 32 mi/gal or better.

34. What speed yields the best fuel economy?

ONGOING ASSESSMENT

35. Open-ended Problem Experiment with different values of *a*, *b*, or *c* to answer parts (a)–(f). For some parts, there may be more than one possible answer.

a. Find b if the graph of $y = x^2 + bx - 2$ has an x-intercept of 1.

b. Find c if the graph of $y = 4x^2 + 4x + c$ is tangent to the x-axis.

c. Find a if the graph of $y = ax^2 + 8x + 2$ is tangent to the x-axis.

d. Find a if the graph of $y = ax^2 + x + 1$ has two x-intercepts.

e. Find b if the graph of $y = x^2 + bx + 3$ has two x-intercepts.

f. Find c if the graph of $y = 3x^2 - 6x + c$ has no x-intercept.

SPIRAL REVIEW

Use the method of completing the square to find the vertex of the graph of each equation. *(Section 5.4)*

36. $y = x^2 - 8x + 1$ **37.** $y = 3x^2 + 6x$ **38.** $y = x^2 + x - \frac{3}{4}$

Solve each equation. *(Section 5.1)*

39. $2x^2 - 50 = 0$ **40.** $\frac{1}{3}x^2 + 1 = 4$ **41.** $-\frac{1}{8}x^2 = -18$

Solve each equation. Round your answers to the nearest hundredth. *(Section 4.4)*

42. $5^{2x} = 12$ **43.** $e^{2x - 1} = 37$ **44.** $\log_2(x + 1) - \log_2(x - 1) = 1$

32. 13.4 mi/h and 70.3 mi/h
33. 30.3 mi/h to 53.3 mi/h
34. 41.8 mi/h
35. a. 1
   b. 1
   c. 8
   d. $a < \frac{1}{4}$
   e. $b > \sqrt{12}$ or $b < -\sqrt{12}$
   f. $c > 3$
36. (4, −15)
37. (−1, −3)
38. $\left(-\frac{1}{2}, -1\right)$
39. ±5
40. ±3
41. ±12
42. 0.77
43. 2.31
44. 3

Section

5.6 Factoring Quadratics

Learn how to...

- factor quadratic expressions
- solve quadratic equations using factoring

So you can...

- find the maximum speed a car can be traveling to be able to stop within a given distance, for example

To find the x-intercepts of the graph of a quadratic function such as $y = x^2 + 7x + 12$, it would be helpful to rewrite the function as a product of factors. The Exploration will show you how this can be done.

EXPLORATION
COOPERATIVE LEARNING

Using Algebra Tiles to Factor

Work with a partner.
You will need:
- algebra tiles

1 Use tiles to model the expression $x^2 + 7x + 12$.

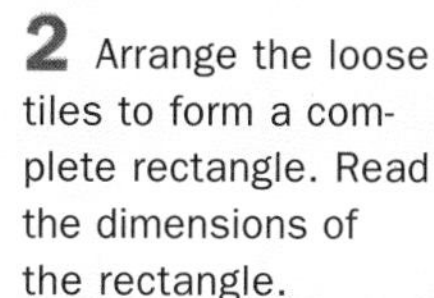
You will need one x^2-tile, seven x-tiles, and twelve 1-tiles.

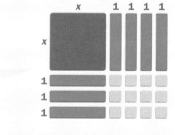

2 Arrange the loose tiles to form a complete rectangle. Read the dimensions of the rectangle.

length: $x + 4$

width: $x + 3$

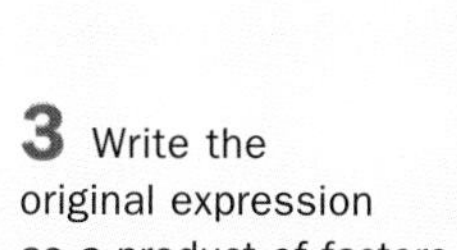

3 Write the original expression as a product of factors.

$$x^2 + 7x + 12 = (x + 3)(x + 4)$$

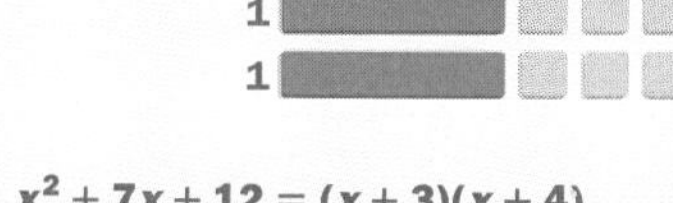
The area of a rectangle ...

... equals the product of its width and length.

Rewrite each expression as a product of factors.

1. $x^2 + 5x + 4$ **2.** $x^2 + 6x + 8$ **3.** $x^2 + 4x + 4$

Exploration Note

Purpose
The purpose of this Exploration is to have students discover how to factor a quadratic expression by using algebra tiles.

Materials/Preparation
Each pair of students needs 1 x^2-tile, 7 x-tiles, and 12 1-tiles.

Procedure
Students work with a partner and follow Steps 1–3 to factor the expression $x^2 + 7x + 12$. They then repeat this process for three other expressions.

Closure
Write the four quadratic expressions and their factored forms on the board. Ask students if they can see any relationships among the coefficients of the original expression and the constant terms of the factored expression. This approach leads into a discussion of the material presented at the top of page 222 and Example 1.

Explorations Lab Manual
See the Manual for more commentary on this Exploration.

For answers to the Exploration, see following page.

Plan⇔Support

Objectives

- Factor quadratic expressions.
- Solve quadratic equations using factoring.
- Solve real-world problems using factoring.

Recommended Pacing

Core and Extended Courses
Section 5.6: 2 day

Block Schedule
Section 5.6: 1 block

Resource Materials

Lesson Support
Lesson Plan 5.6
Warm-Up Transparency 5.6
Overhead Visuals:
Folder B: Algebra Tiles
Practice Bank: Practice 31
Study Guide: Section 5.6
Exploration Lab Manual:
Diagram Masters 2, 3, 4
Assessment Book: Test 18

Technology
Graphing Calculator
McDougal Littell Mathpack
Function Investigator
Internet:
http://www.hmco.com

Warm-Up Exercises

List all the factors of the given number.

1. 20 1 and 20, –1 and –20, 2 and 10, –2 and –10, 4 and 5, –4 and –5

2. –33 1 and –33, –1 and 33, 3 and –11, –3 and 11

3. 76 1 and 76, –1 and –76, 2 and 38, –2 and –38, 4 and 19, –4 and –19

Multiply each expression.

4. $(x + 3)(x - 2)$ $x^2 + x - 6$

5. $(x - 7)(x - 11)$ $x^2 - 18x + 77$

6. $(3x + 4)(2x - 3)$ $6x^2 - x - 12$

Teach⇔Interact

Learning Styles: Kinesthetic

Students with a kinesthetic learning style will benefit from the Exploration. While working through the Exploration, students generate a physical model for factoring a quadratic expression. Students can then see how the original expression and factored form are equivalent representations of the same quantity. Later, when students factor quadratic expressions without the aid of tiles, they should be encouraged to visualize this physical process.

Section Note

Communication: Discussion
When discussing the paragraph at the top of this page with the class, write several specific examples on the board. Students should study the examples and identify the relationships among the constants in the factored and unfactored expressions.

About Example 1

Alternate Approach
Some students may prefer to use a *guess-and-check* strategy to factor quadratic expressions.

Additional Example 1

Write $y = x^2 - 12x + 27$ as a product of factors.

To write $x^2 - 12x + 27$ as $(x + m)(x + n)$, you need to find integers m and n such that $mn = 27$ and $m + n = -12$. Make a table of possibilities.

Factors of 27	Sum of factors
1, 27	28
−1, −27	−28
3, 9	12
−3, −9	−12

The factors −3 and −9 give you the sum you want. A factorization of $y = x^2 - 12x + 27$ is $y = (x - 3)(x - 9)$.

When **factoring** $x^2 + bx + c$, your job is to find (if possible) integers m and n such that $x^2 + bx + c = (x + m)(x + n)$. One way is to use algebra tiles as you did in the Exploration. Another, more general way is to consider relationships among b, c, m, and n:

$$x^2 + bx + c = (x + m)(x + n)$$
$$= x^2 + mx + nx + mn$$
$$= x^2 + (m + n)x + mn$$

Compare corresponding coefficients.

THINK AND COMMUNICATE

1. Complete each statement with either "b" or "c."
   a. Whatever m and n are, their product must equal _?_.
   b. Whatever m and n are, their sum must equal _?_.
2. In the Exploration you saw that $x^2 + 7x + 12 = (x + 3)(x + 4)$. Do the relationships in Question 1 hold in this case?

EXAMPLE 1

Write the function $y = x^2 + 2x - 24$ as a product of factors.

SOLUTION

To write $x^2 + 2x - 24$ as $(x + m)(x + n)$, you need to find integers m and n such that $mn = -24$ and $m + n = 2$. Make a table of possibilities as shown at the right.

Factors of −24	Sum of factors
−1, 24	23
1, −24	−23
−2, 12	10
2, −12	−10
−3, 8	5
3, −8	−5
−4, 6	2
4, −6	−2

These factors give you the sum you want.

A factorization of $y = x^2 + 2x - 24$ is $y = (x - 4)(x + 6)$.

THINK AND COMMUNICATE

3. In Example 1, why would it be difficult to use algebra tiles to factor $x^2 + 2x - 24$?
4. a. **Technology** Use a graphing calculator or graphing software to graph $y = x^2 + 2x - 24$ and $y = (x - 4)(x + 6)$ in the same coordinate plane. What do you notice?
   b. If you rewrite $y = (x - 4)(x + 6)$ as $y = (x - 4)(x - (-6))$, you know that the x-intercepts of the function's graph are _?_ and _?_. Does your graph of $y = (x - 4)(x + 6)$ from part (a) confirm this?

ANSWERS Section 5.6

Exploration

1. $(x + 1)(x + 4)$
2. $(x + 4)(x + 2)$
3. $(x + 2)(x + 2)$

Think and Communicate

1. a. c
   b. b
2. Yes; $3 \cdot 4 = 12$ and $3 + 4 = 7$.
3. If you used 1-tiles to represent the −24, it would not be possible then to interpret the model as a rectangle whose area could be described as $(x + m)(x + n)$ for two numbers.
4. a.

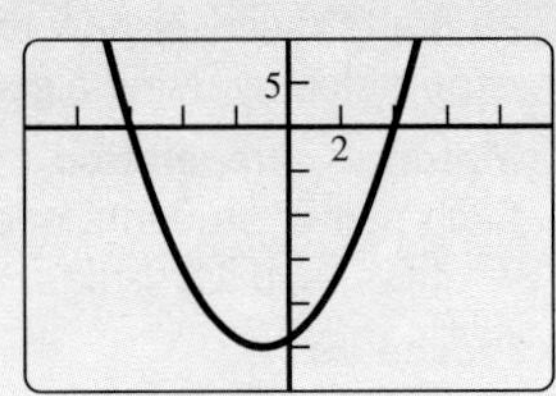

The graphs are the same.

   b. 4; −6; Yes.

To factor $ax^2 + bx + c$ when $a \neq 1$, you need to find (if possible) integers k, l, m, and n where:

$$\begin{aligned} ax^2 + bx + c &= (kx + m)(lx + n) \\ &= klx^2 + knx + lmx + mn \\ &= klx^2 + (kn + lm)x + mn \end{aligned}$$

Note that kn and lm are factors of $klmn$, which is the product of kl (the coefficient of x^2) and mn (the constant).

THINK AND COMMUNICATE

5. Use "a," "b," and "c" to complete this statement: If $ax^2 + bx + c$ can be written as the product of two factors, then _?_ must be the sum of factors of the product of _?_ and _?_.

6. Does the statement you completed in Question 5 apply to $ax^2 + bx + c$ when $a = 1$? Explain.

EXAMPLE 2

Write $y = 4x^2 - 20x + 21$ as a product of factors.

SOLUTION

Step 1 Find the product of the coefficient of x^2 and the constant:

$$4 \cdot 21 = 84$$

List the factors of the product and select the pair whose sum is the coefficient of x.

Factors of 84	Sum of factors
1, 84	85
2, 42	44
3, 28	31
4, 21	25
6, 14	20
7, 12	19

Factors of 84	Sum of factors
−1, −84	−85
−2, −42	−44
−3, −28	−31
−4, −21	−25
−6, −14	−20
−7, −12	−19

The pair of factors −6 and −14 gives the sum you want.

Step 2 Write the function as a product of factors.

$$\begin{aligned} y &= 4x^2 - 20x + 21 \\ &= 4x^2 - 6x - 14x + 21 \\ &= 2x(2x - 3) - 7(2x - 3) \\ &= (2x - 3)(2x - 7) \end{aligned}$$

Separate into two groups of two terms and factor each group.

Pull out the common factor between the two groups.

Written as a product of factors, the function $y = 4x^2 - 20x + 21$ becomes $y = (2x - 3)(2x - 7)$.

Think and Communicate

5. b; a; c

6. Yes; the product of 1 and c is c, and b is the sum of factors of c.

Teach⇔Interact

Think and Communicate

Question 4 leads students to two different ways of using a graphing calculator to check the factoring of a quadratic function. They can either graph both the factored and unfactored expression of the function to see that their graphs coincide, or they can graph the unfactored form to see if the graph has the correct x-intercepts. Question 6 helps students realize that the procedure used to factor an expression of the form $x^2 + bx + c$ is a special case of the procedure discussed on this page for all quadratic expressions.

Additional Example 2

Write $y = 6x^2 + x - 15$ as a product of factors.

Step 1 Find the product of the coefficient of x^2 and the constant: $6(-15) = -90$. List the factors of the product and select the pair whose sum is the coefficient of x.

Factors of −90	Sum of factors
1, −90	−89
−1, 90	89
2, −45	−43
−2, 45	43
3, −30	−27
−3, 30	27
5, −18	−13
−5, 18	13
6, −15	−9
−6 ,15	9
9, −10	−1
−9, 10	1

The pair of factors −9 and 10 gives the sum you want.

Step 2 To write the function as a product of factors, separate into two groups of two terms and factor each group. Finally, pull out the common factor between the two groups.

$$\begin{aligned} y &= 6x^2 + x - 15 \\ &= 6x^2 - 9x + 10x - 15 \\ &= 3x(2x - 3) + 5(2x - 3) \\ &= (3x + 5)(2x - 3) \end{aligned}$$

Written as a product of factors, the function $y = 6x^2 + x - 15$ becomes $y = (3x + 5)(2x - 3)$.

Teach⇔Interact

Additional Example 3

Suppose that on a new type of race track surface, the distance d in feet needed to stop a race car with specially designed tires traveling at a speed s in miles per hour is given by this quadratic function $d = 0.02s^2 + 1.1s$. What is the maximum speed that the car can be traveling in order to come to a stop within 156 ft?

Let $d = 156$ and solve for s.

$156 = 0.02s^2 + 1.1s$

$0 = 0.02s^2 + 1.1s - 156$

$0 = 0.02(s^2 + 55s - 7800)$

$0 = 0.02(s + 120)(s - 65)$

$s + 120 = 0$ or $s - 65 = 0$

$s = -120$ $s = 65$

The car should be traveling at 65 mi/h or less to stop within 156 ft.

Checking Key Concepts

Teaching Tip

When factoring functions, such as the one in question 3, encourage students to factor out any factor common to all three terms before factoring the expression into two binomials.

Closure Question

Describe the process for factoring an equation of the form $y = ax^2 + bx + c$.

Multiply a and c. Make a list of the factors for this product. Find the sum of each pair of factors. Pick the pair of factors whose sum is b. Write the function as $ax^2 + dx + ex + c$, where d and e are the factors found whose sum is b. Separate this expression into two groups of two terms and factor each group. Finally, pull out the common factor between the two groups.

THINK AND COMMUNICATE

7. Would it make any difference in Step 2 of Example 2 if you wrote $4x^2 - 20x + 21$ as $4x^2 - 14x - 6x + 21$? Show what happens if you do.
8. What are the x-intercepts of the graph of the function in Example 2?

EXAMPLE 3 Application: Driving

On dry asphalt, the distance d in feet needed to stop a car traveling at speed s in miles per hour is given by this quadratic function:

$$d = 0.05s^2 + 1.1s$$

The car travels **$0.05s^2$** feet after the brakes are applied.

The car travels **$1.1s$** feet during the time it takes the driver to react to an emergency and apply the brakes.

Suppose a driver has just rounded a curve and sees a stop sign 78 ft ahead. At what maximum speed should the car be traveling in order to come to a stop in time?

78 ft

SOLUTION

Let $d = 78$ and solve for s.

$78 = 0.05s^2 + 1.1s$

$0 = 0.05s^2 + 1.1s - 78$

$0 = 0.05(s^2 + 22s - 1560)$

$0 = 0.05(s + 52)[s + (-30)]$

$s + 52 = 0$ or $s - 30 = 0$

$s = -52$ or $s = 30$

You need to find a factor pair whose product is -1560 and whose sum is 22. Use **52** and **-30**.

The car should be traveling at 30 mi/h or less to stop within 78 ft.

CHECKING KEY CONCEPTS

Write each quadratic function as a product of factors.

1. $y = x^2 - 18x + 81$
2. $y = 2x^2 + 7x + 3$
3. $y = 6x^2 + 10x - 4$
4. What are the x-intercepts of the graph of the function $y = 5x^2 - 21x + 18$?
5. In Example 3, at what maximum speed should the car be traveling in order to come to a stop within 36 ft?

Think and Communicate

7. No; $4x^2 - 14x - 6x + 21 = 2x(2x - 7) - 3(2x - 7) = (2x - 7)(2x - 3) = (2x - 3)(2x - 7)$.
8. 1.5; 3.5

Checking Key Concepts

1. $y = (x - 9)(x - 9)$
2. $y = (2x + 1)(x + 3)$
3. $y = 2(3x - 1)(x + 2)$
4. 1.2; 3
5. 18 mi/h

Exercises and Applications

1. $y = (x - 4)(x - 1)$
2. $y = (x + 8)(x - 3)$
3. $y = (x + 1)(x - 4)$
4. $y = (2x + 3)(x - 5)$
5. $y = (3x + 2)(3x + 2)$
6. $y = 3(2x - 3)(x + 7)$
7. $y = (4x - 5)(x + 2)$
8. $y = (6x - 1)(4x + 3)$
9. $y = 4(3x - 5)(3x + 5)$
10. 20 mi/h
11. a. Let x = the radius in inches and y = the area in square inches; $y = \pi x^2 - \pi(0.375)^2 = \pi(x^2 - (0.375)^2) = \pi(x - 0.375)(x + 0.375)$.
    b. If the tape were completely unrolled, it is clear the area of the top of the tape would equal the length of the tape times the thickness of the tape. The area is not changed by rolling up the tape.
    c. Let d = the distance from the center of the spool to the outer edge of the tape and l = the length of the tape; $l = \frac{\pi(d - 0.375)(d + 0.375)}{0.0005}$.

12–20. Answers are rounded to the nearest tenth when necessary.

12. −10; 2
13. −6
14. −6; 15
15. 1; 3
16. −2; 0.75
17. $-4\frac{1}{2}$; $1\frac{2}{3}$
18. −1; 8
19. −5; 0.6
20. 0; 1

5.6 Exercises and Applications

Extra Practice exercises on page 756

Write each quadratic function as a product of factors.

1. $y = x^2 - 5x + 4$ **2.** $y = x^2 + 5x - 24$ **3.** $y = x^2 - 3x - 4$

4. $y = 2x^2 - 7x - 15$ **5.** $y = 9x^2 + 12x + 4$ **6.** $y = 6x^2 + 33x - 63$

7. $y = 4x^2 + 3x - 10$ **8.** $y = 24x^2 + 14x - 3$ **9.** $y = 36x^2 - 100$

10. DRIVING On wet concrete, the distance d in feet needed to stop a car traveling at speed s in miles per hour is given by the function $d = 0.055s^2 + 1.1s$. At what maximum speed should a car be traveling on wet concrete to stop at a traffic light 44 ft ahead?

11. GEOMETRY The photo shows videocassette tape wound on a spool.

a. Write a function giving the area of the top surface of the reel of tape (without the spool). (*Hint:* Find the area between the two dashed circles shown on the photo.) Express your function both in standard form and as a product of factors.

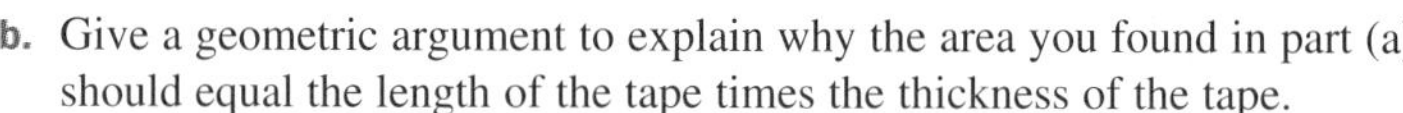

b. Give a geometric argument to explain why the area you found in part (a) should equal the length of the tape times the thickness of the tape.

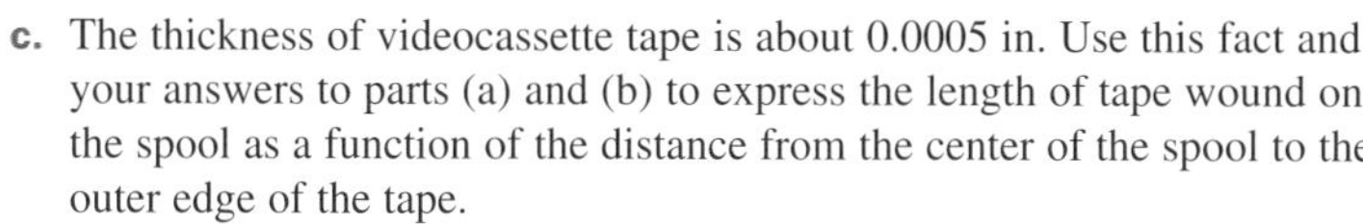

c. The thickness of videocassette tape is about 0.0005 in. Use this fact and your answers to parts (a) and (b) to express the length of tape wound on the spool as a function of the distance from the center of the spool to the outer edge of the tape.

Find the x-intercept(s) of the graph of each function.

12. $y = x^2 + 8x - 20$ **13.** $y = x^2 + 12x + 36$ **14.** $y = x^2 - 9x - 90$

15. $y = 3x^2 - 12x + 9$ **16.** $y = 4x^2 + 5x - 6$ **17.** $y = 6x^2 + 17x - 45$

18. $y = 0.2x^2 - 1.4x - 1.6$ **19.** $y = 25x^2 + 110x - 75$ **20.** $y = 15x - 15x^2$

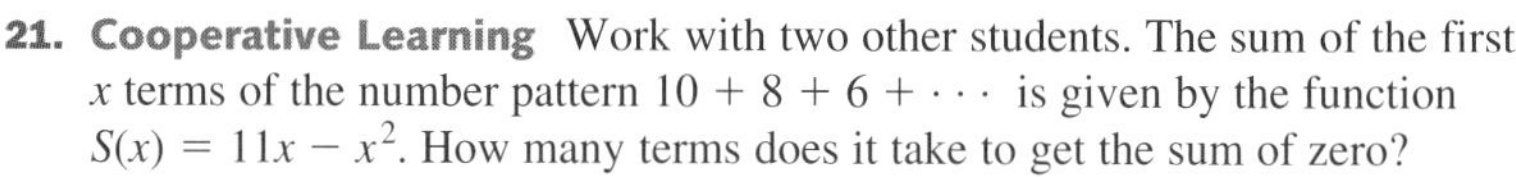

21. Cooperative Learning Work with two other students. The sum of the first x terms of the number pattern $10 + 8 + 6 + \cdots$ is given by the function $S(x) = 11x - x^2$. How many terms does it take to get the sum of zero?

a. One student should use a graph to answer the question. Another student should use algebra. The third student should use arithmetic.

b. Compare the three answers. Which method was easiest to use? Explain.

c. Make a conjecture about a function that gives the sum of the first x terms of the pattern $12 + 10 + 8 + 6 + \cdots$. Test your conjecture using one of the methods in part (a).

22. Challenge Consider the following quadratic functions:

$y = 2x^2 + 3x - 1$ $y = 2x^2 - x - 1$ $y = 8x^2 + 2x - 3$ $y = 3x^2 - 4x + 4$

a. Write each function as a product of factors, if possible.

b. Let $y = 0$ and find the discriminant of each equation. What kind of number is each discriminant?

c. Make a conjecture about the relationship between factorable quadratic functions and discriminants. Test your conjecture using other quadratic functions.

21. a. graph:

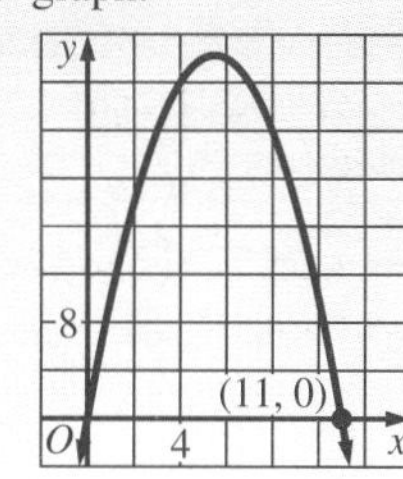

algebra: $0 = 11x - x^2$; $0 = -x(x - 11)$; $x = 0$ or $x = 11$

arithmetic: Add terms until the sum is zero: $10 + 8 + 6 + 4 + 2 + 0 + (-2) + (-4) + (-6) + (-8) + (-10) = 0$; there are 11 terms.

b. The answers are the same; preferences may vary. An example is given. I think the algebraic method was the easiest since the function was so easy to write as a product of factors.

c. Conjecture: The sum of the first x terms of the numbers $12 + 10 + 8 + 6 + \cdots$ is given by the function $S(x) = 13x - x^2$. It takes 13 terms for the function to get the sum of zero.

22. See answers in back of book.

Apply⇔Assess

Suggested Assignment

Core Course
Day 1 Exs. 1–10, 12–21
Day 2 Exs. 26–39, AYP

Extended Course
Day 1 Exs. 1–25
Day 2 Exs. 26–39, AYP

Block Schedule
Day 31 Exs. 1–10, 12–21, 26–39, AYP

Exercise Notes

Cooperative Learning
Exs. 1–9, 12–20 Students should be able to work these problems correctly. You might have students work with a partner to check their answers to these exercises and then work together to redo any for which their answers disagree.

Mathematical Procedures
Ex. 9 This is the first factoring exercise in which the coefficient of x is zero. Point out to students that they can use the same method used for the other exercises. In this case, they are looking for two numbers whose product is 36(–100), or –3600, and whose sum is zero. Students should realize that these numbers must be opposites; in this case, –60 and 60. The rest of the procedure is the same as that for any other quadratic expression.

Second-Language Learners
Ex. 11 Students learning English may not be familiar with the word *spool.* If necessary, explain that it is a plastic reel around which string or tape is wound.

Integrating the Strands
Exs. 11, 21, 23–25 In Exs. 11 and 23–25, students are given the opportunity to work some real-world problems involving both algebra and geometry. Ex. 21 involves concepts from discrete mathematics, functions, algebra, number theory, and arithmetic. The problem is solved using three different methods, each from a different strand.

Reasoning
Exs. 21, 22 In each of these exercises, students are asked to examine some examples in order to find a general pattern, make a conjecture about the pattern, and test the conjecture on other examples. These exercises exemplify the method of inductive reasoning.

Apply⇔Assess

Exercise Notes

Interdisciplinary Problems
Exs. 23–25 In these exercises, students use factoring to solve some astronomy problems involving the construction of a "liquid mirror" telescope.

Career Connection
Exs. 23–25 Astronomers are scientists who study objects in the universe such as stars and planets. As shown in these exercises, astronomers need strong backgrounds in mathematics and science, especially physics and chemistry. Astronomers generally work in colleges or universities as teachers and researchers and usually possess a degree at the doctorate level in astronomy.

Communication: Reading
Exs. 23–25 It may be beneficial to read and discuss this page with students before assigning the problems for homework. In particular, discuss the equation given and how it relates to the picture. You might ask students to list what each letter in the equation represents and whether it is a variable or constant.

Topic Spiraling: Review
Exs. 26–29 These exercises provide students with an opportunity to practice their quadratic factoring at the same time they review the properties of logarithms.

Assessment
Ex. 30 This exercise allows students to review and evaluate the effectiveness of various methods they have learned to find the x-intercepts of the graph of a quadratic equation. For part (a), you might also ask students to comment on how they would find the x-intercepts of the graph of a function given in vertex form.

Connection ASTRONOMY

The surface of a spinning liquid forms a *paraboloid*, a three-dimensional figure created by rotating a parabola about its line of symmetry. By spinning reflective liquids, astronomers have started to make telescopes having "liquid mirrors" with surfaces smoother than glass. A cross section of the surface of a spinning liquid is described by the equation

$$y = \frac{2\pi^2 f^2}{g}x^2 - \frac{\pi^2 f^2 R^2}{g}$$

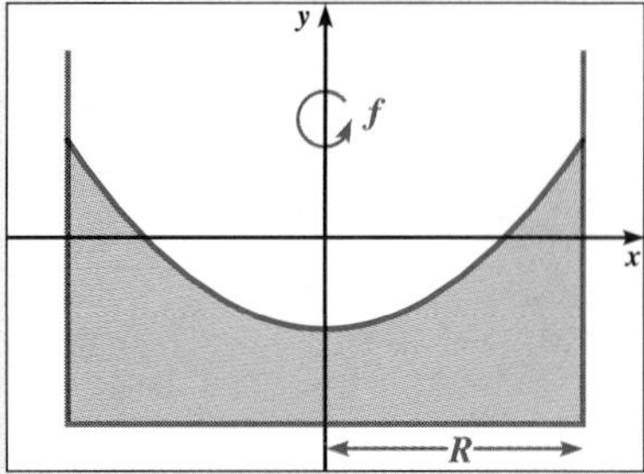

where f is the spinning frequency in revolutions per second, R is the container's radius in meters, and $g = 9.8 \text{ m/s}^2$ (the acceleration due to gravity).

23. What is the equation for the cross-sectional surface of the liquid before it is spun? What does the equation tell you about the placement of the x-axis relative to the liquid?

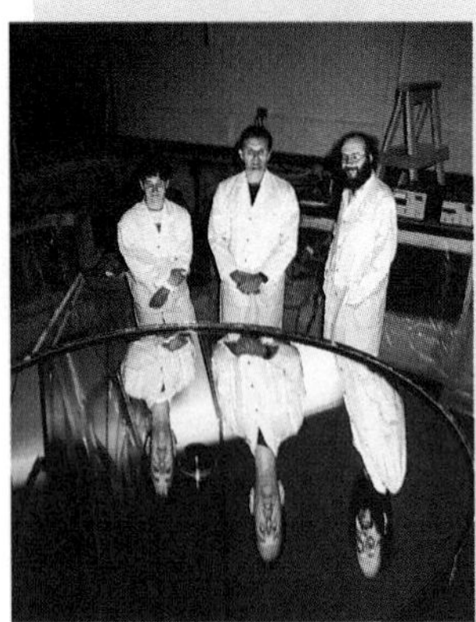

Ph.D. student Luc Girard, Prof. Ermanno F. Borra, and Dr. Robert Content are reflected in a liquid mirror at Université Laval in Quebec City, Canada.

24. Suppose a reflective liquid is put in a container of radius $\sqrt{2}$ m and spun with a frequency of 0.5 revolutions per second.

a. Write an equation describing a cross section of the surface.

b. Rewrite your equation from part (a) as a product of factors.

c. What are the x-intercepts of the graph of the function in part (b)? What are the coordinates of the vertex? Make a sketch of the graph.

d. **Open-ended Problem** Using the same coordinate plane as in part (c), show the cross sections that result when the liquid is spun at other frequencies. What do you notice?

25. Challenge Show that no matter how fast the container is spun, the x-intercepts of the graph of $y = \frac{2\pi^2 f^2}{g}x^2 - \frac{\pi^2 f^2 R^2}{g}$ are $\pm\frac{R}{\sqrt{2}}$.

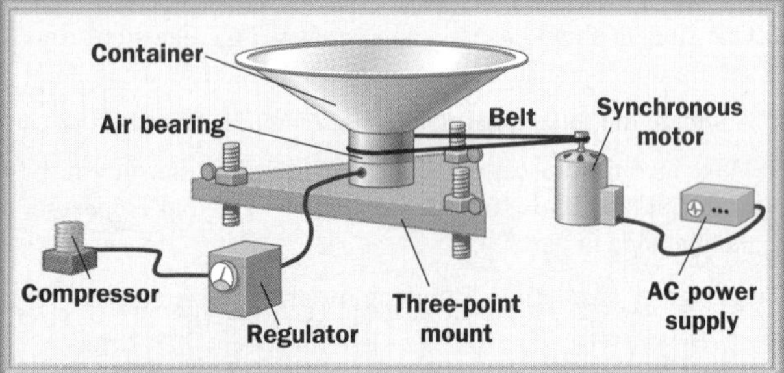

NASA built a liquid mercury mirror based on the design above. It is being used in a telescope for locating space debris that may damage spacecraft or space stations.

23. $y = 0$; The x-axis is at the surface of the liquid before it is spun.
24. See answers in back of book.
25. $y = \frac{2\pi^2 f^2}{g}x^2 - \frac{\pi^2 f^2 R^2}{g} = \frac{2\pi^2 f^2}{g}\left(x^2 - \frac{R^2}{2}\right) = \frac{2\pi^2 f^2}{g}\left(x + \frac{R}{\sqrt{2}}\right)\left(x - \frac{R}{\sqrt{2}}\right)$; The x-intercepts of the graph are $x = \pm\frac{R}{\sqrt{2}}$.
26. −20; 5
27. 0; 6
28. 4
29. $\frac{7}{6}$
30. a. (1) Graph the function. (2) Write the equation in standard form, let $y = 0$, and use the quadratic formula to solve the resulting equation. The solutions are the x-intercepts of the original equation. (3) It may be possible to find the x-intercepts by setting the equation equal to 0 and factoring it. You can determine if this is possible either by observation (if the factors are obvious) or by determining if the discriminant is a perfect square; if so, the function is factorable.
b. Answers may vary. Examples are given. I would use factoring for A, and graphing or the quadratic formula for B, C, and D. The x-intercepts are: A: −2, −3; B: no x-intercepts; C: −2.3, −0.2; D: no x-intercepts
31. 0.4; 7.6
32. $-2; \frac{1}{2}$
33. −0.7; 1.1
34. $-1\frac{1}{2}; -1\frac{1}{3}$
35. about $780

Use factoring and the properties of logarithms to solve for *x*. Be sure to check your answers.

26. $\log(x^2 + 15x) = 2$ **27.** $\log_3(5x^2 - 30x + 45) - \log_3 5 = 2$

28. $\log_2 x + \log_2(x - 2) = 3$ **29.** $\log_5(6x - 1) + \log_5(x + 3) = \log_5 25$

ONGOING ASSESSMENT

30. Cooperative Learning Work with three other students. Each of you should choose one of these four quadratic functions:

A. $y = x^2 + 5x + 6$ **B.** $y = 2x^2 + 5x + 6$

C. $y = 2x^2 + 5x + 1$ **D.** $y = x^2 + 0.8x + 0.3$

a. Writing Describe the methods you could use to find the *x*-intercepts of the graph of *any* quadratic function.

b. Which method would each of you use to find the *x*-intercepts of the graph of the function you chose? Explain your choice.

SPIRAL REVIEW

Solve each equation. *(Section 5.5)*

31. $x^2 - 8x + 3 = 0$ **32.** $2x^2 + 3x - 2 = 0$

33. $5x^2 - 2x - 4 = 0$ **34.** $1.2x^2 + 3.4x + 2.4 = 0$

35. Use a constant growth model to estimate the median weekly earnings for families in 1995. *(Section 1.1)*

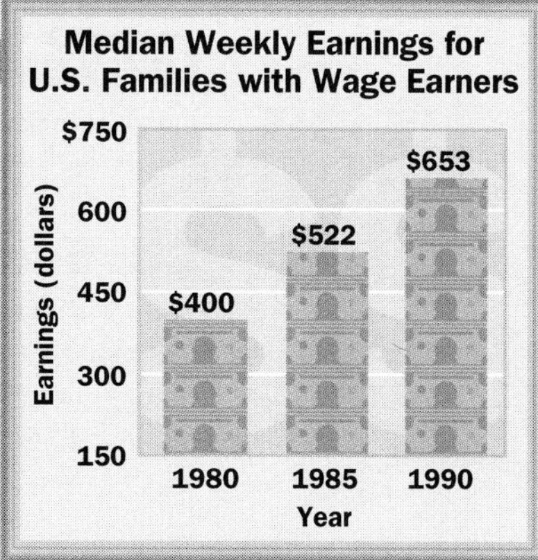

Sketch the graph of each function. *(Section 5.2)*

36. $y = 4(x + 7)^2 - 3$ **37.** $y = -2(x - 1)^2 + 5$

38. $y = \frac{1}{3}(x - 3)^2 - 4$ **39.** $y = -\frac{1}{2}x^2 + 1$

ASSESS YOUR PROGRESS

VOCABULARY

standard form (p. 206) **discriminant** (p. 217)
quadratic formula (p. 216) **factoring** (p. 222)

Write each equation in vertex form. *(Section 5.4)*

1. $y = 15x^2 + 9x$ **2.** $y = x^2 + 12x + 7$ **3.** $y = 4x^2 - 20x - 3$

Find the *x*-intercepts, if any, for the graph of each quadratic function. *(Section 5.5)*

4. $y = 4x^2 + 8x + 3$ **5.** $y = x^2 - 2x + 5$ **6.** $y = 2x^2 - 20$

Write each quadratic function as a product of factors. *(Section 5.6)*

7. $y = x^2 + 7x + 12$ **8.** $y = 3x^2 - 5x - 2$ **9.** $y = 2x^2 + 13x - 7$

10. Journal Explain the advantages and disadvantages of writing quadratic functions in standard, intercept, and vertex form.

Apply⇔Assess

Exercise Notes

Spiral Review: Preview
Ex. 35 This exercise helps to remind students about the idea of modeling data, a major focus of the next chapter.

Assess Your Progress

Journal Entry
While writing the entry for Ex. 10, students should gain a greater appreciation for why there are three forms of quadratic equations.

Progress Check 5.4–5.6

See page 231.

Practice 31 for Section 5.6

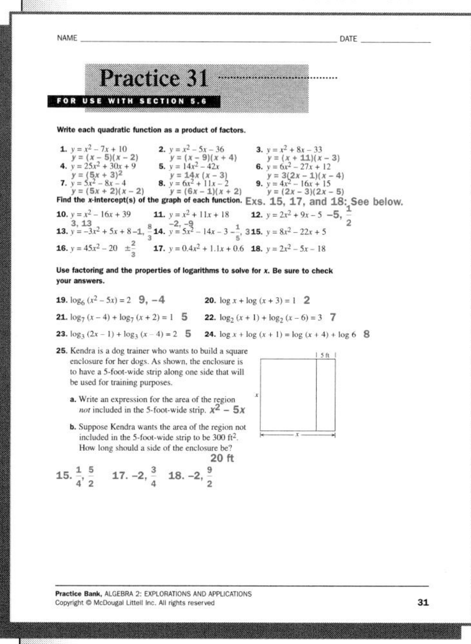

NAME ______________________ DATE __________

Practice 31
FOR USE WITH SECTION 5.6

Write each quadratic function as a product of factors.

1. $y = x^2 - 7x + 10$ $y = (x - 5)(x - 2)$
2. $y = x^2 - 5x - 36$ $y = (x - 9)(x + 4)$
3. $y = x^2 + 8x - 33$ $y = (x + 11)(x - 3)$
4. $y = 25x^2 + 30x + 9$ $y = (5x + 3)^2$
5. $y = 14x^2 - 42x$ $y = 14x(x - 3)$
6. $y = 6x^2 - 27x + 12$ $y = 3(2x - 1)(x - 4)$
7. $y = 5x^2 - 8x - 4$ $y = (5x + 2)(x - 2)$
8. $y = 6x^2 + 11x - 2$ $y = (6x - 1)(x + 2)$
9. $y = 4x^2 - 16x + 15$ $y = (2x - 3)(2x - 5)$

Find the *x*-intercept(s) of the graph of each function. Exs. 15, 17, and 18: See below.

10. $y = x^2 - 16x + 39$ 3, 13
11. $y = x^2 + 11x + 18$ −2, −9
12. $y = 2x^2 + 9x - 5$ $-5, \frac{1}{2}$
13. $y = -3x^2 + 5x + 8$ $-1, \frac{8}{3}$
14. $y = 5x^2 - 14x - 3$ $-\frac{1}{5}, 3$
15. $y = 8x^2 - 22x + 5$
16. $y = 45x^2 - 20$ $\pm\frac{2}{3}$
17. $y = 0.4x^2 + 1.1x + 0.6$
18. $y = 2x^2 - 5x - 18$

Use factoring and the properties of logarithms to solve for *x*. Be sure to check your answers.

19. $\log_6(x^2 - 5x) = 2$ 9, −4
20. $\log x + \log(x + 3) = 1$ 2
21. $\log_7(x - 4) + \log_7(x + 2) = 1$ 5
22. $\log_2(x + 1) + \log_2(x - 6) = 3$ 7
23. $\log_3(2x - 1) + \log_3(x - 4) = 2$ 5
24. $\log x + \log(x + 1) = \log(x + 4) + \log 6$ 8

25. Kendra is a dog trainer who wants to build a square enclosure for her dogs. As shown, the enclosure is to have a 5-foot-wide strip along one side that will be used for training purposes.

a. Write an expression for the area of the region *not* included in the 5-foot-wide strip. $x^2 - 5x$

b. Suppose Kendra wants the area of the region not included in the 5-foot-wide strip to be 300 ft². How long should a side of the enclosure be? 20 ft

15. $\frac{1}{4}, \frac{5}{2}$ **17.** $-2, \frac{3}{4}$ **18.** $-2, \frac{9}{2}$

Practice Bank, ALGEBRA 2: EXPLORATIONS AND APPLICATIONS

31

36. 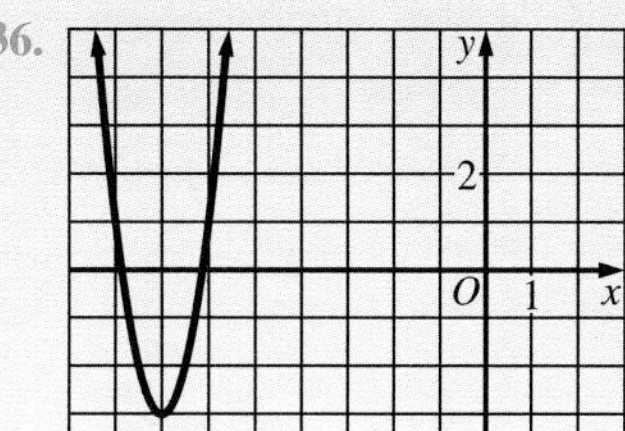

37. 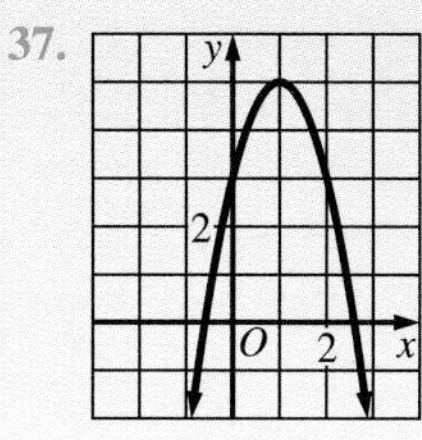

38. 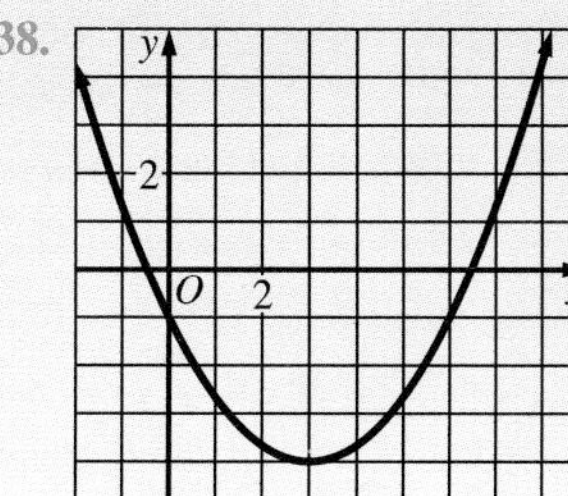

39. 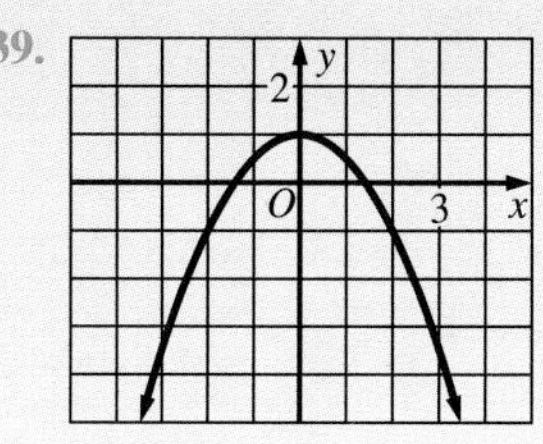

Assess Your Progress

1. $y = 15(x + 0.3)^2 - 1.35$
2. $y = (x + 6)^2 - 29$
3. $y = 4(x - 2.5)^2 - 28$
4. −1.5; −0.5
5. no *x*-intercepts
6. ±3.2
7. $y = (x + 4)(x + 3)$
8. $y = (3x + 1)(x - 2)$
9. $y = (2x - 1)(x + 7)$
10. See answers in back of book.

Project Notes

Mathematical Goals

- Perform an experiment and record data.
- Formulate a quadratic model for the data.
- Display the model and data points on a graph.
- Compare the model with the actual values.

Planning

Materials

- large (metal) can
- nail
- hammer
- metric ruler
- watch
- graphing calculator or graphing software
- paper and pencil

Project Teams

Students can select a partner and perform the experiment together. Then they can decide how to proceed to complete the project.

Guiding Students' Work

Punching the hole in the bottom of the can should be done under teacher supervision. Also, caution students to be careful when using the hammer and nail to create the hole.

This is a good project to illustrate to students the importance of keeping their work organized. An organized table of data values will make it easier to draw a scatter plot and calculate values for the quadratic model. Also, and perhaps more importantly, writing equations in an organized way presents the values so that they are more easily calculated and compared. Students should first simplify the equation so that it contains only the variables t and $h(t)$, then use this simplified equation to substitute t values and find an estimated water height.

Second-Language Learners

Students learning English can benefit from working on the research and writing a report of the project with the help of a peer tutor or aide.

PORTFOLIO PROJECT

Investigating the Flow of Water

Often you will see a town using a tall water tower as a means of creating enough pressure to deliver water throughout the town. The water tower functions on the simple principle that as the height of a column of water increases, the weight of the water in the column increases, which in turn causes the pressure at the base of the column to increase.

You can model this situation using a can filled with water. As you let water flow out of a hole in the bottom, the pressure at the hole decreases. What effect do you think decreasing pressure will have on the time it takes the water to flow out of the can?

PROJECT GOAL **Conduct an experiment to find out how the water level in a can affects the time it takes the water to flow from a hole in the can's bottom.**

Conducting an Experiment

Work with a partner to plan and perform an experiment in which you measure the height of water in a can as the water flows from a hole in the bottom. You will need a large can (like a coffee can), a nail, a hammer, a metric ruler, and a watch. Here are some guidelines for conducting your experiment.

1. **PUNCH** a hole in the bottom of the can.

2. **COVER** the hole in the can's bottom with your finger and fill the can with water.

3. **RECORD** the water's initial height, h_0, in millimeters.

Testing a Quadratic Model

A theoretical model based on physics applies to the situation you have investigated. The model involves the following constants (all measured in millimeters): h_0, the water's initial height; D, the diameter of the can; and d, the diameter of the hole. The model says that the water's height h, in millimeters, is a quadratic function of the time t, in seconds, after the water begins running out of the container:

$$h(t) = \left(\sqrt{h_0} - \frac{70d^2}{D^2}t\right)^2$$

- **MEASURE** the diameters of the can and the hole.
- **SUBSTITUTE** your values of D, d, and h_0 into the model.
- **CALCULATE** $h(t)$ for the t-values in your data table. Compare your calculated water-height values with the actual data from your experiment. Are there differences? What might explain them?
- **GRAPH** the function $h(t)$ along with your data using a graphing calculator or graphing software. Is the graph a good model for your data?

Writing a Report

Write a report about your experiment. Describe your procedures, and present a comparison of your data to the model's predictions. Include the following:

- a statement of the goal of your project
- a data table that compares your experimental water-height data to the values predicted by the quadratic model
- an evaluation of how well the quadratic model predicts water height
- any ideas you may have that could explain differences between the data and the model

To extend your project, you may wish to investigate other factors that could affect the flow of liquid from a container. For example, what if you used a wider can or made a smaller hole? What if you used a different liquid, like pancake syrup?

Self-Assessment

In your report, include a description of any difficulties you had. For example, was it difficult to get precise measurements? How might you alter the experiment to improve the results?

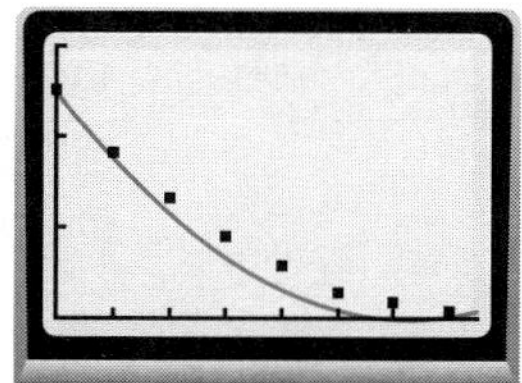

4. **UNCOVER** the hole for 10 s and then cover it with your finger again. Re-measure the height of the water. Repeat this until the can is nearly empty.

5. **ORGANIZE** your data in a table. Then make a scatter plot. Put time on the horizontal axis and water height on the vertical axis.

Project Notes

Guiding Students' Work

Rubric for Chapter Project

4 Students conduct the experiment and record the results in a data table. Students accurately substitute values into the equation and compare the measures and calculated values for experimental and predicted water height. Students graph both a scatter plot and the quadratic model and relate the two. A well-written report is developed that includes a goal statement, a complete data table, an evaluation, the graph, and an explanation, as well as some ideas on the extension of the project. The report shows an understanding of the mathematical content of the project.

3 Students perform the experiment and record the results in a data table. Students measure, substitute, and calculate values for the equation but make at least one mathematical error. Students graph the scatter plot and the quadratic model and compare the two. A report is written that contains all of the required parts and shows an understanding of the mathematics, but an extension of the project is not done.

2 Students complete the experiment and record the data values but have difficulty with the quadratic model. As a result, predicted values are incorrect and the graph of the quadratic model as well as the equation itself are incorrect. Students write a report that relates the mathematical ideas of the project with the experimental values, but some errors in understanding the concepts involved in the project are evident in the report.

1 Students complete the experiment but do not attempt to calculate predicted values. A quadratic model is not found and no attempt is made to relate the experimental data with a model. An incomplete report is written that does not indicate an whole-hearted attempt at doing the project or an understanding of the ideas being explored. Students should be encouraged to speak with the teacher as soon as possible to review their work and to make a new start on the project.

Chapter Support

Course Guide: Chapter 5
Lesson Plans: Chapter 5
Practice Bank: Cumulative Practice 32
Study Guide: Chapter 5 Review
Assessment Book: Chapter Tests 19 and 20; Chapter 5 Alternative Assessment
Test Generator Software
Function Investigator with Matrix Analyzer Activity Book: Function Investigator Activities 13 and 14
Technology Book: Calculator Based Lab 3
Portfolio Project Book
Preparing for College Entrance Tests
Professional Handbook

Progress Check 5.1–5.3

1. The graph below has an equation of the form $y = ax^2$. Find a. *(Section 5.1)* −2

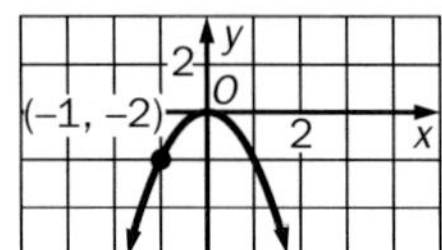

2. Solve $\frac{1}{3}x^2 - 3 = 0$. *(Section 5.1)* ±3
3. Describe the graph of $y = 3(x - 2)^2 - 5$ and then sketch it. *(Section 5.2)* The vertex is (2, −5). The axis of symmetry is $x = 2$. The parabola opens up.

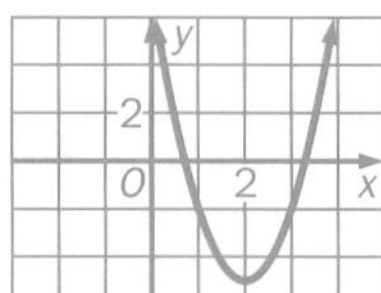

An acorn falls from a 40 ft oak tree. Use $d = 16t^2$ for the distance, in feet, an object falls in t seconds.

4. How many seconds will it take the acorn to hit the ground? *(Section 5.1)* about 1.6 s
5. What is the height of the acorn after 0.5 s? *(Section 5.2)* 36 ft
6. Rewrite $y = (5x + 20)(4 - 2x)$ in intercept form and identify the x-intercepts and the coordinates of the vertex for its graph. *(Section 5.3)* $y = -10(x + 4)(x - 2)$; x-intercepts: −4, 2; vertex: (−1, 90)
7. Find the maximum value of $y = (100 - 5x)(2x - 12)$. *(Section 5.3)* 490

CHAPTER

5 Review

STUDY TECHNIQUE

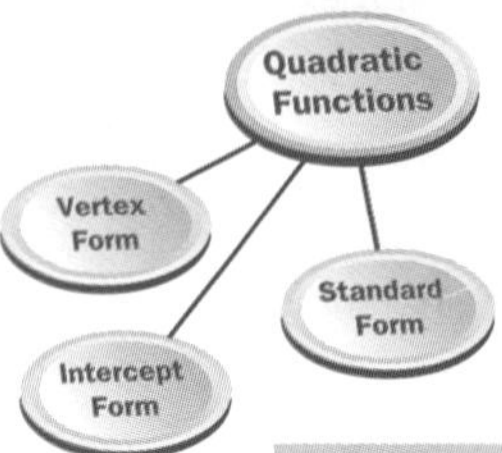

A *concept map* is a diagram that highlights the connections between ideas. Drawing a concept map for a chapter or a section can help you focus on the important ideas and on how they are related. Draw a concept map for this chapter, showing key concepts and vocabulary terms.

VOCABULARY

quadratic function (p. 186)
parabola (p. 186)
line of symmetry (p. 186)
vertex (p. 186)
square root (p. 187)
vertex form (p. 193)
minimum value (p. 195)
maximum value (p. 195)
x-intercept (p. 199)
intercept form (p. 200)
standard form (p. 206)
quadratic formula (p. 216)
discriminant (p. 217)
factoring (p. 222)

SECTIONS 5.1, 5.2, *and* 5.3

The graph of a **quadratic function** is a **parabola**. A quadratic function can be expressed in several forms. In each of the following forms, the value of a affects the width of the parabola and the direction it opens. If $a > 0$, the function has a **minimum value**; if $a < 0$, the function has a **maximum value**.

Vertex form: $y = a(x - h)^2 + k$
vertex: (h, k)
line of symmetry: $x = h$

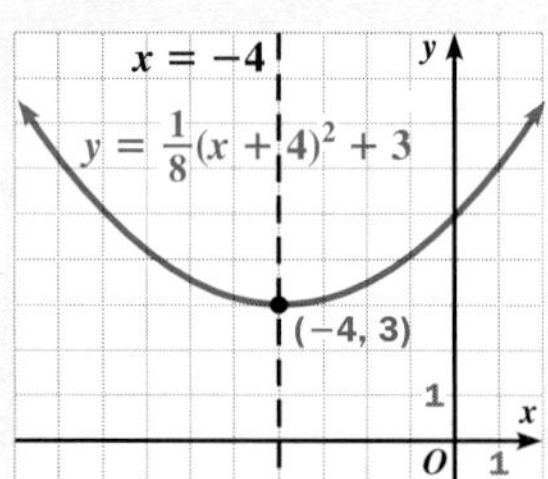

Intercept form: $y = a(x - p)(x - q)$
x-intercepts: p and q
line of symmetry: $x = \frac{p + q}{2}$

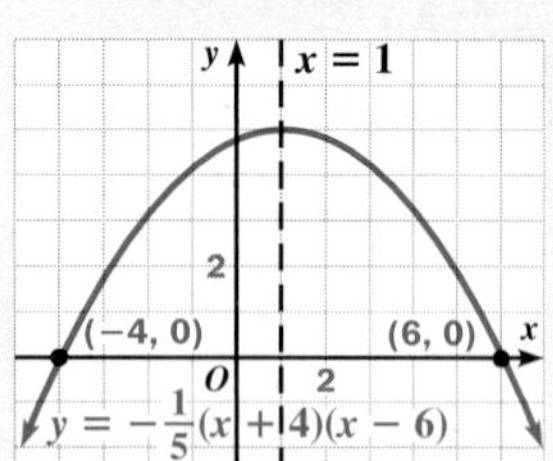

SECTIONS 5.4, 5.5, *and* 5.6

The equation $y = ax^2 + bx + c$ is the **standard form** of a quadratic function. To find the maximum value or minimum value of a quadratic function in standard form, or to find the coordinates of the **vertex** of the function's graph, rewrite the equation in vertex form by completing the square.

$$y = -2x^2 + 12x - 10$$
$$= -2(x^2 - 6x + 9) - 10 + 2(9)$$
$$= -2(x - 3)^2 + 8$$

The vertex is **(3, 8)**.

Add the square of half the coefficient of x.

$\left(\frac{-6}{2}\right)^2 = (-3)^2 = 9$

Add 2(9) to balance **−2(9)**.

To find the ***x*-intercept(s)** of the graph of a quadratic function in standard form or the solutions of a quadratic equation in the form $ax^2 + bx + c = 0$, you can complete the square or use the **quadratic formula**. For example, consider $y = -2x^2 + 12x - 10$.

Method 1: Complete the square. Rewrite $y = -2x^2 + 12x - 10$ as $y = -2(x - 3)^2 + 8$. Let $y = 0$.

$$-2(x - 3)^2 + 8 = 0$$
$$-2(x - 3)^2 = -8$$
$$(x - 3)^2 = 4$$
$$x - 3 = \pm 2$$
$$x = 3 \pm 2 = 1, 5$$

Method 2: Use the quadratic formula. Let $y = 0$.

$$-2x^2 + 12x - 10 = 0$$

Substitute −2 for a, 12 for b, and −10 for c into $x = \dfrac{-b \pm \sqrt{b^2 - 4ac}}{2a}$.

$$x = \frac{-12 \pm \sqrt{12^2 - 4(-2)(-10)}}{2(-2)}$$
$$= \frac{-12 \pm 8}{-4} = 1, 5$$

The **discriminant** of the quadratic formula, $b^2 - 4ac$, tells you how many solutions the equation $ax^2 + bx + c = 0$ has and how many x-intercepts the graph of $y = ax^2 + bx + c$ has, as shown in the table at the right.

$b^2 - 4ac$	Solutions/ x-intercepts
> 0	2
= 0	1
< 0	0

Some quadratic functions of the form $y = ax^2 + bx + c$ can be written as a product of factors by finding two numbers whose product is ac and whose sum is b. For example, to write $y = 3x^2 + 7x - 6$ as a product of factors, find two numbers whose product is $3(-6) = -18$ and whose sum is 7.

Factors of −18	Sum of factors
1, −18	−17
−1, 18	17
2, −9	−7
−2, 9	7
3, −6	−3
−3, 6	3

The factors **−2** and **9** have the needed sum, 7.

$$y = 3x^2 + 7x - 6$$
$$= 3x^2 - 2x + 9x - 6$$
$$= x(3x - 2) + 3(3x - 2)$$
$$= (x + 3)(3x - 2)$$

Write $7x$ as the sum **$-2x + 9x$**.

Factor.

Progress Check 5.4.–5.6

1. Write $y = -3x^2 + 12x - 7$ in vertex form. *(Section 5.4)* $y = -3(x - 2)^2 + 5$
2. A ball is tossed upward from a height of 4 feet with initial velocity 12 ft/s. *(Section 5.4)*
   a. Write an equation to model this situation. $h = -16t^2 + 12t + 4$
   b. What is the maximum height of the ball? 6.25 ft
3. Find the solution(s) of $2x^2 = 4x + 70$ using the quadratic formula. *(Section 5.5)* −5, 7
4. Tell whether $5x^2 - 7x + 11 = 0$ has two solutions, one solution, or no solution. *(Section 5.5)* no solution
5. Find the x-intercepts for the graph of $y = -12x^2 - 4x + 2$. *(Section 5.5)* about −0.61 and 0.27

Factor each expression. *(Section 5.6)*

6. $x^2 + 3x - 18$ $(x + 6)(x - 3)$
7. $15x^2 + 13x + 2$ $(3x + 2)(5x + 1)$

Chapter 5 Assessment
Form A Chapter Test

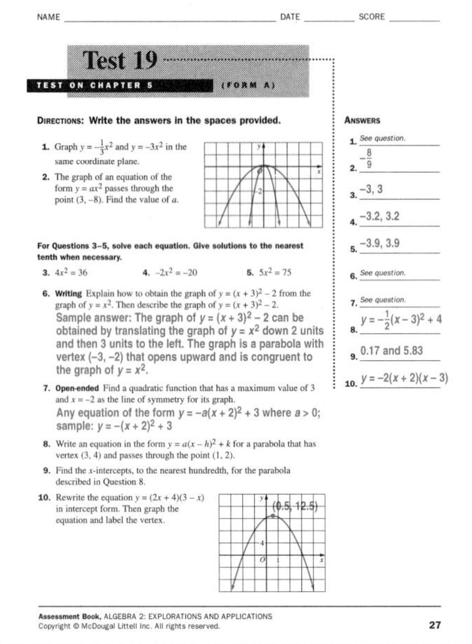

NAME ______ DATE ______ SCORE ______

Test 19
TEST ON CHAPTER 5 (FORM A)

DIRECTIONS: Write the answers in the spaces provided.

1. Graph $y = -\frac{1}{3}x^2$ and $y = -3x^2$ in the same coordinate plane.
2. The graph of an equation of the form $y = ax^2$ passes through the point (3, −8). Find the value of a.

For Questions 3–5, solve each equation. Give solutions to the nearest tenth when necessary.

3. $4x^2 = 36$ 4. $-2x^2 = -20$ 5. $5x^2 = 75$
6. **Writing** Explain how to obtain the graph of $y = (x + 3)^2 - 2$ from the graph of $y = x^2$. Then describe the graph of $y = (x + 3)^2 - 2$.
Sample answer: The graph of $y = (x + 3)^2 - 2$ can be obtained by translating the graph of $y = x^2$ down 2 units and then 3 units to the left. The graph is a parabola with vertex (−3, −2) that opens upward and is congruent to the graph of $y = x^2$.
7. **Open-ended** Find a quadratic function that has a maximum value of 3 and $x = -2$ as the line of symmetry for its graph.
Any equation of the form $y = -a(x + 2)^2 + 3$ where $a > 0$; sample: $y = -(x + 2)^2 + 3$
8. Write an equation in the form $y = a(x - h)^2 + k$ for a parabola that has vertex (3, 4) and passes through the point (1, 2).
9. Find the x-intercepts, to the nearest hundredth, for the parabola described in Question 8.
10. Rewrite the equation $y = (2x + 4)(3 - x)$ in intercept form. Then graph the equation and label the vertex. (0.5, 12.5)

ANSWERS

1. See question.
2. $-\frac{8}{9}$
3. −3, 3
4. −3.2, 3.2
5. −3.9, 3.9
6. See question.
7. See question.
8. $y = -\frac{1}{2}(x - 3)^2 + 4$
9. 0.17 and 5.83
10. $y = -2(x + 2)(x - 3)$

Assessment Book, ALGEBRA 2: EXPLORATIONS AND APPLICATIONS

27

Chapter 5 Assessment
Form B Chapter Test

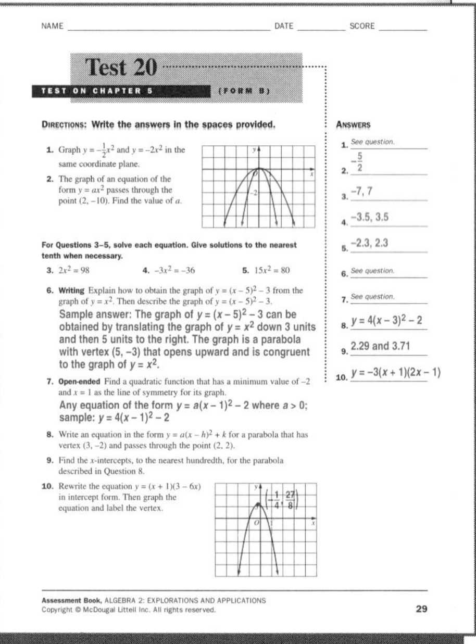

NAME ______ DATE ______ SCORE ______

Test 20
TEST ON CHAPTER 5 (FORM B)

DIRECTIONS: Write the answers in the spaces provided.

1. Graph $y = -\frac{1}{2}x^2$ and $y = -2x^2$ in the same coordinate plane.
2. The graph of an equation of the form $y = ax^2$ passes through the point (2, −10). Find the value of a.

For Questions 3–5, solve each equation. Give solutions to the nearest tenth when necessary.

3. $2x^2 = 98$ 4. $-3x^2 = -36$ 5. $15x^2 = 80$
6. **Writing** Explain how to obtain the graph of $y = (x - 5)^2 - 3$ from the graph of $y = x^2$. Then describe the graph of $y = (x - 5)^2 - 3$.
Sample answer: The graph of $y = (x - 5)^2 - 3$ can be obtained by translating the graph of $y = x^2$ down 3 units and then 5 units to the right. The graph is a parabola with vertex (5, −3) that opens upward and is congruent to the graph of $y = x^2$.
7. **Open-ended** Find a quadratic function that has a minimum value of −2 and $x = 1$ as the line of symmetry for its graph.
Any equation of the form $y = a(x - 1)^2 - 2$ where $a > 0$; sample: $y = 4(x - 1)^2 - 2$
8. Write an equation in the form $y = a(x - h)^2 + k$ for a parabola that has vertex (3, −2) and passes through the point (2, 2).
9. Find the x-intercepts, to the nearest hundredth, for the parabola described in Question 8.
10. Rewrite the equation $y = (x + 1)(3 - 6x)$ in intercept form. Then graph the equation and label the vertex. $\left(-\frac{1}{4}, \frac{27}{8}\right)$

ANSWERS

1. See question.
2. $-\frac{5}{2}$
3. −7, 7
4. −3.5, 3.5
5. −2.3, 2.3
6. See question.
7. See question.
8. $y = 4(x - 3)^2 - 2$
9. 2.29 and 3.71
10. $y = -3(x + 1)(2x - 1)$

Assessment Book, ALGEBRA 2: EXPLORATIONS AND APPLICATIONS

29

CHAPTER

5 Assessment

VOCABULARY QUESTIONS

For Questions 1 and 2, complete each paragraph.

1. The equation $y = 2(x + 1)^2 - 3$ is a __?__ function in __?__ form. The function has a __?__ value of −3. The graph of the equation is a __?__ with vertex (−1, −3) as the __?__ and with $x = -1$ as the __?__ .
2. The equation $y = ax^2 + bx + c$ is the __?__ form of a quadratic function. The expression $b^2 - 4ac$ is called the __?__ , and its value can be used to find the number of __?__ that the graph of $y = ax^2 + bx + c$ has.

SECTIONS 5.1, 5.2, *and* 5.3

3. Graph $y = -0.5x^2$.
4. The graph of an equation of the form $y = ax^2$ passes through the point (−2, 12). Find the value of a.
5. The area of a circular garden is 30 ft². How much fencing is needed to enclose the garden? (*Hint*: Use the formulas $A = \pi r^2$ and $C = 2\pi r$.)
6. **Writing** Explain how to obtain the graph of $y = -(x - 2)^2 + 1$ from the graph of $y = -x^2$. Then describe the graph of $y = -(x - 2)^2 + 1$.
7. **Open-ended Problem** Find a quadratic function that has 0 as its minimum value and $x = -4$ as the line of symmetry for its graph.
8. a. Write an equation in the form $y = a(x - h)^2 + k$ for a parabola that has vertex (−1, −7) and passes through the point (1, 1).
   b. Find the coordinates of the points, to the nearest hundredth, where the parabola from part (a) intersects the x-axis.
9. Rewrite $y = (2 - x)(2 + 2x)$ in intercept form. Then graph the equation and label the vertex.
10. Find an equation for each graph.

a.
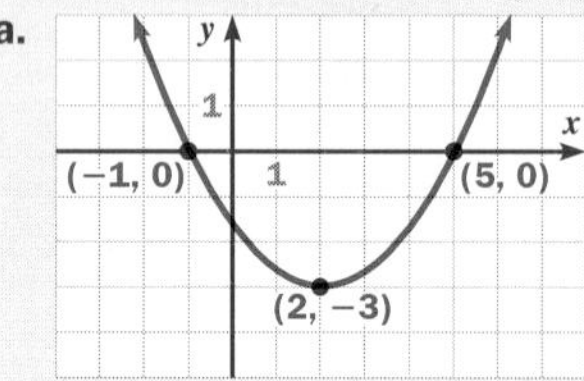

b.
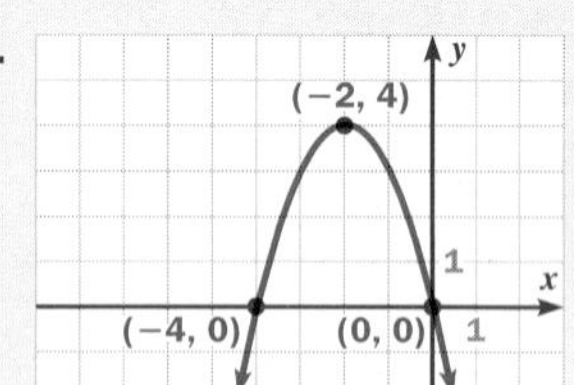

ANSWERS Chapter 5

Assessment

1. quadratic; vertex; minimum; parabola; vertex; line of symmetry
2. general; discriminant; x-intercepts
3. 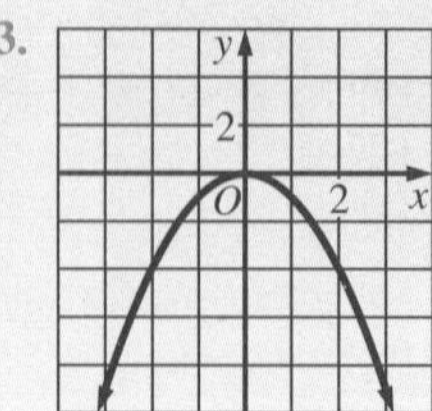

4. 3
5. 19.4 ft
6. Translate the graph 2 units right and up 1 unit; the graph is a parabola that opens down with vertex at (2, 1). It has the same shape as the graph of $y = -x^2$.
7. Answers may vary. An example is given. $y = a(x + 4)^2$ for any positive number a
8. a. $y = 2(x + 1)^2 - 7$
   b. −2.87; 0.87

11. The equation $h = -16t^2 + 20t + 6$ gives the height h, in feet, of a basketball as a function of time t, in seconds.

  a. What is the maximum height the ball reaches?

  b. At what time does the ball hit the ground? What method did you use to find the answer?

SECTIONS 5.4, 5.5, *and* 5.6

Write each function in vertex form.

12. $y = 2x^2 + 9x + 3$ 13. $y = -x^2 + 6x + 3$ 14. $y = \frac{1}{4}x^2 - 2x + 11$

15. **ELECTRICITY** In a 120 volt electrical circuit that has a resistance of 16 ohms, the power P, in watts, is given by $P = 120I - 16I^2$, where I is the amount of current flowing, in amperes.

  a. Write the equation in vertex form.

  b. What are the coordinates of the vertex of the graph of the equation? What do they mean in terms of the situation?

Find the *x*-intercepts, if any, for the graph of each quadratic function.

16. $y = 4x^2 - 20x + 25$ 17. $y = 2x^2 + 3x + 5$ 18. $y = 3x^2 - 16x - 12$

19. **PHYSICS** The equation $y = 0.00139x^2 + 0.0304x + 1.25$ gives the time y, in seconds, that it takes a car to accelerate to a given speed x, in miles per hour. To the nearest integer, what speed can the car reach in 4 s?

Write each quadratic function as a product of factors.

20. $y = x^2 - 6x + 8$ 21. $y = 6x^2 - 7x - 5$ 22. $y = 4x^2 + 27x + 18$

PERFORMANCE TASK

23. The game of Leapfrog involves moving pegs along a line of holes. The game begins with a group of n pegs of the same color at one end of the line, a group of n pegs of another color at the other end, and one hole between the two groups. The goal of the game is to interchange the two groups of pegs by moving individual pegs in one of the following ways:

  (1) Move a peg into an adjacent hole.
  (2) Move a peg of one color over an adjacent peg of the other color to get the peg into a hole on the other side.

Play the game with various numbers of pegs. (You may wish to use coins instead of pegs, and a line of squares drawn on paper instead of a line of holes drilled in wood.) Make a table relating n, the number of pegs of each color, to m, the *minimum* number of moves needed to interchange the colors. Then find a functional relationship between m and n. (*Hint*: m is a quadratic function of n.)

Chapter 5 Assessment
Form C Alternative Assessment

Chapter 5

ALTERNATIVE ASSESSMENT

1. **Open-ended Problem** Write a quadratic equation, if possible, for a parabola that has the following intercepts.

  a. one x-intercept b. two x-intercepts
  c. three x-intercepts d. no x-intercepts
  e. one y-intercept f. two y-intercepts

2. Write a brief summary of all the information that you can get by knowing the value of the discriminant of a quadratic equation.

3. A quadratic equation can be solved by factoring or by using the quadratic formula. Compare these two methods. Discuss the advantages and disadvantages of each method. What factors should you consider when you are trying to decide which method to use?

4. **Performance Task** Find a quadratic equation that has the roots 3 and −6. Find two more quadratic equations that have the same roots. Could a linear equation have these roots? Could a cubic equation have these roots? Explain why.

5. **Project** Use an eye dropper and colored water to make ink blots on a paper towel. Count the number of drops used to make ten blots of different sizes. Place a transparency with a square centimeter grid on each blot and estimate its area. Graph your data, using the number of drops as an independent variable and the area as the dependent variable. Use quadratic regression to find an equation of a line of fit. Compare your equation to the formula for the area of a circle.

6. a. Find an equation of the parabola that passes through (−1, 12), (2, 9), and (1, 6). Show that your equation is correct by graphing the equaiton and the three given points on a graphing calculator.

  b. Is it possible to find three points that would not lie on some parabola? Explain your answer.

118 **Assessment Book,** ALGEBRA 2: EXPLORATIONS AND APPLICATIONS

9. $y = -2(x - 2)(x - (-1))$

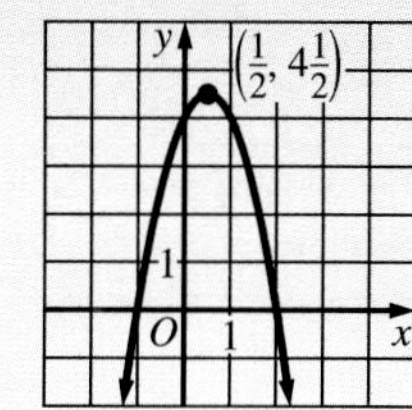

10. a. $y = \frac{1}{3}(x + 1)(x - 5)$

  b. $y = -x(x + 4)$

11. a. 12.25 ft

  b. 1.5 s; Answers may vary. An example is given. I used the quadratic formula to solve the equation $-16t^2 + 20t + 6 = 0$.

12. $y = 2\left(x + \frac{9}{4}\right)^2 - \frac{57}{8}$

13. $y = -(x - 3)^2 + 12$

14. $y = \frac{1}{4}(x - 4)^2 + 7$

15. a. $P = -16(I - 3.75)^2 + 225$

  b. (3.75, 225); The maximum power, 225 watts corresponds to a current of 3.75 amperes.

16. 2.5

17. no x-intercepts

18. $-\frac{2}{3}$; 6

19. 35 mi/h

20. $y = (x - 4)(x - 2)$

21. $y = (3x - 5)(2x + 1)$

22. $y = (4x + 3)(x + 6)$

23. $m = n^2 + 2n$

CHAPTER

6 Investigating Data

Overview

Connecting to Prior and Future Learning

⇔ Different types of data are discussed in Section 6.1. Students will find the review of data displays provided on pages 791–793 of the **Student Resources Toolbox** to be helpful as they study this section.

⇔ Students continue their study of data by learning how to collect data using surveys and samples. The skills students acquire in these areas will be useful to them in future studies of statistics.

⇔ Chapter 6 ends with a study of how to display, analyze, and describe the variation of data. Students also learn how to find the margin of error for a sample. These concepts will enable students to make decisions about the data they collected in Section 6.3.

Chapter Highlights

Interview with Donna Cox: The use of mathematics in computer graphics is illustrated in this interview. Students can explore this relationship further by completing the related exercises on pages 255 and 271.

Explorations can be found in Section 6.2, where students compare ways of asking a question, in Section 6.4, where they use scissors and graph paper to make a box plot, and in Section 6.6, where a calculator with a random number generator is used to make a sample distribution.

The Portfolio Project: Students find a public opinion survey published in a newspaper or magazine and analyze the presentation and conclusions of the survey.

Technology: In Chapter 6, students use a graphing calculator to generate random numbers, draw histograms and box plots, and find the standard deviation of a set of data. Spreadsheets or statistical software can also be used to find the standard deviation.

Objectives

Section	Objectives	NCTM Standards
6.1	• Organize information and classify data. • Analyze and interpret data.	1, 2, 3, 4, 10, 12
6.2	• Write good survey questions. • Judge the accuracy of responses to other people's surveys.	1, 2, 3, 4, 10
6.3	• Choose a representative sample. • Recognize biased samples in surveys.	1, 2, 3, 4, 10
6.4	• Make histograms and box plots.	1, 2, 3, 4, 10, 12
6.5	• Find and interpret the range, interquartile range, and standard deviation of a data set. • Determine the variability of data.	1, 2, 3, 4, 10, 12
6.6	• Find the margin of error for a sample proportion. • Make decisions about opinion poll results.	1, 2, 3, 4, 10, 12

INTEGRATION

Mathematical Connections	6.1	6.2	6.3	6.4	6.5	6.6
algebra	**237–243***	**244–249**	**250–257**	**258–265**	**266–272**	**273–279**
data analysis, probability, discrete math	**237–243**	**244–249**	**250–257**	**258–265**	**266–272**	**273–279**
patterns and functions	243					279
logic and reasoning	237, 238, 241–243	245–249	251–257	261–265	266, 267, 270–272	274, 276–279
Interdisciplinary Connections and Applications						
history and geography				263, 264		
biology and earth science			254		267, 268, 270	
business and economics	243			259, 265		275, 276
arts and entertainment					272	
sports and recreation	241					
political science and government		248	256			
archaeology, medicine, polling, employment, education	239		254		269	275–277

* ***Bold page numbers*** *indicate that a topic is used throughout the section.*

TECHNOLOGY

Section	*opportunities for use with* Student Book	Support Material
6.1	graphing calculator spreadsheet software McDougal Littell Mathpack *Stats!*	**Technology Book:** Spreadsheet Activity 6 **Stats! Activity Book:** Activities 10, 12, and 13
6.2		
6.3	graphing calculator	
6.4	graphing calculator McDougal Littell Mathpack *Stats!*	**Stats! Activity Book:** Activities 5, 7, 17, and 20
6.5	graphing calculator spreadsheet software McDougal Littell Mathpack *Stats!*	**Technology Book:** Spreadsheet Activity 6 **Stats! Activity Book:** Activities 13 and 19
6.6	graphing calculator McDougal Littell Mathpack *Stats!*	**Technology Book:** Calculator Activity 6 **Stats! Activity Book:** Activity 18

Planning Guide

Regular Scheduling (45 min)

Section	Materials Needed	Core Assignment	Extended Assignment	*exercises that feature* Applications	Communication	Technology
6.1	graph paper	1–11, 13–17, 23–31	1–31	8–11, 20–22	10–19, 23	
6.2		1–12, 15–18, 20–26	1–26	12–14, 15	9–11, 13, 15, 20	
6.3	calculator with random number generator	**Day 1:** 1–25 **Day 2:** AYP*	**Day 1:** 1–25 **Day 2:** AYP	6–19	2–6, 10, 12, 19, 20–23	25
6.4	scissors, graph paper	**Day 1:** 1–9, 11, 12 **Day 2:** 14, 15, 17–26	**Day 1:** 1–13 **Day 2:** 14–26	7–10 14–19	1, 13 19, 20	
6.5	graphing calculator or statistical software, graph paper	1–13, 17–19, 23–27	1–27	9–16, 20–22	8, 18, 23	
6.6	calculator with random number generator	**Day 1:** 1–3, 5–11 **Day 2:** 13, 15–20, AYP	**Day 1:** 1–11 **Day 2:** 12–20, AYP	1–3, 7–10	8 12, 15	11 14
Review/ Assess		**Day 1:** 1–11 **Day 2:** 12–20 **Day 3:** Ch. 6 Test	**Day 1:** 1–11 **Day 2:** 12–20 **Day 3:** Ch. 6 Test			
Portfolio Project		Allow 2 days.	Allow 2 days.			

Yearly Pacing (with Portfolio Project)	**Chapter 6 Total**	**Chapters 1–6 Total**	**Remaining**	**Total**
	14 days	80 days	80 days	160 days

Block Scheduling (90 min)

	Day 34	Day 35	Day 36	Day 37	Day 38	Day 39	Day 40	Day 41
Teach/Interact	Ch. 5 Test 6.1	6.2: Exploration, page 244 6.3	Continue with 6.3 6.4: Exploration, page 258	Continue with 6.4 6.5	6.6: Exploration, page 273	Review Port. Proj.	Review Port. Proj.	Ch. 6 Test 7.1
Apply/Assess	**Ch. 5 Test** **6.1:** 1–11, 13–17, 23–31	**6.2:** 1–12, 15–18, 20–26 **6.3:** 1–25	**6.3:** AYP* **6.4:** 1–9, 11, 12	**6.4:** 14, 15, 17–26 **6.5:** 1–13, 17–19, 23–27	**6.6:** 1–3, 5–11, 13, 15–20, AYP	**Review:** 1–11 **Port. Proj.**	**Review:** 12–20 **Port. Proj.**	**Ch. 6 Test** **7.1:** 1–23, 33–38

NOTE: A one-day block has been added for the Portfolio Project—timing and placement to be determined by teacher.

Yearly Pacing (with Portfolio Project)	**Chapter 6 Total**	**Chapters 1–6 Total**	**Remaining**	**Total**
	7 days	$40\frac{1}{2}$ days	$41\frac{1}{2}$ days	82 days

* ***AYP*** *is Assess Your Progress.*

Lesson Support

Section	Practice Bank	Study Guide*	Assessment Book*	Visuals	Explorations Lab Manual	Lesson Plans	Technology Book
6.1	33	6.1		Warm Up 6.1	Masters 1, 2 Add. Expl. 9	6.1	Spreadsheet Act. 6
6.2	34	6.2		Warm-Up 6.2		6.2	
6.3	35	6.3	Test 21	Warm-Up 6.3	Add. Expl. 10	6.3	
6.4	36	6.4		Warm-Up 6.4 Folder 6	Master 1	6.4	
6.5	37	6.5		Warm-Up 6.5 Folder 6	Add. Expl. 11 Master 1	6.5	Spreadsheet Act. 6
6.6	38	6.6	Test 22	Warm-Up 6.6		6.6	Calculator Act. 6
Review Test	39	Chapter Review	Tests 23–25 Alternative Assessment			Review Test	

*__Spanish versions__ of *Study Guide* and *Assessment Book* are available.

Chapter Support

- Course Guide
- Lesson Plans
- Portfolio Project Book
  Additional Project 3: School Survey
- Preparing for College Entrance Tests
- Multi-Language Glossary
- *Test Generator* Software
- Professional Handbook

Software Support

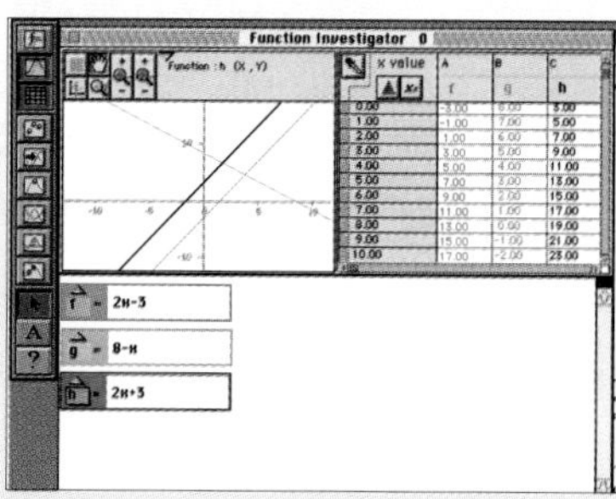

McDougal Littell Mathpack
Stats!

Internet Support

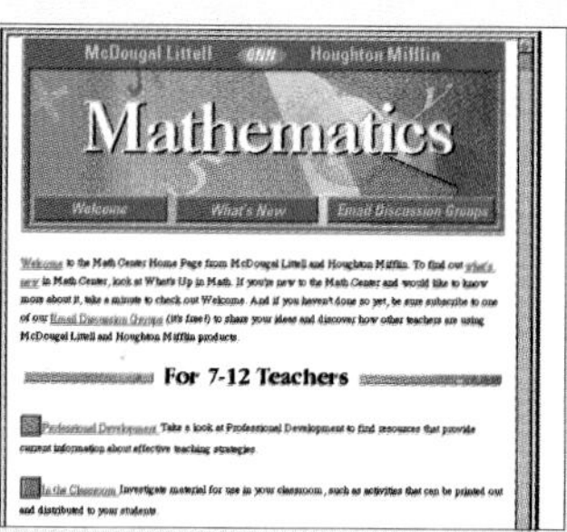

http://www.hmco.com
Next go to McDougal Littell; then the Education Center; then Secondary Math.

Outside Resources

Books, Periodicals

"Information from Surveys," from the *Quantitative Literacy Series*: written by members of the Joint Committee on the Curriculum on Statistics and Probability of the American Statistical Association and the National Council of Teachers of Mathematics. Dale Seymour Publications.

Sanders, Mark. "Technology Tips: Teaching Statistics with Computer Networks." *Mathematics Teacher* (January 1996): pp. 70–72.

Activities, Manipulatives

Somers, Kay, John Dilendik, and Bettie Smolansky. "Class Activities with Student-Generated Data." *Mathematics Teacher* (February 1996): pp. 105–107.

Breuningsen, Chris, Bill Bower, Linda Antinone, and Elisa Breuningsen. "Jump!" *Real-World Math with the CBL System.* Activity 22: pp. 119–123. Texas Instruments, 1995.

Software

Department of Mathematics and Computer Science of the North Carolina School of Science and Mathematics. *Data Analysis.* Materials and software. IBM. Reston, VA: NCTM.

f(g) Scholar. IBM-comp. Southampton, PA: Future Graph.

Videos

Social Studies School Service. "Fact or Fiction?" The 20/20 program attempts to discover the truth behind surveys, statistics, and scientific studies.

Internet

Investigate Internet Math Projects at: http://forum.swarthmore.edu/projects.html

Interview Notes

Background

Computer Graphics
Computer graphics may refer to the images created by a computer or to the process by which computers draw, color, shade, and manipulate images. Either way, computer graphics allow people to gather, display, and understand information more quickly and effectively than in the past.

Computer graphics allows people from many professions to perform tasks and display data in ways that they previously had not been able to do. For example, businesses can prepare sales charts and graphs; engineers can create and test designs for automobiles and airplanes; architects can view building designs from any angle; scientists can track weather systems or describe astronomical phenomena; physicians can see inside the body; and artists can produce cartoons or special effects in movies or video games. The field is continuously growing and new uses are being found all the time.

Donna Cox
Donna Cox and her colleagues based their intricate graphics on 700,000 numbers and 68 years of literature on insect pathology. A run of the data involved more than 1 billion calculations and took about 6.7 minutes. The model itself is highly complex and includes such factors as the 40 life stages of healthy corn borers, movement from plant to plant, the size of the larvae when the disease-causing parasite attacked them, and different modes of transmission.

In addition to producing an image of a corn field, Donna Cox has also used computer graphics to design a thunderstorm simulation. Cox and her students helped to design this simulation to improve weather forecasting. In it, a storm cloud develops according to wind speed, water content, air pressure, and temperature as calculated by the supercomputer. In this simulation, however, she uses symbols to represent variables (wind speed, water content, air pressure, and temperature) and the number or size of the symbol to show the magnitude of the variable.

CHAPTER

6 Investigating Data

Visualizing the data

INTERVIEW **Donna Cox**

"There is natural beauty in mathematics and physics."

As a child growing up in Oklahoma, Donna Cox found herself drawn to the seemingly divergent worlds of art and science. By the time she started college, she still could not decide which subject to pursue. Instead, she "kept bouncing back and forth between the two." In graduate school, she finally discovered computer graphics—a field that draws almost equally on art, science, and mathematics.

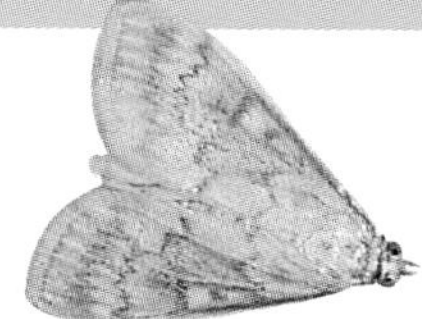

European corn borer adult

Bringing the Data to Life

Now, as an art professor at the University of Illinois and an associate director at the National Center for Supercomputing Applications, Cox specializes in *scientific visualizations*—using pictures or animated movies to display the information contained in raw, numerical data.

She joins forces with scientists to figure out ways of making data more readily accessible. Cox calls these cooperative ventures "Renaissance teams," a term she coined in 1986. The term refers to the Renaissance era, a period of notable intellectual and artistic achievement from the 14th through the 16th centuries.

European corn borer larva

Donna Cox's team created the graphic at the left (*behind Cox*) using mathematics and an experimental computer program. The image that resulted, named *Etruscan Venus*, shows a mathematical figure known as a *Klein bottle* in three-dimensional space.

A Renaissance Team

A good example is the work Cox did with David Onstad, an entomologist at the University of Illinois, and Edward Kornkven, a computer scientist at the South Dakota School of Mines and Technology. Onstad created a mathematical model of the larval and adult forms of the European corn borer, an insect that damages corn plants. He wanted to see how the larval and the adult populations were affected by an insect disease and how this, in turn, affected the corn plants.

Kornkven programmed these interactions on a supercomputer, which produced billions of numbers covering reams and reams of paper. Obviously, there was no shortage of printed output, but extracting useful information from it was a tedious process.

Interview Notes

Background

The European Corn Borer

The European corn borer is the larva of a night-flying moth. It belongs to the snout moth family, Pyralidae, and is also known as *Ostrinia nubilalis.* It came to the United States in 1910 from Europe and spread through the Eastern and Midwestern states. The corn borer eats mostly corn and sorghum, but it also attacks plants such as celery, potatoes, beans, flowers, and weeds. The moths spread by laying their eggs on corn leaves in early June. The larvae will eat the leaves and tassels of the young corn plant and then feed on the stems and ears as they grow larger. In the winter, old corncobs, stems, and stubble provide a home for the insect larvae.

To destroy the corn borers, farmers feed cornstalks to livestock, or shred or burn them in the winter. They may also plant late to avoid the first flight of the moths, and use hybrid plants that are not affected by the corn borer. Some farmers also use insecticides.

Second-Language Learners

Students learning English may not be familiar with a number of terms and expressions used in this interview, such as *seemingly divergent worlds of art and science,* which means "areas of art and science which seem to be different when in fact they have a lot in common." Other terms that may be unfamiliar are *entomologist* and *ream.* If necessary, explain that an entomologist is a person engaged in the scientific study of insects, and a ream of paper is "a measured quantity of paper consisting of 500 sheets." Also, explain that the expression *tedious process* means "a lengthy, slow, and boring task."

Interview Notes

Mathematical Connection

Computer graphics are a good way to display large amounts of data. Yet, even though these large amounts of data are simplified, they still may be too numerous to use in an analysis. In this situation, taking samples of the data and analyzing the samples can be a good way to study the data. This idea is explored in Section 6.3. In this section, students use computer images of corn borer insects and sampling methods to analyze data. Another way of displaying data is in a matrix. In Section 6.5, students use data displayed in a matrix to draw box plots and then analyze the data from the plots.

Explore and Connect

Writing
Students should use the ideas of categorical data and numerical data in their analysis.

Research
If possible, students should interview local farmers or agricultural experts. Also, information may be found in the school or local libraries. Encyclopedias are a good source for information.

Project
Students should use different colors in their visual display to distinguish between temperature and rainfall. A good display for the data would be a bar graph. Once the displays are finished, you may wish to hang some of them in class.

Art Is Communication

The image below is an example of what Cox produced using Onstad and Kornkven's data. The colors of the spikes refer to the health status of larval or adult corn borers, whereas the heights of the spikes refer to the relative numbers of these insects on each plant. In this image, color is used to present *categorical data* and height is used to present *numerical data.*

Cox says, "Artists can make a significant contribution to these technical projects, because art is, in essence, a form of communication. Taking an abstraction such as scientific data and putting it in a visual (or audio) form is something we can do quite well."

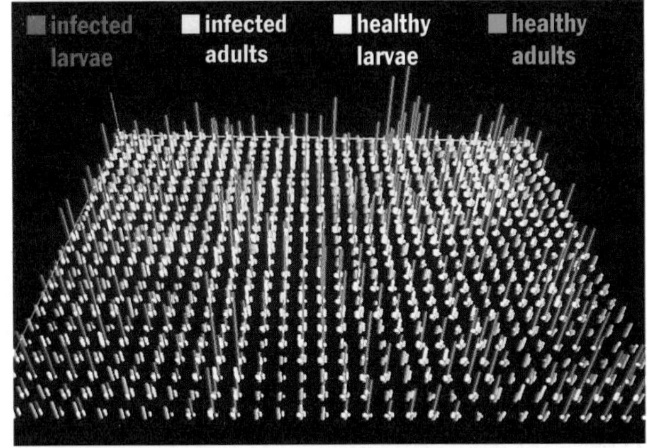

This image is of a cornfield where each group of spikes represents a single corn plant. Disease-infected corn borer larvae and adults do minimal damage to the plants, so only those plants with tall white and blue spikes are being significantly damaged.

"Art is, in essence, a form of communication."

EXPLORE AND CONNECT

Donna Cox works on her computer.

1. Writing Examine the cornfield shown in the computer image above. What do you notice about the distributions of the various categories of corn borers?

2. Research Find out what insects are of concern to farmers near you. What kinds of data do the farmers and agricultural specialists use to make decisions about controlling the insects?

3. Project Use an almanac to find the average monthly temperature and the average monthly rainfall for five cities. Choose cities from different regions of the United States. Transform your numerical data into a visual display.

Mathematics & Donna Cox

In this chapter, you will learn more about ways to describe data used in agriculture.

Related Exercises

Section 6.3
- **Exercises 13–19**

Section 6.5
- **Exercises 14–16**

Section

6.1 Types of Data

Learn how to...

- organize information and classify data

So you can...

- analyze data, such as pizza delivery times

HAYDEN Pizza SHOP
We deliver your pizza in 30 minutes or less
OR YOUR ORDER IS FREE!

Have you ever seen an ad like the one shown? Did you ever wonder how many pizzas are actually given away? The owner of the Hayden Pizza Shop is thinking of placing the ad at the right. The owner uses this method to make a decision:

- The staff keeps a journal like the one shown below.
- The owner makes a graph of all the data gathered by the staff.

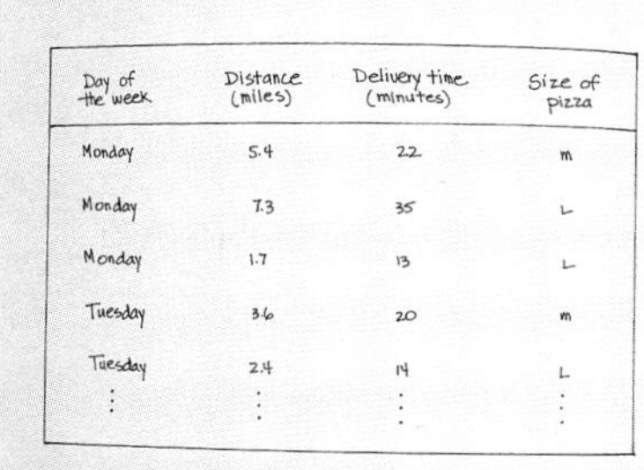

Day of the week	Distance (miles)	Delivery time (minutes)	Size of pizza
Monday	5.4	22	m
Monday	7.3	35	L
Monday	1.7	13	L
Tuesday	3.6	20	m
Tuesday	2.4	14	L
⋮	⋮	⋮	⋮

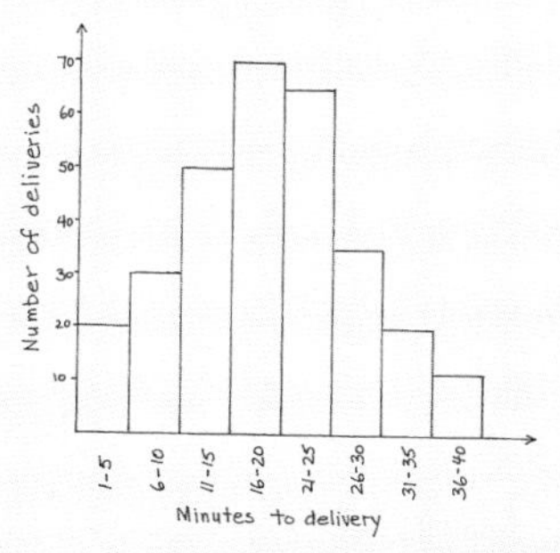

THINK AND COMMUNICATE

1. Look at the journal kept by the staff. Why do you think the owner wants to know the day of delivery? the pizza size?
2. Look at the graph made by the owner. About how long does it take to deliver most pizzas? How many deliveries take longer than 30 minutes?
3. If you were the owner of the Hayden Pizza Shop, would you place the ad shown above? Use the owner's graph to support your answer.

The method used by the owner of the Hayden Pizza Shop can be summarized into five steps. These steps will help you gather data and make conclusions from the data in later sections of this chapter.

Statistical Investigations

1. Consider the problem or question.
2. Collect and record the data.
3. Describe and display the data.
4. Interpret the data.
5. Summarize the findings and make conclusions.

Plan⇔Support

Objectives

- Organize information and classify data.
- Analyze and interpret data.

Recommended Pacing

Core and Extended Courses
Section 6.1: 1 day

Block Schedule
Section 6.1: $\frac{1}{2}$ block
(with Chapter 5 Test)

Resource Materials

Lesson Support
Lesson Plan 6.1
Warm-Up Transparency 6.1
Practice Bank: Practice 33
Study Guide: Section 6.1
Explorations Lab Manual: Additional Exploration 9, Diagram Masters 1, 2

Technology
Technology Book: Spreadsheet Activity 6
Graphing Calculator
Spreadsheet Software
McDougal Littell Mathpack *Stats! Activity Book:* Activities 10, 12, and 13
Internet: http://www.hmco.com

Warm-Up Exercises

1. There are 415 students in the junior class at Richmont High School. If 203 of those students play an intramural sport, what percent of the junior class plays an intramural sport? about 49%

Solve the following proportions.

2. $\frac{x}{25} = \frac{17}{500}$ 0.85
3. $\frac{19}{40} = \frac{475}{x}$ 1000
4. Find the mean of the test scores: 84, 91, 87, 73, and 76. 82.2
5. Find 38% of 45,000. 17,100

ANSWERS Section 6.1

Think and Communicate

1–3. Answers may vary. Examples are given.

1. to determine on which days delivery might take longer than 30 minutes; to determine the cost of pizzas that are delivered late and, therefore, are free
2. Nearly 60% of the deliveries take between 11 and 25 min; 32 deliveries.
3. No; more than 10% of the deliveries take longer than 30 minutes, which would mean a considerable expense for the owner.

Teach⇔Interact

Section Notes

Using Technology
Using spreadsheets, students can easily create charts that display categorical data. The formatting functions for charts allow students to convert the data to different types of charts. By creating several different charts and using available formatting features, students can determine which type of chart is most appropriate for a given set of data.

Teaching Tip
You may want to emphasize the difference between a histogram and a bar graph. In a histogram, the bars touch and the width of each bar represents an equal range of numerical data values. In a bar graph, the bars do not touch and each bar represents a number of nonnumerical things, like days of the week.

Additional Example 1

Tell whether the data that can be gathered about each variable are *categorical* or *numerical.* Then describe the categories or numbers.

a. diameters of trees
The data are numerical. Tree diameters range from a few millimeters when they are seedlings to about 15 m.

b. telephone numbers
The data are categorical. Telephone numbers are number codes.

c. photographic film sizes
The data are categorical. Some standard film sizes are 110 mm, 126 mm, and 35 mm.

Think and Communicate

Students who have difficulty identifying data that appear to be numerical but are actually categorical may find it helpful to ask whether it makes sense to do arithmetic operations with the data. If not, the numbers represent categorical data.

In the journal kept by the staff of the Hayden Pizza Shop, two types of data were gathered. The distances and delivery times are examples of *numerical data.* **Numerical data** are counts or measurements. The days of the week and pizza sizes are examples of *categorical data.* **Categorical data** are names or labels.

Toolbox p. 791
Data Displays

Numerical data are often displayed in a *histogram* or a *box plot.*

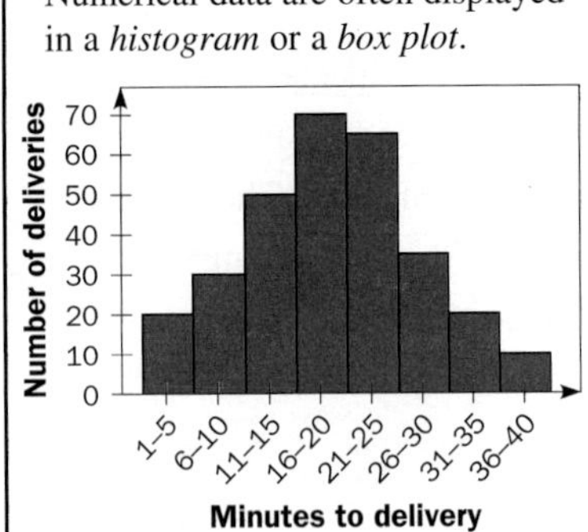

In a histogram, the bars touch and the horizontal axis is a number line.

Categorical data are often displayed in a bar graph or a circle graph.

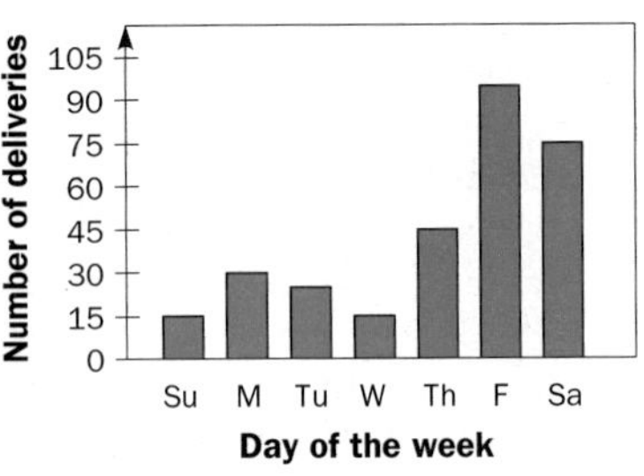

In a bar graph, the bars do not touch and the horizontal axis is not a number line.

The vertical axis for both a histogram and a bar graph is a number line from which you can read the height of each bar.

EXAMPLE 1

Tell whether the data that can be gathered about each variable are *categorical* or *numerical.* Then describe the categories or numbers.

a. blood type **b.** height of an adult **c.** grade point average

SOLUTION

a. The data are *categorical.* Blood types are classified as O+, O−, A+, A−, B+, B−, AB+, and AB−.

b. The data are *numerical.* Most adult heights range from 4 ft to 7 ft.

c. The data are *numerical.* A grade point average usually ranges between 0.0 and 4.0.

Some data may appear to be numerical when they are really categorical. For example, you can describe pizza sizes as 8 in., 12 in., and 16 in., but these numbers are not obtained by actually measuring pizzas. Instead, they represent categories, just as small, medium, and large do.

THINK AND COMMUNICATE

Tell whether the data that can be gathered about each variable are *categorical* or *numerical.* Explain your reasoning.

4. shoe size **5.** words per minute **6.** month of the year

Think and Communicate

4. numerical; Shoe sizes are based on measurement of the foot with average adult sizes, for example, ranging from 3 to 15.
5. numerical; Both the number of words and the number of minutes are counted.
6. categorical; The months are classified by their names.

EXAMPLE 2 Application: Medicine

At the Foley High School blood drive, 200 people each donated a pint of blood. The number of pints of each blood type collected is shown in the table.

Blood type	O+	O−	A+	A−	B+	B−	AB+	AB−
Number of donors	77	6	74	14	11	12	6	0

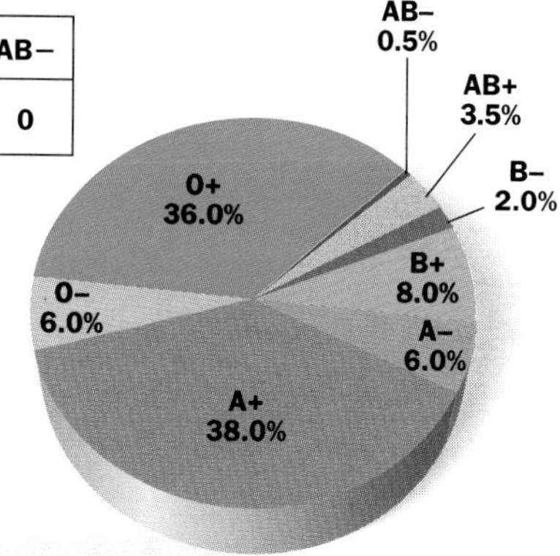

a. Use the results above to estimate the percent of people in the United States population with each blood type.

b. Compare the estimates from part (a) with those made by the American Red Cross, shown at the right. Does the school collection reflect the percent of people in the general population with each blood type? Explain.

SOLUTION

a. Find the percent of the 200 donors with each blood type. A spreadsheet may be helpful.

Blood Donors

	A	B	C	D	E	F	G	H	I
1	Blood type	O+	O–	A+	A–	B+	B–	AB+	AB–
2	Number of donors	77	6	74	14	11	12	6	0
3	Percent of donors	38.5	3	37	7	5.5	6	3	0

b. Almost all of the estimates from the school collection are within 3 percentage points of the estimates for the general population. The school estimate for the number of people in the general population with B− blood was 4 percentage points higher than the estimate made by the American Red Cross.

Since the American Red Cross estimates are based on donations from a much larger group of people, their estimates more closely reflect the figures for the general population. Therefore, the proportion of people with B− blood seems to be higher in the school than in the general population.

BY THE WAY

Dr. Charles Richard Drew was an African-American physician who conducted research in the area of blood plasma and its preservation. The blood banks he organized during World War II helped save millions of lives.

The people who donated at the Foley High School blood drive represent a *sample* of the entire *population*.

A complete group is a **population**. For the American Red Cross data, the population consists of the people who live in the United States.

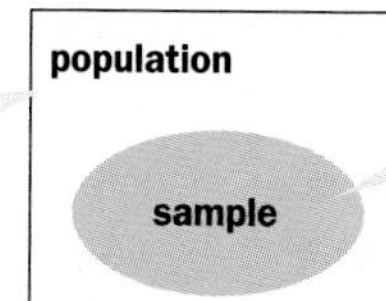

A part of a group is a **sample**. There were 200 people in the sample at the Foley High School blood drive.

Technology Note

Students can use spreadsheets to create a pie chart. By applying both data labels and percents to each pie slice from a format menu, they can create a chart to compare with the chart in Example 2. Students may also choose to create a three-dimensional chart from the chart gallery. They can then use a formatting option to enlarge one slice of the pie in order to emphasize that slice.

Teach⇔Interact

Additional Example 2

The metabolic rate, the rate at which the body consumes energy, was recorded for 150 women athletes at Alverne High School. The rates, which were measured in calories burned per 24 hours, are shown in the table.

Rate	No. of Women Ages 16–18
900–999	11
1000–1099	12
1100–1199	26
1200–1299	29
1300–1399	32
1400–1499	28
1500–1599	12

a. Use the results to estimate the percent of women between the ages of 16 and 18 within each range of metabolic rate.

Find the percent of the 150 participants with each rate.

900–999: $\frac{11}{150} \approx 7.3\%$

1000–1099: $\frac{12}{150} = 8\%$

1100–1199: $\frac{26}{150} \approx 17.3\%$

1200–1299: $\frac{29}{150} \approx 19.3\%$

1300–1399: $\frac{32}{150} \approx 21.3\%$

1400–1499: $\frac{28}{150} \approx 18.7\%$

1500–1599: $\frac{12}{150} = 8\%$

b. Compare your estimates from part (a) to those made by a university from a study of 1000 randomly chosen women between the ages of 16 and 18. Do the school data reflect the percent of people in the general population within each range of metabolic rate as indicated by the university study? Explain.

Rate	% of Population
900–999	15.3
1000–1099	17.6
1100–1199	18.6
1200–1299	16.8
1300–1399	15.0
1400–1499	12.7
1500–1599	4.0

Most of the estimates from the school data are more than 5 percentage points from the estimates of the university study. Since the school study was based on a group of women athletes, their rates may not accurately reflect the rates of the larger group from the university study.

Teach⇔Interact

Checking Key Concepts

Communication: Listening Before answering these questions, ask students to review the attributes of histograms and bar graphs, as well as the definitions of categorical and numerical data. Have students find the definition of the two kinds of data and the text discussion of the similarities and differences between the two kinds of graphs. Choose several students to take turns reading the text while the remaining students listen.

Closure Question

Explain the difference between categorical data and numerical data. Categorical data are names or labels. Numerical data are counts or measurements.

Apply⇔Assess

Suggested Assignment

Core Course
Exs. 1–11, 13–17, 23–31

Extended Course
Exs. 1–31

Block Schedule
Day 34 Exs. 1–11, 13–17, 23–31

Exercise Notes

Student Progress
Exs. 1–4 All students should be able to answer these questions correctly. If not, call upon some students to suggest examples of data and have other students classify the data as numerical or categorical.

Teaching Tip
Exs. 1–4 Remind students that if data represent information that are not normally manipulated arithmetically, then the data are categorical.

CHECKING KEY CONCEPTS

The graphs show the results of a survey on pet ownership. The survey was given to a class of 32 students.

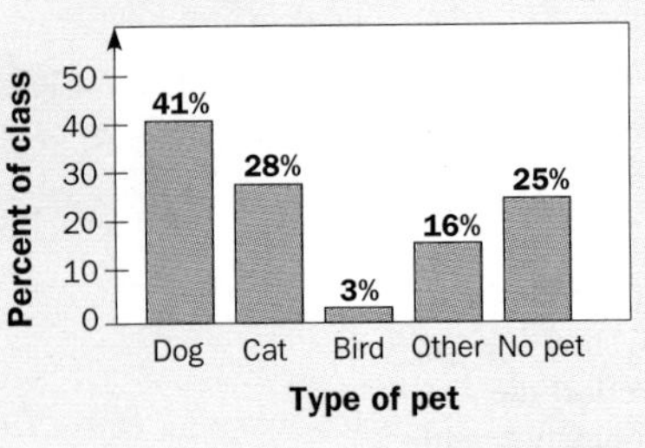

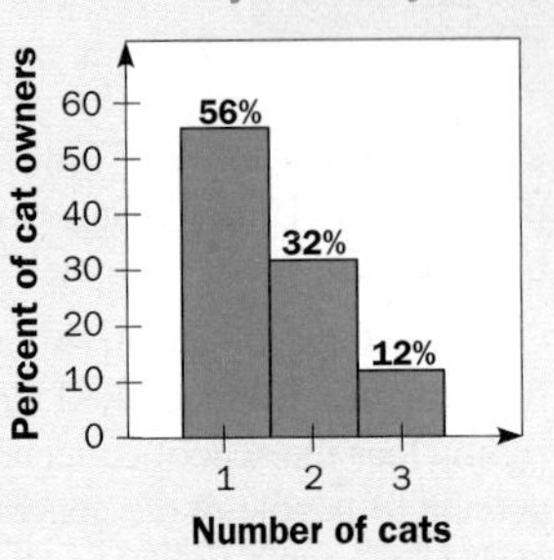

1. Describe each graph and tell whether the data shown are *categorical* or *numerical*.
2. Use the graph of pet ownership to estimate the number of dog owners you might expect to find in a sample of 50,000 households.
3. Repeat Question 2 for the number of cat owners.
4. Why do the percents in the first graph total more than 100%?
5. Use your answer to Question 3 and the graph of cat ownership to estimate the number of cats you might expect to find in a sample of 50,000 households.

6.1 Exercises and Applications

Extra Practice exercises on page 757

Tell whether the data that can be gathered about each variable are *categorical* or *numerical*. Then describe the categories or numbers.

1. the calories in a flavor of yogurt
2. the ZIP code of a city
3. a person's favorite color
4. the scores at a golf tournament

For each graph, tell if it is an appropriate way to display the data. If not, tell what type of display would be better. Explain your reasoning.

5.

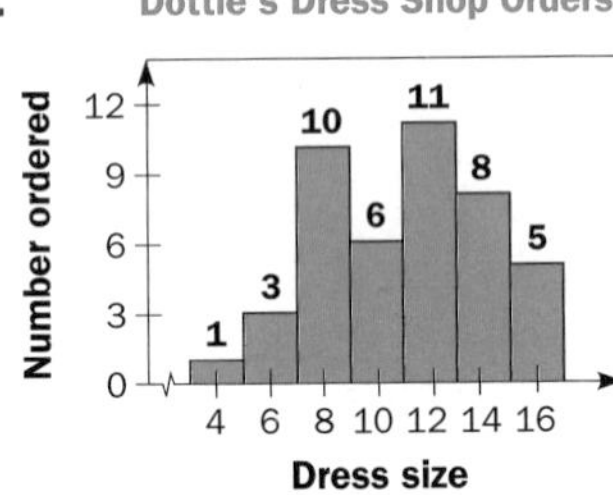

6.

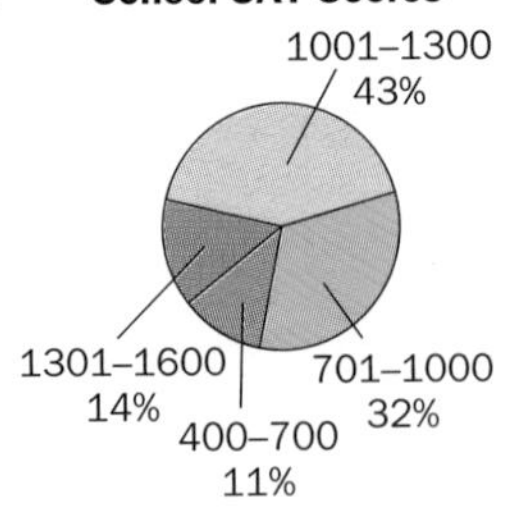

7.

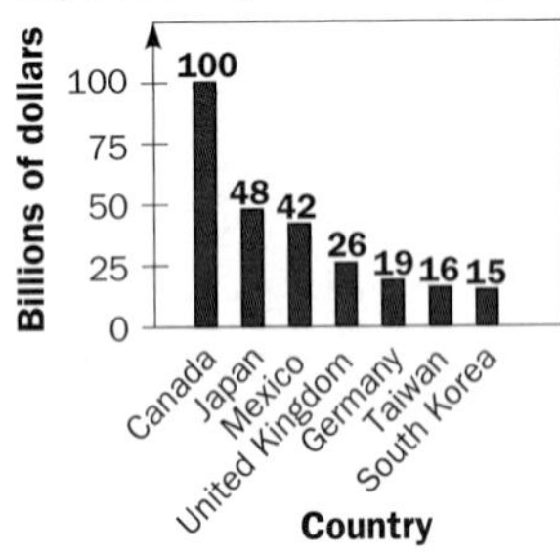

Checking Key Concepts

1. The graph on the left is a bar graph displaying the percents of class members who own a dog, a cat, a bird, another pet, or no pet. The data are categorical. The graph on the right is a histogram displaying the percentages of those students who are cat owners and have 1, 2, or 3 cats. The data are numerical.
2. Estimates may vary; about 20,500 dog owners.
3. Estimates may vary; about 14,000 cat owners.
4. The categories overlap; some people own more than one type of pet.
5. Estimates may vary; about 22,000 cats.

Exercises and Applications

Opinions may vary as to whether some data are numerical or categorical. Classifications other than those given in the following answers should be accepted if they can be reasonably justified.

1. numerical; whole numbers between, say, 100 and 300
2. categorical; five-digit or nine-digit numbers that identify areas
3. categorical; names of colors
4. numerical; whole numbers

Connection SPORTS

Sports teams often provide data about their players. The data tell spectators about each athlete, such as a player's performance statistics, position, and experience. Data for the 1994 Stanford Women's NCAA Volleyball Championship team are given in the table.

Player	Year in school	Position	Height
Lisa Sharpley	1	S-OH	6 ft 0 in.
Cary Wendell	3	S-OH	6 ft 0 in.
Shelly Foster	2	DS	5 ft 5 in.
Marnie Triefenbach	3	OH	6 ft 0 in.
Anne Wicks	4	MB	6 ft 2 in.
Paula McNamee	2	MB-OH	6 ft 0 in.
Barbara Ifejika	1	MB	6 ft 2 in.
Nikki Otto	2	MB	6 ft 0 in.
Eileen Murfee	2	MB	6 ft 2 in.
Maureen McLaren	3	OH	6 ft 1 in.
Catherine Juillard	2	S-DS	5 ft 8 in.
Debbie Lambert	1	OH	6 ft 0 in.
Denise Rotert	4	OH-DS	5 ft 10 in.
Kristin Folkl	1	OH	6 ft 2 in.
Wendy Hromadka	3	OH	5 ft 11 in.

The table lists four positions: outside hitter (OH), middle blocker (MB), setter (S), and defensive specialist (DS).

Stanford's Lisa Sharpley (*left*) sets up a spike for teammate Barbara Ifejika at the NCAA Division 1 Women's Volleyball Championship on December 17, 1994, in Austin, Texas. Stanford defeated UCLA in the best-of-five series.

8. What type of data is represented by a player's year in school? by a player's position? by a player's height? Explain.

9. The graph displays the heights, in inches, of the women on the team.

a. What type of graph is shown?

b. Use the graph to estimate the mean height of the team in inches.

c. Convert the heights in the table to inches. Then calculate the mean of the heights. How accurate was your estimate from part (b)?

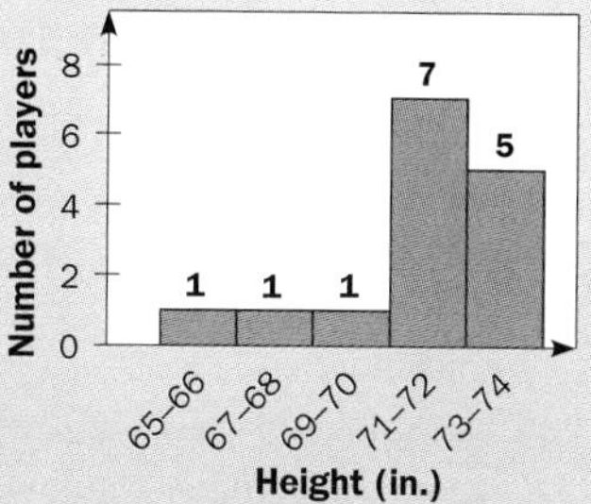

10. a. Make a bar graph to show the number of players at each position.

b. Writing Does it make sense to find the mean of the number of players at each position? Why or why not?

11. Writing What type of graph would you use to display the data for each player's year in school? Why?

Apply⇔Assess

Exercise Notes

Interdisciplinary Problems

Exs. 5–7 These exercises show a use of statistics that benefits retailing, educational administration, and consumer economics. Students should understand that graphs are visual representations of data and, as such, should present the clearest picture of the data. Encourage students to redraw any displays they think would benefit from another kind of graph.

Second-Language Learners

Exs. 8–11 Some students learning English may not be familiar with the terminology of volleyball. Invite volunteers who know the sport to name and explain what each position means.

Alternate Approach

Ex. 9 For part (b), ask students to consider the individual heights as being stretched out in numerical order along a thin rod. The mean is the point at which the rod would balance.

Reasoning

Ex. 10 When working on part (b), students should recognize that because numerical data are counts or measurements, they can be averaged. Since categorical data are only labels, they cannot be averaged.

Journal Entry

Ex. 11 Mention to students that when they are choosing a graph to display data, they should keep in mind that histograms and box plots are often used to display numerical data. Circle graphs and bar graphs are generally suitable for displaying categorical data.

Problem Solving

Ex. 12 This exercise requires students to apply the problem-solving techniques for statistical investigations that were outlined on page 237. Students may discover that more than one graph is suitable for displaying the data.

5. No; the data are categorical, so a bar graph would be more appropriate.

6. No; the data are numerical, so a histogram or a box plot would be more appropriate.

7. Yes; the data are categorical.

8. categorical; The numbers are equivalent to the labels *freshman*, *sophomore*, and so on. categorical; Players are classified according to their function in the game. numerical; Heights, which for this team range from 5 ft 5 in. to 6 ft 2 in., are measured.

9. a. histogram

b. Estimates may vary. about 72 in. (6 ft)

c. heights in inches: 72, 72, 65, 72, 74, 72, 74, 72, 74, 73, 68, 72, 70, 74, and 71; $71\frac{2}{3}$ in. (5 ft $11\frac{2}{3}$ in.); The estimate given in part (b) is fairly close.

10. a.

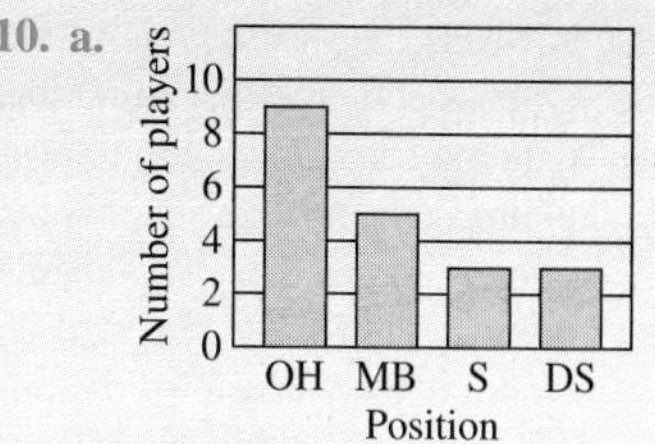

b. No; the mean has no significance for categorical data.

11. bar graph; The data are categorical.

Apply⇔Assess

Exercise Notes

Visual Thinking

Ex. 12 Ask students to sketch the graphs that they would select to display each type of recycling data. Discuss how these graphs could be used to present recycling information to a group. Ask questions such as: Will the graphs help you to explain the data easily? Are there other types of graphs that would make the information clearer? This activity involves the visual skills of *exploration* and *communication*.

Career Connection

Ex. 16 An *actuary* for an insurance company uses statistics to consider the probability of certain risks so the company can determine the cost of its premiums. For example, an actuary can determine the risk of a driver having another car accident if he or she has had three accidents in the last three years. Actuarial training requires the study of advanced mathematics.

Topic Spiraling: Preview

Ex. 19 This exercise may reveal students' intuitive sense of an unbiased sample, a topic introduced in Section 6.3. You might use this opportunity to conduct an informal discussion of the necessity of using unbiased samples to draw reasonable conclusions.

Using Technology

Exs. 20–22 Students can use the list features of a TI-82 graphing calculator to make scatter plots of their data. Using the STAT EDIT menu, enter the whole numbers 1 through 7 into list L1 to represent each orchestra in Aiyana's list. Enter the salaries from her list into list L2. In the STAT CALC menu, select 3:SetUp, select L1 for the Xlist, and select L2 for the Ylist under the 2-Var Stats option. In the STAT PLOTS menu, select 1:Plot1, select ON, then select the scatter plot graphic for Type, L1 for Xlist, and L2 for Ylist. Select any Mark. Make sure an appropriate viewing window is set and view the graph. The same procedure can be repeated using L3 and L4 for Kelly's data and L5 and L6 for Marcus's data. Ask students whether they think the median salary for each list would give a better estimate of each region's salaries.

12. Open-ended Problem Suppose you were doing a survey about people's recycling habits. What numerical data could you collect about the subject? What categorical data could you collect? What type of graph would you use to display each type of data? Explain your choice.

For Exercises 13–17:

a. Describe the population and the sample.

b. Tell whether the data that can be gathered in each situation are *categorical* or *numerical*.

c. Tell what type of graph you would use to display the data.

13. A car manufacturer tests 100 cars of the same model to find the fuel efficiency (in miles per gallon).

14. The cast of the play wants to estimate how many programs to order. Cast members ask students from each grade level if they plan to attend the play.

15. A ratings company polls radio listeners between the ages of 12 and 25 to find the most popular radio station in a city.

16. After a severe storm, an insurance company selects 50 homes to assess the average amount of money each homeowner will need for repairs.

17. Supporters of a political proposal want to know where they need to increase their campaign efforts. On a survey of voters, they ask people to write down their voting district.

18. How many exercises are in the "population" of this exercise set? in the "sample" your teacher assigned?

19. Cooperative Learning Work in a group of four students.

a. Each of you should write one of the variables listed below on a piece of paper so that there is one piece of paper for each variable.

- your eye color
- whether you have a driver's license
- your age
- how many classes you are taking

As a group, discuss whether the data that can be gathered about each variable are *categorical* or *numerical*.

b. Pass each paper to the right until each of you has recorded your data on each piece of paper. When you get your original paper back, summarize the data. Use your summary to predict the number of students in your class who would respond in the same way. Explain how you made your prediction.

c. As a class, use a table like the one shown to record each group's results for each variable. Find the totals and compare them with the predictions you made in part (b). Were your predictions close? Why or why not?

d. Use your class results to make predictions about the entire school population. Do you think your predictions are reasonably accurate? Why or why not?

Class Results for Eye Color

	Blue eyes	Brown eyes	Hazel eyes	Green eyes
Group 1	1	3	0	0
Group 2	1	2	1	0
Group 3	0	3	0	1
Group 4	1	2	1	0
Group 5	2	2	0	0
Group 6	1	2	0	1
Total	6	14	2	2

BY THE WAY

When Hurricane Andrew hit three states in August 1992, the damage amounted to an estimated $15.5 billion in insured losses. This loss is one of the most costly ever recorded in the United States.

12. Answers may vary. Examples are given. I could collect numerical data indicating the number of pounds of paper, glass, or aluminum recycled or the percentage of household waste recycled. I would display this information in a histogram. I could collect categorical data indicating the type of material recycled, such as paper, glass, plastic, or aluminum. I would display this information in a bar graph or a circle graph.

13. a. all cars of a given model; 100 of the cars
b. numerical c. histogram

14. a. all of the students at the school; the students who are surveyed
b. categorical c. bar graph

15. a. all radio listeners in a city between the ages of 12 and 25; those selected for the survey
b. categorical
c. bar graph or circle graph

16. a. all houses that are covered by the insurance company in the area affected by the storm; the 50 houses that are selected
b. numerical
c. histogram

17. a. all eligible voters; the voters who are surveyed
b. categorical
c. bar graph or circle graph

Connection ECONOMICS

Aiyana, Kelly, and Marcus are college students completing a degree in music. Each person contacts some of the orchestras in his or her region of the country to find the average starting salary for musicians in an orchestra.

Aiyana's Results		Kelly's Results		Marcus's Results	
Orchestra	**Salary**	**Orchestra**	**Salary**	**Orchestra**	**Salary**
Dallas	$46,500	Boston	$59,500	Atlanta	$43,750
Houston	$42,000	Buffalo	$32,000	Florida	$22,750
Los Angeles	$58,250	Chicago	$59,500	New Orleans	$16,500
Oregon	$27,500	Cincinnati	$52,250	North Carolina	$29,500
Phoenix	$22,250	Cleveland	$56,500	Louisville	$21,250
Utah	$31,750	Detroit	$52,500	New Jersey	$19,500
San Diego	$25,000	Indianapolis	$39,750	New York Opera	$26,000
Average	$36,179	Average	$50,286	Average	$25,607

20. What type of data did each person collect? What type of graph do you think would be appropriate to display each person's data?

21. From what population are the samples taken? Which person's sample do you think best represents the population? Why?

22. a. Use each person's sample to estimate the percent of orchestras that pay more than $40,000 as a starting salary.

b. Repeat part (a) using the salaries from all three samples. Which estimate do you think is most reliable? Why?

c. There are about 42 major orchestras in the United States. Estimate how many orchestras pay more than $40,000 as a starting salary.

ONGOING ASSESSMENT

23. Cooperative Learning Work in a group of four students. Your group will be conducting a survey about a subject of your choice. In each section of this chapter, you will complete a part of the project.

In this section you should decide on a subject for your survey. Keep in mind that you will need to gather both categorical and numerical data about your subject. Write down the subject and any questions that you would like to consider. Describe the population you want to study.

SPIRAL REVIEW

Find the *x*-intercept(s) of the graph of each function. *(Section 5.6)*

24. $y = 2x^2 + 2x - 12$ **25.** $y = x^2 + 2x - 3$ **26.** $y = 6x^2 + x - 1$ **27.** $y = 36 + 13x + x^2$

Graph each equation. *(Section 5.3)*

28. $y = 2(x + 1)(x + 2)$ **29.** $y = (x - 3)(x + 3)$ **30.** $y = 4(x - 7)(x + 2)$ **31.** $y = -(x + 2)(x - 5)$

Apply⇔Assess

Exercise Notes

Communication: Discussion
Ex. 23 Students should recognize that the difficulty of gathering particular data is an important consideration in determining what survey they choose to conduct. Students might begin this project by making a list of desirable subjects and assigning a level of difficulty for data gathering to each subject. They can then eliminate those subjects for which data would be difficult to obtain. This exercise previews the material covered in Sections 6.2 and 6.3.

Student Study Tip
Exs. 28–31 Point out to students that before they graph a quadratic equation, they should decide whether the equation opens up or down.

Practice 33 for Section 6.1

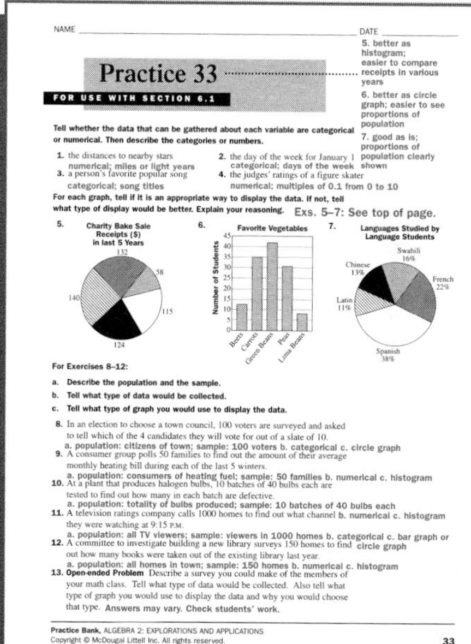
NAME ______ DATE ______

Practice 33
FOR USE WITH SECTION 6.1

5. better as histogram; easier to compare receipts in various years
6. better as circle graph; easier to see proportions of population
7. good as is; proportions of population clearly shown

Tell whether the data that can be gathered about each variable are categorical or numerical. Then describe the categories or numbers.

1. the distances to nearby stars **numerical; miles or light years**
2. the day of the week for January 1 **categorical; days of the week**
3. a person's favorite popular song **categorical; song titles**
4. the judges' ratings of a figure skater **numerical; multiples of 0.1 from 0 to 10**

For each graph, tell if it is an appropriate way to display the data. If not, tell what type of display would be better. Explain your reasoning. Exs. 5–7: See top of page.

For Exercises 8–12:

a. **Describe the population and the sample.**
b. **Tell what type of data would be collected.**
c. **Tell what type of graph you would use to display the data.**

8. In an election to choose a town council, 100 voters are surveyed and asked to tell which of the 4 candidates they will vote for out of a slate of 10. **a. population: citizens of town; sample: 100 voters b. categorical c. circle graph**
9. A consumer group polls 50 families to find out the amount of their average monthly heating bill during each of the last 5 winters. **a. population: consumers of heating fuel; sample: 50 families b. numerical c. histogram**
10. At a plant that produces halogen bulbs, 10 batches of 40 bulbs each are tested to find out how many in each batch are defective. **a. population: totality of bulbs produced; sample: 10 batches of 40 bulbs each b. numerical c. histogram**
11. A television ratings company calls 1000 homes to find out what channel they were watching at 9:15 P.M. **a. population: all TV viewers; sample: viewers in 1000 homes b. categorical c. bar graph or circle graph**
12. A committee to investigate building a new library surveys 150 homes to find out how many books were taken out of the existing library last year. **a. population: all homes in town; sample: 150 homes b. numerical c. histogram**
13. **Open-ended Problem** Describe a survey you could make of the members of your math class. Tell what type of data would be collected. Also tell what type of graph you would use to display the data and why you would choose that type. **Answers may vary. Check students' work.**

Practice Bank, ALGEBRA 2: EXPLORATIONS AND APPLICATIONS

33

18. 31; Answers may vary.

19. a. categorical data: eye color and whether you have a driver's license; numerical data: age and how many classes you are taking

b. Answers may vary. You could predict by writing proportions relating the number of people in your group who share eye color or driver's license status to the number in the class. For age and the number of classes you are taking, you might use the mean in your group to estimate the mean for the class.

c, d. Answers may vary.

20. numerical data; histogram

21. all of the orchestras in the U.S.; Answers may vary. Examples are given. Aiyana: western orchestras, Kelly: eastern and midwestern orchestras, Marcus: eastern orchestras; Although none of the samples represents a geographic cross-section of the country, Aiyana's covers a broader range of salaries. It appears that the symphonies in Kelly's sample have higher-end salaries and those in Marcus's have lower-end salaries.

22–31. See answers in back of book.

Plan⇔Support

Objectives
- Write good survey questions.
- Judge the accuracy of responses to other people's surveys.

Recommended Pacing
Core and Extended Courses
Section 6.2: 1 day

Block Schedule
Section 6.2: $\frac{1}{2}$ block
(with Section 6.3)

Resource Materials
Lesson Support
Lesson Plan 6.2
Warm-Up Transparency 6.2
Practice Bank: Practice 34
Study Guide: Section 6.2

Technology
Internet:
http://www.hmco.com

Warm-Up Exercises
Tell whether the data that can be gathered about each variable are *categorical* or *numerical.*

1. snow depth numerical
2. political party affiliation categorical
3. a car's seating capacity numerical

Describe the population and the sample of each of the following.

4. After a hurricane, an insurance company surveys 60 households to estimate the average amount of money each household will request for repairs. population: the insurance company's policyholders living in the area affected by the hurricane; sample: the 60 households surveyed
5. On election night, a local news station surveys 100 people exiting a polling place to estimate the number of voters who voted for a candidate. population: registered voters who voted at that polling place; sample: the 100 voters surveyed

Section

6.2 Writing Survey Questions

Learn how to...
- write good survey questions

So you can...
- get more accurate responses to your surveys
- judge the accuracy of responses to other people's surveys

What makes a person a good journalist, lawyer, or scientist? The answer may be the way in which people in these fields ask questions. Knowing which questions to ask may help a person get a lead on a story, prevent an innocent person from going to jail, or discover a cure for a life-threatening disease. In this section, you will learn how the way you ask a question can affect the response you receive.

EXPLORATION
COOPERATIVE LEARNING

Comparing Ways of Asking a Question

Work with half the class.

1 One half of the class should answer the questions in Survey I, and the other half of the class should answer the questions in Survey II. The data you collect for this Exploration may be invalid if you read the questions on the other group's survey. To help you avoid this temptation, Survey II has been printed upside down.

Survey I

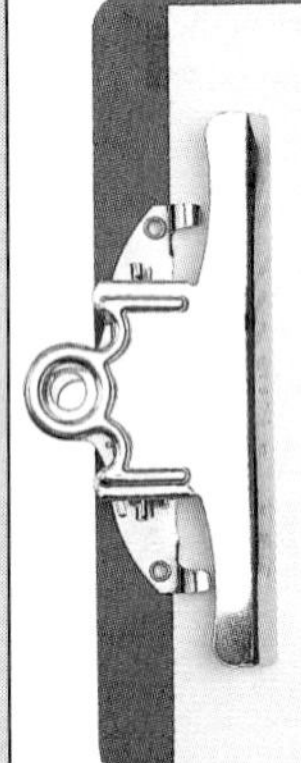

Survey II

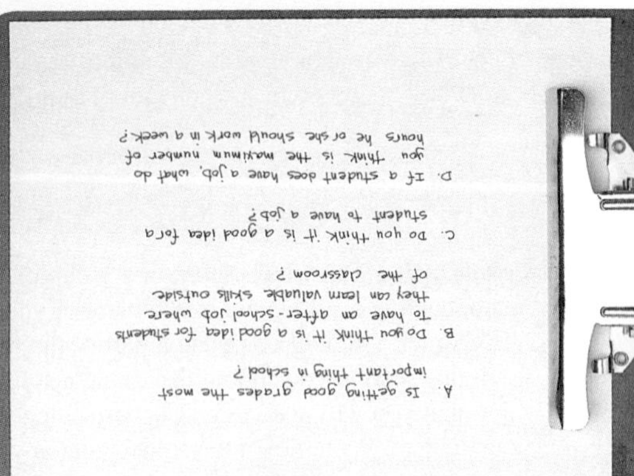

2 Record the percent of people in your group who responded "yes" to survey questions A, B, and C. Also record the percent of people in your group who responded "no" to questions A, B, and C. Finally, find the average of the numbers given in response to question D.

Exploration Note

Purpose
The purpose of this Exploration is to have students discover that the way in which a question is asked can affect the response it receives.

Materials/Preparation
Each group member needs a copy of the group's survey.

Procedure
Each student answers the questions on his or her survey. Each group then records the percent of people in the group who responded "yes" and the percent of people who responded "no" to questions A, B, and C. Each group also records the average of the numbers given in response to question D. The two groups then share their results.

Closure
Students should understand that the way in which a question is asked or the order in which questions are asked can affect the outcome of the survey.

Explorations Lab Manual
See the Manual for more commentary on this Exploration.

Questions

1. Compare the questions on your group's survey with the questions on the other group's survey. Do the questions ask about the same information? Do you think your group's results will be similar to those for the other group? Why or why not?
2. Share the results your group recorded in Step 2 with the other group. In both surveys, question A asked about the importance of good grades. According to the results of the two groups, does it appear that one group thinks good grades are more important than the other group does? Do you think this is really the case? Why or why not?
3. Compare the results for the other three questions. Describe any differences. Why do you think these differences occurred?
4. Suppose questions C and D on each survey in the Exploration were the first two questions on each survey. Do you think the results from these questions would be different? Why or why not?

When a question produces responses that do not accurately reflect the opinions or actions of the respondents, the question is said to be a **biased question**.

A biased question may encourage the respondent to answer in a particular way, may be perceived as too sensitive to respond to truthfully, or may not provide the respondent with enough information to give an accurate opinion. Bias may also be introduced through the order in which the questions are asked.

BY THE WAY

A fictitious law was included in a 1981 study. Thirty percent of the people involved in the study claimed to know of this made-up law.

EXAMPLE 1

Tell why each question may be biased.

a. "Many national parks are being heavily damaged by acid rain. Do you favor government funding to help prevent acid rain?"

b. "Do you agree with the amendments to the Clean Air Act?"

c. Police officers ask mall visitors, "Do you wear your seat belt regularly?"

SOLUTION

a. This is an example of a *leading question*. Respondents may think a "no" response means they are not in favor of supporting the national parks. In this way, the question encourages the respondent to answer "yes."

b. This question assumes that a respondent is familiar with the amendments to the Clean Air Act. Responses by people unfamiliar with the amendments could lead to misleading conclusions.

c. Many motorists may answer untruthfully because a police officer is asking the question, especially if the law requires seat belt use. The data collected might not accurately represent the percent of people who wear seat belts.

Teach⇔Interact

Learning Styles: Verbal

The material in this section on writing survey questions may appeal more to verbal learners than to visual or kinesthetic learners. You may wish to have students work in groups of five to discuss the examples and exercises. Each group should contain at least one verbal learner who can serve as the group's leader.

About Example 1

Reasoning

Example 1 explains several different ways survey questions can be biased. Encourage students to think of other ways questions can be biased and to give examples.

Additional Example 1

Tell why each question may be biased.

a. "Should laws be passed prohibiting the sale of cigarettes or do citizens have the right to decide whether or not they want to smoke?"
This is an example of a leading question. Respondents may think that a "no" response means that they do not support free choice.

b. A television talk show asks viewers, "Do you support regulation of the amount of money presidential candidates can receive from special interest groups? Call the number at the bottom of the screen to register your vote."
Viewers who support such regulation are more likely to call in than those who do not or are undecided. Thus, the results of the survey could be misleading.

c. "Do you agree that the Internet and its Netizens of cyberspace should be censored by cybercops?"
This question assumes that a respondent is familiar with the Internet and its terminology. Responses by people unfamiliar with the subject of the survey question could result in misleading conclusions.

ANSWERS Section 6.2

Exploration

1, 2. Answers may vary. Check students' work.

Questions

1–4. Answers may vary. Examples are given.

1. Questions C and D on the two surveys are identical. Questions A and B ask for similar information in both surveys but might influence different answers. Survey I, with its negative slant on after-school jobs, might prompt negative responses to question C, while Survey II might prompt positive responses.
2. Although it may appear that students answering "no" to question C on Survey I place more importance on good grades than those answering "yes" on Survey II, that assumption is not fair. Survey I encourages a negative response to question C, while Survey II encourages a positive response.
3. The wording of the surveys encourage the following responses to the first three questions. Survey I: (A). Yes. (B) Yes. (C) No. Survey II: (A) No. (B) No. (C) Yes. Numerical responses to question D on Survey I will probably be less than responses to the same question on Survey II.
4. See answers in back of book.

Teach⇔Interact

Section Note

Communication: Discussion Point out that pollsters often report their results with a qualifier like "there is a 3% margin of error." This means that their results may be inaccurate (higher or lower) by as much as 3%.

Think and Communicate

To help avoid writing biased questions, each student can work with a partner who holds opposing views on at least one of the subjects. Two students with different views are likely to be a check on each other's biases as they write their questions.

Additional Example 2

Students who have earned release time from study hall may choose to stay in study hall or to go outside on the school grounds during that time. The student council is considering the establishment of a student center in the school that could be used by students during their release time. The student council would like to know if the student body would provide for the upkeep of the center. Write a group of unbiased questions that will help the council make a decision.

Student Center Survey

1. How many periods of release time do you have each week?
   ___4 periods ___2 or 3 periods
   ___1 period ___no periods
2. Establishment of a student center is dependent on students' willingness to clean it once a day. Would you be willing to donate cleaning time in order to establish a student center or would you rather keep release time activities as they are now and have no cleaning responsibilities?
   ___I would be willing to clean.
   ___I do not want to clean.
3. How much time would you be willing to donate to cleaning a student center if one were established?
   ___15 minutes ___30 minutes
   ___1 hour ___more than 1 hour

THINK AND COMMUNICATE

1. Question B from Survey I of the Exploration is an example of a leading question. What answer does the wording of each version of the question encourage? Why? How can you rewrite the question so that it is not biased?
2. Discuss how you would word an unbiased survey question on each of the following topics:
   a. the need to repave the school's parking lot
   b. the need for a new teen center in your hometown
   c. the need for your state to encourage businesses to hire teens

Some surveys have respondents choose their answer to a question from a list of options. People who conduct surveys, called *pollsters*, have found that results are more accurate when people are given a list of choices than when an open-ended question is asked. Also, because it is easier to place a check mark next to a response than it is to write out a response, more people are likely to complete the survey. A multiple-choice method also makes it easier for the pollster to analyze the results.

EXAMPLE 2 Application: Market Research

Due to rising printing costs, the staff of the school newspaper is considering charging $.50 for each issue of the weekly paper. The staff would like to know how this would affect the number of issues of the paper students read. Write a group of unbiased questions that will help the staff make a decision.

SOLUTION

NEWSPAPER SURVEY

1. How often do you read the weekly school newspaper?
☐ 4 times a month ☐ 2 or 3 times a month
☐ once a month ☐ occasionally
☐ never

Ask this question to determine if the respondent currently reads the paper.

2. Rising printing costs will require that we charge $.50 for each paper if we continue to publish it weekly. Would you pay $.50 for each weekly issue, or would you prefer a monthly paper with no charge?
☐ weekly issue for $.50
☐ monthly issue at no charge

Ask this question to see if students will pay $.50 for the paper.

3. How often would you buy the paper if it is published weekly and costs $.50 a copy?
☐ 4 times a month ☐ 2 or 3 times a month
☐ once a month ☐ occasionally
☐ never

Ask this question to see if the number of issues a student reads will change when a fee is charged.

Think and Communicate

1. Survey I: Yes; it implies that all students who work after school work long hours and therefore have no time and no energy left to study. Survey II: Yes; it implies that jobs give students valuable experience. Answers may vary. An example is given. Do you think that a student can maintain good grades while holding down an after-school job?
2. Answers may vary. Examples are given. In each case, the question should not mention funding (for example, cutting funds from some other program to pay for the proposed plan), positive or negative comments on the present situation, or possible hardships imposed by the implementation of the plan. The question might, however, give details to provide as much information as possible without bias.
   a. Do you think the school's parking lot should be repaved?
   b. Do you think a new teen center providing facilities for athletic and social activities is needed in town?
   c. Do you think the state needs to encourage businesses to hire teens?

CHECKING KEY CONCEPTS

Tell why each question may be biased. Then describe what changes you would make to improve the question.

1. "Which city council candidate's platform do you support?"
2. The last question on a test asks students, "How long did you study for this test?"
3. "Aerosol products containing CFCs damage Earth's ozone layer. Do you think aerosol products containing CFCs should be banned?"
4. "How often do you exercise?"

6.2 Exercises and Applications

Extra Practice exercises on page 757

For Exercises 1–6, tell why each question may be biased. Then describe what changes you would make to improve the question.

1. "How much do you weigh?"
2. "A survey of the voters in the state shows that 85% would support a candidate who favors a tax decrease. Do you favor a tax decrease?"
3. "Which one of the following meals would you eliminate from the menu of our Japanese restaurant?

   ___ yakitori ___ chasoba ___ udon"
4. "Are you in favor of replacing the baseball stadium that has been used by hometown teams since 1905 with a large, new sports complex?"
5. "Do you think the defendant in the Carter case was given a fair trial?"
6. "What is your grade point average?"
7. Look back at the surveys in the Exploration on page 244. Rewrite the questions so that they are not biased.

Do you know which Japanese dish this is?

Use the senior class survey at the right for Exercises 8–11.

8. What information is the survey trying to obtain?
9. **Writing** Which questions are worded well? Which are worded poorly? Are there any additional questions you would ask? Explain your reasoning.
10. Rewrite any questions that you think are poorly worded.
11. Tell what type of data each question will generate. Explain how the data will help the planning committee make decisions about the banquet.

Please complete this questionnaire regarding the senior class banquet and return it to your homeroom teacher by Friday.

- **Are you a male or a female?** ________
- **Do you think a senior banquet is a good idea?**
  ___ yes ___ no ___ no opinion
- **Do you think formal attire should be worn at the banquet?**
  ___ yes ___ no ___ no opinion
- **How much would you be willing to pay to attend the banquet?** ________

Teach⇔Interact

Checking Key Concepts

Teaching Tip
Point out to students that they can test a survey question for bias by asking if it attempts to influence the respondents in any way. A question worded to influence the respondent rather than just inform him or her is biased.

Closure Question

Describe several ways a question can be biased.
A biased question may encourage the respondent to answer in a particular way, may be considered too sensitive to respond to truthfully, or may be lacking sufficient information for the respondent to give an accurate answer. The order in which questions are asked may also cause them to be biased.

Apply⇔Assess

Suggested Assignment

Core Course
Exs. 1–12, 15–18, 20–26

Extended Course
Exs. 1–26

Block Schedule
Day 35 Exs. 1–12, 15–18, 20–26

Exercise Notes

Multicultural Note
Ex. 3 Students may not be familiar with the Japanese foods listed. Explain that *yakitori* is a dish of skewered chicken with teriyaki sauce; *chasoba* is a dish of green-colored noodles made with tea, and *udon* is a dish of big, thick noodles.

Reasoning
Exs. 9, 10 Asking students to explain how they evaluated the questions in the survey may help them rewrite the questions for Ex. 10.

Student Progress
Ex. 11 At this point, all students should be able to distinguish numerical data and categorical data.

Checking Key Concepts

1–4. Answers may vary. Examples are given.

1. The question assumes that the respondent knows all the candidates' platforms and supports one. You could provide a list of the key issues, each with all the candidates' positions given, and ask respondents to check which positions they support.
2. Students may not answer truthfully, since they hope to make a good impression on the teacher. This question would be acceptable if asked after the test scores have been recorded.
3. The first statement encourages the respondent to reply positively. Omit the first sentence.
4. This question is too indefinite. It could be improved by asking, for example, how many times a week the respondent exercises for at least one-half hour.

Exercises and Applications

1–7. Answers may vary. Examples are given.

1. Weight is a sensitive issue and people may not wish to answer or may not answer truthfully. The information should be collected anonymously.
2. This is a leading question. A person might want to be part of the majority. Omit the first sentence.

3–11. See answers in back of book.

Apply⇔Assess

Exercise Notes

Historical Connection

Exs. 12–14 At the time of the 1936 national election, a more scientific sampling method was introduced into politics that correctly predicted Franklin Roosevelt's victory. Since then, only one presidential election has been predicted incorrectly; in 1948, Thomas E. Dewey was predicted to win, but in the final hours, Harry S. Truman won the election. Since 1948, polling methods have improved and the major polling organizations now have only very small errors in their predictions of the national vote.

Communication: Discussion

Exs. 16–19 These exercises can be used to help students learn how to compose a survey sample. In order for a sample to make accurate predictions about a population, the sample must be representative of the population. The *design* of a sample refers to the method used to choose the sample from a population. It should be completely determined before any data are actually collected. This discussion can be used to preview the biased sample designs which are presented in Section 6.3.

Connection POLITICAL SCIENCE

Andrew Jackson

Public polling first became popular in the United States in 1824, when it was used to predict the outcome of the presidential election between Andrew Jackson and John Quincy Adams. Since then, politicians and journalists have questioned people about their political beliefs, candidate preferences, and voting habits.

During the 1992 presidential election, a candidate distributed a questionnaire asking voters about their beliefs. Some of the questions from that survey are shown below.

John Quincy Adams

Please darken box completely.

1. Should the President have the Line Item Veto to eliminate waste?
   YES☐ NO☐
2. Do you want a Constitutional Balanced Budget Amendment, with emergency funding limited exclusively to National Defense?
   YES☐ NO☐
3. Should laws be passed to eliminate all possibilities of special interests giving huge sums of money to candidates?
   YES☐ NO☐

12. **a.** Which questions on this survey do you think are biased? Why?

b. Rewrite the questions so they are not biased.

13. **Writing** In the original survey, 97% of the respondents answered "yes" to question 1. When the question was rewritten to ask "Should the President have the Line Item Veto, or not?", only 57% of the respondents answered "yes."

a. What do you think caused people to support question 1 so strongly in the original questionnaire?

b. What do the results indicate voters are really concerned about? Explain.

14. Question 3 was revised in a new survey that asked, "Should laws be passed to prohibit interest groups from contributing to campaigns, or do groups have a right to contribute to the candidate they support?"

a. Do you think the revised question is biased? Why or why not?

b. Match each set of results with the question you think produced those results. Explain your choice.

Question 3 responses	
YES, pass a law 40%	NO 55%

Question 3 responses	
YES, pass a law 80%	NO 17%

c. Why do you think the results do not add up to 100%?

12. a. All three questions are biased.
   (1) Respondents may not understand what the Line Item Veto is. The question also implies that simply establishing a President's right to a line item veto would eliminate waste.
   (2) Respondents may not fully understand the Balanced Budget Amendment or the implications of "emergency funding."
   (3) The phrase "huge sums" implies that all special interest groups give such sums with the implied result that they buy the candidate's support.

b. (1) Should the President be given the power to veto specific provisions within a budget without vetoing the entire budget?
   (2) Would you vote for a constitutional amendment requiring that the budget be kept balanced at each cycle, with emergency funding allowed to exceed the balance limit only in cases specifically related to national security?
   (3) Should laws be passed that limit the amount of money that a group can contribute to a candidate?

13. a. The phrase "to eliminate waste" biases the question strongly toward a positive response, because few people would argue against eliminating waste.

b. eliminating waste; When the question was changed to eliminate the reference to eliminating waste, many fewer voters supported the question.

14. a. The question is still biased; the phrase "have a right," appeals to the democratic principle of participation in the political process.

b. The original question probably produced the high "yes" response because it implies that politicians are being bought with "huge" sums of money. The revised

15. **Research** Find out about the questioning techniques used by people in the fields of law, journalism, science, or medicine. Give examples of the types of questions used by people in at least one of these fields. Do people in these fields ever use biased questions? Do they always avoid asking open-ended questions? Explain.

Write a survey about each topic. Include at least three unbiased questions on each survey. Tell what population should receive each survey.

16. The manager of a restaurant wants to know whether to add a popular special to the regular menu.

17. The transportation department is trying to decide whether it should expand the number of hours train service is provided.

18. A crafts group is trying to decide what day of the week it should hold its regular meetings so that the most people can attend.

19. A local television station wants to know what percent of viewers watch its programs on Thursday nights.

ONGOING ASSESSMENT

20. **Open-ended Problem** Work with the members of your group from the *Ongoing Assessment* exercise from Section 6.1.
    a. Write a survey to gather information about the topic you chose in Section 6.1. Your survey should include these things:
       - an opening statement that tells the purpose of the survey
       - directions on how to complete the survey
       - at least five well-written questions about the topic, including at least two questions that ask for categorical data and at least two questions that ask for numerical data
    b. After you write your survey, explain what information you want to obtain from each question. Then give your survey to another group to review. Use the comments you receive to improve your survey.

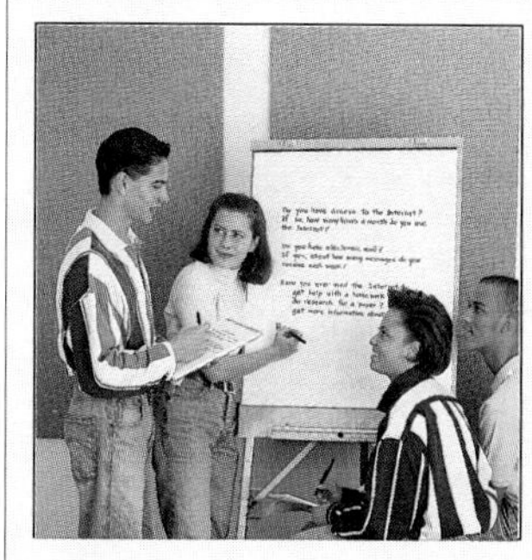

SPIRAL REVIEW

Tell whether the data that can be gathered about each variable are *categorical* or *numerical*. Then describe the categories or numbers. *(Section 6.1)*

21. car mileage
22. gender
23. museum membership

Match each scatter plot with one of the following values of the correlation coefficient r: 0.64, −0.95, 0.25. *(Section 2.5)*

24.
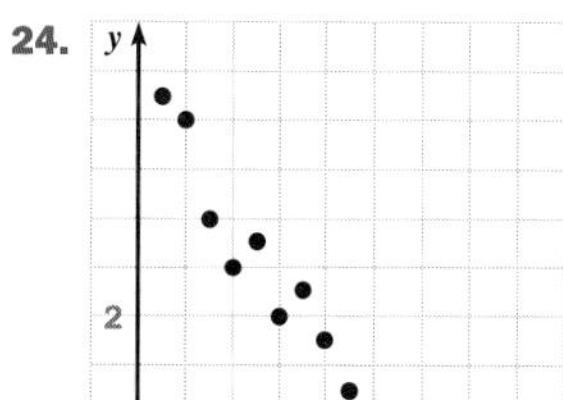

25.
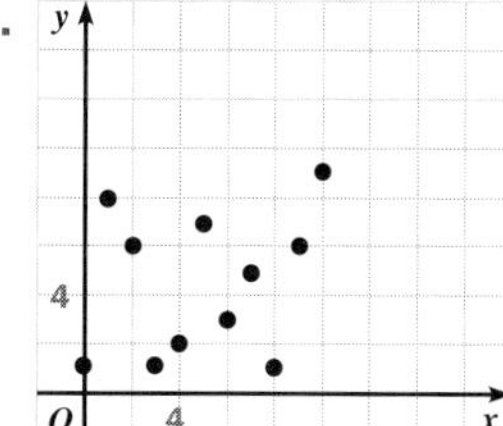

26.
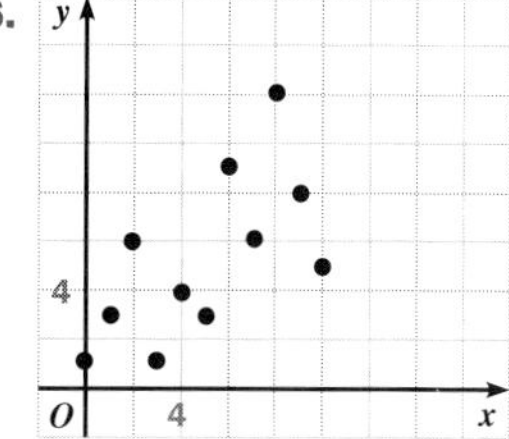

Apply⇔Assess

Exercise Notes

Second-Language Learners
Exs. 16–19 Some students learning English may benefit from working in small groups to brainstorm questions they might ask in their surveys.

Research
Ex. 19 Interested students may want to call a local television station to gather some actual viewing data. They can then base their questions on realistic percentages.

Teaching Tip
Exs. 24–26 Students may need to be reminded that a correlation coefficient close to either 1 or −1 shows a strong fit of the data to a line.

Practice 34 for Section 6.2

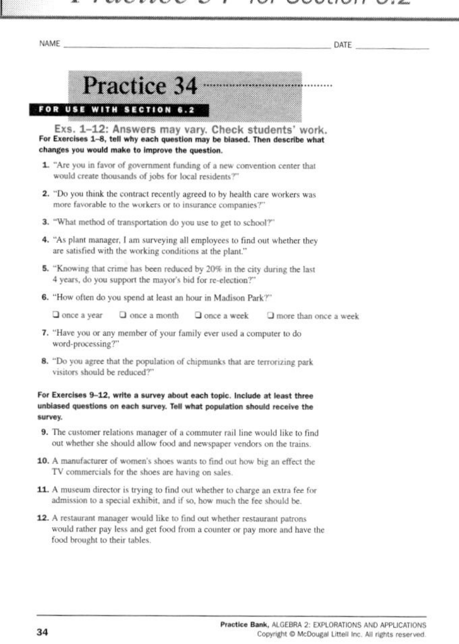

NAME ______ DATE ______

Practice 34

FOR USE WITH SECTION 6.2

Exs. 1–12: Answers may vary. Check students' work.

For Exercises 1–8, tell why each question may be biased. Then describe what changes you would make to improve the question.

1. "Are you in favor of government funding of a new convention center that would create thousands of jobs for local residents?"
2. "Do you think the contract recently agreed to by health care workers was more favorable to the workers or to insurance companies?"
3. "What method of transportation do you use to get to school?"
4. "As plant manager, I am surveying all employees to find out whether they are satisfied with the working conditions at the plant."
5. "Knowing that crime has been reduced by 20% in the city during the last 4 years, do you support the mayor's bid for re-election?"
6. "How often do you spend at least an hour in Madison Park?"
   ❑ once a year ❑ once a month ❑ once a week ❑ more than once a week
7. "Have you or any member of your family ever used a computer to do word-processing?"
8. "Do you agree that the population of chipmunks that are terrorizing park visitors should be reduced?"

For Exercises 9–12, write a survey about each topic. Include at least three unbiased questions on each survey. Tell what population should receive the survey.

9. The customer relations manager of a commuter rail line would like to find out whether she should allow food and newspaper vendors on the trains.
10. A manufacturer of women's shoes wants to find out how big an effect the TV commercials for the shoes are having on sales.
11. A museum director is trying to find out whether to charge an extra fee for admission to a special exhibit, and if so, how much the fee should be.
12. A restaurant manager would like to find out whether restaurant patrons would rather pay less and get food from a counter or pay more and have the food brought to their tables.

34 **Practice Bank,** ALGEBRA 2: EXPLORATIONS AND APPLICATIONS

question probably produced the high "no" response because it says nothing of "huge" gifts and because it implies that groups have a right to support whom they please, which is a basic principle of American democracy.

c. Some respondents either had no opinion, or chose not to give a "yes" or "no" answer.

15. Answers may vary. A lawyer might be inclined to ask questions with very restricted response options to try to produce a desired response. A journalist might vary the types of questions depending on whether he or she is trying to uncover something or just gather information. A scientist needs to ask questions that can be answered specifically, so that responses can be examined by the scientific method. A doctor might use open-ended questions to avoid suggesting the presence of various symptoms.

16–19. See answers in back of book.

20. Answers may vary.

21. numerical; The data might be expressed in miles per gallon, which can vary widely.

22. categorical; The data would be classified as "male" or "female."

23. either categorical (data that classify members as family members, individual members, patrons, students, and so on) or numerical (data that indicate membership totals for given museums)

24. −0.95

25. 0.25

26. 0.64

Plan⇔Support

Objectives

- Choose a representative sample.
- Recognize biased samples in surveys.

Recommended Pacing

Core and Extended Courses
Section 6.3: 2 days

Block Schedule
Section 6.3: 2 half-blocks (with Sections 6.2 and 6.4)

Resource Materials

Lesson Support
Lesson Plan 6.3
Warm-Up Transparency 6.3
Practice Bank: Practice 35
Study Guide: Section 6.3
Explorations Lab Manual: Additional Exploration 10
Assessment Book: Test 21

Technology
Graphing Calculator
Internet: http://www.hmco.com

Warm-Up Exercises

1. Identify the population and the sample.
   A cooking magazine asks its subscribers to send in their favorite chili recipes.
   population: the magazine's subscribers; sample: people who sent in their chili recipes

State whether each survey question is biased or unbiased.

2. "Do you agree with 75% of the school population that two years of mathematics should be required for graduation?" biased
3. "The principal of an elementary school asks his students, 'Do you like school?'" biased
4. "Do you consider the cost of cable television reasonable or unreasonable?" unbiased

Section

6.3 Collecting Data from Samples

Learn how to...

- **choose a representative sample**

So you can...

- **recognize biased samples in surveys**

It is often too difficult, time-consuming, and expensive to survey everyone in a population. As a result, pollsters often gather data from a sample of the population. Some ways of selecting a sample are shown below.

Self-selected sample

Let people volunteer.

Systematic sample

Use a pattern, such as selecting every other person.

Convenience sample

Choose people who are easy to reach, such as those in the front row.

Random sample

Use a method that gives everyone an equally likely chance of being selected, such as drawing names out of a hat.

Stratified random sample

Divide the population into groups and randomly select people from each group.

Cluster sample

Choose people as a group rather than as individuals.

ANSWERS Section 6.3

Think and Communicate

1–7. Answers may vary. Examples are given.

1. The samples least likely to produce a representative sample include self-selected, convenience, and cluster samples. People voluntarily responding to a survey may have stronger feelings about the issue at hand than those in the general population. Convenience and cluster samples may include people from similar or related groups who may share opinions and feelings.
2. You could use a representative sample chosen by an expert or group of experts. You could use a stratified convenience sample. For example, interview the first ten students you see from each of four classes in a high school.
3. I think the method in part (c) is best, because it is most likely to provide a representative sample. The number of people with unlisted numbers or no phones is probably comparatively small.
4. a. You would expect to reach different types of people at different times of day, for example, families with young children early in the day, teens and adults in the evening.

Although there are many ways of sampling a population, a random sample is preferred since it is most likely to produce a representative sample of the population. A sample that overrepresents or underrepresents part of the population is a **biased sample**.

THINK AND COMMUNICATE

1. Which of the sampling methods described on the previous page is *least* likely to produce a representative sample of the population? Explain.
2. Think of some other ways to choose a sample and describe them.

EXAMPLE 1 Application: Market Research

The managers of a movie theater chain want to find out how many movies people in the community usually see in a theater each month. Each manager suggests a method for gathering the data. Identify the type of sample each method describes. Tell if the sample is biased. Explain your reasoning.

a. Have the ticket sellers at each theater survey customers when they purchase their tickets.

b. Place an ad in the local paper and ask people to mail in their responses.

c. Randomly select phone numbers from the phone book and call people to ask their opinions.

SOLUTION

a. This is a convenience sample. The sample is biased because it underrepresents people who seldom or never attend movies in a theater.

b. This is a self-selected sample. The sample is biased because it underrepresents people who do not read the paper. Bias is also introduced because the people who respond to the survey are more likely to enjoy movies than those who do not take the time to respond.

c. This is a random sample. The method of selecting people is not biased, but because people who do not have a phone or who have an unlisted number are not included, the sample is biased.

BY THE WAY

Kinepolis, located in Brussels, Belgium, is the world's largest movie theater complex. It has 24 screens and is able to seat up to 7000 people.

THINK AND COMMUNICATE

3. Which method in Example 1 do you think is best? Why?

For each sample in Example 1, tell how these factors might influence the survey results.

4. time of day the survey is taken
5. day of the week the survey is taken
6. location of the theater
7. time of year the survey is taken

b. no influence

c. If you called homes at any one time of day or on any one day, many people would be underrepresented. For example, if you made all the calls between 8 A.M. and 6 P.M. on weekdays, people working outside the home would be underrepresented.

5. a. A person seeing a movie on a weekday may more likely be a frequent moviegoer than a person seeing a movie on Saturday or Sunday.

b. The Sunday edition of many newspapers is much more widely circulated than the daily edition and so reaches a much broader cross-section of the population.

c. Calling on a weekday would probably result in a much different survey group than calling on a weekend day. Calls should be made on different days and at different times.

6, 7. See answers in back of book.

Teach⇔Interact

Section Notes

Communication: Writing The many different types of samples may confuse some students. Encourage them to relate the type of sample to the adjective used to label it. Have students write the definition of each adjective labeling a sample and give an example of how the adjective is used in a sentence.

Communication: Listening You may wish to call upon some students to explain their understanding of what the word *biased* means. Ask them to relate their understanding of this word to the term *biased sample.*

Additional Example 1

The owner of a small restaurant that serves breakfast and lunch on weekends wants to know if it will be profitable to serve breakfast on weekdays. He outlines three methods for gathering the data. Identify the type of sample each method describes. Tell if the sample is biased. Explain your reasoning.

a. Randomly choose four days in a row during a moderately busy season to survey any customer who visits the restaurant.
This is a random sample. The method of selecting people is not biased.

b. For the next four weekends, survey the people who eat breakfast at the restaurant.
This is a convenience sample. The sample is biased because it underrepresents people who do not eat breakfast regularly at the restaurant.

c. For the next four weekends, ask the party at each table in the restaurant to indicate on the back of the check whether or not they would eat breakfast at the restaurant during the week.
This is a cluster sample. The sample is biased because the answer on the back of each check may not be representative of each individual.

Teach⇔Interact

Additional Example 2

A small company wants to survey employees about setting a date for the spring picnic. Fifty of the 200 employees are to be surveyed. Describe how the company can select a random sample.

There are many ways to select a random sample. Here are two methods.

Method 1 Use a physical model. Write the numbers 1 to 200 on pieces of paper of equal size and place them in a box. Mix the pieces of paper and have each employee draw one number. Repeat the procedure, but this time let one person draw 50 numbers from the box without looking at them. The employees holding these numbers are the sample.

Method 2 Use a random number table. Assign each of the 200 employees a three-digit number 001, 002, 003, and so on to 200, and make a list of these numbers and names. Go into a random number table and examine successive three-digit numbers. Each three-digit number in the table is either an employee on the list or not. Keep examining the random numbers and check off employees' numbers until 50 are selected.

Think and Communicate

A discussion of questions 8 and 9 can clarify the purpose of a stratified random sample. Stratified random sampling allows important groups within a population to be sampled separately first. Then the samples are combined.

EXAMPLE 2

The Science Club at Park High School wants to give a survey about recycling to 50 of the 900 students in the school. The club has a list of names of all the students. Describe how the club can select a random sample.

SOLUTION

There are many ways to select a random sample. Here are two methods.

Method 1 Use a physical model.

Put each name on a piece of paper and place it in a hat. Mix up the pieces of paper and pull out 50 names without looking at them.

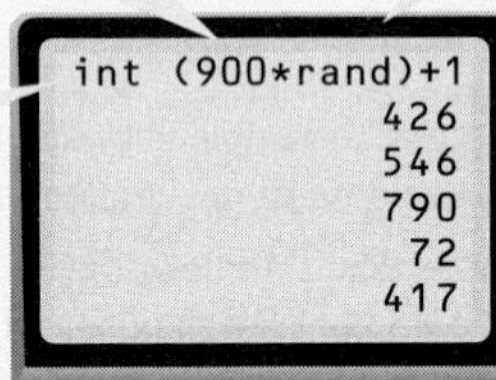

Method 2 Use a calculator.

Number the students' names from 1 to 900. Then use a calculator to generate random numbers from 1 to 900. Most calculators will produce a random number between 0 and 1. To generate other random numbers, you can use a formula as shown on the calculator screen below.

Put the number of people you want to survey here.

This function generates a random number between 0 and 1.

This function cuts off the decimal part of a positive number.

Put the smallest number you want to generate here.

```
int (900*rand)+1
             426
             546
             790
              72
             417
```

Press the ENTER key until you generate 50 different numbers. Then give the survey to the students assigned to the numbers you generate.

In Example 2, the population of the school may be better represented by finding the percent of students in each grade and then choosing a random sample having the same percent of students from each grade.

THINK AND COMMUNICATE

8. Is it possible that the 50 students selected in Example 2 could all be seniors? Do you think that would be a good sample? Why or why not?
9. Suppose the school in Example 2 has 22% freshmen, 28% sophomores, 26% juniors, and 24% seniors. How many students from each grade should be in the sample so that it is a stratified random sample?

Think and Communicate

8. Yes; it is, however, unlikely. Answers may vary. An example is given. I do not think that would be a good sample. I think older students may be more aware of and concerned about environmental problems and more likely to recycle.
9. 11 freshmen; 14 sophomores; 13 juniors; 12 seniors

Checking Key Concepts

1. systematic sample; all customers of the store; The sample is biased unless it is done over a time period covering all the store's hours. Doing the survey at a particular time of day or on a particular day of the week will underrepresent customers who shop at other times.
2. self-selected sample; population of the U.S.; The sample is biased because it is self-selected and because all those who are not viewers of the program are underrepresented.
3. stratified random sample; all students of the school; The sample may be somewhat biased if the number of students in each grade varies too widely. It would be better to find the percent of students in each grade and randomly choose students from the grade to match this percentage.

☑ Checking Key Concepts

Identify the type of sample and describe the population of the survey. Then tell if the sample is biased. Explain your reasoning.

1. A grocery store wants to survey customers about how the store can improve. A survey is given to every fifth person entering the store.
2. A news program asks viewers to phone in a vote for or against increasing funding for the development of electric cars.
3. The student council wants to survey students about school activities. It randomly selects 25 students from each grade to answer a survey.
4. A concert promoter wants to ask people at a jazz concert what other jazz musicians they would like to see perform. The promoter surveys the people sitting in Section D of the concert hall.

BY THE WAY

Call-in surveys became popular after the 1980 presidential debate when a television network asked viewers to decide which candidate had "won" the debate.

6.3 Exercises and Applications

Extra Practice exercises on page 757

1. Suppose the homeroom teachers in a school are asked to send four students to participate on a school festival committee. Here is how some teachers chose the four students.

Ms. Rose mixes the names of all the girls in a box and chooses two without looking. Then she does the same for the boys' names.

Mr. Champine's homeroom is working in groups of four. He sends students from one of the groups.

Mrs. Santanella puts the names of all the students in a box, mixes the names, and pulls out four names without looking.

Mrs. Kim chooses the four students in the first row, closest to where she is standing.

   a. For each homeroom, tell if the teacher used a *random sample*, a *convenience sample*, a *cluster sample*, or a *stratified random sample.*

   b. How would you choose the four students? Why?

Identify the type of sample and describe the population of the survey. Then tell if the sample is biased. Explain your reasoning.

2. The school board wants to find out how voters feel about a proposed addition to the high school. Each board member randomly selects 30 names from the phone book and calls each number during a weekday afternoon.
3. A local sports station wants to find out how many hours per week people in the viewing area watch sporting events on television. The station surveys people at the nearby sports stadium.
4. The alumni club at Pioneer High School wants to set up an employment referral network among alumni. The club mails a survey to all graduates from the past 20 years and asks them to complete and return it.
5. A taxicab company wants to know if its customers are satisfied with the service. Each driver surveys every tenth customer during the day.

Teach⇔Interact

Checking Key Concepts

Student Progress
In order to help students assess their understanding of the concepts involved, review the correct answers in class so they can check their work.

Multicultural Note
Although the music called jazz (originally spelled *jass*) shows many external influences, including European musical forms, it is considered by many to be the only truly American form of music. Early twentieth century New Orleans is considered to be the birthplace of jazz, and many of the early great jazz musicians, such as Louis Armstrong and Jelly Roll Morton, were African American residents of this city.

Closure Question

List and describe six different ways of selecting a sample.
self-selected sample: subjects volunteer to be surveyed; systematic sample: subjects are chosen using a pattern for selection; convenience sample: subjects who are easy to reach are surveyed; random sample: subjects are chosen so that each subject has an equally likely chance of being selected; stratified random sample: the population is divided into groups and subjects from each group are randomly selected; cluster sample: subjects are chosen as a group rather than as individuals.

Apply⇔Assess

Suggested Assignment

Core Course
Day 1 Exs. 1–25
Day 2 AYP

Extended Course
Day 1 Exs. 1–25
Day 2 AYP

Block Schedule
Day 35 Exs. 1–25
Day 36 AYP

4. convenience sample; all the people at the concert; The sample is probably biased, since people sitting together may be acquaintances and share opinions. Also, seating prices may vary by section, so those in Section D may be overrepresentative of a particular economic group.

Exercises and Applications

1. a. Ms. Rose: stratified random sample; Mr. Champine: cluster sample; Mr. Santanella: random sample; Mrs. Kim: convenience sample

   b. Answers may vary. An example is given. I would use Mr. Santanella's method. It is random and gives all students an equal chance of participating.

2. random; voters; Yes; people with no phone or an unlisted number are underrepresented; some people may be called by more than one school board member and be overrepresented; people with daytime jobs are underrepresented; nonvoters may also be overrepresented.

3. convenience or cluster sample; people in the viewing area; Yes; sports fans are overrepresented.

4, 5. See answers in back of book.

Apply⇔Assess

Exercise Notes

Visual Thinking
Exs. 2–6 Ask students to select two of the exercises and prepare diagrams that demonstrate how these sampling processes work. Have them use their diagrams to describe the populations, discuss whether or not each sample is biased, and explain the similarities and differences between the two samples. This activity involves the visual skills of *interpretation* and *generalization.*

Teaching Tip
Ex. 6 Point out to students that it is unlikely that the results from a sample are exactly the same as for the entire population. They should understand that results obtained from a sample population only estimate information about the population.

Second-Language Learners
Ex. 6 Students learning English should be able to complete the writing task in part (b) independently.

Application
Exs. 7–12 In these exercises, students are introduced to an application of sampling to study a past culture. Students should recognize that, unlike many of the populations involved in studies that are related to people, the population in these exercises involves an area of Earth.

Interdisciplinary Problems
Exs. 7–12 Archeology is a field which has links to many other fields and disciplines. For example, radiocarbon dating was developed by physicists; geological dating procedures were developed by geologists; and reconstruction of the archeological site uses techniques developed in sociology, demography, economics, and geography.

Problem Solving
Ex. 12 A discussion of this exercise should help students better understand the problems involved in choosing a sampling method. Students should understand that no sampling method is unconditionally the best method for a particular situation. Rather, the variables in each situation help determine the best method.

6. **BIOLOGY** A biologist wants to estimate the number of deer in a state park. She gathers a random sample of 30 deer and tags each one. Then she releases the tagged deer back into the park. After the deer have mixed with the other deer, she gathers a second random sample of 30 deer and finds that there are 8 tagged deer in this sample.
   a. What percent of the deer in the second sample are tagged? Use this percent to estimate the number of deer in the entire park.
   b. **Writing** Do you think your estimate from part (a) is accurate? Why or why not? What could you do to get a better estimate?

Connection ARCHAEOLOGY

When it becomes too expensive to excavate an entire site, archaeologists resort to sampling portions of the site. Artifacts on the surface of a site may influence where to excavate.

The images below are of an excavation site, where the 42 red squares on each image show the portions to be excavated. For each image, tell what sampling method is suggested.

Archaeological dig at Port au Choix National Historic Park in Newfoundland, Canada

7. 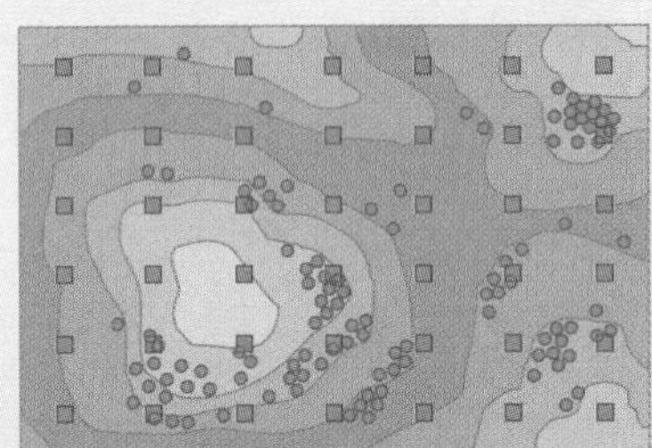

8. 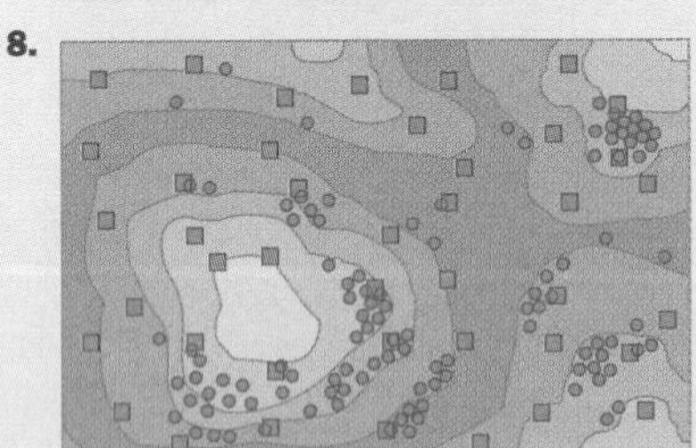

9. 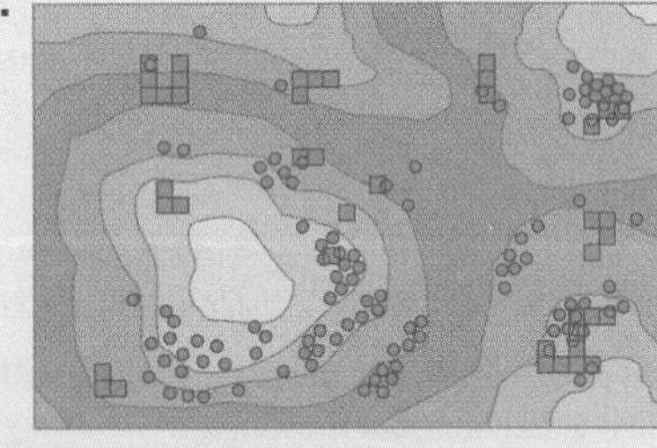

10. **Writing** If there are no artifacts on the surface of the site, which sampling method shown above seems the most reasonable? Why?

11. a. Each blue dot in the images above represents the location of a buried artifact. If you assume that all the artifacts that lie in or partially in the red squares will be found, which sampling method finds the most artifacts?
    b. Do you think this method will always be the best? Why or why not?

12. **Open-ended Problem** Suppose you are in charge of an archaeological dig and must use sampling to stay on budget. Describe the procedure that you would use in your attempt to find the most artifacts. Explain why you think that your procedure is the best.

6. a. $26\frac{2}{3}\%$; about 113 deer
   b. Answers may vary. An example is given. I think the estimate is reasonable. To get a better estimate, I would repeat the second sampling process several times and find an average. I would make sure each time that the sample is taken over the entire area of the park.
7. systematic sampling
8. random sampling
9. cluster sampling
10. systematic sampling; This would allow you to cover the whole site in a systematic manner.
11. a. cluster sampling
    b. Answers may vary. An example is given. I think no one sampling method will always produce the best results; under different circumstances, different methods will produce the most representative samples.
12. Answers may vary. An example is given. I think if there are no visible artifacts, I would use stratified or random sampling, attempting to cover all areas of the site. If there were visible artifacts or other clues as to where artifacts might be found, I would use cluster sampling.

INTERVIEW Donna Cox

Look back at the article on pages 234–236.

Computer images like the ones below, which were created by Donna Cox, are used to study the ecology of crops and the insects that attack them. Each image below shows a cornfield with 800 plants.

The **black squares** show plants with no larvae.

The **green squares** show plants with healthy larvae. These larvae damage crops.

The **red squares** show plants with infected larvae. These larvae do minimal damage to crops.

Other colors show plants with a mixture of healthy and infected larvae.

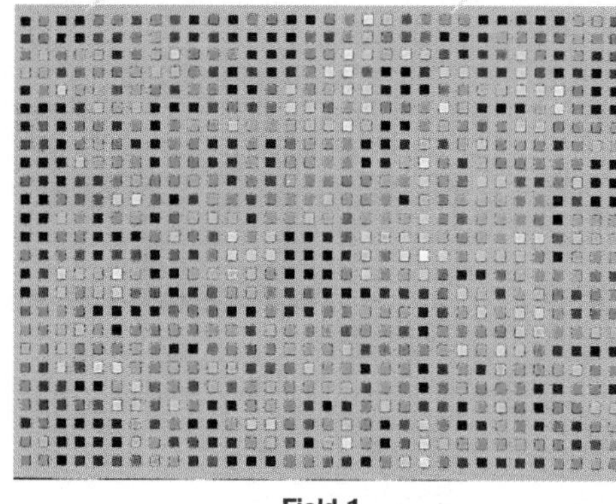

Field 1

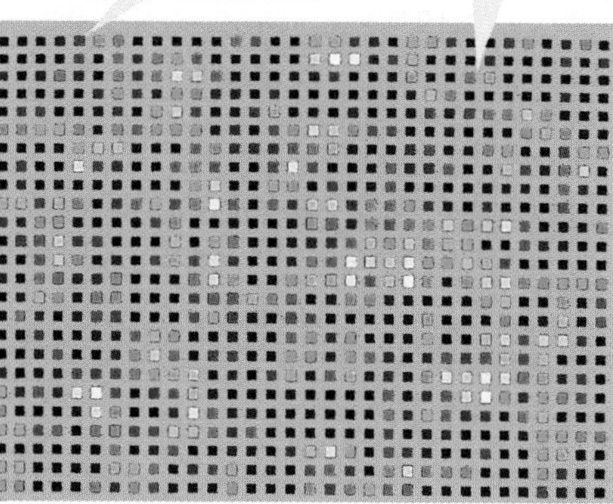

Field 2

13. Visual Thinking For each field, visually estimate the percent of the plants being damaged by healthy larvae. Which field will probably produce more corn? Why?

Use the indicated method to select a sample of 10 plants from each field. Note the color associated with each plant in your samples.

14. cluster **15.** systematic **16.** random

17. a. For each of your samples in Exercises 14–16, count the number of plants being damaged by healthy larvae. Convert the counts to percents.

b. For each field, do your sample percents from part (a) agree with each other? Do they agree with your visual estimate from Exercise 13? If not, explain any discrepancies.

18. Open-ended Problem Obtain another sample from each field. You may want to choose one of the sampling methods in Exercises 14–16 but this time use a sample with more than 10 plants, or you may want to use some other sampling method. Once again calculate the percent of plants being damaged by healthy larvae. Compare the results with those from Exercise 17.

19. Writing Suppose a farmer asks for your help in determining the damage being done to the farmer's corn crop by European corn borer larvae. Describe the steps you would take to judge the health of the crop without examining every plant.

Apply⇔Assess

Exercise Notes

Communication: Discussion

Ex. 13 Have students explain their method of visually estimating the percent of plants being damaged by healthy larvae. Have them tell whether they had used a small sample to estimate the entire field or whether they had other ways of estimating.

Teaching Tip

Ex. 14, 15 An element of randomness can be added to the sample selection for Ex. 14 by choosing the cluster sample randomly once the population is clustered. For Ex. 15, an element of randomness can be added to systematic sampling of this field by using random numbers to pick the starting units.

Interview Note

Ex. 17 Ask students to work in groups to make copies of the images showing the samples they have chosen. Have them use their drawings to make the estimates. This exercise will be helpful to visual learners who may have difficulty estimating populations by simply analyzing data.

Research

Ex. 19 A farmer must be able to limit the number of insects that are potentially destructive to his crop. One means of controlling the insects that damage a crop is by using natural predators. Many beneficial insects that feed on the eggs or larva of other insects are used as a natural control of destructive insects. Both the braconid wasp and the tachinid fly are natural predators of the European corn borer. Ask students to research the ways in which the braconid wasp and the tachinid fly are used as a natural control against the corn borer.

I think visible artifacts would indicate centers of activity and would probably indicate the presence of hidden artifacts.

13. Estimates may vary; field on the left: about 70%; field on the right: about 50%. Field 2; fewer plants are being damaged.

14–16. Answers may vary. Examples are given. Plants are numbered from 1 to 800, starting at the top of the first column and continuing from top to bottom in each column.

14. field on the left: the first ten plants in the tenth column; field on the right: the first ten plants in the ninth column

15. every plant whose number is a multiple of 80

16. Use a calculator to generate random numbers between 1 and 800. Example: plants 9, 737, 669, 205, 581, 198, 81, 585, 178, and 476

17, 18. Answers may vary.

19. Answers may vary. An example is given. I would obtain a computer image of the field like the ones created by Donna Cox. Then I would generate random numbers to determine a sample of say, 25%, of the plants to estimate how many are healthy. I would use my results to determine the health of the crop.

Apply⇔Assess

Exercise Notes

Topic Spiraling: Review

Ex. 20 As students review the description of the population written from the *Ongoing Assessment* in Section 6.1, they should discuss any possible biases that occur because of undercoverage of the population.

Communication: Writing

Ex. 20 As students write their descriptions for part (b), have them also describe how to avoid bias when deciding how they contact people to respond to the survey. Encourage them to take notes on how people respond to help them interpret the results at a later time.

Mathematical Procedures

Ex. 24 Remind students that data should be arranged in numerical order before the median is found.

ONGOING ASSESSMENT

20. Cooperative Learning Work with the members of your group from the *Ongoing Assessment* exercise from Section 6.1.

a. Describe how you will choose a random sample of at least 30 people to take the survey you wrote in Section 6.2.

b. **Writing** Describe how you will contact people to respond to your survey. You should consider these things:

- Will you interview people in person? interview people over the phone? distribute a written questionnaire? use some other method?
- When will you distribute your survey? Where?
- Is there any information you should record about each respondent?
- How will you know when you have responses from enough people? How will you ensure that someone does not take the survey more than once?

c. Distribute your survey from Section 6.2. Keep the surveys or a record of people's responses for your work in later sections.

SPIRAL REVIEW

For Exercises 21–23:

a. Describe any bias in the question. *(Section 6.2)*

b. Tell whether the data that can be gathered from each question are *categorical* or *numerical*. Then tell what type of graph you would use to display the data. *(Section 6.1)*

21. "A recent study says that people who read books regularly have a better vocabulary than people who seldom read books. How many books do you read in a month?"

22. "Do you support the decisions made at last Monday's board meeting?"

23. "What is your favorite flavor of ice cream?"

24. Find the mean, the median, and the mode of the data in the table below. *(Toolbox, page 790)*

City	Bern	Brasília	Buenos Aires	London	Madrid	Mexico City	Paris	Seoul	Tokyo
Price of 1 lb of cheese ($)	6.15	1.28	0.45	2.32	4.23	2.76	2.85	5.40	6.32

25. GOVERNMENT Use the table at the right. *(Section 3.5)*

a. Choose any two periods and write an exponential function to represent the growth. Check your function by comparing its predictions with the other data values in the table.

b. **Technology** Use a graphing calculator or statistical software to perform exponential regression. Compare the equation with your results from part (a).

Period	Number of new stamps issued
1839–1868	88
1869–1898	205
1899–1928	354
1929–1958	476
1959–1988	1277

20. Answers may vary

21–23. Answers may vary. Examples are given.

21. a. The first sentence encourages an exaggerated response because the respondents may want to look good to the pollster.

b. numerical; histogram

22. a. The question assumes the respondent is familiar with the results of the meeting.

b. categorical; bar graph

23. a. The wording of the question implies the respondent eats ice cream and has a favorite flavor. However, the question is not biased if possible responses include "none" and "I don't eat ice cream."

b. categorical; bar graph

24. 3.53; 2.85; no mode

25. a. Answers may vary. An example is given. Let x = the number of 30-year periods since period 1839–1868 and y = the number of new stamps issued. If the first and last data points are used, then the function is $y = 88(1.95)^x$; the equation fits the data reasonably well.

b. $y = 95.5(1.86)^x$; Answers may vary. The fit is similar.

Assess Your Progress

Vocabulary

numerical data (p. 238)
categorical data (p. 238)
population (p. 239)
sample (p. 239)
biased question (p. 245)
self-selected sample (p. 250)
systematic sample (p. 250)
convenience sample (p. 250)
random sample (p. 250)
stratified random sample (p. 250)
cluster sample (p. 250)
biased sample (p. 251)

Tell whether the data that can be gathered about each variable are *categorical* or *numerical.* Then describe the categories or numbers. *(Section 6.1)*

1. savings account balance
2. the month of a person's birthday
3. a person's coat size
4. figure skating scores

For each graph, tell if it is an appropriate way to display the data. If not, tell what type of display would be better and why. *(Section 6.1)*

5. 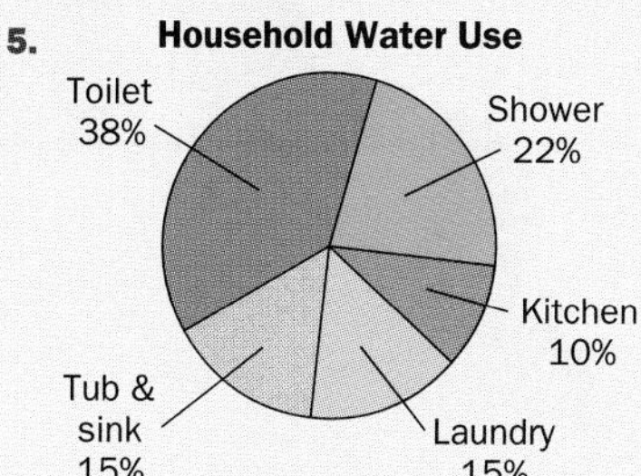

6. 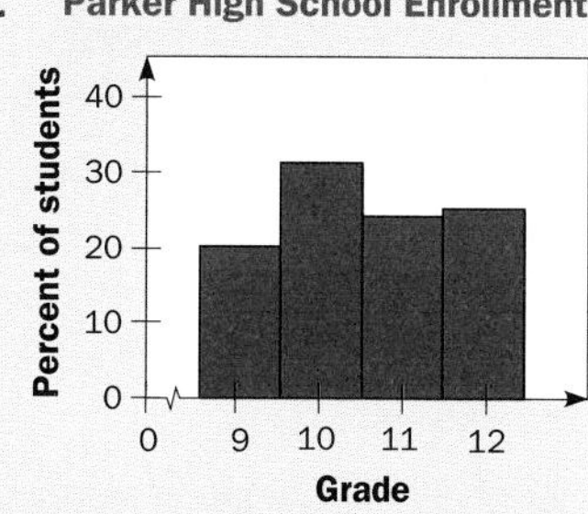

Tell why each question may be biased. Then describe what changes you would make to improve the question. *(Section 6.2)*

7. "College tuition costs are on the rise. Do you favor government spending on student financial aid?"

8. Your dentist asks, "Have you been flossing regularly?"

For Exercises 9 and 10, identify the type of sample and describe the population of the survey. Then tell if the sample is biased. Explain. *(Sections 6.1 and 6.3)*

9. A veterinarian wants to know where to set up a new practice within a county. She randomly telephones 25 people from four towns within the county to ask how many pets are in each household.

10. A department store wants to know customers' opinions of its merchandise and service. The store offers a chance of winning a gift certificate through a random drawing if the customer fills out a survey.

11. Writing Describe two ways to obtain a sample of 50 people from your community to ask about the need for more community parks. *(Section 6.3)*

12. Journal Is it ever appropriate to ask a biased question? Explain.

Apply⇔Assess

Assess Your Progress

Random sampling is a fundamental principle of conducting an unbiased survey. Writing a description of ways to obtain a sample, as in Ex. 11, will help students organize and clarify their understanding of this important procedure.

Journal Entry
You may wish to have students share their answers to Ex. 12 with the class.

Progress Check 6.1–6.3

See page 282.

Practice 35 for Section 6.3

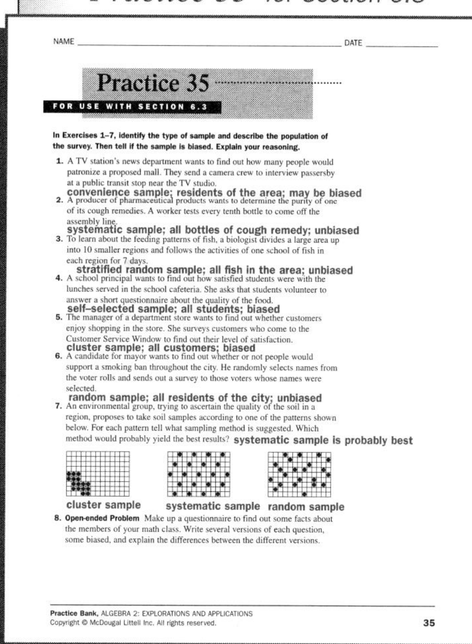
NAME ______________________ DATE __________

Practice 35
FOR USE WITH SECTION 6.3

In Exercises 1–7, identify the type of sample and describe the population of the survey. Then tell if the sample is biased. Explain your reasoning.

1. A TV station's news department wants to find out how many people would patronize a proposed mall. They send a camera crew to interview passersby at a public transit stop near the TV studio.
convenience sample; residents of the area; may be biased

2. A producer of pharmaceutical products wants to determine the purity of one of its cough remedies. A worker tests every tenth bottle to come off the assembly line.
systematic sample; all bottles of cough remedy; unbiased

3. To learn about the feeding patterns of fish, a biologist divides a large area up into 10 smaller regions and follows the activities of one school of fish in each region for 7 days.
stratified random sample; all fish in the area; unbiased

4. A school principal wants to find out how satisfied students were with the lunches served in the school cafeteria. She asks that students volunteer to answer a short questionnaire about the quality of the food.
self-selected sample; all students; biased

5. The manager of a department store wants to find out whether customers enjoy shopping in the store. She surveys customers who come to the Customer Service Window to find out their level of satisfaction.
cluster sample; all customers; biased

6. A candidate for mayor wants to find out whether or not people would support a smoking ban throughout the city. He randomly selects names from the voter rolls and sends out a survey to those voters whose names were selected.
random sample; all residents of the city; unbiased

7. An environmental group, trying to ascertain the quality of the soil in a region, proposes to take soil samples according to one of the patterns shown below. For each pattern tell what sampling method is suggested. Which method would probably yield the best results? **systematic sample is probably best**

cluster sample **systematic sample** **random sample**

8. Open-ended Problem Make up a questionnaire to find out some facts about the members of your math class. Write several versions of each question, some biased, and explain the differences between the different versions.

Practice Bank, ALGEBRA 2: EXPLORATIONS AND APPLICATIONS

35

Assess Your Progress

1. numerical; nonnegative decimal numbers
2. categorical; the names of the month
3. categorical; examples: small, medium, and large
4. numerical; usually numbers between 0.0 and 6.0
5. Yes.
6. No; the data are categorical and should be displayed in a bar graph or a circle graph.

7, 8. Answers may vary. Examples are given.

7. The first sentence encourages a positive response. The first sentence should be omitted.
8. The question is too vague. A better question would be, "How many times a week do you floss between your teeth?"
9. stratified random sample; residents of the county; Yes; people without phones or with unlisted phone numbers are underrepresented.
10. self-selected sample; customers of the store; Yes; only customers (and perhaps people who are not even regular customers) who want to win a gift certificate are represented.
11. Answers may vary. Examples are given. You could call 50 phone numbers chosen randomly or systematically (say every twentieth number) from the phone book. You could divide the community into sections geographically and choose 50 residences randomly but evenly among the sections.
12. Answers may vary. An example is given. There are situations in which it might be appropriate to ask a biased question. A candidate trying to gain support for his or her position might ask leading questions.

Plan⇔Support

Objective

- Make histograms and box plots.

Recommended Pacing

Core and Extended Courses
Section 6.4: 2 days

Block Schedule
Section 6.4: 2 half-blocks
(with Sections 6.3 and 6.5)

Resource Materials

Lesson Support
Lesson Plan 6.4
Warm-Up Transparency 6.4
Overhead Visuals:
Folder 6: Box Plots
Practice Bank: Practice 36
Study Guide: Section 6.4
Explorations Lab Manual:
Diagram Master 1

Technology
Graphing Calculator
McDougal Littell Mathpack
Stats! Activity Book:
Activities 5, 7, 17, and 20
Internet:
http://www.hmco.com

Warm-Up Exercises

Use the data set 5, 8, 12, 17, 20, 21, 28, 33 for Exs. 1–4.

1. Which data are in the top 25% of the data set? 28, 33
2. Which data are in the bottom 25% of the data set? 5, 8
3. Find the number midway between the data in Ex. 1. 30.5
4. Find the number midway between the data in Ex. 2. 6.5
5. What type of graph would you use to display categorical data? bar graph or circle graph

Section

6.4 Displaying and Analyzing Data

Learn how to...

- **make histograms and box plots**

So you can...

- **create appropriate displays for numerical data**

Newspapers and magazines often display data in graphs so that readers can easily understand the information presented. Sometimes a poorly presented display can lead people to draw wrong conclusions about the data. It is important to know how to make good data displays so that you present information accurately.

EXPLORATION
COOPERATIVE LEARNING

Organizing Data in a Box Plot

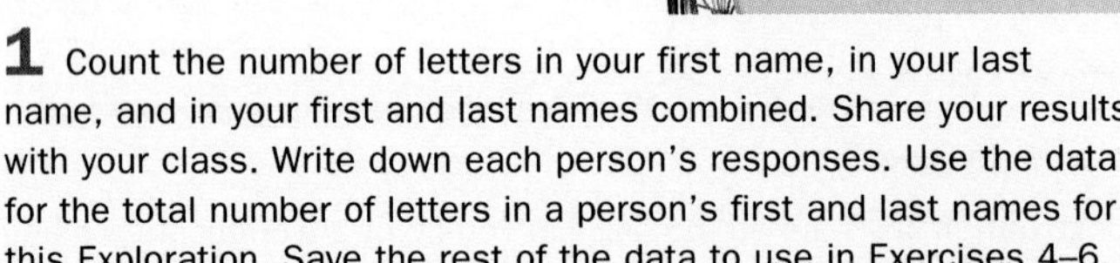

Work with a partner.
You will need:
- scissors
- graph paper

1 Count the number of letters in your first name, in your last name, and in your first and last names combined. Share your results with your class. Write down each person's responses. Use the data for the total number of letters in a person's first and last names for this Exploration. Save the rest of the data to use in Exercises 4–6.

2 List the data in order from least to greatest on a strip of graph paper. Put one number in each square. Then cut the paper so that there are no extra squares to the left or right of the data.

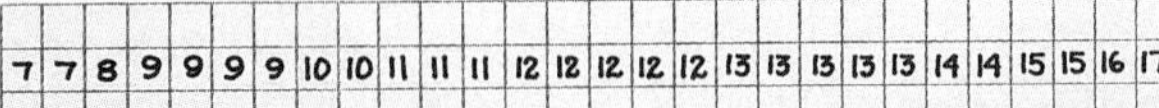

3 Fold the strip of paper in half. Then fold the paper in half again. Open the paper up and draw a line on each fold.

7 7 8 9 9 9 9 | 10 10 11 11 11 12 12 | 12 12 12 13 13 13 13 | 13 14 14 15 15 16 17

Questions

1. What does the middle fold tell you about the data? What do the other folds represent?
2. Below what number of letters do 25% of the names fall? 75% of the names?

Exploration Note

Purpose
The purpose of this Exploration is to have students organize data in a box plot and to see that the data can be divided at the 25%, 50%, and 75% points.

Materials/Preparation
Each pair of students needs scissors and graph paper.

Procedure
Students record the data on a strip of graph paper. The data should be written in numerical order and each number should be written in one square. Students then cut the paper so that there are no extra squares. They fold the strip of graph paper in half and then in half again. Students unfold the strip and draw a line on each fold.

Closure
Students should see that the strip shows the 25%, 50%, and 75% points of their data. The numbers at both ends are the shortest and longest string of letters. The middle fold is the point at which 50% of the data are above or below this point. The first and last folds are the points below which 25% of the data fall and above which 25% of the data fall.

Explorations Lab Manual
See the Manual for more commentary on this Exploration.

In the Exploration you completed the first steps of making a *box-and-whisker plot*, or simply a *box plot*. A **box plot** is a display that shows the median, the *quartiles*, and the *extremes* of a data set.

The **lower extreme** is the least data value.

The **median** divides the data set into two halves.

The **upper extreme** is the greatest data value.

The **lower quartile** is the median of the lower half of the data.

The **upper quartile** is the median of the upper half of the data.

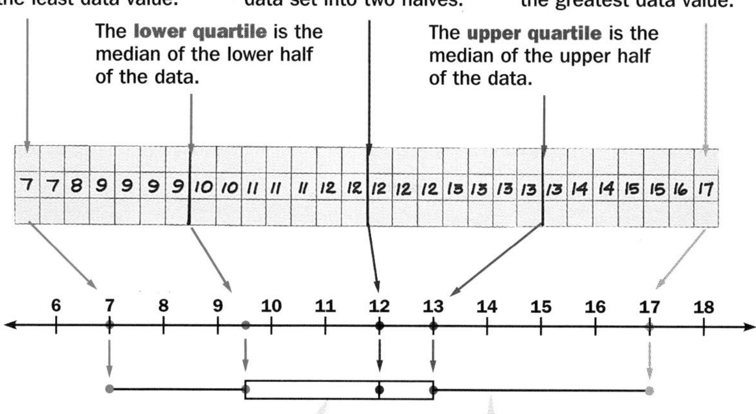

The box in a box plot shows where the **middle 50%** of the data fall.

Each "whisker" in a box plot shows where the **extreme 25%** of the data fall.

EXAMPLE 1 Application: Economics

Make a box plot of the data in the table.

Average 1994 Per Capita Personal Income for Midwestern States	
State	**Income ($)**
Illinois	**23,784**
Indiana	**20,378**
Iowa	**20,265**
Kansas	**20,896**
Michigan	**22,333**
Minnesota	**22,453**
Missouri	**20,717**
Nebraska	**20,488**
North Dakota	**18,546**
Ohio	**20,928**
South Dakota	**19,577**
Wisconsin	**21,019**

SOLUTION

Step 1 Find the median, the quartiles, and the extremes of the data and plot these points just below a number line.

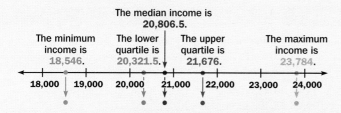

The median income is 20,806.5.

The minimum income is 18,546.

The lower quartile is 20,321.5.

The upper quartile is 21,676.

The maximum income is 23,784.

18,000 19,000 20,000 21,000 22,000 23,000 24,000

Step 2 Make the box plot.

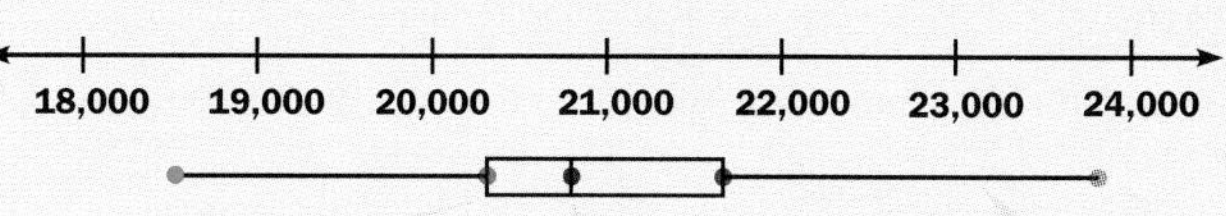

Draw a box from the lower quartile to the upper quartile.

Then draw a vertical line segment through the median.

Draw line segments from the box to the extremes.

Teach⇔Interact

Learning Styles: Visual

The content of this section provides an excellent opportunity for students with visual learning styles to help others with different learning styles. Also, by working in groups of students with varying learning styles, each student can benefit from viewing the material from the perspective of a classmate with a learning style different than his or her own.

Additional Example 1

Make a box plot of the data in the table.

Bushels of corn per acre	
Farm A	150.4
Farm B	118.4
Farm C	120.6
Farm D	135.7
Farm E	139.1
Farm F	164.3
Farm G	134.2
Farm H	138.1
Farm I	119.5
Farm J	150.0

Step 1 Find the median, the quartiles, and the extremes of the data and plot these points just below the number line.

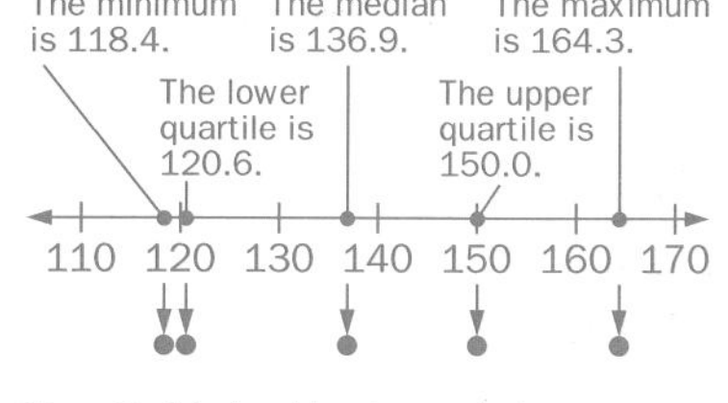

Step 2 Make the box plot.

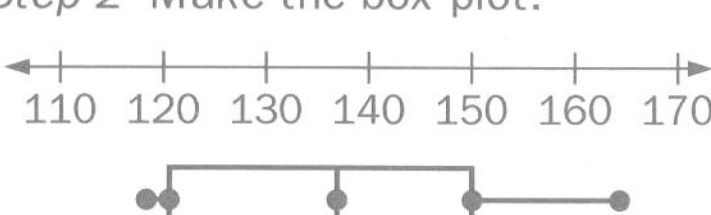

ANSWERS Section 6.4

Exploration

1. The middle fold tells you that half of the data points lie to the left of the fold and half to the right. The left-most fold tells you that one-quarter of the data points lie to the left of the fold and three-quarters lie to the right. The right-most fold tells you that one-quarter of the data points lie to the right of the fold and three-quarters lie to the left.
2. Answers may vary.

Teach⇔Interact

Section Notes

Communication: Discussion Some students may have difficulty understanding the difference between an absolute frequency diagram and a relative frequency diagram. Use the graphs on this page to compare the two. Emphasize the key mathematical difference between the two, namely, that absolute frequency is a *count* while relative frequency is a *percent*.

Using Technology Students can use Activities 17 and 20 in the *Stats! Activity Book* for additional practice creating and analyzing histograms and box plots.

About Example 2

Common Error
Students tend to confuse bar graphs and histograms. Remind students that the bars in a histogram touch each other, but the bars in a bar graph do not.

Additional Example 2

The percents of the popular vote won by the successful candidate in each of the presidential elections from 1948 to 1992 are given below. Display the results in a histogram.

Year	Percent
1948	49.6
1952	55.1
1956	57.4
1960	49.7
1964	61.1
1968	43.4
1972	60.7
1976	50.1
1980	50.7
1984	58.8
1988	53.9
1992	43.2

Step 1 Group the data into equal intervals. The data values range from 43.2 to 61.1. Choose intervals of equal width that will allow you to cover the entire range of values, such as 41–46, 46–51, 51–56, 56–61, and 61–66.

Histograms

A **histogram**, like a box plot, is a graph that displays numerical data. The horizontal axis is a number line divided into intervals of equal width. The vertical axis shows the *frequency* of the data items that fall within each interval.

When the frequency is a count of the data within each interval, the vertical axis shows the **absolute frequency**.

When the frequency gives the count of the data within each interval as a *percent* of all the data, the vertical axis shows the **relative frequency**.

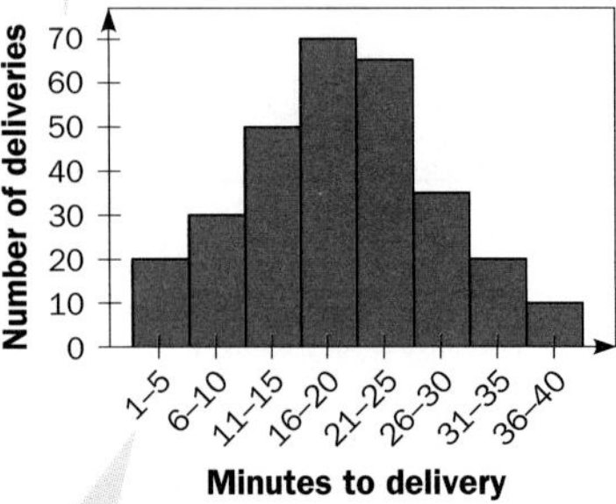

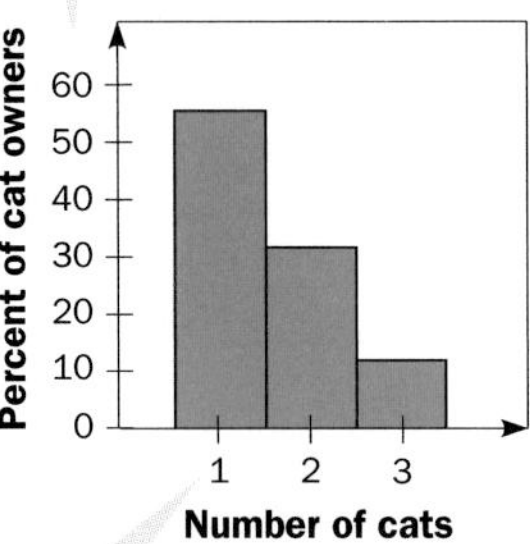

Here, each interval covers a period of 5 min.

Here, each interval includes a single integer.

A histogram displays the overall shape of the distribution of data values. The histogram on the left above shows a **symmetric distribution** because its shape is approximately symmetrical about a vertical line passing through the interval with the greatest frequency. The histogram on the right shows a **skewed distribution** because it is not symmetric.

EXAMPLE 2 Application: Transportation

A company took a survey of its employees to learn the amount of time they spend commuting to work. The survey data (reported to the nearest 5 min) are shown below. Display the results in a histogram.

15, 25, 30, 40, 60, 40, 25, 15, 15, 20, 30, 75, 20, 15, 65, 40,
30, 20, 15, 10, 20, 35, 45, 35, 30, 20, 10, 55, 25, 15, 5, 45, 20,
10, 5, 15, 30, 30, 50, 15, 5, 10, 45, 30, 20, 20, 10, 30, 20, 30

SOLUTION

Step 1 Group the data into equal intervals. (Between 5 and 10 intervals is best.)

The data values range from 5 to 75. Choose intervals of equal width that will allow you to cover the entire range of values, such as 1–10, 11–20, 21–30, 31–40, 41–50, 51–60, 61–70, and 71–80.

Think and Communicate

1. Answers may vary. An example is given. The histogram shows how the data are divided among the nine categories and displays the mode. The box plot shows the median and how closely about the median 50% of the data are grouped.
2. Estimates may vary; about $20,000. On the histogram, there are values far to the right of the median. The box plot has a whisker far to the right of the median.
3. greater than; The values at the far right end of the histogram and to the right of the median in the box plot would increase the mean, perhaps not by much, since so many of the data are grouped closely around the median.

Step 2 Organize the data into a frequency table.

Interval	Tally	Frequency	Relative frequency
1–10	卌 III	8	16%
11–20	卌 卌 卌 II	17	34%
21–30	卌 卌 II	12	24%
31–40	卌	5	10%
41–50	IIII	4	8%
51–60	II	2	4%
61–70	I	1	2%
71–80	I	1	2%
Total		50	100%

Find the relative frequency by dividing each absolute frequency by the total number of data values.

$\frac{2}{50} = 0.04 = 4\%$

Step 3 Draw the histogram.

For each interval, draw a rectangle with a width equal to the interval width and a height equal to the frequency of the data within the interval. You may draw either type of histogram shown below.

Absolute frequency histogram

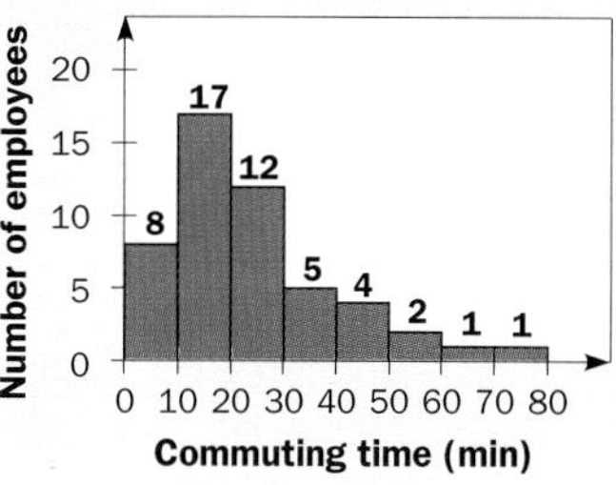

Relative frequency histogram

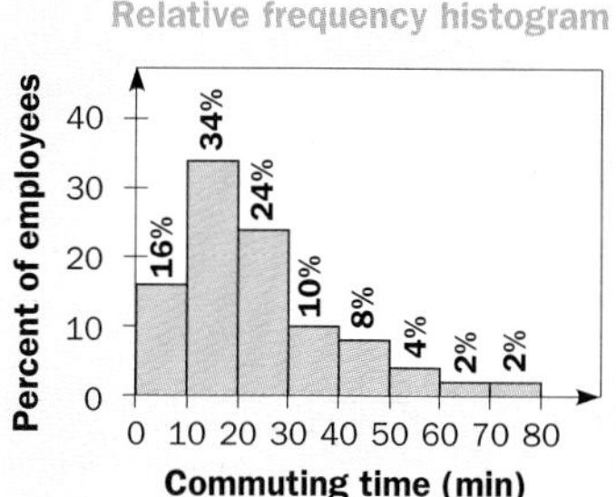

You can use technology to draw histograms and box plots. The displays below both show the same data taken from an auto magazine survey of new car prices.

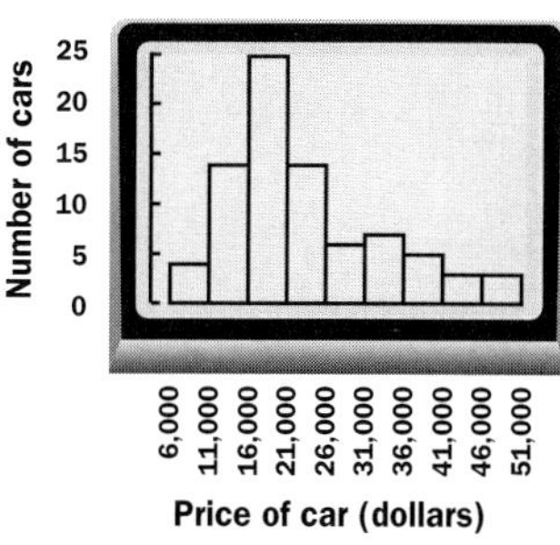

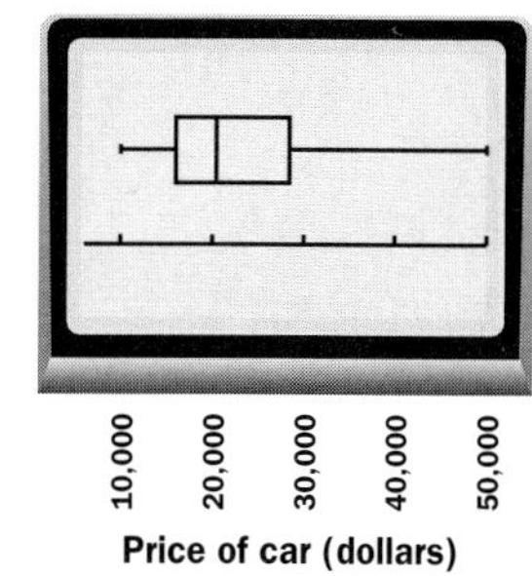

For more information about drawing histograms and box plots, see the *Technology Handbook*, p. 813.

THINK AND COMMUNICATE

1. What visual impressions do you get from each graph of car prices?
2. Estimate the median car price. How does each graph show that there are car prices much greater than the median?
3. Do you think the mean car price is *about the same*, *greater than*, or *less than* the median price? Explain.

Technology Note

To graph a histogram on a TI–82, students must use the frequency values to set Ymin and Ymax in the viewing window. Xscl sets the width of each bar in the window, beginning at Xmin. Students should enter the frequency data in a list from the STAT EDIT menu. In the STAT PLOTS menu, one of the plots can then be turned ON with the histogram chosen for Type and the correct Xlist selected. Freq:1 should also be selected. Then from the ZOOM menu, ZoomStat should be selected so that all x-values will be included on the graph. An appropriate viewing window should also be chosen. The graph is then ready to be viewed.

To graph box plots on a TI–82, Ymin and Ymax are ignored. Xmin and Xmax should be determined to include the whiskers and the box. The procedures for entering the rest of the data are the same as for graphing a histogram except that the plot Type chosen in the STATS PLOT menu is the box plot.

Teach⇔Interact

Additional Example 2 (continued)

Step 2 Organize the data into a frequency table.

Interval	Tally	Freq.	Relative Freq.
41–46	II	2	16.7%
46–51	IIII	4	33.3%
51–56	II	2	16.7%
56–61	III	3	25%
61–66	I	1	8.3%
Total		12	100%

Step 3 Draw the histogram. For each interval draw a rectangle with a width equal to the interval width and a height equal to the frequency of the data within the interval. You may draw either type of histogram shown below.

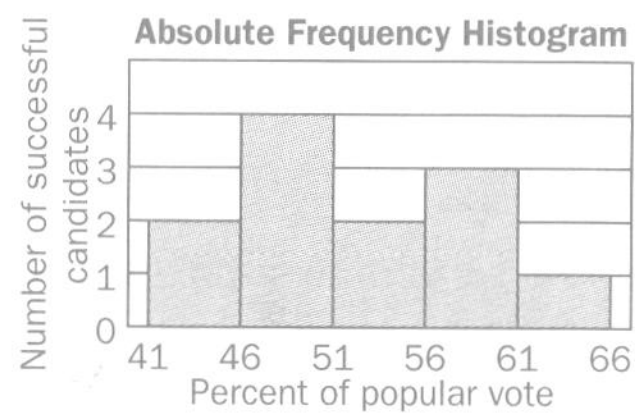

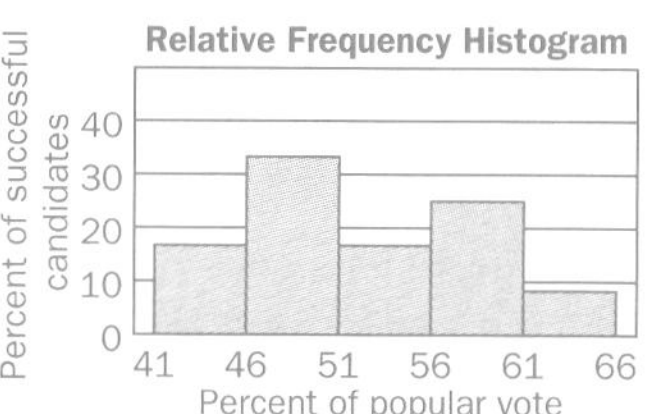

Think and Communicate

Question 1 illustrates an important reason data are displayed in a histogram. A histogram shows the mode of the data.

Teach⇔Interact

Checking Key Concepts

Mathematical Procedures
Remind students to select reasonable intervals so that the histogram does not distort the data by spreading them out too much or by compacting them with too few intervals.

Visual Thinking
For questions 2 and 3, check students' understanding by asking them to label key elements of the two box plots that they have created. The elements that should be identified include the median, the lower and upper extremes, the lower and upper quartiles, and the box and the whiskers. This activity involves the visual skills of *recognition* and *generalization.*

Closure Question

Explain how to draw a box plot and a histogram.

To draw a box plot, put the data in numerical order and find the median, the lower quartile, and the upper quartile. Draw a box from the lower quartile to the upper quartile. Draw a vertical line segment through the median and draw line segments from the box to the extreme values of the data. To draw a histogram, decide on intervals of equal width for a frequency table and enter the data into a table. Choose the *y*-scale to match either the absolute frequency of the data or the relative frequency of the data. Plot the data using the intervals of equal width as the *x*-scale of the graph.

Apply⇔Assess

Suggested Assignment

Core Course
Day 1 Exs. 1–9, 11, 12
Day 2 Exs. 14, 15, 17–26

Extended Course
Day 1 Exs. 1–13
Day 2 Exs. 14–26

Block Schedule
Day 36 Exs. 1–9, 11, 12
Day 37 Exs. 14, 15, 17–26

CHECKING KEY CONCEPTS

The weights, to the nearest quarter pound, of fish gathered in one harvest at a fish farm are listed below. Use the data to answer Questions 1 and 2.

2.25, 1.0, 0.75, 1.5, 2.0, 2.75, 4.0, 1.0, 1.25, 1.75, 1.5, 2.5, 3.25, 5.0, 3.5, 1.0, 2.0, 1.75, 2.25, 1.0, 1.75, 1.25, 2.25, 2.5, 1.25, 3.0, 1.0, 1.5, 2.0, 1.25, 2.75, 1.25, 1.75, 2.25, 1.5, 1.25, 3.75, 1.0, 4.75, 1.5, 3.0, 1.75, 1.0, 3.25, 4.5, 1.25, 2.25, 3.75, 1.5

1. Make a histogram of the data. Does your histogram display *relative frequencies* or *absolute frequencies*? Is the distribution *symmetric* or *skewed*?

2. a. Make a box plot of the data.

b. Above what weight are 50% of the harvest? 25% of the harvest?

The weights of the fish gathered in another harvest are shown below.

1.5, 0.75, 1.25, 1.75, 1.0, 2.25, 1.5, 1.25, 1.0, 1.75, 2.75, 1.25, 1.5, 1.75, 2.0, 1.75, 1.25, 1.5, 1.0, 3.0, 2.0, 2.25, 1.25, 1.5, 1.0, 1.75, 2.5, 1.5, 1.0, 1.25, 1.75, 1.5, 2.0, 1.5, 3.25, 1.75, 1.25, 1.5, 2.25, 1.75, 2.0, 1.25, 1.5, 1.75, 2.5, 2.0, 1.5, 2.0, 1.25, 1.5

3. Make a box plot of the data. Compare this box plot with the box plot you made in Question 2. Which harvest was better? Why?

6.4 Exercises and Applications

Extra Practice exercises on page 758

1. Writing What information does a box plot tell you about a set of data? What information does a histogram tell you?

For each histogram, tell whether the distribution is *symmetric* or *skewed*.

2.

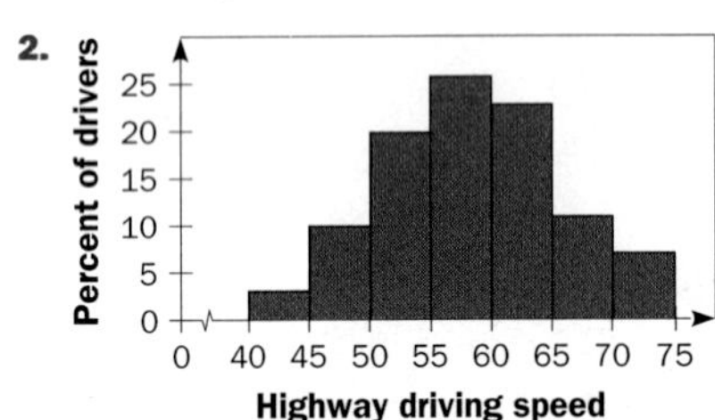

3.

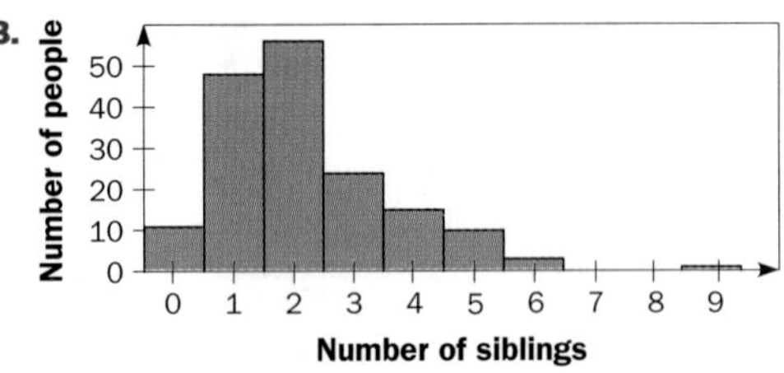

Use the data you gathered in the Exploration on page 258 to answer Exercises 4–6.

4. Make a box plot of the number of letters in the first names of the students in your class. Identify the medians, quartiles, and extremes of the data.

5. Repeat Exercise 4 using the number of letters in the last names of the students in your class.

6. Compare the box plots you made in Exercises 4 and 5. Describe the relationship between the medians, quartiles, and extremes.

Checking Key Concepts

1. Width of intervals may vary. The graph on the left is an absolute frequency histogram. The graph on the right is a relative frequency histogram. The distribution is skewed.

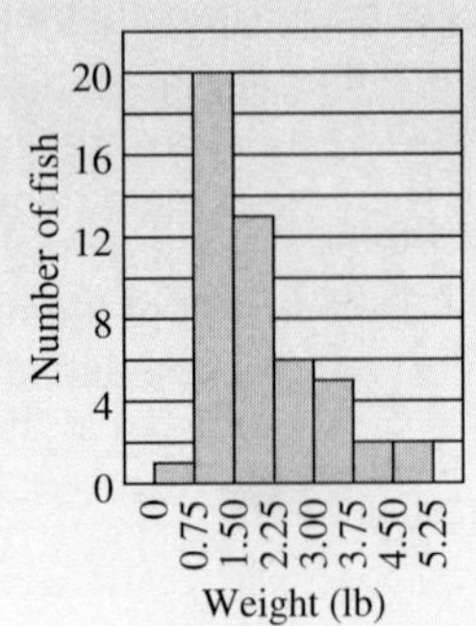

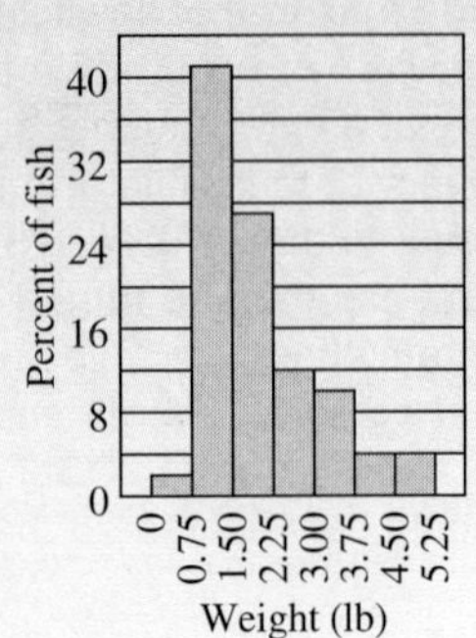

Connection GEOGRAPHY

The highest elevation in the United States is Mt. McKinley, in Alaska. Native Americans living near the mountain call it *Denali*, which means "the great one." The table below gives the highest elevation in each of the 50 states.

Mt. McKinley

HIGHEST ELEVATIONS PER STATE (FEET)

West of the Mississippi						East of the Mississippi					
AK	20,320	MN	2,301	TX	8,749			IN	1,257	NC	6,684
AZ	12,633	MO	1,772	UT	13,528			KY	4,139	OH	1,549
AR	2,753	MT	12,799	WA	14,410			ME	5,267	PA	3,213
CA	14,494	NE	5,426	WY	13,804			MD	3,360	RI	812
CO	14,433	NV	13,140			AL	2,405	MA	3,487	SC	3,560
HI	13,796	NM	13,161			CT	2,380	MI	1,979	TN	6,643
ID	12,662	ND	3,506			DE	442	MS	806	VT	4,393
IA	1,670	OK	4,973			FL	345	NH	6,288	VA	5,729
KS	4,039	OR	11,239			GA	4,784	NJ	1,803	WV	4,861
LA	535	SD	7,242			IL	1,235	NY	5,344	WI	1,951

7. a. Find the median, the upper and lower quartiles, and the extremes of the elevation data for all 50 states.

b. Make a box plot. Between what numbers do 50% of the data fall?

c. If you removed Alaska's highest elevation from the data set, how would the box plot change?

8. a. Make a comparative box plot for the highest elevations east and west of the Mississippi River.

b. The Appalachian Mountains are east of the Mississippi River, and the Rocky Mountains are west. Which mountain range has the higher mountains? How can you tell from the comparative box plot?

9. a. Make a histogram to display the elevation data for all 50 states. Is the distribution *symmetric* or *skewed*?

b. Open-ended Problem Estimate the "middle" of the distribution.

c. Find the mean and median of the elevation data. How do these values compare with your estimate from part (b)?

d. Do you think the *mean* or the *median* is a better measure of the "middle" of the distribution? Why?

10. Research Use an almanac or a set of encyclopedias to find the lowest elevation in each of the 50 states. Display the data in either a box plot or a histogram. Describe what the graph shows about the data. Compare the lowest elevation data with the highest elevation data given here.

BY THE WAY

Denali is a word in a Native American language that is part of a group of related languages called *Athabaskan*. Speakers of Athabaskan languages live in parts of Canada, the United States, and Mexico.

2. a. 0 1 2 3 4 5

b. 1.75 lb; 2.75 lb

3. 0 1 2 3 4

Both box plots have the same lower extreme and lower quartile. The median, upper quartile, and upper extreme of the first data set are greater than those for the second. The first harvest was better. In that harvest, half of the fish were over 1.75 lb, with 25% over 2.75 lb, and the largest fish weighed 5 lb. In the second harvest, 75% of the fish were under 2 lb and the largest fish weighed only 3.25 lb.

Exercises and Applications

1. A box plot gives you the upper and lower extremes, the upper and lower quartiles, and the median of a set of data. It shows where the middle 50% of the data lie. A histogram gives you absolute or relative frequencies corresponding to different intervals. It gives you an idea of how the data are distributed over the intervals and indicates the range and the mode.

2–10. See answers in back of book.

Apply⇔Assess

Exercise Notes

Using Technology
Exs. 4, 5 Students can graph both box plots on the same axes on a TI-82 graphing calculator. The first graph plots in the middle and the second plots at the bottom. Students can use the STAT CALC menu, selecting 3:SetUp, 1-Var Stats, and then the appropriate Xlist to view the median, extremes, and quartiles of each data set.

Multicultural Note
Exs. 7–10 The Athabaskan peoples who live near Denali (Mt. McKinley) include the Ingalik, the Atna, and the Tanaina. All three of these groups utilized the waterways around them to hunt. Both the Ingalik and the Atna people employed complicated systems of nets and basket traps to hunt salmon and other fish in area rivers. The Tanaina people were unique among the Athabaskan groups because of their coastal environment. Using tools like clubs, harpoons, and spears, the Tanaina hunted sea mammals such as seals, sea otters, sea lions, and belugas (white whales).

Mathematical Procedures
Ex. 7 Removing an *outlier* is sometimes used when analyzing data so that more accurate inferences can be drawn from the remaining data sets. Part (c) of this question is an example of that technique.

Reasoning
Ex. 9 Asking students to read their explanations for part (d) aloud will help them to clarify the relationship between the shape of the histogram and the position of the mean in relation to the median on the histogram. In a skewed distribution, extreme values can seriously affect the mean. For this reason, students should see that the median is a better measure of the center of a data set when the data are skewed.

Research
Ex. 10 You may want students to explain why they chose to display the data as a box plot or as a histogram. Have them justify their answers.

Apply⇔Assess

Exercise Notes

Communication: Discussion
Exs. 11, 12 These exercises give students an opportunity to check their understanding of whether a particular graph is an appropriate display for a given set of data. A discussion of students' responses can be instructive to everyone in the class.

Teaching Tip
Ex. 13 Refer students who are having difficulty with this exercise to the graphs in the solution to Example 2 on page 261. Using this specific case, ask students if they can use each type of histogram to create the other type.

Historical Connection
Exs. 14–16 Groups of German-speaking immigrants settled in eastern Pennsylvania in the late 17th and 18th centuries. A mispronunciation of *Deutsch*, the German word meaning *German*, caused the immigrants to become known as Pennsylvania Dutch.

Research
Ex. 16 Some students may be interested in researching the years in which the states in their geographic region achieved statehood. They can also research the reason for any outliers in the data.

Communication: Drawing
Ex. 17 Ask students where the mean and the median appear in the histogram of a nearly symmetric distribution of data (close to each other and near the middle of the tallest bar). Ask students to draw in the mean and median lines of their histograms. They should see that the median divides the area of the histogram in half, whereas the mean can be considered the balance point of the histogram if it were made of a solid material. Students may find it helpful to shade their histograms to demonstrate these ideas.

For each graph, tell why the graph may misrepresent the data. Then suggest a way to improve the graph.

11.

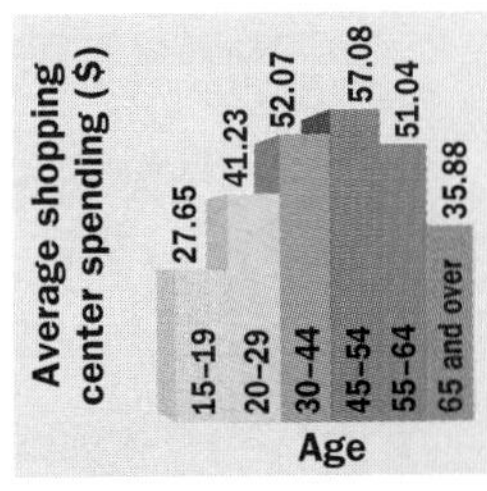

12.

Indianapolis 500 winners
0 10 20 30
33 15 6 3 7 1 5 2 2 2
1–3 4–6 7–9 10–12 13–15 16–18 19–21 22–24 25–27 28–30
Starting position

13. Writing If you are given only an absolute frequency histogram, can you create a relative frequency histogram? Can you create an absolute frequency histogram given only a relative frequency histogram? Explain.

Connection HISTORY

European settlers began arriving in what is now the United States in the sixteenth century. In 1787, Delaware, Pennsylvania, and New Jersey became the first former colonies to achieve statehood. The box plots compare the year in which a present-day state was settled by Europeans and the year in which it achieved statehood.

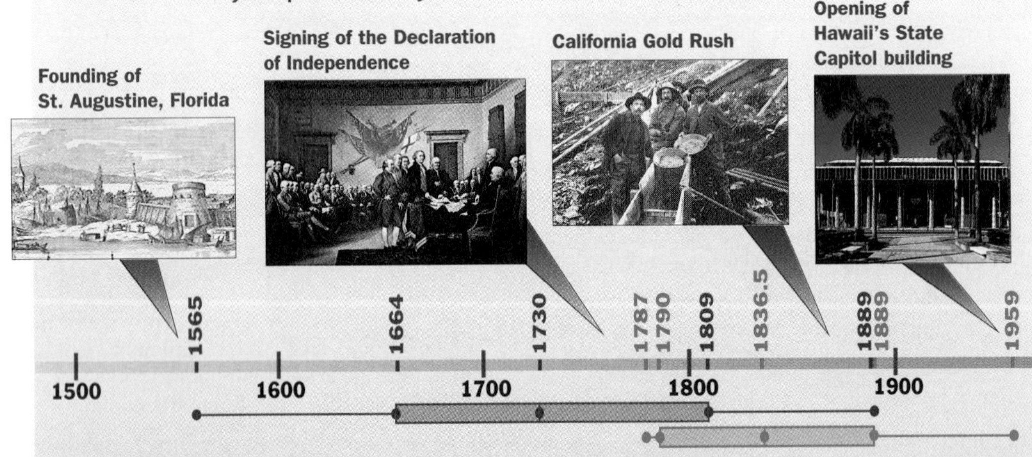

14. By what year did half of the present-day states have European settlements? By what year did half of the present-day states achieve statehood?

15. In 1788, eight former colonies became states—the largest number achieving statehood in one year. This number is twice the next closest number to achieve statehood in one year. What effect does this have on the statehood box plot?

16. Research Find out the year your state was settled by Europeans and the year it achieved statehood. In which quarter of the data is the year your state was settled? In which quarter of the data is the year your state achieved statehood?

BY THE WAY

The oldest city founded by Europeans in what is now the United States is St. Augustine, Florida. It was founded in 1565 by settlers from Spain.

11. The intervals on the horizontal axis are not of equal width. The data should be divided into intervals of equal width.
12. The figures depicting race cars on the top of each bar makes the graph confusing. For example, the bar representing the interval 7–9 should be twice as tall as the bar representing the interval 10–12. The figures should be removed.
13. Yes; divide the number of data points in each interval by the total number of data points to find each percent. No; you can do so only if you are also given the total number of data points, in which case you can find the absolute frequency for each interval by multiplying the percent for that interval by the total number of data points.
14. 1730; 1836.5
15. Since at least 11 territories had become states by 1788, it indicates that the lower quartile is about 1788. Since the lower extreme is 1787, the lower half of the box plot will be very condensed.
16. Answers may vary.

BUSINESS **Use the data below for Exercises 17–19.**

1994 Recording Industry Sales by Age						
Age group (years)	10–14	15–19	20–24	25–29	30–34	35–39
Percent of all sales	7.0	16.8	15.4	12.6	11.8	11.5

17. Make a histogram to display the data in the table from the Recording Industry Association of America. Does the histogram show a *symmetric distribution* or a *skewed distribution*? Explain.

18. Make another histogram of the data. Combine the data so that each interval covers a period of 10 years. How does the choice of intervals change the appearance of the distribution?

19. **Writing** People who are 40 years old or older accounted for 24% of the total recording industry sales in 1994.

a. People who are 10 and older accounted for what percent of the recording industry sales in 1994? Why do you think this total is not equal to 100%?

b. Carly wants to compare spending by people under age 20, people between 20 and 30, and people over 30. Can she use a histogram? Explain.

BY THE WAY

The soundtrack from Walt Disney's "The Lion King" and Ace of Base's "The Sign" tied for the best-selling album in 1994, at about 7 million copies each.

ONGOING ASSESSMENT

20. **Cooperative Learning** Work with the members of your group from the *Ongoing Assessment* exercise from Section 6.1.

a. Organize the data you gathered in Section 6.3. Decide how you will display the data. You should use at least one histogram or box plot and at least one circle graph or bar graph.

b. Make a frequency table for each set of numerical data. Find the mean and median of the data. If you make a histogram, decide if it will display absolute frequencies or relative frequencies.

c. Draw the graphs to display your data. What conclusions can you make about your data?

SPIRAL REVIEW

Identify the type of sample and describe the population of the survey. Then tell if the sample is biased. Explain your reasoning. *(Section 6.3)*

21. The members of the drama club are trying to decide whether they should perform a fall play. The group surveys every fifth person entering the door at the spring play.

22. A sports club wants to find out which football team is the most popular. The manager asks people to call in a vote for their favorite team.

Simplify. *(Toolbox, page 777)*

23. $\sqrt{54}$ **24.** $\sqrt{200}$ **25.** $\sqrt{\frac{300}{3}}$ **26.** $\sqrt{\frac{880}{5}}$

Apply⇔Assess

Exercise Notes

Teaching Tip

Exs. 17, 18 You may want to use these exercises to have students explain how the presentation of data in a graph can convey information. As the interval in the histogram is changed, the appearance of the distribution may lead to different conclusions.

Using Technology

Ex. 18 Using graphing calculators or software, students can quickly make observations of the effects of the width of the intervals on the overall appearance of the histogram.

Assessment Note

Ex. 20 Each group can show their graphs to another group, asking that group to explain what conclusions they draw from the graphs.

Practice 36 for Section 6.4

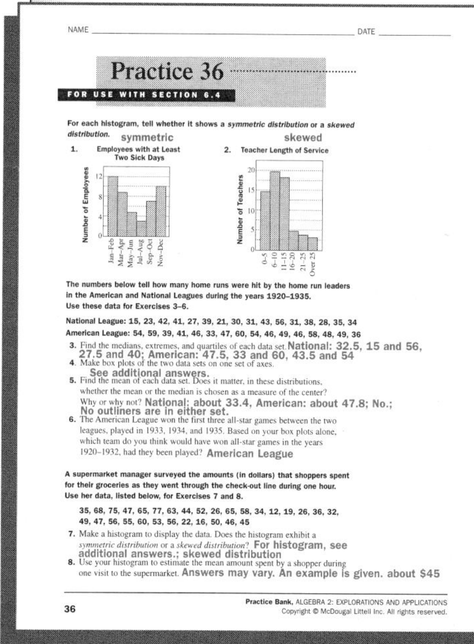

NAME ______ DATE ______

Practice 36

FOR USE WITH SECTION 6.4

For each histogram, tell whether it shows a *symmetric distribution* or a *skewed distribution*.

1. Employees with at Least Two Sick Days — **symmetric**

2. Teacher Length of Service — **skewed**

The numbers below tell how many home runs were hit by the home run leaders in the American and National Leagues during the years 1920–1935. Use these data for Exercises 3–6.

National League: 15, 23, 42, 41, 27, 39, 21, 30, 31, 43, 56, 31, 38, 28, 35, 34

American League: 54, 59, 39, 41, 46, 33, 47, 60, 54, 46, 49, 46, 58, 48, 49, 36

3. Find the medians, extremes, and quartiles of each data set. **National: 32.5, 15 and 56, 27.5 and 40; American: 47.5, 33 and 60, 43.5 and 54**

4. Make box plots of the two data sets on one set of axes. **See additional answers.**

5. Find the mean of each data set. Does it matter, in these distributions, whether the mean or the median is chosen as a measure of the center? Why or why not? **National: about 33.4, American: about 47.8; No.; No outliners are in either set.**

6. The American League won the first three all-star games between the two leagues, played in 1933, 1934, and 1935. Based on your box plots alone, which team do you think would have won all-star games in the years 1920–1932, had they been played? **American League**

A supermarket manager surveyed the amounts (in dollars) that shoppers spent for their groceries as they went through the check-out line during one hour. Use her data, listed below, for Exercises 7 and 8.

35, 68, 75, 47, 65, 77, 63, 44, 52, 26, 65, 58, 34, 12, 19, 26, 36, 32, 49, 47, 56, 55, 60, 53, 56, 22, 16, 50, 46, 45

7. Make a histogram to display the data. Does the histogram exhibit a *symmetric distribution* or a *skewed distribution*? **For histogram, see additional answers.; skewed distribution**

8. Use your histogram to estimate the mean amount spent by a shopper during one visit to the supermarket. **Answers may vary. An example is given. about $45**

36

Practice Bank, ALGEBRA 2: EXPLORATIONS AND APPLICATIONS

17.

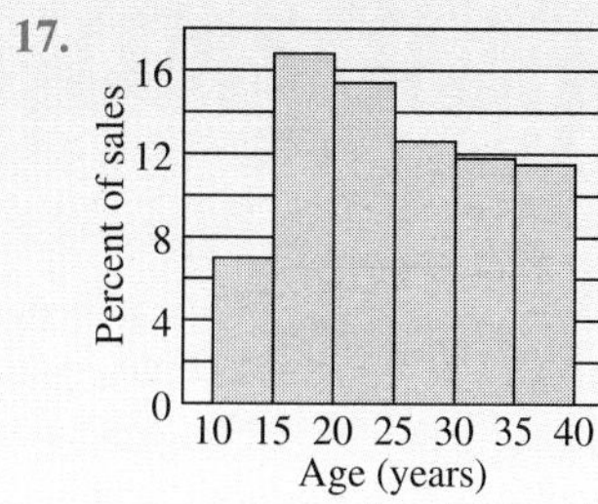

skewed distribution; The shape of the histogram is not symmetric.

18.

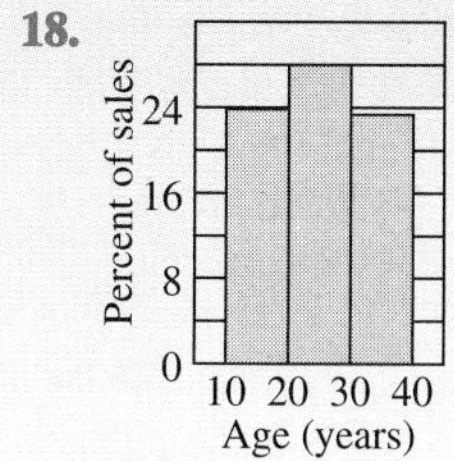

The distribution is approximately symmetric.

19. a. 99.1%; 0.9% of the buyers were under the age of 10.

b. Since the age intervals are not equal, she should not use a histogram. A circle graph might be a good choice for displaying the data.

20. Answers may vary.

21. systematic; people at the spring play; Yes; people who are not at the spring play are not represented.

22. self-selected; Yes; only people with strong feelings one way or the other will probably bother to call.

23. $3\sqrt{6}$

24. $10\sqrt{2}$

25. 10

26. $4\sqrt{11}$

Plan⇔Support

Objectives

- Find and interpret the range, interquartile range, and standard deviation of a data set.
- Determine the variability of data.

Recommended Pacing

Core and Extended Courses
Section 6.5: 1 day

Block Schedule
Section 6.5: $\frac{1}{2}$ block
(with Section 6.4)

Resource Materials

Lesson Support
Lesson Plan 6.5
Warm-Up Transparency 6.5
Overhead Visuals:
Folder 6: Box Plots
Practice Bank: Practice 37
Study Guide: Section 6.5
Explorations Lab Manual:
Additional Exploration 11,
Diagram Master 1

Technology
Technology Book:
Spreadsheet Activity 6
Graphing Calculator
Spreadsheet Software
McDougal Littell Mathpack
Stats! Activity Book:
Activities 13 and 19
Internet:
http://www.hmco.com

Warm-Up Exercises

Use the following set of data.

5.50 5.62 4.88 5.07 5.26
5.57 5.53 5.63 5.29 5.35

1. Find the mean of the data. 5.37
2. Find the extremes in the data set. 4.88 and 5.63
3. How far are the extremes from the mean of the data? 0.49 units and 0.26 units
4. How many of the data are within 0.24 units of the mean? 6
5. Simplify the expression: $\sqrt{\frac{(8-6)^2+(6-6)^2+(4-6)^2}{3}}$ about 1.63

Section

6.5 Describing the Variation of Data

Learn how to...

- **find and interpret the range, interquartile range, and standard deviation of a data set**

So you can...

- **determine the variability of data, such as ozone readings**

Life on Earth would not be possible without the ozone layer in the upper atmosphere filtering the ultraviolet rays of the sun. Recent studies have shown that the thickness of the ozone layer is decreasing. In the tropics, unlike other areas, the change of seasons has little effect on the thickness of the ozone layer. At non-tropical latitudes, the ozone layer is thickest in the spring and thinnest in the fall.

THINK AND COMMUNICATE

The readings below, which give the thickness of the ozone layer in Dobson units (DU), were taken in Fresno, California, for two 20-day periods.

March readings: 427, 466, 372, 299, 293, 284, 298, 284, 314, 302, 286, 296, 306, 308, 320, 318, 344, 345, 354, 381

November readings: 317, 330, 316, 296, 270, 271, 295, 277, 275, 275, 269, 275, 272, 267, 270, 291, 275, 268, 261, 296

The comparative box plot displays these readings.

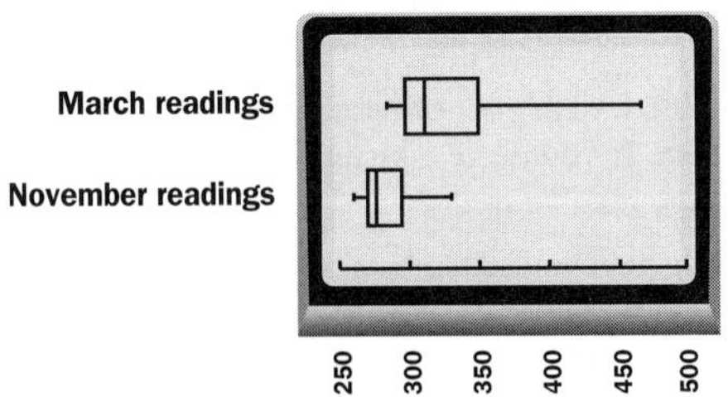

BY THE WAY

The mean ozone reading over the Antarctic dropped from 321 DU in 1956 to 117 DU in 1993. The ozone layer in this area is so thin that it is now referred to as "the Antarctic ozone hole."

1. How does the graph show that the ozone over Fresno is generally thicker in the spring than in the fall?
2. In which month, *March* or *November*, would it be more likely to obtain a reading that is much different from the reading on the previous day? Why?

To determine the variability of a set of data, it is often helpful to find the *range*, the *interquartile range*, and any *outliers* of the data set. The **range** of a data set is the difference between the maximum and minimum data values. The **interquartile range (IQR)** of a data set is the difference between the upper quartile and the lower quartile. A data value can be considered an **outlier** if its distance from the nearer quartile is more than 1.5 times the interquartile range.

ANSWERS Section 6.5

Where necessary, answers are rounded to the nearest tenth, unless otherwise indicated.

Think and Communicate

1. In the comparative box plot the lower quartile for March readings lies above the upper quartile of November readings, so 75% of the March readings lie above the upper quartile for the November readings. Also, at least 25% of the March readings lie above the upper extreme for the November readings.
2. in March; Not only is the range greater, but also the box is longer, meaning that even the middle 50% of the data are more spread out than the middle 50% of the data for November.

EXAMPLE 1 Application: Earth Science

Use the ozone readings on the previous page.

a. Find the range and the interquartile range for each set of readings.

b. Identify the outliers in each data set.

SOLUTION

a. First find the minimum and maximum values (shown in gold below) and the upper and lower quartiles (shown in green) for each data set.

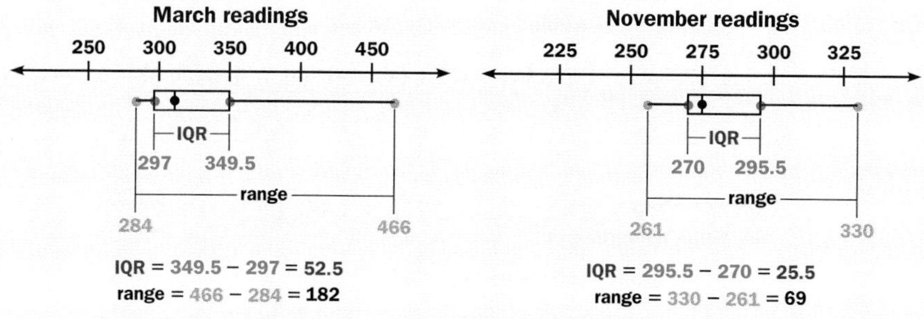

b. Use each IQR from part (a) to set bounds (shown in blue) for the outliers.

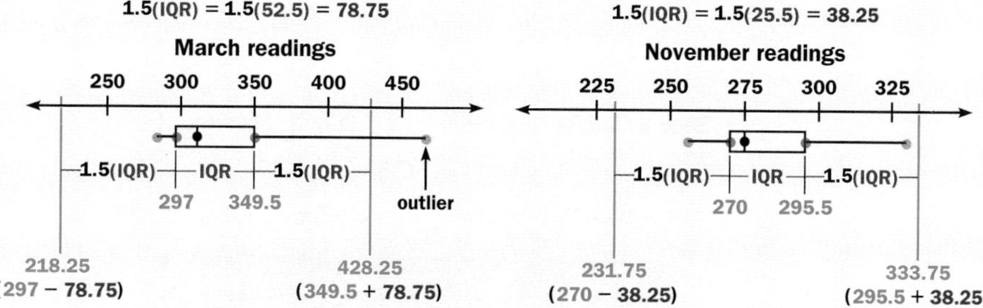

Any values in the March data set that are less than 218.25 or greater than 428.25 are outliers. The ozone reading of 466 is an outlier.

Any values in the November data set that are less than 231.75 or greater than 333.75 are outliers. Since all data values are between 231.75 and 333.75, there are no outliers.

THINK AND COMMUNICATE

3. For each data set in Example 1, compare the range and interquartile range. How do these numbers describe the variability of each data set?

4. A scientist analyzing the March ozone data decides to recalculate the mean, median, range, and interquartile range without the 466 reading. Why do you think the scientist would do this?

Think and Communicate

3. The range of the March data set is more than two-and-a-half times that of the November data set, which means that the March data set varies much more in its overall range. The interquartile range of the March data set is more than twice that of the November data set, which means that the middle half of the data from the March data set varies more than twice as much as the middle half from the November data set.

4. The data point is not only an outlier, but it is also so much greater than any other number in the data set that the scientist may have suspected it was due to a faulty reading or some fluke in the weather.

Teach⇔Interact

Additional Example 1

The following data sets show the home run totals for 10 consecutive years by Babe Ruth and Roger Maris.

Ruth: 54, 59, 35, 41, 46, 25, 47, 60, 54, 46

Maris: 13, 23, 26, 16, 33, 61, 28, 39, 14, 8

a. Find the range and the interquartile range for each set of home runs.

First find the minimum and maximum values and the upper and lower quartiles for each data set.

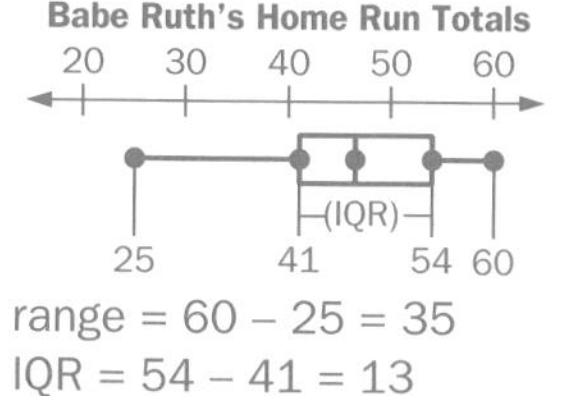

range = 60 – 25 = 35

IQR = 54 – 41 = 13

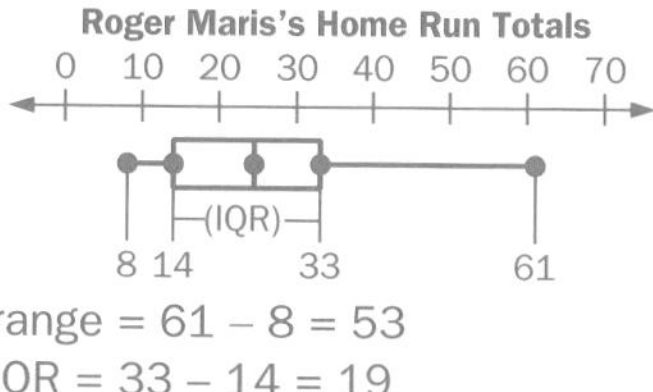

range = 61 – 8 = 53

IQR = 33 – 14 = 19

b. Identify any outliers in each set.

Use the IQR from part (a) to set bounds for the outliers. The outliers are any data values that are more than 1.5 times the interquartile range from the quartiles.

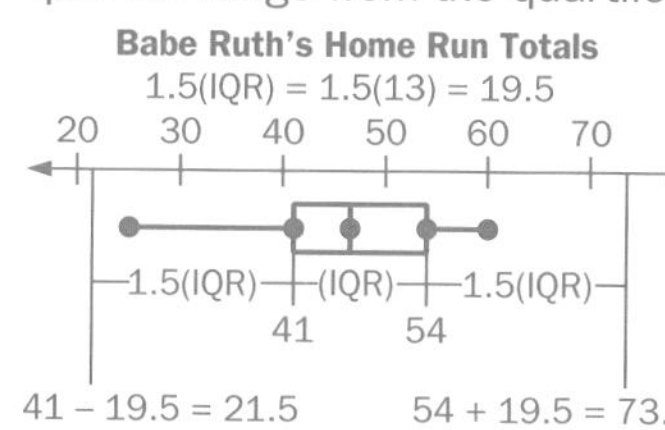

Any values in the Ruth data set that are less than 21.5 or greater than 73.5 are outliers. Since no data values fit these conditions, there are no outliers in this data set.

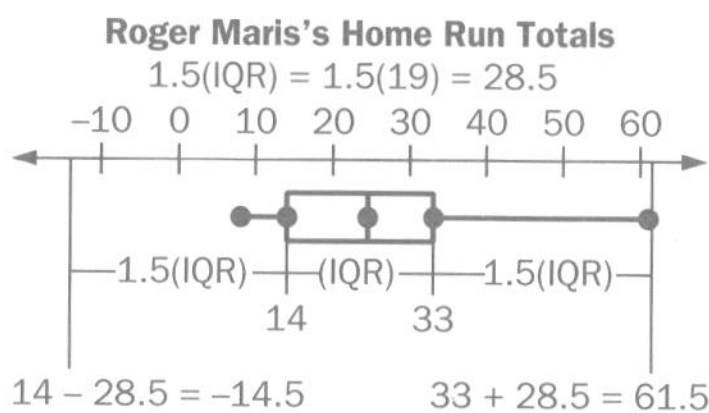

Any values in the Maris data set that are less than –14.5 or greater than 61.5 are outliers. There are no outliers in this data set.

Teach⇔Interact

Section Notes

Reasoning
You may want to point out that if the standard deviation of a set of data is small, then the values are concentrated near the mean, and if the standard deviation is large, the values are scattered greatly about the mean. This means that the proportion of data that must lie within k standard deviations of the mean is at least $1 - \frac{1}{k^2}$, where $k > 1$. See Ex. 12 on page 270.

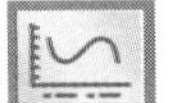
Using Technology
Activities 13 and 19 in the *Stats! Activity Book* offer additional opportunities for students to use statistical software to compute standard deviations and to use standard deviations to reach conclusions.

Additional Example 2

Use the home run totals from Additional Example 1.

a. Find the standard deviation for the Babe Ruth data.
Use a table.
Step 1 Find the mean of the data.
$\bar{x} = \frac{467}{10} = 46.7$
Step 2 Subtract the mean from each data value.
Example: 54 – 46.7 = 7.3
Step 3 Square each difference from Step 2.
Example: $7.3^2 = 53.29$
Step 4 Sum the squares.
sum = 1076.1
Step 5 Divide the sum by the number of data values and take the square root.
$\sigma = \sqrt{\frac{1076.1}{10}} \approx 10.37$

Home Run Total	total – $\bar{x}$	(total – $\bar{x}$)^2
54	7.3	53.29
59	12.3	151.29
35	–11.7	136.89
41	–5.7	32.49
⋮	⋮	⋮
60	13.3	176.89
54	7.3	53.29
46	–0.7	0.49
467		1076.1

The standard deviation is about 10.37.

Standard Deviation

The range and the interquartile range measure the variability of a data set using only the quartiles and extremes of the data. Another measure of variability, called *standard deviation*, uses the square of the deviation of every data value from the mean.

Standard Deviation

For a data set $x_1, x_2, x_3, \ldots, x_n$, where $\bar{x}$ (read "x bar") is the mean of the data values and n is the number of data values, the **standard deviation** σ (read "sigma") is:

$$\sigma = \sqrt{\frac{(x_1 - \bar{x})^2 + (x_2 - \bar{x})^2 + \cdots + (x_n - \bar{x})^2}{n}}$$

EXAMPLE 2 Application: Earth Science

Use the ozone readings on page 266.

a. Find the standard deviation for the March data.

b. Find the standard deviation for the November data.

SOLUTION

You can find the standard deviation of a set of data by using a table or by using a graphing calculator or statistical software.

a. Use a table.

Step 1 Find the mean of the data.
$\bar{x} = \frac{6597}{20} = 329.85$

Step 2 Subtract the mean from each data value.
Example:
427 – 329.85 = 97.15

Step 3 Square each difference from Step 2.
Example:
$(97.15)^2 = 9438.1225$

Ozone reading	reading – $\bar{x}$	(reading – $\bar{x}$)^2
427	97.15	9,438.1225
466	136.15	18,536.8225
372	42.15	1,776.6225
299	–30.85	951.7225
⋮	⋮	⋮
345	15.15	229.5225
354	24.15	583.2225
381	51.15	2,616.3225
6597		46,288.55

Step 4 Sum the squares.

Step 5 Divide the sum by the number of data values and take the square root.

$\sigma = \sqrt{\frac{46{,}288.55}{20}} \approx 48.1$

The standard deviation is about 48.1.

SOLUTION

b. Use a graphing calculator or statistical software.

First enter the data. Then have the calculator or computer find the standard deviation. It is usually shown as σ_x or σ_n.

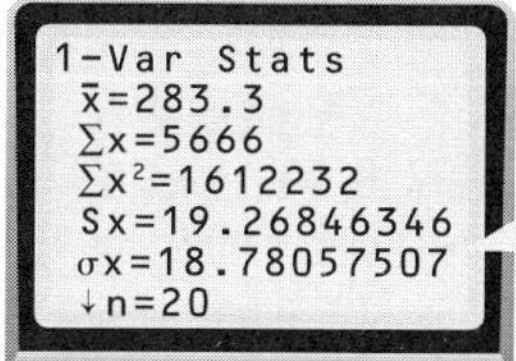

The standard deviation is about **18.8**.

WATCH OUT!

In the display shown, S_x is the standard deviation for a sample. This book will use the standard deviation for a population, σ_x, regardless of whether the data come from a sample or an entire population.

Statisticians often describe a distribution of data by the percent of the data that fall within a certain number of standard deviations of the mean. The histogram below shows the percent of data from the November ozone readings within one and two standard deviations of the mean.

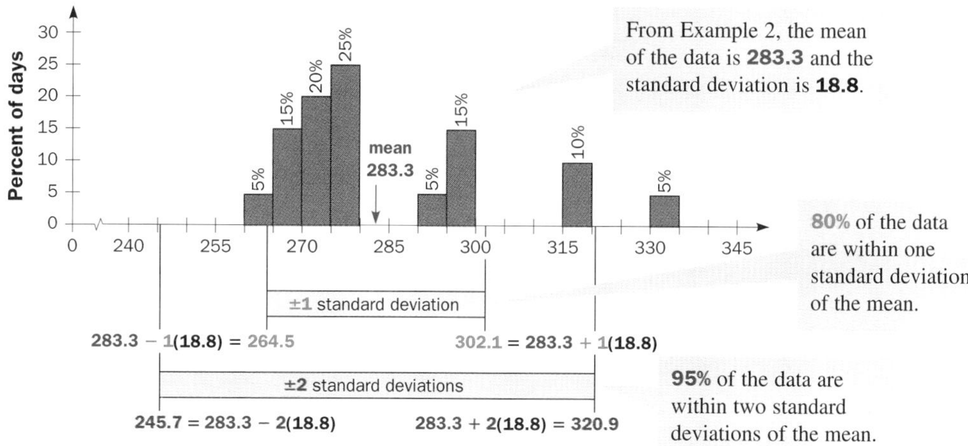

From Example 2, the mean of the data is **283.3** and the standard deviation is **18.8**.

80% of the data are within one standard deviation of the mean.

95% of the data are within two standard deviations of the mean.

CHECKING KEY CONCEPTS

MEDICINE Doctors often take a sample of blood from a patient to help with a diagnosis. The white blood cell counts from 27 patients are shown at the right.

1. Make a box plot of the data. Are there any outliers?
2. Find the mean and the standard deviation of the data.
3. Identify the values that are within one standard deviation of the mean. What percent of the values are within one standard deviation?
4. A white blood cell count greater than 10,000 indicates the possibility of a bacterial infection. How many standard deviations from the mean is 10,000?

White blood cell counts
5620, 5730, 5750, 6210, 6390, 6750, 6900, 7030, 7230, 7450, 7600, 7710, 7730, 7850, 8090, 8370, 8630, 8880, 9060, 9240, 9380, 9440, 9700, 9890, 10,250, 10,900, 11,070

Checking Key Concepts

1. 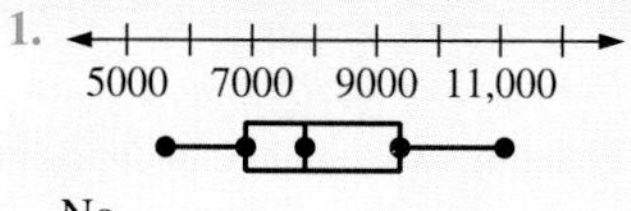

No.

2. 8105.6; 1535.2
3. values between 6570.4 and 9640.8; 63.0%
4. 1.2 standard deviations

Teach⇔Interact

Additional Example 2 (continued)

b. Find the standard deviation for the Roger Maris data.

Use a graphing calculator or statistical software.

First enter the data.

```
L1    L2    L3
13
23
26
16
33
61
28
L1={13,23,26,16...
```

Then have the calculator or computer find the standard deviation. It is usually shown as σ_x or σ_n.

```
1-Var Stats
x̄=26.1
Σx=261
Σx²=9005
Sx=15.60947006
σx=14.80844354
↓n=10
```

The standard deviation is about 14.81.

Checking Key Concepts

Communication: Discussion It may be helpful to discuss the differences between the data summaries obtained from quartiles and from a standard deviation. Students can compare the information obtained from the box plot in question 1 to the information obtained from the mean and the standard deviation in question 2.

Closure Question

Define the following.

a. range of a data set
the difference between the maximum and minimum data values

b. interquartile range of a data set
the difference between the upper quartile and the lower quartile

c. outlier of a data set
a data value whose distance from the nearer quartile of the data set is more than 1.5 times the interquartile range

d. standard deviation
the square root of the quotient of the sum of the squares of the deviation of every data value from the mean and the number of data values

Apply⇔Assess

Suggested Assignment

Core Course
Exs. 1–13, 17–19, 23–27

Extended Course
Exs. 1–27

Block Schedule
Day 37 Exs. 1–13, 17–19, 23–27

Exercise Notes

Reasoning
Exs. 4–7 As students compare their results, ask them if there are data for which this method of estimating a standard deviation will not give reliable results. (Data sets having outliers.)

Student Study Tip
Exs. 9, 10 Point out to students that the quartiles of a data set are appropriate measures of spread when the median is used as a measure of the center of a data set. When the mean is used as a measure of the center, the standard deviation is an appropriate measure of spread.

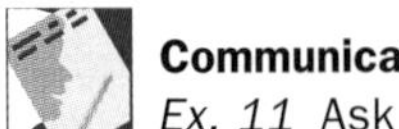

Communication: Discussion
Ex. 11 Ask students to discuss what shape a histogram of the data would have. The more symmetric the histogram is, the more evenly spread out the distribution is about the mean. Students should see that under these conditions, only a very small percent of the data will lie farther than two standard deviations from the mean.

6.5 Exercises and Applications

Extra Practice *exercises on page 758*

The box plot at the right shows the number of minutes a restaurant's customers sit at their table before they place their orders. Use the box plot to find each value asked for in Exercises 1–3.

0 2 4 6 8 10 12 14 16

1. the range **2.** the interquartile range **3.** an outlier

One way to estimate the standard deviation of a set of data is to divide the range by 4. Use this rule to estimate the standard deviation for each set of data. Then find the standard deviation and compare your results.

4. 10, 12, 13, 16, 18 **5.** 15, 22, 30, 40, 55, 75

6. 6, 9, 13, 18, 22, 28, 32 **7.** 1, 5, 15, 25

8. Writing Can a data set have a standard deviation of zero? Explain.

Connection BIOLOGY

In 1995 there were more than 900 plant and animal species protected under the United States Endangered Species Act. The map shows the number of endangered species in each state.

ENDANGERED SPECIES BY STATE as of September 30, 1995

NOTE: Numbers not additive; species often in more than one state

9. Draw a box plot of the data. Find the range and the interquartile range. Identify any outliers.

10. Find the standard deviation of the data.

11. About what percent of the data are within one standard deviation of the mean? within two standard deviations of the mean?

12. *Chebyshev's rule* says that the proportion of data within k standard deviations of the mean, where $k > 1$, is at least $1 - \frac{1}{k^2}$.

a. Use Chebyshev's rule to estimate the percent of data within two standard deviations of the mean. How does the estimate from Chebyshev's rule compare with your answer to the second question in Exercise 11?

b. Use Chebyshev's rule to estimate the percent of data within three standard deviations of the mean.

13. About how many standard deviations from the mean is the data value for California?

BY THE WAY

The bluefin tuna, one of the largest fish in the Atlantic, was placed on the 1994 list of the ten most endangered species. The bluefin tuna population has decreased by 80% since 1974.

Exercises and Applications

1. 12 2. 5 3. There are none.
4. 2; 2.9; The estimate is about 0.7σ.
5. 15; 20.4; The estimate is about 0.7σ.
6. 6.5; 9.0; The estimate is about 0.7σ.
7. 6; 9.3; The estimate is about 0.6σ.
8. Yes; if all the data points are the same, the standard deviation is 0.
9.

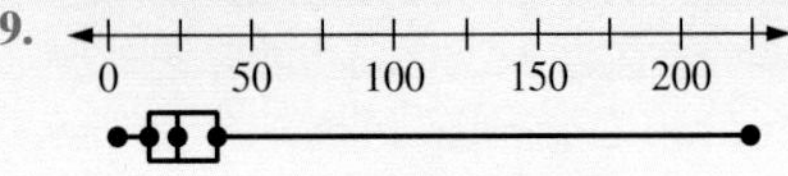

range: 217; IQR: 24; outliers: 79, 89, 97, 161, 224

10. 38.8
11. 90%, 96%
12. a. 75%; The proportion of data within two standard deviations of the mean satisfies Chebyshev's rule.
b. 88.9%
13. about 3.3 standard deviations
14.

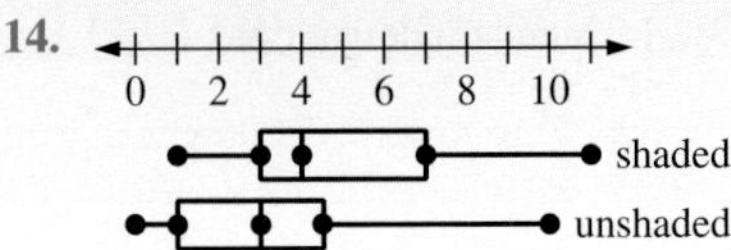

INTERVIEW Donna Cox

Look back at the article on pages 234–236.

The matrix below gives typical cornfield data that Donna Cox might use to create a visual display. Each element of the matrix corresponds to a corn plant and has two values related to corn borer larvae, as described.

	C1		C2		C3		C4		C5		C6		C7		C8		C9	
R1	3	1	7	4	0	0	4	1	8	3	5	3	9	6	8	7	11	5
R2	4	2	0	0	0	0	6	4	4	1	3	2	0	0	6	5	10	10
R3	0	0	0	0	0	0	3	2	3	0	3	3	5	2	1	0	9	7
R4	0	0	0	0	0	0	0	0	1	0	0	0	2	2	0	0	3	3
R5	0	0	0	0	3	0	0	0	0	0	0	0	7	3	9	5	5	1
R6	5	3	6	6	0	0	3	3	6	3	0	0	0	0	0	0	0	0
R7	0	0	0	0	0	0	0	0	0	0	2	0	0	0	0	0	1	0
R8	0	0	0	0	0	0	0	0	0	0	0	0	4	2	8	6	0	0
R9	0	0	0	0	0	0	0	0	0	0	0	0	0	0	3	0	0	0
R10	3	0	0	0	0	0	0	0	0	0	0	0	0	0	0	0	0	0

The numbers in the **shaded column** tell the total number of corn borer larvae on each plant.

The numbers in the **unshaded column** tell how many of the larvae on each plant are infected.

14. Using only the plants that actually have larvae on them, make a comparative box plot of the data in the shaded and unshaded columns of the matrix.

15. Find the median, the range, and the interquartile range of each box plot you made in Exercise 14. How do corresponding values compare with one another?

16. **Challenge** How would your box plots and statistics change if you used all the data?

Use the graphs below to answer Exercises 17–19.

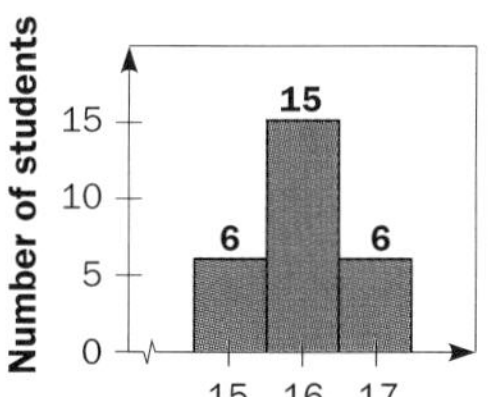

Ages of students in a class

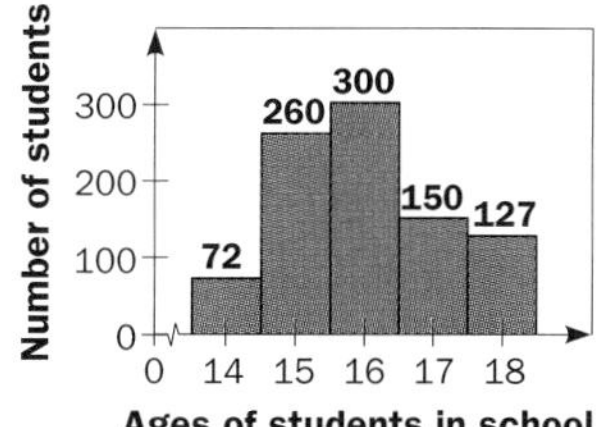

Ages of students in school

Number of people
15
10
5
0
15 8 3 3 3 8 15
0 13 14 15 16 17 18 19

Ages of sports camp participants and staff

17. Find the mean of each data set. How do the means compare?

18. **Visual Thinking** Which set of data do you think has the largest standard deviation? Why?

19. Find the standard deviation of each data set. How do the standard deviations compare with one another?

Apply⇔Assess

Exercise Notes

Using Technology

Exs. 14, 15 Students can create a spreadsheet for each set of data to calculate the necessary measures for these exercises. Many spreadsheets include statistical formulas that will calculate means, medians, maximum and minimum values, and quartiles. For example, the following formulas will return the values in cells B5 to D8 of the spreadsheet below.

B5:=MIN(A2:D4);
B6:=MAX(A2:D4);
B7:=MEDIAN(A2:D4);
D5:=QUARTILE:(A2:D4,1);
D6:=QUARTILE:(A2:D4,3);
D7:=B6–B5;
D8:=D6–D5

	A	B	C	D
1	Total Number of Larvae			
2	3	7	4	8
3	6	4	3	3
4	1	9	7	
5	*Min*	3	*Quartile 1*	3
6	*Max*	9	*Quartile 3*	7
7	*Median*	4	*Range*	8
8			*IQ Range*	4

Common Error

Ex. 16 Some students think that because zero has no value, it does not affect the statistics of a data set. Students can correct this error in their thinking by recalculating the median, range, and interquartile range with one or more zeros added to the data set.

Communication: Discussion

Ex. 18 Point out that a larger standard deviation implies that the data are more widely spread from the mean. Students can visualize which graph has the largest standard deviation.

15. shaded: 4; 10; 4; unshaded: 3; 10; 3.5; The median and interquartile range for the shaded column are higher than for the unshaded column.

16. The box plots would both be "squashed" on their left sides. Because most of the values for both columns are 0, the lower extreme, lower quartile, and median of both box plots would become 0. The upper quartiles would decrease sharply and the upper extremes would remain the same.

17. The mean for each data set is 16.

18. the third data set; Most of its values are farther from the mean, so the squares of the deviations will be large.

19. 0.7; 1.2; 2.5; The standard deviations vary greatly, with that of the third data set much higher than the others.

Apply⇔Assess

Exercise Notes

Assessment Note

Ex. 23 Ask each student to read his or her conclusions to part (c). As students take turns reading their summaries, those listening can review and confirm important facts about the distribution of data. Students should understand that examining graphs gives an overall sense of the data, but measures of center and spread give more precise information about the data.

Practice 37 for Section 6.5

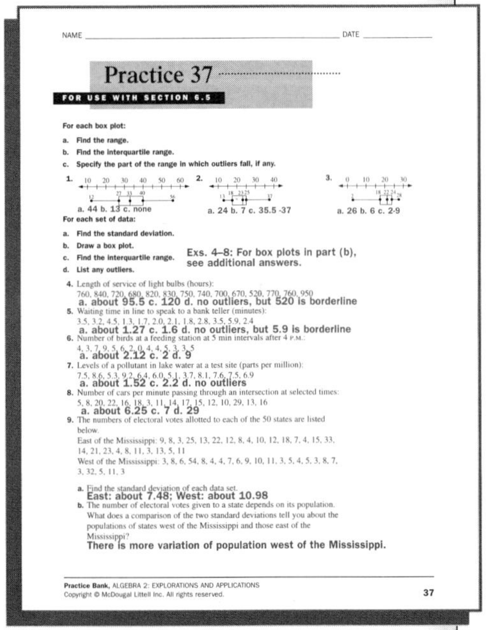
NAME ______ DATE ______

Practice 37

FOR USE WITH SECTION 6.5

For each box plot:

a. Find the range.
b. Find the interquartile range.
c. Specify the part of the range in which outliers fall, if any.

1. (box plot) a. 44 b. 13 c. none
2. (box plot) a. 24 b. 7 c. 35.5 -37
3. (box plot) a. 26 b. 6 c. 2-9

For each set of data:

a. Find the standard deviation.
b. Draw a box plot.
c. Find the interquartile range.
d. List any outliers.

Exs. 4–8: For box plots in part (b), see additional answers.

4. Length of service of light bulbs (hours):
760, 840, 720, 680, 820, 830, 750, 740, 700, 670, 520, 770, 760, 950
a. about 95.5 c. 120 d. no outliers, but 520 is borderline
5. Waiting time in line to speak to a bank teller (minutes):
3.5, 3.2, 4.5, 1.3, 1.7, 2.0, 2.1, 1.8, 2.8, 3.5, 5.9, 2.4
a. about 1.27 c. 1.6 d. no outliers, but 5.9 is borderline
6. Number of birds at a feeding station at 5 min intervals after 4 P.M.:
4, 3, 7, 9, 5, 6, 2, 0, 4, 4, 5, 3, 3, 5
a. about 2.12 c. 2 d. 9
7. Levels of a pollutant in lake water at a test site (parts per million):
7.5, 8.6, 5.3, 9.2, 6.4, 6.0, 5.1, 3.7, 8.1, 7.6, 7.5, 6.9
a. about 1.52 c. 2.2 d. no outliers
8. Number of cars per minute passing through an intersection at selected times:
5, 8, 20, 22, 16, 18, 3, 11, 14, 17, 15, 12, 10, 29, 13, 16
a. about 6.25 c. 7 d. 29
9. The numbers of electoral votes allotted to each of the 50 states are listed below.
East of the Mississippi: 9, 8, 3, 25, 13, 22, 12, 8, 4, 10, 12, 18, 7, 4, 15, 33, 14, 21, 23, 4, 8, 11, 3, 13, 5, 11
West of the Mississippi: 3, 8, 6, 54, 8, 4, 4, 7, 6, 9, 10, 11, 3, 5, 4, 5, 3, 8, 7, 3, 32, 5, 11, 3
a. Find the standard deviation of each data set.
East: about 7.48; West: about 10.98
b. The number of electoral votes given to a state depends on its population. What does a comparison of the two standard deviations tell you about the populations of states west of the Mississippi and those east of the Mississippi?
There is more variation of population west of the Mississippi.

Practice Bank, ALGEBRA 2: EXPLORATIONS AND APPLICATIONS

37

Connection ART

The Pan-African Film and Television Festival was first held in the winter of 1969 to spotlight African-made films. The week-long festival is now held every other year in Burkina Faso, Africa. Countries outside of Africa also participate.

Year	1969	1970	1972	1973	1976	1979	1981	1983
Number of films	23	40	36	51	75	78	78	69
Number of African countries	5	9	18	23	17	16	16	25
Total number of countries	7	9	23	29	26	26	27	37

In downtown Ouagadougou, Burkina Faso, a sculpture representing reels of movie film and camera lenses celebrates the Pan-African Film and Television Festival.

20. a. Make a box plot using the total number of countries participating in each of the first 8 festivals. Find the range and the interquartile range. Are there any outliers?

b. Find the mean and standard deviation of the data. Are any values more than one standard deviation from the mean? more than two standard deviations from the mean?

21. Repeat Exercise 20 for the number of African countries participating in the festival each year.

22. Repeat Exercise 20 for the number of films in the festival.

ONGOING ASSESSMENT

23. Cooperative Learning Work with the members of your group from the *Ongoing Assessment* exercise from Section 6.1. Complete parts (a)–(c) for each set of numerical data you gathered.

a. Draw a box plot of the data. Find the range and the interquartile range. Identify any outliers.

b. Calculate the standard deviation for the data. How many values are within one standard deviation of the mean? within two standard deviations of the mean? within three standard deviations of the mean?

c. What conclusions can you make about your data?

SPIRAL REVIEW

24. A housing committee took a survey of the amount of money people pay to rent an apartment. Use the results of the survey shown in the table at the right to make a histogram of the data. Describe the distribution of the data. *(Section 6.4)*

Apartment rent ($)
295, 600, 225, 280, 430, 290, 310, 200, 300, 725, 350, 575, 260, 375, 450, 400, 380, 250, 175, 325, 420, 425, 130, 475, 300, 220, 340, 550, 375, 625, 330, 525, 180, 325, 360, 350

Solve each inequality. *(Toolbox, p. 787)*

25. $x + 5 \leq 13$ **26.** $-4x - 6 \geq 18$ **27.** $3x + 2 < 8$

20. a. 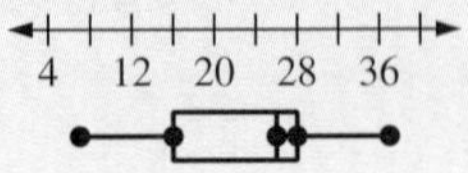
4
12
20
28
36

30; 12; No.

b. 23; 9.5; 7, 9, and 37 are more than one standard deviation from the mean. There are no values more than two standard deviations from the mean.

21. a. 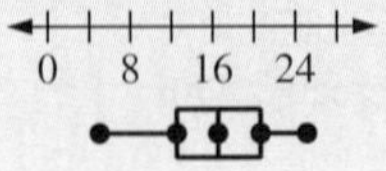
0
8
16
24

20; 8; No.

b. 16.1; 6.2; 5, 9, 23 and 25 are more than one standard deviation from the mean. There are no values more than two standard deviations from the mean.

22. a. 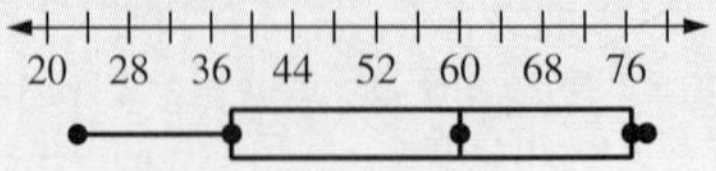
20
28
36
44
52
60
68
76

55; 38.5; No.

b. 56.3; 20.2; 23, 36, 78, and 78 are more than one standard deviation from the mean. There are no values more than two standard deviations from the mean.

23. Answers may vary.

24. 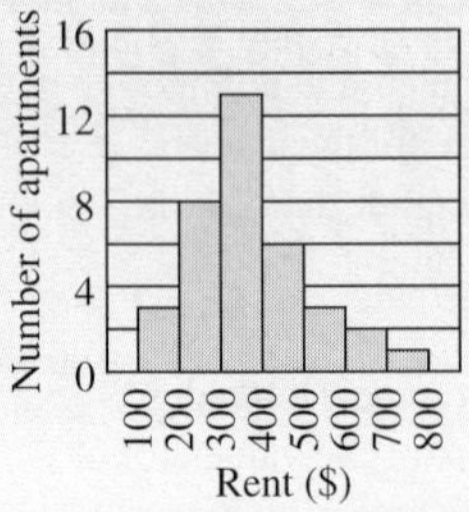
Number of apartments
16
12
8
4
0
100
200
300
400
500
600
700
800
Rent ($)

Answers may vary. An example is given. The distribution is skewed.

25. $x \leq 8$ 26. $x \leq -6$ 27. $x < 2$

Section

6.6 Making Decisions from Samples

Learn how to...

- **find the margin of error for a sample proportion**

So you can...

- **make decisions about opinion poll results**

Although statistics such as mean and standard deviation do not apply to categorical data, you can find the *proportion* of data in a particular category. For example, if 7 out of 10 people respond "yes" to a survey question, then the **sample proportion** of "yes" responses is $\frac{7}{10}$, or 70%. In the Exploration, you will investigate what happens when you take many samples from the same population and look at a distribution of sample proportions.

EXPLORATION
COOPERATIVE LEARNING

Making a Sampling Distribution

Work with a partner.
You will need:
- a calculator with a random number generator

SET UP In 1990, the Consumer Federation of America designed a test of consumer knowledge. The group gave the test to 1139 adults at shopping centers. At the left is one of the questions on the test.

A person who drives 10,000 miles a year in a fuel-efficient car, rather than a gas-guzzler, will save how much in gasoline costs for one year?
A. Under $100
B. Several hundred dollars
C. About $1000
D. Several thousand dollars

1 Write the answer that you think is correct on a piece of paper. Collect all the papers and write the responses on the board, numbering them for easy reference. Your teacher will provide the correct answer.

2 With your partner, use the random number generator on your calculator to choose 6 different random samples of 5 responses from the board. For each sample, find the proportion of correct responses. Write each sample proportion as a percent.

3 In Step 2, did you get the same proportion from each sample? Based on your results from Step 2, would you say that the percent correct for the entire class is closest to *0%, 20%, 40%, 60%, 80%,* or *100%*? Explain.

4 As a class, make a histogram of all the sample proportions. Do they tend to cluster around some value? If so, what is that value?

5 Calculate the *population proportion* (that is, the percent of correct answers for the entire class). Compare this with what you found in Step 4.

Exploration Note

Purpose
The purpose of this Exploration is to investigate the distribution of sample proportions.

Materials/Preparation
Each group needs a calculator.

Procedure
Students record their answers, collect the answers, and write the responses on the board. Provide the correct answer. Each group uses the random number generator to choose 6 random samples of 5 responses. For each sample, the group writes the sample proportion as a percent. The class then makes a histogram and examines it for any clustering. The class calculates the population proportion and compares it with the results of Step 4.

Closure
Students should understand they are simulating a random survey of 6 groups of 5 people and that they are using the results to estimate a population proportion.

Explorations Lab Manual
See the Manual for more commentary on this Exploration.

For answers to the Exploration, see following page.

Plan⇔Support

Objectives

- Find the margin of error for a sample proportion.
- Make decisions about opinion poll results.

Recommended Pacing

Core and Extended Courses
Section 6.6: 2 days
Block Schedule
Section 6.6: 1 block

Resource Materials

Lesson Support
Lesson Plan 6.6
Warm-Up Transparency 6.6
Practice Bank: Practice 38
Study Guide: Section 6.6
Assessment Book: Test 22
Technology
Technology Book: Calculator Activity 6
Graphing Calculator
McDougal Littell Mathpack *Stats! Activity Book:* Activity 18
Internet: http://www.hmco.com

Warm-Up Exercises

Write a formula that a calculator can use to generate random numbers from the given set.

1. between 1 and 10
int(10*rand)+1
2. between 1 and 32
int(32*rand)+1
3. In a random sample, 342 of 1421 people polled answered "Yes." What proportion of the respondents answered "Yes"?
about 24%

Simplify.

4. $\frac{1}{\sqrt{9}}$ $\frac{1}{3}$
5. $-\frac{1}{\sqrt{30}}$ about −0.18

Teach⇔Interact

Section Notes

Communication: Discussion Before beginning this section, it may be helpful to have a discussion of the statistical topics studied so far and to present an overview of the concepts covered in this section. In the first five sections of this chapter, students studied two topics from statistics, data collection and data analysis. They learned methods of producing data that give reliable results. And they learned methods for organizing and describing the data using graphs and numerical summaries. In this section, they will use the data that have been collected to draw conclusions about a larger group. This aspect of statistics, called statistical inference, also provides methods for determining the reliability of any conclusion drawn.

Reasoning
You may want to point out that as n becomes larger and larger (the sample gets larger), the margin of error will become smaller and smaller. However, the gain in reliability is *not* proportional to the increase in sample size. This implies that it may not pay to use a sample size that is very large.

Think and Communicate

These questions prepare students for accommodating the variability inherent in sampling by using the margin of error formula. Discussing their answers to the questions will help students see that using statistics means more than simply manipulating numbers. It means trying to understand what the data say. Therefore, the discussion should highlight the importance of analyzing the results of formulas rather than just performing calculations.

In doing the Exploration, one class made the histogram below. It displays a **sampling distribution** for many samples taken from the same population.

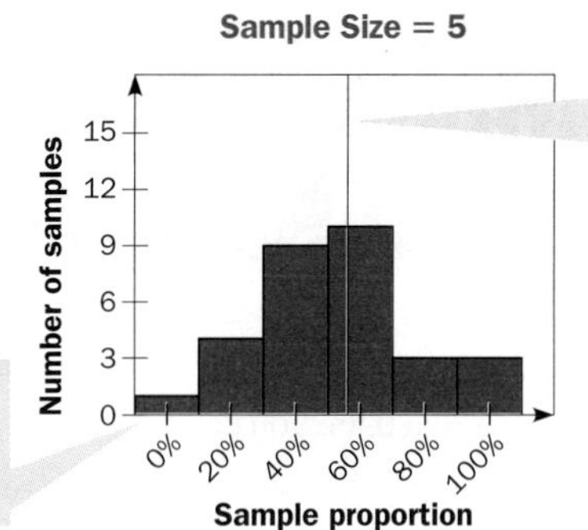

Notice that the results cluster around the class's population proportion of 56%.

Notice the variation in the sample proportions.

Each sample in the histogram above consists of 5 responses, so the **sample size** in this case is 5. The sampling distributions below show what happens when the sample size is increased.

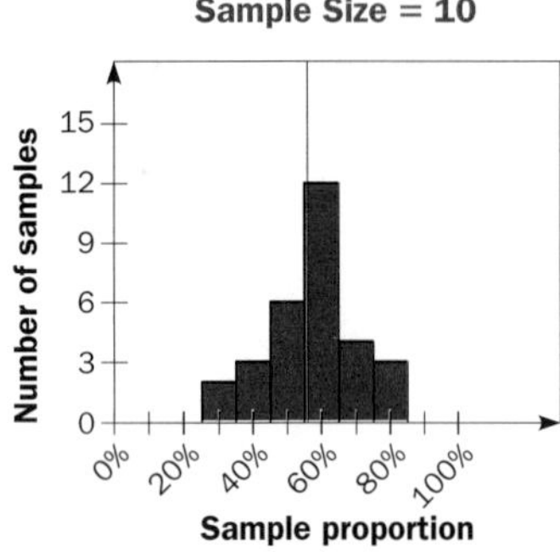

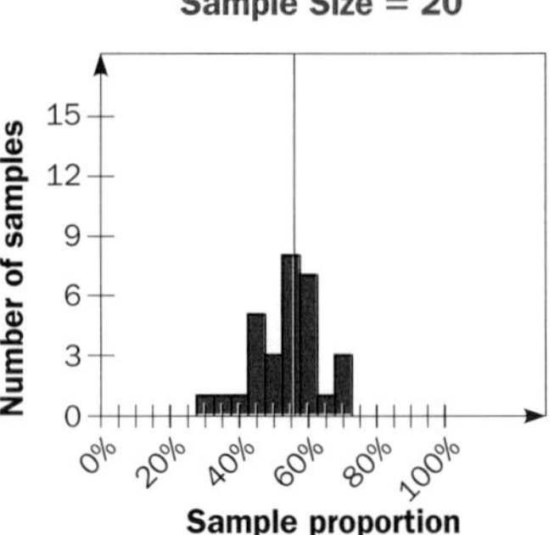

THINK AND COMMUNICATE

1. What do you observe about the clustering of the sample proportions in each histogram?
2. a. What would happen if you used a sample size greater than 20?
   b. What would happen if you made the sample size as large as the population size?

WATCH OUT!
The margin of error formula is most accurate for sample proportions between 40% and 60%. For other proportions, this approximation overestimates the true margin of error.

People often use **margin of error** to indicate the expected variability in sampling. For example, a sample proportion with a margin of error of $\pm 3\%$ is very likely, but not certain, to be within 3 percentage points of the population proportion.

Margin of Error Formula

When a random sample of size n is taken from a large population, the sample proportion has a margin of error approximated by this formula:

$$\textbf{margin of error} \approx \pm\frac{1}{\sqrt{n}}$$

ANSWERS Section 6.6

Margins of error are rounded to whole numbers unless otherwise indicated.

Exploration

Correct answer to test question: B

1–5. Answers may vary.

Think and Communicate

1. The larger the sample size, the more tightly the sample proportions cluster around the population proportion.
2. a. The clustering would be even tighter.
   b. The sample proportion would be the same as the population proportion.

EXAMPLE 1 Application: Consumer Economics

In the test of consumer knowledge conducted by the Consumer Federation of America, 58% of the 1139 adults who participated answered the question stated in the Exploration on page 273 correctly.

a. Find the margin of error for the sample proportion.

b. Find an interval that is likely to contain the proportion of all adult Americans who would give the correct answer to the question.

SOLUTION

a. Find the margin of error using $n = 1139$.

$$\text{margin of error} \approx \pm\frac{1}{\sqrt{1139}} \approx \pm 0.0296 \approx \pm 3\%$$

The margin of error for the sample proportion is about **±3%**.

b.

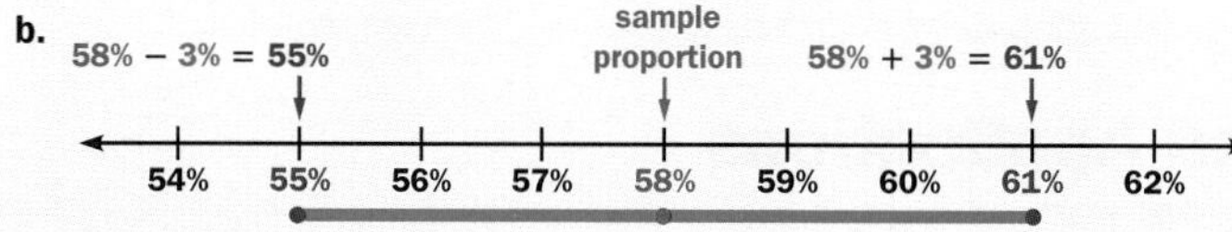

The proportion of people who would answer the question correctly is likely to be between 55% and 61%.

EXAMPLE 2 Application: Polling

Based on the newspaper report at the right, is it reasonable to assume that Costa will win the election?

In a telephone poll, we asked local voters which candidate for mayor they plan to vote for in the upcoming election.

Candidate	Percent of votes
Costa	54%
Kwan	46%

Margin of error: ± 5%

SOLUTION

The diagram shows the range of possible voting results for each candidate.

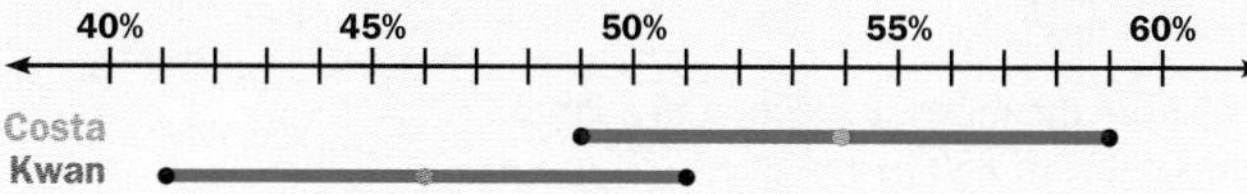

The margin of error for the poll makes it possible that Kwan might get as much as 51% of the vote, and Costa might get as little as 49% of the vote. If this were the case, Kwan would win. Based on this poll, it is not reasonable to assume that Costa will win the election.

Teach⇔Interact

About Example 1

Topic Spiraling: Review

Ask students if the data used are numerical or categorical. (the latter)

Additional Example 1

In a survey of 50 randomly chosen students at Blue Ridge High School, 4% had type AB blood.

a. Find the margin of error for the sample proportion.

Find the margin of error using $n = 50$.

margin of error $\approx \pm\frac{1}{\sqrt{50}} \approx \pm 0.14 \approx \pm 14\%$

The margin of error for the sample proportion is about ±14%.

b. Find an interval that is likely to contain the proportion of all students at the high school who have type AB blood.

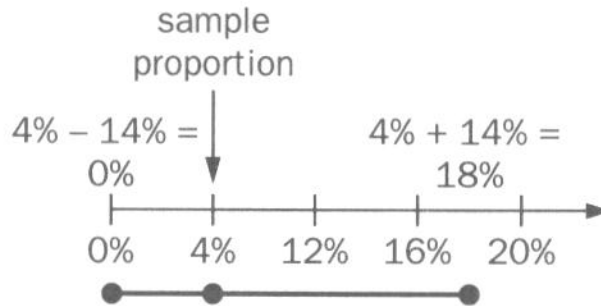

The proportion of students who have blood type AB is likely to be between 0% and 18%.

Additional Example 2

A consumer product company knows that a particular batch of its product has a higher than normal concentration of an active ingredient. Forty-three percent of the batch was estimated to have a concentration higher than normal but less than 0.7434. Fifty-two percent of the batch was estimated to have a concentration higher than 0.7434. Using a 4% margin of error, is it reasonable to conclude that more of the product has a concentration that is higher than 0.7434? The diagram shows the range of possible results for each concentration level.

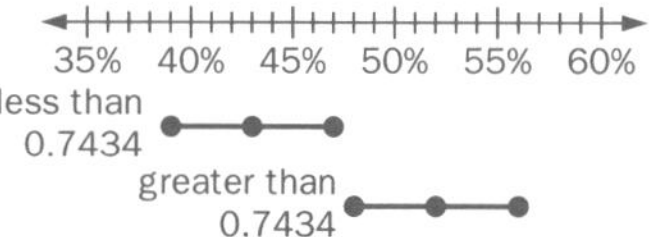

Since the intervals do not overlap, the margin of error makes it reasonable to conclude that more of the product has a concentration that is higher than 0.7434.

Teach⇔Interact

Closure Question

Explain how a sample proportion is related to a population proportion. A sample proportion is a known value that is used to estimate an unknown population proportion.

Apply⇔Assess

Suggested Assignment

Core Course
Day 1 Exs. 1–3, 5–11
Day 2 Exs. 13, 15–20, AYP

Extended Course
Day 1 Exs. 1–11
Day 2 Exs. 12–20, AYP

Block Schedule
Day 38 Exs. 1–3, 5–11, 13, 15–20, AYP

Exercise Notes

Common Error

Exs. 1, 2 A common error made with statistics like those presented in these exercises is to assume that the population proportion can be found in the interval established by the sample proportion and the margin of error. Some students may need to be reminded that the sample proportion simply allows the population proportion to be estimated with a measure of confidence.

Alternate Approach

Ex. 5 In order to demonstrate how statistics can be used to misinform, ask students to also write a report that intentionally misinterprets the data.

Challenge

Ex. 6 Ask students to support their responses to this exercise with a general statement about the relationship between n and $\frac{1}{\sqrt{n}}$. (As n increases, $\sqrt{n}$ increases and $\frac{1}{\sqrt{n}}$ decreases.)

We asked 500 people which candidate they planned to vote for in the upcoming election.

Candidate	Number of votes
Mirdik	280
Wong	220

CHECKING KEY CONCEPTS

POLLING For Questions 1–5, use the poll results at the left.

1. Find the sample proportion for each candidate.
2. Estimate the margin of error for this poll.
3. Find an interval that is likely to contain the percent of *all* voters who would vote for Mirdik.
4. Find an interval that is likely to contain the percent of *all* voters who would vote for Wong.
5. If 1000 people, instead of 500 people, had been included in the sample, what would happen to the margin of error?

6.6 Exercises and Applications

Extra Practice exercises on page 758

1. **EMPLOYMENT** In a survey of 2990 adult Americans, respondents were asked about their jobs. In response to one question, 53.3% said they have a full-time job. Find an interval that is likely to contain the proportion of all adult Americans who have a full-time job.

2. **CONSUMER ECONOMICS** A question from the Consumer Federation of America's test of consumer knowledge is shown at the right. When asked this question, 36% of 1139 adult Americans gave the correct answer. Find an interval that is likely to contain the proportion of all adult Americans who would give the correct answer to this question.

living well & eating right

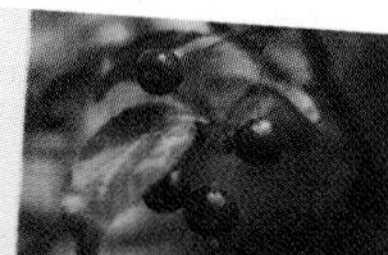

The truth about what's in your food...

TAKE THE FOOD LABEL CHALLENGE

The ingredients on food labels are listed:

A. by nutritional importance, from most to least.
B. by weight, from most to least.
C. alphabetically.
D. in any order the manufacturer chooses.

3. **EDUCATION** A poll reported that 86% of Americans believe that financial need should be a consideration in distributing federal student aid. Is it possible to find the margin of error for this sample proportion? If so, find it. If not, explain why you cannot.

4. Does your class proportion from the Exploration on page 273 fall within the interval found in part (b) of Example 1? If not, what might account for this?

5. **Open-ended Problem** A few days before the mayoral election in a large city, a local newspaper took a random sample of 850 likely voters. Of those surveyed, 453 planned to vote for Timothy Marden and 397 planned to vote for Dipak Johari. If you were a reporter for the newspaper, how would you report the results of this survey?

6. **a.** If the margin of error for a poll is ±4%, find the size of the sample. (*Hint:* Solve for n in the formula for margin of error.)

   b. If the margin of error is ±2%, find the size of the sample. How does this sample size compare with the one you found in part (a)?

Checking Key Concepts

1. Mirdik: 56%; Wong: 44%
2. ±4%
3. between 52% and 60%
4. between 40% and 48%
5. It would be ±3%.

Exercises and Applications

1. between 51.3% and 55.3%
2. between 33% and 39%
3. No; you need to know the sample size.
4. Answers may vary. Differences might be accounted for by many factors, especially the difference in the population sample. Those questioned at the mall were adults.
5. Answers may vary. An example is given. "According to our poll, it appears that if the election were held today, Marden would defeat Johari 53% to 47%. With a margin of error of ±3%, however, the best this poll can predict is that this race is not over yet."
6. **a.** 625
   b. 2500; It is four times as large.
7. The sample is self-selected and, therefore, biased.
8. **a.** ±5%
   b. The results of the voluntary sample fell within the margin of error for the random sample. No; the initial survey was not scientific. The fact that the two

Connection POLLING

Boston Garden was the home sports arena for Boston's professional basketball and hockey teams. On February 26, 1993, the *Boston Globe* published the article below about a deal to build a new arena.

Last Friday and Sunday, Boston Globe readers were asked their opinion of the new Boston Garden deal.... As of noon yesterday the Globe had received 1119 responses from readers. Of those, 841 (approximately 75 percent) said they were in favor of the deal and 278 (25 percent) were against it, though the results are far from scientific.

BY THE WAY

Boston Garden was built in 1928. In 1995, it was replaced by the FleetCenter.

7. Why did the *Boston Globe* say the results of this mail-in survey were "far from scientific"?

8. a. The *Boston Globe* also published the results of a survey based on a random sample of 400 registered voters. In this survey, 71% of the voters favored the plans for the new sports arena. Find the margin of error for this sample proportion.

b. Writing How did the results of the random sample compare with the results of the voluntary sample? Would you expect the two surveys to have the same results? Explain why or why not.

9. The plans for the new sports arena were a popular topic on many talk radio shows in the Boston area. In the survey described above, respondents were also asked whether they listen to talk radio. Of those surveyed, 63% of the men and 54% of the women said "yes." Is it reasonable to conclude that more men than women listen to talk radio, given a $\pm 5\%$ margin of error? Explain.

10. In a poll of 12,300 high school students asked about life in the 22nd century, 80% predicted that Americans will be working in space in the year 2100. Also, 24% predicted that people will learn how to control the weather. Find intervals that are likely to contain the proportion of all high school students who would make these predictions.

11. a. Technology Graph the function $y = \frac{1}{\sqrt{x}}$ on a graphing calculator or graphing software. Use a viewing window with $0 \le x \le 2500$ and $0 \le y \le 0.25$. Use the graph to complete the table.

Sample size	Margin of error
100	?
200	?
400	?
800	?
1600	?
2400	?

b. How are sample size and margin of error related?

c. What sample size is needed for the margin of error to be zero?

12. Writing One organization that regularly conducts public opinion polls uses a random sample of 1000 people for most of its polls. Given that the margin of error decreases as the sample size increases, why do you think the organization does not use a larger sample size?

Apply⇔Assess

Exercise Notes

Student Study Tip

Ex. 7 Mention to students that although a margin of error can be calculated for any survey, the fundamental rules for choosing a random sample cannot be violated if a reasonable inference is to be drawn from the sample. Although some statisticians may disagree about when statistical inference can be used, they all agree that inference is most reliable when the data have been produced by random sampling.

Using Manipulatives

Ex. 9 Students can view the results of the survey by drawing a number line and indicating the intervals given by the margin of error in a segment below the number line. If students use straws of different sizes with a mark indicating the percents obtained from the survey, they can manipulate the straws to see what margins of error, if any, would support the suggested conclusion given in the exercise.

Topic Spiraling: Preview

Ex. 11 You may want to point out that the graph of $y = \frac{1}{\sqrt{2}}$ is the graph of a *radical function*, where $x > 0$ and the line $y = 0$ is an asymptote. Use the graph to emphasize that as x (the sample size) gets very large, there is little change in the y-coordinates. Students will study radical functions in Chapter 8.

Second-Language Learners

Ex. 12 Students learning English may benefit from working with a peer tutor or aide to discuss and complete this writing assignment.

---

agreed may have been coincidental.

9. No. Answers may vary. Examples are given. Even if the intervals did not overlap (and they do), they refer to percents of two different populations. You would have to calculate actual numbers based on population figures. Also, the survey was not scientific.

10. between 79% and 81%; between 23% and 25%

11. a.

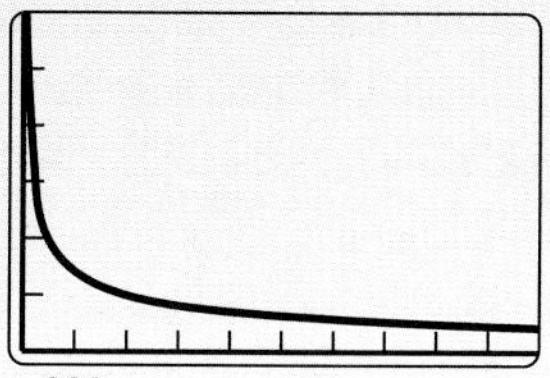

Sample size	Margin of error
100	0.1
200	0.0707
400	0.05
800	0.0354
1600	0.025
2400	0.0204

b. The larger the sample, the smaller the margin of error.

c. Using the margin of error formula, you could not take a large enough sample to have zero error, since there is no number n for which $\frac{1}{\sqrt{n}} = 0$. In actuality, however, the margin of error would be zero if the sample consisted of the entire population.

12. See answers in back of book.

Apply⇔Assess

Exercise Notes

Alternate Approach
Ex. 13 Students can use the spreadsheet to examine what happens to the margin of error if σ is constant and n varies and also when both σ and n vary.

Reasoning
Ex. 14 For part (b), ask students to examine the value of the left member of the equation as $\hat{p}$ (read "p hat") approaches 0.5 from both sides. How does this explain the relationship between a margin of error and the sample proportion? (The margin of error is largest when the sample proportion is 0.50. As the sample proportion increases or decreases, the margin of error gets smaller.)

Cooperative Learning
Ex. 14 Working with a partner to do part (c) may help some students graph the two formulas for the margin of error. Ask them to discuss their interpretations of the graphs with each other.

Assessment Note
Ex. 15 This exercise provides an excellent opportunity for students to review and correlate all the topics studied in this chapter. It is likely that students will need to work through several drafts of their summaries to produce a final draft that clearly and precisely demonstrates their understanding of the material.

13. Sherille Rudbek asked 100 students, "How many hours did you work at a job last week?" The mean number of hours was 18, with a standard deviation of 5.5 hours. In order to find the margin of error for her numerical data, she needed to use a different formula than the one given on page 274. For numerical data, the margin of error for the population mean is given by

$$\text{margin of error} = \pm\frac{2\sigma}{\sqrt{n}}$$

where σ = standard deviation and n = sample size. Find the margin of error for Sherille Rudbek's survey.

14. The formula for the margin of error for a sample proportion given on page 274 can be stated precisely as follows:

$$\text{margin of error} = \pm 2\sqrt{\frac{\hat{p}(1-\hat{p})}{n}}$$

where $\hat{p}$ = sample proportion and n = sample size.

a. Find the margin of error if the sample proportion is 0.20 and the sample size is 500.

b. Show that $2\sqrt{\frac{\hat{p}(1-\hat{p})}{n}} = \frac{1}{\sqrt{n}}$ when $\hat{p} = 0.50$.

c. **Technology** Let $n = 1000$. Use a graphing calculator or graphing software to graph the two formulas for margin of error with $0 \le \hat{p} \le 1$. Describe what the graphs tell you about the relationship between the two formulas.

d. **Challenge** Give a mathematical argument that shows why $\frac{1}{\sqrt{n}} \ge 2\sqrt{\frac{\hat{p}(1-\hat{p})}{n}}$ for all n.

ONGOING ASSESSMENT

15. **Cooperative Learning** Work with the members of your group from the *Ongoing Assessment* exercise from Section 6.1. Use the data you have collected from your survey.

a. Draw conclusions about the categorical data you have collected. Give sample proportions and margins of error.

b. Summarize all of your work in a report. Your report should include the following items:

- a title page
- a copy of your questionnaire
- a description of how you obtained your sample
- the results of the survey, including any graphs that you used to organize and display your data
- a summary of any calculations that you performed, such as mean, median, standard deviation, and margin of error
- a written summary of your findings and conclusions, including comments on any problems that you encountered in doing the project

13. ±1.1 (to two significant digits)

14. a. ±3.6% (to two significant digits)

b. $2\sqrt{\frac{\hat{p}(1-\hat{p})}{n}} = 2\sqrt{\frac{0.5(1-0.5)}{n}} = 2\frac{1}{\sqrt{4n}} = \frac{1}{\sqrt{n}}$

c.

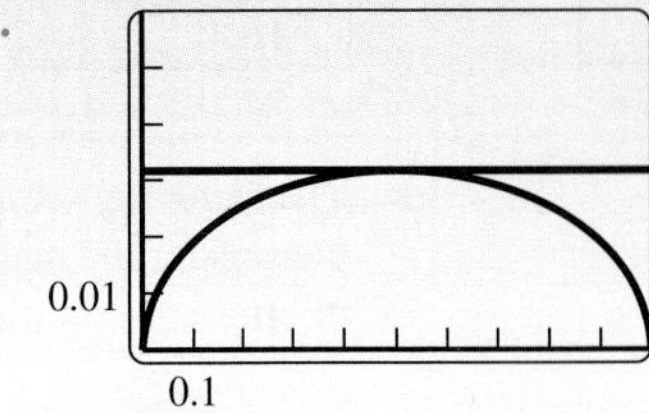

The formulas give the same margin of error for a sample proportion of 0.5. Otherwise, the more complicated formula gives a lower estimate for the margin of error, with the difference between it and the constant value of the simpler formula increasing more rapidly as the sample proportion drops below or rises above one-half.

d. The function $f(\hat{p}) = \hat{p}(1-\hat{p})$ is a quadratic function whose maximum occurs at $\hat{p} = \frac{1}{2}$. Then the maximum value of $2\sqrt{\frac{\hat{p}(1-\hat{p})}{n}}$ is $2\sqrt{\frac{0.5(1-0.5)}{n}} = 2\frac{1}{\sqrt{4n}} = \frac{1}{\sqrt{n}}$.

15. Answers may vary.

16. 12.3; ≈9.98

17. 8.46; ≈2.90

SPIRAL REVIEW

Find the mean and the standard deviation of each data set. *(Section 6.5)*

16. 1, 2, 2, 3, 5, 18, 20, 22, 23, 27

17. 4, 5, 5, 6, 8, 8, 8, 9, 9, 10, 12, 12, 14

Write a point-slope equation of the line passing through the given points. *(Section 2.3)*

18. (6, 3) and (−3, 3) **19.** (−4, 1) and (1, 4) **20.** (−3, 5) and (6, −1)

ASSESS YOUR PROGRESS

VOCABULARY

box plot (p. 259)
lower, upper extremes (p. 259)
lower, upper quartiles (p. 259)
histogram (p. 260)
absolute frequency (p. 260)
relative frequency (p. 260)
symmetric distribution (p. 260)
skewed distribution (p. 260)
range (p. 266)
interquartile range (IQR) (p. 266)
outlier (p. 266)
standard deviation (p. 268)
sample proportion (p. 273)
sampling distribution (p. 274)
sample size (p. 274)
margin of error (p. 274)

The table at the right shows the ages of children attending a summer camp. Use the data for Exercises 1–3. *(Sections 6.4 and 6.5)*

Ages of camp participants			
8	4	10	4
7	8	5	5
6	8	7	9
6	6	10	6
7	5	7	6
5	5	8	6
10	8	8	6
6	9	4	6

1. a. Make a box plot of the data.

b. Above what age are 50% of the children? 25% of the children?

2. a. Writing Suppose four more children register for the camp. Their ages are 11, 9, 8, and 9. Do you think the inclusion of these data will have a significant effect on the distribution of the ages? Why or why not?

b. Make a box plot that includes the new data. Compare this box plot with the one you made in Exercise 1. How did the inclusion of these data affect the distribution of the ages?

3. a. Find the range and the interquartile range for the set of camp participants, including the four children from Exercise 2. Identify any outliers in the data set.

b. Find the standard deviation for the ages of camp participants.

4. A superintendent of a city school system wanted to find out whether the majority of voters in the city favored a proposed building addition to the high school. The superintendent surveyed 500 voters and found that 260 favored the building addition. *(Section 6.6)*

a. Find an interval that would most likely contain the percent of all voters who favor the building addition.

b. Based on the results from part (a), can the superintendent conclude that the voters will pass the building proposal?

5. Journal Explain how you can use the topics you have studied in this chapter to interpret the results of a survey or study that you read about in a newspaper or magazine.

Apply⇔Assess

Assess Your Progress

Journal Entry
A written explanation should help students appreciate more how valuable the use of statistical analysis and inference can be to understand real-world problems. Having a few students read their explanations to the class can provide important feedback for all students.

Progress Check 6.4–6.6

See page 283.

Practice 38 for Section 6.6

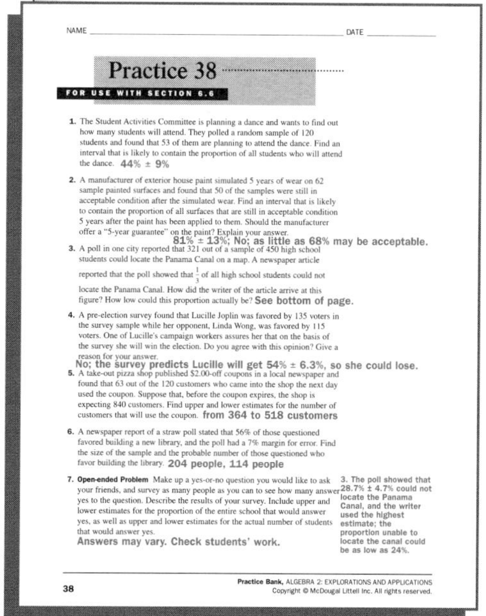
NAME ________ DATE ________

Practice 38
FOR USE WITH SECTION 6.6

1. The Student Activities Committee is planning a dance and wants to find out how many students will attend. They polled a random sample of 120 students and found that 53 of them are planning to attend the dance. Find an interval that is likely to contain the proportion of all students who will attend the dance. **44% ± 9%**

2. A manufacturer of exterior house paint simulated 5 years of wear on 62 sample painted surfaces and found that 50 of the samples were still in acceptable condition after the simulated wear. Find an interval that is likely to contain the proportion of all surfaces that are still in acceptable condition 5 years after the paint has been applied to them. Should the manufacturer offer a "5-year guarantee" on the paint? Explain your answer. **81% ± 13%; No; as little as 68% may be acceptable.**

3. A poll in one city reported that 321 out of a sample of 450 high school students could locate the Panama Canal on a map. A newspaper article reported that the poll showed that $\frac{1}{3}$ of all high school students could not locate the Panama Canal. How did the writer of the article arrive at this figure? How low could this proportion actually be? **See bottom of page.**

4. A pre-election survey found that Lucille Joplin was favored by 135 voters in the survey sample while her opponent, Linda Wong, was favored by 115 voters. One of Lucille's campaign workers assures her that on the basis of the survey she will win the election. Do you agree with this opinion? Give a reason for your answer. **No; the survey predicts Lucille will get 54% ± 6.3%, so she could lose.**

5. A take-out pizza shop published \$2.00-off coupons in a local newspaper and found that 63 out of the 120 customers who came into the shop the next day used the coupon. Suppose that, before the coupon expires, the shop is expecting 840 customers. Find upper and lower estimates for the number of customers that will use the coupon. **from 364 to 518 customers**

6. A newspaper report of a straw poll stated that 56% of those questioned favored building a new library, and the poll had a 7% margin for error. Find the size of the sample and the probable number of those questioned who favor building the library. **204 people, 114 people**

7. **Open-ended Problem** Make up a yes-or-no question you would like to ask your friends, and survey as many people as you can to see how many answer yes to the question. Describe the results of your survey. Include upper and lower estimates for the proportion of the entire school that would answer yes, as well as upper and lower estimates for the actual number of students that would answer yes.
Answers may vary. Check students' work.

3. The poll showed that 28.7% ± 4.7% could not locate the Panama Canal, and the writer used the highest estimate; the proportion unable to locate the canal could be as low as 24%.

38 **Practice Bank,** ALGEBRA 2: EXPLORATIONS AND APPLICATIONS

---

18. $y = 3 + 0(x - 6)$ or
$y = 3 + 0(x + 3)$

19. $y = 4 + \frac{3}{5}(x - 1)$ or
$y = 1 + \frac{3}{5}(x + 4)$

20. $y = -1 - \frac{2}{3}(x - 6)$ or
$y = 5 - \frac{2}{3}(x + 3)$

Assess Your Progress

1. a.

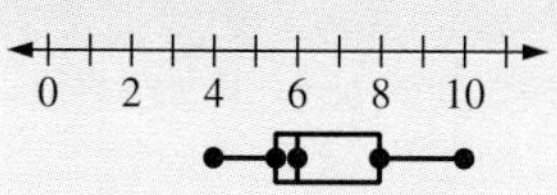

b. 6 years old; 8 years old

2. a. Yes; all of the new values are at or above the upper quartile, so they should have the effect of shifting much of the box plot to the right.

b.

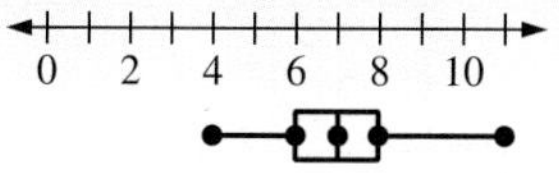

The lower quartile, median, and upper extreme were shifted to the right. The length of the box was shortened somewhat; that is, the interquartile range decreased slightly.

3. a. 7; 2; There are no outliers.

b. 1.8

4. a. between 48% and 56%

b. No.

5. Summaries may vary. Depending on how much information is given, you could determine whether the questions were biased, whether the samples were representative, whether the displays were appropriate or misleading, and whether or not the conclusions were reasonable, taking into account the margin of error.

Project Notes

Mathematical Goals

- Find and analyze a public opinion survey.
- Critique the sampling method of the survey.
- Suggest more accurate sampling methods for the survey.

Planning

Materials

- magazines and newspapers

Project Teams
Students may decide to work separately or in small groups. If students work in small groups, they should meet to determine how to organize their work for the project.

Guiding Students' Work

When students are selecting a survey, they should select one for which they have no bias. If they do have some bias, emphasize the importance of ignoring their own feelings and objectively analyzing the survey. A group discussion is a good way to locate possible problems with a selected survey. If students are working in small groups, have them meet and organize their questions before contacting the sponsors or conductors of the survey.

Second-Language Learners
Students learning English can benefit from working on the research and writing of this project with the help of a peer tutor or aide.

PORTFOLIO PROJECT

Survey Says . . .

Newspapers and magazines often report the results of public opinion surveys. Some surveys are sponsored by independent groups and reported by the press; others are sponsored by news organizations themselves to accompany the articles they publish.

It is important to analyze critically the results of a survey. Does the report fail to note the sampling method used or the sample size? Were the questions worded in ways that might bias the results? Have data been displayed in a misleading manner?

PROJECT GOAL **Analyze the presentation and conclusions of a public opinion survey published in a newspaper or magazine.**

Analyzing a Survey

Skim through some newspapers or magazines to find an opinion survey that interests you. Analyze the presentation and conclusions of the report.

Here are some questions to consider:

- Who conducted the survey? Does the person or group have a stake in its outcome?

Casa Suntuoso, your FAVORITE!!
Which pizza do you prefer?
CASA SUNTUOSO 70%
PIZZA PRIMO 30%
*Based on a survey of 1000 Casa Suntuoso customers

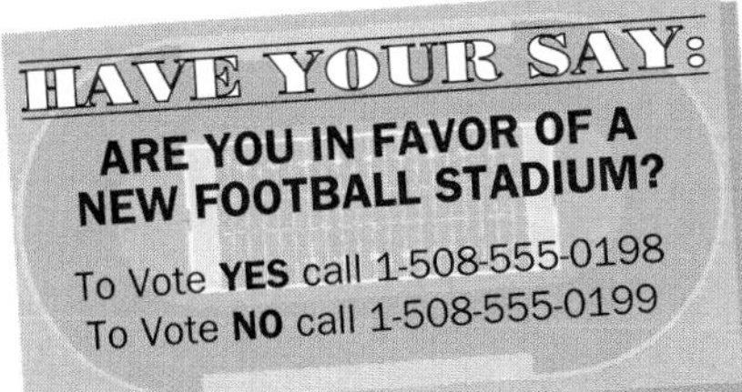

HAVE YOUR SAY:
ARE YOU IN FAVOR OF A NEW FOOTBALL STADIUM?
To Vote YES call 1-508-555-0198
To Vote NO call 1-508-555-0199

• How was the survey done? Does the sampling method seem reasonable? Is the sample representative of the population being studied?

• What is the sample size? Is it large enough to give meaningful results?

• How are the questions worded? Does the wording seem biased? If so, how would you improve the questions?

POLL SHOWS MARCIO IN FRONT IN MAYORAL RACE

Who will get your vote for mayor?

MARCIO	41%
DEHORS	39%
NOT SURE	20%

From a telephone poll of 1000 city residents. Margin of error is +/–3%.

• What is the survey's margin of error? When you take the margin of error into account, are the conclusions drawn from the survey accurate?

• Does the report have any graphs? Are they accurately drawn? Are they in any way misleading?

Writing a Report

Write a report summarizing your work. Include a copy of the survey and your analysis of it. List any information that you think is missing from the survey, and suggest ways the survey can be improved. You may wish to extend your report as follows:

• Write a letter to the sponsor of the survey to ask any questions you still have about it.

• Is the publisher simply reporting the results of a survey from another organization? If so, get the original report from that organization, and compare it with the version that you found.

• In your own community, conduct the same survey that you analyzed. Compare your results with the original findings.

Self-Assessment

Describe any difficulties you had in analyzing your survey. Do you feel that you understand the elements of a well-done public opinion survey? If not, what things are still unclear?

Project Notes

Guiding Students' Work

Rubric for Chapter Project

4 Students select and analyze a survey, including biases, sampling methods, accuracy, and other aspects of the survey. They correctly locate problems and convey them in the report. Students contact the sponsors and find all missing information and settle any questions they have. Students discuss all relevant questions in the report and include ways to improve the problems found. Students conduct their own survey with an appropriate choice of sampling method, display their results, and compare and contrast both surveys.

3 Students select a survey and analyze the various aspects of it. They discuss problems found in conducting the survey but do not locate all of the biases or sampling problems. Students contact the sponsors and ask relevant questions to get unknown facts. Students write a report and suggest solutions to the problems they found. Students perform their own survey and select a sampling method that produces accurate results. They display their results, and compare and contrast the two surveys.

2 Students choose a survey but have problems with the analysis of it. Students only identify some of the biases or problems. Although they make suggestions on correcting the problems found, their analysis is not thorough. Students write a report and conduct their own survey with an appropriate sampling method but repeat some of the mistakes of the original survey. The report contains a display of their data and compares the two surveys.

1 Students select a survey but perform an incomplete analysis of the sampling methods used, biases shown, or displays of the data. Students do not make an effort to critique the survey or ask questions of the people or groups who sponsored it. They do not conduct their own survey, and if a report is turned in, it shows a lack of effort and little understanding of the ideas of the project. Students should be encouraged to speak with the teacher as soon as possible to review their work and to make a new start on the project.

Chapter Support

Course Guide: Chapter 6

Lesson Plans: Chapter 6

Practice Bank:
Cumulative Practice 39

Study Guide: Chapter 6 Review

Assessment Book:
Chapter Tests 23 and 24
Cumulative Test 25
Chapter 6 Alternative Assessment

Test Generator Software

Portfolio Project Book:
Additional Project 3:
School Survey

Preparing for College Entrance Tests

Professional Handbook

Progress Check 6.1–6.3

1. Tell whether the data are *categorical* or *numerical.* *(Section 6.1)*
a person's height at age 16
numerical

2. Describe the population and the sample and tell what type of graph you would use to display the data. *(Section 6.1)*
A used car dealer polls the households in the area to find the most popular class of car.
population: the people who might buy a used car from the dealership; sample: the drivers in the households that were polled; graphs may vary.

3. Write a group of unbiased survey questions for the following topic. *(Section 6.2)*
the need for your town to build and maintain an outdoor ice-skating rink
Answers may vary.

4. Identify the type of sample. Then tell if the sample is biased. *(Section 6.3)*

Conduct a statewide survey to find out if the residents support an increase in the state sales tax. Use a computer to generate 15% of the registered voters' names. At random times of the day, call them and ask "Do you think the state should increase the sales tax?"
random sample; biased

CHAPTER

6 Review

STUDY TECHNIQUE

The steps for performing statistical investigations are listed on page 237. Use what you learned in this chapter to discuss these steps in detail with a partner. If necessary, go back through the chapter to look up any details that are not clear to you.

VOCABULARY

numerical data (p. 238)
categorical data (p. 238)
population (p. 239)
sample (p. 239)
biased question (p. 245)
self-selected sample (p. 250)
systematic sample (p. 250)
convenience sample (p. 250)
random sample (p. 250)
stratified random sample (p. 250)
cluster sample (p. 250)
biased sample (p. 251)
box plot (p. 259)
lower, upper extremes (p. 259)
lower, upper quartiles (p. 259)
histogram (p. 260)
absolute frequency (p. 260)
relative frequency (p. 260)
symmetric distribution (p. 260)
skewed distribution (p. 260)
range (p. 266)
interquartile range (IQR) (p. 266)
outlier (p. 266)
standard deviation (p. 268)
sample proportion (p. 273)
sampling distribution (p. 274)
sample size (p. 274)
margin of error (p. 274)

SECTIONS 6.1, 6.2, *and* 6.3

Suppose you want to gather information about the students in your school. The complete group of students is the **population**. The part of the population that you survey is the **sample**. There are many ways you can choose a sample of the students in your school. For example:

- **self-selected:** Students volunteer.
- **convenience:** Survey the students whose lockers are near yours.
- **stratified random:** Divide student names by grade and randomly choose names from each grade.
- **systematic:** Survey every fourth student in the lunch line.
- **cluster:** Survey the students in your homeroom.
- **random:** Draw names out of a hat and survey the selected students.

You should avoid asking **biased questions** in your survey so that you get responses that accurately reflect the opinions or actions of the respondents.

SECTIONS 6.4 *and* 6.5

Data that are names or labels are **categorical data**. Categorical data are often displayed in a bar graph or a circle graph. Data that are counts or measurements are **numerical data**. Numerical data, such as the fuel efficiency of various models of cars (see table), are often displayed in a *box plot* or a *histogram*.

Fuel efficiency (mi/gal)
30, 25, 41, 29, 22, 29, 18, 23, 26, 23, 29, 24, 25, 28, 25, 20, 17, 19, 17, 17, 18, 20, 13, 23, 12, 17, 21, 17, 18, 26, 29, 29, 26, 22, 17, 22, 26

A **box plot** displays the median, the **quartiles**, and the **extremes** of a data set. The "box" of a box plot shows the middle half of the data.

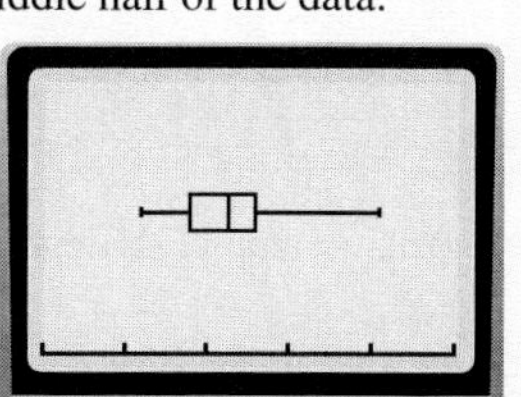

A **histogram** displays the overall shape of a distribution of numerical data. The distribution may be **symmetric** or **skewed**.

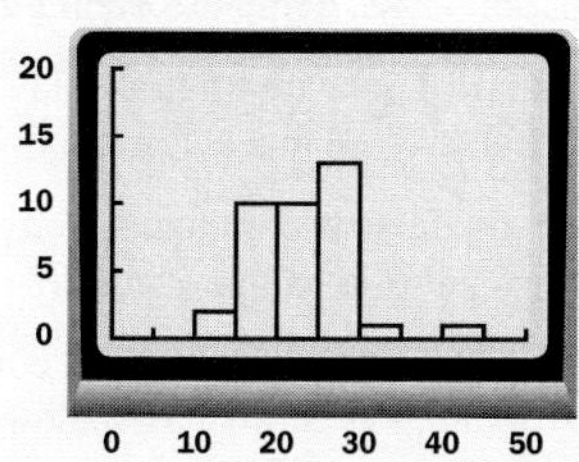

Some measures of variability within a data set are the **range** and **interquartile range**. Another measure of variability is the **standard deviation**. If the data values are $x_1, x_2, \ldots, x_n$, and the mean of the data is $\bar{x}$, then the standard deviation, σ, is given by this formula:

$$\sigma = \sqrt{\frac{(x_1 - \bar{x})^2 + (x_2 - \bar{x})^2 + \cdots + (x_n - \bar{x})^2}{n}}$$

Measures of variability for the fuel efficiency data are shown below.

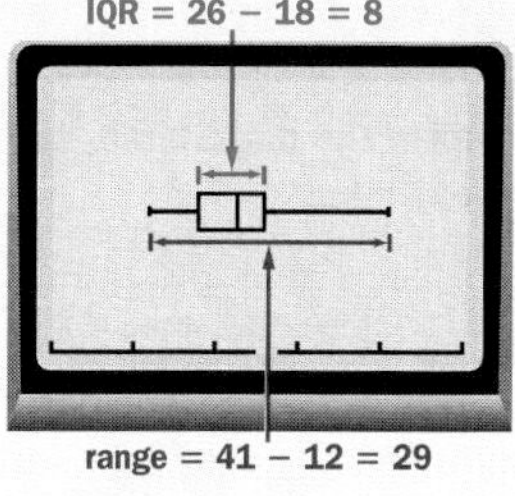

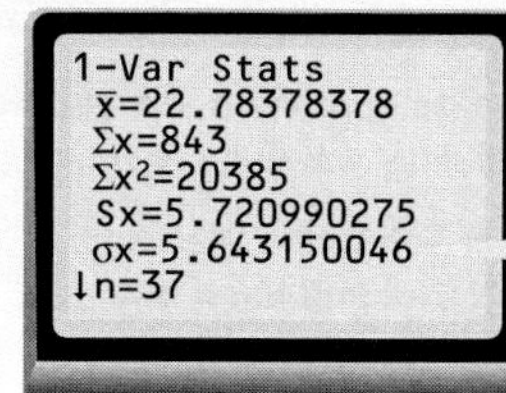

The standard deviation of the fuel efficiency data is about 5.6.

SECTION 6.6

When a random sample of size n is taken from a large population, the **sample proportion**, which is the fraction or percent of responses that fall into a particular category, has a **margin of error** approximated by the formula

$$\textbf{margin of error} \approx \pm\frac{1}{\sqrt{n}}$$

where n is the size of the sample used in the survey. For example, many surveys have a **sample size** of about 1000, which gives a margin of error of about $\pm\frac{1}{\sqrt{1000}} \approx \pm 0.0316$, or $\pm 3\%$.

Progress Check 6.4–6.6

The raw scores of a reading test given to a group of third grade students are listed below. Use the data to answer Exs. 1–4. *(Sections 6.4 and 6.5)*

40 26 39 45 40 31 39 45 14 42
18 25 43 46 27 19 48 21 49 48
52 26 42 26 13 45 52 17 28 50
41 38 19 28 35 29 28 35 29 32
30 46 27 35

1. Make an absolute frequency histogram of the data.

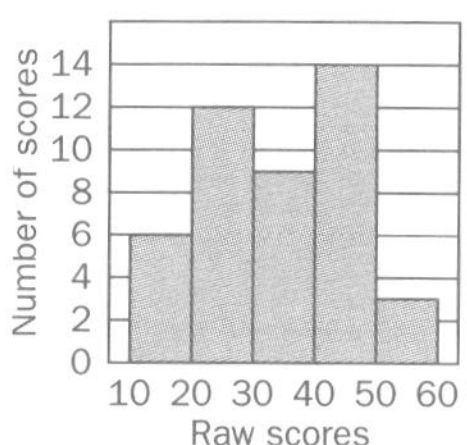

2. Is the distribution symmetric or skewed? symmetric
3. Find the standard deviation of the data. about 10.8
4. **a.** Identify the median, quartiles, and extremes of the data.
   median: 35; 1st quartile: 26.5; 3rd quartile: 44; minimum: 13; maximum: 52

   b. Find the range and the interquartile range of the data.
   range: 39; interquartile range: 17.5
5. **a.** How does a box plot show the distribution of a data set? *(Section 6.4)*
   It uses quartiles to show the spread of the data about the median.

   b. Which measure, mean or median, can be strongly influenced by outliers? *(Section 6.5)* mean

A census of 40,000 American adults revealed that 17,500 of those surveyed had completed 1–3 years of college. *(Section 6.6)*

6. Express the sample proportion as a percent. 43.75%
7. Find the margin of error for this sample. ±0.5%
8. Find an interval that is likely to contain the proportion of all adult Americans who have completed 1–3 years of college. between 43.25% and 44.25%

Chapter 6 Assessment
Form A Chapter Test

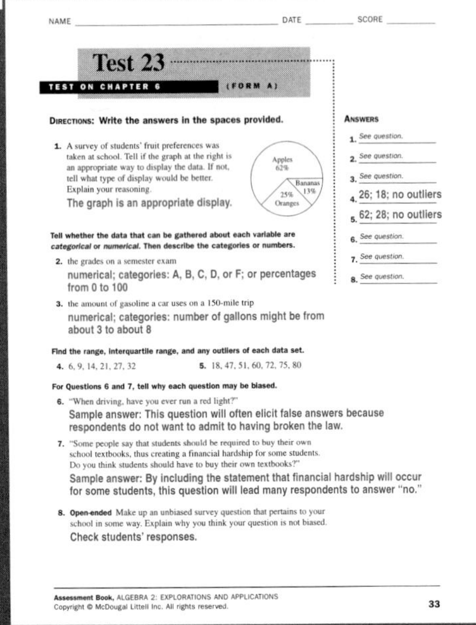

NAME ______ DATE ______ SCORE ______

Test 23

TEST ON CHAPTER 6 (FORM A)

Directions: Write the answers in the spaces provided.

1. A survey of students' fruit preferences was taken at school. Tell if the graph at the right is an appropriate way to display the data. If not, tell what type of display would be better. Explain your reasoning.
   The graph is an appropriate display.

Tell whether the data that can be gathered about each variable are *categorical* or *numerical*. Then describe the categories or numbers.

2. the grades on a semester exam
   numerical; categories: A, B, C, D, or F; or percentages from 0 to 100
3. the amount of gasoline a car uses on a 150-mile trip
   numerical; categories: number of gallons might be from about 3 to about 8

Find the range, interquartile range, and any outliers of each data set.

4. 6, 9, 14, 21, 27, 32
5. 18, 47, 51, 60, 72, 75, 80

For Questions 6 and 7, tell why each question may be biased.

6. "When driving, have you ever run a red light?"
   Sample answer: This question will often elicit false answers because respondents do not want to admit to having broken the law.
7. "Some people say that students should be required to buy their own school textbooks, thus creating a financial hardship for some students. Do you think students should have to buy their own textbooks?"
   Sample answer: By including the statement that financial hardship will occur for some students, this question will lead many respondents to answer "no."
8. **Open-ended** Make up an unbiased survey question that pertains to your school in some way. Explain why you think your question is not biased.
   Check students' responses.

Answers

1. See question.
2. See question.
3. See question.
4. 26; 18; no outliers
5. 62; 28; no outliers
6. See question.
7. See question.
8. See question.

Assessment Book, ALGEBRA 2: EXPLORATIONS AND APPLICATIONS

33

Chapter 6 Assessment
Form B Chapter Test

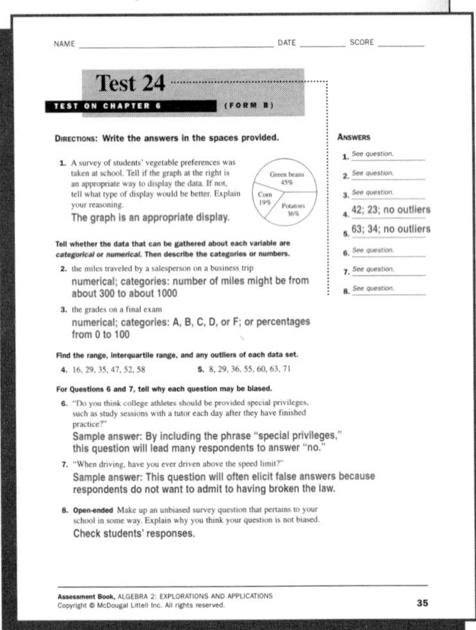

NAME ______ DATE ______ SCORE ______

Test 24

TEST ON CHAPTER 6 (FORM B)

Directions: Write the answers in the spaces provided.

1. A survey of students' vegetable preferences was taken at school. Tell if the graph at the right is an appropriate way to display the data. If not, tell what type of display would be better. Explain your reasoning.
   The graph is an appropriate display.

Tell whether the data that can be gathered about each variable are *categorical* or *numerical*. Then describe the categories or numbers.

2. the miles traveled by a salesperson on a business trip
   numerical; categories: number of miles might be from about 300 to about 1000
3. the grades on a final exam
   numerical; categories: A, B, C, D, or F; or percentages from 0 to 100

Find the range, interquartile range, and any outliers of each data set.

4. 16, 29, 35, 47, 52, 58
5. 8, 29, 36, 55, 60, 63, 71

For Questions 6 and 7, tell why each question may be biased.

6. "Do you think college athletes should be provided special privileges, such as study sessions with a tutor each day after they have finished practice?"
   Sample answer: By including the phrase "special privileges," this question will lead many respondents to answer "no."
7. "When driving, have you ever driven above the speed limit?"
   Sample answer: This question will often elicit false answers because respondents do not want to admit to having broken the law.
8. **Open-ended** Make up an unbiased survey question that pertains to your school in some way. Explain why you think your question is not biased.
   Check students' responses.

Answers

1. See question.
2. See question.
3. See question.
4. 42; 23; no outliers
5. 63; 34; no outliers
6. See question.
7. See question.
8. See question.

Assessment Book, ALGEBRA 2: EXPLORATIONS AND APPLICATIONS

35

CHAPTER

6 Assessment

Vocabulary Questions

For Questions 1–3, complete each paragraph.

1. Numerical data are often displayed in a(n) _?_ or a(n) _?_. _?_ data are often displayed in a bar graph or circle graph.
2. Selecting every tenth name from a list produces a(n) _?_ sample. Drawing names out of a hat produces a(n) _?_ sample.
3. When you can read the actual count of the data within each interval of a histogram, the vertical axis shows _?_. The vertical axis shows _?_ when the count is expressed as a percent of all the data.

Sections 6.1, 6.2, *and* 6.3

Tell whether the data that can be gathered about each variable are *categorical* or *numerical*. Then describe the categories or numbers.

4. the species of animals at a zoo
5. semester grades
6. the amount of electricity a household uses each day

For each graph, tell if it is an appropriate way to display the data. If not, tell what type of display would be better. Explain your reasoning.

7. **Amount of Television Watched by Tenth Graders Each Day**

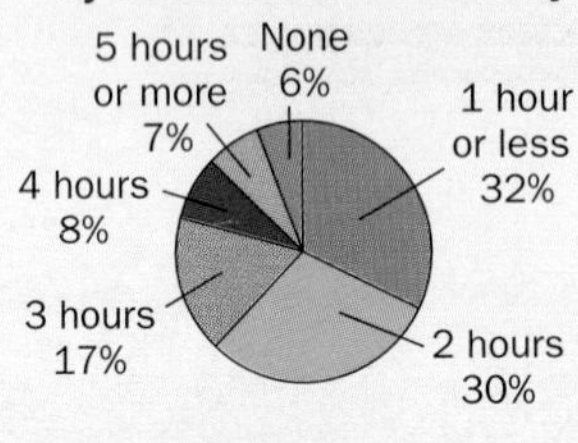

8. **Percent of U.S. Births, by Month, in 1991**

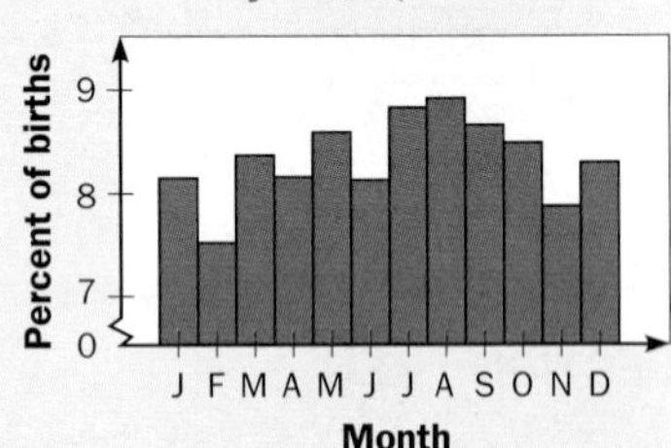

Tell why each question may be biased. Then describe what changes you would make to improve the question.

9. "Have you ever cheated on a test?"
10. "Some people want to censor the material that can be transmitted through the Internet, thus violating people's First Amendment right to freedom of speech. Do you think restrictions should be placed on the material that can be transmitted through the Internet?"

ANSWERS Chapter 6

Assessment

1. histogram; box plot; categorical
2. systematic; random
3. absolute frequencies; relative frequencies
4. categorical; names of the species
5. may be numerical (usually numbers between 0 and 100 or between 0 and 4.0) or categorical (letter grades *A* through *F*)
6. numerical; numbers of kilowatt-hours
7. Yes; the data are categorical.
8. No; the data are categorical. The data should be displayed in a bar graph or circle graph.

9–10. Answers may vary. Examples are given.

9. Unless the question is asked anonymously, it may not elicit an honest response since it involves sensitive personal information. The information should be collected anonymously.
10. The first sentence encourages a negative response, since it appeals to a person's respect for individual freedoms. The first sentence should be eliminated.
11. random; all those in the area covered by the types of numbers generated (for example, with the appropriate area codes); Yes; those without phones are underrepresented.
12. systematic; all students in the class; No (unless seating was not assigned randomly).
13. cluster; all students in the school; Yes; students who are not members of the student council are underrepresented

Identify the type of sample and describe the population of the survey. Then tell if the sample is biased. Explain your reasoning.

11. A marketing group uses a computer to generate phone numbers.

12. A teacher collects the homework of every third student.

13. The principal sends the student council officers to represent the school at a district fundraising event.

SECTIONS 6.4, 6.5, *and* 6.6

SPORTS In 1969, four new teams joined major league baseball. In addition, the height of the pitching mound was lowered, giving batters an advantage. Use the data in the table to answer Questions 14–17.

Average Number of Home Runs per Game

1950–1968	1969–1993
1.23, 1.37,	1.15, 1.28,
1.42, 1.50,	1.36, 1.36,
1.57, 1.66,	1.39, 1.41,
1.67, 1.67,	1.44, 1.46,
1.67, 1.70,	1.47, 1.48,
1.70, 1.72,	1.51, 1.55,
1.78, 1.80,	1.56, 1.57,
1.81, 1.82,	1.60, 1.60,
1.85, 1.85,	1.60, 1.61,
1.91	1.63, 1.71,
	1.73, 1.76,
	1.78, 1.81,
	2.12

14. a. Make two box plots to compare the average number of home runs hit per game for the 1950–1968 seasons and the 1969–1993 seasons.

b. Between what numbers do 50% of the data fall in the box plot of the 1950–1968 data?

c. Repeat part (b) using the box plot of the 1969–1993 data.

15. Make a histogram of each data set. Describe the distribution of the data.

16. a. Find the range, interquartile range, and any outliers of each data set.

b. Find the standard deviation of each data set.

17. Writing Do you think the changes made in 1969 had an effect on the average number of home runs hit per game? Explain your reasoning.

18. The cheerleaders at a high school took a random sample of 80 students and found that 60 students would buy a spirit T-shirt.

a. Find the sample proportion of students who would buy a T-shirt.

b. Find the margin of error. Use the margin of error to find an interval for the proportion of the entire student body that would buy a T-shirt.

19. Writing In what ways can errors in data collection, data displays, and data analysis affect the conclusions made from a set of data?

PERFORMANCE TASK

20. The manager of a restaurant wants to gather data about the restaurant's service, customers, and food. Write a report to the manager suggesting a way to gather, display, and analyze the data. Include these things:

- Give examples of categorical and numerical data that may be collected. Describe how you would display each type of data.
- Write a sample survey of unbiased questions to distribute to customers. Then describe how you would distribute the survey.
- Give examples of how the standard deviation, range, interquartile range, outliers, and margin of error can be used to analyze the data.

Chapter 6 Assessment

Form C Alternative Assessment

Chapter 6

ALTERNATIVE ASSESSMENT

1. **Open-ended Problem** Find a biased question in a newspaper or magazine. Discuss why you think it may be biased.
2. **Open-ended Problem** List several factors that may bias a survey conducted in each manner.
   a. calling people on the telephone
   b. printing the survey in a magazine
   c. using a call-in number during a television show
   d. asking students directly during homeroom in school
3. a. Ask several people working in different types of manufacturing jobs which types of graphs are most useful in their field. Discuss why this is so.
   b. Ask several people working in different types of service industry jobs which types of graphs are most useful in their field. Discuss why this is so.
4. **Open-ended Problem** Find a set of data that has a mean, median, and mode of 10. (Not all the data values can be 10.)
5. **Group Activity** Discuss how the Civil War would have been different if the technology and the knowledge of statistics that exist today had been available in 1861. Make a list of changes discussed by your group.
6. **Performance Task** Use examples to discuss the difference between self-selected samples, systematic samples, random samples, cluster samples, convenience samples, and stratified random samples. Consider the differences within the framework of a school survey.
7. **Open-ended Problem** Compare the quiz grades of the two algebra classes shown in the table by comparing the measures of variation of the two data sets.

First Period	Second Period
10	2
5	10
6	10
5	4
6	2
7	5
8	1
5	10
6	9
6	9
2	7

Assessment Book, ALGEBRA 2: EXPLORATIONS AND APPLICATIONS

119

(while students who are members are overrepresented).

14. a.

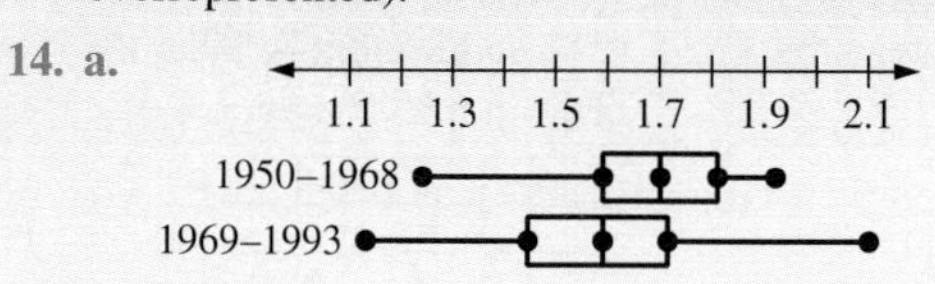

b. 1.23 and 1.7 or 1.57 and 1.81 or 1.7 and 1.91

c. 1.15 and 1.56 or 1.425 and 1.67 or 1.56 and 2.12

15. See answers in back of book

16. a. 1950–1968: 0.68; 0.24; no outliers; 1969–1993: 0.97; 0.25; 2.12

b. 1950–1968: 0.18; 1969–1993: 0.19

17. Answers may vary. An example is given. Since the upper quartile for the later data is about the same as the median for the earlier data, it appears the average number of home runs per game decreased after 1969. I think it is reasonable to assume that the changes described were at least partially responsible.

18. a. 75%

b. ±11%; between 64% and 86%

19. Answers may vary. The information collected can be misleading or uninformative because the questions were biased, the sample was biased, or the sample size was not large enough to produce a reasonable margin of error.

20. Answers may vary.

Cumulative Assessment

CHAPTERS 4–6

CHAPTER 4

For each function:

a. Graph the function and its inverse in the same coordinate plane.

b. Find an equation for the inverse.

1. $f(x) = 6x$ **2.** $g(x) = 4^x$ **3.** $h(x) = -\frac{2}{5}x + \frac{1}{5}$

Write each equation in logarithmic form.

4. $5^0 = 1$ **5.** $81^{-1/2} = \frac{1}{9}$ **6.** $2^5 = 32$

Evaluate each logarithm. Round decimal answers to the nearest hundredth.

7. $\log 0.01$ **8.** $\ln e^{32}$ **9.** $\log_{1/2} 6$

10. Write $\log_4 \frac{p^6}{q^{2/3}}$ in terms of $\log_4 p$ and $\log_4 q$.

11. Write $-\frac{3}{4}\log_a 81 + \frac{1}{2}\log_a 9$ as a logarithm of a single number.

Solve each equation. Round your answers to the nearest hundredth.

12. $4^x = 15$ **13.** $\log_4 x = -3$ **14.** $2 \ln (x + 1) = 6$

Write y as a function of x.

15. $\log y = 0.8x + 1$ **16.** $\ln y = 0.6 - 0.4 \ln x$

17. The table shows the number of households, in millions, in the continental United States with television sets for various years from 1955 through 1985.

Years since 1950	Millions of households
5	32.0
10	45.2
15	53.8
20	60.1
25	69.6
30	76.3
35	84.9

a. Find an exponential function that models the data.

b. Find a power function that models the data.

c. **Writing** Which of the functions from parts (a) and (b) is the better model? Explain your reasoning.

CHAPTER 5

Describe the graph of each function. Make a sketch of each graph.

18. $y = -\frac{1}{2}x^2$ **19.** $y = -\frac{1}{2}(x - 1)^2$ **20.** $y = -\frac{1}{2}(x + 2)(x - 3)$

Solve each equation. Give solutions to the nearest tenth.

21. $-3x^2 = -15$ **22.** $7(x + 4)^2 - 28 = 0$ **23.** $2x^2 - 15x + 21 = 0$

24. **PHYSICS** The equation $h = -4.9(t - 1.1)^2 + 7$ gives the height h of a ball in meters t seconds after it is thrown straight up. When does it hit the ground?

ANSWERS Chapters 4–6

Cumulative Assessment

1. a.

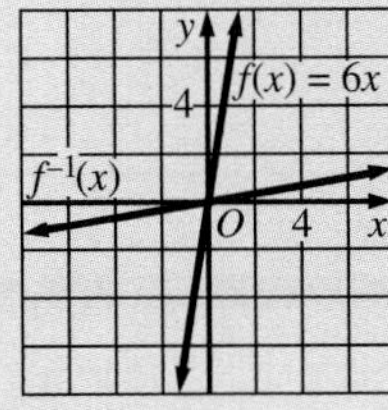

b. $f^{-1}(x) = \frac{1}{6}x$

2. a.

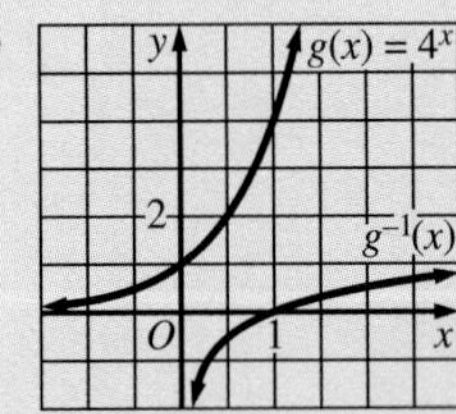

b. $g^{-1}(x) = \log_4 x$

3. a.

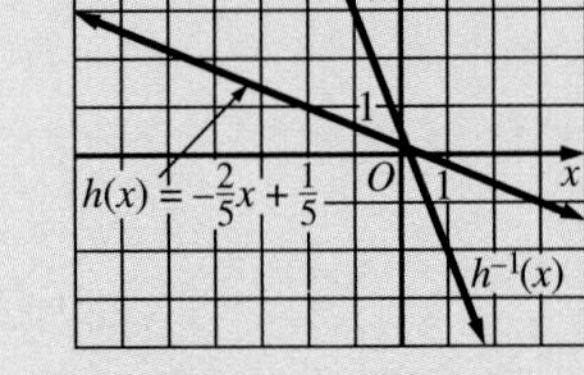

b. $h^{-1}(x) = -\frac{5}{2}x + \frac{1}{2}$

4. $\log_5 1 = 0$ 5. $\log_{81} \frac{1}{9} = -\frac{1}{2}$

6. $\log_2 32 = 5$ 7. –2

8. 32 9. –2.59

10. $6 \log_4 p - \frac{2}{3} \log_4 q$

11. $\log_a \frac{1}{9}$ 12. 1.95

13. 0.02 14. 19.09

15. $y = 10(6.31)^x$

16. $y = 1.82x^{-0.4}$

17. Answers may vary slightly due to differences in technology or to rounding.

a. $y = 31.5(1.03)^x$

b. $y = 14.5x^{0.489}$

c. the power function; The correlation coefficient for the power function is closer to 1.

18–20. Answers may vary.

18. The graph is a parabola that opens down. Its vertex is at the origin and it is wider than the graph of $y = x^2$.

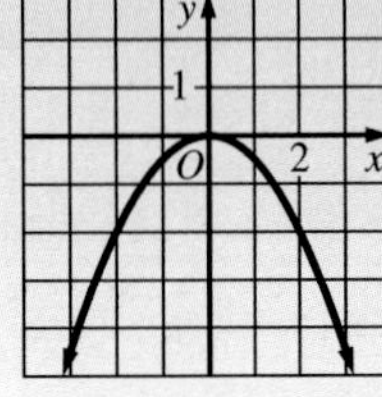

19. The graph is a parabola that opens down. Its vertex is at (1, 0) and it is wider than the graph of $y = x^2$.

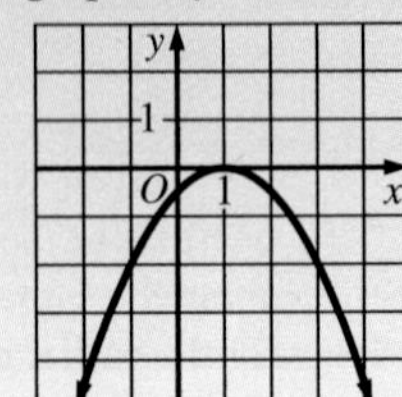

20. The graph is a parabola that opens down. It has x-intercepts –2 and 3 and it is wider than the graph of $y = x^2$.

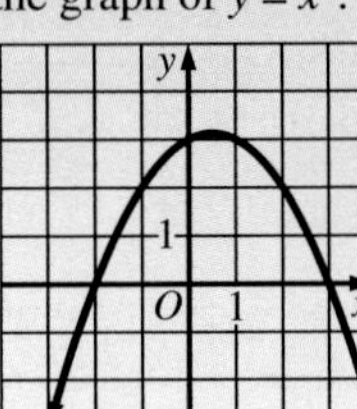

25. Write an equation in the form $y = a(x - h)^2 + k$ for a parabola that has its vertex at $(-4, 1)$ and passes through the point with coordinates $(-2, 9)$.

26. Write $y = -2x^2 - 4x + 6$ in vertex form.

27. Write an equation for the parabola shown at the right.

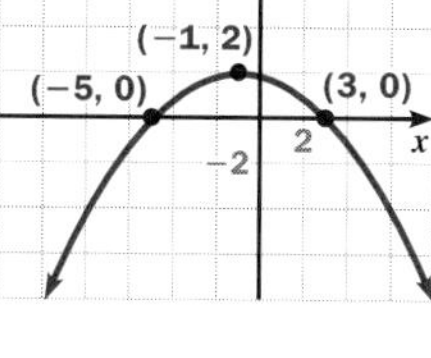

28. **Open-ended Problem** Describe two ways to show that the equation $5x^2 + x + 8 = 0$ has no solution.

29. Find the x-intercepts, if any, for the graph of $y = 0.1x^2 + 0.4x + 0.4$.

30. The equation $y = -0.05x^2 + 2x + 5$, where the distances x and y are measured in feet, describes the path of a stream of water from a fire hose. What is the maximum height of the stream of water?

Write each quadratic function as a product of factors.

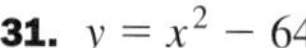

31. $y = x^2 - 64$ 32. $y = 3x^2 + x - 10$ 33. $y = 9x^2 - 24x + 16$

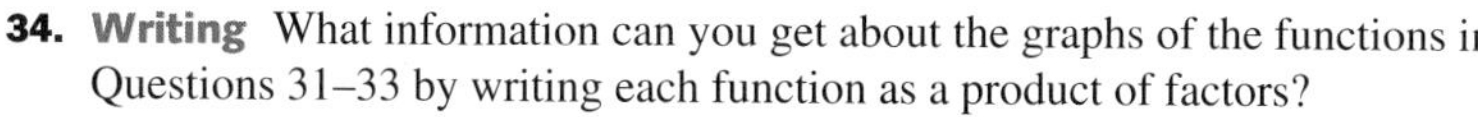

34. **Writing** What information can you get about the graphs of the functions in Questions 31–33 by writing each function as a product of factors?

CHAPTER 6

35. A public radio station sends surveys to its contributors to determine which programs are most popular.
   a. Describe the population of the survey.
   b. Identify the type of sample. Then tell if the sample is biased. Explain.
   c. Tell whether the data that will be gathered are *numerical* or *categorical*.
   d. Tell what type of graph you would use to display the data.

36. **Writing** Describe three ways in which a survey question may be biased. Give an example of each.

37. **Open-ended Problem** Margery Gupta is trying to decide whether to run for mayor. She plans to use an opinion poll to determine the level of support for her candidacy. Explain how she could choose a sample.

SPORTS For Questions 38 and 39, use the table showing the number of wins per team in the National Hockey League in the 1994–1995 season.

Eastern Conference				
27	22	20	19	18
22	15	22	9	28
29	30	17	22	

Western Conference				
16	24	24	17	33
17	16	28	19	21
18	16			

38. a. Make a histogram to display the data for all the teams.
   b. Is the distribution shown on your histogram *symmetric* or *skewed*?

39. a. Identify the median, the quartiles, and the extremes of each data set.
   b. Make a comparative box plot of the data for the two conferences.
   c. For each set of data, find the range and the interquartile range, and identify any outliers.
   d. Find the standard deviation for each set of data. How do the standard deviations compare?

21. ± 2.2 22. $-6; -2$

23. 1.9; 5.6

24. 2.30 s after it is thrown

25. $y = 2(x + 4)^2 + 1$

26. $y = -2(x + 1)^2 + 8$

27. $y = -\frac{1}{8}(x + 5)(x - 3)$

28. Answers may vary. Examples are given. Use a graphing calculator to graph the function and show that the graph has no x-intercepts. Show that the discriminant is negative.

29. -2 30. 25 ft

31. $y = (x + 8)(x - 8)$

32. $y = (3x - 5)(x + 2)$

33. $y = (3x - 4)^2$

34. You can determine the x-intercepts of the graphs of the equations and the solutions of the equation obtained by letting $y = 0$.

35–37. Answers may vary. Examples are given.

35. a. contributors to the station
   b. self-selected; Yes; it under-represents listeners who do not contribute.
   c. numerical (for example, how many listeners listen to a particular type of show); categorical (types of programs that are most popular)
   d. histogram; bar graph; circle graph

36. A biased question may:
   (1) encourage the respondent to answer in a particular way ("Do you plan to vote for Sen. Watson in spite of her voting record on environmental protection?").
   (2) be perceived as too sensitive to respond to truthfully ("If you have ever failed to report income on your federal tax return, please raise your hand.").
   (3) not provide the respondent with enough information to give an accurate opinion ("Which amendment to the United States Constitution do you think had the most significant effect on elections, Article XV, Article XIX, or Article XXVI?").

37. She might obtain the list of voters in the last election and, depending on the size of the group, randomly choose enough voters to get a reasonable sample.

38, 39. See answers in back of book.

CHAPTER

7 Systems

OVERVIEW

Connecting to Prior and Future Learning

- ⇔ The study of systems, begun in Algebra 1, is continued in this chapter. Students first use technology to find points of intersection and then use substitution, adding, or subtracting to solve systems. Students will continue to use these skills throughout this and future courses.
- ⇔ In Section 7.3, the skills students acquired in working with matrices in Chapter 1 are applied to solving linear systems. Students are introduced to identity and inverse matrices in this section.
- ⇔ Systems of inequalities and their use in linear programming are presented in the last two sections of this chapter. A brief review of solving linear inequalities can be found in the **Student Resources Toolbox** on page 787.

Chapter Highlights

Interview with Gina Oliva: Fitness and mathematics come together in this interview with fitness instructor Gina Oliva. Examples and exercises relating to this interview can be found on pages 299, 316, 319, and 326.

Explorations in Chapter 7 use graphing to investigate the ways that lines intersect in Section 7.2 and to explore the meaning of linear inequalities in Section 7.4.

The Portfolio Project: Systems of linear equations and systems of exponential equations are used by students to model Olympic performance data. Students write a report summarizing their results and discuss possible results of future performances.

Technology: Graphing calculators or graphing software provide a convenient way to graph systems of equations and to set up tables of values for systems of equations. Both tools are used in this chapter to solve systems. Graphing calculators are used in Section 7.3 to solve systems using matrices.

OBJECTIVES

Section	Objectives	NCTM Standards
7.1	• Use technology to find points of intersection of graphs. • Use substitution to solve systems of equations. • Use solutions of systems of equations to solve real-world problems.	1, 2, 3, 4, 5
7.2	• Solve linear systems by adding or subtracting equations. • Solve problems involving two unknown variables.	1, 2, 3, 4, 5
7.3	• Solve systems of linear equations using matrices. • Use systems of linear equations to solve real-world problems.	1, 2, 3, 4, 5
7.4	• Write and graph inequalities with two variables. • Solve real-world problems using inequalities with two variables.	1, 2, 3, 4, 5, 8
7.5	• Graph a system of linear inequalities. • Find the system of linear inequalities that describes a given graph. • Represent and interpret situations involving inequalities in real-world problems.	1, 2, 3, 4, 5, 8
7.6	• Find the best solution when several conditions have to be met. • Use linear programming techniques to solve real-world problems.	1, 2, 3, 4, 5

INTEGRATION

Mathematical Connections	7.1	7.2	7.3	7.4	7.5	7.6
algebra	**291–296***	**297–302**	**303–308**	**309–314**	**315–321**	**322–327**
geometry					321	
data analysis, probability, discrete math	291, 292, 294–296	298, 299, 301, 302	307	313, 314	316–321	**322–327**
patterns and functions			308			327
logic and reasoning	291, 292, 294–296	297–299, 301, 302	304–308	**309–314**	315, 317–321	323–327
Interdisciplinary Connections and Applications						
chemistry and physics		301		313		
business and economics	295	302		310	321	322–325
arts and entertainment			307			325
sports and recreation	294	298				
finance	291	302				327
agriculture			307	313		
advertising, social science, cooking, cartography, manufacturing, medicine, geology, meteorology	292, 294		303, 306	313, 314	318, 320	

* ***Bold page numbers** indicate that a topic is used throughout the section.*

TECHNOLOGY

	opportunities for use with	
Section	**Student Book**	**Support Material**
7.1	graphing calculator McDougal Littell Mathpack *Function Investigator*	**Technology Book:** TI-92 Activity 3
7.2	graphing calculator McDougal Littell Mathpack *Function Investigator*	**Function Investigator with Matrix Analyzer Activity Book:** Function Investigator Activities 10 and 11
7.3	graphing calculator McDougal Littell Mathpack *Matrix Analyzer*	**Technology Book:** Calculator Activity 7 Spreadsheet Activity 7
7.4	graphing calculator	
7.5	graphing calculator McDougal Littell Mathpack *Function Investigator*	
7.6	graphing calculator	

PLANNING GUIDE

Regular Scheduling (45 min)

Section	Materials Needed	Core Assignment	Extended Assignment	Applications (exercises that feature)	Communication (exercises that feature)	Technology (exercises that feature)
7.1	graphing calculator or software, graph paper	1–23, 33–38	1, 3, 6, 8–15, 20–38	10, 11, 27–31	26, 32, 33	9, 12–24
7.2	graphing calculator or software	1–18, 22–30	1, 3, 6, 8–11, 15–30	17, 18, 20, 21	19, 22	19
7.3	graphing calculator or software	**Day 1:** 1–14 **Day 2:** 16–22, 27–36, AYP*	**Day 1:** 1–12 **Day 2:** 19–36, AYP	5 23–26	 27	1–15 16–26
7.4	graph paper	1–17, 23–31	1–4, 10–31	18–25	31	
7.5	graph paper	**Day 1:** 1–8, 11–15 **Day 2:** 22–33, 36–47	**Day 1:** 1–3, 7–15 **Day 2:** 22–24, 28–47	9, 14–21 35	9, 17 36	
7.6	graph paper	1–14, 18–27, AYP	1–27, AYP	13–17	22	
Review/ Assess		**Day 1:** 1–20 **Day 2:** Ch. 7 Test	**Day 1:** 1–20 **Day 2:** Ch. 7 Test			
Portfolio Project		Allow 2 days.	Allow 2 days.			

Yearly Pacing (with Portfolio Project)	**Chapter 7 Total** 12 days	**Chapters 1–7 Total** 92 days	**Remaining** 68 days	**Total** 160 days

Block Scheduling (90 min)

	Day 41	Day 42	Day 43	Day 44	Day 45	Day 46	Day 47
Teach/Interact	Ch. 6 Test 7.1	7.2: Exploration, page 297 7.3	Continue with 7.3 7.4: Exploration, page 309	7.5	7.6 Port. Proj.	Review Port. Proj.	Ch. 7 Test 8.1: Exploration, page 337
Apply/Assess	**Ch. 6 Test** **7.1:** 1–23, 33–38	**7.2:** 1–18, 22–30 **7.3:** 1–14	**7.3:** 16–22, 27–36, AYP* **7.4:** 1–17, 23–31	**7.5:** 1–8, 11–15, 22–33, 36–47	**7.6:** 1–14, 18–27, AYP **Port. Proj.**	**Review:** 1–20 **Port. Proj.**	**Ch. 7 Test** **8.1:** 1–7, 11–16, 20–23

NOTE: A one-day block has been added for the Portfolio Project—timing and placement to be determined by teacher.

Yearly Pacing (with Portfolio Project)	**Chapter 7 Total** 6 days	**Chapters 1–7 Total** $46\frac{1}{2}$ days	**Remaining** $35\frac{1}{2}$ days	**Total** 82 days

**AYP* is Assess Your Progress.*

Lesson Support

Section	Practice Bank	Study Guide*	Assessment Book*	Visuals	Explorations Lab Manual	Lesson Plans	Technology Book
7.1	40	7.1		Warm-Up 7.1	Master 1 Add. Expl. 12	7.1	TI-92 Act. 3
7.2	41	7.2		Warm-Up 7.2	Add. Expl. 13	7.2	
7.3	42	7.3	Test 26	Warm-Up 7.3		7.3	Calculator Act. 7 Spreadsheet Act. 7
7.4	43	7.4		Warm-Up 7.4 Folders 7, A	Masters 1, 2	7.4	
7.5	44	7.5		Warm-Up 7.5 Folders 7, A	Masters 1, 2	7.5	
7.6	45	7.6	Test 27	Warm-Up 7.6	Masters 1, 2	7.6	
Review Test	46	Chapter Review	Tests 28, 29 Alternative Assessment			Review Test	Calculator Based Lab 4

***Spanish versions** of *Study Guide* and *Assessment Book* are available.

Chapter Support

- Course Guide
- Lesson Plans
- Portfolio Project Book
- Preparing for College Entrance Tests
- Multi-Language Glossary
- *Test Generator* Software
- Professional Handbook

Software Support

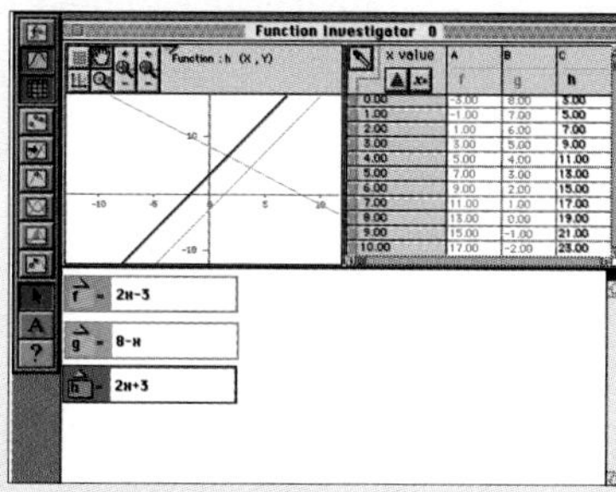

McDougal Littell Mathpack
Function Investigator
Matrix Analyzer

Internet Support

http://www.hmco.com
Next go to McDougal Littell; then the Education Center; then Secondary Math.

Outside Resources

Books, Periodicals

The Madison Project. *Explorations in Math.* "Matrices and Space Capsules," pp. 372–380.

Cooper, Patricia. "Supply & Demand: An Application of Linear Equations." *Mathematics Teacher* (October 1991): pp. 554–559.

Activities, Manipulatives

Breuningsen, Chris, Bill Bower, Linda Antinone, and Elisa Breuningsen. "Meet You at the Intersection." *Real-World Math with the CBL System.* Activity 4: pp. 27–32. Texas Instruments, 1995.

Levine, Maita, Robert Plummer, and Raymond Rolwing. "Using TI-81 to Analyze Sports Data." *Mathematics Teacher* (November 1993): pp. 636–641.

Software

Department of Mathematics and Computer Science of the North Carolina School of Science and Mathematics. *Matrices.* Materials and software. IBM. Reston, VA: NCTM.

Videos

"Juicy Problems." Program No. 4. *For All Practical Purposes.* Arlington, MA: COMAP, 1991.

Video for Unit Project: *Our Human Body.* Episode 10: Nutrition. Agency for Instructional Technology, 1992.

Internet

Subscribe to a newsgroup dedicated to discussions, questions, and answers regarding Texas Instruments graphing calculators by sending a message to:
majordomo@lists.ppp.ti.com
Include no subject; in body of the message, type:
subscribe graph-ti [your e-mail address]

Interview Notes

Background

Sign Language

Sign language is a language of gestures and hand symbols that both deaf and hearing people use to communicate. The earliest known sign language was developed around 1760 in Paris, France by the Abbe Charles Michel de l'Epee and the deaf community who attended his school. Sign language was brought to the United States in 1816 by Laurent Clerc, a graduate of the Paris school, and Dr. Thomas Hopkins Gallaudet. Clerc was named the first head instructor of the American Institution for the Deaf founded in Hartford, Connecticut in 1817, and Dr. Gallaudet's son Edward established Gallaudet College, the world's first college for deaf students, in Washington, DC in 1864.

In the United States, many people use the American Sign Language (ASL), which is the fourth most commonly used language in the United States. The American Sign Language is based on ideas rather than words and uses gestures to express a particular idea or concept. The American manual alphabet, or finger alphabet, is a form of sign language that represents each letter of the alphabet by a different position of the fingers of one hand. Words that do not have an equivalent in the American Sign Language can be spelled out using the American manual alphabet. Together, both the American Sign Language and the American manual alphabet can be used to help the deaf and hearing-impaired communicate information and ideas.

CHAPTER

7 Systems

Seeing how to exercise

(American Manual Alphabet for "exercise")

INTERVIEW

Gina Oliva

When Gina Oliva took her first aerobics class in 1981, she instantly loved it. "When I find something I love, I want to share it with other people," Oliva says. "I also saw that I could do it—that I had a feel for the rhythm." This is no small matter because Oliva is almost completely deaf; she can hear only low-frequency music with a very strong bass line. Yet she picked up the basic aerobics steps quickly and soon decided to "bring the joy of exercise to both deaf and hearing people."

"I've devoted my life to making exercise more accessible to deaf people."

Innovative Hand and Arm Gestures

Oliva was well-positioned to work with deaf people, given that she was already on the staff of Gallaudet University in Washington, D.C., the world's leading learning institution for the deaf. The challenge was coming up with a system to help students follow an aerobics class even when they cannot hear the instructor's commands.

Oliva developed a set of approximately one hundred visual cues—hand and arm gestures that enable teachers to communicate the entire sequence of steps without saying a single word. Like pantomime, most of the visual cues are designed to resemble the dance-exercise movements. She also uses American Sign Language to convey some of the cues not easily represented by gesture.

"Math can help you become more fit."

Visual Cues for Everyone

Equipped with this new "vocabulary," Oliva has taught classes and trained aerobics instructors all over the country. She stresses that the visual cues she invented are for everyone, both hearing and deaf alike.

"Among hearing people, one of the biggest problems is voice injuries to aerobics instructors trying to be heard over the music," she says. "Even when the teacher uses a microphone, many students can't hear the instructions. It helps if you can see a signal at the same time you hear something."

Interview Notes

Background

Gina Oliva

Gina Oliva grew up in New York as the middle child of five children. At the age of 4, it was discovered by her kindergarten teacher that she had a hearing loss, but she still enjoyed music and learned to play the drums and piano at the age of 6. This practice helped her develop an innate sense of rhythm which is crucial to her career today. Despite her hearing loss, she attended public schools from kindergarten through college.

Gina Oliva has produced two workout videotapes, *Sign 'N Sweat* and *Shape Up 'N Sign*. The first is a workout tape appropriate for both hearing and deaf people, and the second is a workout tape for hearing and deaf children ages 6–10 that teaches both movement and sign language. She has also published many papers on the subject of physical fitness in both the deaf and hearing communities, and has won awards for her work, including a selection as one of the ten Healthy American Fitness Leaders for 1989.

Aerobic Exercise

Aerobic exercise is any exercise designed to promote the supply of oxygen in the body. While many people in the United States today refer to aerobics as a combination of exercises and dance steps, there are many other forms of aerobic exercise, such as bicycling, jogging, rowing, skating, swimming, fast walking, and dancing.

The benefits of aerobic exercise are numerous. Physically, aerobic exercise can lessen the risk of heart disease, develop muscle tone, increase strength, help weight loss and weight control, make bones stronger, reduce high blood pressure and cholesterol, and slow down the physical effects of aging. Psychologically, regular exercise contributes to a feeling of well-being and can help relieve stress.

Interview Notes

Second-Language Learners

Make sure all students know that *aerobics* is a form of exercising that combines muscle-strengthening exercises with dance steps. Point out to students learning English that *picked up* in the last sentence means "learned easily."

Mathematical Connection

Many people do various forms of aerobic exercise other than aerobic dance to stay in shape, such as jogging, walking, swimming, skiing, and playing volleyball. In Section 7.2, systems of equations are used to find the time spent walking and jogging during a period of exercise. In Section 7.6, linear programming is used to analyze the effectiveness of various forms of aerobic exercise in terms of calories burned in a certain amount of time. In Section 7.5, systems of inequalities are used to explore the range of a person's target heart rate and also the range of a person's ideal weight.

Explore and Connect

Writing
After students calculate their minimum and maximum heart rates, work with them to write general equations for calculating the minimum and maximum heart rates for a person of any age.

Research
There may also be organizations in your community that are designed for deaf and hearing-impaired athletes. Your local YMCA or other athletic organization may be able to provide information for this project.

Project
Have students engage in physical activities that require constant movement for the entire 5 minutes. Caution students not to overexert themselves.

Getting a Good Workout

In addition to aerobics, Oliva teaches her students how to determine their target heart rate zone—the range of heart rates that indicates a good aerobic workout. Oliva explains, "In a way, math can help you become more fit. It can tell you the amount of exertion you need to get a good workout without overdoing it."

To figure out your target heart rate zone, subtract your age from 220. Multiply that number by 0.60 to get your minimum target heart rate and by 0.80 to get your maximum target heart rate. As a general rule during exercise, try to keep your heart rate between these two values. This requirement can be expressed by a *system of inequalities*, whose graph is shown.

Find your heart rate by counting your pulse for 6 s and multiplying that number by 10. To feel your pulse, place your index and middle fingers on your wrist. Begin counting the beats within 15 s of stopping an activity.

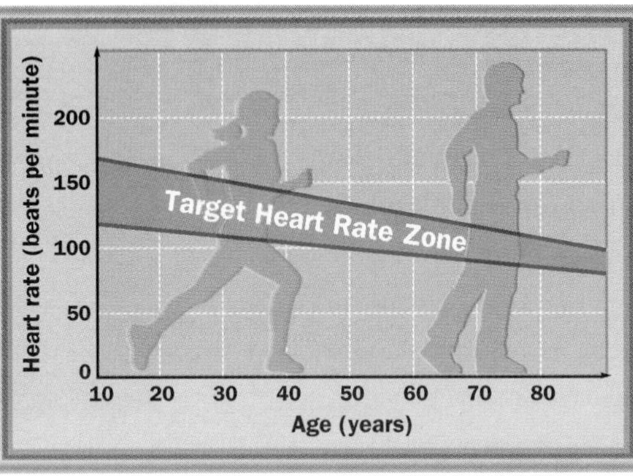

Explore and Connect

Gina Oliva prepares to start an aerobics session.

1. Writing Calculate your own minimum target heart rate and maximum target heart rate using the guidelines above. Then use the graph to explain how these values change as you get older.

2. Research What are some ways that other sports are adapted to make them more accessible to handicapped individuals? You may want to contact the Disabled Sports USA organization in Washington, D.C., for more information. Report your findings to your class.

3. Project Try engaging in different physical activities like walking, running, or bicycling. After 5 min of an activity, find your heart rate. Is this rate within your target heart rate zone? Compare the results for each activity.

Mathematics & Gina Oliva

In this chapter, you will learn more about how mathematics is related to fitness.

Related Examples and Exercises

Section 7.2
- Example 2

Section 7.5
- Example 1
- Exercises 14–18

Section 7.6
- Exercises 15–17

Section

7.1 Systems of Equations

Learn how to...

- use technology to find points of intersection of graphs
- use substitution to solve systems of equations

So you can...

- solve problems about personal finance and advertising, for example

The political party that controls, or has the most members in, the U.S. Senate can more easily accomplish its legislative goals. The graph shows the number of Democrats and Republicans in the U.S. Senate during fifteen Congresses.

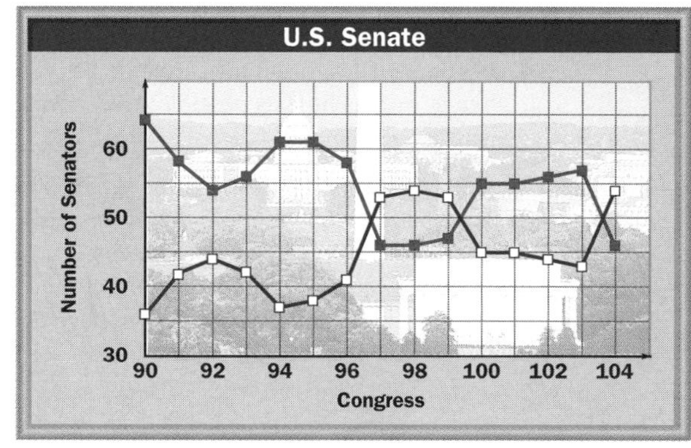

THINK AND COMMUNICATE

Use the graph to answer the following questions.

1. During which Congresses were the Democrats in control? the Republicans in control?
2. How does the graph show when changeovers in control took place? How many changeovers were there?

A point where two graphs intersect often has significance. Sometimes you may want to use technology to find an intersection point.

EXAMPLE 1 Application: Personal Finance

When buying a car, Tom Fiore needs to finance \$12,000 over four years. The car dealer offers him a 2.9% annual interest rate if he finances the full amount *or* a 9.9% annual interest rate if he takes a \$1500 up-front discount. The amount A owed on each loan after p monthly payments is given by:

$$A = 109{,}688 - 97{,}688(1.00242)^p \quad \leftarrow \text{first loan option}$$

$$A = 32{,}218 - 21{,}718(1.00825)^p \quad \leftarrow \text{second loan option}$$

In how many months will the amounts needed to pay off the two loans be equal?

ANSWERS Section 7.1

Think and Communicate

1. 90–96 and 100–103; 97–99 and 104
2. The graphs intersect; 3.

Plan⇔Support

Objectives

- Use technology to find points of intersection of graphs.
- Use substitution to solve systems of equations.
- Use solutions of systems of equations to solve real-world problems.

Recommended Pacing

Core and Extended Courses
Section 7.1: 1 day

Block Schedule
Section 7.1: $\frac{1}{2}$ block
(with Chapter 6 Test)

Resource Materials

Lesson Support
Lesson Plan 7.1
Warm-Up Transparency 7.1
Practice Bank: Practice 40
Study Guide: Section 7.1
Exploration Lab Manual: Additional Exploration 12, Diagram Master 1

Technology
Technology Book: TI-92 Activity 3
Graphing Calculator
McDougal Littell Mathpack *Function Investigator*
Internet: http://www.hmco.com

Warm-Up Exercises

Solve each equation.

1. $15x + 17 = 12x + 7.4$ −3.2
2. $588 = 312 + 12k$ 23
3. $16m^2 + 19 = 1700$ ±10.25

Tell whether the given ordered pair (x, y) is or is not on the graph of the given equation. Answer *Yes* or *No*.

4. (2, −5) $9x + 7y = -27$ No.
5. (−9, 11) $y = -\frac{4}{5}x + \frac{19}{5}$ Yes.
6. (35, 35) $y = \frac{3}{7}x + 20$ Yes.

Teach⇔Interact

Section Note

Using Technology Students using a TI-81 or TI-82 graphing calculator can find intersection points by using the zoom-and-trace approach. Students using the TI-82 have the option of using intersect from the CALC menu.

Additional Example 1

On January 1, Linda Martinez will start withdrawing a fixed amount each month from her retirement account. Her brother will start withdrawing from his retirement account on the same date. The amount Linda will have in her account after m months is given by $A = 700{,}000 - 400{,}000(1.0075)^m$. The amount her brother will have in his account after m months is given by $A = 650{,}000 - 375{,}000(1.007)^m$. After how many months will the amounts remaining be equal?

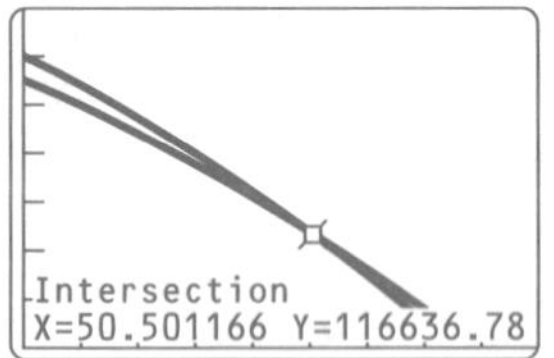

The amounts remaining will be equal after about 51 months.

Additional Example 2

A publisher plans a new edition of its encyclopedia. It will be available in print form and on CD/ROM. The advertising budget is \$8.9 million, with 5.2 times as much going for advertising the CD/ROM as for advertising the print edition. Determine how much money will be spent advertising the two versions.

Step 1 Write a system of equations. Let c = the amount spent on the CD/ROM, in millions, and let p = the amount spent on the print version, in millions.

$c + p = 8.9$

$c = 5.2p$

Step 2 Solve the system using substitution. Substitute $5.2p$ for c.

$5.2p + p = 8.9$

$6.2p = 8.9$

$p \approx 1.44$

Substitute in the second equation of the system to find c.

$c = 5.2p \approx 5.2(1.44) \approx 7.49$

The company will spend about \$1.44 million on advertising the print version and about \$7.49 million on advertising the CD/ROM.

SOLUTION

Use a graphing calculator or graphing software to graph the equations and find where they intersect.

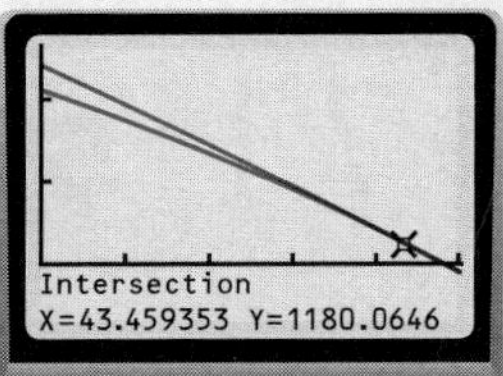

Tom Fiore will owe the same amount on both loans in about **43 months**.

THINK AND COMMUNICATE

3. Compare the two A-values when $p = 0$. What do these numbers represent?

4. If Tom decides to sell his car while he is still making payments on it, he'd like to owe as little as possible. Under which loan plan would Tom owe the least after 43 months?

Two or more equations involving the same variables, such as the equations $A = 109{,}688 - 97{,}688(1.00242)^p$ and $A = 32{,}218 - 21{,}718(1.00825)^p$ from Example 1, form a **system of equations**.

Solving a system means finding the coordinates of the point(s) where the graphs of the equations intersect. When one equation is already solved for one of the variables, you can use substitution to solve the system algebraically.

EXAMPLE 2 Application: Advertising

A footwear company is preparing to market a new kind of sports shoe. The company plans to spend \$28 million on advertising. Past experience shows that the amount spent on television advertising should be 3.5 times as much as the amount spent on magazine advertising. Determine how much money should be spent on each type of advertising.

SOLUTION

Step 1 Write a system of equations. Let t = the amount spent on television advertising, in millions, and let m = the amount spent on magazine advertising, in millions.

$t + m = 28$ ← total money spent on advertising

$t = 3.5m$ ← how money is to be split between television and magazine ads

Step 2 Solve the system using substitution.

Substitute **3.5*m*** for ***t*** and solve for m.

$3.5m + m = 28$

$4.5m = 28$

$m \approx 6.22$

$t = 3.5m$

$\approx 3.5(6.22)$

≈ 21.78

Substitute **6.22** for ***m*** to find t.

The company should spend about \$21.78 million on television advertising and about \$6.22 million on magazine advertising.

Think and Communicate

3. first loan: 12,000; second loan: 10,500; the initial cost of the car

4. the second (first loan: \$1300.47; second loan: \$1296.98)

EXAMPLE 3

Use substitution to solve this system of equations.

$$5x - 2y = 4$$
$$4x - y = 5$$

SOLUTION

Step 1 Solve one equation for one of the variables. In this case, it is convenient to solve the second equation for *y*.

$$4x - y = 5$$
$$y = 4x - 5$$

Step 2 Solve the system using substitution.

$$5x - 2(4x - 5) = 4$$
$$5x - 8x + 10 = 4$$
$$-3x = -6$$
$$x = 2$$

Substitute **4x − 5** for **y** and solve for *x*.

$$y = 4x - 5$$
$$= 4(2) - 5$$
$$= 3$$

Substitute **2** for ***x*** to find *y*.

The solution of the system is $(x, y) = (2, 3)$.

Check

$5(2) - 2(3) \stackrel{?}{=} 4$ $4(2) - 3 \stackrel{?}{=} 5$

$10 - 6 \stackrel{?}{=} 4$ $8 - 3 \stackrel{?}{=} 5$

$4 = 4$ ✓ $5 = 5$ ✓

CHECKING KEY CONCEPTS

1. Explain what it means to solve a system of equations.

Technology Use a graphing calculator or graphing software to graph each pair of equations and find the point(s) of intersection.

2. $y = 2x^2$, $y - x = 6$
3. $y = 3x + 10$, $2y = 12x + 11$
4. $y = 2^x$, $y = x + 5$

Use substitution to solve each system of equations.

5. $y = 13x$, $y + 12x = 75$
6. $2x + y = 14$, $x - 3y = 5$
7. $x + 2y = 11$, $4x + 2y = 17$

7.1 Exercises and Applications

Extra Practice exercises on page 758

Solve each system of equations.

1. $2x - y = 0$, $4x + y = 12$
2. $y = 1.2x$, $y + 2.8x = 5$
3. $x = y - 1$, $3x + y = 13$
4. $-2x + y = 2$, $4x - y = 9$
5. $y + 2x = 15$, $2y - 6x = 10$
6. $3y + x = 10$, $2x + 5y = 19$
7. $y = 2.6x$, $2y + 3x = 12$
8. $3.1x + y = 0$, $3y + 2.5x = 17$

Teach⇔Interact

Additional Example 3

Use substitution to solve this system of equations.

$7x + 3y = 53$

$-x + 4y = -43$

Step 1 Solve one equation for one of the variables.

$-x + 4y = -43$

$-x = -4y - 43$

$x = 4y + 43$

Step 2 Substitute $4y + 43$ for x in the first equation and solve for y.

$7(4y + 43) + 3y = 53$

$28y + 301 + 3y = 53$

$31y = -248$

$y = -8$

Substitute -8 for y to find x.

$x = 4y + 43$

$= 4(-8) + 43$

$= 11$

The solution of the system is $(x, y) = (11, -8)$.

Closure Question

When would you choose to solve a system of equations by graphing rather than by substitution? Give an example. I would use graphing if the coefficients in the system are numbers that would be difficult to use in computations. Example:

$39x - 57.8y = 710$

$-5.13x + 0.9y = 84$

Apply⇔Assess

Suggested Assignment

Core Course
Exs. 1–23, 33–38

Extended Course
Exs. 1, 3, 6, 8–15, 20–38

Block Schedule
Day 41 Exs. 1–23, 33–38

Exercise Notes

Cooperative Learning

Exs. 1–8 You may wish to have students work on these exercises with a partner. One student can solve four of the exercises by graphing, while the other student solves by substitution. Students then trade roles for the remaining four exercises. They should compare results and share the job of checking by substitution.

Checking Key Concepts

1. To solve a system of equations means to find the coordinates of the point(s) where the graphs of the equations intersect (that is, the values of the variables that make all the equations of the system true).
2. (–1.5, 4.5), (2, 8)
3. (1.5, 14.5)
4. (–4.97, 0.032), (3, 8)
5. (3, 39)
6. $\left(\frac{47}{7}, \frac{4}{7}\right)$
7. (2, 4.5)

Exercises and Applications

1. (2, 4)
2. (1.25, 1.5)
3. (3, 4)
4. (5.5, 13)
5. (2, 11)
6. (7, 1)
7. $\left(\frac{60}{41}, \frac{156}{41}\right)$
8. (–2.5, 7.75)

Apply⇔Assess

Exercise Notes

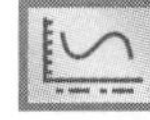

Using Technology

Ex. 9 If students find it confusing to work with two columns of *y*-values, suggest that they add a third column to the table by using Y3=(3X+4)–(4X–1). Ask them to interpret what they observe in that single column of the table. (The value of *x* for which the value of Y3 is 0 gives the *x*-coordinate of the solution pair.)

Research

Ex. 10 Students may consult almanacs or other sources to see if they can find comparable data for the early part of the century. They may also want to think about how the populations of Alaska and Hawaii should be taken into account.

Problem Solving

Ex. 11 Ask for a volunteer to explain how he or she solved this problem. It is possible to solve the problem by setting up and solving a system of equations or by setting up and solving a single equation. Survey the class to see if both approaches were used, and if so, which approach was used more often.

Using Technology

Exs. 12–23 You may want to suggest that students use the *Function Investigator* software for this group of exercises.

Student Study Tip

Ex. 24 Point out to students that they need to experiment with several different graphing windows to find both points of intersection of the graphs.

9.  **Technology** Johanna wanted to solve the system

$$y = 3x + 4$$
$$y = 4x - 1$$

by setting up tables of values. Her calculator displayed the y_1-values (for $y = 3x + 4$) and the y_2-values (for $y = 4x - 1$) for $x = 1, 2, \dots, 7$.

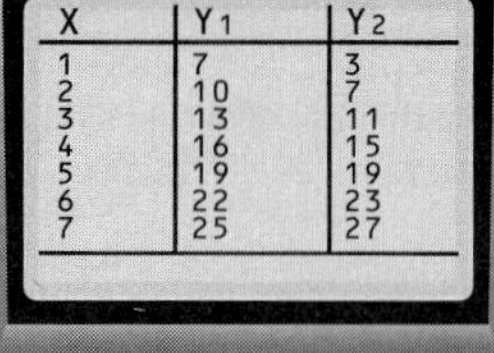

X	Y1	Y2
1	7	3
2	10	7
3	13	11
4	16	15
5	19	19
6	22	23
7	25	27

a. What is the solution of the system? Explain your reasoning.

b. Open-ended Problem Describe other ways that Johanna could solve the system.

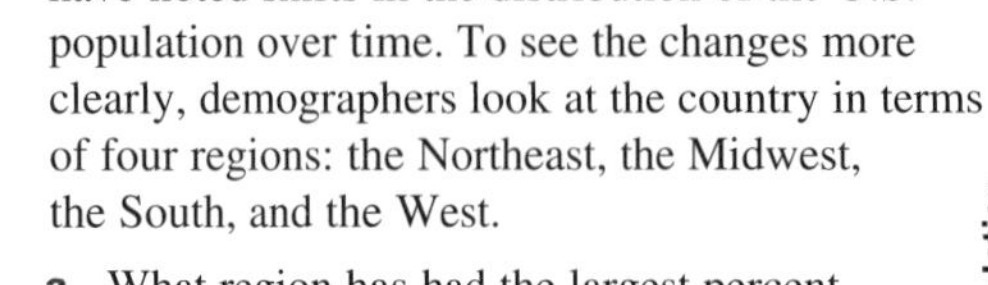

10. SOCIAL SCIENCE People who study demographics have noted shifts in the distribution of the U.S. population over time. To see the changes more clearly, demographers look at the country in terms of four regions: the Northeast, the Midwest, the South, and the West.

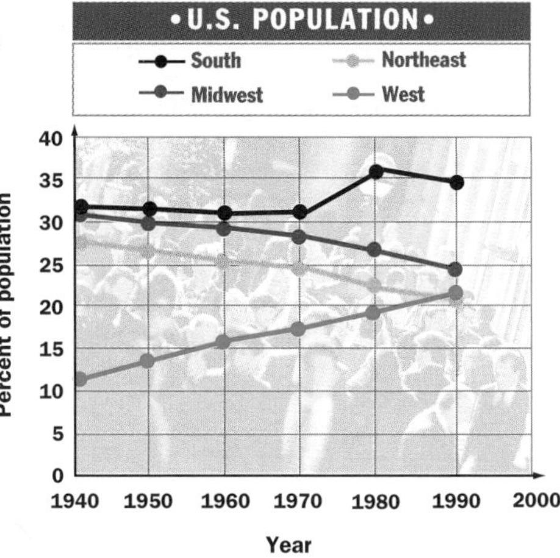

a. What region has had the largest percent of the U.S. population since 1940?

b. What region has been steadily growing since 1940?

c. What was significant about the decade 1980–1990?

d. What predictions might you make about the decade 1990–2000?

11. SPORTS A baseball player's batting average (the number of hits per at-bat) is .280 at the beginning of a game. The player gets 3 hits during 5 at-bats. He ends the game with a .300 batting average. How many times has the player batted this season?

Technology Use a graphing calculator or graphing software to solve each system of equations. (The graphs of the equations may intersect more than once.)

12. $y = 4(2^x)$, $y = 34.6 + 17x$

13. $y = 103(1.08)^x$, $y = 90 + 15x$

14. $y = x^2$, $y = 7$

15. $y = x^2 - 1$, $y = 3x$

16. $y = 3x^2 - 4x + 5$, $y = -5x + 10$

17. $y = x^2 + 2x - 2$, $y = 7 + 10x$

18. $y = x^2 + 2x - 2$, $y = 4x + 4$

19. $y = x^2 - 16$, $y = -x^2 + 6$

20. $y = 2x + 5$, $y = 14x - 17$

21. $y = -2(3^x)$, $y = -53$

22. $y = x^2 - 3x$, $y = 5x + 8$

23. $y = x^2 + x - 1$, $y = 2x^2 - 6$

24. Technology Margot and Claudia are modeling the world's population P, in billions. The world's population in 1990 was 5.33 billion.

- Margot predicts a 1.7% annual increase after 1990, giving the exponential equation $P = 5.33(1.017)^x$, where x = the number of years since 1990.
- Claudia predicts an annual increase of 130 million people, giving the linear equation $P = 5.33 + 0.13x$.

Use a graphing calculator or graphing software to determine the years for which the two models will give the same predicted populations.

9. a. (5, 19); The y_1-value and the y_2-value that correspond to the *x*-value 5 are the same.

b. Answers may vary. Examples are given. She could use substitution. She could also graph the function using graph paper, a graphing calculator, or graphing software and locate the intersection of the graphs.

10. a. the South

b. the West

c. The percentage of the population in the West surpassed that of the Northeast.

d. Answers may vary. An example is given. If present trends continue, I think the percentage of the population in the West will surpass that of the Midwest.

11. 75 times

12. (–1.98, 1.02), (4.88, 118)

13. (2.02, 120), (13.7, 295)

14. (–2.65, 7), (2.65, 7)

15. (–0.303, –0.908), (3.303, 9.908)

16. (–1.47, 17.34), (1.14, 4.32)

17. (–1, –3), (9, 97)

18. (–1.65, –2.58), (3.65, 18.6)

19. (–3.32, –5), (3.32, –5)

25. Challenge Find the time t between successive alignments of the hands of a clock. (*Hint:* The hour hand rotates $d°$ at a rate of 30°/h in time t, while the minute hand rotates $(360 + d)°$ at a rate of 360°/h in the same amount of time.)

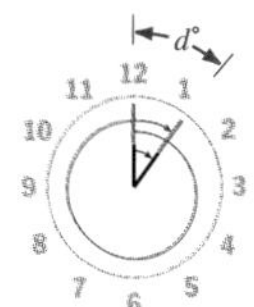

Connection ECONOMICS

Although appliances can be expensive, they often pay for themselves after a certain number of uses.

26. Writing What do you think the phrase "pay for itself" means for an appliance?

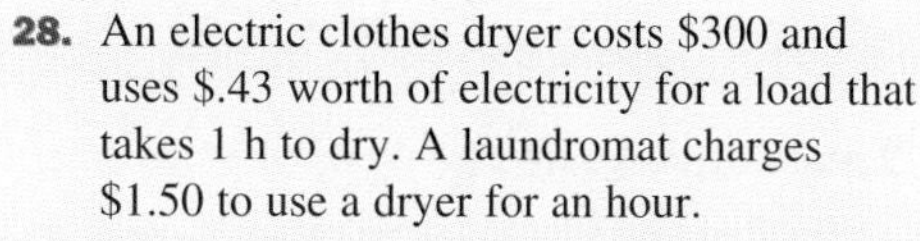

27. A breadmaker costs $199. The ingredients and electricity to make one loaf of bread with the machine cost $.79. A loaf of bread of similar quality costs $1.59 at a grocery store.

a. Write two equations giving the total cost C in terms of the number of loaves l made or purchased for each situation.

b. Solve the system of equations to find the number of loaves of bread you have to make before the breadmaker pays for itself.

28. An electric clothes dryer costs $300 and uses $.43 worth of electricity for a load that takes 1 h to dry. A laundromat charges $1.50 to use a dryer for an hour.

a. Write equations representing the total cost C for drying l loads of laundry in each situation.

b. How many loads of laundry do you have to dry before the dryer pays for itself?

29. A noodle ringer for rolling homemade noodles costs $40. The ingredients for each batch of homemade noodles cost $.25. A batch of store-bought noodles costs $.50.

a. Write equations representing the total cost C for b batches of noodles in each situation.

b. How many batches of noodles do you have to make before the noodle ringer pays for itself?

30. Research Find out how much a home water purifier costs and how much a bottle of purified water costs. Determine how many gallons of water you would have to purify for the purifier to pay for itself.

31. Open-ended Problem Choose an appliance and determine how many times you have to use it before it pays for itself.

Apply⇔Assess

Exercise Notes

Problem Solving

Ex. 25 Students can check the reasonableness of their results by experimenting with a clock or watch. They should keep in mind, however, that an answer that is both exact and reliable requires a careful mathematical analysis of the situation.

Interdisciplinary Problems

Exs. 26–31 These exercises illustrate how systems of equations can be applied to an analysis of the costs involved in using some home appliances.

Communication: Discussion

Exs. 27–31 Ask students what factors they think are significant in determining the cost of using these appliances and what alternatives may be possible. They should not overlook possible maintenance costs.

Second-Language Learners

Ex. 30 Students learning English may not be familiar with home water purifiers. Explain that these devices make drinking water purer by removing chemicals. They may also benefit from working on this research with the help of a peer tutor or aide.

20. (1.83, 8.67)
21. (2.98, –53)
22. (–0.899, 3.51), (8.90, 52.5)
23. (–1.79, 0.417), (2.79, 9.58)
24. 1990, 2031
25. about 65.5 min
26. Answers may vary. An example is given. A new appliance pays for itself when the savings created by using the new appliance are equal to its cost.
27. a. $C = 199 + 0.79l$; $C = 1.59l$
    b. 249 loaves
28. a. $C = 300 + 0.43l$; $C = 1.5l$
    b. 281 loads
29. a. $C = 40 + 0.25b$; $C = 0.5b$
    b. 160 batches
30. Answers may vary. Check students' work.
31. Answers may vary. Check students' work.

Apply⇔Assess

Exercise Notes

Assessment Note

Ex. 33 The requirement in part (a) is intended to ensure that the systems students obtain can be solved easily by substitution. Students should do as much of the work for this exercise as is possible by using paper and pencil. Calculators can be used if some of the systems involve complicated computations.

Topic Spiraling: Preview

Exs. 34–37 These exercises can help prepare students for studying how many solutions are possible for a given system of linear equations, a topic that will be explored in the next section.

Practice 40 for Section 7.1

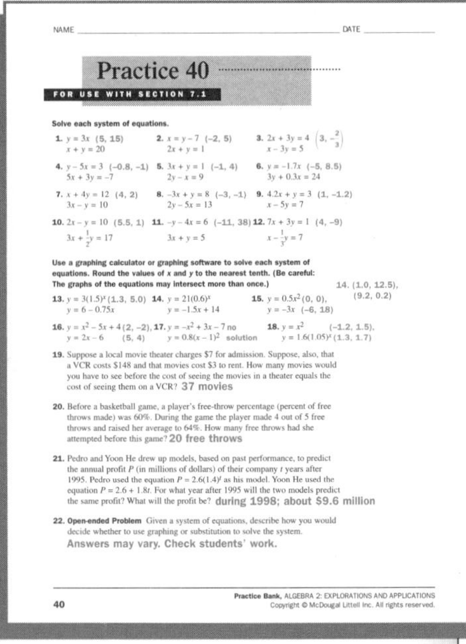

NAME ______ DATE ______

Practice 40

FOR USE WITH SECTION 7.1

Solve each system of equations.

1. $y = 3x$, $x + y = 20$ (5, 15)
2. $x = y - 7$, $2x + y = 1$ (−2, 5)
3. $2x + 3y = 4$, $x - 3y = 5$ $\left(3, -\frac{2}{3}\right)$
4. $y - 5x = 3$, $5x + 3y = -7$ (−0.8, −1)
5. $3x + y = 1$, $2y - x = 9$ (−1, 4)
6. $y = -1.7x$, $3y + 0.3x = 24$ (−5, 8.5)
7. $x + 4y = 12$, $3x - y = 10$ (4, 2)
8. $-3x + y = 8$, $2y - 5x = 13$ (−3, −1)
9. $4.2x + y = 3$, $x - 5y = 7$ (1, −1.2)
10. $2x - y = 10$, $3x + \frac{1}{2}y = 17$ (5.5, 1)
11. $-y - 4x = 6$, $3x + y = 5$ (−11, 38)
12. $7x + 3y = 1$, $x - \frac{1}{3}y = 7$ (4, −9)

Use a graphing calculator or graphing software to solve each system of equations. Round the values of x and y to the nearest tenth. (Be careful: The graphs of the equations may intersect more than once.)

13. $y = 3(1.5)^x$, $y = 6 - 0.75x$ (1.3, 5.0)
14. $y = 21(0.6)^x$, $y = -1.5x + 14$ 14. (1.0, 12.5), (9.2, 0.2)
15. $y = 0.5x^2$, $y = -3x$ (0, 0), (−6, 18)
16. $y = x^2 - 5x + 4$, $y = 2x - 6$ (2, −2), (5, 4)
17. $y = -x^2 + 3x - 7$, $y = 0.8(x - 1)^2$ no solution
18. $y = x^2$, $y = 1.6(1.05)^x$ (−1.2, 1.5), (1.3, 1.7)

19. Suppose a local movie theater charges \$7 for admission. Suppose, also, that a VCR costs \$148 and that movies cost \$3 to rent. How many movies would you have to see before the cost of seeing the movies in a theater equals the cost of seeing them on a VCR? 37 movies

20. Before a basketball game, a player's free-throw percentage (percent of free throws made) was 60%. During the game the player made 4 out of 5 free throws and raised her average to 64%. How many free throws had she attempted before this game? 20 free throws

21. Pedro and Yoon He drew up models, based on past performance, to predict the annual profit P (in millions of dollars) of their company t years after 1995. Pedro used the equation $P = 2.6(1.4)^t$ as his model. Yoon He used the equation $P = 2.6 + 1.8t$. For what year after 1995 will the two models predict the same profit? What will the profit be? during 1998; about \$9.6 million

22. **Open-ended Problem** Given a system of equations, describe how you would decide whether to use graphing or substitution to solve the system. Answers may vary. Check students' work.

40 **Practice Bank,** ALGEBRA 2: EXPLORATIONS AND APPLICATIONS

32. a. Visual Thinking A brainteaser like the one shown below appeared in an ad from a company boasting of its problem-solving abilities. Write an equation for each of the first three pictures.

b. Challenge Use the substitution method several times to determine how many oranges will balance the bunch of grapes in the final picture.

ONGOING ASSESSMENT

33. Cooperative Learning Work with two other students.

a. Each of you should take a piece of paper and write a linear equation using the variables x and y. One of the variables should have the coefficient 1 or −1.

b. Pass your paper to the person on your right. Write a second linear equation on the paper you receive.

c. Pass the paper you have to the person on your right once again. Solve the system of equations on the paper you receive.

d. Pass papers to the right one last time; you should receive your original paper. Check the solution shown by substituting it into the two equations. If it doesn't check, help the person who solved the system find where errors were made.

SPIRAL REVIEW

Solve each equation. If all numbers are solutions, write *all numbers*. If there is no solution, write *no solution*. *(Toolbox, page 786)*

34. $2x + 1 = -2 - 4x$ **35.** $4x - 1 = 4x + 3$ **36.** $2x + 1 = x - 3$ **37.** $x - 1 = x + 1$

38. The table shows the distance from the sun, d, and the period of revolution, p, of five planets. (Distances are given in *astronomical units* (AU), where 1 AU = the distance between Earth and the sun. Periods are given in Earth years.) *(Section 4.5)*

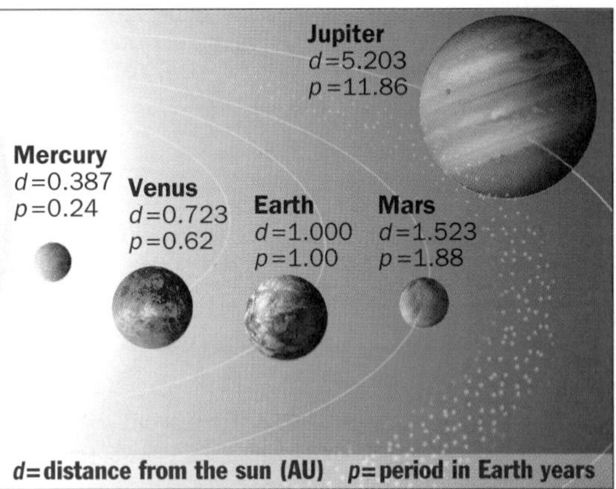

a. Make a scatter plot of the data pairs ($\log d$, $\log p$). What relationship exists between $\log d$ and $\log p$?

b. Find an equation giving p as a function of d.

c. Make a prediction about the period of Saturn if you know that its distance from the sun is 9.541 AU.

32. a. Let o = the number of oranges, g = the number of bunches of grapes, p = the number of pineapples, and b = the number of bananas; $o + g = p$, $o + b = g$, and $2p = 3b$.

b. 5 oranges

33. Answers may vary. Check students' work.

34. $-\frac{1}{2}$

35. no solution

36. −4

37. no solution

38. a.

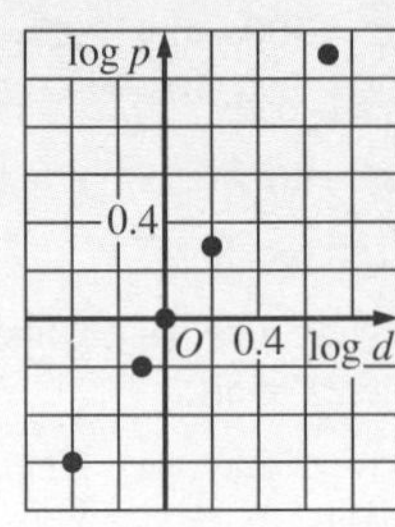

The relationship between $\log d$ and $\log p$ is linear.

b. $p = d^{1.5}$

c. Saturn's period is about 29.5 Earth years.

Section

7.2 Linear Systems

Learn how to...

- solve linear systems by adding equations

So you can...

- solve problems involving two unknown values, such as boat speed and current speed

In Chapter 2, you worked with linear equations in "$y =$" form, such as $y = 2x + 3$. When x and y appear together on the same side of the equation, as in $-2x + y = 3$, the equation is said to be in *standard form*. In the Exploration, you'll investigate systems of linear equations in standard form.

EXPLORATION
COOPERATIVE LEARNING

Investigating Ways that Lines Intersect

Work with a partner.
You will need:

- a graphing calculator or graphing software

Be sure to use parentheses when the coefficient of x is a fraction.

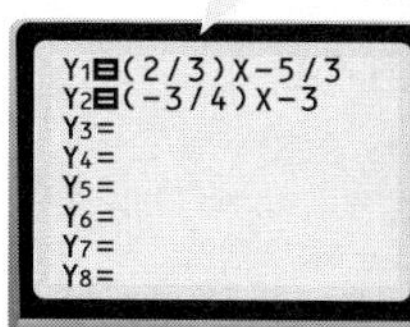

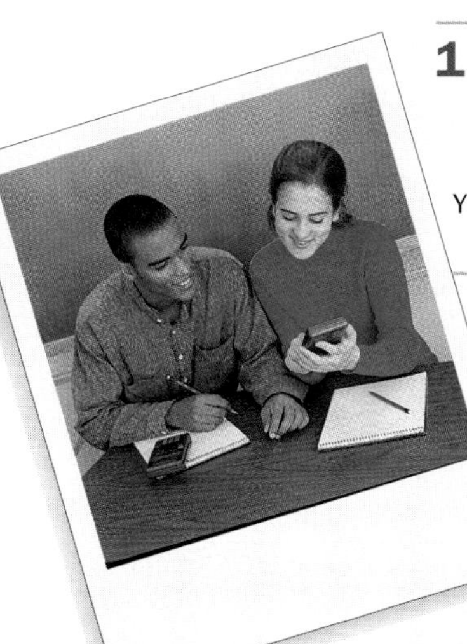

1 Graph this system:

$$2x - 3y = 5$$
$$3x + 4y = -12$$

Your calculator or software may require that you solve each equation for y first.

2 Do the lines intersect in a single point? If so, find the approximate coordinates of the point. If not, describe the geometric relationship between the lines.

3 Repeat Steps 1 and 2 for each of the following systems of equations.

$$2x - 3y = 5 \qquad 2x - 3y = 5$$
$$4x - 6y = 1 \qquad -6x + 9y = -15$$

Questions

1. In what ways can two lines in a plane intersect? What does this mean in terms of the number of solutions that a system of linear equations can have?
2. If a system of linear equations is to have a *unique* solution, what must be true about the slopes of the lines?

When the coefficients of one of the variables in a system of linear equations are the *same*, you can use *subtraction* to solve the system algebraically. Likewise, you can use *addition* when the coefficients of one of the variables are *opposites* of each other.

Exploration Note

Purpose
The purpose of this Exploration is to have students investigate systems of linear equations in standard form in order to discover how the graphs of the equations may be related to the number of solutions of the system.

Materials/Preparation
Each pair of students needs a graphing calculator or graphing software.

Procedure
Students graph a system of equations and describe the geometric relationship between the lines. They repeat this procedure for two more systems.

Closure
Have students share their answers to Questions 1 and 2. All students should understand what the relationship between the graphs of two lines implies for the number of solutions to the related system.

Explorations Lab Manual
See the Manual for more commentary on this Exploration.

For answers to the Exploration, see following page.

Plan⇔Support

Objectives

- Solve linear systems by adding or subtracting equations.
- Solve problems involving two unknown variables.

Recommended Pacing

Core and Extended Courses
Section 7.2: 1 day

Block Schedule
Section 7.2: $\frac{1}{2}$ block
(with Section 7.3)

Resource Materials

Lesson Support
Lesson Plan 7.2
Warm-Up Transparency 7.2
Practice Bank: Practice 41
Study Guide: Section 7.2
Exploration Lab Manual: Additional Exploration 13

Technology
Graphing Calculator
McDougal Littell Mathpack *Function Investigator with Matrix Analyzer Activity Book:* Function Investigator Activities 10 and 11
Internet: http://www.hmco.com

Warm-Up Exercises

Solve for y.

1. $-4x + y = 3$ $y = 4x + 3$
2. $5x - 7y = 8$ $y = \frac{5}{7}x - \frac{8}{7}$
3. $-9x + 3y = -14$ $y = 3x - \frac{14}{3}$

Give the slope and y-intercept for the graph of each equation.

4. $y = -2x + \frac{1}{3}$ $-2, \frac{1}{3}$
5. $6x - 5y = 15$ 1.2, −3
6. $y = -4$ 0, −4

Teach⇔Interact

About Example 1

Second-Language Learners
Several of the terms in this Example may be unfamiliar to students learning English. If necessary, explain that a speedometer is a dial that indicates the speed of a vehicle; nautical miles are units of length measurement used in navigation; a tachometer is an instrument used to measure the number of rotations per minute of a motor.

Additional Example 1

An amusement park has a moving walkway that is popular with children. A child who walks at a certain steady rate can get onto the walkway belt and, in walking in the same direction as the belt, travel 140 ft in 20 s. Walking for 20 s at the same rate but in a direction opposite to that of the belt, the child will travel 52 ft. Use this information to find the rate of motion of the child and the rate of motion of the walkway belt in ft/s.

Step 1 Write a system of equations. The child's speed c is increased by the belt's speed w when the child walks in the direction that the belt moves. When the child walks in the opposite direction, the combined rates give the child a speed of $w - c$ ft/s. Recall that speed × time = distance traveled. When the child walks in the same direction that the belt moves, the equation is $20(w + c) = 140$. When the child walks in the opposite direction, the equation is $20(w - c) = 52$. The system of equations is:

$20w + 20c = 140$
$20w - 20c = 52$

Step 2 Solve the system using addition. Add corresponding sides of the equations to eliminate the variable c.

$$\begin{aligned} 20w + 20c &= 140 \\ \underline{20w - 20c} &= \underline{\ \ 52} \\ 40w \quad\ \ &= 192 \\ w &= 4.8 \end{aligned}$$

Substitute 4.8 for w in the first equation, and then solve for c.

$$\begin{aligned} 20(4.8) + 20c &= 140 \\ 96 + 20c &= 140 \\ 20c &= 44 \\ c &= 2.2 \end{aligned}$$

The speed of the child is 2.2 ft/s. The speed of the walkway belt is 4.8 ft/s.

EXAMPLE 1 Application: Recreation

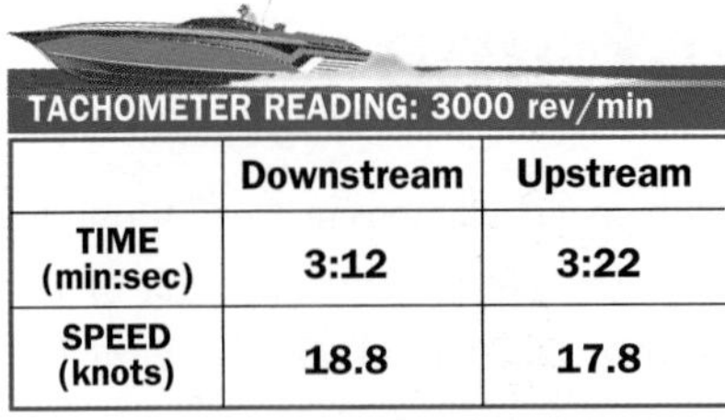

TACHOMETER READING: 3000 rev/min		
	Downstream	**Upstream**
TIME (min:sec)	**3:12**	**3:22**
SPEED (knots)	**18.8**	**17.8**

Jared's boat does not have a speedometer, so he timed how long it took to travel 1 nautical mile (nm) with a river's current (downstream) and 1 nm against it (upstream) while keeping his boat's engine running at a constant 3000 rev/min (as measured on the boat's tachometer).

Use Jared's log to determine the speed of his boat in still water when the engine runs at 3000 rev/min, and find the speed of the current.

The speed of a boat is often measured in nautical miles per hour, or knots: 1 knot ≈ 1.15 mi/h.

SOLUTION

Step 1 Write a system of equations. Use the fact that the boat's speed s is increased by the water's speed w when traveling with the current and decreased by the water's speed when traveling against the current.

$s + w = 18.8$ ⟵ **combined downstream speed**
$s - w = 17.8$ ⟵ **combined upstream speed**

Step 2 Solve the system using addition. Add corresponding sides of each equation to eliminate the variable w.

$$\begin{aligned} s + w &= 18.8 \\ \underline{s - w} &= \underline{17.8} \\ 2s &= 36.6 \\ s &= 18.3 \end{aligned}$$

Substitute **18.3** for s.

$$\begin{aligned} 18.3 + w &= 18.8 \\ w &= 0.5 \end{aligned}$$

The speed of the boat is 18.3 knots. The speed of the current is 0.5 knots.

THINK AND COMMUNICATE

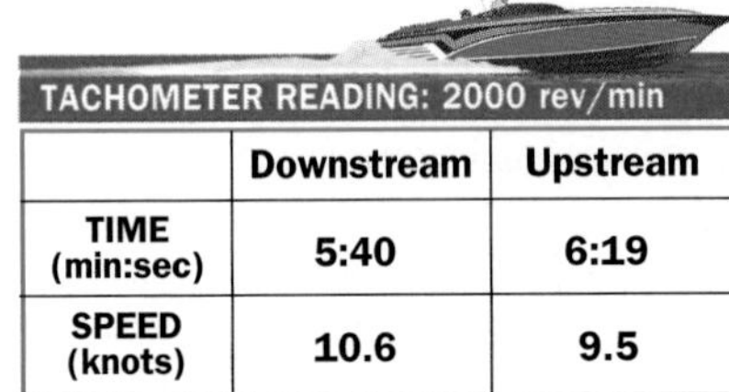

TACHOMETER READING: 2000 rev/min		
	Downstream	**Upstream**
TIME (min:sec)	**5:40**	**6:19**
SPEED (knots)	**10.6**	**9.5**

1. In Example 1, explain how traveling 1 nm in 3:22 translates into a speed of 17.8 knots.

2. Jared repeated his speed experiment, this time keeping his boat's engine running at a constant 2000 rev/min. Use the information shown in Jared's log to determine the speed of his boat in still water when the engine runs at 2000 rev/min.

ANSWERS Section 7.2

Exploration

1.

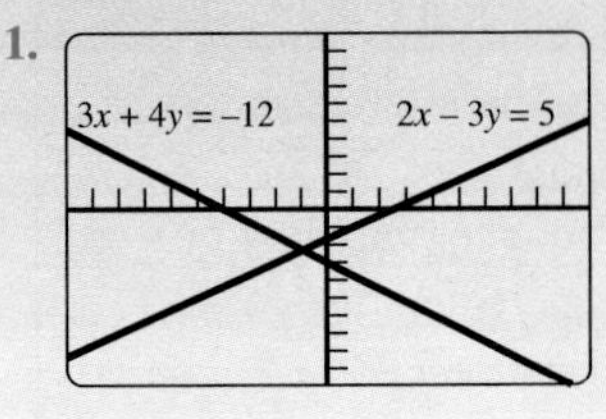

2. Yes; (−0.941, −2.294)

3.

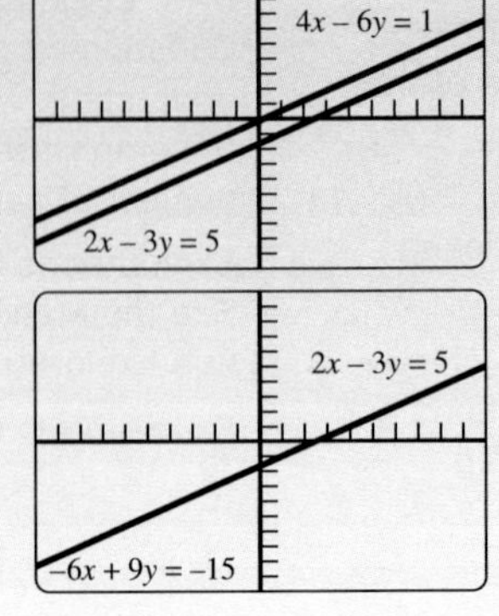

The graphs of the equations $2x - 3y = 5$ and $4x - 6y = 1$ do not intersect; they are parallel. The graphs of the equations $2x - 3y = 5$ and $-6x + 9y = -15$ coincide.

Questions

1. They can intersect in a single point, they can be parallel and not intersect at all, or they can coincide. If they intersect in a

Sometimes you can eliminate a variable by multiplying an equation by a constant before adding or subtracting.

EXAMPLE 2 Interview: Gina Oliva

Gina Oliva likes to combine jogging and walking to exercise. One day she jogged and walked for 1 h and covered 4.2 mi. Her jogging speed was 5 mi/h, and her walking speed was 3 mi/h. Find her time spent jogging and her time spent walking.

SOLUTION

Step 1 Write a system of equations involving jogging time j and walking time w, both in hours.

Step 2 Multiply one of the equations by a constant. In this case, it is convenient to multiply the first equation by 5.

total time ⟶ $j + w = 1 \xrightarrow{\times 5} 5j + 5w = 5$

total distance ⟶ $5j + 3w = 4.2 \longrightarrow 5j + 3w = 4.2$

Step 3 Solve the system using subtraction. Subtract corresponding sides of each equation to eliminate j.

$$\begin{aligned} 5j + 5w &= 5 \\ 5j + 3w &= 4.2 \\ \hline 2w &= 0.8 \\ w &= 0.4 \end{aligned}$$

Substitute **0.4** for w.

$$\begin{aligned} j + 0.4 &= 1 \\ j &= 0.6 \end{aligned}$$

Gina Oliva spent 0.4 h (24 min) walking and 0.6 h (36 min) jogging.

THINK AND COMMUNICATE

3. Describe Sheila's method for solving the system in Example 2.
4. Can the system in Example 2 be solved by the method of substitution described in Section 7.1? Explain.
5. How would the system of equations in Example 2 change if j and w were measured in minutes?

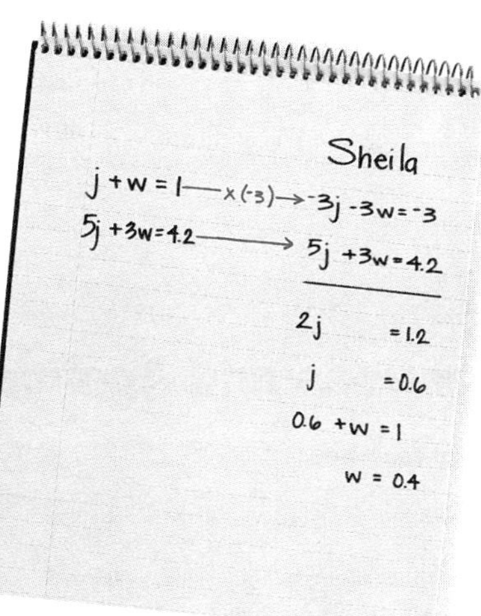

As you saw in the Exploration on page 297, a system of linear equations can have no solution or infinitely many solutions. When a system has *no solution*, it is called an **inconsistent system**. When it has *infinitely many solutions*, it is called a **dependent system**.

Teach⟺Interact

About Example 2

Teaching Tip

You may wish to point out that instead of using subtraction to solve the system, the first equation can be multiplied by −5 and the two equations can be added. Remind students why this is so: subtracting a number is the same as adding its opposite.

Additional Example 2

Bill Chinn has a small motor on a boat that he usually rows. One afternoon, he rowed part of the way across a lake and used the motor to go the rest of the way. His total time was 40 min, and his total distance was 2.45 mi. His rowing speed was 2.6 mi/h, and his speed when using the motor was 6.9 mi/h. Find his time spent rowing and his time spent using the motor.

Step 1 Write a system of equations involving rowing time r and time using the motor m, both in hours.

total time: $r + m = \frac{2}{3}$

total distance: $2.6r + 6.9m = 2.45$

Step 2 Multiply one of the equations by a constant. In this case, it is convenient to multiply the first equation by 6.9.

$6.9r + 6.9m = 4.6$

$2.6r + 6.9m = 2.45$

Step 3 Solve the system using subtraction. Subtract corresponding sides of the equations to eliminate m.

$$\begin{aligned} 6.9r + 6.9m &= 4.6 \\ 2.6r + 6.9m &= 2.45 \\ \hline 4.3r &= 2.15 \\ r &= 0.5 \end{aligned}$$

Substitute 0.5, or $\frac{1}{2}$, in the first equation of the original system.

$\frac{1}{2} + m = \frac{2}{3}$

$m = \frac{1}{6}$

Bill spent $\frac{1}{2}$ h (30 min) rowing and $\frac{1}{6}$ h (10 min) using the motor.

Think and Communicate

You may want to solve the system for question 5 to confirm that $w = 24$ min and $j = 36$ min. Alternatively, simply substitute 24 for w and 36 for j in the changed system.

single point, the system has one solution. If they are parallel, the system has no solution. If they coincide, the system has infinitely many solutions.

2. The slopes must be different or else the slope of exactly one of the lines must be undefined.

Think and Communicate

1. $\frac{1 \text{ nm}}{3\frac{22}{60} \text{ min}} \cdot \frac{60 \text{ min}}{\text{h}} \approx 17.8$ nm/h or 17.8 knots
2. 10.05 knots
3. Sheila multiplied the first equation by −3 to eliminate w instead of j.
4. Yes; use the first equation to solve for w in terms of j or j in terms of w and substitute in the second equation.
5. $j + w = 1$ would become $j + w = 60$, and $5j + 3w = 4.2$ would become $5\left(\frac{j}{60}\right) + 3\left(\frac{w}{60}\right) = 4.2$.

Teach⇔Interact

About Example 3

Reasoning

In part (a) of this Example, point out that by trying to solve the given system, you are trying to find two numbers whose difference is equal to 5 and *at the same time* equal to −3. Obviously, there are no such numbers and the system has no solution. In part (b), suppose there is a number a that can be put in place of x and a number b that can be put in place of y to make $3x + 6y = 9$ true. If $3a + 6b = 9$ is true, then $\frac{2}{3}(3a + 6b) = \frac{2}{3}(9)$ must also be true. Simplifying, $2a + 4b = 6$ must be true, and hence (a, b) is also a solution of $2x + 4y = 6$. The coordinates of any point on the graph of $3x + 6y = 9$ represent a solution of the system. There are infinitely many points on the graph of $3x + 6y = 9$. Therefore, there are infinitely many solutions for the system.

Additional Example 3

Solve each system of equations.

a. $x + y = 7$
$-6x - 6y = 9$

Multiply both sides of the first equation by 6 and add.

$$\begin{array}{r} 6x + 6y = 42 \\ \underline{-6x - 6y = 9} \\ 0 \neq 51 \end{array}$$

Values that satisfy $x + y = 7$ never satisfy the equation $-6x - 6y = 9$. The system is an inconsistent system. It has no solution.

b. $25x + 15y = 40$
$-5x - 3y = -8$

Multiply both sides of the second equation by 5 and add.

$$\begin{array}{r} 25x + 15y = 40 \\ \underline{-25x - 15y = -40} \\ 0 = 0 \end{array}$$

Values that satisfy the equation $25x + 15y = 40$ always satisfy the equation $-5x - 3y = -8$. This is a dependent system. It has infinitely many solutions.

Section Note

Using Technology
Students can use Function Investigator Activity 10 in the *Function Investigator with Matrix Analyzer Activity Book* to reinforce the connection between the number of solutions a system has and the type of graph the system has.

EXAMPLE 3

Solve each system of equations.

a. $x - y = 5$
$x - y = -3$

b. $3x + 6y = 9$
$2x + 4y = 6$

SOLUTION

a. Use subtraction.

$$\begin{array}{r} x - y = 5 \\ \underline{x - y = -3} \\ 0 \neq 8 \end{array}$$

The equation $0 = 8$ is never true.

Values that satisfy the equation $x - y = 5$ never satisfy the equation $x - y = -3$. This is an *inconsistent system*. It has no solution.

b. Multiply both sides of the first equation by 2, and multiply both sides of the second equation by 3. Then subtract.

$$\begin{array}{rcr} 3x + 6y = 9 & \xrightarrow{\times 2} & 6x + 12y = 18 \\ 2x + 4y = 6 & \xrightarrow{\times 3} & \underline{6x + 12y = 18} \\ & & 0 = 0 \end{array}$$

The equation $0 = 0$ is always true.

Values that satisfy the equation $3x + 6y = 9$ always satisfy the equation $2x + 4y = 6$. This is a *dependent system*. It has infinitely many solutions.

CHECKING KEY CONCEPTS

Describe at least one way to solve each system of equations.

1. $3x + 5y = 19$; $5x - 5y = -3$

2. $2x + 4y = 6$; $5x + 4y = 15$

3. $y = 7x + 2$; $y = 2x - 4$

4. $x + 2y = 6$; $3x - 4y = 2$

5. $2x + 4y = 28$; $3x + 6y = 42$

6. $3x - 4y = 55$; $y = 2x$

Write a system of equations that would best be solved by each method.

7. addition

8. subtraction

9. multiplication and addition

7.2 Exercises and Applications

Extra Practice exercises on page 759

Solve each system of equations.

1. $5x - 3y = 13$; $7x + 3y = 11$

2. $12y - 2x = 21$; $3x + 12y = -4$

3. $6x + 2y = 5$; $8x + 2y = 3$

4. $4x + 5y = 12$; $4x - 5y = 6$

5. $y - x = 13$; $2x - 3y = 1$

6. $x + 6y = -1$; $3x - 3y = 6$

7. $3x + 7y = 6$; $2x + 9y = 4$

8. $8x + 9y = 15$; $5x - 2y = 17$

Checking Key Concepts

1–6. Answers may vary. All the systems may be solved by graphing. Examples of algebraic solution methods are given.

1. Add.
2. Subtract.
3. Use substitution.
4. Multiply both sides of the first equation by 2, then add.
5. Multiply both sides of the first equation by 3 and both sides of the second equation by −2, then add.
6. Use substitution.

7–9. Answers may vary. Examples are given.

7. $x - y = 7$, $x + y = 9$
8. $4x + 5y = 3$, $3x + 5y = 1$
9. $2x - 3y = -7$, $3x + 2y = -4$

Exercises and Applications

1. $(2, -1)$
2. $\left(-5, \frac{11}{12}\right)$
3. $\left(-1, 5\frac{1}{2}\right)$
4. $\left(2\frac{1}{4}, \frac{3}{5}\right)$
5. $(-40, -27)$
6. $\left(1\frac{4}{7}, -\frac{3}{7}\right)$
7. $(2, 0)$

Solve each system of equations. State whether the system has *one solution, infinitely many solutions,* or *no solution.*

9. $3x + 7y = 2$, $6x + 14y = 4$

10. $4x + 5y = 3$, $6x + 9y = 9$

11. $6x + 11y = 12$, $12x + 22y = 10$

12. $0.3x + 0.7y = 1.2$, $6x + 14y = 24$

13. $0.4x + 2.1y = 40$, $3.6x + 6.3y = 22$

14. $27x - 42y = 102$, $45x - 70y = 170$

15. $27x - 42y = 102$, $45x - 70y = 150$

16. $9x + 10y = 11$, $10x + 9y = 11$

Connection PHYSICS

A *lever* is a rod that tilts on a *fulcrum*. Any weight applied to a lever creates *torque*, defined as the product of the weight and the distance between the fulcrum and the point where the weight is applied. For a lever *not* to move when force is applied, any torque in a clockwise direction must equal the torque in a counterclockwise direction.

This man creates a counterclockwise torque equal to 60 lb · 8 ft, or 480 ft·lb (read "foot-pounds").
This large stone creates a clockwise torque equal to 120 lb · 4 ft, or 480 ft·lb.
60 lb
120 lb
8 ft
4 ft

Because the counterclockwise and clockwise torques are equal, the lever is in balance.

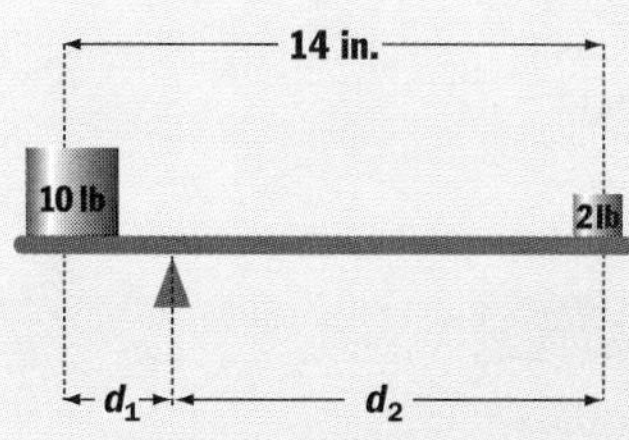

17. In the situation shown at the left, the lever is in balance. Complete parts (a)–(c) to determine the distances d_1 and d_2 between the weights and the fulcrum.

a. Write an equation representing the equal torques.

b. Write an equation for the combined distances.

c. Find d_1 and d_2.

18. Suppose a 100 ft bridge weighs 100 tons and has a support at each end. A 3 ton truck is 10 ft from the right end of the bridge.

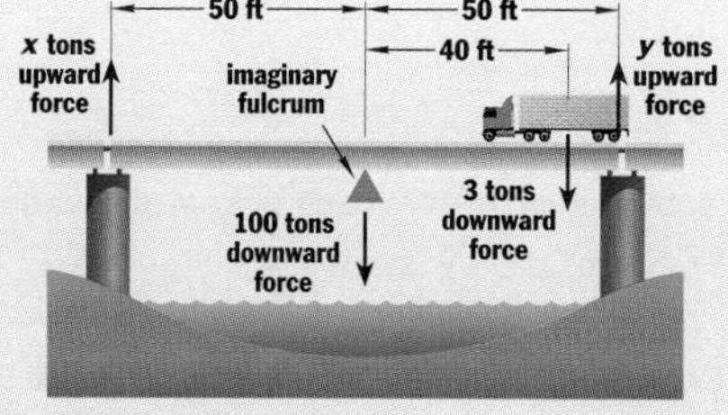

a. You can think of the bridge as a type of lever. Imagine a fulcrum at the midpoint of the bridge, and think of the bridge's weight as being concentrated at the fulcrum. Write an equation representing the balanced torques that occur at each support and at the truck.

b. For an object at rest, the downward forces on the object equal the upward forces. Write an equation representing the balanced vertical forces acting on the bridge. (*Note:* There are no distances involved.)

c. Solve your system of equations to find the force, in tons, that each support must exert to keep the bridge from moving.

Teach⇔Interact

Closure Question

What kind of graph results from a system of equations that has exactly one solution? from a dependent system? from an inconsistent system? two lines that intersect in a single point; a single line; two parallel lines

Apply⇔Assess

Suggested Assignment

Core Course
Exs. 1–18, 22–30

Extended Course
Exs. 1, 3, 6, 8–11, 15–30

Block Schedule
Day 42 Exs. 1–18, 22–30

Exercise Notes

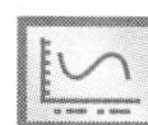

Using Technology
Exs. 1–16 To emphasize ideas about dependent systems, inconsistent systems and systems with a unique solution, suggest that students graph each system. They may use graphing calculators or the *Function Investigator* software.

Communication: Reading
Exs. 17, 18 It is important that students read carefully the paragraph that introduces these two exercises. While doing so, they should refer to the illustration immediately below the paragraph.

Interdisciplinary Problems
Exs. 17, 18 There may be students who do not have a clear understanding of the physical principles involved in these exercises. It may be helpful to invite assistance or suggestions from students who have studied these ideas in a physics course. Or, you may wish to provide some additional background information yourself. As the man pushes down on the end of the lever, he is applying a *force* (weight). This force would cause the lever to rotate in a counterclockwise direction about the fulcrum. However, the rotary effect of a force about a pivotal point depends not only on the magnitude of the force but also upon the distance between the fulcrum and the point where the force is applied. The product of the force and distance is called torque.

---

8. (3, −1)
9. Values that satisfy one equation always satisfy the other; infinitely many.
10. (−3, 3); one solution
11. no solution
12. Values that satisfy one equation always satisfy the other; infinitely many.
13. $\left(-40\frac{5}{6}, 26\frac{52}{63}\right)$; one solution
14. Values that satisfy one equation always satisfy the other; infinitely many.
15. no solution
16. $\left(\frac{11}{19}, \frac{11}{19}\right)$; one solution
17. a. $10d_1 = 2d_2$
b. $d_1 + d_2 = 14$
c. $2\frac{1}{3}$ in.; $11\frac{2}{3}$ in.
18. a. $50x = 50y - 120$
b. $x + y = 103$
c. left support: 50.3 tons; right support: 52.7 tons

Apply⇔Assess

Exercise Notes

Challenge

Ex. 20 Intuitively, it seems reasonable to assume that you should be able to get a mixture with any per-gallon price between \$1.19 and \$1.39. Ask students if they can show algebraically that this is possible.

Visual Thinking

Ex. 21 Ask students to graph the information presented. Have them change their graphs to answer questions such as: How many researchers could be hired if the university could spend up to \$400,000? What if the team were reduced to seven members? Have them use their graphs to explain their responses to the class. This activity involves the visual skills of *recognition* and *exploration*.

Practice 41 for Section 7.2

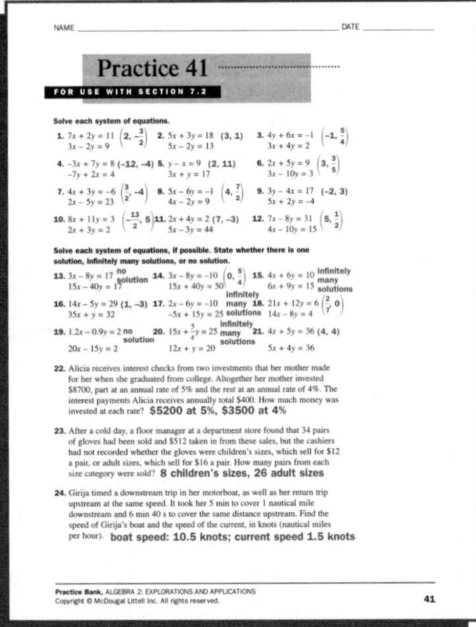
NAME ____________ DATE ________

Practice 41

FOR USE WITH SECTION 7.2

Solve each system of equations.

1. $7x + 2y = 11$, $3x - 2y = 9$ **$\left(2, -\frac{3}{2}\right)$**
2. $5x + 3y = 18$, $5x - 2y = 13$ **(3, 1)**
3. $4y + 6x = -1$, $3x + 4y = 2$ **$\left(-1, \frac{5}{4}\right)$**
4. $-3x + 7y = 8$, $-7y + 2x = 4$ **(−12, −4)**
5. $y - x = 9$, $3x + y = 17$ **(2, 11)**
6. $2x + 5y = 9$, $3x - 10y = 3$ **$\left(3, \frac{3}{5}\right)$**
7. $4x + 3y = -6$, $2x - 5y = 23$ **$\left(\frac{3}{2}, -4\right)$**
8. $5x - 6y = -1$, $4x - 2y = 9$ **$\left(4, \frac{7}{2}\right)$**
9. $3y - 4x = 17$, $5x + 2y = -4$ **(−2, 3)**
10. $8x + 11y = 3$, $2x + 3y = 2$ **$\left(-\frac{13}{2}, 5\right)$**
11. $2x + 4y = 2$, $5x - 3y = 44$ **(7, −3)**
12. $7x - 8y = 31$, $4x - 10y = 15$ **$\left(5, \frac{1}{2}\right)$**

Solve each system of equations, if possible. State whether there is one solution, infinitely many solutions, or no solution.

13. $3x - 8y = 17$, $15x - 40y = 17$ **no solution**
14. $3x - 8y = -10$, $15x + 40y = 50$ **$\left(0, \frac{5}{4}\right)$**
15. $4x + 6y = 10$, $6x + 9y = 15$ **infinitely many solutions**
16. $14x - 5y = 29$, $35x + y = 32$ **(1, −3)**
17. $2x - 6y = -10$, $-5x + 15y = 25$ **infinitely many solutions**
18. $21x + 12y = 6$, $14x - 8y = 4$ **$\left(\frac{2}{7}, 0\right)$**
19. $1.2x - 0.9y = 2$, $20x - 15y = 2$ **no solution**
20. $15x + \frac{5}{4}y = 25$, $12x + y = 20$ **infinitely many solutions**
21. $4x + 5y = 36$, $5x + 4y = 36$ **(4, 4)**

22. Alicia receives interest checks from two investments that her mother made for her when she graduated from college. Altogether her mother invested \$8700, part at an annual rate of 5% and the rest at an annual rate of 4%. The interest payments Alicia receives annually total \$400. How much money was invested at each rate? **\$5200 at 5%, \$3500 at 4%**

23. After a cold day, a floor manager at a department store found that 34 pairs of gloves had been sold and \$512 taken in from these sales, but the cashiers had not recorded whether the gloves were children's sizes, which sell for \$12 a pair, or adult sizes, which sell for \$16 a pair. How many pairs from each size category were sold? **8 children's sizes, 26 adult sizes**

24. Girija timed a downstream trip in her motorboat, as well as her return trip upstream at the same speed. It took her 5 min to cover 1 nautical mile downstream and 6 min 40 s to cover the same distance upstream. Find the speed of Girija's boat and the speed of the current, in knots (nautical miles per hour). **boat speed: 10.5 knots; current speed 1.5 knots**

Practice Bank, ALGEBRA 2: EXPLORATIONS AND APPLICATIONS

41

19. a. **Technology** Graph the equations $-x - y = 3$ and $3x - y = 1$ using a graphing calculator or graphing software. Where do the lines intersect?

b. Writing Add the two equations from part (a), solve for y, and graph the resulting equation on the same screen. Describe what you see.

c. Challenge Use the general equations $Ax + By = C$ and $Dx + Ey = F$ to give an algebraic argument for why the graph of the sum of the two equations passes through the same intersection point as the given equations. (*Hint:* If the coordinates (p, q) satisfy both $Ax + By = C$ and $Dx + Ey = F$, show that they must satisfy the sum of the two equations.)

20. CONSUMER ECONOMICS The octane number of gasoline indicates how well it fights *knock*, the noise an engine makes when fuel is not being burned properly. Octane numbers are based on a standard isooctane-and-heptane reference fuel. For example, 87-octane gasoline fights knock as well as a reference fuel that is 87% isooctane. Alexis decides to mix 87-octane and 93-octane gasolines to get 15 gal of 89-octane gasoline.

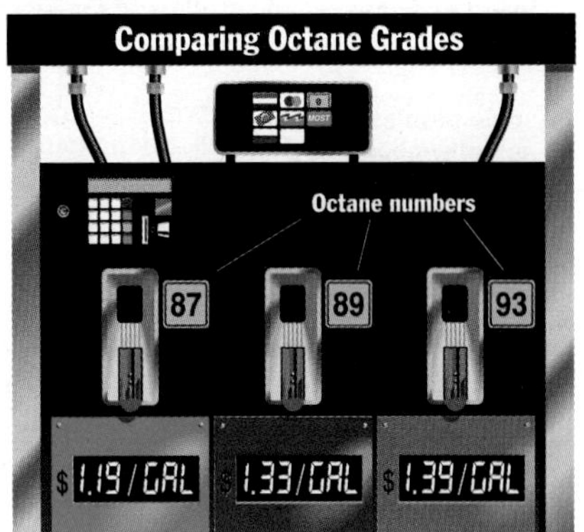

a. Write equations describing the total volume of gasoline and the volume of isooctane in the reference fuel.

b. Solve the system to determine how many gallons of each grade of gasoline are needed.

c. Use the diagram. Is it cheaper for Alexis to buy the 89-octane gasoline directly or to mix the other two grades herself? Explain.

21. FINANCE Suppose a university allots \$335,000 to pay the salaries of a new research team. The team of 10 members will consist of researchers who will be paid \$44,000 and technicians who will be paid \$29,000. Determine how many researchers the university can afford to hire for the team.

ONGOING ASSESSMENT

22. Writing When is the addition method better than the substitution method? When is the substitution method better? Give an example of each.

SPIRAL REVIEW

Use substitution to solve each system of equations. (*Section 7.1*)

23. $y = 7x$, $y + 6x = 39$

24. $x + y = 8$, $2x - 3y = 6$

25. $x + 2y = 11$, $3x + 2y = 17$

26. $x + y = 5.8$, $2x + 3y = 14.4$

For Exercises 27–30, use matrices *A, B, C,* and *D* to evaluate each matrix expression. If an expression is not defined, write *undefined*. (*Section 1.3*)

$$A = \begin{bmatrix} 7 & 11 \\ -15 & 3 \end{bmatrix} \quad B = \begin{bmatrix} 1 & 0 \\ 4 & -\frac{11}{2} \end{bmatrix} \quad C = \begin{bmatrix} 8 & -2 \\ 19 & 6 \\ 1 & -13 \end{bmatrix} \quad D = \begin{bmatrix} 6 & 8 \\ 0 & 12 \\ 10 & -4 \end{bmatrix}$$

27. $A + B$

28. $2A$

29. $D - C$

30. $-4A + 2B$

19. See answers in back of book.

20. a. Let x = the number of gallons of 87-octane gasoline and y = the number of gallons of 93-octane gasoline; $x + y = 15$; $0.87x + 0.93y = 0.89(15) = 13.35$.

b. 10 gal of 87-octane gasoline and 5 gal of 93-octane gasoline

c. It is cheaper for her to mix the grades herself. By mixing the grades herself, she pays \$1.26 per gallon.

21. 3 researchers

22. The addition method is better when the equations are written in standard form and the coefficients of one of the variables are opposites. An example is $3x + 8y = 36$, $5x - 8y = 44$. The substitution method is better when one of the equations is already solved for x or y. An example is $2x + 7y = 8$, $y = x + 8$.

23. (3, 21)

24. (6, 2)

25. (3, 4)

26. (3, 2.8)

27. $\begin{bmatrix} 8 & 11 \\ -11 & -2\frac{1}{2} \end{bmatrix}$

28. $\begin{bmatrix} 14 & 22 \\ -30 & 6 \end{bmatrix}$

29. $\begin{bmatrix} -2 & 10 \\ -19 & 6 \\ 9 & 9 \end{bmatrix}$

30. $\begin{bmatrix} -26 & -44 \\ 68 & -23 \end{bmatrix}$

Section

7.3 Solving Linear Systems with Matrices

Learn how to...

- solve systems of linear equations using matrices

So you can...

- solve problems, such as determining how many batches to prepare of two recipes so that all supplies are used

At the end of the gardening season, the last of the fruits and vegetables are often canned for use during the winter. A system of equations can help you divide your supplies among canning recipes so that you don't waste anything.

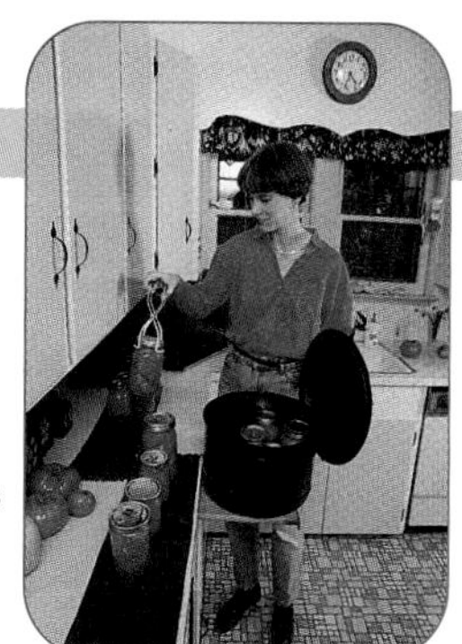

EXAMPLE 1 Application: Cooking

Betsy has harvested her garden before leaving on vacation. She has 42 qt of tomatoes and 6 qt of green peppers. She wants to use up all her tomatoes and peppers by canning tomato sauce and tomato juice.

Use matrices to determine how many batches of sauce and juice she should make so that no tomatoes or peppers are wasted.

Tomato Juice
8 Qt tomatoes
0.5 Qt green peppers
2 onions
1 beet
1 hot pepper
1 garlic clove
celery leaves
Base for Tomato Sauce
4 Qt tomatoes
1 Qt green peppers
1 Qt celery
1 Qt onions

SOLUTION

Step 1 Write a system of equations. Let x = the number of batches of sauce, and let y = the number of batches of juice.

	Number of quarts needed for:	
	x batches of sauce	y batches of juice
Tomatoes	$4x$	$8y$
Peppers	$1x$	$0.5y$

$4x + 8y = 42$ — There are **42 qt** of tomatoes available.

$1x + 0.5y = 6$ — There are **6 qt** of peppers available.

Solution continued on next page.

Plan⇔Support

Objectives

- Solve systems of linear equations using matrices.
- Use systems of linear equations to solve real-world problems.

Recommended Pacing

Core and Extended Courses
Section 7.3: 2 days

Block Schedule
Section 7.3: 2 half-blocks (with Sections 7.2 and 7.4)

Resource Materials

Lesson Support
Lesson Plan 7.3
Warm-Up Transparency 7.3
Practice Bank: Practice 42
Study Guide: Section 7.3
Assessment Book: Test 26

Technology
Technology Book:
Calculator Activity 7
Spreadsheet Activity 7
Graphing Calculator
McDougal Littell Mathpack *Matrix Analyzer*
Internet:
http://www.hmco.com

Warm-Up Exercises

Find the product of each pair of matrices. If a product is not defined, explain why.

$A = \begin{bmatrix} 3 & -5 \\ -1 & 4 \end{bmatrix}$ $B = \begin{bmatrix} 9 & 0 \\ 2 & 10 \end{bmatrix}$

$C = [6\ -4]$ $D = \begin{bmatrix} 5 \\ 3 \end{bmatrix}$

1. AB $\begin{bmatrix} 17 & -50 \\ -1 & 40 \end{bmatrix}$
2. BD $\begin{bmatrix} 45 \\ 40 \end{bmatrix}$
3. AC not defined; A has 2 columns, while C has 1 row.
4. CD $[18]$
5. DC $\begin{bmatrix} 30 & -20 \\ 18 & -12 \end{bmatrix}$
6. BA $\begin{bmatrix} 27 & -45 \\ -4 & 30 \end{bmatrix}$

Teach⇔Interact

Additional Example 1

Adolph Green is baking for a family party. He plans to bake loaves of bread and hamburger buns. Among other ingredients, the recipe for a loaf of bread calls for 2.5 cups of all-purpose flour and 1 cup of milk. The recipe for one batch of hamburger buns calls for 3.5 cups of all-purpose flour and 0.5 cup of milk. Adolph has 17 cups of flour and 5 cups of milk. Use matrices to determine how many loaves of bread and how many batches of buns he can make so that no flour or milk is wasted.

Step 1 Write a system of equations. Let x = the number of loaves of bread, and let y = the number of batches of hamburger buns.

	Number of cups needed for:	
	x loaves of bread	y batches of buns
Flour	$2.5x$	$3.5y$
Milk	$1x$	$0.5y$

The system of equations is

$$\begin{aligned} 2.5x + 3.5y &= 17 \\ 1x + 0.5y &= 5. \end{aligned}$$

Step 2 Rewrite the system as the matrix equation

$$\begin{bmatrix} 2.5 & 3.5 \\ 1 & 0.5 \end{bmatrix}\begin{bmatrix} x \\ y \end{bmatrix} = \begin{bmatrix} 17 \\ 5 \end{bmatrix}.$$

To solve this equation for x and y, multiply both sides of the equation by the inverse of the coefficient matrix: $\begin{bmatrix} x \\ y \end{bmatrix} = \begin{bmatrix} 2.5 & 3.5 \\ 1 & 0.5 \end{bmatrix}^{-1}\begin{bmatrix} 17 \\ 5 \end{bmatrix}.$

Use a graphing calculator or use software with matrix calculation capabilities to find $A^{-1}B$.

```
[A]⁻¹[B]
               [[4]
                [2]]
```

Adolph can bake 4 loaves of bread and 2 batches of hamburger buns.

Section Notes

Using Technology
When using graphing calculators, students should be cautioned to press the x^{-1} key to use the inverse of a matrix in a calculation.

Teaching Tip
Point out that the term *invertible matrix* (used in the Watch Out! note) means any matrix that has an inverse.

SOLUTION *continued*

Step 2 Rewrite the system of equations as a matrix equation.

$$\begin{bmatrix} 4 & 8 \\ 1 & 0.5 \end{bmatrix}\begin{bmatrix} x \\ y \end{bmatrix} = \begin{bmatrix} 42 \\ 6 \end{bmatrix}$$

This equation has the form $A\begin{bmatrix} x \\ y \end{bmatrix} = B$.

To solve this equation for x and y, you multiply both sides of the equation by the *inverse* of the coefficient matrix.

$$\begin{bmatrix} x \\ y \end{bmatrix} = \begin{bmatrix} 4 & 8 \\ 1 & 0.5 \end{bmatrix}^{-1}\begin{bmatrix} 42 \\ 6 \end{bmatrix}$$

This equation has the form $\begin{bmatrix} x \\ y \end{bmatrix} = A^{-1}B$.

For more information about inverse matrices, see the *Technology Handbook*, p. 812.

Use a graphing calculator or software with matrix calculation capabilities to find $A^{-1}B$.

Betsy needs to make **4.5 batches of sauce** and **3 batches of juice**.

THINK AND COMMUNICATE

1. Confirm that $(x, y) = (4.5, 3)$ is the solution of the system of equations in Example 1.
2. In the solution of Example 1, what happens if you try to compute $B(A^{-1})$ instead of $A^{-1}B$? Why doesn't this work?
3. How would you write a matrix equation to represent the system below?

$$\begin{aligned} 15x + 8y &= 9 \\ -3y &= 8 \end{aligned}$$

Square matrices such as $\begin{bmatrix} 1 & 0 \\ 0 & 1 \end{bmatrix}$ and $\begin{bmatrix} 1 & 0 & 0 \\ 0 & 1 & 0 \\ 0 & 0 & 1 \end{bmatrix}$, which have 1's on the diagonal running from upper left to lower right and 0's elsewhere, are called **identity matrices**. An identity matrix I has the special property that

$$AI = IA = A$$

for any matrix A having the same dimensions as I.

The **inverse matrix** of a square matrix A, written A^{-1}, is the matrix that, when multiplied by A, gives an identity matrix:

$$A^{-1}A = A(A^{-1}) = I$$

Not all square matrices have inverses. If a matrix A does not have an inverse, you will get an error message when you try to find A^{-1} on a graphing calculator or computer.

WATCH OUT!
Just as $a \cdot a^{-1} = 1$ for any nonzero number a, $A \cdot A^{-1} = I$ for any invertible matrix A. However, the notation A^{-1} does *not* mean $\frac{1}{A}$.

ANSWERS

Section 7.3

Unless otherwise noted, answers are rounded to three significant digits when necessary.

Think and Communicate

1. $4(4.5) + 8(3) = 42$ and $4.5 + 0.5(3) = 6$
2. $B(A^{-1})$ does not exist, because the dimensions of B are 2×1 and the dimensions of A^{-1} are 2×2.
3. $\begin{bmatrix} 15 & 8 \\ 0 & -3 \end{bmatrix}\begin{bmatrix} x \\ y \end{bmatrix} = \begin{bmatrix} 9 \\ 8 \end{bmatrix}$

THINK AND COMMUNICATE

4. **Technology** Use a graphing calculator or software with matrix calculation capabilities to show that the product of the coefficient matrix and its inverse in Example 1 is an identity matrix.

EXAMPLE 2

Find an equation of the parabola passing through the points $(-3, 1)$, $(4, 4)$, and $(7, -1)$.

SOLUTION

You need to find values of a, b, and c such that the coordinates of each point satisfy $y = ax^2 + bx + c$.

Step 1 Substitute the coordinates of each point into $ax^2 + bx + c = y$.

Using $(-3, 1)$: $a(-3)^2 + b(-3) + c = 1 \longrightarrow 9a - 3b + c = 1$

Using $(4, 4)$: $a(4)^2 + b(4) + c = 4 \longrightarrow 16a + 4b + c = 4$

Using $(7, -1)$: $a(7)^2 + b(7) + c = -1 \longrightarrow 49a + 7b + c = -1$

Step 2 Write the system of equations in matrix form, and solve the system using a graphing calculator or software with matrix calculation capabilities.

$$\begin{bmatrix} 9 & -3 & 1 \\ 16 & 4 & 1 \\ 49 & 7 & 1 \end{bmatrix}\begin{bmatrix} a \\ b \\ c \end{bmatrix} = \begin{bmatrix} 1 \\ 4 \\ -1 \end{bmatrix}$$

$$\begin{bmatrix} a \\ b \\ c \end{bmatrix} = \begin{bmatrix} 9 & -3 & 1 \\ 16 & 4 & 1 \\ 49 & 7 & 1 \end{bmatrix}^{-1}\begin{bmatrix} 1 \\ 4 \\ -1 \end{bmatrix}$$

Rounding the coefficients to two significant digits, you get $\mathbf{y = -0.21x^2 + 0.64x + 4.8}$ as an equation of the parabola.

CHECKING KEY CONCEPTS

Write each system of equations as a matrix equation, and solve using a graphing calculator or software with matrix calculation capabilities.

1. $14x - 13y = -6$
   $-5x - y = -2$
2. $512x + 94y = 318$
   $137x = -89$
3. $81.2x + 17.4y = 0.8$
   $4.6x - 12.8y = 12.8$
4. Find an equation of the parabola passing through the points $(2, 5)$, $(3, -7)$, and $(5, 5)$.
5. Use matrices to solve this system for x, y, and z:

$$3x + 2y + 5z = 9$$
$$2x - 4y + 6z = 10$$
$$5x + 9y - 2z = 15$$

Teach⇔Interact

About Example 2

Alternate Approach

Another way to find an equation of the parabola passing through three given points is to use the statistical features of graphing calculators or software programs. For example, on the TI-82, enter the x-coordinates as list L1 and the y-coordinates as list L2. Then request a quadratic regression equation. The advantage of the method in Example 2 is that it can be extended to find polynomials of degree $n - 1$ given n points on the graph. The size of n is limited only by the size of the matrices the technology can handle.

Additional Example 2

Find an equation of the parabola passing through the points $(2, 3)$, $(4, 8)$, and $(10, 41)$.

Find values of a, b, and c such that the coordinates of each point satisfy $y = ax^2 + bx + c$.

Step 1 Substitute the coordinates of each point into $ax^2 + bx + c = y$.

$a(2)^2 + b(2) + c = 3 \rightarrow$
$4a + 2b + c = 3$

$a(4)^2 + b(4) + c = 8 \rightarrow$
$16a + 4b + c = 8$

$a(10)^2 + b(10) + c = 41 \rightarrow$
$100a + 10b + c = 41$

Step 2 Write the system of equations in matrix form, and solve the system by using a graphing calculator or software with matrix calculation capabilities.

$$\begin{bmatrix} 4 & 2 & 1 \\ 16 & 4 & 1 \\ 100 & 10 & 1 \end{bmatrix}\begin{bmatrix} a \\ b \\ c \end{bmatrix} = \begin{bmatrix} 3 \\ 8 \\ 41 \end{bmatrix}$$

$$\begin{bmatrix} a \\ b \\ c \end{bmatrix} = \begin{bmatrix} 4 & 2 & 1 \\ 16 & 4 & 1 \\ 100 & 10 & 1 \end{bmatrix}^{-1}\begin{bmatrix} 3 \\ 8 \\ 41 \end{bmatrix}$$

```
[A]-1[B]
          [[.375]
           [.25 ]
           [1   ]]
```

An equation of the parabola is $y = 0.375x^2 + 0.25x + 1$.

Think and Communicate

4. If $A = \begin{bmatrix} 4 & 8 \\ 1 & 0.5 \end{bmatrix}$, then $A \cdot A^{-1} = \begin{bmatrix} 1 & 0 \\ 0 & 1 \end{bmatrix}$ and $A^{-1} \cdot A = \begin{bmatrix} 1 & 0 \\ 0 & 1 \end{bmatrix}$.

Checking Key Concepts

1. $\begin{bmatrix} 14 & -13 \\ -5 & -1 \end{bmatrix}\begin{bmatrix} x \\ y \end{bmatrix} = \begin{bmatrix} -6 \\ 2 \end{bmatrix}$; (0.253, 0.734)

2. $\begin{bmatrix} 512 & 94 \\ 137 & 0 \end{bmatrix}\begin{bmatrix} x \\ y \end{bmatrix} = \begin{bmatrix} 318 \\ -89 \end{bmatrix}$; (–0.650, 6.921)

3. $\begin{bmatrix} 81.2 & 17.4 \\ 4.6 & -12.8 \end{bmatrix}\begin{bmatrix} x \\ y \end{bmatrix} = \begin{bmatrix} 0.8 \\ 12.8 \end{bmatrix}$; (0.208, –0.925)

4. $y = 6x^2 - 42x + 65$

5. (4.63, –1.03, –0.567)

Teach⇔Interact

Closure Question

Describe the steps you would follow to solve a linear system of equations by using matrices.
First, rewrite the system of equations as a matrix equation having the form $A\begin{bmatrix} x \\ y \end{bmatrix} = B$. Next, solve this equation for the two variables by multiplying both sides by the inverse of the coefficient matrix. The resulting equation has the form $\begin{bmatrix} x \\ y \end{bmatrix} = A^{-1}B$. Then use a graphing calculator or software with matrix calculation capabilities to find $A^{-1}B$.

Apply⇔Assess

Suggested Assignment

Core Course
Day 1 Exs. 1–14
Day 2 Exs. 16–22, 27–36, AYP

Extended Course
Day 1 Exs. 1–12
Day 2 Exs. 19–36, AYP

Block Schedule
Day 42 Exs. 1–14
Day 43 Exs. 16–22, 27–36, AYP

Exercise Notes

Common Error
Exs. 1–4 Some students may try to solve $A\begin{bmatrix} x \\ y \end{bmatrix} = B$ by evaluating BA^{-1} rather than $A^{-1}B$. You can help these students by reviewing the procedure in Example 1, Step 2 and the related ideas in Think and Communicate question 2.

Using Technology
Exs. 1–4, 7–14, 16–18
You may wish to suggest that students use the *Matrix Analyzer* software for these exercises.

7.3 Exercises and Applications

Extra Practice *exercises on page 759*

Technology **Throughout the exercises, use a graphing calculator or software with matrix calculation capabilities to solve the matrix equations.**

Rewrite each of the systems in matrix form, and solve for x and y.

1. $8x - 3y = 32$
$7x - 9y = 24$

2. $24x - 24y = 89$
$67x - 19y = 22$

3. $5x - 12y = 13$
$14x - 7y = 34$

4. $51x - 25y = 98$
$13x - 19y = 34$

5. COOKING Charlie wants to use up 8 cups of buttermilk and 11 eggs by baking rolls and muffins to freeze. A batch of rolls uses 2 cups of buttermilk and 3 eggs. A batch of muffins uses 1 cup of buttermilk and 1 egg.

a. How many batches should Charlie make of each recipe?

b. How would your answer to part (a) change if Charlie wanted to use up 8 cups of buttermilk and 6 eggs?

6. The graph shows two parallel lines, $y = -2x + 1$ and $y = -2x - 2$.

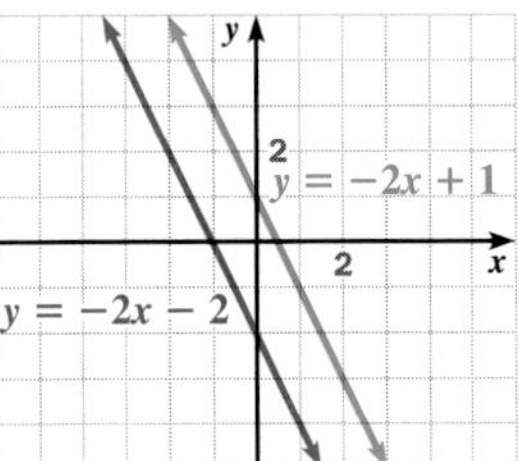

a. Can you solve this system of equations? Explain.

b. Rewrite the equations as a matrix equation of the form $A\begin{bmatrix} x \\ y \end{bmatrix} = B$.

c. Enter the matrices A and B into a graphing calculator or software with matrix calculation capabilities. Try to find $A^{-1}B$. What happens?

d. Open-ended Problem Write two equations that describe the same line. Repeat parts (b) and (c) for this set of equations.

Use matrices to solve each system for x and y.

7. $y = 14x + 2$
$y = 31x - 3$

8. $y = 4.1x - 8.6$
$y = -6.2$

9. $y = 57x + 48$
$y = -12x$

10. $y = -4x + 8.3$
$y = -26x - 2.5$

11. $x = -y + 2$
$y = 2x - 7$

12. $2x - 14y = 57$
$15x + 8y = 18$

13. $31x + 8y = -7$
$2 - 17y = 0$

14. $7.2x - 8.8y = -31.2$
$0.1y - 2.4x = 8.1$

15. Challenge Consider this matrix equation:

$$\begin{bmatrix} a & b \\ c & d \end{bmatrix}\begin{bmatrix} 1 & 2 \\ 3 & 4 \end{bmatrix} = \begin{bmatrix} 1 & 0 \\ 0 & 1 \end{bmatrix}$$

a. Use matrix multiplication to rewrite the left-hand side of the equation as a single matrix.

b. Use your matrices to write four equations involving a, b, c, and d by setting corresponding entries equal. (*Hint:* The first equation is $a + 3b = 1$.)

c. Without using technology, find a, b, c, and d.

d. What is the matrix $\begin{bmatrix} a & b \\ c & d \end{bmatrix}$? Check your answer using technology.

Use matrices to solve each system for x, y, and z.

16. $5x - 4y - 8z = 10$
$7x - 11y - 2z = 13$
$-3x - y - 3z = 17$

17. $6x - z = 10$
$x - 2y - 11z = -3$
$7x - 9z = -1$

18. $3.1x + 0.9y - 8.3z = 1.8$
$0.1x - 0.2y - 1.0z = 4.0$
$8.5x + 2.6y + 5.0z = 5.0$

Exercises and Applications

1. $\begin{bmatrix} 8 & -3 \\ 7 & -9 \end{bmatrix}\begin{bmatrix} x \\ y \end{bmatrix} = \begin{bmatrix} 32 \\ 24 \end{bmatrix}$; (4.24, 0.63)

2. $\begin{bmatrix} 24 & -24 \\ 67 & -19 \end{bmatrix}\begin{bmatrix} x \\ y \end{bmatrix} = \begin{bmatrix} 89 \\ 22 \end{bmatrix}$; (–1.01, –4.72)

3. $\begin{bmatrix} 5 & -12 \\ 14 & -7 \end{bmatrix}\begin{bmatrix} x \\ y \end{bmatrix} = \begin{bmatrix} 13 \\ 34 \end{bmatrix}$; (2.38, –0.90)

4. $\begin{bmatrix} 51 & -25 \\ 13 & -19 \end{bmatrix}\begin{bmatrix} x \\ y \end{bmatrix} = \begin{bmatrix} 98 \\ 34 \end{bmatrix}$; (1.57, –0.71)

5. a. 3 batches of rolls, 2 batches of muffins
b. There is no way Charlie can use up exactly 8 cups of buttermilk and 6 eggs using the given recipes.

6. a. No; if the system had a solution (x, y), the graphs would intersect at that point.
b. $\begin{bmatrix} 2 & 1 \\ 2 & 1 \end{bmatrix}\begin{bmatrix} x \\ y \end{bmatrix} = \begin{bmatrix} 1 \\ -2 \end{bmatrix}$
c. an error message
d. Answers may vary. The results will be the same. A has no inverse.

7. (0.29, 6.12)
8. (0.59, –6.2)
9. (–0.70, 8.35)
10. (–0.49, 10.26)
11. (3, –1)
12. (3.13, –3.62)
13. (–0.26, 0.12)

Find an equation of the parabola passing through each set of points.

19. $(-3, 4.5)$, $(-1.2, -2.0)$, and $(0.8, 1.6)$

20. $(-4, 17)$, $(6, 2)$, and $(13, 15)$

21. $(-1, 5)$, $(7, 2)$, and $(9, 5)$

22. $(-2, 3.5)$, $(3, 8)$, and $(6, 11.5)$

Connection AGRICULTURE

Soil scientists have conducted many experiments to determine how to use various fertilizers. The table shows the yield of hard red spring wheat (in kilograms per hectare) when different amounts of nitrogen fertilizer were applied to a field in the western United States.

Fertilizer (kg/ha)	Wheat yield (kg/ha)
99	1814
155	3562
211	3965
267	4368
323	4032

23. Use the last three entries from the table to find an equation for a parabola passing through those three points. What maximum yield does the model predict for the field?

24. Open-ended Problem Choose another set of three points from the table and find an equation of the parabola passing through them. How well do the two models agree? Is this a good way to fit a quadratic model to data? Why or why not?

25. Technology If you have a calculator or software that will perform quadratic regression, find a quadratic model using all the data. Compare this model with the model you found in Exercise 23. Which model is better? Why?

26. ART Monica is making a mobile. She wants to suspend three objects from a lightweight rod, as shown. Suppose the weights of the objects are $w_1 = 2.5$ oz, $w_2 = 1.2$ oz, and $w_3 = 2.0$ oz.

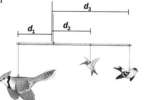

a. Write three equations involving the unknown distances d_1, d_2, and d_3 based on the following facts:

- The length of the rod is 14.5 in.
- To balance the mobile, she must position the objects so that $w_1d_1 = w_2d_2 + w_3d_3$.
- She wants the third object to be 1.5 times as far from the support as the second object is.

b. Write the system of equations as a matrix equation.

c. Find d_1, d_2, and d_3.

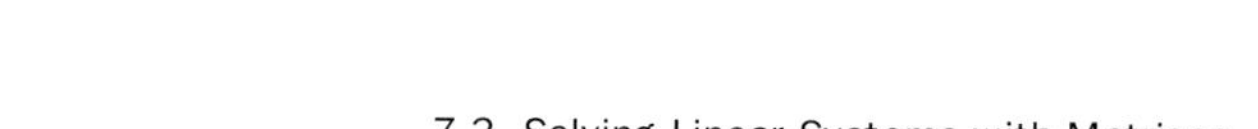

Apply⇔Assess

Exercise Notes

Reasoning

Ex. 6 Students should understand that when they cannot get a matrix solution for $A\begin{bmatrix} x \\ y \end{bmatrix} = B$, there are two possibilities: the original system is inconsistent (no solutions), or the original system is dependent (infinitely many solutions). Further analysis is then needed to discover which of these possibilities hold for the system in question.

Mathematical Procedures

Ex. 15 This exercise demonstrates how to find the inverse of a 2×2 matrix. For 3×3 matrices, the procedure is more complicated. For still larger square matrices, the calculations become unwieldy.

Cooperative Learning

Exs. 19–22 Students can work on these exercises with a partner. One student can use matrices to find the equations, while the other uses the procedure suggested in the Alternate Approach note on page 305.

Problem Solving

Exs. 19–22 In these exercises, suggest that students predict whatever they can about the coefficients a, b, and c in the equation $y = ax^2 + bx + c$.

Research

Exs. 23–25 Students who are interested in agriculture and agricultural research will find it interesting to discover what kinds of mathematical problems are encountered in this field.

Teaching Tip

Ex. 26 To do this exercise, it may help students to review the ideas in Exs. 17 and 18 on page 301.

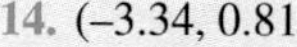

14. $(-3.34, 0.81)$

15. a. $\begin{bmatrix} a+3b & 2a+4b \\ c+3d & 2c+4d \end{bmatrix}$

b. $a + 3b = 1$, $2a + 4b = 0$, $c + 3d = 0$, $2c + 4d = 1$

c. $a = -2$, $b = 1$, $c = \frac{3}{2}$, $d = -\frac{1}{2}$

d. The matrix $\begin{bmatrix} -2 & 1 \\ \frac{3}{2} & -\frac{1}{2} \end{bmatrix}$ is the inverse of $\begin{bmatrix} 1 & 2 \\ 3 & 4 \end{bmatrix}$.

16. $(-3.00, -2.78, -1.74)$

17. $(1.94, -6.43, 1.62)$

18. $(5.89, -17.49, 0.09)$

19–22. Coefficients may vary due to rounding. Examples are given.

19. $y = 1.42x^2 + 2.37x - 1.2$

20. $y = 0.20x^2 - 1.89x + 6.26$

21. $y = 0.19x^2 - 1.5x + 3.31$

22. $y = 0.03x^2 + 0.87x + 5.1$

23. Answers may vary due to rounding. Examples are given. $y = -0.118x^2 + 63.5x - 4191$; about 4350 kilograms per hectare

24. Answers may vary. A model based on the first three points is $y = -0.214x^2 + 85.7x - 4570$. The maximum estimated by this model is about 4010 kilograms per hectare. This is significantly different than the estimate based on the model in Ex. 23. I think this is not a very good method because widely varying results may be produced by using different data points.

25. Answers may vary. An example is given. One model is $y = -0.0949x^2 + 49.4x - 2060$. The maximum estimated by this model is about 4370 kilograms per hectare, which is higher than the estimate in Ex. 24, but is very close to one of the actual data points. Also, the data points all fit this model reasonably well. I think this is a better model.

26. a. $d_1 + d_3 = 14.5$; $2.5d_1 = 1.2d_2 + 2.0d_3$; $d_3 = 1.5d_2$

b. $\begin{bmatrix} 1 & 0 & 1 \\ 2.5 & -1.2 & -2.0 \\ 0 & 1.5 & -1 \end{bmatrix}\begin{bmatrix} d_1 \\ d_2 \\ d_3 \end{bmatrix} = \begin{bmatrix} 14.5 \\ 0 \\ 0 \end{bmatrix}$

c. $d_1 = 7.66$ in., $d_2 = 4.56$ in., $d_3 = 6.84$ in.

Apply⇔Assess

Exercise Notes

Cooperative Learning

Ex. 27 This exercise can be used to review most of the ideas about matrices that have been studied in this section. Have cooperative groups share and discuss their results. A thorough response for this exercise will make a good journal entry.

Assess Your Progress

Journal Entry

A thorough entry will comment on how graphs of dependent systems differ from those of consistent systems. Note, too, whether students have included examples to illustrate their remarks.

Progress Check 7.1–7.3

See page 330.

Practice 42 for Section 7.3

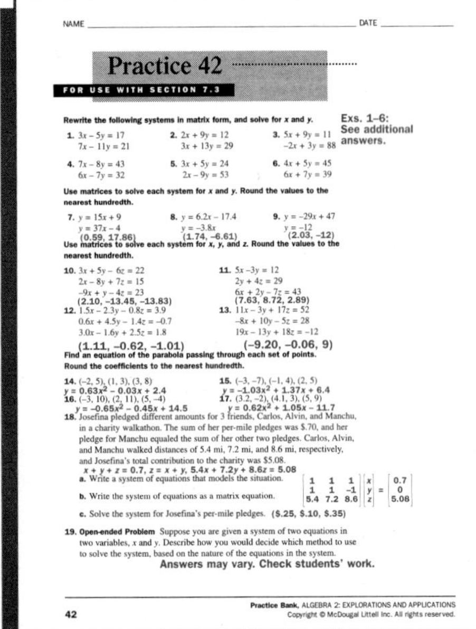

NAME ______________________ DATE __________

Practice 42

FOR USE WITH SECTION 7.3

Rewrite the following systems in matrix form, and solve for *x* and *y*.

Exs. 1–6: See additional answers.

1. $3x - 5y = 17$; $7x - 11y = 21$
2. $2x + 9y = 12$; $3x + 13y = 29$
3. $5x + 9y = 11$; $-2x + 3y = 88$
4. $7x - 8y = 43$; $6x - 7y = 32$
5. $3x + 5y = 24$; $2x - 9y = 53$
6. $4x + 5y = 45$; $6x + 7y = 39$

Use matrices to solve each system for *x* and *y*. Round the values to the nearest hundredth.

7. $y = 15x + 9$; $y = 37x - 4$ **(0.59, 17.86)**
8. $y = 6.2x - 17.4$; $y = -3.8x$ **(1.74, −6.61)**
9. $y = -29x + 47$; $y = -12$ **(2.03, −12)**

Use matrices to solve each system for *x*, *y*, and *z*. Round the values to the nearest hundredth.

10. $3x + 5y - 6z = 22$; $2x - 8y + 7z = 15$; $-9x + y - 4z = 23$ **(2.10, −13.45, −13.83)**
11. $5x - 3y = 12$; $2y + 4z = 29$; $6x + 2y - 7z = 43$ **(7.63, 8.72, 2.89)**
12. $1.5x - 2.3y - 0.8z = 3.9$; $0.6x + 4.5y - 1.4z = -0.7$; $3.0x - 1.6y + 2.5z = 1.8$ **(1.11, −0.62, −1.01)**
13. $11x - 3y + 17z = 52$; $-8x + 10y - 5z = 28$; $19x - 13y + 18z = -12$ **(−9.20, −0.06, 9)**

Find an equation of the parabola passing through each set of points. Round the coefficients to the nearest hundredth.

14. (−2, 5), (1, 3), (3, 8) $y = 0.63x^2 - 0.03x + 2.4$
15. (−3, −7), (−1, 4), (2, 5) $y = -1.03x^2 + 1.37x + 6.4$
16. (−3, 10), (2, 11), (5, −4) $y = -0.65x^2 - 0.45x + 14.5$
17. (3.2, −2), (4.1, 3), (5, 9) $y = 0.62x^2 + 1.05x - 11.7$
18. Josefina pledged different amounts for 3 friends, Carlos, Alvin, and Manchu, in a charity walkathon. The sum of her per-mile pledges was \$.70, and her pledge for Manchu equaled the sum of her other two pledges. Carlos, Alvin, and Manchu walked distances of 5.4 mi, 7.2 mi, and 8.6 mi, respectively, and Josefina's total contribution to the charity was \$5.08.
    a. Write a system of equations that models the situation. $x + y + z = 0.7$, $z = x + y$, $5.4x + 7.2y + 8.6z = 5.08$
    b. Write the system of equations as a matrix equation. $\begin{bmatrix} 1 & 1 & 1 \\ 1 & 1 & -1 \\ 5.4 & 7.2 & 8.6 \end{bmatrix}\begin{bmatrix} x \\ y \\ z \end{bmatrix} = \begin{bmatrix} 0.7 \\ 0 \\ 5.08 \end{bmatrix}$
    c. Solve the system for Josefina's per-mile pledges. **(\$.25, \$.10, \$.35)**
19. **Open-ended Problem** Suppose you are given a system of two equations in two variables, *x* and *y*. Describe how you would decide which method to use to solve the system, based on the nature of the equations in the system. **Answers may vary. Check students' work.**

42 Practice Bank, ALGEBRA 2: EXPLORATIONS AND APPLICATIONS

ONGOING ASSESSMENT

27. **Cooperative Learning** Work with a partner. Each of you should choose three points in the coordinate plane.
    a. Each of you should find an equation of the parabola passing through your three points.
    b. For each set of three points, no two points can be on the same vertical line. Why?
    c. Compare your answers to part (a). Is there always a parabola? What are the values of *a*, *b*, and *c* when the three points lie on a line?

SPIRAL REVIEW

Solve each system of equations. State whether the system has *one solution*, *infinitely many solutions*, or *no solution*. *(Sections 7.1 and 7.2)*

28. $2x - 2y = 6$; $x - y = 1$
29. $2x + 2y = 5$; $x + 3y = 0$
30. $-3x + 4y = 1$; $6x - 8y = -2$

Solve each inequality for *x*. *(Toolbox, page 787)*

31. $2x + 4 \geq 1$
32. $5x - 3 < 7$
33. $12x > 6$

Write each quadratic function as a product of factors. *(Section 5.6)*

34. $y = x^2 - 2x - 8$
35. $y = 2x^2 - 9x - 5$
36. $y = x^2 - 4x + 3$

ASSESS YOUR PROGRESS

VOCABULARY

system of equations (p. 292)
inconsistent system (p. 299)
dependent system (p. 299)
identity matrices (p. 304)
inverse matrix (p. 304)

Solve each system of equations. *(Section 7.1)*

1. $y = 5x + 2$; $y = 2x^2$
2. $3y = 9$; $-5x - y = 6$
3. $3x + 5y = 3$; $14x + 10y = 6$

Solve each system of equations. State whether the system has *one solution*, *infinitely many solutions*, or *no solution*. *(Section 7.2)*

4. $2x + 4y = 7$; $5x - 3y = 1$
5. $8x + 2y = 4$; $-4x - y = 2$
6. $-3x - y = -5$; $x + y = -1$
7. **Technology** Use a graphing calculator or software with matrix calculation capabilities to find an equation of the parabola passing through the points (−11, −7), (−8, 2), and (−3, 3). *(Section 7.3)*
8. **Journal** Explain the difference between an inconsistent system and a dependent system. Can nonlinear systems be inconsistent or dependent? Explain.

27. a. Answers may vary. Check students' work.
    b. A parabola is the graph of a function.
    c. There is not always a parabola. If the points lie on a straight line, $a = 0$, $b =$ the slope of the line, and $c =$ the *y*-intercept.
28. no solution
29. $\left(\frac{15}{4}, -\frac{5}{4}\right)$; one solution
30. Values that satisfy one equation always satisfy the other; infinitely many.
31. $x \geq -\frac{3}{2}$
32. $x < 2$
33. $x > \frac{1}{2}$
34. $y = (x - 4)(x + 2)$
35. $y = (2x + 1)(x - 5)$
36. $y = (x - 3)(x - 1)$

Assess Your Progress

1. (−0.35, 0.25), (2.85, 16.3)
2. $\left(-\frac{9}{5}, 3\right)$
3. $\left(0, \frac{3}{5}\right)$
4. (0.96, 1.27); one solution
5. no solution
6. (3, −4); one solution
7. $y = -0.35x^2 - 3.65x - 4.8$
8. Answers may vary. An example is given. An inconsistent system has no solution. Its graph consists of parallel lines. A dependent system has an infinite number of solutions. The graphs of all the equations are the same. Nonlinear systems can be inconsistent (for example, $y = x^2 + 1$ and $y = -x^2$) or dependent (for example, $y = 4(x + 1)^2$ and $y = 4x^2 + 8x + 4$).

Section

7.4 Inequalities in the Plane

Learn how to...

- write and graph inequalities with two variables

So you can...

- illustrate the relationship between the weight of ground beef and its fat content, for example

For a point to be on the graph of $y = x + 3$, its y-coordinate must be 3 more than its x-coordinate. What do you think must be true about points on the graph of $y > x + 3$ or $y < x + 3$? In the Exploration, you will investigate such inequalities.

EXPLORATION
COOPERATIVE LEARNING

Guessing People's Ages

Work with a group of four or five students.
You will need:
- paper and pencil
- graph paper

1 Each person in your group should estimate the ages of the people shown. (Keep your estimates a secret; it will be more fun.)

2 Get the list of the actual ages from your teacher.

3 Make a scatter plot of your group's data. Use the horizontal axis for a = actual age and the vertical axis for e = estimated age.

Use your scatter plot to answer the following questions.

1. What does each vertical column of dots represent?
2. Do any plotted points fall on the line $e = a$? If so, what does this mean?
3. **a.** What does it mean if a plotted point falls below the line $e = a$? How would you represent this mathematically?
   b. What does it mean if a plotted point falls above the line $e = a$? How would you represent this mathematically?
4. Overall, what pattern do you see in your group's scatter plot? Compare your scatter plot with those from other groups.
5. If people were estimating your age, would you prefer that most points fell above the line $e = a$, below the line, or on the line? Do you think everyone prefers that region?

Exploration Note

Purpose
The purpose of this Exploration is to have students discover how points in a plane above or below a line can be located by using inequalities.

Materials/Preparation
Each group needs paper, pencil, and a sheet of graph paper.

Procedure
Students estimate the ages of the people shown and make a scatter plot of the group's data, including the actual ages you provide. It may help for one student to be in charge of listing all the estimates for his or her group. Students then use their scatter plots to answer five questions about their estimated ages and the actual ages.

Closure
Have groups share their graphs with the class. Students should see how the line $e = a$ separates the plane into two regions and that points above or below the line can be represented by inequalities.

Explorations Lab Manual
See the Manual for more commentary on this Exploration.

For answers to the Exploration, see following page.

Plan⇔Support

Objectives

- Write and graph inequalities with two variables.
- Solve real-world problems using inequalities with two variables.

Recommended Pacing

Core and Extended Courses
Section 7.4: 1 day

Block Schedule
Section 7.4: $\frac{1}{2}$ block
(with Section 7.3)

Resource Materials

Lesson Support
Lesson Plan 7.4
Warm-Up Transparency 7.4
Overhead Visuals:
Folder 7: Linear Inequalities in Two Variables
Folder A: Multi-Use Graphing Packet, Sheets 1–7
Practice Bank: Practice 43
Study Guide: Section 7.4
Exploration Lab Manual: Diagram Masters 1, 2

Technology
Graphing calculator
Internet:
http://www.hmco.com

Warm-Up Exercises

Solve each inequality for x.

1. $2x + 7 \le 15$ $x \le 4$
2. $-3x + 10 > -8$ $x < 6$
3. $9x - 8 \ge 12x + 19$ $x \le -9$
4. $-2.3x < 9.2$ $x > -4$
5. $17.24 + x < 0.5x - 1$ $x < -36.48$

Teach⟺Interact

Learning Styles: Visual

Graphing inequalities in the plane is a highly visual activity. It has important applications for modeling many practical problems that can be solved by using linear programming techniques presented in Section 7.6.

Additional Example 1

A newspaper ad for a clothing store states that sport coats are on sale for at least 30% off the tagged price. Write and graph an inequality that relates the sale price of a sport coat to the tagged price.

Let t = the tagged price of a sport coat, and let s = the sale price of the coat. Then the inequality is $s \le 0.7t$.

Step 1 Draw the line $s = 0.7t$.

Step 2 Shade below the line.

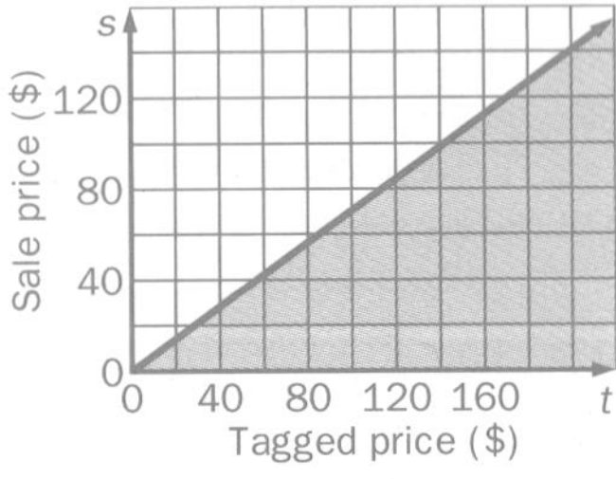

Section Note

Communication: Discussion Have a discussion comparing the conventions for graphing inequalities in the plane (two variables) with those for graphing inequalities on the number line (one variable). Dashed lines in the plane correspond to open dots on the number line. Solid lines in the plane correspond to solid (closed) dots on the number line. In the plane, boundary lines are dashed lines when the inequality symbol is > or <. A solid line is used when the inequality symbol is ≥ or ≤. On the number line, open dots are used when the inequality symbol is > or <. A solid dot is used when the inequality symbol is ≥ or ≤.

EXAMPLE 1 Application: Consumer Economics

The label on a package of ground beef (hamburger) tells you that the beef is "90% lean," which means that it contains no more than 10% fat (by weight). Write and graph an inequality that relates the acceptable amount of fat to the weight of the ground beef being sold.

SOLUTION

Let b = the weight of the ground beef, and let f = the weight of the fat. Then the inequality is $f \le 0.1b$.

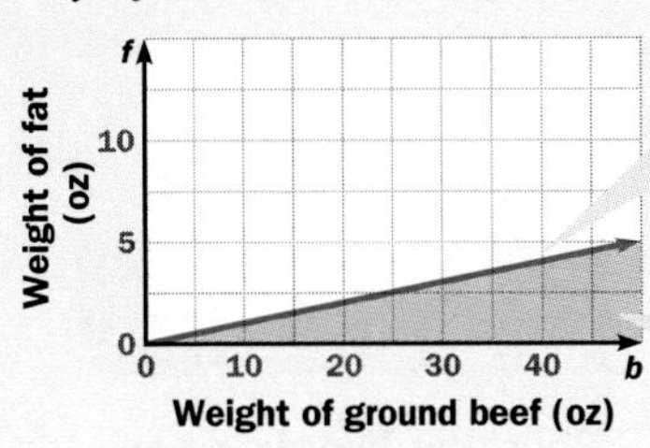

Step 1 Draw the line $f = 0.1b$.

Step 2 Shade below the line.

THINK AND COMMUNICATE

1. Use the graph from Example 1.
   - **a.** For the grocer preparing ground beef, what does the shaded region represent?
   - **b.** What does the region above the line represent? What inequality defines this region?

The graph of $y = ax + b$ divides the coordinate plane into two regions called **half-planes**. Each region is defined by an inequality.

The inequality $y > ax + b$ defines the half-plane **above** the line.

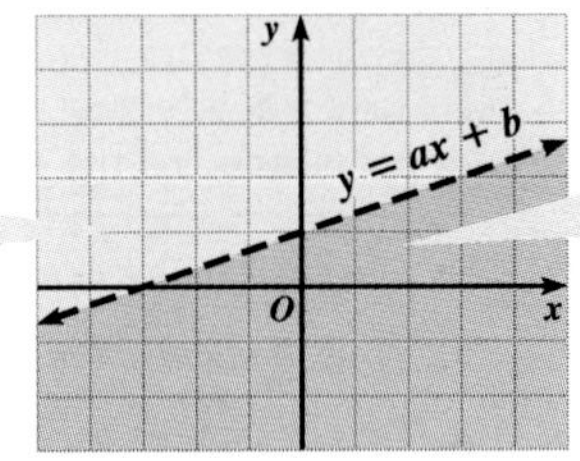

The inequality $y < ax + b$ defines the half-plane **below** the line.

When a line is vertical, it has an equation of the form $x = k$. Vertical lines also define half-planes.

The graph of $x < k$ is the half-plane to the **left** of the line $x = k$.

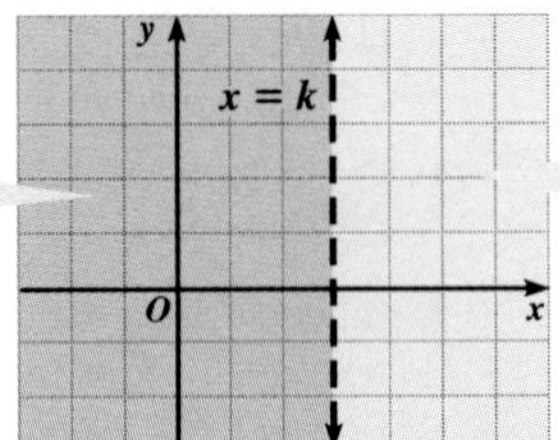

The graph of $x > k$ is the half-plane to the **right** of the line $x = k$.

ANSWERS Section 7.4

Exploration

The ages of people in the photographs are as follows. Philip: 27; Clarice: 38; Alan: 40; Eva: 32; Luke: 41

1. all the guesses for a given actual age
2. Answers may vary. If they do, it means that a guess was correct.
3. a. The guess was less than the actual age; $e < a$.
   b. The guess was greater than the actual age; $e > a$.
4. Answers may vary.
5. Answers may vary. For example, many young people may wish to appear older and so would want most points to fall above the line, while many older people may wish to appear younger and would want most points to fall below the line.

Think and Communicate

1. a. ground beef that is less than 10% fat by weight
   b. ground beef that is more than 10% fat by weight; $f > 0.1b$

EXAMPLE 2

Graph each inequality.

a. $y \geq 2x - 1$ **b.** $y < 2x - 1$

SOLUTION

a.

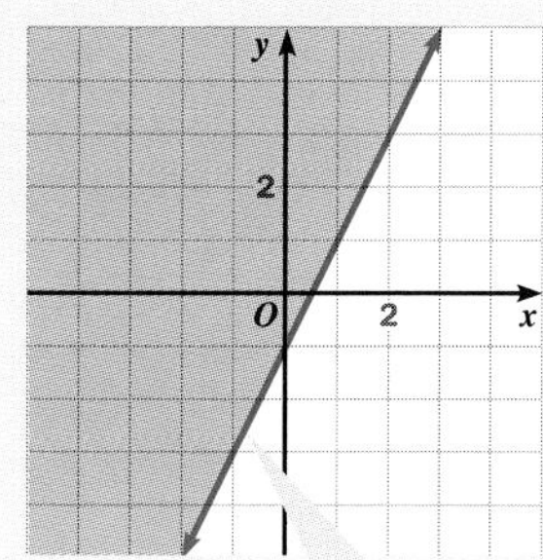

Graph $y = 2x - 1$ as a **solid** line since it is included in the graph of the inequality.

b.

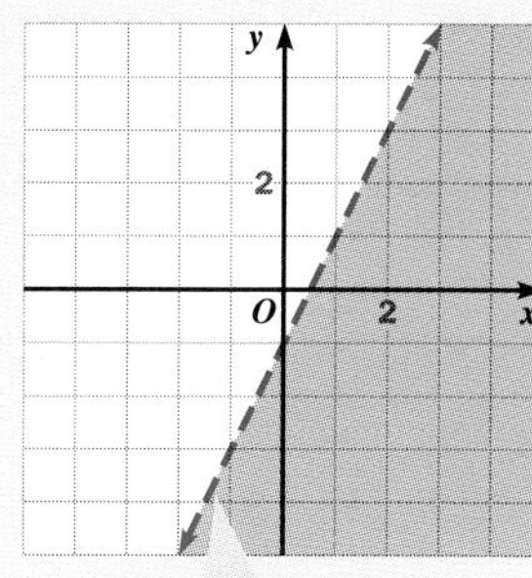

Graph $y = 2x - 1$ as a **dashed** line since it is not included in the graph of the inequality.

EXAMPLE 3

Graph the inequality $2x - 3y > 4$.

SOLUTION

Solve the inequality for y.

$$2x - 3y > 4$$
$$-3y > -2x + 4$$
$$y < \frac{2}{3}x - \frac{4}{3}$$

Remember that an inequality sign **changes direction** when each side is multiplied or divided by a negative number.

Graph $y = \frac{2}{3}x - \frac{4}{3}$ as a dashed line and shade below it.

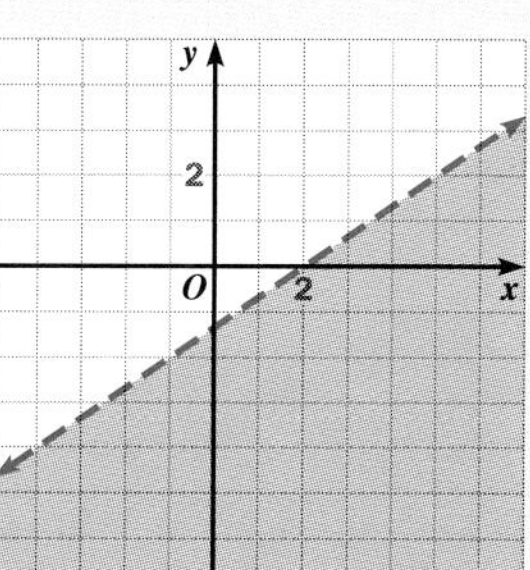

Toolbox p. 787
Solving Linear Inequalities

THINK AND COMMUNICATE

2. a. Do the coordinates $(3, -3)$ satisfy the inequality in Example 3? Do the coordinates $(0, 0)$ satisfy the inequality?

b. How does checking the coordinates of a point help you graph an inequality?

Think and Communicate

2. a. Yes; No.
   b. It helps you to determine which half-plane should be shaded.

Teach⇔Interact

Additional Example 2

Graph each inequality.

a. $y \leq -3x + 2$

Graph $y = -3x + 2$ as a solid line, since it is included in the graph of the inequality.

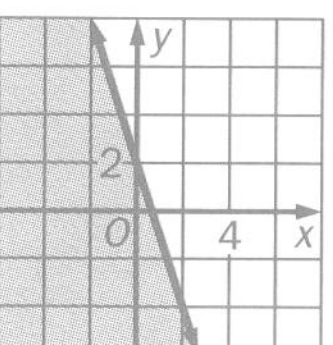

b. $y > -3x + 2$

Graph $y = -3x + 2$ as a dashed line, since it is not included in the graph of the inequality.

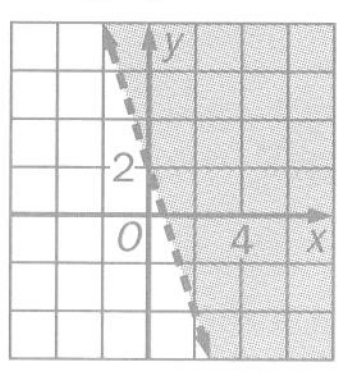

Additional Example 3

Graph the inequality $3x - 5y \leq 15$.

Solve the inequality for y. Remember that an inequality sign changes direction when each side is multiplied or divided by a negative number.

$$3x - 5y \leq 15$$
$$-5y \leq -3x + 15$$
$$y \geq \frac{3}{5}x - 3$$

Graph $y = \frac{3}{5}x - 3$ as a solid line and shade above it.

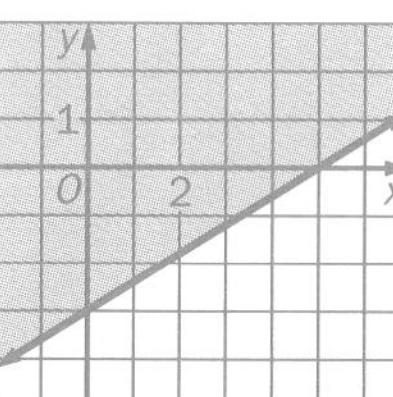

Think and Communicate

For question 2, point out that (0, 0) is always an easy point to check when deciding which half-plane to shade. If (0, 0) is on the boundary line, use another test point that is easy to check. Among the possibilities are (0, 1), (1, 0), (0, –1), and (–1, 0).

Teach⇔Interact

Additional Example 4

Find an inequality that defines the shaded region shown below.

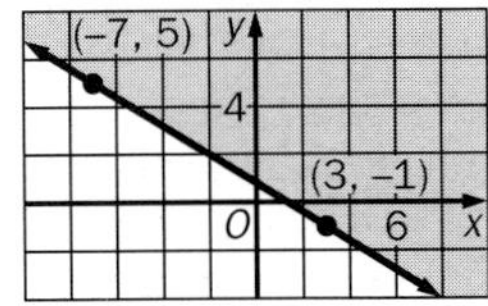

The line through the points (−7, 5) and (3, −1) has the equation $y = \left(\frac{5-(-1)}{-7-3}\right)(x-3) - 1$, or $y = -\frac{3}{5}x + \frac{4}{5}$.
The region is above and includes the line, so the inequality is $y \geq -\frac{3}{5}x + \frac{4}{5}$.

Checking Key Concepts

Reasoning
After graphing each inequality in questions 1–6, ask students how the graph would change if the inequality symbol were changed to each of the other three symbols.

Closure Question

Suppose you have a linear inequality and that you have graphed the line (solid or dashed) that forms the boundary for the graph. How can you decide which half-plane to shade in order to finish the graph? Pick a point whose coordinates you know and which is not on the boundary line. Substitute its coordinates in the inequality. If the resulting inequality is true, shade the half-plane that contains that point. Otherwise, shade the other half-plane.

Apply⇔Assess

Suggested Assignment

Core Course
Exs. 1–17, 23–31

Extended Course
Exs. 1–4, 10–31

Block Schedule
Day 43 Exs. 1–17, 23–31

EXAMPLE 4

Find an inequality that defines the shaded region shown below.

SOLUTION

The line through the points (0, 3) and (4, 1) has the equation $y = \left(\frac{1-3}{4-0}\right)x + 3$, or $y = -\frac{1}{2}x + 3$. The region is below and does not include the line, so the inequality is $y < -\frac{1}{2}x + 3$.

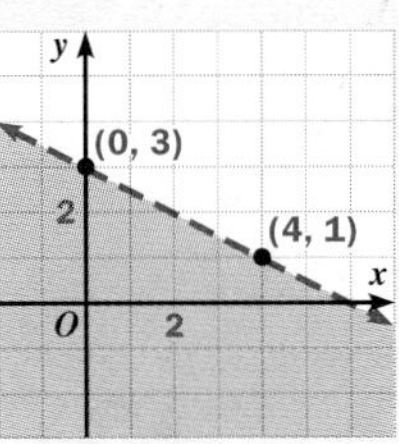

☑ CHECKING KEY CONCEPTS

Graph each inequality.

1. $y < 3x$ **2.** $y > -2x + 4$ **3.** $y \leq 5x + 5$

4. $x < 7$ **5.** $3x + 2y > 1$ **6.** $2y \leq x - 4$

7. During a "red tag" sale, a department store offers discounts of up to 25% on all tagged merchandise. Write and graph an inequality that relates the discounted price to the normal price of tagged merchandise.

7.4 Exercises and Applications

Extra Practice exercises on page 759

1. A day-care center provides at least one supervisor for every three infants. Write and graph an inequality that relates the number of supervisors on duty to the number of infants at the center.

Graph each inequality.

2. $y > 5x$ **3.** $y > 3x - 2$ **4.** $y < -4x + 5$ **5.** $y < -1.2x$

6. $y > 4.5x + 3$ **7.** $x \geq 2.5$ **8.** $4x + 6y > 24$ **9.** $3.1x - 4.6y < 33$

10. $5x + 2y \geq 0$ **11.** $y > 8.5$ **12.** $x + y \geq 6$ **13.** $-5x + 10y < -15$

14. Challenge Graph the quadratic inequality $y > x^2 + 3$.

Find an inequality that defines each shaded region.

15.

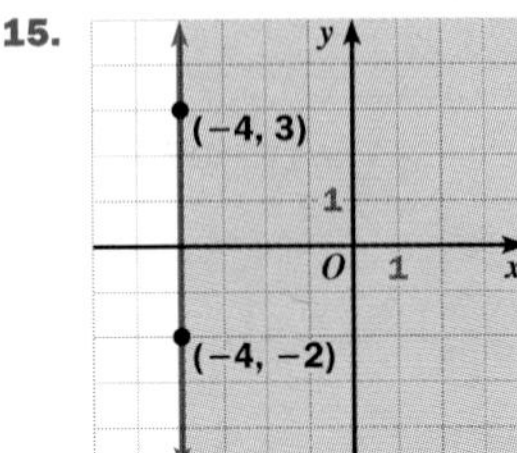

16.

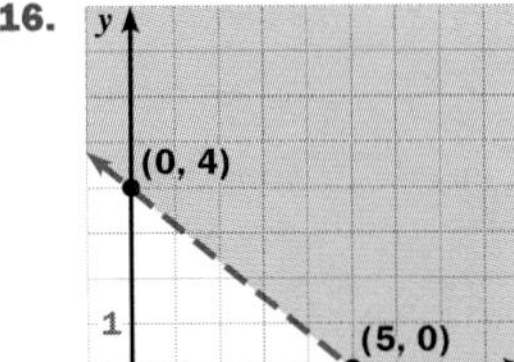

17.

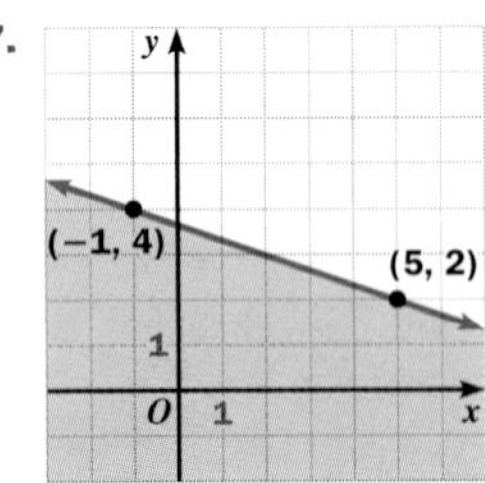

Checking Key Concepts
1–7. See answers in back of book.

Exercises and Applications
1–11. See answers in back of book.

12.

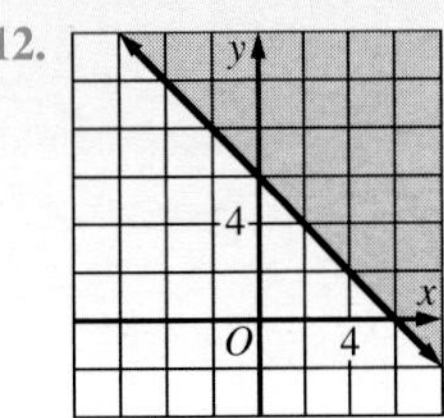

13.

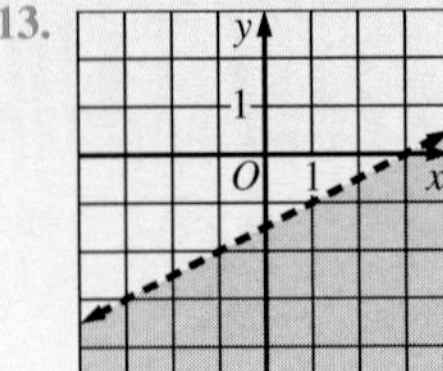

14.

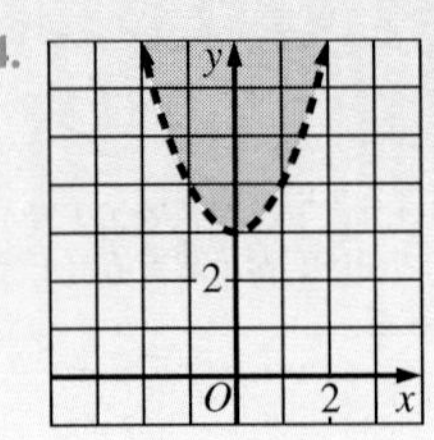

15. $x \geq -4$

16. $y > -\frac{4}{5}x + 4$

17. $y \leq -\frac{1}{3}x + \frac{11}{3}$

Connection PHYSICS

An object immersed in a liquid will be buoyed by a force equal to the weight of the liquid that the object displaces. Since 1 ft^3 of salt water weighs 64 lb, a submerged object having a volume of 1 ft^3 and weighing more than 64 lb will sink, while one having the same volume and weighing less than 64 lb will float.

BY THE WAY

In 1971, a 2400-ton barge was raised in the Gulf of Mexico using the "sphere injection" method.

18. Divers use weight belts to control their buoyancy. In order to float on the surface of the ocean, divers must weigh less than the amount of water they displace. Write and graph an inequality that relates the volume of salt water a floating diver displaces to the weight of the diver.

19. An interesting form of ocean salvage involves pumping hollow spheres into sunken vessels in order to supply buoyancy. To become buoyant, each ton to be raised requires at least 80 specially designed spheres of diameter 11 in. Write and graph an inequality that expresses the relationship between the number of spheres needed to raise a vessel and the weight of the vessel.

20. Lifting bags are sometimes attached to a sunken object and inflated in order to increase the object's buoyancy. Shellie is using a lifting bag to raise an object from the bottom of a freshwater lake. She knows that 1 ft^3 of fresh water weighs 62.4 lb. Write and graph an inequality that relates the volume of water the lifting bag displaces to the weight of the submerged object.

21. **Open-ended Problem** Why do you think that you float more easily in salt water than in fresh water?

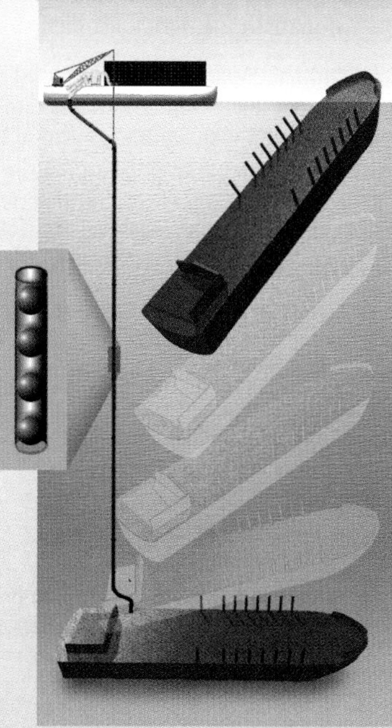

22. **CARTOGRAPHY** In order to make the symbols on maps legible to the average person, cartographers set lower limits for the width and height of the symbols they use. The minimum size of a symbol depends linearly on the distance at which it will be viewed.
    a. The minimum width (in inches) of a legible symbol is 0.007 times the viewing distance (in feet). Write and graph an inequality that relates these two variables.
    b. The minimum height (in inches) of a legible symbol is 0.03 times the viewing distance (in feet). Write and graph an inequality that relates these two variables.

23. **AGRICULTURE** On an ostrich farm, female ostriches lay up to 70 eggs per year. Write and graph an inequality that relates the number of eggs produced per year to the number of female ostriches on the farm.

18. Let w = the weight of the diver in pounds and v = the volume in cubic feet of the water the diver displaces; $w < 64v$.

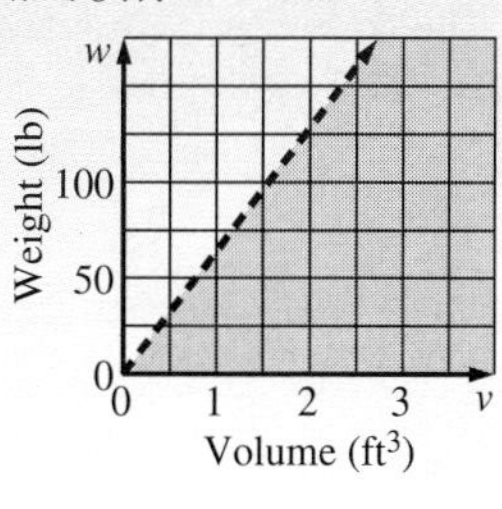

19. Let w = the weight of the vessel in tons and s = the number of spheres; $s \geq 80w$.

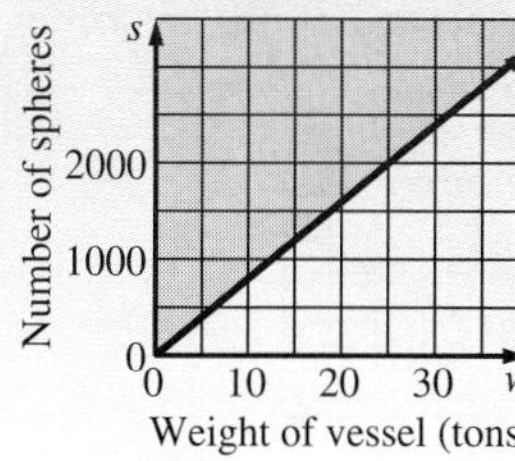

20. Let w = the weight of the object in pounds and v = the volume in cubic feet of water displaced; $v > \frac{w}{62.4}$.

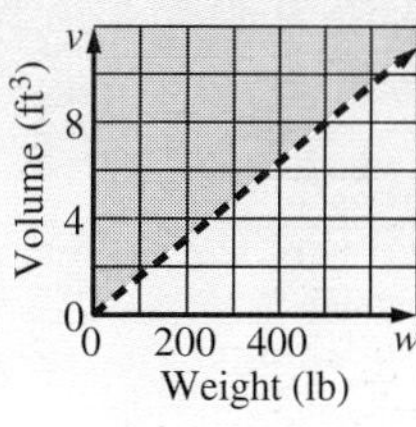

21–23. See answers in back of book.

Apply⇔Assess

Exercise Notes

Application

Ex. 1 Some students may draw graphs that show an entire region of the coordinate plane. Others may show a finite number of isolated points that have whole-number coordinates. Ask students which is more realistic in terms of this application. (the second kind of graph)

Common Error

Exs. 2–13 Some students may draw boundary lines incorrectly or shade the wrong half-plane. Such errors can be corrected by pointing out that if the inequality symbol in the given inequality is $\geq$ or $\leq$, the boundary line should be a solid line. If it is > or <, then the boundary line should be a dashed line. To shade the proper half-plane, substitute coordinates of points that are easy to check.

Challenge

Exs. 15–17 After students have found appropriate inequalities for these regions, ask if anyone can write a linear or nonlinear inequality whose graph is the entire plane. (It is not possible to have a linear inequality whose graph is the entire plane. However, the nonlinear inequality $x^2 + y^2 \geq 0$ is satisfied by the coordinate of every point. Hence, the graph of $x^2 + y^2 \geq 0$ is the entire plane.)

Research

Exs. 18–21 Students may find it interesting to research the salts and minerals in sea water. It is these substances that account for the different weights of a cubic foot of salt water and a cubic foot of fresh water. (Students should note that fresh water is not what a chemist would call *pure* water.)

Visual Thinking

Ex. 22 Assign students to work in groups to collect a variety of maps. Ask them to use the maps and the graphs that they have created to explain how cartographers make certain that the symbols on maps are legible to the average person. This activity involves the visual skills of *interpretation* and *communication.*

Apply⇔Assess

Exercise Notes

Multicultural Note

Ex. 24 According to legend, silk was discovered in 2640 B.C. by the Chinese Empress Xi Ling Shi. China remained the only producer of silk for more than 2000 years. Today, China produces more than half of the world's silkworm cocoons each year.

Topic Spiraling: Preview

Exs. 28–30 Knowing how to solve systems of linear equations is an important goal for this course. The ability to solve such systems is a prerequisite for the work in Sections 7.5 and 7.6.

Practice 43 for Section 7.4

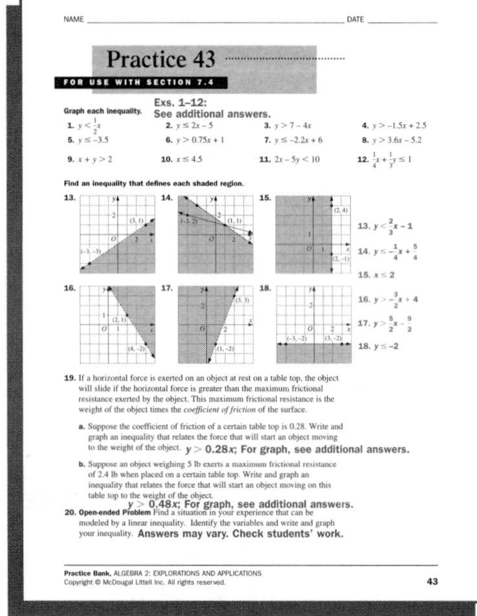

NAME ______________________ DATE __________

Practice 43

FOR USE WITH SECTION 7.4

Graph each inequality. **Exs. 1–12: See additional answers.**

1. $y < \frac{1}{2}x$
2. $y \le 2x - 5$
3. $y > 7 - 4x$
4. $y > -1.5x + 2.5$
5. $y \le -3.5$
6. $y > 0.75x + 1$
7. $y \le -2.2x + 6$
8. $y > 3.6x - 5.2$
9. $x + y > 2$
10. $x \le 4.5$
11. $2x - 5y < 10$
12. $\frac{1}{4}x + \frac{1}{3}y \le 1$

Find an inequality that defines each shaded region.

13. $y < \frac{2}{3}x - 1$
14. $y \le -\frac{1}{4}x + \frac{5}{4}$
15. $x \le 2$
16. $y > -\frac{3}{2}x + 4$
17. $y > \frac{5}{2}x - \frac{9}{2}$
18. $y \le -2$

19. If a horizontal force is exerted on an object at rest on a table top, the object will slide if the horizontal force is greater than the maximum frictional resistance exerted by the object. This maximum frictional resistance is the weight of the object times the *coefficient of friction* of the surface.

   a. Suppose the coefficient of friction of a certain table top is 0.28. Write and graph an inequality that relates the force that will start an object moving to the weight of the object. **$y > 0.28x$; For graph, see additional answers.**

   b. Suppose an object weighing 5 lb exerts a maximum frictional resistance of 2.4 lb when placed on a certain table top. Write and graph an inequality that relates the force that will start an object moving on this table top to the weight of the object. **$y > 0.48x$; For graph, see additional answers.**

20. **Open-ended Problem** Find a situation in your experience that can be modeled by a linear inequality. Identify the variables and write and graph your inequality. **Answers may vary. Check students' work.**

Practice Bank, ALGEBRA 2: EXPLORATIONS AND APPLICATIONS

43

24. **MANUFACTURING** For many farmers in the Tai Lake Valley region of China, silkworm cocoons provide the main source of income. Unfortunately, the industry is being threatened by fluoride pollution from nearby brick factories.

   a. To form its cocoon, each silkworm produces a single filament that stretches up to 1.5 km long. Write an inequality that relates the total possible length of silk filament produced to the number of silkworms.

   b. Mature silkworms, on average, can tolerate 40 micrograms of fluoride in each gram of mulberry leaves they consume. Write an inequality that relates the mass of fluoride a silkworm can safely consume to the mass of mulberry leaves consumed.

BY THE WAY

It takes roughly 700 worms and 15 kg of mulberry leaves to produce the silk to make one blouse.

25. **MEDICINE** A comparison of blood pressures in a patient's arm and ankle can help a doctor find hidden health risks. The doctor may consider it a warning sign if the ankle's systolic blood pressure is less than 90% of the arm's systolic blood pressure. Write and graph an inequality that relates the acceptable blood pressure in the ankle to the blood pressure in the arm. Systolic arm blood pressures, measured in millimeters of mercury (mm Hg), typically range from 100 mm Hg to 130 mm Hg.

26. **SAT/ACT Preview** Which inequality is equivalent to $4x - 3y > 6$?

   A. $y < -\frac{4}{3}x - 2$ **B.** $y > -\frac{4}{3}x + 2$ **C.** $y < \frac{4}{3}x - 2$

   D. $y > \frac{4}{3}x + 2$ **E.** none of these

ONGOING ASSESSMENT

27. **Open-ended Problem** Find an example of an inequality involving two variables in everyday life. Write and graph your inequality.

SPIRAL REVIEW

Solve each system of equations. *(Section 7.2)*

28. $3x + 4y = 17$
$x + 7y = 17$

29. $3.1x - 5.8y = 19$
$2.6x + 1.8y = 51$

30. $23x + 45y = 91$
$26x - 28y = 24$

Use the following data on the number of passengers who entered various stations on two subway lines during one day in Boston. *(Section 6.4)*

Orange Line: 6043; 8368; 5714; 9512; 4743; 2029; 9703; 19,480; 3393; 4886; 15,171; 2769; 5552; 2798; 4167; 2441; 2962; 10,936

Red Line: 9990; 5865; 8744; 14,724; 7877; 8163; 7208; 4636; 16,278; 2960; 3087; 7576; 723; 3777; 1157; 9161; 5114; 4122; 7133; 5920; 5024

31. a. Make two box plots to compare the data for the Orange Line and the Red Line.

   b. **Writing** How do the two plots compare?

24. a. Let s = the number of silkworms and l = the length of silk filament; $l \le 1.5s$.

   b. Let m = the mass in grams of mulberry leaves and f = the mass in micrograms of fluoride; $f \le 40m$.

25. Let x = the blood pressure in the arm and y = the blood pressure in the ankle; $y \ge 0.9x$. (Both x and y are measured in milligrams of mercury.)

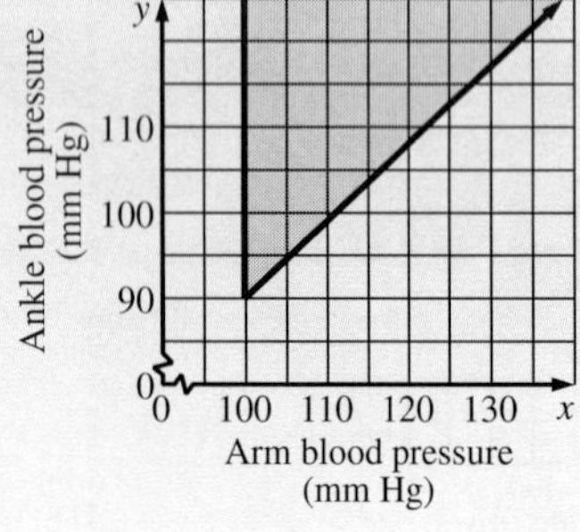

26. C

27. Answers may vary. An example is given. When we make punch for a party, we plan on at least two cups per person. If p = the number of people and c = the number of cups of punch, then $c \ge 2p$.

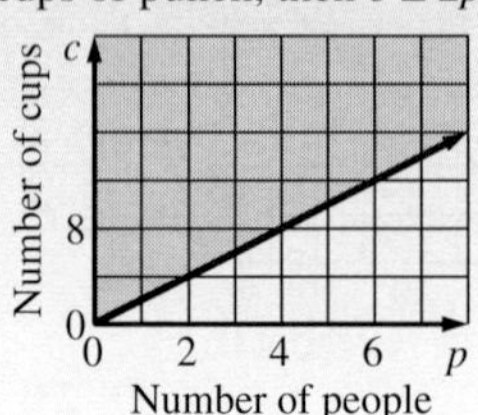

28. (3, 2)

29. (15.97, 5.26)

30. (2, 1)

31. See answers in back of book.

Section

7.5 Systems of Inequalities

Learn how to...

- graph a system of linear inequalities
- find the system of linear inequalities that describes a given graph

So you can...

- represent and interpret situations involving inequalities, such as target heart rate

Hot deserts in North America include three general types: the arid Mojave desert, the grassier Chihuahuan Desert, and the diverse Sonoran Desert. What distinguishes the three deserts climatically is the amount of precipitation that falls in each season. The graph below shows the mean annual precipitation and the percent of precipitation that falls in winter for various sites in the three North American hot deserts.

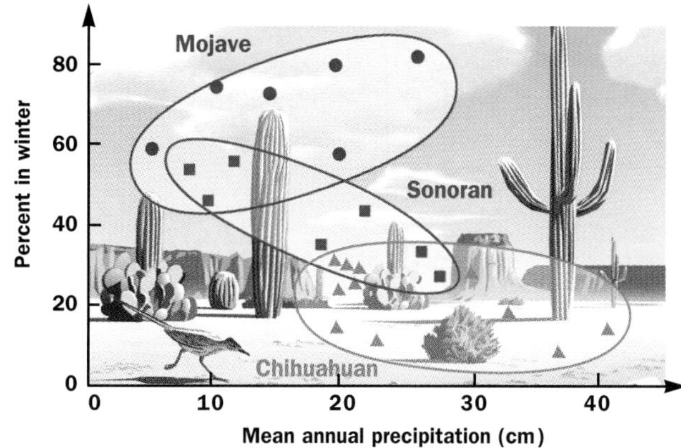

THINK AND COMMUNICATE

Use the graph above to answer the following questions.

1. If a North American hot desert site gets 30 cm of rain, and 20% of that in the winter, which desert do you think it is in? What if the site gets 25 cm of rain, and 30% of that in the winter?
2. How does the graph show that the climate in the Sonoran Desert is sometimes like the climate in the Mojave Desert?
3. Which two deserts are the most different climatically? How does the graph show this?

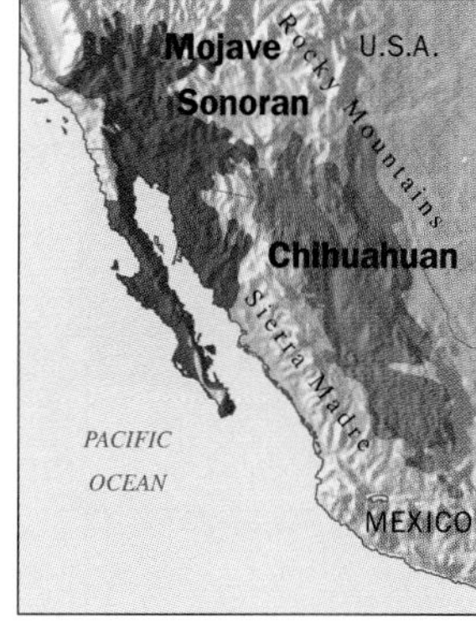

A region of a coordinate plane can be described using a **system of inequalities**. For example, the first quadrant of the *xy*-plane is described by this system of inequalities: $x > 0$ and $y > 0$.

Plan⇔Support

Objectives

- Graph a system of linear inequalities.
- Find the system of linear inequalities that describes a given graph.
- Represent and interpret situations involving inequalities in real-world problems.

Recommended Pacing

Core and Extended Courses
Section 7.5: 2 days

Block Schedule
Section 7.5: 1 block

Resource Materials

Lesson Support
Lesson Plan 7.5
Warm-Up Transparency 7.5
Overhead Visuals:
Folder 7: Linear Inequalities in Two Variables
Folder A: Multi-Use Graphing Packet, Sheets 1–7
Practice Bank: Practice 44
Study Guide: Section 7.5
Exploration Lab Manual: Diagram Masters 1, 2

Technology
Graphing Calculator
McDougal Littell Mathpack *Function Investigator*
Internet:
http://www.hmco.com

Warm-Up Exercises

For each of the given points, write the letter of each inequality whose graph contains the point.

A. $y \le -x + 1$ **B.** $y \ge 0.3x - 4$
C. $y < 2x + 3$ **D.** $y > 2x + 3$
E. $y < 0.3x - 4$

1. (−2, −4) A, B, C
2. (0, −4) A, B, C
3. (1, 6) B, D
4. (5, 3) B, C
5. (7, −2) C, E
6. (−4, −1) A, B, D

ANSWERS Section 7.5

Think and Communicate

1. the Chihuahuan Desert; the Chihuahuan Desert or the Sonoran Desert
2. The regions overlap.
3. the Mojave Desert and the Chihuahuan Desert; The regions do not overlap at all.

Teach⇔Interact

Additional Example 1

An exercise therapist at a health club is often asked by tall members of the club what is a normal range of weights for taller people. The therapist usually advises a minimum weight that can be found by multiplying height in inches by 4 and then subtracting 148. For a maximum weight, she advises multiplying height in inches by 5.4 and then subtracting 205.

a. Write a system of inequalities using h for height in inches and w for weight in pounds.

Inequality 1: w should be at least $4h - 148$.
$w \geq 4h - 148$
Inequality 2: w should be no more than $5.4h - 205$.
$w \leq 5.4h - 205$

b. Graph the system of inequalities.

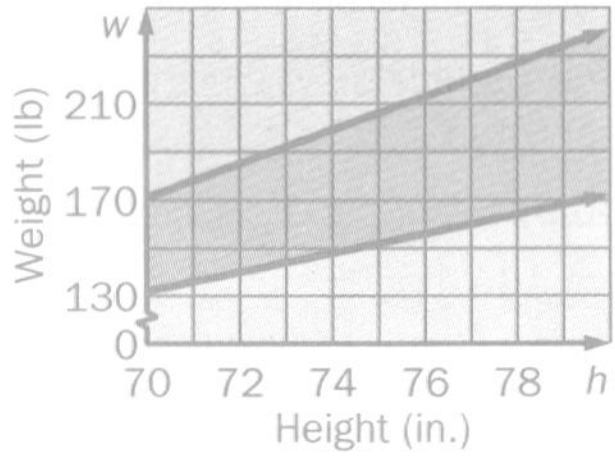

The suggested weights are contained in the overlap of the two shaded regions, that is, on or between the two lines.

Additional Example 2

A newspaper is hiring reporters to cover local and state events. The budget allows for up to 15 people, and management has decided that at least 9 people are needed to get adequate coverage. The editor-in-chief wants at least 4 people for local news and at least 4 people for state news. Graph the region that shows the possible composition of the staff by news area.

Step 1 Write inequalities to represent the restrictions of the problem. Let x = the number of reporters for local events, and let y = the number of reporters for state events.
There should be at most 15 reporters: $x + y \leq 15$.
There should be at least 9 reporters: $x + y \geq 9$.
There should be at least 4 reporters for local events: $x \geq 4$.
There should be at least 4 reporters for state events: $y \geq 4$.

EXAMPLE 1 Interview: Gina Oliva

Look back at the opening interview on pages 288–290. To find a person's target heart rate, subtract the person's age from 220. The target heart rate, in heartbeats per minute, is between 60% and 80% of this number.

a. Write a system of inequalities using a for age and h for target heart rate.

b. Graph the system of inequalities.

SOLUTION

a.

Inequality 1: h should be at least 60% of $(220 - a)$.	**Inequality 2:** h should be no more than 80% of $(220 - a)$.
$h \geq 0.6(220 - a)$, or $h \geq -0.6a + 132$	$h \leq 0.8(220 - a)$, or $h \leq -0.8a + 176$

b.

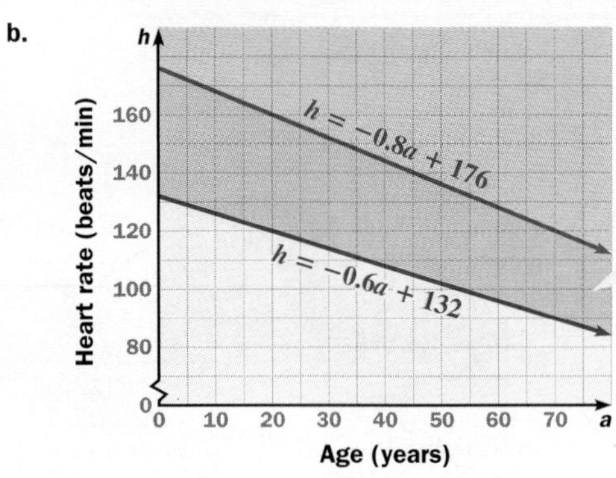

The possible values for the target heart rate are contained in the **overlap** of the red region and the blue region.

EXAMPLE 2

The Junior-Senior Prom Committee must have five to eight representatives drawn from the junior and senior classes. The committee must include at least two members of the junior class and at least two members of the senior class. Write and graph a system of inequalities that represents the possible committees.

SOLUTION

Step 1 Write inequalities to represent the restrictions in the problem. Let x = the number of juniors, and let y = the number of seniors.

The committee has at least five members.	The committee has at most eight members.	There are at least two juniors.	There are at least two seniors.
$x + y \geq 5$, or $y \geq -x + 5$	$x + y \leq 8$, or $y \leq -x + 8$	$x \geq 2$	$y \geq 2$

Think and Communicate

4. No; only the points for which both x and y are integers.

5.

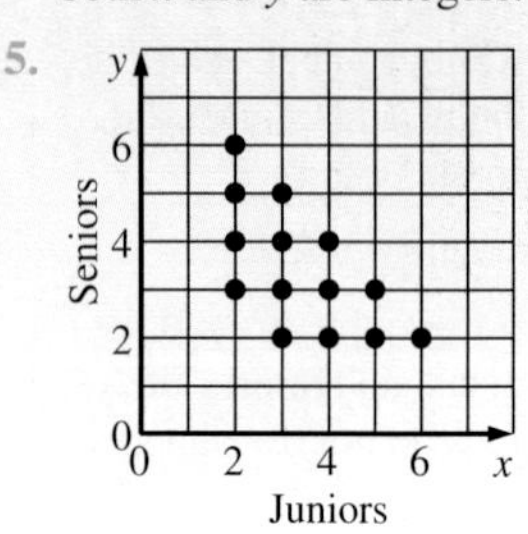

Step 2 Graph the region described by the system of inequalities.

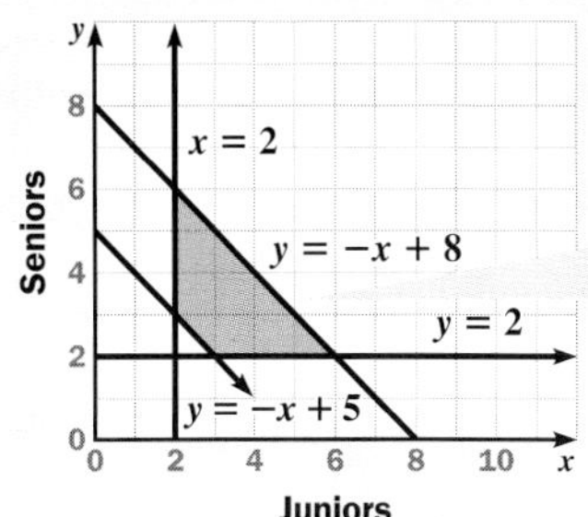

All possible combinations of juniors and seniors on the committee are shown in the shaded region.

THINK AND COMMUNICATE

4. Does *every* point (x, y) in the shaded region in Example 2 represent a committee consisting of x juniors and y seniors? Explain.
5. Redraw the solution of Example 2 using discrete points rather than a shaded region.

EXAMPLE 3

Find a system of inequalities defining the shaded region shown below.

SOLUTION

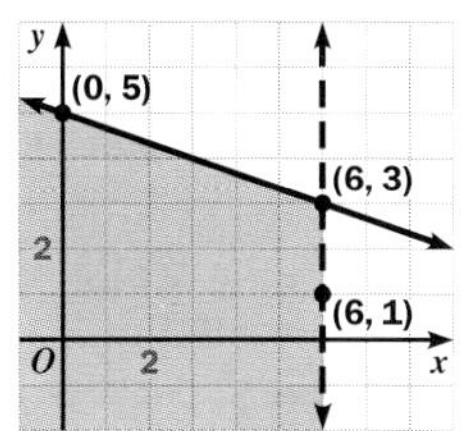

Inequality 1: An equation of the line through the points (0, 5) and (6, 3) is $y = \left(\frac{3-5}{6-0}\right)x + 5$, or $y = -\frac{1}{3}x + 5$. The region is below and includes the line, so the inequality is $y \le -\frac{1}{3}x + 5$.

Inequality 2: An equation of the line through the points (6, 3) and (6, 1) is $x = 6$. The region is to the left of and does not include the line, so an inequality is $x < 6$.

CHECKING KEY CONCEPTS

Graph each system of inequalities.

1. $y \ge x + 3$; $x \le 5$
2. $y > 0$; $y < 10$
3. $y \ge -3x - 3$; $y < x + 1$; $x \le 15$
4. **Writing** What must be true about the coordinates of any point in the graph of a system of inequalities?

Checking Key Concepts

1.

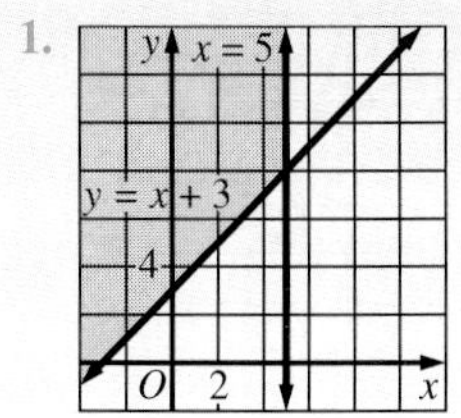

2.

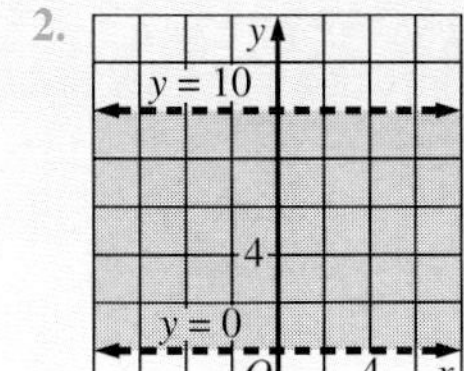

3.

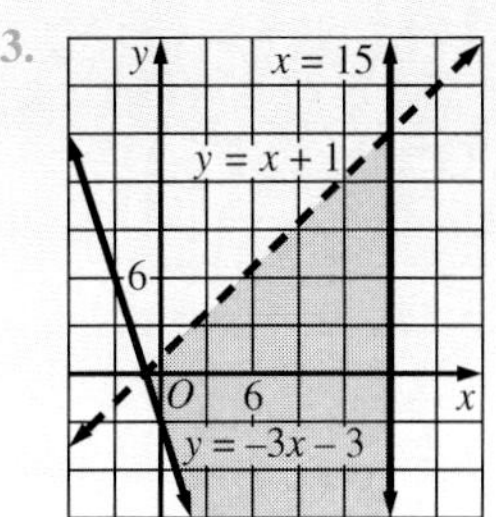

4. They must satisfy each of the inequalities in the system.

Teach⇔Interact

Additional Example 2 (continued)

Step 2 Graph the region described by the system of inequalities.

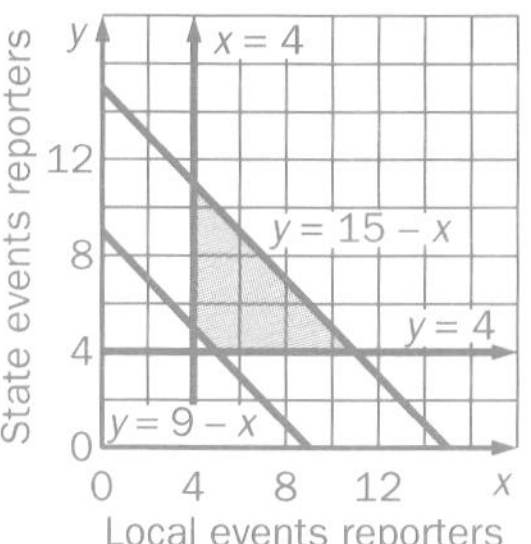

All possible combinations of reporters for local and state events are shown in the shaded region.

Additional Example 3

Find a system of inequalities defining the shaded region shown.

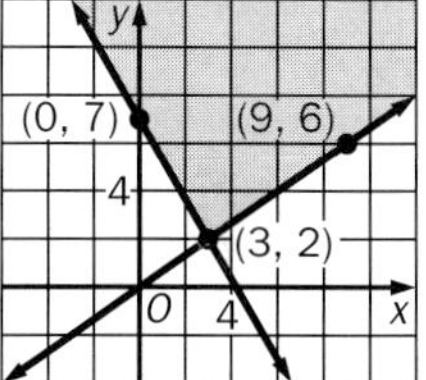

Inequality 1: The line through the points (0, 7) and (3, 2) has the equation $y = \left(\frac{7-2}{0-3}\right)x + 7$, or $y = -\frac{5}{3}x + 7$. The region is above and includes the line, so the inequality is $y \ge -\frac{5}{3}x + 7$.

Inequality 2: The line through the points (9, 6) and (3, 2) has the equation $y = \left(\frac{6-2}{9-3}\right)(x - 3) + 2$, or $y = \frac{2}{3}x$. The region is above and includes the line, so the inequality is $y \ge \frac{2}{3}x$.

Closure Question

Suppose a and b are real numbers and that $(a, 2b)$ is a point belonging to the graph of the system
$2x - y < 7$
$x + 3y > 11$.
Write two inequalities that must be true for the numbers a and b. Explain how you got these inequalities.

$2a - 2b < 7$ and $a + 6b > 11$; Since the point $(a, 2b)$ belongs to the graph of each inequality, its coordinates make each inequality true when x is replaced by a and y is replaced by $2b$.

Apply⇔Assess

Suggested Assignment

Core Course
Day 1 Exs. 1–8, 11–15
Day 2 Exs. 22–33, 36–47

Extended Course
Day 1 Exs. 1–3, 7–15
Day 2 Exs. 22–24, 28–47

Block Schedule
Day 44 Exs. 1–8, 11–15, 22–33, 36–47

Exercise Notes

Application
Ex. 9 Students should be cautious about assuming too much when a model is based on comparatively limited data. The data points in the graph are the result of observations over a brief period of time.

Multicultural Note
Ex. 9 In the late eighteenth century, French traders called the Yellowstone River *la Toche Jaune* (or river of the yellow rock), but many historians are not sure why; the river does not have much yellow rock along it. According to Daniel Old Elk, a member of the Crow people, the French traders did not speak the Crow language very well and as a result, they mistook the Crow name, meaning *elk river* for the Crow word meaning *yellow rock*.

Teaching Tip
Exs. 11–13 Point out to students that exercises such as these are, in a sense, the "opposite" of exercises such as Exs. 1–8.

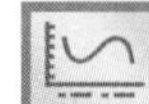

Using Technology
Exs. 14–17 Graphing calculators and the *Function Investigator* software are excellent tools for these exercises.

Research
Exs. 14–18 Students can consult books on medicine and health to find information on recommended ranges for blood pressure, cholesterol, and other health-related factors. Have them share with the class any systems of inequalities that may be suggested by the information they obtain.

7.5 Exercises and Applications

Extra Practice *exercises on page 759*

Graph each system of inequalities.

1. $y \geq -x - 1$, $y \geq 4x - 12$

2. $x \leq 5$, $x \geq -2$

3. $x \leq -1$, $3y \leq -x - 2$

4. $y \geq 1$, $y \leq -3x + 8$

5. $y \geq 5x$, $y \leq 2x - 4$

6. $y \leq 5$, $y \geq \frac{1}{2}x + 1$

7. $y < x + 3$, $y < 4$

8. $y \geq -3x + 3$, $y \leq \frac{1}{4}x + 4$

9. GEOLOGY Although "Old Faithful," a geyser in Yellowstone National Park, is known for its regular eruptions, the lengths of eruptions and the times between eruptions do vary. The scatter plot shows the length of an eruption, l, and the time until the next eruption, t, over a four-day period.

a. Open-ended Problem Imagine a parallelogram that encloses the plotted points. Find the inequalities that describe the interior of the parallelogram.

b. Writing If an eruption lasts 2 min, what range of time would you expect to wait until the next eruption? Explain how you can use your graph from part (a) to answer this question.

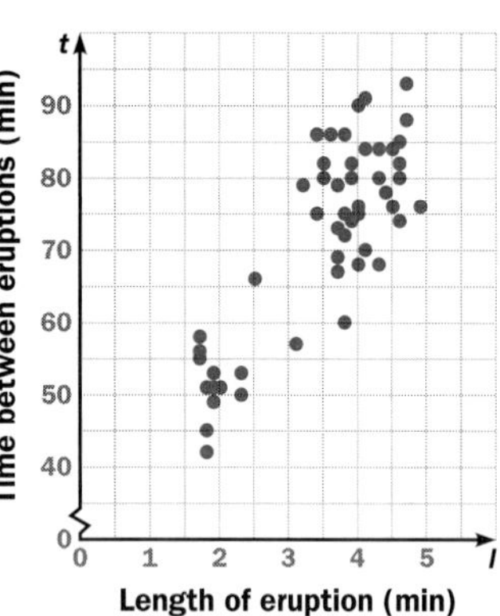

BY THE WAY

A geyser is a spring that intermittently throws hot water up with explosive force. The only geyser group in the United States is in Yellowstone National Park.

10. George has to write a research paper on the work of a well-known author. His bibliography must include at least 12 books or articles, and at least 4 of the references must be original works by the author (primary sources). The other references can be discussions of the author's work or background material (secondary sources).

a. Write inequalities to represent the number of primary sources, p, and the number of secondary sources, s, that George can list in his bibliography.

b. Graph the system of inequalities you found in part (a).

Write a system of inequalities defining each shaded region.

11.

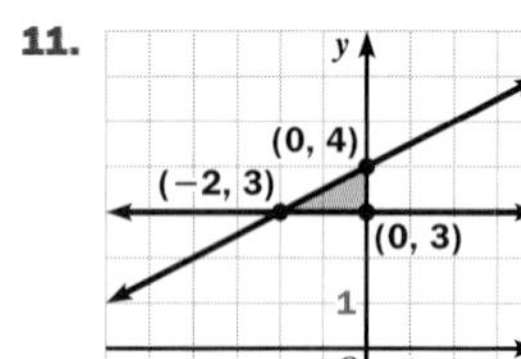

12.

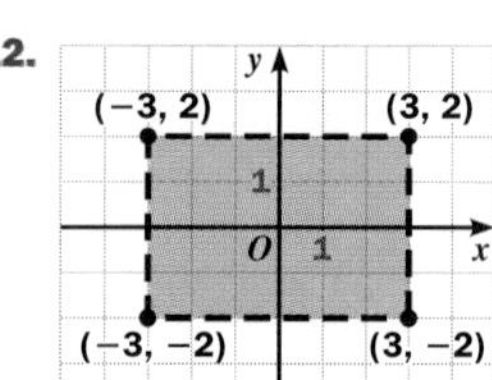

13.

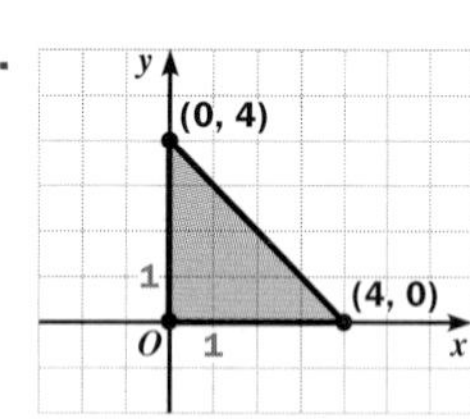

Exercises and Applications

1.

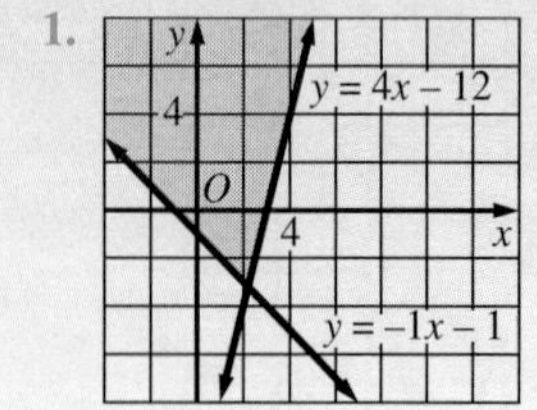

2.

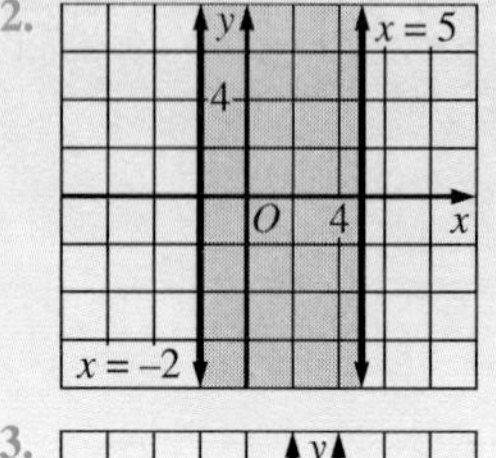

3.

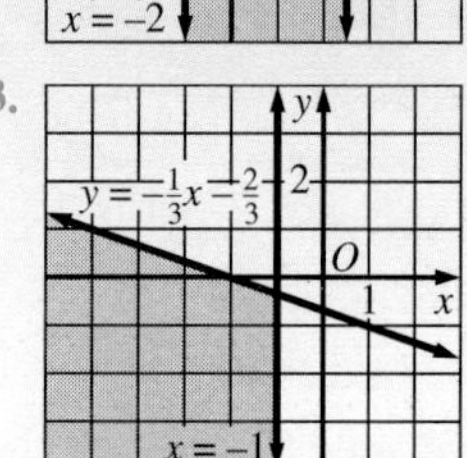

4.

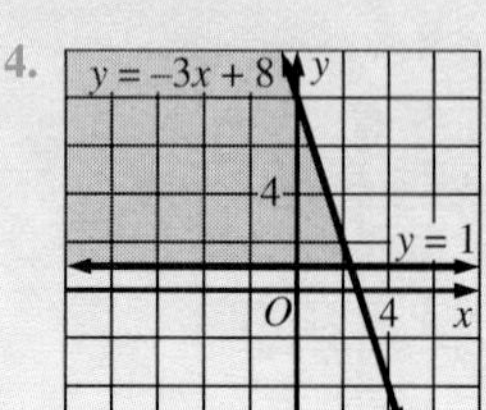

INTERVIEW Gina Oliva

Look back at the article on pages 288–290.

Many people go to aerobics classes like Gina Oliva's to maintain a recommended weight. The table shows the recommended weights for people of various heights and ages.

Healthy Weight Ranges for Men and Women

	Age 19–34 years	Age 35 years and over
Height (in.)	**Weight (lb)**	**Weight (lb)**
60	97–128	108–138
61	101–132	111–143
62	104–137	115–148
63	107–141	119–152
64	111–146	122–157
65	114–150	126–162
66	118–155	130–167
67	121–160	134–172
68	125–164	138–178
69	129–169	142–183
70	132–174	146–188
71	136–179	151–194
72	140–184	155–199

14. Use the data for people aged 19–34. Make a scatter plot with height on the horizontal axis and weight on the vertical axis. For each height, use one color or symbol to plot the lowest recommended weight, and a second color or symbol to plot the highest recommended weight.

15. Using your scatter plot from Exercise 14, draw a line of fit for the lowest recommended weight. Draw a line of fit for the highest recommended weight. Shade the recommended weight region.

16. Write inequalities to describe your graph from Exercise 15.

17. **a.** Repeat Exercises 14–16 using the data for people aged 35 and over.

b. **Writing** Compare the region for people aged 19–34 with the region for people aged 35 and over.

18. **Research** Recommended weight also depends on gender, build, and medical history. Use an almanac or other source to find the recommended weights for a group of people, and repeat Exercises 14–16 for the new data.

Apply⇔Assess

Exercise Notes

Career Connection

Exs. 14–18 Aerobics instructors are members of a growing class of professionals who provide services to people interested in developing and maintaining a healthy lifestyle. The two main components of keeping healthy are exercise and proper nutrition. The careers associated with helping people exercise and eat healthy foods are varied and numerous. For example, a student might be interested in working directly with people as an instructor in a fitness center, or perhaps indirectly by helping to manufacture or sell exercise equipment. A brief discussion by students of careers they are aware of in this area would most likely introduce many different job possibilities.

Visual Thinking

Ex. 18 Check students' understanding by asking them to display the graphs that they created as part of their research assignment. Encourage them to discuss the characteristics of the people who fall within the shaded area of their graphs, and to describe those that would fall outside that area. This activity involves the visual skills of *correlation* and *communication*.

5.

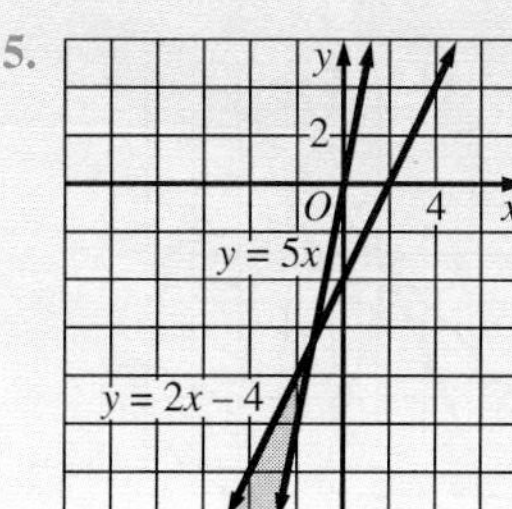

6.

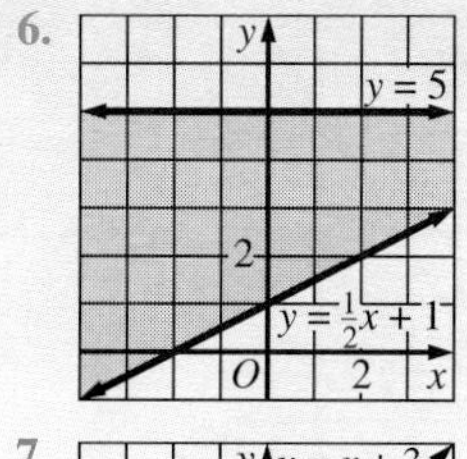

7.

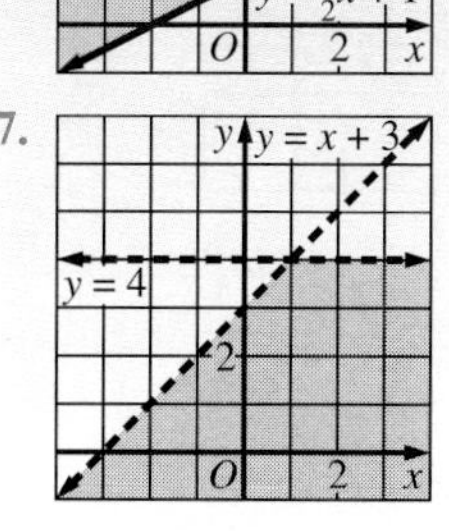

8.

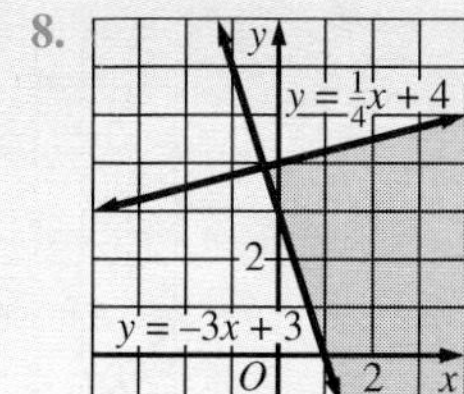

9. Answers may vary. Examples are given.

a. the region that is the solution of the system $t \le \frac{55}{3}l + \frac{185}{6}$, $t \le 95$, $t \le \frac{55}{3}l - \frac{145}{6}$, $t \ge 40$

b. from about 40 min to about 68 min; Determine the t-values inside the parallelogram that correspond to l-values close to 2.

10. a. Let p = the number of primary sources and s = the number of secondary sources; $p + s \ge 12$, $p \ge 4$.

b. Either variable may be the independent variable; this graph uses p as the independent variable. Dots correspond to points with whole-number coordinates that make sense for this situation.

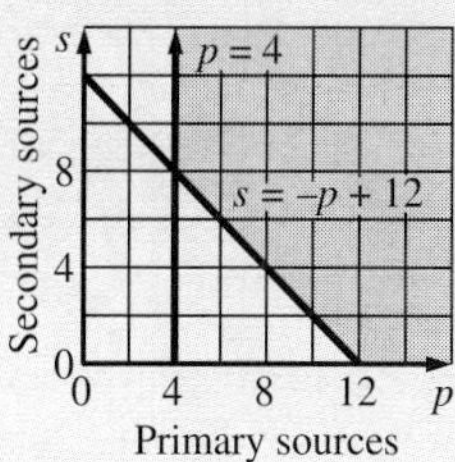

11. $x \le 0, y \ge 3, y \le \frac{1}{2}x + 4$

12. $-3 < x < 3, -2 < y < 2$

13. $x \ge 0, y \ge 0, y \le -x + 4$

14–18. See answers in back of book.

Exercise Notes

Interdisciplinary Problems
Exs. 19–21 The study of weather and climate has only recently been subject to mathematical analysis in a way that holds great promise for understanding and modeling these phenomena at a basic scientific level. The development of the theory of chaos and fractal geometry during the past two decades has given physicists and mathematicians the tools they need to discover the patterns that apply to the physical nature of weather and climate, and to model and analyze these patterns mathematically.

Using Technology
Exs. 22–29 Students may find it helpful to make a rough sketch to help them decide which region to shade. Graphing calculators and the *Function Investigator* software can help. Note that the TI-82 DRAW menu can be used to draw vertical lines in the coordinate plane. The graphing window settings for the x-axis should ensure that the value a is between Xmin and Xmax settings if the vertical line to be drawn is $x = a$. For example, to draw $x = -2$, use the standard settings for the graphing window. Go to a new line on the home screen and display Vertical by pressing [2nd][DRAW]4. Then press [(–)]2[ENTER]. The vertical line will be displayed along with any other graphs that may already be on the screen.

Connection METEOROLOGY

Weather is defined as the condition of the atmosphere during a brief period. *Climate* is the characteristic weather in an area over a long period of time. Climate variables include the variation in temperature from month to month. The table gives the average monthly high and low temperatures in Tokyo, Japan.

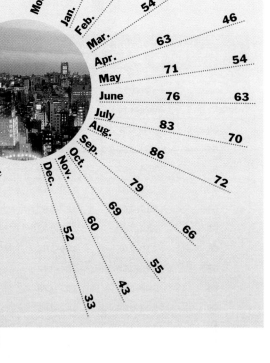

Month	Average high (°F)	Average low (°F)
Jan.	47	29
Feb.	48	31
Mar.	54	36
Apr.	63	46
May	71	54
June	76	63
July	83	70
Aug.	86	72
Sep.	79	66
Oct.	69	55
Nov.	60	43
Dec.	52	33

JAPAN Tokyo

19. Use the data in the table to create a graph showing the range of average temperatures in Tokyo. Put time, measured in months since January, on the horizonal axis, and temperature, measured in degrees Fahrenheit, on the vertical axis.

20. In what months would you expect the temperature to be above 50°? below 60°?

21. **Open-ended Problem** Use an almanac, atlas, or other reference book to find average monthly high and low temperatures for another location. Make a graph, and compare it with your graph from Exercise 19.

BY THE WAY

The climate of the Japanese islands is moderated by the sea. Winters are milder and precipitation is heavier than for locations on the Asian mainland at the same latitude.

Graph each system of inequalities.

22. $y < 7$
$x + y \geq 2$
$y \leq x$

23. $y \leq -2x$
$y \leq \frac{2}{3}x + 5$

24. $y < \frac{3}{2}x + 6$
$y \leq \frac{1}{2}x + 6$
$y \leq x + 8$

25. $x - 2y \geq -2$
$2x + y \geq 0$
$2x - y \leq 8$

26. $y \geq \frac{1}{2}x + 1$
$y \leq -1$

27. $y \leq \frac{3}{2}x + 4$
$y \geq -x - 6$
$x \leq 6$

28. $y \geq x$
$y \leq 2x$
$y \geq -3x - 1$

29. $y \geq \frac{1}{4}x + 3$
$y \leq x + 8$
$y \leq 6$

30. **SAT/ACT Preview** Which one of the following points is included in the region defined by $-2 \leq x < 8$ and $4 < y \leq 12$?

A. $(-2, 7)$ **B.** $(-2, 4)$ **C.** $(-2, -2)$ **D.** $(8, 7)$ **E.** $(8, 12)$

19–23. See answers in back of book.

24.
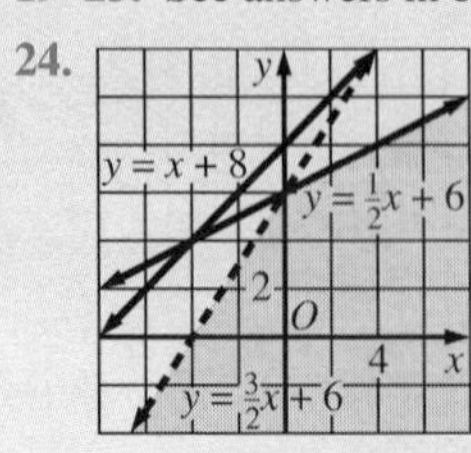

25.
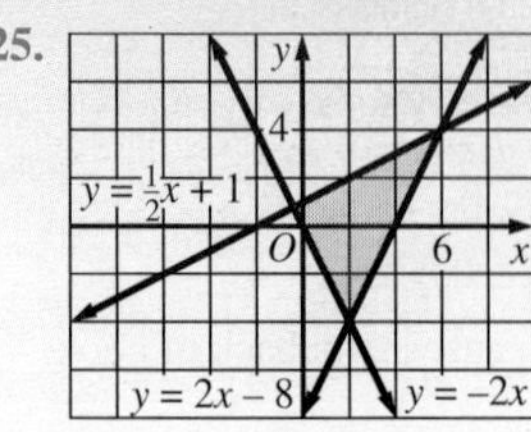

26.
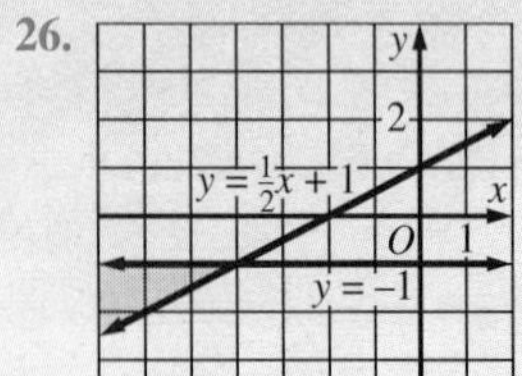

27.
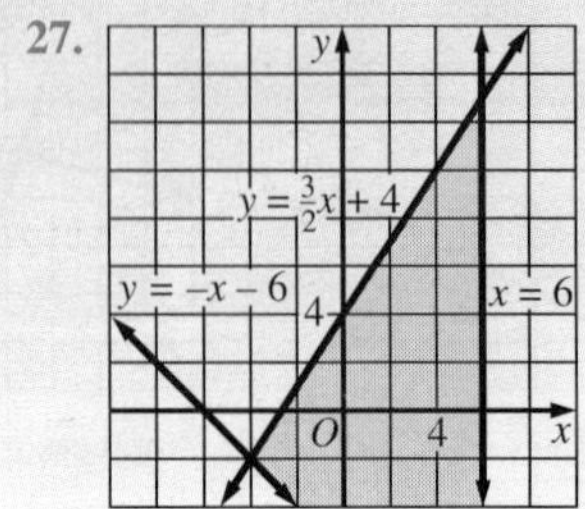

28.
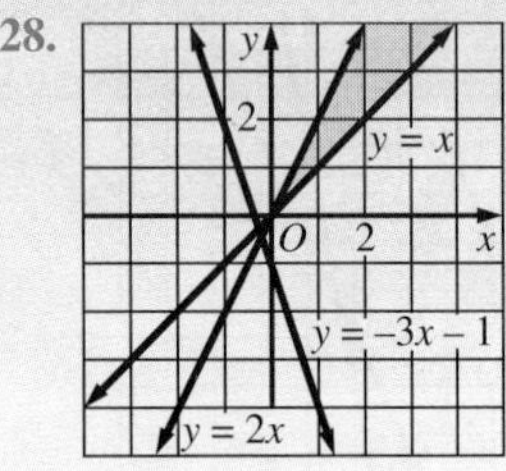

Write a system of inequalities defining each shaded region.

31.

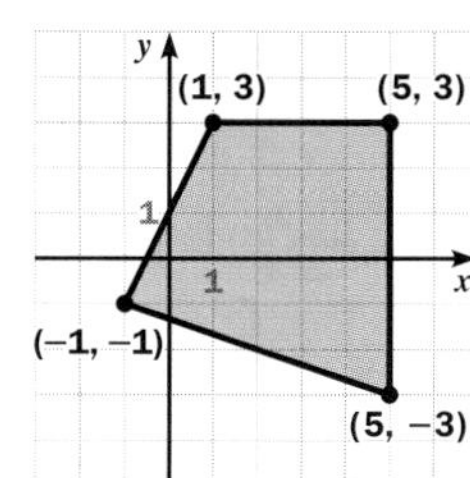

32.

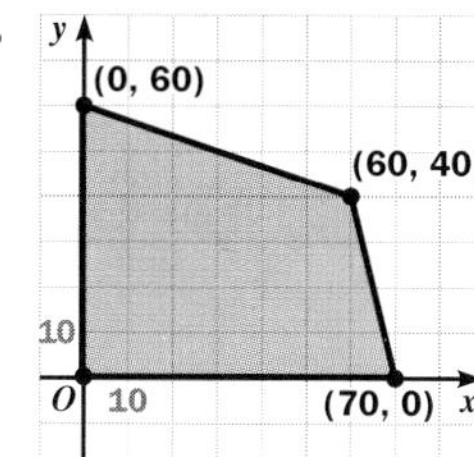

33.

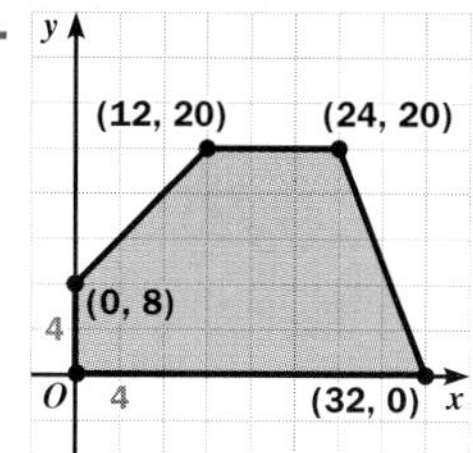

34. Challenge Graph the region defined by the inequalities $y \le \frac{1}{2}x^2 + 6$ and $y \ge 3x^2 - 4$.

35. CONSUMER ECONOMICS An international telephone call from the United States to most countries in Asia costs between \$1.56 and \$5.58 for the first minute. Each additional minute costs between 89% and 95% as much as the first minute.

a. Write inequalities to represent the relationship between the cost of the first minute, f, and the cost of each additional minute, a.

b. Graph the system of inequalities you found in part (a). Put the first-minute cost on the horizontal axis and the additional-minute cost on the vertical axis. Label the lines on your graph.

c. The first minute of a call to Singapore costs \$1.73. What might you expect to pay for each additional minute?

ONGOING ASSESSMENT

36. Cooperative Learning Work with a partner. One of you should use an almanac to find the normal monthly temperature and precipitation for the area where you live. The other should find the normal monthly temperature and precipitation for another area. Each of you should then draw a scatter plot of temperature and precipitation for your chosen area using the same scale. Draw a polygon enclosing the plotted points, and find the inequalities defining the interior of the polygon. Compare the two graphs.

SPIRAL REVIEW

Graph each inequality. *(Section 7.4)*

37. $5x + 3y < 18$ **38.** $x - 4y > 8$ **39.** $-2y \ge 5$ **40.** $-7x - 2y \le 12$

Solve each system of equations. *(Section 7.1)*

41. $2x + y = 15$; $x + y = 10$

42. $y = 3.2x$; $1.5x - 0.5y = 0.8$

43. $-x - y = 11$; $x + 3y = -3$

44. $5x + 4y = 7$; $3x = 9$

Find the *x*-intercepts of the graph of each equation. *(Section 5.3)*

45. $y = 3(x - 1)(x + 4)$ **46.** $y = (6x + 3)(x - 2)$ **47.** $y = (x + 1)(x - 1)$

Apply⇔Assess

Exercise Notes

Integrating the Strands
Exs. 31–33 Students should be able to see that any convex polygonal region can be described by a system of linear inequalities. Ask how the number of inequalities needed to describe such a region is related to the number of sides of the region. (They are equal.)

Practice 44 for Section 7.5

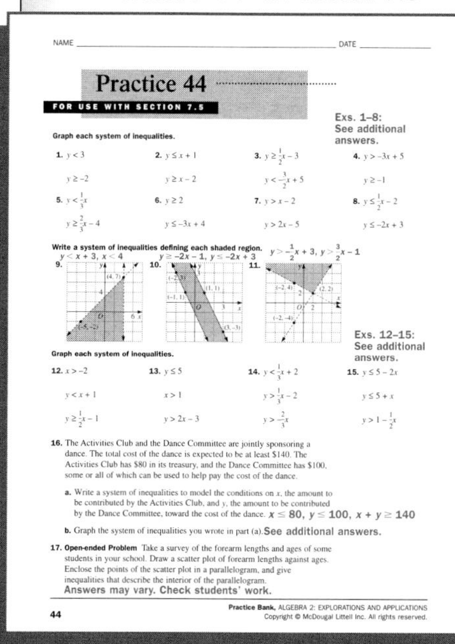

NAME ______ DATE ______

Practice 44

FOR USE WITH SECTION 7.5

Exs. 1–8: See additional answers.

Graph each system of inequalities.

1. $y < 3$, $y \ge -2$ **2.** $y \le x + 1$, $y \ge x - 2$ **3.** $y \ge \frac{1}{2}x - 3$, $y < -\frac{3}{2}x + 5$ **4.** $y > -3x + 5$, $y \ge -1$

5. $y < \frac{1}{3}x$, $y \ge \frac{2}{3}x - 4$ **6.** $y \ge 2$, $y \le -3x + 4$ **7.** $y > x - 2$, $y > 2x - 5$ **8.** $y \le \frac{1}{2}x - 2$, $y \le -2x + 3$

Write a system of inequalities defining each shaded region.

9. $y < x + 3$, $x < 4$ **10.** $y \ge -2x - 1$, $y \le -2x + 3$ **11.** $y > -\frac{1}{2}x + 3$, $y > \frac{3}{2}x - 1$

Exs. 12–15: See additional answers.

Graph each system of inequalities.

12. $x > -2$, $y < x + 1$, $y \ge \frac{1}{2}x - 1$ **13.** $y \le 5$, $x > 1$, $y > 2x - 3$ **14.** $y < \frac{1}{3}x + 2$, $y > \frac{1}{3}x - 2$, $y > -\frac{2}{3}x$ **15.** $y \le 5 - 2x$, $y \le 5 + x$, $y > 1 - \frac{1}{2}x$

16. The Activities Club and the Dance Committee are jointly sponsoring a dance. The total cost of the dance is expected to be at least \$140. The Activities Club has \$80 in its treasury, and the Dance Committee has \$100, some or all of which can be used to help pay the cost of the dance.

a. Write a system of inequalities to model the conditions on x, the amount to be contributed by the Activities Club, and y, the amount to be contributed by the Dance Committee, toward the cost of the dance. $x \le 80$, $y \le 100$, $x + y \ge 140$

b. Graph the system of inequalities you wrote in part (a). See additional answers.

17. Open-ended Problem Take a survey of the forearm lengths and ages of some students in your school. Draw a scatter plot of forearm lengths against ages. Enclose the points of the scatter plot in a parallelogram, and give inequalities that describe the interior of the parallelogram. Answers may vary. Check students' work.

44 **Practice Bank,** ALGEBRA 2: EXPLORATIONS AND APPLICATIONS

29.

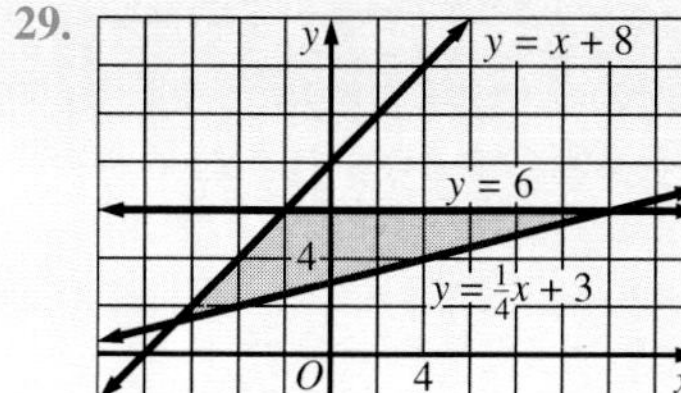

30. A

31. $x \le 5, y \le 3, y \ge -\frac{1}{3}x - \frac{4}{3}, y \le 2x + 1$

32. $x \ge 0, y \ge 0, y \le -\frac{1}{3}x + 60, y \le -4x + 280$

33. $x \ge 0, y \ge 0, y \le 20, y \le x + 8, y \le -\frac{5}{2}x + 80$

34.

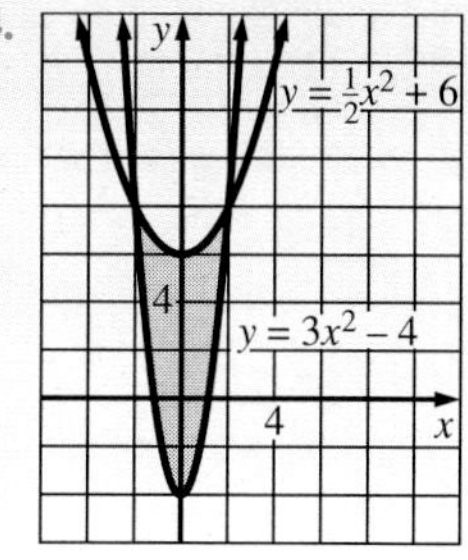

35. a. $a \ge 0.89f, a \le 0.95f$

b.

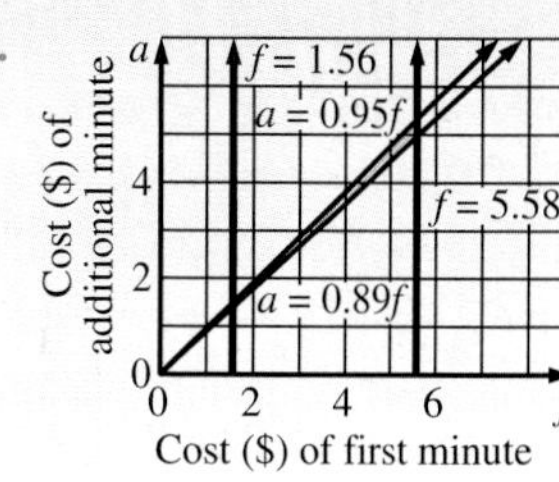

c. between \$1.54 and \$1.64

36. Answers may vary. Check students' work.

37.

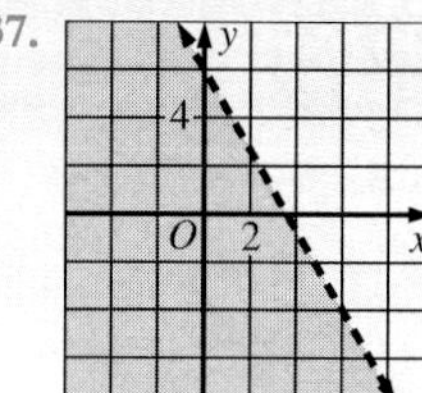

38–40. See answers in back of book.

41. (5, 5) 42. (−8, −25.6)

43. (−15, 4) 44. (3, −2)

45. 1, −4 46. $-\frac{1}{2}$, 2

47. ±1

Plan⇔Support

Objectives

- Find the best solution when several conditions have to be met.
- Use linear programming techniques to solve real-world problems.

Recommended Pacing

Core and Extended Courses
Section 7.6: 1 day

Block Schedule
Section 7.6: $\frac{1}{2}$ block
(with Portfolio Project)

Resource Materials

Lesson Support
Lesson Plan 7.6
Warm-Up Transparency 7.6
Practice Bank: Practice 45
Study Guide: Section 7.6
Exploration Lab Manual: Diagram Masters 1, 2
Assessment Book: Test 27

Technology
Graphing Calculator
Internet: http://www.hmco.com

Warm-Up Exercises

Solve each system of equations.

1. $2x - 3y = 6$
   $-4x + 5y = 10$ (−30, −22)
2. $x = 7$
   $3x - 2y = 11$ (7, 5)
3. $0.5x + 0.8y = 20$
   $2.5x + y = 10$ (−8, 30)

Evaluate $9x + 4y$ for each pair of values of x and y.

4. $x = 7, y = 8$ 95
5. $x = 0, y = -5$ −20
6. $x = -2, y = -3$ −30

Section

7.6 Linear Programming

Learn how to...

- find the best solution when several conditions have to be met

So you can...

- solve problems, such as how many of two types of quilt to stock

Quilting—the art of stitching layers of fabric together—may have originated with the Chinese. Patchwork quilting began in the 1700s when colonial women pieced together scraps of material to make blankets. There are now about 15.5 million quilters and 3000 quilting stores in the United States.

EXAMPLE 1 Application: Business

Elizabeth Ferris owns a quilting store. In addition to quilting supplies, she stocks queen-size and baby quilts. She wants to order up to 50 quilts to display and sell this fall. She knows she should have at least as many queen-size quilts as baby quilts, and she wants to have at least 10 baby quilts. Make a graph to show the possible combinations of quilts she can order.

SOLUTION

Step 1 Express each *constraint* as an inequality.
Let x = the number of queen-size quilts, and let y = the number of baby quilts.

Constraint 1: She will order up to 50 quilts.	**Constraint 2:** She should have at least as many queen-size quilts as baby quilts.	**Constraint 3:** She wants to have at least 10 baby quilts.
$x + y \le 50$, or $y \le -x + 50$	$x \ge y$, or $y \le x$	$y \ge 10$

Step 2 Graph the constraints.

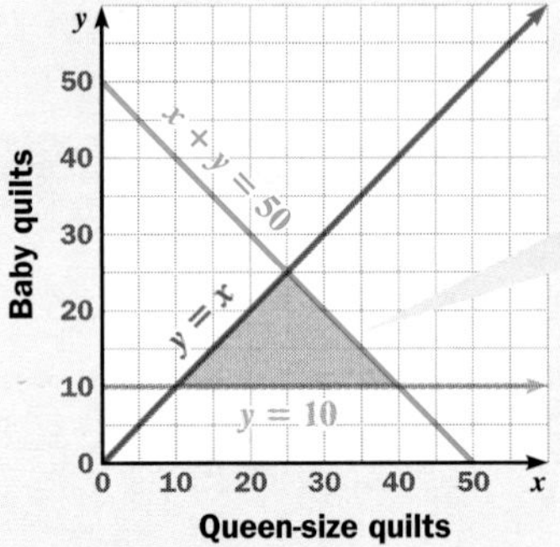

The shaded region contains all possible combinations of quilts that Elizabeth Ferris can order.

In Example 1, suppose Elizabeth Ferris prices her quilts so as to make a profit of \$125 on every queen-size quilt and \$25 on every baby quilt. Then an equation giving her total profit P on the quilts is:

$$P = 125x + 25y$$

The graph from Example 1, which is called a **feasible region**, is shown at the right along with several *profit lines* drawn in red. Each line represents a particular value of P.

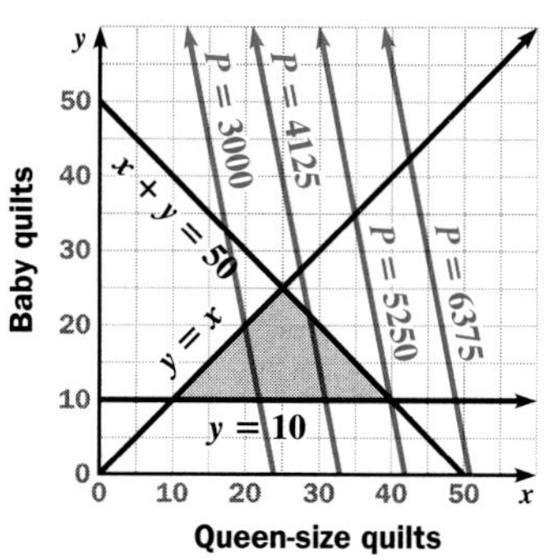

THINK AND COMMUNICATE

1. Confirm that the point (30, 15) is in the feasible region shown above. Which profit line is it on? Name another point in the feasible region that is on the same profit line.
2. For each profit line, name a point with integer coordinates that is on the line and in the feasible region, if possible.
   a. $P = 3000$ b. $P = 5250$ c. $P = 6375$
3. At what point in the feasible region is the profit greatest? Explain.

You often want to find the best solution to a problem, such as the number of quilts that will maximize profit. The **corner-point principle** states that for a polygonal feasible region, the best solution will be at a corner point of the feasible region. The method of finding the best solution is called **linear programming**.

EXAMPLE 2 Application: Business

Suppose Elizabeth Ferris makes a \$100 profit on a queen-size quilt and a \$35 profit on a baby quilt. Given the constraints in Example 1, how many quilts of each type should she order to maximize her profit?

SOLUTION

Step 1 Write an equation describing the profit she earns.

$$P = 100x + 35y$$

Step 2 Check the corner points of the feasible region in the profit equation.

Corner point	Value of P
(10, 10)	P = 100(10) + 35(10) = 1350
(25, 25)	P = 100(25) + 35(25) = 3375
(40, 10)	P = 100(40) + 35(10) = 4350

For the maximum profit, \$4350, she should order **40 queen-size quilts** and **10 baby quilts**.

BY THE WAY

Linear programming was developed after World War II by mathematicians, economists, and others to help choose the best solutions to problems of industry, such as which combination of cost-cutting measures would save the most money.

ANSWERS Section 7.6

Think and Communicate

1. (30, 15) satisfies all three constraints: $15 \le -30 + 50$, $15 \le 30$, and $15 \ge 10$; $P = 4125$; (31, 10) or (29, 20).
2. a. (20, 20), (21, 15), (22, 10)
   b. (40, 10)
   c. not possible
3. (40, 10); The greatest profit is achieved by selling the maximum number of queen-size quilts and the minimum number of baby quilts.

Teach⇔Interact

Additional Example 1

A bedding store places an order with a supplier for king-size sheets. The order is for no more than 90 sheets. The store orders at least 50% as many form-fitted sheets as regular sheets. The store orders 20 or more of the regular sheets. Make a graph to show the possible combinations of king-size sheets the store can order.

Step 1 Express each constraint as an inequality. Let x = the number of regular sheets, and let y = the number of form-fitted sheets.

Constraint 1: The order is for no more than 90 sheets.
$x + y \le 90$, or $y \le -x + 90$

Constraint 2: The store orders at least 50% as many form-fitted sheets as regular sheets. $y \ge 0.5x$

Constraint 3: The order is for 20 or more regular sheets. $x \ge 20$

Step 2 Graph the constraints.

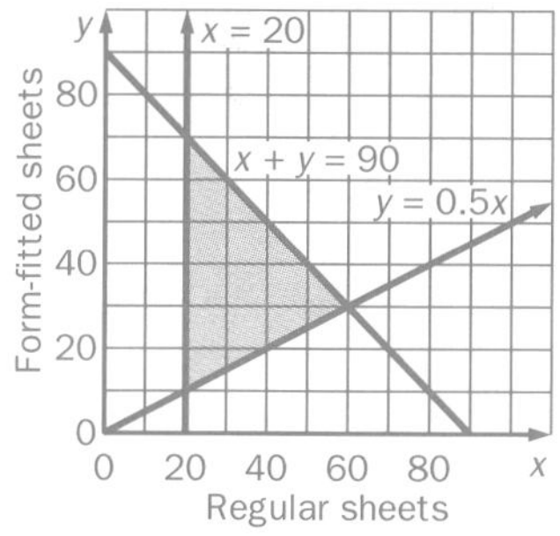

The shaded region contains all possible combinations of sheets that the store can order.

Additional Example 2

Refer to Additional Example 1. Suppose the store makes a profit of \$4 on each regular king-size sheet and \$4.80 on each form-fitted sheet. Given the constraints in Additional Example 1, how many sheets should the store order to maximize its profit?

Step 1 Write an equation describing the profit the store earns.
$P = 4x + 4.8y$

Step 2 Check the corner points of the feasible region in the equation.

Corner point	Value of P
(20, 70)	P = 4(20) + 4.8(70) = 416
(60, 30)	P = 4(60) + 4.8(30) = 384
(20, 10)	P = 4(20) + 4.8(10) = 128

For the maximum profit, \$416, the store should order 20 regular sheets and 70 form-fitted sheets.

Teach⇔Interact

Additional Example 3

Chu and Sally are making a mixture of nuts for a party. They plan to make from 5 lb to 9 lb of the mix. The mixture is to contain almonds and peanuts, with at least $\frac{1}{2}$ as many pounds of almonds as peanuts but not more than $\frac{3}{4}$ as many pounds of almonds as peanuts. They can get peanuts for \$2.75 per pound and almonds for \$4.29 per pound. How many pounds of these kinds of nuts should they buy to minimize the cost of the mixture?

Step 1 Write inequalities to represent the constraints. Let x = the number of pounds of peanuts, and let y = the number of pounds of almonds.

Constraint 1: Make from 5 lb to 9 lb of the mix.

$5 \le x + y \le 9$, or $y \ge 5 - x$ and $y \le 9 - x$

Constraint 2: Use at least $\frac{1}{2}$ as many almonds as peanuts.

$y \ge \frac{1}{2}x$

Constraint 3: Use no more than $\frac{3}{4}$ as many almonds as peanuts.

$y \le \frac{3}{4}x$

Step 2 Graph the feasible region and find the corner points. You can solve a system of two linear equations to find each corner point.

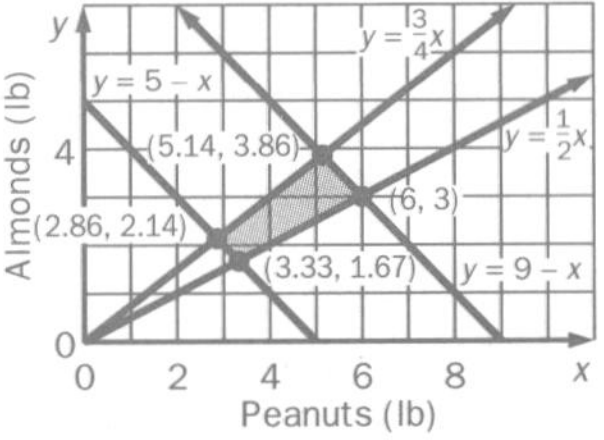

Step 3 Write an expression for the cost function and check the corner points.

$C = 2.75x + 4.29y$

Corner point	Value of C
(2.86, 2.14)	$C = 2.75(2.86) + 4.29(2.14) \approx 17.05$
(5.14, 3.86)	$C = 2.75(5.14) + 4.29(3.86) \approx 30.69$
(6, 3)	$C = 2.75(6) + 4.29(3) = 29.37$
(3.33, 1.67)	$C = 2.75(3.33) + 4.29(1.67) \approx 16.32$

Chu and Sally should purchase 3.33 lb $\left(3\frac{1}{3}\text{ lb}\right)$ of peanuts and 1.67 lb $\left(1\frac{2}{3}\text{ lb}\right)$ of almonds, for a minimum cost of \$16.32.

EXAMPLE 3 Application: Consumer Economics

Julian and Yoshi are making punch for a party using papaya juice and apple juice. They want to make between 12 qt and 15 qt of the punch. They know from experience that there should be at least twice as much, but not more than three times as much, apple juice as papaya juice.

Apple juice costs \$2.49 per quart. Papaya juice costs \$1.53 per quart. How much apple juice and papaya juice should Julian and Yoshi buy to minimize the cost of the punch?

SOLUTION

Step 1 Write inequalities to represent the constraints. Let x = the number of quarts of apple juice, and let y = the number of quarts of papaya juice.

Contraint 1: Make 12 qt to 15 qt of punch.	**Contraint 2:** Use at least twice as much apple juice as papaya juice.	**Contraint 3:** Use no more than three times as much apple juice as papaya juice.
$12 \le x + y \le 15$, or $y \ge 12 - x$ and $y \le 15 - x$	$x \ge 2y$, or $y \le \frac{1}{2}x$	$x \le 3y$, or $y \ge \frac{1}{3}x$

Step 2 Graph the feasible region and find the corner points. You can solve a system of two equations to find each corner point. For example, the graphs of $y = 12 - x$ and $y = \frac{1}{3}x$ intersect at (9, 3).

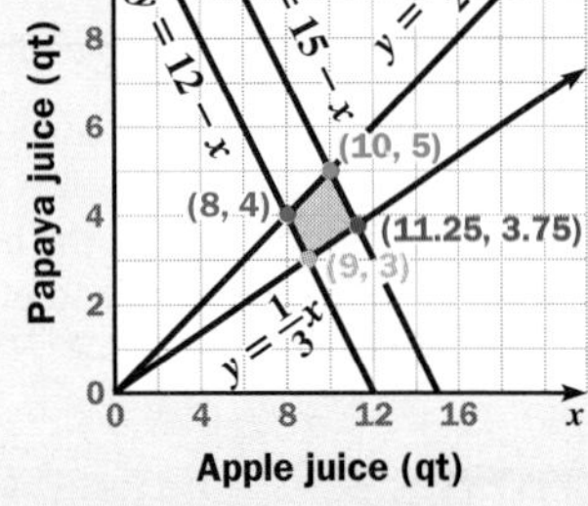

Step 3 Write an expression for the cost function and check the corner points.

$$C = 2.49x + 1.53y$$

Corner point	Value of C
(8, 4)	$C = 2.49(8) + 1.53(4) = 26.04$
(10, 5)	$C = 2.49(10) + 1.53(5) = 32.55$
(11.25, 3.75)	$C = 2.49(11.25) + 1.53(3.75) = 33.75$
(9, 3)	$C = 2.49(9) + 1.53(3) = 27.00$

Julian and Yoshi should purchase 8 qt of apple juice and 4 qt of papaya juice, for a minimum cost of \$26.04.

THINK AND COMMUNICATE

4. In Example 3, suppose the cost lines were parallel to an edge of the feasible region. Does the corner-point principle still work? Explain.

Think and Communicate

4. The best solution occurs at corner-points as well as at other points along one of the edges of the feasible region.

Checking Key Concepts

1. Let x = the number of oatmeal cookies and y = the number of chocolate chip cookies; $x \ge 24$, $y \ge 24$, $y \ge -x + 120$, $y \le -x + 144$.

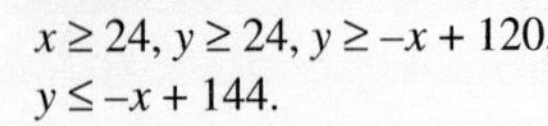

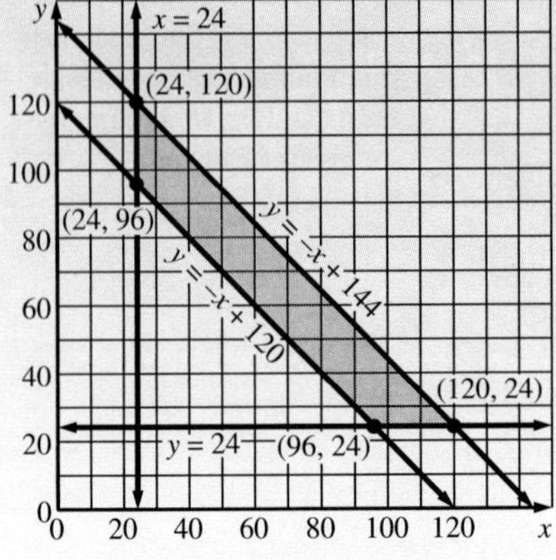

2. 120 oatmeal cookies, 24 chocolate chip cookies
3. 96 oatmeal cookies, 24 chocolate chip cookies

☑ Checking Key Concepts

Jenn and Kate are making oatmeal and chocolate chip cookies for a bake sale. They want to make 120 to 144 cookies. They want at least 24 of each kind.

1. Write constraints for this situation, and graph the feasible region. Label the edges and corner points.

2. Suppose they make a profit of \$.20 per oatmeal cookie and \$.15 per chocolate chip cookie. How many of each type of cookie should they make to maximize their profit?

3. Suppose their cost is \$.20 per oatmeal cookie and \$.25 per chocolate chip cookie. How many of each type of cookie should they make to minimize their cost?

7.6 Exercises and Applications

Extra Practice *exercises on page 760*

Graph each feasible region.

1. $y \le 14$
$y \ge \frac{1}{2}x + 2$
$x \ge 1$

2. $y \ge 2$
$y \le x$
$y \le 22 - 2x$

3. $y \ge 5 - x$
$y \ge \frac{1}{3}x + 1$
$y \le 3 + x$
$y \le 15 - 2x$

4. $y \ge 0$
$y \le 4x$
$y \ge 2x - 14$
$y \le 8$

5–8. For each of the feasible regions you graphed in Exercises 1–4, find the maximum profit for the profit function $P = 5x + 4y$.

9–12. For each of the feasible regions you graphed in Exercises 1–4, find the minimum cost for the cost function $C = 3x + 3y$.

13. BUSINESS A potter is preparing to make serving bowls and plates. A serving bowl uses 5 lb of clay. A plate uses 4 lb of clay. She has 40 lb of clay and wants to make at least 4 serving bowls.

a. Let x = the number of serving bowls, and let y = the number of plates. Write the constraints for this situation in terms of x and y.

b. Graph the feasible region. Label the corner points.

c. If the profit on a serving bowl is \$35 and the profit on a plate is \$30, how many bowls and plates should she make to maximize her profit?

14. ART Laura is making a necklace 15 in. to 18 in. long. She will use two types of beads: brightly painted wooden ovals, 0.75 in. long and \$.65 apiece, and small animal shapes, 0.50 in. long and \$1.00 apiece. She wants to have at least three times as many ovals as animals, but not more than six times as many ovals as animals.

a. Write inequalities to represent the constraints for this situation.

b. Graph the feasible region. Label the corner points.

c. If the cord to string the beads costs \$.75, how many of each type of bead should she buy to minimize her cost?

Exercises and Applications

1.

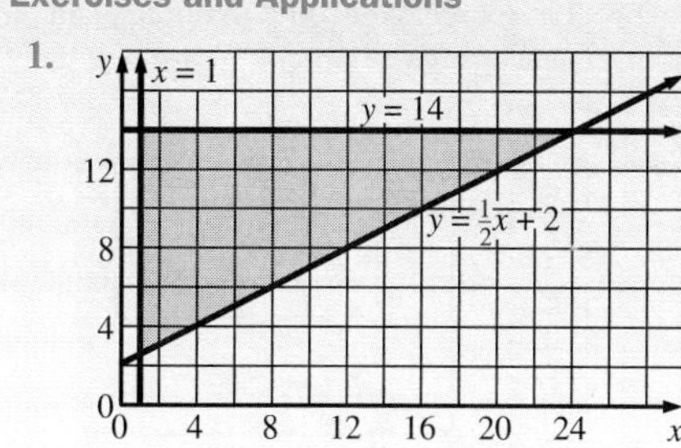

2.

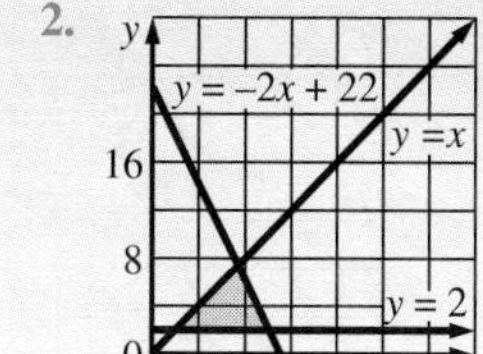

3.

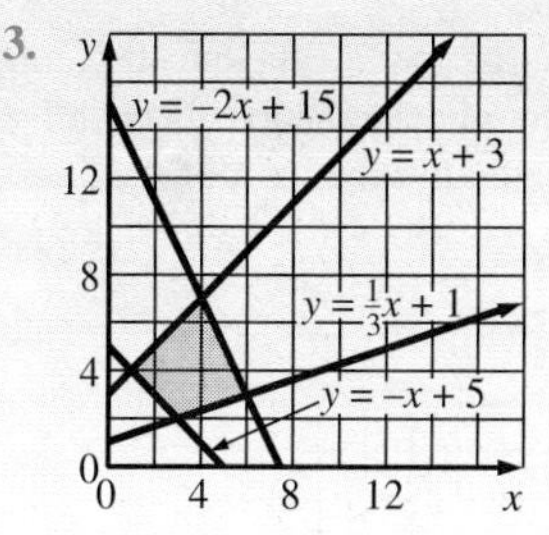

4–14. See answers in back of book.

Teach⇔Interact

Think and Communicate

Question 4 is important because it points out that best solutions may occur at points that are not corner points. They may be solutions that are as good as the corner-point solutions, but they will not be better than the corner-point solutions.

Closure Question

Describe what is involved in solving a linear programming problem. identifying constraints, modeling the constraints with inequalities, and graphing the system of inequalities to determine the feasible region; It is also necessary to identify the profit or cost function and to evaluate it using the coordinates of the corner points of the feasible region.

Apply⇔Assess

Suggested Assignment

Core Course
Exs. 1–14, 18–27, AYP

Extended Course
Exs. 1–27, AYP

Block Schedule
Day 45 Exs. 1–14, 18–27, AYP

Exercise Notes

Cooperative Learning
Exs. 1–12 You may wish to have students work with a partner on these exercises.

Historical Connection
Exs. 1–22 All of these exercises involve concepts of linear programming in one way or another. In essence, linear programming is an organized method of mathematical decision making that was developed after World War II. A key figure in this work was the American mathematician George Dantzig, who discovered what is called the *simplex method.* This method makes it possible to find the best corner point by evaluating only a small fraction of all corners, thus saving an enormous amount of time and computer power.

Apply⇔Assess

Exercise Notes

Common Error

Ex. 13 Some students may forget to include the constraint $y \geq 0$. Suggest that they pick a point with a negative y-coordinate. Ask if it makes sense for this point to be part of the feasible region. Why or why not? (No; you cannot have a negative number of plates.)

Using Technology

Exs. 18–21 Students can use the statistical list features of the TI-82 graphing calculator to do the calculations for these exercises. Enter the x-coordinates for the corner points in list L_1 and the corresponding y-coordinates in list L_2. Use the ▶ and ▲ keys to highlight L_3 at the top of the L_3-column. Then, for Exs. 18 and 20, type $3L_1+5L_2$ and press ENTER. The calculator will compute the value of $3x + 5y$ for the five corner points and display the results in list L_3. The greatest number in L_3 gives the answer for Ex. 18, and the smallest number is the answer for Ex. 20. For Exs. 19 and 21, position the cursor on L_3, type $5L_1+3L_2$, and press ENTER. Look for maximum and minimum values among the new numbers in L_3.

Second-Language Learners

Ex. 22 Students learning English may benefit from being able to respond to this item orally and show examples in order to demonstrate their understanding.

INTERVIEW Gina Oliva

Look back at the article on pages 288–290.

Fitness instructors like Gina Oliva suggest that you vary your workout with different exercises, to work different muscle groups and to avoid boredom with one routine. For Exercises 15–17, use the table below.

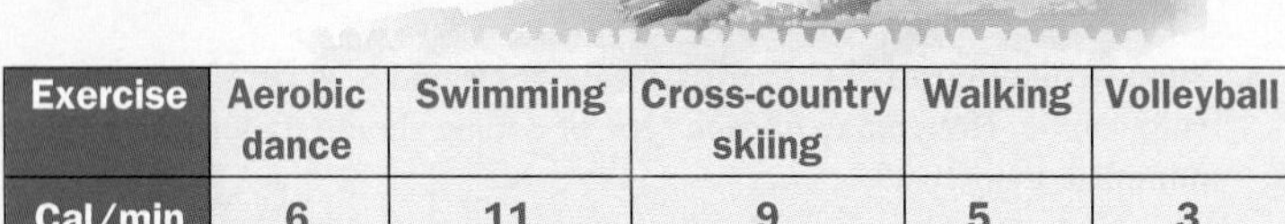

Exercise	Aerobic dance	Swimming	Cross-country skiing	Walking	Volleyball
Cal/min	6	11	9	5	3

15. Aaron wants to combine aerobics and swimming. He wants to work out for 3 h to 5 h (180 min to 300 min) a week. He wants to spend at least 1 h on each activity. He plans to burn at least 1200 Cal a week with this exercise program.

a. Write a system of inequalities to express Aaron's constraints. Let s = the number of minutes spent swimming and a = the number of minutes spent doing aerobics.

b. Graph the feasible region. Label the corner points.

c. How much time should Aaron spend on swimming and on aerobics if he wants to minimize the time spent exercising, but burn as many calories as possible in that time?

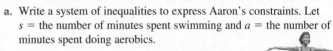

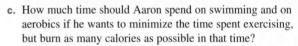

d. How much time should he spend on each exercise if he wants to maximize the number of calories burned?

16. Len wants to combine aerobics and cross-country skiing. He plans to exercise 6 h to 8 h a week this winter. He will spend at least 1 h but no more than 3 h each week doing aerobics and at least 4 h skiing.

a. Write a system of inequalities to express Len's constraints.

b. Graph the feasible region. Label the corner points.

c. How much time should Len spend on each activity if he wants to maximize the number of calories he burns?

17. Open-ended Problem Create a linear programming problem about combining two types of exercise. Then exchange your problem with someone in your class and solve each other's problems.

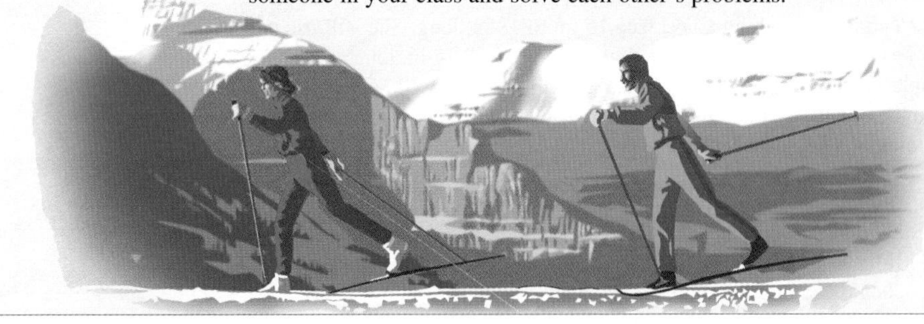

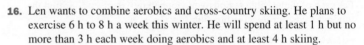

15–22. See answers in back of book.

23. $f^{-1}(x) = \frac{1}{2}x - \frac{7}{2}$

24. $g^{-1}(x) = x$

25. $h^{-1}(x) = \frac{1}{15}x + \frac{1}{5}$

26, 27. Answers may vary. Examples are given.

26. The question is not biased; however, the response may be biased by perceived expectations. To get an unbiased response, I would collect the information anonymously.

27. The first part of the question might encourage a negative answer. I would omit that part of the question as well as the word "another."

Assess Your Progress

1.

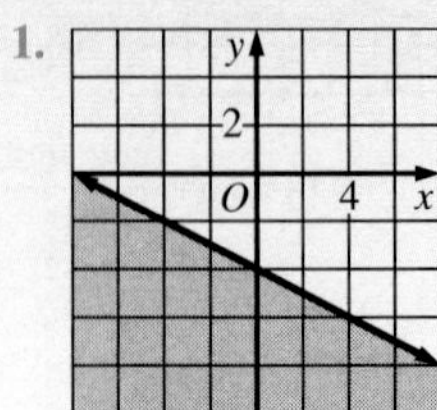

2.

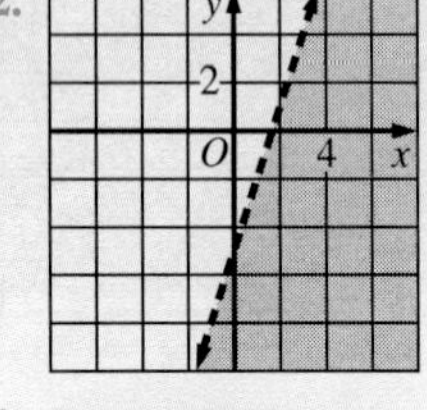

3.

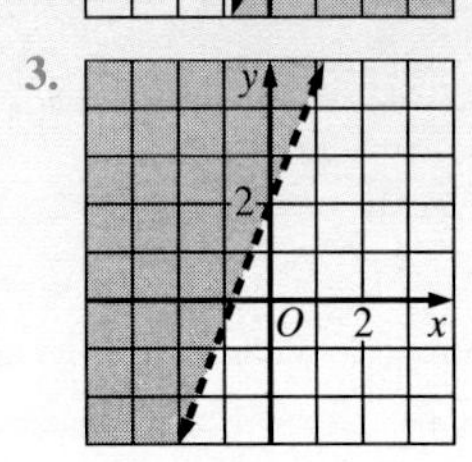

4. Let g = gross income in dollars and m = mortgage amount in dollars; $m \leq 0.28g$.

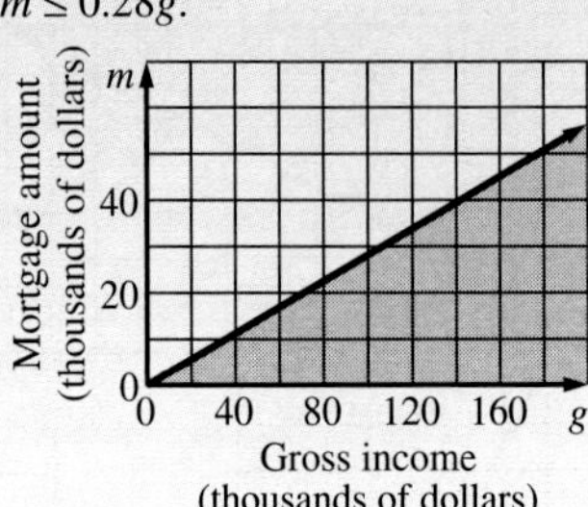

For Exercises 18–21, find the minimum cost or maximum profit for the feasible region shown.

18. $P = 3x + 5y$ **19.** $P = 5x + 3y$

20. $C = 3x + 5y$ **21.** $C = 5x + 3y$

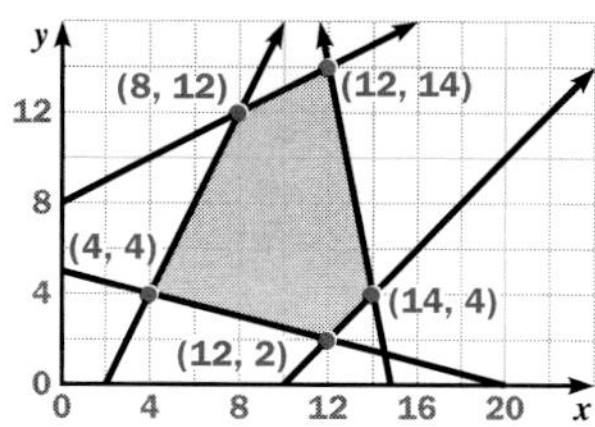

ONGOING ASSESSMENT

22. Writing Explain what a linear programming problem is and how to solve it.

SPIRAL REVIEW

For each function, find the inverse. *(Section 4.1)*

23. $f(x) = 2x + 7$ **24.** $g(x) = x$ **25.** $h(x) = 15x - 3$

Tell why each question might be biased. Then describe what changes you would make to improve the question. *(Section 6.2)*

26. "How much money do you donate to charity each year?"

27. "Given that state taxes were increased twice in the past decade, would you favor another state tax increase?"

ASSESS YOUR PROGRESS

VOCABULARY

half-planes (p. 310) **corner-point principle** (p. 323)
system of inequalities (p. 315) **linear programming** (p. 323)
feasible region (p. 323)

Graph each inequality. *(Section 7.4)*

1. $2x + 4y \le -16$ **2.** $3x - y > 5$ **3.** $-15x + 6y > 12$

4. PERSONAL FINANCE As a general rule, banks allow people to spend up to 28% of their gross incomes on a home mortgage. Write and graph an inequality for this situation. *(Section 7.4)*

Graph each system of inequalities. *(Section 7.5)*

5. $y \ge 2x + 1$
$y \le 4x$

6. $y \ge -2x + 1$
$y < 3x - 6$

7. $y \le 2x + 1$
$y < 4x$

Graph the feasible region and find the maximim profit if $P = 12x + 15y$. *(Section 7.6)*

8. $x \ge 10$
$y \le 18$
$2x - y \le 8$

9. $3x + y \le 13$
$x + 4y \ge 8$
$x \ge 0$

10. $0 \le y \le 12$
$x - y \le 8$
$x \ge 2$

11. Journal How are a system of linear equations and a system of linear inequalities alike? How are they different?

Apply⇔Assess

Exercise Notes

Assessment Note

Ex. 22 Encourage students to illustrate their general descriptions with an example of a linear programming problem. They may wish to use exercises from this section. They should identify constraints and show what function is to be maximized or minimized. A thorough answer to this exercise would make an excellent journal entry.

Progress Check 7.4–7.6

See page 331.

Practice 45 for Section 7.6

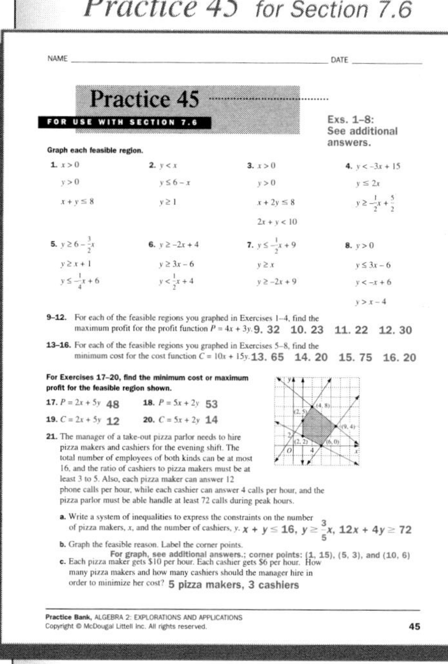

NAME ______________________ DATE __________

Practice 45

FOR USE WITH SECTION 7.6

Exs. 1–8: See additional answers.

Graph each feasible region.

1. $x > 0$
$y > 0$
$x + y \le 8$

2. $y < x$
$y \le 6 - x$
$y \ge 1$

3. $x > 0$
$y > 0$
$x + 2y \le 8$
$2x + y < 10$

4. $y < -3x + 15$
$y \le 2x$
$y \ge -\frac{1}{2}x + \frac{5}{2}$

5. $y \ge 6 - \frac{3}{2}x$
$y \ge x + 1$
$y \le -\frac{1}{4}x + 6$

6. $y \ge -2x + 4$
$y \ge 3x - 6$
$y < \frac{1}{2}x + 4$

7. $y \le -\frac{1}{2}x + 9$
$y \ge x$
$y \ge -2x + 9$

8. $y > 0$
$y \le 3x - 6$
$y < -x + 6$
$y > x - 4$

9–12. For each of the feasible regions you graphed in Exercises 1–4, find the maximum profit for the profit function $P = 4x + 3y$. **9. 32 10. 23 11. 22 12. 30**

13–16. For each of the feasible regions you graphed in Exercises 5–8, find the minimum cost for the cost function $C = 10x + 15y$. **13. 65 14. 20 15. 75 16. 20**

For Exercises 17–20, find the minimum cost or maximum profit for the feasible region shown.

17. $P = 2x + 5y$ **48** **18.** $P = 5x + 2y$ **53**

19. $C = 2x + 5y$ **12** **20.** $C = 5x + 2y$ **14**

(4, 8) (2, 5) (9, 4) (2, 2) (6, 0) 2 O 4 x y

21. The manager of a take-out pizza parlor needs to hire pizza makers and cashiers for the evening shift. The total number of employees of both kinds can be at most 16, and the ratio of cashiers to pizza makers must be at least 3 to 5. Also, each pizza maker can answer 12 phone calls per hour, while each cashier can answer 4 calls per hour, and the pizza parlor must be able handle at least 72 calls during peak hours.

a. Write a system of inequalities to express the constraints on the number of pizza makers, x, and the number of cashiers, y. **$x + y \le 16$, $y \ge \frac{3}{5}x$, $12x + 4y \ge 72$**

b. Graph the feasible reason. Label the corner points. **For graph, see additional answers.; corner points: (1, 15), (5, 3), and (10, 6)**

c. Each pizza maker gets \$10 per hour. Each cashier gets \$6 per hour. How many pizza makers and how many cashiers should the manager hire in order to minimize her cost? **5 pizza makers, 3 cashiers**

Practice Bank, ALGEBRA 2: EXPLORATIONS AND APPLICATIONS

45

5.

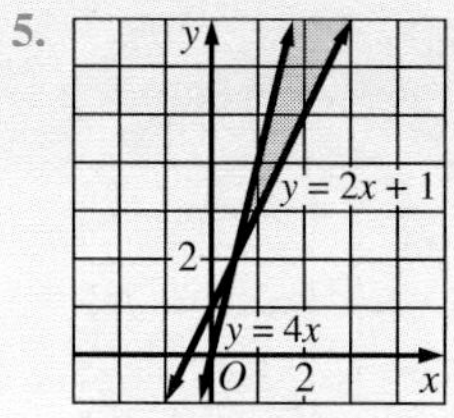

6.

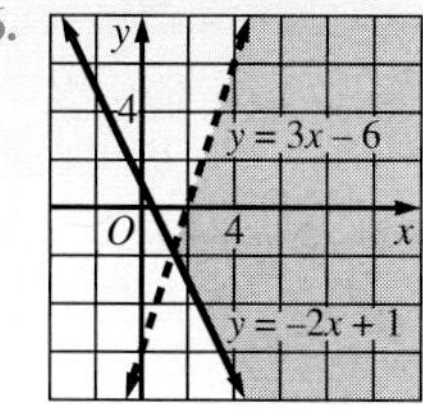

7. y = 2x + 1, y = 4x

8.

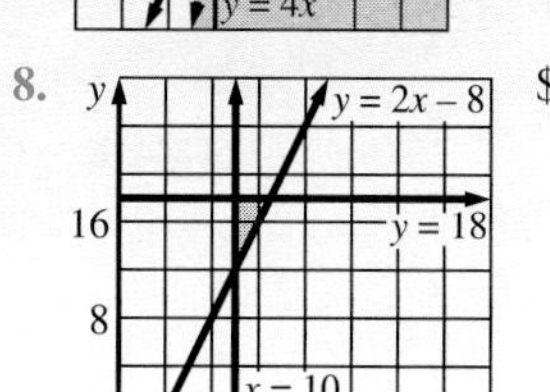

\$426

9. x = 0, y = −3x + 13, $y = -\frac{1}{4}x + 2$

\$195

10.

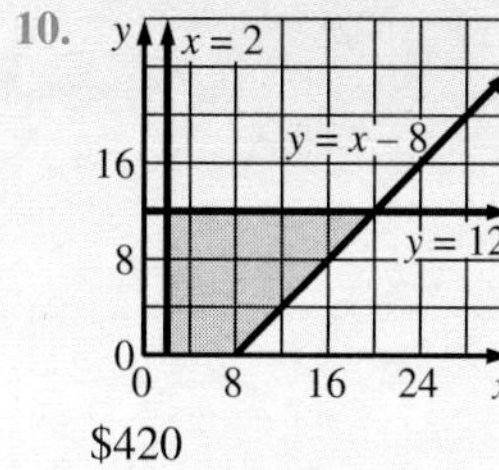

\$420

11. Answers may vary. An example is given. Solving either a system of linear equations or a system of linear inequalities involves finding all ordered pairs that are solutions of every equation or inequality in the system. The solution of a system of linear equations, if it exists, is either a point or a line. The solution of a system of linear inequalities, if it exists, is a region of the plane.

Project Notes

Mathematical Goals

- Collect and organize athletic data for both men and women.
- Display the data in a scatter plot.
- Calculate both linear and exponential regression models of the data.
- Predict future performances based upon the models.

Planning

Materials

- graphing calculator or statistical software
- Olympic data source

Project Teams

Students can decide on the members of their group and then work together to select, draw, and interpret the data. The team members work together to understand the equations and calculations of the project and how to complete the graphs.

Guiding Students' Work

Working together and discussing all aspects of the project can help students to understand the goals of the project and how to complete the necessary calculations. Encourage students to discuss their ideas within the groups. Also, it may help to review the meaning of linear and exponential equations as well as the shapes of their graphs. Discussing both exponential growth and exponential decay may help some students create their exponential model.

Second-Language Learners

The idiomatic expression *compete against the clock* may not be familiar to students learning English. If necessary, explain that it means "to try to accomplish a certain goal speedily, or before there is no time left."

PORTFOLIO PROJECT

Comparing Olympic Performances

In some Olympic events where both men and women compete against the clock, women's times, although not as fast as men's times, have improved more rapidly. This suggests that women may someday surpass men's times in these events. By choosing an appropriate model, you can make reasonable predictions about the future performances of men and women in comparable events.

PROJECT GOAL **Use a system of equations to model men's and women's performance data in an Olympic event.**

Collecting the Data

Work in a group of three students. You will need a graphing calculator or statistical software, and access to an Olympic data source, such as a sports almanac.

Find an event in which both men and women compete. Choose a timed event like swimming or running, or an event like the high jump or long jump, where distance is used to measure performance.

Olympic 100 m Freestyle

	Winning time (s)	
Year	Men	Women
1948	57.3	66.3
1952	57.4	66.8
1956	55.4	62.0
1960	55.2	61.2
1964	53.4	59.5
1968	52.2	60.0
1972	51.2	58.6
1976	50.0	55.7
1980	50.4	54.8
1984	49.8	55.9
1988	48.6	54.9
1992	49.0	54.6

Using Linear Models

Make a scatter plot of your data. Fit a line to each set of points, and find equations of the lines. For example, using data for the 100 m freestyle (see the table) with x = years since 1948 and y = winning times, you get the results shown.

Performing linear regression on the women's data gives $y = -0.283x + 65.4$.

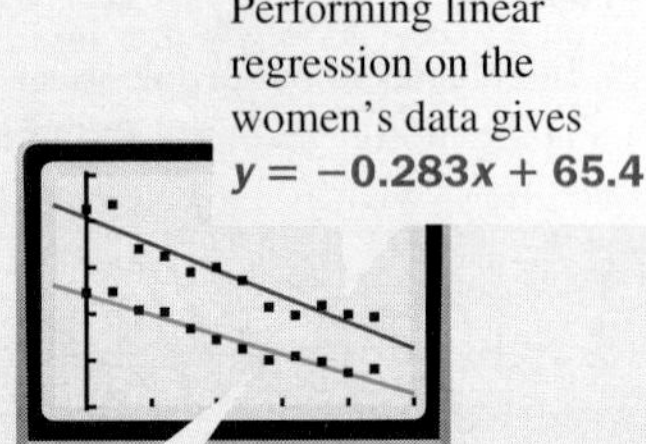

Performing linear regression on the men's data gives $y = -0.214x + 57.2$.

By extending your view of the graphs, you can see that they cross at about x = **120 years**.

Intersection X=119.76531 Y=31.562713

What does your model predict? Why might a linear equation be a poor model for athletic performance?

Using Exponential Models

Suppose you want to model winning times using an exponential decay equation of the form $y = ab^x$ (where $0 < b < 1$). You know y approaches 0 as x increases. Since you wouldn't expect performance times to decrease to 0, it makes sense to choose a reasonable lower bound for the data.

Try using a decay equation of the form $y = ab^x + c$, where the number c is the lower bound. (If you think your data have an upper bound, use a model of the form $y = c - ab^x$.) Suppose, for instance, that you don't expect men's or women's times for the 100 m freestyle to go below 40 s.

STEP 1 Find an exponential model that fits the data points (x, y') where $y' = y - 40$.

STEP 2 In the equation from Step 1, substitute $y - 40$ for y' and solve for y to obtain a model that fits the data points (x, y).

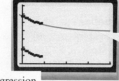

These are the data points (x, y) representing the men's times.

Performing exponential regression on the data points (x, y') gives $y' = 17.7(0.983)^x$.

The function $y = 17.7(0.983)^x + 40$ models the original data points. The graph does not fall below the line $y = 40$.

STEP 3 Repeat Steps 1 and 2 for the women's times, and graph the models for both data sets to see if they intersect.

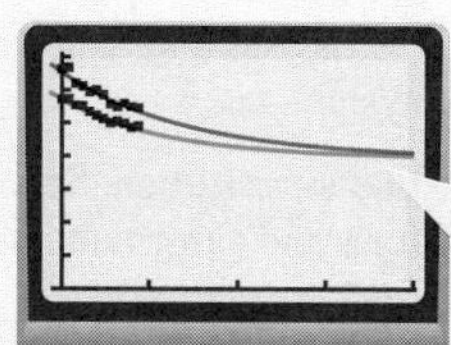

Over a span of 200 years, the graphs don't intersect but do get closer and closer.

Would you get different results if you changed the lower bound? What if you used different lower bounds for men and women?

Writing a Report

Write a report summarizing your results. Include these items: a table and a scatter plot of your data sets, graphs of your linear and exponential models, a comparison of future performances based on your models, and an evaluation of the limitations of each model. To extend your project, you may wish to compare results with other groups, or investigate the topic of human limits in sports.

Self-Assessment

Describe how comfortable you are modeling data and making predictions. What aspects of modeling are you unclear or unsure about?

Project Notes

Guiding Students' Work

Rubric for Chapter Project

4 Students collect and organize data from an Olympic sport and accurately draw scatter plots. The linear and exponential models that are drawn are done accurately. A report is written that includes graphs of the scatter plots and their regression equations, equations for the linear and exponential models and how they were found, and an evaluation of the models. The report is written clearly and an understanding of the mathematics of the project is conveyed. Students also extend the project by comparing results with other groups or discussing human limits.

3 Students collect and organize data and accurately draw scatter plots, but either the linear or the exponential model has at least one error in it. A report is written and an attempt is made to explain the equations and ideas of the project, but some errors are made in the analysis of the models. The report includes the table, scatter plots, graphs, models, and an evaluation of the models. Students also extend the project by comparing their result with other groups or by discussing human limits.

2 Students collect and organize data and draw the scatter plots but do not draw the linear and exponential models correctly or draw only one of the two models. Students write a report, but it does not include all of the graphs, all of the equations, and it only analyzes part of the data. An attempt is made to understand the meaning of the project and to predict future performances, but it is clear from the report that students do not have a good understanding of the key concepts involved. Students also do not extend the project.

1 Students collect data but do not organize it correctly. Both the linear model and the exponential model, if completed at all, are incorrect. If a report is written, it is incomplete because it does not contain graphs that explain the data. No evaluation of limits is done, and the project is not extended in any way. Students should be encouraged to speak with the teacher as soon as possible to review their work and to make a new start on the project.

Chapter Support

Course Guide: Chapter 7
Lesson Plans: Chapter 7
Practice Bank:
Cumulative Practice 46
Study Guide: Chapter 7 Review
Assessment Book:
Chapter Tests 28 and 29
Chapter 7 Alternative Assessment
Test Generator Software
Technology Book:
Calculator Based Lab 4
Portfolio Project Book
Preparing for College Entrance Tests
Professional Handbook

Progress Check 7.1–7.3

1. Use a graphing calculator or graphing software to solve the following system. *(Section 7.1)*
$y = (2.3)^x$
$y = 3x + 1$
(0, 1), (2.62, 8.86)
(Coordinates are rounded to the nearest hundredth.)

2. Use substitution to solve the following system. *(Section 7.1)*
$2x - y = 5$
$7x + y = -2$
$\left(\frac{1}{3}, -\frac{13}{3}\right)$

Tell whether each system has *one solution, infinitely many solutions,* or *no solution.* For those systems that have solutions, name at least one solution. *(Section 7.2)*

3. $-9x + 12y = 3$
$3x - 4y = -1$
infinitely many solutions; possible solution: (5, 4)

4. $-2x + y = 8$
$y = 2x + 7$ no solution

5. $7x + 3y = 21$
$x - 7y = 3$
one solution; (3, 0)

6. Write a matrix equation for the following system. *(Section 7.3)*
$25x - 12y = 30$
$-4x + 7y = 18$
$\begin{bmatrix} 25 & -12 \\ -4 & 7 \end{bmatrix}\begin{bmatrix} x \\ y \end{bmatrix} = \begin{bmatrix} 30 \\ 18 \end{bmatrix}$

7. Use a graphing calculator or software with matrix calculation capabilities to find an equation of the parabola passing through the points (–1, –14), (1, 10), and (3, –22). *(Section 7.3)*
$y = -7x^2 + 12x + 5$

CHAPTER

7 Review

STUDY TECHNIQUE

Form a study group with six members. Have each group member take one section of the chapter and write a "how to" synopsis for the major concepts in that section. Have each member present his or her synopsis to the group, and discuss.

VOCABULARY

system of equations (p. 292)
inconsistent system (p. 299)
dependent system (p. 299)
identity matrices (p. 304)
inverse matrix (p. 304)
half-planes (p. 310)
system of inequalities (p. 315)
feasible region (p. 323)
corner-point principle (p. 323)
linear programming (p. 323)

SECTIONS 7.1, 7.2, *and* 7.3

To solve a **system of equations**, you need to find the values of the variables that make both of the equations true. For example, you can use any of the four methods shown below to solve the system $3x - y = 7$ and $5x + 2y = -3$.

Use technology. Solve each equation for y and graph.

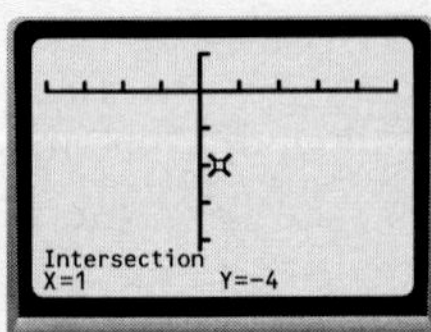

Use substitution. Solve the first equation for y and substitute into the second equation.

$5x + 2(3x - 7) = -3$ $\quad 3(1) - y = 7$
$11x = 11$ $\quad -y = 4$
$x = 1$ $\quad y = -4$

Use addition. Multiply the first equation by 2 and add the two equations.

$6x - 2y = 14$
$5x + 2y = -3$
$11x = 11 \rightarrow x = 1$
$5(1) + 2y = -3 \rightarrow y = -4$

Use matrices. Write the system as a matrix equation and solve using an **inverse matrix**.

$\begin{bmatrix} 3 & -1 \\ 5 & 2 \end{bmatrix}\begin{bmatrix} x \\ y \end{bmatrix} = \begin{bmatrix} 7 \\ -3 \end{bmatrix}$

$\begin{bmatrix} x \\ y \end{bmatrix} = \begin{bmatrix} 3 & -1 \\ 5 & 2 \end{bmatrix}^{-1}\begin{bmatrix} 7 \\ -3 \end{bmatrix} = \begin{bmatrix} 1 \\ -4 \end{bmatrix}$

SECTIONS 7.4, 7.5, *and* 7.6

The graphs of inequalities and **systems of inequalities** are regions of the coordinate plane.

Example: A town's citizens want to plant maple trees and spruce trees in the town square. They want at least 60 trees, and they have enough space for up to 100 trees. To minimize leaf pickup, they want at least 1.5 times as many spruce trees as maples. To provide fall color, they want at least 15 maples. You can follow the steps below to represent this situation graphically.

Step 1 Write a system of inequalities using *s* for the number of spruce trees and *m* for the number of maple trees.

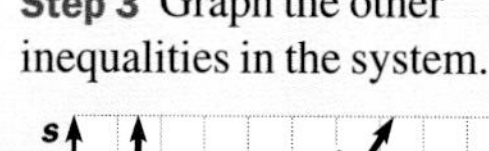

$m + s \leq 100$ $\quad s \geq 1.5m$
$m + s \geq 60$ $\quad m \geq 15$

Step 2 Graph one of the inequalities.

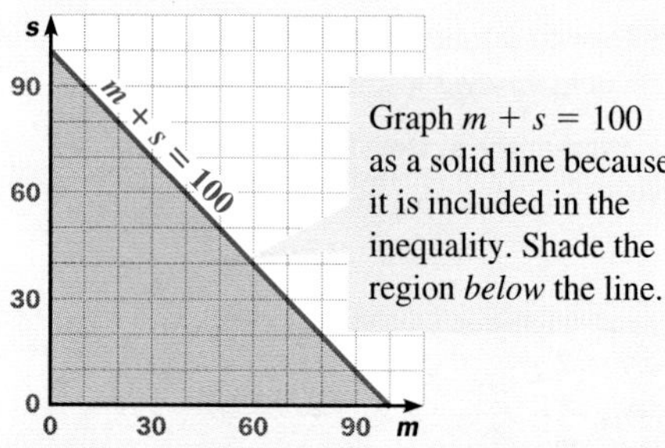

Graph $m + s = 100$ as a solid line because it is included in the inequality. Shade the region *below* the line.

Step 3 Graph the other inequalities in the system.

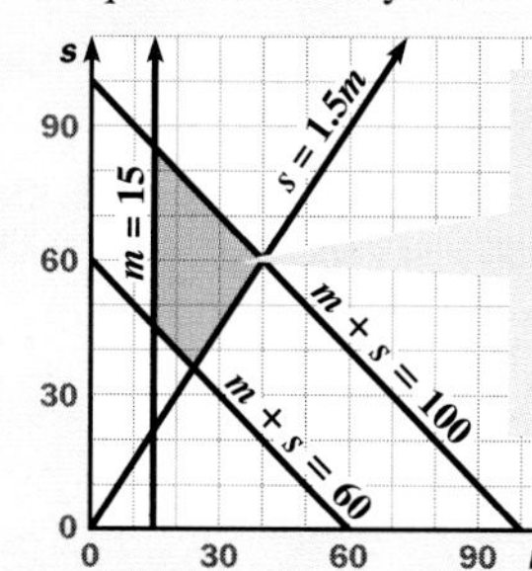

The coordinates of the points in the *overlap* of all the **half-planes** defined by a system of inequalities satisfy the system.

Linear programming involves finding the best solution to a problem subject to various constraints. For example, suppose that in the example above, each maple tree costs \$40 and each spruce tree costs \$60. The town's citizens hope to minimize the cost of the planting.

Step 4 Write a cost function for the situation: $C = 40m + 60s$.

Step 5 Find the corner points of the **feasible region** from Step 3.

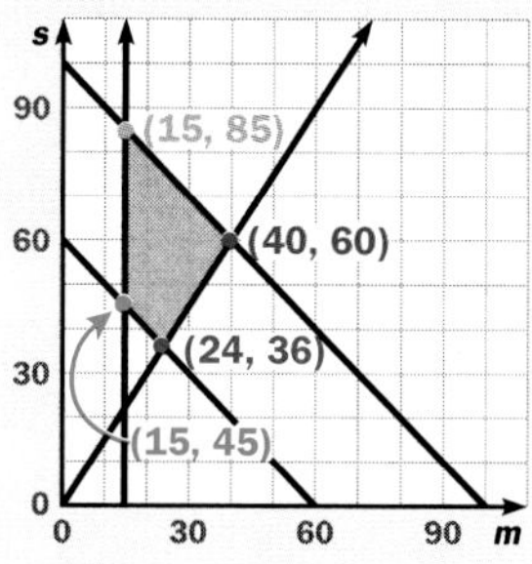

Step 6 Check the coordinates of the corner points of the feasible region in the cost equation.

Corner point	Value of *C*
(15, 45)	C = 40(15) + 60(45) = 3300
(15, 85)	C = 40(15) + 60(85) = 5700
(24, 36)	C = 40(24) + 60(36) = 3120
(40, 60)	C = 40(40) + 60(60) = 5200

For a minimum cost of \$3120, the town's citizens should plant 24 maple trees and 36 spruce trees.

Progress Check 7.4–7.6

1. Graph the inequality $3x + 4y > -16$. *(Section 7.4)*

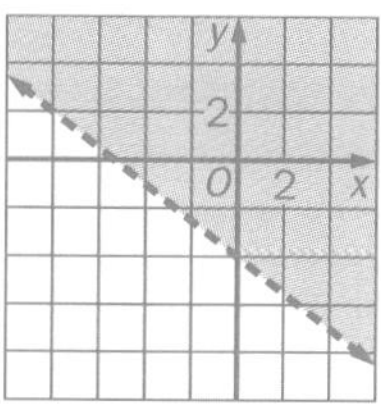

Graph each system of inequalities. *(Section 7.5)*

2. $y > 2x - 1$
$y \leq -3x + 2$

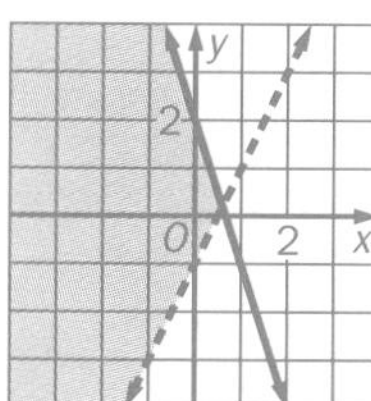

3. $y \leq x + 3$
$y \geq x - 1$

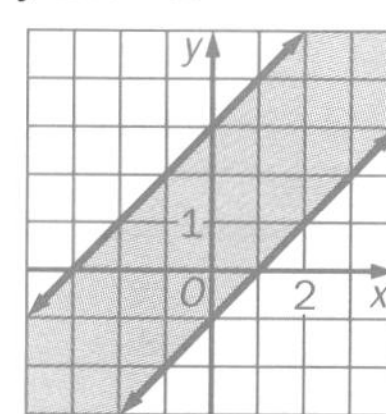

Graph the feasible region and find the minimum cost, in dollars, if $C = 5x + 7y$. *(Section 7.6)*

4. $x + y \geq 4$
$x + y \leq 7$
$x \geq 2$
$y \geq 1$

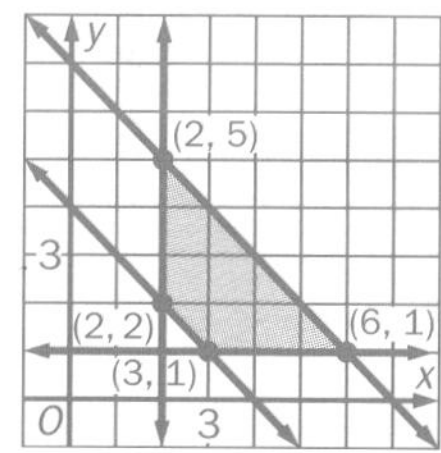

minimum cost: 22, at point (3, 1)

Chapter 7 Assessment
Form A Chapter Test

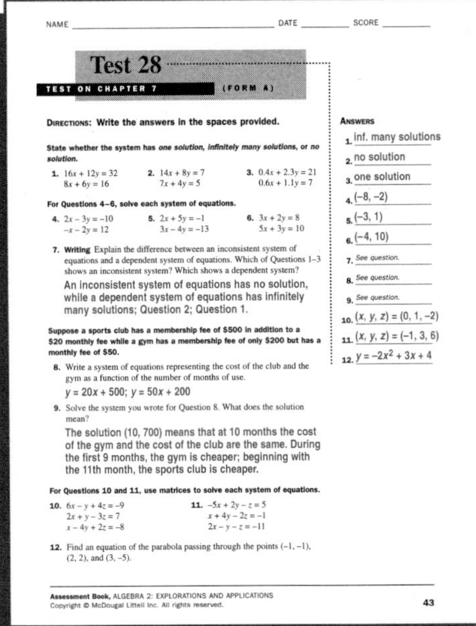
NAME ________ DATE ________ SCORE ________

Test 28

TEST ON CHAPTER 7 (FORM A)

DIRECTIONS: Write the answers in the spaces provided.

State whether the system has *one solution, infinitely many solutions,* or *no solution.*

1. $16x + 12y = 32$, $8x + 6y = 16$
2. $14x + 8y = 7$, $7x + 4y = 5$
3. $0.4x + 2.3y = 21$, $0.6x + 1.1y = 7$

For Questions 4–6, solve each system of equations.

4. $2x - 3y = -10$, $-x - 2y = 12$
5. $2x + 5y = -1$, $3x - 4y = -13$
6. $3x + 2y = 8$, $5x + 3y = 10$
7. **Writing** Explain the difference between an inconsistent system of equations and a dependent system of equations. Which of Questions 1–3 shows an inconsistent system? Which shows a dependent system?
An inconsistent system of equations has no solution, while a dependent system of equations has infinitely many solutions; Question 2; Question 1.

Suppose a sports club has a membership fee of $500 in addition to a $20 monthly fee while a gym has a membership fee of only $200 but has a monthly fee of $50.

8. Write a system of equations representing the cost of the club and the gym as a function of the number of months of use.
$y = 20x + 500$; $y = 50x + 200$
9. Solve the system you wrote for Question 8. What does the solution mean?
The solution (10, 700) means that at 10 months the cost of the gym and the cost of the club are the same. During the first 9 months, the gym is cheaper; beginning with the 11th month, the sports club is cheaper.

For Questions 10 and 11, use matrices to solve each system of equations.

10. $6x - y + 4z = -9$, $2x + y - 3z = 7$, $x - 4y + 2z = -8$
11. $-5x + 2y - z = 5$, $x + 4y - 2z = -1$, $2x - y - z = -11$
12. Find an equation of the parabola passing through the points (–1, –1), (2, 2), and (3, –5).

ANSWERS
1. inf. many solutions
2. no solution
3. one solution
4. (–8, –2)
5. (–3, 1)
6. (–4, 10)
7. See question.
8. See question.
9. See question.
10. $(x, y, z) = (0, 1, -2)$
11. $(x, y, z) = (-1, 3, 6)$
12. $y = -2x^2 + 3x + 4$

Assessment Book, ALGEBRA 2: EXPLORATIONS AND APPLICATIONS

43

Chapter 7 Assessment
Form B Chapter Test

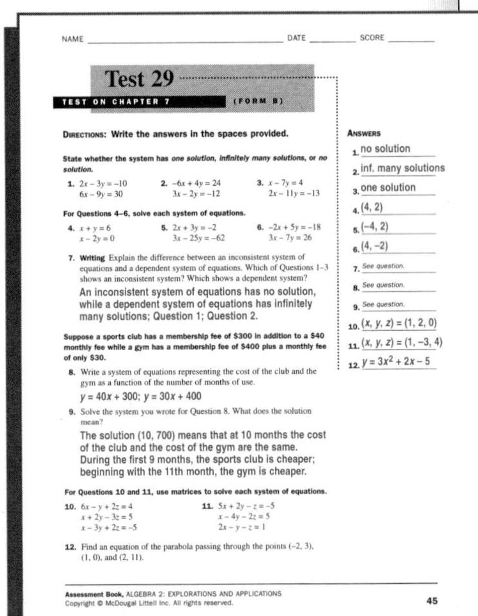
NAME ________ DATE ________ SCORE ________

Test 29

TEST ON CHAPTER 7 (FORM B)

DIRECTIONS: Write the answers in the spaces provided.

State whether the system has *one solution, infinitely many solutions,* or *no solution.*

1. $2x - 3y = -10$, $6x - 9y = 30$
2. $-6x + 4y = 24$, $3x - 2y = -12$
3. $x - 7y = 4$, $2x - 11y = -13$

For Questions 4–6, solve each system of equations.

4. $x + y = 6$, $x - 2y = 0$
5. $2x + 3y = -2$, $3x - 25y = -62$
6. $-2x + 5y = -18$, $3x - 7y = 26$
7. **Writing** Explain the difference between an inconsistent system of equations and a dependent system of equations. Which of Questions 1–3 shows an inconsistent system? Which shows a dependent system?
An inconsistent system of equations has no solution, while a dependent system of equations has infinitely many solutions; Question 1; Question 2.

Suppose a sports club has a membership fee of $300 in addition to a $40 monthly fee while a gym has a membership fee of $400 plus a monthly fee of only $30.

8. Write a system of equations representing the cost of the club and the gym as a function of the number of months of use.
$y = 40x + 300$; $y = 30x + 400$
9. Solve the system you wrote for Question 8. What does the solution mean?
The solution (10, 700) means that at 10 months the cost of the club and the cost of the gym are the same. During the first 9 months, the sports club is cheaper; beginning with the 11th month, the gym is cheaper.

For Questions 10 and 11, use matrices to solve each system of equations.

10. $6x - y + 2z = 4$, $x + 2y - 3z = 5$, $x - 3y + 2z = -5$
11. $5x + 2y - z = -5$, $x - 4y - 2z = 5$, $2x - y - z = 1$
12. Find an equation of the parabola passing through the points (–2, 3), (1, 0), and (2, 11).

ANSWERS
1. no solution
2. inf. many solutions
3. one solution
4. (4, 2)
5. (–4, 2)
6. (4, –2)
7. See question.
8. See question.
9. See question.
10. $(x, y, z) = (1, 2, 0)$
11. $(x, y, z) = (1, -3, 4)$
12. $y = 3x^2 + 2x - 5$

Assessment Book, ALGEBRA 2: EXPLORATIONS AND APPLICATIONS

45

CHAPTER 7

Assessment

VOCABULARY QUESTIONS

For Questions 1 and 2, complete each sentence.

1. A system of linear equations is called a(n) __?__ system if it has no solutions and a(n) __?__ system if it has infinitely many solutions.
2. Multiplying a square matrix by its __?__ , if it has one, gives a(n) __?__ .

SECTIONS 7.1, 7.2, *and* 7.3

3. **Technology** Use technology to estimate the solution of the system of equations $y = 26(1.07)^x$ and $y = 34(1.035)^x$.

Solve each system of equations. State whether the system has *one solution, infinitely many solutions,* or *no solution.*

4. $7x - 3y = -5$
$x + 2y = 4$
5. $4x - 2y = 12$
$2x - y = 3$
6. $0.5x + 3.7y = 2.02$
$0.4x - y = 0$
7. $25x + 10y = 20$
$10x + 4y = 20$
8. $3x + 5y = 2$
$6x - 4y = 11$
9. $2x - 9y = 2$
$7x - 12y = -6$
10. **Writing** Describe what a solution of a system of equations means both in terms of a graph and in terms of algebra.
11. **CONSUMER ECONOMICS** Suppose a 15-watt fluorescent bulb costs $12 to buy and $.0018 per hour to run. A standard bulb with the same light output costs $.50 to buy and $.0072 per hour to run.
   a. Write a system of equations representing the cost of each bulb as a function of hours of use.
   b. Solve the system. What does the solution mean?
   c. **Writing** If a fluorescent bulb lasts 10,000 h and an incandescent bulb lasts 1000 h, will it be cost-effective to use fluorescent bulbs? Explain.
12. **PHYSICS** Estimates of the height of a model rocket for various times after the end of the burn phase are given in the table.

Time after burn(s)	Height (ft)
1	250
3	235
5	100

   a. **Technology** Using matrices and a graphing calculator or software with matrix calculation capabilities, find an equation of the form $h = at^2 + bt + c$ that describes the height h of the rocket t seconds after the burn ends.
   b. What is the rocket's maximum height?

ANSWERS Chapter 7

Assessment

1. inconsistent; dependent
2. inverse; identity matrix
3. (8.07, 44.9)
4. (0.118, 1.94); one solution
5. no solution
6. (1.02, 0.408); one solution
7. no solution
8. $\left(\frac{3}{2}, -\frac{1}{2}\right)$; one solution
9. $\left(-2, -\frac{2}{3}\right)$; one solution
10. Graphically, the solution of a system of equations is the intersection of the graphs of the equations. Algebraically, a solution of a system of two equations is an ordered pair whose coordinates satisfy both equations. If there are three equations, a solution is an ordered triple.
11. a. Let h = the number of hours of use and C = the cost in dollars; $C = 0.0018h + 12$ (fluorescent) and $C = 0.0072h + 0.50$ (incandescent).
   b. (2130, 15.83); If bulbs are used for 2130 h, the cost is the same for fluorescent and incandescent bulbs.
   c. Yes; it would cost a total of $30 to use a single fluorescent bulb for 10,000 h. It would require 10 incandescent bulbs to provide the same number of hours of use at a total cost of $77.
12. a. $h = -15t^2 + 52.5t + 212.5$
   b. about 258 ft
13. 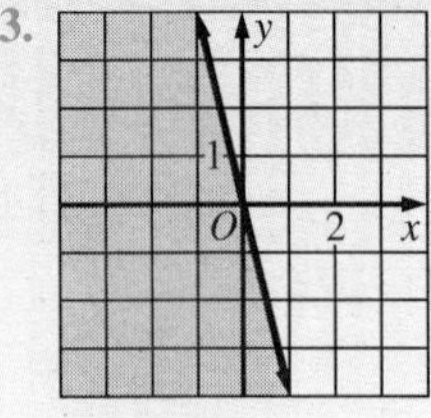

14. 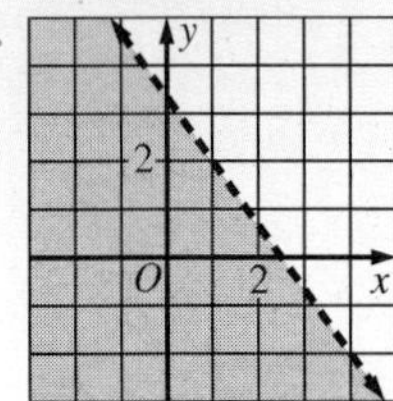

SECTIONS 7.4, 7.5, *and* 7.6

Graph each inequality.

13. $y \le -4x$ **14.** $4x + 3y < 10$ **15.** $3.5x - 7y \ge 1.75$

Write a system of inequalities defining each shaded region.

16.

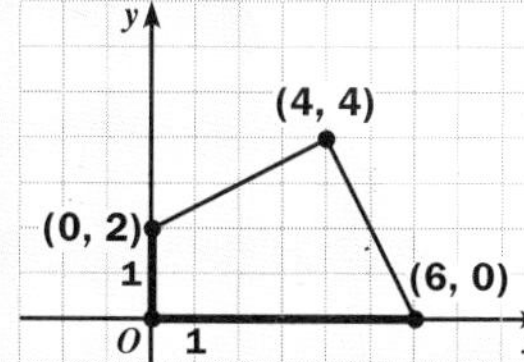

17.

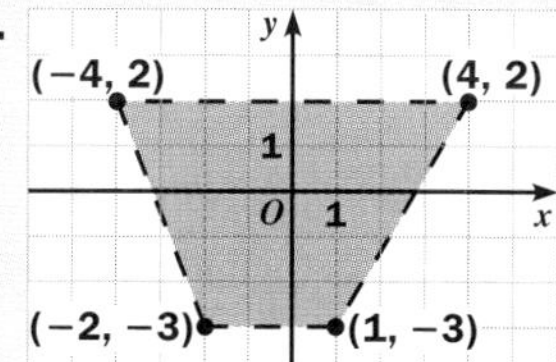

18. FITNESS The President's Council on Physical Fitness program gives awards to children who meet certain standards for a group of exercises. For example, the number of "curl-ups" a boy between the ages of 6 and 14 must complete in a minute to qualify towards an award is modeled by the equations below, where x is age and y is the number of curl-ups.

Presidential Award	National Award
$y = 2.82x + 16.39$	$y = 2.62x + 8.5$

a. Graph the system $y \ge 2.62x + 8.5$ and $y \le 2.82x + 16.39$ for $6 \le x \le 14$.

b. What does the system in part (a) represent for this situation?

19. Mai Wong wants to invest a total of \$12,000 to \$20,000 in two mutual funds: a conservative balanced fund and an aggressive growth fund. She wants at least two-thirds as much money in the balanced fund as in the growth fund, but she wants no more than \$10,000 in the balanced fund.

a. Write inequalities to represent the constraints for this situation.

b. Graph the feasible region. Label each corner point.

c. The balanced fund is expected to grow by 6% and the growth fund by 18% over a year. What investments should Mai Wong make to maximize her gain?

PERFORMANCE TASK

20. From your own experience, describe a situation involving two variables. For example, perhaps you have had to decide how to split your time between two activities, or you belong to a club that has tried to raise funds by making and selling two items.

a. Write a system of equations or a system of inequalities to represent your situation.

b. Show how you can use the system in part (a) to resolve the situation.

Chapter 7 Assessment
Form C Alternative Assessment

Chapter 7

ALTERNATIVE ASSESSMENT

1. **Open-ended Problem** Write three true statements and three false statements about this system of equations.
$-5x + 14 = y$
$7y - 24 = 4x$

2. **Performance Task** Describe three ways that two lines can intersect. What is true about the slopes of the lines in each of the three types of two-equation systems? What can you expect to happen if you solve each type of system algebraically?

3. **Performance Task** Use a table to find the solution of this system of equations. Explain how you identify the solution using a table.
$y = -4x + 3$
$y = 5x - 6$

4. **Performance Task** Write a system of inequalities that are represented by this region.

5. **Open-ended Problem** Make up a system of nonlinear equations that has no solution.

120 **Assessment Book**, ALGEBRA 2: EXPLORATIONS AND APPLICATIONS

15.

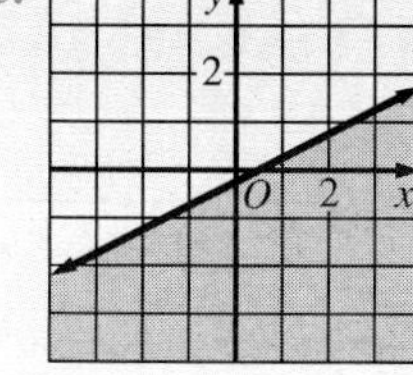

16. $x \ge 0$; $y \ge 0$; $y \le \frac{1}{2}x + 2$; $y \le -2x + 12$

17. $y > -3$; $y < 2$; $y > -\frac{5}{2}x - 8$; $y > \frac{5}{3}x - \frac{14}{3}$

18. a.

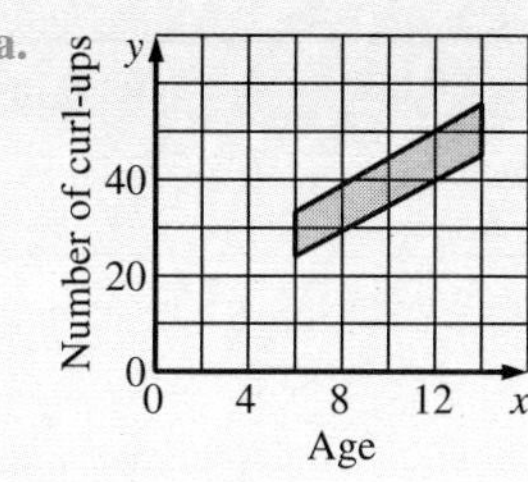

b. With the exception of the upper boundary, the region represents all scores for boys between 6 and 14 that achieve or exceed the limits for the National Award and fall short of the limits for the Presidential Award; every point on the upper boundary represents an age and number of curl-ups that qualify for both awards.

19. a. Let b = the amount in dollars invested in the balanced fund and g = the amount in dollars invested in the growth fund; $b + g \ge 12{,}000$; $b + g \le 20{,}000$; $b \ge \frac{2}{3}g$; $b \le 10{,}000$.

b.

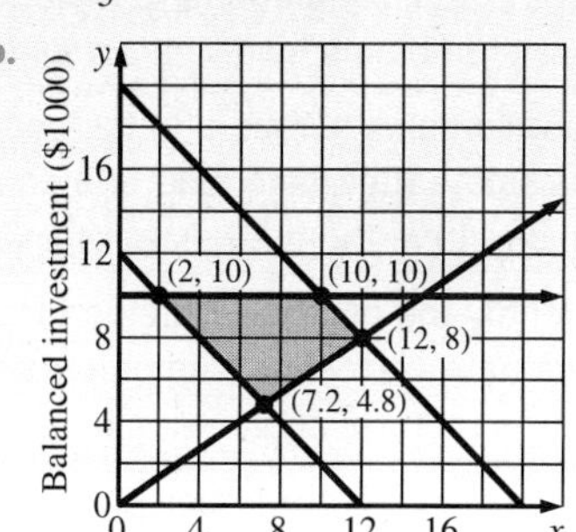

c. \$12,000 in the growth fund and \$8,000 in the balanced fund

20. Answers may vary.

CHAPTER

8 Radical Functions, Number Systems

OVERVIEW

Connecting to Prior and Future Learning

⇔ Students' previous work with square roots is extended in this chapter to square root and radical functions. As students graph functions and solve equations, they develop skills they will use throughout their future studies in mathematics. Students may find the review of square roots on page 777 of the **Student Resources Toolbox** to be helpful.

⇔ In Section 8.4, the number system is extended to include imaginary and complex numbers. Graphing complex numbers in the complex plane is introduced in Section 8.5.

⇔ The chapter closes with a study of number systems and their properties. This study is a foundation for future studies in more advanced mathematics courses. The **Student Resources Toolbox** provides a helpful review of rational numbers and irrational numbers on page 779.

Chapter Highlights

Interview with Ronald Toomer: The role mathematics plays in building amusement park rides is highlighted in this interview. Students can explore this use of mathematics in more detail by completing the exercises on pages 342 and 348.

Explorations can be found in Section 8.1, where students reflect the graph of $y = x^2$ over the line $y = x$, and in Section 8.2, where students reflect the graph of $y = x^n$ over the line $y = x$.

The Portfolio Project: Students conduct an experiment to collect data about the relationship between leg length and walking speed and then compare their results to a model.

Technology: Graphing calculators are used throughout this chapter to solve radical functions by graphing. Spreadsheets are used in Section 8.5 to calculate the critical orbit of a complex number for a given function.

OBJECTIVES

Section	Objectives	NCTM Standards
8.1	• Restrict the domain of a function to obtain an inverse function. • Graph and evaluate square root functions. • Use radical expressions to represent real-world situations.	1, 2, 3, 4, 5, 6
8.2	• Evaluate radical expressions. • Graph radical functions. • Solve problems involving radical functions.	1, 2, 3, 4, 5, 6
8.3	• Solve equations with radical expressions. • Solve problems using radical equations.	1, 2, 3, 4, 5
8.4	• Add, subtract, multiply, and divide complex numbers. • Find complex solutions to equations that have no real solutions. • Solve problems involving complex numbers.	1, 2, 3, 4, 5, 14
8.5	• Plot complex numbers in the complex plane. • Calculate the magnitude of a complex number. • Understand the mathematics used to generate images.	1, 2, 3, 4, 5, 12
8.6	• Identify the number systems to which a number belongs. • Evaluate whether group properties hold for a set and an operation. • See structural similarities in groups.	1, 2, 3, 4, 5, 14

INTEGRATION

Mathematical Connections	8.1	8.2	8.3	8.4	8.5	8.6
algebra	**337–343***	**344–350**	**351–357**	**358–364**	**365–370**	**371–379**
geometry		347		364	**365–370**	
data analysis, probability, discrete math	339, 341, 342	345, 348–350	**351–357**		370	
patterns and functions	**337–343**	**344–350**			**365–370**	377–379
logic and reasoning	**337–343**	344, 346–350	**351–357**	358–360, 362, 364	366–370	371, 375–379
Interdisciplinary Connections and Applications						
biology and earth science		345, 349				
sports and recreation	339	349	356, 357			377, 378
oceanography	341					
astronomy		350				
meteorology			354			
wind chill, electronics, Mandelbrot set			355	363	369	

* ***Bold page numbers*** *indicate that a topic is used throughout the section.*

TECHNOLOGY

Section	*opportunities for use with* Student Book	Support Material
8.1	graphing calculator or software McDougal Littell Mathpack *Function Investigator*	
8.2	graphing calculator or software McDougal Littell Mathpack *Function Investigator*	**Technology Book:** Spreadsheet Activity 8 **Function Investigator with Matrix Analyzer Activity Book:** Function Investigator Activities 19 and 20
8.3	graphing calculator or software McDougal Littell Mathpack *Function Investigator*	
8.4	graphing calculator	**Technology Book:** TI-92 Activity 4
8.5	graphing calculator spreadsheet software McDougal Littell Mathpack *Stats!*	
8.6	graphing calculator	**Technology Book:** Calculator Activity 8 **Function Investigator with Matrix Analyzer Activity Book:** Matrix Analyzer Activity 1

PLANNING GUIDE

Regular Scheduling (45 min)

Section	Materials Needed	Core Assignment	Extended Assignment	*exercises that feature* Applications	Communication	Technology
8.1	graphing calculator, graph paper, MIRA® transparent mirror, tracing paper, ruler	**Day 1:** 1–7, 11–16, 20–23 **Day 2:** 24–50	**Day 1:** 1–23 **Day 2:** 24–50	8–10, 17–19	41	32–39
8.2	graphing calculator, graph paper, round balloon, tape measure	1–28, 40–46	1, 2, 5–21 odd, 22–46	29–35, 38, 39	40	
8.3	graphing calculator, graph paper	**Day 1:** 1–16, 19–34 **Day 2:** 37–40, 45–65, AYP*	**Day 1:** 1–11 odd, 13–22, 23–33 odd **Day 2:** 35–65, AYP	19–22 41–44	18 53	22 37–39
8.4		**Day 1:** 1–35 odd **Day 2:** 2–36 even, 42–47, 50–59	**Day 1:** 1–35 odd **Day 2:** 2–36 even, 37–59	37–41	48	
8.5	graphing calculator, spreadsheet software	1–35, 53–62	1–6, 10–14, 18–22, 25–62	37	36, 47	37–45
8.6		**Day 1:** 1–21, 24–29 **Day 2:** 37–45, 55–64, AYP	**Day 1:** 1–32 **Day 2:** 33–64, AYP	33–36, 39–54	20	
Review/ Assess		**Day 1:** 1–23 **Day 2:** 24–42 **Day 3:** Ch. 8 Test	**Day 1:** 1–23 **Day 2:** 24–42 **Day 3:** Ch. 8 Test			
Portfolio Project		Allow 2 days.	Allow 2 days.			

Yearly Pacing (with Portfolio Project)	Chapter 8 Total	Chapters 1–8 Total	Remaining	Total
	15 days	107 days	53 days	160 days

Block Scheduling (90 min)

	Day 47	Day 48	Day 49	Day 50	Day 51	Day 52	Day 53	Day 54
Teach/Interact	Ch. 7 Test 8.1: Exploration, page 337	Continue with 8.1 8.2: Exploration, page 344	8.3	8.4	8.5 8.6	Continue with 8.6 Review	Review Port. Proj.	Ch. 8 Test Port. Proj.
Apply/Assess	**Ch. 7 Test** **8.1:** 1–7, 11–16, 20–23	**8.1:** 24–50 **8.2:** 1–28, 40–46	**8.3:** 1–16, 19–34, 37–40, 45–65, AYP*	**8.4:** 1–36, 42–47, 50–59	**8.5:** 1–35, 53–62 **8.6:** 1–21, 24–29	**8.6:** 37–45, 55–64, AYP **Review:** 1–23	**Review:** 24–42 **Port. Proj.**	**Ch. 8 Test** **Port. Proj.**

NOTE: A one-day block has been added for the Portfolio Project—timing and placement to be determined by teacher.

Yearly Pacing (with Portfolio Project)	Chapter 8 Total	Chapters 1–8 Total	Remaining	Total
	$7\frac{1}{2}$ days	54 days	28 days	82 days

**AYP* is *Assess Your Progress.*

Lesson Support

Section	Practice Bank	Study Guide*	Assessment Book*	Visuals	Explorations Lab Manual	Lesson Plans	Technology Book
8.1	47	8.1		Warm-Up 8.1 Folders 4, 8	Masters 1, 2	8.1	
8.2	48	8.2		Warm-Up 8.2	Masters 1, 2	8.2	Spreadsheet Act. 8
8.3	49	8.3	Test 30	Warm-Up 8.3	Master 2 Add. Expl. 14	8.3	
8.4	50	8.4		Warm-Up 8.4		8.4	TI-92 Act. 4
8.5	51	8.5		Warm-Up 8.5	Master 1	8.5	
8.6	52	8.6	Test 31	Warm-Up 8.6		8.6	Calculator Act. 8
Review Test	53	Chapter Review	Tests 32, 33 Alternative Assessment			Review Test	

*__Spanish versions__ of *Study Guide* and *Assessment Book* are available.

Chapter Support

- Course Guide
- Lesson Plans
- Portfolio Project Book
- Preparing for College Entrance Tests
- Multi-Language Glossary
- *Test Generator* Software
- Professional Handbook

Software Support

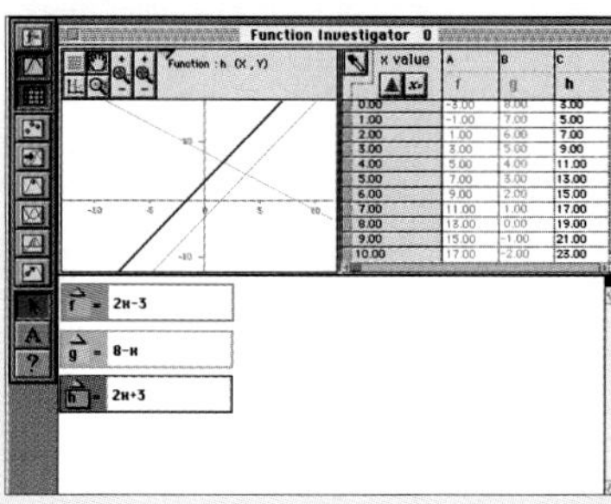

McDougal Littell Mathpack
Stats!
Function Investigator

Internet Support

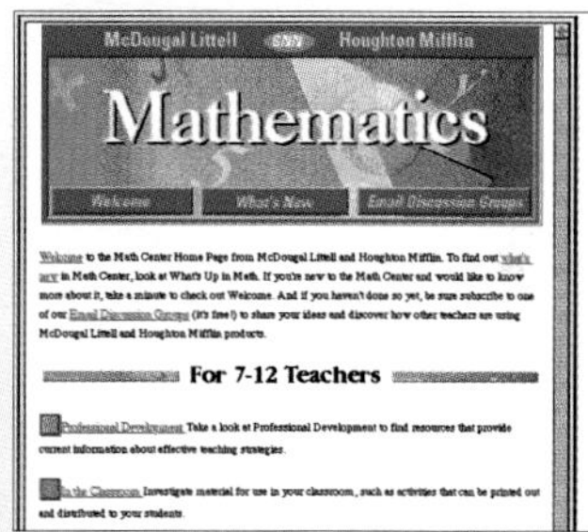

http://www.hmco.com
Next go to McDougal Littell; then the Education Center; then Secondary Math.

Outside Resources

Books, Periodicals

Naraine, Bishnu. "An Alternative Approach to Solving Radical Equations." *Mathematics Teacher* (March 1993): pp. 204–205. (See also: *Mathematics Teacher* (October 1993): p. 608.)

Vonder Embse, Charles. "Graphing Powers and Roots of Complex Numbers." *Mathematics Teacher* (October 1993): pp. 589–597.

Activities, Manipulatives

Disher, Fan. "Activities: Graphing Art." *Mathematics Teacher* (February 1995): pp. 124–128, 134–136.

Software

f(g) Scholar. IBM-comp. Southampton, PA: Future Graph.

Internet

Investigate: Science and mathematics resources—no graphics
http://www-hpcc.astro.washington.edu/scied/science.html

Search for mathematics-related software, teaching materials, other gophers, through
gopher archives.math.utk.edu

Interview Notes

Background

Roller Coasters

The idea for roller coasters came from giant ice slides built in Russia as early as the 1400s. Wooden frames were built 70 feet high and water was poured down the long ramp which froze quickly. Riders had to climb a 70 foot ladder to the top where they got on a sled and raced down the hill. The slides were so fast that the sleds slid at least 600 feet on a straight-away before stopping.

In 1804, the first roller coaster was built in Paris, France. It was named the *Russian Mountains*, after the Russian ice slides. The coasters were small, one-person cars that had few safety precautions. Because the ramp was so high and steep, riders sometimes fell out. The first looping roller coaster was built in 1848 by a French engineer who called it the *Centrifugal Pleasure Railway* and tested it first with sandbags. When the sandbags did not fall out, a workman volunteered to test the ride; he liked it so much he went around again. The first roller coaster in the United States was built in 1884 at Coney Island in New York City by La Marcus Adna Thompson. Today, there are about 175 large, permanent roller coasters in the United States. Two of these coasters have vertical drops of 225 feet: the *Steel Phantom* in West Mifflin, Pennsylvania, and *Desperado* in Jean, Nevada.

CHAPTER

8 Radical Functions, Number Systems

Designs for radical rides

INTERVIEW **Ronald Toomer**

In the fast-moving world of amusement park rides, there's a legend and a classic joke. The legend is Ronald Toomer, the man who has designed and built ninety of the world's biggest and scariest roller coasters. The joke is that Toomer personally never rides these things. He says, "They just don't appeal to me. I have a terrible time with motion sickness and would much rather design them."

"It's just a matter of keeping things moving."

Keeping Things Moving

In the 1960s Toomer was working as a mechanical engineer on missile and rocket systems when he heard about an engineering company that needed help constructing its first roller coaster. He signed up for temporary work on that project. "It's been a long temporary job, lasting about 30 years," says Toomer, who is now president of the company.

"For some crazy reason, people like to be scared."

The company has been so successful that each year about 200 million people ride its coasters. The key to a popular ride is pretty simple, Toomer says. "We try to keep it going as fast as we can, all the time, while throwing in a few changes of pace. Basically, it's just a matter of keeping things moving."

Figuring Out the Fun

His company built *Desperado,* the world's largest roller coaster, located near Las Vegas, Nevada. The ride can theoretically reach a velocity of about 82 mi/h. However, this velocity is reduced both by friction between the wheels and the track and by air resistance against the passenger cars.

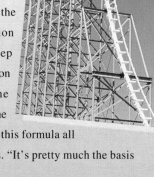

Toomer calculates the velocity of a passenger car at the bottom of a hill by using a formula that includes values for the height of the drop and the acceleration due to gravity. "We keep doing the same calculation for the hills until the coaster finally reaches the end of the line. We use this formula all the time," Toomer says. "It's pretty much the basis of the whole thing."

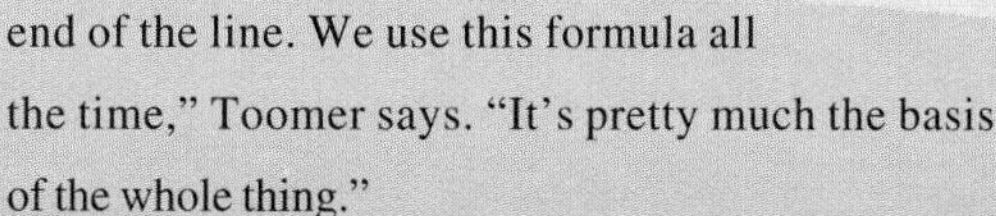

Interview Notes

Background

Ronald Toomer
Ronald Toomer grew up in Pasadena, California during the Great Depression. For entertainment, his family went to an amusement park several times a year. That was where he saw his first roller coaster, the *Cyclone Racer,* and he said he never wanted to go on it. Later, he went to work as a garage mechanic and then earned a mechanical engineering degree in 1961 from the University of Nevada at Reno. After graduation, he worked on the Minuteman missile program, but passed up working on the Space Shuttle program to design amusement-park rides. Today, he is president of his own company, Arrow Dynamics, Inc., which is the largest custom roller coaster producer in the world. Toomer says that when he designs his roller coasters, he purposely uses two-abreast seating so that every passenger can look over the side, and he designs the structural supports beneath the riders so when they look down they see nothing!

Mechanical Engineering
Mechanical engineers are involved with every phase of developing a machine, from the beginning construction of an experimental model to the installation of the finished machine. They design, operate, and test all kinds of machines, from engines that produce power and energy to industrial processing equipment.

Many mechanical engineers work in industry to improve the use of power, transportation, or manufacturing. They also concentrate on research and development because new types of machines are constantly in need. Mechanical engineers are involved in almost every other branch of engineering.

Second-Language Learners

Some second-language learners may not understand the irony in the opening paragraph and may benefit from a class discussion. You may also wish to point out the difference between something that can happen *theoretically* and something that happens in reality.

Interview Notes

Mathematical Connection

In designing roller coasters, Ronald Toomer and his colleagues use the formula $v = \sqrt{2gh}$ repeatedly in their work. The development of this formula and its uses are explained in Sections 8.1 and 8.2. In Section 8.1, students look into the development of the formula by exploring how a roller coaster changes potential energy into kinetic energy. In Section 8.2, students simplify the formula by substituting a standard value for the force of gravity, which results in an equation containing only the variables v and h. Then students explore the velocities of several roller coasters around the United States.

Explore and Connect

Writing
If students do not understand the concepts involved in this question, a demonstration using small model cars may provide some insight.

Project
Remind students that each successive hill should be shorter than the previous one in their sketches.

Research
If students cannot contact an amusement park, they can find the height of the first drop of some roller coasters in a reference book at the library.

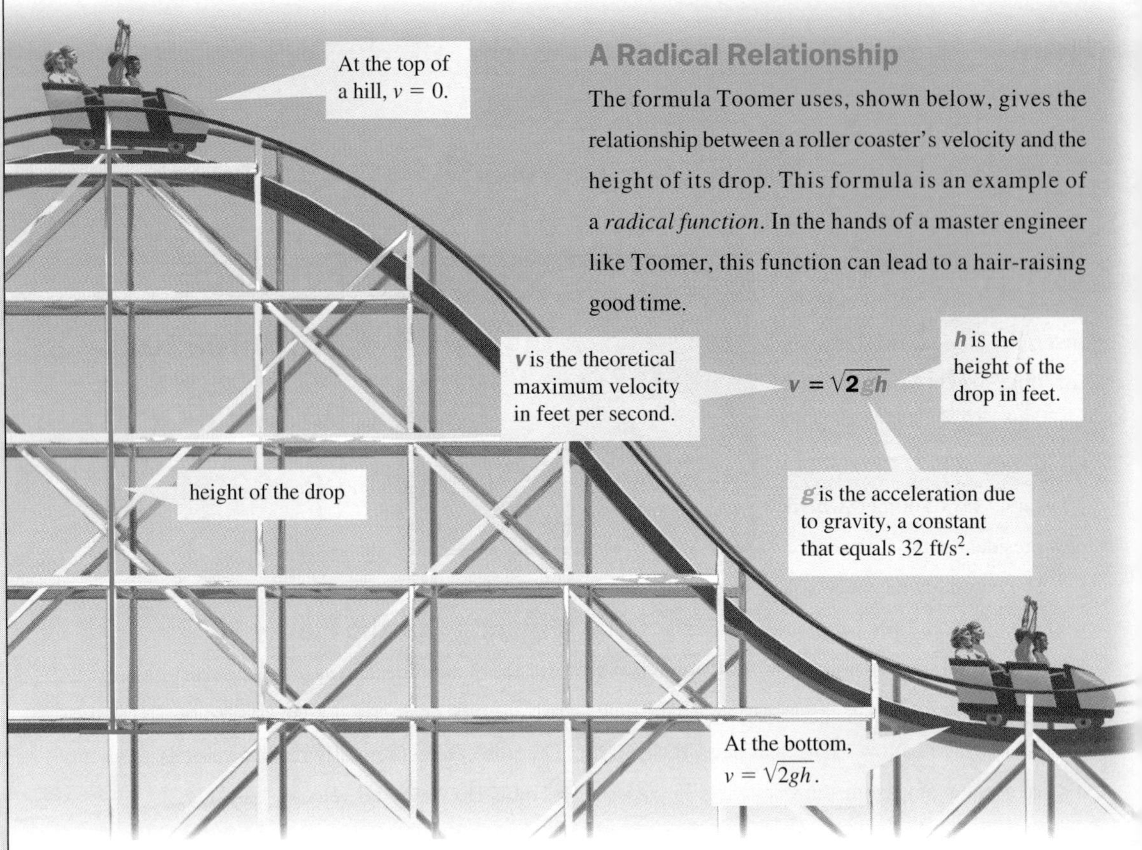
At the top of a hill, $v = 0$.
v is the theoretical maximum velocity in feet per second.
$v = \sqrt{2gh}$
h is the height of the drop in feet.
height of the drop
g is the acceleration due to gravity, a constant that equals 32 ft/s^2.
At the bottom, $v = \sqrt{2gh}$.

A Radical Relationship

The formula Toomer uses, shown below, gives the relationship between a roller coaster's velocity and the height of its drop. This formula is an example of a *radical function*. In the hands of a master engineer like Toomer, this function can lead to a hair-raising good time.

Explore and Connect

Ronald Toomer designs a roller coaster.

1. Writing Roller coasters have motor-driven chains that pull a car up the first and highest hill. The energy a car gains from moving up the first hill keeps it moving for the rest of the ride. Why do you think successive hills must decrease in height?

2. Project Draw a sketch of a roller coaster with three hills. For each hill in your model, measure the height of the drop to the nearest inch. Convert the model heights to life-size heights using 1 in. = 25 ft. Calculate the velocity of a passenger car at the bottom of each drop, assuming that $v = 0$ at the top of each drop.

3. Research Contact a nearby amusement park. Ask for an employee who can tell you the height of the first drop for the park's largest roller coaster. Use the given height to calculate the velocity of the roller coaster at the bottom of the drop.

Mathematics & Ronald Toomer

In this chapter, you will learn more about how mathematics is related to amusement park rides.

Related Exercises

Section 8.1
- Exercises 17–19

Section 8.2
- Exercises 29–32

Section

8.1 Square Root Functions

Learn how to...

- **restrict the domain of a function to obtain an inverse function**
- **graph and evaluate square root functions**

So you can...

- **understand the dynamics of pole vaulting, for example**

In Chapter 4 you saw that reflecting the graph of a function over the line $y = x$ can produce another function, called an *inverse*. For example, the inverse of an exponential function is a logarithmic function.

What happens when the graph of a quadratic function is reflected over the line $y = x$?

EXPLORATION
COOPERATIVE LEARNING

Reflecting the Graph of $y = x^2$ over the Line $y = x$

Work with a partner. You will need:

- graph paper
- a MIRA® transparent mirror, tracing paper, or a ruler

1 Graph $y = x^2$ and $y = x$ in the same coordinate plane.

2 Sketch the reflection of the graph of $y = x^2$ over the line $y = x$. Remember to reflect not only the points above the line but also the points below the line. Here are three ways of reflecting the graph:

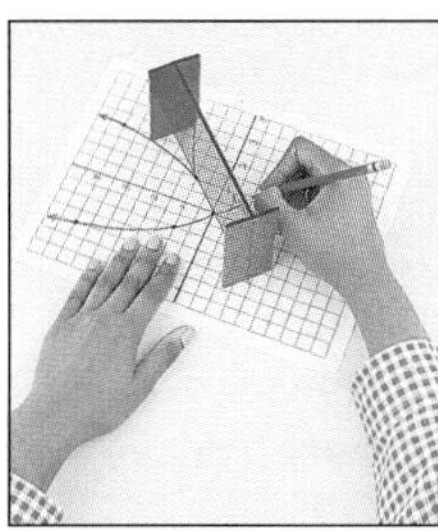

You can place a MIRA® along the line $y = x$ and sketch the reflection you see.

You can fold tracing paper along the line $y = x$ and trace on the other side of the paper.

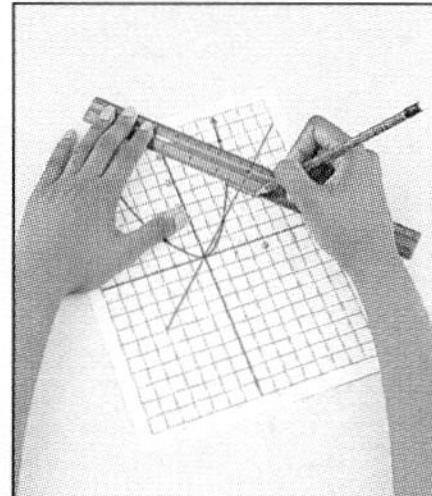

You can hold a ruler perpendicular to the line $y = x$ and measure equal distances on each side.

Questions

1. Does the graph you get represent a function? Explain why or why not.
2. Consider the function $y = x^2$ with a restricted domain of $x \geq 0$. What does the graph look like? Is the reflection of the graph over the line $y = x$ the graph of a function?

Exploration Note

Purpose
The purpose of this Exploration is to have students discover that they need to restrict the domain of the function $y = x^2$ so that its inverse will be a function.

Materials/Preparation
Each pair of students needs graph paper and a MIRA®, tracing paper, or a ruler.

Procedure
Students graph $y = x^2$ and $y = x$ in the same coordinate plane. They then draw the reflection of $y = x^2$ over the line $y = x$. Questions 1 and 2 have students consider whether the reflection is a function and suggest restricting the domain of $y = x^2$ to see if its reflection is a function.

Closure
In answering Questions 1 and 2, students should see that the inverse of the function $y = x^2$ is not a function, but that they can restrict its domain so that the inverse is a function. Ask students if they can think of another way to restrict the domain of $y = x^2$ so that its inverse will be a function.

Explorations Lab Manual
See the Manual for more commentary on this Exploration.

For answers to the Exploration, see following page.

Plan⇔Support

Objectives

- Restrict the domain of a function to obtain an inverse function.
- Graph and evaluate square root functions.
- Use radical expressions to represent real-world situations.

Recommended Pacing

Core and Extended Courses
Section 8.1: 2 days

Block Schedule
Section 8.1: 2 half-blocks (with Chapter 7 Test and Section 8.2)

Resource Materials

Lesson Support
Lesson Plan 8.1
Warm-Up Transparency 8.1
Overhead Visuals:
Folder 4: Functions and Inverses, Sheets 5, 6
Folder 8: Graphs of Square Root Functions
Practice Bank: Practice 47
Study Guide: Section 8.1
Explorations Lab Manual: Diagram Masters 1, 2

Technology
Graphing Calculator
McDougal Littell Mathpack *Function Investigator*
Internet: http://www.hmco.com

Warm-Up Exercises

Simplify each radical expression.

1. $\sqrt{25}$ 5
2. $\sqrt{125}$ $5\sqrt{5}$
3. $\sqrt{200}$ $10\sqrt{2}$

Complete.

4. If (a, b) is on the graph of $y = f(x)$, then __?__ is on the graph of $y = f^{-1}(x)$. (b, a)
5. The graph of $y = f^{-1}(x)$ is the __?__ of the graph of $y = f(x)$ over the line __?__. reflection; $y = x$
6. Find an equation of the inverse of $y = 3x + 2$. $y = \frac{1}{3}x - \frac{2}{3}$

Teach⇔Interact

Section Note

Topic Spiraling: Review
Students first learned the concept of an inverse in Section 4.1. In particular, students need to remember the definition of inverse, the method for obtaining the graph of an inverse by reflecting the function over the line $y = x$, and the method for finding the equation of an inverse. To review the definition of a function, ask students to draw two graphs, one that is the graph of a function and one that is not.

Think and Communicate

Students may be interested in knowing how the graphing calculator in question 1 determines values for the function Y1 = X²(X≥0). If the expression in the parentheses is true, it is assigned a value of 1, and if it is false, it is assigned a value of 0. Thus, when X is positive or zero, Y1 = X², and when X is negative, Y1 = 0. Point out to students that the function obtained this way on a graphing calculator is not identical to the function $y = x^2$, $x \geq 0$ because this function is not defined for $x < 0$, while the graphing calculator function is zero when $x < 0$. Have students trace along the graph or evaluate Y1(X) for negative x-values to verify this fact.

After completing questions 2 and 3, ask students if they can develop a method for determining if a function has an inverse function by examining its graph. Students may be able to come up with a method similar to the horizontal line test, which states that a function has an inverse function if no horizontal line passes through more than one point of its graph.

About Example 1

Second-Language Learners
Students learning English may be unfamiliar with the meaning of the phrase *clearing a bar*. If necessary, explain that it means the athlete was able to "jump over the bar (with the help of the pole) without touching it at all."

To find the inverse of $f(x) = x^2$, you might be tempted to do the following:

$y = x^2$ ← Substitute y for $f(x)$.

$\pm\sqrt{y} = x$ ← Solve for x (see page 187).

$\pm\sqrt{x} = y$ ← Switch x and y.

The equation $y = \pm\sqrt{x}$ does *not* define a function, however, because a positive input results in *two* outputs. For example, if $x = 9$, then $y = \pm\sqrt{9} = \pm 3$.

To get around this problem, you can restrict the domain of $f(x) = x^2$. Two ways of doing this are shown below.

For more information about graphing inverses, see the *Technology Handbook*, p. 809.

1. Restrict the domain of $f(x) = x^2$ to $x \geq 0$.

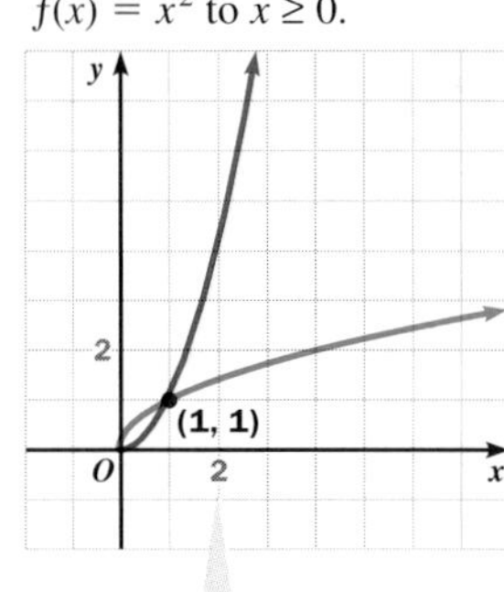

The inverse of $f(x) = x^2$ for $x \geq 0$ is $f^{-1}(x) = \sqrt{x}$.

2. Restrict the domain of $f(x) = x^2$ to $x \leq 0$.

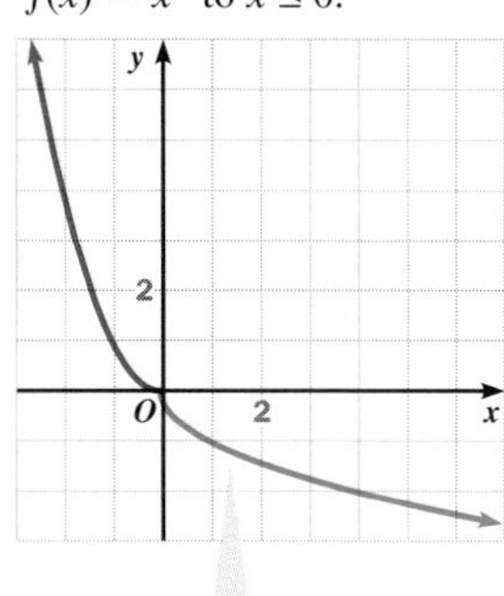

The inverse of $f(x) = x^2$ for $x \leq 0$ is $f^{-1}(x) = -\sqrt{x}$.

THINK AND COMMUNICATE

1. **Technology** Use a graphing calculator or graphing software to graph the equations shown.

a. Use trace to convince yourself that a point (b, a) is on the graph of $y = \sqrt{x}$ whenever the point (a, b) is on the graph of $y = x^2$ for $x \geq 0$.

b. Repeat part (a) using $y = x^2$ for $x \leq 0$ and $y = -\sqrt{x}$.

2. Would you be able to find an inverse of $f(x) = x^2$ if you restricted the domain of f to $-1 \leq x \leq 1$? Explain.

3. Why is it *not* necessary to restrict the domain of a nonconstant linear function or an exponential function when finding its inverse?

ANSWERS

Section 8.1

Note: Answers are rounded to three significant digits when necessary, unless otherwise noted.

Exploration

1, 2.

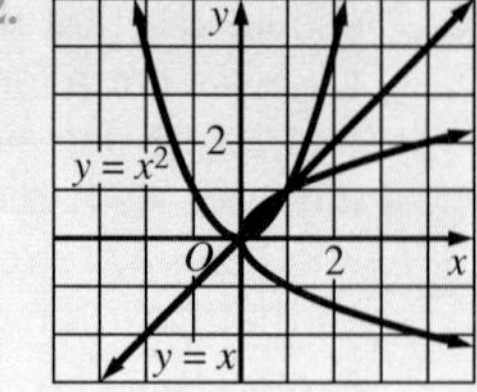

Questions

1. No; for $y = \pm\sqrt{x}$, there are two outputs for every input except 0.
2. a half-parabola; Yes.

Think and Communicate

1. a.

A point (b, a) is on the graph of $y = \sqrt{x}$ whenever the point (a, b) is on the graph of $y = x^2$ for $x \geq 0$.

EXAMPLE 1 Application: Sports

In 1993, Ukrainian athlete Sergei Bubka set a world record in pole vaulting, clearing a bar that was 20 ft 2 in. above the ground. The equation

$$h = \frac{v^2}{64}$$

gives the height h (in feet) that a pole vaulter's center of gravity is raised as a function of running velocity v (in feet per second) at launch.

a. Write an equation for the vaulter's velocity as a function of the height the vaulter's center of gravity is raised.

b. Graph the equation from part (a). State the domain and range.

c. Suppose Bubka's center of gravity while running was 3 ft 2 in. above the ground. Find the running velocity he needed to clear the bar placed 20 ft 2 in. above the ground.

BY THE WAY

There are no length limits on poles for vaulting, but poles longer than 16 ft are considered too heavy to use because they slow down runners.

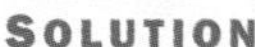

SOLUTION

a. Solve the equation for v.

$$h = \frac{v^2}{64}$$
$$64h = v^2$$
$$\sqrt{64h} = v$$
$$8\sqrt{h} = v$$

Use the **positive square root**, because velocity here must be nonnegative.

An equation is $v = 8\sqrt{h}$.

b.

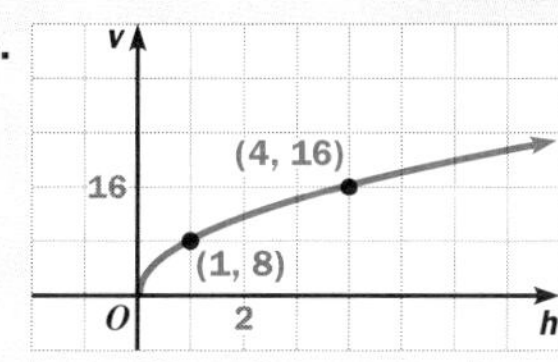

The domain is $h \geq 0$.

The range is $v \geq 0$.

c. Bubka needed to raise his center of gravity from 3 ft 2 in. to 20 ft 2 in. above the ground. The difference in these heights is 17 ft.

$$v = 8\sqrt{17} \quad \longleftarrow \text{Substitute 17 for } h.$$
$$\approx 32.985$$

Bubka needed to run about 33 ft/s for his world record vault.

THINK AND COMMUNICATE

4. **Technology** Verify that $\sqrt{64h} = 8\sqrt{h}$ by using a graphing calculator or graphing software to compare the graphs of $y = \sqrt{64x}$ and $y = 8\sqrt{x}$.

5. Compare the graphs of $y = 8\sqrt{x}$ and $y = \sqrt{x}$. How are they geometrically related?

6. Compare the graphs of $y = 8 + \sqrt{x}$ and $y = \sqrt{x}$. How are they geometrically related?

Toolbox p. 777 *Square Roots*

Teach⇔Interact

About Example 1

Teaching Tip

The formula used in part (a) of this Example is a special case of the formula $v = \sqrt{2gh}$, which relates the change in an object's height to its velocity. This formula was first encountered on page 336, where it was used to determine the maximum velocity possible for a roller coaster at the bottom of a drop of h feet.

Additional Example 1

To win a horse show, Britta's horse must clear a 65 inch fence. The equation $h = \frac{v^2}{768}$ gives the height h (in inches) a horse's center of gravity is raised as a function of running velocity v (in inches per second) as the horse jumps.

a. Write an equation for the horse's velocity as a function of the height its center of gravity is raised.

Solve the equation for v.

$$h = \frac{v^2}{768}$$
$$768h = v^2$$
$$\sqrt{768h} = v$$
$$16\sqrt{3h} = v$$

An equation is $v = 16\sqrt{3h}$.

b. Graph the equation from part (a). State the domain and range.

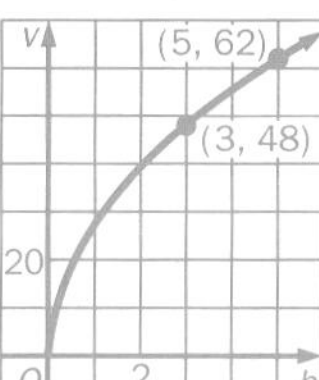

The domain is $h \geq 0$. The range is $v \geq 0$.

c. Suppose the horse's center of gravity while running was 51 in. above the ground. Find the running velocity needed to clear the bar placed at 65 in. above the ground.

The horse needed to raise its center of gravity from a height of 51 in. to a height of 65 in. above the ground. The difference in these heights is 14 in. Substitute 14 for h.

$$v = 16\sqrt{3(14)}$$
$$\approx 103.69$$

Britta's horse needed to run about 104 in./s (5.9 mi/h) to clear the jump.

b. A point (b, a) is on the graph of $y = -\sqrt{x}$ whenever the point (a, b) is on the graph of $y = x^2$ for $x \leq 0$.

2. No; except for $x = 0$, each output corresponds to two inputs. So in attempting to define an inverse, each input corresponds to two outputs.

3. For such functions, each output corresponds to one input, so in defining an inverse, each input corresponds to only one output.

4. The graphs are identical.

5. The graph of $y = 8\sqrt{x}$ is the graph of $y = \sqrt{x}$ stretched vertically by a factor of 8.

6. The graph of $y = 8 + \sqrt{x}$ is the graph of $y = \sqrt{x}$ translated up 8 units.

Teach⇔Interact

Additional Example 2

Compare the graphs of $y = \sqrt{x}$ and $y = 4\sqrt{x + 1}$. How are they geometrically related? State the domain and range of the second function.

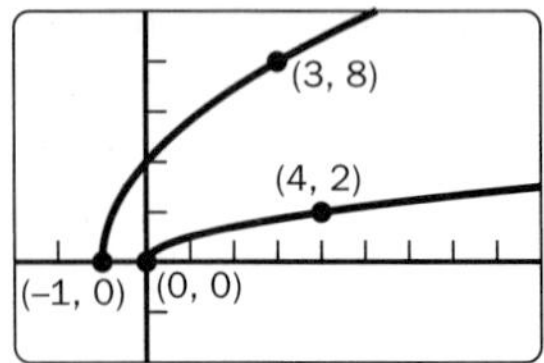

The second graph is the first graph stretched vertically by a factor of 4 and translated 1 unit to the left. The domain of the second function is $x \geq -1$ and the range is $y \geq 0$.

Closure Question

Describe the relationship between the quadratic function $y = x^2$ and the square root function $y = -\sqrt{x}$.
If you restrict the domain of $y = x^2$ to $x \leq 0$, then its inverse is the function $y = -\sqrt{x}$.

Apply⇔Assess

Suggested Assignment

Core Course
Day 1 Exs. 1–7, 11–16, 20–23
Day 2 Exs. 24–50

Extended Course
Day 1 Exs. 1–23
Day 2 Exs. 24–50

Block Schedule
Day 47 Exs. 1–7, 11–16, 20–23
Day 48 Exs. 24–50

Exercise Notes

Reasoning

Ex. 1 Based on the results of this exercise, ask students to make a conjecture about the relationship between the domain and range of a function and the domain and range of its inverse function. Have students test their conjectures on several other translated parabolas.

EXAMPLE 2

Compare the graphs of $y = \sqrt{x}$ and $y = \sqrt{x - 2} + 1$. How are they geometrically related? State the domain and range of the second function.

SOLUTION

WATCH OUT!
Use parentheses on a graphing calculator to indicate what is under the square root sign.

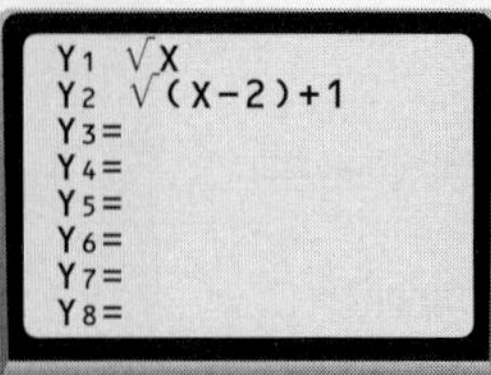

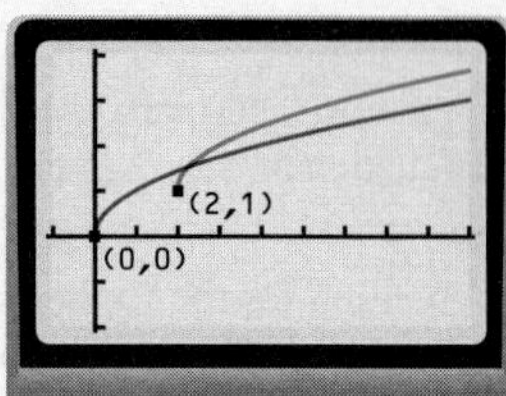

The second graph is the first graph translated 2 units to the right and 1 unit up. The domain of the second function is $x \geq 2$. The range is $y \geq 1$.

CHECKING KEY CONCEPTS

1. If you measure the height that a pole vaulter's center of gravity is raised in meters and the pole vaulter's running velocity in meters per second, an equation relating height and velocity is $h = \frac{v^2}{19.6}$.
   a. Write an equation for the velocity as a function of height.
   b. Graph the equation from part (a). State the domain and range.
   c. Suppose a pole vaulter's center of gravity is 1 m above the ground when running. Find the running velocity needed to clear a bar 5 m above the ground.

Graph each function and compare it with the graph of $f(x) = \sqrt{x}$.

2. $f(x) = 3\sqrt{x}$
3. $f(x) = \sqrt{x} - 3$
4. $f(x) = \sqrt{x - 3}$

8.1 Exercises and Applications

Extra Practice exercises on page 760

1. Use the tools you used in the Exploration on page 337.
   a. Graph $y = x^2 + 1$ and $y = x$ in the same coordinate plane.
   b. Reflect the graph of $y = x^2 + 1$ over the line $y = x$.
   c. Does the graph in part (b) represent a function? Explain.
   d. Write equations for two functions that together create the graph in part (b). What domain restrictions must you apply to the functions?

Checking Key Concepts

1. a. $v = \sqrt{19.6h}$
   b.

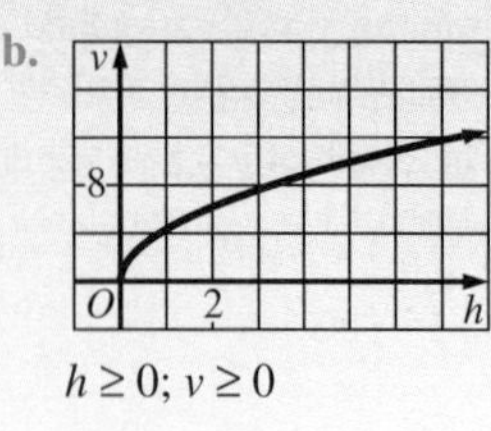

$h \geq 0$; $v \geq 0$

   c. 8.85 m/s

2.

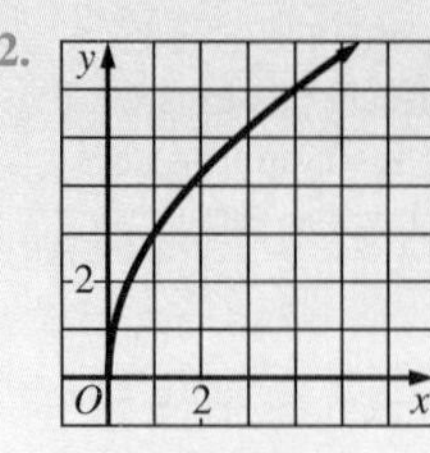

It is the graph of $f(x) = \sqrt{x}$ stretched vertically by a factor of 3.

3.

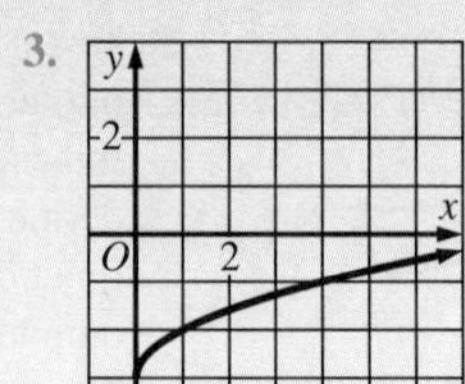

It is the graph of $f(x) = \sqrt{x}$ translated down 3 units.

4.

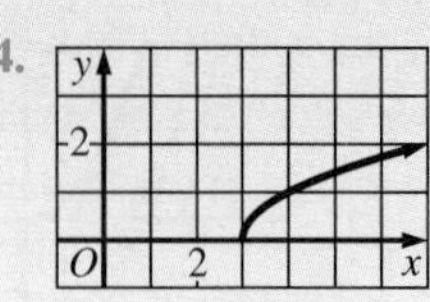

It is the graph of $f(x) = \sqrt{x}$ translated right 3 units.

In Example 1, you saw that you can rewrite $\sqrt{64h}$ as $8\sqrt{h}$ because 64 is a perfect square. Find an equivalent expression for each of the following. Assume $x \geq 0$.

2. $\sqrt{4x}$ **3.** $\sqrt{8x}$ **4.** $\sqrt{16x}$

5. $\sqrt{25x}$ **6.** $\sqrt{50x}$ **7.** $\sqrt{100x}$

Connection OCEANOGRAPHY

When the volcanic island of Krakatau exploded on August 27, 1883, ocean waves caused by the disturbance traveled great distances. The map at the bottom of the page shows how many minutes it took the first wave to reach various places in the Sunda Strait, between Sumatra and Java.

Early photographers captured the Krakatau eruption, one week after the eruptions started.

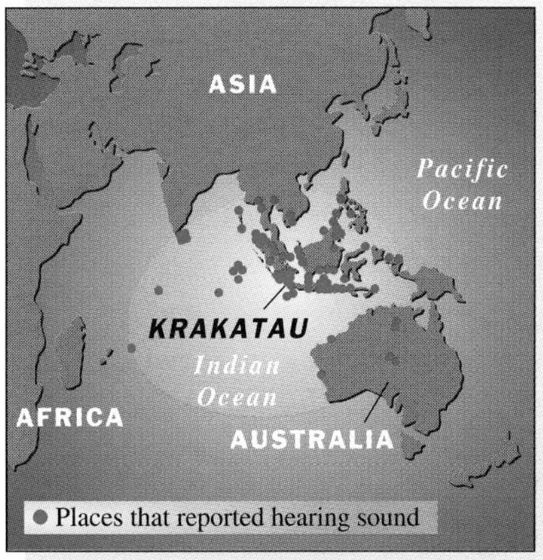

Sounds of the explosions from Krakatau were heard over a region covering 1/14th of the globe.

8. Estimate the speeds of the waves from Krakatau to Princes Island, Telok Betong, and Jakarta. Give your answers in kilometers per minute.

9. The speed of a wave depends on the depth of the water it travels through, according to the equation

$$s = \sqrt{35.28d}$$

where s = the speed of the wave in kilometers per minute and d = the ocean depth in kilometers. Use this equation to estimate the average ocean depths from Krakatau to Princes Island, Telok Betong, and Jakarta. Convert your answers to meters.

10. Waves from Krakatau's eruption reached Port Elizabeth, South Africa, in 15 h 12 min, after traveling 7546 km.

a. Find the average wave speed.

b. Find the average depth of the Indian Ocean between Krakatau and Port Elizabeth.

This 1979 explosion occurred on the island Anak Krakatau ("son of" Krakatau), which was formed after Krakatau exploded.

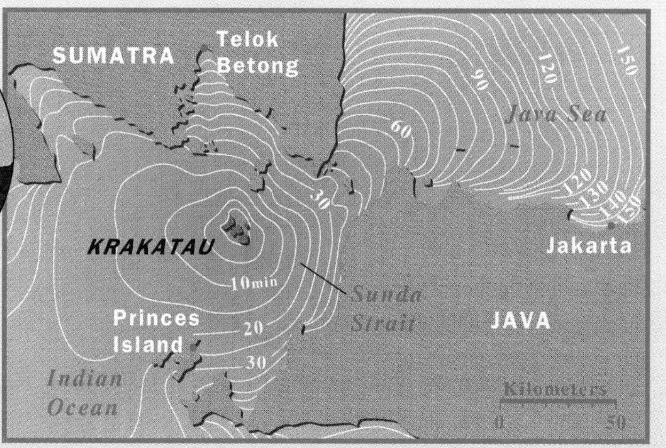

Exercises and Applications

1. a, b.

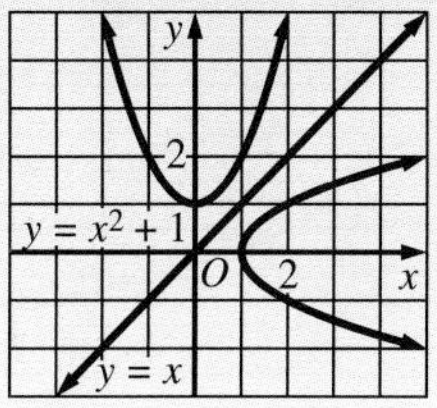

c. No; there are two outputs for each input except 1.

d. $y = \sqrt{x-1}$ and $y = -\sqrt{x-1}$; $x \geq 1$

2. $2\sqrt{x}$

3. $2\sqrt{2x}$

4. $4\sqrt{x}$

5. $5\sqrt{x}$

6. $5\sqrt{2x}$

7. $10\sqrt{x}$

8, 9. Estimates may vary.

8. 2.2 km/min; 1.2 km/min; 1.1 km/min (to the nearest tenth of a kilometer)

9. 137 m; 40.8 m; 34.3 m

10. a. 8.27 km/min

b. 1940 m

Apply⇔Assess

Exercise Notes

Cooperative Learning

Exs. 2–7 These exercises provide an opportunity for some peer tutoring. Pair up students having difficulty with these types of exercises with students who can simplify radicals well. The peer tutor can also make up some additional problems to work on for students who need review.

Interdisciplinary Problems

Exs. 8–10 In doing these exercises, students will see how square root functions can be used in oceanography to estimate average ocean depths from the speed of a wave.

Career Connection

Exs. 8–10 Oceanography is the science of Earth's oceans and their boundaries. The field of oceanography is large and varied, including everything from the study of the sediments at the bottom of oceans to the study of how oceans affect the weather. A person who studies oceanography is called an oceanographer. A strong background in science with an emphasis on physics, chemistry, biology, or geology is generally required to become an oceanographer. Oceanographers usually collect data in small areas of the ocean and study it to answer questions about subjects such as sea life, tides, waves, and the formation of Earth and its continents.

Multicultural Note

Exs. 8–10 More than 13,600 islands make up Indonesia, a country in Southeast Asia. About 60 percent of the total population of Indonesia lives on the island of Java, making it the most heavily populated of the country's islands. Also on Java is the capital of Indonesia, Jakarta (sometimes spelled *Djakarta*). This city is one of the major economic centers of Indonesia; its harbor, Tanjungperiuk, handles over half of Indonesia's foreign trade.

Topic Spiraling: Preview

Exs. 9, 10 Students must solve a radical equation to complete each of these exercises. Solving radical equations will be studied in depth in Section 8.3.

Apply⇔Assess

Exercise Notes

Student Study Tip
Exs. 11–16 Point out to students that the endpoint of the square-root graph can be found easily by noting the similarity between the equation of the square-root function and the quadratic function in vertex form. Remind students that the graph of the equation $y = (x - h)^2 + k$ has vertex (h, k). Similarly, the graph of $y = \sqrt{x - h} + k$ has endpoint (h, k). Once the endpoints are determined, most of the graphs in these problems can be easily identified.

Mathematical Procedures
Exs. 11–16 Encourage students to rewrite the function in Ex. 13 as $f(x) = -\sqrt{-x + 3}$ and to compare it to $g(x) = \sqrt{x + 3}$. To generate the graph of f from the graph of g, you must reflect it about both the x- and y-axes. Also, encourage students to test their answers to all these exercises by picking a point on the graph and substituting it into the chosen equation.

Interview Note
Ex. 17 The equation derived in this exercise was first mentioned in the interview about roller coaster design on page 336. You might ask students to review this page after doing Ex. 17. Students may be surprised to find that the velocity of the car is independent of the mass of the car. Remind them that if air resistance is ignored, all falling bodies near Earth's surface fall at the same rate.

Problem Solving
Ex. 19 Students may find it easier to answer this question if they solve the equation $v = \sqrt{2gh}$ for h and ask what happens to the height if they double the velocity.

Student Progress
Exs. 20–31 Students should be able to make a quick sketch of these graphs by considering transformations of the graph of $y = \sqrt{x}$. You can check this by asking students to describe the transformation(s) they would use for each problem.

Using Technology
Exs. 32–39 You may want to suggest that students use the *Function Investigator* software for this group of exercises.

Match each function with its graph.

11. $f(x) = \sqrt{x - 1} + 2$ **12.** $f(x) = \sqrt{x + 1} - 2$ **13.** $f(x) = -\sqrt{3 - x}$

14. $f(x) = -\sqrt{x - 3}$ **15.** $f(x) = \sqrt{x} + 2$ **16.** $f(x) = \sqrt{x + 2}$

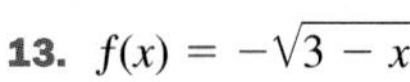

A. 

B. 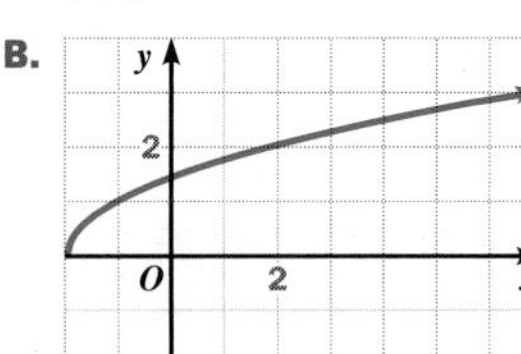

C. 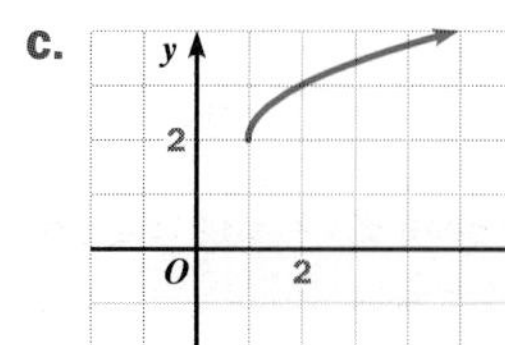

D. 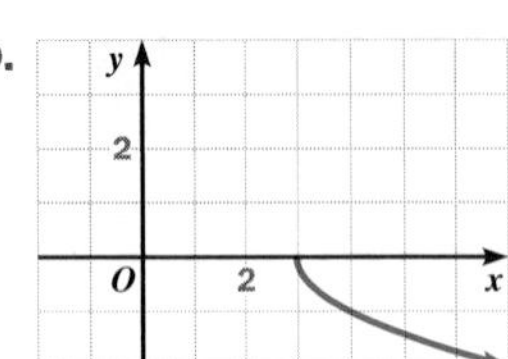

E. 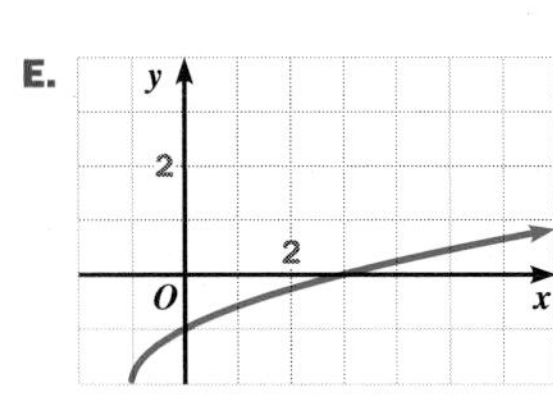

F. 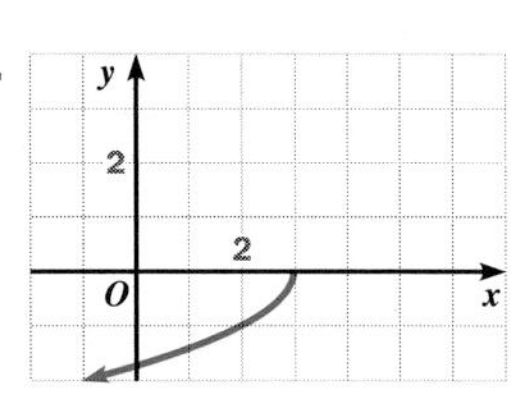

INTERVIEW Ronald Toomer

Look back at the article on pages 334–336.

Amusement park ride designers such as Ronald Toomer use physics to predict the velocity of a ride. In the Demon Drop ride, you plunge dozens of feet before your car follows a curve to slow down. If you ignore friction and air resistance, potential energy at the top is transformed into kinetic energy at the bottom, according to the equation

kinetic energy $\frac{1}{2}mv^2 = mgh$ potential energy

where m is mass, v is velocity, g is the acceleration due to gravity, and h is the height of the drop.

17. Solve the equation for v. Does the Demon Drop car drop faster when people are in it or when it is empty? Explain.

18. The Demon Drop ride involves a drop of 60 ft. Using $g = 32\ \text{ft/s}^2$, find the velocity of the car after it falls 60 ft.

19. By what factor would you have to change the height of the Demon Drop ride in order to double the car's velocity at the bottom?

Over 10 million people have ridden the Demon Drop at Cedar Point in Sandusky, Ohio.

11. C 12. E
13. F 14. D
15. A 16. B
17. $v = \sqrt{2gh}$; The velocity is independent of mass, so the speed is the same whether the car is occupied or not. (Note that the variable m is not involved in the equation for v.)
18. 62.0 ft/s
19. The height must be quadrupled.
20. 

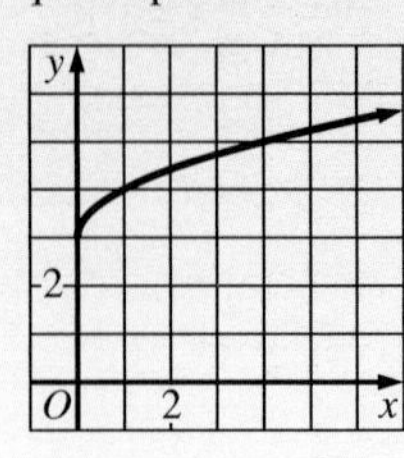

nonnegative numbers; $y \geq 3$

21. 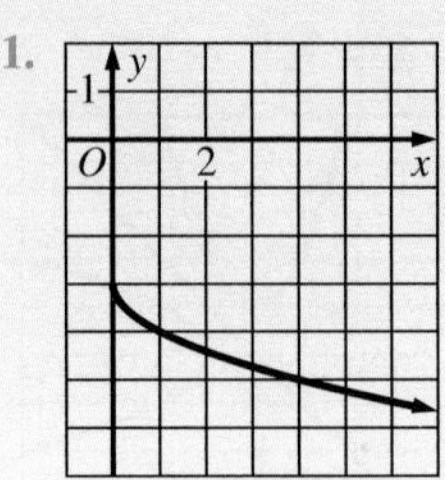

nonnegative numbers; $y \leq -3$

Graph each function. State the domain and range.

20. $y = \sqrt{x} + 3$ **21.** $y = -\sqrt{x} - 3$ **22.** $y = \sqrt{x + 3}$ **23.** $y = -\sqrt{x + 3}$

24. $y = -\sqrt{x}$ **25.** $y = 3 - \sqrt{x}$ **26.** $y = \sqrt{3x}$ **27.** $y = \sqrt{3x - 3}$

28. $y = \sqrt{x - 1}$ **29.** $y = \sqrt{x - 1} + 3$ **30.** $y = \sqrt{x + 1} - 3$ **31.** $y = \sqrt{x + 3} - 1$

Technology **Use a graphing calculator or graphing software to graph the left and right sides of each equation as separate functions. Decide whether each statement is *always true, sometimes true,* or *never true.***

32. $\sqrt{x^2 + 9} = x + 3$ **33.** $\sqrt{x - 2} = \sqrt{x} - 2$ **34.** $\sqrt{x^2} = x$ **35.** $(\sqrt{x})^2 = x$

36. $\sqrt{16x} = 4\sqrt{x}$ **37.** $\sqrt{-x} = -\sqrt{x}$ **38.** $\sqrt{x^4} = x^2$ **39.** $\sqrt{\frac{x}{9}} = \frac{\sqrt{x}}{3}$

40. Writing Is the statement $\sqrt{a + b} = \sqrt{a} + \sqrt{b}$ true for all nonnegative values of a and b? Explain your answer.

ONGOING ASSESSMENT

41. Cooperative Learning Work in a group of four students.

a. Discuss how the values of a, h, and k affect the graph of $y = a(x - h)^2 + k$.

b. Make conjectures about how the values of a, h, and k affect the graph of $y = a\sqrt{x - h} + k$. Then test your conjectures by graphing the equation for several values of a, h, and k, some positive and some negative.

c. Discuss how you can put equations like $y = 3\sqrt{2x - 4} + 1$ and $y = 3 - 2\sqrt{4x + 1}$ in the form of $y = a\sqrt{x - h} + k$ for graphing. Then work out several examples.

SPIRAL REVIEW

42. Chu is ordering T-shirts and sweatshirts for a fundraiser. He wants to order between 200 and 300 shirts altogether, and he wants at least 100 T-shirts and no more than 150 sweatshirts. *(Section 7.6)*

a. Write a system of inequalities and graph the feasible region. Label the coordinates of the vertices of the feasible region.

b. If the T-shirts cost \$5 each and the sweatshirts cost \$9 each, find the number of each that Chu should order to minimize his cost.

c. If the profit on a T-shirt is \$3 and the profit on a sweatshirt is \$6, find the number of each that Chu should order to maximize his profit.

Simplify. *(Section 3.2)*

43. $100^{1/2}$ **44.** $16^{1/4}$ **45.** $125^{1/3}$ **46.** $49^{1/2}$

Graph each inequality. *(Section 7.4)*

47. $y > x$ **48.** $2x - y \geq 1$ **49.** $150x + 40y \leq 200$ **50.** $-5x + 2y < 25$

Apply⇔Assess

Exercise Notes

Challenge

Exs. 32–39 These exercises exemplify many of the properties of radical expressions and disprove other conjectures about radical expressions. Ask students to write a formula for each property or conjecture. For a false conjecture, have them write a property by using the ≠ sign. The conjecture of Ex. 32 is stated and discussed in Ex. 40. Properties of radical expressions will be discussed further on page 346.

Assessment Note

Ex. 41 After completing this exercise, ask each student to write a summary of the group's findings in their notebooks or journals. These summaries would ensure that all students understand the relationship between the equation of a radical function and its graph.

Practice 47 for Section 8.1

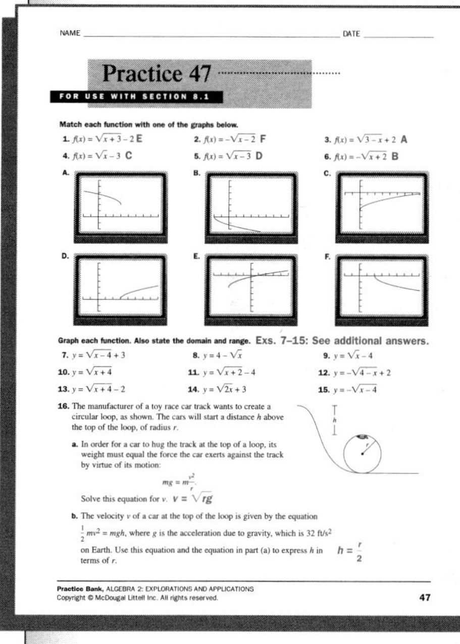

NAME ____________ DATE ________

Practice 47

FOR USE WITH SECTION 8.1

Match each function with one of the graphs below.

1. $f(x) = \sqrt{x + 3} - 2$ E **2.** $f(x) = -\sqrt{x - 2}$ F **3.** $f(x) = \sqrt{3 - x} + 2$ A

4. $f(x) = \sqrt{x} - 3$ C **5.** $f(x) = \sqrt{x - 3}$ D **6.** $f(x) = -\sqrt{x + 2}$ B

A. B. C.

D. E. F.

Graph each function. Also state the domain and range. Exs. 7–15: See additional answers.

7. $y = \sqrt{x - 4} + 3$ **8.** $y = 4 - \sqrt{x}$ **9.** $y = \sqrt{x} - 4$

10. $y = \sqrt{x + 4}$ **11.** $y = \sqrt{x + 2} - 4$ **12.** $y = -\sqrt{4 - x} + 2$

13. $y = \sqrt{x + 4} - 2$ **14.** $y = \sqrt{2x} + 3$ **15.** $y = -\sqrt{x - 4}$

16. The manufacturer of a toy race car track wants to create a circular loop, as shown. The cars will start a distance h above the top of the loop, of radius r.

a. In order for a car to hug the track at the top of a loop, its weight must equal the force the car exerts against the track by virtue of its motion:

$$mg = m\frac{v^2}{r}.$$

Solve this equation for v. $v = \sqrt{rg}$

b. The velocity v of a car at the top of the loop is given by the equation $\frac{1}{2}mv^2 = mgh$, where g is the acceleration due to gravity, which is 32 ft/s² on Earth. Use this equation and the equation in part (a) to express h in terms of r. $h = \frac{r}{2}$

Practice Bank, ALGEBRA 2: EXPLORATIONS AND APPLICATIONS

47

22.

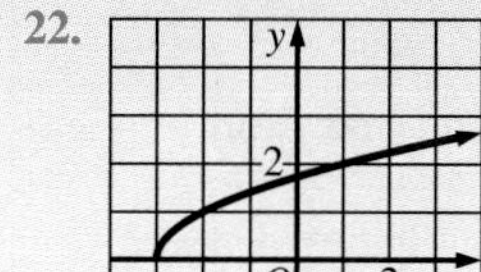

$x \geq -3$; nonnegative numbers

23.

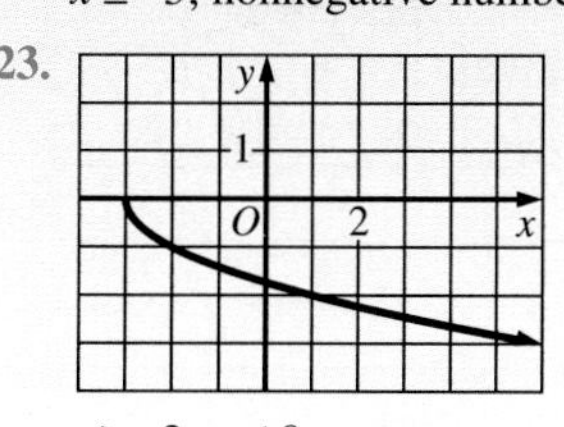

$x \geq -3$; $y \leq 0$

24.

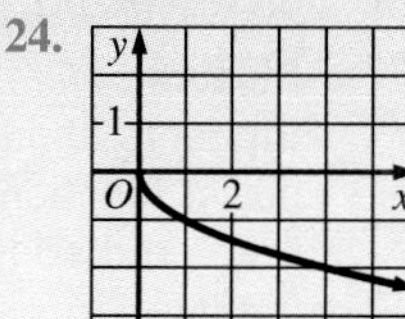

nonnegative numbers; $y \leq 0$

25.

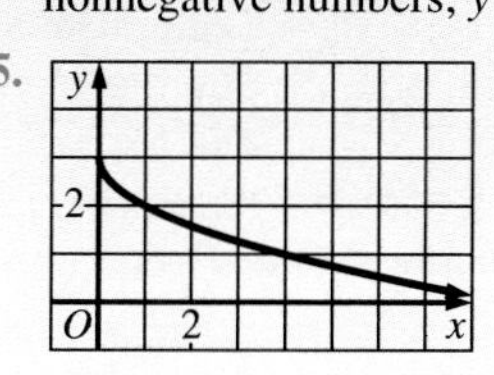

nonnegative numbers; $y \leq 3$

26.

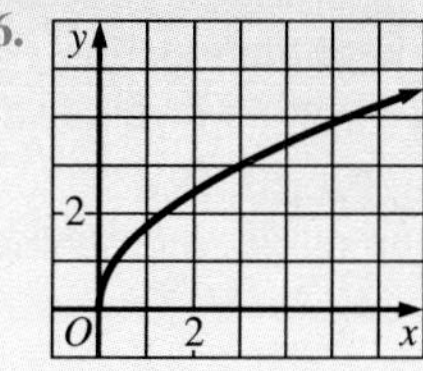

nonnegative numbers; nonnegative numbers

27.

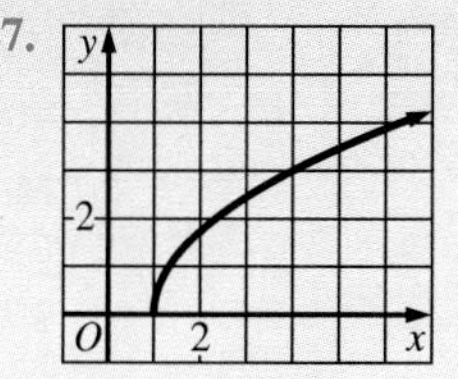

$x \geq 1$; nonnegative numbers

28.

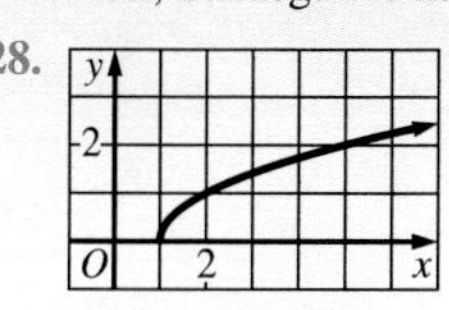

$x \geq 1$; nonnegative numbers

29.

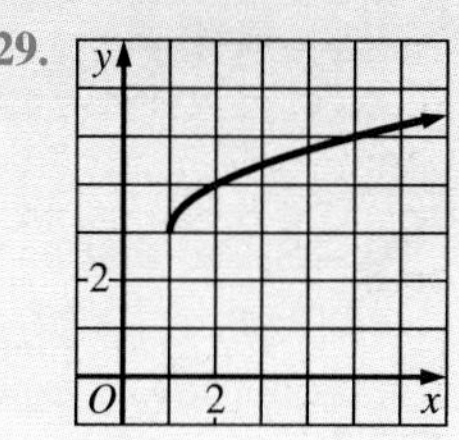

$x \geq 1$; $y \geq 3$

30–50. See answers in back of book.

Plan⇔Support

Objectives

- Evaluate radical expressions.
- Graph radical functions.
- Solve problems involving radical functions.

Recommended Pacing

Core and Extended Courses
Section 8.2: 1 day

Block Schedule
Section 8.2: $\frac{1}{2}$ block
(with Section 8.1)

Resource Materials

Lesson Support
Lesson Plan 8.2
Warm-Up Transparency 8.2
Practice Bank: Practice 48
Study Guide: Section 8.2
Exploration Lab Manual: Diagram Masters 1, 2

Technology
Technology Book: Spreadsheet Activity 8
Graphing Calculator
McDougal Littell Mathpack *Function Investigator with Matrix Analyzer Activity Book:* Function Investigator Activities 19 and 20
Internet: http://www.hmco.com

Warm-Up Exercises

1. Graph $y = (x - 3)^2$ and reflect the function over $y = x$.

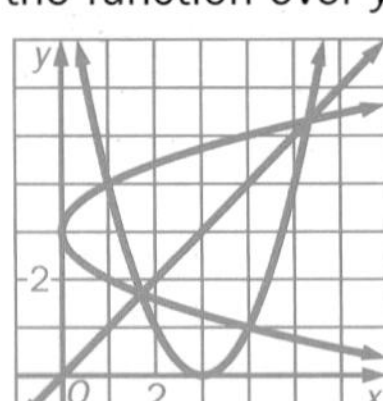

2. Restrict the domain of $y = (x - 3)^2$ so that its inverse is a function. either $x \le 3$ or $x \ge 3$

Use properties of exponents to simplify each expression.

3. $b^{1/2}b^{1/3}$ $b^{5/6}$
4. $(10^{1/3})^3$ 10
5. $\frac{4^{4/5}}{4^{2/5}}$ $4^{2/5}$

Section

8.2 Radical Functions

Learn how to...

- evaluate radical expressions
- graph radical functions

So you can...

- solve problems involving radical functions, such as finding the length of a fish based on its weight

If you restrict the domain of $y = x^2$ to values of $x \ge 0$, then the reflection of the graph of the restricted function over the line $y = x$ represents another function, the inverse $y = \sqrt{x}$. In the Exploration, you will see when you need to restrict the domains of other power functions of the form $y = x^n$ to create **radical functions** of the form $y = \sqrt[n]{x}$.

EXPLORATION
COOPERATIVE LEARNING

Reflecting the Graph of $y = x^n$ over the Line $y = x$

Work with a partner.
You will need:
- a graphing calculator or graphing software

SET UP Adjust the viewing window so that it shows $-3 \le x \le 3$ and $-2 \le y \le 2$.

1 Graph the line $y = x$ and the function $y = x^3$. Then draw the reflection of the function's graph over the line. Does the reflection represent a function?

2 Repeat Step 1 using the functions $y = x^4$, $y = x^5$, and $y = x^6$. (Clear the screen each time.)

Use your calculator or software's draw-inverse feature.

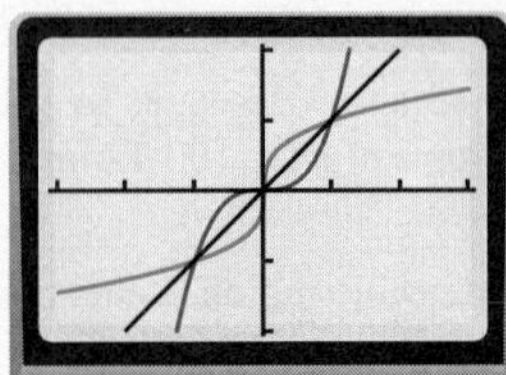

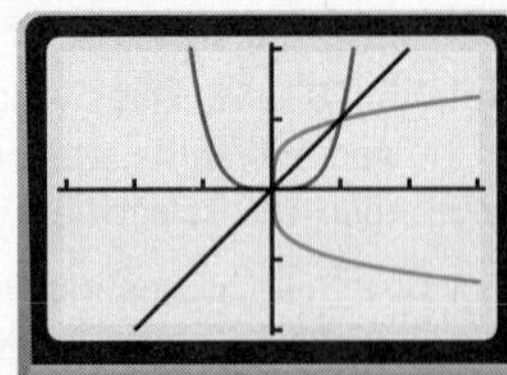

Questions

1. Describe how the graphs of $y = x^n$ when n is odd are different from the graphs when n is even. How are their reflections different?
2. Which functions of the form $y = x^n$ must have their domains restricted in order for the reflections of their graphs over the line $y = x$ to represent functions? How would you restrict the domains of the original functions?

Exploration Note

Purpose
The purpose of this Exploration is to have students discover those values of n for which the domain of the function $y = x^n$ must be restricted so that its inverse will be a function.

Materials/Preparation
Each group needs a graphing calculator or graphing software.

Procedure
For $n = 3$, 4, 5, and 6, students graph $y = x^n$ and $y = x$ in the same coordinate plane and use the draw-inverse feature to reflect $y = x^n$ over the line $y = x$. Questions 1 and 2 ask students to describe how the functions and their reflections differ when n is odd or even, and when the domain of $y = x^n$ should be restricted.

Closure
Students should understand that even power functions must have their domains restricted so that their reflections over the line $y = x$ are functions. Odd power functions need no such restriction.

Explorations Lab Manual
See the Manual for more commentary on this Exploration.

For answers to the Exploration, see answers in back of book.

Radical functions of the form

The positive integer n is called the *index* of the radical. $y = \sqrt[n]{x}$ Read this as "the nth root of x."

have a domain and range of *all* numbers when n is *odd* and a domain and range of *nonnegative* numbers when n is *even*.

Example: $y = \sqrt[3]{x}$ This is a cube root function. Example: $y = \sqrt[4]{x}$ This is a fourth root function.

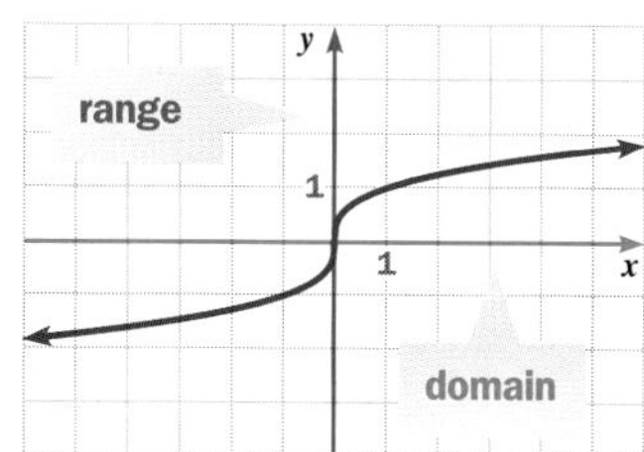

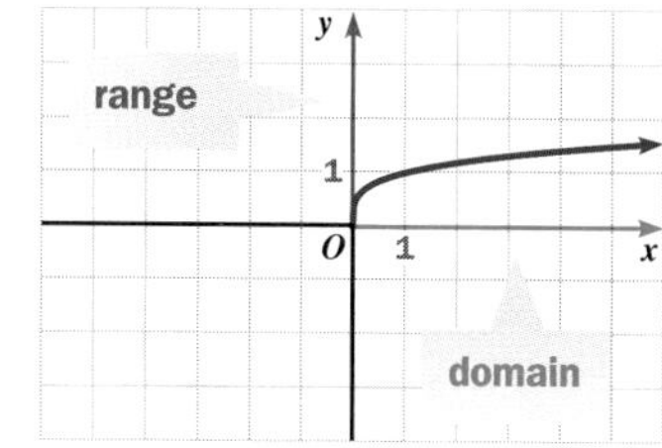

As you saw in the Exploration, a radical function is an inverse of a power function with a suitably restricted domain.

EXAMPLE 1 Application: Biology

A conservationist estimates the weight of a northern pike (in pounds) by cubing its length (in inches) and dividing by 3500.

a. Write an equation for the pike's weight w as a function of its length l.

b. Write an equation for the inverse of the function in part (a).

c. Graph the inverse function. Estimate the length of a 3 lb pike.

SOLUTION

a. $w = \frac{1}{3500}l^3$

b.
$$\frac{1}{3500}l^3 = w$$
$$l^3 = 3500w$$
$$l = \sqrt[3]{3500w}$$

Because weight is a cubic function of length, length is a cube root function of weight.

c.

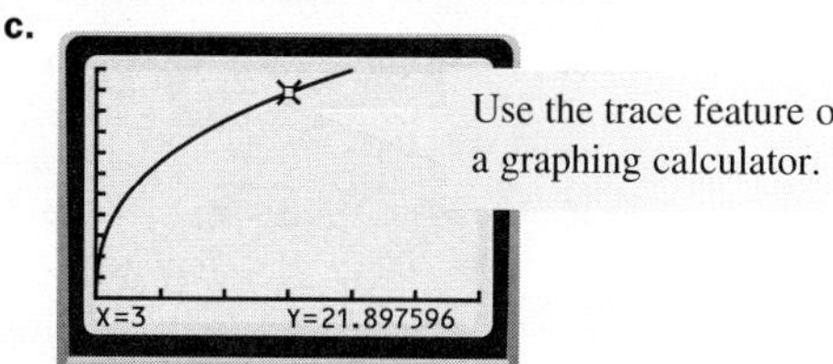

Use the trace feature of a graphing calculator.

A 3 lb northern pike is about 22 in. long.

Technology Note

Many students convert radicals expressions into expressions with fractional exponents before entering them into a calculator. However, students will not see the equivalence of these two types of expressions until page 346. When discussing part (c) of Example 1, the radical expression can be entered into a TI-82 by choosing option 4:$\sqrt[3]{\ }$ from the MATH menu. Similarly, other radicals can be expressed by choosing option 5:$\sqrt[x]{\ }$. In this case, the value of the index must be entered first. Caution students that parentheses must be used around any expression that is to go under the radical sign with the exception of a single number or variable. You may wish to suggest that students use the *Function Investigator* software with this section on radical functions. You can reinforce and extend the ideas of this section by assigning Function Investigator Activities 19 and 20 in the *Function Investigator with Matrix Analyzer Activity Book.*

Teach⇔Interact

Learning Styles: Visual

Encourage visual learners to memorize the graphs at the top of this page as a reminder of the domains and ranges of the various nth root functions.

About Example 1

Second-Language Learners
Students learning English may benefit from a discussion of the role of a *conservationist* (someone who works for a fish or wildlife organization) and the name *northern pike* (a kind of freshwater fish found in the Northern Hemisphere).

Additional Example 1

In an experiment for her biology class, Gwen measured the average swimming speed and tail length of four different fish. Using the data, she predicted that the length of the tail (in inches) of a fish could be found by raising the average swimming speed (in feet per second) to the fourth power and dividing by 333.

a. Write an equation for a fish's tail length t as a function of its average swimming speed s.

$t = \frac{s^4}{333}$

b. Write an equation for the inverse of the function in part (a).

$$\frac{s^4}{333} = t$$
$$s^4 = 333t$$
$$s = \sqrt[4]{333t}$$

Because tail length is a fourth power function of speed, speed is a positive fourth root function of tail length.

c. Graph the inverse function. Estimate the average swimming speed of a fish with a tail 5 inches long.

Use the trace feature of a graphing calculator.

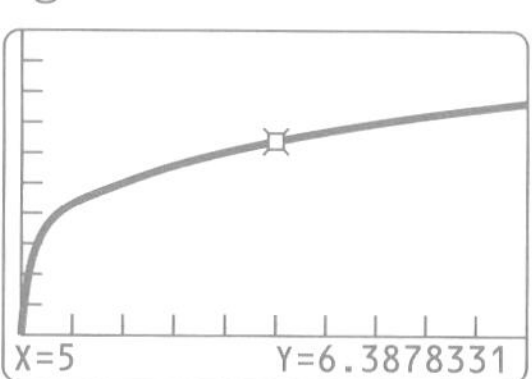

A fish with a 5 in. tail has an average speed of about 6.4 ft/s.

Teach⇔Interact

Think and Communicate

Question 1 calls students' attention to the domain of an nth root function when n is even. Even nth root functions cannot be defined for negative numbers because there is no real number that, when raised to an even power, will result in a negative number. Question 2 leads students to discover the equivalence of the nth root and the $1/n$th power.

Section Notes

Teaching Tip
The properties of radical expressions are presented as an example of the properties of exponents rather than as separate properties. This helps students see that the properties of exponents and radicals are the same even though different notations are used.

About Example 2

Teaching Tip
Remind students that when they simplify square roots, they should look for factors under the radical symbol that are perfect squares. Similarly, to simplify a cube root, they should look for perfect cube factors under the radical symbol, and so on.

Additional Example 2

Simplify each radical expression. Assume $b \geq 0$ in part (c).

a. $\sqrt[4]{\frac{1}{32}}$ $\quad \sqrt[4]{\frac{1}{32}} = \sqrt[4]{\frac{1}{16} \cdot \frac{1}{2}}$

$= \sqrt[4]{\frac{1}{16}} \cdot \sqrt[4]{\frac{1}{2}}$

$= \frac{1}{2}\sqrt[4]{\frac{1}{2}}$

b. $\sqrt[3]{54}$ $\quad \sqrt[3]{54} = \sqrt[3]{27 \cdot 2}$

$= \sqrt[3]{27} \cdot \sqrt[3]{2}$

$= 3\sqrt[3]{2}$

c. $\sqrt[5]{\frac{1}{b^{10}}}$ $\quad \sqrt[5]{\frac{1}{b^{10}}} = \left(\frac{1}{b^{10}}\right)^{1/5}$

$= \frac{1}{(b^{10})^{1/5}}$

$= \frac{1}{b^{10 \cdot 1/5}}$

$= \frac{1}{b^2}$

BY THE WAY

The first documented example of the extraction of higher roots appeared in a book called *Nine Chapters on the Mathematical Art*, compiled in China in the first century B.C. The volume presents many difficult mathematical problems, such as finding the cube root of 1,860,867.

Because radical functions are inverses of power functions, you can use the following definition to help you evaluate $y = \sqrt[n]{x}$ for given values of x.

The Meaning of $\sqrt[n]{b}$

The nth root of a positive number b is the positive number c whose nth power is b.

$\sqrt[n]{b} = c$ **if and only if** $b = c^n$

Examples:

$\sqrt{25} = 5$ because $25 = 5^2$

$\sqrt[3]{343} = 7$ because $343 = 7^3$

$\sqrt[4]{81} = 3$ because $81 = 3^4$

THINK AND COMMUNICATE

1. Can the definition given above be extended to negative values of b if n is even? if n is odd? Explain.
2. Look back at the definition of $b^{1/n}$ given on page 102. What do you notice? What can you conclude about $\sqrt[n]{b}$ and $b^{1/n}$?

You can extend the properties of exponents to radical expressions as long as the values of a and b below are not both negative.

Properties of Exponents	Examples with Fractional Exponents	Examples with Radical Expressions
$(ab)^m = a^m b^m$	$(ab)^{1/2} = a^{1/2} \cdot b^{1/2}$	$\sqrt{ab} = \sqrt{a} \cdot \sqrt{b}$
$\left(\frac{a}{b}\right)^m = \frac{a^m}{b^m}$	$\left(\frac{a}{b}\right)^{1/5} = \frac{a^{1/5}}{b^{1/5}}$	$\sqrt[5]{\frac{a}{b}} = \frac{\sqrt[5]{a}}{\sqrt[5]{b}}$
$(b^m)^n = b^{mn} = b^{nm} = (b^n)^m$	$(b^2)^{1/3} = b^{2/3} = (b^{1/3})^2$	$\sqrt[3]{b^2} = (\sqrt[3]{b})^2$

EXAMPLE 2

Simplify each radical expression. Assume $b \geq 0$ in part (c).

a. $\sqrt[3]{-24}$ **b.** $\sqrt[4]{810}$ **c.** $\sqrt[3]{b^6}$

SOLUTION

a. $\sqrt[3]{-24} = \sqrt[3]{-8 \cdot 3}$
$= \sqrt[3]{-8} \cdot \sqrt[3]{3}$
$= -2\sqrt[3]{3}$

$\sqrt[3]{-8} = -2$ because $(-2)^3 = -8$.

b. $\sqrt[4]{810} = \sqrt[4]{81 \cdot 10}$
$= \sqrt[4]{81} \cdot \sqrt[4]{10}$
$= 3\sqrt[4]{10}$

c. $\sqrt[3]{b^6} = (b^6)^{1/3}$
$= b^{6 \cdot 1/3}$
$= b^2$

ANSWERS Section 8.2

Think and Communicate

1. No; Yes; if n is even and b is negative, $\sqrt[n]{b}$ cannot be defined because there is no real number c for which $b = c^n$. If n is odd and b is negative, $\sqrt[n]{b}$ can be defined because there is a unique real number c for which $b = c^n$.
2. The definition is the same as that for $\sqrt[n]{b}$; $b^{1/n} = \sqrt[n]{b}$.

☑ CHECKING KEY CONCEPTS

State the domain and range of each function.

1. $y = \sqrt[3]{x+2}$
2. $y = \sqrt[4]{x-1}$
3. $y = \sqrt[5]{x} + 7$

Evaluate each radical expression, or state that it is *undefined*.

4. $\sqrt{4}$
5. $\sqrt{-4}$
6. $\sqrt[3]{27}$
7. $\sqrt[3]{-27}$
8. $\sqrt[6]{64}$
9. $\sqrt[3]{-64}$

Simplify each radical expression.

10. $\sqrt{45}$
11. $\sqrt[3]{270}$
12. $\sqrt[4]{2500}$

13. **GEOMETRY** The equation $V = \frac{1}{3}\pi r^3$ gives the volume V of a cone whose height and base radius are both r.
   a. Write an equation for the radius as a function of the volume.
   b. Find the radius of such a cone with a volume of 1000 cm³.

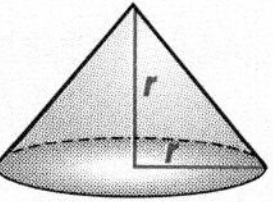

8.2 Exercises and Applications

Extra Practice exercises on page 760

Use the equation from Example 1 to find the length, to the nearest half inch, of a northern pike of each weight.

1. 1 lb
2. 2 lb
3. 4 lb
4. 5 lb

Evaluate each radical expression, or state that it is *undefined*.

5. $\sqrt{-25}$
6. $\sqrt{0}$
7. $\sqrt[3]{1000}$
8. $\sqrt[3]{-27}$
9. $\sqrt[3]{\frac{1}{8}}$
10. $\sqrt[3]{-64}$
11. $\sqrt[4]{625}$
12. $\sqrt[4]{-10{,}000}$
13. $\sqrt[4]{10{,}000}$
14. $\sqrt[5]{-1}$
15. $\sqrt[5]{32}$
16. $\sqrt[5]{-\frac{1}{32}}$

17. **GEOMETRY** The equation $V = \frac{1}{6}\pi d^3$ gives the volume V of a sphere as a function of its diameter d.
   a. Write an equation for the diameter as a function of the volume.
   b. Find the diameter of a sphere with a volume of 1000 cm³.
   c. **Challenge** Use part (a) to show that an equation for the radius r of a sphere as a function of its volume is $r = \sqrt[3]{\frac{3V}{4\pi}}$.

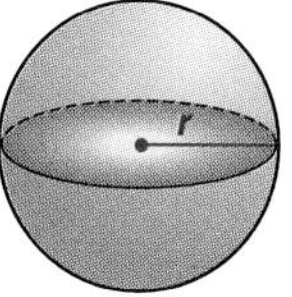

State the domain and range of each function.

18. $y = \sqrt{x+1}$
19. $y = \sqrt[3]{x-2}$
20. $y = \sqrt[4]{x} + 5$
21. $y = 3\sqrt[6]{x}$

Teach⇔Interact

Checking Key Concepts

Teaching Tip

For questions 1–3, encourage students to visualize the graph of each function in order to determine the domain and range. For questions 1 and 3, students should realize that these are translations of odd nth root functions and thus their domains and ranges will be real numbers. For question 2, because the index is even, there will be a restriction of the domain. This is a translation of the fourth root function one unit to the right with endpoint (1, 0).

Alternate Approach

To determine the domain of an even nth root function, such as in question 2, students could solve the inequality formed by setting the expression under the radical greater than or equal to zero.

Integrating the Strands

Question 13 and Exs. 17 and 40 give students an opportunity to see how radical expressions can be used to rewrite some of the common formulas from geometry for the volume of a cone and the volume and circumference of a sphere.

Closure Question

How is the function $y = \sqrt[n]{x}$ related to the function $y = x^n$? How is it related to $y = x^{1/n}$?

If n is odd, $y = \sqrt[n]{x}$ is the inverse of $y = x^n$. If n is even, $y = \sqrt[n]{x}$ is the inverse of $y = x^n$, $x \ge 0$. $y = \sqrt[n]{x}$ is the same function as $y = x^{1/n}$; they are just written in different notation.

Apply⇔Assess

Suggested Assignment

Core Course
Exs. 1–28, 40–46

Extended Course
Exs. 1, 2, 5–21 odd, 22–46

Block Schedule
Day 48 Exs. 1–28, 40–46

Checking Key Concepts

1. all numbers; all numbers
2. $x \ge 1$; nonnegative numbers
3. all numbers; all numbers
4. 2
5. undefined
6. 3
7. –3
8. 2
9. –4
10. $3\sqrt{5}$
11. $3\sqrt[3]{10}$
12. $5\sqrt[4]{4} = 5\sqrt{2}$
13. a. $r = \sqrt[3]{\frac{3V}{\pi}}$ b. 9.85 cm

Exercises and Applications

1. 15 in.
2. 19 in.
3. 24 in.
4. 26 in.
5. undefined
6. 0
7. 10
8. –3
9. $\frac{1}{2}$
10. –4
11. 5
12. undefined
13. 10
14. –1
15. 2
16. $-\frac{1}{2}$
17. a. $d = \sqrt[3]{\frac{6V}{\pi}}$ b. 12.4 cm
   c. $d = \sqrt[3]{\frac{6V}{\pi}}$, so $2r = \sqrt[3]{\frac{6V}{\pi}}$; $r = \frac{1}{2}\sqrt[3]{\frac{6V}{\pi}} = \sqrt[3]{\frac{1}{8}} \cdot \sqrt[3]{\frac{6V}{\pi}} = \sqrt[3]{\frac{1}{8} \cdot \frac{6V}{\pi}} = \sqrt[3]{\frac{3V}{4\pi}}$.
18. $x \ge -1$; nonnegative numbers
19. all numbers; all numbers
20. nonnegative numbers; $y \ge 5$
21. nonnegative numbers; nonnegative numbers

Apply⇔Assess

Exercise Notes

Communication: Listening and Reading

Exs. 22–24 For each exercise, ask a student to first read the expression and then give the answer. This will give students good practice at reading aloud and listening to exponential and radical expressions.

Reasoning

Exs. 25–28 When proving these statements, have students start with the expression on the left-hand side and convert it, using one step and property at a time, into the expression on the right-hand side. Stress to students that they should be able to justify each step of the process by citing a property.

Interview Note

Ex. 29 Students may recognize the formula $v = 8\sqrt{h}$, as it was used in Example 1 on page 339. This formula gives the height a pole vaulter's center of gravity is raised as a function of running velocity at launch. Discuss why the same formula can be used for an object that is rising and one that is falling.

Multicultural Note

Exs. 29–32 In the 1400s, Russians developed the first known "roller coaster"—a wooden slide covered with ice.

In Exercises 22–24, assume all variables are restricted to nonnegative values.

22. Express using fractional exponents.

a. $\sqrt[3]{b}$ b. $\sqrt[4]{d^3}$ c. $\sqrt[5]{s^2}$ d. $\sqrt[6]{t^5}$

23. Express using radical notation.

a. $z^{1/2}$ b. $p^{2/3}$ c. $q^{3/2}$ d. $y^{7/10}$

24. Simplify each expression.

a. $\sqrt{75}$ b. $\sqrt[3]{-432}$ c. $\sqrt[4]{3m^{12}}$ d. $\sqrt[5]{64g^{10}}$

Use fractional exponents and properties of exponents to prove each statement. Assume $x \geq 0$.

25. $\sqrt[6]{x} \cdot \sqrt[6]{x} = \sqrt[3]{x}$ **26.** $\dfrac{\sqrt[3]{x^2}}{\sqrt{x}} = \sqrt[6]{x}$ **27.** $x^{m/n} = \sqrt[n]{x^m}$ **28.** $\left(\sqrt[n]{x}\right)^n = x$

INTERVIEW Ronald Toomer

Look back at the article on pages 334–336.

When designing a roller coaster, Ronald Toomer can use the formula

$$v = \sqrt{2gh}$$

to find the maximum velocity of a roller coaster car, which occurs at the bottom of the first hill. For the exercises, ignore energy losses due to friction and air resistance.

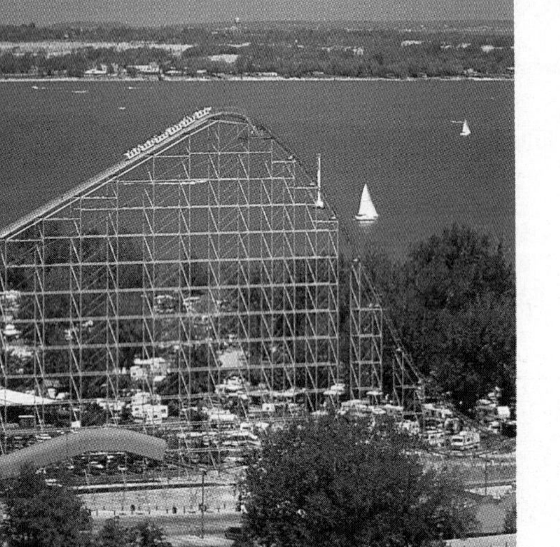

Ronald Toomer designed this roller coaster, the Magnum XL-200 on Lake Erie in Sandusky, Ohio.

29. In customary units, $g = 32$ ft/s^2. Use this fact to show that $v = \sqrt{2gh}$ can be simplified to $v = 8\sqrt{h}$.

30. Find the maximum velocity of each of the roller coaster cars in the table, using the data for the greatest vertical drop distances.

Roller coaster	Drop (ft)
American Eagle (Gurnee, IL)	147
Hercules (Allentown, PA)	157
Magnum XL-200 (Sandusky, OH)	205
Desperado (Jean, NV)	225

31. Use the results of Exercise 30 to graph $v = 8\sqrt{h}$. Identify the points on the graph that represent the roller coasters shown in the table.

32. Suppose you want to design a roller coaster whose maximum velocity is 100 mi/h. Convert this velocity from miles per hour to feet per second. Use your graph or the equation to find the drop for the first hill.

22. a. $b^{1/3}$ b. $d^{3/4}$ c. $s^{2/5}$ d. $t^{5/6}$

23. a. $\sqrt{z}$ b. $\sqrt[3]{p^2}$ c. $\sqrt{q^3}$ d. $\sqrt[10]{y^7}$

24. a. $5\sqrt{3}$ b. $-6\sqrt[3]{2}$ c. $m^3\sqrt[4]{3}$ d. $2g^2\sqrt[5]{2}$

25. $\sqrt[6]{x} \cdot \sqrt[6]{x} = x^{1/6} \cdot x^{1/6} = x^{1/6 + 1/6} = x^{1/3} = \sqrt[3]{x}$

26. $\dfrac{\sqrt[3]{x^2}}{\sqrt{x}} = \dfrac{x^{2/3}}{x^{1/2}} = x^{2/3 - 1/2} = x^{1/6} = \sqrt[6]{x}$

27. $x^{m/n} = x^{m(1/n)} = (x^m)^{1/n} = \sqrt[n]{x^m}$

28. $(\sqrt[n]{x})^n = (x^{1/n})^n = x^{(1/n)n} = x^{n/n} = x^1 = x$

29. $v = \sqrt{2gh} = \sqrt{2(32)h} = \sqrt{64h} = \sqrt{64}\sqrt{h} = 8\sqrt{h}$

30. American Eagle: 97.0 ft/s; Hercules: 100 ft/s; Magnum XL-200: 115 ft; Desperado: 120 ft

31.

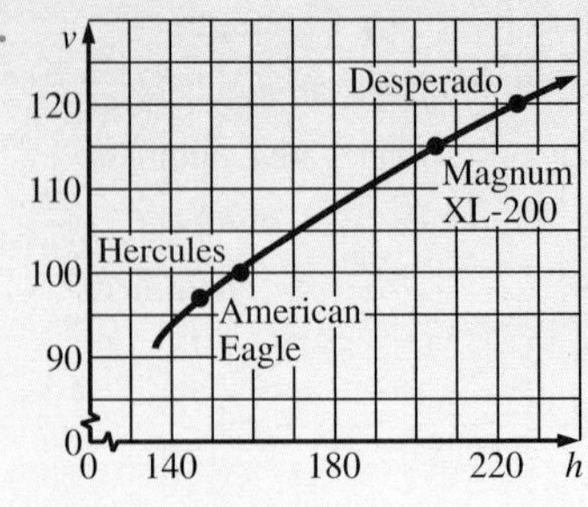

32. 336 ft

Connection SPORTS

A scientist studying rowing found that a crew's speed s (in meters per second) depends on the number of rowers n as follows:

$$s = 4.62\sqrt[9]{n}$$

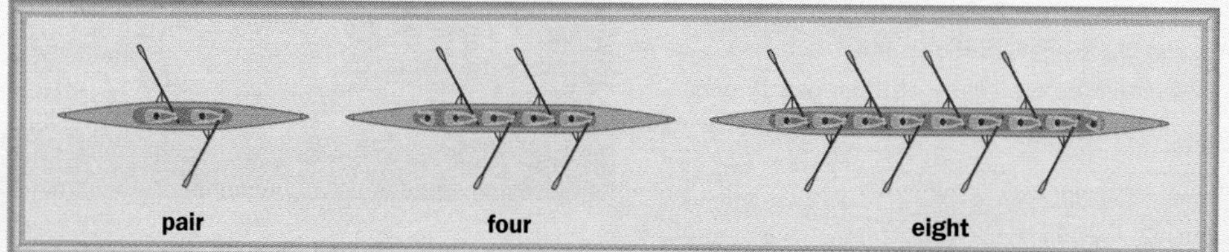

33. For $n = 1, 2, 4,$ and 8 rowers, find s to the nearest hundredth of a meter per second. Does doubling the number of rowers double the speed?

34. Crew races are typically 2000 m long. Find the time in minutes each boat shown above takes to complete a race.

35. Suppose a new kind of boat is designed to accommodate 6 rowers. Predict its speed and race time.

36. Open-ended Problem When is the square root of a number greater than the number itself? Use graphs to show your reasoning. How is the answer different for cube roots? Make a generalization about other even roots and odd roots.

37. SAT/ACT Preview Suppose A $= \sqrt[3]{\sqrt{x}}$ and B $= \sqrt[6]{x}$ for $x \geq 0$. Which of the following is true?

A. A > B **B.** B > A **C.** A = B **D.** relationship cannot be determined

38. BIOLOGY Scientists can compare an animal's metabolic rate M (its daily heat production in Calories) to its weight W (in kilograms). For many species, the results can be modeled by this equation:

$$M = 70\sqrt[4]{W^3}$$

a. Rewrite the equation using a fractional exponent.

b. Calculate the missing metabolic rates in the table.

c. Solve the equation for W.

d. Calculate the missing weights in the table.

Animal	Weight (kg)	Metabolic rate (Cal)
Mouse	?	3.6
Cat	3	?
Rabbit	?	191
Dog	14	?
Goat	?	800
Chimpanzee	38	?
Sheep	?	1160
Pony	253	?
Bull	?	12,100
Elephant	3672	?

Apply⇔Assess

Exercise Notes

Application

Exs. 33–35 In doing these exercises, students learn how a radical function can be used to predict the speed of a crew boat based on the number of rowers. For Ex. 34, students may need to be reminded of the formula $d = st$, which relates distance, speed, and time. They will need to solve the formula for t to get the finish time in seconds and then convert the time to minutes.

Using Technology

Exs. 33, 34, 38 The list features of the TI-82 calculator can be used to simplify the calculations in Exs. 33 and 34. Enter the values of n in L1. Then, moving the cursor onto L2 at the top of the second list, enter the formula for s, replacing n with L1. Use a similar procedure to make L3 the time in seconds (2000/L2), and L4 the time in minutes (L3/60). The function feature of the TI-82 would probably be most useful for Ex. 38. Enter the equation for metabolic rate as Y1 on the Y= list and the equation for weight as Y2. Then on the home screen, enter Y1({3,14,38,253, 3672}). This will give the missing metabolic rates. Enter Y2({3.6, 191,800,1160,12100}) to find the missing weights. (Y1 and Y2 are found under the function option of the Y-VARS menu.)

Problem Solving

Ex. 36 If students have difficulty beginning this exercise, ask them to graph $y = x$ and $y = \sqrt{x}$ and consider the relative sizes of x and $\sqrt{x}$ when the square root graph is above the line and when it is below the line.

33.

Number of rowers	Speed (m/s)
1	4.62
2	4.99
4	5.39
8	5.82

No.

34.

Number of rowers	Time (min)
2	6.68
4	6.18
8	5.73

35. 5.64 m/s; 5.91 min

36. $\sqrt{x} > x$ when $0 < x < 1$; $\sqrt[3]{x} > x$ when $0 < x < 1$ or $x < -1$; if n is even, $\sqrt[n]{x} > x$ when $0 < x < 1$ and if n is odd, $\sqrt[n]{x} > x$ when $0 < x < 1$ or $x < -1$.

y, $y = x$, $y = \sqrt{x}$, 2, O, 2, x

37. C

38. See answers in back of book.

Apply⇔Assess

Exercise Notes

Topic Spiraling: Review

Ex. 39 Students need to recall the method that uses logarithms to find a power function by finding the linear regression equation for points (log d, log t) and then converting this equation into power function form. You may wish to review this topic prior to assigning the exercise.

Mathematical Procedures

Ex. 40 Students should realize in part (d) that C varies directly with the cube root of V and that they are finding a constant of variation. One method to find the constant would be to calculate the ratio of C to $\sqrt[3]{V}$ for each line in the table and then let a be the average. Another method would be to run a linear regression using the values of $\sqrt[3]{V}$ for x and the values of C for y.

Practice 48 for Section 8.2

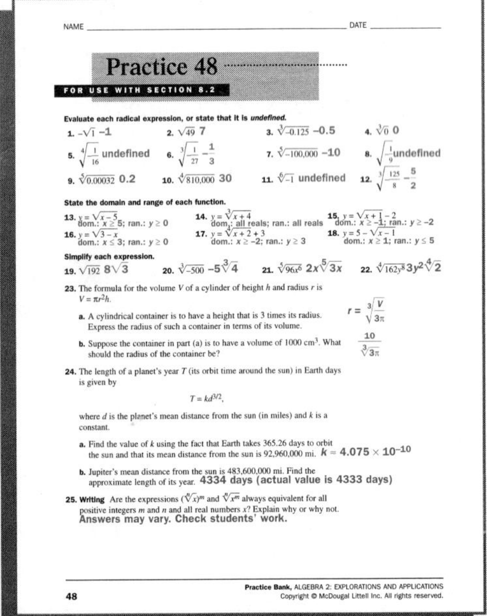

NAME ____________________ DATE __________

Practice 48

FOR USE WITH SECTION 8.2

Evaluate each radical expression, or state that it is *undefined*.

1. $-\sqrt{1}$ **−1**
2. $\sqrt{49}$ **7**
3. $\sqrt[3]{-0.125}$ **−0.5**
4. $\sqrt[3]{0}$ **0**
5. $\sqrt[4]{-\frac{1}{16}}$ **undefined**
6. $\sqrt[3]{-\frac{1}{27}}$ **$-\frac{1}{3}$**
7. $\sqrt[5]{-100{,}000}$ **−10**
8. $\sqrt{-\frac{1}{9}}$ **undefined**
9. $\sqrt[5]{0.00032}$ **0.2**
10. $\sqrt[4]{810{,}000}$ **30**
11. $\sqrt[6]{-1}$ **undefined**
12. $\sqrt[3]{-\frac{125}{8}}$ **$-\frac{5}{2}$**

State the domain and range of each function.

13. $y = \sqrt{x-5}$ **dom.: $x \geq 5$; ran.: $y \geq 0$**
14. $y = \sqrt[3]{x+4}$ **dom.: all reals; ran.: all reals**
15. $y = \sqrt{x+1} - 2$ **dom.: $x \geq -1$; ran.: $y \geq -2$**
16. $y = \sqrt{3-x}$ **dom.: $x \leq 3$; ran.: $y \geq 0$**
17. $y = \sqrt[4]{x+2} + 3$ **dom.: $x \geq -2$; ran.: $y \geq 3$**
18. $y = 5 - \sqrt{x-1}$ **dom.: $x \geq 1$; ran.: $y \leq 5$**

Simplify each expression.

19. $\sqrt{192}$ **$8\sqrt{3}$**
20. $\sqrt[3]{-500}$ **$-5\sqrt[3]{4}$**
21. $\sqrt[5]{96x^6}$ **$2x\sqrt[5]{3x}$**
22. $\sqrt[4]{162y^8}$ **$3y^2\sqrt[4]{2}$**

23. The formula for the volume V of a cylinder of height h and radius r is $V = \pi r^2 h$.
    a. A cylindrical container is to have a height that is 3 times its radius. Express the radius of such a container in terms of its volume. **$r = \sqrt[3]{\frac{V}{3\pi}}$**
    b. Suppose the container in part (a) is to have a volume of 1000 cm^3. What should the radius of the container be? **$\frac{10}{\sqrt[3]{3\pi}}$**

24. The length of a planet's year T (its orbit time around the sun) in Earth days is given by

$$T = kd^{3/2},$$

where d is the planet's mean distance from the sun (in miles) and k is a constant.
    a. Find the value of k using the fact that Earth takes 365.26 days to orbit the sun and that its mean distance from the sun is 92,960,000 mi. **$k \approx 4.075 \times 10^{-10}$**
    b. Jupiter's mean distance from the sun is 483,600,000 mi. Find the approximate length of its year. **4334 days (actual value is 4333 days)**

25. **Writing** Are the expressions $(\sqrt[n]{x})^m$ and $\sqrt[n]{x^m}$ always equivalent for all positive integers m and n and all real numbers x? Explain why or why not. **Answers may vary. Check students' work.**

48 Practice Bank, ALGEBRA 2: EXPLORATIONS AND APPLICATIONS

39. **ASTRONOMY** Saturn has many satellites. The table shows the distances of some of these from Saturn, and the number of days each satellite takes to orbit around Saturn.
    a. Use the methods of Section 4.5 to find a power function that shows how orbit time t is a function of the distance d. Express the power function using a fractional exponent.
    b. Rewrite the power function from part (a) using radical notation.
    c. Predict the orbit time of the moon Phoebe.

Satellite	Distance (100,000 km)	Orbit time (days)
Mimas	1.86	0.942
Enceladus	2.38	1.370
Tethys	2.95	1.888
Dione	3.77	2.737
Rhea	5.27	4.518
Titan	12.22	15.94
Hyperion	14.84	21.28
Iapetus	35.62	79.33
Phoebe	129.30	?

ONGOING ASSESSMENT

40. **Cooperative Learning** Work with another student. You will need a round balloon and a tape measure.
    a. Blow up your balloon using equal shallow breaths. Keep track of the number of breaths you use.
    b. When the balloon looks like a sphere, record its volume V (in number of breaths) and have your partner measure its circumference C. Make a table showing V, C, and $\sqrt[3]{V}$ for your balloon.
    c. Add another breath and measure the balloon's new circumference. Repeat until you have at least four data values in your table.
    d. Find an equation of the form $C = a\sqrt[3]{V}$ that models your data.

SPIRAL REVIEW

Graph each function. State the domain and range. *(Section 8.1)*

41. $y = \sqrt{x} + 1$
42. $y = -\sqrt{2x}$
43. $y = \sqrt{x+4} - 2$

44. An object falls according to the equation $d = 16t^2$, where d is the distance in feet, and t is the time in seconds. *(Section 5.1)*
    a. How far does an object fall in 3 s?
    b. How long does it take an object to fall 210 ft?

Solve. Check to eliminate any extraneous solutions. *(Sections 4.4 and 5.6)*

45. $\log_2(x+6) + \log_2(x-6) = 6$
46. $\log_6(x+1) + \log_6 x = 1$

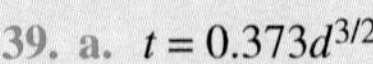

39. a. $t = 0.373d^{3/2}$
    b. $t = 0.373\sqrt{d^3}$
    c. 548 days
40. Answers may vary. Check students' work.
41.

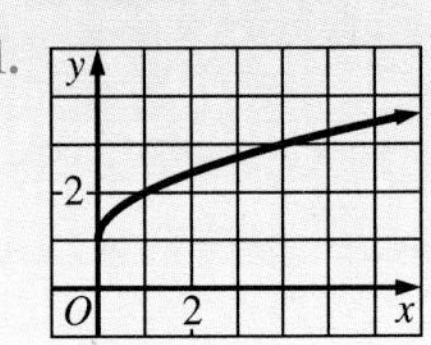

nonnegative numbers; $y \geq 1$

42.

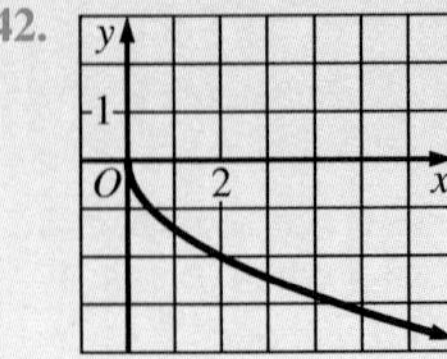

nonnegative numbers; $y \leq 0$

43.

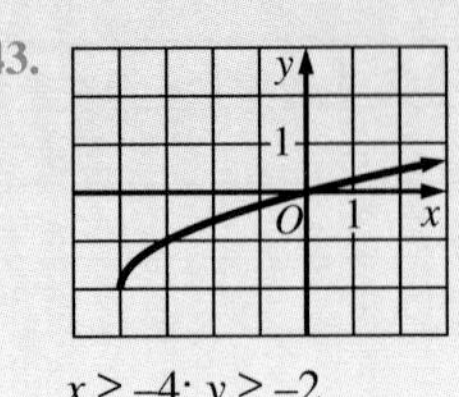

$x \geq -4$; $y \geq -2$

44. a. 144 ft
    b. 3.62 s
45. 10
46. 2

Section

8.3 Solving Radical Equations

Learn how to...

- solve equations with radical expressions

So you can...

- solve problems involving free fall, wind chill, and hurricanes, for example

In *Danny Dunn and the Fossil Cave,* by J. Williams and R. Abrashkin, the young hero Danny Dunn gets separated from his friends while exploring a cave. He finds himself in a narrow passageway near a crevice.

All at once, there was nothing under his right foot.

Nothing! Even as the thought flashed through his mind, his foot went down into empty air and he threw out his right hand to try to find some support. His flashlight went whirling away, and vanished into a huge black space, and in the same instant he realized that he was not going to fall. He was wedged tightly between the rocks by the folds of his jacket, and this held him securely in place.

He was too frightened even to yell. . . . At the same time, his mind was busy counting: *one, two, three, four, five* . . . and then he heard the faint clatter as his flashlight hit bottom somewhere far below.

EXAMPLE 1

It takes time for Danny Dunn's flashlight to fall down the crevice, and it takes time for the sound to travel up to his ears. The combined time is 5 s.

a. Recall from Example 3 on page 187 that the equation $d = 16t^2$ gives the distance d (in feet) that an object drops in t seconds. Use the equation to find the time it takes the flashlight to drop d feet.

b. Write an expression for the time it takes the sound to travel up d feet. Assume the speed of sound in the cave is 1116 ft/s.

c. Combine the results in parts (a) and (b) to write an equation that states the total time is 5 s. Solve the equation to estimate the crevice depth.

SOLUTION

a. Solve $d = 16t^2$ for t.

$$16t^2 = d$$

$$t^2 = \frac{1}{16}d$$

$$t = \sqrt{\frac{1}{16}d}$$

$$t = \frac{1}{4}\sqrt{d}$$

Since time is positive, use the positive square root.

b. Use the formula distance = rate × time:

$$d = 1116t$$

$$\frac{1}{1116}d = t$$

Solution continued on next page.

Plan⇔Support

Objectives

- Solve equations with radical expressions.
- Solve problems using radical equations.

Recommended Pacing

Core and Extended Courses
Section 8.3: 2 days

Block Schedule
Section 8.3: 1 block

Resource Materials

Lesson Support
Lesson Plan 8.3
Warm-Up Transparency 8.3
Practice Bank: Practice 49
Study Guide: Section 8.3
Exploration Lab Manual: Additional Exploration 14, Diagram Master 2
Assessment Book: Test 30

Technology
Graphing Calculator
McDougal Littell Mathpack *Function Investigator*
Internet: http://www.hmco.com

Warm-Up Exercises

1. Multiply $(x - 2)^2$. $x^2 - 4x + 4$

2. Factor $x^2 - 3x - 18$. $(x - 6)(x + 3)$

Evaluate when $x = -3$.

3. $\sqrt{-2(x + 1)}$ 2

4. $\sqrt[3]{14x + 15}$ -3

5. If $d = 16t^2$, find t when $d = 128$. $t = \pm 2\sqrt{2}$

Teach⇔Interact

Additional Example 1

Julie drops a rock from a high bridge into the water below. She hears it hit the water 6 seconds after she dropped it.

a. Use the equation $d = 4.9t^2$, which gives the distance d (in meters) that an object falls in t seconds, to find the time it takes the rock to fall d meters.

Solve $d = 4.9t^2$ for t.

$$4.9t^2 = d$$
$$t^2 \approx 0.2d$$
$$t \approx \sqrt{0.2d}$$
$$t \approx 0.45\sqrt{d}$$

b. Write an expression for the time it takes the sound of the rock hitting the water to travel up d meters. Assume the speed of sound is 342 m/s.

Use the formula distance = rate × time.

$$d = 342t$$
$$\frac{1}{342}d = t$$

c. Combine the results in parts (a) and (b) to write an equation that states the total time is 6 s. Solve the equation to estimate the height of the bridge.

Since the sum of rock travel time and sound travel time is 6, an equation using expressions from parts (a) and (b) is $0.45\sqrt{d} + \frac{1}{342}d = 6$.

Graph $y = 0.45\sqrt{x} + \frac{1}{342}x$ and $y = 6$ on a graphing calculator and find the intersection.

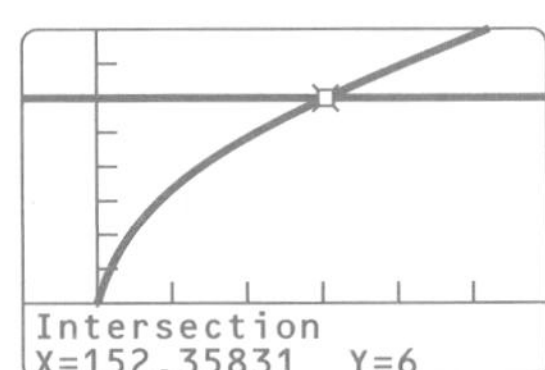

The bridge is about 152 m above the water.

Additional Example 2

Solve $\sqrt{10x + 11} = x + 2$.

Square both sides to eliminate the radical expression. Rearrange terms and solve using factoring.

$$\sqrt{10x + 11} = x + 2$$
$$(\sqrt{10x + 11})^2 = (x + 2)^2$$
$$10x + 11 = x^2 + 4x + 4$$
$$0 = x^2 - 6x - 7$$
$$0 = (x - 7)(x + 1)$$
$$x - 7 = 0 \quad \text{or} \quad x + 1 = 0$$
$$x = 7 \qquad\qquad x = -1$$

SOLUTION *continued*

c. Since the sum of flashlight travel time and sound travel time is 5 s, an equation using the expressions from parts (a) and (b) is $\frac{1}{4}\sqrt{d} + \frac{1}{1116}d = 5$.

Graph $y = \frac{1}{4}\sqrt{x} + \frac{1}{1116}x$ and $y = 5$ on a graphing calculator or graphing software, and find the intersection.

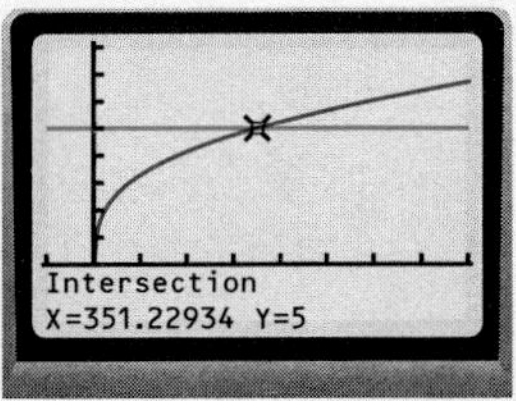

The crevice is about 350 ft deep.

THINK AND COMMUNICATE

1. Explain why the expression $\sqrt{\frac{1}{16}d}$ can be rewritten as $\frac{1}{4}\sqrt{d}$.

2. If the crevice were 700 ft deep instead of 350 ft deep, would Danny Dunn have had to wait 10 s instead of 5 s to hear his flashlight?

Example 1 showed a graphical method for solving radical equations. The following examples show an algebraic method. This method sometimes produces extraneous solutions, however.

EXAMPLE 2

Solve $\sqrt{2x - 3} = x - 3$.

SOLUTION

$$\sqrt{2x - 3} = x - 3$$
$$\left(\sqrt{2x - 3}\right)^2 = (x - 3)^2$$

Square both sides to eliminate the radical expression.

$$2x - 3 = x^2 - 6x + 9$$
$$0 = x^2 - 8x + 12$$
$$0 = (x - 6)(x - 2)$$

Rearrange terms and solve using factoring.

$$x - 6 = 0 \quad \text{or} \quad x - 2 = 0$$
$$x = 6 \quad \text{or} \quad x = 2$$

These are the possible solutions.

Check

$$\sqrt{2 \cdot 6 - 3} \stackrel{?}{=} 6 - 3$$
$$\sqrt{9} \stackrel{?}{=} 3$$
$$3 = 3 \checkmark$$

The solution $x = 6$ does check.

Check

$$\sqrt{2 \cdot 2 - 3} \stackrel{?}{=} 2 - 3$$
$$\sqrt{1} \stackrel{?}{=} -1$$
$$1 \neq -1$$

The solution $x = 2$ does not check.

The solution of the equation $\sqrt{2x - 3} = x - 3$ is 6.

Technology Note

Remind students using a TI-82 graphing calculator that they can find the intersection in Example 1, part (c) easily and quickly using the CALC menu. After graphing both equations and adjusting the window so that the intersection shows, simply choose option 5:intersect.

You may wish to suggest that students use the *Function Investigator* software with this section on solving radicals.

THINK AND COMMUNICATE

3. Which solution in Example 2 is the extraneous solution?
4. **Technology** Solve $\sqrt{2x-3} = x - 3$ using technology. Do you get extraneous solutions when you use technology?
5. Suppose you are solving $\sqrt{2x+3} = x - 3$ instead of $\sqrt{2x-3} = x - 3$. You get the equation $0 = x^2 - 8x + 6$. What would you do next?

EXAMPLE 3

Solve $\sqrt[3]{x-5} + 4 = -3$.

SOLUTION

$\sqrt[3]{x-5} + 4 = -3$ — Isolate the radical on one side.

$\sqrt[3]{x-5} = -7$

$\left(\sqrt[3]{x-5}\right)^3 = (-7)^3$ — Cube both sides of the equation.

$x - 5 = -343$

$x = -338$

Check

$\sqrt[3]{-338-5} + 4 \stackrel{?}{=} -3$

$\sqrt[3]{-343} + 4 \stackrel{?}{=} -3$

$-7 + 4 \stackrel{?}{=} -3$

$-3 = -3$ ✓

CHECKING KEY CONCEPTS

Solve by raising both sides of each equation to the same power.

1. $\sqrt{x-3} = 3$
2. $\sqrt{x+5} = 4$
3. $\sqrt{2x+7} = 1$
4. $\sqrt[3]{5x+4} = 4$
5. $\sqrt[4]{4x+1} = 3$
6. $\sqrt[5]{3x-7} = 2$

Solve. Check to eliminate extraneous solutions.

7. $\sqrt{x+13} = x + 7$
8. $\sqrt{3(x-7)} = x - 7$
9. $\sqrt{3(x+9)} = x + 3$

8.3 Exercises and Applications

Extra Practice exercises on page 761

Solve.

1. $\sqrt{x} = 7$
2. $\sqrt{x-1} = 7$
3. $\sqrt{x+3} = 8$
4. $\sqrt{x+4} = 0$
5. $\sqrt{2x} = 6$
6. $\sqrt{3x-2} = 5$
7. $\sqrt{4x+4} = 8$
8. $\sqrt{5(x+3)} = 10$
9. $5\sqrt{x+3} = 10$
10. $5 + \sqrt{x+3} = 10$
11. $10 + \sqrt{x+3} = 5$
12. $5\sqrt{\frac{x-9}{4}} = 15$

Teach⇔Interact

Additional Example 2 (continued)

Check

$\sqrt{10 \cdot 7 + 11} \stackrel{?}{=} 7 + 2$

$\sqrt{70 + 11} \stackrel{?}{=} 9$

$9 = 9$ ✓

The solution $x = 7$ does check.

$\sqrt{10 \cdot (-1) + 11} \stackrel{?}{=} -1 + 2$

$\sqrt{-10 + 11} \stackrel{?}{=} 1$

$1 = 1$ ✓

The solution $x = -1$ also checks. The solutions of the equation $\sqrt{10x + 11} = x + 2$ are 7 and –1.

Additional Example 3

Solve $-7\sqrt[4]{x+10} = -35$.

Isolate the radical on one side and raise both sides of the equation to the fourth power.

$-7\sqrt[4]{x+10} = -35$

$\sqrt[4]{x+10} = 5$

$\left(\sqrt[4]{x+10}\right)^4 = 5^4$

$x + 10 = 625$

$x = 615$

Check

$-7\sqrt[4]{615+10} \stackrel{?}{=} -35$

$-7\sqrt[4]{625} \stackrel{?}{=} -35$

$-7(5) \stackrel{?}{=} -35$

$-35 = -35$ ✓

Closure Question

Describe two methods for solving a radical equation.

One method is to use a graphing calculator or software and graph the related function for each side of the equation. The x-coordinate(s) of any intersection(s) is a solution. The other method is to use algebra. First, isolate the radical expression and then raise both sides of the equation to the power equal to the index of the radical. Solve the resulting equation. Make sure to check for extraneous solutions.

ANSWERS Section 8.3

Think and Communicate

1. $\sqrt{\frac{1}{16}d} = \sqrt{\frac{1}{16}}\sqrt{d} = \frac{1}{4}\sqrt{d}$
2. No; the equation is not a direct variation.
3. 2
4. 6; No.
5. Solve the equation using the quadratic formula and test both possible solutions.

Checking Key Concepts

1. 12
2. 11
3. –3
4. 12
5. 20
6. 13
7. –4
8. 7; 10
9. 3

Exercises and Applications

1. 49
2. 50
3. 61
4. –4
5. 18
6. 9
7. 15
8. 17
9. 1
10. 22
11. no solution
12. 45

Apply⇔Assess

Suggested Assignment

Core Course
Day 1 Exs. 1–16, 19–34
Day 2 Exs. 37–40, 45–65, AYP

Extended Course
Day 1 Exs. 1–11 odd, 13–22, 23–33 odd
Day 2 Exs. 35–65, AYP

Block Schedule
Day 49 Exs. 1–16, 19–34, 37–40, 45–65, AYP

Exercise Notes

Alternate Approach

Exs. 1–16 Radical equations such as these, where the variable occurs only under the radical, can often be solved mentally or with minimal writing if the following approach is used. First, isolate the radical sign on one side of the equation. Then consider the expression under the radical as a single number and ask yourself what this number would have to be in order for its root (square, cube, or whatever the root is) to equal the number at the right of the equals sign. Then set the variable expression under the radical sign equal to that number and solve for the variable. For example, with Ex. 3, ask: The square root of what number is equal to 8? (64) Thus, the expression under the radical, $x + 3$, must be equal to 64 and so $x = 61$.

Student Progress

Exs. 1–16 Students should be able to do these exercises with minimal errors. Read the answers aloud and have students work briefly with a classmate to review and correct any wrong answers.

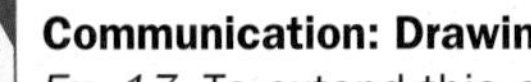

Communication: Drawing

Ex. 17 To extend this exercise, ask students to consider the number of solutions for any radical equation that consists of a radical expression equal to a linear expression. Have them consider the cases where the index of the radical is even and odd. For each case, students should draw graphs to solve the problem and illustrate their solution.

Solve by raising both sides of each equation to the same power.

13. $\sqrt[3]{2x-3} = 7$ **14.** $\sqrt[4]{\frac{1}{2}x+7} = 4$ **15.** $\sqrt[5]{7x+2} = -3$ **16.** $\sqrt[6]{\frac{2}{3}x} - 4 = 0$

17. Visual Thinking The graphs of $y = x - 1$, $y = x$, and $y = x + 1$ are shown with the graph of $y = \sqrt{x}$. Use the graphs to predict the number of solutions of each equation. Then find the solutions.

a. $\sqrt{x} = x - 1$ **b.** $\sqrt{x} = x$ **c.** $\sqrt{x} = x + 1$

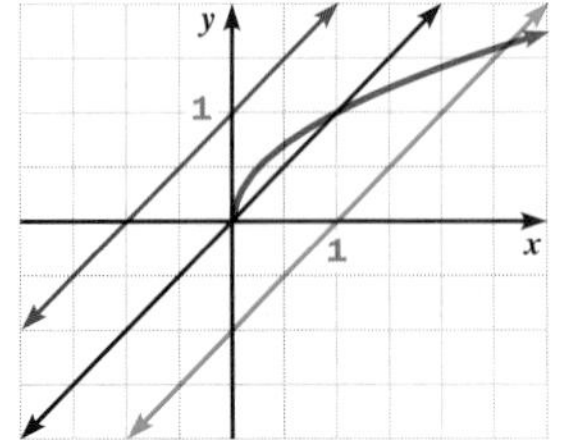

18. Writing Explain why the equation $\sqrt{3x+4} = -5$ has no solution, but the equation $\sqrt[3]{3x+4} = -5$ *does* have a solution.

Connection METEOROLOGY

In a hurricane, the mean sustained wind velocity v, measured in meters per second, is given by

$$v = 6.3\sqrt{1013 - p}$$

where p is the air pressure, measured in millibars (mb), at the center of the hurricane.

The map shows the air pressure and wind velocity at various points along the path of Hurricane Hugo, which passed through the Caribbean in 1989.

19. Estimate the mean sustained wind velocities at 0 h (midnight) on September 13th, 14th, 15th, and 16th.

20. Estimate the air pressures at 12 h (noon) on September 13th, 14th, 15th, and 16th.

21. What happens to wind velocity in a hurricane when air pressure decreases?

13. 173
14. 498
15. –35
16. 6
17. a. 1; $\frac{3+\sqrt{5}}{2}$
b. 2; 0, 1
c. 0; no solution
18. $\sqrt{x} \geq 0$ for every real number x; any number can be written as $\sqrt[3]{x}$ for some number x. The equation $\sqrt[3]{3x+4} = -5$ has solution –43.
19. 27.5 m/s; 33.9 m/s; 45.0 m/s; 59.8 m/s
20. 989 mb; 965 mb; 909 mb; 917 mb
21. Wind velocity increases.
22. a. –10.7°C
b. –3.6°C
c. 28.6 km/h
d. Answers may vary. Examples are given.

T (°C)	*V* (km/h)
–10	34
–20	12.9
–23.2	10
–15.0	20
–11.1	30

23. 7
24. 5
25. –2; –5
26. 1; 9
27. 3
28. –4; –5
29. 9
30. 2; –1

22. WIND CHILL The *wind chill index* gives you an idea of how cold it feels outside when wind increases the rate of heat loss from your skin. The index is the *equivalent temperature* that would have the same cooling effect if the wind velocity were only 6 km/h.

When the wind velocity is less than 100 km/h, an equation for wind chill W (in degrees Celsius), based on temperature T (in degrees Celsius) and wind velocity v (in kilometers per hour), is:

$$W = 33 - (0.0393)(33 - T)(12.36 + 6.13\sqrt{v} - 0.32v)$$

a. Suppose $T = 4$°C and $v = 40$ km/h. Find W.

b. Suppose $W = -15$°C and $v = 20$ km/h. Find T.

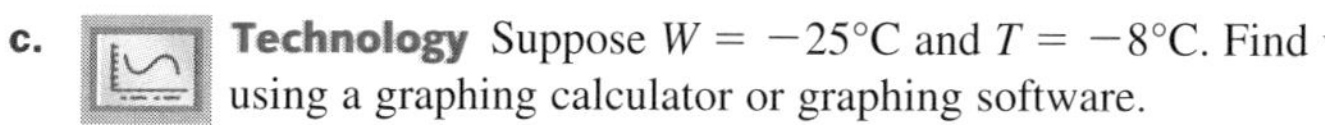

c. **Technology** Suppose $W = -25$°C and $T = -8$°C. Find v using a graphing calculator or graphing software.

d. **Cooperative Learning** Conditions are extremely dangerous for humans when the wind chill index is less than -30°C, because exposed flesh can freeze in less than a minute. Make a wind chill index table that shows several combinations of temperature and wind velocity that produce such dangerous conditions.

BY THE WAY

Working in Antarctica in 1941, Paul Siple and Charles Passel developed a wind chill formula by measuring the time it took a cylinder of water to freeze under various wind and temperature conditions.

Solve by using the method of Example 2. Check to eliminate extraneous solutions.

23. $\sqrt{x - 3} = x - 5$ **24.** $\sqrt{5x + 39} = x + 3$ **25.** $\sqrt{13x + 90} = x + 10$

26. $\sqrt{14x - 5} = x + 2$ **27.** $\sqrt{3x + 7} = x + 1$ **28.** $\sqrt{x + 5} = x + 5$

29. $\sqrt{2x + 7} = x - 4$ **30.** $\sqrt{5x + 6} = x + 2$ **31.** $\sqrt{-(3x + 2)} = x + 4$

32. $\sqrt{9x - 2} = x + 2$ **33.** $2 - x = \sqrt{44 - x}$ **34.** $\sqrt{2x + 15} = x + 6$

35. Open-ended Problem Choose one problem from Exercises 23–34 that has two solutions. Use a graphing calculator or draw sketches to explain why there are two solutions.

36. a. Solve some examples of $\sqrt{x - k} = x - k$ using different values of k.

b. **Challenge** Express the two solutions of $\sqrt{x - k} = x - k$ in terms of k.

Technology **Solve using a graphing calculator or graphing software. Check your answers.**

37. $\sqrt{x} + 2 = x$ **38.** $\sqrt[3]{x - 1} = \frac{1}{4}(x - 1)$ **39.** $\sqrt[4]{x} - 2 = x - 3$

40. a. Lisa tried solving the equation $\sqrt{x - 5} = \sqrt{x} - 1$ algebraically. Part of her work is shown. Complete the solution.

b. Try solving $\sqrt{2x + 1} = \sqrt{x} + 1$ in a similar way.

c. What makes solving equations like those in parts (a) and (b) different from solving the other radical equations in this section?

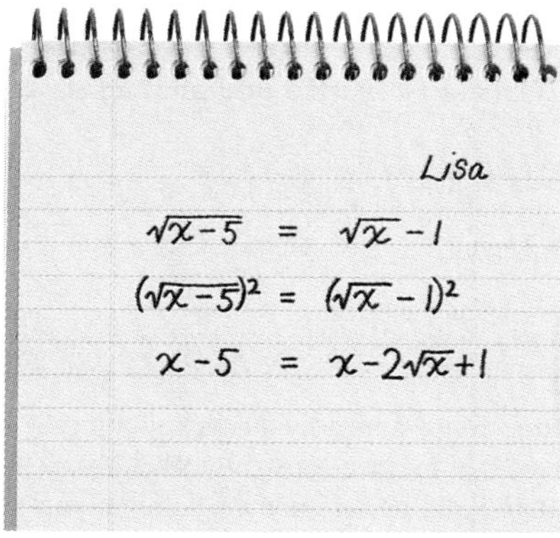

31. –2

32. 2; 3

33. –5

34. –3

35. Selections may vary. In Ex. 25, 26, 28, 30, and 32, the curve represented by the left side of the equation and the line represented by the right side of the equation intersect in two points.

36. a. Answers may vary. Examples are given.

k	Solutions of $\sqrt{x-k} = x-k$
2	2; 3
3	3; 4
5	5; 6
10	10; 11

b. k; $k + 1$

37. 4

38. –7; 1; 9

39. 2.22

40. a. $-5 = -2\sqrt{x} + 1$; $2\sqrt{x} = 6$; $\sqrt{x} = 3$; $x = 9$; checking, $\sqrt{9 - 5} \stackrel{?}{=} \sqrt{9} - 1$; $\sqrt{4} \stackrel{?}{=} 3 - 1$; $2 = 2$

b. $\sqrt{2x + 1} = \sqrt{x} + 1$; $2x + 1 = x + 2\sqrt{x} + 1$; $x = 2\sqrt{x}$; $x^2 = 4x$; $x^2 - 4x = 0$; $x(x - 4) = 0$; $x = 0$ or $x = 4$; checking: $\sqrt{2(0) + 1} \stackrel{?}{=} \sqrt{0} + 1$; $1 = 1$; $\sqrt{2(4) + 1} \stackrel{?}{=} \sqrt{4} + 1$; $3 = 3$

c. The original equation has two square roots so you need to square twice.

Apply⇔Assess

Exercise Notes

Challenge

Ex. 18 Encourage students to write two different responses to this question: one that uses a graphic explanation and one that uses an algebraic explanation.

Interdisciplinary Problems

Exs. 19–22 In doing these exercises, students see how meteorologists can use a radical equation to study the relationship between the air pressure and wind velocity in a hurricane, and how the wind chill index is a radical equation that relates the wind chill with the temperature and wind velocity.

Visual Thinking

Ex. 22 Ask students to work in teams and, using their wind chill index tables, to prepare posters that show the dangers of extremely cold conditions to humans. Encourage them to do additional research, create graphs, and clip photos to make their posters effective and interesting. This activity involves the visual skills of *correlation* and *self-expression.*

Student Study Tip

Exs. 23–34 Mention to students that these problems can be done more quickly if the formulas for squaring a binomial are used.

$(a + b)^2 = a^2 + 2ab + b^2$

$(a - b)^2 = a^2 - 2ab + b^2$

Reasoning

Ex. 36 In doing part (b) of this exercise, students need to make conjectures about the solutions of $\sqrt{x - k} = x - k$ and test these conjectures with the examples they solved in part (a). This is an example of inductive reasoning. This exercise can also be done using deductive reasoning to prove that the solutions to the equation are k and $k + 1$. The proof involves algebra using the quadratic formula, but it is not above the mathematical level of this course and some students may find it interesting.

Mathematical Procedures

Ex. 40 The procedure used for solving an equation with two radical expressions is somewhat involved, so you may want to discuss this exercise as a class activity. You might also ask students to solve the problem by isolating the other radical first to see that the same solutions are achieved.

Apply⇔Assess

Exercise Notes

Application
Exs. 41–44 These exercises engage students in using a radical expression to study various configurations of boats that sail in the America's Cup competition.

Second-Language Learners
Exs. 41–44 Many students learning English may be unfamiliar with the America's Cup competition. If necessary, tell them that it is the most famous international sailing competition, and that it is held on a somewhat irregular basis, usually every three to four years.

Research
Exs. 41–44 You might ask a few students interested in sailing to do some research on the America's Cup race. They could make a short presentation to the class about the race, its history, and the women's team mentioned on this page.

Cooperative Learning
Ex. 44 You might have students work in small groups on this problem. In particular, students should discuss what effects each variable has on the other variable in the expression. For example, if D is increased, must L increase or decrease to keep the rating under 24? Students should also be able to draw on their work for Ex. 22(d) to help them solve this problem.

Common Error
Exs. 45–52 Students often forget to check for extraneous solutions. Point out that even if the directions do not specify checking for extraneous solutions, this is an important part of solving a radical equation or a logarithmic equation.

Assessment Note
Ex. 53 Students who understand how to develop a general method for solving a radical equation, such as the one in this exercise, will have demonstrated a solid understanding of the process for solving a radical equation and an understanding of extraneous solutions.

Spiral Review: Preview
Exs. 58–61 In the next section, students will see that those quadratic equations having no real number solutions will have two complex number solutions. The quadratic formula will be used to motivate the definition of the square root of a negative number.

Connection SPORTS

In the America's Cup competition, boats must meet this rule to qualify:

$$\frac{L + 1.25\sqrt{S} - 9.8\sqrt[3]{D}}{0.679} \leq 24.000$$

where L is boat length (in meters), S is the sail area (in square meters), and D is the displacement (in cubic meters). The expression on the left side of the inequality is the rating of the boat. Boat designers try to make this value as close to 24 as possible.

In 1995, a women's team participated in the America's Cup races. Nearly 700 women asked for applications to try out for the 23 places on the team.

For each boat, find the value of the unknown variable that yields a rating of 24, qualifying the boat for the America's Cup race. Give answers to three decimal places.

	Boat	L (m)	S (m^2)	D (m^3)
41.	Argoknot	?	327.334	24.244
42.	Barnicle	21.870	?	22.440
43.	Coralgrazer	20.950	277.300	?

44. Open-ended Problem Suppose you are designing a boat for the America's Cup.

a. Using the table below, find values for L, S, and D that meet the qualifying rule and fall within the ranges given. Give answers to three decimal places.

b. How close to 24 is the rating of the boat you have designed?

Variable	Minimum	Your boat	Maximum
L (m)	20.600	?	22.000
S (m^2)	250.000	?	330.000
D (m^3)	15.610	?	24.390

Solve. You may need to use the quadratic formula. Check to eliminate extraneous solutions.

45. $\sqrt{6x + 4} = x + 2$ **46.** $\sqrt{5x + 16} = x + 4$ **47.** $\sqrt{x} = x$ **48.** $\sqrt{x} = -x$

49. $\sqrt{3 - 3x} = x$ **50.** $x - 3 = \sqrt{8 - 2x}$ **51.** $\sqrt{3x + 5} = 2x$ **52.** $x = \sqrt{x + 1}$

ONGOING ASSESSMENT

53. Writing Show a step-by-step solution to the equation $\sqrt{ax + b} = c$, where a, b, and c are constants. What restrictions must you place on a, b, or c in order to be sure that the solution you get is not extraneous?

41. 22.044 m
42. 311.673 m^2
43. 17.554 m^3
44. Answers may vary. Examples are given.
 a. $L = 21.000$; $S = 275.000$; $D = 17.479$
 b. 23.9998
45. 0; 2
46. –3; 0
47. 0; 1
48. 0
49. 0.791
50. 3.73
51. 1.55
52. 1.62
53. $\sqrt{ax + b} = c$; $ax + b = c^2$; $ax = c^2 - b$; $x = \frac{c^2 - b}{a}$; $a \neq 0$ and c is nonnegative.
54. $3\sqrt[3]{12}$
55. $5\sqrt{2}$
56. $2\sqrt[3]{9}$
57. $2\sqrt[4]{25} = 2\sqrt{5}$
58. no solution
59. two solutions
60. one solution
61. no solution
62. (–1, 0); (0, 1)
63. $\left(-\frac{4}{7}, -\frac{2}{7}\right)$
64. $\left(-\frac{1}{4}, \frac{3}{4}\right)$
65. (1, 3); (–1, 5)

SPIRAL REVIEW

Simplify each radical expression. *(Section 8.2)*

54. $\sqrt[3]{324}$ **55.** $\sqrt[2]{50}$ **56.** $\sqrt[3]{72}$ **57.** $\sqrt[4]{400}$

Tell whether each equation has *two solutions, one solution,* or *no solution.* *(Section 5.5)*

58. $5x^2 + 3x + 2 = 0$ **59.** $-x^2 + 3x - 2 = 0$ **60.** $2x^2 + 6x + \frac{9}{2} = 0$ **61.** $-4x^2 - 3x - 5 = 0$

Solve each system of equations. *(Section 7.1)*

62. $y = x^2 + 2x + 1$, $y = x + 1$

63. $y = 4x + 2$, $y = \frac{1}{2}x$

64. $y = 5x + 2$, $y = -3x$

65. $y = x^2 - x + 3$, $y = 4 - x$

ASSESS YOUR PROGRESS

VOCABULARY

radical functions (p. 344)

1. a. Graph the function $y = (x - 1)^2$ and its reflection over the line $y = x$. *(Section 8.1)*

b. Write equations for two functions that together create the reflected graph you found in part (a). Include the domain restrictions.

c. Why must you use two functions to describe the graph?

SPORTS **In the sport of powerlifting, the weights lifted in the squat, bench press, and deadlift are combined. Senior men's world record data can be modeled by the equation**

$$T = 47.3\sqrt[3]{w^2}$$

where *T* is the total weight lifted and *w* is the athlete's weight class, both measured in kilograms. *(Section 8.2)*

2. Rewrite the equation using a fractional exponent.

3. Complete the table of the total weights predicted by the equation. Give answers to the nearest half kilogram.

Weight class *w* (kg)	52	56	60	67.5	75	82.5	90	100	110	125
Predicted total *T* (kg)	?	?	?	?	?	?	?	?	?	?

Solve. *(Section 8.3)*

4. $\sqrt{4(-x + 2)} = x - 3$ **5.** $\sqrt{3x + 1} = x$ **6.** $\sqrt{x^2 - 8} = 4$

7. Journal Explain how the relationship between power functions and radical functions is similar to and different from the relationship between exponential functions and logarithmic functions.

BY THE WAY

Egypt, Japan, and Turkey had weightlifting competitions before the twentieth century. Weightlifting was added to the Olympic Games in 1920.

Apply⇔Assess

Assess Your Progress

Journal Entry

This exercise is an excellent way to review the relationship between power functions and radical functions and the relationship between exponential functions and logarithmic functions. You might wish to have several students read their entries to the class while one student lists the differences and one student lists the similarities at the board. This will ensure that all students understand the concepts involved.

Progress Check 8.1–8.3

See page 382.

Practice 49 for Section 8.3

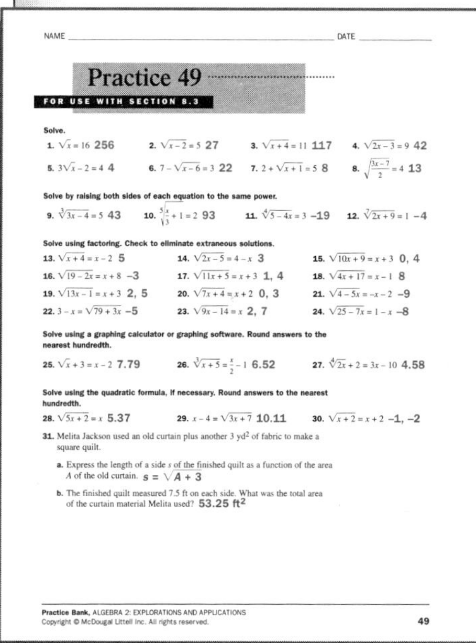

NAME ____________ DATE ________

Practice 49

FOR USE WITH SECTION 8.3

Solve.

1. $\sqrt{x} = 16$ 256 **2.** $\sqrt{x-2} = 5$ 27 **3.** $\sqrt{x+4} = 11$ 117 **4.** $\sqrt{2x-3} = 9$ 42

5. $3\sqrt{x} - 2 = 4$ 4 **6.** $7 - \sqrt{x-6} = 3$ 22 **7.** $2 + \sqrt{x+1} = 5$ 8 **8.** $\sqrt{\frac{3x-7}{2}} = 4$ 13

Solve by raising both sides of each equation to the same power.

9. $\sqrt[3]{3x-4} = 5$ 43 **10.** $\sqrt[5]{\frac{x}{3}} + 1 = 2$ 93 **11.** $\sqrt[4]{5-4x} = 3$ −19 **12.** $\sqrt[7]{2x+9} = 1$ −4

Solve using factoring. Check to eliminate extraneous solutions.

13. $\sqrt{x+4} = x - 2$ 5 **14.** $\sqrt{2x-5} = 4 - x$ 3 **15.** $\sqrt{10x+9} = x + 3$ 0, 4

16. $\sqrt{19-2x} = x + 8$ −3 **17.** $\sqrt{11x+5} = x + 3$ 1, 4 **18.** $\sqrt{4x+17} = x - 1$ 8

19. $\sqrt{13x-1} = x + 3$ 2, 5 **20.** $\sqrt{7x+4} = x + 2$ 0, 3 **21.** $\sqrt{4-5x} = -x - 2$ −9

22. $3 - x = \sqrt{79+3x}$ −5 **23.** $\sqrt{9x-14} = x$ 2, 7 **24.** $\sqrt{25-7x} = 1 - x$ −8

Solve using a graphing calculator or graphing software. Round answers to the nearest hundredth.

25. $\sqrt{x} + 3 = x - 2$ 7.79 **26.** $\sqrt[3]{x+5} = \frac{x}{2} - 1$ 6.52 **27.** $\sqrt[4]{2x} + 2 = 3x - 10$ 4.58

Solve using the quadratic formula, if necessary. Round answers to the nearest hundredth.

28. $\sqrt{5x+2} = x$ 5.37 **29.** $x - 4 = \sqrt{3x+7}$ 10.11 **30.** $\sqrt{x+2} = x + 2$ −1, −2

31. Melita Jackson used an old curtain plus another 3 yd^2 of fabric to make a square quilt.

a. Express the length of a side *s* of the finished quilt as a function of the area *A* of the old curtain. $s = \sqrt{A + 3}$

b. The finished quilt measured 7.5 ft on each side. What was the total area of the curtain material Melita used? 53.25 ft^2

Practice Bank, ALGEBRA 2: EXPLORATIONS AND APPLICATIONS

49

Assess Your Progress

1. a.

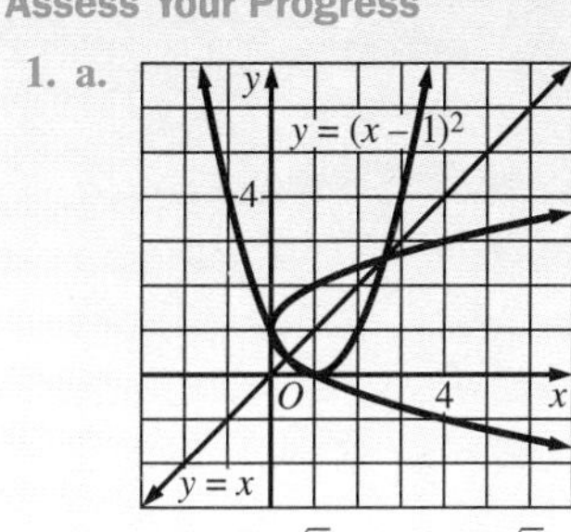

b. $y = 1 + \sqrt{x}$; $y = 1 - \sqrt{x}$, $x \geq 0$

c. The reflection of the graph is not the graph of a function since each input value has two output values. The reflection is the combined graph of two functions.

2. $T = 47.3w^{2/3}$

3. 659; 692.5; 725; 784; 841; 896.5; 950; 1019; 1086; 1182.5

4. no solution

5. $\frac{3 + \sqrt{13}}{2} \approx 3.30$

6. $\pm 2\sqrt{6} \approx \pm 4.90$

7. Summaries may vary. The relationships between power functions and radical functions and between exponential functions and logarithmic functions are both inverse relationships. For every $b > 0$, $b \neq 1$, $y = b^x$ and $y = \log_b x$ are inverse functions. However, $y = x^n$ and $y = \sqrt[n]{x}$ are inverse functions only if n is odd. If n is even, the reflection of $y = x^n$ across the line $y = x$ is not a function. In order to define the inverse of $y = x^n$ for n even, you must restrict the domain to either $x \geq 0$ or $x \leq 0$.

Plan⇔Support

Objectives

- Add, subtract, multiply, and divide complex numbers.
- Find complex solutions to equations that have no real solutions.
- Solve problems involving complex numbers.

Recommended Pacing

Core and Extended Courses
Section 8.4: 2 days

Block Schedule
Section 8.4: 1 block

Resource Materials

Lesson Support
Lesson Plan 8.4
Warm-Up Transparency 8.4
Practice Bank: Practice 50
Study Guide: Section 8.4

Technology
Technology Book:
TI-92 Activity 4
Graphing Calculator
Internet:
http://www.hmco.com

Warm-Up Exercises

1. Add. $(3x - 4) + (-2x + 7)$ $x + 3$
2. Subtract. $(-2x + 6) - (5x - 1)$ $-7x + 7$
3. Multiply. $(2x + 1)(3x - 5)$ $6x^2 - 7x - 5$
4. Expand $(3x - 7)^2$. $9x^2 - 42x + 49$
5. Solve $y = -x^2 + 7x - 4$ using the quadratic formula. 0.6, 6.4
6. How many real solutions does the equation $x^2 - 3x + 1 = 0$ have? two

Section

8.4 Working with Complex Numbers

Learn how to...

- add, subtract, multiply, and divide complex numbers
- find complex solutions to equations that have no real solutions

So you can...

- solve problems involving complex numbers, such as finding the impedance of an electrical circuit

Calvin and Hobbes by Bill Watterson

So far in this book the numbers you have used for the domain and range of functions have been **real numbers**, which include both rational numbers, such as 3 and $-\frac{1}{2}$, and irrational numbers, such as $\sqrt{2}$ and π. In this section you will see how to extend this set of numbers to include other numbers, called *imaginary numbers*, which together with real numbers form the system of *complex numbers.*

THINK AND COMMUNICATE

Use the graphs to answer Questions 1–3.

$y = x^2 - 1$

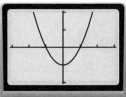

$y = x^2 + 1$

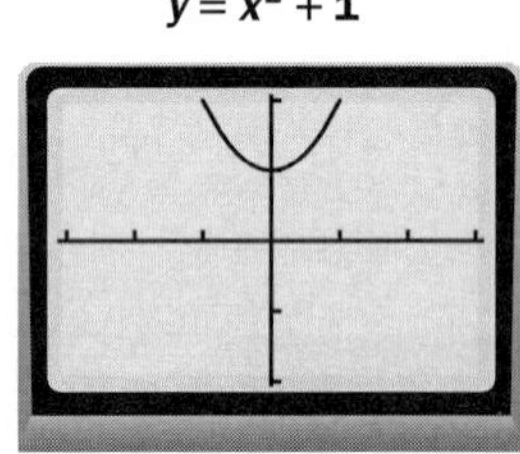

1. What does the graph on the left tell you about solutions of $x^2 - 1 = 0$?
2. What does the graph on the right tell you about solutions of $x^2 + 1 = 0$?
3. Apply the quadratic formula to the equations in Questions 1 and 2. What do you notice?

The equation $x^2 + 1 = 0$ has no *real* solutions, but it does have solutions if you define the square roots of negative numbers.

ANSWERS Section 8.4

Think and Communicate

1. There are two solutions.
2. There are no solutions.
3. The discriminant for the equation on the left is positive; the discriminant for the equation on the right is negative.

Imaginary Numbers

The letter i is used for the square root of -1, the fundamental unit in the system of *imaginary numbers*.

$$i = \sqrt{-1}$$
$$i^2 = -1$$

The square root of any negative number is called a **pure imaginary number**. Every pure imaginary number can be written in the form bi, where $b \neq 0$.

Examples:

$$\sqrt{-2} = \sqrt{(-1)(2)} = \sqrt{-1}\sqrt{2} = i\sqrt{2}$$
$$\sqrt{-3} = \sqrt{(-1)(3)} = \sqrt{-1}\sqrt{3} = i\sqrt{3}$$
$$\sqrt{-4} = \sqrt{(-1)(4)} = \sqrt{-1}\sqrt{4} = 2i$$

WATCH OUT!

The rule $\sqrt{ab} = \sqrt{a}\sqrt{b}$ applies only when a and b are both nonnegative, or when one is nonnegative and the other is negative. The rule does *not* apply if a and b are *both* negative.

EXAMPLE 1

Show that i is a solution of $x^2 + 1 = 0$.

SOLUTION

Substitute i for x in $x^2 + 1 = 0$.

$$i^2 + 1 \stackrel{?}{=} 0$$
$$\left(\sqrt{-1}\right)^2 + 1 \stackrel{?}{=} 0$$
$$-1 + 1 = 0 \ ✔$$

THINK AND COMMUNICATE

4. Show that $-i$ is also a solution of $x^2 + 1 = 0$.

5. Find the pure imaginary solutions of $x^2 + 4 = 0$.

6. Express each radical in the form bi.

a. $\sqrt{-9}$ **b.** $\sqrt{-25}$ **c.** $\sqrt{-200}$

Complex Numbers

Any number of the form $a + bi$ (where a and b are real numbers and $b \neq 0$) is an **imaginary number**. The imaginary numbers together with the real numbers form the set of **complex numbers**.

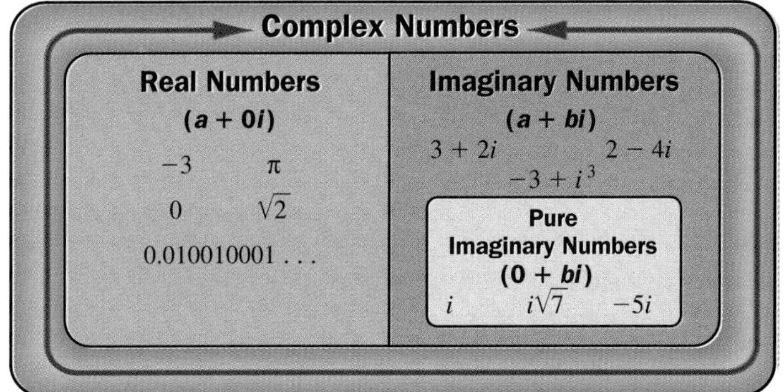

Think and Communicate

4. $(-i)^2 + 1 = (-\sqrt{-1})^2 + 1 = -1 + 1 = 0$

5. $\pm 2i$

6. a. $3i$

b. $5i$

c. $10i\sqrt{2}$

Teach⟺Interact

Section Note

Communication: Drawing
Ask students to illustrate the first sentence of the opening paragraph on page 358 by drawing a Venn diagram to represent the real, rational, and irrational numbers. This diagram can then be added to as students learn about the pure imaginary, imaginary, and complex numbers in this section.

Think and Communicate

Question 3 leads students to recall that when the number under the radical in the quadratic formula is positive, the quadratic equation has two real solutions, and when the number under the radical is negative, the quadratic equation has no real solutions. Students should remember that the expression under the radical in the quadratic formula is called the *discriminant.*

Section Notes

Reasoning
Draw students' attention to the Watch Out! box on this page by having them show, with an example, why this rule does not hold in the case that both a and b are negative numbers.

Teaching Tip
The definition of an imaginary number in this book may be different from one seen by students previously. Some books define an imaginary number as one of the form bi, where b is nonzero. These books make no distinction between pure imaginary numbers and imaginary numbers. In this case, a number of the form $a + bi$, where a is nonzero, would be considered a complex number but not an imaginary number as it is in this text.

Additional Example 1

Show that $-3i$ is a solution to $x^2 + 9 = 0$.

Substitute $-3i$ for x in $x^2 + 9 = 0$.

$$(-3i)^2 + 9 \stackrel{?}{=} 0$$
$$(-3)^2 i^2 + 9 \stackrel{?}{=} 0$$
$$9\left(\sqrt{-1}\right)^2 + 9 \stackrel{?}{=} 0$$
$$9(-1) + 9 \stackrel{?}{=} 0$$
$$-9 + 9 = 0$$

Teach⇔Interact

Learning Styles: Verbal

Encourage verbal learners to write the rules for adding, subtracting, and multiplying complex numbers in words and then to think of the rules as they perform the operations. For example, to multiply complex numbers, a student can think: "Multiply the first terms, the inner terms, the outer terms, and then the last terms. Simplify by replacing i^2 with –1 and collecting like terms."

Section Note

Mathematical Procedures

The procedures used to add, subtract, multiply, and square complex numbers are identical to those used for binomial expressions in one variable. If students are having trouble doing these operations on complex numbers, have them remember the rules used for these operations on binomials.

Additional Example 2

Perform the indicated operation.

a. $(-2 + 3i) + (2 - 7i)$

$(-2 + 3i) + (2 - 7i)$
$= (-2 + 2) + (3i - 7i)$
$= 0 + -4i$
$= -4i$

b. $(3 + 4i) - (-2 + 6i)$

$(3 + 4i) - (-2 + 6i)$
$= (3 - (-2)) + (4i - 6i)$
$= 5 - 2i$

Additional Example 3

Perform the indicated operation.

a. $(-4 + 2i)(3 - i)$

$(-4 + 2i)(3 - i)$
$= -12 + 6i + 4i - 2i^2$
$= -12 + 10i + 2$
$= -10 + 10i$

b. $(4 - 7i)^2$

$(4 - 7i)^2$
$= (4 - 7i)(4 - 7i)$
$= 16 - 28i - 28i + 49i^2$
$= 16 - 56i - 49$
$= -33 - 56i$

c. $(-2 + i)(-2 - i)$

$(-2 + i)(-2 - i)$
$= 4 - 2i + 2i - i^2$
$= 4 + 1$
$= 5$

Adding and Subtracting Complex Numbers

You can add or subtract complex numbers by adding or subtracting their real parts and their pure imaginary parts.

$$(a + bi) + (c + di) = (a + c) + (b + d)i$$
$$(a + bi) - (c + di) = (a - c) + (b - d)i$$

EXAMPLE 2

Perform the indicated operation.

a. $(3 - 4i) + (5 + 9i)$ **b.** $(7 + 2i) - (8 - 3i)$

SOLUTION

a. $(3 - 4i) + (5 + 9i) = (3 + 5) + (-4 + 9)i = 8 + 5i$

b. $(7 + 2i) - (8 - 3i) = (7 - 8) + (2 - (-3))i = -1 + 5i$

Multiplying Complex Numbers

You can multiply complex numbers as you would multiply binomials.

$$\begin{aligned}(a + bi)(c + di) &= ac + bci + adi + bdi^2 \\ &= ac + (bc + ad)i - bd \\ &= (ac - bd) + (bc + ad)i\end{aligned}$$

THINK AND COMMUNICATE

7. Justify each step in the multiplication of $a + bi$ and $c + di$ shown above.
8. Write a rule for multiplying $(a - bi)(c - di)$.

EXAMPLE 3

Perform the indicated operation.

a. $(3 + 2i)(4 + 7i)$ **b.** $(6 + 5i)^2$ **c.** $(2 + 9i)(2 - 9i)$

SOLUTION

a. $(3 + 2i)(4 + 7i) = 12 + 8i + 21i + 14i^2$
$= 12 + 29i - 14$
$= -2 + 29i$

b. $(6 + 5i)^2 = (6 + 5i)(6 + 5i)$
$= 36 + 30i + 30i + 25i^2$
$= 36 + 60i - 25$
$= 11 + 60i$

Use the definition of squaring.

Think and Communicate

7. distributive property; distributive property and definition of i; commutative property of addition and associative property of addition
8. $(a - bi)(c - di) = (ac - bd) - (bc + ad)i$

c. $(2 + 9i)(2 - 9i) = 4 + 18i - 18i - 81i^2$
$= 4 + 81$
$= 85$

In this case, multiplying two complex numbers results in a real number.

Whenever you multiply numbers of the form $a + bi$ and $a - bi$, called **complex conjugates**, the result is a real number. You can use complex conjugates to divide complex numbers.

EXAMPLE 4

Use complex conjugates to write $\frac{3+i}{2-3i}$ in $a + bi$ form.

SOLUTION

$$\frac{3+i}{2-3i} = \frac{3+i}{2-3i} \cdot \frac{2+3i}{2+3i}$$

Multiply the numerator and denominator by the complex conjugate of the denominator.

$$= \frac{(3+i)(2+3i)}{(2-3i)(2+3i)}$$

$$= \frac{6 + 2i + 9i + 3i^2}{4 - 6i + 6i - 9i^2}$$

$$= \frac{6 + 11i - 3}{4 + 9}$$

The denominator is a real number.

$$= \frac{3 + 11i}{13}$$

$$= \frac{3}{13} + \frac{11}{13}i$$

Rewrite the fraction in $a + bi$ form.

EXAMPLE 5

Solve the equation $x^2 - 10x + 41 = 0$. Check the solutions.

SOLUTION

$$x = \frac{-b \pm \sqrt{b^2 - 4ac}}{2a}$$

Substitute **1** for a, **−10** for b, and **41** for c.

$$= \frac{-(-10) \pm \sqrt{(-10)^2 - 4(1)(41)}}{2(1)}$$

$$= \frac{10 \pm \sqrt{100 - 164}}{2}$$

$$= \frac{10 \pm \sqrt{-64}}{2}$$

$$= \frac{10 \pm 8i}{2}$$

$$= 5 \pm 4i$$

Check

Substitute $5 + 4i$ for x in $x^2 - 10x + 41 = 0$.

$(5 + 4i)^2 - 10(5 + 4i) + 41 \stackrel{?}{=} 0$

$(25 + 40i + 16i^2) - 50 - 40i + 41 \stackrel{?}{=} 0$

$25 + 40i - 16 - 50 - 40i + 41 \stackrel{?}{=} 0$

$(25 - 16 - 50 + 41) + (40i - 40i) \stackrel{?}{=} 0$

$0 + 0 = 0$ ✓

The check of the other solution is left to you.

Teach⇔Interact

About Example 4

Teaching Tip

Point out to students that the purpose of using complex conjugates when simplifying complex fractions is to make the denominator a real number. This allows you to rewrite the final version in $a + bi$ form.

Additional Example 4

Use complex conjugates to write $\frac{7-i}{-2+5i}$ in $a + bi$ form.

Multiply the numerator and denominator by the complex conjugate of the denominator, transforming the denominator into a real number.

$$\frac{7-i}{-2+5i} = \frac{7-i}{-2+5i} \cdot \frac{-2-5i}{-2-5i}$$

$$= \frac{(7-i)(-2-5i)}{(-2+5i)(-2-5i)}$$

$$= \frac{-14 + 2i - 35i + 5i^2}{4 - 10i + 10i - 25i^2}$$

$$= \frac{-14 - 33i - 5}{4 + 25}$$

$$= \frac{-19 - 33i}{29}$$

$$= -\frac{19}{29} - \frac{33}{29}i$$

Additional Example 5

Solve the equation $x^2 + 4x + 29 = 0$. Check the solutions.

Substitute 1 for a, 4 for b, and 29 for c.

$$x = \frac{-b \pm \sqrt{b^2 - 4ac}}{2a}$$

$$= \frac{-4 \pm \sqrt{4^2 - 4(1)(29)}}{2(1)}$$

$$= \frac{-4 \pm \sqrt{16 - 116}}{2}$$

$$= \frac{-4 \pm \sqrt{-100}}{2}$$

$$= \frac{-4 \pm 10i}{2}$$

$$= -2 \pm 5i$$

Check

Substitute $-2 + 5i$ for x in $x^2 + 4x + 29 = 0$.

$(-2 + 5i)^2 + 4(-2 + 5i) + 29 \stackrel{?}{=} 0$

$(4 - 20i + 25i^2) - 8 + 20i + 29 \stackrel{?}{=} 0$

$4 - 20i - 25 - 8 + 20i + 29 \stackrel{?}{=} 0$

$(4 - 25 - 8 + 29) + (-20i + 20i) \stackrel{?}{=} 0$

$0 + 0 = 0$ ✓

Substitute $-2 - 5i$ for x in $x^2 + 4x + 29 = 0$.

$(-2 - 5i)^2 + 4(-2 - 5i) + 29 \stackrel{?}{=} 0$

$(4 + 20i + 25i^2) - 8 - 20i + 29 \stackrel{?}{=} 0$

$4 + 20i - 25 - 8 - 20i + 29 \stackrel{?}{=} 0$

$(4 - 25 - 8 + 29) + (20i - 20i) \stackrel{?}{=} 0$

$0 + 0 = 0$ ✓

Teach⇔Interact

Think and Communicate

Ask volunteers to explain their answers to question 10. While most students may use the solution to Example 5, others may use the discriminant. Some students may not understand why these solutions do not appear on the graph. Remind them that the coordinate plane they have been using has only real numbers on the x- and y-axes. Imaginary numbers are not represented on this plane. Students will study the complex number plane in the next section.

Checking Key Concepts

Common Error

Students who get –33 as the solution to question 12 are most likely squaring both terms of the expression. Encourage these students to write out the square as the product of two factors before doing the multiplication.

Closure Question

Describe the two types of numbers in the system of complex numbers and how each type can be found as the solution to a quadratic equation. The complex number system consists of the imaginary numbers and the real numbers. The imaginary numbers have $i = \sqrt{-1}$ in their expression, while real numbers do not. A quadratic equation has real solutions if the expression under the radical sign in the quadratic formula is positive or zero, and it has imaginary solutions if the expression under the radical sign is negative.

Apply⇔Assess

Suggested Assignment

Core Course
Day 1 Exs. 1–35 odd
Day 2 Exs. 2–36 even, 42–47, 50–59

Extended Course
Day 1 Exs. 1–35 odd
Day 2 Exs. 2–36 even, 37–59

Block Schedule
Day 50 Exs. 1–36, 42–47, 50–59

THINK AND COMMUNICATE

9. Show that the solution $5 - 4i$ in Example 5 also checks.

10. Without graphing, tell whether the graph of $y = x^2 - 10x + 41$ intersects the x-axis. How do you know?

CHECKING KEY CONCEPTS

Express in *bi* form.

1. $\sqrt{-7}$ **2.** $\sqrt{-32}$ **3.** $\sqrt{-300}$ **4.** $\sqrt{-16}$

Perform the indicated operation.

5. $(5 - 3i) + (9 + 8i)$ **6.** $(1 + 6i) + (7 - 12i)$

7. $(10 + 9i) - (5 + 3i)$ **8.** $(9 + 6i) - (10 + 7i)$

9. $4(3 + 2i)$ **10.** $6i(5 + 3i)$

11. $(7 + i)(2 + 3i)$ **12.** $(4 + 7i)^2$

13. $\frac{1 + 2i}{3 + 4i}$ **14.** $\frac{1 + 3i}{2 + 4i}$

Solve and check the solutions.

15. $x^2 - 14x + 53 = 0$ **16.** $x^2 - 2x + 26 = 0$

8.4 Exercises and Applications

Extra Practice exercises on page 761

Express in *bi* form.

1. $\sqrt{-36}$ **2.** $\sqrt{-40}$ **3.** $-\sqrt{-28}$ **4.** $\sqrt{-64} + \sqrt{-121}$

Write a quadratic equation with the given solutions.

5. $4i, -4i$ **6.** $10i, -10i$ **7.** $i\sqrt{2}, -i\sqrt{2}$ **8.** $3i\sqrt{3}, -3i\sqrt{3}$

Add or subtract.

9. $(5 + 3i) + (7 + 2i)$ **10.** $(4 + 3i) + (2 + i)$ **11.** $(12 + 5i) + (6 - 2i)$ **12.** $(3 + 7i) + (-2 - i)$

13. $(2 + i) + (-4 + 3i)$ **14.** $(4 - 2i) + (3 - 7i)$ **15.** $(12 + 7i) - (2 + 3i)$ **16.** $(1 + 3i) - (3 + i)$

17. $(-7 + 5i) - (4 + i)$ **18.** $(5 + 11i) - (3 - 2i)$ **19.** $(-8 - i) - (-8 - i)$ **20.** $(2 - 5i) - (-3 + i)$

Multiply.

21. $4(5i)$ **22.** $-2i(6i)$ **23.** $5(2 + 3i)$ **24.** $-8(12 - 3i)$

25. $-9(5 - 2i)$ **26.** $6i(-2 + 4i)$ **27.** $(3 + i)(7 + 2i)$ **28.** $(5 + 2i)(7 - 3i)$

29. $(2 - i)(3 + 4i)$ **30.** $(10 - 5i)(5 - 2i)$ **31.** $(3 + 4i)(3 - 4i)$ **32.** $(6 - 3i)^2$

Use complex conjugates to write each quotient in $a + bi$ form.

33. $\frac{2 + 5i}{8 + i}$ **34.** $\frac{3 - 2i}{2 - i}$ **35.** $\frac{-3 + i}{4 + 3i}$ **36.** $\frac{2 - 5i}{i}$

Think and Communicate

9. $(5 - 4i)^2 - 10(5 - 4i) + 41 = 25 - 40i + 16i^2 - 50 + 40i + 41 = (25 - 16 - 50 + 41) - 40i + 40i = 0$

10. No; the solutions of the equation are not real numbers.

Checking Key Concepts

1. $i\sqrt{7}$
2. $4i\sqrt{2}$
3. $10i\sqrt{3}$
4. $4i$
5. $14 + 5i$
6. $8 - 6i$
7. $5 + 6i$
8. $-1 - i$
9. $12 + 8i$
10. $-18 + 30i$
11. $11 + 23i$
12. $-33 + 56i$
13. $\frac{11}{25} + \frac{2}{25}i$
14. $\frac{7}{10} + \frac{1}{10}i$
15. $7 \pm 2i$
16. $1 \pm 5i$

Exercises and Applications

1. $6i$
2. $2i\sqrt{10}$
3. $-2i\sqrt{7}$
4. $19i$

5–8. Answers may vary. Examples are given.

5. $x^2 + 16 = 0$
6. $x^2 + 100 = 0$
7. $x^2 + 2 = 0$

Connection ELECTRONICS

Electrical circuit components such as resistors, inductors, and capacitors oppose the flow of current in different ways. The opposition is called *impedance*, denoted by Z. The value of Z is a real number R for a resistor of R ohms (Ω), a pure imaginary number Li for an inductor of L ohms, and a pure imaginary number $-Ci$ for a capacitor of C ohms. (See the table at the right for examples.)

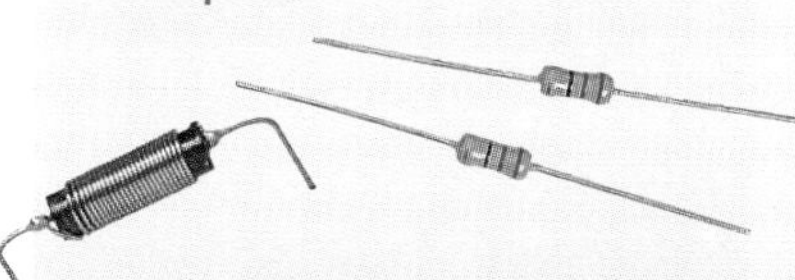

Component	Symbol	Z
resistor	4 Ω	4
inductor	5 Ω	$5i$
capacitor	6 Ω	$-6i$

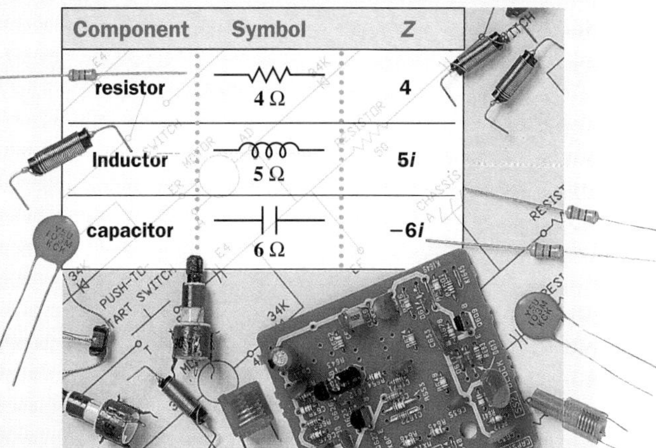

Impedance in a Series Circuit	Impedance in a Parallel Circuit
A series circuit has only one pathway. To find the impedance Z in a series circuit, add the impedance for each component in the circuit.	A parallel circuit has more than one pathway that includes the alternating current source. To find the impedance Z in a parallel circuit with two pathways, find the impedance in each pathway, Z_1 and Z_2, and then apply this formula: $Z = \frac{Z_1Z_2}{Z_1 + Z_2}$
Example: 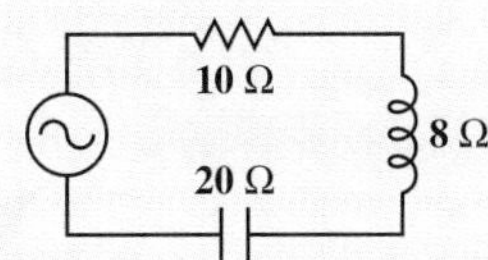 This is the symbol for an alternating current source.	Example: 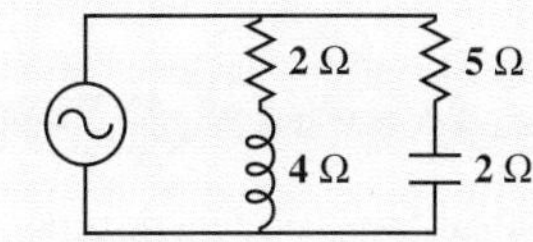
In the series circuit shown, the impedance is: $Z = 10 + 8i - 20i$ $= 10 - 12i$	In the parallel circuit shown, the impedance is: $Z = \frac{(2 + 4i)(5 - 2i)}{(2 + 4i) + (5 - 2i)}$ $= \frac{158}{53} + \frac{76}{53}i$

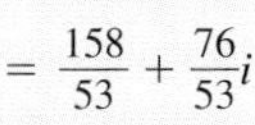

37. Justify that $\frac{(2 + 4i)(5 - 2i)}{(2 + 4i) + (5 - 2i)} = \frac{158}{53} + \frac{76}{53}i$. Show each step.

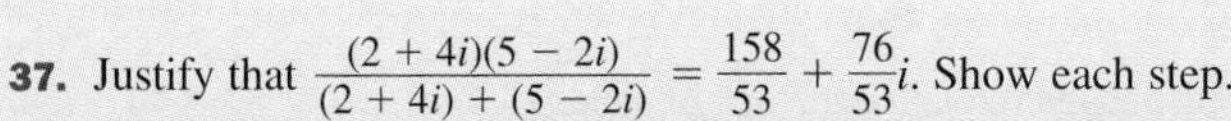

Find the impedance Z for each circuit.

38. series circuit

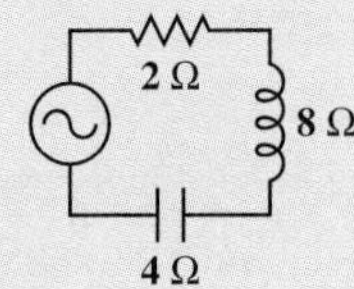

39. parallel circuit

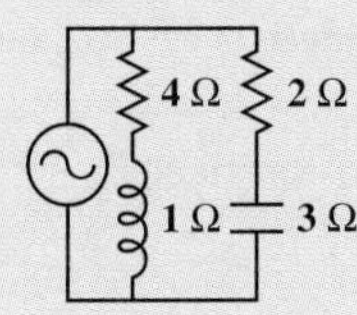

40. series circuit

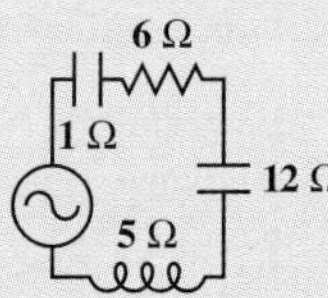

41. parallel circuit

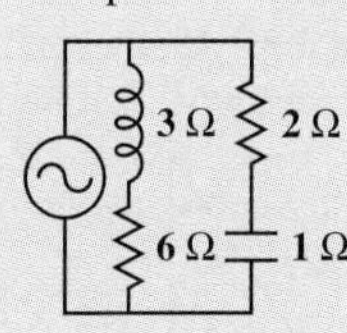

8. $x^2 + 27 = 0$
9. $12 + 5i$ 10. $6 + 4i$
11. $18 + 3i$ 12. $1 + 6i$
13. $-2 + 4i$ 14. $7 - 9i$
15. $10 + 4i$ 16. $-2 + 2i$
17. $-11 + 4i$ 18. $2 + 13i$
19. 0 20. $5 - 6i$
21. $20i$ 22. 12
23. $10 + 15i$ 24. $-96 + 24i$
25. $-45 + 18i$ 26. $-24 - 12i$
27. $19 + 13i$ 28. $41 - i$
29. $10 + 5i$ 30. $40 - 45i$
31. 25 32. $27 - 36i$
33. $\frac{21}{65} + \frac{38}{65}i$ 34. $\frac{8}{5} - \frac{1}{5}i$
35. $-\frac{9}{25} + \frac{13}{25}i$ 36. $-5 - 2i$
37. $\frac{(2 + 4i)(5 - 2i)}{(2 + 4i) + (5 - 2i)} = \frac{10 + 16i + 8}{7 + 2i} = \frac{18 + 16i}{7 + 2i} = \frac{18 + 16i}{7 + 2i} \cdot \frac{7 - 2i}{7 - 2i} = \frac{126 + 76i + 32}{49 + 4} = \frac{158 + 76i}{53} = \frac{158}{53} + \frac{76}{53}i$
38. $2 + 4i$
39. $\frac{43}{20} - \frac{19}{20}i$
40. $6 - 8i$
41. $\frac{30}{17} - \frac{15}{34}i$

Apply⇔Assess

Exercise Notes

Student Progress

Exs. 1–36 These exercises involve straightforward computations. Most students should be able to complete them with a high degree of accuracy. You might wish to go around the room asking each student to read one answer while others correct their work. If some students disagree with the answer read, they can raise their hands, and if many hands are raised, the problem can be reviewed.

Application

Exs. 37–41 In doing these exercises, students will see how complex numbers are used to calculate the impedance in an electrical circuit. It is important for students to understand that these imaginary numbers have very real applications, and to realize that these applications were not even foreseen when imaginary numbers were defined.

Communication: Reading

Exs. 37–41 The introduction to these exercises contains a great deal of information and two examples that students need to understand to do the problems. In addition, students must be able to interpret the table and circuit diagrams. Most students would benefit from having this material read aloud and discussed in class.

Second-Language Learners

Exs. 37–41 Students learning English may find the language and concepts in these exercises challenging. Encourage them to work with a peer tutor or aide or in small groups to discuss the information in the introductory paragraph, table, and diagrams before they begin work on the exercises.

Research

Exs. 37–41 Since much of the material in these exercises may be unfamiliar to students, you might wish to ask a few students to do some research about electrical circuits including resistors, inductors, capacitors, impedance, and circuit diagrams. These students could then present their information to the class the day the problems are assigned. With the help of a science teacher, they might even be able to build a small circuit to bring to class as an example.

Apply⇔Assess

Exercise Notes

Topic Spiraling: Preview

Ex. 49 The process of iteration used in this exercise is vital for Exs. 37–47 of Section 8.5, which examine the Mandelbrot set.

Using Technology

Ex. 49 The real-valued iterations in parts (a) and (b) can be done easily on a TI-82 graphing calculator. For example, to do the iteration in part (a), type the starting value 0 and press [ENTER]. Then press [x2] [+] 2 [ENTER] to get the value of the first iteration. Continue pressing [ENTER] for further iterations.

Assessment

Ex. 50 In completing this problem, students will demonstrate their understanding of all the concepts presented in this section.

Practice 50 for Section 8.4

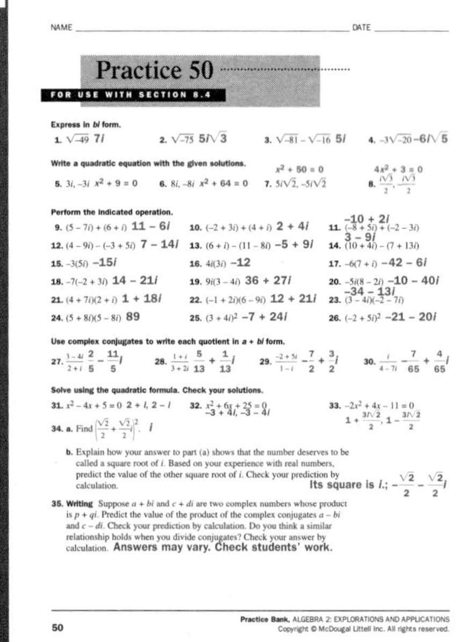

NAME ____________________ DATE ________

Practice 50

FOR USE WITH SECTION 8.4

Express in *bi* form.

1. $\sqrt{-49}$ **7*i***
2. $\sqrt{-75}$ **$5i\sqrt{3}$**
3. $\sqrt{-81} - \sqrt{-16}$ **5*i***
4. $-3\sqrt{-20}$ **$-6i\sqrt{5}$**

Write a quadratic equation with the given solutions.

5. $3i, -3i$ **$x^2 + 9 = 0$**
6. $8i, -8i$ **$x^2 + 64 = 0$**
7. $5i\sqrt{2}, -5i\sqrt{2}$ **$x^2 + 50 = 0$**
8. $\frac{i\sqrt{3}}{2}, -\frac{i\sqrt{3}}{2}$ **$4x^2 + 3 = 0$**

Perform the indicated operation.

9. $(5 - 7i) + (6 + i)$ **$11 - 6i$**
10. $(-2 + 3i) + (4 + i)$ **$2 + 4i$**
11. $(-8 + 5i) + (-2 - 3i)$ **$-10 + 2i$**
12. $(4 - 9i) - (-3 + 5i)$ **$7 - 14i$**
13. $(6 + i) - (11 - 8i)$ **$-5 + 9i$**
14. $(10 + 4i) - (7 + 13i)$ **$3 - 9i$**
15. $-3(5i)$ **$-15i$**
16. $4i(3i)$ **-12**
17. $-6(7 + i)$ **$-42 - 6i$**
18. $-7(-2 + 3i)$ **$14 - 21i$**
19. $9i(3 - 4i)$ **$36 + 27i$**
20. $-5i(8 - 2i)$ **$-10 - 40i$**
21. $(4 + 7i)(2 + i)$ **$1 + 18i$**
22. $(-1 + 2i)(6 - 9i)$ **$12 + 21i$**
23. $(3 - 4i)(-2 - 7i)$ **$-34 - 13i$**
24. $(5 + 8i)(5 - 8i)$ **89**
25. $(3 + 4i)^2$ **$-7 + 24i$**
26. $(-2 + 5i)^2$ **$-21 - 20i$**

Use complex conjugates to write each quotient in *a* + *bi* form.

27. $\frac{3 - 4i}{2 + i}$ **$\frac{2}{5} - \frac{11}{5}i$**
28. $\frac{1 + i}{3 + 2i}$ **$\frac{5}{13} + \frac{1}{13}i$**
29. $\frac{-2 + 5i}{1 - i}$ **$-\frac{7}{2} + \frac{3}{2}i$**
30. $\frac{i}{4 - 7i}$ **$-\frac{7}{65} + \frac{4}{65}i$**

Solve using the quadratic formula. Check your solutions.

31. $x^2 - 4x + 5 = 0$ **$2 + i, 2 - i$**
32. $x^2 + 6x + 25 = 0$ **$-3 + 4i, -3 - 4i$**
33. $-2x^2 + 4x - 11 = 0$ **$1 + \frac{3i\sqrt{2}}{2}, 1 - \frac{3i\sqrt{2}}{2}$**
34. a. Find $\left(\frac{\sqrt{2}}{2} + \frac{\sqrt{2}}{2}i\right)^2$. ***i***

    b. Explain how your answer to part (a) shows that the number deserves to be called a square root of i. Based on your experience with real numbers, predict the value of the other square root of i. Check your prediction by calculation. **Its square is *i*.; $-\frac{\sqrt{2}}{2} - \frac{\sqrt{2}}{2}i$**
35. **Writing** Suppose $a + bi$ and $c + di$ are two complex numbers whose product is $p + qi$. Predict the value of the product of the complex conjugates $a - bi$ and $c - di$. Check your prediction by calculation. Do you think a similar relationship holds when you divide conjugates? Check your answer by calculation. **Answers may vary. Check students' work.**

50 Practice Bank, ALGEBRA 2: EXPLORATIONS AND APPLICATIONS

Solve using the quadratic formula. Check your solutions.

42. $x^2 - 4x + 13 = 0$ **43.** $x^2 - 14x + 50 = 0$ **44.** $x^2 + 6x + 11 = 0$

45. $x^2 + 12 = -6x$ **46.** $2x^2 + 20x = -82$ **47.** $x(x + 1) = -1$

48. Writing If a quadratic equation with real coefficients has two imaginary solutions, are they always complex conjugates? Explain your answer.

49. To *iterate* a function, you evaluate the function for a given starting value, then take the result and evaluate the function for that value, repeating this process over and over. For example, if you iterate the function $f(z) = z^2 + 1$ using a starting z-value of 0, you get these results:

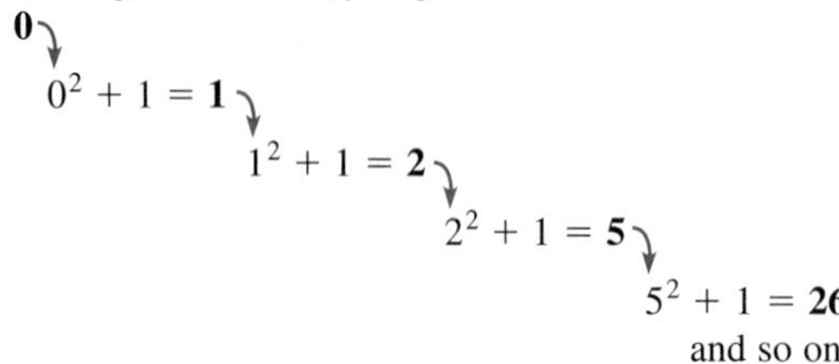

0

$0^2 + 1 = \mathbf{1}$

$1^2 + 1 = \mathbf{2}$

$2^2 + 1 = \mathbf{5}$

$5^2 + 1 = \mathbf{26}$

and so on.

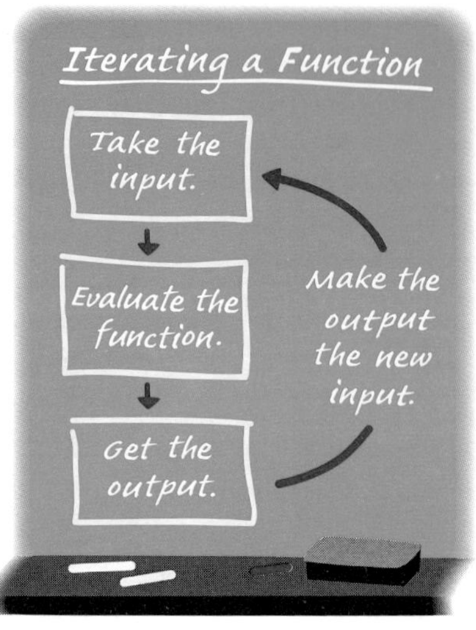

Iterating a Function
Take the input.
Evaluate the function.
Get the output.
Make the output the new input.

a. Iterate the function $f(z) = z^2 + 2$ starting with $z = 0$.

b. Iterate the function $f(z) = z^2 - 1$ starting with $z = 0$.

c. Iterate the function $f(z) = z^2 + i$ starting with $z = 0$.

d. Iterate the function $f(z) = z^2 + 1 + i$ starting with $z = 0$.

e. **Open-ended Problem** Choose any complex number c and iterate the function $f(z) = z^2 + c$ three times starting with $z = 0$. Write the results in $a + bi$ form.

ONGOING ASSESSMENT

50. Open-ended Problem Write a quadratic equation that has complex solutions of the form $a + bi$, where a and b are nonzero. Solve the equation. Show by substitution that the solutions check.

SPIRAL REVIEW

Solve. *(Section 8.3)*

51. $\sqrt{x - 2} = 8$ **52.** $\sqrt[3]{x + 2} = 3$ **53.** $\sqrt[4]{2x - 5} = 2$

Write an exponential function whose graph passes through each pair of points. *(Section 3.5)*

54. (2, 5), (3, 10) **55.** (4, 8), (5, 10) **56.** (2, 2), (5, 6)

Use the Pythagorean theorem to find the length of the hypotenuse of each right triangle. *(Toolbox, page 801)*

57.

7
5

58.

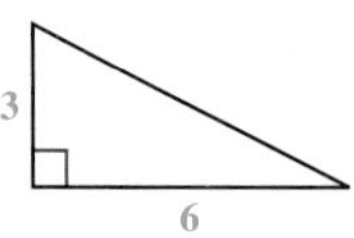

3
6

59.

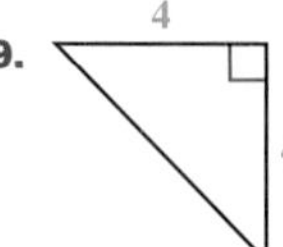

4
4

42. $2 \pm 3i$
43. $7 \pm i$
44. $-3 \pm i\sqrt{2}$
45. $-3 \pm i\sqrt{3}$
46. $-5 \pm 4i$
47. $-\frac{1}{2} \pm \frac{\sqrt{3}}{2}i$
48. Yes; the solutions of the equation $y = ax^2 + bx + c = 0$ are $-\frac{b}{2} \pm \frac{\sqrt{b^2 - 4ac}}{2}$. If the solutions are imaginary, that is, if $b^2 - 4ac < 0$, the solutions are $-\frac{b}{2} \pm \frac{\sqrt{4ac - b^2}}{2a}i$, which are complex conjugates.
49. a. 0; 2; 6; 38; 1446; …
    b. 0; –1; 0; –1; 0; …
    c. 0; i; $-1 + i$; $-i$; $-1 + i$; …
    d. 0; $1 + i$; $1 + 3i$; $-7 + 7i$; $1 - 97i$
    e. Answers may vary. An example is given. The results for $c = i$ are: i, $-1 + i$, $-i$, and $-1 + i$.
50. Answers may vary. Check students' work.
51. 66
52. 25
53. $\frac{21}{2}$
54. $y = 1.25(2^x)$
55. $y = 3.28(1.25)^x$
56. $y = 0.961(1.44)^x$
57. $\sqrt{74} \approx 8.60$
58. $3\sqrt{5} \approx 6.71$
59. $4\sqrt{2} \approx 5.66$

Section

8.5 Visualizing the Complex Plane

Learn how to...

- **plot complex numbers in the complex plane**
- **calculate the magnitude of a complex number**

So you can...

- **understand the mathematics used to generate images of the Mandelbrot set, for example**

Some of the most astonishing images that computers have created are representations of the *Mandelbrot set*, a closed region of the plane with an extremely complicated border. In this section you will learn how repeatedly adding and multiplying complex numbers creates these images.

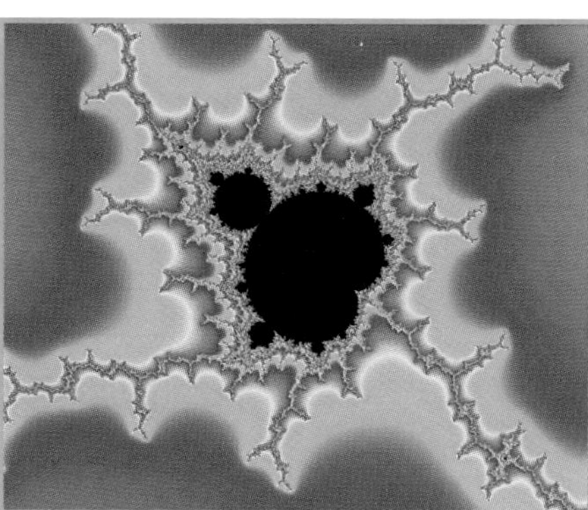

Images of the Mandelbrot set are plotted in the **complex plane**, where each point (a, b) represents the complex number $a + bi$.

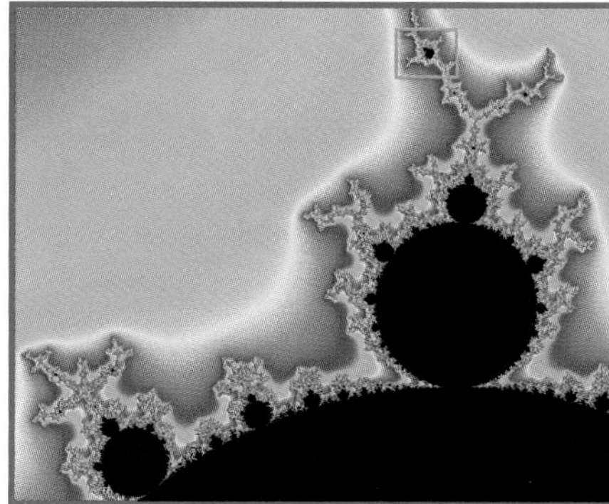

This point is the graph of $-1 + i$.

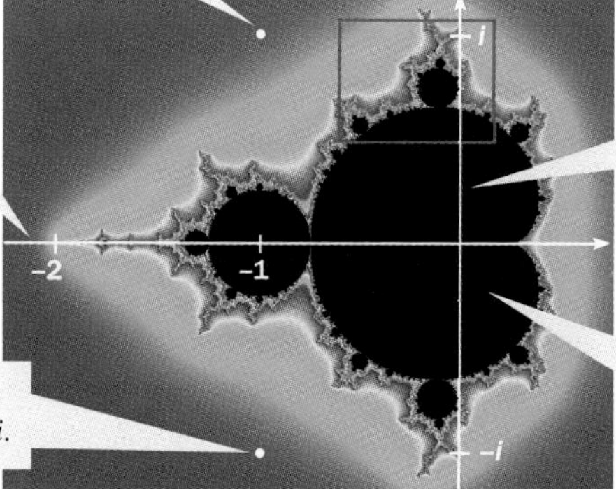

The complex plane has a horizontal **real axis**.

This point is the graph of $-1 - i$.

The complex plane has a vertical **imaginary axis**.

The **Mandelbrot set** is shown in black. It lies inside a circle of radius 2 centered at the origin of the complex plane.

Plan⇔Support

Objectives

- Plot complex numbers in the complex plane.
- Calculate the magnitude of a complex number.
- Understand the mathematics used to generate images.

Recommended Pacing

Core and Extended Courses
Section 8.5: 1 day

Block Schedule
Section 8.5: $\frac{1}{2}$ block
(with Section 8.6)

Resource Materials

Lesson Support
Lesson Plan 8.5
Warm-Up Transparency 8.5
Practice Bank: Practice 51
Study Guide: Section 8.5
Exploration Lab Manual: Diagram Master 1

Technology
Graphing Calculator
Spreadsheet Software
McDougal Littell Mathpack *Stats!*
Internet: http://www.hmco.com

Warm-Up Exercises

Perform the indicated operation.

1. $(2 - 3i) + (-1 + i)$ $1 - 2i$

2. $(5 + 2i) - (-4 + 3i)$ $9 - i$

3. $(6 - i)(2 + 3i)$ $15 + 16i$

4. $2i(-3 - i)$ $2 - 6i$

5. Give the conjugate of $-3 - 4i$. $-3 + 4i$

6. Find the length of the hypotenuse of a right triangle with legs of lengths 3 and 5. $\sqrt{34}$

Teach⇔Interact

Section Note

Historical Connection
Students may be interested in knowing that the Mandelbrot set is a recent discovery. It was found in the early 1980s by Benoit Mandelbrot, a mathematician and scientist. Mandelbrot found the set with the use of a computer program. The Mandelbrot set is an example of a *fractal*, a geometric shape that exhibits the property of self-similarity; that is, within the outer shape, there are smaller versions with exactly the same detail. Fractal geometry is part of a new branch of mathematics called *chaos theory*, which is vital to modeling natural phenomena such as coastlines, trees, clouds, and weather patterns.

Think and Communicate

Question 3 motivates the definition of the magnitude of a complex number. Ask several students to share their responses with the class. Some may have drawn right triangles and used the Pythagorean theorem to demonstrate that the distances are equal.

Additional Example 1

Plot $-4 + 2i$, $1 - 3i$, and their sum in the complex plane. Find their magnitudes.

First find the sum algebraically.
$(-4 + 2i) + (1 - 3i) = -3 - i$
The numbers are plotted in the complex plane below.

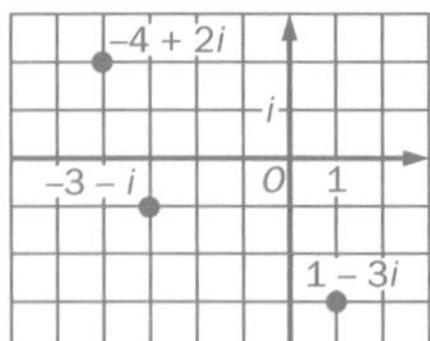

Then calculate the magnitudes.

$|-4 + 2i| = \sqrt{(-4)^2 + 2^2}$
$= \sqrt{16 + 4} = \sqrt{20}$
$|1 - 3i| = \sqrt{1^2 + (-3)^2}$
$= \sqrt{1 + 9} = \sqrt{10}$
$|-3 - i| = \sqrt{(-3)^2 + (-1)^2}$
$= \sqrt{9 + 1} = \sqrt{10}$

THINK AND COMMUNICATE

1. In what quadrant would you plot the complex number $3 - 4i$?
2. In what quadrant would you plot the complex number $-4 + 3i$?
3. Which is farther from the origin, the graph of $3 - 4i$ or $-4 + 3i$? Explain.

WATCH OUT!
For a real number $a + 0i$, the magnitude is $|a + 0i| = |a|$.

Magnitude of a Complex Number

The **magnitude** of a complex number $a + bi$, denoted by $|a + bi|$, is the distance from (0, 0) to (a, b).

You can use the Pythagorean theorem to find this distance:

$$|a + bi| = \sqrt{a^2 + b^2}$$

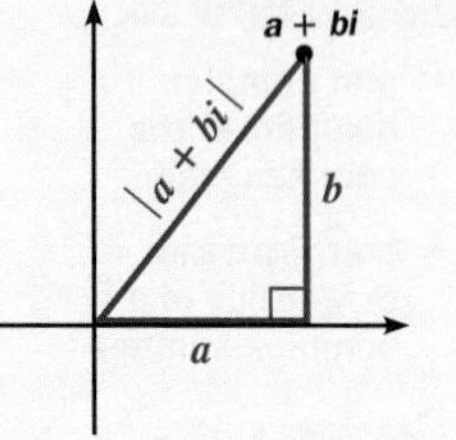

EXAMPLE 1

Plot $3 + i$, $2 + 3i$, and their sum in the complex plane. Find their magnitudes.

SOLUTION

First find the sum algebraically:

$$(3 + i) + (2 + 3i) = 5 + 4i$$

The numbers are plotted in the complex plane at the right.

Then calculate the magnitudes:

$$|3 + i| = \sqrt{3^2 + 1^2} = \sqrt{9 + 1} = \sqrt{10}$$

$$|2 + 3i| = \sqrt{2^2 + 3^2} = \sqrt{4 + 9} = \sqrt{13}$$

$$|5 + 4i| = \sqrt{5^2 + 4^2} = \sqrt{25 + 16} = \sqrt{41}$$

THINK AND COMMUNICATE

4. a. Draw a quadrilateral whose vertices are (0, 0) and the three points in the diagram from Example 1. What is special about the quadrilateral?
   b. Repeat the construction with the sum of two different complex numbers of your own choice. Is the same kind of geometric figure formed?
5. a. In Example 1, is $|(3 + i) + (2 + 3i)| = |3 + i| + |2 + 3i|$ true?
   b. In general, is the *magnitude of the sum* of two complex numbers equal to the *sum of the magnitudes* of the complex numbers?
6. Under what circumstances is the magnitude of the sum of two complex numbers equal to the sum of the magnitudes of the complex numbers?

ANSWERS Section 8.5

Think and Communicate

1. quadrant IV 2. quadrant II
3. neither; (3, −4) and (−4, 3) are both 5 units from the origin.
4. a.

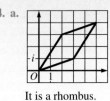

It is a rhombus.

b. Choices of complex numbers may vary. If the points are not collinear, the figure formed is a parallelogram.

5. a. No. b. No.
6. $|(a + bi) + (c + di)| = |a + bi| + |c + di|$ if (0, 0), (a, b), and (c, d) are collinear and (a, b) and (c, d) are in the same quadrant.
7.

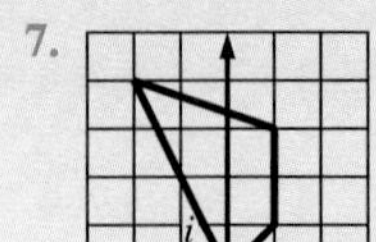

No; No.

8. a. The magnitude of the product is equal to the product of the magnitudes.
   b. The magnitude of the product of two complex numbers is equal to the product of the magnitudes. Answers may vary. An example is

EXAMPLE 2

Plot $1 + i$, $1 + 3i$, and their product in the complex plane. Find their magnitudes.

SOLUTION

First find the product algebraically:

$$(1 + i)(1 + 3i) = 1 + i + 3i + 3i^2$$
$$= 1 + 4i - 3$$
$$= -2 + 4i$$

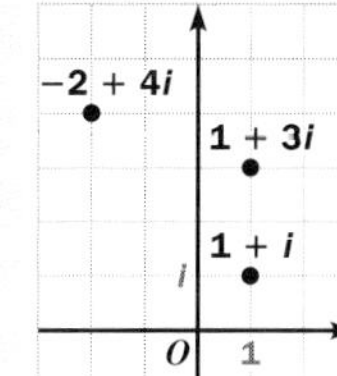

The numbers are plotted in the complex plane at the right.

Then calculate the magnitudes:

$$|1 + i| = \sqrt{1^2 + 1^2} = \sqrt{2}$$

$$|1 + 3i| = \sqrt{1^2 + 3^2} = \sqrt{10}$$

$$|-2 + 4i| = \sqrt{(-2)^2 + 4^2} = \sqrt{4 + 16} = \sqrt{20}$$

THINK AND COMMUNICATE

7. Draw a quadrilateral whose vertices are (0, 0) and the three points in the diagram from Example 2. Is there anything special about it? Is this the same as the type of figure formed in *Think and Communicate* Question 4?

8. a. How are the magnitudes of the complex numbers in Example 2 related?
   b. In general, do you think the *magnitude of the product* of two complex numbers is equal to the *product of the magnitudes* of the complex numbers? Test your conjecture with another pair of complex numbers.

CHECKING KEY CONCEPTS

Plot each pair of complex numbers and their sum in the complex plane. Find their magnitudes.

1. $1, i$
2. $1 + i, -1 + i$
3. $2 + 3i, 3 + 2i$
4. $-4 - 3i, 2 - i$
5. $3 + 5i, 3 - 5i$
6. $-4 + 7i, 4 - 7i$

Plot each pair of complex numbers and their product in the complex plane. Find their magnitudes.

7. $1 + i, 2i$
8. $3i, -2 + 2i$
9. $-4, 2 + i$
10. $3i, -2i$
11. $3 + 4i, 3 - 4i$
12. $-1 + 3i, 4 - i$
13. Plot the numbers $i, i^2, i^3, i^4, i^5, \ldots$ in the complex plane. What do you notice?

Teach⇔Interact

Think and Communicate

The mathematics in this section is a meshing of geometry, algebra, and the theory of complex numbers. Questions 4–8 encourage students to use the geometric representation of complex numbers to examine some of their algebraic properties. The property of question 8 is proven algebraically in Ex. 51.

After drawing the polygons for questions 4 and 7, students can work in groups on questions 5, 6, and 8. A group discussion of these questions should help students understand the ideas contained in them. Encourage the groups to draw figures to aid them with their work, especially when answering question 6.

Additional Example 2

Plot $-1 + i$, $2 - 3i$, and their product in the complex plane. Find their magnitudes.

First find the product algebraically.

$$(-1 + i)(2 - 3i) = -2 + 2i + 3i - 3i^2$$
$$= -2 + 5i + 3$$
$$= 1 + 5i$$

The numbers are plotted in the complex plane below.

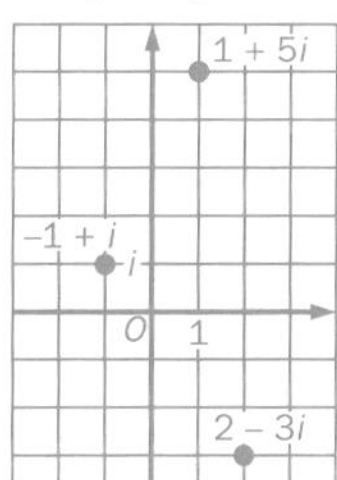

Then calculate the magnitudes.

$$|-1 + i| = \sqrt{(-1)^2 + 1^2} = \sqrt{1 + 1} = \sqrt{2}$$

$$|2 - 3i| = \sqrt{2^2 + (-3)^2} = \sqrt{4 + 9} = \sqrt{13}$$

$$|1 + 5i| = \sqrt{1^2 + 5^2} = \sqrt{1 + 25} = \sqrt{26}$$

Checking Key Concepts

Topic Spiraling: Preview

In question 13, the powers of i yield a repeating cycle: $i, -1, -i, 1, i, -1, -i, 1, \ldots$. In Section 8.6, students will see that the multiplication of these four complex numbers is an example of a *group*.

given. $|(2 + i)(3 - i)| = |7 + i| = \sqrt{50} = 5\sqrt{2}$; $|2 + i| \cdot |3 - i| = \sqrt{5}\sqrt{10} = \sqrt{50} = 5\sqrt{2}$

Checking Key Concepts

1. $1; 1; \sqrt{2}$

2. $\sqrt{2}; \sqrt{2}; 2$

3.

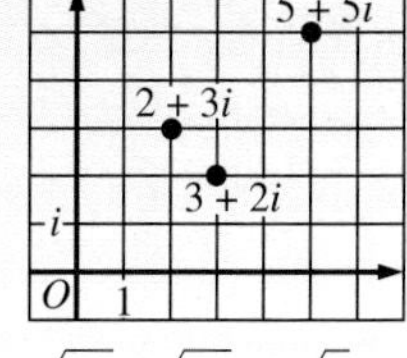

$\sqrt{13}; \sqrt{13}; 5\sqrt{2}$

4.

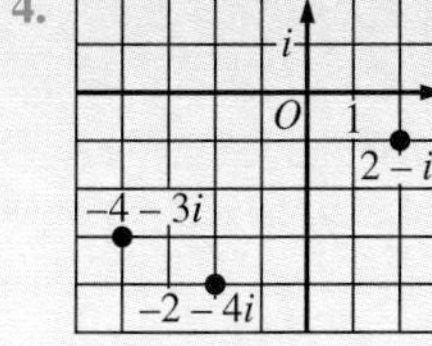

$5; \sqrt{5}; 2\sqrt{5}$

5–13. See answers in back of book.

Teach⇔Interact

Closure Question

Describe how to represent a complex number $a + bi$ geometrically and how to calculate its magnitude.

Draw a complex plane where the horizontal axis represents real numbers and the vertical axis represents pure imaginary numbers. Plot (a, b) to represent $a + bi$. To calculate the magnitude, find $\sqrt{a^2 + b^2}$.

Apply⇔Assess

Suggested Assignment

Core Course
Exs. 1–35, 53–62

Extended Course
Exs. 1–6, 10–14, 18–22, 25–62

Block Schedule
Day 51 Exs. 1–35, 53–62

Exercise Notes

Communication: Discussion
Exs. 20–28 Students may recall the term *opposite* as it applies to real numbers. Here, the concept of opposite is being extended to other complex numbers. When a definition is extended, it is important that its meaning remains the same. Another example of a definition being extended from the real numbers to the complex numbers is that of absolute value. The term absolute value applies only to real numbers, but the idea that the absolute value is the distance from the number to zero on the number line is used to define the magnitude of a complex number. The magnitude is the distance of the number from the origin in the complex plane.

Journal Entry
Exs. 20–36 Students may confuse opposite and conjugate. Suggest that they set aside two facing pages in their journals, one for opposites and one for conjugates. On each page, they should give a definition, some examples with illustrations, and the properties discovered in Exs. 26, 28, 35, and 36.

8.5 Exercises and Applications

Extra Practice exercises on page 761

Name the two complex numbers plotted in each complex plane. Each pair has the same magnitude. Find that magnitude and name two other numbers that have the same magnitude.

1.
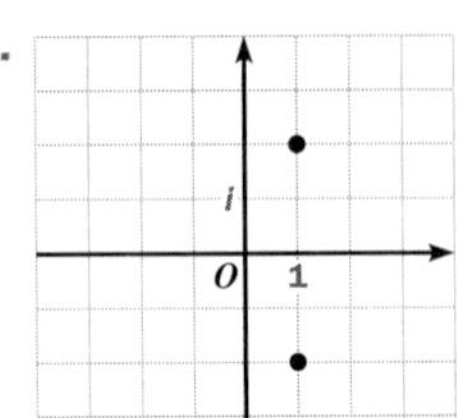

2.
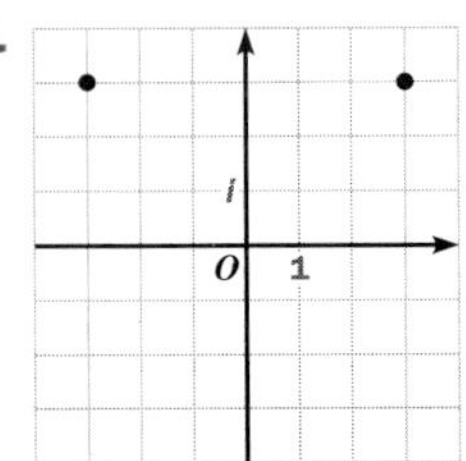

3.
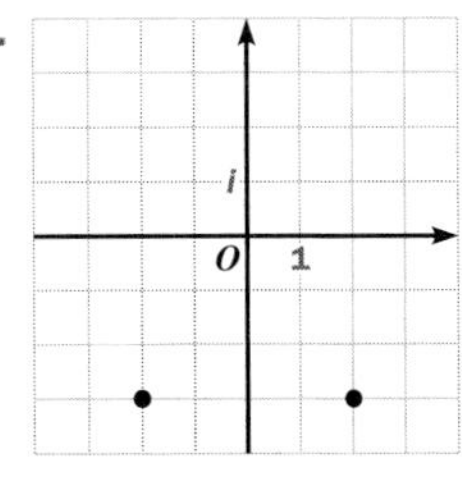

Plot each pair of complex numbers and their sum in the complex plane. Find their magnitudes.

4. $3, -2i$ **5.** $-4 + 3i, -3 + 4i$ **6.** $7i, -3i$ **7.** $2 + i, -2 + i$

8. $-3 + i, 2 - 4i$ **9.** $2 + i, -5 + 3i$ **10.** $3 + 2i, -5 - 3i$ **11.** $8 - 4i, -8 + 4i$

Plot each pair of complex numbers and their product in the complex plane. Find their magnitudes.

12. $i, 4 + 2i$ **13.** $-2i, 3 + 5i$ **14.** $-4i, -2i$ **15.** $5i, -i$

16. $-2, 1 + i$ **17.** $3 + i, 3 + 2i$ **18.** $5 - 6i, 5 + 6i$ **19.** $i, -1$

A complex number and its *opposite* have a sum of 0. Find the opposite of each complex number.

20. $4 + i$ **21.** $-5 + 2i$ **22.** $3 - 6i$

23. $5 + 6i$ **24.** $-1 - i$ **25.** $-(4 - 7i)$

26. Plot the complex numbers and their opposites from Exercises 20–25. How are opposites geometrically related in the complex plane?

27. What is the opposite of the complex number $a + bi$?

28. **a.** **Open-ended Problem** Choose a complex number and its opposite. Square both numbers and plot the results in the complex plane. What do you notice?

b. How are the magnitudes of a complex number and its square related?

c. What do you think is true of the *square roots* of a complex number?

Recall that the *complex conjugate* of $a + bi$ is $a - bi$. Plot each number and its complex conjugate in the complex plane. Then plot their sum.

29. $5 + 3i$ **30.** $6 - 2i$ **31.** $5i$

32. $-3 - 4i$ **33.** $-1 + 7i$ **34.** -8

35. How are complex conjugates geometrically related in the complex plane?

36. **Visual Thinking** Show that the square of any complex number and the square of its conjugate are also conjugates. Use a complex plane diagram to illustrate.

Exercises and Applications

1–3. Choices of numbers with the same magnitude may vary. Examples are given.

1. $1 + 2i$ and $1 - 2i$; $\sqrt{5}$; $-1 + 2i$ and $-1 - 2i$

2. $-3 + 3i$ and $3 + 3i$; $3\sqrt{2}$; $-3 - 3i$ and $3 - 3i$

3. $-2 - 3i$ and $2 - 3i$; $\sqrt{13}$; $2 + 3i$ and $-2 + 3i$

4.
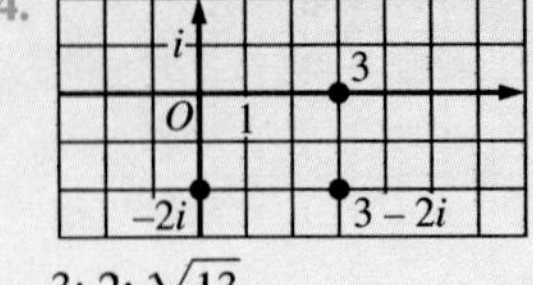

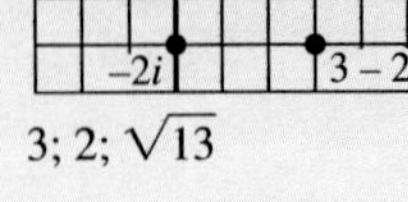

$3; 2; \sqrt{13}$

5.
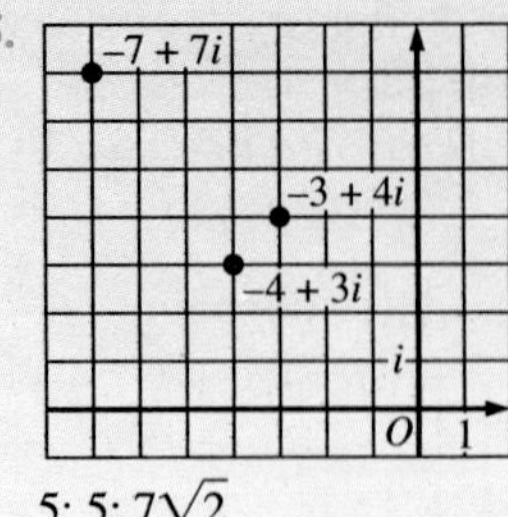

$5; 5; 7\sqrt{2}$

6–36. See answers in back of book.

Connection MANDELBROT SET

In Exercise 49 of Section 8.4, you iterated the function $f(z) = z^2 + c$ for a complex value of c, starting with $z = 0$. If you plot the results in the complex plane, they form the *critical orbit* for that number c.

The complex numbers whose critical orbits settle into predictable cycles (visiting the same point or points in the complex plane repeatedly) form the *Mandelbrot set*. Complex numbers whose critical orbits escape to infinity (eventually moving far away from the origin) are *not* in the Mandelbrot set.

37. **Spreadsheets** You can use a spreadsheet to calculate the critical orbit of a complex number c for the function $f(z) = z^2 + c$.

a. Explain why the cell formulas shown are correct.

b. Create a spreadsheet like the one shown, and use the fill-down feature to iterate the function 1500 times. Test your spreadsheet by entering the c-values from Exercise 49 of Section 8.4. Do the results agree with your earlier work?

c. What happens in your spreadsheet when the magnitudes of numbers in a critical orbit become very large? That is, how does the spreadsheet tell you a value of c is *not* in the Mandelbrot set?

Enter these cell formulas, then fill down:
B8 = B7^2 – D7^2 + B4
D8 = 2*B7*D7 + D4

Mandelbrot Set Calculator

B18 | =B17^2–D17^2+B4

	A	B	C	D	E
1	Mandelbrot Set Calculator: *f*(*z*)=*z*^2 + *c*				
2					
3			*c*		
4		1	+	1	*i*
5					
6	Iteration		*z*		
7	0	0	+	0	*i*
8	1	1	+	1	*i*
9	2	1	+	3	*i*
10	3	–7	+	7	*i*
11	4	1	+	–97	*i*
12	5	–9407	+	–193	*i*
13	6	88454401	+	3631103	*i*
14	7	7.811E+15	+	6.42E+14	*i*
15	8	6.0599E+31	+	1E+31	*i*
16	9	3.5715E+63	+	1.22E+63	*i*
17	10	1.128E+127	+	8.7E+126	*i*
18	11	5.169E+253	+	2E+254	*i*
19	12	#NUM!	+	#NUM!	*i*

The spreadsheet shows that $1 + i$ is *not* in the Mandelbrot set.

Use your spreadsheet to tell whether each complex number *is* or *is not* in the Mandelbrot set.

38. 0 **39.** i **40.** $-1 + 0.5i$ **41.** $-1.2 + 0.1i$

42. $-0.1 + 0.8i$ **43.** $-0.3 + 0.7i$ **44.** $-0.6 - 0.5i$ **45.** $-0.5 - 0.6i$

46. **Investigation** You may have noticed that some critical orbits eventually settle into a cycle that repeatedly visits one, two, three, or more points in the complex plane. Examine these cycles.

a. Look at the orbits of these c-values, which lie in region A: $-0.5 + 0i$, $0.1 + 0.4i$, $-0.6 + 0.2i$. What do you notice?

b. Look at the orbits of these c-values, which lie in region B: $-1 + 0i$, $-0.9 + 0.1i$, $-1.1 + 0.2i$. What do you notice?

c. Look at the orbits of these c-values, which lie in region C: $-0.2 + 0.7i$, $-0.05 + 0.75i$. What do you notice?

d. **Open-ended Problem** Find a point in the Mandelbrot set that has a critical orbit with a cycle that visits 4 or more points.

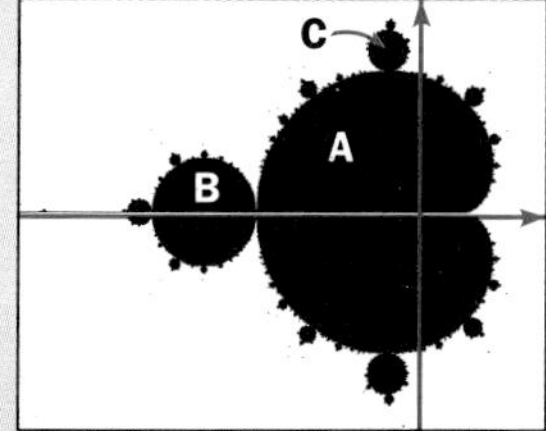

Look back at the graph of the Mandelbrot set on page 365.

47. **Visual Thinking** Examine the critical orbit of any complex number and the critical orbit of its complex conjugate. How do the results help explain the symmetry of the Mandelbrot set?

37. a. $Z_0 = B7 + D7i$ and $C = B4 + D4i$ so $Z_1 = (Z_0)^2 + C = (B7 + D7i)^2 + B4 + D4i = (B7)^2 + 2 \cdot B7 \cdot D7i + (D7)^2i^2 + B4 + D4i = ((B7)^2 - (D7)^2 + B4) + (2 \cdot B7 \cdot D7 + D4)i$ so $B8 = (B7)^2 - (D7)^2 + B4 =$ B7^2 – D7^2 + B4 and $D8 = 2 \cdot B7 \cdot D7 + D4 =$ 2*B7*D7 + D4.

b. Yes.

c. An error message such as #NUM! appears.

38. Yes. 39. Yes.

40. No. 41. Yes.

42. Yes. 43. No.

44. No. 45. Yes.

46. a. The critical orbit settles into a cycle of one point.

b. The critical orbit settles into a cycle of two points.

c. The critical orbit settles into a cycle of three points.

d. Answers may vary. An example is given. $-0.5 - 0.6i$

47. The critical orbit of the complex conjugate of a complex number consists of the complex conjugates of the critical orbit. The Mandelbrot set should be symmetric about the real axis.

Apply⇔Assess

Exercise Notes

Teaching Tip

Ex. 28 Point out to students that the property found in part (b) of this exercise is a special case of the property found in Think and Communicate question 8 on page 367; namely, that the magnitude of the product of two complex numbers is equal to the product of their magnitudes. In this case, the two factors are the same, so that the magnitude of the square is equal to the square of the magnitude of the base.

Research

Exs. 37–47 A great deal has been written recently about chaos and fractals. Some of this material can be understood by high school students. The self-similarity of the Mandelbrot set, in particular, is demonstrated magnificently by a series of photographs in the book *Turbulent Mirror* by John Briggs and F. David Peat. Any library resources should be made available to students for reading and research. Some software packages are also available for studying fractals.

Communication: Reading

Exs. 37–47 The introductory paragraphs for these exercises contain several new concepts and vocabulary words. It would be worthwhile to read and discuss these paragraphs as a class activity before assigning the exercises.

Using Technology

Exs. 37–47 These exercises can be worked on in a computer lab. Bring the class to the lab and have two students work together at each computer. When assigning partners, try to put together students with different computer abilities. Have the student with the least computer experience do the typing while the other student helps. If your school has the equipment, demonstrate how to set up the spreadsheet using an LCD display on an overhead projector. Each pair of students should then set up their own spreadsheet and work through all of the exercises. If time allows, you could finish the lab with some fractal drawing software, ideally, one that has the Mandelbrot set in color and allows the user to zoom in on smaller and smaller parts of it.

You may want to suggest that students use the *Stats!* software for this group of exercises.

Apply⇔Assess

Exercise Notes

Challenge

Ex. 48 After studying several numerical examples, ask students if they can write a formal geometric proof for this exercise when a and b are both positive.

Topic Spiraling: Preview

Exs. 59–62 These exercises demonstrate that matrix addition is commutative while matrix multiplication is not. In the next section, students will examine matrices and other number systems for this and other properties.

48. Writing Show that multiplying a complex number $a + bi$ by the complex number i is equivalent to rotating the point that represents $a + bi$ in the complex plane by 90° counterclockwise around the origin. Use an algebraic argument to explain why this is so.

49. Open-ended Problem Plot two complex numbers and their product in the complex plane. Draw line segments connecting the points to the origin. Measure the angles these segments make with the positive real axis. What do you notice? Make a conjecture and test it.

50. SAT/ACT Preview What is the square of $3 + 2i$?

A. $9 + 4i$ **B.** $25i^2$ **C.** 5 **D.** $5 + 12i$ **E.** $13 + 12i$

51. Challenge Prove that the magnitude of the product of two complex numbers is equal to the product of the magnitudes of the two numbers. That is, show that this equation is always true:

$$|(a + bi)(c + di)| = |a + bi| \cdot |c + di|$$

ONGOING ASSESSMENT

52. a. Open-ended Problem Find two complex numbers whose sum is a pure imaginary number.

b. Open-ended Problem Find two complex numbers whose product is a pure imaginary number.

SPIRAL REVIEW

Use the information in the table to find the mean and standard deviation of each data set. *(Section 6.5)*

53. recycled waste

54. incinerated waste

55. landfilled waste

Solid Waste Management in New England (1991)

State	Recycled (%)	Incinerated (%)	Landfilled (%)
Connecticut	15	65	20
Maine	17	45	38
Massachusetts	29	47	24
New Hampshire	5	23	72
Rhode Island	15	0	85
Vermont	20	8	72

Write a quadratic equation with the given solutions. *(Section 8.4)*

56. $i, -i$

57. $3i, -3i$

58. $i\sqrt{13}, -i\sqrt{13}$

Evaluate each matrix expression. *(Sections 1.3 and 1.4)*

59. $\begin{bmatrix} 1 & 4 \\ 10 & 7 \end{bmatrix} + \begin{bmatrix} 6 & -1 \\ 3 & -2 \end{bmatrix}$

60. $\begin{bmatrix} 6 & -1 \\ 3 & -2 \end{bmatrix} + \begin{bmatrix} 1 & 4 \\ 10 & 7 \end{bmatrix}$

61. $\begin{bmatrix} 1 & 5 \\ 9 & 2 \end{bmatrix}\begin{bmatrix} 6 & 3 \\ 1 & 0 \end{bmatrix}$

62. $\begin{bmatrix} 6 & 3 \\ 1 & 0 \end{bmatrix}\begin{bmatrix} 1 & 5 \\ 9 & 2 \end{bmatrix}$

Practice 51 for Section 8.5

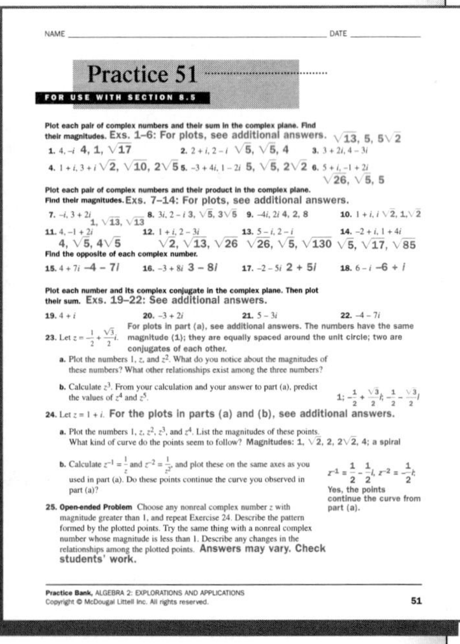

NAME ____________ DATE ____________

Practice 51

FOR USE WITH SECTION 8.5

Plot each pair of complex numbers and their sum in the complex plane. Find their magnitudes. Exs. 1–6: For plots, see additional answers.

1. $4, -i$ $4, 1, \sqrt{17}$
2. $2 + i, 2 - i$ $\sqrt{5}, \sqrt{5}, 4$
3. $3 + 2i, 4 - 3i$ $\sqrt{13}, 5, 5\sqrt{2}$
4. $1 + i, 3 + i$ $\sqrt{2}, \sqrt{10}, 2\sqrt{5}$
5. $-3 + 4i, 1 - 2i$ $5, \sqrt{5}, 2\sqrt{2}$
6. $5 + i, -1 + 2i$ $\sqrt{26}, \sqrt{5}, 5$

Plot each pair of complex numbers and their product in the complex plane. Find their magnitudes. Exs. 7–14: For plots, see additional answers.

7. $-i, 3 + 2i$ $1, \sqrt{13}, \sqrt{13}$
8. $3i, 2 - i$ $3, \sqrt{5}, 3\sqrt{5}$
9. $-4i, 2i$ $4, 2, 8$
10. $1 + i, i$ $\sqrt{2}, 1, \sqrt{2}$
11. $4, -1 + 2i$ $4, \sqrt{5}, 4\sqrt{5}$
12. $1 + i, 2 - 3i$ $\sqrt{2}, \sqrt{13}, \sqrt{26}$
13. $5 - i, 2 - i$ $\sqrt{26}, \sqrt{5}, \sqrt{130}$
14. $-2 + i, 1 + 4i$ $\sqrt{5}, \sqrt{17}, \sqrt{85}$

Find the opposite of each complex number.

15. $4 + 7i$ $-4 - 7i$
16. $-3 + 8i$ $3 - 8i$
17. $-2 - 5i$ $2 + 5i$
18. $6 - i$ $-6 + i$

Plot each number and its complex conjugate in the complex plane. Then plot their sum. Exs. 19–22: See additional answers.

19. $4 + i$
20. $-3 + 2i$
21. $5 - 3i$
22. $-4 - 7i$

23. Let $z = -\frac{1}{2} + \frac{\sqrt{3}}{2}i$. For plots in part (a), see additional answers. The numbers have the same magnitude (1); they are equally spaced around the unit circle; two are conjugates of each other.
   a. Plot the numbers 1, z, and z^2. What do you notice about the magnitudes of these numbers? What other relationships exist among the three numbers?
   b. Calculate z^3. From your calculation and your answer to part (a), predict the values of z^4 and z^5. $1; -\frac{1}{2} + \frac{\sqrt{3}}{2}i; -\frac{1}{2} - \frac{\sqrt{3}}{2}i$

24. Let $z = 1 + i$. For the plots in parts (a) and (b), see additional answers.
   a. Plot the numbers 1, z, z^2, z^3, and z^4. List the magnitudes of these points. What kind of curve do the points seem to follow? Magnitudes: $1, \sqrt{2}, 2, 2\sqrt{2}, 4$; a spiral
   b. Calculate $z^{-1} = \frac{1}{z}$ and $z^{-2} = \frac{1}{z^2}$, and plot these on the same axes as you used in part (a). Do these points continue the curve you observed in part (a)? $z^{-1} = \frac{1}{2} - \frac{1}{2}i, z^{-2} = -\frac{1}{2}i$; Yes, the points continue the curve from part (a).

25. **Open-ended Problem** Choose any nonreal complex number z with magnitude greater than 1, and repeat Exercise 24. Describe the pattern formed by the plotted points. Try the same thing with a nonreal complex number whose magnitude is less than 1. Describe any changes in the relationships among the plotted points. Answers may vary. Check students' work.

Practice Bank, ALGEBRA 2: EXPLORATIONS AND APPLICATIONS

51

48. Answers may vary. An example is given. $(a + bi)i = ai + bi^2 = -b + ai$; $|a + bi| = |-b + ai|$ and the segments joining $a + bi$ and $-b + ai$ to the origin are perpendicular. (One has slope $\frac{b}{a}$ and one has slope $-\frac{a}{b}$.) Then multiplying a complex number by i is equivalent to rotating it 90° clockwise about the origin.

49. Choices of points may vary. The measure of the angle that the product makes with the real axis is the sum of the measures of the angles the two segments make with the positive real axis.

50. D

51. $|(a + bi)(c + di)| = |(ac - bd) + (bc + ad)i| =$
$\sqrt{(ac - bd)^2 + (bc + ad)^2} =$
$\sqrt{a^2c^2 - 2abcd + b^2d^2 + b^2c^2 + 2abcd + a^2d^2} =$
$\sqrt{a^2c^2 + b^2c^2 + a^2d^2 + b^2d^2} = \sqrt{c^2(a^2 + b^2) + d^2(a^2 + b^2)} =$
$\sqrt{(a^2 + b^2)(c^2 + d^2)} = \sqrt{(a^2 + b^2)}\sqrt{(c^2 + d^2)} = |a + bi| \cdot |c + di|$

52. Answers may vary. Examples are given.
   a. $2 + 3i$ and $-2 + 5i$
   b. $2 + 3i$ and $3 + 2i$

53. 16.8; 7.1

54. 31.3; 23.0

55. 51.8; 25.5

56–58. Answers may vary. Examples are given.

56. $x^2 + 1 = 0$

57. $x^2 + 9 = 0$

58. $x^2 + 13 = 0$

59. $\begin{bmatrix} 7 & 3 \\ 13 & 5 \end{bmatrix}$

60. $\begin{bmatrix} 7 & 3 \\ 13 & 5 \end{bmatrix}$

61. $\begin{bmatrix} 11 & 3 \\ 56 & 27 \end{bmatrix}$

62. $\begin{bmatrix} 33 & 36 \\ 1 & 5 \end{bmatrix}$

Section

8.6 Properties of Number Systems

Learn how to...

- identify the number systems to which a number belongs
- evaluate whether group properties hold for a set and an operation

So you can...

- see structural similarities in groups, such as country dancing, puzzle cube rotations, and the multiplication of complex roots of 1

In Peter Høeg's book *Smilla's Sense of Snow,* Smilla Qaavigaaq Jaspersen compares the development of number systems to human development while her neighbor, a mechanic, prepares dinner:

> ". . . the number system is like human life. First you have the natural numbers. The ones that are whole and positive. The numbers of a small child. But human consciousness expands. The child discovers a sense of longing, and do you know what the mathematical expression is for longing?"
>
> He adds cream and several drops of orange juice to the soup.
>
> "The negative numbers. The formalization of the feeling that you are missing something. And human consciousness expands and grows even more, and the child discovers the in between spaces. Between stones, between pieces of moss on the stones, between people. And between numbers. And do you know what that leads to? It leads to fractions. Whole numbers plus fractions produce rational numbers. And human consciousness doesn't stop there. It wants to go beyond reason. It adds an operation as absurd as the extraction of roots. And produces irrational numbers."
>
> He warms French bread in the oven and fills the pepper mill.
>
> "It's a form of madness. Because the irrational numbers are infinite. They can't be written down. They force human consciousness out beyond the limits. And by adding irrational numbers to rational numbers, you get real numbers."
>
> I've stepped into the middle of the room to have more space. It's rare that you have a chance to explain yourself to a fellow human being. Usually you have to fight for the floor. And this is important to me.
>
> "It doesn't stop. It never stops. Because now, on the spot, we expand the real numbers with imaginary square roots of negative numbers. These are numbers we can't picture, numbers that normal human consciousness cannot comprehend. And when we add the imaginary numbers to the real numbers, we have the complex number system. The first number system in which it's possible to explain satisfactorily the crystal formation of ice. It's like a vast, open landscape. The horizons. You head toward them and they keep receding."

THINK AND COMMUNICATE

Toolbox p. 779
Rational Numbers and Irrational Numbers

1. Why does Smilla call the natural numbers "the numbers of a small child"?
2. How does a mathematician use the word *irrational*? What other meaning of *irrational* does Smilla use?
3. Mathematicians sometimes develop number systems in order to solve equations. For each equation, tell what type of number solves it.

   a. $x + 6 = 2$ **b.** $2x = 3$ **c.** $x^2 = 2$ **d.** $x^2 + x + 1 = 0$
4. Diagram the relationships among the sets of numbers that Smilla mentions.

Plan⇔Support

Objectives

- Identify the number systems to which a number belongs.
- Evaluate whether group properties hold for a set and an operation.
- See structural similarities in groups.

Recommended Pacing

Core and Extended Courses
Section 8.6: 2 days

Block Schedule
Section 8.6: 2 half-blocks (with Section 8.5 and Review)

Resource Materials

Lesson Support
Lesson Plan 8.6
Warm-Up Transparency 8.6
Practice Bank: Practice 52
Study Guide: Section 8.6
Assessment Book: Test 31

Technology
Technology Book: Calculator Activity 8
Graphing Calculator
McDougal Littell Mathpack *Function Investigator with Matrix Analyzer Activity Book:* Matrix Analyzer Activity 1
Internet: http://www.hmco.com

Warm-Up Exercises

Give an example of each type of number.

1. real number $2, -9, \sqrt{7}$
2. rational number $\frac{3}{8}, 0, -7$
3. irrational number $\pi, \sqrt{5}$
4. complex number $6, 4 - 2i$
5. integer $6, 0, -12$
6. whole number $0, 11$
7. imaginary number $2 + 3i, -7i$
8. pure imaginary number $i, 6i$

ANSWERS Section 8.6

Think and Communicate

1–2. Answers may vary. Examples are given.

1. Positive integers are the first numbers a small child learns about.
2. A mathematician uses the term irrational to describe a number that cannot be written as the ratio of an integer to a nonzero integer. Smilla also uses the term to mean lacking in reason or illogical.
3. Answers may vary. Examples are given.

   a. whole number b. rational number
   c. irrational number d. complex number
4. 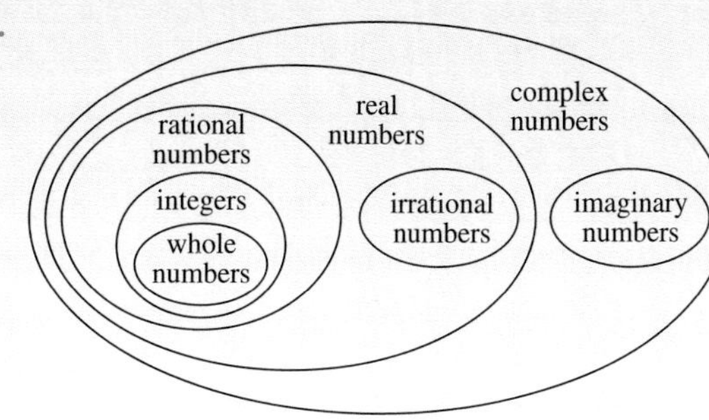

Teach⇔Interact

Learning Styles: Verbal/Visual

The reading that opens this section is a wonderful description of the most common number systems. This description should be an aid to verbal learners. Have them read the paragraph several times and write a journal entry that defines all the systems and gives examples for each one.

The diagram that students draw in Think and Communicate question 4 and the one at the top of this page should be a great aid to visual learners. Encourage these students to add examples of each type of number on the diagram they draw. Students should keep a copy of this diagram with examples in their journals for future reference.

Section Note

Second-Language Learners
The language, concepts, and imagery in the excerpt from *Smilla's Sense of Snow* may prove challenging for many students learning English. You may wish to allow time for a class discussion of the ideas in the passage, making sure that all students understand how each of the analogies works. Then, encourage English-learning students to work with a partner or in small groups to complete Think and Communicate questions 1–4.

Additional Example 1

Identify the number systems to which each number belongs.

a. $-\frac{3}{5}$ rational numbers, real numbers, complex numbers

b. $\sqrt{9}$ whole numbers, integers, rational numbers, real numbers, complex numbers

c. π irrational numbers, real numbers, complex numbers

Additional Example 2

If possible, give an example of a number that satisfies each description.

a. rational but not an integer $\frac{2}{3}$

b. whole and irrational not possible

c. complex but not imaginary 7

Smilla Jaspersen's description of the complex number system shows that it includes real and imaginary numbers. The real number system, in turn, includes rational and irrational numbers.

The diagram below is one way to show how various important number systems are "nested." A number in any box belongs to the number systems of all the boxes that surround it.

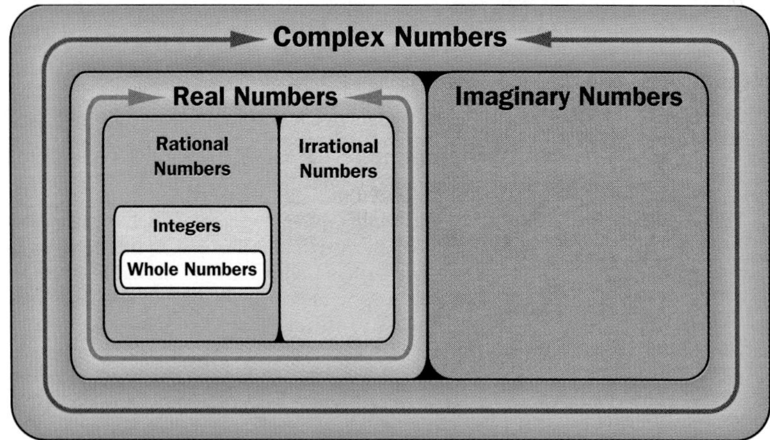

EXAMPLE 1

Identify the number systems to which each number belongs.

a. $3 + 4i$ **b.** $\sqrt{3}$ **c.** -56

SOLUTION

a. imaginary numbers, complex numbers

b. irrational numbers, real numbers, complex numbers

c. integers, rational numbers, real numbers, complex numbers

EXAMPLE 2

If possible, give an example of a number that satisfies each description.

a. integral and real **b.** integral but not whole **c.** rational and imaginary

SOLUTION

a. 3 **b.** -5 **c.** not possible

Mathematicians sometimes study number systems by investigating whether certain properties hold when numbers are combined using operations. For example, if an operation that combines two numbers from a number set yields a number that is also in the set, the number set is *closed* under that operation.

A number set and operation form a **group** if the set is closed under the operation and the *identity, inverse,* and *associative properties* hold. If the *commutative property* also holds for a group, it is called a **commutative group**. The group properties are summarized in the tables on this page and the next.

Group Properties of Real Numbers under Addition

Property	Definition	Example	Summary
Closure	**If *a* and *b* are real numbers, then *a* + *b* is a real number.**	4 and -12 are real; $4+(-12)=-8$, which is also real	**If you add any two real numbers, the sum is also a real number.**
Identity	**If *a* is a real number, then $a+0=0+a=a$. 0 is the identity for addition.**	$5+0=5$ $0+\pi=\pi$	**If you add 0 to any real number, the result is the number you started with.**
Inverse	**If *a* is a real number, then there is a real number $-a$ such that $a+(-a)=0$. Note: $-a$ is called the *opposite*, or *additive inverse*, of *a*.**	$7+(-7)=0$ $-\sqrt{2}+\sqrt{2}=0$	**Every real number has an opposite that is also real. When you add a number and its opposite, the sum is 0, the identity for addition.**
Associative	**If *a, b,* and *c* are real numbers, then $(a+b)+c=a+(b+c)$.**	$(3+7)+4=3+(7+4)$ $\left(1+\frac{1}{3}\right)+\frac{2}{3}=1+\left(\frac{1}{3}+\frac{2}{3}\right)$	**When you add three real numbers, you get the same answer whether you add the sum of the first and second to the third or you add the first to the sum of the second and third.**
Commutative	**If *a* and *b* are real numbers, then $a+b=b+a$.**	$3+4=4+3$ $e+1=1+e$	**The order in which you add two real numbers does not change the sum.**

EXAMPLE 3

Tell whether each number set is closed under the given operation. If so, tell whether the number set and operation *form a group, form a commutative group,* or *do not form a group.* Explain.

a. odd integers; + **b.** even integers; +

SOLUTION

a. The set of odd integers is *not closed* under addition because $1+1=2$, which is not an odd integer. Because the closure property does not hold, the set of odd integers *does not form a group* under the operation of addition.

b. The set of even integers is *closed* under addition, because when you add two even numbers, the result is always even.

Because the *identity, inverse, associative,* and *commutative properties* all hold as well, the set of even integers *forms a commutative group* under addition.

Teach⇔Interact

Section Note

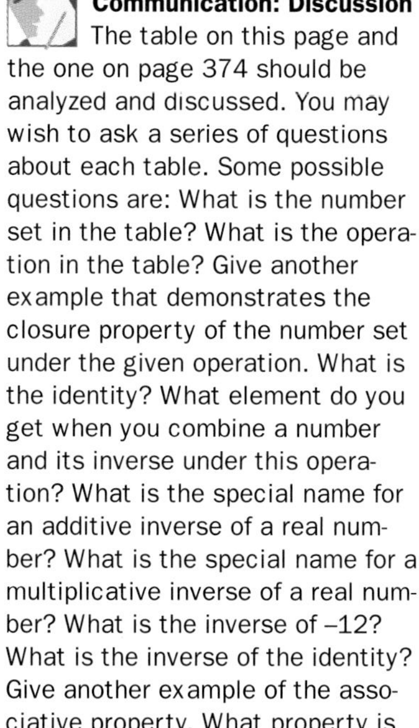

Communication: Discussion The table on this page and the one on page 374 should be analyzed and discussed. You may wish to ask a series of questions about each table. Some possible questions are: What is the number set in the table? What is the operation in the table? Give another example that demonstrates the closure property of the number set under the given operation. What is the identity? What element do you get when you combine a number and its inverse under this operation? What is the special name for an additive inverse of a real number? What is the special name for a multiplicative inverse of a real number? What is the inverse of –12? What is the inverse of the identity? Give another example of the associative property. What property is not necessary for the set and operation to form a group?

Additional Example 3

Tell whether each number set is closed under the given operation. If so, tell whether the number set and operation *form a group, form a commutative group,* or *do not form a group.* Explain.

a. multiples of 4; +
The set of multiples of 4 is *closed* under addition, because when you add two multiples of 4, the result is always a multiple of 4. Because the *identity, inverse, associative,* and *commutative* properties all hold as well, the set of multiples of 4 *forms a commutative group* under addition.

b. pure imaginary numbers; +
The set of pure imaginary numbers is *not closed* under addition because $i+(-i)=0$, which is not a pure imaginary number. Because the closure property *does not hold,* the set of pure imaginary numbers *does not form a group* under the operation of addition. Note that the identity property also does not hold for the set of pure imaginary numbers under addition as 0 is not a pure imaginary number.

Teach⇔Interact

Section Notes

Challenge
Ask students to determine if the real numbers are closed under the operations of division and subtraction. (In both cases, the associative property does not hold.)

Using Technology
Students can use Matrix Analyzer Activity 1 in the *Function Investigator with Matrix Analyzer Activity Book* to discover properties of operations with matrices, and to compare properties of matrices with properties of real numbers.

About Example 4

Reasoning
For part (b), ask students to give examples that demonstrate the identity, inverse, associative, and commutative properties of the positive rational numbers under multiplication.

Additional Example 4

Tell whether each number set is closed under the given operation. If so, tell whether the number set and operation *form a group, form a commutative group,* or *do not form a group.* Explain.

a. multiples of 4; ×
The set of multiples of 4 *is closed* under multiplication because the product of any two multiples of 4 is a multiple of 4. However, the *inverse property does not hold* for the set of multiples of 4 under multiplication, because the multiplicative inverse of 4 is $\frac{1}{4}$, for example, which is not a multiple of 4. Because the inverse property does not hold, the set of multiples of 4 *does not form a group* under multiplication.

b. pure imaginary numbers; ×
The set of pure imaginary numbers is *not closed* under multiplication because $i \times 2i = -2$, which is not a pure imaginary number. Because the closure property does not hold, the set of pure imaginary numbers *does not form a group* under the operation of multiplication.

Group Properties of Real Numbers under Multiplication

Property	Definition	Example	Summary
Closure	**If *a* and *b* are real numbers, then *ab* is a real number.**	4 and -12 are real; $4(-12) = -48$, which is also real	**If you multiply any two real numbers, the product is also a real number.**
Identity	**If *a* is a real number, then $a \cdot 1 = 1 \cdot a = a$. 1 is the identity for multiplication.**	$5 \cdot 1 = 5$ $1 \cdot \pi = \pi$	**If you multiply any real number by 1, the result is the number you started with.**
Inverse	**If *a* is a nonzero real number, then there is a real number $\frac{1}{a}$ such that $a\left(\frac{1}{a}\right) = 1$. Note: $\frac{1}{a}$ is called the *reciprocal*, or *multiplicative inverse*, of *a*.**	$7\left(\frac{1}{7}\right) = 1$ $-\sqrt{2}\left(-\frac{1}{\sqrt{2}}\right) = 1$	**Every nonzero real number has a reciprocal. When you multiply a number and its reciprocal, the product is 1, the identity for multiplication.**
Associative	**If *a, b,* and *c* are real numbers, then $(ab)c = a(bc)$.**	$(3 \cdot 7) \cdot 4 = 3 \cdot (7 \cdot 4)$ $\left(\frac{1}{2} \cdot \frac{3}{4}\right) \cdot \frac{4}{3} = \frac{1}{2} \cdot \left(\frac{3}{4} \cdot \frac{4}{3}\right)$	**When you multiply three real numbers, you get the same answer whether you multiply the product of the first and second by the third or you multiply the first by the product of the second and third.**
Commutative	**If *a* and *b* are real numbers, then $ab = ba$.**	$3 \cdot 4 = 4 \cdot 3$ $2 \cdot e = e \cdot 2$	**The order in which you multiply two real numbers does not change the product.**

EXAMPLE 4

Tell whether each number set is closed under the given operation. If so, tell whether the number set and operation *form a group, form a commutative group,* or *do not form a group.* Explain.

a. positive integers; × **b.** positive rational numbers; ×

SOLUTION

a. The set of positive integers is *closed* under multiplication, because the product of any two positive integers is a positive integer.
However, the *inverse property does not hold* for the set of positive integers under multiplication, because the multiplicative inverse of 3, for example, is $\frac{1}{3}$, which is not an integer. Because the inverse property does not hold, the set of positive integers *does not form a group* under multiplication.

b. The set of positive rational numbers is *closed* under multiplication, because the product of any two positive rational numbers is a positive rational number.
All the other group properties also hold, so the set of positive rational numbers *forms a commutative group* under multiplication.

Think and Communicate

5. No; the set is not closed under addition. Yes; the set is closed under multiplication and all of the properties hold.
6. Yes; the set is closed under addition and all the properties hold. No; the inverse property does not hold.

Checking Key Concepts

1. whole numbers, integers, rational numbers, real numbers, complex numbers
2. integers, rational numbers, real numbers, complex numbers
3. imaginary numbers, complex numbers
4. Answers may vary. An example is given. 4
5. not possible
6. not possible
7. Answers may vary. An example is given. π
8. Yes; form a commutative group; the set is closed under addition and all the group properties as well as the commutative property hold.
9. Yes; do not form a group; there is no identity and the inverse property does not hold.
10. Yes; form a commutative group; the set is closed under multiplication and all the group properties as well as the commutative property hold.

THINK AND COMMUNICATE

5. Does the set of numbers −1, 0, and 1 form a commutative group under addition? under multiplication? Explain.
6. Does the set of integers form a commutative group under addition? under multiplication? Explain.

CHECKING KEY CONCEPTS

Identify the number systems to which each number belongs.

1. 0
2. $-\sqrt{16}$
3. $5i$

If possible, give an example of a number that satisfies each description.

4. rational and even
5. negative and imaginary
6. irrational and odd
7. real and complex

Tell whether each number set is closed under the given operation. If so, tell whether the number set and operation *form a group, form a commutative group,* or *do not form a group.* Explain.

8. multiples of 3; +
9. multiples of 3; ×
10. −1, 1; ×
11. −2, 0, 2; +

8.6 Exercises and Applications

Extra Practice exercises on page 762

Identify the number systems to which each number belongs.

1. $5 - i\sqrt{2}$
2. $\frac{4}{7}$
3. $-\pi$

If possible, give an example of a number that satisfies each description.

4. rational and odd
5. real and imaginary
6. irrational and even
7. whole and complex

Tell whether each number set is closed under the given operation. If so, tell whether the number set and operation *form a group, form a commutative group,* or *do not form a group.* Explain.

8. negative integers; +
9. negative integers; ×
10. rational numbers; +
11. rational numbers; ×
12. irrational numbers; +
13. irrational numbers; ×
14. complex numbers; +
15. complex numbers; ×
16. multiples of 7; +
17. multiples of 7; ×
18. −10, −5, 0, 5, 10; +
19. −10, −5, 0, 5, 10; ×

11. No; do not form a group; the set is not closed under addition.

Exercises and Applications

1. imaginary numbers, complex numbers
2. rational numbers, real numbers, complex numbers
3. irrational numbers, real numbers, complex numbers
4. Answers may vary. An example is given. 3
5. not possible
6. not possible
7. Answers may vary. An example is given. 7
8. Yes; do not form a group; there is no identity and the inverse property does not hold.
9. No; do not form a group; the set is not closed under multiplication.
10. Yes; form a commutative group; the set is closed under addition and all the group properties as well as the commutative property hold.
11. Yes; form a commutative group; the set is closed under multiplication and all the group properties as well as the commutative property hold.
12. No; do not form a group; the set is not closed under addition.

13–19. See answers in back of book.

Teach⇔Interact

Think and Communicate

Students may forget to add or multiply each number in the group by itself when checking for closure. In the case of addition, the set of numbers in question 5 is not closed, because 1 + 1 = 2, which is not an element of the set. To avoid this error when working with a finite set, encourage students to make a table such as the one in Ex. 23, where each element is written in both the first column and first row. The remaining entries of the table are found by combining the corresponding first column and first row entries under the given operation. This is also a good opportunity to introduce these types of tables, as they are used extensively in the exercises of this section.

Closure Question

Describe the criteria necessary for a number set and operation to be a group. Give an example of a group and demonstrate how this group meets the criteria. A number set and operation form a group if the set is closed under the operation and the identity, inverse, and associative properties hold. The numbers −1 and 1 form a group under multiplication. The set is closed under multiplication. The identity is 1 and each element has an inverse; −1 and 1 are their own inverses. The associative property holds because −1 and 1 are real numbers and the associative property holds for all real numbers.

Apply⇔Assess

Suggested Assignment

Core Course
Day 1 Exs. 1–21, 24–29
Day 2 Exs. 37–45, 55–64, AYP

Extended Course
Day 1 Exs. 1–32
Day 2 Exs. 33–64, AYP

Block Schedule
Day 51 Exs. 1–21, 24–29
Day 52 Exs. 37–45, 55–64, AYP

Apply⇔Assess

Exercise Notes

Communication: Writing
Ex. 20 Ask several students to read their responses to this question in class. Then ask how their answers can be used to show that a given set forms a group or commutative group under a given operation. (The associative and commutative properties are true of a subset whenever they are true of the set itself. Thus, any subset of the set of real numbers is associative and commutative.) Caution students that the closure, identity, and inverse properties do not necessarily hold for a subset when they hold for the set itself. These properties must be verified each time.

Using Technology
Ex. 23 Here is a short program for the TI-82 graphing calculator that will do the clock addition in this exercise when the two numbers are input as A and B.

```
PROGRAM: CLOCKADD
:Prompt A
:Prompt B
:A+B–iPart((A+B)/7)*7→C
:Disp C
```

Ask students to explain why the program works. You could also have them write a program that calculates the military time when two military times are added.

Problem Solving
Exs. 23–32, 34–36, 38, 48–55 The tables used in these exercises are very helpful when verifying the group properties of a small set. Closure is easily seen by noting that no element is present in the table which is not in the first column. As Ex. 31 points out, commutativity is also easily seen by noting the symmetry of the table about the diagonal which goes from the upper-left to lower-right corner. You might ask students to develop methods for verifying the identity and inverse properties using the table as well.

20. Writing A *subset* is a set whose members are in another set. If a group property holds for a set of numbers, does it hold for any subset of those numbers? Use the results of Exercises 8–19 to explain your answer.

21. Consider the set of 2×2 matrices whose elements are real numbers.

a. What group properties hold for this set under matrix addition?

b. What group properties hold for this set under matrix multiplication?

22. Consider the operation of exponentiation using the set of real numbers.

a. Is exponentiation commutative? That is, for all real values of a and b (not both zero), is $a^b = b^a$? Explain.

b. Is exponentiation associative? That is, for all real values of a, b, and c (not all zero), is $\left(a^b\right)^c = a^{(bc)}$? Explain.

23. Suppose you number the days of the week with the digits 0 through 6, as shown on the blue "clock" above. You can define *7-day clock addition* to describe the repetitive cycle of days.

a. Suppose today is Friday (day 5). What day will it be six days from now? In 7-day clock addition, $5 + 6 = \underline{?}$.

b. Copy and complete the addition table for the 7-day clock. What group properties hold for this table?

Sun. 0, Mon. 1, Tue. 2, Wed. 3, Thu. 4, Fri. 5, Sat. 6

+	0	1	2	3	4	5	6
0	0	1	2	3	4	5	6
1	1	2	3	4	5	6	0
2	2	3	4	5	6	0	1
3	?	?	?	?	?	?	?
4	?	?	?	?	?	?	?
5	?	?	?	?	?	?	?
6	?	?	?	?	?	?	?

For each table, tell whether the given group property holds. If the group property does not hold, give a counterexample.

24. closure

+	0	2	4	6
0	0	2	4	6
2	2	4	6	8
4	4	6	8	0
6	6	8	0	2

25. identity

†	α	β	γ	δ
α	β	γ	δ	α
β	δ	α	β	γ
γ	γ	δ	α	β
δ	α	β	γ	δ

26. commutativity

*	W	X	Y	Z
W	W	X	Z	Y
X	X	Z	Y	W
Y	Z	Y	W	X
Z	Y	W	X	Z

Copy and complete each table so that it shows the group properties of closure, identity, inverse, and commutativity.

27.

×	A	B	C	D
A	D	C	B	A
B	C	?	A	?
C	B	A	?	?
D	A	?	?	?

28.

+	0	1	2	3
0	0	1	2	3
1	?	2	?	0
2	?	3	0	1
3	?	?	?	2

29.

@	□	●	▽	▲
□	?	?	●	?
●	□	●	▽	▲
▽	●	?	?	?
▲	?	?	?	●

30. Name the identity for each table in Exercises 27–29.

31. Use your answers to Exercises 27–29. How is symmetry across the diagonal of a table related to whether the operation is commutative?

32. Open-ended Problem Do you think the associative property holds for the tables you completed in Exercises 27–29? Give examples or counterexamples to support your conjecture.

20. Not necessarily; for example, the rational numbers under addition form a group and the negative integers are a subset of the rational numbers. But the negative integers under addition do not form a group.

21. a. closure, identity, inverse, associative, commutative

b. closure, identity, associative

22. a. No; for example, $2^3 \neq 3^2$.

b. No; $(a^b)^c = a^{bc}$, which is not the same as $a^{(b^c)}$. For example, $(2^3)^2 = 2^6 = 64$ while $2^{(3^2)} = 2^9 = 512$.

23. See answers in back of book.

24, 25. Counterexamples may vary.

24. No; $2 + 6 = 8$ and 8 is not in the set.

25. No; $\alpha \dagger \delta = \alpha$ but $\gamma \dagger \alpha = \gamma$, so there is not a unique identity.

26. Yes.

27.

×	A	B	C	D
A	D	C	B	A
B	C	D	A	B
C	B	A	D	C
D	A	B	C	D

28.

+	0	1	2	3
0	0	1	2	3
1	1	2	3	0
2	2	3	0	1
3	3	0	1	2

Connection DANCE

In Western square dancing, the pattern known as "right and left grand" is performed by four couples as shown below.

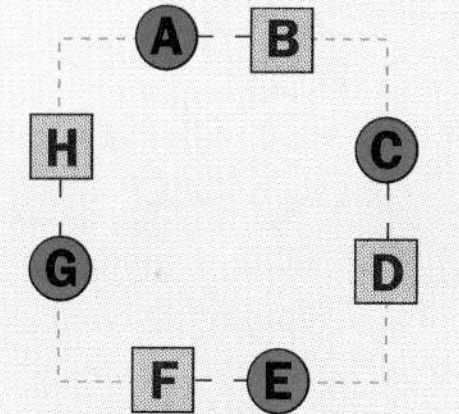

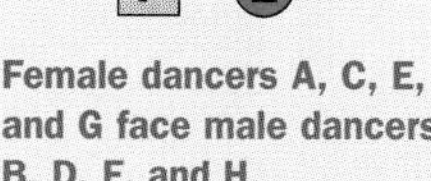

Female dancers A, C, E, and G face male dancers B, D, F, and H.

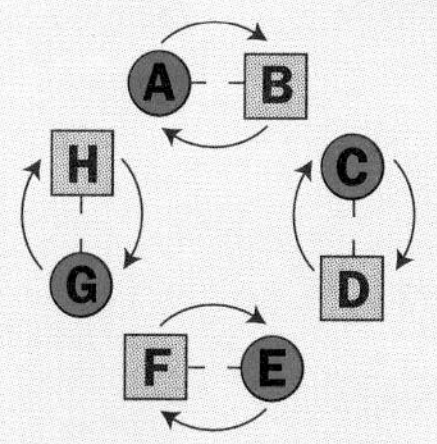

Partners join right hands and pull by each other to switch positions.

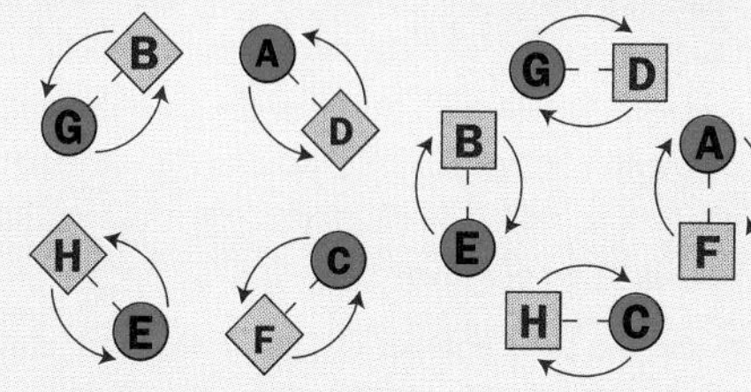

Dancers continue along the circle, giving a left hand to the next person, a right hand to the next, a left hand to the next, and so on.

33. a. Have a group of eight students demonstrate the "right and left grand" pattern twice. Imagine viewing the dancers from above. Which dancers move clockwise? counterclockwise?

b. How many dancers does dancer A pass before she meets her partner again? before she returns to her original place?

The original order of dancers is ABCDEFGH.
After 1 move the order is BADCFEHG.
After 2 moves the order is GDAFCHEB.

	⊕	0	1	2	3	4	5	6	7
ABCDEFGH	0	0	1	2	3	4	5	6	7
BADCFEHG	1	1	2	3	4	5	6	7	0
GDAFCHEB	2	2	3	4	5	6	7	0	1
	3								
	4								
	5								
	6								
	7								

34. Copy and complete the left column of the table, showing how the order changes as dancers move around the circle.

35. Let 0–7 correspond to the 8 orders of dancers you listed in Exercise 34. Using the digits 0–7 for the orders, complete the table.

36. What group properties does this addition table have?

A *field* consists of a set of elements and two operations (such as × and +) that satisfy all the commutative group properties as well as the distributive property:

$$a \times (b + c) = (a \times b) + (a \times c)$$

37. Tell whether each set of numbers forms a field under the operations of standard multiplication and addition.

a. integers **b.** rational numbers **c.** real numbers

38. Do the numbers 0 and 1, and the operations of multiplication and addition, as defined in the tables at the right, form a field? Explain.

×	0	1
0	0	0
1	0	1

+	0	1
0	0	1
1	1	0

Apply⇔Assess

Exercise Notes

Application
Exs. 33–36, 39–54 The group concept can be used to study varied phenomena. In this first set of exercises, the group properties are used to study the movement of dancers in square dancing. Some of the different moves possible on a puzzle cube are examined using group properties in the second set of exercises.

Multicultural Note
Exs. 33–36 The style of American folk dancing, called square dancing, grew out of folk dances brought to the United States by early immigrants from Ireland, England, and Scotland. Square dancing is especially popular in the rural areas of the U.S., and a variety of unique regional styles have evolved.

Teaching Tip
Exs. 37, 38 It is important that students understand that in a field, the set of elements must form a group with each operation separately, and then satisfy the distributive property with both operations. Encourage students to work with each operation separately, verifying each property, and then verifying the distributive property.

29.

@	□	●	▽	▲
□	▲	□	●	▽
●	□	●	▽	▲
▽	●	▽	▲	□
▲	▽	▲	□	●

30. (27) D; (28) 0; (29) ●

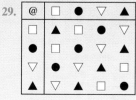

31. The operation is commutative if, and only if, the table is symmetric across the diagonal from the top left to bottom right.

32. The associative property seems to hold for all three tables. Examples are given. A × (B × C) = (A × B) × C = D; 1 + (2 + 3) = (1 + 2) + 3 = 2; □ @ (● @ ▽) = (□ @ ●) @ ▽ = ●

33. a. females; males
b. 4; 8

34, 35. See answers in back of book.

36. closure, identity, inverse, associative, and commutative

37. a. No.
b. Yes.
c. Yes.

38. Yes; the set together with each operation forms a commutative group and the distributive property holds.

Apply⇔Assess

Exercise Notes

Cooperative Learning
Exs. 39–54 These exercises can be done in class in groups of two or three students. A small group discussion of these questions will help all students reach a better understanding of the major concepts of this section.

Second-Language Learners
Exs. 39–54 Some students learning English may be unfamiliar with a puzzle cube. You may wish to bring a puzzle cube to class and give students a chance to experiment with it before starting these exercises. It may also be helpful to read aloud the directions while demonstrating how to complete some of the patterns listed in the exercises.

Topic Spiraling: Review
Exs. 48–54 Students have most likely seen the operation of composition used with sets of functions and sets of geometric transformations. You might ask students to find a set of functions that form a group under composition. One example is $f(x) = x$, $g(x) = x^3$, and $h(x) = \sqrt[3]{x}$.

Assessment
Ex. 55 This exercise will help to assess how clearly students understand the ideas in this section. Have several students volunteer to complete the table at the board and verify all the group properties.

Connection GAMES

If you start with an unmixed puzzle cube and restrict your moves to turning the middle layers 180°, you will always get one of the 8 cube patterns shown.

If you denote 180° turns of the middle layers parallel to the top, right, and front faces of the cube by T, R, and F, then the 8 cube patterns can be named I, T, R, F, TR, TF, RF, and TRF. (The letter I, for identity, names the unmixed cube.)

Use a puzzle cube to complete the following exercises. If you don't have a cube, you can follow the paths that connect cubes in the diagram to see the result of any combination of moves.

Name the cube pattern you get if you start at I and follow each sequence of moves.

39. F, R
40. R, F
41. T, R, F
42. T, R, F, T, F
43. T, R, F, T, R, F
44. T, T
45. T, R, T, F
46. T, T, R, T, F, F, R, R, R, R, T, T, F

47. Writing Based on your answers to Exercises 39–46, describe how you can name the cube pattern you get after *any* sequence of T's, R's, and F's.

BY THE WAY

Hungarian design professor Ernő Rubik invented the puzzle cube. There are more than 4.3×10^{19} possible cube patterns.

You can think of I, T, R, F, TR, TF, RF, and TRF as naming the 8 cube patterns or as naming sequences of moves. *Composition* (∘) is the operation that combines one sequence of moves with another sequence of moves. For example, T ∘ TR = R because if you start at I and move T followed by T and R, you end up at the cube pattern R.

48. Copy and complete the table.

∘	I	T	R	F	TR	TF	RF	TRF
I	I	T	R	F	TR	TF	RF	TRF
T	T	I	TR	TF	R	F	TRF	RF
R	R	TR	I	RF	?	?	?	?
F	F	TF	RF	I	?	?	?	?
TR	?	?	?	?	?	?	?	?
TF	?	?	?	?	?	?	?	?
RF	?	?	?	?	?	?	?	?
TRF	?	?	?	?	?	?	?	?

49. Is the set of 8 elements closed under the operation of composition? Explain.

50. Is there an identity element? If so, what is it?

51. Which elements in the set have an inverse element? What do you notice about their inverses?

52. Does the associative property hold? Give examples or counterexamples to support your conjecture.

53. Does the commutative property hold? Give examples or counterexamples to support your answer.

54. Based on your answers to Exercises 49–53, do the 8 cube patterns form a group under the operation of composition?

39. RF 40. RF 41. TRF
42. R 43. I 44. I
45. RF 46. TRF

47. Answers may vary. An example is given. If an even number of any given single letter appears in the combination, then that letter does not appear in the resulting pattern. If an odd number of any single given letter appear in the combination, then that letter does appear in the resulting pattern. If all three appear an even number of times, the resulting pattern is I. For example, if the pattern is TFFTFTTRTRRRTFF, then the cube pattern is RF.

48. See answers in back of book.

49. Yes; the result of operating on any two elements in the set is in the set.

50. Yes; I.

51. Every element has an inverse; each element is its own inverse.

52, 53. Answers may vary. An example is given. Yes; considering the method described in the answer to Ex. 47, it does not matter how the steps are ordered or grouped.

54. Yes.

55. a.

×	1	i	-1	$-i$
1	1	i	-1	$-i$
i	i	-1	$-i$	1
-1	-1	$-i$	1	i
$-i$	$-i$	1	i	-1

b. Yes; the set is closed under multiplication and all the group properties as well as the commutative property hold.

56.

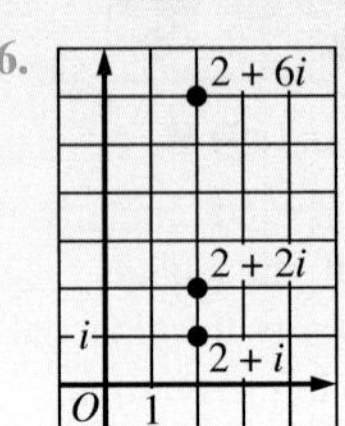

ONGOING ASSESSMENT

55. There are four solutions to the equation $x^4 = 1$ in the complex number system. The solutions are $1, i, -1$, and $-i$.

×	1	i	−1	$-i$
1	1	?	?	?
i	?	−1	?	?
−1	?	?	?	?
$-i$	?	?	?	?

a. Copy and complete the table showing the multiplication of these four complex numbers.

b. Do the numbers $1, i, -1$, and $-i$ form a commutative group under multiplication? Explain.

SPIRAL REVIEW

Plot each pair of complex numbers and their product in the complex plane. *(Section 8.5)*

56. $2 + i, 2 + 2i$ **57.** $-3 + i, -2 - i$ **58.** $5i, -6i$

Write each quadratic function as a product of factors. *(Section 5.6)*

59. $y = 2x^2 + x - 15$ **60.** $y = 16x^2 - 4x - 12$ **61.** $y = 4x^2 - 49$

Simplify. *(Toolbox, page 781)*

62. $(4x + 5) + (6x - 9)$ **63.** $(9x - 7) - (4x + 8)$ **64.** $(x + 2) - (7 - x)$

ASSESS YOUR PROGRESS

VOCABULARY

real number (p. 358)
imaginary number (p. 359)
pure imaginary number (p. 359)
complex number (p. 359)
complex conjugate (p. 361)
complex plane (p. 365)
imaginary axis (p. 365)
real axis (p. 365)
magnitude (p. 366)
group (p. 373)
commutative group (p. 373)
closure property (pp. 373, 374)
identity property (pp. 373, 374)
inverse property (pp. 373, 374)
associative property (pp. 373, 374)
commutative property (pp. 373, 374)

Perform the indicated operation. *(Section 8.4)*

1. $(4 + 2i) - (5 - 3i)$ **2.** $(7 + 3i)(7 - 3i)$ **3.** $(8 - 5i)^2$

4. Plot two numbers that are complex conjugates in the complex plane. Find their sum and their product and plot them as well. *(Section 8.5)*

5. Find the magnitude of $-12 + 3i$.

Tell whether each number set is closed under the given operation. If so, tell whether the number set and operation *form a group, form a commutative group,* or *do not form a group.* Explain. *(Section 8.6)*

6. negative real numbers; + **7.** real numbers a where $0 \le a \le 1$; ×

8. Journal Compare the operations of addition, subtraction, multiplication, and division on real numbers and complex numbers.

Apply⇔Assess

Exercise Notes

Topic Spiraling: Preview
Exs. 59–64 These exercises review students' knowledge of factoring, and adding and subtracting polynomial expressions. Polynomial functions is one of the topics of the next chapter.

Assess Your Progress

Journal Entry
Writing about the operations of addition, subtraction, multiplication, and division of real and complex numbers will help students solidify their understanding of these ideas.

Progress Check 8.4–8.6

See page 383.

Practice 52 for Section 8.6

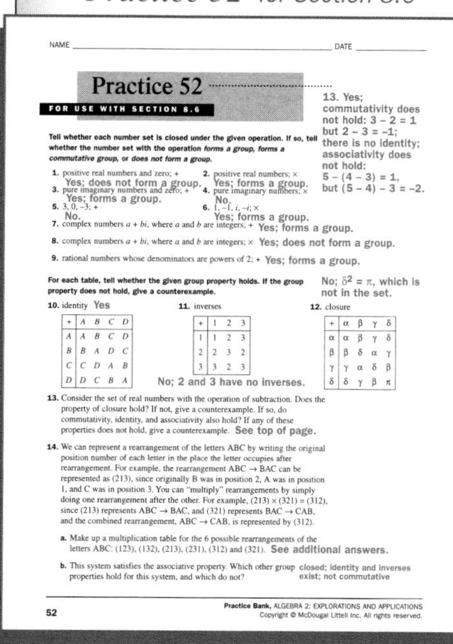
NAME ______ DATE ______

Practice 52
FOR USE WITH SECTION 8.6

13. Yes; commutativity does not hold: $3 - 2 = 1$ but $2 - 3 = -1$; there is no identity; associativity does not hold: $5 - (4 - 3) = 1$, but $(5 - 4) - 3 = -2$.

Tell whether each number set is closed under the given operation. If so, tell whether the number set with the operation *forms a group, forms a commutative group,* or *does not form a group.*

1. positive real numbers and zero; + Yes; does not form a group.
2. positive real numbers; × Yes; forms a group.
3. pure imaginary numbers and zero; + Yes; forms a group.
4. pure imaginary numbers; × No.
5. 3, 0, −3; + No.
6. $1, -1, i, -i$; × Yes; forms a group.
7. complex numbers $a + bi$, where a and b are integers; + Yes; forms a group.
8. complex numbers $a + bi$, where a and b are integers; × Yes; does not form a group.
9. rational numbers whose denominators are powers of 2; + Yes; forms a group.

For each table, tell whether the given group property holds. If the group property does not hold, give a counterexample.

10. identity Yes

+	A	B	C	D
A	A	B	C	D
B	B	A	D	C
C	C	D	A	B
D	D	C	B	A

11. inverses

+	1	2	3
1	1	2	3
2	2	3	2
3	3	2	3

No; 2 and 3 have no inverses.

12. closure No; $\delta^2 = \pi$, which is not in the set.

+	α	β	γ	δ
α	α	β	γ	δ
β	β	δ	α	γ
γ	γ	α	δ	β
δ	δ	γ	β	π

13. Consider the set of real numbers with the operation of subtraction. Does the property of closure hold? If not, give a counterexample. If so, do commutativity, identity, and associativity also hold? If any of these properties does not hold, give a counterexample. See top of page.

14. We can represent a rearrangement of the letters ABC by writing the original position number of each letter in the place the letter occupies after rearrangement. For example, the rearrangement ABC → BAC can be represented as (213), since originally B was in position 2, A was in position 1, and C was in position 3. You can "multiply" rearrangements by simply doing one rearrangement after the other. For example, (213) × (321) = (312), since (213) represents ABC → BAC, and (321) represents BAC → CAB, and the combined rearrangement, ABC → CAB, is represented by (312).

a. Make up a multiplication table for the 6 possible rearrangements of the letters ABC: (123), (132), (213), (231), (312) and (321). See additional answers.

b. This system satisfies the associative property. Which other group properties hold for this system, and which do not? closed; identity and inverses exist; not commutative

52 **Practice Bank,** ALGEBRA 2: EXPLORATIONS AND APPLICATIONS

57.
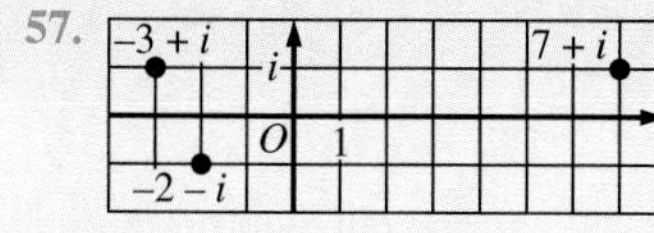

58.
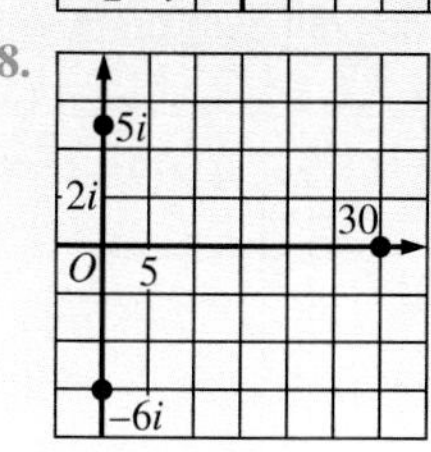

59. $y = (2x - 5)(x + 3)$

60. $y = 4(4x + 3)(x - 1)$

61. $y = (2x - 7)(2x + 7)$

62. $10x - 4$ 63. $5x - 15$ 64. $2x - 5$

Assess Your Progress

1. $-1 + 5i$ 2. 58 3. $39 - 80i$

4. Answers may vary. An example is given.

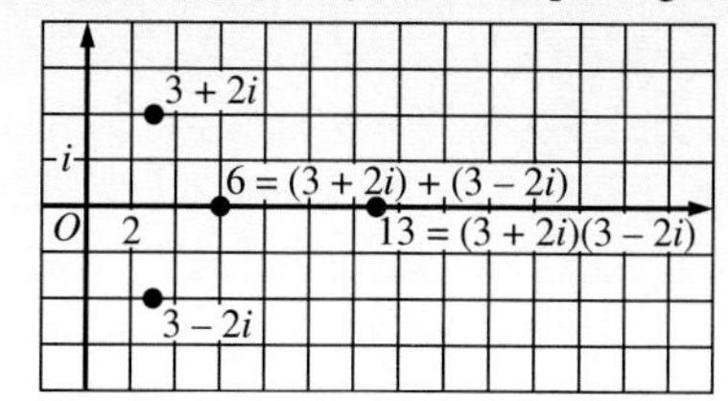

5. $3\sqrt{17}$

6. Yes; do not form a group; there is no identity and the inverse property does not hold.

7. Yes; do not form a group; the inverse property does not hold.

8. Answers may vary. Operations on complex numbers that are not real numbers are based on real-number operations. For example, to add or subtract imaginary numbers that are not real, add or subtract the real number parts and the imaginary parts separately. To multiply complex numbers that are not real, multiply as though the numbers were binomials. To divide, express the division as a fraction and multiply the numerator and denominator by the complex conjugate of the denominator.

Project Notes

Mathematical Goals

- Time and record a person as he or she walks a 60 ft course.
- Calculate each person's actual speed and predicted maximum speed.
- Graph and compare the actual and predicted maximum speeds.

Planning

Materials
- tape measure
- stopwatch

Project Teams
Students can select their partner and come up with a list of possible subjects to participate in the experiment together. Then the partners can decide how they want to divide up the remaining project tasks.

Guiding Students' Work

Make sure that students' measurements are accurate. For example, students are to measure only the leg length that begins at the bottom of the hip where the leg starts and ends at the bottom of the heel where the leg ends. Also, if students are using a track as a course, they should walk along a part of the track that has been measured to be 60 feet. When students are plotting actual and predicted speeds, make sure they select the correct variables. Leg length is the independent variable for both graphs, so it is plotted on the horizontal axis. Speed is the dependent variable for both graphs (actual speed for the first graph and predicted speed for the second graph), so it is plotted on the vertical axis.

Second-Language Learners
Students learning English may benefit from conducting their experiments and working on their reports with the help of a peer tutor or aide.

PORTFOLIO PROJECT

Walk This Way

Try walking across a flat, open space at an increasing speed. You'll notice that at some point you'll feel the urge to switch to a run. Your body recognizes when it's more efficient to run than to walk.

Now think about a small child trying to keep up with an adult who's walking briskly. With shorter legs than the adult, the child often has to run. Obviously, how fast someone can walk (without breaking into a run) has something to do with leg length.

PROJECT GOAL **Examine the relationship between leg length and walking speed.**

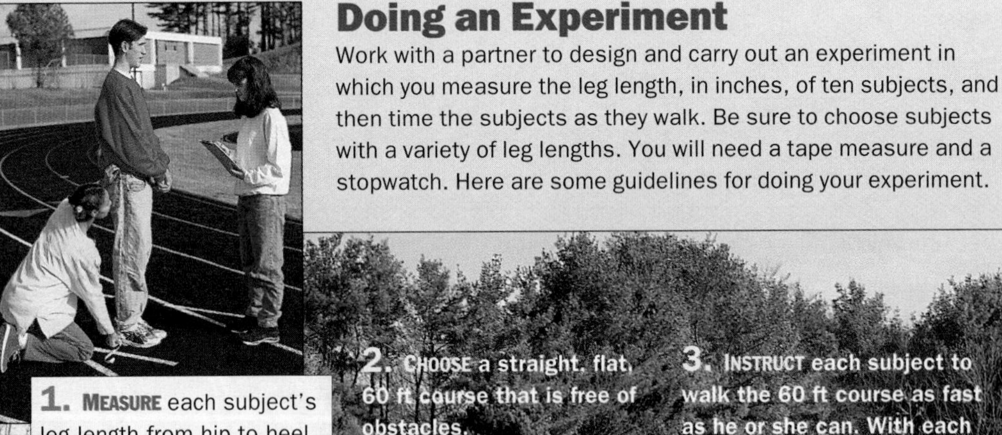

Doing an Experiment

Work with a partner to design and carry out an experiment in which you measure the leg length, in inches, of ten subjects, and then time the subjects as they walk. Be sure to choose subjects with a variety of leg lengths. You will need a tape measure and a stopwatch. Here are some guidelines for doing your experiment.

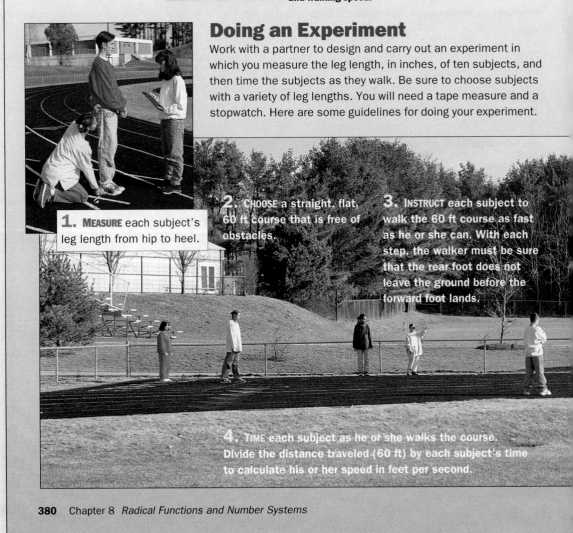

1. MEASURE each subject's leg length from hip to heel.

2. CHOOSE a straight, flat, 60 ft course that is free of obstacles.

3. INSTRUCT each subject to walk the 60 ft course as fast as he or she can. With each step, the walker must be sure that the rear foot does not leave the ground before the forward foot lands.

4. TIME each subject as he or she walks the course. Divide the distance traveled (60 ft) by each subject's time to calculate his or her speed in feet per second.

Using a Model

A person's maximum walking speed s, in feet per second, is given by the following model:

$$s = \frac{\sqrt{gl}}{12}$$

where g, the acceleration due to gravity, is 384 in./s^2, and l is leg length in inches.

- **CALCULATE** the predicted maximum speed for each of your subjects according to the model. You may find a spreadsheet helpful.

Walking Speed Data

	A	B	C	D	E
1	Name	Leg length (in.)	Time (s)	Actual speed (ft/s)	Predicted speed (ft/s)
2	Lester	40	5.9	10.2	10.33
3	Cindy	34	6.4	9.38	9.52
4	Gena	31	6.7	8.96	9.09

- **PLOT** the predicted speeds on the same graph as the actual values. Use different colors or symbols for each data set.
- **COMPARE** the actual values with the predicted values. Are there differences? What might explain them?

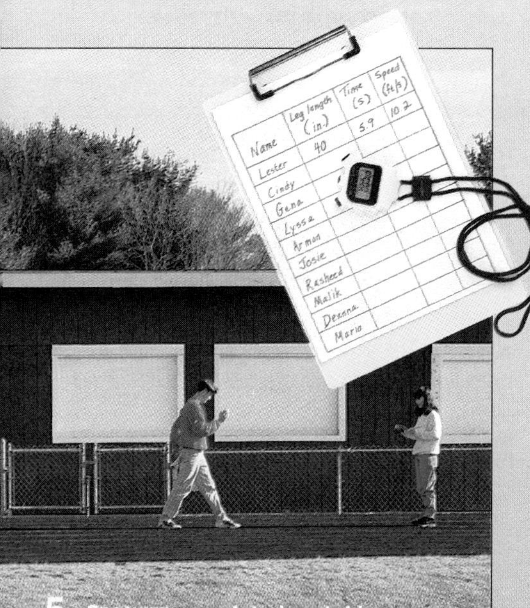

5. ORGANIZE your data in a table, and then make a scatter plot. Put leg length on the horizontal axis and walking speed on the vertical axis.

Writing a Report

Write a report summarizing your results. It should include all your data as well as a comparison of the data with the predictions of the model. You can also extend your project to explore one of these ideas:

- The acceleration due to gravity varies from planet to planet. Suppose you were walking on Mars or on the moon. How fast could you walk? How does gravity affect walking speed?
- The model above is based on the assumption that the *centripetal force* acting on a walker's hips cannot exceed the walker's weight.

  Centripetal force is given by $\frac{ms^2}{l}$, and weight is given by mg, where m is mass. Derive the model from this information. (You'll need to take units of measurement into account.)

Self-Assessment

In your report, describe any difficulties you had gathering data for this project. Did you feel that your data were in agreement with the model? If not, what might have gone wrong?

Project Notes

Guiding Students' Work

Rubric for Chapter Project

4 Students select ten subjects and make accurate measurements. Each subject is timed and a chart showing all times and speeds is completed. Actual and predicted speeds are calculated correctly, and students plot these values on a graph. A well-written report is done that includes all of the requested information. The comparison of the graphs in the report indicates an understanding of the mathematics of the project. Students also extend the project by exploring walking speeds on other planets or by discussing the centripetal force formula.

3 Students select ten subjects and accurately take measurements. The subjects are timed and a chart is completed showing all times and speeds. Some calculations are incorrect, or minor errors are made in plotting the points. Students submit a report that includes all the charts, graphs, and explanations, and that shows an understanding of the concepts of the project. Students also extend the project by looking into walking speeds on other planets or by exploring the centripetal force formula.

2 Students select ten subjects, accurately take measurements and complete the experiment, but they do not display their results correctly. Values are miscalculated and the graphs of the data are incorrect. Students do make an attempt to understand the project. Their report contains all the project data and information requested, but it is clear that some of the main ideas are not completely understood. Students attempt to extend the project, but it is not completed.

1 Students select only a few subjects for the experiment and complete some of the measurements. However, they do not record all of the lengths and speeds and do not calculate correctly the actual or predicted speeds. Students plot only one graph or do not complete a graph at all. If a report is written, it is incomplete and does not show an understanding of the project. Students should be encouraged to speak with the teacher as soon as possible to review their work and to make a new start on the project.

Chapter Support

Course Guide: Chapter 8
Lesson Plans: Chapter 8
Practice Bank: Cumulative Practice 53
Study Guide: Chapter 8 Review
Assessment Book: Chapter Tests 32 and 33; Chapter 8 Alternative Assessment
Test Generator Software
Portfolio Project Book
Preparing for College Entrance Tests
Professional Handbook

Progress Check 8.1–8.3

1. Graph $y = x^2 + 4$ and its reflection over the line $y = x$. *(Section 8.1)*

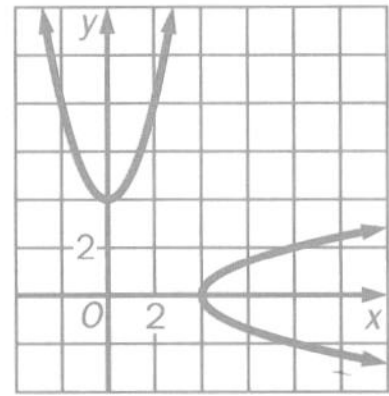

2. Restrict the domain of $y = x^2 + 4$ so that its inverse is a function. Write an equation for the inverse. *(Section 8.1)* $x \geq 0$ or $x \leq 0$; $y = \sqrt{x-4}$ or $y = -\sqrt{x-4}$

State the domain and range of each function. *(Section 8.2)*

3. $y = \sqrt[7]{x+2} + 5$ all real numbers; all real numbers

4. $f(x) = \sqrt[4]{x} - 1$ $x \geq 0$, $y \geq -1$

Simplify each radical expression. *(Section 8.2)*

5. $\sqrt[4]{160}$ $2\sqrt[4]{10}$

6. $\sqrt[3]{-500}$ $-5\sqrt[3]{4}$

7. Solve $\sqrt{-7x-5} = x - 1$. *(Section 8.3)* no solution

CHAPTER

8 Review

STUDY TECHNIQUE

Look through the chapter and identify your trouble spots: questions that you had difficulty answering or solved incorrectly on homework assignments. Review the material covering these topics, and then try to answer the questions again. If you're still having difficulty, seek the help of another student or your teacher.

VOCABULARY

SECTIONS 8.1, 8.2, *and* 8.3

Radical functions of the form $y = \sqrt[n]{x}$ have a domain and range of all real numbers when n is odd, and a domain and range of nonnegative numbers when n is even.

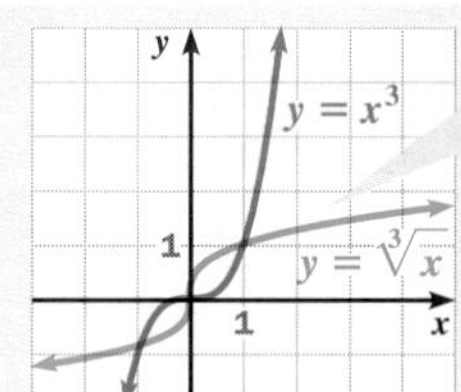

$y = \sqrt[3]{x}$ is the inverse of $y = x^3$.

You can solve radical equations by graphing or by using algebra. For example, you can use the equation $t = \sqrt{\frac{h}{4}}$, which gives the time (in seconds) of a jump that reaches a height of h ft, to find the height of a jump that lasts 0.95 s.

Graph $y = 0.95$ and $y = \sqrt{\frac{x}{4}}$ in the same coordinate plane.

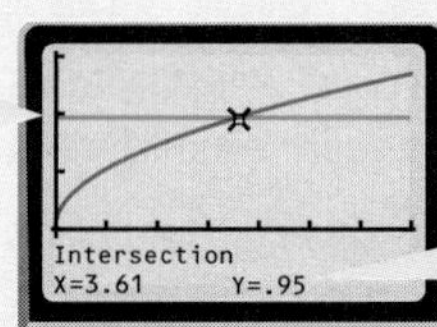

A jump that lasts 0.95 s reaches a height of about 3.6 ft.

$$0.95 = \sqrt{\frac{h}{4}}$$
$$(0.95)^2 = \frac{h}{4}$$
$$4(0.95)^2 = h$$
$$3.6 \approx h$$

SECTIONS 8.4 *and* 8.5

The letter i is used for the square root of -1, the fundamental unit in the system of **imaginary numbers**: $i = \sqrt{-1}$.

A **pure imaginary number** has the form bi where $b \neq 0$.

Examples:

$\sqrt{-4} = 2i$

$5 + \sqrt{-9} = 5 + 3i$

Any number of the form $a + bi$ where a and b are real numbers and $b \neq 0$ is an **imaginary number**.

The real numbers and imaginary numbers together form the set of **complex numbers**, which can be graphed in a **complex plane**.

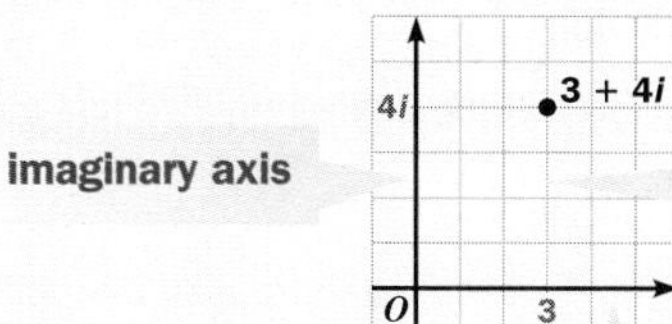

imaginary axis

real axis

The **magnitude** of a complex number is its distance from (0, 0). The magnitude of $3 + 4i$ is 5.

Examples of Operations with Complex Numbers

Operation	Example
Addition or subtraction	$(7 + 5i) - (4 + 2i) = (7 - 4) + (5i - 2i)$ $= 3 + 3i$
Multiplication	$(2 + 3i)(5 + 2i) = (2 \cdot 5) + (2 \cdot 2i) + (3i \cdot 5) + (3i \cdot 2i)$ $= 10 + 4i + 15i + (-6) = 4 + 19i$
Division	$\frac{2 - i}{1 + 3i} = \frac{2 - i}{1 + 3i} \cdot \frac{1 - 3i}{1 - 3i}$ $= \frac{2 - 6i - i + 3i^2}{10} = \frac{2 - 7i - 3}{10} = -\frac{1}{10} - \frac{7}{10}i$

Add or subtract complex numbers by adding or subtracting their real parts and their imaginary parts.

Multiply complex numbers as you would multiply binomials.

Divide complex numbers using a **complex conjugate** to get an answer in $a + bi$ form.

SECTION 8.6

The diagram shows the relationships among the various number systems. A number belongs to all the number systems linked to its left.

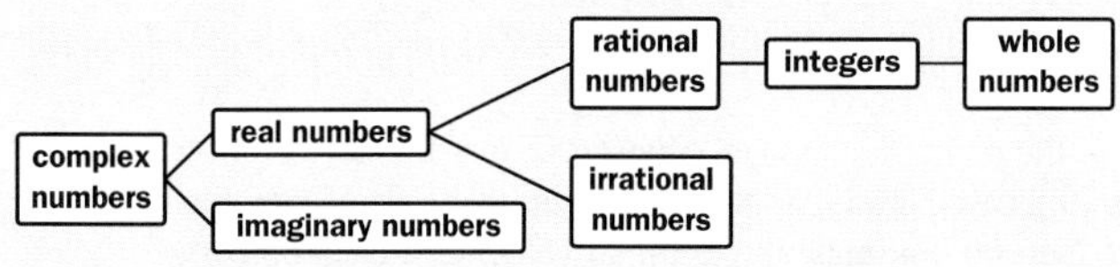

The properties of **closure**, **identity**, **inverse**, **associativity**, and **commutativity** are **group properties**. These properties apply to addition and multiplication on the set of real numbers (see pages 373 and 374).

Progress Check 8.4–8.6

1. Subtract. $(3 - 2i) - (-3 + 4i)$ *(Section 8.4)* $6 - 6i$
2. Write in $a + bi$ form. $\frac{3 + 2i}{4 - i}$ *(Section 8.4)* $\frac{10}{17} + \frac{11}{17}i$
3. Solve $x^2 + 4x + 13 = 0$. *(Section 8.4)* $-2 \pm 3i$
4. Describe how to represent a complex number $a + bi$ geometrically. *(Section 8.5)* Draw a complex plane where the horizontal axis represents real numbers and the vertical axis represents pure imaginary numbers. Plot (a, b) to represent $a + bi$.
5. Find the magnitudes of $-2 + i$, $1 - 5i$, and their product. *(Section 8.5)* $\sqrt{5}$, $\sqrt{26}$, $\sqrt{130}$
6. Define the term *group* as it is used in mathematics. *(Section 8.6)* A group is a set together with an operation that is closed under the operation and for which the identity, inverse, and associative properties hold.
7. Use the table below to determine if a, b, and c form a group under $*$. Explain. *(Section 8.6)*

*	a	b	c
a	b	c	a
b	a	c	b
c	b	a	c

No; there is no identity element.

Chapter 8 Assessment
Form A Chapter Test

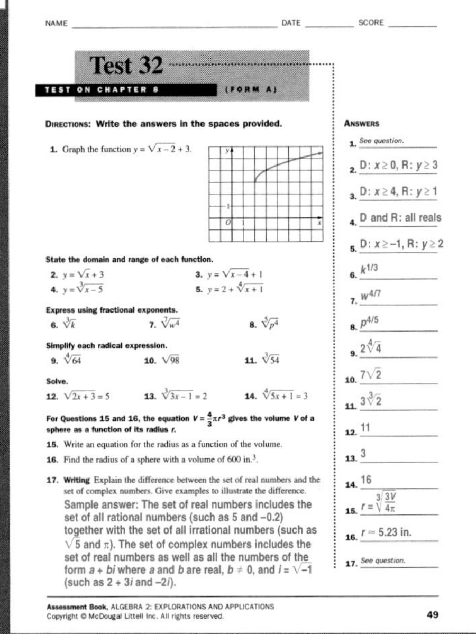

NAME ________ DATE ________ SCORE ________

Test 32
TEST ON CHAPTER 8 (FORM A)

DIRECTIONS: Write the answers in the spaces provided.

1. Graph the function $y = \sqrt{x-2} + 3$.

State the domain and range of each function.

2. $y = \sqrt{x} + 3$
3. $y = \sqrt{x-4} + 1$
4. $y = \sqrt[3]{x-5}$
5. $y = 2 + \sqrt[4]{x+1}$

Express using fractional exponents.

6. $\sqrt[3]{k}$
7. $\sqrt[7]{w^4}$
8. $\sqrt[5]{p^4}$

Simplify each radical expression.

9. $\sqrt[4]{64}$
10. $\sqrt{98}$
11. $\sqrt[3]{54}$

Solve.

12. $\sqrt{2x+3} = 5$
13. $\sqrt[3]{3x-1} = 2$
14. $\sqrt[4]{5x+1} = 3$

For Questions 15 and 16, the equation $V = \frac{4}{3}\pi r^3$ gives the volume V of a sphere as a function of its radius r.

15. Write an equation for the radius as a function of the volume.
16. Find the radius of a sphere with a volume of 600 in.^{3}.
17. **Writing** Explain the difference between the set of real numbers and the set of complex numbers. Give examples to illustrate the difference.
Sample answer: The set of real numbers includes the set of all rational numbers (such as 5 and −0.2) together with the set of all irrational numbers (such as $\sqrt{5}$ and π). The set of complex numbers includes the set of real numbers as well as all the numbers of the form $a + bi$ where a and b are real, $b \neq 0$, and $i = \sqrt{-1}$ (such as $2 + 3i$ and $-2i$).

ANSWERS

1. *See question.*
2. D: $x \geq 0$, R: $y \geq 3$
3. D: $x \geq 4$, R: $y \geq 1$
4. D and R: all reals
5. D: $x \geq -1$, R: $y \geq 2$
6. $k^{1/3}$
7. $w^{4/7}$
8. $p^{4/5}$
9. $2\sqrt[4]{4}$
10. $7\sqrt{2}$
11. $3\sqrt[3]{2}$
12. 11
13. 3
14. 16
15. $r = \sqrt[3]{\frac{3V}{4\pi}}$
16. $r \approx 5.23$ in.
17. *See question.*

Assessment Book, ALGEBRA 2: EXPLORATIONS AND APPLICATIONS

49

Chapter 8 Assessment
Form B Chapter Test

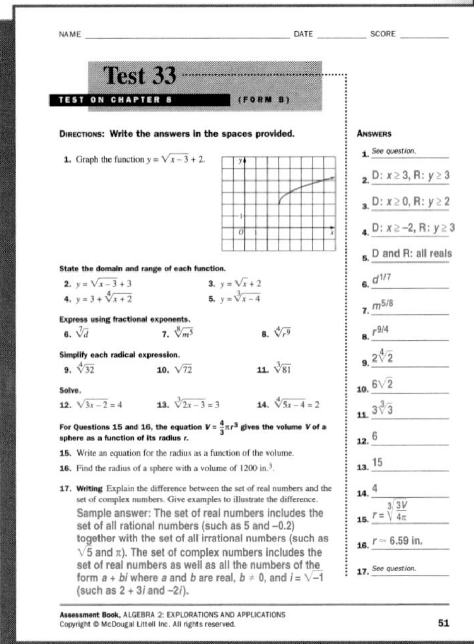

NAME ________ DATE ________ SCORE ________

Test 33
TEST ON CHAPTER 8 (FORM B)

DIRECTIONS: Write the answers in the spaces provided.

1. Graph the function $y = \sqrt{x-3} + 2$.

State the domain and range of each function.

2. $y = \sqrt{x-3} + 3$
3. $y = \sqrt{x} + 2$
4. $y = 3 + \sqrt[4]{x+2}$
5. $y = \sqrt[3]{x-4}$

Express using fractional exponents.

6. $\sqrt[7]{d}$
7. $\sqrt[8]{m^5}$
8. $\sqrt[4]{r^9}$

Simplify each radical expression.

9. $\sqrt[4]{32}$
10. $\sqrt{72}$
11. $\sqrt[3]{81}$

Solve.

12. $\sqrt{3x-2} = 4$
13. $\sqrt[3]{2x-3} = 3$
14. $\sqrt[4]{5x-4} = 2$

For Questions 15 and 16, the equation $V = \frac{4}{3}\pi r^3$ gives the volume V of a sphere as a function of its radius r.

15. Write an equation for the radius as a function of the volume.
16. Find the radius of a sphere with a volume of 1200 in.^{3}.
17. **Writing** Explain the difference between the set of real numbers and the set of complex numbers. Give examples to illustrate the difference.
Sample answer: The set of real numbers includes the set of all rational numbers (such as 5 and −0.2) together with the set of all irrational numbers (such as $\sqrt{5}$ and π). The set of complex numbers includes the set of real numbers as well as all the numbers of the form $a + bi$ where a and b are real, $b \neq 0$, and $i = \sqrt{-1}$ (such as $2 + 3i$ and $-2i$).

ANSWERS

1. *See question.*
2. D: $x \geq 3$, R: $y \geq 3$
3. D: $x \geq 0$, R: $y \geq 2$
4. D: $x \geq -2$, R: $y \geq 3$
5. D and R: all reals
6. $d^{1/7}$
7. $m^{5/8}$
8. $r^{9/4}$
9. $2\sqrt[4]{2}$
10. $6\sqrt{2}$
11. $3\sqrt[3]{3}$
12. 6
13. 15
14. 4
15. $r = \sqrt[3]{\frac{3V}{4\pi}}$
16. $r \approx 6.59$ in.
17. *See question.*

Assessment Book, ALGEBRA 2: EXPLORATIONS AND APPLICATIONS

51

CHAPTER

8 Assessment

VOCABULARY QUESTIONS

For Questions 1–3, complete each statement.

1. The inverse of a power function is a(n) _?_.
2. A complex number can be graphed in the _?_. The distance of a complex number from the origin is called its _?_.
3. The _?_ property of addition says that the sum of any two real numbers is another real number.

SECTIONS 8.1, 8.2, *and* 8.3

State the domain and range of each function.

4. $y = \sqrt{x} - 4$
5. $y = 1 + \sqrt{x}$
6. $y = \sqrt{x-2} + 3$
7. $y = \sqrt[3]{x-3}$
8. $y = 1 + \sqrt[4]{x+5}$
9. $y = \sqrt[6]{x} + 2$

Express using fractional exponents.

10. $\sqrt[9]{z}$
11. $\sqrt[5]{y^3}$
12. $\sqrt[4]{y^7}$

Express using radical notation.

13. $t^{4/3}$
14. $b^{3/8}$
15. $p^{5/7}$

Solve.

16. $\sqrt{4x-7} = 3$
17. $\sqrt[3]{2x+1} = 3$
18. $\sqrt[4]{x-1} = 1$
19. **Writing** Show the steps you would use to solve $\sqrt[n]{ax+b} = c$, where a, b, and c are nonnegative real numbers. What is an expression for x in terms of a, b, and c?

A falling object reaches **terminal velocity** when the force of gravity is balanced by the drag force of the medium through which it falls.

20. **PHYSICS** The *terminal velocity* v (in meters per second) of a spherical object falling through the air can be estimated by using the equation

$$v = \sqrt{\frac{2mg}{0.6A}}$$

where m = the object's mass (in kilograms), g = the acceleration due to gravity (9.8 m/s^2), and A = the cross-sectional area of the object (in square meters). Estimate the terminal velocity of each object.

a. baseball: $m = 0.145$ kg, $A = 0.0042$ m^2
b. golf ball: $m = 0.046$ kg, $A = 0.0014$ m^2
c. hailstone: $m = 0.00048$ kg, $A = 0.000079$ m^2

ANSWERS Chapter 8

Assessment

1. radical
2. complex plane; magnitude
3. closure
4. nonnegative numbers; $y \geq -4$
5. nonnegative numbers; $y \geq 1$
6. $x \geq 2$; $y \geq 3$
7. all real numbers; all real numbers
8. $x \geq -5$; $y \geq 1$
9. nonnegative numbers; $y \geq 2$
10. $z^{1/9}$
11. $y^{3/5}$
12. $y^{7/4}$
13. $\sqrt[3]{t^4}$
14. $\sqrt[8]{b^3}$
15. $\sqrt[7]{p^5}$
16. 4
17. 13
18. 2
19. Raise both sides to the nth power, subtract b from both sides, and divide both sides by a; $x = \frac{c^n - b}{a}$.
20. a. 33.6 m/s b. 32.8 m/s c. 14.1 m/s
21. $3 - 2i$
22. $9 - 3i$
23. 15
24. $12 - 9i$
25. $-2 + 14i$
26. $\frac{38}{29} - \frac{21}{29}i$

27–32. Examples of numbers with the same magnitude may vary.

4 + 4i
1 + 4i
−7 + 3i
−3 + i
i
5 + 2i
O
1
2 − i

27. $\sqrt{17}$; $1 - 4i$ and $-4 + i$
28. $\sqrt{58}$; $7 - 3i$ and $-7 + 3i$
29. $4\sqrt{2}$; $4 - 4i$ and $-4 - 4i$
30. $\sqrt{5}$; $2 + i$ and $-2 + i$
31. $\sqrt{29}$; $5 - 2i$ and $-5 - 2i$
32. $\sqrt{10}$; $3 - i$ and $-3 - i$

SECTIONS 8.4 *and* 8.5

Perform the indicated operation.

21. $(6 + 2i) + (-3 - 4i)$ **22.** $(7 - 6i) + (2 + 3i)$ **23.** $-3i(5i)$

24. $(4 - 5i) - (-8 + 4i)$ **25.** $(3 - i)(-2 + 4i)$ **26.** $\dfrac{8 - i}{5 + 2i}$

Plot each number in the complex plane and find its magnitude. Give two other numbers that have the same magnitude.

27. $1 + 4i$ **28.** $7 + 3i$ **29.** $4 + 4i$

30. $2 - i$ **31.** $5 + 2i$ **32.** $-3 + i$

Plot each pair of numbers and their product in the complex plane.

33. $1 + 4i, 2 - i$ **34.** $7 + 3i, 5 + 2i$ **35.** $-3 + i, 4 + 4i$

SECTION 8.6

Tell whether each number set is closed under the given operation. If so, tell whether the number set and operation *form a group, form a commutative group,* or *do not form a group.* Explain.

36. multiples of 5; + **37.** $-1, 0, 1$; + **38.** $2, 1, \frac{1}{2}$; ×

For each table, tell whether the given group property holds. If the group property does not hold, give a counterexample.

39. commutativity

ǂ	#	@	*	%
#	#	@	%	*
@	@	%	*	#
*	%	*	#	&
%	*	#	&	%

40. identity

Δ	A	B	C	D
A	A	D	C	B
B	C	B	A	D
C	D	C	B	A
D	B	A	D	C

41. closure

×	3	5	7	9
3	9	5	1	7
5	5	5	5	5
7	1	5	9	3
9	7	5	3	1

PERFORMANCE TASK

42. a. Draw a three-dimensional coordinate system with two horizontal x-axes, one real and the other imaginary, and one vertical y-axis.

b. Graph $y = x^2 + 1$ using real values of x as the domain. Then graph $y = x^2 + 1$ using only pure imaginary values of x as the domain.

c. Explain how you can use your graph to solve equations like $x^2 + 1 = 0$, $x^2 + 1 = 5$, and $x^2 + 1 = -8$.

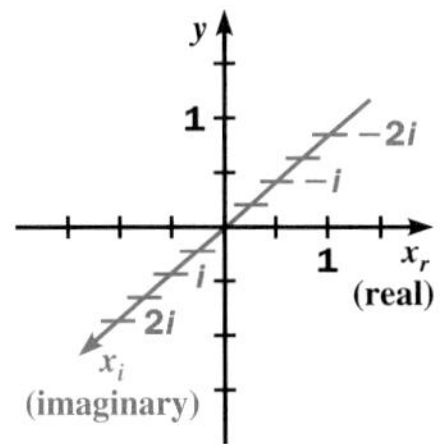

Chapter 8 Assessment
Form C Alternative Assessment

Chapter 8
ALTERNATIVE ASSESSMENT

1. **Performance Task** Find examples of each of the following numbers, if they exist. If such numbers do not exist, explain how you know.
   a. two rational numbers between 3 and 4
   b. two irrational numbers between 3 and 4
   c. two complex numbers between 3 and 4
2. **Group Activity** Make up a list of numbers from the set of real numbers. (Use a variety of numbers like $0, \frac{3}{4}, 0.25, \sqrt{7}, 1, -4, \pi$, and so on.) Write the numbers on index cards. Use tape, string, or clothes pins to attach a card to the back of each student's shirt. Do not let the student see the number. The goal of this activity is that each student will guess their own number. The students can get clues by asking other students questions about their number. The questions must be questions that can be answered with "yes" or "no."
3. **Performance Task** Find the domain and range for $y = \sqrt{x - 2}$. Explain why the domain cannot contain the numbers in the interval $0 < x < 2$.
4. **Performance Task** When dividing two complex numbers, the answer never has an i in the denominator. How is this similar to dividing two radical expressions?
5. **Open-ended Problem** Is there an expression, other than a compex conjugate, that can be used to eliminate the i in the denominator of a rational expression involving two complex numbers? If so, explain.
6. **Performance Task** Simplify $i^1, i^2, i^3, i^4, \ldots, i^{12}$. Describe the pattern in the values for the powers of i.
7. **Open-ended Problem** Subtraction and division do not form a group with most sets of numbers. Find a set of numbers that forms a group with subtraction. Find a set of numbers that forms a group with division.
8. **Open-ended Problem** Some operations on finite sets of numbers are defined by tables. Clock arithmetic addition is an example of an operation that can be defined using a table. Discuss some shortcuts that can be used when you are using a table to check the following properties.
   a. closure property
   b. commutative property
   c. identity property
9. **Performance Task** Multiplying a complex number by i results in a 90° counterclockwise rotation of its graph. What could you do to a complex number to cause its graph to rotate 90° in a clockwise direction?

Assessment Book, ALGEBRA 2: EXPLORATIONS AND APPLICATIONS

121

33.
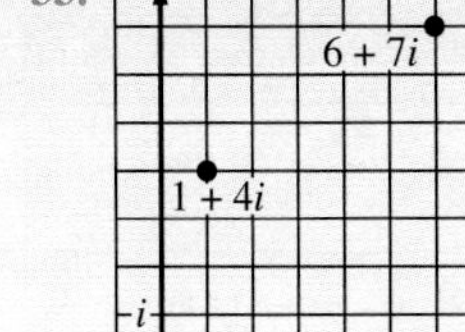

34.
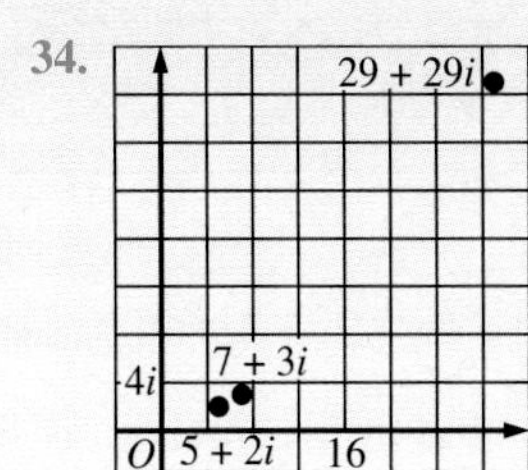

35.
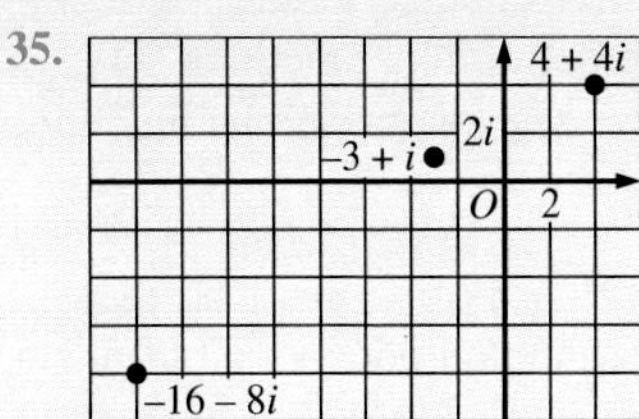

36. Yes; form a commutative group; the set is closed under addition and all the group properties as well as the commutative property hold.

37. No.

38. No.

39. Yes.

40. No; for example A Δ C = C, but C Δ A = D.

41. No; for example, 3 × 7 = 1, which is not in the set.

42. See answers in back of book.

CHAPTER

9 Polynomial and Rational Functions

OVERVIEW

Connecting to Prior and Future Learning

⇔ The chapter opens by reviewing polynomials. Students then use the properties of exponents they studied in Chapter 3 to multiply and divide polynomials.

⇔ Students continue to explore polynomial functions through graphing and finding zeros. Section 9.4 focuses on solving cubic equations.

⇔ The study of rational functions begins with a review of inverse variation which leads to simple and general rational functions. Chapter 9 ends by solving rational equations. Students may find the review of rational expressions on page 783 of the **Student Resources Toolbox** helpful.

Chapter Highlights

Interview with Vera Rubin: Astronomers use mathematics in many aspects of their work. Related exercises on pages 430 and 446 allow students to explore this application even further.

Explorations in Chapter 9 begin in Section 9.3 where students look for patterns in graphs. Section 9.5 starts with an exploration that involves counting real zeros. Students investigate translated hyperbolas in Section 9.7 and vertical asymptotes in Section 9.8.

The Portfolio Project: Students work in groups to design a cylindrical can that will meet certain requirements for volume but minimize material costs.

Technology: Graphing calculators are used to graph polynomial and rational functions throughout this chapter. Students use spreadsheets in Section 9.4 to solve cubic equations and in Section 9.6 to explore inverse variation.

OBJECTIVES

Section	Objectives	NCTM Standards
9.1	• Recognize, evaluate, add, and subtract polynomials. • Understand numeration systems using polynomials.	1, 2, 3, 4, 5
9.2	• Multiply and divide polynomials. • Use multiplication and division of polynomials in real-world problems.	1, 2, 3, 4, 5
9.3	• Recognize graphs of polynomial functions and describe their important features. • Use polynomial functions to solve real-world problems.	1, 2, 3, 4, 5, 6
9.4	• Solve cubic equations. • Find equations for graphs of cubic functions and find zeros of cubic functions. • Solve real-world problems involving cubic functions and equations.	1, 2, 3, 4, 5, 6
9.5	• Find zeros of higher-degree polynomial functions. • Solve problems using higher-degree polynomial functions.	1, 2, 3, 4, 5, 6
9.6	• Recognize inverse variation. • Write and use inverse variation.	1, 2, 3, 4, 5
9.7	• Identify important features and find equations of translated hyperbolas. • Interpret real-world situations using models whose graphs are translated by hyperbolas.	1, 2, 3, 4, 5
9.8	• Identify important features of graphs of rational functions. • Describe the end behavior of rational functions.	1, 2, 3, 4, 5, 6
9.9	• Solve rational equations. • Solve real-world problems by using rational equations.	1, 2, 3, 4, 5

Integration

Mathematical Connections	9.1	9.2	9.3	9.4	9.5	9.6	9.7	9.8	9.9
algebra	**389–396***	**397–403**	**404–411**	**412–418**	**419–425**	**426–432**	**433–440**	**441–448**	**449–453**
geometry	395	397, 402	410, 411					447, 448	451
data analysis, probability, discrete math	394, 396	402	407, 409	412, 417	424	**426–432**	436, 438, 439	444, 446	452, 453
patterns and functions	**389–396**	**397–403**	**404–411**	**412–418**	**419–425**	**426–432**	**433–440**	**441–448**	**449–453**
logic and reasoning	389, 392, 394–396	**397–403**	405, 406, 408–411	414–418	421, 423–425	427–432	434, 436–440	443, 445–448	450–453
Interdisciplinary Connections and Applications									
history and geography		401		418					
biology and earth science	394		407			428			
chemistry and physics					424			448	452
business and economics							439		
sports and recreation		398	409	417		426	438		453
cooking, home repair, gardening		402				429		448	
hydraulics and engineering						432		447	
space travel and meteorology				412		431	439		
psychology, architecture, packaging, education, energy, population growth, dentistry, metallurgy	396		410, 411	417	425		436	444	449, 451

* ***Bold page numbers*** *indicate that a topic is used throughout the section.*

Technology

Section	*opportunities for use with* Student Book	Support Material
9.1–9.3	graphing calculator McDougal Littell Mathpack *Function Investigator*	**Technology Book:** Calculator Activity 9 for Section 9.3 **Function Investigator with Matrix Analyzer Activity Book:** Function Investigator Activity 17 for Section 9.2 and Activities 18, 21, and 22 for Section 9.3
9.4	graphing calculator spreadsheet software McDougal Littell Mathpack *Function Investigator*	
9.5	graphing calculator McDougal Littell Mathpack *Function Investigator*	**Technology Book:** Spreadsheet Activity 9
9.6	graphing calculator spreadsheet software McDougal Littell Mathpack *Function Investigator*	**Function Investigator with Matrix Analyzer Activity Book:** Function Investigator Activity 24
9.7–9.9	graphing calculator McDougal Littell Mathpack *Function Investigator*	**Function Investigator with Matrix Analyzer Activity Book:** Function Investigator Activity 28 for Section 9.7 and Activities 25–27, 35, and 36 for Section 9.8

PLANNING GUIDE

Regular Scheduling (45 min)

Section	Materials Needed	Core Assignment	Extended Assignment	*exercises that feature* Applications	Communication	Technology
9.1		**Day 1:** 1–16, 20–22 **Day 2:** 23–31, 35, 36, 39–51	**Day 1:** 1–22 **Day 2:** 23–51	17 38	33, 34, 37	34
9.2	graphing calculator	1–12, 18–30, 32–45	1–45	16–17	15, 31	37–39
9.3	graphing calculator	**Day 1:** 1–12 **Day 2:** 18–25, AYP*	**Day 1:** 1–15 **Day 2:** 16–25, AYP	13–15 16, 17	1 17	3–13 16
9.4	graphing calculator, spreadsheet software	**Day 1:** 1–9, 13–20 **Day 2:** 21–26, 29–40	**Day 1:** 1–20 **Day 2:** 21–40	10–12 21	27	1–6, 10 21
9.5	graphing calculator	1–18, 21–29, AYP	1–17 odd, 19–29, AYP	19, 20	23	1–8, 20, 22
9.6	graphing calculator, spreadsheet software	1–12, 19, 20, 22–30	1–30	13–18, 21	14, 21, 22	1–8
9.7	graphing calculator	1–4, 6–11, 16–23, 25–33, AYP	1–33, AYP	12–15, 24	15	12, 14
9.8	graphing calculator	1–4, 7–12, 16–25	1–25	5, 6, 13–16	14, 15	5, 13, 15
9.9	graphing calculator	1–12, 16–21, AYP	1–21, AYP	14, 15	14	20, 21
Review/ Assess		**Day 1:** 1–10 **Day 2:** 11–20 **Day 3:** Ch. 9 Test	**Day 1:** 1–10 **Day 2:** 11–20 **Day 3:** Ch. 9 Test			
Portfolio Project		Allow 2 days.	Allow 2 days.			

Yearly Pacing (with Portfolio Project)	Chapter 9 Total	Chapters 1–9 Total	Remaining	Total
	17 days	124 days	36 days	160 days

Block Scheduling (90 min)

	Day 55	Day 56	Day 57	Day 58	Day 59	Day 60	Day 61	Day 62	Day 63
Teach/ Interact	9.1	9.2 9.3: Exploration, page 404	Continue with 9.3 9.4	Continue with 9.4 9.5: Exploration, page 419	9.6 9.7: Exploration, page 433	9.8: Exploration, page 441 9.9	Review Port. Proj.	Review Port. Proj.	Ch. 9 Test 10.1
Apply/ Assess	**9.1:** 1–16, 20–31, 35, 36, 39–51	**9.2:** 1–12, 18–30, 32–45 **9.3:** 1–12	**9.3:** 18–25, AYP* **9.4:** 1–9, 13–20	**9.4:** 21–26, 29–40 **9.5:** 1–18, 21–29, AYP	**9.6:** 1–12, 19, 20, 22–30 **9.7:** 1–4, 6–11, 16–23, 16–21, 25–34, AYP	**9.8:** 1–4, 7–12, 16–25 **9.9:** 1–12, AYP	**Review:** 1–10 **Port. Proj.**	**Review:** 11–20 **Port. Proj.**	**Ch. 9 Test** **10.1:** 1–14, 16–35

NOTE: A one-day block has been added for the Portfolio Project—timing and placement to be determined by teacher.

Yearly Pacing (with Portfolio Project)	Chapter 9 Total	Chapters 1–9 Total	Remaining	Total
	$8\frac{1}{2}$ days	$62\frac{1}{2}$ days	$19\frac{1}{2}$ days	82 days

* ***AYP*** *is Assess Your Progress.*

Lesson Support

Section	Practice Bank	Study Guide*	Assessment Book*	Visuals	Explorations Lab Manual	Lesson Plans	Technology Book
9.1	54	9.1		Warm-Up 9.1	Add. Expl. 15	9.1	
9.2	55	9.2		Warm-Up 9.2	Master 2	9.2	
9.3	56	9.3	Test 34	Warm-Up 9.3	Master 1	9.3	Calculator Act. 9
9.4	57	9.4		Warm-Up 9.4 Folder 9	Master 1	9.4	
9.5	58	9.5	Test 35	Warm-Up 9.5 Folder 9		9.5	Spreadsheet Act. 9
9.6	59	9.6		Warm-Up 9.6	Masters 1, 2	9.6	
9.7	60	9.7	Test 36	Warm-Up 9.7	Master 2	9.7	
9.8	61	9.8		Warm-Up 9.8		9.8	
9.9	62	9.9	Test 37	Warm-Up 9.9		9.9	
Review Test	63	Chapter Review	Tests 38, 39 Alternative Assessment			Review Test	

*__Spanish versions__ of *Study Guide* and *Assessment Book* are available.

Chapter Support

- Course Guide
- Lesson Plans
- Portfolio Project Book
- Preparing for College Entrance Tests
- Multi-Language Glossary
- *Test Generator* Software
- Professional Handbook

Software Support

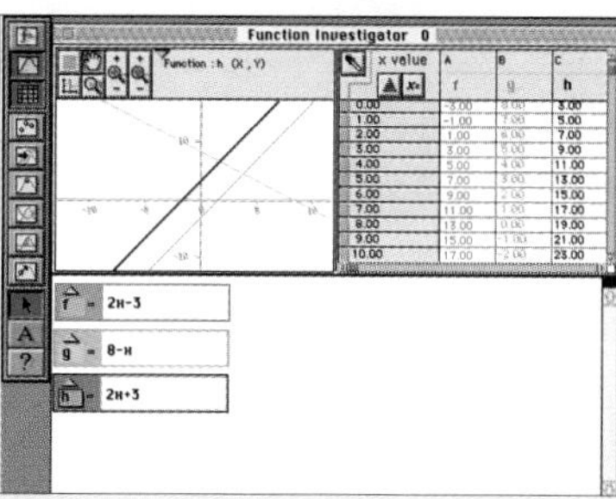

McDougal Littell Mathpack
Function Investigator

Internet Support

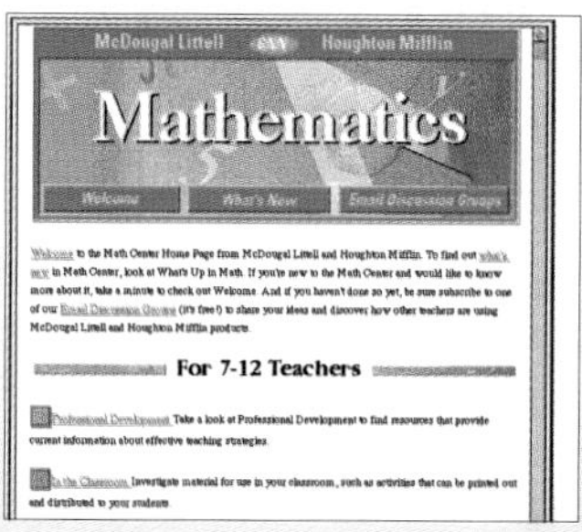

http://www.hmco.com
Next go to McDougal Littell; then the Education Center; then Secondary Math.

Outside Resources

Books, Periodicals

Binder, Margery. "A Calculator Investigation of an Interesting Polynomial." *Mathematics Teacher* (October 1995): pp. 558–560.

Hunt, William J. "Spreadsheets—A Tool for the Mathematics Classroom." *Mathematics Teacher* (December 1995): pp. 774–777.

Isdell, Wendy. *A Gebra Named Al.* With teacher's guide "Using a Gebra Named Al in the Classroom." Minneapolis, MN: Free Spirit Publishing.

Software

f(*g*) *Scholar.* IBM-comp. Southampton, PA: Future Graph.

Keir, Marilyn and Roy A. Keir. *AlgeBrush.* IBM PS/2 and comp. Salt Lake City, UT: MareWare.

Videos

Apostol, Tom. *Polynomials.* Videotape and guide. Reston, VA: NCTM, 1991.

Internet

From an internet account, access lesson plans for K–12 mathematics, compiled by the Eisenhower Network, using the command:
gopher enc.org

Interview Notes

Background

Voyager 2

Identical to its twin, *Voyager 1*, the space probe *Voyager 2* was launched from Cape Canaveral, Florida on August 20, 1977. The probe took pictures and recorded huge amounts of data from four planets. It flew past and photographed Jupiter in 1979, Saturn in 1981, Uranus in 1986, and Neptune in 1989. One discovery from the space probe pictures indicates that geysers are present on Triton, a moon of Neptune.

Because of the remote possibility that one or both of the *Voyager* probes would be recovered by extraterrestrial beings (if they exist), a gold-coated phonograph record was attached to each craft. Each record contains 117 pictures of our planet and human beings, greetings in 54 different languages, a selection of "the sounds of Earth," and a 90-minute selection of the world's music.

Vera Rubin

Vera Rubin built her first telescope with the help of her father when she was 14 years old. Today, she uses some of the most scientifically advanced telescopes in the world. Her career in astronomy has brought her many awards, degrees, and accomplishments, including election into the National Academy of Sciences in 1981, where she is one of only 75 women among the 3508 researchers. Throughout her career, Dr. Rubin has worked at such places as The University of California at San Diego, Georgetown University, and the Carnegie Institution's Department of Terrestrial Magnetism in Washington, DC. Today, much of Dr. Rubin's time is spent giving professional talks, meeting with visitors, and working for groups such as the National Academy's Committee on Human Rights. She is also interested in recruiting young people, especially girls, into science.

CHAPTER

9 Polynomial and Rational Functions

Stellar mysteries in the data

INTERVIEW **Vera Rubin**

"Sometimes all it takes to solve a puzzle is to look at it through new eyes."

As a child, Vera Rubin used to lie awake studying the stars outside her bedroom window. She picked out the constellations and the planets, and saw the night sky as a "great mystery waiting to be solved." She decided to become an astronomer, even though, she says, "teachers in those days just didn't know what to do with girls who were interested in science." Rubin was not discouraged, however, and went on to make an important discovery about the universe.

Vera Rubin holds a picture of herself as a young woman using her college telescope.

386

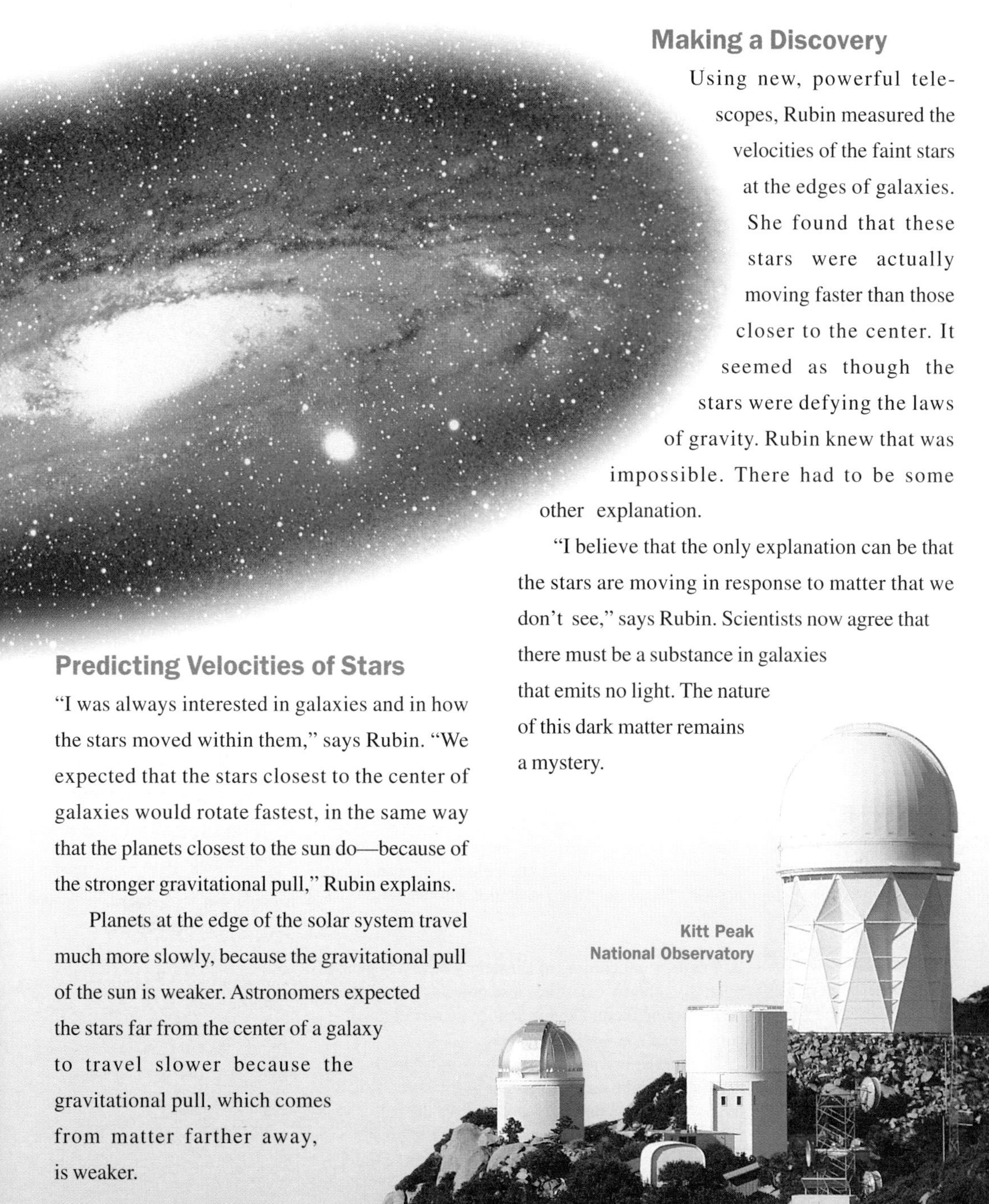

Making a Discovery

Using new, powerful telescopes, Rubin measured the velocities of the faint stars at the edges of galaxies. She found that these stars were actually moving faster than those closer to the center. It seemed as though the stars were defying the laws of gravity. Rubin knew that was impossible. There had to be some other explanation.

"I believe that the only explanation can be that the stars are moving in response to matter that we don't see," says Rubin. Scientists now agree that there must be a substance in galaxies that emits no light. The nature of this dark matter remains a mystery.

Predicting Velocities of Stars

"I was always interested in galaxies and in how the stars moved within them," says Rubin. "We expected that the stars closest to the center of galaxies would rotate fastest, in the same way that the planets closest to the sun do—because of the stronger gravitational pull," Rubin explains.

Planets at the edge of the solar system travel much more slowly, because the gravitational pull of the sun is weaker. Astronomers expected the stars far from the center of a galaxy to travel slower because the gravitational pull, which comes from matter farther away, is weaker.

Kitt Peak National Observatory

Interview Notes

Background

Albert Einstein
Albert Einstein (1879–1955) is recognized as one of the true geniuses of the twentieth century. When Einstein was fifteen, he and his family moved from his birthplace in Germany to Milan, Italy. During the next year, Einstein took an examination that would have enabled him to attend the Zurich Polytechnic Institute and earn a degree in electrical engineering. He failed the test, but entered the school the following year, graduating in 1900 with a teaching degree. Only five years later, at the age of 26, Einstein published three scientific papers that had a profound impact on the world, including his paper on the "special theory of relativity."

Second-Language Learners

Students learning English may not understand what is meant by the term *stellar mysteries* in the title. If necessary, tell them that *stellar* is an adjective meaning "of the stars"; it also is used figuratively to mean "truly excellent."

Interview Notes

Mathematical Connection

Astronomers like Vera Rubin are concerned with many different aspects of the universe. In our solar system, astronomers study the relationship between the nine known planets and the sun. These relationships are explored in Sections 9.6 and 9.8 of this chapter. In Section 9.6, students are introduced to the idea of apparent size and how to calculate the apparent size of the sun from each of the nine planets. Then, students discover how to find the apparent size of the sun from any location in the solar system by writing an inverse variation equation. In Section 9.8, the force exerted by Earth on an object as it travels through the solar system is discussed using the space probe *Voyager 2*. Students examine the relationship between force, mass, and distance on an object as it travels through the solar system.

Explore and Connect

Research
In addition to encyclopedias, science magazines or periodicals can be used to research what astronomers believe is happening to the universe.

Writing
After students complete this writing activity, have them discuss their answers. If students suggest different factors for the decrease in gravitational force, have them pursue their reasoning until a single factor is agreed to by all students.

Project
This activity can be done by the entire class. A class discussion of the results can help solidify the concepts of Einstein's theory in students' minds.

Gravitational Force

Central to Rubin's work is the idea that any two objects exert a gravitational force on each other. This force, typically measured in newtons (N), changes with the distance between the objects.

Consider, for example, the probe *Voyager 2*, launched by NASA in the late 1970s to explore the outer planets of our solar system. Although the gravitational force between the probe and Earth was 9143 N at Earth's surface, the force decreased as the probe moved away from Earth. As the graph shows, gravitational force is a function of distance. This relationship is an example of a *rational function*.

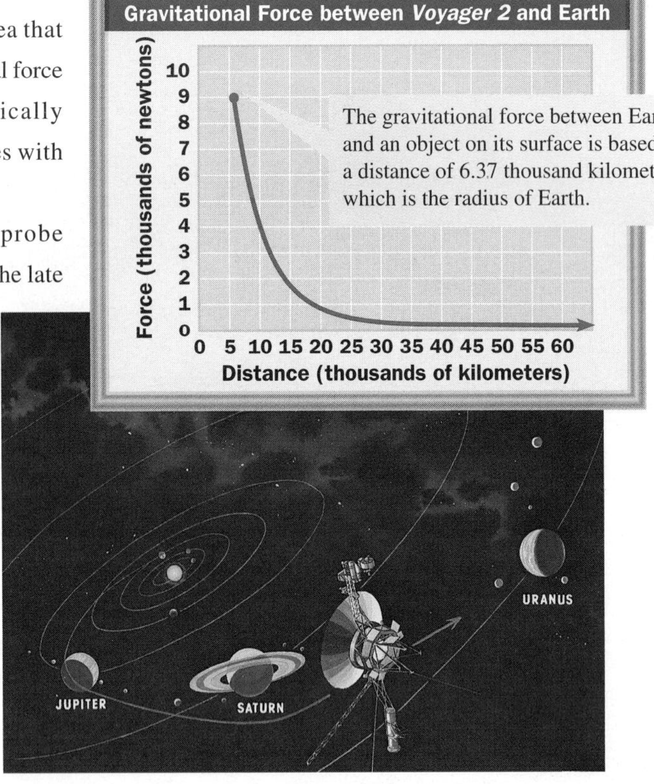

EXPLORE AND CONNECT

Vera Rubin accepts the National Medal of Science from President Clinton.

1. Research Use an encyclopedia to find out if astronomers expect the universe to continue expanding, or if they expect the gravitational forces acting on matter to eventually stop or reverse the expansion.

2. Writing Use the graph shown above to describe the change in force as *Voyager 2* moves from 10 thousand kilometers to 20 thousand kilometers away from Earth. By what factor does the gravitational force between them decrease?

3. Project According to Einstein's theory of general relativity, gravity exists because objects "curve" the space around them. Do this activity to model the curvature of space by an object:

Hold a flat bed sheet above the ground and have a person pull tightly at each corner of the sheet. Roll a tennis ball across the sheet and observe the ball's movement. Next, place a basketball on the sheet's center and roll the tennis ball across one more time. What happens to the movement of the tennis ball?

Mathematics & Vera Rubin

In this chapter, you will learn more about how mathematics is related to astronomy.

Related Exercises

Section 9.6
- **Exercises 13 and 14**

Section 9.8
- **Exercises 5 and 6**

Section

9.1 Polynomials

Learn how to...

- recognize, evaluate, add, and subtract polynomials

So you can...

- understand numeration systems of ancient cultures, for example

Many early civilizations used drawings of everyday objects to represent numbers. Some of the symbols used by the Egyptian and Aztec peoples are shown below. Notice that the Egyptian numeration system was based on powers of 10, while the Aztec system was based on powers of 20.

This Egyptian painting is from the tomb of Tashep-Khonsu.

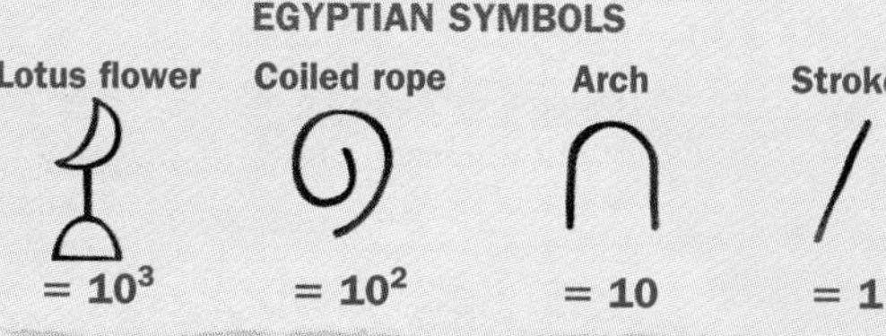

AZTEC SYMBOLS

Maize doll $= 20^3$; Maize plant $= 20^2$; Flag $= 20$; Maize seed $= 1$

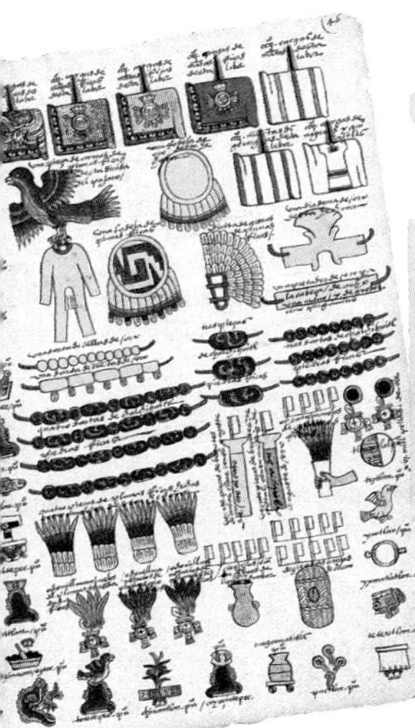

This page is from an Aztec *codex*, an illustrated history of their culture.

You can use the information above to find the decimal form of the numbers represented by these groups of symbols:

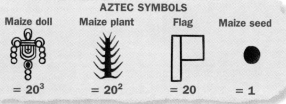

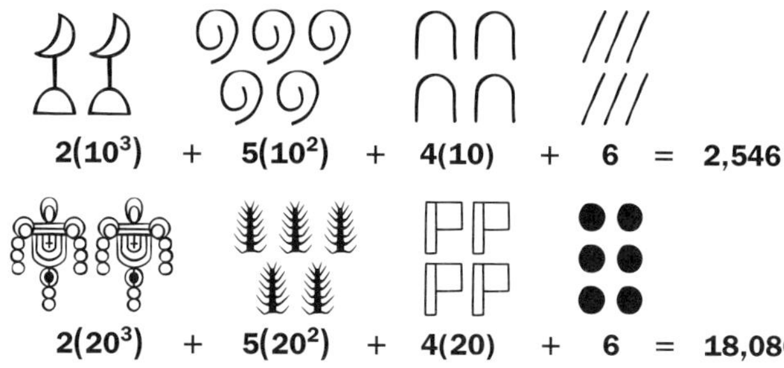

THINK AND COMMUNICATE

1. Compare the Egyptian numeration system with the decimal system used in the United States today. How are the systems alike? How are they different?
2. What is the maximum number of maize seeds that can be used in the Aztec representation of a number? Explain.

Plan⇔Support

Objectives

- Recognize, evaluate, add, and subtract polynomials.
- Understand numeration systems using polynomials.

Recommended Pacing

Core and Extended Courses
Section 9.1: 2 days

Block Schedule
Section 9.1: 1 block

Resource Materials

Lesson Support
Lesson Plan 9.1
Warm-Up Transparency 9.1
Practice Bank: Practice 54
Study Guide: Section 9.1
Explorations Lab Manual: Additional Exploration 15

Technology
Graphing Calculator
McDougal Littell Mathpack *Function Investigator*
Internet: http://www.hmco.com

Warm-Up Exercises

Find the value of each expression.

1. $8 \cdot 10^2 + 9 \cdot 10 + 2$ 892
2. $3 \cdot 10^3 + 7 \cdot 10 + 5$ 3075
3. $2 \cdot 10^4 + 8 \cdot 10^3 + 4 \cdot 10^2 + 4 \cdot 10 + 7$ 28,447
4. $4(-2)^2 + 5(-2)$ 6
5. $7(-1)^3 + 0(-1)^2 + 4(-1) + 8$ −3
6. $12 \cdot 3^4 - 6 \cdot 3^2 + 2 \cdot 3$ 924

ANSWERS Section 9.1

Think and Communicate

1. Summaries may vary. The Egyptian system is based on the number 10, which makes it more like our system than the Aztec system is. Both systems use pictures instead of position to determine values. In the decimal system, the position of a digit determines its value. In the Egyptian and Aztec systems, the order of the pictures is not important.
2. 19; 20 maize seeds equal 1 flag.

Teach⇔Interact

Think and Communicate

After discussing questions 1 and 2, ask students what advantages the decimal system has over the Egyptian and Aztec systems.

Section Note

Research
Students may find it interesting to investigate the development of the notation we use for polynomials.

Additional Example 1

Tell whether each expression is a polynomial. If so, write the polynomial in standard form and state its degree. If not, explain why not.

a. $-5x^2 + 0x^4 - 5x^3 + 2$
The expression is a polynomial. The standard form of the polynomial is $-5x^3 - 5x^2 + 2$. The degree of the polynomial is 3.

b. $7x^0 + 9x^5 - \frac{3}{4}x + \frac{(x^3 + 2)}{5}$
The expression is a polynomial. The standard form of the polynomial is $9x^5 + \frac{1}{5}x^3 - \frac{3}{4}x + \frac{37}{5}$. The degree of the polynomial is 5.

c. $3a^2 + 4a^5 - 3\sqrt{a} + a$
The expression is not a polynomial because $\sqrt{a} = a^{1/2}$ and $\frac{1}{2}$ is not a whole number.

d. $9x^2 + 7x - 5x^{-3}$
The expression is not a polynomial, because $-5x^{-3}$ is not the product of a real number and a whole-number power of x.

The expressions

$$2(10^3) + 5(10^2) + 4(10) + 6$$

and

$$2(20^3) + 5(20^2) + 4(20) + 6$$

on the previous page are similar in form to the *polynomial* below:

A **polynomial** is a term or a sum of terms like $2x^3$. Each term is the product of a real-number coefficient and a variable with a whole-number exponent.

The exponent of this variable is 1, since $x^1 = x$.

$$2x^3 + 5x^2 + 4x + 6$$

The **degree** of a polynomial is the greatest exponent. This polynomial has degree 3.

This term is called the **constant term** of the polynomial. Note that 6 is the same as $6x^0$, since $x^0 = 1$.

A polynomial whose exponents decrease from left to right is said to be in **standard form**. For example, the polynomial shown above is in standard form.

EXAMPLE 1

Tell whether each expression is a polynomial. If so, write the polynomial in standard form and state its degree. If not, explain why not.

a. $x^2 - 8x^3 - 2 + 3x^4 + 7x$ **b.** $4x^{-3} + 9x + x^{1/2} + 10$

c. $7r^4 - 2r^3 + 5r^2 + \frac{1}{r+1}$ **d.** $-6.4m + \frac{2}{7}m^3 - \pi m^5 + \sqrt{3}m^2$

SOLUTION

a. The expression *is* a polynomial.
The standard form of the polynomial is $3x^4 - 8x^3 + x^2 + 7x - 2$.
The degree of the polynomial is 4.

b. The expression *is not* a polynomial, because the exponents -3 and $\frac{1}{2}$ are not whole numbers.

c. The expression *is not* a polynomial, because $\frac{1}{r+1}$ is not the product of a real number and a whole-number power of r.

d. The expression *is* a polynomial.
The standard form of the polynomial is $-\pi m^5 + \frac{2}{7}m^3 + \sqrt{3}m^2 - 6.4m$.
The degree of the polynomial is 5.

Synthetic Substitution

Look at the group of Aztec symbols shown.

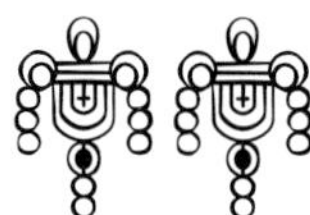  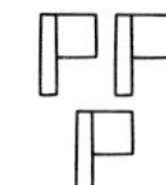 

You can find the decimal form of the number these symbols represent by evaluating the polynomial

$$2x^3 + x^2 + 3x + 7$$

when $x = 20$. One way to do this is simply to substitute 20 for x in the polynomial. However, it is more efficient to first rewrite the polynomial this way:

$$2x^3 + x^2 + 3x + 7 = (2x^2 + x + 3)x + 7$$
$$= ((2x + 1)x + 3)x + 7$$

To evaluate the polynomial for any value of x, follow these steps:

$$((2x + 1)x + 3)x + 7$$

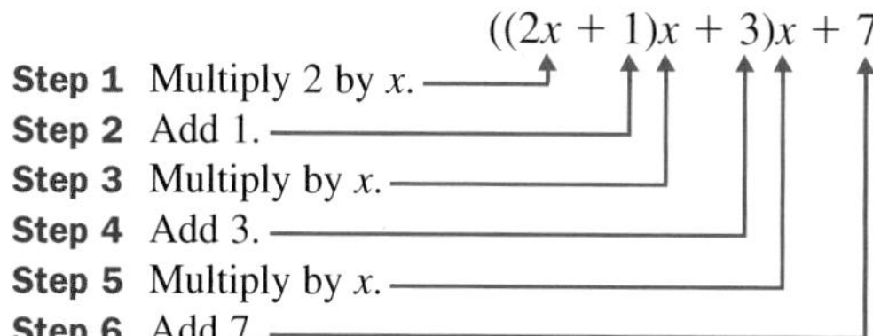

Step 1 Multiply 2 by x.
Step 2 Add 1.
Step 3 Multiply by x.
Step 4 Add 3.
Step 5 Multiply by x.
Step 6 Add 7.

Steps 1–6 illustrate a procedure called **synthetic substitution**. The next example shows an easy way to perform this procedure.

EXAMPLE 2

Evaluate $2x^3 + x^2 + 3x + 7$ when $x = 20$.

SOLUTION

Write the polynomial's coefficients in a row. Perform Steps 1–6 above.

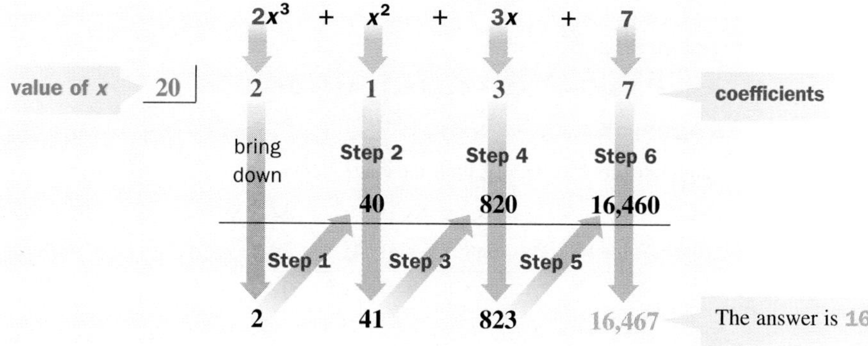

BY THE WAY

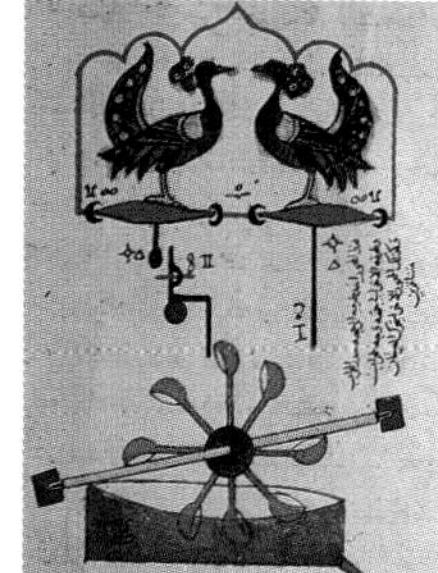

The numeration system used in the United States and most other countries is called the *Hindu-Arabic* system. Hindu mathematicians developed the system between 300 B.C. and A.D. 600. Arabic peoples introduced the system in Europe during the twelfth century A.D.

Teach⇔Interact

Section Notes

Historical Connection

The Hindu-Arabic system is a *positional numeral system* with base 10. The number 407, for example, can be written as $4 \cdot 10^2 + 0 \cdot 10^1 + 7 \cdot 10^0$. In this type of system, a symbol is needed for zero to indicate any missing powers of the base. The earliest versions of the Hindu-Arabic system did not contain zero and were not positional. It is thought, however, that a zero and positional value must have been introduced in India before 800 A.D., because such a Hindu system is described for the first time in a book written in 825 A.D. by the Persian mathematician al-Khowârizmi.

Mathematical Procedures

Synthetic substitution is useful for paper-and-pencil calculations. In Section 9.4, students will see that the usefulness of the synthetic substitution algorithm goes beyond evaluating polynomials. They will see how the algorithm yields valuable information about situations in which a polynomial is divided by $x - k$, where k is a real number.

Teaching Tip

In discussing the six steps for evaluating the polynomial $2x^3 + x^2 + 3x + 7$, point out that the alternate form for the polynomial was obtained by working from right to left, each time factoring out x by means of the distributive property.

Additional Example 2

Evaluate $-7x^4 + 5x^3 + x^2 - 6x + 8$ when $x = -2$.

Write the polynomial's coefficients in a row and use the synthetic substitution algorithm.

$$-7x^4 + 5x^3 + x^2 - 6x + 8$$

−2	−7	5	1	−6	8
		14	−38	74	−136
	−7	19	−37	68	−128

The answer is −128.

Teach⇔Interact

Section Notes

Using Technology
The synthetic substitution algorithm is easily performed with a calculator. For instance, in Example 2, with the TI-81 or TI-82 graphing calculator, first use 20 → X to store 20 as the value of X. Then use appropriate keystrokes to display the following on the home screen.

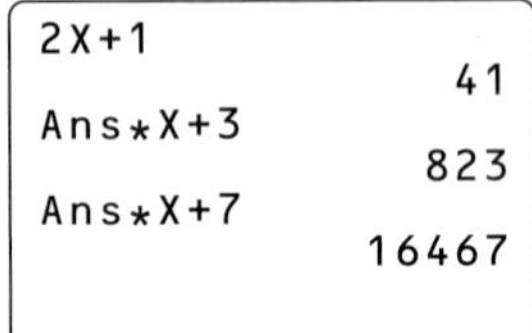

Common Error
Students may forget to include a 0 coefficient for the missing terms when using synthetic substitution. Stress to students that they must include any zeros needed to perform the synthetic substitution correctly. Think and Communicate question 3 will help to solidify this concept.

Integrating the Strands
Using ancient number systems to introduce the addition of polynomials helps students see the connection between the two. Remind students that polynomials are algebraic representations of numbers. Students may also see the similarity of operations with polynomials to operations with complex numbers.

Think and Communicate

In discussing question 3, students should perform the algorithm to see what result they get if the zeros are left out.

About Example 3

Teaching Tip
Students should note that the coefficient of x^2 in the first polynomial is –1. Some students may want to rewrite the polynomial as $3x^4 + 8x^3 - 1x^2 - 2x + 6$ before adding.

Additional Example 3

Add $-7x^5 + 8x^3 - 2x^2 + x - 9$ and $-5x^3 - 12x + 2$.

Align like terms vertically, then add coefficients.

$$\begin{array}{r} -7x^5 + 8x^3 - 2x^2 + x - 9 \\ - 5x^3 \qquad\quad - 12x + 2 \\ \hline -7x^5 + 3x^3 - 2x^2 - 11x - 7 \end{array}$$

The sum is $-7x^5 + 3x^3 - 2x^2 - 11x - 7$.

When writing the row of coefficients in synthetic substitution, you must include a zero coefficient for each "missing" term of the polynomial. For example, you would write these coefficients for the polynomial $4x^5 - 7x^3 - x^2 + 12$:

4 0 –7 –1 0 12

This zero represents the missing x^4-term.

This zero represents the missing x-term.

THINK AND COMMUNICATE

3. Use synthetic substitution to evaluate $4x^5 - 7x^3 - x^2 + 12$ when $x = -2$. Would you get the same answer if you left out the zeros?

Adding and Subtracting Polynomials

Look at the two Egyptian numbers shown. You can find the decimal form of their sum by adding coefficients of terms containing the same power of 10.

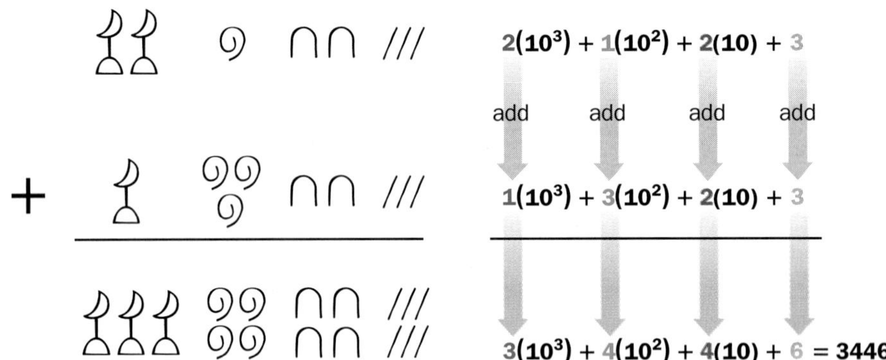

To add two polynomials, you add coefficients of terms containing the same power of the variable, such as $2x^3$ and $5x^3$. Such terms are called **like terms**.

EXAMPLE 3

Add $3x^4 + 8x^3 - x^2 - 2x + 6$ and $9x^4 - 4x^2 + 2x - 1$.

SOLUTION

Align like terms vertically, then add coefficients.

$$\begin{array}{r} 3x^4 + 8x^3 - x^2 - 2x + 6 \\ 9x^4 \qquad\quad - 4x^2 + 2x - 1 \\ \hline 12x^4 + 8x^3 - 5x^2 + 0x + 5 \end{array}$$

The sum is $12x^4 + 8x^3 - 5x^2 + 5$.

Think and Communicate

3.

$\underline{-2}\rfloor$	4	0	–7	–1	0	12
		–8	16	–18	38	–76
	4	–8	9	–19	38	–64

The value of $4x^5 - 7x^3 - x^2 + 12$ when $x = -2$ is –64. No.

Subtracting polynomials is similar to subtracting real numbers. To subtract a real number, you add the opposite of the number. To subtract a polynomial, you add the opposite of each term of the polynomial.

EXAMPLE 4

Subtract $-3x^2 - 5x - 7 + 6x^3$ from $7x^3 - 5x^2 + x - 4$.

SOLUTION

Align like terms vertically.

$$\begin{array}{r} 7x^3 - 5x^2 + x - 4 \\ -(6x^3 - 3x^2 - 5x - 7) \\ \hline \end{array}$$

Write the **opposite** of each term, then add.

$$\begin{array}{r} 7x^3 - 5x^2 + x - 4 \\ -6x^3 + 3x^2 + 5x + 7 \\ \hline x^3 - 2x^2 + 6x + 3 \end{array}$$

The difference is $x^3 - 2x^2 + 6x + 3$.

CHECKING KEY CONCEPTS

Tell whether each expression is a polynomial. If so, write the polynomial in standard form and state its degree. If not, explain why not.

1. $1 + x^{-1} + x^{-2} + x^{-3}$
2. $5 + 2x^4 - 4x^3 + x - 9x^2$
3. $-1.6y^2 - 3.8y^4 + 0.22y - 0.94y^5$
4. $4t^{4/5} + 3t^3 + 2t^2 + t$

Use synthetic substitution to evaluate each polynomial for the given value of the variable.

5. $x^3 - 4x^2 + 6x - 2; x = 3$
6. $-2t^5 + 3t^3 + 5t^2 - 8; t = -1$

Find each polynomial sum or difference.

7. $(6x^3 + x^2 + 3x + 4) + (2x^3 + 7x^2 + x + 5)$
8. $(-9n^5 + 3n^4 - 2n^2 - 2n - 8) - (-n^5 + 5n^4 + 10n^3 - 2n^2 + 4)$

9.1 Exercises and Applications

Extra Practice exercises on page 762

Tell whether each expression is a polynomial. If so, write the polynomial in standard form and state its degree. If not, explain why not.

1. $5x^{2.9}$
2. $5x^3$
3. $6t + t^2 - 1$
4. $6t + \frac{1}{t^2} - 1$
5. $-\frac{3}{2}u^4 + 5u^7 + \pi - 0.6u^{10} - \sqrt{2}u^8$
6. $-9u^2 + 7u + 2\sqrt[3]{u} + 4$
7. $1 + 2^r + 3^r + 4^r$
8. $1 + r^2 + r^3 + r^4$

Teach⇔Interact

Section Note

Teaching Tip

If students have difficulty subtracting polynomials, it may be due to not aligning the terms properly. Suggest that they write out the terms with zero coefficients.

Additional Example 4

Subtract $-3x^3 + 5x - 4$ from $2x^2 - 7x + 8$.

Align like terms vertically.

$$\begin{array}{r} 2x^2 - 7x + 8 \\ -(-3x^3 \qquad + 5x - 4) \\ \hline \end{array}$$

Write the opposite of each term, then add.

$$\begin{array}{r} 2x^2 - 7x + 8 \\ 3x^3 \qquad - 5x + 4 \\ \hline 3x^3 + 2x^2 - 12x + 12 \end{array}$$

The difference is $3x^3 + 2x^2 - 12x + 12$.

Closure Question

Describe the procedures used to add and subtract polynomials.

To add two polynomials, align the like terms vertically. Then add the coefficients. To subtract a polynomial from another polynomial, align the like terms vertically. Then add the opposite of each term of the polynomial that is being subtracted.

Apply⇔Assess

Suggested Assignment

Core Course

Day 1 Exs. 1–16, 20–22

Day 2 Exs. 23–31, 35, 36, 39–51

Extended Course

Day 1 Exs. 1–22

Day 2 Exs. 23–51

Block Schedule

Day 55 Exs. 1–16, 20–31, 35, 36, 39–51

Checking Key Concepts

1. No; the exponents are not whole numbers.
2. Yes; $2x^4 - 4x^3 - 9x^2 + x + 5$; 4
3. Yes; $-0.94y^5 - 3.8y^4 - 1.6y^2 + 0.22y$; 5
4. No; the exponent $\frac{4}{5}$ is not a whole number.
5. 7
6. −4
7. $8x^3 + 8x^2 + 4x + 9$
8. $-8n^5 - 2n^4 - 10n^3 - 2n - 12$

Exercises and Applications

1. No; the exponent 2.9 is not a whole number.
2. Yes; $5x^3$; 3
3. Yes; $t^2 + 6t - 1$; 2
4. No; $\frac{1}{t^2} = t^{-2}$ and the exponent −2 is not a whole number.
5. Yes; $-0.6u^{10} - \sqrt{2}u^8 + 5u^7 - \frac{3}{2}u^4 + \pi$; 10
6. No; $2\sqrt[3]{u} = 2u^{1/3}$ and the exponent $\frac{1}{3}$ is not a whole number.
7. No; 2^r is not the product of a real number and a whole-number power of r.
8. Yes; $r^4 + r^3 + r^2 + 1$; 4

Apply⇔Assess

Exercise Notes

Common Error

Exs. 9–16, 20–31 Some students forget that it is necessary to write a zero for the coefficient of each missing term in order to apply the synthetic substitution algorithm correctly. Working a few problems of this type as examples for students should help them to understand this procedure. For Exs. 20–31, students make errors for a variety of reasons. Urge students to briefly review their work to eliminate simple arithmetical errors. Some errors may be the result of aligning terms incorrectly. These errors can be corrected by checking that the same powers of the variable are being aligned. Point out that it is always a good idea to write both polynomials in standard form before adding or subtracting them. Doing so can help students reduce the number of alignment errors.

Application

Ex. 17 Ask students if it would be reasonable to substitute 80 for l in the equation $w = 0.0005l^3$. Why or why not? (No; a rainbow trout never grows to so great a length.)

Research

Exs. 18, 19 Ask some students to research the system used by the Sumerians around 2100 B.C. This system was a place value system using base 60. There are symbols for 1, 60, 3600, 60^{-1}, and 60^{-2}. A number such as 61 would be written using two of the 1 symbols to represent $1 \cdot 60 + 1 \cdot 1 = 61$. You might have students make up some problems such as those in these exercises using this system.

Use synthetic substitution to evaluate each polynomial for the given value of the variable.

9. $x^2 + 4x + 3;\ x = 10$

10. $2x^3 - 3x^2 - 7x + 5;\ x = 2$

11. $t^4 + t^3 - 4t^2 - 8t - 10;\ t = -2$

12. $-3x^4 + 6x^3 + x^2 + 5x - 2;\ x = -3$

13. $2n^3 - 6n + 1;\ n = 5$

14. $-w^5 - 7w^4 - 2w^2 - 9;\ w = -4$

15. $6x^4 + 7x^3 + 5x - 3;\ x = \frac{1}{3}$

16. $-5u^7 + 9u^5 - 8u^2 + 4u + 11;\ u = \sqrt{2}$

17. **BIOLOGY** The average weight w of a rainbow trout l in. long can be modeled by the equation

$$w = 0.0005l^3$$

where w is measured in pounds.

a. Is $0.0005l^3$ a polynomial? If so, what is the polynomial's degree?

b. Find the average weight of a rainbow trout 20 in. long.

c. By what factor does average weight for a rainbow trout increase when length doubles?

Use the information about Egyptian and Aztec numeration systems on page 389 to find the decimal form of each number.

18. 

19. 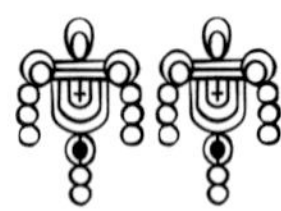   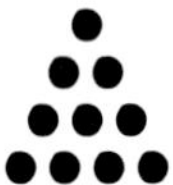

Add.

20. $(4x^2 + x + 3) + (3x^2 + 5x + 1)$

21. $(6x^3 + 8x^2 - 2x + 4) + (10x^3 + x^2 + 11x + 9)$

22. $(-t^4 + 7t^3 + 18t - 13) + (4t^4 - 2t^3 - 6t^2 + 5t - 20)$

23. $(3.9y^5 + 0.5y^4 - 4.8y^3 - 2y^2 + 7.6y) + (5.2y^5 - 2.7y^4 - 8.8y^2 + 0.9y + 4)$

24. $\left(\frac{3}{8}x^2 - \frac{5}{3}x^3 + \frac{7}{12} + \frac{1}{2}x^4 - \frac{2}{5}x\right) + \left(-\frac{1}{6}x^3 + \frac{2}{9} - \frac{5}{2}x^4 + \frac{11}{4}x^2\right)$

25. $\left(3\sqrt{2}m^3 - 6\sqrt{5}m^2 + 9m + \sqrt{3}\right) + \left(-\sqrt{3} + 14m - \sqrt{5}m^2 - 7\sqrt{2}m^3\right)$

Subtract.

26. $(2x^2 + 5x + 3) - (x^2 + 2x + 1)$

27. $(8x^3 - 4x^2 - 7x + 2) - (9x^3 + 6x^2 - 7x - 2)$

28. $(-3u^4 - 10u^3 + 19u^2 - 12) - (-16u^3 + 14u^2 + 5u + 4)$

29. $\left(\frac{3}{4}v - \frac{2}{7}v^2 + \frac{1}{3}v^3 - \frac{11}{12}\right) - \left(\frac{4}{3}v^3 - \frac{3}{5}v + \frac{13}{18} - \frac{5}{14}v^2\right)$

30. $\left(4\sqrt{3} + 5y - 10\sqrt{7}y^2 + 9y^4 - 3\sqrt{3}y^6\right) - \left(4\sqrt{7}y^2 - \sqrt{3}y^6 + 5\sqrt{3} - y\right)$

31. $(0.65t^5 - 0.7t^4 + 1.09t^2 - 0.3) - (-3.82 - 1.5t^2 + 2.1t + 0.36t^5 + 0.08t^4)$

9. 143
10. −5
11. −2
12. −413
13. 221
14. −809
15. −1
16. −5
17. a. Yes; 3.
 b. 4 lb
 c. 8
18. 4268
19. 17,250
20. $7x^2 + 6x + 4$
21. $16x^3 + 9x^2 + 9x + 13$
22. $3t^4 + 5t^3 - 6t^2 + 23t - 33$
23. $9.1y^5 - 2.2y^4 - 4.8y^3 - 10.8y^2 + 8.5y + 4$
24. $-2x^4 - \frac{11}{6}x^3 + \frac{25}{8}x^2 - \frac{2}{5}x + \frac{29}{36}$
25. $-4\sqrt{2}m^3 - 7\sqrt{5}m^2 + 23m$
26. $x^2 + 3x + 2$
27. $-x^3 - 10x^2 + 4$
28. $-3u^4 + 6u^3 + 5u^2 - 5u - 16$
29. $-v^3 + \frac{1}{14}v^2 + \frac{27}{20}v - \frac{59}{36}$
30. $-2\sqrt{3}y^6 + 9y^4 - 14\sqrt{7}y^2 + 6y - \sqrt{3}$
31. $0.29t^5 - 0.78t^4 + 2.59t^2 - 2.1t + 3.52$

Connection GEOMETRY

For each positive integer *n*, you can make a type of pyramid called a *tetrahedron* with *n* layers of dots. The number of dots *T(n)* in the *n*th tetrahedron is called the *nth tetrahedral number.*

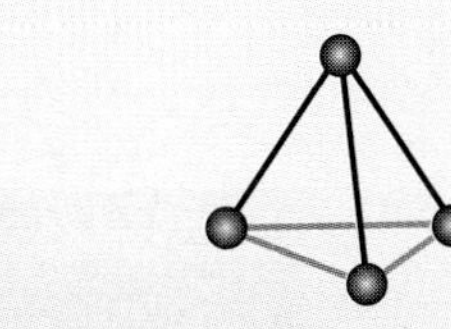

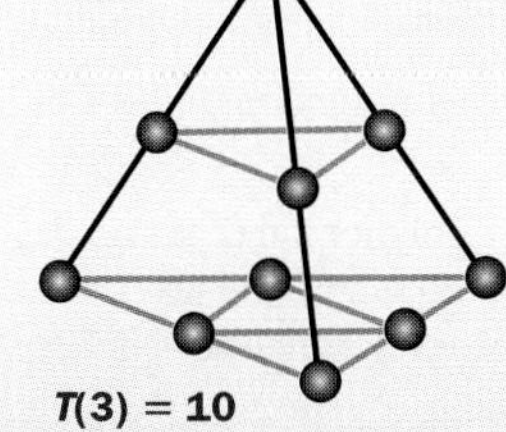

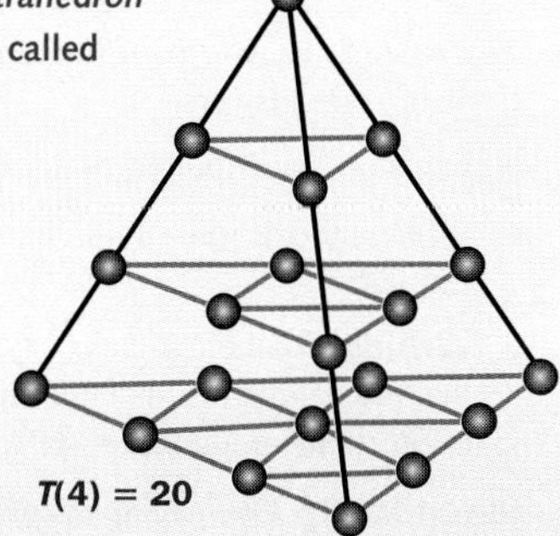

$T(1) = 1$ $T(2) = 4$ $T(3) = 10$ $T(4) = 20$

In Exercises 32–34, you will develop an equation for *T(n)*.

32. Consider the polynomial function $f(x) = x^2 + 3x + 1$. The diagram below shows how to compute *1st differences* and *2nd differences* for $f(x)$.

x	1	2	3	4	5	6
$f(x)$	5	11	19	?	?	?

11 − 5 19 − 11

1st differences	6	8	?	?	?

8 − 6

2nd differences	2	?	?	?

a. Copy and complete the diagram.

b. What do you notice about the 2nd differences?

33. Now consider these polynomial functions:

$$g(x) = 3x^3 - 5x^2 + x + 10$$
$$h(x) = 2x^4 - 7x^3 - 4x^2 - 6x + 1$$

a. Find 1st, 2nd, 3rd, and 4th differences for $g(x)$ and $h(x)$ using the x-values 1, 2, 3, . . . , 7.

b. **Writing** Based on your results from part (a) and Exercise 32, what can you say about the *k*th differences for a polynomial function of degree *k*?

34. a. **Visual Thinking** Show that the 5th and 6th tetrahedral numbers, $T(5)$ and $T(6)$, are 35 and 56, respectively.

b. It can be shown that $T(n)$ is a polynomial function of *n*. Use the values of $T(1)$, $T(2)$, . . . , $T(6)$ to compute various differences for $T(n)$. Based on your results, what is the degree of $T(n)$?

c. **Technology** Look back at Section 7.3, where you used matrices to fit quadratic functions to data. Use an extension of this matrix method to find an equation for $T(n)$. Verify that the first six tetrahedral numbers given by your equation are correct.

Apply⇔Assess

Exercise Notes

Integrating the Strands
Exs. 32–34 The concepts involved in these exercises integrate the strands of number theory, geometry, algebra, and functions.

Using Technology
Ex. 33 A TI-82 graphing calculator can be used to calculate quickly the values of $g(x)$ and $h(x)$ for x-values 1, 2, 3, ..., 7. Various approaches are feasible. Here is one possibility. Clear lists L1 and L2 of the STAT data lists. Press [Y=] and enter the polynomial for $g(x)$ on the first line. The Y= list will show Y1=3X^3–5X²+X+10. Next, press [STAT] 1 to go to the statistical data edit screen. Enter 1, 2, 3, ..., 7 in list L1. Place the cursor on L2 at the very top of the L2-column. Press [2nd] <Y-VARS> <Function><Y1> [(] [2nd] [L1] [)] [ENTER]. The values for $g(x)$ will be displayed in the L2-column.

Second-Language Learners
Ex. 33 Students learning English may benefit from orally discussing the results of Exs. 32 and 33(a) with a peer tutor or aide before writing their explanations for part (b).

Research
Ex. 34 Polynomials and finite differences can be a fascinating topic for research by interested students. Books on finite mathematics are a good source of information. The proofs of the results in such books may be too difficult for most students, but the results themselves are interesting and easy to confirm for specific polynomials.

32. a. $f(x)$: 5, 11, 19, 29, 41, 55; 1st differences: 6, 8, 10, 12, 14; 2nd differences: 2, 2, 2, 2

b. They are all the same.

33. a. 1st differences for $g(x)$: 7, 33, 77, 139, 219, 317; 2nd differences for $g(x)$: 26, 44, 62, 80, 98; 3rd differences for $g(x)$: 18, 18, 18, 18; 4th differences for $g(x)$: 0, 0, 0; 1st differences for $h(x)$: –37, –29, 57, 269, 655, 1263; 2nd differences for $h(x)$: 8, 86, 212, 386, 608; 3rd differences for $h(x)$: 78, 126, 174, 222; 4th differences for $h(x)$: 48, 48, 48

b. They are all the same.

34. a. The 5th layer has 15 dots and the 6th has 21, making the 5th and 6th tetrahedral numbers 35 and 56.

b. 1st differences: 3, 6, 10, 15, 21; 2nd differences: 3, 4, 5, 6; 3rd differences: 1, 1, 1; The degree of $T(n)$ is 3.

c. $T(n) = \frac{1}{6}n^3 + \frac{1}{2}n^2 + \frac{1}{3}n$; $T(1) = 1$; $T(2) = 4$; $T(3) = 10$; $T(4) = 20$; $T(5) = 35$; $T(6) = 56$

Apply⇔Assess

Exercise Notes

Assessment Note

Ex. 39 This exercise can help students see some interesting connections between algebra and geometry. Many polynomials that arise in geometry involve more than one variable.

Challenge

Exs. 40–43 You may wish to have students consider whether polynomials with real-number coefficients have any of the group properties with respect to addition and subtraction.

Practice 54 for Section 9.1

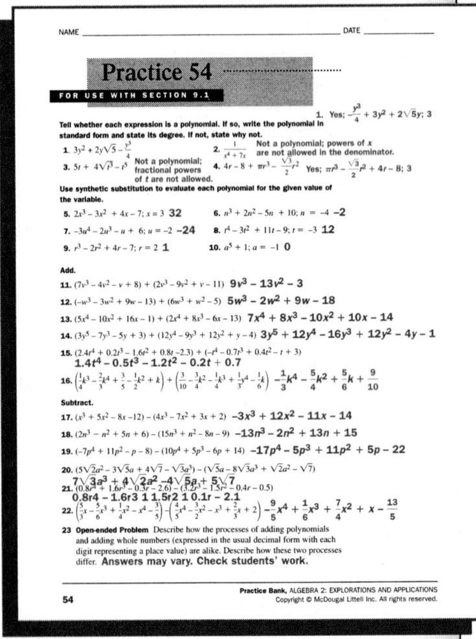

NAME ______________________ DATE __________

Practice 54

FOR USE WITH SECTION 9.1

Tell whether each expression is a polynomial. If so, write the polynomial in standard form and state its degree. If not, state why not.

1. $3y^2 + 2y\sqrt{5} - \frac{y^3}{4}$ **Yes; $-\frac{y^3}{4} + 3y^2 + 2\sqrt{5}y$; 3**
2. $\frac{1}{x^4 + 7x}$ **Not a polynomial; powers of x are not allowed in the denominator.**
3. $5t + 4\sqrt{t^3} - t^5$ **Not a polynomial; fractional powers of t are not allowed.**
4. $4r - 8 + \pi r^3 - \frac{\sqrt{3}}{2}r^2$ **Yes; $\pi r^3 - \frac{\sqrt{3}}{2}r^2 + 4r - 8$; 3**

Use synthetic substitution to evaluate each polynomial for the given value of the variable.

5. $2x^3 - 3x^2 + 4x - 7;\ x = 3$ **32**
6. $n^3 + 2n^2 - 5n + 10;\ n = -4$ **−2**
7. $-3u^4 - 2u^3 - u + 6;\ u = -2$ **−24**
8. $t^4 - 3t^2 + 11t - 9;\ t = -3$ **12**
9. $r^3 - 2r^2 + 4r - 7;\ r = 2$ **1**
10. $a^5 + 1;\ a = -1$ **0**

Add.

11. $(7v^3 - 4v^2 - v + 8) + (2v^3 - 9v^2 + v - 11)$ **$9v^3 - 13v^2 - 3$**
12. $(-w^3 - 3w^2 + 9w - 13) + (6w^3 + w^2 - 5)$ **$5w^3 - 2w^2 + 9w - 18$**
13. $(5x^4 - 10x^2 + 16x - 1) + (2x^4 + 8x^3 - 6x - 13)$ **$7x^4 + 8x^3 - 10x^2 + 10x - 14$**
14. $(3y^5 - 7y^3 - 5y + 3) + (12y^4 - 9y^3 + 12y^2 + y - 4)$ **$3y^5 + 12y^4 - 16y^3 + 12y^2 - 4y - 1$**
15. $(2.4t^4 + 0.2t^3 - 1.6t^2 + 0.8t - 2.3) + (-t^4 - 0.7t^3 + 0.4t^2 - t + 3)$ **$1.4t^4 - 0.5t^3 - 1.2t^2 - 0.2t + 0.7$**
16. $\left(\frac{1}{4}k^3 - \frac{2}{3}k^4 + \frac{3}{5} - \frac{1}{2}k^2 + k\right) + \left(\frac{3}{10} - \frac{3}{4}k^2 - \frac{1}{4}k^3 + \frac{1}{3}y^4 - \frac{1}{6}k\right)$ **$-\frac{1}{3}k^4 - \frac{5}{4}k^2 + \frac{5}{6}k + \frac{9}{10}$**

Subtract.

17. $(x^3 + 5x^2 - 8x - 12) - (4x^3 - 7x^2 + 3x + 2)$ **$-3x^3 + 12x^2 - 11x - 14$**
18. $(2n^3 - n^2 + 5n + 6) - (15n^3 + n^2 - 8n - 9)$ **$-13n^3 - 2n^2 + 13n + 15$**
19. $(-7p^4 + 11p^2 - p - 8) - (10p^4 + 5p^3 - 6p + 14)$ **$-17p^4 - 5p^3 + 11p^2 + 5p - 22$**
20. $(5\sqrt{2}a^2 - 3\sqrt{5}a + 4\sqrt{7} - \sqrt{3}a^3) - (\sqrt{5}a - 8\sqrt{3}a^3 + \sqrt{2}a^2 - \sqrt{7})$ **$7\sqrt{3}a^3 + 4\sqrt{2}a^2 - 4\sqrt{5}a + 5\sqrt{7}$**
21. $(0.8r^4 + 1.6r^3 - 0.3r - 2.6) - (3.2r^3 - 1.5r^2 - 0.4r - 0.5)$ **0.8r4 – 1.6r3 1 1.5r2 1 0.1r – 2.1**
22. $\left(\frac{5}{3}x - \frac{5}{6}x^3 + \frac{1}{4}x^2 - x^4 - \frac{3}{5}\right) - \left(\frac{4}{5}x^4 - \frac{3}{2}x^2 - x^3 + \frac{2}{3}x + 2\right)$ **$-\frac{9}{5}x^4 + \frac{1}{6}x^3 + \frac{7}{4}x^2 + x - \frac{13}{5}$**

23 **Open-ended Problem** Describe how the processes of adding polynomials and adding whole numbers (expressed in the usual decimal form with each digit representing a place value) are alike. Describe how these two processes differ. **Answers may vary. Check students' work.**

54 Practice Bank, ALGEBRA 2: EXPLORATIONS AND APPLICATIONS

35. SAT/ACT Preview If A $= x^4 + x^2 + 1$, B $= x^5 + x^3 + x$, and $0 < x < 1$, then:

A. A > B **B.** B > A **C.** A = B **D.** relationship cannot be determined

36. Tell whether each type of function is a polynomial function. Explain your answers.

a. linear **b.** exponential **c.** logarithmic **d.** quadratic **e.** radical

37. a. What is the degree of a nonzero constant polynomial like 2 or −3? Explain.

b. Writing Mathematicians say that the polynomial 0 has no degree, or that the degree of 0 is undefined. Explain why this makes sense. (*Hint:* Is there a unique whole number n for which $0 = 0x^n$?)

38. PSYCHOLOGY One social scientist found that the percent P of people who describe themselves as "highly annoyed" by noise from traffic, airports, and other sources can be modeled by the equation

$$P = 0.8553L - 0.0401L^2 + 0.00047L^3$$

where L is the average loudness of the noise in decibels (dB).

a. Is $0.8553L - 0.0401L^2 + 0.00047L^3$ a polynomial? If so, write the polynomial in standard form and state its degree.

b. The noise level of city traffic is typically about 70 dB. At this level, what percent of people are highly annoyed by the noise?

BY THE WAY

Highway barriers protect nearby areas from traffic noise. A typical barrier reduces noise levels by up to 20 dB. The amount of noise reduction depends on the barrier's material, height, and length.

ONGOING ASSESSMENT

39. Open-ended Problem Recall from geometry that the formula $V = \frac{4}{3}\pi r^3$ gives the volume V of a sphere as a polynomial function of its radius r. List three other polynomial formulas from geometry, and tell what information each formula gives.

SPIRAL REVIEW

Tell whether each set of numbers is closed under the given operation. If so, tell whether the number set and operation *form a group, form a commutative group,* or *do not form a group.* Explain. (*Section 8.6*)

40. whole numbers; +

41. whole numbers; ×

42. real numbers; +

43. odd numbers; ×

Evaluate each logarithm. (*Section 4.2*)

44. $\log_2 8$ **45.** $\log 10{,}000$ **46.** $\log_5 \frac{1}{25}$ **47.** $\log_9 27$

Multiply. (*Toolbox, page 782*)

48. $(x + 1)(x + 2)$ **49.** $(x - 3)(x + 4)$ **50.** $(2y + 5)(2y - 5)$ **51.** $(6n + 1)^2$

35. A

36. a. Yes; it has degree 1.

b. No; ab^x is not the product of a real number and a whole-number power of x.

c. No; $\log_b x$ is not a whole-number power of x.

d. Yes; it has degree 2.

e. No; $\sqrt[n]{x} = x^{1/n}$ is not a whole number power of x unless $n = 1$.

37. a. 0; $2 = 2 \cdot 1 = 2x^0$

b. $0x^n = 0$ for every n, so there is no unique n for which $0 = 0x^n$.

38. a. Yes; $0.00047L^3 - 0.0401L^2 + 0.8553L$; 3

b. about 24.6%

39. Answers may vary. Examples are given.
$A = s^2$ (the area of a square);
$V = s^3$ (the volume of a cube);
$A = \pi r^2$ (the area of a circle)

40. Yes; do not form a group; the inverse property does not hold.

41. Yes; do not form a group; the inverse property does not hold.

42. Yes; form a commutative group; the set is closed under addition and all the group properties as well as the commutative property hold.

43. Yes; do not form a group; the inverse property does not hold.

44. 3

45. 4

46. −2

47. $\frac{3}{2}$

48. $x^2 + 3x + 2$

49. $x^2 + x - 12$

50. $4y^2 - 25$

51. $36n^2 + 12n + 1$

Section

9.2 Multiplying and Dividing Polynomials

Learn how to...

- multiply and divide polynomials

So you can...

- find the power needed to maintain a given speed on a bicycle, for example

The diagrams show the same rectangular prism drawn two different ways. These drawings suggest two methods for computing the prism's volume V.

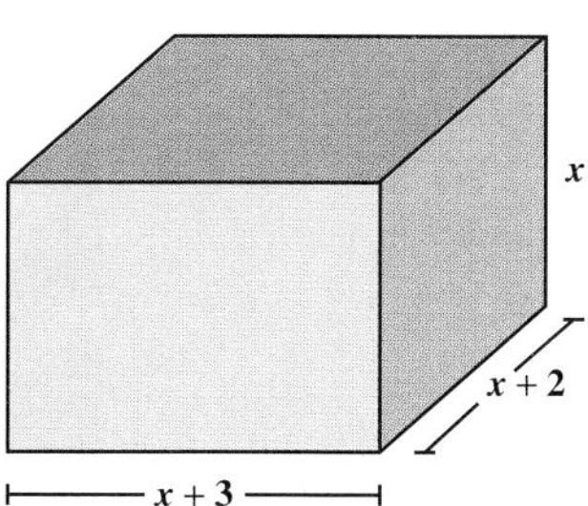

V = length × width × height
$= (x + 3)(x + 2)x$

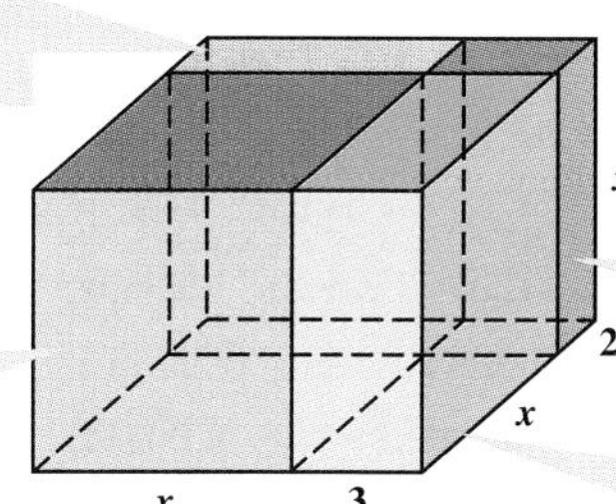

V_2 = volume of this piece

V_1 = volume of this piece

$V = V_1 + V_2 + V_3 + V_4$

V_4 = volume of this piece

V_3 = volume of this piece

THINK AND COMMUNICATE

1. Express the volume of each piece of the prism shown above in terms of x.
   a. V_1 b. V_2 c. V_3 d. V_4
2. Use your results from Question 1 to complete this equation with a polynomial in standard form:
$$(x + 3)(x + 2)x = \underline{\ ?\ }$$
3. a. Use the distributive property to find the product $(x + 3)(x + 2)$.
   b. Use the distributive property again to find the product of x and your answer to part (a).
   c. How does your answer to part (b) compare with your answer to Question 2?

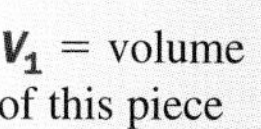
Toolbox p. 774 *Properties of Real Numbers*

Plan⇔Support

Objectives

- Multiply and divide polynomials.
- Use multiplication and division of polynomials in real-world problems.

Recommended Pacing

Core and Extended Courses
Section 9.2: 1 day

Block Schedule
Section 9.2: $\frac{1}{2}$ block
(with Section 9.3)

Resource Materials

Lesson Support
Lesson Plan 9.2
Warm-Up Transparency 9.2
Practice Bank: Practice 55
Study Guide: Section 9.2
Explorations Lab Manual: Diagram Master 2

Technology
Graphing Calculator
McDougal Littell Mathpack *Function Investigator with Matrix Analyzer Activity Book:* Function Investigator Activity 17
Internet: http://www.hmco.com

Warm-Up Exercises

Simplify.

1. $(5x^2)(6x^3)$ $30x^5$
2. $(-2x^5)(2x^4)$ $-4x^9$
3. $(-4x)(-7x^{10})$ $28x^{11}$

Use the distributive property to find each product.

4. $5x(x + 3)$ $5x^2 + 15x$
5. $-7x(-9x + 1)$ $63x^2 - 7x$
6. $2x(5x - 4)$ $10x^2 - 8x$

ANSWERS Section 9.2

Think and Communicate

1. a. x^3
   b. $2x^2$
   c. $3x^2$
   d. $6x$
2. $x^3 + 5x^2 + 6x$
3. a. $x^2 + 5x + 6$
   b. $x^3 + 5x^2 + 6x$
   c. They are the same.

Teach⇔Interact

Learning Styles: Visual

The diagrams on page 397 showing the same rectangular prism drawn two different ways should help visual learners see how the volume of the prism can be found. This can help them to make the transition to understanding how to compute the volume algebraically by multiplying polynomials.

Think and Communicate

Questions 1–3 lead students to the concept of polynomial multiplication by using a visual geometric model. Students should come to the realization that polynomial multiplication can be accomplished through a repeated use of the distributive property.

Additional Example 1

An engineer is testing materials to be used in the manufacture of diving boards. In one test, she determined that if a weight of 165 lb is placed x ft from the support, the far end of a 10 ft board will be deflected vertically by d ft, where $d = 0.00003x^2(240 - x)$.

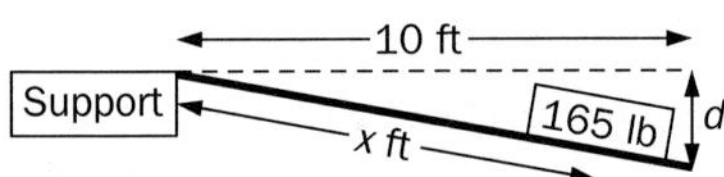

a. Write d as a polynomial function of x.

Use the equation for d.

$d = 0.00003x^2(240 - x)$

$= 0.00003x^2(240) - 0.00003x^2(x)$

$= 0.0072x^2 - 0.00003x^3$

$= -0.00003x^3 + 0.0072x^2$

b. By how much will the end of the board be deflected if the weight is placed 9.75 ft from the support?

Substitute 9.75 for x in the equation from part (a).

$d = -0.00003(9.75)^3 + 0.0072(9.75)^2$

≈ 0.657

The end of the board will be deflected by about 0.657 ft.

Polynomial Multiplication

You can use the distributive property to write a product of polynomials as a single polynomial.

EXAMPLE 1 Application: Bicycling

The power P, measured in horsepower (hp), required to keep a certain kind of bicycle moving at s mi/h is given by

$$P = 0.00267sF$$

where F is the force of road and air resistance in pounds. On level ground, this force is a quadratic function of speed:

$$F = 0.0116s^2 + 0.789$$

a. Write P as a polynomial function of s alone.

b. How much power must a cyclist provide to keep the bicycle moving at 15 mi/h on level ground?

SOLUTION

a. Use the equation for P.

$$P = 0.00267sF$$
$$= 0.00267s(\mathbf{0.0116s^2 + 0.789})$$ (Substitute **$0.0116s^2 + 0.789$** for **F**.)
$$= 0.00267s(0.0116s^2) + 0.00267s(0.789)$$ (Use the distributive property.)
$$\approx 0.0000310s^3 + 0.00211s$$ (Simplify.)

The function is $P = 0.0000310s^3 + 0.00211s$.

b. Substitute 15 for s in the equation from part (a).

$$P = 0.0000310(15^3) + 0.00211(15)$$
$$\approx 0.136$$

A cyclist must provide about 0.136 hp.

THINK AND COMMUNICATE

4. a. How much power must a cyclist provide to keep the bicycle in Example 1 moving at 30 mi/h on level ground?

b. Compare your answer to part (a) with the answer to part (b) of Example 1. Does it take twice as much power to double speed? Explain.

c. For a certain moped, the power from the engine must be 1.35 hp in order to travel 30 mi/h on level ground. Which vehicle—the moped or the bicycle in Example 1—uses energy more efficiently? Explain.

Think and Communicate

4. a. 0.900 hp

b. $0.900 \approx 6.62(0.136)$; No; answers may vary. An example is given. If doubling speed required doubling power, you would expect the function to be linear rather than cubic.

c. the bicycle; It requires less power to move the bicycle at 30 mi/h than to move the moped.

When you use the distributive property to multiply two polynomials, you multiply each term of one polynomial by each term of the other polynomial. You can do this easily using the vertical format shown in Example 2.

EXAMPLE 2

Multiply $2x - 3$ and $4x^2 - 5x + 8$.

SOLUTION

Align like terms vertically.

$4x^2 - 5x + 8$	Write the polynomial with more terms first.
$2x - 3$	
$8x^3 - 10x^2 + 16x$	← This is $2x(4x^2 - 5x + 8)$.
$- 12x^2 + 15x - 24$	← This is $-3(4x^2 - 5x + 8)$.
$8x^3 - 22x^2 + 31x - 24$	← Add like terms.

The product is $8x^3 - 22x^2 + 31x - 24$.

THINK AND COMMUNICATE

5. a. Multiply $2x - 3$ and $4x^2 - 5x + 8$ using a horizontal format:

$$(2x - 3)(4x^2 - 5x + 8) = 2x(4x^2 - 5x + 8) - 3(4x^2 - 5x + 8) = \underline{\ ?\ }$$

b. Compare the horizontal format in part (a) with the vertical format in Example 2. Which format do you prefer? Why?

Polynomial Division

In elementary school, you learned how to use long division to write a fraction like $\frac{173}{3}$ as a mixed number:

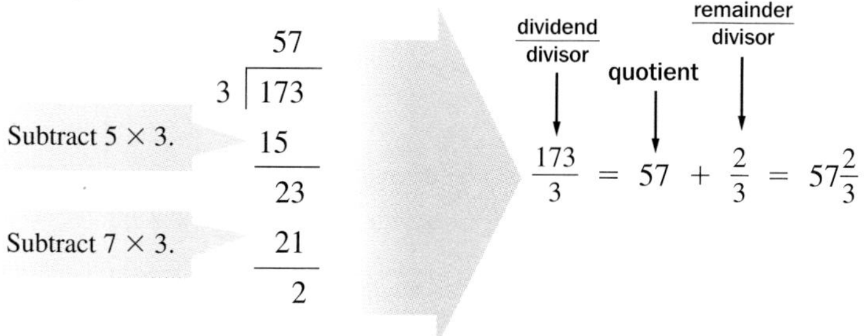

You can use a similar procedure to divide one polynomial by another as shown in Example 3 on the next page.

Think and Communicate

5. a. $8x^3 - 10x^2 + 16x - 12x^2 + 15x - 24 = 8x^3 - 22x^2 + 31x - 24$

b. Answers may vary.

Teach⇔Interact

Additional Example 2

Multiply $-3x^2 + x - 4$ and $7x^2 - 6x + 5$.

Align like terms vertically. Write the partial products $7x^2(-3x^2 + x - 4)$, $-6x(-3x^2 + x - 4)$, and $5(-3x^2 + x - 4)$. Add like terms.

$$\begin{array}{r} -3x^2 + x - 4 \\ 7x^2 - 6x + 5 \\ \hline -21x^4 + 7x^3 - 28x^2 \qquad\qquad \\ 18x^3 - 6x^2 + 24x \qquad \\ -15x^2 + 5x - 20 \\ \hline -21x^4 + 25x^3 - 49x^2 + 29x - 20 \end{array}$$

The product is $-21x^4 + 25x^3 - 49x^2 + 29x - 20$.

Think and Communicate

In connection with question 5, ask students whether they could start with the far-right terms when multiplying polynomials. (Yes, this is possible.) Doing so makes the procedure more closely resemble the way whole numbers are multiplied when using pencil and paper. The format used in Example 2 has the advantage, however, that it is easier to align the partial products.

Section Note

Using Technology Students can use the table features of graphing calculators or the *Function Investigator* software to check products obtained by multiplying polynomials. For instance, the answer for Example 2 can be checked on the TI-82 as follows. Press [Y=] and enter Y1=(2X–3)(4X²–5X+8) and Y2=8X³–22X²+31X–24. Press [2nd][TblSet] and choose values for TblMin and △Tbl. (You might use, for example, TblMin = 0, △Tbl = .5.) Finally, press [2nd][TABLE] to display values for Y1 and Y2. Since the numbers in the Y1-column and Y2-column are exactly the same, the product appears to be correct.

Teach⇔Interact

About Example 3

Mathematical Procedures
Polynomials can be multiplied without writing the factors in standard form, although having the factors in standard form helps organize the calculations. However, students should not try to divide polynomials that are not in standard form. In discussing Example 3, call attention to the step in which $0x$ is written for the missing x-term. Also, students should see that the answer is not a polynomial but is the sum of a polynomial and a nonpolynomial expression.

Additional Example 3

Divide $-12x^4 - 4x^3 + 6x^2 - 13x + 6$ by $3x + 1$.

$$\begin{array}{r|l} & -4x^3 \qquad\quad + 2x - 5 \\ 3x+1 & -12x^4 - 4x^3 + 6x^2 - 13x + 6 \\ & -12x^4 - 4x^3 \\ \hline & \qquad\qquad 6x^2 - 13x \\ & \qquad\qquad 6x^2 + 2x \\ \hline & \qquad\qquad\quad -15x + 6 \\ & \qquad\qquad\quad -15x - 5 \\ \hline & \qquad\qquad\qquad\quad 11 \end{array}$$

Therefore,
$$\frac{-12x^4 - 4x^3 + 6x^2 - 13x + 6}{3x+1} = -4x^3 + 2x - 5 + \frac{11}{3x+1}.$$

Think and Communicate

After discussing question 6, call on a volunteer to show a check of the answer in Example 3 at the board.

Checking Key Concepts

Teaching Tip
Use these multiplication and division problems to stress the importance of organizing calculations carefully. Call upon individual students to work each problem at the board.

Closure Question

Ask students to select a multiplication exercise and a division exercise from the Checking Key Concepts questions and explain the procedures they used to solve each exercise. Answers may vary.

EXAMPLE 3

Divide $6x^3 - 5x^2 + 9$ by $2x + 1$.

SOLUTION

$\frac{6x^3}{2x} = 3x^2$ $\qquad \frac{-8x^2}{2x} = -4x$ $\qquad \frac{4x}{2x} = 2$

$$\begin{array}{r|l} & 3x^2 - 4x + 2 \\ 2x+1 & 6x^3 - 5x^2 + 0x + 9 \\ \text{Subtract } 3x^2(2x+1). \rightarrow & 6x^3 + 3x^2 \\ \hline & \quad -8x^2 + 0x \\ \text{Subtract } -4x(2x+1). \rightarrow & \quad -8x^2 - 4x \\ \hline & \qquad\quad 4x + 9 \\ \text{Subtract } 2(2x+1). \rightarrow & \qquad\quad 4x + 2 \\ \hline & \qquad\qquad\quad 7 \end{array}$$

Insert the "missing" x-term by using 0 as the coefficient.

Stop when either the degree of the remainder is less than the degree of the divisor, or the remainder is 0.

Therefore, $\frac{6x^3 - 5x^2 + 9}{2x + 1} = 3x^2 - 4x + 2 + \frac{7}{2x+1}$.

$$\frac{\textbf{dividend}}{\textbf{divisor}} = \textbf{quotient} + \frac{\textbf{remainder}}{\textbf{divisor}}$$

THINK AND COMMUNICATE

6. a. Explain why the word equation at the end of Example 3 is equivalent to this equation:

dividend = divisor × quotient + remainder

b. Use the word equation in part (a) to check the answer to Example 3.

CHECKING KEY CONCEPTS

Multiply.

1. $(3x - 5)(x + 2)$

2. $(x^3 + 2x)(4x^4 - 7x^2)$

3. $(u + 1)(u^2 - 5u - 1)$

4. $(2t - 9)(-3t^2 + 8t + 2)$

Divide.

5. $\frac{x^2 + 6x + 10}{x + 2}$

6. $\frac{6w^2 + 7w + 5}{3w - 1}$

7. $\frac{y^3 + 4y^2 - 20y + 1}{y - 3}$

8. $\frac{8x^3 - 12x - 7}{2x + 3}$

Think and Communicate

6. a. The equation given here is the result of "multiplying" the equation at the end of Example 3 by the divisor.

b. $6x^3 - 5x^2 + 9 = (2x + 1)(3x^2 - 4x + 2) + 7$; $6x^3 - 5x^2 + 9$

Checking Key Concepts

1. $3x^2 + x - 10$
2. $4x^7 + x^5 - 14x^3$
3. $u^3 - 4u^2 - 6u - 1$
4. $-6t^3 + 43t^2 - 68t - 18$
5. $x + 4 + \frac{2}{x+2}$
6. $2w + 3 + \frac{8}{3w-1}$
7. $y^2 + 7y + 1 + \frac{4}{y-3}$
8. $4x^2 - 6x + 3 - \frac{16}{2x+3}$

Exercises and Applications

1. $8x^2 + 26x - 7$
2. $25x^2 + 30x + 9$
3. $-6t^5 + 12t^4 + 2t^3$
4. $y^3 - 3y^2 + 2y$
5. $u^3 + 2u^2 + 2u + 1$
6. $-v^3 + 10v^2 - 25v + 18$
7. $6m^3 - 31m^2 - 40m + 16$
8. $x^3 + 6x^2 + 12x + 8$
9. $40n^3 + 66n^2 - 31n - 42$

9.2 Exercises and Applications

Extra Practice exercises on page 762

Multiply.

1. $(4x-1)(2x+7)$ **2.** $(5x+3)^2$ **3.** $-2t^3(3t^2-6t-1)$

4. $y(y-1)(y-2)$ **5.** $(u+1)(u^2+u+1)$ **6.** $(v-2)(-v^2+8v-9)$

7. $(3m+4)(2m^2-13m+4)$ **8.** $(x+2)^3$ **9.** $(-10n^2+n+6)(-4n-7)$

10. $(2x-1)(3x+5)(4x+3)$ **11.** $(w-3)^4$ **12.** $(r^2-5r+2)(2r^2+8r-3)$

Connection HISTORY

According to Chinese legend, an emperor named Yu found a mystical turtle while walking along the Lo River. A drawing of the turtle is shown. The square array gives the number of dots on each plate of the turtle's shell. The array is called a *magic square* because the entries in each row, column, and diagonal add up to the same number, called a *magic constant*.

The entries in this 3×3 magic square are the integers from 1 to 9. In general, the entries in an $n \times n$ magic square are the integers from 1 to n^2.

2	9	4
7	5	3
6	1	8

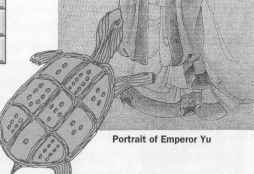

13. a. Verify that the square array shown is a magic square. What is the magic constant?

b. In general, the magic constant M for an $n \times n$ magic square is given by this equation:

$$M = \frac{1}{2}n^3 + \frac{1}{2}n$$

Find M when $n = 3$, and compare this value with the magic constant you found in part (a).

Portrait of Emperor Yu

BY THE WAY

The 3×3 magic square on the shell of Emperor Yu's turtle is called the *Lo shu*, which means "Lo River writing."

14. a. Use the equation for M in Exercise 13 to find an equation giving the sum S of *all* the entries in an $n \times n$ magic square.

$\left(\textit{Hint:}\text{ Note that } S = \text{sum of the numbers in each row} \times \text{number of rows}\right)$

b. Use your equation from part (a) to find the sum of all the entries in the given 3×3 magic square. Check your answer by computing the sum directly.

15. Cooperative Learning Work with a partner to create a 4×4 magic square. Verify that the equation for M in Exercise 13 applies to your square.

10. $24x^3 + 46x^2 + x - 15$

11. $w^4 - 12w^3 + 54w^2 - 108w + 81$

12. $2r^4 - 2r^3 - 39r^2 + 31r - 6$

13. a. The entries in each row, column, and diagonal add up to 15, which is the magic constant.

b. $M = \frac{27}{2} + \frac{3}{2} = 15$, which is the sum found in part (a).

14. a. $S = \frac{1}{2}n^4 + \frac{1}{2}n^2$

b. $S = \frac{1}{2}(81) + \frac{1}{2}(9) = 45 = 1 + 2 + 3 + 4 + 5 + 6 + 7 + 8 + 9$

15. Answers may vary. An example is given.

1	15	4	14
12	6	9	7
13	3	16	2
8	10	5	11

$M = \frac{1}{2}(64) + \frac{1}{2}(4) = 34$

Apply⇔Assess

Suggested Assignment

Core Course
Exs. 1–12, 18–30, 32–45

Extended Course
Exs. 1–45

Block Schedule
Day 56 Exs. 1–12, 18–30, 32–45

Exercise Notes

Common Error

Exs. 1–12 Students can make many types of errors when multiplying polynomials, most due to carelessness. Stress the importance of checking each product to see that all the correct terms have been multiplied; that all signs and exponents are correct; that like terms have been combined correctly; and that there are no arithmetic mistakes.

Reasoning

Ex. 13 The formula in part (b) tells what the magic constant must be, *provided* there actually is an arrangement of the n^2 integers for which all row, column, and diagonal sums are equal. Establishing the existence of such an arrangement is a separate problem.

Problem Solving

Exs. 13, 14 In Chapter 10, students will see that the sum of the first k positive whole numbers is equal to $\frac{k(k+1)}{2}$. Some students may wish to see if they can use this fact to deduce the formula in Ex. 13, part (b). One approach is to substitute n^2 for k in $\frac{k(k+1)}{2}$. This gives an expression for the sum S in Ex. 14, part (a). Divide this expression by n to find the magic constant M.

Visual Thinking

Ex. 15 Encourage students to create visual models of their 4×4 magic squares and present them to the class. Ask them to explain how their models work and to discuss such questions as: "Are other solutions possible?" and "Can a 5×5 magic square be created?" This activity involves the visual skills of *generalization* and *communication*.

Apply⇔Assess

Exercise Notes

Application
Exs. 16, 17 This group of exercises shows how algebra and geometry can be used to find the number of cups of meringue needed to cover a hemisphere of ice cream.

Second-Language Learners
Exs. 16, 17 Students learning English may be unfamiliar with the word *meringue*. If necessary, explain that this is a mixture of whipped egg whites and sugar.

Student Study Tip
Ex. 16 In part (b), students need to expand the expression $(r+1)^3$. Point out that they should first write the expression as a product, multiply two of the factors, and then multiply the result by the third factor. Students will see how to do this expansion quickly in Chapter 12 when studying binomial expansion.

Mathematical Procedures
Ex. 17 The expression for part (c) is an example of a 2-variable polynomial. Discuss how to define the term *degree* for polynomials with two or more variables.

Connection COOKING

The diagram shows one version of a dessert called Baked Alaska. You make this dessert by covering a hemisphere-shaped mound of ice cream with a 1 in. layer of meringue and then browning the meringue in an oven. The amount of meringue needed depends on the radius of the ice cream mound.

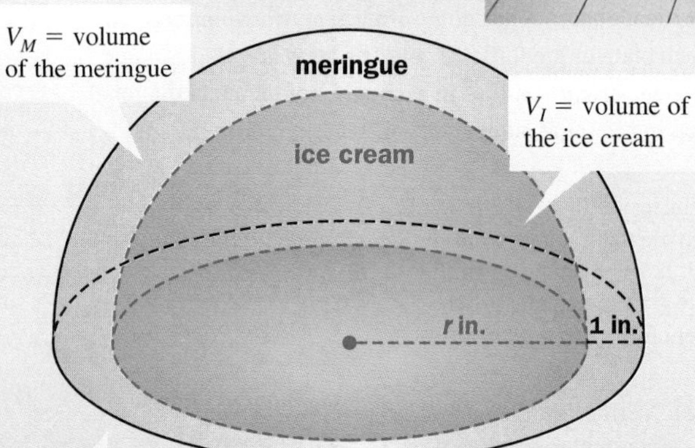

16. Use the diagram.

a. Express V_I as a polynomial function of r. (*Hint*: Recall from geometry that the volume of a sphere is given by $V = \frac{4}{3}\pi r^3$. Use 3.14 for π.)

b. Express V_{I+M} as a polynomial function of r.

c. Use your results from parts (a) and (b) to express V_M as a polynomial function of r. Write the polynomial in standard form.

d. How much meringue is needed to make Baked Alaska if the radius of the ice cream mound is 4 in.?

V_M = volume of the meringue

meringue

ice cream

V_I = volume of the ice cream

r in.

1 in.

V_{I+M} = combined volume of the ice cream and meringue

BY THE WAY

Baked Alaska was invented by Thomas Jefferson, the third President of the United States.

17. The function you found in part (c) of Exercise 16 gives the volume of meringue needed in cubic inches. It is more convenient to measure meringue in cups.

a. Use the fact that 1 in.$^3 \approx 0.0693$ cups to express V_M in cups as a function of r in inches.

b. How many cups of meringue are needed if the radius of the ice cream mound is 5 in.?

c. **Challenge** Suppose you want the thickness of the meringue to be something other than 1 in. Express the number of cups of meringue needed in terms of r and the desired meringue thickness h (where r and h are in inches).

16. a. $V_I \approx 2.09r^3$
b. $V_{I+M} \approx 2.09(r+1)^3$
c. $V_M \approx 6.27r^2 + 6.27r + 2.09$
d. about 127.5 in.^{3}

17. a. $V_M \approx 0.435r^2 + 0.435r + 0.145$
b. about 13.2 cups
c. $V_M = 0.435r^2h + 0.435rh^2 + 0.145h^3$

18. E

19. $x + 2 + \frac{4}{x+1}$

20. $2x + 4 + \frac{17}{x-3}$

21. $3t + 13 + \frac{26}{t-2}$

22. $u - 2 + \frac{8}{u+2}$

23. $y^2 + 4y - 2 + \frac{5}{y+4}$

24. $2v^2 + 3v - 8 - \frac{3}{2v+5}$

25. $6x^2 + 9x + 4$

26. $w^2 - 5w - 5 - \frac{4}{w-1}$

27. $2x^3 - 5x^2 + 11x - 12 + \frac{18}{x+2}$

28. $-2x^3 + 3x^2 - x + 4$

29. $4n - 1 + \frac{12n+4}{2n^2+7n-5}$

30. $r^2 - 3r - 8 + \frac{8r+37}{r^2-r+3}$

31. Let ax^m and bx^n, respectively, be the terms of highest degree of $p(x)$ and $d(x)$.
a. $m + n$; The term of highest degree in the product would be abx^{m+n}.

18. **SAT/ACT Preview** Suppose $(2x + a)(bx^2 + cx + d) = 2x^3 + 7x^2 - 19x + 6$. Which of the following equals $ac + 2d - 3b$?

A. 13 **B.** 14 **C.** 16 **D.** −19 **E.** −22

Divide.

19. $\frac{x^2 + 3x + 6}{x + 1}$
20. $\frac{2x^2 - 2x + 5}{x - 3}$
21. $\frac{3t^2 + 7t}{t - 2}$
22. $\frac{u^2 + 4}{u + 2}$
23. $\frac{y^3 + 8y^2 + 14y - 3}{y + 4}$
24. $\frac{4v^3 + 16v^2 - v - 43}{2v + 5}$
25. $\frac{12x^3 - 19x - 12}{2x - 3}$
26. $\frac{w^3 - 6w^2 + 1}{w - 1}$
27. $\frac{2x^4 - x^3 + x^2 + 10x - 6}{x + 2}$
28. $\frac{-6x^4 + 7x^3 + 11x + 4}{3x + 1}$
29. $\frac{8n^3 + 26n^2 - 15n + 9}{2n^2 + 7n - 5}$
30. $\frac{r^4 - 4r^3 - 2r^2 + 7r + 13}{r^2 - r + 3}$

31. **Writing** Let $p(x)$ and $d(x)$ be polynomials with degrees m and n, respectively, where $m > n$.
   a. What is the degree of $p(x) \cdot d(x)$? How do you know?
   b. Suppose that when $p(x)$ is divided by $d(x)$, the quotient is $q(x)$ and the remainder is $r(x)$. What is the degree of $q(x)$? What can you say about the degree of $r(x)$? Explain.

ONGOING ASSESSMENT

32. **Open-ended Problem** Write two polynomials $p(x)$ and $d(x)$ such that $\frac{p(x)}{d(x)} = 4x^2 - x + 6$. Explain how you found your polynomials, and verify that they satisfy the given equation.

SPIRAL REVIEW

Use synthetic substitution to evaluate each polynomial for the given value of the variable. *(Section 9.1)*

33. $2x^2 + 6x + 11;\ x = 3$
34. $x^3 - 9x^2 - 5x + 5;\ x = 2$
35. $3y^4 + 2y^3 - 7y + 1;\ y = -2$
36. $-4m^3 + 13m^2 + 5m - 10;\ m = \frac{1}{4}$

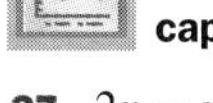
Technology Use a graphing calculator or software with matrix calculation capabilities to solve each system of equations. *(Section 7.3)*

37. $2x - y = 5$
    $x + 3y = 6$
38. $-70x + 10y = 19$
    $-10x + 2y = 3$
39. $3x + 5y - 9z = 26$
    $4x + 7y + 2z = -7$
    $6x - 9y - 8z = 3$

Graph each equation. *(Sections 2.2 and 5.4)*

40. $y = 3x - 2$
41. $y = -2x + 6$
42. $y = \frac{1}{3}x + 4$
43. $y = x^2 - 6x + 4$
44. $y = -3x^2 - 12x - 8$
45. $y = 2x^2 - 5x - 7$

Apply⇔Assess

Exercise Notes

Cooperative Learning
Exs. 19–30 Students may find it helpful to work on these exercises with a partner.

Using Technology
Exs. 19–30 Students may wish to explore how they can check their answers to these exercises using a graphing calculator or the *Function Investigator* software. See the Using Technology note on page 399 and discuss with students how the idea in this note can be adapted to check these answers.

Assessment Note
Ex. 32 Discuss this exercise with the class to bring out the idea that there are infinitely many pairs of polynomials $p(x)$ and $d(x)$ that satisfy the conditions of the problem.

Practice 55 for Section 9.2

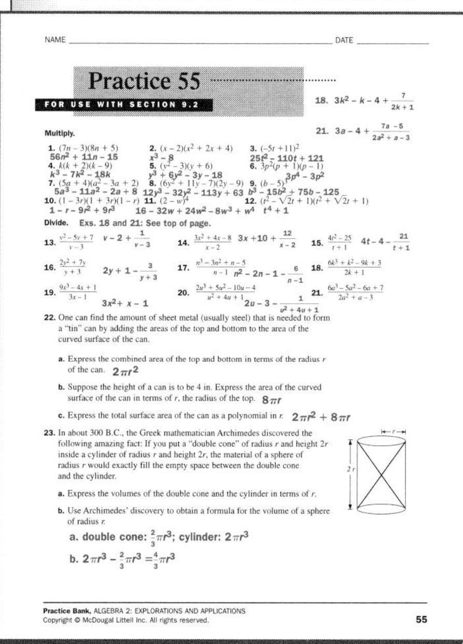

b. $m - n$; It is less than n; the term of highest degree in the quotient would be $\frac{ax^m}{bx^n} = \frac{a}{b}x^{m-n}$ and the degree of the remainder is always less than the degree of the divisor.

32. Answers may vary. An example is given. Let $p(x) = 4x^3 - 5x^2 + 7x - 6$ and $d(x) = x - 1$; I chose the binomial $x - 1$ to be $d(x)$ and multiplied it by $4x^2 - x + 6$ to get $p(x)$.

33. 47
34. −33
35. 47
36. −8
37. (3, 1)
38. $\left(-\frac{1}{5}, \frac{1}{2}\right)$
39. (−2, 1, −3)
40.
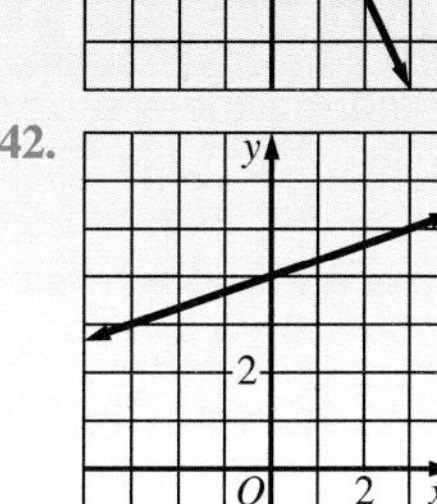

41.
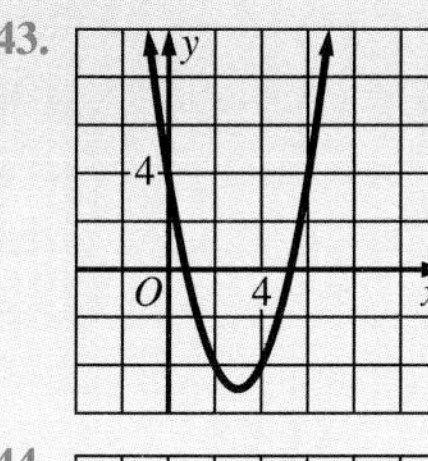

42.
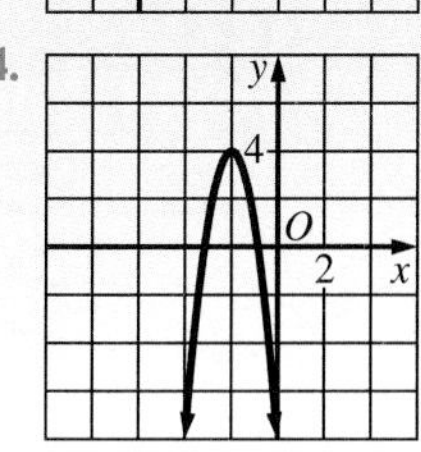

43.

44.

45.
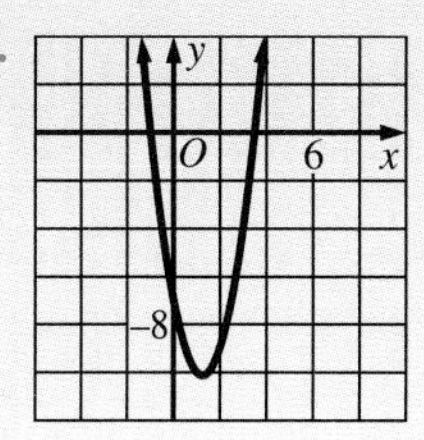

Plan⇔Support

Objectives

- Recognize graphs of polynomial functions and describe their important features.
- Use polynomial functions to solve real-world problems.

Recommended Pacing

Core and Extended Courses
Section 9.3: 2 days

Block Schedule
Section 9.3: 2 half-blocks (with Sections 9.2 and 9.4)

Resource Materials

Lesson Support
Lesson Plan 9.3
Warm-Up Transparency 9.3
Practice Bank: Practice 56
Study Guide: Section 9.3
Explorations Lab Manual: Diagram Master 1
Assessment Book: Test 34

Technology
Technology Book: Calculator Activity 9
Graphing Calculator
McDougal Littell Mathpack *Function Investigator with Matrix Analyzer Activity Book:* Function Investigator Activities 18, 21, and 22
Internet: http://www.hmco.com

Warm-Up Exercises

For each equation, find the *y*-value for the given *x*-value.

1. $y = 3x^2$, $x = 100$ 30,000
2. $y = -3x^2$, $x = -100$ −30,000
3. $y = 2x^3$, $x = 20$ 16,000
4. $y = 2x^3$, $x = -20$ −16,000
5. $y = 4x^4 + 5x^3 - 17$, $x = -10$ 34,983
6. $y = -3x^5 + 4x^4 - x^2$, $x = 10$ −260,100

Section

9.3 Exploring Graphs of Polynomial Functions

Learn how to...

- recognize graphs of polynomial functions and describe their important features

So you can...

- understand how the volume of air in a person's lungs changes as the person breathes, for example

You already know what the graphs of some polynomial functions look like. For example, the graph of $y = 2x + 5$ is a line, and the graph of $y = 4x^2 - 8x - 3$ is a parabola. You'll look at graphs of higher-degree polynomial functions in the Exploration.

EXPLORATION
COOPERATIVE LEARNING

Looking for Patterns in Graphs

Work with a partner.
You will need: • a graphing calculator or graphing software

SET UP Adjust the viewing window for your calculator or software so that the intervals $-5 \le x \le 5$ and $-20 \le y \le 20$ are shown on the axes. Complete Steps 1 and 2 for each of these functions:

- $y = 4x^2 - 8x - 3$
- $y = x^3 - x^2 - 3x + 1$
- $y = x^4 + 2x^3 - 5x^2 - 7x + 3$
- $y = 2x^5 + 6x^4 - 2x^3 - 14x^2 + 5$
- $y = 3x^6 - 13x^4 + 15x^2 + x - 17$
- $y = x^7 - 8x^5 + 18x^3 - 6x$

1 Describe what happens to the graph as *x* takes on large positive and large negative values.

Example: $y = 2x^2 - 2x - 11$

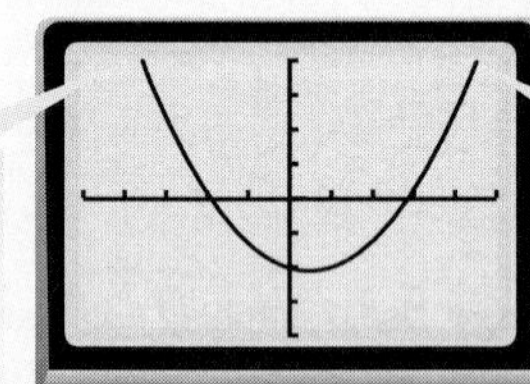

As *x* takes on large negative values, the graph rises.

As *x* takes on large positive values, the graph rises.

2 Find the number of *turning points.* Give the approximate coordinates of each turning point.

Example: $y = 2x^2 - 2x - 11$

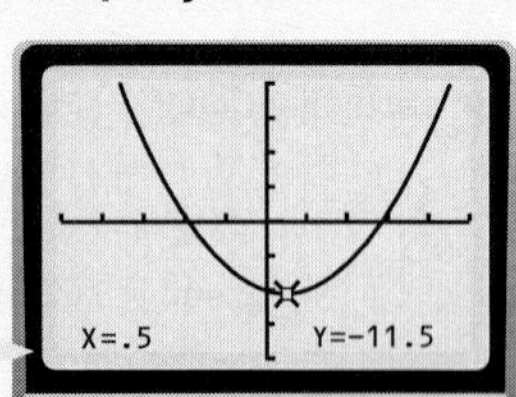

The graph has one turning point. Its coordinates are (0.5, −11.5).

Exploration Note

Purpose
The purpose of this Exploration is to have students discover relationships between a polynomial's degree, its end behavior, and the number of turning points it has.

Materials/Preparation
Each pair of students needs a graphing calculator or graphing software.

Procedure
Suggest that partners sketch each graph they obtain when using technology. The sketches do not need to be exact but should indicate where turning points occur and how the graphs behave for large positive and negative values of *x*. Students should write down their answers to Questions 1–3.

Closure
Call upon different groups to present their answers to Questions 1–3. Students should understand the effects that the degree of a polynomial function has on its graph.

Explorations Lab Manual
See the Manual for more commentary on this Exploration.

Questions

1. How do the graphs of the even-degree polynomial functions behave as x takes on large positive values? as x takes on large negative values?
2. How do the graphs of the odd-degree polynomial functions behave as x takes on large positive values? as x takes on large negative values?
3. a. How is the number of turning points on the graph of each function in the Exploration related to the function's degree?
   b. Decide if the relationship from part (a) is always true by graphing other polynomial functions. If necessary, modify your statement of the relationship so that it applies to *all* polynomial functions.

An important characteristic of a polynomial function f is its *end behavior*. The phrase **end behavior** refers to what happens to $f(x)$ as x takes on large positive and large negative values.

You can describe end behavior with the symbols $+\infty$ and $-\infty$, which stand for "positive infinity" and "negative infinity," respectively. The diagram explains how these symbols are used.

To indicate that x takes on larger and larger negative values, you write "$x \to -\infty$."

To indicate that $f(x)$ takes on larger and larger positive values, you write "$f(x) \to +\infty$."

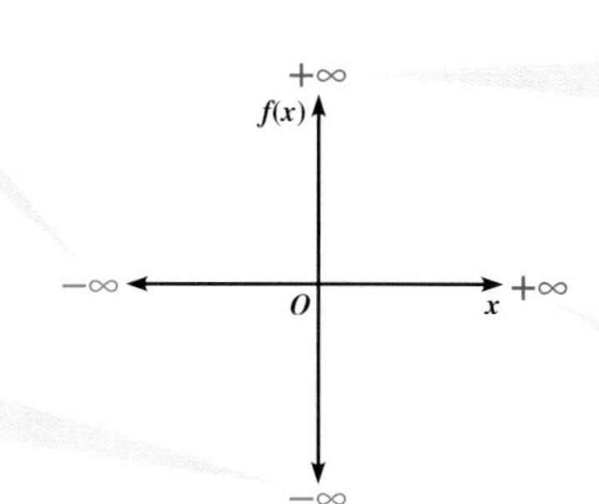

To indicate that $f(x)$ takes on larger and larger negative values, you write "$f(x) \to -\infty$."

To indicate that x takes on larger and larger positive values, you write "$x \to +\infty$."

EXAMPLE 1

Graph $f(x) = 2x^3 + x^2 - 3x + 4$ using a graphing calculator or graphing software. Describe the end behavior of f.

SOLUTION

The graph of f is shown. Use infinity notation to describe the end behavior.

As $x \to -\infty$, $f(x) \to -\infty$. This sentence is read, "As x approaches negative infinity, $f(x)$ approaches negative infinity."

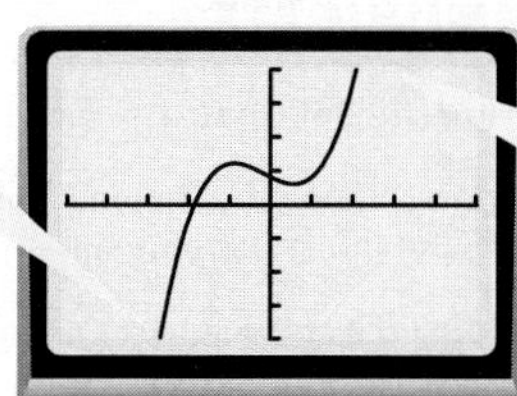

As $x \to +\infty$, $f(x) \to +\infty$.

ANSWERS Section 9.3

Exploration

See answers in back of book.

Teach⇔Interact

Section Note

Mathematical Procedures
Point out that it is necessary to write $x \to +\infty$ and $x \to -\infty$, not $x = +\infty$ or $x = -\infty$ in discussing end behavior. The symbols $+\infty$ and $-\infty$ do not represent actual numbers, thus, the equality symbol is not appropriate.

About Example 1

Teaching Tip
Point out to students that since f is a 3rd degree polynomial, it can have at most 2 turning points, both of which are in the given graph. This should help students realize that the function will not change behavior for values of x beyond the given graph.

Using Technology
The graph shown in the Solution uses Xmin = –5, Xmax = 5 as the settings for the x-axis. You can reinforce students' understanding of the end behavior of f by suggesting that they use their graphing calculators to display a table of values for $f(x)$ when x is a very large value.

Additional Example 1

Graph $f(x) = 0.003x^4 - 99x^3 - 9x^2 - 12x - 124$ using a graphing calculator or graphing software. Describe the end behavior of f.

The graph of f is shown. Use infinity notation to describe the end behavior.

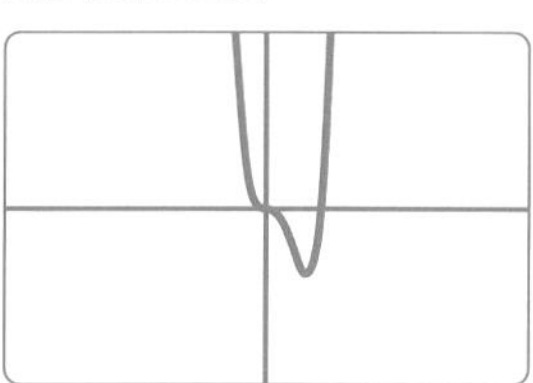

As $x \to -\infty$, $f(x) \to +\infty$.
As $x \to +\infty$, $f(x) \to +\infty$.

The end behavior of a polynomial function depends on the function's degree and the sign of its *leading coefficient*. The **leading coefficient** is the coefficient of the highest power of the variable. For example, the leading coefficient of $f(x) = 2x^3 + x^2 - 3x + 4$ is **2**. The graphs below illustrate the four types of end behavior for nonconstant polynomial functions.

Type 1: As $x \to +\infty$, $f(x) \to +\infty$.
As $x \to -\infty$, $f(x) \to -\infty$.

Example:

$f(x) = \mathbf{4}x^3 - 2x^2 - 5x + 3$

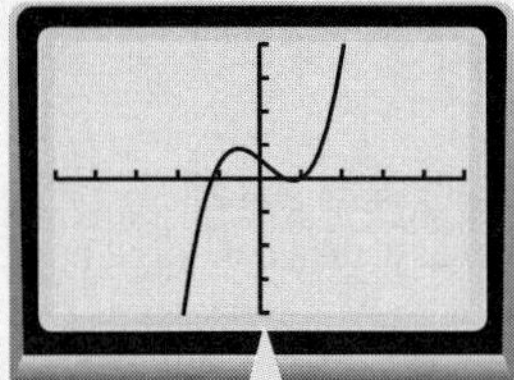

This type applies to functions whose **degree** is **odd** and whose **leading coefficient** is **positive**.

Type 2: As $x \to +\infty$, $f(x) \to -\infty$.
As $x \to -\infty$, $f(x) \to +\infty$.

Example:

$f(x) = \mathbf{-1}x^5 + 8x^3 - 10x$

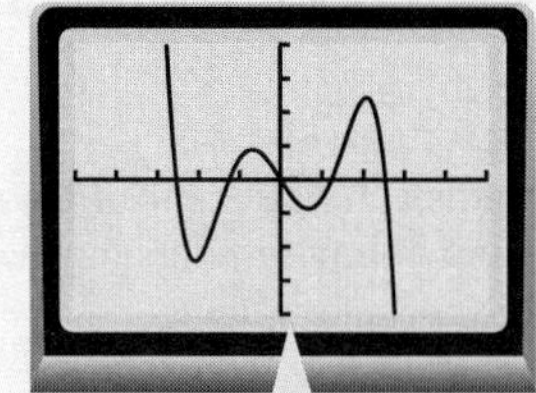

This type applies to functions whose **degree** is **odd** and whose **leading coefficient** is **negative**.

Type 3: As $x \to +\infty$, $f(x) \to +\infty$.
As $x \to -\infty$, $f(x) \to +\infty$.

Example:

$f(x) = \mathbf{1}x^4 + 3x^3 + x^2 - 7x - 10$

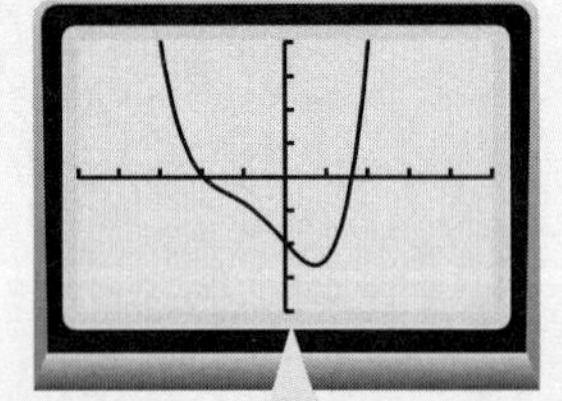

This type applies to functions whose **degree** is **even** and whose **leading coefficient** is **positive**.

Type 4: As $x \to +\infty$, $f(x) \to -\infty$.
As $x \to -\infty$, $f(x) \to -\infty$.

Example:

$f(x) = \mathbf{-2}x^6 + 9x^4 - 9x^2 + x + 7$

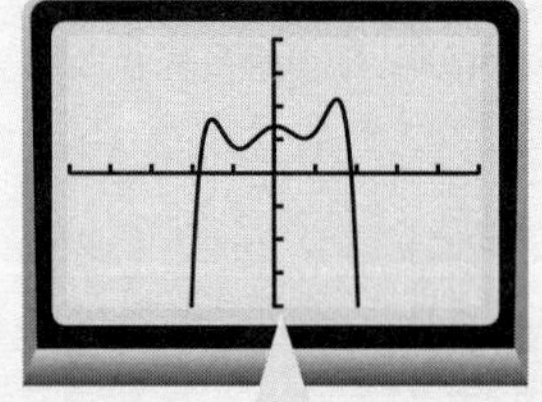

This type applies to functions whose **degree** is **even** and whose **leading coefficient** is **negative**.

THINK AND COMMUNICATE

1. What is the leading coefficient of $f(x) = 7x^3 - x^2 + 2x - 5x^4 + 4$?

2. a. Use infinity notation to describe the end behavior of the function $h(x) = 3x^6 - 6x^4 + 5x - 7$ *without* actually graphing it.

b. **Technology** Graph $y = h(x)$ using a graphing calculator or graphing software. Does the graph support your answer to part (a)?

Teach⇔Interact

Section Notes

Reasoning
Students may think that if they make the coefficient of the term of highest degree very small and use large coefficients of the opposite sign for the coefficients of all other terms, then they can get a polynomial for which the term of highest degree does not dominate the end behavior. Use Additional Example 1 to show this is not true.

Problem Solving
Rather than memorizing the end behavior given for the four different cases on this page, students should be encouraged to mentally substitute a number with the appropriate sign in the leading term of the polynomial to determine the sign of that term. For example, with the function $f(x) = -2x^6 + 9x^4 - 9x^2 + x + 7$, to determine the behavior as $x \to +\infty$, think of substituting a positive number. The sixth power of a positive number is positive, and when this is multiplied by –2, the result will be a negative number. Since this is true for all positive values of x, it will be true for all the numbers as $x \to +\infty$ and, thus, $f(x) \to -\infty$.

Communication: Writing
Students should be encouraged to write polynomials in standard form. This will help them see quickly what the end behavior of the function is. The infinity notation used here is an alternative to the limit notation commonly used. In limit notation, "As $x \to +\infty$, $f(x) \to +\infty$" would be written "$\lim_{x \to +\infty} f(x) = +\infty$." The infinity notation is preferable in this section because the formal concept of limit has not been introduced. Note that ∞ is also written in some texts as an alternative to $+\infty$.

Think and Communicate

For part (b) of question 2, a good graphing window is one that uses Xmin = –3, Xmax = 3, Ymin = –25, Ymax = 50. Another good window might be one that uses Xmin = –5, Xmax = 5, Ymin = –20, Ymax = 20.

Think and Communicate

1. –5
2. a. As $x \to +\infty$, $h(x) \to +\infty$.
As $x \to -\infty$, $h(x) \to +\infty$.

b.

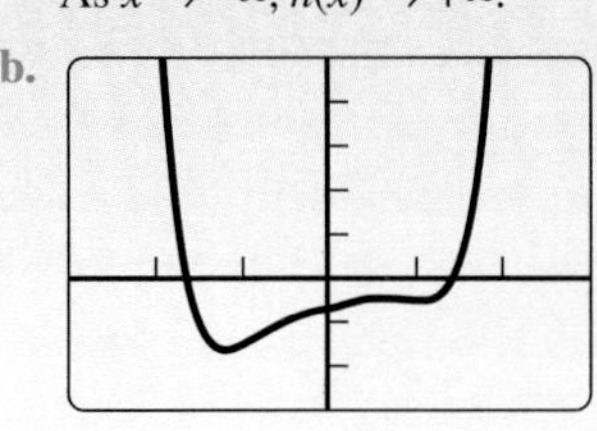

Yes.

A **turning point** of the graph of a function is a point that is higher or lower than all nearby points. The graph of $g(x) = -2x^3 - 3x^2 + 12x + 5$, shown below, has two turning points. In general, the graph of a polynomial function of degree n has at most $n - 1$ turning points.

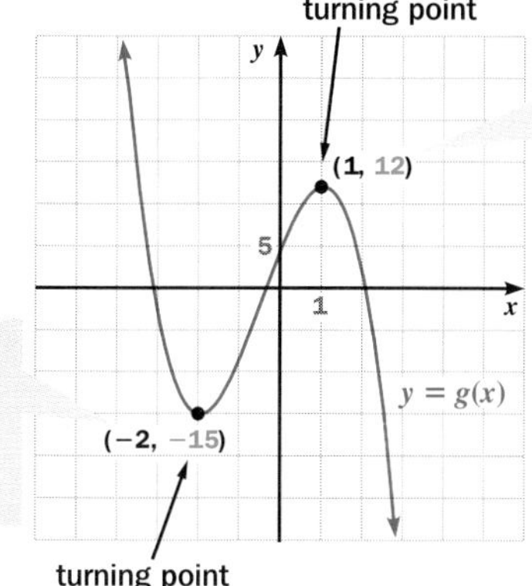

The *y*-coordinate of a turning point *higher* than all nearby points is a **local maximum**.

The *y*-coordinate of a turning point *lower* than all nearby points is a **local minimum**.

EXAMPLE 2 Application: Physiology

During a normal five-second respiratory cycle in which a person inhales and then exhales, the volume V of air brought into the person's lungs can be modeled by the function

$$V = 0.027t^3 - 0.27t^2 + 0.675t$$

where volume is measured in liters and t is the number of seconds after the person starts inhaling. When is this volume at a maximum? What is the maximum volume?

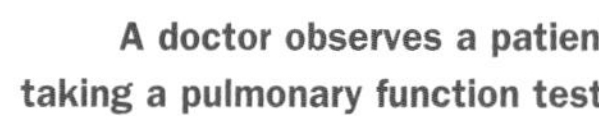

A doctor observes a patient taking a pulmonary function test.

SOLUTION

Use a graphing calculator or graphing software to graph the function $y = 0.027x^3 - 0.27x^2 + 0.675x$. Adjust the viewing window so that the interval $0 \leq x \leq 5$ is shown on the x-axis.

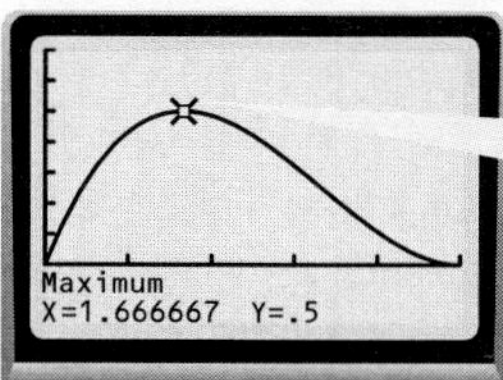

The volume reaches a **maximum** at this turning point. The coordinates of the point are about (1.7, 0.5).

The volume of air brought into the lungs is at a maximum after about 1.7 s. The maximum volume is about 0.5 L.

Teach⇔Interact

About Example 2

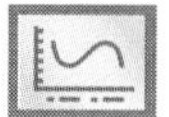

Using Technology
Students using TI-82 graphing calculators can quickly identify turning points by selecting maximum or minimum from the CALCULATE menu. Students using TI-81 graphing calculators can use the zoom-and-trace approach.

Additional Example 2

A bird watcher estimated the size of a bird population on a small island during the first ten years after measures were introduced to protect the species from hunters. Her data can be modeled by the function $P = 0.68x^4 - 20.7x^3 + 226x^2 - 949x + 2388$, where x is the number of years after the measures were introduced and P is the total estimated population. According to this function, what was the minimum population before the protection measures were introduced? When was the population at a minimum?

Use a graphing calculator or graphing software to graph the function $y = 0.68x^4 - 20.7x^3 + 226x^2 - 949x + 2388$. Adjust the viewing window so that the interval $0 \leq x \leq 10$ is shown on the x-axis and the interval $-1000 \leq y \leq 4000$ is shown on the y-axis.

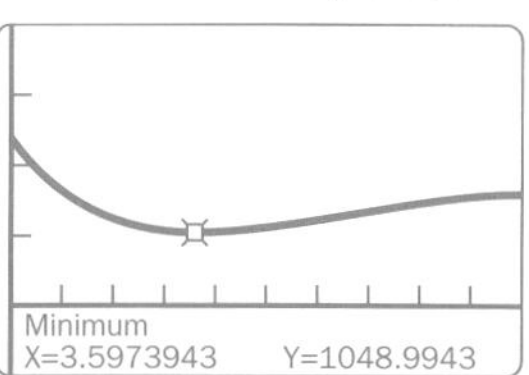

The population reached a minimum at the turning point located at about (3.6, 1050). The population reached a minimum of about 1050. The minimum occurred about 3.6 years after the protection measures were introduced.

Teach⇔Interact

Checking Key Concepts

Reasoning
After discussing questions 1–4, you may wish to give students the graph of a polynomial function and ask them some questions about it. Include some turning points and tell students to assume that all turning points are shown. Ask whether the graph is the graph of a polynomial of even or odd degree. Then have students tell whether the leading coefficient is positive or negative. Be sure students can explain their answers.

Closure Question

What can you tell about the end behavior of a polynomial function $f(x)$ if you know only that it has even degree and that the leading coefficient is negative? Explain.
$f(x) \to -\infty$ as $x \to -\infty$, and $f(x) \to -\infty$ as $x \to +\infty$. If the degree is even and equal to n, then for very large negative values of x, x^n will be a very large positive number. Multiplying the value of x^n by the leading coefficient will give a very large negative number. So, $f(x) \to -\infty$ as $x \to -\infty$. For large positive values of x, x^n will be a very large positive number. When you multiply x^n by the leading coefficient, you get a large negative number. So, $f(x) \to -\infty$ as $x \to +\infty$.

Apply⇔Assess

Suggested Assignment

Core Course
Day 1 Exs. 1–12
Day 2 Exs. 18–25, AYP

Extended Course
Day 1 Exs. 1–15
Day 2 Exs. 16–25, AYP

Block Schedule
Day 56 Exs. 1–12
Day 57 Exs. 18–25, AYP

THINK AND COMMUNICATE

3. In Example 2, the volume given by $V = 0.027t^3 - 0.27t^2 + 0.675t$ does not include the "residual volume" of air present in the lungs even after a person exhales. This volume is about 2.5 L.

a. Write an equation for the total volume of air in the lungs during a five-second respiratory cycle.

b. Find the maximum total volume of air in the lungs *without* graphing the function from part (a). Explain how you obtained your answer.

CHECKING KEY CONCEPTS

For Questions 1–4, use what you know about end behavior to match each polynomial function with its graph.

1. $f(x) = 3x^4 - 10x^2 - x + 3$ **2.** $f(x) = -2x^3 + 3x^2 + 2x + 3$

3. $f(x) = x^5 - 6x^3 + 5x + 3$ **4.** $f(x) = -4x^6 + 7x^5 + 8x^2 - 8x + 3$

A.
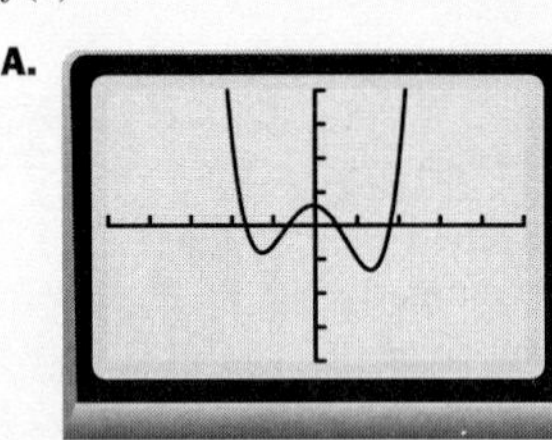

B.
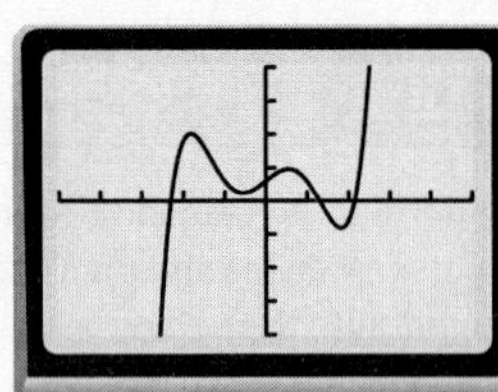

C.
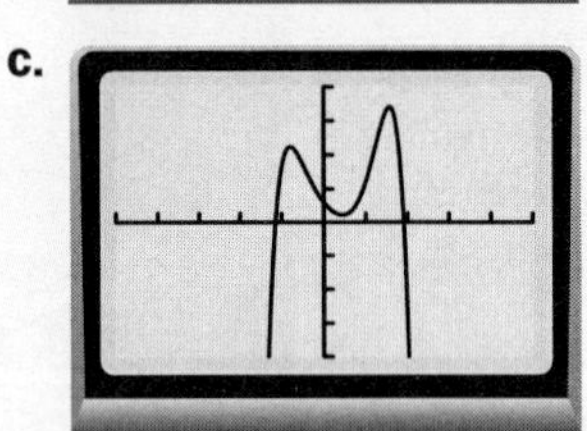

D.
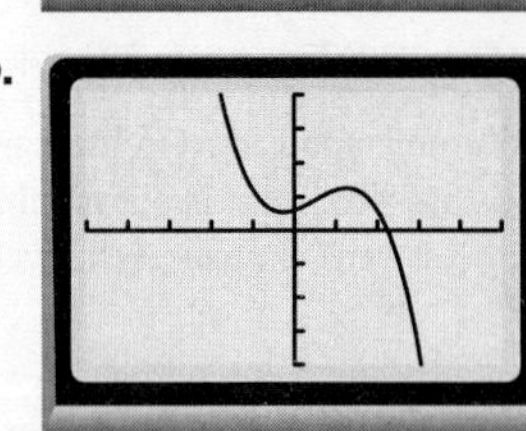

5–8. **Technology** Use a graphing calculator or graphing software to graph each function in Questions 1–4. Find all local maximums and local minimums.

9.3 Exercises and Applications

Extra Practice *exercises on page 762*

1. Writing In your own words, explain what is meant by the "end behavior" of a polynomial function. How is the end behavior related to the function's degree and leading coefficient?

2. Look back at the graph of $g(x) = -2x^3 - 3x^2 + 12x + 5$ near the top of the previous page. Why does it make sense to call 12 a "local maximum," but not a "maximum," of the function g?

Think and Communicate

3. a. $V = 0.027t^3 - 0.27t^2 + 0.675t + 2.5$

b. 3 L; This is the maximum amount inhaled plus the residual air in the lung.

Checking Key Concepts

1–8. See answers in back of book.

Exercises and Applications

Answers that involve approximations may vary due to rounding. Examples are given.

1. Answers may vary. An example is given. The end behavior of a function f is what happens to $f(x)$ as x takes on very large negative values and very large positive values. If the degree of f is odd and the leading coefficient is positive, then as $x \to +\infty, f(x) \to +\infty$ and as $x \to -\infty, f(x) \to -\infty$. If the degree of f is odd and the leading coefficient is negative, then as $x \to +\infty, f(x) \to -\infty$ and as $x \to -\infty, f(x) \to +\infty$. If the degree of f is even (and positive) and the leading coefficient is positive, then as $x \to +\infty, f(x) \to +\infty$ and as $x \to -\infty, f(x) \to +\infty$. If the degree of f is even (and positive) and the leading coefficient is negative, then as $x \to +\infty, f(x) \to -\infty$ and as $x \to -\infty, f(x) \to -\infty$.

Technology **Use a graphing calculator or graphing software to graph each polynomial function. For each function:**

a. Describe the end behavior using infinity notation.

b. Find all local maximums and local minimums.

3. $f(x) = 3x^2 + 6x - 4$

4. $f(x) = -5x^2 + 19x - 8$

5. $f(x) = -x^3 + x^2 - 5x + 2$

6. $f(x) = 2x^3 - 7x$

7. $g(x) = 4x^5 + 7x^4 - 9x^3 - 15x^2$

8. $g(x) = x^4 + x^3 - 6x^2 - x + 4$

9. $h(x) = -2x^4 + 5$

10. $h(x) = -2x^5 + 3x^4 + x^2 + 6$

11. $f(x) = x^6 - 7x^4 + 8x^2 + 7$

12. $f(x) = -5x^6 + x^5 + 21x^4 - 3x^3 - 22x^2 + 9$

Connection SWIMMING

The graph shows how the speed of a swimmer doing the breaststroke changes over the course of one complete stroke. An equation for the graph is:

$$s = -241.0t^7 + 1062t^6 - 1871t^5 + 1647t^4 - 737.4t^3 + 143.9t^2 - 2.432t$$

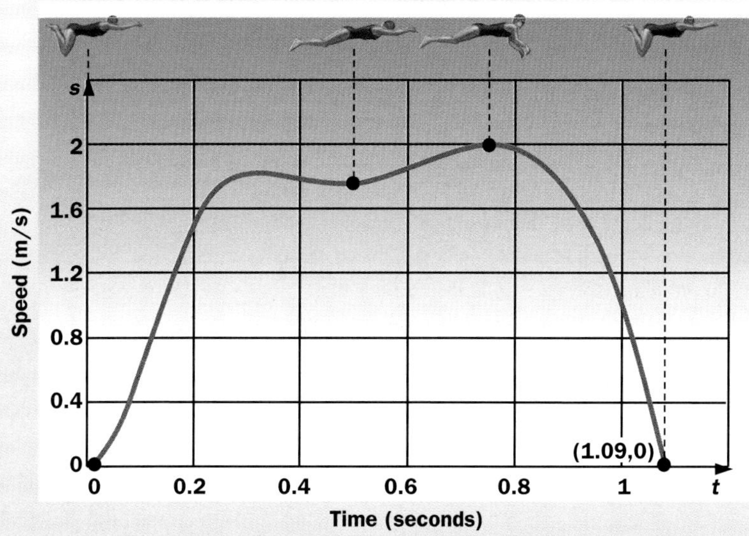

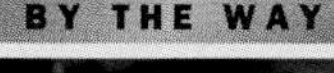

Swim meets are held in both long-course pools, which are 50 m in length, and short-course pools, which are 25 m in length. Dai Guohong of China holds the women's short-course world records in the 100 m and 200 m breaststroke.

13. a. Use the graph. Estimate the coordinates of all turning points. For each turning point, tell whether a *local maximum* or *local minimum* occurs at the point.

b. **Technology** Use a graphing calculator or graphing software to graph the given equation and check your answers to part (a). At what time is the swimmer's speed greatest? What is the greatest speed?

14. Visual Thinking Extend the graph shown so that it gives the speed of the swimmer over the course of three complete strokes.

15. Open-ended Problem What do you think is the average speed of the swimmer during one complete stroke? Explain your reasoning.

Apply⇔Assess

Exercise Notes

Communication: Writing

Ex. 2 Some mathematicians use the terms *absolute maximum* and *local maximum.* After discussing this exercise, ask students how they could write an inequality to explain what it means to say that *M* is the absolute maximum of the function *f*. (For all real numbers *x*, $f(x) \le M$.)

Using Technology

Exs. 3–12 Students can use graphing calculators or the *Function Investigator* software for these exercises. Students using a TI-81 or TI-82 calculator can use the zoom-and-trace approach. With the TI-82, it is more efficient to use minimum and maximum from the CALCULATE menu.

Application

Exs. 13–15 Polynomial functions can be used to model many real-world phenomena. In Exs. 13–15, students will work with a 7th degree polynomial that models the speed of a swimmer doing the breaststroke. You might ask a student who has had some swimming experience to explain the components of the breaststoke to the class. This should help all students understand the physical interpretation of the graph.

Topic Spiraling: Preview

Ex. 14 The graph created in this exercise is an example of a periodic graph. Students will see additional periodic graphs and functions in Chapter 15.

2. The y-coordinate is greater than those of other nearby points, but not of all other points.

3.

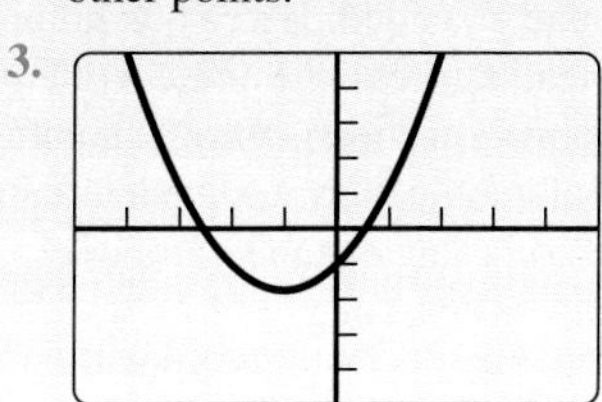

a. As $x \to +\infty, f(x) \to +\infty$ and as $x \to -\infty, f(x) \to +\infty$.

b. local minimum: –7

4.

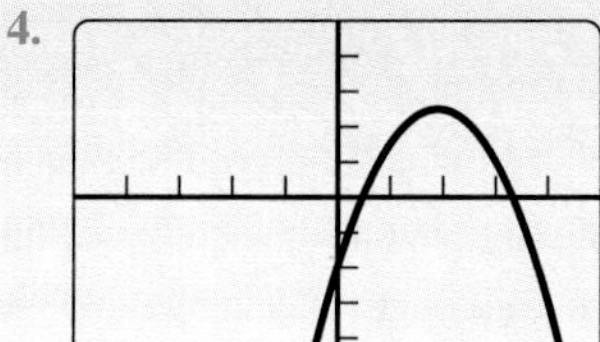

a. As $x \to +\infty, f(x) \to -\infty$ and as $x \to -\infty, f(x) \to -\infty$.

b. local maximum: 10.05

5.

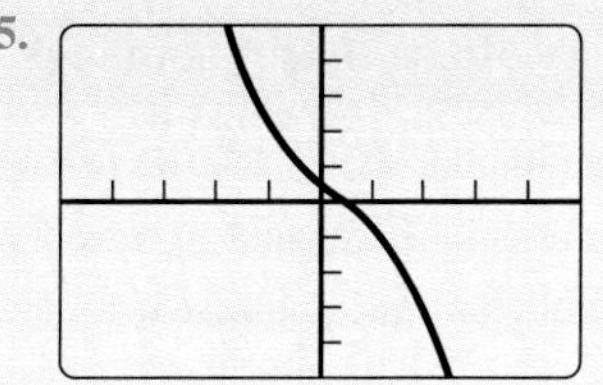

a. As $x \to +\infty, f(x) \to -\infty$ and as $x \to -\infty, f(x) \to +\infty$.

b. no local maximum, no local minimum

6.

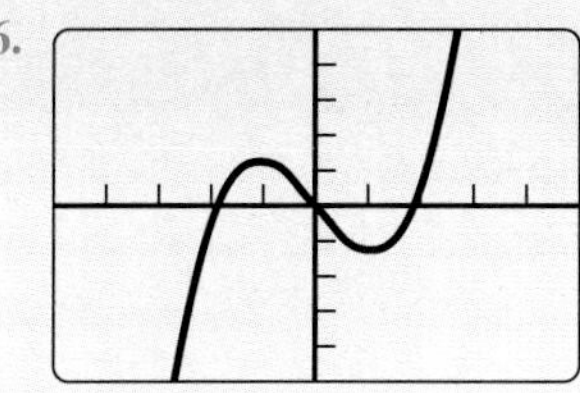

a. As $x \to +\infty, f(x) \to +\infty$ and as $x \to -\infty, f(x) \to -\infty$.

b. local maximum: 5.04; local minimum: –5.04

7–15. See answers in back of book.

Apply⇔Assess

Exercise Notes

Visual Thinking

Ex. 16 To find the radius that maximizes the volume, students will need to locate the turning points of their graphs. Encourage them to discuss the meaning of the turning points in relation to the diagram of the neesquttow. Have them explain in their own words what is happening in regard to the neesquttow at these points. This activity involves the visual skills of *identification* and *communication.*

Multicultural Note

Ex. 16 The Wampanoag were one of the first Native American groups encountered by the Pilgrims after they landed at Plymouth in 1620 and were in large part responsible for the early survival of the colony. Massasoit, the Wampanoag chief at that time, was kind and generous to the settlers and promised that as long as he was alive, there would be peace between the Wampanoag and the Pilgrims.

Integrating the Strands

Exs. 16, 17 By applying the concepts of this section to situations involving the surface area of cylinders, these exercises integrate algebraic and geometric concepts.

Reasoning

Ex. 18 Students should be able to support their answers for part (b) with plausible arguments.

Assessment Note

Ex. 19 This exercise provides a good opportunity for students to demonstrate an understanding of all the major ideas of this section. Note that in order for the polynomial to have turning points, it should not be possible to express it in the form $f(x) = (x - a)^5$.

Connection ARCHITECTURE

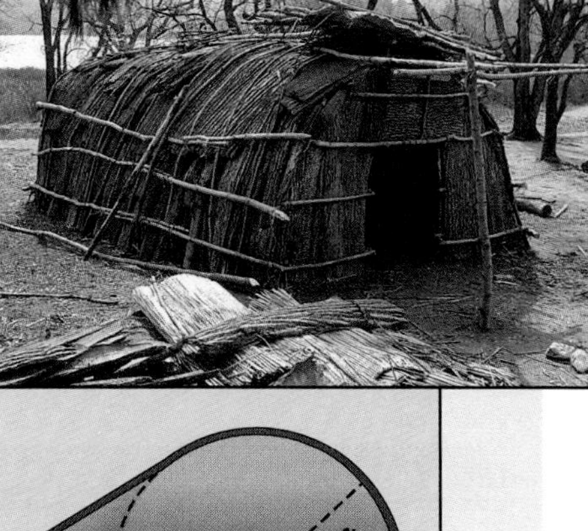

The photograph shows a reconstruction of a dwelling called a *neesquttow* once built by the Wampanoag, a Native American people. A *neesquttow* was constructed with poles covered by strips of bark.

16. Challenge Suppose 600 ft^2 of bark is to be used to make a *neesquttow* in the shape of a half-cylinder. You can find the dimensions of the roomiest *neesquttow* that can be built with this amount of bark.

a. Use the diagram. Write an equation giving the surface area S of the *neesquttow* in terms of its radius r and length l. Assume the *neesquttow* has a rectangular opening at one end as shown. (*Hint:* The surface area of a cylinder is given by the formula $S = 2\pi r^2 + 2\pi rl$.)

4 ft
5 ft
r ft
l ft

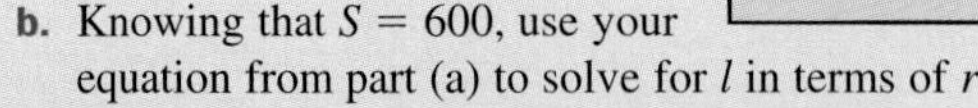

b. Knowing that $S = 600$, use your equation from part (a) to solve for l in terms of r.

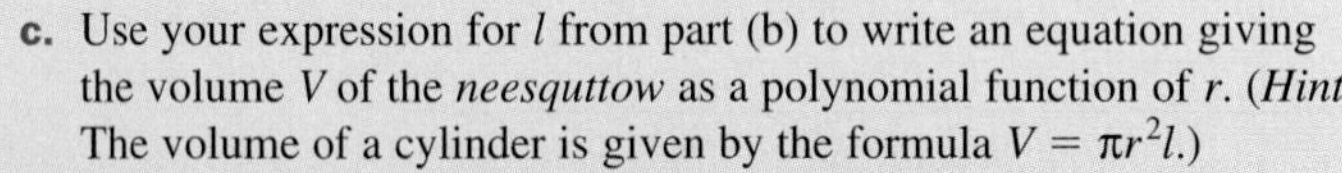

c. Use your expression for l from part (b) to write an equation giving the volume V of the *neesquttow* as a polynomial function of r. (*Hint:* The volume of a cylinder is given by the formula $V = \pi r^2 l$.)

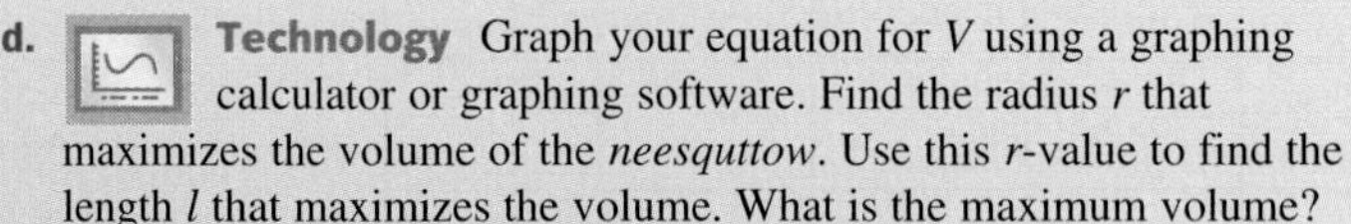

d. **Technology** Graph your equation for V using a graphing calculator or graphing software. Find the radius r that maximizes the volume of the *neesquttow*. Use this r-value to find the length l that maximizes the volume. What is the maximum volume?

17. Writing Is the "best" *neesquttow* that can be built with a given amount of bark necessarily the one having maximum volume? What factors besides volume do you think the Wampanoag considered when building a *neesquttow*?

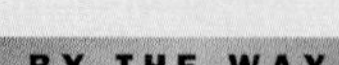

BY THE WAY

The Wampanoag live in what is now Rhode Island and southern Massachusetts. They shared the first Thanksgiving feast with the Pilgrims in 1621.

18. a. Find the domain and range of each function whose graph is shown on page 406.

b. In general, what can you say about the domain and range of an odd-degree polynomial function? of an even-degree polynomial function?

ONGOING ASSESSMENT

19. Open-ended Problem Write a polynomial function f of degree 5 such that $f(x) \to -\infty$ as $x \to +\infty$. Use a graphing calculator or graphing software to find the coordinates of each turning point on the function's graph.

16. a. $S = \pi r^2 + \pi rl - 20$

b. $l = \dfrac{620 - \pi r^2}{\pi r}$ c. $V = 310r - \dfrac{1}{2}\pi r^3$

d. about 8.11 ft; about 16.2 ft; about 1676 ft^3

17. Answers may vary. An example is given. I think the "best" *neesquttow* may depend on many factors. For example, depending on how the building is to be used, it may be better to maximize the floor area instead of the radius (and consequently, the height) of the building.

18. a. (1) all real numbers; all real numbers
(2) all real numbers; all real numbers
(3) all real numbers; $f(x) \geq -13.14$
(4) all real numbers; $f(x) \leq 11.14$

b. The domain and range of an odd-degree polynomial function are the real numbers; the domain of an even-degree polynomial function is the real numbers, but the range of the function is some interval of the real numbers (or a single number if the polynomial is a nonzero constant).

19. Answers may vary. An example is given. $f(x) = -2x^5 + 6x^3$; turning points at $(-1.34, -5.80)$ and about $(1.34, 5.80)$

20. $y = 1 + 2(x - 2)$ or $y = 3 + 2(x - 3)$

SPIRAL REVIEW

Write a point-slope equation of the line passing through the given points. *(Section 2.3)*

20. (2, 1) and (3, 3) **21.** (−5, −3) and (4, 0) **22.** (6, 6) and (8, −2)

Solve each equation. *(Sections 5.5 and 5.6)*

23. $x^2 - 7x + 12 = 0$ **24.** $6x^2 + x - 15 = 0$ **25.** $2x^2 + 3x - 1 = 0$

ASSESS YOUR PROGRESS

VOCABULARY

polynomial (p. 390)	**like terms** (p. 392)
degree (p. 390)	**end behavior** (p. 405)
constant term (p. 390)	**leading coefficient** (p. 406)
standard form (p. 390)	**turning point** (p. 407)
synthetic substitution (p. 391)	**local maximum/minimum** (p. 407)

1. Write the polynomial $-13x + 8x^4 + 9 + x^3 - 2x^5$ in standard form and state its degree. *(Section 9.1)*
2. Use synthetic substitution to evaluate $3t^4 - 4t^2 + t + 7$ when $t = 2$. *(Section 9.1)*
3. Add $2x^3 - 5x^2 - 6x + 3$ and $x^3 + 2x^2 + 6x - 10$. *(Section 9.1)*
4. Subtract $3y^4 - 5y^3 + y + 8$ from $7y^4 + 4y^3 - 8y^2 - 1$. *(Section 9.1)*
5. Multiply $4u + 7$ and $2u^2 - u + 5$. *(Section 9.2)*
6. Divide $4x^3 - x^2 + 8$ by $x + 2$. *(Section 9.2)*
7. Describe the end behavior of $f(x) = -7x^4 + 3x^3 - 5x + 1$ using infinity notation. *(Section 9.3)*
8. **PACKAGING** A manufacturer of packaging materials makes open-top boxes by cutting an x in. by x in. square from each corner of an 18 in. by 24 in. piece of cardboard as shown. *(Section 9.3)*

   a. Write an equation giving the volume V of the box as a function of x.

   b. **Technology** Use a graphing calculator or graphing software to graph the equation from part (a). Find the value of x that maximizes the volume of the box. What is the maximum volume?

9. **Journal** Choose two of the four basic operations—addition, subtraction, multiplication, and division—and explain how to perform the operations on polynomials.

Apply⇔Assess

Assess Your Progress

Journal Entry

In their responses to Ex. 9, students should use examples to illustrate their general statements about how the operations are performed. To provide for a more meaningful response from students, you may wish to suggest that they choose either addition or subtraction and then either multiplication or division. You may also wish to extend the exercise by having students explain all four operations. This will provide an excellent check on their understanding of these procedures.

Progress Check 9.1–9.3

See page 456.

Practice 56 for Section 9.3

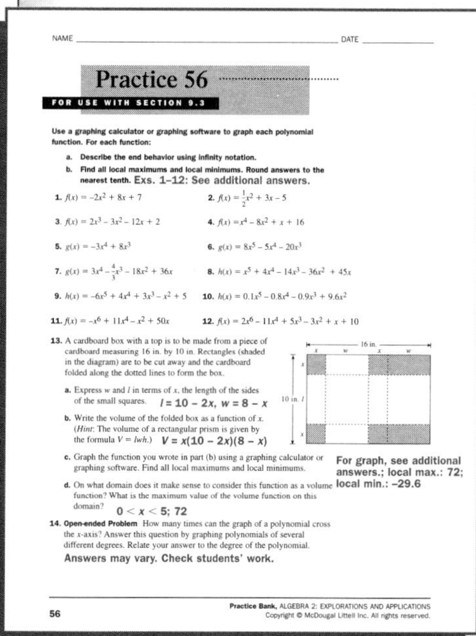

NAME ______________________ DATE __________

Practice 56

FOR USE WITH SECTION 9.3

Use a graphing calculator or graphing software to graph each polynomial function. For each function:

a. **Describe the end behavior using infinity notation.**

b. **Find all local maximums and local minimums. Round answers to the nearest tenth.** **Exs. 1–12: See additional answers.**

1. $f(x) = -2x^2 + 8x + 7$ **2.** $f(x) = \frac{1}{2}x^2 + 3x - 5$

3. $f(x) = 2x^3 - 3x^2 - 12x + 2$ **4.** $f(x) = x^4 - 8x^2 + x + 16$

5. $g(x) = -3x^4 + 8x^3$ **6.** $g(x) = 8x^5 - 5x^4 - 20x^3$

7. $g(x) = 3x^4 - \frac{4}{3}x^3 - 18x^2 + 36x$ **8.** $h(x) = x^5 + 4x^4 - 14x^3 - 36x^2 + 45x$

9. $h(x) = -6x^5 + 4x^4 + 3x^3 - x^2 + 5$ **10.** $h(x) = 0.1x^5 - 0.8x^4 - 0.9x^3 + 9.6x^2$

11. $f(x) = -x^6 + 11x^4 - x^2 + 50x$ **12.** $f(x) = 2x^6 - 11x^4 + 5x^3 - 3x^2 + x + 10$

13. A cardboard box with a top is to be made from a piece of cardboard measuring 16 in. by 10 in. Rectangles (shaded in the diagram) are to be cut away and the cardboard folded along the dotted lines to form the box.

a. Express w and l in terms of x, the length of the sides of the small squares. **$l = 10 - 2x$, $w = 8 - x$**

b. Write the volume of the folded box as a function of x. (*Hint:* The volume of a rectangular prism is given by the formula $V = lwh$.) **$V = x(10 - 2x)(8 - x)$**

c. Graph the function you wrote in part (b) using a graphing calculator or graphing software. Find all local maximums and local minimums. **For graph, see additional answers.; local max.: 72; local min.: –29.6**

d. On what domain does it make sense to consider this function as a volume function? What is the maximum value of the volume function on this domain? **$0 < x < 5$; 72**

14. Open-ended Problem How many times can the graph of a polynomial cross the x-axis? Answer this question by graphing polynomials of several different degrees. Relate your answer to the degree of the polynomial. **Answers may vary. Check students' work.**

56 **Practice Bank,** ALGEBRA 2: EXPLORATIONS AND APPLICATIONS

21. $y = -3 + \frac{1}{3}(x + 5)$ or $y = \frac{1}{3}(x - 4)$
22. $y = -4(x - 6) + 6$ or $y = -2 - 4(x - 8)$
23. 3; 4
24. $-\frac{5}{3}; \frac{3}{2}$
25. $-\frac{3}{4} \pm \frac{\sqrt{17}}{4}$

Assess Your Progress

1. $-2x^5 + 8x^4 + x^3 - 13x + 9$; 5
2. 41
3. $3x^3 - 3x^2 - 7$
4. $4y^4 + 9y^3 - 8y^2 - y - 9$
5. $8u^3 + 10u^2 + 13u + 35$
6. $4x^2 - 9x + 18 - \frac{28}{x + 2}$
7. As $x \to +\infty$, $f(x) \to -\infty$ and as $x \to -\infty$, $f(x) \to -\infty$.
8. a. $V = x(24 - 2x)(18 - 2x)$

   b.

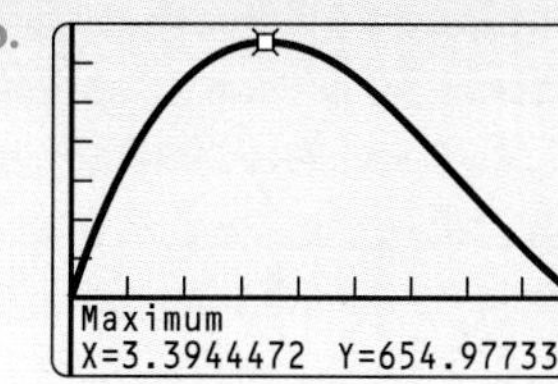

about 3.39 in.; about 655 in.^{3}

9. Answers may vary. Examples are given. To add polynomials, add the coefficients of like terms. To subtract a polynomial, add the opposite of each term of the polynomial. To multiply polynomials, use the distributive property to multiply each term of one polynomial by each term of the other. To divide one polynomial by another, use a process similar to long division in arithmetic, stopping when either the degree of the remainder is less than the degree of the divisor or the remainder is zero.

Plan⇔Support

Objectives

- Solve cubic equations.
- Find equations for graphs of cubic functions.
- Find zeros of cubic functions.
- Solve real-world problems involving cubic functions and equations.

Recommended Pacing

Core and Extended Courses
Section 9.4: 2 days

Block Schedule
Section 9.4: 2 half-blocks (with Sections 9.3 and 9.5)

Resource Materials

Lesson Support
Lesson Plan 9.4
Warm-Up Transparency 9.4
Overhead Visuals:
Folder 9: Graphs of Cubic Functions
Practice Bank: Practice 57
Study Guide: Section 9.4
Explorations Lab Manual:
Diagram Master 1

Technology
Graphing Calculator
Spreadsheet Software
McDougal Littell Mathpack
Function Investigator
Internet:
http://www.hmco.com

Warm-Up Exercises

Use synthetic substitution to find the value of each polynomial function for the given value of x.

1. $f(x) = 5x^3 - x^2 + 4x + 7$, $x = 2$ 51
2. $g(x) = -2x^3 + x - 8$, $x = 5$ −253
3. $h(x) = -4x^3 + 17x^2 + 80x - 21$, $x = -3$ 0

Find exact solutions for each quadratic equation.

4. $x^2 + 2x - 35 = 0$ −7, 5
5. $x^2 + 5x - 1 = 0$ $\frac{-5 \pm \sqrt{29}}{2}$
6. $3x^2 + 7x + 2 = 0$ $-2, -\frac{1}{3}$

Section

9.4 Solving Cubic Equations

Learn how to...

- solve cubic equations
- find equations for graphs of cubic functions
- find zeros of cubic functions

So you can...

- analyze the flight of the space shuttle after launch, for example

The space shuttle is powered by three liquid-fuel motors in its tail and two external solid-fuel booster rockets. When the shuttle reaches a speed of about 3000 mi/h, the booster rockets fall off and return to Earth. The liquid-fuel motors continue firing and propel the shuttle into orbit.

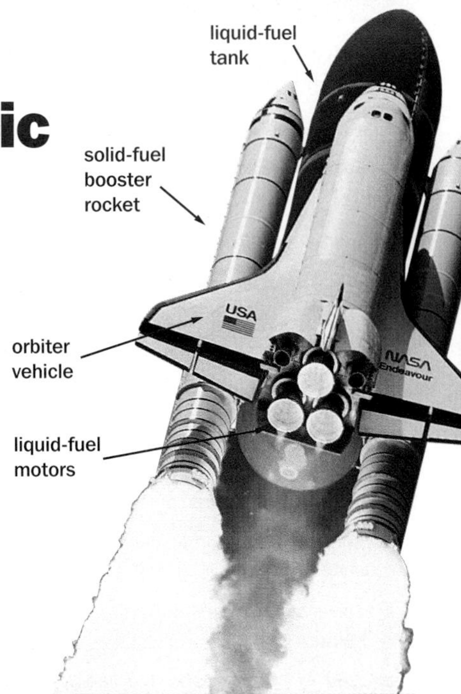

EXAMPLE 1 Application: Space Travel

The speed of the space shuttle t seconds after launch can be approximated using the function

$$s(t) = 0.000559t^3 + 0.0313t^2 + 13.6t$$

where $s(t)$ is measured in miles per hour. How long after launch do the booster rockets fall off?

SOLUTION

The booster rockets fall off when $s(t) = 3000$, so you need to solve this equation:

$$0.000559t^3 + 0.0313t^2 + 13.6t = 3000$$

Method 1 Use a graphing calculator or graphing software.

Graph $y = 0.000559x^3 + 0.0313x^2 + 13.6x$ and $y = 3000$ in the same coordinate plane.

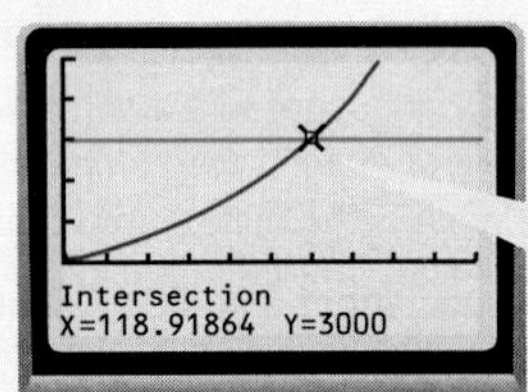

Find the x-coordinate of the point where the graphs intersect. The x-coordinate is about **119**.

The booster rockets fall off about 119 s (or about 2 min) after launch.

Method 2 Use a spreadsheet.

Step 1 Have the spreadsheet calculate the shuttle's speed at ten-second intervals from $t = 50$ to $t = 150$.

	A	B	C	D	E	F	G	H	I	J	K	L
	Shuttle Speeds 1											
1	*t*	50	60	70	80	90	100	110	120	130	140	150
2	*s*(*t*)	828	1049	1297	1575	1885	2232	2619	3049	3525	4051	4631

Since 3000 is between these two numbers, the desired t-value is between **110** and **120**.

Step 2 Have the spreadsheet calculate the shuttle's speed at one-second intervals from $t = 110$ to $t = 120$.

	A	B	C	D	E	F	G	H	I	J	K	L
	Shuttle Speeds 2											
1	*t*	110	111	112	113	114	115	116	117	118	**119**	120
2	*s*(*t*)	2619	2660	2701	2743	2785	2828	2871	2915	2959	3004	3049

Notice that **s(119)** $\approx$ **3000**.

The booster rockets fall off about 119 s (or about 2 min) after launch.

The function $s(t) = 0.000559t^3 + 0.0313t^2 + 13.6t$ in Example 1 is called a *cubic function.* A **cubic function** is a polynomial function of degree 3.

Cubic functions may be written in standard form, $f(x) = ax^3 + bx^2 + cx + d$, or in *intercept form*, $f(x) = a(x - p)(x - q)(x - r)$. The next example shows why the phrase "intercept form" is used.

EXAMPLE 2

Let $f(x) = 2(x - 1)(x - 3)(x - 4)$. Find the x-intercepts of the graph of f.

SOLUTION

To find the x-intercepts, solve the equation $f(x) = 0$.

$$2(x - 1)(x - 3)(x - 4) = 0$$

Divide both sides by 2.

$$(x - 1)(x - 3)(x - 4) = 0$$

$$x - 1 = 0 \quad \text{or} \quad x - 3 = 0 \quad \text{or} \quad x - 4 = 0$$

$$x = 1 \quad \text{or} \quad x = 3 \quad \text{or} \quad x = 4$$

The product of the factors equals 0 if and only if at least one of the factors equals 0.

The x-intercepts are 1, 3, and 4.

If f is a polynomial function, then each solution of the equation $f(x) = 0$ is called a **zero** of f. For cubic functions in the form $f(x) = a(x - p)(x - q)(x - r)$, the zeros are p, q, and r. Each zero is an x-intercept of the graph of f provided the zero is a real number.

Teach⇔Interact

Additional Example 1

A manufacturer stores fuel in spherical tanks with a radius of 30 ft. When the fuel has a depth of d ft, the approximate amount of fuel (in cubic feet) can be found by $V(d) = -1.047198d^3 + 94.24778d^2$. The manufacturer prefers that the amount of fuel not be less than 20,000 ft^3. What must be the depth of the fuel in the tank to meet this requirement? The least amount of fuel there should be is 20,000 ft^3, so you need to solve $-1.047198d^3 + 94.24778d^2 = 20{,}000$.

Method 1 Use a graphing calculator or graphing software. Graph $y = -1.047198x^3 + 94.24778x^2$ and $y = 20{,}000$ in the same coordinate plane. Find the x-coordinate where the graphs intersect.

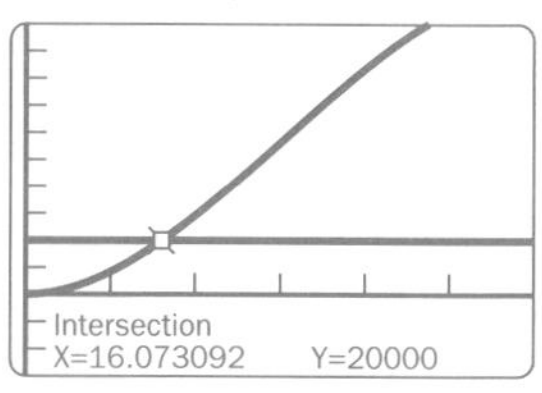

The x-coordinate is about 16.1. The depth should be at no less than 16.1 ft.

Method 2 Use a spreadsheet.

Step 1 Calculate the volume at one-foot intervals. (See spreadsheet below.) Since 20,000 is between 19,838 and 22,093, the desired d value is between 16 and 17.

Step 2 Calculate the volume at intervals of one-tenth of a foot from $d = 16$ to $d = 16.5$. (See spreadsheet below.) Notice that $V(16.1) \approx 20{,}000$. The depth should be about 16.1 ft.

Additional Example 2

Let $f(x) = -5(x - 2)(x - 6)(x - 7)$. Find the x-intercepts of the graph of f. To find the x-intercepts, solve the equation $f(x) = 0$.

$-5(x + 2)(x - 6)(x + 7) = 0$

$(x + 2)(x - 6)(x + 7) = 0$

$x + 2 = 0$ or $x - 6 = 0$ or $x + 7 = 0$

$x = -2$ or $x = 6$ or $x = -7$

The intercepts are −7, −2, and 6.

For Additional Example 1

Step 1

	A	B	C	D	E	F	G
	Fuel Volume						
1	*d*	12	13	14	15	16	17
2	*v*(*d*)	11762	13627	15599	17671	19838	22093

Step 2

	A	B	C	D	E	F	G
	Fuel Volume						
1	*d*	16.0	16.1	16.2	16.3	16.4	16.5
2	*v*(*d*)	19838	20060	20282	20506	20730	20955

Teach⇔Interact

Additional Example 3

Find an equation for the cubic function h whose graph is shown.

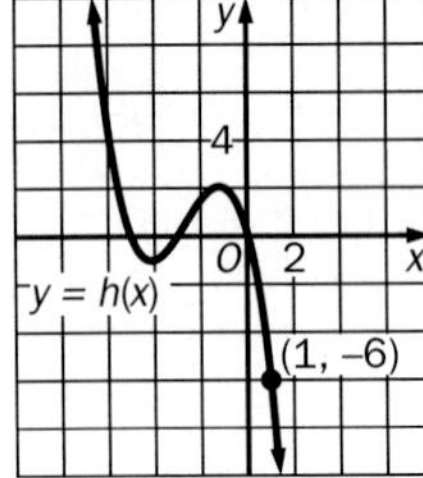

The x-intercepts of the graph of h are -5, -3, and 0. Therefore, an equation for h has the form $h(x) = a(x - (-5))(x - (-3))(x - 0)$ or $h(x) = ax(x + 5)(x + 3)$.
To find the value of a, use the fact that $h(1) = -6$.

$$a \cdot 1(1 + 5)(1 + 3) = -6$$
$$24a = -6$$
$$a = -\frac{1}{4}$$

An equation for h is $h(x) = -\frac{1}{4}x(x + 5)(x + 3)$.

Think and Communicate

When discussing question 3, have students refer to question 2. Ask how the results for these two questions are related logically. (They are converses of one another.) Remind students that when two statements "if p, then q" and "if q, then p" are both true, they can be combined into the single statement "p if and only if q." Relate this idea to the statement of the factor theorem.

Section Note

Reasoning
Point out that the factor theorem and what is known about the end behavior of cubic functions make it possible to deduce a number of things about zeros of cubic polynomials. First, every cubic function has at least one real zero. A cubic function goes from $-\infty$ to the left of $x = 0$ to $+\infty$ to the right of $x = 0$, or else it goes from $+\infty$ to the left of $x = 0$ and to $-\infty$ to the right of $x = 0$. In each case, the function changes sign, and so it must have the value 0 somewhere along the x-axis.

THINK AND COMMUNICATE

1. What are the zeros of $f(x) = (x - 1)(x^2 + 1)$? What are the x-intercepts of the graph of f? Is every zero an x-intercept? Explain.

2. In general, if $x - k$ is a factor of a polynomial $f(x)$, what can you say about $f(k)$?

EXAMPLE 3

Find an equation for the cubic function g whose graph is shown.

SOLUTION

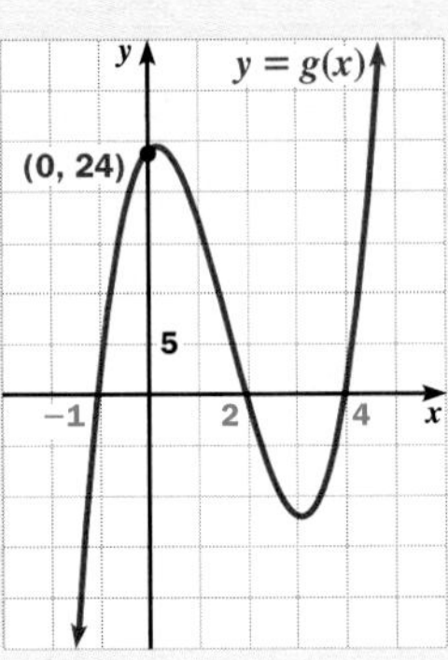

The x-intercepts of the graph of g are $-1, 2,$ and 4. Therefore, an equation for g has the form

$$g(x) = a(x - (-1))(x - 2)(x - 4),$$

or

$$g(x) = a(x + 1)(x - 2)(x - 4).$$

To find the value of a, use the fact that $g(0) = 24$:

$$a(0 + 1)(0 - 2)(0 - 4) = 24$$
$$8a = 24$$
$$a = 3$$

An equation for g is $g(x) = 3(x + 1)(x - 2)(x - 4)$.

THINK AND COMMUNICATE

3. In general, if $f(x)$ is a polynomial and $f(k) = 0$, what can you say about $x - k$?

Examples 2 and 3 illustrate the following result, known as the *factor theorem.*

Factor Theorem

Let $f(x)$ be a polynomial.
Then $x - k$ is a factor of $f(x)$ if and only if $f(k) = 0$.

Suppose you want to find the zeros of $f(x) = ax^3 + bx^2 + cx + d$, a cubic function given in standard form. The factor theorem says that if you can find one zero k, then you can write $f(x)$ in the form

$$f(x) = (x - k) \cdot q(x)$$

where $q(x) = \frac{f(x)}{x - k}$ is a polynomial of degree 2. You can then find the remaining zeros by solving the quadratic equation $q(x) = 0$, since any zero of q is also a zero of f.

ANSWERS Section 9.4

Think and Communicate

1. 1, i, and $-i$; 1; No; i and $-i$, which are not real numbers, are not x-intercepts.
2. $f(k) = 0$
3. $x - k$ is a factor of $f(x)$.

There is a simple test you can use to determine whether a given polynomial function has any integral zeros.

Finding Integral Zeros of Polynomial Functions

Let f be a polynomial function with integral coefficients. Then the only possible integral zeros of f are the divisors of the constant term.

Example: Let $f(x) = 3x^3 + x^2 - 19x + \mathbf{10}$. The possible integral zeros of f are the divisors of **10**: ±1, ±2, ±5, and ±10.

THINK AND COMMUNICATE

4. a. Use the function $f(x) = 3x^3 + x^2 - 19x + 10$. Show that f has only one integral zero by using synthetic substitution to evaluate $f(x)$ at each divisor of the constant term.
   b. Let k = the integral zero you found in part (a). Divide $f(x)$ by $x - k$. Compare the coefficients of the quotient with the numbers in the last row of the synthetic substitution array for k. What do you notice?

EXAMPLE 4

Find the zeros of $f(x) = x^3 + 4x^2 - 5x - 8$.

SOLUTION

Step 1 List the possible integral zeros of f.
The possible integral zeros are the divisors of -8: ±1, ±2, ±4, and ±8.

Step 2 Use synthetic substitution to find one zero from among the divisors.

Try $x = 1$:

1	1	4	−5	−8
		1	5	0
	1	5	0	**−8**

Since $f(1) = -8$, 1 *is not* a zero.

Try $x = -1$:

−1	1	4	−5	−8
		−1	−3	8
	1	**3**	**−8**	**0**

Since $f(-1) = \mathbf{0}$, -1 *is* a zero.

Step 3 Find the quotient polynomial $q(x) = \dfrac{f(x)}{x-(-1)} = \dfrac{f(x)}{x+1}$.

As you saw in *Think and Communicate* Question 4, the coefficients of $q(x)$ are the red numbers in the synthetic substitution array for $x = -1$:

$$q(x) = \mathbf{1}x^2 + \mathbf{3}x - \mathbf{8}$$

Step 4 Find the remaining zeros of f by solving the equation $q(x) = 0$.

$$x^2 + 3x - 8 = 0$$

$$x = \frac{-3 \pm \sqrt{3^2 - 4(1)(-8)}}{2(1)} = \frac{-3 \pm \sqrt{41}}{2}$$

Use the quadratic formula.

The zeros of f are -1, $\dfrac{-3+\sqrt{41}}{2}$, and $\dfrac{-3-\sqrt{41}}{2}$.

Think and Communicate

4. a. $f(1) = -5$; $f(-1) = 27$;
$f(2) = 0$; $f(-2) = 28$;
$f(5) = 315$; $f(-5) = -245$;
$f(10) = 2920$; $f(-10) = -2700$
   b. $3x^2 + 7x - 5$; They are the same.

Teach⇔Interact

Section Note

Communication: Reading
It is important that students be clear on the meaning of the result about integral zeros at the top of this page. The statement does not claim that there *are* any integral zeros, only that *if* there are such zeros, then they divide the constant term. The class should look at examples to see why the integer-coefficients condition is important.

Additional Example 4

Find the zeros of $f(x) = 10x^3 - 33x^2 + 23x + 6$.

Step 1 List the possible integral zeros of f. The possible integral zeros are the divisors of 6: ±1, ±2, ±3, and ±6.

Step 2 Use synthetic substitution to find one zero from among the divisors.

Try $x = 1$.

1	10	−33	23	6
		10	−23	0
	10	−23	0	6

Since $f(1) = 6$, 1 *is not* a zero.

Try $x = -1$.

−1	10	−33	23	6
		−10	43	−66
	10	−43	66	−60

Since $f(-1) = -60$, -1 *is not* a zero.

Try $x = 2$.

2	10	−33	23	6
		20	−26	−6
	10	−13	−3	0

Since $f(2) = 0$, 2 *is* a zero.

Step 3 Find the quotient polynomial $q(x) = \dfrac{f(x)}{x-2}$. Use the first three numbers in the last row of the synthetic substitution array for $x = 2$: $q(x) = 10x^2 - 13x - 3$.

Step 4 Find the remaining zeros of f by solving the equation $q(x) = 0$.

$$10x^2 - 13x - 3 = 0$$
$$(2x - 3)(5x + 1) = 0$$
$$2x - 3 = 0 \text{ or } 5x + 1 = 0$$
$$x = \frac{3}{2} \text{ or } x = -\frac{1}{5}$$

The zeros of f are $-\frac{1}{5}$, $\frac{3}{2}$, and 2.

Teach⇔Interact

Closure Question

Suppose $f(x)$ is a cubic polynomial with integral coefficients and you know that there is exactly one integral zero of $f(x)$. Describe how you would identify that zero and use it to find any other zeros that the function might have.

List the factors of the constant term. Check the factors one at a time until you find the factor that makes the value of the function zero. Suppose the factor is k. Divide $f(x)$ by $x - k$. Then find all zeros of the quotient polynomial $q(x)$, which is a quadratic polynomial. The zeros of $q(x)$, together with k, are all the zeros of $f(x)$.

Apply⇔Assess

Suggested Assignment

Core Course
Day 1 Exs. 1–9, 13–20
Day 2 Exs. 21–26, 29–40

Extended Course
Day 1 Exs. 1–20
Day 2 Exs. 21–40

Block Schedule
Day 57 Exs. 1–9, 13–20
Day 58 Exs. 21–26, 29–40

Exercise Notes

Reasoning

Exs. 7–9 After discussing these exercises, ask students how many cubic functions have a given set of 3 different real numbers as zeros. Ask them to also explain their thinking. (infinitely many; If the three zeros are a, b, and c, then $f(x) = (x - a)(x - b)(x - c)$ is a cubic function. However, if k is a nonzero real number, then $g(x) = k(x - a)(x - b)(x - c)$ also has a, b, and c as zeros.) You might also ask them how varying the value of k affects the graph.

THINK AND COMMUNICATE

5. Use a calculator to verify that $\frac{-3 + \sqrt{41}}{2}$ and $\frac{-3 - \sqrt{41}}{2}$ are zeros of the function $f(x) = x^3 + 4x^2 - 5x - 8$ in Example 4.

6. Write the function f in Example 4 in intercept form.

7. Can you use the procedure described in Example 4 to find the zeros of $f(x) = x^3 + x^2 - 9x + 2$? Why or why not? How can you approximate the zeros using a graphing calculator or graphing software? using a spreadsheet?

CHECKING KEY CONCEPTS

Technology Use technology to approximate all real solutions of each equation to the nearest hundredth.

1. $x^3 + x^2 + 5x = 4$

2. $-2.28x^3 + 9.14x^2 - 5.07 = 0$

Find an equation for the cubic function whose graph has the given x-intercepts and passes through the given point.

3. x-intercepts: 1, 2, 3; (0, −12)

4. x-intercepts: −2, −1, 2; (1, 18)

Find the zeros of each function.

5. $f(x) = (x - 2)(x - 4)(x - 5)$

6. $f(x) = 4(x + 9)(x + 3)(x - 3)$

7. $g(x) = x^3 - 7x^2 + 4x + 12$

8. $h(x) = 3x^3 + 6x^2 - 22x + 8$

9.4 Exercises and Applications

Extra Practice exercises on page 762

Technology Use technology to approximate all real solutions of each equation to the nearest hundredth.

1. $x^3 - 2x^2 + x = 12$

2. $-x^3 + x^2 + x = 10$

3. $2x^3 + 3x^2 - 7x - 3 = 0$

4. $3x^3 - 5x^2 - 66x + 120 = 0$

5. $-0.5x^3 + 2.1x^2 + 1.3x = 4$

6. $0.45x^3 - 2.07x^2 + 0.12x + 3.84 = 0$

Find an equation for each cubic function whose graph is shown.

7.

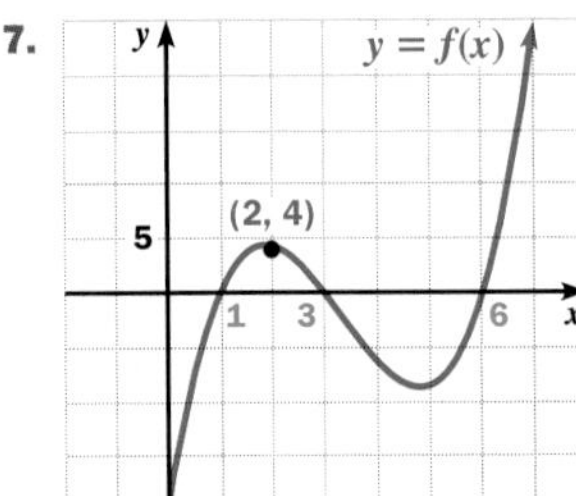

8.

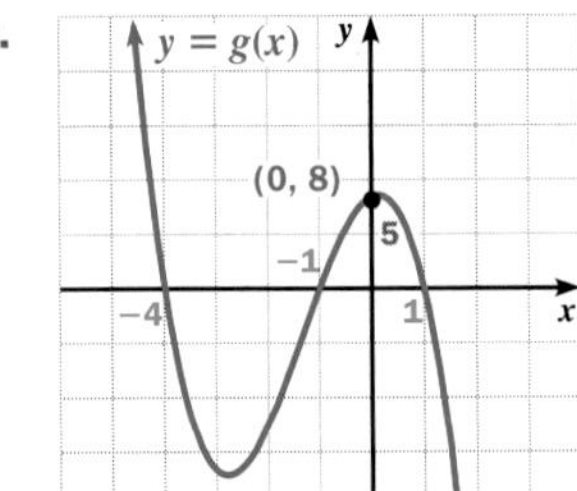

9.

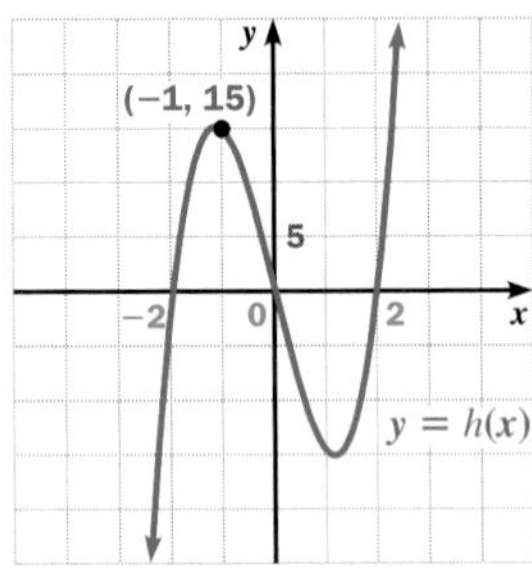

Think and Communicate

5. The calculator will show that $\frac{-3 + \sqrt{41}}{2}$ and $\frac{-3 - \sqrt{41}}{2}$ are zeros of the function.

6. $f(x) = (x + 1)\left(x + \frac{3 - \sqrt{41}}{2}\right)\left(x + \frac{3 + \sqrt{41}}{2}\right)$

7. No; $f(x)$ has no integral zeros. Use a graphing calculator to graph the function and use the trace or root function to find the zeros, or graph $y = 0$ on the same screen and use the intersect function to find the zeros. To use a spreadsheet, calculate the values of the functions over intervals of the domain, narrowing the intervals until the desired level of accuracy is achieved.

Checking Key Concepts

Answers that involve approximations may vary due to rounding. Examples are given.

1. 0.66
2. −0.69; 0.84; 3.86
3. $f(x) = 2(x - 1)(x - 2)(x - 3)$
4. $f(x) = -3(x + 2)(x + 1)(x - 2)$
5. 2; 4; 5
6. −9; −3; 3
7. −1; 2; 6
8. $-4; \frac{3 + \sqrt{3}}{3}; \frac{3 - \sqrt{3}}{3}$

Connection BOATING

Unlike cars, most motorboats do not have speedometers that tell how fast they are moving. Instead, the speed of a motorboat can be estimated using the boat's *tachometer* and a graph called a *speed curve*. A tachometer shows the propeller speed in revolutions per minute (rev/min). A speed curve gives the boat's speed as a function of the propeller speed.

10. **Technology** A speed curve for a certain motorboat traveling in still water is shown. An equation of the curve is:

$$b = 0.00547p^3 - 0.224p^2 + 3.60p - 11.0$$

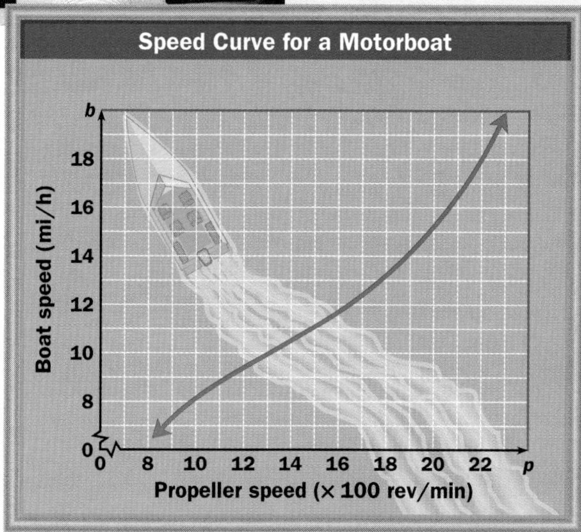

a. Suppose the boat's tachometer reads 1400 rev/min. How fast is the boat traveling?

b. At what tachometer reading does the boat's speed reach 15 mi/h?

c. What minimum tachometer reading should be maintained in order to cross a lake 2 mi wide in at most 10 min?

11. Suppose the motorboat in Exercise 10 is traveling against a current of 3 mi/h. Sketch a new speed curve for the boat, and find an equation of this curve. How is the new speed curve geometrically related to the speed curve shown?

12. **Open-ended Problem** Describe how you would go about creating a speed curve for a motorboat.

Find the zeros of each function.

13. $f(x) = 2(x - 3)(x + 10)(x - 7)$

14. $f(x) = (3x + 1)(2x - 5)(x + 8)$

15. $g(x) = x^3 - 4x^2 - x + 4$

16. $g(x) = x^3 + 8x^2 + 17x + 10$

17. $h(x) = x^3 + 8x^2 - 10x - 9$

18. $h(x) = -x^3 + 6x^2 - 2x$

19. $f(x) = 2x^3 - 16x^2 + 19x + 30$

20. $f(x) = 3x^3 + 17x^2 + 16x - 24$

21. **EDUCATION** For the years from 1985 to 1991, the number of women (in thousands) enrolled in graduate school in the United States can be modeled by the function

$$W = 1.97t^3 - 14.1t^2 + 50.4t + 701$$

where t is the number of years since 1985. Use technology to estimate the year when the number of female graduate students reached 803 thousand.

BY THE WAY

The first college in the United States to admit women was Oberlin Collegiate Institute in Oberlin, Ohio. The college opened in 1833 with 44 students, 15 of whom were women.

Write each function in intercept form.

22. $f(x) = (2x + 4)(x - 5)(x - 6)$

23. $g(x) = (2x + 3)(3x + 4)(4x + 5)$

24. $h(x) = x^3 - 3x^2 - 4x + 12$

25. $f(x) = x^3 + 5x^2 - 18x - 28$

Exercises and Applications

Answers that involve approximations may vary due to rounding. Examples are given.

1. 3
2. −2
3. −2.62; −0.38; 1.50
4. −4.75; 1.85; 4.57
5. −1.44; 1.27; 4.38
6. −1.19; 1.79; 4.00
7. $f(x) = (x - 1)(x - 3)(x - 6)$
8. $f(x) = -2(x + 4)(x + 1)(x - 1)$
9. $f(x) = 5x(x + 2)(x - 2)$
10. a. about 10.5 mi/h
    b. about 1987 rev/min
    c. about 1652 rev/min
11, 12. See answers in back of book.
13. −10; 3; 7
14. $-8; -\frac{1}{3}; \frac{5}{2}$
15. −1; 1; 4
16. −5; −2; −1
17. $-9; \frac{1-\sqrt{5}}{2}; \frac{1+\sqrt{5}}{2}$
18. $0; 3-\sqrt{7}; 3+\sqrt{7}$
19. $6; \frac{2-\sqrt{14}}{2}; \frac{2+\sqrt{14}}{2}$
20. $-3; \frac{-4-2\sqrt{10}}{3}; \frac{-4+2\sqrt{10}}{3}$
21. 1989
22. $f(x) = 2(x + 2)(x - 5)(x - 6)$
23. $g(x) = 24\left(x + \frac{3}{2}\right)\left(x + \frac{4}{3}\right)\left(x + \frac{5}{4}\right)$
24. $h(x) = (x + 2)(x - 2)(x - 3)$
25. $f(x) = (x + 7)(x - 1 - \sqrt{5})(x - 1 + \sqrt{5})$

Apply⇔Assess

Exercise Notes

Challenge

Exs. 7–9 Ask students how many points (not necessarily intercepts) it is necessary to know on the graph of a cubic function before you can find the function. If you know that number of points, how can you find the function? (four points; Substitute the x- and y-coordinates of the four points into the equation $y = ax^3 + bx^2 + cx + d$ to obtain a system of four linear equations in four unknowns. Solve the system by using matrices to find a, b, c, and d.)

Cooperative Learning

Exs. 13–20 Since students may be prone to making computational errors, you may want them to work in pairs. Have one person work a part of the solution, while the other checks the computations. Have them switch roles frequently. Being able to discuss these problems as they work will help students solidify their understanding of the methods for finding the zeros of cubic equations.

Problem Solving

Exs. 13–20 Several of these exercises can be solved in different ways. Students should realize that Exs. 13 and 14 can be most easily done using the method of Example 2. The method of Example 4 can be used for all the others. One way to make this method more efficient is for students to use graphing technology to help them guess the integral root and then use synthetic substitution to confirm their guess and find the polynomial $q(x)$. You may want to point out that Ex. 15 can be solved by grouping terms and factoring, and that Ex. 18 can be solved by factoring out the common x-term.

Common Error

Exs. 22, 23 Some students may think that these functions are already in intercept form. Refer these students to the definition of intercept form on page 413. Point out that the definition requires that each factor be of the form $x - p$ (with 1 as the coefficient of x).

Apply⇔Assess

Exercise Notes

Historical Connection

Ex. 27 Some students may wish to study the history of attempts by mathematicians to solve the general cubic equation. In particular, they should look for information about the contributions of the sixteenth century mathematicians Niccolo Tartaglia and Girolamo Cardano (also spelled Cardan).

Assessment Note

Ex. 29 This exercise provides a good opportunity to assess students' understanding of the ideas of this section. Unless the integers selected have large absolute values, students can do this exercise without using technology.

Practice 57 for Section 9.4

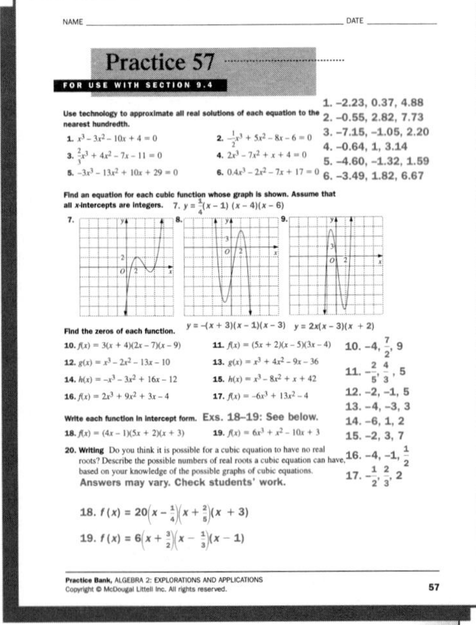

NAME ______ DATE ______

Practice 57

FOR USE WITH SECTION 9.4

Use technology to approximate all real solutions of each equation to the nearest hundredth.

1. $x^3 - 3x^2 - 10x + 4 = 0$
2. $-\frac{1}{2}x^3 + 5x^2 - 8x - 6 = 0$
3. $\frac{2}{3}x^3 + 4x^2 - 7x - 11 = 0$
4. $2x^3 - 7x^2 + x + 4 = 0$
5. $-3x^3 - 13x^2 + 10x + 29 = 0$
6. $0.4x^3 - 2x^2 - 7x + 17 = 0$

1. –2.23, 0.37, 4.88
2. –0.55, 2.82, 7.73
3. –7.15, –1.05, 2.20
4. –0.64, 1, 3.14
5. –4.60, –1.32, 1.59
6. –3.49, 1.82, 6.67

Find an equation for each cubic function whose graph is shown. Assume that all x-intercepts are integers.

7. $y = \frac{1}{4}(x - 1)(x - 4)(x - 6)$

8. $y = -(x + 3)(x - 1)(x - 3)$

9. $y = 2x(x - 3)(x + 2)$

Find the zeros of each function.

10. $f(x) = 3(x + 4)(2x - 7)(x - 9)$
11. $f(x) = (5x + 2)(x - 5)(3x - 4)$
12. $g(x) = x^3 - 2x^2 - 13x - 10$
13. $g(x) = x^3 + 4x^2 - 9x - 36$
14. $h(x) = -x^3 - 3x^2 + 16x - 12$
15. $h(x) = x^3 - 8x^2 + x + 42$
16. $f(x) = 2x^3 + 9x^2 + 3x - 4$
17. $f(x) = -6x^3 + 13x^2 - 4$

10. $-4, \frac{7}{2}, 9$
11. $-\frac{2}{5}, \frac{4}{3}, 5$
12. –2, –1, 5
13. –4, –3, 3
14. –6, 1, 2
15. –2, 3, 7
16. $-4, -1, \frac{1}{2}$
17. $-\frac{1}{2}, \frac{2}{3}, 2$

Write each function in intercept form. Exs. 18–19: See below.

18. $f(x) = (4x - 1)(5x + 2)(x + 3)$
19. $f(x) = 6x^3 + x^2 - 10x + 3$

20. **Writing** Do you think it is possible for a cubic equation to have no real roots? Describe the possible numbers of real roots a cubic equation can have, based on your knowledge of the possible graphs of cubic equations.
Answers may vary. Check students' work.

18. $f(x) = 20\left(x - \frac{1}{4}\right)\left(x + \frac{2}{5}\right)(x + 3)$

19. $f(x) = 6\left(x + \frac{3}{2}\right)\left(x - \frac{1}{3}\right)(x - 1)$

Practice Bank, ALGEBRA 2: EXPLORATIONS AND APPLICATIONS

57

26. **SAT/ACT Preview** Suppose $x - 2$ is a factor of $f(x) = 3x^3 - x^2 + cx - 8$. What is the value of c?

    A. -18 **B.** -6 **C.** 0 **D.** 3 **E.** 4

27. **HISTORY** Bhaskara was an Indian mathematician who lived during the twelfth century A.D. In his book *Siddhanta Siromani* (*The Gem of Mathematics*), he solved the cubic equation $x^3 - 6x^2 = -12x + 35$ using these steps:

$$x^3 - 6x^2 + 12x - 8 = 27$$

Add $12x - 8$ to both sides so that each side becomes a perfect cube.

$$(x - 2)^3 = 3^3$$

Write each side as the cube of a number or an expression.

$$x - 2 = 3$$

Take the cube root of both sides.

$$x = 5$$

BY THE WAY

Bhaskara was the first mathematician to recognize that a fraction like $\frac{3}{0}$ whose denominator is zero is not defined in the real number system.

   a. Use Bhaskara's method to solve the equation $x^3 + 3x^2 = -3x + 7$.

   b. **Writing** Can Bhaskara's method be used to solve *any* cubic equation? If not, what types of equations are solvable with his method?

   c. **Research** Find a book on the history of mathematics. Discuss some of the contributions made by mathematicians of various cultures to solving polynomial equations.

28. Let $f(x) = ax^3 + bx^2 + cx + d$ where a, b, c, and d are integers.

   a. **Challenge** Prove that if k is an integral zero of f, then k is a divisor of the constant term, d. (*Hint:* Start with the equation $f(k) = 0$. Solve this equation for d, and show that d can be written as the product of k and another integer.)

   b. Prove or disprove the converse of the statement in part (a): If k is a divisor of d, then k is a zero of f.

ONGOING ASSESSMENT

29. **Cooperative Learning** Work with a partner. One person should choose three integers and write a cubic function f in standard form whose zeros are the chosen integers. The other person should use the methods from this section to "discover" the zeros of f. Reverse roles with your partner and repeat the activity.

SPIRAL REVIEW

Describe the end behavior of each function using infinity notation. *(Section 9.3)*

30. $f(x) = 2x^3$ 31. $g(x) = -x^4 + 2x^3$ 32. $h(x) = -8x^7 - x^6 + 2x^5 - 14x^2$

Simplify. *(Section 3.2)*

33. 5^{-2} 34. 2^{-5} 35. $9^{1/2}$ 36. $32^{4/5}$

Solve using the quadratic formula. Check your solutions. *(Section 8.4)*

37. $x^2 + 16 = 0$ 38. $x^2 - 2x + 2 = 0$ 39. $x^2 = 4x - 13$ 40. $3x^2 + 6x = -7$

26. B

27. a. $x^3 + 3x^2 = -3x + 7$; $x^3 + 3x^2 + 3x + 1 = 8$; $(x + 1)^3 = 8$; $x + 1 = 2$; $x = 1$

   b. No; only when the coefficients of the powers of x match the coefficients of the same powers of x in a perfect cube; that is, only when the cube can be completed.

   c. Answers may vary. Possible topics include the works of Ahmes, Diophantus, Al-Khwarizmi, and Omar Khayyam.

28. a. If k is an integral zero of f, then $f(k) = 0$ and $0 = ak^3 + bk^2 + ck + d$. So $ak^3 + bk^2 + ck = -d$, $-k(ak^2 + bk + c) = d$. Since a, b, c, and k are integers, $ak^2 + bk + c$ is an integer and k is a divisor of d.

   b. The converse is not true. For example, consider $f(x) = x^2 + 24$. The whole number 2 is a divisor of 24, but is not a zero of f. (Also consider that a quadratic function has at most two real zeros. If the converse were true, every quadratic function would have as many real zeros as its constant term has factors.)

29. Answers may vary. Check students' work.

30. As $x \to +\infty$, $f(x) \to +\infty$ and as $x \to -\infty$, $f(x) \to -\infty$.

31. As $x \to +\infty$, $g(x) \to -\infty$ and as $x \to -\infty$, $g(x) \to -\infty$.

32. As $x \to +\infty$, $h(x) \to -\infty$ and as $x \to -\infty$, $h(x) \to +\infty$.

33. $\frac{1}{25}$ 34. $\frac{1}{32}$

35. 3 36. 16

37. $\pm 4i$ 38. $1 \pm i$

39. $2 \pm 3i$ 40. $\frac{-3 \pm 2i\sqrt{3}}{3}$

Section

9.5 Finding Zeros of Polynomial Functions

Learn how to...

- find zeros of higher-degree polynomial functions

So you can...

- solve problems about beam deflection, for example

You can find the number of real zeros of a polynomial function by counting the x-intercepts of the function's graph. As you'll see in the Exploration, the number of real zeros is related to the function's degree.

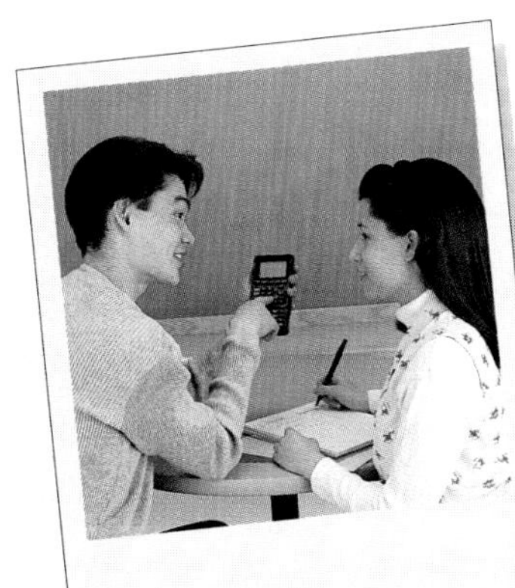

EXPLORATION
COOPERATIVE LEARNING

Counting Real Zeros

Work with a partner.
You will need:

- a graphing calculator or graphing software

SET UP Adjust the viewing window for your calculator or software so that the intervals $-5 \le x \le 5$ and $-20 \le y \le 20$ are shown on the axes.

1 Graph the following functions. Tell how many real zeros each function has.

- $y = x^3 + x^2 - 7x - 5$
- $y = x^4 - 10x^2 + 9$
- $y = x^5 - 7x^3 - x^2 + 8x$
- $y = x^6 - 11x^4 + 29x^2 - x - 8$

◀ **Example: $y = x^3 - 2x^2 - 5x + 6$**

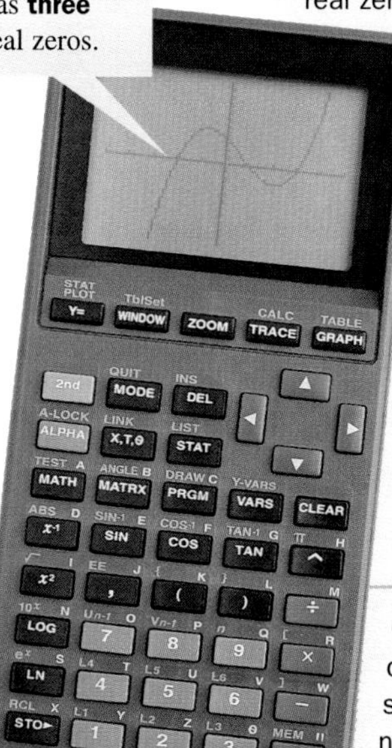

This function has **three** real zeros.

2 For each function in Step 1, how is the number of real zeros related to the function's degree?

3 Repeat Step 1 using these functions:

- $y = x^3 + x^2 + 3x + 5$
- $y = x^4 + 2x^3 - x^2 + 6x + 7$
- $y = x^5 - x^4 - 6x^3 + 4x^2 + 8x$
- $y = x^6 - 3x^4 - 7x^2 + 10$

Is the relationship you found in Step 2 true for these functions? Explain.

4 Graph several other polynomial functions having different degrees. If necessary, modify your answer for Step 2 so that the stated relationship between a polynomial function's degree and its number of real zeros is *always* true.

Exploration Note

Purpose
The purpose of this Exploration is to have students discover that the degree of a polynomial function indicates the maximum number of real zeros the function can have.

Materials/Preparation
Each pair of students will need a graphing calculator or graphing software.

Procedure
Students graph four functions and observe how the number of real zeros is related to the function's degree. They then check their observations using four other functions. After graphing several more functions, they state the relationship they think is true.

Closure
Call on several partners to describe their results and share their conjectures. Students should understand that the degree of a polynomial function sets an upper limit on the possible number of real zeros. It is not necessary to say what the values of the zeros actually are, only *how many* there are.

Explorations Lab Manual
See the Manual for more commentary on this Exploration.

For answers to the Exploration, see answers in back of book.

Plan⇔Support

Objectives

- Find zeros of higher-degree polynomial functions.
- Solve problems using higher degree polynomial functions.

Recommended Pacing

Core and Extended Courses
Section 9.5: 1 day

Block Schedule
Section 9.5: $\frac{1}{2}$ block
(with Section 9.4)

Resource Materials

Lesson Support
Lesson Plan 9.5
Warm-Up Transparency 9.5
Overhead Visuals:
Folder 9: Graphs of Conic Sections
Practice Bank: Practice 58
Study Guide: Section 9.5
Assessment Book: Test 35

Technology
Technology Book:
Spreadsheet Activity 9
Graphing Calculator
McDougal Littell Mathpack
Function Investigator
Internet:
http://www.hmco.com

Warm-Up Exercises

Use synthetic substitution to evaluate each polynomial when x has the given value.

1. $2x^3 - x^2 - 23x - 20$, $x = 4$ 0
2. $-3x^4 + 8x^3 + 9x + 1$, $x = 10$ −21,909
3. $16x^3 + 8x^2 - 12x + 7$, $x = -1$ 11

List all of the integer divisors of each number.

4. 54 ±1, ±2, ±3, ±6, ±9, ±18, ±27, ±54
5. 100 ±1, ±2, ±4, ±5, ±10, ±20, ±25, ±50, ±100

Teach⇔Interact

Section Notes

Challenge
Ask students if they can present a logical argument in support of the idea that a polynomial function of degree n can have at most n real zeros. (Use the idea that $x - k$ is a factor of $f(x)$ if and only if $f(k) = 0$. If there were more than n real zeros, then there would be more than n linear factors of $f(x)$. This would imply that $f(x)$ has degree $n + 1$ or greater, contradicting the assumption that the degree of $f(x)$ is n.)

Teaching Tip
Continue to emphasize that information about zeros provides information about factors and vice versa.

About Example 1

Reasoning
If the coefficients of a polynomial are real numbers, then any imaginary zeros will come in conjugate pairs. This Example should help students understand why this is so.

Additional Example 1

Find the zeros of $f(x) = x^3 - 3x^2 + 3x + 7$.

Find one zero of f by checking the divisors of the constant term, 7. One zero is -1:

$$\begin{array}{r|rrrr} -1 & 1 & -3 & 3 & 7 \\ & & -1 & 4 & -7 \\ \hline & 1 & -4 & 7 & 0 \end{array}$$

The quotient polynomial is $q(x) = \frac{f(x)}{x+1} = 1x^2 - 4x + 7$. Find the remaining zeros of f by solving the equation $q(x) = 0$.

$x^2 - 4x + 7 = 0$

$$x = \frac{-(-4) \pm \sqrt{(-4)^2 - 4(1)(7)}}{2(1)}$$

$$= \frac{4 \pm \sqrt{-12}}{2} = \frac{4 \pm 2i\sqrt{3}}{2} = 2 \pm i\sqrt{3}$$

The zeros of f are -1, $2 + i\sqrt{3}$, and $2 - i\sqrt{3}$.

The following result gives a maximum for the number of real zeros of a polynomial function.

Real Zeros of Polynomial Functions

A polynomial function of degree n has at most n real zeros.

For example, a cubic function may have 1, 2, or 3 real zeros, as shown below.

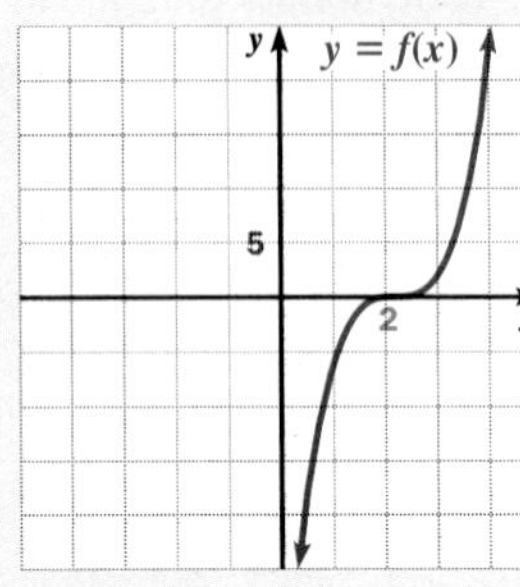

$f(x) = 4(x - 2)^3$
Number of real zeros: 1

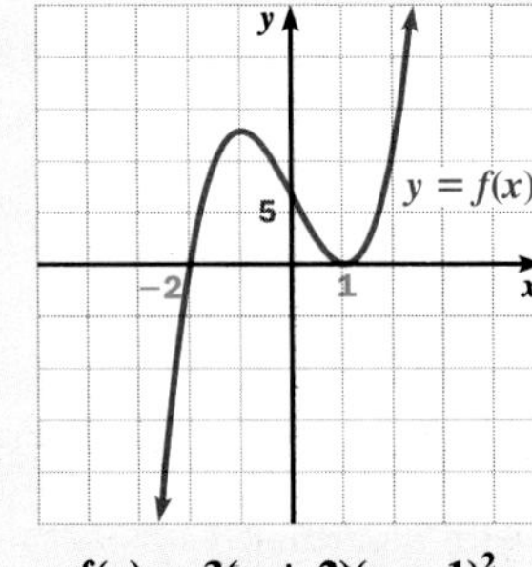

$f(x) = 3(x + 2)(x - 1)^2$
Number of real zeros: 2

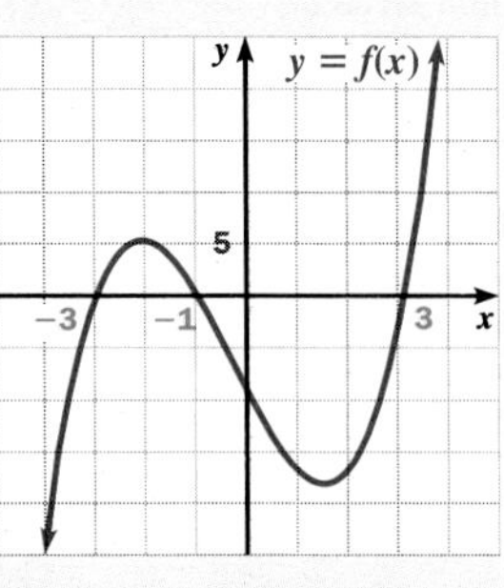

$f(x) = (x + 3)(x + 1)(x - 3)$
Number of real zeros: 3

The second of the three functions, $f(x) = 3(x + 2)(x - 1)^2$, has a *double zero* at $x = 1$. A number k is a **double zero** of a polynomial function f if $(x - k)^2$ is a factor of $f(x)$. Similarly, k is a **triple zero** of f if $(x - k)^3$ is a factor of $f(x)$. The first function, $f(x) = 4(x - 2)^3$, has a triple zero at $x = 2$.

The next example shows that a polynomial function may have imaginary zeros as well as real zeros.

EXAMPLE 1

Find the zeros of $f(x) = x^3 - x^2 - 7x + 15$.

SOLUTION

Find one zero of f by checking the divisors of the constant term, 15. You will find that -3 is a zero:

$$\begin{array}{r|rrrr} -3 & 1 & -1 & -7 & 15 \\ & & -3 & 12 & -15 \\ \hline & 1 & -4 & 5 & 0 \end{array}$$

The quotient polynomial is $q(x) = \frac{f(x)}{x+3} = 1x^2 - 4x + 5$. Find the remaining zeros of f by solving the equation $q(x) = 0$:

$$x^2 - 4x + 5 = 0$$

Use the quadratic formula. $$x = \frac{-(-4) \pm \sqrt{(-4)^2 - 4(1)(5)}}{2(1)} = \frac{4 \pm \sqrt{-4}}{2} = \frac{4 \pm 2i}{2} = 2 \pm i$$

The zeros of f are -3, $2 + i$, and $2 - i$.

ANSWERS Section 9.5

Exploration

1–4. See answers in back of book.

In Example 1, you saw that the cubic function $f(x) = x^3 - x^2 - 7x + 15$ has three zeros in the set of complex numbers. The next theorem, known as the *fundamental theorem of algebra*, extends this result.

Fundamental Theorem of Algebra

A polynomial function of degree n has exactly n complex zeros, provided each double zero is counted as 2 zeros, each triple zero is counted as 3 zeros, and so on.

BY THE WAY

The fundamental theorem of algebra was first proved in 1799 by Carl Friedrich Gauss, one of the greatest mathematicians of all time. Gauss, a child prodigy, is said to have detected an error in his father's bookkeeping when he was only three years old.

THINK AND COMMUNICATE

1. Explain why the fundamental theorem of algebra implies the property stated on the previous page: A polynomial function of degree n has at most n real zeros.

Rational Zeros of Polynomial Functions

In Section 9.4, you learned a procedure for finding the possible integral zeros of a polynomial function. The following more general result lets you find all possible rational zeros (that is, zeros of the form $\frac{p}{q}$ where p and q are integers and $q \neq 0$).

Rational Zeros Theorem

Let f be a polynomial function with integral coefficients. Then the only possible rational zeros of f are $\frac{p}{q}$ where p is a divisor of the constant term of $f(x)$ and q is a divisor of the leading coefficient.

For example, you can use the theorem above to find the possible rational zeros of $f(x) = 4x^3 - 7x^2 + 2x - 3$:

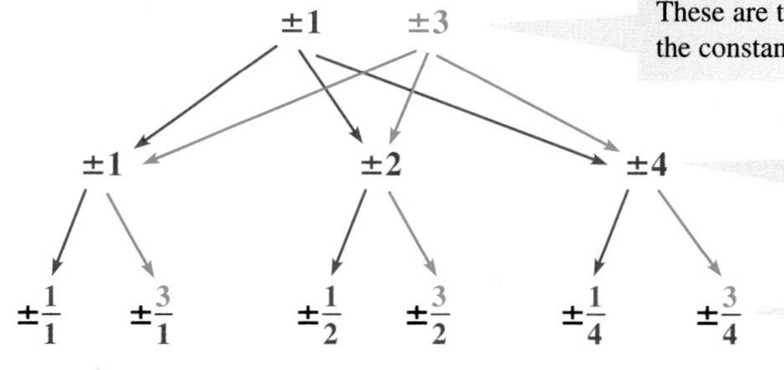

These are the divisors of the constant term, −3.

These are the divisors of the leading coefficient, 4.

These are the possible rational zeros of f.

Note that the possible rational zeros include the possible integral zeros: **±1** and **±3**.

Think and Communicate

1. If a polynomial function has exactly n complex zeros altogether, it has n real zeros if all the zeros are real and fewer than n real zeros if some of the zeros are imaginary.

Teach⇔Interact

Section Note

Historical Connection

Carl Friedrich Gauss (1777–1855) was born in the German city of Brunswick. His elementary school teachers were impressed by his intelligence. Later, he was supported financially by Duke Karl Wilhelm, who paid for Gauss to attend higher schools and the University of Göttingen. Gauss's first major mathematical work was his doctoral thesis in which he proved the fundamental theorem of algebra. Gauss made many contributions to algebra, complex functions, geometry, physics, and astronomy. During his lifetime, he was considered the greatest mathematician since Newton, and by the time of his death in 1855, he was called the "prince of mathematicians."

Think and Communicate

You may wish to point out to students that mathematicians would say that the fundamental theorem of algebra is a *stronger* result than the property stated on page 420 because it tells exactly how many zeros there are.

Section Note

Teaching Tip

The rational zeros theorem assumes that the fractions $\frac{p}{q}$ are in lowest terms. For example, if the value $\frac{14}{35}$ is used for x in the polynomial $10x^2 - 19x + 6 = 0$, it can be shown that $\frac{14}{35}$ is a zero of the polynomial, even though 14 is *not* a divisor of 6, and 35 is *not* a divisor of 10. However, $\frac{14}{35}$ is not in lowest terms. When $\frac{14}{35}$ is written in lowest terms as $\frac{2}{5}$, 2 is a divisor of 6, and 5 is a divisor of 10.

Teach⇔Interact

Additional Example 2

Find the zeros of $g(x) = 9x^4 + 21x^3 - 23x^2 - 85x - 50$.

Step 1 Find the possible rational zeros of g. The divisors of the constant term are ±1, ±2, ±5, ±10, ±25, and ±50. The divisors of 9 are ±1, ±3, and ±9. So the possible rational zeros of g are $\pm1, \pm2, \pm5, \pm10, \pm25, \pm50, \pm\frac{1}{3}, \pm\frac{2}{3}, \pm\frac{5}{3}, \pm\frac{10}{3}, \pm\frac{25}{3}, \pm\frac{50}{3}, \pm\frac{1}{9}, \pm\frac{2}{9}, \pm\frac{5}{9}, \pm\frac{10}{9}, \pm\frac{25}{9}$, and $\pm\frac{50}{9}$.

Step 2 Use the short form of synthetic substitution to find one zero of g.

	9	21	−23	−85	−50	←coeff. of $g(x)$
1	9	30	7	−78	−128	←$g(1) = -128$
−1	9	12	−35	−50	0	←$g(-1) = 0$

One zero of g is −1.

Step 3 Find a second zero of g by finding a zero of the quotient polynomial $q_1(x) = \frac{g(x)}{x+1} = 9x^3 + 12x^2 - 35x - 50$. Continue checking the possible rational zeros of g listed in Step 1, starting with −1 (since −1 may be a double zero).

	9	12	−35	−50	←coeff. of $q_1(x)$
−1	9	3	−38	−12	←$q_1(-1) = -12$
2	9	30	25	0	←$q_1(2) = 0$

So 2 is a zero of g.

Step 4 Find the remaining zeros of g by solving the equation $q_2(x) = 0$, where $q_2(x)$ is the new quotient polynomial: $q_2(x) = \frac{q_1(x)}{x-1} = 9x^2 + 30x + 25$. The equation is a quadratic equation that can be solved by factoring.

$$9x^2 + 30x + 25 = 0$$
$$(3x + 5)(3x + 5) = 0$$
$$3x + 5 = 0$$
$$x = -\frac{5}{3}$$

Since the factor $3x + 5$ occurs twice in the solution of the quadratic equation, $-\frac{5}{3}$ is a double zero of g.

The zeros of g are $-\frac{5}{3}$ (a double zero), −1, and 2.

Think and Communicate

When discussing question 3, ask students why, in Step 3, the search for zeros resumes with $\frac{1}{3}$ instead of starting over again with 1, −1, and so on.

EXAMPLE 2

Find the zeros of $f(x) = 9x^4 - 6x^3 + 19x^2 - 12x + 2$.

SOLUTION

Step 1 Find the possible rational zeros of f.
The divisors of the constant term, 2, are ±1 and ±2.
The divisors of the leading coefficient, 9, are ±1, ±3, and ±9.
So the possible rational zeros of f are $\pm1, \pm2, \pm\frac{1}{3}, \pm\frac{2}{3}, \pm\frac{1}{9}$, and $\pm\frac{2}{9}$.

Step 2 Use synthetic substitution to find one zero of f from the list in Step 1. To save space, you can use a "short form" of synthetic substitution in which the coefficients of $f(x)$ are listed only once and all additions are done mentally.

possible rational zeros	9	−6	19	−12	2	← coefficients of $f(x)$
1	9	3	22	10	12	← $f(1) = 12$
−1	9	−15	34	−46	48	← $f(-1) = 48$
2	9	12	43	74	150	← $f(2) = 150$
−2	9	−24	67	−146	294	← $f(-2) = 294$
$\frac{1}{3}$	9	−3	18	−6	0	← $f\left(\frac{1}{3}\right) = 0$

One zero of f is $\frac{1}{3}$.

Step 3 Find a second zero of f by finding a zero of the quotient polynomial $q_1(x) = \frac{f(x)}{x - \frac{1}{3}} = 9x^3 - 3x^2 + 18x - 6$.

Continue checking the possible rational zeros of f listed in Step 1, starting with $\frac{1}{3}$ $\left(\text{since } \frac{1}{3} \text{ may be a double zero}\right)$.

	9	−3	18	−6	← Use the coefficients of $q_1(x)$.
$\frac{1}{3}$	9	0	18	0	← $q_1\left(\frac{1}{3}\right) = 0$

So $\frac{1}{3}$ is a double zero of f.

Step 4 Find the remaining zeros of f by solving the equation $q_2(x) = 0$ where $q_2(x)$ is the new quotient polynomial: $q_2(x) = \frac{q_1(x)}{x - \frac{1}{3}} = 9x^2 + 0x + 18$.

$$9x^2 + 18 = 0$$
$$9x^2 = -18$$
$$x^2 = -2$$
$$x = \pm\sqrt{-2} = \pm i\sqrt{2}$$

The zeros of f are $\frac{1}{3}$ (a double zero), $i\sqrt{2}$, and $-i\sqrt{2}$.

Think and Communicate

2. $q(x)$ is the quotient of a fourth-degree polynomial divided by a first-degree polynomial, so its degree is $4 - 1 = 3$.

3. $f(x) = \left(x - \frac{1}{3}\right) \cdot q_1(x)$, so any zero of q must be a zero of f.

4. a. $f(x) = \left(x - \frac{1}{3}\right)^2 \cdot q_2(x)$; If k is a zero of q_2, then $q_2(k) = 0$ and $f(k) = \left(k - \frac{1}{3}\right)^2 \cdot 0 = 0$.

b. $f(i\sqrt{2}) = 9(4) + 6i(2\sqrt{2}) - 19(2) - 12i\sqrt{2} + 2 = 0$; $f(-i\sqrt{2}) = 9(4) - 6i(2\sqrt{2}) - 19(2) + 12i\sqrt{2} + 2 = 0$

5. No; the fundamental theorem of algebra does not require that the zeros be distinct, rather that double zeros be counted twice, triple zeros three times, and so on.

6. The possible rational zeros are $\pm1, \pm5, \pm\frac{1}{3}$, and $\pm\frac{5}{3}$. Checking all of these using synthetic substitution shows that none is a zero of h. This does not mean h has no real zeros. One way to check this is to use a graphing calculator. The real zeros can be approximated using the trace function or the root function.

THINK AND COMMUNICATE

2. In Example 2, how do you know that the quotient polynomial $q_1(x)$ has degree 3?
3. In Step 3 of Example 2, it was assumed that any rational zero of q_1 must be in the list of possible rational zeros of f. Explain why this is true.
4. Refer to the functions f and q_2 in Example 2.
   a. Complete this equation: $f(x) = \underline{?} \cdot q_2(x)$. How does the equation show that each zero of q_2 is also a zero of f?
   b. Verify directly that the zeros of q_2 are zeros of f by evaluating $f(x)$ when $x = i\sqrt{2}$ and when $x = -i\sqrt{2}$.
5. Does the degree of the function f in Example 2 equal the number of *distinct* complex zeros of f? Does this contradict the fundamental theorem of algebra? Explain.
6. Use the rational zeros theorem to show that $h(x) = 3x^3 + 4x^2 + x - 5$ has no rational zeros. Does this mean that h has no *real* zeros? If not, how can you approximate any real zeros of h?

CHECKING KEY CONCEPTS

Technology Use a graphing calculator or graphing software to approximate all real zeros of each function to the nearest hundredth.

1. $y = x^3 - 5x^2 + 11x - 10$
2. $y = x^4 + 4x^3 - 2x^2 - 11x + 1$
3. $y = 2x^5 + 7x^4 - 3x^3 - 19x^2 + 7$
4. $y = -3x^6 + 12x^4 - 4x^2 - 8$

For each function in Questions 5–8:

a. List the possible rational zeros of the function.

b. Find all real and imaginary zeros of the function. Identify any double or triple zeros.

5. $f(x) = 2x^3 - 5x^2 - 4x + 3$
6. $f(x) = 4x^3 + 3x^2 + 12x + 9$
7. $g(x) = x^4 - 6x^3 + 14x^2 - 16x + 8$
8. $h(x) = 3x^4 + 8x^3 + 6x^2 - 1$

9.5 Exercises and Applications

Extra Practice *exercises on page 763*

Technology Use a graphing calculator or graphing software to approximate all real zeros of each function to the nearest hundredth.

1. $y = x^3 + x^2 - 9x + 3$
2. $y = 2x^3 + 3x^2 - 3x + 5$
3. $y = x^4 - x^3 - 4x^2 + 2x - 12$
4. $y = 3x^4 - 14x^2 - 2x + 7$
5. $y = -6x^5 - 18x^4 + x^3 + 27x^2 - 4$
6. $y = x^5 - 6x^4 + 5x^3 + 15x^2 - 9x - 8$
7. $y = 2x^6 - 16x^4 + 32x^2 - x - 10$
8. $y = -4x^6 + 13x^4 + x^3 - 11x^2 - 2x + 10$

Teach⇔Interact

Section Note

Using Technology

A TI-82 graphing calculator can be used to find rational zeros quickly. For example, suppose you want to find the rational zeros, if any, of $f(x) = 343x^3 - 833x^2 + 28x + 192$. Enter the equation $y = 343x^3 - 833x^2 + 28x + 192$ on the Y= list and choose the standard viewing window. When the graph is displayed, it is not useful, but by changing the interval for y to $-500 \le y \le 500$, you get a fairly good graph. Use root from the CALCULATE menu to approximate the rightmost x-intercept. The display at the bottom of the window shows X=2.2857143. Press [2nd] [QUIT] to go to the home screen. Then press [X,T,θ] [MATH] <▶Frac> [ENTER]. The calculator converts the x-value to a fraction and displays the exact value for the intercept, $\frac{16}{7}$. (Note that this approach does not work if the absolute value of the denominator is too large.)

Closure Question

Describe the procedure you would use to find all the zeros of a polynomial function with integer coefficients. Find all divisors of the constant term and leading coefficient of $f(x)$. Use these divisors and the rational zero theorem to obtain a list of all possible rational zeros. Check systematically to see which of these numbers are zeros. If a number $\frac{a}{b}$ is a zero, divide $f(x)$ by $x - \frac{a}{b}$ and resume the check on the quotient polynomial starting with $\frac{a}{b}$. Continue in this way until the quotient polynomial is a quadratic function. Solve this quadratic to find the remaining zeros.

Checking Key Concepts

Answers that involve approximations may vary due to rounding. Examples are given.

1. 2.00
2. −3.72; −1.92; 0.09; 1.56
3. −2.98; −1.87; −0.70; 0.63; 1.42
4. ±1.84; ±1.12
5. a. $\pm1; \pm3; \pm\frac{1}{2}; \pm\frac{3}{2}$
   b. $-1; \frac{1}{2}; 3$
6. a. $\pm1; \pm3; \pm9; \pm\frac{1}{2}; \pm\frac{1}{4}; \pm\frac{3}{2}; \pm\frac{3}{4}; \pm\frac{9}{2}; \pm\frac{9}{4}$
   b. $-\frac{3}{4}; \pm i\sqrt{3}$
7. a. ±1; ±2; ±4; ±8
   b. 2 (double zero); $1 \pm i$
8. a. $\pm1; \pm\frac{1}{3}$
   b. −1 (triple zero); $\frac{1}{3}$

Exercises and Applications

Answers that involve approximations may vary due to rounding. Examples are given.

1. −3.67; 0.35; 2.32
2. −2.50
3. −2.24; 2.76
4. −1.92; −0.86; 0.67; 2.11
5. −2 (double zero); −0.41; 0.41; 1
6. −1.23; −0.55; 1.12; 2.70; 3.95
7. −2.21; −1.67; −0.59; 0.64; 1.57; 2.26
8. −1.59; 1.61

Apply⇔Assess

Suggested Assignment

Core Course
Exs. 1–18, 21–29, AYP

Extended Course
Exs. 1–17 odd, 19–29, AYP

Block Schedule
Day 58 Exs. 1–18, 21–29, AYP

Apply⇔Assess

Exercise Notes

Teaching Tip

Exs. 9–18 To help students remember where the divisors of the constant term and leading coefficient are used when applying the rational zeros theorem, you can suggest the following procedure. Imagine that a polynomial $f(x)$ with integer coefficients has been factored completely: $f(x) = (A_1x - B_1)(A_2x - B_2) \dots (A_nx - B_n)$. When these binomial factors are multiplied, the product is $(A_1A_2 \dots A_n)x^n + \dots \pm (B_1B_2 \dots B_n)$. The zeros are found by setting the factors equal to zero and solving for x; hence, the zeros are $\frac{B_1}{A_1}, \frac{B_2}{A_2}, \dots, \frac{B_n}{A_n}$. Note that $B_1, B_2, \dots$, and B_n are in the numerators and appear in the expression for the constant term. The numbers $A_1, A_2, \dots, A_n$ are in the denominators and appear in the expression for the leading coefficient. Every polynomial can be factored in the way suggested above, although some or all of the numbers $A_1, \dots, A_n$ and $B_1, \dots, B_n$ may be imaginary.

Interdisciplinary Problems

Exs. 19, 20 These exercises provide students with an opportunity to see how the mathematical properites of polynomial functions can be used to examine the deflection of a beam. In particular, students will examine the deflection of a diving board and the deflection of a bookshelf. In Ex. 19, students should use the length of the dividing board to help them define a good viewing window. They can solve this problem by graphing both $y = d(x)$ and $y = 1$ and looking for the intersection, or by graphing $y = d(x) - 1$ and looking for the zero. A similar option is available for part (c) of Ex. 20.

For each function in Exercises 9–18:

a. List the possible rational zeros of the function.

b. Find all real and imaginary zeros of the function. Identify any double or triple zeros.

9. $f(x) = x^3 - 5x^2 + 17x - 13$

10. $f(x) = 4x^3 + 12x^2 + x + 3$

11. $f(x) = 3x^3 + 10x^2 + 4x - 8$

12. $f(x) = 30x^3 - x^2 - 6x + 1$

13. $g(x) = x^4 - 2x^3 - 21x^2 + 22x + 40$

14. $g(x) = 4x^4 - 4x^3 - 11x^2 + 6x + 9$

15. $g(x) = 16x^4 + 8x^3 + 17x^2 + 8x + 1$

16. $g(x) = -6x^4 + 19x^3 - 6x^2 - 41x + 20$

17. $h(x) = -x^5 + 2x^4 + 10x^3 + 8x^2 - x - 2$

18. $h(x) = 2x^5 - 5x^4 - 40x^3 - x^2 - 52x - 84$

Connection PHYSICS

A *beam* is an oblong piece of metal, wood, or other material used as a horizontal support. When a force is applied to a beam, the beam is *deflected*, or bent, from its original position. The deflection can often be modeled by a polynomial function.

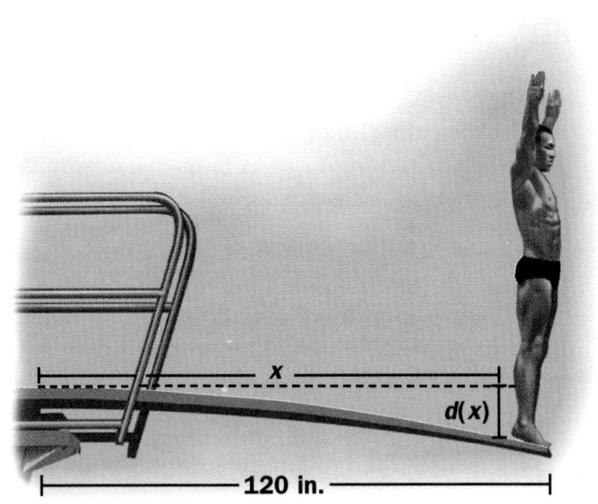

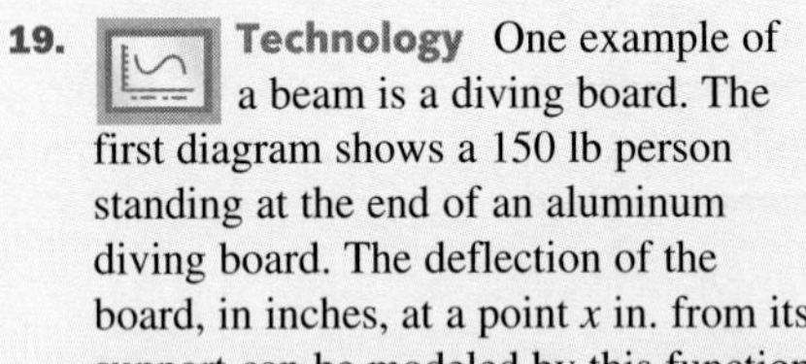

19. **Technology** One example of a beam is a diving board. The first diagram shows a 150 lb person standing at the end of an aluminum diving board. The deflection of the board, in inches, at a point x in. from its support can be modeled by this function:

$$d(x) = (-4.752 \times 10^{-7})x^3 + (1.711 \times 10^{-4})x^2$$

Use a graphing calculator or graphing software to find the portion of the diving board where the deflection exceeds 1 in.

20. **Technology** Another example of a beam is a bookshelf. The second diagram shows a wooden bookshelf loaded with 180 lb of books. The deflection of the bookshelf, in inches, at a point x in. from its left end can be modeled by this function:

$$d(x) = (2.724 \times 10^{-7})x^4 - (3.269 \times 10^{-5})x^3 + (9.806 \times 10^{-4})x^2$$

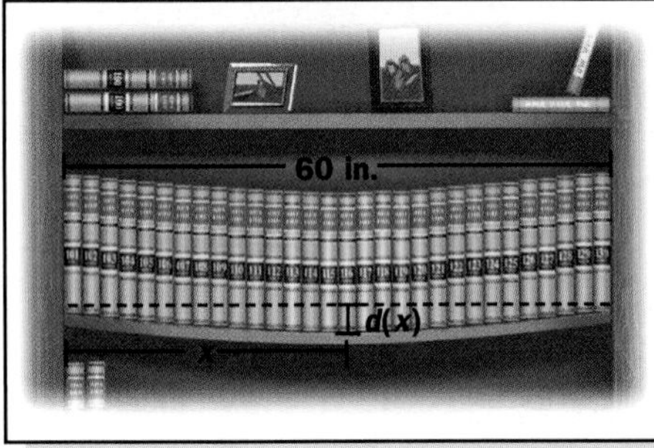

a. Use a graphing calculator or graphing software to graph $y = d(x)$. Approximate all real zeros of $y = d(x)$ in the interval $0 \le x \le 60$. Explain why your answers make sense in the given situation.

b. At what point on the bookshelf is the deflection at a maximum? What is the maximum deflection?

c. Find the portion of the bookshelf where the deflection exceeds 0.1 in.

9. a. $\pm 1; \pm 13$
   b. $1, 2 \pm 3i$

10. a. $\pm 1; \pm 3; \pm\frac{1}{2}; \pm\frac{1}{4}; \pm\frac{3}{2}; \pm\frac{3}{4}$
    b. $-3; \pm\frac{i}{2}$

11. a. $\pm 1; \pm 2; \pm 4; \pm 8; \pm\frac{1}{3}; \pm\frac{2}{3}; \pm\frac{4}{3}; \pm\frac{8}{3}$
    b. $\frac{2}{3}$; -2 (double zero)

12. a. $\pm 1; \pm\frac{1}{2}; \pm\frac{1}{3}; \pm\frac{1}{5}; \pm\frac{1}{6}; \pm\frac{1}{10}; \pm\frac{1}{15}; \pm\frac{1}{30}$
    b. $-\frac{1}{2}; \frac{1}{3}; \frac{1}{5}$

13. a. $\pm 1; \pm 2; \pm 4; \pm 5; \pm 8; \pm 10; \pm 20; \pm 40$
    b. $-4; -1; 2; 5$

14. a. $\pm 1; \pm 3; \pm 9; \pm\frac{1}{2}; \pm\frac{1}{4}; \pm\frac{3}{2}; \pm\frac{3}{4}; \pm\frac{9}{2}; \pm\frac{9}{4}$
    b. -1 (double zero); $\frac{3}{2}$ (double zero)

15. a. $\pm 1; \pm\frac{1}{2}; \pm\frac{1}{4}; \pm\frac{1}{8}; \pm\frac{1}{16}$
    b. $-\frac{1}{4}$ (double zero); $\pm i$

16. a. $\pm 1; \pm 2; \pm 4; \pm 5; \pm 10; \pm 20; \pm\frac{1}{2}; \pm\frac{5}{2}; \pm\frac{1}{3}; \pm\frac{2}{3}; \pm\frac{4}{3}; \pm\frac{5}{3}; \pm\frac{10}{3}; \pm\frac{20}{3}; \pm\frac{1}{6}; \pm\frac{5}{6}$
    b. $-\frac{4}{3}; \frac{1}{2}; 2 \pm i$

21. SAT/ACT Preview Let $f(x) = 15x^4 + lx^3 + mx^2 + nx + 12$ where l, m, and n are integers. Which of the following numbers *cannot* be a zero of f?

A. $\frac{12}{5}$ **B.** 6 **C.** $-\frac{2}{3}$ **D.** $\frac{5}{4}$ **E.** $-\frac{1}{15}$

22. Investigation For parts (a)–(d), identify the double zeros of each function.

a. $f(x) = (x + 2)(x - 2)^2$ **b.** $g(x) = -3(x + 2)^2(x - 1)$

c. $h(x) = x^2(x - 3)^2$ **d.** $f(x) = -(x - 1)^2(x - 4)^2$

e. **Technology** Use a graphing calculator or graphing software to graph the functions in parts (a)–(d). Describe the behavior of each graph when the value of x is close to a double zero.

ONGOING ASSESSMENT

23. a. Writing Explain why an odd-degree polynomial function must have at least one real zero.

b. Open-ended Problem Give an example of an even-degree polynomial function that has no real zeros.

SPIRAL REVIEW

Find an equation for the cubic function whose graph has the given x-intercepts and passes through the given point. *(Section 9.4)*

24. x-intercepts: 1, 2, 4; $(0, -8)$ **25.** x-intercepts: -7, -5, 0; $(-1, 48)$

For each equation, tell whether y varies directly with x. *(Section 2.1)*

26. $y = 3x$ **27.** $y = 3x + 3$ **28.** $y = \frac{x}{3}$ **29.** $y = \frac{3}{x}$

ASSESS YOUR PROGRESS

VOCABULARY

cubic function (p. 413) **double zero** (p. 420)
zero (p. 413) **triple zero** (p. 420)

1. ENERGY For a certain windmill, the power produced when the wind speed is s m/s can be modeled by the function

$$P(s) = -0.699s^3 + 19.8s^2 - 120s + 220$$

where $P(s)$ is in kilowatts (kW) and $4 \le s \le 16$. At what wind speed does the power produced by the windmill reach 300 kW? *(Section 9.4)*

Find the zeros of each function. *(Sections 9.4 and 9.5)*

2. $f(x) = 5(x - 1)(x + 2)(x - 3)$ **3.** $f(x) = x^3 - 14x + 8$

4. $g(x) = 6x^3 - x^2 - 9x - 10$ **5.** $h(x) = 2x^4 - 5x^3 + 5x^2 - 20x - 12$

6. Journal Discuss the relationship between the factors and zeros of a polynomial function f. Explain how the rational zeros theorem can help you find the zeros of f.

BY THE WAY

The first windmills were probably built in Iran in the seventh century A.D. These windmills had sails that revolved around a vertical axis. They were used mainly to grind grain.

Apply⇔Assess

Exercise Notes

Assessment Note

Ex. 23 Students should discuss their answers to these exercises to ensure that they understand the concepts involved.

Assess Your Progress

Journal Entry

This journal entry offers students a good opportunity to demonstrate their understanding of the important concepts and results of Sections 9.1 through 9.5. If they mention any of the theorems from this section or preceding sections, be sure they state and use them correctly.

Progress Check 9.4–9.5

See page 456.

Practice 58 for Section 9.5

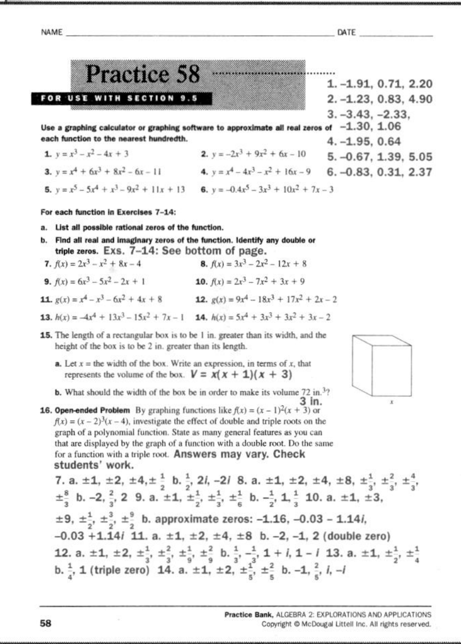

NAME ______ DATE ______

Practice 58

FOR USE WITH SECTION 9.5

Use a graphing calculator or graphing software to approximate all real zeros of each function to the nearest hundredth.

1. $y = x^3 - x^2 - 4x + 3$ 2. $y = -2x^3 + 9x^2 + 6x - 10$
3. $y = x^4 + 6x^3 + 8x^2 - 6x - 11$ 4. $y = x^4 - 4x^3 - x^2 + 16x - 9$
5. $y = x^5 - 5x^4 + x^3 - 9x^2 + 11x + 13$ 6. $y = -0.4x^5 - 3x^3 + 10x^2 + 7x - 3$

1. –1.91, 0.71, 2.20 2. –1.23, 0.83, 4.90 3. –3.43, –2.33, –1.30, 1.06 4. –1.95, 0.64 5. –0.67, 1.39, 5.05 6. –0.83, 0.31, 2.37

For each function in Exercises 7–14:

a. **List all possible rational zeros of the function.**

b. **Find all real and imaginary zeros of the function. Identify any double or triple zeros.** Exs. 7–14: See bottom of page.

7. $f(x) = 2x^3 - x^2 + 8x - 4$ 8. $f(x) = 3x^3 - 2x^2 - 12x + 8$
9. $f(x) = 6x^3 - 5x^2 - 2x + 1$ 10. $f(x) = 2x^3 - 7x^2 + 3x + 9$
11. $g(x) = x^4 - x^3 - 6x^2 + 4x + 8$ 12. $g(x) = 9x^4 - 18x^3 + 17x^2 + 2x - 2$
13. $h(x) = -4x^4 + 13x^3 - 15x^2 + 7x - 1$ 14. $h(x) = 5x^4 + 3x^3 + 3x^2 + 3x - 2$

15. The length of a rectangular box is to be 1 in. greater than its width, and the height of the box is to be 2 in. greater than its length.

a. Let x = the width of the box. Write an expression, in terms of x, that represents the volume of the box. $V = x(x + 1)(x + 3)$

b. What should the width of the box be in order to make its volume 72 in.^{3}? 3 in.

16. **Open-ended Problem** By graphing functions like $f(x) = (x - 1)^2(x + 3)$ or $f(x) = (x - 2)^3(x - 4)$, investigate the effect of double and triple roots on the graph of a polynomial function. State as many general features as you can that are displayed by the graph of a function with a double root. Do the same for a function with a triple root. Answers may vary. Check students' work.

7. a. $\pm1, \pm2, \pm4, \pm\frac{1}{2}$ b. $\frac{1}{2}, 2i, -2i$ 8. a. $\pm1, \pm2, \pm4, \pm8, \pm\frac{1}{3}, \pm\frac{2}{3}, \pm\frac{4}{3}, \pm\frac{8}{3}$ b. $-2, \frac{2}{3}, 2$ 9. a. $\pm1, \pm\frac{1}{2}, \pm\frac{1}{3}, \pm\frac{1}{6}$ b. $-\frac{1}{2}, 1, \frac{1}{3}$ 10. a. $\pm1, \pm3, \pm9, \pm\frac{1}{2}, \pm\frac{3}{2}, \pm\frac{9}{2}$ b. approximate zeros: $-1.16, -0.03 - 1.14i, -0.03 + 1.14i$ 11. a. $\pm1, \pm2, \pm4, \pm8$ b. $-2, -1, 2$ (double zero) 12. a. $\pm1, \pm2, \pm\frac{1}{3}, \pm\frac{2}{3}, \pm\frac{1}{9}, \pm\frac{2}{9}$ b. $\frac{1}{3}, -\frac{1}{3}, 1 + i, 1 - i$ 13. a. $\pm1, \pm\frac{1}{2}, \pm\frac{1}{4}$ b. $\frac{1}{4}$, 1 (triple zero) 14. a. $\pm1, \pm2, \pm\frac{1}{5}, \pm\frac{2}{5}$ b. $-1, \frac{2}{5}, i, -i$

58 Practice Bank, ALGEBRA 2: EXPLORATIONS AND APPLICATIONS

---

17. a. $\pm1; \pm2$
b. -1 (triple zero); $\frac{5 \pm \sqrt{17}}{2}$

18. a. $\pm1; \pm2; \pm3; \pm4; \pm6; \pm7; \pm12; \pm14; \pm21; \pm28; \pm42; \pm84; \pm\frac{1}{2}; \pm\frac{3}{2}; \pm\frac{7}{2}; \pm\frac{21}{2}$
b. $-\frac{7}{2}; -1; 6; \frac{1 \pm i\sqrt{7}}{2}$

19. from about 87.9 in. to 120 in.

20. a. 0; 60; There should be no deflection at the ends where the bookshelf is supported.
b. 30 in. in from the left end; about 0.221 in.
c. at a point about 12.9 in. in from the left end to a point about 47.1 in. in from the left end

21. D

22. a. 2 b. -2
c. 0; 3 d. 1; 4
e. Answers may vary. An example is given. When the value of x approaches a double zero from either direction, the graph approaches the x-axis. At the double zero, the graph changes direction. In the area of the double zero, the graph is U-shaped.

23. Answers may vary. Examples are given.
a. Recall the end behavior of the graph of an odd-degree polynomial function. The graph is on opposite sides of the x-axis as $x \to +\infty$ and as $x \to -\infty$, so the graph must cross the x-axis at some point and, therefore, have a real zero.
b. $f(x) = x^2 + 1$

24. $f(x) = (x - 1)(x - 2)(x - 4)$

25. $f(x) = -2x(x + 7)(x + 5)$

26. Yes. 27. No.

28. Yes. 29. No.

Assess Your Progress

1–6. See answers in back of book.

Plan⇔Support

Objectives

- Recognize inverse variation.
- Write and use inverse variation equations.

Recommended Pacing

Core and Extended Courses
Section 9.6: 1 day

Block Schedule
Section 9.6: $\frac{1}{2}$ block
(with Section 9.7)

Resource Materials

Lesson Support
Lesson Plan 9.6
Warm-Up Transparency 9.6
Practice Bank: Practice 59
Study Guide: Section 9.6
Explorations Lab Manual: Diagram Masters 1, 2

Technology
Graphing Calculator
Spreadsheet Software
McDougal Littell Mathpack *Function Investigator with Matrix Analyzer Activity Book:* Function Investigator Activity 24
Internet: http://www.hmco.com

Warm-Up Exercises

Suppose the product of x and y is 180. Use the given value of one variable to find the value of the other variable.

1. $x = 45$, $y =$ __?__ 4
2. $y = -60$, $x =$ __?__ −3
3. $x = -100$, $y =$ __?__ −1.8
4. $y = \frac{1}{9}$, $x =$ __?__ 1620

Tell whether the graph of each equation does or does not contain the given point. Answer *Yes* or *No.*

5. $y = \frac{14}{x}$, $(-2, 7)$ No.
6. $3xy = 100$, $\left(1, \frac{100}{3}\right)$ Yes.

Section

9.6 Inverse Variation

Learn how to...

- recognize inverse variation
- write and use inverse variation equations

So you can...

- find the pressure that a person exerts when walking on snow, for example

If you've ever walked through deep snow, you can appreciate the value of snowshoes. Snowshoes prevent you from sinking by distributing your weight over a large area. This reduces the pressure you exert on the snow.

EXAMPLE 1 Application: Physics

For a 120 lb person, the average pressure P exerted on the snow beneath the person's footwear is given by

$$P = \frac{120}{A}$$

where A is the area of the bottom of the footwear in square feet and P is measured in pounds per square foot.

a. Use the pictures below. Compare the pressure that a 120 lb person exerts when wearing the snowshoes with the pressure the person exerts when wearing the boots.

b. Graph $P = \frac{120}{A}$ using a graphing calculator or graphing software. Describe what happens to pressure as area gets very small and very large.

The area A of the soles of a pair of boots is $A = 0.4$ ft^2.

The area A of a pair of snowshoes is $A = 4$ ft^2.

SOLUTION

a. For the boots, $P = \frac{120}{0.4} = 300$ lb/ft^2.

For the snowshoes, $P = \frac{120}{4} = 30$ lb/ft^2.

The pressure exerted when the snowshoes are worn is only one tenth of the pressure exerted when the boots are worn.

b.

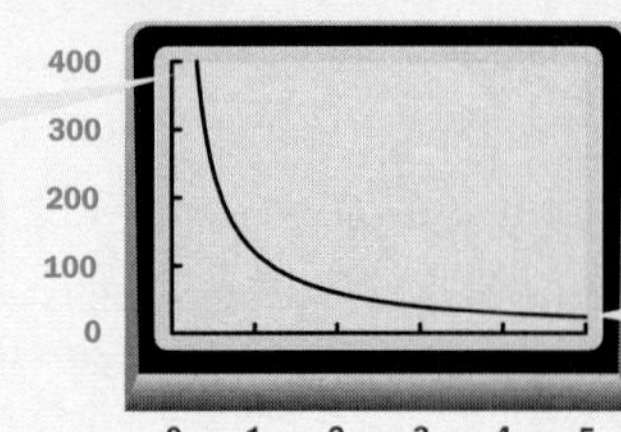

As **area** gets very **small**, **pressure** gets very **large**.

As **area** gets very **large**, **pressure** gets very **small**.

THINK AND COMMUNICATE

1. **a.** In Example 1, what happens to the pressure exerted on the snow when the area of the footwear is doubled? tripled? increased by a factor of n?
   b. What can you say about the product of pressure and area?

2. 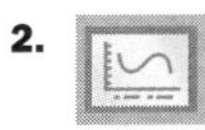 **Technology** Use a graphing calculator or graphing software to graph $y = \frac{120}{x}$. Adjust the viewing window so that the intervals $-5 \le x \le 5$ and $-400 \le y \le 400$ are shown on the axes. Compare this graph with the graph in Example 1. What do you notice?

Two variables x and y are said to show **inverse variation** if they are related by an equation of this form:

You can say that y **varies inversely** with x, or that y is **inversely proportional** to x.

$$y = \frac{a}{x}$$

The number a, which must be nonzero, is called the **constant of variation**.

In Example 1, for instance, the equation $P = \frac{120}{A}$ indicates that P varies inversely with A. The constant of variation is 120.

The graph of an inverse variation equation is called a **hyperbola**. A hyperbola consists of two pieces, called *branches*. For $a > 0$, the graph of $y = \frac{a}{x}$ looks like this:

The y-axis is a vertical *asymptote* of the graph. An **asymptote** is a line that a graph approaches more and more closely.

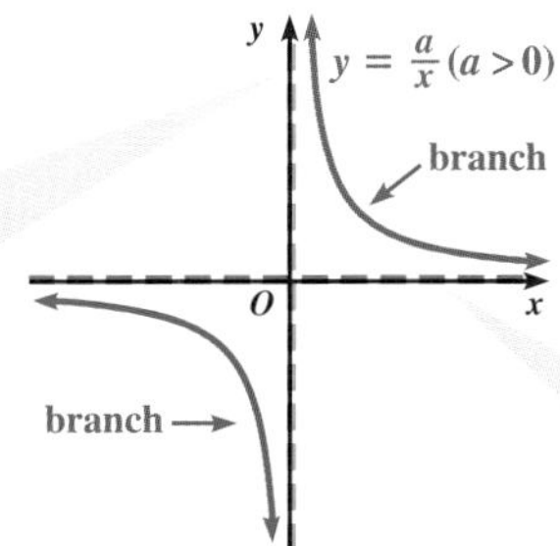

The **x-axis** is a horizontal asymptote of the graph.

Note that if x and y represent positive quantities (such as area and pressure), the graph of $y = \frac{a}{x}$ is only the first-quadrant branch of a hyperbola.

THINK AND COMMUNICATE

3. 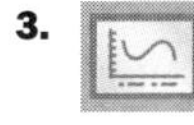 **Technology** Use a graphing calculator or graphing software to graph $y = \frac{a}{x}$ for several negative values of a. Compare the graph of $y = \frac{a}{x}$ where $a < 0$ with the graph of $y = \frac{a}{x}$ where $a > 0$.

4. Explain why the graph of $y = \frac{a}{x}$ (where $a \ne 0$) never crosses either the x- or y-axis.

BY THE WAY

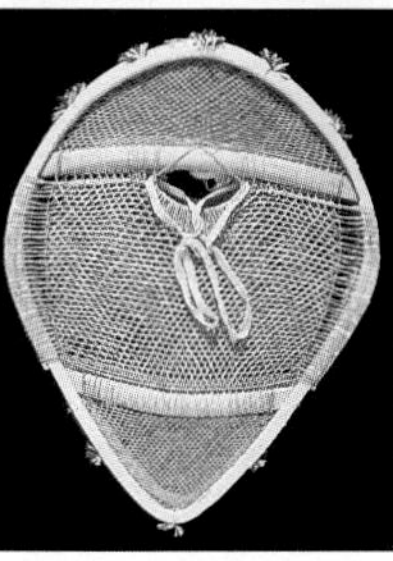

Snowshoes were first used by Native Americans. The snowshoe above was made by the Montagnais people from Canada. This snowshoe is in a bear paw style, with colored wool strings decorating the rim.

WATCH OUT!

The asymptotes of a graph are *not* part of the graph.

ANSWERS Section 9.6

Think and Communicate

1. a. It is multiplied by $\frac{1}{2}$; it is multiplied by $\frac{1}{3}$; it is multiplied by $\frac{1}{n}$.
   b. It is constant.

2. 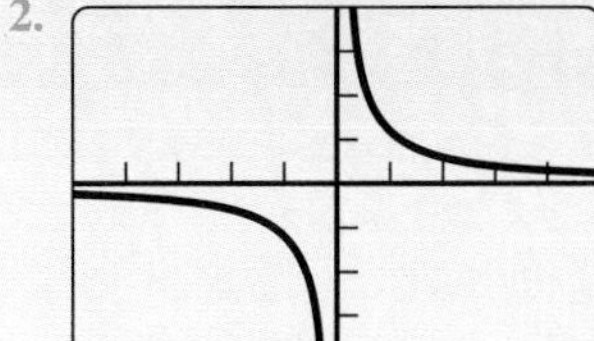
   There are two sections of the graph; the first-quadrant portion is the graph in Example 1.

3. If $a < 0$, the graph is in the second and fourth quadrants; if $a > 0$, the graph is in the first and third quadrants.

4. Since $y = \frac{a}{x}$ and $\frac{a}{x}$ is not defined for $x = 0$, the graph never crosses the y-axis. Since $a \ne 0$, there is no value of x for which $\frac{a}{x} = 0$, so the graph never crosses the x-axis.

Teach⇔Interact

Additional Example 1

According to Boyle's law, the relation between the pressure P of a gas at a constant temperature and the volume V of the gas can be modeled by an equation of the form $P = \frac{k}{V}$, where k is a constant.

a. Suppose that at a particular temperature, $P = \frac{48}{V}$ is the equation that relates P (lb/in.^{2}) to V (in.^{3}). Compare the pressure of the gas when the volume is 6 in.^{3} to the pressure of the gas when the volume is 8 in.^{3}.

When the volume is 6 in.^{3}, $P = \frac{48}{6} = 8$ lb/in.^{2}.
When the volume is 8 in.^{3}, $P = \frac{48}{8} = 6$ lb/in.^{2}.
When the volume is 6 in.^{3}, the pressure is $\frac{8}{6}$, or $\frac{4}{3}$ times as great as the pressure when the volume is 8 in.^{3}.

b. Graph $P = \frac{48}{V}$ using a graphing calculator or graphing software. Describe what happens to the volume as the pressure gets very large or very small.

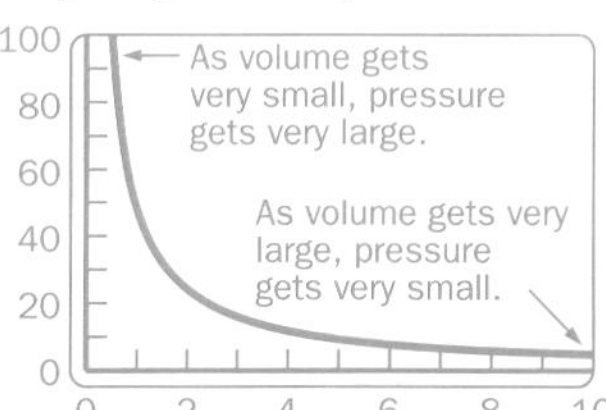

Section Notes

Reasoning

You can use the graph of the inverse variation $y = \frac{a}{x}$ $(a > 0)$ to demonstrate visually that as one of the variables x or y increases in either a positive or negative direction, the other variable decreases. The equation $xy = a$ at the top of page 428 can also be used to demonstrate this relationship.

Using Technology

Students can use Function Investigator Activity 24 in the *Function Investigator with Matrix Analyzer Activity Book* to explore the characteristics of graphs of equations of the form $y = \frac{a}{x}$.

An inverse variation equation $y = \frac{a}{x}$ can also be written this way:

$$xy = a$$

Therefore, y varies inversely with x if and only if the product of x and y is constant.

EXAMPLE 2 Application: Biology

The rate r at which a bird flaps its wings depends on the bird's wing length l. The pictures show values of l and r for different species of birds.

a. Show that r varies approximately inversely with l.

b. Write an equation giving r as a function of l.

Merganser

$l = 28.9$ cm
$r = 4.6$ flaps per second

Widgeon

$l = 26.2$ cm
$r = 5.1$ flaps per second

Peregrine falcon

$l = 30.9$ cm
$r = 4.3$ flaps per second

Egyptian vulture

$l = 49.5$ cm
$r = 2.7$ flaps per second

SOLUTION

a. Find the product lr for each species of bird. A spreadsheet may be helpful.

Bird Data

	A	B	C
1	l	r	lr
2	26.2	5.1	133.6
3	28.9	4.6	132.9
4	30.9	4.3	132.9
5	49.5	2.7	133.7

Each product is about 133.

Since the products lr are nearly constant, r varies approximately inversely with l.

b. From part (a), the relationship between l and r can be modeled by the equation $lr = 133$, or $r = \frac{133}{l}$.

THINK AND COMMUNICATE

5. The wing length of a cormorant is about 35.0 cm. About how many times per second does a cormorant flap its wings?

Teach⇔Interact

About Example 2

Teaching Tip

Point out to students that y varies inversely with x if and only if x varies inversely with y. That is, in this Example, as the product lr is constant, l can also be expressed using inverse variation, $l = \frac{133}{r}$. Using this equation, the wing length of a bird can be estimated when the rate at which it flaps its wings is known.

Additional Example 2

The diet of a particular caterpillar consists exclusively of leaves from a certain type of plant. A biologist studying this caterpillar collected data on how many hours it took for caterpillar populations of different sizes to consume 1 kg of leaves. The data are shown in the following table.

Population (p)	Hours (h) to consume 1 kg
150	41.4
200	31.0
120	51.7
170	36.3

a. Show that h varies approximately inversely with p.

Find the product ph for each data pair. Use a calculator.

$150 \times 41.4 = 6210$
$200 \times 31.0 = 6200$
$120 \times 51.7 = 6204$
$170 \times 36.3 = 6171$

Each product is about 6200. Since the products ph are nearly constant, h varies approximately inversely with p.

b. Write an equation giving h as a function of p.

From part (a), the relationship between h and p can be modeled by the equation $hp = 6200$, or $h = \frac{6200}{p}$.

Think and Communicate

5. about 3.8 times per second

EXAMPLE 3 Application: Home Repair

The force F needed to loosen a bolt with a wrench is inversely proportional to the turning radius r.

The wrench shown can loosen the bolt with 250 lb of force. How much force is needed when a wrench with a 10 in. turning radius is used?

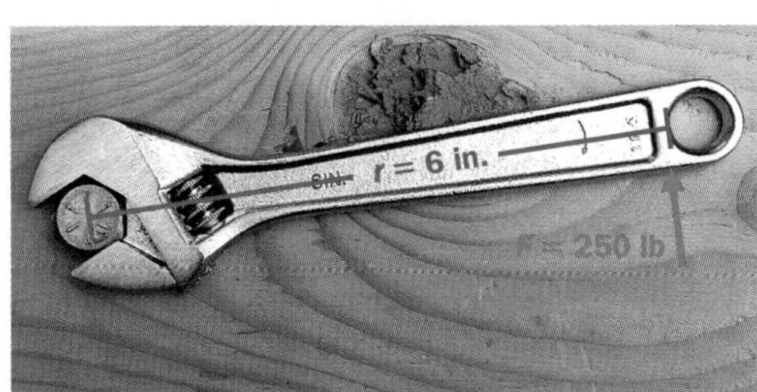

SOLUTION

The product of turning radius and force is constant. Therefore:

$$10 \text{ in.} \times \text{force needed with a 10 in. radius} = 6 \text{ in.} \times \text{force needed with a 6 in. radius}$$

F is the unknown force.

$$10F = 6 \cdot 250$$
$$10F = 1500$$
$$F = 150$$

The force needed is 150 lb.

THINK AND COMMUNICATE

6. Write an equation giving the force F needed to loosen the bolt in Example 3 as a function of the wrench's turning radius r.

CHECKING KEY CONCEPTS

Tell whether y varies inversely with x. If so, state the constant of variation.

1. $y = \frac{5}{x}$ **2.** $y = \frac{x}{5}$ **3.** $\frac{y}{x} = -0.3$ **4.** $xy = -0.3$

Tell whether y varies inversely with x. If so, write an equation giving y as a function of x.

5.

x	y
1	3
2	6
3	9
4	12

6.

x	y
1	24
2	12
3	8
4	6

7.

x	y
1	4
2	3
3	2
4	1

8. The time it takes Teri Petersen to drive home from college is inversely proportional to her average speed. When Teri drives at an average speed of 45 mi/h, it takes her 6 h to get home. How fast must Teri drive if she wants to get home in 5 h?

Teach⇔Interact

Additional Example 3

The length of a straight driveway that can be covered to a given depth by 1 ton of gravel varies inversely with the width of the driveway. If 1 ton of gravel can cover a length of 28 ft when the width of the driveway is 6 ft, how many feet can the gravel cover if the driveway were $5\frac{1}{2}$ ft wide?

The product of length and width is constant. Let L be the length for a driveway that is $5\frac{1}{2}$ ft ft wide. Then

$$L \times 5\frac{1}{2} = 28 \times 6$$
$$5\frac{1}{2}L = 168$$
$$L = 30\frac{6}{11}, \text{ or about } 30\frac{1}{2}.$$

The length that can be covered is about $30\frac{1}{2}$ ft.

Checking Key Concepts

Topic Spiraling: Review

While discussing these questions, ask whether any of the equations or tables show *direct* variation. (Yes; questions 2, 3, and 5 show direct variation.)

Closure Question

Suppose you are given a graph of a function in the coordinate plane. The graph has two branches, one in the first quadrant and the other in the third quadrant. Each branch has the x- and y-axes as asymptotes. How can you tell if the graph shows inverse variation or not? Estimate the coordinates of several points on the graph. Select points on both branches. Find the product of the x- and y-coordinates for each of the points. If the products are constant or very nearly constant, then y varies approximately inversely with x.

Think and Communicate

6. $F = \frac{1500}{l}$

Checking Key Concepts

1. Yes; 5.
2. No.
3. No.
4. Yes; –0.3.
5. No.
6. Yes; $y = \frac{24}{x}$.
7. No.
8. 54 mi/h

Apply⇔Assess

Suggested Assignment

Core Course
Exs. 1–12, 19, 20, 22–30

Extended Course
Exs. 1–30

Block Schedule
Day 59 Exs. 1–12, 19, 20, 22–30

Exercise Notes

Common Error
Ex. 5 Some students think that an equation of the form $y = \frac{a}{x + b}$ shows inverse variation because the variable x is in the denominator of the fraction and the graph of the equation is a hyperbola with vertical and horizontal asymptotes. Refer these students to the definition of inverse variation on page 427 to correct this error. Remind them that the product of the x- and y-coordinates of points on the graph of an inverse variation equation must be constant. Have them test some points on the graph of $y = \frac{2}{x + 1}$ to see that they fail this test.

Communication: Discussion
Exs. 9–12 These questions may help verbal learners to recognize situations in which inverse variation may occur. Note that in all these problems, as one quantity increases, the other decreases and vice versa. Point out to students that this alone is not enough to indicate an inverse relationship. In each case, they must verify that the product of the two quantities is a constant. For example in Ex. 10, it must be assumed that the total rent of the apartment is constant regardless of the number of students living in it. Ask students how this problem would change if the rent increased slightly for a larger number of students.

Interview Note
Exs. 13, 14 Students will see in Ex. 13, that the apparent size of the sun from the planets in our solar system varies directly with the distance of that planet from the sun. In Ex. 14, students will have the opportunity to draw a model to help them visualize how large the sun appears from the planets in our solar system.

9.6 Exercises and Applications

Extra Practice exercises on page 763

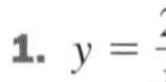

Technology For each equation, tell whether y varies inversely with x. If so, state the constant of variation and graph the equation using a graphing calculator or graphing software.

1. $y = \frac{2}{x}$
2. $xy = 10$
3. $y = -\frac{7}{2}x$
4. $y = -\frac{7}{2x}$
5. $y = \frac{2}{x + 1}$
6. $x = -\frac{1.6}{y}$
7. $3xy = 15$
8. $xy^3 = 15$

Tell whether you think the two quantities show inverse variation. Explain your reasoning.

9. the possible length and width of a rectangular garden having an area of 72 ft^2
10. the number of college students sharing an apartment, and each student's rent
11. the distance driven in a car and the amount of gasoline used
12. the possible radius and height of a can holding 1 L of juice

INTERVIEW Vera Rubin

Look back at the article on pages 386–388.

Astronomers like Vera Rubin use a measurement called **apparent size** *to describe how big an object in space looks from different locations. For example, the apparent size of the sun from the planet Mercury is about 2.58. This means that the sun appears about 2.58 times as wide on Mercury as it does on Earth.*

Planet	Mean distance from sun (millions of miles)	Apparent size of sun
Mercury	36.0	2.58
Venus	67.2	1.38
Earth	93.0	1.00
Mars	142	0.655
Jupiter	484	0.192
Saturn	885	0.105
Uranus	1780	0.0522
Neptune	2790	0.0333
Pluto	3660	0.0254

13. Use the table.
    a. Show that the apparent size of the sun varies inversely with the distance from where it is observed.
    b. Write an equation giving the sun's apparent size s from a point d million miles away.
    c. Ceres is an asteroid located between the orbits of Mars and Jupiter. The mean distance of Ceres from the sun is about 257 million miles. What is the apparent size of the sun from Ceres?

14. **Visual Thinking** Draw a circle representing the sun as seen from Earth. Use the diameter of this circle and the apparent sizes in the table to draw eight more circles representing the sun as seen from each of the other eight planets.

Exercises and Applications

Answers that involve approximations may vary due to rounding. Examples are given.

1. Yes; 2.

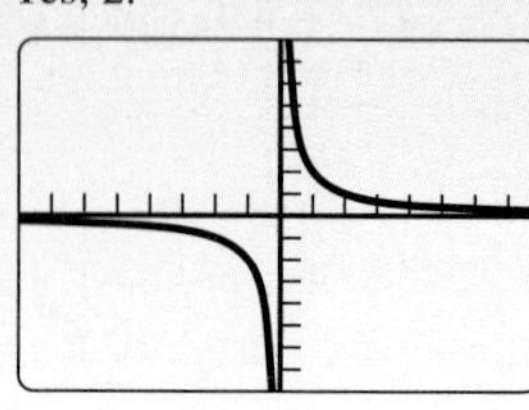

2. Yes; 10.

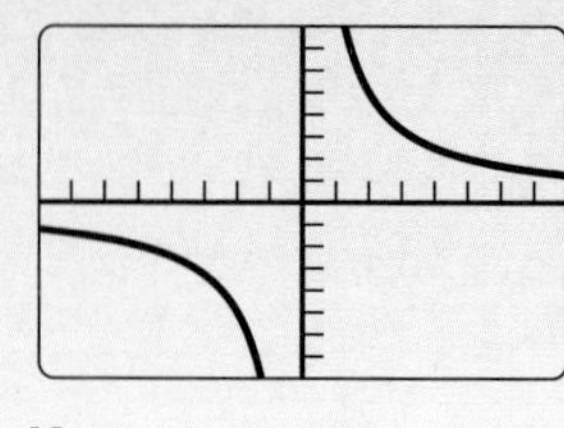

3. No.

4. Yes; $-\frac{7}{2}$.

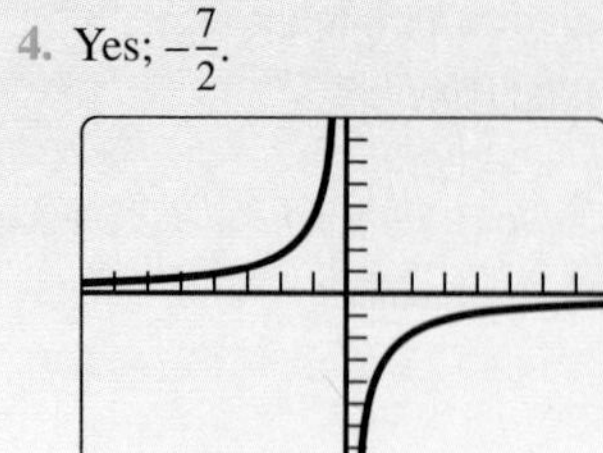

5. No.

6–14. See answers in back of book.

Connection METEOROLOGY

On May 11, 1970, a powerful tornado struck Lubbock, Texas. The diagram below shows the horizontal cross section of the tornado at ground level. Notice that a tornado consists of a central **core region** and a surrounding free vortex region.

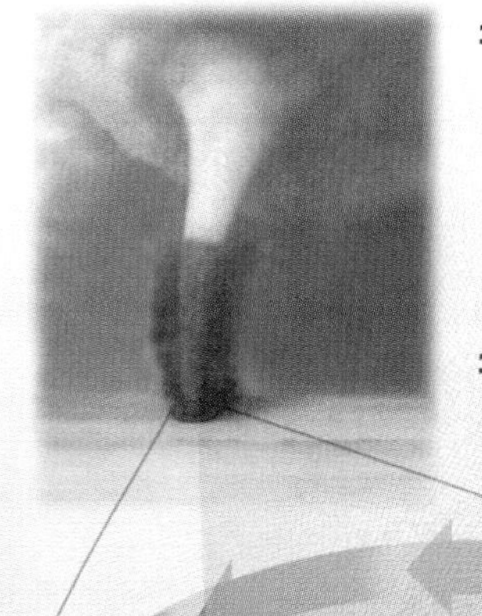

The May 11, 1970, tornado caused major damage to Lubbock, Texas.

15. In the core region, the rotational wind speed s varies *directly* with the distance d from the tornado's center. Use the information in the diagram to estimate the rotational wind speed 700 ft from the center of the Lubbock tornado.

16. In the free vortex region, s varies *inversely* with d. Use the information in the diagram to estimate the rotational wind speed 2000 ft from the center of the Lubbock tornado.

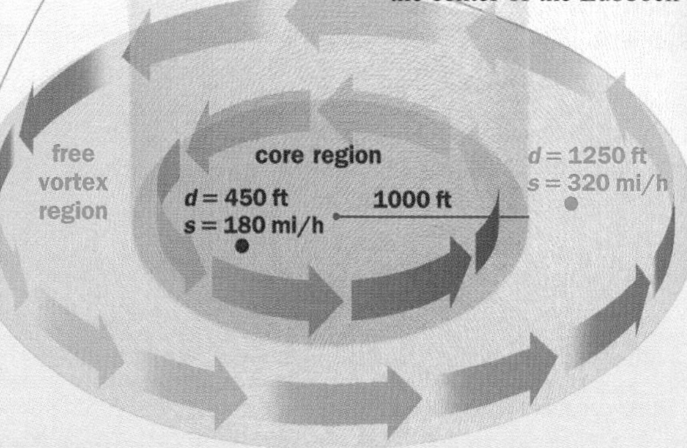

17. a. Write an equation giving s as a function of d in the core region of the Lubbock tornado. What is the function's domain?

b. Write an equation giving s as a function of d in the free vortex region of the Lubbock tornado. What is this function's domain?

c. Challenge Draw a graph showing s as a function of d for $0 \le d \le 4000$.

BY THE WAY

Tornadoes rotate counterclockwise in the Northern Hemisphere and clockwise in the Southern Hemisphere.

18. Use the equations and graph from Exercise 17.

a. How far from the center of the Lubbock tornado did the maximum rotational wind speed occur? What is the significance of this distance? What was the maximum speed?

b. How far from the center of the tornado was the rotational wind speed 300 mi/h?

19. Open-ended Problem Draw a graph of a function having $x = 3$ as a vertical asymptote and $y = -2$ as a horizontal asymptote.

20. SAT/ACT Preview Suppose $A = \frac{a}{x}$ and $B = ax$ where a is a positive constant. Which of the following is true for all positive values of x?

A. $A > B$ **B.** $B > A$ **C.** $A = B$ **D.** relationship cannot be determined

15. 280 mi/h

16. 200 mi/h

17. a. $s = 0.4d$; $0 \le d \le 1000$

b. $s = \frac{400{,}000}{d}$; $d \ge 1000$

c.

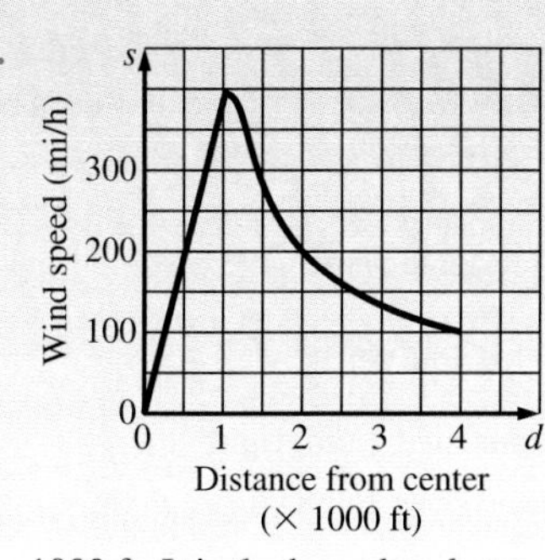

18. a. 1000 ft; It is the boundary between the core region and the free vortex region; 400 mi/h.

b. 750 ft and 1333 ft

19. Answers may vary. An example is given.

20. D

Apply⇔Assess

Exercise Notes

Communication: Reading
Exs. 15–18 Students should read these exercises carefully and study the diagram to be sure they understand the information they must use to solve the problems.

Research
Exs. 15–18 Some students may find it interesting to research tornadoes, hurricanes, and related weather phenomena to see how they are defined by meteorologists. A brief report on these types of storms, illustrated with diagrams, could be of interest to the entire class.

Using Technology
Exs. 17, 18 To help students draw the graph in part (c) of Ex. 17, you might suggest that they graph the equations from parts (a) and (b) simultaneously using a graphing calculator or graphing software. They should realize that the graph they need is the graph from part (a) until $d = 1000$ and the one from part (b) after $d = 1000$. Students using a TI-82 can make use of the intersect and evaluate options from the CALC menu when answering Ex. 18.

Problem Solving
Ex. 19 One approach students can use to draw the graph is to translate the graph of an equation of the form $y = \frac{k}{x} (k > 0)$ 3 units to the right and 2 units down. This method also yields an equation for the function $f(x) = \frac{k}{x-3} - 2$. In this case, the graph is a hyperbola. Students should see that curves other than hyperbolas are possible. For example, the graph of any function of the form $f(x) = \frac{k}{(x-3)^n} - 2$, where k is a nonzero constant and n is a whole number greater than 1, satisfies the conditions of the problem.

Reasoning
Ex. 20 Ask a few volunteers to explain the reasoning behind their choice of answer. This can help all students gain a greater understanding of the relationships between direct and indirect variation equations.

Apply⇔Assess

Exercise Notes

Application

Ex. 21 Part (c) provides an excellent opportunity for students to apply reasoning and mathematics to a real-world situation. Encourage students to comment on any difficulties they encounter and also comment on how they overcame each difficulty.

Visual Thinking

Ex. 22 Ask students to create sketches of the graphs for each real-life example. Encourage them to show their graphs to the class and to explain how they demonstrate inverse and direct variation. Ask them to explain what the graphs mean in relation to the real-life situations they represent. This activity involves the visual skills of *recognition* and *communication*.

Practice 59 for Section 9.6

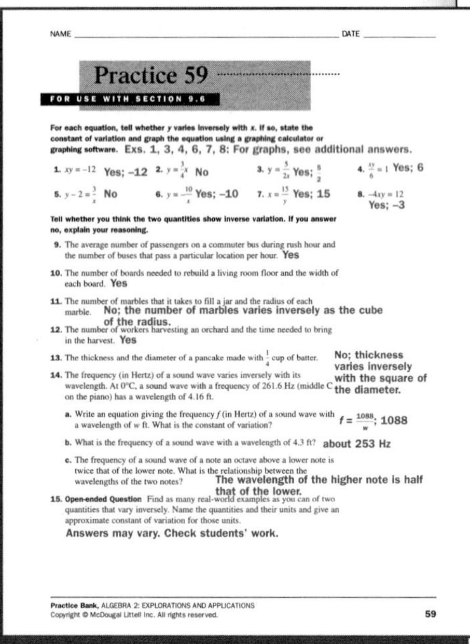

NAME ______________________ DATE __________

Practice 59

FOR USE WITH SECTION 9.6

For each equation, tell whether y varies inversely with x. If so, state the constant of variation and graph the equation using a graphing calculator or graphing software. Exs. 1, 3, 4, 6, 7, 8: For graphs, see additional answers.

1. $xy = -12$ Yes; -12
2. $y = \frac{3}{4}x$ No
3. $y = \frac{5}{2x}$ Yes; $\frac{5}{2}$
4. $\frac{xy}{6} = 1$ Yes; 6
5. $y - 2 = \frac{3}{x}$ No
6. $y = -\frac{10}{x}$ Yes; -10
7. $x = \frac{15}{y}$ Yes; 15
8. $-4xy = 12$ Yes; -3

Tell whether you think the two quantities show inverse variation. If you answer no, explain your reasoning.

9. The average number of passengers on a commuter bus during rush hour and the number of buses that pass a particular location per hour. Yes
10. The number of boards needed to rebuild a living room floor and the width of each board. Yes
11. The number of marbles that it takes to fill a jar and the radius of each marble. No; the number of marbles varies inversely as the cube of the radius.
12. The number of workers harvesting an orchard and the time needed to bring in the harvest. Yes
13. The thickness and the diameter of a pancake made with $\frac{1}{4}$ cup of batter. No; thickness varies inversely with the square of the diameter.
14. The frequency (in Hertz) of a sound wave varies inversely with its wavelength. At 0°C, a sound wave with a frequency of 261.6 Hz (middle C on the piano) has a wavelength of 4.16 ft.
    a. Write an equation giving the frequency f (in Hertz) of a sound wave with a wavelength of w ft. What is the constant of variation? $f = \frac{1088}{w}$; 1088
    b. What is the frequency of a sound wave with a wavelength of 4.3 ft? about 253 Hz
    c. The frequency of a sound wave of a note an octave above a lower note is twice that of the lower note. What is the relationship between the wavelengths of the two notes? The wavelength of the higher note is half that of the lower.
15. **Open-ended Question** Find as many real-world examples as you can of two quantities that vary inversely. Name the quantities and their units and give an approximate constant of variation for those units.
    Answers may vary. Check students' work.

Practice Bank, ALGEBRA 2: EXPLORATIONS AND APPLICATIONS
 59

21. **HYDRAULICS** The time t required to fill a container with water from a hose is given by the equation

$$t = \frac{V}{As}$$

where V is the volume of the container, A is the cross-sectional area of the hose, and s is the speed of the water from the hose.

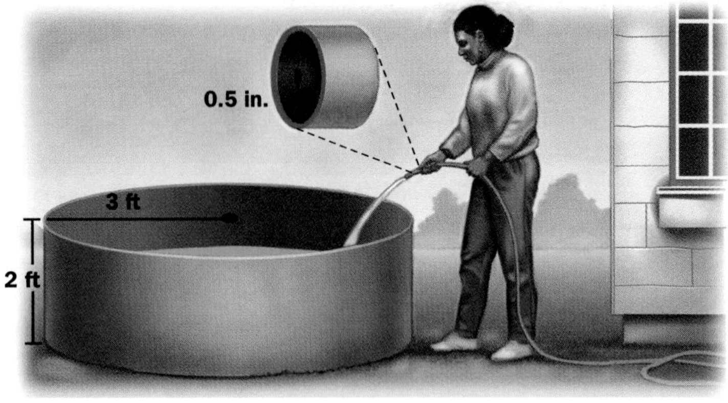

a. Use the diagram. Find an equation giving the time required to fill the swimming pool as a function of the water speed. How long will it take to fill the pool if the water speed is 10 ft/s? (*Hint:* First convert the units of the hose radius to feet.)

b. In part (a), does time vary inversely with water speed? If so, what is the constant of variation?

c. **Cooperative Learning** Working with a partner, devise an experiment in which you use the equation $t = \frac{V}{As}$ to estimate the speed of water from a garden hose. If you have a garden hose available, perform the experiment and write a report summarizing your results.

ONGOING ASSESSMENT

22. **Writing** Explain how inverse variation differs from direct variation. Give real-life examples illustrating each type of variation.

SPIRAL REVIEW

Find all real and imaginary zeros of each function. Identify any double or triple zeros. *(Section 9.5)*

23. $f(x) = 6x^3 - 23x^2 - 5x + 4$
24. $g(x) = 15x^3 - x^2 + 3x - 2$
25. $h(x) = -4x^4 - 12x^3 - 5x^2 + 16x - 5$
26. $f(x) = 18x^5 + 63x^4 + 82x^3 + 48x^2 + 12x + 1$

Describe how the graph of each equation is related to the graph of $y = 5x^2$. *(Section 5.2)*

27. $y = 5(x - 1)^2$
28. $y = 5x^2 + 3$
29. $y = 5(x - 1)^2 + 3$
30. $y = 5(x + 1)^2 - 3$

21. a. Let t = time in seconds and s = speed in ft/s; $t = \frac{10{,}368}{s}$; 1036.8 s ≈ 17.3 min
    b. Yes; 10,368.
    c. Fill a container with water. Determine the volume of the container and the cross-sectional area of the hose. Record how long it takes to fill the container, then use the formula $s = \frac{V}{tA}$ to estimate the speed of the water from the hose.
22. Answers may vary. An example is given. If two quantities vary inversely, as one increases, the other decreases. For example, as the number of people sharing a pizza equally increases, the number of slices per person decreases. If two quantities vary directly, as one increases, the other also increases. For example, if each person in a group is to be given two slices of pizza, as the size of the group increases, the number of slices increases as well.
23. $\frac{1}{3}; -\frac{1}{2}; 4$
24. $\frac{2}{5}; \frac{-1 \pm i\sqrt{11}}{6}$
25. $\frac{1}{2}$ (double zero); $-2 \pm i$
26. -1 (triple zero); $-\frac{1}{3}; -\frac{1}{6}$
27. translated 1 unit right
28. translated 3 units up
29. translated 1 unit right and 3 units up
30. translated 1 unit left and 3 units down

Section

9.7 Working with Simple Rational Functions

Learn how to...

- **identify important features of translated hyperbolas**
- **find equations of translated hyperbolas**

So you can...

- **understand population models, for example**

In Chapter 5, you learned how to translate parabolas by introducing new constants in their equations. As you'll see in the Exploration, you can do the same thing with hyperbolas.

EXPLORATION
COOPERATIVE LEARNING

Investigating Translated Hyperbolas

Work with a partner.
You will need: • a graphing calculator or graphing software

1 Graph the following equations. For each graph, identify the asymptotes and tell how the graph is related to the graph of $y = \frac{1}{x}$.

- $y = \frac{1}{x-1}$
- $y = \frac{1}{x-3}$
- $y = \frac{1}{x+2}$
- $y = \frac{1}{x+4}$

Example: $y = \frac{1}{x-2}$

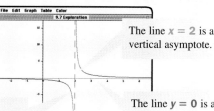

The line $x = 2$ is a vertical asymptote.

The line $y = 0$ is a horizontal asymptote.

The graph is the hyperbola $y = \frac{1}{x}$ translated 2 units right.

2 Repeat Step 1 using these equations.

- $y = \frac{1}{x} + 1$
- $y = \frac{1}{x} + 3$
- $y = \frac{1}{x} - 2$
- $y = \frac{1}{x} - 4$

3 Predict how the graph of each equation below is related to the graph of $y = \frac{2}{x}$. Use your calculator or software to check your predictions.

- $y = \frac{2}{x-4}$
- $y = \frac{2}{x+1}$
- $y = \frac{2}{x} + 3$
- $y = \frac{2}{x} - 5$

Exploration Note

Purpose
The purpose of this Exploration is to have students discover how the graph of an equation of the form $y = \frac{a}{x-h} + k$ is related geometrically to the graph of $y = \frac{a}{x}$.

Materials/Preparation
Each pair of students needs a graphing calculator or graphing software.

Procedure
As students complete Steps 1–3, they may find it helpful to sketch each graph including asymptotes. They should write the equation alongside the sketch. This procedure can facilitate a class discussion and provide students with a record of their work.

Closure
Discuss the graphs and asymptotes for all equations. Students should be able to predict accurately how the graph of any equation of the form $y = \frac{a}{x-h} + k$ is related to the graph of $y = \frac{a}{x}$.

Explorations Lab Manual
See the Manual for additional commentary on this Exploration.

For answers to the Exploration, see answers in back of book.

Plan⇔Support

Objectives

- Identify important features of translated hyperbolas.
- Find equations of translated hyperbolas.
- Interpret real-world situations using models whose graphs are translated hyperbolas.

Recommended Pacing

Core and Extended Courses
Section 9.7: 1 day

Block Schedule
Section 9.7: $\frac{1}{2}$ block
(with Section 9.6)

Resource Materials

Lesson Support
Lesson Plan 9.7
Warm-Up Transparency 9.7
Practice Bank: Practice 60
Study Guide: Section 9.7
Explorations Lab Manual: Diagram Master 2
Assessment Book: Test 36

Technology
Graphing Calculator
McDougal Littell Mathpack *Function Investigator with Matrix Analyzer Activity Book:* Function Investigator Activity 28

Internet:
http://www.hmco.com

Warm-Up Exercises

Tell how the graph of each equation is related to the graph of $y = x^2$ or the graph of $y = -x^2$.

1. $y = (x - 3)^2 + 5$
graph of $y = x^2$ translated 3 units right and 5 units up

2. $y = -(x + 2)^2 + 8$
graph of $y = -x^2$ translated 2 units left and 8 units up

3. $y = \left(x + \frac{3}{2}\right)^2 - 4$
graph of $y = x^2$ translated $\frac{3}{2}$ units left and 4 units down

4. $y = -(x - 7)^2 - 1$
graph of $y = -x^2$ translated 7 units right and 1 unit down

Teach⇔Interact

Think and Communicate

After discussing questions 1 and 2 and the paragraph that follows question 2, generalize what has been discovered. Ask students how the graph of $y = f(x - h) + k$ is related to the graph of $y = f(x)$. (The graph of $y = f(x - h) + k$ is just like the graph of $y = f(x)$, except that it has been translated horizontally $|h|$ units and vertically $|k|$ units.)

About Example 1

Mathematical Procedures

Students have just seen how to obtain the graph of $y = \frac{a}{x - h} + k$ by translating the graph of $y = \frac{a}{x}$. In this Example, students are given the graph, asymptotes, and one point on the graph and asked to find the equation. Knowing the asymptotes gives the values of h and k. Knowing the coordinates of one point on the graph makes it possible to use substitution to find the value of a.

Additional Example 1

Find an equation for the function g whose graph is the hyperbola shown.

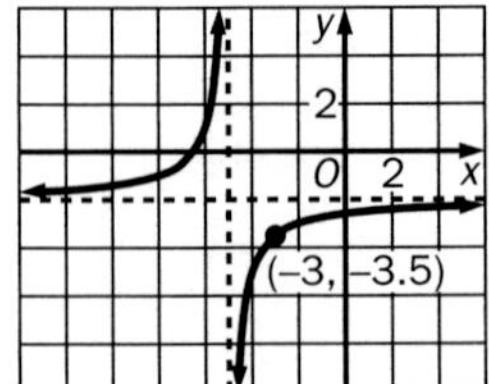

A vertical asymptote of the graph is $x = -5$, and a horizontal asymptote is $y = -2$. Therefore, an equation for g has this form:

$g(x) = \frac{a}{x + 5} - 2$

To find the value of a, use the fact that $f(-3) = -3.5$:

$$\frac{a}{-3 + 5} - 2 = -3.5$$
$$\frac{a}{2} = -1.5$$
$$a = -3$$

An equation for g is $g(x) = \frac{-3}{x + 5} - 2$.

THINK AND COMMUNICATE

1. How is the graph of $y = \frac{a}{x - h}$ geometrically related to the graph of $y = \frac{a}{x}$ when h is positive? when h is negative?

2. How is the graph of $y = \frac{a}{x} + k$ geometrically related to the graph of $y = \frac{a}{x}$ when k is positive? when k is negative?

You can obtain the graph of $y = \frac{a}{x - h} + k$ by translating the graph of $y = \frac{a}{x}$ **horizontally h units** and **vertically k units.**

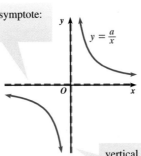

horizontal asymptote: $y = 0$

$y = \frac{a}{x}$

vertical asymptote: $x = 0$

horizontal asymptote: $y = k$

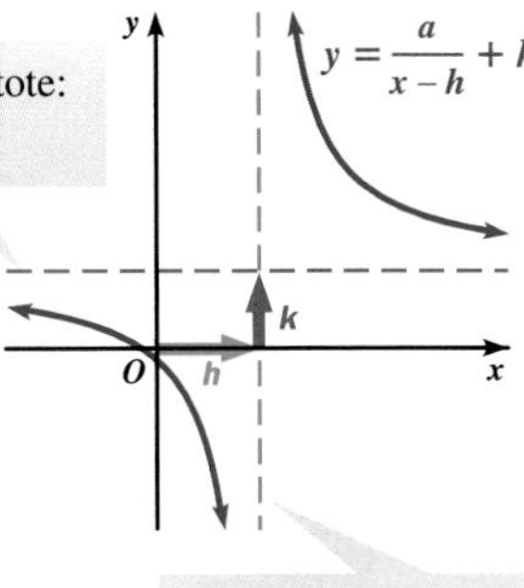

vertical asymptote: $x = h$

EXAMPLE 1

Find an equation for the function f whose graph is the hyperbola shown.

SOLUTION

A vertical asymptote of the graph is $x = 3$, and a horizontal asymptote is $y = 1$. Therefore, an equation for f has this form:

$$f(x) = \frac{a}{x - 3} + 1$$

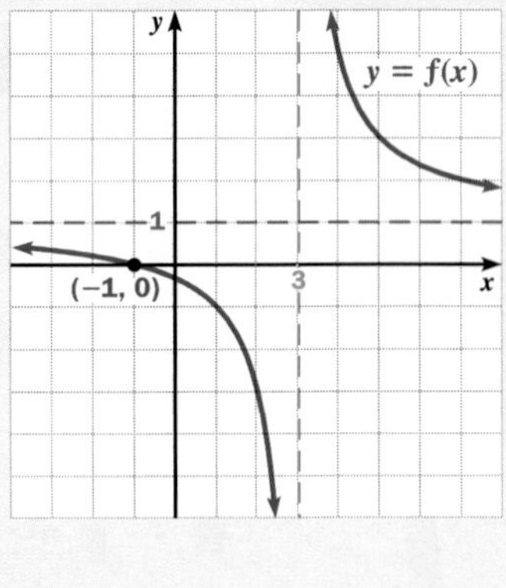

To find the value of a, use the fact that $f(-1) = 0$:

$$\frac{a}{-1 - 3} + 1 = 0$$
$$\frac{a}{-4} = -1$$
$$a = 4$$

An equation for f is $f(x) = \frac{4}{x - 3} + 1$.

ANSWERS Section 9.7

Think and Communicate

1. It is the graph of $y = \frac{a}{x}$ translated h units right; it is the graph of $y = \frac{a}{x}$ translated $|h|$ units left.

2. It is the graph of $y = \frac{a}{x}$ translated k units up; it is the graph of $y = \frac{a}{x}$ translated $|k|$ units down.

Rational Functions

Notice that the function $f(x) = \frac{4}{x-3} + 1$ in Example 1 can be written this way:

Toolbox p. 783
Adding and Subtracting Rational Expressions

$$\begin{aligned} f(x) &= \frac{4}{x-3} + 1 \\ &= \frac{4}{x-3} + \frac{x-3}{x-3} \\ &= \frac{4 + x - 3}{x-3} \\ &= \frac{x+1}{x-3} \begin{array}{l} \leftarrow \text{polynomial} \\ \leftarrow \text{polynomial} \end{array} \end{aligned}$$

This is an example of a *rational function.* A **rational function** has the form $f(x) = \frac{p(x)}{q(x)}$ where $p(x)$ and $q(x)$ are polynomials. In this section, $p(x)$ and $q(x)$ will always be constant or first-degree polynomials.

EXAMPLE 2

Let $g(x) = \frac{-6x+7}{2x+1}$. Find the asymptotes of the graph of g, and tell how the graph is related to a hyperbola with equation of the form $y = \frac{a}{x}$.

SOLUTION

Step 1 Divide the denominator of $g(x)$ into the numerator.

$$\begin{array}{r} -3 \\ 2x+1 \enclose{longdiv}{-6x+7} \\ \underline{-6x-3} \\ 10 \end{array}$$

So $\frac{-6x+7}{2x+1} = -3 + \frac{10}{2x+1}$.

Step 2 Write $g(x)$ in the form $g(x) = \frac{a}{x-h} + k$.

$$\begin{aligned} g(x) &= \frac{-6x+7}{2x+1} \\ &= \frac{10}{2x+1} - 3 \\ &= \frac{\frac{1}{2}}{\frac{1}{2}} \cdot \frac{10}{2x+1} - 3 \\ &= \frac{5}{x+\frac{1}{2}} - 3 \\ &= \frac{5}{x-\left(-\frac{1}{2}\right)} + (-3) \end{aligned}$$

A vertical asymptote is $x = -\frac{1}{2}$. A horizontal asymptote is $y = -3$. The graph of g is the graph of $y = \frac{5}{x}$ translated $\frac{1}{2}$ unit left and 3 units down.

Teach⇔Interact

About Example 2

Reasoning

Ask students how they can tell that $x = -\frac{1}{2}$ is a vertical asymptote, even before $g(x)$ is written in the form $g(x) = \frac{a}{x-h} + k$. (The value of the denominator, $2x + 1$, is 0 when $x = -\frac{1}{2}$. As x approaches $-\frac{1}{2}$, the quotient $\frac{-6x+7}{2x+1}$ therefore becomes very large in absolute value.)

Additional Example 2

Let $f(x) = \frac{-6x+11}{-2x+5}$. Find the asymptotes of the graph of f, and tell how the graph is related to a hyperbola with equation of the form $y = \frac{a}{x}$.

Step 1 Divide the denominator of $f(x)$ into the numerator.

$$\begin{array}{r} 3 \\ -2x+5 \enclose{longdiv}{-6x+11} \\ \underline{-6x+15} \\ -4 \end{array}$$

So $\frac{-6x+11}{-2x+5} = 3 + \frac{-4}{-2x+5} = 3 + \frac{4}{2x-5}$.

Step 2 Write $f(x)$ in the form $f(x) = \frac{a}{x-h} + k$.

$$\begin{aligned} f(x) &= \frac{4}{2x-5} + 3 \\ &= \frac{\frac{1}{2}}{\frac{1}{2}} \cdot \frac{4}{2x-5} + 3 \\ &= \frac{2}{x-\frac{5}{2}} + 3 \end{aligned}$$

A vertical asymptote is $x = \frac{5}{2}$.

A horizontal asymptote is $y = 3$.

The graph of f is the graph of $y = \frac{2}{x}$ translated $\frac{5}{2}$ units right and 3 units up.

Teach⇔Interact

About Example 3

Second-Language Learners
Students learning English may wish to discuss their ideas about the *doomsday model* with an English-proficient partner and then complete the rest of the writing task independently.

Additional Example 3

A company specializing in environmental clean-up claimed to be able to remove a certain chemical pollutant from the soil in a $\frac{1}{4}$-acre area for a cost described by the equation $C = \frac{1425}{100 - x}$, where C is in dollars and x is the number ($0 < x < 100$) that tells what percent of the pollutant is removed.

a. Graph the cost equation using a graphing calculator or graphing software. What is the predicted cost of removing 99% of the pollutant from a $\frac{1}{4}$-acre area?

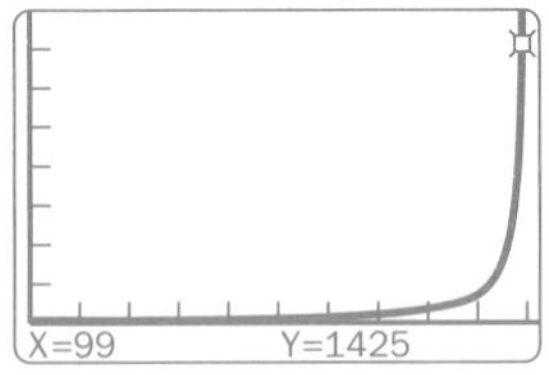

The vertical asymptote is $x = 100$. The predicted cost for removing 99 percent of the pollutant is $1425.

b. According to the company's cost equation, how does the clean-up cost change as the percent of the pollutant to be removed approaches 100 percent? Would it be reasonable for a developer who wants to build houses on 100 half-acre plots to buy the land and sign a contract for having 99.95% of the pollutant removed before building begins? The clean-up cost increases very rapidly as the percent of the pollutant to be removed approaches 100%. The clean-up cost being considered by the housing developer would be $28,500 for each $\frac{1}{4}$ acre, or $57,000 for each of the 100 half-acre plots. Thus, because of this high cost, it would not be reasonable for the developer to sign a contract to remove 99.95% of the pollutant.

THINK AND COMMUNICATE

3. In Example 2, how do you know that the direction of the translation is left and down?

4. a. What are the domain and range of the function g in Example 2?

b. Generalize your answer to part (a) by identifying the domain and range of any function that can be written in the form $f(x) = \frac{a}{x - h} + k$.

EXAMPLE 3 Application: Population Growth

In 1960, three scientists from the University of Illinois published a paper claiming that the world population P in some future year t can be predicted with the equation

$$P = \frac{179}{2026 - t}$$

where P is in billions and $t < 2026$. This population model is often called the "doomsday" model.

a. Graph the scientists' equation using a graphing calculator or graphing software. What is the predicted world population in the year 2000?

b. Writing According to the doomsday model, what general trend in world population will occur in future years? Why is the name "doomsday" appropriate?

SOLUTION

a.

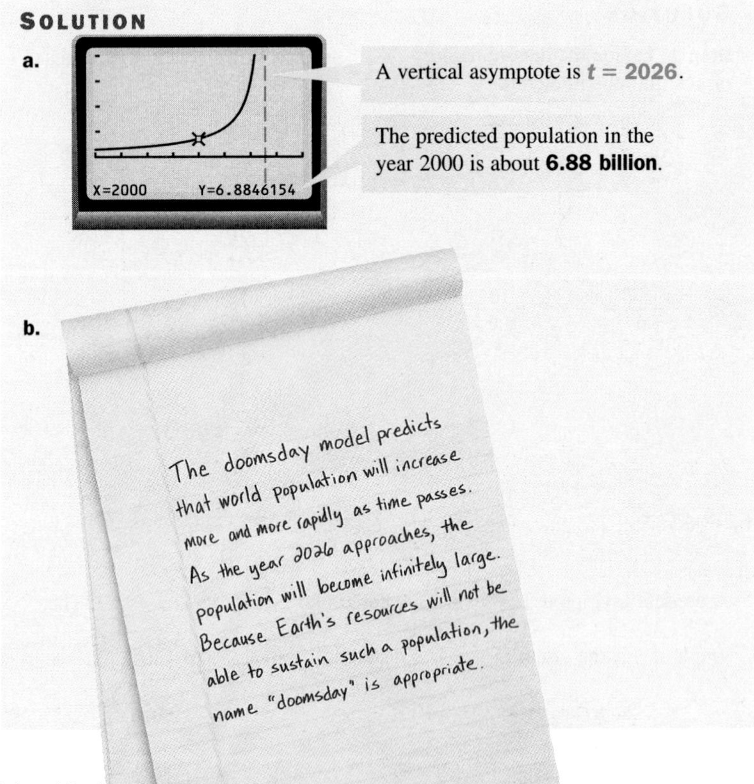

X=2000 Y=6.8846154
A vertical asymptote is t = 2026.
The predicted population in the year 2000 is about 6.88 billion.
b.
The doomsday model predicts that world population will increase more and more rapidly as time passes. As the year 2026 approaches, the population will become infinitely large. Because Earth's resources will not be able to sustain such a population, the name "doomsday" is appropriate.

Think and Communicate

3. h and k are both negative.

4. a. all real numbers except $-\frac{1}{2}$; all real numbers except -3

b. all real numbers except h; all real numbers except k

THINK AND COMMUNICATE

5. Do you think the doomsday model is realistic? Why or why not?

6. The world population in 1990 was about 5.29 billion. How does this figure compare with the prediction from the doomsday model?

CHECKING KEY CONCEPTS

Match each equation with its graph.

1. $y = \frac{3}{x+4} + 2$ **2.** $y = \frac{3}{x+4} - 2$

3. $y = \frac{3}{x-4} + 2$ **4.** $y = \frac{3}{x-4} - 2$

A.
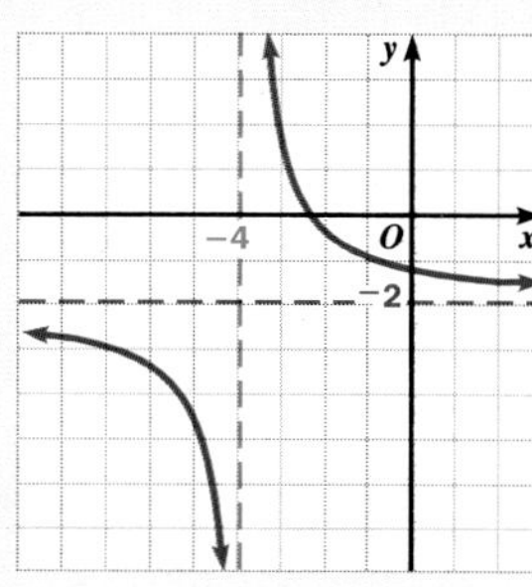

B.
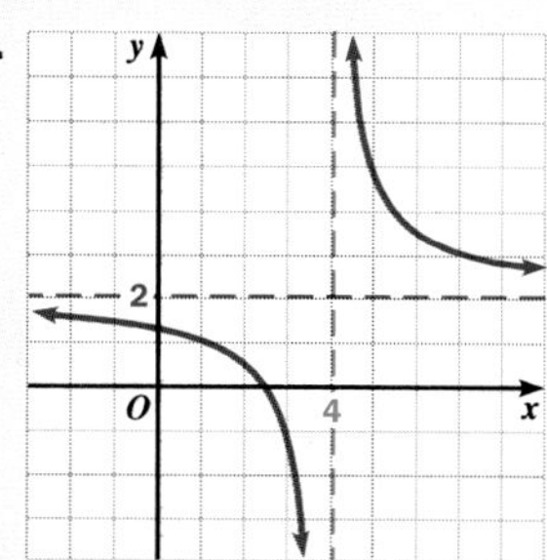

C.
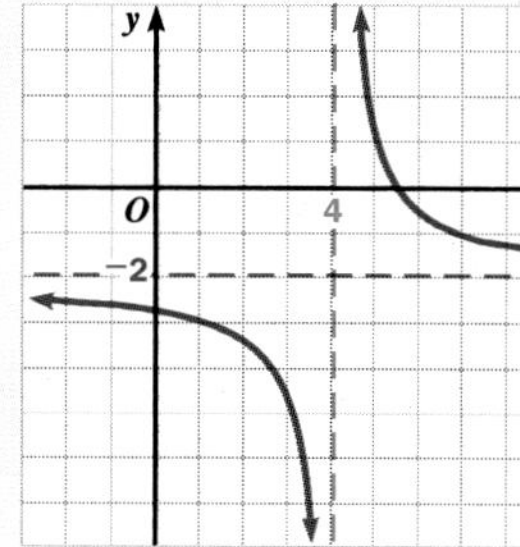

D.
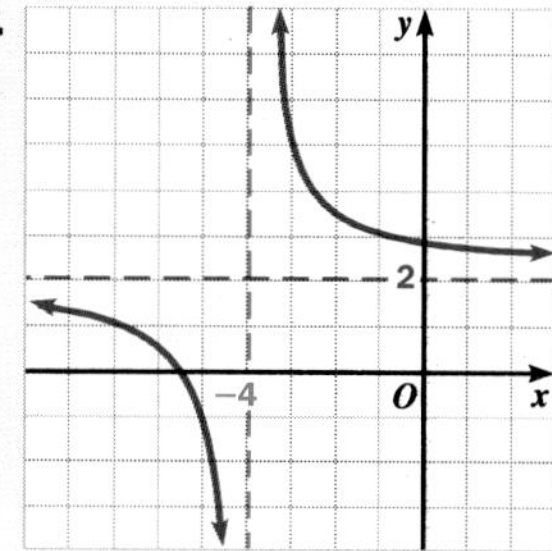

9.7 Exercises and Applications

Extra Practice exercises on page 763

Tell whether each function is a rational function. Explain your answers.

1. $f(x) = \frac{2x+3}{5x-1}$ **2.** $f(x) = \frac{\sqrt{x-4}}{x}$

3. $g(t) = \frac{t^2+1}{2^t+1}$ **4.** $h(t) = \frac{t^2+1}{6t^3 - t^2 - 12}$

5. Challenge Show that the equation $xy - x + 2y = 5$ defines a rational function.

Teach⇔Interact

Think and Communicate

Questions 5 and 6 point out the danger of *extrapolation*, or making predictions beyond the scope of the data upon which a model is based. The doomsday model was created in 1960 and thus based on data prior to that time. Students should use extrapolation with caution, being aware that predictions may be far from actual values.

Closure Question

Is the function $y = \frac{a}{x-h} + k$ an example of inverse variation? If so, give the constant of variation. If not, explain how it is related to an inverse variation function.

If h and k are both zero, then this is inverse variation with constant of variation a. Otherwise, the function is not inverse variation, but the graph of this function is the graph of the inverse variation function $y = \frac{a}{x}$ translated $|h|$ units horizontally and $|k|$ units vertically.

Apply⇔Assess

Suggested Assignment

Core Course
Exs. 1–4, 6–11, 16–23, 25–33, AYP

Extended Course
Exs. 1–33, AYP

Block Schedule
Day 59 Exs. 1–4, 6–11, 16–23, 25–34, AYP

Exercise Notes

Reasoning
Ex. 4 Ask whether it is possible to say, without graphing the function, whether the graph of $h(t)$ is a hyperbola. Students should be able to explain their reasoning. (No; the graph is not a hyperbola. The denominator of $h(t)$ can be factored to obtain $(3t + 4)(2t - 3)$. Hence, there are two vertical asymptotes. A hyperbola cannot have more than one vertical asymptote.)

Think and Communicate

5. Answers may vary. An example is given. I do not think the doomsday model is realistic. As births increase, so will deaths. Throughout history, population growth has been limited by disease, natural disasters, war, and the inability of limited resources to support unlimited growth.

6. It is more than the 4.97 billion predicted by the model.

Checking Key Concepts

1. D
2. A
3. B
4. C

Exercises and Applications

Answers that involve approximations may vary due to rounding. Examples are given.

1. Yes; $2x + 3$ and $5x - 1$ are both polynomials.
2. No; $\sqrt{x-4}$ is not a polynomial.
3. No; $2^t + 1$ is not a polynomial.
4. Yes; $t^2 + 1$ and $6t^3 - t^2 - 12$ are both polynomials.
5. $xy - x + 2y = 5$ is equivalent to $y = \frac{x+5}{x+2}$.

Apply⇔Assess

Exercise Notes

Cooperative Learning
Exs. 6–11 You may wish to have students work on these exercises with a partner.

Mathematical Procedures
Exs. 6–11 Students having difficulty with these exercises should refer back to Example 1 on page 434. You may want to have these students use this Example to write the procedure for finding the equation of a hyperbola as a series of steps. These steps could then be used when completing these exercises.

Student Study Tip
Exs. 6–11 Remind students that when the branches of the hyperbola are above and to the right of the asymptotes, and below and to the left of the asymptotes, the constant a in $y = \frac{a}{x-h} + k$ is positive, and when the branches of the hyperbola are below and to the right of the asymptotes, and above and to the left of the asymptotes, the constant a is negative. Have students use this fact to check their equations.

Application
Exs. 12–15 Students examine a rational function in Exs. 12 and 13 that gives the percent of oxygen a diver needs to breath at a given depth. In Exs. 14 and 15, a rational function is used to study the relationship between air temperature, distance, and the time it takes for thunder to be heard.

Research
Exs. 12, 13 The introductory paragraph for these exercises states that oxygen under high pressure can be toxic to the human body. Students may be interested to find out why this is so.

Find an equation of each hyperbola.

6. 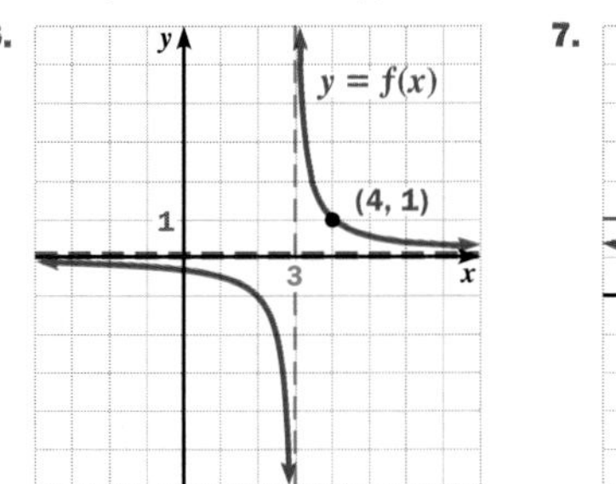

7. 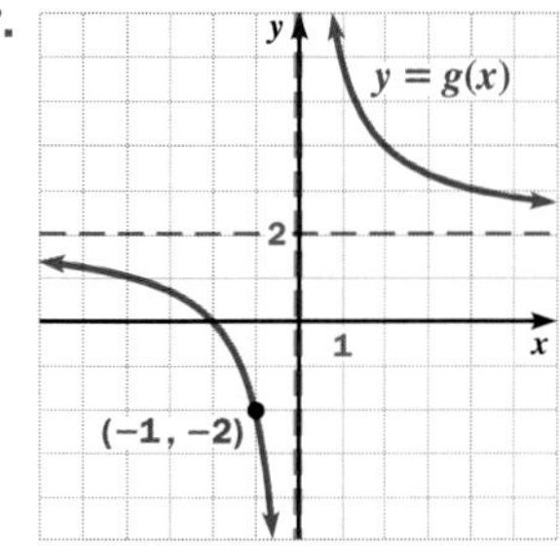

8. 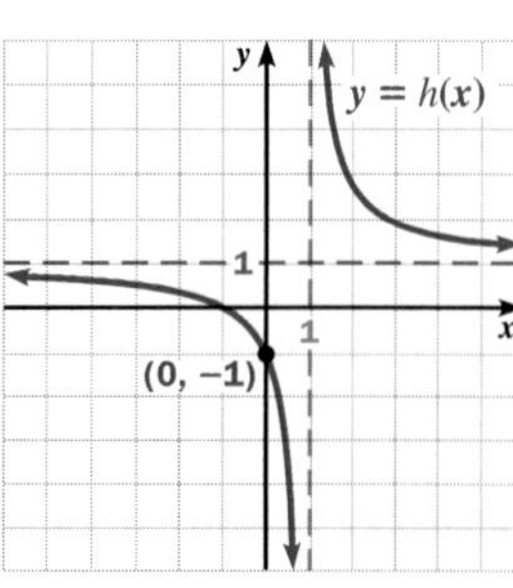

9. 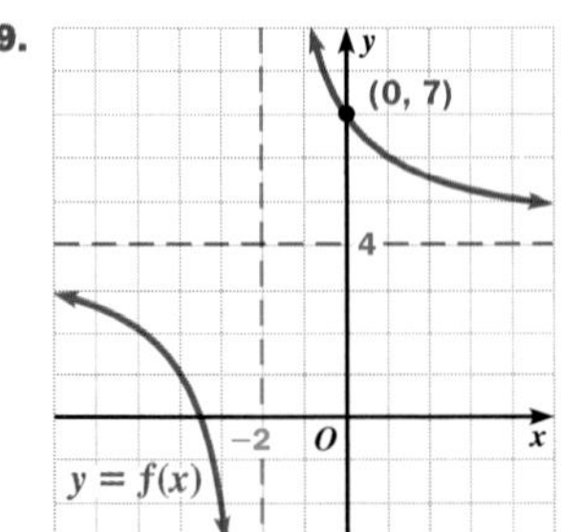

10. 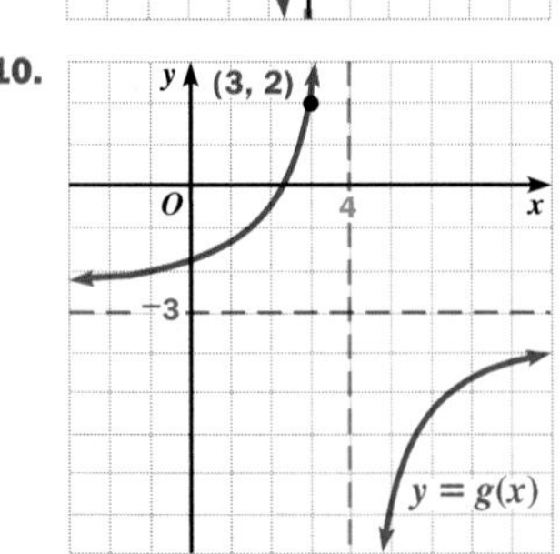

11. 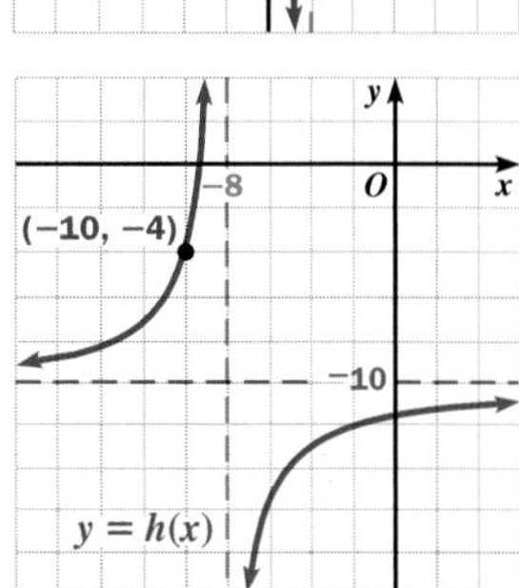

Connection SCUBA DIVING

You may be surprised to learn that oxygen under high pressure can be toxic to the human body. This is a concern for scuba divers, who often experience high pressure caused by the weight of the water above them. In order to breathe safely, divers working in deep water must fill their scuba tanks with air containing less oxygen than Earth's atmosphere.

BY THE WAY

Earth's atmosphere consists mainly of oxygen and nitrogen. However, because nitrogen under high pressure produces confusion and giddiness, divers working in deep water often breathe a mixture of oxygen and helium.

12. The recommended percent p of oxygen (by volume) in the air a diver breathes is given by

$$p = \frac{660}{d + 33}$$

where d is the depth (in feet) at which the diver is working.

a. 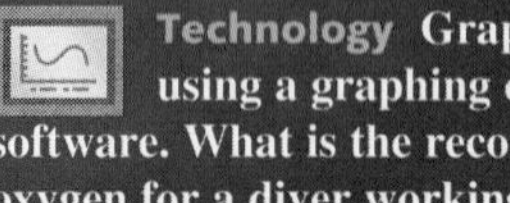 **Technology** Graph the given equation using a graphing calculator or graphing software. What is the recommended percent of oxygen for a diver working at a depth of 100 ft? 200 ft? 300 ft?

b. What value does the recommended percent of oxygen approach as a diver's depth increases?

13. Research Find out the percent of oxygen (by volume) in Earth's atmosphere. Compare this figure with the one given by the equation in Exercise 12 for a diver at the surface.

6. $f(x) = \frac{1}{x-3}$

7. $f(x) = \frac{4}{x} + 2$

8. $f(x) = \frac{2}{x-1} + 1$

9. $f(x) = \frac{6}{x+2} + 4$

10. $f(x) = \frac{-5}{x-4} - 3$

11. $f(x) = \frac{-12}{x+8} - 10$

12. a. 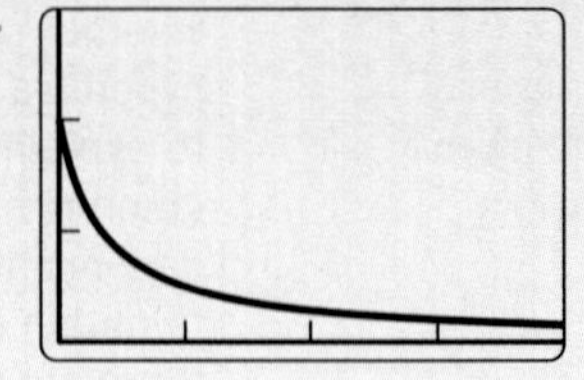

about 4.96%; about 2.83%; about 1.98%

b. 0

13. Answers may vary. Earth's atmosphere is about 21% oxygen (by volume). This is slightly greater than the percent recommended by the equation for a diver at the surface (20%).

14. a. $t = \frac{11{,}220}{1.09T + 1050}$

b. 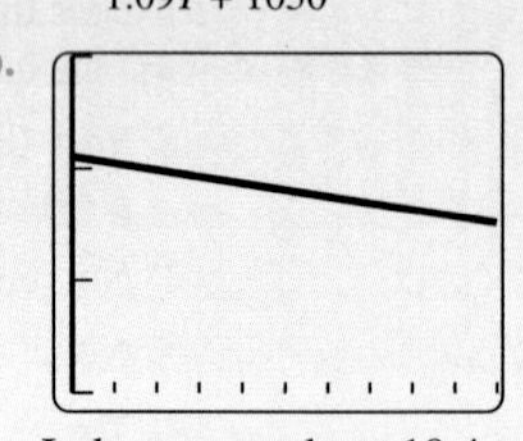

It decreases; about 10.4 s; about 10.1 s; about 9.8 s.

15. If you know the temperature (in °F), count the seconds between the time you see the flash and the time you hear the thunder. Use the formula $d = t(1.09T + 1050)$ to estimate how far away (in feet) the lightning is.

16. a. $x = 1$; $y = 2$

b. It is the graph of $y = \frac{7}{x}$ translated 1 unit right and 2 units up.

17. a. $x = \frac{3}{2}$; $y = 4$

b. It is the graph of $y = \frac{13}{x}$ translated $\frac{3}{2}$ units right and 4 units up.

Connection METEOROLOGY

When a bolt of lightning strikes, you don't hear thunder right away. The time t (in seconds) it takes for the thunder to reach your ears is given by

$$t = \frac{d}{1.09T + 1050}$$

where d is your distance from the lightning (in feet) and T is the air temperature (in degrees Fahrenheit).

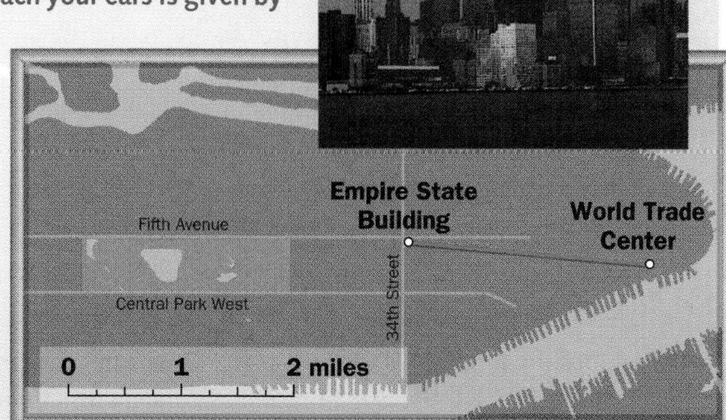

14. The World Trade Center in New York City is often struck by lightning during storms. Suppose you are at the Empire State Building when lightning strikes the World Trade Center. (See the map.)

a. Write an equation giving the time it takes to hear the thunder as a function of air temperature. (*Note:* 1 mi = 5280 ft)

b. **Technology** Graph the equation from part (a) using a graphing calculator or graphing software. How does the time it takes to hear the thunder change as air temperature increases? What is this time when the temperature is 30°F? 60°F? 90°F?

15. Writing Explain how you can use the equation $t = \frac{d}{1.09T + 1050}$ to estimate how far away lightning is.

For each function:

a. Find the asymptotes of the function's graph.

b. Tell how the function's graph is related to a hyperbola with equation of the form $y = \frac{a}{x}$.

16. $f(x) = \frac{2x + 5}{x - 1}$ **17.** $f(x) = \frac{8x + 14}{2x - 3}$ **18.** $f(x) = \frac{21x + 16}{3x + 7}$ **19.** $f(x) = \frac{3x + 11}{x + 2}$

20. $g(x) = \frac{-x + 5}{x + 4}$ **21.** $g(x) = \frac{-20x + 3}{4x + 1}$ **22.** $h(x) = \frac{-4x + 19}{7x - 21}$ **23.** $h(x) = \frac{-30x - 1}{5x - 4}$

24. CONSUMER ECONOMICS Tanya wants to order several pizzas from Marco's Pizza Palace and have them delivered to her graduation party. The price of each pizza is \$8, and there is a \$3 delivery charge regardless of the number of pizzas ordered.

a. Write an equation giving Tanya's total cost C if she orders p pizzas.

b. Use your answer from part (a) to write an equation giving the *average* cost A of a pizza if p pizzas are ordered.

c. Copy and complete the table. What value does the average cost of a pizza approach as the number of pizzas ordered increases? What does this value tell you about the effect of the delivery charge on large orders?

d. Write your equation from part (b) in the form $A = \frac{a}{p - h} + k$. How does this form show you what value the average cost approaches as the number of pizzas ordered increases?

p	A
2	?
5	?
10	?
20	?
50	?
100	?

Apply⇔Assess

Exercise Notes

Interdisciplinary Problems

Exs. 14, 15 Ask students to share any ideas they have about how physics might predict that the value of t (time) decreases as the value of T (temperature) increases.

Communication: Writing

Ex. 15 Students should realize that when one of the three variables in this expression is replaced by a constant, the result is a function of two variables. Ask students to determine which variable is the constant, which is the dependent variable, and which is the independent variable. Also, ask students what type of function results. (a direct variation function with independent variable t and dependent variable d)

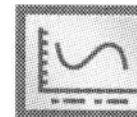

Using Technology

Exs. 16–23 Students can use a graphing calculator or the *Function Investigator* software to confirm whether they have correctly identified the asymptotes. Have students describe their methods to the class. The table features of a TI-82 or spreadsheet software can also be used to check the asymptotes. Again, ask students who use this approach to describe their methods to the class.

Challenge

Exs. 16–23 After completing these exercises, ask students if they can find a pattern that allows them to find the horizontal asymptote of the graph without doing the division. (For an equation of the form $y = \frac{ax + b}{cx + d}$, the horizontal asymptote is $y = \frac{a}{c}$.) This formula can also be developed by using division on the general equation $y = \frac{ax + b}{cx + d}$. This result is included in a more general result about the end behavior of rational functions on page 444.

18. a. $x = -\frac{7}{3}$; $y = 7$

b. It is the graph of $y = -\frac{11}{x}$ translated $\frac{7}{3}$ units left and 7 units up.

19. a. $x = -2$; $y = 3$

b. It is the graph of $y = \frac{5}{x}$ translated 2 units left and 3 units up.

20. a. $x = -4$; $y = -1$

b. It is the graph of $y = \frac{9}{x}$ translated 4 units left and 1 unit down.

21. a. $x = -\frac{1}{4}$; $y = -5$

b. It is the graph of $y = \frac{2}{x}$ translated $\frac{1}{4}$ unit left and 5 units down.

22. a. $x = 3$; $y = -\frac{4}{7}$

b. It is the graph of $y = \frac{1}{x}$ translated 3 units right and $\frac{4}{7}$ unit down.

23. a. $x = \frac{4}{5}$; $y = -6$

b. It is the graph of $y = -\frac{5}{x}$ translated $\frac{4}{5}$ unit right and 6 units down.

24. a. $C = 8p + 3$

b. $A = \frac{8p + 3}{p} = \frac{3}{p} + 8$

c.

p	A
2	\$9.50
5	\$8.60
10	\$8.30
20	\$8.15
50	\$8.06
100	\$8.03

\$8.00; It is negligible.

d. $A = \frac{3}{p} + 8$; As $p \to +\infty$, $\frac{3}{p} \to 0$, so $A \to 8$.

Apply⇔Assess

Exercise Notes

Assessment Note

Ex. 25 Some students may wish to discuss a particular function (with specific values for *a*, *b*, *c*, and *d*) and then go on to consider the general case.

Assess Your Progress

Journal Entry

When students answer this question, be sure they can state the conditions that *a*, *h*, and *k* must satisfy in those cases where $y = \frac{a}{x-h} + k$ does model inverse variation.

Progress Check 9.6–9.7

See page 457.

Practice 60 for Section 9.7

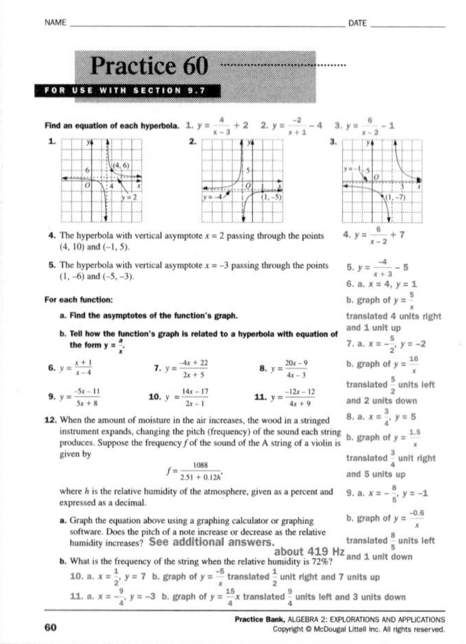

NAME ______________________ DATE ____________

Practice 60

FOR USE WITH SECTION 9.7

Find an equation of each hyperbola. 1. $y = \frac{4}{x-3} + 2$ 2. $y = \frac{-2}{x+1} - 4$ 3. $y = \frac{6}{x-2} - 1$

4. The hyperbola with vertical asymptote $x = 2$ passing through the points (4, 10) and (−1, 5). 4. $y = \frac{6}{x-2} + 7$

5. The hyperbola with vertical asymptote $x = -3$ passing through the points (1, −6) and (−5, −3). 5. $y = \frac{-4}{x+3} - 5$

For each function:

a. **Find the asymptotes of the function's graph.**

b. **Tell how the function's graph is related to a hyperbola with equation of the form $y = \frac{a}{x}$.**

6. $y = \frac{x+1}{x-4}$ 7. $y = \frac{-4x+22}{2x+5}$ 8. $y = \frac{20x-9}{4x-3}$

9. $y = \frac{-5x-11}{5x+8}$ 10. $y = \frac{14x-17}{2x-1}$ 11. $y = \frac{-12x-12}{4x+9}$

12. When the amount of moisture in the air increases, the wood in a stringed instrument expands, changing the pitch (frequency) of the sound each string produces. Suppose the frequency *f* of the sound of the A string of a violin is given by

$$f = \frac{1088}{2.51 + 0.12h},$$

where *h* is the relative humidity of the atmosphere, given as a percent and expressed as a decimal.

a. Graph the equation above using a graphing calculator or graphing software. Does the pitch of a note increase or decrease as the relative humidity increases? See additional answers.

b. What is the frequency of the string when the relative humidity is 72%? about 419 Hz

6. a. $x = 4$, $y = 1$ b. graph of $y = \frac{5}{x}$ translated 4 units right and 1 unit up

7. a. $x = -\frac{5}{2}$, $y = -2$ b. graph of $y = \frac{16}{x}$ translated $\frac{5}{2}$ units left and 2 units down

8. a. $x = \frac{3}{4}$, $y = 5$ b. graph of $y = \frac{1.5}{x}$ translated $\frac{3}{4}$ unit right and 5 units up

9. a. $x = -\frac{8}{5}$, $y = -1$ b. graph of $y = \frac{-0.6}{x}$ translated $\frac{8}{5}$ units left and 1 unit down

10. a. $x = \frac{1}{2}$, $y = 7$ b. graph of $y = \frac{-5}{x}$ translated $\frac{1}{2}$ unit right and 7 units up

11. a. $x = -\frac{9}{4}$, $y = -3$ b. graph of $y = \frac{15}{4}x$ translated $\frac{9}{4}$ units left and 3 units down

60 Practice Bank, ALGEBRA 2: EXPLORATIONS AND APPLICATIONS

ONGOING ASSESSMENT

25. **Open-ended Problem** Write a rational function of the form $f(x) = \frac{ax+b}{cx+d}$. Find the asymptotes of the graph of your function, and make a sketch of the graph.

SPIRAL REVIEW

Tell whether *y* varies inversely with *x*. If so, state the constant of variation. *(Section 9.6)*

26. $\frac{y}{x} = 3$ 27. $y = \frac{1.5}{x}$ 28. $4xy = -28$ 29. $x^2y = 2$

30. Ben took 11 tests in his algebra class. The scores he received were 75, 83, 84, 66, 75, 70, 84, 91, 80, 84, and 77. Find each statistic for Ben's test data. *(Sections 6.4 and 6.5)*

a. mean b. median c. range d. standard deviation

Divide. *(Section 9.2)*

31. $\frac{x^2+8x+10}{x+1}$ 32. $\frac{3x^2+x+7}{x^2-3x+2}$ 33. $\frac{8x^3-6x^2-x-4}{2x-1}$

ASSESS YOUR PROGRESS

VOCABULARY

inverse variation (p. 427)
constant of variation (p. 427)
hyperbola (p. 427)
asymptote (p. 427)
rational function (p. 435)

1. Use the table. Tell whether *y* varies inversely with *x*. If so, write an equation giving *y* as a function of *x*. *(Section 9.6)*

x	−2	−0.5	1.5	3
y	6	24	−8	−4

2. The time it takes a person to type a handwritten essay varies inversely with the person's typing rate. Matt, who types at a rate of 40 words per minute, can type a five-page essay in 30 min. If Matt takes a typing class that increases his rate to 60 words per minute, how long will it take him to type a five-page essay? *(Section 9.6)*

3. Find an equation of the hyperbola that has asymptotes $x = 1$ and $y = -3$ and that passes through the point with coordinates $(0, -5)$. *(Section 9.7)*

4. Let $g(x) = \frac{6x+13}{3x+2}$. Find the asymptotes of the graph of *g*, and tell how the graph is related to a hyperbola with equation of the form $y = \frac{a}{x}$. *(Section 9.7)*

5. **Journal** If $y = \frac{a}{x-h} + k$, can you say that *y* varies inversely with *x*? Explain.

25. Answers may vary. An example is given. Let $f(x) = \frac{3x-1}{x-1}$. The asymptotes are $x = 1$ and $y = 3$.

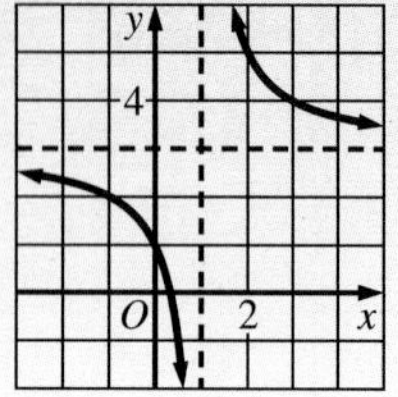

26. No. 27. Yes; 1.5

28. Yes; −7 29. No.

30. a. 79 b. 80

c. 25 d. about 6.89

31. $x + 7 + \frac{3}{x+1}$

32. $3 + \frac{10x+1}{x^2-3x+2}$

33. $4x^2 - x - 1 - \frac{5}{2x-1}$

Assess Your Progress

1. Yes; $y = \frac{-12}{x}$. 2. 20 min

3. $f(x) = \frac{2}{x-1} - 3$

4. $x = -\frac{2}{3}$; $y = 2$; It is the graph of $y = \frac{3}{x}$ translated $\frac{2}{3}$ unit left and 2 units up.

5. If *h* and *k* are not both zero, *y* does not vary inversely with *x*. Explanations may vary. The asymptotes of the graph of the given function are $x = h$ and $y = k$. The asymptotes of an inverse variation are the axes.

Section

9.8 Working with General Rational Functions

Learn how to...

- identify important features of graphs of rational functions
- describe the end behavior of rational functions

So you can...

- analyze how the acidity level in your mouth changes when you eat sugary foods, for example

So far, the graph of each rational function you have seen has been a hyperbola with a single vertical asymptote. In the Exploration, you'll work with rational functions having more complicated graphs.

EXPLORATION
COOPERATIVE LEARNING

Investigating Vertical Asymptotes

Work with a partner.
You will need:
- a graphing calculator or graphing software

One vertical asymptote is $x = -2$.

A second vertical asymptote is $x = 3$.

1 Graph the following functions. Identify the vertical asymptotes of each graph.

Example: $y = \frac{5}{(x+2)(x-3)}$

- $y = \frac{1}{(x-1)(x-3)}$
- $y = \frac{x-1}{(x-2)(x-5)}$
- $y = \frac{2x^2 - 3x - 7}{(x+1)(x-4)}$
- $y = \frac{4}{x(x+3)(x-3)}$
- $y = \frac{-x^2 - 6x + 2}{(x+3)(x+1)(x-4)}$
- $y = \frac{x^4 + x^3 - 5x^2}{(x+1)^2(x-2)}$

2 Predict the vertical asymptotes of the graph of each function below. Use your calculator or software to check your predictions.

- $y = \frac{-3x^2}{(x+3)(x-2)}$
- $y = \frac{7x+2}{(x+5)(x+1)(x-3)}$
- $y = \frac{x^3}{x^2 - 16}$

3 Based on your results from Steps 1 and 2, how are the vertical asymptotes of a rational function's graph related to the function's equation?

Exploration Note

Purpose
The purpose of this Exploration is to have students discover relationships between the equation for a rational function and the vertical asymptotes of the function.

Materials/Preparation
Each pair of students needs a graphing calculator or graphing software.

Procedure
As students graph their functions using technology, suggest that they sketch each of the graphs they obtain. The sketches can be used to facilitate a class discussion of students' results and provide them with a record of their work.

Closure
Discuss each graph with the class. Students should understand that each vertical asymptote corresponds to a zero of the denominator of the given function.

Explorations Lab Manual
See the Manual for more commentary on this Exploration.

For answers to the Exploration, see answers in back of book.

Plan⇔Support

Objectives

- Identify important features of graphs of rational functions.
- Describe the end behavior of rational functions.

Recommended Pacing

Core and Extended Courses
Section 9.8: 1 day

Block Schedule
Section 9.8: $\frac{1}{2}$ block
(with Section 9.9)

Resource Materials

Lesson Support
Lesson Plan 9.8
Warm-Up Transparency 9.8
Practice Bank: Practice 61
Study Guide: Section 9.8

Technology
Graphing Calculator
McDougal Littell Mathpack
Function Investigator with Matrix Analyzer Activity Book: Function Investigator Activities 25–27, 35, and 36

Internet:
http://www.hmco.com

Warm-Up Exercises

Find the solutions of each equation.

1. $(x-3)(x+3)(2x-1) = 0$ $-3, \frac{1}{2}, 3$
2. $x(4x-7)(-3x+1) = 0$ $0, \frac{1}{3}, \frac{7}{4}$
3. $x^2(5x+11)^2 = 0$ $-\frac{11}{5}, 0$

Use factoring to find all real-number solutions of each equation.

4. $x^3 + x^2 + x = 0$ 0
5. $x^2 - 13x + 36 = 0$ $4, 9$

Teach⇔Interact

Section Notes

Mathematical Procedures
When discussing the statement about finding vertical asymptotes, call attention to the assumption that $p(x)$ and $q(x)$ have no common factors. When students use algebra to find vertical asymptotes, they should factor both the numerator and the denominator of the rational function. They should *not* divide out common polynomial factors, unless those factors are constants or are polynomials with no real zeros. The reasons for not eliminating common factors that have real zeros will become clear when students discuss Think and Communicate question 1.

Using Technology
The behavior of rational functions can be complicated. It is possible to study their behavior algebraically, without examining their graphs. Most students, however, may find it helpful to use graphing calculators or graphing software, such as the *Function Investigator* software, to graph all of the functions in this section.

Students can use Function Investigator Activity 27 in the *Function Investigator with Matrix Analyzer Activity Book* to examine the behavior of the graphs of rational functions. In Function Investigator Activity 36, students explore the limits of rational functions.

Additional Example 1

Find the vertical asymptotes of the graph of $g(x) = \frac{x^2 - 5x + 6}{2x^3 + 11x^2 + 15x}$.

Step 1 Write the numerator and denominator of $g(x)$ as a product of factors to see if they have common factors.

$$g(x) = \frac{x^2 - 5x + 6}{2x^3 + 11x^2 + 15x} = \frac{(x-2)(x-3)}{x(2x+5)(x+3)}$$

The numerator and denominator have no common factors. Therefore, the vertical asymptotes occur where the denominator is 0.

Step 2 Solve the equation $x(2x + 5)(x + 3) = 0$.

$x(2x + 5)(x + 3) = 0$

$x = 0$ or $2x + 5 = 0$ or $x + 3 = 0$

$x = 0$ or $x = -\frac{5}{2}$ or $x = -3$

The vertical asymptotes are $x = 0$, $x = -\frac{5}{2}$, and $x = -3$.

The following result makes it easy to identify the vertical asymptotes of a rational function's graph.

Finding Vertical Asymptotes for Rational Functions

Let $p(x)$ and $q(x)$ be polynomials having no common factors. Then the graph of $f(x) = \frac{p(x)}{q(x)}$ has a vertical asymptote at each real solution of $q(x) = 0$.

For example, you can use the summary above to find the vertical asymptotes for the graph of $f(x) = \frac{x+2}{(x+3)(x-1)}$.

The solutions of $(x + 3)(x - 1) = 0$ are -3 and 1.

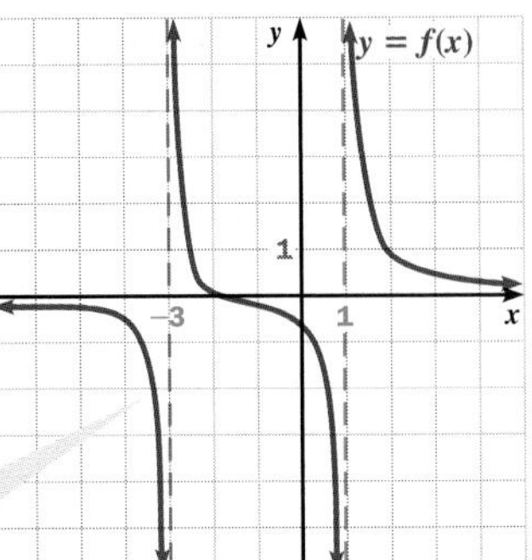

The vertical asymptotes are $x = -3$ and $x = 1$.

EXAMPLE 1

Find the vertical asymptotes of the graph of $f(x) = \frac{x^2 - 1}{2x^2 + 5x - 12}$.

SOLUTION

Step 1 Write the numerator and denominator of $f(x)$ as a product of factors to see if they have common factors.

$$f(x) = \frac{x^2 - 1}{2x^2 + 5x - 12} = \frac{(x-1)(x+1)}{(2x-3)(x+4)}$$

The numerator and denominator have **no common factors**. Therefore, the vertical asymptotes occur where the denominator is 0.

Step 2 Solve the equation $(2x - 3)(x + 4) = 0$.

$$(2x - 3)(x + 4) = 0$$

$$2x - 3 = 0 \quad \text{or} \quad x + 4 = 0$$

$$x = \frac{3}{2} \quad \text{or} \quad x = -4$$

The vertical asymptotes are $x = \frac{3}{2}$ and $x = -4$.

THINK AND COMMUNICATE

1. a.  **Technology** Graph $g(x) = \dfrac{x+3}{x^2+x-6}$ using a graphing calculator or graphing software. Identify the vertical asymptotes.

b. Solve the equation $x^2 + x - 6 = 0$. Does the graph of g have a vertical asymptote at each solution? If not, does this contradict the boxed statement at the top of the previous page? Explain.

2. a.  **Technology** Graph $h(x) = \dfrac{5}{x^2+1}$ using a graphing calculator or graphing software. Identify any vertical asymptotes.

b. Solve the equation $x^2 + 1 = 0$. How do your solutions support your answer to part (a)?

End Behavior of Rational Functions

As x takes on large positive and negative values, a rational function $y = f(x)$ can behave in one of several different ways. You can describe this end behavior using the infinity notation introduced in Section 9.3.

EXAMPLE 2

Use infinity notation to describe the end behavior of each function.

a. $f(x) = \dfrac{2x-1}{x^2-4}$ **b.** $g(x) = \dfrac{6x^2+2x-13}{3x^2+x-10}$ **c.** $h(x) = \dfrac{-x^3+5}{x^2}$

SOLUTION

Graph each function using a graphing calculator or graphing software.

a.

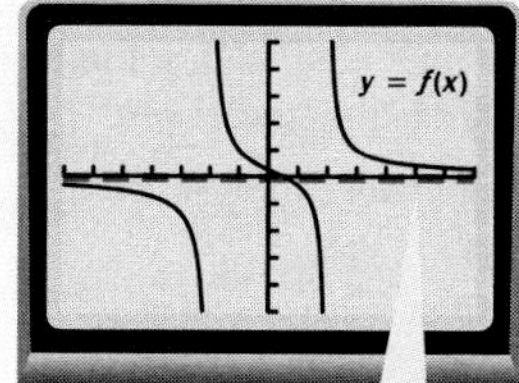

As $x \to \pm\infty$, $f(x) \to 0$.

b.

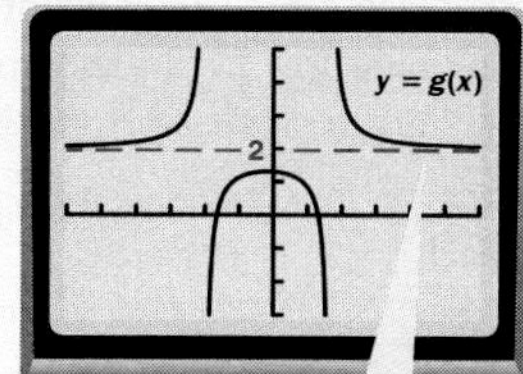

As $x \to \pm\infty$, $g(x) \to 2$.

c.

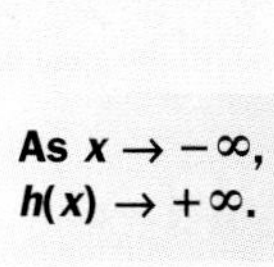

As $x \to -\infty$, $h(x) \to +\infty$.

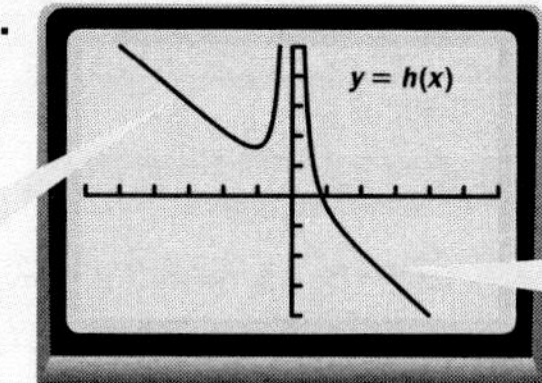

As $x \to +\infty$, $h(x) \to -\infty$.

ANSWERS Section 9.8

Think and Communicate

1. a.

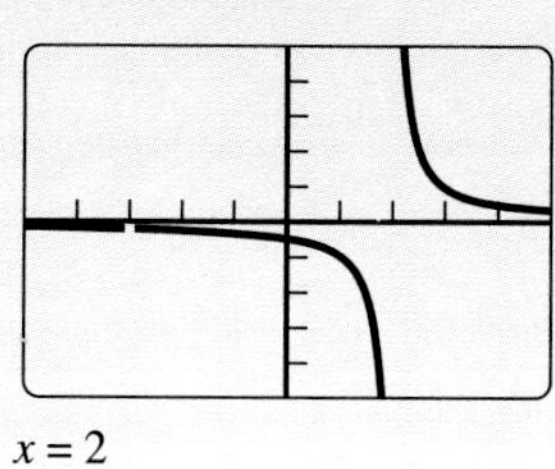

$x = 2$

b. $x^2 + x - 6 = (x+3)(x-2)$; No; $x + 3$ and $x^2 + x - 6$ have a common factor, $x + 3$, and the boxed statement makes no claim about whether there is or is not a vertical asymptote associated with that factor.

2. a.

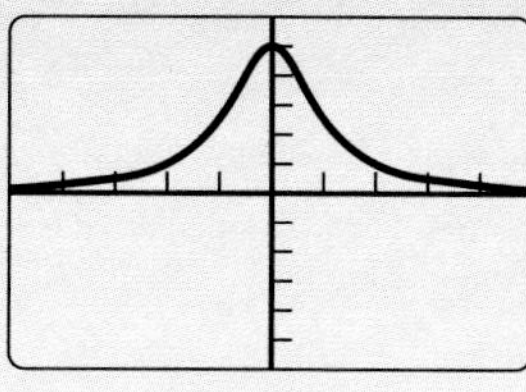

no vertical asymptotes

b. $x = \pm i$; The denominator has no real solutions, so the graph of the function has no vertical asymptotes.

Teach⇔Interact

Think and Communicate

When discussing question 1, ask students if -3 is in the domain of $g(x)$. (No.) Is it correct to say that the graph of $g(x) = \dfrac{x+3}{x^2+x-6}$ is identical to the graph of $h(x) = \dfrac{1}{x-2}$? Why? (No. $g(x)$ is not defined for $x = -3$, but $h(x)$ is defined for $x = -3$.) Point out to students that the graph of $y = g(x)$ will have an open circle when $x = -3$.

About Example 2

Problem Solving

The graph of $h(x)$ for part (c) has a slant, or *oblique*, asymptote. You may wish to show students how they can find an equation for this asymptote. Notice that $\dfrac{-x^3+5}{x^2} = -x + \dfrac{5}{x^2}$. Since $\dfrac{5}{x^2} \to 0$ as $x \to \pm\infty$, the graph of $y = -x + \dfrac{5}{x^2}$ approaches the graph of $y = -x$ as $x \to \pm\infty$. Thus, an equation for the slant asymptote is $y = -x$.

Additional Example 2

Use infinity notation to describe the end behavior of each function.

a. $f(x) = \dfrac{4x^2}{2x^3+x^2-2}$

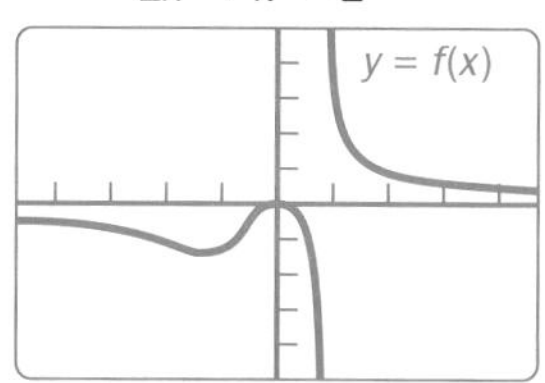

As $x \to \pm\infty$, $f(x) \to 0$.

b. $g(x) = \dfrac{5x^3+1}{x^3-1}$

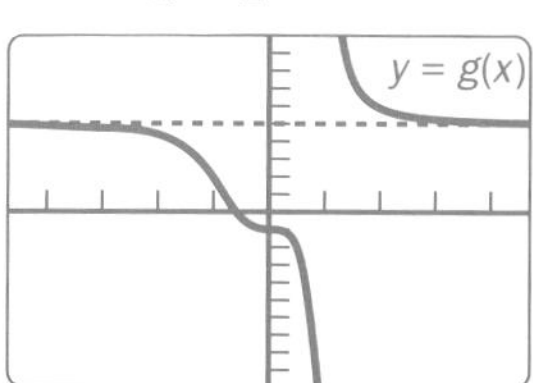

As $x \to \pm\infty$, $g(x) \to 5$.

c. $h(x) = \dfrac{2x^4+6}{3x^3-x^2+4}$

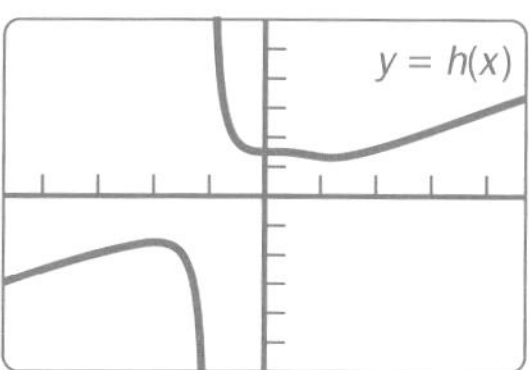

As $x \to -\infty$, $h(x) \to -\infty$.
As $x \to +\infty$, $h(x) \to +\infty$.

Teach⇔Interact

Section Note

Alternate Approach

Another approach to determining the end behavior of a rational function $f(x) = \frac{p(x)}{q(x)}$ is to divide the numerator and denominator by the leading term of the denominator. For $x \neq 0$, the values of the resulting expression are the same as those of $\frac{p(x)}{q(x)}$. The end behavior of this expression is easy to describe and is the same as the end behavior of $\frac{p(x)}{q(x)}$. In Example 2, part (a) on page 443, consider the values of $\frac{\frac{1}{x^2}(2x-1)}{\frac{1}{x^2}(x^2-4)}$, or $\frac{\frac{2}{x}-\frac{1}{x^2}}{1-\frac{4}{x^2}}$. As $x \to \pm\infty$, all the fractions in this last expression will have values that approach 0. Therefore, the value of the expression approaches $\frac{0-0}{1-0}$, or 0.

Additional Example 3

A radio station that changed its program format a year ago conducted weekly surveys of its listeners to see how many minutes per day, on the average, they listened to the station. When they plotted the data as a graph, they obtained a graph that can be modeled by the function $T(x) = \frac{120x^3 + 1080x^2 + 3000x + 2280}{4x^3 + 276x + 106}$, where x is the number of weeks since the format change, and $T(x)$ is listening time per day in minutes.

a. About how many weeks after the format change did listening time reach its maximum? What is this listening time? Graph the equation using a graphing calculator or graphing software. Find the coordinates of the highest point on the graph.

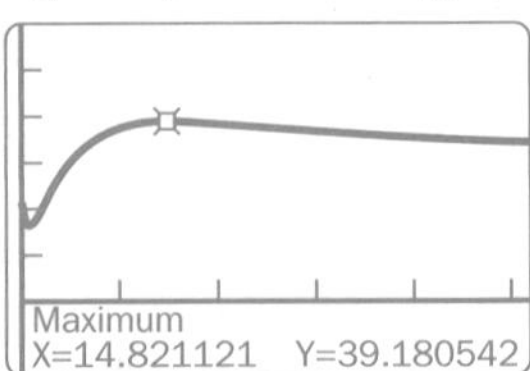

The coordinates of the highest point are about (14.82, 39.18). Listening time per day reached a maximum about 15 weeks after the format change. The maximum listening time is about 39 min.

You can also determine a rational function's end behavior by comparing the degrees of the numerator and denominator.

Determining the End Behavior of Rational Functions

Let $f(x) = \frac{p(x)}{q(x)}$ where $p(x)$ is a polynomial of degree m and $q(x)$ is a polynomial of degree n. Let a be the leading coefficient of $p(x)$, and let b be the leading coefficient of $q(x)$.

- If $m < n$, then $f(x) \to 0$ as $x \to \pm\infty$. The graph of f has a horizontal asymptote at $y = 0$.
- If $m = n$, then $f(x) \to \frac{a}{b}$ as $x \to \pm\infty$. The graph of f has a horizontal asymptote at $y = \frac{a}{b}$.
- If $m > n$, then f has the same end behavior as the polynomial function $y = \frac{a}{b}x^{m-n}$. The graph of f has no horizontal asymptote.

EXAMPLE 3 Application: Dentistry

When you eat sugary foods, the amount of acid in your mouth temporarily increases. This causes your mouth's pH to decrease. (See page 140 for a description of the pH scale.) The pH level t minutes after eating can be approximated with this equation:

$$\text{pH} = \frac{65t^2 - 204t + 2340}{10t^2 + 360}$$

a. Tooth decay can occur if the pH level in your mouth falls too low (and remains low for a period of time). How long after sugary foods are eaten is the lowest pH level reached? What is the lowest pH?

b. As time passes (that is, as $t \to +\infty$), what value does the pH level approach?

BY THE WAY

The pH model in Example 3 was developed by Ron Lancaster, a mathematics teacher in Ontario, Canada. Mr. Lancaster's students used litmus paper to verify the model's accuracy.

SOLUTION

a. Graph the equation using a graphing calculator or graphing software.

Find the coordinates of the lowest point on the graph. These coordinates are **(6, 4.8)**.

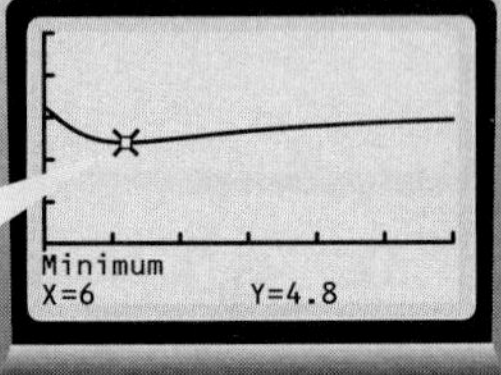

The lowest pH level is reached after 6 min. The lowest pH is 4.8.

b. In the equation for pH, the numerator $\mathbf{65t^2} - 204t + 2340$ has the same degree as the denominator $\mathbf{10t^2} + 360$. So the pH level approaches $\frac{\mathbf{65}}{\mathbf{10}}$, or 6.5, as time passes.

Think and Communicate

3. Use the equation $\text{pH} = \dfrac{65t^2 - 204t + 2340}{10t^2 + 360}$ to find the pH level when $t = 0$. What do you notice? Explain why your answer makes sense.
4. Show that the rules for end behavior at the top of the previous page apply to the rational functions in Example 2.
5. Find a polynomial function that has the same end behavior as the function $f(x) = \dfrac{-15x^4 + x^3 + 4x^2 + 4}{3x^2 + 11x + 1}$.

Checking Key Concepts

For each function:

a. Find the vertical asymptotes of the function's graph.

b. Describe the function's end behavior using infinity notation.

1. $f(x) = \dfrac{2}{(x + 7)(x - 4)}$
2. $f(x) = \dfrac{8x^2 + 2x - 15}{2x^2 + 13x + 11}$
3. $g(x) = \dfrac{3x^2}{4x - 17}$
4. $h(x) = \dfrac{-x^5 + x + 1}{x^3 - 2x}$

9.8 Exercises and Applications

Extra Practice exercises on page 764

Use what you know about vertical asymptotes to match each function with its graph.

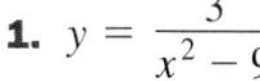

1. $y = \dfrac{3}{x^2 - 9}$
2. $y = \dfrac{x - 8}{x^2 - 2x - 3}$
3. $y = \dfrac{-2x}{x^2 + 2x - 3}$

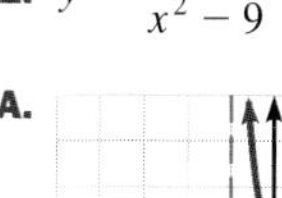

A.

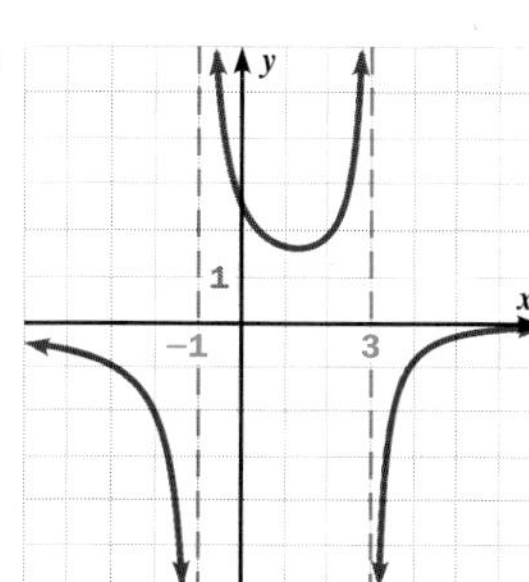

B.

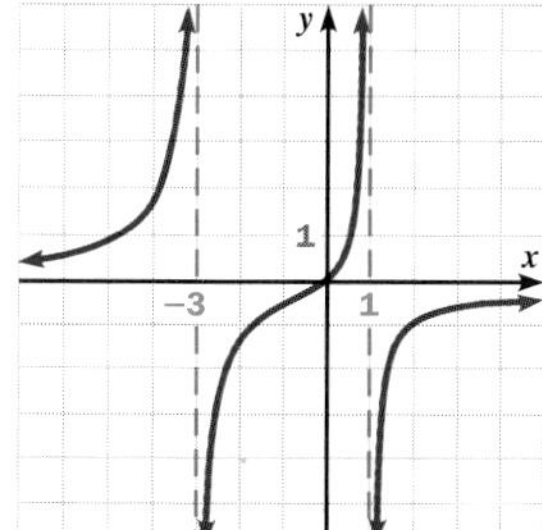

C.

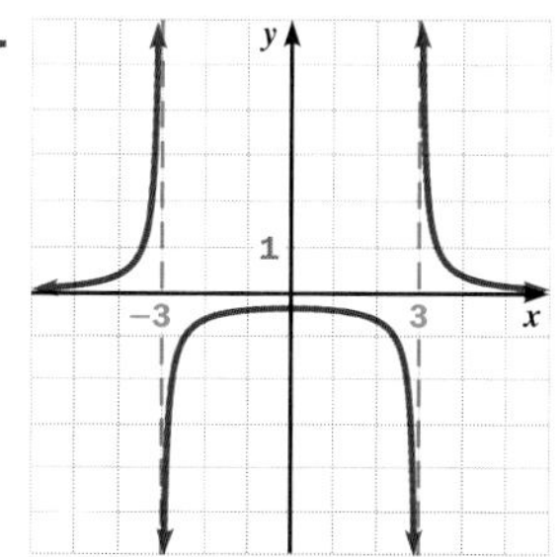

4. **SAT/ACT Preview** Let $f(x) = \dfrac{p(x)}{q(x)}$ be a rational function such that $f(x) \to 3$ as $x \to \pm\infty$. If A = the degree of $p(x)$ and B = the degree of $q(x)$, then:

A. A > B **B.** B > A **C.** A = B **D.** relationship cannot be determined

Think and Communicate

3. 6.5; This is the same as the pH that is approached as t gets very large. That makes sense because it is the pH level when a person has not eaten for awhile.
4. a. $m < n$ and $f(x) \to 0$ as $x \to \pm\infty$.
   b. $m = n$ and $g(x) \to \frac{a}{b} = \frac{6}{3} = 2$ as $x \to \pm\infty$.
   c. $m > n$ and $h(x)$ has the same end behavior as $y = -x$; that is, as $x \to -\infty$, $h(x) \to +\infty$ and as $x \to +\infty$, $h(x) \to -\infty$.
5. $g(x) = -5x^2$

Checking Key Concepts

1. a. $x = -7$; $x = 4$
   b. $f(x) \to 0$ as $x \to \pm\infty$.
2. a. $x = -\frac{11}{2}$; $x = -1$
   b. $f(x) \to 4$ as $x \to \pm\infty$.
3. a. $x = \frac{17}{4}$
   b. $g(x) \to -\infty$ as $x \to -\infty$ and $g(x) \to +\infty$ as $x \to +\infty$.
4. a. $x = 0$; $x = \pm\sqrt{2}$
   b. $h(x) \to -\infty$ as $x \to \pm\infty$.

Exercises and Applications

Answers that involve approximations may vary due to rounding. Examples are given.

1. C 2. A
3. B 4. C

Teach⇔Interact

Additional Example 3 (continued)

b. According to the function model, as time passes, what value does the listening time per day approach?
In the equation for $T(x)$, the numerator and denominator are polynomials of the same degree. So the listening time per day approaches about $\frac{120}{4}$, or about 30 minutes.

Think and Communicate

When discussing question 5, use a graphing calculator (or graphing software) to graph the given function along with $y = -5x^2$. On a TI-81 or TI-82, set the zoom factors to have XFact = 2, YFact = 10. (Choose 4:SetFactors from the ZOOM MEMORY menu.) Zoom out three or four times and note the change in appearance of the graphs of the two functions. Ask students whether the change in appearance is consistent with the prediction of the rules of end behavior at the top of page 444. (Yes, since the graphs eventually appear to coincide.)

Checking Key Concepts

Teaching Tip
When discussing the asymptotes for the functions in these questions, remind students that it is a good idea to check the numerator and denominator for common factors. Students may wish to graph these functions using technology to see if the graphs support their answers.

Closure Question

Suppose you are given a rational function $\frac{p(x)}{q(x)}$, where $p(x)$ and $q(x)$ are polynomials of degree m and n, respectively, with $m < n$. What would you do to determine if there are any vertical asymptotes? Write the numerator and denominator as a product of factors to see if they have any common factors. Eliminate the common factors, and then find all the real zeros of the remaining factors in the denominator. The function will have a vertical asymptote $x = h$ for each h that is a zero.

Apply⇔Assess

Suggested Assignment

Core Course
Exs. 1–4, 7–12, 16–25

Extended Course
Exs. 1–25

Block Schedule
Day 60 Exs. 1–4, 7–12, 16–25

Exercise Notes

Interdisciplinary Problems
Exs. 5, 6 These problems allow students to see how rational functions can be used by astronomers to study the force of Earth's gravity on objects such as a space probe. You might ask students who have had some introductory physics to explain to the class the concepts of force and gravity.

Research
Exs. 5, 6 Students who are interested in space and astronomy might want to research *Voyager 2* and report their findings to the class.

Problem Solving
Ex. 5 Ask students how the hint in part (b) is helpful. (Using the hint can help when deciding on the viewing window settings. One such setting is: Xmin = 10^6, Xmax = 10^9, Xscl = 10^8, Xmin = 0, Ymax = 50, Yscl = 10.)

Communication: Discussion
Exs. 5, 6 After students have read these exercises, discuss the concepts involved so that all students understand them and can see the types of complicating factors space scientists must deal with when applying the law of gravity to planning space probes.

INTERVIEW Vera Rubin

Look back at the article on pages 386–388.

As an astronomer, Vera Rubin uses gravity to help explain the motion of planets, stars, and other objects in space. Of all these objects, Earth is the source of the gravity that affects you most directly. The gravitational force F, measured in newtons (N), that Earth exerts on a relatively small object is given by

$$F = \frac{(4.0 \times 10^{14})m}{d^2}$$

where m is the object's mass (in kilograms) and d is the distance (in meters) of the object from the center of Earth.

5. The space probe *Voyager 2* was launched in 1977 to explore the outer planets of the solar system. In order to leave Earth, *Voyager 2* had to overcome Earth's gravity. The mass of *Voyager 2* is 930 kg.

a. Write an equation giving Earth's gravitational force on *Voyager 2* as a function of the distance d of *Voyager 2* from Earth's center.

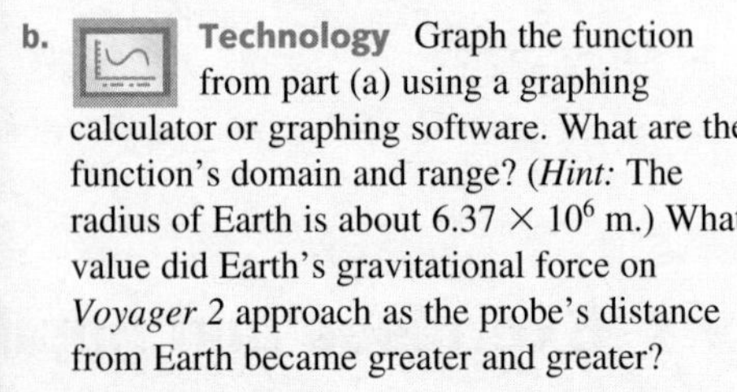

b. **Technology** Graph the function from part (a) using a graphing calculator or graphing software. What are the function's domain and range? (*Hint:* The radius of Earth is about 6.37×10^6 m.) What value did Earth's gravitational force on *Voyager 2* approach as the probe's distance from Earth became greater and greater?

c. The average distance between the center of Earth and the center of the moon is about 3.84×10^8 m. When *Voyager 2* was this far from Earth's center, what was Earth's gravitational force on the probe? How does this force compare with the force required to lift an algebra book (about 20 N)?

6. Suppose two space probes, probe A and probe B, are traveling away from Earth. The mass of probe B is twice the mass of probe A, and probe B is twice as far from Earth's center as probe A is. On which probe is the pull of Earth's gravity greater? Give a mathematical justification for your answer.

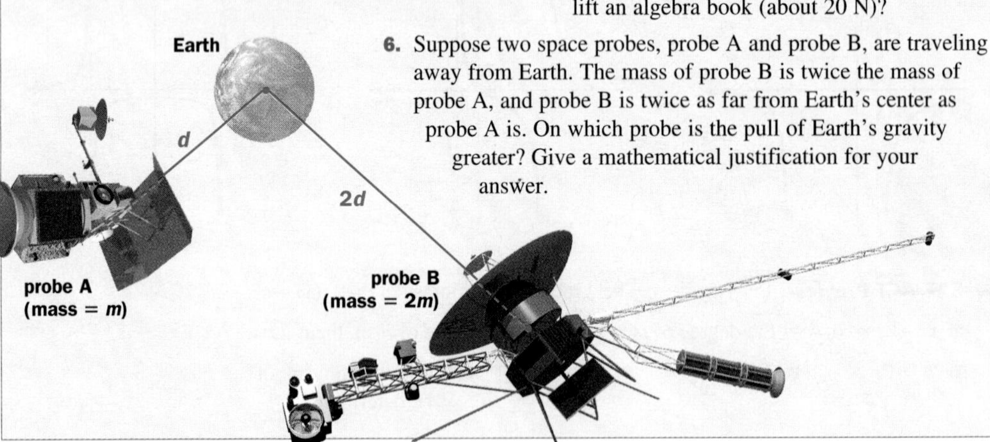

5. a. $F = \frac{(4.0 \times 10^{14})930}{d^2} = \frac{3.72 \times 10^{17}}{d^2}$

b. 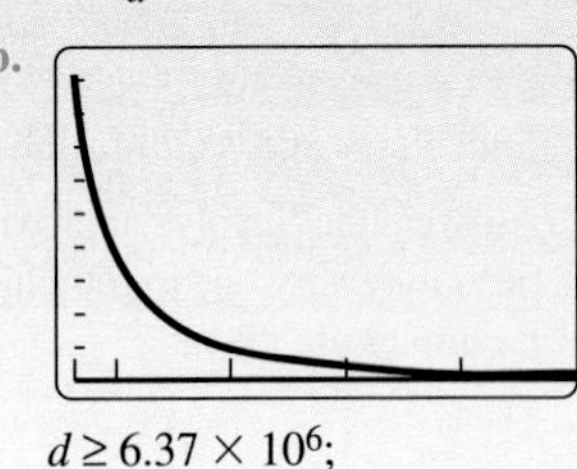

$d \geq 6.37 \times 10^6$; $0 < F \leq 9167.8$; 0

c. 2.52 N; It is about one-eighth the force required to lift an algebra book.

6. probe A; The force on probe A is $\frac{(4.0 \times 10^{14})m}{d^2}$. The force on probe B is $\frac{(4.0 \times 10^{14})2m}{(2d)^2} = \frac{(4.0 \times 10^{14})m}{2d^2}$, which is half the force on probe A.

7. a. $x = -5$; $x = 2$

b. $f(x) \to 0$ as $x \to \pm\infty$.

8. a. $x = -\frac{1}{3}$; $x = \frac{1}{4}$

b. $g(x) \to \frac{1}{3}$ as $x \to \pm\infty$.

9. a. $x = \frac{9}{5}$

b. $h(x) \to +\infty$ as $x \to -\infty$ and $h(x) \to -\infty$ as $x \to +\infty$.

10. a. $x = 0$; $x = 1$; $x = 4$

b. $f(x) \to 0$ as $x \to \pm\infty$.

11. a. $x = -4$; $x = 2$; $x = 3$

b. $g(x) \to -1$ as $x \to \pm\infty$.

For each function:

a. Find the vertical asymptotes of the function's graph.

b. Describe the function's end behavior using infinity notation.

7. $f(x) = \dfrac{1}{(x+5)(x-2)}$

8. $g(x) = \dfrac{4x^2}{12x^2 + x - 1}$

9. $h(x) = \dfrac{-10x^2 - 13x + 3}{5x - 9}$

10. $f(x) = \dfrac{x-5}{x^3 - 5x^2 + 4x}$

11. $g(x) = \dfrac{-x^3 + 1}{x^3 - x^2 - 14x + 24}$

12. $h(x) = \dfrac{6x^4 + 11}{x^2 + 1}$

Connection ENGINEERING

An *impluvium* is a type of dwelling found in western Africa. One version of an impluvium consists of several houses that surround a courtyard and share a single funnel-shaped roof. The roof is used to direct rainwater into a concrete tank or into clay jars.

13. Challenge Suppose you are making a cylindrical tank for the courtyard of an impluvium. You have 100 ft^3 of concrete. The sides and base of the tank should be 1 ft thick. The tank's inner radius r and inner height h should be chosen so that the tank holds as much water as possible. You can determine the values of r and h as follows.

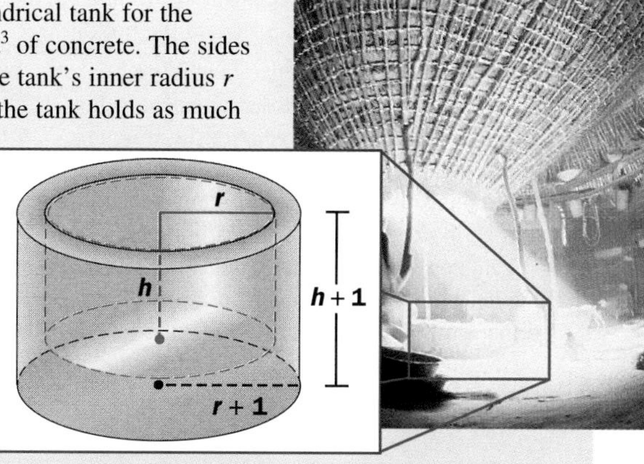

The photo above, taken by Jean-Paul Bourdier, shows an impluvium courtyard.

a. Write an equation giving the volume V_c of concrete needed to make the tank in terms of r and h. (*Hint:* The volume of concrete is the difference of the volumes of two cylinders.)

b. Knowing that $V_c = 100$, use your equation from part (a) to solve for h in terms of r.

c. Use your expression for h from part (b) to write an equation giving the tank's water capacity V_w as a rational function of r.

d. **Technology** Graph your equation for V_w using a graphing calculator or graphing software. Find the radius r that maximizes the tank's water capacity. Use this r-value to find the height h that maximizes the water capacity. What is the maximum capacity?

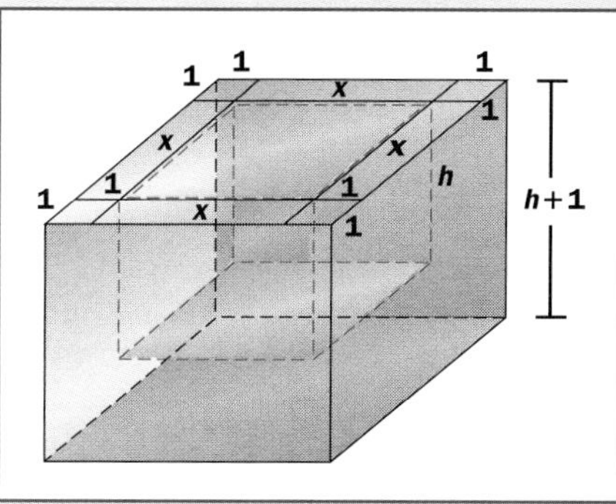

14. Cooperative Learning Suppose you want to make a tank like the one at the left for an impluvium. Work with a partner to find the dimensions x and h that maximize the tank's water capacity, assuming you can use 100 ft^3 of concrete. Compare the maximum capacity for this type of tank with the maximum capacity for a cylindrical tank. Which type of tank is better for storing water? Explain.

Apply⇔Assess

Exercise Notes

Cooperative Learning

Exs. 7–12 Both parts of these exercises can be done by working with the algebraic expression of the function, or by using a graphing calculator or the *Function Investigator* software. Students should be adept at using both methods. You might have students work in pairs on these exercises and have one student use each method. They can then check each other's work and switch methods for the next exercise.

Challenge

Ex. 13 This challenge problem requires quite a bit of algebraic manipulation, with many opportunities for error. Encourage students to organize their work and to work slowly and carefully. You might have students check their answer to each part before beginning the next part.

Topic Spiraling: Preview

Exs. 13, 14 Students may be interested to know that problems such as these are often encountered in calculus couses.

Integrating the Strands

Exs. 13, 14 These exercises show how algebra, geometry, and functions can be used together in studying real-world situations.

12. a. no vertical asymptotes

b. $h(x) \to +\infty$ as $x \to \pm\infty$.

13. a. $V_c = \pi(r+1)^2(h+1) - \pi r^2 h$

b. $h = \dfrac{100 - \pi(r+1)^2}{\pi(2r+1)}$

c. $V_w = \dfrac{r^2(100 - \pi(r+1)^2)}{(2r+1)}$

d.

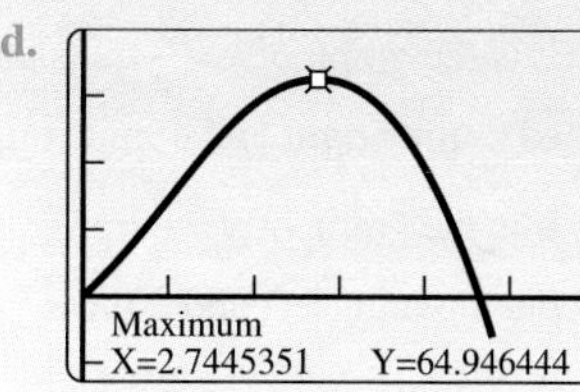

2.74 ft; 2.75 ft; 64.9 ft^3

14. 4.74 ft; 2.38 ft; The maximum capacity is 53.4 ft^3, which means that, using 100 ft^3 of concrete, you can build a cylindrical tank that holds more water than any rectangular tank you can build with that same volume of concrete.

Apply⇔Assess

Exercise Notes

Using Technology

Ex. 15 In part (a), students using graphing calculators may find it helpful to store each of the constants $k_0, k_1, \ldots, k_6$ in a variable before entering the equation for the function. For example, on the TI-82, they might use 9.9983952*10²→A, 1.6945176*10→B, and so on, through 1.6879850*10^-2→G. Then enter the equation on the Y= list by using Y1=(A+BX+CX²+DX^3+EX^4+FX^5)/(1+GX). This procedure is useful because it simplifies the equation on the Y= list.

Assessment Note

Ex. 17 Question students about the thinking they used when they constructed their functions. Their answers can help you to assess how well they have understood the important ideas of this section.

Practice 61 for Section 9.8

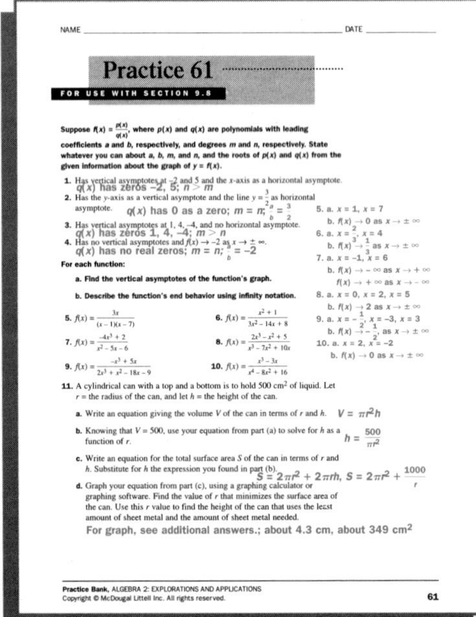

NAME ______________________ DATE __________

Practice 61

FOR USE WITH SECTION 9.8

Suppose $f(x) = \frac{p(x)}{q(x)}$, where $p(x)$ and $q(x)$ are polynomials with leading coefficients a and b, respectively, and degrees m and n, respectively. State whatever you can about a, b, m, and n, and the roots of $p(x)$ and $q(x)$ from the given information about the graph of $y = f(x)$.

1. Has vertical asymptotes at -2 and 5 and the x-axis as a horizontal asymptote. $q(x)$ has zeros -2, 5; $n > m$
2. Has the y-axis as a vertical asymptote and the line $y = \frac{3}{2}$ as horizontal asymptote. $q(x)$ has 0 as a zero; $m = n$; $\frac{a}{b} = \frac{3}{2}$
3. Has vertical asymptotes at 1, 4, -4, and no horizontal asymptote. $q(x)$ has zeros 1, 4, -4; $m > n$
4. Has no vertical asymptotes and $f(x) \to -2$ as $x \to \pm\infty$. $q(x)$ has no real zeros; $m = n$; $\frac{a}{b} = -2$

For each function:

a. Find the vertical asymptotes of the function's graph.

b. Describe the function's end behavior using infinity notation.

5. $f(x) = \frac{3x}{(x-1)(x-7)}$
6. $f(x) = \frac{x^2+1}{3x^2-14x+8}$
7. $f(x) = \frac{-4x^3+2}{x^2-5x-6}$
8. $f(x) = \frac{2x^3-x^2+5}{x^3-7x^2+10x}$
9. $f(x) = \frac{-x^3+5x}{2x^3+x^2-18x-9}$
10. $f(x) = \frac{x^3-3x}{x^4-8x^2+16}$

5. a. $x = 1$, $x = 7$
 b. $f(x) \to 0$ as $x \to \pm\infty$
6. a. $x = \frac{2}{3}$, $x = 4$
 b. $f(x) \to \frac{1}{3}$ as $x \to \pm\infty$
7. a. $x = -1$, $x = 6$
 b. $f(x) \to -\infty$ as $x \to +\infty$
 $f(x) \to +\infty$ as $x \to -\infty$
8. a. $x = 0$, $x = 2$, $x = 5$
 b. $f(x) \to 2$ as $x \to \pm\infty$
9. a. $x = -\frac{1}{2}$, $x = -3$, $x = 3$
 b. $f(x) \to -\frac{1}{2}$, as $x \to \pm\infty$
10. a. $x = 2$, $x = -2$
 b. $f(x) \to 0$ as $x \to \pm\infty$

11. A cylindrical can with a top and a bottom is to hold 500 cm² of liquid. Let r = the radius of the can, and let h = the height of the can.
 a. Write an equation giving the volume V of the can in terms of r and h. $V = \pi r^2 h$
 b. Knowing that $V = 500$, use your equation from part (a) to solve for h as a function of r. $h = \frac{500}{\pi r^2}$
 c. Write an equation for the total surface area S of the can in terms of r and h. Substitute for h the expression you found in part (b). $S = 2\pi r^2 + 2\pi rh$, $S = 2\pi r^2 + \frac{1000}{r}$
 d. Graph your equation from part (c), using a graphing calculator or graphing software. Find the value of r that minimizes the surface area of the can. Use this r value to find the height of the can that uses the least amount of sheet metal and the amount of sheet metal needed. For graph, see additional answers.; about 4.3 cm, about 349 cm²

Practice Bank, ALGEBRA 2: EXPLORATIONS AND APPLICATIONS

61

15. **CHEMISTRY** The density of water depends on the temperature of the water. The density D and temperature T are related by the equation

$$D = \frac{k_0 + k_1T + k_2T^2 + k_3T^3 + k_4T^4 + k_5T^5}{1 + k_6T}$$

where D is measured in kilograms per cubic meter, T is measured in degrees Celsius, and the constants $k_0, k_1, \ldots, k_6$ are given in the table.

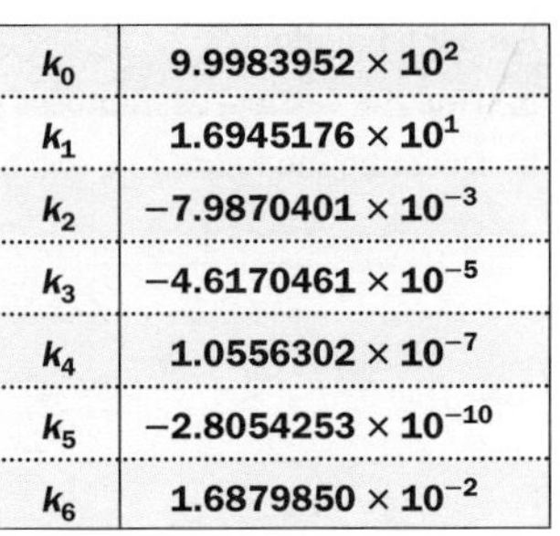

k_0	9.9983952×10^2
k_1	1.6945176×10^1
k_2	$-7.9870401 \times 10^{-3}$
k_3	$-4.6170461 \times 10^{-5}$
k_4	1.0556302×10^{-7}
k_5	$-2.8054253 \times 10^{-10}$
k_6	1.6879850×10^{-2}

a. **Technology** Graph the equation for D using a graphing calculator or graphing software. Adjust the viewing window so that the intervals $0 \le T \le 10$ and $999.5 \le D \le 1000.5$ are shown on the axes. At what temperature is water most dense? What is the maximum density?

b. **Writing** For a typical liquid, density increases steadily as the temperature decreases to the liquid's freezing point. Is this true for water? Explain. (*Note:* The freezing point of water is 0°C.)

c. **Writing** How does your answer to part (b) explain why lakes freeze at the top, rather than at the bottom?

16. **GARDENING** Brian Keegan is making a rectangular vegetable garden adjacent to his house as shown. The vegetables he wants to plant require 200 ft² of earth in which to grow. Brian plans to put fencing on three sides of the garden to keep out animals that might eat the vegetables. To save money, he wants to use the least amount of fencing possible. Find the dimensions x and y that minimize the length of fencing needed. (*Hint:* Use a procedure similar to the one given in Exercise 13.)

ONGOING ASSESSMENT

17. **Open-ended Problem** Write a rational function $f(x)$ so that the graph of f has vertical asymptotes at $x = -1$ and $x = 2$, and so that $f(x) \to -\infty$ as $x \to \pm\infty$.

SPIRAL REVIEW

Tell how the graph of each function is related to a hyperbola with equation of the form $y = \frac{a}{x}$. (*Section 9.7*)

18. $f(x) = \frac{1}{x-2} + 4$
19. $f(x) = \frac{5}{x+6} - 3$
20. $g(x) = \frac{2x+9}{x+1}$
21. $h(x) = \frac{-14x+3}{2x-1}$

Solve each proportion. (*Toolbox, page 785*)

22. $\frac{x}{10} = \frac{3}{2}$
23. $\frac{3}{8} = \frac{9}{4x}$
24. $\frac{4}{3k-4} = \frac{7}{2}$
25. $\frac{2n-3}{5} = \frac{n+2}{6}$

15. a.

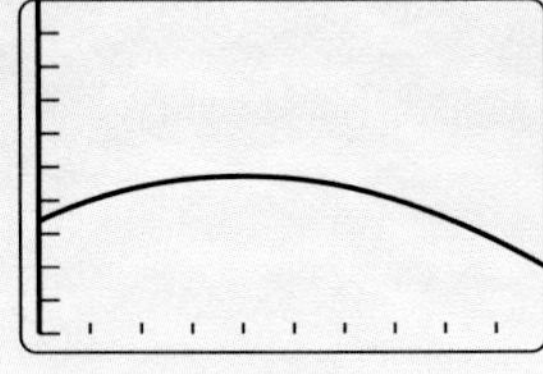

3.98°C; 999.97 kg/m³

b. No; the density of water decreases as the temperature decreases from about 4°C to 0°C.

c. Since the density of water decreases as water approaches its freezing point, the freezing water would be at the top, over the denser water below.

16. 10 ft; 20 ft

17. Answers may vary. An example is given. $f(x) = \frac{-x^4}{(x+1)(x-2)}$

18. It is the graph of $y = \frac{1}{x}$ translated 2 units right and 4 units up.

19. It is the graph of $y = \frac{5}{x}$ translated 6 units left and 3 units down.

20. It is the graph of $y = \frac{7}{x}$ translated 1 unit left and 2 units up.

21. It is the graph of $y = \frac{-2}{x}$ translated $\frac{1}{2}$ unit right and 7 units down.

22. 15

23. 6

24. $\frac{12}{7}$

25. 4

Section

9.9 Solving Rational Equations

Learn how to...

- solve rational equations

So you can...

- solve problems that involve mixing metals, for example

When two or more metals are melted, mixed together, and allowed to harden, the material they form is called an *alloy*. Some alloys, such as steel (a mixture of iron and carbon), are used mainly for industrial purposes. Other alloys, including those made from gold and silver, are highly valued for their beauty.

yellow gold
white gold
red gold

EXAMPLE 1 Application: Metallurgy

"Green gold," a gold-and-silver alloy named for its greenish color, is used to make jewelry. Green gold is 75% gold and 25% silver by weight. A naturally occurring alloy called *electrum* is 80% gold and 20% silver by weight. How much pure silver should be mixed with 12 oz of electrum to make green gold?

SOLUTION

Let x = the weight in ounces of pure silver to be mixed with the electrum. The value of x should be chosen so that this equation is satisfied:

This is the fraction of the electrum-and-silver mixture that is gold.

$$\frac{\text{weight of gold in mixture}}{\text{total weight of mixture}} = 0.75$$

Green gold is 75% gold.

The weight of the gold is 80% of the electrum's weight.

$$\frac{(0.80)(12)}{12 + x} = 0.75$$

The total weight is the weight of the electrum, 12, plus the weight x of the silver added.

$$\frac{9.6}{12 + x} = 0.75$$

Multiply both sides by $12 + x$.

$$(12 + x)\left(\frac{9.6}{12 + x}\right) = (12 + x)(0.75)$$

$$9.6 = 9 + 0.75x$$

$$0.6 = 0.75x$$

$$0.8 = x$$

The amount of silver that should be mixed with the electrum is 0.8 oz.

Plan⇔Support

Objectives

- Solve rational equations.
- Solve real-world problems by using rational equations.

Recommended Pacing

Core and Extended Courses
Section 9.9: 1 day

Block Schedule
Section 9.9: $\frac{1}{2}$ block
(with Section 9.8)

Resource Materials

Lesson Support
Lesson Plan 9.9
Warm-Up Transparency 9.9
Practice Bank: Practice 62
Study Guide: Section 9.9
Assessment Book: Test 37

Technology
Graphing Calculator
McDougal Littell Mathpack
Function Investigator
Internet:
http://www.hmco.com

Warm-Up Exercises

Solve each equation.

1. $17x + 19 = 155$ 8
2. $2(m - 2) - 7(m + 1.5) = 2.5$ −3.4
3. $42(n + 9) + 35(n - 3) = 306$ $\frac{3}{7}$

Solve by factoring or by using the quadratic formula.

4. $x^2 + 12x + 32 = 21$ −11, −1
5. $(x - 3)(x + 10) = 90$ −15, 8

Teach⇔Interact

Additional Example 1

The weight of a salt water solution is 60 lb. By weight, the solution is 2% salt. How much water must evaporate for the solution to be 5% salt?

Let x = the number of pounds of water that must evaporate. The weight of the salt is 2% of the original weight of the solution. The final weight of the solution is the original weight minus the weight x of water evaporated.

$$\frac{\text{weight of salt}}{\text{weight of solu. after evaporation}} = 0.05$$

$$\frac{(0.02)(60)}{60 - x} = 0.05$$

$$(60 - x)\left(\frac{1.2}{60 - x}\right) = (60 - x)(0.05)$$

$$1.2 = 3 - 0.05x$$

$$-1.8 = -0.05x$$

$$36 = x$$

For the solution to be 5% salt, 36 lb of water must evaporate.

Additional Example 2

Solve the rational equation $\frac{4}{x - 10} + \frac{x}{x + 2} = \frac{24}{(x - 10)(x + 2)}$.

Step 1 Multiply each side of the equation by the LCD of the fractions. Solve for x.

$$(x - 10)(x + 2)\left(\frac{4}{x - 10} + \frac{x}{x + 2}\right) = (x - 10)(x + 2)\left(\frac{24}{(x - 10)(x + 2)}\right);$$

$$(x - 10)(x + 2)\left(\frac{4}{x - 10}\right) + (x - 10)(x + 2)\left(\frac{x}{x + 2}\right) = 24$$

$$4(x + 2) + x(x - 10) = 24$$

$$4x + 8 + x^2 - 10x = 24$$

$$x^2 - 6x + 8 = 24$$

$$x^2 - 6x - 16 = 0$$

$$(x + 2)(x - 8) = 0$$

$$x + 2 = 0 \quad \text{or} \quad x - 8 = 0$$

$$x = -2 \quad \text{or} \quad x = 8$$

Step 2 Check the solutions. Use the original equation.

Check $x = -2$:

$$\frac{4}{-2 - 10} + \frac{-2}{-2 + 2} \stackrel{?}{=} \frac{24}{(-2 - 10)(-2 + 2)}$$

$$\frac{4}{-12} + \frac{-2}{0} \stackrel{?}{=} \frac{24}{0}$$

You cannot divide by 0, so –2 is not a solution.

Check $x = 8$:

$$\frac{4}{8 - 10} + \frac{8}{8 + 2} \stackrel{?}{=} \frac{24}{(8 - 10)(8 + 2)}$$

$$\frac{4}{-2} + \frac{8}{10} \stackrel{?}{=} \frac{24}{(-2)(10)}$$

$$-2 + \frac{4}{5} \stackrel{?}{=} -\frac{6}{5}$$

$$-\frac{6}{5} = -\frac{6}{5} \checkmark$$

The only solution is 8.

THINK AND COMMUNICATE

1. **a.** Solve the problem in Example 1 by starting with this equation:

$$\frac{\text{weight of silver in mixture}}{\text{total weight of mixture}} = 0.25$$

   b. How does this method compare with the method shown in Example 1?

The equation $\frac{9.6}{12 + x} = 0.75$ in Example 1 is called a *rational equation.* A **rational equation** contains only polynomials or quotients of polynomials.

EXAMPLE 2

Solve the rational equation $\frac{x}{x + 1} + \frac{2}{x - 3} = \frac{4}{(x + 1)(x - 3)}$.

SOLUTION

Step 1 Multiply each side of the equation by the least common denominator (LCD) of the fractions. Solve the resulting equation for x.

The LCD is $(x + 1)(x - 3)$.

$$(x + 1)(x - 3)\left(\frac{x}{x + 1} + \frac{2}{x - 3}\right) = (x + 1)(x - 3)\left(\frac{4}{(x + 1)(x - 3)}\right)$$

$$x(x - 3) + 2(x + 1) = 4$$

$$(x + 1)(x - 3)\left(\frac{x}{x + 1}\right) = x(x - 3) \qquad (x + 1)(x - 3)\left(\frac{2}{x - 3}\right) = 2(x + 1)$$

$$x^2 - 3x + 2x + 2 = 4$$

$$x^2 - x - 2 = 0$$

$$(x + 1)(x - 2) = 0$$

$$x + 1 = 0 \quad \text{or} \quad x - 2 = 0$$

$$x = -1 \quad \text{or} \quad x = 2$$

Step 2 Check the solutions. Use the *original* equation.

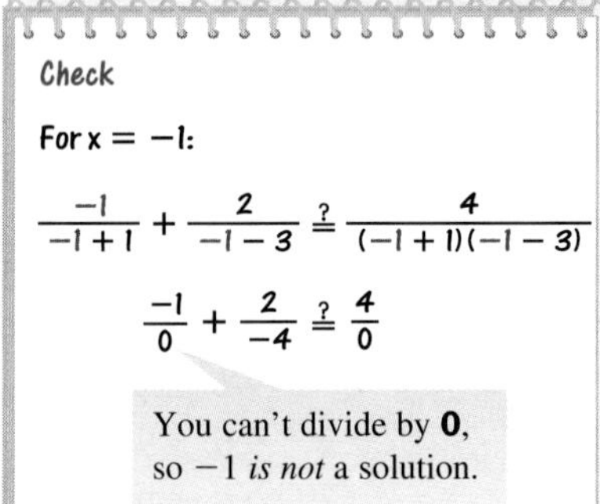

Check
For x = −1:
$\frac{-1}{-1 + 1} + \frac{2}{-1 - 3} \stackrel{?}{=} \frac{4}{(-1 + 1)(-1 - 3)}$
$\frac{-1}{0} + \frac{2}{-4} \stackrel{?}{=} \frac{4}{0}$
You can't divide by **0**, so −1 *is not* a solution.

Check

For x = 2:

$$\frac{2}{2 + 1} + \frac{2}{2 - 3} \stackrel{?}{=} \frac{4}{(2 + 1)(2 - 3)}$$

$$\frac{2}{3} + \frac{2}{-1} \stackrel{?}{=} \frac{4}{(3)(-1)}$$

$$-\frac{4}{3} = -\frac{4}{3} \checkmark$$

The only solution is 2.

ANSWERS **Section 9.9**

Think and Communicate

1. a. $\frac{0.2(12) + x}{12 + x} = 0.25$; $\frac{2.4 + x}{12 + x} = 0.25$; $2.4 + x = 0.25(12) + 0.25x$; $0.75x = 0.6$; $x = 0.8$

   b. The methods are similar. In this case, since silver is being added, the variable occurs in the numerator of the fraction as well as in the denominator.

Example 2 shows that rational equations can have extraneous solutions. You should always check your solutions to eliminate any that are extraneous.

THINK AND COMMUNICATE

2. Solve the equation $\frac{6x-1}{2x+5} = 3$. What happens? What does this tell you about the equation's solutions?
3. What happens when you try to solve $\frac{3x-6}{x-2} = 3$? What are the solutions of this equation? (*Hint:* Be careful. Remember that a solution must satisfy the *original* equation.)

CHECKING KEY CONCEPTS

Solve each equation.

1. $\frac{20+x}{4+x} = 3$
2. $\frac{2}{x-1} = \frac{x}{6}$
3. $\frac{3}{t} + \frac{8}{t-5} = 1$
4. $\frac{u}{u-3} + \frac{1}{u-4} = \frac{1}{(u-3)(u-4)}$
5. **METALLURGY** *Sterling silver* and *jewelry silver* are both silver-and-copper alloys. Sterling silver is 92.5% silver and 7.5% copper by weight. Jewelry silver is 80% silver and 20% copper by weight. How much pure silver must be mixed with 10 oz of jewelry silver to make sterling silver?

9.9 Exercises and Applications

Extra Practice exercises on page 764

Solve each equation.

1. $\frac{16-x}{5+x} = 2$
2. $\frac{1}{x} + \frac{8}{5x} = 1$
3. $\frac{x}{x-3} = 4 + \frac{3}{x-3}$
4. $\frac{6}{x(x-6)} - \frac{1}{x-6} = 0$
5. $\frac{1}{x-2} - \frac{8}{x-1} = -3$
6. $\frac{2}{x-10} + \frac{3x}{x-4} = 5$
7. $\frac{t^2-7t-8}{(3t-4)(t+2)} + \frac{3t+7}{t+2} = 3$
8. $\frac{4}{y+1} - \frac{1}{y} = 1$
9. $\frac{u-2}{(u+3)(u+2)} = \frac{3}{u+3} - \frac{2}{u+2}$
10. $\frac{v-3}{v-5} + \frac{4}{10-2v} = 1$
11. $\frac{5}{r-1} - \frac{3}{r+1} = \frac{2}{r}$
12. $\frac{s+1}{s} + \frac{14}{s-7} = \frac{3s-7}{s^2-7s}$
13. **Challenge** The ancient Greeks thought that the rectangles with the most pleasing appearance were those for which the ratio of the width w to the length l equals the ratio of l to $l + w$. The ratio of width to length for these "ideal" rectangles is called the *golden ratio*. Find the value of the golden ratio. (*Hint:* Write a proportion involving w and l. Express your proportion in terms of $x = \frac{w}{l}$, and solve the proportion for x.)

Teach⇔Interact

Think and Communicate

When discussing question 2, ask students to describe the graph of $y = \frac{6x-1}{2x+5}$. Students should be able to apply what they have learned in previous sections to see that the graph is a hyperbola with horizontal asymptote $y = 3$. The fact that a hyperbola never intersects its asymptotes is consistent with the fact that the equation in question 2 has no solutions.

Closure Question

What is the procedure for a paper-and-pencil solution of a rational equation? Why is it important to check all solutions in the original equation?

Multiply both sides of the equation by the least common denominator of the fractions. Solve the resulting equation, and check each solution in the original equation. Checking solutions of rational equations is important, because multiplying both sides of the original equation by the LCD of the fractions may introduce extraneous solutions.

Apply⇔Assess

Suggested Assignment

Core Course
Exs. 1–12, 16–21, AYP

Extended Course
Exs. 1–21, AYP

Block Schedule
Day 60 Exs. 1–12, 16–21, AYP

Exercise Notes

Cooperative Learning
Exs. 1–12 If students work on these exercises with a partner, they can share their solutions as a check on the accuracy of their work.

Common Error
Exs. 1–12 If students list extraneous solutions along with other solutions, stress the importance of checking each solution. In so doing, students can identify extraneous solutions and may also find other errors due to mistakes in computation.

Think and Communicate

2. You get $-1 = 15$. The equation has no solution.
3. You get $0 = 0$, which is true for all real numbers. However, since substituting 2 in the original equation produces division by 0, all real numbers *except* 2 are solutions.

Checking Key Concepts

1. 4
2. –3; 4
3. 1; 15
4. –1
5. $16\frac{2}{3}$ oz

Exercises and Applications

1. 2
2. $\frac{13}{5}$
3. no solution
4. no solution
5. $\frac{7}{3}$; 3
6. 8; 13
7. 6
8. 1
9. no solution
10. all real numbers except 5
11. $-\frac{1}{4}$
12. –5
13. $\frac{-1+\sqrt{5}}{2} \approx 0.618$

Apply⇔Assess

Exercise Notes

Research

Ex. 13 Some students may wish to research the golden ratio and write a report about the various contexts in which the golden ratio is found both in mathematics and in real-world applications.

Communication: Reading

Exs. 14, 15 Students who are not familiar with the concepts from chemistry that are used in these exercises should read the introductory passage carefully and analytically.

Interdisciplinary Problems

Exs. 14, 15 In doing these exercises, students will see how a rational equation can be solved to determine the amounts of hydrogen, iodine, and hydrogen iodide gases present in a mixture in a state of equilibrium. You may want to ask students who have studied chemistry to discuss some of the concepts present in these exercises such as, *reverse reaction*, *equilibrium*, and the *law of mass action*.

Career Connection

Exs. 14, 15 People who work in the field of chemistry are called chemists. Most chemists work to develop and produce chemicals used in industry. Some products and chemicals produced by chemists include plastics, resins, rubber, synthetic fibers, pharmaceutical drugs, gasoline and other petroleum products, fertilizers, pesticides, and detergents. Chemists have college degrees and often have completed graduate work in the field.

Connection CHEMISTRY

H_2 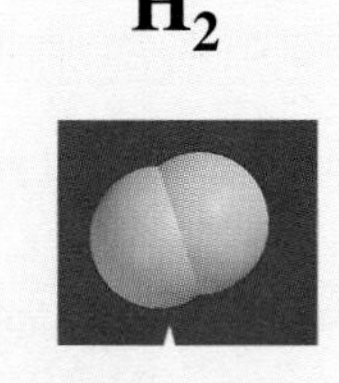 + I_2 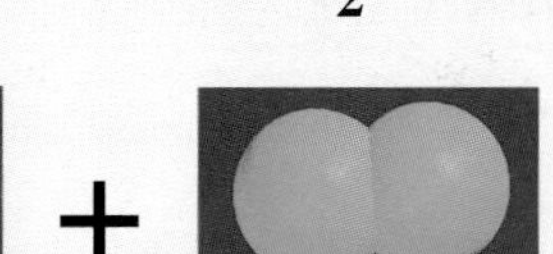 ⇌ $2HI$

The double arrow means that the reaction occurs in both directions.

An H_2 molecule consists of two hydrogen atoms.

An I_2 molecule consists of two iodine atoms.

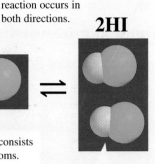

An HI molecule consists of one hydrogen atom and one iodine atom.

When you mix hydrogen and iodine gases together, a third gas, called *hydrogen iodide*, is formed. Each molecule of hydrogen (H_2) reacts with one molecule of iodine (I_2) to produce two molecules of hydrogen iodide (HI). Simultaneously, a "reverse reaction" occurs, in which the newly-formed hydrogen iodide breaks down into hydrogen and iodine. This process is illustrated above.

Eventually, the amounts of the three gases stabilize. When this happens, the gas mixture is said to be in *equilibrium*. At equilibrium, the amounts of the gases are related by the *law of mass action*,

$$\frac{(\text{amount of HI})^2}{(\text{amount of } H_2)(\text{amount of } I_2)} = K$$

where K is a constant that depends on the temperature of the gases.

14. Suppose 1 mole of hydrogen gas reacts with 2 moles of iodine gas to produce hydrogen iodide. (*Note:* 1 mole $\approx 6.023 \times 10^{23}$ molecules) The temperature of the gas mixture is 430°C. Assume the mixture has reached equilibrium.

a. **Writing** Let x = the amount of hydrogen (in moles) that is converted into hydrogen iodide. Use the illustration above to explain why x moles of iodine are also converted.

b. Copy the table and replace each "?" with an expression involving x. (All amounts are in moles.)

Molecule	H_2	I_2	HI
Initial amount	1	2	0
Net change	$-x$	$-x$	?
Equilibrium amount	?	?	?

c. At a temperature of 430°C, the value of K is 54.3 in the law of mass action. Use this K-value and the equilibrium amounts in the table to find the value of x. (*Hint:* You will obtain two x-values, but only one of the values makes sense in this situation.) At equilibrium, how much of the gas mixture is hydrogen? iodine? hydrogen iodide?

The bottle shown contains iodine in its gaseous and solid forms.

15. Suppose 5 moles of hydrogen react with 3 moles of iodine at 430°C. Find the equilibrium amounts of hydrogen, iodine, and hydrogen iodide.

14. a. A molecule of hydrogen iodide consists of one molecule of hydrogen and one molecule of iodine, so for each molecule of hydrogen that is converted, one molecule of iodine is also converted. Then if one mole of hydrogen is converted, one mole of iodine is converted.

b.

Molecules	H_2	I_2	HI
Initial amount	1	2	0
Net change	$-x$	$-x$	$2x$
Equilibrium amount	$1-x$	$1-x$	$2x$

c. 0.061 moles; 1.061 moles; 1.878 moles

15. 2.25 moles; 0.25 moles; 5.50 moles
16. Answers may vary.
17. $13 - 4i$
18. $-4 + 12i$
19. $8 + 6i$
20. $y = 2x^2$
21. $y = x^2 - x$

ONGOING ASSESSMENT

16. **Open-ended Problem** Choose a real-world application of a rational function from Section 9.7 or 9.8. Write a question based on this application that can be answered by solving a rational equation. Exchange questions with a classmate, and answer your classmate's question.

SPIRAL REVIEW

Perform the indicated operation. *(Section 8.4)*

17. $(5 - 7i) + (8 + 3i)$ 18. $(2 + i) - (6 - 11i)$ 19. $(3 + i)^2$

Technology Use a graphing calculator or software with matrix calculation capabilities to find an equation of the parabola passing through each set of points. *(Section 7.3)*

20. (1, 2), (2, 8), and (3, 18) 21. (1, 0), (2, 2), and (3, 6)

ASSESS YOUR PROGRESS

VOCABULARY

rational equation (p. 450)

For each function:

a. Find the vertical asymptotes of the function's graph.

b. Describe the function's end behavior using infinity notation. *(Section 9.8)*

1. $f(x) = \dfrac{2x - 7}{(x - 1)(x - 5)}$
2. $g(x) = \dfrac{-x^2 - 2x + 8}{x + 3}$
3. $h(x) = \dfrac{x^4}{x^2 + 3x - 10}$
4. $f(x) = \dfrac{3x^3 + 1}{x^3 - 9x}$

Solve each equation. *(Section 9.9)*

5. $\dfrac{5}{3} + \dfrac{2x + 9}{x + 4} = 4$
6. $\dfrac{w}{w - 5} = \dfrac{2}{w - 3} + \dfrac{4}{(w - 3)(w - 5)}$
7. **SPORTS** The 1994 major league baseball season ended prematurely because of a players' strike. When the season ended, Tony Gwynn of the San Diego Padres had the league's highest batting average (the ratio of hits to at-bats). Gwynn had 165 hits in 419 at-bats, for an average of .394. If there had been no strike and Gwynn had continued playing, how many consecutive hits would he have needed to raise his average to .400? *(Section 9.9)*

Tony Gwynn

8. **Journal** Explain how the end behavior of a rational function $f(x) = \dfrac{p(x)}{q(x)}$ is related to the degrees of the polynomials $p(x)$ and $q(x)$.

BY THE WAY

If Tony Gwynn had batted at least .400 in 1994, he would have been the first major league player to do so since 1941, when Ted Williams batted .406.

Apply⇔Assess

Exercise Notes

Assessment Note

Ex. 16 A good response to this exercise would be a question that shows how mathematics can be used to provide an understanding of a situation. Check to see if students give some indication of why the question to be answered might be of interest to someone working with the application.

Assess Your Progress

Students should be able to show that they can answer Exs. 1–4 by using two approaches: an algebraic approach and a graphic approach.

Progress Check 9.8–9.9

See page 457.

Practice 62 for Section 9.9

NAME ______ DATE ______

Practice 62

FOR USE WITH SECTION 9.9

Solve each equation.

1. $\frac{5}{x-1} + \frac{4}{x-1} = 3$ 4
2. $\frac{2}{y+2} - \frac{3}{y} = 5$ $-\frac{6}{5}, -1$
3. $\frac{1}{x(x-3)} - \frac{4}{x-3} = \frac{7}{x}$ 2
4. $\frac{x-2}{x+5} - \frac{4}{x+1} = \frac{1}{x+1}$ −3, 9
5. $\frac{6}{x-2} + \frac{3x+1}{x+3} = 4$ 5, −8
6. $\frac{5}{r-1} + \frac{r+2}{(r-4)(r-1)} = 3$ 2, 5
7. $\frac{t}{t-2} - \frac{t+1}{t^2-4} = \frac{11}{t+2}$ 3, 7
8. $\frac{u+4}{u^2+3u} - \frac{u}{u+3} = 1$ 1, −2
9. $\frac{x}{(x-2)(x-5)} + \frac{x+2}{x-2} = 1$ 4
10. $\frac{x}{x+4} + \frac{x-3}{(x+4)(x+1)} = \frac{-12}{x+1}$ −5, −9
11. $\frac{3}{v+4} - \frac{4}{v-3} = \frac{-2}{v}$ −1, 24
12. $\frac{3}{x-7} + \frac{9}{x^2-5x-14} = \frac{x+1}{x+2}$ 11
13. $\frac{3}{x+2} - \frac{10x}{x^2-x-6} = \frac{x-1}{x-3}$ −1, −7
14. $\frac{1}{x(x-4)} + \frac{5-2x}{(x+1)(x-4)} = \frac{5}{x(x+1)}$ $-3, \frac{7}{2}$
15. A manufacturer of pancake syrup has 28 gal of a blend of syrups that is 10% pure maple syrup. She wants to mix enough pure maple syrup to the blend so that the resulting mixture is 16% maple syrup. How much pure maple syrup should she add? 2 gal
16. A sailboat captain calculated that her boat's average speed on a certain day in the Charles River in Boston was 4.5 knots (nautical miles per hour) without taking the river's current into account. The boat traveled 2 nautical miles downstream (with the current) and the same distance back upstream (against the current) in 1 h. What was the speed of the current? 1.5 knots
17. When Donya went to buy a car, the salesman told her that the more expensive of two models got 4 mi/gal better fuel economy than the cheaper model. On test drives, she found that a 40-mile drive in the more expensive model and a 21-mile drive in the cheaper model used a total of 2 gal of gasoline. What were the fuel economies of the two cars (in mi/gal)? 28 mi/gal, 32 mi/gal

Assess Your Progress

1. a. $x = 1; x = 5$
   b. $f(x) \to 0$ as $x \to \pm\infty$.
2. a. $x = -3$
   b. $g(x) \to +\infty$ as $x \to -\infty$ and $g(x) \to -\infty$ as $x \to +\infty$.
3. a. $x = -5; x = 2$
   b. $h(x) \to +\infty$ as $x \to \pm\infty$.
4. a. $x = 0; x = -3; x = 3$
   b. $f(x) \to 3$ as $x \to \pm\infty$.
5. −1
6. 2
7. 5 consecutive hits
8. Let $f(x) = \frac{p(x)}{q(x)}$, where $p(x)$ is a polynomial of degree m and $q(x)$ is a polynomial of degree n. Let a be the leading coefficient of $p(x)$ and b be the leading coefficient of $q(x)$. If $m < n$, then $f(x) \to 0$ as $x \to \pm\infty$. If $m = n$, then $f(x) \to \frac{a}{b}$ as $x \to \pm\infty$. If $m > n$, then f has the same end behavior as the polynomial function $y = \frac{a}{b}x^{m-n}$.

Project Notes

Mathematical Goals

- Express the surface area of a cylinder as a function of its base radius only.
- Find the dimensions of the cylinder that minimize surface area.
- Investigate other shapes and minimize their surface area for the given volume.
- Select the best geometric shape with minimum surface area for the given volume.

Planning

Materials
- graphing calculator or graphing software
- spreadsheet software

Project Teams
Students can select their groups and all three members can work together on the ideas and calculations for the project. Students may find it helpful to have group discussions to understand the ideas of the project and to decide how to solve the problems.

Guiding Students' Work

Some students may need a review of the formulas for surface area and volume. If this is the case, review the formulas for the basic shapes, such as cylinders, cubes, and prisms. Also, the project involves writing the formula for surface area in terms of base radius only. If students are having difficulty with this, guide them in their solution by explaining the hint given in Step 1. If students cannot understand one type of solution method, have them try a different method. For example, if students are using a graph to try to minimize surface area but do not know how to read the graph, have them use a spreadsheet. Then have them relate the spreadsheet to the graph.

Second-Language Learners
Students learning English can benefit from working in small groups to design their containers and write their reports.

PORTFOLIO PROJECT

The Shape of Things

Grocery store shelves are filled with containers having many different sizes and shapes, and made of many different materials. When designing containers, food packagers must consider serving size, the cost of materials, and other factors. Packages must be durable, attractive, and easy to handle.

For liquid products that can assume the shape of any container, selecting the shape with the least surface area for a specified volume may reduce the cost of the material used to make the container.

PROJECT GOAL **Determine the best shape and dimensions of a container that will hold a specified volume of liquid.**

Designing an Efficient Container

Work in groups of three. Suppose you are part of a package design team assigned to design a container that will hold 500 cm^3 of tomato sauce. The most desirable package design will minimize material costs, yielding the most efficient container. Here's a possible plan of action for your team:

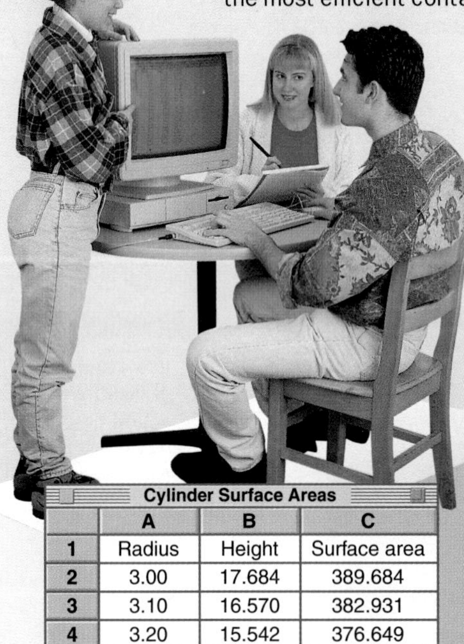

1. CONSIDER a cylindrical can. Since many food containers are cylindrical, this is a logical shape to consider first. Your team must determine the radius and height of a 500 cm^3 cylindrical can with the least surface area.

STEP 1 Express surface area as a function of base radius only. (*Hint:* The fact that the volume must be 500 cm^3 will allow you to replace h with an expression involving r.)

STEP 2 Find the dimensions of the cylinder that minimizes surface area. One method is to examine a graph using a graphing calculator or graphing software. Another method is to examine a table of values using a spreadsheet like the one shown.

Cylinder Surface Areas

	A	B	C
1	Radius	Height	Surface area
2	3.00	17.684	389.684
3	3.10	16.570	382.931
4	3.20	15.542	376.649
5	3.30	14.615	371.266

2. **INVESTIGATE** other shapes for your tomato sauce container. Are there other geometric shapes that can hold 500 cm^3 using less surface area than the cylinder you identified in Part 1?

With other team members, discuss the advantages and disadvantages of using unusual container shapes, such as a cone, pyramid, prism, or sphere. Analyze their surface areas as you did in Part 1.

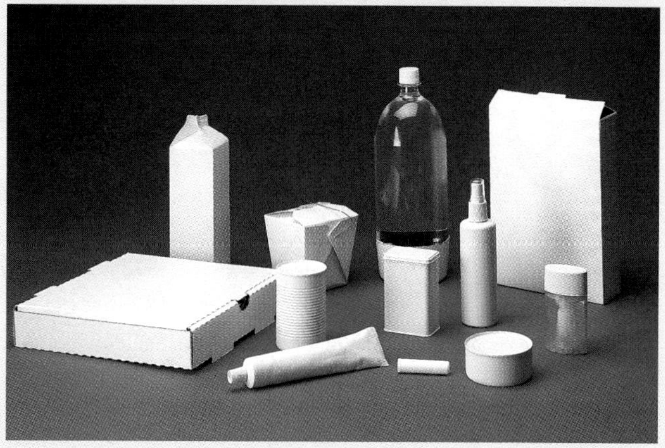

Presenting a Report

Present your report as a recommendation to the management of your company. Describe your procedures, and provide data that support the conclusions you make. You may also wish to extend your project. Here are a few ideas:

- Based on your findings for a cylindrical can with a volume of 500 cm^3, make a generalization about the ratio of radius to height for a cylindrical can of any specified volume if the surface area is to be minimized.
- Visit a grocery store. Can you find cylindrical containers that have the radius-to-height ratio described above? Based on the examples you see in the store, what are some other factors that should be considered in packaging a product?
- Consider the problems of packaging objects with unusual shapes. How would you package a football, a boomerang, or a dozen coat hangers?
- Most products are shipped in large quantities after packaging. What issues need to be considered when making larger packages from smaller ones? For example, how would you package 24 cans of tomato sauce so that they can be shipped?

Self-Assessment

How did the team divide up the tasks? What problems, if any, did you have setting up and analyzing the equations you used for the project? In what ways, if any, did the results of the project surprise you?

Project Notes

Guiding Students' Work

Rubric for Chapter Project

4 Students correctly express the surface area of a cylinder as a function of base radius only and accurately find the dimensions that minimize surface area. Students then investigate other shapes and minimize their surface areas. Students make a well-organized presentation of their report, describing their procedures and using data to support their conclusions. The report includes all of the information requested. Students also extend the project by completing at least one of the suggestions given.

3 Students write the surface area as a function of base radius only and accurately find the dimensions that minimize the surface area of the cylinder. Students' investigate other shapes and minimize their surface areas as well, but make errors in some of the calculations. The presentation is made and conclusions are drawn, but there are some errors in logic or conclusions are not supported by data. Students extend the project using one of the suggestions given.

2 Students make an attempt to write the surface area as a function of base radius but make an error in their calculations and do not minimize the surface area correctly. Students make the same mistake when investigating other shapes. Students try to make a presentation, but it is not done well because of the lack of understanding of the ideas of the project. Students draw incorrect conclusions and try to support them with data. The written report is not clear. Students do not extend the project.

1 Students do not correctly minimize the surface area of a cylinder because they do not write the surface area as a function of base radius only. Students do not investigate other shapes for possible containers and therefore no comparison is made to find the best shape. If a presentation is made, it demonstrates a poor attempt at completing the project. Students should be encouraged to speak with the teacher as soon as possible to review their work and to make a new start on the project.

Chapter Support

Course Guide: Chapter 9
Lesson Plans: Chapter 9
Practice Bank: Cumulative Practice 63
Study Guide: Chapter 9 Review
Assessment Book: Chapter Tests 38 and 39; Chapter 9 Alternative Assessment
Test Generator Software
Portfolio Project Book
Preparing for College Entrance Tests
Professional Handbook

Progress Check 9.1–9.3

1. Write the polynomial $10x^3 + x^4 - x - x^2 + 6$ in standard form and state its degree. *(Section 9.1)* $x^4 + 10x^3 - x^2 - x + 6$; 4
2. Subtract $4x^4 + 3x^3 - 5x - 8$ from $-9x^3 + x^2 + 7x + 10$. *(Section 9.1)* $-4x^4 - 12x^3 + x^2 + 12x + 18$
3. Multiply $-3x^3 + 2x + 1$ by $2x^2 - 5x + 6$. *(Section 9.2)* $-6x^5 + 15x^4 - 14x^3 - 8x^2 + 7x + 6$
4. Divide $-15x^3 - x^2 + 27x + 20$ by $3x + 2$. $-5x^2 + 3x + 7 + \frac{6}{3x + 2}$
5. Describe the end behavior of $f(x) = 7x^5 - x^4 + 3x^2 + 1$ using infinity notation. *(Section 9.3)* $f(x) \to -\infty$ as $x \to -\infty$. $f(x) \to +\infty$ as $x \to +\infty$.

Progress Check 9.4–9.5

1. Find an equation for the cubic function whose graph has –2, 1.5, and 6 as x-intercepts and 126 as a y-intercept. Write the equation in intercept form. *(Section 9.4)* $y = 7(x + 2)(x - 1.5)(x - 6)$

Find the zeros of each function.

2. $f(x) = 8(x + 7)\left(x - \frac{1}{2}\right)(x - 5)$ *(Section 9.4)* $-7, \frac{1}{2}, 5$
3. $g(x) = 18x^3 + 51x^2 + 20x - 25$ *(Section 9.4)* $\frac{1}{2}, -\frac{5}{3}$ (double zero)
4. $r(x) = 2x^3 - 19x^2 + 58x + 34$ *(Section 9.4)* $-\frac{1}{2}, 5 \pm 3i$
5. $h(x) = 2x^4 + x^3 - 7x^2 - 2x + 6$ *(Section 9.5)* $-\frac{3}{2}, 1, \pm\sqrt{2}$

CHAPTER

9 Review

STUDY TECHNIQUE

Reread the chapter, writing down questions about the ideas you don't understand. If any questions are still unanswered at the end of the chapter, ask a parent, friend, or teacher to help you with them.

VOCABULARY

polynomial (p. 390)	**cubic function** (p. 413)
degree (p. 390)	**zero** (p. 413)
constant term (p. 390)	**double zero** (p. 420)
standard form (p. 390)	**triple zero** (p. 420)
synthetic substitution (p. 391)	**inverse variation** (p. 427)
like terms (p. 392)	**constant of variation** (p. 427)
end behavior (p. 405)	**hyperbola** (p. 427)
leading coefficient (p. 406)	**asymptote** (p. 427)
turning point (p. 407)	**rational function** (p. 435)
local maximum/minimum (p. 407)	**rational equation** (p. 450)

SECTIONS 9.1 *and* 9.2

A **polynomial** is a term or a sum of terms such that each term is the product of a real-number coefficient and a variable with a whole-number exponent. An example of a polynomial is $3x^4 - 5x^2 + 6x - 1$. The **degree**, or greatest exponent, of this polynomial is 4. The polynomial is in **standard form** because its exponents decrease from left to right.

You can use **synthetic substitution** to evaluate a polynomial for a given value of the variable. An example is shown at the right.

Evaluate $3x^4 - 5x^2 + 6x - 1$ when $x = 2$.

3	0	−5	6	−1
	6	12	14	40
3	6	7	20	39 ← answer

Repeatedly **multiply by 2**, and then add.

You can also add, subtract, multiply, and divide polynomials. An example is shown at the right.

Add $x^4 + 2x^3 + 4x^2 - 7$ and $3x^4 - 5x^2 + 6x - 1$.

$$\begin{array}{r} x^4 + 2x^3 + 4x^2 - 7 \\ 3x^4 - 5x^2 + 6x - 1 \\ \hline 4x^4 + 2x^3 - x^2 + 6x - 8 \end{array}$$ ← answer

Add **like terms**.

SECTIONS 9.3, 9.4, *and* 9.5

Four important features of a polynomial function and its graph are **end behavior**, **turning points**, **local maximums and minimums**, and **zeros**.

You can obtain information about a polynomial function's zeros using the *factor theorem*, the *fundamental theorem of algebra*, and the *rational zeros theorem*.

$f(x) = 3(x + 1)(x - 2)^2$

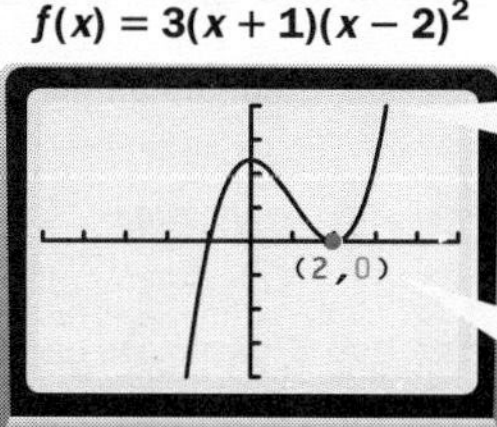

As $x \to +\infty$, $f(x) \to +\infty$.
As $x \to -\infty$, $f(x) \to -\infty$.

This is a **turning point** of the graph. Its x-coordinate is a **zero** of the function. Its y-coordinate is a **local minimum**.

SECTIONS 9.6 *and* 9.7

Two variables x and y show **inverse variation** if $y = \frac{a}{x}$, or $xy = a$, for some nonzero number a. For example, the time it takes to travel a fixed distance varies inversely with speed.

The graph of $y = \frac{a}{x}$ is a **hyperbola**. The graph of $y = \frac{a}{x - h} + k$ is the graph of $y = \frac{a}{x}$ translated h units horizontally and k units vertically.

A **rational function** has the form $f(x) = \frac{p(x)}{q(x)}$ where $p(x)$ and $q(x)$ are polynomials.

$y = \frac{1}{x - 2} + 3$

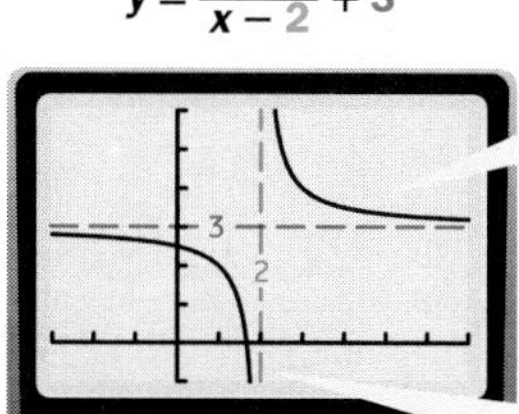

This graph is the graph of $y = \frac{1}{x}$ shifted **2 units right** and **3 units up**.

The **asymptotes** of the graph are $x = 2$ and $y = 3$.

SECTIONS 9.8 *and* 9.9

A rational function $f(x) = \frac{p(x)}{q(x)}$ has a vertical asymptote at each real solution of $q(x) = 0$, provided $p(x)$ and $q(x)$ have no common factors.

The end behavior of $f(x) = \frac{p(x)}{q(x)}$ is determined by the degrees and leading coefficients of $p(x)$ and $q(x)$.

$f(x) = \frac{2x^2}{x^2 - 1}$

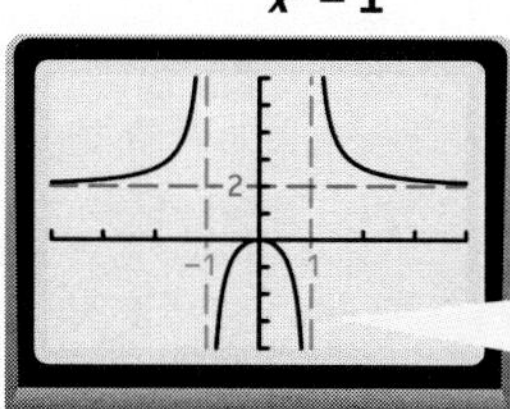

The numerator $2x^2$ has the same **degree** as the denominator $1x^2 - 1$. So $f(x) \to \frac{2}{1} = 2$ as $x \to \pm\infty$.

The graph has vertical asymptotes where $x^2 - 1 = 0$: at $x = 1$ and $x = -1$.

A **rational equation**, such as $\frac{3x}{x - 4} + \frac{2}{x - 10} = 5$, contains only polynomials or quotients of polynomials. When solving a rational equation, be sure to check your solutions to eliminate any that are extraneous.

Progress Check 9.6–9.7

1. Use the table. Tell whether y varies inversely with x. If so, write an equation giving y as a function of x. If not, explain why not. *(Section 9.6)*

x	y
2	50
8	12.5
2.4	40
–10	10

No; the product of x and the corresponding value of y is not constant.

2. Let $f(x) = \frac{-8x + 61}{x - 7}$. Find the asymptotes of the graph of f, and tell how the graph is related to a hyperbola of the form $y = \frac{a}{x}$. *(Section 9.7)*
The asymptotes are $x = 7$ and $y = -8$. The graph of f is the graph of $y = \frac{5}{x}$ translated 7 units right and 8 units down.

3. Find an equation for the function h whose graph is a hyperbola with asymptotes $y = 0$ and $x = 4$ and passing through the point (6, 5). *(Section 9.7)*
$h(x) = \frac{10}{x - 4}$

Progress Check 9.8–9.9

1. Find the vertical and horizontal asymptotes, if any, for the function $f(x) = \frac{x^3 + 5x^2}{x^2 - 9}$. *(Section 9.8)*
$x = 3$, $x = -3$; no horizontal asymptotes

Describe the end behavior of each function. *(Section 9.8)*

2. $r(x) = \frac{x^3 - x^2 + 2}{(x + 1)(x + 10)}$
$r(x) \to -\infty$ as $x \to -\infty$.
$r(x) \to +\infty$ as $x \to +\infty$.

3. $s(x) = \frac{7x^3 + 5x}{2x^3 - 8}$
$s(x) \to \frac{7}{2}$ as $x \to \pm\infty$.

Solve each equation. *(Section 9.9)*

4. $\frac{x}{x + 1} - \frac{2}{x - 6} = \frac{7}{(x + 1)(x - 6)}$ 9

5. $\frac{4}{x - 1} + \frac{7}{x + 3} = \frac{-6}{(x - 1)(x + 3)}$ –1

Chapter 9 Assessment
Form A Chapter Test

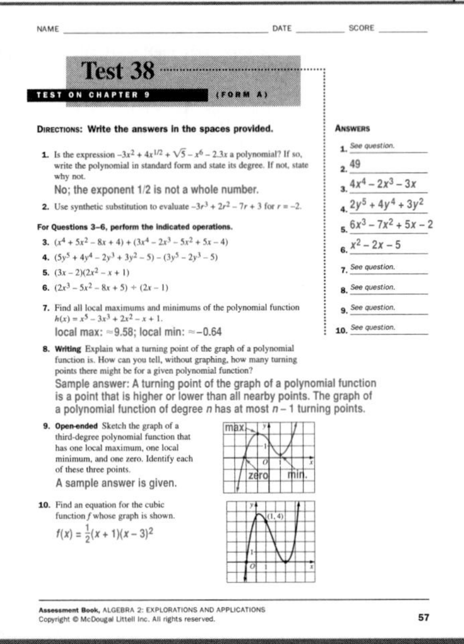

NAME ______ DATE ______ SCORE ______

Test 38

TEST ON CHAPTER 9 (FORM A)

DIRECTIONS: **Write the answers in the spaces provided.**

1. Is the expression $-3x^2 + 4x^{1/2} + \sqrt{5} - x^6 - 2.3x$ a polynomial? If so, write the polynomial in standard form and state its degree. If not, state why not.
   No; the exponent 1/2 is not a whole number.
2. Use synthetic substitution to evaluate $-3r^3 + 2r^2 - 7r + 3$ for $r = -2$.

For Questions 3–6, perform the indicated operations.

3. $(x^4 + 5x^2 - 8x + 4) + (3x^4 - 2x^3 - 5x^2 + 5x - 4)$
4. $(5y^5 + 4y^4 - 2y^3 + 3y^2 - 5) - (3y^5 - 2y^3 - 5)$
5. $(3x - 2)(2x^2 - x + 1)$
6. $(2x^3 - 5x^2 - 8x + 5) \div (2x - 1)$
7. Find all local maximums and minimums of the polynomial function $h(x) = x^5 - 3x^3 + 2x^2 - x + 1$.
   local max: ≈ 9.58; local min: ≈ -0.64
8. **Writing** Explain what a turning point of the graph of a polynomial function is. How can you tell, without graphing, how many turning points there might be for a given polynomial function?
   Sample answer: A turning point of the graph of a polynomial function is a point that is higher or lower than all nearby points. The graph of a polynomial function of degree n has at most $n - 1$ turning points.
9. **Open-ended** Sketch the graph of a third-degree polynomial function that has one local maximum, one local minimum, and one zero. Identify each of these three points.
   A sample answer is given.
10. Find an equation for the cubic function f whose graph is shown.
   $f(x) = \frac{1}{2}(x + 1)(x - 3)^2$

ANSWERS

1. See question.
2. 49
3. $4x^4 - 2x^3 - 3x$
4. $2y^5 + 4y^4 + 3y^2$
5. $6x^3 - 7x^2 + 5x - 2$
6. $x^2 - 2x - 5$
7. See question.
8. See question.
9. See question.
10. See question.

Assessment Book, ALGEBRA 2: EXPLORATIONS AND APPLICATIONS

57

Chapter 9 Assessment
Form B Chapter Test

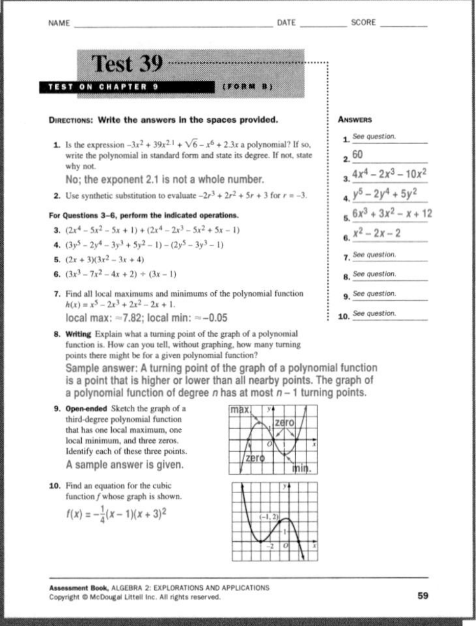

NAME ______ DATE ______ SCORE ______

Test 39

TEST ON CHAPTER 9 (FORM B)

DIRECTIONS: **Write the answers in the spaces provided.**

1. Is the expression $-3x^2 + 39x^{2.1} + \sqrt{6} - x^6 + 2.3x$ a polynomial? If so, write the polynomial in standard form and state its degree. If not, state why not.
   No; the exponent 2.1 is not a whole number.
2. Use synthetic substitution to evaluate $-2r^3 + 2r^2 + 5r + 3$ for $r = -3$.

For Questions 3–6, perform the indicated operations.

3. $(2x^4 - 5x^2 - 5x + 1) + (2x^4 - 2x^3 - 5x^2 + 5x - 1)$
4. $(3y^5 - 2y^4 - 3y^3 + 5y^2 - 1) - (2y^5 - 3y^3 - 1)$
5. $(2x + 3)(3x^2 - 3x + 4)$
6. $(3x^3 - 7x^2 - 4x + 2) \div (3x - 1)$
7. Find all local maximums and minimums of the polynomial function $h(x) = x^5 - 2x^3 + 2x^2 - 2x + 1$.
   local max: ≈ 7.82; local min: ≈ -0.05
8. **Writing** Explain what a turning point of the graph of a polynomial function is. How can you tell, without graphing, how many turning points there might be for a given polynomial function?
   Sample answer: A turning point of the graph of a polynomial function is a point that is higher or lower than all nearby points. The graph of a polynomial function of degree n has at most $n - 1$ turning points.
9. **Open-ended** Sketch the graph of a third-degree polynomial function that has one local maximum, one local minimum, and three zeros. Identify each of these three points.
   A sample answer is given.
10. Find an equation for the cubic function f whose graph is shown.
   $f(x) = -\frac{1}{4}(x - 1)(x + 3)^2$

ANSWERS

1. See question.
2. 60
3. $4x^4 - 2x^3 - 10x^2$
4. $y^5 - 2y^4 + 5y^2$
5. $6x^3 + 3x^2 - x + 12$
6. $x^2 - 2x - 2$
7. See question.
8. See question.
9. See question.
10. See question.

Assessment Book, ALGEBRA 2: EXPLORATIONS AND APPLICATIONS

59

CHAPTER

9 Assessment

VOCABULARY QUESTIONS

For Questions 1 and 2, complete each paragraph.

1. Let $f(x) = 5x^3 - x^2 + 9x - 7$. For this function, the _?_ is 3, the _?_ is 5, and the _?_ is -7.

2. If $y = \dfrac{16}{x}$, then x and y show _?_. The number 16 is called the _?_.

SECTIONS 9.1 *and* 9.2

3. Use synthetic substitution to evaluate $-2t^4 + 5t^3 + t - 8$ when $t = 3$.

4. Add $x^3 + 6x^2 - 10x + 3$ and $-3x^3 + 7x^2 - x - 2$.

5. Subtract $9u^2 - 8u^3 + 20 + 4u^4$ from $-1 + 11u + 2u^2 - 5u^3 + 8u^4$.

6. Multiply $3y - 1$ and $4y^2 - 2y + 7$.

7. Divide $10x^3 + 3x^2 - 5$ by $2x + 3$.

SECTIONS 9.3, 9.4, *and* 9.5

8. Describe the end behavior of $f(x) = -4x^5 + 2x^4 + 6x^2 + 12$ using infinity notation.

9. Find all local maximums and minimums of $g(x) = x^4 - 7x^2 - 4x + 9$.

10. **Writing** Explain why odd-degree polynomial functions can have only *local* maximums and *local* minimums, but even-degree polynomial functions can have maximums and minimums.

11. **POPULATION** For the period 1890–1990, the American Indian, Eskimo, and Aleut population P (in thousands) can be modeled by

$$P = 0.00496t^3 - 0.432t^2 + 11.3t + 212$$

where t is the number of years since 1890. In what year did the population reach 502 thousand?

12. Find an equation for the cubic function whose graph is shown at the left.

Find all real and imaginary zeros of each function.

13. $f(x) = x^3 - 10x^2 + 25x - 4$ **14.** $g(x) = 4x^4 + 4x^3 + x^2 + 30x + 36$

ANSWERS Chapter 9

Assessment

1. degree; leading coefficient; constant term
2. inverse variation; constant of variation
3. -32
4. $-2x^3 + 13x^2 - 11x + 1$
5. $4u^4 + 3u^3 - 7u^2 + 11u - 21$
6. $12y^3 - 10y^2 + 23y - 7$
7. $5x^2 - 6x + 9 - \dfrac{32}{2x + 3}$
8. As $x \to +\infty, f(x) \to -\infty$ and as $x \to -\infty, f(x) \to +\infty$.
9. local minimums: 3.92 and -11; local maximum: 9.58
10. Answers may vary. The end behavior at $+\infty$ and $-\infty$ are opposite for odd-degree polynomials. Thus, the values of the polynomial grow without bound on one end and decrease without bound on the other and so there can be no maximum or minimum value of the function. Even-degree polynomials have identical end behavior at $+\infty$ and $-\infty$. If the function approaches $+\infty$ at both ends, then it will have a minimum, and if it approaches $-\infty$, then it will have a maximum.
11. 1956
12. $f(x) = 2x(x + 2)(x - 3)$
13. 4; $3 + 2\sqrt{2} \approx 5.83$; $3 - 2\sqrt{2} \approx 0.17$
14. $-\frac{3}{2}$ (double zero); $1 \pm i\sqrt{3}$
15. a. No; the products dl are not constant.

SECTIONS 9.6 *and* 9.7

15. HOME REPAIR On some tubes of caulking, the diameter of the round nozzle opening from which the caulking flows can be adjusted by the user. As the diameter d increases, the length l of caulking obtained from the tube decreases, as shown in the table.

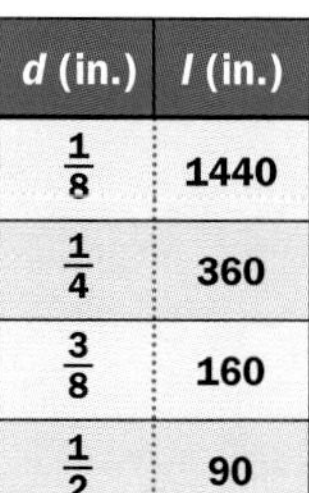

d (in.)	l (in.)
$\frac{1}{8}$	1440
$\frac{1}{4}$	360
$\frac{3}{8}$	160
$\frac{1}{2}$	90

a. Does l vary inversely with d? Explain.

b. Find the area A of each nozzle opening whose diameter is given in the table. Does l vary inversely with A? Explain.

c. Write an equation giving l as a function of A.

16. Let $f(x) = \frac{30x + 17}{6x + 1}$. Find the asymptotes of the graph of f, and tell how the graph is related to a hyperbola with equation of the form $y = \frac{a}{x}$.

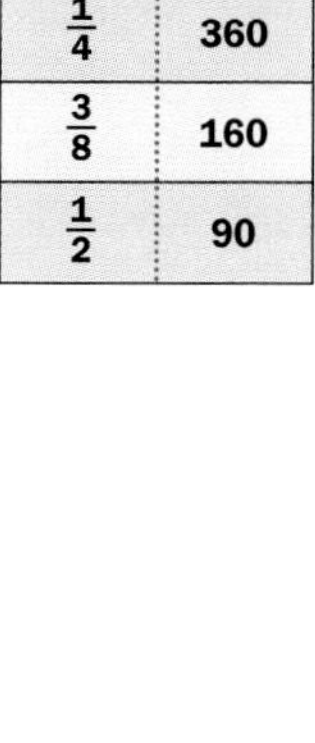

SECTIONS 9.8 *and* 9.9

17. Let $h(x) = \frac{x^4}{4x^2 + 11x - 3}$. Find the vertical asymptotes of the graph of h, and describe the end behavior of h using infinity notation.

18. SPORTS Janice is a member of her high school's varsity basketball team. So far this season, she has made 26 of the 80 three-point shots she has attempted, for a shooting percentage of 32.5%. How many consecutive three-point shots must she make to raise her shooting percentage to 40%?

19. Solve the equation $\frac{2}{x(x - 2)} - \frac{1}{x - 2} = 1$.

PERFORMANCE TASK

20. This problem was posed 700 years ago by the Chinese mathematician Ch'in Chiu-shao:

There is a circular town of unknown diameter having four gates. A tall tree stands 3 li from the north gate as shown. When you exit the south gate and turn east, you must walk 9 li before you can see the tree. Find the diameter of the town. (*Note:* 1 li ≈ 0.33 mi)

a. Show that the radius r of the town must satisfy the polynomial equation $4r^4 + 12r^3 + 9r^2 - 486r - 729 = 0$.

b. Find all real and imaginary solutions of the equation in part (a). Explain why only one of the solutions makes sense in the given situation. What is this solution? What is the diameter of the town?

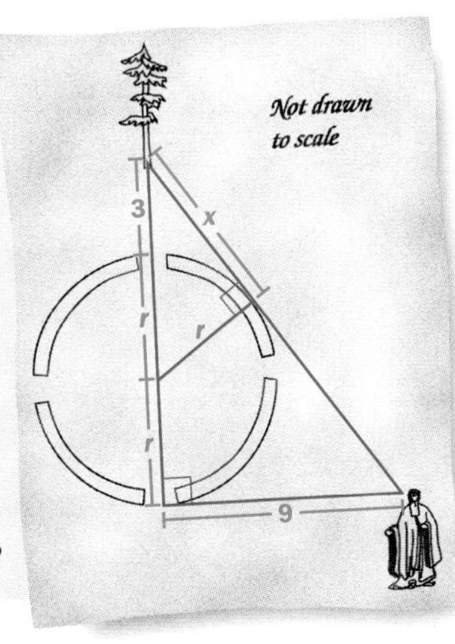

Assessment **459**

Chapter 9 Assessment
Form C Alternative Assessment

Chapter 9
ALTERNATIVE ASSESSMENT

1. **Open-ended Problem** Make up five different expressions that are *not* polynomials. Explain why each expression is not a polynomial.

2. **Group Activity** Arrange the following groups of expressions in order from greatest to least. Use a graphing calculator (for example, graph $y = x$, $y = x + 2$, and $y = x - 2$ for part (a)) to confirm your answer.
   - a. $x, x + 2, x - 2$
   - b. $2x, 2x + 1, 2x - 1$
   - c. $|x|, x, x^2$
   - d. $3x - 2, 2x - 3, 2x + 3$

3. a. Compare the multiplication of two whole numbers to the multiplication of two polynomials. Explain how the procedures are similar and how they are different.
   b. Compare the division of whole numbers to the division of polynomials. Explain how the procedures are similar and how they are different.

4. **Project** Think of something important to remember when you are simplifying rational expressions. Draw a cartoon that will help students remember your idea.

5. **Performance Task** Does $\frac{x-2}{2-x} = -1$ for every value for x? Why or why not?

6. **Performance Task** Make up a polynomial equation that has the following possible rational roots.
   $\frac{6}{1}, \frac{3}{1}, \frac{2}{1}, \frac{1}{1}, \frac{6}{2}, \frac{3}{2}, \frac{2}{2}, \frac{1}{2}, -\frac{6}{1}, -\frac{3}{1}, -\frac{2}{1}, -\frac{1}{1}, -\frac{6}{2}, -\frac{3}{2}, -\frac{2}{2}, -\frac{1}{2}$

7. If you solve a polynomial equation by graphing, can you tell if your answers are double or triple zeroes? If so, explain how.

8. **Project** A company that makes pasta is working on packaging for a new product line. The box will be cut from a piece of cardboard that is 15 in. long by 10 in. wide. Find an equation that represents the maximum volume of the box made by cutting squares from each corner and folding. Do not make a lid for the box from this cardboard. Graph your equation on a graphing calculator. Use the graph to find the maximum volume of the box. Explain why some parts of your graph can be ignored when answering this question.

9. **Research Project** Design a can for a 355 mL juice drink that uses the least amount of aluminum. Write a letter to a bottling company explaining why your design is better than the can they are presently using.

122 **Assessment Book,** ALGEBRA 2: EXPLORATIONS AND APPLICATIONS

---

b.

d (in.)	l (in.)	A (in.^{2})
$\frac{1}{8}$	1440	$\frac{\pi}{256}$
$\frac{1}{4}$	360	$\frac{\pi}{64}$
$\frac{3}{8}$	160	$\frac{9\pi}{256}$
$\frac{1}{2}$	90	$\frac{\pi}{16}$

Yes; for all corresponding values of l and A, $lA = \frac{45\pi}{8}$.

c. $l = \frac{45\pi}{8A}$

16. horizontal asymptote: $y = 5$; vertical asymptote: $x = -\frac{1}{6}$; The graph is the graph of $y = \frac{2}{x}$ translated $\frac{1}{6}$ unit left and 5 units up.

17. $x = -3$ and $x = \frac{1}{4}$; As $x \to \pm\infty$, $h(x) \to +\infty$.

18. at least 10 three-point shots

19. -1

20. a. Since tangents to a circle from a given point are equal in length, the distance from the person's eye to the east gate is also 9 li. Using the large and small right triangles indicated, you get $x^2 + r^2 = (r + 3)^2$ or $x^2 = 6r + 9$, and $(2r + 3)^2 + 9^2 = (x + 9)^2$ or $x^2 + 18x = 4r^2 + 12r + 9$. Substituting, $6r + 9 + 18\sqrt{6r + 9} = 4r^2 + 12r + 9$. Then $4r^2 + 6r = 18\sqrt{6r + 9}$. Dividing by 2 and squaring both sides, $4r^4 + 12r^3 + 9r^2 = 486r + 729$, and $4r^4 + 12r^3 + 9r^2 - 486r - 729 = 0$.

b. $4.5; -1.5; -3 \pm 3i\sqrt{2}$; The solution represents a distance and must be a positive real number; 45; 9 li or about 2.97 mi.

Cumulative Assessment

CHAPTERS 7–9

CHAPTER 7

Solve each system of equations. Tell whether the system has *one solution*, *infinitely many solutions*, or *no solution*.

1. $3x - 2y = 12$
$5x + 2y = 4$

2. $7x + y = 7$
$5x + 3y = 15$

3. $x - y = 6$
$2x - 3y = 8$

4. **Open-ended Problem** Write a system of equations that has infinitely many solutions and another system that has no solution. Use graphs and algebraic methods to show that your choices of systems are correct.

5. Rewrite the system at the right in matrix form and solve for x, y, and z.

$$\begin{aligned} 0.3x - 2.1y + 4.5z &= 12.4 \\ 5.2x + 0.2y - 3.7z &= 4.0 \\ -1.8x - 3.0y - 1.6z &= 9.2 \end{aligned}$$

6. **Writing** Explain how to decide which half-plane to shade when you graph an inequality such as $7x + 5y > 35$.

Graph each inequality or system of inequalities.

7. $y \leq -0.4x + 0.6$

8. $x > -4$
$y \leq 5$

9. $2x - 3y \geq -12$
$y < -\frac{2}{3}x - 2$

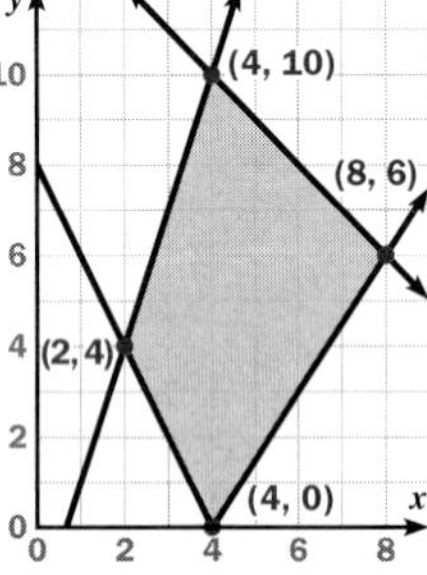

10. a. Write a system of inequalities defining the feasible region shown.

b. For the feasible region in part (a), find the minimum cost for the cost function $C = 5x + 8y$.

11. Daniel can spend up to \$100 on tapes and CDs. He wants to buy at least as many CDs as tapes, and at least 1 tape. A tape costs \$10 and a CD costs \$15. How many CDs and how many tapes should Daniel buy to maximize the number of items bought?

CHAPTER 8

12. **OCEANOGRAPHY** The equation $l = \frac{\pi v^2}{4.9}$ shows how the velocity v of a deep water wave (in meters per hour) is related to the wavelength l (in meters).

a. Write an equation for the velocity as a function of the wavelength.

b. Find the domain and range of the function you wrote in part (a).

c. Find the velocity of a deep water wave with wavelength 10 m.

13. Graph the function $y = -\sqrt{x - 2} + 2$. State the domain and range.

14. **Writing** Explain why $\sqrt[3]{-729}$ is a real number, but $\sqrt{-729}$ is not.

ANSWERS Chapters 7–9

Cumulative Assessment

1. $(2, -3)$; one solution
2. $\left(\frac{3}{8}, 4\frac{3}{8}\right)$; one solution
3. $(10, 4)$; one solution
4. Answers may vary. Examples are given. The system $x + y = 1$, $x + y = 2$ has no solution. This can be shown algebraically by solving simultaneously. The result is $1 = 2$, which is never true. It is shown geometrically by the fact that the graphs of the two equations do not intersect. The system $2x + y = 5$, $4x + 2y = 10$ has infinitely many solutions. This can be shown algebraically by solving simultaneously. The result is $0 = 0$, which is always true. It is shown geometrically by the fact the graphs of the two equations are the same line.
5. 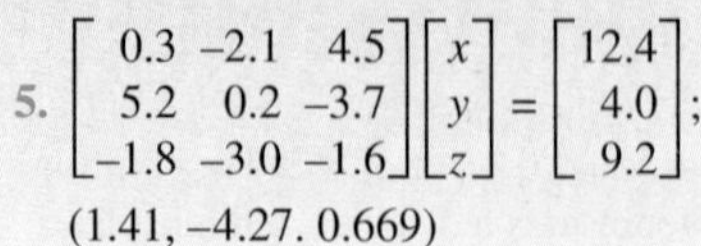
$\begin{bmatrix} 0.3 & -2.1 & 4.5 \\ 5.2 & 0.2 & -3.7 \\ -1.8 & -3.0 & -1.6 \end{bmatrix}\begin{bmatrix} x \\ y \\ z \end{bmatrix} = \begin{bmatrix} 12.4 \\ 4.0 \\ 9.2 \end{bmatrix}$;
$(1.41, -4.27. 0.669)$
6. Write the inequality with y alone on one side. Since the inequality is $y > -\frac{7}{5}x + 7$, shade above the graph. If the inequality sign were $<$ or $\leq$, you would shade below the graph.
7. 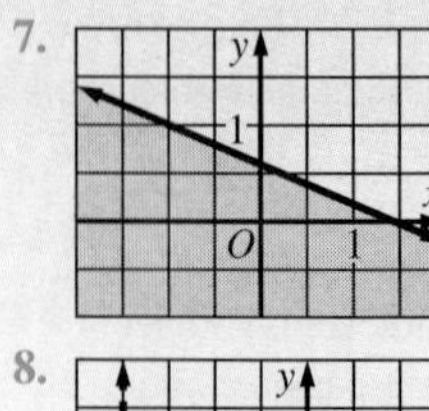
8. 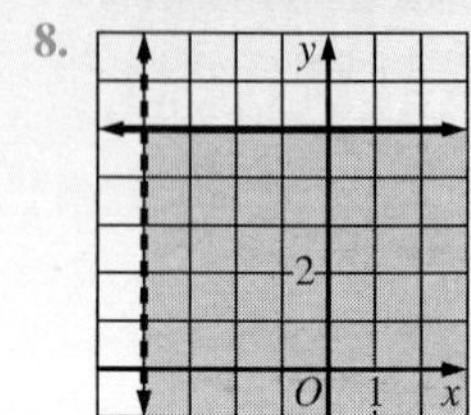
9. 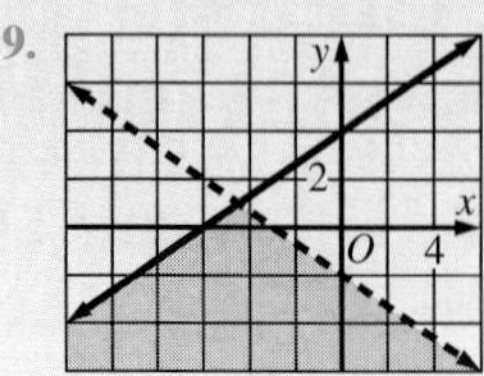
10. a. 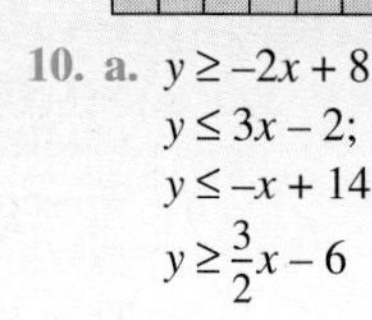
$y \geq -2x + 8$;
$y \leq 3x - 2$;
$y \leq -x + 14$;
$y \geq \frac{3}{2}x - 6$
b. 20
11. 4 tapes; 4 CDs
12. a. 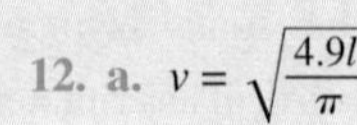 $v = \sqrt{\frac{4.9l}{\pi}}$
b. nonnegative real numbers; nonnegative real numbers
c. 3.95 m/h
13. a. 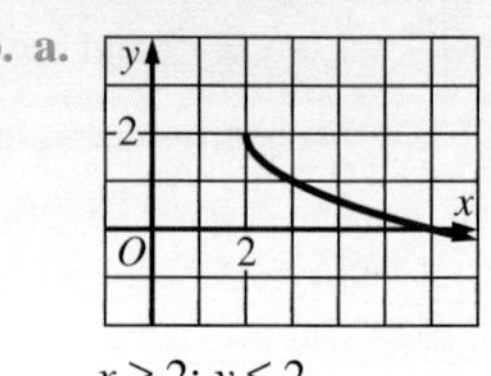
$x \geq 2$; $y \leq 2$
14. $\sqrt[3]{-729}$ is a real number because there is a real number, namely -9, such that $(-9)^3 = -729$. $\sqrt{-729}$ is not a real number because there is no real number whose square is negative.
15. 5
16. 108.5
17. $\pm 4i\sqrt{2}$

Solve. Check to eliminate extraneous solutions.

15. $2\sqrt{2x-1}=6$ **16.** $\sqrt[3]{2x-1}=6$ **17.** $x^2+32=0$

Perform the indicated operation. Plot the result in the complex plane and find its magnitude.

18. $(7+i)+(1-i)$ **19.** $(7+i)-(1-i)$ **20.** $(7+i)(1-i)$

21. Use complex conjugates to write the quotient $\frac{7+i}{7-i}$ in $a+bi$ form.

22. Tell which group properties hold for the whole numbers under multiplication. If a group property does not hold, give a counterexample.

23. **Open-ended Problem** Give an example of a geometry problem whose solution is an irrational number.

CHAPTER 9

Perform the indicated operation.

24. $\frac{2v^2+6v-5}{v-2}$ **25.** $(1.2x^3+3x-5)+(-x^2+7x-8)$

26. Use synthetic substitution to evaluate $8x^3-7x^2+3x-5$ for $x=\sqrt{3}$.

27. **Writing** Explain why $2\sqrt{x}$ is not a polynomial, but $x\sqrt{2}$ is a polynomial.

28. **Technology** Use a graphing calculator or graphing software to graph $f(x)=x^4-5x^3+5x^2$.

a. Describe its end behavior.

b. Find all local maximums and local minimums of the function.

29. Find an equation for the cubic function whose graph has x-intercepts -2, 1, and 4 and passes through the point with coordinates (2, 4).

For each function, list the possible rational zeros, and find all real and imaginary zeros. Identify any double or triple zeros.

30. $f(x)=6x^3-23x^2+9x+5$ **31.** $g(x)=x^4-2x^3-3x^2-50x-50$

32. **Open-ended Problem** Find a polynomial function that has degree 4 and a triple zero at $x=1$.

33. **GEOMETRY** Does the base length of a triangle with area 20 cm^2 vary inversely with the triangle's height? If so, state the constant of variation.

For each function, find the vertical asymptotes of the function's graph, and describe the function's end behavior using infinity notation.

34. $f(x)=\frac{2x-1}{x+5}$ **35.** $g(x)=\frac{x}{x^2+x-6}$

36. Find an equation of the hyperbola that has asymptotes $x=2$ and $y=-3$ and that passes through the point with coordinates (1, 5).

37. Solve the rational equation $\frac{1}{x-1}+\frac{2}{x+1}=1$.

18–20.

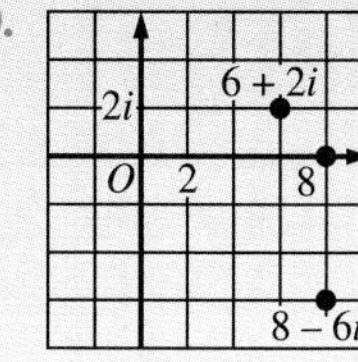

18. 8; 8

19. $6+2i$; $2\sqrt{10}$

20. $8-6i$; 10

21. $\frac{24}{25}+\frac{7}{25}i$

22. closure, identity, associative, and commutative; The inverse property does not hold. No whole number except 1 has a multiplicative inverse that is a whole number.

23. Answers may vary. An example is given. Find the radius of a circle with area 4 m^2.

24. $2v+10+\frac{15}{v-2}$

25. $1.2x^3-x^2+10x-13$

26. $27\sqrt{3}-26$

27. $2\sqrt{x}=2x^{1/2}$, so the exponent of x is not a whole number. The coefficient in $x\sqrt{2}$ is real and the exponent of x, 1, is a whole number.

28.

a. As $x\to\pm\infty$, $f(x)\to+\infty$.

b. minimums: 0 and –9.17; maximum: 1.06

29. $y=-\frac{1}{2}(x+2)(x-1)(x-4)$

30. possible rational zeros: ±1, ±5; $\pm\frac{1}{2}$; $\pm\frac{1}{3}$; $\pm\frac{1}{6}$; $\pm\frac{5}{2}$; $\pm\frac{5}{3}$; $\pm\frac{5}{6}$; zeros: $\frac{5}{6}$; 0.833, 3.30

31. possible rational zeros: ±1; ±2; ±5; ±10; ±25; ±50; zeros: –1; 5; $-1\pm3i$

32. Answers may vary. An example is given. $f(x)=x(x-1)^3$

33. Yes; 40.

34. $x=-5$; As $x\to\pm\infty$, $f(x)\to2$.

35. $x=-3$; $x=2$; As $x\to\pm\infty$, $g(x)\to0$.

36. $f(x)=\frac{-8}{x-2}-3$

37. 0; 3

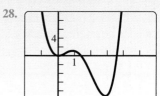

CHAPTER

10 Sequences and Series

OVERVIEW

Connecting to Prior and Future Learning

⇔ Chapter 10 begins with a presentation of sequences. Students study both arithmetic and geometric sequences and examine how they can be used. A review of ratios and proportions on page 785 of the **Student Resources Toolbox** should help students recall some of the concepts from Algebra 1 that they can use in their study of sequences.

⇔ In Section 10.3, students explore recursion. Both explicit and recursive formulas are discussed. These concepts will be used by students in their future studies in mathematics.

⇔ Arithmetic and geometric series are introduced and will be used by students as they pursue higher levels of mathematical studies.

Chapter Highlights

Interview with Jhane Barnes: The relationship between fashion and mathematics is emphasized in this interview, with related exercises on pages 471, 479, and 502.

Explorations in Chapter 10 involve investigating sequences by paper cutting in Section 10.2, investigating the process of recursion by using repeated addition and multiplication in Section 10.3, and investigating how to find the sum of an infinite geometric series in Section 10.5.

The Portfolio Project: Students collect data about the cost to attend various colleges of their choice. They use these data and a spreadsheet to project future college costs and to explore different methods of saving for college.

Technology: Students use a graphing calculator in Section 10.3 to investigate repeated addition and multiplication. Spreadsheets are used in Sections 10.4 and 10.5 to find the sums of arithmetic and geometric series.

OBJECTIVES

Section	Objectives	NCTM Standards
10.1	• Find and use formulas for sequences. • Determine whether a sequence is finite or infinite. • Use sequences to make predictions.	1, 2, 3, 4, 5, 6, 12
10.2	• Determine whether a sequence is arithmetic or geometric. • Find an arithmetic or geometric mean. • Use sequences to solve real-life problems.	1, 2, 3, 4, 5, 6, 12
10.3	• Find the terms of a sequence defined recursively. • Write a recursive formula for a sequence. • Apply the recursive formula for a sequence to a real-world situation.	1, 2, 3, 4, 5, 12
10.4	• Use a formula to find the sum of an arithmetic series. • Use sigma notation for series.	1, 2, 3, 4, 5, 12
10.5	• Use a formula to find the sum of a finite geometric series. • Use a formula to find the sum of an infinite geometric series, if the sum exists. • Apply the formula for the sum of a finite geometric series to real-world situations.	1, 2, 3, 4, 5, 12

Integration

Mathematical Connections	10.1	10.2	10.3	10.4	10.5
algebra	**465–471***	**472–479**	**480–487**	**488–494**	**495–503**
geometry	471	477–479	485	492	502
data analysis, probability, discrete math	467, 470	476		494	
patterns and functions	**465–471**	**472–479**	**480–487**	**488–494**	**495–503**
logic and reasoning	465, 468–471	472, 473, 477–479	481, 484–487	488, 489, 491–494	495–498, 500–503
Interdisciplinary Connections and Applications					
reading and language arts			484		
biology and earth science	468	477	485		
chemistry and physics	467	476		494	
business and economics					501
arts and entertainment	469	475			
sports and recreation		478			496
astronomy, graphic design, architecture, medicine, photography, quilting	470	477, 478	482, 486	489, 492	

* ***Bold page numbers*** *indicate that a topic is used throughout the section.*

Technology

Section	*opportunities for use with* Student Book	Support Material
10.1	scientific calculator graphing calculator	**Geometry Inventor Activity Book:** Activity 9
10.2	scientific calculator graphing calculator	**Technology Book:** Spreadsheet Activity 10
10.3	scientific calculator graphing calculator	**Technology Book:** Calculator Activity 10 Spreadsheet Activity 10
10.4	graphing calculator spreadsheet software McDougal Littell Mathpack *Stats!* *Matrix Analyzer*	
10.5	graphing calculator spreadsheet software McDougal Littell Mathpack *Stats!*	

PLANNING GUIDE

Regular Scheduling (45 min)

Section	Materials Needed	Core Assignment	Extended Assignment	*exercises that feature* Applications	Communication	Technology
10.1	graph paper	**Day 1:** 1–14, 16–35 **Day 2:** 43–48, 55–67	**Day 1:** 1–9, 14, 16–35 odd **Day 2:** 36–67	14 40–42, 49–53	14, 15	
10.2	scissors, graph paper	1–14, 17–25, 35–43	1–43	15, 25–29, 32–37	16, 34, 38	
10.3	graphing calculator or software	**Day 1:** 1–8, 10–21 **Day 2:** 22, 27, 28, 32–41, AYP*	**Day 1:** 1–21 **Day 2:** 22–41, AYP	9 23–26	22	
10.4	spreadsheet software, graphing calculator	1–6, 8–21, 25–27, 29–34	1–21 odd, 22–34	22, 23, 28	7, 29	26
10.5	scissors, spreadsheet software	**Day 1:** 1–21 **Day 2:** 25–27, 37, 40–53, AYP	**Day 1:** 1–21 odd, 22–24 **Day 2:** 25–35 odd, 37–53, AYP	23 38, 39	24 37	22
Review/ Assess		**Day 1:** 1–31 **Day 2:** Ch. 10 Test	**Day 1:** 1–31 **Day 2:** Ch. 10 Test			
Portfolio Project		Allow 2 days.	Allow 2 days.			

Yearly Pacing (with Portfolio Project)	**Chapter 10 Total** 12 days	**Chapters 1–10 Total** 136 days	**Remaining** 24 days	**Total** 160 days

Block Scheduling (90 min)

	Day 63	Day 64	Day 65	Day 66	Day 67	Day 68	Day 69
Teach/Interact	Ch. 9 Test 10.1	Continue with 10.1 10.2: Exploration, page 472	10.3: Exploration, page 480	10.4 10.5: Exploration, page 497	Continue with 10.5 Port. Proj.	Review Port. Proj.	Ch. 10 Test 11.1: Exploration, page 513
Apply/Assess	**Ch. 9 Test** **10.1:** 1–14, 16–35	**10.1:** 43–48, 55–67 **10.2:** 1–14, 17–25, 35–43	**10.3:** 1–8, 10–22, 27, 28, 32–41, AYP*	**10.4:** 1–6, 8–21, 25–27, 29–34 **10.5:** 1–21	**10.5:** 25–27, 37, 40–53, AYP **Port. Proj.**	**Review:** 1–31 **Port. Proj.**	**Ch. 10 Test** **11.1:** 1–13

NOTE: A one-day block has been added for the Portfolio Project—timing and placement to be determined by teacher.

Yearly Pacing (with Portfolio Project)	**Chapter 10 Total** 6 days	**Chapters 1–10 Total** $68\frac{1}{2}$ days	**Remaining** $13\frac{1}{2}$ days	**Total** 82 days

* ***AYP*** *is Assess Your Progress.*

LESSON SUPPORT

Section	Practice Bank	Study Guide*	Assessment Book*	Visuals	Explorations Lab Manual	Lesson Plans	Technology Book
10.1	64	10.1		Warm-Up 10.1	Masters 1, 2	10.1	
10.2	65	10.2		Warm-Up 10.2	Masters 1, 12	10.2	Spreadsheet Act. 10
10.3	66	10.3	Test 40	Warm-Up 10.3	Add. Expl. 16	10.3	Calculator Act. 10 Spreadsheet Act. 10
10.4	67	10.4		Warm-Up 10.4		10.4	
10.5	68	10.5	Test 41	Warm-Up 10.5 Folder 10	Master 13	10.5	
Review Test	69	Chapter Review	Tests 42, 43 Alternative Assessment			Review Test	Calculator Based Lab 5

***Spanish versions** of *Study Guide* and *Assessment Book* are available.

Chapter Support

- Course Guide
- Lesson Plans
- Portfolio Project Book:
  Additional Project 4: Music Theory
- Preparing for College Entrance Tests
- Multi-Language Glossary
- *Test Generator* Software
- Professional Handbook

Software Support

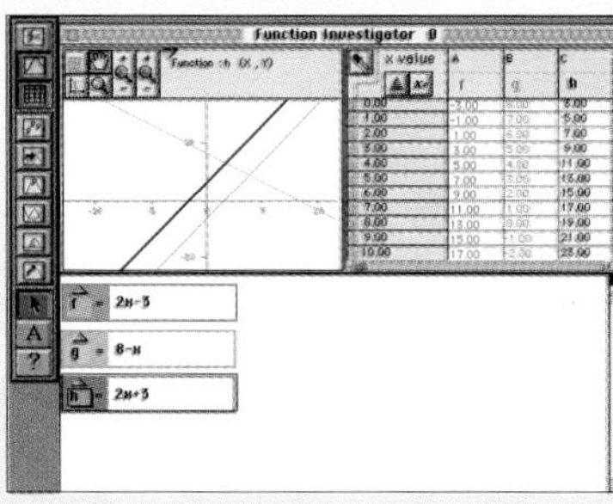

McDougal Littell Mathpack
Stats!
Matrix Analyzer

Internet Support

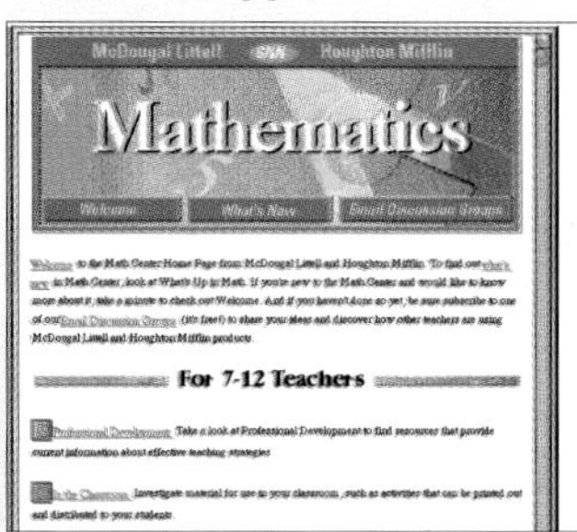

http://www.hmco.com
Next go to McDougal Littell; then the Education Center; then Secondary Math.

OUTSIDE RESOURCES

Books, Periodicals

Sandefur, J. *Discrete Mathematics with Difference Equations.* Lecture notes available through the Mathematics Department of Georgetown University, Washington, D.C.

Schielack, Vincent P., Jr. "Tournaments and Geometric Sequences." *Mathematics Teacher* (February 1993): pp. 127–129.

HiMAP Module 2. *Recurrence Relations.* Section 3: "Counting Backwards: Carbon Dating": pp. 8–12.

Activities, Manipulatives

Dion, Gloria S. "Fibonacci Meets the TI-82." *Mathematics Teacher* (February 1995): pp. 101–105.

Software

Meridian Creative Group. "Exploration: Modeling Geometric Sequences I" and "Exploration: Modeling Geometric Sequences II." Worksheets and software included for use with the CBL.

Explorations in Precalculus for the TI-82: pp. 61–72.

Internet

Explore a webcrawler site (use Webcrawler software to explore the Web)
http://webcrawler.cs.washington.edu/web

A Web site for "Multimedia Math" is located at
http://www.sa.ua.edu/m3func.htm

Interview Notes

Background

Geometric Style
The term *geometric style* refers to many different ancient and modern art styles. These styles are characterized by distinctive abstract, or geometric, designs. The ancient Greeks applied this style to their ceramics as early as 900 B.C. Native Americans used geometric styles in making baskets, clothing, pottery, and paintings. Women of the Blackfoot, Cheyenne, Crow, Kiowa, Pawnee, and Sioux tribes traditionally produced art that was geometric. Most of their art was painted with porcupine quills. When glass beads became available, they were used to embroider clothing. Twined baskets and featherwork done in a geometric style are also characteristic of the Hupa and Yurok tribes. A variety of twining techniques were used to make different geometric patterns on the watertight baskets.

Jhane Barnes
Jhane Barnes founded Jhane Barnes, Inc. in 1979. Since founding her company, she has used computers in many ways to help her design fabrics. In addition to producing fractal patterns with computers, Ms. Barnes also produces symmetrical designs and other intricate patterns. Computers help her work faster in dealing with questions regarding color arrangements on the loom and woven fabric structure. They also help her to see the fabric on a shirt before it is made so she can decide on the right proportion for the design. What makes her patterns appealing, however, is Ms. Barnes's understanding of fabrics and her eye for color.

CHAPTER

10 Sequences and Series

Fractals for fashions

INTERVIEW **Jhane Barnes**

Imagine starting your own business while still a college student. That's what Jhane Barnes did while she was enrolled in New York's Fashion Institute of Technology. Starting with a loan from a professor, she built a business that has grown to become both a commercial and a critical success. In 1980, at the age of 25, Barnes was the youngest person, and the first female, to win a Coty American Fashion Critics' Award for menswear.

"Everything I create is ultimately visual. And fractals are very visual."
"Trilobyte"
"Mandelbrot cloud"
"Koch curve snowflake"
462

The design in the background is the pattern Jhane Barnes calls "Dragon."

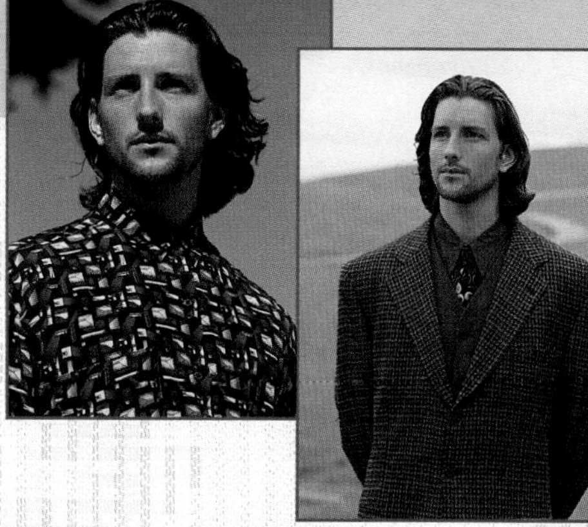

Working with several programmers, Barnes develops the computer code that creates her most intricate designs. "Instead of thinking of patterns only in a visual sense, I now break them down to their simplest elements and think of them mathematically."

Creating Magical Patterns

Barnes designs textiles that are immediately recognizable for their unique use of color and texture. Many of her designs start with a basic element, to which geometric transformations such as rotations and translations are repeatedly applied. In some cases the basic element is as simple as a swirl; in other cases it is a complex figure known as a *fractal*. A close look at a fractal reveals patterns that repeat on an ever-diminishing scale.

Designing with Mathematics

Barnes can go through hundreds of pattern ideas a day for a tie, a jacket, or a pair of socks. Her creative genius shines through in her color choices, and in knowing which patterns will adapt to which fabrics. The outcome, she says, is like magic.

"I realized that if patterns can be generated mathematically, so can the arrangement of color. Many of my designs demonstrate how the sequence of color enhances the movement and flow of the pattern," says Barnes.

Programs with Fashion Finesse

Barnes uses computer programs to transfer her designs to a weaving loom. Only certain designs are transferable, because the loom's frame must hold vertical and horizontal threads known as *warp* and *weft*. "I'm constrained by fabrics," Barnes says. "I can only do things that can be made with warp and weft thread."

Using a computer to control the loom is what led Barnes to discover that a computer could also help her with fabric design.

"Symmetry star"

463

Interview Notes

Background

Waclaw Sierpinski

During a sixty-year career spanning much of the twentieth century, Polish mathematician Waclaw Sierpinski (1882–1969) published over seven hundred papers on mathematics. His new geometrical constructions, theorems, and findings inspired mathematicians throughout the world, including those studying fractals. Fractals have helped scientists to understand vegetation growth and weather patterns, and artists have used fractals to create colorful images.

Second-Language Learners

The meaning of the opening statement *Everything I create is ultimately visual* may be somewhat difficult for students learning English to understand. If necessary, explain that in the fashion industry, the visual appeal of the product is of primary importance. Point out to students that the phrase *a commercial and critical success* means that the designer's fashions sold well, and were also well liked by the experts in the field. Explain that the phrase *I'm constrained by fabrics* means "the designer is limited to the kinds of designs she can create because of the nature of the fabrics she uses."

Interview Notes

Mathematical Connection

There are many different ways that Jhane Barnes can create fabric pattern designs. Three of these ways are explored in this chapter. In Section 10.1, students see how to make a pattern with a transformation. In Section 10.2, they study the Sierpinski triangle to decide if the area and perimeter of the triangles formed are sequential. In Section 10.5, a fractal designed by the Swedish mathematician Helge von Koch is explored to discover the relationship between its area and perimeter.

Explore and Connect

Project
Students may use the stages to create a design of their own. Students can make a poster using different colors and sizes of the square.

Writing
Students can improve their description after someone has attempted to follow their instructions to draw the figure. Encourage them to rewrite their description until it is clear and others can draw the figure by reading their instructions.

Research
Students may find the ideas of warp and weft in a discussion of textiles.

A Sierpinski Pattern

Barnes designed one textile using a fractal known as a *Sierpinski triangle*, named after Waclaw Sierpinski, the Polish mathematician who introduced it in 1916. This fractal is developed in stages, beginning with a solid triangle. At the next stage, the original triangle is divided into four congruent, smaller triangles and the center one is removed. This process is repeated to create successive stages involving smaller and smaller triangles. As the illustration shows, these stages form a *sequence* of triangles, each more elaborate than the one before.

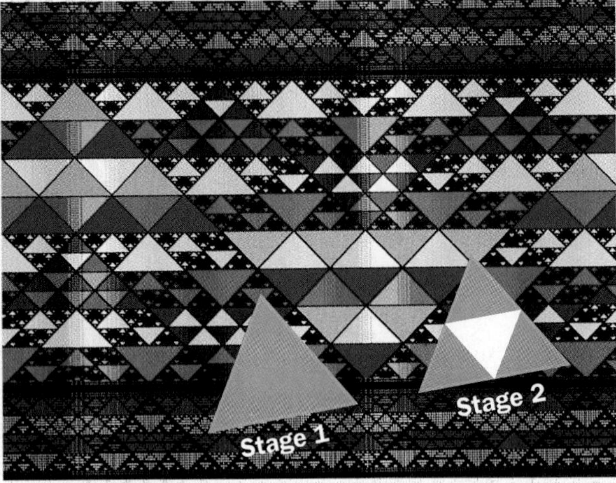

Fabrics created by Jhane Barnes are also used in interior design.

Mathematics & Jhane Barnes

In this chapter, you will learn more about how mathematics is related to fashion.

Related Exercises

Section 10.1
- Exercises 52 and 53

Section 10.2
- Exercises 35–37

Section 10.5
- Exercises 38 and 39

EXPLORE AND CONNECT

Jhane Barnes works on one of her fabric designs using a computer.

1. Project A Sierpinski carpet is like a Sierpinski triangle except that it starts with a shaded square. At each stage of creating the carpet, you must divide all shaded squares from the previous stage into nine congruent, smaller squares and shade all but the middle square. Show three stages of this sequence.

2. Writing Give a written description of the stages in this sequence:

  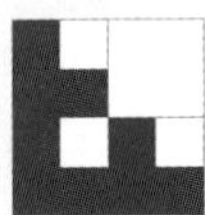

Ask someone who has not seen this sequence to try drawing it from your description of it.

3. Research Use an encyclopedia to find out about weaving. Describe the use of warp and weft in plain and satin weaves. How do the patterns of warp and weft differ in each of the weaves?

Section

10.1 Sequences

Learn how to...

- find and use formulas for sequences
- determine whether a sequence is finite or infinite

So you can...

- predict the maximum number of electrons in an energy level of an atom, for example

The sights and sounds of everyday life are full of interesting patterns. Some are easy to detect, while others demand careful study. Here are just a few examples. Think about what they have in common.

A. Part of a Philippine folk dance

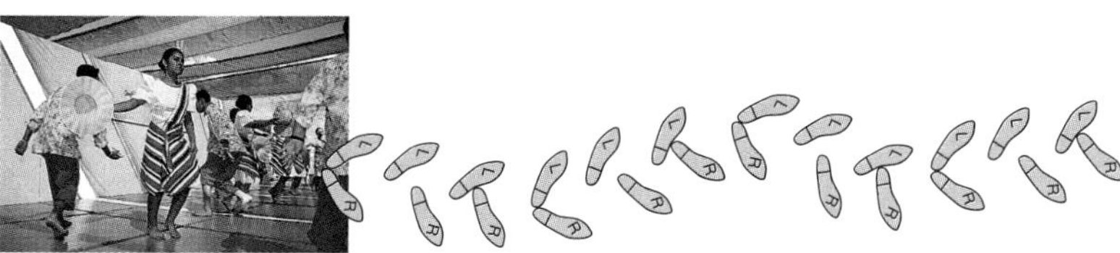

B. A pattern completion exercise from a standardized test

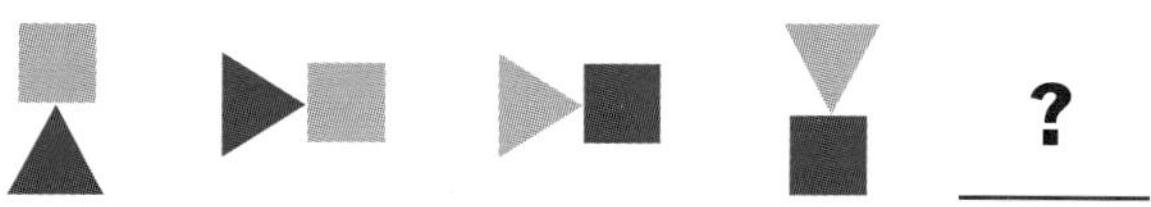
?

C. Part of the sheet music for the national anthem of Mexico, words by Francisco González Bocanegra and music by Jaime Nunó

ce - ro a - pre-stad y el bri - dón.

Think and Communicate

1. Explain why order is important in each situation shown.
2. Describe any patterns you see. Use them to predict what comes next when the pattern is continued.

In each situation shown above, the elements are ordered in time or space. Each set of ordered elements forms a **sequence**, or ordered list. The elements of a sequence are called its **terms**. In mathematics, the terms of a sequence are numbers or geometric figures.

ANSWERS Section 10.1

Think and Communicate

1. In each case, the pattern would be changed by arranging the individual elements differently.
2. Answers may vary. Examples are given. The first three positions in the dance form a pattern and are repeated to complete the dance. The figure in the pattern-completion exercise is rotated 90° clockwise and the colors of the triangle and square are switched. This pattern is repeated. In the music, the pattern of the first three notes, moved up a step, is repeated.

Plan⇔Support

Objectives

- Find and use formulas for sequences.
- Determine whether a sequence is finite or infinite.
- Use sequences to make predictions.

Recommended Pacing

Core and Extended Courses
Section 10.1: 2 days

Block Schedule
Section 10.1: 2 half-blocks (with Chapter 9 Test and Section 10.2)

Resource Materials

Lesson Support

Lesson Plan 10.1

Warm-Up Transparency 10.1

Practice Bank: Practice 64

Study Guide: Section 10.1

Exploration Lab Manual: Diagram Masters 1, 2

Technology

Scientific Calculator

Graphing Calculator

McDougal Littell Mathpack *Geometry Inventor Activity Book:* Activity 9

Internet: http://www.hmco.com

Warm-Up Exercises

1. Find the next two numbers in the pattern −8, −20, −32, −44, ... −56, −68
2. Find the next two numbers in the pattern 2, 4, 8, 16, 32, ... 64, 128
3. Evaluate $3^2 - 3$. 6
4. If $t = 3n - 4$, find t when $n = 7$. 17
5. Describe how to get the next number in the pattern 1000, 100, 10, 1, 0.1, ... Divide the previous number by 10.

Teach⇔Interact

Think and Communicate

For question 2, stress the idea that patterns can be used to make predictions.

About Example 1

Common Error
A common error in making predictions from real-life patterns is to assume that if a pattern exists, then it will continue. This may or may not happen.

Additional Example 1

Find the next 4 terms of each sequence, if possible. If not, explain why.

a. $2, 1, \frac{1}{2}, \frac{1}{4}, \ldots$ $\frac{1}{8}, \frac{1}{16}, \frac{1}{32}, \frac{1}{64}$

b. the number of sea turtles that hatch on a beach each year: 2000, 2650, 2100, 2900, 3350, ...
The number of turtles that hatch on a beach each year depends on many factors, which cannot be predicted. Therefore, it is not possible to find the next 4 terms in the sequence.

Section Note

Reasoning
The choice of a notation to represent mathematical ideas can often influence the ease or difficulty of using the idea. To represent the *n*th term of a sequence, the subscript notation is much more descriptive than the standard notation for a function.

Additional Example 2

Find the first 4 terms of the sequence $t_n = 3n - 4$.
Substitute 1, 2, 3, and 4 for *n* in the formula.

$t_1 = 3(1) - 4 = -1$ $\quad t_2 = 3(2) - 4 = 2$

$t_3 = 3(3) - 4 = 5$ $\quad t_4 = 3(4) - 4 = 8$

The first four terms of the sequence are –1, 2, 5, 8.

EXAMPLE 1

Find the next 4 terms of each sequence, if possible. If not, explain why.

a. 13, 10, 7, 4, . . .

b. daily closing prices of a stock:

$18\frac{3}{8}, 18\frac{1}{4}, 17, 17\frac{5}{8}, \ldots$

SOLUTION

a. 13, 10, 7, 4, . . . (differences: −3, −3, −3)

Notice that each term after the first is 3 less than the term before it. The next 4 terms of the sequence are 1, −2, −5, −8.

b. The price of a stock at the close of the trading day depends on many factors, most of which cannot be predicted. Therefore, it is not possible to find the next 4 terms of the sequence.

When a number sequence has a pattern, you may be able to write a rule for the relationship between the terms of a sequence and their position in the sequence. For example, consider the sequence of positive even numbers:

terms of the sequence: 2, 4, 6, 8, . . .
↕ ↕ ↕ ↕
position in the sequence: 1 2 3 4 . . .

Each term *t* is simply 2 times its position number *n*. This means that *t* is a function of *n*, and you can write:

$$t(n) = 2n$$

For any sequence, the **domain** consists of the **counting numbers** 1, 2, 3, ...

For any sequence, the **range** consists of the **terms** of the sequence.

To distinguish sequences from other functions, formulas for sequences are written in *subscript notation*, where t_n represents the *n*th term of the sequence:

$$t_n = 2n$$

Read this as "*t* sub *n*."

EXAMPLE 2

Find the first 4 terms of the sequence $t_n = n^2 - 3$.

SOLUTION

Substitute 1, 2, 3, and 4 for *n* in the formula.

$t_1 = 1^2 - 3 = -2$ $\quad t_2 = 2^2 - 3 = 1$ $\quad t_3 = 3^2 - 3 = 6$ $\quad t_4 = 4^2 - 3 = 13$

The first 4 terms of the sequence are −2, 1, 6, 13.

EXAMPLE 3 Application: Chemistry

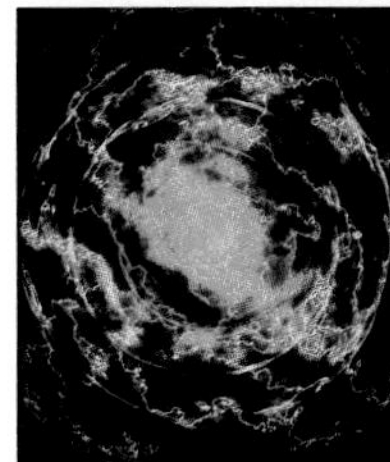
Representation of the energy levels of a helium atom

In an atom, the electrons can be found in energy levels outside the nucleus. An atom can have from 1 to 7 energy levels, and there is a maximum number of electrons that each level can hold. For the first 4 levels, these numbers are 2, 8, 18, 32.

a. Write a formula for the *n*th term of the sequence.

b. Determine the maximum number of electrons in the outermost energy level.

SOLUTION

a. Look for a relationship between each term and its position number.

Energy level	1	2	3	4
Maximum number of electrons	2	8	18	32
	1 · 2	2 · 4	3 · 6	4 · 8

Each term t_n is the product of n and a positive even number, so:

$$t_n = n \cdot 2n$$
$$= 2n^2$$

b. Level 7 is the outermost level. Find t_7.

$$t_7 = 2 \cdot 7^2$$
$$= 98$$

Substitute 7 for *n*.

The maximum number of electrons in the outermost energy level is 98.

A sequence like 2, 8, 18, 32, . . . , 98 that has a last term is a **finite sequence**. A sequence like 2, 4, 6, 8, . . . that continues without end is an **infinite sequence**.

EXAMPLE 4

Tell whether each sequence is *finite* or *infinite*. Explain.

a. the nonnegative multiples of 3

b. the dates of all Sundays this month

SOLUTION

a. The nonnegative multiples of 3 are produced by multiplying 3 by the whole numbers:

$$3 \cdot 0 = 0, \quad 3 \cdot 1 = 3, \quad 3 \cdot 2 = 6, \quad \ldots, \quad 3n, \quad \ldots$$

This sequence continues without end, so it is infinite.

b. No month has more than 5 Sundays, so the sequence is always finite.

Teach⇔Interact

About Example 3

Interdisciplinary Problems

This example illustrates how mathematical ideas can be used to describe physical reality. Mention to students that although the maximum number of electrons at each energy level forms a sequence, each energy level differs from the other levels with regard to their size and orientation around the nucleus.

Additional Example 3

In a certain college lecture hall, there are 8 rows with seats arranged so that each row after the first is higher than the previous row. For the first 4 rows, the number of seats is 13, 17, 21, 25.

a. Write a formula for the *n*th term of the sequence.

Look for a relationship between each term and its position number.

Row Number	1	2	3	4
Number of seats	13	17	21	25
	4·1+9	4·2+9	4·3+9	4·4+9

Each term t_n is the sum of 4 times the term number and 9, so $t_n = 4 \cdot n + 9 = 4n + 9$.

b. Determine the number of seats in the last row.

Row 8 is the last row. Find t_8.

$$t_8 = 4 \cdot 8 + 9$$
$$= 41$$

The number of seats in the last row is 41.

Additional Example 4

Tell whether each sequence is *finite* or *infinite*. Explain.

a. the leap years during the 21st century

The leap years will start with the year 2000 and proceed to the year 2096. There are 25 leap years, so the sequence is finite.

b. the positive powers of 10

The positive powers of 10 are produced by finding each whole number power of 10 greater than 0.

$10^1, 10^2, 10^3, \ldots, 10^n, \ldots$

This sequence continues without end, so it is infinite.

Teach⇔Interact

Checking Key Concepts

Communication: Discussion Writing a formula for the nth term of a sequence is difficult for many students. Ask students to share their formulas for questions 1–3.

Closure Question

Explain how to find and use a formula for a sequence.
Look for a pattern between each term and the position number of the term. Write t_n as a function of n to represent that pattern. Use t_n by substituting a value for n in the function.

Apply⇔Assess

Suggested Assignment

Core Course
Day 1 Exs. 1–14, 16–35
Day 2 Exs. 43–48, 55–67

Extended Course
Day 1 Exs. 1–9, 14, 16–35 odd
Day 2 Exs. 36–67

Block Schedule
Day 63 Exs. 1–14, 16–35
Day 64 Exs. 43–48, 55–67

Exercise Notes

Challenge
Ex. 8 This is an example of a sequence that does not increase or decrease. Instead, the values alternate between –7 and 7. Interested students can try to find an equation for this sequence using $(-1)^{n-1}$, where n is the term number.

Reasoning
Exs. 10, 12 These exercises illustrate that even if real-life sequences of numbers follow a pattern, students should not assume that the next number in the pattern is a valid term of the sequence. Just because there is a pattern of rainy days in June, for example, it is invalid to assume it will rain on the next day in the pattern.

CHECKING KEY CONCEPTS

For each sequence:

a. Find the next 4 terms. **b. Write a formula for t_n.**

1. $1, \frac{1}{4}, \frac{1}{9}, \frac{1}{16}, \ldots$ **2.** 8, 14, 20, 26, . . . **3.** $1, \sqrt{2}, \sqrt{3}, 2, \ldots$

Find the 7th term of each sequence.

4. $t_n = 2n + 3$ **5.** $t_n = 4^{n+1}$ **6.** $t_n = 5n^3 - 6$

10.1 Exercises and Applications

Extra Practice *exercises on page 764*

Find the next 4 terms of each sequence.

1. 4, 16, 64, 256, . . .
2. 800, 400, 200, 100, . . .
3. 1, 9, 3, 18, 5, 36, . . .
4. 1, 0, −1, 5, 1, 10, . . .
5. 5, 55, 555, 5555, . . .
6. 121, 11,311, 1,114,111, . . .
7. 43, 63, 93, 133, . . .
8. 7, −7, 7, −7, . . .
9. $\sqrt{3}, 3, \sqrt{27}, 9, \ldots$

Tell whether it is possible to find the next 4 terms of each sequence. Explain.

10. balances in a checkbook: \$500.00, \$467.32, \$421.84, \$471.84, . . .

11. Presidential election years in the U.S. since 1790: 1792, 1796, 1800, 1804, . . .

12. days on which it rains in June: 3, 7, 8, 15, . . .

13. house numbers along one side of a street: 901, 903, 905, 907, . . .

14. BIOLOGY Frogs belong to a class of animals called *amphibians.* After tadpoles hatch from frogs' eggs, they undergo a sequence of changes in body structure, called *metamorphosis,* before becoming adult frogs. The major stages in the metamorphosis of a frog are shown out of order.

A.
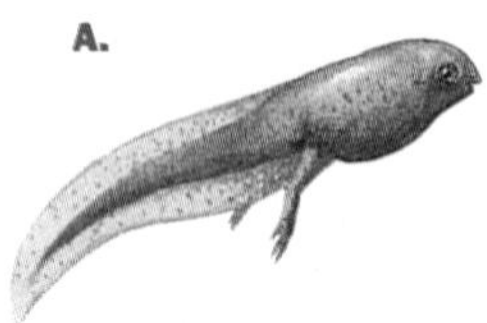
B.

C.
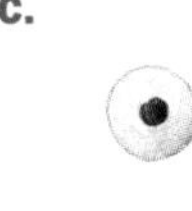
D.

E.
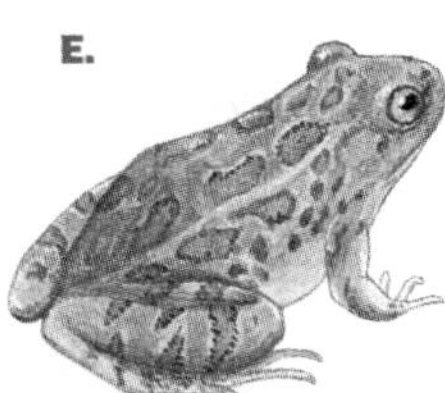
F.

a. Put the pictures in the correct order to show the sequence of events that occurs during the metamorphosis of a frog.

b. Writing Describe the sequence of events shown.

Checking Key Concepts

1. a. $\frac{1}{25}, \frac{1}{36}, \frac{1}{49}, \frac{1}{64}$
   b. $t_n = \frac{1}{n^2}$
2. a. 32, 38, 44, 50
   b. $t_n = 6n + 2$
3. a. $\sqrt{5}, \sqrt{6}, \sqrt{7}, 2\sqrt{2}$
   b. $t_n = \sqrt{n}$
4. 17
5. 65,536
6. 1709

Exercises and Applications

1. 1024, 4096, 16,384, 65,536
2. 50, 25, 12.5, 6.25
3. 7, 72, 9, 144
4. –1, 15, 1, 20
5. 55,555, 555,555, 5,555,555, 55,555,555
6. 111,151,111, 11,111,611,111, 1,111,117,111,111, 111,111,181,111,111
7. 183, 243, 313, 393
8. 7, –7, 7, –7
9. $\sqrt{243}, 27, \sqrt{2187}, 81$
10. No; the balance in a checkbook is affected by both deposits and withdrawals, which are often not predictable.
11. Yes; U.S. presidential elections are held every four years.
12. No; it is not possible to predict the weather with certainty.
13. Yes; on most streets, house lots are numbered by consecutive even or consecutive odd integers. (There are exceptions to this rule.)

15. Cooperative Learning Work with another student.

a. Each of you should cut out a comic strip from a newspaper. (Do not show your comic strip to your partner.)

b. Cut out the individual frames of the strip, mix them up, and pass them to your partner.

c. Each of you should try to put the frames in the correct sequence.

Find the 2nd, 5th, and 12th terms of each sequence.

16. $t_n = 2n + 1$ **17.** $t_n = n^3$ **18.** $t_n = \frac{3}{n}$ **19.** $t_n = \frac{10n}{3n + 2}$

20. $t_n = 3n - 2^n$ **21.** $t_n = (3n - 2)^2$ **22.** $t_n = 2n(5 - n)$ **23.** $t_n = (2n)^2$

24. $t_n = (4n)^{1/n}$ **25.** $t_n = (-1)^{n+3}$ **26.** $t_n = 6^{n-2}$ **27.** $t_n = \log 8n$

Write a formula for t_n.

28. 1, 8, 15, 22, . . . **29.** 43, 49, 55, 61, . . . **30.** 7, 49, 343, 2401, . . . **31.** $\frac{3}{5}, \frac{3}{25}, \frac{3}{125}, \frac{3}{625}, \ldots$

32. $-3, 3, -3, 3, \ldots$ **33.** 103, 97, 91, 85, . . . **34.** $1, 2\sqrt{2}, 3\sqrt{3}, 8, \ldots$ **35.** $1, -\frac{1}{3}, \frac{1}{5}, -\frac{1}{7}, \frac{1}{9}, \ldots$

36. 12, 42, 92, 162, . . . **37.** $\frac{\sqrt{2}}{2}, 1, \sqrt{2}, 2, \ldots$ **38.** 6, 30, 150, 750, . . . **39.** 84, 42, 21, 10.5, . . .

Connection MUSIC

The *acoustics,* or sound quality, in a concert hall depends in part on the stereo effect that occurs when sound reaches listeners' ears from the sides of the hall. To improve acoustics, architects often install *diffusers* with wells of varying depths that scatter sound in all directions.

In one type of diffuser, the depths of the wells are the terms of this sequence: the remainders when n^2 (for $n = 0, 1, 2, \ldots$) is divided by a prime number p. A shorthand way to write this is $t_n = (n^2)_{\text{mod } p}$.

BY THE WAY

In an earlier age, the high, narrow shape of concert halls ensured good acoustics. Today's concert halls are built wider to accommodate more people, more comfortably.

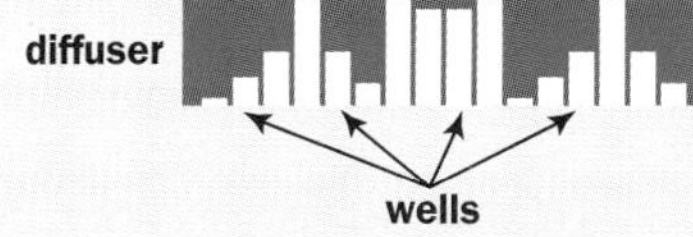

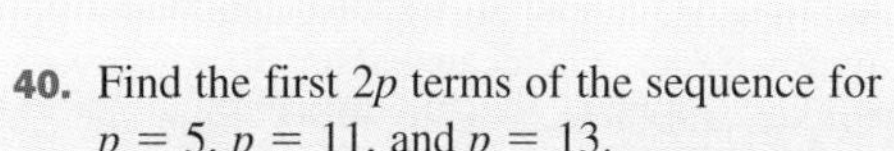

The Michael Fowler Centre, Wellington, New Zealand

40. Find the first $2p$ terms of the sequence for $p = 5$, $p = 11$, and $p = 13$.

41. Use your results in Exercise 40 to make a conjecture about the terms of $t_n = (n^2)_{\text{mod } p}$.

42. The Michael Fowler Centre opened in 1983 in Wellington, New Zealand. It has diffusers with well depths $t_n = (n^2)_{\text{mod } 7}$. Find the terms of the sequence and sketch the pattern of the wells in a diffuser.

Apply⇔Assess

Exercise Notes

Cooperative Learning

Ex. 14 Working in pairs will allow students to compare their answers to this exercise. Have each group explain why the concept of order is important in real-life phenomena.

Research

Ex. 14 After discussing this exercise, some students may be interested in finding other examples of real-life sequences. Have them describe each sequence of events in writing.

Using Technology

Exs. 16–27 Students can use a TI-82 calculator to generate the terms of these sequences. For Ex. 16, choose 5:seq(under the LIST menu and enter 2*X+1,X,1,12,1) to see the first twelve numbers in the sequence $t_n = 2n + 1$.

Topic Spiraling: Preview

Exs. 28, 29, 33 These exercises are examples of arithmetic sequences, which are discussed in the next section.

Alternate Approach

Exs. 28–33 You may wish to have students find a common difference or a common ratio between terms. Use the pattern $t_n = t_1 + d(n - 1)$ for sequences with a common difference between the terms, where t_1 is the first term and d is the common difference. For sequences whose terms have a common ratio, use the pattern $t_n = t_1 \cdot r^{n-1}$, where t_1 is the first term and r is the common ratio.

Topic Spiraling: Preview

Exs. 30–32 These exercises are examples of geometric sequences, which are discussed in the next section.

Problem Solving

Exs. 40, 41 For these exercises, encourage students to make a table like the one in Example 3 on page 467 for each value of p. This will help them to keep their data organized.

Student Study Tip

Ex. 41 If students do not understand the concept of "mod p," point out that a clock is "mod 12." The numbers on a clock face repeat themselves after 12. Likewise, the numbers in a modular (mod) system repeat themselves.

14. a. C, F, A, D, B, E

b. Summaries may vary. A tiny tadpole hatches from an egg. The tadpole develops rear legs, then front legs. It develops into a frog and finally reaches adulthood.

15. Answers may vary. Check students' work.

16. 5; 11; 25 17. 8; 125; 1728

18. $\frac{3}{2}; \frac{3}{5}; \frac{1}{4}$ 19. $\frac{5}{2}; \frac{50}{17}; \frac{60}{19}$

20. 2; –17; –4060

21. 16; 169; 1156

22. 12; 0; –168

23. 16, 100, 576

24. $2\sqrt{2}; \sqrt[5]{20}; \sqrt[12]{48}$

25. –1; 1; –1

26. 1; 216; 60,466,176

27. Logs are rounded to two decimal places. $\log 16 \approx 1.20$; $\log 40 \approx 1.60$; $\log 96 \approx 1.98$

28. $t_n = 7n - 6$

29. $t_n = 6n + 37$ 30. $t_n = 7^n$

31. $t_n = \frac{3}{5^n}$ 32. $t_n = 3(-1)^n$

33. $t_n = 109 - 6n$ 34. $t_n = n\sqrt{n}$

35. $t_n = \frac{1}{2n - 1}(-1)^{n+1}$

36. $t_n = 10n^2 + 2$

37. $t_n = \frac{\sqrt{2}}{2}(\sqrt{2})^{n-1}$

38. $t_n = 6 \cdot 5^{n-1}$

39. $t_n = 84\left(\frac{1}{2}\right)^{n-1}$

40–42. See answers in back of book.

Apply⇔Assess

Exercise Notes

Historical Connection

Exs. 49–51 Astronomy is the oldest of all sciences and has been responsible for the development of many new mathematical ideas and discoveries. Ancient astronomers developed the basic concepts of trigonometry to do astronomical calculations. In the sixteenth and seventeenth centuries, a good deal of the interest in algebra was motivated by the need to solve equations and work with identities in making trigonometry tables. In 1594, John Napier developed logarithms to help do astronomical calculations. Napier's work was guided by a growing body of knowledge that was developing at this time about arithmetic and geometric sequences and series.

Alternate Approach

Ex. 51 An alternative way to generate Bode's law is to write the sequence 0, 3, 6, 12, 24, 48, 96, ... (Each number after the second is twice the previous number.) Then add 4 to each number in the sequence and divide by 10. Generated this way, students can see that Neptune does not seem to fit the sequence, while Pluto does fit the sequence.

Interview Note

Ex. 52 This exercise helps students make a connection between the transformations they studied in geometry and the use of sequences in algebra.

Assessment Note

Ex. 55 This open-ended problem has many solutions. Encourage students to work backward from a pattern to the terms of the sequence to help them find more solutions.

Tell whether each sequence is *finite* or *infinite*. Explain.

43. a pattern around the rim of a pottery vase

44. the digits in the decimal form of π

45. the positive odd numbers

46. the squares of the counting numbers

47. the digits in the decimal form of $\frac{1}{128}$

48. the positive integer powers of 10 less than one billion

Connection ASTRONOMY

In 1766 the German astronomer Johann Daniel Titius announced a rule for computing the approximate distances of the planets from the sun in astronomical units (AU). One astronomical unit is the mean distance between Earth and the sun. Today, Titius's rule is known as *Bode's law*, for Johann Elert Bode, who made it famous.

According to Bode's law, the distances of the first four planets from the sun are given by the sequence in the table.

Johann Elert Bode

BY THE WAY

VIRGO

The first planetary system outside our solar system was discovered in 1992. It is in the constellation Virgo, where three planets are in orbit around a pulsar.

Planet	Distance (AU)
Mercury	0.4 = 4/10
Venus	0.7 = (4 + 3)/10
Earth	1.0 = (4 + 3 · 2)/10
Mars	1.6 = (4 + 3 · 2 · 2)/10

49. Write a formula for Bode's law. (*Hint:* Begin with Venus.)

50. a. When Bode's law was developed, only six of the nine planets had been discovered. The other two known planets were Jupiter, about 5 AU from the sun, and Saturn, about 10 AU from the sun. Use these distances to find each planet's position in the sequence for Bode's law.

b. In 1781, the planet Uranus was discovered. Uranus fits into Bode's sequence immediately after Saturn. Use Bode's law to find the approximate distance of Uranus from the sun in astronomical units.

c. Research Find the 5th term in Bode's sequence. Use an encyclopedia or astronomy book to confirm that there is no planet at this distance from the sun. What astronomical object *is* located at this distance? When was it discovered?

51. The last two planets, Neptune and Pluto, were not discovered until 1846 and 1930, respectively. Neptune is located about 30 AU from the sun and Pluto is about 39 AU from the sun. Do Neptune and Pluto fit into Bode's sequence? Explain.

43. finite; The circumference of the vase is finite, so the pattern can only contain a finite number of terms.
44. infinite; π is irrational, so the sequence 3, 1, 4, 1, 5, 9, 2, 6, 5, 4 continues without end.
45. infinite; The sequence 1, 3, 5, 7, 9, ... continues without end.
46. infinite; The sequence of counting numbers is infinite, so the sequence of the squares of the counting numbers continues without end.
47. finite; $\frac{1}{128} = 0.0078125$
48. finite; There are eight such numbers ($10^1, 10^2, 10^3, 10^4, 10^5, 10^6, 10^7$, and 10^8).
49. Number the planets in order of their distance from the sun: Mercury = 1, Venus = 2, Earth = 3, and Mars = 4. Then for $t_1 = 0.4$ and for $n \geq 2$, $t_n = \frac{4 + 3 \cdot 2^{n-2}}{10}$.
50. a. Solving $5 = \frac{4 + 3 \cdot 2^{n-2}}{10}$, Jupiter's position is 6. Similarly, Saturn's is 7.
    b. 19.6 A.U.
    c. 2.8; The asteroid belt, consisting of about 30,000 pieces of rocky debris called asteroids or planetoids, is located at about that distance. The first and largest of the asteroids, Ceres, was discovered in 1801.
51. Solving the formula, Neptune's position would be 8.6, which does not seem to fit. Pluto's position would be 9, which does fit.
52. The upper left-hand square is rotated 90° clockwise about its lower right-hand corner; call it point *P*. Each resulting square is then rotated 90° clockwise about *P*.
53. Answers may vary. Check students' work.

INTERVIEW Jhane Barnes

Look back at the article on pages 462–464.

Jhane Barnes created the fabric design shown. Examine the pattern inside the four small squares outlined in this design.

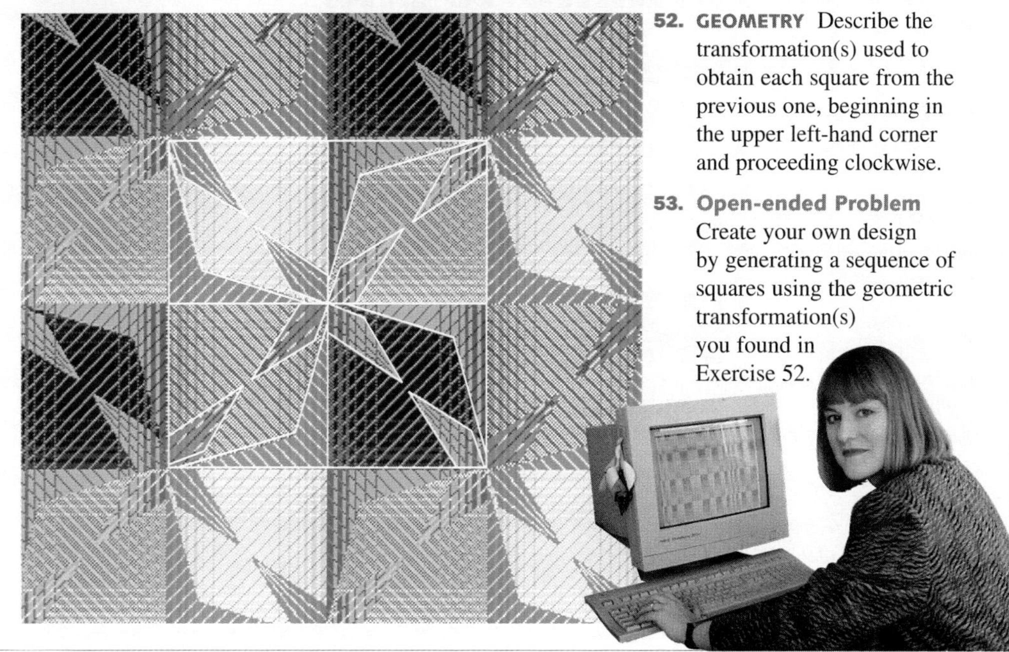

52. **GEOMETRY** Describe the transformation(s) used to obtain each square from the previous one, beginning in the upper left-hand corner and proceeding clockwise.

53. **Open-ended Problem** Create your own design by generating a sequence of squares using the geometric transformation(s) you found in Exercise 52.

54. **Challenge** Is 22,211 a term of the sequence 11, 15, 19, 23, . . . ? If so, what is its position in the sequence? Explain how you found your answer.

ONGOING ASSESSMENT

55. **Open-ended Problem** Think of as many sequences as you can that begin with 2, 4, . . . and have a pattern. For each sequence, write the next 4 terms and, if possible, write a formula for the nth term of the sequence. If it is not possible to write a formula, describe the pattern in words.

SPIRAL REVIEW

Solve each equation. *(Section 9.9)*

56. $\frac{1}{x} = \frac{5}{x+2}$ 57. $\frac{y}{4} = \frac{7}{y-3}$ 58. $\frac{t-1}{t} + \frac{6}{5t} = 2$ 59. $\frac{1}{x-4} = \frac{2}{x^2-16}$

Simplify each radical expression. *(Section 8.2)*

60. $\sqrt{4^8}$ 61. $\sqrt[3]{9^9}$ 62. $\sqrt[3]{128}$ 63. $\sqrt[5]{96}$

Sketch the graph of each equation. *(Sections 2.2 and 3.3)*

64. $y = -3x + 5$ 65. $y = 3^x$ 66. $y = 6x - 14$ 67. $y = (1.5)^x$

Apply⇔Assess

Exercise Notes

Using Technology

Ex. 55 Students can use the regression equation features of a graphing calculator to study this situation. Enter the coordinates (1, 2), (2, 4), (3, 5), and (4, 4) as statistical data. (The first two points correspond to the first two terms of a sequence that begins with 2, 4, ...) Have the calculator generate a cubic regression equation. The resulting equation leads to the sequence formula $t_n = -\frac{1}{6}n^3 + \frac{1}{2}n^2 + \frac{5}{3}n$. Does the sequence begin with 2, 4, ...? (Yes.) Is there a pattern? (Yes, the pattern given by the formula.) Now change (4, 4) to, say, (4, 3) and find a new cubic regression equation. (It yields a different sequence formula, $t_n = -\frac{1}{3}n^3 + \frac{3}{2}n^2 + -\frac{1}{6}n + 1$. Again, the sequence begins with 2, 4, ..., and there is a pattern.)

Practice 64 for Section 10.1

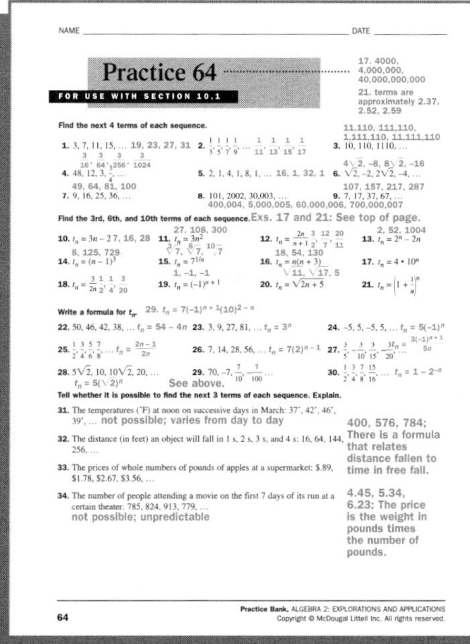

NAME ______ DATE ______

Practice 64

FOR USE WITH SECTION 10.1

17. 4000, 4,000,000, 40,000,000,000
21. terms are approximately 2.37, 2.52, 2.59

Find the next 4 terms of each sequence.

1. 3, 7, 11, 15, … 19, 23, 27, 31
2. $\frac{1}{3}, \frac{1}{5}, \frac{1}{7}, \frac{1}{9}, \ldots$ $\frac{1}{11}, \frac{1}{13}, \frac{1}{15}, \frac{1}{17}$
3. 10, 110, 1110, … 11,110, 111,110, 1,111,110, 11,111,110
4. 48, 12, 3, $\frac{3}{4}$, … $\frac{3}{16}, \frac{3}{64}, \frac{3}{256}, \frac{3}{1024}$
5. 2, 1, 4, 1, 8, 1, … 16, 1, 32, 1
6. $\sqrt{2}, -2, 2\sqrt{2}, -4, \ldots$ $4\sqrt{2}, -8, 8\sqrt{2}, -16$
7. 9, 16, 25, 36, … 49, 64, 81, 100
8. 101, 2002, 30,003, … 400,004, 5,000,005, 60,000,006, 700,000,007
9. 7, 17, 37, 67, … 107, 157, 217, 287

Find the 3rd, 6th, and 10th terms of each sequence. Exs. 17 and 21: See top of page.

10. $t_n = 3n - 2$ 7, 16, 28
11. $t_n = 3n^2$ 27, 108, 300
12. $t_n = \frac{2n}{n+1}$ $\frac{3}{2}, \frac{12}{7}, \frac{20}{11}$
13. $t_n = 2^n - 2n$ 2, 52, 1004
14. $t_n = (n-1)^3$ 8, 125, 729
15. $t_n = 7^{1/n}$ $\sqrt[3]{7}, \sqrt[6]{7}, \sqrt[10]{7}$
16. $t_n = n(n+3)$ 18, 54, 130
17. $t_n = 4 \cdot 10^n$
18. $t_n = \frac{3}{2n}$ $\frac{1}{2}, \frac{1}{4}, \frac{3}{20}$
19. $t_n = (-1)^{n+1}$ 1, –1, –1
20. $t_n = \sqrt{2n+5}$ $\sqrt{11}, \sqrt{17}, 5$
21. $t_n = \left(1 + \frac{1}{n}\right)^n$

Write a formula for t_n. 29. $t_n = 7(-1)^{n+1}(10)^{2-n}$

22. 50, 46, 42, 38, … $t_n = 54 - 4n$
23. 3, 9, 27, 81, … $t_n = 3^n$
24. –5, 5, –5, 5, … $t_n = 5(-1)^n$
25. $\frac{1}{2}, \frac{3}{4}, \frac{5}{6}, \frac{7}{8}, \ldots$ $t_n = \frac{2n-1}{2n}$
26. 7, 14, 28, 56, … $t_n = 7(2)^{n-1}$
27. $\frac{3}{5}, -\frac{3}{10}, \frac{3}{15}, -\frac{3}{20}, \ldots$ $t_n = \frac{3(-1)^{n+1}}{5n}$
28. $5\sqrt{2}, 10, 10\sqrt{2}, 20, \ldots$ $t_n = 5(\sqrt{2})^n$
29. $70, -7, \frac{7}{10}, -\frac{7}{100}, \ldots$ See above.
30. $\frac{1}{2}, \frac{3}{4}, \frac{7}{8}, \frac{15}{16}, \ldots$ $t_n = 1 - 2^{-n}$

Tell whether it is possible to find the next 3 terms of each sequence. Explain.

31. The temperatures (°F) at noon on successive days in March: 37°, 42°, 46°, 39°, … not possible; varies from day to day
32. The distance (in feet) an object will fall in 1 s, 2 s, 3 s, and 4 s: 16, 64, 144, 256, … 400, 576, 784; There is a formula that relates distance fallen to time in free fall.
33. The prices of whole numbers of pounds of apples at a supermarket: \$.89, \$1.78, \$2.67, \$3.56, … 4.45, 5.34, 6.23; The price is the weight in pounds times the number of pounds.
34. The number of people attending a movie on the first 7 days of its run at a certain theater: 785, 824, 913, 779, … not possible; unpredictable

64 **Practice Bank,** ALGEBRA 2: EXPLORATIONS AND APPLICATIONS

54. Yes; it is t_{5551}. I found the formula $t_n = 4n + 7$ and solved the equation $4n + 7 = 22,211$ for n.

55. Answers may vary. Examples are given.

Terms	Formula or Description
2, 4, 6, 8, 10, …	$t_n = 2n$
2, 4, 8, 16, 32, 64, …	$t_n = 2^n$
2, 4, 7, 11, 16, 22, …	The first term is 2. Each term is the previous term plus the position number.
2, 4, 2, 4, 2, 4, …	The terms 2 and 4 repeat without end.

56. $\frac{1}{2}$ 57. –4; 7

58. $\frac{1}{5}$ 59. –2

60. 256 61. 729

62. $4\sqrt[3]{2}$ 63. $2\sqrt[5]{3}$

64.

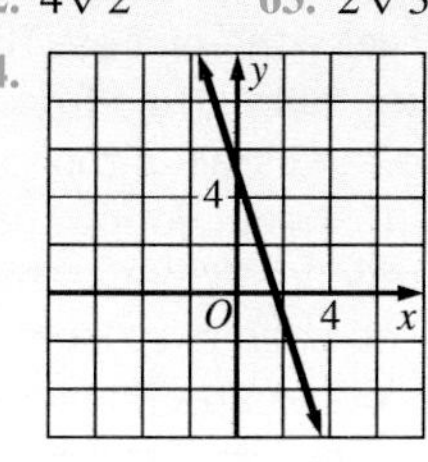

65.

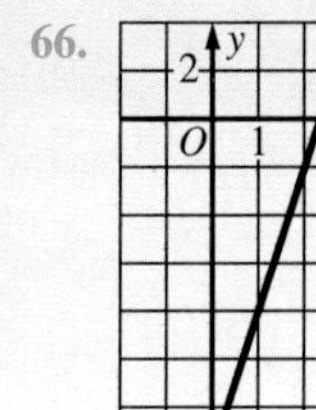

66.

67.

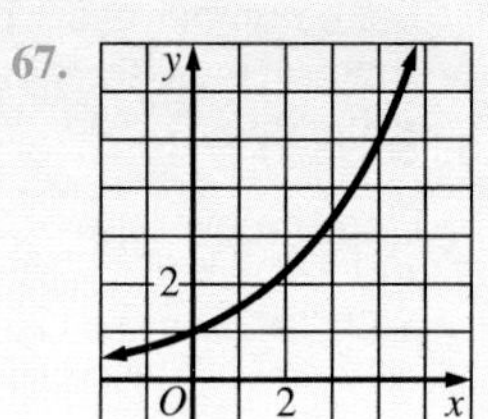

Plan⇔Support

Objectives

- Determine whether a sequence is arithmetic or geometric.
- Find an arithmetic or geometric mean.
- Use sequences to solve real-life problems.

Recommended Pacing

Core and Extended Courses
Section 10.2: 1 day

Block Schedule
Section 10.2: $\frac{1}{2}$ block
(with Section 10.1)

Resource Materials

Lesson Support
Lesson Plan 10.2
Warm-Up Transparency 10.2
Practice Bank: Practice 65
Study Guide: Section 10.2
Exploration Lab Manual:
Diagram Masters 1, 12

Technology
Technology Book:
Spreadsheet Activity 10
Scientific Calculator
Graphing Calculator
Internet:
http://www.hmco.com

Warm-Up Exercises

1. What is the common difference in the numbers 18, 21, 24, 27, and 30? 3
2. What is the ratio of 8 to 24 and of 6 to 18? 1 to 3
3. Describe the pattern in the sequence 15, 21, 27, 33, ... Each number is six more than the previous number.
4. Solve $x^2 = 49$. 7 or –7
5. Find the average of 36 and 72. 54

Section

10.2 Arithmetic and Geometric Sequences

Learn how to...

- **determine whether a sequence is arithmetic or geometric**
- **find an arithmetic or geometric mean**

So you can...

- **find a slide position on a trombone, for example**

When cutting up vegetables, experienced cooks usually begin by making a sequence of parallel cuts. Then they gather all the pieces and make another sequence of cuts through all the pieces at once. Do you know why?

EXPLORATION
COOPERATIVE LEARNING

Investigating Methods for Cutting Paper

Work with a partner.
You will need:

- scissors

SET UP Turn an $8\frac{1}{2}$ in. by 11 in. sheet of paper sideways. Cut the paper into strips about a half inch wide. One of you should use Method A, and the other should use Method B. Each time you make a cut, record the total number of cuts and the total number of pieces of paper in a table.

	No. of pieces	
No. of cuts	A	B
1	2	2
2	4	3
⋮	⋮	⋮

Method A

1. Cut the paper in half.
2. Stack the halves. Cut the stack in half.
3. Continue stacking and cutting.

Method B

1. Cut a thin strip from one end.
2. Continue cutting off strips.

Answer Questions 1–3 for each method.

1. How does the number of pieces change with each cut after the first?
2. Find the next 3 terms of the sequence of the number of pieces.
3. Write a formula for the nth term of the sequence in terms of n and the first term of the sequence.

Exploration Note

Purpose
The purpose of this Exploration is to have students explore arithmetic and geometric sequences.

Materials/Preparation
Each group needs paper and scissors.

Procedure
One student of a pair cuts strips of paper in half repeatedly and stacks the pieces. The other student in the pair cuts strips from one end of a piece of paper and stacks the strips. The first student should notice that cutting strips in half repeatedly generates a tall stack of papers. The second student should notice that his or her stack increases slowly because only one piece is added at a time. Students are asked to describe how their stacks increase and write a formula for the sequence generated.

Closure
Students should understand that they have generated two different types of sequences: one which has a common ratio and another which has a common difference.

Explorations Lab Manual
See the Manual for more commentary on this Exploration.

Diagram Master 12

For answers to the Exploration, see answers in back of book.

The sequence formed by Method A in the Exploration is a *geometric sequence*. In a **geometric sequence**, the ratio of any term to the term before it is constant. The constant ratio is called the **common ratio**. For example, consider this sequence:

The common ratio is 5.

2, 10, 50, 250, . . . , $2 \cdot 5^{n-1}$
×5 ×5 ×5

To find the *n*th term, **multiply** the first term by the common ratio $(n - 1)$ times.

The sequence formed by Method B in the Exploration is an *arithmetic sequence*. In an **arithmetic sequence**, the difference between any term and the term before it is constant. The constant difference is called the **common difference**. For example, consider this sequence:

The common difference is 5.

2, 7, 12, 17, . . . , $2 + 5(n - 1)$
+5 +5 +5

To find the *n*th term, **add** the common difference to the first term $(n - 1)$ times.

Arithmetic and Geometric Sequences

Arithmetic Sequence

General formula:

$$t_n = t_1 + d(n-1)$$

common difference

Example: $t_n = 2 + 5(n - 1)$

Geometric Sequence

General formula:

$$t_n = t_1 \cdot r^{n-1}$$

common ratio

Example: $t_n = 2 \cdot 5^{n-1}$

THINK AND COMMUNICATE

1. How are arithmetic sequences and linear functions related? Which elements of the general formula for an arithmetic sequence correspond to the slope, the dependent variable, and the independent variable of a linear function? Explain.
2. How are geometric sequences and exponential functions related? Which elements of the general formula for a geometric sequence correspond to the proportional growth factor, the dependent variable, and the independent variable of an exponential function? Explain.
3. What effect does a common ratio *between 0 and 1* have on a geometric sequence? on its graph?
4. What effect does a *negative* common difference have on an arithmetic sequence? on its graph?

Technology Note

Graphing calculators and software with parametric graphing capability can be used to display graphs of sequences. For example, on the TI-82, graph X1T=T and Y1T=2(5)^(T–1) to obtain a graph like the one shown at the right on this page. Be sure the calculator is set to parametric mode and that the graphing mode is dot mode rather than connected mode. For the graphing window, use Tmin = 1 and Tstep = 1. Students may experiment to find a suitable setting for the graphing window.

Teach⇔Interact

Learning Styles: Verbal

The presentation on this page summarizes the intuitive results of the Exploration on page 472. The new terms and formulas introduced may not present any difficulties to verbal learners but may be somewhat difficult for visual or kinesthetic learners to absorb at first. The alternative approach given below may be helpful for all students, as would be the use of other specific examples.

Section Notes

Alternate Approach

Students may see the patterns in the sequences more easily if they arrange the terms in a table.

geometric

n	terms	pattern
1	2	
2	2 × 5	2×5^1
3	2 × (5 × 5)	2×5^2
4	2 × (5 × 5 × 5)	2×5^3
n	2 × (5 × 5 × 5 ...)	$2 \times 5^{n-1}$

arithmetic

n	terms	pattern
1	2	
2	2 + 5	2 + 5 × 1
3	2 + (5 + 5)	2 + 5 × 2
4	2 + (5 + 5 + 5)	2 + 5 × 3
n	2 + (5 + 5 + 5 ...)	2 + 5 × (*n* – 1)

Student Progress

Students should recognize that the graph of the arithmetic sequence $t_n = 2 + 5(n - 1)$ is linear, where 5 represents the slope and –3 is the *y*-intercept. They should also recognize that the graph of the geometric sequence $t_n = 2 \cdot 5^{n-1}$ is similar to the graph of an exponential function with base 5.

Think and Communicate

For question 3, have students graph their conjectures with a calculator to verify that a ratio between 0 and 1 gives a sequence whose terms are decreasing. For question 4, have them graph their conjectures with a calculator to verify that a negative difference also gives a sequence whose terms are decreasing.

For answers to Think and Communicate, see answers in back of book.

Teach⇔Interact

About Example 1

Teaching Tip
You may wish to point out that for a geometric sequence, the common difference of two terms has the same ratio as the terms themselves. This is not the case with arithmetic sequences or sequences that do not have a geometric pattern.

Additional Example 1

Tell whether each sequence is *arithmetic, geometric,* or *neither.*

a. $\frac{1}{2}, \frac{1}{3}, \frac{1}{4}, \frac{1}{5}, \ldots$
Look for a common difference or a common ratio.
$\frac{1}{2}, \frac{1}{3}, \frac{1}{4}, \frac{1}{5}, \ldots$
differences: $-\frac{1}{6}$ $-\frac{1}{12}$ $-\frac{1}{20}$
ratios: $\frac{2}{3}$ $\frac{3}{4}$ $\frac{4}{5}$
There is neither a common difference nor a common ratio. The sequence is neither geometric nor arithmetic.

b. –1, –8, –64, –512, ...
–1, –8, –64, –512,...
differences: –7 –56 –448
ratios: 8 8 8
The common ratio is 8. The sequence is geometric.

c. 5, 8, 11, 14, ...
5, 8, 11, 14,...
differences: 3 3 3
ratios: 1.6 1.38 1.27
The common difference is 3. The sequence is arithmetic.

Additional Example 2

Find the 9th term of each sequence.

a. the sequence in part (b) of Additional Example 1
The sequence –1, –8, –64, –512, ... is geometric with a common ratio of 8.
Step 1 Write a formula for t_n.
$t_n = -1(8^{n-1})$
Step 2 Find t_9.
$t_9 = -1(8^{9-1})$
$= -16{,}777{,}216$
The 9th term of the sequence is –16,777,216.

EXAMPLE 1

Tell whether each sequence is *arithmetic, geometric,* or *neither.*

a. −14, −42, −126, −378, . . . **b.** 1, 19, 49, 91, . . . **c.** 31, 27, 23, 19, . . .

SOLUTION

Look for a common difference or a common ratio.

a. −14, −42, −126, −378, . . .
differences: −28 −84 −252
ratios: 3 3 3
The common ratio is 3. The sequence is geometric.

b. 1, 19, 49, 91, . . .
differences: 18 30 42
ratios: 19 2.579 1.857
There is neither a common difference nor a common ratio. The sequence is neither geometric nor arithmetic.

c. 31, 27, 23, 19, . . .
differences: −4 −4 −4
ratios: 0.871 0.852 0.826
The common difference is −4. The sequence is arithmetic.

EXAMPLE 2

Find the 11th term of each sequence.

a. the sequence in part (a) of Example 1
b. the sequence in part (c) of Example 1

SOLUTION

a. The sequence −14, −42, −126, −378, . . . is geometric with a common ratio of 3.

Step 1 Write a formula for t_n.

$$t_n = -14(3^{n-1})$$

Substitute **−14** for t_1 and **3** for r in the general formula.

Step 2 Find t_{11}.

$$t_{11} = -14(3^{11-1}) = -14(59{,}049) = -826{,}686$$

Substitute **11** for n.

The 11th term of the sequence is −826,686.

b. The sequence 31, 27, 23, 19, . . . is arithmetic with a common difference of −4.

Step 1 Write a formula for t_n.

$$t_n = 31 - 4(n - 1)$$

Substitute **31** for t_1 and **−4** for d in the general formula.

Step 2 Find t_{11}.

$$t_{11} = 31 - 4(11 - 1)$$
$$= 31 - 40 = -9$$

Substitute 11 for n.

The 11th term of the sequence is −9.

EXAMPLE 3 Application: Music

To play different notes on a trombone, you change the *effective length* of the tube by moving the slide in and out. The sequence of lengths at different slide positions is geometric. Find the effective length at 4th position on this trombone.

In 1st position, the slide is fully drawn in.

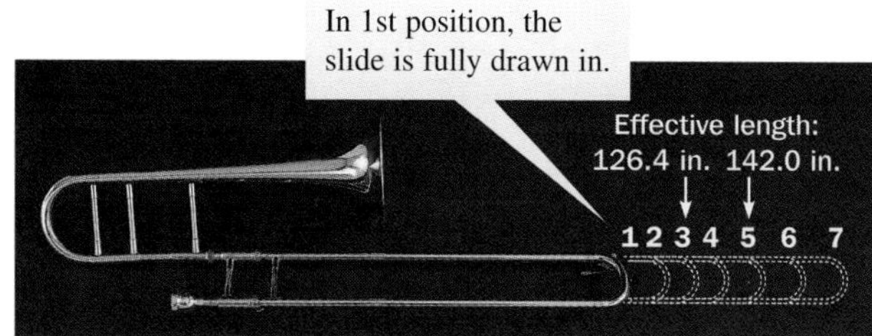

SOLUTION

Use the fact that a geometric sequence has a common ratio.

The ratios of successive terms are equal.

$$\frac{t_4}{t_3} = \frac{t_5}{t_4}$$

Substitute **126.4** for t_3 and **142.0** for t_5.

Toolbox p. 785 *Ratios and Proportions*

$$\frac{t_4}{126.4} = \frac{142.0}{t_4}$$

Use the means-extremes property.

$$(t_4)^2 = 126.4 \cdot 142.0 = 17{,}948.8$$

$$t_4 = \sqrt{17{,}948.8} \approx 134.0$$

Since length is always positive, find the **positive square root**.

The effective length at 4th position is about 134.0 in.

The term between any two terms of a geometric sequence is called the **geometric mean** of the given terms. In Example 3, you found the geometric mean of two positive terms. In general, for any two positive terms a and b:

geometric mean of a and b = $\sqrt{ab}$

The term between any two terms of an arithmetic sequence is called the **arithmetic mean** of the given terms. The arithmetic mean is simply the *average* or *mean* of the given terms. For any two terms a and b:

arithmetic mean of a and b = $\frac{a+b}{2}$

Teach⇔Interact

Additional Example 2 (continued)

b. the sequence in part (c) of Additional Example 1

The sequence 5, 8, 11, 14, ... is arithmetic with a common difference of 3.

Step 1 Write a formula for t_n.

$t_n = 5 + 3(n - 1)$

Step 2 Find t_9.

$t_9 = 5 + 3(9 - 1)$
$= 29$

The 9th term of the sequence is 29.

About Example 3

Reasoning

Ask students to explain why the equation $\frac{t_4}{t_3} = \frac{t_5}{t_4}$ is true for any geometric sequence.

Additional Example 3

A superball bounces to three-fourths its previous height with each bounce. The sequence of heights is geometric. Find the height on the 6th bounce if the height of the 5th bounce is 9 ft and the height of the 7th bounce is 5 ft.

Use the fact that a geometric sequence has a common ratio. Substitute 9 for t_5 and 5 for t_7.

$\frac{t_6}{t_5} = \frac{t_7}{t_6}$

$\frac{t_6}{9} = \frac{5}{t_6}$

$(t_6)^2 = 9 \cdot 5$

$t_6 = \sqrt{45} \approx 6.7$

The height on the 6th bounce is about 6.7 ft.

Section Note

Integrating the Strands

Students have already found the geometric mean in geometry when they solved an equation like $\frac{6}{x} = \frac{x}{9}$. For example, the altitude to the hypotenuse of a right triangle is the geometric mean between the two segments of the hypotenuse.

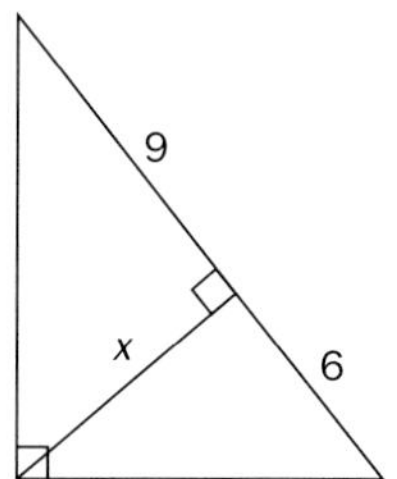

Teach⇔Interact

Checking Key Concepts

Reasoning

For questions 7–9, point out that the geometric mean of two positive numbers is always positive, although solving the equivalent algebraic expression $x^2 = ab$ has two solutions, $\sqrt{ab}$ and $-\sqrt{ab}$.

Closure Question

Explain how to determine whether a sequence is arithmetic or geometric. Find the difference and the ratio of each term and the term before it. If the difference is constant, the sequence is arithmetic. If the ratio is constant, the sequence is geometric.

Apply⇔Assess

Suggested Assignment

Core Course
Exs. 1–14, 17–25, 35–43

Extended Course
Exs. 1–43

Block Schedule
Day 64 Exs. 1–14, 17–25, 35–43

Exercise Notes

Communication: Discussion
Exs. 1–12 The answers to these exercises can be reviewed quickly in class. Ask students to justify their responses.

Teaching Tip
Exs. 13, 14 Remind students that they can use the formulas on page 473 to help them find the patterns for the sequences.

Topic Spiraling: Review
Ex. 15 Students should understand that the formula for the nth term of the sequence they found in part (b) is also an exponential equation. Encourage students to solve part (c) by using logarithms.

Checking Key Concepts

For each sequence:

a. Tell whether it is *arithmetic, geometric,* or *neither.*

b. Find the 11th term.

1. 11, 15, 19, 23, . . . **2.** 3, 15, 75, 375, . . . **3.** 53, 47, 41, 35, . . .

4. 63, 21, 7, $2\frac{1}{3}$, . . . **5.** 1, 4, 9, 16, . . . **6.** −3, −1, 1, 3, . . .

For each pair of numbers:

a. Find the arithmetic mean. **b. Find the geometric mean.**

7. 4, 25 **8.** 5, 45 **9.** $\sqrt{2}$, $3\sqrt{2}$

10.2 Exercises and Applications

Extra Practice exercises on page 764

Tell whether each sequence is *arithmetic, geometric,* or *neither.*

1. 8, 16, 24, 32, . . . **2.** −4, −16, −64, −256, . . . **3.** 36, 49, 64, 81, . . .

4. 32, 33, 34, 35, . . . **5.** 243, 81, 27, 9, . . . **6.** 1, −1, 1, −1, . . .

7. 0, 14, 52, 254, . . . **8.** 129, 88, 47, 6, . . . **9.** $\frac{1}{3}, \frac{1}{9}, \frac{1}{27}, \frac{1}{81}, \ldots$

10. $\sqrt{3}$, $2\sqrt{3}$ $4\sqrt{3}$, $7\sqrt{3}$, . . . **11.** −0.61, −0.64, −0.67, −0.70, . . . **12.** i, −2, $-4i$, 8, . . .

13. For each arithmetic sequence in Exercises 1–12, find:
 a. the common difference b. the 14th term

14. For each geometric sequence in Exercises 1–12, find:
 a. the common ratio b. the 14th term

15. BIOCHEMISTRY When scientists analyze DNA for medical diagnosis, criminal law, evolutionary biology, and gene research, they often need larger samples than they are given. The *polymerase chain reaction (PCR)* process amplifies a tiny sample of DNA by making multiple copies of the DNA molecules. In each cycle of the process, the number of DNA molecules doubles.

a. Is the sequence of the number of DNA molecules after each cycle *arithmetic, geometric,* or *neither*? Explain.

b. Write a formula for the nth term of the sequence, beginning with one molecule of DNA.

c. How many cycles are needed to produce a billion copies?

d. One cycle takes about 6.5 min. About how long does it take to produce a billion copies?

BY THE WAY

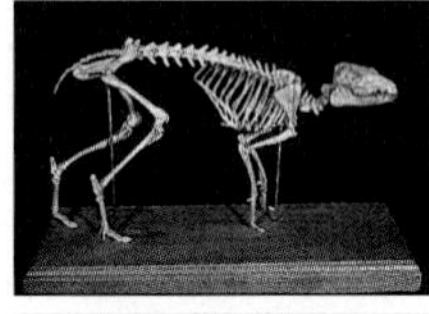

The PCR process was discovered in 1983 by biochemist Kary Mullis. One of its uses has been to study the genes of dinosaurs and other animals that lived millions of years ago.

ANSWERS Section 10.2

Checking Key Concepts

1. a. arithmetic b. 51
2. a. geometric b. 29,296,875
3. a. arithmetic b. −7
4. a. geometric b. $\frac{63}{3^{10}} \approx 0.00107$
5. a. neither b. 121
6. a. arithmetic b. 17
7. a. 14.5 b. 10
8. a. 25 b. 15
9. a. $2\sqrt{2}$ b. $\sqrt{6}$

Exercises and Applications

1. arithmetic 2. geometric
3. neither 4. arithmetic
5. geometric 6. geometric
7. neither 8. arithmetic
9. geometric 10. neither
11. arithmetic 12. geometric

13. a. (1) 8; (4) 1; (8) −41; (11) −0.03
 b. (1) 112; (4) 45; (8) −404; (11) −1

14. a. (2) 4; (5) $\frac{1}{3}$; (6) −1; (9) $\frac{1}{3}$; (12) $2i$
 b. (2) −268,435,456; (5) $\frac{1}{3^8} \approx 0.000152$; (6) −1; (9) $\frac{1}{3^{14}} \approx 0.000000209$; (12) $-8192i$

16. Use the information about trombone slide positions in Example 3 on page 475.
   a. Find the common ratio of the sequence of effective lengths.
   b. **Writing** Explain what the common ratio means in this situation.

For each pair of numbers:

a. Find the arithmetic mean. **b. Find the geometric mean.**

17. 9, 64
18. 5, 19
19. 16, 5
20. $\frac{7}{8}, \frac{1}{7}$
21. 3.7, 2.3
22. 7, 33
23. 4, 0.04
24. $8i$, $32i$

25. **BIOLOGY** Domestic honeybees construct a honeycomb by adding ring after ring of hexagonal cells around a single hexagonal cell.
   a. **Visual Thinking** Is the sequence of the number of cells in each ring *arithmetic, geometric,* or *neither*? Explain.
   b. Write a formula for the nth term of the sequence.

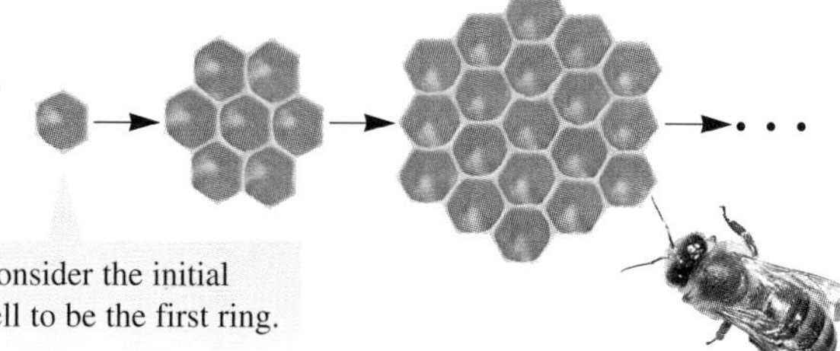

Consider the initial cell to be the first ring.

Connection GRAPHIC DESIGN

The *weight* of a typeface refers to the thickness of the strokes used in the letters. Sometimes type designers use numbers to represent typeface weights. The weights of the typeface in the table form a geometric sequence.

Light 215	The quick brown fox jumps over the lazy red dog.
Regular _?_	The quick brown fox jumps over the lazy red dog.
Semibold 422	The quick brown fox jumps over the lazy red dog.
Bold _?_	**The quick brown fox jumps over the lazy red dog.**
Black 830	**The quick brown fox jumps over the lazy red dog.**

26. Complete the table of typeface weights.
27. What is the common ratio for this geometric sequence?
28. For this typeface, suppose you want to create an Ultra Light weight one step lighter in weight than Light. What number should you use to describe it?
29. For this typeface, suppose you want to create an Extra Black weight one step heavier than Black. What number should you use to describe it?

Apply⇔Assess

Exercise Notes

Second-Language Learners

Ex. 15 Students learning English may not know what the medical term *DNA* means. If necessary, tell them that DNA molecules are a part of the cells of living organisms that contain the genetic material and essentially determine the nature of the organism. You may also wish to discuss such terms as *medical diagnosis, evolutionary biology,* and *gene research.*

Research

Ex. 26 Interested students can research what typefaces are available for a personal computer and which ones are commonly used. Have them prepare a report including examples.

Application

Exs. 26–29 These exercises are examples of how mathematics applies to real-world situations. Have students discuss how finding the previous term of a geometric sequence or the next term of a sequence is related to finding the size of a typeface.

Career Connection

Exs. 26–29 The field of graphic design offers many employment opportunities to students having the necessary training and experience. Independent schools of graphic design and departments of graphic design in colleges and universities provide opportunities for students to learn the basic skills. Job opportunities exist in all businesses that have to design and create products. For example, a graphic designer was responsible for the design of this textbook, including the selection, size, and weights of the typefaces used.

15. a. geometric; Since each term is double the preceding term, the sequence is geometric with common ratio 2.
   b. $t_n = 2^{n-1}$
   c. 30 cycles
   d. about 195 min or 3 h 15 min
16. a. 1.060
   b. The effective length at each position is 1.060 times the effective length at the preceding position.
17. a. 36.5 b. 24
18. a. 12 b. $\sqrt{95}$
19. a. 10.5 b. $4\sqrt{5}$
20. a. $\frac{57}{112}$ b. $\sqrt{\frac{1}{8}}$
21. a. 3 b. $\sqrt{8.51}$
22. a. 20 b. $\sqrt{231}$
23. a. 2.02 b. 0.4
24. a. $20i$ b. $16i$
25. a. neither; The sequence is 1, 6, 12, 18, 24, ….
   b. $t_1 = 1$ and for $n \geq 2$, $t_n = 6(n - 1)$.
26. regular: 301; bold: 592
27. 1.40
28. 154
29. 1162

Apply⇔Assess

Exercise Notes

Problem Solving

Exs. 30, 31 Students having difficulty doing these exercises can try considering the last term given as the *n*th term and then work backward to solve for the first term of the sequence.

Using Technology

Exs. 30, 31 A graphing calculator can be used for these and similar exercises. For Ex. 30, use the information about the terms as statistical data. Enter the coordinates (3, 5) and (9, –7) into lists L1 and L2. Find a linear regression equation and use it to write a formula for the sequence. For Ex. 31, have students find an exponential regression equation.

Integrating the Strands

Ex. 32 In this exercise, students use the 30°-60°-90° triangle ratios from geometry.

Multicultural Note

Ex. 32 The Navajo, who call themselves *Dinneh*, or "the people," are one of the largest Native American groups in the United States. More than 150,000 Navajo live on the Navajo reservation, located on 16 million acres in New Mexico, Arizona, and Utah. Traditionally, the Navajo lived in extended families and farmed the dry region that is now the southwestern United States. Today, many Navajo still farm or raise sheep, but others are teachers, miners, engineers, or craft workers. Important Navajo crafts include jewelry and woven blankets and rugs.

Application

Ex. 33, 34 Ask students who are on the track team at school to explain their answers to these sports applications.

Challenge **Write a formula for each sequence.**

30. arithmetic sequence: $t_3 = 5$ and $t_9 = -7$

31. geometric sequence: $t_4 = 15$ and $t_7 = 120$

32. ARCHITECTURE The Navajo people of the western United States used a technique called *corbeling* to build dome-shaped log houses. In a corbeled house, or *hogan*, the logs are stacked in layers. In each layer, the ends of the logs meet at the midpoints of the logs below.

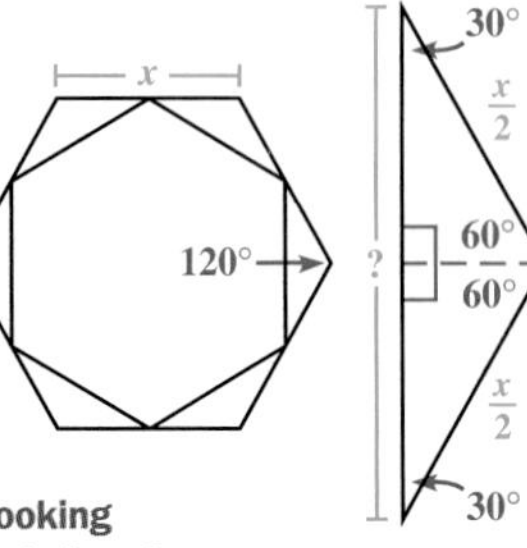

This is the view looking down onto a corbeled roof.

a. Suppose the logs in each layer of a corbeled roof form a regular hexagon. Write a formula for the *n*th term of the sequence of log lengths, starting with *x*.

b. Is the sequence of log lengths *arithmetic, geometric,* or *neither*? Explain.

Connection SPORTS

The track shown is shaped like a rectangle with semicircles on either end. The track has 8 lanes that are each 1.22 m wide. The lanes are numbered in ascending order from the inside. In a 400 m race, the runners must stay in their lanes. (Assume that the distance around each lane is measured along the center of the lane.)

finish line
83.4 m
36.5 m

33. a. Each red radius on the diagram is the radius of the curve for one of the lanes. Tell whether the sequence formed by these radii is *arithmetic, geometric,* or *neither.*

b. Write a formula for the sequence in part (a).

c. According to the rules of the International Amateur Athletic Federation (IAAF), world records can be set only on tracks that have a curve radius of at most 50 m in the outside lane. Does the track shown meet the IAAF requirement?

34. a. On the track shown, each lane except the first is longer than 400 m. For each lane, find how much longer.

b. Tell whether the sequence in part (a) is *arithmetic, geometric,* or *neither.*

c. Writing Explain why the IAAF requires that the starting line be staggered when 400 m races are run on a track like the one shown.

30. $t_n = 9 + (-2)(n - 1)$ or $t_n = -2n + 11$

31. $t_n = \frac{15}{8} \cdot 2^{n-1}$

32. a. $t_n = x \cdot \left(\frac{\sqrt{3}}{2}\right)^{n-1}$

b. geometric; The formula describes a geometric sequence with common ratio $\frac{\sqrt{3}}{2}$.

33. a. arithmetic

b. $t_n = 37.11 + 1.22(n - 1)$ or $t_n = 1.22n + 35.89$

c. Yes; the curve radius of the outside lane is 45.65 m.

34. a. Answers may vary due to rounding. 7.63 m; 15.30 m; 22.97 m; 30.63 m; 38.30 m; 45.96 m; 53.63 m

b. arithmetic

c. The starting lines are staggered so that the runner in each lane runs exactly 400 m and all the runners finish at the same finish line.

35. a. 1, 3, 9, 27

b. geometric; Each term is 3 times the preceding term.

c. $t_n = 3^{n-1}$

36. a. 3; 4.5; 6.75; 10.125

b. geometric; Each term is $\frac{3}{2}$ times the preceding term.

INTERVIEW Jhane Barnes

Look back at the article on pages 462–464.

Jhane Barnes named this fabric pattern "Sierpinski triangle" for the fractal shape it contains. A **fractal** *is an irregular shape that has the property of* **self-similarity,** *so that any part looks like the whole fractal.*

The first four stages in the construction of a Sierpinski triangle are shown. The initial triangle is equilateral with sides 1 unit long. The cut-out triangles are formed by connecting the midpoints of the sides of the shaded triangles.

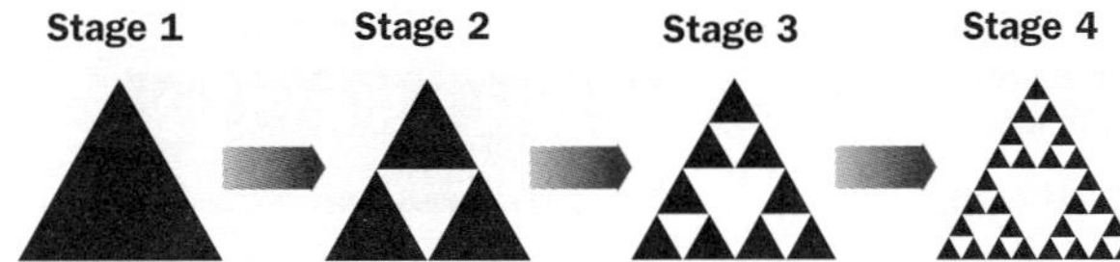

For each sequence described in Exercises 35–37:

a. Find the first 4 terms.

b. Tell whether the sequence is *arithmetic, geometric,* or *neither.* Explain.

c. Write a formula for the *n*th term.

35. the number of shaded triangles at each stage

36. the perimeter of the figure that remains at each stage (*Hint:* Add the perimeters of all the shaded triangles at each stage.)

37. the area of the figure that remains at each stage (*Hint:* Add the areas of all the shaded triangles at each stage.)

ONGOING ASSESSMENT

38. a. Graph the sequences $t_n = 3(2)^{n-1}$ and $u_n = 3(-2)^{n-1}$ in the same coordinate plane using two different symbols, such as "X" and "O," for your plotted points.

b. Writing How are the graphs alike? How are they different?

c. Writing Explain how the two sequences suggest the need to restrict the base b to positive numbers (other than 1) for an exponential function of the form $y = ab^x$.

SPIRAL REVIEW

Find the 15th term of each sequence. *(Section 10.1)*

39. $\sqrt{3}, 2\sqrt{3}, 3\sqrt{3}, 4\sqrt{3}, \ldots$ **40.** 0.9, 0.09, 0.009, 0.0009, . . . **41.** 1, 6, 15, 28, . . .

Find the zeros of each function. *(Section 9.4)*

42. $f(x) = 6(x + 2)(x + 1)(x - 1)$ **43.** $f(x) = x^3 - 5x^2 - 2x + 10$

Apply⇔Assess

Exercise Notes

Historical Connection

Exs. 35–37 The Sierpinski triangle is named after the famous Polish mathematician W. Sierpinski, who made important contributions to set theory, mathematical logic, and topology during the first part of the twentieth century.

Challenge

Ex. 37 Ask students to explain why each successive Sierpinski triangle is three-fourths the area of the previous triangle.

Using Technology

Ex. 38 Students can generate a table of values for each sequence by entering the sequences as functions into a graphing calculator and then choosing TABLE. A comparison of the tables shows how the values for the sequences are alike and how they are different. The signs alternate for integral table values for the powers of –2.

Practice 65 for Section 10.2

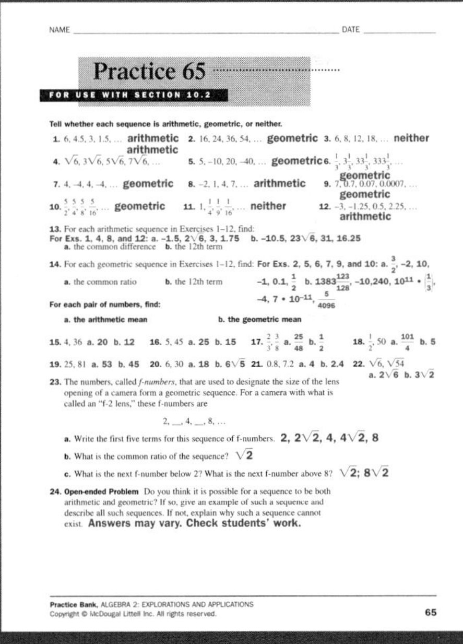

NAME ______________________ DATE __________

Practice 65

FOR USE WITH SECTION 10.2

Tell whether each sequence is arithmetic, geometric, or neither.

1. 6, 4.5, 3, 1.5, … **arithmetic** **2.** 16, 24, 36, 54, … **geometric** **3.** 6, 8, 12, 18, … **neither**

4. $\sqrt{6}, 3\sqrt{6}, 5\sqrt{6}, 7\sqrt{6}, \ldots$ **arithmetic** **5.** 5, –10, 20, –40, … **geometric** **6.** $\frac{1}{3}, 3\frac{1}{3}, 33\frac{1}{3}, 333\frac{1}{3}, \ldots$ **geometric**

7. 4, –4, 4, –4, … **geometric** **8.** –2, 1, 4, 7, … **arithmetic** **9.** 7, 0.7, 0.07, 0.0007, … **geometric**

10. $\frac{5}{2}, \frac{5}{4}, \frac{5}{8}, \frac{5}{16}, \ldots$ **geometric** **11.** $1, \frac{1}{4}, \frac{1}{9}, \frac{1}{16}, \ldots$ **neither** **12.** –3, –1.25, 0.5, 2.25, … **arithmetic**

13. For each arithmetic sequence in Exercises 1–12, find:
For Exs. 1, 4, 8, and 12: a. –1.5, $2\sqrt{6}$, 3, 1.75 b. –10.5, $23\sqrt{6}$, 31, 16.25
a. the common difference **b.** the 12th term

14. For each geometric sequence in Exercises 1–12, find: **For Exs. 2, 5, 6, 7, 9, and 10: a. $\frac{3}{2}$, –2, 10, –1, 0.1, $\frac{1}{2}$ b. $1383\frac{123}{128}$, –10,240, $10^{11} \cdot \left(\frac{1}{3}\right)$, –4, $7 \cdot 10^{-11}$, $\frac{5}{4096}$**
a. the common ratio **b.** the 12th term

For each pair of numbers, find:

a. the arithmetic mean **b. the geometric mean**

15. 4, 36 **a. 20 b. 12** **16.** 5, 45 **a. 25 b. 15** **17.** $\frac{2}{3}, \frac{3}{8}$ **a. $\frac{25}{48}$ b. $\frac{1}{2}$** **18.** $\frac{1}{2}$, 50 **a. $\frac{101}{4}$ b. 5**

19. 25, 81 **a. 53 b. 45** **20.** 6, 30 **a. 18 b. $6\sqrt{5}$** **21.** 0.8, 7.2 **a. 4 b. 2.4** **22.** $\sqrt{6}, \sqrt{54}$ **a. $2\sqrt{6}$ b. $3\sqrt{2}$**

23. The numbers, called *f-numbers*, that are used to designate the size of the lens opening of a camera form a geometric sequence. For a camera with what is called an "f-2 lens," these f-numbers are

2, __, 4, __, 8, …

a. Write the first five terms for this sequence of f-numbers. **2, $2\sqrt{2}$, 4, $4\sqrt{2}$, 8**

b. What is the common ratio of the sequence? **$\sqrt{2}$**

c. What is the next f-number below 2? What is the next f-number above 8? **$\sqrt{2}$; $8\sqrt{2}$**

24. Open-ended Problem Do you think it is possible for a sequence to be both arithmetic and geometric? If so, give an example of such a sequence and describe all such sequences. If not, explain why such a sequence cannot exist. **Answers may vary. Check students' work.**

Practice Bank, ALGEBRA 2: EXPLORATIONS AND APPLICATIONS

65

---

c. $t_n = 3\left(\frac{3}{2}\right)^{n-1}$

37. a. $\frac{\sqrt{3}}{4}, \frac{3\sqrt{3}}{16}, \frac{9\sqrt{3}}{64}, \frac{27\sqrt{3}}{256}$

b. geometric; Each term is $\frac{3}{4}$ times the preceding term.

c. $t_n = \frac{\sqrt{3}}{4}\left(\frac{3}{4}\right)^{n-1}$

38. a.

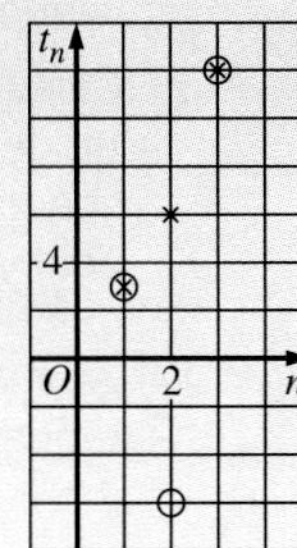

b. For n odd, the points on the graph are the same. For n even, the points of one graph are the images of the other reflected over the n-axis.

c. Summaries may vary. It is convenient to have the family of exponential functions have graphs that fit the same pattern. If the base of such a function is negative, its graph is not an exponential curve but a discontinuous collection of points.

39. $15\sqrt{3}$

40. 0.000000000000009

41. 435

42. –2; ±1

43. $\pm\sqrt{2}$; 5

Plan⇔Support

Objectives

- Find the terms of a sequence defined recursively.
- Write a recursive formula for a sequence.
- Apply the recursive formula for a sequence to a real-world situation.

Recommended Pacing

Core and Extended Courses
Section 10.3: 2 days

Block Schedule
Section 10.3: 1 block

Resource Materials

Lesson Support
Lesson Plan 10.3
Warm-Up Transparency 10.3
Practice Bank: Practice 66
Study Guide: Section 10.3
Exploration Lab Manual:
Additional Exploration 16
Assessment Book: Test 40

Technology
Technology Book:
Calculator Activity 10
Spreadsheet Activity 10
Scientific Calculator
Graphing Calculator
Internet:
http://www.hmco.com

Warm-Up Exercises

Find the 4th term of each sequence.

1. $t_n = 6n - 2$ 22
2. $t_n = 6^{n-2}$ 36
3. a sequence with a common difference of 8 if $t_3 = -14$ −6
4. Write a formula for the sequence in Ex. 3 above. $t_n = 8n - 38$
5. Write a formula for the sequence $\frac{3}{1}, \frac{3}{4}, \frac{3}{16}, \frac{3}{64}, \ldots$. $3\left(\frac{1}{4}\right)^{n-1}$

Section

10.3 Exploring Recursion

Learn how to...

- **find the terms of a sequence defined recursively**
- **write a recursive formula for a sequence**

So you can...

- **find out what happens to the level of medication in the bloodstream over time, for example**

In Sections 10.1 and 10.2, you wrote formulas for the nth term, t_n, of a sequence. These formulas, which express t_n as a function of its position number n, are called **explicit formulas**. In this section you will explore another way to write a formula for the nth term of a sequence.

EXPLORATION
COOPERATIVE LEARNING

Investigating Repeated Addition and Multiplication

Work with a partner.
You will need:
- graphing calculators

1 The calculator screen shows the procedure you should follow to generate a sequence. Choose various starting values and various constants (both positive and negative), and write down the first 6 terms of the sequences you generate.

Choose a starting value.

Add a constant to the "last answer."

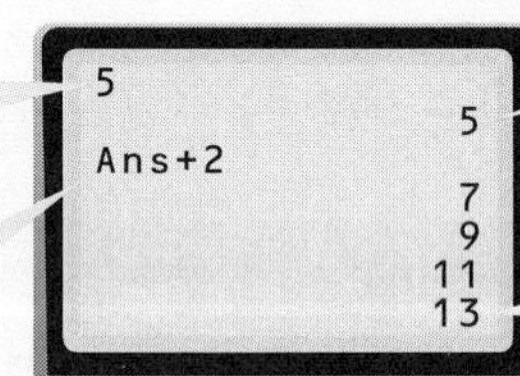

Press ENTER.

By repeatedly pressing ENTER, you will continue to add the constant to the "last answer."

2 Repeat Step 1, but this time *multiply* by a constant instead of adding a constant. For example, use Ans * 2 instead of Ans + 2.

Questions

1. In each sequence, what is the relationship between any term (except the first) and the term before it?
2. Tell whether each sequence formed by repeatedly adding a constant is *arithmetic, geometric,* or *neither.*
3. Tell whether each sequence formed by repeatedly multiplying by a constant is *arithmetic, geometric,* or *neither.*

Exploration Note

Purpose
The purpose of this Exploration is to have students generate a sequence by using repeated addition or multiplication.

Materials/Preparation
A graphing calculator is needed.

Procedure
Students enter a starting value and then add a constant to the last value by using the last-answer key. They then keep pressing ENTER to see the same constant added repeatedly. This generates an arithmetic sequence. They repeat the procedure by multiplying by a constant. This generates a geometric sequence. Encourage students to experiment with various starting values and various constants.

Closure
Students should conclude that an arithmetic sequence can be generated by repeatedly adding the common difference to the previous term. They also should conclude that a geometric sequence can be generated by repeatedly multiplying the previous term by the common ratio.

Explorations Lab Manual
See the Manual for more commentary on this Exploration.

In the Exploration you generated the terms of a sequence by entering a starting value and repeatedly applying the same operation to your last answer. This process is called **recursion**.

Recursion gives you another way to write a formula for the nth term of a sequence. A **recursive formula** for a sequence has two parts. The first part assigns a starting value. The second part is a *recursion equation* for t_n as a function of t_{n-1}, the term before it.

Recursive Formulas

Arithmetic Sequence

General formula:

t_1 = starting value

$t_n = t_{n-1} + d$

Example: $t_1 = 5$

$t_n = t_{n-1} + 2$

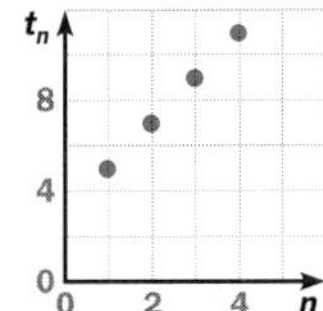

Geometric Sequence

General formula:

t_1 = starting value

$t_n = r(t_{n-1})$

Example: $t_1 = 5$

$t_n = 2(t_{n-1})$

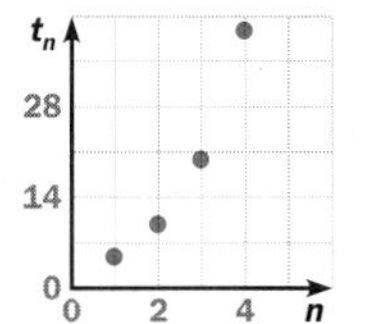

THINK AND COMMUNICATE

1. How are the general explicit formula (see page 473) and the general recursive formula for arithmetic sequences alike? different?
2. Answer Question 1 for geometric sequences.

EXAMPLE 1

Find the first 4 terms of each sequence.

a. $t_1 = -3; t_n = t_{n-1} + 8$

b. $t_1 = 1875; t_n = \frac{1}{5}(t_{n-1})$

SOLUTION

Each time you find a term of the sequence, use it to find the next term.

a. $t_1 = -3$

$t_2 = t_1 + 8 = -3 + 8 = 5$

$t_3 = t_2 + 8 = 5 + 8 = 13$

$t_4 = t_3 + 8 = 13 + 8 = 21$

The sequence is −3, 5, 13, 21, ...

b. $t_1 = 1875$

$t_2 = \frac{1}{5}(t_1) = \frac{1}{5}(1875) = 375$

$t_3 = \frac{1}{5}(t_2) = \frac{1}{5}(375) = 75$

$t_4 = \frac{1}{5}(t_3) = \frac{1}{5}(75) = 15$

The sequence is 1875, 375, 75, 15, ...

Teach⇔Interact

Section Notes

Communication: Discussion A thorough discussion of the concept of recursion and the recursive formulas for an arithmetic sequence and a geometric sequence would help students to generalize their understanding of the Exploration on page 480. Be sure to ask students to describe the two graphs shown on this page.

Teaching Tip
Emphasize the importance of including the first term when writing a recursive formula for a sequence.

Think and Communicate

Ask students to discuss the advantages and disadvantages of giving an explicit or recursive formula for a sequence.

About Example 1

Teaching Tip
After discussing the Solution to this Example, you can emphasize the nature of recursion by asking students to find t_{10} for each sequence. They will not be able to find this term quickly since it is necessary to find all the previous terms first.

Additional Example 1

Find the first 4 terms of each sequence.

a. $t_1 = 132; t_n = t_{n-1} - 6$

Each time you find a term of the sequence, use it to find the next term.

$t_1 = 132$

$t_2 = t_1 - 6 = 132 - 6 = 126$

$t_3 = t_2 - 6 = 126 - 6 = 120$

$t_4 = t_3 - 6 = 120 - 6 = 114$

The sequence is 132, 126, 120, 114, ...

b. $t_1 = -12; t_n = -4(t_{n-1})$

$t_1 = -12$

$t_2 = -4(t_1) = -4(-12) = 48$

$t_3 = -4(t_2) = -4(48) = -192$

$t_4 = -4(t_3) = -4(-192) = 768$

The sequence is −12, 48, −192, 768, ...

ANSWERS Section 10.3

Exploration

1, 2. Answers may vary. Examples are given.

1. For a starting value of 5 with 3 added, the calculator produces the sequence 5, 8, 11, 14, 17, 20, ...
2. For a starting value of 5 multiplied by 3, the calculator produces the sequence 5, 15, 45, 135, 405, 1215, ...

Questions

1. If a sequence is produced by adding a constant, consecutive terms have a common difference, the constant. If a sequence is produced by multiplying by a constant, consecutive terms have a common ratio, the constant.
2. arithmetic
3. geometric

Think and Communicate

1. The formulas have the same form. The general explicit formula gives a value for t_n in terms of t_1 and the common difference d. The general recursive formula indicates the value of t_1 and gives the value for t_n in terms of t_{n-1} and the common difference d.
2. See answers in back of book.

Teach⇔Interact

About Example 2

Teaching Tip
For part (b), the sequence is neither arithmetic nor geometric, although a recursive formula can still be written. Stress the importance of being able to identify a pattern in order to write a recursive formula.

Additional Example 2

Write a recursive formula for each sequence.

a. $76, 19, \frac{19}{4}, \frac{19}{16}, \ldots$
The sequence is geometric with a common ratio of $\frac{1}{4}$. Use the general recursive formula for a geometric sequence.
$t_1 = 76 \quad t_n = \frac{1}{4}(t_{n-1})$

b. 2, 5, 14, 41, ...
The sequence is neither arithmetic nor geometric. Look for a relationship between each term after the first term and the term before it.
2, 5, 14, 41, ...
$3 \cdot 2 - 1 \quad 3 \cdot 5 - 1 \quad 3 \cdot 14 - 1$
$t_1 = 2 \quad t_n = 3(t_{n-1}) - 1$

c. 8, 8.05, 8.1, 8.15, ...
The sequence is arithmetic with a common difference of 0.05. Use the general recursive formula for an arithmetic sequence.
$t_1 = 8 \quad t_n = t_{n-1} + 0.05$

About Example 3

Teaching Tip
While explicit formulas are usually preferred because you can calculate values directly, this Example shows that certain biological processes do not have an associated explicit formula.

Additional Example 3

Drinking water is processed in a home filtration system. The water originally contains 3 mg of impurities. Every hour the water is passed through filters until 48% of these impurities remain. How long does it take before the water contains less than 0.01 mg of impurities?
Write a recursive formula to show the level of impurities in the water after each filtration.
$t_1 = 3 \quad t_n = 0.48(t_{n-1})$

EXAMPLE 2

Write a recursive formula for each sequence.

a. 1, 2.5, 6.25, 15.625, . . . **b.** 2, 5, 26, 677, . . . **c.** 4, 2.8, 1.6, 0.4, . . .

SOLUTION

a. The sequence is geometric with a common ratio of 2.5. Use the general recursive formula for a geometric sequence.

Substitute **1** for the starting value. $t_1 = 1 \quad t_n = 2.5(t_{n-1})$ Substitute **2.5** for *r*.

b. The sequence is neither arithmetic nor geometric. Look for a relationship between each term after the first and the term before it:

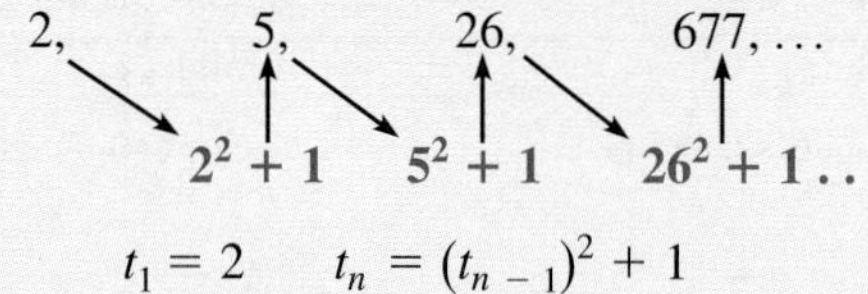

$t_1 = 2 \quad t_n = (t_{n-1})^2 + 1$

c. The sequence is arithmetic with a common difference of −1.2. Use the general recursive formula for an arithmetic sequence.

Substitute **4** for the starting value. $t_1 = 4 \quad t_n = t_{n-1} + (-1.2)$ Substitute **−1.2** for *d*.

EXAMPLE 3 Application: Medicine

One medication used to treat high blood pressure is supplied in 1.25 mg tablets. Patients take one tablet at the same time every day. By the time a patient takes the next dose, only about 32% of the medication is left in the bloodstream. What happens to the level of medication in a patient's bloodstream over time?

SOLUTION

Write a recursive formula to model the level of medication in the bloodstream after each dose.

$t_1 = 1.25$

$t_n = 0.32(t_{n-1}) + 1.25$

Method 1 Use the last-answer key on a calculator.

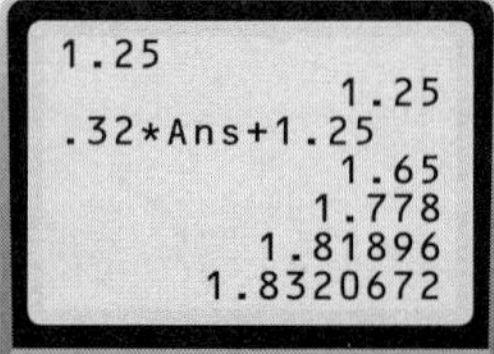

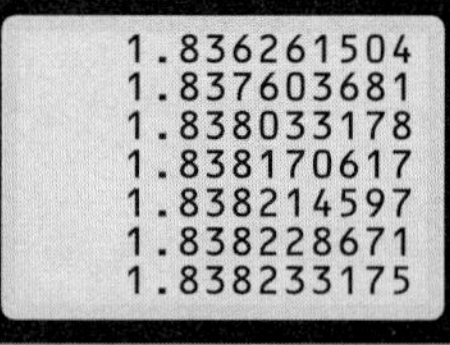

After several days, the amount of medication in the bloodstream levels off at about 1.84 mg.

Method 2 Use a graph.

Be sure your calculator is set for sequence graphing in dot mode.

Step 1 Enter the recursion equation. **Step 2** Set the viewing window.

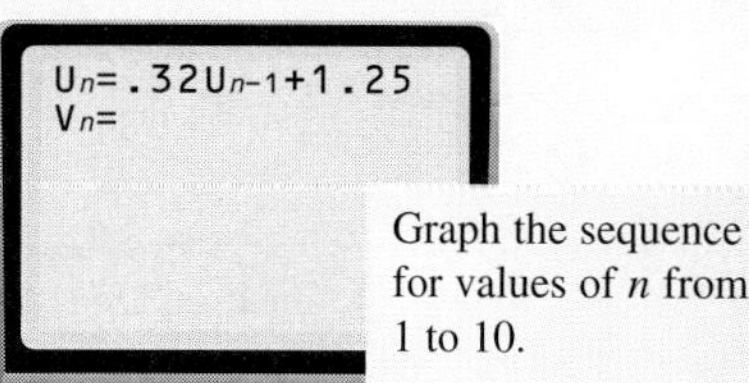

Graph the sequence for values of n from 1 to 10.

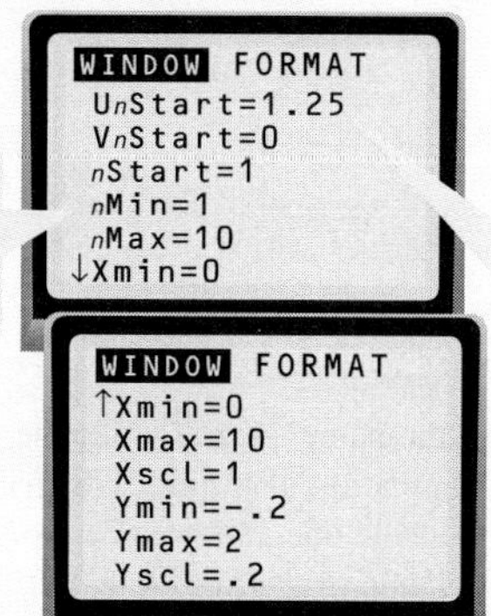

Enter the starting value of the sequence.

Step 3 Graph the recursive formula.

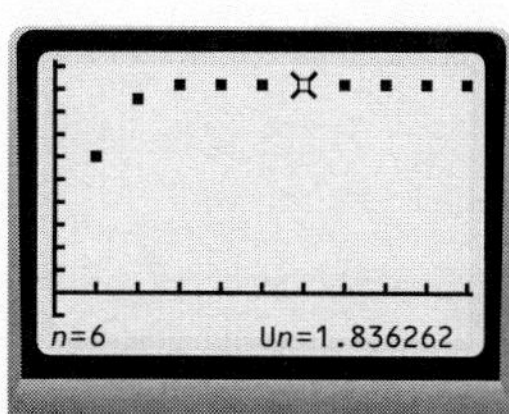

After several days, the amount of medication in the bloodstream levels off at about 1.84 mg.

For more information about graphing sequences, see the *Technology Handbook*, p. 810.

THINK AND COMMUNICATE

3. Doctors sometimes begin a course of medication with a "booster" dose of 1.6 times the regular dose. What is the booster dose for the medication in Example 3? How does it compare with the solution to Example 3?

CHECKING KEY CONCEPTS

Find the first 4 terms of each sequence.

1. $t_1 = 0$, $t_n = t_{n-1} + 6$
2. $t_1 = 8$, $t_n = -2(t_{n-1})$
3. $t_1 = -2$, $t_n = t_{n-1} - 5$
4. $t_1 = 3$, $t_n = 4(t_{n-1}) + 0.3$
5. $t_1 = 6.2$, $t_n = \frac{1}{2}(t_{n-1})$
6. $t_1 = 5\frac{1}{4}$, $t_n = 2(t_{n-1})^2 + 1$

Write a recursive formula for each sequence.

7. 4, 28, 196, 1372, ...
8. 7, −14, 28, −56, ...
9. 19, 13, 7, 1, ...
10. 384, 96, 24, 6, ...
11. $\frac{1}{3}, \frac{1}{9}, \frac{1}{81}, \frac{1}{6561}, \ldots$
12. 0.66, 0.69, 0.72, 0.75, ...

Think and Communicate

3. 2 mg; It is slightly higher than the eventual level, 1.84 mg.

Checking Key Concepts

1. 0, 6, 12, 18
2. 8, −16, 32, −64
3. −2, −7, −12, −17
4. 3, 12.3, 49.5, 198.3
5. 6.2, 3.1, 1.55, 0.775
6. 5.25, 56.125, 6301.03125, 79,405,990.62696
7. $t_1 = 4$; $t_n = 7(t_{n-1})$
8. $t_1 = 7$; $t_n = -2(t_{n-1})$
9. $t_1 = 19$; $t_n = t_{n-1} - 6$
10. $t_1 = 384$; $t_n = \frac{1}{4}(t_{n-1})$
11. $t_1 = \frac{1}{3}$; $t_n = (t_{n-1})^2$
12. $t_1 = 0.66$; $t_n = t_{n-1} + 0.03$

Teach⇔Interact

Additional Example 3 (continued)

Method 1 Use the last-answer key.

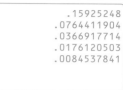

.15925248
.0764411904
.0366917714
.0176120503
.0084537841

After 8 hours, the amount of impurities left is less than 0.01 mg.

Method 2 Use a graph. Be sure the calculator is set for sequence graphing in dot mode.

Step 1 Enter the recursion equation.

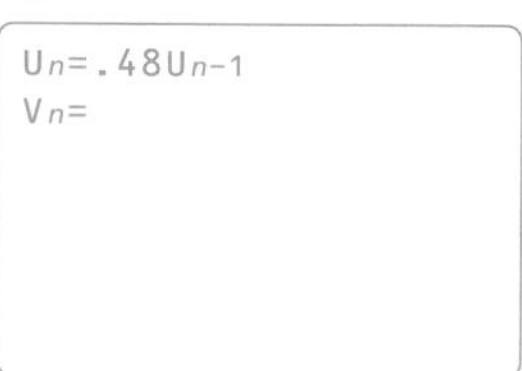

Step 2 Set the viewing window.

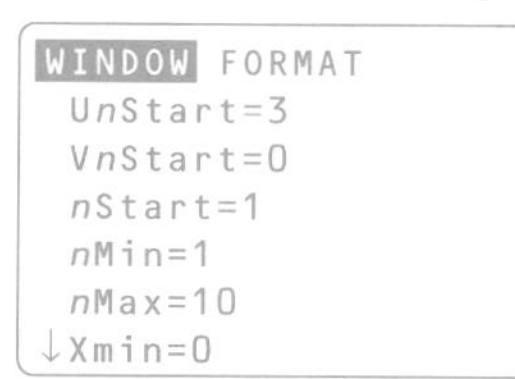

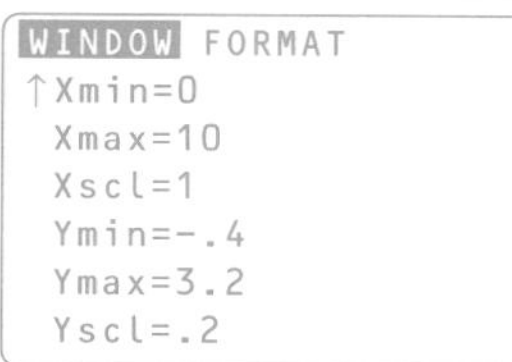

Step 3 Graph the recursive formula.

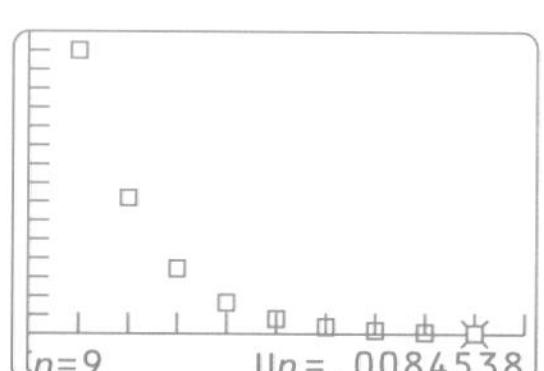

At the beginning of the 9th hour, that is, after 8 hours, the amount of impurities left in the water is less than 0.01 mg.

Section Note

Using Technology
Activity 9 in the *Geometry Inventor Activity Book* can provide additional practice with recursion, as well as a connection to geometry. In this activity, students explore the number of diagonals in a polygon with n sides.

Teach⇔Interact

Closure Question

Describe how to write a recursive formula for a sequence.

First classify the sequence as arithmetic, geometric, or neither and identify the first term as t_1. If the sequence is arithmetic with common difference d, complete the general recursive formula $t_n = t_{n-1} + d$. If the sequence is geometric with common ratio r, complete the general recursive formula with $t_n = r(t_{n-1})$. If the sequence is "neither," complete the formula using a relationship between each term after the first and the term before it.

Apply⇔Assess

Suggested Assignment

Core Course
Day 1 Exs. 1–8, 10–21
Day 2 Exs. 22, 27, 28, 32–41, AYP

Extended Course
Day 1 Exs. 1–21
Day 2 Exs. 22–41, AYP

Block Schedule
Day 65 Exs. 1–8, 10–22, 27, 28, 32–41, AYP

Exercise Notes

Mathematical Procedures
Exs. 1–8 Students should not try to write the explicit formula until they find the first 4 terms.

Interdisciplinary Problems
Ex. 9 This exercise relates the concept of mathematical recursion to the linguistic concept of recursion as nested grammatical sentence patterns.

Second-Language Learners
Ex. 9 Some of this language might prove challenging for students learning English. Some students may not be familiar with the word *linguist*. If necessary, explain that linguists are people who study the structure of language. Be sure students understand how the rhyme and sequence of the language work.

10.3 Exercises and Applications

Extra Practice exercises on page 765

For each sequence:
a. Find the first 4 terms.
b. Write an explicit formula.

1. $t_1 = 4$, $t_n = -\frac{1}{2}(t_{n-1})$

2. $t_1 = 7$, $t_n = t_{n-1} + 3$

3. $t_1 = 4.5$, $t_n = 6(t_{n-1})$

4. $t_1 = -1$, $t_n = t_{n-1} - 6$

5. $t_1 = 9$, $t_n = t_{n-1}$

6. $t_1 = -3$, $t_n = t_{n-1} + 6$

7. $t_1 = 0$, $t_n = t_{n-1} + 6$

8. $t_1 = 2i$, $t_n = i(t_{n-1})$

9. **LANGUAGE ARTS** Linguists use the name *recursion* for the property of mental grammar that permits the creation of infinitely many new sentences by nesting grammatical patterns inside one another. Verna Aardema uses linguistic recursion in her book *Bringing the Rain to Kapiti Plain,* a retelling of a folktale of the Nandi culture of Kenya, in Africa. How is the linguistic recursion Verna Aardema used like mathematical recursion?

Here are the first four verses of *Bringing the Rain to Kapiti Plain*:

This is the cloud, all heavy with rain, that
shadowed the ground on Kapiti Plain.

This is the grass, all brown and dead, that
needed the rain from the cloud overhead—
the big, black cloud, all heavy with rain, that
shadowed the ground on Kapiti Plain.

These are the cows, all hungry and dry, who
mooed for the rain to fall from the sky; to
green-up the grass, all brown and
dead, that needed the rain
from the cloud overhead—
the big, black cloud, all
heavy with rain, that
shadowed the ground
on Kapiti Plain.

This is Ki-pat, who watched his herd as he stood
on one leg, like the big stork bird;
Ki-pat, whose cows were
so hungry and dry, they
mooed for the rain to
fall from the sky;
to green-up the grass,
all brown and dead,
that needed the rain
from the cloud
overhead—the big,
black cloud, all
heavy with rain,
that shadowed
the ground on
Kapiti Plain.

Write a recursive formula for each sequence.

10. $-8, -10, -12, -14, \ldots$

11. $1.1, 2.2, 3.3, 4.4, \ldots$

12. $71, 65, 59, 53, \ldots$

13. $50, 25, 12.5, 6.25, \ldots$

14. $7, \frac{14}{3}, \frac{28}{9}, \frac{56}{27}, \ldots$

15. $3, 9, 81, 6561, \ldots$

16. $1.3, -2.7, -6.7, -10.7, \ldots$

17. $3, 3\sqrt{5}, 15, 15\sqrt{5}, \ldots$

18. $572, 57.2, 5.72, 0.572, \ldots$

19. $t_n = 3n - 1$

20. $t_n = 2^n$

21. $t_n = n$

Exercises and Applications

1. a. $4, -2, 1, -\frac{1}{2}$
   b. $t_n = 4\left(-\frac{1}{2}\right)^{n-1}$
2. a. 7, 10, 13, 16
   b. $t_n = 7 + 3(n-1)$ or $t_n = 3n + 4$
3. a. 4.5, 27, 162, 972
   b. $t_n = 4.5 \cdot 6^{n-1}$
4. a. −1, −7, −13, −19
   b. $t_n = -1 - 6(n-1)$ or $t_n = -6n + 5$
5. a. 9, 9, 9, 9
   b. $t_n = 9$
6. a. −3, 3, 9, 15
   b. $t_n = -3 + 6(n-1)$ or $t_n = 6n - 9$
7. a. 0, 6, 12, 18
   b. $t_n = 6(n-1)$ or $t_n = 6n - 6$
8. a. $2i, -2, -2i, 2$
   b. $t_n = 2i(i^{n-1})$
9. Summaries may vary. The poem begins with an initial verse, analogous to the first term of a sequence, t_1. Each subsequent verse contains all of the preceding verses along with new lines.
10. $t_1 = -8$; $t_n = t_{n-1} - 2$
11. $t_1 = 1.1$; $t_n = t_{n-1} + 1.1$
12. $t_1 = 71$; $t_n = t_{n-1} - 6$
13. $t_1 = 50$; $t_n = \frac{1}{2}(t_{n-1})$
14. $t_1 = 7$; $t_n = \frac{2}{3}(t_{n-1})$

22. Suppose a tree farm has 3800 trees. Each year the farmer harvests 20% of the trees and plants 900 new trees.

a. Write a recursive formula for the total number of trees each year.

b. What happens to the total number of trees over time?

c. **Writing** What other factors may affect the number of trees on the farm? How do these factors affect the formula you wrote in part (a)?

Connection BIOLOGY

The chambered nautilus is a cone-shaped shellfish. To protect its soft body, it makes a shell that contains many chambers, or compartments. Each successive chamber is larger than the previous one but has the same shape as the animal's body. As the nautilus grows, it moves into the next larger chamber. The cross section of a nautilus shell is an *equiangular spiral* in which all the walls between the chambers intersect the outer boundary at congruent angles.

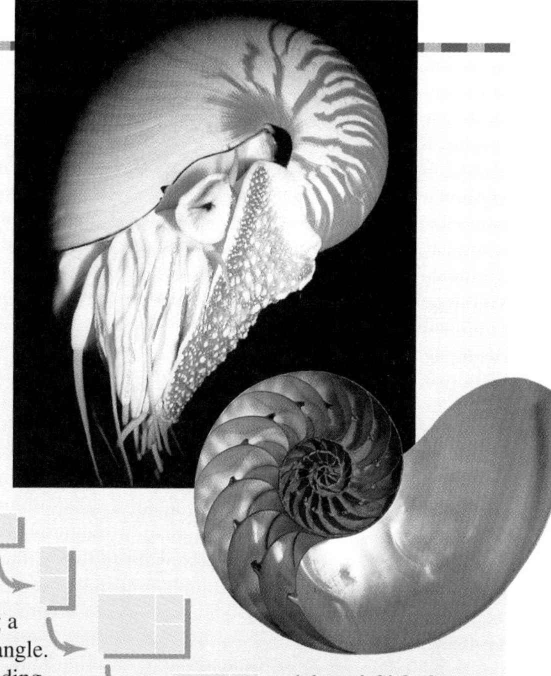

***(above left)* A chambered nautilus *(above)* Cross section of a chambered nautilus shell, showing the chamber walls intersecting the outside of the shell at congruent angles**

23. a. GEOMETRY Follow these steps to construct a figure that approximates an equiangular spiral.

Step 1 Begin with a small unit square.

Step 2 Form a rectangle by adding another unit square above it.

Step 3 Form a larger rectangle by adding a square on the left side of the previous rectangle.

Step 4 Form a still larger rectangle by adding a square below the previous rectangle.

Step 5 Continue adding squares in this way, moving in a counterclockwise direction, until there are nine squares in all.

Step 6 Starting at the lower right-hand vertex of the *second* square, draw a spiral that passes through one pair of opposite vertices of each square after the first.

b. In terms of the length of a side of the unit square, find the lengths of the sides of the nine squares you drew in part (a).

24. The numbers you wrote in part (b) of Exercise 23 are the first 9 terms of an infinite sequence called the *Fibonacci sequence.* Write a recursive formula for this sequence. (*Hint:* Each term after the second is a function of the *two* terms immediately before it.)

BY THE WAY

"Fibonacci" is the pen name of mathematician Leonardo of Pisa. Almost 800 years ago, he introduced the Western world to both Hindu-Arabic numerals and the mathematics developed in China, Arabia, and India.

Apply⇔Assess

Exercise Notes

Common Error

Exs. 10–21 Some students may forget to include t_1 in their recursive formula for each sequence. Stress that t_1 is part of the definition.

Using Technology

Ex. 22 Students who use the graphing method in Example 3 on page 482 should use the iPart function of a graphing calculator in their definition of U*n* to represent the fact that a whole tree will be harvested. On a TI-82 calculator, iPart is located under the MATH NUM menu.

Integrating the Strands

Exs. 23, 24 These two exercises integrate concepts from algebra, geometry, and discrete mathematics.

Problem Solving

Ex. 24 Students should conclude that if t_{n-1} is the term preceding a term, t_{n-2} is the term preceding the term t_{n-1}.

Historical Connection

Ex. 24 Leonardo Fibonacci (1170?–1250?) was a talented mathematician of the Middle Ages. He was born in the city of Pisa, Italy, and as a boy traveled widely throughout Egypt, Sicily, Greece, and Syria with his father, who was in the mercantile business. He developed an early interest in arithmetic and was impressed with the practical superiority of Hindu-Arabic notation for writing numerals and doing calculations. In 1228, he published his famous book called the *Liber abaci.* It was in this book that the following problem appeared, which gives rise to the Fibonacci sequence: How many pairs of rabbits can be produced from a single pair in a year if every month each pair begets a new pair, which from the second month on, becomes productive?

Research

Ex. 24 Ask students to find other examples of the Fibonacci sequence in nature.

15. $t_1 = 3$; $t_n = (t_{n-1})^2$

16. $t_1 = 1.3$; $t_n = t_{n-1} - 4$

17. $t_1 = 3$; $t_n = \sqrt{5}(t_{n-1})$

18. $t_1 = 572$; $t_n = 0.1(t_{n-1})$

19. $t_1 = 2$; $t_n = t_{n-1} + 3$

20. $t_1 = 2$; $t_n = 2(t_{n-1})$

21. $t_1 = 1$; $t_n = t_{n-1} + 1$

22. a. $t_1 = 3800$; $t_n = 0.8(t_{n-1}) + 900$

b. The number of trees levels off at 4500.

c. Answers may vary. Examples are given. A disease or poor weather conditions might destroy some of the trees so that the coefficient of t_{n-1} might be less than 0.8. Economic conditions might force the farmer to plant fewer trees, which would decrease the constant term.

23. a.

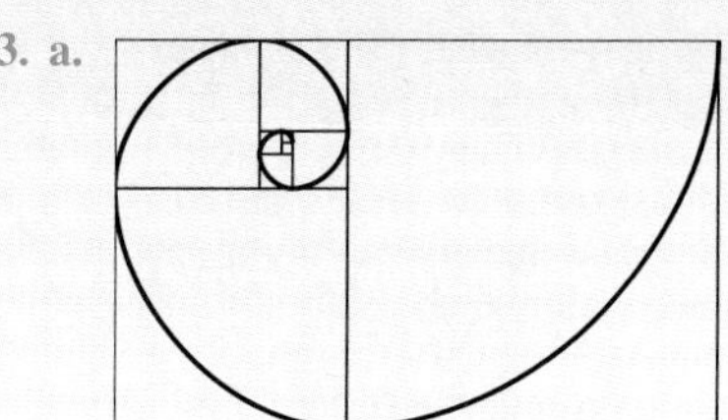

b. 1, 1, 2, 3, 5, 8, 13, 21, 34

24. $t_1 = 1$; $t_2 = 1$; $t_n = t_{n-1} + t_{n-2}$

Exercise Notes

Application
Exs. 25, 26 In these exercises, the mathematics of recursion is applied to camera settings.

Second-Language Learners
Exs. 25, 26 The intricacies and workings of a camera might not be familiar to some students learning English. You may wish to review such terms as *manual 35 mm camera, camera settings, film speed,* and *flash guide.* Invite volunteers familiar with cameras to tell the class what each of these terms means.

Reasoning
Ex. 25 Students need to organize their work and reason carefully from step to step to do this exercise. First they must note that the f-stop itself and the light the f-stop lets in are separate quantities. Encourage students to start with the hint and remember what they need to do, namely, to find the relationship between f_n and f_{n-1}.

Topic Spiraling: Review
Ex. 30 Some students may need to be reminded how to find the factorial of a number.

Visual Thinking
Ex. 32 Ask students to sketch their own visual models that demonstrate recursion. These could take a variety of forms, including abstract designs, nature sketches, and visual stories. Encourage students to discuss their models with the class. This activity involves the visual skills of *generalization* and *self-expression.*

Connection PHOTOGRAPHY

Photographers who use manual 35mm cameras must adjust various camera settings to account for different lighting conditions.

25. To allow more (or less) light to enter the camera, the photographer changes the camera setting that increases (or decreases) the area of a circular opening. The camera settings, called *f-stops,* are inversely proportional to the diameter of the opening. Each f-stop lets in half as much light as the previous one. The amount of light is proportional to the area of the opening.

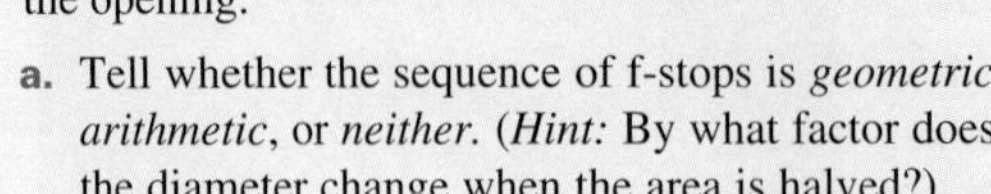

a. Tell whether the sequence of f-stops is *geometric, arithmetic,* or *neither.* (*Hint:* By what factor does the diameter change when the area is halved?)

b. Use 2 as the first f-stop. Find the next 7 f-stops.

c. Write a recursive formula for the sequence of f-stops.

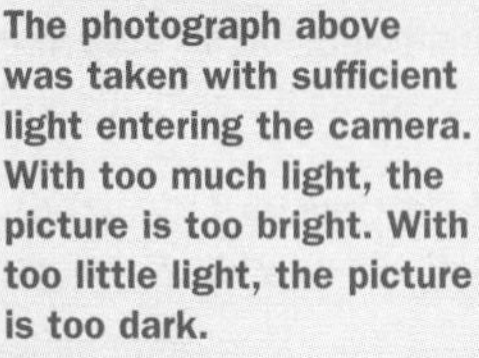

The photograph above was taken with sufficient light entering the camera. With too much light, the picture is too bright. With too little light, the picture is too dark.

26. The table gives factors for modifying the flash guide number (G) and the film speed number (ASA) in order to set the correct f-stop and exposure time for flash photography. Use the table to write a recursive formula for each sequence of factors:

a. G—available light is principal source
b. G—flash is principal source
c. ASA—available light is principal source
d. ASA—flash is principal source

Principal Light Source

	Available light		Flash	
Lighting ratio	Factor for G	Factor for ASA	Factor for G	Factor for ASA
2:1	$\sqrt{2}$	2	$\sqrt{2}$	2
3:1	$\sqrt{3}$	$\frac{3}{2}$	$\sqrt{\frac{3}{2}}$	3
4:1	$\sqrt{4}$	$\frac{4}{3}$	$\sqrt{\frac{4}{3}}$	4
5:1	$\sqrt{5}$	$\frac{5}{4}$	$\sqrt{\frac{5}{4}}$	5

For each sequence in Exercises 27 and 28, any term t_n after the second term is a function of the two terms, t_{n-1} and t_{n-2}, immediately before it in the sequence. Write a recursive formula for each sequence.

27. 6, 7, 26, 66, 184, 500, ...

28. 2, 3, 2, 2, $\frac{4}{3}$, $\frac{8}{9}$, ...

Challenge **Write a recursive formula that defines each operation.**

29. a number raised to a positive integral power

30. a factorial of a positive integer

25. a. geometric (The diameter changes by $\frac{1}{\sqrt{2}}$ or $\frac{\sqrt{2}}{2}$ when the area is halved.)
b. 2, $2\sqrt{2}$, 4, $4\sqrt{2}$, 8, $8\sqrt{2}$, 16, $16\sqrt{2}$
c. $t_1 = 2$; $t_n = \sqrt{2}(t_{n-1})$

26. a. $t_1 = \sqrt{2}$; $t_n = \left(\sqrt{\frac{n+1}{n}}\right)t_{n-1}$
b. $t_1 = \sqrt{2}$; $t_n = \left(\sqrt{\frac{n^2-1}{n^2}}\right)t_{n-1}$
c. $t_1 = 2$; $t_n = \left(\frac{n^2-1}{n^2}\right)t_{n-1}$
d. $t_1 = 2$; $t_n = t_{n-1} + 1$

27. $t_1 = 6$; $t_2 = 7$; $t_n = 2(t_{n-1} + t_{n-2})$

28. $t_1 = 2$; $t_2 = 3$; $t_n = \frac{1}{3}(t_{n-1} \cdot t_{n-2})$

29. Let k be any number and let $t_n = k^n$; $t_1 = k$; $t_n = k \cdot t_{n-1}$.

30. Let p be any positive integer and let $t_p = p!$; $t_1 = 1$; $t_n = n \cdot t_{n-1}$.

31. C

32. Answers may vary. An example is given. Let $t_1 = 1$, $t_2 = 2$, $t_3 = 3$, and $t_n = t_{n-1} + t_{n-2} + t_{n-3}$. The first 12 terms are 1, 2, 3, 6, 11, 20, 37, 68, 125, 230, 423, and 778.

33. $12\sqrt{3}$

34. $2\sqrt{78}$

35. $2\sqrt{6\sqrt{10}}$

36–38. Answers may vary. Justifications are given for both.

31. **SAT/ACT Preview** The third term of a sequence is 143. The recursion equation for the sequence is $t_n = 2(t_{n-1}) - 3$ for $n = 2, 3, \ldots$. What is the starting value for the sequence?

A. 140 **B.** 283 **C.** 38 **D.** 73 **E.** none of these

ONGOING ASSESSMENT

32. **Open-ended Problem** Create a sequence in which each term after the third term is a function of the three terms immediately before it. Write a recursive formula for your sequence and use it to find the first 12 terms.

SPIRAL REVIEW

Find the geometric mean of each pair of numbers. *(Section 10.2)*

33. 72, 6 34. 12, 26 35. $8\sqrt{6}, \sqrt{15}$

Tell whether the data that can be gathered about each variable are *categorical* or *numerical*. Then describe the categories or numbers. *(Section 6.1)*

36. class size 37. clothing size 38. pets

Evaluate each expression when $a = 7$ and $b = 2$. *(Toolbox, page 780)*

39. $5(a + 4b)$ 40. $2a + 3(a - b)$ 41. $\frac{-5(a + 2b)}{ab}$

ASSESS YOUR PROGRESS

VOCABULARY

sequence (p. 465)
term (p. 465)
finite sequence (p. 467)
infinite sequence (p. 467)
geometric sequence (p.473)
common ratio (p. 473)
arithmetic sequence (p. 473)
common difference (p. 473)
geometric mean (p. 475)
arithmetic mean (p. 475)
explicit formula (p. 480)
recursion (p. 481)
recursive formula (p. 481)

For each sequence, find the next three terms and tell whether the sequence is *arithmetic, geometric,* or *neither.* *(Sections 10.1 and 10.2)*

1. $-5, -7, -10, -14, \ldots$ 2. $8, -16, 32, -64, \ldots$ 3. $4, 13, 22, 31, \ldots$

Find the arithmetic mean and the geometric mean for each pair of numbers. *(Section 10.2)*

4. 8, 22 5. 0.9, 62.5 6. $\sqrt{3}, 5\sqrt{3}$

Write an explicit formula and a recursive formula for each sequence. *(Sections 10.2 and 10.3)*

7. $-7, -13, -19, -25, \ldots$ 8. $3, 2, \frac{4}{3}, \frac{8}{9}, \ldots$ 9. $19, 22, 25, 28, \ldots$

10. **Journal** What are some of the advantages and disadvantages of explicit formulas for sequences? of recursive formulas for sequences?

Apply⇔Assess

Exercise Notes

Assessment Note

Ex. 32 Discuss this problem in class. Having three or four students show their sequences to the class would provide useful feedback for all students to evaluate what they have written.

Assess Your Progress

Journal Entry

Encourage students to give examples of sequences that can be written both explicitly and recursively. Then have students give examples of sequences which can be written recursively but not explicitly. This should help them answer the questions in this journal activity.

Progress Check 10.1–10.3

See page 506.

Practice 66 for Section 10.3

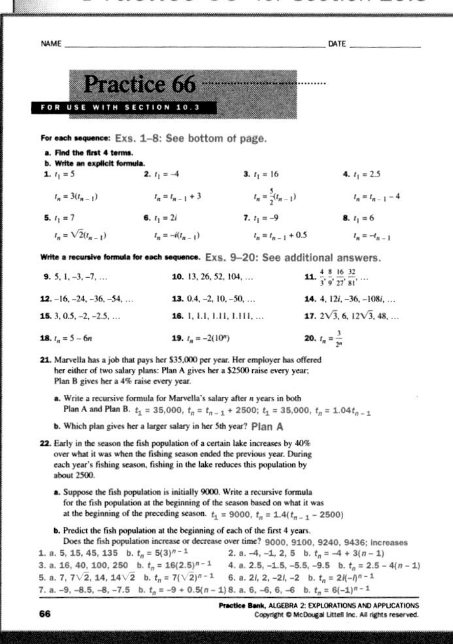

NAME ______ DATE ______

Practice 66

FOR USE WITH SECTION 10.3

For each sequence: Exs. 1–8: See bottom of page.

a. Find the first 4 terms.
b. Write an explicit formula.

1. $t_1 = 5$, $t_n = 3(t_{n-1})$ 2. $t_1 = -4$, $t_n = t_{n-1} + 3$ 3. $t_1 = 16$, $t_n = \frac{5}{2}(t_{n-1})$ 4. $t_1 = 2.5$, $t_n = t_{n-1} - 4$

5. $t_1 = 7$, $t_n = \sqrt{2}(t_{n-1})$ 6. $t_1 = 2i$, $t_n = -i(t_{n-1})$ 7. $t_1 = -9$, $t_n = t_{n-1} + 0.5$ 8. $t_1 = 6$, $t_n = -t_{n-1}$

Write a recursive formula for each sequence. Exs. 9–20: See additional answers.

9. $5, 1, -3, -7, \ldots$ 10. $13, 26, 52, 104, \ldots$ 11. $\frac{4}{3}, \frac{8}{9}, \frac{16}{27}, \frac{32}{81}, \ldots$

12. $-16, -24, -36, -54, \ldots$ 13. $0.4, -2, 10, -50, \ldots$ 14. $4, 12i, -36, -108i, \ldots$

15. $3, 0.5, -2, -2.5, \ldots$ 16. $1, 1.1, 1.11, 1.111, \ldots$ 17. $2\sqrt{3}, 6, 12\sqrt{3}, 48, \ldots$

18. $t_n = 5 - 6n$ 19. $t_n = -2(10^n)$ 20. $t_n = \frac{3}{2^n}$

21. Marvella has a job that pays her \$35,000 per year. Her employer has offered her either of two salary plans: Plan A gives her a \$2500 raise every year; Plan B gives her a 4% raise every year.
   a. Write a recursive formula for Marvella's salary after n years in both Plan A and Plan B. $t_1 = 35{,}000$, $t_n = t_{n-1} + 2500$; $t_1 = 35{,}000$, $t_n = 1.04t_{n-1}$
   b. Which plan gives her a larger salary in her 5th year? Plan A

22. Early in the season the fish population of a certain lake increases by 40% over what it was when the fishing season ended the previous year. During each year's fishing season, fishing in the lake reduces this population by about 2500.
   a. Suppose the fish population is initially 9000. Write a recursive formula for the fish population at the beginning of the season based on what it was at the beginning of the preceding season. $t_1 = 9000$, $t_n = 1.4(t_{n-1} - 2500)$
   b. Predict the fish population at the beginning of each of the first 4 years. Does the fish population increase or decrease over time? 9000, 9100, 9240, 9436; increases

1. a. 5, 15, 45, 135 b. $t_n = 5(3)^{n-1}$ 2. a. −4, −1, 2, 5 b. $t_n = -4 + 3(n-1)$
3. a. 16, 40, 100, 250 b. $t_n = 16(2.5)^{n-1}$ 4. a. 2.5, −1.5, −5.5, −9.5 b. $t_n = 2.5 - 4(n-1)$
5. a. 7, $7\sqrt{2}$, 14, $14\sqrt{2}$ b. $t_n = 7(\sqrt{2})^{n-1}$ 6. a. $2i$, 2, $-2i$, −2 b. $t_n = 2i(-i)^{n-1}$
7. a. −9, −8.5, −8, −7.5 b. $t_n = -9 + 0.5(n-1)$ 8. a. 6, −6, 6, −6 b. $t_n = 6(-1)^{n-1}$

66 **Practice Bank,** ALGEBRA 2: EXPLORATIONS AND APPLICATIONS

36. Numerical data may involve numbers of students. Categorical data may involve classes described as small or large.
37. Numerical data may involve measurement such as with shirts sized by neck measurements. Categorical data may involve clothing such as sweaters grouped into sizes such as small, medium, and large.
38. Numerical data may involve sizes or numbers of pets. Categorical data may involve types of pets (cat, dog, turtle, and so on).
39. 75
40. 29
41. $-\frac{55}{14} \approx -3.93$

Assess Your Progress

1. −19, −25, −32; neither
2. 128, −256, 512; geometric
3. 40, 49, 58; arithmetic
4. 15; $4\sqrt{11}$
5. 31.7; 7.5
6. $3\sqrt{3}$; $\sqrt{15}$
7. explicit: $t_n = -7 + (-6)(n - 1)$ or $t_n = -6n - 1$; recursive: $t_1 = -7$; $t_n = t_{n-1} - 6$
8. explicit: $t_n = 3 \cdot \left(\frac{2}{3}\right)^{n-1}$; recursive: $t_1 = 3$; $t_n = \frac{2}{3}(t_{n-1})$
9. explicit: $t_n = 19 + 3(n - 1)$ or $t_n = 3n + 16$; recursive: $t_1 = 19$; $t_n = t_{n-1} + 3$
10. Summaries may vary. With an explicit formula it is simple to find any term of the sequence by substituting the position number in the formula. With a recursive formula, that is not possible. A recursive formula makes it simple to generate a sequence using a graphing calculator.

Plan⇔Support

Objectives

- Use a formula to find the sum of an arithmetic series.
- Use sigma notation for series.

Recommended Pacing

Core and Extended Courses
Section 10.4: 1 day

Block Schedule
Section 10.4: $\frac{1}{2}$ block
(with Section 10.5)

Resource Materials

Lesson Support
Lesson Plan 10.4
Warm-Up Transparency 10.4
Practice Bank: Practice 67
Study Guide: Section 10.4

Technology
Graphing Calculator
Spreadsheet Software
McDougal Littell Mathpack
Stats!
Matrix Analyzer
Internet:
http://www.hmco.com

Warm-Up Exercises

1. Find the 20th term of the arithmetic sequence $t_n = -18n + 2$. −358
2. How many terms are in the sequence 2, 4, 6, ..., 150? 75
3. Find the sum of the numbers in the sequence 1, 2, 3, ..., 10. 55
4. Tell whether the sequence −18, −14, −10, −6, ... is *arithmetic*, *geometric*, or *neither*. arithmetic
5. Give a formula for the sequence in Ex. 2 above. $t_n = 2n$

Section

10.4 Sums of Arithmetic Series

Learn how to...

- use a formula to find the sum of an arithmetic series
- use sigma notation for series

So you can...

- find the number of seats in a theater, for example

Imagine the amazement of his elementary school teacher when young Carl Friedrich Gauss announced that he had added the integers from 1 to 100 in his head! This happened over 200 years ago in Germany. Gauss was a child prodigy who grew up to be one of the greatest mathematicians of all time. To see how Gauss did it, look at this:

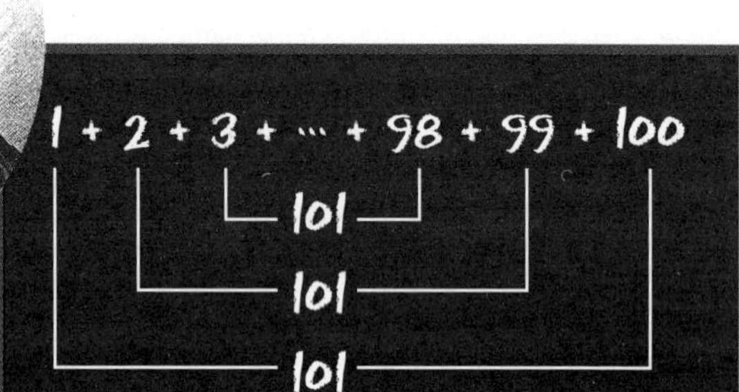

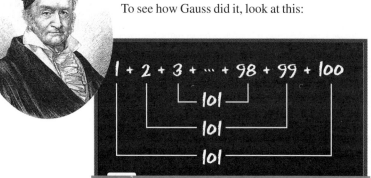

THINK AND COMMUNICATE

1. a. How many sums of 101 are there in the sum of the integers from 1 to 100? How do you know?
   b. Find the sum of the integers from 1 to 100. Explain how you found your answer.
2. Is the sequence 1, 2, 3, . . . , 100 *arithmetic, geometric,* or *neither*? Is it *finite* or *infinite*?
3. Use Gauss's method to find the sum of the terms of each sequence. Check your answers with a calculator.
   a. 1, 2, 3, . . . , 200
   b. 5, 10, 15, . . . , 100
   c. 60, 58, 56, . . . , 0
   d. −12, −11, −10, . . . , 12

When the terms of a sequence are added, the indicated sum is called a **series**:

sequence: 1, 2, 3, . . . , 100

series: 1 + 2 + 3 + · · · + 100

The indicated sum of an arithmetic sequence is called an **arithmetic series**.

ANSWERS Section 10.4

Think and Communicate

1. a. 50 sums; The 100 numbers are divided into pairs.
   b. 5050; 50(101) = 5050
2. arithmetic; finite
3. a. 20,100
   b. 1050
   c. 930
   d. 0

The method Gauss used to find the sum of the integers from 1 to 100 leads to a general formula for the sum S_n of any finite arithmetic series with n terms.

Sum of a Finite Arithmetic Series

The general formula for the sum S_n of a finite arithmetic series $t_1 + t_2 + t_3 + t_4 + \cdots + t_n$ is:

$$S_n = \frac{n}{2}(t_1 + t_n)$$

sum of n terms

half the number of terms first term last term

THINK AND COMMUNICATE

4. Describe Gauss's formula in words.
5. Derive an alternative formula for the sum of a finite arithmetic series by substituting the explicit formula $t_1 + d(n - 1)$ for t_n.

EXAMPLE 1 Application: Architecture

The seats in the theater shown are staggered to improve visibility. How many seats are there in the section shown? Use the information that accompanies the photo.

The last row has 45 seats.

The first row has 22 seats.

Each row after the first has one more seat than the row before it.

SOLUTION

To find the total number of seats, find the sum of the series whose terms are the number of seats in each row.

Step 1 Find the related sequence.
The first four terms of the sequence are: 22, 23, 24, 25.
The sequence is arithmetic with a common difference of 1.

Step 2 Find the number of terms in the related sequence.
Use the formula for the nth term of an arithmetic sequence.

$$45 = 22 + 1(n - 1)$$ Substitute **45** for t_n, **22** for t_1, and **1** for d.

Solve for n.

$$45 = 22 + n - 1$$

$$24 = n$$

Step 3 Find the sum of the series.
Use the formula for the sum of a finite arithmetic series.

$$S_{24} = \frac{24}{2}(22 + 45)$$ Substitute 24 for n, **22** for t_1, and **45** for t_n.

$$= 12 \cdot 67 = 804$$

There are 804 seats in the section shown.

Teach⇔Interact

Section Note

Communication: Discussion
When discussing the formula for the sum of a finite arithmetic series, ask students to identify the value of n, t_1, and t_n for the series $1 + 2 + 3 + \ldots + 99 + 100$, and then find the value of S_n.

About Example 1

Reasoning
To find the sum of the terms of an arithmetic sequence, it is necessary to identify how many terms are in the sequence in order to apply the sum formula for the series.

Additional Example 1

In another theater, the first row has 23 seats and each row after the first has three seats more than the previous row. If the last row has 77 seats, how many seats are there all together?

Step 1 Find the related sequence. The first 4 terms of the sequence are: 23, 26, 29, 32. The sequence is arithmetic with a common difference of 3.

Step 2 Find the number of terms in the related sequence. Use the formula for the nth term of an arithmetic sequence. Substitute 77 for t_n, 23 for t_1, and 3 for d.
$77 = 23 + 3(n - 1)$
Solve for n.
$77 = 23 + 3n - 3$
$57 = 3n$
$19 = n$

Step 3 Find the sum of the series. Use the formula for the sum of a finite arithmetic series. Substitute 19 for n, 23 for t_1, and 77 for t_n.
$S_{19} = \frac{19}{2}(23 + 77)$
$= 9.5 \cdot 100 = 950$
There are 950 seats in this theater.

Think and Communicate

4. Summaries may vary. The sum of a finite arithmetic series is the sum of the first and the last terms multiplied by half the number of terms.
5. $S_n = \frac{n}{2}(2t_1 + d(n - 1))$

Teach⇔Interact

Section Notes

Historical Connection
The use of the summation symbol in mathematics can be attributed to the Swiss mathematician Leonhard Euler (1707–1783), who was the key figure in eighteenth century mathematics.

Reasoning
After students have learned the meaning of sigma notation, ask them what the coefficient of n represents in the explicit formula $3n + 10$ for the related sequence. (The coefficient of n is the common difference in the related sequence.)

About Example 2

Mathematical Procedures
Expressing a series in sigma notation is especially useful when there are many terms in a series or when the series has a simple, short explicit formula. Point out that using sigma notation in Example 2 simplifies a discussion of the pattern in the terms.

Additional Example 2

A grocery clerk can create a display of canned items if the cans are stacked so that each row has two more cans than the previous row. Use sigma notation to write the series that represents the number of cans in 20 rows.

Step 1 Write an explicit formula for the sequence.
From the top, the rows contain the following number of cans:

$$1, \quad 3, \quad 5, \quad \ldots, \quad 39.$$
$$2(1)-1, \; 2(2)-1, \; 2(3)-1, \; \ldots, \; 2(20)-1$$

Each term is equal to one less than twice its position number, so the explicit formula is $t_n = 2n - 1$.

Step 2 Write the series in sigma notation.

$$\sum_{n=1}^{20} (2n-1)$$

Sigma Notation

You can write a series in compact form by using the summation symbol Σ, which is the capital Greek letter *sigma*. For example, the expression below represents the sum of the values of $3n + 10$ for integer values of n from 1 to 9.

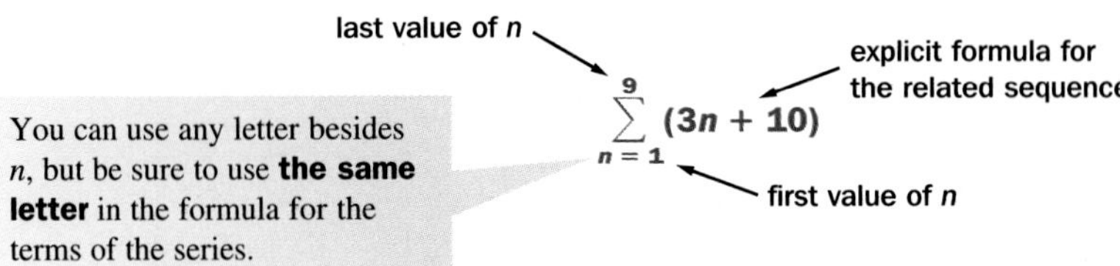

You can use any letter besides n, but be sure to use **the same letter** in the formula for the terms of the series.

To write a series in *expanded form,* substitute each value of n into the formula. For the expression above, you have:

$$\sum_{n=1}^{9} (3n + 10) = (3 \cdot 1 + 10) + (3 \cdot 2 + 10) + \cdots + (3 \cdot 9 + 10)$$
$$= 13 + 16 + \cdots + 37$$

EXAMPLE 2

In Korea it is a tradition to celebrate three special birthdays over a person's lifetime: the *paegil* or hundredth day after birth, the *tol* or first birthday, and the *hwan'gap* or 60th birthday. For these celebrations, tables are set with stacks of fruit, rice cakes, sweetened bean cakes, and other special foods.

Jae Joon Hyun's mother stacks oranges for his *tol.* Use sigma notation to write the series that represents the number of oranges in the stack.

6 6

There are 6 layers.

SOLUTION

Step 1 Write an explicit formula for the sequence.
From the top, the layers contain these numbers of oranges:

$$1, \; 4, \; 9, \; \ldots, 36$$
$$1^2, \; 2^2, \; 3^2, \; \ldots, 6^2$$

Each term is the square of its position number, so the explicit formula is:

$$t_n = n^2$$

Step 2 Write the series in sigma notation.

$$\sum_{n=1}^{6} n^2$$

CHECKING KEY CONCEPTS

For each series:

a. Find the number of terms.

b. Find the sum.

1. $0 + 6 + 12 + 18 + \cdots + 120$ **2.** $1 + 9 + 17 + 25 + \cdots + 97$

3. $15 + 12 + 9 + 6 + \cdots + (-21)$ **4.** $100 + 99 + 98 + 97 + \cdots + 5$

Expand each series.

5. $\sum_{n=1}^{6} 4n^3$ **6.** $\sum_{k=1}^{7} (k + 5)$ **7.** $\sum_{n=1}^{5} (3n - 2)$ **8.** $\sum_{s=1}^{8} (2s^2 + 7)$

Write each series in sigma notation.

9. $3 + 12 + 48 + 192$ **10.** $(-14) + (-9) + (-4) + 1 + 6$

11. $26 + 24 + 22 + 20 + \cdots + 10$ **12.** $4 + 9 + 16 + 25 + \cdots + 81$

10.4 Exercises and Applications

Extra Practice exercises on page 765

Find the sum of each series.

1. $1 + 2 + 3 + 4 + \cdots + 80$ **2.** $0 + 4 + 8 + 12 + \cdots + 52$

3. $200 + 199 + 198 + 197 + \cdots + 100$ **4.** $20 + 14 + 8 + 2 + \cdots + (-76)$

5. $58 + 61 + 64 + 67 + \cdots + 349$ **6.** $11 + 4 + (-3) + (-10) + \cdots + (-73)$

7. Visual Thinking Cut out a block of unit squares from a piece of graph paper to represent the terms of a finite arithmetic series. Then cut out an identical block and put it together with the first block to form a rectangle.

Example: $4 + 7 + 10 + 13$

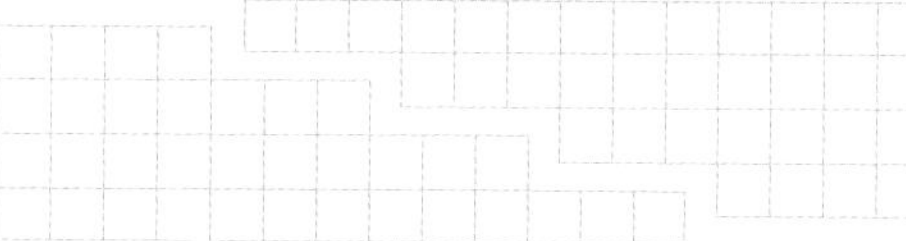

Repeat this procedure for three other arithmetic series. Then answer these questions:

a. What feature of the series does the length of each rectangle represent?

b. What feature of the series does the width of each rectangle represent?

c. What is the relationship between the area of each rectangle and the sum of the series?

The first four terms and the sum of a finite arithmetic series are given. Find the last term of each series.

8. $1 + 1\frac{1}{4} + 1\frac{1}{2} + 1\frac{3}{4} + \cdots + \underline{?};\ S_n = 51$ **9.** $1 + 2 + 3 + 4 + \cdots + \underline{?};\ S_n = 1275$

10. $0 + 11 + 22 + 33 + \cdots + \underline{?};\ S_n = 396$ **11.** $90 + 88 + 86 + 84 + \cdots + \underline{?};\ S_n = 948$

Teach⇔Interact

Checking Key Concepts

Student Progress

Students should be able to expand each series in questions 5–8 correctly before proceeding with the Exercises and Applications.

Closure Question

Explain how to find the sum of the terms of a finite arithmetic series. Identify the first and last terms of the series, t_1 and t_n, and the number of terms, n. Then apply the formula $S_n = \frac{n}{2}(t_1 + t_n)$.

Apply⇔Assess

Suggested Assignment

Core Course
Exs. 1–6, 8–21, 25–27, 29–34

Extended Course
Exs. 1–21 odd, 22–34

Block Schedule
Day 66 1–6, 8–21, 25–27, 29–34

Exercise Notes

Communication: Discussion
Exs. 1–6 To ensure that all students can apply the general formula for the sum of a finite arithmetic series, ask six students to write their solutions to these exercises on the board.

Integrating the Strands
Ex. 7 This exercise illustrates an application of arithmetic sequences to a geometric situation.

Teaching Tip
Exs. 8–11 These exercises are challenging as both n and t_n are unknown. Students can use the formula $t_n = t_1 + (n - 1)d$ to find an expression for t_n in terms of n and then substitute this into $S_n = \frac{n}{2}(t_1 + t_n)$ to solve for n.

Checking Key Concepts

1. a. 21 terms b. 1260
2. a. 13 terms b. 637
3. a. 13 terms b. –39
4. a. 96 terms b. 5040
5. $4 + 32 + 108 + 256 + 500 + 864$
6. $6 + 7 + 8 + 9 + 10 + 11 + 12$
7. $1 + 4 + 7 + 10 + 13$
8. $9 + 15 + 25 + 39 + 57 + 79 + 105 + 135$
9. $\sum_{n=1}^{4} 3 \cdot 4^{n-1}$
10. $\sum_{n=1}^{5} (5n - 19)$
11. $\sum_{n=1}^{9} (28 - 2n)$
12. $\sum_{n=1}^{8} (n + 1)^2$

Exercises and Applications

1. 3240 2. 364
3. 15,150 4. –476
5. 19,943 6. –403
7. a. the sum of the first term and the last term
b. the number of terms
c. The area of the rectangle is twice the sum of the series.
8. 5 9. 50
10. 88 11. 68 or –66

Apply⇔Assess

Exercise Notes

Cooperative Learning

Exs. 12–21 You may wish to have students work in small groups to do these exercises. Each group can then present its solution to one exercise so that the other groups can check their work. Remind students that they need to find the number of terms before they can find the sum.

Teaching Tip

Exs. 13, 14 When expanding the sigma notation for an arithmetic or geometric series, encourage students to determine the common difference or ratio and the first term and then use recursion to find the remaining terms rather than substituting each value of the index. With arithmetic series, the common difference is the coefficient of n and with geometric series, the common ratio is the base of the power. For example, in Ex. 14, the first term is $6(1) + 1 = 7$. The common difference is 6, so the next terms are $7 + 6 = 13$, $13 + 6 = 19$, $19 + 6 = 25$, and so on.

Application

Exs. 22, 23 Patterns are very commonly used in quilt designs. These exercises illustrate how sequences and series can be used to analyze two particular patterns called an *around the world block* and a *log cabin*.

Expand each series.

12. $\sum_{m=1}^{4} 9m^2$ **13.** $\sum_{n=1}^{6} (n - 12)$ **14.** $\sum_{n=1}^{5} (6n + 1)$ **15.** $\sum_{k=1}^{7} (-2k^3 + 5)$

Write each series in sigma notation and find the sum.

16. $13 + 25 + 37 + 49 + \cdots + 217$

17. $0 + 2.3 + 4.6 + 6.9 + \cdots + 92$

18. $164 + 113 + 62 + 11 + \cdots + (-499)$

19. $8 + 11\frac{1}{2} + 15 + 18\frac{1}{2} + \cdots + 67\frac{1}{2}$

20. $(-184) + (-170) + (-156) + (-142) + \cdots + (-72)$

21. $7.1 + 6.9 + 6.7 + 6.5 + \cdots + 0.7$

Connection QUILTING

Quilters may create original patterns or choose from a variety of traditional patterns, including those used in the two quilts shown. A portion of each quilt has been enlarged to make the pattern easier to see.

22. The pattern shown at the right is called *around the world,* because the quilter sews groups of squares around a central square.

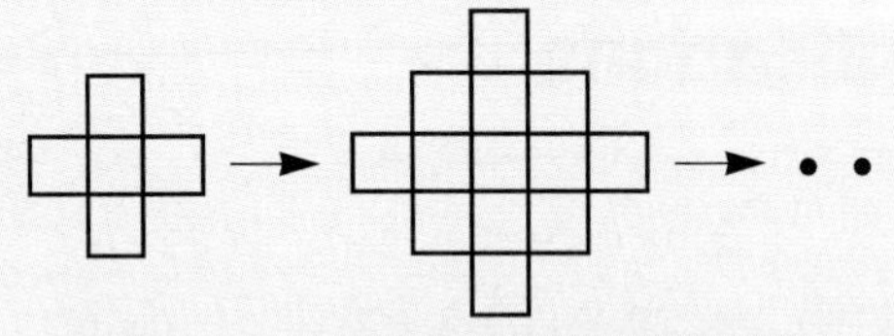

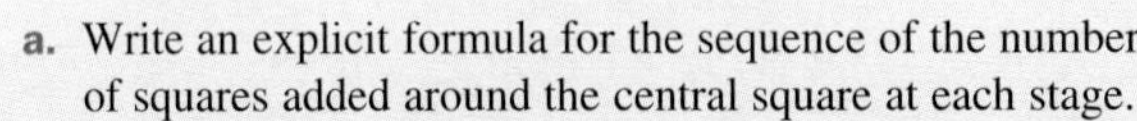

a. Write an explicit formula for the sequence of the number of squares added around the central square at each stage.

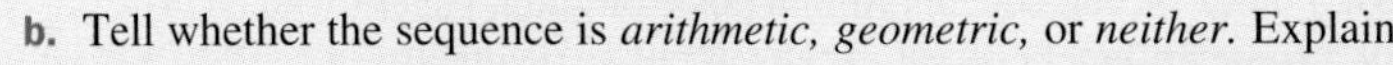

b. Tell whether the sequence is *arithmetic, geometric,* or *neither.* Explain.

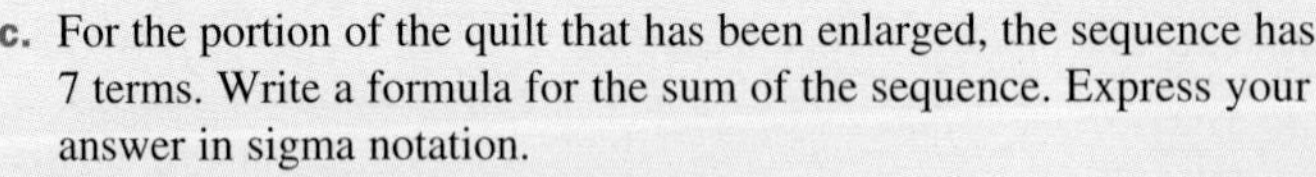

c. For the portion of the quilt that has been enlarged, the sequence has 7 terms. Write a formula for the sum of the sequence. Express your answer in sigma notation.

d. Suppose the pattern were continued. How many stages would be needed to bring the total number of squares added around the central square to 480?

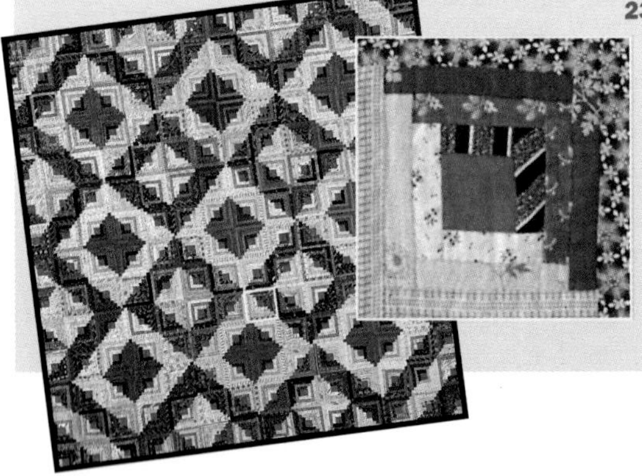

23. The pattern shown at the left is called a *log cabin.* It is created by cutting pieces from a long strip of cloth. All the pieces are 1 unit wide. The central piece is a unit square. Write a formula for the total length of cloth used as a function of the number of pieces cut from the cloth strip. Express your answer in sigma notation.

12. $9 + 36 + 81 + 144$

13. $-11 + (-10) + (-9) + (-8) + (-7) + (-6)$

14. $7 + 13 + 19 + 25 + 31$

15. $3 + (-11) + (-49) + (-123) + (-245) + (-427) + (-681)$

16. $\sum_{n=1}^{18} (12n + 1)$; 2070

17. $\sum_{n=1}^{41} (2.3n - 2.3)$; 1886

18. $\sum_{n=1}^{14} (-51n + 215)$; -2345

19. $\sum_{n=1}^{18} \left(\frac{7}{2}n + \frac{9}{2}\right)$; $679\frac{1}{2}$

20. $\sum_{n=1}^{9} (14n - 198)$; -1152

21. $\sum_{n=1}^{33} (7.3 - 0.2n)$; 128.7

22. a. $t_n = 4n$

b. arithmetic; Each term is 4 more than the preceding term.

c. Let k = the number of stages; $\sum_{n=1}^{k} 4n$.

For $k = 7$, $\sum_{n=1}^{7} 4n$.

d. 15 stages

23. Let k = the number of pieces cut from the strip; $\sum_{n=1}^{k} (2n - 1)$.

24. The *Fibonacci sequence* appears in many natural patterns (see Exercises 23 and 24 on page 485). The first and second terms of the sequence are both 1. Each term after the second term is the sum of the two terms immediately before it.

a. Write the first 15 terms of the Fibonacci sequence.

b. Find the sum of n terms of the Fibonacci sequence when $n = 2, 3, 4, \ldots, 13$.

c. Add 1 to each sum you found in part (b) and compare the results with the terms of the Fibonacci sequence. Make a conjecture about the sum of the terms of the Fibonnaci sequence.

d. **Challenge** Make a conjecture about the sum of any 10 consecutive terms of the Fibonacci sequence.

For Exercises 25 and 26, refer to Example 2 on page 490.

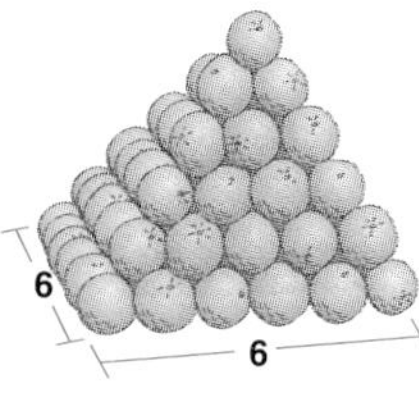

25. a. How many oranges are in the stack?

b. Suppose the stack of oranges had 12 layers. Use sigma notation to write the series that represents the number of oranges in the stack.

26. **Technology** Use a spreadsheet in part (a) and a graphing calculator or software with matrix calculation capabilities in part (b).

a. Evaluate $\sum_{k=1}^{n} k^2$ for $n = 1, 2, 3, \ldots, 20$. These values form a *sequence of partial sums.*

Sums of Squares

C8 | =C7 + B8

	A	B	C
1	Position number	Term	Partial sum
2	1	1	1
3	2	4	5
4	3	9	14
5	4	16	30
6	5	25	55
7	6	36	91
8	7	49	140

b. You know that $\sum_{k=1}^{n} k = \frac{n(n+1)}{2}$, which means that the sum of the first n positive integers is a *quadratic* function of n. Using the steps below, test the conjecture that the sum of the squares of the first n positive integers is a *cubic* function of n.

Step 1 Write a system of four equations of this form using $n = 1, 2, 3, 4$:

$$\sum_{k=1}^{n} k^2 = an^3 + bn^2 + cn + d, \text{ where } a, b, c, \text{ and } d \text{ are constants}$$

Step 2 Solve the system using matrices (see Section 7.3).
Step 3 Check to see whether the values you found for a, b, c, and d produce the correct sums when $n = 5, 6, 7, \ldots, 20$. (See part (a).)

BY THE WAY

In gold, nickel, and some other metals, the atoms are stacked in layers like oranges. The atoms in each layer fit into the gaps between the atoms in the layer underneath. In 1990 Wu-Yi Hsiang proved that this is one of the densest ways to pack spheres.

Apply⇔Assess

Exercise Notes

Challenge

Ex. 24 Encourage students to use their solutions to part (c) when answering part (d). If the sum of the first n terms is F_{n+2}, the sum of any 10 consecutive terms ending with the nth term is the sum of the first n terms minus the sum of the first $n - 10$ terms, which equals $F_{n+2} - F_{(n-10)+2} = F_{n+2} - F_{n-8}$.

Using Technology

Ex. 26 The table features and list features of a TI-82 graphing calculator can be used to generate a table very much like the spreadsheet in part (a). Press [Y=] and for the first two equations in the Y= list, use Y1= and X². Next, enter Y2=sum seq(A²,A,1,X,1). Use [2nd][LIST] to access sum and seq. For sum, use the LIST MATH menu. For seq, use the LIST OPS menu. Having entered the equations on the Y= list, press [2nd][TblSet]. Set TblMin equal to 1 and ΔTbl equal to 1. Then press [2nd][TABLE] to view the table. For part (b), use the first four rows of columns 1 and 3 of the table as statistical data. (Clear list L1 and L2 before you do so.) Enter 1,2,3,4 in L1, and enter 1,5,14,30 in L2. Then press [STAT][▶]7[ENTER] to get coefficients for a cubic regression equation. The calculator will display four coefficients on the home screen. These coefficients suggest the following fractional values: $a = \frac{1}{3}$, $b = \frac{1}{2}$, $c = \frac{1}{6}$, $d = 0$. These, in turn, lead to the formula $\sum_{k=1}^{n} k^2 = \frac{1}{3}n^3 + \frac{1}{2}n^2 + \frac{1}{6}n$. You may want to suggest that students use the *Stats!* and *Matrix Analyzer* software for this exercise.

24. a. 1, 1, 2, 3, 5, 8, 13, 21, 34, 55, 89, 144, 233, 377, 610

b. 2, 4, 7, 12, 20, 33, 54, 88, 143, 232, 376, 609

c. 3, 5, 8, 13, 21, 34, 55, 89, 144, 233, 377, 610; Let f_n be the nth term of the Fibonacci sequence; for $n \geq 2$, $\sum_{k=1}^{n} f_k = f_{n+2} - 1$.

d. Let $f_j, f_{j+1}, \ldots, f_{j+9}$ be any 10 consecutive terms of the Fibonacci sequence; $\sum_{n=j}^{j+9} f_j = f_{j+11} - f_{j+1}$.

25. a. 91 oranges

b. $\sum_{n=1}^{12} n^2$

26. a. 1, 5, 14, 30, 55, 91, 140, 204, 285, 385, 506, 650, 819, 1015, 1240, 1496, 1785, 2109, 2470, 2870

b. Step 1: The equations are $a + b + c + d = 1$, $8a + 4b + 2c + d = 5$, $27a + 9b + 3c + d = 14$, and $64a + 16b + 4c + d = 30$. Step 2: The solution is $\left(\frac{1}{3}, \frac{1}{2}, \frac{1}{6}, 0\right)$, so for positive integers n, $\sum_{k=1}^{n} k^2 = \frac{1}{3}n^3 + \frac{1}{2}n^2 + \frac{1}{6}n$. Step 3: The values found in Step 2 do produce the correct sums.

Apply⇔Assess

Exercise Notes

Problem Solving

Ex. 28 You may want to suggest that students make a table in order to organize the terms of the sequence.

Assessment Note

Ex. 29 As students work on this exercise in their groups, you can assess their understanding of the concepts involved by joining each group for a short period of time. The activities involved provide an excellent review of the key concepts in this section.

Practice 67 for Section 10.4

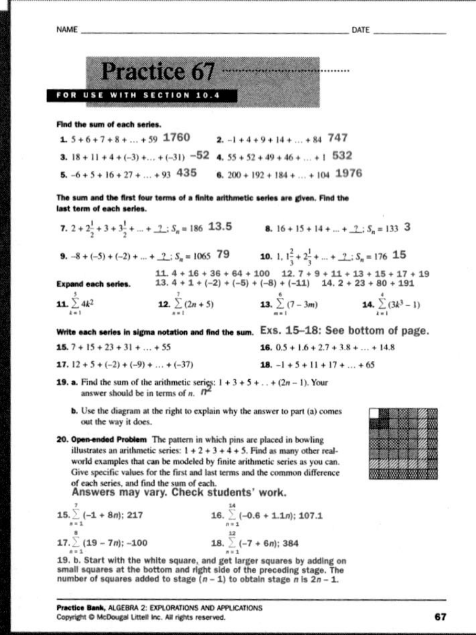

NAME ______________ DATE ______

Practice 67

FOR USE WITH SECTION 10.4

Find the sum of each series.

1. $5 + 6 + 7 + 8 + \ldots + 59$ 1760
2. $-1 + 4 + 9 + 14 + \ldots + 84$ 747
3. $18 + 11 + 4 + (-3) + \ldots + (-31)$ −52
4. $55 + 52 + 49 + 46 + \ldots + 1$ 532
5. $-6 + 5 + 16 + 27 + \ldots + 93$ 435
6. $200 + 192 + 184 + \ldots + 104$ 1976

The sum and the first four terms of a finite arithmetic series are given. Find the last term of each series.

7. $2 + 2\frac{1}{2} + 3 + 3\frac{1}{2} + \ldots + \underline{?}$; $S_n = 186$ 13.5
8. $16 + 15 + 14 + \ldots + \underline{?}$; $S_n = 133$ 3
9. $-8 + (-5) + (-2) + \ldots + \underline{?}$; $S_n = 1065$ 79
10. $1, 1\frac{2}{3} + 2\frac{1}{3} + \ldots + \underline{?}$; $S_n = 176$ 15

Expand each series.

11. $4 + 16 + 36 + 64 + 100$ 12. $7 + 9 + 11 + 13 + 15 + 17 + 19$
13. $4 + 1 + (-2) + (-5) + (-8) + (-11)$ 14. $2 + 23 + 80 + 191$

11. $\sum_{k=1}^{5} 4k^2$
12. $\sum_{n=1}^{7} (2n + 5)$
13. $\sum_{m=1}^{6} (7 - 3m)$
14. $\sum_{k=1}^{4} (3k^3 - 1)$

Write each series in sigma notation and find the sum. Exs. 15–18: See bottom of page.

15. $7 + 15 + 23 + 31 + \ldots + 55$
16. $0.5 + 1.6 + 2.7 + 3.8 + \ldots + 14.8$
17. $12 + 5 + (-2) + (-9) + \ldots + (-37)$
18. $-1 + 5 + 11 + 17 + \ldots + 65$

19. a. Find the sum of the arithmetic series: $1 + 3 + 5 + \ldots + (2n - 1)$. Your answer should be in terms of n. n^2

 b. Use the diagram at the right to explain why the answer to part (a) comes out the way it does.

20. **Open-ended Problem** The pattern in which pins are placed in bowling illustrates an arithmetic series: $1 + 2 + 3 + 4 + 5$. Find as many other real-world examples that can be modeled by finite arithmetic series as you can. Give specific values for the first and last terms and the common difference of each series, and find the sum of each.
 Answers may vary. Check students' work.

15. $\sum_{n=1}^{7} (-1 + 8n)$; 217
16. $\sum_{n=1}^{14} (-0.6 + 1.1n)$; 107.1
17. $\sum_{n=1}^{8} (19 - 7n)$; −100
18. $\sum_{n=1}^{12} (-7 + 6n)$; 384

19. b. Start with the white square, and get larger squares by adding on small squares at the bottom and right side of the preceding stage. The number of squares added to stage $(n - 1)$ to obtain stage n is $2n - 1$.

Practice Bank, ALGEBRA 2: EXPLORATIONS AND APPLICATIONS

67

27. **SAT/ACT Preview** What is the relationship between A and B for all values of n?

$$A = \sum_{k=1}^{n} (k + 4) \qquad B = \sum_{k=1}^{n} (k + 6)$$

 A. A = B **B.** A < B **C.** A > B **D.** relationship cannot be determined

28. **PHYSICS** When an object is in free fall, it experiences a constant acceleration of 32 ft/s^2 due to gravity. This means that the object's speed increases by 32 ft/s each second.

 a. Starting with an initial speed of 0 ft/s, write the sequence of speeds for an object in free fall after 0, 1, 2, 3, . . . , t seconds.

 b. You can find the distance that the object falls during any second by multiplying the *average* speed by the time. For example, during the first second of fall, the object's average speed is $\frac{0 + 32}{2} = 16$ ft/s, so the object falls (16 ft/s)(1 s) = 16 ft. Write the sequence of distances the object falls during each of the t seconds.

 c. The total distance that an object falls is the sum of the terms of the sequence you wrote in part (b). Find the total distance. How does this compare with the function given in Example 3 on page 187?

ONGOING ASSESSMENT

29. **Cooperative Learning** Work with four other students.

 For each step described below, each of you should write your answer on the paper you have and then pass it to the student on your right. When you have completed all the steps, pass the papers to the right one last time. Check the work on the paper you receive and correct any errors.

 a. **Open-ended Problem** Make up a finite arithmetic sequence and write the terms of the sequence at the top of a blank sheet of paper.

 b. Write an explicit formula for the nth term of the sequence on the paper you receive.

 c. Using sigma notation, write an expression for the series related to the sequence on the paper you next receive.

 d. Find the sum of the series on the paper you next receive.

SPIRAL REVIEW

Write a recursive formula for each sequence. *(Section 10.3)*

30. $12, 12i, -12, -12i, \ldots$

31. $14, 6, -2, -10, \ldots$

Multiply. *(Section 1.4)*

32. $\begin{bmatrix} 5 & -1 & 6 \\ 0 & 0 & 2 \\ -1 & 1 & 0 \end{bmatrix}\begin{bmatrix} 3 & 8 \\ 4 & 7 \\ -2 & 0 \end{bmatrix}$

33. $\begin{bmatrix} 8 & -3 \\ 7 & -2 \\ 1 & -3 \end{bmatrix}\begin{bmatrix} -3 & 0 & -1 \\ 6 & 9 & -9 \end{bmatrix}$

34. $\begin{bmatrix} 0 & 1 \\ 1 & 1 \end{bmatrix}\begin{bmatrix} 0 \\ 1 \end{bmatrix}$

27. B

28. a. 0, 32, 64, 96, 128, 160, 192, 224, . . . , $32t$

 b. 16, 48, 80, 112, 144, 176, 208, . . . , $32t - 16$

 c. $S_t = \frac{t}{2}(16 + 32t - 16) = 16t^2$; It is the same function.

29. Answers may vary. Check students' work.

30. $t_1 = 12$; $t_n = i \cdot t_{n-1}$

31. $t_1 = 14$; $t_n = t_{n-1} - 8$

32. $\begin{bmatrix} -1 & 33 \\ -4 & 0 \\ 1 & -1 \end{bmatrix}$

33. $\begin{bmatrix} -42 & -27 & 19 \\ -33 & -18 & 11 \\ -21 & -27 & 26 \end{bmatrix}$

34. $\begin{bmatrix} 1 \\ 1 \end{bmatrix}$

Section

10.5 Sums of Geometric Series

Learn how to...

- **use a formula to find the sum of a finite geometric series**
- **use a formula to find the sum of an infinite geometric series, if the sum exists**

So you can...

- **find the number of games in an elimination tournament, for example**

Many companies set up a telephone tree to notify employees when the business is shut down due to a hurricane, snow storm, or other emergency. Suppose a company sets up a telephone tree so that each employee (except those in the last level of the tree) calls three other employees. The first three levels of the tree look like this:

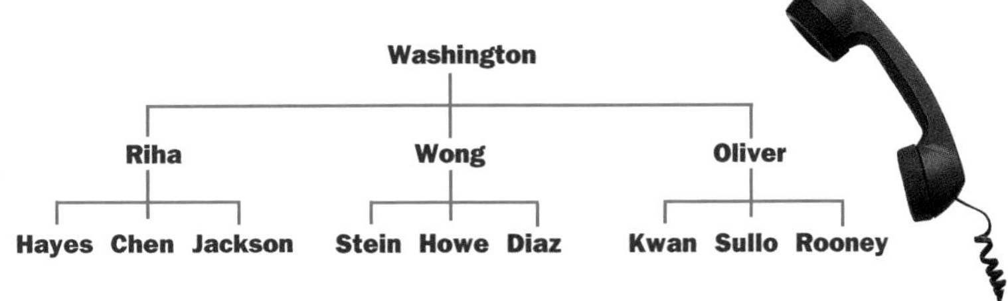

THINK AND COMMUNICATE

1. a. Write the first 4 terms of the sequence for the number of employees at each level of the tree.
   b. Tell whether the sequence is *arithmetic, geometric,* or *neither.* Explain.
   c. Write an explicit formula for the sequence.
2. a. What does the related series represent in this situation?
   b. Write the related series in sigma notation, using k as the variable and n as the number of terms.

The indicated sum of a geometric sequence is called a **geometric series**. To find a general formula for the sum of a finite geometric series, first multiply the expanded series by the common ratio and then subtract the product from the original series:

$$\begin{aligned} S_n &= t_1 + t_1 \cdot r + t_1 \cdot r^2 + t_1 \cdot r^3 + \cdots + t_1 \cdot r^{n-1} \\ r \cdot S_n &= \qquad t_1 \cdot r + t_1 \cdot r^2 + t_1 \cdot r^3 + \cdots + t_1 \cdot r^{n-1} + t_1 \cdot r^n \\ \hline S_n - r \cdot S_n &= t_1 \qquad\qquad\qquad\qquad\qquad\qquad - t_1 \cdot r^n \end{aligned}$$

Subtract $r \cdot S_n$ from S_n.

THINK AND COMMUNICATE

3. Solve the equation above for S_n. Then check the formula using $n = 3$, $t_1 = 1$, and $r = 3$ to see whether it gives the number of employees shown in the telephone tree at the top of the page.

Plan⇔Support

Objectives

- Use a formula to find the sum of a finite geometric series.
- Use a formula to find the sum of an infinite geometric series, if the sum exists.
- Apply the formula for the sum of a finite geometric series to real-world situations.

Recommended Pacing

Core and Extended Courses
Section 10.5: 2 days

Block Schedule
Section 10.5: 2 half-blocks (with Section 10.4 and Portfolio Project)

Resource Materials

Lesson Support
Lesson Plan 10.5
Warm-Up Transparency 10.5
Overhead Visuals:
Folder 10: Sum of an Infinite Geometric Series
Practice Bank: Practice 68
Study Guide: Section 10.5
Exploration Lab Manual:
Diagram Master 13
Assessment Book: Test 41

Technology
Graphing Calculator
Spreadsheet Software
McDougal Littell Mathpack *Stats!*
Internet:
http://www.hmco.com

Warm-Up Exercises

1. Give the formula for the nth term of a geometric sequence. $t_n = t_1 \cdot r^{n-1}$
2. Tell whether the series $\sum_{n=1}^{8} 6\left(\frac{1}{2}\right)^{n-1}$ is *finite* or *infinite*. finite
3. Give the number of terms in the sequence 3, 9, 27, ..., 729. 6
4. Find t_8 for the series $\sum_{n=1}^{8} (6n - 3)$. 45

ANSWERS Section 10.5

Think and Communicate

1. a. 1, 3, 9, 27
   b. geometric; Each term is three times the preceding term.
   c. $t_n = 3^{n-1}$
2. a. The sum of n terms of the series represents the total number of employees called after n levels.
   b. $\sum_{k=1}^{n} (3^{k-1})$
3. $S_n = \frac{t_1(1 - r^n)}{1 - r}$; $S_3 = \frac{1(1 - 3^3)}{1 - 3} = 13$; The formula gives the number of employees shown in the table at the top of page 495.

Teach⇔Interact

Section Note

Challenge

You may want to challenge students to derive the formula for the sum S_n of a finite arithmetic series using a method similar to the one shown on page 495 for the sum of a finite geometric series. Students should begin by writing S_n two different ways: one starting with t_1, whose 2nd term is $t_1 + d$ and so on, and one starting with t_n, whose 2nd term is $t_n - d$ and so on.

Think and Communicate

For part (b) of question 2 on page 495, students should see that a geometric series is expressed in sigma notation the same way that an arithmetic series is, except that a geometric pattern is involved instead of an arithmetic pattern. Question 3 leads to the general formula for the sum S_n of a finite geometric series, which is discussed at the top of this page. Question 5 asks students to think about why the sum of a geometric series cannot be determined if the common ratio is 1. Ask students to make up a simple series with common ratio 1 and find the sum.

Additional Example 1

A family tree includes you, your parents, your four grandparents, their eight parents, and so on. Each level in a tree is called a *generation*, and each person in a previous generation is called your *ancestor*.

a. How many generations will you need to go back to find one with 512 ancestors?

Sketch a tree diagram of a simple family tree, such as one with you and two generations of ancestors.

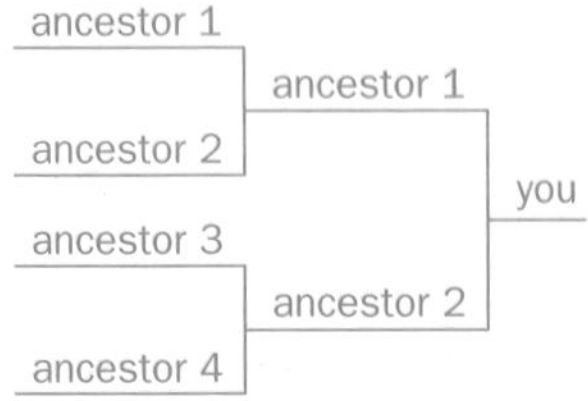

In each generation, the number of ancestors is twice the previous number of ancestors. The number of ancestors in each generation forms a geometric sequence with common ratio 2.

Sum of a Finite Geometric Series

The general formula for the sum S_n of the finite geometric series $t_1 + t_1 \cdot r + t_1 \cdot r^2 + \cdots + t_1 \cdot r^{n-1}$ where $r \neq 1$ is:

$$S_n = \frac{t_1(1 - r^n)}{1 - r}$$

THINK AND COMMUNICATE

4. Describe the general formula for the sum of a finite geometric series in your own words.

5. Why does the general formula for the sum of a finite geometric series apply only to series that have a common ratio that is not equal to 1? What sum would you get if $r = 1$?

EXAMPLE 1 Application: Sports

The four grand-slam tennis tournaments played each year are *elimination tournaments.* In both the men's and women's divisions, 128 players enter the first round of each tournament. In each round, pairs of players compete in matches. Players who win their matches move on to the next round. Players who lose are eliminated from the tournament.

a. How many rounds are there in a grand-slam tournament?

b. How many matches are played in a grand-slam tournament?

SOLUTION

Sketch a tree diagram of a simpler tournament, such as one with only four players.

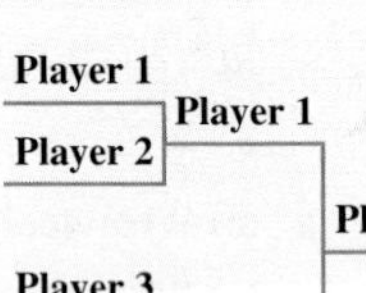

In the final round, only **1** match is played.

In each round, the number of matches is **half** the number of players.

The number of matches in each round forms a geometric sequence with common ratio $\frac{1}{2}$. The number of matches in the first round is $128\left(\frac{1}{2}\right) = \mathbf{64}$.

a. Use the formula for the nth term of a geometric sequence: $t_n = t_1 \cdot r^{n-1}$.

$$1 = 64\left(\frac{1}{2}\right)^{n-1}$$

Substitute **1** for t_n, **64** for t_1, and $\frac{1}{2}$ for r.

$$\frac{1}{64} = \left(\frac{1}{2}\right)^{n-1}$$

$$\left(\frac{1}{2}\right)^6 = \left(\frac{1}{2}\right)^{n-1}$$

$$6 = n - 1$$

The exponents are equal.

$$7 = n$$

There are **7 rounds** in a grand-slam tournament.

BY THE WAY

The four grand-slam tournaments are the Australian Open, the French Open, Wimbledon, and the U.S. Open. Steffi Graf of Germany won all four in 1988.

Think and Communicate

4. Summaries may vary. To find the sum S_n of a finite geometric series with common ratio $r \neq 1$, first find r^n. Subtract the result from 1 and multiply by the first term. Finally, divide the result by $1 - r$.

5. If $r = 1$, $\frac{t_1(1 - r^n)}{1 - r}$ is not defined. If $r = 1$, the sequence is $t_1, t_1, t_1, \ldots$ and $S_n = nt_1$.

Exploration

1. a. $\frac{1}{2} + \frac{1}{4} + \frac{1}{8} + \frac{1}{16} + \ldots$

   b. infinite; geometric; Every term is half the preceding term and there is no last term.

2. 1; If you could halve the leftover piece forever, the area of the amount leftover would approach 0, so the combined area of the pieces put aside would approach 1.

b. The total number of matches is the sum of the number of matches in each round. Use the formula for the sum of a finite geometric series with $r \neq 1$.

$$S_7 = \frac{64\left(1 - \left(\frac{1}{2}\right)^7\right)}{1 - \frac{1}{2}} = 127$$

Substitute **64** for t_1, $\frac{1}{2}$ for r, and 7 for n.

In a grand-slam tournament, 127 matches are played.

Infinite Geometric Series

An infinite geometric series has the general form:

$$t_1 + t_1 \cdot r + t_1 \cdot r^2 + t_1 \cdot r^3 + t_1 \cdot r^4 + \cdots$$

Do you think it is possible for a series like this to have a sum?

EXPLORATION
COOPERATIVE LEARNING

Investigating an Infinite Geometric Series

Work with a partner.
You will need:

- scissors

SET UP Begin with a large square piece of paper. Let the length of a side of the square be 1 unit.

1 Cut the square in half to form two rectangles. Put one half aside and record its area as shown in the table below.

2 Cut the leftover piece in half to form two rectangles. Again put one half aside, placing it next to the piece set aside in the previous step. Express the combined area of the pieces put aside as an indicated sum.

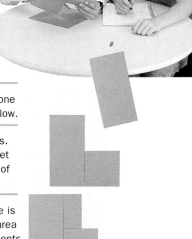

3 Repeat Step 2 over and over until the leftover piece is too small to cut in half. Each time be sure to add the area of the piece put aside to the indicated sum that represents the combined area of the pieces put aside.

Questions

1. Suppose you can halve the leftover piece forever.
   a. Write the series that represents the combined area of the pieces put aside. Express your answer in expanded form.
   b. Is the series in part (a) *finite* or *infinite*? Is it *arithmetic*, *geometric*, or *neither*? Explain.
2. What would you say is the sum of the series in Question 1? Explain your reasoning.

Pieces put aside	Combined area
1	$\frac{1}{2}$
2	$\frac{1}{2} + ?$
3	$\frac{1}{2} + ? + ?$
⋮	$\frac{1}{2} + ? + ? + \cdots$

Exploration Note

Purpose
The purpose of this Exploration is to have students create an infinite geometric sequence and to investigate its properties.

Materials/Preparation
Each group needs paper and scissors.

Procedure
Each pair of students starts with a square of paper and cuts the square in half to form two rectangles. They put aside one rectangle and record its area. They then repeat the procedure until the leftover rectangle is too small to cut. They keep a total of the area each time one rectangle is put aside.

Closure
Students should see that the series of rectangles put aside forms a geometric sequence with ratio $\frac{1}{2}$ and that the series is actually infinite, although the paper gets too small to cut. Students should also see that the total area of the pieces approaches a limiting value of 1, the original area of the square.

Explorations Lab Manual
See the Manual for more commentary on this Exploration.

Diagram Master 13

Teach⇔Interact

Additional Example 1 (continued)

The number of ancestors one generation before you in the tree is 2.
Use the formula for the nth term of a geometric sequence:
$t_n = t_1 \cdot r^{n-1}$.
$512 = 2(2)^{n-1}$
$256 = 2^{n-1}$
$2^8 = 2^{n-1}$
$8 = n - 1$
$9 = n$
You will have 512 ancestors in the 9th generation before your own.

b. What is the total number of ancestors in 20 generations before you?
The total number of ancestors is the sum of the number of ancestors in each generation. Use the formula for the sum of a finite geometric series with $r \neq 1$.
$S_n = \frac{2(1 - (2)^{20})}{1 - 2} = 2,097,150$
In 20 generations, you had 2,097,150 ancestors.

Section Notes

Teaching Tip
Before students do the Exploration, you may want to show them some examples of infinite geometric series. Have students conjecture whether the sum of each series can be found. For series that may have a sum, ask for the ratio. Then have students conjecture what a requirement for the ratio may be.

Using Technology
You can have students use the list and sequence graphing features of a TI-82 graphing calculator to continue the Exploration a bit further. To gain another visual picture of the sum of the combined area, you can have students graph the sequence of partial sums. Set the calculator to sequence and dot mode. On the Y= list, type Un=sum seq(0.5^A,A,1,X,1) and enter window settings U_nStart = 0.5, V_nStart = 0, nStart = 1, nMin = 1, nMax = 20, Xmin = 0, Xmax = 20, Ymin = 0.4, and Ymax = 1.1. (Sum is found under the LIST MATH menu and seq(is found under the LIST OPS menu.) You might also have students graph the sequence of terms of the series to see what happens to them.

Teach⇔Interact

Think and Communicate

For questions 6 and 7, you may need to explain how to read the notation used.

Section Note

Reasoning

If some students cannot understand how the sum of an infinite geometric series formula is arrived at, go back to the sum formula for a finite geometric series and examine the term r^n for larger and larger values of n, with r a number whose absolute value is less than 1, such as $\frac{1}{3}, \frac{1}{4}, \frac{1}{5}$, and so on. Students should begin to see that for very large values of n with $|r| < 1$, r^n gets closer and closer to zero and thus can be set equal to zero as n becomes infinite. This process yields the formula $S = \frac{t_1}{1-r}$.

About Example 2

Student Progress

This example has students find the sum of a geometric series expressed both in sigma notation and in expanded form. Students should be able to understand each form at this point.

Additional Example 2

Find the sum of each infinite geometric series, if the sum exists.

For each series, first find the common ratio r. If $|r| < 1$, use the general formula to find the sum of the series.

a. $\sum_{n=1}^{\infty} 6\left(\frac{1}{4}\right)^{n-1}$

The common ratio is $\frac{1}{4}$. Since $|r| < 1$, the series has a sum:

$$S = \frac{6}{1-\frac{1}{4}} = \frac{6}{\frac{3}{4}} = 6\left(\frac{4}{3}\right) = 8$$

b. $-2 + 8 + (-32) + 128 - \ldots$

The common ratio is -4. Since $|r| > 1$, the series does not have a sum.

In the Exploration you saw that the infinite geometric series

$$\frac{1}{2} + \frac{1}{4} + \frac{1}{8} + \frac{1}{16} + \cdots$$

appears to have a sum of 1. To see why this is so, consider the sum of the first n terms of this series:

$$S_n = \frac{\frac{1}{2}\left(1 - \left(\frac{1}{2}\right)^n\right)}{1 - \frac{1}{2}} = 1 - \left(\frac{1}{2}\right)^n$$

Substitute $\frac{1}{2}$ for t_1 and $\frac{1}{2}$ for r in the general formula for the sum of a finite geometric series.

THINK AND COMMUNICATE

6. a. What happens to $\left(\frac{1}{2}\right)^n$ as $n \to \infty$? b. What happens to S_n as $n \to \infty$?
7. a. Describe what happens to $S_n = \frac{t_1(1 - r^n)}{1 - r}$ as $n \to \infty$ for *any* geometric series with $|r| < 1$.
   b. Why is your conclusion in part (a) *not* true when $|r| > 1$?

Sum of an Infinite Geometric Series

The general formula for the sum S of the infinite geometric series $t_1 + t_1 \cdot r + t_1 \cdot r^2 + t_1 \cdot r^3 + t_1 \cdot r^4 + \cdots$ where $|r| < 1$ is:

$$S = \frac{t_1}{1 - r}$$

You can use sigma notation to express the sum of an infinite geometric series. For example, you can write:

$$\frac{1}{2} + \frac{1}{4} + \frac{1}{8} + \frac{1}{16} + \cdots = \sum_{n=1}^{\infty} \frac{1}{2^n}$$

Use the symbol for **infinity** in place of the last value of n.

EXAMPLE 2

Find the sum of each infinite geometric series, if the sum exists.

a. $\sum_{n=1}^{\infty} -21(2)^{n-1}$ **b.** $5 + \frac{5}{3} + \frac{5}{9} + \frac{5}{27} + \cdots$

SOLUTION

For each series, first find the common ratio r. If $|r| < 1$, use the general formula to find the sum of the series.

a. The common ratio r is 2. Since $|r| > 1$, the series does not have a sum.

Think and Communicate

6. a. $\left(\frac{1}{2}\right)^n \to 0$
   b. $S_n \to 1$
7. a. If $|r| < 1$, as $n \to \infty$, $r^n \to 0$, so $\frac{t_1(1 - r^n)}{1 - r} \to \frac{t_1}{1-r}$.
   b. If $|r| > 1$, the absolute value of r^n increases without limit.

b. The common ratio r is $\frac{1}{3}$. Since $|r| < 1$, the series has a sum:

$$S = \frac{5}{1 - \frac{1}{3}} = \frac{5}{\frac{2}{3}} = 5\left(\frac{3}{2}\right) = 7.5$$

Substitute **5** for t_1 and $\frac{1}{3}$ for r in the general formula for the sum of an infinite geometric series.

EXAMPLE 3

Express 0.4545. . . as a fraction.

SOLUTION

Express the repeating decimal as an infinite series:

$$0.4545\ldots = 0.45 + 0.0045 + 0.000045 + 0.00000045 + \cdots$$

This series is geometric with $r = 0.01$. Since $|r| < 1$, the series has a sum:

$$S = \frac{0.45}{1 - 0.01} = \frac{0.45}{0.99} = \frac{45}{99} = \frac{5}{11}$$

Substitute **0.45** for t_1 and **0.01** for r in the general formula for the sum of an infinite geometric series.

CHECKING KEY CONCEPTS

Find the sum of each geometric series, if the sum exists. If a series does not have a sum, explain why not.

1. $2 + 6 + 18 + 54 + \cdots + 486$

2. $1 + 5 + 25 + 125 + \cdots$

3. $3 + (-6) + 12 + (-24) + \cdots + (-384)$

4. $200 + 50 + 12.5 + 3.125 + \cdots$

5. $\sum_{n=1}^{\infty} 4^{n-1}$

6. $\sum_{n=1}^{8} 7\left(\frac{1}{2}\right)^{n-1}$

7. $\sum_{n=1}^{\infty} \left(\frac{1}{3}\right)^{n-1}$

8. $\sum_{n=1}^{5} 3(-2)^{n-1}$

9–12. Write each series in Questions 1–4 in sigma notation.

Teach⇔Interact

About Example 3

Mathematical Procedures

This example shows students how to use the formula for the sum of the terms of an infinite geometric series to express a repeating decimal as a fraction. Point out to students that a value of r such that $|r| < 1$ can be found for any repeating decimal.

Additional Example 3

Express 3.248248... as a fraction.

First express the number as 3 + 0.248248.... Then express the repeating decimal as an infinite series:

$0.248248\ldots = 0.248 + 0.000248 + 0.000000248 + \ldots$.

This series is geometric with $r = 0.001$. Since $|r| < 1$, the series has a sum:

$$S = 3 + \frac{0.248}{1 - 0.001} = 3 + \frac{0.248}{0.999} = 3\frac{248}{999}$$

Checking Key Concepts

Common Error

For questions 5–8, students may make errors in finding the common ratio for the series. If so, suggest that they write out the first few terms of the expanded form of the series and then find the common ratio of successive terms.

Closure Question

Describe how to find the sum of the terms of a finite or infinite geometric series.

Find the first term and the common ratio of terms for either series. If the series is finite, use the formula $S_n = \frac{t_1(1 - r^n)}{1 - r}$. If the series is infinite and if $|r| < 1$, use the formula $S = \frac{t_1}{1 - r}$.

Checking Key Concepts

1. 728
2. The sum does not exist because the series is infinite with $r = 5$ and $|5|$ is not less than 1.
3. −255
4. $266\frac{2}{3}$
5. The sum does not exist because the series is infinite with $r = 4$ and $|4|$ is not less than 1.
6. $\frac{1785}{128} \approx 13.9$
7. 1.5
8. 33
9. $\sum_{n=1}^{6} 2 \cdot 3^{n-1}$
10. $\sum_{n=1}^{\infty} 5^{n-1}$
11. $\sum_{n=1}^{8} 3(-2)^{n-1}$
12. $\sum_{n=1}^{\infty} 200\left(\frac{1}{4}\right)^{n-1}$

Apply⇔Assess

Suggested Assignment

Core Course
Day 1 Exs. 1–21
Day 2 Exs. 25–27, 37, 40–53, AYP

Extended Course
Day 1 Exs. 1–21 odd, 22–24
Day 2 Exs. 25–35 odd, 37–53, AYP

Block Schedule
Day 66 Exs. 1–21
Day 67 Exs. 25–27, 37, 40–53, AYP

Exercise Notes

Student Progress
Exs. 1–6 A review of the answers to these exercises will help to ensure that all students can find the correct sums.

Common Error
Exs. 7–10, 18–21 Some students may not be able to find r or the first term for series expressed in sigma notation. Suggest to these students that they begin by writing the first few terms of the expanded series.

Using Technology
Ex. 22 A spreadsheet is a good tool to show how the partial sums of the terms of an infinite geometric series can approach the limiting value for the sum, the value obtained by using the formula $S = \frac{t_1}{1-r}$. It also can show how the series does not have a sum in the case of $|r| > 1$. Any geometric series can be expressed the same way for any number of terms. All a student has to do is make the first term and the common ratio easily changeable. You may want to suggest that students use the *Stats!* software for this exercise.

10.5 Exercises and Applications

Extra Practice exercises on page 765

Find the sum of each series.

1. $4 + 8 + 16 + 32 + \cdots + 256$
2. $125 + 75 + 45 + 27 + \cdots + 5.832$
3. $2 + 2 \times 10 + 2 \times 10^2 + \cdots + 2 \times 10^{12}$
4. $\pi + 3\pi + 9\pi + 27\pi + \cdots + 6561\pi$
5. $12 + (-6) + 3 + \left(-\frac{3}{2}\right) + \cdots + \left(-\frac{3}{128}\right)$
6. $0.31 + 0.93 + 2.79 + 8.37 + \cdots + 225.99$
7. $\sum_{n=1}^{9} 43(0.3)^{n-1}$
8. $\sum_{n=1}^{5} -1\left(\frac{1}{3}\right)^{n-1}$
9. $\sum_{n=1}^{30} 5^{n-1}$
10. $\sum_{n=1}^{22} -2(-3)^{n-1}$
11. Suppose the company telephone tree on page 495 has 12 levels when it is completed. How many employees does the company have?

Find the sum of each series, if the sum exists. If a series does not have a sum, explain why not.

12. $300 + 150 + 75 + 37.5 + \cdots$
13. $6 + 4 + \frac{8}{3} + \frac{16}{9} + \cdots$
14. $-0.18 + 0.36 + (-0.72) + 1.44 + \cdots$
15. $88 + 22 + 5.5 + 1.375 + \cdots$
16. $1 + \left(-\frac{1}{3}\right) + \frac{1}{9} + \left(-\frac{1}{27}\right) + \cdots$
17. $12 + 16 + 21\frac{1}{3} + 28\frac{4}{9} + \cdots$
18. $\sum_{n=1}^{\infty} \frac{1}{2}(3)^{n-1}$
19. $\sum_{n=1}^{\infty} 5(0.2)^{n-1}$
20. $\sum_{n=1}^{\infty} -14\left(-\frac{1}{5}\right)^{n-1}$
21. $\sum_{n=1}^{\infty} \frac{5}{4}(-1)^{n-1}$
22.  **Spreadsheets** Use a spreadsheet to explore the *sequence of partial sums*

$$S_1 = t_1,\quad S_2 = t_1 + t_2,\quad S_3 = t_1 + t_2 + t_3,\quad S_4 = t_1 + t_2 + t_3 + t_4,\quad \ldots$$

for an infinite geometric series $t_1 + t_2 + t_3 + t_4 + \cdots$.

Infinite Geometric Series

C7 | = C6 + B7

	A	B	C
1	First term =	0.5	
2	Common ratio =	0.5	
3			
4	Position number	Sequence	Partial sum
5	1	0.5	0.5
6	2	0.25	0.75
7	3	0.125	0.875

Set up your spreadsheet so that you can easily change the first term and the common ratio.

a. Do the partial sums of the series you investigated in the Exploration on page 497 appear to get closer and closer to any single number? If so, what is the number and how does it compare with the sum of the series?

b. Investigate the sequence of partial sums for three other infinite geometric series where $|r| < 1$. What do you notice?

c. Investigate the sequence of partial sums for three infinite geometric series where $|r| > 1$. What do you notice?

Exercises and Applications

Rounded answers are given to three significant digits.

1. 508
2. 303.752
3. 2,222,222,222,222
4. 9841π
5. 7.9921875
6. 338.83
7. 61.4
8. $-1\frac{40}{81}$
9. 2.33×10^{20}
10. about 1.57×10^{10}
11. 265,720 employees
12. 600
13. 18
14. The sum does not exist because the series is infinite with $r = -2$ and $|-2|$ is not less than 1.
15. $117\frac{1}{3}$
16. 0.75
17. The sum does not exist because the series is infinite with $r = 1\frac{1}{3}$ and $\left|1\frac{1}{3}\right|$ is not less than 1.
18. The sum does not exist because the series is infinite with $r = 3$ and $|3|$ is not less than 1.
19. 6.25
20. $-11\frac{2}{3}$
21. The sum does not exist because the series is infinite with $r = -1$ and $|-1|$ is not less than 1.
22. a. Yes; the partial sums appear to get closer and closer to 1, which is the sum of the series.
b. The partial sums get closer and closer to $\frac{t_1}{1-r}$.
c. The absolute values of the partial sums increase without limit.
23. a. $1 + 0.805 + (0.805)^2 + (0.805)^3$
b. about 2.97
c. M$24.21
24. a. Instead of cutting the paper into halves, cut it into thirds. Put one strip into a pile labeled A and a second into a pile labeled B. Record the area of each piece. Divide the third piece into thirds. Again put one strip in pile A and a second into pile B. Express the combined area of the pieces put

23. ECONOMICS In 1974 the Malaysian Tourist Development Corporation (MTDC) estimated the economic benefit of tourist brochures. The MTDC based its analysis on these assumptions:

- Each person or organization will spend 80.5% of each *ringgit,* or Malaysian dollar (M$), received in payment for goods or services.

- Malaysian dollars remain in circulation in Malaysia.
- Each tourist receiving a brochure about the capital city Kuala Lumpur will spend an additional M$4.72 per person while in Malaysia.

a. Write the first 4 terms of the series that represents the total spending generated from each additional ringgit spent by a tourist. (*Hint:* The person who receives M$1 spends 80.5% of M$1, the person who receives 80.5% of M$1 spends 80.5% of 80.5% of M$1, and so on.)

b. Find the sum of the series you wrote in part (a).

c. What is the total additional spending that the MTDC predicted from each additional tourist brochure?

BY THE WAY

The Southeast Asian nation of Malaysia lies partly on the Malay peninsula and partly on the island of Borneo. In 1992 tourists spent about $1.8 billion in Malaysia.

24. Investigation You can use a procedure like the one described in the Exploration on page 497 to find the sum of any series of the form

$$\sum_{n=1}^{\infty} \frac{1}{c^n} \text{ where } c = 2, 3, 4, \ldots.$$

a. **Writing** Describe how to modify the procedure in the Exploration for finding the sum of this series:

$$\frac{1}{3} + \frac{1}{9} + \frac{1}{27} + \frac{1}{81} + \cdots$$

b. Carry out the procedure you described in part (a) and explain why it shows that the sum of the series is $\frac{1}{2}$.

c. Modify the procedure once again to find $\sum_{n=1}^{\infty} \frac{1}{4^n}$.

d. **Challenge** Make a conjecture about an explicit formula for this sequence: $\sum_{n=1}^{\infty} \frac{1}{2^n}, \sum_{n=1}^{\infty} \frac{1}{3^n}, \sum_{n=1}^{\infty} \frac{1}{4^n}, \ldots$. Explain your reasoning.

Express each repeating decimal as a fraction.

25. 0.888. . .	**26.** 0.181818. . .	**27.** 0.040404. . .
28. 0.3823823823. . .	**29.** 3.717171. . .	**30.** 0.13121312. . .
31. 1.059059. . .	**32.** 0.005005005. . .	**33.** −0.0453453. . .
34. 4.002626. . .	**35.** −10.8787. . .	**36.** 3.0005555. . .

Apply⇔Assess

Exercise Notes

Multicultural Note

Ex. 23 Many tourists are attracted to Kuala Lumpur, the capital and largest city of Malaysia. The National Mosque with its 48 domes is one outstanding site; one dome forms the shape of an 18-pointed star, with the points representing the five Pillars of Islam and the 13 states of Malaysia. Other popular sites include the National Museum; the Lake Gardens, which contain tropical forests and formal gardens; and the Parliament House, where the Malaysia legislature meets.

Alternate Approach

Ex. 24 To verify that the sum of the series is $\frac{1}{2}$, students can rewrite the series shown in part (a) as $\sum_{n=1}^{\infty} \frac{1}{3}\left(\frac{1}{3}\right)^{n-1}$ and then substitute into the formula for the sum of an infinite geometric series. They can then use this technique to solve the challenge problem in part (d).

Teaching Tip

Exs. 25–36 Those students having difficulty with these exercises should refer to Example 3 on page 499.

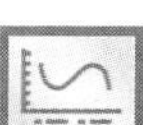

Using Technology

Exs. 25–36 All of these exercises can be done by using a TI-82 graphing calculator. For example, for Ex. 26, enter 0.1818181818181818 on the home screen. (Use about 16 digits after the decimal point.) Then press MATH 1 ENTER. The calculator will display the result 2/11. Challenge students to show how they can modify the procedure to find a fraction for the decimal in Ex. 36.

aside as an indicated sum. Repeat until the pieces are too small to divide into thirds.

b. The combined area of the strips in pile A is $\frac{1}{3} + \frac{1}{9} + \frac{1}{27} + \frac{1}{81} + \ldots$. The combined area of the strips in both piles is 1, so $\sum_{n=1}^{\infty} \left(\frac{1}{3}\right)^n = \frac{1}{2}$.

c. Repeat the process described above, but divide each strip into fourths, place three of the strips into piles labeled A, B, and C and use the fourth as the "leftover" strip. The combined area of the strips in a single pile is $\frac{1}{4} + \frac{1}{16} + \frac{1}{64} + \frac{1}{256} + \ldots$. The combined area of the strips in all three piles is 1, so $\sum_{n=1}^{\infty} \left(\frac{1}{4}\right)^n = \frac{1}{3}$.

d. An explicit formula for the sequence is $t_n = \frac{1}{n}$. The method described in the Exploration on page 497 can be modified to show that for any $k \geq 2$, $\sum_{n=1}^{\infty} \left(\frac{1}{k}\right)^n = \frac{1}{k-1}$.

25. $\frac{8}{9}$ 26. $\frac{2}{11}$

27. $\frac{4}{99}$ 28. $\frac{382}{999}$

29. $\frac{368}{99}$ 30. $\frac{1312}{9999}$

31. $\frac{1058}{999}$ 32. $\frac{5}{999}$

33. $-\frac{151}{3330}$ 34. $\frac{19{,}813}{4950}$

35. $-\frac{359}{33}$ 36. $\frac{5401}{1800}$

Apply⇔Assess

Exercise Notes

Cooperative Learning
Ex. 37 It is usually quite surprising to students that any integer can be expressed as a repeating decimal. Have students go over their descriptions in groups to refine the ideas they are discussing.

Historical Connection
Exs. 38, 39 The concept of fractional dimension, or *fractal*, was developed by Benoit Mandelbrot, a Fellow of the Thomas J. Watson Research Center, IBM Corporation, in order to describe the shapes of natural objects. Mandelbrot first published his ideas on fractal theory in 1975.

Interview Note
Ex. 38 Remind students that if they know a series is geometric, they only need to know a few terms in order to find the pattern $t_n = t_1 \cdot r^{n-1}$. Students should divide two consecutive terms to get the common ratio.

Using Technology
Ex. 38 For each geometric series in this exercise, students can trace the related exponential function $y = t_1 \cdot r^{x-1}$, where x replaces n, graphed under the ZOOM 8:ZInteger feature to get successive terms of the geometric series.

Second-Language Learners
Ex. 39 After students learning English have written their formulas for Exs. 38 and 39(a) independently, they may wish to write their explanations for Ex. 39(b) with the help of an English-proficient partner.

Visual Thinking
Exs. 40–43 Ask students to work in groups to create tree diagrams of real-life or fictional situations. Have the groups explain their diagrams as geometric series. This activity involves the visual skills of *interpretation* and *communication.*

37. a. Express the repeating decimal 0.9999. . . as a fraction. Are you surprised by the result?

b. Writing Use your result in part (a) to explain why any integer can be expressed as a repeating decimal.

INTERVIEW Jhane Barnes

Look back at the article on pages 462–464.

Jhane Barnes likes to use fractals in her fabric designs. The fractal shown below is one of a class of Koch constructions, named for the Swedish mathematician Helge von Koch. It is formed by beginning with a unit square and repeatedly replacing the middle third of each boundary segment with a small square that is missing one side:

Stage 1

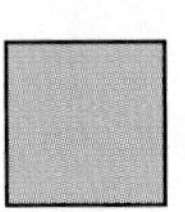

Stage 2

Stage 3

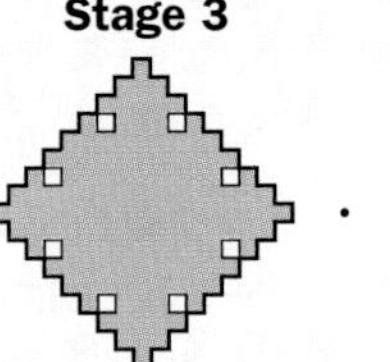

. . .

38. a. Consider the *number* of new squares added at each stage of this Koch construction. Write an explicit formula for the nth term of this sequence. (*Hint:* The sequence is geometric. Express t_n in the form $t_1 \cdot r^{n-1}$.)

b. Consider the *area* of *each* of the new squares added at each stage of this Koch construction. Write an explicit formula for the nth term of this sequence. See the hint for part (a).

c. Consider the *combined area* of *all* the new squares added at each stage of this Koch construction. Write an explicit formula for the nth term of this sequence. See the hint for part (a).

d. Writing Find the total area enclosed by this Koch construction after infinitely many stages. Explain how you found your answer.

39. a. Consider the length added to the perimeter of this Koch construction at each stage. Write an explicit formula for the nth term of this sequence. See the hint for part (a) of Exercise 38.

b. Writing Explain why the perimeter of this Koch construction after infinitely many stages is infinitely long.

"Koch curve snowflake"

37. a. 1; Answers may vary.

b. For every integer n, $n - 1$ is also an integer and n can be written as $(n - 1) + 1 = n - 1 + 0.999\ldots$, a repeating decimal.

38. a. $t_n = 4 \cdot 5^{n-1}$

b. $t_n = \frac{1}{9}\left(\frac{1}{9}\right)^{n-1}$

c. $t_n = \frac{4}{9}\left(\frac{5}{9}\right)^{n-1}$

d. The total area is the original area plus $\sum_{n=1}^{\infty} \frac{4}{9}\left(\frac{5}{9}\right)^{n-1} =$

$1 + \dfrac{\frac{4}{9}}{1 - \frac{5}{9}} = 1 + 1 = 2.$

39. a. $t_n = \frac{8}{3}\left(\frac{5}{3}\right)^{n-1}$

b. The perimeter of the construction after infinitely many stages is the original perimeter plus $\sum_{n=1}^{\infty} \frac{8}{3}\left(\frac{5}{3}\right)^{n-1}$. Since for this series $r = \frac{5}{3}$ and $\left|\frac{5}{3}\right| > 1$, the sum of the series does not exist. The perimeter increases without limit and is infinitely long.

40–43. Answers may vary. Examples are given.

40. $\sum_{n=1}^{10} 2^{n-1} = 1023$ (Every finite series has a sum.)

41. not possible; Every finite series has a sum.

ONGOING ASSESSMENT

Open-ended Problem Make up a geometric series of each type. Write each series in sigma notation. If the series has a sum, tell what the sum is.

40. finite series that has a sum

41. finite series that does not have a sum

42. infinite series that has a sum

43. infinite series that does not have a sum

SPIRAL REVIEW

Write each series in sigma notation and find the sum. *(Section 10.4)*

44. $0.6 + 0.67 + 0.74 + 0.81 + \cdots + 1.58$

45. $183 + 169 + 155 + 141 + \cdots + (-391)$

Solve each system of equations. *(Section 7.2)*

46. $2x - 3y = 9$
$x + y = 2$

47. $4y + 2x = -1$
$3x + 8y = 12$

48. $3x + 6y = -9$
$4x + 8y = -12$

49. $6x - y = 2$
$x + 5y = 16$

Find the slope of the line passing through each pair of points. *(Section 2.2)*

50. (0, 0) and (7, 9)

51. (0, 2) and (5, 1)

52. (3, −2) and (−1, 4)

53. (−1, −5) and (1, 5)

ASSESS YOUR PROGRESS

VOCABULARY

series (p. 488)
arithmetic series (p. 488)
geometric series (p. 495)

For each series:

a. Tell whether it is *arithmetic* or *geometric*.

b. Find the number of terms.

c. Find the sum. *(Sections 10.4 and 10.5)*

1. $640 + 320 + 160 + 80 + \cdots + 5$

2. $\left(-\frac{3}{8}\right) + \left(-\frac{1}{4}\right) + \left(-\frac{1}{8}\right) + 0 + \cdots + \frac{1}{2}$

3. $6 + 30 + 150 + 750 + \cdots + 93{,}750$

Find the sum of each geometric series, if the sum exists. *(Section 10.5)*

4. $\frac{1}{64} + \frac{1}{16} + \frac{1}{4} + 1 + \cdots$

5. $100 + 20 + 4 + 0.8 + \cdots$

Write each series in sigma notation and find the sum, if the sum exists. *(Sections 10.4 and 10.5)*

6. $114 + 97 + 80 + 63 + \cdots + (-5)$

7. $3 + 2 + \frac{4}{3} + \frac{8}{9} + \cdots$

Expand each series and find the sum. *(Sections 10.4 and 10.5)*

8. $\sum_{n=1}^{9} (5n + 1)$

9. $\sum_{n=1}^{7} 0.3(7)^{n-1}$

10. $\sum_{n=1}^{\infty} (0.8)^n$

11. Journal Explain why infinite arithmetic series never have a sum but infinite geometric series sometimes have a sum.

Apply⇔Assess

Exercise Notes

Assessment Note
Exs. 40–43 These are excellent exercises to assess what students know about geometric series. You may wish to call upon various students to write their examples on the board. This will help you and the class to assess how well the concepts of this section have been learned.

Assess Your Progress

Journal Entry
You may want students to share their explanations about why infinite arithmetic series have no sum but infinite geometric series may have a sum.

Progress Check 10.4–10.5

See page 507.

Practice 68 for Section 10.5

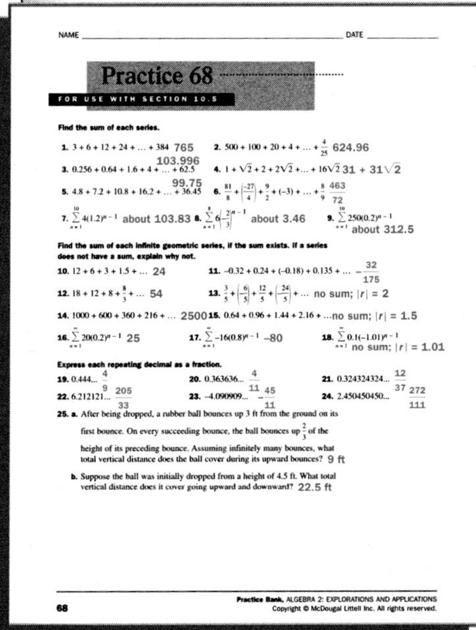
NAME ____________ DATE ________

Practice 68

FOR USE WITH SECTION 10.5

Find the sum of each series.

1. $3 + 6 + 12 + 24 + \ldots + 384$ 765
2. $500 + 100 + 20 + 4 + \ldots + \frac{4}{25}$ 624.96
3. $0.256 + 0.64 + 1.6 + 4 + \ldots + 62.5$ 103.996
4. $1 + \sqrt{2} + 2 + 2\sqrt{2} + \ldots + 16\sqrt{2}$ $31 + 31\sqrt{2}$
5. $4.8 + 7.2 + 10.8 + 16.2 + \ldots + 36.45$ 99.75
6. $\frac{81}{8} + \left(\frac{-27}{4}\right) + \frac{9}{2} + (-3) + \ldots + \frac{8}{9}$ $\frac{463}{72}$
7. $\sum_{n=1}^{10} 4(1.2)^{n-1}$ about 103.83
8. $\sum_{n=1}^{8} 6\left(-\frac{2}{3}\right)^{n-1}$ about 3.46
9. $\sum_{n=1}^{10} 250(0.2)^{n-1}$ about 312.5

Find the sum of each infinite geometric series, if the sum exists. If a series does not have a sum, explain why not.

10. $12 + 6 + 3 + 1.5 + \ldots$ 24
11. $-0.32 + 0.24 + (-0.18) + 0.135 + \ldots$ $-\frac{32}{175}$
12. $18 + 12 + 8 + \frac{8}{3} + \ldots$ 54
13. $\frac{3}{5} + \left(-\frac{6}{5}\right) + \frac{12}{5} + \left(-\frac{24}{5}\right) + \ldots$ no sum; $|r| = 2$
14. $1000 + 600 + 360 + 216 + \ldots$ 2500
15. $0.64 + 0.96 + 1.44 + 2.16 + \ldots$ no sum; $|r| = 1.5$
16. $\sum_{n=1}^{\infty} 20(0.2)^{n-1}$ 25
17. $\sum_{n=1}^{\infty} -16(0.8)^{n-1}$ −80
18. $\sum_{n=1}^{\infty} 0.1(-1.01)^{n-1}$ no sum; $|r| = 1.01$

Express each repeating decimal as a fraction.

19. 0.444... $\frac{4}{9}$
20. 0.363636... $\frac{4}{11}$
21. 0.324324324... $\frac{12}{37}$
22. 6.212121... $\frac{205}{33}$
23. −4.090909... $-\frac{45}{11}$
24. 2.450450450... $\frac{272}{111}$

25. a. After being dropped, a rubber ball bounces up 3 ft from the ground on its first bounce. On every succeeding bounce, the ball bounces up $\frac{2}{3}$ of the height of its preceding bounce. Assuming infinitely many bounces, what total vertical distance does the ball cover during its upward bounces? 9 ft

b. Suppose the ball was initially dropped from a height of 4.5 ft. What total vertical distance does it cover going upward and downward? 22.5 ft

68

Practice Bank, ALGEBRA 2: EXPLORATIONS AND APPLICATIONS

---

42. $\sum_{n=1}^{\infty} \left(\frac{1}{2}\right)^{n-1} = 2$

43. $\sum_{n=1}^{\infty} 2^{n-1}$

44. $\sum_{n=1}^{15} (0.07n + 0.53) = 16.35$

45. $\sum_{n=1}^{42} (-14n + 197) = -4368$

46. $(3, -1)$

47. $\left(-14, 6\frac{3}{4}\right)$

48. all ordered pairs (x, y) such that $x + 2y = -3$

49. $\left(\frac{26}{31}, \frac{94}{31}\right)$ 50. $\frac{9}{7}$

51. $-\frac{1}{5}$ 52. $-\frac{3}{2}$ 53. 5

Assess Your Progress

1. a. geometric b. 8 terms
c. 1275

2. a. arithmetic b. 8 terms
c. $\frac{1}{2}$

3. a. geometric
b. 7 terms
c. 117,186

4. does not exist

5. 125

6. $\sum_{n=1}^{8} (-17n + 131)$; 436

7. $\sum_{n=1}^{\infty} 3\left(\frac{2}{3}\right)^{n-1}$; 9

8. $6 + 11 + 16 + 21 + 26 + 31 + 36 + 41 + 46$; 234

9. $0.3 + 2.1 + 14.7 + 102.9 + 720.3 + 5042.1 + 35{,}294.7$; 41,177.1

10. $0.8 + 0.64 + 0.512 + 0.4096 + \ldots$; 4

11. Answers may vary. An example is given. Infinite arithmetic series never have a sum because the absolute value of each term of the sequence is greater than the one before, so the sum of the terms continues to grow, in absolute value, without limit. A geometric series has a sum if the absolute value of the common ratio of the sequence is less than one. The terms of the sequence get closer and closer to 0 as n gets larger, so eventually adding terms has no effect and the series has a sum.

Project Notes

Mathematical Goals

- Gather data on college costs.
- Organize the data in a spreadsheet.
- Analyze the data using different scenarios of saving for college.
- Compare the scenarios and discuss their feasibility.

Planning

Materials

- sources to find data on college costs
- spreadsheet software

Project Teams

Students can select the members of their group and can work together on all aspects of the project, including gathering the data, deciding how to analyze the data, completing the calculations, and writing the report.

Guiding Students' Work

If students decide to select data from different types of colleges, make sure that they do not mix the data selections and compare the colleges or universities incorrectly. If students do not have access to sources for data, you can provide the following information for them.

Average College Costs		
Year	Public Institutions	Private Institutions
1979–80	$2,165	$4,912
1980–81	$2,373	$5,470
1981–82	$2,663	$6,166
1982–83	$2,945	$6,920
1983–84	$3,156	$7,508
1984–85	$3,408	$8,202
1985–86	$3,571	$8,885
1986–87	$3,805	$9,676
1987–88	$4,050	$10,512
1988–89	$4,274	$11,189
1989–90	$4,504	$12,018
1990–91	$4,757	$12,910
1991–92	$5,135	$13,907
1992–93	$5,379	$14,634
1993–94	$5,695	$15,532

Second-Language Learners

Students learning English will benefit from working with a peer tutor or in small groups to create the spreadsheets and write their reports for this project.

PORTFOLIO PROJECT

The Future Is Now

A college education can be expensive, and the cost continues to rise as time passes. Planning ahead makes the task of providing the money needed to pay for college much easier. Paying for a college education can be difficult if you wait too long before starting a savings program.

Based on today's costs, you can estimate the cost of a college education at some point in the future. You can then explore various savings and investment plans to see how they can help pay for college.

PROJECT GOAL **Estimate the future cost of a college education, and test the effectiveness of several plans for saving this amount of money.**

Collecting the Data

Suppose you had 20 years to save enough money to pay for a college education. How could you determine the best method for achieving this goal?

Work in groups of three to collect the data necessary to develop good models. Each group member should do research in one of these areas:

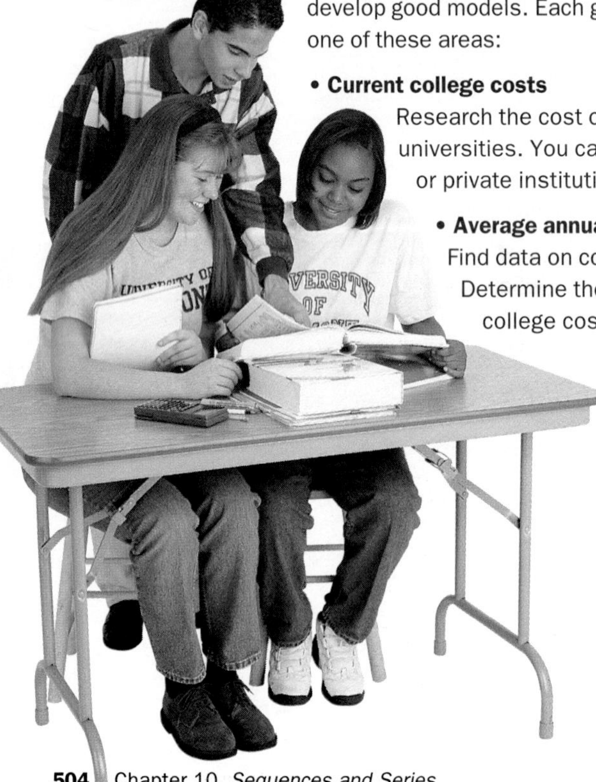

- **Current college costs**
  Research the cost of going to various 4-year colleges and universities. You can use either an average cost for all public or private institutions or the cost for a college of your choice.
- **Average annual percent increase in college costs**
  Find data on college costs over the past 10–15 years. Determine the average annual percent increase in total college costs.
- **Interest rates for various investments**
  Find out the average annual rate of return for various investments, such as savings accounts, certificates of deposit, and mutual funds.

Two good sources for college data are the *Digest of Educational Statistics* and the *Statistical Abstract of the United States*. For investment data, try calling a bank or brokerage firm.

Projecting Costs and Savings

Use a spreadsheet to project future college costs and to explore different methods of saving for college. For example, the spreadsheet below shows college costs growing at 5.9% per year. It also shows a savings plan in which annual deposits of $1200 earn 10% annual interest. Will this plan yield enough savings to pay for college in 20 years?

College Planning

	A	B	C
1	Year	College cost	Savings
2	1	24876	1200
3	2	B2∗1.059	C2∗1.1+120

Enter the **amount of money saved annually** in the first cell of this column. Use a recursive formula to calculate the effect of annual savings and deposits.

Enter the **current college cost** in the first cell of this column. Use a recursive formula to calculate costs for future years.

Use a spreadsheet to investigate scenarios like the one above, as well as other savings and investment programs, including those that involve:

- making only a large initial investment
- increasing your annual investment each year
- waiting until the final few years before starting to save

You can extend your project by investigating the cost differences between public and private institutions.

Writing a Report

Write a report summarizing your results. Include the following:

- all your data, estimates, and sources
- a description of each scenario you investigated and a printout of each spreadsheet you created
- a discussion of the feasibility of each scenario and which one makes the most sense for most people

Self-Assessment

What do you think are the limitations of the models you used in this project? Did you have any trouble setting up your spreadsheet models? What insights did you gain from this project?

Project Notes

Guiding Students' Work

Rubric for Chapter Project

4 Students collect data on college costs for various colleges and universities and use a spreadsheet correctly to project future costs and savings. Students investigate several scenarios for saving and extend the project by comparing both private and public institutions. Students complete a report and convey their data and findings clearly and concisely.

3 Students collect data about various colleges and universities and use a spreadsheet to project future costs and savings. Some of the calculations are incorrect, however, and, therefore, the analysis is somewhat inaccurate. Students write a report that contains all of the data and information used in the project. The report is well written and complete.

2 Students collect college cost data and attempt to analyze the data in a spreadsheet but make mistakes both in their calculations and in the analysis of the results. Students only look at a few ways to save for college and do not explore other possibilities for comparison. Students write a report, but it is poorly organized and does not contain a good discussion of the feasibility of the scenarios.

1 Students gather some cost data on colleges and universities but do not correctly analyze it in a spreadsheet. An analysis of the data is not completed and only one, if any, scenario is investigated. If a report is written, it is incomplete and does not indicate that any effort was made to complete the project. Students should be encouraged to speak with the teacher as soon as possible to review their work and to make a new start on the project.

Chapter Support

Course Guide: Chapter 10
Lesson Plans: Chapter 10
Practice Bank:
Cumulative Practice 69
Study Guide: Chapter 10 Review
Assessment Book:
Chapter Tests 42 and 43
Chapter 10 Alternative Assessment
Test Generator Software
Technology Book:
Calculator Based Lab 5
Portfolio Project Book:
Additional Project 4: Music Theory
Preparing for College Entrance Tests
Professional Handbook

Progress Check 10.1–10.3

1. Find the next four terms of the sequence 1, –2, 4, –8, ... *(Section 10.1)*
16, –32, 64, –128
2. Tell whether the sequence of days on which there is a full moon in a year is finite or infinite. *(Section 10.1)* finite
3. Find the 4th and 9th terms of the sequence $t_n = 5^{n-2}$. *(Section 10.1)* 25; 78,125

Tell whether each sequence is *arithmetic*, *geometric*, or *neither*. Write a formula for each sequence. *(Section 10.2)*

4. 2, 11, 101, 1001, ...
neither; $t_n = 10^{n-1} + 1$
5. –5, –9, –13, –17, ...
arithmetic; $t_n = -4n - 1$
6. $-\frac{1}{4}, \frac{3}{4}, -\frac{9}{4}, \frac{27}{4}, \ldots$ geometric; $t_n = -\frac{1}{4}(-3)^{n-1}$
7. Find the arithmetic mean and the geometric mean of 12 and 18. *(Section 10.2)*
15; $\sqrt{216} = 6\sqrt{6}$
8. Find the first 4 terms of the sequence.
$t_1 = -3$
$t_n = t_{n-1} - 8$ *(Section 10.3)*
–3, –11, –19, –27
9. Write a recursive formula for the sequence 18, 9, 4.5, 2.25, ... *(Section 10.3)*
$t_1 = 18$; $t_n = \frac{1}{2}(t_{n-1})$

CHAPTER

10 Review

STUDY TECHNIQUE

Taking a practice test is a good way to determine how ready you are for an actual test. One way to prepare for a test is to write one yourself. Writing a test requires you to review the chapter and become familiar with it. Create a test for this chapter, then exchange it with a classmate. After taking the tests, correct and discuss them. You should study any topics that gave you difficulty.

VOCABULARY

SECTIONS 10.1 *and* 10.2

A **sequence** is an ordered list of numbers or figures, called **terms**.

Arithmetic sequence

$t_n = t_1 + d(n-1)$

nth term — first term — common difference

An **explicit formula** gives t_n in terms of n.

Geometric sequence

$t_n = t_1 \cdot r^{n-1}$

nth term — first term — common ratio

Examples of arithmetic and geometric sequences are given below.

A sequence is **finite** if it has a last term, and it is **infinite** if it continues without end.

Sequence	50, 47, 44, 41, . . .	1000, 100, 10, 1, . . . , 0.00000001
Type	infinite arithmetic	finite geometric
Explicit formula	$t_n = 50 + (-3)(n-1)$	$t_n = 1000(0.1)^{n-1}$
10th term	$t_{10} = 50 + (-3)(9) = 23$	$t_{10} = 1000(0.1)^9 = 0.000001$

Suppose a, x, and b are consecutive terms of a sequence.

- If the sequence is arithmetic, then $x = \frac{a+b}{2}$.

  Example: 50, 47, 44, 41, 38, 35, . . . ⟵ $41 = \frac{44+38}{2}$

- If the sequence is geometric, then $x = \sqrt{ab}$.

  Example: 1000, 100, 10, 1, 0.1, 0.01, . . . ⟵ $1 = \sqrt{10 \cdot 0.1}$

x is the **arithmetic mean** of a and b.

x is the **geometric mean** of a and b.

SECTION 10.3

Recursion is the process of obtaining each term of a sequence from the previous one by repeatedly performing the same operation(s). You can find **recursive formulas** for arithmetic and geometric sequences.

Arithmetic sequence	Geometric sequence
t_1 = starting value	**t_1 = starting value**
$t_n = t_{n-1} + d$	$t_n = r(t_{n-1})$
Example: The sequence 50, 47, 44, 41, . . . is defined recursively as $t_1 = 50$ and $t_n = t_{n-1} - 3$.	Example: The sequence 1000, 100, 10, 1, . . . is defined recursively as $t_1 = 1000$ and $t_n = 0.1(t_{n-1})$.

SECTIONS 10.4 *and* 10.5

When the terms of a sequence are added, as in $t_1 + t_2 + \cdots + t_n$, the indicated sum is called a **series**. A series in *expanded form* can be expressed more compactly by using *sigma notation*. You can use formulas to find the sum S_n of a finite arithmetic series and a finite geometric series, each with n terms.

Arithmetic series	Geometric series ($r \neq 1$)
$S_n = \frac{n}{2}(t_1 + t_n)$	$S_n = \frac{t_1(1 - r^n)}{1 - r}$
Example: $9 + 22 + 35 + 48 + \cdots + 152$	Example: $1 + 2 + 4 + 8 + \cdots + 256$
$= \sum_{n=1}^{12} [9 + 13(n - 1)]$	$= \sum_{n=1}^{9} 1(2^{n-1})$
$= \frac{12}{2}(9 + 152) = 966$	$= \frac{1(1 - 2^9)}{1 - 2} = 511$

If a geometric series is infinite and $|r| < 1$, then the sum is $S = \frac{t_1}{1 - r}$.

Example: $32 + 16 + 8 + 4 + \cdots = \sum_{n=1}^{\infty} 32(0.5)^{n-1} = \frac{32}{1 - 0.5} = 64$

Progress Check 10.4–10.5

Find the sum of each arithmetic series. *(Section 10.4)*

1. $72 + 76 + 80 + 84 + \ldots + 116$
1128

2. $20 + 11 + 2 + (-7) + \ldots + (-97)$
−539

3. The sum of the following finite arithmetic series is −28. Find the last term. *(Section 10.4)*
$7 + 4 + 1 + (-2) + \ldots +$ __?__
−14

4. Write the arithmetic series $30.5 + 29 + 27.5 + 26 + \ldots + 18.5$ in sigma notation. *(Section 10.4)*
$\sum_{n=1}^{9} (-1.5n + 32)$

Find the sum of each geometric series. *(Section 10.5)*

5. $3 + 12 + 48 + 192 + \ldots + 49{,}152$ 65,535

6. $\sum_{n=1}^{8} 6^{n-1}$ 335,923

Find the sum of each infinite geometric series, if the sum exists. If the series does not have a sum, explain why not. *(Section 10.5)*

7. $1 + \left(-\frac{1}{2}\right) + \frac{1}{4} + \left(-\frac{1}{8}\right) + \ldots$ $\frac{2}{3}$

8. $\sum_{n=1}^{\infty} \frac{2}{3}(-1)^{n-1}$
There is no sum because $|r| = 1$.

Chapter 10 Assessment
Form A Chapter Test

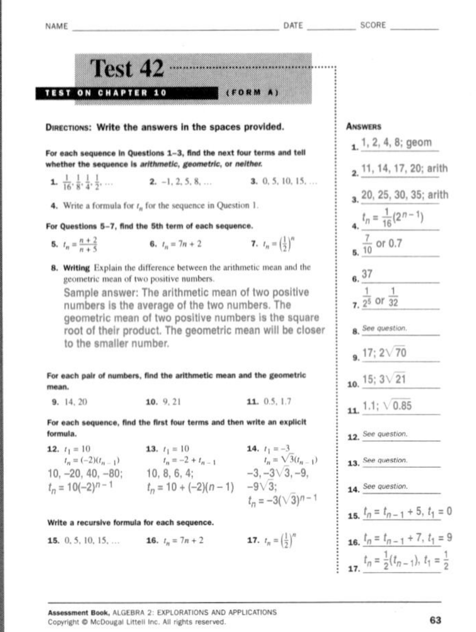

NAME ______ DATE ______ SCORE ______

Test 42

TEST ON CHAPTER 10 (FORM A)

DIRECTIONS: Write the answers in the spaces provided.

For each sequence in Questions 1–3, find the next four terms and tell whether the sequence is *arithmetic, geometric,* or *neither.*

1. $\frac{1}{16}, \frac{1}{8}, \frac{1}{4}, \frac{1}{2}, \ldots$ 2. $-1, 2, 5, 8, \ldots$ 3. $0, 5, 10, 15, \ldots$

4. Write a formula for t_n for the sequence in Question 1.

For Questions 5–7, find the 5th term of each sequence.

5. $t_n = \frac{n+2}{n+5}$ 6. $t_n = 7n + 2$ 7. $t_n = \left(\frac{1}{2}\right)^n$

8. **Writing** Explain the difference between the arithmetic mean and the geometric mean of two positive numbers.
Sample answer: The arithmetic mean of two positive numbers is the average of the two numbers. The geometric mean of two positive numbers is the square root of their product. The geometric mean will be closer to the smaller number.

For each pair of numbers, find the arithmetic mean and the geometric mean.

9. 14, 20 10. 9, 21 11. 0.5, 1.7

For each sequence, find the first four terms and then write an explicit formula.

12. $t_1 = 10$, $t_n = (-2)(t_{n-1})$
10, −20, 40, −80; $t_n = 10(-2)^{n-1}$

13. $t_1 = 10$, $t_n = -2 + t_{n-1}$
10, 8, 6, 4; $t_n = 10 + (-2)(n-1)$

14. $t_1 = -3$, $t_n = \sqrt{3}(t_{n-1})$
$-3, -3\sqrt{3}, -9, -9\sqrt{3}$; $t_n = -3(\sqrt{3})^{n-1}$

Write a recursive formula for each sequence.

15. $0, 5, 10, 15, \ldots$ 16. $t_n = 7n + 2$ 17. $t_n = \left(\frac{1}{2}\right)^n$

ANSWERS

1. 1, 2, 4, 8; geom
2. 11, 14, 17, 20; arith
3. 20, 25, 30, 35; arith
4. $t_n = \frac{1}{16}(2^{n-1})$
5. $\frac{7}{10}$ or 0.7
6. 37
7. $\frac{1}{2^5}$ or $\frac{1}{32}$
8. See question.
9. $17; 2\sqrt{70}$
10. $15; 3\sqrt{21}$
11. $1.1; \sqrt{0.85}$
12. See question.
13. See question.
14. See question.
15. $t_n = t_{n-1} + 5, t_1 = 0$
16. $t_n = t_{n-1} + 7, t_1 = 9$
17. $t_n = \frac{1}{2}(t_{n-1}), t_1 = \frac{1}{2}$

Assessment Book, ALGEBRA 2: EXPLORATIONS AND APPLICATIONS

63

Chapter 10 Assessment
Form B Chapter Test

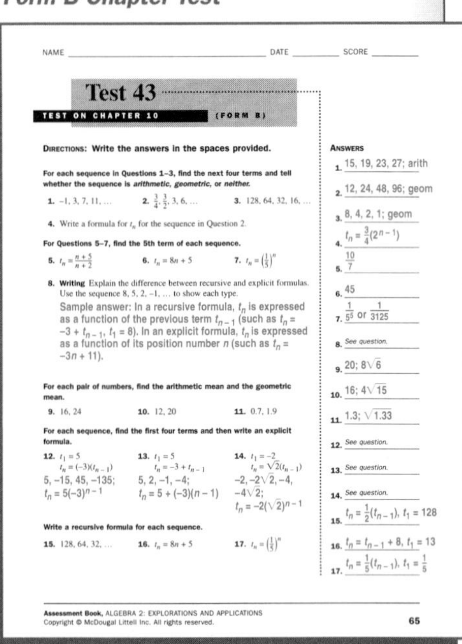

NAME ______ DATE ______ SCORE ______

Test 43

TEST ON CHAPTER 10 (FORM B)

DIRECTIONS: Write the answers in the spaces provided.

For each sequence in Questions 1–3, find the next four terms and tell whether the sequence is *arithmetic, geometric,* or *neither.*

1. $-1, 3, 7, 11, \ldots$ 2. $\frac{3}{4}, \frac{3}{2}, 3, 6, \ldots$ 3. $128, 64, 32, 16, \ldots$

4. Write a formula for t_n for the sequence in Question 2.

For Questions 5–7, find the 5th term of each sequence.

5. $t_n = \frac{n+5}{n+2}$ 6. $t_n = 8n + 5$ 7. $t_n = \left(\frac{1}{5}\right)^n$

8. **Writing** Explain the difference between recursive and explicit formulas. Use the sequence 8, 5, 2, −1, … to show each type.
Sample answer: In a recursive formula, t_n is expressed as a function of the previous term t_{n-1} (such as $t_n = -3 + t_{n-1}, t_1 = 8$). In an explicit formula, t_n is expressed as a function of its position number n (such as $t_n = -3n + 11$).

For each pair of numbers, find the arithmetic mean and the geometric mean.

9. 16, 24 10. 12, 20 11. 0.7, 1.9

For each sequence, find the first four terms and then write an explicit formula.

12. $t_1 = 5$, $t_n = (-3)(t_{n-1})$
5, −15, 45, −135; $t_n = 5(-3)^{n-1}$

13. $t_1 = 5$, $t_n = -3 + t_{n-1}$
5, 2, −1, −4; $t_n = 5 + (-3)(n-1)$

14. $t_1 = -2$, $t_n = \sqrt{2}(t_{n-1})$
$-2, -2\sqrt{2}, -4, -4\sqrt{2}$; $t_n = -2(\sqrt{2})^{n-1}$

Write a recursive formula for each sequence.

15. $128, 64, 32, \ldots$ 16. $t_n = 8n + 5$ 17. $t_n = \left(\frac{1}{5}\right)^n$

ANSWERS

1. 15, 19, 23, 27; arith
2. 12, 24, 48, 96; geom
3. 8, 4, 2, 1; geom
4. $t_n = \frac{3}{4}(2^{n-1})$
5. $\frac{10}{7}$
6. 45
7. $\frac{1}{5^5}$ or $\frac{1}{3125}$
8. See question.
9. $20; 8\sqrt{6}$
10. $16; 4\sqrt{15}$
11. $1.3; \sqrt{1.33}$
12. See question.
13. See question.
14. See question.
15. $t_n = \frac{1}{2}(t_{n-1}), t_1 = 128$
16. $t_n = t_{n-1} + 8, t_1 = 13$
17. $t_n = \frac{1}{5}(t_{n-1}), t_1 = \frac{1}{5}$

Assessment Book, ALGEBRA 2: EXPLORATIONS AND APPLICATIONS

65

CHAPTER

10 Assessment

VOCABULARY QUESTIONS

For Questions 1 and 2, complete each paragraph.

1. When you divide any term by its preceding term in a(n) _?_ sequence, the quotient is a constant called the _?_. The term between any two terms in a(n) _?_ sequence is the average, or _?_, of the given terms.
2. If a sequence has a last term, it is a(n) _?_ sequence. When you add the terms of a sequence, the indicated sum is called a(n) _?_.

SECTIONS 10.1 *and* 10.2

For each sequence, find the next 4 terms and tell whether the sequence is *arithmetic, geometric,* or *neither.*

3. 1, 8, 27, 64, . . .
4. 1, −1, 1, −1, . . .
5. 7, 2, −3, −8, . . .

6–8. Write a formula for t_n for each sequence in Questions 3–5.

Find the 12th term of each sequence.

9. $t_n = \frac{n}{n+1}$
10. $t_n = \left(\frac{1}{3}\right)^n$
11. $t_n = 5n - 2$

For each pair of numbers:

a. Find the arithmetic mean. **b. Find the geometric mean.**

12. 16, 25
13. 7, 11
14. 0.3, 1.2

SECTION 10.3

For each sequence:

a. Find the first 4 terms. **b. Write an explicit formula.**

15. $t_1 = 100$, $t_n = -0.1(t_{n-1})$
16. $t_1 = 100$, $t_n = -0.1 + t_{n-1}$
17. $t_1 = 3$, $t_n = \sqrt{2}(t_{n-1})$

Write a recursive formula for the sequence given in each question.

18. Question 5
19. Question 10
20. Question 11
21. **GENEALOGY** Assume that a person has two parents, four grandparents, and so on. Write an explicit formula and a recursive formula for the number of ancestors the person has when you go back n generations. Is the sequence *arithmetic, geometric,* or *neither*? Explain.

ANSWERS Chapter 10

Assessment

1. geometric; common ratio; arithmetic; arithmetic mean
2. finite; series
3. 125, 216, 343, 512; neither
4. 1, −1, 1, −1; geometric
5. −13, −18, −23, −28; arithmetic
6. $t_n = n^3$
7. $t_n = (-1)^{n-1}$
8. $t_n = 7 - 5(n-1)$ or $t_n = -5n + 12$
9. $\frac{12}{13}$
10. $\frac{1}{531{,}441}$
11. 58
12. a. 20.5
    b. 20
13. a. 9
    b. $\sqrt{77}$
14. a. 0.75
    b. 0.6
15. a. 100, −10, 1, −0.1
    b. $t_n = 100(-0.1)^{n-1}$
16. a. 100, 99.9, 99.8, 99.7
    b. $t_n = 100 - 0.1(n-1)$ or $t_n = -0.1n + 100.1$
17. a. $3, 3\sqrt{2}, 6, 6\sqrt{2}$
    b. $t_n = 3(\sqrt{2})^{n-1}$
18. $t_1 = 7; t_n = t_{n-1} - 5$
19. $t_1 = \frac{1}{3}; t_n = \frac{1}{3}(t_{n-1})$
20. $t_1 = 3; t_n = t_{n-1} + 5$
21. $t_n = 2^n; t_1 = 2, t_n = 2(t_{n-1})$; geometric; Each term is twice the preceding term.

22. **Writing** Explain how to find the 20th term when given a recursive formula for an arithmetic sequence and for a geometric sequence.

SECTIONS 10.4 *and* 10.5

For each series:

a. Tell whether it is *arithmetic* or *geometric*.

b. Find the number of terms.

c. Write the series in sigma notation.

d. Find the sum.

23. $20 + 12 + 4 + \cdots + (-100)$

24. $2 + 10 + 50 + \cdots + 156{,}250$

25. $8 + (-4) + 2 + \cdots + \left(-\frac{1}{4}\right)$

26. $7 + 7.5 + 8 + \cdots + 15$

27. ARCHITECTURE A concert hall has seating for 800 people. There are 21 seats in the first row. Each of the other rows has two more seats than the row in front of it. How many rows are in the concert hall? How many seats are in the last row?

28. Expand each series and find the sum.

a. $\sum_{n=1}^{5} \frac{1}{2}n^3$ **b.** $\sum_{n=1}^{\infty} 1000\left(-\frac{1}{4}\right)^{n-1}$

29. Open-ended Problem Write a geometric series that does not have a sum. Explain why the series has no sum.

30. Express 0.0175175175. . . as a fraction.

PERFORMANCE TASK

31. The Tower of Hanoi puzzle involves moving disks from one post to another. The puzzle begins with a group of n disks of increasing diameter (from top to bottom) on one of three wooden posts. The goal of the puzzle is to transfer all the disks from this starting post to one of the other two posts following these rules:

(1) Only one disk can be moved at a time.

(2) No disk can be placed on top of a smaller disk.

Solve the puzzle with various numbers of disks. (You may wish to use coins of different sizes instead of disks, and three locations A, B, and C on a sheet of paper to represent the three posts.) Make a table relating n, the number of disks, to M_n, the *minimum* number of moves needed to transfer all the disks to another post. Then find an explicit formula for M_n. (*Hint:* Try applying a recursive procedure to moving the disks.)

Chapter 10 Assessment
Form C Alternative Assessment

Chapter 10
ALTERNATIVE ASSESSMENT

1. **Open-ended Problem** Make up your own patterned sequence of numbers. Give your sequence to a partner. Your partner should write a description of the pattern in your sequence, and then predict more terms for your sequence. Would graphing the numbers on a number line help a person predict the rule for a sequence?

2. **Performance Task** Is it possible to find a sequence of numbers that has an explicit rule, but does not have a recursive rule? Is it possible to find a sequence of numbers that has a recursive rule, but does not have an explicit rule? Support your answers with examples.

3. **Research Project** Find out how sequences can be represented using a web graph. The manual for your graphing calculator would be a good place to find information about this type of graph. Describe the process for creating a web graph. Create a web graph from a sequence of numbers. Interpret your graph.

4. **Open-ended Problem** Find a real-world example of an arithmetic sequence. State the common difference and list the first five terms of your sequence.

5. **Open-ended Problem** Find a real-world example of a geometric sequence. State the common ratio and list the first five terms of your sequence.

6. **Group Activity** Build a model of a structure (like a pyramid) with sugar cubes. Complete at least ten layers. Use what you know about sequences to predict the number of cubes in the one-hundredth layer from the top. Find the total number of cubes needed to enlarge your structure so it has 100 layers.

22. The procedure is the same for either an arithmetic sequence or geometric sequence. Start with the given 1st term. Substitute it into the formula for t_{n-1} to find the 2nd term. Substitute the 2nd term into the formula for t_{n-1} to find the 3rd term. Continue this process until you reach the 20th term.

23. a. arithmetic
 b. 16 terms
 c. $\sum_{n=1}^{16} (-8n + 28)$
 d. −640

24. a. geometric
 b. 8 terms
 c. $\sum_{n=1}^{8} 2 \cdot 5^{n-1}$
 d. 195,312

25. a. geometric
 b. 6 terms
 c. $\sum_{n=1}^{6} 8\left(-\frac{1}{2}\right)^{n-1}$
 d. $5\frac{1}{4}$

26. a. arithmetic
 b. 17 terms
 c. $\sum_{n=1}^{17} (0.5n + 6.5)$
 d. 187

27. 20 rows; 59 seats

28. a. $\frac{1}{2} + 4 + \frac{27}{2} + 32 + \frac{125}{2}$; $112\frac{1}{2}$
 b. $1000 - 250 + 62.5 - 15.625 + \ldots$; 800

29. Answers may vary. An example is given. $\sum_{n=1}^{\infty} 2\left(-\frac{4}{3}\right)^{n-1}$ does not have a sum because the series is infinite with $r = -\frac{4}{3}$ and $\left|-\frac{4}{3}\right|$ is not less than 1.

30. $\frac{35}{1998}$

31. $M_n = 2^n - 1$

CHAPTER

11 Analytic Geometry

OVERVIEW

Connecting to Prior and Future Learning

⇔ In Chapter 11, students expand their study of geometry from previous mathematics courses to include analytic geometry. Section 11.1 reviews the basic concepts of finding the distance between two points and finding the midpoint of a line segment.

⇔ Properties and equations of parabolas, circles, ellipses, and hyperbolas are discussed in Sections 11.2 to 11.5. Students will use these concepts in the future as they continue to study mathematics.

⇔ The chapter closes with an introduction to identifying conics. This topic will be covered in detail in more advanced mathematics courses.

Chapter Highlights

Interview with Kija Kim: This interview highlights the relationship between mathematics and mapmaking, with related exercises on pages 517 and 532.

Explorations in Chapter 11 begin in Section 11.1, with students using graph paper and a ruler to find distances and midpoints and using graph paper, a protractor, and a straightedge to explore slopes of perpendicular lines. In Section 11.2, students use focus-directrix paper to draw a parabola. Students use graph paper, string, push pins, and cardboard to draw an ellipse in Section 11.4.

The Portfolio Project: Students use a graphing calculator or graphing software to design a logo composed of conic sections.

Technology: In this chapter, graphing calculators are used to graph the equations of parabolas, circles, ellipses, and hyperbolas.

OBJECTIVES

Section	Objectives	NCTM Standards
11.1	• Find the distance between two points in a coordinate plane. • Find the coordinates of the midpoint of a line segment in a coordinate plane. • Use the distance formula to solve real-world problems.	1, 2, 3, 4, 5, 8
11.2	• Find the focus and directrix of a parabola. • Write an equation of a parabola. • Use parabolas to solve real-world problems.	1, 2, 3, 4, 5, 8
11.3	• Write an equation of a circle. • Graph an equation of a circle. • Locate points on a circle.	1, 2, 3, 4, 5, 8
11.4	• Write an equation of an ellipse. • Graph an equation of an ellipse. • Describe an elliptical object.	1, 2, 3, 4, 5, 8
11.5	• Write an equation of a hyperbola. • Graph an equation of a hyperbola. • Describe an object with a hyperbolic shape.	1, 2, 3, 4, 5, 8
11.6	• Find conics by taking a cross section of a double cone. • Describe how a conic section can be used in applications.	1, 2, 3, 4, 5, 8

INTEGRATION

Mathematical Connections	11.1	11.2	11.3	11.4	11.5	11.6
algebra	**513–519***	**520–527**	**528–534**	**535–542**	**543–549**	**550–553**
geometry	**513–519**	**520–527**	**528–534**	**535–542**	**543–549**	**550–553**
patterns and functions			534		549	
logic and reasoning	513, 517–519	521, 526, 527	528, 531–534	535, 540–542	543, 547–549	550, 552, 553
Interdisciplinary Connections and Applications						
history and geography			534			
reading and language arts	518					552
arts and entertainment				540		
architecture	514			539		
astronomy		523, 525				
aviation						552
flag making, acoustics, communications, medicine, carpentry, navigation	518	526	533	541	548	

****Bold page numbers** indicate that a topic is used throughout the section.*

TECHNOLOGY

	opportunities for use with	
Section	Student Book	Support Material
11.1	graphing calculator	**Technology Book:** Spreadsheet Activity 11 **Geometry Inventor Activity Book:** Activity 15
11.2	graphing calculator McDougal Littell Mathpack *Function Investigator*	
11.3	graphing calculator	**Technology Book:** TI-92 Activity 5 **Geometry Inventor Activity Book:** Activity 19
11.4	graphing calculator	**Technology Book:** Spreadsheet Activity 11
11.5	graphing calculator	**Technology Book:** Calculator Activity 11
11.6	graphing calculator	

PLANNING GUIDE

Regular Scheduling (45 min)

Section	Materials Needed	Core Assignment	Extended Assignment	*exercises that feature* Applications	Communication	Technology
11.1	metric ruler, centimeter graph paper, protractor, graph paper	**Day 1:** 1–21 **Day 2:** 25, 26, 29, 33–40	**Day 1:** 1–19 odd, 20–25 **Day 2:** 26–40	22–24 27, 28	27	
11.2	focus directrix paper, graph paper	**Day 1:** 1–20 **Day 2:** 21–26, 30–35, AYP*	**Day 1:** 1–20 **Day 2:** 21–35, AYP	11–13 27, 28	1 26	14
11.3	graph paper	**Day 1:** 1–21 **Day 2:** 24, 27–30, 33–35, 40–46	**Day 1:** 1–9 odd, 10–14, 15–23 odd **Day 2:** 24–46	10–14 25, 26, 37–39	31, 40	38
11.4	graph paper, 10 in. piece of string, push pins, piece of cardboard	1–9, 20–31, AYP	1–9 odd, 10–31, AYP	10–18, 24	19	
11.5	bifocal conic paper, graph paper	1–9, 11–14, 18–25, 28–34	1–9 odd, 10–34	15–17	18–20, 28	17
11.6	graph paper, flashlight	1–6, 12–23, AYP	1–23, AYP	7–11	18	
Review/ Assess		**Day 1:** 1–26 **Day 2:** Ch. 11 Test	**Day 1:** 1–26 **Day 2:** Ch. 11 Test			
Portfolio Project		Allow 2 days.	Allow 2 days.			

Yearly Pacing (with Portfolio Project)	Chapter 11 Total	Chapters 1–11 Total	Remaining	Total
	13 days	149 days	11 days	160 days

Block Scheduling (90 min)

	Day 69	Day 70	Day 71	Day 72	Day 73	Day 74	Day 75	Day 76
Teach/Interact	Ch. 10 Test 11.1: Exploration, page 513	Continue with 11.1: Exploration, page 515	11.2: Exploration, page 520	11.3	11.4: Exploration, page 535 11.5	11.6 Review	Review Port. Proj.	Ch. 11 Test Port. Proj.
Apply/Assess	**Ch. 10 Test** **11.1:** 1–13	**11.1:** 14–21, 25, 26, 29, 33–40	**11.2:** 1–26, 30–35, AYP*	**11.3:** 1–21, 24, 27–30, 33–35, 40–46	**11.4:** 1–9, 20–31, AYP **11.5:** 1–9, 11–14, 18–25, 28–34	**11.6:** 1–6, 12–23, AYP **Review:** 1–15	**Review:** 16–26 **Port. Proj.**	**Ch. 11 Test** **Port. Proj.**

NOTE: A one-day block has been added for the Portfolio Project—timing and placement to be determined by teacher.

Yearly Pacing (with Portfolio Project)	Chapter 11 Total	Chapters 1–11 Total	Remaining	Total
	$7\frac{1}{2}$ days	76 days	6 days	82 days

****AYP*** *is Assess Your Progress.*

Lesson Support

Section	Practice Bank	Study Guide*	Assessment Book*	Visuals	Explorations Lab Manual	Lesson Plans	Technology Book
11.1	70	11.1		Warm-Up 11.1 Folder A	Masters 1, 2, 14	11.1	Spreadsheet Act. 11
11.2	71	11.2	Test 44	Warm-Up 11.2 Folder 11	Masters 2, 15	11.2	
11.3	72	11.3		Warm-Up 11.3 Folder 11	Master 2	11.3	TI-92 Act. 5
11.4	73	11.4	Test 45	Warm-Up 11.4 Folder 11	Master 2	11.4	Spreadsheet Act. 11
11.5	74	11.5		Warm-Up 11.5 Folder 11	Masters 2, 16	11.5	Calculator Act. 11
11.6	75	11.6	Test 46	Warm-Up 11.6 Folder 11	Master 2 Add. Expl. 17	11.6	
Review Test	76	Chapter Review	Tests 47, 48 Alternative Assessment			Review Test	

***Spanish versions** of *Study Guide* and *Assessment Book* are available.

Chapter Support

- Course Guide
- Lesson Plans
- Portfolio Project Book
- Preparing for College Entrance Tests
- Multi-Language Glossary
- *Test Generator* Software
- Professional Handbook

Software Support

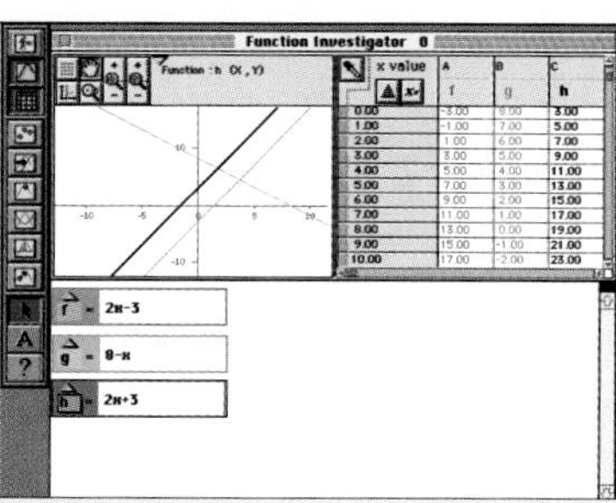

McDougal Littell Mathpack
Function Investigator

Internet Support

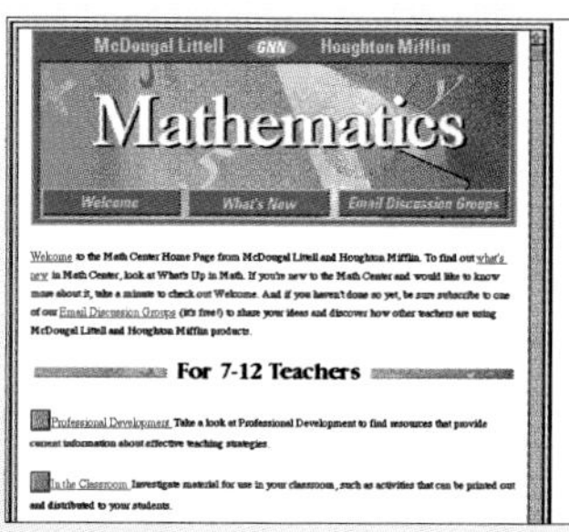

http://www.hmco.com
Next go to McDougal Littell; then the Education Center; then Secondary Math.

Outside Resources

Books, Periodicals

Germain-McCarthy, Yvelyne. "Circular Graphs: Vehicles for Conic and Polar Connections." *Mathematics Teacher* (January 1995): pp. 26–29.

Edwards, Thomas G. "Exploring Quadratic Functions: From *a* to *c*." *Mathematics Teacher* (February 1996): pp. 144–146.

Metz, James. "Sharing Teaching Ideas: Seeing *b* in $y = ax^2 + bx + c$." *Mathematics Teacher* (January 1994): pp. 23–25.

Amick, H. Louise. "A Unique Slope for a Parabola." Sharing Teaching Ideas. *Mathematics Teacher* (January 1995): pp. 38–39.

Activities, Manipulatives

Breuningsen, Chris, Bill Bower, Linda Antinone, and Elisa Breuningsen. "Swinging Ellipses." *Real-World Math with the CBL System.* Activity 17: pp. 91–94. Texas Instruments, 1995.

Software

f(*g*) *Scholar.* IBM-comp. Southampton, PA: Future Graph.

Internet

A source of NSF press releases and publications is found on the web at:
http://nsf.gov/

Interview Notes

Background

Ancient Maps and Mapmaking
The oldest known map was made in Babylonia around 2500 B.C. and shows a settlement in a mountain-lined river valley. The Babylonians also developed the system of dividing a circle into 360 degrees, which is the basis for the latitudinal and longitudinal system used today. The Egyptians made maps as early as 1300 B.C., probably to chart property boundaries since the Nile River flooded every year and these boundaries needed to be redrawn after each flood. The Greeks had an influence on mapmaking with their advances in geometry and surveying. Many Greeks also thought that the world was a sphere, including the mathematician Eratosthenes, who calculated the circumference of Earth with remarkable accuracy around the year 250 B.C.

Kija Kim
Kija Kim's mapmaking company, HDM, can provide a wide variety of information, both pictoral and numerical. For example, HDM helped an electric company map out its wires, poles, and manholes so that a person could select a specific pole and find its height, type of wood, and last inspection date. These data replaced many manual checking systems and paper maps scattered throughout the company. Kim's company has also put together a package that contains flood maps for flood-prone Florida counties. With it, real estate agents, mortgage lenders, and insurance agents can instantly know the flood risk of a piece of property. Many other companies and the government are using the maps and information from Kija Kim's company.

CHAPTER

11 Analytic Geometry

Taking maps into the FUTURE

INTERVIEW **Kija Kim**

Kija Kim starts her day with three games of table tennis and ends her day with sleep. In between, there are maps—lots and lots of maps. Kim runs a computerized-mapping company, Harvard Design & Mapping (HDM), which is one of the world leaders in a new technology called "Geographic Information Systems." It is a very competitive arena, which explains her morning pastime. "Table tennis builds up my competitiveness and gets me mentally ready," she says.

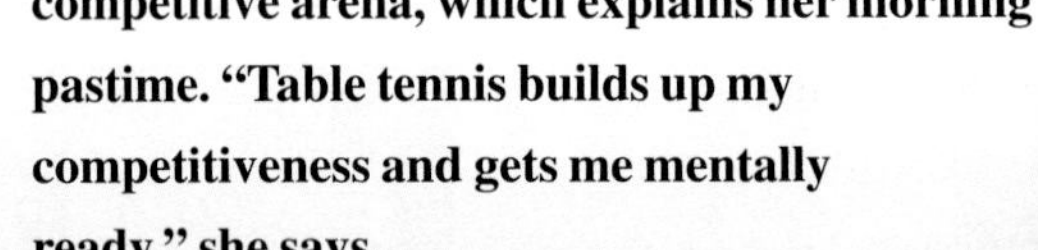

"Ten years ago, we relied on rulers to measure distances. Now computers do the job instantly."

> "Geography combines the other subjects I like—physical science, social science, and mathematics."

Leading the Field

Kim began considering a profession in geography early in life, while she was still a high school student in South Korea. She did not work with computers until many years later, after she had moved to the United States. "During my first course in computer programming, I started thinking about how to automate mapmaking processes," she recalls. "That's when it all began."

Now her company designs maps for a variety of government and business applications. "When I first began applying computers to *cartography*, or mapmaking, people didn't think you could build a business around that idea. Now it's one of the fastest-growing areas in the whole computer field."

Creating Thinking Maps

HDM's computerized maps can display a vast amount of information compared with normal maps. For example, one of HDM's "intelligent maps" provides a detailed picture of all the water flowing into a Massachusetts reservoir, shows the industrial activity in the region, locates the toxic waste sites, and determines how both affect the water quality.

Another computerized map, shown on the next page, predicts what would happen if a break occurred in the Sudbury Dam, which is located on the Sudbury River near Framingham, Massachusetts. "Our software can tell you in 30 seconds whether you're in a potential flood zone," Kim notes. "Your insurance company could take two weeks getting that information."

Interview Notes

Background

Ptolemy

Claudius Ptolemy, c. 85–165 A.D., was perhaps the most influential mapmaker of ancient times. He was a Greek astronomer and geographer who wrote two major works in his lifetime. One was the *Almagest,* in which he describes mathematically the motions of the sun, moon, and planets. His other major work, the *Geography,* consisted of eight volumes that included a list of about 8000 places along with their latitude and longitude, several maps, and a description of various techniques of mapmaking. Years later, in the 800s, Ptolemy's *Geography* was translated into Arabic and used by Arab scholars to develop methods of determining latitude and longitude. In the 1400s, it was translated into Latin and used to develop methods of map projection and how to locate places systematically.

Second-Language Learners

Point out to students learning English that the phrase *competitive arena* in the fourth sentence means that "many companies in the computer-mapping business are trying hard to develop the best products and win the most sales, and so win out over their rivals." The word *superimpose* may also be unfamiliar to students learning English. If necessary, explain that this means "to place one layer over another."

Interview Notes

Mathematical Connection

Many different mathematical skills are needed for both making and reading maps. One of the most common skills is finding distance. In Section 11.1, students are asked to find and estimate the distance between given points. In Section 11.3, students calculate the regions covered by a fire station for different times given a specific response time for a calculated distance. Students are also asked to write an equation for the circular region that a fire station covers and analyze its response area.

Explore and Connect

Writing
If students are having difficulty with this problem, guide them by drawing a right triangle and help them find the distances of the legs. Then have them find the length of the hypotenuse.

Research
A winding road may not be indicated on a map because of the large reduction in scale. When students are reading the maps, have them pay attention to the small numbers (usually in black) alongside the roads that indicate the number of miles of a section of highway or state route. Discuss with students why paying attention to these numbers is important when looking at distance traveled.

Project
Students should divide the project tasks equally and take part in each step of the project. You may wish to display the maps in class when they are completed.

Automating Calculations

Computers can provide information more quickly and accurately than old-fashioned cartographic techniques. "Just ten years ago, we relied on rulers and other tools to determine the position of objects on maps and to measure the distance between them. Now computers do that job instantly," Kim says.

A useful technique for identifying points and finding distances is to superimpose a coordinate grid onto a map. For example, the coordinate grid shown below allows you to locate a point on Sudbury Dam and another point on a nearby office building using coordinates. To calculate the distance between these two points, you can use the *distance formula*, which you will learn about in this chapter.

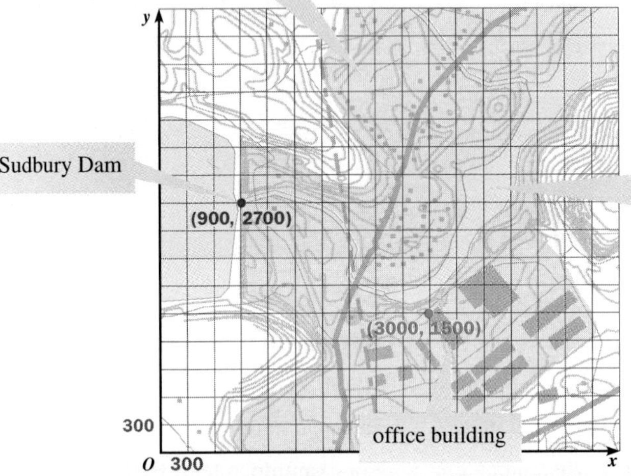

Explore and Connect

Kija Kim and an assistant review the details of a map.

1. Writing Using the coordinates, given in feet, on the map above, find the east-west distance and the north-south distance between Sudbury Dam and the office building. How can you use these distances to calculate the straight-line distance between the dam and the building? Explain.

2. Research Find two maps of your state and compare them. How does each map represent the cities and landmarks? What scale does each map use for distances? Why is a scale important for reading a map?

3. Project Work in a group of four students to make a map of a public area, such as a local park or the school grounds. Identify at least 10 landmarks in the area. Choose a scale and make a map of the area, identifying the landmarks using coordinates.

Mathematics & Kija Kim

In this chapter, you will learn more about how mathematics is related to making maps.

Related Exercises

Section 11.1
- **Exercises 22 and 23**

Section 11.3
- **Exercises 10–14**

Section

11.1 Distances, Midpoints, and Lines

Learn how to...

- find the distance between two points in a coordinate plane
- find the coordinates of the midpoint of a line segment in a coordinate plane

So you can...

- find the distance between two points on a map, for example

Given any right triangle with sides of length a, b, and c (with c being the length of the hypotenuse), the Pythagorean theorem states that $c^2 = a^2 + b^2$. Ancient writings from Egypt, China, Babylonia, India, and Greece show that this relationship was known and used by many civilizations thousands of years ago.

EXPLORATION
COOPERATIVE LEARNING

Finding Distances and Midpoints

Work with a partner.
You will need:
- centimeter graph paper
- a metric ruler

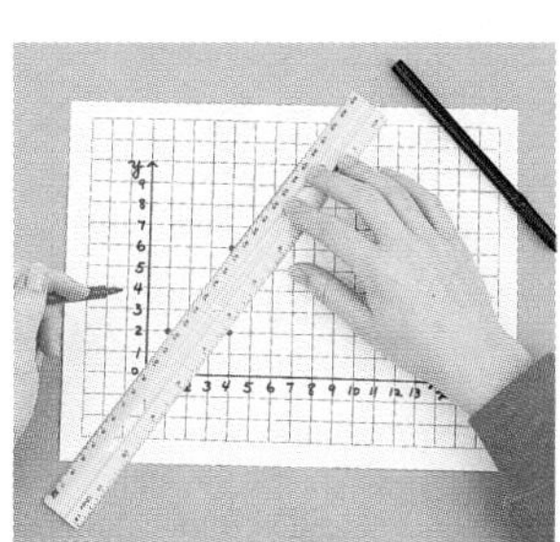

1 Draw coordinate axes on a piece of graph paper. Plot the points (1, 2), (4, 2), and (4, 6). Find the distance in centimeters between (1, 2) and (4, 6). Verify your measurement using the Pythagorean theorem applied to the right triangle formed by the three plotted points.

2 Explain how you can use the Pythagorean theorem to find the distance between any two points in a coordinate plane. Then choose two points and find the distance between them.

3 On the same graph, use your ruler to estimate the coordinates of the point halfway between (1, 2) and (4, 6).

4 Find the mean of the x-coordinates of the points (1, 2) and (4, 6). Also find the mean of the y-coordinates of these points. How are these means related to the coordinates of the point you found in Step 3?

5 Explain how you can find the midpoint of the line segment connecting any two points in a coordinate plane. Then choose two points and find the coordinates of the midpoint of the line segment connecting them.

Exploration Note

Purpose
The purpose of this Exploration is to have students discover the distance and midpoint formulas.

Materials/Preparation
Each pair of students needs centimeter graph paper and a metric ruler.

Procedure
Students find the distance between two points and then verify their measurements by applying the Pythagorean theorem. Next, students use a ruler to estimate the midpoint of a segment. They find the arithmetic means of the x- and y-coordinates of the endpoints of the segments, and check to see how they are related to the midpoint.

Closure
Students should be able to explain how to find the distance between any two points in a coordinate plane. They should understand that the coordinates of the midpoint are found by using the coordinates of the endpoints of the segment and taking the means of the x- and y-coordinates.

Explorations Lab Manual
See the Manual for more commentary on this Exploration.

For answers to the Exploration, see answers in back of book.

Plan⇔Support

Objectives

- Find the distance between two points in a coordinate plane.
- Find the coordinates of the midpoint of a line segment in a coordinate plane.
- Use the distance formula to solve real-world problems.

Recommended Pacing

Core and Extended Courses
Section 11.1: 2 days

Block Schedule
Section 11.1: $1\frac{1}{2}$ blocks
(with Chapter 10 Test)

Resource Materials

Lesson Support
Lesson Plan 11.1
Warm-Up Transparency 11.1
Overhead Visuals:
Folder A: Multi-Use Graphing Packet, Sheets 1–3
Practice Bank: Practice 70
Study Guide: Section 11.1
Explorations Lab Manual:
Diagram Masters 1, 2, 14

Technology
Technology Book:
Spreadsheet Activity 11
Graphing Calculator
McDougal Littell Mathpack
Geometry Inventor Activity Book: Activity 15
Internet:
http://www.hmco.com

Warm-Up Exercises

1. The legs of a right triangle have lengths of 16 cm and 20 cm. Find the approximate length of the hypotenuse. about 25.6 cm
2. Find the distance between (–5, 10) and (–5, –12). 22
3. Find k so that the distance between (6, –4) and (k, –4) is 14. –8 or 20
4. Find the slope of the line through (3, 8) and (7, 5). $-\frac{3}{4}$
5. Find the slope of the line through (3, –2) and (7, –2). 0

Teach⇔Interact

Section Notes

Student Progress
The distance and midpoint formulas may be familiar to students from a previous course.

Common Error
Caution students that the midpoint formula results in an x-value and a y-value, and not a single value as in the distance formula.

About Example 1

Communication: Discussion
Ask students to explain why including coordinates on a map or floor plan improves how the map or floor plan can be used. Ask them to give examples of maps they have used that include coordinates.

Additional Example 1

Washington, D.C. is located along the Potomac River between Maryland and Virginia.

a. Find the distance between Annandale, VA at point (5.8, 3.5) and New Carrollton, MD at point (22.6, 12.1) on a map of the Washington, D.C. area.

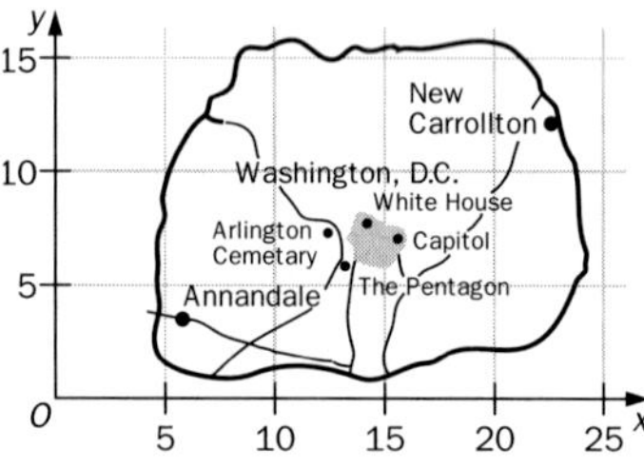

Use the distance formula.

$d = \sqrt{(x_2 - x_1)^2 + (y_2 - y_1)^2}$

$= \sqrt{(22.6 - 5.8)^2 + (12.1 - 3.5)^2}$

$= \sqrt{16.8^2 + 8.6^2}$

$= \sqrt{356.2} \approx 18.9$

The distance between Annandale and New Carrollton is about 19 mi.

b. Find the midpoint of the line segment joining these towns. What is located on the map at the approximate midpoint of this line segment?

Use the midpoint formula.

$\left(\frac{x_1 + x_2}{2}, \frac{y_1 + y_2}{2}\right)$

$= \left(\frac{5.8 + 22.6}{2}, \frac{3.5 + 12.1}{2}\right)$

$= (14.2, 7.8)$

The midpoint of the line connecting Annandale and New Carrollton is (14.2, 7.8). This point is the approximate location of the White House.

BY THE WAY

Zi jin cheng, the Forbidden City, was home to the Chinese imperial court from 1421 to 1911. It is now open to the public and is known as both *Gugong*, the Old Palace, and the Palace Museum.

You can find the distance between any two points in a coordinate plane by finding the length of the hypotenuse of a right triangle. You can find the coordinates of the midpoint of a segment by finding the mean of the x-coordinates and the mean of the y-coordinates of the segment's endpoints.

Distance and Midpoint Formulas

The distance d between two points (x_1, y_1) and (x_2, y_2) is:

$$d = \sqrt{(x_2 - x_1)^2 + (y_2 - y_1)^2}$$

The midpoint of the line segment joining the points (x_1, y_1) and (x_2, y_2) has coordinates:

$$\left(\frac{x_1 + x_2}{2}, \frac{y_1 + y_2}{2}\right)$$

EXAMPLE 1 Application: Architecture

The Palace Museum is located in the city of Beijing in China. A moat and a 35 ft high wall completely surround nearly 1000 buildings with 9000 rooms.

a. Find the distance between the watchtowers at point $A(220, 220)$ and point $B(2580, 3240)$, where the coordinates are given in feet, on the map of the Palace Museum shown below.

b. Find the coordinates of the midpoint of the line segment joining A and B. What is located on the map at the midpoint of this line segment?

SOLUTION

a. Use the distance formula.

$$\begin{aligned} d &= \sqrt{(x_2 - x_1)^2 + (y_2 - y_1)^2} \\ &= \sqrt{(2580 - 220)^2 + (3240 - 220)^2} \\ &= \sqrt{2360^2 + 3020^2} \\ &= \sqrt{14{,}690{,}000} \\ &\approx 3833 \end{aligned}$$

The distance between A and B is about 3830 ft.

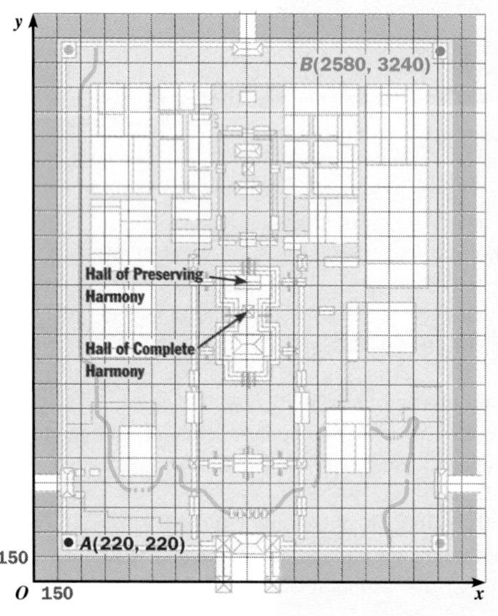

b. Use the midpoint formula.

$$\left(\frac{x_1 + x_2}{2}, \frac{y_1 + y_2}{2}\right) = \left(\frac{220 + 2580}{2}, \frac{220 + 3240}{2}\right) = (1400, 1730)$$

The midpoint of the segment connecting A and B has coordinates (1400, 1730). This point is the center of the terrace between the Hall of Preserving Harmony and the Hall of Complete Harmony.

ANSWERS Section 11.1

Exploration

1–4. Check students' work. Tables may vary. An example is given.

Slope of line	Slope of perpendicular line
1	-1
2	$-\frac{1}{2}$
$-\frac{1}{3}$	3
$\frac{2}{3}$	$-\frac{3}{2}$

5. Answers may vary. Examples are given. The product of the slopes of two perpendicular lines is -1. If the slope of a line is m, then the slope of a line perpendicular to the line is $-\frac{1}{m}$.

Parallel and Perpendicular Lines

When you solved systems of linear equations in Chapter 7, you found that some systems of equations are inconsistent because the graphs of the equations are parallel lines. Parallel lines never intersect because they have the same slope.

Parallel Lines

If line l has slope a_1 and line k has slope a_2, lines l and k are parallel if and only if $a_1 = a_2$.

Example:

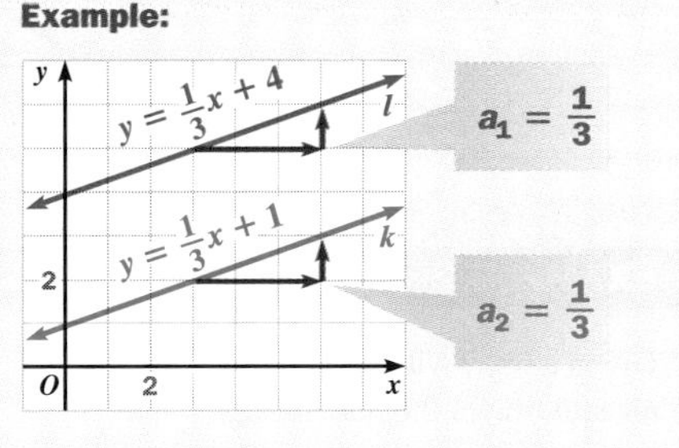

EXPLORATION
COOPERATIVE LEARNING

Exploring Slopes of Perpendicular Lines

Work with a partner.
You will need:
- graph paper
- a straightedge
- a protractor

1 Draw a line with a slope of 1. Choose a point on the line and label the point.

2 Use a protractor to draw another line perpendicular to the first line at the labeled point.

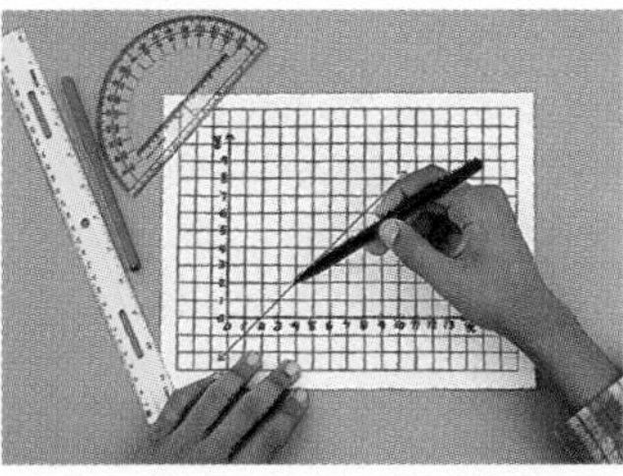

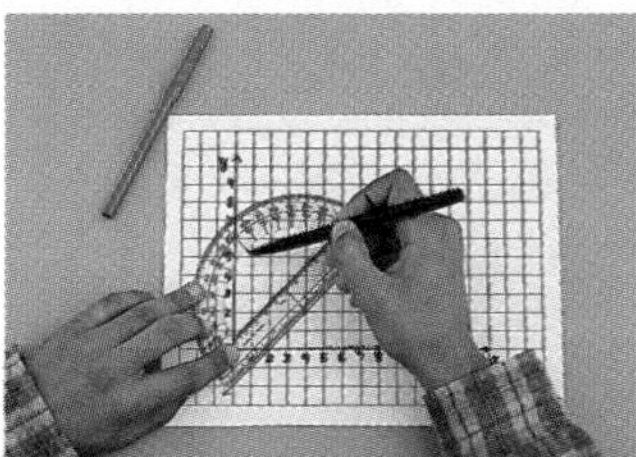

3 Find the slope of the perpendicular line and record it in a table like the one shown.

4 Repeat Steps 1–3 at least three more times, each time starting with a new line having a different slope, such as 2.

5 Make a conjecture about the slopes of perpendicular lines. Compare your conjecture with those of other groups.

Slope of line	Slope of perpendicular line
1	?
2	?
$-\frac{1}{3}$	?

Exploration Note

Purpose
The purpose of this Exploration is to have students discover that the slopes of two perpendicular lines are negative reciprocals.

Materials/Preparation
Each pair of students needs graph paper, a protractor, and a straightedge.

Procedure
Students draw a line with slope of 1, label a point, and then use a protractor to draw another line perpendicular to the first line at the labeled point. They find the slope of the perpendicular line and record it. They then repeat the procedure several times until they can make a conjecture about the relationship between the slopes of perpendicular lines.

Closure
Students should understand that the slopes of two perpendicular lines are negative reciprocals of each other.

Explorations Lab Manual
See the Manual for more commentary on this Exploration.

Diagram Master 14

For answers to the Exploration, see page 514.

Teach⇔Interact

Learning Styles: Visual

The use of a coordinate plane integrates the strands of geometry and algebra and helps visual learners to better understand algebraic concepts. Here, the coordinate plane is used to view an inconsistent linear system as a pair of parallel lines in the plane.

Section Notes

Topic Spiraling: Review
Students should understand that parallel lines have the same slope but have different *y*-intercepts. Two lines with the same slope and the same *y*-intercept are coincident and form a consistent system.

Historical Connection
The development of analytic geometry by René Descartes (1596–1650) was a momentous event in the development of mathematics. Until the seventeenth century, geometry and algebra developed as separate branches of mathematics, and there was no way to solve geometrical problems algebraically. The establishment of a correspondence between ordered pairs of real numbers and points in the plane made possible a correspondence between lines and curves in a plane and equations of two variables. This correspondence could then be used to shift the task of proving a theorem in geometry to that of proving a corresponding theorem in algebra or analysis.

Using Technology
As an alternative way to have students explore the slopes of parallel and perpendicular lines, you may wish to assign Activity 15 in the *Geometry Inventor Activity Book*.

Teach⇔Interact

Section Note

Teaching Tip
Make sure students understand that if a line is vertical, then a line perpendicular to it has slope 0.

About Example 2

Alternate Approach
Students can also find the equation of a line through point $P(x_1, y_1)$ parallel to or perpendicular to a given line by using $y - y_1 = a(x - x_1)$.

Additional Example 2

Given a line l with equation $y = \frac{1}{3}x - 4$ and point P with coordinates $(-3, 1)$, find an equation of the line through P and:

a. parallel to l

A line parallel to l will also have a slope of $\frac{1}{3}$. An equation of the line through P and parallel to l is $y = ax + b$, where:

$1 = \frac{1}{3}(-3) + b$
$1 = -1 + b$
$2 = b$

An equation for the line through P and parallel to l is $y = \frac{1}{3}x + 2$.

b. perpendicular to l

A line perpendicular to l will have a slope of -3. An equation of the line through P and perpendicular to l is $y = ax + b$, where:

$1 = -3(-3) + b$
$1 = 9 + b$
$-8 = b$

An equation for the line through P and perpendicular to l is $y = -3x - 8$.

Closure Question

Explain how to find the length and the midpoint of a segment in a coordinate plane.

To find the length of the segment connecting (x_1, y_1) and (x_2, y_2), use the distance formula $d = \sqrt{(x_2 - x_1)^2 + (y_2 - y_1)^2}$. To find the midpoint, use the midpoint formula $\left(\frac{x_1 + x_2}{2}, \frac{y_1 + y_2}{2}\right)$.

Perpendicular Lines

If line l has slope a_1 and line k has slope a_2, lines l and k are perpendicular if and only if $a_1 = -\frac{1}{a_2}$ (that is, the slopes are negative reciprocals).

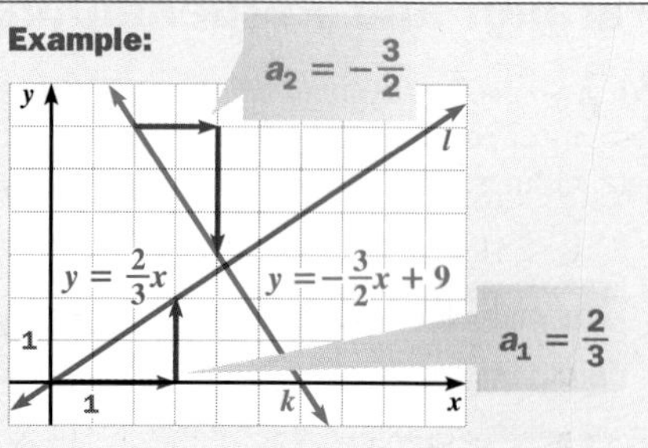

EXAMPLE 2

Given a line l with equation $y = 3x + 2$ and point P with coordinates $(4, 2)$, find an equation of the line through P and:

a. parallel to l **b.** perpendicular to l

SOLUTION

a. Line l has a slope of 3. A line parallel to l will also have a slope of 3. An equation of the line through P and parallel to l is $y = ax + b$ where:

Substitute **4** for x, **2** for y, and **3** for a.

$$2 = 3 \cdot 4 + b$$
$$2 = 12 + b$$
$$-10 = b$$

An equation of the line through P and parallel to l is $y = 3x - 10$.

b. A line perpendicular to l will have a slope of $-\frac{1}{3}$. An equation of the line through P and perpendicular to l is $y = ax + b$ where:

Substitute **4** for x, **2** for y, and $-\frac{1}{3}$ for a.

$$2 = -\frac{1}{3} \cdot 4 + b$$
$$2 = -\frac{4}{3} + b$$
$$\frac{10}{3} = b$$

An equation of the line through P and perpendicular to l is $y = -\frac{1}{3}x + \frac{10}{3}$.

CHECKING KEY CONCEPTS

For each pair of points, find:

a. the distance between the points

b. the coordinates of the midpoint of the line segment connecting the points

1. (1, 5), (4, 8) **2.** (−2, 6), (1, 4) **3.** (−3, −2), (−6, 0)

4. Given a line l with equation $y = 4x - 3$ and point P with coordinates $(-3, 5)$, find an equation of the line through P and:

a. parallel to l **b.** perpendicular to l

Checking Key Concepts

1. a. $3\sqrt{2} \approx 4.24$
   b. (2.5, 6.5)
2. a. $\sqrt{13} \approx 3.61$
   b. $\left(-\frac{1}{2}, 5\right)$
3. a. $\sqrt{13} \approx 3.61$
   b. (−4.5, −1)
4. a. $y = 4x + 17$
   b. $y = -\frac{1}{4}x + \frac{17}{4}$

Exercises and Applications

1. a. (1.5, −4)
   b. $\frac{4}{5}$ c. $-\frac{5}{4}$
   d. $y + 4 = -\frac{5}{4}(x - 1.5)$ or $y = -\frac{5}{4}x - \frac{17}{8}$
2. a. $\sqrt{194} \approx 13.93$
   b. $y = -\frac{5}{13}x + \frac{206}{13}$
3. a. 2.5
   b. $y = \frac{3}{4}x - \frac{21}{16}$
4. a. 5
   b. $y = -\frac{3}{4}x - \frac{23}{8}$
5. a. $\sqrt{197} \approx 14.04$
   b. $y = -14x - 60.5$
6. a. $\sqrt{58} \approx 7.62$
   b. $y = \frac{7}{3}x + \frac{5}{3}$
7. a. $\sqrt{106} \approx 10.30$
   b. $y = -\frac{5}{9}x + \frac{2}{9}$
8. a. $\sqrt{47.77} \approx 6.91$

11.1 Exercises and Applications

Extra Practice *exercises on page 766*

1. **a.** Find the midpoint M of segment $\overline{GH}$ for $G(4, -2)$ and $H(-1, -6)$.
   b. What is the slope of $\overline{GH}$?
   c. What is the slope of a line perpendicular to $\overline{GH}$ at M?
   d. Find an equation of the perpendicular bisector of $\overline{GH}$.

For each pair of points, find:

a. the distance between the points

b. an equation of the perpendicular bisector of the line segment connecting the points

2. (1, 8), (6, 21) **3.** (5, 4), (6.5, 2) **4.** (−8, 0), (−5, 4) **5.** (−11, −5), (3, −4)

6. (−4, 2), (3, −1) **7.** (−3, −4), (2, 5) **8.** (3.4, 4.6), (−2.5, 8.2) **9.** (4.3, 0.4), (−6.4, −7.9)

10. (−12, 15), (4, −16) **11.** $\left(-4, \frac{1}{2}\right), \left(\frac{3}{4}, \frac{5}{8}\right)$ **12.** $\left(\frac{1}{3}, \frac{5}{6}\right), \left(-\frac{2}{5}, \frac{8}{3}\right)$ **13.** (−3.5, 12.4), (−1.8, −9)

14. GEOMETRY Verify that $\triangle ABC$ with vertices $A(1, 7)$, $B(6, 2)$, and $C(8, 6)$ is isosceles.

Find the value(s) of k so that the given points are n units apart.

15. $(6, k), (9, 11), n = 5$ **16.** $(-3, 2), (5, k), n = 17$ **17.** $(3, k), (-7, -10), n = 26$

18. $(3.5, 1.4), (k, 2.8), n = 5$ **19.** $(-3, 6), (k, 4), n = 6$ **20.** $(-4, 1), (k, 6), n = 12$

21. GEOMETRY Show that $Q(-5, 7)$, $R(0, 12)$, $S(5, 7)$, and $T(0, 2)$ are the vertices of a square. Explain your reasoning.

INTERVIEW Kija Kim

Look back at the article on pages 510–512.

Storm drains in the city shown carry rainwater from the streets to a local river. The water may pick up contaminants before reaching the river. To determine the impact that this water has on local recreation areas, city engineers start by finding the distance between a discharge pipe and a nearby recreation area.

22. Find the distance between the discharge pipe and the closest point of the recreation area shown on the map, where the coordinates are given in feet.

23. Estimate the distance between the discharge pipe and the farther end of the recreation area. Explain how you found your estimate.

Apply⇔Assess

Suggested Assignment

Core Course
Day 1 Exs. 1–21
Day 2 Exs. 25, 26, 29, 33–40

Extended Course
Day 1 Exs. 1–19 odd, 20–25
Day 2 Exs. 26–40

Block Schedule
Day 69 Exs. 1–13
Day 70 Exs. 14–21, 25, 26, 29, 33–40

Exercise Notes

Reasoning
Exs. 2–13 For each exercise, students first need to find the midpoint and slope of each segment. They can then use the midpoint with the negative reciprocal slope to find an equation of the perpendicular bisector of the segment.

Using Technology
Exs. 2–13 Students can use the TI-82 graphing calculator program below to find the distance between the points.

```
:Input "X1",A:Input "Y1",B
:Input "X2",C:Input "Y2",D
:√((C–A)^2+(D–B)^2 )→E
:Disp "DISTANCE IS ":Disp E
```

Integrating the Strands
Exs. 14, 21 You may need to remind students that an isosceles triangle has two equal sides. They can use the distance formula to find the length of the sides. After discussing Ex. 21, some students may wish to think about how the distance formula or the midpoint formula can be used to classify other polygons whose vertices are points in a coordinate plane.

Cooperative Learning
Exs. 15–20 Students would benefit from working in pairs to compare their answers to these exercises. Have each pair verify why each exercise has two possible solutions by drawing sketches.

Interview Note
Exs. 22, 23 These exercises exemplify how the distance formula can be applied to maps. Using the scale given on a map and a ruler can also help students to find estimates of distances.

b. $y = \frac{59}{36}x + \frac{453}{80}$

9. a. $\sqrt{183.38} \approx 13.54$
   b. $y = -\frac{107}{83}x - \frac{2118}{415}$

10. a. $\sqrt{1217} \approx 34.89$
    b. $y = \frac{16}{31}x + \frac{97}{62}$

11. a. $\frac{\sqrt{1445}}{8} \approx 4.75$
    b. $y = -38x - \frac{979}{16}$

12. a. $\frac{\sqrt{31,581}}{90} \approx 1.97$
    b. $y = -\frac{2}{5}x + \frac{529}{300}$

13. a. $\sqrt{460.85} \approx 21.47$
    b. $y = \frac{17}{214}x + \frac{8177}{4280}$

14. $AB = 5\sqrt{2}$ and $AC = 5\sqrt{2}$

15. 7 or 15 16. −13 or 17

17. 14 or −34 18. −1.3 or 8.3

19. $-3 \pm 4\sqrt{2}$; about 2.66 or −8.66

20. $-4 \pm \sqrt{119}$; about 6.91 or −14.91

21. Answers may vary. An example is given. $QR = RS = ST = TQ = 5\sqrt{2}$. The slope of $\overline{QR}$ and $\overline{ST}$ is 1, and the slope of $\overline{RS}$ and $\overline{TQ}$ is −1. The adjacent sides of $QRST$ are perpendicular. $QRST$ is a square.

22. ≈430 ft

23. ≈744 ft; Explanations may vary. The coordinates of the farther end of the recreation area are about (500,950).

Apply⇔Assess

Exercise Notes

Research

Ex. 24 Some students may be interested in researching the rules for creating their own state flag. Have them confirm the positions of the elements of the flag mathematically.

Integrating the Strands

Exs. 25, 26 When discussing part (b) of Ex. 25, emphasize that proofs involving coordinates can be done algebraically with the actual lengths of sides or slopes, rather than synthetically, as is usually done in a geometry course. Students may recognize that both Ex. 25(b) and Ex. 26 can be done with slopes or by using the distance formula. Have them review the properties of right triangles and parallelograms that apply to these exercises.

Interdisciplinary Problems

Exs. 27, 28 These problems allow students to see an example of how mathematics can be used to better understand and interpret a passage from literature.

Cooperative Learning

Exs. 27, 28 Reading and using maps is difficult for some students. You may want students to discuss their answers to these exercises in small groups.

Second-Language Learners

Exs. 27, 28 Some of the language and idiomatic expressions in the excerpt from *To Kill a Mockingbird* might be challenging for students learning English. They may benefit from reading and discussing the passage with an English-proficient partner.

Integrating the Strands

Ex. 31 Point out to students that for coordinate geometry proofs, drawing a figure in a convenient position on a coordinate plane is allowed and desirable as long as the diagram does not add any information that is not given. In this case, using the coordinates listed in the hint gives the midpoint of the hypotenuse as (x, y).

24. FLAG MAKING According to the rules for creating California's state flag, the length of the bear is measured from the nose tip N to the rear of the right-hand paw P.

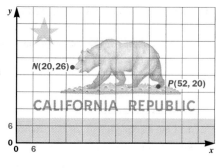

a. Find the length of the bear using the flag shown, where the coordinates are given in inches. Confirm that this length is about two thirds the height of the flag.

b. The midpoint M of $\overline{NP}$ should also be the midpoint of a segment drawn horizontally across the flag. Find the coordinates of M. Confirm that the bear is correctly positioned horizontally.

N(20, 26)
P(52, 20)
CALIFORNIA REPUBLIC

25. a. GEOMETRY Find the perimeter of quadrilateral $ABCD$ with vertices $A(-2, 5)$, $B(3, 8)$, $C(3, 3)$, and $D(-2, 0)$.

b. Prove that $ABCD$ is a parallelogram.

26. GEOMETRY Show that $\triangle PQR$ with vertices $P(3, 12)$, $Q(6, 5)$, and $R(-1, 2)$ is a right triangle. Which side is the hypotenuse? Which sides are perpendicular?

Connection LITERATURE

Harper Lee's novel *To Kill a Mockingbird* is set in the fictional town of Maycomb, Alabama, the county seat of Maycomb County. In the book, the exact location of the county seat is credited to Mr. Sinkfield, who ran an inn near the geographic center of the county.

> Business was excellent when Governor William Wyatt Bibb, with a view to promoting the newly created county's domestic tranquillity, dispatched a team of surveyors to locate its exact center and there establish its seat of government. The surveyors, Sinkfield's guests, . . . showed him the probable spot where the county seat would be built. Had not Sinkfield made a bold stroke to preserve his holdings, Maycomb would have sat in the middle of Winston Swamp. . . . Sinkfield . . . induced them to bring forward their maps and charts, lop off a little here, add a bit there, and adjust the center of the county to meet his requirements.

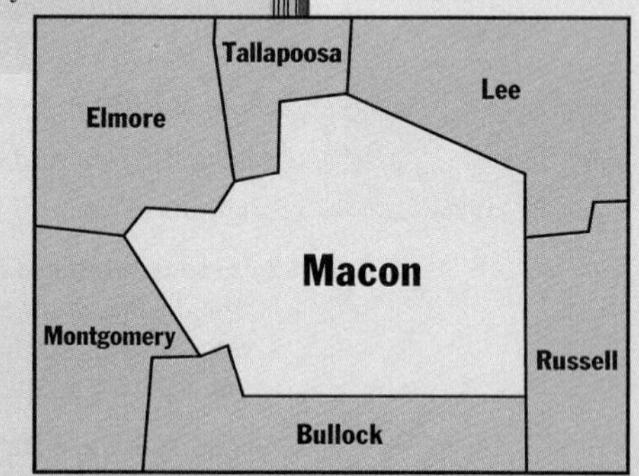

27. Writing Explain how you can approximate the center of a county using a coordinate map. Why does it make sense to locate the county seat at the center of the county?

28. Open-ended Problem Maycomb County is fictional. Use the map of Macon County in Alabama. Sketch a copy of the map. Locate the spot that you think is the center of the county. Explain your reasoning.

24. a. $NP \approx 32.6$ in., which is about two thirds of the height of the flag (48 in.).

b. $M = (36, 23)$, which is the midpoint of the horizontal line segment that is 23 in. from the bottom edge of the flag.

25. a. $10 + 2\sqrt{34} \approx 21.66$

b. Proofs may vary. An example is given. $AB = CD = \sqrt{34}$, and $BC = AD = 5$. Since both pairs of opposite sides are congruent, $ABCD$ is a parallelogram.

26, 27. See answers in back of book.

28. Answers may vary. Check students' work.

29. a. (1.5, 2.5), (2.5, 6.5), (6.5, 5.5), (5.5, 1.5)

b. (2, 4.5), (4.5, 6), (6, 3.5), (3.5, 2)

c. $4\sqrt{34}$; $4\sqrt{17}$; $4\sqrt{8.5}$; The perimeter of the next square should be $4\sqrt{4.25}$. To check the prediction, find two adjacent vertices of the next square, for example, (3.25, 5.25) and (5.25, 4.75). These give a side length of $\sqrt{4.25}$, and a perimeter of $4\sqrt{4.25}$, as predicted.

29. The pattern at the right was created by connecting the midpoints of the sides of one square to form a new square.

a. Find the coordinates of the vertices of the red square.

b. Find the coordinates of the vertices of the green square.

c. Find the perimeters of the blue square, the red square, and the green square. Predict the perimeter of the next square in the pattern. Check your prediction.

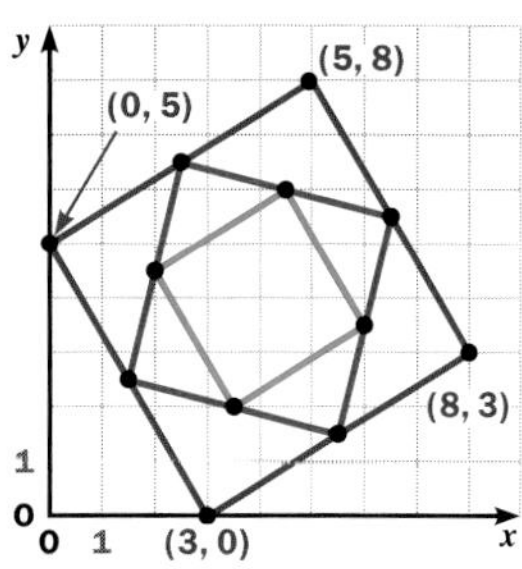

30. The second pattern at the right was created by connecting the points one quarter of the way along the sides of one square to form a new square.

a. Explain how you can use the midpoint formula to find the coordinates of the vertices for each new square in this pattern.

b. Find the coordinates of the vertices of the red square and the green square.

c. Find the areas of the blue square, the red square, and the green square. Predict the area of the next square in the pattern. Check your prediction.

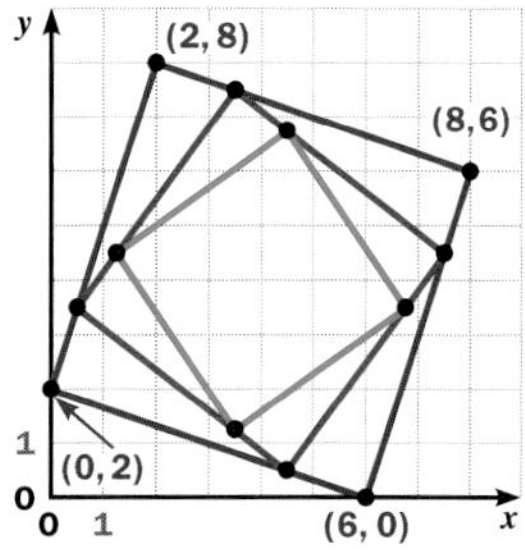

31. GEOMETRY Show that the midpoint of the hypotenuse of any right triangle is equidistant from each vertex. (*Hint:* Any right triangle can be placed on the coordinate plane so that its vertices are (0, 0), (2*x*, 0), and (0, 2*y*).)

32. Challenge Use the distance formula to show that $M\left(\frac{x_1 + x_2}{2}, \frac{y_1 + y_2}{2}\right)$ is the same distance from point $P(x_1, y_1)$ as it is from point $Q(x_2, y_2)$.

ONGOING ASSESSMENT

33. Open-ended Problem Work in a group of 2 or 3 students. Create a game or puzzle that involves finding the lengths and midpoints of line segments and finding parallel and perpendicular lines. Write the rules for your game or puzzle. Suggest strategies for winning your game or provide a solution for your puzzle.

SPIRAL REVIEW

Write each geometric series in sigma notation. Find the sum of each series, if the sum exists. If a series does not have a sum, explain why not. *(Section 10.5)*

34. $512 + (-128) + 32 + (-8) + \cdots$

35. $6 + 18 + 54 + 162 + \cdots + 4374$

For Exercises 36 and 37:

a. Graph the feasible region.

b. Find the maximum profit for the profit function $P = 3x + 5y$.

c. Find the minimum cost for the cost function $C = 0.5x + 3y$. *(Section 7.6)*

36. $y \le 5$
$y \ge \frac{4}{5}x - \frac{3}{5}$
$y \le 3x - 5$

37. $y \ge 3x + 1$
$y \le \frac{1}{2}x + 8$
$y \ge -4x + 9$

Describe the graph of each function. Make a sketch of each graph. *(Section 5.2)*

38. $y = 2(x - 5)^2$ **39.** $y = 0.2(x - 3)^2 - 4$ **40.** $y = -3(x + 3)^2 - 3$

Apply⇔Assess

Exercise Notes

Visual Thinking

Ex. 33 Ask students to create models of their games and to explain how they work to the class. Encourage them to show how finding the lengths and midpoints of line segments, and parallel and perpendicular lines, are integral to playing the game. This activity involves the visual skills of *correlation* and *communication*.

Assessment Note

Ex. 33 Creating a game or puzzle should help students organize and clarify their thoughts about the concepts involved. You may wish to call upon some groups to share their game or puzzle with the class.

Practice 70 for Section 11.1

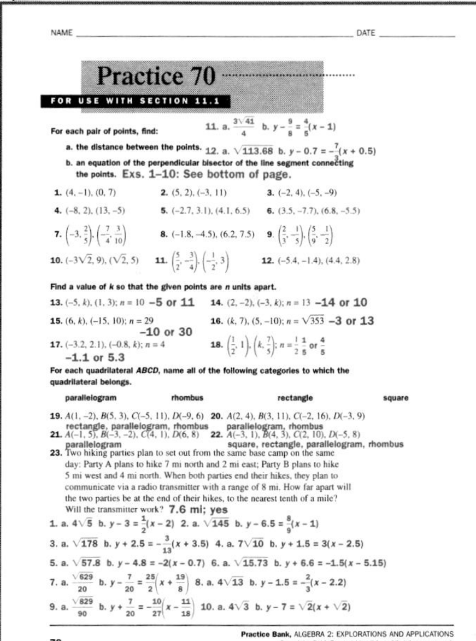
NAME ______ DATE ______

Practice 70

FOR USE WITH SECTION 11.1

For each pair of points, find:

a. the distance between the points.

b. an equation of the perpendicular bisector of the line segment connecting the points. Exs. 1–10: See bottom of page.

11. a. $\frac{3\sqrt{41}}{4}$ b. $y - \frac{9}{8} = \frac{4}{5}(x - 1)$

12. a. $\sqrt{113.68}$ b. $y - 0.7 = -\frac{7}{3}(x + 0.5)$

1. (4, –1), (0, 7) **2.** (5, 2), (–3, 11) **3.** (–2, 4), (–5, –9)

4. (–8, 2), (13, –5) **5.** (–2.7, 3.1), (4.1, 6.5) **6.** (3.5, –7.7), (6.8, –5.5)

7. $\left(-3, \frac{2}{5}\right), \left(-\frac{7}{4}, \frac{3}{10}\right)$ **8.** (–1.8, –4.5), (6.2, 7.5) **9.** $\left(\frac{2}{3}, -\frac{1}{5}\right), \left(\frac{5}{9}, -\frac{1}{2}\right)$

10. $(-3\sqrt{2}, 9), (\sqrt{2}, 5)$ **11.** $\left(\frac{5}{2}, -\frac{3}{4}\right), \left(-\frac{1}{2}, 3\right)$ **12.** (–5.4, –1.4), (4.4, 2.8)

Find a value of *k* so that the given points are *n* units apart.

13. (–5, *k*), (1, 3); $n = 10$ –5 or 11 **14.** (2, –2), (–3, *k*); $n = 13$ –14 or 10

15. (6, *k*), (–15, 10); $n = 29$ –10 or 30 **16.** (*k*, 7), (5, –10); $n = \sqrt{353}$ –3 or 13

17. (–3.2, 2.1), (–0.8, *k*); $n = 4$ –1.1 or 5.3 **18.** $\left(\frac{1}{2}, 1\right), \left(k, \frac{7}{5}\right)$; $n = \frac{1}{2}$ $\frac{1}{5}$ or $\frac{4}{5}$

For each quadrilateral *ABCD*, name all of the following categories to which the quadrilateral belongs.

parallelogram rhombus rectangle square

19. *A*(1, –2), *B*(5, 3), *C*(–5, 11), *D*(–9, 6) rectangle, parallelogram, rhombus

20. *A*(2, 4), *B*(3, 11), *C*(–2, 16), *D*(–3, 9) parallelogram, rhombus

21. *A*(–1, 5), *B*(–3, –2), *C*(4, 1), *D*(6, 8) parallelogram

22. *A*(–3, 1), *B*(4, 3), *C*(2, 10), *D*(–5, 8) square, rectangle, parallelogram, rhombus

23. Two hiking parties plan to set out from the same base camp on the same day: Party A plans to hike 7 mi north and 2 mi east; Party B plans to hike 5 mi west and 4 mi north. When both parties end their hikes, they plan to communicate via a radio transmitter with a range of 8 mi. How far apart will the two parties be at the end of their hikes, to the nearest tenth of a mile? Will the transmitter work? 7.6 mi; yes

1. a. $4\sqrt{5}$ b. $y - 3 = \frac{1}{2}(x - 2)$ 2. a. $\sqrt{145}$ b. $y - 6.5 = \frac{8}{9}(x - 1)$
3. a. $\sqrt{178}$ b. $y + 2.5 = -\frac{3}{13}(x + 3.5)$ 4. a. $7\sqrt{10}$ b. $y + 1.5 = 3(x - 2.5)$
5. a. $\sqrt{57.8}$ b. $y - 4.8 = -2(x - 0.7)$ 6. a. $\sqrt{15.73}$ b. $y + 6.6 = -1.5(x - 5.15)$
7. a. $\frac{\sqrt{629}}{20}$ b. $y - \frac{7}{20} = \frac{25}{2}\left(x + \frac{19}{8}\right)$ 8. a. $4\sqrt{13}$ b. $y - 1.5 = -\frac{2}{3}(x - 2.2)$
9. a. $\frac{\sqrt{829}}{90}$ b. $y + \frac{7}{20} = -\frac{10}{27}\left(x - \frac{11}{18}\right)$ 10. a. $4\sqrt{3}$ b. $y - 7 = \sqrt{2}(x + \sqrt{2})$

70 **Practice Bank,** ALGEBRA 2: EXPLORATIONS AND APPLICATIONS

30. a. If $\overline{AB}$ is a side of a square, find the midpoint *M* of $\overline{AB}$, and then find the midpoint *N* of $\overline{AM}$. *N* is one fourth of the way from *A* to *B*.

b. red square: (4.5, 0.5), (0.5, 3.5), (3.5, 7.5), (7.5, 4.5); green square: (3.5, 1.25), (1.25, 4.5), (4.5, 6.75), (6.75, 3.5)

c. 40; 25; $15.625 = \frac{125}{8}$; The area of the next square should be $\frac{625}{64} = 9.765625$. To check the prediction, find two consecutive vertices of the next square, for example, (2.9375, 2.0625) and (2.0625, 5.0625). These give a side length of $\sqrt{9.765625}$, and an area of 9.765625, as predicted.

31. Place the right triangle on the coordinate plane with right angle *C* at the origin, *A* on the *x*-axis at (2*x*, 0), and *B* on the *y*-axis at (0, 2*y*). The midpoint *M* of the hypotenuse, $\overline{AB}$, is (*x*, *y*). $AM = BM = CM = \sqrt{x^2 + y^2}$, so *M* is equidistant from *A*, *B*, and *C*.

32–38. See answers in back of book.

39. parabola that opens up and has vertex (3, –4)

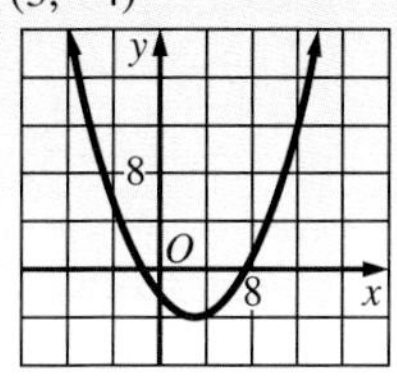

40. parabola that opens down and has vertex (–3, –3)

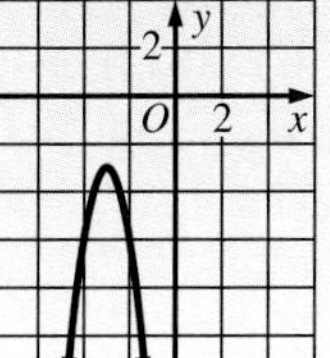

Plan⇔Support

Objectives

- Find the focus and directrix of a parabola.
- Write an equation of a parabola.
- Use parabolas to solve real-world problems.

Recommended Pacing

Core and Extended Courses
Section 11.2: 2 days

Block Schedule
Section 11.2: 1 block

Resource Materials

Lesson Support
Lesson Plan 11.2
Warm-Up Transparency 11.2
Overhead Visuals:
Folder 11: Conic Sections
Practice Bank: Practice 71
Study Guide: Section 11.2
Explorations Lab Manual:
Diagram Masters 2, 15
Assessment Book: Test 44

Technology
Graphing Calculator
McDougal Littell Mathpack
Function Investigator
Internet:
http://www.hmco.com

Warm-Up Exercises

1. Find the distance between the points (−3, 1) and (5, 7) in a coordinate plane. 10
2. Find the midpoint of the line segment connecting (4, 8) and (−6, −2). (−1, 3)
3. Tell whether the parabola $y = -4x^2$ opens up or down. down
4. Find the vertex of the parabola $y + 2 = (x - 4)^2$. (4, −2)
5. Find the vertex of the parabola $y = x^2 + 2x$ by completing the square. (−1, −1)

Section

11.2 Parabolas

Learn how to...

- find the focus and directrix of a parabola
- write an equation of a parabola

So you can...

- locate the antenna of a radio telescope dish, for example

Many antennas are shaped so that a cross section of the antenna is a parabola. This shape improves the ability of the antenna to focus the signal that it is sending or receiving. The sender or receiver for the antenna is located at a point called the *focus* of the parabolic cross section. You can define a parabola using this point and a line called the *directrix*.

EXPLORATION
COOPERATIVE LEARNING

Using Focus-Directrix Paper to Draw a Parabola

Work with a partner.
You will need:

- focus-directrix paper

SET UP The circles on focus-directrix paper share a common center and the parallel lines are tangent to the circles.

1 Label as d the line that is tangent to the circle with radius 4 units and is *below* the center of the circle. Mark the single point where the line that is 2 units above d intersects the circle with radius 2 units.

2 Mark the two points where the line 3 units above d intersects the circle with radius 3 units.

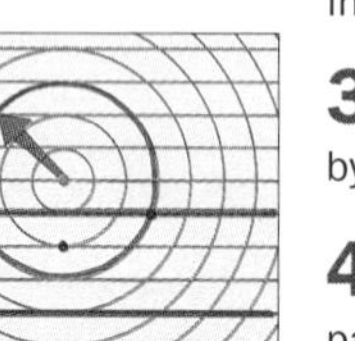

3 Continue marking points as in Step 2, each time increasing by 1 unit the distance above d and the radius of the circle.

4 All the points you marked in Steps 2 and 3 should lie on a parabola. Sketch the parabola.

Questions

1. What is special about the point you marked in Step 1?
2. Line d is the *directrix* of the parabola, and the common center of the circles is the *focus* of the parabola. Based on Steps 2 and 3, how is a point on a parabola related to its focus and directrix?
3. What happens if you start with a directrix that lies *above* the focus?

Exploration Note

Purpose
The purpose of this Exploration is to have students discover that each point on a parabola is an equal distance from the focus and the directrix.

Materials/Preparation
Each group needs focus-directrix paper.

Procedure
Students label a directrix line below the center of a circle. The center of the circle is the focus of the parabola, and the vertex is the midpoint of the perpendicular line segment joining the directrix to the focus. They then use the concentric circles to label two points whose distance from the center of the circle is equal to their distance from the directrix line. They continue to mark points on concentric circles with larger radii until they have sketched the points of a parabola defined by the given directrix line and focus.

Closure
Students should understand that each point of a parabola is an equal distance from the focus and the directrix.

Explorations Lab Manual
See the Manual for more commentary on this Exploration.
Diagram Master 15

You can define a parabola geometrically in terms of a focus and directrix.

A **parabola** is the set of all points in a plane that are the same distance from a point F, called the **focus**, and a line d, called the **directrix**.

The distance between a point on the parabola and the directrix is measured along a line perpendicular to the directrix.

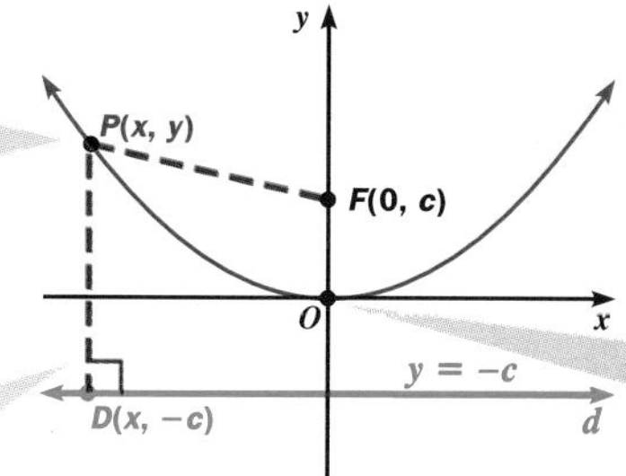

The **vertex** of the parabola lies halfway between the focus and the directrix.

In the graph shown above, the focus is the point $F(0, c)$ and the directrix is the horizontal line $y = -c$. If $P(x, y)$ is any point on the parabola, then you can derive an equation of the parabola as follows:

$$PF = PD$$

$D(x, -c)$ is the point of intersection of d and the line through P that is perpendicular to d.

$$\sqrt{(x-0)^2 + (y-c)^2} = y - (-c)$$

$$x^2 + (y-c)^2 = (y+c)^2$$

Square both sides.

$$x^2 + y^2 - 2cy + c^2 = y^2 + 2cy + c^2$$

$$x^2 = 4cy$$

$$\frac{1}{4c}x^2 = y$$

THINK AND COMMUNICATE

1. What type of function does the equation $y = \frac{1}{4c}x^2$ represent? Does this surprise you? Explain.
2. What would happen to the equation of the parabola shown above if the parabola were translated h units horizontally and k units vertically?

Equation of a Parabola that Opens Up or Down

The graph of $y - k = \frac{1}{4c}(x - h)^2$ is a parabola that:

- has vertex at (h, k) and focus at $(h, k + c)$
- has directrix with equation $y = k - c$
- has a line of symmetry with equation $x = h$
- opens up if $c > 0$ and opens down if $c < 0$

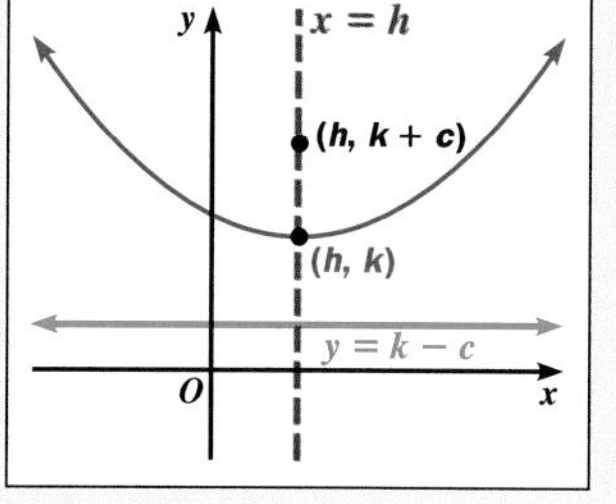

Teach⇔Interact

Section Notes

Application
There are many examples of real objects that are shaped like parabolas, such as the antenna used to introduce this section. Ask students to name some other familiar examples.

Student Progress
The graph of a parabola should be familiar to most students from Chapter 5. Thus, students should recognize that a parabola has a vertex and can open up or down.

Reasoning
The algebraic derivation of an equation of the parabola depends on the distance formula, as does the derivation of the other conic sections. Make sure students understand this algebraic connection to the geometric definition of a parabola in terms of its focus and directrix and can follow the reasoning involved in the derivation of an equation of the parabola.

Think and Communicate

Question 2 is important to students' understanding of parabolas because it makes a connection between quadratic functions studied in Chapter 5 and the way parabolas are defined in this chapter.

Section Note

Communication: Discussion
The text at the bottom of this page should be thoroughly discussed to help students organize the information they should know about parabolas. Encourage students to refer to this page when doing the exercises.

ANSWERS Section 11.2

Exploration

1–4. Check students' work.

Questions

1. It is the vertex of the parabola.
2. The distance between any point on a parabola and its focus is equal to the perpendicular distance from the point to the directrix of the parabola.
3. The parabola opens down.

Think and Communicate

1. a quadratic function; No; the Exploration suggests that the set of points that are the same distance from a point and a line is a parabola and the equation for a parabola is quadratic.
2. An equation of the translated parabola is $y - k = \frac{1}{4c}(x - h)^2$.

Teach⇔Interact

About Example 1

Teaching Tip

Point out that the graph of a parabola never crosses its directrix. This is helpful in finding the direction the parabola opens. It is always away from the directrix.

Additional Example 1

A parabola has the point (3, –2) as its focus and the line $y = -6$ as its directrix.

a. Write an equation of the parabola.

Since the directrix is a horizontal line that lies below the focus, the parabola will open up.

Step 1 Find the vertex of the parabola. The vertex (h, k) is at the midpoint of the line segment connecting the focus (3, –2) and the point (3, –6) on the directrix.

$$(h, k) = \left(\frac{3+3}{2} + \frac{-2+(-6)}{2}\right)$$
$$= (3, -4)$$

Step 2 Find the value of c. Use the fact that the directrix is the line $y = k - c$.

$$-6 = -4 - c$$
$$c = 2$$

Step 3 Find an equation of the parabola using the general form of an equation of a parabola that opens up or down.

$$y - k = \frac{1}{4c}(x - h)^2$$
$$y - (-4) = \frac{1}{4(2)}(x - 3)^2$$
$$y + 4 = \frac{1}{8}(x - 3)^2$$

An equation of the parabola is $y + 4 = \frac{1}{8}(x - 3)^2$.

b. Graph the equation.

Plot the vertex and at least one other point. Use symmetry to sketch the graph.

The vertex is (3, –4).

When $x = -1$,

$y = \frac{1}{8}(-1 - 3)^2 - 4 = -2.$

Plot the point (–1, –2).

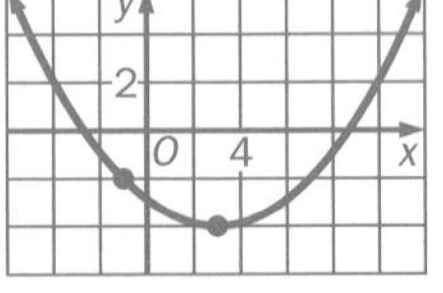

EXAMPLE 1

A parabola has the point (2, 1) as its focus and the line $y = 5$ as its directrix.

a. Write an equation of the parabola. **b.** Graph the equation.

SOLUTION

a. Since the directrix is a horizontal line that lies above the focus, the parabola will open down.

Step 1 Find the vertex of the parabola. The vertex (h, k) is at the midpoint of the line segment connecting the focus (2, 1) and the point (2, 5) on the directrix.

$$(h, k) = \left(\frac{2+2}{2}, \frac{1+5}{2}\right)$$
$$= (2, 3)$$

Use the midpoint formula.

Step 2 Find the value of c. Use the fact that the directrix is the line $y = k - c$.

Substitute **5** for y and **3** for k.

$$5 = 3 - c$$
$$-2 = c$$

Note that a negative c-value agrees with the fact that the parabola opens down.

Check:

The focus is located at $(h, k + c)$.

$$(2, 1) \stackrel{?}{=} (2, 3 + (-2))$$
$$1 \stackrel{?}{=} 3 + (-2)$$
$$1 = 1 ✔$$

Step 3 Find an equation of the parabola using the general form of an equation of a parabola that opens up or down.

$$y - k = \frac{1}{4c}(x - h)^2$$

Substitute **2** for h, **3** for k, and **–2** for c.

$$y - 3 = \frac{1}{4(-2)}(x - 2)^2$$
$$y - 3 = -\frac{1}{8}(x - 2)^2$$

An equation of the parabola is $y - 3 = -\frac{1}{8}(x - 2)^2$.

b. Plot the vertex and at least one other point. Use symmetry to sketch the graph.

The vertex is the point (2, 3).

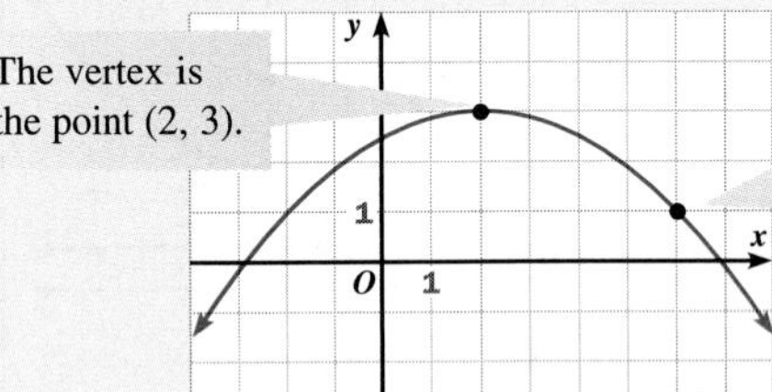

When $x = 6$,

$y = -\frac{1}{8}(6 - 2)^2 + 3 = 1.$

Plot the point (6, 1).

EXAMPLE 2 Application: Astronomy

Astronomers use radio telescopes to collect data from space. A main component of a radio telescope is a large dish with a parabolic cross section. The dish captures radio waves from space and reflects them to a receiver at the focus of the dish. The dish for the Fortaleza radio telescope in Brazil has a width of 14.2 m and a depth of 2.16 m. How far from the vertex is the receiver located?

SOLUTION

Step 1 Sketch a parabolic cross section of the radio telescope in a coordinate plane so that the vertex of the parabola is at (0, 0). Then an equation of the parabola is

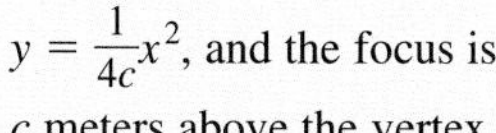

$y = \frac{1}{4c}x^2$, and the focus is c meters above the vertex.

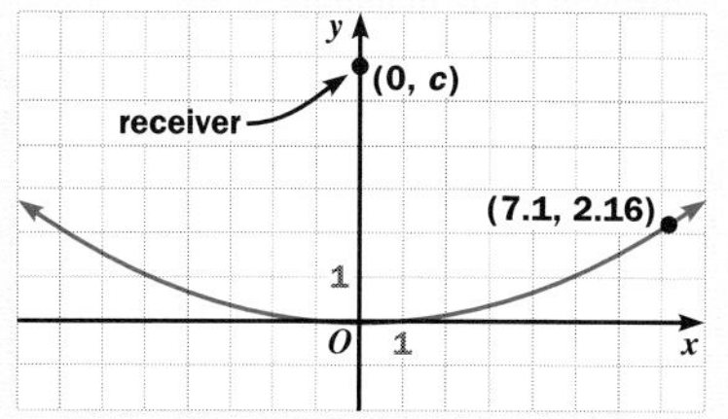

Step 2 Choose a point on the rim of the dish, as shown. The x-coordinate of this point is equal to half the diameter of the dish, 7.1 m. The y-coordinate of this point is equal to the depth of the dish, 2.16 m.

$$2.16 = \frac{1}{4c}(7.1)^2$$

Substitute **7.1** for x and **2.16** for y into the equation of the parabola.

$$2.16c = \frac{1}{4}(50.41)$$

$$2.16c = 12.6025$$

$$c \approx 5.83$$

Multiply both sides by c and solve for c.

The receiver is about 5.8 m from the vertex of the dish.

BY THE WAY

To express the location of objects in space, modern astronomers use a form of the equatorial system that the Chinese developed about 2400 B.C. This system divides the sky into 28 sections.

Parabolas that Open Right or Left

So far, you have seen parabolas that open either up or down. If a parabola opens to the left or right, its equation changes, as stated below.

Equation of a Parabola that Opens Right or Left

The graph of $x - h = \frac{1}{4c}(y - k)^2$ is a parabola that:

- has vertex at (h, k) and focus at $(h + c, k)$
- has directrix with equation $x = h - c$
- has a line of symmetry with equation $y = k$
- opens right if $c > 0$ and opens left if $c < 0$

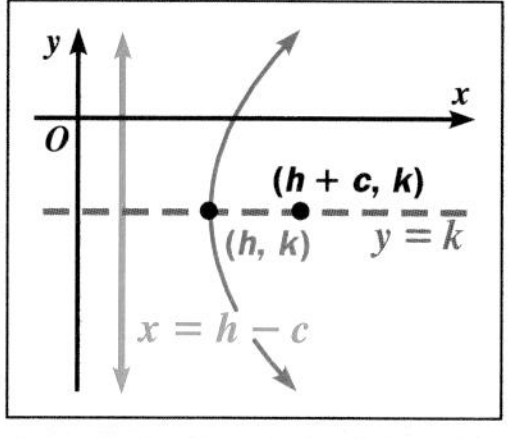

Teach⇔Interact

About Example 2

Teaching Tip

Example 2 shows students how to find the equation of a parabola that models the cross section of a radio telescope. Stress the importance of sketching the parabola at a convenient location in a coordinate plane, the location with vertex (0, 0). This simplifies finding the focus. Point out that the parabola for the cross section must be identical to this model.

Additional Example 2

A beam projector is a portable lighting instrument used around theatrical stages. Light rays emanating from the bulb at the focus of a parabolic reflector are projected out as parallel rays. A typical size for the parabolic reflector has a width of 8 in. and a depth of 1 in. How far from the vertex is the bulb located?

Step 1 Sketch a parabolic cross section of the parabolic reflector in a coordinate plane so that the vertex of the parabola is at (0, 0). Then an equation of the parabola is $y = \frac{1}{4c}x^2$, and the focus is c inches above the vertex.

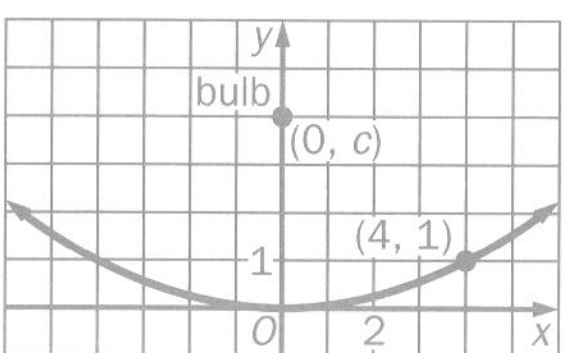

Step 2 Choose a point on the parabolic reflector of the projector, as shown. The x-coordinate of this point is equal to half the diameter of the reflector, 4 in. The y-coordinate of this point is equal to the depth of the reflector, 1 in.

$$1 = \frac{1}{4c}(4)^2$$

$$c = \frac{1}{4} \cdot 16$$

$$c = 4$$

The bulb is 4 in. from the vertex of the reflector.

Teach⇔Interact

Additional Example 3

An equation for a parabola is $x = \frac{1}{8}(y - 2)^2 - 8$. Find the vertex, focus, and directrix of the parabola. Sketch the parabola with its focus and directrix.

Step 1 The given equation involves squaring y, so its graph is a parabola that opens left or right.

Step 2 Find the values of h and k. You can rewrite the equation as $x - (-8) = \frac{1}{8}(y - 2)^2$, so $h = -8$ and $k = 2$.

Step 3 Find the value of c.

$$\frac{1}{4c} = \frac{1}{8}$$
$$\frac{1}{4} = \frac{c}{8}$$
$$2 = c$$

Because $c > 0$, the parabola opens to the right.

Step 4 Find the vertex, focus, and directrix.

Vertex: $(h, k) = (-8, 2)$

Focus: $(h + c, k) = (-8 + 2, 2) = (-6, 2)$

Directrix: $x = h - c = -8 - 2 = -10$

Step 5 Sketch the graph. When $y = -4$, $x = -3.5$, so the parabola passes through $(-3.5, -4)$.

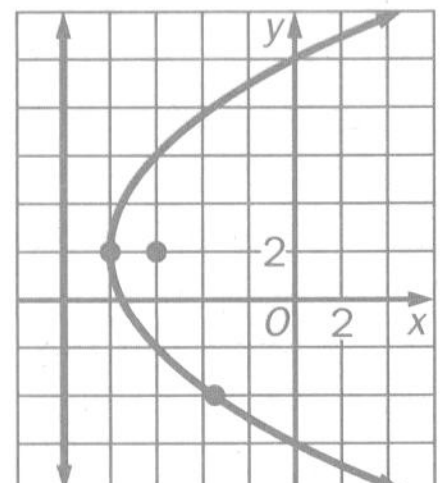

Closure Question

Explain how to find an equation of a parabola given its focus and directrix.

Sketch the focus and directrix on a coordinate plane. Determine the general form of the parabola to use from their position. Find the vertex by finding the midpoint of the perpendicular segment joining the focus and directrix. Find the value of c using the appropriate general form and substitute the vertex and c into the same general form.

EXAMPLE 3

An equation for a parabola is $x = -\frac{1}{3}(y + 1)^2 + 2$. Find the vertex, focus, and directrix of the parabola. Sketch the parabola with its focus and directrix.

SOLUTION

Step 1 Recognize the general form of the equation. The given equation involves squaring y, so its graph is a parabola that opens left or right.

Step 2 Find the values of h and k. You can rewrite the equation as

$$x - 2 = -\frac{1}{3}(y - (-1))^2,$$

so $h = 2$ and $k = -1$.

Step 3 Find the value of c.

$$\frac{1}{4c} = -\frac{1}{3}$$
$$\frac{1}{4} = -\frac{1}{3}c$$
$$-\frac{3}{4} = c$$

Multiply both sides by c and solve for c.

Because $c < 0$, the parabola opens to the left.

Step 4 Find the vertex, focus, and directrix.

Vertex: $(h, k) = (2, -1)$

Focus: $(h + c, k) = \left(2 + \left(-\frac{3}{4}\right), -1\right) = \left(1\frac{1}{4}, -1\right)$

Directrix: $x = h - c = 2 - \left(-\frac{3}{4}\right) = 2\frac{3}{4}$

Step 5 Sketch the parabola with its focus and directrix.

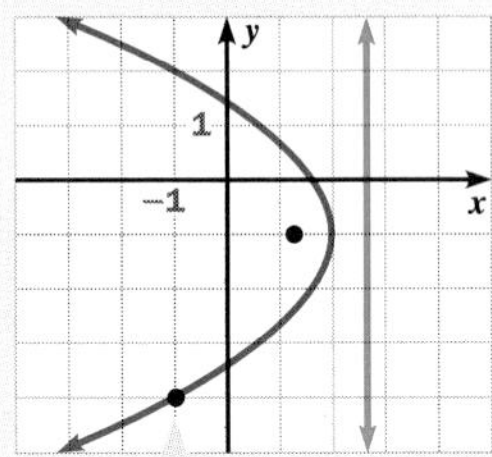

When $y = -4$, $x = -1$, so the parabola passes through the point $(-1, -4)$.

CHECKING KEY CONCEPTS

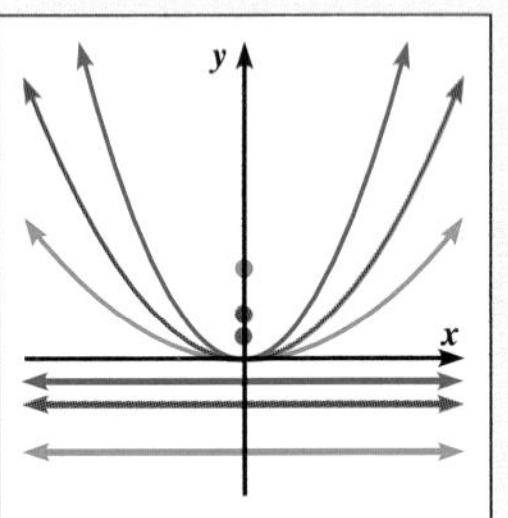

1. The parabolas at the right all have the same vertex. Each parabola is the same color as its directrix and focus. Explain how the distance between the focus and the directrix affects the shape of the parabola.

Find an equation of the parabola with focus *F* and directrix *d*.

2. $F(0, 4); d: y = -2$

3. $F(1, 3); d: y = 0$

4. $F(4, -3); d: y = 5$

5. $F(-2, -1); d: x = 4$

Checking Key Concepts

1. The farther apart the directrix and focus are, the wider the parabola becomes.
2. $y - 1 = \frac{1}{12}x^2$
3. $y - \frac{3}{2} = \frac{1}{6}(x - 1)^2$
4. $y - 1 = -\frac{1}{16}(x - 4)^2$
5. $x - 1 = -\frac{1}{12}(y + 1)^2$

Exercises and Applications

1. Answers may vary. An example is given. The vertex is the midpoint of the perpendicular segment connecting the focus and the directrix. If the directrix is horizontal and the focus is above it, the parabola opens up. If the directrix is horizontal and the focus is below it, the parabola opens down. If the directrix is vertical and the focus is to its right, the parabola opens to the right. If the directrix is vertical and the focus is to its left, the parabola opens to the left.
2. C
3. A
4. B
5. a. $y = -\frac{1}{12}x^2$

   b. 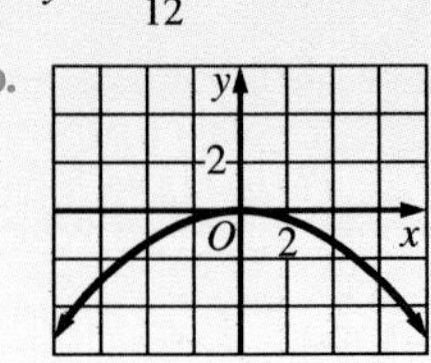

6. a. $y = \frac{1}{8}(x + 3)^2$

11.2 Exercises and Applications

Extra Practice exercises on page 766

1. **Writing** Describe how you can find the vertex of a parabola and tell whether the parabola opens up, down, left, or right if you know its focus and directrix.

Match each equation with its graph.

2. $y - 4 = 2(x - 1)^2$
3. $x - 1 = \frac{1}{4}(y - 2)^2$
4. $x - 1 = 2(y - 2)^2$

A. 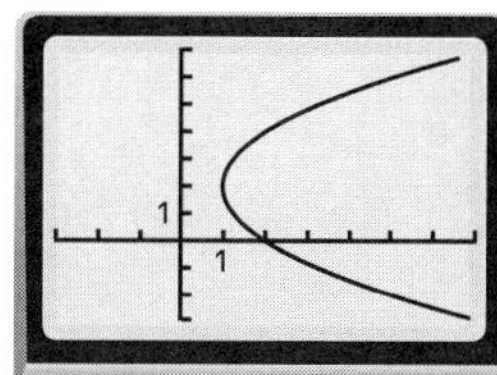

B. 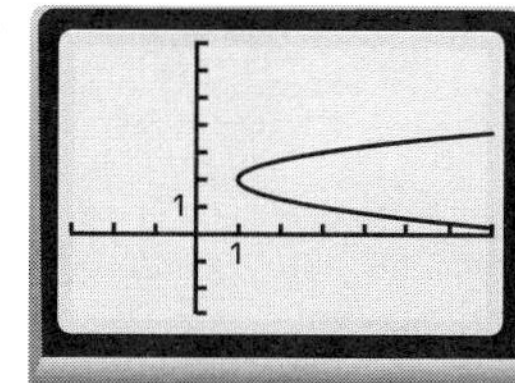

C. 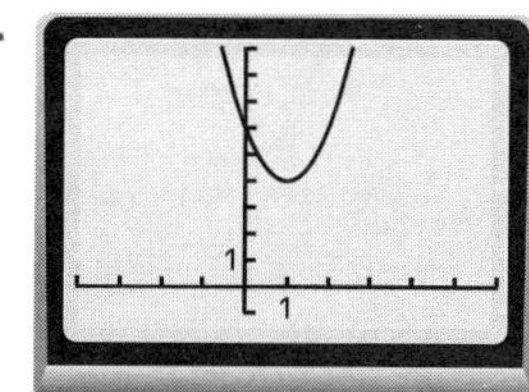

For Exercises 5–10:

a. Find an equation of the parabola with the given characteristics.

b. Graph your equation from part (a).

5. vertex (0, 0); directrix $y = 3$
6. vertex $(-3, 0)$; directrix $y = -2$
7. vertex (0, 0); directrix $y = -1$
8. vertex (0, 0); focus $(0, -4)$
9. focus $(-4, 6)$; vertex (2, 6)
10. vertex $(-3, 3)$; directrix $x = 5$

ASTRONOMY Draw a sketch of a cross section of each radio telescope dish in a coordinate plane. Find the distance of the receiver from the vertex of each dish.

11. The submillimeter telescope at Steward Observatory in Tucson, Arizona, has a width of 10 m and a depth of 3.5 m.
12. The Hobart telescope in Tasmania has a width of 85 ft and a depth of 12.96 ft.
13. The Parkes telescope at the Australia Telescope National Facility has a width of 64 m and a depth of 9.2 m.
14. a. Solve the equation $x = 2(y + 1)^2 - 3$ for y.
    b. **Technology** Graph the equation from part (a) using a graphing calculator or graphing software. Explain what you must do to obtain a graph of the entire parabola.

The Parkes telescope is located in the state of New South Wales in southeastern Australia.

Name the vertex, focus, and directrix of a parabola with the given equation. Sketch the parabola with its focus and directrix. You may want to use a graphing calculator or graphing software to check your work.

15. $y = 3x^2$
16. $x = -\frac{1}{2}y^2$
17. $y = \frac{1}{8}(x + 1)^2 + 4$
18. $y = \frac{1}{16}(x - 3)^2 - 2$
19. $x = 4(y - 1)^2 + 2$
20. $x = \frac{1}{12}(y - 4)^2$
21. $x^2 = y - 2$
22. $x = 2y^2 + 5$
23. $y^2 = 4x + 3$

Apply⇔Assess

Suggested Assignment

Core Course
Day 1 Exs. 1–20
Day 2 Exs. 21–26, 30–35, AYP

Extended Course
Day 1 Exs. 1–20
Day 2 Exs. 21–35, AYP

Block Schedule
Day 71 Exs. 1–26, 30–35, AYP

Exercise Notes

Communication: Discussion
Ex. 1 Ask some students to read their answers to this exercise. Discussing the answers may help clear up any misconceptions.

Problem Solving
Exs. 2–4 Encourage students to analyze the general form of a parabola so they can eliminate possibilities for these exercises.

Teaching Tip
Exs. 11–13 Remind students to draw a diagram of each cross section with the vertex at (0, 0). Then the focal length is the depth of the telescope.

Multicultural Note
Exs. 11–13 The largest radio telescope in the world is in a deep, natural valley near Arecibo, Puerto Rico. Frequently used to study pulsars, this telescope has a reflector dish with a diameter of 1000 feet. Since late 1991, Uruguayan-born Dr. Daniel Altschuler has been the director of the observatory; he is the first director from a Latin American country. Many of the engineers and computer scientists who work for the observatory are from Puerto Rico; Dr. Sixto Gonzalez, who studies atmospheric physics, is the first Puerto Rican member of the scientific staff.

Using Technology
Exs. 14, 16, 19, 20, 22
With a graphing calculator, students need to graph the parabola in two parts if it is not a function. In general, the parts are $Y_1 = k + \sqrt{4c(x - h)}$ and $Y_2 = k - \sqrt{4c(x - h)}$.

b. 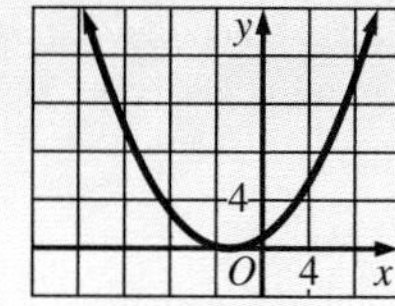

7. a. $y = \frac{1}{4}x^2$

b. 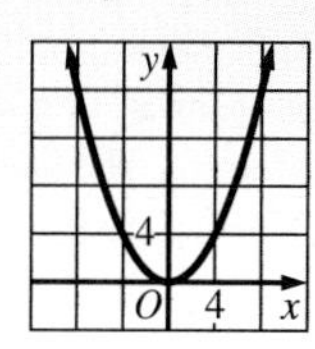

8. a. $y = -\frac{1}{16}x^2$

b. 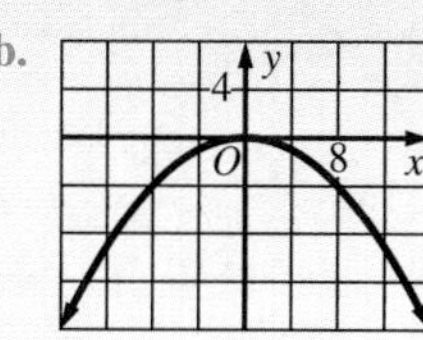

9. a. $x - 2 = -\frac{1}{24}(y - 6)^2$

b. 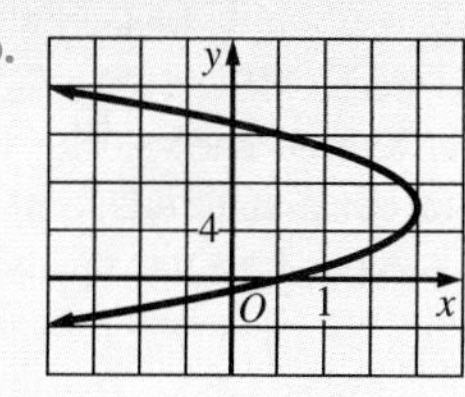

10–23. See answers in back of book.

Apply⇔Assess

Exercise Notes

Using Manipulatives

Ex. 24 Students fold paper in such a way as to physically create a parabola. In part (b), students explain why the figure they have created is a parabola.

Application

Ex. 25 This exercise shows students that lines or rays reflect from a parabola to the focus, thus demonstrating an application of a parabola as a reflector.

Communication: Writing

Ex. 26 Have students review the definition of a function and use it in their explanation of why some parabolas (and later, other quadratics) are not functions.

Communication: Discussion

Ex. 28 The parabolic dishes need to face each other directly because sound waves reflecting off the dishes are essentially parallel to each other. Ask students to explain their reasoning in class.

24. a. Investigation Complete the following steps and describe the results.

Step 1 Use a sheet of lined paper. Turn the paper so that the lines are vertical. Mark a point near the center of the paper.

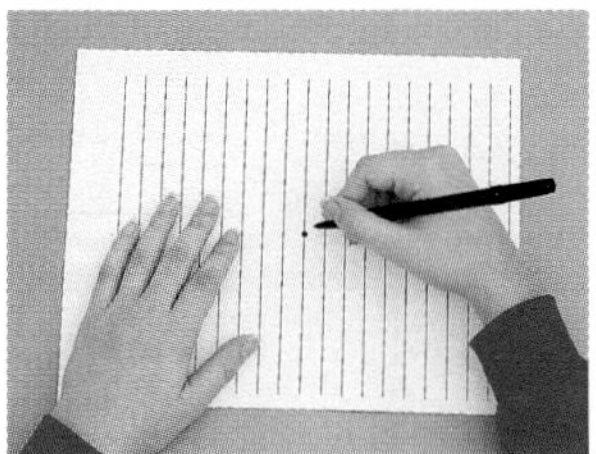

Step 2 Fold the paper up from the bottom so that the bottom edge passes through the point you marked in Step 1.

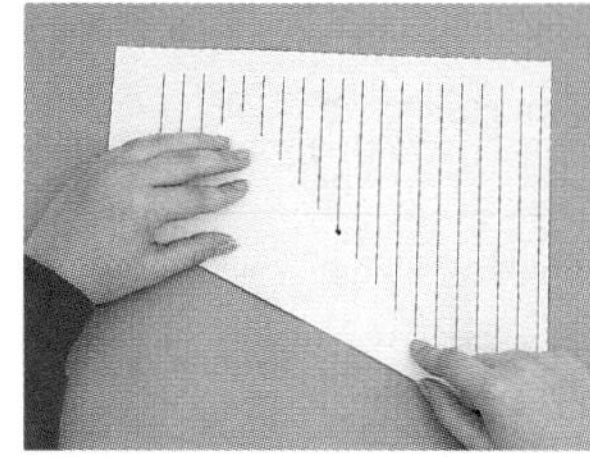

Step 3 Repeat Step 2 about 20 times, each time matching a different point along the bottom edge with the marked point.

b. Explain how you know that the figure traced out by the folds is a parabola. What are the focus and directrix of the parabola?

25. a. Challenge Use your parabola from Exercise 24. Trace along one of the tangent lines you folded. Mark the point where this line is tangent to the parabola. On both sides of the paper, draw a vertical line through the marked point. Fold the paper on the tangent. What happens to the part of the vertical line below the parabola?

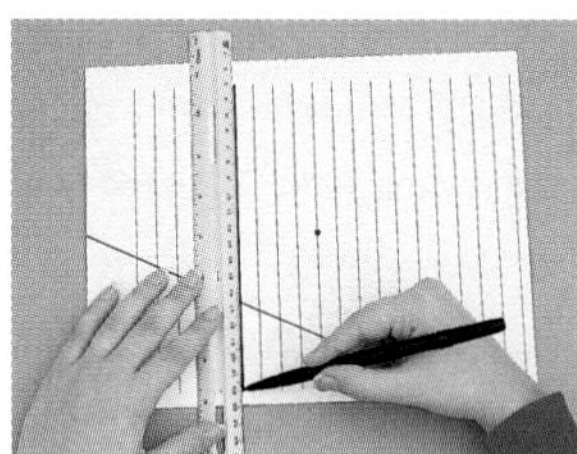

b. The vertical line in part (a) is reflected in the same way that a light ray would be reflected by the parabola. How does this help to explain how the focus got its name?

26. Writing Explain why an equation whose graph is a parabola that opens up or down is a function. Explain why an equation whose graph is a parabola that opens left or right is not a function.

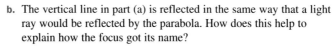

Connection ACOUSTICS

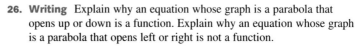

Many science museums have an exhibit that allows two people to whisper to each other over long distances. One person speaks into a ring attached to a parabolic dish while another person listens at the ring attached to a second dish across the room from the first.

27. At the Ann Arbor Hands-On Museum in Ann Arbor, Michigan, the "whisper dishes" are 49 in. wide and 8 in. deep. On graph paper, sketch a cross section of one of the parabolic dishes. Put the vertex of the dish at the origin. Show the location of the ring on your sketch, and explain why it is in the correct position.

28. When the dishes directly face each other, they can be used to transmit messages. Use your results from Exercise 25 to draw a sketch that shows how messages are transmitted. Explain your reasoning.

24. a. Check students' work.

b. Answers may vary. An example is given. Each time you fold the paper, the distance from the marked point to the point on the fold is equal to the distance from the point on the fold to the bottom edge of the paper. By the definition of a parabola given on page 521, the figure is a parabola with focus at the marked point and directrix the bottom edge of the paper.

25. a. The lower part of the vertical line now passes through the marked point, which is the focus of the parabola.

b. Answers may vary. An example is given. Light rays entering a parabolic mirror on a path that is parallel to the axis of the parabola are reflected to the focus, just like the vertical lines on the piece of paper.

26. Answers may vary. An example is given. A parabola that opens up or down has an equation of the form $y = \frac{1}{4c}(x - h)^2 + k$. For a parabola like this, each value of x is associated with exactly one value of y, and so an equation of this form is a function. A parabola that opens left or right has an equation of the form $x = \frac{1}{4c}(y - k)^2 + h$. For a parabola like this, except at the vertex, each value of x is associated with two values of y, and so an equation of this form is not a function.

27.

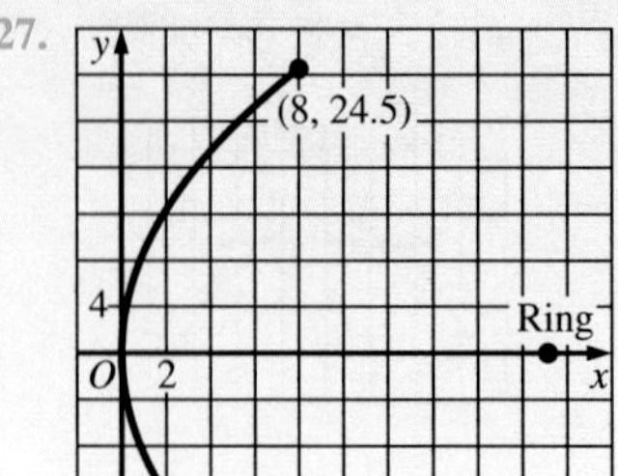

The ring is located at the focus of the parabola, which is approximately at the point (18.8, 0). Point (8, 24.5) is on the rim of the dish, since the dish is 49 in. wide and 8 in. deep. Since $x = \frac{1}{4c}y^2$, $8 = \frac{1}{4c}(24.5)^2$; $c \approx 18.8$ in.

28. See answers in back of book.

29. a. Answers may vary. An example is given. Let (3, 6) be the focus and $y = 2$ be the directrix. The vertex of the resulting parabola is (3, 4), and its axis of symmetry is the line $x = 3$. The

ONGOING ASSESSMENT

29. a. Open-ended Problem Choose the coordinates of a point and an equation of a horizontal or vertical line. Give all of the information that you can about the parabola with the chosen point as its focus and the chosen line as its directrix.

b. Write an equation of the parabola from part (a). Explain the steps you need to take in order to sketch a graph of your equation. Include a sketch in your explanation.

SPIRAL REVIEW

For each pair of points, find:

a. the distance between the points

b. the coordinates of the midpoint of the line segment connecting the points *(Section 11.1)*

30. $(9, -1), (14, -5)$ **31.** $(-8, 5), (16, -11)$ **32.** $(7, 2), (-3, 8)$

Solve each system of equations. *(Section 7.1)*

33. $x - 3y = 0$, $x + y = 5$

34. $y = 4.5x$, $y + 4.5x = 18$

35. $2x = y - 4$, $x + y = 15$

ASSESS YOUR PROGRESS

VOCABULARY

parabola (p. 521) **directrix** (p. 521)
focus (p. 521) **vertex** (p. 521)

For each pair of points, find:

a. the distance between the points

b. the coordinates of the midpoint of the line segment connecting the points *(Section 11.1)*

1. $(-3, 8), (5, -1)$ **2.** $(-6, -3), (7, 1)$

3. Given a line l with equation $y = -\frac{4}{5}x - 6$ and point P with coordinates $(5, -3)$, find an equation of the line through P and:

a. parallel to l **b.** perpendicular to l *(Section 11.1)*

Name the vertex, focus, and directrix of each parabola with the given equation. *(Section 11.2)*

4. $x + 3 = (y - 1)^2$ **5.** $y + 6 = \frac{1}{20}(x + 5)^2$

Find an equation of each parabola with the given characteristics. *(Section 11.2)*

6. vertex $(3, 1)$; directrix $y = -2$ **7.** vertex $(-2.5, 0)$; directrix $x = 1$

8. Journal How are the new ideas about lines and parabolas in Sections 11.1 and 11.2 related to what you already knew about lines and parabolas when you started studying this chapter?

Apply⇔Assess

Exercise Notes

Assessment Note

Ex. 29 Asking students to summarize what they know about constructing a parabola from a given focus and directrix may help you assess and clarify any misunderstandings they may have.

Progress Check 11.1–11.2

See page 556.

Practice 71 for Section 11.2

NAME ______ DATE ______

Practice 71

FOR USE WITH SECTION 11.2

For Exercises 1–8: Exs. 1–8: For graphs in part (b), see additional answers.

a. Find an equation of the parabola with the given characteristics.

b. Graph your equation from part (a).

1. vertex $(0, 0)$; directrix $y = -1$ a. $y = \frac{1}{4}x^2$

2. vertex $(0, 0)$; directrix $x = \frac{1}{2}$ a. $x = -\frac{1}{2}y^2$

3. vertex $(2, 0)$; directrix $y = 2$ a. $y = -\frac{1}{8}(x - 2)^2$

4. vertex $(0, 4)$; focus $(0, 5)$ a. $y - 4 = \frac{1}{4}x^2$

5. focus $\left(\frac{5}{4}, 3\right)$; directrix $x = \frac{3}{4}$ a. $x - 1 = (y - 3)^2$

6. focus $\left(4, -\frac{1}{2}\right)$; vertex $(4, -1)$ a. $y + 1 = \frac{1}{2}(x - 4)^2$

7. focus $\left(-3, -\frac{7}{8}\right)$; directrix $y = -\frac{9}{8}$ a. $y + 1 = 2(x + 3)^2$

8. focus $(-5, 1)$; vertex $(-2, 1)$ a. $x + 2 = -\frac{1}{12}(y - 1)^2$

Name the vertex, focus, and directrix of a parabola with the given equation. Then sketch the parabola. You may want to use a graphing calculator or graphing software to check your work.

Exs. 9–17: See bottom of page. For graphs, see additional answers.

9. $y = \frac{1}{4}x^2$ **10.** $x = -2y^2 + 1$ **11.** $y = -\frac{1}{8}(x - 4)^2$

12. $y = \frac{1}{2}(x - 3)^2 + 1$ **13.** $y^2 = 3 - x$ **14.** $x = 2(y + 1)^2 - 4$

15. $y = -x^2 + 5$ **16.** $x = \frac{1}{12}(y - 1)^2 - 4$ **17.** $y = -4(x + 2)^2 + 3$

18. Sealed-beam automobile headlights have a rear silvered reflector whose cross-section has the shape of a parabola. In order to insure that emerging reflected rays of light will be parallel, the filament (the tiny wire that produces light) is located at the focus of this parabola. For each diameter and depth measurement, find the distance of the filament from the vertex of the reflector.

a. diameter = 4.8 in.; depth = 2.4 in. 0.6 in. **b.** diameter = 6.4 in.; depth = 4 in. 0.64 in.

19. Writing Suppose a parabola has focus (p, q) and directrix $y = -q$. Choose several values of p and q and graph the parabolas that correspond to these values of p and q. Describe, as specifically as you can, how changing these values affects the corresponding parabolas. In particular, describe the relationship between $|q|$ and the shape of the parabola.
Answers may vary. Check students' work.

9. $(0, 0), (0, 1), y = -1$ 10. $(1, 0), \left(\frac{7}{8}, 0\right), x = \frac{9}{8}$ 11. $(4, 0), (4, -2), y = 2$
12. $(3, 1), \left(3, \frac{3}{2}\right), y = \frac{1}{2}$ 13. $(3, 0), \left(\frac{11}{4}, 0\right), x = \frac{13}{4}$ 14. $(-4, -1), \left(-\frac{31}{8}, -1\right), x = -\frac{33}{8}$
15. $(0, 5), \left(0, \frac{19}{4}\right), y = \frac{21}{4}$ 16. $(-4, 1), (-1, 1), x = -7$ 17. $(-2, 3), \left(-2, \frac{47}{16}\right), y = \frac{49}{16}$

Practice Bank, ALGEBRA 2: EXPLORATIONS AND APPLICATIONS

71

parabola opens up, and has no x-intercepts.

b. Since $c = 2$, an equation of the parabola is $y - 4 = \frac{1}{8}(x - 3)^2$. To sketch the parabola, plot the vertex. Then calculate and plot a few more points, such as $(7, 6)$ and $(-1, 6)$. Connect the points with a smooth curve.

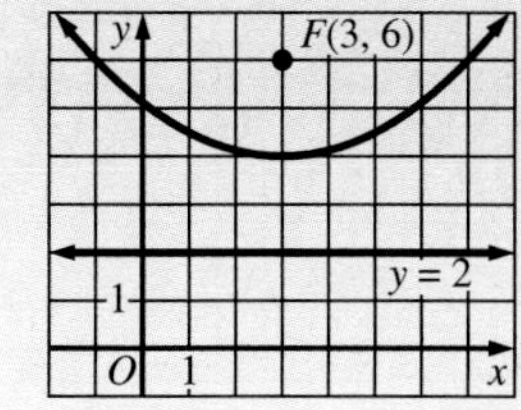

30. a. $\sqrt{41} \approx 6.4$ b. $\left(\frac{23}{2}, -3\right)$

31. a. $8\sqrt{13} \approx 28.84$ b. $(4, -3)$

32. a. $\sqrt{136} \approx 11.7$ b. $(2, 5)$

33. $(3.75, 1.25)$

34. $(2, 9)$

35. $\left(\frac{11}{3}, \frac{34}{3}\right)$

Assess Your Progress

1. a. $\sqrt{145} \approx 12.04$ b. $(1, 3.5)$

2. a. $\sqrt{185} \approx 13.60$ b. $(0.5, -1)$

3. a. $y = -\frac{4}{5}x + 1$ b. $y = \frac{5}{4}x - \frac{37}{4}$

4. vertex: $(-3, 1)$; focus: $\left(-2\frac{3}{4}, 1\right)$; directrix: $x = -3\frac{1}{4}$

5. vertex: $(-5, -6)$; focus: $(-5, -1)$; directrix: $y = -11$

6. $y - 1 = \frac{1}{12}(x - 3)^2$

7. $x + 2\frac{1}{2} = -\frac{1}{14}y^2$

8. See answers in back of book.

Plan⇔Support

Objectives

- Write an equation of a circle.
- Graph an equation of a circle.
- Locate points on a circle.

Recommended Pacing

Core and Extended Courses
Section 11.3: 2 days

Block Schedule
Section 11.3: 1 block

Resource Materials

Lesson Support
Lesson Plan 11.3
Warm-Up Transparency 11.3
Overhead Visuals:
Folder 11: Conic Sections
Practice Bank: Practice 72
Study Guide: Section 11.3
Explorations Lab Manual:
Diagram Master 2

Technology
Technology Book:
TI-92 Activity 5
Graphing Calculator
McDougal Littell Mathpack
Geometry Inventor Activity Book: Activity 19
Internet:
http://www.hmco.com

Warm-Up Exercises

1. Write the formula for the distance d between points (x_1, y_1) and (x_2, y_2).
$d = \sqrt{(x_2 - x_1)^2 + (y_2 - y_1)^2}$
2. Find the distance between the points (–8, 5) and (–12, 2). 5
3. Solve the equation $(y - 2)^2 = 36$ for y. 8; –4
4. Solve the equation $x^2 + y^2 = 25$ for y. $y = \sqrt{25 - x^2}$ or $y = -\sqrt{25 - x^2}$
5. Explain why the graph of a parabola that opens right or left cannot be the graph of a single function.
For each x-value other than the x-coordinate of the vertex, there are two y-values.

Section

11.3 Circles

Learn how to...

- **write an equation of a circle**
- **graph an equation of a circle**

So you can...

- **locate points on a circle, such as the location of a ship just within range of a lighthouse beam**

The Pharos of Alexandria, Egypt, is thought to be the first lighthouse ever constructed. Built in about 280 B.C., this lighthouse is considered one of the Seven Wonders of the Ancient World. A fire burning at the top of the lighthouse provided light. The use of reflectors increased the visibility of the light so that it could be seen from a distance of 35 miles. Modern lighthouses, like Boston Light shown above, are powered by electricity.

THINK AND COMMUNICATE

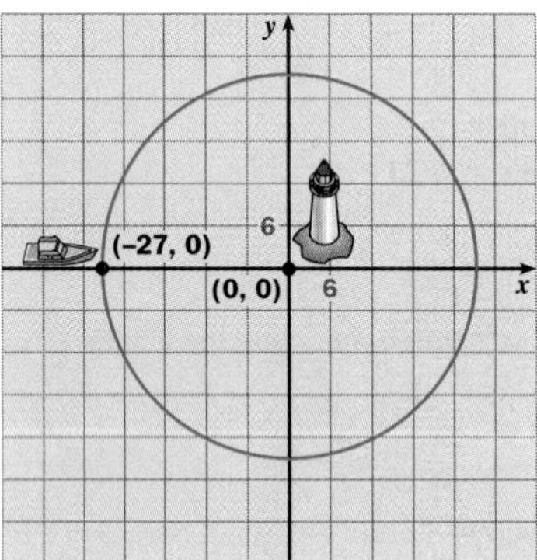

1. The ship shown on the map, where the coordinates are given in miles, has just come within range of the beam from Boston Light. Find the range of the beam.
2. Use the distance formula to find an equation for the distance between the lighthouse and any ship with coordinates (x, y) just within range of the beam.
3. Can a ship with coordinates (24, 12) see the beam? Explain.

The set of points that lie at the outer range of the lighthouse beam form a *circle*. A **circle** is the set of all points in a plane that are the same distance from the **center** of the circle. The **radius** of a circle is the distance between the center of the circle and any point on the circle. Any line containing the center of a circle is a line of symmetry for the circle.

The distance between the center $C(h, k)$ and any point $P(x, y)$ on the circle shown is r. You can use the distance formula to express this relationship as an equation:

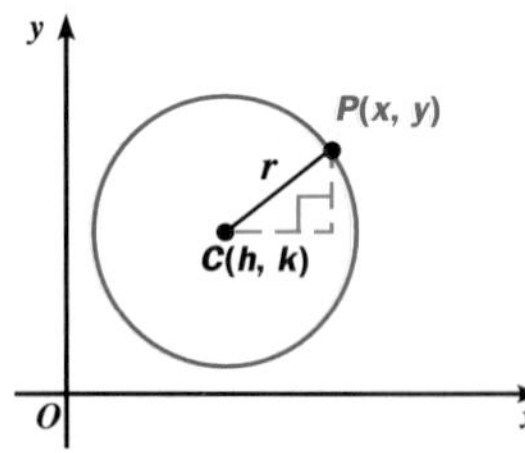

$$r = \sqrt{(x - h)^2 + (y - k)^2}$$
$$r^2 = (x - h)^2 + (y - k)^2$$

ANSWERS Section 11.3

Think and Communicate

1. 27 miles
2. $27 = \sqrt{x^2 + y^2}$
3. Yes.

Standard Form of an Equation of a Circle

An equation of a circle with center $C(h, k)$ and radius r is:

$$(x - h)^2 + (y - k)^2 = r^2$$

EXAMPLE 1

Write an equation of the circle with center (1, −4) and radius 5.

SOLUTION

$$(x - h)^2 + (y - k)^2 = r^2$$

Use the standard form of an equation of a circle.

$$(x - 1)^2 + (y - (-4))^2 = 5^2$$

$$(x - 1)^2 + (y + 4)^2 = 25$$

Substitute **1** for h, **−4** for k, and **5** for r.

EXAMPLE 2

Graph the circle with equation $(x + 3)^2 + (y - 2)^2 = 36$.

SOLUTION

Method 1 Use graph paper.

The center is $(-3, 2)$. ← In the given equation, $h = -3$ and $k = 2$.

The radius is 6. ← In the equation, $r^2 = 36$, so $r = 6$.

Step 1 Plot the center point $(-3, 2)$.

Step 2 The radius is 6. Plot the points that are 6 units above, below, to the left of, and to the right of the center.

Step 3 Sketch the circle through the four points.

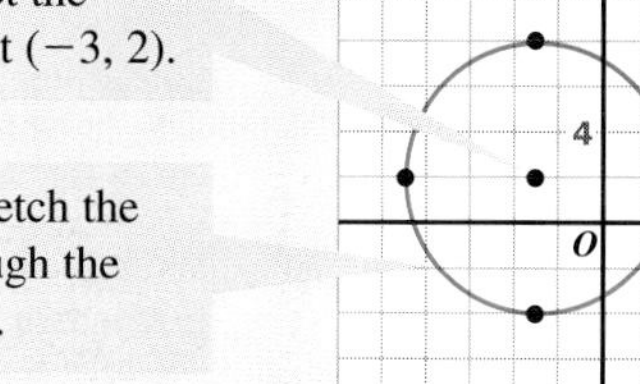

Method 2 Use a graphing calculator or graphing software.

Solve the equation for y.

$$(x + 3)^2 + (y - 2)^2 = 36$$

$$(y - 2)^2 = 36 - (x + 3)^2$$

$$y - 2 = \pm\sqrt{36 - (x + 3)^2}$$

$$y = 2 \pm \sqrt{36 - (x + 3)^2}$$

There are two equations to graph: $y = 2 + \sqrt{36 - (x + 3)^2}$ represents the top half of the circle, and $y = 2 - \sqrt{36 - (x + 3)^2}$ represents the bottom half.

Solution continued on next page.

Technology Note

The term "squaring the viewing window" means to adjust the window so that the units on the x- and y-axis are equal. When the screen is not squared, graphs may look distorted. This is especially necessary when viewing perpendicular lines and circles.

Students may square the window manually on their graphing calculators by setting the ratio of the ranges of the x-values to the y-values to be 3:2. For example, if Xmax − Xmin = 15, then Ymax − Ymin = 10.

Another way to square the window on a TI-82 is to set a window which contains the circle. Then select 5:ZSquare from the ZOOM menu. This will readjust the variables in one direction to square the window.

Teach⇔Interact

About Example 1

Common Error

When negative values are involved, students tend to make sign errors writing the equation of a circle in standard form. Caution students to be careful about this, as shown in Example 1.

Additional Example 1

Write an equation of the circle with center $(-\frac{5}{3}, -2)$ and radius 12.

$$(x - h)^2 + (y - k)^2 = r^2$$

$$\left(x - \left(-\frac{5}{3}\right)\right)^2 + (y - (-2))^2 = 12^2$$

$$\left(x + \frac{5}{3}\right)^2 + (y + 2)^2 = 144$$

Additional Example 2

Graph the circle with equation $(x - 4)^2 + (y + 2)^2 = 25$.

Method 1 Use graph paper. The center is $(4, -2)$. The radius is 5.

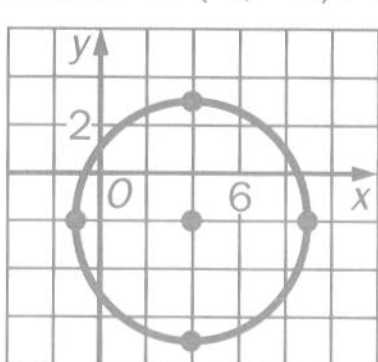

Method 2 Use a graphing calculator or graphing software. Solve the equation for y.

$$(x - 4)^2 + (y + 2)^2 = 25$$

$$(y + 2)^2 = 25 - (x - 4)^2$$

$$y + 2 = \pm\sqrt{25 - (x - 4)^2}$$

$$y = -2 \pm \sqrt{25 - (x - 4)^2}$$

There are two equations to graph: $y = -2 + \sqrt{25 - (x - 4)^2}$ represents the top half of the circle, and $y = -2 - \sqrt{25 - (x - 4)^2}$ represents the bottom half.

Enter the equations into the calculator or graphing software. Set an appropriate square window and graph the circle.

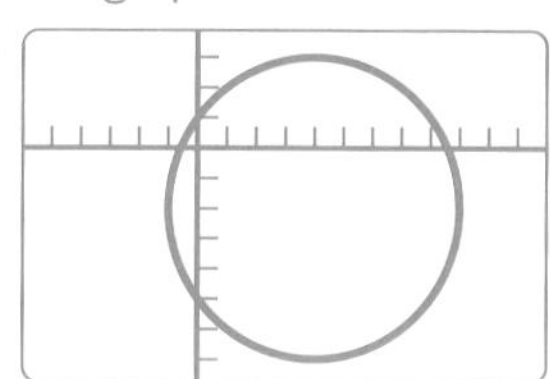

Teach⇔Interact

Additional Example 3

Find the points of intersection of the graphs of $(x + 3)^2 + (y - 2)^2 = 16$ and $(x - 4)^2 + (y - 3)^2 = 36$.

Method 1 Solve the system of equations.

Step 1 Expand both equations and subtract like terms.

$$\begin{aligned} x^2 + 6x + 9 + y^2 - 4y + 4 &= 16 \\ x^2 - 8x + 16 + y^2 - 6y + 9 &= 36 \\ \hline 14x - 7 + 2y - 5 &= -20 \end{aligned}$$

Solve for y.

$14x + 8 = -2y$

$-7x - 4 = y$

Step 2 Substitute the equation from Step 1 into one of the original equations and solve.

$$\begin{aligned} (x + 3)^2 + (-7x - 4 - 2)^2 &= 16 \\ x^2 + 6x + 9 + 49x^2 + 84x + 36 &= 16 \\ 50x^2 + 90x + 29 &= 0 \end{aligned}$$

$x = \dfrac{-90 \pm \sqrt{90^2 - 4(50)(29)}}{2(50)}$

$x = \dfrac{-9 \pm \sqrt{23}}{10}$

$x \approx -0.42, -1.38$

Step 3 Find the coordinates of the points of intersection.

When $x = \dfrac{-9 + \sqrt{23}}{10}$,

$y = -7\left(\dfrac{-9 + \sqrt{23}}{10}\right) - 4 \approx -1.06.$

When $x = \dfrac{-9 - \sqrt{23}}{10}$,

$y = -7\left(\dfrac{-9 - \sqrt{23}}{10}\right) - 4 \approx 5.66.$

The circles intersect at about (−0.42, −1.06) and at about (−1.38, 5.66).

Method 2 Use a graphing calculator or graphing software.

Step 1 Solve the equations for y.

```
Y1=2+√(16-(X+3)2
)
Y2=2-√(16-(X+3)2
)
Y3=3+√(36-(X-4)2
)
Y4=3-√(36-(X-4)2
)
```

Step 2 Find the point(s) of intersection.

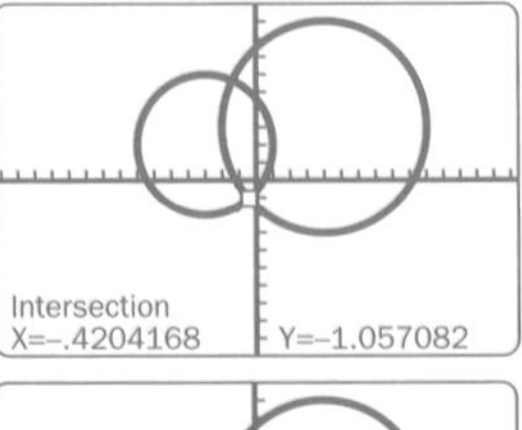

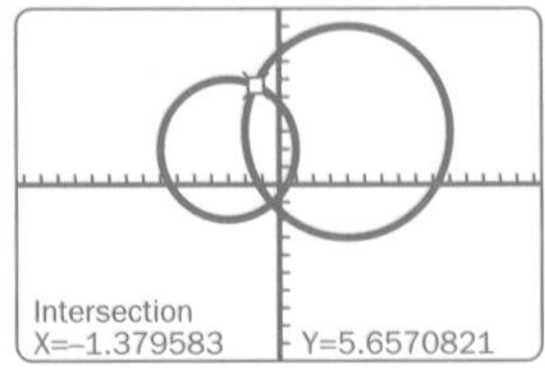

The graphs intersect at about (−0.42, −1.06) and at about (−1.38, 5.66).

For more information about squaring the viewing window, see the *Technology Handbook*, p. 805.

SOLUTION *continued*

Enter the equations into a graphing calculator or graphing software.

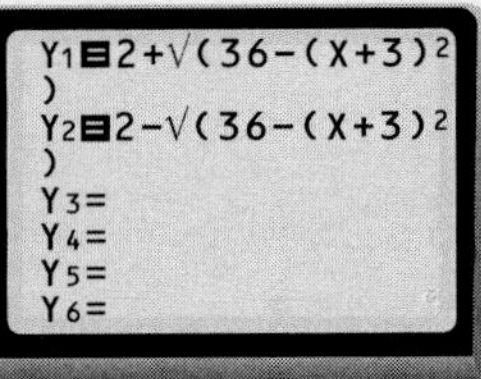

Set an appropriate square window and graph the circle.

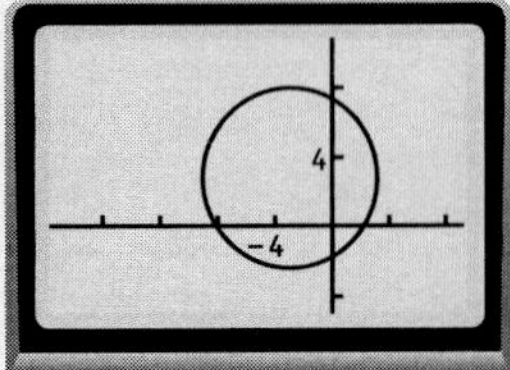

EXAMPLE 3

Find the points of intersection of the graphs of $(x + 4)^2 + (y - 1)^2 = 25$ and $(x - 2)^2 + (y - 3)^2 = 9$.

SOLUTION

Method 1 Solve the system of equations.

Step 1 Expand both equations and subtract like terms.

$$\begin{aligned} x^2 + 8x + 16 + y^2 - 2y + 1 &= 25 \\ x^2 - 4x + 4 + y^2 - 6y + 9 &= 9 \\ \hline 12x + 12 \quad + 4y - 8 &= 16 \\ 12x - 12 &= -4y \\ -3x + 3 &= y \end{aligned}$$

Solve for y.

Step 2 Substitute the equation from Step 1 into one of the original equations and solve.

Substitute $-3x + 3$ for y.

$$\begin{aligned} (x - 2)^2 + (-3x + 3 - 3)^2 &= 9 \\ x^2 - 4x + 4 + 9x^2 &= 9 \\ 10x^2 - 4x - 5 &= 0 \end{aligned}$$

Write the equation in standard form.

Use the quadratic formula.

$$x = \frac{-(-4) \pm \sqrt{(-4)^2 - 4(10)(-5)}}{2(10)}$$

$$x = \frac{2 \pm 3\sqrt{6}}{10}$$

$$x \approx 0.93, -0.53$$

Step 3 Find the coordinates of the points of intersection.

When $x = \dfrac{2 + 3\sqrt{6}}{10}$, $y = -3\left(\dfrac{2 + 3\sqrt{6}}{10}\right) + 3 \approx 0.20.$

When $x = \dfrac{2 - 3\sqrt{6}}{10}$, $y = -3\left(\dfrac{2 - 3\sqrt{6}}{10}\right) + 3 \approx 4.60.$

The circles intersect at about (0.93, 0.20) and at about (−0.53, 4.60).

Method 2 Use a graphing calculator or graphing software.

Step 1 Solve the equations for y. Enter the equations into the calculator or software.

```
Y1=1+√(25-(X+4)²)
Y2=1-√(25-(X+4)²)
Y3=3+√(9-(X-2)²)
Y4=3-√(9-(X-2)²)
```

Step 2 Find the point(s) of intersection.

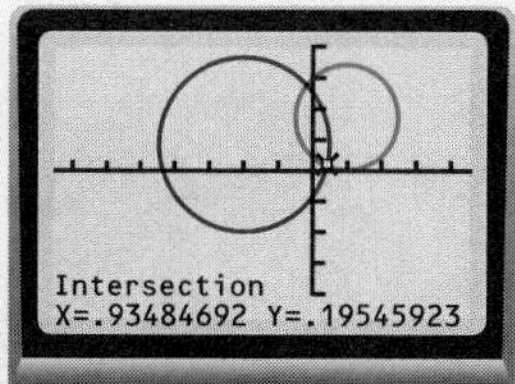

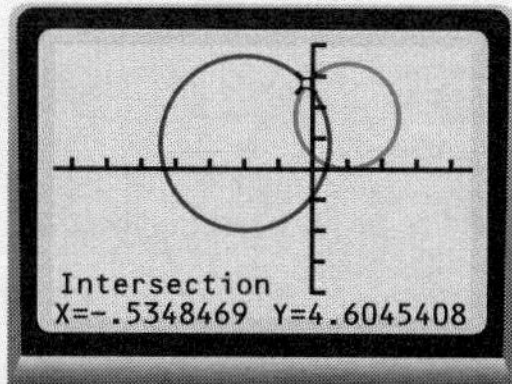

The graphs intersect at about (0.93, 0.20) and at about (−0.53, 4.60).

THINK AND COMMUNICATE

4. Why do you have to enter two equations to graph a circle on a graphing calculator?
5. The circles in Example 3 intersect at two points. Can two circles intersect at only one point? at more than two points? Explain.

CHECKING KEY CONCEPTS

For each equation of a circle, identify the center and the radius.

1. $(x-6)^2+(y-4)^2=144$
2. $(x+1)^2+(y-3)^2=10$
3. $(x+5)^2+y^2=36$
4. $(x+3)^2+(y+2)^2=18$

For each graph, write an equation of the circle.

5. 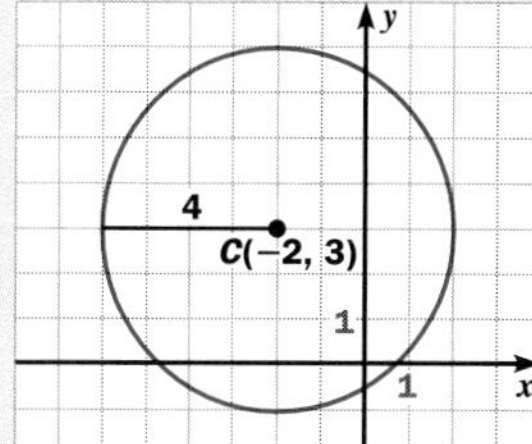

6. 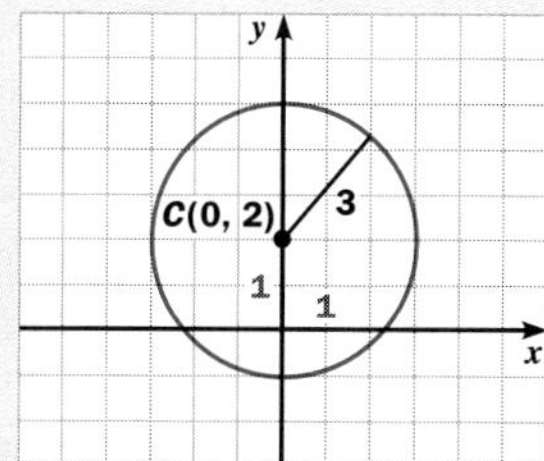

Graph each equation.

7. $(x-5)^2+(y+1)^2=4$
8. $x^2+(y-3)^2=25$
9. Find the point(s) of intersection of the graphs of $x^2+(y-1)^2=9$ and $x^2+y^2=4$.

Teach⇔Interact

Think and Communicate

Question 4 can be used to review the concept of function. Question 5 can be used to introduce the concepts of consistent and dependent systems in relation to circles. You may wish to extend the discussion by asking about circles that do not intersect.

Checking Key Concepts

Common Error

For questions 5 and 6, some students may make errors in finding the equation of a circle when its center is not (0, 0). Encourage them to practice translating the circle $x^2+y^2=r^2$ and then writing the new equation of the circle using the standard form.

Closure Question

Describe how to write an equation of a circle given its center and radius.

Use the standard form of the equation of a circle, $(x-h)^2+(y-k)^2=r^2$, where (h, k) is the center, and r is the radius.

Think and Communicate

4. Answers may vary. An example is given. You have to enter two equations because the equation of a circle is not a function. The upper half and lower half of a circle do represent functions. The upper half involves a positive square root, and the lower half involves a negative square root.
5. Yes; two circles can intersect at one point if they are tangent to one another. No; two circles cannot intersect at more than two points.

Checking Key Concepts

1. $C(6, 4)$; $r=12$
2. $C(-1, 3)$; $r=\sqrt{10}\approx 3.16$
3. $C(-5, 0)$; $r=6$
4. $C(-3, -2)$; $r=3\sqrt{2}\approx 4.24$
5. $(x+2)^2+(y-3)^2=16$
6. $x^2+(y-2)^2=9$
7. 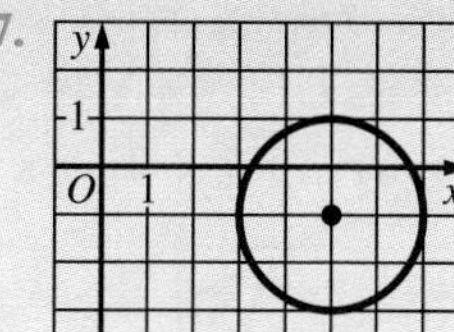

8. 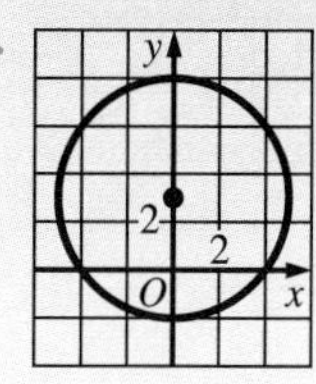

9. (0, −2)

Apply⇔Assess

Suggested Assignment

Core Course
Day 1 Exs. 1–21
Day 2 Exs. 24, 27–30, 33–35, 40–46

Extended Course
Day 1 Exs. 1–9 odd, 10–14, 14–23 odd
Day 2 Exs. 24–46

Block Schedule
Day 72 Exs. 1–21, 24, 27–30, 33–35, 40–46

Exercise Notes

Student Progress
Exs. 1–7, 15–23 Students should be able to complete these exercises quickly and accurately. You may want to have students check their answers with a partner before moving on to the rest of the exercises.

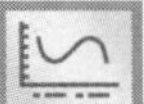

Using Technology
Exs. 1–9 Students may want to check their hand-drawn graphs with a graphing calculator. For Ex. 8, note that with technology, the completing-the-square method is not necessary. In Ex. 9, however, completing the square must be used to solve the equation for *y*.

Topic Spiraling: Review
Exs. 8, 9 Students can find the equation of a circle in standard form for these exercises by completing the square (see Section 5.4). From the standard form, they can identify the center and radius.

Interview Note
Exs. 10–12 These exercises illustrate a typical application of circles to define regions of a map. You may want to point out that the area of each concentric circle gets larger by the square of the radius. This may have implications for the adequacy of having only one fire station.

Visual Thinking
Ex. 24 Ask students to sketch the circle on a piece of graph paper. Then, ask them to sketch the circles that have these two endpoints as their radius and to find the equations of those two circles. This activity involves the visual skills of *recognition* and *exploration.*

11.3 Exercises and Applications

Extra Practice exercises on page 766

For each equation of a circle, identify the center and the radius. Then graph the equation.

1. $(x-4)^2+(y-2)^2=25$
2. $x^2+y^2=16$
3. $x^2+(y-1)^2=4$
4. $(x-3)^2+(y+5)^2=6^2$
5. $(x+1)^2+(y+1)^2=10$
6. $(x+6)^2+y^2=1$
7. $x^2+(y+3)^2=2$
8. $x^2-10x+25+y^2=4$
9. $x^2+y^2-2y+1=8$

INTERVIEW Kija Kim

Look back at the article on pages 510–512.

The maps that Kija Kim's company makes are used by a variety of other companies and organizations. Kim's company created the computerized map below for a Massachusetts fire department. The map illustrates the estimated fire engine response time from the fire station, which is located at the origin.

A fire engine leaving the station can reach a fire that is within 0.6 mi from the station within 3 min.

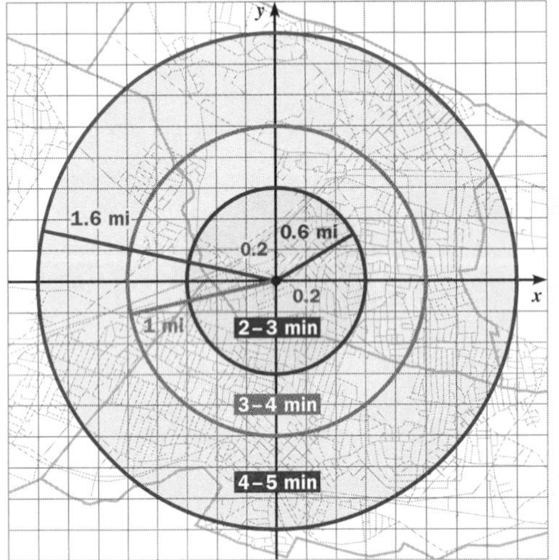

10. Write an equation of the circle defining the region that a fire engine leaving the station can reach within 3 min. What is the radius of this circle?

11. Repeat Exercise 10 for the circle defining the region that a fire engine can reach within 4 min.

12. Repeat Exercise 10 for the circle defining the region that a fire engine can reach within 5 min.

13. Suppose the station receives an alarm from a building located at the map coordinates (0.9, 1.2). The fire chief says that the building is near the outer edge of the ring defining the 4–5 min response time. Do you agree? Explain your reasoning.

14. Suppose a second fire station located at (1.6, 1.6) sometimes sends fire engines to help the fire department referred to in Exercises 10–13.

a. Suppose the second station assists with fires within a 1 mi radius. Write an equation of the circle defining the region that the second station covers.

b. Would the second station assist with a fire at the building in Exercise 13?

Exercises and Applications

1. $C(4, 2)$; $r = 5$

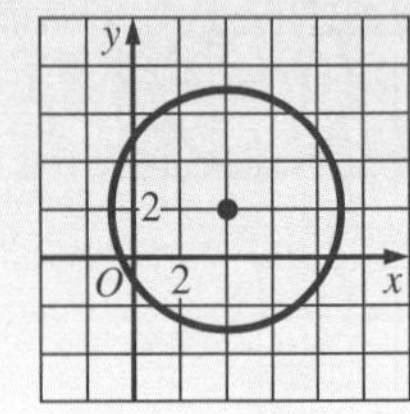

2. $C(0, 0)$; $r = 4$

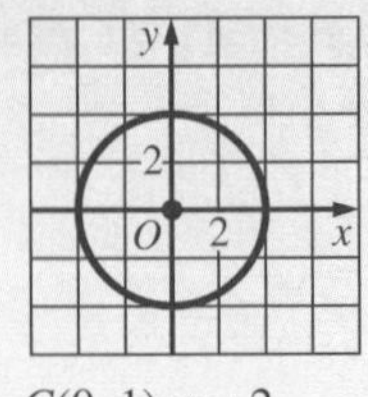

3. $C(0, 1)$; $r = 2$

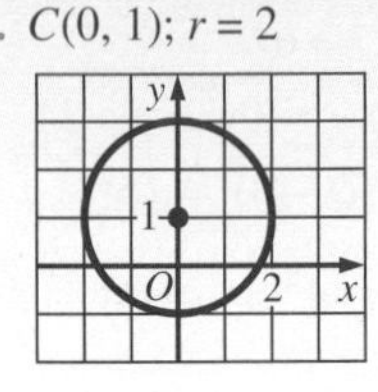

4. $C(3, -5)$; $r = 6$

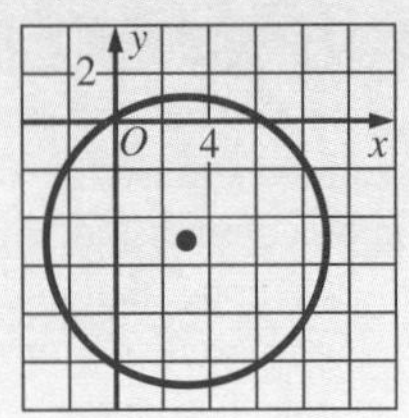

5–14. See answers in back of book.

Write an equation of the circle with the given center *C* and radius *r*.

15. $C(0, 0); r = 7$ **16.** $C(2, 5); r = 3$ **17.** $C(-1, 6); r = 3$

18. $C(-3, 2); r = \sqrt{6}$ **19.** $C(0, \sqrt{3}); r = 11$ **20.** $C(-4, -2\sqrt{2}); r = 3$

21. $C(a, b); r = 11$ **22.** $C(a, -b); r = 2k$ **23.** $C(-a, 2b); r = c^2$

24. A diameter of a circle has endpoints $(-2, -3)$ and $(4, 1)$. Find an equation of the circle.

Connection COMMUNICATIONS

A cellular phone network uses towers like the one shown to transmit calls. When a person makes a call, a tower forwards the signal to a switching station. The switching station transfers the call to the regional telephone office, which then processes the call. The range of a tower can be defined by a circle.

25. Suppose a cellular phone service company is installing towers in a town. The current tower locations are shown on the map, where the coordinates are given in miles.

a. Tower A is located at the origin and has a range of 2 mi. Find an equation of the circle that defines the tower's range.

b. Write an equation of the circle that defines the range of tower B.

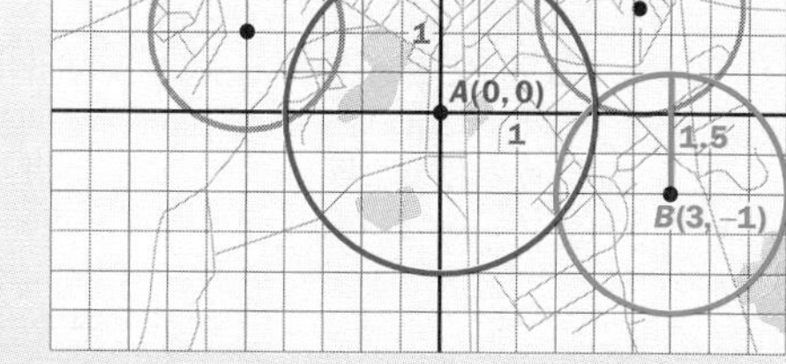

26. The company determines that additional towers should be installed to provide more complete service in the town.

a. **Open-ended Problem** Select a location for a new tower in a region that does not have coverage. Determine a reasonable range for the tower.

b. Write an equation of the circle that defines the tower's range.

Use substitution to determine if each given point is on the circle with equation $(x + 1)^2 + (y - 5)^2 = 16$.

27. $(-1, 1)$ **28.** $(3, 4)$ **29.** $(-5, 5)$ **30.** $(-3.4, 1.8)$

31. Writing Is the center of a circle a point on the circle? Why or why not?

32. Challenge The point $(-4, -1)$ is on the circle with equation $(x - 1)^2 + (y + 3)^2 = 29$. Write an equation of the line that is tangent to the circle and passes through the point $(-4, -1)$. (*Hint:* The tangent is perpendicular to the radius with an endpoint at $(-4, -1)$.)

33. Challenge Find an equation of the circle that passes through the points $(3, 6)$, $(-1, -2)$, and $(6, 5)$.

Find the point(s) of intersection of each pair of circles with the given equations.

34. $(x - 3)^2 + y^2 = 24$, $x^2 + y^2 = 9$

35. $x^2 + (y - 1)^2 = 16$, $x^2 + (y + 1)^2 = 28$

36. $(x + 3)^2 + (y + 2)^2 = 18$, $(x + 2)^2 + (y + 1)^2 = 11$

Apply⇔Assess

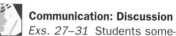

Exercise Notes

Teaching Tip

Ex. 24 Students need to find the radius of the circle and the mid-point of the diameter to find the equation of this circle. You may wish to review the appropriate formulas.

Application

Exs. 25, 26 Some students may have used their family's cellular phone to make calls and can discuss billing rates with their parents. If possible, ask these students to explain how cellular phone rates are determined, including what "roaming" charges are and how they relate to these exercises.

Second-Language Learners

Exs. 25, 26 Students learning English may not be familiar with some of the terms in the first paragraph, such as *cellular phones* (a kind of cordless phone that can be used within a certain radius), *beacons* (transmitters that receive and send radio signals), and *switching station* (the place where the calls are processed). You may also wish to point out that the word *forwards* means "moves a message along to another place or destination."

Communication: Discussion

Exs. 27–31 Students sometimes forget that if a point is a solution to an equation that it can be substituted for the *x*- and *y*-values in the equation. Discuss the concepts of the interior of a circle, the circle itself, and the exterior of a circle. Ask students to explain why the interior points of a circle are not points of the circle. (Refer them to the definition of a circle.) This discussion will help students to answer Ex. 31.

Problem Solving

Ex. 33 Students should use the distance formula to find an expression for the distance from each point to the center of the circle, (x, y). They can then set the distances equal to each other to form two equations and solve for x and y. Substituting these values along with one of the original points into the equation of the circle will enable them to find r.

15. $x^2 + y^2 = 49$
16. $(x - 2)^2 + (y - 5)^2 = 9$
17. $(x + 1)^2 + (y - 6)^2 = 9$
18. $(x + 3)^2 + (y - 2)^2 = 6$
19. $x^2 + (y - \sqrt{3})^2 = 121$
20. $(x + 4)^2 + (y + 2\sqrt{2})^2 = 9$
21. $(x - a)^2 + (y - b)^2 = 121$
22. $(x - a)^2 + (y + b)^2 = 4k^2$
23. $(x + a)^2 + (y - 2b)^2 = c^4$
24. $(x - 1)^2 + (y + 1)^2 = 13$
25. a. $x^2 + y^2 = 4$
    b. $(x - 3)^2 + (y + 1)^2 = 2.25$
26. Answers may vary. An example is given.
    a. Place a tower at (1, 3) with a range of 1.5 mi.
    b. $(x - 1)^2 + (y - 3)^2 = 2.25$
27. Yes. 28. No.
29. Yes. 30. Yes.
31. No. Explanations may vary. An example is given. Every point on a circle is a positive distance r from the center. The center is at a distance of 0 from the center.
32. $y = \frac{5}{2}x + 9$
33. $(x - 3)^2 + (y - 1)^2 = 25$
34. about (–1, 2.83) and (–1, –2.83)
35. about (3.46, 3) and (–3.46, 3)
36. about (1.23, –1.73) and (–2.73, 2.23)

Apply⇔Assess

Exercise Notes

Interdisciplinary Problems
Exs. 37–39 These exercises allow students to see how the mathematics of circles can be used in geography to determine the location of a hiker.

Cooperative Learning
Exs. 37–39 You may wish to have students work on Exs. 37–39 with a partner.

Assessment Note
Ex. 40 Having three or four groups show their work to the class would provide useful feedback for all students to evaluate the equations they have written.

Practice 72 for Section 11.3

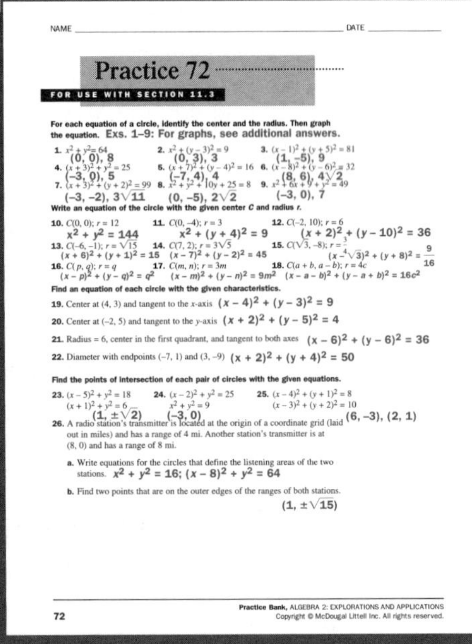

NAME ______________________ DATE ____________

Practice 72

FOR USE WITH SECTION 11.3

For each equation of a circle, identify the center and the radius. Then graph the equation. Exs. 1–9: For graphs, see additional answers.

1. $x^2 + y^2 = 64$ (0, 0), 8
2. $x^2 + (y - 3)^2 = 9$ (0, 3), 3
3. $(x - 1)^2 + (y + 5)^2 = 81$ (1, −5), 9
4. $(x + 3)^2 + y^2 = 25$ (−3, 0), 5
5. $(x + 7)^2 + (y - 4)^2 = 16$ (−7, 4), 4
6. $(x - 8)^2 + (y - 6)^2 = 32$ (8, 6), $4\sqrt{2}$
7. $(x + 3)^2 + (y + 2)^2 = 99$ (−3, −2), $3\sqrt{11}$
8. $x^2 + y^2 + 10y + 25 = 8$ (0, −5), $2\sqrt{2}$
9. $x^2 + 6x + 9 + y^2 = 49$ (−3, 0), 7

Write an equation of the circle with the given center *C* and radius *r*.

10. $C(0, 0)$; $r = 12$ $x^2 + y^2 = 144$
11. $C(0, -4)$; $r = 3$ $x^2 + (y + 4)^2 = 9$
12. $C(-2, 10)$; $r = 6$ $(x + 2)^2 + (y - 10)^2 = 36$
13. $C(-6, -1)$; $r = \sqrt{15}$ $(x + 6)^2 + (y + 1)^2 = 15$
14. $C(7, 2)$; $r = 3\sqrt{5}$ $(x - 7)^2 + (y - 2)^2 = 45$
15. $C(\sqrt{3}, -8)$; $r = \frac{3}{4}$ $(x - \sqrt{3})^2 + (y + 8)^2 = \frac{9}{16}$
16. $C(p, q)$; $r = q$ $(x - p)^2 + (y - q)^2 = q^2$
17. $C(m, n)$; $r = 3m$ $(x - m)^2 + (y - n)^2 = 9m^2$
18. $C(a + b, a - b)$; $r = 4c$ $(x - a - b)^2 + (y - a + b)^2 = 16c^2$

Find an equation of each circle with the given characteristics.

19. Center at (4, 3) and tangent to the *x*-axis $(x - 4)^2 + (y - 3)^2 = 9$
20. Center at (−2, 5) and tangent to the *y*-axis $(x + 2)^2 + (y - 5)^2 = 4$
21. Radius = 6, center in the first quadrant, and tangent to both axes $(x - 6)^2 + (y - 6)^2 = 36$
22. Diameter with endpoints (−7, 1) and (3, −9) $(x + 2)^2 + (y + 4)^2 = 50$

Find the points of intersection of each pair of circles with the given equations.

23. $(x - 5)^2 + y^2 = 18$, $(x + 1)^2 + y^2 = 6$ $(1, \pm\sqrt{2})$
24. $(x - 2)^2 + y^2 = 25$, $x^2 + y^2 = 9$ (−3, 0)
25. $(x - 4)^2 + (y + 1)^2 = 8$, $(x - 3)^2 + (y + 2)^2 = 10$ (6, −3), (2, 1)
26. A radio station's transmitter is located at the origin of a coordinate grid (laid out in miles) and has a range of 4 mi. Another station's transmitter is at (8, 0) and has a range of 8 mi.
    a. Write equations for the circles that define the listening areas of the two stations. $x^2 + y^2 = 16$; $(x - 8)^2 + y^2 = 64$
    b. Find two points that are on the outer edges of the ranges of both stations. $(1, \pm\sqrt{15})$

72 **Practice Bank,** ALGEBRA 2: EXPLORATIONS AND APPLICATIONS

Connection GEOGRAPHY

The NAVSTAR Global Positioning System (GPS) enables people, such as hikers, to locate their position on Earth using a handheld device that communicates with a group of satellites. Suppose three satellites are transmitting signals to a receiver held by a hiker. Each satellite determines a circle on which the hiker is positioned. The intersection of the three circles gives the coordinates of the hiker's location.

37. Write an equation of each circle shown on the map.

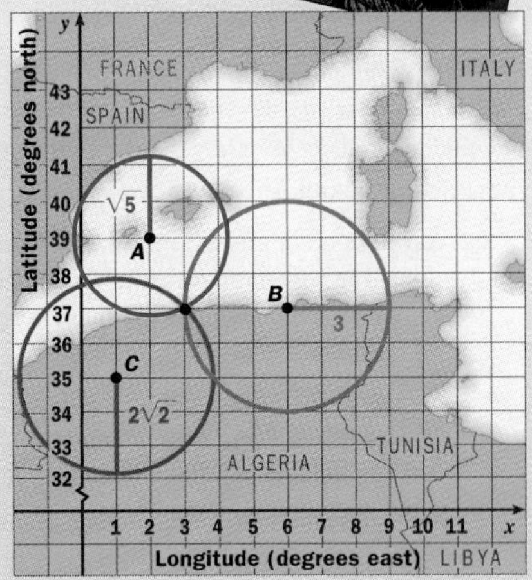

38. **Technology** Use a graphing calculator or graphing software.

a. Find the points of intersection of circles *A* and *B*.

b. Find the points of intersection of circles *A* and *C*.

c. Find the points of intersection of circles *B* and *C*.

39. a. Use your answers to Exercise 38 to find the coordinates for the position of the hiker.

b. **Challenge** Suppose the hiker lives near Barcelona, Spain, at coordinates (3, 42). Use the fact that the hiker's current location and the hiker's home are at the same longitude to find how far from home, in miles, the hiker is. (*Hint:* The radius of Earth is about 3963 mi.)

ONGOING ASSESSMENT

40. **Cooperative Learning** Work with a partner. You will need graph paper.

a. Each of you should draw axes on a piece of graph paper and sketch a circle so that its center is not at the origin. Label the coordinates of the center and a point on your circle. Give your paper to your partner.

b. Write an equation of the circle you receive. Then graph your equation. Does your graph match the one your partner sketched? If not, check your work and correct any errors.

SPIRAL REVIEW

Find an equation of each parabola with the given focus and directrix. Sketch the parabola with its focus and directrix. ***(Section 11.2)***

41. focus (3, 1)
directrix $y = -1$

42. focus $(-2, -1)$
directrix $x = 2$

43. focus $(4, -3)$
directrix $y = 1$

Graph each function. State the domain and range. ***(Section 8.1)***

44. $y = \sqrt{x + 2}$

45. $y = \sqrt{x - 1} + 4$

46. $y = \sqrt{x} + 3$

37. circle *A*: $(x - 2)^2 + (y - 39)^2 = 5$;
circle *B*: $(x - 6)^2 + (y - 37)^2 = 9$;
circle *C*: $(x - 1)^2 + (y - 35)^2 = 8$

38. a. (3, 37) and (4.2, 39.4)
b. (3, 37) and about (0.18, 37.71)
c. (3, 37) and about (3.83, 35)

39. a. (3, 37)
b. about 346 mi

40. Answers may vary. Check students' work.

41. $y = \frac{1}{4}(x - 3)^2$

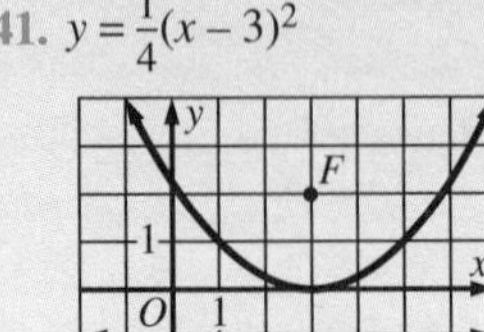

42. $x = -\frac{1}{8}(y + 1)^2$

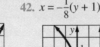

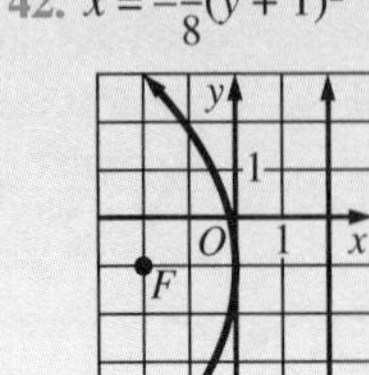

43. $y + 1 = -\frac{1}{8}(x - 4)^2$

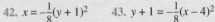

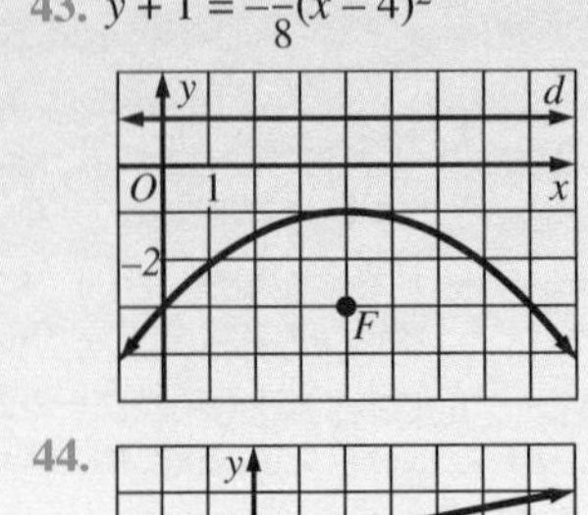

44.

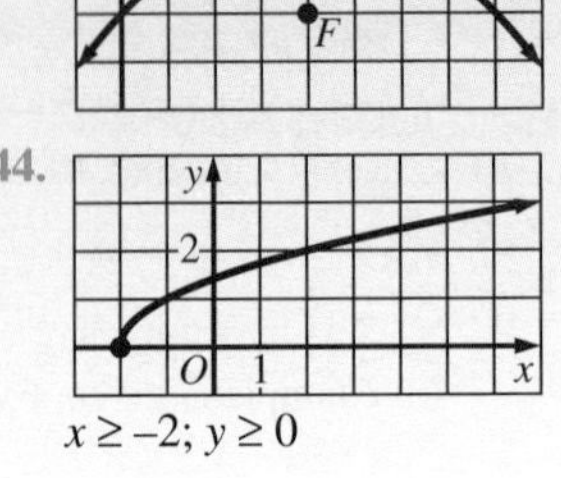

$x \geq -2$; $y \geq 0$

45.

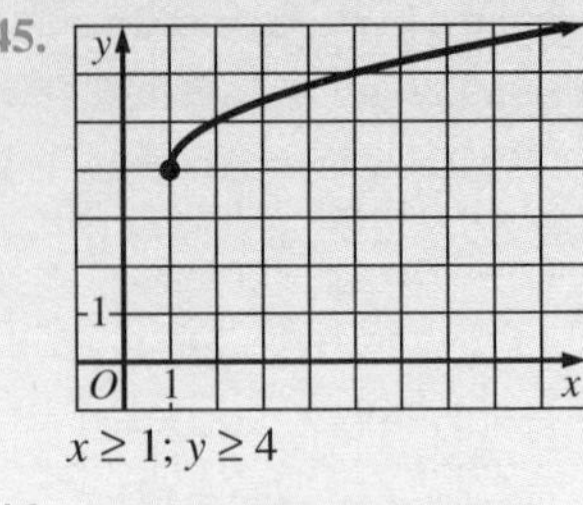

$x \geq 1$; $y \geq 4$

46.

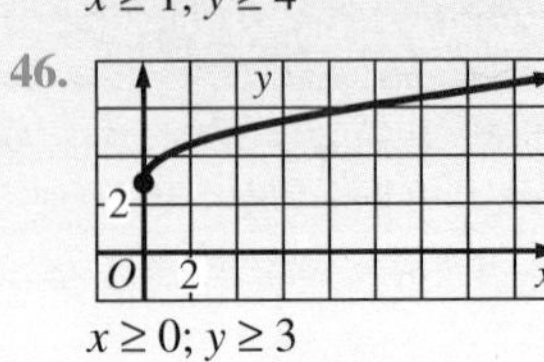

$x \geq 0$; $y \geq 3$

Section

11.4 Ellipses

Learn how to...

- write an equation of an ellipse
- graph an equation of an ellipse

So you can...

- describe an elliptical object, such as the Oval Office in the White House

At one time, astronomers thought that the orbit of a planet around the sun was a circle or a combination of circles. In 1609, the German astronomer Johannes Kepler realized that the orbit of a planet follows an elliptical path, with the sun located at one focus of the elliptical orbit. In the Exploration, you will see how to draw an *ellipse*.

EXPLORATION
COOPERATIVE LEARNING

Drawing an Ellipse

Work with a partner.
You will need:

- graph paper
- a 10-inch piece of string
- two pushpins
- a piece of cardboard

1 Draw axes on a piece of graph paper. Label the points (–4, 0) and (4, 0). Place the graph paper on the cardboard. Use the pins to hold one end of the string at (–4, 0) and the other end at (4, 0).

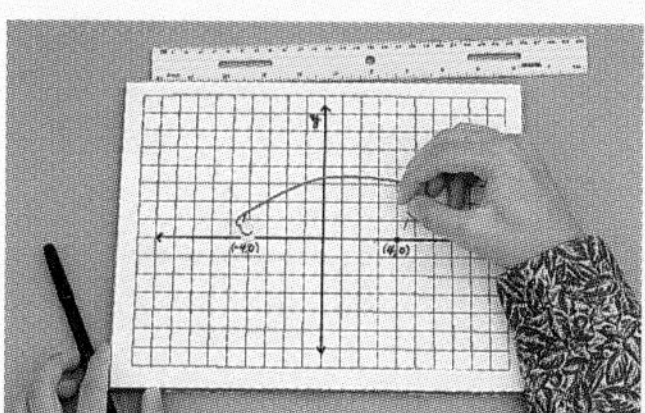

2 Use a pencil to pull the string until it is taut. Move the pencil above and below the x-axis, keeping the string taut, until you have sketched a closed geometric figure called an *ellipse*.

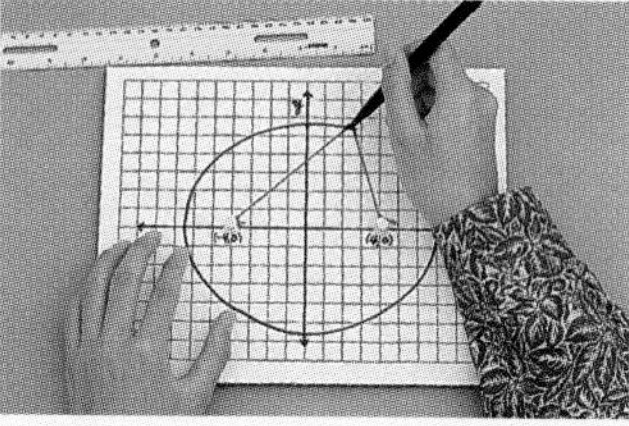

Questions

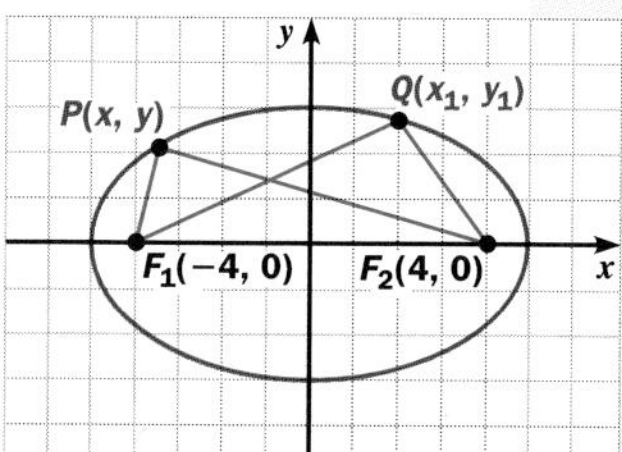

1. If P and Q are any two points on the figure you drew, how do the lengths F_1P and F_2P compare with the lengths F_1Q and F_2Q?
2. Give a definition for the figure you drew using the relationship between F_1P, F_2P, and the length of the string.
3. Move the pins closer together and repeat Steps 1 and 2. How did the ellipse change? What do you think happens to the shape of the ellipse when the pins are moved even closer together?

Plan⇔Support

Objectives

- Write an equation of an ellipse.
- Graph an equation of an ellipse.
- Describe an elliptical object.

Recommended Pacing

Core and Extended Courses
Section 11.4: 1 day

Block Schedule
Section 11.4: $\frac{1}{2}$ block
(with Section 11.5)

Resource Materials

Lesson Support
Lesson Plan 11.4
Warm-Up Transparency 11.4
Overhead Visuals:
Folder 11: Conic Sections
Practice Bank: Practice 73
Study Guide: Section 11.4
Explorations Lab Manual:
Diagram Master 2
Assessment Book: Test 45

Technology
Technology Book:
Spreadsheet Activity 11
Graphing Calculator
Internet:
http://www.hmco.com

Warm-Up Exercises

1. Find the distance between the points (–2, 5) and (6, 9) in a coordinate plane. $4\sqrt{5}$
2. Solve $\sqrt{x^2 + 4} = 7$. $\pm 3\sqrt{5}$
3. Solve $\sqrt{x} + 2 = \sqrt{x + 8}$. 1
4. Express the circle $x^2 + y^2 + 2x - 4y = 31$ in standard form. $(x + 1)^2 + (y - 2)^2 = 36$
5. Solve $(x - 1)^2 + (y - 3)^2 = 25$ for y. $3 \pm \sqrt{25 - (x - 1)^2}$

Exploration Note

Purpose
The purpose of this Exploration is to have students discover that an ellipse is all points in a plane such that the sum of the distances from two fixed points is constant.

Materials/Preparation
Each pair of students needs graph paper, a 10-inch piece of string, two push pins, and a piece of cardboard.

Procedure
Students draw a coordinate axes on graph paper, plot two points, and pin the graph paper to the cardboard at those points. They attach a string to each of the points and use a pencil to pull it taut. By moving the pencil above and below the axis, they trace out an ellipse. Students then repeat the activity with the fixed points in different positions.

Closure
Students should be able to define an ellipse as the set of points in a plane whose distance from two fixed points is constant.

Explorations Lab Manual
See the Manual for more commentary on this Exploration.

For answers to the Exploration, see following page.

Teach⇔Interact

Section Notes

Communication: Discussion The derivation of the equation of the ellipse having foci $F_1(4, 0)$ and $F_2(-4, 0)$ should be discussed with all students in class. Although there are many steps in the derivation, they are straightforward and students should be able to follow the algebra involved. This discussion will enable students to understand how the general formula for an ellipse can be found. Notice that the derivation depends on students being able to solve a radical equation.

Teaching Tip
Some students may need extra help corresponding the geometric parts of an ellipse to its related equation. You could have these students draw a graph like the one at the bottom of the page using the particular values for the ellipse $\frac{x^2}{25} + \frac{y^2}{9} = 1$. Have them label the vertices, foci, center, major axis, and minor axis of their sketch. Make sure students understand that an ellipse has only two vertices, not four, and that they are the endpoints of the major axis.

Communication: Discussion Discuss with students how the standard forms for the equation of an ellipse are generated by translating a horizontal or a vertical ellipse centered at (0, 0). Explain the magnitude of the translation as h units horizontally and k units vertically.

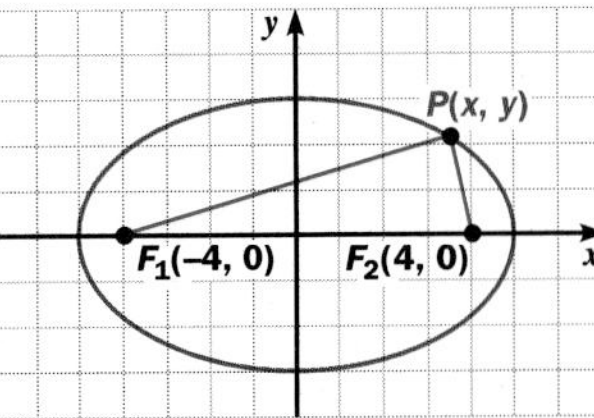

An **ellipse** is the set of all points in a plane such that the sum of the distances between any point and two fixed points F_1 and F_2, called the **foci**, is constant. (*Foci* is the plural of *focus*.) The **center** of the ellipse is the midpoint of $\overline{F_1F_2}$.

You can use the definition of an ellipse to find an equation of an ellipse. The ellipse in the Exploration has its center at the origin and has foci $F_1(-4, 0)$ and $F_2(4, 0)$. The string is 10 inches long. Let $P(x, y)$ be any point on the ellipse.

$$PF_1 + PF_2 = 10$$

Use the distance formula.

$$\sqrt{(x+4)^2 + y^2} + \sqrt{(x-4)^2 + y^2} = 10$$

$$\sqrt{(x-4)^2 + y^2} = 10 - \sqrt{(x+4)^2 + y^2}$$

Square both sides of the equation.

$$(x-4)^2 + y^2 = 100 - 20\sqrt{(x+4)^2 + y^2} + (x+4)^2 + y^2$$

$$(x^2 - 8x + 16) + y^2 = 100 - 20\sqrt{(x+4)^2 + y^2} + (x^2 + 8x + 16) + y^2$$

$$20\sqrt{(x+4)^2 + y^2} = 100 + 16x$$

$$5\sqrt{(x+4)^2 + y^2} = 25 + 4x$$

Square both sides of the equation.

$$25\left[(x+4)^2 + y^2\right] = (25 + 4x)^2$$

$$25\left(x^2 + 8x + 16 + y^2\right) = 625 + 200x + 16x^2$$

Combine like terms and simplify.

$$9x^2 + 25y^2 = 225$$

Divide each term by 225.

$$\frac{x^2}{25} + \frac{y^2}{9} = 1$$

So an equation of the ellipse is $\frac{x^2}{25} + \frac{y^2}{9} = 1$. This equation can be generalized as:

$$\frac{x^2}{a^2} + \frac{y^2}{b^2} = 1$$

The foci are at $(c, 0)$ and $(-c, 0)$.

The points $(a, 0)$ and $(-a, 0)$ are the *vertices* of the ellipse. The **vertices** are the points of intersection of the ellipse and the line containing its foci.

(0, b) (−a, 0) (a, 0) (−c, 0) O (c, 0) (0, −b)

The **major axis** of an ellipse is the line segment that contains the foci and has the two vertices as its endpoints.

The segment with endpoints $(0, b)$ and $(0, -b)$ is the **minor axis** of this ellipse.

The lines containing the axes of an ellipse are lines of symmetry for the ellipse. Also, the values of a, b, and c are related by the equation $c^2 = a^2 - b^2$. You will investigate this relationship in the exercises.

ANSWERS Section 11.4

Exploration

1, 2. Check students' work.

3. The ellipse became more and more round. The ellipse looks more and more like a circle.

Questions

1. $F_1P + F_2P = F_1Q + F_2Q$
2. The figure I drew is the set of all points in a plane such that, for all points, the sum of the distances to the two points F_1 and F_2 is the same.

The summary below describes ellipses whose vertices are aligned either vertically or horizontally and whose centers are at $C(h, k)$.

Standard Forms of an Equation of an Ellipse

An equation of a *horizontal ellipse* (an ellipse with a horizontal major axis) with center $C(h, k)$ is:

$$\frac{(x-h)^2}{a^2} + \frac{(y-k)^2}{b^2} = 1 \quad (a > b > 0)$$

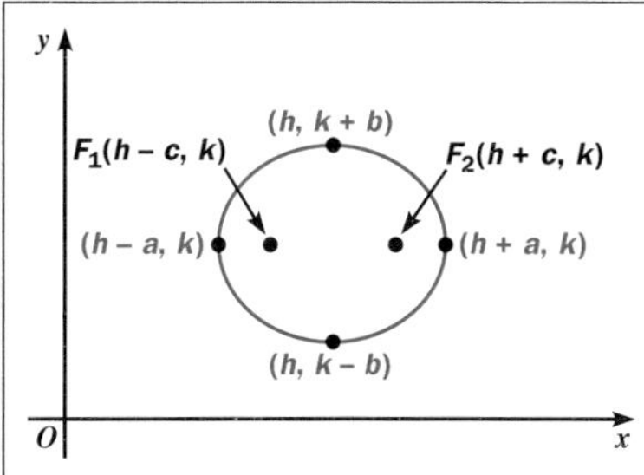

Foci: $F_1(h-c, k)$ and $F_2(h+c, k)$ where $c > 0$ and $c^2 = a^2 - b^2$

Length of major axis: $2a$

Length of minor axis: $2b$

An equation of a *vertical ellipse* (an ellipse with a vertical major axis) with center $C(h, k)$ is:

$$\frac{(x-h)^2}{b^2} + \frac{(y-k)^2}{a^2} = 1 \quad (a > b > 0)$$

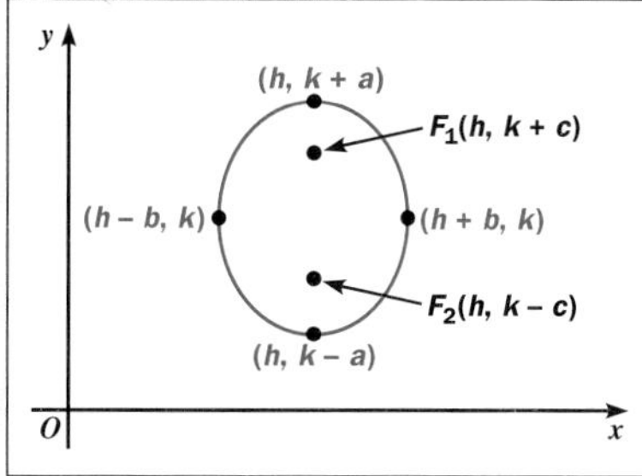

Foci: $F_1(h, k+c)$ and $F_2(h, k-c)$ where $c > 0$ and $c^2 = a^2 - b^2$

Length of major axis: $2a$

Length of minor axis: $2b$

EXAMPLE 1

Graph the equation $\frac{(x+2)^2}{16} + \frac{(y-3)^2}{25} = 1$.

SOLUTION

Method 1 Use graph paper.

In the given equation, $h = -2$ and $k = 3$. The center is $C(-2, 3)$.

Since the denominator of the term involving y is greater than the denominator of the term involving x, the ellipse is vertical. The major axis is parallel to the y-axis.

Since $a^2 = 25$, $a = \sqrt{25} = 5$. The major axis has length $2a = 10$.

Since $b^2 = 16$, $b = \sqrt{16} = 4$. The minor axis has length $2b = 8$.

Step 1 Plot the center point $C(-2, 3)$.

Step 2 Plot the endpoints of the major axis. They are **5 units** above and below the center.

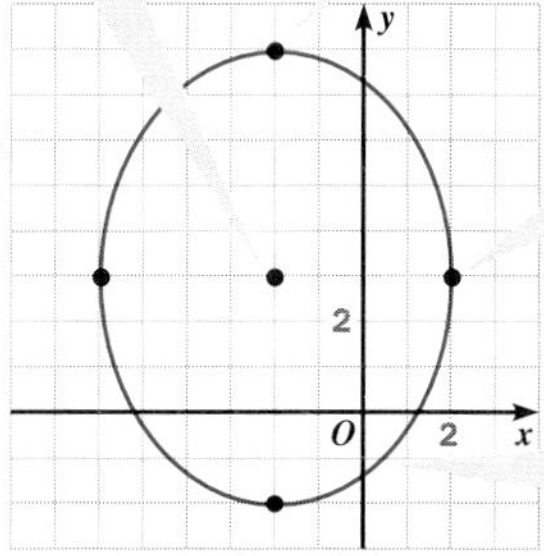

Step 3 Plot the endpoints of the minor axis. They are **4 units** to the left and right of the center.

Step 4 Sketch the ellipse through the four points plotted in Steps 2 and 3.

Solution continued on next page.

Teach⇔Interact

Additional Example 1

Graph the equation $\frac{(x+2)^2}{25} + \frac{(y+4)^2}{9} = 1$.

Method 1 Use graph paper. In the given equation $h = -2$ and $k = -4$. The center is $C(-2, -4)$. Since the denominator of the term involving x is greater than the denominator of the term involving y, the ellipse is horizontal. The major axis is parallel to the x-axis. Since $a^2 = 25$, $a = \sqrt{25} = 5$. The major axis has length $2a = 10$. Since $b^2 = 9$, $b = \sqrt{9} = 3$. The minor axis has length $2b = 6$.

Step 1 Plot the center point $C(-2, -4)$.

Step 2 Plot the endpoints of the major axis. They are 5 units to the right and to the left of the center.

Step 3 Plot the endpoints of the minor axis. They are 3 units above and below the center.

Step 4 Sketch the ellipse through the four points from Steps 2 and 3.

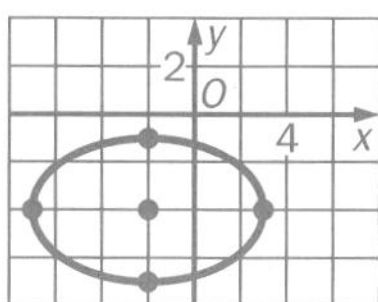

Method 2 Use a graphing calculator or graphing software.

Solve the equation for y.

$$\frac{(x+2)^2}{25} + \frac{(y+4)^2}{9} = 1$$

$$\frac{(y+4)^2}{9} = 1 - \frac{(x+2)^2}{25}$$

$$(y+4)^2 = 9 - \frac{9(x+2)^2}{25}$$

$$y + 4 = \pm\sqrt{9 - \frac{9(x+2)^2}{25}}$$

$$y = -4 \pm \sqrt{9 - \frac{9(x+2)^2}{25}}$$

Enter the equations into a graphing calculator or graphing software.

```
Y1=-4+√(9-9(X+2)
²/25)
Y2=-4-√(9-9(X+2)
²/25)
Y3=
Y4=
Y5=
```

Set an appropriate square window and graph the ellipse.

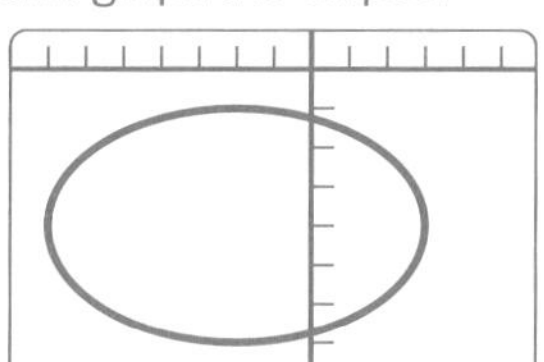

Teach⇔Interact

About Example 2

Mathematical Procedures
Step 1 of the Solution has students first determine if the ellipse is horizontal or vertical. This is very important so that the vertices and foci are located correctly. The other values necessary to use the standard form for an ellipse can then be calculated. Note also that when finding the value of b, the equation $b^2 = a^2 - c^2$ is used. This expression is equivalent to $c^2 = a^2 - b^2$, which was given on page 537.

Communication: Drawing
Some students, especially those with visual learning styles, will benefit from sketching the ellipse described before finding its equation. Begin by having students plot the vertices and foci. Students will also be able to use the drawing to find the values of a and c. Their relative orientation to each other should help students decide if the ellipse is horizontal or vertical. If the points are located on a vertical line, the ellipse is vertical. If the vertices and foci are located on a horizontal line, the ellipse is horizontal.

Additional Example 2

Write an equation of the ellipse with center (–1, 2), one vertex at (–1, 7), and one focus at (–1, 5).

Step 1 Determine if the ellipse is *horizontal* or *vertical*. Since the x-coordinates of the center, vertex, and focus are the same, the ellipse is vertical.

Step 2 Find a.
$a = 7 - 2 = 5$

Step 3 Find c.
$c = 5 - 2 = 3$

Step 4 Find b.
$b^2 = 5^2 - 3^2 = 16$
$b = 4$

Step 5 Substitute the values for h, k, a, and b into the standard form for an equation of a vertical ellipse.
$$\frac{(x+1)^2}{16} + \frac{(y-2)^2}{25} = 1$$

SOLUTION *continued*

Method 2 Use a graphing calculator or graphing software.

Solve the equation for y.

$$\frac{(x+2)^2}{16} + \frac{(y-3)^2}{25} = 1$$
$$\frac{(y-3)^2}{25} = 1 - \frac{(x+2)^2}{16}$$
$$(y-3)^2 = 25 - \frac{25(x+2)^2}{16}$$
$$y - 3 = \pm\sqrt{25 - \frac{25(x+2)^2}{16}}$$
$$y = 3 \pm\sqrt{25 - \frac{25(x+2)^2}{16}}$$

There are two equations to graph. The equation with + represents the top half of the ellipse, and the equation with – represents the bottom half.

Enter the equations into a graphing calculator or graphing software.

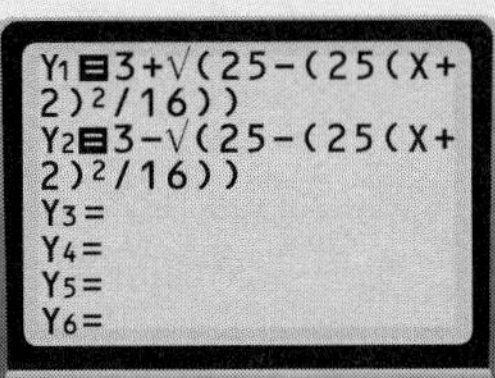

Set an appropriate square window and graph the ellipse.

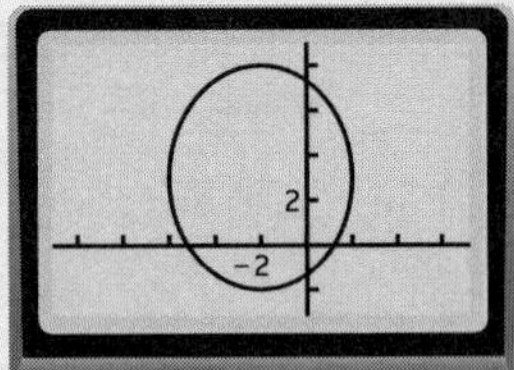

EXAMPLE 2

Write an equation of the ellipse with center (6, 8), one vertex at (1, 8), and one focus at (2, 8).

SOLUTION

Step 1 Determine if the ellipse is *horizontal* or *vertical*. Since the y-coordinates of the center, vertex, and focus are the same, the ellipse is horizontal.

Step 2 Find a.
$a = 6 - 1 = 5$

Subtract the x-coordinates of the center and the vertex to find a.

Step 3 Find c.
$c = 6 - 2 = 4$

The distance between the center and the focus is c.

Step 4 Find b.
$b^2 = 5^2 - 4^2 = 9$
$b = 3$

Substitute the values of a and c into the equation $b^2 = a^2 - c^2$.

Step 5 Substitute the values for h, k, a, and b into the standard form for an equation of a horizontal ellipse.

$$\frac{(x-6)^2}{25} + \frac{(y-8)^2}{9} = 1$$

Substitute 6 for h, 8 for k, 5 for a, and 3 for b.

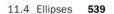

EXAMPLE 3 Application: Architecture

The Oval Office in the White House is in the shape of an ellipse. Write an equation of the ellipse.

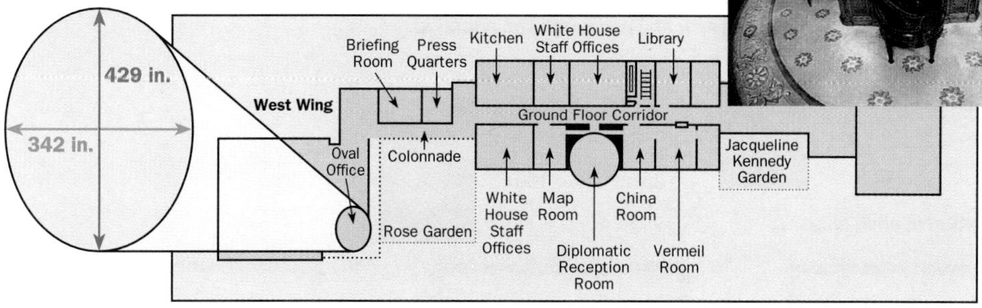

SOLUTION

Step 1 Find a.

$2a = 429$

$a = 214.5$

The major axis is 429 in. long.

Step 2 Find b.

$2b = 342$

$b = 171$

The minor axis is 342 in. long.

Step 3 Write an equation of the ellipse. Suppose the center of the Oval Office is positioned at the origin and the major axis lies on the y-axis.

$$\frac{x^2}{b^2} + \frac{y^2}{a^2} = 1 \longrightarrow \frac{x^2}{29{,}241} + \frac{y^2}{46{,}010.25} = 1$$

Use the equation of a vertical ellipse.

The coordinates of points on this ellipse are measured in inches.

CHECKING KEY CONCEPTS

Use the ellipse shown to answer Questions 1–3.

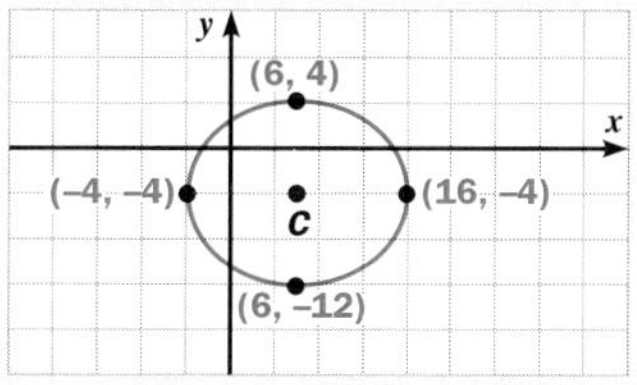

1. What are the coordinates of the center? the foci?
2. What is the length of the major axis? the minor axis?
3. Write an equation of the ellipse.

Identify the center, the vertices, and the foci of each ellipse with the given equation. Tell whether the ellipse is *horizontal* or *vertical*. Then graph the equation.

4. $\dfrac{(x-6)^2}{16} + \dfrac{(y-4)^2}{49} = 1$
5. $\dfrac{(x+1)^2}{64} + \dfrac{(y-3)^2}{4} = 1$
6. Discuss the graph of $\dfrac{x^2}{a^2} + \dfrac{y^2}{b^2} = 1$ when $a = b$. What can you say about the foci of this ellipse?

Checking Key Concepts

1. (6, –4); (0, –4) and (12, –4)
2. 20; 16
3. $\dfrac{(x-6)^2}{100} + \dfrac{(y+4)^2}{64} = 1$
4. center: (6, 4); vertices: (6, 11) and (6, –3); foci: $(6, 4 + \sqrt{33})$ and $(6, 4 - \sqrt{33})$; vertical

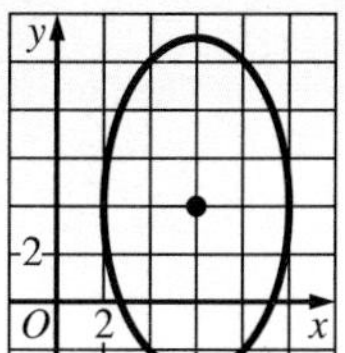

5. center: (–1, 3); vertices: (–9, 3) and (7, 3); foci: $(-1 - 2\sqrt{15}, 3)$ and $(-1 + 2\sqrt{15}, 3)$; horizontal

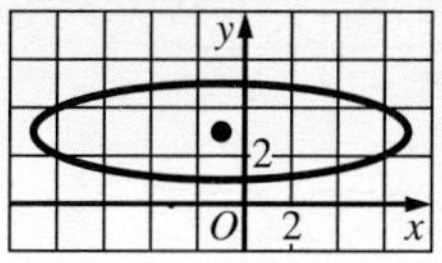

6. When $a = b$, the equation $\frac{x^2}{a^2} + \frac{y^2}{b^2} = 1$ becomes $x^2 + y^2 = a^2$, which is the equation of a circle with center at the origin and radius a. Since $c^2 = a^2 - b^2 = 0$, the foci "collapse" together at the center.

Teach⇔Interact

Additional Example 3

The Yellow Oval Room on the second floor of the White House is an ellipse. Write an equation for the shape of the Yellow Oval Room.

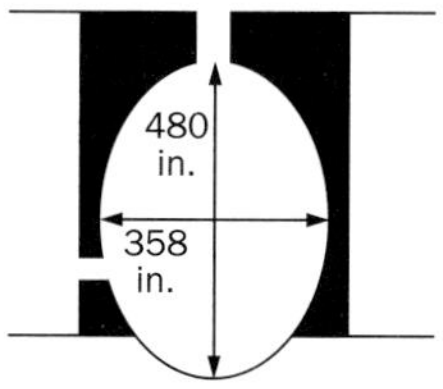

Step 1 Find a.
The major axis is 480 in. long.
$2a = 480$
$a = 240$
Step 2 Find b.
The minor axis is 358 in. long.
$2b = 358$
$b = 179$
Step 3 Write an equation of the ellipse. Suppose the center of the Yellow Oval Room is positioned at the origin and the major axis lies on the y-axis. Use the equation of a vertical ellipse.

$$\frac{x^2}{b^2} + \frac{y^2}{a^2} = 1$$

$$\frac{x^2}{32{,}041} + \frac{y^2}{57{,}600} = 1$$

Checking Key Concepts

Topic Spiraling: Preview
Question 6 allows students to see a relationship between circles and ellipses. This relationship will be expanded in Section 11.6, when students study parabolas, circles, ellipses, and hyperbolas as various types of conic sections.

Closure Question

Explain how to write an equation of an ellipse given the center, a vertex point, and a focus point.
Determine if the ellipse is horizontal or vertical. From the given points, calculate the values of a, c, and b. Substitute these values into the appropriate standard form for the equation of an ellipse along with the values of h and k from the center point.

Apply⇔Assess

Suggested Assignment

Core Course
Exs. 1–9, 20–31, AYP

Extended Course
Exs. 1–9 odd, 10–31, AYP

Block Schedule
Day 73 Exs. 1–9, 20–31, AYP

Exercise Notes

Teaching Tip
Exs. 1–6 Remind students that the foci of an ellipse may have irrational coordinates.

Common Error
Exs. 2, 5 Some students may identify the vertices incorrectly for these ellipses by thinking that the first denominator (for the x^2-term) represents the a^2-term for an ellipse in standard form. Remind them that a^2 is the largest denominator, not necessarily the first denominator.

Cooperative Learning
Exs. 10–13 You may wish to have students work on these exercises in groups. In so doing, they can reinforce their understanding of how ellipses are constructed.

Teaching Tip
Exs. 10–13 Ex. 11 will help students understand how to find an equation of an ellipse whose center is not (0, 0). For Ex. 12, have students generalize the answer to be $2c$, the distance between the foci. For Ex. 13, the barrier in place represents a necessary path of $2a$, the length of the major axis.

Challenge
Exs. 10–13 Have students consider the path of the ball if a focus is not the starting point. The paths would form an envelope of tangents around an inner ellipse with the same foci.

Multicultural Note
Exs. 10–13 The first women's international golfing tournament, held in 1976, was won by Sally Little of South Africa. The prize was \$10,000. Two year later, Nancy Lopez, a 21-year-old United States golfer who is of Mexican-American ancestry, won nine tournaments (five of which were consecutive) and earned \$189,814, setting a record for earnings for a rookie professional golfer.

11.4 Exercises and Applications

Extra Practice *exercises on page 766*

Identify the center, the vertices, and the foci of each ellipse with the given equation. Tell whether the ellipse is *horizontal* or *vertical*. Then graph the equation.

1. $\dfrac{x^2}{64} + \dfrac{y^2}{25} = 1$ **2.** $\dfrac{x^2}{4} + \dfrac{y^2}{49} = 1$ **3.** $\dfrac{x^2}{8} + \dfrac{(y-1)^2}{2} = 1$

4. $\dfrac{(x-3)^2}{16} + \dfrac{(y+3)^2}{9} = 1$ **5.** $\dfrac{(x+8)^2}{25} + \dfrac{(y+1)^2}{100} = 1$ **6.** $\dfrac{(x-5)^2}{81} + y^2 = 1$

Write an equation of each ellipse.

7.
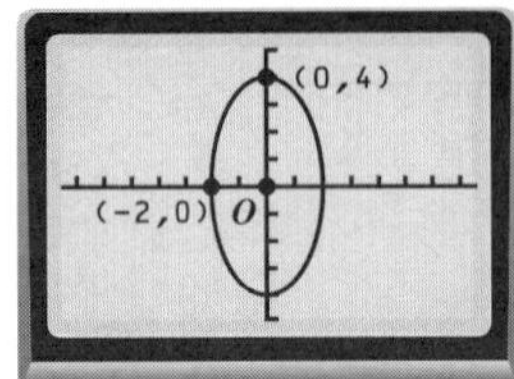

8.

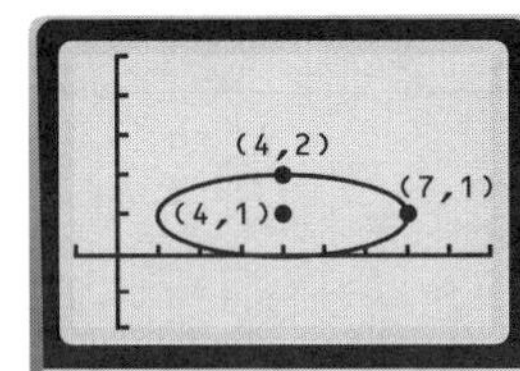

9.
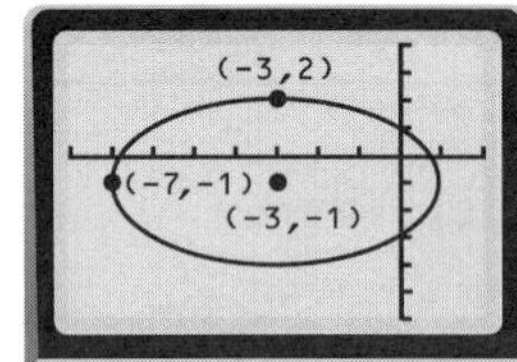

Connection ART

Artist William Wainwright used mathematics to design a miniature golf hole in the shape of an ellipse for a museum exhibit entitled "Strokes of Genius: Mini Golf by Artists." The starting position of the ball is at one focus and the cup is at the other focus. Wainwright positioned a barrier at the center of the ellipse.

10. The length of the major axis of the elliptical frame of the hole is 12 ft. The length of the minor axis is 8 ft.
 a. Find an equation of the elliptical frame. Assume the ellipse is centered at the origin and the major axis is horizontal.
 b. Estimate the location of the foci.

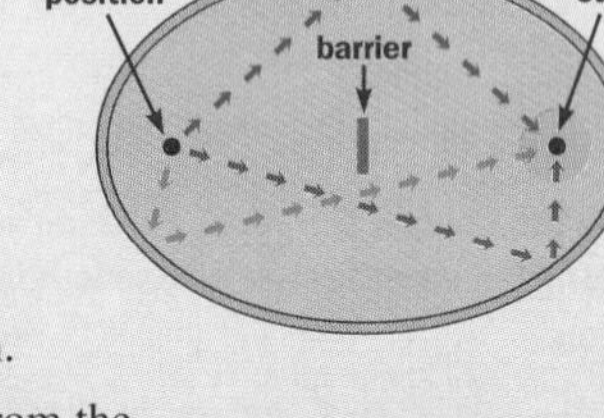

11. Suppose you position the ellipse in a coordinate plane so that the origin is located at the starting position of the ball.
 a. What are the coordinates of the center of the barrier?
 b. What are the coordinates of the cup?
 c. Write an equation for the elliptical frame in this position.

12. Suppose the barrier is removed. What is the shortest path from the starting position of the ball to the cup? What is the length of the path?

13. With the barrier in place, what is the shortest path with a single putt from the starting position of the ball to the cup? What is the length of the path?

Exercises and Applications

1. center: (0, 0); vertices: (8, 0) and (–8, 0); foci: $(\sqrt{39}, 0)$ and $(-\sqrt{39}, 0) \approx (\pm 6.24, 0)$; horizontal

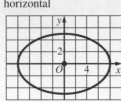

2. center: (0, 0); vertices: (0, 7) and (0, –7); foci: $(0, 3\sqrt{5})$ and $(0, -3\sqrt{5}) \approx (0, \pm 6.71)$; vertical

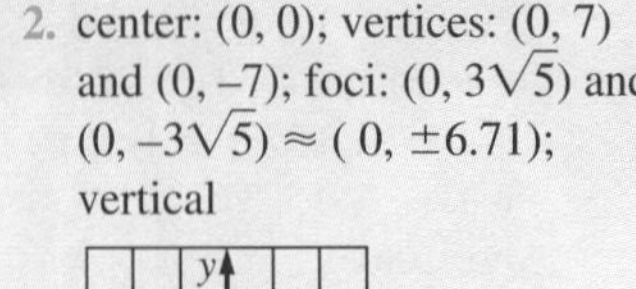
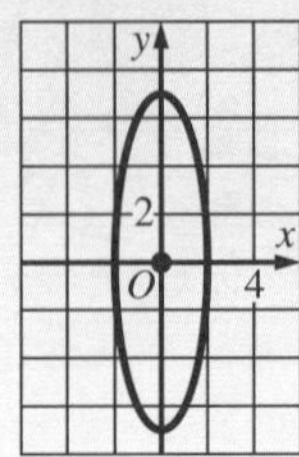

3. center: (0, 1); vertices: $(2\sqrt{2}, 1)$ and $(-2\sqrt{2}, 1) \approx (\pm 2.83, 1)$; foci: $(\sqrt{6}, 1)$ and $(-\sqrt{6}, 1) \approx (\pm 2.45, 1)$; horizontal

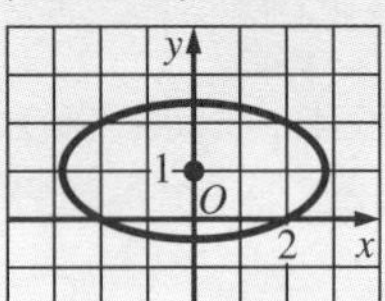

4–11. See answers in back of book.

Connection MEDICINE

A *lithotripter* is an elliptical medical instrument used to disintegrate a kidney stone within a patient's body. Shock waves, initiated by a generator located at one focus of the instrument, reflect off the sides of a three-dimensional elliptical reflecting dish and converge at the other focus. The patient is positioned so that the kidney stone is at the other focus.

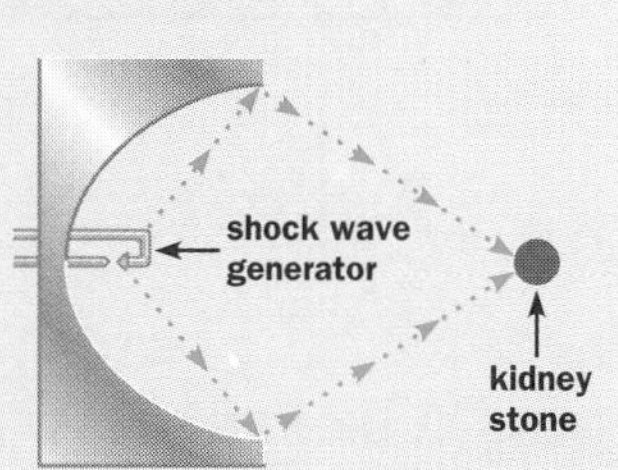

The shape of a cross section of the reflecting dish on a certain lithotripter can be represented by the equation $\frac{x^2}{196} + \frac{y^2}{63.6804} = 1$, where x and y are measured in centimeters.

14. What are the coordinates of the shock wave generator? of the kidney stone?

15. Graph the equation.

Scientists have determined that a deeper reflecting dish operates more effectively. A different model lithotripter is 1.5 times as wide (along the x-axis) as the model discussed in Exercises 14 and 15.

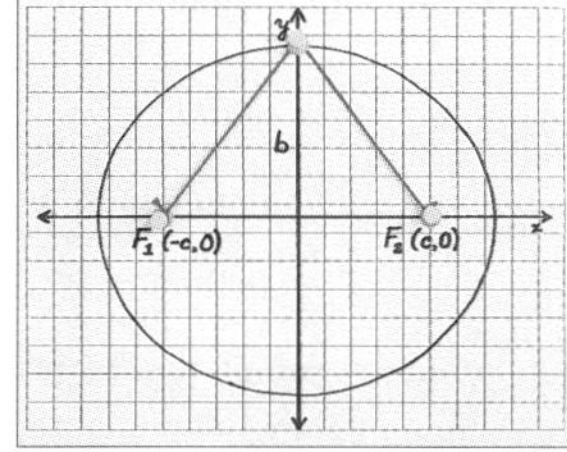

16. Write an equation of a cross section of this reflecting dish. How is this equation different from the equation you used in Exercises 14 and 15?

17. Graph the equation you wrote in Exercise 16.

18. For this lithotripter, what are the coordinates of the shock wave generator? of the kidney stone?

19. Writing The ends of the string in the figure at the right are at $(-c, 0)$ and $(c, 0)$. Suppose that the length of the string is $2a$.

a. Explain why the vertices of the ellipse must be $(-a, 0)$ and $(a, 0)$.

b. Explain why the endpoints of the minor axis must be $(0, -b)$ and $(0, b)$, where $b^2 = a^2 - c^2$.

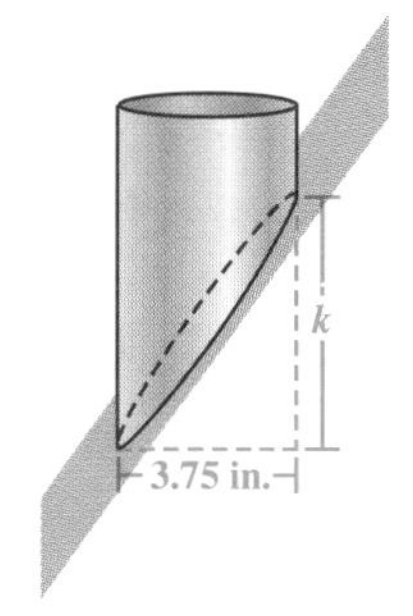

Write an equation of the ellipse with the given characteristics.

20. center $(0, 0)$; vertex $(-4, 0)$; y-intercept -3

21. center $(5, -3)$; $a = 3$; $b = 2$; major axis is vertical

22. center $(6, 2)$; vertex $(6, 7)$; focus $(6, -1)$

23. center $(-2, 4)$; vertex $(-9, 4)$; minor axis is 6 units long

24. CARPENTRY A standard vent pipe with a diameter of 3.75 in. needs to be installed on a roof with a slope of $\frac{4}{3}$. The hole that needs to be cut into the roof to fit the pipe is in the shape of an ellipse.

a. Use the slope of the roof to find k.

b. What is the length of the major axis? the minor axis?

c. Write an equation of the ellipse that should be cut from the roof to fit the pipe.

k
3.75 in.

12. The shortest path is the straight path from the starting position to the cup. The length of the path is the same as the distance between the two foci. The length of the path is about 9 ft.

13. The shortest path is the path that hits the elliptical rim once. The length of this path is $2a = 12$ ft.

14. about $(-11.50, 0)$; about $(11.50, 0)$

15.

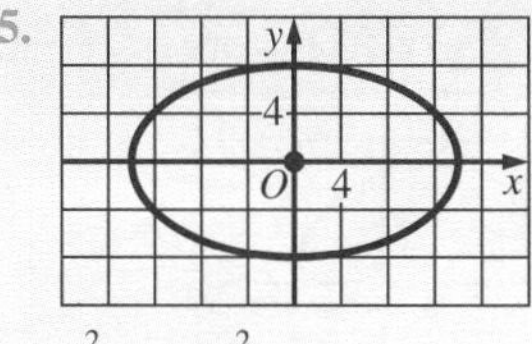

16. $\frac{x^2}{441} + \frac{y^2}{63.6804} = 1$; The value of a^2 is $(1.5)^2 =$ 2.25 times the previous value.

17.

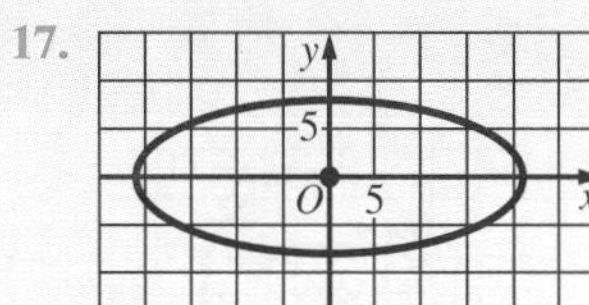

18. about $(-19.42, 0)$; about $(19.42, 0)$

19–24. See answers in back of book.

Apply⇔Assess

Exercise Notes

Second-Language Learners
Exs. 14–18 Students learning English may gain a clearer understanding of the procedure of how kidney stones are dissolved by working with English-proficient partners who can help them orally review the process described in the introductory paragraph for these exercises.

Communication: Discussion
Exs. 14–18 Discuss the significance of placing a generator at a focus of the ellipse. This means that the shock waves will all converge at one point, the other focus. This has the effect of greatly magnifying the shock wave. The shock wave is strong enough to sound like a pounding hammer when the machine is in operation.

Career Connection
Exs. 14–18 There are many career opportunities for trained medical technicians to help doctors and nurses administer treatment to patients that requires the use of sophisticated medical instruments and equipment. One of the first such jobs available involved using equipment to take chest X-rays. Ask students to suggest any experiences they are familiar with that would involve the use of other medical technology.

Problem Solving
Exs. 20–23 Students need to work in an organized manner to find the equation of the ellipse with the given characteristics. Have them find a, b, c, h, and k in each case and then use the standard form for the equation of an ellipse.

Reasoning
Ex. 24 In part (b), the minor axis of the ellipse is the width of the pipe, 3.75 in. The major axis is the hypotenuse of the right triangle with 3.75 in. as a leg and k as a leg, where $k{:}3.75 = 4{:}3$.

Apply⇔Assess

Exercise Notes

Assessment Note

Ex. 25 This exercise reviews the major concepts of this section. To summarize these concepts, have students compare their choices of ellipses and check each other's work.

Assess Your Progress

Journal Entry

Have students share ideas about how circles and ellipses are alike and different. Emphasize that a circle is an ellipse with $a = b$.

Progress Check 11.3–11.4

See page 557.

Practice 73 for Section 11.4

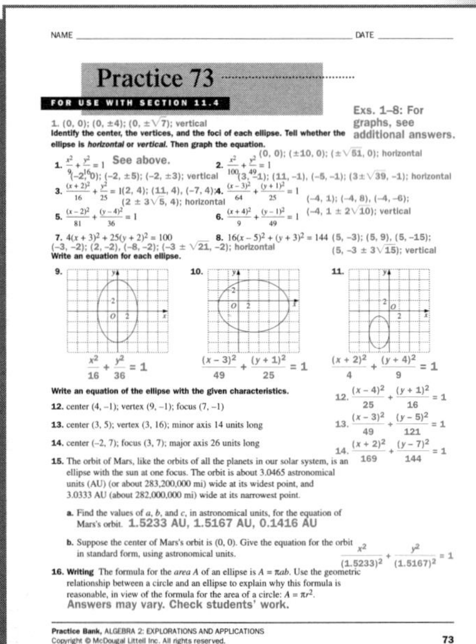

NAME ______________________ DATE __________

Practice 73

FOR USE WITH SECTION 11.4

Exs. 1–8: For graphs, see additional answers.

Identify the center, the vertices, and the foci of each ellipse. Tell whether the ellipse is *horizontal* or *vertical*. Then graph the equation.

1. $\frac{x^2}{9} + \frac{y^2}{16} = 1$ See above. [1. $(0, 0)$; $(0, \pm 4)$; $(0, \pm\sqrt{7})$; vertical]
2. $\frac{x^2}{100} + \frac{y^2}{49} = 1$ $(0, 0)$; $(\pm 10, 0)$; $(\pm\sqrt{51}, 0)$; horizontal
3. $\frac{(x+2)^2}{16} + \frac{y^2}{25} = 1$ $(-2, 0)$; $(-2, \pm 5)$; $(-2, \pm 3)$; vertical
4. $\frac{(x-3)^2}{64} + \frac{(y+1)^2}{25} = 1$ $(3, -1)$; $(11, -1)$, $(-5, -1)$; $(3 \pm \sqrt{39}, -1)$; horizontal
5. $\frac{(x-2)^2}{81} + \frac{(y-4)^2}{36} = 1$ $(2, 4)$; $(11, 4)$, $(-7, 4)$; $(2 \pm 3\sqrt{5}, 4)$; horizontal
6. $\frac{(x+4)^2}{9} + \frac{(y-1)^2}{49} = 1$ $(-4, 1)$; $(-4, 8)$, $(-4, -6)$; $(-4, 1 \pm 2\sqrt{10})$; vertical
7. $4(x+3)^2 + 25(y+2)^2 = 100$ $(-3, -2)$; $(2, -2)$, $(-8, -2)$; $(-3 \pm \sqrt{21}, -2)$; horizontal
8. $16(x-5)^2 + (y+3)^2 = 144$ $(5, -3)$; $(5, 9)$, $(5, -15)$; $(5, -3 \pm 3\sqrt{15})$; vertical

Write an equation for each ellipse.

9. $\frac{x^2}{16} + \frac{y^2}{36} = 1$
10. $\frac{(x-3)^2}{49} + \frac{(y+1)^2}{25} = 1$
11. $\frac{(x+2)^2}{4} + \frac{(y+4)^2}{9} = 1$

Write an equation of the ellipse with the given characteristics.

12. center $(4, -1)$; vertex $(9, -1)$; focus $(7, -1)$ 12. $\frac{(x-4)^2}{25} + \frac{(y+1)^2}{16} = 1$
13. center $(3, 5)$; vertex $(3, 16)$; minor axis 14 units long 13. $\frac{(x-3)^2}{49} + \frac{(y-5)^2}{121} = 1$
14. center $(-2, 7)$; focus $(3, 7)$; major axis 26 units long 14. $\frac{(x+2)^2}{169} + \frac{(y-7)^2}{144} = 1$
15. The orbit of Mars, like the orbits of all the planets in our solar system, is an ellipse with the sun at one focus. The orbit is about 3.0465 astronomical units (AU) (or about 283,200,000 mi) wide at its widest point, and 3.0333 AU (about 282,000,000 mi) wide at its narrowest point.
 a. Find the values of a, b, and c, in astronomical units, for the equation of Mars's orbit. 1.5233 AU, 1.5167 AU, 0.1416 AU
 b. Suppose the center of Mars's orbit is $(0, 0)$. Give the equation for the orbit in standard form, using astronomical units. $\frac{x^2}{(1.5233)^2} + \frac{y^2}{(1.5167)^2} = 1$
16. **Writing** The formula for the *area A* of an ellipse is $A = \pi ab$. Use the geometric relationship between a circle and an ellipse to explain why this formula is reasonable, in view of the formula for the area of a circle: $A = \pi r^2$. Answers may vary. Check students' work.

Practice Bank, ALGEBRA 2: EXPLORATIONS AND APPLICATIONS

73

ONGOING ASSESSMENT

25. **Open-ended Problem** Choose the length of the major and minor axes of an ellipse. Write an equation of your ellipse. Graph the equation. Find the vertices and foci.

SPIRAL REVIEW

Write an equation of the circle with the given center *C* and radius *r*. Then sketch its graph. *(Section 11.3)*

26. $C(3, 0)$; $r = 5$
27. $C(-2, -1)$; $r = 6$
28. $C(4, -3)$; $r = \sqrt{8}$

Find the slope-intercept equation of each line. *(Section 2.2)*

29.

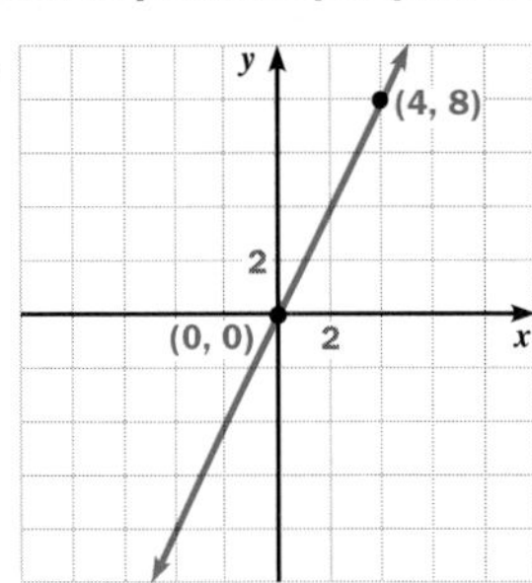

30.

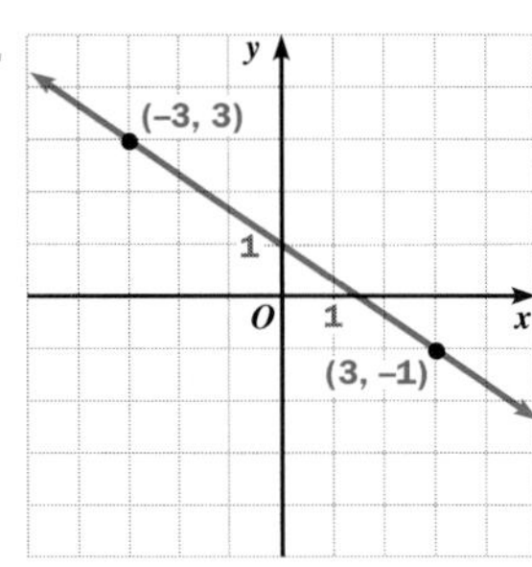

31.

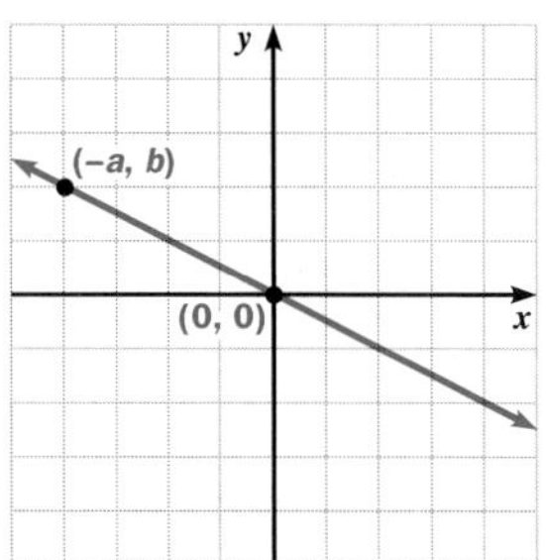

ASSESS YOUR PROGRESS

VOCABULARY

circle (p. 528)	**foci** (p. 536)
center (pp. 528, 536)	**vertices** (p. 536)
radius (p. 528)	**major axis** (p. 536)
ellipse (p. 536)	**minor axis** (p. 536)

For each equation of a circle, identify the center and the radius. *(Section 11.3)*

1. $(x + 1)^2 + (y - 5)^2 = 16$
2. $(x - 3)^2 + y^2 = 25$
3. Find the point(s) of intersection of the equations in Exercises 1 and 2. *(Section 11.3)*

Identify the center, the vertices, and the foci of each ellipse with the given equation. Tell whether the ellipse is *horizontal* or *vertical*. *(Section 11.4)*

4. $\frac{x^2}{64} + \frac{(y-5)^2}{36} = 1$
5. $\frac{(x+3)^2}{100} + \frac{(y-5)^2}{196} = 1$

Write an equation of each figure described. Then graph the equation. *(Sections 11.3 and 11.4)*

6. circle with center $(0, -5)$ and radius 4
7. ellipse with center $(-1, 2)$, a vertex at $(-1, -3)$, and a focus at $(-1, 6)$
8. **Journal** Discuss the geometric and algebraic relationships between circles and ellipses.

25. Answers may vary. An example is given. If an ellipse has a horizontal major axis of length 10, a minor axis of length 4, and center at the origin, its equation is $\frac{x^2}{25} + \frac{y^2}{4} = 1$. The coordinates of its vertices are $(\pm 5, 0)$. The coordinates of its foci are $(\pm\sqrt{21}, 0) \approx (4.58, 0)$.

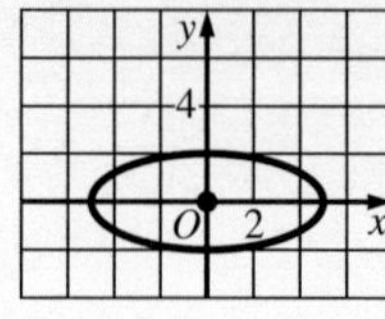

26. $(x - 3)^2 + y^2 = 25$

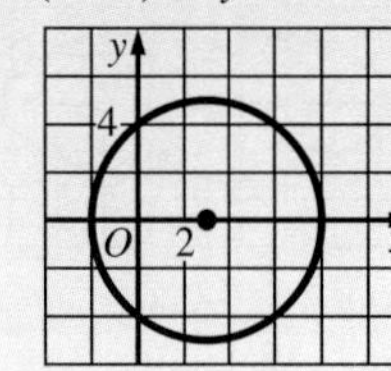

27. $(x + 2)^2 + (y + 1)^2 = 36$

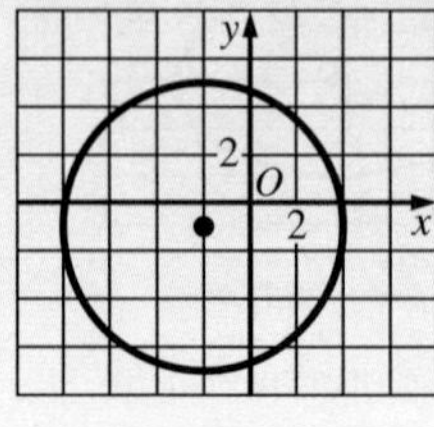

28. $(x - 4)^2 + (y + 3)^2 = 8$

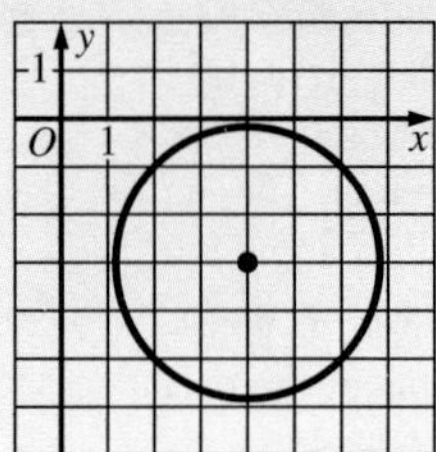

29. $y = 2x$ 30. $y = -\frac{2}{3}x + 1$ 31. $y = -\frac{b}{a}x$

Assess Your Progress

1–8. See answers in back of book.

Section

11.5 Hyperbolas

Learn how to...

- write an equation of a hyperbola
- graph an equation of a hyperbola

So you can...

- describe an object with a hyperbolic shape, such as the *Acqua Alle Funi* sculpture

The Fermi National Accelerator Laboratory in Batavia, Illinois, is the site of several large geometric sculptures. One of these, named *Acqua Alle Funi*, is a three-sided sculpture in the shape of a *hyperbola*.

THINK AND COMMUNICATE

By superimposing a coordinate system on one side of the sculpture, as shown at the left, you can describe the vertical edges of the sculpture by the equation $x^2 - \frac{3y^2}{169} = 1$, where x and y are measured in feet.

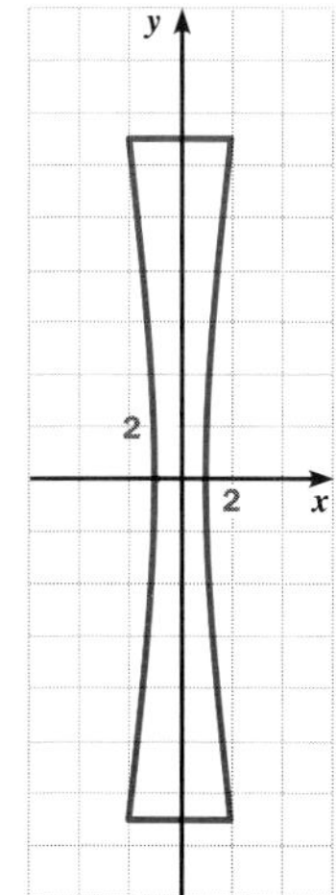

1. How is the equation like the equation of an ellipse? How is it different?
2. a. The narrowest part of the sculpture corresponds to the x-intercepts of the graph of the given equation. What are the x-intercepts?
   b. How wide is the sculpture at its narrowest width?
3. The height of the sculpture is 26 ft. How wide is the sculpture at the top?

A **hyperbola** is the set of all points in a plane such that the difference of the distances between a point and each of two **foci**, F_1 and F_2, is constant. The **center** of the hyperbola is the midpoint of $\overline{F_1F_2}$.

The sculpture's equation is of this form:

$$\frac{x^2}{a^2} - \frac{y^2}{b^2} = 1$$

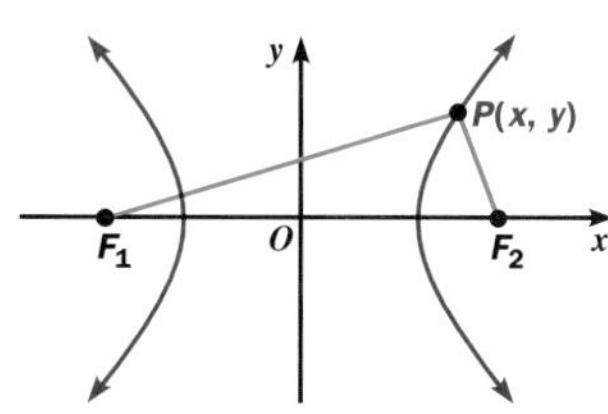

$|F_1P - F_2P|$ = a constant

The hyperbola has the lines $y = \frac{b}{a}x$ and $y = -\frac{b}{a}x$ as asymptotes. The asymptotes contain the diagonals of a rectangle with dimensions $2a$ and $2b$.

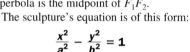

The **major axis** has endpoints at the vertices, so its length is $2a$.

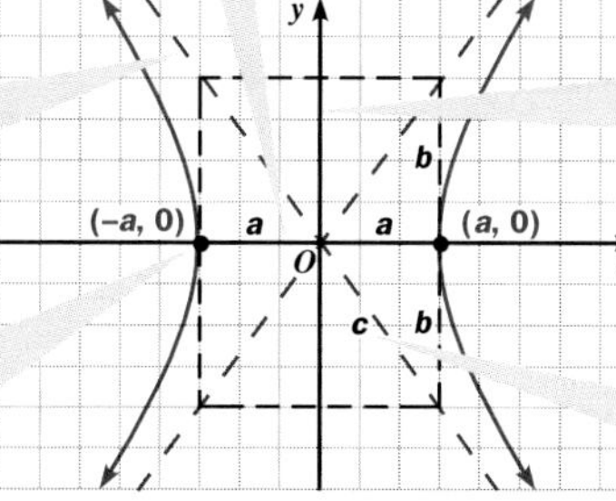

The **minor axis** is perpendicular to the major axis and has length $2b$.

The **vertices** of the hyperbola are $(a, 0)$ and $(-a, 0)$.

By the Pythagorean theorem, $c^2 = a^2 + b^2$.

The lines containing the axes of a hyperbola are lines of symmetry.

Plan⇔Support

Objectives

- Write an equation of a hyperbola.
- Graph an equation of a hyperbola.
- Describe an object with a hyperbolic shape.

Recommended Pacing

Core and Extended Courses
Section 11.5: 1 day

Block Schedule
Section 11.5: $\frac{1}{2}$ block
(with Section 11.4)

Resource Materials

Lesson Support
Lesson Plan 11.5
Warm-Up Transparency 11.5
Overhead Visuals:
Folder 11: Conic Sections
Practice Bank: Practice 74
Study Guide: Section 11.5
Explorations Lab Manual:
Diagram Masters 2, 16

Technology
Technology Book:
Calculator Activity 11
Graphing Calculator
Internet:
http://www.hmco.com

Warm-Up Exercises

1. Given points $F_1(5, 0)$, $F_2(-5, 0)$, and $P(8, 3\sqrt{3})$, find $|F_1P - F_2P|$. 8
2. Find the length of the major axis of the ellipse $\frac{x^2}{16} + \frac{y^2}{25} = 1$. 10
3. Give the center, vertices, and foci of the ellipse $\frac{(x+1)^2}{25} + \frac{(y-2)^2}{16} = 1$. center: (–1, 2); vertices: (4, 2), (–6, 2); foci: (2, 2), (–4, 2)
4. Solve the equation $\frac{(x+1)^2}{25} + \frac{(y+4)^2}{16} = 1$ for y. $y = -4 \pm \sqrt{16 - \frac{16(x+1)^2}{25}}$

ANSWERS Section 11.5

Think and Communicate

1. Both the equation for an ellipse and the equation for a hyperbola include squared terms of x and y on one side and 1 on the other. The equation of an ellipse involves the sum of the squared terms, while that of a hyperbola involves the difference of the terms.
2. a. 1 and –1
   b. 2 ft
3. 4 ft

Teach⇔Interact

Section Note

Topic Spiraling: Review
Review the procedure used to find the values of a and b for the standard form of the equation of an ellipse before doing this section.

Think and Communicate

For question 1, the similarities and differences in the equations of an ellipse and a hyperbola are important connections for students to make.

Section Note

Teaching Tip
Emphasize that the equation of a hyperbola, as with the equation of an ellipse, depends on the distances of each point on the hyperbola to the foci. In a hyperbola however, the constant is the *difference* of these distances, not the sum of them. Point out that a does not have to be larger than b in the equation of a hyperbola; instead, a can be identified because a^2 is the denominator of the positive term.

About Example 1

Using Technology
Students may graph the rectangle used in this Example with the following additional graphs. The logical operators are found under the TEST menu.
Y5 = 6(–6 ≤ X)(X ≤ 4)
Y6 = 0(–6 ≤ X)(X ≤ 4)
Line(–6, 6, –6, 0)
Line(4, 6, 4, 0)

Additional Example 1

Graph the hyperbola with equation $\frac{(y-1)^2}{16} + \frac{(x+2)^2}{25} = 1$.

Method 1 Use graph paper.
The center is $C(-2, 1)$. Since the term involving y is positive, the hyperbola is vertical. The major axis is parallel to the y-axis. Since $a^2 = 16$, $a = \sqrt{16} = 4$. The major axis has length $2a = 8$. Since $b^2 = 25$, $b = \sqrt{25} = 5$. The minor axis has length $2b = 10$. The vertices are $V_1(-2, 5)$ and $V_2(-2, -3)$. The hyperbola has asymptotes $y = \frac{4}{5}(x+2) + 1$ and $y = -\frac{4}{5}(x+2) + 1$.

The summary below describes hyperbolas whose vertices are aligned either vertically or horizontally and whose centers are at $C(h, k)$.

Standard Forms of an Equation of a Hyperbola

An equation of a *horizontal hyperbola* (a hyperbola with a horizontal major axis) with center $C(h, k)$ is:

$$\frac{(x-h)^2}{a^2} - \frac{(y-k)^2}{b^2} = 1 \quad (a > 0, b > 0)$$

$y = -\frac{b}{a}(x-h) + k$ $\quad$ $y = \frac{b}{a}(x-h) + k$

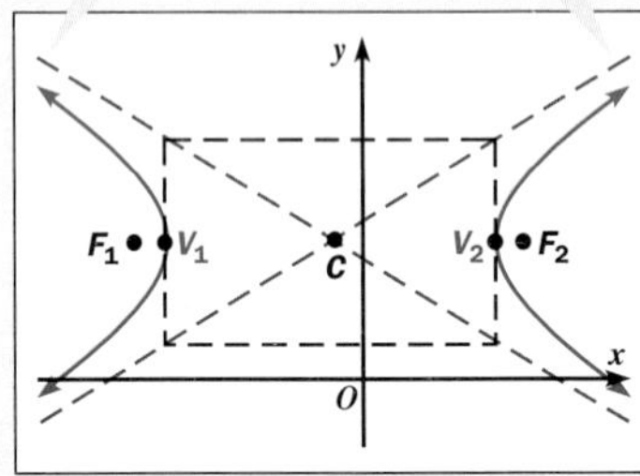

Foci: $F_1(h-c, k)$ and $F_2(h+c, k)$ where $c > 0$ and $c^2 = a^2 + b^2$
Vertices: $V_1(h-a, k)$ and $V_2(h+a, k)$
Asymptotes: $y = \pm\frac{b}{a}(x-h) + k$
Length of major axis: $2a$
Length of minor axis: $2b$

An equation of a *vertical hyperbola* (a hyperbola with a vertical major axis) with center $C(h, k)$ is:

$$\frac{(y-k)^2}{a^2} - \frac{(x-h)^2}{b^2} = 1 \quad (a > 0, b > 0)$$

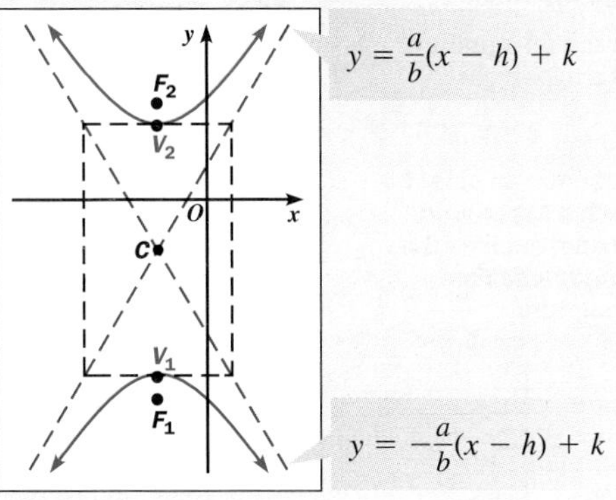

$y = \frac{a}{b}(x-h) + k$

$y = -\frac{a}{b}(x-h) + k$

Foci: $F_1(h, k-c)$ and $F_2(h, k+c)$ where $c > 0$ and $c^2 = a^2 + b^2$
Vertices: $V_1(h, k-a)$ and $V_2(h, k+a)$
Asymptotes: $y = \pm\frac{a}{b}(x-h) + k$
Length of major axis: $2a$
Length of minor axis: $2b$

EXAMPLE 1

Graph the hyperbola with equation $\frac{(x+1)^2}{25} - \frac{(y-3)^2}{9} = 1$.

SOLUTION

Method 1 Use graph paper.
The center is $C(-1, 3)$.

In the given equation, $h = -1$ and $k = 3$.

Since the term involving y is subtracted from the term involving x, the hyperbola is horizontal. The major axis is parallel to the x-axis.

Since $a^2 = 25$, $a = \sqrt{25} = 5$. The major axis has length $2a = 10$.

Since $b^2 = 9$, $b = \sqrt{9} = 3$. The minor axis has length $2b = 6$.

The vertices are $V_1(-6, 3)$ and $V_2(4, 3)$.

The hyperbola has asymptotes $y = -\frac{3}{5}(x+1) + 3$ and $y = \frac{3}{5}(x+1) + 3$.

WATCH OUT!
The major axis of a hyperbola is $\overline{V_1V_2}$. It is not necessarily the longer axis.

Step 1 Plot the center $C(-1, 3)$ and the vertices $V_1(-6, 3)$ and $V_2(4, 3)$.

Step 2 Draw the rectangle centered at $(-1, 3)$ and having a horizontal dimension of 10 and a vertical dimension of 6.

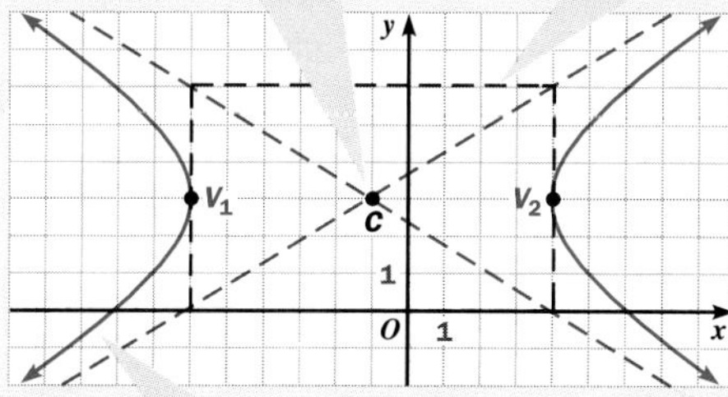

Step 3 Sketch the asymptotes $y = -\frac{3}{5}(x + 1) + 3$ and $y = \frac{3}{5}(x + 1) + 3$, which are the diagonals of the rectangle.

Step 4 Sketch the hyperbola through the vertices, extending toward the asymptotes.

Method 2 Use a graphing calculator or graphing software.

Solve the equation for y.

$$\frac{(x + 1)^2}{25} - \frac{(y - 3)^2}{9} = 1$$

$$\frac{(y - 3)^2}{9} = \frac{(x + 1)^2}{25} - 1$$

$$(y - 3)^2 = \frac{9(x + 1)^2}{25} - 9$$

$$y - 3 = \pm\sqrt{\frac{9(x + 1)^2}{25} - 9}$$

$$y = 3 \pm \sqrt{\frac{9(x + 1)^2}{25} - 9}$$

There are two equations to graph: $y = 3 + \sqrt{\frac{9(x + 1)^2}{25} - 9}$ represents the top half of the hyperbola, and $y = 3 - \sqrt{\frac{9(x + 1)^2}{25} - 9}$ represents the bottom half of the hyperbola.

Enter the equations into a graphing calculator or graphing software.

```
Y1=3+√((9(X+1)²/
25)-9)
Y2=3-√((9(X+1)²/
25)-9)
Y3=
Y4=
Y5=
Y6=
```

Set an appropriate square window and graph the hyperbola.

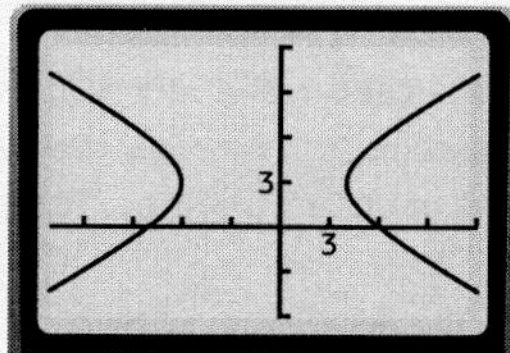

Teach⇔Interact

Additional Example 1 (continued)

Step 1 Plot the center $C(-2, 1)$ and the vertices $V_1(-2, 5)$ and $V_2(-2, -3)$.

Step 2 Draw the rectangle centered at $(-2, 1)$ having a horizontal dimension of 10 and a vertical dimension of 8.

Step 3 Sketch the asymptotes $y = \frac{4}{5}(x + 2) + 1$ and $y = -\frac{4}{5}(x + 2) + 1$, which are the diagonals of the rectangle.

Step 4 Sketch the hyperbola through the vertices, extending toward the asymptotes.

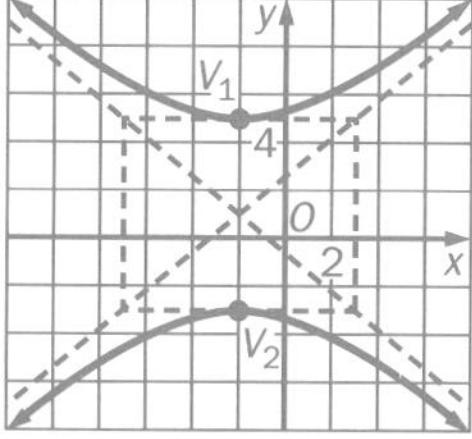

Method 2 Use a graphing calculator or graphing software.

Solve the equation for y.

$$\frac{(y - 1)^2}{16} - \frac{(x + 2)^2}{25} = 1$$

$$\frac{(y - 1)^2}{16} = 1 + \frac{(x + 2)^2}{25}$$

$$(y - 1)^2 = 16 + \frac{16(x + 2)^2}{25}$$

$$y - 1 = \pm\sqrt{16 + \frac{16(x + 2)^2}{25}}$$

$$y = 1 \pm \sqrt{16 + \frac{16(x + 2)^2}{25}}$$

There are two equations to graph: $y = 1 + \sqrt{16 + \frac{16(x + 2)^2}{25}}$ represents the top branch of the hyperbola, and $y = 1 - \sqrt{16 + \frac{16(x + 2)^2}{25}}$ represents the bottom branch of the hyperbola.

Enter the equations into a graphing calculator or graphing software.

```
Y1=1+√(16+16(X+2)
²/25)
Y2=1-√(16+16(X+2)
²/25)
Y3=
Y4=
Y5=
```

Set an appropriate square window and graph the hyperbola.

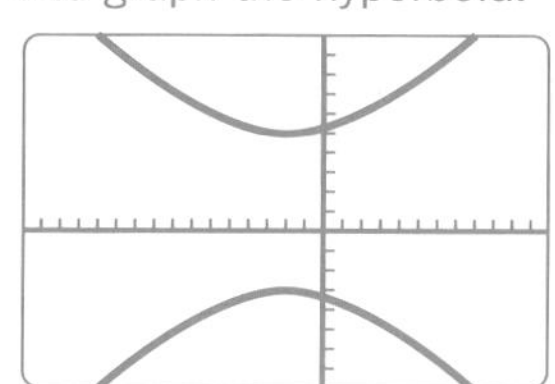

Teach⇔Interact

About Example 2

Teaching Tip

This Example has students find the equation of a hyperbola given its center, a vertex, and one focus. Have students sketch these points, if necessary, to help them see the orientation of the hyperbola.

Additional Example 2

Write an equation of the hyperbola with center (1, –3), one vertex at (6, –3), and one focus at $(1 + \sqrt{41}, -3)$.

Step 1 Determine if the hyperbola is *horizontal* or *vertical*. Since the y-coordinates of the center, vertex, and focus are the same, the major axis is parallel to the x-axis. The hyperbola is horizontal.

Step 2 Find a.
$a = 6 - 1 = 5$.

Step 3 Find c.
$c = 1 + \sqrt{41} - 1 = \sqrt{41}$

Step 4 Find b.
$(\sqrt{41})^2 = 5^2 + b^2$
$b^2 = 41 - 25 = 16$
$b = 4$

Step 5 Substitute the values for h, k, a, and b into the standard form of an equation of a horizontal hyperbola.

$$\frac{(x-h)^2}{a^2} - \frac{(y-k)^2}{b^2} = 1$$
$$\frac{(x-1)^2}{25} - \frac{(y+3)^2}{16} = 1$$

Closure Question

Describe how to find the equation of a hyperbola given its center, a vertex point, and a focus point.

Determine whether the hyperbola is horizontal or vertical by determining if the x-coordinates or if the y-coordinates are the same in the given points. If the x-coordinates are the same, the hyperbola is vertical. If the y-coordinates are the same, the hyperbola is horizontal. Find the values of a and c from the points. Then use the equation $c^2 = a^2 + b^2$ to find b. Substitute the values of h, k, a, and b into the standard form of the equation of a hyperbola.

EXAMPLE 2

Write an equation of the hyperbola with center $(-2, 4)$, one vertex at $(-2, 1)$, and one focus at $(-2, 9)$.

SOLUTION

Step 1 Determine if the hyperbola is horizontal or vertical.
Since the x-coordinates of the center, vertex, and focus are the same, the major axis is parallel to the y-axis. The hyperbola is vertical.

Step 2 Find a.
$a = 4 - 1 = 3$

The distance between the center and a vertex is a.

Step 3 Find c.
$c = 9 - 4 = 5$

The distance between the center and a focus is c.

Step 4 Find b.
$5^2 = 3^2 + b^2$
$b^2 = 25 - 9 = 16$
$b = 4$

Use the fact that $c^2 = a^2 + b^2$.

Step 5 Substitute the values for h, k, a, and b into the standard form of an equation of a vertical hyperbola.

$$\frac{(y-k)^2}{a^2} - \frac{(x-h)^2}{b^2} = 1$$
$$\frac{(y-4)^2}{9} - \frac{(x+2)^2}{16} = 1$$

Substitute -2 for h, 4 for k, 3 for a, and 4 for b.

CHECKING KEY CONCEPTS

Use the hyperbola shown to answer Questions 1–3.

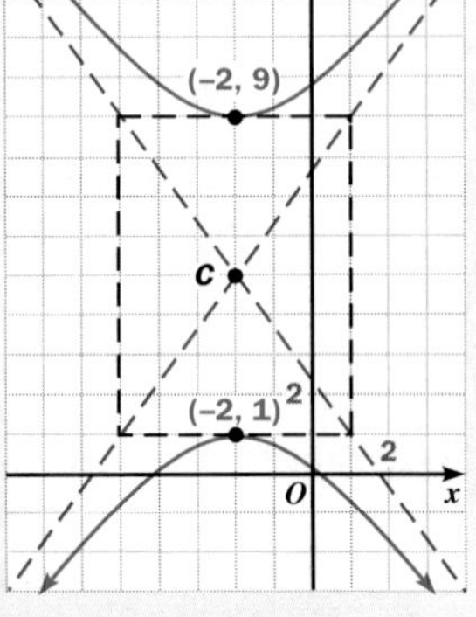

1. What are the coordinates of the center? the foci?
2. Find equations of the asymptotes.
3. Write an equation of the hyperbola.
4. Graph $\frac{(x-2)^2}{4} - \frac{(y+1)^2}{25} = 1$ and $\frac{(x-2)^2}{25} - \frac{(y+1)^2}{4} = 1$. How are the graphs alike? How are they different?

Identify the center, the vertices, and the foci of each hyperbola with the given equation. Tell whether the hyperbola is *horizontal* or *vertical*. Find equations of the asymptotes. Then graph the equation.

5. $\frac{(x-6)^2}{16} - \frac{(y-4)^2}{49} = 1$
6. $\frac{(x+1)^2}{64} - \frac{(y-3)^2}{4} = 1$

Checking Key Concepts

1. (–2, 5); (–2, 10) and (–2, 0)
2. $y = \pm\frac{4}{3}(x + 2) + 5$
3. $\frac{(y-5)^2}{16} - \frac{(x+2)^2}{9} = 1$
4.

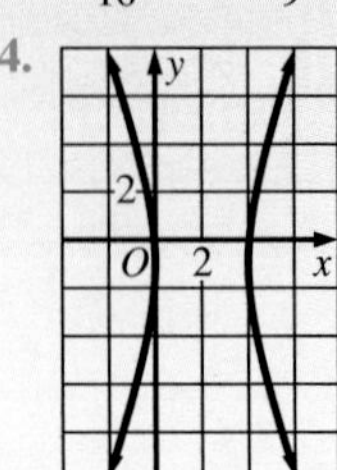

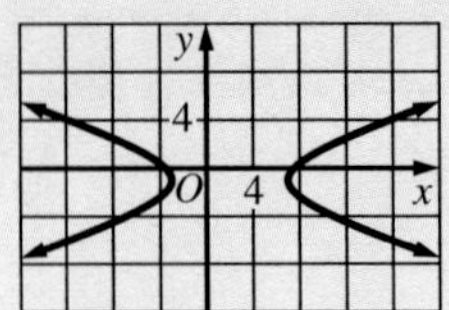

The graphs are alike since they have the same center, and both are horizontal hyperbolas. They are different since their vertices differ, as do the equations for their asymptotes. Also, the first graph is much steeper than the second graph.

5. center: (6, 4); vertices: (2, 4) and (10, 4); foci: $(6 \pm \sqrt{65}, 4) \approx$ (14.06, 4) and (–2.06, 4); horizontal; asymptotes: $y = \pm\frac{7}{4}(x - 6) + 4$

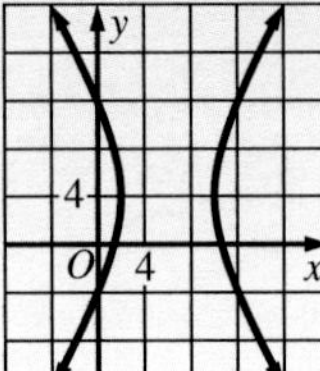

11.5 Exercises and Applications

Extra Practice exercises on page 767

Identify the center, the vertices, and the foci of each hyperbola with the given equation. Tell whether the hyperbola is *horizontal* or *vertical*. Find equations of the asymptotes. Then graph the equation.

1. $\dfrac{x^2}{36} - \dfrac{y^2}{25} = 1$ **2.** $\dfrac{y^2}{4} - \dfrac{x^2}{49} = 1$ **3.** $\dfrac{x^2}{81} - \dfrac{(y-3)^2}{9} = 1$

4. $\dfrac{(x-1)^2}{16} - \dfrac{(y+3)^2}{64} = 1$ **5.** $\dfrac{(y+7)^2}{25} - \dfrac{(x-5)^2}{100} = 1$ **6.** $\dfrac{(x-2)^2}{81} - y^2 = 1$

7. $\dfrac{y^2}{3^2} - \dfrac{(x+4)^2}{7^2} = 1$ **8.** $4x^2 - 25y^2 = 100$ **9.** $\dfrac{y^2}{36} - \dfrac{x^2}{25} = 3$

10. Investigation Use a piece of bifocal paper. Let the centers of each set of concentric circles represent the foci, F_1 and F_2, of a hyperbola.

a. Plot points on the paper that satisfy the equation $|F_1P - F_2P| = 2$. Then connect the points to sketch the curve.

b. Use a different color pen to plot points on the same paper that satisfy the equation $|F_1P - F_2P| = 4$. Then connect the points to sketch the curve.

c. How does the shape of the hyperbola change as you increase the value of $|F_1P - F_2P|$?

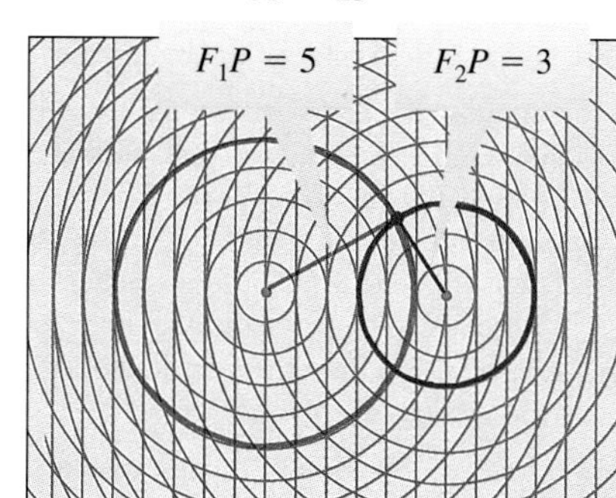

Match each graph with its equation.

11. $\dfrac{(x+3)^2}{9} - \dfrac{(y-1)^2}{25} = 1$ **12.** $\dfrac{(y+3)^2}{9} - \dfrac{(x-1)^2}{25} = 1$

13. $\dfrac{(x+3)^2}{25} - \dfrac{(y-1)^2}{9} = 1$ **14.** $\dfrac{(y+3)^2}{25} - \dfrac{(x-1)^2}{9} = 1$

A.

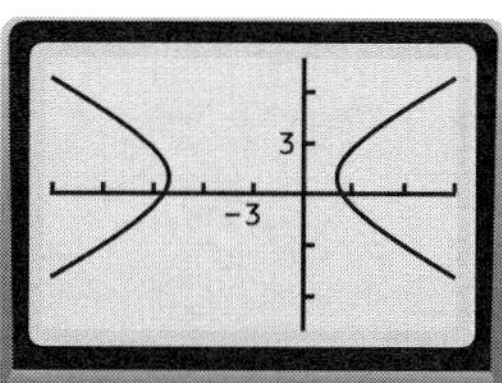

B.

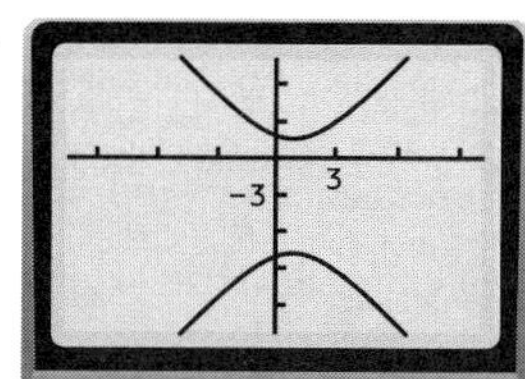

C.

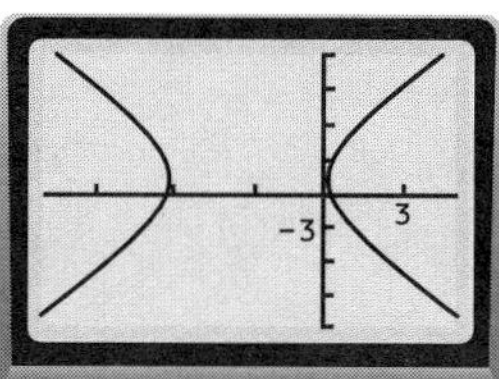

D.

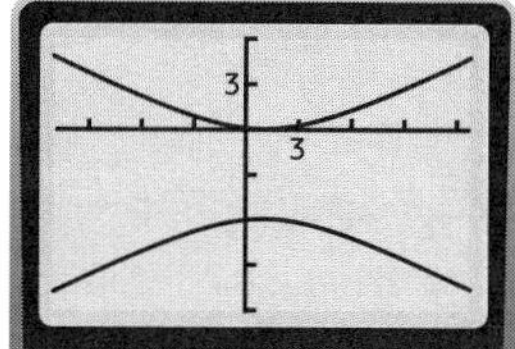

Apply⇔Assess

Suggested Assignment

Core Course
Exs. 1–9, 11–14, 18–25, 28–34

Extended Course
Exs. 1–9 odd, 10–34

Block Schedule
Day 73 Exs. 1–9, 11–14, 18–25, 28–34

Exercise Notes

Student Progress
Exs. 1, 2, 8, 9 Students should be able to tell that the center for each of these hyperbolas is (0, 0) by inspection.

Common Error
Exs. 2, 4, 5, 7 Some students may think that the x^2-term of a hyperbola written in standard form always gives them the value of *a*. Emphasize the difference in the standard forms to help correct this error. For Exs. 5 and 7, make sure students recognize that the vertex for a vertical hyperbola is still (h, k), with h the value in the x^2-term. Some students may list the wrong vertex for a vertical hyperbola.

Student Study Tip
Exs. 8, 9 Remind students that one side of the standard form equation for a hyperbola must be 1. In these exercises, students must first divide by the constant on the right side of the equation.

Using Manipulatives
Ex. 10 Diagram Master 16 can be used for this exercise. Students should see that the hyperbola becomes narrower as $|F_1P - F_2P|$ increases. Have students verify their answers by making up a new set of exercises.

Communication: Drawing
Exs. 11–14 Students should be able to match the orientation of the hyperbola, horizontal or vertical, with the form of the equation by inspection. If not, encourage them to draw the rectangle with dimensions $2a$ and $2b$ to help them match the actual equation with the graph.

6. center: (–1, 3); vertices: (–9, 3) and (7, 3); foci: $(-1 \pm 2\sqrt{17}, 3) \approx$ (7.25, 3) and (–9.25, 3); horizontal; asymptotes: $y = \pm\frac{1}{4}(x+1) + 3$

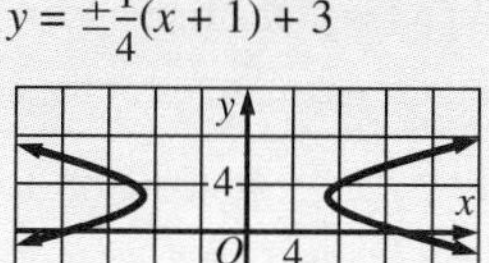

Exercises and Applications

1. center: (0, 0); vertices: $(\pm 6, 0)$; foci: $(\pm\sqrt{61}, 0) \approx (\pm 7.81, 0)$; horizontal; asymptotes: $y = \pm\frac{5}{6}x$

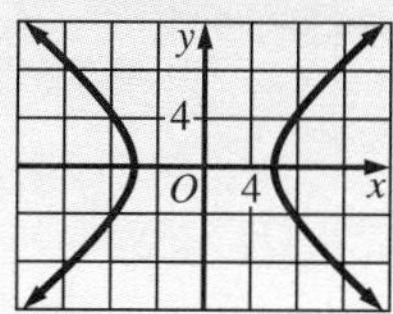

2. center: (0, 0); vertices: $(0, \pm 2)$; foci: $(0, \pm\sqrt{53}) \approx (0, \pm 7.28)$; vertical; asymptotes: $y = \pm\frac{2}{7}x$

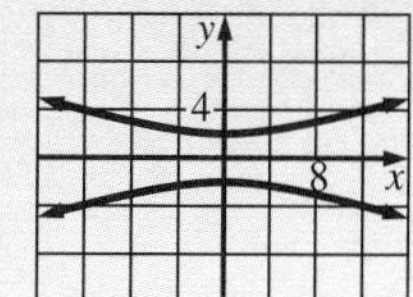

3–10. See answers in back of book.

11. C 12. D

13. A 14. B

Apply⇔Assess

Exercise Notes

Teaching Tip

Ex. 15–17 Point out that Exs. 15 and 16 must be completed before the location of the ship can actually be found in Ex. 17. With one set of signals, the ship's location is anywhere along a branch of the hyperbola. The second hyperbola pinpoints its location. For part (b) of Ex. 15, remind students that the difference in the distances between a ship and the transmitters represents the distance $2a$, where a is one-half the major axis.

Communication: Writing

Ex. 18 This exercise can be used for summary and review. Encourage students to explain their answers with sketches, showing all the key points of a hyperbola. You may want students to give examples to verify that the asymptotes are not the same.

Using Technology

Ex. 20 Have students generate each graph with a graphing calculator to see how the rectangular hyperbola differs in its equation than one in standard form. This can help students review the concept of standard form.

Problem Solving

Exs. 21, 22 Students should be able to tell by inspection that these hyperbolas have center (0, 0). It is incorrect for them to assume, however, that the values of a and b can be read from the asymptote, since the equation of an asymptote represents the ratio of a to b or b to a. Students can still write an equation as long as they recognize there may be more than one equation.

Connection NAVIGATION

Hyperbolic navigation allows ships to locate their position using radio transmitters. A master transmitter, M, and two secondary transmitters, T_1 and T_2, send signals to a receiver on the ship. The time it takes for the ship to receive each signal determines the ship's position along two hyperbolas. The intersection of the two hyperbolas gives the coordinates of the ship's location. All three transmitters are located at foci of the hyperbolas.

15. Complete parts (a)–(c) for the hyperbola with foci $M(2, 1)$ and $T_1(8, 1)$.

a. What is the center of the hyperbola?

b. Suppose the difference in the distances between a ship and the transmitters at M and T_1 is 2. (This means that the ship received a signal from one of the transmitters 2 units of time sooner than it received a signal from the other transmitter.) Find the coordinates of the vertices of the hyperbola.

c. Write an equation of the hyperbola.

16. Repeat Exercise 15 for the hyperbola with foci M and $T_2(2, 9)$. In part (b), suppose that the difference in the distances between the ship and the transmitters at M and T_2 is 6.

17. **Technology** Graph the equations you wrote in part (c) of Exercises 15 and 16. The ship is located at the point where the branch of the hyperbola in Exercise 15 that is closer to T_1 intersects the branch of the hyperbola in Exercise 16 that is closer to T_2. Find the coordinates of the ship.

18. **Writing** How is the hyperbola with equation $\frac{x^2}{a^2} - \frac{y^2}{b^2} = 1$ like the hyperbola with equation $\frac{y^2}{a^2} - \frac{x^2}{b^2} = 1$? How are they different? How do equations of the asymptotes of each graph compare?

19. **Cooperative Learning** Work in a group of three students.

a. Graph each equation.

$\frac{x^2}{4} - \frac{y^2}{4} = 1$ $\quad \frac{x^2}{9} - \frac{y^2}{4} = 1$ $\quad \frac{x^2}{16} - \frac{y^2}{4} = 1$

b. What happens to the graph of each equation as the value of a increases?

c. What do you think will happen to the graph of each equation as the value of a stays the same and the value of b increases? Graph three equations with the same value for a and different values for b to test your prediction.

15. a. (5, 1)
 b. (4, 1) and (6, 1)
 c. $(x-5)^2 - \frac{(y-1)^2}{8} = 1$

16. a. (2, 5)
 b. (2, 2) and (2, 8)
 c. $\frac{(y-5)^2}{9} - \frac{(x-2)^2}{7} = 1$

17.

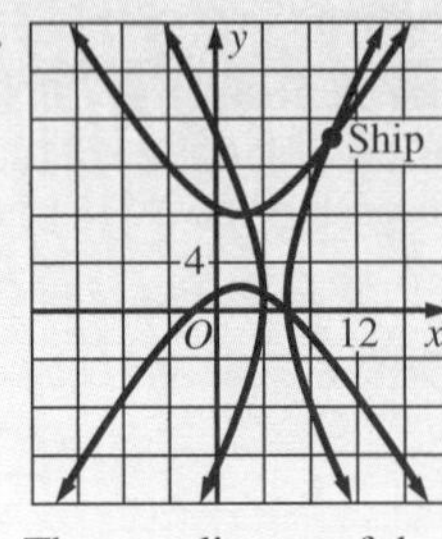

The coordinates of the ship are approximately (9.83, 14.38).

18. Answers may vary. An example is given. The hyperbolas have the same shape, and both are centered at the origin. They are 90° rotations of each other, so that the first is a horizontal hyperbola, while the second is a vertical hyperbola. The asymptotes of the first hyperbola are $y = \pm\frac{b}{a}x$, and the asymptotes of the second hyperbola are $y = \pm\frac{a}{b}x$. Both pairs of asymptotes contain the origin. One asymptote of each hyperbola is perpendicular to an asymptote of the other. For example, $y = \frac{b}{a}x$ is perpendicular to $y = -\frac{a}{b}x$.

20. a. Graph the equation $xy = 4$.

b. What is the center of the graph? What are equations of the asymptotes?

c. Writing How is the hyperbola with equation $xy = 4$ like the other hyperbolas in this section? How is it different?

Write an equation for each hyperbola with the given characteristics. Then sketch its graph.

21. asymptotes: $y = \pm\frac{4}{3}x$; focus $(5, 0)$

22. asymptotes: $y = \pm\frac{5}{12}x$; focus $(0, 13)$

23. center $(0, 0)$; vertex $(6, 0)$; focus $(10, 0)$

24. center $(-5, 2)$; vertex $(-5, -1)$; focus $(-5, 7)$

25. center $(1, -4)$; vertex $(7, -4)$; focus $(-9, -4)$

26. a. Show that $\frac{x^2}{a^2} - \frac{y^2}{b^2} = 1$ can be rewritten as $y = \pm\frac{b}{a}\sqrt{x^2 - a^2}$.

b. Explain why $\pm\frac{b}{a}\sqrt{x^2 - a^2} \approx \pm\frac{b}{a}x$ for large positive values of x.

c. How does your answer to part (b) explain what happens to the graph of $\frac{x^2}{a^2} - \frac{y^2}{b^2} = 1$ as $x \to \infty$?

27. Challenge Suppose that the foci of a hyperbola are $F_1(-5, 0)$ and $F_2(5, 0)$ and the difference in the distances between any point $P(x, y)$ and these foci is 8. Derive an equation of the form $\frac{x^2}{a^2} - \frac{y^2}{b^2} = 1$ to describe this hyperbola.

ONGOING ASSESSMENT

28. Writing Graph the equations $\frac{x^2}{16} - \frac{y^2}{9} = 1$ and $\frac{x^2}{16} + \frac{y^2}{9} = 1$ in the same coordinate plane. How are the graphs alike? How are they different?

SPIRAL REVIEW

Write an equation of the ellipse with the given characteristics. Then sketch its graph. *(Section 11.4)*

29. center $(-6, -1)$; $a = 4$; $b = 1$; major axis is horizontal

30. center $(1, 2)$; vertex $(1, -4)$; focus $(1, 7)$

31. center $(2, -3)$; vertex $(15, -3)$; focus $(-10, -3)$

Write the equation of each function in vertex form and in intercept form. *(Sections 5.4 and 5.5)*

32. $y = x^2 - 8x - 20$ **33.** $y = x^2 - 12x + 32$ **34.** $y = 2x^2 - 5x - 3$

Apply⇔Assess

Exercise Notes

Assessment Note

Ex. 28 This exercise will help you to assess how well students understand the ideas in Sections 11.4 and 11.5. Students should know that the ellipse has the same vertices and is internally tangent to the rectangle with dimensions $2a$ and $2b$.

Topic Spiraling: Review

Exs. 32–34 These exercises should help students review the method of completing the square for finding the vertex of a parabola. This previews how to identify a conic, the topic of the next section.

Practice 74 for Section 11.5

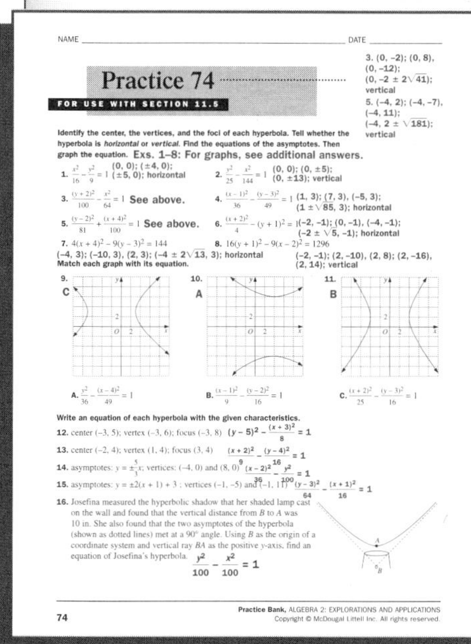

NAME ____________ DATE ____________

Practice 74

FOR USE WITH SECTION 11.5

Identify the center, the vertices, and the foci of each hyperbola. Tell whether the hyperbola is *horizontal* or *vertical*. Find the equations of the asymptotes. Then graph the equation. Exs. 1–8: For graphs, see additional answers.

1. $\frac{x^2}{16} - \frac{y^2}{9} = 1$ (0, 0); (±4, 0); (±5, 0); horizontal
2. $\frac{y^2}{25} - \frac{x^2}{144} = 1$ (0, 0); (0, ±5); (0, ±13); vertical
3. $\frac{(y+2)^2}{100} - \frac{x^2}{64} = 1$ See above.
4. $\frac{(x-1)^2}{36} - \frac{(y-3)^2}{49} = 1$ (1, 3); (7, 3), (–5, 3); $(1 \pm \sqrt{85}, 3)$; horizontal
5. $\frac{(y-2)^2}{81} + \frac{(x+4)^2}{100} = 1$ See above.
6. $\frac{(x+2)^2}{4} - (y+1)^2 = 1$ (–2, –1); (0, –1), (–4, –1); $(-2 \pm \sqrt{5}, -1)$; horizontal
7. $4(x+4)^2 - 9(y-3)^2 = 144$ (–4, 3); (–10, 3), (2, 3); $(-4 \pm 2\sqrt{13}, 3)$; horizontal
8. $16(y+1)^2 - 9(x-2)^2 = 1296$ (–2, –1); (2, –10), (2, 8); (2, –16), (2, 14); vertical

3. (0, –2); (0, 8), (0, –12); $(0, -2 \pm 2\sqrt{41})$; vertical
5. (–4, 2); (–4, –7), (–4, 11); $(-4, 2 \pm \sqrt{181})$; vertical

Match each graph with its equation.

9. C 10. A 11. B

A. $\frac{y^2}{36} - \frac{(x-4)^2}{49} = 1$ B. $\frac{(x-1)^2}{9} - \frac{(y-2)^2}{16} = 1$ C. $\frac{(x+2)^2}{25} - \frac{(y-3)^2}{16} = 1$

Write an equation of each hyperbola with the given characteristics.

12. center (–3, 5); vertex (–3, 6); focus (–3, 8) $(y-5)^2 - \frac{(x+3)^2}{8} = 1$
13. center (–2, 4); vertex (1, 4); focus (3, 4) $\frac{(x+2)^2}{9} - \frac{(y-4)^2}{16} = 1$
14. asymptotes: $y = \pm\frac{5}{3}x$; vertices: (–4, 0) and (8, 0) $\frac{(x-2)^2}{36} - \frac{y^2}{100} = 1$
15. asymptotes: $y = \pm 2(x+1) + 3$; vertices (–1, –5) and (–1, 11) $\frac{(y-3)^2}{64} - \frac{(x+1)^2}{16} = 1$
16. Josefina measured the hyperbolic shadow that her shaded lamp cast on the wall and found that the vertical distance from B to A was 10 in. She also found that the two asymptotes of the hyperbola (shown as dotted lines) met at a 90° angle. Using B as the origin of a coordinate system and vertical ray BA as the positive y-axis, find an equation of Josefina's hyperbola. $\frac{y^2}{100} - \frac{x^2}{100} = 1$

74 **Practice Bank,** ALGEBRA 2: EXPLORATIONS AND APPLICATIONS
Copyright © McDougal Littell Inc. All rights reserved.

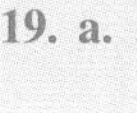

19. a.

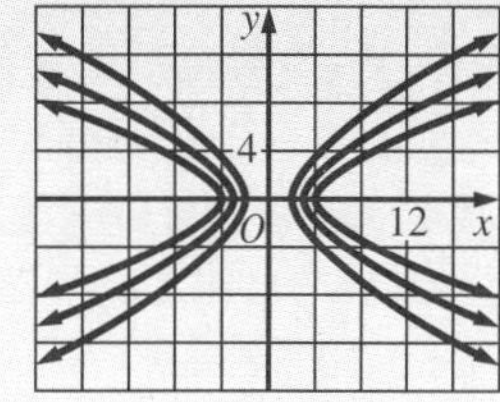

b. The vertices move out farther from the center (the origin), and the hyperbola becomes narrower.

c. Answers may vary. An example is given, using the equations $\frac{x^2}{4} - \frac{y^2}{4} = 1$, $\frac{x^2}{4} - \frac{y^2}{9} = 1$, and $\frac{x^2}{4} - \frac{y^2}{16} = 1$.

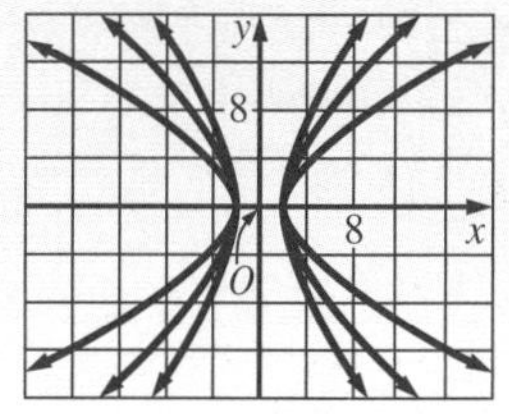

As the value of b increases, the hyperbola widens, while the vertices remain fixed.

20. a.

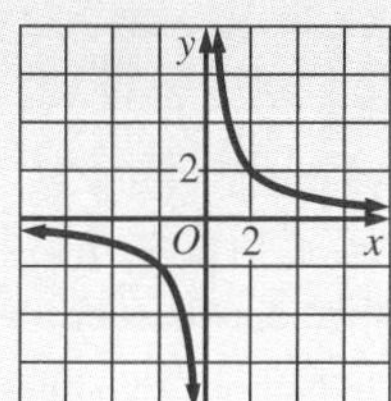

b. the origin; the axes, $x = 0$ and $y = 0$

c. Answers may vary. An example is given. The hyperbola with equation $xy = 4$ has the same shape as other hyperbolas in this section, with vertices, center, and foci all on one line, and asymptotes intersecting at the center of the hyperbola. However, the hyperbola $xy = 4$ is oriented differently from other hyperbolas in this section, with its major axis rotated 45° counterclockwise from the x-axis rather than being parallel or perpendicular to the x-axis. Its asymptotes are the x-axis and the y-axis, unlike the other hyperbolas.

21–34. See answers in back of book.

Plan⇔Support

Objectives

- Find conics by taking a cross section of a double cone.
- Describe how a conic section can be used in applications.

Recommended Pacing

Core and Extended Courses
Section 11.6: 1 day

Block Schedule
Section 11.6: $\frac{1}{2}$ block (with Review)

Resource Materials

Lesson Support
Lesson Plan 11.6
Warm-Up Transparency 11.6
Overhead Visuals: Folder 11: Conic Sections
Practice Bank: Practice 75
Study Guide: Section 11.6
Explorations Lab Manual: Additional Exploration 17, Diagram Master 2
Assessment Book: Test 46

Technology
Graphing Calculator
Internet: http://www.hmco.com

Warm-Up Exercises

1. Factor $9x^2 - y^2$. $(3x - y)(3x + y)$
2. Find k so that $x^2 + 12x + k$ is a perfect square. 36
3. Identify the graph of $\frac{(x-3)^2}{9} - \frac{(y+1)^2}{16} = 1$ and give its center and vertices. horizontal hyperbola with center $(3, -1)$, vertices $(6, -1)$, $(0, -1)$
4. Give the radius of the circle $6x^2 + 6y^2 = 60$. $\sqrt{10}$
5. Express the equation $\frac{x^2}{16} - \frac{y^2}{25} = 0$ in the form $Ax^2 + By^2 + C = 0$. $25x^2 - 16y^2 = 0$

Section

11.6 Identifying Conics

Learn how to...

- find conics by taking a cross section of a double cone

So you can...

- describe the shock wave generated by a supersonic aircraft, for example

Consider two cones placed together so that they have a common vertex and axis of symmetry. If you cut across the cones with a plane at different places and different angles, you can produce all of the curves that you have studied in this chapter. The name **conic section**, or simply *conic*, applies to these curves because they are all slices of a double cone.

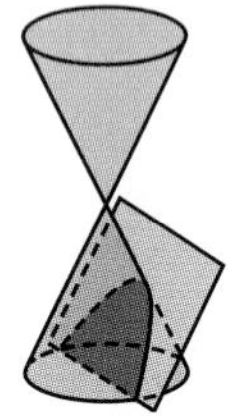
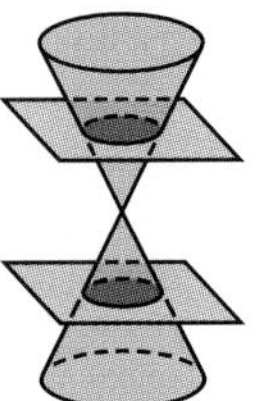
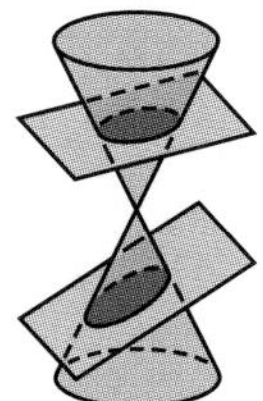
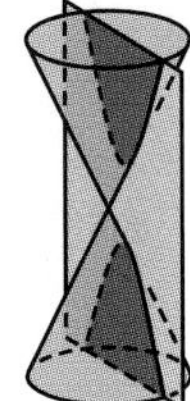

Think and Communicate

1. For each plane, use the diagrams above to tell what conic you get when the plane intersects the double cone and does *not* pass through the vertex. Also tell what conic you get when the plane *does* pass through the vertex.
   a. a horizontal plane **b.** a vertical plane

The general equation for all conics is $Ax^2 + Bxy + Cy^2 + Dx + Ey + F = 0$. For any conic that has an axis of symmetry parallel to the x-axis or the y-axis, $B = 0$. In this section, you will work only with cases where $B = 0$.

Example 1

For each equation of a conic, rewrite the equation in standard form, identify the conic, and state its important characteristics.

a. $x^2 + y^2 + 6x - 4y - 12 = 0$ **b.** $x^2 - y^2 + 6x + 8y - 32 = 0$

Solution

a. Complete the square in x and y.

$$x^2 + y^2 + 6x - 4y - 12 = 0$$
$$(x^2 + 6x + \underline{?}) + (y^2 - 4y + \underline{?}) = 12 + \underline{?} + \underline{?}$$
$$(x^2 + 6x + 9) + (y^2 - 4y + 4) = 12 + 9 + 4$$
$$(x + 3)^2 + (y - 2)^2 = 25$$

This is an equation of the circle with center $C(-3, 2)$ and radius 5.

ANSWERS Section 11.6

Think and Communicate

1. a. a circle; a point, the vertex of the double cone
   b. a hyperbola; two lines that intersect at the vertex of the double cone

b. Complete the square in x and y.

$$x^2 - y^2 + 6x + 8y - 32 = 0$$
$$(x^2 + 6x + \underline{?}) - (y^2 - 8y + \underline{?}) = 32 + \underline{?} - \underline{?}$$
$$(x^2 + 6x + 9) - (y^2 - 8y + 16) = 32 + 9 - 16$$
$$(x + 3)^2 - (y - 4)^2 = 25$$
$$\frac{(x + 3)^2}{25} - \frac{(y - 4)^2}{25} = 1$$

This is an equation of the horizontal hyperbola with center $C(-3, 4)$, vertices $V_1(-8, 4)$ and $V_2(2, 4)$, and asymptotes $y = \pm 1(x + 3) + 4$.

You can use the coefficients of the x^2- and y^2-terms in the general equation of a conic to identify the conic without completing the square.

Identifying Conics

Provided $B = 0$, the graph of $Ax^2 + Bxy + Cy^2 + Dx + Ey + F = 0$ is:

- a parabola if $A = 0$ or $C = 0$, but not both
- a circle if $A = C$ and neither A nor C is 0
- an ellipse if A and C have the same sign ($AC > 0$), and $A \neq C$
- a hyperbola if A and C have opposite signs ($AC < 0$)

Degenerate Conics

The graph of the equation

$$Ax^2 + Bxy + Cy^2 + Dx + Ey + F = 0$$

can be the graph of either a conic section or a *degenerate conic*. A **degenerate conic** occurs when a plane intersects a double cone at its vertex.

Conic	Degenerate case
Ellipse or circle	Point
Parabola	Line
Hyperbola	Pair of intersecting lines

EXAMPLE 2

Identify and graph the degenerate conic $4x^2 - y^2 = 0$.

SOLUTION

$$4x^2 - y^2 = 0$$
$$(2x - y)(2x + y) = 0$$
$$2x - y = 0 \quad \text{or} \quad 2x + y = 0$$
$$y = 2x \quad \text{or} \quad y = -2x$$

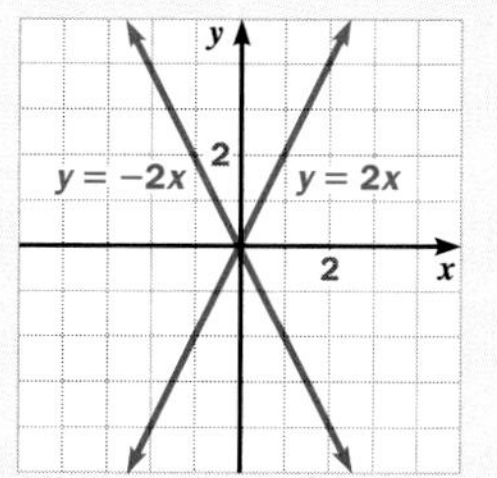

The graphs of these equations are intersecting lines. They represent a degenerate hyperbola.

Teach⇔Interact

Additional Example 1

For each equation of a conic, rewrite the equation in standard form, identify the conic, and state its important characteristics.

a. $x^2 - 2x - 12y + 49 = 0$

Complete the square in x and y.

$$x^2 - 2x - 12y + 49 = 0$$
$$(x^2 - 2x + \underline{?}) - 12y + 49 = 0$$
$$(x^2 - 2x + 1) - 12y + 49 = 1$$
$$(x^2 - 2x + 1) = 12y - 48$$
$$(x - 1)^2 = 12(y - 4)$$
$$\frac{1}{12}(x - 1)^2 = y - 4$$

This is an equation of a parabola that opens up with vertex (1, 4), directrix $y = 1$, and focus (1, 7).

b. $16x^2 + 9y^2 - 64x + 18y - 71 = 0$

Complete the square in x and y.

$$16(x^2 - 4x + \underline{?}) + 9(y^2 + 2y + \underline{?}) = 71 + \underline{?} + \underline{?}$$
$$16(x^2 - 4x + 4) + 9(y^2 + 2y + 1) = 71 + 64 + 9$$
$$16(x - 2)^2 + 9(y + 1)^2 = 144$$
$$\frac{(x - 2)^2}{9} + \frac{(y + 1)^2}{16} = 1$$

This is an equation of a vertical ellipse with center (2, –1) and vertices (2, 3) and (2, –5).

Section Note

Alternate Approach

Another way to identify a conic is to calculate $B^2 - 4AC$ for the general form $Ax^2 + Bxy + Cy^2 + Dx + EY + F = 0$. If the graph is not degenerate and A, B, and C are not all 0, then: if $B^2 - 4AC < 0$, the conic is an ellipse or a circle; if $B^2 - 4AC = 0$, the conic is a parabola; if $B^2 - 4AC > 0$, the conic is a hyperbola.

Additional Example 2

Identify and graph the degenerate conic $4x - y + 8 = 0$.

$$4x - y + 8 = 0$$
$$4x + 8 = y$$

This is a graph of a line. It represents the graph of a degenerate parabola.

Teach⇔Interact

Checking Key Concepts

Communication: Discussion For question 3, emphasize that students should still find the value of *A*, *B*, and *C*. They can identify the conic as an ellipse according to the rules on page 551; then identify the conic as degenerate because, in standard form, the sum of the terms in $x^2 + \frac{y^2}{4} = 0$ is 0, not 1.

Closure Question

Describe how to identify a conic written in general form $Ax^2 + Bxy + Cy^2 + Dx + Ey + F = 0$.

If $A = 0$ or $C = 0$ but not both, the equation represents a parabola. If $A = C$ and neither is 0, the equation is a circle. If $A \neq C$ and $AC > 0$, the equation represents an ellipse. If $AC < 0$, the equation represents a hyperbola.

Apply⇔Assess

Suggested Assignment

Core Course
Exs. 1–6, 12–23, AYP

Extended Course
Exs. 1–23, AYP

Block Schedule
Day 74 Exs. 1–6, 12–23, AYP

Exercise Notes

Student Progress
Exs. 1–6 Students should be able to identify the conics in these exercises before rewriting the conic in standard form. Encourage them to use the rules on page 551.

Reasoning
Ex. 7 You may want students to generalize how to find the values of *s* and *t*. The slant angle *s* is the measure of the acute angle between the edge of the cone and a plane intersecting the cone parallel to the base. The tilt angle *t* is the angle between a plane that cuts the cone and a plane intersecting the cone parallel to the base. Generalize that $t < s < 90°$ for an ellipse, $s = t$ for a parabola, and $s < t$ for a hyperbola.

CHECKING KEY CONCEPTS

For each equation of a conic, rewrite the equation in standard form, identify the conic, and state its important characteristics.

1. $x^2 + y^2 - 6x + y + 3 = 0$

2. $12x^2 + 6y^2 - 36 = 0$

3. Identify and graph the degenerate conic $4x^2 + y^2 = 0$.

11.6 Exercises and Applications

Extra Practice *exercises on page 767*

For each equation of a conic, rewrite the equation in standard form, identify the conic, and state its important characteristics.

1. $9x^2 - 4y^2 + 90x - 16y + 173 = 0$

2. $4x^2 + 8y^2 - 8x - 32 = 0$

3. $-6x^2 - 12x + y - 6 = 0$

4. $9x^2 + y^2 - 18x + 6y - 12 = 0$

5. $4y^2 + 4x^2 + 12y - 16 = 0$

6. $y^2 - 4x^2 - 8x - 18y + 17 = 0$

7. LANGUAGE ARTS The slant angle *s* of a cone and the tilt angle *t* of a plane that cuts the cone determine which conic is created.

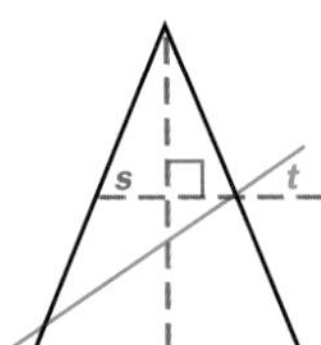

a. If *t* (where $0° < t < 90°$) "falls short" of *s* (that is, $t < s$), an ellipse is created. What is created when *t* equals *s*? What is created when *t* exceeds *s*?

b. Research Look up the definitions of "hyperbole," "parable," and "ellipsis." Describe how these words relate to exceeding, equaling, or falling short of reality.

Connection AVIATION

When an aircraft flies at the speed of sound or faster, it leaves behind a conical pressure wave, or shock wave. This wave causes a loud "sonic boom" that is heard simultaneously at all points of intersection of the cone with the ground.

Major Charles E. Yeager, USAF pilot, was the first person to fly faster than the speed of sound.

8. Consider an aircraft flying parallel to the ground and faster than the speed of sound. What conic section is generated when the shock wave intersects the ground?

9. What conic section(s) can be generated if the plane is ascending?

10. Suppose the sonic boom is heard along a circle on the ground. In what direction is the aircraft traveling?

11. Is it possible for the sonic boom to be heard along a straight line? Explain.

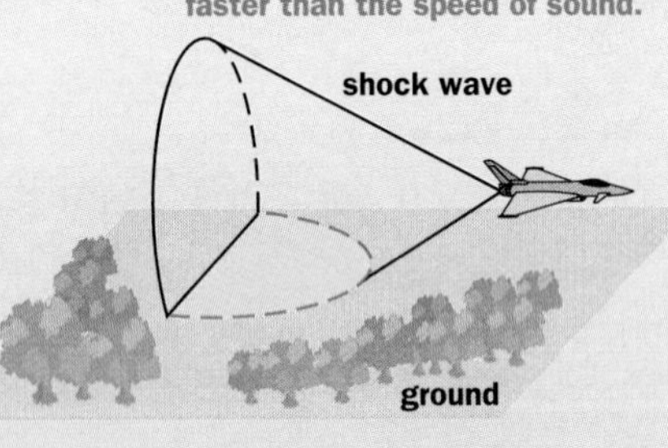

Checking Key Concepts

1. $(x-3)^2 + \left(y + \frac{1}{2}\right)^2 = \frac{25}{4}$; a circle with center $\left(3, -\frac{1}{2}\right)$ and radius $\frac{5}{2}$

2. $\frac{x^2}{3} + \frac{y^2}{6} = 1$; a vertical ellipse with center (0, 0), vertices $(0, \pm\sqrt{6}) \approx (0, \pm 2.45)$, and minor axis with endpoints $(\pm\sqrt{3}, 0) \approx (\pm 1.73, 0)$

3. a point, (0, 0), representing a degenerate ellipse; The graph is the point (0, 0).

Exercises and Applications

1. $\frac{(x+5)^2}{4} - \frac{(y+2)^2}{9} = 1$; a horizontal hyperbola with center (–5, –2), vertices (–3, –2) and (–7, –2), and asymptotes $y = \pm\frac{3}{2}(x+5) - 2$

2. $\frac{(x-1)^2}{9} + \frac{y^2}{4.5} = 1$; a horizontal ellipse with center (1, 0), vertices (4, 0) and (–2, 0), and minor axis with endpoints $(1, \pm\sqrt{4.5}) \approx (1, \pm 2.12)$

3. $y = 6(x+1)^2$; a parabola, opening upward, with vertex (–1, 0) and axis $x = -1$

4. $\frac{(x-1)^2}{\frac{10}{3}} + \frac{(y+3)^2}{30} = 1$; a vertical ellipse with center (1, –3), vertices $(1, -3 \pm \sqrt{30}) \approx$ (1, 2.48) and (1, –8.48), and minor axis with endpoints $\left(1 \pm \sqrt{\frac{10}{3}}, -3\right) \approx (2.83, -3)$ and (–0.83, –3)

5–11. See answers in back of book.

Identify and graph each equation of a degenerate conic.

12. $x^2 + y^2 = 0$ **13.** $(x + y)^2 = 0$ **14.** $-3x^2 + y^2 - 30x - 6y - 66 = 0$

15. $x^2 + 3y^2 - 4x + 4 = 0$ **16.** $x^2 - 9y^2 = 0$ **17.** $-x^2 + 4y^2 - 2x - 8y + 3 = 0$

ONGOING ASSESSMENT

18. **Cooperative Learning** Work with a partner. Hold a flashlight perpendicular to a wall and observe the shape of the light. Change the angle of the flashlight and describe the new shape of the light. What other conics can you create with a flashlight?

SPIRAL REVIEW

Identify the center, the vertices, and the foci of each hyperbola with the given equation. Tell whether the hyperbola is *horizontal* or *vertical*. Find equations of the asymptotes. Then graph the equation. *(Section 11.5)*

19. $\frac{(x + 1)^2}{49} - \frac{(y - 3)^2}{25} = 1$ **20.** $\frac{y^2}{16} - \frac{(x + 2)^2}{36} = 1$

Graph each system of inequalities. *(Section 7.5)*

21. $y \geq 2x$
$y \leq -3x - 1$

22. $y \leq 3x + 2$
$y \geq -2x + 7$

23. $y < -0.5x + 1$
$2y \leq -x + 6$

ASSESS YOUR PROGRESS

VOCABULARY

hyperbola (p. 543)
foci (p. 543)
center (p. 543)
vertices (p. 543)
major axis (p. 543)
minor axis (p. 543)
conic section (p. 550)
degenerate conic (p. 551)

Identify the center, the vertices, and the foci of each hyperbola with the given equation. Tell whether the hyperbola is *horizontal* or *vertical*. Find equations of the asymptotes. Then graph the equation. *(Section 11.5)*

1. $\frac{x^2}{81} - \frac{(y - 5)^2}{36} = 1$ **2.** $\frac{(y - 2)^2}{25} - \frac{(x + 2)^2}{4} = 1$

For each equation of a conic, rewrite the equation in standard form, identify the conic, and state its important characteristics. *(Section 11.6)*

3. $x^2 + y^2 + 14x - 8y + 1 = 0$ **4.** $x^2 - y^2 + 4x + 10y - 30 = 0$

Identify and graph each equation of a degenerate conic. *(Section 11.6)*

5. $x^2 + 36y^2 = 0$ **6.** $x^2 - 36y^2 = 0$

7. **Journal** Compare the conic sections from this chapter. Describe any similarities you notice in the graphs or equations. Which conics do you think are most closely related? Why?

Apply⇔Assess

Exercise Notes

Communication: Discussion
Exs. 12, 14–17 If you ask students to generalize the common characteristics of each of these conics in standard form, they should notice that in each case the square terms are still expressed in the equation, but the equation equals 0, not 1.

Assessment Note
Ex. 18 This exercise will help you assess how well students understand the shapes formed when a plane intersects a cone.

Progress Check 11.5–11.6

See page 557.

Practice 75 for Section 11.6

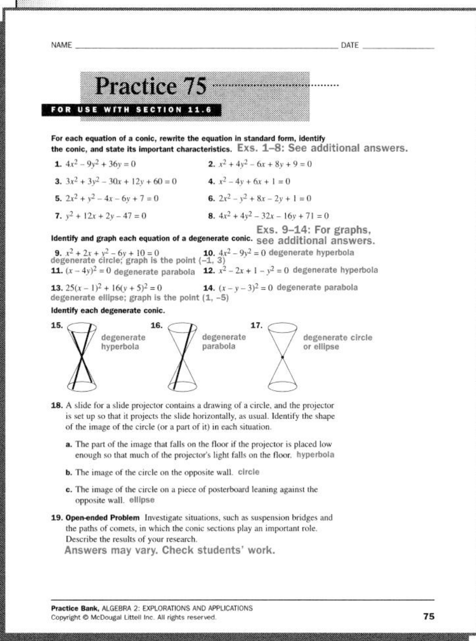

NAME ______ DATE ______

Practice 75

FOR USE WITH SECTION 11.6

For each equation of a conic, rewrite the equation in standard form, identify the conic, and state its important characteristics. Exs. 1–8: See additional answers.

1. $4x^2 - 9y^2 + 36y = 0$ **2.** $x^2 + 4y^2 - 6x + 8y + 9 = 0$

3. $3x^2 + 3y^2 - 30x + 12y + 60 = 0$ **4.** $x^2 - 4y + 6x + 1 = 0$

5. $2x^2 + y^2 - 4x - 6y + 7 = 0$ **6.** $2x^2 - y^2 + 8x - 2y + 1 = 0$

7. $y^2 + 12x + 2y - 47 = 0$ **8.** $4x^2 + 4y^2 - 32x - 16y + 71 = 0$

Identify and graph each equation of a degenerate conic. Exs. 9–14: For graphs, see additional answers.

9. $x^2 + 2x + y^2 - 6y + 10 = 0$ degenerate circle; graph is the point (–1, 3)
10. $4x^2 - 9y^2 = 0$ degenerate hyperbola
11. $(x - 4y)^2 = 0$ degenerate parabola
12. $x^2 - 2x + 1 - y^2 = 0$ degenerate hyperbola
13. $25(x - 1)^2 + 16(y + 5)^2 = 0$ degenerate ellipse; graph is the point (1, –5)
14. $(x - y - 3)^2 = 0$ degenerate parabola

Identify each degenerate conic.

15. degenerate hyperbola **16.** degenerate parabola **17.** degenerate circle or ellipse

18. A slide for a slide projector contains a drawing of a circle, and the projector is set up so that it projects the slide horizontally, as usual. Identify the shape of the image of the circle (or a part of it) in each situation.

a. The part of the image that falls on the floor if the projector is placed low enough so that much of the projector's light falls on the floor. hyperbola

b. The image of the circle on the opposite wall. circle

c. The image of the circle on a piece of posterboard leaning against the opposite wall. ellipse

19. **Open-ended Problem** Investigate situations, such as suspension bridges and the paths of comets, in which the conic sections play an important role. Describe the results of your research.
Answers may vary. Check students' work.

Practice Bank, ALGEBRA 2: EXPLORATIONS AND APPLICATIONS

75

---

12. a degenerate circle; the point (0, 0)

13. a degenerate parabola; the line $y = -x$

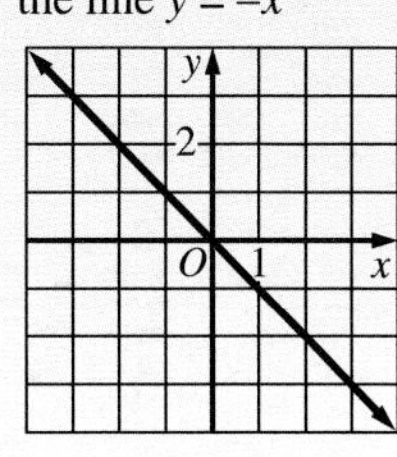

14. a degenerate hyperbola; the intersecting lines $y = 3 \pm \sqrt{3}(x + 5)$

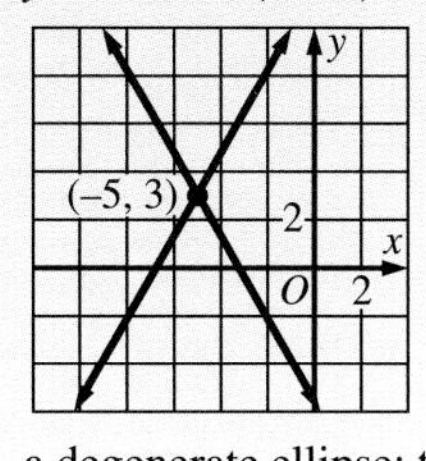

15. a degenerate ellipse; the point (2,0)

16. a degenerate hyperbola; the intersecting lines $y = \frac{x}{3}$ and $y = -\frac{x}{3}$

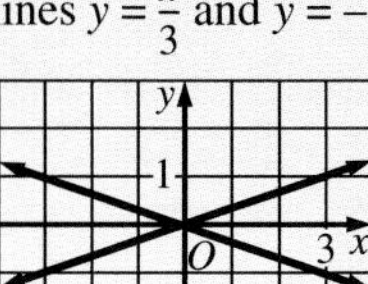

17. a degenerate hyperbola; the intersecting lines $y = \frac{1}{2}x + \frac{3}{2}$ and $y = -\frac{1}{2}x + \frac{1}{2}$

18–23. See answers in back of book.

Assess Your Progress

1. center: (0, 5); vertices: (±9, 5); foci: $(\pm 3\sqrt{13}, 5) \approx (10.82, 5)$; horizontal; asymptotes: $y = \pm\frac{2}{3}x + 5$

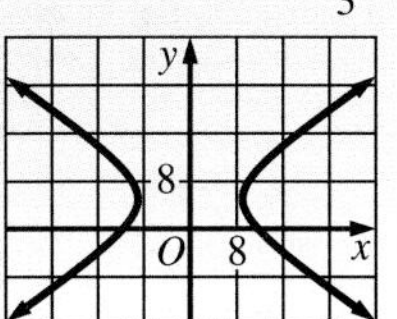

2–7. See answers in back of book.

Project Notes

Mathematical Goal

- Design a logo composed of conic sections for an organization.

Planning

Materials

- graphing calculator or graphing software
- posterboard
- markers or other coloring tools

Project Teams
Students can work alone or with a partner to complete the project. If students work with a partner, they should work together on all phases of the project, including selecting the organization, choosing the conic sections for the logo, and designing the logo.

Guiding Students' Work

Students may have difficulty writing the equations for their conic sections. If so, work through the logic with them so they can translate their ideas into the mathematical equations. Also, if students have difficulty selecting an organization, you may want to develop a list of possible companies with names that lend themselves to logos containing conic sections.

Second-Language Learners
Students learning English should be able to design and present their logos to the class independently, but they might benefit from discussing with English-proficient students what each of the signs, represented by the logos or icons, means.

PORTFOLIO PROJECT

Designing a Logo

Do you recognize the logo shown? It's the symbol for the Olympic Games, which include thousands of participants from around the world.

The five rings symbolize Africa, Asia, Australia, Europe, and the Americas.

The interlocking of the rings symbolizes the meeting of athletes from all over the world.

At least one of the colors of the Olympic logo appears in the flag of every nation.

A *logo* is a symbol that visually represents an organization, such as a club, a charity, or a company. A good logo will often suggest or explain something about the organization it represents. In this project, you will design a logo for an organization that interests you.

PROJECT GOAL **Use a graphing calculator or graphing software to design a logo composed of conic sections.**

Drawing with a Graphing Calculator

Paloma is designing a logo for her mother's company, *Orbital Transmissions,* which sells satellite dishes. For her design, Paloma decides to feature a planet with an orbiting satellite. These are the steps she takes to make her logo.

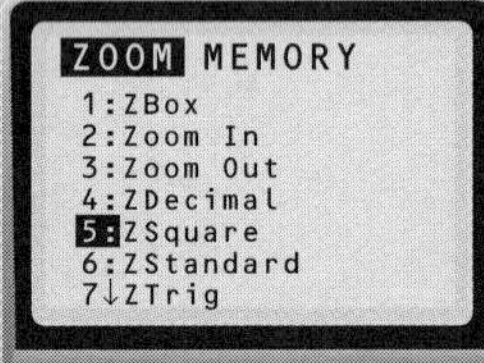

1. USE a square window so that circles are not distorted.

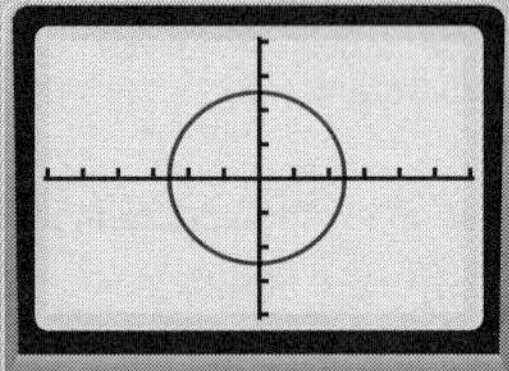

2. ENTER equations to draw a circle, leaving enough room for the rest of the drawing.

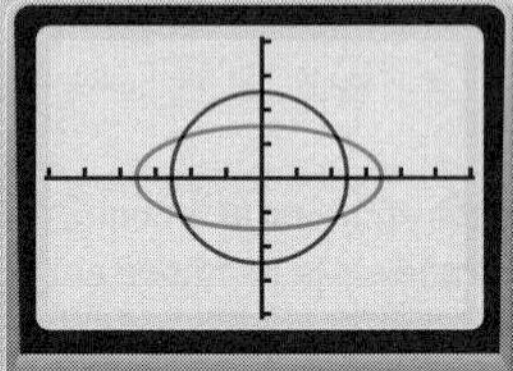

3. ENTER equations to draw an ellipse with a minor axis that is shorter than the radius of the circle and a major axis that is longer.

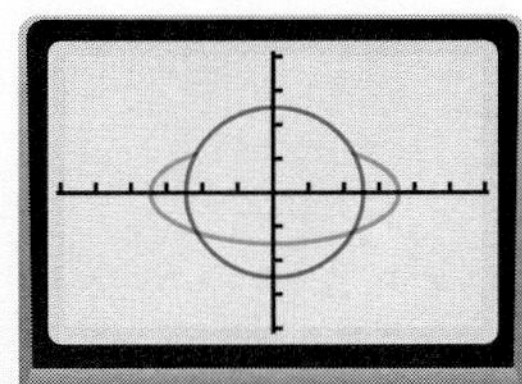

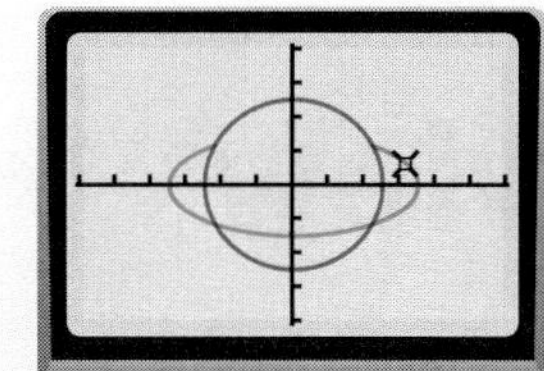

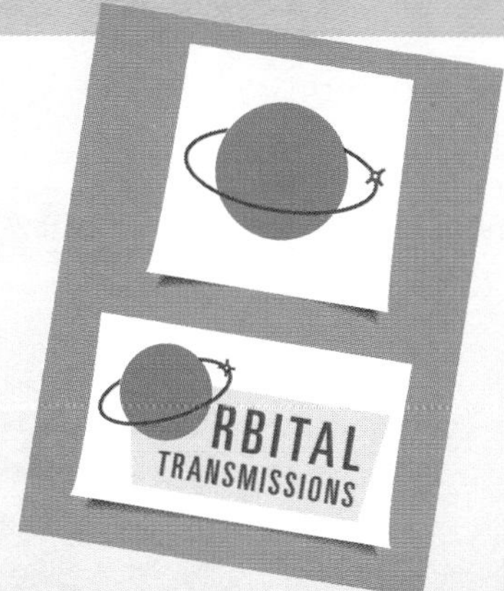

4. RESTRICT the domain of the equation for the upper half of the ellipse so that the ellipse appears to go behind the circle.

5. PLOT a point on the ellipse to give the appearance of an orbiting satellite.

Making Your Own Logo

Describe the organization for which you will design a logo. Then use a graphing calculator or graphing software to create the logo using conic sections. Include points and lines if you wish. Your logo should meet these criteria:

- The logo should be simple and easy to read.
- It should be distinctive and easily understood.
- It should feature information about the organization's functions, services, or products.

Transfer your logo to a poster. Use color to highlight parts of your logo.

Presenting Your Design

Present your logo to the class. Explain what the parts of your logo represent and how you created your design. Show all the equations you used as well as any restrictions on the variables. Explain how your logo meets the criteria given above.

You can extend your project by exploring these ideas:

- Try using a graphing calculator or graphing software to duplicate an actual logo for which conic sections are part of the design.
- Make a list of places where simple visual symbols like the ones shown are used as a substitute for words.
- Try using a graphing calculator or graphing software to draw simple pictures (not necessarily logos) that include conic sections.

Self-Assessment
Describe any problems that you had while creating your logo. How has your understanding of conic sections improved as a result of doing this project?

Project Notes

Guiding Students' Work

Rubric for Chapter Project

4 Students select an organization and design an interesting logo using conic sections. The logo is appropriate for the organization and contains all of the features listed in the text. The poster of the logo is well drawn and neatly done. Students make an interesting presentation of their design to the class and include all of the aspects of designing the logo in the presentation. Students extend the project in one of the three ways listed.

3 Students select an organization and design a logo using conic sections. The logo is appropriate for the organization and the poster is neatly done. Students make a presentation of their logo to the class, but the presentation is incomplete because students do not explain all aspects of the project, such as the equations and their restrictions or how the logo relates to the organization. Students extend the project in one of the three ways listed.

2 Students select an organization and design a logo using conic sections, but the design selected does not relate to the organization or is not easily understood. A neatly drawn poster is made of the logo. Students make a presentation but do not explain all of the parts of the project. No attempt is made to extend the project.

1 Students select an organization and come up with an idea for a logo but do not draw the logo with a graphing calculator or graphing software. No poster is made of a logo, and if a presentation is made, it is incomplete because it does not include an explanation of the equations used in the project or how the logo was designed. Students should be encouraged to speak with the teacher as soon as possible to review their work and to make a new start on the project.

Chapter Support

Course Guide: Chapter 11
Lesson Plans: Chapter 11
Practice Bank:
Cumulative Practice 76
Study Guide: Chapter 11 Review
Assessment Book:
Chapter Tests 47 and 48
Chapter 11 Alternative Assessment
Test Generator Software
Portfolio Project Book
Preparing for College Entrance Tests
Professional Handbook

Progress Check 11.1–11.2

1. Given points (–2, 3) and (–6, 0), find the distance between the points, the midpoint of the line segment connecting the points, and an equation of the perpendicular bisector of this segment. *(Section 11.1)*
5; (–4, 1.5); $y = -\frac{4}{3}x - \frac{23}{6}$

2. Given a line with equation $y = 5x + 2$ and point P with coordinates (–3, –6), find an equation of the line through P and:
a. parallel to l $y = 5x + 9$
b. perpendicular to l *(Section 11.1)*
$y = -\frac{1}{5}x - \frac{33}{5}$

Give the vertex, focus, and directrix of each parabola. *(Section 11.2)*

3. $y - 4 = \frac{1}{12}(x + 3)^2$
vertex: (–3, 4); focus: (–3, 7); directrix: $y = 1$

4. $x + 4 = -\frac{1}{4}(y - 2)^2$
vertex: (–4, 2); focus: (–5, 2); directrix: $x = -3$

5. Find an equation of the parabola with the given characteristic. Then graph the equation.
vertex: (–2, 3); directrix: $y = 4$ *(Section 11.2)*
$y - 3 = -\frac{1}{4}(x + 2)^2$

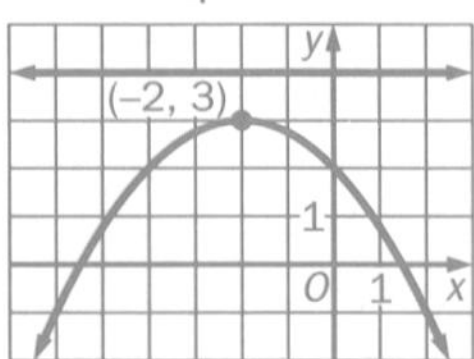

CHAPTER

11 Review

STUDY TECHNIQUE

Without looking at the book or your notes, write a list of important concepts from this chapter. Then look through the chapter and your notes and compare them with your list. Did you miss anything?

VOCABULARY

parabola (p. 521)
focus, foci (pp. 521, 536, 543)
directrix (p. 521)
vertex, vertices (pp. 521, 536, 543)
circle (p. 528)
center (pp. 528, 536, 543)
radius (p. 528)
ellipse (p. 536)
major axis (pp. 536, 543)
minor axis (pp. 536, 543)
hyperbola (p. 543)
conic section (p. 550)
degenerate conic (p. 551)

SECTION 11.1

For any two points $A(x_1, y_1)$ and $B(x_2, y_2)$:

- the distance AB between the two points is $AB = \sqrt{(x_2 - x_1)^2 + (y_2 - y_1)^2}$
- the midpoint M of $\overline{AB}$ has coordinates $\left(\frac{x_1 + x_2}{2}, \frac{y_1 + y_2}{2}\right)$

For example, if A has coordinates (3, 1) and B has coordinates (–2, 5), then:

$$AB = \sqrt{(-2 - 3)^2 + (5 - 1)^2} = \sqrt{41} \approx 6.4$$

$$M = \left(\frac{3 + (-2)}{2}, \frac{1 + 5}{2}\right) = \left(\frac{1}{2}, 3\right)$$

Two lines are parallel if and only if they have the same slope and different y-intercepts. Two lines are perpendicular if and only if the slope of one line is the negative reciprocal of the slope of the other line.

SECTIONS 11.2, 11.3, 11.4, *and* 11.5

A **parabola** with **vertex** $V(h, k)$ and a distance of $2c$ between its **focus** F and **directrix** d has equation $y - k = \frac{1}{4c}(x - h)^2$ if it opens up or down, and has equation $x - h = \frac{1}{4c}(y - k)^2$ if it opens left or right.

A **circle** with **center** (h, k) and **radius** r has equation $(x - h)^2 + (y - k)^2 = r^2$.

An equation of an **ellipse** with **major axis** of length $2a$, **minor axis** of length $2b$, and center $C(h, k)$ is written in one of two ways:

- A *horizontal* ellipse has equation

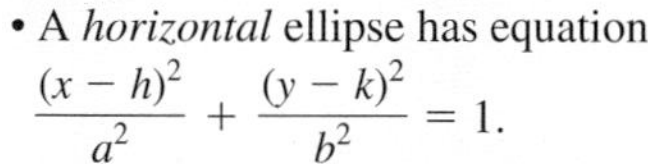

$$\frac{(x-h)^2}{a^2} + \frac{(y-k)^2}{b^2} = 1.$$

- A *vertical* ellipse has equation $\frac{(x-h)^2}{b^2} + \frac{(y-k)^2}{a^2} = 1.$

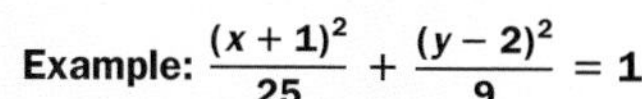

Example: $\frac{(x+1)^2}{25} + \frac{(y-2)^2}{9} = 1$

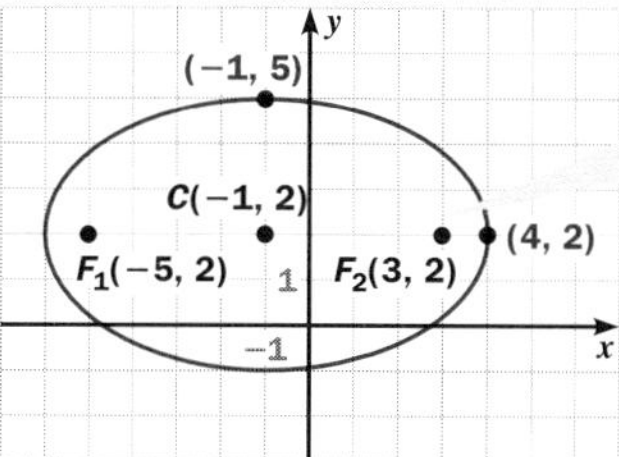

The foci lie on the major axis c units from the center. For an ellipse, $c = \sqrt{a^2 - b^2}$.

An equation of a **hyperbola** with major axis of length $2a$, minor axis of length $2b$, and center $C(h, k)$ is written in one of two ways:

- A *horizontal* hyperbola has equation $\frac{(x-h)^2}{a^2} - \frac{(y-k)^2}{b^2} = 1$ with asymptotes $y = \pm\frac{b}{a}(x - h) + k$.
- A *vertical* hyperbola has equation $\frac{(y-k)^2}{a^2} - \frac{(x-h)^2}{b^2} = 1$ with asymptotes $y = \pm\frac{a}{b}(x - h) + k$.

Example: $\frac{(y-3)^2}{4} - \frac{(x-5)^2}{9} = 1$

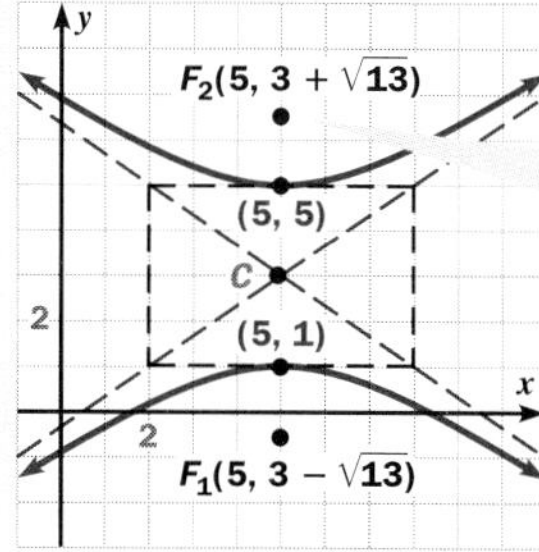

The foci lie on the major axis c units from the center. For a hyperbola, $c = \sqrt{a^2 + b^2}$.

SECTION 11.6

The general form of equations of all **conic sections**, including **degenerate conics**, is $Ax^2 + Bxy + Cy^2 + Dx + Ey + F = 0$. If $B = 0$, you can use the method of completing the square to write the equation in standard form. For example, you know the graph of $16x^2 - 25y^2 - 64x - 150y - 561 = 0$ is a hyperbola because the coefficients of x^2 and y^2 have opposite signs.

$$16x^2 - 25y^2 - 64x - 150y - 561 = 0$$
$$16(x^2 - 4x + \underline{?}) - 25(y^2 + 6y + \underline{?}) = 561 + \underline{?} + \underline{?}$$

Complete the square in x and y.

$$16(x^2 - 4x + 4) - 25(y^2 + 6y + 9) = 561 + 64 - 225$$
$$16(x - 2)^2 - 25(y + 3)^2 = 400$$
$$\frac{(x-2)^2}{25} - \frac{(y+3)^2}{16} = 1$$

Add 64 and subtract 225 on both sides of the equation.

This is an equation of the horizontal hyperbola with center $C(2, -3)$, vertices $V_1(-3, -3)$ and $V_2(7, -3)$, and asymptotes $y = \pm\frac{4}{5}(x - 2) - 3$.

Progress Check 11.3–11.4

1. Write an equation of a circle with center (−2, 5) and radius 4. *(Section 11.3)*
$(x + 2)^2 + (y - 5)^2 = 16$
2. Graph the circle $(x + 3)^2 + (y + 1)^2 = 25$. *(Section 11.3)*

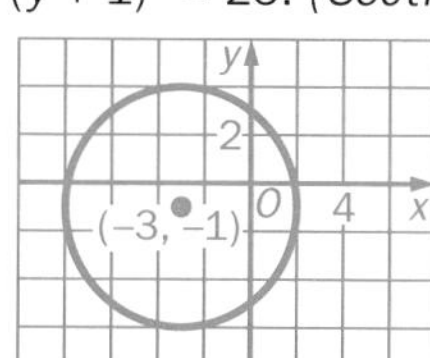

3. Solve the system of equations $(x + 2)^2 + (y - 1)^2 = 16$ and $(x - 4)^2 + (y - 3)^2 = 9$. *(Section 11.3)*
(2, 0.77), (1.06, 3.6)
4. Identify the center, the vertices, and the foci of the ellipse $\frac{(x+2)^2}{16} + \frac{(y-4)^2}{9} = 1$ and tell whether it is *horizontal* or *vertical*. Then graph the equation. *(Section 11.4)*
center: (−2, 4); vertices: (2, 4), (−6, 4); foci: $(-2 + \sqrt{7}, 4)$, $(-2 - \sqrt{7}, 4)$; horizontal

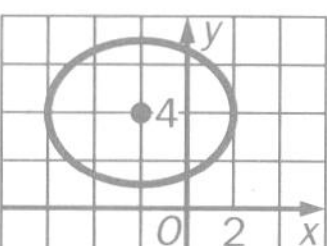

Progress Check 11.5–11.6

1. Graph $\frac{(x-1)^2}{16} - \frac{(y+3)^2}{9} = 1$. *(Section 11.5)*

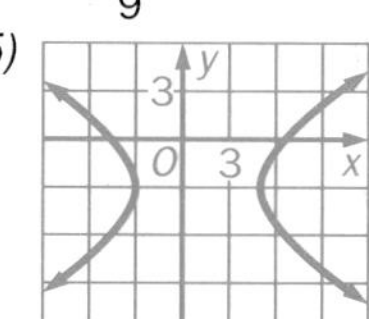

2. Write an equation of a hyperbola with center (−2, −4), one vertex at (−2, −1), and one focus at (−2, 1). *(Section 11.5)*
$\frac{(y+4)^2}{9} - \frac{(x+2)^2}{16} = 1$
3. Identify the center, asymptotes, vertices, and foci of the hyperbola $4x^2 - 9y^2 = 36$. *(Section 11.5)*
asymptotes: $y = \pm\frac{2}{3}x$;
center: (0,0);
vertices: (3, 0), (−3, 0);
foci: $(\sqrt{13}, 0)$, $(-\sqrt{13}, 0)$
4. Given $y^2 + 8x + 4y + 20 = 0$, rewrite the equation in standard form. *(Section 11.6)*
$x + 2 = -\frac{1}{8}(y + 2)^2$
5. Identify $12x^2 + 25y^2 + 24x - 150y - 163 = 0$. *(Section 11.6)*
ellipse
6. Identify the degenerate conic $9x^2 - 16y^2 = 0$. *(Section 11.6)*
degenerate hyperbola

CHAPTER

11 Assessment

VOCABULARY QUESTIONS

For Questions 1–3, complete each paragraph.

1. When two lines have the same slope and different y-intercepts, the lines are _?_. Two lines are _?_ to each other when their slopes are negative reciprocals.
2. A _?_ consists of all points that are the same distance from a point F, called the _?_, and a line d, called the _?_.
3. The _?_ of an ellipse is the line segment that contains the foci and has the two _?_ as its endpoints.

SECTION 11.1

For each pair of points, find:

a. the distance between the points

b. the coordinates of the midpoint of the line segment connecting the points

4. $(1, 5), (2, 4)$
5. $(3, 2), (7, 5)$
6. $(-7, 0), (-3, 1)$
7. Given the point $P(-2, 6)$ and the line l with equation $y = \frac{3}{5}x - 4$, find an equation of the line through P and:
   a. parallel to l
   b. perpendicular to l
8. a. **GEOMETRY** If the endpoints of a diameter of circle P are $(-3, 4)$ and $(2, -2)$, find the coordinates of the center of P.
   b. What is the length of the diameter of circle P? What is the radius?
   c. Find an equation of the line that is tangent to the circle at $(-3, 4)$.

SECTIONS 11.2, 11.3, 11.4, *and* 11.5

9. Find an equation of the parabola with focus $F(4, -2)$ and directrix $d{:}x = -3$. Then graph the equation.

For each equation, identify the conic and state its important characteristics. Then graph the equation.

10. $(x + 2)^2 + (y - 5)^2 = 16$
11. $\frac{(x - 1)^2}{9} - \frac{(y - 6)^2}{16} = 1$
12. $\frac{(x + 2)^2}{36} + \frac{(y - 1)^2}{64} = 1$
13. $4(x + 1)^2 + 25(y + 7)^2 = 100$

Chapter 11 Assessment

Form A Chapter Test

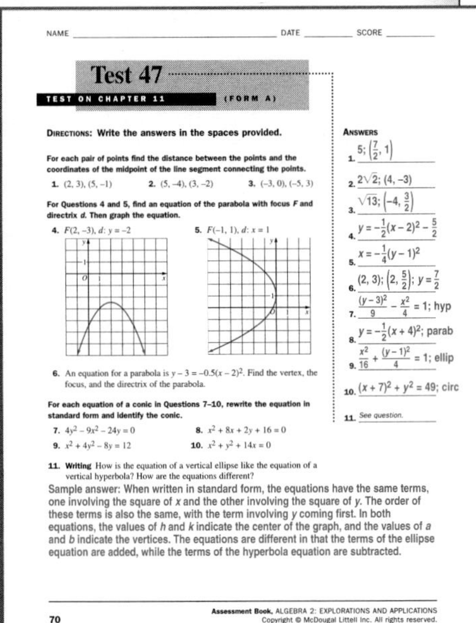

NAME ________ DATE ________ SCORE ________

Test 47

TEST ON CHAPTER 11 (FORM A)

DIRECTIONS: Write the answers in the spaces provided.

For each pair of points find the distance between the points and the coordinates of the midpoint of the line segment connecting the points.

1. $(2, 3), (5, -1)$
2. $(5, -4), (3, -2)$
3. $(-3, 0), (-5, 3)$

For Questions 4 and 5, find an equation of the parabola with focus F and directrix d. Then graph the equation.

4. $F(2, -3)$, d: $y = -2$
5. $F(-1, 1)$, d: $x = 1$

6. An equation for a parabola is $y - 3 = -0.5(x - 2)^2$. Find the vertex, the focus, and the directrix of the parabola.

For each equation of a conic in Questions 7–10, rewrite the equation in standard form and identify the conic.

7. $4y^2 - 9x^2 - 24y = 0$
8. $x^2 + 8x + 2y + 16 = 0$
9. $x^2 + 4y^2 - 8y = 12$
10. $x^2 + y^2 + 14x = 0$

11. **Writing** How is the equation of a vertical ellipse like the equation of a vertical hyperbola? How are the equations different?

Sample answer: When written in standard form, the equations have the same terms, one involving the square of x and the other involving the square of y. The order of these terms is also the same, with the term involving y coming first. In both equations, the values of h and k indicate the center of the graph, and the values of a and b indicate the vertices. The equations are different in that the terms of the ellipse equation are added, while the terms of the hyperbola equation are subtracted.

ANSWERS

1. $5; \left(\frac{7}{2}, 1\right)$
2. $2\sqrt{2}; (4, -3)$
3. $\sqrt{13}; \left(-4, \frac{3}{2}\right)$
4. $y = -\frac{1}{2}(x - 2)^2 - \frac{5}{2}$
5. $x = -\frac{1}{4}(y - 1)^2$
6. $(2, 3); \left(2, \frac{5}{2}\right); y = \frac{7}{2}$
7. $\frac{(y - 3)^2}{9} - \frac{x^2}{4} = 1$; hyp
8. $y = -\frac{1}{2}(x + 4)^2$; parab
9. $\frac{x^2}{16} + \frac{(y - 1)^2}{4} = 1$; ellip
10. $(x + 7)^2 + y^2 = 49$; circ
11. *See question.*

70

Assessment Book, ALGEBRA 2: EXPLORATIONS AND APPLICATIONS

Chapter 11 Assessment

Form B Chapter Test

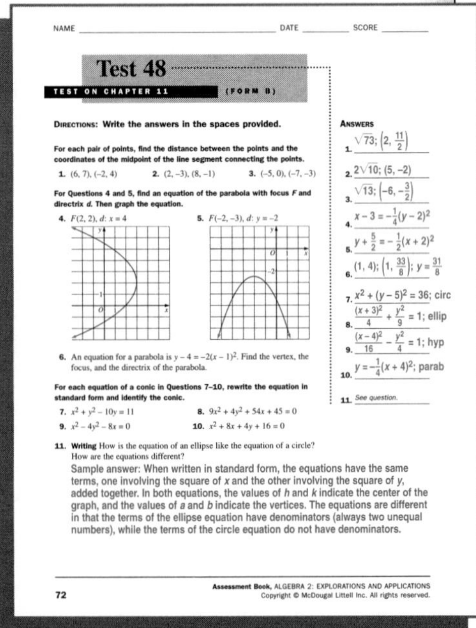

NAME ________ DATE ________ SCORE ________

Test 48

TEST ON CHAPTER 11 (FORM B)

DIRECTIONS: Write the answers in the spaces provided.

For each pair of points, find the distance between the points and the coordinates of the midpoint of the line segment connecting the points.

1. $(6, 7), (-2, 4)$
2. $(2, -3), (8, -1)$
3. $(-5, 0), (-7, -3)$

For Questions 4 and 5, find an equation of the parabola with focus F and directrix d. Then graph the equation.

4. $F(2, 2)$, d: $x = 4$
5. $F(-2, -3)$, d: $y = -2$

6. An equation for a parabola is $y - 4 = -2(x - 1)^2$. Find the vertex, the focus, and the directrix of the parabola.

For each equation of a conic in Questions 7–10, rewrite the equation in standard form and identify the conic.

7. $x^2 + y^2 - 10y = 11$
8. $9x^2 + 4y^2 + 54x + 45 = 0$
9. $x^2 - 4y^2 - 8x = 0$
10. $x^2 + 8x + 4y + 16 = 0$

11. **Writing** How is the equation of an ellipse like the equation of a circle? How are the equations different?

Sample answer: When written in standard form, the equations have the same terms, one involving the square of x and the other involving the square of y, added together. In both equations, the values of h and k indicate the center of the graph, and the values of a and b indicate the vertices. The equations are different in that the terms of the ellipse equation have denominators (always two unequal numbers), while the terms of the circle equation do not have denominators.

ANSWERS

1. $\sqrt{73}; \left(2, \frac{11}{2}\right)$
2. $2\sqrt{10}; (5, -2)$
3. $\sqrt{13}; \left(-6, -\frac{3}{2}\right)$
4. $x - 3 = -\frac{1}{4}(y - 2)^2$
5. $y + \frac{5}{2} = -\frac{1}{2}(x + 2)^2$
6. $(1, 4); \left(1, \frac{33}{8}\right); y = \frac{31}{8}$
7. $x^2 + (y - 5)^2 = 36$; circ
8. $\frac{(x + 3)^2}{4} + \frac{y^2}{9} = 1$; ellip
9. $\frac{(x - 4)^2}{16} - \frac{y^2}{4} = 1$; hyp
10. $y = -\frac{1}{4}(x + 4)^2$; parab
11. *See question.*

72

Assessment Book, ALGEBRA 2: EXPLORATIONS AND APPLICATIONS

ANSWERS Chapter 11

Assessment

1. parallel; perpendicular
2. parabola; focus; directrix
3. major axis; vertices
4. a. $\sqrt{2} \approx 1.41$ b. $(1.5, 4.5)$
5. a. 5 b. $(5, 3.5)$
6. a. $\sqrt{17} \approx 4.12$ b. $(-5, 0.5)$
7. a. $y = \frac{3}{5}x + \frac{36}{5}$
   b. $y = -\frac{5}{3}x + \frac{8}{3}$
8. a. $\left(-\frac{1}{2}, 1\right)$
   b. $\sqrt{61} \approx 7.81$; $\frac{1}{2}\sqrt{61} \approx 3.91$
   c. $y = \frac{5}{6}x + \frac{13}{2}$
9. $x - \frac{1}{2} = \frac{1}{14}(y + 2)^2$

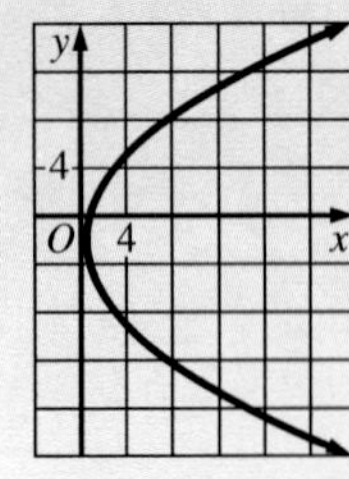

10. a circle with center $(-2, 5)$ and radius 4

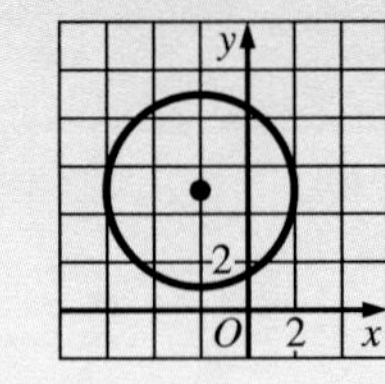

11. a horizontal hyperbola with center $(1, 6)$, vertices $(4, 6)$ and $(-2, 6)$, foci $(6, 6)$ and $(-4, 6)$, and asymptotes $y = \pm\frac{4}{3}(x - 1) + 6$

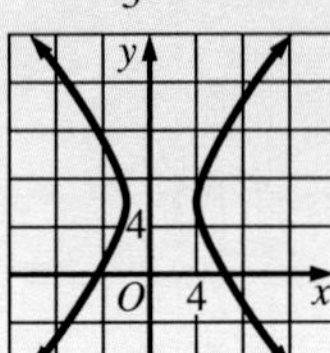

14. ASTRONOMY Halley's comet travels around the sun in a path that is an ellipse. Assume that the center of the ellipse is placed at the origin of a coordinate plane and the foci are located on the x-axis. The distance between the vertices is 35.6 astronomical units (AU), and the distance between the foci is 34.4 AU.

a. The sun is located at one of the foci of the ellipse. What are the possible coordinates of the sun on this graph?

b. Write an equation of the ellipse that models the path of the comet. Then graph the equation.

15. Writing How is the equation of a horizontal hyperbola different from the equation of a vertical hyperbola? How are the equations alike?

Write an equation of each conic with the given characteristics.

16. a parabola with focus (3, 2) and directrix $y = 6$

17. a circle with center (0, 8) and radius 2

Write an equation of each graph.

18.

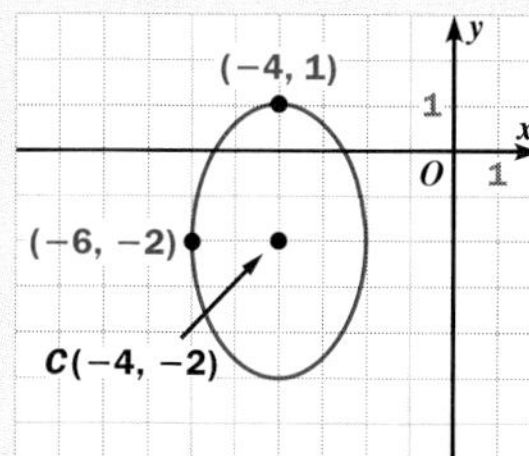

19.

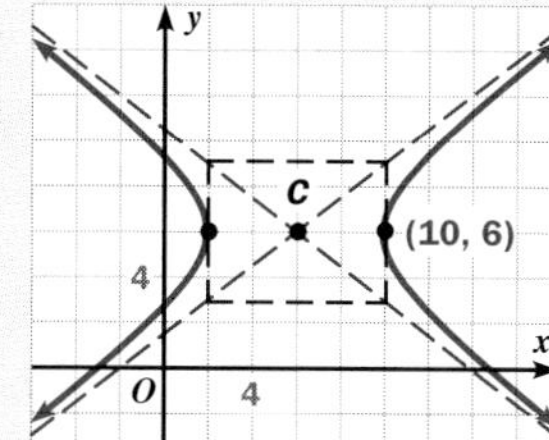

SECTION 11.6

For each equation of a conic, rewrite the equation in standard form and identify the conic.

20. $y^2 - 6y + 16x + 25 = 0$

21. $x^2 - 4y^2 - 2x + 16y - 19 = 0$

22. $x^2 + y^2 - 12y + 25 = 0$

23. $x^2 + 25y^2 - 6x - 100y + 84 = 0$

Identify and graph each equation of a degenerate conic.

24. $25x^2 - y^2 = 0$

25. $x^2 + y^2 + 2x + 2y + 2 = 0$

PERFORMANCE TASK

26. Make a poster to display the conic sections from this chapter. Highlight the important characteristics of each conic and give its equation. Describe the similarities and differences of the equations of the conics and of their graphs.

Assessment **559**

Chapter 11 Assessment
Form C Alternative Assessment

Chapter 11
ALTERNATIVE ASSESSMENT

1. **a.** Draw a line segment on graph paper that has (2, 1) and (14, 6) as its endpoints.
   b. Find the length of the line segment.
   c. Draw four more line segments that have the same length as the line segment in part (a).
   d. Use the endpoints of your new line segments and the distance formula to show that all the lengths are the same.
   e. Without measuring or using the distance formula, describe how you can tell that your segments are congruent to the original line segment.
2. **Open-ended Problem** Draw an isosceles triangle on graph paper. Use the grids to estimate the area of your triangle. Now use the formula for area of a triangle, $A = \frac{1}{2}bh$, to find the exact area of your triangle.
3. **a.** Write a program for a graphing calculator that calculates the distance between two points input by the user.
   b. Write a program for a graphing calculator that calculates the midpoint of a line segment when the user enters the endpoints of the line segment.
4. **Project** Bend a wire to model the path of the water from a drinking fountain. To determine if the path is parabolic, lay your wire on graph paper and find some points on the curve. Use these points to try to find a quadratic equation for your model.
5. **Performance Task** Compare and contrast and the standard form equations of the four types of conic sections.
6. **Group Activity** Make a picture using the graphs of conic sections. Write the equations for the graphs and any limits on the domains on a separate sheet of paper. Exchange your equation papers with another group. Try to reproduce the intended drawings.
7. **Open-ended Problem** How does string art reflect what you have learned about conics? Consider the definitions of each conic section in your answer.

124 **Assessment Book,** ALGEBRA 2: EXPLORATIONS AND APPLICATIONS

12. a vertical ellipse with center (−2, 1), vertices (−2, −7) and (−2, 9), and minor axis with endpoints (−8, 1) and (4, 1)

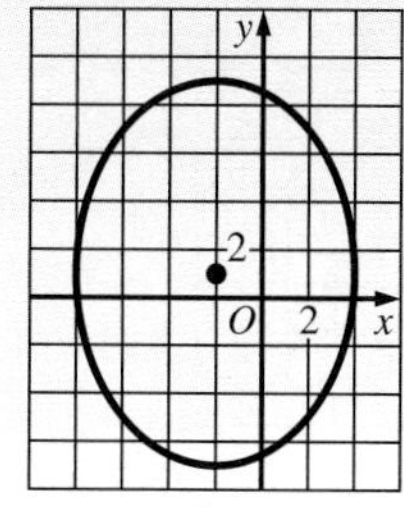

13. a horizontal ellipse with center (−1, −7), vertices (−6, −7) and (4, −7), and minor axis with endpoints (−1, −9) and (−1, −5)

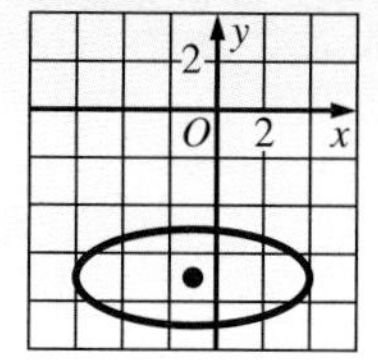

14. a. (17.2, 0) or (−17.2, 0)

b. $\frac{x^2}{316.84} + \frac{y^2}{21} = 1$

15. Answers may vary. An example is given. The equations are different in that a horizontal hyperbola has a positive x-term coefficient and a negative y-term coefficient, while a vertical hyperbola has a negative x-term coefficient and a positive y-term coefficient. The equations are alike in that each has the form $\frac{(\text{quantity})^2}{a^2} - \frac{(\text{quantity})^2}{b^2} = 1$.

16. $y - 4 = -\frac{1}{8}(x - 3)^2$

17. $x^2 + (y - 8)^2 = 4$

18–26. See answers in back of book.

CHAPTER

12 Discrete Mathematics

OVERVIEW

Connecting to Prior and Future Learning

⇔ This chapter opens with a discussion of how to color graphs with as few colors as possible, building on the study of discrete mathematics begun in Algebra 1. Directed graphs and their representations with matrices are presented in Section 12.2.

⇔ Permutations and combinations were introduced in Algebra 1 and are discussed again in Sections 12.3 and 12.4. Both of these concepts will be used in Chapter 13, when students study probability.

⇔ Pascal's triangle is introduced in Section 12.5 and will be used again in Chapter 13. Students discover how Pascal's triangle can be used when expanding binomials.

Chapter Highlights

Interview with Maria Rodriguez: The use of mathematics in electrical engineering is emphasized in this interview. The related exercises on pages 574 and 582 allow students to explore this application even further.

Explorations in Chapter 12 focus on how to color a map using as few colors as possible in Section 12.1, investigating the multiplication counting principle in Section 12.3, and investigating ways of choosing winners in Section 12.4.

The Portfolio Project: Students choose a real-world situation that has many options and make a flip book to show all the possible combinations of the options.

Technology: Graphing calculators are used to perform matrix calculations and to find the number of permutations or combinations. In Section 12.4, an exercise on using computers to color graphs is provided.

OBJECTIVES

Section	Objectives	NCTM Standards
12.1	• Represent situations with graphs. • Color graphs, trying to use as few colors as possible. • Solve problems that require sorting items into groups.	1, 2, 3, 4, 12
12.2	• Represent situations with directed graphs. • Represent directed graphs with matrices. • Analyze situations involving direction.	1, 2, 3, 4, 5, 12
12.3	• Use the multiplication counting principle. • Find the number of permutations of the elements of a set. • Count the possibilities in real-world situations.	1, 2, 3, 4, 5, 12
12.4	• Find the number of combinations of the elements of a set. • Count choices in which order is not important.	1, 2, 3, 4, 5, 12
12.5	• Apply the concept of combinations to Pascal's triangle. • Easily count the outcomes of sequences of events for which there are exactly two possibilities.	1, 2, 3, 4, 5, 12

INTEGRATION

Mathematical Connections	12.1	12.2	12.3	12.4	12.5
algebra	**563–568***	**569–576**	**577–584**	**585–592**	**593–599**
geometry	568				
data analysis, probability, discrete math	**563–568**	**569–576**	**577–584**	**585–592**	**593–599**
logic and reasoning	564–568	**569–576**	577, 578, 582–584	588–592	**593–599**
Interdisciplinary Connections and Applications					
history and geography					597
reading and language arts			583		
biology and earth science	567				
chemistry and physics	568				
business and economics		576			
arts and entertainment			583		
sports and recreation		575	579, 583, 584	589–591	596
medicine, forensics, genetics		573	582, 584		
scheduling, manufacturing, industry	565		583	586, 588, 592	
cartography, ecology, fashion, computers, advertising, fundraising	564, 567	570, 572, 573, 576	578, 581	590, 591	

**Bold page numbers* indicate that a topic is used throughout the section.*

TECHNOLOGY

Section	*opportunities for use with* Student Book	Support Material
12.1	scientific calculator	
12.2	graphing calculator McDougal Littell Mathpack *Matrix Analyzer*	**Technology Book:** Spreadsheet Activity 12
12.3	graphing calculator	**Technology Book:** Calculator Activity 12 Spreadsheet Activity 12
12.4	graphing calculator	**Technology Book:** Spreadsheet Activity 12
12.5	graphing calculator	**Technology Book:** Spreadsheet Activity 12

Planning Guide

Regular Scheduling (45 min)

Section	Materials Needed	Core Assignment	Extended Assignment	*exercises that feature* Applications	Communication	Technology
12.1	colored pens or pencils	1–5, 9–12, 17–27	1–27	6, 9–12, 16	8	
12.2	graphing calculator or software, colored pencils	1–10, 20–23, 26–30, AYP*	1–30, AYP	1, 6–15, 20–24	17	8, 9, 16
12.3	4 different colors of paper, scissors	1–11, 20–23, 25–28, 31, 33–36, 41–45	1–21, 25–32, 35–37, 39–45	11–19, 27, 29, 30, 37–40, 42	41	
12.4	graphing calculator	**Day 1:** 1–20, 22–25 **Day 2:** 29–34, 36, 38, 41–56	**Day 1:** 1–25 **Day 2:** 26–56	13, 22, 23, 25 27–36, 38	15, 16 35, 40, 41	
12.5	graphing calculator or software	**Day 1:** 1–19 **Day 2:** 29, 30, 32–35, 37–46, AYP	**Day 1:** 1–24 **Day 2:** 25–46, AYP	21–24 25–27	3 29, 31, 33, 38	32
Review/ Assess		**Day 1:** 1–27 **Day 2:** Ch. 12 Test	**Day 1:** 1–27 **Day 2:** Ch. 12 Test			
Portfolio Project		Allow 2 days.	Allow 2 days.			

Yearly Pacing (with Portfolio Project)	**Chapter 12 Total** 11 days	**Chapters 1–12 Total** 160 days	**Remaining** 0 days	**Total** 160 days

Block Scheduling (90 min)

	Day 77	Day 78	Day 79	Day 80	Day 81	Day 82
Teach/Interact	12.1: Exploration, page 563 12.2	12.3: Exploration, page 577 12.4: Exploration, page 585	Continue with 12.4 12.5	Continue with 12.5 Review	Review Port. Proj.	Ch. 12 Test Port. Proj.
Apply/Assess	**12.1:** 1–5, 9–12, 17–27 **12.2:** 1–10, 20–23, 26–30, AYP*	**12.3:** 1–11, 20–23, 25–28, 31, 33–36 **12.4:** 1–20	**12.4:** 22–25, 29–34, 36, 38, 41–56 **12.5:** 1–19	**12.5:** 29, 30, 32–35, 37–46, AYP **Review:** 1–10	**Review:** 11–27 **Port. Proj.**	**Ch. 12 Test** **Port. Proj.**

NOTE: A one-day block has been added for the Portfolio Project—timing and placement to be determined by teacher.

Yearly Pacing (with Portfolio Project)	**Chapter 12 Total** 6 days	**Chapters 1–12 Total** 82 days	**Remaining** 0 days	**Total** 82 days

**AYP* is *Assess Your Progress.*

Lesson Support

Section	Practice Bank	Study Guide*	Assessment Book*	Visuals	Explorations Lab Manual	Lesson Plans	Technology Book
12.1	77	12.1		Warm-Up 12.1	Master 17	12.1	
12.2	78	12.2	Test 49	Warm-Up 12.2 Folder 12	Add. Expl. 18	12.2	Spreadsheet Act. 12
12.3	79	12.3		Warm-Up 12.3		12.3	Calculator Act. 12 Spreadsheet Act. 12
12.4	80	12.4		Warm-Up 12.4		12.4	Spreadsheet Act. 12
12.5	81	12.5	Test 50	Warm-Up 12.5		12.5	Spreadsheet Act. 12
Review Test	82	Chapter Review	Tests 51–53 Alternative Assessment			Review Test	

*__Spanish versions__ of *Study Guide* and *Assessment Book* are available.

Chapter Support

- Course Guide
- Lesson Plans
- Portfolio Project Book
- Preparing for College Entrance Tests
- Multi-Language Glossary
- *Test Generator* Software
- Professional Handbook

Software Support

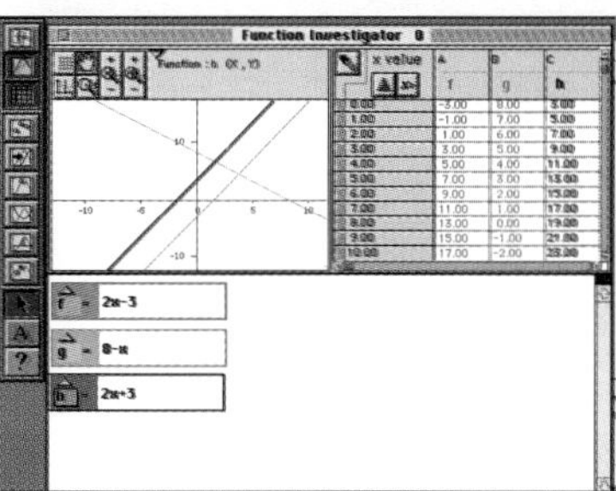

McDougal Littell Mathpack
Matrix Analyzer

Internet Support

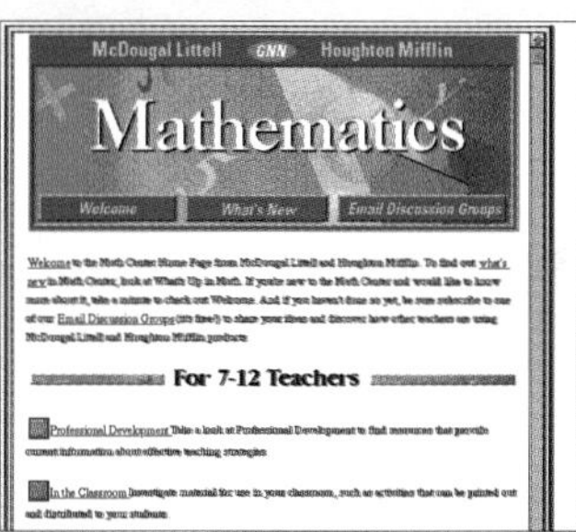

http://www.hmco.com
Next go to McDougal Littell; then the Education Center; then Secondary Math.

Outside Resources

Books, Periodicals

Perham, Bernadette H. and Arnold E. Perham. "Discrete Mathematics and Historical Analysis: A Study of Magellan." *Mathematics Teacher* (February 1995): pp. 106–112.

Copes, Wayne, William Sacco, Clifford Sloyer, and Robert Stark. "Queues." *Contemporary Applied Mathematics*, Graph Theory, Part IV, "Variable Arrivals and Service Times—Simulation": pp. 16–34. Janson Publications.

Copes, Wayne, William Sacco, Clifford Sloyer, and Robert Stark. "Reachability." *Contemporary Applied Mathematics*, Graph Theory: pp. 47–51. Janson Publications.

Activities, Manipulatives

Bredlau, C. L. "The Towers of Hanoi: Recursion, Induction, and the Microcomputer." *The New Jersey Mathematics Teacher*, 41(1): pp. 13–22.

Software

Kemeny, J. *Discrete Mathematics.* Hanover, NH: TrueBASIC.

Videos

Management Science. Show No. 1: "Street Smarts: Street Networks." Arlington, MA: COMAP.

Management Science. Show No. 2: "Trains, Planes, and Critical Paths." Arlington, VA: COMAP.

Internet

FTP Math Shareware Sites:
garbo.uwasa.fi
oak.oakland.edu

Interview Notes

Background

Conserving Energy
Most of the electricity we use today comes from the burning of fossil fuels, such as coal, oil, and natural gas. As a result, there are several environmental challenges that society faces. Three of the most prominent include (1) developing new energy sources; (2) improving energy production, transportation, and use; and (3) conserving energy. The first challenge can be met by continuing to explore renewable sources of energy, such as solar energy. Engineers can help achieve the second challenge by designing and building more efficient power plants and engines. The third challenge can be met by using energy wisely, including turning off unneeded lights, setting thermometer settings higher in summer and lower in winter, and recycling products made of paper, aluminum, glass, and plastic.

Maria Rodriguez
Maria Rodriguez attended Hunter College before entering Columbia University. While at Hunter, Ms. Rodriguez was involved in many Hispanic community activities. Some of these activities included being a member of the Hispanic Literary Academy, co-producer of the monthly Hispanic Literary magazine, and director of the first and second Annual Hispanic Folklore Festival at Hunter College. As a woman in a highly technological field, Ms. Rodriguez has some advice to offer women entering the work force today: "Be flexible, ready to learn different job skills, and keep up with the new technological trends."

12 Discrete Mathematics

Delivering POWER

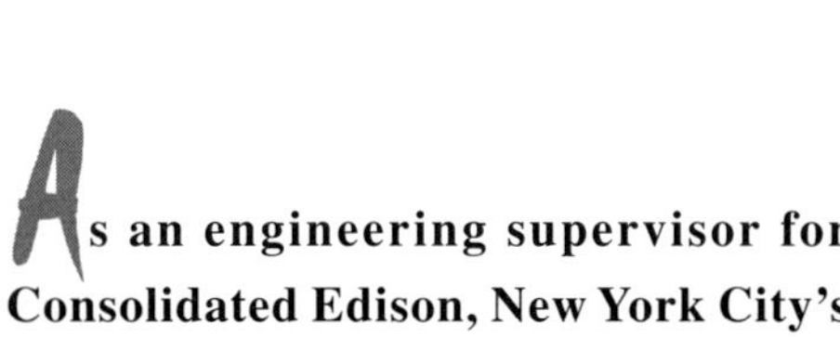

INTERVIEW **Maria Rodriguez**

As an engineering supervisor for Consolidated Edison, New York City's electric utility, Maria Rodriguez is responsible for providing and maintaining all the power lines coming into the borough of Queens. The major facilities supplied include John F. Kennedy Airport, LaGuardia Airport, *The New York Times*, Riker's Island prison, hospitals, schools, factories, apartment buildings, and homes. "My main concern is that no one loses power," Rodriguez says. "If for some reason the lights go out, it's my job to get them back on as quickly as possible."

"My main concern is that no one loses power."

560

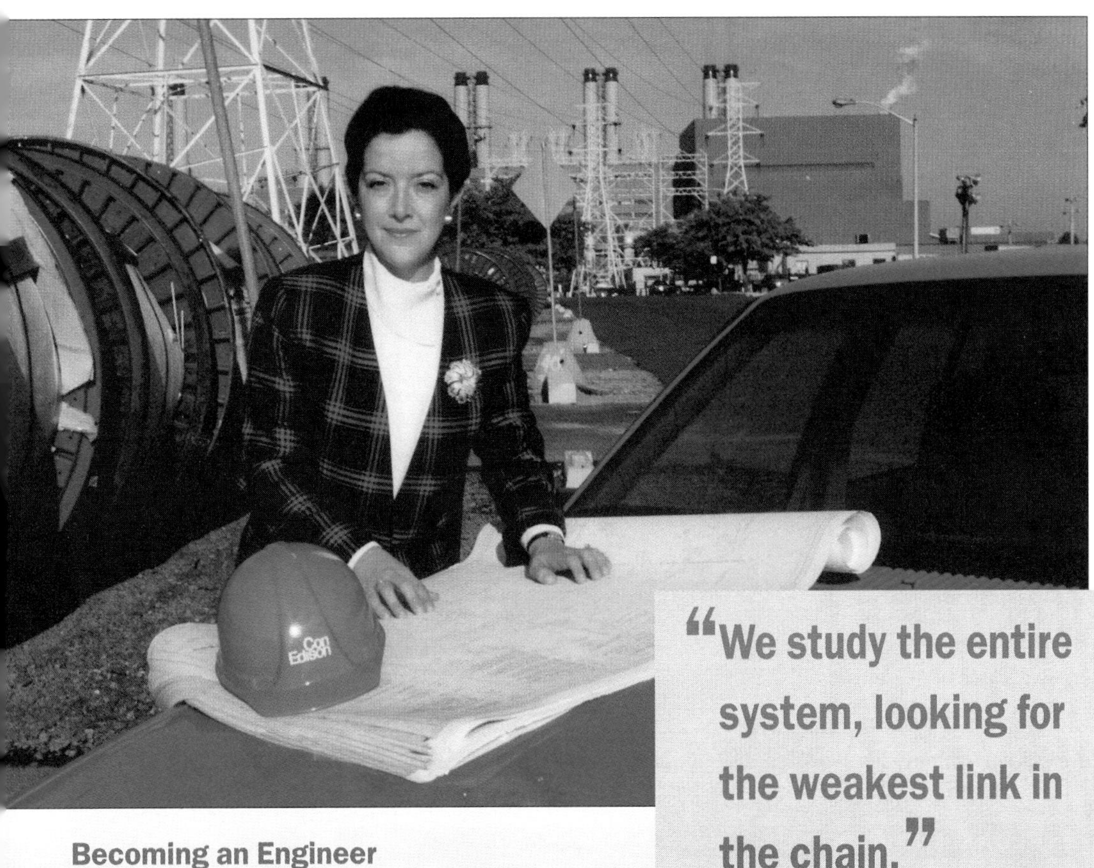

"We study the entire system, looking for the weakest link in the chain."

Becoming an Engineer

Rodriguez has traveled far to reach a position of such great responsibility. Born in England and raised in Spain, she moved to the United States in 1980 and studied English for a year. Fluent in both Spanish and French, she had planned to become a translator until she entered college, when she discovered engineering.

After graduating from Columbia University in 1987 with a degree in electrical engineering, she participated in a management training program at Consolidated Edison. Five years and several promotions later, she was put in charge of delivering electric power to Queens.

Determining Peak Capacity

To ensure reliable electrical service, she has to anticipate peak power demands a year or more in advance. "We study the entire system, looking for the weakest link in the chain," she says. "Then we try to determine whether that link will be capable of withstanding the projected electrical load."

Too much electric current can overload power lines. "Pipes will burst if you try to make them carry too much water. Electrical equipment fails, too, when pushed beyond the limit," she explains.

Interview Notes

Background

Queens, New York

Queens is the largest in area of the five boroughs, or districts, of New York City. It covers 126 square miles. It has a population of almost 2 million people, and is second in population only to the borough of Brooklyn. The other three boroughs are Manhattan, The Bronx, and Staten Island. Queens was settled by the Dutch in 1635 and became part of the British province of New York in 1683. It was named for England's Catherine of Braganza, Queen of Charles II. In 1898, Queens became part of New York City.

Second-Language Learners

Point out to students learning English that the work *borough* in the first sentence means "a self-governing district located within a city." Explain that the city in the United States best known for its boroughs is New York City. Also, students may not be familiar with the idiomatic expressions to *anticipate peak power demands* and *the weakest link in the chain.* If necessary, explain that the first expression means "to try to predict ahead of time when the most power might be needed and to be prepared for it." The second expression means "in a structured sequence of a process, each part must be as strong as the others; otherwise, a weak spot could cause the entire process to collapse."

Interview Notes

Mathematical Connection

Electrical engineers have to understand and follow the connections and paths of electricity as it flows from one place to another. There are many mathematical areas that can be used to complete this job. One of those areas is using directed graphs, and this idea is explored in Section 12.2. In this section, students investigate the flow of electricity and how it travels through five stations. Another component in studying the flow of electricity involves resistors. In Section 12.3, students use permutations to look at the resistance of the flow of electricity and the meanings of the bands on the resistors.

Explore and Connect

Writing
Students should apply the multiplication counting principle described in the article.

Research
Students can also talk to electric companies about ways families or other users of electricity can reduce their costs.

Project
The wattage of an appliance may be given on a metal tag on the appliance, or students may be able to find it in the manual for the appliance.

Distributing Electricity

Electricity reaching the area stations in Queens is generated at a power plant and transmitted through high-voltage cables to switching stations, where the voltage is "stepped down," or reduced. The electricity then goes to area stations, where the voltage is reduced again. The area stations feed a network of transformers that distribute electricity to residential and commercial customers.

An Easy Way to Count

If Rodriguez needs to count the number of transformers branching out from the area stations in the diagram shown, she can use the *multiplication counting principle* to simplify her calculations. First, she counts the lines that leave the power plant. Then she counts the lines that leave each switching station. Next, she counts the lines that leave each area station. Finally, she multiplies these numbers together to determine the total number of transformers. This knowledge, in turn, helps her decide whether the system has the capacity to meet the growing need for electricity.

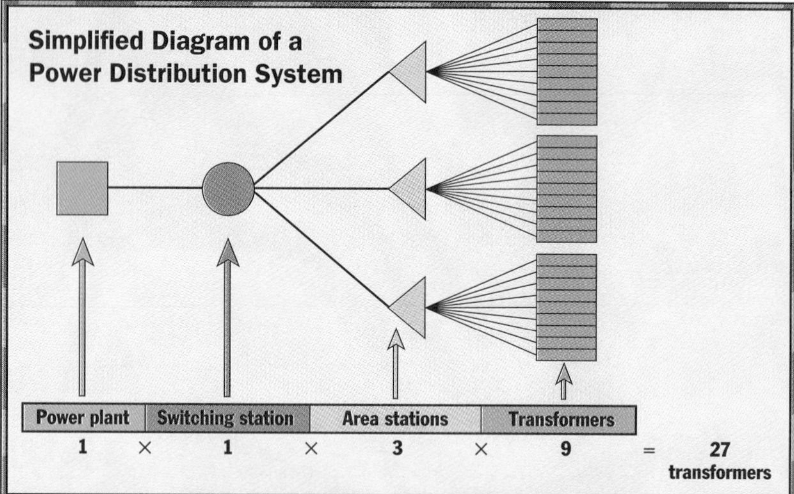

Explore and Connect

Maria Rodriguez stands at the Astoria Generating Station in Queens, New York.

1. Writing Suppose a second switching station is connected to the power plant in the diagram shown above. If the second switching station has the same subsequent connections as the first, how many transformers would there be? Explain.

2. Research Contact your local utility company to find out the cost of one kilowatt-hour of electricity for your school and for your home. Are these costs different? If so, what is the reason for this cost difference?

3. Project Determine the cost for using some household appliance during one year. Find out the wattage of the appliance and divide that figure by 1000 to convert it to kilowatts. Multiply your answer by the hours of usage for a year. Use the cost per kilowatt-hour from Question 2 to figure out the total cost for one year.

Mathematics & Maria Rodriguez

In this chapter, you will learn more about how mathematics is used in electrical engineering.

Related Exercises

Section 12.2
- Exercises 11–15

Section 12.3
- Exercises 12–18

Section

12.1 Coloring a Graph

Learn how to...

- represent situations with graphs
- color graphs, trying to use as few colors as possible

So you can...

- solve problems that require sorting items into groups, such as coloring maps, scheduling exams, and planning aquariums

Cartographers (mapmakers) are careful to use different colors to distinguish adjacent countries on a map. However, each additional color adds to the cost of printing the map. How can a cartographer color a map using as few colors as possible?

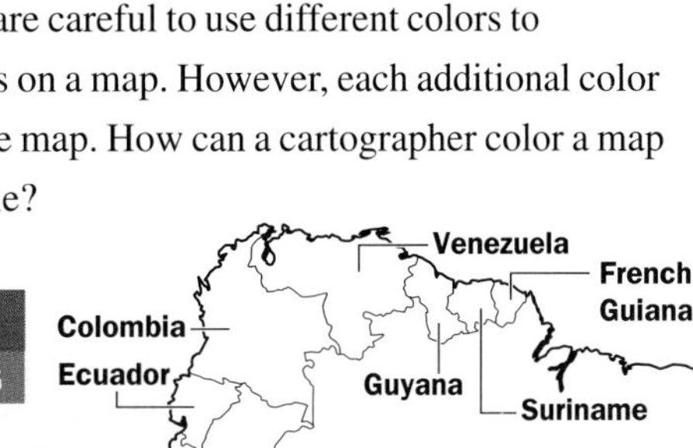

EXPLORATION
COOPERATIVE LEARNING

Investigating How to Color a Map

Work with a partner.
You will need:

- colored pens or pencils

1 Trace or sketch the map of South America.

2 Using as few colors as possible, color the map so that no two countries sharing a border have the same color. How many colors do you need?

3 Compare your results with those of others in the class. Is there more than one way to group countries by color?

WATCH OUT!

In a graph, the position of each vertex does not matter. For example, these two graphs are equivalent:

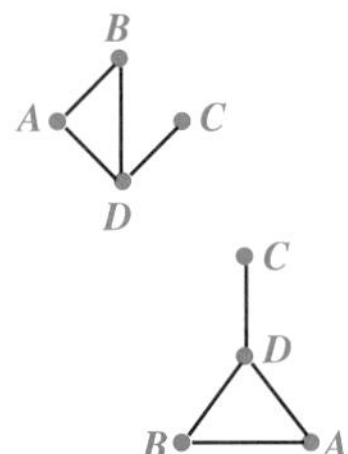

To simplify a map like the one of South America above, you can represent it with a **graph** consisting of **vertices** (plural of *vertex*) connected by **edges**.

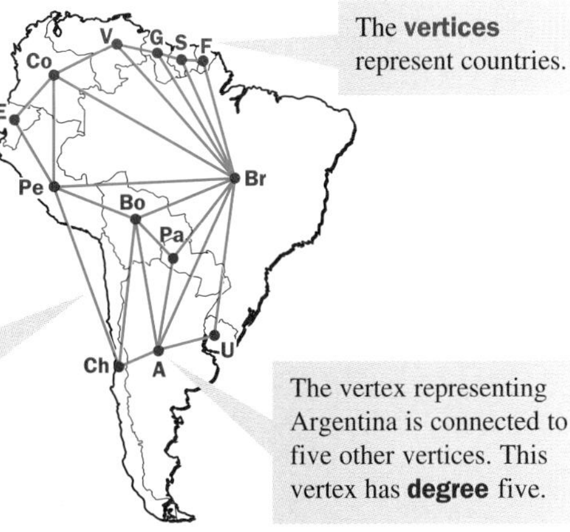

The **vertices** represent countries.

An **edge** connects two vertices if the corresponding countries share a border.

The vertex representing Argentina is connected to five other vertices. This vertex has **degree** five.

Exploration Note

Purpose
The purpose of this Exploration is to have students investigate how a map can be colored using as few colors as possible.

Materials/Preparation
Each group needs colored pens or pencils.

Procedure
Students sketch the map of South America. Then using as few colors as possible, they color the map so that no two countries sharing a border have the same color.

Closure
Students should understand that four colors are needed, but that there is more than one way to group countries by color.

Explorations Lab Manual
See the Manual for more commentary on this Exploration.

Diagram Master 17

For answers to the Exploration, see following page.

Plan⇔Support

Objectives

- Represent situations with graphs.
- Color graphs, trying to use as few colors as possible.
- Solve problems that require sorting items into groups.

Recommended Pacing

Core and Extended Courses
Section 12.1: 1 day

Block Schedule
Section 12.1: $\frac{1}{2}$ block
(with Section 12.2)

Resource Materials

Lesson Support
Lesson Plan 12.1
Warm-Up Transparency 12.1
Practice Bank: Practice 77
Study Guide: Section 12.1
Explorations Lab Manual: Diagram Master 17

Technology
Internet: http://www.hmco.com

Warm-Up Exercises

Identify the number of vertices in each polygon.

1. hexagon 6
2. decagon 10

For the given number of non-collinear points, determine the number of segments that can be drawn from each point to every other point.

3. 3 points 2
4. 6 points 5
5. 8 points 7

Teach⇔Interact

Additional Example 1

Color a map of Central America. Try to use as few colors as possible.

Use a graph to represent Central America. Label each vertex with its degree.

Step 1 Choose a color, such as green (①), and color green the vertex of highest degree (Honduras). Then color green the vertex of the next highest degree that does *not* share an edge with the first vertex. Continue until all vertices are either green or adjacent to a green vertex.

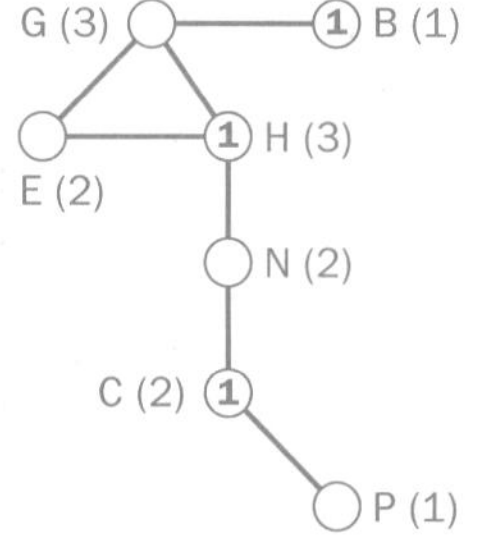

Step 2 Choose another color, such as white (②), and follow the procedure established in Step 1. Color white the uncolored vertex of highest degree (Guatemala). Continue coloring as in Step 1 until you can no longer use white.

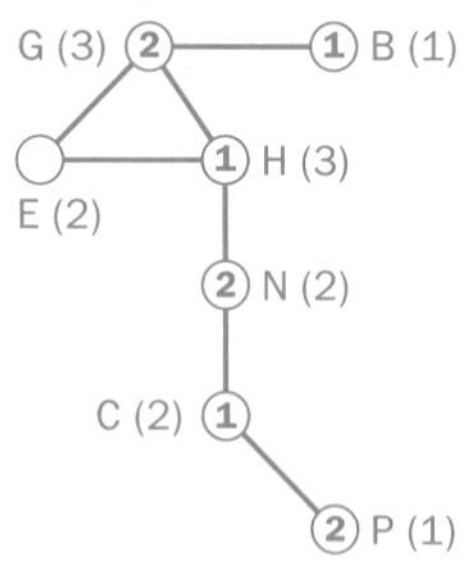

Step 3 Repeat using blue (③).

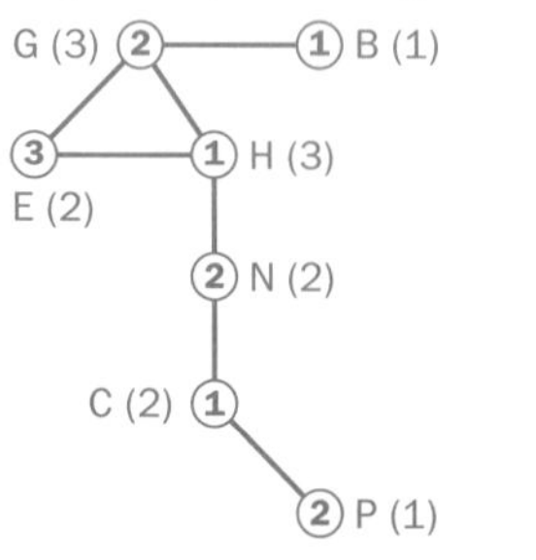

EXAMPLE 1 Application: Cartography

Color a map of South America. Try to use as few colors as possible.

SOLUTION

Use a graph to represent South America. Label each vertex with its degree.

Step 1 Choose a color, such as red, and apply it as shown.

First color **red** the vertex of highest degree (Brazil).

Then color **red** the vertex of next highest degree that does *not* share an edge with the first vertex.

Continue until all vertices are either red or adjacent to a red vertex.

Step 2 Choose another color, such as blue, and follow the procedure established in Step 1.

Color **blue** the uncolored vertex of highest degree. (Since there are three vertices of degree 5, choose one—say, Peru.)

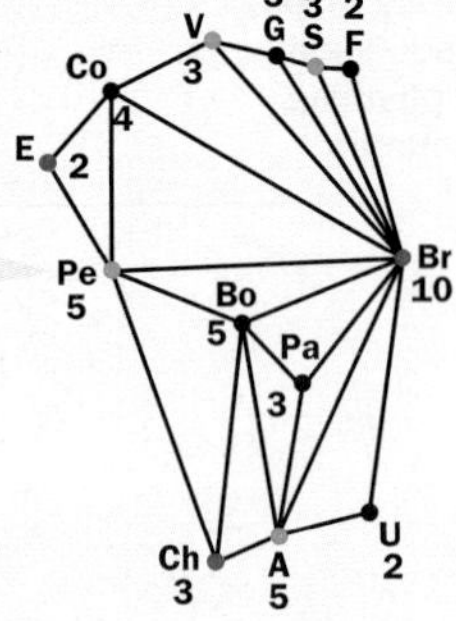

Step 3 Repeat using green and then yellow.

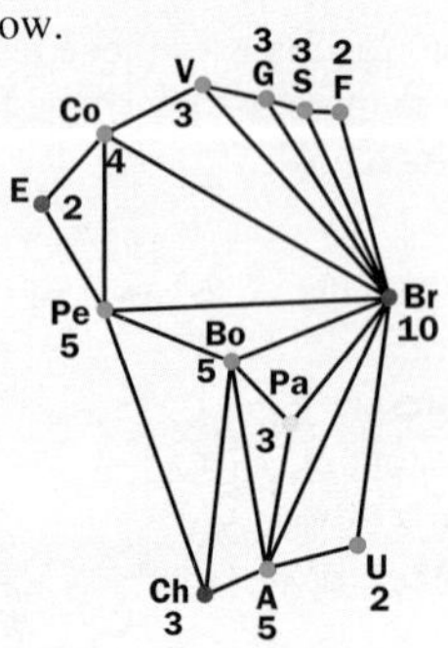

Step 4 Use the colored graph to color the map.

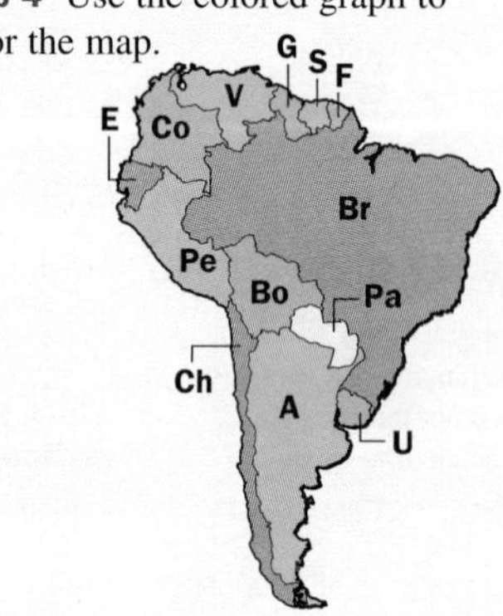

THINK AND COMMUNICATE

1. Describe another strategy for coloring a graph when you are trying to use as few colors as possible. Compare your strategy with the one in Example 1. Which is easier to use? Does one strategy use fewer colors than the other?

ANSWERS Section 12.1

Exploration

1, 2. Check students' work. Four colors are needed.

3. Yes.

Think and Communicate

1. Answers may vary. An example is given. Choose any country and color it. Color the countries next to it, using a new color only when necessary. The strategy in Example 1 is easier because it indicates where to begin coloring and tends to require the fewest number of colors.

EXAMPLE 2 Application: Scheduling

Rosa Perez schedules final exams for Carver High School's summer school. The table lists the students in each class. Rosa Perez wants to ensure that no student is scheduled for more than one test at a time. How can she set up a testing schedule with the fewest time slots? Which exams can be given at the same time?

Algebra	Biology	Civics	Driver's Ed.	English	French
Marie-Ange	Jamarl	Ansu	Ansu	Eduardo	Eduardo
Marlon	James	Jamarl	Nancy	Marie-Ange	Marie-Ange
Mbuyi	Kiyana	James	Rachel	Parker	Melinda
Micah	Melinda	Marlon	Washington	Rachel	Micah
Tish	Wayne		Wayne	Tish	Washington

SOLUTION

Use a graph to represent the situation.

Represent each **class** by a **vertex**.

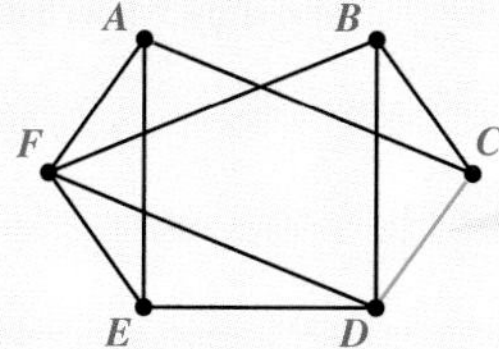

Draw **edges** between classes that **share one or more students**. The exams for these classes must *not* be given at the same time.

Use a different color to represent each testing period. Use the procedure from Example 1 to color the graph.

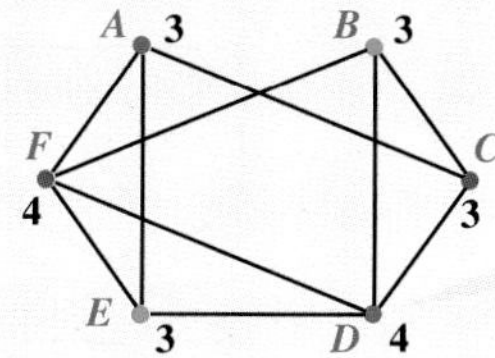

Two vertices have degree 4. Start by choosing one and coloring it **red**.

So, three testing times are needed: one for **Driver's Education** and **Algebra**, one for **French** and **Civics**, and one for **Biology** and **English**.

WATCH OUT!

Sometimes edges in a graph may cross each other. The intersections are not vertices.

THINK AND COMMUNICATE

2. Explain why the two graphs at the right are equivalent.
3. Describe how to move the vertices of the graph in Example 2 so that no edges cross.

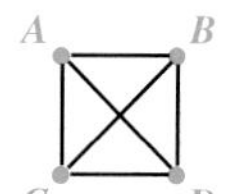

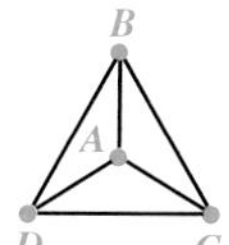

Think and Communicate

2. Each graph has four vertices of degree 3.
3. Choose a vertex. Move that vertex to the interior of a triangle formed by three other vertices that are connected by an edge to the chosen vertex. Continue the process until no edges cross. For example, move F to the interior of $\triangle ABE$. Then move B to the interior of $\triangle FDC$.

Teach⇔Interact

Additional Example 1 (continued)

Step 4 Use the colored graph to color the map.

Additional Example 2

Ky Dinh schedules meetings for his school's clubs. He wants to ensure that no student is scheduled for more than one meeting at a time. How can he set up a meeting schedule with the fewest time slots? Which meetings can be held at the same time?

Business Club	Chemistry Club	Drama Club
Darrell	Camilla	Abby
Becky	Dylan	Becky
Fletcher	Carlos	Carlos
Sheila	Ramon	Pedro
Yolanda	Han	Ramon

Fine Arts Club	Math Team	Student Council
Angelica	Greg	Becky
Jim	Sue	Sheila
Sayumi	Han	Fletcher
Maria	Fran	Ky Dinh
Sue	Abby	Lamar

Use a graph. Represent each club by a vertex. Draw edges between groups that share one or more students. The meetings for these groups must not be at the same time.

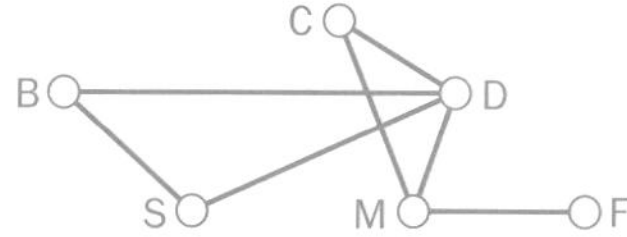

Use a different color to represent each meeting time. Use the procedure from Example 1 to color the graph. Start by coloring D white (①).

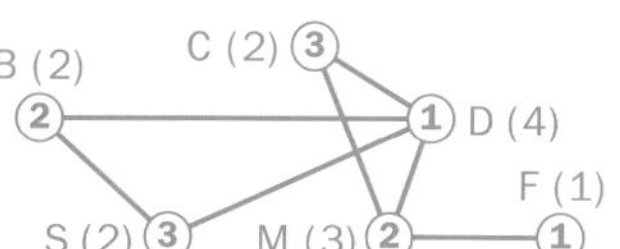

So, three meeting times are needed: one for Student Council and Chemistry Club, one for Business Club and Math Team, and one for Fine Arts Club and Drama Club.

Teach⇔Interact

Checking Key Concepts

Communication: Reading Students would benefit from hearing some of their classmates read their explanations to these exercises. Each reading can then serve as a review of the essential concepts that have been introduced in this section.

Closure Question

How would you attempt to solve a problem that requires sorting items into groups?
Represent the situation with a graph using vertices and edges. Label each vertex with its degree. Use a different color to represent each group and color the graph using the procedure from Example 1 on page 564. The number of colors used gives the solution to the problem.

Apply⇔Assess

Suggested Assignment

Core Course
Exs. 1–5, 9–12, 17–27

Extended Course
Exs. 1–27

Block Schedule
Day 77 Exs. 1–5, 9–12, 17–27

Exercise Notes

Common Error
Ex. 1 Some students may have trouble distinguishing between vertices and points of intersection that are not vertices. You can use these graphs to emphasize the difference between a vertex and a point of intersection.

Alternate Approach
Exs. 2–6, 10, 11, 14, 16 You may want to let students use a good algorithm from Think and Communicate question 1, instead of the method of Example 1, to do these exercises. In this case, the answers may vary somewhat from those given.

CHECKING KEY CONCEPTS

1. Which of the following graphs are equivalent? Explain your answer.

A. **B.** **C.**

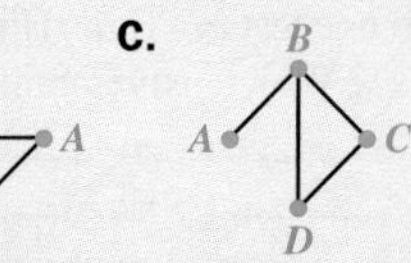

D.

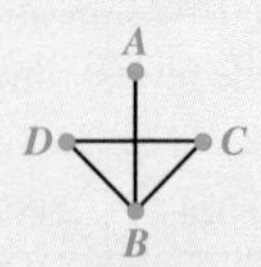

For Questions 2–4, look back at the graph on page 563. Give the degree of the vertex that represents each country.

2. Bolivia **3.** Uruguay **4.** Argentina

5. If you drew a graph like the one on page 563 for the United States, what would be the degree of your state?

6. A group of friends are making plans for the weekend. Tracy and Kim want to spend a day hiking; Tim wants to spend a day with his family; Tracy and Manuel want to spend a day working on a project for school; and Tim, Manuel, and Kim want to spend a day shopping.

a. Explain how the graph at the left represents this situation.

b. In two days, can everyone do what they want? How can you tell from the graph?

12.1 Exercises and Applications

Extra Practice exercises on page 767

1. Which two graphs are equivalent? How do you know?

A.

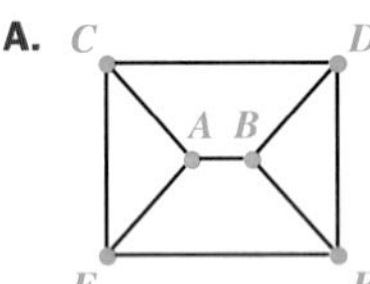

B.

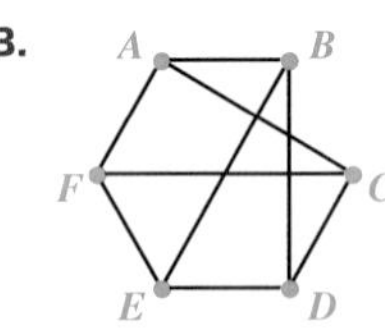

C.

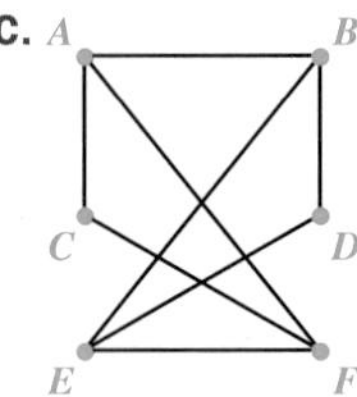

For each graph in Exercises 2–5:

a. Find the degree of each vertex.

b. Draw an equivalent graph.

c. Color your graph so that vertices that share an edge are different colors. Try to use as few colors as possible.

2.

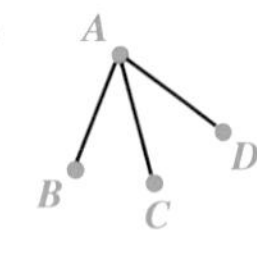

3.

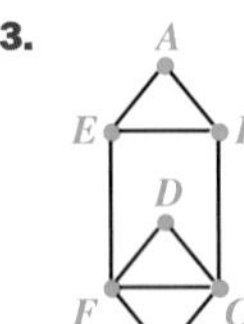

4.

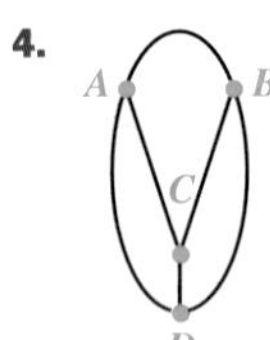

5.

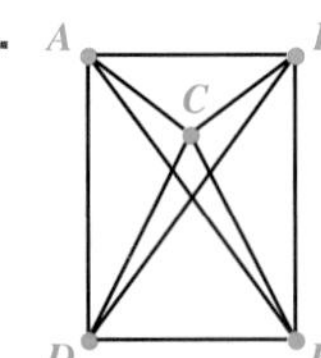

Checking Key Concepts

1. Graphs A and B are equivalent. Answers may vary. An example is given. Each graph consists of edges connecting A and B, B and C, C and D, and D and A. Graphs C and D are equivalent.

2. 5 3. 2

4. 5

5. Answers may vary. An example is given. Vermont would have degree 3.

6. a. Each vertex represents an activity and each edge represents the person who would like to participate in the activities associated with the vertices of that edge.

b. No; three colors would be needed to color the graph, so three days would be needed for everyone to do what they want.

Exercises and Applications

1. A and B; In both graphs, each vertex is connected in the same way to the other vertices.

2–5. Answers to parts (b) and (c) may vary. Examples are given.

2. a. A, 3; B, 1; C, 1; D, 1

b, c. Two colors are needed.

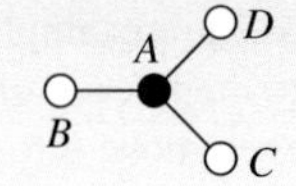

6. a. CARTOGRAPHY Use a graph to decide how to color the map at the right. Try to use as few colors as possible. Tasmania does not border any states or territories, so do not include it in your graph.

b. Trace or sketch the map of Australia and use your results from part (a) to color it. You can color Tasmania any color.

7. Research Suppose you represent a map of the United States with a graph as in Example 1 on page 564. Which vertex (or vertices) would have the highest degree?

8. Writing Look back at Example 2 on page 565.

a. If one student were enrolled in all six of the courses being offered, how many testing periods would be needed?

b. Explain why the graph at the right describes the situation in part (a). What happens if you try to use the method in Example 1 to color the graph?

Australia
Northern Territory
Western Australia
Queensland
South Australia
New South Wales
Australian Capital Territory
Victoria
Tasmania

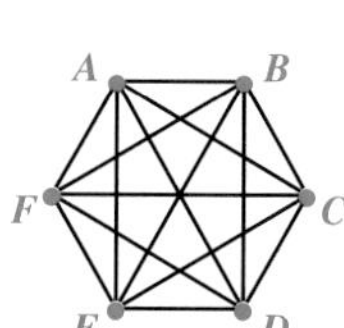

Connection MARINE BIOLOGY

Aquariums and pet stores must be careful when putting fish together in the same tank. For example, to minimize territorial disputes, a highly territorial species like the damselfish should be introduced to tanks after the other fish have had a chance to become established. Certain fish should never share a tank. The table below shows which fish are compatible and which are not.

9. a. Draw a graph in which each vertex represents a type of fish, and edges connect fish that are *incompatible.*

b. A graph is *planar* if it can be drawn without any edges crossing. Is your graph planar? Justify your answer.

10. Color your graph to decide how many tanks you need if you have all six types of fish named in the table. What grouping of fish does your graph represent?

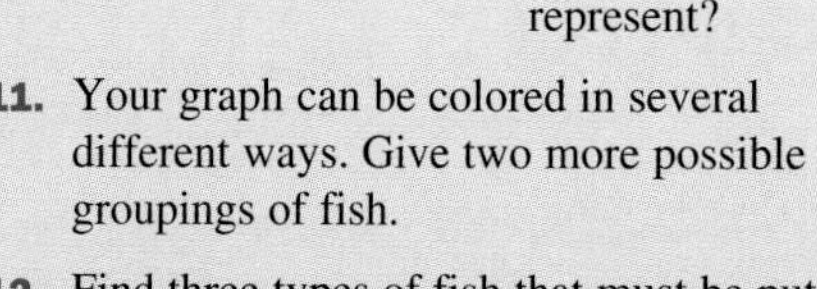

11. Your graph can be colored in several different ways. Give two more possible groupings of fish.

12. Find three types of fish that must be put into three different tanks. (*Hint:* Look for a triangle formed by three edges and three vertices in your graph.)

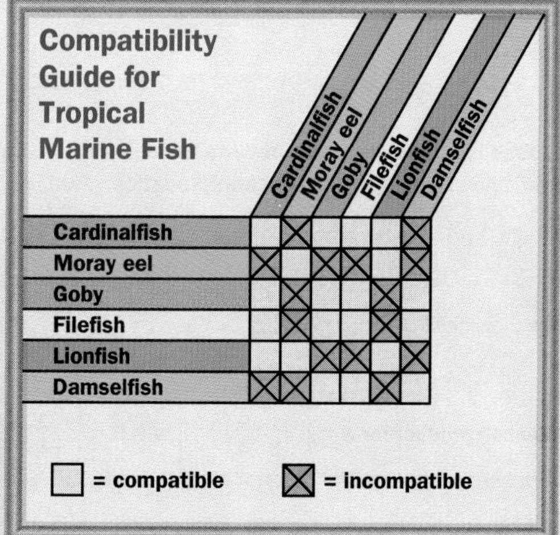

Compatibility Guide for Tropical Marine Fish

	Cardinalfish	Moray eel	Goby	Filefish	Lionfish	Damselfish
Cardinalfish		☒				☒
Moray eel	☒		☒	☒		☒
Goby		☒			☒	
Filefish		☒			☒	
Lionfish			☒	☒		☒
Damselfish	☒	☒			☒	

☐ = compatible ☒ = incompatible

Apply⇔Assess

Exercise Notes

Application
Ex. 6 The field of cartography or map making has changed dramatically with the introduction of geographical information systems, or *GIS.* These systems use large databases of cartographic information to electronically produce maps.

Challenge
Ex. 8 The graph in this exercise is an example of a *complete* graph, which is a graph having each pair of vertices connected by exactly one edge. Ask students what the maximum degree of each vertex of a complete graph with n vertices would be. ($n - 1$)

Second-Language Learners
Exs. 9–12 It may be helpful to explain that *territorial* in the phrase *highly territorial species* means "a kind of animal that does not let others of its kind come into the area in which it lives."

Research
Ex. 9 Some students may be interested in finding out why some fish listed in the table are incompatible. Ask them to report their findings to the class.

Reasoning
Ex. 9 Students should be able to use their justifications to formulate a general rule about the relationship between vertices of a planar graph.

Student Progress
Ex. 11 Students should understand that although the vertices of their graphs may be colored differently, the graphs are equivalent because they have the same vertices and edges.

Reasoning
Ex. 12 Ask students if a quadrilateral formed by four edges and four vertices would necessarily represent four types of fish that must be put into four different tanks. (No.)

3. a. A, 2; B, 3; C, 4; D, 2; E, 3; F, 4; G, 2

b, c. Three colors are needed.

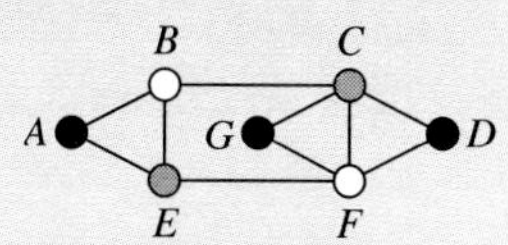

4. a. A, 3; B, 3; C, 3; D, 3

b, c. Four colors are needed.

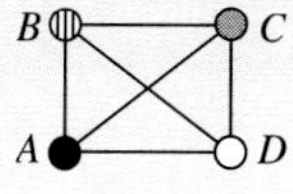

5. a. A, 4; B, 4; C, 4; D, 4; E, 4

b, c. Five colors are needed.

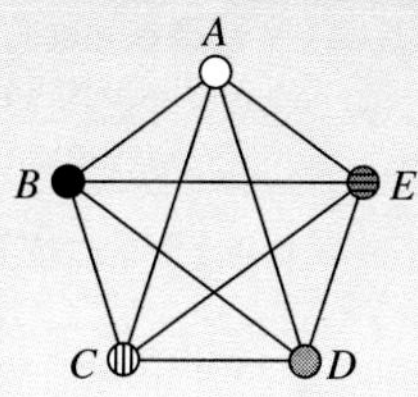

6. See answers in back of book.

7. Tennessee and Missouri

8. a. 6

b. Every class shares a student with every other class. Since there is an edge connecting every vertex with every other vertex, 6 colors would be needed.

9–12. See answers in back of book.

Apply⇔Assess

Exercise Notes

Cooperative Learning

Ex. 14 By working in small groups to produce several of each kind of map, students can share their creative approaches with each other.

Interdisciplinary Problems

Ex. 16 Graph theory can be applied to a wide variety of other disciplines because of its general nature. Point out that some household cleaners, such as ammonia and bleach, are incompatible and should not be mixed together.

Visual Thinking

Ex. 17 Ask students to create displays of their maps and the graphs that they used to color them. Encourage them to explain how they used the graphs to color the maps. This activity involves the visual skills of *interpretation* and *communication.*

Practice 77 for Section 12.1

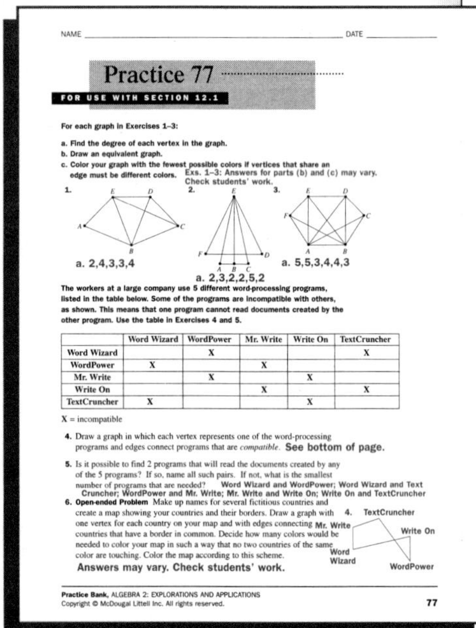

NAME ______________________ DATE __________

Practice 77

FOR USE WITH SECTION 12.1

For each graph in Exercises 1–3:

a. Find the degree of each vertex in the graph.
b. Draw an equivalent graph.
c. Color your graph with the fewest possible colors if vertices that share an edge must be different colors. Exs. 1–3: Answers for parts (b) and (c) may vary. Check students' work.

1. a. 2,4,3,3,4
2. a. 2,3,2,2,5,2
3. a. 5,5,3,4,4,3

The workers at a large company use 5 different word-processing programs, listed in the table below. Some of the programs are incompatible with others, as shown. This means that one program cannot read documents created by the other program. Use the table in Exercises 4 and 5.

	Word Wizard	WordPower	Mr. Write	Write On	TextCruncher
Word Wizard		X			X
WordPower	X		X		
Mr. Write		X		X	
Write On			X		X
TextCruncher	X			X	

X = incompatible

4. Draw a graph in which each vertex represents one of the word-processing programs and edges connect programs that are *compatible*. See bottom of page.

5. Is it possible to find 2 programs that will read the documents created by any of the 5 programs? If so, name all such pairs. If not, what is the smallest number of programs that are needed? Word Wizard and WordPower; Word Wizard and TextCruncher; WordPower and Mr. Write; Mr. Write and Write On; Write On and TextCruncher

6. **Open-ended Problem** Make up names for several fictitious countries and create a map showing your countries and their borders. Draw a graph with one vertex for each country on your map and with edges connecting countries that have a border in common. Decide how many colors would be needed to color your map in such a way that no two countries of the same color are touching. Color the map according to this scheme. Answers may vary. Check students' work.

4. TextCruncher, Mr. Write, Write On, Word Wizard, WordPower

Practice Bank, ALGEBRA 2: EXPLORATIONS AND APPLICATIONS

77

13. Research Look up the history of the *four-color problem.* Write a brief summary of your findings, including the origin of the problem and attempts to prove that all maps on a sphere can be colored with four or fewer colors.

14. Open-ended Problem Make up a map having six regions that can be colored using exactly:

a. two colors **b.** three colors **c.** four colors

15. Challenge Make up a map that corresponds to the graph in Exercise 3. The vertices represent countries and the edges connect countries that share a border.

16. CHEMISTRY Many chemicals can be dangerous if allowed to come into contact with each other. For this reason, it is good practice to store incompatible chemicals in separate areas to avoid accidents. The table shows which chemicals in a high school lab are incompatible.

a. Use a graph to decide how to store the chemicals. How many different storage areas are needed?

b. If the lab did not have halogens and ethers, how many storage areas would be needed?

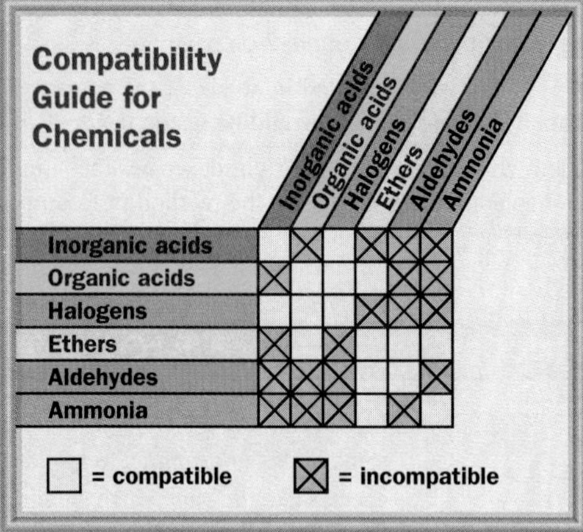

Compatibility Guide for Chemicals

	Inorganic acids	Organic acids	Halogens	Ethers	Aldehydes	Ammonia
Inorganic acids		X		X	X	X
Organic acids	X				X	X
Halogens				X	X	X
Ethers	X		X			
Aldehydes	X	X	X			X
Ammonia	X	X	X		X	

☐ = compatible ☒ = incompatible

ONGOING ASSESSMENT

17. Open-ended Problem Choose a map showing several countries, states, or other subdivisions. Use a graph to color your map. Try to use as few colors as possible. If possible, show another way to color your map and explain how you found it.

SPIRAL REVIEW

For each equation of a conic, rewrite the equation in standard form, identify the conic, and state its important characteristics. *(Section 11.6)*

18. $y^2 + 2x^2 - 2y = 8$ **19.** $x^2 + y^2 + 3x = 4$

20. $2x^2 - 3y^2 + 6y = 9$ **21.** $y^2 + 10y + 2x = -24$

Solve. *(Section 8.3)*

22. $\sqrt{x} = 2$ **23.** $\sqrt{x - 3} = 6$ **24.** $\sqrt{4x - 1} = 5$

Find each product for $A = \begin{bmatrix} -1 & 0 & 2 \\ 3 & 5 & -4 \end{bmatrix}$ **and** $B = \begin{bmatrix} 2 & 1 \\ -1 & 0 \end{bmatrix}$**. If the matrices cannot be multiplied, state that the product is *undefined.*** *(Section 1.4)*

25. AB **26.** BA **27.** B^2

13. Answers may vary. An example is given. The four-color theorem was first proposed in the mid-19th century by Francis Guthrie. The theorem stated that, for any map on which countries sharing a border were colored with different colors, no more than four colors would be needed to color the map. Although proofs of the conjecture were attempted earlier, it was not until 1976 that Kenneth Appel and Wolfgang Haken finally proved the theorem using a modem computer.

14–17. See answers in back of book.

18. $\frac{(y-1)^2}{9} + \frac{x^2}{4.5} = 1$; an ellipse with center (0, 1), foci $(0, 1 + \sqrt{4.5})$ and $(0, 1 - \sqrt{4.5})$, and vertices (0, 4) and (0, –2)

19. $\left(x + \frac{3}{2}\right)^2 + y^2 = \frac{25}{4}$; a circle with center $\left(-\frac{3}{2}, 0\right)$ and radius $\frac{5}{2}$

20. $\frac{x^2}{3} - \frac{(y-1)^2}{2} = 1$; a hyperbola with center (0, 1), foci $(\sqrt{5}, 1)$ and $(-\sqrt{5}, 1)$, and vertices $(\sqrt{3}, 1)$ and $(-\sqrt{3}, 1)$

21. $x - \frac{1}{2} = -\frac{1}{2}(y + 5)^2$; a parabola with vertex $\left(\frac{1}{2}, -5\right)$, focus (0, –5), and directrix $x = 1$

22. 4

23. 39

24. $\frac{13}{2}$

25. undefined

26. $\begin{bmatrix} 1 & 5 & 0 \\ 1 & 0 & -2 \end{bmatrix}$

27. $\begin{bmatrix} 3 & 2 \\ -2 & -1 \end{bmatrix}$

Section

12.2 Directed Graphs and Matrices

Learn how to...

- represent situations with directed graphs
- represent directed graphs with matrices

So you can...

- analyze situations involving direction, such as predator-prey relationships and electricity transfers

A *food web* shows how energy, in the form of food, is transferred through an ecosystem. The illustration below shows a food web for polar seas.

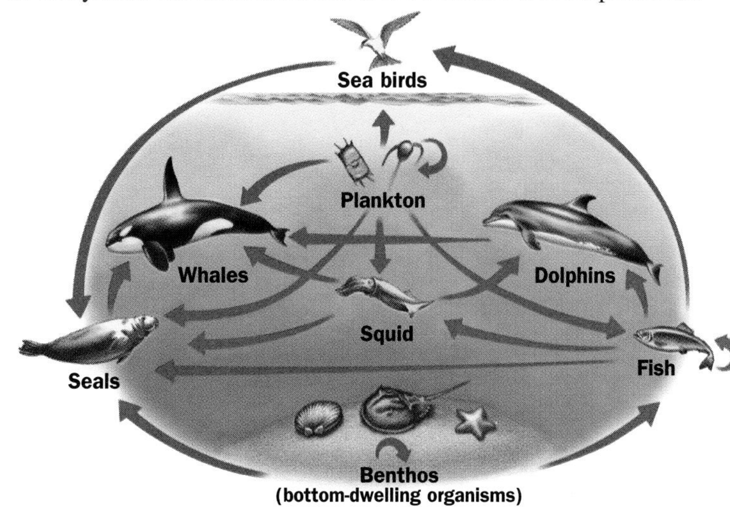

BY THE WAY

There are several kinds of whales in the polar seas. Bottlenose whales and sperm whales eat squid, killer whales eat dolphins and seals, and whalebone whales eat plankton.

THINK AND COMMUNICATE

1. In the food web above, what do the arrows represent?
2. Why is there an arrow from fish back to fish?
3. Which animals eat squid?
4. Seals eat benthos directly. They also eat benthos indirectly by eating fish that have eaten benthos. How else do seals eat benthos indirectly?
5. List the direct and indirect food sources for dolphins.
6. How would squid be affected if all of the plankton died?
7. Which animal is the most important food source for the rest of the animals? Explain.

You can represent the food web by a *directed graph* as shown at the right. A **directed graph** is a graph in which the **edges** are **arrows**.

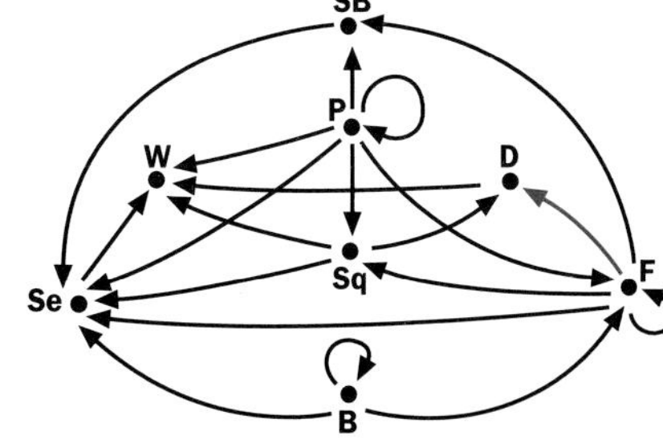

Plan⇔Support

Objectives

- Represent situations with directed graphs.
- Represent directed graphs with matrices.
- Analyze situations involving direction.

Recommended Pacing

Core and Extended Courses
Section 12.2: 1 day

Block Schedule
Section 12.2: $\frac{1}{2}$ block
(with Section 12.1)

Resource Materials

Lesson Support
Lesson Plan 12.2
Warm-Up Transparency 12.2
Overhead Visuals:
Folder 12: Matrices and Directed Graphs
Practice Bank: Practice 78
Study Guide: Section 12.2
Explorations Lab Manual:
Additional Exploration 18
Assessment Book: Test 49

Technology
Technology Book:
Spreadsheet Activity 12
Graphing Calculator
McDougal Littell Mathpack
Matrix Analyzer
Internet:
http://www.hmco.com

Warm-Up Exercises

Let $A = \begin{bmatrix} -1 & 3 \\ 0 & 2 \end{bmatrix}$ and $B = \begin{bmatrix} 2 & -1 \\ 1 & 0 \\ 3 & -2 \end{bmatrix}$.

1. Is the sum $A + B$ defined? No.
2. Is the product BA defined? Yes.
3. Define the size of the matrix BA. 3 rows × 2 columns
4. Find A^2. $\begin{bmatrix} 1 & 3 \\ 0 & 4 \end{bmatrix}$
5. Find $A + A^2$. $\begin{bmatrix} 0 & 6 \\ 0 & 6 \end{bmatrix}$

ANSWERS Section 12.2

Think and Communicate

1. The arrows point from prey to predator.
2. Some fish eat other fish.
3. dolphins, whales, and seals
4. Answers may vary. Examples are given. Seals eat benthos indirectly by eating squid that have eaten fish that have eaten benthos, or by eating sea birds that have eaten fish that have eaten benthos.
5. The direct food sources are squid and fish. The indirect food sources are benthos, plankton, and fish (through squid).
6. The squid could eat only fish, whose population would be reduced since fish eat plankton.
7. plankton; It is a direct food source for itself and five other animals, and is an indirect food source for another animal.

Teach⇔Interact

Section Note

Problem Solving
The directed graphs and matrices presented in this section are an important problem-solving tool that provide a visual means for determining the relationships.

Using Technology
You might suggest that students use the *Matrix Analyzer* software to perform calculations like those at the bottom of page 571 and in Example 3 on page 572. Students can use the Matrix Analyzer Activity 1 in the *Function Investigator with Matrix Analyzer Activity Book.*

About Example 1

Teaching Tip
Call students' attention to the fact that the loop beginning and ending at the insect vertex indicates that insects are both predators and prey of other insects.

Additional Example 1

Part of a field ecosystem is described below. Draw a directed graph to represent the food web.

- Plants are eaten by aphids, rabbits, birds, foxes, and mice.
- Aphids are eaten by birds.
- Praying mantises are eaten by birds and by other praying mantises.
- Birds are eaten by foxes.
- Rabbits are eaten by foxes.
- Mice are eaten by foxes.

The vertices represent plants and animals. The arrows point from prey to predator.

Think and Communicate

The direction of the arrows determine whether a vertex is a predator or prey. Thus, students should see that the number of edges with arrows pointing away from the plant vertex determine the number of ways in which plants are a direct food source for the animals represented in the graph.

BY THE WAY

Earthworms eat decaying plant and animal matter in the soil. As they burrow through the ground, earthworms loosen and mix the soil, providing a better environment for plants.

EXAMPLE 1 Application: Ecology

Part of a forest ecosystem is described below. Draw a directed graph to represent the food web.

- Plants are eaten by insects, deer, earthworms, and songbirds.
- Insects are eaten by amphibians (frogs and toads), songbirds, raptors (hawks and owls), and each other.
- Amphibians are eaten by raptors.
- Earthworms are eaten by songbirds.
- Songbirds are eaten by raptors.

SOLUTION

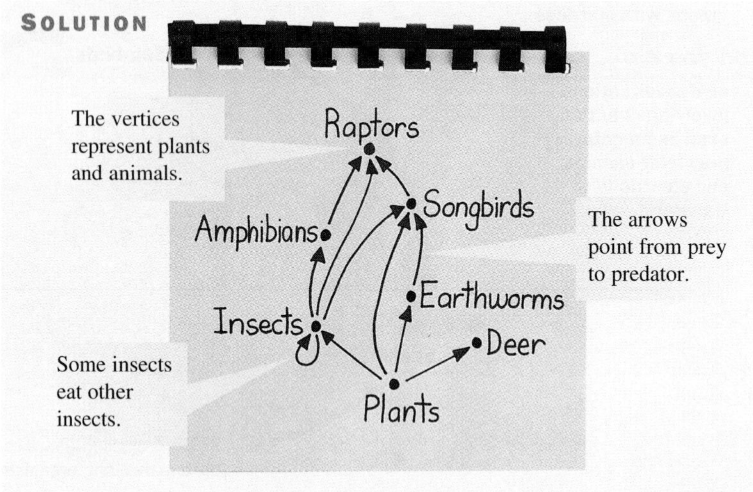

THINK AND COMMUNICATE

8. In Example 1, are there any plants or animals that are not direct or indirect food sources for raptors? If so, what are they?

9. Plants (P) are an indirect food source for raptors (R) in several ways. Two are shown at the right.

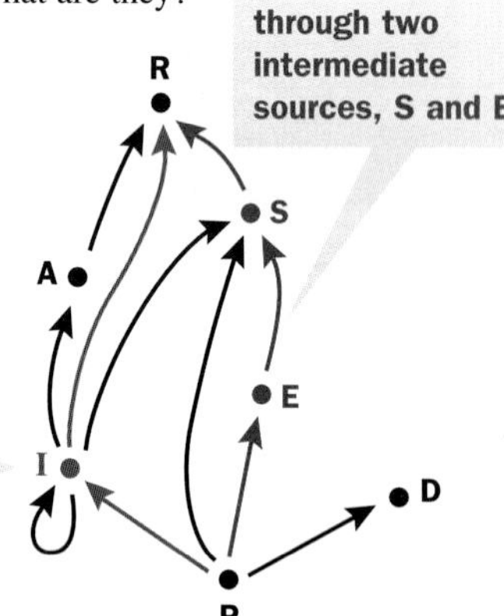

a. Describe at least two more ways that plants are a food source for raptors through two intermediate sources.

b. Describe three ways that insects are a food source for raptors through one intermediate source. **through one intermediate source**

c. Are there any ways that plants are a food source for raptors through three intermediate sources?

Think and Communicate

8. Yes; deer.
9. Answers may vary. Examples are given.
   a. plants → insects → songbirds → raptors; plants → insects → amphibians → raptors
   b. insects → insects → raptors; insects → songbirds → raptors; insects → amphibians → raptors
   c. Yes; plants → insects → insects → amphibians → raptors and plants → insects → insects → songbirds → raptors.

If you represent a directed graph with a matrix, you can use a calculator or a computer to find direct and indirect relationships between vertices.

EXAMPLE 2

Represent the directed graph of the food web in Example 1 with a matrix.

SOLUTION

Let the rows and columns represent the vertices of the graph. Let an element of the matrix be 1 if the vertex represented by the row has an arrow going to the vertex represented by the column. Otherwise, let the element be 0.

Predator

$$\begin{array}{c|ccccccc} & D & R & A & I & E & S & P \\ \hline D & 0 & 0 & 0 & 0 & 0 & 0 & 0 \\ R & 0 & 0 & 0 & 0 & 0 & 0 & 0 \\ A & 0 & 1 & 0 & 0 & 0 & 0 & 0 \\ I & 0 & 1 & 1 & 1 & 0 & 1 & 0 \\ E & 0 & 0 & 0 & 0 & 0 & 1 & 0 \\ S & 0 & 1 & 0 & 0 & 0 & 0 & 0 \\ P & 1 & 0 & 0 & 1 & 1 & 1 & 0 \end{array} = M$$

Prey

There is no arrow from deer to plants.

There is an arrow from plants to deer.

THINK AND COMMUNICATE

Questions 10 and 11 refer to both the matrix in Example 2 and the food web in Example 1.

10. In terms of the food web, what does it mean when an element of the matrix is 1? when an element is 0?

11. Find the sum of the elements in the last row of the matrix. What does the sum represent in terms of the food web? What does the sum of the elements in the second column represent?

The square of the matrix above tells you how many ways each plant or animal is a food source for the others through one intermediate source.

$$M \times M = M^2$$

$$\begin{array}{c|ccccccc} & D & R & A & I & E & S & P \\ \hline D & 0&0&0&0&0&0&0 \\ R & 0&0&0&0&0&0&0 \\ A & 0&1&0&0&0&0&0 \\ I & 0&1&1&1&0&1&0 \\ E & 0&0&0&0&0&1&0 \\ S & 0&1&0&0&0&0&0 \\ P & 1&0&0&1&1&1&0 \end{array} \times \begin{array}{c|ccccccc} & D & R & A & I & E & S & P \\ \hline D & 0&0&0&0&0&0&0 \\ R & 0&0&0&0&0&0&0 \\ A & 0&1&0&0&0&0&0 \\ I & 0&1&1&1&0&1&0 \\ E & 0&0&0&0&0&1&0 \\ S & 0&1&0&0&0&0&0 \\ P & 1&0&0&1&1&1&0 \end{array} = \begin{array}{c|ccccccc} & D & R & A & I & E & S & P \\ \hline D & 0&0&0&0&0&0&0 \\ R & 0&0&0&0&0&0&0 \\ A & 0&0&0&0&0&0&0 \\ I & 0&3&1&1&0&1&0 \\ E & 0&1&0&0&0&0&0 \\ S & 0&0&0&0&0&0&0 \\ P & 0&2&1&1&0&2&0 \end{array}$$

Plants are eaten by **insects** and **earthworms**.

Insects and **earthworms** are eaten by **songbirds**.

$0 + 0 + 0 + \mathbf{1} + 1 + 0 + 0 = \mathbf{2}$
Plants are eaten by songbirds indirectly in **two** ways, through **insects** and through **earthworms**.

Think and Communicate

10. When an element is 1, the plant or animal in the element row is a food source for the plant or animal in the element column. When an element is 0, the plant or animal in the element row is not a food source for the plant or animal in the element column.

11. 4; Plants are a direct food source for four of the plants or animals. The sum represents the number of plants and animals that are direct food sources for raptors.

Teach⇔Interact

About Example 2

Teaching Tip
Some students have difficulty setting up their matrices because they are confused by what both a 0 and a 1 represent. Remind them that in a base 2 system, a 1 often represents a true value, while a 0 represents a false value. Thus, it makes sense to use a 1 when a row item is the prey of a column item.

Additional Example 2

Represent the directed graph in Additional Example 1 with a matrix.

Let the rows and columns represent the vertices of the graph. Let an element of the matrix be 1 if the vertex represented by the row has an arrow going to the vertex represented by the column. Otherwise, let the element be 0.

$$\begin{array}{c|ccccccc} & A & B & F & M & P & PM & R \\ \hline A & 0&1&0&0&0&0&0 \\ B & 0&0&1&0&0&0&0 \\ F & 0&0&0&0&0&0&0 \\ M & 0&0&1&0&0&0&0 \\ P & 1&1&1&1&0&0&1 \\ PM & 0&1&0&0&0&1&0 \\ R & 0&0&1&0&0&0&0 \end{array}$$

Section Note

Communication: Discussion
Students would benefit from a discussion of the material at the bottom of this page. Have them use the matrices to see how the 2 in row 7, column 6 of the M^2 matrix is calculated. Students, especially those who are visual learners, should also correlate the calculation of this matrix element with the paths in the directed graph on page 570, which show plants as a food source for songbirds through one intermediate source. Students need to realize that the numbers in M^2 represent the number of ways the prey and predator are connected through *one* intermediate source, not the number of intermediate sources that connect the two species. Thus, students should also see that a 0 in any one location does not mean that the prey is not a food source for the predator through fewer or more than one intermediate source. Think and Communicate questions 12 and 13 will also help students make some of these connections.

Teach⇔Interact

Additional Example 3

Using the matrix in Additional Example 2, find a matrix that gives the total number of ways each plant or animal is a food source for the others directly or through one intermediate source.

Because M gives the number of ways that each plant or animal is a food source for the others directly, and M^2 gives the number of ways that each plant or animal is a food source for the others through one intermediate source, the totals are given by $M + M^2$.

```
      A B F M P PM R
A   [[0 1 1 0 0 0 0...
B    [0 0 1 0 0 0 0...
F    [0 0 0 0 0 0 0...
M    [0 0 1 0 0 0 0...
P    [1 2 4 1 0 0 1...
PM   [0 2 1 0 0 2 0...
R    [0 0 1 0 0 0 0...
```

$M + M^2$

Think and Communicate

You may wish to extend the discussion of these questions with students to include the fact that the matrix M^3 shows how many 2-step paths there are from any row vertex to a column vertex. Ask students what the matrix M^4 would show. (how many 3-step paths there are from any row vertex to a column vertex)

Closure Question

If matrix M represents the ways in which each plant or animal is a food source of the others in a food web, explain what the matrix M^2 and the matrix $M + M^2$ represent.
M^2: how many ways each plant or animal is a food source for the others in the food web through one intermediate source; $M + M^2$: the total number of ways each plant or animal is a food source for the others in the web directly or through one intermediate source

EXAMPLE 3 Application: Ecology

Using the matrix in Example 2, find a matrix that gives the total number of ways each plant or animal is a food source for the others directly or through one intermediate source.

SOLUTION

Because M gives the number of ways that each plant or animal is a food source for the others directly, and M^2 gives the number of ways that each plant or animal is a food source for the others through one intermediate source, the totals are given by $M + M^2$.

Enter M into a graphing calculator or computer software with matrix calculation capabilities. Find $M + M^2$.

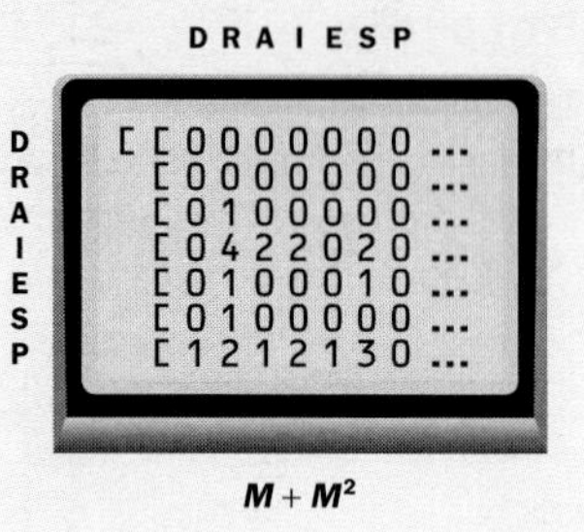

D R A I E S P
D R A I E S P
[[0 0 0 0 0 0 0 ...
[0 0 0 0 0 0 0 ...
[0 1 0 0 0 0 0 ...
[0 4 2 2 0 2 0 ...
[0 1 0 0 0 1 0 ...
[0 1 0 0 0 0 0 ...
[1 2 1 2 1 3 0 ...

$M + M^2$

THINK AND COMMUNICATE

12. In how many ways are plants eaten by raptors directly or through one intermediate source? How can you tell from the matrix above? List the ways.

13. List the songbirds' food sources that are direct or through one intermediate source. How can you tell from the matrix above?

CHECKING KEY CONCEPTS

Some of the mathematics courses offered at Fremont High School are listed below with their direct prerequisites.

Math C requires Math A.	**Math D requires Math A or B.**
Math E requires Math B, C, or D.	**Math F requires Math D or E.**

1. Which of the following directed graphs represents this situation?

A., **B.**
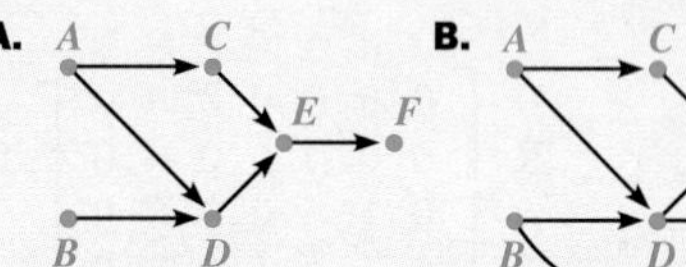

C.
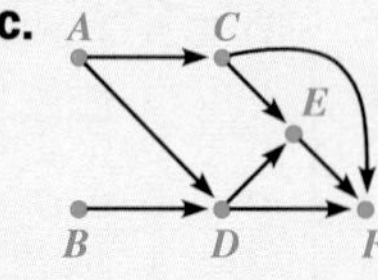

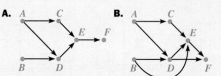

2. What do the arrows represent?

3. List the ways to take Math F with just two courses preceding it.

4. Write a matrix that represents the directed graph.

5. Square the matrix from Question 4. What do the elements represent?

Think and Communicate

12. 2; 2 is the element in row P, column R; plants → insects → raptors and plants → songbirds → raptors.

13. insects, earthworms, and plants; The nonzero elements in the songbird column occur in the rows for these food sources.

Checking Key Concepts

1. B

2. The arrows point from a prerequisite course to a course that requires it.

3. $A \to D \to F$, $B \to D \to F$, and $B \to E \to F$

4.

	A	B	C	D	E	F
A	0	0	1	1	0	0
B	0	0	0	1	1	0
C	0	0	0	0	1	0
D	0	0	0	0	1	1
E	0	0	0	0	0	1
F	0	0	0	0	0	0

5.

	A	B	C	D	E	F
A	0	0	0	0	2	1
B	0	0	0	0	1	2
C	0	0	0	0	0	1
D	0	0	0	0	0	1
E	0	0	0	0	0	0
F	0	0	0	0	0	0

the number of ways a course can be taken with two prerequisites

12.2 Exercises and Applications

Extra Practice exercises on page 768

1. **MEDICINE** Six students at Ouray High School have come down with infectious mononucleosis in the past week. The school nurse, who suspects that the disease entered the school through a single student, is trying to trace the spread so that the disease can be contained. The nurse gathered the information at the right by interviewing each of the six students.
   a. Draw a directed graph representing possible transmissions from one student to another.
   b. List all possible pathways that the disease could have spread from Barbara to Katrina.
   c. Is it possible that the disease entered the school with only one of these six students and then spread to the others? Explain.

Student	Could have passed mononucleosis to:
Barbara	Ezra, Fatima, Ilana
Ezra	Fatima
Fatima	Katrina
Geoff	Ezra, Fatima
Ilana	Barbara, Fatima
Katrina	Ezra

Write a matrix representing each directed graph. Let an element be 1 if the row vertex has an edge directed to the column vertex. Otherwise, let the element be 0.

2. 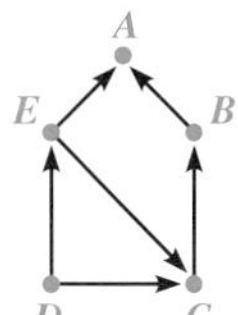

3. 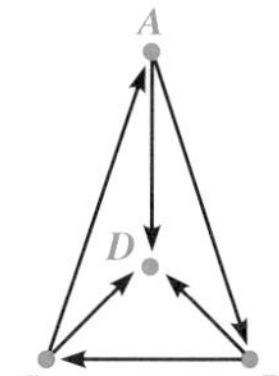

4. 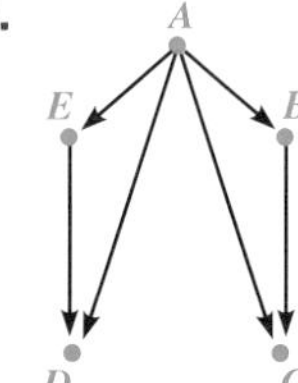

5. 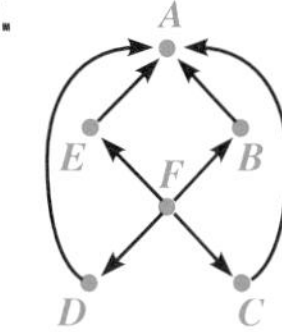

ECOLOGY The graph on page 570 represents part of a forest ecosystem. For Exercises 6–10, use the following additional information about the ecosystem.

- Plants are eaten by foxes, raccoons, and mice.
- Insects are eaten by snakes and mice.
- Snakes are eaten by raptors, foxes, and each other.
- Mice are eaten by raccoons, foxes, and snakes.
- Raccoons are eaten by foxes.

6. Using the information above, copy and extend the graph on page 570.
7. Represent the directed graph in Exercise 6 with a matrix.
8. **Technology** Find a matrix that gives the number of ways that each plant or animal is a food source for the others through one intermediate source.
9. **Technology** Find a matrix that gives the number of ways that each plant or animal is a food source for the others directly or through one intermediate source.
10. Which animal would be most affected if all of the mice died? Why? How can you tell from the matrix in Exercise 9?

Apply⇔Assess

Suggested Assignment

Core Course
Exs. 1–10, 20–23, 26–30, AYP

Extended Course
Exs. 1–30, AYP

Block Schedule
Day 77 Exs. 1–10, 20–23, 26–30, AYP

Exercise Notes

Application
Ex. 1 This exercise can be used to initiate a discussion of the usefulness of mathematics to solve a real-world problem. You might ask some students to research and report to the class on infectious mononucleosis and how the disease is spread. The exercise may motivate some students to research the role of the Center of Disease Control in Atlanta, GA in gathering data about the occurrence of diseases in the United States. Research of the 1976 outbreak of Legionnaire's disease in Philadelphia may further interest some students.

Cooperative Learning
Exs. 2–5 As a means of checking their matrices, have students work with a partner. One student can write the first matrix for Ex. 2. Without looking at the graph, the second student can then read from his or her partner's matrix the number of directed edges that are indicated from each vertex to every other vertex. The information in the matrix should match the graph. The partners can take turns writing the matrix and reading the information in the matrix for Exs. 3–5.

Using Technology
Exs. 8–10 These exercises show the advantage of using technology to solve problems that involve a finite number of steps. Because a graph is a finite, nonempty set of vertices, together with a set of edges, the greatest possible number of edges between any two vertices is a finite number. Thus, using technology to perform the matrix calculations allows students to focus on interpreting the results of the calculations.

Exercises and Applications

1. a. 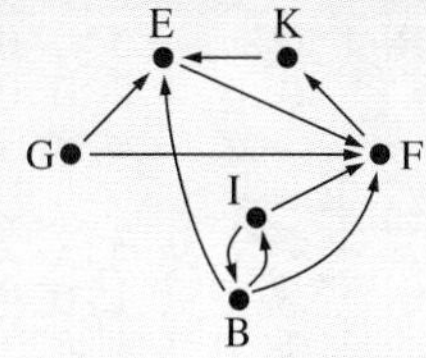

b. Barbara → Fatima → Katrina, Barbara → Ezra → Fatima → Katrina, Barbara → Ilana → Fatima → Katrina

c. No. Explanations may vary. An example is given. Ilana could have gotten the disease only from Barbara, and Barbara could have gotten it only from Ilana. Since Geoff could not have gotten the disease from either Barbara or Ilana, it is not possible that the disease began with one of these six students.

2. $\begin{array}{c} \\ A \\ B \\ C \\ D \\ E \end{array} \begin{array}{c} A\ B\ C\ D\ E \\ \begin{bmatrix} 0 & 0 & 0 & 0 & 0 \\ 1 & 0 & 0 & 0 & 0 \\ 0 & 1 & 0 & 0 & 0 \\ 0 & 0 & 1 & 0 & 1 \\ 1 & 0 & 1 & 0 & 0 \end{bmatrix} \end{array}$

3. $\begin{array}{c} \\ A \\ B \\ C \\ D \end{array} \begin{array}{c} A\ B\ C\ D \\ \begin{bmatrix} 0 & 1 & 0 & 1 \\ 0 & 0 & 1 & 1 \\ 1 & 0 & 0 & 1 \\ 0 & 0 & 0 & 0 \end{bmatrix} \end{array}$

4–10. See answers in back of book.

Apply⇔Assess

Exercise Notes

Interview Note

Exs. 11–15 For Exs. 12 and 13, students should use their graphs as a visual aid to support their answers. Students can discuss the advantages of the different routes for Ex. 15. They should understand that in the application of graph theory to real-life applications, finding a route often involves satisfying a set of conditions so that cost and resources are minimized. Such a route may not necessarily be the most direct route.

Career Connection

Ex. 17 The term *environmentalist* encompasses many job fields. Most commonly, environmentalists are concerned with the effects of pollution on Earth's resources, plant life, and animal life. Environmental scientists use other branches of science, such as biology or chemistry, to study the physical effects of pollution on the environment. Environmental lobbyists publicize environmental dangers and work for action or laws to prevent, repair, or alleviate these dangers and their effects. Environmental lawyers write laws to protect the environment from further pollution and uphold these laws by suing violators.

Communication: Writing

Ex. 17 In order for mathematics to be useful, people must be able to interpret the results of a mathematical procedure in the context of a real-world problem. This writing exercise should help students learn how to use the matrix they produced in Ex. 16 to answer a real-world question. It would be a good idea to have several students read their responses to this question aloud so that all students can gain greater insight into how to interpret these types of matrices.

Reasoning

Ex. 18 By asking students to explain what the vertices and arrows represent, this exercise highlights the importance of identifying all the variables in a problem. If possible, ask students to work in groups that include a student with an interest in biology or a related science.

INTERVIEW Maria Rodriguez

Look back at the article on pages 560–562.

Electrical systems throughout the United States and Canada are connected together in large networks. During emergencies or times of high energy demand, one company may sell power to another company in the network. Rather than flowing directly from the source to the receiving station, the power divides and follows several paths of the network. Engineers such as Maria Rodriguez must monitor the transfer to make sure that no parts of the network are overloaded.

Suppose 100 megawatts (MW) of electricity must be sent from station A to station B within a network. Some constraints of the network are noted in the table.

Station	Can receive power from
B	A, C, D, E
C	E
D	A
E	D

11. Draw a directed graph that represents the possible paths of power transfer in the network.

12. If only 60 MW is transmitted directly from A to B, where does the additional 40 MW of electricity go initially?

13. If 10 MW flows from D to E and 5 MW flows from E to C, how much power flows from C to B, D to B, and E to B? Label your graph with all of the power transfers. Assume that no power is lost.

14. What is the most indirect path that power takes to get from A to B? How much power takes this path?

15. In how many ways can power flow from A to B through one intermediate station? through two? through three?

For Exercises 16 and 17, refer to the food web and graph on page 569.

16. **Technology** Find a matrix that gives the number of ways each plant or animal is a food source for the others directly or through one intermediate source. Use a graphing calculator or computer software with matrix calculation capabilities.

17. **Writing** Some environmentalists have been concerned about the impact of whaling on marine ecosystems. Explain how you can use the matrix from Exercise 16 to study how the polar marine ecosystem would be affected if the whale population were significantly reduced.

18. **Research** Draw a directed graph that summarizes the *hydrologic* (water) cycle. Explain what the vertices and arrows represent. Make a matrix to represent the information in your graph.

11. Graphs may vary. An example is given.

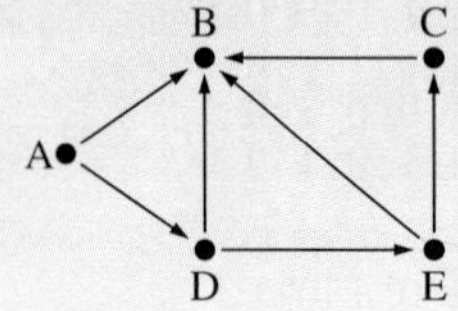

12. to D

13. 5 MW; 30 MW; 5 MW

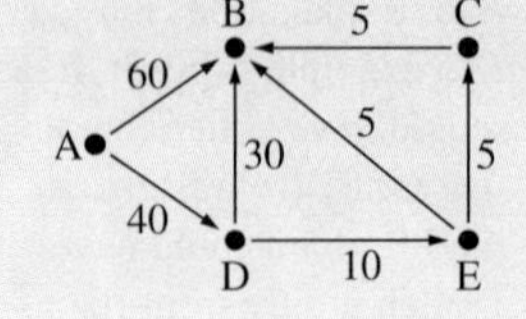

14. A → D → E → C → B; 5 MW

15. 1; 1; 1

16. In the following matrix, the rows represent the prey and the columns represent the predators.

	Se	SB	W	P	D	F	Sq	B
Se	0	0	1	0	0	0	0	0
SB	1	0	1	0	0	0	0	0
W	0	0	0	0	0	0	0	0
P	5	3	4	2	2	3	3	0
D	0	0	1	0	0	0	0	0
F	4	2	3	0	3	2	2	0
Sq	1	0	3	0	1	0	0	0
B	3	1	1	0	1	3	1	2

17. Answers may vary. An example is given. Whales are direct or indirect predators of all

19. a. Challenge Represent the map of one-way streets with a 9×9 matrix M. The rows and columns should represent street intersections. What do the elements represent?

b. Use a graphing calculator or computer software with matrix calculation capabilities to find M^2, M^3, and M^4. What do they represent?

c. Can you drive from the intersection of First and C to the intersection of First and A without leaving the area shown in the map? How can you use matrices to find out?

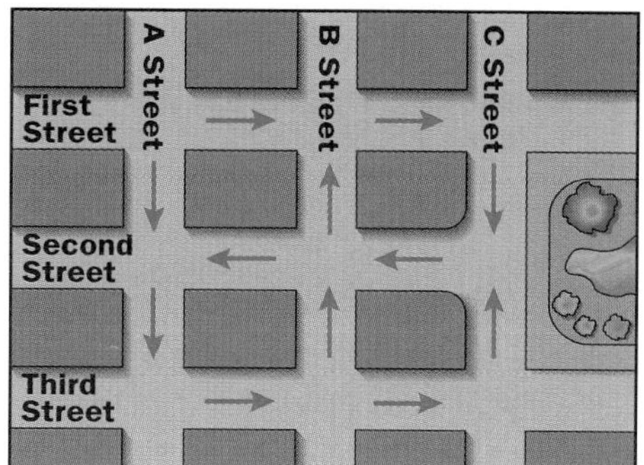

Connection SPORTS

To avoid ties when determining the regional champion for high school football, the Ohio High School Athletic Association rates Division I teams that play each other with a system similar to the following: Award the team one point for each win. Then add an additional point for each team that a defeated team has beaten.

BY THE WAY

Each school in the Ohio High School Athletic Association is assigned to a division based on the number of male students enrolled in grades 9–11.

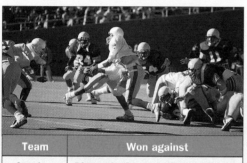

Team	Won against
Angels	Diamonds, Cougars, Flames, Giants
Bison	Angels, Elks, Giants
Cougars	Diamonds, Bison, Elks
Diamonds	Bison
Elks	Angels, Diamonds, Flames, Giants
Flames	Diamonds, Cougars, Bison
Giants	Diamonds, Cougars, Flames

Suppose the results for a small region are shown in the table. The Diamonds receive 1 point for beating the Bison and 1 point for each of the Bison's wins, for a total of 4 points.

20. Draw a directed graph representing the information in the table. Can you tell who the champion would be? Explain.

21. a. Write a matrix M to represent the results for all 7 teams, using a 1 if the row team defeated the column team and a 0 otherwise.

b. What do the row sums of M represent?

22. a. Find M^2. What does it represent?

b. What do the row sums of M^2 represent?

23. a. Find $M + M^2$. What do the row sums represent?

b. Give the number of points awarded each team. Which team wins the championship?

24. Open-ended Problem Describe another method for choosing a regional champion. How can you use a graph or matrix with your method? Does your method give the same results as the one described above?

Apply⇔Assess

Exercise Notes

Problem Solving

Ex. 19 Students may have difficulty creating the matrix directly from the map. Encourage them to make a directed graph first. The street intersections should be the vertices. Choose a simple labeling system, such as 1A to represent the corner of First Street and A Street. For part (c), students must realize that for there to be a way to drive from 1C to 1A, the entry in the third row and first column of one of the powers of M must be nonzero. Ask students how many powers of M they must test before they can answer the question conclusively. (8: M through M^7, as there are 7 possible intermediate intersections)

Challenge

Ex. 21 Ask students to give a general rule that shows the relationship in matrix M between an entry in row j, column i, where $j \neq i$, and the entry in row i, column j and to explain why this is so. (If the entry in row j, column $i = 0$, then the entry in row i, column $j = 1$. If the entry in row j, column $i = 1$, then the entry in row i, column $j = 0$. This is so because, in each game, one team must be the winner and the other team must be the loser.)

Reasoning

Ex. 22 Encourage students to pick a particular entry of M^2 and use the directed graph to determine the meaning of that particular entry. This should help them understand what the matrix as a whole represents. For example, students should see that the number 1 in the second row, first column means that there is one team the Bison defeated who defeated the Angels (the Elks).

Reasoning

Ex. 23 Students should understand that each element in $M + M^2$ represents a total, as does each row sum and each column sum. Ask them to explain how the totals are different. This question will help them to clarify their understanding of the relationship of a graph to its matrix.

plant and animal life mentioned in the food web for polar seas. If the whale population were significantly reduced, the seal, dolphin, and squid populations might increase significantly, causing depletion of shared food sources such as plankton and fish. This, in turn, could produce a decline in the population of most plants and animals in the ecosystem.

18. Answers may vary. An example is given. In the graph, the vertices represent ways in which water moves through the environment; the arrows indicate the direction of the movement. In the matrix, a "1" shows that there is an arrow from the row vertex to the column vertex.

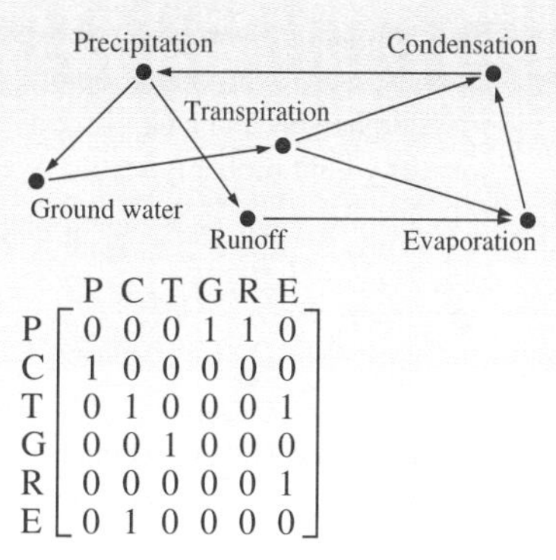

$$\begin{array}{c} \\ P \\ C \\ T \\ G \\ R \\ E \end{array} \begin{array}{c} \begin{array}{cccccc} P & C & T & G & R & E \end{array} \\ \begin{bmatrix} 0 & 0 & 0 & 1 & 1 & 0 \\ 1 & 0 & 0 & 0 & 0 & 0 \\ 0 & 1 & 0 & 0 & 0 & 1 \\ 0 & 0 & 1 & 0 & 0 & 0 \\ 0 & 0 & 0 & 0 & 0 & 1 \\ 0 & 1 & 0 & 0 & 0 & 0 \end{bmatrix} \end{array}$$

19–24. See answers in back of book.

Apply⇔Assess

Exercise Notes

Using Manipulatives

Ex. 25 Students can use toothpicks and small buttons to model their graphs. This way they can remodel the graph to remove a plant or animal.

Assess Your Progress

Journal Entry

Ask students to be specific about what kinds of algebraic graphs they are discussing. In addition to algebraic graphs, students can discuss similarities and differences between directed graphs and the graphs they use in statistics.

Progress Check 12.1–12.2

See page 602.

Practice 78 for Section 12.2

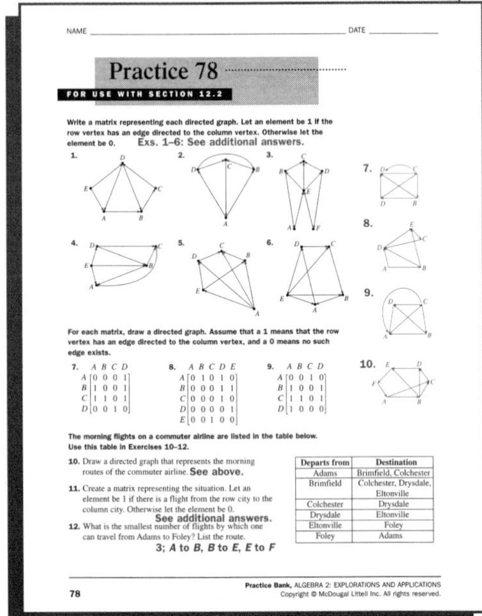
NAME ______________________ DATE __________

Practice 78

FOR USE WITH SECTION 12.2

Write a matrix representing each directed graph. Let an element be 1 if the row vertex has an edge directed to the column vertex. Otherwise let the element be 0. Exs. 1–6: See additional answers.

1. 2. 3. 4. 5. 6.

For each matrix, draw a directed graph. Assume that a 1 means that the row vertex has an edge directed to the column vertex, and a 0 means no such edge exists.

7. $\begin{matrix} & A & B & C & D \\ A & 0 & 0 & 0 & 1 \\ B & 1 & 0 & 0 & 1 \\ C & 1 & 1 & 0 & 1 \\ D & 0 & 0 & 1 & 0 \end{matrix}$

8. $\begin{matrix} & A & B & C & D & E \\ A & 0 & 1 & 0 & 1 & 0 \\ B & 0 & 0 & 0 & 1 & 1 \\ C & 0 & 0 & 0 & 1 & 0 \\ D & 0 & 0 & 0 & 0 & 1 \\ E & 0 & 0 & 1 & 0 & 0 \end{matrix}$

9. $\begin{matrix} & A & B & C & D \\ A & 0 & 0 & 1 & 0 \\ B & 1 & 0 & 0 & 1 \\ C & 1 & 1 & 0 & 1 \\ D & 1 & 0 & 0 & 0 \end{matrix}$

The morning flights on a commuter airline are listed in the table below. Use this table in Exercises 10–12.

10. Draw a directed graph that represents the morning routes of the commuter airline. **See above.**

11. Create a matrix representing the situation. Let an element be 1 if there is a flight from the row city to the column city. Otherwise let the element be 0. **See additional answers.**

12. What is the smallest number of flights by which one can travel from Adams to Foley? List the route. **3; *A* to *B*, *B* to *E*, *E* to *F***

Departs from	Destination
Adams	Brimfield, Colchester
Brimfield	Colchester, Drysdale, Eltonville
Colchester	Drysdale
Drysdale	Eltonville
Eltonville	Foley
Foley	Adams

78

Practice Bank, ALGEBRA 2: EXPLORATIONS AND APPLICATIONS

ONGOING ASSESSMENT

25. **Open-ended Problem** Draw a food web. Use a directed graph and matrices to analyze your food web. What would happen if one of the animals or plants were removed from the web?

SPIRAL REVIEW

26. Pete Morrill has several things he must do before leaving on a trip. Pete can't eat, pack, go to the store, or talk to his friend while he's showering. He can't pack or eat as he goes to the store. He also can't talk to his friend while he plans what to pack. Draw and color a graph to decide which of the six things Pete should do at the same time so that he can be ready to leave as soon as possible. *(Section 12.1)*

Find the sum of each series, if the sum exists. If a series does not have a sum, explain why not. *(Section 10.5)*

27. $3 + 9 + 27 + 81 + \cdots + 6561$

28. $\frac{1}{12} + \frac{1}{6} + \frac{1}{3} + \frac{2}{3} + \cdots + \frac{16}{3}$

29. $250 + (-50) + 10 + (-2) + \cdots$

30. $0.125 + 0.25 + 0.5 + 1 + \cdots$

ASSESS YOUR PROGRESS

VOCABULARY

graph (p. 563)
vertices (p. 563)
edges (p. 563)
degree of a vertex (p. 563)
directed graph (p. 569)

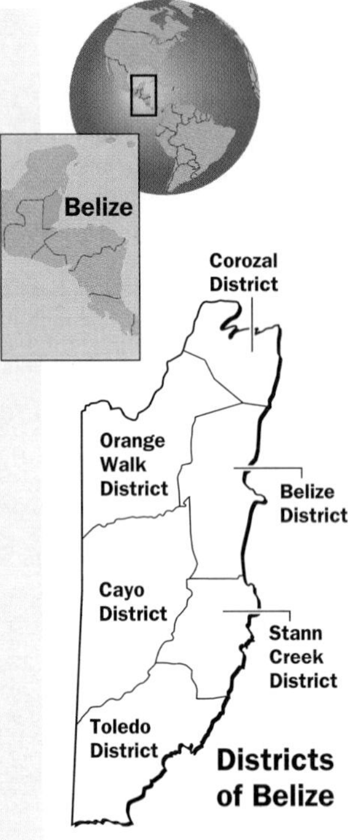

1. Draw a graph representing the map of Belize, a country in Central America. Label the degree of each vertex. Use your graph to color the map. Try to use as few colors as possible. *(Section 12.1)*

2. **BUSINESS** A company is planning meeting times for six committees. The following committees share members: 1 and 2, 1 and 3, 1 and 4, 2 and 5, 4 and 5, and 5 and 6. Use a graph to find the smallest number of meeting times that can be scheduled without conflict. *(Section 12.1)*

3. a. **ECOLOGY** Along the protected rocky shore of southern New England, plankton is eaten by mussels and barnacles; mussels are eaten by crabs, fish, and carnivorous snails; barnacles are eaten by crabs and fish; and crabs and carnivorous snails are eaten by fish. Draw a directed graph to represent this food web. *(Section 12.2)*

   b. Write a matrix M to represent the graph, using a 1 if the row animal is a food source for the column animal, and a 0 otherwise.

   c. Find M^2 and $M + M^2$. What does each matrix represent? Give specific examples.

4. **Journal** How are the graphs used in discrete mathematics different from the graphs you use in algebra? How are they alike?

25. Answers may vary. An example, based on a pond food web, is given.

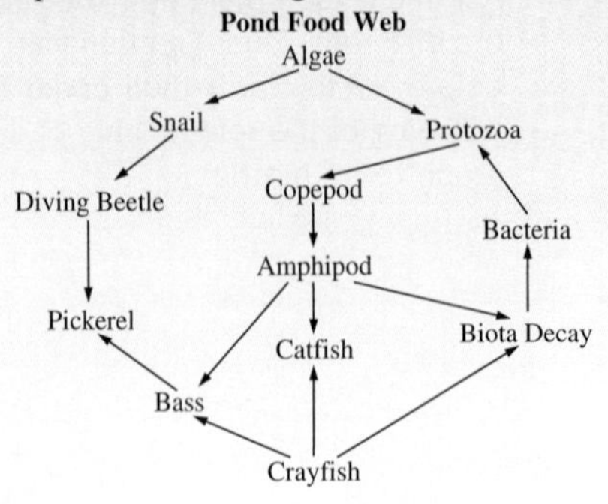

If an animal or plant were removed from the web, overpopulation of some organisms and lack of food sources would probably result.

26.

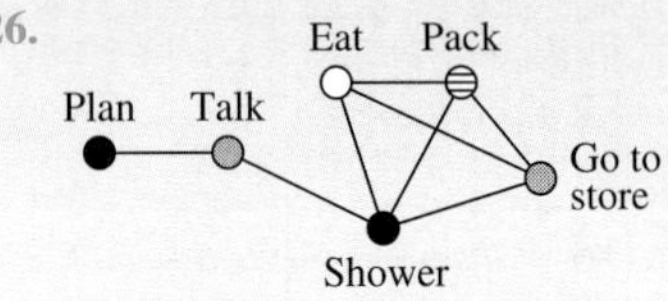

He should (1) shower and plan what to pack, (2) go to the store and talk to his friend, (3) eat, and (4) pack.

27. 9840

28. $\frac{127}{12} \approx 10.58$

29. $\frac{625}{3} \approx 208.33$

30. There is no sum since $r > 1$.

Assess Your Progress

1. Answers may vary. An example is given.

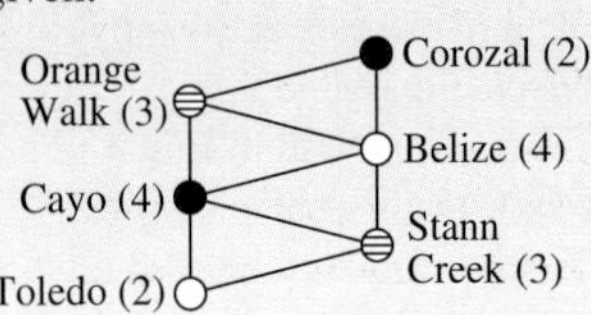

Three colors are needed. Color groupings: Corozal and Cayo, Belize and Toledo, Orange Walk and Stann Creek.

2–4. See answers in back of book.

Section

12.3 Permutations

Learn how to...

- use the multiplication counting principle
- find the number of permutations of the elements of a set

So you can...

- count the possibilities in situations such as relay race strategy and protein synthesis

Will musicians ever run out of new tunes? In how many ways can the components of DNA be arranged? The answers to such questions involve counting large numbers of possibilities. As you will see in the Exploration, even designing the simplest of flags leads to numerous alternatives.

EXPLORATION
COOPERATIVE LEARNING

Investigating the Multiplication Counting Principle

Work with a partner.
You will need:

- paper of four different colors
- scissors

Indonesia

1 Cut a 1 in. by 3 in. strip from each of three colors of paper. Arrange the strips to form flags with two horizontal bars like those at the right.
List all of the different flags you can make. How many are there? This number is the product of what two consecutive integers?

Burkina Faso

2 Cut a 1 in. by 3 in. strip of the fourth color. List all of the two-bar flags you can make with four colors. How many are there? This number is the product of what two consecutive integers?

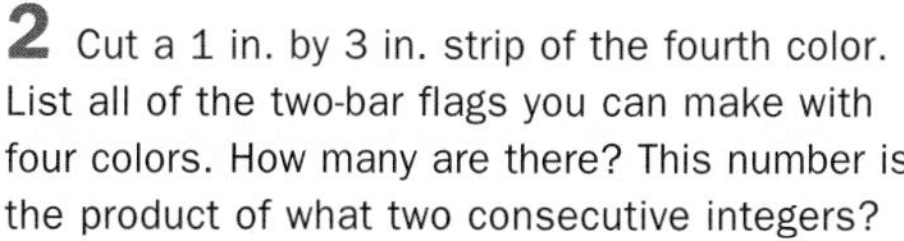

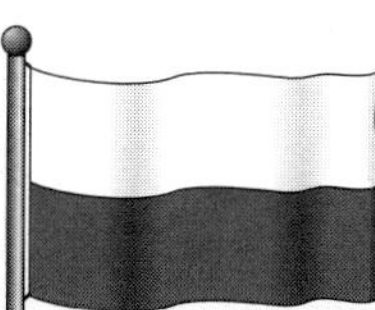

Poland

3 Look for a pattern in Steps 1 and 2. If you had one strip of each of ten different colors, how many different two-bar flags could you make?

Morocco

4 Cut another strip of each of the three original colors. How many different two-bar flags can you make from three colors if the bars may be the same color (as on the Moroccan flag) or different colors? Explain. This number is the square of what integer?

5 How many two-bar flags do you think can be made from four colors if repetition of colors is allowed? Explain.

Exploration Note

Purpose
The purpose of this Exploration is to have students discover the multiplication counting principle.

Materials/Preparation
Each group needs paper of four different colors and scissors.

Procedure
Students use colored strips of paper to determine the number of flags that can be made with two horizontal bars when three colors are used with no repetition, three colors are used with repetition, and four colors are used without repetition. Students count the number of flags and try to find patterns between that number and the number of colors used. They then use these patterns to predict the number of flags possible when ten colors are used without repetition and when four colors are used with repetition.

Closure
Students should intuitively understand the multiplication counting principle.

Explorations Lab Manual
See the Manual for more commentary on this Exploration.

For answers to the Exploration, see following page.

Plan⇔Support

Objectives

- Use the multiplication counting principle.
- Find the number of permutations of the elements of a set.
- Count the possibilities in real-world situations.

Recommended Pacing

Core and Extended Courses
Section 12.3: 1 day

Block Schedule
Section 12.3: $\frac{1}{2}$ block
(with Section 12.4)

Resource Materials

Lesson Support
Lesson Plan 12.3
Warm-Up Transparency 12.3
Practice Bank: Practice 79
Study Guide: Section 12.3

Technology
Technology Book:
Calculator Activity 12
Spreadsheet Activity 12
Graphing Calculator
Internet:
http://www.hmco.com

Warm-Up Exercises

1. List the possible outcomes of two tosses of a fair coin. HH, HT, TH, TT

Simplify.

2. $6 \cdot 5 \cdot 4 \cdot 3 \cdot 2 \cdot 1$ 720
3. $10 \cdot 10 \cdot 10 \cdot 9 \cdot 9 \cdot 8$ 648,000
4. $\frac{6 \cdot 5 \cdot 4 \cdot 3 \cdot 2 \cdot 1}{4 \cdot 3 \cdot 2 \cdot 1}$ 30
5. $\frac{9 \cdot 8 \cdot 7 \cdot 6 \cdot 5 \cdot 4 \cdot 3 \cdot 2 \cdot 1}{5 \cdot 4 \cdot 3 \cdot 2 \cdot 1}$ 3024

Teach⇔Interact

Learning Styles: Visual

Tree diagrams are a useful device to illustrate the multiplication counting principle. As students study the examples and discuss the questions, encourage them to draw tree diagrams to verify the counting that is necessary to solve each problem.

Additional Example 1

A family restaurant offers 4 different lunches on its children's menu: a hamburger, a hot dog, spaghetti, and a tuna fish sandwich. Each lunch comes with one of two desserts, a hot fudge sundae or strawberry shortcake. How many different meals can a child have for lunch?

Make a tree diagram to list the possibilities.

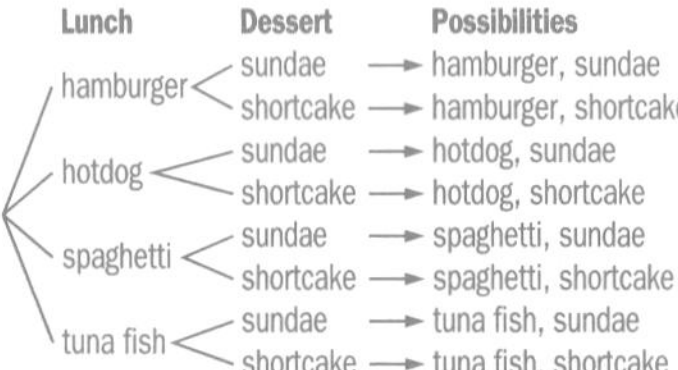

For each of the four choices of lunch, there are two choices of dessert. So there are $4 \times 2 =$ 8 possible meals.

Section Note

Communication: Writing
To solidify their understanding of the multiplication counting principle, ask students to rewrite the principle using the situation of Example 1. (If you must choose among 3 watch faces and for each face you must choose among 4 watchbands, then you have a total of 3×4 possible watches.)

Think and Communicate

Ask students to consider the possibilities when colors may be repeated and when they may not be repeated. Students can discuss why a country might not want to allow colors to be repeated.

EXAMPLE 1 Application: Fashion

How many different *Furry Fashions* watches can you make?

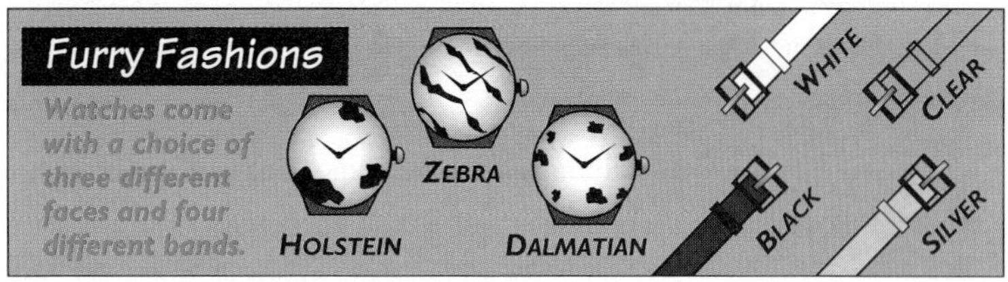

SOLUTION

Make a *tree diagram* to list the possibilities.

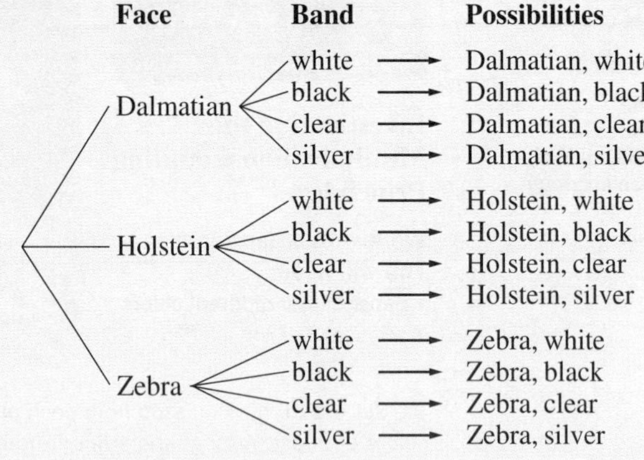

For each of the three choices of faces, there are four choices of bands. So there are $3 \times 4 = 12$ possible watches.

Example 1 suggests the following general principle.

Multiplication Counting Principle

If you must choose among m things, and for each of these you must choose among n more things, then you have a total of $m \times n$ possibilities.

THINK AND COMMUNICATE

1. How can you use the multiplication counting principle to tell how many two-bar flags you can make with 10 strips of different-colored paper?
2. How can you extend the multiplication counting principle to tell how many three-bar flags you can make with the 10 strips of paper?

ANSWERS Section 12.3

Exploration

1–4. Choice of colors may vary. Examples, using red, blue, white, and green are given.

1. Red/Blue, Red/White, Blue/Red, Blue/White, White/Red, White/Blue; 6; 2×3
2. Red/Blue, Red/White, Red/Green, Blue/Red, Blue/White, Blue/Green, White/Red, White/Blue, White/Green, Green/Red, Green/Blue, Green/White; 12; 3×4
3. The pattern is number of flags = (number of colors) × (number of colors – 1); $10 \times 9 = 90$.
4. 9; Answers may vary. An example is given. There are 3 choices of color for the top bar and 3 choices of color for the bottom bar, or $3 \times 3 = 9$ flags in all; $9 = 3^2$.
5. 16; There are 4 choices of color for the top bar and 4 choices of color for the bottom bar, so $4 \times 4 = 16$ flags can be made.

EXAMPLE 2 Application: Sports

Short-track speed skaters occasionally compete in relay races. In planning the race strategy, a team's coach can arrange the four team members to skate in any order. How many arrangements are possible?

SOLUTION

Use the multiplication counting principle.

For the **first** skater, there are **4** choices.

For the **third** skater, 2 choices remain.

$$4 \times 3 \times 2 \times 1 = 24$$

For the **second** skater, 3 choices remain.

Only **1** choice remains for the **fourth** skater.

There are 24 possible orders in which the team can compete.

BY THE WAY

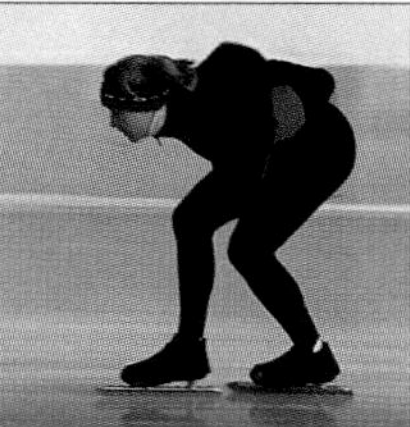

Short-track speed skaters skate around a 111 m track on a hockey or Olympic-size rink. Some skaters can reach a speed of 35 mi/h and take turns leaning at a 65° angle.

You can write $4 \times 3 \times 2 \times 1$ as 4! (read "four factorial"). In general, for any positive integer n, $n!$ is the product of all the integers from n down to 1.

$$n! = n(n-1)(n-2)\cdots(2)(1)$$

For consistency, 0! is defined to be equal to 1.

Permutations

An arrangement of the elements of a set is called a **permutation**. A set of four objects, such as a four-person relay race team, has 4! permutations. In general, the number of permutations of a set of n objects is given by $n!$.

EXAMPLE 3 Application: Sports

If the coach in Example 2 had filled the four slots from a pool of seven eligible skaters, how many four-member arrangements are possible?

SOLUTION

The **first** skater can be any of the **7**.

For the **third** skater, 5 choices remain.

$$7 \times 6 \times 5 \times 4 = 840$$

For the **second** skater, 6 choices remain.

For the **fourth** skater, **4** choices remain.

There are 840 permutations of four skaters chosen from seven.

Think and Communicate

1. a. There are 10 choices for the top bar, and for each of these, there are 9 choices for the bottom bar, which gives a total of $10 \times 9 = 90$ different two-bar flags.
   b. For each of the 90 choices of two bars, there are 8 choices for the third bar, which gives a total of $10 \times 9 \times 8 = 720$ different three-bar flags.

Teach⇔Interact

Additional Example 2

Five members of the senior class were chosen to address the class during graduation. The principal can choose the order in which the seniors address their class. How many arrangements of the speakers are possible?

Use the multiplication counting principle.

For the first speaker, there are 5 choices. For the second speaker, 4 choices remain. For the third speaker, 3 choices remain. For the fourth speaker, 2 choices remain. Only 1 choice remains for the fifth speaker.

$5 \times 4 \times 3 \times 2 \times 1 = 120$

There are 120 possible orders in which the five seniors can address their class.

Section Note

Reasoning

Some students may have difficulty understanding the definition of 0!. Use a few numerical examples with $n \geq 2$ to show them that $n[(n-1)!] = n!$. Then point out that in order for the statement to be true for $n = 1$, 0! must be defined as being equal to 1.

About Example 3

Communication: Discussion
Students can discuss the relationship they see between the arrangement of the skaters in Example 3 and the Exploration in which two stripes of four colors were chosen for a flag and colors were not allowed to be repeated.

Additional Example 3

If the principal in Additional Example 2 had chosen the five speakers from a group of eight possible speakers, how many five-member arrangements are possible?

The first speaker can be any of the 8. For the second speaker, 7 choices remain. For the third speaker, 6 choices remain. For the fourth speaker, 5 choices remain. For the fifth speaker, 4 choices remain.

$8 \times 7 \times 6 \times 5 \times 4 = 6720$.

There are 6720 permutations of five speakers chosen from the eight possible speakers.

Teach⇔Interact

Additional Example 4

How many distinguishable permutations are there of the letters in CONNECTICUT?

Use the formula for permutations with repeated elements. There are 11 letters, so $n = 11$. There are three C's ($q_1 = 3$), one O ($q_2 = 1$), two N's ($q_3 = 2$), one E ($q_4 = 1$), two T's ($q_5 = 2$), one I ($q_6 = 1$), and one U ($q_7 = 1$).

$$\frac{11!}{3!\,1!\,2!\,1!\,2!\,1!\,1!}$$

$$= \frac{11 \cdot 10 \cdot 9 \cdot 8 \cdot 7 \cdot 6 \cdot 5 \cdot 4 \cdot 3 \cdot 2 \cdot 1}{3 \cdot 2 \cdot 1 \cdot 1 \cdot 2 \cdot 1 \cdot 1 \cdot 2 \cdot 1 \cdot 1 \cdot 1}$$

$$= 11 \cdot 10 \cdot 9 \cdot 8 \cdot 7 \cdot 6 \cdot 5$$

$$= 1{,}663{,}200$$

There are 1,663,200 distinguishable permutations of the letters in CONNECTICUT.

Section Note

Common Error

Students often find counting problems difficult. These difficulties generally stem from an inability to distinguish the important characteristics of the situation. Thus, students may apply the wrong formula. To help students avoid these errors, have them make a flow chart that they can use to solve the counting problems in this section. This can be done individually or as a class activity. For example, the first question in the flow chart could be: "Are the elements of the arrangement being chosen from one set?" After each question in the flow chart, there should be two subbranches, one for *Yes* and one for *No*. Each of these branches should contain either another question or the method to determine the solution for that case. For the question above, a *No* answer should lead to a summary of the method of Example 1, while a *Yes* may lead to the question "Can the elements of the set be used more than once?" A *No* answer to this question would lead to the method of Example 2, while a *Yes* response should lead to another question. The flow chart should be continued until each branch ends in a method for counting the number of arrangements.

In Example 3, the answer is the product of the first four factors of 7!. This is the same as $\frac{7!}{3!} = \frac{7 \cdot 6 \cdot 5 \cdot 4 \cdot \cancel{3} \cdot \cancel{2} \cdot \cancel{1}}{\cancel{3} \cdot \cancel{2} \cdot \cancel{1}}$ and suggests the following result.

Permutations

The number of permutations of r objects taken from a set of n objects is given by ${}_nP_r$ where

$${}_nP_r = \frac{n!}{(n-r)!}.$$

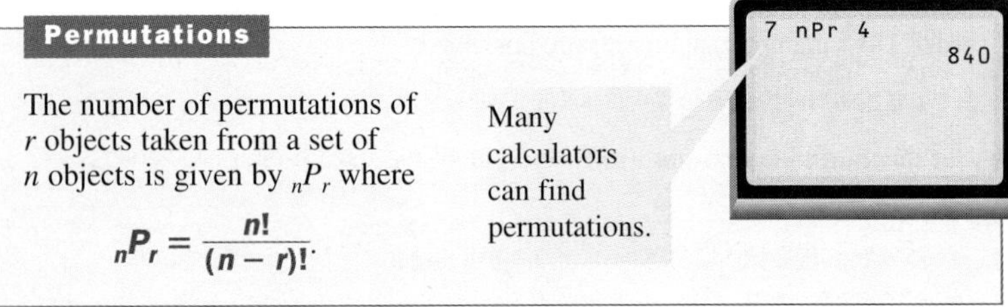

Many calculators can find permutations.

Distinguishable Permutations

When a set contains elements that are identical, some of the permutations look the same. For example, how many permutations are there of the letters in DAD? If you color the D's, you can arrange the 3 letters in 3! different ways.

DAD	ADD	DDA
DAD	ADD	DDA

If you ignore the colors, you see **2!** copies of each "word" because there are **2!** ways to arrange the two D's.

So without color there are only $\frac{3!}{2!} = 3$ different permutations: DAD, ADD, and DDA.

Permutations with Repeated Elements

If a set of n elements has one element repeated q_1 times, another element repeated q_2 times, . . . , and the last element repeated q_k times, then the number of distinguishable permutations of the n elements is given by

$$\frac{n!}{q_1!q_2! \ldots q_k!}.$$

EXAMPLE 4

How many distinguishable permutations are there of the letters in MISSISSIPPI?

SOLUTION

Use the formula for permutations with repeated elements above. There are 11 letters, so $n = 11$. There are **one** M ($q_1 = 1$), four I's ($q_2 = 4$), **four** S's ($q_3 = 4$), and **two** P's ($q_4 = 2$).

$$\frac{11!}{1!4!4!2!} = \frac{11 \cdot 10 \cdot 9 \cdot \cancel{8} \cdot 7 \cdot \cancel{6} \cdot 5 \cdot \cancel{4} \cdot \cancel{3} \cdot \cancel{2} \cdot 1}{1 \cdot \cancel{4} \cdot \cancel{3} \cdot \cancel{2} \cdot 1 \cdot \cancel{4} \cdot \cancel{3} \cdot \cancel{2} \cdot 1 \cdot \cancel{2} \cdot 1}$$

$$= 11 \cdot 10 \cdot 9 \cdot 7 \cdot 5$$

$$= 34{,}650$$

There are 34,650 distinguishable permutations of the letters in MISSISSIPPI.

Checking Key Concepts

1. a. spaghetti/corn/pie, spaghetti/corn/pudding, spaghetti/beans/pie, spaghetti/beans/pudding, spaghetti/squash/pie, spaghetti/squash/pudding, chicken/corn/pie, chicken/corn/pudding, chicken/beans/pie, chicken/beans/pudding, chicken/squash/pie, chicken/squash/pudding
   b. There are 2 choices of entrees, 3 choices of vegetables, and 2 choices of desserts, so there are $2 \times 3 \times 2 = 12$ choices of meals.
2. a. 10!
   b. 10^{10}
3. 6
4. 40,320
5. 6720
6. 336
7. D
8. 6!
9. $\frac{6!}{2!\,2!\,1!\,1!}$
10. $\frac{6!}{3!\,2!\,1!}$

CHECKING KEY CONCEPTS

1. a. List the different meals you can choose from the menu at the right.
   b. Explain how you can use the multiplication counting principle to find the number of possible meals.
2. Write an expression for the number of ten-digit numerical codes that can be formed if:
   a. the digits must all be different   b. the digits may repeat

Simplify each expression.

3. $3!$ 4. $8!$ 5. ${}_8P_5$ 6. ${}_8P_3$

7. Which of the following represents the number of ways that eight runners can fill the top three slots in a race?

   A. $3!$ B. $8!$ C. $\frac{8!}{3!}$ D. $\frac{8!}{5!}$ E. ${}_8P_5$

Write an expression for the number of distinguishable permutations of the letters in each of the following six-letter words.

8. MATRIX   9. DIVIDE   10. BAOBAB

Dinner $6.99
Entrees (choose one)
Spaghetti with Meatballs
Chicken Enchiladas
Vegetables (choose one)
Corn on the Cob
Green Beans
Butternut Squash
Desserts (choose one)
Apple Pie
Bread Pudding

12.3 Exercises and Applications

Extra Practice exercises on page 768

Calculate each expression in Exercises 1–3.

1. a. $6!$ b. $(2!)(3!)$
2. a. $7!$ b. $2! + 5!$
3. a. $4!$ b. $4(3!)$

Tell whether each statement is true for all positive integers *m* and *n*. Explain your answer.

4. $(m!)(n!) = (mn)!$ 5. $m! + n! = (m + n)!$ 6. $m! = m \cdot (m - 1)!$

Simplify.

7. $\frac{7!}{6!}$ 8. $\frac{8!}{4!4!}$ 9. $\frac{7!}{3!2!2!}$ 10. $\frac{(n+2)!}{(n+1)!}$

11. **FASHION** Janice has the clothes shown below, all of which coordinate. In how many ways can she create an outfit by choosing one of each item?

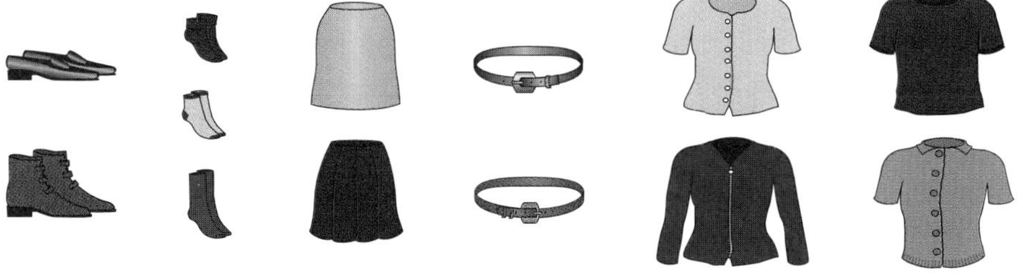

Teach⇔Interact

Checking Key Concepts

Alternate Approach
One way to calculate permutations quickly is to think of ${}_nP_r$ as representing the first r factors of $n!$. For example, in question 6, ${}_8P_3$ is $8 \cdot 7 \cdot 6 = 336$.

Closure Question

What are the formulas for finding the number of permutations of a set of n objects? of r objects taken from a set of n objects?

$n! = n(n-1)(n-2) \cdot \ldots \cdot (2)(1)$; ${}_nP_r = \frac{n!}{(n-r)!}$

Apply⇔Assess

Suggested Assignment

Core Course
Exs. 1–11, 20–23, 25–28, 31, 33–36, 41–45

Extended Course
Exs. 1–21, 25–32, 35–37, 39–45

Block Schedule
Day 78 Exs. 1–11, 20–23, 25–28, 31, 33–36, 41–45

Exercise Notes

Reasoning
Exs. 4–6 Call upon three students to discuss their answers to one of these exercises. Specific examples may help some students to understand why a particular statement is true or false.

Exercises and Applications

1. a. 720
   b. 12
2. a. 5040
   b. 122
3. a. 24
   b. 24

4–6. Answers may vary. Examples are given.

4. No; $(2!)(3!) \neq 6!$.
5. No; $2! + 5! \neq 7!$.
6. Yes; $m! = m(m-1)(m-2) \cdots (2)(1) = m[(m-1)(m-2) \cdots (2)(1)] = m \cdot (m-1)!$.
7. 7
8. 70
9. 210
10. $n + 2$
11. 96 ways

Exercise Notes

Interview Note
Exs. 12–14 The directed graph that students drew in Ex. 11 on page 574 represented the possible paths of power transfer in the network. One of the constraints of a network such as this is the amount of power lost along each path. Electrical engineers can perform resistance tests to determine the power loss for each path.

Research
Exs. 12–18 Students may recall that resistors were first examined in a series of problems from Chapter 8 on page 363. You might ask students to research how the impedance of a resistor is related to its resistance. Students may also be interested to know why resistors are needed in electrical circuits.

Mathematical Procedures
Exs. 15–17 Ask students to explain how they found their answers to these exercises. Their explanations may help clarify and reinforce the procedures used for solving different kinds of problems that involve permutations.

Application
Ex. 19 This exercise introduces an interesting application of permutations. Some students may be interested in researching the development of the Photo-FIT system.

Using Technology
Exs. 20–23 These exercises can be completed quickly on a graphing calculator by choosing 2:nPr from the MATH:PRB menu. Note that the value for n must be entered first and the value for r entered after the $_nP_r$. Once the entry is calculated, it can be recalled by pressing [2nd][ENTER]. Then the arrow keys can be used to replace the values of n and r in the formula.

INTERVIEW Maria Rodriguez

Look back at the article on pages 560–562.

Electrical engineers like Maria Rodriguez know that the components of an electrical circuit resist the flow of electricity. One such component, called a resistor, is shown below. The amount of resistance, measured in ohms, is indicated on the resistor using bands of color.

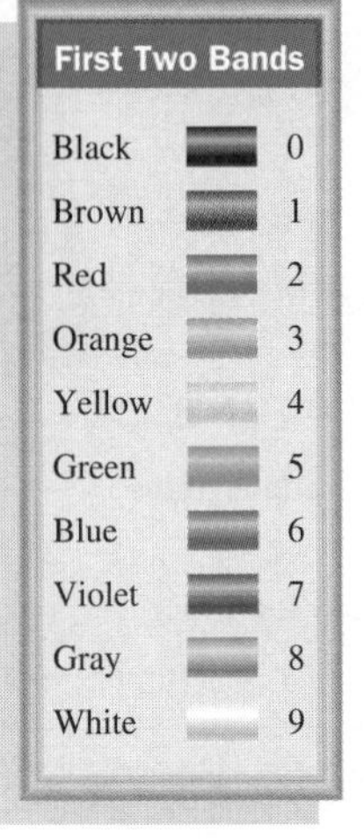

First Two Bands	
Black	0
Brown	1
Red	2
Orange	3
Yellow	4
Green	5
Blue	6
Violet	7
Gray	8
White	9

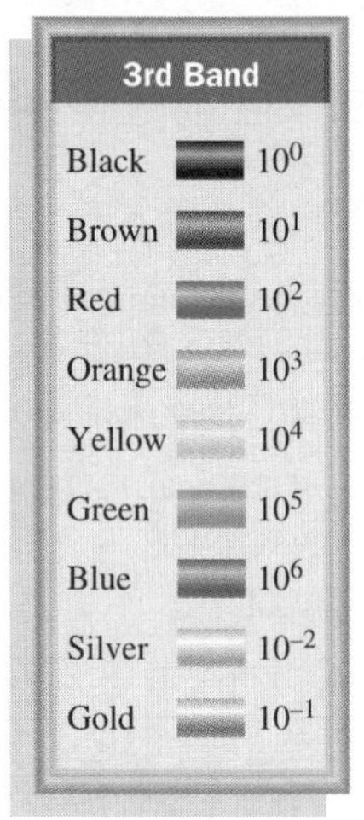

3rd Band	
Black	10^0
Brown	10^1
Red	10^2
Orange	10^3
Yellow	10^4
Green	10^5
Blue	10^6
Silver	10^{-2}
Gold	10^{-1}

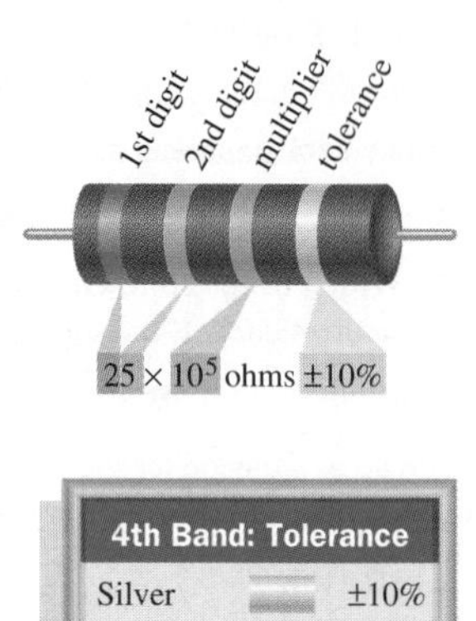

4th Band: Tolerance	
Silver	±10%
Gold	±5%
No Band	±20%

BY THE WAY

There are resistors inside electronics such as radios, answering machines, and computers.

Find the resistance of each resistor.

12.  **13.**  **14.** 

15. How many different possible color sequences are there for the first three bands on a resistor?

16. How many different possible color sequences are there if tolerances are included? (Count "no band" as a color, since it still gives information.)

17. On some resistors a 5th color band indicates reliability. If the 5th band can be any of four colors, how many different color patterns can the resistor have?

18. **Challenge** Do all of the possible color patterns correspond to different resistance ratings? Give examples to explain your answer.

19. **FORENSICS** In the United Kingdom, a system called Photo-FIT (Facial Identification Technique) has been used by police to identify suspects. The basic five-section kit contains 195 hairlines, 99 eyes and eyebrows, 89 noses, 105 mouths, and 74 chins and cheeks. How many different faces can be constructed from the basic kit?

Simplify.

20. $_9P_5$ **21.** $_8P_1$ **22.** $_6P_2$ **23.** $_6P_4$

12. $49 \cdot 10^2 \pm 10\%$
13. $68 \cdot 10^5 \pm 5\%$
14. $25 \cdot 10^6 \pm 20\%$
15. 900 sequences
16. 2700 sequences
17. 10,800 sequences
18. No. Examples may vary. An example is given. Color bands of (Black, Brown, Red, Silver) correspond to a resistance rating of 01×10^2 ohms $\pm 10\% =$ 100 ohms $\pm 10\%$, and color bands of (Brown, Black, Brown, Silver) correspond to a resistance rating of 10×10^1 ohms $\pm 10\% =$ 100 ohms $\pm 10\%$.
19. 13,349,986,650 different faces
20. 15,120
21. 8
22. 30
23. 360
24. Answers may vary. Examples are given.
    a. finding the number of ways in which the runners of an 8-person race can place
    b. finding the number of ways in which the winner of an 8-person race can be decided
25. 259 other cars
26. 720 ways

24. a. **Open-ended Problem** Describe a situation in which you would find ${}_8P_8$.

b. Describe a situation in which you would find ${}_8P_1$.

25. Adam saw a car speeding away from an accident scene. The car had a license plate from his state, which uses three letters followed by three digits. He remembers the first two letters and the first two digits. How many other cars could possibly share these same letters and digits?

26. You and five friends go to the movies. In how many ways can you line up to buy tickets?

27. **SPORTS** In baseball, the coach must plan the order in which the nine players will bat. How many possible batting orders are there?

28. Suppose 22 students are entering a classroom that has 22 desks.

a. How many possible seating arrangements are there? (*Note:* Assume that no one moves the desks.)

b. Some scientists estimate that the universe is about 10 billion years old. If the class had put itself into one of the seating arrangements each second since the beginning of the universe, when would the possibilities have been exhausted, or how much longer would it take?

29. **SPORTS** If 18 gymnasts are competing in an event, in how many ways can the gold, silver, and bronze medals be awarded?

30. **SCHEDULING** Jeanne Remsen, a health inspector, is making random checks of local restaurants. She has time today to visit four of nine restaurants on her list. In how many possible ways can her trips for the day be ordered?

31. **SAT/ACT Preview** In how many ways can a president, vice-president, treasurer, and recorder be chosen from a club of 11 members?

A. 14,641 **B.** 24 **C.** 7920 **D.** 1,663,200 **E.** 66

32. a. How many permutations of n objects are there?

b. What happens when you let $r = n$ in the formula for ${}_nP_r$?

c. What must be true for your answers to parts (a) and (b) to agree?

Give an expression for the number of distinguishable permutations of the letters of each word. Then evaluate the expression.

33. ROTOR 34. HUMDRUM 35. SYZYGY 36. ONOMATOPOEIA

37. a. **MUSIC** In the key of G, the first 14 notes of "Twinkle, Twinkle, Little Star" include three G's, two A's, two B's, two C's, three D's, and two E's. Write and then evaluate an expression for the number of different melodies that can be formed from these 14 notes.

b. The first two measures of J. S. Bach's "Musette" from *English Suite No. 3* are shown at the right. How many different melodies can be formed from these 9 notes?

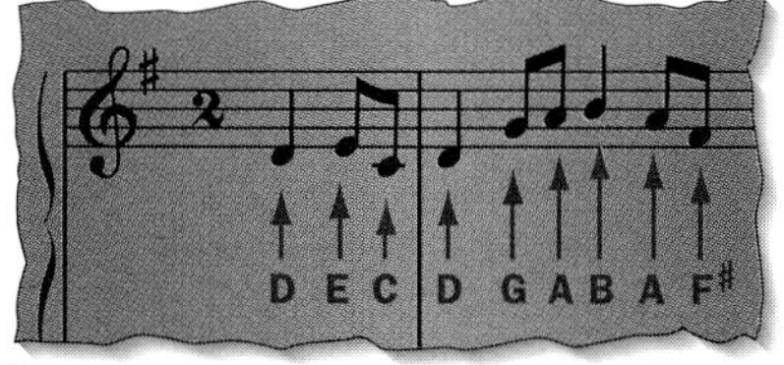

38. **LITERATURE** French poet Raymond Queneau's book *Cent mille milliards de poèmes* contains a sonnet on each of its 10 pages. The pages are cut so that any of the 14 lines of the sonnet can be turned independently, allowing any line of one sonnet to be combined with any line of another sonnet. How many different sonnets can result if the lines are structured so that all the possibilities make sense?

Apply⇔Assess

Exercise Notes

Cooperative Learning

Exs. 24–38 Because there is a minimal amount of written work needed to solve these types of problems, students who get an incorrect answer often cannot find their mistake. Allowing students the opportunity to go over and correct these exercises in small groups can help them better understand the methods used to categorize and solve these problems.

Student Progress

Ex. 24 Students should be able to use their examples to explain the difference between the values ${}_8P_8$ and ${}_8P_1$.

Challenge

Ex. 28 Ask students to find the possible number of seating arrangements of 22 people around a circle of desks if only the order is considered, not the particular desk in which the person is sitting. (21!)

Reasoning

Ex. 32 Students should realize after completing this exercise that the formula for the permutation of n objects ($n!$) is a special case of the formula for the number of permutations of r objects taken from a set of n objects (${}_nP_r$), the case when $n = r$.

Interdisciplinary Problems

Ex. 38 Some students may be interested in creating their own versions of Raymond Queneau's book. They may find it easier to include poems that have fewer than 14 lines. For example, they may wish to have each page of the book contain a *haiku*, which is an unrhymed Japanese verse form of three lines containing 5, 7, and 5 syllables, respectively.

27. 362,880 batting orders

28. a. $22! \approx 1.124 \times 10^{21}$

b. It would take about 35,600 billion more years.

29. 4896 ways

30. 3024 ways

31. C

32. a. $n!$

b. You get ${}_nP_n = \frac{n!}{0!}$.

c. 0! must be equal to 1.

33. $\frac{5!}{2!\,2!\,1!} = 30$

34. $\frac{7!}{2!\,2!\,1!\,1!\,1!} = 1260$

35. $\frac{6!}{3!\,1!\,1!\,1!} = 120$

36. $\frac{12!}{4!\,2!\,1!\,1!\,1!\,1!\,1!\,1!} = 9{,}979{,}200$

37. a. $\frac{14!}{3!\,2!\,2!\,2!\,3!\,2!} =$ 151,351,200 different melodies

b. 90,720 different melodies

38. 10^{14} sonnets

Apply⇔Assess

Exercise Notes

Historical Connection

Exs. 39, 40 An enzyme involved in the synthesis of RNA was discovered by the biochemist Severo Ochoa. He used his discovery to synthesize RNA in a test tube. The results of his work earned him the 1959 Nobel Prize for medicine, which he shared with Arthur Kornberg.

Assessment Note

Ex. 41 Ask students to consider why ${}_{10}P_{10} = {}_{10}P_9$. (In the case of ${}_{10}P_{10}$ the last, or 10th choice, has already been determined by the first nine choices and therefore, gives no additional possibilities.)

Practice 79 for Section 12.3

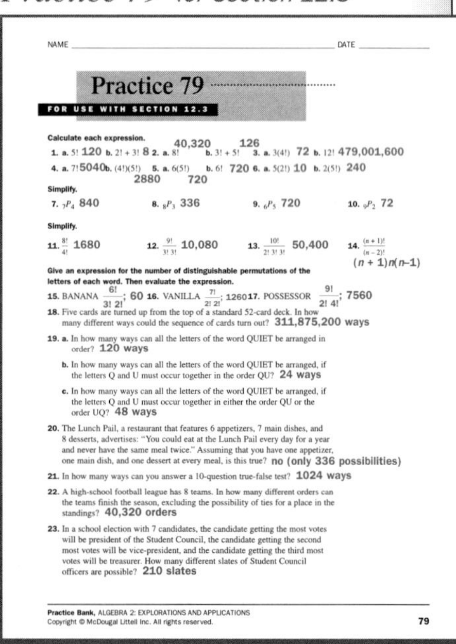

NAME ______________________ DATE __________

Practice 79

FOR USE WITH SECTION 12.3

Calculate each expression.

1. a. 5! 120 b. 2! + 3! 8 2. a. 8! 40,320 b. 3! + 5! 126 3. a. 3(4!) 72 b. 12! 479,001,600
4. a. 7! 5040 b. (4!)(5!) 2880 5. a. 6(5!) 720 b. 6! 720 6. a. 5(2!) 10 b. 2(5!) 240

Simplify.

7. ${}_7P_4$ 840 8. ${}_8P_3$ 336 9. ${}_6P_5$ 720 10. ${}_9P_2$ 72

Simplify.

11. $\frac{8!}{4!}$ 1680 12. $\frac{9!}{3!\,3!}$ 10,080 13. $\frac{10!}{2!\,3!\,3!}$ 50,400 14. $\frac{(n+1)!}{(n-2)!}$ $(n+1)n(n-1)$

Give an expression for the number of distinguishable permutations of the letters of each word. Then evaluate the expression.

15. BANANA $\frac{6!}{3!\,2!}$; 60 16. VANILLA $\frac{7!}{2!\,2!}$; 1260 17. POSSESSOR $\frac{9!}{2!\,4!}$; 7560

18. Five cards are turned up from the top of a standard 52-card deck. In how many different ways could the sequence of cards turn out? 311,875,200 ways

19. a. In how many ways can all the letters of the word QUIET be arranged in order? 120 ways
   b. In how many ways can all the letters of the word QUIET be arranged, if the letters Q and U must occur together in the order QU? 24 ways
   c. In how many ways can all the letters of the word QUIET be arranged, if the letters Q and U must occur together in either the order QU or the order UQ? 48 ways

20. The Lunch Pail, a restaurant that features 6 appetizers, 7 main dishes, and 8 desserts, advertises: "You could eat at the Lunch Pail every day for a year and never have the same meal twice." Assuming that you have one appetizer, one main dish, and one dessert at every meal, is this true? no (only 336 possibilities)

21. In how many ways can you answer a 10-question true-false test? 1024 ways

22. A high-school football league has 8 teams. In how many different orders can the teams finish the season, excluding the possibility of ties for a place in the standings? 40,320 orders

23. In a school election with 7 candidates, the candidate getting the most votes will be president of the Student Council, the candidate getting the second most votes will be vice-president, and the candidate getting the third most votes will be treasurer. How many different slates of Student Council officers are possible? 210 slates

Practice Bank, ALGEBRA 2: EXPLORATIONS AND APPLICATIONS

79

Connection GENETICS

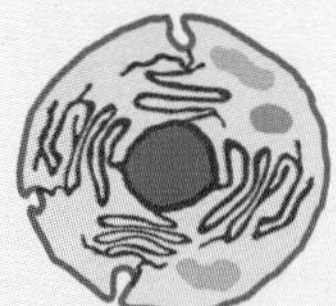

Special molecules called *RNA* tell your cells how to make proteins out of smaller molecules called *amino acids*. The RNA is made up of four kinds of *nucleotides*, which are often represented by the letters U, C, A, and G. Nucleotides are like letters in a sentence. Three nucleotides make a word, or *codon*. Each codon tells the cell to do something.

RNA

"Start making the protein. The first amino acid should be methionine."

"Add an alanine amino acid to the protein."

"Add a glycine amino acid."

"The protein is done. Stop."

AUG GCC UCU UGC AAA GGC UAU AGU AGU UAG

The cell always starts assembly with an AUG codon. This codon may also appear later in the sequence.

(Each codon except UAA, UGA, and UAG tells the cell to add an amino acid to the protein.)

The codons UAA, UGA, and UAG tell the cell to stop assembly.

39. How many possible codons are there? (*Note:* Nucleotides may repeat within a codon.)

40. Write an expression for the theoretical number of sequences of codons that can code for a protein 20 amino acids long. Remember that each sequence must begin with the "start" codon and must end with one of the "stop" codons, which do *not* add an amino acid.

ONGOING ASSESSMENT

41. Writing Explain why the number of ways to choose the top three winners from 10 contestants is $\frac{10!}{7!}$.

SPIRAL REVIEW

42. HOBBIES Five friends use ham radios to keep in touch. Api can receive broadcasts only from Destin. Ben can receive broadcasts from both Api and Destin. Charma can receive broadcasts from Api and Eddie. Destin can receive broadcasts from Eddie. And Eddie can receive broadcasts only from Ben. ***(Section 12.2)***

a. Draw a directed graph representing this situation.

b. Can anyone send a broadcast that can be passed to all of the others with at most one intermediate stage? How you can use a matrix to find out?

Find the asymptotes of the graph of each function. Tell how the graph is related to a hyperbola with equation of the form $y = \frac{a}{x}$. ***(Section 9.7)***

43. $f(x) = \frac{2}{x-1} + 3$ **44.** $g(x) = \frac{x}{x+2}$ **45.** $h(x) = \frac{2x+1}{x-4}$

39. 64 codons

40. $1 \times 61^{19} \times 3 \approx 2.50 \times 10^{34}$ sequences

41. Answers may vary. An example is given. There are 10 choices for first place. For each first-place choice, there are 9 choices for second place. When first and second place winners have been chosen, there are 8 ways to choose a third place winner. Therefore, there are $10 \times 9 \times 8$ ways to choose the top three winners. Also, $\frac{10!}{7!} = \frac{10 \times 9 \times 8 \times 7 \times 6 \times 5 \times 4 \times 3 \times 2 \times 1}{7 \times 6 \times 5 \times 4 \times 3 \times 2 \times 1} = 10 \times 9 \times 8$.

42. a.

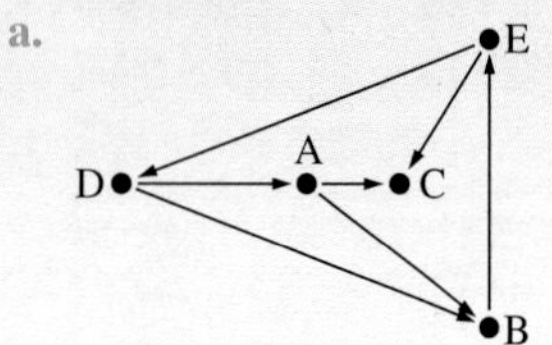

b. Yes; Eddie and Destin can both transmit to everyone else with at most one intermediate stage. To show this, write a matrix M to represent the graph in part (a), using a 1 if the row person can send directly to the column person, and a 0 otherwise. Find $M^2 + M$, and see if there is a row which has a nonzero number in every position except the position in which the row person is the same as the column person.

43. $x = 1$ and $y = 3$; $y = \frac{2}{x}$ is translated 1 unit right and 3 units up.

44. $x = -2$ and $y = 1$; $y = -\frac{2}{x}$ is translated 2 units left and 1 unit up.

45. $x = 4$ and $y = 2$; $y = \frac{9}{x}$ is translated 4 units right and 2 units up.

Section

12.4 Combinations

Lucky 3 Drawing
Three lucky winners will receive $3000 each!
Grand Sweepstakes
1st prize: $30,000
2nd prize: $300
3rd prize: $30

Learn how to...

- find the number of combinations of the elements of a set

So you can...

- count choices in which order is not important, such as the number of ways to plan a figure skating program

In which contest would you rather be a winner? In the Lucky 3 Drawing, it matters only that you're in the top three, while in the Grand Sweepstakes, your exact position in the top three is important. There are more ways to award prizes for the Grand Sweepstakes than for the Lucky 3 Drawing. In the Exploration you'll see why.

EXPLORATION
COOPERATIVE LEARNING

Investigating Ways of Choosing Winners

Work in a group of 4 or 5 students.

Suppose only the members of your group enter the Grand Sweepstakes.

1 Each of you should list all of the ways that others in your group can win the second and third prizes if you are the first-prize winner.

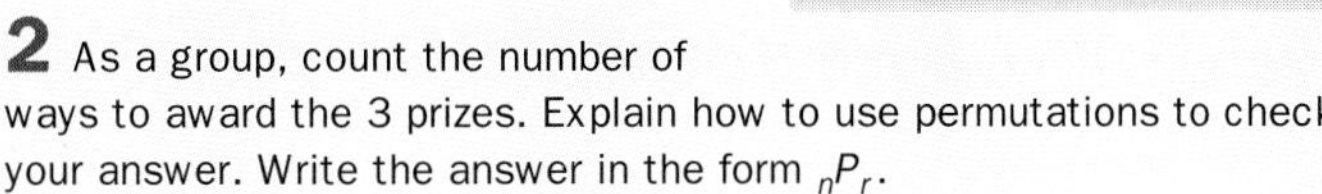

2 As a group, count the number of ways to award the 3 prizes. Explain how to use permutations to check your answer. Write the answer in the form ${}_nP_r$.

Suppose only the members of your group enter the Lucky 3 Drawing.

3 Since each winner receives the same amount of money, the group of winners "John, Kia, and Ann," for example, is the same as "Ann, John, and Kia." In your lists from Step 1, how many times is each group of three winners listed? Explain why you can write your answer in the form $r!$.

4 Use the lists from Step 1 to count the number of ways to award the Lucky 3 prizes. How is your answer related to the answer in Step 2? Express the number of ways to choose the Lucky 3 winners in the form $\frac{{}_nP_r}{r!}$.

A selection made from a set when position is not important, as in the Lucky 3 Drawing, is called a **combination**. The symbol ${}_nC_r$ stands for the number of combinations of r items that can be chosen from a set of n items. The number of combinations of **three** Lucky 3 winners from a group of **five** entrants is ${}_5C_3$.

Exploration Note

Purpose
The purpose of this Exploration is to have students explore the concept of combinations.

Procedure
Students work in groups of 4 or 5. Each member assumes he or she is the first-prize winner in the Grand Sweepstakes and lists the ways the others in the group can win the second- and third-place prizes. The group then combines these lists, counts the total number of ways to award the 3 prizes, and writes their answer in the form ${}_nP_r$. Next, each group uses the lists to count the number of ways the three winners can be chosen for the Lucky 3 Drawing. They express this total in the form $\frac{{}_nP_r}{r!}$

Closure
Students should understand that when groups of n elements of a set are taken r at a time and position is not important, the $r!$ groups are not unique. Therefore, dividing the total number of permutations of the set, ${}_nP_r$, by $r!$ will give the combinations of the set, ${}_nC_r$.

Explorations Lab Manual
See the Manual for more commentary on this Exploration.

For answers to the Exploration, see following page.

Plan⇔Support

Objectives

- Find the number of combinations of the elements of a set.
- Count choices in which order is not important.

Recommended Pacing

Core and Extended Courses
Section 12.4: 2 days

Block Schedule
Section 12.4: 2 half-blocks (with Sections 12.3 and 12.5)

Resource Materials

Lesson Support
Lesson Plan 12.4
Warm-Up Transparency 12.4
Practice Bank: Practice 80
Study Guide: Section 12.4

Technology
Technology Book: Spreadsheet Activity 12
Graphing Calculator
Internet: http://www.hmco.com

Warm-Up Exercises

Simplify.

1. ${}_5P_3$ 60
2. $\frac{{}_5P_3}{3!}$ 10
3. $\frac{5!}{(5-3)!3!}$ 10
4. $\frac{{}_5P_3}{3!} \cdot \frac{{}_5P_4}{4!}$ 50
5. $\frac{6!}{(6-4)!4!} + \frac{4!}{(4-1)!1!}$ 19

Teach⇔Interact

Learning Styles: Visual

Since many combination problems lend themselves to tree diagrams, it may be helpful to students who are visual learners to use them to list the possible arrangements. In this way, they can see which arrangements are not unique and must be disregarded.

Section Note

Student Study Tip
Point out to students the *Watch Out!* note on this page. Understanding and remembering this note can help students decide whether permutations or combinations should be used in a particular situation. You might suggest that students copy this note in their notebooks or journals.

About Example 1

Mathematical Procedures
Students should notice in Method 1 of this example that the expression for $(n - r)!$ in the denominator of ${}_nC_r$ was not written out as factors but was canceled with the equivalent factorial in the numerator. Note that the numerator was written as factors only until the remaining factorial was equal to the factorial in the denominator. Students should use this same procedure to simplify other combinations when doing them without a graphing calculator.

Additional Example 1

A college administrator must select 4 candidates to interview for a teaching position from a group of 12 candidates. How many possible sets of candidates can be chosen from the set of 12?

The order in which the candidates are chosen is not important.

Use the combination formula with $n = 12$ and $r = 4$.

$${}_{12}C_4 = \frac{12!}{(12-4)!4!}$$

$$= \frac{\overset{3}{\cancel{12}} \cdot 11 \cdot \overset{5}{\cancel{10}} \cdot \overset{3}{\cancel{9}} \cdot \cancel{8!}}{\cancel{8!} \cdot \cancel{4} \cdot \cancel{3} \cdot \cancel{2} \cdot 1} = 495$$

There are 495 sets of four candidates from a set of 12.

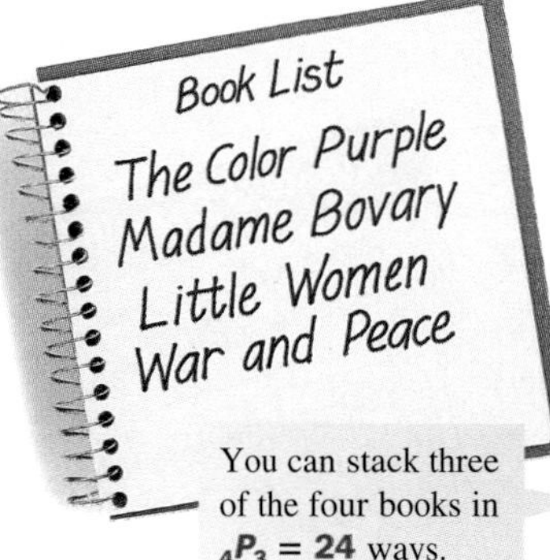

You can stack three of the four books in ${}_4P_3 = 24$ ways.

Suppose you are at a bookstore and have enough money to buy only three of the four books on your list. In how many ways can you choose three books? To find out, you could start by listing all the ways in which three of the four books can be stacked on the checkout counter.

T	T	M	M	L	L	M	M	L	L	W	W
M	L	L	T	T	M	L	W	W	M	M	L
L	M	T	L	M	T	W	L	M	W	L	M
L	L	W	W	T	T	W	W	T	T	M	M
W	T	T	L	L	W	T	M	M	W	W	T
T	W	L	T	W	L	M	T	W	M	T	W

Since it doesn't matter how you stack the books you buy, all of the permutations of ***W***, ***T***, and ***M*** are the same. There are ${}_3P_3 = 6$ permutations of ***W***, ***T***, and ***M***.

So there are $\frac{{}_4P_3}{{}_3P_3} = \frac{24}{6} = 4$ combinations of three books that you can buy.

The example above suggests the following general formula.

WATCH OUT!
A **permutation** is an arrangement of items in which **position is important**, and a **combination** is a selection of items in which **position is not important**.

Combinations

The number of combinations of r objects taken from a set of n objects is given by ${}_nC_r$ where

$${}_nC_r = \frac{{}_nP_r}{{}_rP_r} = \frac{n!}{(n-r)!r!}$$

EXAMPLE 1 Application: Manufacturing

A quality control engineer in a manufacturing plant is inspecting five of each set of 100 resistors to make sure that they are within tolerance limits. How many possible sets of five resistors can be chosen from a set of 100?

SOLUTION

Note that the order in which the resistors are chosen is not important.

Method 1 Use the combinations formula with $n = 100$ and $r = 5$.

$${}_{100}C_5 = \frac{100!}{(100-5)!5!}$$

$$= \frac{\overset{5}{\cancel{100}} \cdot \overset{33}{\cancel{99}} \cdot \overset{49}{\cancel{98}} \cdot 97 \cdot 96 \cdot \cancel{95!}}{\cancel{95!} \cdot \cancel{5} \cdot \cancel{4} \cdot \cancel{3} \cdot \cancel{2} \cdot 1}$$

$$= 75{,}287{,}520$$

Method 2 Use a calculator to find ${}_{100}C_5$.

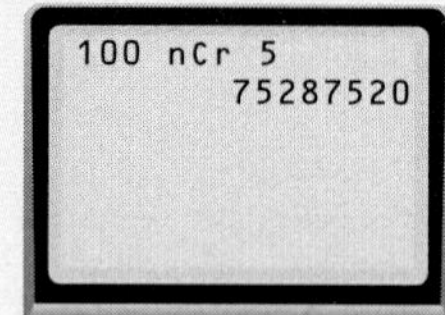

There are 75,287,520 sets of five resistors that can be chosen from a set of 100.

ANSWERS Section 12.4

Exploration

1–4. Results may vary, depending on the number of students in each group.

1. For a four-member group, there are $3 \times 2 = 6$ ways; for a five-member group, there are $4 \times 3 = 12$ ways.
2. For a four-member group, there are $4 \times 3 \times 2 = {}_4P_3$ ways; for a five-member group, there are $5 \times 4 \times 3 = {}_5P_3$ ways. If there are n people in the group, there are $n(n-1)(n-2)$ ways to award the three prizes, that is, ${}_nP_3$ ways.
3. 6 times; You can write 6 as 3! because there are 3! ways to order the names in each group of three.
4. For a four-member group, there are 4 ways, or $\frac{{}_4P_3}{6}$ ways, to award the prizes; for a five-member group, there are 10 ways, or $\frac{{}_5P_3}{6}$ ways, to award the prizes. The list in Step 2 gives the number of ways to award the prizes when the order of the names is important. If the order of the names is not important, as in Step 4, then the number of ways to award the prizes is equal to the answer in Step 2 divided by 3!.

Suppose on a rainy afternoon you can choose to see one of 5 movies **or** read one of 3 books. You have a total of $5 + 3 = 8$ activities to choose from. But if you want to see one of 5 movies **and then** read one of 3 books, you have $5 \cdot 3 = 15$ possible ways to plan the afternoon.

When you have more than one choice to make, you must decide whether to use the multiplication counting principle or whether to add the possibilities.

EXAMPLE 2

Jasmine wants to buy at least 4 bagels of different flavors. How many different combinations of bagels can she buy?

SOLUTION

Jasmine can buy 4 *or* 5 bagels, so add the possibilities. You must consider purchases of 4 bagels and 5 bagels separately.

Step 1 Suppose Jasmine buys 4 different bagels.

$${}_5C_4 = \frac{5!}{1!4!} = 5$$

Jasmine has **5** combinations of 4 flavors to choose from.

Step 2 Suppose Jasmine buys 5 different bagels.

$${}_5C_5 = \frac{5!}{0!5!} = \mathbf{1}$$

Jasmine has **1** *additional* combination to choose from.

Step 3 Jasmine has $\mathbf{5} + \mathbf{1} = 6$ different combinations of bagel flavors to choose from.

WATCH OUT!

Use the multiplication counting principle only when, for each choice from *m* things, you must choose from *n* more things. You shouldn't multiply in Step 3 because Jasmine will make one choice or the other, not both.

EXAMPLE 3

The school chorus has been invited to an international competition in Spain. How many different five-member fundraising committees can be formed from the 16 seniors, 14 juniors, and 10 sophomores in the chorus if the committee must have 2 seniors, 2 juniors, and 1 sophomore?

SOLUTION

The chorus must choose 2 seniors *and* 2 juniors *and* 1 sophomore for the committee, so use the multiplication counting principle.

The two seniors can be chosen in ${}_{16}C_2$ ways. | The two juniors can be chosen in ${}_{14}C_2$ ways. | The sophomore can be chosen in ${}_{10}C_1$ ways.

$$\begin{aligned} {}_{16}C_2 \cdot {}_{14}C_2 \cdot {}_{10}C_1 &= \frac{16!}{14!2!} \cdot \frac{14!}{12!2!} \cdot \frac{10!}{9!1!} \\ &= 120 \cdot 91 \cdot 10 \\ &= 109{,}200 \end{aligned}$$

There are 109,200 possible committees.

Teach⇔Interact

Additional Example 2

Don wants to ski on 2 or 3 of the 4 Saturdays in February. On how many different combinations of Saturdays can he ski?

Don can ski on 2 or 3 Saturdays, so add the possibilities. You must consider skiing on 2 Saturdays and 3 Saturdays separately.

Step 1 Suppose Don skis on 2 Saturdays. ${}_4C_2 = \frac{4!}{2!\,2!} = 6$, so he has 6 combinations of 2 Saturdays to choose from.

Step 2 Suppose Don skis on 3 Saturdays. ${}_4C_3 = \frac{4!}{1!\,3!} = 4$, so he has 4 additional combinations to choose from.

Step 3 Don has $6 + 4 = 10$ different combinations of Saturdays to choose from.

About Example 3

Problem Solving

Unlike the choices in Example 2, the choices in Example 3 are not "either or" choices. Instead, one choice follows another choice. When this is the case, the multiplication counting principle is required. Point out the use of the word *and* in the solution. If students can reword a problem using the word *and,* then the multiplication counting principle should be used. If the problem can be reworded using the word *or,* addition should be used.

Additional Example 3

A paint company is printing a card with chips of 12 different paints. How many different combinations of paint chips can be made from 6 shades of green chips, 7 shades of blue chips, and 5 shades of gray chips if the company wishes to use 4 different green chips, 5 different blue chips, and 3 different gray chips?

The paint company must choose 4 of the green chips *and* 5 of the blue chips *and* 3 of the gray chips for the card, so use the multiplication counting principle.

The 4 green chips can be chosen in ${}_6C_4$ ways. The 5 blue chips and be chosen in ${}_7C_5$ ways. The 3 gray chips can be chosen in ${}_5C_3$ ways.

$$\begin{aligned} {}_6C_4 \cdot {}_7C_5 \cdot {}_5C_3 &= \frac{6!}{2!\,4!} \cdot \frac{7!}{2!\,5!} \cdot \frac{5!}{2!\,3!} \\ &= 15 \cdot 21 \cdot 10 \\ &= 3150 \end{aligned}$$

There are 3150 possible cards.

Teach⇔Interact

Checking Key Concepts

Student Study Tip

For questions 1–3, point out to students that *arrangement* and *order* are often clue words for permutations. Similarly, *group* and *committee* are often clue words for combinations. Encourage students to try to reword each of the questions using one of these clue words.

Closure Question

Explain how you can use the formula ${}_nP_r$ to find the number of combinations of r objects from a set of n objects.

${}_nC_r = \frac{{}_nP_r}{{}_rP_r} = \frac{n!}{(n-r)!r!}$

Apply⇔Assess

Suggested Assignment

Core Course

Day 1 Exs. 1–20, 22–25

Day 2 Exs. 29–34, 36, 38, 41–56

Extended Course

Day 1 Exs. 1–25

Day 2 Exs. 26–56

Block Schedule

Day 78 Exs. 1–20

Day 79 Exs. 22–25, 29–34, 36, 38, 41–56

Exercise Notes

Challenge

Exs. 5–8 After examining the results of these exercises, ask students to make a prediction about ${}_8C_2$ and ${}_8C_6$. (${}_8C_2 = {}_8C_6$) This property is examined further in Ex. 24.

Reasoning

Exs. 13, 14 Students should be able to explain, using examples, why the situations in these exercises are examples of combinations rather than permutations.

Checking Key Concepts

For Questions 1–3, tell whether each situation describes *permutations* or *combinations*. Represent each number by an expression of the form ${}_nC_r$ or ${}_nP_r$.

1. the number of different three-filling burritos that you can choose if there are 9 fillings to choose from
2. the number of ways you can choose the four questions that you will answer on the test below
3. the number of orders in which you can answer four of the questions on the test below

> Name ______________________ Date ________
>
> **Answer four of the following six questions.**
>
> 1. Describa las comidas que le gustan.

4. Write an expression for the number of ways that you can choose one or two puppies from a litter of seven.
5. Write an expression for the number of committees of four girls and four boys that can be chosen from a class of 10 girls and 13 boys.
6. In the combinations formula on page 586, why does $\frac{{}_nP_r}{{}_rP_r} = \frac{n!}{(n-r)!r!}$?
7. Simplify each expression.

   a. ${}_7C_5$ b. ${}_7C_4$ c. ${}_7C_3$ d. ${}_{100}C_{98}$

12.4 Exercises and Applications

Extra Practice exercises on page 768

Simplify each expression.

1. ${}_{12}C_5$
2. ${}_8C_4$
3. ${}_{16}C_4$
4. ${}_{16}C_{12}$
5. ${}_8C_0$
6. ${}_8C_8$
7. ${}_8C_1$
8. ${}_8C_7$
9. ${}_5C_3 + {}_5C_2$
10. ${}_6C_2 + {}_8C_2$
11. ${}_7C_5 \cdot {}_7C_2$
12. ${}_nC_1$
13. **MANUFACTURING** A quality control engineer is inspecting 10 of each set of 500 computers to make sure that they work properly. How many possible sets of 10 computers can be chosen from a set of 500?
14. Tyrone has four kinds of lettuce growing in his garden. How many different salads can he make using two of the types of lettuce?

Visual Thinking For Exercises 15 and 16, suppose eight pegs are arranged in a circle on a board. You can stretch a rubber band around the pegs as shown.

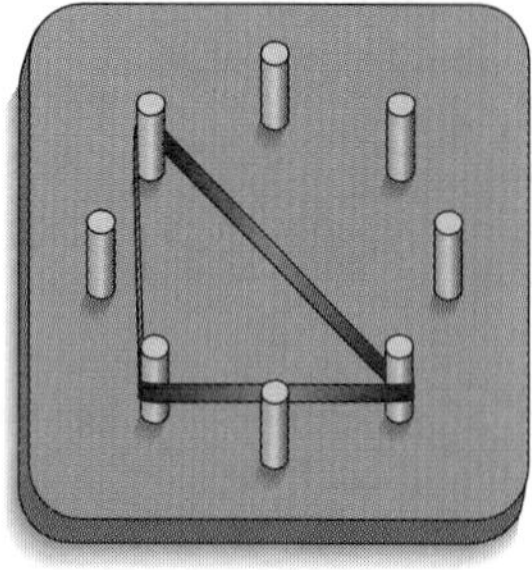

15. How many different triangles can be created? (*Note:* If two triangles have the same shape but are in different orientations, then consider them to be different triangles.)
16. How many quadrilaterals can be created?

Checking Key Concepts

1. combinations; ${}_9C_3$
2. combinations; ${}_6C_4$
3. permutations; ${}_6P_4$
4. ${}_7C_1 + {}_7C_2$
5. ${}_{10}C_4 \cdot {}_{13}C_4$
6. ${}_nP_r = \frac{n!}{(n-r)!}$ and ${}_rP_r = \frac{r!}{(r-r)!} = \frac{r!}{1} = r!$, so $\frac{{}_nP_r}{{}_rP_r} = \frac{n!}{(n-r)!} \div r! = \frac{n!}{(n-r)!} \cdot \frac{1}{r!} = \frac{n!}{(n-r)r!}$
7. a. 21
   b. 35
   c. 35
   d. 4950

Exercises and Applications

1. 792
2. 70
3. 1820
4. 1820
5. 1
6. 1
7. 8
8. 8
9. 20
10. 43
11. 441
12. n
13. ${}_{500}C_{10} \approx 2.458 \times 10^{20}$ sets
14. 6 different salads
15. 56 triangles

17. Karl wants either to go to the movies or to go bowling this evening. There are four movies and two kinds of bowling (tenpins and candlepins) to choose from. In how many different ways can Karl spend his evening?

18. Andrea wants to go bowling early in the evening and then see a movie later. She has two kinds of bowling and four movies to choose from. In how many ways can Andrea spend her evening?

19. Five bands that Sean likes have each released new music on both compact disc (CD) and cassette tape. Since tapes are less expensive than CDs, Sean plans to buy either three of the cassette tapes or two of the CDs. How many possible purchases does he have to choose from?

20. Trinh has roses, day lilies, clematis, lady's slippers, daisies, and poppies growing in her flower garden. She is making a bouquet to take to a neighbor. How many combinations of flowers can she have for the bouquet if she wants to include *at least* four different types of flowers?

21. **Open-ended Problem** For each expression, describe a situation in which the expression would be useful. Evaluate each expression and explain what the answer means in terms of the situation.

a. ${}_6C_4 + {}_8C_4$ **b.** ${}_6C_4 \cdot {}_8C_4$ **c.** ${}_6P_4$

Connection SPORTS

Singles figure skating competitions are composed of two parts: the *short program*, which consists of required moves, and the *long program*, which skaters choreograph themselves. Each skater in a Ladies' Intermediate Freeskating competition must include all of the following moves in her short program:

Jumps:
(1) an axel
(2) a double loop or double salchow
(3) a jump combination: a single jump and a double jump, or two double jumps

Spins:
(4) a camel, sit, or upright spin
(5) a spin combination

Step Sequences:
(6) a straight line, circular, or serpentine step sequence

22. Skaters typically satisfy requirement (3) by choosing from six kinds of jumps: the salchow, axel, loop, toe loop, lutz, and flip. Each of these may be either a single or a double jump, but the double jump used to fulfill requirement (2) may not be used again in the jump combination. If the order of the jumps does not matter, and two double jumps of the same type may be used, how many different jump combinations may a skater use in her program to satisfy requirement (3)?

23. Suppose a skater has 36 choices for the spin combination to satisfy requirement (5). In how many ways can she choose all of the moves to include in her short program?

Apply⇔Assess

Exercise Notes

Student Progress
Exs. 17–20 These exercises are straightforward. However, students having difficulty should identify each problem as being similar to either Example 2 or Example 3 and then follow that example as a model.

Cooperative Learning
Exs. 17–20 Students can work in groups to correct these exercises, beginning each one with a discussion of whether possibilities should be added or the multiplication counting principle used. In such a discussion, students will be able to reinforce their own correct reasoning or identify and correct any invalid reasoning.

Communication: Writing
Ex. 21 Asking students to write problems that can be solved using a particular expression involving combinations and permutations can help them solidify their understanding of the connections between words and these mathematical expressions.

Teaching Tip
Ex. 22 This exercise is more difficult than the others on this page because students must use both the multiplication counting principle and addition to calculate the number of jump combinations. You may want to give students the hint that they must consider the two types of jump combinations (single and double or two doubles) separately and then add the number of possibilities for each case. Ask students why the number of double jump combinations is $5 \cdot 5$ rather than ${}_5C_2$. (The jumps may be repeated.)

Communication: Listening
Exs. 22, 23 Students who may have studied or practiced figure skating can be an interesting resource for their classmates. For example, these students may know which jumps are done most often, or they may be able to report some special accomplishments that have been made by amateur skaters.

16. 70 quadrilaterals
17. 6 ways
18. 8 ways
19. 20 possible purchases
20. 22 combinations of flowers
21. Answers may vary. Examples are given.
    a. Choose 4 employees either from a group of 6 candidates or from another group of 8 candidates; 85. There are 85 ways to choose 4 employees either from a group of 6 candidates or from another group of 8 candidates.
    b. Choose 4 employees from a group of 6 candidates and choose another 4 employees from another group of 8 candidates; 1050. There are 1050 ways to choose 4 employees from a group of 6 candidates and choose another 4 employees from another group of 8 candidates.
    c. Choose the left fielder, right fielder, center fielder, and shortstop from a group of 6 baseball players; 360. There are 360 ways to assign the positions to 4 players chosen from a group of 6.
22. 55 different jump combinations
23. 35,640 ways

Apply⇔Assess

Exercise Notes

Alternate Approach

Ex. 24 Encourage students to use algebraic manipulations to show how ${}_nC_{n-r} = \frac{n!}{r!(n-r)!} = {}_nC_r$.

Common Error

Ex. 25 In part (c), students may not realize that as the players are being chosen from a single set of classmates, the number n in the combination will decrease appropriately as the positions are filled. Thus, the answer is ${}_{20}C_1 \cdot {}_{19}C_2 \cdot {}_{18}C_2$ rather than ${}_{20}C_1 \cdot {}_{20}C_2 \cdot {}_{20}C_2$.

Interdisciplinary Problems

Ex. 27 This exercise highlights the usefulness of using computers to perform tasks having a large finite number of steps. It also shows students how mathematical writing techniques can be used to estimate the time it takes for a computer to perform such a task. If students have difficulty with parts (a) and (b), have them consider the case where $n = 4$ and draw sample graphs.

Career Connection

Ex. 27 There are numerous job opportunities in fields associated with computers. Some people work directly in the production of computers, either as hardware engineers who design the computer, or software engineers who design and program software applications. Other people are involved with computer education. Still others use computer applications to aid them with their work. Some of these fields include automotive designers, who use computer-aided design (CAD), graphics designers, statisticians, accountants, writers, editors, and many others.

Topic Spiraling: Preview

Ex. 28 Students should be able to explain why the combination ${}_{11}C_0$ must be considered. This value previews the ways rows and row elements are numbered in Pascal's triangle, which is introduced in the next section.

24. Explain how ${}_nC_r$ is related to ${}_nC_{n-r}$. Give an example.

25. **SPORTS** Suppose your physical education class has 20 students.

a. In how many ways can a six-member volleyball team be chosen if players rotate so that all the positions are equivalent?

b. How many baseball teams can be arranged given that each of the nine players plays a different position on the field?

c. A basketball team consists of a center, two equivalent guards, and two equivalent forwards. In how many ways can a basketball team be chosen and positions assigned?

26. To open a combination lock like the one at the right, you dial a sequence of three numbers called the lock's *combination*. You turn the dial clockwise for the first number, counter-clockwise for the second, and clockwise again for the third.

a. How many *combinations* are possible for a lock like the one shown?

b. Explain why a "combination" lock is misnamed.

27. **COMPUTERS** Before giving a computer a task, such as coloring a graph like those on page 566, computer scientists may predict how long it might take.

a. In how many ways can you color a graph with n vertices using 4 colors? (For now, ignore the edges. Each vertex may be any color.)

b. At most, how many edges must you check to be sure that no vertices that share an edge are the same color? (Consider the worst-case scenario in which each pair of the n vertices is connected by an edge. Assume that no two vertices are connected by more than one edge.)

c. At most, how many edges must you check if you check every edge of every possible 4-color coloring of a graph with n vertices? What does it mean if each possible coloring of the graph has an edge that joins two vertices of the same color?

d. Suppose a computer can check 1000 edges each second. At most, how long will the computer take to check every edge of every 4-color coloring of a graph with 10 vertices? 17 vertices?

28. **ADVERTISING** A national pizza chain had the following promotion: If you buy two pizzas, you may receive up to five toppings on each pizza free. In the television advertisement for this promotion, a child and an adult are waiting to buy pizza.

If the pizza chain offers 11 toppings, how did the child get the answer 1,048,576? Do you think the child is right? Explain your answer.

CHILD: The possibilities are endless.

ADULT 1: Five plus five is ten.

(The child does some quick calculations on a notepad.)

CHILD: Actually, there are 1,048,576.

ADULT 1: Well, ten was just a ballpark figure.

ADULT 2: You got that right! Some ballpark!

24. ${}_nC_r = {}_nC_{n-r}$; Examples may vary. An example is given. ${}_5C_3 = 10 = {}_5C_2$

25. a. 38,760 ways
 b. ${}_{20}P_9 \approx 6.095 \times 10^{10}$ ways
 c. 465,120 ways

26. a. 64,000 combinations
 b. The order of the three numbers is important, so the choice of numbers is a permutation, not a combination.

27. a. 4^n
 b. ${}_nC_2 = \frac{n(n-1)}{2}$
 c. $4^n \cdot \frac{n(n-1)}{2}$; It means that more colors are necessary to color the graph so that no edge connects vertices of the same color.
 d. about 47,186 s, or 13.1 h; about 2,336,000,000 s, or about 74 years

28. The number of possibilities for one pizza are ${}_{11}C_0 + {}_{11}C_1 + {}_{11}C_2 + {}_{11}C_3 + {}_{11}C_4 + {}_{11}C_5 =$ 1024, so the child figured that the possibilities for two pizzas are $1024 \cdot 1024 = 1{,}048{,}576$. This would give the number of possibilities if the order of the pizzas mattered. Since the order does not matter, the actual number of possibilities is 524,288.

Connection GAMES

A standard deck of playing cards contains 52 cards, with 13 cards in each of four suits: clubs, diamonds, hearts, and spades. Many games begin with one player dealing out a *hand* of cards to each player. The order of the cards within a hand is unimportant.

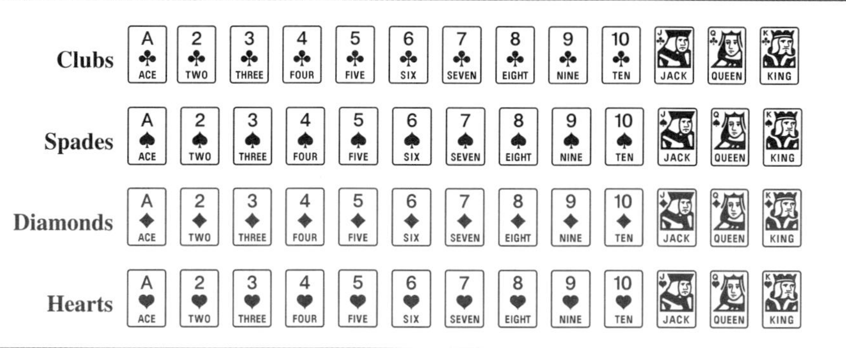

29. How many different 13-card hands can be dealt from a standard deck of cards?

30. In how many different orders can you play four cards from your 13-card hand?

31. How many 5-card hands contain a pair of aces? (*Hint:* The dealer must deal two of the four aces in the deck *and* three cards that are not aces.)

32. How many 5-card hands contain a pair of aces and three kings?

33. A hand with two cards of one value and three of another, such as in Exercise 32, is sometimes called a *full house*. A full house with 3 jacks and 2 fives is considered to be different from a full house with 3 fives and 2 jacks, but the suits of the cards are unimportant. How many different kinds of full houses are there?

34. How many 5-card hands contain a full house? Explain your answer. (*Note:* The two hands at the right are different.)

35. Writing How many different 5-card hands can be dealt from a standard deck of cards? Do you think it is likely that one of the 5-card hands in a 4-player game will be a full house? Explain.

36. FUNDRAISING Many states sell lottery tickets to raise money for education and other state expenses. If the numbers on your ticket match the numbers the state draws at random, then you win. In some lottery games you get to choose six numbers from 1 to 40 with no numbers repeated. The order in which you choose the numbers does not matter. How many possible number combinations can you choose? Do you think it is likely that you will choose the same six numbers that the state will draw?

Apply⇔Assess

Exercise Notes

Communication: Discussion
Exs. 29, 30 A discussion of the wording of these exercises should reveal that one exercise is asking for *different combinations* and the other exercise is asking for *different orders*. Once students recognize this distinction, they should see that one exercise involves combinations and the other involves permutations.

Problem Solving
Exs. 33, 34 When completing these exercises, students should be encouraged to write down a series of steps they would take to choose any one full house. For example, in Ex. 33, the first step might be: Choose the two values that make up the full house. The second step could be: Pick one of the two values to be the 3-card value. These steps can then be translated into combinations which should be multiplied to find the solution; in the case of Ex. 33, ${}_{13}C_2 \cdot {}_2C_1$.

29. 635,013,559,600 different hands

30. 17,160 different orders

31. 103,776 hands

32. 24 hands

33. 156 different kinds of full houses

34. 3744 different hands; Answers may vary. An example is given. There are ${}_{13}C_1$ ways to choose a value such as an ace, and ${}_4C_3$ ways to choose 3 of the 4 cards of that value from the deck. There are ${}_{12}C_1$ ways to choose another value, and ${}_4C_2$ ways to choose 2 of the 4 cards of that value from the deck. By the multiplication counting principle, there are ${}_{13}C_1 \cdot {}_4C_3 \cdot {}_{12}C_1 \cdot {}_4C_2 = 3744$ hands that contain a full house.

35. 2,598,960 different 5-card hands; It is not likely that a player will be dealt a full house, because, for example, the probability that the first hand dealt is a full house is $\frac{3744}{2,598,960} \approx 0.00144$.

36. 3,838,380 combinations; No.

Exercise Notes

Communication: Reading
Ex. 37 Before students begin this exercise, they should clarify the meaning of "some of the dishes may be the same."

Student Study Tip
Ex. 38 In part (d), point out to students that asking for the number of samples containing at least one defective chip is the same as asking for all the samples except those that contain only good chips.

Assessment Note
Ex. 41 If students trade papers with a partner and check each other's work, they will have an opportunity to clarify their understanding of the relationship between permutations and combinations.

Practice 80 for Section 12.4

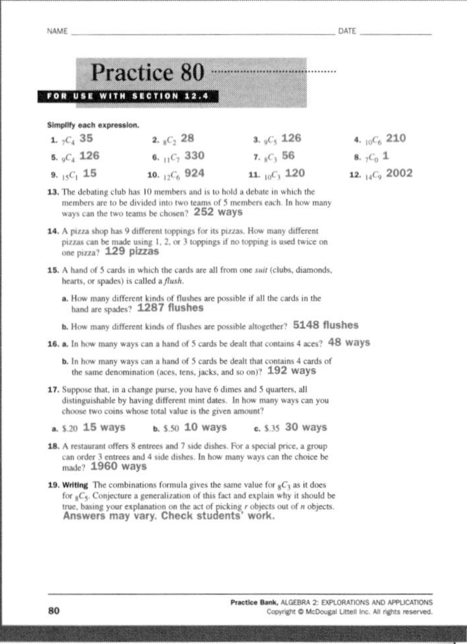

NAME ________________ DATE ________

Practice 80

FOR USE WITH SECTION 12.4

Simplify each expression.

1. $_7C_4$ 35
2. $_8C_2$ 28
3. $_9C_5$ 126
4. $_{10}C_6$ 210
5. $_9C_4$ 126
6. $_{11}C_7$ 330
7. $_8C_3$ 56
8. $_7C_0$ 1
9. $_{15}C_1$ 15
10. $_{12}C_6$ 924
11. $_{10}C_3$ 120
12. $_{14}C_9$ 2002
13. The debating club has 10 members and is to hold a debate in which the members are to be divided into two teams of 5 members each. In how many ways can the two teams be chosen? **252 ways**
14. A pizza shop has 9 different toppings for its pizzas. How many different pizzas can be made using 1, 2, or 3 toppings if no topping is used twice on one pizza? **129 pizzas**
15. A hand of 5 cards in which the cards are all from one *suit* (clubs, diamonds, hearts, or spades) is called a *flush*.
    a. How many different kinds of flushes are possible if all the cards in the hand are spades? **1287 flushes**
    b. How many different kinds of flushes are possible altogether? **5148 flushes**
16. a. In how many ways can a hand of 5 cards be dealt that contains 4 aces? **48 ways**
    b. In how many ways can a hand of 5 cards be dealt that contains 4 cards of the same denomination (aces, tens, jacks, and so on)? **192 ways**
17. Suppose that, in a change purse, you have 6 dimes and 5 quarters, all distinguishable by having different mint dates. In how many ways can you choose two coins whose total value is the given amount?
    a. \$.20 **15 ways** b. \$.50 **10 ways** c. \$.35 **30 ways**
18. A restaurant offers 8 entrees and 7 side dishes. For a special price, a group can order 3 entrees and 4 side dishes. In how many ways can the choice be made? **1960 ways**
19. **Writing** The combinations formula gives the same value for $_8C_3$ as it does for $_8C_5$. Conjecture a generalization of this fact and explain why it should be true, basing your explanation on the act of picking r objects out of n objects. **Answers may vary. Check students' work.**

80 Practice Bank, ALGEBRA 2: EXPLORATIONS AND APPLICATIONS

37. Challenge Some friends are at an Ethiopian restaurant that offers the specials shown. In how many ways can they choose four dishes to share if some of the dishes may be the same?

Today's Specials
1. Minchet Abish
2. T'ibs We't
3. Yesiga T'ibs
4. Dullet
5. Kitfo
6. Ye'assa We't
7. Doro We't
8. Yenqulal We't
9. Dinich We't
10. Ye'atakilt Alich'a

38. INDUSTRY Of 20 computer chips, a sample of two will be tested. Suppose that two of the 20 are defective and the other 18 are good.

a. How many different samples can be selected?

b. How many samples contain only good chips?

c. How many samples contain only defective chips?

d. How many samples contain at least one defective chip? (*Hint:* Use your answers to parts (a) and (b).)

39. Challenge Prove that $_{n+1}C_r = {}_nC_{r-1} + {}_nC_r$. (*Hint:* Rewrite the right-hand side of the equation with a common denominator and then factor the numerator.)

40. Writing Describe how to tell, for a given situation, when to add numbers of combinations and when to multiply numbers of combinations to determine the total number of possibilities.

ONGOING ASSESSMENT

41. a. Writing How are combinations and permutations similar? How are they different? Use examples in your answer.

b. Explain what $_nC_r$ and $_nP_r$ represent. How can you find $_nC_r$ if you know $_nP_r$? Use an example in your answer.

SPIRAL REVIEW

Simplify each expression. *(Section 12.3)*

42. $_5P_3$ **43.** $_6P_3$ **44.** $_{30}P_2$

45. $_8P_8$ **46.** $_8P_0$ **47.** $_{12}P_{10}$

Multiply. *(Section 9.2)*

48. $(5x + 2)(2x - 1)$ **49.** $(2t - 1)^2$ **50.** $(m - 3)(m^2 - 6m + 9)$

51. $(n - 2)^3$ **52.** $(n - 2)^4$ **53.** $(y + 1)^4$

Find an equation for each cubic function whose graph is shown. *(Section 9.4)*

54.

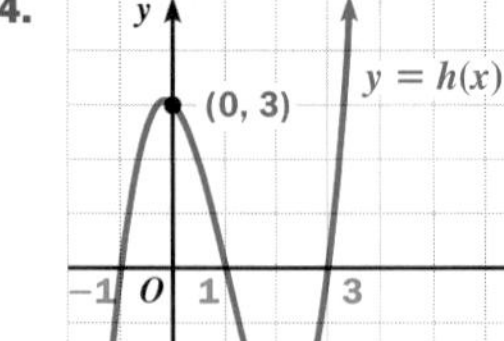

55.

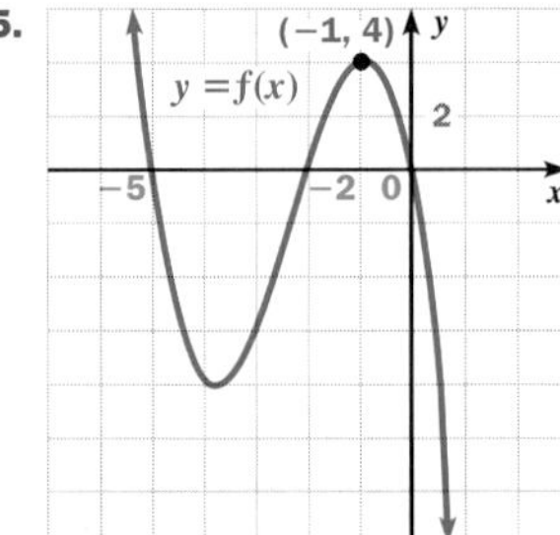

56.

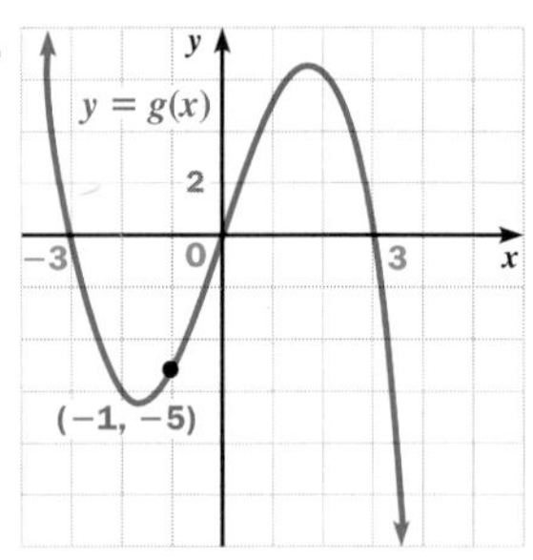

37. 715 ways

38. a. 190 samples b. 153 samples
c. 1 sample d. 37 samples

39. $_nC_{r-1} + {}_nC_r = \dfrac{n!}{(r-1)!(n-(r-1))!} + \dfrac{n!}{r!(n-r)!} =$

$\dfrac{n!}{(r-1)!(n-r+1)!} + \dfrac{n!}{r!(n-r)!} =$

$\dfrac{n!}{(r-1)!(n-r+1)(n-r)!} + \dfrac{n!}{r(r-1)!(n-r)!} =$

$\dfrac{r \cdot n! + (n-r+1)\cdot n!}{r(r-1)!(n-r+1)(n-r)!} = \dfrac{(r+n-r+1)\cdot n!}{r(r-1)!(n-r+1)(n-r)!} =$

$\dfrac{(n+1)\cdot n!}{r(r-1)!(n-r+1)(n-r)!} =$

$\dfrac{(n+1)!}{r!(n-r+1)!} = \dfrac{(n+1)!}{r!(n+1-r)!} =$

$_{n+1}C_r$

40. Answers may vary. An example is given. Combinations should be added when the problem involves one choice or another, but not both. When successive choices are made, combinations should be multiplied.

41. See answers in back of book.

42. 60

43. 120

44. 870

45. 40,320

46. 1

47. 239,500,800

48. $10x^2 - x - 2$

49. $4t^2 - 4t + 1$

50. $m^3 - 9m^2 + 27m - 27$

51. $n^3 - 6n^2 + 12n - 8$

52. $n^4 - 8n^3 + 24n^2 - 32n + 16$

53. $y^4 + 4y^3 + 6y^2 + 4y + 1$

54. $h(x) = (x + 1)(x - 1)(x - 3)$

55. $f(x) = -x(x + 5)(x + 2)$

56. $g(x) = -\frac{5}{8}x(x + 3)(x - 3)$

Section

12.5 Pascal's Triangle

Learn how to...

- apply the concept of combinations to Pascal's triangle

So you can...

- easily count the outcomes of sequences of events for which there are exactly two possibilities, such as left and right turns

In the nineteenth century a device similar to the one shown was used to illustrate ideas about probability. A marble dropped into the device hits a sequence of pins and then falls into a slot at the bottom. The marble is equally likely to turn left or right when it hits each pin.

THINK AND COMMUNICATE

1. a. Express the paths of the red marble and the blue marble as sequences of left turns (L) and right turns (R).
   b. Make a list of all possible paths that lead to the slot containing each of the two marbles shown.
   c. For a falling marble to reach the slot that the blue marble is in, it must make exactly two right turns at any combination of two out of the six rows of pins that it falls through. What is ${}_6C_2$? How does this relate to your list from part (b)? What values of n and r in ${}_nC_r$ apply to the paths that lead to the slot where the red marble is?
   d. Find the number of paths that lead to each of the other slots.

If you count the number of paths that a marble can follow to reach each pin, you get this result:

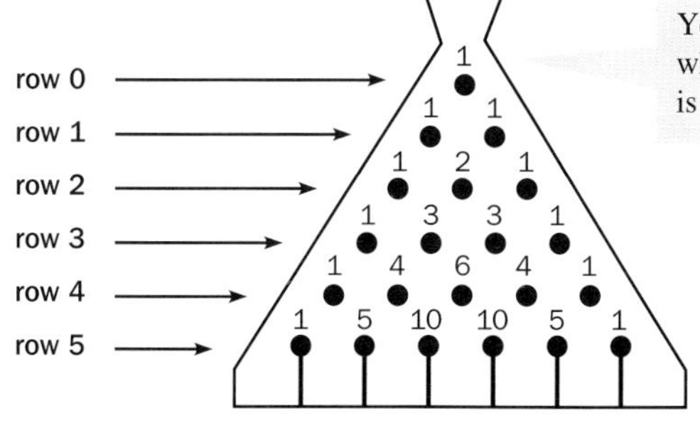

You will see later why the first row is called row 0.

This triangular display of numbers is called **Pascal's triangle**, named after French mathematician Blaise Pascal (1623–1662), who published it in 1653.

BY THE WAY

Pascal's remarkable achievements included publishing his first geometric theorem at the age of 12 and inventing an adding machine at the age of 19.

Plan⇔Support

Objectives

- Apply the concept of combinations to Pascal's triangle.
- Easily count the outcomes of sequences of events for which there are exactly two possibilities.

Recommended Pacing

Core and Extended Courses
Section 12.5: 2 days

Block Schedule
Section 12.5: 2 half-blocks (with Section 12.4 and Review)

Resource Materials

Lesson Support
Lesson Plan 12.5
Warm-Up Transparency 12.5
Practice Bank: Practice 81
Study Guide: Section 12.5
Assessment Book: Test 50

Technology
Technology Book: Spreadsheet Activity 12
Graphing Calculator
Internet: http://www.hmco.com

Warm-Up Exercises

Find the number of combinations.

1. ${}_8C_2$ 28
2. ${}_8C_6$ 28

Expand each polynomial.

3. $(x-2)^2$ x^2-4x+4
4. $(3a+2b)^2$ $9a^2+12ab+4b^2$
5. $(x+3)^3$ $x^3+9x^2+27x+27$

ANSWERS Section 12.5

Think and Communicate

1. a. red marble: R-R-R-R-R-R; blue marble: L-L-R-L-L-R
   b. red: R-R-R-R-R-R; blue: R-R-L-L-L-L, R-L-R-L-L-L, R-L-L-R-L-L, R-L-L-L-R-L, R-L-L-L-L-R, L-R-R-L-L-L, L-R-L-R-L-L, L-R-L-L-R-L, L-R-L-L-L-R, L-L-R-R-L-L, L-L-R-L-R-L, L-L-R-L-L-R, L-L-L-R-R-L, L-L-L-R-L-R, L-L-L-L-R-R
   c. 15; It is the number of paths for the blue marble; $n = 6$ and $r = 6$.
   d.

Slot no.	1	2	3	4	5	6	7
No. of paths	${}_6C_0 = 1$	${}_6C_1 = 6$	${}_6C_2 = 15$	${}_6C_3 = 20$	${}_6C_4 = 15$	${}_6C_5 = 6$	${}_6C_6 = 1$

Additional Example 1

Generate row 8 of Pascal's triangle.

Method 1 Calculate $_nC_r$ for each term.

$_8C_0$ $_8C_1$ $_8C_2$ $_8C_3$ $_8C_4$ $_8C_5$ $_8C_6$ $_8C_7$ $_8C_8$
1 8 28 56 70 56 28 8 1

Method 2 Add pairs of terms from the previous row.

1 7 21 35 35 21 7 1
1 8 28 56 70 56 28 8 1

Think and Communicate

Questions 4 and 5 prepare students for the statement of the binomial theorem. In anticipation of this, you might ask students why they think using Pascal's triangle would not be an efficient method of expanding $(a + b)^{15}$.

Section Note

Alternate Approach
For some students, the connection between $_nC_r$ and the coefficients of terms in a binomial expansion may not be obvious. These students might consider the following explanation. In the expansion $(a + b)^3 = (a + b)(a + b)(a + b)$, the variable a occurs 3 times, once in each factor. The first term of the expansion comes from the product of these 3 a's. The product a^3 can occur in just this one way, by taking an a from each factor in the expansion. In contrast, the term a^2b can occur in more than one way. For example, a^2b might be the product of the first two a's and the last b, or it might be the product of the last two a's and the first b. Since we are choosing 2 a's from the expansion of 3 factors, there are $_3C_2$ a^2b terms. Thus, the second term in the expansion of $(a + b)^3$ is $3a^2b$.

About Example 2

Common Error
When the coefficient of a term in a binomial to be expanded is not equal to one, students often forget to consider the coefficient when expanding the binomial. In part (b) of the Example, call students' attention to the fact that $3p$ and $-2q$, and not just p and q, must be raised to each power.

As you saw in *Think and Communicate* Question 1, term r in row n of Pascal's triangle, where r and n start at 0, is given by $_nC_r$.

THINK AND COMMUNICATE

2. Look closely at Pascal's triangle on the previous page to complete this statement: $_5C_4 = {_4C_3} + {_4C_{?}}$. Then verify that the statement is true.

3. Use the pattern in Question 2 to write any term $_nC_r$ $(0 < r < n)$ of Pascal's triangle as a sum of terms from the preceding row. (*Hint:* The term $_nC_r$ is in row n of the triangle, so the preceding row is row $n - 1$.)

EXAMPLE 1

Generate row 7 of Pascal's triangle.

SOLUTION

Remember, r starts at 0.

Method 1 Calculate $_nC_r$ for each term.

$_7C_0$	$_7C_1$	$_7C_2$	$_7C_3$	$_7C_4$	$_7C_5$	$_7C_6$	$_7C_7$
1	7	21	35	35	21	7	1

These are the numbers you found in part (d) of *Think and Communicate* Question 1.

Method 2 Add pairs of terms from the previous row.

1 6 15 20 15 6 1
1 7 21 35 35 21 7 1

Each row begins and ends with 1.

Pascal's triangle is useful in *expanding*, or multiplying out, powers of binomials. Here are some expansions of the powers of $(a + b)$:

$$(a + b)^0 = 1$$
$$(a + b)^1 = a + b$$
$$(a + b)^2 = a^2 + 2ab + b^2$$
$$(a + b)^3 = a^3 + 3a^2b + 3ab^2 + b^3$$
$$(a + b)^4 = a^4 + 4a^3b + 6a^2b^2 + 4ab^3 + b^4$$
$$(a + b)^5 = a^5 + 5a^4b + 10a^3b^2 + 10a^2b^3 + 5ab^4 + b^5$$

THINK AND COMMUNICATE

4. What do you notice about the coefficients of the terms in the expansions? What do you notice about the exponents of a and b?

5. What do you think the expansion of $(a + b)^6$ is?

The patterns in the expansions of $(a + b)^0$, $(a + b)^1$, $(a + b)^2$, and so on are summarized by the **binomial theorem**, which allows you to expand any binomial without using the distributive property repeatedly.

Binomial Theorem

For any positive integer n,

$$(a + b)^n = {_nC_0}a^n + {_nC_1}a^{n-1}b + {_nC_2}a^{n-2}b^2 + \cdots + {_nC_{n-1}}ab^{n-1} + {_nC_n}b^n$$

Think and Communicate

2. 4; $_5C_4 = 5$, $_4C_3 = 4$, and $_4C_4 = 1$
3. $_nC_r = {_{n-1}C_{r-1}} + {_{n-1}C_r}$
4. The coefficients of $(a + b)^n$ are the terms of the nth row of Pascal's triangle. The exponents of a begin with n and decrease to 0. The exponents of b begin with 0 and increase to n. The sum of the exponents in each term is n.
5. $a^6 + 6a^5b + 15a^4b^2 + 20a^3b^3 + 15a^2b^4 + 6ab^5 + b^6$

EXAMPLE 2

Expand each binomial.

a. $(a + b)^7$ **b.** $(3p - 2q)^3$

SOLUTION

a. Use the binomial theorem. Let $n = 7$.

$$(a + b)^7 = {}_7C_0a^7 + {}_7C_1a^6b + {}_7C_2a^5b^2 + {}_7C_3a^4b^3 + {}_7C_4a^3b^4 + {}_7C_5a^2b^5 + {}_7C_6ab^6 + {}_7C_7b^7$$
$$= a^7 + 7a^6b + 21a^5b^2 + 35a^4b^3 + 35a^3b^4 + 21a^2b^5 + 7ab^6 + b^7$$

b. Use the binomial theorem. Let $a = 3p$, $b = -2q$, and $n = 3$.

$$(3p - 2q)^3 = {}_3C_0(3p)^3 + {}_3C_1(3p)^2(-2q) + {}_3C_2(3p)(-2q)^2 + {}_3C_3(-2q)^3$$
$$= 3^3p^3 + 3(3^2)(-2)p^2q + 3(3)(-2)^2pq^2 + (-2)^3q^3$$
$$= 27p^3 - 54p^2q + 36pq^2 - 8q^3$$

CHECKING KEY CONCEPTS

1. Row 7 of Pascal's triangle is 1, 7, 21, 35, 35, 21, 7, 1. Find row 8.
2. Terms 7 and 8 of row 14 of Pascal's triangle are 3432 and 3003, respectively. Use these numbers to find ${}_{15}C_8$.
3. Suppose you toss a coin 10 times and get this sequence of heads (H) and tails (T): HTTHHHTTTH. How many such sequences of ten coin tosses show exactly five heads and five tails?

Give the values of *a*, *b*, and *n* that you would substitute in the binomial theorem to expand each power of a binomial.

4. $(x - y)^6$
5. $(4x + y)^{12}$
6. $(\pi p - 4q)^m$

12.5 Exercises and Applications

Extra Practice exercises on page 768

1. Find row 9 of Pascal's triangle.
2. Suppose you toss a coin 12 times and get this sequence of heads (H) and tails (T): HTTHHHHTTTHHT. How many such sequences of twelve coin tosses show exactly six heads and six tails?

3. **Writing** Look back at the device shown at the top of page 593. What do you notice about the number of paths that involve one right turn out of six turns and the number of paths that involve five right turns out of six turns? Explain why Pascal's triangle has this symmetry. Compare paths with r right turns and $6 - r$ right turns.

Teach⇔Interact

Additional Example 2

Expand each binomial.

a. $(c - d)^6$

Use the binomial theorem. Let $a = 1c$, $b = -1d$, and $n = 6$.

$(c - d)^6$
$= {}_6C_0\,c^6 + {}_6C_1\,c^5(-1d) + {}_6C_2\,c^4(-1d)^2 + {}_6C_3\,c^3(-1d)^3 + {}_6C_4\,c^2(-1d)^4 + {}_6C_5\,c(-1d)^5 + {}_6C_6\,(-1d)^6$
$= c^6 - 6c^5d + 15c^4d^2 - 20c^3d^3 + 15c^2d^4 - 6cd^5 + d^6$

b. $(2r + 4s)^4$

Use the binomial theorem. Let $a = 2r$, $b = 4s$, and $n = 4$.

$(2r + 4s)^4$
$= {}_4C_0\,(2r)^4 + {}_4C_1\,(2r)^3(4s) + {}_4C_2\,(2r)^2(4s)^2 + {}_4C_3\,(2r)(4s)^3 + {}_4C_4\,(4s)^4$
$= 2^4r^4 + 4(2^3)(4)r^3s + 6(2^2)(4^2)r^2s^2 + 4(2)(4^3)rs^3 + 4^4s^4$
$= 16r^4 + 128r^3s + 384r^2s^2 + 512rs^3 + 256s^4$

Closure Question

How can you use Pascal's triangle or the binomial theorem to find the coefficient of the fifth term in the binomial expansion of $(x + y)^{12}$?

In Pascal's triangle, the fifth term in row 12 of the triangle, ${}_{12}C_4$, is the coefficient of the fifth term in the expansion.

Apply⇔Assess

Suggested Assignment

Core Course
Day 1 Exs. 1–19
Day 2 Exs. 29, 30, 32–35, 37–46, AYP

Extended Course
Day 1 Exs. 1–24
Day 2 Exs. 25–46, AYP

Block Schedule
Day 79 Exs. 1–19
Day 80 Exs. 29, 30, 32–35, 37–46, AYP

Exercise Notes

Second-Language Learners

Ex. 3 Students learning English may benefit from discussing their findings with a peer tutor before writing their explanations.

Checking Key Concepts

1. 1, 8, 28, 56, 70, 56, 28, 8, 1
2. 6435
3. 252 sequences
4. $a = x$; $b = -y$; $n = 6$
5. $a = 4x$; $b = y$; $n = 12$
6. $a = \pi p$; $b = -4q$; $n = m$

Exercises and Applications

1. 1, 9, 36, 84, 126, 126, 84, 36, 9, 1
2. 924 sequences
3. Each is equal to 6. Pascal's triangle is symmetric because ${}_nC_r = {}_nC_{n-r}$. The number of paths with r right turns is the same as the number of paths with $6 - r$ right turns: ${}_6C_r = \dfrac{6!}{(6-r)!r!}$ and ${}_6C_{6-r} = \dfrac{6!}{(6-(6-r))!(6-r)!} = \dfrac{6!}{(6-r)!r!} = {}_6C_r$.

Apply⇔Assess

Exercise Notes

Reasoning

Exs. 4–7 Students should recognize that for any term of a binomial expansion ${}_nC_r a^x b^y$, $x + y = n$. They should also be able to explain why this is true.

Student Study Tip

Exs. 8–11, 15, 17 Mention to students that when the coefficients of a and b in the binomial $(a + b)^n$ are both 1, and n is reasonably small (for example, less than 10), they can use Pascal's triangle to find the coefficients of the binomial expansion of $(a + b)^n$.

Challenge

Ex. 18 Ask students to explain why ${}_nC_r = {}_nC_{n-r}$. (There are just as many ways of choosing a group of r from a set of n as there are of choosing a group of $n - r$ from the set of n to be left behind.)

Integrating the Strands

Ex. 20 Part (c) of this exercise can be used to relate the concept of combinations to the number of subsets a set has in set theory. The sum of the elements in each row of Pascal's triangle can be expressed as a power of two. For example, in row 4 of the triangle, $1 + 4 + 6 + 4 + 1 = 16 = 2^4$. If students were to list the subsets of a set having 4 elements, they would discover that the number of subsets having 0 elements is 1, the number of subsets having 1 element is 4, the number of subsets having 2 elements is 6, the number of subsets having 3 elements is 4, and the number of subsets having 4 elements is 1. Thus, the total number of distinct subsets of a set having 4 elements is 2^4 or 16.

Research

Exs. 21–24 The ancient Greeks were intrigued by numbers that represented geometric shapes. These numbers are called *figurate numbers. Square* numbers are one kind of figurate number. Ask students to research square numbers and then find where in the grid for Exs. 21–24 the sum of groups of numbers are square numbers. (For example, $1 + 3 = 2^2$, $3 + 6 = 3^2$, $6 + 10 = 4^2$. These pairs of numbers can be found in both row C and column 3 of the grid, which represent the third diagonal of Pascal's triangle.)

Use the binomial theorem to find the value of *k* for each of the following terms of a binomial expansion.

4. ${}_6C_4a^{6-k}b^k$ **5.** ${}_5C_ka^3b^2$ **6.** ${}_kC_4a^7b^4$ **7.** ${}_{14}C_6a^kb^{14-k}$

8. Find the coefficient of a^5b^4 in the expansion of $(a + b)^9$.

9. Find the coefficient of x^3y^8 in the expansion of $(x + y)^{11}$.

Use the binomial theorem to expand each power of a binomial.

10. $(x + y)^8$ **11.** $(x - y)^5$ **12.** $(3p + 4q)^4$ **13.** $(2y - 3z)^5$

14. $(a + 5b)^4$ **15.** $(m^2 - n)^7$ **16.** $(y - 2z^2)^5$ **17.** $(x^2 + y^3)^6$

18. a. Find the first three terms in the expansion of $(m + n)^{20}$.

b. Find the last three terms in the expansion of $(m + n)^{20}$.

19. One term in the expansion of $(a + b)^n$ is $66a^{10}b^2$.

a. What is the value of n?

b. Express 66 in the form ${}_nC_r$.

c. What is the next term of the expansion?

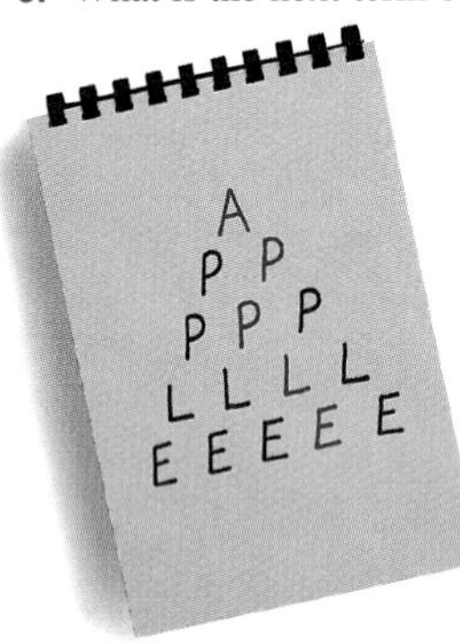

20. You can make a triangle out of a word by repeating the letters as shown in the arrangement of APPLE at the left. Then you can reconstruct the word by beginning at the top and choosing a letter in each row as you follow a path downward (such as the one shown in red) by turning right or left at each row.

a. Explain why this situation is related to Pascal's triangle.

b. How many ways are there to reconstruct APPLE?

c. Express the number in part (b) as a sum of combinations.

d. Make a word triangle from SUBTRACT. How many ways are there to reconstruct the word? To what sum of combinations does this correspond?

RECREATION Every day, Anchara jogs from her house at the intersection of 1st Street and Avenue A to the park entrance at the intersection of 6th Street and Avenue F. She runs so that she is always getting closer to the park, but she likes to take a different path as often as possible.

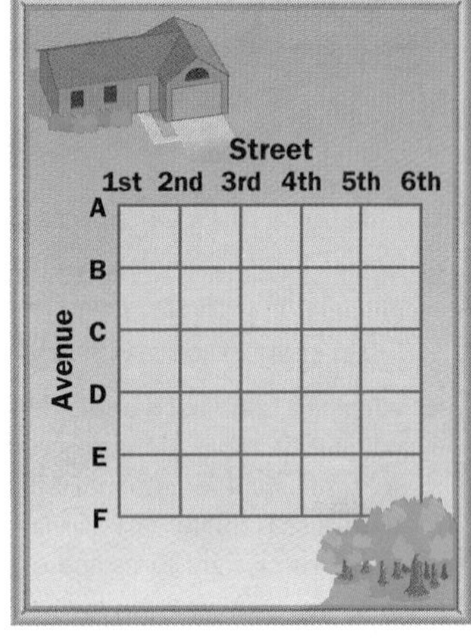

21. There are two paths from Anchara's house to 2nd and B: one passes through 1st and B, and one passes through 2nd and A. There are three paths from her house to 3rd and B: two paths pass through 2nd and B, and one passes through 3rd and A. Give a similar argument for the number of paths to 3rd and C.

22. Sketch a copy of the grid, and use the method in Exercise 21 to label each intersection with the number of different paths from Anchara's house to that intersection. What do you notice?

23. The distance from Anchara's house to the park is 10 blocks. To get to the park, how many times must she travel to a higher-numbered street, and how many times must she travel to a higher-lettered avenue? Use these facts to express the number of different paths to the park as a combination.

24. If Anchara also jogs back home from the park, how many different round-trip paths can she choose from? Explain.

4. 4 5. 2
6. 11 7. 8
8. 126 9. 165
10. $x^8 + 8x^7y + 28x^6y^2 + 56x^5y^3 + 70x^4y^4 + 56x^3y^5 + 28x^2y^6 + 8xy^7 + y^8$
11. $x^5 - 5x^4y + 10x^3y^2 - 10x^2y^3 + 5xy^4 - y^5$
12. $81p^4 + 432p^3q + 864p^2q^2 + 768pq^3 + 256q^4$
13. $32y^5 - 240y^4z + 720y^3z^2 - 1080y^2z^3 + 810yz^4 - 243z^5$
14. $a^4 + 20a^3b + 150a^2b^2 + 500ab^3 + 625b^4$
15. $m^{14} - 7m^{12}n + 21m^{10}n^2 - 35m^8n^3 + 35m^6n^4 - 21m^4n^5 + 7m^2n^6 - n^7$
16. $y^5 - 10y^4z^2 + 40y^3z^4 - 80y^2z^6 + 80yz^8 - 32z^{10}$
17. $x^{12} + 6x^{10}y^3 + 15x^8y^6 + 20x^6y^9 + 15x^4y^{12} + 6x^2y^{15} + y^{18}$
18. a. $m^{20} + 20m^{19}n + 190m^{18}n^2$
b. $190m^2n^{18} + 20mn^{19} + n^{20}$
19. a. 12
b. ${}_{12}C_2$
c. $220a^9b^3$
20. a. Answers may vary. An example is given. The number of possible paths is the same as the numbers in Pascal's triangle.
b. 16 ways
c. ${}_4C_0 + {}_4C_1 + {}_4C_2 + {}_4C_3 + {}_4C_4$

Connection HISTORY

Some 350 years before Pascal, Chinese mathematicians Yang Hui and Chu Shih Chieh gave a detailed treatment of what is now called Pascal's triangle. Their work was based on another Chinese work from about A.D. 1050.

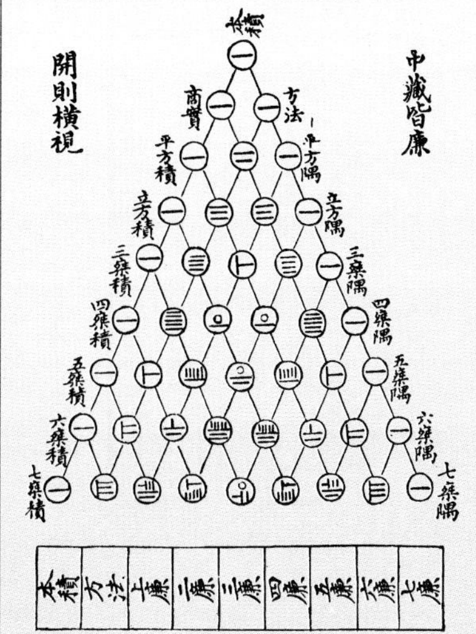

25. Examine the symbols used in the Chinese triangle.

a. What number does the symbol ⊢ represent?

b. What number does the symbol ⊥ represent?

c. How do you think the number 77 is written using the Chinese symbols? Explain your reasoning.

The Chinese discovered how to use the terms of the triangle to find roots of large numbers. Use the following version of the ancient Chinese method to complete Exercises 26 and 27.

To find an approximation for $\sqrt{465}$, first note that $21^2 < 465 < 22^2$, so $21 < \sqrt{465} < 22$. Then let $\sqrt{465} = 21 + d$, where d is the decimal part of the root $(0 < d < 1)$. So:

$$(21 + d)^2 = 465$$

Use Pascal's triangle to expand $(21 + d)^2$.

$$21^2 + 2(21)d + d^2 = 465$$

$$441 + 42d + d^2 = 465$$

$$42d + d^2 = 24$$

$$d(d + 42) = 24$$

$$d = \frac{24}{d + 42}$$

Solve for d in terms of itself.

If you let $d = 0$, you get $\sqrt{465} \approx 21 + \frac{24}{0 + 42} = 21\frac{4}{7}$, an overestimate of $\sqrt{465}$.

If you let $d = 1$, you get $\sqrt{465} \approx 21 + \frac{24}{1 + 42} = 21\frac{24}{43}$, an underestimate of $\sqrt{465}$.

26. Find an approximation for $\sqrt{301}$.

27. Challenge Find an approximation for $\sqrt[3]{623}$.

28. A diagonal of Pascal's triangle consists of the numbers ${}_nC_r$, where r is constant and n varies. You can generate one diagonal from another.

a. Describe the numbers on the diagonal given by the terms ${}_nC_1$.

b. The numbers on the diagonal given by the terms ${}_nC_2$ are called the *triangular numbers* because they give the total number of objects in a triangular arrangement of objects. Describe how to find the triangular numbers as sums of the terms in the diagonal given by the terms ${}_nC_1$.

c. Find the sum of the first n terms along any diagonal. Where can that sum be found in Pascal's triangle? Do you think this always works? Why?

Apply⇔Assess

Exercise Notes

Multicultural Note

Ex. 25 Ancient Chinese mathematicians discovered, explored, and employed a number of mathematical principles and concepts long before Western mathematicians began considering them. Pascal's triangle is one such concept. Other examples include the Chinese use of the decimal system in the fourteenth century B.C.; their use of an empty place for zero, dating as far back as the fourth century B.C.; and their advances in computing the value of pi in the third century A.D.

Historical Connection

Ex. 25 Blaise Pascal, together with another mathematician, Pierre de Fermat, developed the field of mathematics called *probability theory,* which applies to all games of chance. Probability theory, in turn, has led to the development of *game theory,* which is used in economics, military strategy, and voting systems.

Challenge

Exs. 26, 27 There are several places in the Chinese triangle where square numbers can be found. One of the places is along a diagonal. For example ${}_2C_0 + {}_3C_1 = 2^2$, ${}_3C_1 + {}_4C_2 = 3^2$, ${}_4C_2 + {}_5C_3 = 4^2$, and ${}_5C_3 + {}_6C_4 = 5^2$. Ask students to study this pattern and find a place in the Chinese triangle where they could find 21^2 and 22^2. (${}_{21}C_{19} + {}_{22}C_{20} = 21^2$; ${}_{22}C_{20} + {}_{23}C_{21} = 22^2$)

Communication: Discussion

Ex. 28 In part (c), a discussion of why the sum can always be found in the same position relative to its addends may help those students who do not understand the logic of the pattern.

d.

S
U U
B B B
T T T T
R R R R R
A A A A A A
C C C C C C C
T T T T T T T T

There are 128 ways to reconstruct the word: ${}_7C_0 + {}_7C_1 + {}_7C_2 + {}_7C_3 + {}_7C_4 + {}_7C_5 + {}_7C_6 + {}_7C_7$.

21. Answers may vary. An example is given. There are 3 paths to 3rd and B. By symmetry, there are 3 paths to 2nd and C. From each of these points, there is only one way to get to 3rd and C, so there are 6 paths altogether.

22.

	1st	2nd	3rd	4th	5th	6th
A		1	1	1	1	1
B	1	2	3	4	5	6
C	1	3	6	10	15	21
D	1	4	10	20	35	56
E	1	5	15	35	70	126
F	1	6	21	56	126	252

The numbers on the grid are the same as the numbers in Pascal's triangle when Anchara's house is at position ${}_0C_0$.

23. 5 times; 5 times; ${}_{10}C_5 = 252$

24. 63,504 paths; There are 252 possible paths each way, so by the multiplication counting principle, there are 252 • 252 round-trip paths.

25. a. 6 b. 10 c. [illegible]

26. $17\frac{12}{35} < \sqrt{301} < 17\frac{6}{17}$

27. $8\frac{111}{217} < \sqrt[3]{623} < 8\frac{37}{64}$

28. See answers in back of book.

Apply⇔Assess

Exercise Notes

Application
Ex. 30 Both botanists and geneticists use mathematics extensively in their fields. Fibonacci numbers have applications to both fields.

Integrating the Strands
Exs. 30, 32 Pascal's triangle is extremely versatile and can be related to nearly every strand of mathematics. It is related to number theory because of the triangular and square numbers it contains. It is related to discrete mathematics and probability because its elements are combinations, and because it can be used to generate sequences such as the Fibonacci sequence. Its connection to the binomial theorem relates the triangle to algebra, and Ex. 32 shows how the triangle can be related to functions.

Problem Solving
Ex. 35 Encourage students to look at several binomial expansions for a pattern that can suggest a general formula for finding any term of a binomial expansion.

Cooperative Learning
Ex. 36 Students may benefit from discussing their ideas to this open-ended problem in small groups before writing a response. In this way, they can get feedback from others and refine their ideas.

Student Progress
Ex. 37 Students should quickly recognize that the sum of the exponents in choice D is not equal to 7.

Assessment Note
Ex. 38 This assessment exercise can be used to evaluate a student's ability to see patterns and relationships.

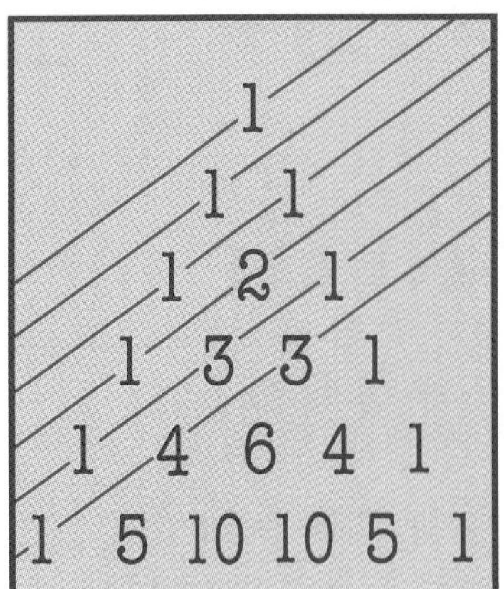

29. Writing Explain why the first and last numbers in a row of Pascal's triangle are always 1's.

30. The sums of the terms along the diagonals shown give a sequence called the *Fibonacci numbers.* Use Pascal's triangle to give the first 10 Fibonacci numbers.

31. Research Find out more about Pascal and Pascal's triangle. Describe some patterns that appear in the triangle that haven't been discussed in your class. Share your discoveries with your class.

BY THE WAY

Fibonacci numbers appear in many patterns in nature, such as in the spirals of seeds in a sunflower.

32. a. Technology Use a graphing calculator or graphing software to graph $y = x^4$ and $y = x^4 + 4x^3 + 6x^2 + 4x + 1$. How is the second graph geometrically related to the first? Why?

b. Repeat part (a) using $y = x^3$ and $y = x^3 - 6x^2 + 12x - 8$.

33. Writing Find the sum of the numbers in each of the first few rows of Pascal's triangle. Then give a formula for the sum of the numbers in row *n*. Explain why the sum of the numbers in a row doubles from row to row.

34. Substitute $a = 1$ and $b = 1$ into the binomial theorem. How do the results relate to the answers to Exercise 33?

35. a. Challenge Use the binomial theorem to find a general formula for term *r* (where $0 \le r \le n$) in the expansion of $(a + b)^n$.

b. Use your formula to find term 11 of $(a - 3b)^{13}$.

c. Use your formula to find term 6 of $(2m + n)^{12}$.

36. Open-ended Problem Think of a situation that you think can be modeled by Pascal's triangle. Explain your reasoning.

37. SAT/ACT Preview Which term is *not* part of the expansion of $(x + y)^7$?

A. x^7 **B.** $21x^5y^2$ **C.** $7xy^6$ **D.** $35x^3y^3$ **E.** y^7

ONGOING ASSESSMENT

38. Cooperative Learning Work in groups of three or four. Each group should have several copies of the first 20 rows or so of Pascal's triangle. Begin by circling or coloring all of the odd numbers. The pattern represents a portion of the fractal called a *Sierpenski triangle.* (See page 464 for more information about a Sierpenski triangle.)

a. On one copy of the triangle, circle or color all of the multiples of 3. What do you notice?

b. On a different copy of the triangle, circle or color all of the multiples of 5. What do you notice?

c. Repeat this process with multiples of other numbers. Does a pattern always seem to appear? If so, describe it.

29. The first term is ${}_nC_0 = 1$ and the last term is ${}_nC_n = 1$.

30. 1, 1, 2, 3, 5, 8, 13, 21, 34, 55

31. Answers may vary. An example is given. The sum of the numbers in row *n* of Pascal's triangle is equal to 2^n. The sum of all the terms in row 0 through row *n* is equal to $2^{n+1} - 1$. If you choose a number that is surrounded by 6 other numbers and you find the product of the 6 numbers, the product will be a perfect square.

32. a. The second graph is obtained by translating the first graph 1 unit to the left, because the second equation can be written as $y = (x + 1)^4$.

b. The second graph is obtained by translating the first graph 2 units to the right, because the second equation can be written as $y = (x - 2)^3$.

33. 1, 2, 4, 8, 16; 2^n; The sum doubles from row to row because each number in row *n* is used twice to get the numbers in row $n + 1$.

34. $(1 + 1)^n = 2^n$, which is the sum of the numbers in row *n*

35. a. ${}_nC_r \cdot a^{n-r}b^r$

b. $-13{,}817{,}466a^2b^{11}$

c. $59{,}136m^6n^6$

36. Answers may vary. An example is given. If you toss a coin once, there are two possibilities: H or T. If you toss a coin twice, there are four possibilities with 2 the same: HH, HT and TH, or TT. If you toss a coin three times, there are eight possibilities with two sets of 3 the same: HHH, HHT and HTH and THH, HTT and THT and TTH, or TTT. This pattern can be continued, and the number of outcomes for *n* tosses are given by the terms in row *n* of Pascal's triangle.

37. D 38. See answers in back of book.

SPIRAL REVIEW

Simplify each expression. *(Section 12.4)*

39. ${}_1C_0$ **40.** ${}_7C_1$ **41.** ${}_7C_6$ **42.** ${}_{15}C_{12}$

43. The radius of a basketball is 12.0 cm. Find its surface area. *(Toolbox, page 802)*

For Exercises 44–46, write a point-slope equation of the line passing through the given point and having the given slope. *(Section 2.3)*

44. point: (1, 1)
slope = 2

45. point: (3, −2)
slope = 6

46. point: (−5, 5)
slope = −2

ASSESS YOUR PROGRESS

VOCABULARY

permutation (p. 579)
combination (p. 585)
Pascal's triangle (p. 593)

For Questions 1–6, simplify each expression. *(Sections 12.3 and 12.4)*

1. ${}_9C_1$ **2.** ${}_9P_1$ **3.** ${}_{10}C_4$

4. ${}_{10}P_4$ **5.** ${}_6C_3 + {}_6C_4$ **6.** ${}_5P_2 - {}_4P_2$

For Questions 7–9, give and evaluate an expression of the form ${}_nC_r$ or ${}_nP_r$ for each situation. *(Sections 12.3 and 12.4)*

7. the number of ways that five different jobs can be filled from among 40 employees

8. the number of different vanilla frozen-yogurt sundaes that can be created by choosing three of eight possible toppings

9. the number of ways that you can place 5 of your 20 compact discs into the slots of your stereo's CD changer

10. If you think of a rainbow as having seven different colors, how many possible rainbows can you paint if you painted the color bands in any order? *(Section 12.3)*

11. Niti plans to take two of four possible social science classes, one of three possible mathematics classes, one of four possible English classes, and two of six possible electives. How many possible combinations of classes can she take? *(Section 12.4)*

12. Find term 8 of row 11 of Pascal's triangle. *(Section 12.5)*

Use the binomial theorem to expand each power of a binomial. *(Section 12.5)*

13. $(j + k)^6$ **14.** $(2p - q)^5$ **15.** $(p + 0.5q)^3$

16. Journal Describe as many situations as you can in which you could use permutations to count possibilities during the course of a typical day. Then describe as many situations as you can in which you could use combinations.

Apply⇔Assess

Exercise Notes

Using Technology
Exs. 44–46 If students convert their equations to slope-intercept form, they can graph the equations on a graphing calculator. Then they can use the 1:value option of the CALC menu to verify that the point given in the exercise is on the graph.

Assess Your Progress

Journal Entry
This exercise not only allows students to relate permutations and combinations to real life, but also affords an opportunity to check whether students fully understand these two concepts.

Progress Check 12.3–12.5

See page 603.

Practice 81 for Section 12.5

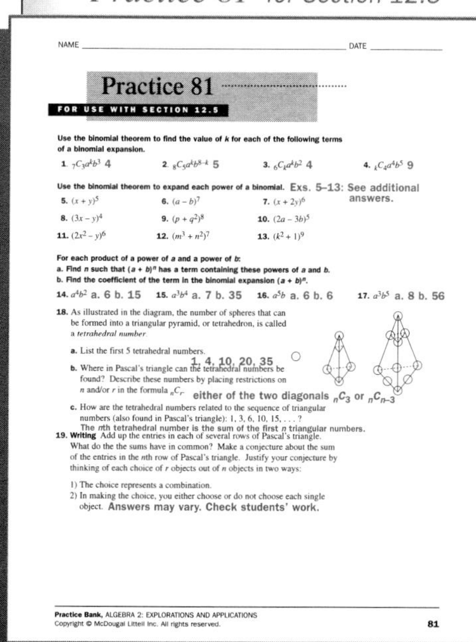

NAME ________ DATE ________

Practice 81

FOR USE WITH SECTION 12.5

Use the binomial theorem to find the value of k for each of the following terms of a binomial expansion.

1. ${}_7C_3a^kb^3$ 4 **2.** ${}_8C_5a^kb^{8-k}$ 5 **3.** ${}_6C_ka^kb^2$ 4 **4.** ${}_kC_4a^4b^5$ 9

Use the binomial theorem to expand each power of a binomial. Exs. 5–13: See additional answers.

5. $(x + y)^5$ **6.** $(a - b)^7$ **7.** $(x + 2y)^6$

8. $(3x - y)^4$ **9.** $(p + q^2)^8$ **10.** $(2a - 3b)^5$

11. $(2x^2 - y)^6$ **12.** $(m^3 + n^2)^7$ **13.** $(k^2 + 1)^9$

For each product of a power of a and a power of b:
a. Find n such that $(a + b)^n$ has a term containing these powers of a and b.
b. Find the coefficient of the term in the binomial expansion $(a + b)^n$.

14. a^4b^2 a. 6 b. 15 **15.** a^3b^4 a. 7 b. 35 **16.** a^5b a. 6 b. 6 **17.** a^3b^5 a. 8 b. 56

18. As illustrated in the diagram, the number of spheres that can be formed into a triangular pyramid, or tetrahedron, is called a *tetrahedral number*.

a. List the first 5 tetrahedral numbers. 1, 4, 10, 20, 35

b. Where in Pascal's triangle can the tetrahedral numbers be found? Describe these numbers by placing restrictions on n and/or r in the formula ${}_nC_r$. either of the two diagonals ${}_nC_3$ or ${}_nC_{n-3}$

c. How are the tetrahedral numbers related to the sequence of triangular numbers (also found in Pascal's triangle): 1, 3, 6, 10, 15, . . . ? The nth tetrahedral number is the sum of the first n triangular numbers.

19. Writing Add up the entries in each of several rows of Pascal's triangle. What do the the sums have in common? Make a conjecture about the sum of the entries in the nth row of Pascal's triangle. Justify your conjecture by thinking of each choice of r objects out of n objects in two ways:

1) The choice represents a combination.
2) In making the choice, you either choose or do not choose each single object. Answers may vary. Check students' work.

Practice Bank, ALGEBRA 2: EXPLORATIONS AND APPLICATIONS

81

---

39. 1 40. 7
41. 7 42. 455
43. $576\pi \approx 1810$ cm^2
44. $y - 1 = 2(x - 1)$
45. $y + 2 = 6(x - 3)$
46. $y - 5 = -2(x + 5)$

Assess Your Progress

1. 9 2. 9
3. 210 4. 5040
5. 35 6. 8
7. ${}_{40}P_5 = 78{,}960{,}960$
8. ${}_8C_3 = 56$
9. ${}_{20}P_5 = 1{,}860{,}480$
10. 5040 possible rainbows
11. 1080 possible combinations of classes
12. 165
13. $j^6 + 6j^5k + 15j^4k^2 + 20j^3k^3 + 15j^2k^4 + 6jk^5 + k^6$
14. $32p^5 - 80p^4q + 80p^3q^2 - 40p^2q^3 + 10pq^4 - q^5$
15. $p^3 + 1.5p^2q + 0.75pq^2 + 0.125q^3$
16. Answers may vary. An example is given. Ten people want to bat in a softball game. In how many ways can the first 3 batters be chosen? (${}_{10}P_3 = 720$ ways) In how many ways can a group of three batters be chosen? (${}_{10}C_3 = 120$ ways)

Project Notes

Mathematical Goals

- Select a real-world situation to model with a flip book.
- Design and create a flip book modeling the situation.
- Calculate the number of combinations the flip book shows.

Planning

Materials

- pad of paper
- posterboard
- scissors
- photos from magazines or coloring utensils

Project Teams

Students can work individually to make their flip books, but if they decide to work in pairs, every part of the project should be completed as a team.

Guiding Students' Work

The description of the flip book and how it works will be better understood by students if they can see an example before trying to make their own. If possible, put together a real flip book for students to see, such as the planning a dinner flip book described in the project. Also, if students are having difficulty coming up with an idea for a flip book, it may help to have a list of areas where they may find ideas, such as selecting classes for the coming school year, clothing combinations, color combinations for parts of a design or map, or paths they can take home from school.

Second-Language Learners

The professions *police sketch artist, landscaper,* and *fashion designer* may not be familiar to some students learning English. If necessary, explain that a police sketch artist is "an artist who helps police track down criminals by using witness accounts to draw composite pictures of the criminals." A landscaper is "someone who designs and builds gardens, or decorates grounds." A fashion designer is "someone who designs fashions that are made into clothes." Students learning English may also wish to work on creating their flip books with the help of a peer tutor or aide.

PORTFOLIO PROJECT

Exploring the Possibilities

The average day presents many situations that call for choices. Various options and combinations present themselves. "Does the blue sweater go with this outfit?" "Would I rather have the entree with a baked potato or rice? And what type of dressing for the salad?" "Should I buy the two-door or the four-door model of this car?"

You can examine your options by writing them all down, but sometimes it helps to see the possibilities. By making a flip book, you can visualize all the combinations of choices.

PROJECT GOAL **Make a flip book to show all the possibilities for a real-world situation involving choices.**

Making Your Own Flip Book

Choose a real-world situation that has many options and where the options are divided into at least four categories. Make a flip book that displays every possible combination of the options. You will need a pad of paper, posterboard, and scissors.

The flip book below illustrates the choices for someone planning a dinner. Assemble your flip book in a similar manner.

1. USE a pad of paper or several sheets of paper stapled together.

2. DIVIDE the pad by cutting the paper so that there is a section for each category of options. Be sure to leave a border on the side where the sheets of paper are bound together.

3. MOUNT the pad on posterboard and put the name of each category on the side opposite the binding.

4. DRAW pictures (or clip and paste photos from magazines) to illustrate the options for each category. Remove any unused pieces of paper.

Writing a Report

Write a report about your flip book. Consider these questions:

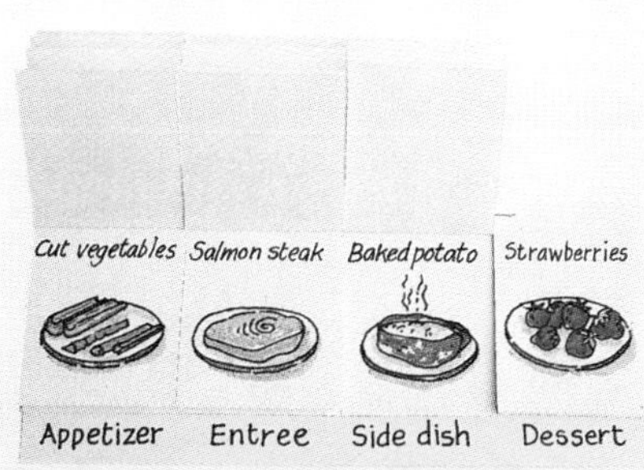

- How many different combinations of options does your flip book display? How do you know?
- What are the advantages and drawbacks of using a flip book to display options?
- How might other people, such as a police sketch artist, a landscaper, or a fashion designer, use a flip book? Can you think of other areas in which a flip book might be useful?

Extend your investigation by doing related projects that involve combinations. For example:

- Make drawings on strips of paper that you can put next to each other and slide back and forth to change, say, the expressions on a face or the clothes on a model.
- Make drawings on clear plastic overlays to explore options such as toppings on a pizza or disguises for a face.
- Create a colorful toy or game that a young child can play with to explore combinations.

Self-Assessment

Describe your understanding of combinations. In what ways do you feel that you are better able to recognize and examine situations involving choices?

Project Notes

Guiding Students' Work

Rubric for Chapter Project

4 Students produce a flip book that is original and innovative. It is neatly drawn and well organized. A report is made that contains all of the information relating to the production of the book including the number of combinations in the book. The report is easy to read and discusses how other people in different professions can use a flip book for their job. Students extend the project by completing one or more of the suggestions given in the project.

3 Students produce a flip book that displays an interesting idea and is neatly drawn and put together. A report is made, but the report is either missing information relating to producing the book or it is not clear nor well written. The report does contain a discussion of how flip books can be used in careers, and students extend the project using one of the suggestions given in the project.

2 Students produce a flip book that displays an idea, but it is poorly put together. A report is made and contains some of the information relating to the book, but it is incomplete and is not well written. Students discuss some of the ideas about the flip book in their report. Students do not make any attempt to extend the project.

1 Students produce a flip book, but it is very poorly done and is incomplete. The flip book shows a lack of effort for the project. If a report is done, it is incomplete and contains only one or two of the pieces of information relating to the book. The report also shows a lack of effort to complete the project. Students make no attempt to extend the project. Students should be encouraged to speak with the teacher as soon as possible to review their work and to make a new start on the project.

Chapter Support

Course Guide: Chapter 12
Lesson Plans: Chapter 12
Practice Bank:
Cumulative Practice 82
Study Guide: Chapter 12 Review
Assessment Book:
Chapter Tests 51 and 52
Cumulative Test 53
Chapter 12 Alternative Assessment
Test Generator Software
Portfolio Project Book
Preparing for College Entrance Tests
Professional Handbook

Progress Check 12.1–12.2

1. Draw and color a graph representing the six New England states: Maine, New Hampshire, Vermont, Massachusetts, Rhode Island, and Connecticut. Label the degree of each vertex. *(Section 12.1)*
Graphs may vary. An example is given.

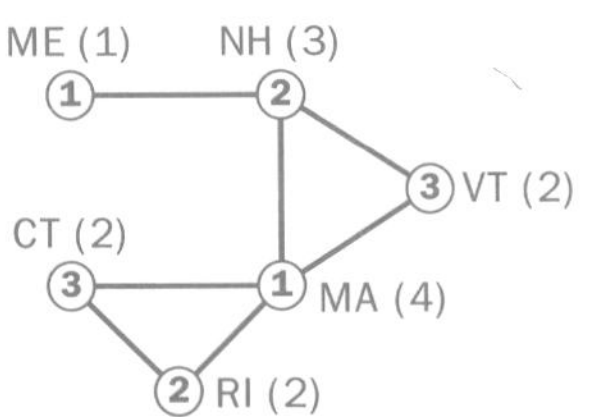

2. Use your graph from Ex. 1 to color the map. Try to use as few colors as possible. *(Section 12.1)*
Maps may vary. An example is given using the graph above.

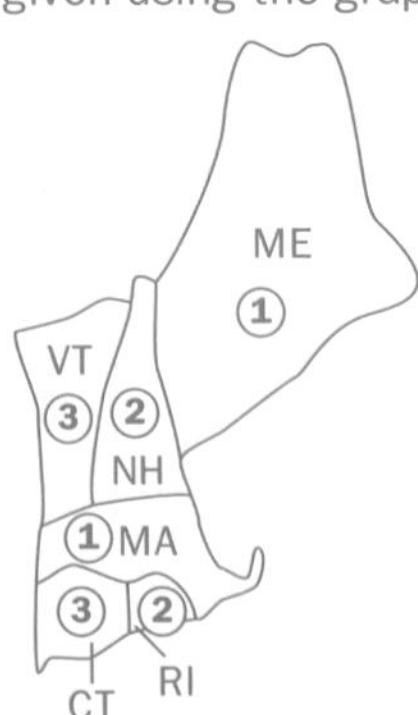

3. Would adding the state of New York to the map require the use of four colors? *(Section 12.1)*
Four colors would not be required for the graph given in Ex. 1 because NY could be colored the same color as NH and RI.

CHAPTER

12 Review

STUDY TECHNIQUE

Describe any exercises or ideas in this chapter that you found difficult but eventually understood. How did you resolve your difficulties? Now describe some things you still don't understand. Can you use some of the methods you used before to help you resolve these difficulties?

VOCABULARY

graph (p. 563)
vertices (p. 563)
edges (p. 563)
degree of a vertex (p. 563)
directed graph (p. 569)
permutation (p. 579)
combination (p. 585)
Pascal's triangle (p. 593)

SECTIONS 12.1 *and* 12.2

You can use a **graph** to solve problems that require sorting items into groups, such as coloring a map of New England. The graph shown consists of **vertices,** representing states, connected by **edges,** representing a shared border. Each vertex is labeled with its **degree.**

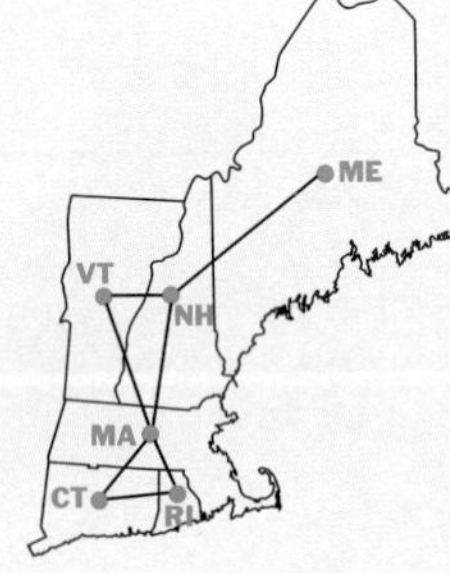

Using a procedure like that in Example 1 on page 564, you can assign colors.

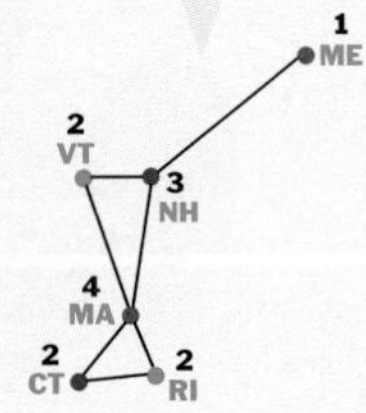

You can use **directed graphs** and matrices to solve problems involving one-way relationships, such as predator-prey relationships.

The arrows point from prey to predator.

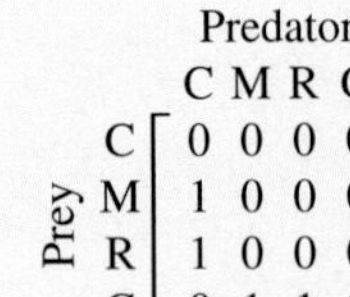

Matrix M tells you how many ways each plant or animal is a direct food source for the others.

Predator (columns C M R G), Prey (rows C M R G):

$$\begin{array}{c} \\ C \\ M \\ R \\ G \end{array}\begin{array}{c} C\ M\ R\ G \\ \begin{bmatrix} 0 & 0 & 0 & 0 \\ 1 & 0 & 0 & 0 \\ 1 & 0 & 0 & 0 \\ 0 & 1 & 1 & 0 \end{bmatrix} \end{array} = M$$

Matrix M^2 tells you how many ways each plant or animal is a food source for the others through one intermediate source.

$$\begin{array}{c} \\ C \\ M \\ R \\ G \end{array}\begin{array}{c} C\ M\ R\ G \\ \begin{bmatrix} 0 & 0 & 0 & 0 \\ 0 & 0 & 0 & 0 \\ 0 & 0 & 0 & 0 \\ 2 & 0 & 0 & 0 \end{bmatrix} \end{array} = M^2$$

SECTIONS 12.3 and 12.4

The multiplication counting principle states that if you must choose among m things, and for each of these you must choose among n more things, then you have a total of $m \times n$ possibilities. The multiplication counting principle leads to two important counting formulas.

	Permutation	Combination
Definition	an arrangement of items in which position is important	a selection of items in which position is *not* important
Formula	${}_nP_r = \frac{n!}{(n-r)!}$	${}_nC_r = \frac{{}_nP_r}{{}_rP_r} = \frac{n!}{(n-r)!r!}$
Example	In a club with 20 members, the number of ways to select 4 officers is: ${}_{20}P_4 = \frac{20!}{16!} = 116{,}280$	In a club with 20 members, the number of ways to select a committee of 4 is: ${}_{20}C_4 = \frac{20!}{16!4!} = 4845$

For any positive integer n, $n!$ is the product of all the integers from n down to 1.

When a set contains elements that are identical, there are fewer distinguishable permutations. For example, the number of distinguishable arrangements of the letters in the word EEL is given by:

Divide by 2! because there are two E's.

$$\frac{3!}{2!} = 3$$

The three arrangements are: EEL, ELE, LEE.

SECTION 12.5

Pascal's triangle is formed by calculating ${}_nC_r$ $(0 \le r \le n)$ for term r in row n, or by adding pairs of terms from the previous row.

Pascal's triangle gives the coefficients when you expand powers of binomials.

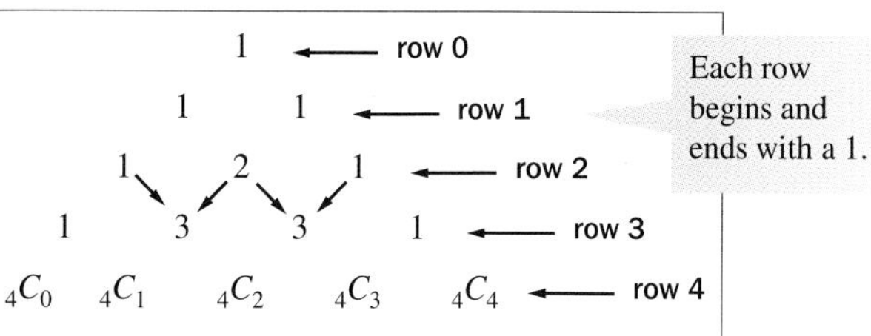

Each row begins and ends with a 1.

Binomial Theorem

For any positive integer n,

$$(a + b)^n = {}_nC_0a^n + {}_nC_1a^{n-1}b + {}_nC_2a^{n-2}b^2 + \cdots + {}_nC_{n-1}ab^{n-1} + {}_nC_nb^n$$

The binomial theorem gives the coefficients when you expand any power of any binomial, such as $(x + 3y)^4$.

$$\begin{aligned}(x + 3y)^4 &= {}_4C_0x^4 + {}_4C_1x^3(3y) + {}_4C_2x^2(3y)^2 + {}_4C_3x(3y)^3 + {}_4C_4(3y)^4 \\ &= x^4 + 4(3)x^3y + 6(3^2)x^2y^2 + 4(3^3)xy^3 + 3^4y^4 \\ &= x^4 + 12x^3y + 54x^2y^2 + 108xy^3 + 81y^4\end{aligned}$$

Progress Check 12.1–12.2 (cont.)

Matrix M represents the direct connections a bus company makes between five towns. *(Section 12.2)*

	A	B	C	D	E
A	0	0	1	0	1
B	0	0	1	1	0
C	0	0	0	1	0
D	1	1	0	0	1
E	0	1	0	1	0

4. What does the 1 in row 3, column 4 represent? the zero in row 5, column 3? There is a direct connection from C to D. There is no direct connection from E to C.

5. Draw a directed graph.

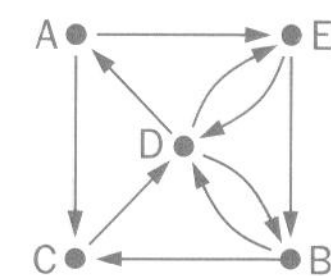

6. Find M^2. What do the numbers in this matrix represent?

$M^2 =$

	A	B	C	D	E
A	0	1	0	2	0
B	1	1	0	1	1
C	1	1	0	0	1
D	0	1	2	2	1
E	1	1	1	1	1

the number of routes from the town in the row to the town in the column that include exactly one intermediate stop

Progress Check 12.3–12.5

Simplify each expression. *(Sections 12.3 and 12.4)*

1. ${}_7C_2$ 21

2. ${}_7P_2$ 42

3. ${}_{15}P_5 + {}_{10}P_5$ 390,600

4. ${}_5C_3 \cdot {}_8C_4$ 700

Give and evaluate an expression of the form ${}_nC_r$ or ${}_nP_r$ for each situation. *(Sections 12.3 and 12.4)*

5. Five cards are dealt from a standard 52-card deck. How many possible hands have only face cards? ${}_{12}C_5 = 792$

6. In how many ways can an employer choose 5 employees from a pool of 12 for 5 different job assignments? ${}_{12}P_5 = 95{,}040$

Use the binomial theorem to expand each power of a binomial. *(Section 12.5)*

7. $(v - w)^5$
$v^5 - 5v^4w + 10v^3w^2 - 10v^2w^3 + 5vw^4 - w^5$

8. $(r^3 + 2p)^6$
$r^{18} + 12r^{15}p + 60r^{12}p^2 + 160r^9p^3 + 240r^6p^4 + 192r^3p^5 + 64p^6$

Chapter 12 Assessment
Form A Chapter Test

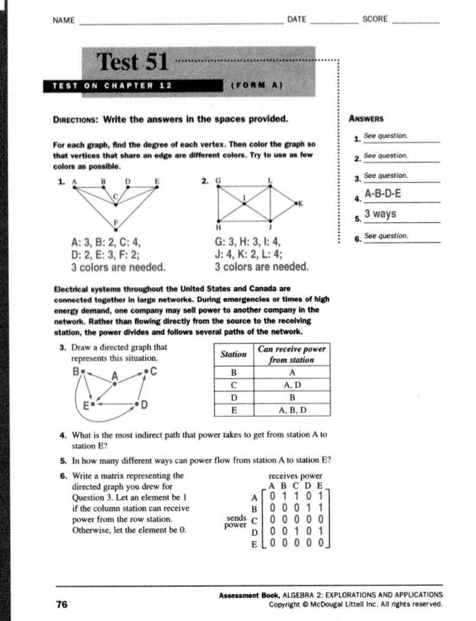

NAME ______ DATE ______ SCORE ______

Test 51

TEST ON CHAPTER 12 (FORM A)

DIRECTIONS: Write the answers in the spaces provided.

For each graph, find the degree of each vertex. Then color the graph so that vertices that share an edge are different colors. Try to use as few colors as possible.

1. A: 3, B: 2, C: 4, D: 2, E: 3, F: 2; 3 colors are needed.

2. G: 3, H: 3, I: 4, J: 4, K: 2, L: 4; 3 colors are needed.

Electrical systems throughout the United States and Canada are connected together in large networks. During emergencies or times of high energy demand, one company may sell power to another company in the network. Rather than flowing directly from the source to the receiving station, the power divides and follows several paths of the network.

3. Draw a directed graph that represents this situation.

Station	Can receive power from station
B	A
C	A, D
D	B
E	A, B, D

4. What is the most indirect path that power takes to get from station A to station E?
5. In how many different ways can power flow from station A to station E?
6. Write a matrix representing the directed graph you drew for Question 3. Let an element be 1 if the column station can receive power from the row station. Otherwise, let the element be 0.

receives power (columns A B C D E); sends power (rows A B C D E)

$$\begin{bmatrix} 0 & 1 & 1 & 0 & 1 \\ 0 & 0 & 0 & 1 & 1 \\ 0 & 0 & 0 & 0 & 0 \\ 0 & 0 & 1 & 0 & 1 \\ 0 & 0 & 0 & 0 & 0 \end{bmatrix}$$

ANSWERS
1. See question.
2. See question.
3. See question.
4. A-B-D-E
5. 3 ways
6. See question.

76

Assessment Book, ALGEBRA 2: EXPLORATIONS AND APPLICATIONS

Chapter 12 Assessment
Form B Chapter Test

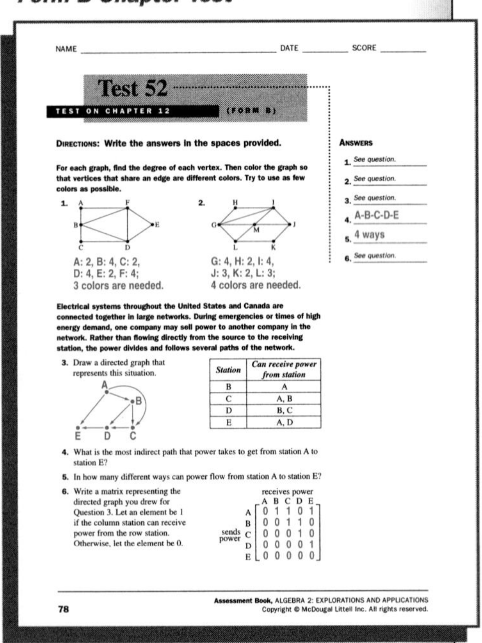

NAME ______ DATE ______ SCORE ______

Test 52

TEST ON CHAPTER 12 (FORM B)

DIRECTIONS: Write the answers in the spaces provided.

For each graph, find the degree of each vertex. Then color the graph so that vertices that share an edge are different colors. Try to use as few colors as possible.

1. A: 2, B: 4, C: 2, D: 4, E: 2, F: 4; 3 colors are needed.

2. G: 4, H: 2, I: 4, J: 3, K: 2, L: 3; 4 colors are needed.

Electrical systems throughout the United States and Canada are connected together in large networks. During emergencies or times of high energy demand, one company may sell power to another company in the network. Rather than flowing directly from the source to the receiving station, the power divides and follows several paths of the network.

3. Draw a directed graph that represents this situation.

Station	Can receive power from station
B	A
C	A, B
D	B, C
E	A, D

4. What is the most indirect path that power takes to get from station A to station E?
5. In how many different ways can power flow from station A to station E?
6. Write a matrix representing the directed graph you drew for Question 3. Let an element be 1 if the column station can receive power from the row station. Otherwise, let the element be 0.

receives power (columns A B C D E); sends power (rows A B C D E)

$$\begin{bmatrix} 0 & 1 & 1 & 0 & 1 \\ 0 & 0 & 1 & 1 & 0 \\ 0 & 0 & 0 & 1 & 0 \\ 0 & 0 & 0 & 0 & 1 \\ 0 & 0 & 0 & 0 & 0 \end{bmatrix}$$

ANSWERS
1. See question.
2. See question.
3. See question.
4. A-B-C-D-E
5. 4 ways
6. See question.

78

Assessment Book, ALGEBRA 2: EXPLORATIONS AND APPLICATIONS

CHAPTER

12 Assessment

VOCABULARY QUESTIONS

For Questions 1–3, complete each sentence.

1. On a graph, the _?_ of a vertex is five if the vertex is connected to five other vertices.
2. A(n) _?_ is a graph in which the edges are arrows.
3. A(n) _?_ is an arrangement of items in which position is important, and a(n) _?_ is a selection of items in which position is *not* important.

SECTIONS 12.1 *and* 12.2

For each graph in Questions 4–6:

a. Find the degree of each vertex.

b. Draw an equivalent graph.

c. Color your graph so that vertices that share an edge are different colors. Try to use as few colors as possible.

4. A B C D E F

5.

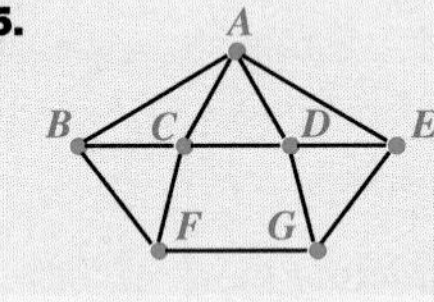

6.

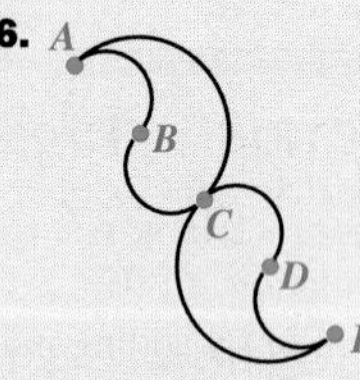

GAMES Some of the characters in a game are listed below along with their required companions.

- A healer requires an interpreter.
- A merchant requires a scribe.
- An advisor requires a merchant or an interpreter.
- A knight requires a healer.
- A prince requires a knight or an advisor.

7. Draw a directed graph to represent this situation.
8. What do the arrows in the graph represent?
9. Write a matrix that represents the directed graph.
10. Square the matrix from Question 9. What do the elements represent?

ANSWERS Chapter 12

Assessment

1. degree
2. directed graph
3. permutation; combination

4. a. *A*, 3; *B*, 3; *C*, 2; *D*, 3; *E*, 3; *F*, 2

b, c. Answers may vary. An example is given.

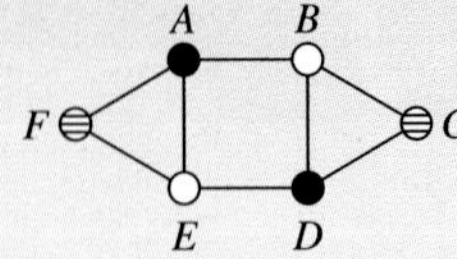

5. a. *A*, 4; *B*, 3; *C*, 4; *D*, 4; *E*, 3; *F*, 3; *G*, 3

b, c. Answers may vary. An example is given.

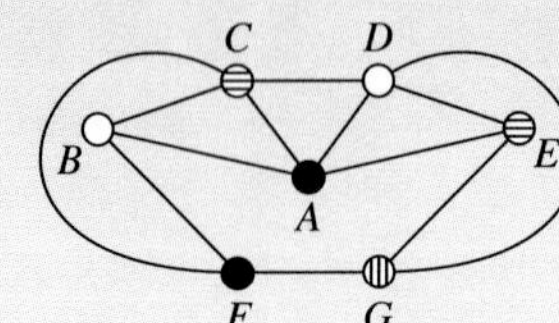

6. a. *A*, 2; *B*, 2; *C*, 4; *D*, 2; *E*, 2

b, c. Answers may vary. An example is given.

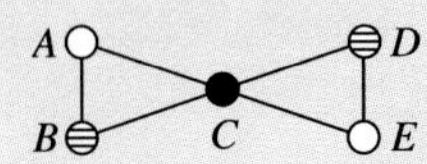

7–10. Answers may vary. Examples are given.

7. In the diagram, H = healer, I = interpreter, M = merchant, S = scribe, A = advisor, K = knight, and P = prince. The arrows are drawn from a person who requires another to a person who is required, although

Sections 12.3 *and* 12.4

Simplify each expression.

11. ${}_7P_4$ **12.** ${}_7P_1$ **13.** ${}_6P_2$ **14.** ${}_6P_4$

15. ${}_7C_4$ **16.** ${}_7C_1$ **17.** ${}_5C_3 + {}_5C_4$ **18.** ${}_9C_3 + {}_9C_4$

19. GOVERNMENT Your Social Security number consists of nine digits from 0 to 9. The first three encode the location of your birth. For example, numbers beginning "001" originate in a certain area of New York. If "000" is not allowed for the first three digits, and "000000" is not allowed for the last six digits, how many possible nine-digit Social Security numbers are there? (*Hint:* Don't subtract the case of all 0 digits twice.)

20. Open-ended Problem Write two words with five letters each. One of your words should have a single letter repeated. Find the number of different five-letter permutations that can be made for each word. Are the numbers of permutations the same? Why or why not?

Sections 12.5

21. Find term 6 of row 12 of Pascal's triangle.

22. Find the coefficient of a^6b^2 in the expansion of $(a + b)^8$.

Use the binomial theorem to expand each power of a binomial.

23. $(x - y)^3$ **24.** $(2a + 2b)^4$ **25.** $(2y + z^2)^5$

26. One term in the expansion of $(a + b)^n$ is $78a^{11}b^2$. What is the next term of the expansion?

Performance Task

27. Draw a map of a fictitious continent with at least seven territories. Draw a graph that represents your map and use it to color your map so that territories that border each other are different colors. Write a story about the continent you created. Demonstrate what you have learned in this chapter by including mathematical explanations in your story, such as:

- a *directed graph* to show which territories export to other territories
- a *matrix* to represent your directed graph
- a *squared matrix* to show which territories export to the others through one intermediate territory
- the number of ways that all of the leaders of your territories can line up for a group photograph (a *permutation*)
- the number of handshakes if all the leaders shake hands after the group photograph (a *combination*)

Chapter 12 Assessment
Form C Alternative Assessment

Chapter 12
ALTERNATIVE ASSESSMENT

1. **Open-ended Problem** Using a map of the United States, draw an outline around 7 adjoining states. Use the rules of map coloring to color your states. What is the minimum number of colors that you can use?

2. **Research Project** Look up some more information about Fibonacci and his numbers. Make a poster about Fibonacci. Include some real examples of Fibonacci numbers that occur in nature.

3. **Open-ended Problem** If a coach wanted to count how many different ways his players could be placed into positions for a sport, he might use permutations or combinations. Name three sports for which a coach would use permutations and three sports for which a coach would use combinations.

4. **Open-ended Problem** Look up the word *discrete* in the dictionary. How does the word discrete relate to the mathematics content of Chapter 12?

5. **Research Project** Find an illustration of a food chain in an encyclopedia or science book. Draw a directed graph of your food chain. Use matrices to model the total number of primary food sources and the total number of secondary food sources.

6. **Open-ended Problem** Explain the difference between a directed graph and a graph that is not directed. Use examples to clarify your explanation.

Assessment Book, ALGEBRA 2: EXPLORATIONS AND APPLICATIONS

125

---

the arrows can be drawn in the opposite direction.

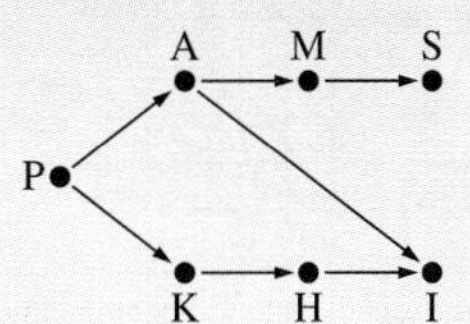

8. The arrows indicate which characters are required by others.

9, 10. See answers in back of book.

11. 840

12. 7

13. 30

14. 360

15. 35

16. 7

17. 15

18. 210

19. $10^9 - 10^6 - 10^3 + 1 = 998{,}999{,}001$

20. Answers may vary. An example is given. A 5-letter word with no letters repeated, such as SMILE, has ${}_5C_5 = 5! = 120$ permutations. A 5-letter word with just one letter repeated, such as MILLS, has $\frac{5!}{2!\,1!\,1!\,1!} = 60$ permutations. The numbers of permutations are different because you cannot distinguish between the repeated letters.

21. ${}_{12}C_6 = 924$

22. 28

23. $x^3 - 3x^2y + 3xy^2 - y^3$

24. $16a^4 + 64a^3b + 96a^2b^2 + 64ab^3 + 16b^4$

25. $32y^5 + 80y^4z^2 + 80y^3z^4 + 40y^2z^6 + 10yz^8 + z^{10}$

26. ${}_{13}C_3a^{10}b^3 = 286a^{10}b^3$

27. Answers may vary. Check students' work.

Cumulative Assessment

CHAPTERS 10–12

CHAPTER 10

Tell whether each sequence is *arithmetic, geometric,* or *neither*. Then write a formula for t_n and find the 10th term of the sequence.

1. $\frac{1}{2}, \frac{2}{3}, \frac{3}{4}, \frac{4}{5}, \ldots$ **2.** $2, 1, 0.5, 0.25, \ldots$ **3.** $5\sqrt{2}, 4\sqrt{2}, 3\sqrt{2}, 2\sqrt{2}, \ldots$

4. Find the arithmetic mean and the geometric mean of 0.9 and 1.6.

5. Write a recursive formula for each sequence in Questions 2 and 3.

Find the first 4 terms of each sequence.

6. $t_n = n(n + 1)$ **7.** $t_1 = 2;\ t_n = 2(t_{n-1}) + 1$

8. Open-ended Problem Make up an arithmetic sequence with a starting value of -8. Write a recursive formula and an explicit formula for your sequence.

Find the sum of each series, if the sum exists.

9. $2 + (-6) + 18 + (-54) + \cdots + 13{,}122$

10. $\sum_{n=1}^{\infty} 100(0.2)^{n-1}$ **11.** $\sum_{n=1}^{\infty} (2n - 3)$

12. a. Write the series of the first 100 positive even integers in sigma notation and find the sum.

b. Find the value of n for which the sum of the first n positive even integers is 342.

13. Writing Explain how to write 0.727272. . . as a fraction in simplest form.

CHAPTER 11

14. For the points $A(-4, -9)$ and $B(1, 3)$, find the distance between the points and an equation of the perpendicular bisector of $\overline{AB}$.

15. Show that the points $P(-5, -2)$, $Q(1, 2)$, $R(3, -1)$, and $S(-3, -5)$ are the vertices of a rectangle. Explain your reasoning.

For Questions 16–19, write an equation of each figure and then graph the equation.

16. a parabola with focus $(1, 4)$ and directrix $x = 3$

17. a circle with center $(-4, 0)$ and radius $2\sqrt{2}$

18. an ellipse with center $(1, -3)$, one vertex at $(1, 7)$, and one focus at $(1, 5)$

19. a hyperbola with center $(0, 0)$, one vertex at $(-3, 0)$, and one focus at $(-5, 0)$

ANSWERS Chapters 10–12

Cumulative Assessment

1. neither; $t_n = \frac{n}{n+1}$; $\frac{10}{11}$
2. geometric; $t_n = 2\left(\frac{1}{2}\right)^{n-1}$; $\frac{1}{256}$
3. arithmetic; $t_n = (6 - n)\sqrt{2}$; $-4\sqrt{2}$
4. 1.25; 1.2
5. (2) $t_1 = 2;\ t_n = \frac{1}{2}t_{n-1}$;
   (3) $t_1 = 5\sqrt{2};\ t_n = t_{n-1} - \sqrt{2}$
6. 2, 6, 12, 20
7. 2, 5, 11, 23
8. Answers may vary. An example is given. $-8, -7, -6, -5, \ldots$; $t_1 = -8,\ t_n = t_{n-1} + 1;\ t_n = n - 9$
9. 9842
10. 125
11. no sum
12. a. $\sum_{n=1}^{100} 2n = 10{,}100$
    b. 18
13. Express 0.727272… as the infinite series $0.72 + 0.0072 + 0.000072 + \ldots$. This series is geometric with $r = 0.01$. Since $|r| < 1$, the sum of the series is $\frac{0.72}{1 - 0.01} = \frac{0.72}{0.99} = \frac{8}{11}$.
14. 13; $y = -\frac{5}{12}x - \frac{29}{8}$
15. $\overleftrightarrow{PQ}$ and $\overleftrightarrow{SR}$ both have slope $\frac{2}{3}$; $\overleftrightarrow{PS}$ and $\overleftrightarrow{QR}$ both have slope $-\frac{3}{2}$. Then $\overleftrightarrow{PQ} \perp \overleftrightarrow{QR}$, $\overleftrightarrow{PQ} \perp \overleftrightarrow{PS}$, $\overleftrightarrow{SR} \perp \overleftrightarrow{PS}$, and $\overleftrightarrow{SR} \perp \overleftrightarrow{QR}$, so $\angle P$, $\angle Q$, $\angle R$, and $\angle S$ are right angles and $PQRS$ is a rectangle.

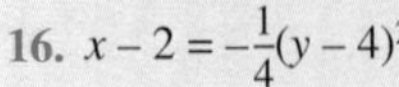

16. $x - 2 = -\frac{1}{4}(y - 4)^2$

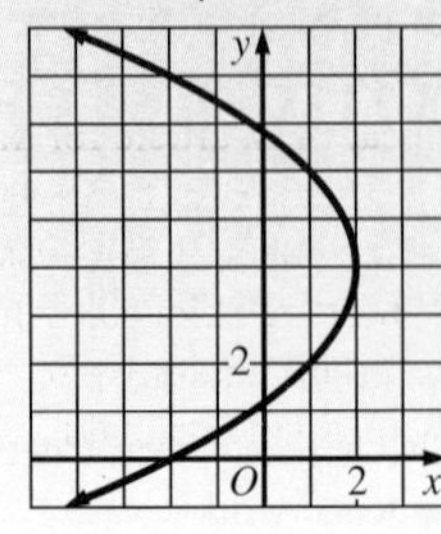

17. $(x + 4)^2 + y^2 = 8$

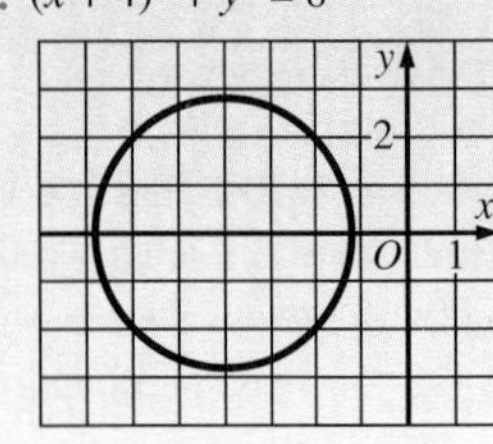

18. $\frac{(x-1)^2}{36} + \frac{(y+3)^2}{100} = 1$

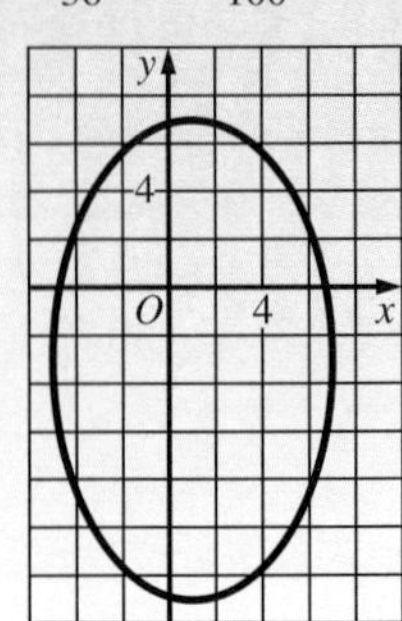

19. $\frac{x^2}{9} - \frac{y^2}{16} = 1$

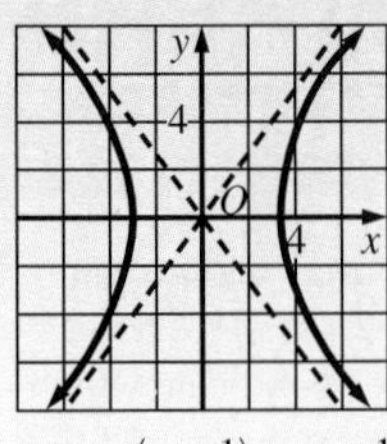

20. $(3, 5)$; $\left(3, 5\frac{1}{2}\right)$; $y = 4\frac{1}{2}$
21. $(4, 0)$; 2

20. Identify the vertex, the focus, and the directrix of the parabola with equation $y = \frac{1}{2}(x - 3)^2 + 5$.

21. Identify the center and the radius of the circle with equation $x^2 - 8x + 16 + y^2 = 4$.

22. Identify the center, the vertices, and the foci of the hyperbola with equation $\frac{(x-2)^2}{64} - \frac{y^2}{225} = 1$. Find equations of the asymptotes.

23. Open-ended Problem Write an equation of a degenerate conic. Identify the conic and graph your equation.

24. Rewrite the equation $x^2 + 4y^2 + 6x + 5 = 0$ in standard form. Identify the conic, and state its important characteristics.

25. Writing Describe the placement and angle of a plane that intersects a double cone to form each of the conic sections.

CHAPTER 12

26. For the graph at the right, find the degree of each vertex, draw an equivalent graph, and color your graph so that vertices that share an edge must be different colors. Try to use as few colors as possible.

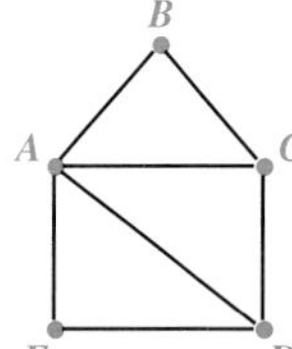

27. Writing Suppose a group of students have signed up to participate in volunteer activities, and some students want to participate in more than one activity. Describe how a graph can be used to find the least number of meeting times required to plan the volunteer activities.

28. COMPUTERS The table describes the connections in a computer network.

a. Draw a directed graph to represent the computer network.

b. Represent the directed graph in part (a) with a matrix.

c. **Technology** Use a graphing calculator or software with matrix calculation capabilities to determine which computers can send data to all the others either directly or through an intermediate computer.

Transmitting computer	Receiving computer(s)
A	B, C
B	D, E
C	A, D
D	E
E	B, C

Calculate each expression.

29. $\frac{9!}{5!}$ **30.** ${}_7P_4$ **31.** ${}_7C_4$ **32.** $\frac{n!}{(n-2)!}$

33. In how many ways can five cards be randomly chosen and turned face up in a row from a standard deck of 52 cards?

34. A person who buys a punch card at an amusement center can choose 8 rides out of 12 and two activities from the following: miniature golf, bowling, and a water slide. How many different choices are possible?

35. Open-ended Problem Describe situations in which you would find ${}_4P_2$ and ${}_4C_2$. How are the two situations similar? How are they different?

36. Find the coefficients of x^7y^7 and x^9y^5 in the expansion of $(x + y)^{14}$.

27. Draw a graph with each vertex representing an activity. Draw an edge connecting two vertices if one or more students signed up for both activities. The least number of colors needed to color the graph (as described in Example 1 on page 564) is the least number of meeting times required.

28. a.

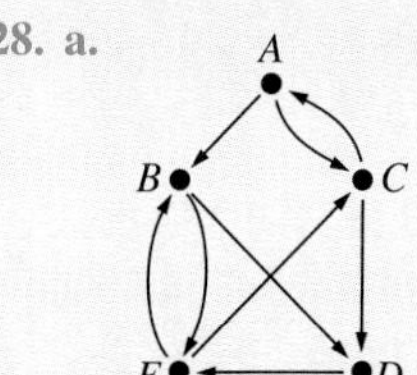

b.

	A	B	C	D	E
A	0	1	1	0	0
B	0	0	0	1	1
C	1	0	0	1	0
D	0	0	0	0	1
E	0	1	1	0	0

c. A, C, and E

29. 3024

30. 840

31. 35

32. $n(n - 1)$

33. 311,875,200 ways

34. 1485 choices

22. (2, 0); (–6, 0), (10, 0); (–15, 0), (19, 0); $y = \pm\frac{15}{8}(x - 2)$

23. Answers may vary. An example is given. $x^2 + y^2 = 0$ is a degenerate circle. Its graph is the point (0, 0).

24. $\frac{(x+3)^2}{4} + y^2 = 1$; an ellipse with center (–3, 0) and vertices (–5, 0) and (–1, 0)

25. When a plane intersects a double cone (and does not pass through the vertex), the result is a circle if the plane is horizontal, a hyperbola if the plane is vertical, a parabola if the plane has the same slant as the double cone, and an ellipse otherwise. Degenerate conics (points, lines, or pairs of intersecting lines) are produced if the plane intersects the vertex of the double cone.

26. *A*: 4; *B*: 2; *C*: 3; *D*: 3; *E*: 2; Drawings may vary. Color vertex *A* one color, *B* and *D* a second color, and *C* and *E* a third color.

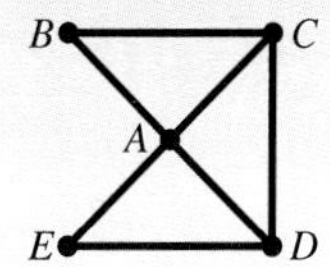

35. Answers may vary. An example is given. If a class has a four-member executive committee and plans to choose two to serve as chair and co-chair, there are ${}_4P_2$ ways to choose the members. If the class plans to choose two of the members to appear in an article for the school paper, there are ${}_4C_2$ ways to choose the members. In both cases, two out of four are being chosen. However, in the first case order matters, while in the second it does not.

36. 3432; 2002

CHAPTER

13 Probability

OVERVIEW

Connecting to Prior and Future Learning

⇔ In this chapter, the study of probability, begun in Algebra 1, is continued with an exploration of the meanings of experimental, theoretical, and geometric probabilities.

⇔ Probabilities involving independent, mutually exclusive, and complementary events are studied in Section 13.2. Conditional probabilities are explored in Section 13.3. Students will encounter these topics again as they continue to study mathematics.

⇔ The study of Pascal's triangle in Chapter 12 is applied in this chapter as students study binomial distributions. Normal distributions are also presented. As students study normal distributions, the review of mean, median, and mode on page 790 in the **Student Resources Toolbox** may prove helpful.

Chapter Highlights

Interview with Robert Ward: The use of mathematics in legal issues is highlighted in this interview with Robert Ward, an attorney who teaches criminal law. This application is emphasized in the related example and exercises on pages 628, 630, and 637.

Explorations involve conducting a movie survey in Section 13.3 and finding the probabilities of certain outcomes when guessing on a true/false test in Section 13.4.

The Portfolio Project: Students perform an experiment to determine whether people can distinguish a generic beverage from one or more brand-name beverages. Students evaluate their results using probability concepts developed in this chapter.

Technology: The uses of graphing calculators in this chapter include generating random numbers, finding combinations, and graphing the standard normal curve. Spreadsheets are used in Section 13.4 to find probabilities.

OBJECTIVES

Section	Objectives	NCTM Standards
13.1	• Calculate experimental, theoretical, and geometric probabilities. • Use probability to make predictions in real-world situations.	1, 2, 3, 4, 5, 11
13.2	• Find probabilities involving independent, mutually exclusive, and complementary events. • Use probability to solve problems in real-world situations.	1, 2, 3, 4, 5, 11
13.3	• Find conditional probabilities. • Use probability to make judgments in real-world situations.	1, 2, 3, 4, 5, 11
13.4	• Find the probability distribution for a binomial experiment. • Make predictions using binomial probability.	1, 2, 3, 4, 5, 10, 11
13.5	• Recognize a normal distribution. • Find probabilities involving normally distributed data. • Find probabilities in real-world situations.	1, 2, 3, 4, 5, 10, 11

INTEGRATION

Mathematical Connections	13.1	13.2	13.3	13.4	13.5
algebra	**611–617***	**618–624**	**625–631**	**632–638**	**639–645**
geometry	614–617	624	631		645
data analysis, probability, discrete math	**611–617**	**618–624**	**625–631**	**632–638**	**639–645**
patterns and functions	617				
logic and reasoning	611, 614–617	618, 620–624	626–631	632, 633, 635–638	639, 640, 642–645
Interdisciplinary Connections and Applications					
history and geography			631		643
reading and language arts		622	629		
biology and earth science					641
chemistry and physics	617				
arts and entertainment		621			
sports and recreation	612, 613, 615, 616	620, 622, 624			
manufacturing and marketing	615		631	636, 638	643, 645
health and medicine			630		644
astronomy, meteorology, electronics, architecture, driver's education, botany, agriculture	614, 615	621, 623		634, 636	642, 645

* ***Bold page numbers*** *indicate that a topic is used throughout the section.*

TECHNOLOGY

Section	*opportunities for use with* Student Book	Support Material
13.1	graphing calculator	**Technology Book:** Spreadsheet Activity 13 **Probability Constructor Activity Book:** Activities 11–17
13.2	scientific calculator	**Probability Constructor Activity Book:** Activities 18–21
13.3	graphing calculator	
13.4	graphing calculator spreadsheet software	**Technology Book:** Calculator Activity 13
13.5	graphing calculator McDougal Littell Mathpack *Stats!*	**Stats! Activity Book:** Activities 9, 15, and 19

PLANNING GUIDE

Regular Scheduling (45 min)

See Alternative Pacing Charts, page T42.

Section	Materials Needed	Core Assignment	Extended Assignment	*exercises that feature* Applications	Communication	Technology
13.1	graphing calculator, cup, thumbtack	1–12, 17–24	1–24	13, 14, 16	14, 15, 17, 18	15
13.2		1–8, 11–13, 15, 17–20	1–20	2–12, 14, 15	1, 12, 16	
13.3		1–8, 12–17, 19–25, AYP*	1–25, AYP	9–11, 18, 19		
13.4	coins, graphing calculator or software, graph paper, plastic cups, spreadsheet software	1–12, 15–18, 28–34	1–34	13, 14, 21–27	1, 11, 19, 24, 28	21, 22
13.5	graphing calculator or software, graph paper	1, 2, 6–9, 12–22, AYP	1–22, AYP	2–6, 12–14	5, 7, 15	10
Review/ Assess		**Day 1:** 1–20 **Day 2:** Ch. 13 Test	**Day 1:** 1–20 **Day 2:** Ch. 13 Test			
Portfolio Project		Allow 2 days.	Allow 2 days.			
Pacing (with Portfolio Project)		**Chapter 13 Total** 9 days				

Block Scheduling (90 min)

	Day 1	Day 2	Day 3	Day 4	Day 5
Teach/Interact	13.1 13.2	13.3: Exploration, page 625 13.4: Exploration, page 632	13.5 Port. Proj.	Review Port. Proj.	Ch. 13 Test
Apply/Assess	**13.1:** 1–12, 17–24 **13.2:** 1–8, 11–13, 15, 17–20	**13.3:** 1–8, 12–17, 19–25, AYP* **13.4:** 1–12, 15–18, 28–34	**13.5:** 1, 2, 6–9, 12–22, AYP **Port. Proj.**	**Review:** 1–20 **Port. Proj.**	**Ch. 13 Test**
NOTE: A one-day block has been added for the Portfolio Project—timing and placement to be determined by teacher.					
Pacing (with Portfolio Project)	**Chapter 13 Total** $4\frac{1}{2}$ days				

* ***AYP*** *is Assess Your Progress.*

Lesson Support

Section	Practice Bank	Study Guide*	Assessment Book*	Visuals	Explorations Lab Manual	Lesson Plans	Technology Book
13.1	83	13.1		Warm-Up 13.1		13.1	Spreadsheet Act. 13
13.2	84	13.2		Warm-Up 13.2	Add. Expl. 19	13.2	
13.3	85	13.3	Test 54	Warm-Up 13.3	Master 18	13.3	
13.4	86	13.4		Warm-Up 13.4	Master 1	13.4	Calculator Act. 13
13.5	87	13.5	Test 55	Warm-Up 13.5 Folder 13	Master 1	13.5	
Review Test	88	Chapter Review	Tests 56, 57 Alternative Assessment			Review Test	

**Spanish versions* of *Study Guide* and *Assessment Book* are available.

Chapter Support

- Course Guide
- Lesson Plans
- Portfolio Project Book
- Preparing for College Entrance Tests
- Multi-Language Glossary
- *Test Generator* Software
- Professional Handbook

Software Support

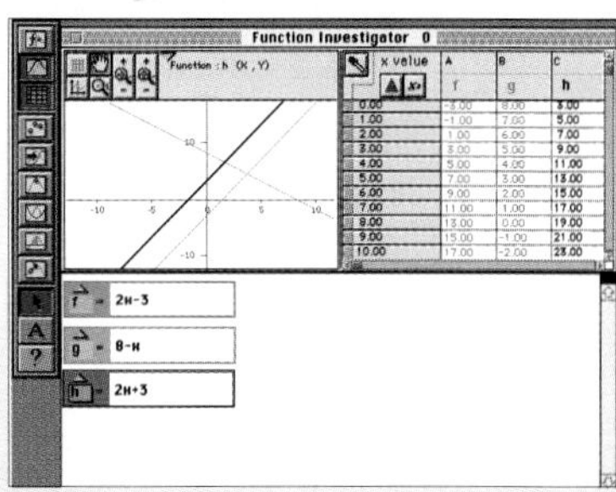

McDougal Littell Mathpack
Stats!

Internet Support

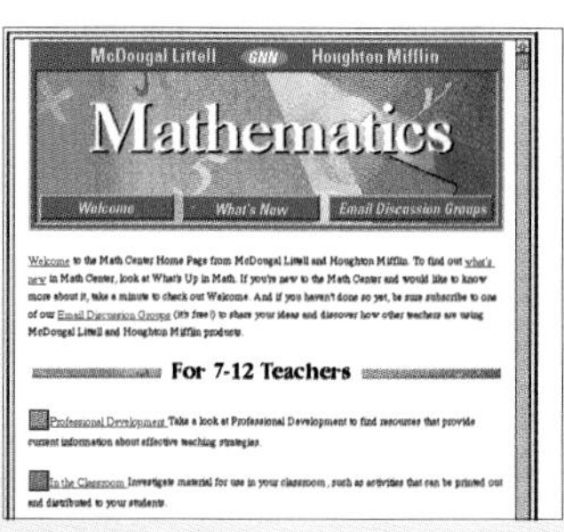

http://www.hmco.com
Next go to McDougal Littell; then the Education Center; then Secondary Math.

Outside Resources

Books, Periodicals

Watson, Jane M. "Conditional Probability: Its Place in the Mathematics Curriculum." *Mathematics Teacher* (January 1995): pp. 12–17.

Ekeland, Ivar. *The Broken Dice and Other Mathematical Tales of Chance.* University of Chicago Press, 1993.

Buckhiester, Philip G. "Probability, Problem-Formulation, and Two-Player Games." *Mathematics Teacher* (March 1994): pp. 154–159.

Floyd, Jeffrey K. "A Discrete Analysis of Final Jeopardy." *Mathematics Teacher* (May 1994): pp. 328–331.

Sloyer, Clifford W. and Richard J. Crouse. "Mathematics and Medical Indexes: a Life-saving Connection." *Mathematics Teacher* (November 1993): pp. 624–626.

Software

Kemeny, J. *Probability.* Hanover, NH: TrueBASIC.

Videos

"What is Probability?" Show No. 15 from *Against All Odds.* Arlington, MA: COMAP.

Internet

Web site:
"Carl Miller's Math Problem of the Day"
http://mmm.mbhs.edu/~cmiller/MPOTD

Chance data base. Available on the World Wide Web at
http://www.geom.umn.edu/locate/chance

Interview Notes

Background

DNA Testing
DNA testing can be applied to many areas, both inside and outside the courtroom. Inside the courtroom, DNA testing has already been used to prove a person's innocence rather than their guilt. Outside the courtroom, DNA testing may be used in such fields as medicine and anthropology. In medicine, researchers can use DNA to look for sources and cures of genetically inherited diseases, while anthropologists can use DNA profiling to examine the bones of humans and animals from ancient civilizations to see how their gene structure compares to that of humans and animals alive today.

Robert Ward
Robert Ward's perspective on teaching and practicing law has been shaped by the unusual path he has taken into law. Ward came to Boston from Philadelphia after graduating from high school to work for the Prudential Insurance Co. While taking undergraduate night courses at Northeastern University, one of his English teachers encouraged him to pursue a career in law. After graduating from Northeastern, Ward attended and graduated from Suffolk University Law School in Boston. He then joined the faculty at Suffolk to teach legal writing and research. At the New England School of Law, Ward teaches criminal procedure, evidence, and prisoners' rights.

CHAPTER

13 Probability

INTERVIEW **Robert Ward**

If you like drama, says attorney Robert Ward, there's nothing like a trial. "Courtrooms are a great place to study human behavior. You can see a full range of emotions in a single trial. For me, the attraction of trying cases is that it forces me to use the ideas and strategies I teach in the classroom. I get a firsthand chance to see how these principles actually play out in the courtroom."

"Someday, hopefully, DNA testing will make it easier for us to get criminals off the streets."

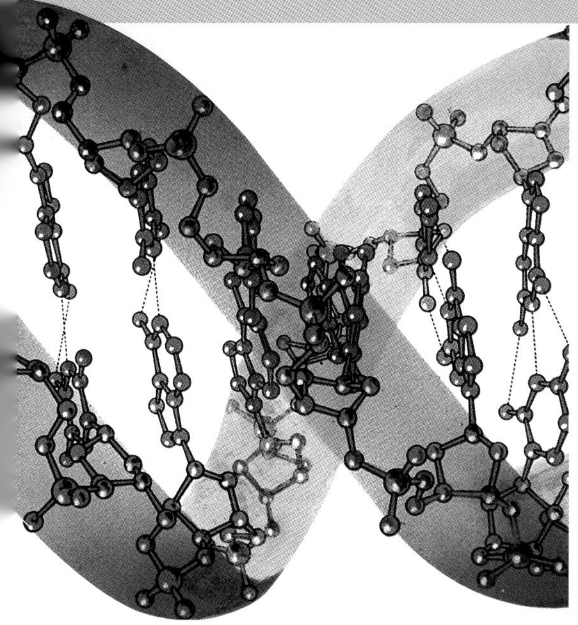

In Search of the Perfect Match

Scientists can analyze blood or hair samples by isolating the DNA contained within a cell nucleus. DNA is considered to be the chemical basis of heredity because it makes up the genes. Everyone in the world, except for identical twins, has a unique set of genes. The DNA found at a crime scene is compared with a DNA sample taken from the suspect to see if they match.

The DNA samples are compared at several well-known sites on the DNA chain. The genetic sequence found at each site is called a *marker*. If all the markers in the two samples "match up" (that is, look the same), then the two samples just may come from the same person.

Relevant Evidence

When he's not defending clients in court, Ward can be found teaching students the nuances of criminal law at New England School of Law. One of his classes is a course on evidence, which explores, among other things, the circumstances in which laboratory findings can be admitted in a trial.

In his class, Ward discusses the pros and cons of DNA testing, a relatively new technique hailed by some crime fighters as the greatest technological advance since fingerprinting. This method can be used when samples of, say, a perpetrator's blood or hair are found at the scene of a crime.

DNA Testing on Trial

DNA testing is based on the idea that the markers occur independently of one another in a strand of DNA. "Some critics have argued that DNA markers are not necessarily independent," Ward notes. "Questions have also been raised about the reliability of the tests themselves. In some cases, lab technicians may do the tests properly but still misread the results."

Given present uncertainties about the technique, Ward believes that DNA testing is most useful right now for upholding someone's innocence rather than proving his or her guilt. In time, he says, the technique should become more dependable as testing procedures improve and scientists reach a broader consensus on the methodology.

Interview Notes

Background

The Law Profession
An attorney or lawyer is a person who is licensed to represent people in a court of law or to counsel them on matters of law. Once a person becomes a lawyer, there are many avenues that their career can take. They may decide to go into areas of law dealing with defense, prosecution, corporations, or administration, just to name a few. The law profession is also the most common profession of people in Congress, state legislatures, and the administrative agencies. Almost all judges have been lawyers and about two-thirds of the presidents of the United States were lawyers.

Second-Language Learners

Some students learning English may not understand what is meant by the phrase *nuances of criminal law.* If necessary, explain that nuances are "small differences or fine points" that are likely to be overlooked or misunderstood by anyone who is not an expert. Also, some second-language learners may not be familiar with the idiomatic expression *pros and cons.* If necessary, explain that this means "arguments for a certain subject, and arguments against that subject"—in this case, DNA testing.

Interview Notes

Mathematical Connection

In Section 13.3, students explore the ideas behind the uniqueness of fingerprints and how they were proven to be unique. Another aspect of the criminal system is the use of a jury and how the members decide to either convict or acquit a person on trial. Two models that explore the likelihood of a conviction, an acquittal, or whether the jury would not be able to make a decision are explained in Section 13.4. In this section, students use the two models to explore the probability of a conviction with both 6-member and 12-member juries, and then compare the probabilities.

Explore and Connect

Writing
Students may find it helpful to relate this probability to the probability of different population samples. For example, a large crowd at a sporting event is often around 50,000 people, so about 50 of them could have both of these markers. In a large city, the population could be 1 million people, and thus 1000 of them could have these markers.

Research
Students can use the library as a source for information, or if they know someone who is an attorney, they may want to ask them these questions and discuss the ideas about evidence.

Project
If students ask an attorney about the research information, they may want to also discuss this project with the attorney. Doing this could lead to a broader understanding of the ideas and types of evidence used in trials.

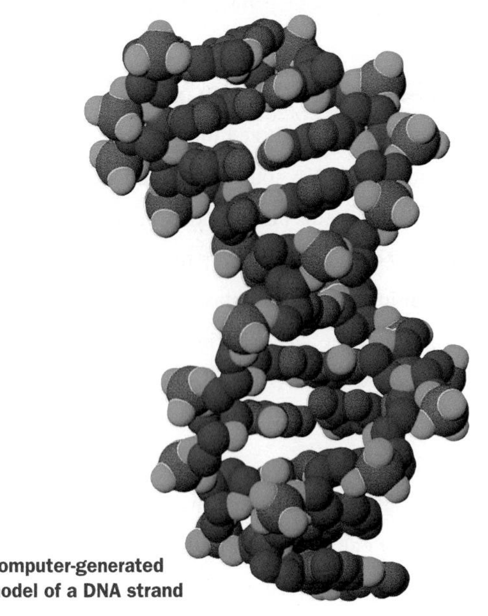

Computer-generated model of a DNA strand

Determining the Probability

A single strand of DNA contains 3 billion markers. It is impossible to compare so many markers between two DNA samples using present technology. Typically, 3 to 6 distinguishing markers are compared. In order to determine the probability that two different DNA samples come from the same person, the probabilities associated with each of the markers are multiplied, assuming that the markers are *independent* of one another.

Suppose, for example, that marker A occurs in 2% of the population and marker B occurs in 5% of the population. The probability of two DNA samples having both markers is 1 in 1000:

$$\text{probability of having both markers} = \frac{2}{100} \times \frac{5}{100} = \frac{1}{1000}$$

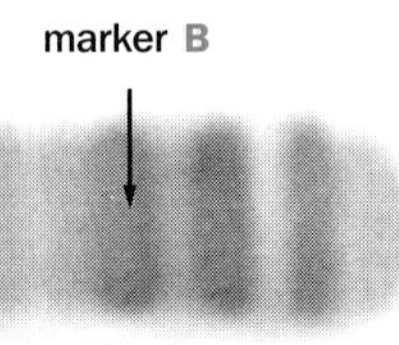

EXPLORE AND CONNECT

Robert Ward reads a legal brief in his office.

1. Writing Do you think the probability calculated above is low enough to convict a suspect if a match is made between the suspect's DNA and the DNA found at a crime scene? What do you think happens to the probability of a match if more markers are compared?

2. Research DNA is just one type of identifying evidence used in court trials. Find out what other kinds of identifying evidence are admitted in trials. Are there any problems associated with them?

3. Project Using information from your research, rank the different types of identifying evidence allowed in trials, based on your estimates of the probability of their accuracy. Explain your ranking.

Mathematics & Robert Ward

In this chapter, you will learn more about how mathematics is related to legal issues.

Related Examples and Exercises

Section 13.3
- **Example 3**
- **Exercises 10 and 11**

Section 13.4
- **Exercises 21–24**

Section

13.1 Exploring Probability

Learn how to...

- **calculate experimental, theoretical, and geometric probabilities**

So you can...

- **predict the likelihood that a person will make a free throw, for example**

During a Chicago Bulls basketball game on April 14, 1993, spectator Don Calhoun got a chance to win \$1 million. All he had to do was make a 71 ft, three-quarter-court shot from the opposite free-throw line. Amazingly, Calhoun made the shot and became a rich man.

THINK AND COMMUNICATE

1. If Don Calhoun had taken 100 shots at a distance of 71 ft from the basket, how many do you think he would have made? How would you express Calhoun's chances of making a 71 ft shot?
2. A regulation free throw is shot 15 ft from the basket. What do you think your chances are of making a regulation free throw?

The **probability** of an event A, denoted $P(A)$, is the fraction of the time that A is expected to happen. For example, if you make 3 out of every 4 free throws you attempt, then the probability of your making a free throw is $\frac{3}{4}$, or 0.75.

For any event A, $0 \le P(A) \le 1$. The closer $P(A)$ is to 0, the *less* likely A is to happen. The closer $P(A)$ is to 1, the *more* likely A is to happen.

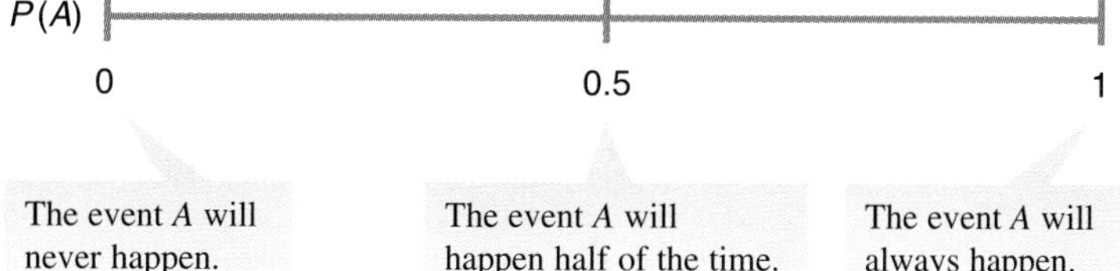

The event A will never happen.

The event A will happen half of the time.

The event A will always happen.

You can estimate your probability of making a free throw by performing an experiment in which you shoot, say, 50 free throws and record how many you make. The ratio of the number of free throws made to the number attempted is your *experimental probability* of making a free throw.

Plan⇔Support

Objectives

- Calculate experimental, theoretical, and geometric probabilities.
- Use probability to make predictions in real-world situations.

Recommended Pacing

Core and Extended Courses
Section 13.1: 1 day

Block Schedule
Section 13.1: $\frac{1}{2}$ block
(with Section 13.2)

Resource Materials

Lesson Support
Lesson Plan 13.1
Warm-Up Transparency 13.1
Practice Bank: Practice 83
Study Guide: Section 13.1

Technology
Technology Book:
Spreadsheet Activity 13
Graphing Calculator
McDougal Littell Mathpack
Probability Constructor
Activity Book:
Activities 18–21
Internet:
http://www.hmco.com

Warm-Up Exercises

1. Give the equation of a circle of radius 1 centered at the origin. $x^2 + y^2 = 1$
2. What is the area of the circle in Ex. 1? π
3. What is the area of a circle with radius 3? 9π
4. Give the ratio of the areas of the circles in Exs. 1 and 3. $\frac{1}{9}$
5. Use the formula S.A. = $4\pi r^2$ which gives the surface area of a sphere with radius r. Calculate the surface area of a sphere with radius 14. $784\pi \approx 2463$

ANSWERS Section 13.1

Think and Communicate

1. Answers may vary. You can express the chances as the fraction $\frac{\text{shots made}}{\text{shots attempted}}$ or its decimal equivalent.
2. Answers may vary.

Teach⇔Interact

Learning Styles: Visual

Students with a visual learning style should be encouraged to take a visual note of the probability number line illustration on page 611. This illustration can help them remember that $0 \le P(A) \le 1$ and that events are more likely to happen as $P(A)$ gets closer to 1.

Section Notes

Historical Connection
Probability began as a study of games of chance. In 1654, two French mathematicians, Pascal and Fermat, corresponded in a series of letters about a dice game. These letters form the basis for probability theory. Interest in problems involving probability was generated at that time due to the establishment of insurance.

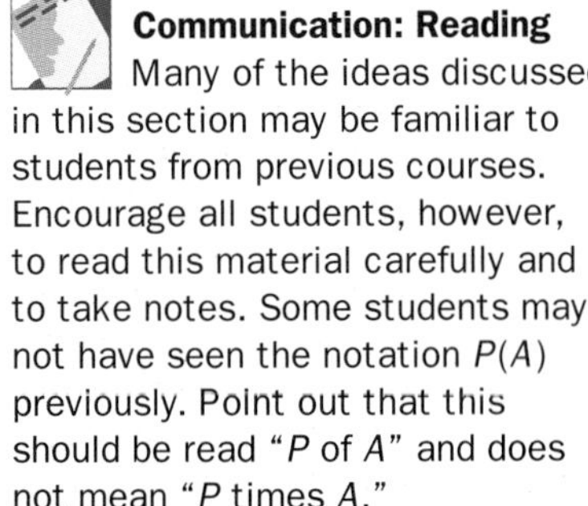

Communication: Reading
Many of the ideas discussed in this section may be familiar to students from previous courses. Encourage all students, however, to read this material carefully and to take notes. Some students may not have seen the notation $P(A)$ previously. Point out that this should be read "P of A" and does not mean "P times A."

Additional Example 1

While playing for her high school softball team, Lena got 17 hits during 43 times at bat. What was the probability of Lena getting a hit?

Use the formula for experimental probability.

$$P(\text{Lena gets a hit}) = \frac{\text{number of hits}}{\text{number of at bats}} = \frac{17}{43} \approx 0.395$$

The probability of Lena getting a hit was about 0.395.

Section Note

Teaching Tip
Point out to students that a theoretical probability is a prediction about what will happen while an experimental probability is based on events that have already happened.

Experimental Probability

Suppose you perform an experiment consisting of a certain number of *trials* (such as free-throw attempts). The **experimental probability** of an event A (such as making a free throw) is given by:

$$P(A) = \frac{\text{number of trials where } A \text{ happens}}{\text{total number of trials}}$$

EXAMPLE 1 Application: Sports

Lorri Bauman (55) prepares to take a shot.

Lorri Bauman holds the National Collegiate Athletic Association (NCAA) Division 1 women's basketball record for the most free throws made. While at Drake University during 1981–1984, she made **907** free throws in **1090** attempts. What was the probability of Lorri Bauman making a free throw?

SOLUTION

Use the formula for experimental probability.

$$P\left(\begin{matrix}\text{Lorri Bauman}\\ \text{makes a free throw}\end{matrix}\right) = \frac{\text{number of free throws made}}{\text{number of free throws attempted}} = \frac{907}{1090} \approx 0.832$$

The probability of Lorri Bauman making a free throw was about 0.832.

Theoretical Probability

Suppose you want to find the probability of getting an odd number when you roll a six-sided die. There are 6 possible outcomes:

Of these outcomes, 3 correspond to getting an odd number:

Since the possible outcomes are equally likely to happen, it is reasonable to expect that the fraction of the time you will roll an odd number is the ratio of the number of odd outcomes to the number of possible outcomes:

$$P(\text{rolling an odd number}) = \frac{3}{6} = 0.5$$

This value, 0.5, is the *theoretical probability* of rolling an odd number.

Theoretical Probability

Suppose a procedure (such as rolling a die) can result in n equally likely outcomes. If an event A (such as rolling an odd number) corresponds to m of these outcomes, then the **theoretical probability** of A is given by:

$$P(A) = \frac{m}{n}$$

EXAMPLE 2 Application: Sports

Every spring, the National Basketball Association (NBA) holds a lottery to determine the order in which its teams choose new members from among college players. In the 1993 lottery, each team that did not make the playoffs was assigned a certain number of lottery entries. Teams with poorer records received more entries.

Team (with 1992–93 win-loss record)	Number of entries
Dallas (11–71)	11
Minnesota (19–63)	10
Washington (22–60)	9
Sacramento (25–57)	8
Philadelphia (26–56)	7
Milwaukee (28–54)	6
Golden State (34–48)	5
Denver (36–46)	4
Miami (36–46)	3
Detroit (40–42)	2
Orlando (41–41)	1

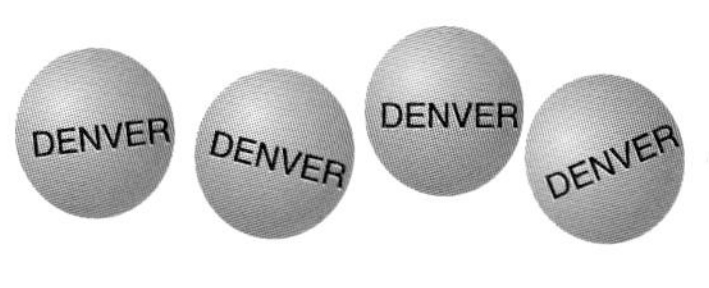

Each of a team's entries was represented by a ball with the team's name on it.

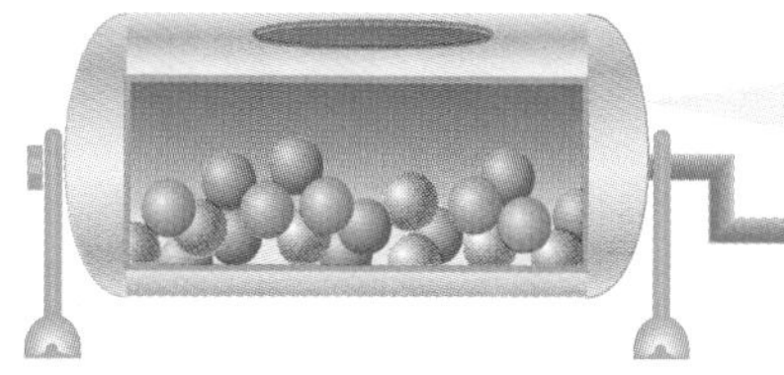

The balls were placed in a cylinder and were mixed well. The first ball drawn determined the team that chose first from the pool of college players.

What was the probability that Denver got to choose first in 1993?

SOLUTION

Each ball in the cylinder had an equally likely chance of being drawn first. Therefore:

$$P\begin{pmatrix}\text{Denver got}\\ \text{to choose first}\end{pmatrix} = \frac{\text{number of balls with Denver's name on them}}{\text{total number of balls}}$$

$$= \frac{4}{1 + 2 + 3 + \cdots + 11}$$

$$= \frac{4}{66}$$

$$= \frac{2}{33}, \text{ or about } 0.0606$$

The probability that Denver got to choose first was about 0.0606.

Teach⇔Interact

Section Note

Common Error

When calculating theoretical probabilities, it is essential that the outcomes being counted are equally likely. Students often forget this fact. For example, when tossing a coin twice, a student might say that the probability of getting two heads is $\frac{1}{3}$ as there are 3 possible outcomes: one head, two heads, or no heads. However, this is incorrect because these three outcomes are not equally likely. There are two ways to get one head, a head on the first toss or a head on the second toss, and thus this outcome is more likely than the other two. The correct equally likely outcomes are TT, HT, TH, and HH, one of which is two heads, and thus the correct probability is $\frac{1}{4}$.

Additional Example 2

In a regional cross-country race, seven teams are competing. The number of students running from each team is given in the table below. Because the race course is narrow, it is decided that the start of the race will be staggered. To decide the starting order, numbers are drawn. What is the probability that the first runner to start the race will be from Oxbridge?

Team	Number of runners
Wilton	9
Brownsville	7
Colton	9
Urbane	12
Oxbridge	3
River Bend	6
Briarwood	5

Each runner has an equal chance of drawing number 1. Therefore:

P(Oxbridge starts first)

$$= \frac{\text{number of runners from Oxbridge}}{\text{total number of runners}}$$

$$= \frac{3}{9 + 7 + 9 + 12 + 3 + 6 + 5}$$

$$= \frac{3}{51}$$

$$= \frac{1}{17}, \text{ or about } 0.0588$$

Teach⇔Interact

Think and Communicate

Question 4 focuses students' attention on those situations where the theoretical probability can be determined and on those where it cannot be determined, but a simulation or previous data can be used to estimate the theoretical probability. Point out to students that in the lottery example, there were a finite number of equally likely possibilities. This allowed the theoretical probability to be calculated. In the basketball player example, the outcomes, a basket or no basket, are not equally likely, and thus a theoretical probability cannot be calculated.

Section Notes

Integrating the Strands
Geometric probability is an excellent example of how two seemingly unrelated parts of mathematics, geometry and probability, can be used together to solve problems that cannot be solved otherwise.

Using Technology
The *Probability Constructor Activity Book* provides additional opportunities for students to explore experimental, theoretical, and geometric probabilities.

Additional Example 3

The area of the Atlantic Ocean is about 106,450,000 km^2. If a meteorite were to strike Earth, what is the probability that it would land in the Atlantic Ocean? Use that fact that Earth's radius is about 6378 km.

It is reasonable to assume that each point on Earth has an equally likely chance of being hit by a meteorite. Therefore, the probability that a meteorite lands in the Atlantic Ocean is just the fraction of Earth's surface area that the Atlantic Ocean occupies. Using the formula S.A. = $4\pi r^2$,

P(a meteorite lands in the Atlantic)

$$= \frac{\text{area of the Atlantic}}{\text{surface area of Earth}} = \frac{106,450,000}{4\pi(6378)^2} \approx \frac{106,450,000}{511,185,932} \approx 0.20824$$

The probability that a meteorite striking Earth lands in the Atlantic Ocean is about 0.20824.

THINK AND COMMUNICATE

3. Orlando won the first pick in the NBA lottery in 1993 and chose Chris Webber of the University of Michigan. What was the probability of Orlando's winning?

4. a. Why is it possible to compute a team's probability of winning the first pick in the NBA lottery without first performing an experiment in which the lottery drawing is simulated many times?

b. Why *can't* you estimate a basketball player's probability of making a free throw without first having the player shoot many free throws?

Geometric Probability

Sometimes the probability of an event can be found by comparing the areas of geometric figures. The study of probability based on length, area, or volume is called **geometric probability**.

BY THE WAY

Mare is a Latin word meaning "sea." The Italian scientist Galileo first used the word to name the moon's plains because he mistakenly believed they were covered with water. The name *Mare Serenitatis* means "Sea of Serenity."

EXAMPLE 3 Application: Astronomy

Each year, millions of meteorites strike the moon, producing craters in the moon's surface. Many of these meteorites land in a *mare*—a large, flat plain that is seen as a dark patch from Earth. What is the probability that a meteorite striking the moon lands in Mare Serenitatis?

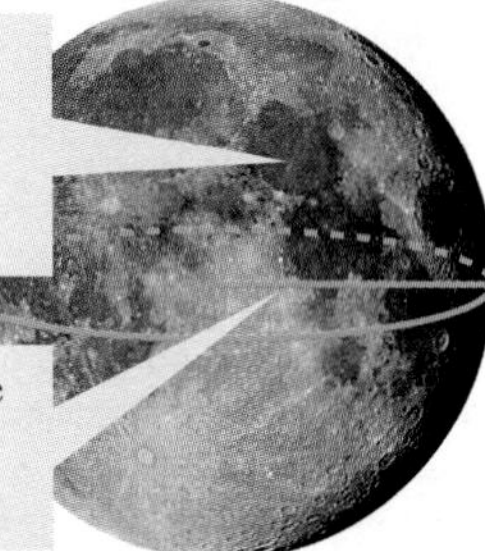

The area of Mare Serenitatis is approximately 125,000 mi^2.

The radius of the moon is about 1080 mi.

SOLUTION

It is reasonable to assume that each point on the moon has an equally likely chance of being hit by a meteorite. Therefore, the probability that a meteorite lands in Mare Serenitatis is just the fraction of the moon's surface area that the mare occupies:

$$P\left(\begin{matrix}\text{a meteorite lands}\\ \text{in Mare Serenitatis}\end{matrix}\right) = \frac{\text{area of Mare Serenitatis}}{\text{surface area of moon}} \approx \frac{125,000}{4\pi(1080)^2} \approx \frac{125,000}{14,657,415} \approx 0.00853$$

Use the formula **S.A. = $4\pi r^2$** for the surface area of a sphere.

The probability that a meteorite striking the moon lands in Mare Serenitatis is about 0.00853.

Think and Communicate

3. $\frac{1}{66}$, or about 0.0152

4. a. There are 66 possible equally likely outcomes, one for each ball in the cylinder. Because each event (a team winning the lottery) corresponds to a known number of outcomes, you can compute the theoretical probability.

b. because you are not dealing with equally likely outcomes but outcomes that will vary with each player

THINK AND COMMUNICATE

5. Out of every 1 million meteorites that strike the moon, how many would you expect to land in Mare Serenitatis? Explain.

CHECKING KEY CONCEPTS

1. MANUFACTURING A quality control engineer randomly selects 250 hard disk drives from the assembly line of a computer manufacturing plant. The engineer finds that 8 of the drives are defective. Estimate the probability that a drive on the assembly line is defective.

2. Suppose you roll 2 six-sided dice, one white and one red. Find the probability that the number showing on the red die is greater than the number showing on the white die.

3. ASTRONOMY The area of the United States is about 3,618,770 mi^2. What is the probability that a meteorite striking Earth lands in the United States? (*Hint:* The radius of Earth is about 3963 mi.)

13.1 Exercises and Applications

Extra Practice exercises on page 769

GAMES A standard deck of playing cards consists of 52 cards, with 13 cards in each of 4 *suits:* clubs, spades, diamonds, and hearts. *Face cards* are jacks, queens, and kings. Find the probability of choosing each type of card at random from a standard deck.

Clubs ♣	ace, 2, 3, 4, 5, 6, 7, 8, 9, 10, jack, queen, king
Spades ♠	ace, 2, 3, 4, 5, 6, 7, 8, 9, 10, jack, queen, king
Diamonds ♦	ace, 2, 3, 4, 5, 6, 7, 8, 9, 10, jack, queen, king
Hearts ♥	ace, 2, 3, 4, 5, 6, 7, 8, 9, 10, jack, queen, king

1. a 9 of diamonds
2. an ace
3. a heart
4. a face card
5. a red card
6. not a 3 of spades
7. not a club
8. a red or black card
9. an 11 of clubs

The targets shown are an equilateral triangle, a square, and a circle. Suppose a randomly thrown dart hits each target. Find the probability that the dart hits the target's shaded region. (The shaded regions are also formed from equilateral triangles, squares, and circles.)

10. 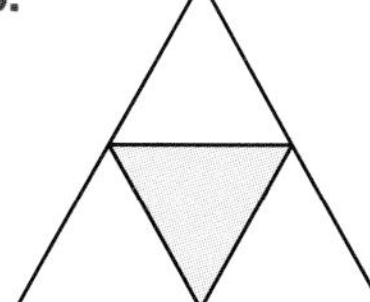

11. 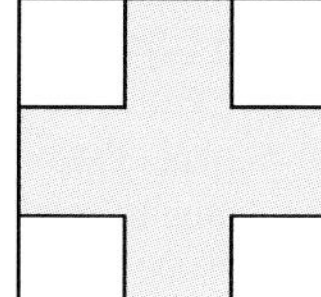

12. 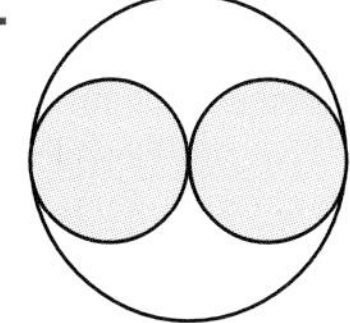

Teach⇔Interact

Checking Key Concepts

Problem Solving
When determining the set of equally likely possibilities, such as for question 2, it is important for students to organize their work so that no possibilities are forgotten. In this case, a table can be constructed where the outcomes on the white die are written in the first column and the outcomes of the red die are in the first row. Another way to organize the work is by making a systematic list. This is usually done by holding all but one item constant while that one changes and then by changing each of the others in turn.

Student Progress
After students have answered these questions ask them which probabilities are theoretical and which are experimental.

Closure Question

Describe three different types of probabilities and how each one is found.

Theoretical probability is the likelihood that an event will occur. In a situation where there are a fixed number of equally likely outcomes, it is found by dividing the number of outcomes that make up the event by the total number of possible outcomes. Geometric probability is a type of theoretical probability found by comparing the areas (or other geometric measurements) of geometric figures. Experimental probability is related to performing an experiment and is found by dividing the number of times an event occurs by the total number of trials.

Think and Communicate

5. about 8530; The decimal 0.00853 is equivalent to the fraction $\frac{8530}{1{,}000{,}000}$, which means that this ratio, interpreted as 8530 meteorites landing in Mare Serenitatis out of 1,000,000 hitting the moon, expresses the same probability that was determined geometrically.

Checking Key Concepts

1. $\frac{8}{250}$, or 0.032
2. $\frac{15}{36}$, or about 0.417
3. $\frac{3{,}618{,}770}{4\pi(3963)^2}$, or about 0.0183

Exercises and Applications

1. $\frac{1}{52}$, or about 0.0192
2. $\frac{1}{13}$, or about 0.0769
3. $\frac{1}{4}$, or 0.25
4. $\frac{3}{13}$, or about 0.231
5. $\frac{1}{2}$, or 0.5
6. $\frac{51}{52}$, or about 0.981
7. $\frac{3}{4}$, or 0.75
8. 1
9. 0
10. $\frac{1}{4}$, or 0.25
11. $\frac{5}{9}$, or about 0.556
12. $\frac{1}{2}$, or 0.5

Apply⇔Assess

Suggested Assignment

Core Course
Exs. 1–12, 17–24

Extended Course
Exs. 1–24

Block Schedule
Day 1 Exs. 1–12, 17–24

Apply⇔Assess

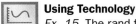

Exercise Notes

Multicultural Note

Exs. 13, 14 People around the world have played games with dice for millennia. Materials for making dice have included fruit pits, animal bones or teeth, pebbles, pottery, walnut shells, and a variety of seashells. Dice were first mentioned in print in the *Mahabharata*, a sacred epic poem written in India more than 2000 years ago.

Using Technology

Ex. 15 The rand feature of the TI-82, found under the MATH PRB menu, will give a random number greater than 0 and less than 1. Students can enter rand on the home screen and then repeatedly press ENTER to generate their random numbers. They will need 40, 20 *x*'s and 20 *y*'s. Another way to use the calculator in this exercise is to enter the random coordinates as lists and then use the list features to calculate the sum of the squares of the coordinates. To do this, type on the home screen

```
seq(rand,X,1,20,1)→L1
seq(rand,X,1,20,1)→L2
L1²+L2²→L3
```

This will generate a sequence of 20 random numbers in L1 for the *x*-coordinates and a sequence of 20 random numbers in L2 for the *y*-coordinates. Then it will calculate the sum of the squares of each pair of coordinates and enter the result in L3. To see how many of the 20 points lie in the quarter circle, scan L3 and count how many of these numbers are less than or equal to 1. This procedure would be an easy way for students to test what happens when more points are used.

Alternate Approach

Ex. 15 As an alternative to testing all 20 points algebraically to see if they fit in the quarter circle, students could draw a large diagram of the square and quarter circle on graph paper and plot each random point to see if it lies in the quarter circle. Most points will be clearly in or out of the circle. Those that lie near the boundary should also be tested algebraically as a check. This method will help visual learners better understand this exercise.

Connection GAMES

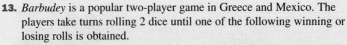

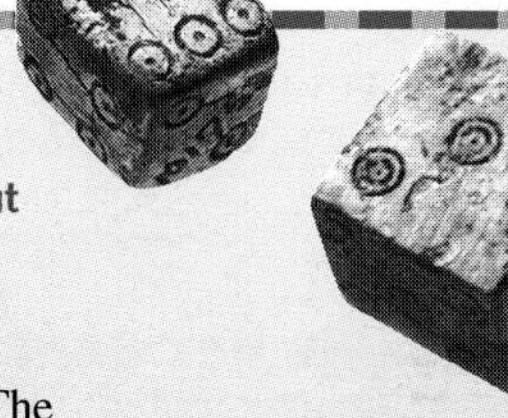

Ancient Egyptian dice

Did you know that the six-sided dice used in the United States today are almost identical to those used in China about 600 B.C. and in Egypt about 2000 B.C.? Games involving dice have been developed in virtually every part of the world.

13. *Barbudey* is a popular two-player game in Greece and Mexico. The players take turns rolling 2 dice until one of the following winning or losing rolls is obtained.

Winning rolls:

Losing rolls:

Find the probability of getting each type of roll.

a. a winning roll **b.** a losing roll **c.** neither a winning nor a losing roll

14. The Chinese game *kon mín yéung* is played with 6 six-sided dice. Players score points by rolling at least three of a kind.

a. Cooperative Learning Work with a partner. You need 6 six-sided dice. Roll the dice 100 times. For each roll, record whether you get at least 3 of a kind. What is your experimental probability of getting at least 3 of a kind?

b. Challenge Show that the theoretical probability of getting at least 3 of a kind when rolling 6 dice is $\frac{119}{324}$. How does this value compare with the experimental probability you obtained in part (a)?

15. Cooperative Learning You can use probability to approximate π. Work in a group of five students. You each need a graphing calculator.

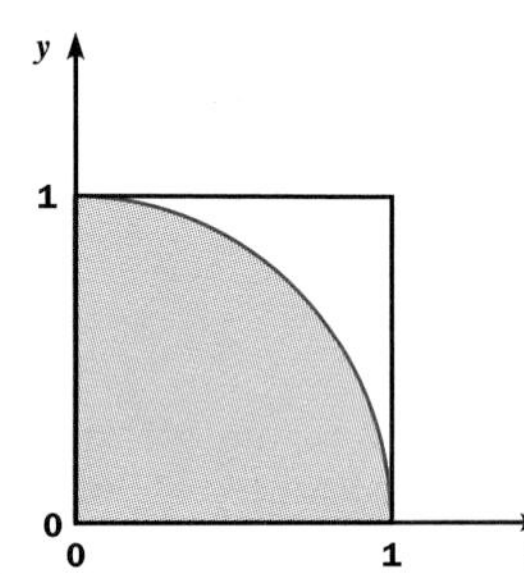

a. Suppose you randomly choose a point in the square region at the right. Show that:

$$P\left(\begin{matrix}\text{the point lies in the}\\ \text{shaded quarter circle}\end{matrix}\right) = \frac{\pi}{4}$$

b. Technology You can generate a random point (x, y) in the square region by having your calculator produce two random numbers x and y between 0 and 1. Each of you should generate 20 such points and determine whether each point lies in the quarter circle. (*Hint:* Note that (x, y) lies in the quarter circle if and only if $x^2 + y^2 \le 1$.)

c. Based on the 100 points your group generated in part (b), what is the experimental probability that a random point (x, y) lies in the quarter circle? Use this probability and the equation in part (a) to approximate π. How close is your approximation to 3.14, the value of π to two decimal places? How would using more random points affect your approximation?

13. a. $\frac{5}{36}$, or about 0.139

b. $\frac{5}{36}$, or about 0.139

c. $\frac{13}{18}$, or about 0.722

14. See answers in back of book.

15. a. The geometric probability that the point lies in the shaded region is given by $\frac{\text{shaded area}}{\text{area of square}} = \frac{\frac{1}{4} \cdot \pi \cdot 1^2}{1^2} = \frac{\pi}{4}$.

b. Answers may vary. On average, about 16 of the 20 points will lie in the quarter circle.

c. The experimental probabilities will vary. Students should multiply their experimental probability by four to obtain their estimate of π. The more random points used, the more accurate the approximation.

16. a. $R_2 + R_3$

b. $\frac{\pi(R_2 + R_3)^2}{\pi R_1^2} = \frac{(R_2 + R_3)^2}{R_1^2}$

c. about 3.05×10^{-9}

d. 328,000,000

16. PHYSICS A *neutron* is a subatomic particle that has no electric charge. When a beam of neutrons is directed at a piece of lead foil, almost all of the neutrons pass straight through the foil. This is because the lead atoms that make up the foil are mostly empty space. Sometimes, however, a neutron will strike a *nucleus*—the positively charged central region of an atom—and be deflected from its original path.

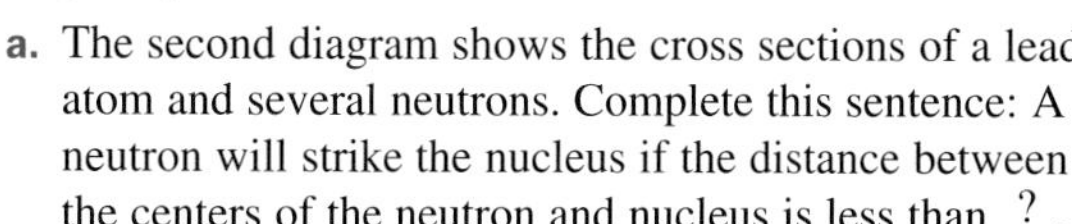

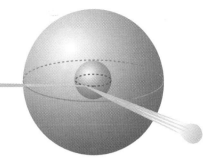

a. The second diagram shows the cross sections of a lead atom and several neutrons. Complete this sentence: A neutron will strike the nucleus if the distance between the centers of the neutron and nucleus is less than _?_.

b. In terms of R_1, R_2, and R_3, what is the probability that a neutron passing through a lead atom strikes the atom's nucleus? (*Hint:* Assume the path of the neutron's center intersects the atom's cross section at a random point.)

c. The values (in meters) of R_1, R_2, and R_3 are as follows:

$$R_1 = 1.75 \times 10^{-10}$$
$$R_2 = 8.27 \times 10^{-15}$$
$$R_3 = 1.40 \times 10^{-15}$$

Use these values to compute the probability expressed in part (b).

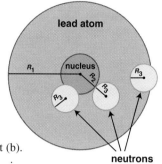

d. Complete this sentence: About 1 out of every _?_ neutrons passing through a lead atom will strike the atom's nucleus.

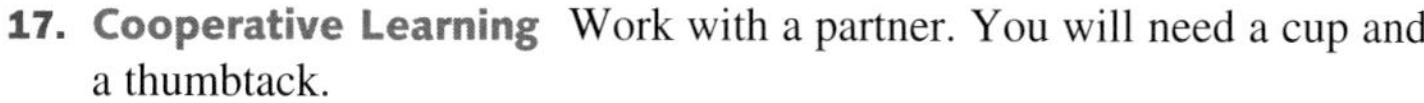

17. Cooperative Learning Work with a partner. You will need a cup and a thumbtack.

a. One of you should put the thumbtack into the cup, shake the cup, and "pour" the thumbtack onto a flat surface. The other should record whether the thumbtack lands "point up" or "point down." Repeat this process at least 20 times. What is the experimental probability that the thumbtack lands "point up"?

b. How do you think the probability in part (a) would change if the thumbtack had a wider head? a longer point?

ONGOING ASSESSMENT

18. Writing In your own words, define experimental, theoretical, and geometric probability. Give examples of real-world situations where each type of probability is used.

SPIRAL REVIEW

Use the binomial theorem to expand each power of a binomial. *(Section 12.5)*

19. $(a + b)^3$ **20.** $(x - y)^5$ **21.** $(2u + 3v)^4$ **22.** $\left(u^2 + v^3\right)^4$

Find all real and imaginary zeros of each function. Identify any double or triple zeros. *(Section 9.5)*

23. $f(x) = x^3 - 3x^2 - 7x + 12$ **24.** $g(x) = 4x^4 - 4x^3 + 5x^2 - 4x + 1$

Exercise Notes

Second-Language Learners

Ex. 16 The information in this exercise may prove challenging for some students learning English. If necessary, have students work in small groups to discuss the facts and then work in pairs on parts (a) through (d). Also, it may be helpful to explain that the word *lead* refers to the metallic element and is pronounced with a short *e*.

Communication: Writing

Ex. 18 This exercise provides students with an excellent opportunity to review and summarize the main concepts of this section. Challenge students to think of new real-world examples. You may want to ask several students to read their responses to the class so that all students can check what they have written.

Practice 83 for Section 13.1

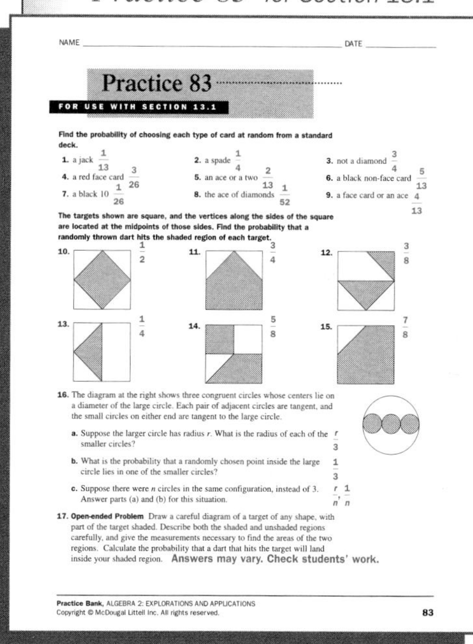

NAME ______ DATE ______

Practice 83

FOR USE WITH SECTION 13.1

Find the probability of choosing each type of card at random from a standard deck.

1. a jack $\frac{1}{13}$
2. a spade $\frac{1}{4}$
3. not a diamond $\frac{3}{4}$
4. a red face card $\frac{3}{26}$
5. an ace or a two $\frac{2}{13}$
6. a black non-face card $\frac{5}{13}$
7. a black 10 $\frac{1}{26}$
8. the ace of diamonds $\frac{1}{52}$
9. a face card or an ace $\frac{4}{13}$

The targets shown are square, and the vertices along the sides of the square are located at the midpoints of those sides. Find the probability that a randomly thrown dart hits the shaded region of each target.

10. $\frac{1}{2}$
11. $\frac{3}{4}$
12. $\frac{3}{8}$
13. $\frac{1}{4}$
14. $\frac{5}{8}$
15. $\frac{7}{8}$

16. The diagram at the right shows three congruent circles whose centers lie on a diameter of the large circle. Each pair of adjacent circles are tangent, and the small circles on either end are tangent to the large circle.

a. Suppose the larger circle has radius r. What is the radius of each of the smaller circles? $\frac{r}{3}$

b. What is the probability that a randomly chosen point inside the large circle lies in one of the smaller circles? $\frac{1}{3}$

c. Suppose there were n circles in the same configuration, instead of 3. Answer parts (a) and (b) for this situation. $\frac{r}{n}, \frac{1}{n}$

17. **Open-ended Problem** Draw a careful diagram of a target of any shape, with part of the target shaded. Describe both the shaded and unshaded regions carefully, and give the measurements necessary to find the areas of the two regions. Calculate the probability that a dart that hits the target will land inside your shaded region. **Answers may vary. Check students' work.**

Practice Bank, ALGEBRA 2: EXPLORATIONS AND APPLICATIONS

83

17. a. Answers may vary.

b. Answers may vary. An example is given. A wider head might increase the probability, while a longer point might decrease the possibility.

18. Answers may vary. Examples are given. An experimental probability is estimated by conducting a real-world experiment of many trials, and then finding the ratio of the number of successes to the number of trials. A batting average is an experimental probability. A theoretical probability is calculated by finding the ratio of the number of outcomes representing an event to the total number of equally likely outcomes. You can find the theoretical probability of winning a prize in a contest or lottery. A geometric probability is calculated for random events in a region by finding the ratio of the length, area, or volume of a target portion of the region to the total length, area, or volume. You might use a geometric probability to describe the likelihood of a given area receiving rainfall from randomly scattered summer showers.

19. $a^3 + 3a^2b + 3ab^2 + b^3$

20. $x^5 - 5x^4y + 10x^3y^2 - 10x^2y^3 + 5xy^4 - y^5$

21. $16u^4 + 96u^3v + 216u^2v^2 + 216uv^3 + 81v^4$

22. $u^8 + 4u^6v^3 + 6u^4v^6 + 4u^2v^9 + v^{12}$

23. $4, \frac{-1 - \sqrt{13}}{2}, \frac{-1 + \sqrt{13}}{2}$

24. $\frac{1}{2}$ (double zero), $i, -i$

Plan⇔Support

Objectives

- Find probabilities involving independent, mutually exclusive, and complementary events.
- Use probability to solve problems in real-world situations.

Recommended Pacing

Core and Extended Courses
Section 13.2: 1 day

Block Schedule
Section 13.2: $\frac{1}{2}$ block
(with Section 13.1)

Resource Materials

Lesson Support
Lesson Plan 13.2
Warm-Up Transparency 13.2
Practice Bank: Practice 84
Study Guide: Section 13.2
Explorations Lab Manual: Additional Exploration 19

Technology
Scientific Calculator
McDougal Littell Mathpack *Probability Constructor*
Activity Book: Activities 11–17
Internet: http://www.hmco.com

Warm-Up Exercises

Find the probability of choosing each type of card at random from a standard deck of 52 playing cards.

1. a king $\frac{1}{13}$
2. a queen of spades $\frac{1}{52}$
3. not a 7 $\frac{48}{52} = \frac{12}{13}$
4. a black card $\frac{1}{2}$

Find the probability of each sum when two six-sided die are thrown.

5. two $\frac{1}{36}$
6. eight $\frac{5}{36}$

Section

13.2 Working with Multiple Events

Learn how to...

- **find probabilities involving independent, mutually exclusive, and complementary events**

So you can...

- **solve problems about genetics, for example**

Many traits of plants and animals are passed from one generation to another through *genes*. Genes occur in pairs, and each pair controls a particular trait. One kind of pea plant has a pair of "height genes" that determines whether the plant is tall or short.

Gene pair:	TT	TS	SS
Plant height:	tall	tall	short

Use T for a "tall gene" and S for a "short gene."

Notice that a TS pair—a tall gene and a short gene—produces a tall plant. The tall gene is said to be *dominant* because it "overpowers" the short gene. The short gene is said to be *recessive*.

BY THE WAY

In 1866, Gregor Mendel, an Austrian monk, described how pea plants inherit certain traits. Mendel is considered the founder of modern genetics.

THINK AND COMMUNICATE

Suppose two pea plants, plant 1 and plant 2, both have a TS gene pair. The plants can be crossed to produce a new "child" plant. The height genes of the child consist of one height gene from each parent. A parent is equally likely to transmit its T or S gene.

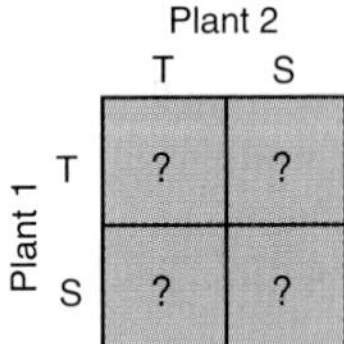

Plant 1 \ Plant 2	T	S
T	?	?
S	?	?

1. Copy and complete the table showing the possible gene pairs for the child plant.
2. Find each probability.
   a. $P\left(\begin{array}{c}\text{plant 1}\\\text{transmits S gene}\end{array}\right)$ **b.** $P\left(\begin{array}{c}\text{plant 2}\\\text{transmits S gene}\end{array}\right)$ **c.** $P\left(\begin{array}{c}\text{child}\\\text{is short}\end{array}\right)$
   d. How are the probabilities in parts (a)–(c) related?
3. Find each probability.
   a. $P\left(\begin{array}{c}\text{child has a}\\\text{TT gene pair}\end{array}\right)$ **b.** $P\left(\begin{array}{c}\text{child has a}\\\text{TS gene pair}\end{array}\right)$ **c.** $P\left(\begin{array}{c}\text{child}\\\text{is tall}\end{array}\right)$
   d. How are the probabilities in parts (a)–(c) related?
4. How can you use *P*(child is short) to find *P*(child is tall)?

ANSWERS Section 13.2

Think and Communicate

1.

Plant 1 \ Plant 2	T	S
T	TT	TS
S	TS	SS

2. a. $\frac{1}{2}$, or 0.5
   b. $\frac{1}{2}$, or 0.5
   c. $\frac{1}{4}$, or 0.25
   d. The probability in part (c) is the product of the probabilities in parts (a) and (b).
3. a. $\frac{1}{4}$, or 0.25
   b. $\frac{1}{2}$, or 0.5
   c. $\frac{3}{4}$, or 0.75
   d. The probability in part (c) is the sum of the probabilities in parts (a) and (b).
4. Since a plant is either short or tall, *P*(child is short) + *P*(child is tall) = 1. So, *P*(child is tall) = 1 – *P*(child is short) = $1 - \frac{1}{4} = \frac{3}{4}$.

Think and Communicate Questions 2–4 illustrate the following types of events.

Multiple Events

Definition	Rule	Example: Suppose two pea plants, each with TS height genes, are crossed to produce a third plant.
Events A and B are **independent** if the occurrence of A does not affect whether B happens.	**$P(A \text{ and } B) = P(A) \cdot P(B)$**	A: One parent plant transmits an S gene. B: The other parent plant transmits an S gene. $P(A \text{ and } B) = \frac{1}{2} \cdot \frac{1}{2} = \frac{1}{4}$
Events A and B are **mutually exclusive** if they cannot both happen.	**$P(A \text{ or } B) = P(A) + P(B)$**	A: The child plant gets a TT gene pair. B: The child plant gets a TS gene pair. $P(A \text{ or } B) = \frac{1}{4} + \frac{1}{2} = \frac{3}{4}$
Events A and B are **complementary** if they are mutually exclusive and one of the events must happen.	**$P(B) = 1 - P(A)$**	A: The child plant is short. B: The child plant is tall. $P(B) = 1 - \frac{1}{4} = \frac{3}{4}$

EXAMPLE 1

Suppose a card is drawn at random from a standard deck of 52 playing cards. (See page 615 for a description of such a deck.) The card is put back in the deck, and a card is again drawn at random. Find the probability of each event.

a. The first card is a 7 of hearts, and the second card is a 2 of spades.

b. The first card is either a face card or an ace.

SOLUTION

a. Find $P(A \text{ and } B)$ where A is the event "the first card is a 7 of hearts" and B is the event "the second card is a 2 of spades."

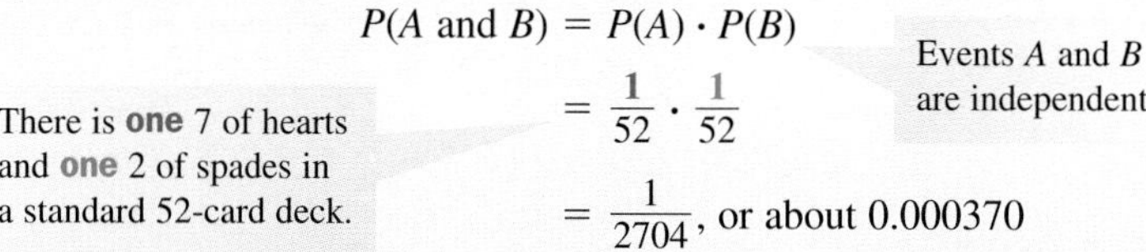

$$P(A \text{ and } B) = P(A) \cdot P(B)$$

Events A and B are independent.

$$= \frac{1}{52} \cdot \frac{1}{52}$$

There is **one** 7 of hearts and **one** 2 of spades in a standard 52-card deck.

$$= \frac{1}{2704}, \text{ or about } 0.000370$$

b. Find $P(A \text{ or } B)$ where A is the event "the first card is a face card" and B is the event "the first card is an ace."

$$P(A \text{ or } B) = P(A) + P(B)$$

Events A and B are mutually exclusive.

$$= \frac{12}{52} + \frac{4}{52}$$

There are **12** face cards and **4** aces in a standard 52-card deck.

$$= \frac{16}{52}$$

$$= \frac{4}{13}, \text{ or about } 0.308$$

Teach⇔Interact

Section Notes

Common Error

Students often have difficulty distinguishing independent and mutually exclusive events. Stress that these terms do not apply to the same type of situation. When considering a situation where event A *and* event B can occur together, it must be verified that these events are independent before using the multiplication rule. When considering a situation where either event A *or* event B can occur, it must be verified that events A and B are mutually exclusive before applying the addition rule.

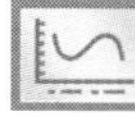

Using Technology Activities 20 and 21 of the *Probability Constructor Activity Book* offer additional opportunities for students to investigate mutually exclusive events and complementary events.

Additional Example 1

Suppose one card is drawn at random from a standard deck of 52 playing cards. The card is then put back in the deck and a card is again drawn at random. Find the probability of each event.

a. The first card is a 6 of clubs, and the second card is a black face card.

Find $P(A \text{ and } B)$, where A is the event "the first card is a 6 of clubs" and B is the event "the second card is a black face card." Events A and B are independent. There is one 6 of clubs and there are six black face cards in a standard 52-card deck.

$P(A \text{ and } B) = P(A) \cdot P(B) = \frac{1}{52} \cdot \frac{6}{52} = \frac{6}{2704} = \frac{3}{1352}$, or about 0.00222

b. The first card is either a diamond or a black 10.

Find $P(A \text{ or } B)$ where A is the event "the first card is a diamond" and B is the event "the first card is a black 10." Events A and B are mutually exclusive. There are 13 diamonds and 2 black 10's in a standard 52-card deck.

$$P(A \text{ or } B) = P(A) + P(B)$$

$$= \frac{13}{52} + \frac{2}{52}$$

$$= \frac{15}{52}, \text{ or about } 0.288$$

Teach⇔Interact

Think and Communicate

Question 5 has students discover that the rule given for adding probabilities on page 619 only works when the events are mutually exclusive. See Ex. 16 on page 624 for an addition rule for the case where events are not mutually exclusive. Question 6 has students discover that the draws from the deck are not independent when the card is not replaced, and that the multiplication rule does not apply in this case. Students will learn how to work with this situation when they learn about conditional probability in Section 13.3.

Additional Example 2

On a game show, contestants spin a wheel to determine the amount of money they can win. Sometimes a part of the wheel is marked with the name of a prize in addition to the money. Suppose that the wheel has 24 sections, one of which contains a prize. What is the probability that a spin will land on the prize at least once in 4 spins?

The desired probability is P(landing on prize in 4 spins). Let S_i be the event "the prize is landed on during the ith spin," and let N_i be the event "the prize is not landed on the ith spin." Because landing on the prize and not landing on the prize are complementary events and because each spin is an independent event:

P(landing on prize in 4 spins)
$= 1 - P$(not landing on prize in 4 spins)
$= 1 - P(N_1 \text{ and } N_2 \text{ and } N_3 \text{ and } N_4)$
$= 1 - P(N_1) \cdot P(N_2) \cdot P(N_3) \cdot P(N_4)$

On a given spin, there are 24 possible outcomes, 1 of which is the prize. Therefore, as S_i and N_i are complementary events:

$P(S_i) = \frac{1}{24}$ and $P(N_i) = 1 - P(S_i) = \frac{23}{24}$

It follows that:

P(landing on prize in 4 spins)
$= 1 - \left(\frac{23}{24}\right)\left(\frac{23}{24}\right)\left(\frac{23}{24}\right)\left(\frac{23}{24}\right)$
$= 1 - \frac{279{,}841}{331{,}776}$
$= \frac{51{,}935}{331{,}776}$, or about 0.157

The probability of landing on the prize in 4 spins is about 0.157.

THINK AND COMMUNICATE

5. Look back at Example 1. Let A be the event "the first card is a 2," and let B be the event "the first card is a diamond." Find $P(A \text{ or } B)$ and $P(A) + P(B)$. Are these expressions equal? If not, why not?

6. Suppose Example 1 is changed so that the first card drawn is *not* put back in the deck before the second card is drawn. What effect, if any, does this have on the answer to part (a) of Example 1?

EXAMPLE 2 Application: Games

In the game of Monopoly®, a player sometimes has to "go to jail." The player can get out of jail by rolling doubles with a pair of six-sided dice. ("Doubles" is a roll such that both dice show the same number.) If the player doesn't roll doubles in three tries, the player must pay $50 to get out of jail. What is the probability of getting out of jail without paying $50?

SOLUTION

The desired probability is the same as P(rolling doubles in 3 tries). Let D_i be the event "doubles is rolled on the ith try," and let N_i be the event "doubles is not rolled on the ith try." Note that:

$$P\binom{\text{rolling doubles}}{\text{in 3 tries}} = 1 - P\binom{\text{not rolling doubles}}{\text{in 3 tries}}$$

Rolling and not rolling doubles in 3 tries are complementary events.

$$= 1 - P(N_1 \text{ and } N_2 \text{ and } N_3)$$
$$= 1 - P(N_1) \cdot P(N_2) \cdot P(N_3)$$

N_1, N_2, and N_3 are independent events.

On a given roll, there are 6 possible outcomes for each of the two dice. So there are $6 \cdot 6 = 36$ possible outcomes for the roll. Of these 36 outcomes, 6 are doubles (two 1's, two 2's, . . . , two 6's). Therefore:

$$P(D_i) = \frac{6}{36} = \frac{1}{6} \text{ and } P(N_i) = 1 - P(D_i) = \frac{5}{6}$$

D_i and N_i are complementary events.

It follows that:

$$P\binom{\text{rolling doubles}}{\text{in 3 tries}} = 1 - \left(\frac{5}{6}\right)\left(\frac{5}{6}\right)\left(\frac{5}{6}\right)$$
$$= 1 - \frac{125}{216}$$
$$= \frac{91}{216}, \text{ or about } 0.421$$

The probability of getting out of jail without paying $50 is about 0.421.

Think and Communicate

5. $P(A \text{ or } B) = \frac{16}{52}$, and $P(A) + P(B) = \frac{17}{52}$; No; Events A and B are not mutually exclusive.

6. There is still a probability of $\frac{1}{52}$ that the first card is a 7 of hearts, but now there is a $\frac{1}{51}$ chance that the second card is a 2 of spades, so $P(A \text{ and } B) = \frac{1}{52} \cdot \frac{1}{51} = \frac{1}{2652}$, or about 0.000377.

7. a. $\frac{1}{6}$, or about 0.167

b. $\frac{5}{36}$, or about 0.139

c. $\frac{25}{216}$, or about 0.116

d. $\frac{91}{216}$; The sum is the probability of rolling doubles in three tries. This makes sense because the probability of rolling doubles in one of three tries is the sum of the mutually exclusive events of rolling doubles on the first try (D_1), rolling doubles on the second of two tries (N_1 and D_2), and rolling doubles on the third of three tries (N_1 and N_2 and D_3).

THINK AND COMMUNICATE

7. Look back at Example 2. Find each probability.
   a. $P(D_1)$ b. $P(N_1 \text{ and } D_2)$ c. $P(N_1 \text{ and } N_2 \text{ and } D_3)$
   d. Find the sum of the probabilities in parts (a)–(c). What do you notice? Explain why this result makes sense.

CHECKING KEY CONCEPTS

Suppose a card is drawn at random from a standard deck of 52 playing cards. (See page 615 for a description of such a deck.) The card is put back in the deck, and a card is again drawn at random. Find the probability of each event.

1. The first card is a club, and the second card is a spade.
2. The first card is a heart or an 8 of spades.
3. The cards are not both jacks.
4. The first card is black, and the second card is a queen of hearts or a 2.
5. At least one of the cards is a diamond.
6. Exactly one of the cards is a diamond.

13.2 Exercises and Applications

Extra Practice exercises on page 769

1. **Writing** Explain the conditions under which each rule applies.
   a. $P(A \text{ and } B) = P(A) \cdot P(B)$ b. $P(A \text{ or } B) = P(A) + P(B)$ c. $P(B) = 1 - P(A)$

METEOROLOGY A meteorologist giving the weekend weather forecast says that there is a 30% chance of rain on Saturday and a 40% chance of rain on Sunday. Assuming these chances are correct and Sunday's weather is independent of Saturday's weather, find the probability of each event.

2. It doesn't rain on Saturday.
3. It rains on both Saturday and Sunday.
4. It rains on Saturday and doesn't rain on Sunday.
5. It doesn't rain at all over the weekend.
6. It rains on at least one day of the weekend.
7. It rains on exactly one day of the weekend.

8. **MOVIES** In the 1943 film *Casablanca*, casino owner Rick Blaine tells a roulette player to choose number 22. Incredibly, number 22 comes up twice in a row. What is the probability of this happening? (*Note:* In roulette, a ball spins around a wheel and lands in a slot on the wheel's edge. There are 36 slots numbered 1 through 36, plus two special slots labeled 0 and 00. The ball is equally likely to land in any slot.)

Teach⇔Interact

Think and Communicate

Question 7 provides students with an alternative approach to Example 2. This method does not use the rule for complements, but calculates the probability directly. Studying this question will allow students to verify the complement rule and help them feel more confident about using it.

Checking Key Concepts

Alternate Approach
Like many probability problems, questions 3–6 can be done in several ways. For example, question 3 can be done by adding the probability that only the first card is a jack to the probability that only the second card is a jack to the probability that neither card is a jack. Or, it can be found by subtracting the probability that both cards are jacks from 1. You may want to have 4 volunteers write their solutions to these questions on the board and then ask for alternative solutions from other students. It should help all students to see these various approaches.

Closure Question

Describe three types of multiple events.
Independent events are those in which the occurrence of one event does not affect the occurrence of the other. Mutually exclusive events are those that cannot both happen. Complementary events are mutually exclusive events where one of them must happen.

Apply⇔Assess

Suggested Assignment

Core Course
Exs. 1–8, 11–13, 15, 17–20

Extended Course
Exs. 1–20

Block Schedule
Day 1 Exs. 1–8, 11–13, 15, 17–20

Checking Key Concepts

1. $\frac{1}{16}$, or 0.0625
2. $\frac{7}{26}$, or about 0.269
3. $\frac{168}{169}$, or about 0.994
4. $\frac{5}{104}$, or about 0.0481
5. $\frac{7}{16}$, or 0.4375
6. $\frac{3}{8}$, or 0.375

Exercises and Applications

1. a. *A* and *B* are independent events.
   b. *A* and *B* are mutually exclusive events.
   c. *A* and *B* are complementary events, that is, they are mutually exclusive and one of the events must occur.
2. 0.7
3. 0.12
4. 0.18
5. 0.42
6. 0.58
7. 0.46
8. $\frac{1}{1444}$, or about 0.000693

Apply⇔Assess

Exercise Notes

Application
Exs. 2–12, 14, 15 Probability has applications to many aspects of life. These exercises introduce students to some ways in which probability is used in meteorology, games, literature, sports, and electronics.

Communication: Discussion
Exs. 9, 10 Students would benefit from a discussion of these exercises. You may want to assign Exs. 9, 10(a), and 10(b) for homework and then ask a volunteer or pair of students to explain their solution to the class. The next day, all students could work on Ex. 10(c) with a basic understanding of the problem.

Teaching Tip
Exs. 10, 13 In part (c) of Ex. 10 and in Ex. 13, remind students that they can use either logarithms or graphing technology to solve for the exponents in these problems.

Problem Solving
Exs. 10, 14, 15 These problems are very complex. If students have difficulty, they should review Example 2 and use that solution as a model. Students may benefit from having additional time to think about these problems.

Mathematical Procedures
Exs. 11, 14, 15 When asked for the probability that at least one success was achieved, it is usually easier to find the probability of the complement (no success was achieved) and subtract that from 1.

Challenge
Ex. 11 In some foul situations in basketball, players are given one free throw and if they make that throw, they get to take a second shot. You might ask students to determine the possible outcomes and probabilities for each outcome in this situation.

Connection LITERATURE

In Douglas Hofstadter's short story *The Tale of Happiton*, a mischievous demon writes a letter to the residents of a town called Happiton. The letter begins like this:

> I've got some bad news and some good news for you. The bad first. You know your bell [in the courthouse clock] that rings every hour on the hour? Well, I've set it up so that each time it rings, there is exactly one chance in a hundred thousand—that is, $\frac{1}{100,000}$—that a Very Bad Thing will occur. The way I determine if that Bad Thing will occur is, I have this robot arm fling five dice and see if they all land with "7" on top.

9. The demon's dice are 20-sided. The sides of each die are numbered 0 through 9, with each number appearing on two opposite sides.

a. What is the probability that a given die lands with "7" showing?

b. Use your answer from part (a) to prove that the probability of all five dice landing with "7" showing is $\frac{1}{100,000}$, as stated in the passage.

10. In Hofstadter's story, Nellie Doobar, the mathematics teacher at Happiton High School, says that "the chances we'll make it through any eight-year period [without the Very Bad Thing happening] are almost exactly fifty-fifty."

a. How many times will the dice be rolled in an eight-year period? (Assume 1 year = 365.25 days.)

b. Find the probability that the Very Bad Thing happens during an eight-year period. Is your answer consistent with Nellie Doobar's remark?

c. **Challenge** The "good news" referred to in the passage is that the demon will make the clock bell ring less often than once an hour provided the residents of Happiton write him postcards. If the demon receives p postcards on a given day, then the time t (in hours) between rings the next day will be:

$$t = (1.00001)^p$$

The population of Happiton is 20,000. How many postcards must each resident write per day for there to be only a 5% chance of the Very Bad Thing happening during an eight-year period?

11. **SPORTS** Elena, a member of her high school's basketball team, makes 80% of the free throws she attempts. Suppose Elena is fouled and gets to shoot a pair of free throws. Find the probability of each event.

a. Elena makes both free throws.

b. Elena makes at least 1 free throw.

c. Elena makes exactly one free throw.

d. Elena misses both free throws.

9. a. $\frac{1}{10}$

b. Since the throws are independent, the probability that all five land with "7" showing is $\left(\frac{1}{10}\right)^5 = \frac{1}{100,000}$.

10. a. 70,128

b. about 0.504; This is consistent with Nellie Doobar's remark.

c. about 13.1 postcards; Thus, if each person writes 14 postcards per day, there will be less than a 5% chance of the Very Bad Thing happening.

11. a. 0.64

b. 0.96

c. 0.32

d. 0.04

12. In some state lotteries, you buy a ticket and choose 6 numbers from among the integers 1, 2, 3, . . . , 49. (The order in which you choose the numbers doesn't matter.) You win all or a portion of the lottery if your numbers are the ones selected in that day's lottery drawing.

a. If you buy a single lottery ticket, what is your probability of winning?

b. Suppose you buy a lottery ticket every day for 60 years. What is your probability of winning at least once during that time? (Assume 1 year = 365.25 days.)

c. **Writing** Based on your answer, do you think playing this type of lottery is a good idea? Explain.

13. **SAT/ACT Preview** What is the minimum number of times you must roll a six-sided die to have at least a 50% chance of getting a 1?

A. 3 **B.** 4 **C.** 5 **D.** 6 **E.** 7

14. **ELECTRONICS** An automatic garage door opener includes a transmitter that you carry in your car and a receiver that you attach to your garage. On one brand of opener, both the transmitter and the receiver have 8 switches that you can set to either "on" or "off." The switches on the transmitter and the receiver must be set identically for the garage door to open.

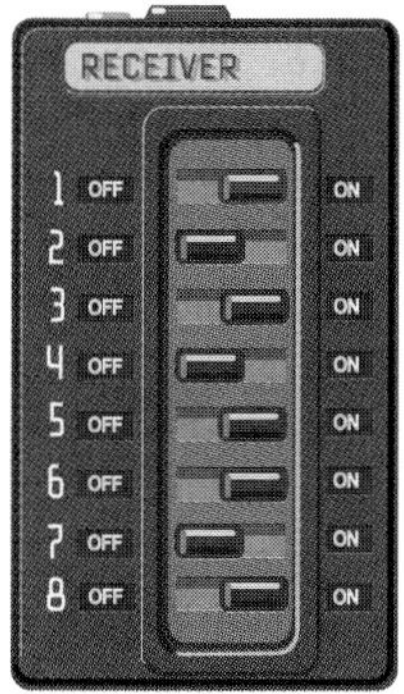

Setting 1

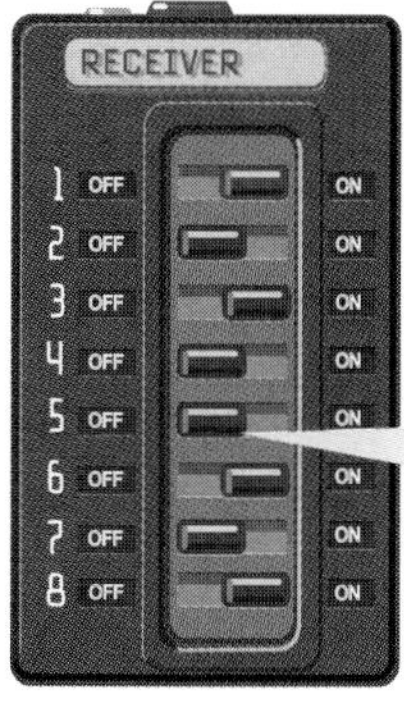

Setting 2

The position of switch 5 is different for the transmitter and the receiver, so the garage door *will not* open.

Setting 1 matches the transmitter's setting, so the garage door *will* open.

a. Kaya bought the brand of garage door opener described above. In how many ways can she set the switches on the transmitter and the receiver so that the door opens?

b. Suppose 10 houses on Kaya's street have her brand of garage door opener (including Kaya's house). What is the probability that at least one other opener has the same switch settings as Kaya's? Should Kaya feel confident that a neighbor won't accidentally open her garage door? Explain.

c. Consider the probability that *any* 2 or more of the 10 houses in part (b) have the same switch settings for their garage door openers. Would you expect this probability to be higher or lower than the probability you found in part (b)? Check your answer by calculating this probability.

d. **Challenge** Find the probability that at least 3 of the 10 houses in part (b) have the same switch settings for their garage door openers.

Apply⇔Assess

Exercise Notes

Topic Spiraling: Review

Ex. 12 Students must use combinations to solve this problem.

Research

Ex. 14 The switch settings in this problem are an example of a binary code. Computers also store their information using a binary code. Students may want to research how binary codes work and how computers use them to store information.

Career Connection

Ex. 14 There are varied career opportunities in the field of electronics. Most electrical or battery-operated items used today have electronic components. Some examples are computers, phone systems, sound recording equipment, airplane controls, and many medical instruments. People who research, develop, and manufacture electronic equipment are generally electrical engineers. This field requires a college degree, with a major in electrical engineering, including a strong emphasis on mathematics and the physical sciences.

12. a. $\frac{1}{13{,}890{,}000}$, or about 7.20×10^{-8}

b. about 0.00158; Students should see that they have very little chance (about one in 633) of winning the lottery even after the purchase of almost 22,000 tickets.

13. B

14. a. $2^8 = 256$

b. $1 - \left(\frac{255}{256}\right)^9 \approx 0.0346$; Students' answers to whether this probability, which is about 1 out of 30, will help Kaya to be confident will vary.

c. Students should see that the probability of any two or more houses having the same setting is higher, since another house's settings matching Kaya's is a subset of any two houses having matching settings. The probability is about 0.163.

d. about 0.0106

Apply⇔Assess

Exercise Notes

Integrating the Strands

Ex. 16 In this exercise, students must use geometric reasoning to demonstrate a rule of probability, namely, the addition rule for two events which are not mutually exclusive. This approach to the problem should be especially helpful for students with visual learning styles.

Assessment Note

Ex. 17 This open-ended problem gives students the chance to review and solidify the concepts learned in this section. You may want to have several students share their work with the class to give students some measure for evaluating their own work.

Second-Language Learners

Ex. 17 Students learning English may benefit from first discussing the examples they want to list, and then completing the writing independently.

Practice 84 for Section 13.2

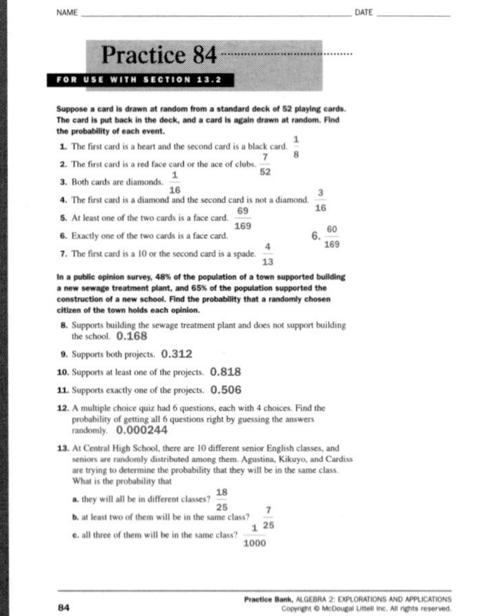

NAME ______ DATE ______

Practice 84

FOR USE WITH SECTION 13.2

Suppose a card is drawn at random from a standard deck of 52 playing cards. The card is put back in the deck, and a card is again drawn at random. Find the probability of each event.

1. The first card is a heart and the second card is a black card. $\frac{1}{8}$
2. The first card is a red face card or the ace of clubs. $\frac{7}{52}$
3. Both cards are diamonds. $\frac{1}{16}$
4. The first card is a diamond and the second card is not a diamond. $\frac{3}{16}$
5. At least one of the two cards is a face card. $\frac{69}{169}$
6. Exactly one of the two cards is a face card. 6. $\frac{60}{169}$
7. The first card is a 10 or the second card is a spade. $\frac{4}{13}$

In a public opinion survey, 48% of the population of a town supported building a new sewage treatment plant, and 65% of the population supported the construction of a new school. Find the probability that a randomly chosen citizen of the town holds each opinion.

8. Supports building the sewage treatment plant and does not support building the school. 0.168
9. Supports both projects. 0.312
10. Supports at least one of the projects. 0.818
11. Supports exactly one of the projects. 0.506
12. A multiple choice quiz had 6 questions, each with 4 choices. Find the probability of getting all 6 questions right by guessing the answers randomly. 0.000244
13. At Central High School, there are 10 different senior English classes, and seniors are randomly distributed among them. Agustina, Kikuyo, and Cardiss are trying to determine the probability that they will be in the same class. What is the probability that
    a. they will all be in different classes? $\frac{18}{25}$
    b. at least two of them will be in the same class? $\frac{7}{25}$
    c. all three of them will be in the same class? $\frac{1}{1000}$

84 **Practice Bank,** ALGEBRA 2: EXPLORATIONS AND APPLICATIONS

15. SPORTS In 1941, baseball player Joe DiMaggio of the New York Yankees got at least one hit in 56 consecutive games. What is the probability of this happening in any given sequence of 56 games? Assume DiMaggio batted an average of 4 times per game and had a 32.5% chance of getting a hit during each at-bat.

16. Visual Thinking The rectangular target shown contains two overlapping colored regions—the red square *PQSU* and the blue triangle *QRT*. Suppose a randomly thrown dart hits the target. Let *A* be the event "the dart hits square *PQSU*," and let *B* be the event "the dart hits triangle *QRT*."

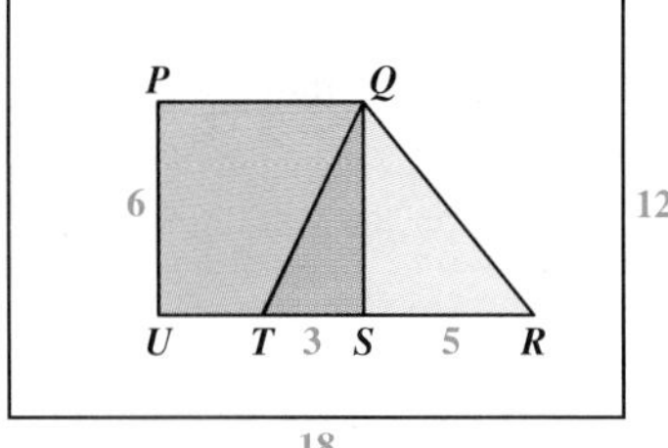

a. **Writing** Explain why the rule

$$P(A \text{ or } B) = P(A) + P(B)$$

on page 619 does not apply in this situation.

b. Find $P(A \text{ or } B)$ by calculating the ratio of the area of quadrilateral *PQRU* to the area of the entire rectangular target. Show that:

$$P(A \text{ or } B) = P(A) + P(B) - P(A \text{ and } B)$$

c. Suppose a six-sided die is rolled. Let *A* be the event "the number showing on the die is even," and let *B* be the event "the number showing on the die is greater than 4." Find $P(A \text{ or } B)$, $P(A)$, $P(B)$, and $P(A \text{ and } B)$. Show that these probabilities satisfy the equation in part (b), but not the equation in part (a).

d. **Writing** Explain why the equation in part (b) is a generalization of the equation in part (a).

ONGOING ASSESSMENT

17. Open-ended Problem Give examples of independent, mutually exclusive, and complementary events. Write and solve at least two probability problems involving the events you chose.

SPIRAL REVIEW

18. The sequence of numbers and letters on a California license plate has the form

d–LLL–DDD

where *d* is a digit from 1 to 9, each *L* is any letter except "O," and each *D* is any digit from 0 to 9. Find the probability that the three letters on a California license plate spell "CAT." *(Section 13.1)*

Find the sum of each series. *(Sections 10.4 and 10.5)*

19. $2 + 7 + 12 + 17 + \cdots + 87$

20. $\sum_{n=1}^{\infty} 17(-0.25)^{n-1}$

15. about 0.000002192

16. a. The rule does not apply because events *A* and *B* are not mutually exclusive; if the dart lands in triangle *QST*, then events *A* and *B* have occurred simultaneously.

b. $\frac{17}{72}$, or about 0.236; $P(A) + P(B) - P(A \text{ and } B) = \frac{36}{216} + \frac{24}{216} - \frac{9}{216} = \frac{51}{216} = \frac{17}{72} = P(A \text{ or } B)$

c. $P(A \text{ or } B) = \frac{2}{3}$, $P(A) = \frac{1}{2}$, $P(B) = \frac{1}{3}$, $P(A \text{ and } B) = \frac{1}{6}$; Using the equation in part (b), $\frac{2}{3} = \frac{1}{2} + \frac{1}{3} - \frac{1}{6}$ because $\frac{4}{6} = \frac{3+2-1}{6}$. Using the equation in part (a), $\frac{2}{3} \neq \frac{1}{2} + \frac{1}{3}$ because $\frac{4}{6} \neq \frac{3+2}{6}$.

d. Answers may vary. An example is given. The equation in part (a) holds only for mutually exclusive events *A* and *B*. This case is included in part (b), which holds for any events *A* and *B*, because $P(A \text{ and } B) = 0$ for mutually exclusive events.

17. Answers may vary. Check students' work.

18. $\frac{1}{15,625} = 0.000064$

19. 801

20. 13.6

Section

13.3 Using Conditional Probability

Learn how to...

- find conditional probabilities

So you can...

- make judgments about evidence presented in court cases, for example

Most movies are screened by test audiences before they open at your local theater. This helps movie studios predict whether their movies will be successful. It also allows studios to target advertising toward groups (such as men or women) that seem most receptive to a particular movie.

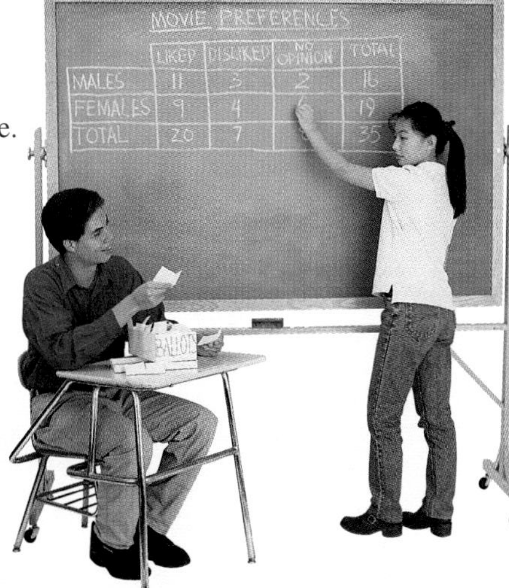

EXPLORATION
COOPERATIVE LEARNING

Conducting a Movie Survey

Work as an entire class.
You will need:

- small slips of paper

1 Your class should choose a movie that most of you have seen.

2 Each of you should have a slip of paper. On your slip, write whether you are male or female and whether you liked or disliked the movie. If you haven't seen the movie, write "no opinion."

3 Collect all the slips of paper. Tally the results of the survey in a table like the one shown in the photo.

Questions

1. Suppose a student is randomly selected from your class. Let A be the event "the student is female," and let B be the event "the student liked the movie."
   a. Find $P(A \text{ and } B)$ and $P(A) \cdot P(B)$.
   b. Are the events A and B independent? Explain.
2. Suppose a student is randomly selected from the *males* in your class. What is the probability that the student liked the movie? disliked the movie? had no opinion about the movie?

Exploration Note

Purpose
The purpose of this Exploration is to have students examine a situation in which events are not independent.

Materials/Preparation
Each student needs a small slip of paper.

Procedure
The class chooses a movie that most students have seen. Each student writes whether they are male or female and whether they liked or disliked the movie. If the student did not see the movie, they write "no opinion." The slips are collected and a table is made.

Closure
Students should understand that the events *male* or *female* and *liked movie* or *disliked movie* are not independent, and that the probability that a person liked or disliked the movie is different when considering males or females than it is when you consider the whole class.

Explorations Lab Manual
See the Manual for more commentary on this Exploration.

Diagram Master 18

For answers to the Exploration, see following page.

Plan⇔Support

Objectives

- Find conditional probabilities.
- Use probability to make judgments in real-world situations.

Recommended Pacing

Core and Extended Courses
Section 13.3: 1 day

Block Schedule
Section 13.3: $\frac{1}{2}$ block
(with Section 13.4)

Resource Materials

Lesson Support
Lesson Plan 13.3
Warm-Up Transparency 13.3
Practice Bank: Practice 85
Study Guide: Section 13.3
Explorations Lab Manual: Diagram Master 18
Assessment Book: Test 54

Technology
Graphing Calculator

Internet:
http://www.hmco.com

Warm-Up Exercises

1. What are independent events? events where the occurrence of one does not affect the occurrence of the other
2. Complete the rule for independent events: $P(A \text{ and } B) = \underline{\ ?\ }$. $P(A) \cdot P(B)$

A drawer contains 4 blue socks, 7 black socks, and 2 white socks. One sock is drawn at random, replaced, and then a second sock is drawn. Find the probability of each event.

3. The first sock is black and the second sock is blue. $\frac{28}{169}$
4. At least one of the socks is white. $\frac{48}{169}$
5. Both socks are the same color. $\frac{69}{169}$

Teach⇔Interact

About Example 1

Teaching Tip

When data are given in a table, as in this Example, encourage students to circle or underline in pencil the row or column to which the probabilities are being restricted. For instance, when doing part (a), the probability is restricted to males, so students should circle the row headed "Males." This method can help them to focus on the data needed to answer the question.

Additional Example 1

A poll was taken in a small town to see how people felt about a plan to to build a new cafeteria in town. People were classified as retired (R), nonretired (N), or high school student (S), and asked if they liked or disliked the plan. The results are summarized in the table below.

	Liked	Disliked	Total
R	105	67	172
N	49	97	146
S	17	12	29
Total	171	176	347

Find each probability for a randomly selected person in the poll.

a. $P(R|\text{liked the plan})$

171 of those who were polled liked the plan. Of those people, 105 are retired. Therefore:

$P(R \mid \text{liked the plan}) = \frac{105}{171}$, or about 0.614.

b. $P(\text{disliked the plan} \mid S)$

29 of those who were polled are high school students. Of those students, 12 disliked the plan. Therefore:

$P(\text{disliked the plan}|S) = \frac{12}{29}$, or about 0.414.

Think and Communicate

Question 1 leads students to discover that the rule for independent events can be generalized to the rule for conditional probabilities in the case that the events are not independent. This rule is summarized at the top of page 627.

It is often important to know the probability of an event under restricted conditions. For example, a physician may need to know the probability that a patient has the flu even though the patient has no fever. You can write this probability as:

$$P\left(\begin{array}{c}\text{the patient}\\\text{has the flu}\end{array}\middle|\begin{array}{c}\text{the patient}\\\text{has no fever}\end{array}\right)$$

You read this as "the probability that the patient has the flu, **given that** the patient has no fever."

In general, the **conditional probability** $P(B|A)$ is the probability of event B, given that event A has occurred.

EXAMPLE 1

Susan's class performed the Exploration on the previous page. Each student was asked whether he or she liked the movie *Jurassic Park*. The results of the survey are shown in the table.

	Liked	Disliked	No opinion	Total
Males	11	3	2	16
Females	9	4	6	19
Total	20	7	8	35

Find each probability for a randomly selected student in Susan's class.

a. $P\left(\begin{array}{c}\text{student liked}\\\textit{Jurassic Park}\end{array}\middle|\begin{array}{c}\text{student}\\\text{is male}\end{array}\right)$ **b.** $P\left(\begin{array}{c}\text{student}\\\text{is female}\end{array}\middle|\begin{array}{c}\text{student disliked}\\\textit{Jurassic Park}\end{array}\right)$

SOLUTION

a. There are 16 male students. Of these students, 11 liked *Jurassic Park*. Therefore:

$$P\left(\begin{array}{c}\text{student liked}\\\textit{Jurassic Park}\end{array}\middle|\begin{array}{c}\text{student}\\\text{is male}\end{array}\right) = \frac{11}{16}\text{, or about } 0.688$$

b. There are 7 students who disliked *Jurassic Park*. Of these students, 4 are female. Therefore:

$$P\left(\begin{array}{c}\text{student}\\\text{is female}\end{array}\middle|\begin{array}{c}\text{student disliked}\\\textit{Jurassic Park}\end{array}\right) = \frac{4}{7}\text{, or about } 0.571$$

THINK AND COMMUNICATE

1. Suppose a student is randomly selected from Susan's class. Let A be the event "the student liked *Jurassic Park*," and let B be the event "the student is male."

a. Find $P(A \text{ and } B)$ and $P(A) \cdot P(B)$. Are the events A and B independent? Explain.

b. Find $P(A) \cdot P(B|A)$. What do you notice?

ANSWERS Section 13.3

Exploration

Answers may vary. Examples are given based on the table on page 626.

1. a. $P(A \text{ and } B) = \frac{9}{35}$,

$P(A) \cdot P(B) = \frac{19}{35} \cdot \frac{4}{7} = \frac{76}{245}$

b. No; since $P(A \text{ and } B) \neq P(A) \cdot P(B)$, events A and B are not independent.

2. $\frac{11}{16}, \frac{3}{16}, \frac{1}{8}$

Think and Communicate

1. a. $P(A \text{ and } B) = \frac{11}{35}$,

$P(A) \cdot P(B) = \frac{4}{7} \cdot \frac{16}{35} = \frac{64}{245}$;

No; since $P(A \text{ and } B) \neq P(A) \cdot P(B)$, events A and B are not independent.

b. $\frac{11}{35}$; $P(A) \cdot P(B|A) = P(A \text{ and } B)$

In Section 13.2, you saw that $P(A \text{ and } B) = P(A) \cdot P(B)$ provided the events A and B are independent. Using conditional probability, you can generalize this equation so that it applies to *any* events A and B.

Conditional Probability

For any events A and B:

$$P(A \text{ and } B) = P(A) \cdot P(B|A) \quad \text{and} \quad P(B|A) = \frac{P(A \text{ and } B)}{P(A)}$$

EXAMPLE 2

Royce randomly chooses a marble from the jar shown. He places the marble in his pocket, then randomly chooses a second marble from the jar. Find the probability that the first marble is red and the second marble is green.

SOLUTION

Let R be the event "a red marble is chosen" and G be the event "a green marble is chosen."

Step 1 Make a *probability tree diagram* showing the possible outcomes of each stage of the experiment and the probabilities of these outcomes.

P(R) → $\frac{5}{8}$; $\frac{3}{8}$ — 3 of the 8 marbles are green.

first choice → R, G

From R: $\frac{4}{7}$ → R, $\frac{3}{7}$ ← P(G|R) → G

From G: $\frac{5}{7}$ → R, $\frac{2}{7}$ → G — If the first marble chosen is green, then 2 of the remaining 7 marbles are green.

second choice → R, G, R, G

Step 2 Find $P(R \text{ and } G)$ using a formula for conditional probability.

$$\begin{aligned} P(R \text{ and } G) &= P(R) \cdot P(G|R) \\ &= \frac{5}{8} \cdot \frac{3}{7} \\ &= \frac{15}{56}, \text{ or about } 0.268 \end{aligned}$$

Read these probabilities from the tree diagram.

The probability of choosing red first and green second is about 0.268.

THINK AND COMMUNICATE

2. Suppose an outcome of an experiment, such as choosing two marbles from a jar, corresponds to a path through a probability tree diagram. How is the probability of the outcome related to the probabilities assigned to the branches of the path?

Think and Communicate

2. The probability of the outcome is the product of the probabilities assigned to the branches.

Teach⇔Interact

Section Notes

Teaching Tip

Point out to students that the two formulas for conditional probability are equivalent. They only need to memorize the equation on the left, as it can be transformed into the one on the right by dividing both sides of the equation by $P(A)$.

Reasoning

Ask students to find the value of $P(B|A)$ when B and A are independent. ($P(B)$) This shows that the rule for independent events is a special case of the property of conditional probabilities

Additional Example 2

A math team orders 3 pizzas before its next meet. One large pizza is pepperoni and one large and one small pizza are plain cheese. Each large pizza has eight slices and each small pizza has four slices. The team captain picks two pieces at random for the coach. Find the probability that both pieces are cheese.

Let P be the event "a slice is pepperoni" and C be the event "a slice is cheese."

Step 1 Make a probability tree diagram showing the possible outcomes of each stage of the experiment and the probabilities of these outcomes. Use the fact that when the first slice is taken, 8 of the 20 pieces are pepperoni, and if the first slice taken is pepperoni, 12 of the remaining 19 slices are cheese.

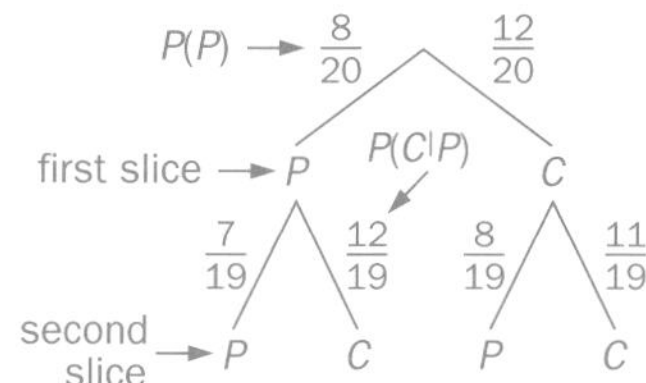

Step 2 Find $P(C \text{ and } C)$ using a formula for conditional probability. Read these probabilities from the tree diagram.

$$\begin{aligned} P(C \text{ and } C) &= P(C) \cdot P(C \mid C) \\ &= \frac{12}{20} \cdot \frac{11}{19} \\ &= \frac{33}{95}, \text{ or about } 0.347 \end{aligned}$$

The probability of choosing cheese first and cheese second is about 0.347.

Teach⇔Interact

About Example 3

Common Error
Students may not understand why the answer is not 80%. Point out that this is the probability of the witness saying the cab was blue given that the cab was blue; or $P(B_w|B)$. The probability that is asked for is $P(B|B_w)$. The order of the arguments in a conditional probability is not generally reversible.

Additional Example 3

You are a juror in a case involving a store robbery. The suspect has a mustache. You are told that:

- The clerk said the man did not have a mustache.
- 9% of men have a mustache.
- In reenactments, the clerk correctly identified whether the man had a mustache or not 87% of the time.

What is the probability that the robber has a mustache, given the clerk's statement?

Step 1 Use symbols to represent the different events.

M: The robber had a mustache.
N: The robber did not have a mustache.
M_c: The clerk says the robber had a mustache.
N_c: The clerk says the robber did not have a mustache.

Step 2 Make a tree diagram.

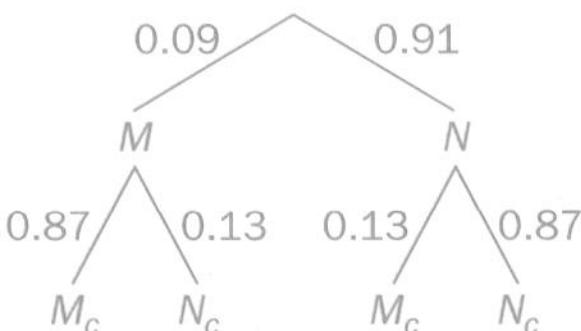

Step 3 Find $P(M|N_c)$, the probability that the man had a mustache given that the clerk says he did not have a mustache. First use a formula for conditional probability. Then find the outcome of each of the two paths that lead to N_c, and add these probabilities together.

$$P(M|N_c) = \frac{P(M \text{ and } N_c)}{P(N_c)}$$

$$= \frac{P(M \text{ and } N_c)}{P(M \text{ and } N_c) + P(N \text{ and } N_c)}$$

$$= \frac{(0.09)(0.13)}{(0.09)(0.13) + (0.91)(0.87)}$$

$$\approx 0.0146$$

The probability that the man had a mustache, given the clerk's statement, is only 0.0146.

BY THE WAY

The problem in Example 3 was first posed by psychologists Amos Tversky (*left*) and Daniel Kahneman, who specialize in the study of "cognitive illusions."

EXAMPLE 3 Interview: Robert Ward

Attorneys like Robert Ward have to make sure that jurors understand how to interpret evidence based on probability. Consider the following problem:

You are a juror in a case involving a nighttime hit-and-run accident by a taxicab. Two cab companies, one with green cabs and one with blue cabs, operate in your city. You are told that:

- Of the cabs in your city, 85% are green and 15% are blue.
- A witness identified the cab as blue.
- In reenactments of the accident, the witness correctly identified the color of the cab 80% of the time.

What is the probability that the cab involved in the accident was blue, given the witness's statement?

SOLUTION

Step 1 Use symbols to represent the different events.

G: The cab was green. G_w: The witness says the cab was green.
B: The cab was blue. B_w: The witness says the cab was blue.

Step 2 Make a probability tree diagram representing the situation.

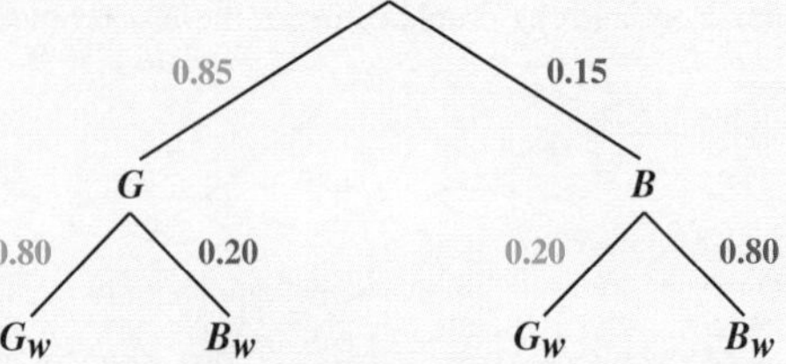

Step 3 Find $P(B|B_w)$, the probability that the cab involved in the accident was blue, given that the witness says it was blue.

$$P(B|B_w) = \frac{P(B \text{ and } B_w)}{P(B_w)}$$

Use a formula for conditional probability.

$$= \frac{P(B \text{ and } B_w)}{P(G \text{ and } B_w) + P(B \text{ and } B_w)}$$

There are two paths through the tree diagram that lead to B_w. **Add** the probabilities associated with each path.

$$= \frac{(0.15)(0.80)}{(0.85)(0.20) + (0.15)(0.80)}$$

$$\approx 0.414$$

The probability that the cab was blue, given the witness's statement, is about 0.414.

THINK AND COMMUNICATE

3. The problem in Example 3 is sometimes called the *juror's fallacy*. Why do you think this name is used?

Think and Communicate

3. Answers may vary. An example is given. An attorney might expect that a juror will believe that if witnesses correctly identify the color 80% of the time, then there is an 80% chance that the cab was blue, given that the witness identified it as such. This is a fallacy, because the actual chance in this case is only about 41%.

CHECKING KEY CONCEPTS

Suppose two marbles are taken at random from the jar shown. The first marble is *not* put back in the jar before the second marble is taken. Find the probability of each event.

1. The second marble is green, given that the first marble is red.
2. The first marble is red, and the second marble is green.
3. The first marble is green, and the second marble is red.
4. Both marbles are red.
5. Both marbles are green.
6. The second marble is red.

13.3 Exercises and Applications

Extra Practice exercises on page 769

The table gives the majors of students at a small technical college.

	Freshmen	Sophomores	Juniors	Seniors	Total
Architecture	50	30	40	25	145
Business	60	55	45	30	190
Engineering	40	35	50	55	180
Total	150	120	135	110	515

Suppose a student from the college is selected at random. Find each probability.

1. P(sophomore)
2. P(architecture major)
3. P(engineering major | freshman)
4. P(architecture major | senior)
5. P(freshman | engineering major)
6. P(junior | business major)
7. P(business major | junior or senior)
8. P(business or engineering major | sophomore)
9. **LITERATURE** In Michael Crichton's novel *Congo*, a team of scientists searches for the legendary Lost City of Zinj in Africa. At one point, the scientists consider parachuting into the jungles of Zaire.

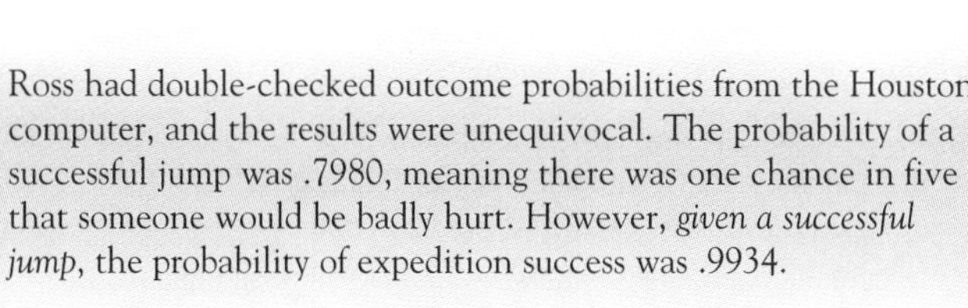

> Ross had double-checked outcome probabilities from the Houston computer, and the results were unequivocal. The probability of a successful jump was .7980, meaning there was one chance in five that someone would be badly hurt. However, *given a successful jump*, the probability of expedition success was .9934.

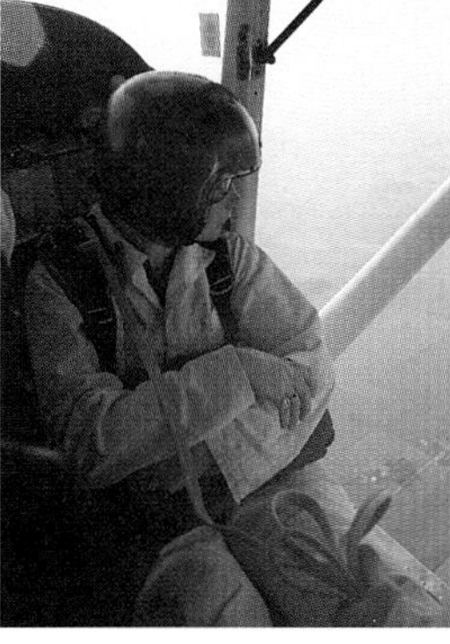

What is the probability that the scientists make a successful jump and then complete their expedition successfully?

Teach⇔Interact

Checking Key Concepts

Communication: Drawing Encourage students to make a probability tree diagram for this situation before answering the questions.

Closure Question

Consider the events A and B. Describe how to find $P(A \text{ and } B)$ when A and B are independent and when A and B are not independent. If A and B are independent, then find $P(A)$ and $P(B)$ and multiply these quantities to get $P(A \text{ and } B)$. If A and B are not independent, then calculate $P(A)$ and $P(B|A)$, the probability of B given A, and multiply these quantities to get $P(A \text{ and } B)$.

Apply⇔Assess

Suggested Assignment

Core Course
Exs. 1–8, 12–17, 19–25, AYP

Extended Course
Exs. 1–25. AYP

Block Schedule
Day 2 Exs. 1–8, 12–17, 19–25, AYP

Exercise Notes

Student Progress
Exs. 1–8 Students should be able to complete these problems with few errors before continuing on to the rest of the exercises. Students having difficulty with this work may need to work with a peer tutor.

Using Technology
Exs. 1–8, 12–17 The TI-82 can reduce the fractions in these exercises quickly and easily. Enter the fraction using the division key and press MATH ENTER ENTER. Or, when other operations need to be done as in Ex. 17, enter the entire expression, in this case, 4/52*3/51+48/52*4/51, on the home screen and then press MATH ENTER ENTER. The calculator will give the answer in reduced fraction form.

Checking Key Concepts

1. $\frac{4}{9}$, or about 0.444
2. $\frac{4}{15}$, or about 0.267
3. $\frac{4}{15}$, or about 0.267
4. $\frac{1}{3}$, or about 0.333
5. $\frac{2}{15}$, or about 0.133
6. $\frac{3}{5}$, or 0.6

Exercises and Applications

1. $\frac{24}{103}$, or about 0.233
2. $\frac{29}{103}$, or about 0.282
3. $\frac{4}{15}$, or about 0.267
4. $\frac{5}{22}$, or about 0.227
5. $\frac{2}{9}$, or about 0.222
6. $\frac{9}{38}$, or about 0.237
7. $\frac{15}{49}$, or about 0.306
8. $\frac{3}{4}$, or 0.75
9. 0.793

Apply⇔Assess

Exercise Notes

Interdisciplinary Problems
Ex. 9 As is shown in this example, a basic knowledge of probability is important for understanding and interpreting some literature and writing in general.

Application
Exs. 10, 11, 18, 19 These exercises allow students to see some of the many areas to which probability can be applied to help make judgments. Here students see examples from law, health, and market research.

Using Manipulatives
Exs. 12–17 Suggest that students having difficulty with conditional probabilities use an actual deck of cards to help them with these exercises. Have them remove the given card and then calculate the probabilities for the second pick.

Challenge
Ex. 18 Have students calculate the probability that a randomly selected person in North America is HIV-negative, given that the person tests positive on two ELISA tests.

Research
Ex. 18 Students may want to find out the false-positive probabilities for certain types of medical testing.

Problem Solving
Exs. 18, 19 Encourage students having difficulty with these problems to use Example 3 on page 628 as a model and to make a probability tree diagram.

INTERVIEW Robert Ward

Look back at the article on pages 608–610.

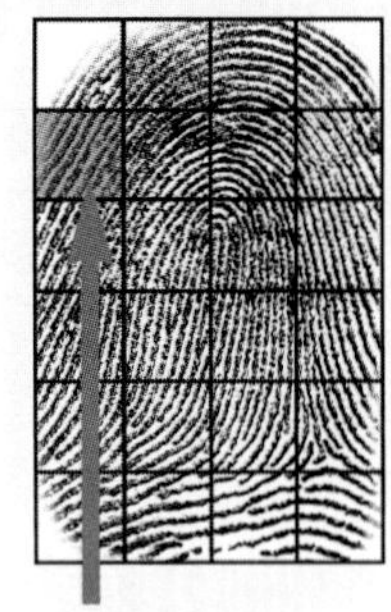

corresponding squares

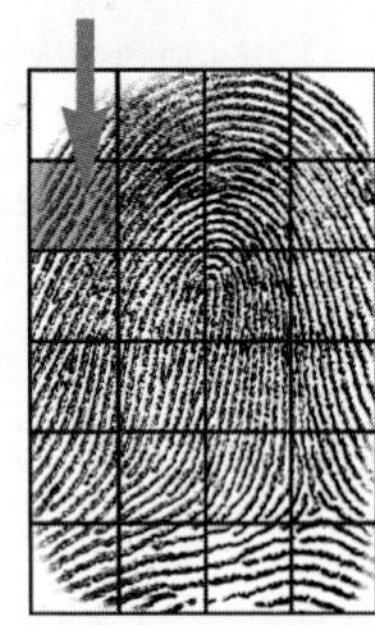

Attorneys like Robert Ward often rely on fingerprints to place a suspect at the scene of a crime. Fingerprint evidence is considered very strong because each person's fingerprints are assumed to be unique. This was established mathematically by the British scientist Sir Francis Galton in 1892.

Galton wanted to estimate the probability that two fingerprints from two randomly chosen people match. He divided each of two hypothetical fingerprints into 24 square sections. He then defined these three events:

A: The same number of ridges pass through two corresponding squares.

B: The ridges adjacent to two corresponding squares have the same "general course."

C: Two corresponding squares match.

10. Galton claimed that:

$$P(C) = P(C \mid (A \text{ and } B)) \cdot P(A \text{ and } B)$$

He assumed A and B are independent events, and estimated that $P(A) = \frac{1}{2^{1/3}}$, $P(B) = \frac{1}{2^{1/6}}$, and $P(C \mid (A \text{ and } B)) = \frac{1}{2}$. Find the probability that two corresponding squares match.

11. Galton also assumed that matches between corresponding squares are independent events. Based on this assumption and your answer to Exercise 10, what is the probability of a complete match between two fingerprints from two randomly chosen people?

Suppose two cards are drawn at random from a standard deck of 52 playing cards. (See page 615 for a description of such a deck.) The first card is *not* put back in the deck before the second card is drawn. Find the probability of each event.

12. The second card is a 4, given that the first card is an ace.

13. The second card is a king, given that the first card is a king.

14. The first card is a club, and the second card is a diamond.

15. Both cards are black.

16. The first card is a queen, and the second card is a face card.

17. The second card is a 9.

18. **HEALTH** The ELISA test is used for diagnosing HIV (the human immunodeficiency virus, which causes AIDS). The accuracy of this test is such that 99.3% of people who are HIV-positive will test positive, and 99.99% of people who are HIV-negative will test negative. It has been estimated that about 0.7365% of the population in North America is HIV-positive. What is the probability that a randomly selected person in North America is HIV-negative, given that the person tests positive?

BY THE WAY

Juan Vucetich, a police officer from Argentina, published one of the first systems of fingerprint classification in 1904. Vucetich's system is still used in many countries.

10. $\frac{1}{(\sqrt{2})^3}$, or about 0.354
11. $\left(\frac{1}{(\sqrt{2})^3}\right)^{24}$, or about 1.46×10^{-11}
12. $\frac{4}{51}$, or about 0.0784
13. $\frac{1}{17}$, or about 0.0588
14. $\frac{13}{204}$, or about 0.0637
15. $\frac{25}{102}$, or about 0.245
16. $\frac{11}{663}$, or about 0.0166
17. $\frac{1}{13}$, or about 0.0769
18. about 0.0134
19. about 0.620
20. Answers may vary. An example is given. What is the probability that a student is male, given that the student disliked the movie? Answer: $\frac{3}{7}$
21. a. $\frac{1}{3}$, or about 0.333
    b. $\frac{1}{12}$, or about 0.0833
    c. $\frac{11}{12}$, or about 0.917
22. $x^4 + 4x^3y + 6x^2y^2 + 4xy^3 + y^4$
23. $x^3 - 3x^2y + 3xy^2 - y^3$
24. $a^5 + 10a^4b + 40a^3b^2 + 80a^2b^3 + 80ab^4 + 32b^5$
25. $s^6 + 3s^4t^4 + 3s^2t^8 + t^{12}$

19. MARKET RESEARCH In a focus group designed to test the appeal of a new TV sitcom, 80% of the women and 60% of the men said they plan to watch the sitcom. Women made up 55% of the focus group. If you assume the composition and preferences of the focus group are representative of the TV audience as a whole, what is the probability that a person watching the sitcom is female?

ONGOING ASSESSMENT

20. Open-ended Problem Write a conditional probability exercise based on the table in Example 1. Then solve your exercise.

SPIRAL REVIEW

21. Suppose a six-sided die is rolled twice. Find the probability of each event. *(Section 13.2)*

a. The first roll is a 1 or a 4.

b. The first roll is a 5, and the second roll is an even number.

c. The sum of the two rolls is less than 11.

Use the binomial theorem to expand each power of a binomial. *(Section 12.5)*

22. $(x + y)^4$ **23.** $(x - y)^3$ **24.** $(a + 2b)^5$ **25.** $\left(s^2 + t^4\right)^3$

ASSESS YOUR PROGRESS

VOCABULARY

probability (p. 611)
experimental probability (p. 612)
theoretical probability (p. 613)
geometric probability (p. 614)
independent events (p. 619)
mutually exclusive events (p. 619)
complementary events (p. 619)
conditional probability (p. 626)

1. Suppose a randomly thrown dart hits the circular target shown. Find the probability that the dart hits the target's square shaded region. *(Section 13.1)*

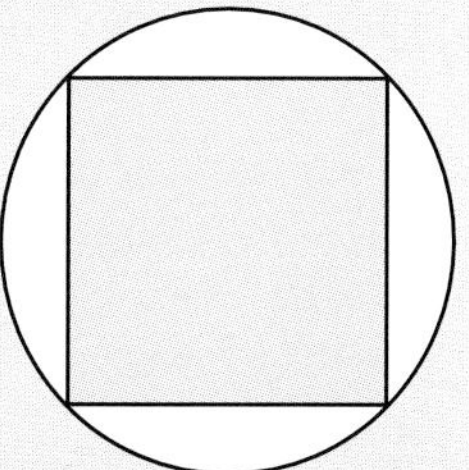

2. HISTORY In a game popular in France during the seventeenth century, a player tried to roll a six-sided die four times without getting a 6. What is the probability that a player wins this game? *(Section 13.2)*

3. MANUFACTURING Seven percent of the welds made on an automobile assembly line are defective. An X-ray machine correctly rejects 90% of the defective welds and correctly accepts 95% of the good welds. Find the probability that an accepted weld is defective. *(Section 13.3)*

4. Journal Explain why the equation $P(A \text{ and } B) = P(A) \cdot P(B|A)$ is a generalization of the equation $P(A \text{ and } B) = P(A) \cdot P(B)$. Give an example of two events A and B that satisfy the first equation but not the second.

Apply⇔Assess

Exercise Notes

Assessment Note
Ex. 20 Students would benefit from hearing other's responses to this exercise. Ask a few volunteers to read their responses or pair students up and have them critique each other's writing.

Topic Spiraling: Preview
Exs. 22–25 The techniques of binomial expansion will be used in Section 13.4 when studying binomial probability.

Assess Your Progress

Journal Entry
Writing this entry should help students compare and contrast probabilities for independent events with those for conditional events.

Progress Check 13.1–13.3

See page 648.

Practice 85 for Section 13.3

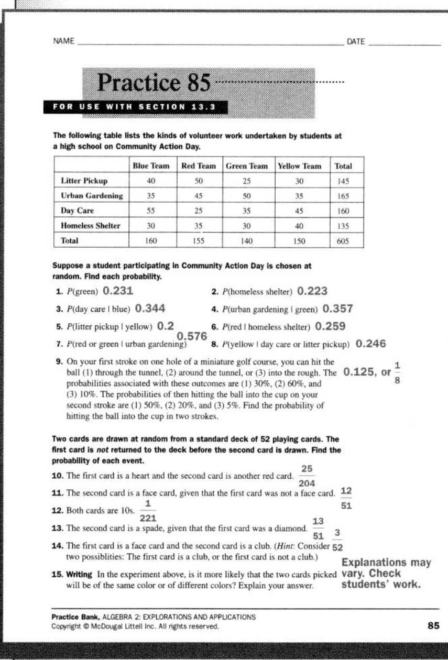

NAME ____________ DATE ____________

Practice 85
FOR USE WITH SECTION 13.3

The following table lists the kinds of volunteer work undertaken by students at a high school on Community Action Day.

	Blue Team	Red Team	Green Team	Yellow Team	Total
Litter Pickup	40	50	25	30	145
Urban Gardening	35	45	50	35	165
Day Care	55	25	35	45	160
Homeless Shelter	30	35	30	40	135
Total	160	155	140	150	605

Suppose a student participating in Community Action Day is chosen at random. Find each probability.

1. $P(\text{green})$ 0.231
2. $P(\text{homeless shelter})$ 0.223
3. $P(\text{day care} \mid \text{blue})$ 0.344
4. $P(\text{urban gardening} \mid \text{green})$ 0.357
5. $P(\text{litter pickup} \mid \text{yellow})$ 0.2
6. $P(\text{red} \mid \text{homeless shelter})$ 0.259
7. $P(\text{red or green} \mid \text{urban gardening})$ 0.576
8. $P(\text{yellow} \mid \text{day care or litter pickup})$ 0.246
9. On your first stroke on one hole of a miniature golf course, you can hit the ball (1) through the tunnel, (2) around the tunnel, or (3) into the rough. The probabilities associated with these outcomes are (1) 30%, (2) 60%, and (3) 10%. The probabilities of then hitting the ball into the cup on your second stroke are (1) 50%, (2) 20%, and (3) 5%. Find the probability of hitting the ball into the cup in two strokes. 0.125, or $\frac{1}{8}$

Two cards are drawn at random from a standard deck of 52 playing cards. The first card is *not* returned to the deck before the second card is drawn. Find the probability of each event.

10. The first card is a heart and the second card is another red card. $\frac{25}{204}$
11. The second card is a face card, given that the first card was not a face card. $\frac{12}{51}$
12. Both cards are 10s. $\frac{1}{221}$
13. The second card is a spade, given that the first card was a diamond. $\frac{13}{51}$
14. The first card is a face card and the second card is a club. (*Hint*: Consider two possiblities: The first card is a club, or the first card is not a club.) $\frac{3}{52}$
15. **Writing** In the experiment above, is it more likely that the two cards picked will be of the same color or of different colors? Explain your answer. Explanations may vary. Check students' work.

Practice Bank, ALGEBRA 2: EXPLORATIONS AND APPLICATIONS

85

Assess Your Progress

1. $\frac{2}{\pi}$, or about 0.637
2. $\frac{625}{1296}$, or about 0.482
3. about 0.00786
4. Answers may vary. If events A and B are independent, then $P(B)$ is not influenced by whether or not A has occurred, so $P(B|A) = P(B)$, which gives the more restricted result that $P(A \text{ and } B) = P(A) \cdot P(B)$ when events A and B are independent. An example is given. Roll two dice. Let A be the event "doubles is rolled," and B be the event "the sum of the roll is 6." Then $P(A \text{ and } B) = P(A) \cdot P(B|A)$ gives $\frac{1}{36} = \frac{1}{6} \cdot \frac{1}{6}$, which is true, while $P(A \text{ and } B) = P(A) \cdot P(B)$ gives $\frac{1}{36} = \frac{1}{6} \cdot \frac{5}{36}$, which is false.

Plan⇔Support

Objectives

- Find the probability distribution for a binomial experiment.
- Make predictions using binomial probability.

Recommended Pacing

Core and Extended Courses
Section 13.4: 1 day

Block Schedule
Section 13.4: $\frac{1}{2}$ block
(with Section 13.3)

Resource Materials

Lesson Support
Lesson Plan 13.4
Warm-Up Transparency 13.4
Practice Bank: Practice 86
Study Guide: Section 13.4
Explorations Lab Manual:
Diagram Master 1

Technology
Technology Book:
Calculator Activity 13
Graphing Calculator
Spreadsheet Software
Internet:
http://www.hmco.com

Warm-Up Exercises

Suppose you flip a coin 3 times.

1. What is the probability that the coin will land on heads in any one flip? $\frac{1}{2}$
2. List the equally likely outcomes of this experiment. TTT, TTH, THT, HTT, THH, HTH, HHT, HHH
3. Find the probability that the coin lands on heads 2 or 3 times. $\frac{1}{2}$

Suppose the coin is weighted so that the probability that it lands on heads in any particular flip is 0.4.

4. Find the probability that the coin lands on heads all three times. 0.064
5. Find the probability that the coin lands on heads 2 or 3 times. 0.352

Section

13.4 Binomial Distributions

Learn how to...

- find the probability distribution for a binomial experiment

So you can...

- make predictions in cases where there are many trials, each with two possible outcomes, such as guessing answers on a true/false test

Do you think you could pass a true/false test in a language you can't read? Suppose you are taking a six-question true/false test and decide to guess each answer by flipping a coin. How likely are you to get all of the questions right? How likely are you to get more than half of the questions right? You will examine this situation experimentally in the Exploration.

EXPLORATION
COOPERATIVE LEARNING

Guessing on a True/False Test

Work with your class.
You will need:

- a coin

1 Flip a coin to guess the answer to each question on the test shown. If the coin comes up heads, answer that question *True*. If it comes up tails, answer *False*.

Esperanto Geography Test
Tell whether each statement is *True* or *False*.

1. Pli da homoj loĝas en Kalifornio ol en la tuta Aŭstralio.
2. La ĉefa rikoltaĵo en Ukrajno estas tritiko.
3. Kenjo iĝis sendependa en 1963.
4. Kalkuto estas la ĉefurbo de Bharato.
5. Ŝanhajo estas lando en Azio.
6. Malavio troviĝas sur la okcidenta bordo de granda laĝo.

2 What is the theoretical probability of answering the first question correctly? of answering all six correctly?

3 Your teacher will tell you the answers to the test. Find your score by counting the number of questions you answered correctly.

BY THE WAY

Esperanto is an international language invented by Ludwik Lazar Zamenhof, a Polish opthamologist, in 1887. It is the most widely used of the more than 200 universal languages that have been proposed since the 1600s.

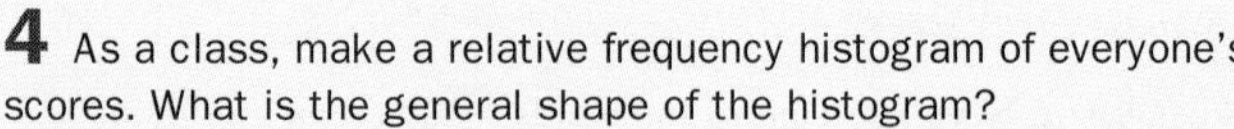

4 As a class, make a relative frequency histogram of everyone's scores. What is the general shape of the histogram?

5 Based on the histogram, what is the probability of answering all of the questions correctly? Does this agree with the probability you calculated in Step 2? Why or why not?

6 Based on the histogram, what is the probability of answering four or more questions correctly? of answering three or fewer questions correctly? Is one situation theoretically more likely than the other? Why?

Exploration Note

Purpose
The purpose of this Exploration is to conduct a binomial experiment and to introduce students to a binomial distribution.

Materials/Preparation
Each student in the class needs a coin.

Procedure
Each student simulates random guessing by flipping a coin and answering *True* if heads comes up and *False* if tails comes up. Students find the number of questions they answered correctly. (The translation and answers are in the answers to the Exploration.) A relative frequency histogram is made. Students use this histogram to find experimental probabilities. They also calculate theoretical probabilities and compare them with the experimental ones.

Closure
Ask students to copy the histogram into their journals. Have each student turn to a classmate and ask a question that can be answered using the histogram.

Explorations Lab Manual
See the Manual for more commentary on this Exploration.

For answers to the Exploration, see answers in back of book.

In Steps 5 and 6 of the Exploration, you used your class results to calculate probabilities experimentally. You can also calculate the probabilities theoretically, as Example 1 illustrates.

EXAMPLE 1

Find the probability of getting exactly 4 out of 6 right on the following two types of six-question tests if you guess the answer to each question randomly.

a. a true/false test

b. a multiple-choice test, with 5 possible responses to each question

SOLUTION

a. The probability of getting any particular question right (or wrong) is $\frac{1}{2}$, so the probability of guessing a particular sequence of answers—such as right, right, wrong, right, wrong, right—is $\left(\frac{1}{2}\right)^6$. There are ${}_6C_4$ different sequences that contain exactly 4 right answers.

$$P(\text{4 out of 6 right}) = {}_6C_4 \cdot \left(\frac{1}{2}\right)^6$$
$$\approx 0.234$$

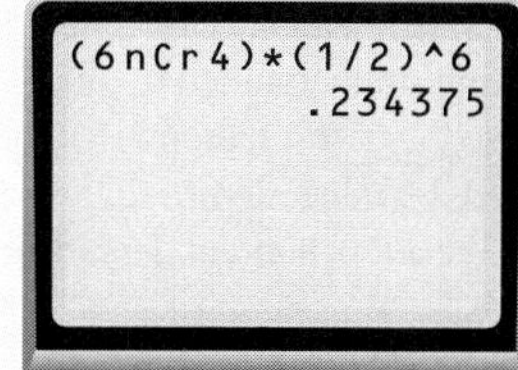

b. The probability of getting a particular question right is $\frac{1}{5}$. The probability of getting a particular question wrong is $1 - \frac{1}{5} = \frac{4}{5}$.

$$P(\text{4 out of 6 right}) = {}_6C_4 \cdot \left(\frac{1}{5}\right)^4 \cdot \left(\frac{4}{5}\right)^2$$
$$= 15 \cdot \frac{1}{625} \cdot \frac{16}{25}$$
$$\approx 0.0154$$

There are ${}_6C_4$ different sequences that contain exactly 4 right answers.

The probability of guessing a particular sequence of **4 right** and **2 wrong** answers is $\left(\frac{1}{5}\right)^4 \cdot \left(\frac{4}{5}\right)^2$.

THINK AND COMMUNICATE

1. For each test in Example 1, what is P(2 out of 6 *wrong*)? Explain.

2. Based on part (b) of Example 1, explain how to find P(5 out of 6 right).

3. Based on parts (a) and (b) of Example 1, what can you say about P(4 out of 6 right) on a six-question test as the number of answer choices increases?

ANSWERS Section 13.4

Think and Communicate

1. about 0.234 for part (a) and 0.0154 for part (b); These probabilities are the same as those found in Example 1 because getting 2 out of 6 wrong is the same as getting 4 out of 6 right.

2. There are ${}_6C_5$ different sequences that contain exactly 5 right answers. The probability of guessing a particular sequence of 5 right and 1 wrong answer is $\left(\frac{1}{5}\right)^5 \cdot \left(\frac{4}{5}\right)^1$, so P(5 out of 6 right) = ${}_6C_5 \cdot \left(\frac{1}{5}\right)^5 \cdot \left(\frac{4}{5}\right)^1 = 6 \cdot \frac{1}{3125} \cdot \frac{4}{5} = \frac{24}{15{,}625} \approx 0.00154$.

3. It decreases.

Teach⇔Interact

About Example 1

Communication: Discussion This Example utilizes many concepts learned previously. It is important to discuss it in detail and continually ask students to justify what is being done at each step. Challenge students to point out which events are being considered independent, mutually exclusive, or complementary, and where the corresponding properties are being used to solve each part. Also, ask students to explain why there are ${}_6C_4$ different sequences of exactly 4 right answers. (You are choosing 4 of the 6 questions to be correct. Students may want to list these possibilities the first time through.) A thorough discussion of this Example should help students realize that binomial probability is not entirely new but is a combination of previously learned concepts.

Additional Example 1

Suppose that you are going to spin a spinner with 8 sections labeled 1 through 8. Find the probability of each event.

a. The spinner lands on an even number exactly 3 out of 10 times.

The probability that any one spin lands on an even number is $\frac{1}{2}$, so the probability of getting any particular sequence of even or odd numbers—such as even, odd, odd, even, odd, even, even, even, odd, odd—is $\left(\frac{1}{2}\right)^{10}$. There are ${}_{10}C_3$ different sequences that contain exactly 3 out of 10 even numbers.

$$P(\text{3 out of 10 even}) = {}_{10}C_3 \cdot \left(\frac{1}{2}\right)^{10} \approx 0.117$$

b. The spinner lands on the number 5 three out of ten times.

The probability of landing on a 5 in any one spin is $\frac{1}{8}$. The probability of landing on any other number is $1 - \frac{1}{8} = \frac{7}{8}$. There are ${}_{10}C_3$ different sequences that contain exactly three 5's, and the probability of spinning a particular sequence of three 5's and seven non-5's is $\left(\frac{1}{8}\right)^3 \cdot \left(\frac{7}{8}\right)^7$.

$$P(\text{three 5's out of 10}) = {}_{10}C_3 \cdot \left(\frac{1}{8}\right)^3 \cdot \left(\frac{7}{8}\right)^7 \approx 0.0920$$

Teach⇔Interact

Section Note

Student Study Tip

Mention the similarity of the binomial probability formula with that for the kth term of a binomial expansion. In particular, students should note that the exponents should add to n. Also, in the case of probability, the bases add to one. Remind them that Pascal's triangle can be used to find some of the combinations, especially those where n and k are equal or differ by 1.

Additional Example 2

A middle school is purchasing new lockers for its students. The lockers have a coat-storage section at the bottom and a book storage section at the top. The top section is 4.5 feet from the floor. The locker company did a study and found that, in general, 2% of middle school students cannot reach their books out of such a locker. The school intends to purchase some shorter lockers for the students. How many of these shorter lockers are needed so that each of the shorter students in the school population of 340 can have one?

a. Write a formula and make a histogram giving the binomial distribution for this situation.
Substitute $n = 340$ and $p = 0.02$ into the formula for a binomial distribution.
$P(k \text{ shorter students}) = {}_{340}C_k \cdot (0.02)^k \cdot (0.98)^{340-k}$
Find $P(k$ shorter students) for each k and make a histogram.
The probabilities for $k \geq 17$ are less than 10^{-3}.

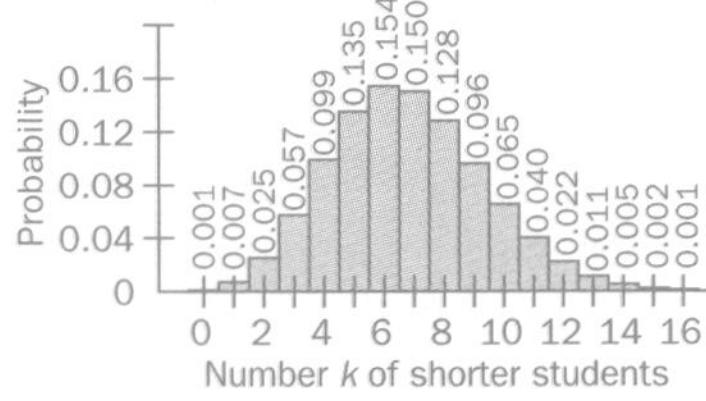

b. Find the probability that more than 12 students in the school will need the shorter lockers.
The probability that more than 12 students in the school will need the shorter lockers is the sum of the probabilities that 13, 14, 15, 16, ..., or 340 students will need the shorter lockers.
$P(\text{more than } 12) \approx 0.011 + 0.005 + 0.002 + 0.001 \approx 0.019$
The probability that more than 12 students will need shorter lockers is about 0.019.

Flipping a coin is an example of a *binomial experiment* because there are two possible results, heads or tails. In a binomial experiment, the two mutually exclusive outcomes are often called *success* and *failure*. For each trial, $P(\text{success}) = p$ and $P(\text{failure}) = 1 - p$. The probability of k successes in n independent trials is given by the following formula.

$$P(k \text{ successes in } n \text{ trials}) = {}_nC_k \cdot p^k \cdot (1 - p)^{n-k}$$

There are ${}_nC_k$ different sequences of trials that contain exactly k successes.

The probability of a particular sequence of k successes and $n - k$ failures is $p^k \cdot (1 - p)^{n-k}$.

This formula determines a theoretical **binomial distribution**. The distribution of your class's scores in the Exploration is an experimental binomial distribution.

EXAMPLE 2 Application: Architecture

An architect is designing a new lecture hall for a university. Each seat will have a writing desk attached to one of the arms. Left-handed students usually prefer to have the desk on their left. The architect wants to know the number of left-handed desks needed so that each left-handed student in a class of 180 students can have a left-handed desk. About 10% of the general population is left-handed.

a. Write a formula and make a histogram giving the binomial distribution for this situation.

b. Find the probability that more than 25 students in the class will be left-handed.

SOLUTION

a. Substitute $n = 180$ and $p = 0.1$ into the formula for a binomial distribution.

$$P(k \text{ left-handed students}) = {}_{180}C_k \cdot (0.1)^k \cdot (0.9)^{180-k}$$

Find $P(k$ left-handed students) for each possible value of k and make a histogram.

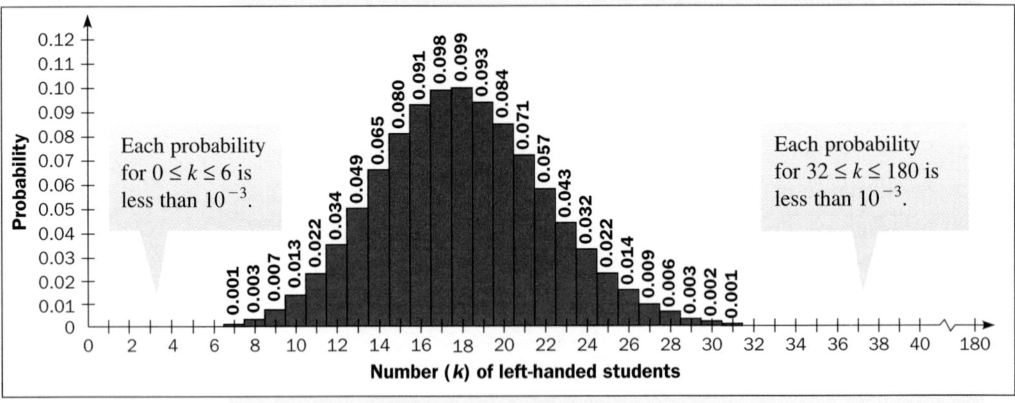

Checking Key Concepts

1. true/false test; On a true/false test, the probability of getting any question right is $\frac{1}{2}$, while on a multiple-choice test, the probability of guessing any question right from among 3 answers is $\frac{1}{3}$, from among 4 answers is $\frac{1}{4}$, and so on.
2. $\frac{5}{16}$, or 0.3125
3. $\frac{15}{1024}$, or about 0.0146
4. The probability of at least 8 successes is the sum of the probabilities of the mutually exclusive events of exactly 8 successes, exactly 9 successes, and exactly 10 successes, so $P(\text{at least 8 successes}) = {}_{10}C_8(0.4)^8(0.6)^2 + {}_{10}C_9(0.4)^9(0.6)^1 + {}_{10}C_{10}(0.4)^{10}(0.6)^0$.
5. For each trial, there are two possible mutually exclusive results: correct or incorrect. Each trial is also independent, so the distribution is binomial; ${}_6C_k\left(\frac{1}{2}\right)^6$, where k is the number correct.

b. The probability that more than 25 students in the class will be left-handed is the sum of the probabilities that 26, 27, 28, 29, . . . , or 180 students will be left-handed.

$$P(\text{more than } 25) \approx 0.014 + 0.009 + 0.006 + 0.003 + 0.002 + 0.001 = 0.035$$

The probability that more than 25 students will be left-handed is about 0.035.

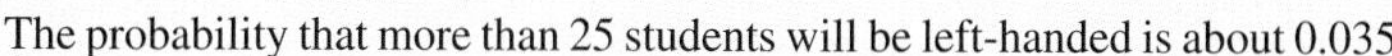

CHECKING KEY CONCEPTS

1. Are you likely to guess more questions correctly on a 20-question true/false test or on a 20-question multiple-choice test? Why?
2. What is the probability of answering exactly 3 questions correctly on a 5-question true/false test if you guess each answer randomly?
3. Each of the 5 questions on a multiple-choice test has 4 possible answers. Find P(exactly 4 correct) if you guess each answer randomly.
4. A binomial experiment has 10 trials where each trial has a 40% chance of success. Explain how to find the probability of at least 8 successes.
5. Why is the distribution of your class's scores in the Exploration on page 632 an example of a binomial distribution? Write a formula to describe the theoretical distribution.

13.4 Exercises and Applications

Extra Practice exercises on page 770

1. **Writing** Write the binomial probability formula, and explain how each part of the formula relates to a binomial experiment.

Write a formula to describe the probability distribution for a binomial experiment with *n* trials, each with probability *p* of success. Then find the probability that the experiment will have exactly *k* successful trials.

2. $n = 5, p = 0.4; k = 2$
3. $n = 10, p = \frac{1}{2}; k = 7$
4. $n = 9, p = 0.02; k = 1$
5. $n = 7, p = 10\%; k = 6$
6. $n = 12, p = \frac{1}{4}; k = 0$
7. $n = 1, p = 70\%; k = 1$
8. Suppose you flip a coin to answer each of the ten questions on a true/false quiz. What is the probability that you will get:
   a. all the answers right?
   b. half the answers right?
   c. 9 or 10 answers wrong?
9. About 10% of the population is left-handed. What is the probability that a class of 30 students will have:
   a. no left-handed students?
   b. exactly 1 left-handed student?
   c. exactly 2 left-handed students?
   d. more than 2 left-handed students?

Teach⇔Interact

Closure Question

What are the characteristics of a binomial experiment?
There are two possible results which are mutually exclusive. There are a given number of independent trials. The probability of success is the same for every trial.

Apply⇔Assess

Suggested Assignment

Core Course
Exs. 1–12, 15–18, 28–34

Extended Course
Exs. 1–34

Block Schedule
Day 2 Exs. 1–12, 15–18, 28–34

Exercise Notes

Second-Language Learners
Ex. 1 Some students learning English may benefit from working with a peer tutor or aide on this writing task.

Using Technology
Exs. 2–7 After students have successfully completed these exercises and have had practice using the binomial probability formula, you may want to challenge them to write a program for a graphing calculator to find binomial probabilities when given *n*, *p*, and *k*. Here is a short one written for the TI-82 graphing calculator.

```
:Prompt N
:Prompt P
:Prompt K
:(N nCr K)*P^K*(1–P)^(N–K)→B
:Disp B
```

Exercises and Applications

1. $P(k \text{ successes in } n \text{ trials}) = {}_nC_k \cdot p^k \cdot (1-p)^{n-k}$; ${}_nC_k$ gives the number of ways that k successes can result from n independent trials, and that p^k gives the probability of success multiplied together the number of times that the trials succeed, and that $(1-p)^{n-k}$ gives the probability of failure multiplied together the number of times that the trials fail.
2. ${}_5C_k(0.4)^k(0.6)^{5-k}$; about 0.346
3. ${}_{10}C_k(0.5)^{10}$; about 0.117
4. ${}_9C_k(0.02)^k(0.98)^{9-k}$; about 0.153
5. ${}_7C_k(0.1)^k(0.9)^{7-k}$; 6.3×10^{-6}
6. ${}_{12}C_k(0.25)^k(0.75)^{12-k}$; about 0.0317
7. ${}_1C_k(0.7)^k(0.3)^{1-k}$; 0.7
8. a. $\frac{1}{1024}$, or about 0.000977
   b. $\frac{63}{256}$, or about 0.246
   c. $\frac{11}{1024}$, or about 0.0107
9. a. about 0.0424
   b. about 0.141
   c. about 0.228
   d. about 0.589

Apply⇔Assess

Exercise Notes

Problem Solving

Exs. 8–10, 12–14 Students having difficulty with these problems are most likely having trouble identifying *n, p,* and *k* or making errors when substituting these values into the binomial probability formula. Make sure these students can describe what each variable represents and then encourage them to write down the values for each variable and the formula itself before substituting.

Application

Exs. 13, 14, 20–27

Binomial probability theory has applications to many areas of real-life. These exercises expose students to applications in marketing, driver's education, sports, and the judicial system.

Using Manipulatives

Ex. 19 This exercise gives students an opportunity to use coins and a shaker to develop their own experimental probability distribution. The physical act of generating the probabilities and creating the histogram can help students to gain greater insight into binomial probability distributions.

Visual Thinking

Ex. 19 Ask students to present their histograms to the class. Encourage them to point out various conditions that are represented by the histograms and to respond to questions from the class. This activity involves the visual skills of *recognition* and *communication.*

10. Look back at the description of a standard deck of playing cards on page 615.

a. If you draw a card at random from a standard deck of cards, what is the probability of getting a heart? a non-heart?

b. Suppose you draw a card at random from each of two standard decks of cards. Copy and complete the table.

Number of hearts	0	1	2	3 or more
Probability	?	?	?	?

11. Writing Suppose you roll a die 3 times and count how many fives occur. Is this a binomial experiment? Explain.

12. a. If you toss a coin twice, what is the probability of getting 2 heads? 1 head? 0 heads?

b. If the coin from part (a) is slightly bent so that it comes up heads 60% of the time, what is the probability of getting 2 heads? 1 head? 0 heads?

13. MARKETING A telemarketer has found that 20% of the calls made result in a sale.

a. If the telemarketer makes 10 calls, what is the probability that half of them result in a sale?

b. What minimum number of calls is needed for the probability of at least one sale to be greater than 99%?

14. DRIVER'S EDUCATION The written test for a regular Massachusetts driver's license has 20 multiple-choice questions. You must answer at least 14 questions correctly to pass the test. If you can always narrow the choices down so that you have a 50% chance of guessing the answer to each question correctly, what is the probability that you will pass? Do you think the minimum score for passing the test should be changed? Explain.

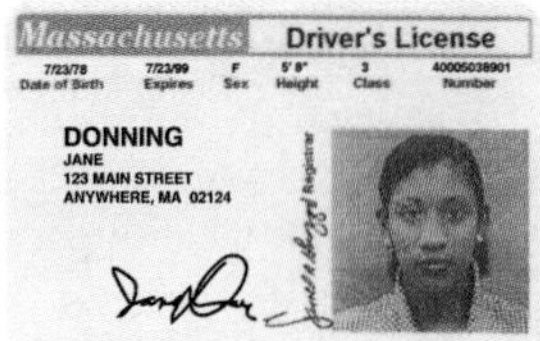

Describe how *P*(*k* successes in *n* trials, each with probability *p* of success) is related to *P*(*j* successes in *m* trials, each with probability *q* of success) for each set of values of *n, k, p, m, j,* and *q*. Explain.

15. $n = 7, k = 4, p = 0; m = 7, j = 2, q = 0$

16. $n = 9, k = 3, p = \frac{1}{2}; m = 9, j = 6, q = \frac{1}{2}$

17. $n = 5, k = 1, p = \frac{1}{5}; m = 5, j = 4, q = \frac{4}{5}$

18. $n = 8, k \geq 4, p = 0.1; m = 8, j \leq 4, q = 0.9$

19. Cooperative Learning Work with a partner. One of you should do part (a) and the other should do part (b). Work together on part (c).

a. Shake 4 coins in a plastic cup or other container and "pour" them out. Write down how many of the 4 coins come up heads. Repeat the experiment 20 times and make a relative frequency histogram.

b. Make a histogram of a theoretical binomial distribution with $n = 4$ and $p = 0.5$.

c. Writing Compare your results from parts (a) and (b). Discuss the reasons behind any similarities and differences.

20. Challenge In the 1991 National Hockey League (NHL) playoffs, Minnesota beat Chicago in a best-of-seven series. Suppose Minnesota had a 45% chance of beating Chicago in any particular game of the series. The winner of the series is the first team to win four games. What was the probability that Minnesota would win the series?

10. a. $\frac{1}{4}$, or 0.25; $\frac{3}{4}$, or 0.75

b. Probability: $\frac{9}{16}$, or 0.5625; $\frac{3}{8}$, or 0.375; $\frac{1}{16}$, or 0.0625; 0

11. Yes; consider rolling a 5 a success, and not rolling a 5 a failure, then the experiment is binomial, with $n = 3, p = \frac{1}{6}$, and k = the number of 5's rolled.

12. a. $\frac{1}{4}$, or 0.25; $\frac{1}{2}$, or 0.5; $\frac{1}{4}$, or 0.25

b. 0.36; 0.48; 0.16

13. a. about 0.0264

b. 21

14. about 0.0577; Answers may vary. An example is given. With the given conditions, about 1 out of 17 people could pass with very little knowledge, which is unacceptable. Also, a person who knew only 5 answers but had a 50% chance of guessing the rest would have nearly a one-in-three chance of passing, which seems too high.

15. P(k successes) and P(j successes) both equal 0, since k and j are both positive, and there is a 0% chance of success (certain failure) on every trial of either experiment.

16. P(k successes) and P(j successes) are the same (about 0.164). This is because each experiment has a 50% chance of failure or success in each trial. So, the probability of 3 successes (and 6 failures) in 9 trials is the same as the probability of 6 successes (and 3 failures) in 9 trials.

17. P(k successes) and P(j successes) are the same (about 0.410). This is because the first experiment asks for the probability of 1 success in 5 trials with a 20% chance of success. You can restate this as the probability of 4 failures in 5 trials with an 80% chance of failure. This is what the second experiment describes, redefining "failure" as "success."

18. P(k successes) and P(j successes) are the same (about 0.00502). This is because the first experiment asks for the probability of at least 4 successes in 8 trials with a 0.1 chance of success. You can restate this

INTERVIEW Robert Ward

In criminal court cases, which require a unanimous decision, it is often difficult for a jury to reach a decision. Suppose the probability that any given jury member thinks the defendant is guilty is a constant, p. This constant could be called the "appearance of guilt" based on the evidence and arguments given by lawyers like Robert Ward. You will use two models to explore how this probability and the jury size affect the chances of a guilty verdict.

Look back at the article on pages 608–610.

BY THE WAY

Until the 1950s, most criminal cases that went to a jury trial were required by law to use a 12-member jury.

21. **Technology** The *Friedman* model of jury decision-making assumes that each jury member votes independently and that the defendant is declared guilty if all members vote for conviction or not guilty if all members vote for acquittal. Otherwise, there is a "hung jury."

a. Suppose the vote of each jury member is a binomial trial with probability p of a vote for conviction. Express the probabilities of conviction by a 6-member jury and by a 12-member jury as functions of p.

b. Graph both functions from part (a) using a graphing calculator or graphing software. Compare the probability of a conviction for a 12-person jury with that for a 6-person jury.

22. **Spreadsheets** The *Walbert* model assumes that the majority will convince the minority. If the jury is initially split evenly, the *Walbert* model states that there is a 50% chance of eventual conviction.

a. According to the *Walbert* model, a 6-member jury will always vote for conviction if 4, 5, or 6 members initially think the defendant is guilty, and half the time if just 3 members believe the defendant is guilty. If $p = 0.9$, what is the probability of a conviction?

b. Use the *Walbert* model to express the probability of a conviction as a function of p if a jury has 6 members. Use a spreadsheet to find this probability for $p = 0, 0.1, 0.2, \ldots, 1$.

c. If a jury has 12 members, use the *Walbert* model to express the probability of a conviction as a function of p. Use a spreadsheet to find this probability for the same values of p as in part (b).

d. Compare the results of parts (b) and (c). Which size jury is more likely to convict the defendant for each value of p?

23. **Open-ended Problem** The *Walbert* model is more realistic than the *Friedman* model, but it still differs from an actual jury. How is the *Walbert* model unrealistic? Can you suggest any improvements? How do you think your changes would affect the probability of a conviction?

24. **Writing** Do you think juries in criminal cases should have 6 members or 12 members? Why?

Apply⇔Assess

Exercise Notes

Interview Note

Exs. 21–24 Students may need some help reading and interpreting these exercises. You could ask students to read the exercises in class and set up the problems with a partner while you answer any questions. Some hints may be helpful. For example, in part (a) of Ex. 21, since the decision must be unanimous, $n = k$. Also, when writing the models in parts (b) and (c) of Ex. 22, students will need to consider two cases, one in which the jury is evenly divided, and one in which the majority favors conviction.

Integrating the Strands

Exs. 21, 22 These exercises use ideas from discrete mathematics, probability, and function theory.

Cooperative Learning

Exs. 21–24 These exercises are appropriate for small group work. Set aside enough time in class for students to work together. They could either work in the computer lab and complete the exercises in one day, or they could work on Exs. 21 and 22 (excluding the spreadsheet work) one day, do the spreadsheet work for homework, and discuss the remainder of the questions the next day in class.

Using Technology

Ex. 21 Students will need to think about an appropriate viewing window for part (b) of this exercise, keeping in mind that the probabilities are between 0 and 1.

Student Progress

Ex. 22 The formulas needed for the spreadsheets are fairly complex. You may want to check students' equations before they work at the computer.

---

as the probability of at most 4 failures in 8 trials with an 0.9% chance of failure. This is what the second experiment describes, redefining "failure" as "success."

19. a. Answers may vary. Check students work.

b.

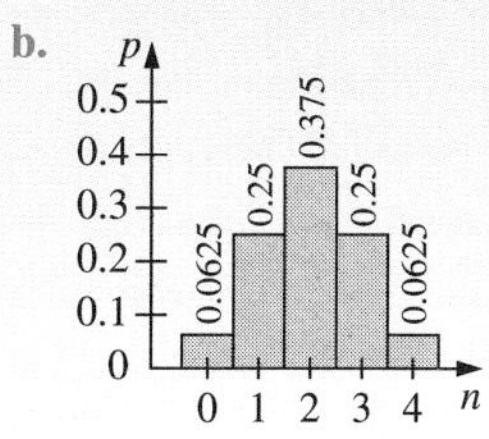

c. Students' histograms should have the same general shape as the theoretical histogram, with the results clustered around obtaining 2 heads. Because of the small number of trials, however, results could vary significantly.

20. about 0.392

21. a. 6-member: p^6; 12-member: p^{12}

b.

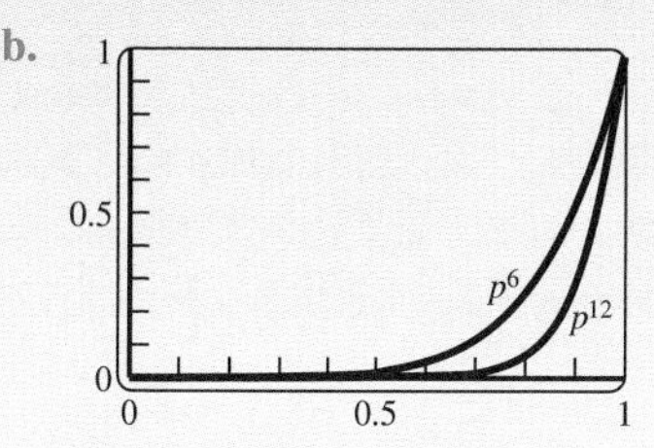

For $0 < p < 1$, it is always less likely for a 12-person jury to vote for conviction than for a 6-person jury to vote for conviction.

22, 23. See answers in back of book.

24. Answers may vary. An example is given. Juries should have 12 members because the unanimity requirement means that an innocent person is less likely to be convicted.

Apply⇔Assess

Exercise Notes

Mathematical Procedures

Exs. 25–27 These exercises provide students with an informal introduction to hypothesis testing. In hypothesis testing, probability is used to show that an outcome different from what is hypothesized is not due to chance alone, hence, invalidating the hypothesis. In Ex. 26, for example, the hypothesis would be that people have no preference between two types of cola. The outcome is that 11 out of 15 people said they preferred the cola. Assuming the hypothesis is true, it is shown that the probability of 11 or more people out of 15 people preferring the one cola is less than 6%. Since the outcome of the taste test is theoretically so unlikely, it is concluded that the hypothesis is false and that people do indeed prefer the new cola.

Practice 86 for Section 13.4

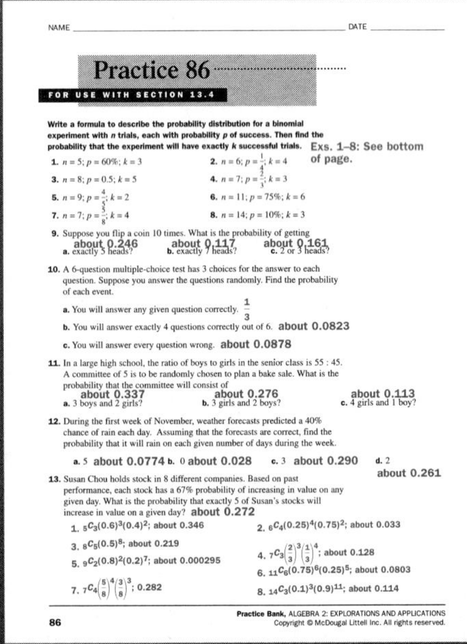
NAME ______________________ DATE ____________

Practice 86

FOR USE WITH SECTION 13.4

Write a formula to describe the probability distribution for a binomial experiment with n trials, each with probability p of success. Then find the probability that the experiment will have exactly k successful trials. Exs. 1–8: See bottom of page.

1. $n = 5$; $p = 60\%$; $k = 3$
2. $n = 6$; $p = \frac{1}{4}$; $k = 4$
3. $n = 8$; $p = 0.5$; $k = 5$
4. $n = 7$; $p = \frac{2}{3}$; $k = 3$
5. $n = 9$; $p = \frac{4}{5}$; $k = 2$
6. $n = 11$; $p = 75\%$; $k = 6$
7. $n = 7$; $p = \frac{5}{8}$; $k = 4$
8. $n = 14$; $p = 10\%$; $k = 3$

9. Suppose you flip a coin 10 times. What is the probability of getting
   a. exactly 5 heads? about 0.246
   b. exactly 7 heads? about 0.117
   c. 2 or 3 heads? about 0.161

10. A 6-question multiple-choice test has 3 choices for the answer to each question. Suppose you answer the questions randomly. Find the probability of each event.
    a. You will answer any given question correctly. $\frac{1}{3}$
    b. You will answer exactly 4 questions correctly out of 6. about 0.0823
    c. You will answer every question wrong. about 0.0878

11. In a large high school, the ratio of boys to girls in the senior class is 55 : 45. A committee of 5 is to be randomly chosen to plan a bake sale. What is the probability that the committee will consist of
    a. 3 boys and 2 girls? about 0.337
    b. 3 girls and 2 boys? about 0.276
    c. 4 girls and 1 boy? about 0.113

12. During the first week of November, weather forecasts predicted a 40% chance of rain each day. Assuming that the forecasts are correct, find the probability that it will rain on each given number of days during the week.
    a. 5 about 0.0774 b. 0 about 0.028 c. 3 about 0.290 d. 2 about 0.261

13. Susan Chou holds stock in 8 different companies. Based on past performance, each stock has a 67% probability of increasing in value on any given day. What is the probability that exactly 5 of Susan's stocks will increase in value on a given day? about 0.272

1. ${}_5C_3(0.6)^3(0.4)^2$; about 0.346
2. ${}_6C_4(0.25)^4(0.75)^2$; about 0.033
3. ${}_8C_5(0.5)^8$; about 0.219
4. ${}_7C_3\left(\frac{2}{3}\right)^3\left(\frac{1}{3}\right)^4$; about 0.128
5. ${}_9C_2(0.8)^2(0.2)^7$; about 0.000295
6. ${}_{11}C_6(0.75)^6(0.25)^5$; about 0.0803
7. ${}_7C_4\left(\frac{5}{8}\right)^4\left(\frac{3}{8}\right)^3$; 0.282
8. ${}_{14}C_3(0.1)^3(0.9)^{11}$; about 0.114

86 **Practice Bank,** ALGEBRA 2: EXPLORATIONS AND APPLICATIONS

Connection MARKETING

Before launching a new product, a manufacturing company will conduct various marketing surveys to see how well the product will compete in the existing market. For example, a new food product will be taste-tested, consumer reactions to different marketing approaches will be charted, and so on.

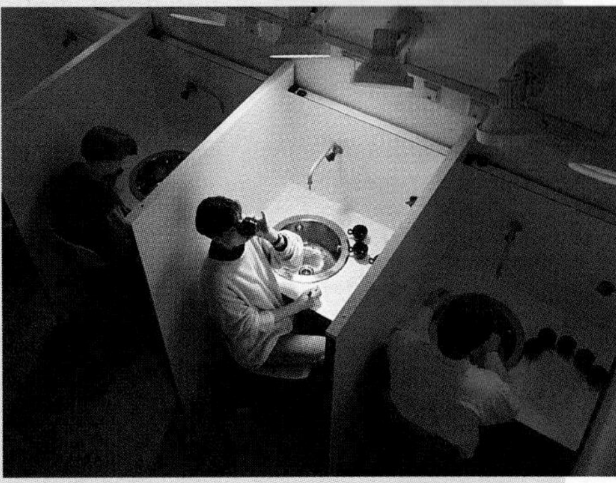

25. Suppose people at a shopping mall are asked to look at an advertisement and then answer some questions about it. Out of the first ten shoppers surveyed independently, 7 said they thought the claims seemed believable. What is the probability of getting this result if the proportion of the general population believing the ad is 50%?

26. In a blind taste test, 11 out of 15 people said they preferred a new formulation for a company's diet cola. What is the probability of getting at least this many positive responses if the average consumer really has no preference between the two?

27. a. Write an expression for the probability of getting k out of 15 positive responses to a new diet cola if you assume that people in general have no preference.

b. Of the 15 people polled, let $x =$ the number who indicated a preference for the new cola. Use the expression from part (a) to find $P(x \geq k)$ for each value of k from 0 to 15. Make a histogram of these probabilities as a function of k.

c. In part (b), for what value of k is $P(x \geq k)$ first less than 10%? first less than 5%? At what point do you think it is safe to conclude that people really prefer the new formulation to the old?

ONGOING ASSESSMENT

28. Writing Explain what a binomial experiment is and how to find binomial distributions. Include some examples.

SPIRAL REVIEW

29. Suppose $P(A) = 0.8$ and $P(A \text{ and } B) = 0.32$. Find $P(B \mid A)$. *(Section 13.3)*

Find the mean and the standard deviation of each data set. *(Section 6.5)*

30. 125, 170, 250, 270, 144, 185, 176, 100, 220, 160

31. 0.18, 0.25, 0.63, 0.40, 0.26, 0.25, 0.20

Graph each inequality. *(Section 7.4)*

32. $y < 3x + 6$

33. $5x - 4y \geq 10$

34. $y > 2.7x - 1.3$

25. about 0.117

26. about 0.0592

27. a. ${}_{15}C_k\left(\frac{1}{2}\right)^{15}$

b. $P(x \geq 0)$: 1; $P(x \geq 1)$: 0.999969; $P(x \geq 2)$: 0.999512; $P(x \geq 3)$: 0.996307; $P(x \geq 4)$: 0.982422; $P(x \geq 5)$: 0.940765; $P(x \geq 6)$: 0.849121; $P(x \geq 7)$: 0.696381; $P(x \geq 8)$: 0.5; $P(x \geq 9)$: 0.303619; $P(x \geq 10)$: 0.150879; $P(x \geq 11)$: 0.059235; $P(x \geq 12)$: 0.017578; $P(x \geq 13)$: 0.003693; $P(x \geq 14)$: 0.000488; $P(x \geq 15)$: 0.000031

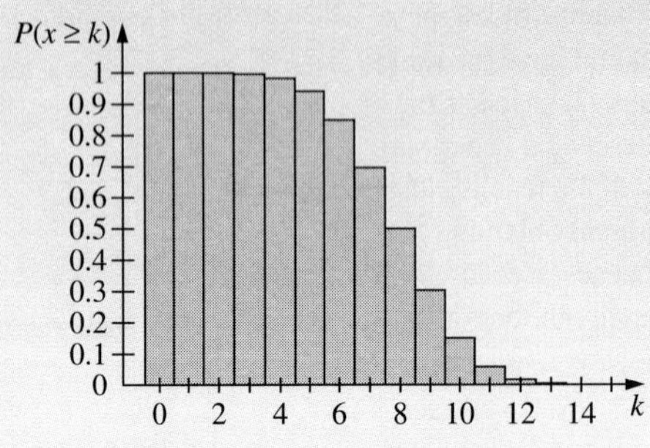

c. $k \geq 11$; $k \geq 12$; Answers may vary, though 10% and 5% tests of statistical significance are often used.

28. Answers may vary. An example is given. A binomial experiment consists of a particular number of independent trials with the outcome either success, with probability p, or failure, with probability $1 - p$. The probability of k successes in p trials is then given by ${}_nC_k \cdot p^k \cdot (1 - p)^{n-k}$. For example, if a basketball player averages making 65% of free-throw attempts, then the probability that that player makes 14 out of 20 attempts is ${}_{20}C_{14}(0.65)^{14}(0.35)^6$.

29. 0.4

30–34. See answers in back of book.

Section

13.5 Normal Distributions

Learn how to...

- recognize a normal distribution
- find probabilities involving normally distributed data

So you can...

- find probabilities involving the heights of women, for example

Chances are that most of the women you see during a typical week are all close to the same height, and that very few are either much shorter or much taller. This is due to the fact that women's heights are *normally* distributed. The histogram below shows the distribution of the heights of 1000 women.

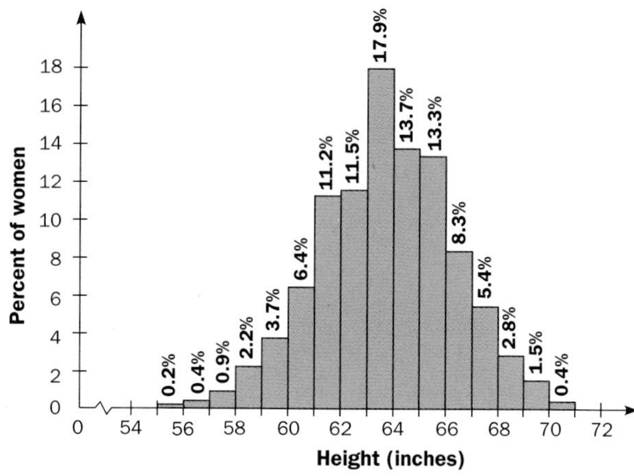

THINK AND COMMUNICATE

1. Does the histogram show a *symmetric distribution* or a *skewed distribution*?
2. Estimate the mean, median, and mode of the data in the histogram.
3. The height data have a mean of 64 in. and a standard deviation of 2.5 in. Estimate the percent of the data that lies within:
   a. one standard deviation of the mean (that is, between $64 - 2.5 = 61.5$ in. and $64 + 2.5 = 66.5$ in.)
   b. two standard deviations of the mean
   c. three standard deviations of the mean
4. Imagine a smooth curve that comes reasonably close to passing through the midpoints of the tops of the bars in the histogram. What shape does the curve have?

Toolbox p. 790
Mean, Median, and Mode

Rather than working with a large set of data, mathematicians sometimes use a curve that approximates the shape of a distribution. In particular, when a histogram has a **normal distribution** like the one above, it is modeled with an equation whose graph is a bell-shaped curve.

The *area under the curve* (that is, the area between the curve and the horizontal axis) for some interval on the horizontal axis gives an approximation of the percent of data that lies within the interval.

Plan⇔Support

Objectives

- Recognize a normal distribution.
- Find probabilities involving normally distributed data.
- Find probabilities in real-world situations.

Recommended Pacing

Core and Extended Courses
Section 13.5: 1 day

Block Schedule
Section 13.5: $\frac{1}{2}$ block
(with Portfolio Project)

Resource Materials

Lesson Support
Lesson Plan 13.5
Warm-Up Transparency 13.5
Overhead Visuals:
Folder 13: The Normal Curve
Practice Bank: Practice 87
Study Guide: Section 13.5
Explorations Lab Manual:
Diagram Master 1
Assessment Book: Test 55

Technology
Graphing Calculator
McDougal Littell Mathpack
Stats! Activity Book:
Activities 9, 15, and 19
Internet:
http://www.hmco.com

Warm-Up Exercises

Use the following set of data.
29.1, 33.2, 27.7, 31.6, 30.0, 27.8, 29.6, 31.3, 30.2, 28.6, 29.9, 30.7, 30.2, 32.4, 35.0, 27.6, 31.1, 25.4

1. Find the mean and standard deviation of the data. 30.1; 2.2
2. What percent of the data lies below the mean? 50%
3. What percent of the data lies below 32? 83%
4. What percent of the data lies above 32? 17%
5. Complete. If x is a real number, then $P(x > 5) = 1 -$ __?__ . $P(x \le 5)$

ANSWERS Section 13.5

Think and Communicate

1. symmetric
2. Answers may vary. Examples are given: mean: about 64 in.; median: about 64 in.; mode: between 63 and 64 in.
3. Answers may vary. Examples are given.
   a. about 70%
   b. about 95%
   c. almost 100%
4. bell-shaped

Teach⇔Interact

Learning Styles: Visual

The normal curve can help visual learners understand and solve problems involving normal distributions. Suggest to students that they sketch normal curves for each problem. Have them include the mean on the horizontal axis as well as the x-value(s) that define the interval. They can then shade the area which is equal to the probability they are seeking.

Think and Communicate

Questions 1–4 lead students through several properties of a normal distribution, namely, that it is symmetric; its mean, median, and modes are equal; it has a bell shape; and a fixed percentage of data lies within one, two, or three standard deviations of the mean.

Section Note

Communication: Drawing Have students sketch the normal distribution from this page into their journals. In addition to the labels on this page, students should calculate the percent of data that lies in each portion of the curve and place those values in the appropriate section.

Additional Example 1

Use the data on page 639. Find the theoretical probability that the height of a randomly selected woman is between 56.5 in. and 66.5 in. Use the fact that the mean of the data is 64 in. and the standard deviation is 2.5 in.

Step 1 Find how far above or below the mean the endpoints of the given interval are.

56.5 in. – 64 in. = –7.5 in.

66.5 in. – 64 in. = 2.5 in.

The lower endpoint is 3 standard deviations below the mean and the upper endpoint is 1 standard deviation above the mean.

Step 2 Use the mean to rewrite the interval as two separate intervals. Let x be the height in inches of a randomly selected woman.

$P(56.5 \text{ in.} < x < 66.5 \text{ in.})$

$= P(56.5 \text{ in.} < x < 64 \text{ in.}) + P(64 \text{ in.} < x < 66.5 \text{ in.})$

$= 0.495 + 0.34$

$= 0.835$

The probability is 0.835.

The graph shown below is a normal distribution with mean $\bar{x}$ and standard deviation σ.

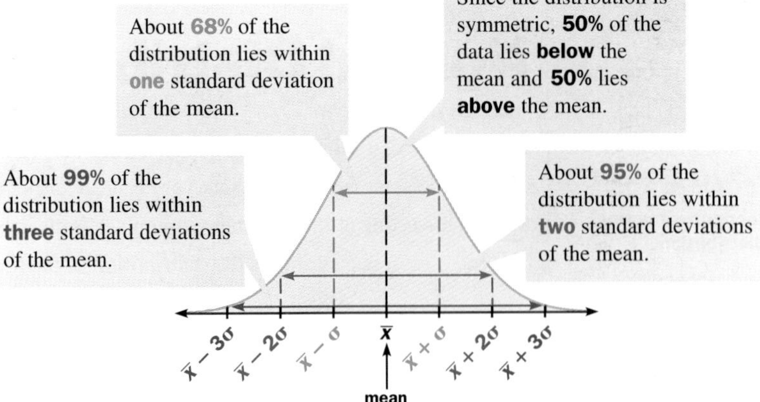
About 68% of the distribution lies within one standard deviation of the mean.
Since the distribution is symmetric, 50% of the data lies below the mean and 50% lies above the mean.
About 99% of the distribution lies within three standard deviations of the mean.
About 95% of the distribution lies within two standard deviations of the mean.
$\bar{x}-3\sigma$
$\bar{x}-2\sigma$
$\bar{x}-\sigma$
$\bar{x}$
$\bar{x}+\sigma$
$\bar{x}+2\sigma$
$\bar{x}+3\sigma$
mean

THINK AND COMMUNICATE

5. What percent of the data in a normal distribution lies between the mean and one standard deviation above the mean? Explain.

You can interpret the percent of data that lies within a given interval as a probability. For example, your chance of selecting any one data value from a set of normally distributed data and having it lie within one standard deviation of the mean is about 68%, or 0.68.

Because a normal curve is a *model* for a histogram of actual data, probabilities obtained from the curve are theoretical, and probabilities obtained from the data are experimental.

EXAMPLE 1

Find the theoretical probability that the height of a randomly selected woman is between 61.5 in. and 69 in. tall. Use the fact that the mean of the data is 64 in. and the standard deviation is 2.5 in.

SOLUTION

Step 1 Find how far above or below the mean the endpoints of the given interval are.

61.5 in. − 64 in. = **−2.5 in.** ⟵ 1 standard deviation below the mean

69 in. − 64 in. = **5 in.** ⟵ 2 standard deviations above the mean

Step 2 Use the mean to rewrite the interval as two separate intervals. Let x be the height in inches of a randomly selected woman.

$$P(61.5 \text{ in.} < x < 69 \text{ in.}) = P(61.5 \text{ in.} < x < 64 \text{ in.}) + P(64 \text{ in.} < x < 69 \text{ in.})$$
$$= 0.34 + 0.475$$
$$= 0.815$$

The probability that a woman's height is between 61.5 in. and 69 in. is 0.815.

Think and Communicate

5. about 34%; About 68% of the data lies within one standard deviation of the mean. Because the curve is symmetric about the mean, this means that half of the data within one standard deviation lies on either side of the mean.

The Standard Normal Distribution

The **standard normal distribution** is the normal distribution with mean 0 and standard deviation 1. You can use the *standard normal table* shown below to find the probability that a randomly selected data value is less than a given number of standard deviations from the mean. The table is a refinement of the percents given on the previous page.

To use the table, you need to convert a given x-value from a normal distribution with mean $\bar{x}$ and standard deviation σ to a ***z*-score** given by:

Subtract the mean from the given value.

$$z = \frac{x - \bar{x}}{\sigma}$$

Divide by the standard deviation.

A z-score gives the number of standard deviations that x lies from the mean. Obtaining a z-score for a data value is called *standardizing* the value.

z	.0	.1	.2	.3	.4	.5	.6	.7	.8	.9
−3	.0013	.0010	.0007	.0005	.0003	.0002	.0002	.0001	.0001	.0000+
−2	.0228	.0179	.0139	.0107	.0082	.0062	.0047	.0035	.0026	.0019
−1	.1587	.1357	.1151	.0968	.0808	.0668	.0548	.0446	.0359	.0287
−0	.5000	.4602	.4207	.3821	.3446	.3085	.2743	.2420	.2119	.1841
0	.5000	.5398	.5793	.6179	.6554	.6915	.7257	.7580	.7881	.8159
1	.8413	.8643	.8849	.9032	.9192	.9332	.9452	.9554	.9641	.9713
2	.9772	.9821	.9861	.9893	.9918	.9938	.9953	.9965	.9974	.9981
3	.9987	.9990	.9993	.9995	.9997	.9998	.9998	.9999	.9999	1.0000−

This means "slightly more than 0."

The probability that a data value is less than 1.9 standard deviations above the mean is **0.9713**.

This means "slightly less than 1."

EXAMPLE 2 Application: Biology

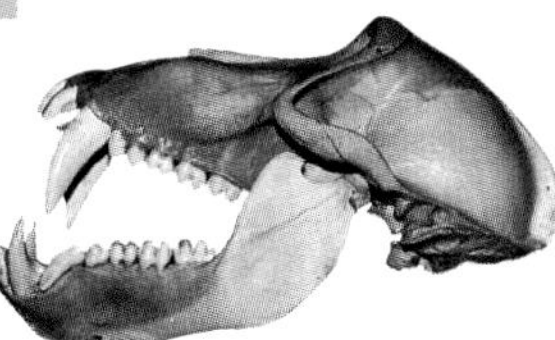

Baboon skulls can be classified by their dental structure. A fossilized baboon skull with a third premolar of length 9.0 mm was discovered in Angola. In genus *Papio*, the third premolar has a mean length of 8.18 mm with a standard deviation of 0.47 mm. If you assume molar lengths are normally distributed, what is the probability of one having a length of at least 9.0 mm? What does this probability suggest about the baboon?

SOLUTION

Standardize the observed value so you can use the standard normal table.

$$\begin{aligned} P(\text{length} \geq 9.0) &= 1 - P(\text{length} < 9.0) \\ &= 1 - P\left(z < \frac{9.0 - 8.18}{0.47}\right) \\ &\approx 1 - P(z < 1.7) \\ &\approx 1 - 0.9554 \\ &= 0.0446 \end{aligned}$$

Use the complementary probability.

Standardize the value 9.0.

Use the standard normal table.

The probability that an animal of this genus has a third premolar at least 9.0 mm long is about 0.045. This small probability suggests that the premolar may not be from genus *Papio*.

Teach⇔Interact

Section Notes

Alternate Approach

Students who are comfortable with transformations may understand this material better if the relationship between the standard normal distribution and other normal distributions is examined in terms of scale change and translation. Have them think of the standard normal distribution as the basic distribution from which all other normal distributions can be generated by using a scale change followed by a translation in the horizontal direction.

Communication: Reading

Students may never have used a table such as this one before and may need help in reading it. Make sure students understand that the numbers along the left and top represent z-scores and that the numbers within the table are probabilities.

Teaching Tip

Point out that a negative z-score indicates that the x-value lies below the mean, while a positive z-score indicates that the x-value lies above the mean.

Additional Example 2

Suppose that a bus company is considering reducing its rate for senior citizens (over 65) and children 10 and under. If the age of a typical bus rider is normally distributed with a mean of 37 and a standard deviation of 12.2, what is the probability that a rider will get a reduced fare? A rider will get a reduced fare if he or she is over 65 or 10 and under. This is a compound probability with mutually exclusive events so that they can be added. Then use complements to find the probability that the rider is over 65. Standardize the endpoint values so that you can use a standard normal table.

$P(\text{reduced fare}) = P(x > 65) + P(x \leq 10)$

$= [1 - P(x \leq 65)] + P(x \leq 10)$

$= \left[1 - P\left(z \leq \frac{65 - 37}{12.2}\right)\right] + P\left(z \leq \frac{10 - 37}{12.2}\right)$

$= [1 - P(z \leq 2.2951)] + P(z \leq -2.213)$

$\approx [1 - 0.9893] + 0.0139$

$= 0.0246$

The probability that a rider will get a reduced fare is about 0.0246.

Teach⇔Interact

Think and Communicate

After students complete questions 6 and 7, have them write down the strategies suggested by these problems. Question 6 gives them an alternative method to the one used in Example 2 for finding the probability that a value lies above a given value, namely, that after converting to z-scores, $P(z > k) = P(z < -k)$. This can be seen clearly by drawing a standard normal curve and labeling k and $-k$. Students should realize that the corresponding areas are equal. In completing question 7, students will need to develop a method for the type of problem where the interval has two given endpoints. In question 8, students use the standard normal table to verify the 68, 95, and 99 percentages for data within 1, 2, and 3 deviations of the mean, respectively.

Closure Question

Describe a normal distribution.
A normal distribution is bell-shaped and symmetric about the mean. The area under the curve for some interval on the horizontal axis is equal to the percent of data that lies within the interval. About 68% of the data lies within 1 standard deviation of the mean, about 95% within two standard deviations, and about 99% within three standard deviations.

THINK AND COMMUNICATE

6. Look at the standard normal table on the previous page. How is $P(z < k)$ related to $P(z < -k)$ for some positive constant k? Explain this relationship in terms of the area under the standard normal curve.

7. Describe how you can use the standard normal table to solve Example 1.

8. Use the standard normal table to show that about 68% of the data in a normal distribution lies within one standard deviation of the mean, about 95% lies within two standard deviations, and about 99% lies within three standard deviations.

CHECKING KEY CONCEPTS

1. At a large university, the heights of male students are normally distributed with mean 175 cm and standard deviation 10 cm.

a. Sketch a normal curve to show the distribution of male student heights. Label the mean. Also label the heights that are one, two, and three standard deviations from the mean.

b. About what percent of the male students have heights between 155 cm and 195 cm? between 165 cm and 205 cm?

2. The following values are from a normal distribution with mean 25 and standard deviation 15. Standardize each value.

a. 25 **b.** 10 **c.** 55 **d.** 30 **e.** 50

3. Suppose the scores on a national mathematics test are normally distributed with mean 500 and standard deviation 100. What percent of students who took the test scored below 420? above 690?

13.5 Exercises and Applications

Extra Practice exercises on page 770

1. Find the probability that a randomly chosen x-value lies within each interval if x comes from a normal distribution with mean $\bar{x}$ and standard deviation σ. Explain how you found each answer.

a. $\bar{x} - 2\sigma < x < \bar{x}$ **b.** $\bar{x} < x < \bar{x} + 3\sigma$ **c.** $\bar{x} - 2\sigma < x < \bar{x} + 3\sigma$

2. BOTANY The *guayule* plant, which grows in the southwestern United States and in Mexico, is one of several plants that can be used as a source of rubber. Suppose that in a large group of *guayule* plants, the heights of the plants are normally distributed with mean 12 in. and standard deviation 2 in.

a. What percent of the *guayule* plants are taller than 16 in.?

b. What percent of the *guayule* plants are between 10 in. and 14 in. tall?

c. What percent of the *guayule* plants are between 8 in. and 12 in. tall?

d. What percent of the *guayule* plants are no taller than 6 in.?

Apply⇔Assess

Suggested Assignment

Core Course
Exs. 1, 2, 6–9, 12–22, AYP

Extended Course
Exs. 1–22, AYP

Block Schedule
Day 3 Exs. 1, 2, 6–9, 12–22, AYP

Think and Communicate

6. $P(z < k) = 1 - P(z < -k)$; Because of the symmetry of the standard normal curve, $P(z < -k) = P(z > k)$. Now, $P(z < k) + P(z > k)$ accounts for all of the area under the curve, so $P(z < k) + P(z > k) = 1$, or $P(z < k) = 1 - P(z < -k)$.

7. Answers may vary. An example is given. Since you are already given that you are looking between 1 standard deviation below the mean and 2 standard deviations above the mean, you know that the relevant z-scores are −1 and 2. From the table, $P(z < -1) = 0.1587$ and $P(z < 2) = 0.9772$. So, $P(z < 2) - P(z < -1) = P(-1 < z < 2) = 0.9772 - 0.1587 = 0.8185$, since $P(z < -1)$ is included as part of $P(z < 2)$.

8. $P(-1 < z < 1) = P(z < 1) - P(z < -1) = 0.8413 - 0.1587 = 0.6826$, or about 68%; $P(-2 < z < 2) = P(z < 2) - P(z < -2) = 0.9772 - 0.0228 = 0.9544$, or about 95%; $P(-3 < z < 3) = P(z < 3) - P(z < -3) = 0.9987 - 0.0013 = 0.9974$, or about 99%

Checking Key Concepts

1–3. See answers in back of book.

Connection HISTORY

Adolphe Quetelet was a Belgian scholar who demonstrated that normal distributions can be used to model many different kinds of data. In one of his studies he measured the chests of 5738 Scottish soldiers. To check whether the data were normally distributed, Quetelet compared theoretical and experimental probabilities.

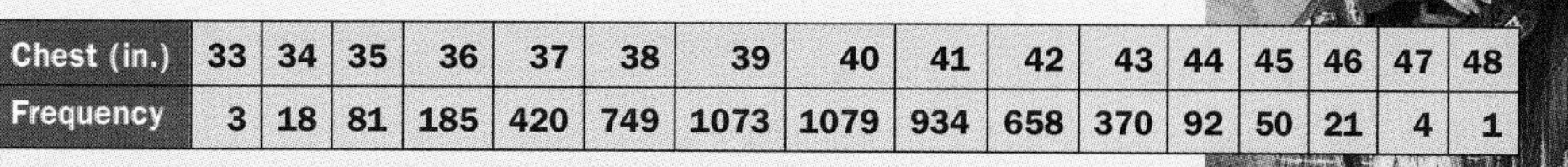

Chest (in.)	33	34	35	36	37	38	39	40	41	42	43	44	45	46	47	48
Frequency	3	18	81	185	420	749	1073	1079	934	658	370	92	50	21	4	1

3. The mean of the data is $\bar{x} = 39.8$ in. and the standard deviation is $\sigma = 2.05$ in. If you assume that the data are normally distributed, what is the theoretical probability for each interval?

a. $\bar{x} - \sigma < x < \bar{x}$ **b.** $\bar{x} < x < \bar{x} + 2\sigma$ **c.** $\bar{x} - 3\sigma < x < \bar{x} - \sigma$

4. Use the table to find the experimental probability for each interval in Exercise 3.

5. Writing Do you think the data are normally distributed? Why or why not? You may want to test other intervals before deciding.

6. MANUFACTURING Boxes of cereal are filled by a machine. Tests of the machine's accuracy show that the amount of cereal in each box varies. The weights are normally distributed with mean 20 oz and standard deviation 0.25 oz.

a. Sketch a normal curve to show the distribution of weights. Label the mean. Also label the weights that are one, two, and three standard deviations from the mean.

b. What percent of the boxes weigh more than 20 oz?

c. What percent of the boxes weigh less than 19.5 oz?

7. Cooperative Learning Work with a partner.

a. Ask at least 30 people to each draw a line segment 5 in. long without using a ruler. Measure each line segment to the nearest $\frac{1}{16}$ in. and display the data in a histogram.

b. Find the mean and the standard deviation of the data. What percent of your data values are within one standard deviation of the mean? within two standard deviations of the mean? within three standard deviations of the mean? Do you think the data are normally distributed? Explain.

8. The following values are from a normal distribution with mean 17 and standard deviation 4. Standardize each value.

a. 17 **b.** 21 **c.** 15 **d.** 26 **e.** 30

9. The scores on a statewide mathematics test were normally distributed with mean 400 and standard deviation 40. Each score is reported as a *percentile*, the percent of people who scored less than or equal to that score. José's score was 550. What was his percentile score?

Apply⇔Assess

Exercise Notes

Mathematical Procedures
Exs. 3–5, 7 These exercises illustrate a method that can be used to determine if data are normally distributed. Students look at different intervals of data and compare the experimental probability of being in that interval with the theoretical probability based on the normal distribution.

Interdisciplinary Problems
Exs. 3–5, 12–14 Normal distributions occur in nearly every discipline. These exercises introduce students to problems involving the normal distribution from history and medicine.

Application
Ex. 6 This exercise and *Assess Your Progress* Ex. 4 allow students to see how normal distributions are used in manufacturing.

Cooperative Learning
Ex. 7 Students will need time to gather their data outside of class. You might have each group member collect some of the data and bring them to class prior to working with their partner on the rest of the exercise. After each group completes the exercise, discuss the results together and account for any differences.

Challenge
Ex. 9 Students may be familiar with percentiles as they are generally given for standardized tests such as the SAT. Some students may want to know what score they need to achieve on a standardized test to be at a particular percentile. As an example, have students find the score (to the nearest ten) that a student would have to get to be at the 60th percentile on the test in this exercise. (410)

Exercises and Applications

1. Answers may vary. Examples are given.
   a. about 0.475; Because about 95% of the distribution lies within two standard deviations of the mean, half of that lies between the mean and two standard deviations below the mean.
   b. about 0.495; Because about 99% of the distribution lies within three standard deviations of the mean, half of that lies between the mean and three standard deviations above the mean.
   c. about 0.976; The probability can be found by subtracting the probability for a z-score of up to –2 from the probability for a z-score of up to 3: 0.9987 – 0.0228 = 0.9759.
2. a. about 2.28%
   b. about 68.3%
   c. about 47.7%
   d. about 0.13%
3. a. about 0.341
   b. about 0.477
   c. about 0.157
4. (a) about 0.317; (b) about 0.530; (c) about 0.123
5. Answers may vary. An example is given. The data are fairly close to the theoretical probabilities, so they seem to be normally distributed.
6. See answers in back of book.
7. a, b. Answers may vary. Check students' work.
8. a. 0 b. 1
   c. –0.5 d. 2.25
   e. 3.25
9. 99.99%

Apply⇔Assess

Exercise Notes

Using Technology
Ex. 10 To integrate on a TI-82 graphing calculator, choose 9:fnInt(from the MATH MATH menu. Then complete the entry on the home screen so that it reads fnInt(*expression,variable,lower, upper*). The expression is the function that is being integrated, the variable is the variable that was used in the expression, and lower and upper are the endpoints of the x-interval for which the integral is being calculated. If the function to be integrated is already on the Y= list, Y1 can be used for the expression. For example, after entering the function from part (a) on the Y= list as Y1, enter fnInt(Y1,X,–4,4) on the home screen to answer part (b).

Integrating the Strands
Ex. 10 This exercise gives students some insight into how probability, exponential functions, and calculus are intimately linked in the study of normal distributions.

Visual Thinking
Ex. 11 Ask students to sketch graphs of their distributions and show the graphs to the class. Encourage them to explain the graphs in relation to the real-life data that they represent. This activity involves the visual skills of *correlation* and *communication.*

Assessment Note
Ex. 11 This problem could be made into a project and used to assess students' understanding of this section. In addition to the report, students can be asked to make a poster showing their data and illustrating their conclusion.

Communication: Discussion
Exs. 12–14 Students may need some help interpreting the information given in these problems. Stress to students that each age (each column) forms its own normal distribution. You might have students do a sketch of two of these distributions, making sure to give them a title and label the axis. The graph students make in Ex. 12 uses the first two rows of the given data. Students may need to think about why the data do not make a normal curve. You might also ask students to look at the standard deviations as the age increases. What does this say about the weights of girls as they age?

10. a. 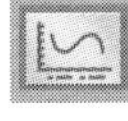 **Technology** Use a graphing calculator or graphing software to graph $f(x) = \frac{1}{\sqrt{2\pi}}e^{-x^2/2}$. Use a viewing window with $-4 \le x \le 4$ and $0 \le y \le 1$.

b. The graph in part (a) is the standard normal curve. What should the total area under this curve be? Why? Some calculators and graphing software can *integrate* to find the area under the curve, denoted by $\int f(x)\,dx$. Check your answer by finding $\int f(x)\,dx$ for $-4 \le x \le 4$.

c. What value would you expect for $\int f(x)\,dx$ when x is between -1 and 1? between -2 and 2? Explain. Check your answers.

11. Open-ended Problem Think of a real-world variable that might have a normal distribution. Gather at least 30 data values. Are the data normally distributed? Write a brief report about your project. Be sure to justify your conclusion.

Connection MEDICINE

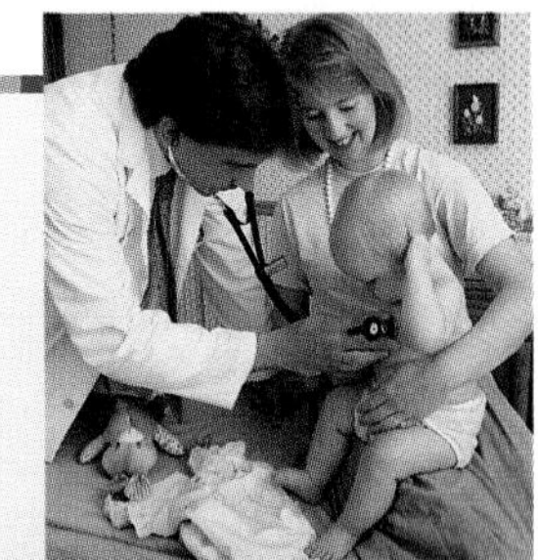

Pediatricians check a child's height and weight against standard tables to track his or her rate of growth. Measurements of weight for each age group tend to be normally distributed, and each child tends to remain at the same place within the distribution as she or he grows. If a child's location within the distribution changes significantly, the pediatrician may want to find out why.

Distributions of Weights for Girls Aged 2 Years to 13 Years

Age (years)	2	3	4	5	6	7	8	9	10	11	12	13
Weight (kg)	11.9	14.1	15.9	17.7	19.5	21.8	24.8	28.3	32.5	36.9	41.4	45.9
Standard deviation	1.34	1.88	2.37	2.98	3.77	4.68	5.90	7.36	8.81	10.33	11.79	12.83

12. Make a scatter plot of mean weight versus age. Over which year does the average girl gain the most weight?

13. About what percent of each group weighs more than 19 kg?

a. 5-year-old girls **b.** 6-year-old girls **c.** 7-year-old girls

14. a. The table at the right gives Cathy's weights at different ages. At each age, what percent of girls in Cathy's age group weigh less than she does?

b. At 10 years old Cathy weighed 31.9 kg, and at 13 years old she weighed 47.8 kg. Compare her growth through age 13 with the typical growth pattern of girls.

Age (years)	Weight (kg)
2	11.55
3	13.43
4	14.35
5	15.87

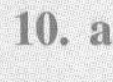

10. a.

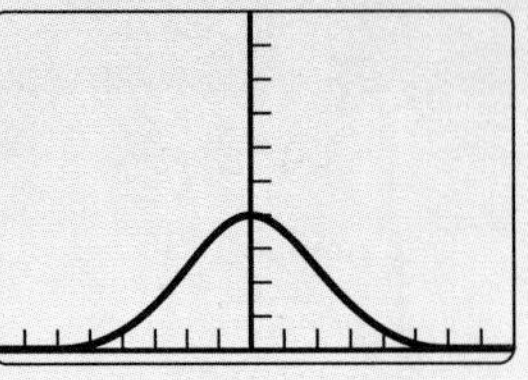

b. 1; The total area represents the sum of the probabilities, which is 1. The result of integrating is about 0.9999, which is very close to 1.

c. about 0.6827; about 0.9545; These expressions represent the probabilities that a data value lies within one standard deviation of the mean and within two standard deviations of the mean, respectively. The answers check closely with the table values of 0.6826 and 0.9544.

11. Answers may vary. Check students' work.

12.

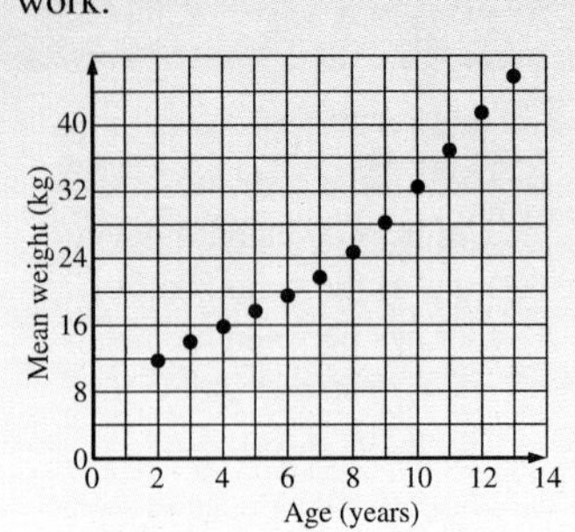

during her 13th year

ONGOING ASSESSMENT

15. Writing How is the standard normal distribution like other normal distributions? What is special about the standard normal distribution?

SPIRAL REVIEW

16. What is the probability that you will answer exactly 3 questions correctly on a 4-question true/false test if you guess each answer randomly? *(Section 13.4)*

17. A company sells *FunBags* that each contain a random assortment of 10 colored erasers. If 20% of the erasers manufactured for the *FunBags* are green, what is the probability that a given *FunBag* will contain at least 3 green erasers? *(Section 13.4)*

Simplify each expression. *(Section 12.4)*

18. ${}_5C_1$ **19.** ${}_7C_4$ **20.** ${}_{22}C_0$ **21.** ${}_{19}C_{17}$

22. Are $\triangle ABC$ and $\triangle ADE$ similar? Explain. *(Toolbox, page 801)*

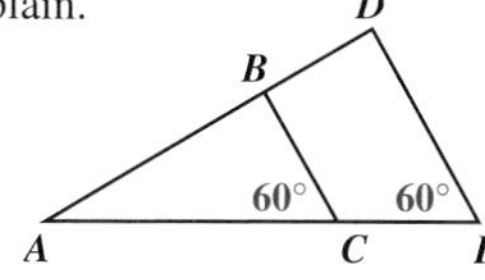

ASSESS YOUR PROGRESS

VOCABULARY

binomial distribution (p. 634)
normal distribution (p. 639)
standard normal distribution (p. 641)
z-score (p. 641)

1. Suppose you roll a die 6 times and count how many times the die shows a one. Find P(exactly 3 ones) and P(at least 2 ones). *(Section 13.4)*

2. MANUFACTURING There is a 7% chance that any given capacitor is defective. If an inspector examines 20 capacitors, what is the probability that at least one will be defective? *(Section 13.4)*

3. AGRICULTURE The pumpkins that a farmer harvested varied in size, having diameters that were normally distributed with mean diameter 12 in. and standard deviation 3 in. *(Section 13.5)*

a. About what percent of the pumpkins were larger than 15 in. in diameter?

b. About what percent were smaller than 6 in. in diameter?

4. MANUFACTURING A manufacturer produces bags of pretzels marked "Net weight 16 oz." The machine is set to fill each bag with 16.3 oz, but the actual weights are normally distributed with mean 16.3 oz and standard deviation 0.18 oz. What is the probability that a given bag will contain less than 16 oz of pretzels? *(Section 13.5)*

5. Journal Compare the binomial and normal distributions. How are they alike? How are they different? Give examples of each.

Apply⇔Assess

Assess Your Progress

Have students review the vocabulary and write down the formulas for the binomial probability function and z-score. Ask students to consider how they would distinguish a binomial distribution problem from a normal distribution problem.

Journal Entry
This entry will help students organize their ideas about normal and binomial distributions. You might ask some students to share their examples with the class so that all students can judge their own work.

Progress Check 13.4–13.5

See page 649.

Practice 87 for Section 13.5

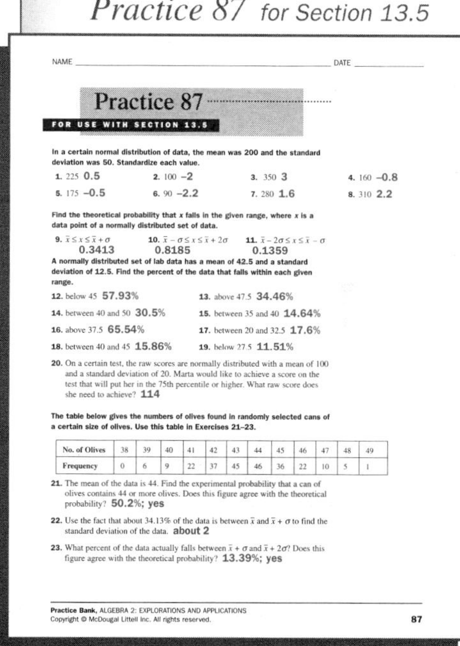

NAME ______ DATE ______

Practice 87

FOR USE WITH SECTION 13.5

In a certain normal distribution of data, the mean was 200 and the standard deviation was 50. Standardize each value.

1. 225 **0.5** **2.** 100 **–2** **3.** 350 **3** **4.** 160 **–0.8**

5. 175 **–0.5** **6.** 90 **–2.2** **7.** 280 **1.6** **8.** 310 **2.2**

Find the theoretical probability that x falls in the given range, where x is a data point of a normally distributed set of data.

9. $\bar{x} \le x \le \bar{x} + \sigma$ **0.3413** **10.** $\bar{x} - \sigma \le x \le \bar{x} + 2\sigma$ **0.8185** **11.** $\bar{x} - 2\sigma \le x \le \bar{x} - \sigma$ **0.1359**

A normally distributed set of lab data has a mean of 42.5 and a standard deviation of 12.5. Find the percent of the data that falls within each given range.

12. below 45 **57.93%** **13.** above 47.5 **34.46%**

14. between 40 and 50 **30.5%** **15.** between 35 and 40 **14.64%**

16. above 37.5 **65.54%** **17.** between 20 and 32.5 **17.6%**

18. between 40 and 45 **15.86%** **19.** below 27.5 **11.51%**

20. On a certain test, the raw scores are normally distributed with a mean of 100 and a standard deviation of 20. Marta would like to achieve a score on the test that will put her in the 75th percentile or higher. What raw score does she need to achieve? **114**

The table below gives the numbers of olives found in randomly selected cans of a certain size of olives. Use this table in Exercises 21–23.

No. of Olives	38	39	40	41	42	43	44	45	46	47	48	49
Frequency	0	6	9	22	37	45	46	36	22	10	5	1

21. The mean of the data is 44. Find the experimental probability that a can of olives contains 44 or more olives. Does this figure agree with the theoretical probability? **50.2%; yes**

22. Use the fact that about 34.13% of the data is between $\bar{x}$ and $\bar{x} + \sigma$ to find the standard deviation of the data. **about 2**

23. What percent of the data actually falls between $\bar{x} + \sigma$ and $\bar{x} + 2\sigma$? Does this figure agree with the theoretical probability? **13.39%; yes**

Practice Bank, ALGEBRA 2: EXPLORATIONS AND APPLICATIONS

87

13, 14. See answers in back of book.

15. Answers may vary. An example is given. Like other normal distributions, the standard normal distribution is a bell-shaped curve that is symmetric about its maximum at $x = \bar{x}$. It also shares the fact that the area between the curve and the x-axis, its asymptote, is 1. The special characteristics of the standard normal distribution are that it has a mean of 0 and a standard deviation of 1.

16. 0.25

17. about 0.322

18. 5

19. 35

20. 1

21. 171

22. Yes; both share $\angle A$, and $\angle C$ and $\angle E$ are congruent, since they both measure 60°. Thus, the triangles are similar by the Angle-Angle theorem.

Assess Your Progress

1. about 0.0536; about 0.263

2. about 0.766

3. a. 15.9%

b. 2.28%

4. about 0.0446

5. Answers may vary. An example is given. The binomial distribution involves countable independent trials, and is represented by a histogram, with the sum of the heights of the columns equal to 1. The histogram is bell-shaped, like the smooth curve representing the normal distribution, which has an area of 1 between it and the x-axis. Both distributions are symmetric about the mean, but the binomial distribution refers to trials where success or failure are the only possible outcomes, while the normal distribution often refers to measurements that can vary widely. Examples may vary.

Project Notes

Mathematical Goals

- Survey at least twenty students on the taste of generic and name brand beverages.
- Evaluate the results of the survey.
- Analyze the evaluations to see if people can distinguish between generic beverages.

Planning

Materials
- generic version of a drink and one or more brand name versions of the same drink
- small paper cups for beverage sampling

Project Teams
Students can perform the survey and complete the project individually or in pairs. If students work in pairs, they should complete all phases of the project together, including calculating the various probabilities.

Guiding Students' Work

Students may need help understanding the meaning of the two formulas used in the project. If necessary, take the time to go over the variables and ideas involved in the equations and how they were developed.

Second-Language Learners
Students learning English should benefit from working cooperatively with a partner on their experiments. They might also wish to work with their partners to conduct the research and write the report.

PORTFOLIO PROJECT

A Matter of Taste

If you've ever gone grocery shopping, you know that generic foods are generally less expensive than their brand-name counterparts. However, many shoppers believe that "buying generic" means sacrificing quality. Is there really a difference? You can find out by doing some market research.

PROJECT GOAL **Determine whether people can distinguish a generic beverage from one or more brand-name beverages.**

Performing an Experiment

1. **CHOOSE** a beverage, such as cola or orange juice, that you want to test. Buy a generic version and one or more brand-name versions of the beverage.

2. **FIND** a person to survey. Have the person taste each beverage *without* letting him or her see the beverage containers. You may want to use small paper cups to hold beverage samples.

3. **ASK** the person to identify the generic beverage. Record whether the person is correct.

4. **REPEAT** Steps 2 and 3 with other people. You should survey a total of 20 people.

Evaluating Your Results

If the responses of the people you surveyed were totally random (that is, if the people were only guessing when they tried to identify the generic beverage), then the probability that *r* of the 20 people responded correctly is

$$P(r) = {}_{20}C_r \cdot p^r \cdot (1 - p)^{20 - r}$$

where *p* is the probability of a correct guess.

1. What kind of probability distribution does the formula for $P(r)$ determine?

2. What is the value of *p* for your experiment? How is this value related to the number of beverages you tested? Explain.

In a binomial experiment like this one (with more than just a few trials), the value of $P(r)$ is small for any value of *r*. So $P(r)$ alone is not a good indicator of the likelihood that the experiment's outcome is due to chance. A better indicator is the probability of getting *at least r* correct responses:

$$P(\text{at least } r) = P(r) + P(r + 1) + \cdots + P(20)$$

3. Calculate the value of $P(\text{at least } r)$ for the value of *r* from your experiment. Does this probability suggest that chance determined the outcome of the experiment? Explain your reasoning.

Being very cautious, scientists and mathematicians *presume* that chance determines the outcome of an experiment unless there is strong evidence to the contrary. In this case, "strong evidence" might mean that the value of $P(\text{at least } r)$ is less than 0.1 or even 0.05 (for extra-cautious types).

4. Given the requirement for strong evidence stated above, can you conclude that people really can distinguish the generic beverage from the brand-name beverage(s)?

Writing a Report

Write a report summarizing your results. Include the data from your experiment and your answers to the questions above. If you wish, you can extend your project by conducting a survey that compares generic and brand-name versions of a different product, such as peanut butter or paper towels.

Self-Assessment

After completing this project, how comfortable are you with the ideas of probability as they relate to a scientific experiment? What ideas do you understand well? What ideas are still unclear?

Project Notes

Guiding Students' Work

Rubric for Chapter Project

4 Students conduct the survey of twenty people and include their results in a report. The report is well written, neat, and contains the correct answers to the questions. The report also shows a solid understanding of the mathematical ideas being discussed in the project. Students extend the project by conducting another survey with another product and calculate the probabilities for that product.

3 Students conduct the survey of twenty people and include their results in a report. The report is neat and clear, but one or two of the calculations are not done correctly or the answers to some of the questions are not complete. The report indicates a good understanding of the mathematical ideas of the project. Students extend the project by conducting another survey with another product and calculate the probabilities for that product.

2 Students conduct the survey of twenty people and include their results in a report. While the report shows an effort to complete the project and understand the ideas, it does not contain all of the answers to the questions and does not indicate a good grasp of the meaning of the probabilities involved in the project. Students do not attempt to extend the survey.

1 Students conduct a survey but survey only a few people. Students may or may not complete a report, but if one is done, it does not contain all of the answers to the questions and is poorly written. It is clear that students do not have an understanding of the probabilities involved in the project. Students should be encouraged to speak with the teacher as soon as possible to review their work and to make a new start on the project.

Chapter Support

Course Guide: Chapter 13
Lesson Plans: Chapter 13
Practice Bank: Cumulative Practice 88
Study Guide: Chapter 13 Review
Assessment Book: Chapter Tests 56 and 57; Chapter 13 Alternative Assessment
Test Generator Software
Portfolio Project Book
Preparing for College Entrance Tests
Professional Handbook

Progress Check 13.1–13.3

Suppose that in a new game, on every turn, the player rolls a pair of tetrahedral (four-sided) dice.

1. Find the probability that a player rolls a sum of 7 on any one roll. *(Section 13.1)* $\frac{1}{8}$
2. Find the probability that the player rolls doubles on any one roll. *(Section 13.1)* $\frac{1}{4}$
3. Consider one roll of the dice. Give an example of two mutually exclusive events and an example of two independent events. *(Section 13.2)* Answers may vary. Examples are given. mutually exclusive: rolling a sum of 7 or rolling doubles; independent: rolling a 2 on the first die and a 1 on the second die
4. Find the probability that a player gets a sum of 5 on any of his first 4 turns. *(Section 13.2)* about 0.684

In the game described above, suppose that if the player rolls doubles on any one turn, she spins a spinner with 3 blue sections and 1 red section. If the player does not roll doubles, she spins a spinner with 2 blue sections and 2 red sections.

CHAPTER

13 Review

STUDY TECHNIQUE

Read the title of each section in the chapter. Without looking at the section's contents, write a summary of the section and at least three questions that test the section's main objectives.

VOCABULARY

SECTION 13.1

The **probability** of an event A, denoted $P(A)$, is the fraction of the time that A is expected to happen. There are several types of probability.

For example, the table shows the results of an experiment in which a six-sided die is rolled 50 times. You can use these results to find the **experimental probability** of rolling a number less than 3:

Number on die	Number of times rolled
1	8
2	11
3	9
4	5
5	7
6	10

$$P\left(\begin{matrix}\text{rolling a number}\\ \text{less than 3}\end{matrix}\right) = \frac{8+11}{50} = \frac{19}{50}\text{, or about } 0.380$$

To find the corresponding **theoretical probability**, compare the number of possible rolls with the number of rolls less than 3:

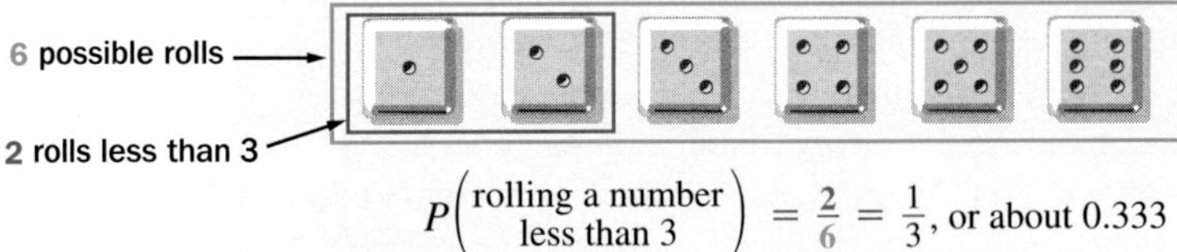

$$P\left(\begin{matrix}\text{rolling a number}\\ \text{less than 3}\end{matrix}\right) = \frac{2}{6} = \frac{1}{3}\text{, or about } 0.333$$

A **geometric probability** is a probability based on length, area, or volume. For example, suppose a randomly thrown dart hits the target shown. Since the area of each wedge is one sixth the area of the circle,

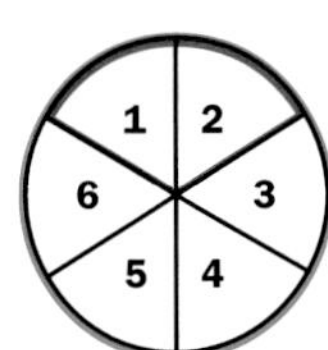

$$P\left(\begin{matrix}\text{dart hitting}\\ \text{wedge 1 or wedge 2}\end{matrix}\right) = \frac{2}{6} = \frac{1}{3}\text{, or about } 0.333$$

SECTIONS 13.2 *and* 13.3

Two events, A and B, are:

- **independent** if the occurrence of A does not affect whether B happens.
- **mutually exclusive** if A and B cannot both happen.
- **complementary** if A and B are mutually exclusive and either A or B *must* happen.

To illustrate these definitions, suppose that two marbles are chosen at random—one from jar 1 and one from jar 2—and consider the following pairs of events and corresponding rules.

Jar 1

A: The marble chosen from jar 1 is **red**.
B: The marble chosen from jar 2 is **red**.

These events are independent.

$P(A \text{ and } B) = P(A) \cdot P(B) = \frac{4}{9} \cdot \frac{2}{7} = \frac{8}{63}$, or about 0.127

C: The marble chosen from jar 1 is **green**.
D: The marble chosen from jar 1 is **blue**.

These events are mutually exclusive.

$P(C \text{ or } D) = P(C) + P(D) = \frac{3}{9} + \frac{2}{9} = \frac{5}{9}$, or about 0.556

Jar 2

E: The marble chosen from jar 2 is **green**.
F: The marble chosen from jar 2 is *not* **green**.

These events are complementary.

$P(F) = 1 - P(E) = 1 - \frac{4}{7} = \frac{3}{7}$, or about 0.429

The **conditional probability** $P(B|A)$ is the probability of event B, given that event A has occurred. For example, suppose two marbles are removed one at a time from jar 1. Let A be the event "the first marble is **blue**," and let B be the event "the second marble is **blue**." Then $P(B|A) = \frac{1}{8}$, or 0.125.

SECTIONS 13.4 *and* 13.5

A *binomial experiment* consists of n independent trials, each of which has two possible outcomes, *success* and *failure*. If $P(\text{success}) = p$ and $P(\text{failure}) = 1 - p$, then the probability that k of the n trials are successes is given by the formula:

$$P(k \text{ successes in } n \text{ trials}) = {}_nC_k \cdot p^k \cdot (1-p)^{n-k}$$

This formula defines a **binomial distribution**.

For example, the probability of getting 4 threes in 20 rolls of a six-sided die is $P(4) = {}_{20}C_4 \cdot \left(\frac{1}{6}\right)^4 \cdot \left(\frac{5}{6}\right)^{16}$, or about 0.202.

A **normal distribution** is characterized by a bell-shaped curve. In a normal distribution, about **68%** of the data lies within **1** standard deviation of the mean, about **95%** of the data lies within **2** standard deviations of the mean, and about **99%** of the data lies within **3** standard deviations of the mean.

Progress Check 13.1–13.3 (cont.)

5. Draw a probability tree diagram to represent the possible outcomes of one player's turn. *(Section 13.2)*
Let D represent the event "The player rolled doubles," N the event "The player did not roll doubles," B the event "The player spun a blue section," and R the event "The player spun a red section."

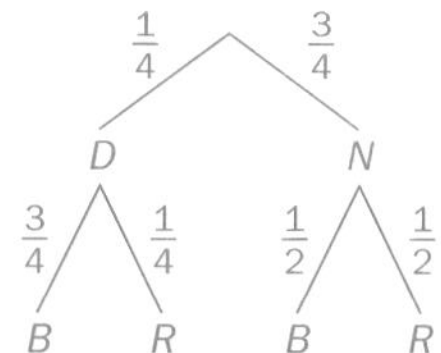

6. Find the probability that a player who landed on a blue section rolled doubles. *(Section 13.3)* $\frac{1}{3}$

Progress Check 13.4–13.5

A history test contains 10 true-false questions and 8 multiple-choice questions with 4 possible responses to each question. Find the probability of getting each outcome if the student guesses the answer to each question randomly.

1. A student gets exactly 8 out of 10 true-false questions correct. *(Section 13.4)* 0.044

2. A student gets at least one multiple-choice question correct. *(Section 13.4)* 0.900

3. The teacher gives a retest if a student scores below 60. The probability of any one student needing a retest is 4%. A class of 32 students takes the test (not random guessing). What is the probability that more than 3 students need a retest? *(Section 13.4)* about 0.038

After giving the test to 32 students, the teacher found their scores to be normally distributed with a mean of 71 and a standard deviation of 5.6. Use this information to estimate the following. *(Section 13.5)*

4. the percent of students between 65.4 and 76.6 68%

5. the percent of students between 65.4 and 82.2 81.5%

6. the percent and number of students who scored below 60 about 2.3%; 1 student

7. the percent of students who scored above 80 about 5.5%

Chapter 13 Assessment
Form A Chapter Test

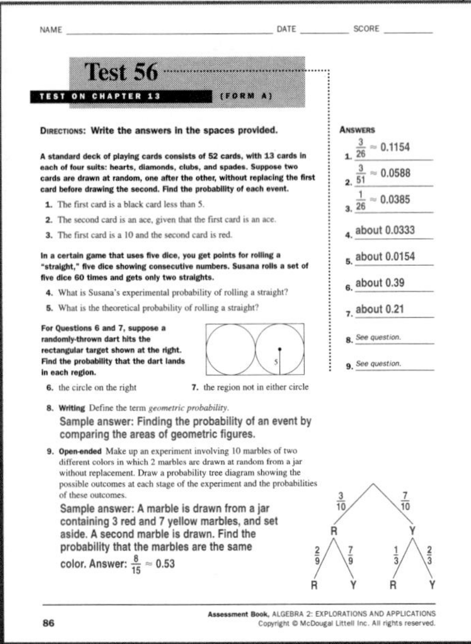

NAME ______ DATE ______ SCORE ______

Test 56

TEST ON CHAPTER 13 (FORM A)

DIRECTIONS: **Write the answers in the spaces provided.**

A standard deck of playing cards consists of 52 cards, with 13 cards in each of four suits: hearts, diamonds, clubs, and spades. Suppose two cards are drawn at random, one after the other, without replacing the first card before drawing the second. Find the probability of each event.

1. The first card is a black card less than 5.
2. The second card is an ace, given that the first card is an ace.
3. The first card is a 10 and the second card is red.

In a certain game that uses five dice, you get points for rolling a "straight," five dice showing consecutive numbers. Susana rolls a set of five dice 60 times and gets only two straights.

4. What is Susana's experimental probability of rolling a straight?
5. What is the theoretical probability of rolling a straight?

For Questions 6 and 7, suppose a randomly-thrown dart hits the rectangular target shown at the right. Find the probability that the dart lands in each region.

6. the circle on the right
7. the region not in either circle

8. **Writing** Define the term *geometric probability*.
Sample answer: Finding the probability of an event by comparing the areas of geometric figures.

9. **Open-ended** Make up an experiment involving 10 marbles of two different colors in which 2 marbles are drawn at random from a jar without replacement. Draw a probability tree diagram showing the possible outcomes at each stage of the experiment and the probabilities of these outcomes.
Sample answer: A marble is drawn from a jar containing 3 red and 7 yellow marbles, and set aside. A second marble is drawn. Find the probability that the marbles are the same color. Answer: $\frac{8}{15} \approx 0.53$

ANSWERS
1. $\frac{3}{26} \approx 0.1154$
2. $\frac{3}{51} \approx 0.0588$
3. $\frac{1}{26} \approx 0.0385$
4. about 0.0333
5. about 0.0154
6. about 0.39
7. about 0.21
8. *See question.*
9. *See question.*

86

Assessment Book, ALGEBRA 2: EXPLORATIONS AND APPLICATIONS

Chapter 13 Assessment
Form B Chapter Test

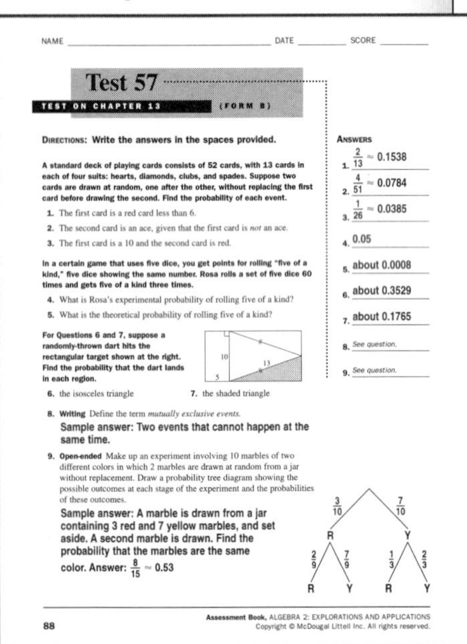

NAME ______ DATE ______ SCORE ______

Test 57

TEST ON CHAPTER 13 (FORM B)

DIRECTIONS: **Write the answers in the spaces provided.**

A standard deck of playing cards consists of 52 cards, with 13 cards in each of four suits: hearts, diamonds, clubs, and spades. Suppose two cards are drawn at random, one after the other, without replacing the first card before drawing the second. Find the probability of each event.

1. The first card is a red card less than 6.
2. The second card is an ace, given that the first card is *not* an ace.
3. The first card is a 10 and the second card is red.

In a certain game that uses five dice, you get points for rolling "five of a kind," five dice showing the same number. Rosa rolls a set of five dice 60 times and gets five of a kind three times.

4. What is Rosa's experimental probability of rolling five of a kind?
5. What is the theoretical probability of rolling five of a kind?

For Questions 6 and 7, suppose a randomly-thrown dart hits the rectangular target shown at the right. Find the probability that the dart lands in each region.

6. the isosceles triangle
7. the shaded triangle

8. **Writing** Define the term *mutually exclusive events*.
Sample answer: Two events that cannot happen at the same time.

9. **Open-ended** Make up an experiment involving 10 marbles of two different colors in which 2 marbles are drawn at random from a jar without replacement. Draw a probability tree diagram showing the possible outcomes at each stage of the experiment and the probabilities of these outcomes.
Sample answer: A marble is drawn from a jar containing 3 red and 7 yellow marbles, and set aside. A second marble is drawn. Find the probability that the marbles are the same color. Answer: $\frac{8}{15} \approx 0.53$

ANSWERS
1. $\frac{2}{13} \approx 0.1538$
2. $\frac{4}{51} \approx 0.0784$
3. $\frac{1}{26} \approx 0.0385$
4. 0.05
5. about 0.0008
6. about 0.3529
7. about 0.1765
8. *See question.*
9. *See question.*

88

Assessment Book, ALGEBRA 2: EXPLORATIONS AND APPLICATIONS

CHAPTER

13 Assessment

VOCABULARY QUESTIONS

For Questions 1–3, complete each sentence.

1. If events A and B are _?_, then $P(A \text{ and } B) = P(A) \cdot P(B)$.
2. The _?_ distribution has mean 0 and standard deviation 1.
3. If a procedure can result in n equally likely outcomes, and an event A corresponds to m of these outcomes, then $\frac{m}{n}$ is called the _?_ of A.

SECTION 13.1

Find the probability of choosing each type of card at random from a standard deck of 52 playing cards. (See page 615 for a description of such a deck.)

4. a 6 of hearts
5. a 6
6. a heart
7. a black card
8. a red face card
9. a 15 of clubs

10. **GAMES** In a certain game that uses 5 six-sided dice, you get points for rolling a "full house." A full house consists of three dice showing one number and two dice showing a different number (for example, three 1's and two 4's as shown).
    a. Jorge rolls a set of 5 six-sided dice 80 times and gets 4 full houses. What is his experimental probability of rolling a full house?
    b. What is the theoretical probability of rolling a full house?

Suppose a randomly thrown dart hits the circular target shown. Find the probability that the dart lands in each region.

11. the blue region
12. the white region
13. the red region

SECTIONS 13.2 *and* 13.3

14. A soda manufacturer puts a message on the underside of 10% of its bottle caps indicating that the purchaser has won a free bottle of soda. Yolanda buys two bottles of the manufacturer's soda. Find the probability that Yolanda wins the following:
    a. two bottles of soda
    b. no bottles of soda
    c. at least one bottle of soda
    d. exactly one bottle of soda

ANSWERS Chapter 13

Assessment

1. independent
2. standard normal distribution
3. probability
4. $\frac{1}{52}$, or about 0.0192
5. $\frac{1}{13}$, or about 0.0769
6. $\frac{1}{4}$, or 0.25
7. $\frac{1}{2}$, or 0.5
8. $\frac{3}{26}$, or about 0.115
9. 0
10. a. $\frac{1}{20}$, or 0.05
    b. $\frac{25}{648}$, or about 0.0386
11. $\frac{1}{2}$, or 0.5
12. $\frac{24}{25\pi}$, or about 0.306
13. $0.5 - \frac{24}{25\pi}$, or about 0.194
14. a. 0.01
    b. 0.81
    c. 0.19
    d. 0.18

15. SPORTS On July 28, 1991, Dennis Martinez of the Montreal Expos pitched a "perfect game" against the Los Angeles Dodgers. This means that he prevented all 27 batters he faced from getting on base. If you assume each batter had a 30% chance of getting on base, what was the probability of Martinez pitching a perfect game that day?

16. POLITICS Two candidates, one Democratic and one Republican, are running for a seat in the United States Senate. Recent polls predict that the Republican will receive 53% of the vote. Of voters who say they will vote Republican, 82% support a balanced-budget amendment to the Constitution. Of voters who say they will vote Democratic, only 59% support such an amendment. What is the probability that a voter who supports a balanced-budget amendment votes Republican?

SECTIONS 13.4 *and* 13.5

17. Matt is taking a history quiz that consists of 6 multiple-choice questions. Each question has 4 choices. Since Matt didn't study for the quiz, he is forced to guess each answer randomly. What is the probability that he gets:

a. all the answers right? **b.** none of the answers right?

c. exactly half the answers right? **d.** at least half the answers right?

18. Open-ended Problem A company hires 10 engineers—8 men and 2 women—from a large pool of equally qualified applicants. Of the applicants, 65% are men and 35% are women. Do you think the company's hiring decisions indicate a bias against women? Give a mathematical justification for your answer.

19. MANUFACTURING A manufacturer makes ball bearings that must have diameters between 20.0 mm and 21.0 mm. Quality control tests indicate that the diameters of the bearings produced are normally distributed with mean 20.4 mm and standard deviation 0.4 mm.

a. What percent of the ball bearings have diameters less than 20.0 mm?

b. What percent of the ball bearings have diameters greater than 21.0 mm?

c. What percent of the ball bearings have diameters that are *not* between 20.0 mm and 21.0 mm?

PERFORMANCE TASK

20. In Section 13.1, you found probabilities associated with the dice games *barbudey* and *kon mín yéung*. Choose another game and analyze some probabilities associated with playing this game. Your analysis should incorporate ideas from at least three of the five sections in this chapter.

Chapter 13 Assessment
Form C Alternative Assessment

Chapter 13

ALTERNATIVE ASSESSMENT

1. **Project** Design an experiment to find the probability of some real-world event. Use an event that has no theoretical probability. Describe your experiment and its results.

2. **Open-ended Problem** Count the number of As, Bs, Cs, and so on in a newspaper or magazine article. For each letter of the alphabet, calculate the probability that the letter might occur in a given body of text. Would the probabilities change significantly if the language of the text was Spanish? How could this information be used in deciphering an encoded message?

3. **Open-ended Problem** Describe several sets of real data that might have a normal distribution.

4. **Open-ended Problem** Some teachers believe that students' grades should form a normal distribution. Why do you think that they believe this? Ask several teachers what "grading on the curve" means? Do you think that students should be "graded on the curve?" Explain your answer.

5. **Project** Draw a bull's-eye target on a posterboard. Place the target on the floor and randomly drop small pebbles on the target. Count the number of pebbles that landed in each region. Use this information to discuss the area of each region of the target. If you were blindfolded and randomly threw darts at the target, what would be the probability of hitting each region? Discuss the connection between probability and area.

6. **a.** Describe an event that has a probability of 0.
b. Describe an event that has a probability of 1.

126 Assessment Book, ALGEBRA 2: EXPLORATIONS AND APPLICATIONS

15. about 6.57×10^{-5}
16. about 0.610
17. a. about 0.000244
    b. about 0.178
    c. about 0.132
    d. about 0.169

18. Answers may vary. The probability of picking at least 8 men at random is about 0.261, meaning that if you drew names from the pool at random, you would select at least 8 males a little more than $\frac{1}{4}$ of the time.

19. a. about 15.9%
    b. about 6.68%
    c. about 22.6%

20. Answers may vary. Check students' work.

CHAPTER

14 Triangle Trigonometry

OVERVIEW

Connecting to Prior and Future Learning

⇔ In this chapter, students expand their study of triangles to include triangle trigonometry. The basic trigonometric functions of sine, cosine, and tangent are introduced in the first two sections of the chapter. Students can review triangle relationships on page 801 in the **Student Resources Toolbox**.

⇔ The study of trigonometry continues with a discussion of finding the trigonometric functions of different angles of rotation and using the sine function to find the area of a triangle. Students will encounter these topics again in future mathematics courses.

⇔ The chapter concludes with a study of the law of sines and the law of cosines. These laws provide a foundation that students can build upon in their future studies of mathematics.

Chapter Highlights

Interview with Johnpaul Jones: The relationship between mathematics and designing animal parks is emphasized in this interview. Related exercises can be found on pages 661 and 668.

Explorations in Chapter 14 begin in Section 14.1, with students using graph paper and protractors to find ratios of side lengths in similar right triangles. In Section 14.2, students use a ruler, yardstick, or meterstick to estimate the distance across their classroom. Students use a ruler, a protractor, and a compass to construct a triangle in Section 14.5.

The Portfolio Project: Students work with a partner to determine the height of a tall object of their choice using two trigonometric methods.

Technology: In Chapter 14, graphing calculators are used to find the sine, cosine, and tangent of various angles and their inverses.

OBJECTIVES

Section	Objectives	NCTM Standards
14.1	• Find the sine and cosine of an acute angle. • Use the sine or cosine of an acute angle to solve problems.	1, 2, 3, 4, 5, 9
14.2	• Find the tangent of an acute angle. • Use the tangent ratio to solve problems.	1, 2, 3, 4, 5, 9
14.3	• Find the trigonometric functions of an angle of rotation. • Convert course angles to angles of rotation. • Use the angles of rotation and course angles to solve problems.	1, 2, 3, 4, 5, 9
14.4	• Find the area of a triangle given two side lengths and the measure of the included angle. • Find the area of a sector.	1, 2, 3, 4, 5, 9
14.5	• Use the law of sines to solve a triangle. • Use the law of sines to solve real-world problems.	1, 2, 3, 4, 5, 9
14.6	• Use the law of cosines to solve a triangle. • Use the law of cosines to solve real-world problems.	1, 2, 3, 4, 5, 9

INTEGRATION

Mathematical Connections	14.1	14.2	14.3	14.4	14.5	14.6
algebra	**655–662***	**663–670**	**671–677**	**678–683**	**684–690**	**691–697**
geometry	**655–662**	**663–670**	**671–677**	**678–683**	**684–690**	**691–697**
data analysis, probability, discrete math	662		677			
patterns and functions	662	670				
logic and reasoning	655, 658–662	663, 664, 667–670	671, 674–677	**678–683**	686, 688–690	691, 692, 695–697
Interdisciplinary Connections and Applications						
history and geography				683	690	
reading and language arts					689	
chemistry and physics	660, 661					695
arts and entertainment			675			696
sports and recreation	659, 662	670				696
architecture and landscaping			675	680		
aviation and navigation		666, 667, 670	673, 676		684	
archaeology, astronomy, surveying, meteorology, robotics, cooking, paleontology	656, 657, 661	664, 668, 669	676	681, 682		693

* ***Bold page numbers*** *indicate that a topic is used throughout the section.*

TECHNOLOGY

Section	*opportunities for use with* Student Book	Support Material
14.1	graphing calculator McDougal Littell Mathpack *Function Investigator*	**Technology Book:** Spreadsheet Activity 14 TI-92 Activity 6 **Function Investigator with Matrix Analyzer Activity Book:** Function Investigator Activity 31
14.2	graphing calculator McDougal Littell Mathpack *Function Investigator*	**Technology Book:** Calculator Activity 14 TI-92 Activity 6
14.3	graphing calculator	**Geometry Inventor Activity Book:** Activity 20
14.4	scientific calculator	
14.5	graphing calculator	
14.6	graphing calculator	**Technology Book:** Spreadsheet Activity 14

PLANNING GUIDE

Regular Scheduling (45 min)

See Alternative Pacing Charts, page T42.

Section	Materials Needed	Core Assignment	Extended Assignment	*exercises that feature* Applications	Communication	Technology
14.1	graph paper, protractor, graphing calculator	1–17, 20–22, 28–36	1–9, 11, 13, 15, 17–36	7, 18, 19, 23–26	29	19
14.2	ruler, yardstick, or meterstick	1–13, 15, 22–24, AYP*	1–24, AYP	4, 11, 12, 14, 16–20, 22	13, 21	
14.3	protractor, compass, graph paper	1–21, 27–30	1–30	7, 8, 21, 23, 24	17, 25, 27	
14.4	graph paper	1–9, 12–20, 26–33, AYP	1–6, 10–15, 18–33, AYP	10, 11, 21–25	24	
14.5	ruler, protractor, compass, graphing calculator	1–7, 9–17, 26–36	1–7, 9–36	18–25	8, 20, 25	
14.6	graphing calculator	**Day 1:** 1–9, 14–17 **Day 2:** 18–22, 24, 25, 28–40, AYP	**Day 1:** 1–20 **Day 2:** 21–40, AYP	11–13 23–26	10	
Review/ Assess		**Day 1:** 1–25 **Day 2:** Ch. 14 Test	**Day 1:** 1–25 **Day 2:** Ch. 14 Test			
Portfolio Project		Allow 2 days.	Allow 2 days.			
Pacing (with Portfolio Project)		**Chapter 14 Total** 11 days				

Block Scheduling (90 min)

	Day 1	Day 2	Day 3	Day 4	Day 5	Day 6
Teach/Interact	14.1: Exploration, page 655 14.2: Exploration, page 663	14.3 14.4	14.5: Exploration, page 685 14.6	Continue with 14.6 Port. Proj.	Review Port. Proj.	Ch. 14 Test
Apply/Assess	**14.1:** 1–17, 20–22, 28–36 **14.2:** 1–13, 15, 22–24, AYP*	**14.3:** 1–21, 27–30 **14.4:** 1–9, 12–20, 26–33, AYP	**14.5:** 1–7, 9–17, 26–36 **14.6:** 1–9, 14–17	**14.6:** 18–22, 24, 25, 28–40, AYP **Port. Proj.**	**Review:** 1–25 **Port. Proj.**	**Ch. 14 Test**
NOTE: A one-day block has been added for the Portfolio Project—timing and placement to be determined by teacher.						
Pacing (with Portfolio Project)	**Chapter 14 Total** $5\frac{1}{2}$ days					

****AYP*** *is Assess Your Progress.*

Lesson Support

Section	Practice Bank	Study Guide*	Assessment Book*	Visuals	Explorations Lab Manual	Lesson Plans	Technology Book
14.1	89	14.1		Warm-Up 14.1 Folder 14	Master 1	14.1	Spreadsheet Act. 14 TI-92 Act. 6
14.2	90	14.2	Test 58	Warm-Up 14.2 Folder 14	Add. Expl. 20	14.2	Calculator Act. 14 TI-92 Act. 6
14.3	91	14.3		Warm-Up 14.3 Folder A	Master 2	14.3	
14.4	92	14.4	Test 59	Warm-Up 14.4	Masters 1, 2	14.4	
14.5	93	14.5		Warm-Up 14.5	Master 19	14.5	
14.6	94	14.6	Test 60	Warm-Up 14.6		14.6	Spreadsheet Act. 14
Review Test	95	Chapter Review	Tests 61, 62 Alternative Assessment			Review Test	

***Spanish versions** of *Study Guide* and *Assessment Book* are available.

Chapter Support

- Course Guide
- Lesson Plans
- Portfolio Project Book
- Preparing for College Entrance Tests
- Multi-Language Glossary
- *Test Generator* Software
- Professional Handbook

Software Support

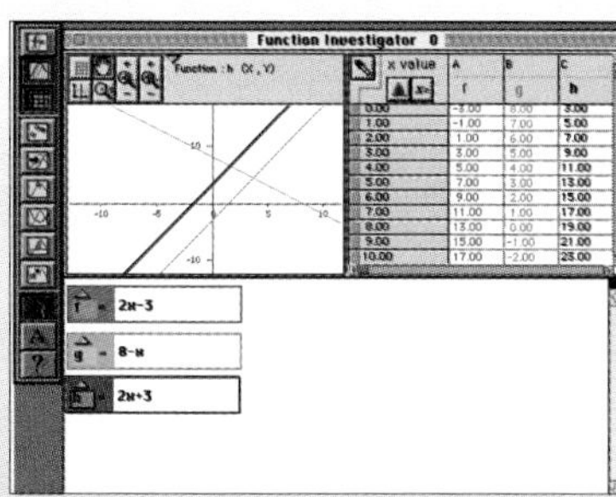

McDougal Littell Mathpack
Function Investigator

Internet Support

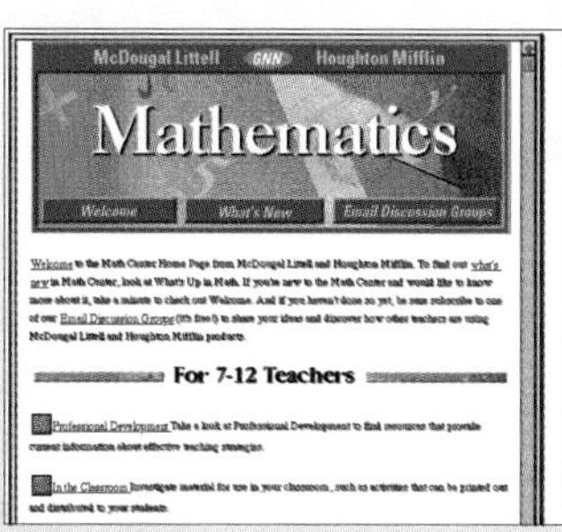

http://www.hmco.com
Next go to McDougal Littell; then the Education Center; then Secondary Math.

Outside Resources

Books, Periodicals

Ren, Guanshen. "Match Geometric Figures with Trigonometric Identities." *Mathematics Teacher* (January 1995): pp. 24–25.

Thoemke, Sharon S., Pamela A. Shriver, and Joby M. Anthony. "A Mathematical Model for the Height of a Satellite." *Mathematics Teacher* (October 1993): pp. 563–565.

Zerger, Monte. "Hidden Treasures in Students' Assumptions." *Mathematics Teacher* (October 1993): pp. 567–569.

Activities, Manipulatives

Breuningsen, Chris, Bill Bower, Linda Antinone, and Elisa Breuningsen. "Stay Tuned." *Real-World Math with the CBL System.* Activity 15: pp. 79–84. Texas Instruments, 1995.

Hurwitz, Marsha. "Discovering the Law of Sines." *Mathematics Teacher* (November 1991): pp. 634–635.

Levine, Bernard S. "The Taming of the Ambiguous Case." *Mathematics Teacher* (March 1992): p. 200.

Software

de Lange, Jan. *Gliding.* Macintosh., Pleasantville, NY: Wings for Learning/Sunburst, 1992.

Videos

Apostol, Tom M. *The Theorem of Pythagoras.* Reston, VA: NCTM.

Internet

Investigate: "More than Anyone Wants to Know About Pi"
http://uts.cc.utexas.edu/~joeting/math/pitime.html

Teacher's Place: Classroom Materials
http://forum.swarthmore.edu/classroom.html

Interview Notes

Background

Landscape Architecture
A landscape architect is concerned with the design and development of land for both use and enjoyment. The design for a project is usually broken down into three parts: land planning, site design, and land use management. In the land planning phase, architects study the climate, water supply, vegetation, soil composition, and the slope of the land to determine land use to avoid flooding, erosion, and air and water pollution. The site design phase involves setting the location of the structures so that they best fit the use of the physical features of the land and the natural features such as sunlight, shade, breezes, or existing trees. During the land use management phase, architects help administrators develop ways to preserve and conserve the productivity and beauty of the land and its architecture.

Johnpaul Jones
Johnpaul Jones is a Native American from the Cherokee and Choctaw tribes. He grew up in Oklahoma where he had the opportunity to develop an interest in the natural world. This was the beginning of his interest in landscapes and landscape architecture. Jones and his company have also designed other zoological exhibits that give people a chance to view animals in their natural habitat. Two of these exhibits are the Arctic and Northwest Coast Exhibits in Tacoma, Washington, and the Seattle Zoo in Seattle, Washington. In addition to helping in the design of animal habitat exhibits like Tiger River, Jones has also been selected by the Smithsonian Institution to assist in the design of the National Museum for Native Americans, which is scheduled to open in 2001.

CHAPTER

14 Triangle Trigonometry

It's a JUNGLE *in there*

INTERVIEW **Johnpaul Jones**

Deep in the still greenness of *Tiger River*, a three-acre replica of a Southeast Asian jungle at the San Diego Zoo, live tapirs, pythons, crocodiles, purple herons, gold-crested mynas, and rare Sumatran tigers. The exhibit was designed to feel like a patch of tropical rain forest in Indonesia. Johnpaul Jones, a senior partner in the architectural and landscape design firm of Jones & Jones, and his partners pioneered the design of exhibits like *Tiger River*.

"The idea we started, putting wild animals and wild nature together, has traveled to zoos around the world."

Knowing the Animals

In addition to examining the site, the staff studied each of the animal species destined for *Tiger River*. "We figured out the height, length, and weight of every animal and came up with diagrams showing what each wild animal was capable of doing in terms of jumping, running, and leaping," Jones says.

Creating Safe Barriers

With information about the site and the animals, Jones could design safe, natural barriers to keep the tigers and other animals within their new habitats. "We could put a pool in the foreground of the tigers' space and make it deep enough so that a running tiger couldn't put its feet on the bottom and leap out. Then we could make the barrier above the water line lower so people could get closer. At all of the various barriers around the tiger exhibit, trigonometry played a major role in creating a better place for the wild animal and for the visitor," explains Jones.

Developing Natural Environments

Designing each animal exhibit for *Tiger River* meant solving two basic problems: creating an environment that looks and feels like the animals' natural wild habitat, and providing a safe, close viewing space for visitors. In the case of *Tiger River*, Jones and his colleagues first studied the zoo site, which had been a canyon occupied by African birds and antelope. Because the existing canyon would eventually house rain forest creatures, he decided to install pools of water that would seem to flow continuously throughout the canyon, representing the Southeast Asian rain forest floor.

Interview Notes

Background

Zoos
The oldest existing public zoo is the Zoological Society of London, England, founded in 1826. As of 1993, the zoo held 18,128 specimens of animals on 577 acres in both Regent's Park and Whipsnade Park in England. In the United States, the oldest and most attended zoo is Lincoln Park Zoo in Chicago, Illinois. The zoo opened its doors to the public in 1868 with the donation of two swans from Central Park.

Second-Language Learners

Some students may not be familiar with some of the animals named in this passage, such as the *tapir,* a short, heavy, hoofed animal that is rarely found in the wild anymore; the *heron,* a kind of long, thin, aquatic bird; and the *gold-crested myna,* a bird about the size of a robin that has a gold-colored crest on its head.

Interview Notes

Mathematical Connection

The design of Tiger River involved the use of many concepts from trigonometry. Safe dimensions for the moat surrounding the exhibit and the height of the wall that separates the animals from the visitors can be calculated using the trigonometric functions of sine and cosine. In Section 14.1, students examine the trigonometric formulas used to calculate the dimensions of both the moat and the wall using realistic values for a tiger's speed and jump. Other design features of Tiger River include an enclosure for birds. In Section 14.2, students use the tangent function to investigate the construction dimensions and design locations for an aviary.

Explore and Connect

Writing
Ask students to explain why, as the angle measures increase, the vertical velocity is increasing and the horizontal velocity is decreasing.

Research
Students may find sources for the information in the school or local library. Most or all of the information may be contained in encyclopedias.

Project
When students have completed their exhibits, have them describe to the class how they designed and put their exhibit together. You may wish to display the zoo exhibits in the classroom.

Analyzing a Tiger's Leap

Jones analyzed the leaping capabilities of the tiger to design the barriers for its habitat. A tiger's initial leaping velocity v_0 can be broken down into vertical and horizontal components. The tiger's vertical velocity, $v_0 \sin \theta$ (θ is pronounced THAY *tuh*), and its horizontal velocity, $v_0 \cos \theta$, correspond to the tiger's upward and forward motion in the diagram.

These velocity components are used in parametric equations that define the path of the tiger as it leaps through the air. The factors $\sin \theta$ and $\cos \theta$ are examples of *trigonometric ratios* that you will study in this chapter.

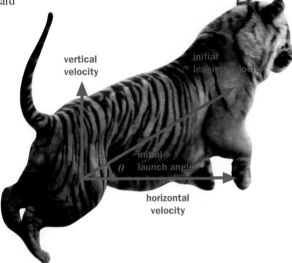

EXPLORE AND CONNECT

Johnpaul Jones reviews the plans for the *Tiger River* exhibit.

1. Writing Suppose a tiger's initial leaping velocity is 1.5 ft/s. Use a calculator to evaluate $1.5 \sin \theta$ and $1.5 \cos \theta$ for a launch angle θ of 25°, 35°, and 45°. What happens to the vertical velocity, $1.5 \sin \theta$, and the horizontal velocity, $1.5 \cos \theta$, as the launch angle increases?

2. Research Choose an animal from the wild. Find out about the animal's natural habitat and its physical capabilities. Write a paragraph describing the habitat's climate, landscape, and vegetation. Write another paragraph about the animal's height, weight, and body structure, along with any leaping, flying, or running abilities.

3. Project Describe and sketch a zoo exhibit for the wild animal you chose to research in Question 2. Share your exhibit with your classmates and find ways to combine exhibits to house various wild animals.

Mathematics & Johnpaul Jones

In this chapter, you will learn more about how mathematics is related to designing animal parks.

Related Exercises

Section 14.1
- **Exercises 24 and 25**

Section 14.2
- **Exercises 16 and 17**

Section

14.1 Using Sine and Cosine

Learn how to...

- find the sine and cosine of an acute angle

So you can...

- find the length of a ladder needed to reach a given height, for example

Trigonometry is a branch of mathematics used in fields such as astronomy, architecture, surveying, aviation, and navigation. The term "trigonometry" comes from two Greek words, *trigonos* and *metron,* meaning "triangle measurement." Trigonometry can be used to find an unknown angle measure or an unknown length of a side of a triangle.

EXPLORATION
COOPERATIVE LEARNING

Finding Ratios of Side Lengths in Similar Right Triangles

Work with a partner.
You will need:

- graph paper
- a protractor

1 Draw a segment along a horizontal grid line on your graph paper. Label one endpoint *A*.

2 Use a protractor to draw a 60° angle at *A*.

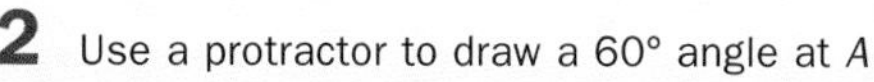

3 Draw segments along vertical grid lines to form right triangles, such as $\triangle ABC$, $\triangle ADE$, and $\triangle AFG$ shown.

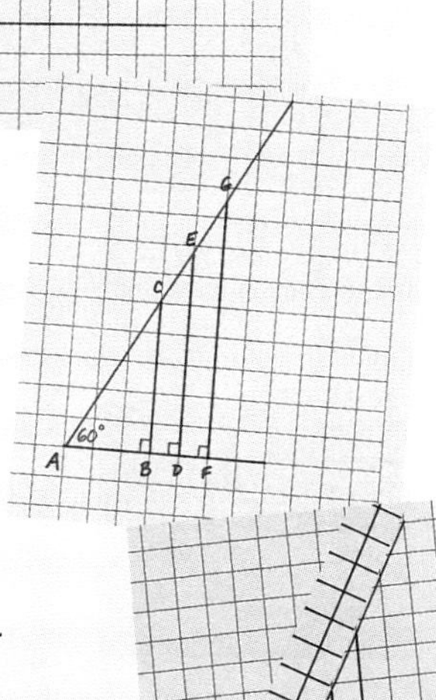

Questions

Toolbox p. 801 *Triangle Relationships*

1. Why are $\triangle ABC$, $\triangle ADE$, and $\triangle AFG$ similar triangles?
2. Cut a strip of graph paper to use as a ruler. Use your ruler to find the value of the following ratio for $\triangle ABC$ and for $\triangle ADE$:
$$\frac{\text{length of leg opposite } \angle A}{\text{length of hypotenuse}}$$
What do you notice?
3. Make a conjecture about the value of the ratio in Question 2 for any triangle similar to $\triangle ABC$. To test your conjecture, find the value of this ratio for $\triangle AFG$.

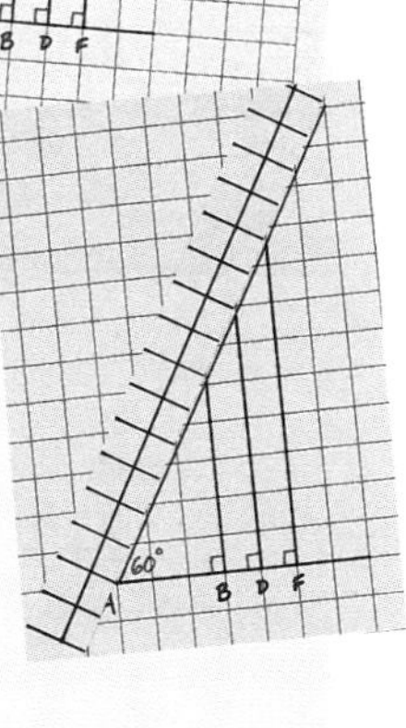

Exploration Note

Purpose
The purpose of this Exploration is to have students discover that the corresponding ratios of side lengths in similar triangles do not depend on the lengths of the sides.

Materials/Preparation
Each pair of students needs graph paper and a protractor.

Procedure
Students draw right triangles on graph paper, each with a 60° angle and discuss why the triangles they have drawn are similar. They then find the ratio of the length of the leg opposite the 60° angle to the length of the hypotenuse. They should determine that the ratio is the same for each triangle even though the lengths of the sides are different.

Closure
Students should understand that the corresponding ratios of side lengths in similar triangles do not depend on the lengths of the sides, but they do depend on the measures of the acute angles.

Explorations Lab Manual
See the Manual for more commentary on this Exploration.

For answers to the Exploration, see following page.

Plan⇔Support

Objectives

- Find the sine and cosine of an acute angle.
- Use the sine or cosine of an acute angle to solve problems.

Recommended Pacing

- **Core and Extended Courses**
  Section 14.1: 1 day
- **Block Schedule**
  Section 14.1: $\frac{1}{2}$ block
  (with Section 14.2)

Resource Materials

Lesson Support
Lesson Plan 14.1
Warm-Up Transparency 14.1
Overhead Visuals:
  Folder 14: Trigonometric Ratios
Practice Bank: Practice 89
Study Guide: Section 14.1
Explorations Lab Manual:
  Diagram Master 1

Technology
Technology Book:
  Spreadsheet Activity 14
  TI-92 Activity 6
Graphing Calculator
McDougal Littell Mathpack
  Function Investigator with Matrix Analyzer Activity Book: Function Investigator Activity 31
Internet:
  http://www.hmco.com

Warm-Up Exercises

1. If $\triangle ABC$ is a right triangle with right angle *C*, name the hypotenuse and the side opposite $\angle A$. hyp: $\overline{AB}$; opp side: $\overline{BC}$
2. If $\triangle XYZ$ is a right triangle with legs 5 and 12, find the length of the hypotenuse. 13
3. Solve the equation $0.01654 = \frac{500}{x}$ for *x*. Give the answer to the nearest whole number. 30,230
4. The graph of a function passes through the points (–3, 4), (5, 2), and (–2, 7). Find three points that lie on the graph of the inverse of this function. (4, –3), (2, 5), (7, –2)

Teach⇔Interact

Section Notes

Communication: Discussion You may want to have students repeat the Exploration on page 655 with the measure of $\angle A = 30°$ to verify that the sine ratio depends on the measure of the angle, not on the lengths of the sides.

Mathematical Procedures The angle symbol is usually not written when using the trigonometric functions.

Additional Example 1

For $\triangle ABC$ below, find sin A and cos A.

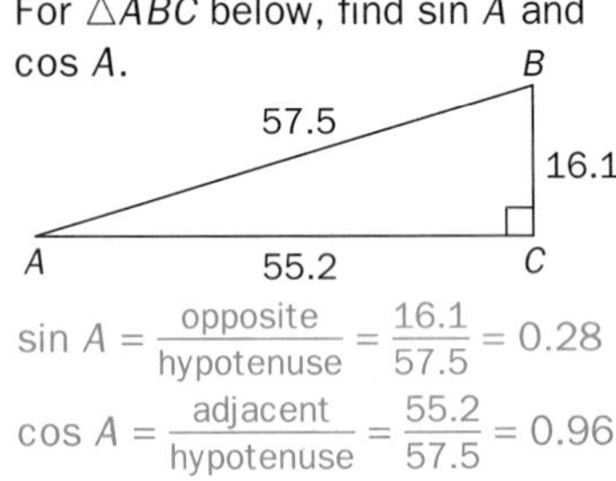

$\sin A = \frac{\text{opposite}}{\text{hypotenuse}} = \frac{16.1}{57.5} = 0.28$

$\cos A = \frac{\text{adjacent}}{\text{hypotenuse}} = \frac{55.2}{57.5} = 0.96$

Additional Example 2

Some archeologists believe that the ancient Egyptians may have built the Great Pyramid at Giza by dragging the massive stones up a ramp. Suppose the ramp shown is on the ground and extends one third the way up the pyramid, or 46 meters. How long is the ramp?

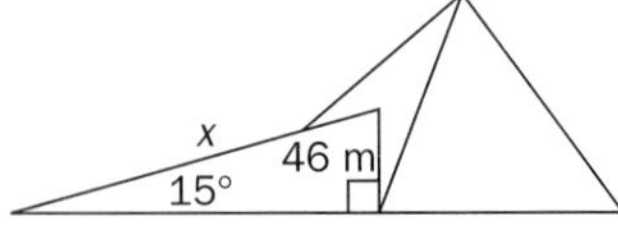

Use the sine ratio to find the unknown side length.

$\sin 15° = \frac{\text{opposite}}{\text{hypotenuse}} = \frac{46}{x}$

$x = \frac{46}{\sin 15°} \approx \frac{46}{0.2588} \approx 177.7$

The ramp is about 178 m long.

In the Exploration you found that corresponding ratios of side lengths in similar right triangles do *not* depend on the lengths of the sides. These ratios depend *only* on the shape of the triangles as determined by the measures of the acute angles. These constant ratios are so important they are given names.

WATCH OUT!

Sometimes the lengths of sides of triangles are named with the lowercase form of the capital letter at the vertex of the opposite angle.

Sine and Cosine of an Angle

In right $\triangle ABC$, the **sine** of $\angle A$, which is written "sin A," is given by

$$\sin A = \frac{\text{length of leg opposite } \angle A}{\text{length of hypotenuse}} = \frac{a}{c}$$

and the **cosine** of $\angle A$, which is written "cos A," is given by

$$\cos A = \frac{\text{length of leg adjacent to } \angle A}{\text{length of hypotenuse}} = \frac{b}{c}$$

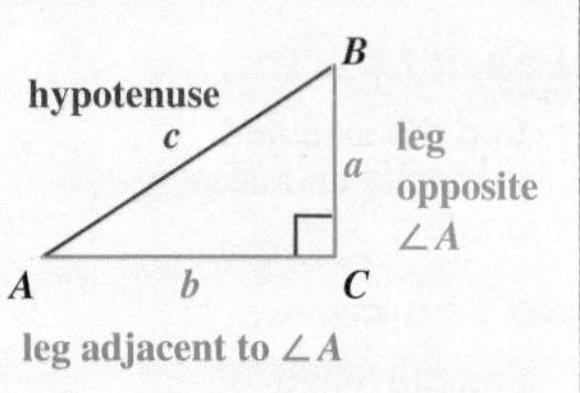

EXAMPLE 1

For $\triangle ABC$ below, find sin A and cos A.

SOLUTION

$\sin A = \frac{\text{opposite}}{\text{hypotenuse}} = \frac{36}{39} \approx 0.9231$

$\cos A = \frac{\text{adjacent}}{\text{hypotenuse}} = \frac{15}{39} \approx 0.3846$

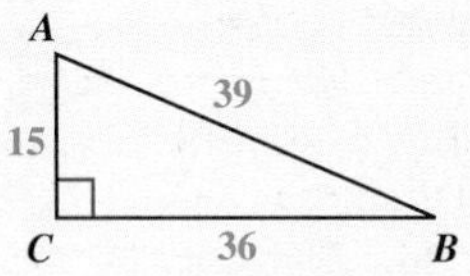

EXAMPLE 2 Application: Archaeology

Between A.D. 1000 and 1300, the Anasazi people lived in cliff dwellings in the southwestern part of the United States. The doors to the cliff dwellings opened onto balconies that were reached by climbing ladders.

Suppose the ladder shown rests on the ground and extends 3 ft above the balcony. How long is the ladder?

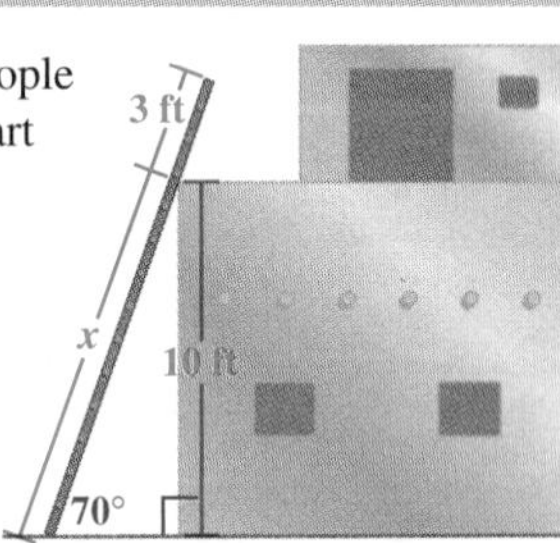

BY THE WAY

The Anasazi built and inhabited a cliff dwelling with almost 200 rooms in what is now Colorado. Known as "Cliff Palace," it is located in Mesa Verde National Park, a United Nations world heritage site.

SOLUTION

Use the sine ratio to find the unknown side length.

$\sin 70° = \frac{\text{opposite}}{\text{hypotenuse}} = \frac{10}{x}$

$x = \frac{10}{\sin 70°} \approx \frac{10}{0.9397} \approx 11$

The ladder is about 11 + 3 = 14 ft long.

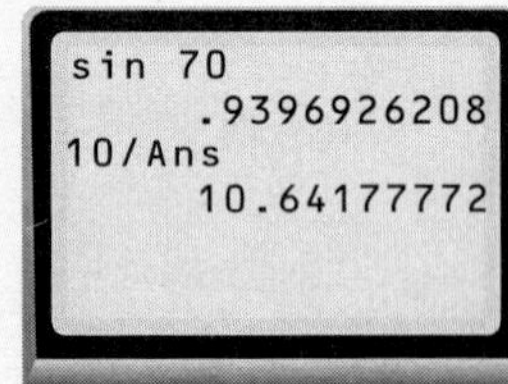

ANSWERS Section 14.1

Exploration

1–3. Check students' work.

Questions

1. The corresponding angles of the three triangles are congruent.
2. The ratio is about 0.87 for both triangles.
3. The ratio will be the same for any similar triangle. The ratio for $\triangle AFG$ is the same.

All angles do not have measures that are an integral number of degrees, such as the 70° angle in Example 2. The measures of some angles involve fractions of a degree, such as 74.3125°. Angle measures written in this form are said to be in *decimal degrees*.

Angle measures can also be written using *degrees* (°), *minutes* (′), *and seconds* (″), such as 74°18′45″. Just as a circle is divided into 360 degrees, a degree is divided into 60 minutes, and a minute is divided into 60 seconds.

$$1° = 60' \qquad 1' = 60''$$

You can use the equations above to convert an angle measure from degrees, minutes, seconds to decimal degrees. For example, consider 74°18′45″:

$$74°18'45'' = 74° + 18\left(\frac{1}{60}^{\circ}\right) + 45\left(\frac{1}{3600}^{\circ}\right) = 74.3125°$$

Each minute is $\frac{1}{60}$ of a degree.

Each second is $\frac{1}{60}\left(\frac{1}{60}\right) = \frac{1}{3600}$ of a degree.

BY THE WAY

About 4000 years ago, the Babylonians developed a numeration system based on 60. We still use this system in measuring angles and in telling time.

EXAMPLE 3 Application: Astronomy

Thousands of years before the invention of telescopes, satellites, computers, and other tools of modern astronomy, people used trigonometry to calculate the sizes and distances of the moon and the sun.

Use the diagram below to estimate the distance between the surface of Earth and the moon using trigonometry. In the diagram, the moon is directly above point P and is just visible from point Q. Points P and Q are on the equator. Use 3963 mi for the radius of Earth at the equator.

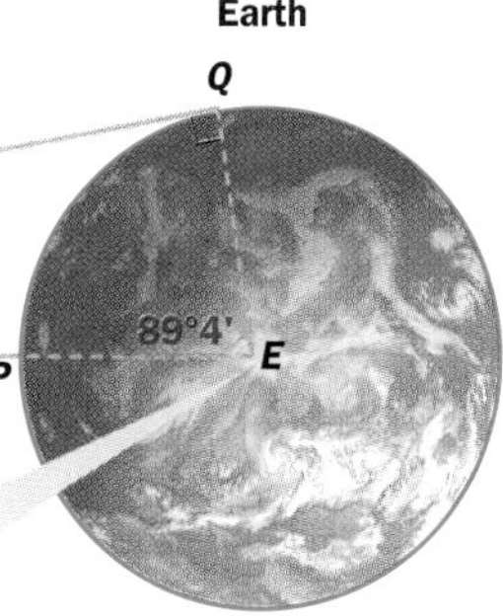

The measure of $\angle E$ is the difference in longitude between points P and Q on the surface of Earth.

SOLUTION

Find the distance between E, the center of Earth, and M, the moon. Then subtract the radius of Earth.

$$\cos 89°4' = \frac{\text{adjacent}}{\text{hypotenuse}}$$

$$= \frac{3963}{EM}$$

Substitute **3963** for ***EQ***, the radius of Earth at the equator.

$$EM = \frac{3963}{\cos 89°4'}$$

$$\approx \frac{3963}{0.01629}$$

$$\approx 243{,}000$$

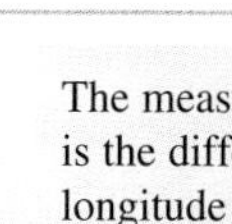

```
cos 89°4'
        .0162890193
3963/Ans
        243292.7321
```

The distance between the surface of Earth and the moon is about 243,000 − 3,963 ≈ 239,000 mi.

Teach⇔Interact

About Example 3

Reasoning

Students should understand that the cosine ratio is used instead of the sine ratio because the information given in the figure includes the adjacent side of the angle, not the opposite side. Point out that a choice of ratios must always be made when using trigonometry.

Using Technology On a TI-82 graphing calculator, the expression cos 89°4′ must be entered as cos 89′4′, where the ′ symbol is found on the Angle menu. In general, angles given in degrees, minutes, and seconds must be entered as degrees′ minutes′ seconds′.

Additional Example 3

Many telecommunication satellites are placed in geostationary orbits around Earth so that the satellite's speed matches that of the planet. Use the diagram below to estimate the height of the orbit above Earth using trigonometry. In the diagram, the satellite is directly above point P and is just visible from point Q. Points P and Q are on the equator. Use 3963 mi for the radius of Earth at the equator.

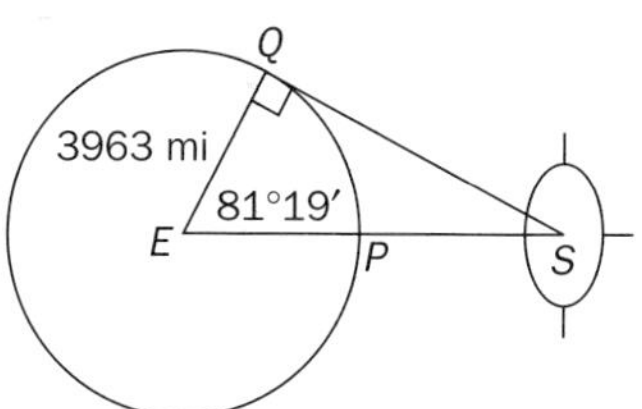

Find the distance between E, the center of Earth, and S, the satellite. Then subtract the radius of Earth.

$$\cos 81°19' = \frac{\text{adjacent}}{\text{hypotenuse}} = \frac{3963}{ES}$$

$$ES = \frac{3963}{\cos 81°19'}$$

$$ES \approx \frac{3963}{0.1510}$$

$$\approx 26{,}250$$

The distance above the surface of Earth is about 26,250 − 3963 ≈ 22,287 mi.

Teach⇔Interact

Section Note

Topic Spiraling: Review

It is important that students have a thorough understanding of the concepts of domain, range, and inverse of a function before proceeding with this section.

Additional Example 4

Find θ when $\cos \theta = 0.6235$. Express the answer in two ways:

a. in decimal degrees, to the nearest tenth of a degree

To find θ, use inverse cosine.

$\theta = \cos^{-1} 0.6235$

A calculator gives θ as 51.4° to the nearest tenth of a degree.

b. in degrees, minutes, and seconds, to the nearest second

```
cos⁻¹ 0.6235
        51.42782406
Ans▶DMS
      51°25'40.167"
```

Converting θ to degrees, minutes, and seconds, you get 51°25′40″, to the nearest second.

Inverse Sine and Cosine Functions

For a right triangle, sine and cosine are functions whose domain consists of all angle measures between 0° and 90°. What do you think the range of each function is?

THINK AND COMMUNICATE

1. **a.** What is the longest side of a right triangle?

   b. What does your answer to part (a) tell you about the ratios $\frac{\text{opposite}}{\text{hypotenuse}}$ and $\frac{\text{adjacent}}{\text{hypotenuse}}$?

   c. What does your answer to part (b) tell you about the range of sine and cosine?

The inverses of the sine and cosine functions are written "$\sin^{-1}$" and "$\cos^{-1}$." The inverse sine function is formed by reversing the domain and range of the sine function. Similarly, the inverse cosine function is formed by reversing the domain and range of the cosine function.

For a right triangle, the **domain** of sine and cosine consists of **angle measures between 0° and 90°**, and the **range** consists of **ratios between 0 and 1**.

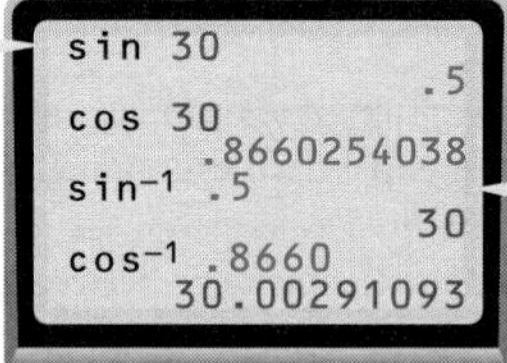

For a right triangle, the **domain** of inverse sine and inverse cosine consists of **ratios between 0 and 1**, and the **range** consists of **angle measures between 0° and 90°**.

You can use the inverse sine or cosine to find the measure of an angle whose sine or cosine is known. In this and certain other situations, it is convenient to refer to the sine or cosine of an angle without reference to its vertex label. In such cases, the Greek letter θ (read "theta") is often used to name the angle.

EXAMPLE 4

Find θ when $\sin \theta = 0.4731$. Express the answer in two ways:

a. in decimal degrees, to the nearest tenth of a degree

b. in degrees, minutes, and seconds, to the nearest second

SOLUTION

To find θ, use inverse sine.

$$\theta = \sin^{-1} 0.4731$$

a. A calculator gives θ as 28.2° to the nearest tenth of a degree.

b. Converting θ to degrees, minutes, and seconds, you get 28°14′9″ to the nearest second.

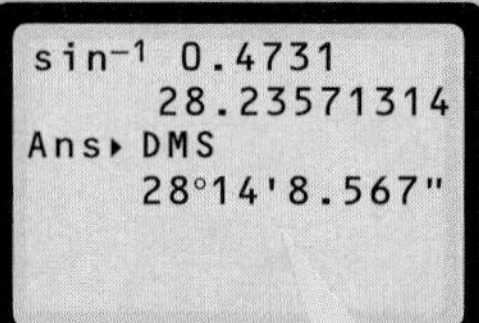

Many calculators will convert decimal degrees to degrees, minutes, and seconds (DMS).

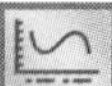

Technology Note

Point out that the reciprocal key of a calculator (usually labeled x^{-1}) is never used to find the inverse sine or inverse cosine. Emphasize that the x^{-1} key is used for finding the reciprocal of a number or expression, but that the $\sin^{-1}$ key is not the reciprocal of the sine, and the $\cos^{-1}$ key is not the reciprocal of the cosine. The $\sin^{-1}$ key returns the value of the inverse sine, or an angle between –90° and 90°, inclusive. Likewise, the $\cos^{-1}$ key returns the value of the inverse cosine, or an angle between 0° and 180°, inclusive. For the exercises in this section, only values between 0° and 90° are used.

CHECKING KEY CONCEPTS

1. For $\triangle ABC$ with right $\angle C$, write $\sin B$ and $\cos B$ in terms of the side lengths a, b, and c.
2. Find θ to the nearest tenth of a degree when $\cos \theta = 0.7859$.
3. Given $\theta = 32°19'$, find $\sin \theta$ and $\cos \theta$ to four decimal places.
4. A kite is on a 50 yard string that makes a 50° angle with the ground.
   a. Make a sketch to represent the situation described. Label the sides of the triangle formed by the kite, the string, and the ground.
   b. Find the height of the kite above the ground.

14.1 Exercises and Applications

Extra Practice exercises on page 770

Find sin *A* and cos *A* for each right triangle. Express answers in decimal form.

1. Triangle: 15 m, 20 m, 25 m (A, B, C)

2. 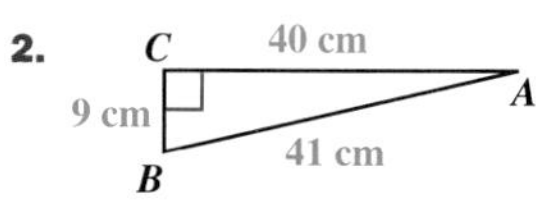

3. 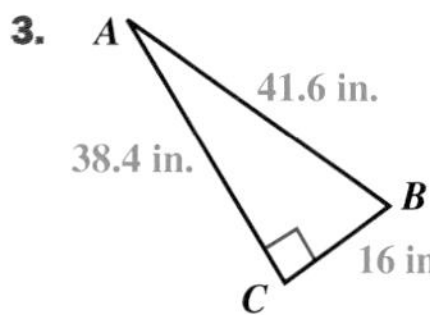

4–6. Find $\sin B$ and $\cos B$ for each of the triangles in Exercises 1–3. Compare these ratios with the ones you found in Exercises 1–3. What do you notice?

7. **EXERCISING** When Carol read these directions for an exercise, she thought, "How can I know when I have lifted my shoulders 30°? Wouldn't it be easier if I knew how *high* to lift my shoulders instead?"
   Suppose Carol's shoulder-to-hip length is 28 in. How high should she lift her shoulders to do this exercise correctly?

CRUNCHES: Legs bent; arms back, supporting head and neck; chin up. Lift shoulders and back about 30°. (Lifting up more than 30° activates back muscles, thus negating abdominal work.)

x
30°

For each trigonometric ratio, find θ. Express the answer in two ways:

a. in decimal degrees, to the nearest tenth of a degree

b. in degrees, minutes, and seconds, to the nearest second

8. $\sin \theta = 0.9532$
9. $\cos \theta = 0.7248$
10. $\sin \theta = 0.2361$
11. $\cos \theta = 0.2854$
12. $\sin \theta = 0.5377$
13. $\sin \theta = 0.4444$
14. $\cos \theta = 0.7800$
15. $\sin \theta = 0.8580$

Complete each statement using <, >, =, ≤, or ≥. Assume that ∠*A* and ∠*B* are both acute angles.

16. If $\angle A > \angle B$, then $\sin A \underline{\ ?\ } \sin B$.
17. If $\angle A > \angle B$, then $\cos A \underline{\ ?\ } \cos B$.

Teach⇔Interact

Closure Question

Explain how to find the sine or cosine of an acute angle of a right triangle.

The sine of an acute angle is the ratio of the length of the side opposite the angle to the length of the hypotenuse of the triangle. The cosine of the angle is the ratio of the length of the side adjacent to the angle to the length of the hypotenuse.

Apply⇔Assess

Suggested Assignment

Core Course
Exs. 1–17, 20–22, 28–36

Extended Course
Exs. 1–9, 11, 13, 15, 17–36

Block Schedule
Day 1 Exs. 1–17, 20–22, 28–36

Exercise Notes

Common Error
Exs. 1–3 Some students may confuse the sine and cosine ratios. Refer these students to the definitions on page 656.

Problem Solving
Exs. 4–6 Students should notice that $\sin A = \cos B$. Have them express this relationship in terms of complementary angles.

Student Study Tip
Exs. 8–15 Point out to students that they can use these exercises to support their conjecture for Exs. 4–6 by finding the sine or cosine of the complementary angle.

Teaching Tip
Exs. 16, 17 You can use these exercises to point out that the sine increases as the angle increases from 0° to 90°, while the cosine decreases in the same interval.

Think and Communicate

1. a. the hypotenuse
   b. The ratios must be between 0 and 1.
   c. For all angle measures between 0° and 90°, the range of each function consists of ratios between 0 and 1.

Checking Key Concepts

1. $\sin B = \frac{b}{c}$; $\cos B = \frac{a}{c}$

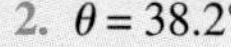

2. $\theta = 38.2°$
3. $\sin \theta = 0.5346$; $\cos \theta = 0.8451$
4. a.

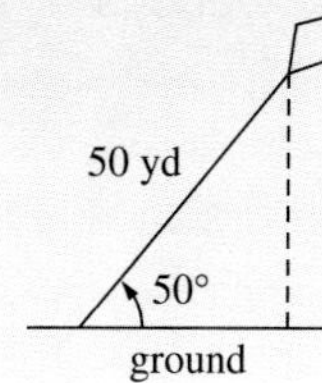

   b. about 38 yd

Exercises and Applications

Angle measures are rounded to the nearest 0.1 degree and trigonometric ratios are rounded to four significant digits.

1. $\sin A = 0.8000$; $\cos A = 0.6000$
2. $\sin A = 0.2195$; $\cos A = 0.9756$
3. $\sin A = 0.3846$; $\cos A = 0.9231$
4. $\sin B = 0.6000$; $\cos B = 0.8000$
5. $\sin B = 0.9756$; $\cos B = 0.2195$
6. $\sin B = 0.9231$; $\cos B = 0.3846$

4–6. $\sin B = \cos A$ and $\cos B = \sin A$.

7–17. See answers in back of book.

Apply⇔Assess

Exercise Notes

Topic Spiraling: Review

Exs. 18, 19 You may want to review some of the quadratic modeling concepts from Chapter 5, before assigning these exercises. For Ex. 18, remind students that the number of seconds the ball is in the air corresponds to the number of seconds it takes for the y-coordinate to "become" 0. Students not using technology can solve the given y-equation as a quadratic with $y = 0$. Students using technology can use trace and zoom to find the x-intercepts of the graph. (See the Using Technology note below for Ex. 18.)

Student Study Tip

Exs. 18, 19 Remind students that parametric equations are often used to express a relationship between two variables (x and y) in terms of another variable (such as t, in this case).

Using Technology

Exs. 18, 19 To graph the equations for Ex. 18, set the calculator to parametric mode (Par), degree mode (Degree), replace θ with 45 in each equation, and enter the equations using the [Y=] key. A convenient window to use is Tmin = 0, Tmax = 10, Xmin = 0, Xmax = 600, Ymin = –1, Ymax = 200. For Ex. 19(a), you can enter the equations directly into the calculator by using parametric mode and the variable θ in each equation. Before graphing, store a value for θ. This eliminates the need to reenter the equation as θ is varied.

Function Investigator Activity 31 of the *Function Investigator with Matrix Analyzer Activity Book* provides further opportunities for students to explore projectile motion using parametric equations. In this activity, students can examine how the initial velocity and launch angle of a projectile affect the path of the projectile through the air.

Cooperative Learning

Ex. 19 You may wish to have students work on this exercise in groups so they can help each other find a good viewing window for the graph of the path of the jumper and develop their solutions together.

Connection PHYSICS

Recall from Chapter 5 that a *projectile* is an object that is thrown or launched. The horizontal position x, in feet, and the vertical position y, in feet, of a projectile t seconds after it is launched at an angle θ can be found using the parametric equations

$$x = (v_0 \cos \theta)t + x_0$$
$$y = -16t^2 + (v_0 \sin \theta)t + y_0$$

where v_0 is the launch velocity in feet per second, and x_0 and y_0 are the initial horizontal and vertical positions, respectively, in feet. This model has its limitations, since it does not take into account factors such as air resistance, which affect a projectile's movement.

BY THE WAY

Golfers use two types of clubs, woods and irons. Both woods and irons are numbered 1 through 9 depending on the angle of the surface that contacts the ball. This angle affects how far away the ball lands.

In Exercises 18 and 19, assume the origin of the coordinate system is at (x_0, y_0).

18. Suppose a golfer hits a golf ball 167 yd using a 7-iron. If the launch angle is 45° and the launch velocity is 127 ft/s, for how many seconds is the golf ball in the air?

19. In a court case, a track coach was called as an expert witness. The coach testified that it was impossible for the defendant to have jumped the 13 ft gap between the roof of one building and the roof of another building that was one story (about 10 ft) lower. The reason given was that the world record for the standing broad jump is 12 ft.

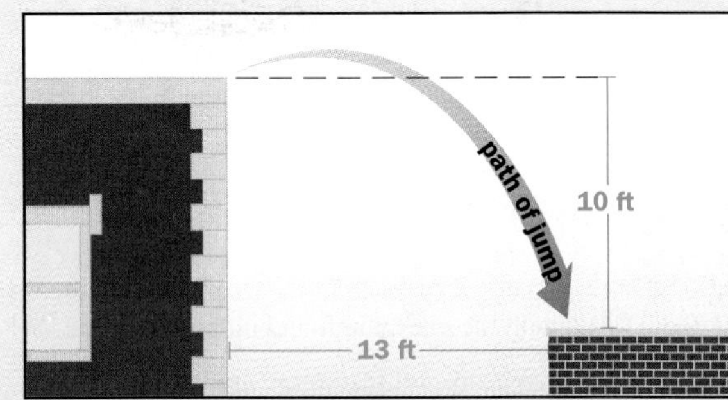

a. **Technology** Graph the parametric equations $x = (v_0 \cos \theta)t$ and $y = -16t^2 + (v_0 \sin \theta)t$ on a graphing calculator. Experiment to find the value of θ that maximizes the horizontal distance that a projectile travels for any given value of v_0. (*Hint:* Choose any value for v_0 and then vary the value of θ.)

b. Estimate the launch velocity for the world record standing broad jump by completing Steps 1–3.

Step 1 Express the time in the air for a standing broad jump in terms of v_0 and θ. (*Hint:* Solve $y = -16t^2 + (v_0 \sin \theta)t$ for t when $y = 0$. Why?)

Step 2 Express the length of a broad jump in terms of v_0 and θ. (*Hint:* Substitute the nonzero value of t from Step 1 into $x = (v_0 \cos \theta)t$.)

Step 3 Express the length of the longest possible standing broad jump as a function of the launch velocity. (*Hint:* Use the value of θ you found in part (a).)

c. **Technology** On a graphing calculator, graph the path of the jumper in the situation described. Determine whether the conclusion drawn by the expert witness is correct. (*Hint:* Is $x > 13$ when $y = -10$?)

18. about 5.6 s

19. a. Check students' work. $\theta = 45°$

b. Step 1: $t = \frac{v_0}{16} \sin \theta$

Step 2: $x = \frac{v_0^2}{16} \sin \theta \cos \theta$

Step 3: $x_{max} = \frac{v_0^2}{32}$; The launch velocity for the world record standing broad jump is about 19.6 ft/s.

c. Check students' work. The expert witness is incorrect; $x \approx 18.6$ when $y = -10$.

Find the measure of ∠ A in decimal degrees, to the nearest tenth of a degree.

20.

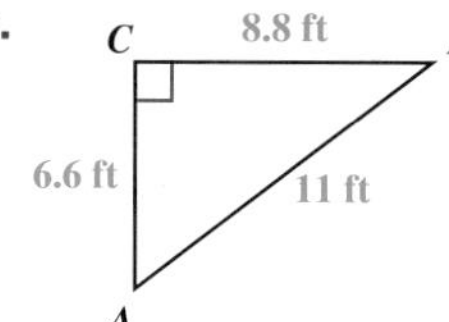

21.

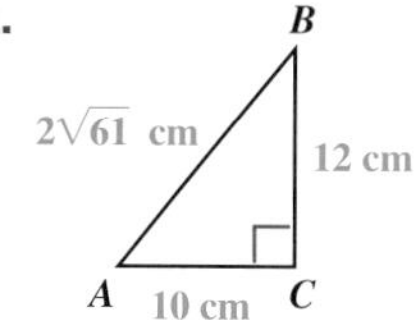

22.

B
45.5 in.
17.5 in.
C
42 in.
A

23. ASTRONOMY For the triangle formed by Earth E, Venus V, and the sun S, $\angle E$ reaches its maximum value, 47°, when $\overline{EV} \perp \overline{VS}$. Use the diagram to find each distance. (The distance between Earth and the sun is about 93,000,000 mi.)

a. distance between Venus and the sun

b. distance between Venus and Earth

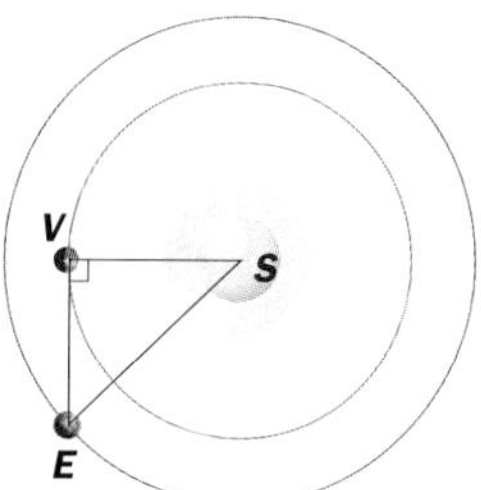

INTERVIEW Johnpaul Jones

Look back at the article on pages 652–654.

The moat that surrounds the Sumatran tiger habitat in Johnpaul Jones's Tiger River exhibit benefits both the human visitors to the zoo and the tigers. In warm climates like that of San Diego, tigers may spend hours lying in the water to escape the heat and the flies. Combined with a wall that slants inward toward the base, the moat helps to keep the tigers inside the simulated jungle.

Use the parametric equations for projectile motion given on page 660.

24. Suppose that the observation deck for the Sumatran tiger habitat were on the same level as the water in the moat. If the launch angle and velocity for the *longest* possible leap by a Sumatran tiger are 20° and 19 mi/h, respectively, what should be the minimum width of the moat?

25. Suppose that the tiger habitat were not surrounded by a moat, and that the wall below the observation deck were straight, not slanted. If the launch angle and velocity for the *highest* possible leap by a Sumatran tiger are 30° and 24 mi/h, respectively, what should be the minimum height of the wall?

Apply⇔Assess

Exercise Notes

Reasoning

Exs. 20–22 Students should realize that they can use either the sine ratio or the cosine ratio for each of these exercises.

Historical Connection

Ex. 23 Nicholaus Copernicus, a Polish astronomer, devised this straightforward method of determining the distance from the inner planets in the early 1500s. Because the orbits of the planets are elliptical, these distances are calculated when the planet is at its greatest distance from the sun, a point known as *aphelion.*

Interview Note

Ex. 24 You may want to suggest that students use the technology method discussed in Ex. 19, part (b), Step 1 for this exercise.

Using Technology

Ex. 25 If students graph the parametric equations to find the answer to Ex. 24, ask them to find the maximum value by zooming and tracing the graph.

Second-Language Learners

Ex. 26 The sports of hang-gliding and water-skiing may not be familiar to some students learning English. If necessary, explain that in hang-gliding a person is strapped into a harness, which comprises an aluminum frame covered with synthetic material. The hang-glider then jumps off the side of a hill and the wind carries him or her through the air. Water-skiing is a sport in which a power boat pulls a skier, who glides along the surface of the water on skis.

20. 53.1°
21. 50.2°
22. 22.6°
23. a. about 68,000,000 mi
    b. about 63,000,000 mi
24. about 16 ft
25. about 4.8 ft

Apply⇔Assess

Exercise Notes

Topic Spiraling: Preview

Ex. 27 Students are asked to find special values of sine and cosine related to 30°-60°-90° triangles and 45°-45°-90° triangles. They will need to know these values for the next chapter.

Assessment Note

Ex. 29 Thinking about how the value of the sine function changes as the measure of an angle increases will help students understand the presentation of the sine function when they discuss the unit circle in the next chapter. You may wish to call upon two or three students to explain their answers to the class.

Practice 89 for Section 14.1

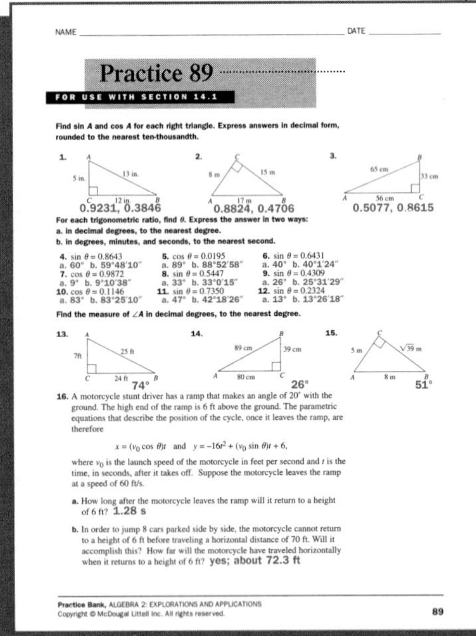

NAME ______ DATE ______

Practice 89

FOR USE WITH SECTION 14.1

Find sin *A* and cos *A* for each right triangle. Express answers in decimal form, rounded to the nearest ten-thousandth.

1. (A, C, B; 5 in., 13 in., 12 in.) 0.9231, 0.3846
2. (C, A, B; 8 m, 15 m, 17 m) 0.8824, 0.4706
3. (B, A, C; 65 cm, 33 cm, 56 cm) 0.5077, 0.8615

For each trigonometric ratio, find θ. Express the answer in two ways:
a. in decimal degrees, to the nearest degree.
b. in degrees, minutes, and seconds, to the nearest second.

4. $\sin\theta = 0.8643$ a. 60° b. 59°48′10″
5. $\cos\theta = 0.0195$ a. 89° b. 88°52′58″
6. $\sin\theta = 0.6431$ a. 40° b. 40°1′24″
7. $\cos\theta = 0.9872$ a. 9° b. 9°10′38″
8. $\sin\theta = 0.5447$ a. 33° b. 33°0′15″
9. $\sin\theta = 0.4309$ a. 26° b. 25°31′29″
10. $\cos\theta = 0.1146$ a. 83° b. 83°25′10″
11. $\sin\theta = 0.7350$ a. 47° b. 42°18′26″
12. $\sin\theta = 0.2324$ a. 13° b. 13°26′18″

Find the measure of ∠*A* in decimal degrees, to the nearest degree.

13. (A, C, B; 7ft, 25 ft, 24 ft) 74°
14. (B, A, C; 89 cm, 39 cm, 80 cm) 26°
15. (C, A, B; 5 m, $\sqrt{39}$ m, 8 m) 51°

16. A motorcycle stunt driver has a ramp that makes an angle of 20° with the ground. The high end of the ramp is 6 ft above the ground. The parametric equations that describe the position of the cycle, once it leaves the ramp, are therefore

$$x = (v_0 \cos\theta)t \quad \text{and} \quad y = -16t^2 + (v_0 \sin\theta)t + 6,$$

where v_0 is the launch speed of the motorcycle in feet per second and t is the time, in seconds, after it takes off. Suppose the motorcycle leaves the ramp at a speed of 60 ft/s.

a. How long after the motorcycle leaves the ramp will it return to a height of 6 ft? 1.28 s

b. In order to jump 8 cars parked side by side, the motorcycle cannot return to a height of 6 ft before traveling a horizontal distance of 70 ft. Will it accomplish this? How far will the motorcycle have traveled horizontally when it returns to a height of 6 ft? yes; about 72.3 ft

Practice Bank, ALGEBRA 2: EXPLORATIONS AND APPLICATIONS

89

26. PARASAILING This sport is a cross between hang-gliding and water-skiing. The rider wears a harness fitted with a special parachute that acts like an airplane wing. As the tow line connecting the rider to the boat is let out, the parachute lifts the rider up into the air.

Suppose the tow line is let out to its maximum length of 800 ft. What is the measure of the angle between the tow line and the surface of the water when the rider is at a height of 400 ft?

Not drawn to scale

800 ft

400 ft

27. Challenge Without using a calculator, find sin θ and cos θ for each value of θ. Explain how you found your answers.

a. $\theta = 45°$ (*Hint:* Sketch a right triangle with a 45° angle. How are the lengths of the two shorter sides related?)

b. $\theta = 30°$ (*Hint:* Sketch an equilateral triangle and draw an altitude.)

c. $\theta = 60°$

28. SAT/ACT Preview For right $\triangle ABC$, which equation can you use to solve for c?

A. $\cos 72.8° = \dfrac{34.2}{c}$ **B.** $\sin 72.8° = \dfrac{34.2}{c}$

C. $\sin 72.8° = \dfrac{c}{34.2}$ **D.** $\cos 72.8° = \dfrac{c}{34.2}$

E. none of these

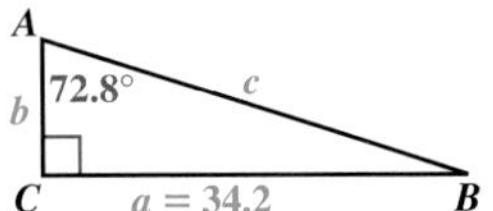

ONGOING ASSESSMENT

29. Writing Use the diagram and a calculator to describe how the value of sin A changes with the measure of $\angle A$ in a right triangle when the length of the hypotenuse remains constant.

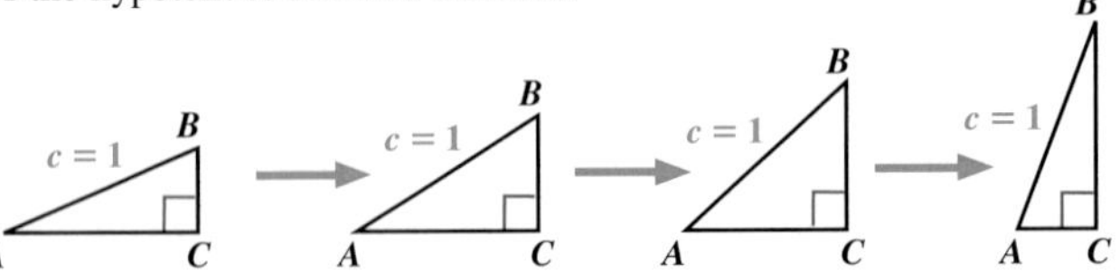

SPIRAL REVIEW

30. Suppose the donations received at a charity event form a normal distribution. If the mean donation was \$55 and the standard deviation was \$10, what percent of donations were below \$50? above \$75? *(Section 13.5)*

Find the geometric mean for each pair of numbers. *(Section 10.2)*

31. 17, 6 **32.** 1.8, 0.5 **33.** $\frac{1}{2}, \frac{3}{8}$

Solve each proportion. *(Toolbox, p. 785)*

34. $\frac{x}{16} = \frac{3}{4}$ **35.** $\frac{18}{x} = \frac{x}{4}$ **36.** $\frac{2}{3} = \frac{10}{x+1}$

26. 30°

27. a–c. Explanations may vary.

a. $\sin 45° = \cos 45° = \frac{1}{\sqrt{2}} = \frac{\sqrt{2}}{2}$; If each leg has length x, then by the Pythagorean theorem, the hypotenuse must have length $\sqrt{2}x$. Thus, $\sin 45° = \frac{x}{\sqrt{2}x} = \frac{1}{\sqrt{2}}$. The same is true for cos 45°.

b. $\sin 30° = \frac{1}{2}$; $\cos 30° = \frac{\sqrt{3}}{2}$

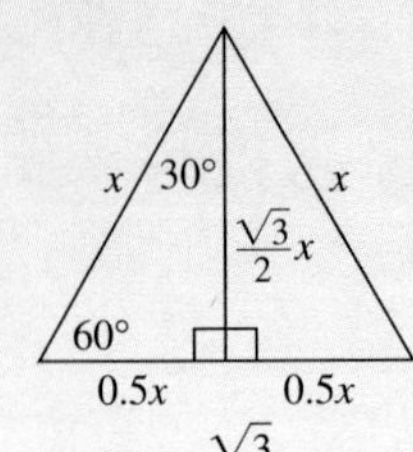

c. $\sin 60° = \frac{\sqrt{3}}{2}$; $\cos 60° = \frac{1}{2}$; See diagram in part (b).

28. B

29. The value of sin A increases from 0 to 1 as the measure of $\angle A$ increases from 0° to 90°.

30. 30.85%; 2.28%

31. 10.1

32. 0.95

33. 0.43

34. $x = 12$

35. $x = \pm 6\sqrt{2}$

36. $x = 14$

Section

14.2 Using Tangent

Learn how to...
- find the tangent of an acute angle

So you can...
- find the length of a runway when you know the height of a plane and the angle of descent, for example

In addition to the sine and cosine ratios, you can use the *tangent* ratio to find the measures of the sides and angles of a right triangle.

Tangent of an Angle

In right $\triangle ABC$, the **tangent** of $\angle A$, which is written "tan A," is given by

$$\tan A = \frac{\text{length of leg opposite } \angle A}{\text{length of leg adjacent to } \angle A} = \frac{a}{b}$$

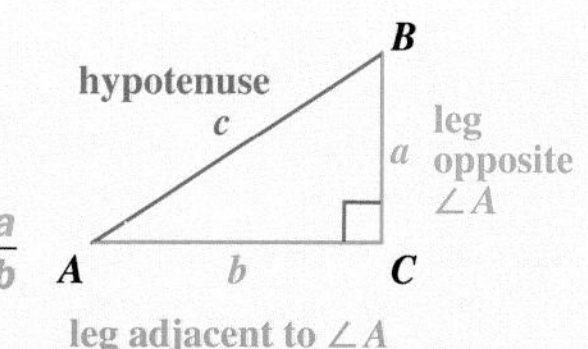

EXPLORATION
COOPERATIVE LEARNING

Estimating the Distance Across Your Classroom

Work with a partner.
You will need:
- a ruler, yardstick, or meterstick

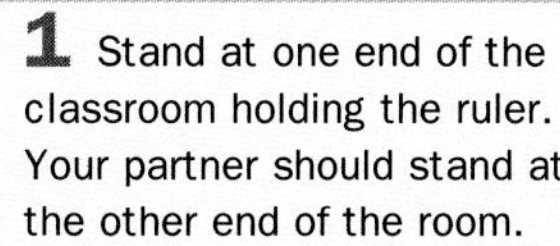

1 Stand at one end of the classroom holding the ruler. Your partner should stand at the other end of the room.

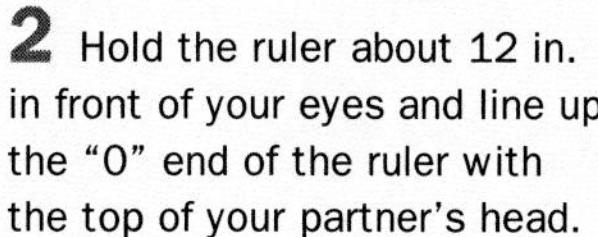

2 Hold the ruler about 12 in. in front of your eyes and line up the "0" end of the ruler with the top of your partner's head.

3 Spot the floor at your partner's feet along the ruler. What are your partner's "ruler height" and actual height?

4 Copy and complete the diagram below. Include the measurements from Step 3.

Questions

1. What is the value of tan θ using the small triangle?
2. Let d be the unknown distance across the room. Using the large triangle, write an expression involving d for the value of tan θ.

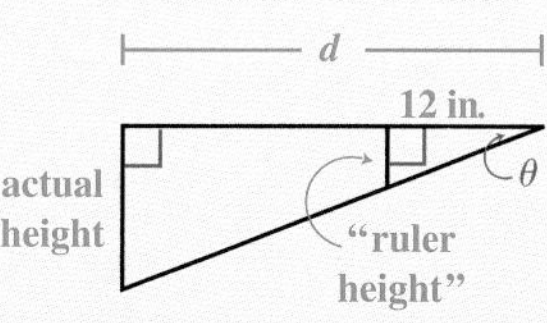

3. Use your answers from Questions 1 and 2 to write an equation. Solve for d. (You may want to use the ruler to find the actual distance across the room and compare it with your calculated result.)

Exploration Note

Purpose
The purpose of this Exploration is to have students use the tangent ratio and similar triangles to find the distance across a room.

Materials/Preparation
Each pair of students needs a ruler, a yardstick, or a meterstick.

Procedure
One student stands across the room while the other lines up a ruler vertically 12 in. in front of his or her eyes. This student then uses the ruler to line up the partner within his or her line of sight. The students then set up similar triangles by using the actual height and ruler height of the partner. The tangent ratio is then calculated for each triangle and is used to find the estimated distance across the room.

Closure
Students should understand that the tangent ratio can be used to find distances that are difficult to measure.

Explorations Lab Manual
See the Manual for more commentary on this Exploration.

For answers to the Exploration, see following page.

Plan⇔Support

Objectives
- Find the tangent of an acute angle.
- Use the tangent ratio to solve problems.

Recommended Pacing
Core and Extended Courses
Section 14.2: 1 day

Block Schedule
Section 14.2: $\frac{1}{2}$ block
(with Section 14.1)

Resource Materials
Lesson Support
Lesson Plan 14.2
Warm-Up Transparency 14.2
Overhead Visuals:
Folder 14: Trigonometric Ratios
Practice Bank: Practice 90
Study Guide: Section 14.2
Explorations Lab Manual:
Additional Exploration 20
Assessment Book: Test 58

Technology
Technology Book:
Calculator Activity 14
TI-92 Activity 6
Graphing Calculator
McDougal Littell Mathpack
Function Investigator
Internet:
http://www.hmco.com

Warm-Up Exercises
1. What is the measure of $\angle A$ in $\triangle ABC$ with right angle C if $a = 3$ and $b = 4$? about 36.9°
2. What is the ratio of the opposite side of $\angle A$ to the adjacent side for the triangle in Ex. 1? $\frac{3}{4}$
3. Solve the equation $\frac{x}{48} = 2.096$ for x. 100.608
4. Solve the equation $0.2 = \frac{1}{2 + x}$ for x. 3
5. Find the measures of the acute angles of a right triangle with side lengths 5, 12, and 13. 22.6°; 67.4°

Teach⇔Interact

Think and Communicate

When discussing question 2, students should realize that the range of the tangent function increases without bound as the value of the angle approaches 90°.

Section Note

Teaching Tip

Students may need to be reminded that the x^{-1} key on a calculator is used for finding the reciprocal of a number or expression, but that the $\tan^{-1}$ key is not the reciprocal of the tangent. The $\tan^{-1}$ key gives the value of the inverse tangent, or an angle between –90° and 90°. For the exercises in this section, only values between 0° and 90° are given.

Additional Example 1

Two buildings are located as shown in the diagram. Find the distance between the buildings.

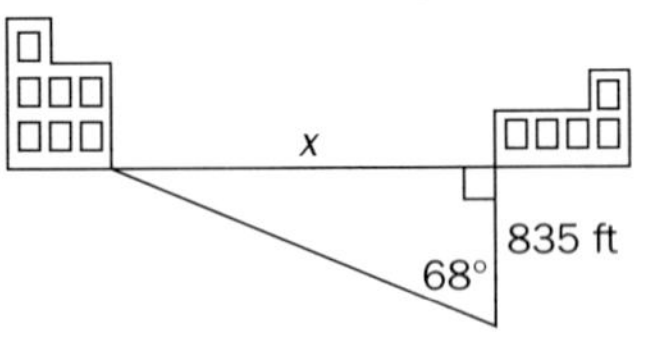

Let x = the distance in feet between the buildings. Write an equation involving x and a trigonometric ratio.

$$\frac{x}{835} = \tan 68°$$

$$x = 835 \tan 68°$$

$$x \approx 835(2.4751) \approx 2066.7$$

The distance between the two buildings is about 2067 ft.

Section Note

Problem Solving

You can help students organize their method of solving a triangle by listing these steps:

1. Read the problem carefully.
2. Draw and label the triangle, if necessary.
3. Determine the trigonometric ratio than can be used to find an unknown side or angle.
4. Set up a trigonometric equation and solve it.
5. Repeat Steps 3 and 4 until all the angles and sides are known.

THINK AND COMMUNICATE

Use the definition of tangent given at the top of the previous page.

1. What is the value of tan A for a right $\triangle ABC$ with the given leg lengths?

a. $a = 1, b = 10$ **b.** $a = 10, b = 10$ **c.** $a = 100, b = 10$

2. For a right triangle, tangent is a function whose domain consists of all angle measures between 0° and 90°. What do you think is the range of the tangent function for a right triangle?

The inverse of the tangent function is written "$\tan^{-1}$." The inverse tangent function is formed by reversing the domain and range of the tangent function.

For a right triangle, the **domain** of tangent consists of **angle measures between 0° and 90°**, and the **range** consists of **all nonnegative numbers.**

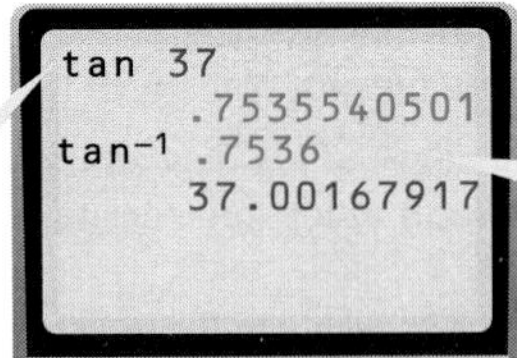

For a right triangle, the **domain** of inverse tangent consists of **all nonnegative numbers**, and the **range** consists of **angle measures between 0° and 90°**.

BY THE WAY

Surveyors sometimes use a professional instrument called a *transit* to measure angles and then use trigonometry to calculate distances.

EXAMPLE 1 Application: Surveying

Surveyors use trigonometry to calculate distances that would otherwise be difficult to find, such as the distance between two houses located across a lake from each other, as shown in the diagram. What is this distance?

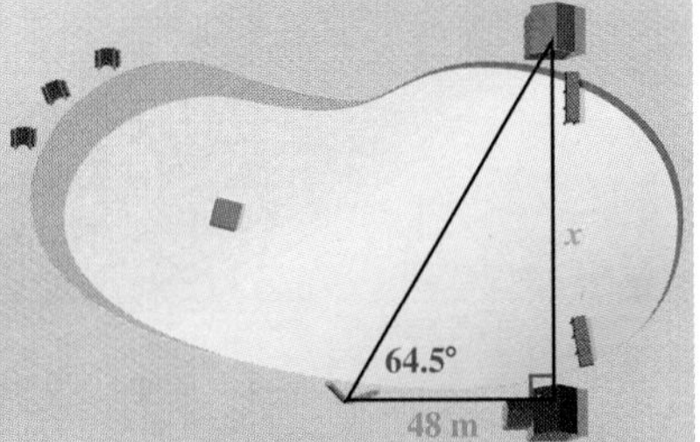

SOLUTION

Let x = the distance in meters across the lake. Write an equation involving x and a trigonometric ratio.

```
tan 64.5
        2.096543599
Ans*48
        100.6340928
```

$$\frac{x}{48} = \tan 64.5°$$

$$x = 48 \tan 64.5°$$

$$x \approx 48(2.0965) \approx 100.6$$

The distance between the two houses is about 101 m.

THINK AND COMMUNICATE

3. In Example 1, what would you do to find the other acute angle in the diagram? What would you do to find the hypotenuse?

4. Use a calculator to find $\tan^{-1}\left(\frac{101}{48}\right)$. Does this agree with the angle given in the diagram for Example 1?

ANSWERS Section 14.2

Exploration

1–4. Check students' work.

Questions

1. Answers may vary. Students' ratios should be $\frac{\text{ruler height}}{12 \text{ in.}}$.
2. Answers may vary. Students' ratios should be $\frac{\text{actual height}}{d}$.
3. Answers may vary. Students' equations should be $d = \frac{12 \cdot \text{actual height}}{\text{ruler height}}$.

Think and Communicate

1. a. $\tan A = 0.1$
   b. $\tan A = 1$
   c. $\tan A = 10$
2. all positive real numbers
3. To find the other angle, subtract 64.5 from 90. To find the hypotenuse, use either the Pythagorean theorem or the definition of sine or cosine.
4. 64.6°; Yes, the measures are approximately equal.

Finding the lengths of *all* sides and the measures of *all* angles of a triangle is called **solving a triangle**.

EXAMPLE 2

Solve right $\triangle ABC$.

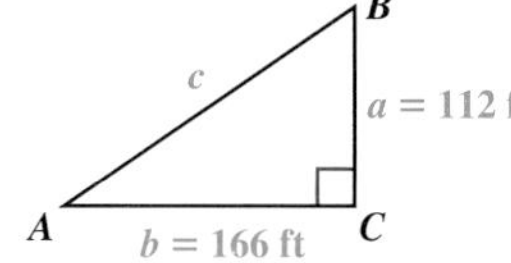

SOLUTION

Step 1 The lengths of the two legs of the triangle are known. To find $\angle A$, use inverse tangent.

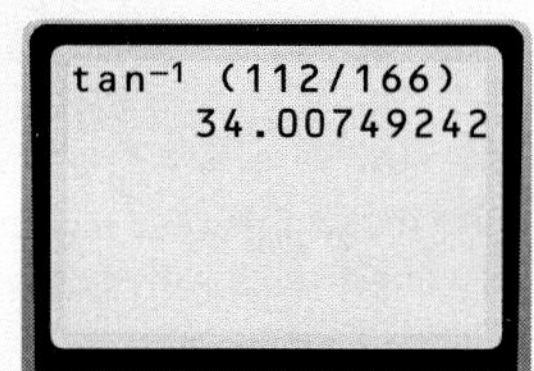

$$\tan A = \frac{\text{opposite}}{\text{adjacent}} = \frac{112}{166}$$

$$\angle A = \tan^{-1}\left(\frac{112}{166}\right) \approx 34°$$

Step 2 To find $\angle B$, use the fact that the sum of angle measures in a triangle is 180°.

$$\angle A + \angle B + \angle C = 180°$$
$$34° + \angle B + 90° \approx 180°$$
$$\angle B \approx 56°$$

Step 3 Now use the Pythagorean theorem (or a trigonometric ratio) to find c, the length of the hypotenuse.

$$c^2 = 166^2 + 112^2$$
$$c = \sqrt{40{,}100}$$
$$c \approx 200.2$$

Therefore, $\angle A \approx 34°$, $c \approx 200$ ft, and $\angle B \approx 56°$.

Toolbox p. 801
Triangle Relationships

Angles of Elevation and Depression

In the Exploration on page 663, you measured an **angle of depression** from your eye level to your partner's feet. There is an angle of equal measure, called the **angle of elevation**, from your partner's feet up to your eyes.

Angles of depression and elevation are often used with trigonometry to calculate distances in navigation, aviation, and surveying.

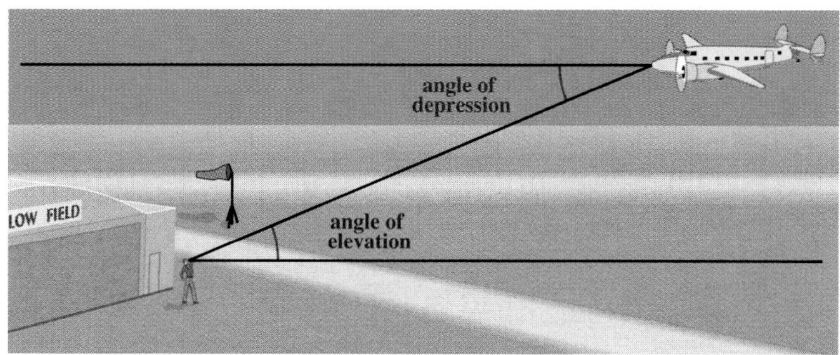

Teach⇔Interact

About Example 2

Alternate Approach

Step 3 of the solution to this Example uses the Pythagorean theorem to find the length of the hypotenuse. Point out that either the sine or the cosine ratio could also have been used to find the hypotenuse once $\angle A$ is known.

Additional Example 2

Solve right $\triangle ABC$.

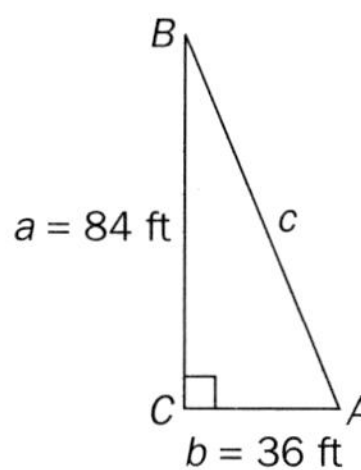

Step 1 The length of the two legs of the triangle are known. To find $\angle A$, use inverse tangent.

$$\tan A = \frac{\text{opposite}}{\text{adjacent}} = \frac{84}{36}$$

$$\angle A = \tan^{-1}\left(\frac{84}{36}\right) \approx 67°$$

Step 2 To find $\angle B$, use the fact that the sum of angle measures in a triangle is 180°.

$$\angle A + \angle B + \angle C = 180°$$
$$67° + \angle B + 90° \approx 180°$$
$$\angle B \approx 23°$$

Step 3 Now use the Pythagorean theorem to find c, the length of the hypotenuse.

$$c^2 = 84^2 + 36^2$$
$$c = \sqrt{8352}$$
$$c \approx 91.4$$

Therefore, $\angle A \approx 67°$, $c \approx 91.4$ ft, and $\angle B \approx 23°$.

Section Note

Teaching Tip

Labeling an angle of elevation or an angle of depression is difficult for some students. Point out that each type of angle is measured from the horizontal.

Teach⇔Interact

About Example 3

Communication: Drawing In this Example, students solve trigonometric equations. Point out the importance of drawing a good diagram to represent the information.

Additional Example 3

A ship traveling at sea passes a marker known to be 1000 m from a lighthouse on a cliff as measured along the water.

a. If the observation point at the top of the lighthouse is 96 m above the sea, what is the angle of depression from the lighthouse to the ship?

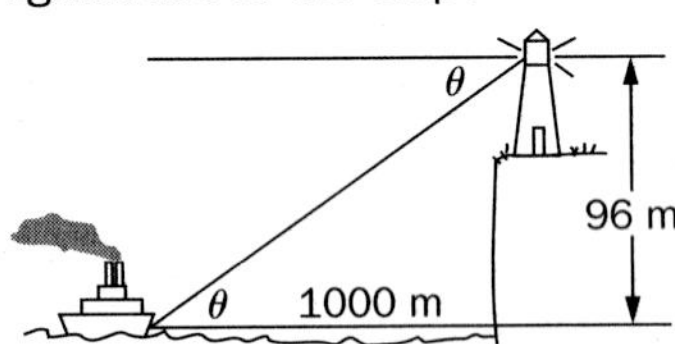

Use the fact that the measures of the angle of elevation and the angle of depression are the same to find the angle of depression from the lighthouse to the ship using the tangent ratio.

$\tan\theta = \frac{96}{1000}$

$\theta = \tan^{-1}\left(\frac{96}{1000}\right)$

$\approx 5.5°$

The angle of depression is about 5.5°.

b. How far has the ship traveled if the angle of depression from the lighthouse to the ship is 2.9°?

The distance to the ship is 1000 + x, where x is the distance the ship traveled.

$\tan 2.9° = \frac{96}{1000 + x}$

$0.0507 \approx \frac{96}{1000 + x}$

$1000 + x \approx \frac{96}{0.0507}$

$x \approx 1893 - 1000$

$= 893$ m

The ship has traveled about 893 m.

BY THE WAY

The longest runway at O'Hare International Airport in Chicago, the busiest airport in the United States, is 13,000 ft long.

EXAMPLE 3 Application: Aviation

Suppose an airplane begins its descent to a runway from an altitude of 10,000 ft.

a. If the airplane is still 75,000 ft (about 14 mi) from the runway as measured along the ground, what is its angle of descent (the angle of depression)?

b. What is the length of the runway if the angle of elevation from the far end of the runway to the airplane is 6.6°?

SOLUTION

Not drawn to scale

a.

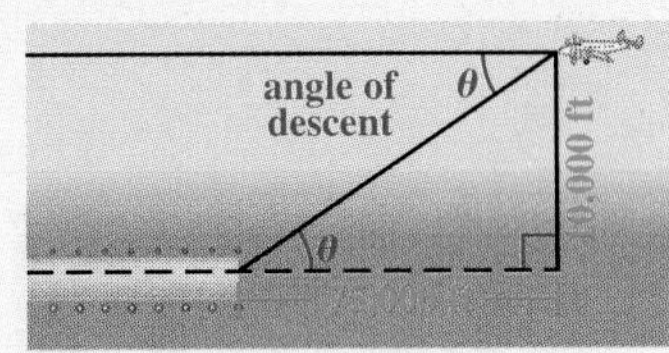

Use the fact that the measures of the angle of elevation and the angle of depression are the same to find the airplane's angle of descent using the tangent ratio.

$\tan\theta = \frac{10{,}000}{75{,}000}$

$\theta = \tan^{-1}\left(\frac{10{,}000}{75{,}000}\right)$

$\approx 7.59°$

The angle of descent is about 7.6°.

b.

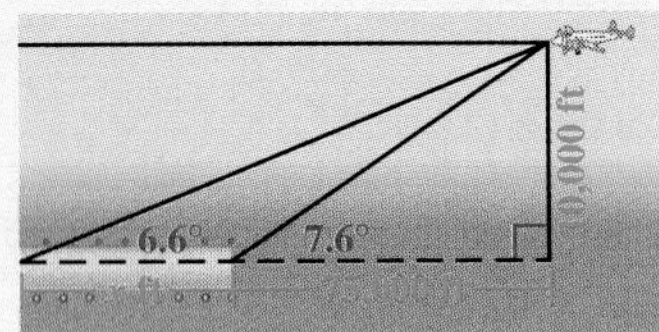

The distance from the airplane to the far end of the runway as measured along the ground is 75,000 + x, where x is the length in feet of the runway.

$\tan 6.6° = \frac{10{,}000}{75{,}000 + x}$

$0.1157 \approx \frac{10{,}000}{75{,}000 + x}$

$75{,}000 + x \approx \frac{10{,}000}{0.1157}$

Multiply both sides by 75,000 + x and divide both sides by 0.1157.

$x \approx 86{,}430 - 75{,}000$

$= 11{,}430$

The runway is about 11,000 ft long.

CHECKING KEY CONCEPTS

1. a. For right △ABC shown, state an equation you can use to solve for a.

b. Solve △ABC.

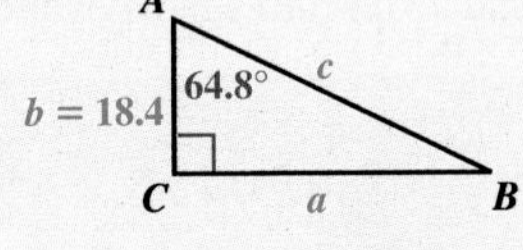

2. In right △RST, ∠R is a right angle, t = 3 ft, and s = 4 ft. Find ∠S.

3. Use your knowledge of geometry to explain why the angle of elevation (∠1) and the angle of depression (∠2) between two objects at different heights are equal in measure.

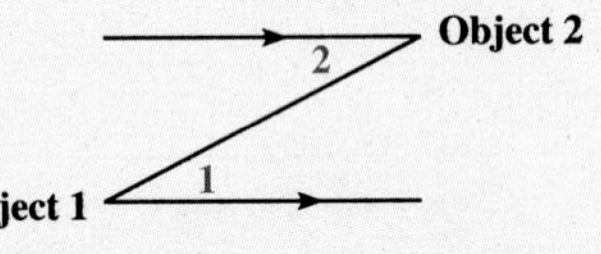

Checking Key Concepts

1. a. $\tan 64.8° = \frac{a}{18.4}$

   b. ∠B = 25.2 °; $a \approx 39.1$; $c \approx 43.2$

2. about 53.1°

3. They are alternate interior angles formed by parallel lines and a transversal and are therefore congruent.

Exercises and Applications

Angle measures are rounded to the nearest 0.1 degree and trigonometric ratios are rounded to four significant digits.

1. a. $\frac{3}{4}$

   b. $\frac{4}{3}$

   c. They are reciprocals.

2. a. 0.6445

   b. 0.5471

   c. 63.66

   d. 1.000

3. A

4. a. about 957 m

   b. about 764 m

5–10. Side lengths are rounded to the nearest tenth.

5. ∠B = 71°; a = 1.2; c = 3.7

6. ∠A = 18°; a = 9.9; b = 30.4

7. ∠A = 52.0°; ∠B = 38°; a = 15.4

14.2 Exercises and Applications

Extra Practice exercises on page 770

1. Refer to $\triangle XYZ$.

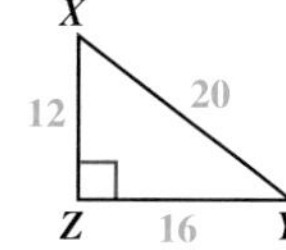

   a. Find $\tan Y$. b. Find $\tan X$.

   c. How are your answers to parts (a) and (b) related?

2. Find each value.

   a. $\tan 32.8°$ b. $\tan 28°41'$ c. $\tan 89.1°$ d. $\tan 45°$

3. **SAT/ACT Preview** For right $\triangle ABC$, which equation can you use to solve for a?

   A. $\tan 68.2° = \frac{24.8}{a}$ B. $\tan 68.2° = \frac{a}{24.8}$

   C. $\tan 21.8° = \frac{24.8}{a}$ D. $\sin 21.8° = \frac{a}{24.8}$

   E. none of these

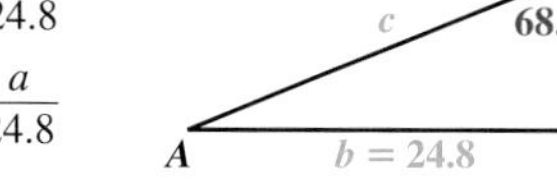

4. **AVIATION** The pilot of an airplane finds that the angle of depression to the start of a runway is 52°34′.

   a. If the altitude of the airplane is 1250 m, how far is the airplane from the runway as measured along the ground?

   b. If the angle of depression to the other end of the runway is 36°, how long is the runway to the nearest meter?

Solve $\triangle ABC$ using the given measures.

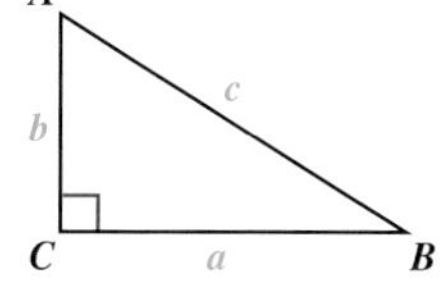

5. $\angle A = 19°$, $b = \sqrt{12}$ 6. $c = 32$, $\angle B = 72°$

7. $b = 12$, $c = 19.5$ 8. $a = 114$, $\angle B = 39.3°$

9. $\tan A = \frac{1}{3}$, $a = 7$ 10. $\angle B = 62.8°$, $c = 18.4$

11. **ASTRONOMY** On the moon, when the sun is at an angle of 30° to the horizon, the rim of a crater casts a 120 ft long shadow at the bottom of the crater. Assuming the sides of the crater are vertical, find the depth of the crater.

12. **GEOMETRY** For any line that has a positive slope and intersects the x-axis, let θ be the angle that the line makes with the x-axis. (Measure θ counterclockwise from the x-axis to the line, so that $0° < \theta < 90°$.) Discuss how θ is related to the slope of the line.

13. **Cooperative Learning** Work with a partner. You may want to use graph paper.

   a. Find the perimeter of rectangle $ABCD$ with vertices $A(3, 7)$, $B(5, 5)$, $C(1, 1)$, and $D(-1, 3)$.

   b. Find the measures of $\angle DAC$ and $\angle CAB$ formed by drawing diagonals $\overline{AC}$ and $\overline{BD}$. Describe your method.

   c. Find the measure of the acute angle formed by the intersection of the diagonals. Describe your method.

8. $\angle A = 50.7°$; $b = 93.3$; $c = 147.3$

9. $\angle A = 18.4°$; $\angle B = 71.6°$; $b = 21$; $c = 22.1$

10. $\angle A = 27.2°$; $a = 8.4$; $b = 16.4$

11. about 69 ft

12. The value of $\tan \theta$ is the slope of the line.

13. a. $12\sqrt{2} \approx 17$

   b, c. Methods may vary. Examples are given.

   b. $\angle DAC = 26.6°$; $\angle CAB = 63.4°$; $\triangle DAC$ and $\triangle CAB$ are right triangles with right angles at D and B. So, $\angle DAC = \tan^{-1}\left(\frac{\sqrt{8}}{\sqrt{32}}\right) = 26.6°$ and $\angle CAB = \tan^{-1}\left(\frac{\sqrt{32}}{\sqrt{8}}\right) = 63.4°$.

   c. 53.2°; By continuing the process in part (b), you can find that two of the angles of each acute triangle have measures of 63.4°. The sum of the angles of a triangle is 180°, so the measure of the third angle is 53.2°.

Teach⇔Interact

Closure Question

Explain when to use each trigonometric ratio to help solve a triangle.

Use the sine ratio when the calculations involve the length of the leg opposite the angle and the length of the hypotenuse. Use the cosine ratio when the calculations involve the length of the leg adjacent to the angle and the length of the hypotenuse. Use the tangent ratio when the calculations involve the length of the leg opposite the angle and the length of the leg adjacent to the angle.

Apply⇔Assess

Suggested Assignment

Core Course
Exs. 1–13, 15, 22–24, AYP

Extended Course
Exs. 1–24, AYP

Block Schedule
Day 1 Exs. 1–13, 15, 22–24, AYP

Exercise Notes

Reasoning
Ex. 1 This exercise points out the relationship between the tangents of complementary angles.

Visual Thinking
Ex. 4 Encourage students to make a sketch that represents the situation described. Ask them to label the angles, the sides, and the angles of depression of the two triangles. This activity involves the visual skills of *identification* and *generalization*.

Communication: Discussion
Exs. 5–10 These exercises can be done in a variety of ways. Students may benefit from a discussion of each other's methods.

Integrating the Strands
Ex. 12 This exercise integrates concepts from geometry and trigonometry. The angle a line makes with the x-axis, in this case θ, is called the *angle of inclination*. The tangent of θ is equal to the slope of the line.

Apply⇔Assess

Exercise Notes

Career Connection

Ex. 14 Students interested in aviation can pursue many career opportunities with commercial airlines or in military service. Many pilots received their initial training in flying airplanes in one of the four branches of the armed services. Others learn to fly by taking private lessons. Pilots of large commercial airliners generally have two or more years of college education and many of them have bachelor or masters degrees in a large variety of fields. Being the pilot of a jetliner today is a complex job that requires both analytical and visual skills in the cockpit. Modern jetliners are highly computerized and pilots need to be able to read and understand the visual displays in the cockpit that provide critical information about the flight of the aircraft.

Challenge

Ex. 15 You may wish to ask students if they can determine the degree measure of angles L and M. ($\angle L = 30°$ and $\angle M = 60°$)

Interview Note

Ex. 17 Students can use similar triangles to solve this exercise. They should notice that $\angle 1 = \angle 2$ because they are alternate interior angles for two parallel lines.

14. **METEOROLOGY** Pilots flying at night need to know the *cloud ceiling*, the lowest height of a mass of clouds. Pilots cannot take off or land if this ceiling is too low. To determine the cloud ceiling at night, a meteorologist shines a searchlight, located a known distance from a weather station, directly up at the clouds. The meteorologist then measures the angle of elevation to the spot of light and finds the height of the clouds. Find the height of the clouds shown in the diagram.

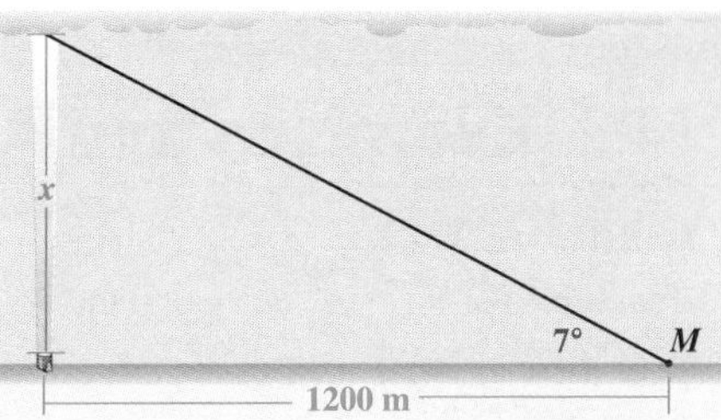
x
7°
M
1200 m
Not drawn to scale

15. Refer to $\triangle LMN$. Name an angle whose tangent is $\sqrt{3}$. Name an angle whose tangent is $\frac{1}{\sqrt{3}}$. Is there an angle of $\triangle LMN$ whose tangent is $\frac{1}{2}$? Explain.

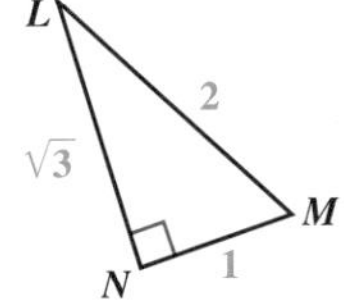
L
2
√3
M
N
1

INTERVIEW **Johnpaul Jones**

Look back at the article on pages 652–654.

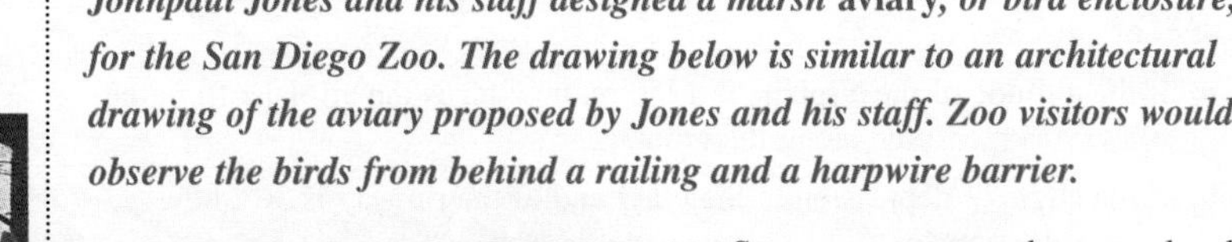

Johnpaul Jones and his staff designed a marsh **aviary,** *or bird enclosure, for the San Diego Zoo. The drawing below is similar to an architectural drawing of the aviary proposed by Jones and his staff. Zoo visitors would observe the birds from behind a railing and a harpwire barrier.*

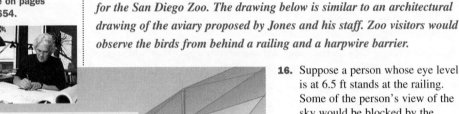

The roof is designed to be high enough to give birds flying room, but not so high that birds can fly out of sight from viewers.
6 ft
1
2
x
9 ft
6.5 ft
y
Not drawn to scale
Tubular steel posts hold up the nylon netting roof.
Harpwire is strung between roof supports.

16. Suppose a person whose eye level is at 6.5 ft stands at the railing. Some of the person's view of the sky would be blocked by the overhang as shown.

a. What would be the vertical distance between the person's eyes and the overhang?

b. What should be the measure of $\angle 1$?

17. Suppose the person in Exercise 16 is 60 ft from a point where a tubular steel post is to be placed, and suppose that the bottom of the post is at the person's eye level.

a. What should be the measure of $\angle 2$ (the angle of elevation determined by the overhang)?

b. How tall should the post be for the person to just see the top of it?

c. Would someone shorter still find this an acceptable height for the pole? Explain.

14. about 147 m
15. $\angle M$; $\angle L$; No; the tangent is the ratio of the legs of a right triangle, but $\overline{LM}$ is the hypotenuse.
16. a. 2.5 ft
    b. 22.6°
17. a. 22.6°
    b. 25 ft
    c. Yes; A shorter person would have a greater angle of elevation from eye to the overhang, and thus a higher line of sight.

Connection ASTRONOMY

Throughout history, people have constructed sundials (or "shadow clocks") to tell time. In the 13th century, a Moroccan scientist named Abu'l Hasan first introduced the idea of pointing the *gnomon*, or shadow pole, toward the North Star. In this way, the hour lines on a sundial could be positioned so that consecutive lines measure equal intervals of elapsed time.

A sundial can be constructed using the directions shown.

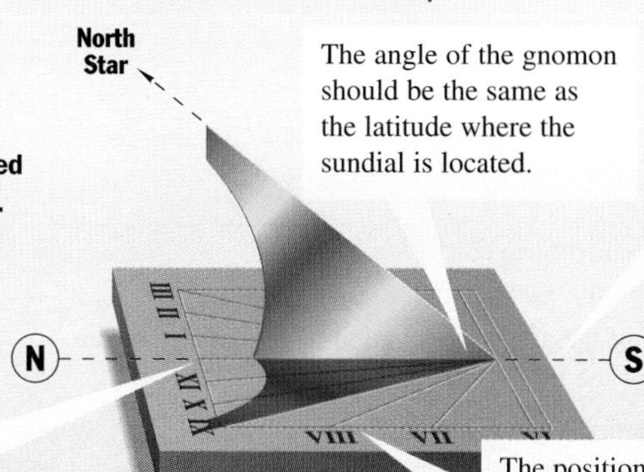
North Star
The angle of the gnomon should be the same as the latitude where the sundial is located.
Place the gnomon on the north-south line.
The noon shadow points directly north. Place the mark for 12:00 noon on the north-south line. The hour angle for noon is 0°.
N
S
The positions of the hour marks around the sundial depend on the latitude where the sundial is located.

18. To position the hour marks around the sundial, you need to determine the shadow angles that correspond to each hour. These angles depend on the sundial's latitude and can be found by this formula:

$$\text{tan (shadow angle)} = \text{sin (latitude)} \times \text{tan (hour angle)}$$

a. Each hour is $\frac{1}{24}$ of a day. Each day, Earth makes a revolution of 360°, so *each hour* Earth turns through an angle of $\frac{1}{24} \cdot 360°$, or 15°. At 2:00 P.M., the hour angle is 30°. What is the hour angle for 5:00 P.M.?

b. What shadow angle corresponds to 5:00 P.M. in New York City, which has a latitude of 41°?

19. Research What is the latitude where you live? If you build a sundial like the one shown, what is the shadow angle on your sundial at 5:00 P.M.?

BY THE WAY

Before Abu'l Hasan, the gnomon of a sundial pointed directly up into the sky. Hours measured on these sundials were called *temporary hours* because their lengths changed with the day of the year.

20. Challenge From point A, at an elevation of about 11,750 ft above sea level, suppose the angle of elevation to the top of Mt. Everest is about 8.7°. From point B, at the same elevation as A but 18,325 ft farther away from the mountain than A, suppose the angle of elevation is about 7.5°. How high is Mt. Everest?

Not drawn to scale
B
7.5°
A
8.7°
18,325 ft

18. a. 75°
 b. 67.8°
19. Answers will vary.
20. about 29,030 ft

Apply⇔Assess

Exercise Notes

Communication: Discussion
Ex. 18 Students may have difficulty understanding the "hour angle" for this exercise. Point out that part (a) establishes an angle of 15° for each hour. Thus, an hour angle equals the number of hours past 12:00 times 15°.

Research
Exs. 18, 19 Students interested in astronomy may wish to research the origins of the sundial and the role it has played throughout history to tell time. The oldest known sundial was made in Egypt about 1500 B.C.

Historical Connection
Ex. 20 Mt. Everest is named for George Everest who helped complete the Great Trigonometrical Survey of India in the 19th century. See page 690 more more information on the Great Trigonometrical Survey.

Apply⇔Assess

Exercise Notes

Assessment Note

Ex. 21 Asking students to show algebraically that the tangent of an angle is the same as the ratio of the sine of the angle to the cosine of the angle can help them to assess their understanding of these ratios and how they are related.

Teaching Tip

Ex. 21 A common mnemonic to help students remember the sine, cosine, and tangent ratios is *sohcahtoa*: **s**ine is **o**pposite over **h**ypotenuse, **c**osine is **a**djacent over **h**ypotenuse, and **t**angent is **o**pposite over **a**djacent.

Progress Check 14.1–14.2

See page 700.

Practice 90 for Section 14.2

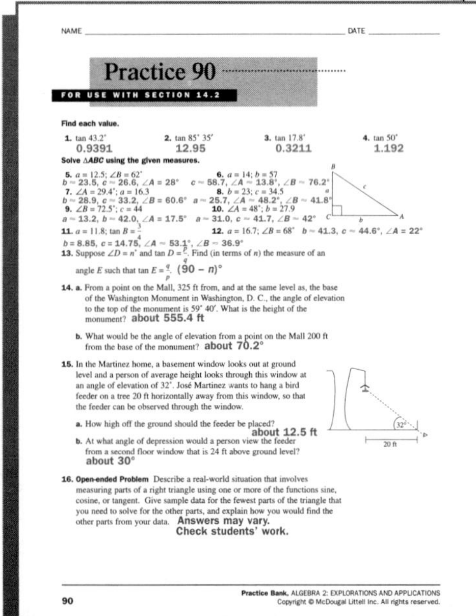

NAME ______________________ DATE __________

Practice 90

FOR USE WITH SECTION 14.2

Find each value.

1. tan 43.2° **0.9391**
2. tan 85° 35′ **12.95**
3. tan 17.8° **0.3211**
4. tan 50° **1.192**

Solve △ABC using the given measures.

5. $a = 12.5$; $\angle B = 62°$ **$b \approx 23.5$, $c \approx 26.6$, $\angle A = 28°$**
6. $a = 14$; $b = 57$ **$c \approx 58.7$, $\angle A \approx 13.8°$, $\angle B \approx 76.2°$**
7. $\angle A = 29.4°$; $a = 16.3$ **$b \approx 28.9$, $c \approx 33.2$, $\angle B = 60.6°$**
8. $b = 23$; $c = 34.5$ **$a \approx 25.7$, $\angle A \approx 48.2°$, $\angle B \approx 41.8°$**
9. $\angle B = 72.5°$; $c = 44$ **$a \approx 13.2$, $b \approx 42.0$, $\angle A = 17.5°$**
10. $\angle A = 48°$; $b = 27.9$ **$a \approx 31.0$, $c \approx 41.7$, $\angle B \approx 42°$**
11. $a = 11.8$; $\tan B = \frac{3}{4}$ **$b = 8.85$, $c = 14.75$, $\angle A \approx 53.1°$, $\angle B \approx 36.9°$**
12. $a = 16.7$; $\angle B = 68°$ **$b \approx 41.3$, $c \approx 44.6°$, $\angle A = 22°$**
13. Suppose $\angle D = n°$ and $\tan D = \frac{p}{q}$. Find (in terms of n) the measure of an angle E such that $\tan E = \frac{q}{p}$. **$(90 - n)°$**
14. a. From a point on the Mall, 325 ft from, and at the same level as, the base of the Washington Monument in Washington, D. C., the angle of elevation to the top of the monument is 59° 40′. What is the height of the monument? **about 555.4 ft**

    b. What would be the angle of elevation from a point on the Mall 200 ft from the base of the monument? **about 70.2°**
15. In the Martinez home, a basement window looks out at ground level and a person of average height looks through this window at an angle of elevation of 32°. José Martinez wants to hang a bird feeder on a tree 20 ft horizontally away from this window, so that the feeder can be observed through the window.

    a. How high off the ground should the feeder be placed? **about 12.5 ft**

    b. At what angle of depression would a person view the feeder from a second floor window that is 24 ft above ground level? **about 30°**
16. **Open-ended Problem** Describe a real-world situation that involves measuring parts of a right triangle using one or more of the functions sine, cosine, or tangent. Give sample data for the fewest parts of the triangle that you need to solve for the other parts, and explain how you would find the other parts from your data. **Answers may vary. Check students' work.**

90 **Practice Bank,** ALGEBRA 2: EXPLORATIONS AND APPLICATIONS

ONGOING ASSESSMENT

21. **Writing** Use ratios to help you explain this alternative definition of tan A:

$$\tan A = \frac{\sin A}{\cos A}$$

SPIRAL REVIEW

22. **SPORTS** It takes only about 5 min to travel up the *Extreme Access* chair lift at the Copper Mountain Resort ski area in Colorado. *(Section 14.1)*

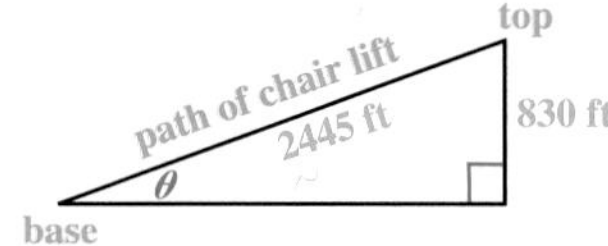

    a. How fast does the chair lift travel in feet per minute? in miles per hour?

    b. Find θ, the measure of the angle of elevation from the base of the chair lift to the top of the lift.

23. Tell whether each sequence is *geometric* or *arithmetic*. State the common difference or common ratio for each sequence. *(Section 10.2)*

    a. 35, 125, 215, 305, . . .   b. 0.5234, −0.5234, 0.5234, −0.5234, . . .

24. Name the vertex and the line of symmetry for the graph of the equation $y = 3x^2 + 6x + 2$. *(Section 5.2)*

ASSESS YOUR PROGRESS

VOCABULARY

sine (p. 656)	**solving a triangle** (p. 665)
cosine (p. 656)	**angle of depression** (p. 665)
tangent (p. 663)	**angle of elevation** (p. 665)

1. For $\triangle ABC$ with right $\angle C$, find cos A if $a = 5$ m and $b = 12$ m. *(Section 14.1)*
2. **NAVIGATION** An observer at the top of a lighthouse 100 m above sea level spots a boat at sea. *(Section 14.2)*

    a. The angle of depression of the boat is 28°. How far is the boat from the shore?

    b. Suppose a second boat, in line with the observer and the first boat, is sighted far away at an angle of depression of 12°. How far apart are the two boats?

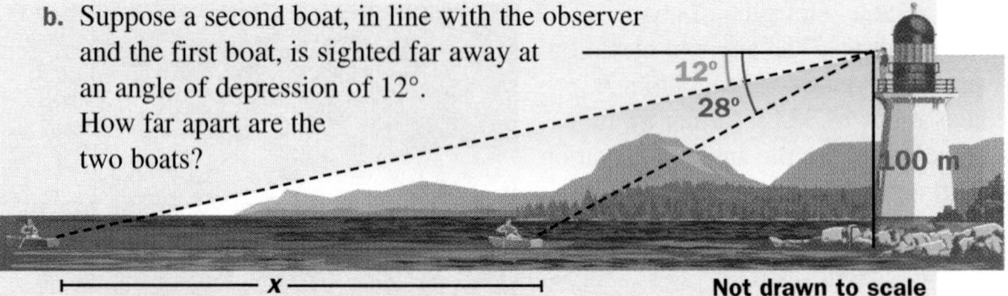

3. **Journal** Give specific examples of situations where you can use trigonometry to find unknown side lengths or angle measures of right triangles.

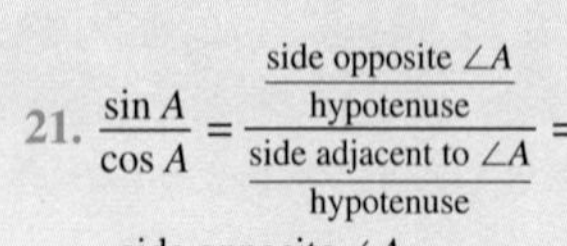

21. $\dfrac{\sin A}{\cos A} = \dfrac{\frac{\text{side opposite } \angle A}{\text{hypotenuse}}}{\frac{\text{side adjacent to } \angle A}{\text{hypotenuse}}} = \dfrac{\text{side opposite } \angle A}{\text{side adjacent to } \angle A} = \tan A$

22. a. 489 ft/min; about 5.6 mi/h

    b. 19.8°

23. a. arithmetic; common difference = 90

    b. geometric; common ratio = −1

24. (−1, −1); $x = -1$

Assess Your Progress

1. $\frac{12}{13}$
2. a. about 188 m

    b. about 282 m
3. Answers may vary. An example is given. You could use the tangent ratio to calculate the height of a building or the angle of descent toward a runway.

Section

14.3 Angles of Rotation

Learn how to...

- **find the trigonometric functions of an angle of rotation**

So you can...

- **convert courses to angles of rotation and find the coordinates of a plane's position, for example**

The direction that an airplane or ship travels relative to north is called its *course*. Courses can be any angle between 0° and 360°. The navigator of a plane or ship measures angles in a *clockwise* direction from north.

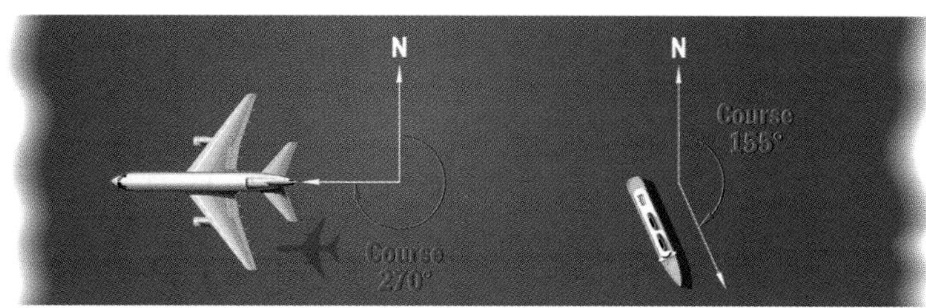

THINK AND COMMUNICATE

1. a. What course corresponds to a ship traveling northwest? southeast?

b. In general, what mathematical relationship holds for the courses of two ships traveling in opposite directions?

2. If a plane flies from Chicago to St. Louis on a course of 205° as shown, what is its return course? Explain.

Chicago
205°
N
?
St. Louis

Mathematicians create angles between 0° and 360° in the coordinate plane by rotating a ray in a *counterclockwise* direction from the positive x-axis, which is called the **initial side** of an angle. The other ray that forms the angle is its **terminal side**. This is known as the **standard position** of an angle.

This is a **Quadrant III angle** because the terminal side lies in Quadrant III.

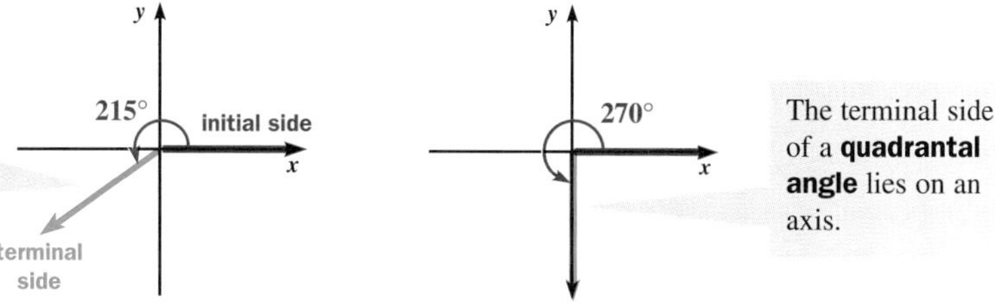

The terminal side of a **quadrantal angle** lies on an axis.

THINK AND COMMUNICATE

3. Name a Quadrant II angle and a Quadrant IV angle.

4. Name a quadrantal angle other than the one shown above.

Plan⇔Support

Objectives

- Find the trigonometric functions of an angle of rotation.
- Convert course angles to angles of rotation.
- Use angles of rotation and course angles to solve problems.

Recommended Pacing

Core and Extended Courses
Section 14.3: 1 day

Block Schedule
Section 14.3: $\frac{1}{2}$ block
(with Section 14.4)

Resource Materials

Lesson Support
Lesson Plan 14.3
Warm-Up Transparency 14.3
Overhead Visuals:
Folder A: Multi-Use Graphing Packet, Sheet 1
Practice Bank: Practice 91
Study Guide: Section 14.3
Explorations Lab Manual:
Diagram Master 2

Technology
Graphing Calculator
McDougal Littell Mathpack
Geometry Inventor Activity Book: Activity 20
Internet:
http://www.hmco.com

Warm-Up Exercises

1. Name the quadrant containing each of the following points: (−3, 6), (4, −2), (−7, −1)
II; IV; III

2. Find sin A, cos A, and tan A if $A = 20°$.
0.3420; 0.9397; 0.3640

3. Find a value of θ if $\sin \theta = 0.8660$. 60°

4. Find a value of θ if $\cos \theta = 0.7660$. 40°

5. Find a value of θ if $\tan \theta = 1$. 45°

ANSWERS Section 14.3

Think and Communicate

1. a. 315°; 135°
   b. They differ by 180°.
2. 25°; The plane is going in the opposite direction, so its course must differ by 180° from the original course.
3. Answers may vary. Acceptable ranges are given; any angle between 90° and 180°; any angle between 270° and 360°.
4. 0°, 90°, 180°, or 360°

Teach⇔Interact

Section Notes

Communication: Discussion Some students may have difficulty remembering that rotating a ray *counterclockwise* gives a positive angle, while a direction angle corresponding to courses is positive when it is measured *clockwise* from the north. A thorough discussion of the diagrams on this page should help alleviate this problem.

Teaching Tip
Point out that the legs of a triangle really do not have lengths that are negative or 0, but since the coordinates of the point $P(x, y)$ corresponds to a leg of length x and a leg of length y, it is convenient to give these "lengths" a sign.

Additional Example 1

Find the sine, cosine, and tangent of each angle.

a. The terminal side of θ_1 passes through $(3, -3\sqrt{3})$.

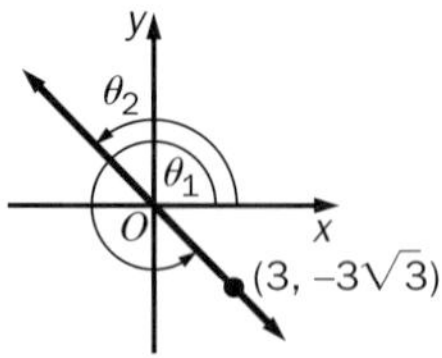

Find the value of r and then use the definitions of the trigonometric functions.

$r = \sqrt{3^2 + (-3\sqrt{3})^2} = \sqrt{36} = 6$

$\sin \theta_1 = \frac{y}{r} = \frac{-3\sqrt{3}}{6} = \frac{-\sqrt{3}}{2} \approx -0.8660$

$\cos \theta_1 = \frac{x}{r} = \frac{3}{6} = \frac{1}{2} = 0.5$

$\tan \theta_1 = \frac{y}{x} = \frac{-3\sqrt{3}}{3} = -\sqrt{3}$

b. The measure of θ_2 is 145°.
Use a calculator set in degree mode.

```
sin 145
     .5735764364
cos 145
    -.8191520443
tan 145
    -.7002075382
```

Therefore:
sin 145° ≈ 0.5736
cos 145° ≈ −0.8192
tan 145° ≈ −0.7002

Trigonometric Functions

The *trigonometric functions* of an angle θ between 0° and 360° are defined by placing θ in standard position in a coordinate plane and choosing a point P on the terminal side of θ. P must be a point other than the origin.

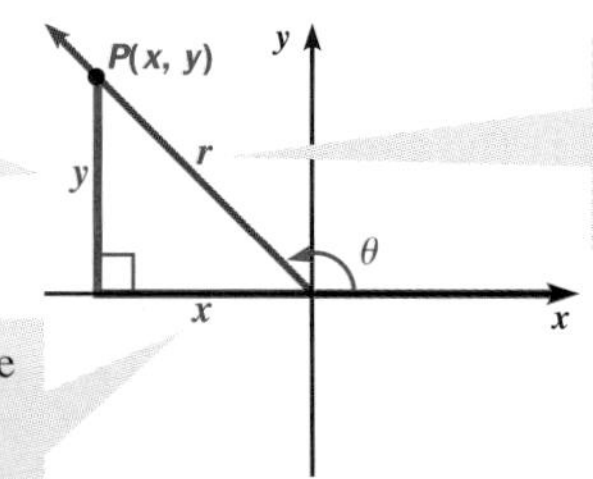

Dropping a perpendicular from P to the x-axis forms a right triangle.

The hypotenuse has length $r = \sqrt{x^2 + y^2}$, which is always positive.

The legs of the triangle have "lengths" that can be positive, negative, or 0.

The trigonometric functions of θ are defined in terms of the x- and y-coordinates of P together with r, the distance between P and the origin.

WATCH OUT!
The ratio $\frac{y}{x}$ is undefined when $x = 0$, so $\tan \theta$ is undefined when $\theta = 90°$ or $\theta = 270°$.

Trigonometric Functions

If θ is an angle in standard position with the point $P(x, y)$ on its terminal side and if r is the distance between P and the origin, then:

$$\sin \theta = \frac{y}{r} \qquad \cos \theta = \frac{x}{r} \qquad \tan \theta = \frac{y}{x}$$

EXAMPLE 1

Find the sine, cosine, and tangent of each angle.

a. The terminal side of θ_1 passes through $(-2, 2)$.

b. The measure of θ_2 is 295°.

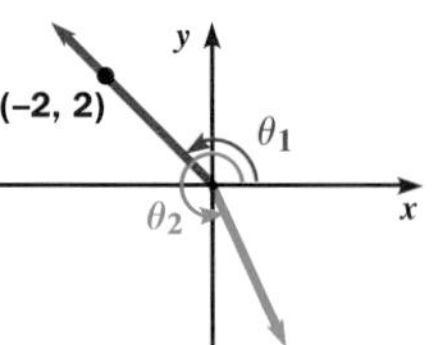

SOLUTION

a. Find the value of r and then use the definitions given above.

$r = \sqrt{(-2)^2 + 2^2} = \sqrt{8} = 2\sqrt{2}$

$\sin \theta_1 = \frac{y}{r} = \frac{2}{2\sqrt{2}} = \frac{1}{\sqrt{2}} \approx 0.7071$

$\cos \theta_1 = \frac{x}{r} = \frac{-2}{2\sqrt{2}} = \frac{-1}{\sqrt{2}} \approx -0.7071$

$\tan \theta_1 = \frac{y}{x} = \frac{2}{-2} = -1$

b. Use a calculator set in degree mode.

```
sin 295
    -.906307787
cos 295
     .4226182617
tan 295
    -2.144506921
```

Therefore:

sin 295° ≈ −0.9063

cos 295° ≈ 0.4226

tan 295° ≈ −2.145

Notice that the signs of the x- and y-coordinates of the point in part (a) of Example 1 determined whether the sine, cosine, and tangent values of the angle were positive or negative. The trigonometric functions can be positive or negative depending on the quadrant in which the terminal side lies.

The table at the right and the coordinate plane below show which trigonometric functions are positive in each quadrant.

I	II	III	IV
all	sine	tangent	cosine

Quadrant II:
$x < 0$, $y > 0$
$90° < \theta < 180°$
$\sin \theta > 0$
$\cos \theta < 0$
$\tan \theta < 0$

Quadrant I:
$x > 0$, $y > 0$
$0° < \theta < 90°$
$\sin \theta > 0$
$\cos \theta > 0$
$\tan \theta > 0$

Quadrant III:
$x < 0$, $y < 0$
$180° < \theta < 270°$
$\sin \theta < 0$
$\cos \theta < 0$
$\tan \theta > 0$

Quadrant IV:
$x > 0$, $y < 0$
$270° < \theta < 360°$
$\sin \theta < 0$
$\cos \theta > 0$
$\tan \theta < 0$

You may find it helpful to remember this sentence:
All scholars take calculus.

The first letter of each word tells you which trigonometric functions are positive in each quadrant:
A for all (Q. I)
s for sine (Q. II)
t for tangent (Q. III)
c for cosine (Q. IV)

EXAMPLE 2 Application: Aviation

After leaving an airport, a plane travels on a course of 160° for 300 mi.

a. What angle of rotation is equivalent to the plane's course?

b. Write and solve two equations relating the plane's x- and y-coordinates to the angle of rotation found in part (a).

c. How far has the plane traveled in an east-west direction? in a north-south direction?

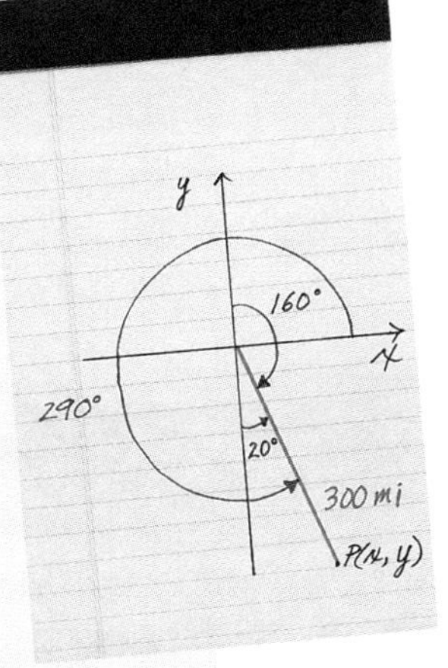

SOLUTION

a. Imagine the airport at the origin of a coordinate plane. The positive y-axis represents north, and the positive x-axis represents east. Make a sketch of the course and the corresponding angle of rotation as shown. The terminal side of a **160° course** is in Quadrant IV. The **angle of rotation** is $270° + 20° = 290°$.

b. $\cos 290° = \dfrac{x}{300}$ $\qquad$ $\sin 290° = \dfrac{y}{300}$

$$x = 300 \cos 290° \qquad y = 300 \sin 290°$$
$$\approx 300(0.3420) \qquad \approx 300(-0.9397)$$
$$\approx 102.6 \qquad \approx -281.9$$

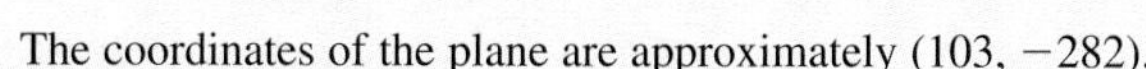

The coordinates of the plane are approximately $(103, -282)$.

c. The plane has traveled about 103 mi east and about 282 mi south.

Teach⇔Interact

Section Note

Communication: Discussion Students need to make the connection between the coordinates of a point on the terminal side of the angle and the sign of the trigonometric functions. Discuss this thoroughly. Make sure students understand that the cosine function corresponds to the x-value of the point, while the sine function corresponds to the y-value of the point. Therefore, the sign of the function comes from the sign of the coordinate it corresponds to.

Additional Example 2

After leaving an airport, a plane travels on a course of 290° for 150 mi.

a. What angle of rotation is equivalent to the plane's course?
Imagine the airport at the origin of a coordinate plane. The positive y-axis represents north, and the positive x-axis represents east. Make a sketch of the course and the corresponding angle of rotation as shown. The terminal side of a 290° course is in Quadrant II. The angle of rotation is $180° - 20° = 160°$.

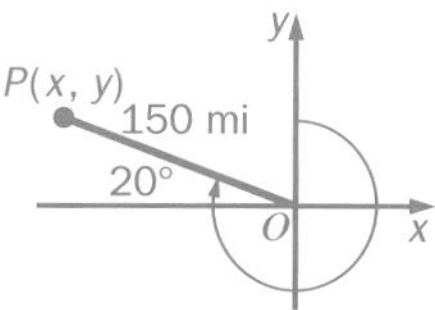

b. Write and solve two equations relating the plane's x- and y-coordinates to the angle of rotation found in part (a).

$$\cos 160° = \frac{x}{150}$$
$$x = 150 \cos 160°$$
$$\approx 150(-0.9397)$$
$$\approx -141$$
$$\sin 160° = \frac{y}{150}$$
$$y = 150 \sin 160°$$
$$\approx 150\,(0.3420)$$
$$\approx 51.3$$

The coordinates of the plane are approximately $(-141, 51)$.

c. How far has the plane traveled in an east-west direction? in a north-south direction?
The plane has traveled about 141 mi west and 51 mi north.

Teach⇔Interact

Section Note

Teaching Tip

Emphasize that the trigonometric values repeat themselves except for their sign just as the *x*- and *y*-coordinates repeat themselves except for their sign. The trigonometric values depend on the coordinate or angle, while the sign of the value depends on the quadrant containing the angle.

About Example 3

Reasoning

Students should understand that Quadrant II contains another angle with the same sine value because Quadrant II contains another point with the same *y*-coordinate. Point out that every second quadrant point has the same *y*-coordinate but the opposite *x*-coordinate as its reflection point in the *y*-axis (a first quadrant point).

Additional Example 3

Find all angles θ between 0° and 360° such that $\cos\theta = 0.7193$.

Find a Quadrant I angle using the inverse cosine on a calculator.

$\theta = \cos^{-1} 0.7193 \approx 44°$

Another angle with the same cosine value is in Quadrant IV.

$360° - \theta \approx 360° - 44° = 316°$

So, $\theta \approx 44°$ or $\theta \approx 316°$.

Closure Question

Describe how to find the trigonometric functions of an angle of rotation. Choose a point (x, y) on the terminal side of the angle of rotation. Construct a perpendicular from the point to the *x*-axis and find the hypotenuse of the triangle, the *r*-value. Then the sine of the angle is $\frac{y}{r}$, the cosine of the angle is $\frac{x}{r}$, and the tangent of the angle is $\frac{y}{x}$.

THINK AND COMMUNICATE

5. In part (a) of Example 2, why was 20° added to 270° to get the angle of rotation?

Consider an angle θ with its terminal side in Quadrant I. If you reflect that terminal side over the axes, you will get other angles whose trigonometric values are related to those for θ, as shown below.

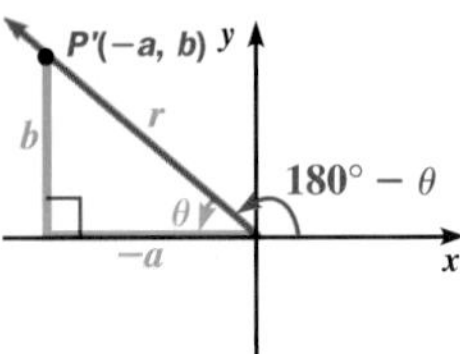

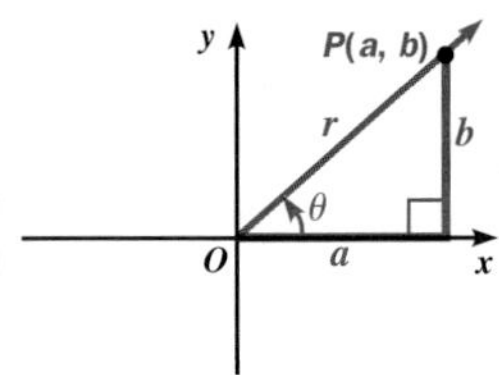

The *x*- and *y*-coordinates of each reflected point ***P′*** are the **same as or the opposite of** the coordinates of ***P***.

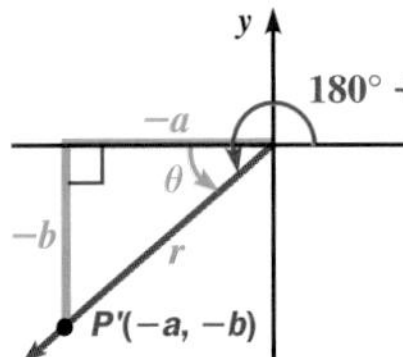

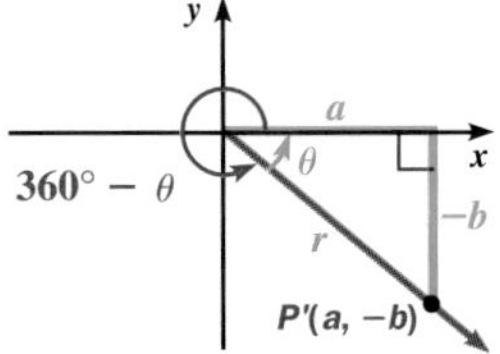

EXAMPLE 3

Find all angles θ between 0° and 360° such that $\sin\theta = 0.5878$.

SOLUTION

Find a Quadrant I angle using inverse sine on a calculator.

$\theta = \sin^{-1} 0.5878 \approx 36°$

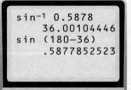

```
sin⁻¹ 0.5878
     36.00104446
sin (180−36)
     .5877852523
```

Another angle with the same sine value is in Quadrant II.

$180° - \theta \approx 180° - 36° = 144°$

So, $\theta \approx 36°$ or $\theta \approx 144°$.

CHECKING KEY CONCEPTS

1. In what quadrant is the terminal side of each angle?

a. 150° **b.** 252° **c.** 76°

2. An angle θ is in standard position with the given point P on its terminal side. Find $\sin\theta$, $\cos\theta$, and $\tan\theta$.

a. $P(-12, -5)$ **b.** $P(1, \sqrt{2})$ **c.** $P(24, -7)$

3. Find all angles θ between 0° and 360° such that $\tan\theta = 2.246$.

Think and Communicate

5. because a course of 160° is 20° east of due south

Checking Key Concepts

1. a. II
   b. III
   c. I
2. a. −0.3846; −0.9231; 0.4167
   b. 0.8165; 0.5774; 1.414
   c. −0.28; 0.96; −0.2917
3. about 66°, about 246°

Exercises and Applications

Angle measures are rounded to the nearest 0.1 degree and trigonometric ratios are rounded to four significant digits.

1.

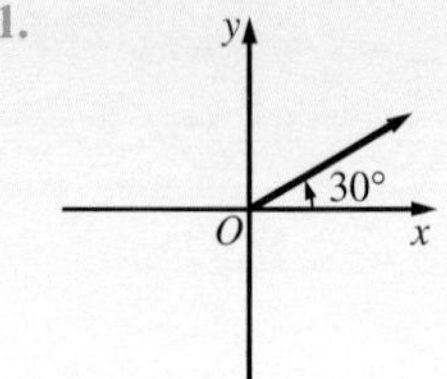

2.

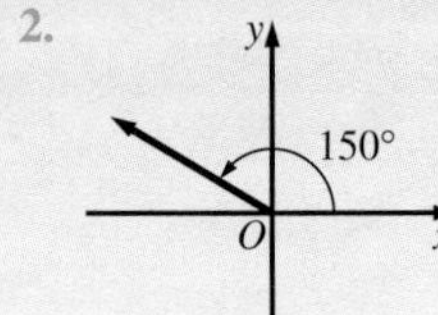

3.

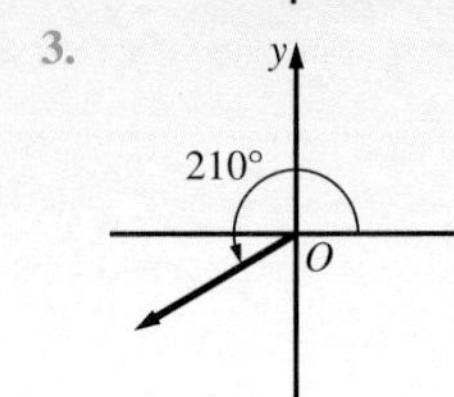

14.3 Exercises and Applications

Extra Practice exercises on page 770

Sketch each angle of rotation in a coordinate plane.

1. $30°$ 2. $150°$ 3. $210°$ 4. $330°$

5. A snorkeler swam away from an anchored boat on a course of 25°. What course should the snorkeler use to swim back to the boat?

6. a. Name all the quadrantal angles from 0° to 360°.
   b. What quadrantal angles correspond to courses that are exactly north, south, east, and west?
   c. Find the sine and cosine of each quadrantal angle you listed in part (a).

7. **DANCING** In ballroom dancing, some of the moves are described as "quarter turns."
   a. What is the measure of a quarter turn?
   b. If a pair of dancers has completed three quarter turns, through what angle have they rotated?

8. **ARCHITECTURE** The Space Needle is a 605 ft tall tower built for the 1962 World's Fair in Seattle, Washington. The dining section of a circular restaurant near the top of the Space Needle rotates, completing one revolution each hour.

   a. Suppose your seat in the restaurant is near a window. Through what angle have you rotated in 40 min when you move from a view of Mt. Rainier to a view of Puget Sound?
   b. Suppose your seat at the restaurant has rotated through an angle of 84°. How long have you been sitting there?

An angle θ is in standard position with the given point P on its terminal side. Find $\sin\theta$, $\cos\theta$, and $\tan\theta$.

9. $P(8, 15)$ 10. $P(3, -3\sqrt{3})$ 11. $P(-4, -3)$ 12. $P(-100, 100)$

For each angle θ, use a calculator to find $\sin\theta$, $\cos\theta$, and $\tan\theta$.

13. $\theta = 192°$ 14. $\theta = 302°$ 15. $\theta = 111.6°$ 16. $\theta = 24°16'45''$

17. **Writing** Suppose an angle θ has its terminal side in Quadrant IV. Explain in your own words how you know that $\tan\theta$ is negative.

18. What is the relationship between $\sin\theta$ and $\sin(180° + \theta)$? Explain.

19. Find two values of θ for each trigonometric value below. Round your answers to the nearest degree.
   a. $\sin\theta = 0.8660$ b. $\cos\theta = 0.7660$ c. $\tan\theta = 2.475$

20. Use the results of Exercise 19 to predict two values of θ for each trigonometric value given. Check your answers with a calculator.
   a. $\sin\theta = -0.8660$ b. $\cos\theta = -0.7660$ c. $\tan\theta = -2.475$

Apply⇔Assess

Suggested Assignment

Core Course
Exs. 1–21, 27–30

Extended Course
Exs. 1–30

Block Schedule
Day 2 Exs. 1–21, 27–30

Exercise Notes

Application
Ex. 8 Part (a) illustrates that when a real-life situation using angles is considered, a student has to decide if the angle is measured counterclockwise or clockwise. In this case, the restaurant could rotate either way and the angle is still $\frac{2}{3}$ of 360°, or 240°.

Cooperative Learning
Exs. 17–20 These exercises can be used to provide a review and summary of the basic concepts of this section. Have students work in groups to complete the exercises and discuss their answers.

Visual Thinking
Ex. 18 Ask students to create a sketch in a coordinate plane of this situation. Ask them to identify the two locations described on their sketch. Encourage them to discuss additional questions that can be answered by using their sketches. This activity involves the visual skills of *identification* and *exploration.*

4.

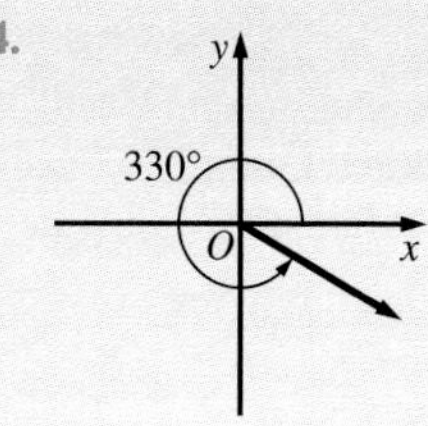

5. 205°

6. a. 0° (or 360°), 90°, 180°, 270°
   b. 90°; 270°; 0° (or 360°); 180°
   c. $\sin 0° = 0$, $\cos 0° = 1$; $\sin 90° = 1$, $\cos 90° = 0$; $\sin 180° = 0$, $\cos 180° = -1$; $\sin 270° = -1$, $\cos 270° = 0$

7. a. 90° b. 270°

8. a. 240° b. 14 min

9. 0.8824; 0.4706; 1.875

10. –0.8660; 0.5; –1.732

11. –0.6; –0.8; 0.75

12. 0.7071; –0.7071; –1

13. –0.2079; –0.9781; 0.2126

14. –0.8480; 0.5299; –1.600

15. 0.9298; –0.3681; –2.526

16. 0.4112; 0.9116; 0.4511

17. A point on the terminal side of an angle in Quadrant IV will have a positive x-coordinate and a negative y-coordinate, so $\tan\theta$ is negative.

18. $\sin(180° + \theta) = -\sin\theta$; Adding 180° to θ reverses the direction, which reverses the sign of the y-coordinate, reversing the value of $\sin\theta$.

19. a. 60°; 120°
   b. 40°; 320°
   c. 68°; 248°

20. a. 240°; 300°
   b. 140°; 220°
   c. 112°; 292°

Apply⇔Assess

Exercise Notes

Teaching Tip

Ex. 22 Remind students that a ship uses course angles (measured clockwise) to set its direction. To express the direction in terms of quadrants, students need to give the equivalent angle of rotation.

Application

Ex. 23 The science of robotics has grown increasingly important since the first industrial robot was invented in 1976 by NASA for use aboard the space shuttle. Robots are used in many areas that would pose some danger to a human being. They are also used in industrial processes to increase both productivity and quality control. Stress the importance of specifying a set of mathematically correct instructions to a computer-driven robot. A computer is normally programmed to know that a counterclockwise direction corresponds to a positive angle of rotation. An air traffic controller, on the other hand, uses computers that consider course angles (clockwise) to be positive.

Mathematical Procedures

Ex. 23 The instructions in this exercise are an algorithm. They must be followed in the exact order for the given result to occur. Ask students to give other examples of algorithms, either mathematical or real-world.

Problem Solving

Ex. 24 Ask students to describe how they found the total distance traveled in part (b), and explain any examples they used to the class.

21. NAVIGATION After leaving a harbor, a ship travels on a course of 190° for 98 mi.

a. What angle of rotation is equivalent to the ship's course?

b. Write and solve two equations relating the ship's x- and y-coordinates to the angle of rotation found in part (a).

c. How far has the boat traveled in an east-west direction? in a north-south direction?

22. Challenge A ship travels 38 mi on a course of θ degrees for 1 h. Find formulas for the distance d traveled north or south as a function of θ. Consider different quadrants.

Connection ROBOTICS

Robots are useful in such diverse fields as manufacturing, office automation, and space and underwater exploration. A robot is sometimes controlled by a human operator, who gives the robot instructions to follow. These instructions can take the form of turn angles and distances.

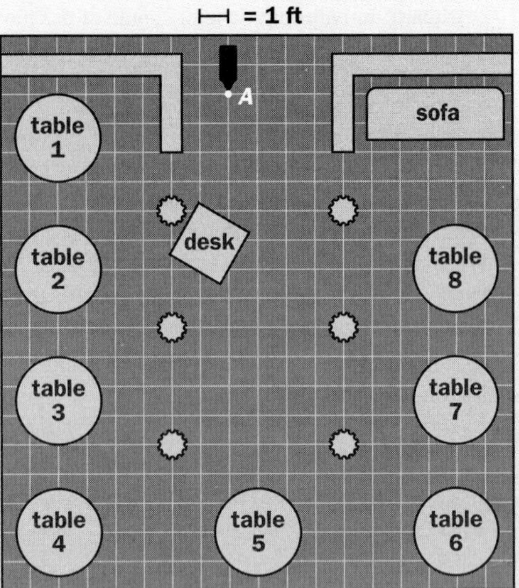

Suppose a robot can move only forward and must first turn its body to face the direction it needs to travel. The diagram shows the obstacles that the robot must avoid while moving around a room.

23. Suppose the robot travels from its current location A to another location B by following the given set of instructions. Each of the turns described is *counterclockwise* from the direction that the robot faces at any given time.

a. Copy the diagram. Sketch the robot and obstacles. Determine the robot's location after carrying out the instructions.

b. If the robot can make only *clockwise* turns, how would the instructions to get from A to B change?

```
From location A
turn 27°, move 4.5 ft;
turn 333°, move 4.0 ft;
turn 288°, move 6.5 ft;
  to reach location B.
```

24. Sometimes it is important for a robot to complete instructions in the least amount of time or to travel the shortest distance possible.

a. **Open-ended Problem** Choose what you think is a shorter path from location A to location B. Write a set of instructions using counterclockwise turns to get the robot from A to B.

b. Find the total distance traveled for your path in part (a) and the path described in Exercise 23. Is your new path shorter than the one in Exercise 23?

21. a. 260°

b. $\cos 260° = \frac{x}{98}$; about –17; $\sin 260° = \frac{y}{98}$; about –97

c. about 17 mi; about 97 mi

22. Let θ represent the course of the ship. Then in Quadrant I, $d = 38 \sin (90° - \theta)$, and in Quadrants II, III, and IV, $d = 38 \sin (450° - \theta)$. If d is positive, the distance is northward, and if d is negative, the distance is southward. (Or, $d = 38 \cos \theta$ for all four quadrants.)

23. a. on the right side of Table 3

b. All the angles would change. To find the new angles, subtract the previous angles from 360°.

24. a. Answers may vary. An example is given. From location A, turn 320°, walk 73 ft; turn 40°, walk 4 ft.

b. 15 ft, 11.3 ft; Yes.

25. Cooperative Learning Work in a group of three students. You will design a treasure map using a protractor and a compass.

a. Find or draw a map of some open area such as the schoolyard, the town green, a golf course, or a local park.

b. Put a starting point on your map and secretly choose a location for the treasure.

c. Write directions to the treasure in terms of courses and distances with no fewer than four moves and no more than eight moves.

d. Exchange your map and directions with another group. Use the process shown in Example 2 on page 673 to separate each course and distance you receive into north-south and east-west components. Then combine all the north-south components and all the east-west components to find out the location of the treasure on the map you receive.

e. Compare your group's answer with the treasure's location chosen by the group that created the map. Were you successful in finding the treasure?

26. Research Find out about a famous journey in history or a treasure map in literature. Discuss how to get to the goal using courses and distances. Then separate these courses and distances into north-south and east-west components.

ONGOING ASSESSMENT

27. Writing Describe how you can find all angles θ between 0° and 360° whose tangent is $\frac{3}{4}$. Include sketches of the angles in a coordinate plane.

SPIRAL REVIEW

28. Solve each triangle. *(Section 14.2)*

a.

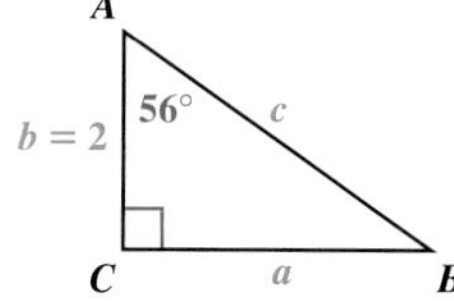

b.

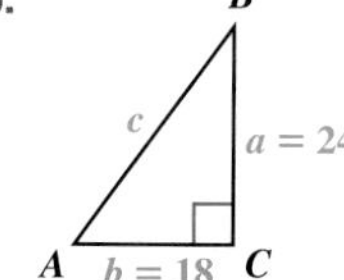

29. Find the area of each triangle described. *(Toolbox, page 802)*

a. base length = 17 ft; height = 24 ft

b. base length = $3\sqrt{2}$ cm; height = $\sqrt{6}$ cm

30. Suppose you roll a six-sided die three times. What is the probability of getting at least one 6? *(Section 13.2)*

Apply⇔Assess

Exercise Notes

Reasoning

Ex. 25 Ask students to explain why adding all the north-south components and all the east-west components in part (d) can give the mathematical location of the treasure.

Application

Ex. 25 This exercise illustrates how trigonometry can be used in navigation to locate reference points.

Assessment Note

Ex. 27 Having three or four students show their sketches to the class would provide useful feedback for all students to evaluate what they have written.

Practice 91 for Section 14.3

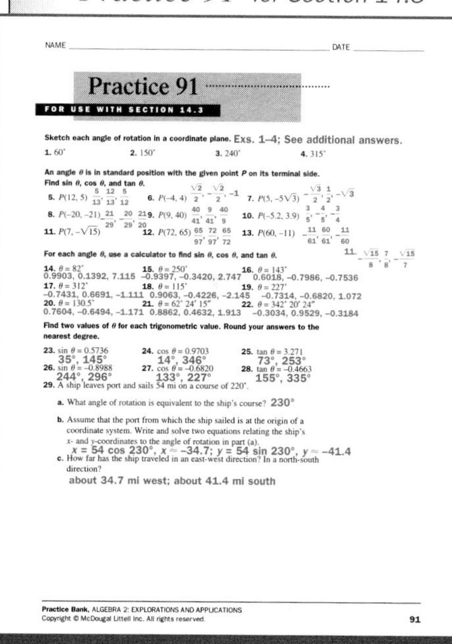

NAME ______ DATE ______

Practice 91

FOR USE WITH SECTION 14.3

Sketch each angle of rotation in a coordinate plane. Exs. 1–4; See additional answers.

1. 60° 2. 150° 3. 240° 4. 315°

An angle θ is in standard position with the given point P on its terminal side. Find $\sin\theta$, $\cos\theta$, and $\tan\theta$.

5. $P(12, 5)$ $\frac{5}{13}, \frac{12}{13}, \frac{5}{12}$ 6. $P(-4, 4)$ $\frac{\sqrt{2}}{2}, -\frac{\sqrt{2}}{2}, -1$ 7. $P(5, -5\sqrt{3})$ $-\frac{\sqrt{3}}{2}, \frac{1}{2}, -\sqrt{3}$

8. $P(-20, -21)$ $-\frac{21}{29}, -\frac{20}{29}, \frac{21}{20}$ 9. $P(9, 40)$ $\frac{40}{41}, \frac{9}{41}, \frac{40}{9}$ 10. $P(-5.2, 3.9)$ $\frac{3}{5}, -\frac{4}{5}, -\frac{3}{4}$

11. $P(7, -\sqrt{15})$ 12. $P(72, 65)$ $\frac{65}{97}, \frac{72}{97}, \frac{65}{72}$ 13. $P(60, -11)$ $-\frac{11}{61}, \frac{60}{61}, -\frac{11}{60}$

11. $-\frac{\sqrt{15}}{8}, \frac{7}{8}, -\frac{\sqrt{15}}{7}$

For each angle θ, use a calculator to find $\sin\theta$, $\cos\theta$, and $\tan\theta$.

14. $\theta = 82°$ 0.9903, 0.1392, 7.115 15. $\theta = 250°$ −0.9397, −0.3420, 2.747 16. $\theta = 143°$ 0.6018, −0.7986, −0.7536

17. $\theta = 312°$ −0.7431, 0.6691, −1.111 18. $\theta = 115°$ 0.9063, −0.4226, −2.145 19. $\theta = 227°$ −0.7314, −0.6820, 1.072

20. $\theta = 130.5°$ 0.7604, −0.6494, −1.171 21. $\theta = 62° 24' 15''$ 0.8862, 0.4632, 1.913 22. $\theta = 342° 20' 24''$ −0.3034, 0.9529, −0.3184

Find two values of θ for each trigonometric value. Round your answers to the nearest degree.

23. $\sin\theta = 0.5736$ 35°, 145° 24. $\cos\theta = 0.9703$ 14°, 346° 25. $\tan\theta = 3.271$ 73°, 253°

26. $\sin\theta = -0.8988$ 244°, 296° 27. $\cos\theta = -0.6820$ 133°, 227° 28. $\tan\theta = -0.4663$ 155°, 335°

29. A ship leaves port and sails 54 mi on a course of 220°.

a. What angle of rotation is equivalent to the ship's course? 230°

b. Assume that the port from which the ship sailed is at the origin of a coordinate system. Write and solve two equations relating the ship's x- and y-coordinates to the angle of rotation in part (a).
$x = 54\cos 230°$, $x \approx -34.7$; $y = 54\sin 230°$, $y \approx -41.4$

c. How far has the ship traveled in an east-west direction? In a north-south direction?
about 34.7 mi west; about 41.4 mi south

Practice Bank, ALGEBRA 2: EXPLORATIONS AND APPLICATIONS

91

25. Check students' work.

26. Check students' work.

27. $\tan^{-1}\left(\frac{3}{4}\right) \approx 37°$; The other angle is $(180 + 37)°$, or about 217°.

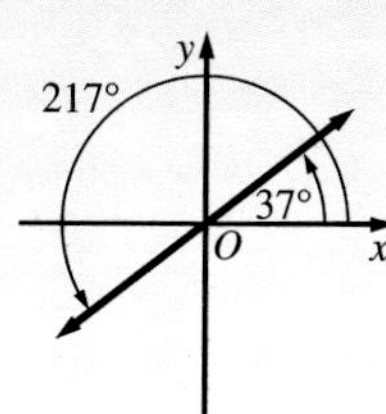

28. a. $\angle B = 34°$; $a \approx 3.0$; $c \approx 3.6$

b. $\angle A = 53.1°$; $\angle B = 36.9°$; $c = 30$

29. a. 204 ft²

b. $3\sqrt{3}$ cm², or about 5.2 cm²

30. about 0.42

Plan⇔Support

Objectives

- Find the area of a triangle given two side lengths and the measure of the included angle.
- Find the area of a sector.

Recommended Pacing

Core and Extended Courses
Section 14.4: 1 day

Block Schedule
Section 14.4: $\frac{1}{2}$ block
(with Section 14.3)

Resource Materials

Lesson Support
Lesson Plan 14.4
Warm-Up Transparency 14.4
Practice Bank: Practice 92
Study Guide: Section 14.4
Explorations Lab Manual: Diagram Masters 1, 2
Assessment Book: Test 59

Technology
Scientific Calculator
Internet: http://www.hmco.com

Warm-Up Exercises

1. Find the area of a triangle with height 62.4 in. and base 74.3 in. about 2318 in.^{2}
2. Find K if $K = \frac{1}{2}(6.2)(7.1) \sin 20°$. about 7.5
3. Solve $12.2 = \frac{1}{2}(8)(4) \sin \theta$ for $\sin \theta$. $\sin \theta = 0.7625$
4. Find the area of a circle whose radius is 4 cm. 16π cm^2
5. Given $\triangle ABC$ with $b = 18$, $\angle A = 42°$, and right angle C, find the length of the altitude to side AB. about 12

Section

14.4 Finding the Area of a Triangle

Learn how to...

- find the area of a triangle given two side lengths and the measure of the included angle
- find the area of a sector

So you can...

- find the area of the Bermuda Triangle, for example

The map below shows the Bermuda Triangle, a large region of the Atlantic Ocean where many ships and planes have mysteriously disappeared. A trigonometric ratio can help you to find the area of this region.

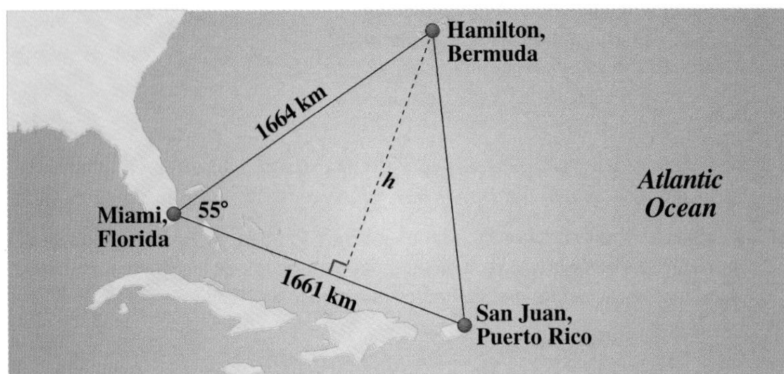

THINK AND COMMUNICATE

1. a. Write and solve an equation involving a trigonometric ratio to find the height h of the Bermuda Triangle.
   b. Use the value of h you found in part (a) to find the area of the Bermuda Triangle.
2. Write an expression for the area of $\triangle XYZ$ in terms of x, y, and the measure of $\angle Z$. Use a diagram to explain your reasoning.

You can find the area of any triangle if you know the lengths of two sides and the measure of the included angle.

Area of a Triangle

The area K of $\triangle ABC$ is given by any of the following formulas:

$$K = \frac{1}{2}bc \sin A$$

$$K = \frac{1}{2}ac \sin B$$

$$K = \frac{1}{2}ab \sin C$$

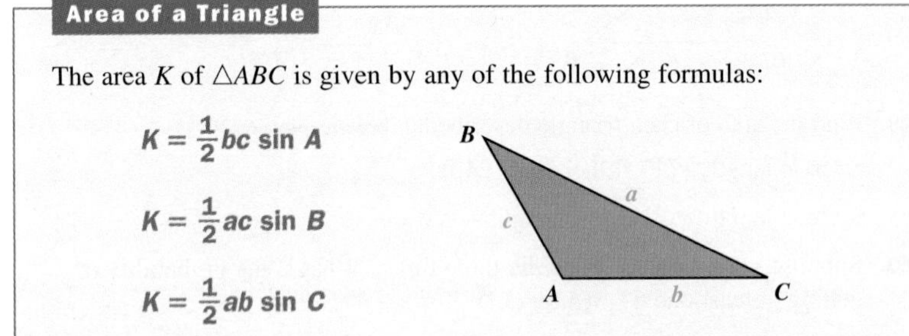

ANSWERS Section 14.4

Think and Communicate

1. a. $\sin 55° = \frac{h}{1664}$; about 1363 km
   b. about 1,132,000 km^2
2. $h = x \sin Z$; $A = \frac{1}{2}bh = \frac{1}{2}xy \sin Z$

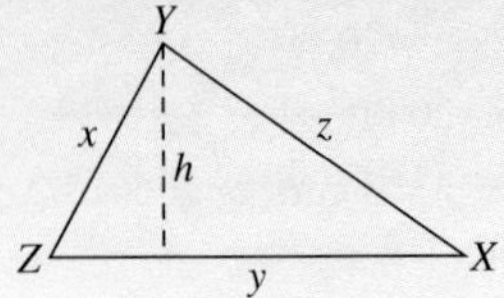

EXAMPLE 1

Find the area of $\triangle XYZ$.

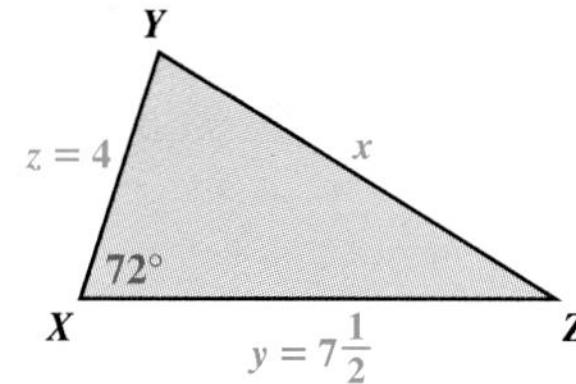

SOLUTION

Since you know the lengths of two sides and the measure of the included angle, you can find the area using $K = \frac{1}{2}yz \sin X$.

$$K = \frac{1}{2}\left(7\frac{1}{2}\right)(4) \sin 72^\circ$$

$$\approx 14.3$$

Substitute $7\frac{1}{2}$ for y, **4** for z, and **72°** for X.

The area of $\triangle XYZ$ is about 14.3 square units.

THINK AND COMMUNICATE

3. Complete parts (a)–(c) to show that the area formula given on the previous page applies to an obtuse triangle as well as an acute triangle.

a. Express the height h of $\triangle ABC$ in terms of one or more of the labeled parts of $\triangle ABD$.

b. Use your answer to part (a) to express the area K of $\triangle ABC$ in terms of the labeled parts of the diagram.

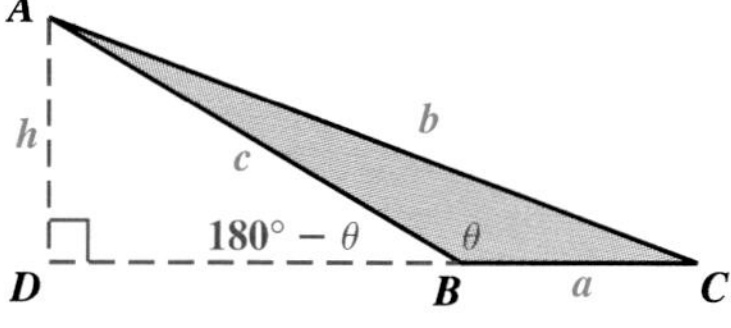

c. What is the relationship between $\sin (180^\circ - \theta)$ and $\sin \theta$? Use your answer to simplify the area formula you wrote for part (b).

EXAMPLE 2

Find the area of $\triangle RST$ where $t = 11.4$, $r = 8.6$, and $\angle S = 125°$.

SOLUTION

Step 1 Make a sketch. You can see that the given angle is between the two given sides.

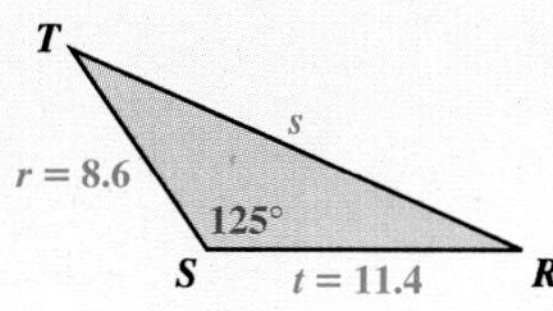

Step 2 Use an area formula.

$$K = \frac{1}{2}tr \sin S$$

$$= \frac{1}{2}(11.4)(8.6) \sin 125^\circ$$

$$\approx 40.2$$

The area of $\triangle RST$ is **about 40.2 square units.**

Think and Communicate

3. a. $h = c \sin (180^\circ - \theta)$

b. $K = \frac{1}{2}ac \sin (180^\circ - \theta)$

c. $\sin (180^\circ - \theta) = \sin \theta$, so $K = \frac{1}{2}ac \sin \theta$.

Teach⇔Interact

Section Notes

Problem Solving

You may want to have students verify one of the area formulas. For example, students can draw the altitude from A to $\overline{BC}$ and let its length be h. Then $\sin C = \frac{h}{b}$, or $h = b \sin C$. Use the standard formula for finding area, namely, $\frac{1}{2}$ base times height. Then, $K = \frac{1}{2}ab \sin C$.

Teaching Tip

The area formulas may be summarized as follows: the area of a triangle equals one-half the product of two sides times the sine of their included angle. If the included angle is a right angle, its sine equals one and the formula becomes one-half base times height.

Additional Example 1

Find the area of $\triangle RST$.

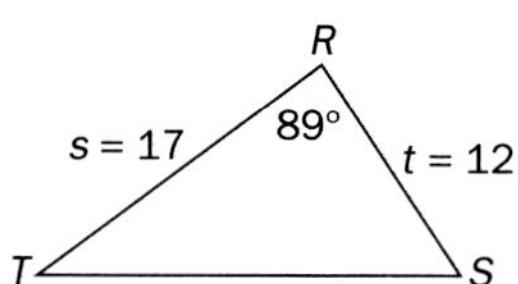

Since you know the lengths of two sides and the measure of the included angle, you can find the area using $K = \frac{1}{2}ts \sin R$.

$K = \frac{1}{2}(12)(17)\sin 89^\circ$

≈ 102

The area of $\triangle RST$ is about 102 square units.

Additional Example 2

Find the area of $\triangle ABC$, where $a = 62.3$, $c = 82.6$, and $\angle B = 132°$.

Step 1 Make a sketch. You can see that the given angle is between the two given sides.

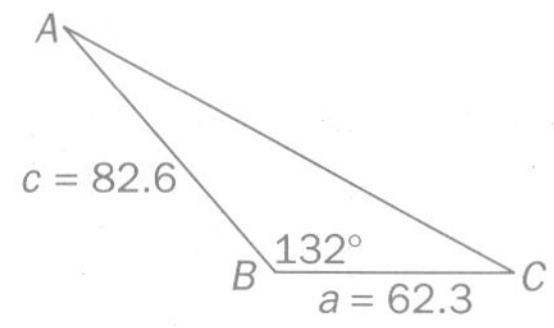

Step 2 Use an area formula.

$K = \frac{1}{2}ac \sin B$

$K = \frac{1}{2}(62.3)(82.6) \sin 132^\circ$

≈ 1912.1

The area of $\triangle ABC$ is about 1912.1 square units.

Teach⇔Interact

Section Note

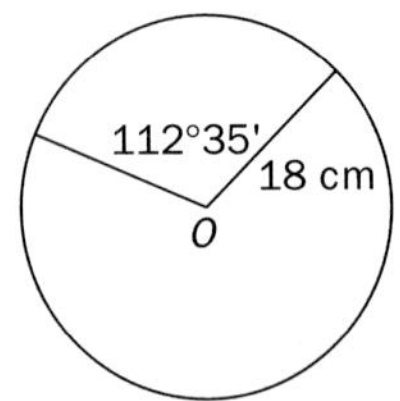

Communication: Reading Students should interpret the first formula given on this page as saying that the ratio of the area of a sector to the area of the circle is the same as the ratio of the central angle of the circle to the total circle, or 360°.

Think and Communicate

Question 5 requires students to think about possible restrictions when using area formulas. For a triangle, the area formulas were restricted to angles less than 180°. For the area of a sector, however, the angles can be greater than 180° but not more than 360°.

Additional Example 3

Find the area of the sector of a circle if $\angle O = 112°35'$ and the radius of the circle is 18 cm.

112°35'
18 cm
O

Substitute 112°35' for θ and 18 for r into the formula for the area of a sector.

$$\text{area of sector} = \frac{\theta}{360}\pi r^2$$
$$= \left(\frac{112 + \frac{35}{60}}{360}\right)\pi(18)^2$$
$$= 101.325\pi$$
$$\approx 318$$

The area is about 318 cm².

Area of a Sector

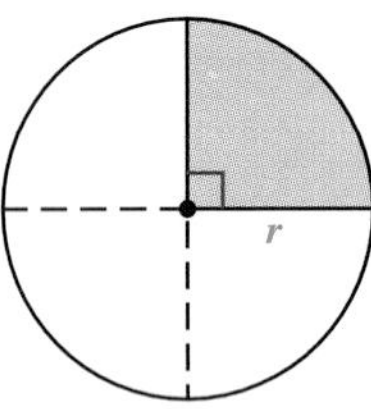

Recall from geometry that the region enclosed by a central angle of a circle and the intercepted arc is called a *sector*. The central angle of the shaded sector shown is 90°. You can see that the area of the sector is $\frac{90}{360}$, or $\frac{1}{4}$, of the total area of the circle.

In general, for any sector having radius r and central angle θ:

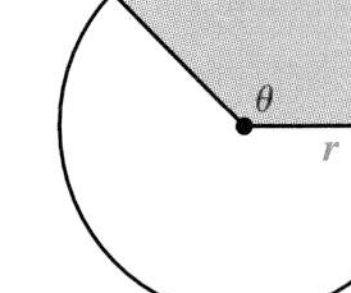

$$\frac{\text{area of the sector}}{\text{area of the circle}} = \frac{\theta}{360}$$

$$\text{area of the sector} = \frac{\theta}{360} \cdot (\text{area of the circle})$$

$$\textbf{area of the sector} = \frac{\theta}{360} \cdot \pi r^2$$

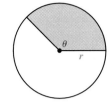

THINK AND COMMUNICATE

4. Explain the significance of the "360" in the formula above.

5. What values of θ can be substituted into the formula?

EXAMPLE 3 Application: Landscaping

A water sprinkler positioned in the corner of a yard is set to rotate only 120° in order to keep the pavement around the grass dry. Water from the sprinkler reaches as far as 15 ft away. What is the area of the yard that is watered by the sprinkler?

SOLUTION

Substitute 120° for θ and 15 for r into the formula for the area of a sector.

$$\text{area of sector} = \frac{\theta}{360}\pi r^2$$

$$\text{area of yard watered} = \frac{120}{360}\pi(15)^2$$
$$= 75\pi$$
$$\approx 240$$

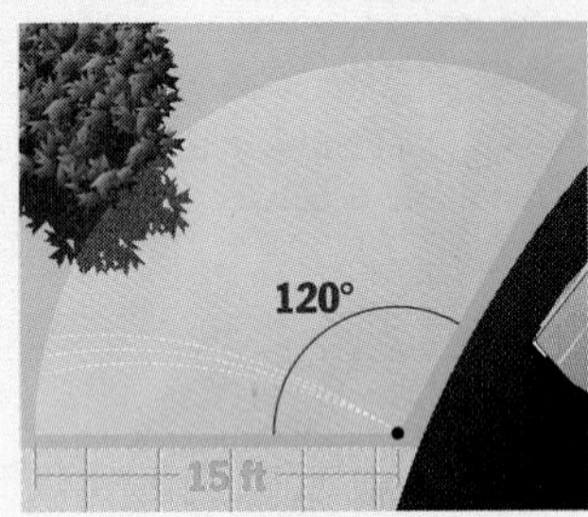

About 240 ft² of the yard is watered by the sprinkler.

Think and Communicate

4. The central angle corresponding to a circle measures 360°.
5. $0 \le \theta \le 360°$

CHECKING KEY CONCEPTS

Find the area of each figure.

1.

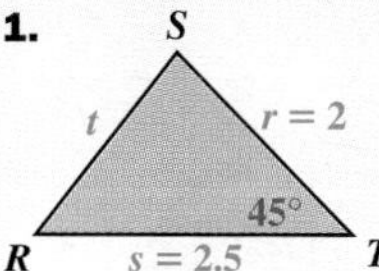

2.

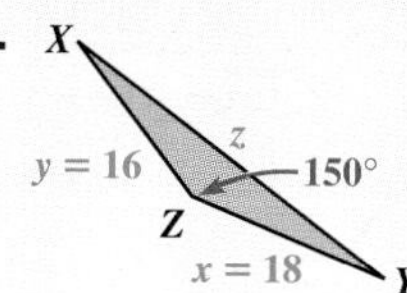

3.

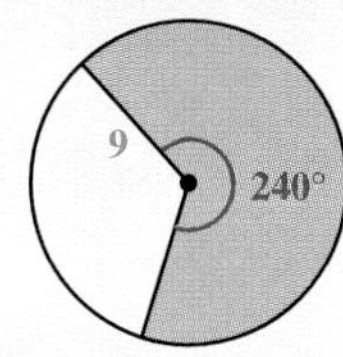

Find the area of △*ABC*.

4. $a = \frac{1}{2}, b = 12, \angle C = 40°$ **5.** $b = 10, c = 3.9, \angle A = 145°$

Find the area of each sector having central angle θ and radius *r*.

6. $\theta = 60°, r = \frac{3}{5}$ **7.** $\theta = 144°, r = \sqrt{6}$ **8.** $\theta = 110°, r = 1.5$

9. Writing In your own words, state the formula for the area of a triangle given on page 678.

14.4 Exercises and Applications

Extra Practice exercises on page 771

For Exercises 1–9, find the area of △*DEF*. If there is not enough information, explain.

1.

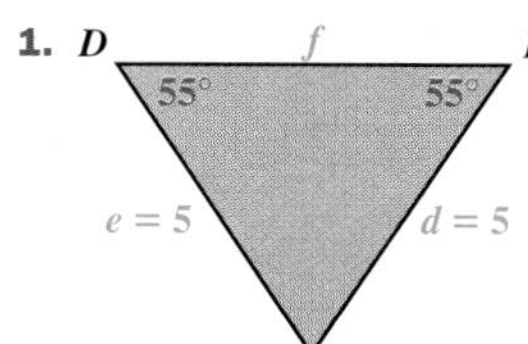

2.

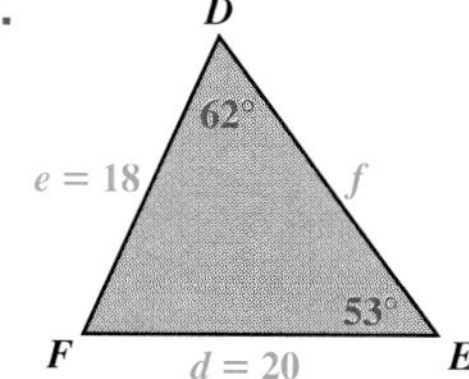

3.

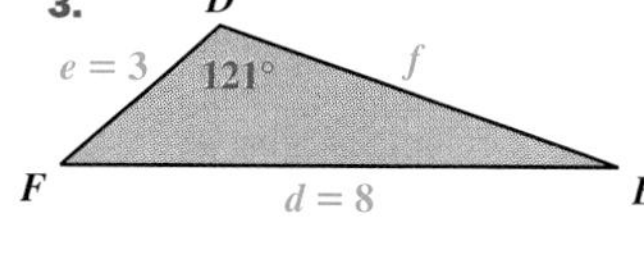

4. $e = 18.2, d = 9.4, \angle F = 91°$ **5.** $f = 32, e = 40, \angle D = 115°$ **6.** $d = 6, e = 11, \angle E = 30°$

7. $d = 9, f = 8\frac{1}{2}, \angle E = 57°$ **8.** $f = 13, e = 4, \angle F = 135°40'$ **9.** $f = 10, d = 9.7, \angle D = 85°$

COOKING One kind of pastry is shaped by rolling a triangular piece of dough from the wide end, as shown. The long sides of the triangle measure about $4\frac{1}{2}$ in. each, and the angle between them measures $22\frac{1}{2}°$.

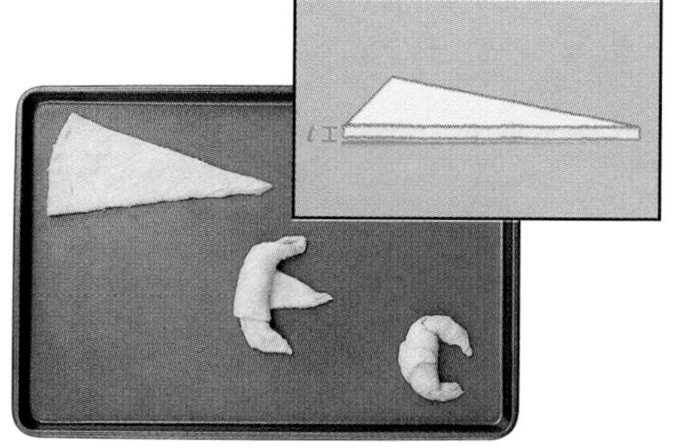

10. Let t = the thickness of a triangular piece of dough before it is rolled. Express the volume of the dough in terms of t.

11. Challenge How many pieces of pastry can be made from $1\frac{1}{2}$ cups of dough when the dough is $\frac{1}{8}$ in. thick? (*Hint:* 1 cup $\approx$ 14.4 in.^{3})

Teach⇔Interact

Checking Key Concepts

Student Progress
Question 9 provides an opportunity to check on how well students understand this formula. You may wish students to do the same thing for the area of a sector.

Closure Question

When can you use the area of a triangle formula that involves sine to find the area of a triangle? What is the formula for finding the area of the sector of a circle?
You can use the area of a triangle formula when the lengths of two sides and the measure of their included angle are known.
area of a sector = $\frac{\theta}{360} \cdot \pi r^2$

Apply⇔Assess

Suggested Assignment

Core Course
Exs. 1–9, 12–20, 26–33, AYP

Extended Course
Exs. 1–6, 10–15, 18–33, AYP

Block Schedule
Day 2 Exs. 1–9, 12–20, 26–33, AYP

Exercise Notes

Communication: Discussion
Exs. 1–9 These exercises can help students understand exactly when the area formulas apply. Call upon various students to explain their answers.

Teaching Tip
Ex. 2 Students may think they cannot find the area of the triangle for this exercise. Remind them they may need to find the missing angle first.

Problem Solving
Ex. 10 Remind students that the volume of a triangular prism, the shape of the dough, is the area of the base times the height.

Checking Key Concepts

1. about 1.8 square units
2. 72 square units
3. about 170 square units
4. about 1.9 square units
5. about 11.2 square units
6. about 0.19 square units
7. about 7.5 square units
8. about 2.2 square units
9. The area of a triangle is equal to one-half the product of the lengths of two sides and the sine of the included angle.

Exercises and Applications

1. about 11.7 square units
2. about 163 square units
3. not enough information; need the measure of included angle
4. about 85.5 square units
5. about 580 square units
6. not enough information; need the measure of included angle
7. about 32.1 square units
8. not enough information; need the measure of included angle
9. not enough information; need the measure of included angle
10. $V \approx 3.87t$
11. about 45 pieces

Apply⇔Assess

Exercise Notes

Common Error
Exs. 13, 17 Some students may think that if the central angle is greater than 180°, they must find the area of the "smaller sector." Emphasize that a sector can represent more than half a circle.

Teaching Tip
Ex. 21 Remind students that an arc of a circle is expressed with at least three reference points on the arc so that the "direction" is clear. A major arc has measure greater than 180° and a minor arc has measure less than 180°.

Interdisciplinary Problems
Exs. 21–24 The disciplines of astromony and mathematics have been linked together from the beginning of civilization. Geometry and trigonometry provided the first mathematical tools to understand the relationships among the sun, Earth, and moon. Later, algebra and calculus provided astronomers with the tools they needed to study the universe.

Second-Language Learners
Ex. 24 Students learning English may benefit from working on this writing task with the help of a peer tutor or aide.

Multicultural Note
Ex. 24 Accra, the capital and largest city of Ghana, is located on the Gulf of Guinea. It is the site of the University of Ghana, as well as a center of trade and manufacturing, producing clothing, timber and plywood, and processed foods.

Find the area of each sector having central angle θ and radius r.

12. $\theta = 180°, r = \frac{5}{8}$ **13.** $\theta = 280°, r = 3\frac{1}{4}$ **14.** $\theta = 72°, r = 2.5$

15. $\theta = 22°30', r = 12$ **16.** $\theta = 3°36', r = 5$ **17.** $\theta = 225°, r = 1.6$

Find the area of each shaded sector.

18.
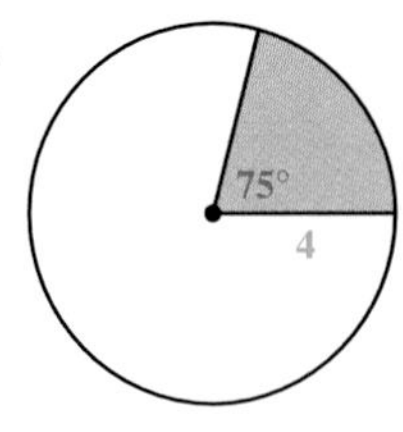

19.
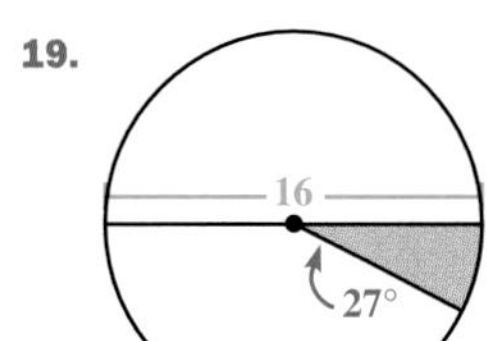

20.
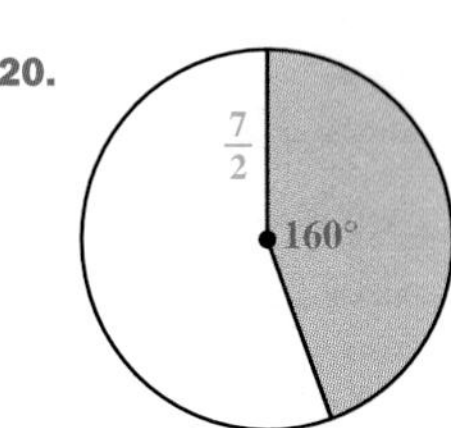

Connection ASTRONOMY

During a solar eclipse, the moon blocks some or all of your view of the sun. You can use the areas of sectors and triangles to find the fraction of the sun that is blocked by the moon at any given moment during an eclipse.

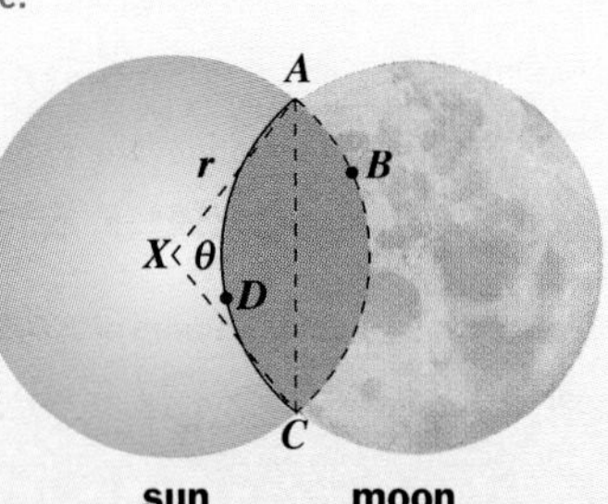

21. a. Write an expression for the area of sector *XABC*.
b. Write an expression for the area of $\triangle XAC$.
c. The region bounded by $\overline{AC}$ and arc *ABC* is called a *segment* of the circle with center *X*. Write an expression for the area of segment *ABC*. Explain your reasoning.

22. a. From Earth, the sun and the moon appear to be circular disks of about the same size. What does this tell you about the areas of segments *ADC* and *ABC*?
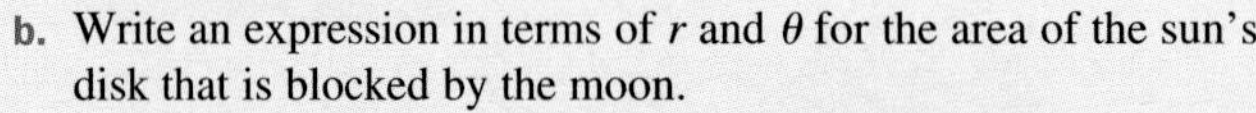
b. Write an expression in terms of r and θ for the area of the sun's disk that is blocked by the moon.
c. Write an expression in terms of θ for the *fraction* of the sun's disk blocked by the moon. Explain how you found your answer.

23. On March 29, 2006, there will be a solar eclipse over Africa. Use the expression you wrote in part (c) of Exercise 22 to calculate the fraction of the sun's disk blocked at 10 A.M. (universal time) at each location. Write your answers as percents.
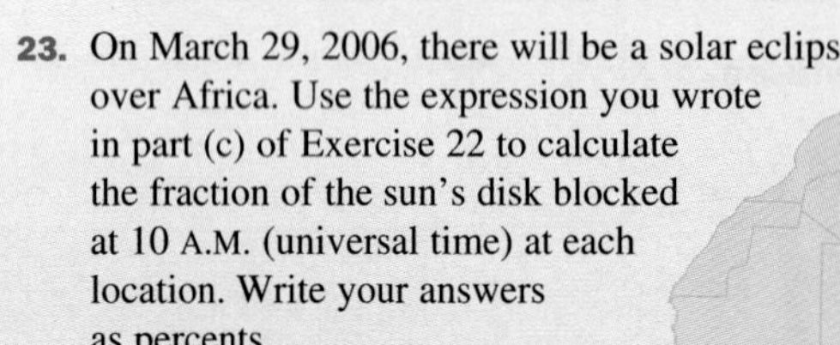
a. Lagos, Nigeria
b. Lake Chad in Chad
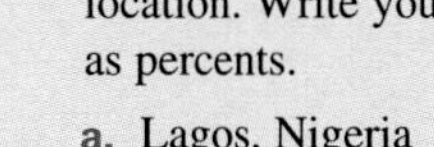
c. Cairo, Egypt

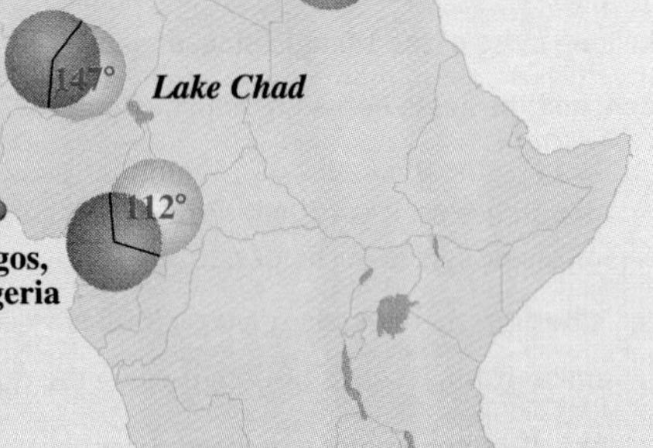

24. Writing On March 29, 2006, at about 9:15 A.M. (universal time), the moon will completely block the sun in Accra, Ghana. What will the value of θ be in this case? Explain your answer.
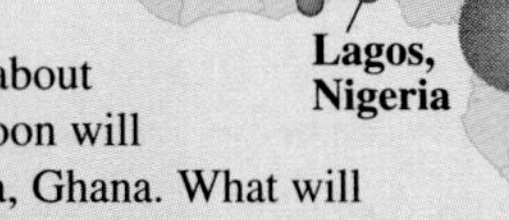

12. about 0.61 square units
13. about 25.8 square units
14. about 3.9 square units
15. about 28.3 square units
16. about 0.79 square units
17. about 5 square units
18. about 10.5 square units
19. about 15.1 square units
20. about 17.1 square units
21. a. $\left(\frac{\theta}{360}\right)\pi r^2$ b. $\frac{1}{2}r^2 \sin\theta$
c. area of segment = $\left(\frac{\theta}{360}\right)\pi r^2 - \frac{1}{2}r^2\sin\theta = \frac{r^2}{2}\left(\frac{\theta\pi}{180} - \sin\theta\right)$
22. a. The areas are the same.
b. $r^2\left(\frac{\theta\pi}{180} - \sin\theta\right)$
c. $\frac{\theta}{180} - \frac{\sin\theta}{\pi}$; Find the ratio of the shaded area to the total area.
23. a. about 33%
b. about 64%
c. about 21%
24. $\theta = 180°$; When $\theta = 180°$, $\frac{\theta}{180} = 1$, and $\sin\theta = 0$, so, $\frac{\theta}{180} - \frac{\sin\theta}{\pi} = 1$, indicating that 100% of the sun is covered.
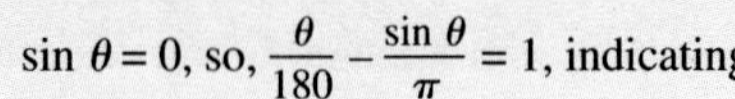
25. a. about 20 ft
b. $16\pi \approx 50.3$ ft
c. 144°
d. about 503 ft²
26. Check students' work.
27. 55°, 125°
28. 73°, 253°
29. 26°, 334°
30. Yes; –5.
31. Yes; 6.
32. No.
33. Yes; 12.

25. **HISTORY** All the Native American people of the Great Plains used *tipis* for shelter. Tipis were made of wooden poles slanted toward a center point and covered by buffalo hides sewn together in a shape similar to a sector.

a. What is the length x from the ground to where the poles meet? This is the radius of the buffalo hide.

b. Find the length of the circular edge of the buffalo hide.

c. **Challenge** Find the central angle θ of the buffalo hide. $\left(\textit{Hint: } \frac{\text{length of circular edge of the sector}}{\text{circumference of the circle}} = \frac{\theta}{360}\right)$

d. Find the surface area of the tipi.

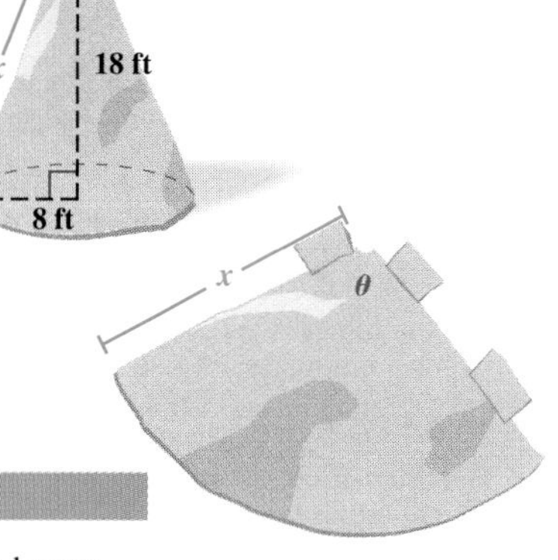

ONGOING ASSESSMENT

26. **Open-ended Problem** Draw a triangle and a sector where each has an area of 40 square units.

SPIRAL REVIEW

Find two measures of $\angle A$ for each trigonometric value. *(Section 14.3)*

27. $\sin A = 0.8192$
28. $\tan A = 3.271$
29. $\cos A = 0.8988$

Tell whether y varies inversely with x. If so, state the constant of variation. *(Section 9.6)*

30. $y = -\frac{5}{x}$
31. $xy = 6$
32. $y = \frac{8}{x - 4}$
33. $x = \frac{12}{y}$

ASSESS YOUR PROGRESS

VOCABULARY

initial side of an angle (p. 671)
terminal side of an angle (p. 671)
standard position of an angle (p. 671)

Sketch each angle of rotation in a coordinate plane. Then use a calculator to find $\sin\theta$, $\cos\theta$, and $\tan\theta$. *(Section 14.3)*

1. $\theta = 40°$
2. $\theta = 225°$
3. $\theta = 308°$
4. $\theta = 95°$

Each point P lies on the terminal side of an angle θ in standard position. Find $\sin\theta$, $\cos\theta$, and $\tan\theta$. *(Section 14.3)*

5. $P(6, 2)$
6. $P(4, -1)$
7. $P(-7, 9)$
8. $P(-5, -8)$

Find the area of $\triangle JKL$. *(Section 14.4)*

9. $j = 22, l = 17, \angle K = 38°$
10. $k = 0.75, l = 1.2, \angle J = 165°$

Find the area of each sector having central angle θ and radius r. *(Section 14.4)*

11. $\theta = 330°, r = 60$
12. $\theta = 200°, r = \sqrt{3}$
13. $\theta = 54°, r = 15$

14. **Journal** List the important ideas you learned in Sections 14.3 and 14.4. Tell whether each was easy or difficult for you to understand.

Apply⇔Assess

Exercise Notes

Using Manipulatives
Ex. 25 Students can model this exercise by cutting a small sector from a circle then folding it to be a cone. They should label the model as is done in the diagram, "unwrap" it, and find the measurements. This will make part (c) easier. For part (d), you may need to point out that the surface area of the hide is the same as the area of the sector.

Visual Thinking
Ex. 26 Ask students to trace the sector over the triangle and to discuss the relationship between the areas that do not overlap. This activity involves the visual skills of *recognition* and *generalization*.

Progress Check 14.3–14.4

See page 701.

Practice 92 for Section 14.4

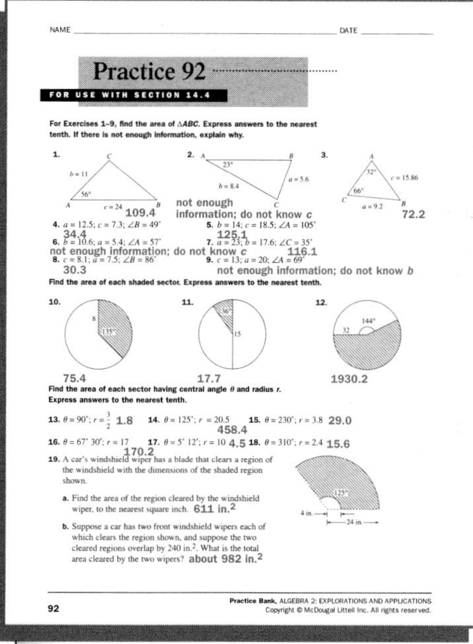
NAME ______ DATE ______

Practice 92

FOR USE WITH SECTION 14.4

For Exercises 1–9, find the area of $\triangle ABC$. Express answers to the nearest tenth. If there is not enough information, explain why.

1. $b = 11$, $\angle A = 56°$, $c = 24$ 109.4
2. $\angle A = 23°$, $b = 8.4$, $a = 5.6$ not enough information; do not know c
3. $\angle A = 32°$, $c = 15.86$, $\angle C = 66°$, $a = 9.2$ 72.2
4. $a = 12.5$; $c = 7.3$; $\angle B = 49°$ 34.4
5. $b = 14$; $c = 18.5$; $\angle A = 105°$ 125.1
6. $b = 10.6$; $a = 5.4$; $\angle A = 57°$ not enough information; do not know c
7. $a = 23$; $b = 17.6$; $\angle C = 35°$ 116.1
8. $c = 8.1$; $a = 7.5$; $\angle B = 86°$ 30.3
9. $c = 13$; $a = 20$; $\angle A = 69°$ not enough information; do not know b

Find the area of each shaded sector. Express answers to the nearest tenth.

10. 8, 135° 75.4
11. 36°, 15 17.7
12. 144°, 32 1930.2

Find the area of each sector having central angle θ and radius r. Express answers to the nearest tenth.

13. $\theta = 90°$; $r = \frac{3}{2}$ 1.8
14. $\theta = 125°$; $r = 20.5$ 458.4
15. $\theta = 230°$; $r = 3.8$ 29.0
16. $\theta = 67°\,30'$; $r = 17$ 170.2
17. $\theta = 5°\,12'$; $r = 10$ 4.5
18. $\theta = 310°$; $r = 2.4$ 15.6

19. A car's windshield wiper has a blade that clears a region of the windshield with the dimensions of the shaded region shown. (125°, 4 in., 24 in.)

a. Find the area of the region cleared by the windshield wiper, to the nearest square inch. 611 in.^{2}

b. Suppose a car has two front windshield wipers each of which clears the region shown, and suppose the two cleared regions overlap by 240 in.^{2}. What is the total area cleared by the two wipers? about 982 in.^{2}

92 **Practice Bank,** ALGEBRA 2: EXPLORATIONS AND APPLICATIONS

Assess Your Progress

1. 0.6428; 0.7660; 0.8391

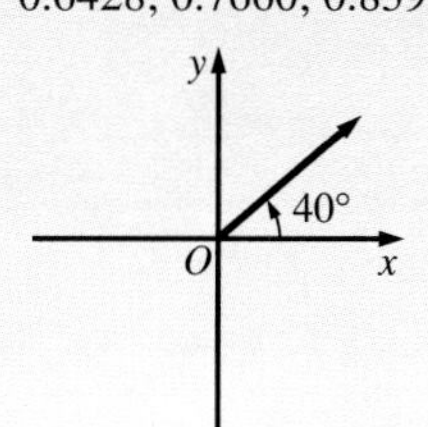

2. –0.7071; –0.7071; 1

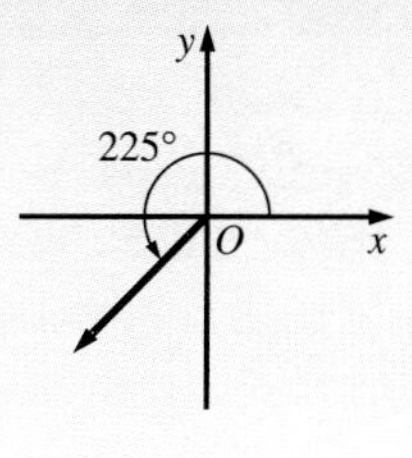

3. –0.7880; 0.6157; –1.280

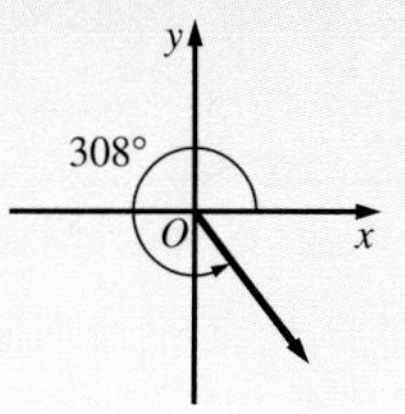

4. 0.9962; –0.0872; –11.43

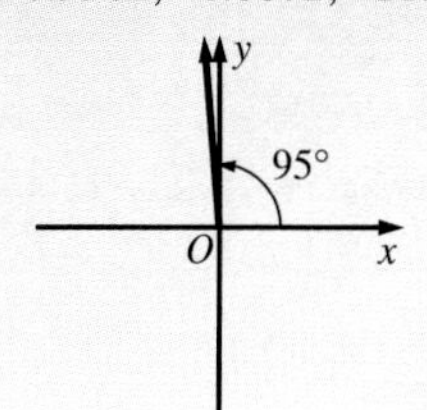

5. 0.3162; 0.9487; 0.3333
6. –0.2425; 0.9701; –0.25
7. 0.7894; –0.6139; –1.286
8. –0.8480; –0.5300; 1.6
9. about 115 square units
10. about 0.12 square units
11. about 10,367 square units
12. about 5.2 square units
13. about 106 square units
14. Answers may vary.

Plan⇔Support

Objectives

- Use the law of sines to solve a triangle.
- Use the law of sines to solve real-world problems.

Recommended Pacing

Core and Extended Courses
Section 14.5: 1 day

Block Schedule
Section 14.5: $\frac{1}{2}$ block
(with Section 14.6)

Resource Materials

Lesson Support
Lesson Plan 14.5
Warm-Up Transparency 14.5
Practice Bank: Practice 93
Study Guide: Section 14.5
Explorations Lab Manual:
Diagram Master 19

Technology
Graphing Calculator
Internet:
http://www.hmco.com

Warm-Up Exercises

1. Solve for b. $\frac{0.1226}{16} = \frac{0.6872}{b}$ about 89.7
2. Solve for b in terms of a, sin A, and sin B. $\frac{\sin A}{a} = \frac{\sin B}{b}$ $b = a\frac{\sin B}{\sin A}$
3. In $\triangle ABC$, $\angle C = 90°$ and $\angle A = 32°$. If $c = 48$ in., find the length of a. about 25.4 in.
4. List the four ways triangles can be proved congruent. side-side-side, side-angle-side, angle-side-angle, angle-angle-side
5. Suppose you are given sides of lengths 8 and 12 and a non-included angle with measure 62°. Do they form a unique triangle? Explain. No. There is more than one triangle with those same side lengths and nonincluded angle measure (side-side-angle).

Section

14.5 The Law of Sines

Learn how to...

- use the law of sines to solve a triangle

So you can...

- find the distance of a ship from the Rock of Gibraltar, for example

In his book *Shakl al-qita*, the Persian mathematician al-Tusi (1201–1274) described a relationship that is true for any triangle. To derive this law, first set the three expressions for area shown below equal to each other.

$$K = \frac{1}{2}bc \sin A \qquad K = \frac{1}{2}ac \sin B \qquad K = \frac{1}{2}ab \sin C$$

$$\frac{1}{2}bc \sin A = \frac{1}{2}ac \sin B = \frac{1}{2}ab \sin C$$

Multiply by $\frac{2}{abc}$.

$$\frac{2}{abc}\left(\frac{1}{2}bc \sin A\right) = \frac{2}{abc}\left(\frac{1}{2}ac \sin B\right) = \frac{2}{abc}\left(\frac{1}{2}ab \sin C\right)$$

$$\frac{\sin A}{a} = \frac{\sin B}{b} = \frac{\sin C}{c}$$

Notice that each ratio involves only one angle and its opposite side.

Law of Sines

For any $\triangle ABC$:

$$\frac{\sin A}{a} = \frac{\sin B}{b} = \frac{\sin C}{c}$$

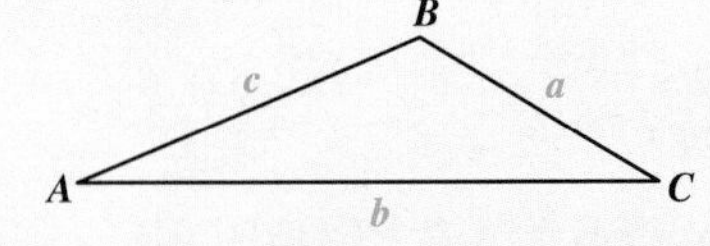

EXAMPLE 1 Application: Navigation

A ship at S is sighted simultaneously from Point Almina at A and from the Rock of Gibraltar at G, as shown. Find the ship's distance from the Rock of Gibraltar.

BY THE WAY

The Rock of Gibraltar is located on the coast of Spain but owned by Great Britain. Point Almina is located on the coast of Morocco but owned by Spain.

SOLUTION

You need to find $\angle S$ before you can use the law of sines.

$$\angle S = 180° - \angle A - \angle G$$
$$= 180° - 32° - 37° = 111°$$

Use the fact that $\frac{\sin S}{s} = \frac{\sin A}{a}$.

$$\frac{\sin 111°}{15} = \frac{\sin 32°}{a}$$

$$a = \frac{15 \sin 32°}{\sin 111°}$$

$$a \approx 8.5$$

The ship is about 8.5 mi from the Rock of Gibraltar.

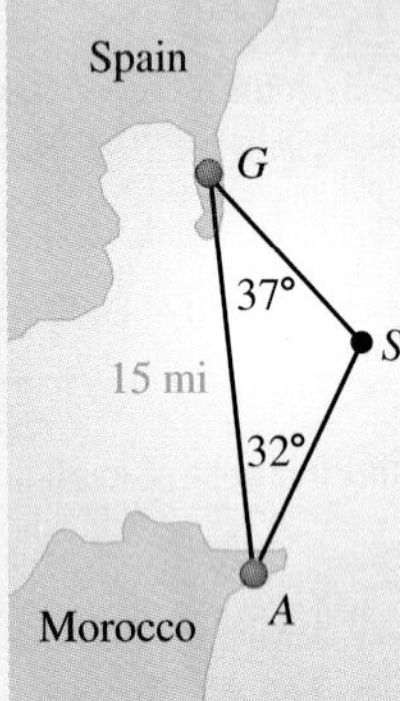

You can use the law of sines when you know two side lengths of a triangle and the measure of a non-included angle. In the Exploration below, you will investigate whether this information determines a unique triangle.

EXPLORATION
COOPERATIVE LEARNING

Constructing a Triangle

Work with a partner.
You will need:
- a ruler
- a protractor
- a compass

1 Draw $\overrightarrow{AX}$ at least 15 cm long. Then draw $\overline{AC}$ so that $\angle CAX = 30°$ and $AC = b = 8$ cm.

2 Place the compass point at C. Choose a radius from the table below and draw an arc as shown.

3 Each point (if any) where the arc meets $\overrightarrow{AX}$ should be labeled B to complete a $\triangle ABC$. Determine the number of such triangles that can be drawn with the given radius.

4 Copy and complete the table by repeating Steps 1–3 for each radius r.

Radius r (cm)	Number of triangles
2	?
4	?
6	?
8	?
10	?

Questions

1. Find the height h of $\triangle ABC$ when $\angle A = 30°$ and $b = 8$ cm. Do all the triangles you constructed have the same height? Explain.

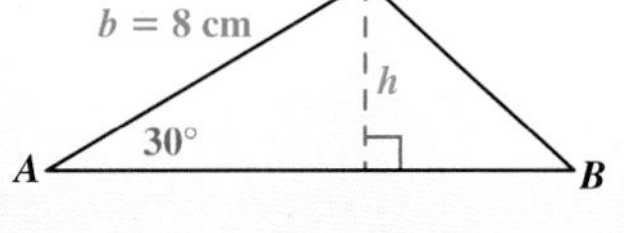

2. How many triangles are possible when $\overrightarrow{AX}$ is tangent to the arc drawn with the compass? For which value(s) of r does this happen?
3. **a.** For which value(s) of r does the arc intersect $\overrightarrow{AX}$ twice?
   b. Express your answer to part (a) as an inequality in terms of b, h, and r.
4. How many triangles are possible when the arc intersects $\overrightarrow{AX}$ exactly once? For which value(s) of r does this happen?
5. Repeat Steps 1–4 above using $\angle A = 120°$. Under what circumstances do you get a triangle? don't you get a triangle?

Exploration Note

Purpose
The purpose of this Exploration is to have students investigate whether two side lengths of a triangle and the measure of a non-included angle are enough information to determine a unique triangle.

Materials/Preparation
Each pair of students needs a ruler, a protractor, and a compass.

Procedure
Students construct a 30° angle along a ray and then mark off an 8 cm segment along the side of the angle. From the endpoint of the side, they draw arcs with different radii. They then determine which radii complete a triangle with the given angle and given side. This is the side-side-angle condition studied in geometry.

Closure
Students should conclude that two sides and a non-included angle do not always determine a unique triangle.

Explorations Lab Manual
See the Manual for more commentary on this Exploration.

Diagram Master 19

For answers to the Exploration, see following page.

Teach⇔Interact

Section Notes

Student Study Tip
Sometimes the law of sines is written as the reciprocal relationship $\frac{a}{\sin A} = \frac{b}{\sin B} = \frac{c}{\sin C}$.

Teaching Tip
Point out to students that the law of sines applies to triangles with three known parts (side-side-angle, angle-side-angle, or angle-angle-side).

Additional Example 1

From point A, a surveyor locates a tree at point C across the river at an angle of 68.6°. From point B on the same side of the river and 120 m from point A, the surveyor locates a tree at an angle of 41°, as shown. Find the distance to the tree from point A.

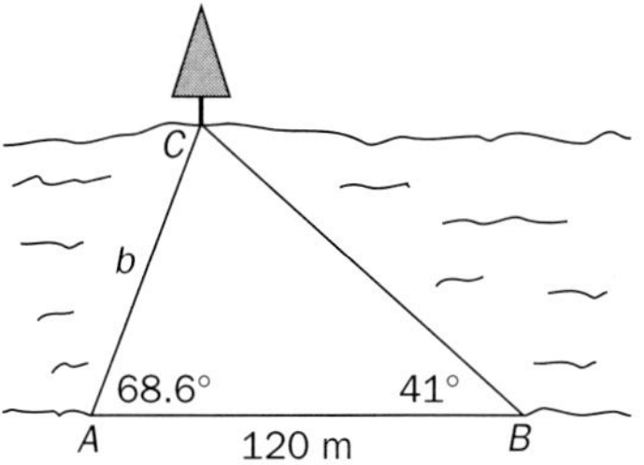

You need to find $\angle C$ before you can use the law of sines.

$\angle C = 180° - \angle A - \angle B$
$= 180° - 68.6° - 41° = 70.4°$

Use the fact that $\frac{\sin C}{c} = \frac{\sin B}{b}$.

$$\frac{\sin 70.4°}{120} = \frac{\sin 41°}{b}$$
$$b = \frac{120 \sin 41°}{\sin 70.4°}$$
$$b \approx 83.6$$

The tree is about 83.6 m from point A.

Teach⇔Interact

Section Note

Communication: Discussion The *ambiguous case* may be confusing for some students because of the many possible combinations of angle measures and side lengths. Emphasize that the ambiguous case occurs only when the given conditions are side-side-angle or two given sides with the non-included angle. Remind students that in geometry, side-side-angle is *not* a postulate or congruence theorem for triangles.

Think and Communicate

Students should see that $b \sin A$ is the height of each triangle. Therefore, the given side a is being compared to the height. Side a is either less than, equal to, or greater than the height. It is this relationship that helps determine the number of possible triangles.

In considering question 5, students should see that since a triangle can only have one obtuse angle, if the angle given is obtuse, the ambiguous case either gives one solution or no solution. If students are not aware of this, a second solution may give angle measures that add up to more than 180°.

The Exploration showed that two sides and a non-included angle do not always determine a unique triangle. This is called the *ambiguous case*. All the possible outcomes of the ambiguous case are shown below.

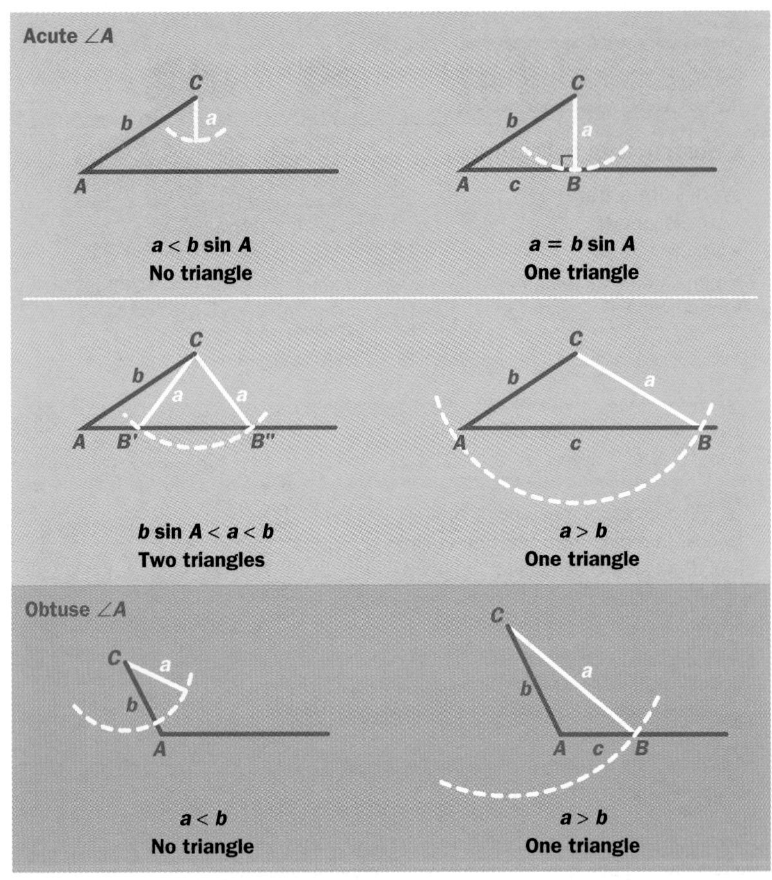

THINK AND COMMUNICATE

1. What does the expression $b \sin A$ represent for the triangles above?
2. When $\angle A$ is acute and $a < b \sin A$, why can't any triangles be formed?
3. What kind of triangle is formed when $\angle A$ is acute and $a = b \sin A$?
4. How are $\angle AB'C$ and $\angle AB''C$ related in the diagram for the situation where $\angle A$ is acute and $b \sin A < a < b$?
5. **a.** How many obtuse angles can a triangle have?
   b. What must be true about the length of the side opposite an obtuse angle of a triangle in comparison with the lengths of the other two sides? How does this explain the conditions and diagrams for the situation where $\angle A$ is obtuse?
6. Based on the diagrams above, explain why the case where the measures of two sides and a non-included angle are known is called the *ambiguous* case.

ANSWERS Section 14.5

Exploration

Radius r (cm)	Number of Triangles
2	0
4	1
6	2
8	1
10	1

Questions

1. 4 cm; Yes; the height is always the length of the perpendicular dropped from the vertex at C to the line containing the opposite side. Since AC and $\angle A$ are the same for all triangles, h is the same for all triangles.
2. 1; 4 cm
3. a. 6 cm b. $h < r < b$
4. 1; 4 cm, 8 cm, 10 cm
5. $r > b$; $r \leq b$

Think and Communicate

1. the height of the triangle
2. The shortest distance from C to the base is the perpendicular, whose length is $b \sin A$. If $a < b \sin A$, the arc will not intercept the base, and no triangle will be formed.
3. a right triangle
4. They are supplementary angles.

EXAMPLE 2

For $\triangle ABC$, $\angle A = 38°$, $a = 14$, and $b = 21$. Find the measure of $\angle B$.

SOLUTION

Step 1 Check to see how many triangles are possible.

$b \sin A = 21 \sin 38°$
≈ 12.9

$12.9 < 14 < 21$

Compare a with $b \sin A$ and with b.

Since $b \sin A < a < b$, two triangles are possible: $\triangle ACB'$ and $\triangle ACB''$.

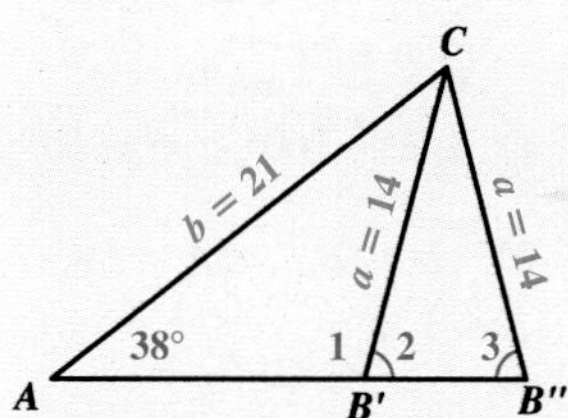

Note that $\triangle B'CB''$ is isosceles, so $\angle 2 = \angle 3$. Therefore, since $\angle 1$ and $\angle 2$ are supplementary, $\angle 1$ and $\angle 3$ are also supplementary.

Step 2 Use the fact that $\frac{\sin A}{a} = \frac{\sin B}{b}$.

$$\frac{\sin 38°}{14} = \frac{\sin B}{21}$$

$$\frac{21 \sin 38°}{14} = \sin B$$

$$0.9235 \approx \sin B$$

$\angle B \approx 67.4°$ or $\angle B \approx 180° - 67.4° = 112.6°$

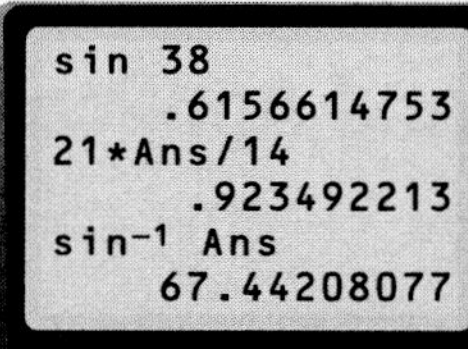

EXAMPLE 3

For $\triangle JKL$, $\angle J = 128°$, $j = 32$, and $k = 20$. Find the measure of $\angle K$.

SOLUTION

Step 1 Check to see how many triangles are possible. Since $\angle J$ is obtuse and $j > k$, one triangle is possible.

Note that $\angle K$ must be acute since $\angle J$ is obtuse and a triangle can have only one obtuse angle.

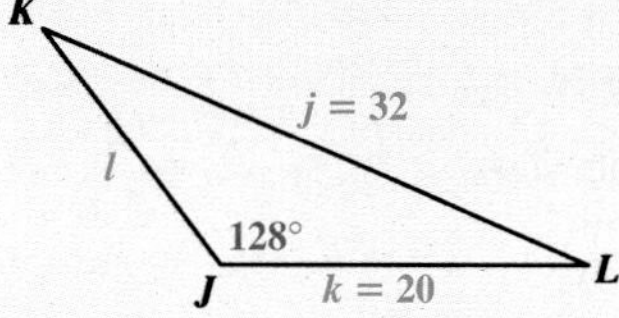

Step 2 Use the fact that $\frac{\sin K}{k} = \frac{\sin J}{j}$.

$$\frac{\sin K}{20} = \frac{\sin 128°}{32}$$

$$\sin K = \frac{20 \sin 128°}{32}$$

$$\sin K \approx 0.4925$$

$$\angle K \approx 29.5°$$

5. a. 1
   b. It must be longer than either of the other sides. If it is not longer than the given side, no triangle can be formed, and if it is longer, then one triangle can be formed.
6. Two triangles are possible, one with an acute angle at B and one with an obtuse angle at B.

Teach⇔Interact

Additional Example 2

For $\triangle ABC$, $\angle A = 62°$, $a = 20$, and $b = 22$. Find the measure of $\angle B$.

Step 1 Check to see how many triangles are possible.
$b \sin A = 22 \sin 62°$
≈ 19.4
Since $b \sin A < a < b$, two triangles are possible: $\triangle ACB'$ and $\triangle ACB''$.

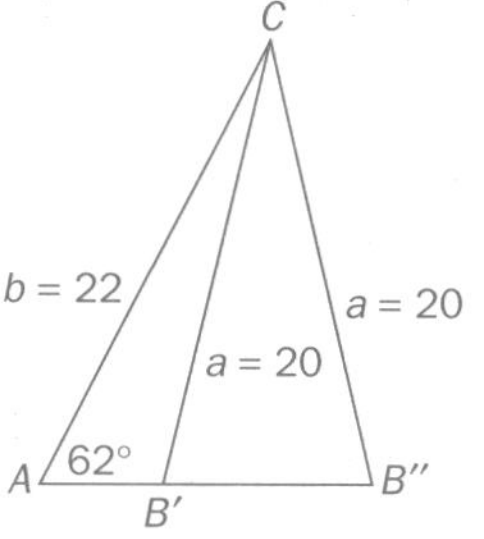

Step 2 Use the fact that $\frac{\sin A}{a} = \frac{\sin B}{b}$.

$$\frac{\sin 62°}{20} = \frac{\sin B}{22}$$

$$\frac{22 \sin 62°}{20} = \sin B$$

$$0.9712 \approx \sin B$$

$\angle B \approx 76.2°$ or
$\angle B \approx 180° - 76.2° = 103.8°$

Additional Example 3

For $\triangle RST$, $\angle R = 111°$, $r = 43$, and $s = 32$. Find the measure of $\angle S$.

Step 1 Check to see how many triangles are possible. Since $\angle R$ is obtuse and $r > s$, one triangle is possible.

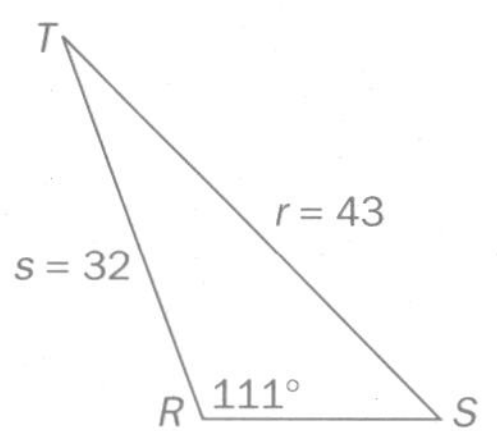

Step 2 Use the fact that $\frac{\sin S}{s} = \frac{\sin R}{r}$.

$$\frac{\sin S}{32} = \frac{\sin 111°}{43}$$

$$\sin S = \frac{32 \sin 111°}{43}$$

$$\sin S \approx 0.6948$$

$$\angle S \approx 44.0°$$

Teach⇔Interact

Checking Key Concepts

Communication: Drawing Have students use a ruler and compass to draw the "triangles" listed in questions 4–7. They should then be able to see which sides and angles do not form triangles (no solution).

Closure Question

Describe how to use the law of sines to solve a triangle.
Check to see how many triangles are possible. If the given conditions are angle-side-angle or angle-angle-side, one triangle is possible. If the given conditions are the ambiguous case, compare $b \sin A$ to a and b as is shown on page 686. Then apply the law of sines to each triangle possible.

Apply⇔Assess

Suggested Assignment

Core Course
Exs. 1–7, 9–17, 26–36

Extended Course
Exs. 1–7, 9–36

Block Schedule
Day 3 Exs. 1–7, 9–17, 26–36

Exercise Notes

Teaching Tip
Exs. 1–7 Remind students that to solve a triangle they have to find all the missing sides and angles. Students may find it useful to make a list of the sides and angles they need to find so as not to miss any.

Cooperative Learning
Ex. 8 After groups have played the game provided by this exercise, it would be worthwhile to have a class discussion of the strategies developed by the various groups.

Checking Key Concepts

Solve each $\triangle ABC$. If there are two solutions, give both. If there is no solution, explain why.

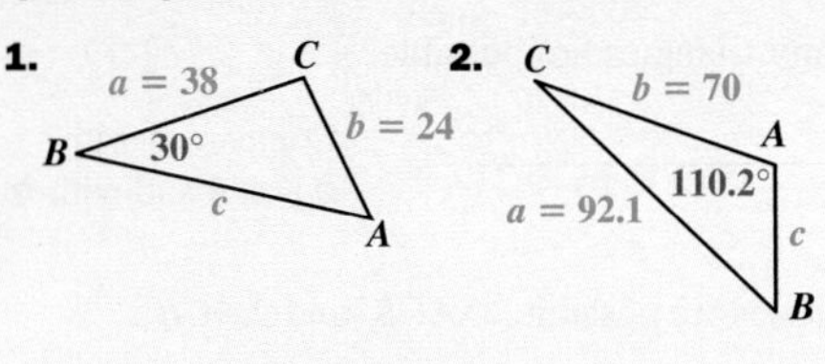

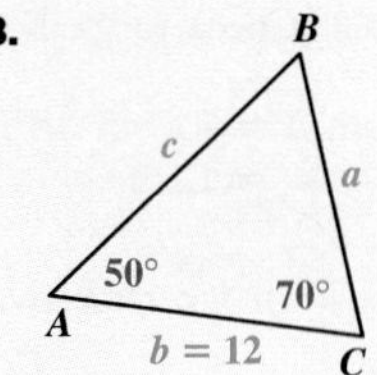

4. $a = 16, b = 19.2, \angle A = 81°$

5. $b = 14, \angle A = 100°, \angle C = 32°$

6. $a = 7.9, b = 9.9, \angle A = 53°$

7. $a = 21\frac{2}{5}, b = 24\frac{1}{3}, \angle A = 59°$

14.5 Exercises and Applications

Extra Practice exercises on page 771

Solve each triangle.

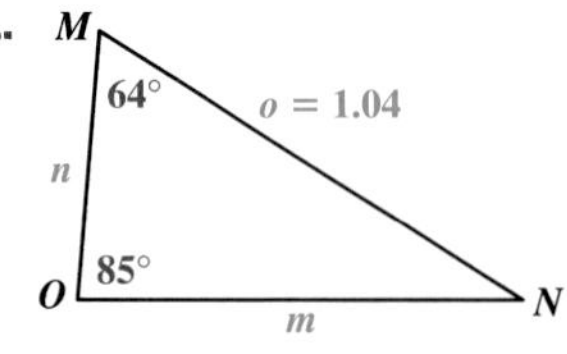

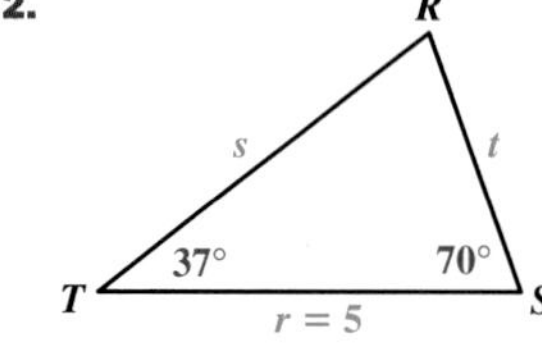

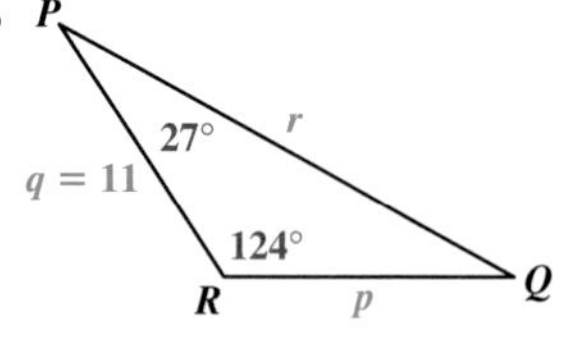

4. $\triangle DEF$ with $f = 30, \angle D = 50°, \angle E = 25°$

5. $\triangle XYZ$ with $y = 18, \angle X = 84°, \angle Z = 39°$

6. $\triangle JKL$ with $j = 7.25, \angle K = 116°, \angle L = 44°$

7. $\triangle ABC$ with $a = 4\frac{3}{5}, \angle B = 28°, \angle C = 76°$

8. Cooperative Learning Play three rounds of the following game with two other students. The object is to make the greatest number of triangles.

- Have the person to your left choose a measure for $\angle A$.
- Have the person to your right choose a length for b.
- You should choose a length for a. You earn one point for each $\triangle ABC$ that can be made with the given side lengths and angle measure.

a. What are the possible scores you can get in one turn?

b. What kind of angle should the person to your left choose to make sure that you will not earn more than one point during your turn?

c. How should you choose a when $\angle A$ is obtuse? when $\angle A$ is acute?

d. Writing Describe a strategy for the game that maximizes the points you earn. Try out your strategy by playing two or three more rounds with your partners. Did your partners use similar strategies? What happens when you all use the same strategy?

Checking Key Concepts

1. $\angle A \approx 52.3°$; $\angle C \approx 97.7°$; $c \approx 47.6$
2. $\angle B \approx 45.5°$; $\angle C \approx 24.3°$; $c \approx 40.4$
3. $\angle B = 60°$; $a \approx 10.6$; $c \approx 13.0$
4. no solution; $a < b \sin A$
5. $\angle B = 48°$; $a \approx 18.6$; $c \approx 10.0$
6. no solution; $a < b \sin A$
7. $\angle B \approx 77.1°$, $\angle C \approx 43.9°$; $c \approx 17.3$; $\angle B \approx 102.9°$; $\angle C \approx 18.1°$; $c \approx 7.8$

Exercises and Applications

1. $\angle N = 31°$; $m \approx 0.94$; $n \approx 0.54$
2. $\angle R = 73°$; $s \approx 4.9$; $t \approx 3.1$
3. $\angle Q = 29°$; $r \approx 19$; $p \approx 10$
4. $\angle F = 105°$; $d \approx 24$; $e \approx 13$
5. $\angle Y = 57°$; $x \approx 21$; $z \approx 14$
6. $\angle J = 20°$; $k \approx 19$; $l \approx 15$
7. $\angle A = 76°$; $b \approx 2.2$; $c = 4.6$
8. a. 0, 1, or 2
   b. obtuse angle
   c. If $\angle A$ is obtuse, choose any a longer than b; if $\angle A$ is acute, choose a shorter than b but longer than $b \sin A$.
   d. Answers may vary. An example is given. If it is your turn to pick an angle, choose an obtuse angle. It does not matter what you pick for b when it is your turn to pick b. When choosing a for an obtuse

Solve each $\triangle ABC$. If there are two solutions, give both. If there is no solution, explain why.

9.

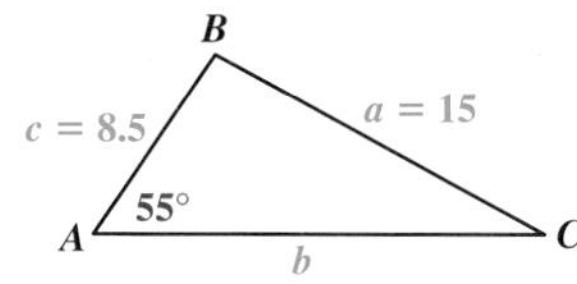

10.

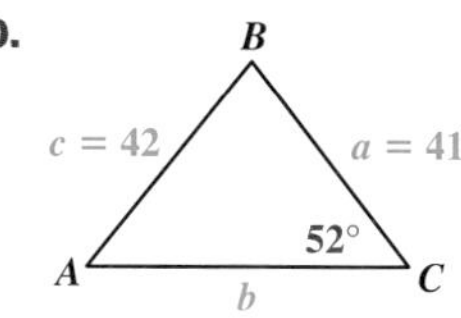

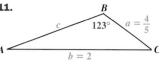

11. B, c, 123°, $a = \frac{4}{5}$, A, $b = 2$, C

12. $a = 34, b = 42, \angle A = 47°$

13. $\angle A = 56°, a = 20, b = 12$

14. $\angle A = 102°, b = 27, a = 41$

15. $\angle C = 137.1°, b = 97.2, c = 72$

16. $a = 12, b = 8, \angle B = 44°$

17. $a = 17.4, c = 21.3, \angle C = 92°$

Connection LITERATURE

In the C.S. Forester novel *The African Queen,* the mechanic Mr. Allnut and his employer Miss Rose travel along a river on a dilapidated boat during World War I. In this passage from the novel, the enemy captain tries to capture the boat as it travels past him.

> They were right in the eye of the sun now, and the glare off the water made the foresight indistinct. It was very easy to lose sight of the white awning of the boat as he aimed.
>
> A thousand metres was a long range for a Martini rifle with worn rifling. He fired, reloaded, fired again, and again, and again.

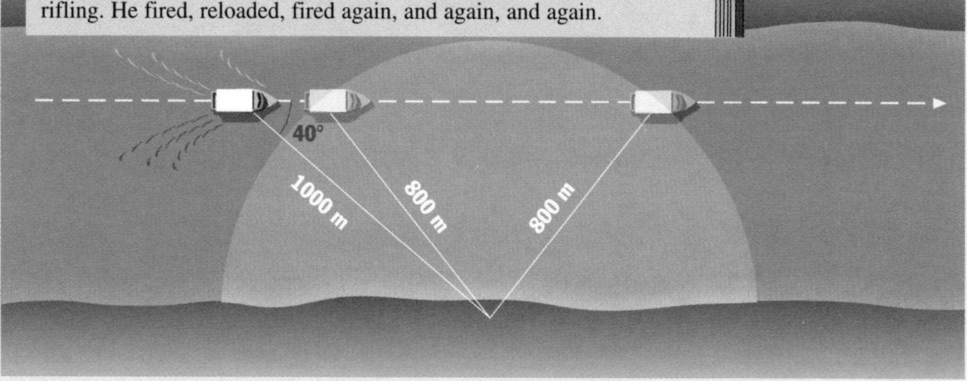

Suppose that when the boat was 1000 m from the captain, the angle formed by the direction of the boat and the line of sight from the boat to the enemy captain was 40°. Assume the range of the Martini rifle was 800 m.

18. How far did the boat travel before the captain's fire could reach it?

19. What distance did the boat have to travel through the captain's fire?

20. **Visual Thinking** Where should the boat travel to minimize its exposure to the gunfire? Explain.

21. **Challenge** Suppose the boat traveled a path 100 m closer to the bank of the river where the captain stood. How far would the boat have to travel through the captain's fire?

Apply⇔Assess

Exercise Notes

Common Error

Ex. 12 Students sometimes forget to find the second solution when there are two. Remind students to check for the value of $b \sin A$ in this triangle. Since $b \sin A < a < b$ and $\angle A$ is acute, there are two solutions.

Reasoning

Exs. 15, 16 Students may assume that if they are given two sides and an angle that there is a unique triangle determined by those conditions. For these exercises, there is no triangle that can be constructed.

Multicultural Note

Exs. 22–25 India possesses one of the world's most varied landscapes, including fertile plains, arid deserts, tropical forests, and the soaring Himalayan Mountains. Some Indians live in villages that have not changed much over the centuries, while others live in modern cities such as Bombay, which has more than 10 million residents. Most Indians are followers of the Hindu faith, but there are also many millions who are followers of Islam, as well as smaller groups of Christians, Sikhs, Buddhists, and Jains. Dating back to the third millennium B.C., Indian culture has produced a great many works of outstanding literature, including the sacred *Mahabharata*, impressive architecture such as the Taj Mahal, and countless works of art and music.

angle, choose any a longer than b. For an acute angle, choose a just slightly shorter than b. If everyone uses the same strategy, the scores will be about the same.

9. $\angle B \approx 97.3°$; $\angle C \approx 27.7°$; $b \approx 18.2$

10. $\angle A \approx 50.3°$; $\angle B \approx 77.7°$; $b \approx 52$

11. $\angle A \approx 19.6°$; $\angle C \approx 37.4°$; $c \approx 1.4$

12. $\angle B \approx 64.6°$, $\angle C \approx 68.4°$, $c \approx 43.2$; $\angle B \approx 115.4°$, $\angle C \approx 17.6°$, $c \approx 14.1$

13. $\angle B \approx 29.8°$; $\angle C \approx 94.2°$; $c \approx 24$

14. $\angle B \approx 40.1°$; $\angle C \approx 37.9°$; $c \approx 26$

15. no solution; The side opposite the obtuse angle is not the longest.

16. no solution; Since $\angle B$ is acute and $b < a \sin b$, no triangle can be formed.

17. $\angle A = 54.7°$; $\angle B = 33.3°$; $b \approx 11.7$

18. about 290 m

19. about 953 m

20. as close to the opposite bank as possible, since the intersection of the path of the boat with the arc representing danger is shorter the farther you get from the captain

21. about 1175 m

Apply⇔Assess

Exercise Notes

Historical Connection
Exs. 22–25 The angles of triangles constructed during a survey by triangulation are measured with a *theodolite*, which is part protractor and part telescope. For the survey of India, the theodolite measured the horizontal angles, and the law of sines gave the horizontal distance from a vertex of the triangle (called a *station*) to an inaccessible point below the summit of a mountain. To measure the height of a mountain, the surveyors took an average of several calculations.

Assessment Note
Ex. 26 A discussion of this exercise will show how well students understand the ideas in this section.

Practice 93 for Section 14.5

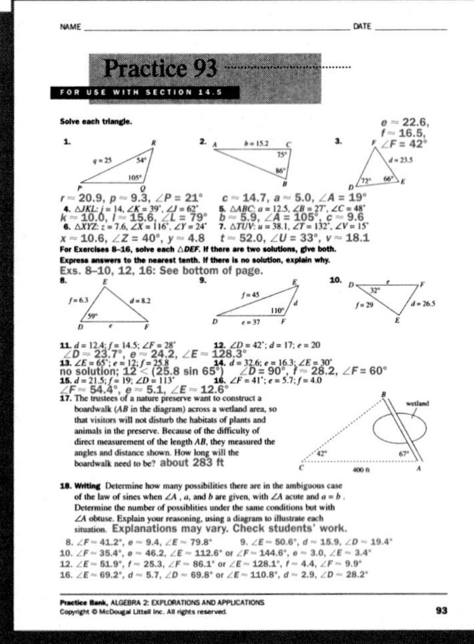

NAME ______ DATE ______

Practice 93

FOR USE WITH SECTION 14.5

Solve each triangle.

1. (triangle PQR: $q = 25$, $54°$, $105°$) $r \approx 20.9$, $p \approx 9.3$, $\angle P = 21°$
2. (triangle ABC: $b = 15.2$, $75°$, $86°$) $c \approx 14.7$, $a \approx 5.0$, $\angle A = 19°$
3. (triangle DEF: $d = 23.5$, $72°$, $66°$) $e \approx 22.6$, $f \approx 16.5$, $\angle F = 42°$
4. $\triangle JKL$: $j = 14$, $\angle K = 39°$, $\angle J = 62°$ $k \approx 10.0$, $l \approx 15.6$, $\angle L = 79°$
5. $\triangle ABC$: $a = 12.5$, $\angle B = 27°$, $\angle C = 48°$ $b \approx 5.9$, $\angle A = 105°$, $c \approx 9.6$
6. $\triangle XYZ$: $z = 7.6$, $\angle X = 116°$, $\angle Y = 24°$ $x \approx 10.6$, $\angle Z = 40°$, $y \approx 4.8$
7. $\triangle TUV$: $u = 38.1$, $\angle T = 132°$, $\angle V = 15°$ $t \approx 52.0$, $\angle U = 33°$, $v \approx 18.1$

For Exercises 8–16, solve each $\triangle DEF$. If there are two solutions, give both. Express answers to the nearest tenth. If there is no solution, explain why.
Exs. 8–10, 12, 16: See bottom of page.

8. ($f = 6.3$, $d = 8.2$, $59°$)
9. ($f = 45$, $e = 37$, $110°$)
10. ($f = 29$, $d = 26.5$, $32°$)
11. $d = 12.4$; $f = 14.5$; $\angle F = 28°$ $\angle D \approx 23.7°$, $e \approx 24.2$, $\angle E \approx 128.3°$
12. $\angle D = 42°$; $d = 17$; $e = 20$
13. $\angle E = 65°$; $e = 12$; $f = 25.8$ no solution; $12 < (25.8 \sin 65°)$
14. $d = 32.6$; $e = 16.3$; $\angle E = 30°$ $\angle D = 90°$, $f \approx 28.2$, $\angle F = 60°$
15. $d = 21.5$; $f = 19$; $\angle D = 113°$ $\angle F \approx 54.4°$, $e \approx 5.1$, $\angle E \approx 12.6°$
16. $\angle F = 41°$; $e = 5.7$; $f = 4.0$
17. The trustees of a nature preserve want to construct a boardwalk (AB in the diagram) across a wetland area, so that visitors will not disturb the habitats of plants and animals in the preserve. Because of the difficulty of direct measurement of the length AB, they measured the angles and distance shown. How long will the boardwalk need to be? about 283 ft (wetland; $42°$, $67°$, 400 ft)
18. **Writing** Determine how many possibilities there are in the ambiguous case of the law of sines when $\angle A$, a, and b are given, with $\angle A$ acute and $a = b$. Determine the number of possiblities under the same conditions but with $\angle A$ obtuse. Explain your reasoning, using a diagram to illustrate each situation. Explanations may vary. Check students' work.

8. $\angle F \approx 41.2°$, $e \approx 9.4$, $\angle E \approx 79.8°$ 9. $\angle E \approx 50.6°$, $d \approx 15.9$, $\angle D \approx 19.4°$
10. $\angle F \approx 35.4°$, $e \approx 46.2$, $\angle E \approx 112.6°$ or $\angle F \approx 144.6°$, $e \approx 3.0$, $\angle E \approx 3.4°$
12. $\angle E \approx 51.9°$, $f \approx 25.3$, $\angle F \approx 86.1°$ or $\angle E \approx 128.1°$, $f \approx 4.4$, $\angle F \approx 9.9°$
16. $\angle E \approx 69.2°$, $d \approx 5.7$, $\angle D \approx 69.8°$ or $\angle E \approx 110.8°$, $d \approx 2.9$, $\angle D \approx 28.2°$

Practice Bank, ALGEBRA 2: EXPLORATIONS AND APPLICATIONS
 93

Connection HISTORY

The map below represents part of the Great Trigonometrical Survey of India in the nineteenth century. The surveyors began with one known distance and used triangle trigonometry to find other distances. Complete Exercises 22–25 to see how measurements were found from these surveys.

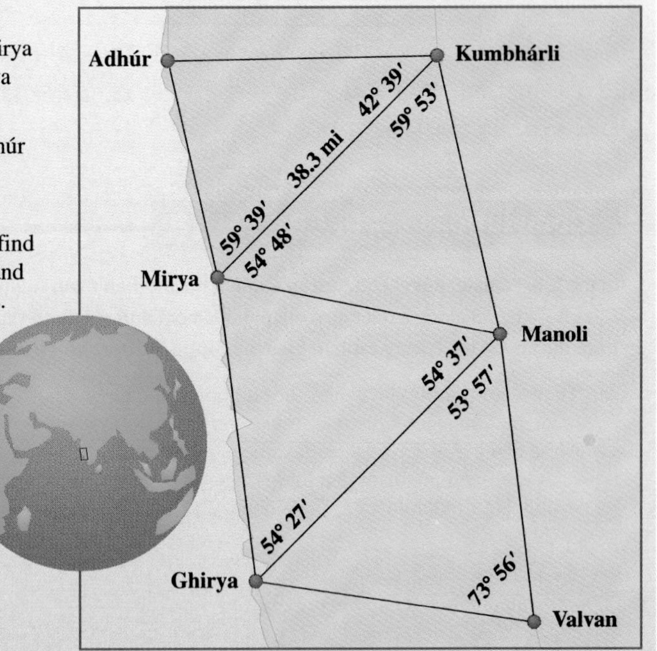

22. Find the distances between Ghirya and Manoli and between Ghirya and Valvan.

23. Find the distance between Adhúr and Kumbhárli and between Manoli and Kumbhárli.

24. Use two different triangles to find the distance between Manoli and Mirya. Compare your answers.

25. a. Find the distance between Ghirya and Adhúr going through Mirya.

b. Writing Do you think the distance you found in part (a) is a good approximation of the straight-line distance between Ghirya and Adhúr? Explain.

ONGOING ASSESSMENT

26. Open-ended Problem Suppose $a = 8$ and $\angle A = 62°$ in $\triangle ABC$. Find a value for b so that $\angle B$ has:

a. two possible values **b.** one possible value **c.** no possible values

SPIRAL REVIEW

Find the area of $\triangle XYZ$. *(Section 14.4)*

27. $x = 4, y = 6, \angle Z = 50°$

28. $y = 22, z = 38, \angle X = 160°$

29. $x = 5.6, z = 9.2, \angle Y = 97°$

30. $x = \frac{8}{9}, y = \frac{3}{4}, \angle Z = 34°$

Write as a logarithm of a single number or expression. *(Section 4.3)*

31. $3 \log_a 4$

32. $5 \log_b 6 - \log_b 12$

33. $-2 \log_b 5$

34. $\log_a 8 + 2 \log_a 4$

35. $2 \ln m^2 - \ln m^4$

36. $\log_c 3 + \log_c 9$

22. about 42.4 mi; about 35.6 mi
23. about 33.8 mi; about 34.4 mi
24. Both estimates round to 36.5 mi.
25. a. about 63.1 mi
 b. Yes; the angle from Adhúr to Ghirya centered at Mirya is 185°, so it is very nearly a straight line.
26. a. $8 < b < \frac{8}{\sin 62°} (\approx 9.06)$
 b. $0 < b \le 8$, or $b = \frac{8}{\sin 62°}$
 c. $b > \frac{8}{\sin 62°}$
27. about 9.2 square units
28. about 143 square units
29. about 25.6 square units
30. about 0.19 square units
31. $\log_a 64$
32. $\log_b 648$
33. $\log_b \frac{1}{25}$
34. $\log_a 128$
35. $\ln 1 = 0$
36. $\log_c 27$

Section

14.6 The Law of Cosines

Learn how to...

- use the law of cosines to solve a triangle

So you can...

- find the step angle of dinosaur tracks, for example

You can think of the hands of a clock as two sides of a triangle whose third side connects the ends of the hands. The shape of the triangle changes as the hands move and the angle they form changes.

Los Angeles

Chicago

New York

THINK AND COMMUNICATE

On each clock shown above, the minute hand is 4 in. long and the hour hand is 2.5 in. long. Use this information for Questions 1–4.

1. What is the measure of the angle formed by the hands of the Chicago clock? How far apart are the ends of the hands?
2. Compare the Los Angeles clock with the Chicago clock.
   a. Is the measure of the angle formed by the hands *greater than* or *less than* the measure of the angle for the Chicago clock?
   b. Is the distance between the ends of the hands *greater than* or *less than* the distance for the Chicago clock?
3. Compare the New York clock with the Chicago clock by repeating parts (a) and (b) of Question 2.
4. Can you find the distance between the ends of the hands of the Los Angeles clock or the New York clock? Why or why not?

You already know how to use the Pythagorean theorem to find the third side of a right triangle. The *law of cosines* is a general rule you can apply to any triangle.

Law of Cosines

For any $\triangle ABC$:

$$a^2 = b^2 + c^2 - 2bc \cos A$$
$$b^2 = a^2 + c^2 - 2ac \cos B$$
$$c^2 = a^2 + b^2 - 2ab \cos C$$

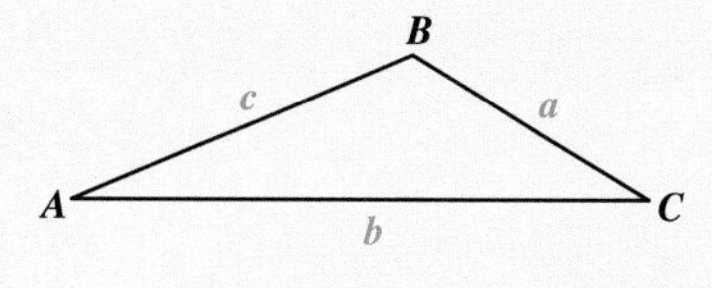

ANSWERS Section 14.6

Think and Communicate

1. 90°; about 4.7 in.
2. a. less than
   b. less than
3. a. greater than
   b. greater than
4. No; they are not right triangles, so you cannot use the Pythagorean theorem. Also, the length of the side opposite the known angle is not given, so you cannot use the law of sines.

Plan⇔Support

Objectives

- Use the law of cosines to solve a triangle.
- Use the law of cosines to solve real-world problems.

Recommended Pacing

Core and Extended Courses
Section 14.6: 2 days

Block Schedule
Section 14.6: 2 half-blocks (with Section 14.5 and Portfolio Project)

Resource Materials

Lesson Support
Lesson Plan 14.6
Warm-Up Transparency 14.6
Practice Bank: Practice 94
Study Guide: Section 14.6
Assessment Book: Test 60

Technology
Technology Book: Spreadsheet Activity 14
Graphing Calculator
Internet: http://www.hmco.com

Warm-Up Exercises

1. If $\triangle ABC$ has $a = 18$, $b = 27$, and $\angle C = 114.5°$, find the area of the triangle. 221.1 square units
2. Describe when the ambiguous case of the law of sines applies. when two sides and the non-included angle are given
3. Solve the equation $a^2 = 18^2 + 27^2 - 2(-201)$ for a. Express the answer to the nearest tenth. ± 38.1
4. The law of sines can be used if the given information in a triangle fits the angle-side-angle, angle-angle-side, or side-side-angle pattern. Which combinations of angles and sides are not included in this list. side-side-side, side-angle-side angle-angle-angle
5. For $\triangle ABC$, $a = 42.5$, $b = 18$, and $\angle A = 124°$. Solve the triangle. $\angle B \approx 20.6°$; $\angle C \approx 35.4°$; $c \approx 29.7$

Teach⟺Interact

Section Notes

Communication: Discussion
Point out that the three forms of the law of cosines are all equivalent. They each have the following pattern:
square of one side = sum of squares of the other two sides minus twice the product of these two sides times the cosine of the angle between them

Reasoning
Ask students to explain why the law of cosines is the same as the Pythagorean theorem when the triangle is a right triangle.

Additional Example 1

Find the missing side length c for $\triangle ABC$ below.

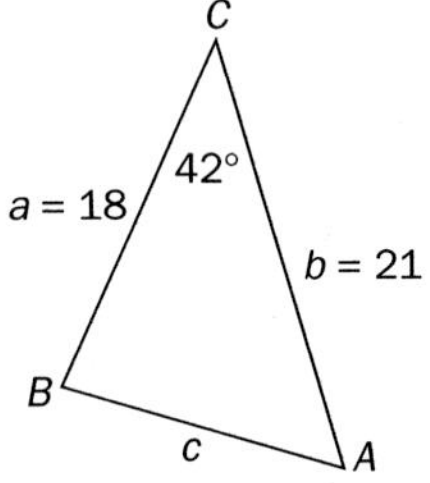

You know the lengths of two sides and the measure of the included angle, so you can use the law of cosines.

$c^2 = a^2 + b^2 - 2ab \cos C$

$c^2 = 18^2 + 21^2 - 2(18)(21) \cos 42°$

$c = \sqrt{18^2 + 21^2 - 2(18)(21) \cos 42°}$

≈ 14.3

The length c is about 14.3 units.

To derive the law of cosines, draw the altitude $\overline{CD}$ of any $\triangle ABC$ and let $CD = h$. Then let $AD = x$. It follows that $DB = c - x$.

Use the Pythagorean theorem to find two different expressions for h^2.

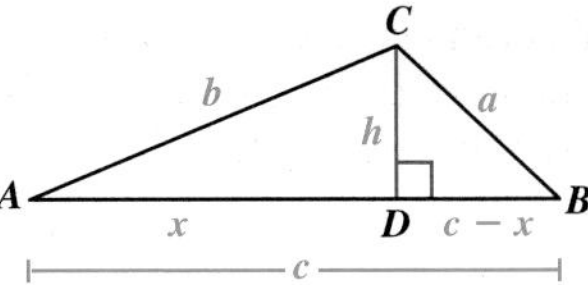

Right $\triangle ACD$

$$x^2 + h^2 = b^2$$
$$h^2 = b^2 - x^2$$

Right $\triangle BCD$

$$h^2 + (c - x)^2 = a^2$$
$$h^2 = a^2 - (c - x)^2$$

Equate the two expressions for h^2.

$$b^2 - x^2 = a^2 - (c - x)^2$$
$$b^2 - x^2 = a^2 - c^2 + 2cx - x^2$$
$$b^2 = a^2 - c^2 + 2cx$$
$$a^2 = b^2 + c^2 - 2cx$$
$$a^2 = b^2 + c^2 - 2c(b \cos A)$$
$$a^2 = b^2 + c^2 - 2bc \cos A$$

Since $\cos A = \frac{x}{b}$, $x = b \cos A$. Substitute **$b \cos A$** for **x**.

THINK AND COMMUNICATE

5. What happens to the law of cosines when the included angle is 90°?

6. Look back at the clocks shown on page 691. Use the law of cosines to find the distance between the ends of the hands of the Los Angeles clock and between the ends of the hands of the New York clock.

EXAMPLE 1

Find the missing side length e for $\triangle DEF$ below.

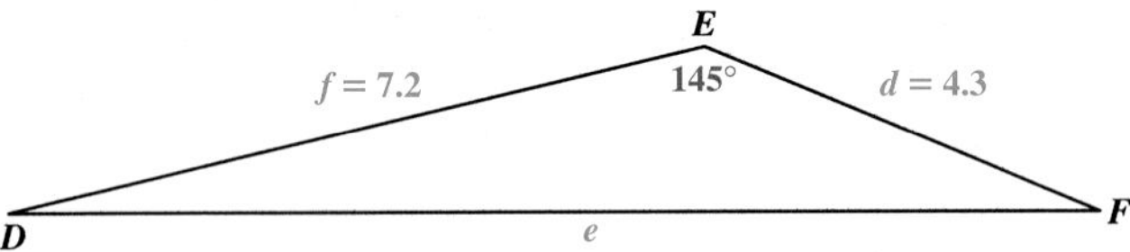

SOLUTION

You know the lengths of two sides and the measure of the included angle, so you can use the law of cosines.

Substitute **4.3** for **d**, **7.2** for **f**, and **145°** for **E**.

$$e^2 = d^2 + f^2 - 2df \cos E$$
$$e^2 = (4.3)^2 + (7.2)^2 - 2(4.3)(7.2) \cos 145°$$
$$e = \sqrt{(4.3)^2 + (7.2)^2 - 2(4.3)(7.2) \cos 145°}$$
$$\approx 11.0$$

The length e is about 11 units.

```
cos 145
     -.8191520443
√(4.3²+7.2²-
2(4.3*7.2*Ans))
      11.00235859
```


Think and Communicate

5. It becomes the Pythagorean theorem.
6. about 2.2 in.; about 5.7 in.

You can also use the law of cosines to find angle measures when only the side lengths of a triangle are known. Start with the formula $a^2 = b^2 + c^2 - 2bc \cos A$ and solve for $\cos A$:

$$a^2 = b^2 + c^2 - 2bc \cos A$$
$$a^2 - b^2 - c^2 = -2bc \cos A$$
$$\frac{b^2 + c^2 - a^2}{2bc} = \cos A$$

You can derive the following formulas for $\cos B$ and $\cos C$ in the same way.

$$\frac{a^2 + c^2 - b^2}{2ac} = \cos B \qquad \frac{a^2 + b^2 - c^2}{2ab} = \cos C$$

EXAMPLE 2 Application: Paleontology

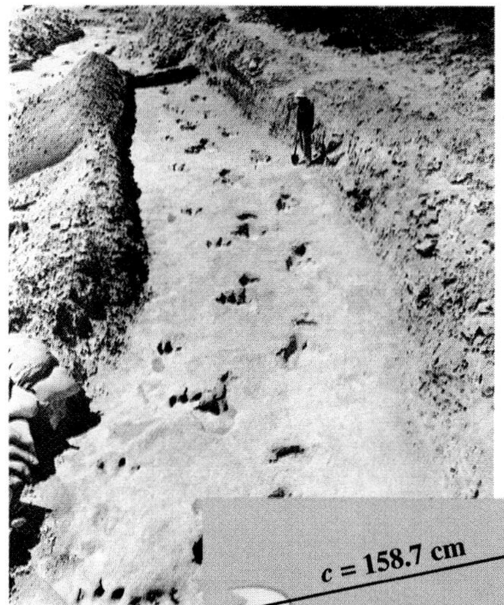

The tracks shown at the left were made by a carnivorous dinosaur during the Cretaceous period in what is now Texas. The tracks are on display at the American Museum of Natural History in New York City.

The *pace lengths*, a and c, and the *stride length*, b, were measured from the tracks themselves. Use the measurements shown in the diagram to find $\angle B$, the *step angle*.

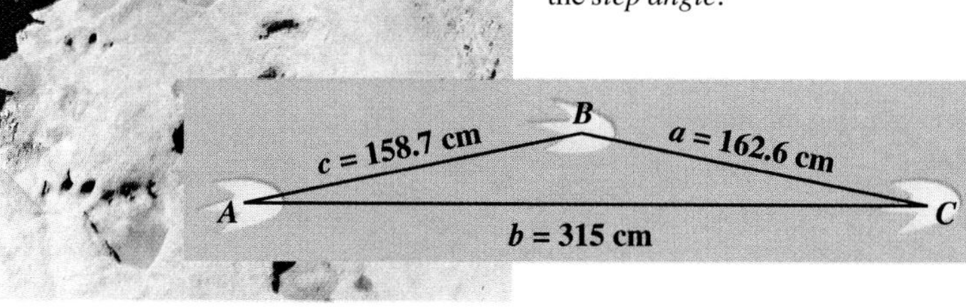

SOLUTION

You know the three side lengths, so you can use the law of cosines to find the angle.

$$\cos B = \frac{a^2 + c^2 - b^2}{2ac}$$

$$\cos B = \frac{(162.6)^2 + (158.7)^2 - 315^2}{2(162.6)(158.7)}$$

Substitute **162.6** for a, **158.7** for c, and **315** for b.

$$\cos B \approx -0.9223$$
$$\angle B \approx \cos^{-1}(-0.9223)$$
$$\angle B \approx 157°$$

The step angle is about 157°.

BY THE WAY

Step angles are important to paleontologists because they indicate roughly how fast an animal is moving: The step angle approaches 180° as the animal moves faster.

Teach⇔Interact

Section Note

Teaching Tip

The alternative form of the law of cosines given on this page should be used when no angle is given. Another way to express the measure of angle A is $\cos^{-1}\left(\frac{b^2 + c^2 - a^2}{2bc}\right)$. The inverse cosine function key on a calculator returns a value between 0° and 180°. This corresponds to the possible angle measures for a triangle.

Additional Example 2

The Great Pyramid at Giza in Egypt had square bases with sides 756 ft long and was originally 481 ft high. Use the measurements shown in the diagram to find $\angle B$, the apex angle of a face of the pyramid.

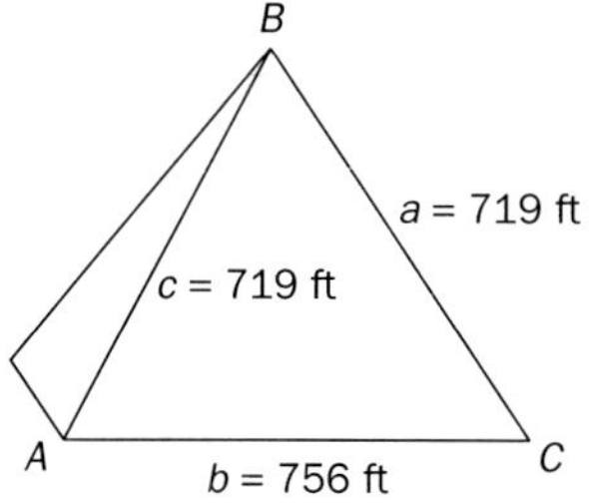

You know the three side lengths, so you can use the law of cosines to find the angle.

$$\cos B = \frac{a^2 + c^2 - b^2}{2ac}$$
$$\cos B = \frac{719^2 + 719^2 - 756^2}{2(719)(719)}$$
$$\cos B \approx 0.4472$$
$$\angle B \approx \cos^{-1}(0.4472)$$
$$\angle B \approx 63.4°$$

The apex angle was about 63.4°.

Teach⇔Interact

Section Notes

Reasoning
Students should understand that a combination of the law of sines and the law of cosines can be used to solve a triangle. For example, once an angle is found for the side-side-side case of the law of cosines, then the law of sines can be used to find the remaining angles of the triangle.

Using Technology
The following program can be used to find ∠*C* *or* side *c* in △*ABC* using the law of cosines, given side-side-side or side-angle-side information for the triangle.

```
PROGRAM:LAWCOS
:Lbl 1
:Input "0=SIDE-SIDE-SIDE 1=SIDE-
   ANGLE-SIDE ",G
:If G>1:Goto 1
:Input "ENTER SIDE A ",A
:Input "ENTER SIDE B ",B
:If G=0:Then:Input "SIDE C ",C
:Disp "ANGLE C = ",cos-1
   ((A2+B2-C2)/(2AB))
:Else:Input "ANGLE C ",C
:Disp "SIDE C= ",√(A2+B2
   -2ABcosC)
:End
```

Checking Key Concepts

Common Error
Some students have difficulty using the law of cosines if the angles and sides are not labeled *A*, *B*, *C*, and *a*, *b*, *c*, respectively. These students may find it useful to rewrite the law of cosines using the labels given on the triangle before trying to solve each exercise.

Closure Question

Explain when to use the law of sines and when to use the law of cosines to solve a given triangle.
See the chart on this page.

You can use the table below to help you remember when to use the law of sines and when to use the law of cosines.

Information given	Law to use	Information to find
angle-angle-side	law of sines	remaining sides*
side-side-angle	law of sines	remaining side and angles
angle-side-angle	law of sines	remaining sides*
side-angle-side	law of cosines	remaining side and one angle*
side-side-side	law of cosines	all three angles

Remember: This is the ambiguous case, where the given information may lead to 0, 1, or 2 triangles.

*You can find the remaining angle by using the fact that the sum of the measures of the angles of a triangle is 180°.

CHECKING KEY CONCEPTS

Find the missing side length of each triangle.

1. *H*, *i* = 13, *g*, 40°, *G*, *h* = 16, *I*

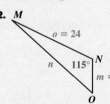

2. *M*, *o* = 24, *N*, *n*, 115°, *m* = 9, *O*

3.

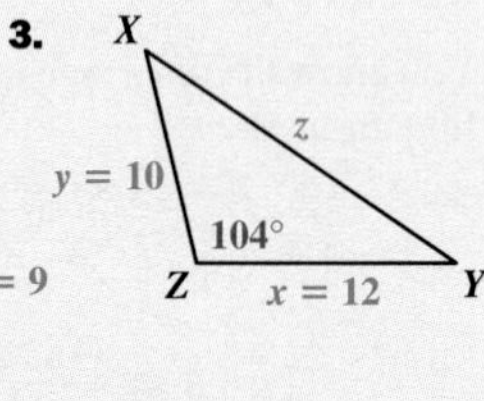

Find the angle measures of each triangle.

4. *R*, *s* = 6, *t* = 15, *T*, *r* = 16, *S*

5. *J*, *l* = 15, *K*, *k* = 12, *j* = 21, *L*

6.

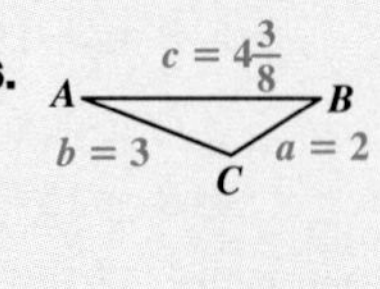

Checking Key Concepts

1. about 10
2. about 29
3. about 17
4. $\angle R \approx 88.4°$; $\angle S \approx 22.0°$; $\angle T \approx 69.6°$
5. $\angle J \approx 101.5°$; $\angle K \approx 34.0°$; $\angle L \approx 44.4°$
6. $\angle A \approx 23.1°$; $\angle B \approx 36.1°$; $\angle C \approx 120.8°$

14.6 Exercises and Applications

Extra Practice exercises on page 771

For Exercises 1–9, find the missing side length of each triangle.

1. 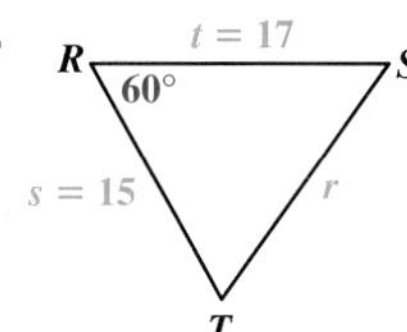

2. 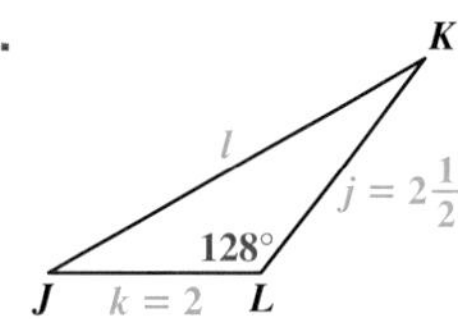

3.
K
l
$j = 2\frac{1}{2}$
128°
J
k = 2
L

4. $\triangle MNO$ with $m = 1$, $o = 7$, $\angle N = 75°$

5. $\triangle XYZ$ with $x = 10$, $y = 11$, $\angle Z = 83°$

6. $\triangle ABC$ with $a = 0.47$, $b = 0.47$, $\angle C = 96°$

7. $\triangle LKJ$ with $k = 20.5$, $l = 29.2$, $\angle J = 15°$

8. $\triangle RST$ with $s = \frac{8}{3}$, $t = \frac{3}{4}$, $\angle R = 172°$

9. $\triangle DEF$ with $d = 6\frac{1}{2}$, $f = 9\frac{1}{4}$, $\angle E = 154°$

10. Writing Chandra used the law of cosines to solve $\triangle ABC$ as shown at the right. Describe Chandra's reasoning. Then apply the same reasoning to side-side-angle cases that lead to two triangles and no triangle.

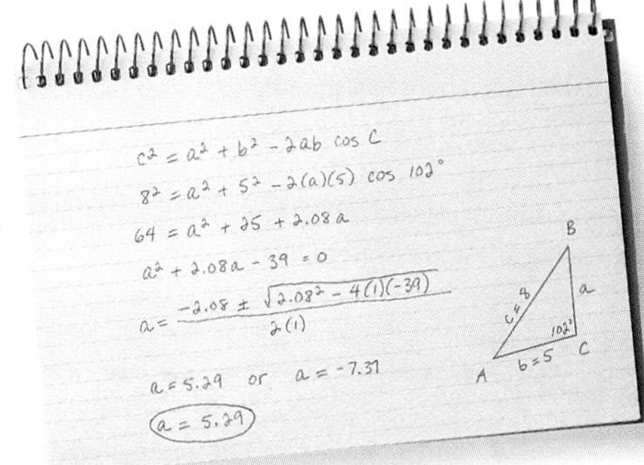

Connection CHEMISTRY

A water molecule is made up of one atom of oxygen and two atoms of hydrogen. The bonds between the oxygen atom and each hydrogen atom form an angle that makes the molecule triangular. As water freezes, the shape of the triangle changes. (*Note:* All measurements are given in picometers (pm), with 1 pm = 10^{-12} m.)

liquid water

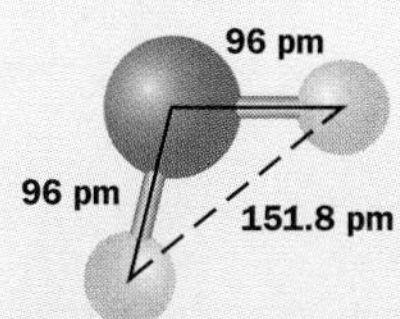

ice

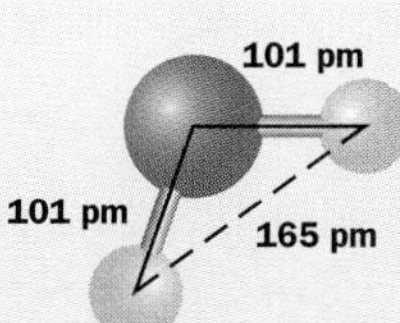

11. Find the angle formed by the two bonds in a molecule of liquid water.

12. Find the angle formed by the two bonds in a molecule of ice.

13. Research Find out which is less dense, *liquid water* or *ice*. How do your answers to Exercises 11 and 12 support this fact?

Apply⇔Assess

Suggested Assignment

Core Course
Day 1 Exs. 1–9, 14–17
Day 2 Exs. 18–22, 24, 25, 28–40, AYP

Extended Course
Day 1 Exs. 1–20
Day 2 Exs. 21–40, AYP

Block Schedule
Day 3 Exs. 1–9, 14–17
Day 4 Exs. 18–22, 24, 25, 28–40, AYP

Exercise Notes

Communication: Discussion *Exs. 1–9, 14–22* You can use these exercises to strengthen students' understanding of when the law of cosines or the law of sines is used. Exs. 1–9 involve the law of cosines only. Exs. 14–22 require an initial use of the law of cosines, followed by at least one application of the law of sines.

Communication: Writing *Ex. 10* Students should be able to list the steps in solving a triangle using the law of cosines, as is done in this exercise. Students should discuss their reasoning in the cases mentioned.

Application *Exs. 11, 12* This application of the law of cosines to chemistry helps to illustrate the wide applicability of trigonometry to science. Point out that the atoms in a molecule are not actually connected to form a triangle, but since they are bonded by physical forces, the distance between the atoms can be measured.

Exercises and Applications

1. about 16
2. about 20
3. about 4.05
4. about 7
5. about 14
6. about 0.70
7. about 10.8
8. about 3.41
9. about 15.4
10. Answers may vary. An example is given. Chandra used the law of cosines equation for the 102° angle. This formed a quadratic equation in the unknown variable. She then used the quadratic formula to solve for the unknown, discarding the negative solution. Cases that lead to two triangles will have two positive solutions to the quadratic equation and cases that lead to no triangle will have no positive solutions.
11. 104.5°
12. 109.5°
13. Ice is less dense. Answers may vary. An example is given. You can see from the dimensions of the drawing that a frozen molecule of water occupies more space than a liquid molecule, thus making ice less dense.

Apply⇔Assess

Exercise Notes

Communication: Discussion
Exs. 14–22 Discuss when to use the law of sines or the law of cosines as they apply to these exercises. After the first angle is found with the law of cosines, students can switch to the law of sines to find the other angles.

Teaching Tip
Exs. 14–22 Remind students that the largest angle in a triangle is opposite the longest side.

Problem Solving
Ex. 23 Students need to complete a triangle before solving for one of the sides in this application.

Second-Language Learners
Ex. 23 Some students may not be familiar with the game of softball. Invite volunteers familiar with the rules of the game to explain such terms as *first base* and *home plate.* Second-language learners may benefit from working on part (c) with the help of a peer tutor or aide.

Visual Thinking
Ex. 23 Ask students to work in groups to create a large diagram of a baseball field with all the dimensions and angles labeled. Have them discuss how they determined all of the measurements represented. This activity involves the visual skills of *identification* and *correlation.*

Application
Exs. 24–26 The use of geometry and symmetry is apparent in many works of art. You may wish to ask students to share their drawings for Ex. 26.

Teaching Tip
Ex. 27 Remind students that a course angle is measured from north in a clockwise direction.

For Exercises 14–22, find the angle measures of each triangle.

14.

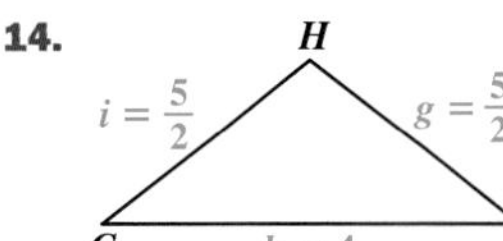

15.

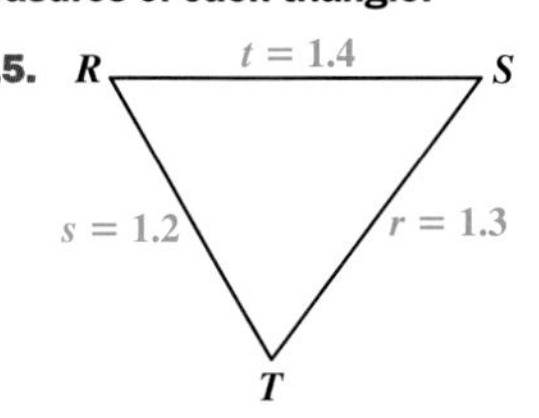

16.

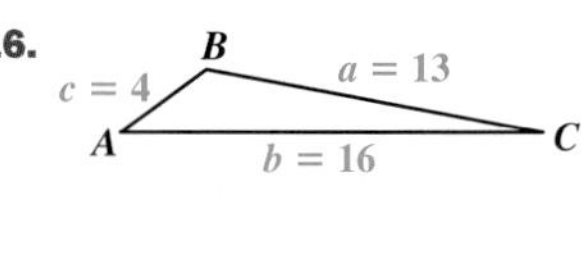

17. $\triangle XYZ$ with $x = 20$, $y = 19$, $z = 2$

18. $\triangle MNO$ with $m = 245$, $n = 720$, $o = 595$

19. $\triangle PQR$ with $p = 8.0$, $q = 14.6$, $r = 6.7$

20. $\triangle DEF$ with $d = 7.5$, $e = 6.3$, $f = 11.9$

21. $\triangle JKL$ with $j = 4\frac{2}{9}$, $k = 5\frac{1}{3}$, $l = 2\frac{7}{8}$

22. $\triangle GHI$ with $g = \frac{17}{2}$, $h = \frac{13}{3}$, $i = \frac{34}{5}$

23. SPORTS The three bases and home plate of a slow-pitch softball field form a 65 ft by 65 ft square. The pitcher's mound lies between home plate and second base, 46 ft from home plate.

a. Find the distance between the pitcher's mound and second base.

b. Find the distance between the pitcher's mound and first base.

c. **Writing** How do the distances between the pitcher's mound and each base differ? Do you think this affects the play of the game? Explain.

d. **Research** Find out the dimensions of a baseball field. Repeat parts (a)–(c) using these measurements.

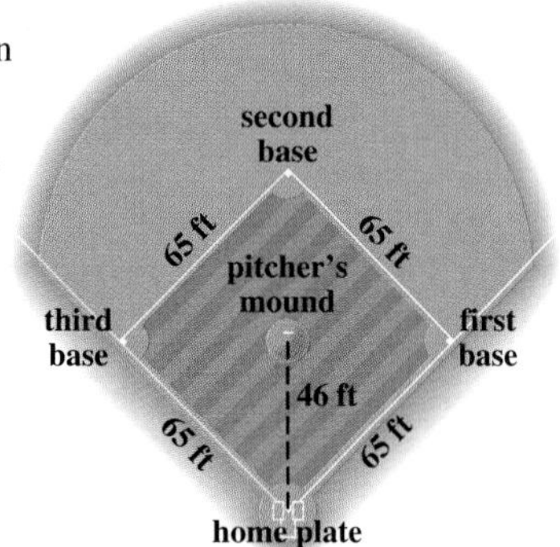

Connection ART

In 1974, the town of Vegreville in Alberta, Canada, honored the Ukrainian settlers of the region by building a giant *pysanki*, or Ukrainian decorated egg. The egg is made of 2208 equilateral triangles and 524 three-pointed stars. Each star consists of three isosceles triangles with 12 in. legs and bases that form equilateral triangles of different sizes.

24. Suppose the point of a three-pointed star is 20°. How long is each side of the equilateral triangle in the center of the star?

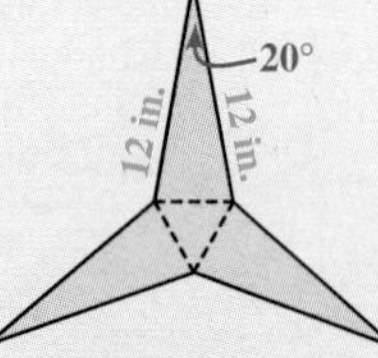

25. To make a three-pointed star so that the equilateral triangle in the center is 8 in. on each side, what must be the angle measure of each point of the star?

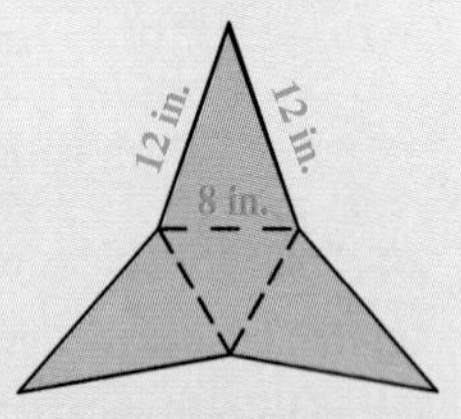

26. Open-ended Problem Draw a star like the ones shown. Give lengths and angle measures for all four triangles that make up the star.

14. $\angle G \approx 36.9°$; $\angle H \approx 106.3°$; $\angle I \approx 36.9°$
15. $\angle R \approx 59.4°$; $\angle S \approx 52.6°$; $\angle T \approx 68.0°$
16. $\angle A \approx 36.4°$; $\angle B \approx 133.1°$; $\angle C \approx 10.5°$
17. $\angle X \approx 117.4°$; $\angle Y \approx 57.5°$; $\angle Z \approx 5.1°$
18. $\angle M \approx 18.5°$; $\angle N \approx 111.0°$; $\angle O \approx 50.5°$
19. $\angle P \approx 7.3°$; $\angle Q \approx 166.6°$; $\angle R \approx 6.1°$
20. $\angle D \approx 33.5°$; $\angle E \approx 27.6°$; $\angle F \approx 118.9°$
21. $\angle J \approx 52.0°$; $\angle K \approx 95.6°$; $\angle L \approx 32.4°$
22. $\angle G \approx 97.0°$; $\angle H \approx 30.4°$; $\angle I \approx 52.6°$

23. a. about 46 ft b. about 46 ft
 c. The pitcher's mound is closest to home plate, and a little closer to first and third bases than it is to second base. Explanations of play may vary.
 d. The bases are 90 ft apart, and the center of the pitcher's mound is 60.5 ft from home plate. about 67 ft; about 64 ft; The pitcher is closest to home plate and farthest from second base.

24. about 4.2 in. 25. 38.9°

26. Answers may vary. Check students' work.

27. 48°; about 1790 km

28. Answers may vary. An example is given. Use the law of cosines to find the angle measure between two of the sides. Draw the first side, then, using a protractor, draw the second side to form the calculated angle. Draw the third side by connecting the other endpoints.

29. $\angle C = 84°$; $a \approx 5.0$; $c \approx 8.5$

30. $\angle B = 48°$; $a \approx 4$; $b \approx 3$

31. $\angle B = 90°$; $\angle C = 60°$; $c \approx 7.36$

32. $\angle A \approx 20.2°$; $\angle B = 10.8°$; $a \approx 7$

27. Challenge A pilot flies a plane 3580 km on a 146° course to get from Caracas to Brasília. On a second flight, the pilot flies 3780 km on a 298° course to get from Brasília to Quito. What course and distance must the pilot fly to return directly to Caracas from Quito?

ONGOING ASSESSMENT

28. Writing Explain how you can use the law of cosines to draw accurately a triangle with side lengths 5, 6, and 7.

SPIRAL REVIEW

Find each missing side length and angle measure of $\triangle ABC$. *(Section 14.5)*

29. $b = 7.4$, $\angle A = 36°$, $\angle B = 60°$ **30.** $c = 2$, $\angle A = 104°$, $\angle C = 28°$

31. $a = 4.25$, $b = 8.5$, $\angle A = 30°$ **32.** $b = 4$, $c = 11$, $\angle C = 149°$

Find $\sin\theta$ and $\cos\theta$ for each value of θ. *(Section 14.3)*

33. $\theta = 135°$ **34.** $\theta = 120°$ **35.** $\theta = 90°$ **36.** $\theta = 45°$

37. $\theta = 330°$ **38.** $\theta = 270°$ **39.** $\theta = 225°$ **40.** $\theta = 210°$

ASSESS YOUR PROGRESS

Solve $\triangle XYZ$. If there are two solutions, give both. If there is no solution, explain why. *(Sections 14.5 and 14.6)*

1. 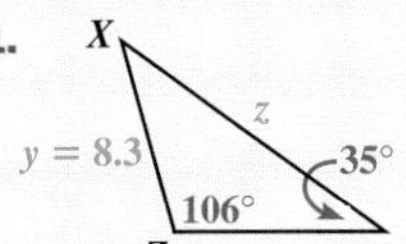

2. 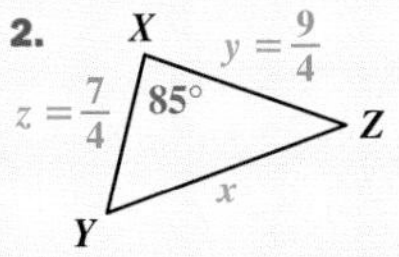

3. 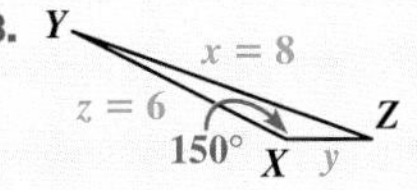

4. $y = 9.75$, $z = 11$, $\angle Y = 57°$ **5.** $x = 40$, $y = 29$, $z = 58$

6. $x = 9$, $z = 7$, $\angle Z = 165°$ **7.** $y = 13.7$, $\angle Y = 43°$, $\angle Z = 28°$

8. $x = 8.3$, $y = 6.05$, $z = 2.6$ **9.** $x = 14.5$, $y = 8$, $\angle Y = 35°$

10. $x = 13$, $y = 19$, $\angle Y = 25°$ **11.** $y = 12.5$, $z = 5$, $\angle X = 124°$

12. $x = \frac{3}{8}$, $z = \frac{3}{4}$, $\angle X = 30°$ **13.** $x = 6\frac{2}{5}$, $z = 6\frac{3}{5}$, $\angle X = 72°$

14. Journal Is the *law of sines* or the *law of cosines* easier for you to remember and use? Explain.

Apply⇔Assess

Exercise Notes

Multicultural Note

Ex. 27 In 1960, Brasilia became the capital of Brazil. President Juscelino Kubitschek hoped that the new capital would spur development in Brazil's sparsely populated interior. Brazilian architect Lúcio Costa planned the city, which includes business and cultural centers, government buildings, apartment buildings, and an artificial lake. During construction, materials were flown to the site; roads connecting Brasilia with the rest of the nation were built later. Today, Brasilia's population is growing rapidly; many residents work in industries such as commerce and construction, or for the Brazilian government.

Progress Check 14.5–14.6

See page 701.

Practice 94 for Section 14.6

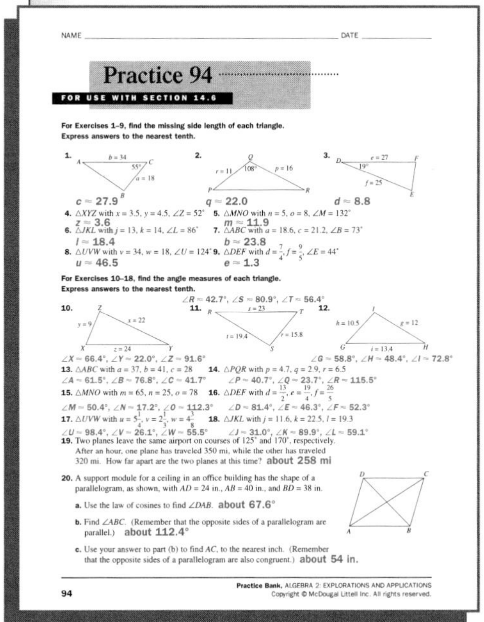
NAME ______ DATE ______

Practice 94

FOR USE WITH SECTION 14.6

For Exercises 1–9, find the missing side length of each triangle. Express answers to the nearest tenth.

1. $b = 34$, $\angle C = 55°$, $a = 18$ — $c \approx 27.9$
2. $r = 11$, $\angle Q = 108°$, $p = 16$ — $q \approx 22.0$
3. $e = 27$, $\angle D = 19°$, $f = 25$ — $d \approx 8.8$

4. $\triangle XYZ$ with $x = 3.5$, $y = 4.5$, $\angle Z = 52°$ — $z \approx 3.6$
5. $\triangle MNO$ with $n = 5$, $o = 8$, $\angle M = 132°$ — $m \approx 11.9$
6. $\triangle JKL$ with $j = 13$, $k = 14$, $\angle L = 86°$ — $l \approx 18.4$
7. $\triangle ABC$ with $a = 18.6$, $c = 21.2$, $\angle B = 73°$ — $b \approx 23.8$
8. $\triangle UVW$ with $v = 34$, $w = 18$, $\angle U = 124°$ — $u \approx 46.5$
9. $\triangle DEF$ with $d = \frac{7}{4}$, $f = \frac{9}{5}$, $\angle E = 44°$ — $e \approx 1.3$

For Exercises 10–18, find the angle measures of each triangle. Express answers to the nearest tenth.

10. $y = 9$, $x = 22$, $z = 24$ — $\angle X \approx 66.4°$, $\angle Y \approx 22.0°$, $\angle Z \approx 91.6°$
11. $s = 23$, $t = 19.4$, $r = 15.8$ — $\angle R \approx 42.7°$, $\angle S \approx 80.9°$, $\angle T \approx 56.4°$
12. $h = 10.5$, $g = 12$, $i = 13.4$ — $\angle G \approx 58.8°$, $\angle H \approx 48.4°$, $\angle I \approx 72.8°$
13. $\triangle ABC$ with $a = 37$, $b = 41$, $c = 28$ — $\angle A \approx 61.5°$, $\angle B \approx 76.8°$, $\angle C \approx 41.7°$
14. $\triangle PQR$ with $p = 4.7$, $q = 2.9$, $r = 6.5$ — $\angle P \approx 40.7°$, $\angle Q \approx 23.7°$, $\angle R \approx 115.5°$
15. $\triangle MNO$ with $m = 65$, $n = 25$, $o = 78$ — $\angle M \approx 50.4°$, $\angle N \approx 17.2°$, $\angle O \approx 112.3°$
16. $\triangle DEF$ with $d = \frac{13}{2}$, $e = \frac{19}{4}$, $f = \frac{26}{5}$ — $\angle D \approx 81.4°$, $\angle E \approx 46.3°$, $\angle F \approx 52.3°$
17. $\triangle UVW$ with $u = 5\frac{1}{4}$, $v = 2\frac{1}{3}$, $w = 4\frac{3}{8}$ — $\angle U \approx 98.4°$, $\angle V \approx 26.1°$, $\angle W \approx 55.5°$
18. $\triangle JKL$ with $j = 11.6$, $k = 22.5$, $l = 19.3$ — $\angle J \approx 31.0°$, $\angle K \approx 89.9°$, $\angle L \approx 59.1°$

19. Two planes leave the same airport on courses of 125° and 170°, respectively. After an hour, one plane has traveled 350 mi, while the other has traveled 320 mi. How far apart are the two planes at this time? **about 258 mi**

20. A support module for a ceiling in an office building has the shape of a parallelogram, as shown, with $AD = 24$ in., $AB = 40$ in., and $BD = 38$ in.

a. Use the law of cosines to find $\angle DAB$. **about 67.6°**

b. Find $\angle ABC$. (Remember that the opposite sides of a parallelogram are parallel.) **about 112.4°**

c. Use your answer to part (b) to find AC, to the nearest inch. (Remember that the opposite sides of a parallelogram are also congruent.) **about 54 in.**

94 **Practice Bank,** ALGEBRA 2: EXPLORATIONS AND APPLICATIONS

33. $\frac{\sqrt{2}}{2} \approx 0.7071$; $-\frac{\sqrt{2}}{2} \approx -0.7071$

34. $\frac{\sqrt{3}}{2} \approx 0.8660$; -0.5

35. 1; 0

36. $\frac{\sqrt{2}}{2} \approx 0.7071$; $\frac{\sqrt{2}}{2} \approx 0.7071$

37. -0.5; $\frac{\sqrt{3}}{2} \approx 0.8660$

38. -1; 0

39. $-\frac{\sqrt{2}}{2} \approx -0.7071$; $-\frac{\sqrt{2}}{2} \approx -0.7071$

40. -0.5; $-\frac{\sqrt{3}}{2} \approx -0.8660$

Assess Your Progress

1. $\angle X = 39°$; $x \approx 9.1$; $z \approx 13.9$

2. $\angle Y \approx 55.3°$; $\angle Z \approx 39.7°$; $x \approx 2.73$

3. $\angle Y \approx 8.0°$; $\angle Z \approx 22.0°$; $y \approx 2$

4. $\angle X \approx 51.9°$, $\angle Z \approx 71.1°$, $x \approx 9.15$; $\angle X \approx 14.1°$, $\angle Z \approx 108.9°$, $x \approx 2.83$

5. $\angle X \approx 39.3°$, $\angle Y \approx 27.3°$; $\angle Z \approx 113.4°$

6. no solution; The side opposite the obtuse angle is not the longest side.

7. $\angle X \approx 109°$; $x \approx 19.0$; $z \approx 9.4$

8. $\angle X = 144.2°$, $\angle Y = 25.2°$; $\angle Z = 10.5°$

9. no solution; $y < x \sin y$, so no triangle can be formed.

10. $\angle X \approx 16.8°$, $\angle Z \approx 138.2°$, $z \approx 30$

11. $\angle Y \approx 40.8°$; $\angle Z \approx 15.2°$; $x \approx 15.8$

12. $\angle Y = 60°$; $\angle Z = 90°$; $y = \frac{3\sqrt{3}}{8} \approx 0.65$

13. $\angle Y = 29.3°$, $\angle Z = 78.7°$, $y \approx 3.3$; $\angle Y = 6.7°$, $\angle Z = 101.3°$, $y \approx 0.79$

14. Answers may vary.

Project Notes

Mathematical Goals

- Measure and record an angle of elevation using a transit.
- Measure and record distances from a tall object.
- Apply trigonometric ratios to calculate the height of a tall object using the angle of elevation and distance measurements.

Planning

Materials

- protractor
- drinking straw
- a piece of string
- small weight (such as a key or a washer)
- masking tape
- tape measure

Project Teams

Students can select their own partners and can work together to make the transit. Then they can use the transit with both methods to find the height of a tall object. Students should also work together to compile a report of their project.

Guiding Students' Work

It may be difficult for students to understand how to use a transit. If this is the case, make your own transit before explaining the project and demonstrate how it is used. Also, the steps listed in calculating the height of a tall object using the indirect approach involve the use of geometry. Students may find it helpful to redraw the figure given in the project pages on a piece of paper and label the parts of the triangles (the angles and lengths) being calculated at each step. This may help in understanding the logic behind the approach.

Second-Language Learners

Students learning English may benefit from working with a peer tutor or aide to write their summary reports.

PORTFOLIO PROJECT

How High Is Up?

A surveyor can determine positions and elevations with the help of a tool called a *transit*, which can be used to measure angles. After taking some measurements, a surveyor uses trigonometry to calculate distances that would be impractical or impossible to measure directly.

In this project, you will make a type of transit. Then you will explore and evaluate two trigonometric methods of calculating the height of a tall object.

PROJECT GOAL **Determine the height of a tall object using two trigonometric methods.**

Tape a straw to the base of a protractor.

Tie a string to a weight. Attach the string to the center mark (or hole) on the base of the protractor.

Making and Using a Transit

Work with a partner. You will need a protractor, a drinking straw, a piece of string, a small weight (such as a key or a washer), and masking tape. Assemble your transit as shown.

To measure the angle of elevation to some high point, view the point through the straw. Have your partner record the angle of elevation while you view the point.

For each of the following methods, you will need a tape measure.

The angle of elevation is equal to the measure of the angle formed by the string and the 90° mark.

Method 1: The Direct Approach

Choose a tall object whose height you wish to measure. Stand some distance away from the object, and look through your transit at the top of the object. Draw a diagram of the situation, and record your data on the diagram.

1. MEASURE and record the angle of elevation of the top of the object.

2. MEASURE and record your distance from the bottom of the object.

3. USE a trigonometric ratio to calculate the object's height above eye level.

4. ADD the height at eye level to the height in Step 3 to find the object's total height.

Method 2: The Indirect Approach

Even when it's not possible to measure the distance between observer and object, it's still possible to measure the height of the object. To do this, you must measure the angle of elevation from two points, *A* and *B*, that lie on a line running directly to the object. Draw a diagram like the one shown, and record your data on the diagram.

1. MEASURE and record the angle of elevation of the top of the object from each of points *A* and *B*.

2. MEASURE and record the distance between *A* and *B*.

3. CALCULATE the measure of $\angle ABC$.

4. CALCULATE the measure of $\angle ACB$.

5. USE the law of sines to find *BC*.

6. USE a trigonometric ratio to calculate *CD*, and add the height at eye level to find the total height of the object.

Writing a Report

Write a report summarizing your results. Include your data, calculations, and a completed diagram for each method used. Compare your results from the first and second methods, and try to account for any differences. Why might someone prefer to use one method rather than the other?

Here are some ideas for extending your project:

- Give a geometric argument to explain why the angle between the 90° mark on the protractor and the position of the string on your transit is the same as the angle of elevation.
- Apply the methods of this project to calculate the height of a tall object whose height is already known from some reliable source, and compare your results with the known height.

Self-Assessment

Describe any problems that you had completing the project. How well were you able to apply the trigonometric formulas needed to calculate some of the distances? What formulas, if any, still give you trouble?

Project Notes

Guiding Students' Work

Rubric for Chapter Project

4 Students make their transit and measure the angle of elevation and distances accurately. Students then calculate the height of the tall object correctly using both methods and assemble their measurements and results in a report. The report contains all of the project information and conveys this information so that it is clear and easily understood. The project is extended by completing one of the suggestions given in the text.

3 Students make their transit correctly. They measure the angle of elevation accurately as well as the distances needed to complete the calculations. Students try to calculate the height of a tall object but make an error in applying and using the trigonometric ratios. A report is written and contains all of the data and calculations completed. The report also presents the ideas clearly. Students accomplish or attempt to extend the project by completing one of the suggestions given in the text.

2 Students make their transit correctly but do not use it properly to measure the angle of elevation. Their calculations of the height of the tall object using trigonometric ratios are not accurate, but students make an attempt at completing the project. Students write a report and it contains most of the information from the project. Students do not attempt to extend the project.

1 Students do not make their transit correctly and, as a result, do not have an accurate measurement for an angle of elevation. Students do not attempt to calculate the height of the tall object using either method and, if a report is written, it is incomplete and shows little effort to complete the project. Students should be encouraged to speak with the teacher as soon as possible to review their work and to make a new start on the project.

Chapter Support

Course Guide: Chapter 14
Lesson Plans: Chapter 14
Practice Bank:
Cumulative Practice 95
Study Guide: Chapter 14 Review
Assessment Book:
Chapter Tests 61 and 62
Chapter 14 Alternative Assessment
Test Generator Software
Portfolio Project Book
Preparing for College Entrance Tests
Professional Handbook

Progress Check 14.1–14.2

1. For $\triangle ABC$ below, find sin A and cos A. *(Section 14.1)* 0.8; 0.6

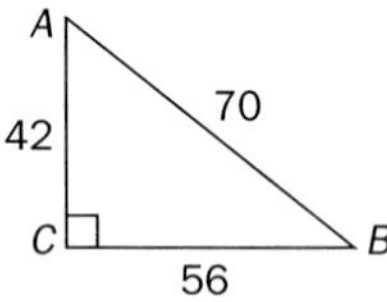

2. Sin $A = 0.7682$ for $\triangle ABC$ with right angle C. Find the measure of $\angle A$ to the nearest tenth of a degree. *(Section 14.1)* 50.2°
3. Find the value of tan A for right $\triangle ABC$ with leg lengths $a = 2$ and $b = 8$. *(Section 14.2)* 0.25
4. The angle of elevation of a flagpole is 32° as shown below. How high is the flagpole? *(Section 14.2)* about 34.4 ft

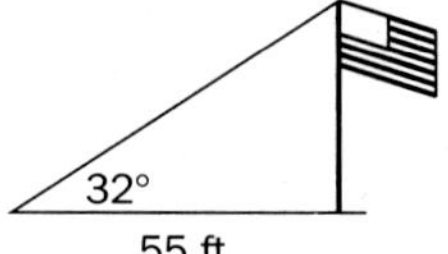

5. Solve $\triangle RST$ with right angle T, $\angle S = 42°$, and $r = \sqrt{18}$. *(Section 14.2)* $t \approx 5.7$, $\angle R = 48°$, $s \approx 3.8$

CHAPTER

14 Review

STUDY TECHNIQUE

Explain a section of the chapter to someone who is not in the class: a parent, a brother or sister, or a friend. Trying to teach someone else what you know often helps to clarify your own understanding.

VOCABULARY

sine (p. 656)
cosine (p. 656)
tangent (p. 663)
solving a triangle (p. 665)
angle of depression (p. 665)
angle of elevation (p. 665)
initial side of an angle (p. 671)
terminal side of an angle (p. 671)
standard position of an angle (p. 671)

SECTIONS 14.1 *and* 14.2

You can find the unknown lengths of sides or measures of angles in a right triangle by using the **sine**, **cosine**, and **tangent** ratios.

Sine, Cosine, and Tangent Ratios

In right $\triangle ABC$, the sine, cosine, and tangent of $\angle A$ are given by:

$$\sin A = \frac{\text{length of leg opposite } \angle A}{\text{length of hypotenuse}}$$

$$\cos A = \frac{\text{length of leg adjacent to } \angle A}{\text{length of hypotenuse}}$$

$$\tan A = \frac{\text{length of leg opposite } \angle A}{\text{length of leg adjacent to } \angle A}$$

Example:

hypotenuse $c = 17$; leg opposite $\angle A$ $a = 8$; $b = 15$ leg adjacent to $\angle A$

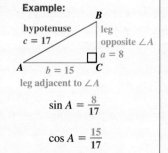

$\sin A = \frac{8}{17}$

$\cos A = \frac{15}{17}$

$\tan A = \frac{8}{15}$

You can measure angles in decimal degrees or in degrees, minutes, and seconds. For example, the angle measure 54.68° is the same as the measure 54°40′48″.

You can find the measure of an angle θ in a right triangle if you know its sine, cosine, or tangent value by using the inverse trigonometric functions. For example, if $\tan \theta = 1.256$, then $\theta = \tan^{-1} 1.256 \approx 51.5°$.

SECTION 14.3

If θ is an angle in **standard position** with the point $P(x, y)$ on its **terminal side** and if r is the distance between P and the origin, then the **trigonometric functions** of θ are defined by:

$\sin\theta = \frac{y}{r}$ $\quad \cos\theta = \frac{x}{r}$ $\quad \tan\theta = \frac{y}{x}$

For any Quadrant I angle θ, there are three other angles between 90° and 360° having the same trigonometric values as θ or opposite trigonometric values.

Examples:

$r = \sqrt{4^2 + 3^2} = 5$

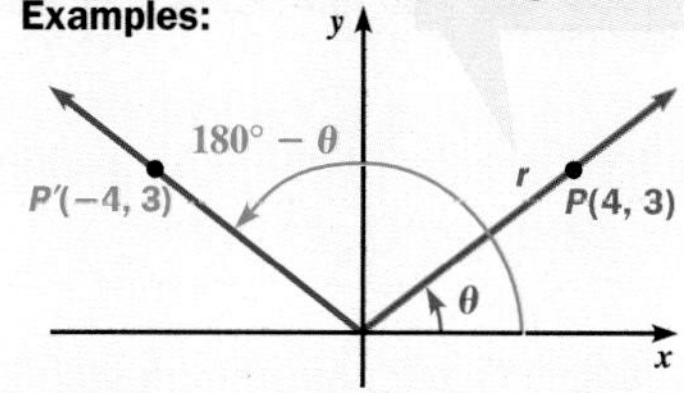

$\sin\theta = \frac{3}{5}$ $\quad \sin(180° - \theta) = \frac{3}{5}$

$\cos\theta = \frac{4}{5}$ $\quad \cos(180° - \theta) = -\frac{4}{5}$

$\tan\theta = \frac{3}{4}$ $\quad \tan(180° - \theta) = -\frac{3}{4}$

SECTIONS 14.4, 14.5, *and* 14.6

You can find the area K of $\triangle ABC$ by using any of these formulas:

$K = \frac{1}{2}bc\sin A$ $\quad K = \frac{1}{2}ac\sin B$ $\quad K = \frac{1}{2}ab\sin C$

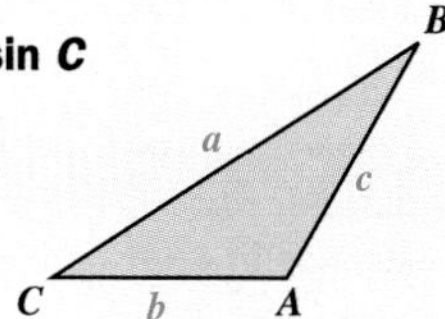

Example: The area of $\triangle ABC$ where $b = 5$, $c = 6$, and $\angle A = 120°$ is:

$K = \frac{1}{2}(5)(6)\sin 120° \approx 15(0.8660) \approx 13$

When you know the measures of some sides or angles in $\triangle ABC$, you can use the *law of sines* or the *law of cosines* to **solve the triangle** to find any unknown measures.

Law of Sines

$$\frac{\sin A}{a} = \frac{\sin B}{b} = \frac{\sin C}{c}$$

Example: If $a = 10$, $b = 8$, and $\angle A = 42°$ for $\triangle ABC$, use the law of sines to find the measure of $\angle B$.

$$\frac{\sin 42°}{10} = \frac{\sin B}{8}$$

$$\sin B = \frac{8\sin 42°}{10} \approx 0.5353$$

$$\angle B = \sin^{-1} 0.5353 \approx 32.4°$$

Law of Cosines

$$a^2 = b^2 + c^2 - 2bc\cos A$$

$$b^2 = a^2 + c^2 - 2ac\cos B$$

$$c^2 = a^2 + b^2 - 2ab\cos C$$

Example: If $a = 5$, $b = 12$, and $\angle C = 125°$ for $\triangle ABC$, use the law of cosines to find c.

$$c^2 = 5^2 + 12^2 - 2(5)(12)\cos 125°$$

$$\approx 237.8$$

$$c \approx \sqrt{237.8} \approx 15.4$$

By solving these formulas for $\cos A$, $\cos B$, and $\cos C$, you can find $\angle A$, $\angle B$, and $\angle C$ when a, b, and c are known.

Progress Check 14.3–14.4

1. Sketch an angle of rotation for $\theta = 315°$. Then use a calculator to find $\sin\theta$, $\cos\theta$, and $\tan\theta$. *(Section 14.3)*

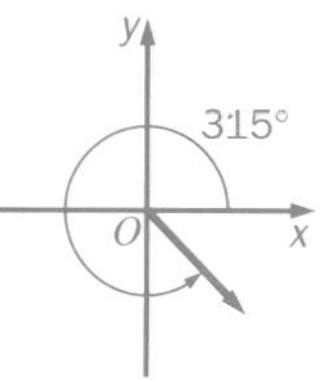

−0.7071; 0.7071; −1

Each point lies on the terminal side of an angle θ in standard position. Find $\sin\theta$, $\cos\theta$, and $\tan\theta$. *(Section 14.3)*

2. (−3, 6) 0.8944; −0.4472; −2
3. (2, −5) −0.9285; 0.3714; −2.5
4. (−6, −8) −0.8; −0.6; 1.3333

Find the area of $\triangle ABC$. *(Section 14.4)*

5. $a = 28.5$, $b = 34.6$, $\angle C = 78°$ about 482.3 square units
6. $b = 35.7$, $c = 48.6$, $\angle A = 118°32'$. about 762.1 square units

Find the area of each sector having central angle θ and radius r. *(Section 14.4)*

7. $\theta = 112°$, $r = \sqrt{12}$ about 11.7 square units
8. $\theta = 314°$, $r = 8.5$ about 198 square units

Progress Check 14.5–14.6

Solve $\triangle ABC$. If there are two solutions, give both. If there is no solution, explain why. *(Sections 14.5, 14.6)*

1. B; $c = 12$; 82°; $a = 21$; A; b; C

   $\angle A \approx 66.4°$, $\angle C \approx 31.6°$, $b \approx 22.7$

2. a; B; C; 127°; $c = 14.36$; $b = 38$; A

   $\angle A \approx 35.4°$, $\angle C \approx 17.6°$; $a \approx 27.6$

3. $a = 3.2$, $b = 5.1$, $\angle C = 53°$ $\angle A \approx 38.8°$, $\angle B \approx 88.2°$, $c \approx 4.1$
4. $a = 10$, $b = 12.8$, $\angle A = 98°$ no solution because $\angle A$ is obtuse and $a < b$

Chapter 14 Assessment
Form A Chapter Test

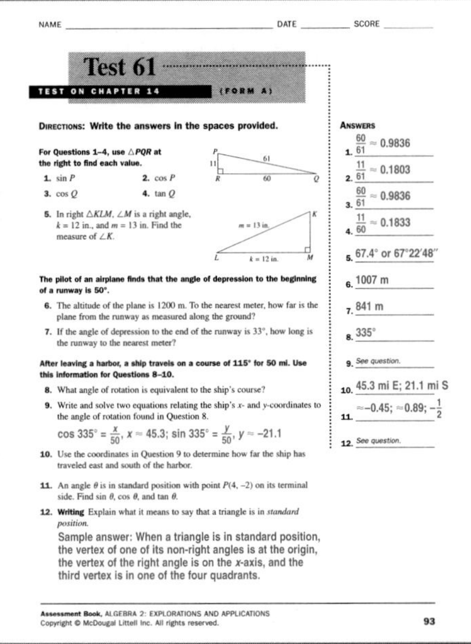

NAME ______________ DATE ________ SCORE ________

Test 61

TEST ON CHAPTER 14 (FORM A)

DIRECTIONS: Write the answers in the spaces provided.

For Questions 1–4, use $\triangle PQR$ at the right to find each value.

1. $\sin P$
2. $\cos P$
3. $\cos Q$
4. $\tan Q$
5. In right $\triangle KLM$, $\angle M$ is a right angle, $k = 12$ in., and $m = 13$ in. Find the measure of $\angle K$.

The pilot of an airplane finds that the angle of depression to the beginning of a runway is 50°.

6. The altitude of the plane is 1200 m. To the nearest meter, how far is the plane from the runway as measured along the ground?
7. If the angle of depression to the end of the runway is 33°, how long is the runway to the nearest meter?

After leaving a harbor, a ship travels on a course of 115° for 50 mi. Use this information for Questions 8–10.

8. What angle of rotation is equivalent to the ship's course?
9. Write and solve two equations relating the ship's x- and y-coordinates to the angle of rotation found in Question 8.
   $\cos 335° = \frac{x}{50}$, $x \approx 45.3$; $\sin 335° = \frac{y}{50}$, $y \approx -21.1$
10. Use the coordinates in Question 9 to determine how far the ship has traveled east and south of the harbor.
11. An angle θ is in standard position with point $P(4, -2)$ on its terminal side. Find $\sin \theta$, $\cos \theta$, and $\tan \theta$.
12. **Writing** Explain what it means to say that a triangle is in *standard position.*
    Sample answer: When a triangle is in standard position, the vertex of one of its non-right angles is at the origin, the vertex of the right angle is on the x-axis, and the third vertex is in one of the four quadrants.

ANSWERS

1. $\frac{60}{61} \approx 0.9836$
2. $\frac{11}{61} \approx 0.1803$
3. $\frac{60}{61} \approx 0.9836$
4. $\frac{11}{60} \approx 0.1833$
5. 67.4° or 67°22′48″
6. 1007 m
7. 841 m
8. 335°
9. *See question.*
10. 45.3 mi E; 21.1 mi S
11. ≈ -0.45; ≈ 0.89; $-\frac{1}{2}$
12. *See question.*

Assessment Book, ALGEBRA 2: EXPLORATIONS AND APPLICATIONS

93

Chapter 14 Assessment
Form B Chapter Test

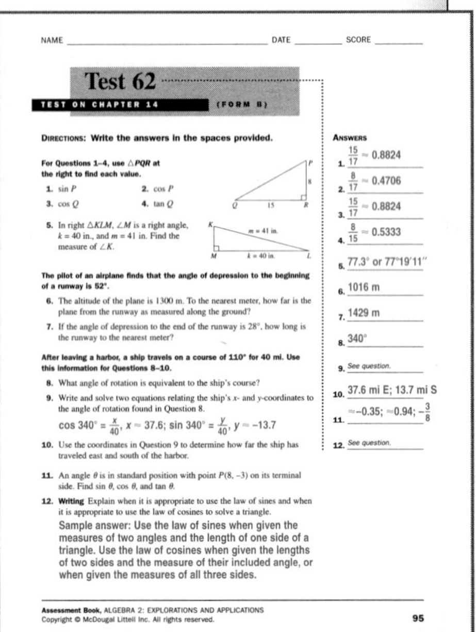

NAME ______________ DATE ________ SCORE ________

Test 62

TEST ON CHAPTER 14 (FORM B)

DIRECTIONS: Write the answers in the spaces provided.

For Questions 1–4, use $\triangle PQR$ at the right to find each value.

1. $\sin P$
2. $\cos P$
3. $\cos Q$
4. $\tan Q$
5. In right $\triangle KLM$, $\angle M$ is a right angle, $k = 40$ in., and $m = 41$ in. Find the measure of $\angle K$.

The pilot of an airplane finds that the angle of depression to the beginning of a runway is 52°.

6. The altitude of the plane is 1300 m. To the nearest meter, how far is the plane from the runway as measured along the ground?
7. If the angle of depression to the end of the runway is 28°, how long is the runway to the nearest meter?

After leaving a harbor, a ship travels on a course of 110° for 40 mi. Use this information for Questions 8–10.

8. What angle of rotation is equivalent to the ship's course?
9. Write and solve two equations relating the ship's x- and y-coordinates to the angle of rotation found in Question 8.
   $\cos 340° = \frac{x}{40}$, $x \approx 37.6$; $\sin 340° = \frac{y}{40}$, $y \approx -13.7$
10. Use the coordinates in Question 9 to determine how far the ship has traveled east and south of the harbor.
11. An angle θ is in standard position with point $P(8, -3)$ on its terminal side. Find $\sin \theta$, $\cos \theta$, and $\tan \theta$.
12. **Writing** Explain when it is appropriate to use the law of sines and when it is appropriate to use the law of cosines to solve a triangle.
    Sample answer: Use the law of sines when given the measures of two angles and the length of one side of a triangle. Use the law of cosines when given the lengths of two sides and the measure of their included angle, or when given the measures of all three sides.

ANSWERS

1. $\frac{15}{17} \approx 0.8824$
2. $\frac{8}{17} \approx 0.4706$
3. $\frac{15}{17} \approx 0.8824$
4. $\frac{8}{15} \approx 0.5333$
5. 77.3° or 77°19′11″
6. 1016 m
7. 1429 m
8. 340°
9. *See question.*
10. 37.6 mi E; 13.7 mi S
11. ≈ -0.35; ≈ 0.94; $-\frac{3}{8}$
12. *See question.*

Assessment Book, ALGEBRA 2: EXPLORATIONS AND APPLICATIONS

95

CHAPTER 14 Assessment

VOCABULARY QUESTIONS

For Questions 1 and 2, complete each sentence.

1. In a right $\triangle ABC$, the ratio $\dfrac{\text{length of leg opposite } \angle A}{\text{length of leg adjacent to } \angle A}$ is the value of the __?__ of $\angle A$.
2. When you __?__, you find the measures of all the sides and all the angles of the triangle.

SECTIONS 14.1 *and* 14.2

For $\triangle ABC$ shown at the right, find each value.

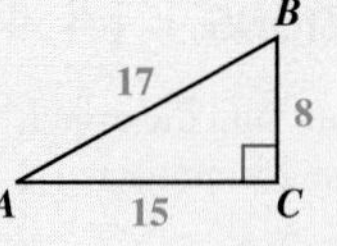

3. $\sin A$
4. $\cos A$
5. $\tan B$
6. In right $\triangle DEF$, $\angle F$ is a right angle, $d = 12$ in., and $f = 24$ in. Find the measure of $\angle D$.
7. **BOTANY** Suppose you want to estimate the height of a several-hundred-year-old redwood in a redwood forest in northern California. The angle of elevation is 82.4° from a point 35 ft from the base of the tree. How tall is the redwood?

SECTION 14.3

8. **NAVIGATION** After leaving an airport, a plane travels on a course of 250° for 100 mi.
   a. What angle of rotation is equivalent to the plane's course?
   b. Write and solve equations relating the plane's x- and y-coordinates to the angle of rotation found in part (a).
   c. How far has the plane traveled in an east-west direction? in a north-south direction?
9. An angle θ is in standard position with point $P(-4, -5)$ on its terminal side. Find $\sin \theta$, $\cos \theta$, and $\tan \theta$.
10. Find all the angles θ between 0° and 360° such that $\cos \theta = 0.2345$.

ANSWERS Chapter 14

Assessment

1. tangent
2. solve a triangle
3. $\frac{8}{17}$
4. $\frac{15}{17}$
5. $\frac{15}{8}$
6. 30°
7. about 262 ft
8. a. 200°
   b. $\cos 200° = \frac{x}{100}$; $x \approx -94$;
   $\sin 200° = \frac{y}{100}$; $y \approx -34$
   c. about 94 mi west; about 34 mi south
9. –0.7809; –0.6247; 1.25
10. 76.4°, 283.6°

Find the area of each triangle.

11.

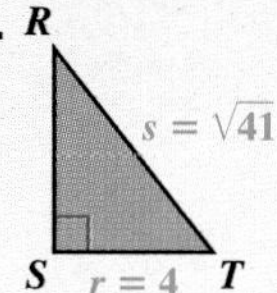

12.

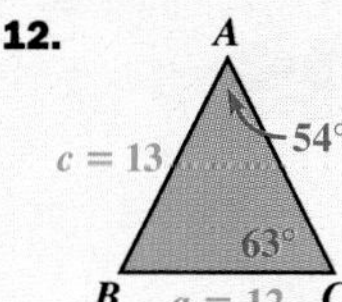

13.

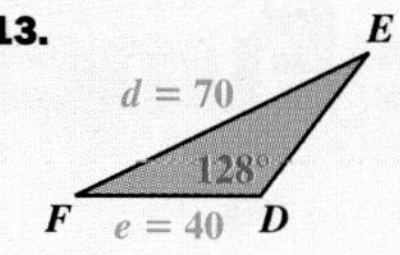

14. **GEOMETRY** Find the area of the stop sign, which is a regular octagon. (*Hint:* Divide the sign into eight triangles. What is the area of each triangle?)

Find the area of each sector.

15.

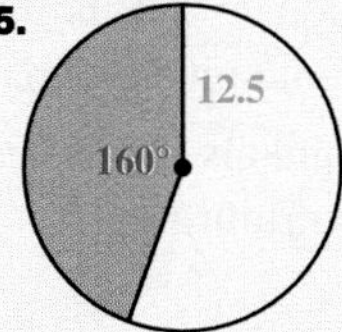

16.

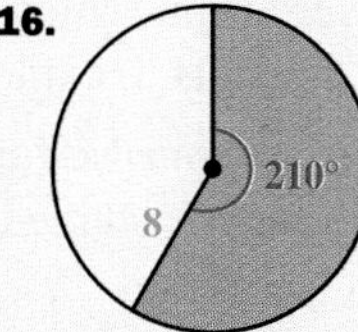

17.

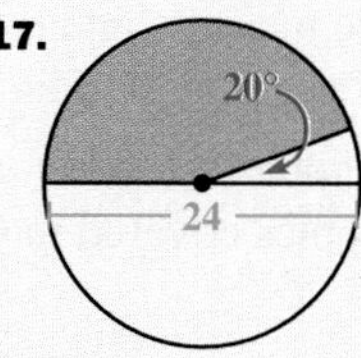

Solve each $\triangle ABC$. If there are two solutions, give both. If there is no solution, explain why.

18. $a = 40, b = 32, \angle A = 115°20'$

19. $a = 71, b = 84, \angle A = 160°$

20. $a = 6, b = 6\sqrt{2}, \angle C = 45°$

21. $b = 25, c = 50, \angle B = 15°$

22. $a = 54, c = 112, \angle C = 120°$

23. $a = 60, b = 50, c = 66$

24. **Writing** Describe the situations in which you can use the law of sines to solve a triangle. Also, describe the situations in which you can use the law of cosines to solve a triangle.

PERFORMANCE TASK

25. Choose two angle measures and a side length, or choose two side lengths and an angle measure. Indicate whether the single side (or angle) chosen is included between the two angles (or sides) chosen.

a. How many triangles can you form with your chosen values?

b. If your answer to part (a) was "none," choose new values until you can form at least one triangle.

c. Solve your triangle(s).

d. Find the area(s) of your triangle(s).

Chapter 14 Assessment
Form C Alternative Assessment

Chapter 14
ALTERNATIVE ASSESSMENT

1. Project Cut a 3 in. by 5 in. index card into three right triangles as shown below.

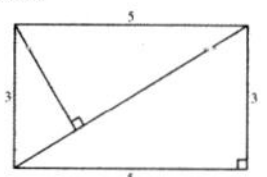

a. Use what you learned in geometry to prove that the triangles are similar.

b. Measure the sides and angles of each triangle.

c. Use what you know about trigonometry to compare the three triangles. Discuss the trigonometric ratios for the angles of each triangle and how their relationships confirm the angle measures.

d. How would the situation change if you used a 4 in. by 6 in. index card instead?

2. Open-ended Problem Suppose the entrance ramp for a restaurant must have a 25° incline. Find some reasonable vertical and horizontal measurements that would result in the correct incline.

3. Open-ended Problem The building code on most stairs requires that the slope be between $\frac{5}{10}$ and $\frac{7}{10}$. Use trigonometry to find the possible angles of elevation for a set of stairs that meets the code.

4. Research Talk to an airplane pilot or a hot air balloonist about how they use *triangulation.*

5. Open-ended Problem Solve the equation $\cos 30° = \frac{20}{x}$ using two different methods.

6. Research Ask a surveyor, an appraisor, or an engineer to describe the device that they use to measure large distances when they do not use triangulation.

Assessment Book, ALGEBRA 2: EXPLORATIONS AND APPLICATIONS

127

11. 10 square units
12. about 69 square units
13. about 597 square units
14. about 742 in.^{2}
15. about 218 square units
16. about 117 square units
17. about 201 square units
18. $\angle B \approx 46°18'$; $\angle C \approx 18°22'$; $c \approx 14$
19. no solution; The side opposite the obtuse angle is not the longest side.
20. $\angle A = 45°$; $\angle B = 90°$; $c = 6$
21. $\angle A \approx 133.8°$, $\angle C \approx 31.2°$, $a \approx 70$; $\angle A \approx 16.2°$, $\angle C \approx 148.8°$, $a \approx 27$
22. $\angle A \approx 24.7°$; $\angle B \approx 35.3°$; $b \approx 75$
23. $\angle A \approx 60.4°$; $\angle B \approx 46.5°$; $\angle C \approx 73.1°$
24. You can use the law of sines when you know the measures of two angles and a side or two sides and an opposite angle. You can use the law of cosines when you know the measures of two sides and their included angle or three sides.
25. Check students' work.

CHAPTER

15 Trigonometric Functions

OVERVIEW

Connecting to Prior and Future Learning

⇔ Students expand their study of trigonometry in this chapter as they study trigonometric functions. The sine, cosine, and tangent functions are presented. A review of dimensional analysis on page 793 of the **Student Resources Toolbox** will be helpful as students complete the exercises in Section 15.1.

⇔ Converting between degree measure and radian measure is presented in Section 15.2. Finding the sine and cosine of angles given in radians allows students to explore periodic phenomena.

⇔ The amplitude, period, and phase shifts of graphs of trigonometric functions are also explored in this chapter. All of the topics covered in Chapter 15 will be studied again in future mathematics courses.

Chapter Highlights

Interview with Joe Lopez: The interview and related example and exercises on pages 715, 718, and 724 emphasize the use of mathematics in the recording industry.

Explorations in Chapter 15 provide students with the opportunity to graph sine and cosine functions in Section 15.1 and explore radian measure by making a radian string in Section 15.2.

The Portfolio Project: Students study simple harmonic motion as they suspend a weight from a spring and find an equation to model the motion of the weight when it is pulled down and released.

Technology: In Section 15.1, scientific calculators are used to complete a table of values for the sine and cosine functions. The uses of graphing calculators throughout this chapter include graphing the sine and cosine functions, exploring amplitude, period, and phase shifts of the graphs of the sine and cosine functions, and converting between degree measure and radian measure.

OBJECTIVES

Section	Objectives	NCTM Standards
15.1	• Work with sine and cosine as functions. • Extend the domain of sine and cosine. • Use sine and cosine functions to solve problems.	1, 2, 3, 4, 5, 6, 9
15.2	• Convert between degree measure and radian measure. • Find the sine and cosine of angles given in radians. • Explore models of periodic phenomena.	1, 2, 3, 4, 5, 6, 9
15.3	• Graph equations of these forms: $y = a \sin bx + k$ and $y = a \cos bx + k$. • Understand periodic phenomena.	1, 2, 3, 4, 5, 6, 9
15.4	• Graph equations of these forms: $y = a \sin b(x - h) + k$ and $y = a \cos b(x - h) + k$. • Write equations for sine and cosine graphs that have a phase shift. • Describe periodic phenomena that have a phase shift.	1, 2, 3, 4, 5, 6, 9
15.5	• Graph the tangent function. • Solve problems using a tangent function.	1, 2, 3, 4, 5, 6, 9

INTEGRATION

Mathematical Connections	15.1	15.2	15.3	15.4	15.5
algebra	**707–712***	**713–718**	**719–726**	**727–734**	**735–739**
geometry	**707–712**	**713–718**	**719–726**	**727–734**	**735–739**
data analysis, probability, discrete math		717, 718		732, 733	
patterns and functions	**707–712**	714–718	**719–726**	**727–734**	**735–739**
logic and reasoning	708–712	713, 715–718	719–721, 723–726	**727–734**	735, 736, 738, 739
Interdisciplinary Connections and Applications					
arts and entertainment				733	
sports and recreation					735, 738
amusement park rides	710, 711		722		
surveying and architecture			725	732	
solar power		717			
automotive mechanics				731	
astronomy					739

* ***Bold page numbers*** *indicate that a topic is used throughout the section.*

TECHNOLOGY

	opportunities for use with	
Section	**Student Book**	**Support Material**
15.1	scientific calculator graphing calculator	
15.2	graphing calculator McDougal Littell Mathpack *Function Investigator*	
15.3	graphing calculator McDougal Littell Mathpack *Function Investigator*	**Function Investigator with Matrix Analyzer Activity Book:** Function Investigator Activities 32 and 34 **Geometry Inventor Activity Book:** Activities 22 and 23
15.4	graphing calculator McDougal Littell Mathpack *Function Investigator*	**Technology Book:** Calculator Activity 15 **Function Investigator with Matrix Analyzer Activity Book:** Function Investigator Activity 33 **Geometry Inventor Activity Book:** Activity 22
15.5	graphing calculator McDougal Littell Mathpack *Function Investigator*	**Technology Book:** Spreadsheet Activity 15 **Function Investigator with Matrix Analyzer Activity Book:** Function Investigator Activity 37

PLANNING GUIDE

Regular Scheduling (45 min)

See Alternative Pacing Charts, page T42.

Section	Materials Needed	Core Assignment	Extended Assignment	*exercises that feature* Applications	Communication	Technology
15.1	scientific calculator, graph paper, graphing calculator or software	1–20, 27–44	1–44	21–26		35
15.2	cylinder, ruler, string, marker, tape, protractor, graphing calculator, graph paper	**Day 1:** 1–20 **Day 2:** 21–38, 46–51	**Day 1:** 1–20 **Day 2:** 21–51	39–45	39, 43	13 41, 50
15.3	graphing calculator, graph paper	1–23, 33, 37–46, AYP*	1–19 odd, 20–46, AYP	24–32	29, 37	23
15.4	graphing calculator, graph paper	**Day 1:** 1–17, 23–25 **Day 2:** 26–30, 32–48	**Day 1:** 1–25 **Day 2:** 26–48	17–22 31	22 39, 40	24, 25 26–38
15.5	graphing calculator, graph paper	1–14, 19–27, AYP	1–27, AYP	15–18	13, 14, 18, 22	
Review/ Assess		**Day 1:** 1–28 **Day 2:** Ch. 15 Test	**Day 1:** 1–28 **Day 2:** Ch. 15 Test			
Portfolio Project		Allow 2 days.	Allow 2 days.			

Pacing (with Portfolio Project) — **Chapter 15 Total** 11 days

Block Scheduling (90 min)

	Day 1	Day 2	Day 3	Day 4	Day 5	Day 6
Teach/Interact	15.1: Exploration, page 707 15.2: Exploration, page 713	Continue with 15.2 15.3	15.4	15.5 Port. Proj.	Review Port. Proj.	Ch. 15 Test
Apply/Assess	**15.1:** 1–20, 27–44 **15.2:** 1–20	**15.2:** 21–38, 46–51 **15.3:** 1–23, 33, 37–46, AYP*	**15.4:** 1–17, 23–30, 32–48	**15.5:** 1–14, 19–27, AYP **Port. Proj.**	**Review:** 1–28 **Port. Proj.**	**Ch. 15 Test**

NOTE: A one-day block has been added for the Portfolio Project—timing and placement to be determined by teacher.

Pacing (with Portfolio Project) — **Chapter 15 Total** $5\frac{1}{2}$ days

AYP is *Assess Your Progress.*

Lesson Support

Section	Practice Bank	Study Guide*	Assessment Book*	Visuals	Explorations Lab Manual	Lesson Plans	Technology Book
15.1	96	15.1		Warm-Up 15.1	Masters 1, 2, 20	15.1	
15.2	97	15.2		Warm-Up 15.2	Master 1	15.2	
15.3	98	15.3	Test 63	Warm-Up 15.3 Folder 15	Master 2	15.3	
15.4	99	15.4		Warm-Up 15.4 Folder 15	Masters 1, 2 Add. Expl. 21	15.4	Calculator Act. 15
15.5	100	15.5	Test 64	Warm-Up 15.5	Master 2	15.5	Spreadsheet Act. 15
Review Test	101	Chapter Review	Tests 65–68 Alternative Assessment			Review Test	Calculator Based Lab 6

***Spanish versions** of *Study Guide* and *Assessment Book* are available.

Chapter Support

- Course Guide
- Lesson Plans
- Portfolio Project Book
- Preparing for College Entrance Tests
- Multi-Language Glossary
- *Test Generator* Software
- Professional Handbook

Software Support

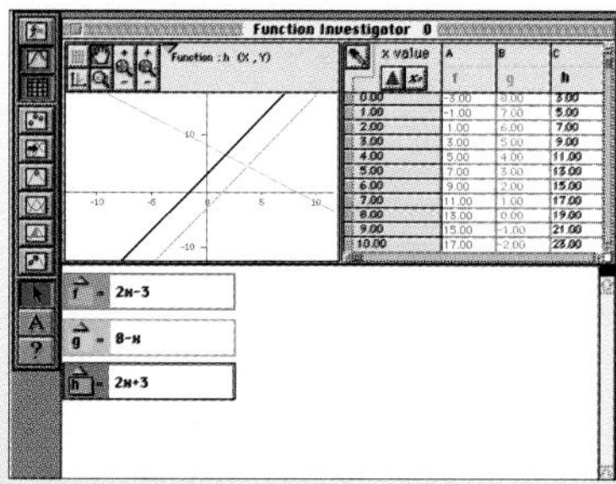

McDougal Littell Mathpack
Function Investigator

Internet Support

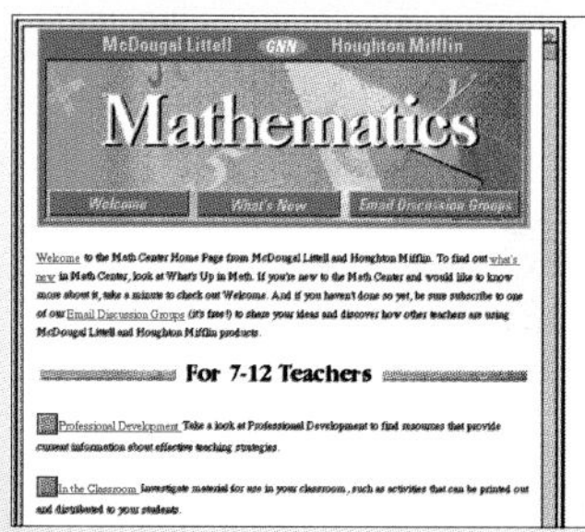

http://www.hmco.com
Next go to McDougal Littell; then the Education Center; then Secondary Math.

Outside Resources

Books, Periodicals

Germain-McCarthy, Yvelne. "Circular Graphs: Vehicles for Conic and Polar Connections." *Mathematics Teacher* (January 1995): pp. 26–28.

Edwards, Thomas. "Building Mathematical Models of Simple Harmonic and Damped Motion." *Mathematics Teacher* (January 1995): pp. 18–22.

Activities, Manipulatives

Moody, Mally. "Sharing Teaching Ideas: Trig Skits." *Mathematics Teacher* (December 1994): p. 702.

Schultz, Harris S. and Martin V. Bonsangue. "Time for Trigonometry." *Mathematics Teacher* (May 1995): pp. 393–396.

Software

f(*g*) *Scholar.* IBM-comp. Southampton, PA: Future Graph.

de Lange, Jan. *Gliding.* Macintosh. Pleasantville, NY: Wings for Learning/Sunburst, 1992.

Videos

Apostol, Tom M. *Sines and Cosines*, Parts 1 and 2. Reston, VA: NCTM.

Internet

With a modem, access the FAA Aviation Education menu for information about Aircraft Owners and Pilots Association's (AOPA) aviation education program. AOPA distributes *A Teacher's Guide to Aviation.*

Interview Notes

Background

Acoustics
The study of acoustics looks at how sounds are created, transmitted, and received. It may also refer to the quality of the sound heard. In the area of acoustics, there are two major fields: architectural and environmental. Architectural acoustics deals with making buildings and rooms quiet, and for ensuring good conditions for listening to sound, such as speeches or music. Environmental acoustics deals with decreasing noise pollution. Other fields of acoustics are physiological acoustics (the way we hear sounds), psychological acoustics (the way we interpret sounds), musical acoustics (the way instruments and voices produce sounds), and speech communication (the way we produce and hear speech). Sound is measured in decibels, abbreviated dB. Decibel levels increase dramatically. For example, 0 dB is the threshold of hearing, 30 dB is a bedroom at night, 60 dB is a conversation at one meter, 90 dB is a subway train moving through a station, 110 dB is a rock band, and 140 dB is the threshold of pain.

Joe Lopez
Joe Lopez started his business in 1968. Today, his business has grown so much he has opened satellite offices in California and Mexico. More than three dozen of his family members are involved in the company, including his daughter who is president, and his son-in-law who is in charge of the Mexico offices. Lopez looks at his family business as a tree and considers himself the trunk of that tree. Everyone works together to help the business, and the business helps them all to succeed.

CHAPTER

15 Trigonometric Functions

Making beautiful music

INTERVIEW **Joe Lopez**

In 1962 Joe Lopez started Zaz Recording Studio when he purchased a one-track tape recorder and a small, rundown building on the outskirts of San Antonio, Texas. He'd already spent countless hours in studios as a guest bass player, and he was ready to use what he'd learned to go into business for himself. Today, Lopez heads a family-run operation that is the largest Hispanic-owned recording complex in the Southwest.

"Every engineer or studio producer relies on what he hears. It's all mathematical, yes, but in the end, it all depends on your ears."

704

A Blend of Musical Worlds

Lopez, a San Antonio native, is no stranger to success. In the mid-1950s his band, "Los Guadalupanos," was one of the most popular *conjunto* groups in south Texas. *Conjunto* music features the accordion and *bajo sexto,* a Mexican twelve-string guitar.

This musical form developed in the late 19th century along the Texas-Mexico border when German, Czech, and Polish immigrants brought their accordions, along with polkas, waltzes, and mazurkas, to local dances. *Conjunto* is now regarded as one of this country's purest forms of folk music, and in San Antonio, it's as popular as chili.

Changing Tracks

Lopez gradually moved from creating music to recording it. He now owns Joey Records International, which occupies six acres of land and includes two 24-track recording studios, cassette and vinyl production facilities, a talent agency, and several record labels featuring more than 300 artists.

But Lopez still spends most of his time in the studio. "It takes a lot of coordination to create a beautiful sound," he says. "My heart's always in what I'm doing, and I never have a dull day in the studio. You get in there with the musicians and you create something. That's why music is so beautiful. You try to do the best you can."

The Science behind the Sound

When Lopez started in the recording business, he hung heavy curtains and tacked thousands of empty egg cartons to the walls of his makeshift studio for soundproofing. "What we used for acoustics then is nothing like what we use today. But when I hear those tapes, they still sound great," he says.

The old building was eventually torn down, and Lopez consulted architectural acoustics experts from around the country to design the studios with higher ceilings, concrete walls and floors, and foam insulation that is shaped like egg cartons. Thick glass divides the musicians' booths and allows them to see and keep time with each other. The result, Lopez says, is state-of-the-art.

Nick Villarreal (accordion) and Rogelio Valdéz (guitar) prepare to record with Joe Lopez.

Interview Notes

Background

Conjunto Music

Conjunto, Tejano, and Tex-Mex all describe a type of music with roots in the early twentieth century near the Texas-Mexico border. Conjunto includes elements of German polka, Mexican mariachi, and American folk music. Don Leonardo Jimenez, also known as *Flaco,* or "the skinny one," is one of the greatest Tejano accordion players. His grandfather learned the accordion from German neighbors in Texas, and his father carried on the accordion-playing tradition. The biggest conjunto festival in the world is staged by the Guadalupe Cultural Arts Center in San Antonio, Texas. Over 40,000 people attend each year.

Second-Language Learners

Students learning English may not be familiar with the idiomatic compound words *rundown, outskirts,* and *countless.* If necessary, explain that the word *rundown* means "something that is no longer in good condition," *outskirts* means "an area at the very edge of a town," and *countless* means "too many to count." Also, point out that the word *makeshift* means "something that is temporary, not permanent," and that the idiomatic expression *state-of-the-art* means "equipment, a method, or a device that is of top quality and at its highest stage of development at that particular time."

Interview Notes

Mathematical Connection

The properties of the graphs of trigonometric functions have direct applications and connections to sound and music. Some of these applications and connections are explored in this chapter. In Section 15.2, students use radian measure to examine a sound wave graph for a soft concert A in an Example, and then apply these ideas to the graph of a soft concert C in the exercises. In Section 15.3, students explore the idea of harmonics and how the mathematical equations representing sound pressures can be applied to tuning a guitar. Students use fundamental equations for each string to calculate the corresponding pressure equations at the 12th, 7th, and 5th frets.

Explore and Connect

Writing
Have students describe the waves of the graph in terms of height, width, number of maximum and minimum values, intercepts, and number of wave repetitions. This will introduce them to the ideas of amplitude, period, and phase shift.

Project
It may be difficult for students to find a vinyl record. If so, you may wish to find a recording of a song on all three formats and play it in class.

Research
The school or local library are good sources for information. The information may be found in encyclopedias.

Music to the Ears

Sound waves are generated by any object that is vibrating. Sound waves move in all directions, bouncing off the floor, ceiling, and walls. The ear receives a mixture of direct sound waves and their reflections. Acoustics is used in the music studio to minimize unwanted reflections and other noise in recordings.

The image of a sound wave can be displayed by an electronic instrument called an *oscilloscope*. Pictured below are oscilloscope traces of a C-note played at three different volumes on a guitar. Each sound wave establishes a repetitive pattern, which can be modeled by *sine functions*. You will learn more about the graphs of sine, cosine, and tangent functions in this chapter.

Eva Ybarra, known as "the queen of the accordion," plays *conjunto* music in San Antonio, Texas.

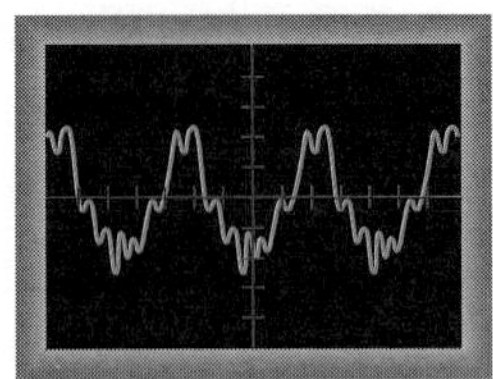

Guitar C-note at low volume

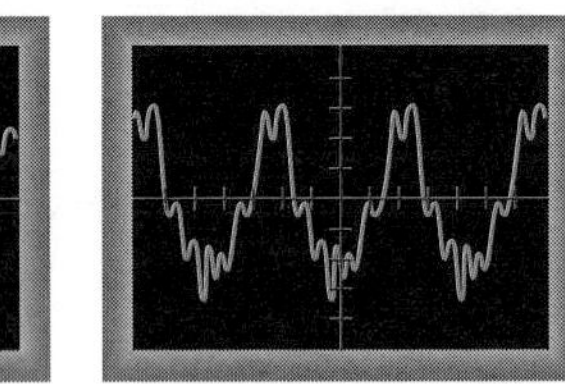

Guitar C-note at normal volume

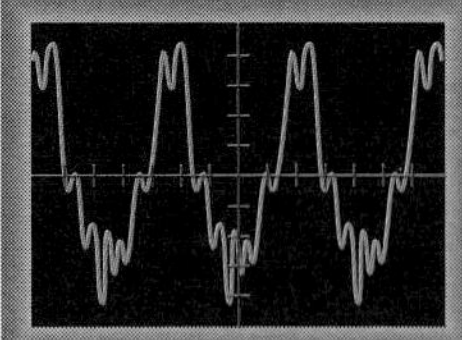

Guitar C-note at high volume

EXPLORE AND CONNECT

Joe Lopez plays his *bajo sexto* at his recording studio.

1. Writing Describe the similarities and differences you see between the three oscilloscope traces of the guitar's C-note. What happens to the wave as the note is played louder? softer? Do you think a C-note played by another instrument would look the same on the oscilloscope? Why or why not?

2. Project Find a recording of one song, by the same artist, on a vinyl record, cassette tape, and compact disc (CD). Listen to the three recordings, and rank them according to their sound quality. Explain your ranking. Describe the similarities and differences you notice between the three different recordings.

3. Research There are two methods of recording sound—analog and digital. Find out how each of these methods works. For each method, what is a sound wave converted to and how is it stored? Which recording formats are typically used for analog recordings? for digital recordings?

Mathematics & Joe Lopez

In this chapter, you will learn more about how mathematics is related to music and recording.

Related Examples and Exercises

Section 15.2
- **Example 3**
- **Exercises 43–45**

Section 15.3
- **Exercises 24–27**

Section

15.1 Sine and Cosine on the Unit Circle

Learn how to...

- work with sine and cosine as functions
- extend the domain of sine and cosine

So you can...

- solve problems, such as finding the height of a seat on a Ferris wheel as a function of time

In Chapter 14 you used sine and cosine to find missing parts of triangles. The sine and cosine functions are not restricted to triangles, however. They are also used to describe many types of *periodic*, or repeating, phenomena, such as the motion of a Ferris wheel.

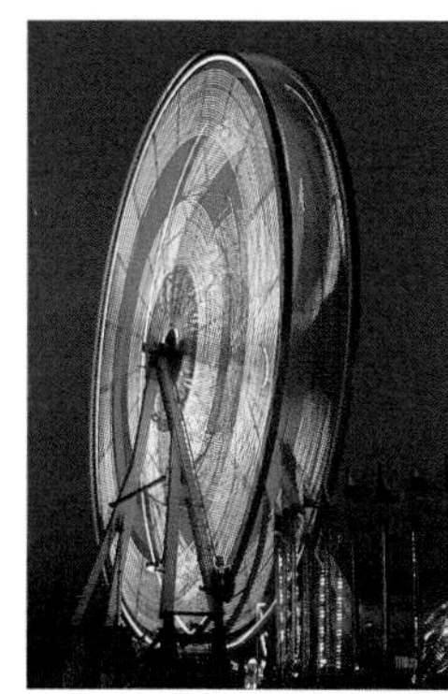

EXPLORATION
COOPERATIVE LEARNING

Graphing Sine and Cosine as Functions

Work with a partner.
You will need:

- a scientific calculator
- graph paper

Angle θ	sin θ	cos θ
0°	?	?
30°	?	?
60°	?	?
90°	?	?
120°	?	?
150°	?	?
180°	?	?
210°	?	?
240°	?	?
270°	?	?
300°	?	?
330°	?	?
360°	?	?

1 Copy the table. Use a calculator to complete the table.

2 Use the table to graph the sine function $y = \sin \theta$ for $0° \le \theta \le 360°$. With θ on the horizontal axis and y on the vertical axis, plot the points (θ, y). Then connect the points with a smooth curve.

3 Use the table to graph the cosine function $y = \cos \theta$ for $0° \le \theta \le 360°$. With θ on the horizontal axis and y on the vertical axis, plot the points (θ, y). Connect the points with a smooth curve.

Questions

1. How are the two graphs alike? How are they different?
2. How many values of θ for $0° \le \theta \le 360°$ satisfy each of the following?
   a. $\sin \theta = 0.5$
   b. $\cos \theta = 0.5$
   c. $\cos \theta = -1$
   d. $\sin \theta = -0.8$

Exploration Note

Purpose
The purpose of this Exploration is to have students plot the sine and cosine functions and analyze them.

Materials/Preparation
Each pair of students needs a scientific calculator and graph paper.

Procedure
Students use a calculator to complete a table of values for $y = \sin \theta$ and $y = \cos \theta$ for θ between 0° and 360°. They plot points from the table to graph both functions. Students compare the two graphs and find the number of angles on the graph that produce a given sine or cosine value.

Closure
Discuss questions 1 and 2 in class. For question 1, students should point out the shape of the curves, the maximum and minimum values, and the θ-intercepts. For question 2, students can see how many values satisfy each equation by examining the graphs.

Explorations Lab Manual
See the Manual for more commentary on this Exploration.

Diagram Master 20

For answers to the Exploration, see answers in back of book.

Plan⇔Support

Objectives

- Work with sine and cosine as functions.
- Extend the domain of sine and cosine.
- Use sine and cosine functions to solve problems.

Recommended Pacing

Core and Extended Courses
Section 15.1: 1 day

Block Schedule
Section 15.1: $\frac{1}{2}$ block
(with Section 15.2)

Resource Materials

Lesson Support
Lesson Plan 15.1
Warm-Up Transparency 15.1
Practice Bank: Practice 96
Study Guide: Section 15.1
Explorations Lab Manual: Diagram Masters 1, 2, 20

Technology
Scientific Calculator
Graphing Calculator
McDougal Littell Mathpack *Function Investigator*
Internet: http://www.hmco.com

Warm-Up Exercises

1. In what quadrant is the terminal side of an angle of rotation with measure 243°? Quadrant III
2. An angle θ is in standard position with the point $P(5, -12)$ on its terminal side. Find $\sin \theta$, $\cos \theta$, and $\tan \theta$. −0.9231, 0.3846, −2.4
3. Find all angles θ between 0° and 360° such that $\cos \theta = 0.6423$. 50° and 310°

Evaluate each expression.

4. sin 135° 0.7071
5. cos 298° 0.4695

Teach⟺Interact

Learning Styles: Verbal

Students with a verbal learning style may need help in understanding the sine and cosine functions as modeling the movement of an object counterclockwise around a unit circle. Have these students focus on either the sine or cosine function and work quadrant by quadrant relating the unit circle to that portion of the function graph. For example, with the sine function in the first quadrant, ask students what is happening to the height of the point (the sine value) as it moves from (1, 0) to (0, 1). They should see how this corresponds to the increase in the sine function from 0 to 1 as the angle of rotation changes from 0° to 90°.

Think and Communicate

Question 2 directs students attention to the link between the unit circle and the values of the sine and cosine function. This question could be extended by asking students to use the unit circle to explain other characteristics of the sine and cosine functions, such as why the sine function decreases from 90° to 270°.

Additional Example 1

Find all angles θ for $0° \le \theta \le 360°$ that satisfy each equation. Give answers to the nearest degree.

a. $\sin \theta = -0.3241$

Use a graphing calculator or graphing software set in degree mode. Graph $y = \sin x$ and $y = -0.3241$, and find the intersections of the two graphs.

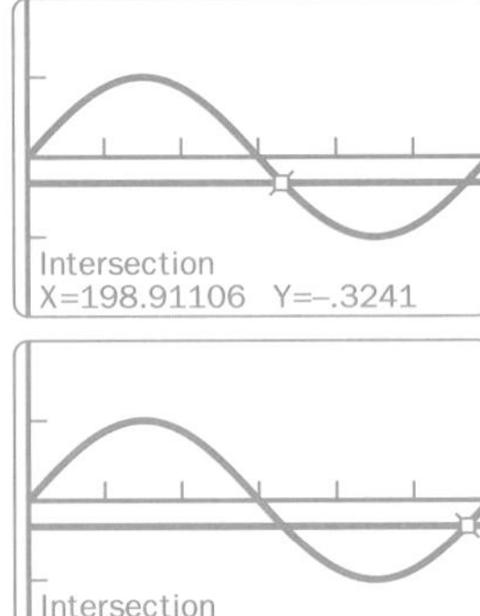

$\sin \theta = -0.3241$ for $\theta \approx 199°$ and $\theta \approx 341°$.

When an object moves on a circular path, you can use **sine** and **cosine** to determine the object's position. For example, consider an object moving counterclockwise around a *unit circle* (a circle of radius 1, centered at the origin).

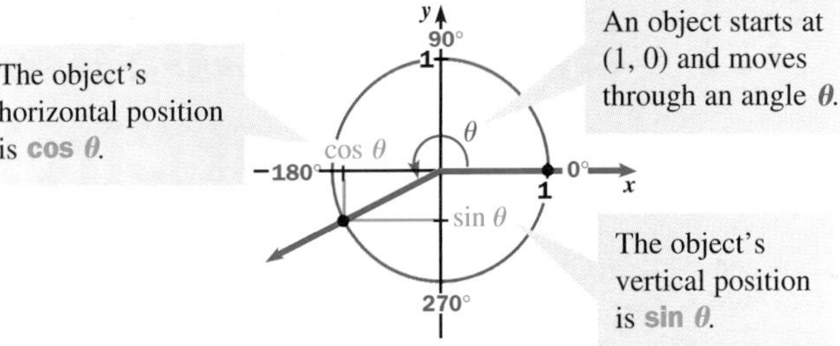

Think and Communicate

1. Use the definitions of sine and cosine on page 672 to explain why the coordinates of the object in the diagram above are $(\cos \theta, \sin \theta)$.
2. How does the unit circle tell you the ranges of the sine and cosine functions?

Graphs of Sine and Cosine for $0° \le \theta \le 360°$

The graphs of $y = \sin \theta$ and $y = \cos \theta$ for $0° \le \theta \le 360°$ are shown below.

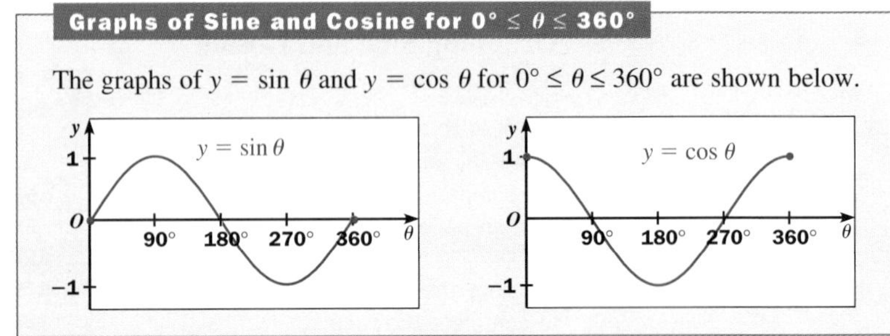

Example 1

Find all angles θ for $0° \le \theta \le 360°$ that satisfy each equation. Give answers to the nearest degree.

a. $\sin \theta = 0.7000$ **b.** $\cos \theta = -0.4518$

Solution

Use a graphing calculator or graphing software set in degree mode.

a. Graph $y = \sin x$ and $y = 0.7000$, and find the intersections of the two graphs.

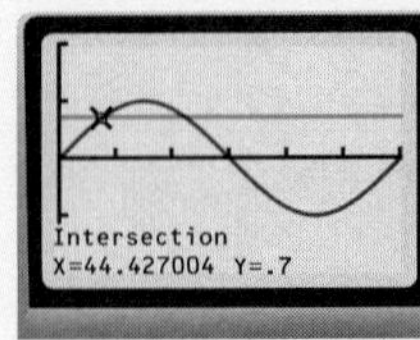

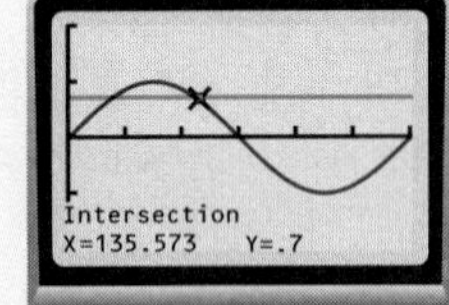

$\sin \theta = 0.7000$ for $\theta \approx$ **44°** and $\theta \approx$ **136°**.

ANSWERS

Section 15.1

Answers are rounded to the nearest degree.

Think and Communicate

1. Points on a circle of radius r have coordinates $(r \cos \theta, r \sin \theta)$. The graphs of $y = r \sin \theta$ and $y = r \cos \theta$ are like the graphs of $y = \sin \theta$ and $y = \cos \theta$ but have amplitude r.
2. The sine and cosine functions both have range $-1 \le y \le 1$ because the values of the sine and cosine correspond to y- and x-coordinates of points on the unit circle, which by definition are between −1 and 1.

b. Graph $y = \cos x$ and $y = -0.4518$, and find the intersections of the two graphs.

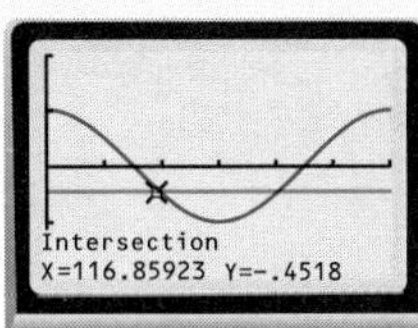

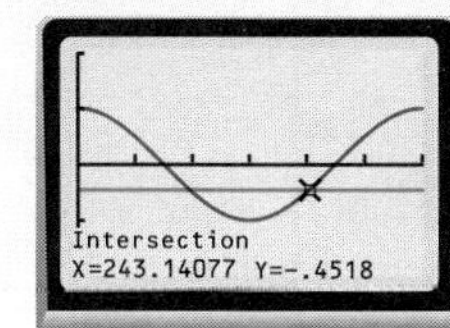

$\cos \theta = -0.4518$ for $\theta \approx$ **117°** and $\theta \approx$ **243°**.

THINK AND COMMUNICATE

3. Use a calculator to find $\sin^{-1} 0.7000$ and $\cos^{-1}(-0.4518)$. How do these values relate to the solutions found in Example 1?

4. Why does the equation $\sin \theta = 2$ have no solution?

Extending Sine and Cosine's Domain

An object moving on the unit circle might keep moving beyond one revolution counterclockwise ($\theta > 360°$). It might also move clockwise ($\theta < 0°$).

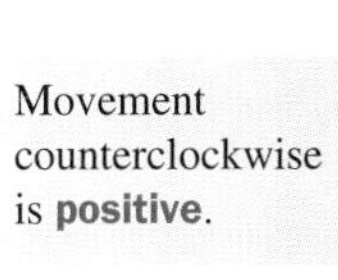

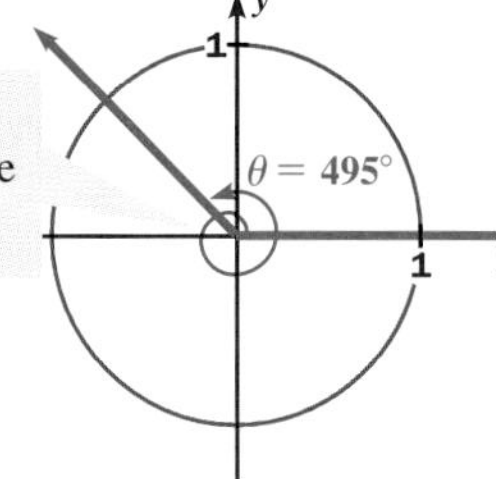

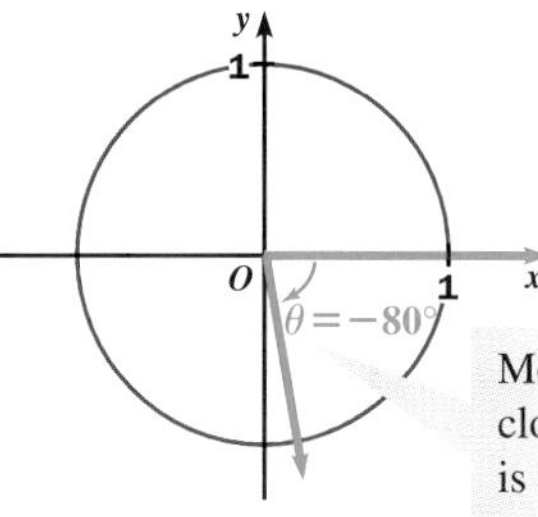

You can extend the domain of the sine and cosine functions to include any angle, positive or negative.

Graphs of Sine and Cosine for All Angles θ

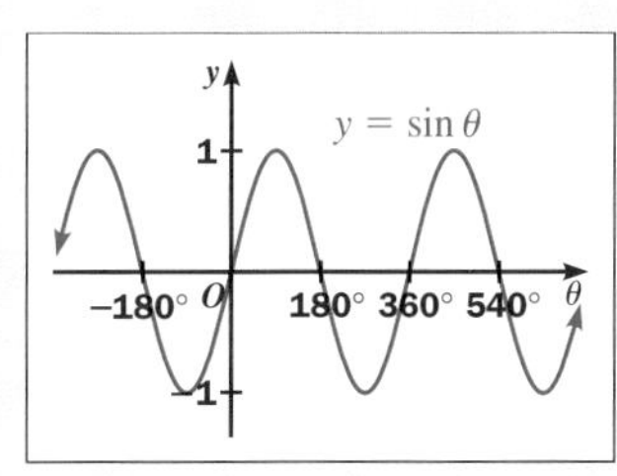

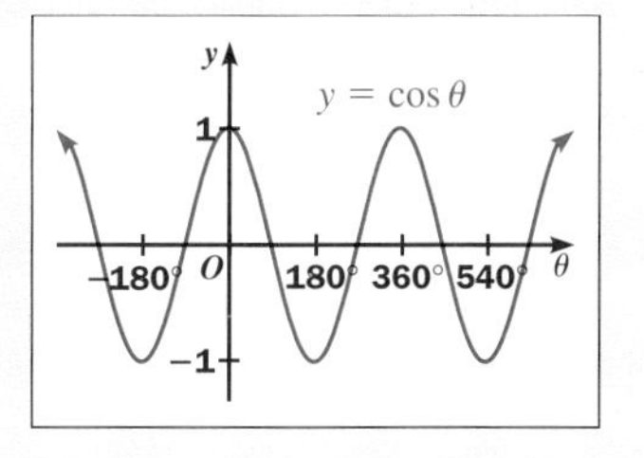

THINK AND COMMUNICATE

5. Compare the complete graphs of sine and cosine shown above to the graphs for $0° \le \theta \le 360°$ shown on the previous page. What do you notice?

Teach⇔Interact

Additional Example 1 (continued)

b. $\cos \theta = 0.8274$

Graph $y = \cos x$ and $y = 0.8274$, and find the intersections of the two graphs.

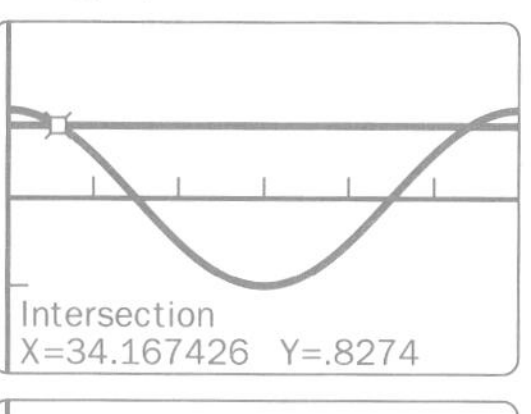

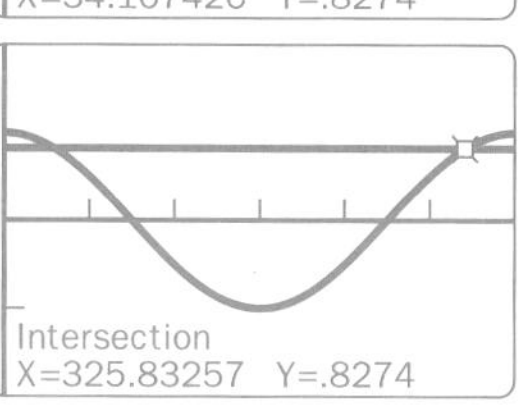

$\cos \theta = 0.8274$ for $\theta \approx 34°$ and $\theta \approx 326°$.

Section Notes

Mathematical Procedures
Point out to students that it is simply a mathematical convention that angles formed by rotation in the clockwise direction are negative and angles formed by rotation in the counterclockwise direction are positive.

Using Technology
When using the TI-82, students can choose 7:ZTrig from the ZOOM menu to get a "friendly" window for many trigonometric functions. This window is set so that the trace feature will include the common angles such as 45° and 180°.

Think and Communicate

3. $\sin^{-1} 0.7000 \approx 44°$ and $\cos^{-1}(-0.4518) \approx 117°$; In each case, these values agree with one of the two values found in Example 1.

4. Because the range of the sine function is $-1 \le y \le 1$, $\sin \theta = 2$ has no solution.

5. Each graph repeats itself every 360°.

Teach⇔Interact

About Example 2

Multicultural Note

The largest Ferris wheel in the world is the Cosmoclock 21, located in Yokohama City, Japan. Standing 344.5 ft tall and measuring 328 ft across, the wheel accommodates 480 riders at a time. Laser beams light the wheel, and synthesizers provide sound effects. At the hub of the wheel is a 42.5 ft electric clock; 60 radiating spokes not only hold the wheel's seats but also serve as second hands of the clock.

Additional Example 2

Jason is jumping rope. The center of the jump rope is rotating at a constant speed of 420°/s. The height h, in inches, of the center of the rope above the ground is given as a function of time t, in seconds, by $h = 36 + 33 \sin (420° \cdot t)$. Find h when:

a. $t = 0.5$

$h = 36 + 33 \sin (420° \cdot 0.5)$
$= 36 + 33 \sin (210°)$
$= 36 + 33(-0.5)$
$= 19.5$

After 0.5 s, the center of the rope is at a height of 19.5 in.

b. $t = 2$

$h = 36 + 33 \sin (420° \cdot 2)$
$= 36 + 33 \sin (840°)$
$\approx 36 + 33(0.8660)$
≈ 65

After 2 s, the center of the rope is at a height of about 65 in.

Think and Communicate

For question 7, students should realize that the Ferris wheel graph is a vertical stretch and a vertical translation of the sine function. Vertical stretches correspond to amplitude changes. Amplitude variations and vertical translations will be studied in Section 15.3.

Closure Question

How can the sine and cosine functions be used to describe the movement of an object around the unit circle? The coordinates of the object are given by $(\cos \theta, \sin \theta)$, where θ is the angle of rotation measured counterclockwise from the positive x-axis.

BY THE WAY

The modern Ferris wheel developed from the pleasure wheel, which probably originated in Eastern Europe or the Near East. The first pleasure wheel may have been a water wheel children clung to for rides.

EXAMPLE 2 Application: Amusement Park Rides

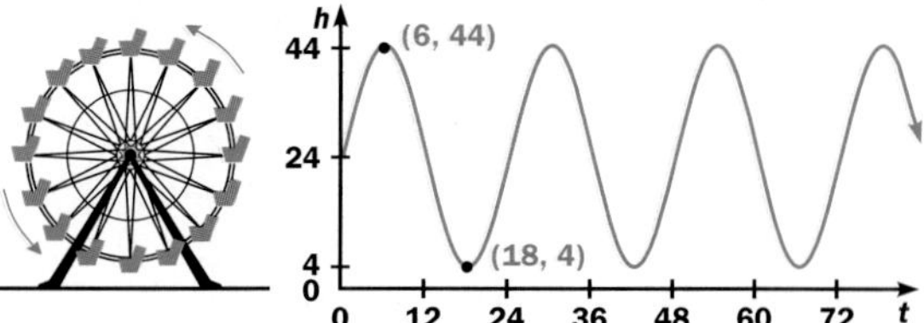

The Ferris wheel shown has a radius of 20 ft. The center of the wheel is 24 ft above the ground, and the wheel rotates at a constant speed of 15°/s. The height h, in feet, of the red seat above the ground is given as a function of time t, in seconds, by:

$$h = 24 + 20 \sin (15° \cdot t)$$

When $t = 0$, the red seat is in the position shown in the diagram above.

Find h when:

a. $t = 1$ **b.** $t = 30$

SOLUTION

Evaluate the function for each value of t.

Use a calculator to find a sine or cosine value you don't know.

a. $h = 24 + 20 \sin (15° \cdot 1)$
$= 24 + 20 \sin (15°)$
$\approx 24 + 20(0.2588)$
≈ 29

After 1 s, the red seat is at a height of about 29 ft.

b. $h = 24 + 20 \sin (15° \cdot 30)$
$= 24 + 20 \sin (450°)$
$= 24 + 20(1)$
$= 44$

After 30 s, the red seat is at a height of 44 ft.

THINK AND COMMUNICATE

6. In Example 2, can you get the same h-value for two different t-values? Explain.

7. How is the graph of the height function in Example 2 like the graph of the sine function shown on the previous page? How is it different?

CHECKING KEY CONCEPTS

Find all angles θ for $0° \le \theta \le 360°$ that satisfy each equation. Give answers to the nearest degree.

1. $\sin \theta = 0.8244$ **2.** $\sin \theta = -0.4257$ **3.** $\cos \theta = 0.2209$

4. $\cos \theta = -0.3000$ **5.** $\sin \theta = \sin (-135°)$ **6.** $\cos \theta = \sin 395°$

Use the height function in Example 2 to find the height of the red seat for each value of t.

7. $t = 5$ **8.** $t = 45$ **9.** $t = 60$

Think and Communicate

6. Yes; each turn of the Ferris wheel brings you once to the maximum and minimum heights and twice to every height in between.

7. The graph of the height function has the same repeating wave-like shape as the sine function, but its range is $4 \le h \le 44$ instead of $-1 \le y \le 1$, and the height function repeats every 24 units (seconds), while the sine function repeats every 360 units (degrees).

Checking Key Concepts

1. 56°, 124°
2. 205°, 335°
3. 77°, 283°
4. 107°, 253°
5. 225°, 315°
6. 55°, 305°
7. about 43 ft
8. about 10 ft
9. 24 ft

15.1 Exercises and Applications

Extra Practice exercises on page 771

Find all angles θ for $0° \le \theta \le 360°$ that satisfy each equation. Give answers to the nearest degree.

1. $\sin\theta = -1$ **2.** $\cos\theta = -1$ **3.** $\sin\theta = 0$ **4.** $\cos\theta = 0$

5. $\sin\theta = 0.5000$ **6.** $\cos\theta = -0.5000$ **7.** $\sin\theta = 0.4580$ **8.** $\cos\theta = 0.3223$

9. $\sin\theta = -0.5000$ **10.** $\sin\theta = 1$ **11.** $\cos\theta = -0.3250$ **12.** $\cos\theta = 0.9140$

Evaluate each expression.

13. $\sin 390°$ **14.** $\cos 450°$ **15.** $\cos(-75°)$ **16.** $\sin(-69°)$

17. $\sin 1350°$ **18.** $\cos(-480°)$ **19.** $\cos 1000°$ **20.** $\sin(-10°)$

Connection CAROUSELS

For a typical carousel, or merry-go-round, a horse makes a complete up-and-down movement 4.8 times for each revolution of the carousel, and the horse is raised 6 in. above and 6 in. below its center position. The up-and-down motion of the horse can be modeled by the function

$$h = 6\cos(4.8\theta)$$

where h is the horse's displacement from its center position and θ is the angle turned by the carousel.

The brown horse started at its highest position at $\theta = 0°$.

Find the value of h for each value of θ.

21. $\theta = 90°$ **22.** $\theta = 360°$ **23.** $\theta = 400°$

24. Open-ended Problem Do you think that the function $h = 6\cos(4.8\theta)$ holds for $\theta < 0°$? What would a negative value of θ indicate about the motion of the carousel?

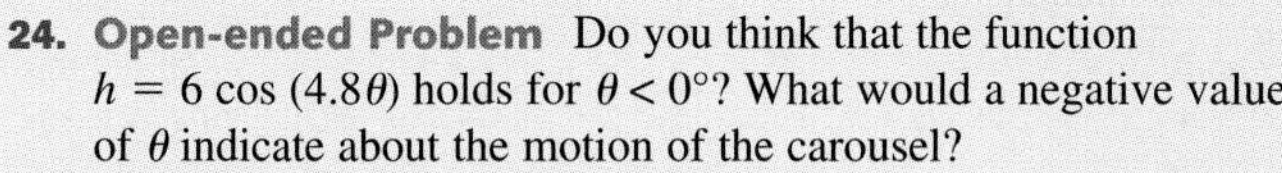

The height of the horse can also be expressed as a function of time, t, in minutes, assuming the speed of the carousel is constant. If the carousel makes 4 revolutions per minute, then:

Toolbox p. 793 Dimensional Analysis

$$\theta = \frac{360°}{1\text{ revolution}} \cdot \frac{4\text{ revolutions}}{1\text{ min}} \cdot t$$

25. Write an equation for h as a function of t.

26. Use the height function you found in Exercise 25 to find the value of h for each value of t.

a. $t = 1$ **b.** $t = 5$

c. $t = 0.5$ **d.** $t = 0.25$

e. $t = 2$ **f.** $t = -1$

Apply⇔Assess

Suggested Assignment

Core Course
Exs. 1–20, 27–44

Extended Course
Exs. 1–44

Block Schedule
Day 1 Exs. 1–20, 27–44

Exercise Notes

Student Progress
Exs. 1–4, 10 Students should commit the cosine and sine function graphs to memory and be able to visualize or quickly sketch the graphs to solve these exercises.

Mathematical Procedures
Exs. 13, 14, 17, 18 Students should be able to do these exercises without the aid of a calculator. They need to use the fact that the values of the sine and cosine function repeat every 360°. Thus, for Ex. 17, since 1350 = 360(3) + 270, sin 1350° = sin 270° = –1.

Application
Exs. 21–26 These exercises and Example 2 of this section show that the sine and cosine functions are often used to model circular motion.

Reasoning
Ex. 25 Students should analyze the conversion used above this exercise to rewrite θ in terms of t in the equation for h as a function of t. There are 360° in one revolution and the carousel makes 4 revolutions per minute, so there are 360(4) = 1440° turned per minute. As θ is the number of degrees turned in t minutes, $\theta = 1440t$.

Exercises and Applications

1. 270°
2. 180°
3. 0°, 180°, 360°
4. 90°, 270°
5. 30°, 150°
6. 120°, 240°
7. 27°, 153°
8. 71°, 289°
9. 210°, 330°
10. 90°
11. 109°, 251°
12. 24°, 336°
13. 0.5
14. 0
15. 0.2588
16. –0.9336
17. –1
18. –0.5
19. 0.1736
20. –0.1736
21. about 2 in.
22. about 2 in. 23. –3 in.
24. The same function holds for $\theta < 0°$, but it represents motion of the carousel in the opposite direction.
25. $h = 6\cos(6912t)$
26. a. about 2 in.
 b. 6 in.
 c. about –5 in.
 d. about 2 in.
 e. about –5 in.
 f. about 2 in.

Apply⇔Assess

Exercise Notes

Topic Spiraling: Preview

Exs. 27–34 These exercises introduce students to some sine and cosine curves that have varying amplitudes, periods, and vertical translations. These topics will be explored in Section 15.3.

Challenge

Ex. 36 This exercise is an example of the property which states that for all angles, $\sin(-\theta) = -\sin\theta$. This property can be demonstrated using the unit circle by observing that the height of a point rotated on the unit circle in one direction is the opposite of the height of a point rotated the same magnitude in the opposite direction.

Practice 96 for Section 15.1

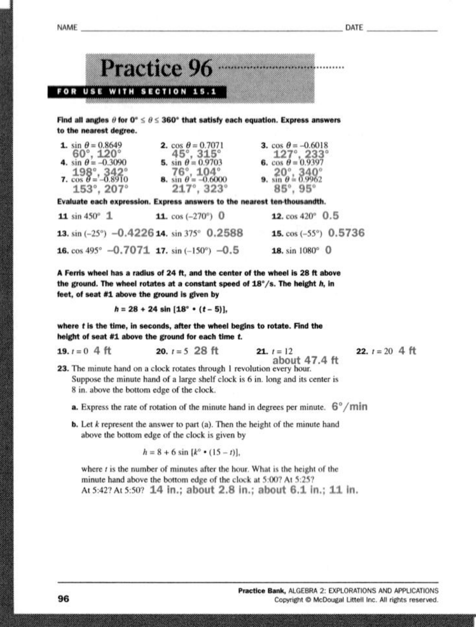

NAME ______________________ DATE ____________

Practice 96

FOR USE WITH SECTION 15.1

Find all angles θ for $0° \le \theta \le 360°$ that satisfy each equation. Express answers to the nearest degree.

1. $\sin\theta = 0.8649$ 60°, 120°
2. $\cos\theta = 0.7071$ 45°, 315°
3. $\cos\theta = -0.6018$ 127°, 233°
4. $\sin\theta = -0.3090$ 198°, 342°
5. $\sin\theta = 0.9703$ 76°, 104°
6. $\cos\theta = 0.9397$ 20°, 340°
7. $\cos\theta = -0.8910$ 153°, 207°
8. $\sin\theta = -0.6000$ 217°, 323°
9. $\sin\theta = 0.9962$ 85°, 95°

Evaluate each expression. Express answers to the nearest ten-thousandth.

11 $\sin 450°$ 1
11. $\cos(-270°)$ 0
12. $\cos 420°$ 0.5
13. $\sin(-25°)$ −0.4226
14. $\sin 375°$ 0.2588
15. $\cos(-55°)$ 0.5736
16. $\cos 495°$ −0.7071
17. $\sin(-150°)$ −0.5
18. $\sin 1080°$ 0

A Ferris wheel has a radius of 24 ft, and the center of the wheel is 28 ft above the ground. The wheel rotates at a constant speed of 18°/s. The height h, in feet, of seat #1 above the ground is given by

$$h = 28 + 24\sin[18° \cdot (t - 5)],$$

where t is the time, in seconds, after the wheel begins to rotate. Find the height of seat #1 above the ground for each time t.

19. $t = 0$ 4 ft
20. $t = 5$ 28 ft
21. $t = 12$ about 47.4 ft
22. $t = 20$ 4 ft
23. The minute hand on a clock rotates through 1 revolution every hour. Suppose the minute hand of a large shelf clock is 6 in. long and its center is 8 in. above the bottom edge of the clock.
    a. Express the rate of rotation of the minute hand in degrees per minute. 6°/min
    b. Let k represent the answer to part (a). Then the height of the minute hand above the bottom edge of the clock is given by
    $$h = 8 + 6\sin[k° \cdot (15 - t)],$$
    where t is the number of minutes after the hour. What is the height of the minute hand above the bottom edge of the clock at 5:00? At 5:25? At 5:42? At 5:50? 14 in.; about 2.8 in.; about 6.1 in.; 11 in.

96

Practice Bank, ALGEBRA 2: EXPLORATIONS AND APPLICATIONS

If $y = 4 - \cos\theta$, find the value of y for each value of θ.

27. 60° **28.** 120° **29.** 420° **30.** −180°

If $y = 3 + \cos 3\theta$, find the value of y for each value of θ.

31. 60° **32.** 120° **33.** 420° **34.** −180°

35. **Technology** Graph $y = 3 + 2\cos\theta$ on a graphing calculator or graphing software.

a. What is the maximum value of the function?

b. What is the minimum value of the function?

c. What values of θ for $0 \le \theta \le 360°$ give the maximum and minimum values of the function?

36. SAT/ACT Preview If A = sin (−40°) and B = −sin 40°, then:

A. A > B **B.** A = B **C.** A < B

D. relationship cannot be determined

ONGOING ASSESSMENT

37. a. Open-ended Problem Name a positive angle and a negative angle that have the same sine value and the same cosine value. How are the two angles related to each other?

b. Are there other positive or negative angles that have the same sine value and cosine value as the two angles you named? Explain.

SPIRAL REVIEW

Solve each triangle. *(Section 14.6)*

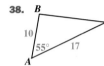

38. **39.**

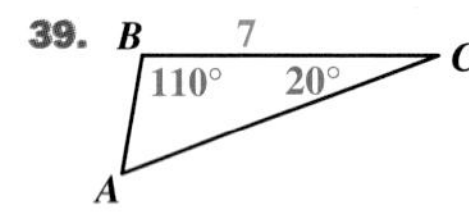

40.

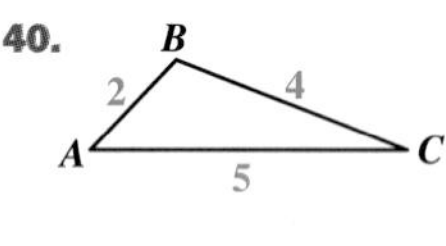

41. A circle has a radius of 1 in. *(Toolbox, page 802)*

a. What is the circumference of the circle?

b. An arc of the circle is intercepted by a 270° central angle. What is the length of the arc?

Find the area of each shaded region. *(Section 14.4)*

42.

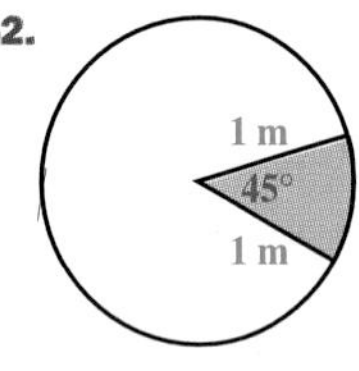

43.

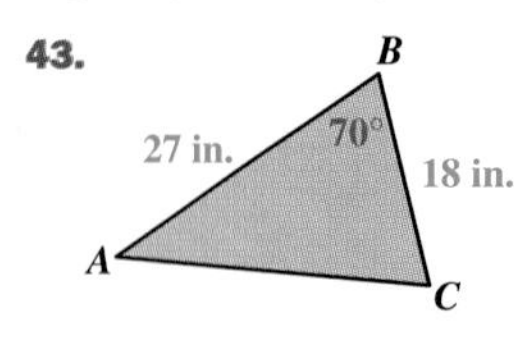

44.

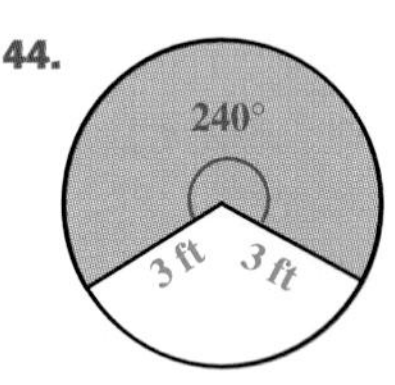

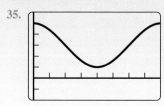

27. 3.5
28. 4.5
29. 3.5
30. 5
31. 2
32. 4
33. 2
34. 2
35. a. 5
    b. 1
    c. maximum: 0°, 360°; minimum: 180°
36. B
37. Answers may vary. An example is given.
    a. 60° and −300° have the same sine and cosine values. These two angles represent the same point on the unit circle.
    b. Any other angles that differ from these two angles by integral multiples of 360° will have the same sine and cosine.
38. $BC \approx 14$; $\angle B \approx 89°$; $\angle C \approx 36°$
39. $\angle A = 50°$; $AB \approx 3$; $AC \approx 9$
40. $\angle A \approx 50°$; $\angle B \approx 108°$; $\angle C \approx 22°$
41. a. 2π in.
    b. $\frac{3\pi}{2}$ in.
42. $\frac{\pi}{8}$ m²
43. 228 in.²
44. 6π ft²

Section

15.2 Measuring in Radians

Learn how to...

- convert between degree measure and radian measure
- find the sine and cosine of angles given in radians

So you can...

- explore models of periodic phenomena, such as music

So far you have measured angles in degrees. People who use the sine and cosine functions for modeling often use *radians,* rather than degrees, to measure angles.

EXPLORATION
COOPERATIVE LEARNING

Making a Radian String

Work with a partner.
You will need:

- a cylinder
- a string
- tape
- a ruler
- a marker
- a protractor

SET UP Measure the diameter of a circular cross section of the cylinder and then calculate the radius. Cut a piece of string so that it is about 14 times as long as the radius. Mark the midpoint of the string, and make marks at one radius (one *radian*) intervals to the left and right of the midpoint. You may want to use a different color in each direction.

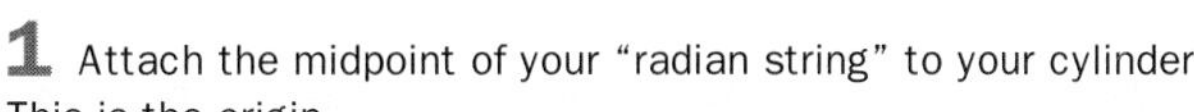

1 Attach the midpoint of your "radian string" to your cylinder. This is the origin.

2 Wrap the string around the cylinder. The counterclockwise direction is positive, and the clockwise direction is negative.

Questions

Use your string and cylinder to answer the following questions.

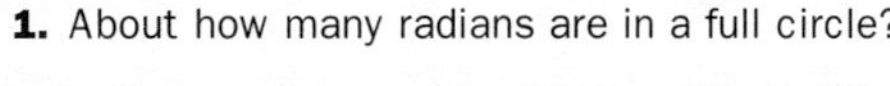

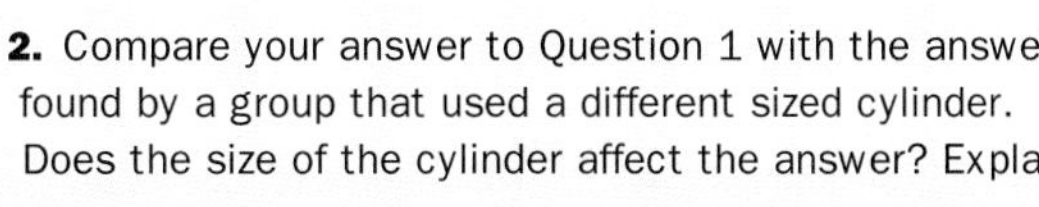

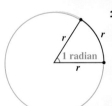

1. About how many radians are in a full circle?
2. Compare your answer to Question 1 with the answer found by a group that used a different sized cylinder. Does the size of the cylinder affect the answer? Explain.
3. How are the points at 2 radians and −2 radians geometrically related to each other? Does this relationship hold for any two points that are at $\pm r$ radians from the origin?

Exploration Note

Purpose
The purpose of this Exploration is to introduce students to the concept of radian measure.

Materials/Preparation
Each group needs a cylinder, string, tape, a ruler, a marker, and a protractor.

Procedure
Students determine the radius of their cylinder by measuring its diameter. They then mark the midpoint of the string and repeatedly mark lengths equal to one radius in each direction from the midpoint. The string is then taped to the cylinder at the midpoint (representing the origin) and wrapped around it, with the counterclockwise direction representing positive and the clockwise direction representing negative. Students then answer questions about their model.

Closure
Discuss questions 1–3 in class. Students should understand that a radian is the *measure* of a central angle that intercepts an arc whose length is equal to the radius.

Explorations Lab Manual
See the Manual for more commentary on this Exploration.

For answers to the Exploration, see following page.

Plan⇔Support

Objectives

- Convert between degree measure and radian measure.
- Find the sine and cosine of angles given in radians.
- Explore models of periodic phenomena.

Recommended Pacing

Core and Extended Courses
Section 15.2: 2 days

Block Schedule
Section 15.2: 2 half-blocks (with Sections 15.1 and 15.3)

Resource Materials

Lesson Support
Lesson Plan 15.2
Warm-Up Transparency 15.2
Practice Bank: Practice 97
Study Guide: Section 15.2
Explorations Lab Manual: Diagram Master 1

Technology
Graphing Calculator
McDougal Littell Mathpack *Function Investigator*
Internet: http://www.hmco.com

Warm-Up Exercises

Evaluate each expression.

1. sin 90° 1
2. cos −45° 0.7071
3. sin 210° −0.5
4. Give the maximum value, minimum value, and θ-intercepts of the graph of $y = \cos\theta$. 1; −1; odd multiples of $\frac{\pi}{2}$
5. What is the formula for the circumference of a circle? $C = 2\pi r$

Teach⇔Interact

Section Notes

Communication: Discussion Students should discuss what they learned from the Exploration and relate it to the definition of radian measure. Ask students how they would determine the number of radians that are in any given central angle. (Measure the arc cut out by the angle and divide this measurement by the radius of the circle.) Ask students why they might prefer to use the unit circle to find radian measure. (The radius of the unit circle is 1 so that the radian measure of the angle is equal to the length of the intercepted arc.)

Common Error
Some students may be confused as to when an angle is given in radians and when it is given in degrees. Point out that when no units are given, the angle is in radians. Also point out that π is not necessary in the expression for an angle to be measured in radians.

Additional Example 1

a. Convert 135° to radians.

$135° = 135 \cdot \frac{\pi}{180}$ radians
$= 0.75\pi$ radians
≈ 2.36

b. Convert $-\frac{5\pi}{6}$ radians to degrees.

$-\frac{5\pi}{6} = -\frac{5\pi}{6} \cdot \frac{180°}{\pi}$
$= -\frac{900°}{6}$
$= -150°$

Additional Example 2

a. Find sin (-3π).
Since -3π represents one and a half times around the unit circle in the clockwise direction, sin (-3π) is the y-coordinate of the point $(-1, 0)$. sin $(-3\pi) = 0$

b. Find cos 4.2.
Set a calculator to radian mode and find cos 4.2.
cos 4.2 ≈ −0.4903

The **radian** measure of an angle is the length of the unit circle's arc intercepted by the angle. Since the circumference of the unit circle is 2π, you have the following relationship between radians and degrees:

In this diagram, $\theta = 135° = \frac{3\pi}{4}$ radians.

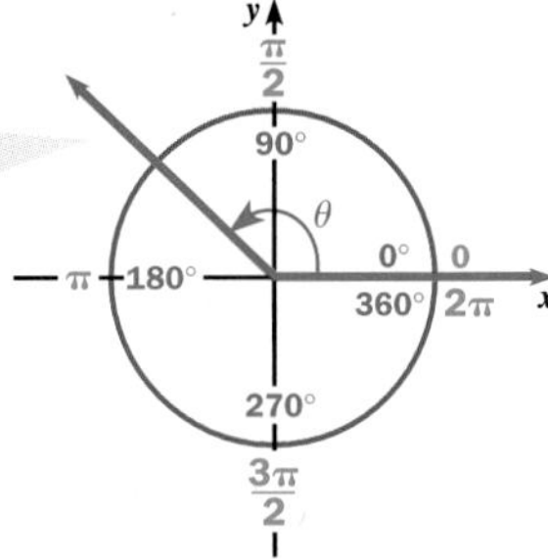

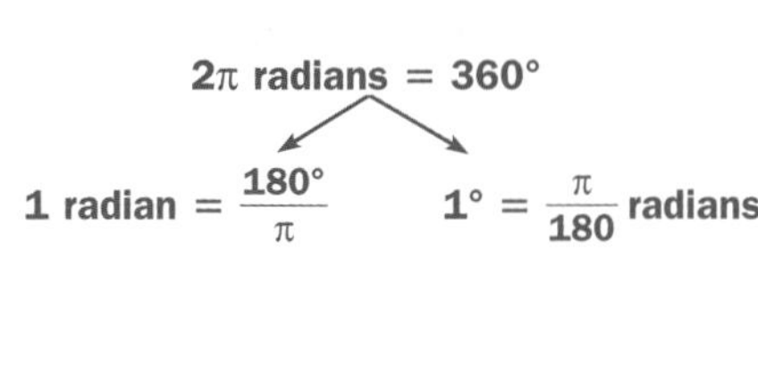

EXAMPLE 1

a. Convert 236° to radians. **b.** Convert $\frac{\pi}{4}$ radians to degrees.

SOLUTION

a. $236° = 236 \cdot \frac{\pi}{180}$ radians
$\approx 1.31\pi$ radians
≈ 4.12

b. $\frac{\pi}{4} = \frac{\pi}{4} \cdot \frac{180°}{\pi}$
$= \frac{180°}{4}$
$= 45°$

If an angle measure does not have a degree symbol, it is understood to be in radians.

The sine and cosine functions can be defined for radians as well as degrees.

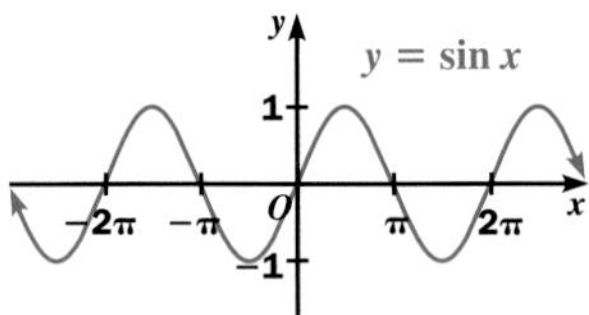

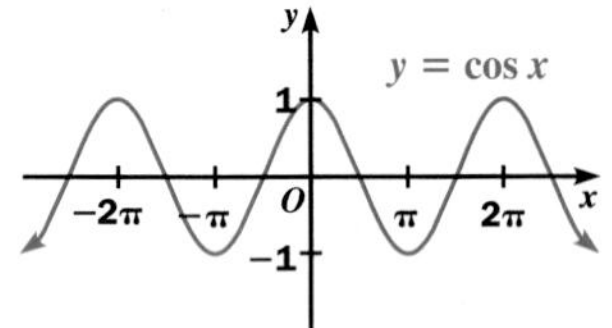

In general, this book will use x to represent angles measured in radians and θ to represent angles measured in degrees.

EXAMPLE 2

a. Find cos $\frac{\pi}{2}$. **b.** Find sin 3.

SOLUTION

a. Since $\frac{\pi}{2}$ represents a quarter of the way around the unit circle, cos $\frac{\pi}{2}$ is the x-coordinate of the point $(0, 1)$.

$$\cos \frac{\pi}{2} = 0$$

b.

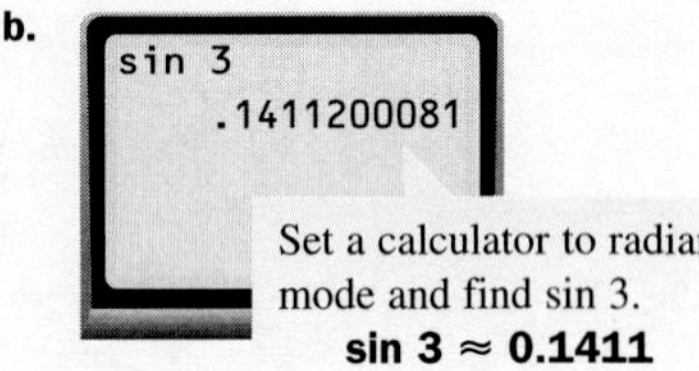

Set a calculator to radian mode and find sin 3.
sin 3 ≈ 0.1411

WATCH OUT!
In part (b), if your calculator gives 0.0523, it is set in degree mode. Remember: sin 3 ≠ sin 3°

ANSWERS Section 15.2

Exploration

1. about 6.3
2. No; because the ratio of the circumference to the radius is the same for all circles, the size of the cylinder does not matter.
3. They are the same distance from the origin but in different directions. Yes.

Periodic Phenomena

The tides rising and falling, the moon waxing and waning, and a pendulum swinging back and forth are all examples of *periodic* phenomena, because the events repeat themselves in a regular cycle.

Many periodic phenomena can be modeled with sine or cosine functions. Radians are usually used when describing or discussing periodic phenomena that are not obviously related to angles, such as musical tones.

EXAMPLE 3 Interview: Joe Lopez

When Joe Lopez records music, he is recording small pressure changes in the air. If the atmospheric pressure is 15.000 pounds per square inch (psi), a loud noise might cause the pressure to vary between 14.999 psi and 15.001 psi.

A pure musical tone will cause the air pressure to vary in a way that can be modeled with a sine function. The variation in pressure for a soft Concert A is given by

$$P = (5 \cdot 10^{-4}) \sin (2\pi \cdot 440t)$$

where P is the deviation from the average atmospheric pressure in psi, and t is the time in seconds.

a. Graph the function. Find the maximum deviation in pressure.

b. Find the time between consecutive maximum P-values.

SOLUTION

a. Graph $y = (5 \cdot 10^{-4}) \sin (2\pi \cdot 440x)$ using a graphing calculator or graphing software, and use trace to find the maximum value of y.

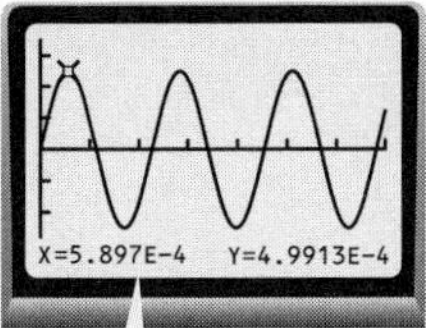

The maximum deviation from the average pressure is about **$5 \cdot 10^{-4}$ psi**.

b. Use trace to find the values of x at the first and second maximum values of y, and then subtract the two x-values.

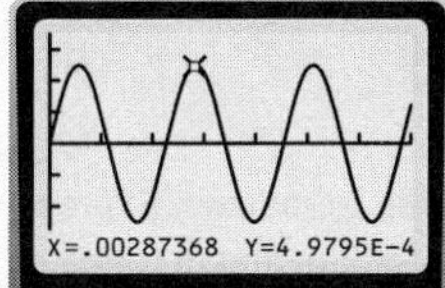

Find the difference between the x-values shown here and on the screen in part (a).

$$x_2 - x_1 \approx 0.0029 - 0.0006$$
$$\approx 0.0023$$

The time between consecutive maximum P-values is about 0.0023 s.

BY THE WAY

Technicians use pure tones to test stereo speakers. They feed a test tone into the speaker and check the tone coming out for any distortion.

THINK AND COMMUNICATE

1. Maximum P-values occur at about 0.0006 s and 0.0029 s. When does the next maximum P-value occur?

Think and Communicate

1. at about 0.0052 s

Teach⇔Interact

About Example 3

Using Technology Students need to set the viewing window of their graphing calculators or graphing software carefully in order to see the sinusoidal nature of the curve. These settings may be used: 0≤X≤0.007 and −6*10^−4≤Y≤6*10^−4.

Additional Example 3

A weather station recorded the number of hours of daylight, N, in each day beginning on March 20. They found that N could be modeled with the sine function $N = 4.1 \sin \frac{\pi}{183}x + 11.8$, where x is the number of days after March 20.

a. Graph the function. Find the maximum number of hours of daylight.

Graph $y = 4.1 \sin \frac{\pi}{183}x + 11.8$ using a graphing calculator or graphing software, and use trace to find the maximum value of y.

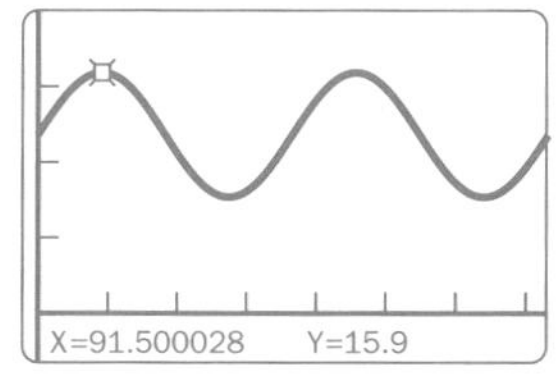

The maximum number of hours of daylight is 15.9.

b. Find the number of days between consecutive maximum N-values.

Use trace to find the values of x at the first and second maximum values of y, and then subtract the two x-values.

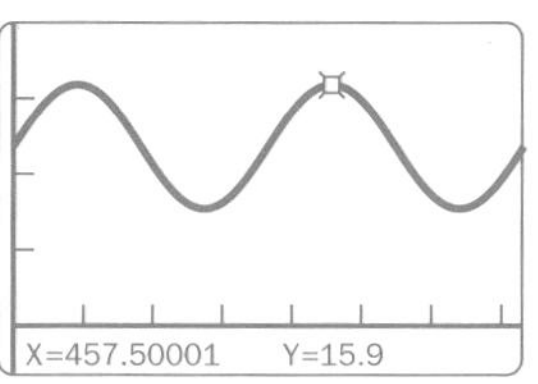

$$x_2 - x_1 \approx 457.5 - 91.5$$
$$\approx 366$$

The number of days between consecutive maximum N-values is about 366.

Teach⇔Interact

Checking Key Concepts

Student Progress
Students should be able to complete these problems with little or no difficulty. You may want to discuss the answers in class so that students can check their work before beginning the exercises.

Closure Question

What is radian measure and how does it relate to degree measure? Radian measure of an angle is the measure defined as the length of the unit circle's arc intercepted by the angle. 180° is equivalent to π radians.

Apply⇔Assess

Suggested Assignment

Core Course
Day 1 Exs. 1–20
Day 2 Exs. 21–38, 46–51

Extended Course
Day 1 Exs. 1–20
Day 2 Exs. 21–51

Block Schedule
Day 1 Exs. 1–20
Day 2 Exs. 21–38, 46–51

Exercise Notes

Alternate Approach
Exs. 2, 3, 5–7, 10, 11 When converting a radian measure expressed in terms of π to a degree measure, students can substitute 180° for π in the expression and simplify.

Topic Spiraling: Review
Ex. 13 This exercise reviews parametric equations and, at the same time, allows students to see a dynamic presentation of a point moving around the unit circle. In part (a), the second graph shows that the *y*-coordinate of such a point traces out a sine function. In part (b), the second graph shows that the *x*-coordinate of such a point traces out a cosine function.

CHECKING KEY CONCEPTS

Convert each angle measure from radians to degrees (to the nearest degree).

1. $\frac{5\pi}{2}$ **2.** $-\frac{3\pi}{4}$ **3.** 3.86 **4.** 8.95

Convert each angle measure from degrees to radians (to the nearest hundredth of a radian).

5. $-200°$ **6.** 540° **7.** 30° **8.** 311°

Evaluate each expression.

9. $\sin \frac{3\pi}{2}$ **10.** $\cos(-3\pi)$ **11.** $\cos \frac{4\pi}{3}$ **12.** $\sin \frac{\pi}{2}$

15.2 Exercises and Applications

Extra Practice exercises on page 772

Convert each angle measure from radians to degrees (to the nearest degree).

1. 1 **2.** $\frac{5\pi}{8}$ **3.** $\frac{5\pi}{4}$ **4.** -2.5

5. 3π **6.** $-\pi$ **7.** 7π **8.** -6.5

9. 5.89 **10.** 7.25π **11.** -0.5π **12.** -13.27

13. **Technology** Use a graphing calculator set in parametric, simultaneous, and radian mode. Use $0 \le t \le 2\pi$, $-2 \le x \le 7$, and $-3 \le y \le 3$. Have the calculator increment t by 0.05. Graph each pair of parametric equations below. Describe what happens on your calculator screen. Explain the two graphs in terms of what you learned in this section.

a. $x_1 = \cos t$ $\quad x_2 = t$
$y_1 = \sin t$ $\quad y_2 = \sin t$

b. $x_1 = \cos t$ $\quad x_2 = t$
$y_1 = \sin t$ $\quad y_2 = \cos t$

Convert each angle measure from degrees to radians. Express each answer as a decimal to the nearest hundredth of a radian, and in terms of π.

14. 45° **15.** $-180°$ **16.** 210° **17.** 720°

18. $-90°$ **19.** $-65°$ **20.** 20° **21.** 390°

22. $-700°$ **23.** 87° **24.** $-35°$ **25.** 405°

26. **GEOMETRY** Steve's racing bicycle has tires with a diameter of 27 in. Joan's mountain bike has tires with a diameter of 24 in. If both bikes are wheeled forward 4π radians, how far (in inches) will each bike travel? Explain why the two bikes travel different distances, even though the tires move through the same number of radians.

racing bicycle

mountain bicycle

Checking Key Concepts

1. 450° 2. –135° 3. 221° 4. 513° 5. –3.49 6. 9.42 7. 0.52 8. 5.43 9. –1 10. –1 11. –0.5 12. 1

Exercises and Applications

1. 57° 2. 113° 3. 225° 4. –143° 5. 540° 6. –180° 7. 1260° 8. –372° 9. 337° 10. 1305° 11. –90° 12. –760°

13. a. The first pair of equations graphs the unit circle counterclockwise from (1, 0). As each point is plotted, the second set of equations plots the sine value of the corresponding angle.

b. The first pair of equations again plots the unit circle counterclockwise from (1, 0). As each point is plotted on the unit circle, the second pair of equations plots the cosine function of the corresponding angle.

14. 0.79; $\frac{\pi}{4}$ 15. –3.14; –π

16. 3.67; $\frac{7\pi}{6}$ 17. 12.57; 4π

Evaluate each expression.

27. $\sin\left(-\frac{\pi}{2}\right)$ **28.** $\cos\frac{7\pi}{4}$ **29.** $\sin 8\pi$ **30.** $\cos\left(-\frac{2\pi}{3}\right)$

31. $\sin\pi$ **32.** $\cos 3\pi$ **33.** $\cos 0.82$ **34.** $\sin(-2.18)$

35. $\cos(-5.34)$ **36.** $\cos 2.8\pi$ **37.** $\sin 2$ **38.** $\sin 6.32$

Connection SOLAR POWER

The amount of sunlight striking Earth at a particular latitude changes through the year because the orientation of Earth's axis relative to the sun changes. The table shows the variation in the average solar radiation falling on a solar panel lying flat on the ground at latitude 40° north.

The heliostat field and solar tower at Sandia National Laboratories in NM (above) and the Solar One Power Plant in Dagget, CA (below) collect solar energy.

Month	Average daily solar radiation (BTU/ft^2)
March	1182
April	1584
May	1862
June	1920
July	1862
August	1584
September	1182
October	838
November	524
December	463
January	524
February	838

39. Make a scatter plot of the data. Put time t, in months since March, on the horizontal axis, and the average daily solar radiation s, in BTU per square foot, on the vertical axis. Connect the points with a smooth curve.

a. What function does your graph resemble? What are the maximum and minimum values?

b. **Visual Thinking** Extend your graph one year backward (letting t be negative) and one year forward. Explain what you did.

40. Open-ended Problem How do you think your graph in Exercise 39 would change for a location closer to the equator? Explain your reasoning.

41.  **Technology** The data in the table can be modeled by this equation:

$$s = 1190 + 730\sin\left(\frac{\pi}{6}t\right)$$

Use a graphing calculator or graphing software to graph the function and compare it with the graph you made in Exercise 39. How well does the model fit the data?

42. Open-ended Problem Given what you learned above, when are solar collectors most useful at latitude 40° north? What might someone do to gather more energy from sunlight?

Apply⇔Assess

Exercise Notes

Mathematical Procedures
Exs. 27–32, 41 Students should know how to do Exs. 27–32 without the aid of a calculator. (Follow the Solution to Example 2(a) on page 714.) In fact, if students do use a calculator for Ex. 29, they will get $-4 \cdot 10^{-13}$ as the answer. This is very close to the correct answer of 0; the difference is due to rounding error. For Ex. 41, encourage students to use the maximum and minimum values from the table to determine an appropriate viewing window.

Interdisciplinary Problems
Exs. 39–42, 43–45 Trigonometry is used extensively in physics. In these exercises, trigonometric functions are used to model solar radiation and sound waves.

Second-Language Learners
Exs. 39–42 Students learning English may benefit from a discussion of such phrases as *orientation of Earth's axis* and *variation in the average solar radiation.* These students may also benefit from discussing their ideas with an English-proficient partner before completing Ex. 42.

Historical Connection
Exs. 39–42 While the idea of harnessing the sun's radiation for heat and hot water goes back to ancient times, it was not until the 1970s that scientists began to seriously consider the possibility of solar power as an alternative to other energy sources to meet the needs of modern society. This renewed interest in solar power came after the 1973 OPEC oil embargo, during which time the price of oil increased nearly 1000% and people became aware that access to unlimited amounts of petroleum could not be taken for granted.

Research
Exs. 39–42 You might ask students to research other periodic phenomena that can be modeled using a sine or cosine function. Students can find data in an almanac or newspaper and then make a scatter plot using the data.

18. $-1.57; -\frac{\pi}{2}$
19. $-1.13; -\frac{13\pi}{36}$
20. $0.35; \frac{\pi}{9}$
21. $6.81; \frac{13\pi}{6}$
22. $-12.22; -\frac{35\pi}{9}$
23. $1.52; \frac{29\pi}{60}$
24. $-0.61; -\frac{7\pi}{36}$
25. $7.07; \frac{9\pi}{4}$
26. Steve's bicycle: 170 in.; Joan's bicycle: 151 in. Because 4π radians just means two revolutions, each bicycle will travel twice as far as the circumference of its tires, so the bicycle with the larger tires will travel farther.
27. −1
28. 0.7071
29. 0
30. −0.5
31. 0
32. −1
33. 0.6822
34. −0.8201
35. 0.5872
36. −0.8090
37. 0.9093
38. 0.0368
39–42. See answers in back of book.

Apply⇔Assess

Exercise Notes

Visual Thinking

Ex. 45 Encourage students to make sketches of what they think the graphs of pressure over time will look like when the music is twice as loud and half as loud. Ask them to also sketch what they think the graphs of voltage over time will look like in both cases. This activity involves the visual skills of *interpretation* and *correlation.*

Assessment Note

Ex. 46 Students can extend this activity to create a reference chart for themselves. Have students draw a large unit circle on a piece of paper and put points on the circle to represent all angles of rotation that are multiples of 30° or 45° from 0° to 360°. Label each point with its degree measure, its radian measure, and its coordinates, representing the cosine and sine of the angles, respectively.

Practice 97 for Section 15.2

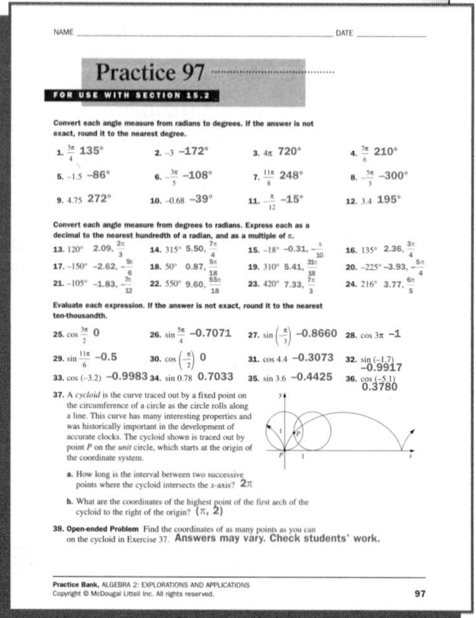

NAME ______ DATE ______

Practice 97

FOR USE WITH SECTION 15.2

Convert each angle measure from radians to degrees. If the answer is not exact, round it to the nearest degree.

1. $\frac{3\pi}{4}$ **135°** 2. -3 **−172°** 3. 4π **720°** 4. $\frac{7\pi}{6}$ **210°**

5. -1.5 **−86°** 6. $-\frac{3\pi}{5}$ **−108°** 7. $\frac{11\pi}{8}$ **248°** 8. $-\frac{5\pi}{3}$ **−300°**

9. 4.75 **272°** 10. -0.68 **−39°** 11. $-\frac{\pi}{12}$ **−15°** 12. 3.4 **195°**

Convert each angle measure from degrees to radians. Express each as a decimal to the nearest hundredth of a radian, and as a multiple of π.

13. 120° **2.09, $\frac{2\pi}{3}$** 14. 315° **5.50, $\frac{7\pi}{4}$** 15. −18° **−0.31, $-\frac{\pi}{10}$** 16. 135° **2.36, $\frac{3\pi}{4}$**

17. −150° **−2.62, $-\frac{5\pi}{6}$** 18. 50° **0.87, $\frac{5\pi}{18}$** 19. 310° **5.41, $\frac{31\pi}{18}$** 20. −225° **−3.93, $-\frac{5\pi}{4}$**

21. −105° **−1.83, $-\frac{7\pi}{12}$** 22. 550° **9.60, $\frac{55\pi}{18}$** 23. 420° **7.33, $\frac{7\pi}{3}$** 24. 216° **3.77, $\frac{6\pi}{5}$**

Evaluate each expression. If the answer is not exact, round it to the nearest ten-thousandth.

25. $\cos \frac{3\pi}{2}$ **0** 26. $\sin \frac{5\pi}{4}$ **−0.7071** 27. $\sin\left(-\frac{\pi}{3}\right)$ **−0.8660** 28. $\cos 3\pi$ **−1**

29. $\sin \frac{11\pi}{6}$ **−0.5** 30. $\cos\left(-\frac{\pi}{2}\right)$ **0** 31. $\cos 4.4$ **−0.3073** 32. $\sin(-1.7)$ **−0.9917**

33. $\cos(-3.2)$ **−0.9983** 34. $\sin 0.78$ **0.7033** 35. $\sin 3.6$ **−0.4425** 36. $\cos(-5.1)$ **0.3780**

37. A *cycloid* is the curve traced out by a fixed point on the circumference of a circle as the circle rolls along a line. This curve has many interesting properties and was historically important in the development of accurate clocks. The cycloid shown is traced out by point P on the *unit* circle, which starts at the origin of the coordinate system.

a. How long is the interval between two successive points where the cycloid intersects the x-axis? **2π**

b. What are the coordinates of the highest point of the first arch of the cycloid to the right of the origin? **(π, 2)**

38. **Open-ended Problem** Find the coordinates of as many points as you can on the cycloid in Exercise 37. **Answers may vary. Check students' work.**

Practice Bank, ALGEBRA 2: EXPLORATIONS AND APPLICATIONS

97

INTERVIEW Joe Lopez

Look back at the article on pages 704–706.

When Joe Lopez records music, the varying pressure of the sound wave against the microphone is transformed into varying electrical voltage. He then records this electrical signal.

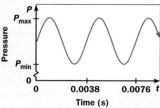

P
P_{max}
Pressure
P_{min}
0
0 0.0038 0.0076 t
Time (s)

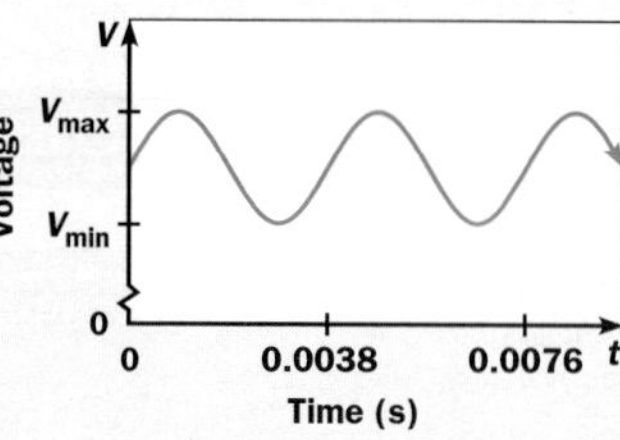

43. **Visual Thinking** The graphs show the air pressure over time of a pure musical tone, Concert C, and the voltage over time as the tone is transformed into an electrical signal. Describe the similarities and differences between the two graphs.

44. Look at the graph of pressure over time. How much time passes before the pattern repeats itself?

45. **Open-ended Problem** What do you think happens to the graph of pressure over time when the music gets louder? softer? What do you think happens to the graph of voltage over time?

ONGOING ASSESSMENT

46. **Open-ended Problem** Choose four angles, one in each of the four quadrants. Give the radian measure, the sine value, and the cosine value of each angle.

SPIRAL REVIEW

Find all angles θ for $0° \le \theta \le 360°$ that satisfy each equation. Give answers to the nearest degree. *(Section 15.1)*

47. $\sin \theta = 0.5218$ 48. $\cos \theta = -0.9900$ 49. $\sin \theta = 0.4050$

50.  **Technology** Use a graphing calculator or graphing software to graph the function $y = 5 + 2 \sin \theta$. *(Section 15.1)*

a. What is the maximum value of the function?

b. What is the minimum value of the function?

51. Suppose you roll a 20-sided die, marked with the numbers 1 through 20. What is the probability of getting a 1 or a 2? *(Section 13.1)*

43. Both of the graphs are periodic and have the same shape. The minimum and maximum values of the voltage graph vary less than the minimum and maximum values of the pressure graph.

44. 0.0038 s

45. Answers may vary. An example is given. For both graphs, louder music would result in a greater amplitude, with the same mean; softer sounds would result in a smaller amplitude.

46. Answers may vary. An example is given. Quadrant I: $\frac{\pi}{4}$ radians or 45°, $\sin \approx 0.7071$, $\cos \approx 0.7071$; Quadrant II: $\frac{2\pi}{3}$ radians or 120°, $\sin \approx 0.8660$, $\cos = -0.5$; Quadrant III: $\frac{7\pi}{6}$ or 210°, $\sin = -0.5$, $\cos \approx -0.8660$; Quadrant IV: $\frac{7\pi}{4}$ or 315°, $\sin \approx -0.7071$, $\cos \approx 0.7071$

47. 31°; 149°

48. 172°; 188°

49. 24°; 156°

50. 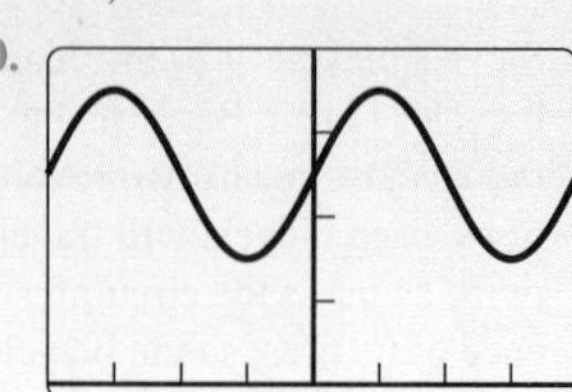

a. $y = 7$

b. $y = 3$

51. $\frac{1}{10}$

Section

15.3 Exploring Amplitude and Period

Learn how to...

- **graph equations of these forms:** $y = a \sin bx + k$ / $y = a \cos bx + k$

So you can...

- **understand periodic phenomena, such as the pure-tone sounds used in hearing tests**

An audiologist tests your hearing by changing the pressure of pure-tone sounds to find the lowest pressure you can hear at various frequencies.

An audiologist uses a frequency of 1000 cycles per second (cps) first, increasing the sound pressure level until you can hear the tone. The graphs below show the changes in air pressure associated with tones at sound pressure levels of 20, 25, and 30 decibels (db).

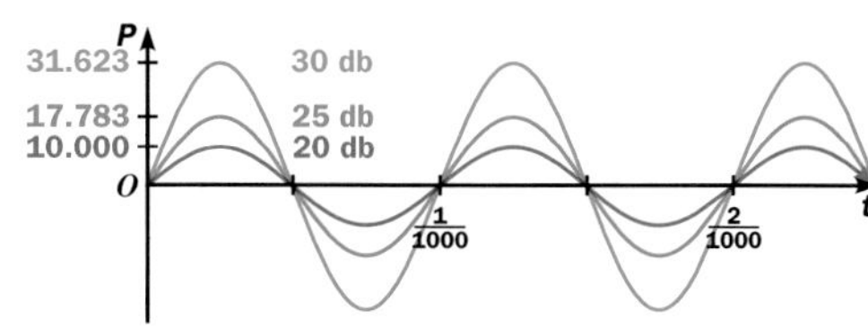

After using the frequency of 1000 cps, the audiologist may use other frequencies, such as 500 cps (a lower pitch) and 2000 cps (a higher pitch).

The unit of sound pressure on the vertical axis is the smallest sound pressure audible to most humans, approximately 3.53×10^{-9} atmosphere.

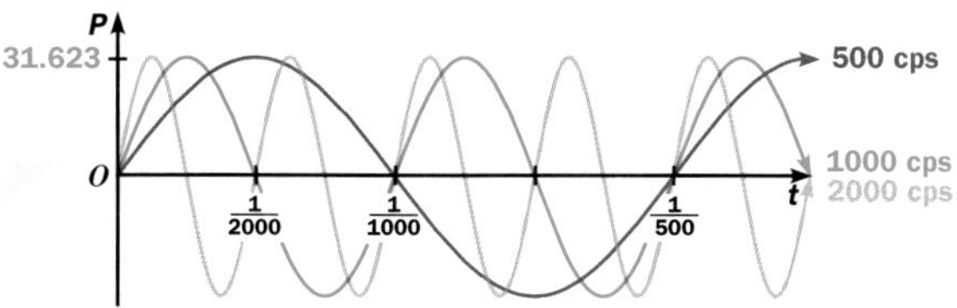

THINK AND COMMUNICATE

1. In the first set of graphs, the frequency of the tone is constant but the pressure level varies. How are the graphs alike? How are they different?
2. The equations $P = 10.000 \sin 2000\pi t$, $P = 17.783 \sin 2000\pi t$, and $P = 31.623 \sin 2000\pi t$ describe the first set of graphs. How does changing the value of a affect the graph of the equation $y = a \sin bx$?
3. In the second set of graphs, the pressure level is constant but the frequency varies. How are the graphs alike? How are they different?
4. The equations $P = 31.623 \sin 1000\pi t$, $P = 31.623 \sin 2000\pi t$, and $P = 31.623 \sin 4000\pi t$ describe the second set of graphs. How does changing the value of b affect the graph of the equation $y = a \sin bx$?

Plan⇔Support

Objectives

- Graph equations of these forms: $y = a \sin bx + k$ / $y = a \cos bx + k$.
- Understand periodic phenomena.

Recommended Pacing

Core and Extended Courses
Section 15.3: 1 day

Block Schedule
Section 15.3: $\frac{1}{2}$ block (with Section 15.2)

Resource Materials

Lesson Support
Lesson Plan 15.3
Warm-Up Transparency 15.3
Overhead Visuals: Folder 15: Graphing $y = a \cos bx + k$
Practice Bank: Practice 98
Study Guide: Section 15.3
Explorations Lab Manual: Diagram Master 2
Assessment Book: Test 63

Technology
Graphing Calculator
McDougal Littell Mathpack *Function Investigator with Matrix Analyzer Activity Book:* Function Investigator Activities 32 and 34
Geometry Inventor Activity Book: Activities 22 and 23
Internet: http://www.hmco.com

Warm-Up Exercises

Convert each expression from radians to degrees.

1. $\frac{3\pi}{2}$ 270°
2. $-\frac{5\pi}{6}$ −150°

Evaluate each expression.

3. $\cos \frac{\pi}{4}$ 0.7071
4. $\sin(-5\pi)$ 0

Consider the graph of the function $y = \sin x$.

5. What are the maximum and minimum values of the function? 1 and −1
6. How many radians are there between consecutive maximum values? 2π

ANSWERS Section 15.3

Think and Communicate

1. The graphs all have the same horizontal intercepts, and they all reach maximum and minimum values for the same t-values, but the distances between their maximum and minimum values vary.
2. The larger the (absolute) value of a, the greater the difference between minimum and maximum values of the graph of $y = a \sin bx$.
3. The graphs have the same maximum and minimum values, but the distance before each "wave" repeats itself varies.
4. As the (absolute) value of b increases, the degree of horizontal compression of the graph increases.

Teach⇔Interact

Section Notes

Communication: Reading
Students should read the material at the top of this page carefully and take notes. You may then want to ask volunteers to give a definition of one of the concepts discussed and to illustrate how to find its value from an equation and from a graph.

Teaching Tip
Absolute value notation is used to describe the amplitude and period of the graphs of $y = a \sin bx$ and $y = a \cos bx$ because amplitude and period are generally given as positive numbers, regardless of whether the values of a and b are positive or negative.

Using Technology
You may want to suggest that students use *Function Investigator* software to graph equations like the one in Example 3. Also, students can use Function Investigator Activities 32 and 34 in the *Function Investigator with Matrix Analyzer Activity Book* to discover the relationship between the values of a and b in the equation $y = a \sin bx$ and the amplitude and period of the function.

Additional Example 1

Graph each equation.

a. $y = \frac{1}{2} \cos \frac{1}{2}\pi x$

The amplitude is $\frac{1}{2}$.

The period is $\frac{2\pi}{\frac{1}{2}\pi} = 4$.

The graph rises $\frac{1}{2}$ unit above and falls $\frac{1}{2}$ unit below the x-axis. The graph completes one cycle in 4 units.

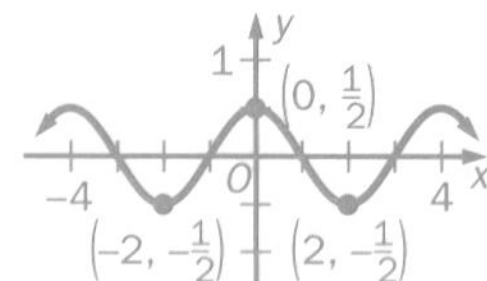

b. $y = 2.5 \sin 1.5x$

The amplitude is 2.5.

The period is $\frac{4\pi}{3}$.

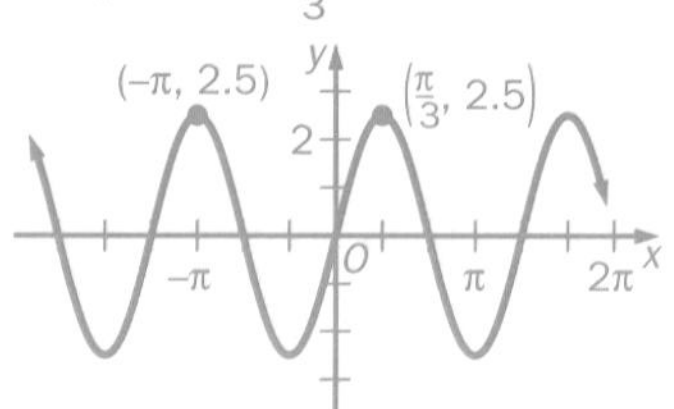

The values of a and b affect the *amplitude* and *period* of the graphs of $y = a \sin bx$ and $y = a \cos bx$ as shown below.

The **amplitude** is $\frac{M - m}{2}$ where M and m are the maximum and minimum values of the function.

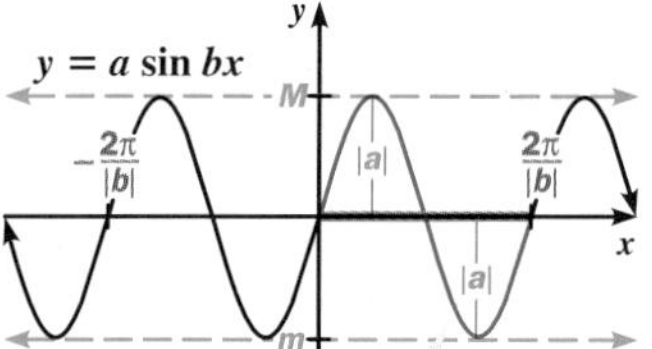

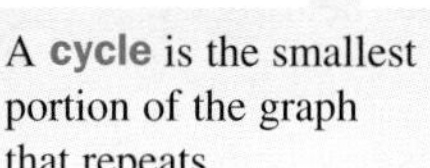

A **cycle** is the smallest portion of the graph that repeats.

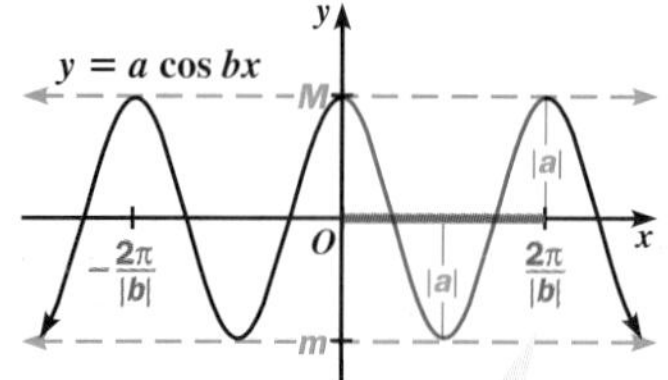

The **period** is the length of one cycle.

In general, the graphs of $y = a \sin bx$ and $y = a \cos bx$ have an amplitude of $|a|$ and a period of $\frac{2\pi}{|b|}$. Throughout this chapter, only positive values of b will be used.

Think and Communicate

5. Suppose an audiologist uses a 10 db pure-tone sound with the sound pressure equation $P = 3.162 \sin 16{,}000\pi t$. What are the amplitude and period of the sound?
6. If the test subject cannot hear the sound in Question 5, how should the audiologist change it? What effect would this have on the equation?

Example 1

Graph each equation.

a. $y = 2 \sin 4\pi x$

b. $y = 3.5 \cos 0.5x$

Solution

a. The amplitude is 2.

The period is $\frac{2\pi}{4\pi} = \frac{1}{2}$.

The graph rises 2 units above and falls 2 units below the x-axis.

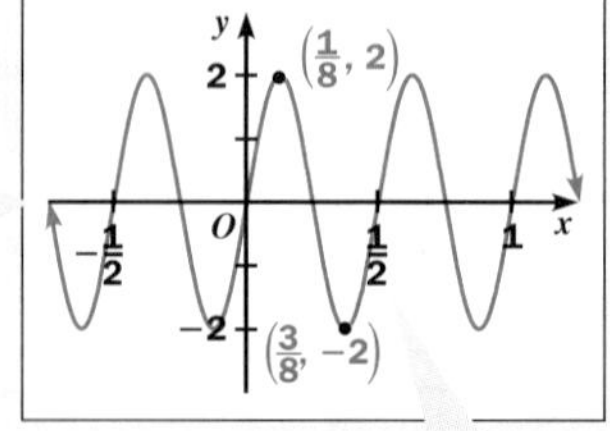

The graph completes one cycle in $\frac{1}{2}$ unit.

b. The amplitude is 3.5.

The period is $\frac{2\pi}{0.5} = 2\pi \cdot 2 = 4\pi$.

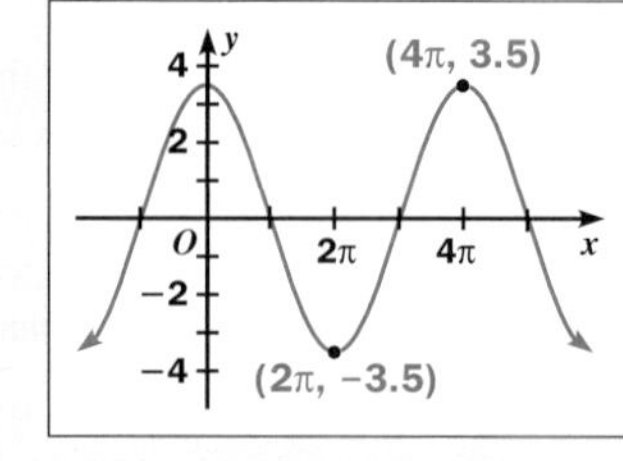

Think and Communicate

5. amplitude: 3.162 (times 3.53×10^{-9} atm); period: $\frac{1}{8000}$ s
6. Answers may vary. An example is given. Increase the sound pressure level, which would increase the amplitude in the sound pressure equation (that is, the coefficient of sin), or alter the frequency (a higher frequency giving a higher tone), which would change the coefficient of t in the argument of sin.

Think and Communicate

7. In the same coordinate plane, graph each pair of functions. How are the graphs geometrically related?

   a. $y = \sin x$
      $y = -\sin x$

   b. $y = \cos x$
      $y = -\cos x$

8. Suppose the value of b in the equation $y = a \sin bx$ is between 0 and 1. Is the period of the graph *greater than* or *less than* 2π?

Example 2

Find an equation for each graph.

a.

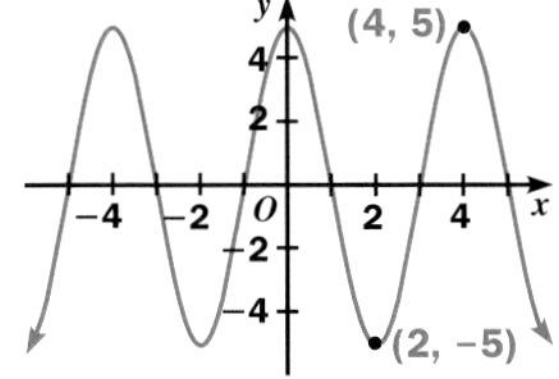

b.

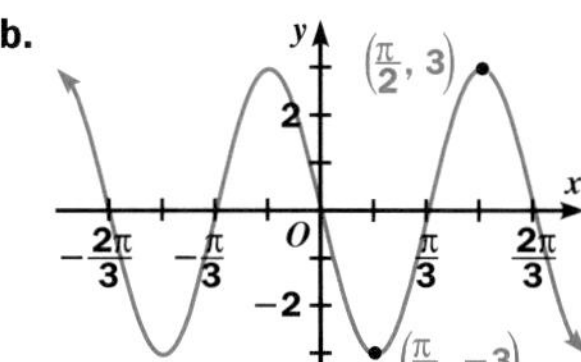

Solution

a. The curve resembles the graph of a cosine function.

The amplitude is 5, so $a = 5$.

The period is 4, so $\frac{2\pi}{b} = 4$ and $b = \frac{2\pi}{4} = \frac{\pi}{2}$.

An equation for the graph is $y = 5 \cos \frac{\pi}{2}x$.

b. The curve resembles the graph of a sine function reflected over the x-axis.

The amplitude is 3, so $a = -3$.

Use a negative value for a because of the reflection.

The period is $\frac{2\pi}{3}$, so $\frac{2\pi}{b} = \frac{2\pi}{3}$ and $b = 3$.

An equation for the graph is $y = -3 \sin 3x$.

Vertical Translation

If you translate the graph of a sine or cosine function vertically by k units, its equation will have the form $y = a \sin bx + k$ or $y = a \cos bx + k$.

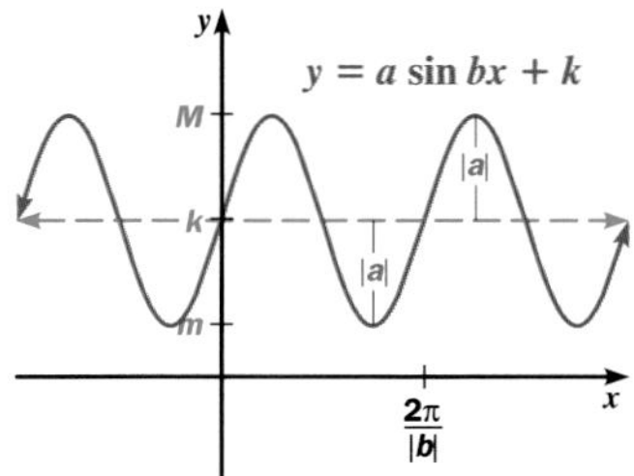

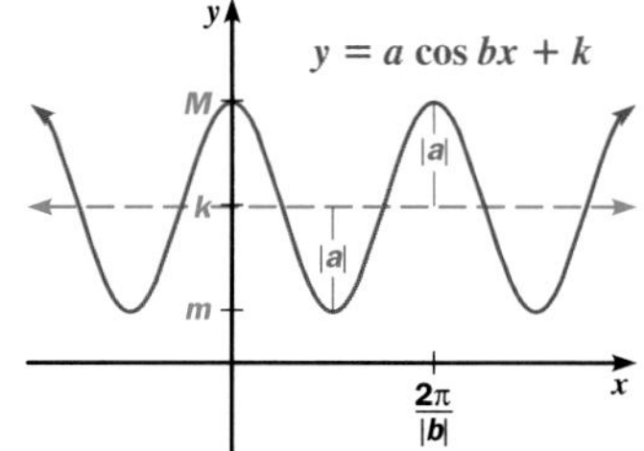

The line $y = k$ is called the *axis*. The value of k is the average of M and m.

Think and Communicate

7. a.

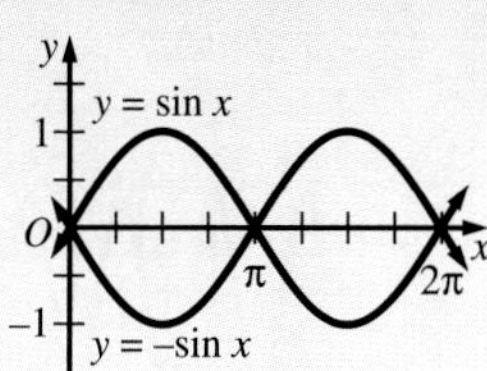

b. The graphs are reflections over the x-axis.

8. greater than 2π

Teach⇔Interact

Think and Communicate

After completing question 7, students should realize that the graph of $y = -a \sin bx$ is a reflection over the x-axis of the graph of $y = a \sin bx$ and that the same holds true for the graphs of $y = a \cos bx$ and $y = -a \cos bx$. Question 8 leads students to discover that when the value of b in the equation $y = a \sin bx$ is between 0 and 1, the period of the graph is greater than 2π.

Additional Example 2

Find an equation for each graph.

a.

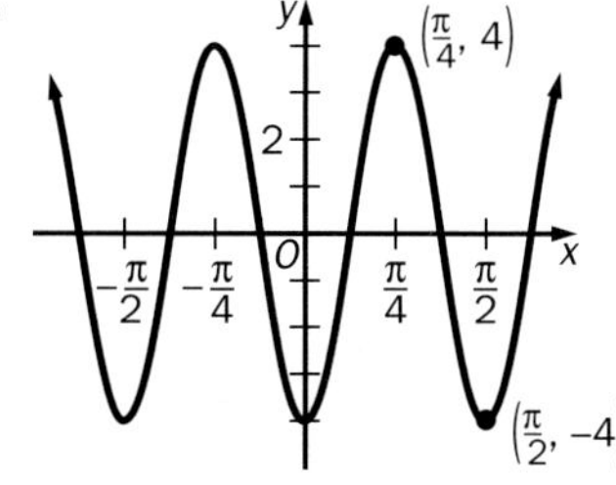

The curve resembles the graph of a cosine function reflected over the x-axis.

The amplitude is 4, so $a = -4$.

The period is $\frac{\pi}{2}$, so $\frac{2\pi}{b} = \frac{\pi}{2}$ and $b = 4$.

An equation for the graph is $y = -4 \cos 4x$.

b.

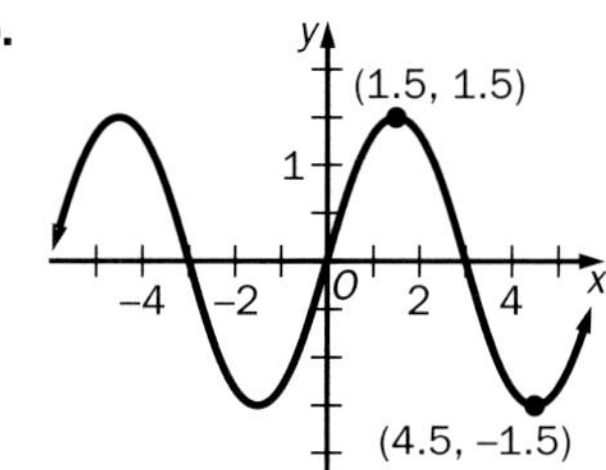

The curve resembles the graph of a sine function.

The amplitude is 1.5, so $a = 1.5$, or $\frac{3}{2}$.

The period is 6, so $\frac{2\pi}{b} = 6$ and $b = \frac{2\pi}{6} = \frac{\pi}{3}$.

An equation for the graph is $y = \frac{3}{2} \sin \frac{\pi}{3}x$.

Teach⇔Interact

Additional Example 3

Consider a Ferris wheel with an axle that is 55 ft above the ground with a wheel radius of 49 ft. After the passengers are seated, the wheel makes one complete counterclockwise turn in 160 s.

a. Suppose you are in the seat at the far left when the wheel begins to turn. Draw a graph of your height above the ground as a function of time.

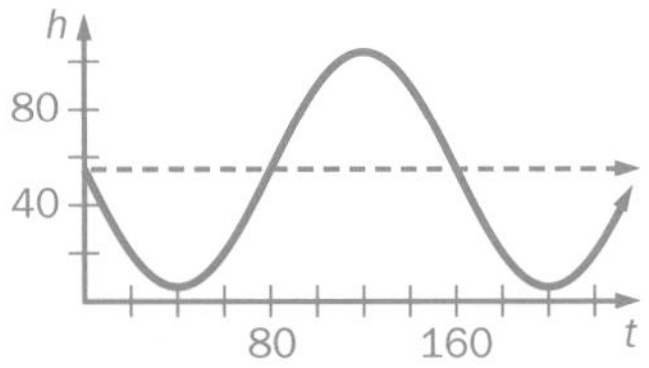

b. Write an equation for your height above the ground as a function of time.

The radius is 49 ft, the period is 160 s, and the axle is 55 ft above the ground.

$h = -49 \sin \frac{2\pi}{160}t + 55$

c. Find the first two times you are 90 ft above the ground.

Method 1 Use a graphing calculator or graphing software.

Graph $y = -49 \sin \frac{2\pi}{160}x + 55$ and $y = 90$ and find the points of intersection of the two graphs.

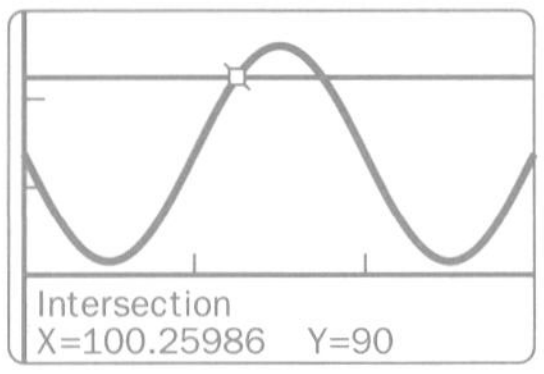

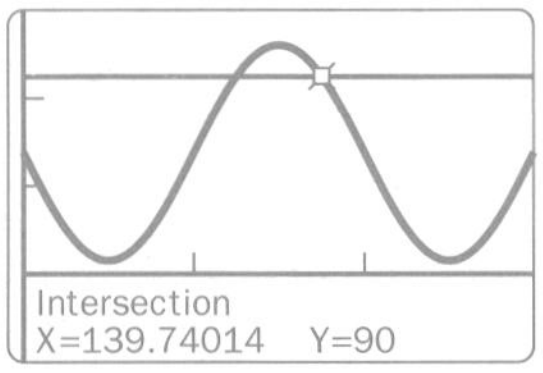

You are at 90 ft about 100 s and about 140 s after the wheel begins to turn.

Method 2 Solve the equation using algebra.

$$-49 \sin \frac{2\pi}{160}t + 55 = 90$$

$$-49 \sin \frac{2\pi}{160}t = 35$$

$$\sin \frac{2\pi}{160}t = -\frac{35}{49}$$

$$\frac{2\pi}{160}t = \sin^{-1}\left(-\frac{35}{49}\right)$$

$$t = \frac{160}{2\pi} \sin^{-1}\left(-\frac{35}{49}\right)$$

$$t \approx -20.3$$

EXAMPLE 3 Application: Amusement Park Rides

The Navy Pier Ferris wheel in Chicago, Illinois, has an axle that is 80 ft above the ground. The wheel has a radius of 70 ft. After passengers are seated, the wheel makes one complete turn in 210 s.

a. Suppose you are in the seat at the bottom when the wheel begins to turn. Draw a graph of your height above the ground as a function of time.

b. Write an equation for your height above the ground as a function of time.

c. Find the first two times you are 100 ft above the ground.

SOLUTION

a.

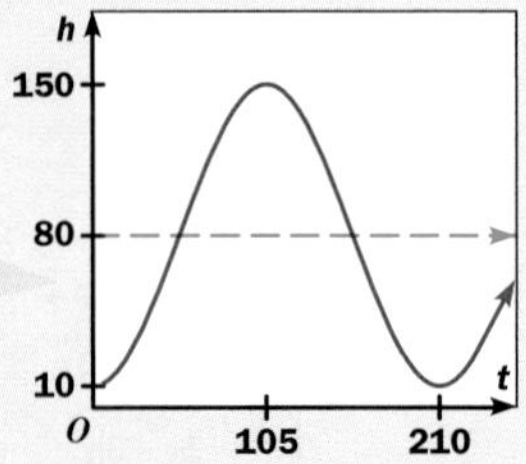

This graph looks like the graph of a cosine function reflected over its axis.

b. $h = -70 \cos \frac{2\pi}{210}t + 80$

The radius is 70 ft. The period is 210 s. The axle is 80 ft above the ground.

c. **Method 1** Use a graphing calculator or graphing software.

Graph $y = -70 \cos \frac{2\pi}{210}x + 80$ and $y = 100$ and find the points of intersection of the graphs.

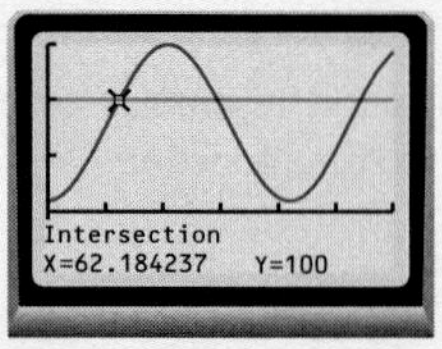

You are at 100 ft about 62 s and about 148 s after the wheel begins to turn.

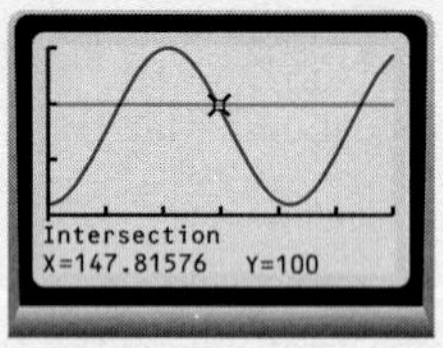

Method 2 Solve the equation using algebra.

$$-70 \cos \frac{2\pi}{210}t + 80 = 100$$

$$-70 \cos \frac{2\pi}{210}t = 20$$

$$\cos \frac{2\pi}{210}t = -\frac{20}{70}$$

$$\frac{2\pi}{210}t = \cos^{-1}\left(-\frac{20}{70}\right)$$

$$t = \frac{210}{2\pi} \cos^{-1}\left(-\frac{20}{70}\right)$$

$$t \approx 62.2$$

The first time is 43 s *before* the peak at 105 s, so the second time is 43 s *after* the peak, at about 148 s.

You reach 100 ft when $t \approx 62$ s and when $t \approx 148$ s.

Checking Key Concepts

1. a.

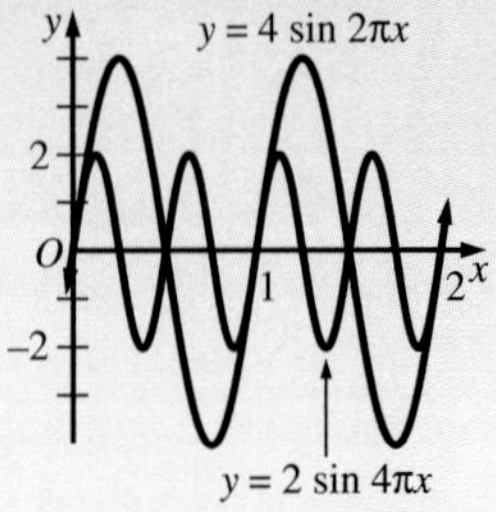

b. $y = 2 \sin 4\pi x$ has amplitude 2 and period 0.5; $y = 4 \sin 2\pi x$ has amplitude 4 and period 1.

2. Answers may vary. An example is given.

a.

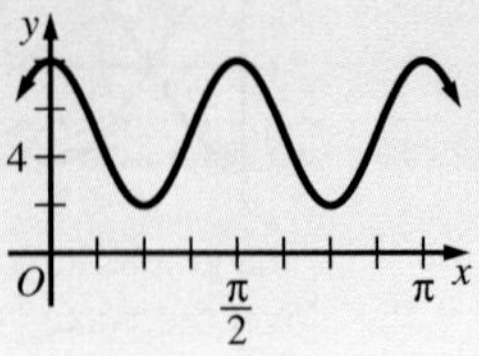

b. $y = 3 \cos 4x + 5$

Exercises and Applications

1. amplitude: 2; period: 1

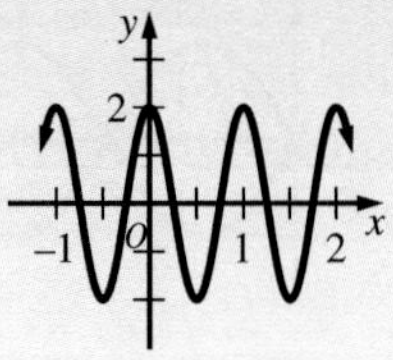

✓ Checking Key Concepts

1. a. In the same coordinate plane, graph $y = 2 \sin 4\pi x$ and $y = 4 \sin 2\pi x$.
   b. Compare the amplitudes and periods of the two graphs.
2. a. Draw the graph of a cosine function that has period $\frac{\pi}{2}$, amplitude 3, and a vertical translation of 5 units.
   b. What is an equation for the graph in part (a)?

15.3 Exercises and Applications

Extra Practice exercises on page 772

Graph each function. State the amplitude and period of each graph.

1. $y = 2 \cos 2\pi x$
2. $y = \frac{1}{2} \sin \frac{1}{2}x$
3. $y = -\frac{1}{2} \sin 3x$
4. $y = 2 \cos \frac{3\pi}{2}x$
5. $y = -\cos \frac{\pi}{2}x$
6. $y = 4 \sin 4\pi x$
7. $y = 4 \sin 4x$
8. $y = -\frac{1}{3} \cos 2x$
9. $y = -\frac{1}{3} \sin \frac{1}{4}x$
10. $y = 3 \cos \pi x$
11. $y = 2 \sin 2.5x$
12. $y = 2 \cos 4x$

Write an equation for each graph.

13.

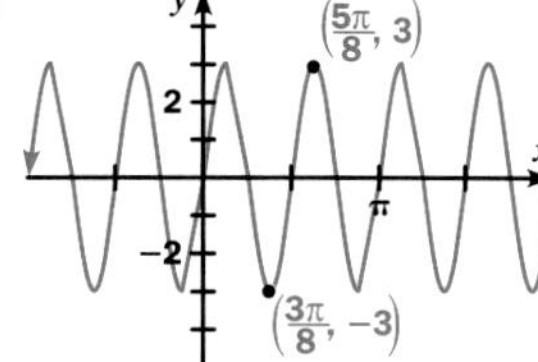

14.

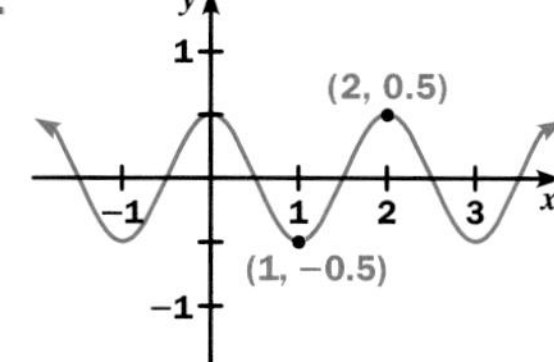

15.

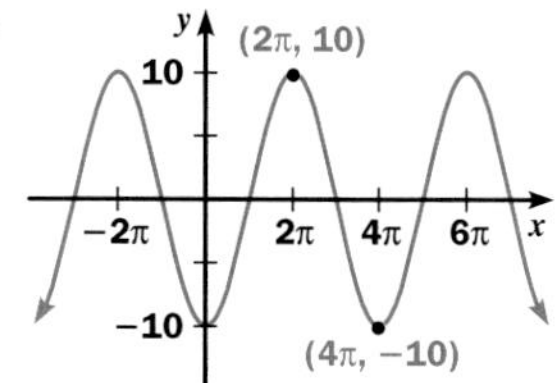

Write a sine function with the given maximum *M*, minimum *m*, and period.

16. $M = 5, m = -5$, period $= \frac{1}{4}$
17. $M = 7, m = 3$, period $= 3\pi$
18. $M = 6, m = -9$, period $= 1$
19. $M = -4, m = -10$, period $= \frac{\pi}{2}$

Match each equation with its graph.

20. $y = 2 + \frac{1}{2} \cos 2x$
21. $y = -1 - 3 \sin 2x$
22. $y = 2 - \cos \frac{\pi}{2}x$

A.

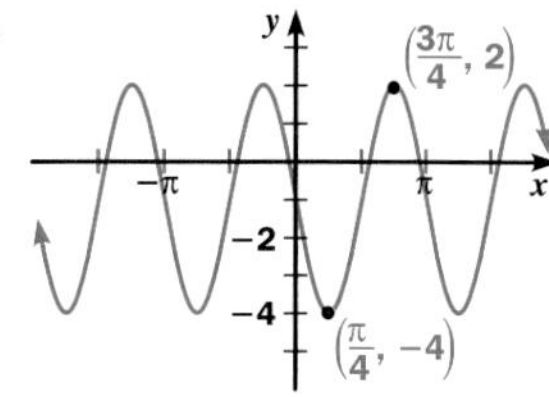

B.

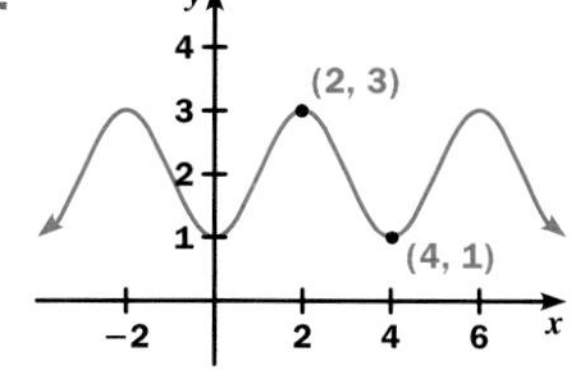

C.

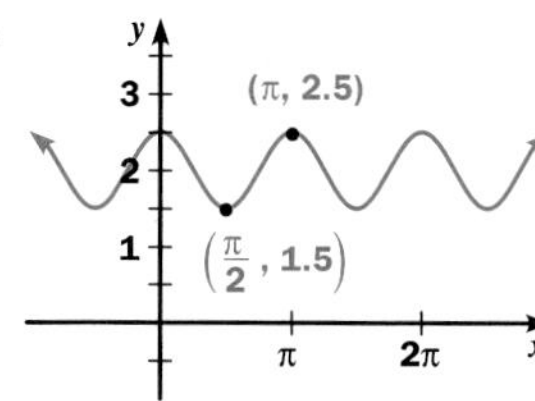

2. amplitude: $\frac{1}{2}$; period: 4π

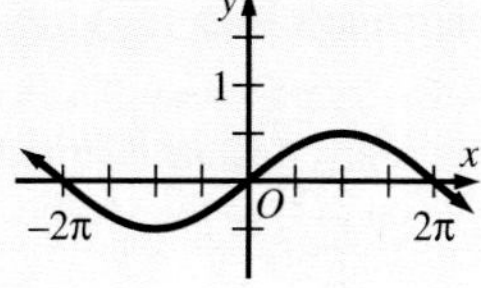

3. amplitude: $\frac{1}{2}$; period: $\frac{2\pi}{3}$

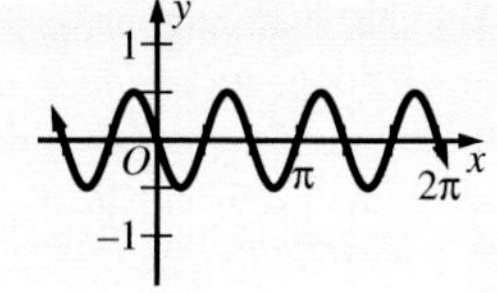

4–12. See answers in back of book.

13. $y = 3 \sin 4x$
14. $y = \frac{1}{2} \cos \pi x$
15. $y = -10 \cos \frac{1}{2}x$
16. $y = 5 \sin 8\pi x$
17. $y = 2 \sin \frac{2}{3}x + 5$
18. $y = 7.5 \sin 2\pi x - 1.5$
19. $y = 3 \sin 4x - 7$
20. C
21. A
22. B

Teach⇔Interact

Additional Example 3 (continued)

If the Ferris wheel had been rotating before the time started, you would have been at a height of 90 ft about 20 s before reaching the left-most position. This corresponds to a positive time of 140 s. The peak is reached at 120 s, 20 s before the wheel reaches 90 ft the second time. Thus, the wheel reaches 90 ft the first time 20 s before the peak, or at about 100 s.

Closure Question

Give the period, amplitude, and vertical translation for the graph of $y = a \cos bx + k$.

The amplitude is $|a|$. The period is $\frac{2\pi}{|b|}$ and the vertical translation is $|k|$ units up if k is positive and $|k|$ units down if k is negative.

Apply⇔Assess

Suggested Assignment

Core Course
Exs. 1–23, 33, 37–46, AYP

Extended Course
Exs. 1–19 odd, 20–46, AYP

Block Schedule
Day 2 Exs. 1–23, 33, 37–46, AYP

Exercise Notes

Communication: Drawing

Exs. 1–12 With practice, sine and cosine curves can be drawn by hand easily when the amplitude, period, and vertical translation are known. Students sometimes have difficulty scaling the axes using π. Recommend that they start by marking off one period, then work backward, dividing the period interval in half and marking that with half of the period. Then divide each of these intervals in half and mark these with one fourth of the period. Students can check their graphs by substituting the coordinates for one maximum and one minimum point into the equation.

Apply⇔Assess

Exercise Notes

Reasoning

Ex. 23 After completing this exercise, ask students to explain their findings. They should realize that since the period is $\frac{2\pi}{b}$, b is 2π divided by the period, or the number of cycles that fit into 2π.

Interview Note

Exs. 24–27 Some of the terms related to the playing of stringed instruments, such as *fret*, *open string*, and *harmonic* may be unfamiliar to many students. If possible, invite volunteers familiar with these instruments to explain or demonstrate what the terms mean.

Application

Exs. 24–27 These exercises allow students to see how considerations of the period and amplitude of a sine function can be used to study the tuning of a guitar. Point out to students that the intervals (ratios of frequencies) between the 1st and 2nd, the 2nd and 3rd, 3rd and 4th, and 5th and 6th strings of a guitar are all the same. However, the interval between the 4th and 5th strings is different. For this reason, the patterns in the table are interrupted between the 4th and 5th rows. Also, for this reason, the 6th string can be tuned in relation to the 1st string, because it is exactly two octaves higher. The 5th string can then be tuned in relation to the 6th string using harmonics.

Multicultural Note

Exs. 24–27 Stringed instruments have been played around the world for thousands of years. Music scholars debate the origins of the guitar, with some tracing it to the ancient Greek kithara, some to the long-necked lutes of Mesopotamia, and others to Coptic lutes discovered in Egypt. It is unclear whether the guitar developed independently in Europe, or whether it was brought there by Arabs during the Middle Ages. Different types of guitars are made and played in various parts of the Americas, the Pacific world, Africa, and Asia.

23. **Technology** Use a graphing calculator or graphing software. Set it to radian mode and use a viewing window that shows $0 \le x \le 2\pi$ and $-1.5 \le y \le 1.5$.

a. Graph $y = \sin x$, $y = \sin 1.5x$, $y = \sin 2x$, and $y = \sin 2.5x$. How many cycles appear between 0 and 2π along the x-axis for each graph?

b. How many cycles do you expect will appear between 0 and 2π in the graphs of $y = \sin 1.25x$, $y = \sin 1.75x$, and $y = \sin 2.25x$? Check your predictions by graphing.

c. How many cycles will appear between 0 and 2π in a graph with equation of the form $y = \sin bx$? in a graph with equation of the form $y = \cos bx$?

Joe Lopez tunes his *bajo sexto* using harmonics.

INTERVIEW Joe Lopez

Look back at the article on pages 704–706.

One way to tune a guitar is to use **harmonics**, *the ringing tones you hear if you pluck a string while lightly touching the string at points one half, one third, or one fourth of the way along its length. To find the exact points to touch, you can use the 12th, 7th, and 5th frets as guides.*

The plucked string vibrates this way.

String (frequency)	Fundamental (open string)	1st harmonic (12th fret)	2nd harmonic (7th fret)	3rd harmonic (5th fret)
E $\left(82\frac{1}{2}\text{ cps}\right)$	$P = \sin 165\pi t$	$P = \sin 330\pi t$	$P = \sin 495\pi t$	$P = \sin 660\pi t$
A (110 cps)	$P = \sin 220\pi t$	$P = \sin 440\pi t$	$P = \sin 660\pi t$	$P = \sin 880\pi t$
D $\left(146\frac{2}{3}\text{ cps}\right)$	$P = \sin 293\frac{1}{3}\pi t$	$P = \sin 586\frac{2}{3}\pi t$	$P = \sin 880\pi t$	?
G $\left(195\frac{5}{9}\text{ cps}\right)$	$P = \sin 391\frac{1}{9}\pi t$	$P = \sin 782\frac{2}{9}\pi t$	?	?
B $\left(247\frac{1}{2}\text{ cps}\right)$	$P = \sin 495\pi t$	?	?	?
E (330 cps)	?	?	?	?

5th fret

7th fret

12th fret

The table shows sound pressure equations for the harmonic tones of a guitar tuned in the key of A.

24. Look at the first two rows of the table. How do the equations of the 1st, 2nd, and 3rd harmonics compare with the fundamental equations?

25. Copy and complete the table.

26. Strings that vibrate according to the same equation sound the same. Explain how the first four strings can be tuned by comparing the 3rd harmonic of one string with the 2nd harmonic of the next string.

27. **Open-ended Problem** Explain how you might tune the last two strings based on the harmonics of one of the first four strings.

23. a.

1 cycle; 1.5 cycles; 2 cycles; 2.5 cycles

b. 1.25 cycles; 1.75 cycles; 2.25 cycles

c. $y = \sin bx$: $|b|$ cycles; $y = \cos bx$: $|b|$ cycles

24. If the fundamental equation is represented by $P = \sin b\pi t$, then the first harmonic is $P = \sin 2b\pi t$, the second harmonic is $P = \sin 3b\pi t$, and the third harmonic is $P = \sin 4b\pi t$.

25. See answers in back of book.

26. Once the E at 82.5 cps is set, tune the second harmonic of A to match the third harmonic of E. Then match the second harmonic of D to the third harmonic of A and the second harmonic of G to the third harmonic of D.

27. Answers may vary. An example is given. Tune the fundamental of B to the second harmonic of low E and the fundamental of high E to the third harmonic of low E.

28. between 860 m and 880 m

29. You know the distance traveled by the wave within one

Connection SURVEYING

Electronic Distance Measurement (EDM) is a surveying technique. Because of its accuracy, it is used to monitor dams so that small but potentially disastrous changes can be recognized early and repaired. EDM is also used for tunneling through mountains or below water. To measure the distance between two points *A* and *B* by EDM, an electromagnetic wave of known period is transmitted from *A* and reflected back from a prism at *B*. A receiver at *A* analyzes the amplitude of the wave when it returns.

28. Suppose previous measurements show that the distance from *A* to *B* is between 430 m and 440 m. If you transmit a wave from *A* to *B* and back to *A*, how far might it travel?

29. Writing Explain how using a wave with a period of 20 m can help you find the exact distance between *A* and *B*.

30. Write an equation for a sine wave with period 20 m and amplitude a.

31. Suppose that when a sine wave with period 20 m returns to *A*, the receiver detects that the wave height is $0.6a$, as shown in the diagram at the right. What are the two possible distances the wave has traveled when going from *A* to *B* and back to *A*?

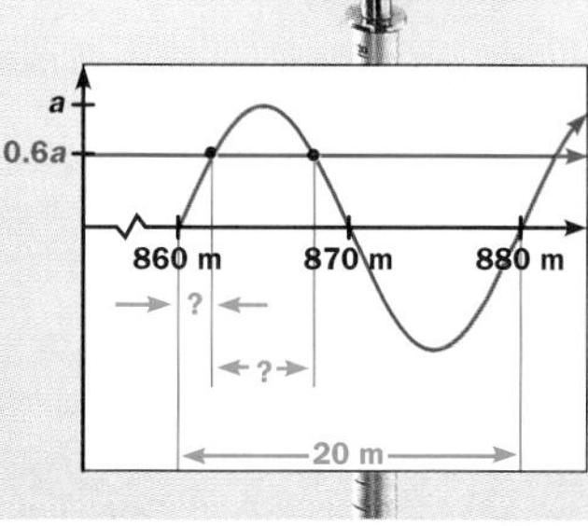

32. Suppose the receiver also detects that the wave height is *decreasing* when the wave reaches the receiver. Which of your two answers from Exercise 31 is the actual distance the wave has traveled? What is the actual distance between *A* and *B*?

33. SAT/ACT Preview A sine function has a maximum value of 20, a minimum value of -10, and period 4π. Which equation describes this function?

A. $y = 20 - 10 \sin 4x$ **B.** $y = 5 - 15 \sin \frac{1}{2}x$ **C.** $y = 15 \sin \frac{1}{2}x - 5$

D. $y = 20 - 10 \sin \frac{1}{2}x$ **E.** none of these

34. Consider the function $y = a \sin bx + k$.

a. Give expressions in terms of a, b, and k for M and m, the maximum and minimum values of the function.

b. Use your results from part (a) to show that the definition of amplitude applies to vertically translated graphs of sine functions. That is,

$$\frac{M - m}{2} = |a|$$

35. Open-ended Problem In the same coordinate plane, sketch the graphs of two sine functions, one with twice the amplitude and half the period of the other. Write equations for the two graphs.

36. Challenge Find three different cosine equations whose graphs include the points (0, 2) and (3, 0).

Apply⇔Assess

Exercise Notes

Problem Solving
Exs. 30, 31 Students having difficulty with these problems should review the Solution to Example 3 on page 722. To solve the equation for Ex. 31, after dividing both sides of the equation by *a*, a graphing calculator or graphing software can be used. Enter $y = 0.6$ and $y = \sin \frac{\pi}{10}x$ and find the two intersections.

Teaching Tip
Ex. 34 You might want to give students a hint for part (a): they should use the fact that the values of sin *bx* are between –1 and 1. Students should then be able to determine that the maximum value of $a \sin bx$ is $|a|$, and thus the maximum value of y is $|a| + k$. Likewise, they should see that the minimum value will be $-|a| + k$.

Student Progress
Ex. 35 Completing this open-ended problem will ensure an understanding of the concepts of period and amplitude and how they are represented in the graph's equation.

Visual Thinking
Ex. 35 Ask students to explain their equations in relation to their sketches. Encourage them to change a variable of the equation and then sketch the sine curve of the new equation. This activity involves the visual skills of *recognition* and *communication.*

Challenge
Ex. 36 This problem can be done in many ways. The points, for example, could be the maximum and minimum values of the function in a single period, the maximum and minimum values of consecutive periods, or one point could be on the axis and the other a maximum or minimum. The variations are infinite. Ask students to share their results with the class so that the variety of solutions can be seen.

cycle. By finding the height of the returning wave, you can determine the possible points in that cycle (at most, three points) at which the receiver intercepts the wave.

30. $y = a \sin \frac{\pi}{10}x$

31. 862 m or 868 m

32. 868 m; 434 m

33. B

34. a. $M = |a| + k$; $m = -|a| + k$

b. $\frac{M - m}{2} = \frac{|a| + k - (-|a| + k)}{2} = \frac{2|a|}{2} = |a|$

35. Answers may vary. An example is given. $y = \sin x$; $y = 2 \sin 2x$

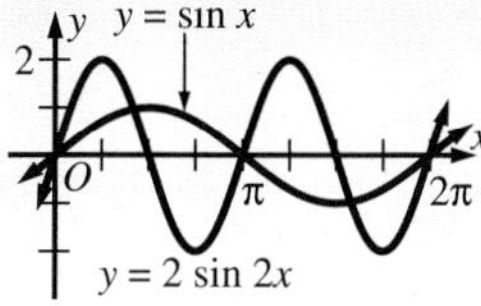

36. Answers may vary. An example is given. $y = 2 \cos \frac{\pi}{2}x$; $y = \cos \frac{\pi}{3}x + 1$; $y = 2 \cos \frac{5\pi}{6}x$

Apply⇔Assess

Exercise Notes

Cooperative Learning

Ex. 37 Students will need some time to work together on this exercise, although it is not extensive. In part (a), all of the graphs should have period 2π with the same x-intercepts as the graph of $y = \cos x$. Suggest that each group use both positive and negative values of a. In part (b), all the graphs should have amplitude 1 with varying periods. These graphs will be easier to draw if the periods are chosen so that the two smaller periods divide the larger period. The groups could share their summary statement with the entire class.

Progress Check 15.1–15.3

See page 742.

Practice 98 for Section 15.3

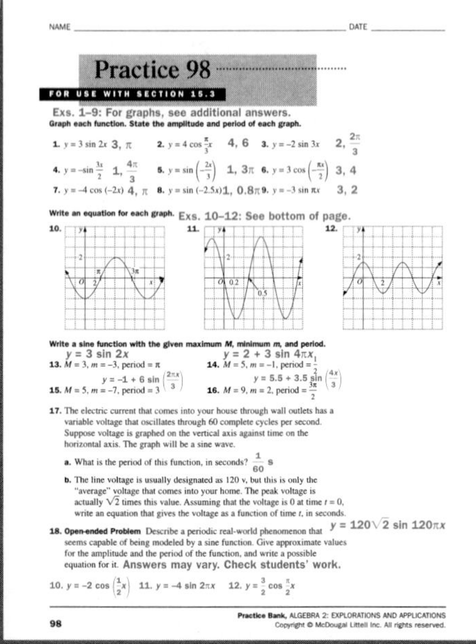

NAME ______ DATE ______

Practice 98

FOR USE WITH SECTION 15.3

Exs. 1–9: For graphs, see additional answers.

Graph each function. State the amplitude and period of each graph.

1. $y = 3 \sin 2x$ 3, π
2. $y = 4 \cos \frac{\pi}{3}x$ 4, 6
3. $y = -2 \sin 3x$ 2, $\frac{2\pi}{3}$
4. $y = -\sin \frac{3x}{2}$ 1, $\frac{4\pi}{3}$
5. $y = \sin\left(-\frac{2x}{3}\right)$ 1, 3π
6. $y = 3 \cos\left(-\frac{\pi x}{2}\right)$ 3, 4
7. $y = -4 \cos(-2x)$ 4, π
8. $y = \sin(-2.5x)$ 1, 0.8π
9. $y = -3 \sin \pi x$ 3, 2

Write an equation for each graph. Exs. 10–12: See bottom of page.

10. 11. 12.

Write a sine function with the given maximum M, minimum m, and period.

13. $M = 3, m = -3$, period $= \pi$ $y = 3 \sin 2x$
14. $M = 5, m = -1$, period $= \frac{1}{2}$ $y = 2 + 3 \sin 4\pi x$
15. $M = 5, m = -7$, period $= 3$ $y = -1 + 6 \sin\left(\frac{2\pi x}{3}\right)$
16. $M = 9, m = 2$, period $= \frac{3\pi}{2}$ $y = 5.5 + 3.5 \sin\left(\frac{4x}{3}\right)$

17. The electric current that comes into your house through wall outlets has a variable voltage that oscillates through 60 complete cycles per second. Suppose voltage is graphed on the vertical axis against time on the horizontal axis. The graph will be a sine wave.
    a. What is the period of this function, in seconds? $\frac{1}{60}$ s
    b. The line voltage is usually designated as 120 v, but this is only the "average" voltage that comes into your home. The peak voltage is actually $\sqrt{2}$ times this value. Assuming that the voltage is 0 at time $t = 0$, write an equation that gives the voltage as a function of time t, in seconds. $y = 120\sqrt{2} \sin 120\pi x$

18. **Open-ended Problem** Describe a periodic real-world phenomenon that seems capable of being modeled by a sine function. Give approximate values for the amplitude and the period of the function, and write a possible equation for it. Answers may vary. Check students' work.

10. $y = -2 \cos\left(\frac{1}{2}x\right)$ 11. $y = -4 \sin 2\pi x$ 12. $y = \frac{3}{2} \cos \frac{\pi}{2}x$

98 **Practice Bank,** ALGEBRA 2: EXPLORATIONS AND APPLICATIONS

ONGOING ASSESSMENT

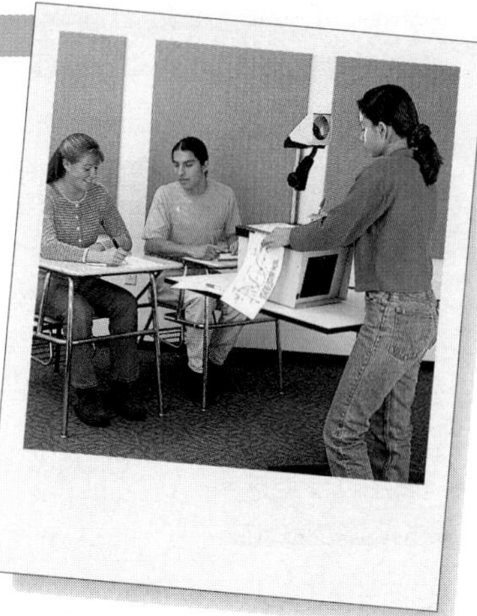

37. Cooperative Learning Work in a group of four students. Prepare your results on a poster or overhead transparency.

a. In the same coordinate plane, graph $y = a \cos x$ for three different positive values of a, showing at least one period. In a different color, graph three functions with negative values of a. Label the intersections of the graphs with the y-axis.

b. In the same coordinate plane, graph $y = \cos bx$ for three different positive values of b, showing at least one period. Label the intersections of each graph with the x-axis.

c. Write a summary statement that describes the effects of varying a and b in the equations $y = a \cos x$ and $y = \cos bx$.

SPIRAL REVIEW

Convert each angle measure from radians to degrees. *(Section 15.2)*

38. $\frac{\pi}{6}$ **39.** -3π **40.** $\frac{9}{4}\pi$

Multiply. *(Section 8.4)*

41. $(4 + i)(2 + 3i)$ **42.** $(6 - 3i)^2$ **43.** $(1 + i)(1 - i)$

Graph each function. *(Section 9.7)*

44. $y = \frac{1}{x - 1}$ **45.** $y = \frac{1}{x + 2} + 4$ **46.** $y = \frac{1}{x + 7} - 3$

ASSESS YOUR PROGRESS

VOCABULARY

sine (p. 708)	**cycle** (p. 720)
cosine (p. 708)	**period** (p. 720)
radian (p. 714)	**amplitude** (p. 720)

Find all angles θ for $0° \le \theta \le 360°$ that satisfy each equation. Give answers to the nearest degree. *(Section 15.1)*

1. $\sin \theta = 0.7589$ **2.** $\cos \theta = 0.8234$ **3.** $\sin \theta = -0.5239$

Evaluate each expression. *(Section 15.2)*

4. $\cos 4.1$ **5.** $\cos(-1.5)$ **6.** $\sin(-315°)$

Graph each function. State the amplitude and period of each graph. *(Section 15.3)*

7. $y = 2 \sin \frac{\pi}{2}x + 4$ **8.** $y = -\frac{1}{2} \cos \frac{1}{2}x$

9. Journal Explain how to write an equation of the form $y = a \sin bx + k$ if you know the maximum and minimum values, the amplitude, and the period of a sine function.

---

37. a, b. Answers may vary. Check students' work.

c. Answers may vary. An example is given. Varying a changes the amplitude of the waves (amplitude $= |a|$). Varying b changes the period of the waves $\left(\text{period} = \frac{2\pi}{b}\right)$.

38. 30° 39. –540°

40. 405° 41. $5 + 14i$

42. $27 - 36i$ 43. 2

44.

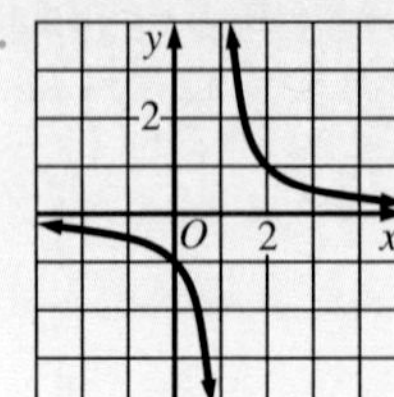

45.

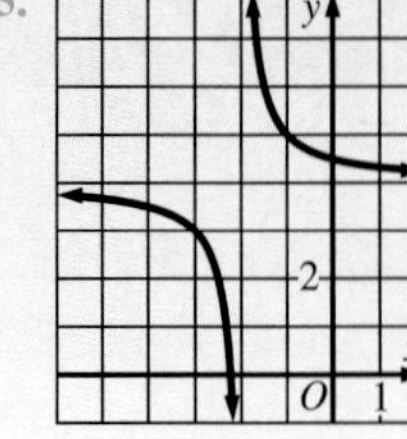

46.

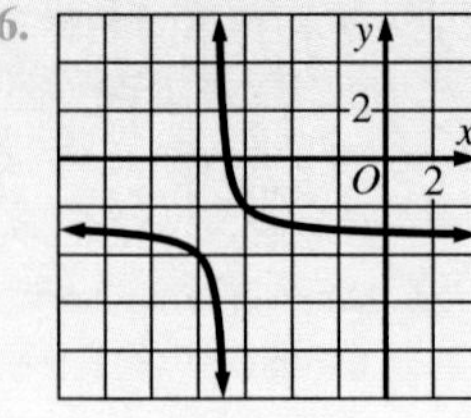

Assess Your Progress

1. 49°, 131°
2. 35°, 325°
3. 328°, 212°
4. –0.5748
5. 0.0707
6. 0.7071

7–9. See answers in back of book.

Section

15.4 Exploring Phase Shifts

Learn how to...

- graph equations of these forms: $y = a \sin b(x - h) + k$, $y = a \cos b(x - h) + k$
- write equations for sine and cosine graphs that have a phase shift

So you can...

- describe periodic phenomena that have a phase shift, such as the motions of pistons in an engine

The combustion of fuel causes the pistons of a car engine to move up and down. Connecting rods link the pistons to the throws of a crankshaft, transforming the vertical motion of the pistons into a circular motion that powers the wheels.

1 2 3 4 5 6
piston
connecting rod
throw
crankshaft

Because the throws are placed at 120° angles around the crankshaft, the pistons must be fired "out of phase" (that is, at different times) for the engine to run smoothly.

The graphs below show the heights of the pistons above the crankshaft in a 6-cylinder engine as a function of time. There are only three graphs, because the pistons are paired. Pistons **1** and **6** move together, as do pistons **2** and **5**, and pistons **3** and **4**.

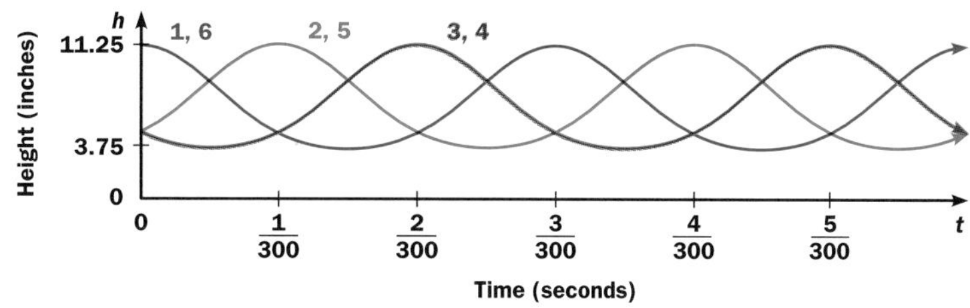

BY THE WAY

To help the crankshaft rotate smoothly, a film of oil about 0.002 inch thick is placed between the crankshaft and the surrounding bearings.

THINK AND COMMUNICATE

1. Based on the graphs, how long does it take the crankshaft to rotate once?
2. At what times are pistons 1 and 6 at their maximum height? pistons 2 and 5? pistons 3 and 4?
3. Describe how the graphs of the piston heights are geometrically related to each other.
4. Suppose the equation $h = f(t)$ describes the height of piston 1 as a function of time t. Complete each sentence.
   a. An equation for piston 2 is $h = f(t - \underline{?})$.
   b. An equation for piston 3 is $h = f(t - \underline{?})$.

Plan⇔Support

Objectives

- Graph equations of these forms: $y = a \sin b(x - h) + k$, $y = a \cos b(x - h) + k$.
- Write equations for sine and cosine graphs that have a phase shift.
- Describe periodic phenomena that have a phase shift.

Recommended Pacing

Core and Extended Courses
Section 15.4: 2 days

Block Schedule
Section 15.4: 1 block

Resource Materials

Lesson Support
Lesson Plan 15.4
Warm-Up Transparency 15.4
Overhead Visuals:
Folder 15: Graphing $y = a \cos bx + k$
Practice Bank: Practice 99
Study Guide: Section 15.4
Explorations Lab Manual:
Additional Exploration 21, Diagram Masters 1, 2

Technology
Technology Book:
Calculator Activity 15
Graphing Calculator
McDougal Littell Mathpack
Function Investigator with Matrix Analyzer Activity Book: Function Investigator Activity 33
Geometry Inventor Activity Book: Activity 22
Internet:
http://www.hmco.com

ANSWERS Section 15.4

Think and Communicate

1. 0.01 s
2. 1 and 6: $\frac{3}{300}$ s, $\frac{6}{300}$ s;
   2 and 5: $\frac{1}{300}$ s, $\frac{4}{300}$ s;
   3 and 4: $\frac{2}{300}$ s, $\frac{5}{300}$ s
3. The graph for pistons 2 and 5 is the graph for pistons 1 and 6 shifted $\frac{1}{300}$ s to the right. The graph for pistons 3 and 4 is the graph for pistons 1 and 6 shifted $\frac{2}{300}$ s to the right.
4. a. $\frac{1}{300}$
   b. $\frac{2}{300}$

Warm-Up Exercises

State the amplitude, period, and vertical translation of each graph.

1. $y = 3 \sin 2\pi x$ 3; 1; none
2. $y = -\frac{1}{2} \cos 3x + 4$ $\frac{1}{2}$; $\frac{2\pi}{3}$; 4 up
3. $y = 5 \cos 0.25x - 5$ 5; 8π; 5 down
4. Give an equation for a cosine function with maximum 13, minimum 5, and period 3π. $y = \pm 4 \cos \frac{2}{3}x + 9$

Teach⇔Interact

Section Note

Challenge
The graphs of the heights of the pistons on page 727 look like cosine curves but they are not. The crank pins and connecting rods transform the circular motion of the crankshaft into an up and down motion that is not exactly sinusoidal. The motion of pistons 1 and 6, for example, can be modeled by the equation $y = 3.75 \cos x + \sqrt{(7.5)^2 - (3.75 \sin x)^2}$, where x is the angle of rotation of the throw measured clockwise from vertical. Ask students to verify this equation using triangle trigonometry. They should draw a diagram using the fact that the throw, and hence the radius of the circle is 3.75 in. and the length of the connecting rod is 7.5 in. The connecting rod goes from a point on the circle to the piston. The height of the piston is measured from the center of the circle to the end of the connecting rod. To find the equation, students need to draw the altitude of the triangle formed by the throw, connecting rod and height of the piston that goes from the point on the circle. Two of the legs of the two right triangles formed make up the height of the piston. To transform the equation into the one graphed on page 727, students need to change the variable from degrees to time using the fact that the throw turns 360° in 0.01 s.

Additional Example 1

Graph each equation.

a. $y = 2 \sin\left(x - \frac{\pi}{4}\right)$

Rewrite the equation in the form $y = a \sin b(x - h)$.

$y = 2 \sin 1\left(x - \frac{\pi}{4}\right)$

$a = 2,\ b = 1,\ h = \frac{\pi}{4}$

Think of shifting the graph of $y = 2 \sin x$ (shown as a dashed curve) to the right $\frac{\pi}{4}$ units.

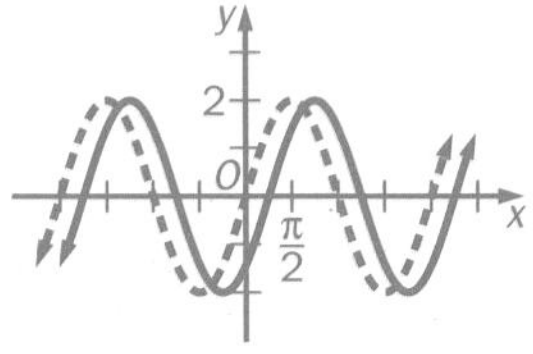

A horizontal translation of a periodic function is called a **phase shift.**

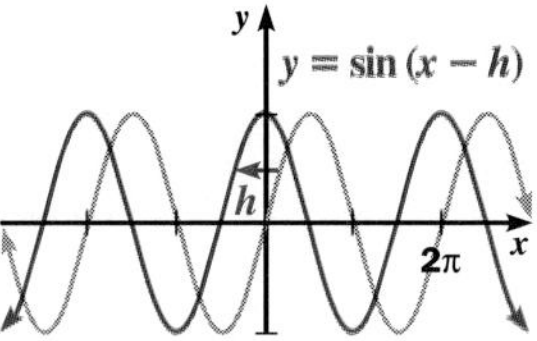

If $h < 0$, the graph is shifted to the **left**.

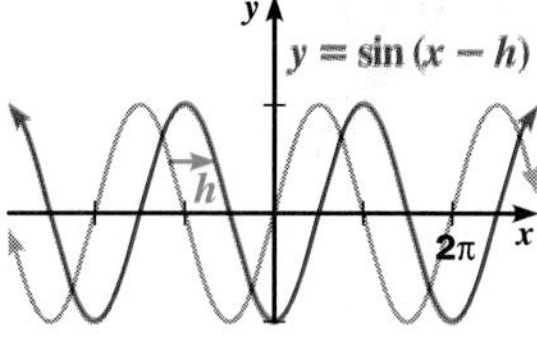

If $h > 0$, the graph is shifted to the **right**.

THINK AND COMMUNICATE

5. Refer to the piston height graphs on the previous page. Compared with piston 1, what is the phase shift for piston 2? for piston 3?

6. a. What is the smallest phase shift that will move a sine graph onto a cosine graph? Express your answer in radians.

b. Complete this equation: $\cos x = \sin (x - \underline{?})$.

EXAMPLE 1

Graph each equation.

a. $y = 3 \cos (x - \pi)$ **b.** $y = \cos\left(3x - \frac{\pi}{2}\right)$

SOLUTION

Rewrite each equation in the form $y = a \cos b(x - h)$.

WATCH OUT!
In part (b), the phase shift is *not* $\frac{\pi}{2}$.

a. $y = 3 \cos (x - \pi)$
$= 3 \cos 1(x - \pi)$

$a = 3 \quad b = 1 \quad h = \pi$

Think of shifting the graph of $y = 3 \cos x$ (shown in gray) to the right π units.

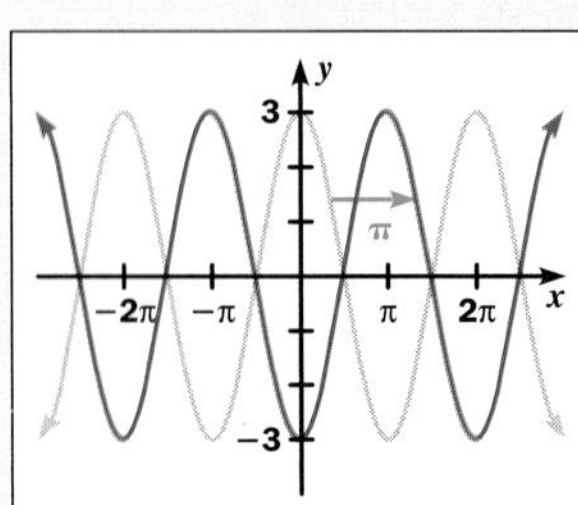

b. $y = \cos\left(3x - \frac{\pi}{2}\right)$
$= 1 \cos 3\left(x - \frac{\pi}{6}\right)$

$a = 1 \quad b = 3 \quad h = \frac{\pi}{6}$

Think of shifting the graph of $y = \cos 3x$ (shown in gray) to the right $\frac{\pi}{6}$ units.

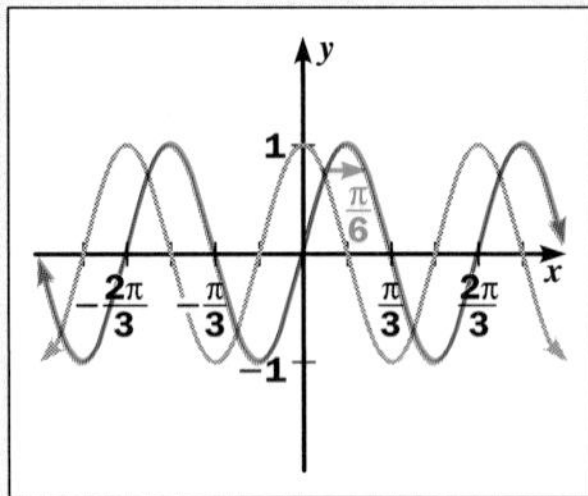

Think and Communicate

5. $\frac{1}{300}$ s to the right; $\frac{2}{300}$ s to the right

6. a. $\frac{\pi}{2}$ radians to the left

b. $-\frac{\pi}{2}$

EXAMPLE 2

Write a sine function for the blue graph shown below.

SOLUTION

Step 1 Find the amplitude and period of the graph.

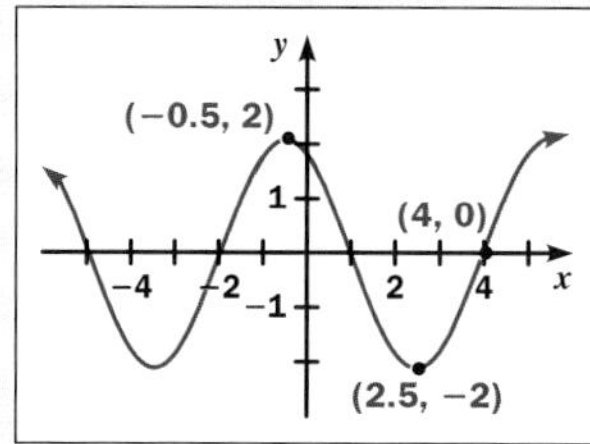

The amplitude is 2, so $a = 2$.

The period is 6, so $\frac{2\pi}{b} = 6$, and $b = \frac{\pi}{3}$.

There is no vertical shift for this graph.

Step 2 Draw a helping graph using a sine function with the values of a and b from Step 1.

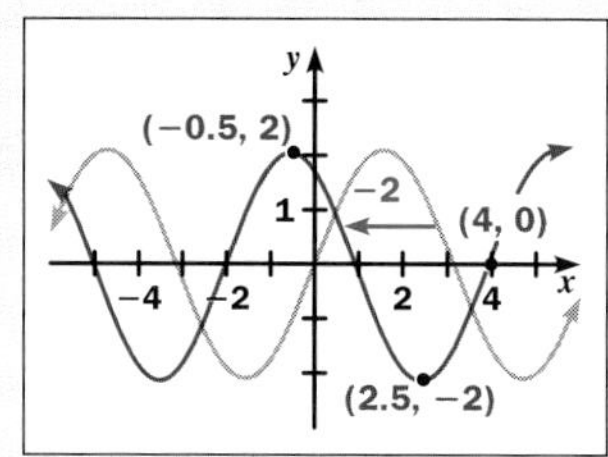

The graph of $y = 2 \sin \frac{\pi}{3}x$ is shown in gray.

The blue graph is shifted 2 units to the left of the gray graph.

A sine function that describes the blue graph is $y = 2 \sin \frac{\pi}{3}(x + 2)$.

Combining Sine and Cosine Functions

When you add sine or cosine functions, the resulting functions can be periodic, but their graphs may be more complicated than the graphs of simple sine and cosine functions.

In general, the **sum** of the functions $y = f(x)$ and $y = g(x)$ is the function $y = f(x) + g(x)$.

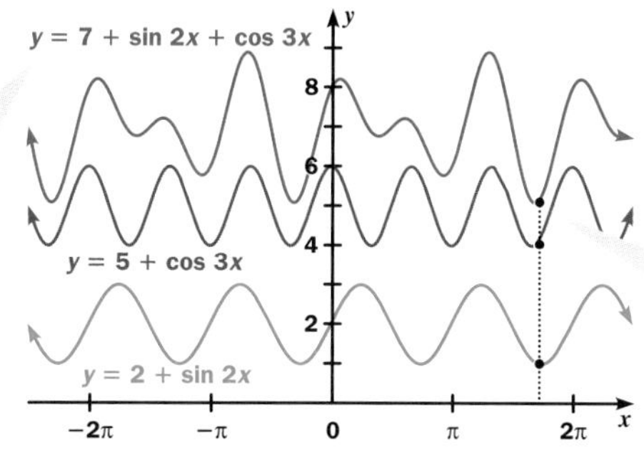

The y-coordinate of this point is the **sum of the y-coordinates** of the points with the same x-coordinate on the green and blue graphs.

THINK AND COMMUNICATE

7. What is the period of each of the three functions shown above?
8. Is the *period of the sum* function (shown in red) equal to the *sum of the periods* of the two original functions (shown in green and blue)?

Think and Communicate

7. $y = 2 + \sin 2x$: π;
$y = 5 + \cos 3x$: $\frac{2\pi}{3}$;
$y = 7 + \sin 2x + \cos 3x$: 2π
8. No.

Teach⇔Interact

Additional Example 1 (continued)

b. $y = \sin(2x + \pi)$

$y = \sin(2x + \pi)$

$= 1 \sin 2\left(x + \frac{\pi}{2}\right)$

$a = 1, b = 2, h = -\frac{\pi}{2}$

Think of shifting the graph of $y = \sin 2x$ (shown as a dashed curve) to the left $\frac{\pi}{2}$ units.

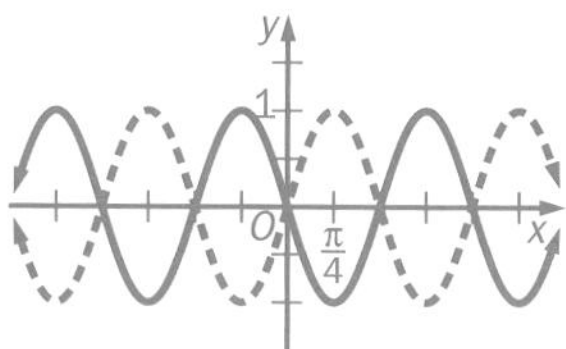

Additional Example 2

Write a cosine function for the graph shown below.

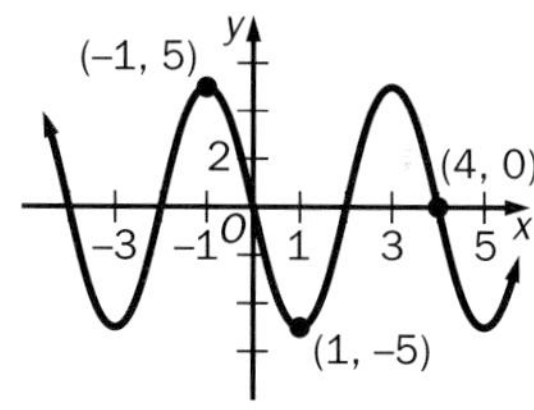

Step 1 Find the amplitude and period of the graph. The amplitude is 5, so $a = 5$. The period is 4, so $\frac{2\pi}{b} = 4$, and $b = \frac{\pi}{2}$. There is no vertical shift for this graph.

Step 2 Draw a helping graph using a cosine function with the values of a and b from Step 1.

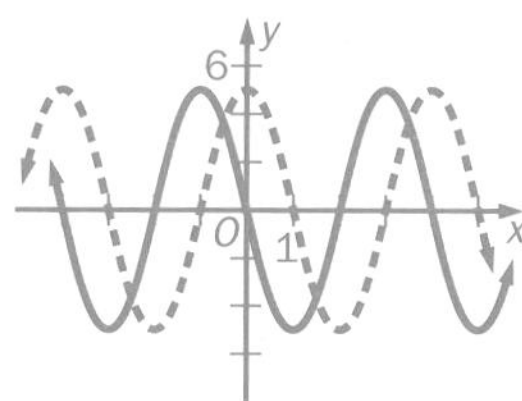

The graph of $y = 5 \cos \frac{\pi}{2}x$ is shown as a dashed curve. The original graph is shifted 1 unit to the left of the dashed graph. A cosine function that describes the original graph is $y = 5 \cos \frac{\pi}{2}(x + 1)$.

Section Note

Reasoning

Ask students to think about how the period of the sum of two periodic functions is related to the periods of the two functions. You may wish to review the concept of "least common multiple" to guide students toward an answer.

Teach⇔Interact

Additional Example 3

In the same coordinate plane, graph the functions $y = 3 \sin \frac{1}{2}x$, $y = \sin \frac{1}{2}(x + 2\pi)$, and their sum. Find the period and range of each function.

Use a graphing calculator or graphing software.

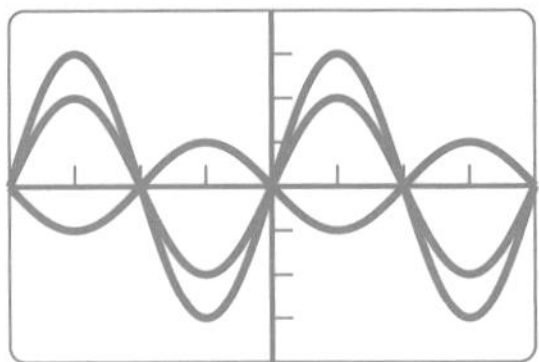

The period of all three functions is 4π. The range of $y = 3 \sin \frac{1}{2}x$ is $-3 \le y \le 3$. The range of $y = \sin \frac{1}{2}(x + 2\pi)$ is $-1 \le y \le 1$. Use a graphing calculator to find the minimum and maximum values of the sum function.

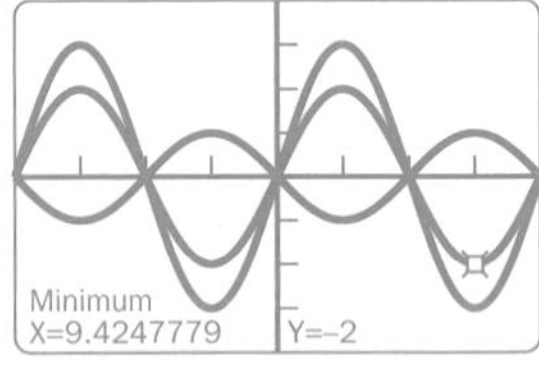

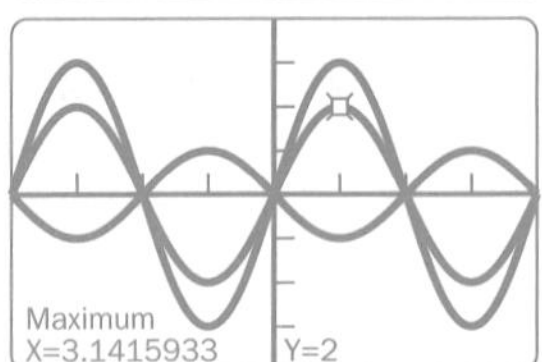

The minimum is exactly –2. The maximum is exactly 2. The range of $y = 3 \sin \frac{1}{2}x + \sin \frac{1}{2}(x + 2\pi)$ is $-2 \le y \le 2$.

Closure Question

Describe the procedure for graphing a sine or cosine curve that has a phase shift.

Rewrite the equation in the form $y = a \sin b(x - h)$ or $y = a \cos b(x - h)$. Identify a, b, and h and draw a helping graph using the values of a and b. Translate the helping graph horizontally $|h|$ units to the left if h is negative and $|h|$ units to the right if h is positive.

EXAMPLE 3

In the same coordinate plane, graph the functions $y = 4 + \cos \frac{2\pi}{3}x$, $y = 2 + \sin \pi x$, and their sum. Find the period and range of each function.

SOLUTION

Use a graphing calculator or graphing software.

The period of $y = 4 + \cos \frac{2\pi}{3}x$ is **3**.

The period of $y = 2 + \sin \pi x$ is **2**.

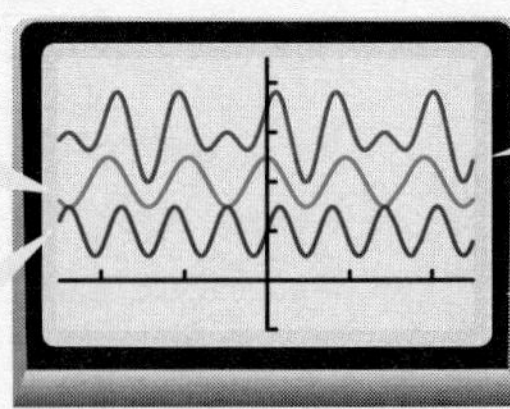

The period of the **sum** of the functions is **6**.

On some calculators, you can enter $y_3 = y_1 + y_2$ to graph the sum of two functions.

The range of $y = 4 + \cos \frac{2\pi}{3}x$ is $3 \le y \le 5$.

The range of $y = 2 + \sin \pi x$ is $1 \le y \le 3$.

Use a graphing calculator or graphing software to find the minimum and maximum values of the sum function.

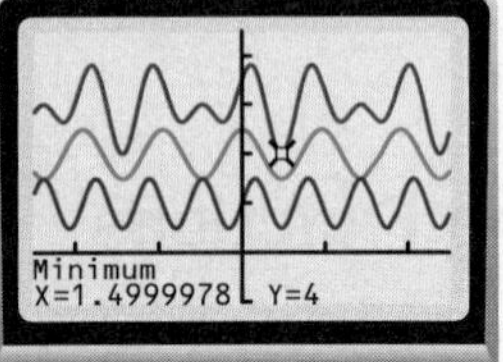

The minimum is exactly 4.

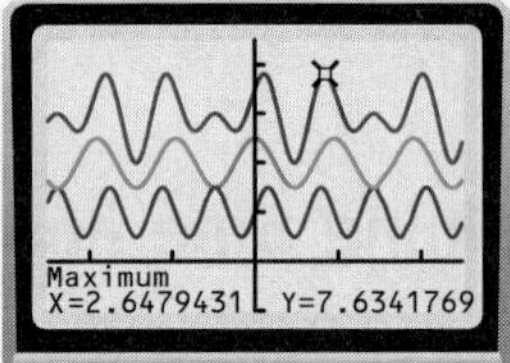

The maximum is about 7.63.

The range of $y = 6 + \sin \pi x + \cos \frac{2\pi}{3}x$ is about $4 \le y \le 7.63$.

THINK AND COMMUNICATE

9. In Example 3, the maximum values of the given functions (graphs shown in green and blue) are 5 and 3, but the maximum value of the sum function (graph shown in red) is *not* $5 + 3 = 8$. Explain why not.

CHECKING KEY CONCEPTS

1. Graph the functions $y = \sin\left(x - \frac{\pi}{2}\right)$ and $y = \sin\left(x + \frac{\pi}{2}\right)$. How do they compare with the graph of $y = \sin x$?

2. Write a cosine function that is equivalent to $y = \sin \pi(x - 3)$.

3. Graph the functions $y = 2 \sin 2x$, $y = \sin (2x - \pi)$, and their sum. Find the period and range of each function.

Think and Communicate

9. The maximum values for the two functions never occur for the same x-values, so the value of the sum function is always less than the sum of the maxima of the two original functions.

Checking Key Concepts

1. $y = \sin\left(x + \frac{\pi}{2}\right)$; $y = \sin\left(x - \frac{\pi}{2}\right)$

The graph of $y = \sin\left(x - \frac{\pi}{2}\right)$ is the graph of $y = \sin x$ shifted $\frac{\pi}{2}$ to the right, while the graph of $y = \sin\left(x + \frac{\pi}{2}\right)$ is the graph of $y = \sin x$ shifted $\frac{\pi}{2}$ to the left.

2. Answers may vary. An example is given. $y = \cos \pi(x - 1.5)$

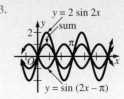

3. $y = 2 \sin 2x$; sum; $y = \sin (2x - \pi)$

15.4 Exercises and Applications

Extra Practice exercises on page 772

Graph each function. Show at least one period.

1. $y = 3\sin\left(x - \frac{\pi}{2}\right)$ **2.** $y = 2\cos\left(x + \frac{\pi}{2}\right)$ **3.** $y = 4\sin\frac{\pi}{4}(x + 5)$ **4.** $y = 10\cos\frac{\pi}{6}(x - 7)$

5. $y = \sin 3\left(x + \frac{\pi}{9}\right)$ **6.** $y = 1.5\cos 2\left(x - \frac{3\pi}{4}\right)$ **7.** $y = -2\sin\pi\left(x - \frac{3}{4}\right)$ **8.** $y = -5\cos\frac{\pi}{2}(x + 2.5)$

Write a sine function for each graph.

9.

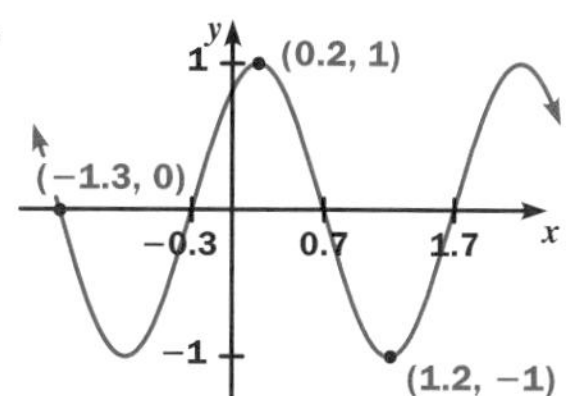

10.

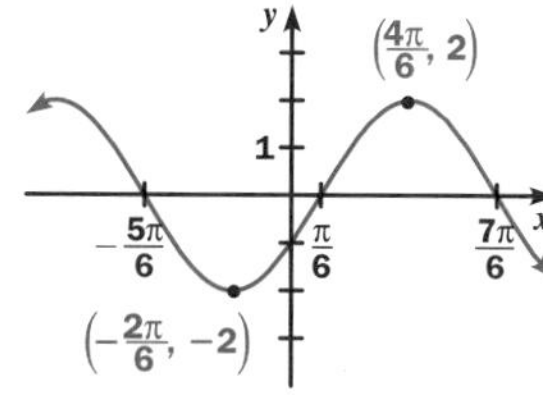

11.

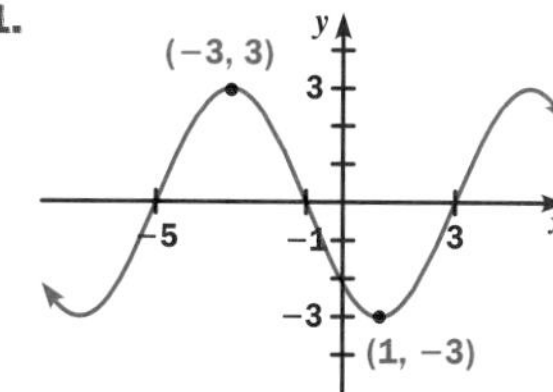

12. Open-ended Problem Write a cosine function for one of the graphs in Exercises 9–11.

Write a cosine function for each graph.

13.

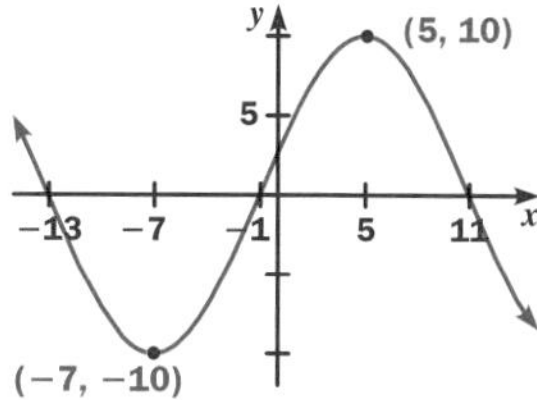

14.

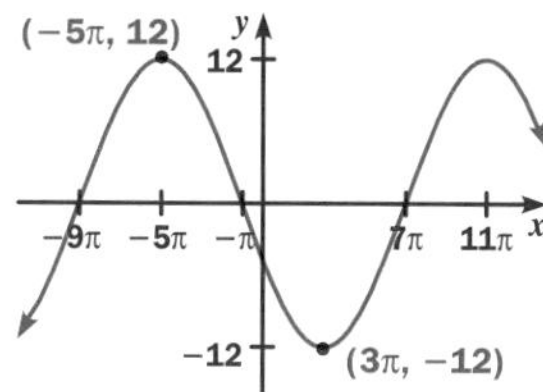

15.

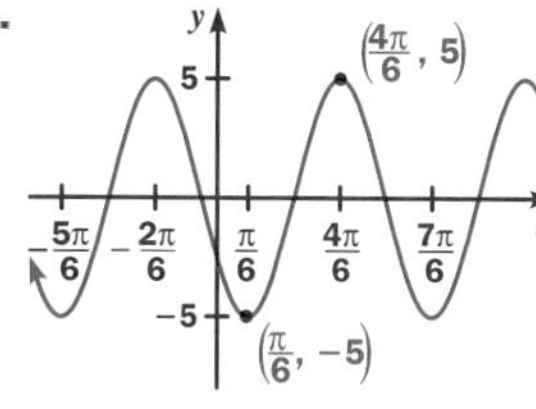

16. Open-ended Problem Write a sine function for one of the graphs in Exercises 13–15.

17. AUTOMOTIVE MECHANICS The pistons in an engine force the crank pins to rotate in a circle around the center of the crankshaft. The graphs show the heights of crankpins 1, 2, and 3 relative to the axle as a function of time.

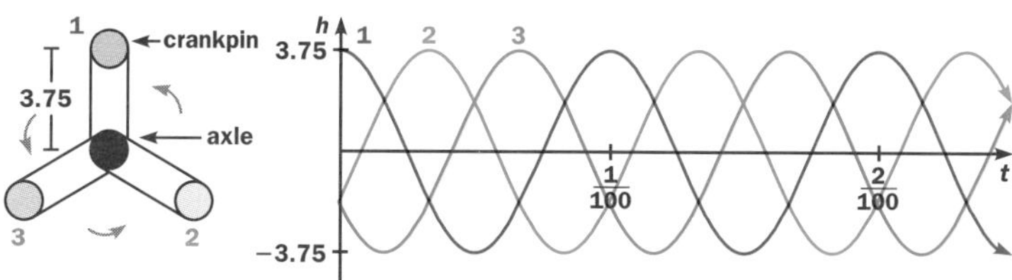

a. Explain why the periods of the graphs are $\frac{1}{100}$ s if the engine speed is 6000 rev/min.

b. Write a cosine function for the height of crank pin 1.

c. Use phase shifts to write equations for crank pins 2 and 3.

Apply⇔Assess

Suggested Assignment

Core Course
Day 1 Exs. 1–17, 23–25
Day 2 Exs. 26–30, 32–48

Extended Course
Day 1 Exs. 1–25
Day 2 Exs. 26–48

Block Schedule
Day 3 Exs. 1–17, 23–30, 32–48

Exercise Notes

Teaching Tip

Exs. 1–8 Students should know how to sketch these graphs by hand. They will find this easier to do if they first draw a helping graph, one of the same type with the same period and amplitude as the one they wish to draw without being translated. For each period of the helping graph, students should graph the minimum, the maximum and the two horizontal intercepts. These points can then be easily translated and connected to form the desired graph.

Communication: Drawing

Exs. 1–8 You may want to assign each of these exercises to different students and ask them to draw a graph on overhead sheets and bring it to class the next day. At the beginning of class, the overheads can be shown, allowing students to quickly correct their own work.

Application

Exs. 17, 18–22, 31 Many phenomena can be modeled by functions that are phase shifts or sums of sine and cosine curves. These exercises provide examples from automotive mechanics, architecture, and music.

Second-Language Learners

Ex. 17 For second-language learners unfamiliar with the function of an automobile engine, this problem may pose difficulties. Consider having these students work cooperatively with English-fluent students on this exercise.

Function	Period	Range
$y = 2\sin 2x$	π	$-2 \le y \le 2$
$y = \sin(2x - \pi)$	π	$-1 \le y \le 1$
sum	π	$-1 \le y \le 1$

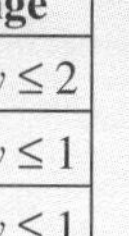

2.

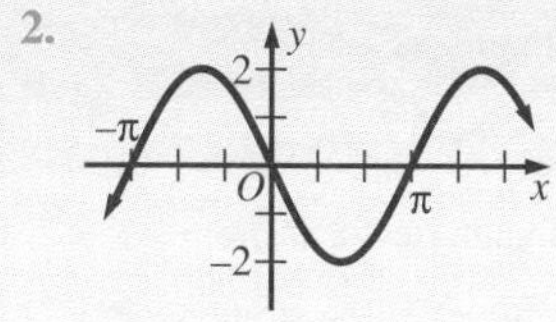

4.

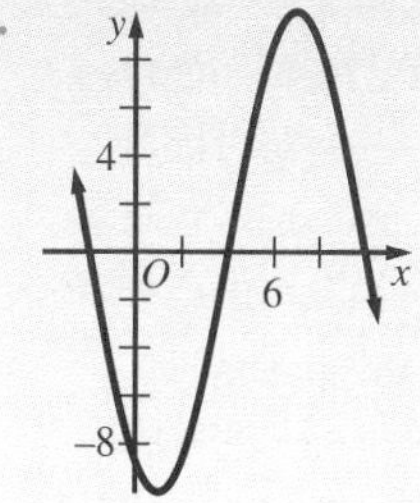

Exercises and Applications

1.

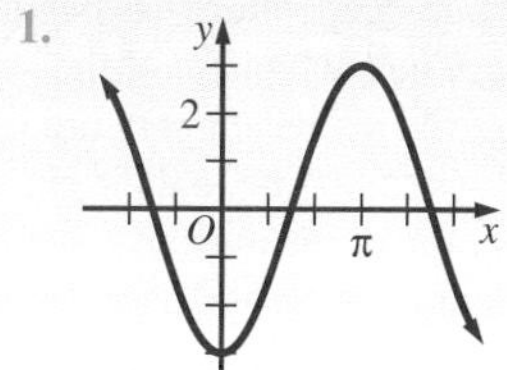

3.

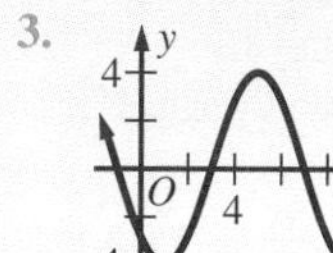

5–17. See answers in back of book.

Apply⇔Assess

Exercise Notes

Career Connection

Exs. 18–22 The design of some buildings and other structures is considered a form of art because of the aesthetic impact the structure has on the surrounding environment. Architects are required to have completed their undergraduate degree in architecture and many have had some graduate work in the field. Architecture majors usually take courses in mechanical drawing and design, as well as in mathematics and science. Hand in hand with the aesthetic design of a building is also the functional design of systems, such as the heating and electrical systems. Architects may work for themselves or in small or large firms.

Multicultural Note

Exs. 18, 19 For thousands of years, the Inuit have inhabited the icy regions in arctic and subarctic North America, Greenland, and eastern Siberia. Inuit means "real people," and it is the word these people use to refer to themselves. The term *eskimo*, which was used in the past by European settlers to refer to the Inuit, means "eaters of raw meat"; it is a name that was used by some other native groups to refer to the Inuit. During the winter, Inuit traditionally lived in igloos, or snow houses, some of which contained several rooms. Smaller igloos were built for shelter during winter hunts. In the spring and summer, the Inuit moved to wood and caribou skin tents. Today, most Inuit in the former Soviet Union, United States, and Canada live in modern frame houses with electricity and running water.

Research

Exs. 18–22 Some students may be interested in learning how other cultures have designed buildings to protect their people from extreme outdoor temperatures. Some good places to start are with the ancient Inca city of Machu Picchu, which was built in a way that would take advantage of solar radiation, or with the ancient Romans, who built raised floors so that heat could circulate underneath them.

Connection ARCHITECTURE

Architecture protects people from environments in which they would not survive unsheltered. An igloo, for instance, maintains inside air temperatures near or above freezing while outside temperatures are well below freezing. An adobe house protects its inhabitants from extreme heat in the daytime (by absorbing the sun's energy) as well as from uncomfortably low temperatures at night (by radiating the absorbed energy).

The diagrams below show the temperature patterns over a 24-hour period inside and outside an igloo and an adobe house.

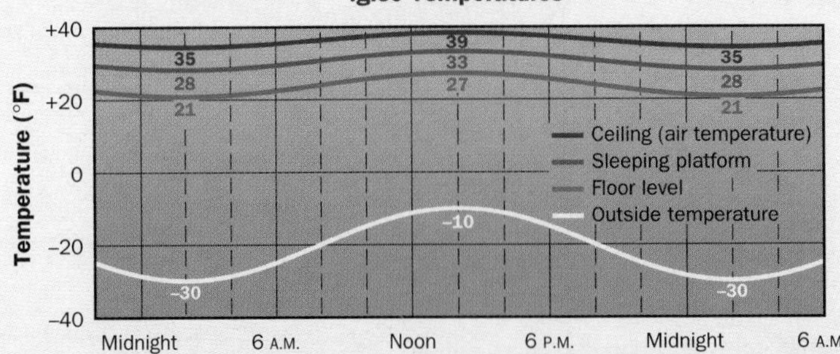

18. Suppose t represents hours after midnight. Explain why the temperature outside the igloo, T_o, can be modeled by this equation:

$$T_o = -20 + 10 \sin \frac{\pi}{12}(t - 8)$$

19. Write sine functions like the one in Exercise 18 for T_c, T_s, and T_f, the temperatures at the ceiling, sleeping platform, and floor of the igloo. How are the equations alike? How are they different?

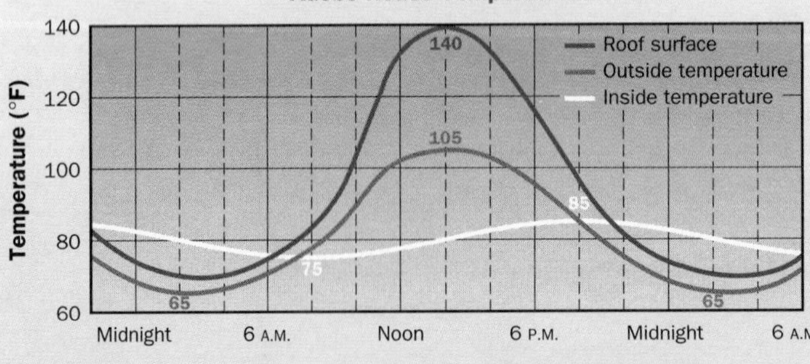

20. Write sine functions to model the temperatures inside and outside the adobe house.

21. Explain why the functions in Exercise 20 have different phase shifts.

22. **Writing** Explain why a sine function is not as good a model for the adobe house roof temperature as it is for the inside temperature.

18. The axis or k-value (vertical translation) is $\frac{-30-10}{2} = -20$. The amplitude, a, is $\frac{-10-(-30)}{2} = 10$. From the graph, the period is 24 hours, so $\frac{2\pi}{b} = 24$, or $b = \frac{\pi}{12}$. The graph looks like a sine function translated 8 hours to the right, so $h = 8$.

19. T_c: $37 + 2 \sin \frac{\pi}{12}(t-8)$; T_s: $30.5 + 2.5 \sin \frac{\pi}{12}(t-8)$; T_f: $24 + 3 \sin \frac{\pi}{12}(t-8)$. The equations all have the same period and phase shift, but they have different amplitudes and vertical shifts.

20. inside: $T = 80 + 5 \sin \frac{\pi}{12}(t-14)$; outside: $T = 85 + 20 \sin \frac{\pi}{12}(t-8)$.

21. Because the adobe walls cool and heat slowly, they keep radiating away heat (and thus cooling the inside) even while the outside temperature is rising. Likewise, they retain heat (and thus continue to warm the inside) even after the outside temperature has begun cooling. Thus, the phase of the graph of the inside temperature is shifted several hours to the right of the graph of the outside temperature.

23. Investigation Draw a graph of $y = \sin x$ using the domain $-4\pi \le x \le 4\pi$. On a piece of tracing paper, draw a graph of $y = \cos x$ using the same domain. Place the cosine graph over the sine graph and shift it until the two graphs coincide. Using different phase shifts, write three cosine functions that are equivalent to $y = \sin x$.

Technology In the same coordinate plane, graph each pair of functions and their sum. Find the period and range of each function.

24. $y = \sin x$
$y = 2 \sin x$

25. $y = 3 + \cos x$
$y = -3 + \cos x$

26. $y = 5 \sin \frac{1}{2}x$
$y = -2 \sin \frac{1}{2}x$

27. $y = 2 + \sin \pi x$
$y = 4 + \sin \frac{\pi}{2}x$

28. $y = 1 + \sin \pi x$
$y = 3 + \sin \frac{\pi}{2}\left(x - \frac{1}{2}\right)$

29. $y = 5 + 2 \cos \frac{2\pi}{3}x$
$y = -5 + 2 \cos \frac{\pi}{2}x$

30. a. **Technology** Use a graphing calculator or graphing software to graph $y = (\cos x)^2$, $y = (\sin x)^2$, and $y = (\cos x)^2 + (\sin x)^2$ in the same coordinate plane.

b. Explain why what you saw in part (a) is reasonable, using what you know about the coordinates of points on the unit circle.

31. MUSIC If you simultaneously play two musical notes that are very close in frequency (measured in cycles per second), you may hear "beats," which are loud-soft pulsations in the sound. The diagram shows how adding two sounds with frequencies of 15 cps and 10 cps creates a wave form that has 5 beats per second.

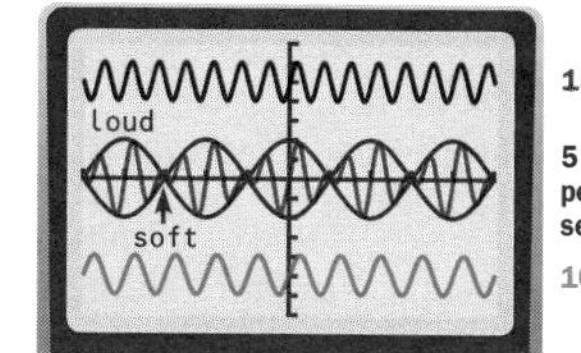

a. **Technology** Copy and complete the table. Use a graphing calculator or graphing software with this viewing window: $-0.5 \le x \le 0.5$ and $-7 \le y \le 7$.

Notes	Equations	Sum	Beats/sec
15 cps 10 cps	$y = \sin 30\pi x + 5$ $y = \sin 20\pi x - 5$	$y = \sin 30\pi x + \sin 20\pi x$	5
14 cps 10 cps	$y = \sin 28\pi x + 5$ $y = \sin 20\pi x - 5$	$y = \sin 28\pi x + \sin 20\pi x$	?
13 cps 10 cps	$y = \sin 26\pi x + 5$ $y = \sin 20\pi x - 5$	?	?
12 cps 10 cps	$y = ?$ $y = \sin 20\pi x - 5$	?	?
11 cps 10 cps	$y = ?$ $y = \sin 20\pi x - 5$	?	?

The constants +5 and −5 are in the equations to help separate the graphs so that you can see the sum better.

b. What happens to the number of beats per second as two musical notes get closer in frequency? How is the number of beats related to the frequencies of the two notes?

BY THE WAY

Piano tuner technician Clarence Koop, who is blind, strikes two notes at the same time and uses the number of beats he hears to tune the strings.

Apply⇔Assess

Exercise Notes

Communication: Discussion
Exs. 24–29 After students complete these exercises, it would be worthwhile to discuss the results. Have students consider how the two given functions are alike or different, and how this affects the sum function.

Using Technology
Exs. 24-30 You may want to suggest that students use the *Function Investigator* software to graph these functions.

Integrating the Strands
Ex. 30 In this exercise, students use functions to demonstrate the Pythagorean Identity. This identity can also be proved using analytic geometry. Have students draw the unit circle and a point on the circle in the first quadrant. The right triangle that is formed by the radius to the point and the horizontal axis has legs of length cos *x* and sin *x*, where *x* is the angle of rotation to the point. Use the Pythagorean theorem on this triangle to prove the identity for acute angles. For other angles, a congruent triangle can be drawn and used.

Communication: Reading
Ex. 31 You may want to discuss this exercise before students attempt to do it. Studying the black curves that enclose the red graph in the illustration may help students see how the resulting sound appears to pulsate as loud-soft-loud-soft. This pulsation of loudness is what is referred to by the term *beats*. Point out to students that *x* in the equations is time measured in seconds. Also, make sure that they understand that the +5 and –5 cancel each other out and thus have no effect on the sum. The viewing window that they use has a width of one second, so that the number of cycles that they see for each graph is the reciprocal of the period (or the frequency). For example, with the first equation in the table, the period is $\frac{2\pi}{30\pi} = \frac{1}{15}$ seconds per cycle. Thus, in one second or one viewing window, 15 cycles will be completed.

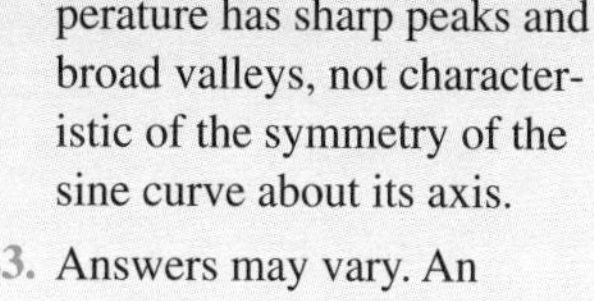

22. The graph of the roof temperature has sharp peaks and broad valleys, not characteristic of the symmetry of the sine curve about its axis.

23. Answers may vary. An example is given.
$y = \cos\left(x - \frac{\pi}{2}\right)$;
$y = \cos\left(x + \frac{3\pi}{2}\right)$;
$y = \cos\left(x - \frac{5\pi}{2}\right)$

24.

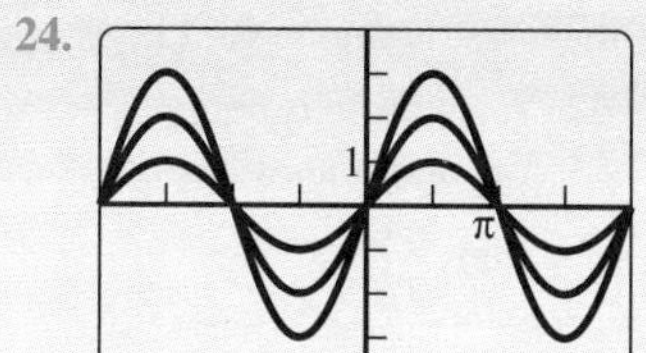

Function	Period	Range
$y = \sin x$	2π	$-1 \le y \le 1$
$y = 2 \sin x$	2π	$-2 \le y \le 2$
sum	2π	$-3 \le y \le 3$

25.

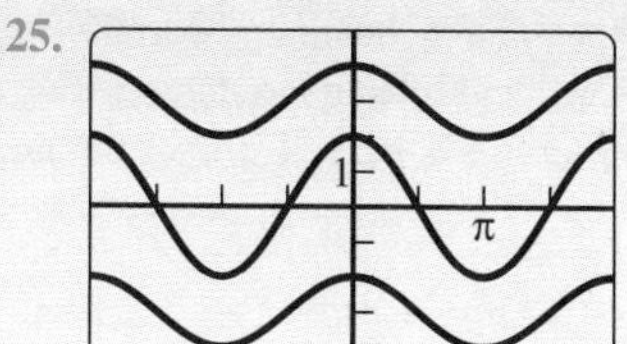

Function	Period	Range
$y = 3 + \cos x$	2π	$2 \le y \le 4$
$y = -3 + \cos x$	2π	$-4 \le y \le -2$
sum	2π	$-2 \le y \le 2$

26–31. See answers in back of book.

Apply⇔Assess

Exercise Notes

Journal Entry

Ex. 39 You might suggest that students respond to this question in their journals so that this information will be available for future reference. They should be sure to include examples where *a*, *h*, or *k* are negative. As an extension, students can investigate what happens when *b* is changed from positive to negative.

Cooperative Learning

Ex. 40 This exercise provides an opportunity for students to evaluate and increase their understanding of this section's material. Each student is responsible for one sine curve. These curves are easy to check since each student's graph should be a phase shift of the other students' graphs.

Practice 99 for Section 15.4

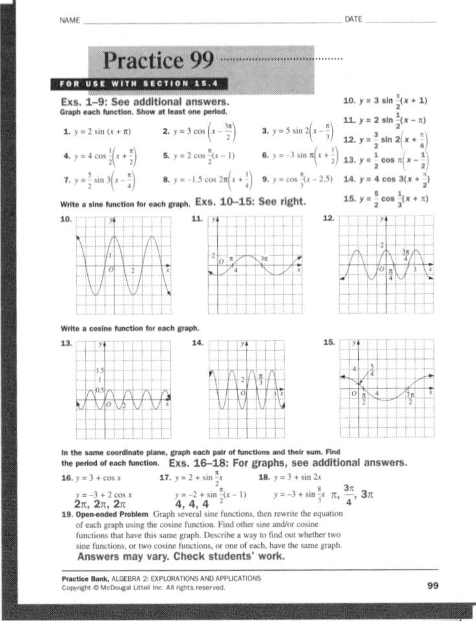

NAME ______ DATE ______

Practice 99

FOR USE WITH SECTION 15.4

Exs. 1–9: See additional answers.

Graph each function. Show at least one period.

1. $y = 2\sin(x + \pi)$
2. $y = 3\cos\left(x - \frac{3\pi}{2}\right)$
3. $y = 5\sin 2\left(x - \frac{\pi}{3}\right)$
4. $y = 4\cos\frac{1}{2}\left(x + \frac{\pi}{2}\right)$
5. $y = 2\cos\frac{\pi}{2}(x - 1)$
6. $y = -3\sin\pi\left(x + \frac{1}{2}\right)$
7. $y = \frac{5}{2}\sin 3\left(x - \frac{\pi}{4}\right)$
8. $y = -1.5\cos 2\pi\left(x + \frac{1}{4}\right)$
9. $y = \cos\frac{\pi}{3}(x - 2.5)$

Write a sine function for each graph. **Exs. 10–15: See right.**

10. $y = 3\sin\frac{\pi}{2}(x + 1)$
11. $y = 2\sin\frac{1}{2}(x - \pi)$
12. $y = \frac{3}{2}\sin 2\left(x + \frac{\pi}{4}\right)$

Write a cosine function for each graph.

13. $y = \frac{1}{2}\cos\pi\left(x - \frac{1}{2}\right)$
14. $y = 4\cos 3\left(x + \frac{\pi}{2}\right)$
15. $y = \frac{5}{2}\cos\frac{1}{3}(x + \pi)$

In the same coordinate plane, graph each pair of functions and their sum. Find the period of each function. **Exs. 16–18: For graphs, see additional answers.**

16. $y = 3 + \cos x$; $y = -3 + 2\cos x$ **2π, 2π, 2π**
17. $y = 2 + \sin\frac{\pi}{2}x$; $y = -2 + \sin\frac{\pi}{2}(x - 1)$ **4, 4, 4**
18. $y = 3 + \sin 2x$; $y = -3 + \sin\frac{8}{3}x$ **π, $\frac{3\pi}{4}$, 3π**
19. **Open-ended Problem** Graph several sine functions, then rewrite the equation of each graph using the cosine function. Find other sine and/or cosine functions that have this same graph. Describe a way to find out whether two sine functions, or two cosine functions, or one of each, have the same graph. **Answers may vary. Check students' work.**

Practice Bank, ALGEBRA 2: EXPLORATIONS AND APPLICATIONS

99

Technology **Use a graphing calculator or graphing software to graph each function. Find its maximum value by using trace.**

32. $y = \sin x + \cos x$
33. $y = 3\sin x + 4\cos x$
34. $y = 4\sin x + 3\cos x$
35. $y = 5\sin x + 12\cos x$
36. $y = 8\sin x + 6\cos x$
37. $y = 2\sin x + 5\cos x$
38. Based on your results in Exercises 32–37, write a conjecture about the maximum value of the function $y = a\sin x + b\cos x$. Check your conjecture by testing other values of *a* and *b*.
39. **Writing** Describe how changing the values of *a*, *b*, *h*, and *k* affects the graph of $y = a\cos b(x - h) + k$. Use drawings in your explanation.

ONGOING ASSESSMENT

40. **Cooperative Learning** Work in a group of four students.
    a. Suppose each of you is in a different seat on the Ferris wheel shown when it begins to turn counterclockwise. It takes 160 s to make one complete turn. Draw a graph of your height above the ground as a function of time.
    b. Write a sine function for your graph in part (a). What is the phase shift for your graph compared with an unshifted sine graph?
    c. Combine your results with those of the others in your group on a single graph, labeling each graph with an appropriate equation.

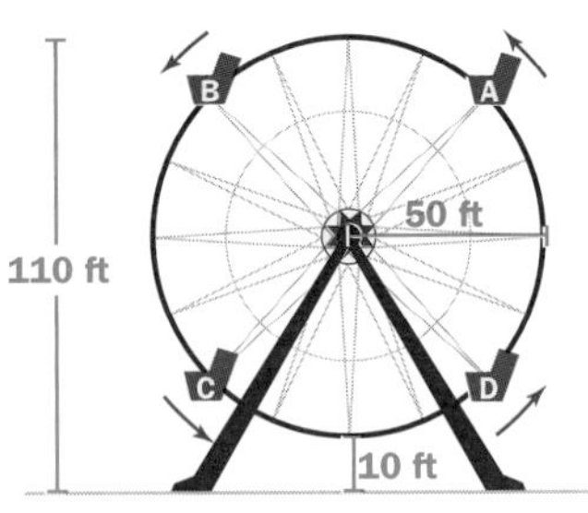

SPIRAL REVIEW

Write a cosine function with the given maximum *M*, minimum *m*, and period. *(Section 15.3)*

41. $M = 8$, $m = 2$, period $= 6$
42. $M = -1$, $m = -5$, period $= \pi$

Find the vertical asymptotes of the graph of each function. *(Section 9.8)*

43. $y = \frac{1}{(x - 2)(x - 5)}$
44. $y = \frac{5}{(x + 3)(x - 4)}$
45. $y = \frac{2}{x^2 - 36}$

Find tan *A* for each right triangle. *(Section 14.2)*

46.

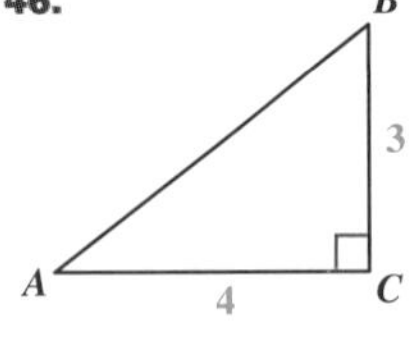

47.

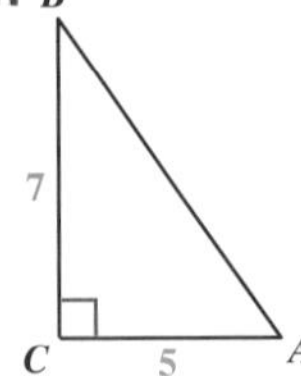

48.

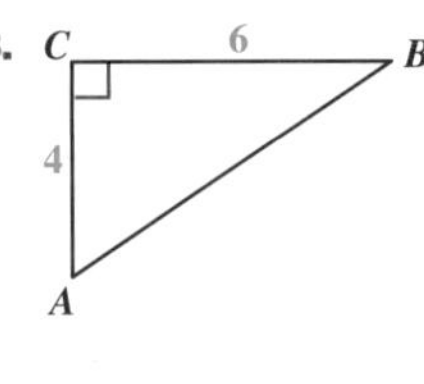

32. 1.41
33. 5
34. 5
35. 13
36. 10
37. 5.39
38. $\sqrt{a^2 + b^2}$
39. Answers may vary. An example is given. Varying the constants in $y = a\cos b(x - h) + k$ will vary the amplitude, period, phase shift, and vertical shift of the graph. The amplitude is given by $|a|$, and the period by $\frac{2\pi}{b}$. *h* represents a phase shift, *h* units to the right for $h > 0$ and $|h|$ units to the left for $h < 0$. *k* represents a vertical shift, *k* units upward for $k > 0$ and $|k|$ units downward for $k < 0$.

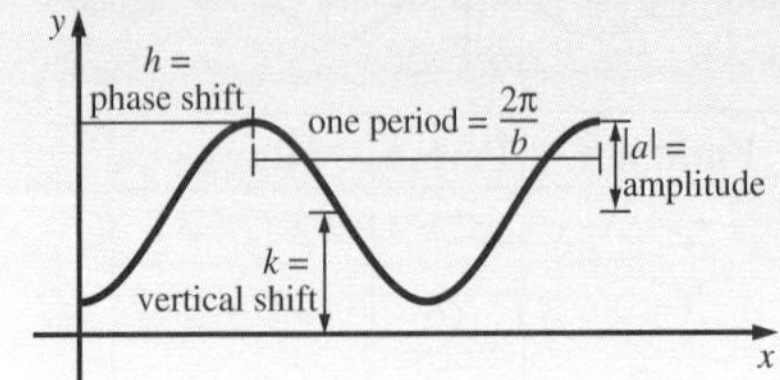

40. See answers in back of book.
41, 42. Answers may vary. Examples are given.
41. $y = 3\cos\frac{\pi}{3}x + 5$
42. $y = 2\cos 2x - 3$
43. $x = 2$; $x = 5$
44. $x = -3$; $x = 4$
45. $x = -6$; $x = 6$
46. 0.75
47. 1.4
48. 1.5

Section

15.5 The Tangent Function

Learn how to...

- **graph the tangent function**

So you can...

- **solve problems, such as finding the height of a balloon**

More than two thousand years ago, Chinese children made the first hot air balloons from empty eggshells. They heated the air inside the eggshells by burning small pieces of dried plants placed inside the eggshells. The air inside a modern hot air balloon is heated with a propane burner.

EXAMPLE 1 Application: Hot Air Ballooning

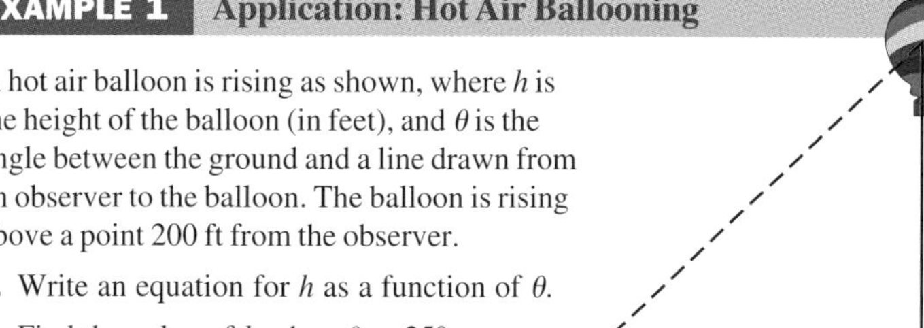

A hot air balloon is rising as shown, where h is the height of the balloon (in feet), and θ is the angle between the ground and a line drawn from an observer to the balloon. The balloon is rising above a point 200 ft from the observer.

a. Write an equation for h as a function of θ.

b. Find the value of h when $\theta = 25°$.

c. Graph h as a function of θ.

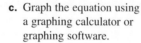

SOLUTION

a. From Section 14.2 you know that $\tan \theta = \frac{h}{200}$. So $h = 200 \tan \theta$.

b. $h = 200 \tan \theta$

$= 200 \tan 25°$

$= 200(0.4663)$

≈ 93

Use a calculator to find a tangent value you don't know.

The balloon is at a height of about 93 ft when $\theta = 25°$.

c. Graph the equation using a graphing calculator or graphing software.

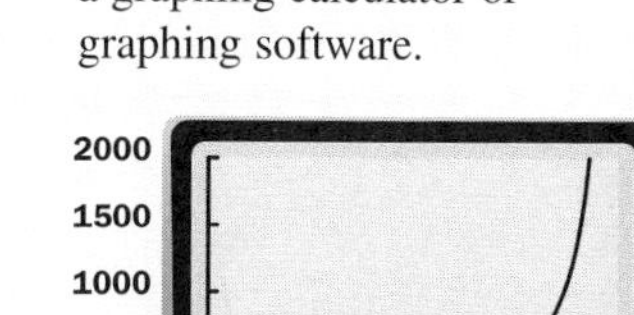

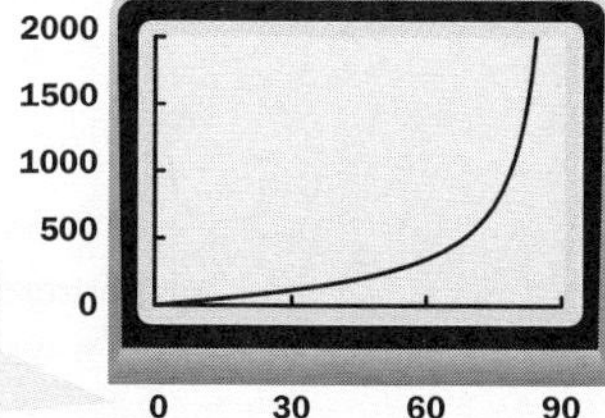

The domain of the height function is $0° \le \theta < 90°$.

THINK AND COMMUNICATE

1. What is the balloon's height when $\theta = 0°$? Why does this make sense?
2. What happens to the height of the balloon as θ approaches 90°? How is this shown in the graph?

Plan⇔Support

Objectives

- Graph the tangent function.
- Solve problems using a tangent function.

Recommended Pacing

Core and Extended Courses
Section 15.5: 1 day

Block Schedule
Section 15.5: $\frac{1}{2}$ block (with Portfolio Project)

Resource Materials

Lesson Support

Lesson Plan 15.5

Warm-Up Transparency 15.5

Practice Bank: Practice 100

Study Guide: Section 15.5

Explorations Lab Manual: Diagram Master 2

Assessment Book: Test 64

Technology

Technology Book: Spreadsheet Activity 15

Graphing Calculator

McDougal Littell Mathpack *Function Investigator with Matrix Analyzer Activity Book:* Function Investigator Activity 37

Internet: http://www.hmco.com

Warm-Up Exercises

1. Evaluate tan 45°. 1
2. Evaluate tan 63°. 1.9626
3. Express $\tan \theta$ in terms of $\sin \theta$ and $\cos \theta$. $\tan \theta = \frac{\sin \theta}{\cos \theta}$

For Exs. 4–6, use $\triangle ABC$.

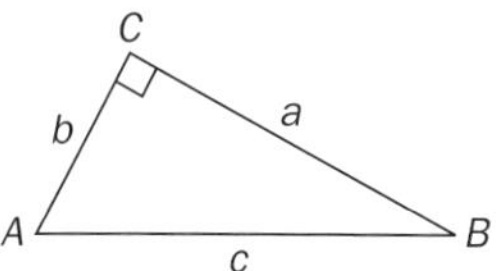

4. Write an expression for tan A. $\tan A = \frac{a}{b}$
5. Find a if $\angle A = 46°$ and $b = 3$. 3.1
6. Find b if $\angle A = 46°$ and $a = 7$. 6.8

ANSWERS Section 15.5

Think and Communicate

1. 0 ft; When $\theta = 0°$, the line of sight to the balloon is horizontal, which means that the balloon is on the ground.
2. As θ approaches 90°, the height of the balloon increases more and more rapidly, shown in the graph by the increasing steepness of the tangent graph near $\theta = 90°$.

Teach⇔Interact

Additional Example 1

An oil spill spreads downstream from an anchored tanker as shown, where d is the distance of the leading edge of the spill from the tanker (in feet), and θ is the angle between the line joining the tanker with the dock and the line joining the dock with the edge of the spill.

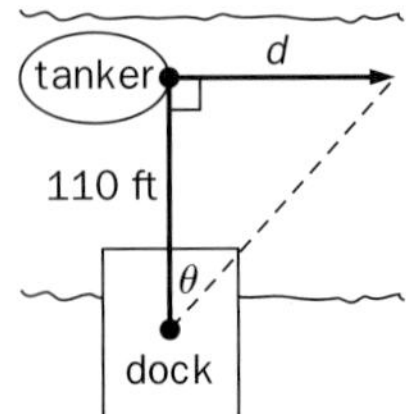

a. Write an equation for d as a function of θ.

$\tan\theta = \frac{d}{110}$

$d = 110\tan\theta$

b. Find the value of d when $\theta = 47°$.

$d = 110\tan\theta$

$= 110\tan 47°$

$\approx 110(1.0724)$

≈ 118

The distance of the leading edge of the spill from the tanker is about 118 ft when $\theta = 47°$.

c. Graph d as a function of θ.

Graph the equation using a graphing calculator or graphing software.

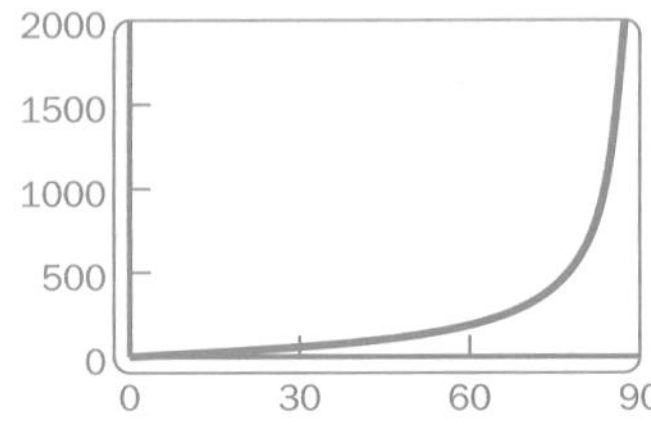

Think and Communicate

For question 2, students can imagine that if the balloon continues to rise, the angle it makes with the horizon gets closer and closer to 90° but is never exactly 90°. This will help them understand that the asymptote of the tangent graph occurs when $\theta = 90°$, or $\frac{\pi}{2}$ radians.

Section Note

Student Study Tip

Students should know the asymptotes and some key points on the tangent function, such as the x-intercepts and points $\left(\frac{\pi}{4}, 1\right)$ and $\left(-\frac{\pi}{4}, -1\right)$. This will help students to graph $y = \tan x$.

As with sine and cosine, you can define **tangent** on the unit circle. The domain of the tangent function can be extended to all angles except those for which tangent is undefined, and the angles can be expressed in degrees or radians.

$\theta = -60° = -\frac{\pi}{3}$

$\tan\theta = \frac{-\frac{\sqrt{3}}{2}}{\frac{1}{2}} \approx -1.732$

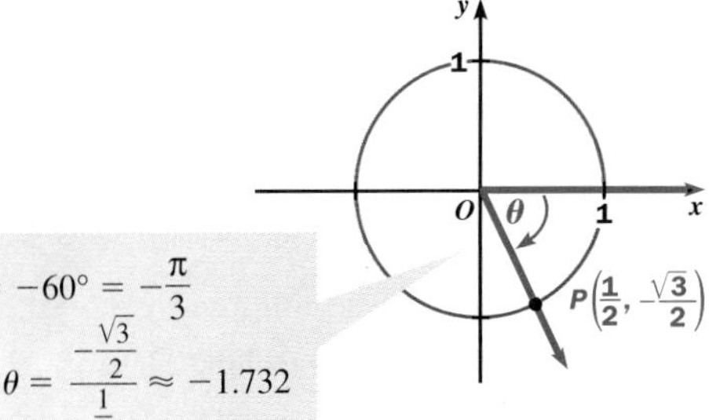

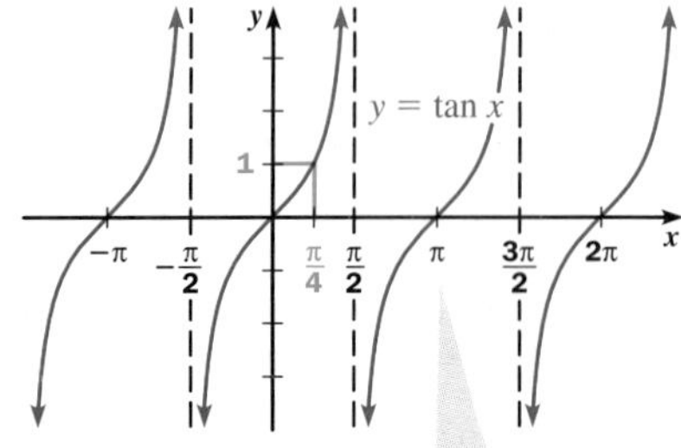

The tangent function has a period of π.

The tangent function is undefined at odd multiples of $\frac{\pi}{2}$ (that is, at $x = \frac{(2n+1)\pi}{2}$, where n is an integer). The graph of $y = \tan x$ has vertical asymptotes at these x-values.

THINK AND COMMUNICATE

3. Explain why $y = \tan x = \frac{\sin x}{\cos x}$ is undefined at odd multiples of $\frac{\pi}{2}$.

4. Compare the period of the tangent function with the periods of the sine and cosine functions. Why is there a difference?

The graph of the tangent function $y = \tan x$ can be shifted and stretched vertically and horizontally. Examples are shown below.

Vertical stretch: $y = 2\tan x$

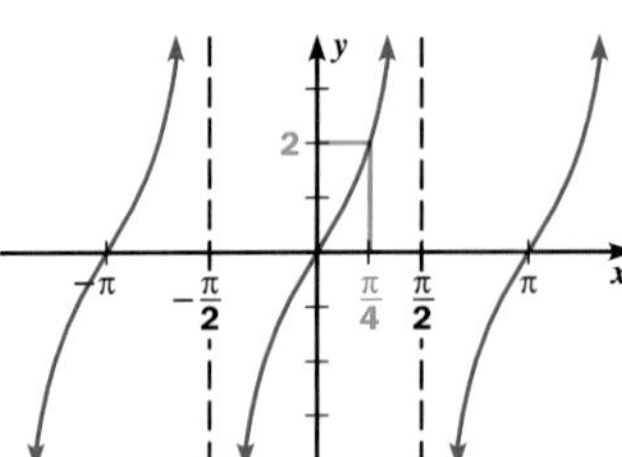

Horizontal stretch: $y = \tan 2x$

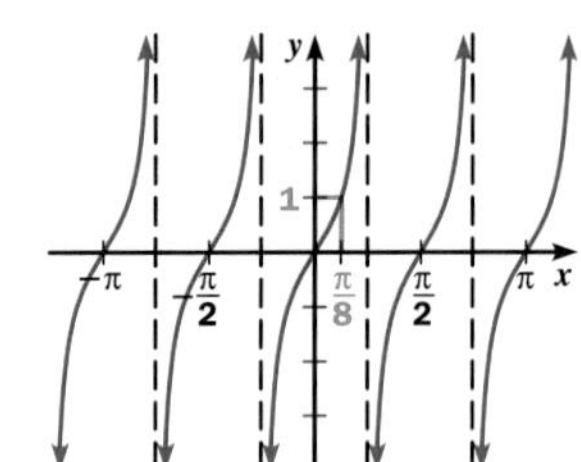

Horizontal shift: $y = \tan\left(x - \frac{\pi}{4}\right)$

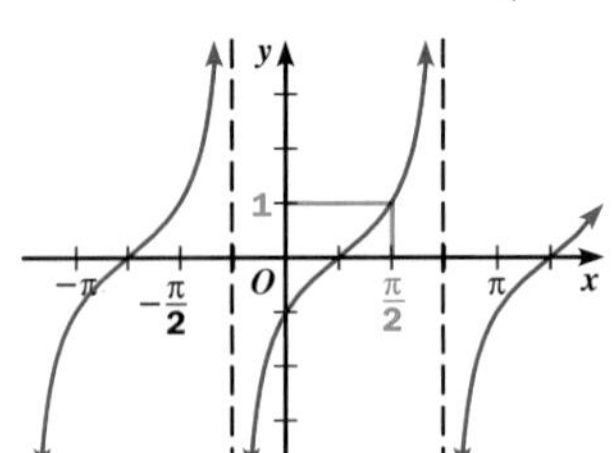

Vertical shift: $y = \tan x + 1$

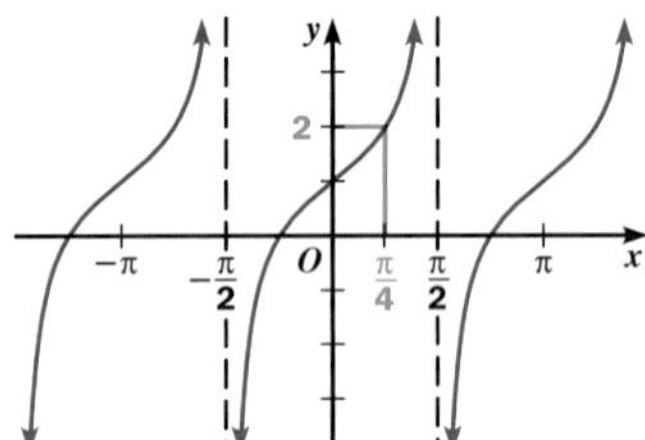

Think and Communicate

3. At odd multiples of $\frac{\pi}{2}$, the cosine function is 0, so $\tan x = \frac{\sin x}{\cos x}$ is undefined.

4. The period of the tangent function is one-half the periods of the sine and cosine functions. This is because, for similar triangles, the tangent values in Quadrant III repeat those in Quadrant I (because sine and cosine are both negative in the third quadrant, yielding a positive tangent), and similarly, the tangent values in Quadrant IV repeat those in Quadrant II.

The general form of the equation for a tangent function is $y = a \tan b(x - h) + k$, where a, b, h, and k determine vertical and horizontal stretches and shifts.

EXAMPLE 2

Graph the function $y = 3 \tan \frac{x}{2}$.

SOLUTION

The graph will involve a vertical stretch and a horizontal stretch of the basic tangent graph.

Step 1 Find the asymptotes of the graph. You want to know when $\frac{x}{2}$ equals odd multiples of $\frac{\pi}{2}$.

$$\frac{x}{2} = \frac{(2n + 1)\pi}{2}$$

$$x = (2n + 1)\pi$$

So asymptotes occur when x is an odd multiple of π.

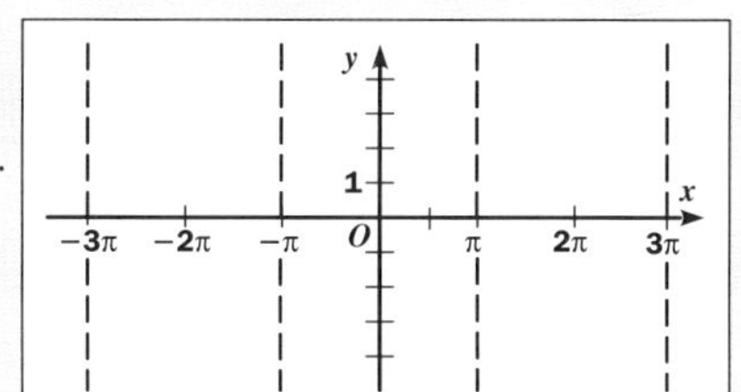

Step 2 The graph is not shifted vertically ($k = 0$), so the x-intercepts of the basic tangent graph are not shifted vertically. The x-intercepts remain halfway between the asymptotes, at even multiples of π.

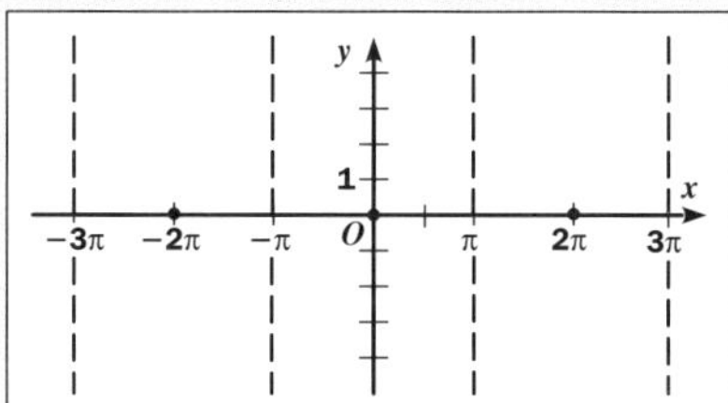

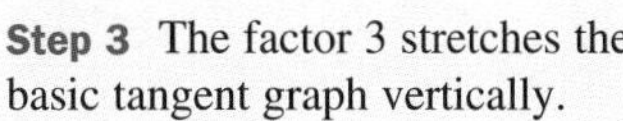

Step 3 The factor 3 stretches the basic tangent graph vertically.

For $y = \tan x$, $y = 1$ when $x = \frac{\pi}{4}$, midway between the x-intercept at 0 and the asymptote at $x = \frac{\pi}{2}$.

For $y = 3 \tan \frac{x}{2}$, $y = 3$ when $x = \frac{\pi}{2}$, midway between the x-intercept at 0 and the asymptote at $x = \pi$.

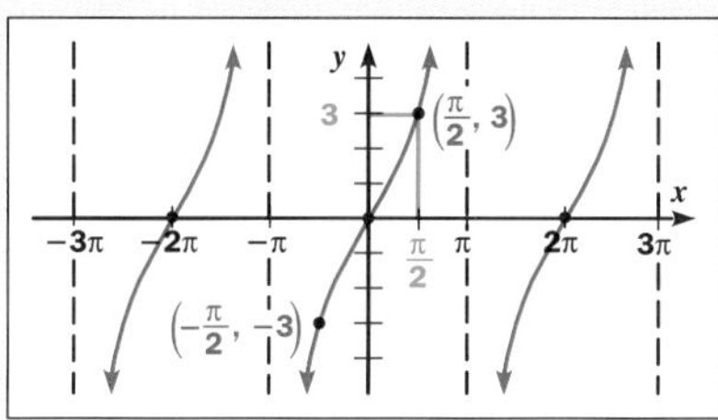

CHECKING KEY CONCEPTS

Evaluate each expression.

1. $\tan 38°$
2. $\tan (-5°)$
3. $\tan 0.75\pi$
4. $\tan \left(-\frac{2\pi}{3}\right)$

Graph each function.

5. $y = -\tan x$
6. $y = \tan x - 2$
7. $y = \tan \left(x + \frac{\pi}{3}\right) + \frac{1}{2}$

Checking Key Concepts

1. 0.7813
2. −0.0875
3. −1
4. 1.7321
5. 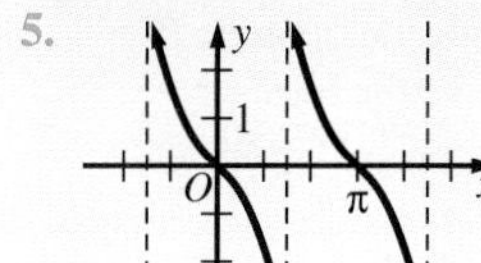

6. 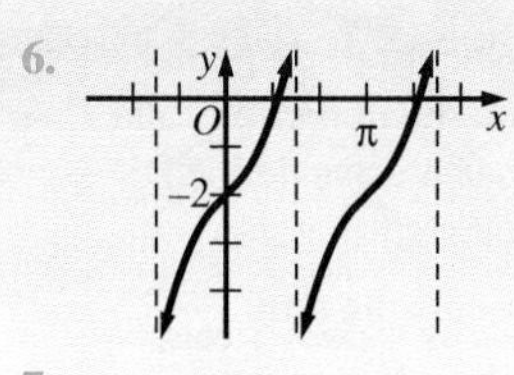

7. 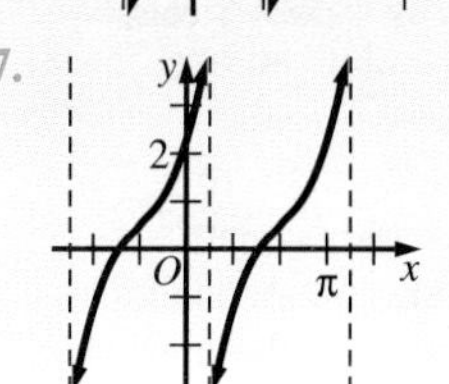

Teach⇔Interact

About Example 2

Alternate Approach

To find the asymptotes of the graph of $y = \tan bx$, rather than using the method in Step 1 of the Solution, students may prefer to find the period of the function. The period of a tangent function of this form is $\frac{\pi}{b}$. In this case, since $b = \frac{1}{2}$, the period is 2π. All graphs of this form are centered at the origin. To find the asymptotes, go half the period to the right of the origin and half the period to the left of the origin. After finding these two asymptotes, continue marking off full periods in either direction.

Additional Example 2

Graph the function

$y = \tan\left(x - \frac{\pi}{2}\right) + 2.$

Step 1 Draw a helping graph. In this case, the helping graph is $y = \tan x$.

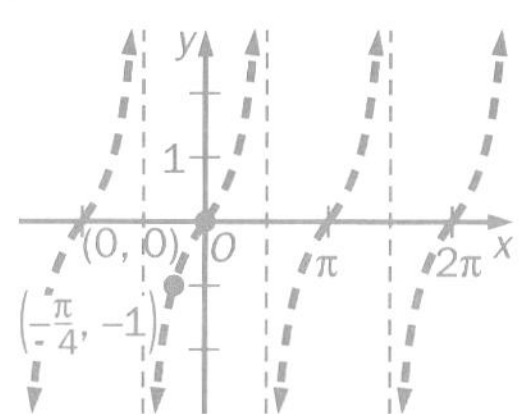

Step 2 Find the asymptotes of the graph by translating the asymptotes of the helping graph. In this case, the asymptotes of the helping graph are translated right $\frac{\pi}{2}$ units.

Step 3 Translate the x-intercepts and other key points of the helping graph by the horizontal and vertical translation. All the points of the helping graph are translated right $\frac{\pi}{2}$ units and up 2 units. For example, the point (0, 0) becomes $\left(\frac{\pi}{2}, 2\right)$ and the point $\left(-\frac{\pi}{4}, -1\right)$ becomes $\left(\frac{\pi}{4}, 1\right)$.

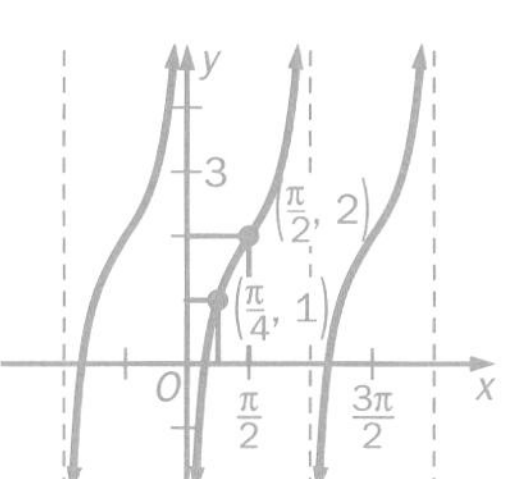

Teach⇔Interact

Closure Question

Describe the graph of $y = \tan x$.

This graph has vertical asymptotes at every odd multiple of $\frac{\pi}{2}$. It has x-intercepts at every multiple of π. The period of the function is π. Between the asymptotes, the graph rises.

Apply⇔Assess

Suggested Assignment

Core Course
Exs. 1–14, 19–27, AYP

Extended Course
Exs. 1–27, AYP

Block Schedule
Day 4 Exs. 1–14, 19–27, AYP

Exercise Notes

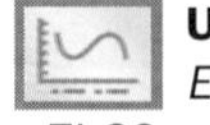

Using Technology

Exs. 1–12 Students using a TI-82 graphing calculator can indicate on an angle-by-angle basis if the measure is in degrees or radians, rather than switching the mode. To do this, enter the expression and immediately following the angle, type [2nd][ANGLE] and choose item 1 if the measure is degrees or item 3 if the angle is in radians. This does not change the mode of the calculator.

Integrating the Strands

Ex. 14 This exercise helps students connect the geometric idea of a tangent line with the tangent function. After answering part (c), students should examine the graph of $y = \tan x$ and see how the variations in the rate of increase of this function are reflected here as well.

Application

Exs. 15–18 This application of sine and cosine functions to bicycle riding is not only one that most students will enjoy, but also one that can help them see some of the mathematics that relates to common things in their lives.

Second-Language Learners

Ex. 18 Students learning English may benefit from discussing their ideas with an English-proficient partner before beginning this writing task.

15.5 Exercises and Applications

Extra Practice *exercises on page 772*

Evaluate each expression.

1. $\tan \pi$ **2.** $\tan \frac{\pi}{6}$ **3.** $\tan \frac{11\pi}{6}$ **4.** $\tan(-30°)$

5. $\tan \frac{3\pi}{8}$ **6.** $\tan\left(-\frac{7\pi}{4}\right)$ **7.** $\tan 520°$ **8.** $\tan \frac{13\pi}{16}$

9. $\tan 0.18$ **10.** $\tan 95°$ **11.** $\tan 0$ **12.** $\tan(-3.10)$

13. Writing Explain why the graph of the tangent function has asymptotes, while the graphs of the sine and cosine functions do not.

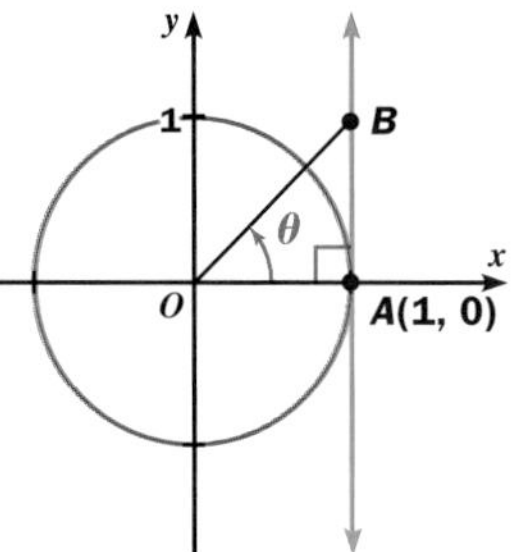

14. GEOMETRY In the diagram, $\overleftrightarrow{AB}$ is tangent to the unit circle at $A(1, 0)$ and $\angle AOB = \theta$.

a. Use trigonometry to express the length of $\overline{AB}$ in terms of θ.

b. Writing Use your answer to part (a) to explain the relationship between the tangent line and the tangent function.

c. If θ increases steadily, does $\tan \theta$ increase steadily? (*Hint:* Find the tangent of 0°, 10°, 20°, and so on.)

Connection SPORTS

The more you turn the handlebars of your bicycle, the smaller the circle followed by each tire. The diagram shows the turning radius f of the front wheel and the turning radius r of the rear wheel. Suppose the length of the bike frame between the centers of the wheels is 3 ft and the handlebars are turned an angle θ.

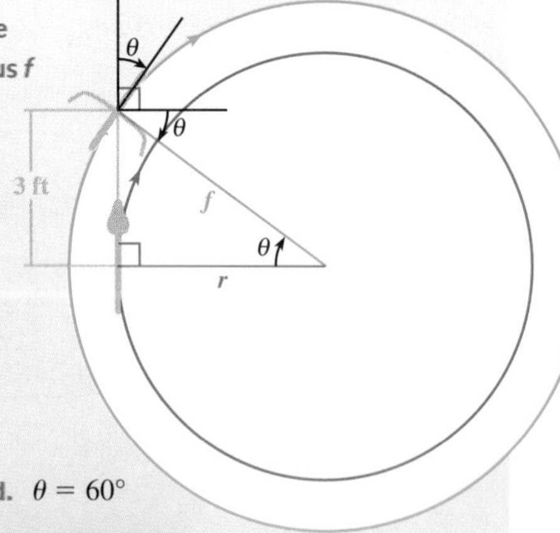

15. a. Use the diagram to write an expression for $\sin \theta$ and an expression for $\tan \theta$.

b. Use the equations you found in part (a) to express f and r as functions of θ. Use these equations for Exercises 16–18.

16. Find the values of f and of r for each turning angle.

a. $\theta = 0°$ **b.** $\theta = 30°$ **c.** $\theta = 45°$ **d.** $\theta = 60°$

17. Challenge For what approximate values of θ will f and r have close to the same value? $\left(\textit{Hint:}\text{ Remember that } \tan \theta = \frac{\sin \theta}{\cos \theta}.\right)$

18. Writing Some trick riders spin their bikes around quickly, with the handlebars at $\theta = 90°$. What are the front and rear turning radii for this value of θ? Why does this make sense? Describe what the bike does.

Exercises and Applications

1. 0
2. 0.5774
3. −0.5774
4. −0.5774
5. 2.4142
6. 1
7. −0.3640
8. −0.6682
9. 0.1820
10. −11.4301
11. 0
12. 0.0416
13. Answers may vary. An example is given. The sine and cosine functions represent horizontal and vertical distances from the origin to points around the unit circle, and are thus always defined. The tangent function represents the ratio $\frac{\sin x}{\cos x}$, so it is undefined for x-values where the denominator is zero. Because the tangent function is unbounded for x approaching these values, it has asymptotes.
14. a. $AB = \tan \theta$
 b. Answers may vary. An example is given. The tangent function gives the signed distance along the tangent line from the x-axis to the point where the terminal side of angle θ intercepts the line.
 c. No.
15. a. $\sin \theta = \frac{3}{f}$; $\tan \theta = \frac{3}{r}$

Graph each function. Show at least one period.

19. $y = 3 \tan \frac{x}{4}$ **20.** $y = \tan \frac{\pi}{2}(x - 1)$ **21.** $y = \tan 4x - 2$

ONGOING ASSESSMENT

22. Writing Write an equation of a tangent function of the form $y = a \tan bx$. Then graph your equation. Write an explanation of each step you take.

SPIRAL REVIEW

23. Write a cosine function that is equivalent to $y = 2 \sin x$. *(Section 15.4)*

Evaluate each expression. *(Section 15.1)*

24. sin 45° **25.** cos 450° **26.** cos (−45°) **27.** sin (−450°)

ASSESS YOUR PROGRESS

VOCABULARY

phase shift (p. 728) **tangent** (p. 736)

Graph each function. Show at least one period. *(Section 15.4)*

1. $y = -3 \sin \left(x + \frac{\pi}{6}\right)$ **2.** $y = \cos \left(x + \frac{\pi}{2}\right) - 1$

3. ASTRONOMY In 1725 James Bradley discovered that there is a small shift in the observed position of a star due to the movement of Earth. This *aberration angle* is greatest when the true direction of the star is perpendicular to the direction of Earth's motion around the sun.

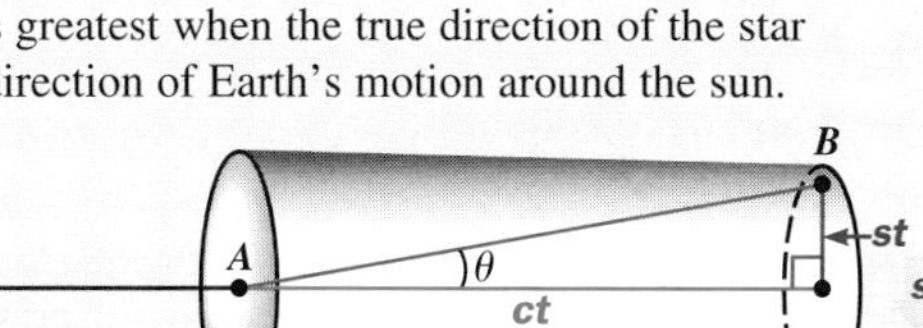

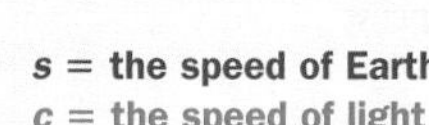

The diagram shows that light enters the telescope at A, but while it travels the distance ct, the telescope has moved a distance st, so the light hits the focal plane at B. The telescope must be tilted along the red line to view the star, and the star appears to be shifted by θ from its true position.

a. Use the diagram to write an expression for s in terms of c and θ.

b. Find s, given that $c = 3.00 \times 10^8$ m/s and $\theta = 0.00569°$. *(Section 15.5)*

Graph each function over two periods. *(Section 15.5)*

4. $y = \tan x - 2$ **5.** $y = 2 \tan 2x$ **6.** $y = \tan (2x - \pi)$

7. Journal Consider the three general equations below:

$y = a \sin b(x - h) + k$ $y = a \cos b(x - h) + k$ $y = a \tan b(x - h) + k$

Explain what effect the values of a, b, h, and k have on the graph of each function.

Apply⇔Assess

Exercise Notes

Assessment Note

Ex. 22 Students having difficulty writing the steps for this exercise should refer to the Solution of Example 2 on page 737.

Assess Your Progress

Journal Entry

Writing this entry can help students organize all of the concepts they have studied in this chapter about the graphs of sine, cosine, and tangent functions. Encourage students to focus on the similarities and differences of the three types of graphs. A general illustration that labels a, b, h, and k for each graph would be useful for future reference.

Progress Check 15.4–15.5

See page 743.

Practice 100 for Section 15.5

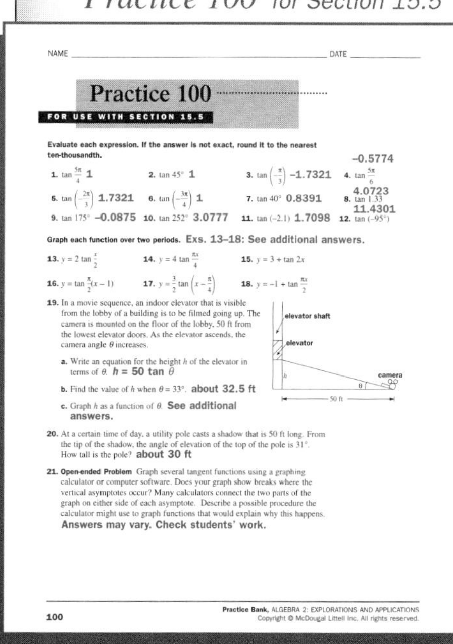

NAME ______ DATE ______

Practice 100

FOR USE WITH SECTION 15.5

Evaluate each expression. If the answer is not exact, round it to the nearest ten-thousandth.

1. $\tan \frac{5\pi}{4}$ **1** **2.** tan 45° **1** **3.** $\tan \left(-\frac{\pi}{3}\right)$ **−1.7321** **4.** $\tan \frac{5\pi}{6}$ **−0.5774**

5. $\tan \left(-\frac{2\pi}{3}\right)$ **1.7321** **6.** $\tan \left(-\frac{3\pi}{4}\right)$ **1** **7.** tan 40° **0.8391** **8.** tan 1.33 **4.0723**

9. tan 175° **−0.0875** **10.** tan 252° **3.0777** **11.** tan (−2.1) **1.7098** **12.** tan (−95°) **11.4301**

Graph each function over two periods. **Exs. 13–18: See additional answers.**

13. $y = 2 \tan \frac{x}{2}$ **14.** $y = 4 \tan \frac{\pi x}{4}$ **15.** $y = 3 + \tan 2x$

16. $y = \tan \frac{\pi}{2}(x - 1)$ **17.** $y = \frac{3}{2} \tan \left(x - \frac{\pi}{4}\right)$ **18.** $y = -1 + \tan \frac{\pi x}{2}$

19. In a movie sequence, an indoor elevator that is visible from the lobby of a building is to be filmed going up. The camera is mounted on the floor of the lobby, 50 ft from the lowest elevator doors. As the elevator ascends, the camera angle θ increases.

elevator shaft
elevator
camera
h
θ
50 ft

a. Write an equation for the height h of the elevator in terms of θ. **$h = 50 \tan \theta$**

b. Find the value of h when $\theta = 33°$. **about 32.5 ft**

c. Graph h as a function of θ. **See additional answers.**

20. At a certain time of day, a utility pole casts a shadow that is 50 ft long. From the tip of the shadow, the angle of elevation of the top of the pole is 31°. How tall is the pole? **about 30 ft**

21. Open-ended Problem Graph several tangent functions using a graphing calculator or computer software. Does your graph show breaks where the vertical asymptotes occur? Many calculators connect the two parts of the graph on either side of each asymptote. Describe a possible procedure the calculator might use to graph functions that would explain why this happens. **Answers may vary. Check students' work.**

100

Practice Bank, ALGEBRA 2: EXPLORATIONS AND APPLICATIONS

---

b. $f = \frac{3}{\sin \theta}$; $r = \frac{3}{\tan \theta}$

16. a. f: undefined; r: undefined (both wheels continue in a straight line)

b. $f = 6$ ft; $r \approx 5$ ft c. $f \approx 4$ ft; $r = 3$ ft

d. $f \approx 3$ ft; $r \approx 2$ ft

17. $f \approx r$ when $\cos \theta \approx 1$, or θ is close to 0°.

18. The front turning radius is 3 ft. The rear turning radius is undefined, but as θ approaches 90°, r approaches 0 (that is, the rear wheel pivots around its point of contact, while the front wheel circles this point).

19.

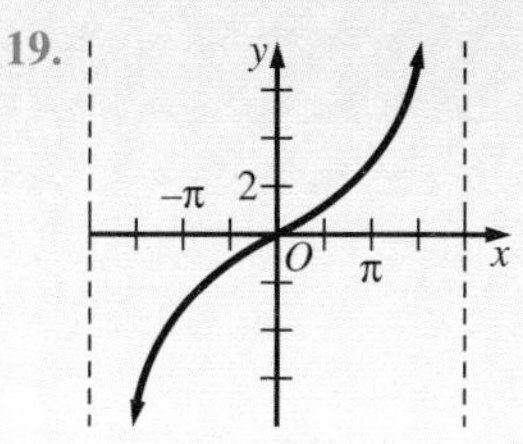

20.

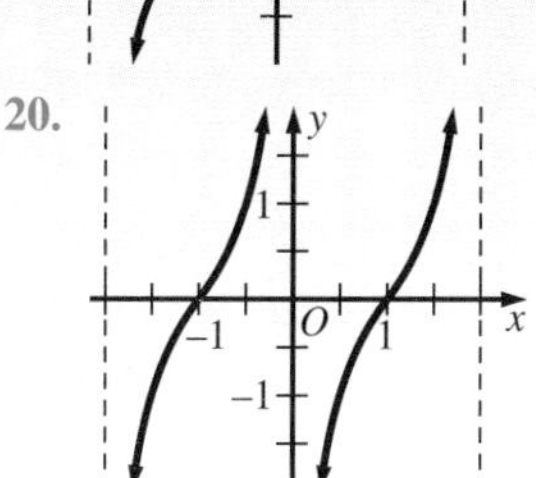

21.

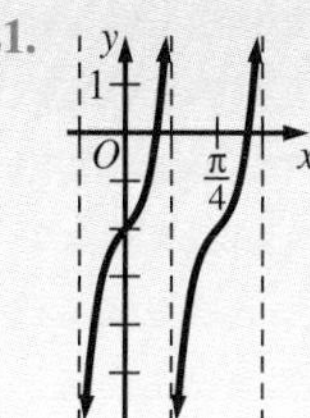

22. Answers may vary. Check students' work.

23. Answers may vary. An example is given. $y = 2 \cos \left(x - \frac{\pi}{2}\right)$

24. 0.7071 25. 0

26. 0.7071 27. −1

Assess Your Progress

1.

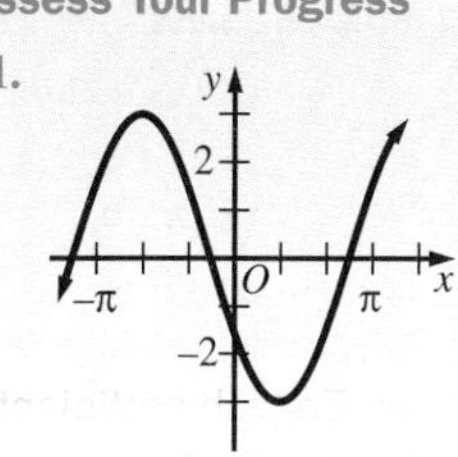

2–7. See answers in back of book.

Project Notes

Mathematical Goals

- Perform an experiment to model simple harmonic motion.
- Relate experimental measurements to values in a trigonometric equation involving cosine.
- Graph the cosine equation.
- Compare and evaluate the results.

Planning

Materials

- a spring
- paper clip
- masking tape
- 1-lb weight
- string
- meterstick
- stopwatch

Project Teams

Students can select their project teams and can work together to gather all of the materials and perform the experiment. Team members should also work together to write and graph the equations and to develop a written report for the group.

Guiding Students' Work

It may be easier for students to find the materials for the project at school and perform the experiment there. Also, before the groups perform the experiment, discuss the ideas of equilibrium position, oscillation, and amplitude, as they apply to the experiment. This will help students when they need to model their oscillation. Also, discuss why simple harmonic motion would be modeled with the cosine function and not the sine function by relating the graph for simple harmonic motion with that of the cosine function.

Second-Language Learners

Students learning English may benefit from writing their reports with the help of a peer tutor or aide.

PORTFOLIO PROJECT

Modeling Ups and Downs

Suppose a weight is attached to a spring, pulled down, and released. As you can see from the diagram below, the position of the weight as it moves up and down is a periodic function of time.

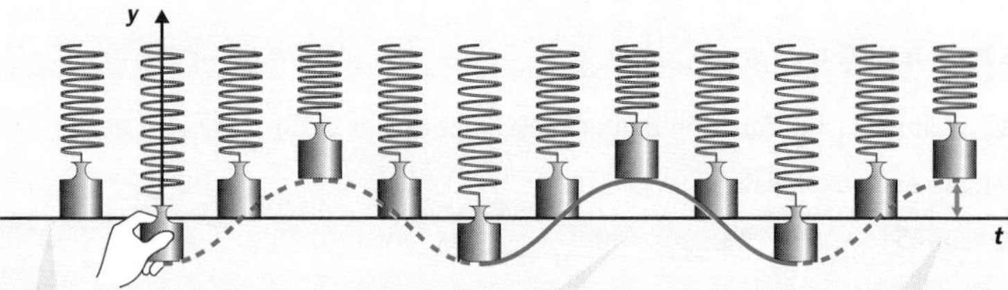

The line $y = 0$ represents the weight's **equilibrium position**, that is, its position before being pulled down and released.

One **oscillation** of the weight is one complete motion up and down.

The **amplitude** of the weight's motion is the maximum displacement of the weight from its equilibrium position.

The motion of the weight is an example of *simple harmonic motion*. An object undergoes simple harmonic motion if its displacement from its equilibrium position can be modeled by a function of the form

$$y = a \cos bt$$

where a and b are constants.

PROJECT GOAL **Use a cosine function to model the motion of a weight as it oscillates on a spring.**

Experimenting with a Spring

Suspend a weight from a spring, and find an equation modeling the motion of the weight when it is pulled down and released. Work in a group of four students. Your group will need a spring, a paper clip, some masking tape, a 1 lb weight, some string, a meterstick, and a stopwatch.

1. **BEND** a paper clip and tape it to the edge of a desk. Hook one end of the spring to the paper clip and use string to tie the weight to the other end.

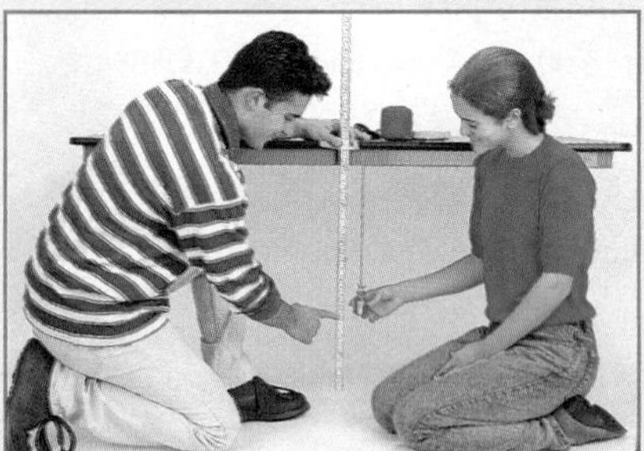

2. **TAPE** the meterstick next to the spring and weight as shown.

3. **STRETCH** the spring some distance d, using the meterstick to measure the distance stretched.

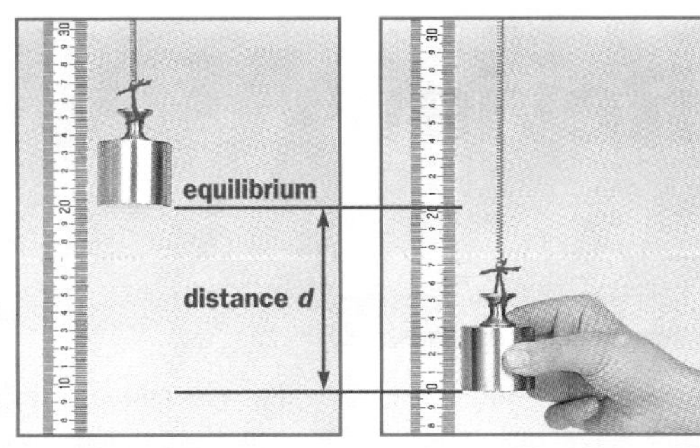

4. **RELEASE** the weight and use the stopwatch to measure the time it takes the weight to complete five oscillations. Divide this time by 5 to calculate the time of one oscillation.

5. **REPEAT** Step 4 two more times using the same stretch distance d. Find T, the average of your three oscillation times.

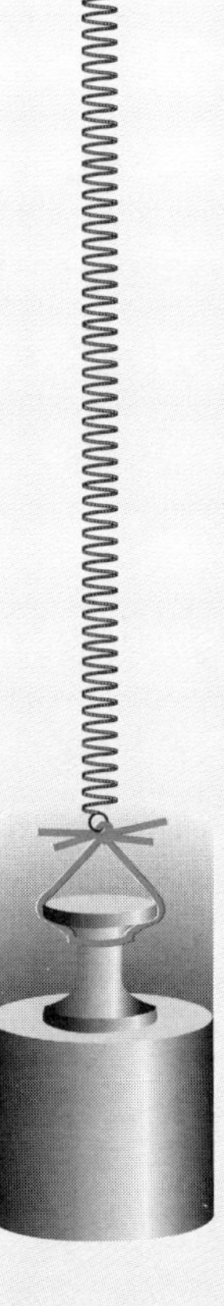

Modeling Oscillation

1. What is the relationship between your values of d and T and the values of a and b in the equation $y = a \cos bt$?

2. Write an equation of the form $y = a \cos bt$ for the oscillating weight.

3. Graph your equation from Question 2, and use your graph to predict the time required for the weight to complete four oscillations. Test your prediction.

4. What does your equation assume about the amplitude of the oscillating weight as time passes? Do you think this assumption is accurate?

Writing a Report

Write a report about your experiment. Describe your procedures, and present your data and the equation of your model. Also include these items:

- a graph of your model
- a comparison of your model with your experimental results
- an evaluation of your model's strengths and weaknesses

An equation of the form $y = ae^{-kt} \cos bt$ (where $k > 0$) more accurately models the weight's displacement as a function of time. This is because the factor e^{-kt} *damps*, or decreases, the amplitude of the oscillations as time passes. You can extend your report by finding an equation of the form $y = ae^{-kt} \cos bt$ for your oscillating weight.

Self-Assessment

Describe your understanding of the usefulness of trigonometric functions in modeling. What is it about the graphs of these functions that makes them useful? What are some other types of situations that might be modeled by trigonometric functions?

Project Notes

Guiding Students' Work

Rubric for Chapter Project

4 Students perform the experiment and make accurate measurements for the values of d and T. They then correctly relate these values to the cosine equation. An accurate graph is made and included in a well-written report, along with the other items listed in the text. Also included in the report is an in-depth comparison and evaluation of the experiment's model. Students also extend the project by addressing the idea of the damping function.

3 Students perform the experiment and make accurate measurements for the values of d and T. They then relate these values to the cosine function but make an error in the calculation of the equation or in the graph of the function. A report is written and contains all of the information relating to the project but does not make a thorough comparison or evaluation of the model. Students extend the project by investigating the damping function.

2 Students perform the experiment and make estimates for d and T but incorrectly relate them to the cosine function. Their graph is also incorrect and the written report shows that students do not understand the main ideas of the project. The report does contain an attempt at an evaluation of the model, but it is not complete or well done. Students do not make an attempt to extend the project.

1 Students make an attempt at performing the experiment but do not calculate the values of d and T correctly and do not make an effort to complete the project. As a result, an equation is not found and a graph is not drawn. If a report is written, it is poorly done and does not indicate an attempt to complete the project or an understanding of the mathematical ideas of the project. Students should be encouraged to speak with the teacher as soon as possible to review their work and to make a new start on the project.

Chapter Support

Course Guide: Chapter 15
Lesson Plans: Chapter 15
Practice Bank: Cumulative Practice 101
Study Guide: Chapter 15 Review
Assessment Book:
Chapter Tests 65 and 66
Cumulative Tests 67 and 68
Chapter 15 Alternative Assessment
Test Generator Software
Technology Book: Calculator Based Lab 6
Portfolio Project Book
Preparing for College Entrance Tests
Professional Handbook

Progress Check 15.1–15.3

1. Find all angles θ for which $0° \leq \theta \leq 360°$ and that satisfy the equation $\cos \theta = -0.2395$. Give your answer to the nearest degree. *(Section 15.1)* 104°, 256°

Evaluate each expression. *(Sections 15.1 and 15.2)*

2. $\sin 630°$ -1
3. $\cos\left(\frac{-3\pi}{4}\right)$ -0.7071
4. $\sin 4$ -0.7568
5. Convert $-279°$ from degrees to radians. *(Section 15.2)* -4.87
6. A Ferris wheel makes one revolution every 64 seconds. The height h, in feet, of the seat that begins at the highest position of this wheel is given as a function of time t, in seconds, by the equation $h = 100 + 90 \cos \frac{\pi}{32}t$. Give the amplitude, period, and vertical translation for the graph of the height function. *(Section 15.3)* 90; 64; up 100 ft
7. Graph the function $y = 2 \sin 0.5x$. *(Section 15.3)*

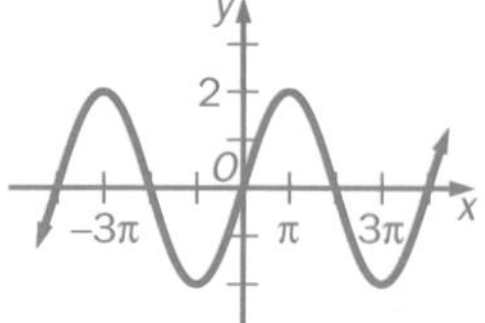

CHAPTER

15 Review

STUDY TECHNIQUE

Make a test study sheet. Write down definitions, equations, diagrams, or hints that will help you prepare for a test. Even if your teacher does not allow you to use your study sheet during the test, you will benefit from having organized and summarized your notes.

VOCABULARY

sine (p. 708)
cosine (p. 708)
radian (p. 714)
amplitude (p. 720)
cycle (p. 720)
period (p. 720)
phase shift (p. 728)
tangent (p. 736)

SECTIONS 15.1 *and* 15.2

You can measure angles in degrees or **radians.**

$$360 \text{ degrees} = 2\pi \text{ radians}$$

$$1 \text{ degree} = \frac{\pi}{180} \text{ radians}$$

$$\frac{180}{\pi} \text{ degrees} = 1 \text{ radian}$$

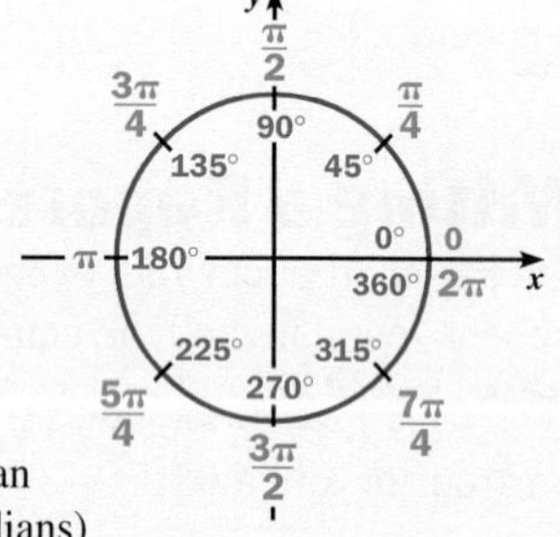

The domain of the **sine** and **cosine** functions can include angles with measures less than 0° (0 radians) or greater than 360° (2π radians).

Graph of $y = \sin \theta$ (θ measured in degrees)

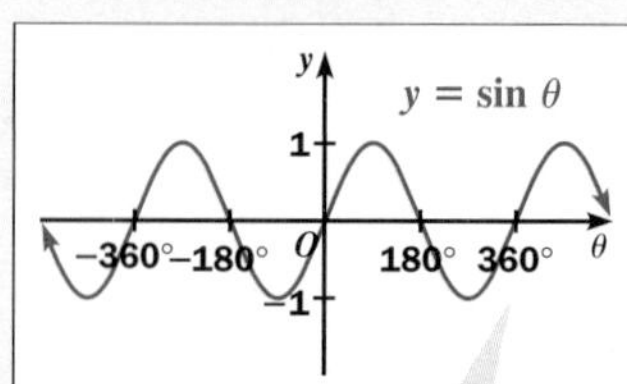

The period is 360°.

Graph of $y = \cos x$ (x measured in radians)

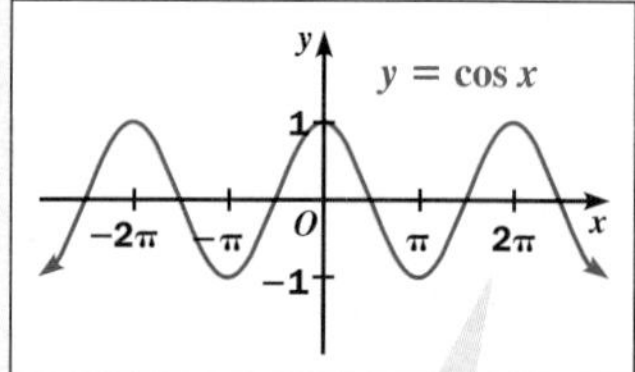

The period is 2π.

SECTIONS 15.3 *and* 15.4

Functions of the form $y = a \sin b(x - h) + k$ and $y = a \cos b(x - h) + k$ have graphs with **amplitude** $|a|$, **period** $\frac{2\pi}{|b|}$, **phase shift** h, and vertical shift k.

Example:

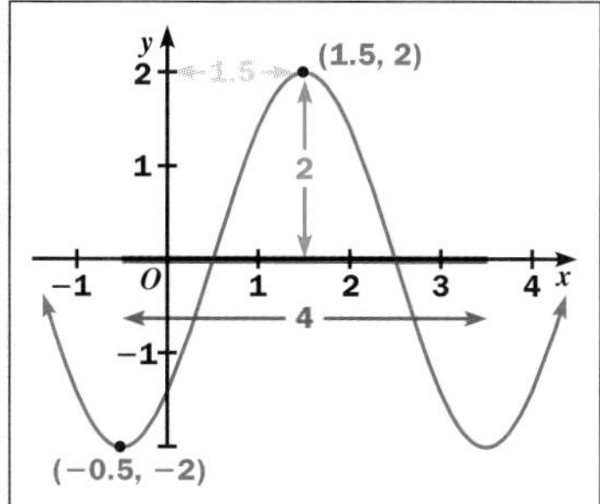

The amplitude is 2.

$y = 2 \cos \frac{2\pi}{4}(x - 1.5)$

The period is 4.

The phase shift is 1.5.

You can use trigonometric functions to model periodic phenomena, such as the mean monthly temperature in Columbia, South Carolina.

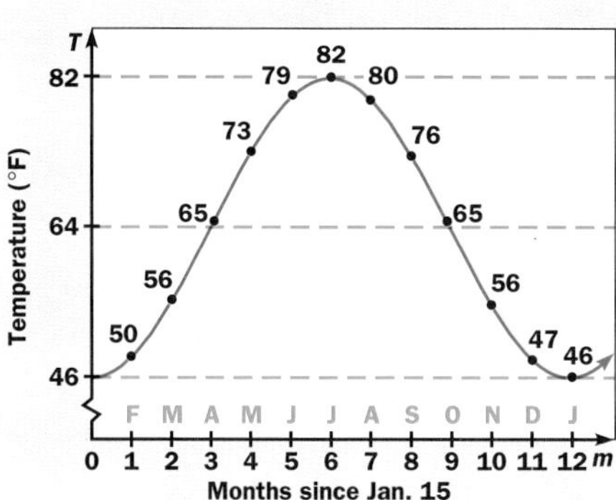

The amplitude is 18.

$T = -18 \cos \frac{2\pi}{12}m + 64$

The period is 12.

The graph has a vertical shift of 64, because the temperature varies around a mean of 64°F.

SECTION 15.5

As with the sine and cosine functions, the domain of the **tangent** function can be extended.

The function $y = \tan x$ is undefined when x is an odd multiple of $\frac{\pi}{2}$.

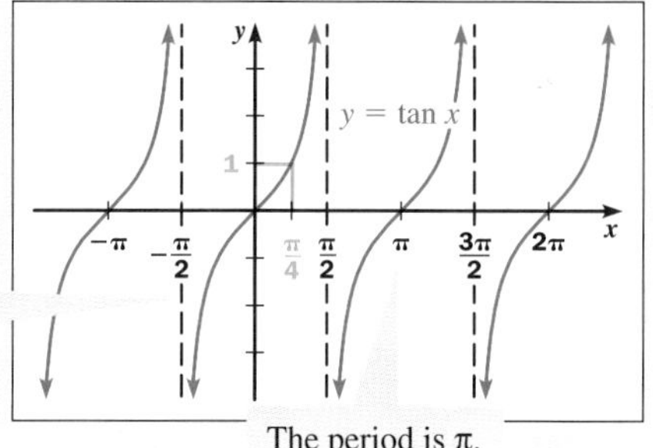

The period is π.

Progress Check 15.4–15.5

1. Graph $y = -3 \cos \left(x + \frac{\pi}{4}\right)$. *(Section 15.4)*

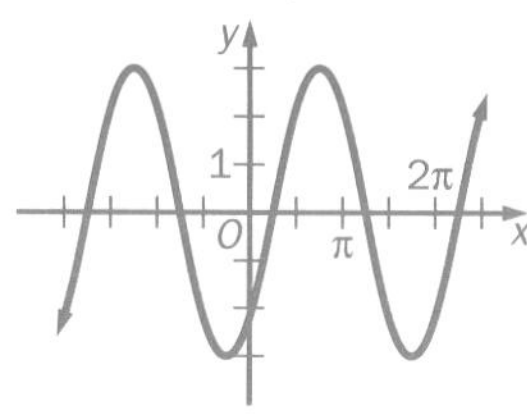

2. Write an equation for a sine function with maximum 12, minimum −2, and period 4. *(Section 15.4)* $y = 7 \sin \frac{\pi}{2}x + 5$

3. Write a cosine function that is equivalent to $y = \sin 2x$. *(Section 15.4)* $y = \cos 2\left(x - \frac{\pi}{4}\right)$

4. Evaluate each expression. *(Section 15.5)*

a. $\tan \left(-\frac{3\pi}{4}\right)$ 1

b. $\tan 415°$ 1.4281

5. Graph $y = \tan \frac{1}{2}x - 1$. *(Section 15.5)*

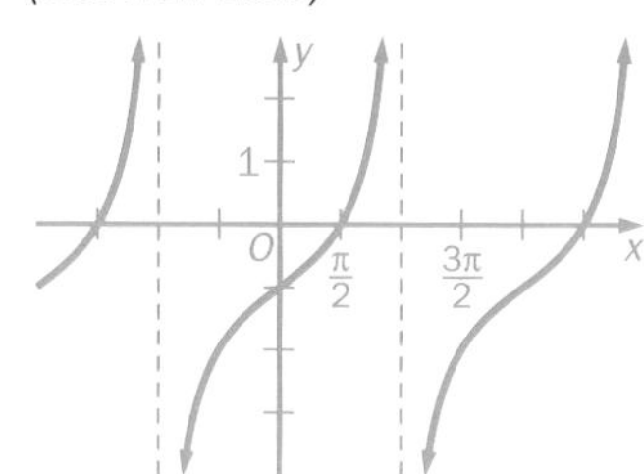

6. A helicopter is lifting off from an airfield. An observer 20 ft from the lift-off point measures the angle θ between the ground and the line connecting himself to the helicopter. *(Section 15.5)*

a. Write the height, h, of the helicopter as a function of θ. $h = 20 \tan \theta$

b. Find the value of h when $\theta = 62°$ about 38 ft

Chapter 15 Assessment
Form A Chapter Test

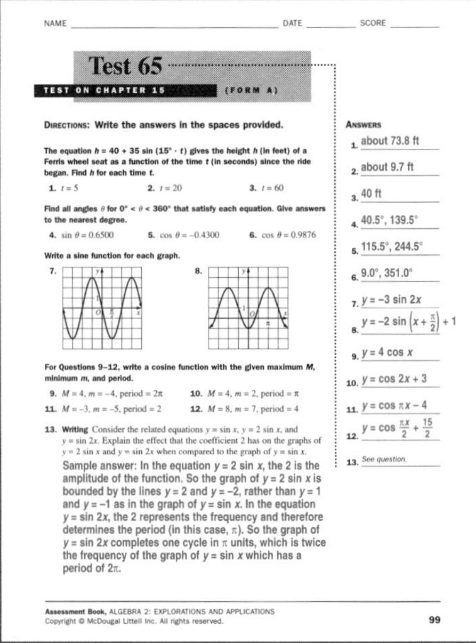
NAME ____________ DATE ______ SCORE ______

Test 65
TEST ON CHAPTER 15 (FORM A)

DIRECTIONS: Write the answers in the spaces provided.

The equation $h = 40 + 35 \sin (15° \cdot t)$ gives the height h (in feet) of a Ferris wheel seat as a function of the time t (in seconds) since the ride began. Find h for each time t.

1. $t = 5$ 2. $t = 20$ 3. $t = 60$

Find all angles θ for $0° < \theta < 360°$ that satisfy each equation. Give answers to the nearest degree.

4. $\sin \theta = 0.6500$ 5. $\cos \theta = -0.4300$ 6. $\cos \theta = 0.9876$

Write a sine function for each graph.

7. 8.

For Questions 9–12, write a cosine function with the given maximum M, minimum m, and period.

9. $M = 4, m = -4$, period $= 2\pi$ 10. $M = 4, m = 2$, period $= \pi$
11. $M = -3, m = -5$, period $= 2$ 12. $M = 8, m = 7$, period $= 4$

13. **Writing** Consider the related equations $y = \sin x$, $y = 2 \sin x$, and $y = \sin 2x$. Explain the effect that the coefficient 2 has on the graphs of $y = 2 \sin x$ and $y = \sin 2x$ when compared to the graph of $y = \sin x$.
Sample answer: In the equation $y = 2 \sin x$, the 2 is the amplitude of the function. So the graph of $y = 2 \sin x$ is bounded by the lines $y = 2$ and $y = -2$, rather than $y = 1$ and $y = -1$ as in the graph of $y = \sin x$. In the equation $y = \sin 2x$, the 2 represents the frequency and therefore determines the period (in this case, π). So the graph of $y = \sin 2x$ completes one cycle in π units, which is twice the frequency of the graph of $y = \sin x$ which has a period of 2π.

ANSWERS
1. about 73.8 ft
2. about 9.7 ft
3. 40 ft
4. 40.5°, 139.5°
5. 115.5°, 244.5°
6. 9.0°, 351.0°
7. $y = -3 \sin 2x$
8. $y = -2 \sin \left(x + \frac{\pi}{2}\right) + 1$
9. $y = 4 \cos x$
10. $y = \cos 2x + 3$
11. $y = \cos \pi x - 4$
12. $y = \cos \frac{\pi x}{2} + \frac{15}{2}$
13. See question.

Assessment Book, ALGEBRA 2: EXPLORATIONS AND APPLICATIONS

99

Chapter 15 Assessment
Form B Chapter Test

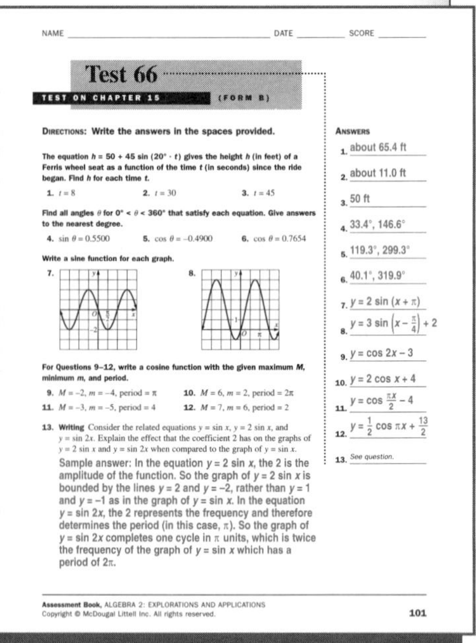
NAME ____________ DATE ______ SCORE ______

Test 66
TEST ON CHAPTER 15 (FORM B)

DIRECTIONS: Write the answers in the spaces provided.

The equation $h = 50 + 45 \sin (20° \cdot t)$ gives the height h (in feet) of a Ferris wheel seat as a function of the time t (in seconds) since the ride began. Find h for each time t.

1. $t = 8$ 2. $t = 30$ 3. $t = 45$

Find all angles θ for $0° < \theta < 360°$ that satisfy each equation. Give answers to the nearest degree.

4. $\sin \theta = 0.5500$ 5. $\cos \theta = -0.4900$ 6. $\cos \theta = 0.7654$

Write a sine function for each graph.

7. 8.

For Questions 9–12, write a cosine function with the given maximum M, minimum m, and period.

9. $M = -2, m = -4$, period $= \pi$ 10. $M = 6, m = 2$, period $= 2\pi$
11. $M = -3, m = -5$, period $= 4$ 12. $M = 7, m = 6$, period $= 2$

13. **Writing** Consider the related equations $y = \sin x$, $y = 2 \sin x$, and $y = \sin 2x$. Explain the effect that the coefficient 2 has on the graphs of $y = 2 \sin x$ and $y = \sin 2x$ when compared to the graph of $y = \sin x$.
Sample answer: In the equation $y = 2 \sin x$, the 2 is the amplitude of the function. So the graph of $y = 2 \sin x$ is bounded by the lines $y = 2$ and $y = -2$, rather than $y = 1$ and $y = -1$ as in the graph of $y = \sin x$. In the equation $y = \sin 2x$, the 2 represents the frequency and therefore determines the period (in this case, π). So the graph of $y = \sin 2x$ completes one cycle in π units, which is twice the frequency of the graph of $y = \sin x$ which has a period of 2π.

ANSWERS
1. about 65.4 ft
2. about 11.0 ft
3. 50 ft
4. 33.4°, 146.6°
5. 119.3°, 299.3°
6. 40.1°, 319.9°
7. $y = 2 \sin (x + \pi)$
8. $y = 3 \sin \left(x - \frac{\pi}{4}\right) + 2$
9. $y = \cos 2x - 3$
10. $y = 2 \cos x + 4$
11. $y = \cos \frac{\pi x}{2} - 4$
12. $y = \frac{1}{2} \cos \pi x + \frac{13}{2}$
13. See question.

Assessment Book, ALGEBRA 2: EXPLORATIONS AND APPLICATIONS

101

CHAPTER

15 Assessment

VOCABULARY QUESTIONS

For Questions 1–3, complete each sentence.

1. There are 360° or 2π __?__ in a full circle.
2. The graph of $y = 3 \sin 4(x - 2)$ has a(n) __?__ of 3, a(n) __?__ of $\frac{\pi}{2}$, and a(n) __?__ of 2.
3. The __?__ and __?__ functions have a period of 2π.

SECTIONS 15.1 *and* 15.2

4. The equation $h = 30 + 25 \sin (2° \cdot t)$ gives the height h (in meters) of a Ferris wheel seat as a function of the time t (in seconds) since the ride began. Find h when:
   a. $t = 10$ **b.** $t = 45$ **c.** $t = 90$

Find all angles θ for $0° \le \theta \le 360°$ that satisfy each equation. Give answers to the nearest degree.

5. $\sin \theta = 0.7315$ 6. $\cos \theta = -0.7315$ 7. $\cos \theta = 0.8250$

Evaluate each expression.

8. $\sin 720°$ 9. $\cos (-270°)$ 10. $\sin (-450°)$
11. $\cos (-\pi)$ 12. $\sin \pi$ 13. $\cos 16\pi$
14. **Writing** Explain how to convert an angle measurement from degrees to radians, and from radians to degrees. Show some examples.

SECTIONS 15.3 *and* 15.4

15. **ENGINEERING** To discourage pigeons from nesting in subway stations, electronic soundmaking devices are installed that emit a pitch that is inaudible to humans yet disturbs pigeons.
   a. At one setting, the device emits a sound with the equation $P = 316{,}200 \sin 44{,}000\pi t$, where P represents pressure and t represents time in seconds. What are the amplitude and period of the equation's graph?
   b. **Writing** Most humans cannot hear sounds with frequencies greater than 16,000 cycles per second. Explain why the sound described in part (a) is inaudible to most humans.

ANSWERS Chapter 15

Assessment

1. radians
2. amplitude, period, phase shift
3. sine, cosine
4. a. 39 m
   b. 55 m
   c. 30 m
5. 47°, 133°
6. 137°, 223° 7. 34°, 326°
8. 0 9. 0
10. –1 11. –1
12. 0 13. 1
14. See answers in back of book.
15. a. amplitude: 316,200; period: $\frac{1}{22{,}000}$
    b. The sound described in part (a) has frequency of 22,000 cps (since period and frequency are reciprocals). Since most humans cannot detect frequencies greater than 16,000 cps, this sound is inaudible.

16–19. Answers may vary. Examples are given.
16. $y = 7 \cos 2x$
17. $y = 5 \cos \pi x - 1$
18. $y = 3 \cos \frac{\pi}{6}(x - 4)$
19. $y = 2 \sin \left(x + \frac{\pi}{3}\right)$
20.

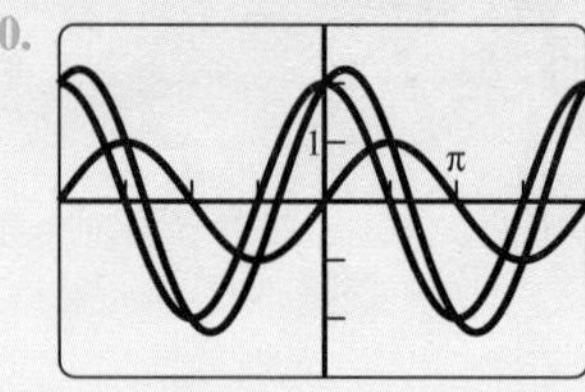

Function	Period	Range
$y = \sin x$	2π	$-1 \le y \le 1$
$y = 2 \cos x$	2π	$-2 \le y \le 2$
sum	2π	$-2.24 \le y \le 2.24$

Write a cosine function with the given maximum *M*, minimum *m*, and period.

16. $M = 7, m = -7$, period $= \pi$

17. $M = 4, m = -6$, period $= 2$

Open-ended Problem **Write a sine or cosine function for each graph.**

18.

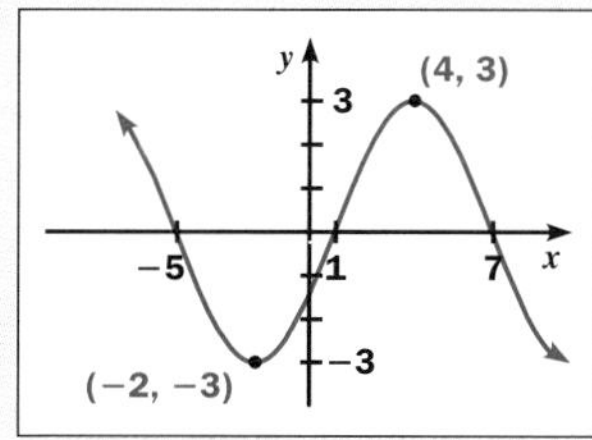

19.

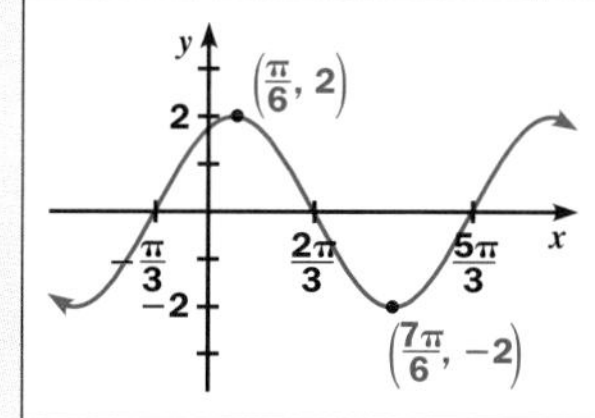

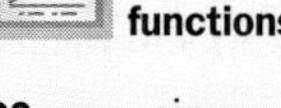

Technology **In the same coordinate plane, graph each pair of functions and their sum. Find the period and range of each function.**

20. $y = \sin x$
$y = 2 \cos x$

21. $y = \cos \frac{1}{2}x$
$y = 2 \sin x$

22. $y = 3 \sin \pi x$
$y = \cos 2\pi x$

SECTION 15.5

Graph each function. Show at least two periods.

23. $y = 4 \tan x$

24. $y = \tan \pi x$

25. $y = 2 \tan \left(x - \frac{\pi}{2}\right)$

26. **Open-ended Problem** Write a tangent function with period 3π and a phase shift of your choice. Graph your function and show its horizontal intercepts and vertical asymptotes.

27. **ASTRONOMY** If you push a stick into the ground so it points directly at the sun at noon, then wait a while, the shadow it casts will point east.

a. Use the diagram to explain why the length l of the shadow cast by a stick of height h is $l = h \tan \theta$.

b. What is the length of the shadow cast by an 18 in. stick when $\theta = 0°$? $\theta = 30°$? $\theta = 60°$? $\theta = 90°$?

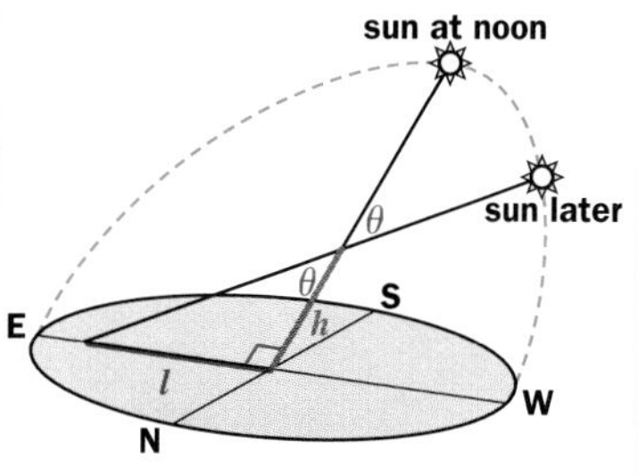

A stick pointed directly at the sun later casts a shadow pointing east.

PERFORMANCE TASK

28. **a.** Research a phenomenon that can be modeled using a sine function.

b. Find the maximum value, minimum value, and period of the function. Write an equation for the function. Graph your equation.

c. Prepare a poster or overhead transparency that explains the phenomenon and shows how your function models it.

Chapter 15 Assessment
Form C Alternative Assessment

Chapter 15
ALTERNATIVE ASSESSMENT

1. **Open-ended Problem** Describe some real-world functions that you think would be modeled by a sine or cosine graph.
2. **Project** Use a microphone, a keyboard or tuning fork, and a data collection device to determine which notes have sound waves that form sine curves.
3. **Project** Draw a large quadrilateral on a piece of paper. Draw one of the diagonals. Take five direct measurements from your drawing using a ruler or protractor. Use the law of sines and the law of cosines to find the other angle and side measures indirectly. Confirm your indirect measurements by direct measurement.
4. **Open-ended Problem** Find some examples of graphs that represent real-world periodic functions. Make a display of your graphs.
5. **a.** Graph $y = \sin x$ and $y = \sin(-x)$ on a graphing calculator or using graphing software.
   b. Graph $y = -\sin x$ and $y = \sin(-x)$ on a graphing calculator or using graphing software.
   c. Graph $y = \cos x$ and $y = \cos(-x)$ on a graphing calculator or using graphing software.
   d. Graph $y = -\cos x$ and $y = \cos(-x)$ on a graphing calculator or using graphing software.
   e. Describe any patterns that you observed in your graphs.

128 **Assessment Book,** ALGEBRA 2: EXPLORATIONS AND APPLICATIONS

21.

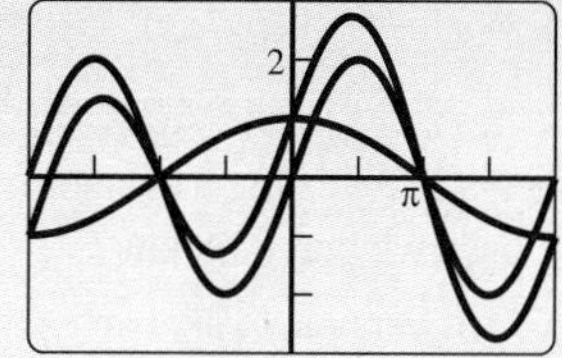

Function	Period	Range
$y = \cos \frac{1}{2}x$	4π	$-1 \le y \le 1$
$y = 2 \sin x$	2π	$-2 \le y \le 2$
sum	4π	$-2.74 \le y \le 2.74$

22.

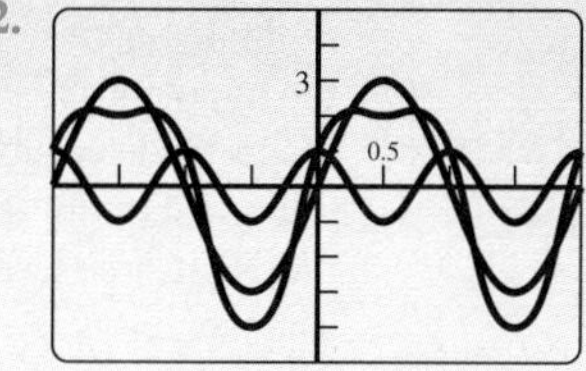

Function	Period	Range
$y = 3 \sin \pi x$	2	$-3 \le y \le 3$
$y = \cos 2\pi x$	1	$-1 \le y \le 1$
sum	2	$-4 \le y \le 2.13$

23.

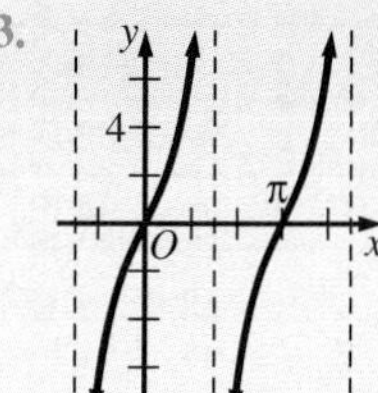

24.

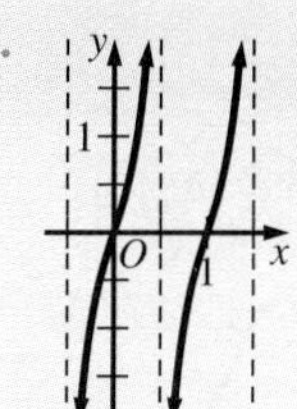

25.

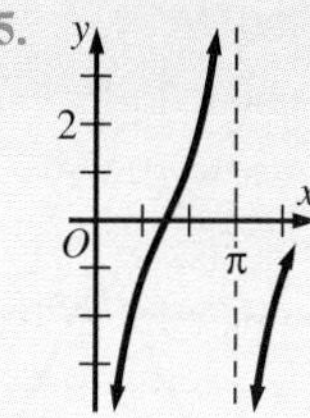

26. Answers may vary. Check students' work.

27. a. By the diagram, $\tan \theta = \frac{l}{h}$, and thus $l = h \tan \theta$.

b. 0, 10.4 in., 31.2 in., and (when $\theta = 90°$) undefined.

28. Check students' work.

Cumulative Assessment

CHAPTERS 13–15

CHAPTER 13

1. **Writing** Describe how you can use experimental probability to find the likelihood that a thumbtack will land "point down" when dropped.

2. A sunken ship is known to be within a circular region of the ocean with a 10 mi radius. Divers search 0.5 mi in all directions from a point directly below a salvage ship anchored within the circular region. What is the probability that the divers find the sunken ship on the first try?

Suppose there is a 2% chance that an accident will occur in a certain factory during any month. Find the probability of each event.

3. at least one accident occurs during three consecutive months

4. accidents occur in exactly two months during one year

For Questions 5 and 6, suppose that a new machine produces 75% of the rivets made in a factory. Of these rivets, 2% are defective. An older machine produces the rest of the rivets. Of these, 5% are defective. Find the probability of each event.

5. a randomly chosen rivet is from the new machine and is not defective

6. a randomly chosen rivet is from the older machine and is defective

7. There are 8 *bits* in a computer *byte*. If each bit in a particular byte is equally likely to be on or off, find the probability that 6 or more bits are on.

8. **Open-ended Problem** Sketch two normal distributions that have the same standard deviation but different means.

9. The scores on a quiz are normally distributed with mean 72.5 and standard deviation 7.5. About what percent of the scores are:

   a. over 80? **b.** under 92? **c.** between 65 and 75?

CHAPTER 14

10. In right $\triangle ABC$, $a = 12$, $b = 35$, and $c = 37$.

    a. Find the sine, cosine, and tangent of $\angle A$ and of $\angle B$. Express your answers in decimal form.

    b. Find the measure of $\angle A$ and of $\angle B$. Express your answers in decimal degrees, to the nearest tenth of a degree, and in degrees, minutes, and seconds, to the nearest second.

11. The wires that support an 800 ft tower extend from the top of the tower to the ground and form 70° angles with the ground. Find the distance from the base of the tower to the point at which a wire is anchored to the ground.

ANSWERS Chapters 13–15

Cumulative Assessment

1. Answers may vary. An example is given. Drop a thumbtack many times, say 100, and count the number of times, n, that it lands point down. For this experiment, the experimental probability is $\frac{n}{100}$.
2. 0.0025
3. about 0.0588
4. about 0.0216
5. 0.735
6. 0.0125
7. about 0.145
8. Answers may vary. An example is given.

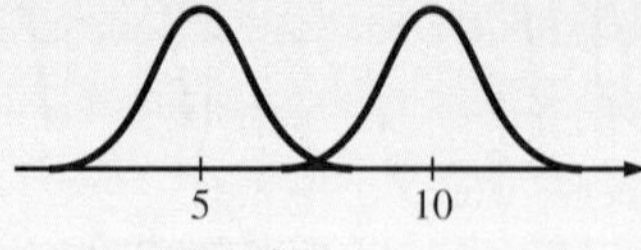

9. a. about 0.1587
   b. about 0.995
   c. about 0.4592

10. a.

Angle	sine	cosine	tangent
A	0.3243	0.9459	0.3429
B	0.9459	0.3243	2.917

    b. $\angle A \approx 18.9°$ or $18°55'29''$; $\angle B \approx 71.1°$ or $71°4'31''$

11. about 291 ft
12. Answers may vary. An example is given. For each acute angle in a right triangle, the sine and cosine are both given by the ratio of the length of a leg to the length of the hypotenuse. Since the hypotenuse is the longest side of any right triangle, both of these ratios must be less than 1. The tangent is the ratio of the lengths of two legs. Since those lengths may be equal or either one may be greater than the other, the tangent may be 1, less than 1, or greater than 1.
13. –0.7071; –0.7071; 1
14. about 53.1° and about 306.9°
15. a. 170°
    b. 49.2 km; 8.68 km
16. Answers may vary. An example is given. If $a = 15$, $b = 30$, and $\angle A = 35°$, then no triangle can be formed because $a < b \sin A$.
17. $\angle C = 83°$; $b \approx 7.18$; $c \approx 10.1$

12. **Writing** Explain why the sine and the cosine of an acute angle must be less than 1, but the tangent of an acute angle may be greater than 1. (*Hint:* Sketch a right triangle.)

13. An angle θ is in standard position with the point $P(-\sqrt{2}, -\sqrt{2})$ on its terminal side. Find $\sin \theta$, $\cos \theta$, and $\tan \theta$.

14. Find all angles θ between 0° and 360° such that $\cos \theta = \frac{3}{5}$.

15. **NAVIGATION** A ship traveled on a course of 280° for 50 km.
   a. What angle of rotation is equivalent to the ship's course?
   b. How far did the ship travel in an east-west direction? in a north-south direction?

16. **Open-ended Problem** Give an example of the ambiguous case for which no triangle can be formed. Explain why no triangle can be formed.

Solve $\triangle ABC$. If there are two solutions, give both. If there is no solution, explain why.

17. $\angle A = 52°, \angle B = 45°, a = 8$ 18. $\angle B = 50°, b = 16, c = 20$

CHAPTER 15

Evaluate each expression.

19. $\tan(-135°)$ 20. $\cos 1.47$ 21. $\sin\left(-\frac{4\pi}{3}\right)$ 22. $\sin 150°$

23. Find the maximum and minimum values of the function $y = 2 - \sin 2\theta$. What values of θ for $0° \le \theta \le 360°$ give the maximum and minimum values?

24. Convert $\frac{10\pi}{9}$ from radians to degrees (to the nearest degree).

25. Convert $-84°$ from degrees to radians. Express your answer as a decimal to the nearest hundredth of a radian, and as a multiple of π.

Graph each function. Show at least one period. State the amplitude and period of each graph.

26. $y = 3\sin(-2x)$ 27. $y = -\frac{3}{2}\cos\frac{\pi}{2}x$ 28. $y = -\cos(x - \pi)$

29. a. Write an equation for a cosine function with period $\frac{\pi}{3}$, maximum value 1, and minimum value -5.
   b. **Open-ended Problem** Choose a phase shift to apply to the graph of the equation you wrote in part (a). Write a function that describes the graph after the phase shift.
   c. How is the graph of the function from part (b) different from the graph of the function from part (a)?

30. **Writing** Explain how to use technology to find the period and the maximum value of a function of the form $y = a\sin x + b\cos x$.

31. Graph at least two periods of the function $y = \tan\frac{1}{2}\left(x - \frac{\pi}{2}\right)$.

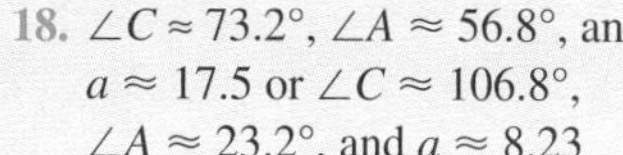

18. $\angle C \approx 73.2°$, $\angle A \approx 56.8°$, and $a \approx 17.5$ or $\angle C \approx 106.8°$, $\angle A \approx 23.2°$, and $a \approx 8.23$

19. 1

20. about 0.1006

21. about 0.8660

22. 0.5

23. 3; 1; maximum: $\theta = 135°$ and $\theta = 315°$; minimum: $\theta = 45°$ and $\theta = 225°$

24. 200°

25. about -1.47 or $-\frac{7\pi}{15}$

26.

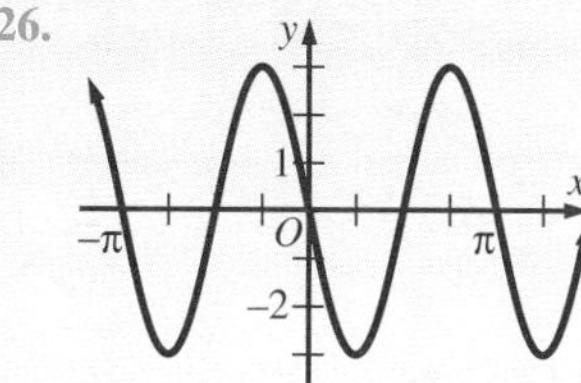

amplitude: 3; period: π

27.

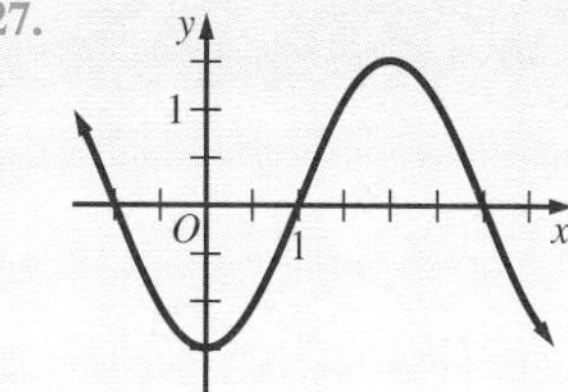

amplitude: $\frac{3}{2}$; period: 4

28.

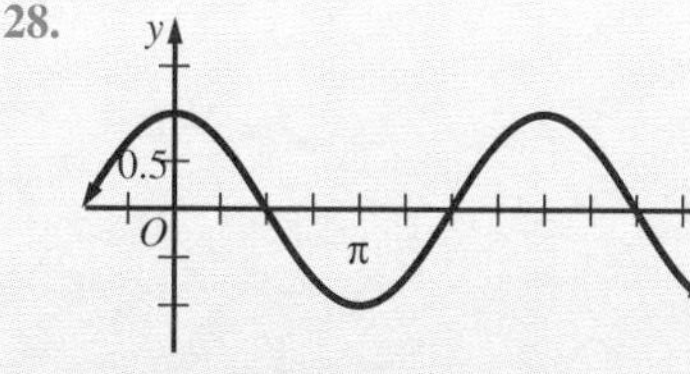

amplitude: 1; period: 2π

29. a. $y = 3\cos 6x - 2$
   b. Answers may vary. An example is given. $\frac{\pi}{2}$; $y = 3\cos 6\left(x - \frac{\pi}{2}\right) - 2$
   c. The graph of the function from part (b) is the graph of the function from part (a) shifted $\frac{\pi}{2}$ units to the right.

30. Answers may vary. Examples are given. Use a graphing calculator or graphing software to graph the sum. Use the maximum feature or the trace feature to find the maximum value. To find the period, use the trace feature to find the difference between consecutive x-values that produce maximums.

31.

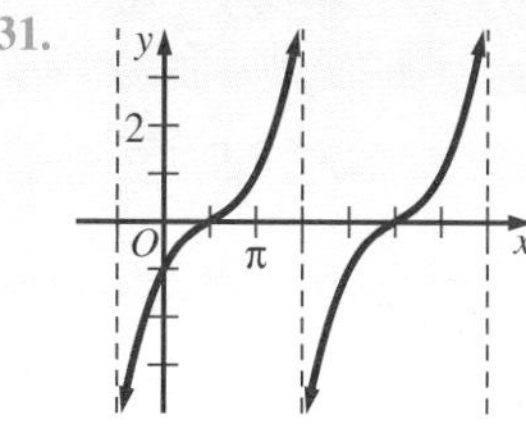

Contents of Student Resources

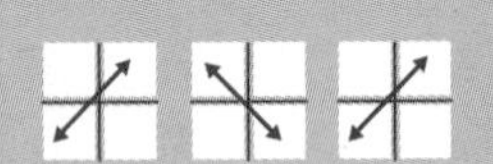

748

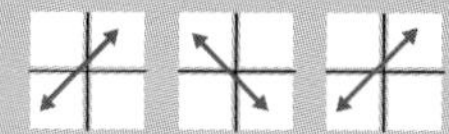

Extra Practice

CHAPTER 1

For Exercises 1–5, use the histogram below, which gives the circulation of Sunday newspapers in the United States in selected years. *Section 1.1*

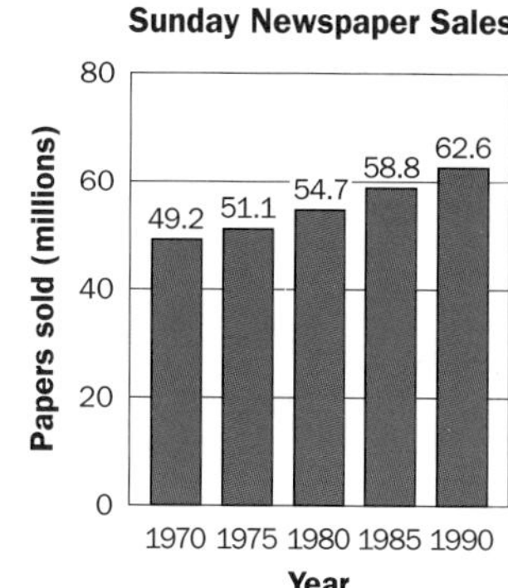

1. Find the average 5-year increase in Sunday newspaper sales between 1970 and 1990.
2. Find the average 5-year percent increase in Sunday newspaper sales between 1970 and 1990.
3. Using a constant growth model, estimate the Sunday newspaper sales in the years 1995 and 2000.
4. Repeat Exercise 3 using a proportional growth model.
5. Which model do you think provides a more accurate estimate? Give a reason for your answer.

For Exercises 6–8, use the table below, which lists the expenditures for public elementary and secondary schools in the United States from 1985 to 1993. *Section 1.2*

Year	1985	1986	1987	1988	1989	1990	1991	1992	1993
Expenditure (billions)	137	149	161	173	193	212	229	243	258

6. a. Write a linear function to model the annual expenditures for public elementary and secondary schools from 1985 to 1993. Let $x =$ the number of years since 1985.
   b. Repeat part (a) with an exponential function.
7. **Technology** Use a graphing calculator or graphing software to graph your functions from Exercise 6 along with the data from the table. Does one model fit the data better? If so, which one?
8. Predict the expenditures for public elementary and secondary schools in 2000 and in 2010 using:
   a. the linear function from part (a) of Exercise 6
   b. the exponential function from part (b) of Exercise 6

Use matrices A and B to evaluate each matrix expression. *Section 1.3*

$$A = \begin{bmatrix} 4 & 0 & -1 \\ 2 & -3 & 1 \end{bmatrix} \qquad B = \begin{bmatrix} 2 & -6 & 10 \\ 0 & 8 & -4 \end{bmatrix}$$

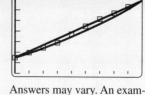

9. $A + B$
10. $B - A$
11. $3A$
12. $2A - B$
13. $A - 2B$
14. $A + \frac{1}{2}B$
15. $-4A + 3B$
16. $A - \frac{3}{2}B$

ANSWERS Extra Practice

Chapter 1

1–5. Answers may vary. Examples are given.

1. 3.35 million
2. about 6.2%
3. 66.0 million; 69.3 million
4. 66.5 million; 70.6 million
5. The constant growth model provides a more accurate estimate, because the lower estimates provided by this model seem to reflect the fact that sales did not increase as much from 1985–1990 as from 1980–1985.
6. a. $y = 137 + 15.1x$
   b. $y = 137(1.08)^x$
7. Answers may vary. An example is given. The linear model appears to fit the data better.
8. a. about \$364 billion; about \$515 billion
   b. about \$435 billion; about \$938 billion
9. $\begin{bmatrix} 6 & -6 & 9 \\ 2 & 5 & -3 \end{bmatrix}$
10. $\begin{bmatrix} -2 & -6 & 11 \\ -2 & 11 & -5 \end{bmatrix}$
11. $\begin{bmatrix} 12 & 0 & -3 \\ 6 & -9 & 3 \end{bmatrix}$
12. $\begin{bmatrix} 6 & 6 & -12 \\ 4 & -14 & 6 \end{bmatrix}$
13. $\begin{bmatrix} 0 & 12 & -21 \\ 2 & -19 & 9 \end{bmatrix}$
14. $\begin{bmatrix} 5 & -3 & 4 \\ 2 & 1 & -1 \end{bmatrix}$
15. $\begin{bmatrix} -10 & -18 & 34 \\ -8 & 36 & -16 \end{bmatrix}$
16. $\begin{bmatrix} 1 & 9 & -16 \\ 2 & -15 & 7 \end{bmatrix}$

Use matrices *P*, *Q*, *R*, and *S* to find each product. If the matrices cannot be multiplied, state that the product is *undefined*. *Section 1.4*

$$P = \begin{bmatrix} 2 & 3 \\ 0 & -1 \\ 4 & 1 \end{bmatrix} \quad Q = \begin{bmatrix} 4 & 3 & -2 \\ 1 & 12 & 0 \end{bmatrix} \quad R = \begin{bmatrix} 3 & 1 & -2 \\ -5 & 0 & 4 \\ -1 & 2 & 6 \end{bmatrix} \quad S = \begin{bmatrix} 3 & -1 \\ 1 & 4 \end{bmatrix}$$

17. PQ **18.** QP **19.** QR **20.** RQ

21. PS **22.** SP **23.** SQ **24.** RP

For Exercises 25–28, describe how you would simulate each situation. *Section 1.5*

25. guessing the right answer to a multiple-choice question with 4 choices

26. finding that a randomly chosen student in a school is left-handed, if 1 out of every 10 people is left-handed

27. getting on base in a given at-bat in baseball, if the player gets on base 2 out of every 5 at-bats

28. having a rainstorm start between 12:00 noon and 3:00 P.M., if the storm arrives at some time during a 24-hour day

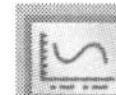
Technology Write a calculator formula that will generate a random number satisfying the given condition. *Section 1.5*

29. a positive integer less than 7

30. one of the integers 5–12

31. one of the integers 2, 3, 4, 5, and 6

32. one of the integers 0–7

CHAPTER 2

For each equation, tell whether *y* varies directly with *x*. If so, graph the equation. *Section 2.1*

1. $y = \frac{x}{2}$ **2.** $y = 4x - 1$ **3.** $y = \frac{4}{3x}$ **4.** $y = -\frac{4}{5}x$

5. $y = 5x$ **6.** $y = 3x - 2$ **7.** $y = 2$ **8.** $y = 4x - 9$

For each table, write a direct variation equation relating the two variables. Use the equation to find the missing value. *Section 2.1*

9.

x	y
3	4.5
4	6
9	?
14	21

10.

x	y
6	2.7
?	3.6
12	5.4
22	9.9

11.

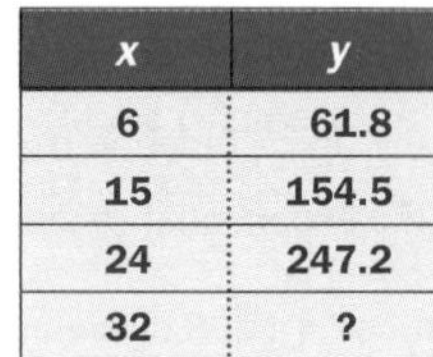

x	y
6	61.8
15	154.5
24	247.2
32	?

Find the slope-intercept equation of the line passing through each pair of points. *Section 2.2*

12. (1, 4) and (3, 8) **13.** (−1, 5) and (1, 11) **14.** (6, −1) and (7, −4)

15. (−3, 2) and (5, −10) **16.** (−2, 1) and (−4, −9) **17.** (−13, −9) and (5, 3)

17. $\begin{bmatrix} 11 & 42 & -4 \\ -1 & -12 & 0 \\ 17 & 24 & -8 \end{bmatrix}$

18. $\begin{bmatrix} 0 & 7 \\ 2 & -9 \end{bmatrix}$

19. $\begin{bmatrix} -1 & 0 & -8 \\ -57 & 1 & 46 \end{bmatrix}$

20. undefined

21. $\begin{bmatrix} 9 & 10 \\ -1 & -4 \\ 13 & 0 \end{bmatrix}$

22. undefined

23. $\begin{bmatrix} 11 & -3 & -6 \\ 8 & 51 & -2 \end{bmatrix}$

24. $\begin{bmatrix} -2 & 6 \\ 6 & -11 \\ 22 & 1 \end{bmatrix}$

25–28. Answers may vary. Examples are given.

25. Draw cards at random from a standard deck of 52 cards, letting each suit represent one of the four answer choices, and choosing a particular suit to represent a right answer.

26. Generate a random number from 1–10 on a calculator by using the formula "int(10*rand)+1." Let "1" represent left-handedness.

27. Generate a random number from 1–5 on a calculator by using the formula "int(5*rand)+1." Let "1" and "2" represent getting on base.

28. Generate a random number from 1–8 on a calculator by using the formula "int(8*rand)+1." Let "1" represent the rainstorm starting in the 3 hr period from 12:00 noon–3:00 P.M.

29. int(6*rand)+1

30. int(8*rand)+5

31. int(5*rand)+2

32. int(8*rand)

Chapter 2

1. Yes.

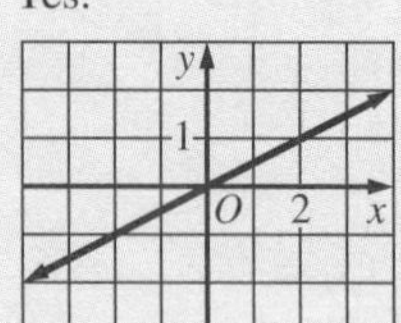

2. No.

3. No.

4. Yes.

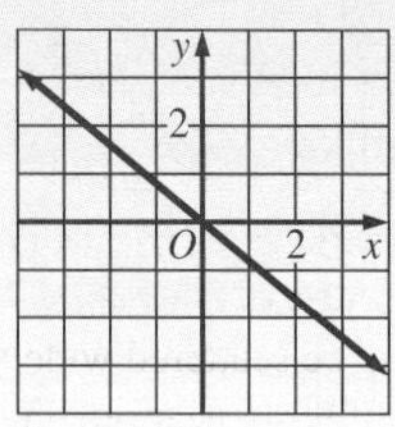

5. Yes.

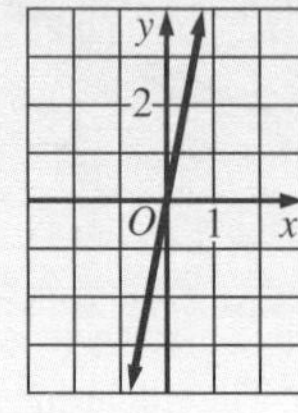

6. No.

7. No.

8. No.

9. $y = 1.5x$; 13.5

10. $y = 0.45x$; 8

11. $y = 10.3x$; 329.6

12. $y = 2x + 2$

13. $y = 3x + 8$

14. $y = -3x + 17$

15. $y = -\frac{3}{2}x - \frac{5}{2}$

16. $y = 5x + 11$

17. $y = \frac{2}{3}x - \frac{1}{3}$

18.

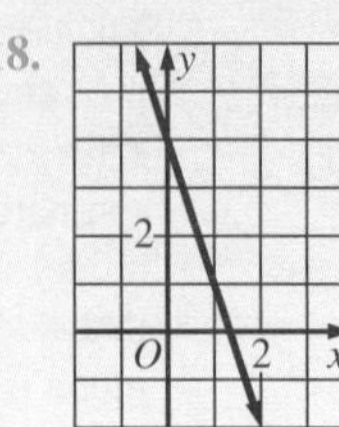

19.

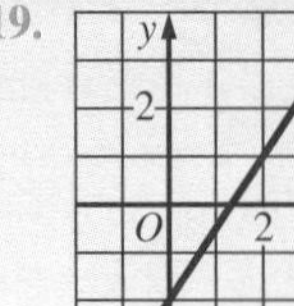

Graph each equation. *Section 2.2*

18. $y = 4 - 3x$ **19.** $y = -2 + \frac{3}{2}x$ **20.** $y = -4$

21. $y = 3 + \frac{1}{3}x$ **22.** $y = 5 - \frac{3}{4}x$ **23.** $y = -1 - \frac{2}{3}x$

For Exercises 24–29:

a. Write a point-slope equation of the line passing through the given point and having the given slope.

b. Graph the equation.

c. Write the equation in slope-intercept form. *Section 2.3*

24. point: (4, 1)
slope $= -2$

25. point: $(-3, 2)$
slope $= \frac{1}{3}$

26. point: $(5, -6)$
slope $= \frac{3}{4}$

27. point: $(6, -2)$
slope $= 2$

28. point: $(-1, 5)$
slope $= -3$

29. point: (9, 2)
slope $= \frac{2}{3}$

Find a point-slope equation of the line passing through each pair of points. *Section 2.3*

30. (2, 4) and (3, 7) **31.** $(-1, 5)$ and (4, 3) **32.** $(7, -2)$ and (5, 4)

33. (1, 6) and $(-3, -2)$ **34.** $(9, -5)$ and $(-1, 3)$ **35.** $(-1, -4)$ and $(-6, 5)$

For Exercises 36–39, use the table below, which lists the number of cable television subscribers in the United States in selected years. *Section 2.4*

Year	1982	1983	1985	1986	1988	1990	1991	1993
Subscribers (millions)	29.3	34.1	39.8	42.2	48.6	54.9	55.8	58.8

36. Make a scatter plot of the data and draw a line of fit.

37. Find an equation of the line. Let $x =$ the number of years since 1982.

38. What does the slope of the line from Exercise 37 represent?

39. Use the equation from Exercise 37 to predict the number of cable television subscribers in the year 2000.

For Exercise 40, use the table below, which lists the 1993 assets and incomes of 8 major life insurance companies in the United States. *Section 2.5*

Assets (billions)	67.5	53.6	51.5	48.4	47.3	44.1	43.7	40.0
Income (billions)	3.2	8.9	6.5	5.3	3.7	5.3	8.4	9.9

40. **Technology** Use a graphing calculator or statistical software to make a scatter plot of the data.

a. Writing Is the correlation between a life insurance company's assets and its income strong or weak? Explain.

b. Find the correlation coefficient for the data.

20.

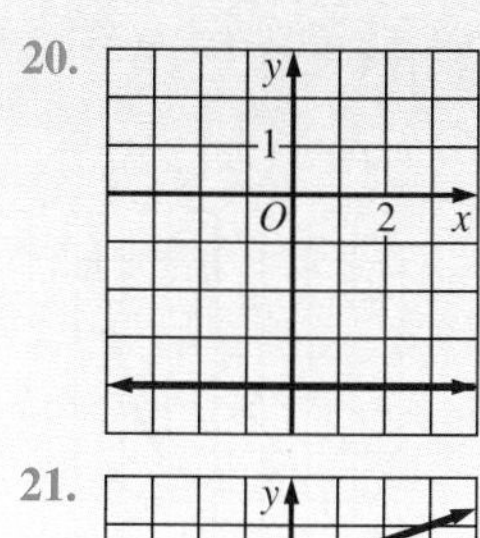

21.

22.

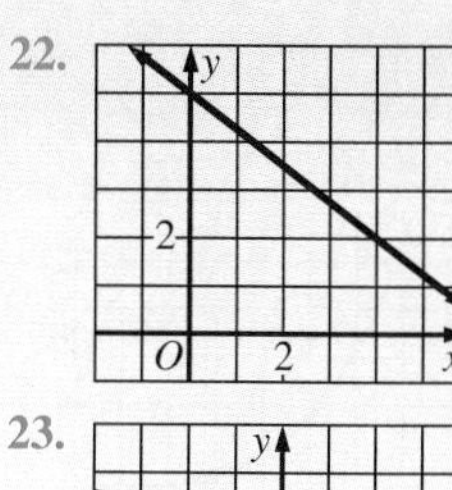

23.

24. a. $y = 1 - 2(x - 4)$

b.

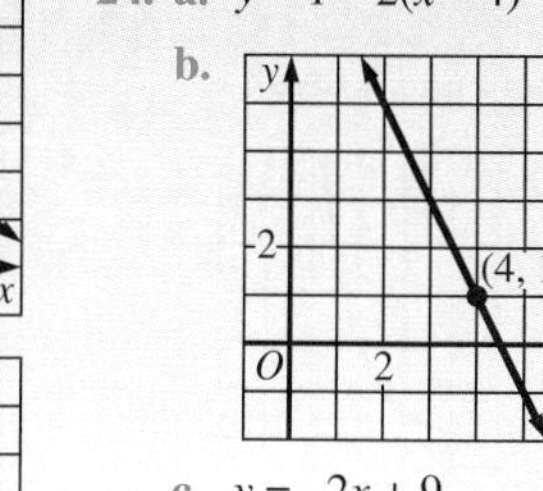

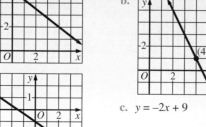

c. $y = -2x + 9$

25. a. $y = 2 + \frac{1}{3}(x + 3)$

b.

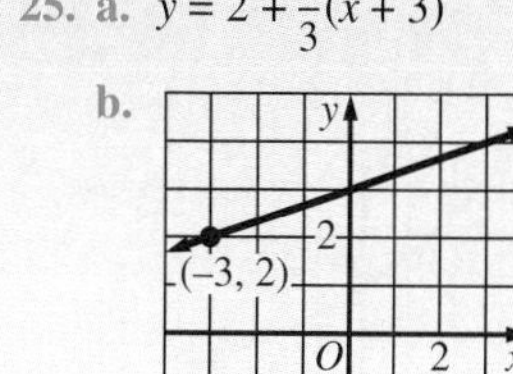

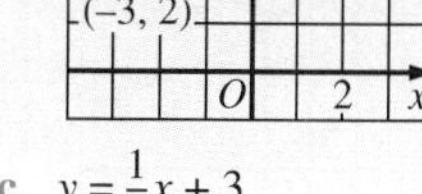

c. $y = \frac{1}{3}x + 3$

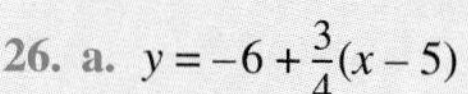

26. a. $y = -6 + \frac{3}{4}(x - 5)$

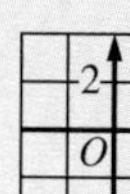

b.

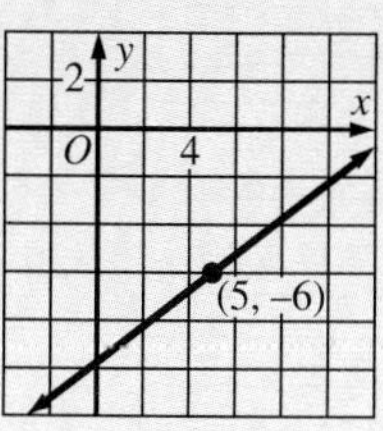

c. $y = \frac{3}{4}x - \frac{39}{4}$

27. a. $y = -2 + 2(x - 6)$

b.

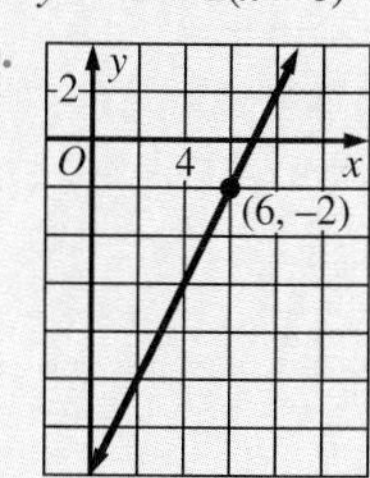

c. $y = 2x - 14$

28. a. $y = 5 - 3(x + 1)$

b.

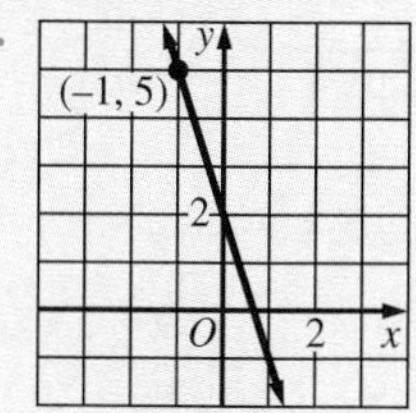

c. $y = -3x + 2$

29. a. $y = 2 + \frac{2}{3}(x - 9)$

b.

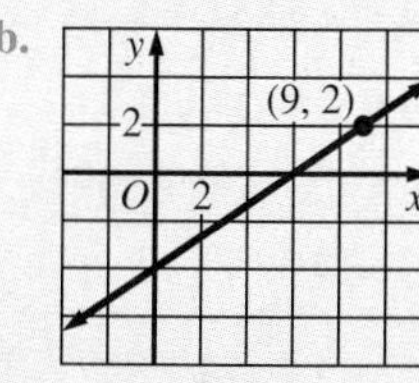

c. $y = \frac{2}{3}x - 4$

30. $y = 4 + 3(x - 2)$ or $y = 7 + 3(x - 3)$

31. $y = 5 - \frac{2}{5}(x + 1)$ or $y = 3 - \frac{2}{5}(x - 4)$

32. $y = -2 - 3(x - 7)$ or $y = 4 - 3(x - 5)$

33. $y = 6 + 2(x - 1)$ or
$y = -2 + 2(x + 3)$

34. $y = -5 - \frac{4}{5}(x - 9)$ or
$y = 3 - \frac{4}{5}(x + 1)$

35. $y = -4 - \frac{9}{5}(x + 1)$ or
$y = 5 - \frac{9}{5}(x + 6)$

36–39. See answers in back of book.

40.

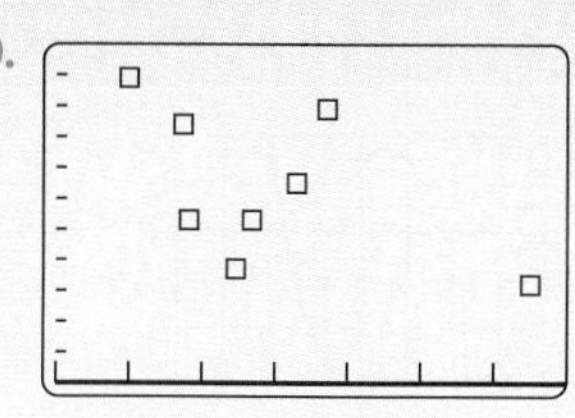

a. Answers may vary. An example is given. weak; The points appear to be scattered widely instead of lying near any line.

b. about -0.54

41. about zero 42. positive

43.

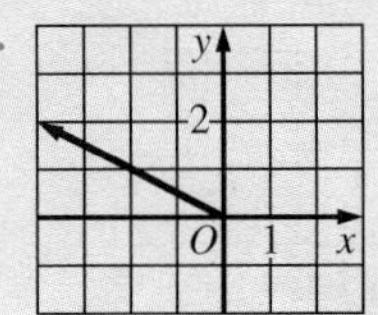

44.

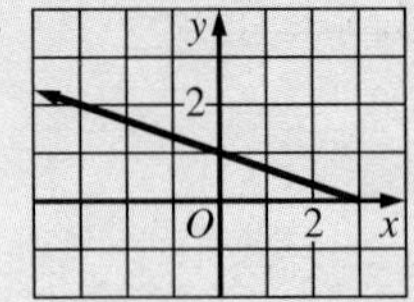

45.

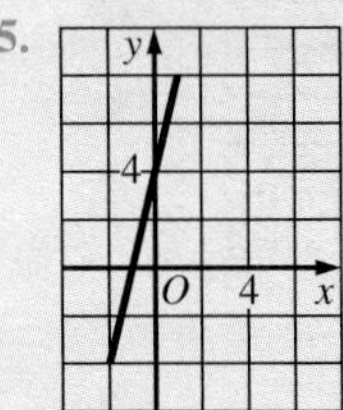

46.

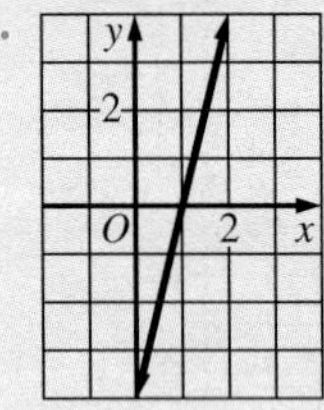

Chapter 3

1. 2^5 2. 2^7
3. 2^8 4. 2^{10}
5. 640 6. 5
7. 8 8. 25
9. exponential 10. neither
11. linear 12. exponential
13. $\frac{1}{64}$ 14. 1
15. $\frac{1}{243}$ 16. $\frac{1}{1296}$
17. 5 18. 27
19. 64 20. 36
21. $\frac{1}{5}$ 22. 4
23. 7 24. $\frac{1}{3}$

25. a. 7
b. exponential growth
c.

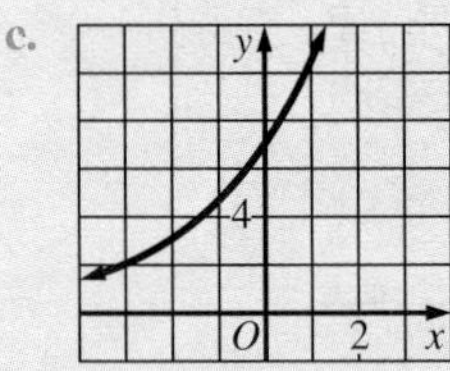

26. a. 1
b. exponential growth
c.

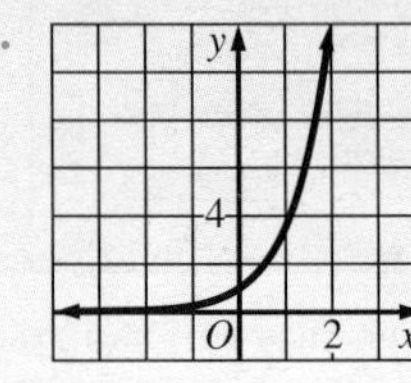

Tell whether you would expect the correlation between the two quantities to be *positive, negative,* or *about zero.* *Section 2.5*

41. a student's travel time to and from school and his or her average grade in Spanish class

42. number of hours of daylight in a certain city in a given month and the average high temperature in that city for that month

Graph each pair of parametric equations for the given restriction on *t*. *Section 2.6*

43. $x = 2t$, $y = -t$, $t \leq 0$

44. $x = -3t$, $y = t + 1$, $t \geq -1$

45. $x = -2 + t$, $y = -4 + 4t$, $0 \leq t \leq 3$

46. $x = 1 + \frac{1}{2}t$, $y = 2t$, no restriction on t

CHAPTER 3

Write each expression as a power of 2. *Section 3.1*

1. $2 \cdot 2 \cdot 2 \cdot 2 \cdot 2$ **2.** $8 \cdot 16$ **3.** $4 \cdot 2^6$ **4.** $2^3 \cdot 2^7$

Evaluate each expression when $x = 6$. *Section 3.1*

5. $10(2^x)$ **6.** $320\left(\frac{1}{2}\right)^x$ **7.** $0.125(2^x)$ **8.** $1600\left(\frac{1}{2}\right)^x$

Tell whether each equation represents growth that is *linear, exponential,* or *neither.* *Section 3.1*

9. $y = 5(2^x)$ **10.** $y = 5x^2$ **11.** $y = (5^2)x$ **12.** $y = 2(5^x)$

Simplify. *Section 3.2*

13. 4^{-3} **14.** 5^0 **15.** 3^{-5} **16.** 6^{-4}

Simplify using the properties of exponents. *Section 3.2*

17. $5^{1/3} \cdot 5^{2/3}$ **18.** $3^{5/2} \cdot 3^{1/2}$ **19.** $4^{7/2} \cdot 4^{-1/2}$ **20.** $6^{13/4} \cdot 6^{-5/4}$

21. $\frac{25^{1/4}}{25^{3/4}}$ **22.** $\left(16^{5/2}\right)^{1/5}$ **23.** $\frac{49^{5/2}}{49^2}$ **24.** $\left(27^{4/3}\right)^{-1/4}$

For each function:

a. Find the *y*-intercept of the graph.

b. Tell whether the graph represents *exponential growth* or *exponential decay.*

c. Sketch the graph. *Section 3.3*

25. $y = 7(1.5)^x$ **26.** $y = (3.5)^x$ **27.** $y = 6(0.9)^x$

28. $y = 4^x$ **29.** $y = 2(0.75)^x$ **30.** $y = 0.5(1.8)^x$

Sketch the graph of $y = 5 \cdot 2^x$ and its reflection in each axis, and give an equation for each reflection. *Section 3.3*

31. the y-axis **32.** the x-axis

27. a. 6
b. exponential decay
c.

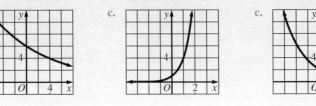

28. a. 1
b. exponential growth
c.

29. a. 2
b. exponential decay
c.

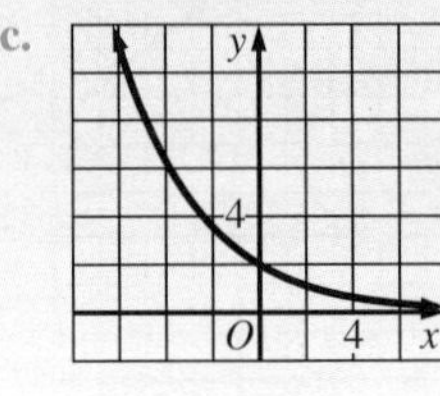

30. a. 0.5
b. exponential growth
c.

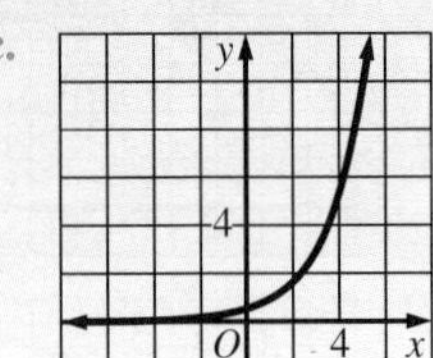

Suppose a bank offers interest compounded continuously. Use the formula $A = Pe^{rt}$ to find the value of \$1000 after 10 years at each interest rate. *Section 3.4*

33. 5% **34.** 7.5% **35.** 4.25% **36.** 9.75%

37. The growth of railways in the United States from 1830 to 1920 can be modeled by the logistic function

$$M(t) = \frac{303.8}{1 + e^{-0.0687(t - 1891.85)}}$$

where $M(t)$ = railway mileage in thousands and t = year. *Section 3.4*

a. **Technology** Use a graphing calculator or graphing software to graph the function.

b. Evaluate the function when $t = 1900$.

c. Find the year when railway mileage reached 162 thousand miles.

Write an exponential function that passes through each pair of points. *Section 3.5*

38. (0, 6), (1, 8) **39.** (0, 4), (2, 9) **40.** (0, 50), (2, 8) **41.** (2, 10), (3, 25)

For Exercises 42–45, use the table, which shows the population growth for the city of Austin, Texas, from 1960 to 1990. *Section 3.5*

Year	1960	1970	1980	1990
Population	186,545	253,539	345,496	465,648

42. Use the population values for 1960 and 1970 to write an exponential function to represent the growth. Let x = number of years since 1960.

43. **Technology** Use a graphing calculator or statistical software to perform an exponential regression. What function do you get?

44. Compare your exponential functions from Exercises 42 and 43. Which function gives the closest values to the data for 1980 and 1990?

45. Use the model you chose in Exercise 44 to predict the population of Austin in the year 2000.

CHAPTER 4

For each function:

a. Graph the function and its inverse in the same coordinate plane.

b. Find an equation for the inverse. *Section 4.1*

1. $f(x) = -4x$ **2.** $g(x) = -2x + 3$ **3.** $y = -5x + 7$ **4.** $h(x) = 3x - 1$

5. $y = -\frac{5}{4}x$ **6.** $y = \frac{2}{5}x$ **7.** $f(x) = -\frac{3}{2}x - 6$ **8.** $y = \frac{1}{4}x + 7$

Write each equation in logarithmic form. *Section 4.2*

9. $5^3 = 125$ **10.** $49^{1/2} = 7$ **11.** $\left(\frac{2}{3}\right)^2 = \frac{4}{9}$ **12.** $\left(\frac{1}{3}\right)^{-1} = 3$

31.
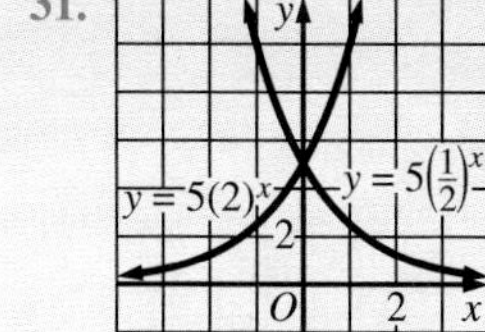

32.
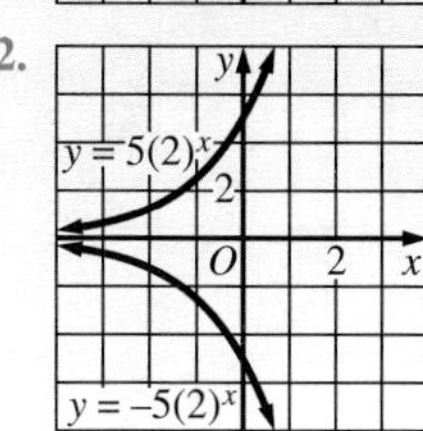

33. \$1648.72 **34.** \$2117.00

35. \$1529.59 **36.** \$2651.17

37. a.
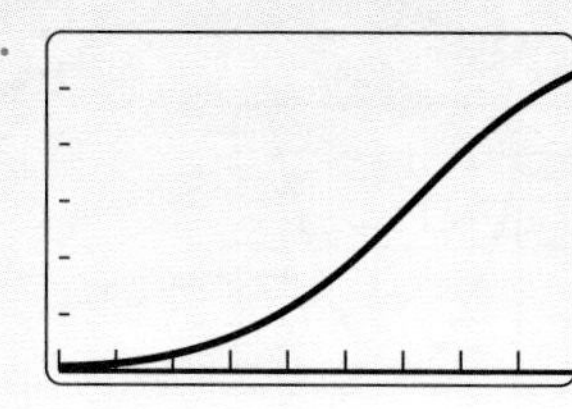

b. about 193,000 mi

c. in 1893, shortly before the turn of 1894

38. $y = 6\left(\frac{4}{3}\right)^x$ **39.** $y = 4\left(\frac{3}{2}\right)^x$

40. $y = 50\left(\frac{2}{5}\right)^x$ **41.** $y = 1.6\left(\frac{5}{2}\right)^x$

42. $y = 186{,}545(1.03116)^x$

43. $y = 186{,}810(1.03101)^x$

44. Answers may vary. The function from Ex. 42 gives the closest value for 1980; the function from Ex. 43 gives the closest value for 1990.

45. Predictions should range from 633,746 to 636,540.

Chapter 4

1. a.
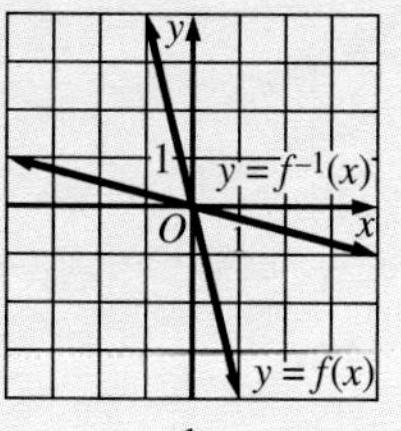

b. $f^{-1}(x) = -\frac{1}{4}x$

2. a.
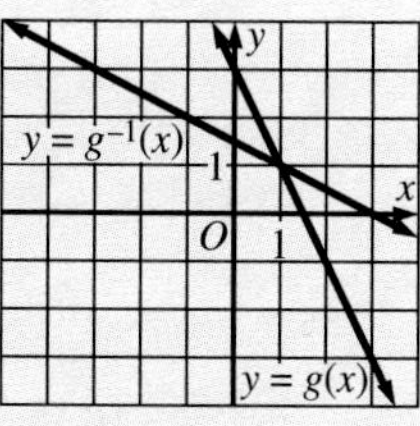

b. $g^{-1}(x) = -\frac{1}{2}x + \frac{3}{2}$

3. a.
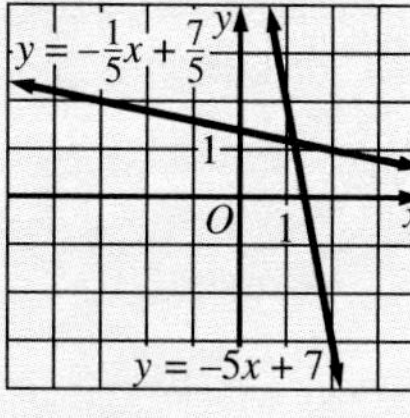

b. $y = -\frac{1}{5}x + \frac{7}{5}$

4. a.
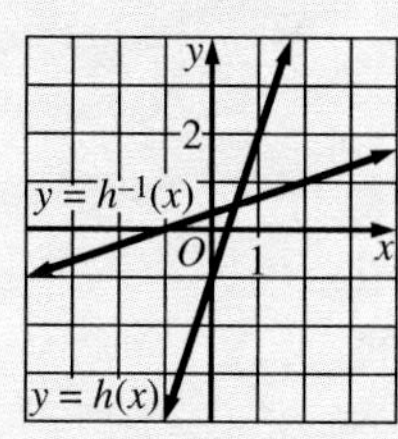

b. $h^{-1}(x) = \frac{1}{3}x + \frac{1}{3}$

5. a.
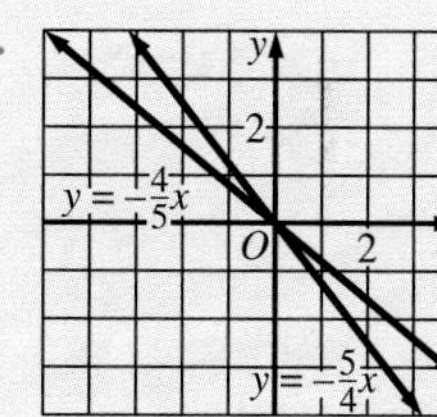

b. $y = -\frac{4}{5}x$

6. a.
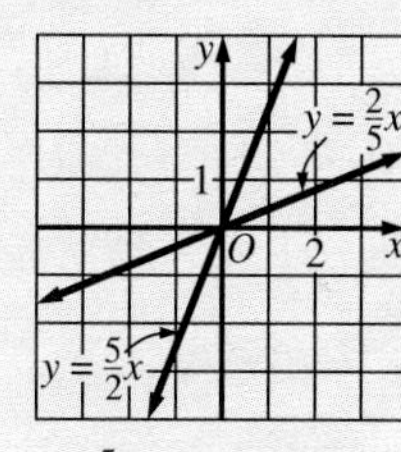

b. $y = \frac{5}{2}x$

7. a.
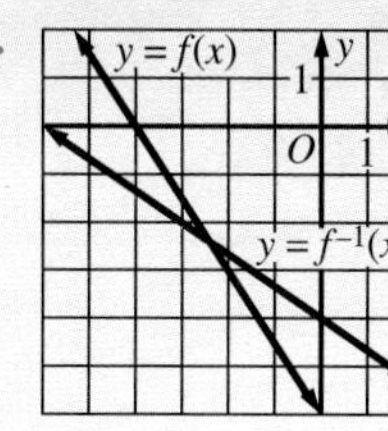

b. $f^{-1}(x) = -\frac{2}{3}x - 4$

8–12. See answers in back of book.

13. $\left(\frac{1}{2}\right)^{-3} = 8$

14. $64^{1/6} = 2$

15. $(0.36)^{1/2} = 0.6$

16. $\left(\frac{3}{2}\right)^3 = \frac{27}{8}$

17. 7 18. –4

19. 0 20. $\frac{1}{2}$

21. $3 \log_5 x - 2 \log_5 y$

22. $7(\log_5 x + \log_5 y)$

23. $\frac{1}{2} \log_5 x + \log_5 y$

24. $-4 \log_5 x - 7 \log_5 y$

25. $\log_2 \frac{x}{y}$ 26. $\log_a 1000$

27. $\log_b 45$ 28. $\log_8 p^4q^3$

29. $\ln \frac{1}{2}$

30. $\log \frac{1}{x}$, or $-\log x$

31. 0.36 32. 3.87

33. 0.86 34. 0.49

35. 1.64 36. 7.39

37. 0.28 38. 0.56

39. 1.89 40. 1.39

41. 2.07 42. –0.26

43. –0.16 44. –17.25

45. –0.79 46. 2.22

47. $y = 251(63.1)^x$

48. $y = 0.0316(2.00)^x$

49. $y = 40.4(0.549)^x$

50. $y = 2.66(66.7)^x$

51. $y = 0.00398x^{0.76}$

52. $y = 50.1x^{-0.2}$

53. $y = 2.5x^{5.8}$

54. $y = 36.6x^{6.2}$

55. $y = 337x^{-1.09}$

56. $y = 0.010x^2$

Chapter 5

1. C
2. B
3. A
4. $\pm 2\sqrt{7} \approx \pm 5.3$
5. ± 3
6. ± 15
7. $\pm 5\sqrt{2} \approx \pm 7.1$
8. $\pm \frac{1}{3}$
9. $\pm \sqrt{15} \approx \pm 3.9$
10. $\pm 2\sqrt{3} \approx \pm 3.5$
11. $\pm \sqrt{30} \approx \pm 5.5$
12. 3
13. –4.5
14. $\frac{2}{3}$
15. minimum value; –2
16. maximum value; 4
17. minimum value; –5
18. maximum value; 13

Write each equation in exponential form. *Section 4.2*

13. $\log_{1/2} 8 = -3$ **14.** $\log_{64} 2 = \frac{1}{6}$ **15.** $\log_{0.36} 0.6 = \frac{1}{2}$ **16.** $\log_{3/2} \frac{27}{8} = 3$

Evaluate each logarithm. *Section 4.2*

17. $\log_2 128$ **18.** $\log_{1/3} 81$ **19.** $\ln 1$ **20.** $\log \sqrt{10}$

Write each expression in terms of $\log_5 x$ and $\log_5 y$. *Section 4.3*

21. $\log_5 \frac{x^3}{y^2}$ **22.** $\log_5 (xy)^7$ **23.** $\log_5 y\sqrt{x}$ **24.** $\log_5 \frac{1}{x^4y^7}$

Write as a logarithm of a single number or expression. *Section 4.3*

25. $\log_2 x - \log_2 y$ **26.** $3 \log_a 10$ **27.** $2 \log_b 3 + \frac{1}{2} \log_b 25$

28. $4 \log_8 p + 3 \log_8 q$ **29.** $\frac{2}{3} \ln 8 - 3 \ln 2$ **30.** $6 \log x^2 - 2 \log x^5 - \log x^3$

Solve each equation. Round each answer to the nearest hundredth. *Section 4.4*

31. $2e^{5x} = 12$ **32.** $e^{0.7x} = 15$ **33.** $10^{x+1} = 73$ **34.** $10^{4x} = 88$

35. $6(2.5)^x = 27$ **36.** $\left(\frac{3}{2}\right)^x = 20$ **37.** $5e^{2x} - 7 = e^{2x}$ **38.** $12 - 3(10^x) = 1$

Evaluate each logarithm. Round each answer to the nearest hundredth. *Section 4.4*

39. $\log_3 8$ **40.** $\log_7 15$ **41.** $\log_5 28$ **42.** $\log_{12} 0.52$

43. $\log_6 \frac{3}{4}$ **44.** $\log_{0.8} 47$ **45.** $\log_{1/4} 3$ **46.** $\log_{5/6} \frac{2}{3}$

Write y as an exponential function of x. *Section 4.5*

47. $\log y = 1.8x + 2.4$ **48.** $\log y = 0.3x - 1.5$ **49.** $\ln y = -0.6x + 3.7$ **50.** $\ln y = 4.2x + 0.98$

Write y as a power function of x. *Section 4.5*

51. $\log y = 0.76 \log x - 2.4$ **52.** $\log y = 1.7 - 0.2 \log x$ **53.** $\log y = 5.8 \log x + 0.4$

54. $\ln y = 6.2 \ln x + 3.6$ **55.** $\ln y = -1.09 \ln x + 5.82$ **56.** $\ln y = -4.6 + 2 \ln x$

CHAPTER 5

Match each graph with its equation. *Section 5.1*

1.

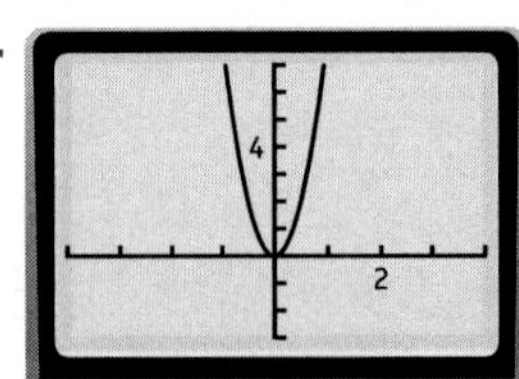

2.

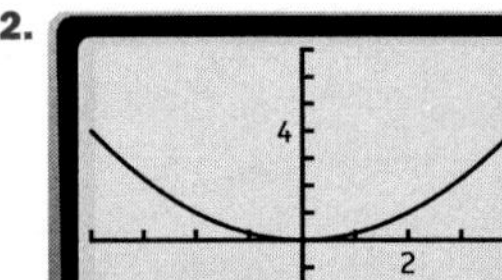

3.

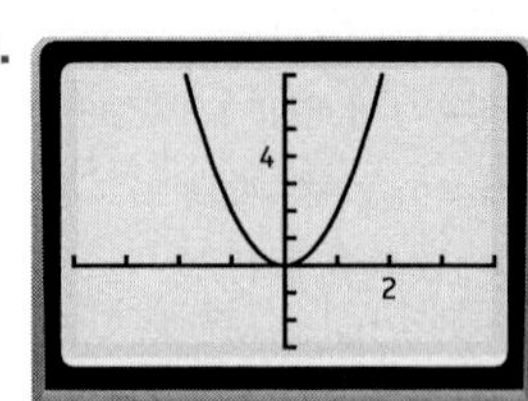

A. $y = 2x^2$ **B.** $y = \frac{1}{4}x^2$ **C.** $y = 8x^2$

19. minimum value; –10

20. maximum value; 7

21. a. $x = -3$
b. $(-3, 0)$
c. opens down
d.

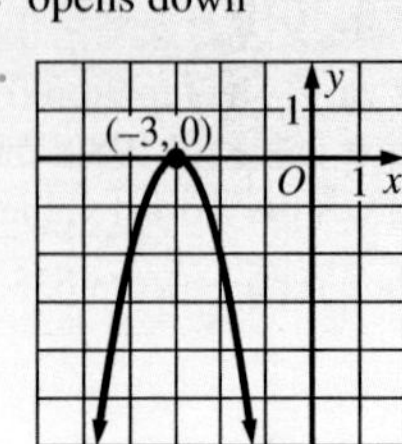

22. a. $x = 5$
b. $(5, 1)$
c. opens up
d.

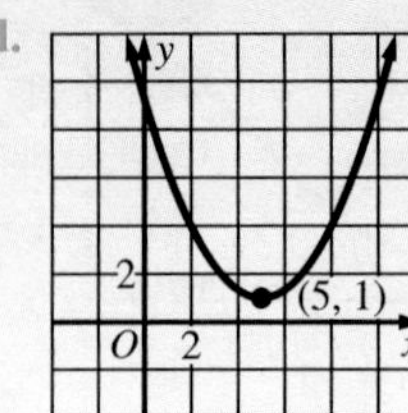

23. a. $x = -1$
b. $(-1, 3)$
c. opens down
d.

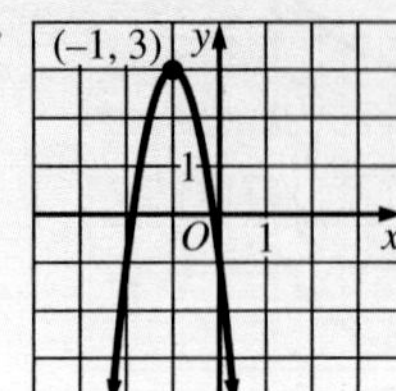

Solve each equation. Give solutions to the nearest tenth when necessary. *Section 5.1*

4. $x^2 = 28$ **5.** $5x^2 = 45$ **6.** $-\frac{1}{3}x^2 = -75$ **7.** $\frac{2}{5}x^2 = 20$

8. $-27x^2 = -3$ **9.** $3x^2 - 45 = 0$ **10.** $-6x^2 = -72$ **11.** $x^2 - 10 = 20$

Each graph has an equation of the form $y = ax^2$. Find a. *Section 5.1*

12.

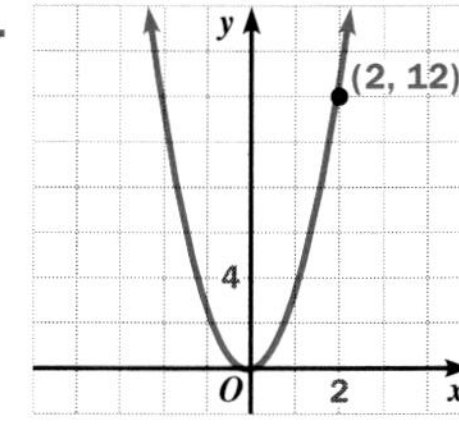

13.

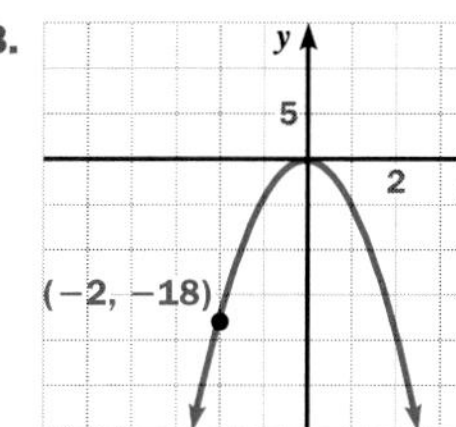

14.

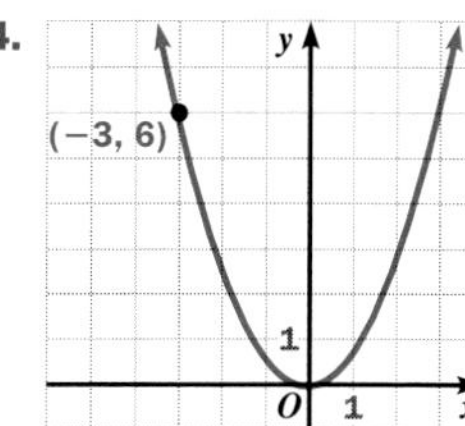

State whether each function has a *maximum value* or a *minimum value*. Then find that value. *Section 5.2*

15. $f(x) = 3(x - 5)^2 - 2$ **16.** $f(x) = -1.7(x + 11)^2 + 4$ **17.** $f(x) = 9(x - 3)^2 - 5$

18. $f(x) = -0.6(x - 8)^2 + 13$ **19.** $f(x) = 5(x + 9)^2 - 10$ **20.** $f(x) = -0.8(x - 4)^2 + 7$

For each function:

a. Find an equation of the line of symmetry for the graph of the function.

b. Find the coordinates of the vertex.

c. Tell whether the graph *opens up* or *opens down*.

d. Sketch the graph. *Section 5.2*

21. $y = -2(x + 3)^2$ **22.** $y = \frac{1}{3}(x - 5)^2 + 1$ **23.** $y = -4(x + 1)^2 + 3$

24. $y = \frac{3}{2}(x - 6)^2 - 5$ **25.** $y = -\frac{2}{3}(x + 4)^2 - 2$ **26.** $y = 3(x + 5)^2 - \frac{5}{2}$

Write an equation in the form $y = a(x - h)^2 + k$ for each parabola shown. *Section 5.2*

27.

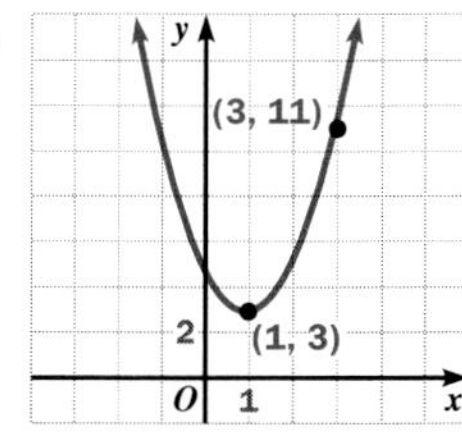

28.

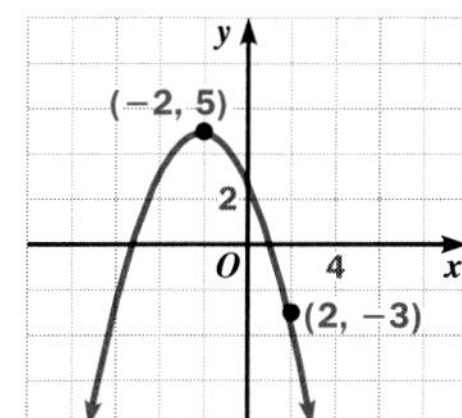

29.

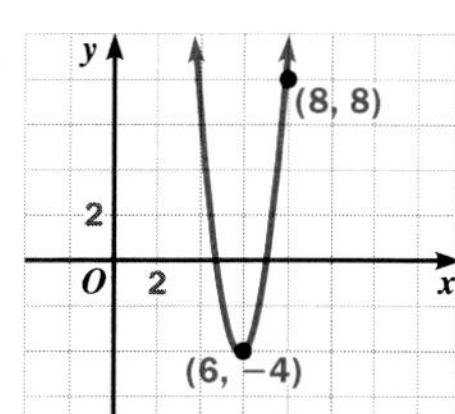

For each equation:

a. Find the x-intercept(s).

b. Find the vertex.

c. Sketch the graph. *Section 5.3*

30. $y = (x - 1)(x + 3)$ **31.** $y = -2(x - 2)(x - 5)$ **32.** $y = \frac{1}{3}(x + 8)(x - 4)$

33. $y = -1.5(x + 3)(x + 7)$ **34.** $y = (2x - 6)(x - 2)$ **35.** $y = (12 - 3x)(x - 2)$

24. a. $x = 6$
b. $(6, -5)$
c. opens up
d.

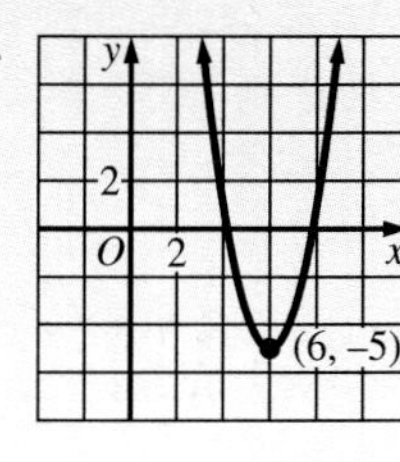

25. a. $x = -4$
b. $(-4, -2)$
c. opens down
d.

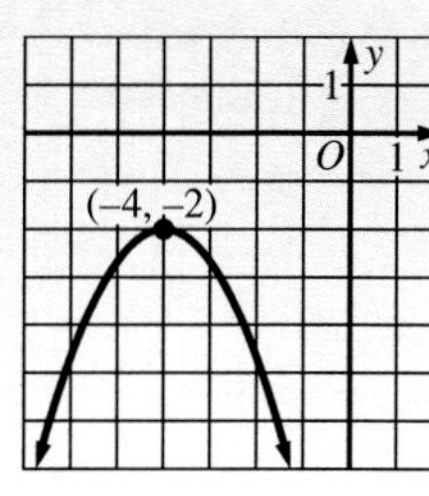

26. a. $x = -5$
b. $\left(-5, -\frac{5}{2}\right)$
c. opens up
d.

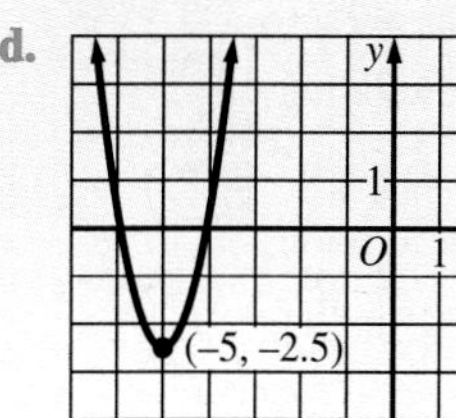

27. $y = 2(x - 1)^2 + 3$

28. $y = -\frac{1}{2}(x + 2)^2 + 5$

29. $y = 3(x - 6)^2 - 4$

30. a. 1, –3
b. (–1, –4)
c.

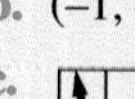

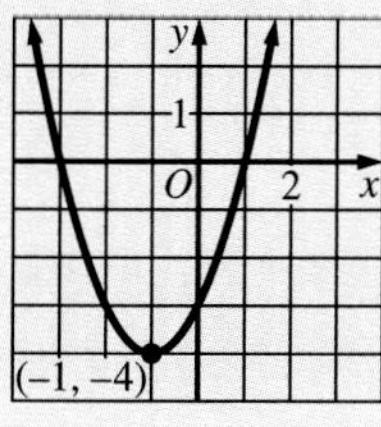

31. a. 2, 5
b. (3.5, 4.5)
c.

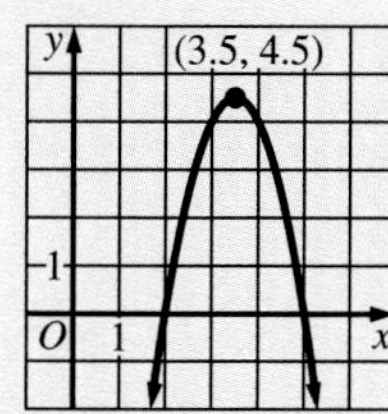

32. a. –8, 4
b. (–2, –12)
c.

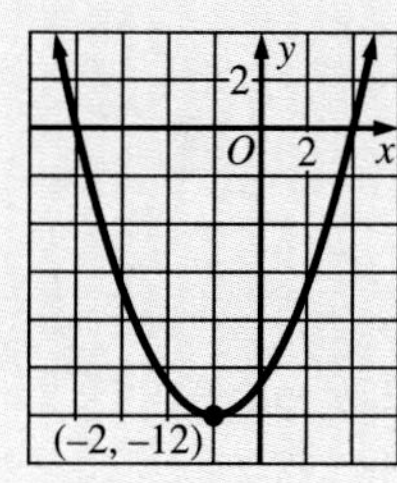

33. a. –3, –7
b. (–5, 6)
c.

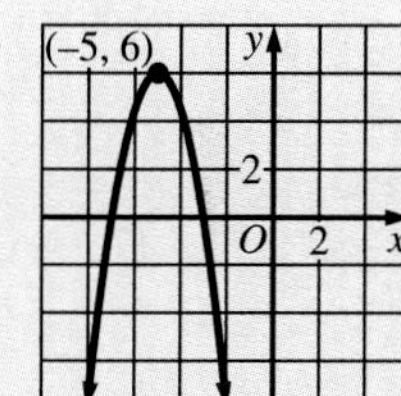

34. a. 3, 2
b. $\left(\frac{5}{2}, -\frac{1}{2}\right)$
c.

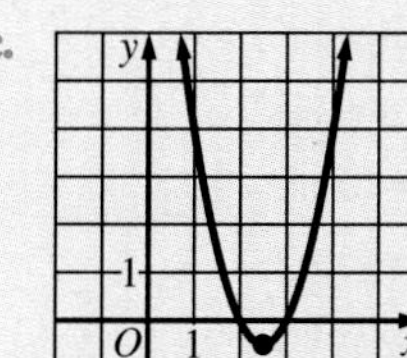

35. a. 2, 4
b. (3, 3)
c.

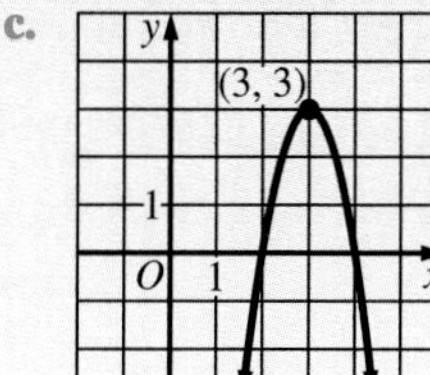

Write an equation for each graph. *Section 5.3*

36.

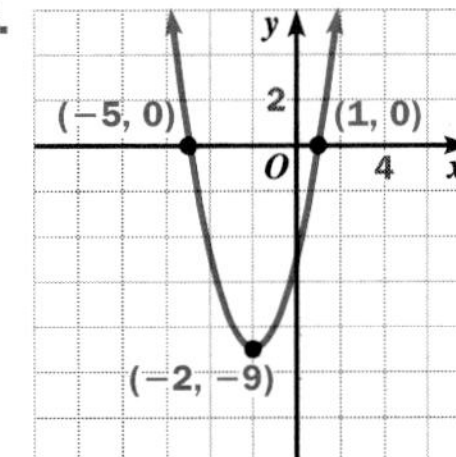

37.

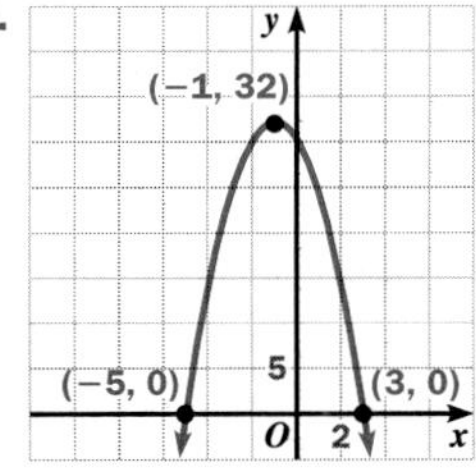

38.

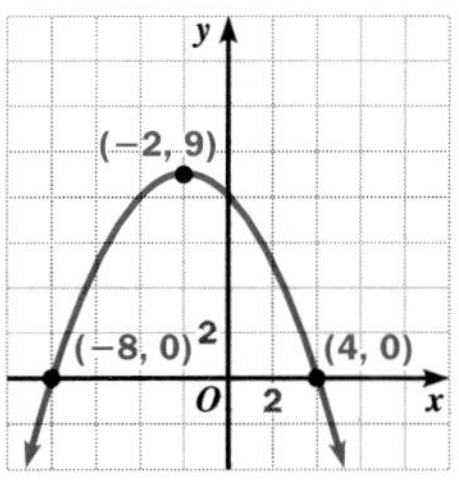

Write each function in vertex form. *Section 5.4*

39. $y = x^2 - 6x + 8$ **40.** $y = x^2 + 10x + 31$ **41.** $y = x^2 - 8x$

42. $y = -x^2 + 3x - 1$ **43.** $y = 2x^2 - 8x + 5$ **44.** $y = 4x^2 + 6x - 7$

State whether each function has a *maximum value* or a *minimum value*. Then find that value. *Section 5.4*

45. $y = 8x^2 - 4x + 3$ **46.** $y = -5x^2 - 10x + 2$ **47.** $y = -x^2 + 6x - 10$

48. $y = 2x^2 + 14x + 5$ **49.** $y = \frac{3}{4}x^2 - 6x$ **50.** $y = -2x^2 - 8x - 1$

Find the solution(s) of each quadratic equation. Use the quadratic formula. *Section 5.5*

51. $x^2 - 4x - 12 = 0$ **52.** $2x^2 + 3x - 2 = 0$ **53.** $5x^2 - 9x + 2 = 0$

54. $3x^2 - 7x + 2 = 0$ **55.** $2x^2 + x - 5 = 0$ **56.** $4x^2 + 12x + 9 = 0$

57. $-x^2 + 4x - 1 = 0$ **58.** $6x^2 - x - 5 = 0$ **59.** $x^2 + 12x + 8 = 0$

Tell whether each equation has *two solutions, one solution,* or *no solution.* *Section 5.5*

60. $x^2 - 3x + 5 = 0$ **61.** $2x^2 - 7x - 4 = 0$ **62.** $x^2 + 6x + 3 = 0$

63. $4x^2 - 20x + 25 = 0$ **64.** $3x^2 - 2x + 5 = 0$ **65.** $\frac{1}{4}x^2 - 3x + 9 = 0$

Write each quadratic function as a product of factors. *Section 5.6*

66. $y = x^2 - 7x - 18$ **67.** $y = 2x^2 - 9x + 10$ **68.** $y = 3x^2 - 2x - 5$

69. $y = 25x^2 - 49$ **70.** $y = 4x^2 - 20x + 9$ **71.** $y = 5x^2 + 17x - 12$

72. $y = 2x^2 + 19x + 35$ **73.** $y = 3x^2 + 16x - 12$ **74.** $y = x^2 - 8x + 12$

Find the *x*-intercept(s) of the graph of each function. *Section 5.6*

75. $y = x^2 - 12x + 35$ **76.** $y = x^2 + 6x - 40$ **77.** $y = 2x^2 - 5x + 3$

78. $y = 0.3x^2 + 1.1x - 0.4$ **79.** $y = 8x - 10x^2$ **80.** $y = 5x^2 - 17x + 6$

Use factoring and the properties of logarithms to solve for *x*. Be sure to check your answers. *Section 5.6*

81. $\log x + \log (x - 3) = 1$ **82.** $\log_3 (x - 6) + \log_3 (x + 2) = 2$

83. $\log_4 (3x + 2) + \log_4 x = 2$ **84.** $\log_2 x + \log_2 (x - 4) = 5$

36. $y = (x - 1)(x + 5)$
37. $y = -2(x + 5)(x - 3)$
38. $y = -\frac{1}{4}(x + 8)(x - 4)$
39. $y = (x - 3)^2 - 1$
40. $y = (x + 5)^2 + 6$
41. $y = (x - 4)^2 - 16$
42. $y = -\left(x - \frac{3}{2}\right)^2 + \frac{5}{4}$
43. $y = 2(x - 2)^2 - 3$
44. $y = 4\left(x + \frac{3}{4}\right)^2 - \frac{37}{4}$
45. minimum value; 2.5
46. maximum value; 7
47. maximum value; –1
48. minimum value; –19.5
49. minimum value; –12
50. maximum value; 7
51. –2, 6
52. $-2, \frac{1}{2}$
53. $\frac{9 \pm \sqrt{41}}{10} \approx 0.26, 1.54$
54. $\frac{1}{3}, 2$
55. $\frac{-1 \pm \sqrt{41}}{4} \approx -1.85, 1.35$
56. $-\frac{3}{2}$
57. $2 \pm \sqrt{3} \approx 0.27, 3.73$
58. $-\frac{5}{6}, 1$
59. $-6 \pm 2\sqrt{7} \approx -11.29, -0.71$
60. no solution
61. two solutions
62. two solutions
63. one solution
64. no solution
65. one solution
66. $y = (x - 9)(x + 2)$
67. $y = (2x - 5)(x - 2)$
68. $y = (3x - 5)(x + 1)$
69. $y = (5x - 7)(5x + 7)$
70. $y = (2x - 9)(2x - 1)$
71. $y = (5x - 3)(x + 4)$
72. $y = (x + 7)(2x + 5)$
73. $y = (3x - 2)(x + 6)$
74. $y = (x - 6)(x - 2)$
75. 5, 7
76. –10, 4
77. $1, \frac{3}{2}$
78. $-4, \frac{1}{3}$
79. $0, \frac{4}{5}$
80. 0.4, 3
81. 5
82. 7
83. 2
84. 8

CHAPTER 6

Tell whether the data that can be gathered about each variable are *categorical* or *numerical*. Then describe the categories or numbers. *Section 6.1*

1. the batting average of a baseball player
2. the country of origin of a car
3. a commuter's mode of transportation
4. a person's annual salary

Each member of the junior class was asked to name his or her favorite subject. The graphs show the results of the survey in two different ways. *Section 6.1*

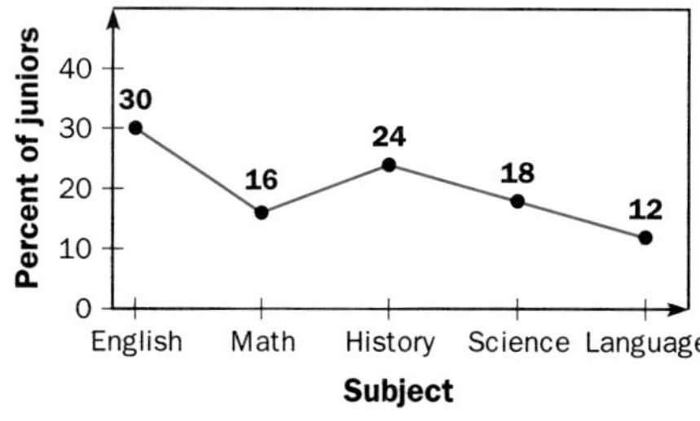

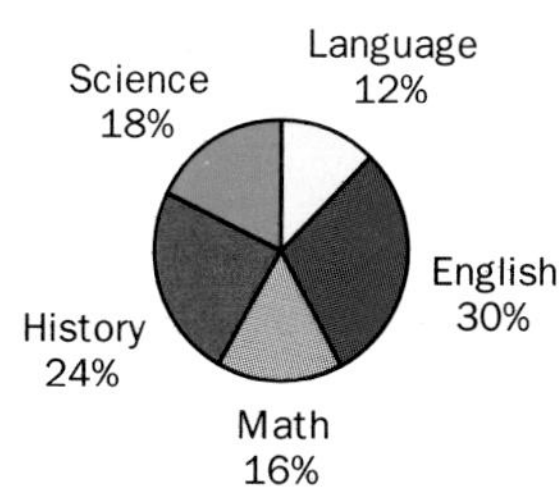

5. Tell which of the two graphs is a better way to display the data. Explain your reasoning.
6. **Open-ended** Display the data using another method. Explain why the graph you created is an appropriate way to organize the data.

For Exercises 7–10, tell why each question may be biased. Then describe what changes you would make to improve the question. *Section 6.2*

7. "Do you think this city should re-elect Mayor Fong, who is in favor of lower taxes and more public recreation facilities?"
8. "Are you in favor of repealing Proposition 17?"
9. "How many times a year do you visit the public library?"
10. "Do you favor building a dam that will bring low-cost electricity to the region?"

For Exercises 11–14, suppose a survey is to be conducted to find out the average number of hours students spend doing homework each night. Identify the type of sample and tell if the sample is biased. Explain your reasoning. *Section 6.3*

11. The survey will be given out to the three homerooms nearest the principal's office.
12. The names of 20 students will be chosen by a computer, which will select students by randomly generating 20 identification numbers.
13. Students will be asked to volunteer to participate in the survey. The first 30 students to volunteer will be selected.
14. Every fourth student on the school computer's list of all students will be selected to participate in the survey.

Chapter 6

1–4. Descriptions may vary. Examples are given.

1. numerical; A batting average can vary from 0.0 to 1.0, with values around 0.2 to 0.25 being common.
2. categorical; Countries might include the U.S., Germany, Japan, and so on.
3. categorical; Some of the modes are train, car, bus, bicycle, and taxi.
4. numerical; A person's annual salary is a positive number that can vary widely.
5. Answers may vary. An example is given. the circle graph; The circle graph allows you to see more easily how each choice is related to the results as a whole.
6. Answers may vary. An example is given. A bar graph is appropriate for categorical data such as these.

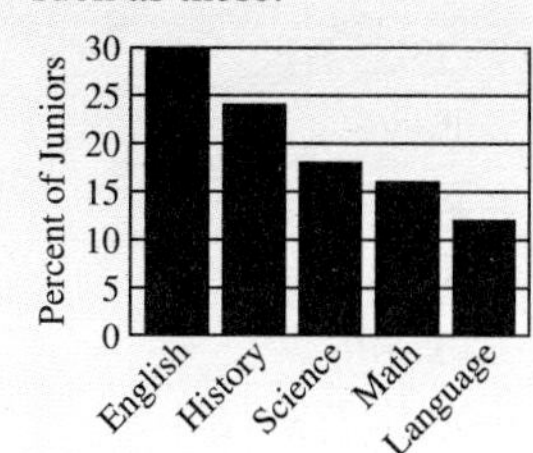

7–10. Answers may vary. Examples are given.

7. This is a leading question, since most people would like lower taxes and public recreation facilities. The question could be ended after "Do you think this city should re-elect Mayor Fong?"
8. Voters might not be familiar with Proposition 17, but might answer anyway to avoid the appearance of ignorance. The results of retaining or repealing Proposition 17 should be discussed objectively.
9. Respondents might inflate their answers to look good in the eyes of the questioner. The question would be best if it could be answered anonymously.
10. This is a leading question, since most people would like low-cost electricity. The issue should be stated as specifically as possible, detailing the location and various impacts of the dam.

11–14. Answers may vary. Examples are given.

11. convenience sample; biased; The students in a few specific homerooms might not represent students as a whole.
12. random sample; unbiased; A random sample is unbiased, though a stratified random sample might be preferred to ensure that students from all grades are well-represented.
13. self-selected sample; biased; Students who volunteer might spend more time doing homework than others.
14. systematic; unbiased; Since students' study habits should not depend on the placing of names in an alphabetical list, this technique should be about as effective as a simple random sample.

A magazine surveyed its subscribers to find out their annual incomes and collected the data listed below (in thousands of dollars). Use the data to answer Exercises 15–17. *Section 6.4*

25, 36, 18, 23.5, 19, 44, 34, 32, 28, 29.5, 24, 26, 29.5, 25, 33, 52, 56, 33, 30, 27

15. Find the median of the data and make a box plot.

16. Find the mean of the data and make a histogram.

17. Is the median or the mean a better measure of the center of the distribution? Explain.

Use the graphs below to answer Exercises 18–20. *Section 6.5*

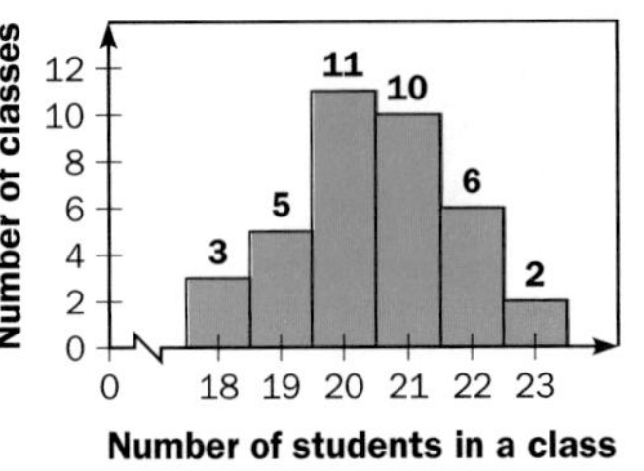

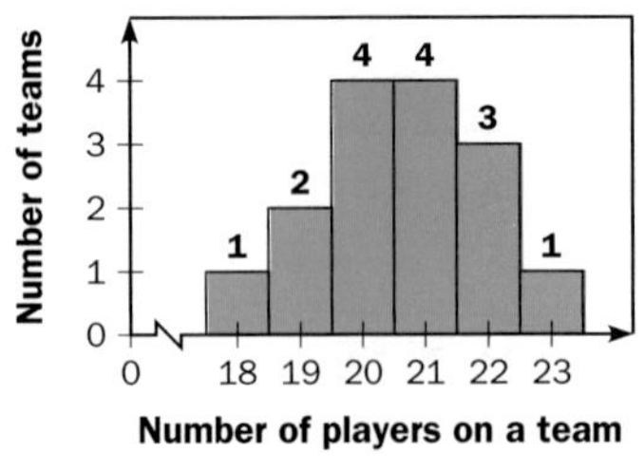

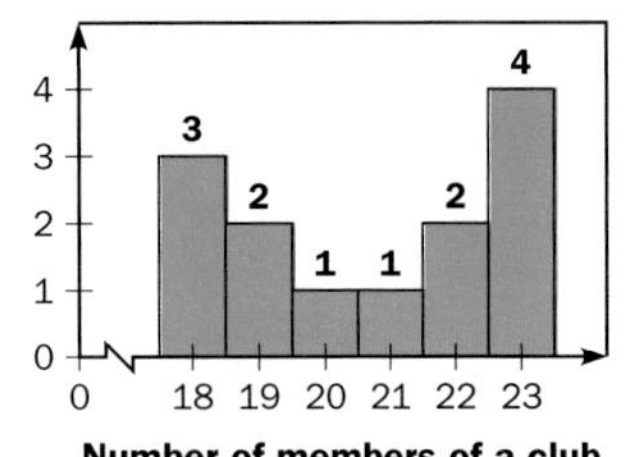

18. Find the mean of each data set. Compare the three means.

19. Find the standard deviation of each data set. Compare the standard deviations.

20. For each data set, what percent of the values fall within one standard deviation of the mean? within two standard deviations of the mean?

In a random telephone survey, only 18% of those questioned could name the current United States Secretary of State. *Section 6.6*

21. a. If 2540 people responded to the survey, what is the margin of error?

b. Find an interval that is likely to contain the proportion of the entire United States population that knows the name of the Secretary of State.

22. Suppose the margin of error of a similar survey is 1.5%, based on the sample size. What is the sample size in this case?

In a poll conducted by a potential advertiser, 324 out of 700 people questioned said they planned to watch the Super Bowl game. *Section 6.6*

23. What percent of the entire population of the United States would be likely to watch the Super Bowl game, based on the sample data?

24. What is the margin of error for this poll?

CHAPTER 7

Solve each system of equations. *Section 7.1*

1. $4x - y = 5$
$2x + y = 13$

2. $-2x + 5y = 9$
$x = y - 3$

3. $6x + 1.4y = 5$
$x + y = -3$

4. $x + 3y = 23$
$5x + 2y = -2$

15. median: $29,500

15 25 35 45 55
Income (thousands of dollars)

16. mean: about $31,200

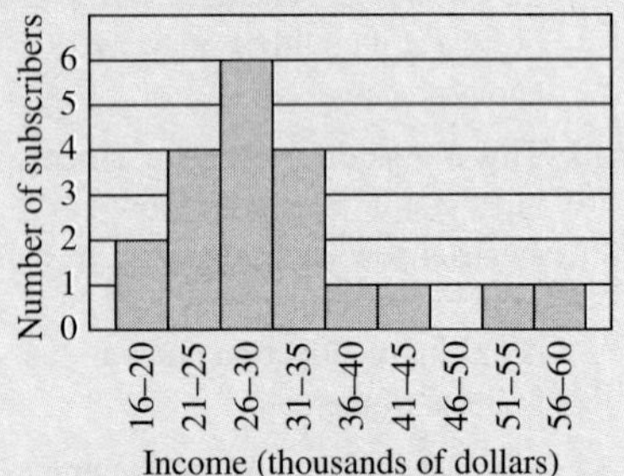

17. Answers may vary. An example is given. The median is a better measure of the center because the two highest incomes skew the mean toward the right, which makes it less representative of most of the incomes.

18. 1st set: about 20.5; 2nd set: about 20.6; 3rd set: about 20.7; The means are almost identical.

19. 1st set: about 1.3; 2nd set: about 1.3; 3rd set: about 2.0; The standard deviation of the 3rd set is much greater.

20. Answers may vary, depending upon how students round and choose columns from the histogram. 1st set: one standard deviation—about 57%; two standard deviations—100%
2nd set: one standard deviation—about 53%; two standard deviations—100%
3rd set: one standard deviation—about 46%; two standard deviations—100%

21. a. ±2%
b. 16–20%

22. about 4444

23. about 46.3%

24. about ±3.8% (42.5%–50.1%)

Chapter 7

1. (3, 7)
2. (–2, 1)
3. (2, –5)
4. (–4, 9)
5. about (–3.54, 7.54) and (2.54, 1.46)
6. about (2.19, 1.77) and (–3.19, –19.77)
7. about (1.65, –2.31) and (–6.65, 14.31)
8. about (1.33, 1.01) and (–4.29, 17.86)
9. about (–1.58, 1.67) and (1.48, 13.90)
10. about (2.55, –2.5) and (–2.55, –2.5)
11. (–1, 4); one solution
12. no solution
13. infinitely many solutions
14. (1, –3); one solution
15. (6, 5); one solution
16. no solution
17. $\begin{bmatrix} 11 & -8 \\ 15 & -14 \end{bmatrix}\begin{bmatrix} x \\ y \end{bmatrix} = \begin{bmatrix} 48 \\ 33 \end{bmatrix}$; (12, 10.5)
18. $\begin{bmatrix} 1.3 & 2.7 \\ 2.5 & -3.5 \end{bmatrix}\begin{bmatrix} x \\ y \end{bmatrix} = \begin{bmatrix} 18 \\ 17 \end{bmatrix}$; about (9.64, 2.03)

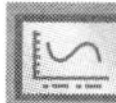

Technology **Use a graphing calculator or graphing software to solve each system of equations.** *Section 7.1*

5. $y = x^2 - 5$
$y = 4 - x$

6. $y = 3x - x^2$
$y = 4x - 7$

7. $y = x^2 + 3x - 10$
$y = 1 - 2x$

8. $y = 2(0.6)^x$
$y = -3x + 5$

9. $y = 5(2)^x$
$y = 4x + 8$

10. $y = 4 - x^2$
$y = x^2 - 9$

Solve each system of equations. State whether the system has *one solution, infinitely many solutions,* or *no solution.* *Section 7.2*

11. $2x + 3y = 10$
$3x + 2y = 5$

12. $5x - 25y = 8$
$4x - 20y = 8$

13. $4x - 2y = 8$
$6x - 3y = 12$

14. $3x - 11y = 36$
$11x - 3y = 20$

15. $2.5x + 3.2y = 31$
$1.5x - 2.6y = -4$

16. $6x + 15y = 28$
$7x + 17.5y = 24$

Rewrite each system in matrix form, and solve for *x* and *y*. *Section 7.3*

17. $11x - 8y = 48$
$15x - 14y = 33$

18. $1.3x + 2.7y = 18$
$2.5x - 3.5y = 17$

19. $5x + 13y = 40.8$
$9x - 4y = 2.2$

20. $7x + 15y = 103$
$21x - 9y = 65$

21. $9x + y = -13.5$
$1.6x + y = 0$

22. $-3x + y = 7.4$
$-6.2x + y = -19$

Use matrices to solve each system for *x*, *y*, and *z*. *Section 7.3*

23. $3x - 2y + 5z = 11$
$7x + 8y - 4z = 10$
$-5x + y + 2z = 7$

24. $y - 6z = 5$
$8x + 3y = 21$
$5x - y + 2z = 18$

25. $2.4x + 1.8y - 3.5z = 6.8$
$0.6x - 2.7y - 4.8z = 3.1$
$7.3x + 5.6y + 3.4z = 9.2$

Graph each inequality. *Section 7.4*

26. $y < -2x + 7$

27. $y \geq x - 4$

28. $y \leq 5$

29. $y < \frac{1}{2}x - 3$

30. $y \geq 1.4x - 2.6$

31. $y > -\frac{4}{3}x + 3$

32. $3x + 2y \leq 6$

33. $2x - 5y < 8$

Find an inequality that defines each shaded region. *Section 7.4*

34.

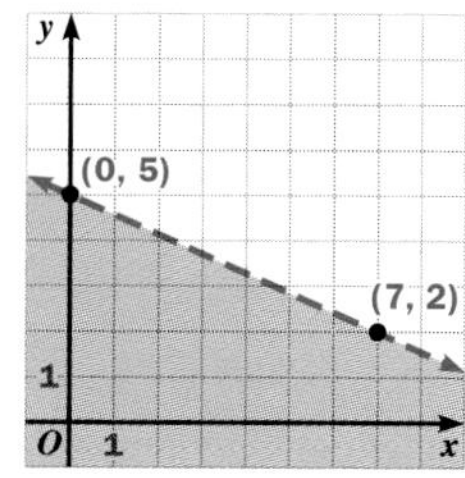

35.

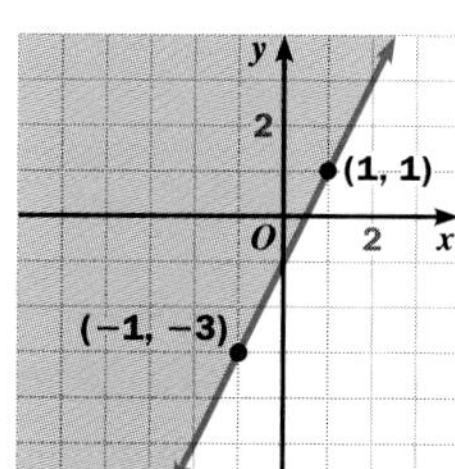

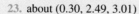

36.

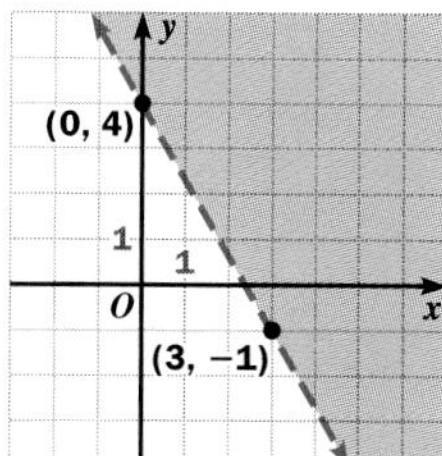

Graph each system of inequalities. *Section 7.5*

37. $y < 3x - 8$
$x < 2$

38. $y \geq -\frac{5}{3}x + 4$
$y < 2x - 7$

39. $y > \frac{1}{2}x + 1$
$y < 5$

40. $y \leq -2x + 9$
$y > \frac{3}{4}x - 2$

29.

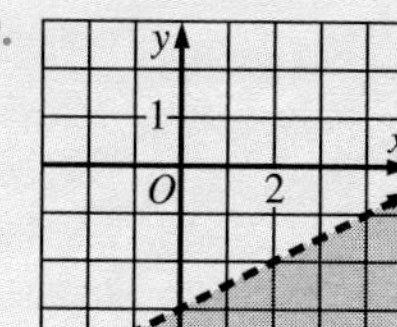

30.

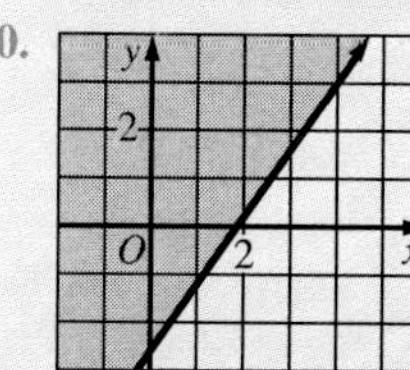

31.

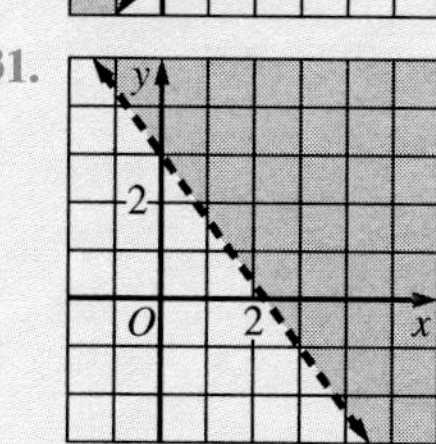

32.

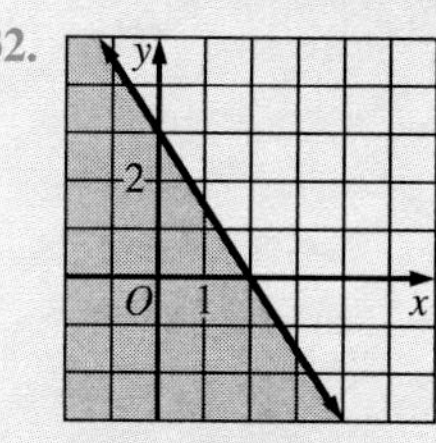

33.

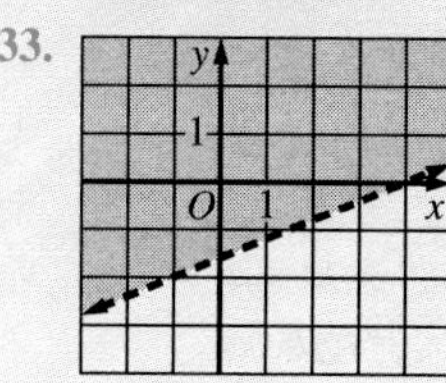

34. $y < -\frac{3}{7}x + 5$ 35. $y \geq 2x - 1$

36. $y > -\frac{5}{3}x + 4$

37.

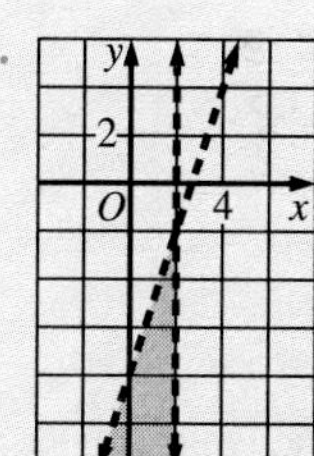

38.

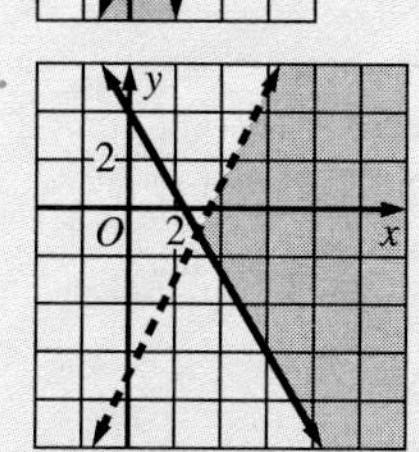

39.

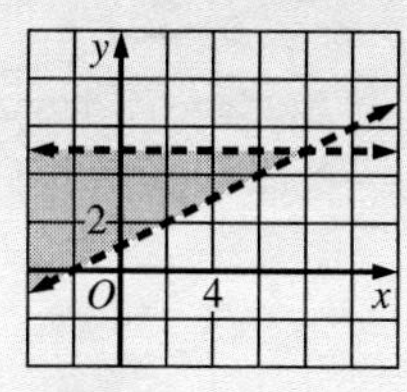

40.

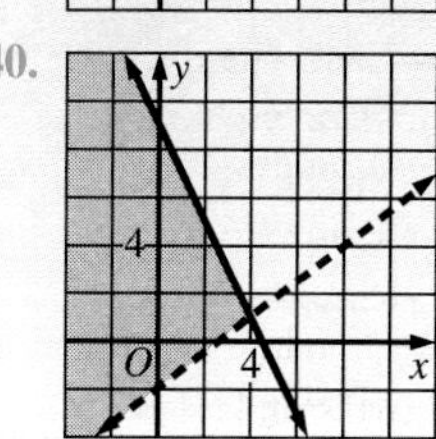

19. $\begin{bmatrix} 5 & 13 \\ 9 & -4 \end{bmatrix}\begin{bmatrix} x \\ y \end{bmatrix} = \begin{bmatrix} 40.8 \\ 2.2 \end{bmatrix}$; (1.4, 2.6)

20. 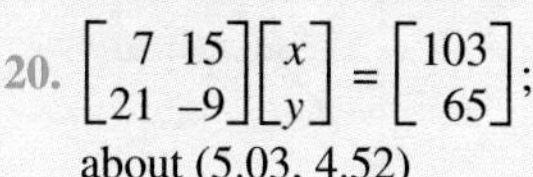

$\begin{bmatrix} 7 & 15 \\ 21 & -9 \end{bmatrix}\begin{bmatrix} x \\ y \end{bmatrix} = \begin{bmatrix} 103 \\ 65 \end{bmatrix}$; about (5.03, 4.52)

21. $\begin{bmatrix} 9 & 1 \\ 1.6 & 1 \end{bmatrix}\begin{bmatrix} x \\ y \end{bmatrix} = \begin{bmatrix} -13.5 \\ 0 \end{bmatrix}$; about (–1.82, 2.92)

22. $\begin{bmatrix} -3 & 1 \\ -6.2 & 1 \end{bmatrix}\begin{bmatrix} x \\ y \end{bmatrix} = \begin{bmatrix} 7.4 \\ -19 \end{bmatrix}$; (8.25, 32.15)

23. about (0.30, 2.49, 3.01)

24. about (3.59, –2.57, –1.26)

25. about (1.20, 0.58, –0.82)

26.

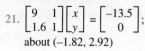

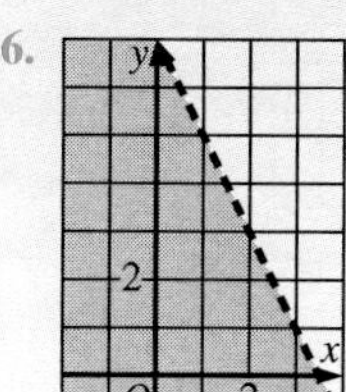

27.

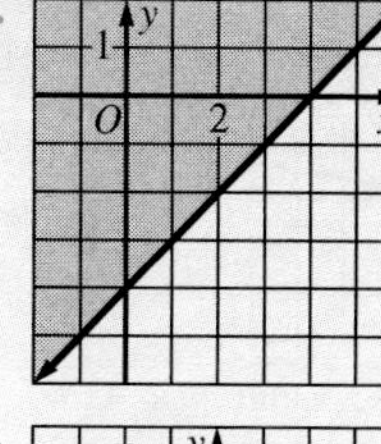

28.

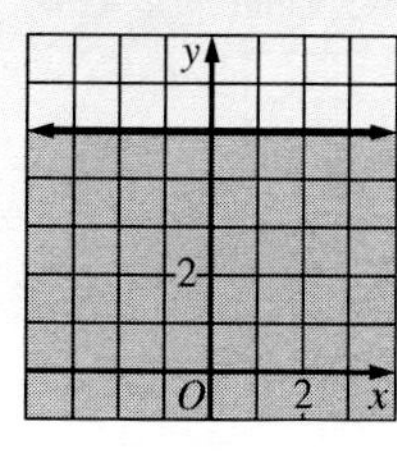

41. $x \geq -2$
$y > 2x - 1$

42. $y \leq 3x + 4$
$y \leq -\frac{5}{2}x + 4$

43. $y < 2x$
$y < -3x + 5$
$y > \frac{1}{3}x - \frac{5}{3}$

44.
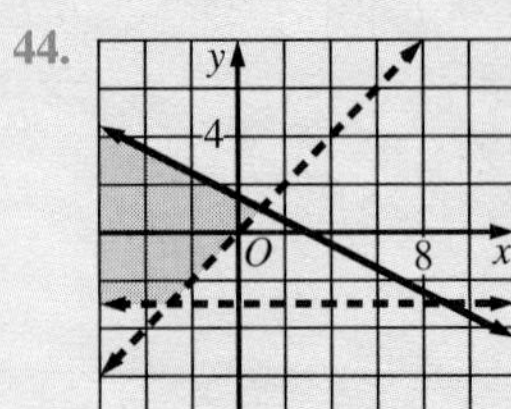

45.
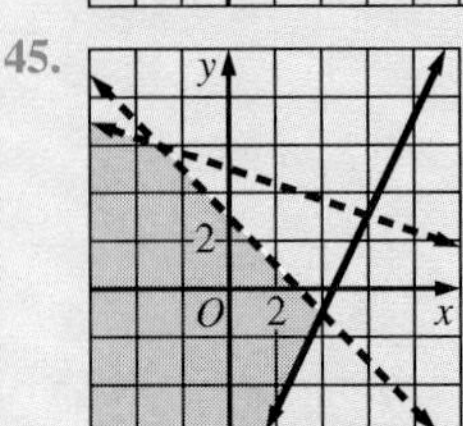

46.
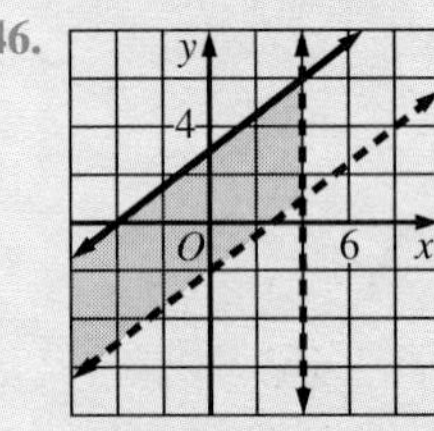

47.
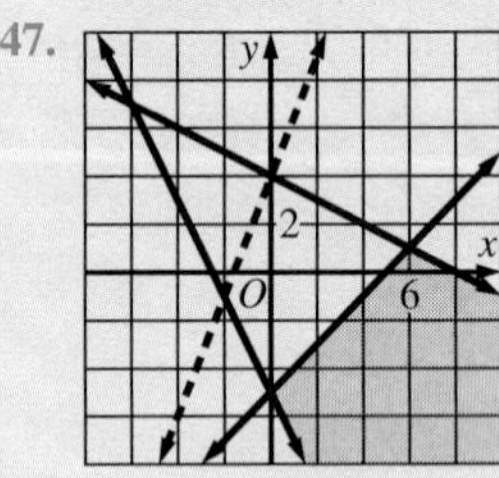

48. a. $x \geq 0$
$y \geq 0$
$0.028x + 0.007y \geq 0.63$,
or $y = -4x + 90$
$0.002x + 0.003y \geq 0.21$,
or $y = -\frac{2}{3}x + 70$

b.
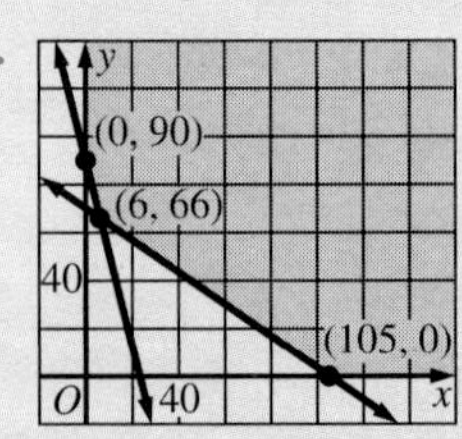

c. 6 kg of seaweed meal and 66 kg of mushroom compost

Write a system of inequalities defining each shaded region. *Section 7.5*

41.
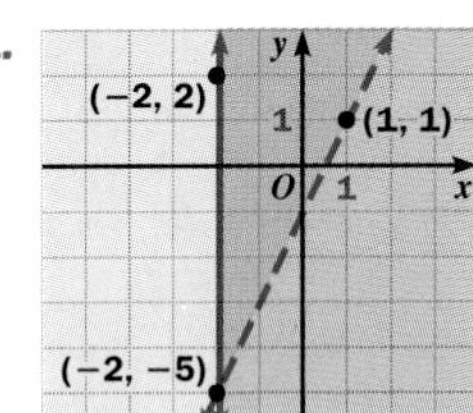

42.
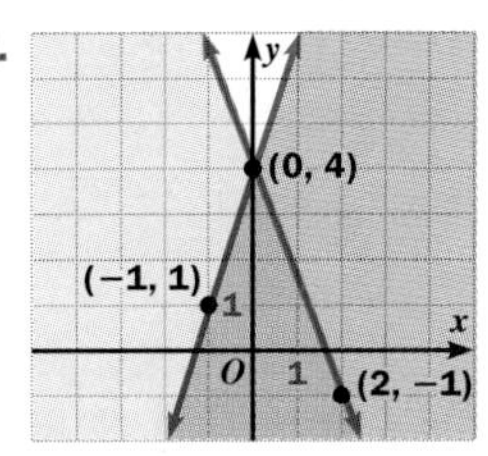

43.
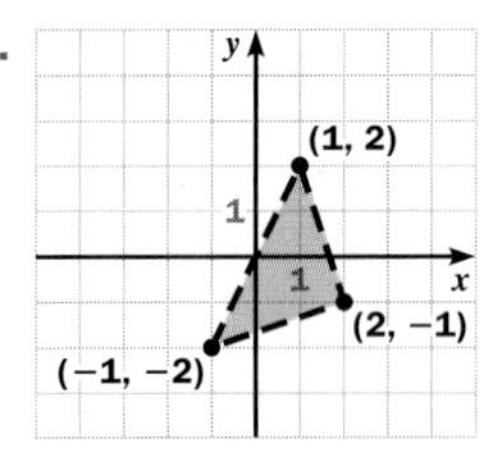

Graph each system of inequalities. *Section 7.5*

44. $y > -3$
$y \leq -\frac{1}{2}x + \frac{3}{2}$
$y > x$

45. $y \geq 2x - 9$
$x + y < 3$
$y < 5 - \frac{1}{3}x$

46. $y \leq \frac{3}{4}x + 3$
$y > \frac{3}{4}x - 2$
$x < 4$

47. $x + 2y \leq 8$
$x - y \geq 5$
$2x + y \geq -5$
$2y - 5x < 8$

48. Ana Lopez wants to mix seaweed meal and mushroom compost to make a fertilizer. These ingredients contain nitrogen (N) and phosphorus (P) as shown in the table. Ana needs a mixture that will contain at least 0.63 kg of N and at least 0.21 kg of P. *Section 7.6*

	Percent N	Percent P
Seaweed meal	2.8	0.2
Mushroom compost	0.7	0.3

a. Let x = number of kilograms of seaweed meal, and let y = number of kilograms of mushroom compost in the mixture. Write a system of inequalities to represent the constraints.

b. Graph the feasible region. Label the corner points.

c. Suppose seaweed meal costs \$.40 per kilogram and mushroom compost costs \$.15 per kilogram. How much of each ingredient should Ana use in order to minimize the cost of the mixture?

CHAPTER 8

Graph each function. State the domain and range. *Section 8.1*

1. $y = \sqrt{x + 2}$
2. $y = \sqrt{2 - x}$
3. $y = \sqrt{x} - 2$
4. $y = \sqrt{x + 2} - 2$
5. $y = 3\sqrt{x} - 3$
6. $y = \sqrt{4 - x} + 2$

Technology Use a graphing calculator or graphing software to graph the left and right sides of each equation as separate functions. Decide whether each statement is *always true, sometimes true,* or *never true.* *Section 8.1*

7. $\sqrt{x^2 - 4} = x - 2$
8. $\sqrt{9x^2} = |3x|$
9. $2\sqrt{x} + 3 = \sqrt{2x + 3}$
10. $\sqrt{25x^6} = 5x^3$
11. $\sqrt{x - 3} = \sqrt{3 - x}$
12. $\sqrt{x^3} = \sqrt{x} \cdot \sqrt{x^2}$

Evaluate each radical expression, or state that it is *undefined.* *Section 8.2*

13. $\sqrt[3]{-1}$
14. $\sqrt{0.0009}$
15. $\sqrt[4]{-16}$
16. $\sqrt[3]{1,000,000}$
17. $\sqrt[5]{-32}$
18. $\sqrt{-\frac{1}{49}}$
19. $\sqrt[4]{\frac{81}{16}}$
20. $\sqrt[3]{\frac{1}{125}}$

Chapter 8

1. domain: $x \geq -2$; range: $y \geq 0$

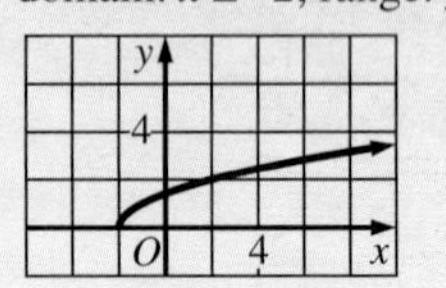

2. domain: $x \leq 2$; range: $y \geq 0$

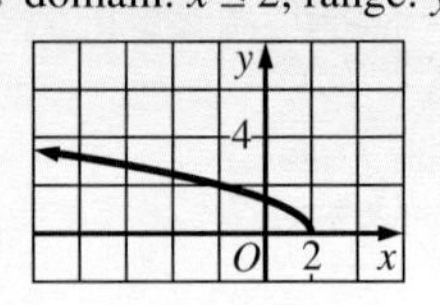

3. domain: $x \geq 0$; range: $y \geq -2$

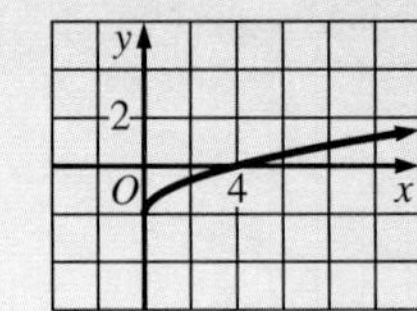

4. domain: $x \geq -2$; range: $y \geq -2$

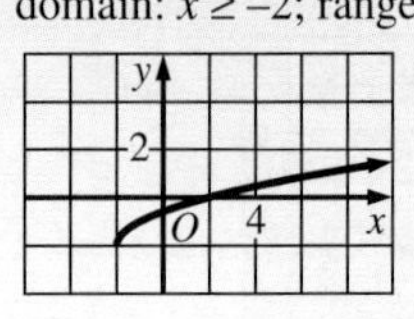

5. domain: $x \geq 0$; range: $y \geq -3$

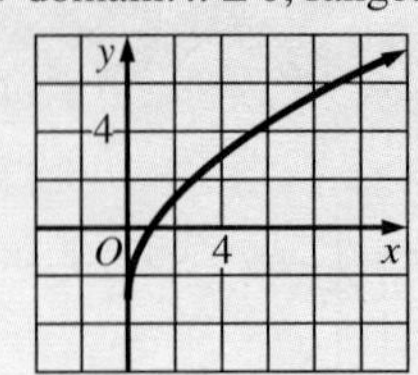

6. domain: $x \leq 4$; range: $y \geq 2$

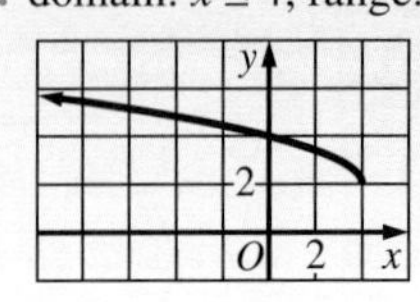

7. sometimes true

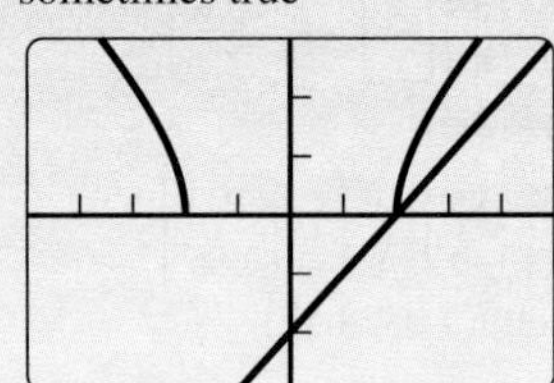

8–20. See answers in back of book.

21. domain: $x \geq 5$; range: $y \geq 0$

22. domain: all real numbers; range: all real numbers

State the domain and range of each function. *Section 8.2*

21. $\sqrt{x - 5}$ **22.** $\sqrt[3]{x + 4}$ **23.** $\sqrt[4]{x} - 3$ **24.** $\sqrt[4]{2x + 8}$

In Exercises 25–32, assume all variables are restricted to nonnegative values. Simplify each expression. *Section 8.2*

25. $\sqrt{18}$ **26.** $\sqrt[3]{-250}$ **27.** $\sqrt[4]{48}$ **28.** $\sqrt[3]{135}$

29. $\sqrt{5x^8}$ **30.** $\sqrt[5]{96x^{15}}$ **31.** $\sqrt{9x^{16}}$ **32.** $\sqrt[3]{8x^{27}}$

Solve. *Section 8.3*

33. $\sqrt{x} = 9$ **34.** $\sqrt{2x - 1} = 5$ **35.** $\sqrt{15 - x} = 6$ **36.** $\sqrt{x + 11} = 8$

37. $\sqrt{\frac{2x}{5}} = 8$ **38.** $3 + \sqrt{x - 2} = 6$ **39.** $\sqrt{\frac{x - 5}{2}} = 7$ **40.** $\sqrt{2x - 3} - 1 = 4$

Solve by raising both sides of each equation to the same power. *Section 8.3*

41. $\sqrt[4]{3x + 5} = 1$ **42.** $\sqrt[3]{5 - 2x} = 5$ **43.** $\sqrt[5]{\frac{1}{3}x + 8} = 2$ **44.** $\sqrt[4]{4x - 7} = 3$

Solve using factoring. Check to eliminate extraneous solutions. *Section 8.3*

45. $\sqrt{x + 2} = x - 4$ **46.** $\sqrt{3x - 8} = x - 6$ **47.** $\sqrt{5x + 9} = x - 1$

48. $\sqrt{89 - 2x} = x + 5$ **49.** $\sqrt{7x + 15} = x + 3$ **50.** $\sqrt{34 - 3x} = 2x - 1$

Perform the indicated operation. *Section 8.4*

51. $(3 - 7i) + (2 + 5i)$ **52.** $(2 + 3i) - (-5 + 6i)$ **53.** $(-11 + 2i) + (-3 - 9i)$

54. $(2 - 5i)(3 + 8i)$ **55.** $(7 - 2i)^2$ **56.** $(-6 + 5i)(-6 - 5i)$

Use complex conjugates to write each quotient in $a + bi$ form. *Section 8.4*

57. $\frac{5 - 2i}{3 + i}$ **58.** $\frac{3 + 8i}{4 - i}$ **59.** $\frac{-6 + 7i}{2 + 3i}$ **60.** $\frac{9 - 5i}{3i}$

Solve using the quadratic formula. Check your solutions. *Section 8.4*

61. $x^2 - 4x + 5 = 0$ **62.** $x^2 + 8x + 20 = 0$ **63.** $-3x^2 + 6x - 30 = 0$

Plot each pair of complex numbers and their sum in the complex plane. Find their magnitudes. *Section 8.5*

64. $2, -5i$ **65.** $3 + 2i, 3 - 2i$ **66.** $-1 + 5i, 1 - 3i$

67. $-3 - 6i, 2 - i$ **68.** $5, -2 + 7i$ **69.** $4 + 2i, -3 + 3i$

Plot each pair of complex numbers and their product in the complex plane. Find their magnitudes. *Section 8.5*

70. $3 + i, 3 - i$ **71.** $-2 + 3i, 1 + i$ **72.** $-3, 4 - i$ **73.** $5 - 3i, 2 - i$

A complex number and its *opposite* have a sum of 0. Find the opposite of each complex number. *Section 8.5*

74. $4i$ **75.** $3 + 2i$ **76.** $5 - 7i$ **77.** $-3 - 2i$

23. domain: $x \ge 0$; range: $y \ge -3$
24. domain: $x \ge -4$; range: $y \ge 0$
25. $3\sqrt{2}$ 26. $-5\sqrt[3]{2}$
27. $2\sqrt[4]{3}$ 28. $3\sqrt[3]{5}$
29. $x^4\sqrt{5}$ 30. $2x^3\sqrt[5]{3}$
31. $3x^8$ 32. $2x^9$
33. 81 34. 13
35. –21 36. 53
37. 160 38. 11
39. 103 40. 14

41. $-\frac{4}{3}$ 42. –60
43. 72 44. 22
45. 7 46. 11
47. 8 48. 4
49. –2, 3 50. 3
51. $5 - 2i$ 52. $7 - 3i$
53. $-14 - 7i$ 54. $46 + i$
55. $45 - 28i$ 56. 61
57. $\frac{13}{10} - \frac{11}{10}i$ 58. $\frac{4}{17} + \frac{35}{17}i$

59. $\frac{9}{13} + \frac{32}{13}i$ 60. $-\frac{5}{3} - 3i$
61. $2 \pm i$ 62. $-4 \pm 2i$
63. $1 \pm 3i$
64. $|2| = 2$; $|-5i| = 5$; $|2 - 5i| = \sqrt{29}$

65. $|3 + 2i| = \sqrt{13}$; $|3 - 2i| = \sqrt{13}$; $|6| = 6$

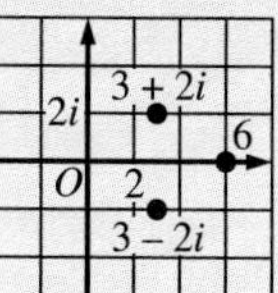

66. $|-1 + 5i| = \sqrt{26}$; $|1 - 3i| = \sqrt{10}$; $|2i| = 2$

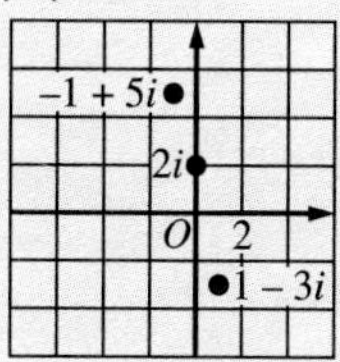

67. $|-3 - 6i| = 3\sqrt{5}$; $|2 - i| = \sqrt{5}$; $|-1 - 7i| = 5\sqrt{2}$

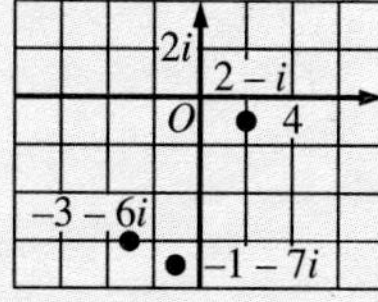

68. $|5| = 5$; $|-2 + 7i| = \sqrt{53}$; $|3 + 7i| = \sqrt{58}$

69. $|4 + 2i| = 2\sqrt{5}$; $|-3 + 3i| = 3\sqrt{2}$; $|1 + 5i| = \sqrt{26}$

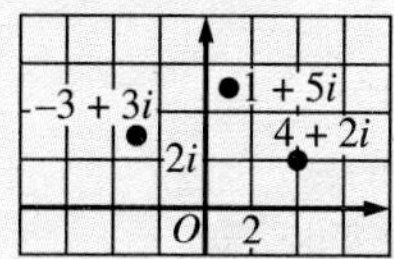

70. $|3 + i| = \sqrt{10}$; $|3 - i| = \sqrt{10}$; $|10| = 10$

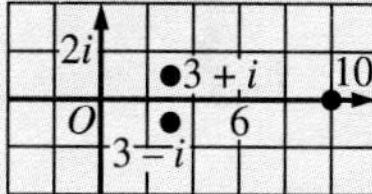

71. $|-2 + 3i| = \sqrt{13}$; $|1 + i| = \sqrt{2}$; $|-5 + i| = \sqrt{26}$

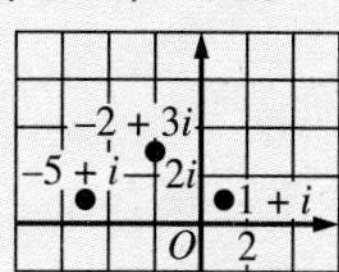

72. $|-3| = 3$; $|4 - i| = \sqrt{17}$; $|-12 + 3i| = 3\sqrt{17}$

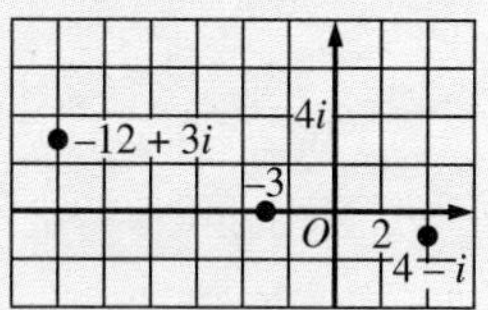

73. $|5 - 3i| = \sqrt{34}$; $|2 - i| = \sqrt{5}$; $|7 - 11i| = \sqrt{170}$

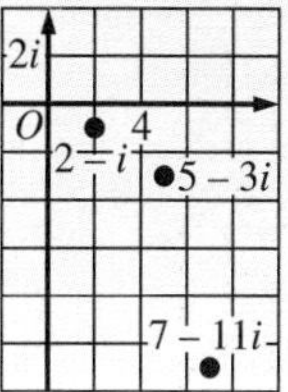

74. $-4i$ 75. $-3 - 2i$
76. $-5 + 7i$ 77. $3 + 2i$

For each table, tell whether the given group property holds. If a group property does not hold, give a counterexample. *Section 8.6*

78. commutative

+	0	1	2	3
0	0	1	2	3
1	1	2	3	0
2	2	3	0	1
3	3	0	1	2

79. closure

×	1	2	3
1	1	3	3
2	2	4	6
3	3	6	9

80. identity

+	α	β	γ	δ
α	α	β	γ	δ
β	β	α	δ	γ
γ	γ	δ	α	β
δ	δ	β	γ	α

If possible, give an example of a number that satisfies each description. *Section 8.6*

81. integral and complex **82.** whole and imaginary **83.** real and rational

CHAPTER 9

Use synthetic substitution to evaluate each polynomial for the given value of the variable. *Section 9.1*

1. $x^3 - 3x^2 + 2x - 5; x = 5$

2. $2x^3 + 4x^2 - 5x + 7; x = -4$

3. $u^4 + 2u^3 - u + 10; u = -2$

4. $-3r^4 + 5r^3 + 12r^2 - 11; r = 3$

Add or subtract, as indicated. *Section 9.1*

5. $(5x^3 - 3x^2 + 4x - 8) + (-2x^3 + 3x^2 - 7x + 15)$

6. $(-4x^5 + x - 10) + (7x^5 + 5x^2 + 4)$

7. $(3x^4 - 5x^3 + 4) - (-4x^4 - 2x^2 + 7)$

8. $(-2x^5 + x^3 - 7x + 8) - (5x^5 - 3x^4 - 5)$

Multiply or divide, as indicated. *Section 9.2*

9. $(5x + 3)(2x - 7)$

10. $(v - 3)(v^2 + 3v + 9)$

11. $(y - 2)(5y^2 - y + 4)$

12. $(n^2 + n - 2)(n^2 - n + 2)$

13. $(a + 2)(a - 1)(a + 1)$

14. $(x - 4)^3$

15. $\dfrac{2x^2 - 5x - 3}{x - 2}$

16. $\dfrac{3k^2 - 5k}{k + 3}$

17. $\dfrac{4c^3 - 3c^2 + c - 1}{c - 3}$

18. $\dfrac{y^3 - 40}{y - 4}$

19. $\dfrac{2b^3 - 9b^2 + 4}{2b - 5}$

20. $\dfrac{4x^3 + 7x^2 - 2x - 8}{x^2 + 3x - 1}$

Technology Use a graphing calculator or graphing software to graph each polynomial function. For each function:

a. Describe the end behavior using infinity notation.

b. Find all local maximums and local minimums. *Section 9.3*

21. $f(x) = -2x^2 + 8x - 5$

22. $f(x) = -x^4 + 4x^3 - 5$

Find the zeros of each function. *Section 9.4*

23. $f(x) = 3(x + 2)(x - 1)(x - 5)$

24. $f(x) = (2x - 5)(x + 2)(x + 7)$

25. $g(x) = x^3 - 4x^2 + x + 6$

26. $g(x) = x^3 + 5x^2 - 4x - 20$

78. Yes.
79. No; for example, $3 \times 2 = 6$, but 6 is not in the set {1, 2, 3}.
80. Yes.
81. any integer
82. not possible
83. any rational number

Chapter 9

1. 55
2. –37
3. 12
4. –11
5. $3x^3 - 3x + 7$
6. $3x^5 + 5x^2 + x - 6$
7. $7x^4 - 5x^3 + 2x^2 - 3$
8. $-7x^5 + 3x^4 + x^3 - 7x + 13$
9. $10x^2 - 29x - 21$
10. $v^3 - 27$
11. $5y^3 - 11y^2 + 6y - 8$
12. $n^4 - n^2 + 4n - 4$
13. $a^3 + 2a^2 - a - 2$
14. $x^3 - 12x^2 + 48x - 64$
15. $2x - 1 - \dfrac{5}{x - 2}$
16. $3k - 14 + \dfrac{42}{k + 3}$
17. $4c^2 + 9c + 28 + \dfrac{83}{c - 3}$
18. $y^2 + 4y + 16 + \dfrac{24}{y - 4}$
19. $b^2 - 2b - 5 - \dfrac{21}{2b - 5}$
20. $4x - 5 + \dfrac{17x - 13}{x^2 + 3x - 1}$

21. X=2 Y=3

a. As $x \to -\infty, f(x) \to -\infty$. As $x \to +\infty, f(x) \to -\infty$.

b. local (and global) maximum: (2, 3)

22. 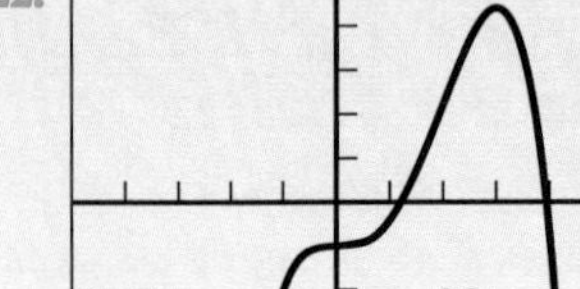

a. As $x \to -\infty, f(x) \to -\infty$. As $x \to +\infty, f(x) \to -\infty$.

b. local (and global) maximum: (3, 22)

23. –2, 1, 5
24. $-7, -2, \frac{5}{2}$
25. –1, 2, 3
26. –5, –2, 2
27. $y = 2x(x + 1)(x - 3)$
28. $y = -\frac{1}{2}(x + 4)(x + 2)(x - 2)$
29. $y = \frac{3}{2}(x + 3)(x + 1)(x - 3)$
30. –1.35, 0.45, 4.90

Find an equation for each cubic function whose graph is shown. *Section 9.4*

27.

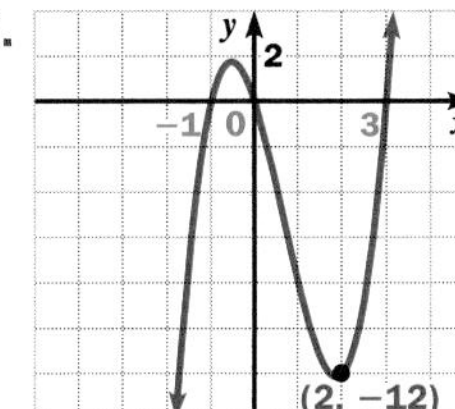

28.

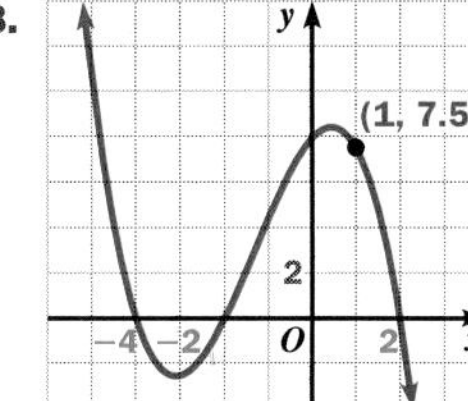

29.

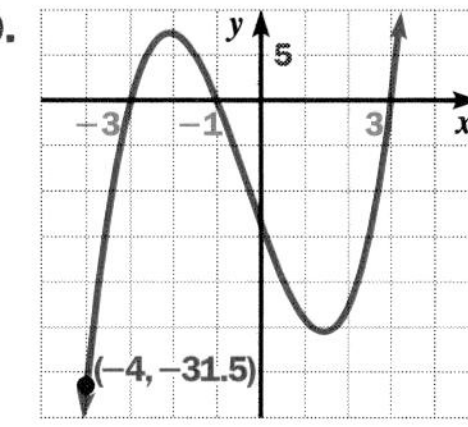

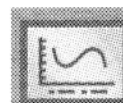

Technology **Use a graphing calculator or graphing software to approximate all real zeros of each function to the nearest hundredth.** *Section 9.5*

30. $f(x) = x^3 - 4x^2 - 5x + 3$

31. $f(x) = x^3 - 6x^2 + 5x + 7$

For each function:

a. List the possible rational zeros of the function.

b. Find all real and imaginary zeros of the function. Identify any double or triple zeros. *Section 9.5*

32. $f(x) = 2x^3 - 11x^2 + 17x - 6$

33. $f(x) = 10x^4 - 33x^3 + 33x^2 - 7x - 3$

Technology **Use a graphing calculator or graphing software to graph each equation. For each equation, tell whether *y* varies inversely with *x*. If so, state the constant of variation.** *Section 9.6*

34. $y = -\dfrac{6}{x}$

35. $xy = 8$

36. $x = \dfrac{4}{y}$

37. $y = -9x$

Tell whether you think the two quantities show inverse variation. Explain your reasoning. *Section 9.6*

38. the time it takes to drive a fixed distance and the average speed for the trip

39. the number of pounds of pears in a grocery bag and the price of the bag

Find an equation of each hyperbola. *Section 9.7*

40.

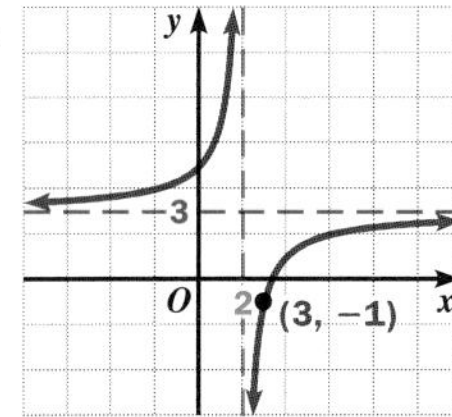

41.

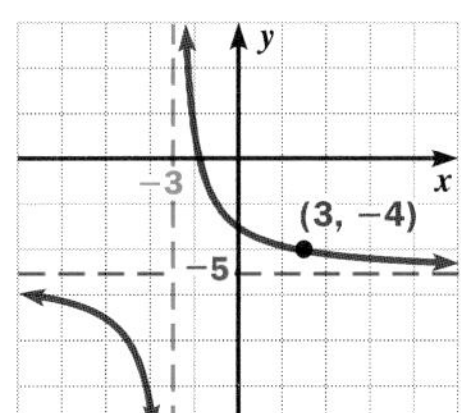

42.

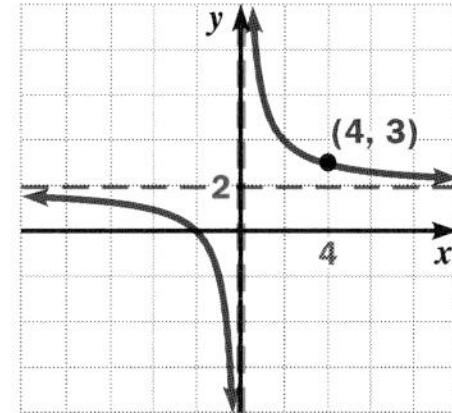

For each function:

a. Find the asymptotes of the function's graph.

b. Tell how the function's graph is related to a hyperbola with equation of the form $y = \frac{a}{x}$. *Section 9.7*

43. $y = \dfrac{2x + 4}{x - 3}$

44. $y = \dfrac{-5x - 2}{x + 2}$

45. $y = \dfrac{3x - 6}{x - 5}$

31. −0.71, 2.14, 4.57

32. a. $\pm1, \pm2, \pm3, \pm6, \pm\frac{1}{2}, \pm\frac{3}{2}$

b. $\frac{1}{2}$, 2, 3

33. a. $\pm1, \pm3, \pm\frac{1}{2}, \pm\frac{3}{2}, \pm\frac{1}{5}, \pm\frac{3}{5}, \pm\frac{1}{10}, \pm\frac{3}{10}$

b. $-\frac{1}{5}, \frac{3}{2}$, 1 (double root)

34. Yes; −6.

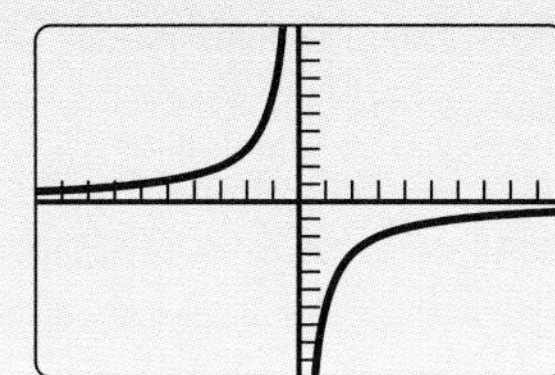

35. Yes; 8.

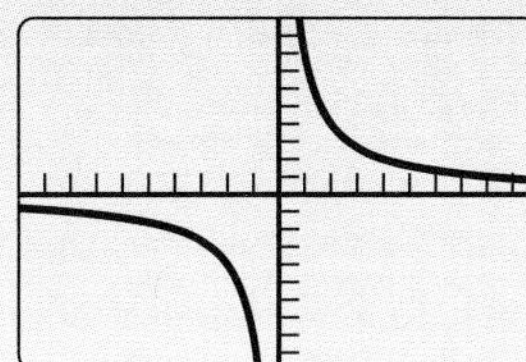

36. Yes; 4.

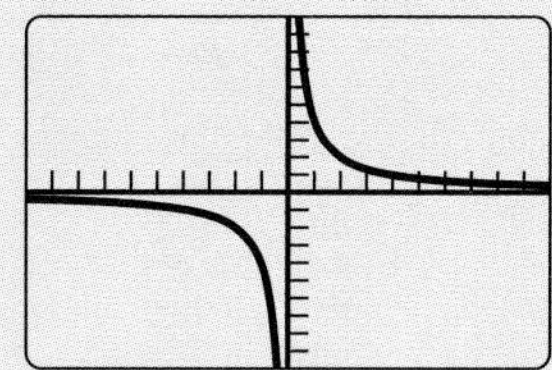

37. No.

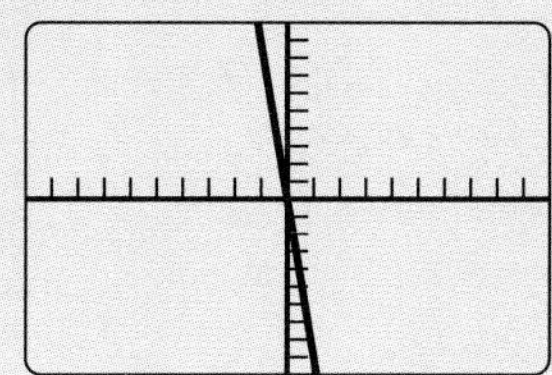

38. Yes; if k is the fixed distance, then $rt = k$, which is an inverse variation equation.

39. No; let n = the number of pounds in the bag, c = the total cost of the bag, and k = the fixed price per pound. Then $nk = c$, or $\frac{c}{n} = k$, which is a direct variation equation.

40. $y = -\dfrac{4}{x-2} + 3$

41. $y = \dfrac{6}{x+3} - 5$

42. $y = \dfrac{4}{x} + 2$

43. a. $x = 3$ and $y = 2$

b. The graph is the graph of $y = \frac{10}{x}$ translated 3 units to the right and 2 units up.

44. a. $x = -2$ and $y = -5$

b. The graph is the graph of $y = \frac{8}{x}$ translated 2 units to the left and 5 units down.

45. a. $x = 5$ and $y = 3$

b. The graph is the graph of $y = \frac{9}{x}$ translated 5 units to the right and 3 units up.

46. degree $p(x) <$ degree $q(x)$; degree $q(x) \geq 2$; roots of $q(x)$: −5, 3

For Exercises 46–48, suppose $f(x) = \frac{p(x)}{q(x)}$, where $p(x)$ and $q(x)$ are polynomials.

State what you can conclude about the degrees and zeros of $p(x)$ and $q(x)$ from the given information about the graph. *Section 9.8*

46. has vertical asymptotes at $x = 3$ and $x = -5$ and a horizontal asymptote at $y = 0$

47. has vertical asymptotes at $x = -2$, $x = 2$, and $x = 4$ and has no horizontal asymptote

48. has a vertical asymptote at the y-axis and the line $y = \frac{2}{3}$ as a horizontal asymptote

For each function:

a. Find the vertical asymptotes of the function's graph.

b. Describe the function's end behavior using infinity notation. *Section 9.8*

49. $f(x) = \frac{-x}{x^2 - 3x - 4}$

50. $f(x) = \frac{x^3 + 5}{15x^2 + 2x - 1}$

51. $g(x) = \frac{-3x + 8}{x^3 - 2x^2 + x}$

52. $g(x) = \frac{6x^3}{x^3 + 4x^2 - x - 4}$

53. $h(x) = \frac{x^2 - 3x - 10}{2x^3 - 3x^2 + 6x - 9}$

54. $h(x) = \frac{5x^4 + 7}{x^2 - 16}$

Solve each equation. *Section 9.9*

55. $\frac{2}{x} + \frac{5}{2x} = 3$

56. $\frac{4}{x} - \frac{3}{x + 1} = 1$

57. $\frac{6}{x} - \frac{2}{x(x - 3)} = 1$

58. $\frac{y + 2}{y - 1} + \frac{2}{y + 1} = 3$

59. $\frac{4}{n - 2} = \frac{4}{(n + 3)(n - 2)}$

60. $\frac{2r}{r - 1} + \frac{5}{r + 5} = 4$

CHAPTER 10

Find the next 4 terms of each sequence. *Section 10.1*

1. $2, -4, 8, -16, \ldots$

2. $5, 8, 11, 14, \ldots$

3. $11, 101, 1001, 10{,}001, \ldots$

4. $1, 3, 6, 10, \ldots$

5. $50, 5, 0.5, 0.05, \ldots$

6. $-15, -8, -1, 6, \ldots$

7. $1, \frac{1}{4}, \frac{1}{9}, \frac{1}{16}, \ldots$

8. $\frac{3}{4}, \frac{4}{5}, \frac{5}{6}, \frac{6}{7}, \ldots$

9. $\frac{1}{9}, 9\frac{1}{9}, 99\frac{1}{9}, 999\frac{1}{9}, \ldots$

Write a formula for t_n. *Section 10.1*

10. $\frac{1}{5}, \frac{3}{10}, \frac{9}{15}, \frac{27}{20}, \ldots$

11. $17, 13, 9, 5, \ldots$

12. $3, 3\sqrt{2}, 6, 6\sqrt{2}, \ldots$

13. $\frac{1}{2}, \frac{1}{4}, \frac{1}{6}, \frac{1}{8}, \ldots$

14. $\sqrt{6}, 3\sqrt{6}, 5\sqrt{6}, 7\sqrt{6}, \ldots$

15. $\frac{4}{5}, -\frac{8}{5}, \frac{16}{5}, -\frac{32}{5}, \ldots$

Tell whether each sequence is *arithmetic*, *geometric*, or *neither*. *Section 10.2*

16. $1, 1.2, 1.04, 1.008, \ldots$

17. $-3, 6, -12, 24, \ldots$

18. $\frac{1}{3}, 1, \frac{5}{3}, \frac{7}{3}, \ldots$

19. $5, 43, 81, 119, \ldots$

20. $-\frac{4}{5}, -\frac{7}{5}, -2, -\frac{13}{5}, \ldots$

21. $0.4, 0.44, 0.444, \ldots$

22. $3, 3i, -3, -3i, \ldots$

23. $6, 1.2, 0.24, 0.048, \ldots$

24. $-2, -3.5, -5, -6.5, \ldots$

47. degree $p(x) >$ degree $q(x)$; degree $q(x) \geq 3$; roots of $q(x)$: –2, 2, 4

48. degree $p(x) =$ degree $q(x)$; degree $q(x) \geq 1$; roots of $q(x)$: 0

49. a. $x = 4$ and $x = -1$
   b. As $x \to \pm\infty, f(x) \to 0$.

50. a. $x = \frac{1}{5}$ and $x = -\frac{1}{3}$
   b. As $x \to -\infty, f(x) \to -\infty$. As $x \to +\infty, f(x) \to +\infty$.

51. a. $x = 0$ and $x = 1$
   b. As $x \to \pm\infty, g(x) \to 0$.

52. a. $x = -4$ and $x = -1$ and $x = 1$
   b. As $x \to \pm\infty, g(x) \to 6$.

53. a. $x = \frac{3}{2}$
   b. As $x \to \pm\infty, h(x) \to 0$.

54. a. $x = 4$ and $x = -4$
   b. As $x \to \pm\infty, h(x) \to +\infty$.

55. $\frac{3}{2}$

56. –2, 2

57. 4, 5

58. $-\frac{1}{2}, 3$

59. –2

60. $-3, \frac{5}{2}$

Chapter 10

1. 32, –64, 128, –256
2. 17, 20, 23, 26
3. 100,001, 1,000,001, 10,000,001, 100,000,001
4. 15, 21, 28, 36
5. 0.005, 0.0005, 0.00005, 0.000005
6. 13, 20, 27, 34
7. $\frac{1}{25}, \frac{1}{36}, \frac{1}{49}, \frac{1}{64}$
8. $\frac{7}{8}, \frac{8}{9}, \frac{9}{10}, \frac{10}{11}$
9. $9999\frac{1}{9}, 99{,}999\frac{1}{9}, 999{,}999\frac{1}{9}, 9{,}999{,}999\frac{1}{9}$,

10–15. Answers may vary. Examples are given.

10. $t_n = \frac{3^{n-1}}{5n}$
11. $t_n = -4n + 21$
12. $t_n = 3(\sqrt{2})^{n-1}$
13. $t_n = \frac{1}{2n}$
14. $t_n = (2n - 1)\sqrt{6}$
15. $t_n = \frac{1}{5}(-2)^{n+1}$
16. neither
17. geometric
18. arithmetic
19. arithmetic
20. arithmetic
21. neither
22. geometric
23. geometric
24. arithmetic
25. a. 62.5
   b. 22

For each pair of numbers:

a. Find the arithmetic mean.

b. Find the geometric mean. *Section 10.2*

25. 4, 121 **26.** 2, 18 **27.** 5, 80 **28.** $\frac{1}{3}$, 75

29. 0.45, 0.05 **30.** 7, 29 **31.** 2, $\frac{81}{2}$ **32.** $6i$, $54i$

Write a recursive formula for each sequence. *Section 10.3*

33. 3, 15, 75, 375, . . . **34.** 16, 9, 2, −5, . . . **35.** $\frac{5}{2}, \frac{5}{4}, \frac{5}{8}, \frac{5}{16}, \ldots$

36. $t_n = -0.5n + 3$ **37.** $t_n = \frac{8}{3} \cdot \frac{1}{2^n}$ **38.** $t_n = 4n - 11$

For each sequence:

a. Find the first 4 terms.

b. Write an explicit formula. *Section 10.3*

39. $t_1 = 0.2$, $t_n = -5t_{n-1}$ **40.** $t_1 = -10$, $t_n = t_{n-1} + 7$ **41.** $t_1 = 32$, $t_n = -\frac{1}{2}t_{n-1}$ **42.** $t_1 = 4$, $t_n = i(t_{n-1})$

Find the sum of each series. *Section 10.4*

43. $-1 + 5 + 11 + 17 + \cdots + 89$ **44.** $38 + 35 + 32 + 29 + \cdots + (-13)$

45. $\frac{3}{2} + 5 + \frac{17}{2} + 12 + \cdots + 33$ **46.** $-5.6 + (-4.8) + (-4) + \cdots + 2.4$

Expand each series. *Section 10.4*

47. $\sum_{n=1}^{7} (n + 3)$ **48.** $\sum_{n=1}^{5} (3n^2 - 6)$ **49.** $\sum_{n=1}^{6} (5 - 10n^3)$

Write each series in sigma notation and find the sum. *Section 10.4*

50. $9 + 13 + 17 + 21 + \cdots + 149$ **51.** $-87 + (-62) + (-37) + \cdots + 113$

52. $28 + 25 + 22 + 19 + \cdots + (-41)$ **53.** $\frac{1}{3} + \frac{7}{3} + \frac{13}{3} + \frac{19}{3} + \cdots + \frac{91}{3}$

Find the sum of each series. *Section 10.5*

54. $162 + 108 + 72 + 48 + \cdots + \frac{128}{9}$ **55.** $\frac{1}{4} + \frac{1}{2} + 1 + 2 + \cdots + 64$

56. $\frac{5}{4}\sqrt{3} + 5\sqrt{3} + 20\sqrt{3} + \cdots + 1280\sqrt{3}$ **57.** $250 + (-50) + 10 + (-2) + \cdots + 0.016$

Find the sum of each series, if the sum exists. If a series does not have a sum, explain why not. *Section 10.5*

58. $54 + 36 + 24 + 16 + \cdots$ **59.** $-6 + 2 + \left(-\frac{2}{3}\right) + \frac{2}{9} + \cdots$ **60.** $\frac{100}{7} + \frac{10}{7} + \frac{1}{7} + \frac{1}{70} + \cdots$

61. $\sum_{n=1}^{\infty} 4\left(\frac{3}{4}\right)^n$ **62.** $\sum_{n=1}^{\infty} 3(1.1)^n$ **63.** $\sum_{n=1}^{\infty} 8\left(-\frac{3}{5}\right)^n$

26. a. 10
 b. 6
27. a. 42.5
 b. 20
28. a. $37\frac{2}{3}$
 b. 5
29. a. 0.25
 b. 0.15
30. a. 18
 b. $\sqrt{203}$
31. a. $21\frac{1}{4}$
 b. 9
32. a. $30i$
 b. $18i$
33. $t_1 = 3$; $t_n = 5(t_{n-1})$
34. $t_1 = 16$; $t_n = t_{n-1} - 7$
35. $t_1 = \frac{5}{2}$; $t_n = \frac{1}{2}(t_{n-1})$
36. $t_1 = 2.5$; $t_n = t_{n-1} - 0.5$
37. $t_1 = \frac{4}{3}$; $t_n = \frac{1}{2}(t_{n-1})$
38. $t_1 = -7$; $t_n = t_{n-1} + 4$
39. a. 0.2, −1, 5, −25
 b. $t_n = 0.2(-5)^{n-1}$
40. a. −10, −3, 4, 11
 b. $t_n = -10 + 7(n-1)$, or $t_n = 7n - 17$
41. a. 32, −16, 8, −4
 b. $t_n = 32\left(-\frac{1}{2}\right)^{n-1}$
42. a. 4, $4i$, −4, $-4i$
 b. $t_n = 4(i)^{n-1}$
43. 704
44. 225
45. 172.5
46. −17.6
47. 4 + 5 + 6 + 7 + 8 + 9 + 10
48. −3 + 6 + 21 + 42 + 69
49. −5 − 75 − 265 − 635 − 1245 − 2155
50–53. Answers may vary. Examples are given.
50. $\sum_{n=1}^{36}(4n + 5) = 2844$
51. $\sum_{n=1}^{9}(25n - 112) = 117$
52. $\sum_{n=1}^{24}(31 - 3n) = -156$
53. $\sum_{n=1}^{16}\left(2n - \frac{5}{3}\right) = 245\frac{1}{3}$
54. $457\frac{5}{9}$
55. $127\frac{3}{4}$
56. $1706\frac{1}{4}\sqrt{3}$
57. 208.336
58. 162
59. −4.5
60. $\frac{1000}{63}$
61. 12
62. no sum; $|r| = 1.1 > 1$
63. −3

Chapter 11

1. a. $6\sqrt{2}$ b. $y = -x + 11$
2. a. $4\sqrt{5}$ b. $y = 2x - 1$
3. a. $6\sqrt{10}$ b. $y = \frac{1}{3}x + \frac{8}{3}$
4. a. $\sqrt{41}$ b. $y = 0.8x + 13.9$
5. 2 or –8
6. 12 or –6
7. 1.5 or 0.1
8. a. $y = \frac{1}{4}(x - 3)^2$

b.

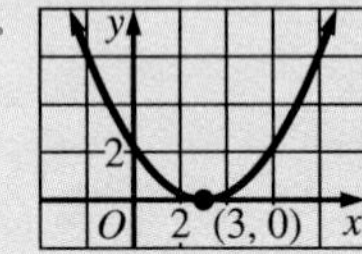

9. a. $x - 2 = -(y + 7)^2$

b.

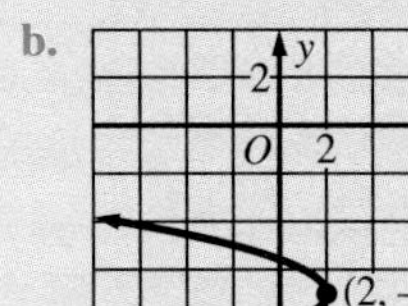

10. a. $y - 3 = \frac{1}{4}(x - 5)^2$

b.

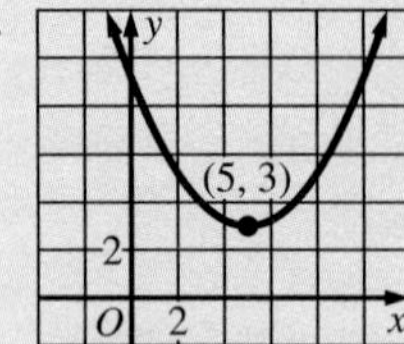

11. vertex: (0, 0); focus: $\left(0, -\frac{1}{8}\right)$; directrix: $y = \frac{1}{8}$

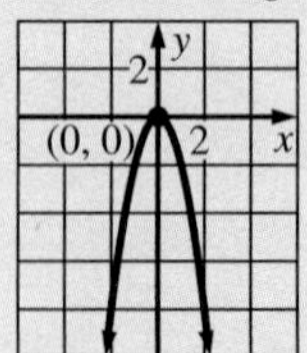

12. vertex: (0, –1); focus: (0, 1); directrix: $y = -3$

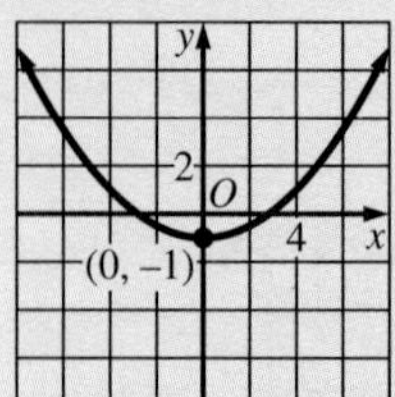

13. vertex: (–4, 3); focus: (–4, 0); directrix: $y = 6$

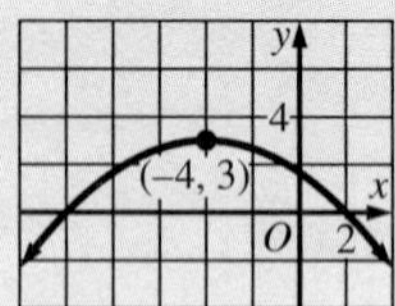

14. vertex: (–2, –3); focus: $\left(-\frac{3}{2}, -3\right)$; directrix: $x = -\frac{5}{2}$

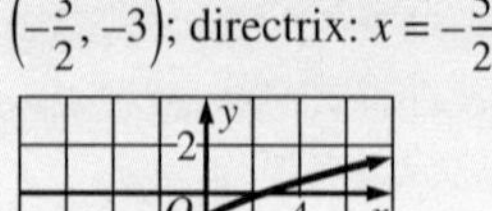

CHAPTER 11

For each pair of points, find:

a. the distance between the points

b. an equation of the perpendicular bisector of the line segment connecting the points *Section 11.1*

1. (2, 3), (8, 9)
2. (−2, 5), (6, 1)
3. (1, −7), (−5, 11)
4. (−1, 9), (−5, 14)

Find the value(s) of *k* so that the given points are *n* units apart. *Section 11.1*

5. (7, −3), (−5, *k*); $n = 13$
6. (4, *k*), (−8, 3); $n = 15$
7. (*k*, 1.9), (0.8, 4.3); $n = 2.5$

For Exercises 8–10:

a. Find an equation of the parabola with the given characteristics.

b. Graph your equation from part (a). *Section 11.2*

8. vertex (3, 0); directrix $y = -1$
9. vertex (2, −7); focus $\left(\frac{7}{4}, -7\right)$
10. directrix $y = 2$; focus (5, 4)

Name the vertex, focus, and directrix of a parabola with the given equation. Sketch the parabola with its focus and directrix. You may want to use a graphing calculator or graphing software to check your work. *Section 11. 2*

11. $y = -2x^2$
12. $y = \frac{1}{8}x^2 - 1$
13. $y = -\frac{1}{12}(x + 4)^2 + 3$
14. $x = \frac{1}{2}(y + 3)^2 - 2$

For each equation of a circle, identify the center and the radius. Then graph the equation. *Section 11.3*

15. $(x + 3)^2 + (y - 5)^2 = 16$
16. $x^2 + (y - 4)^2 = 49$
17. $x^2 - 6x + 9 + y^2 = 10$

Write an equation of the circle with the given center *C* and radius *r*. *Section 11.3*

18. $C(2, 0)$; $r = 3$
19. $C(-1, 8)$; $r = \sqrt{7}$
20. $C(3, -4)$; $r = 2\sqrt{3}$
21. $C(-\sqrt{5}, -1)$; $r = 6$
22. $C(0, k)$; $r = 4$
23. $C(p, q)$; $r = p$

Identify the center, the vertices, and the foci of each ellipse with the given equation. Tell whether the ellipse is *horizontal* or *vertical*. Then graph the equation. *Section 11.4*

24. $\frac{x^2}{36} + \frac{y^2}{81} = 1$
25. $\frac{x^2}{49} + \frac{y^2}{25} = 1$
26. $\frac{(x + 2)^2}{36} + \frac{(y - 5)^2}{100} = 1$

Write an equation of each ellipse. *Section 11.4*

27.

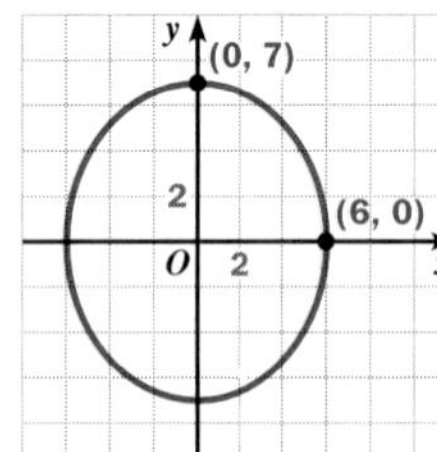

28.

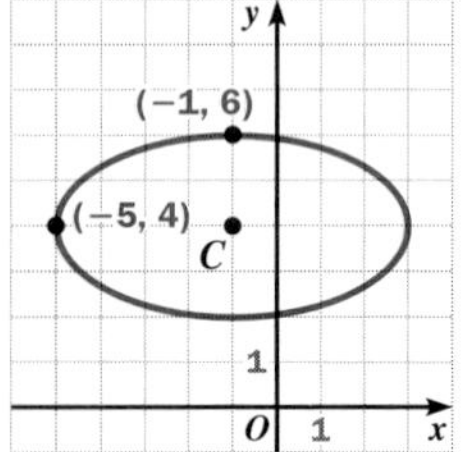

29.

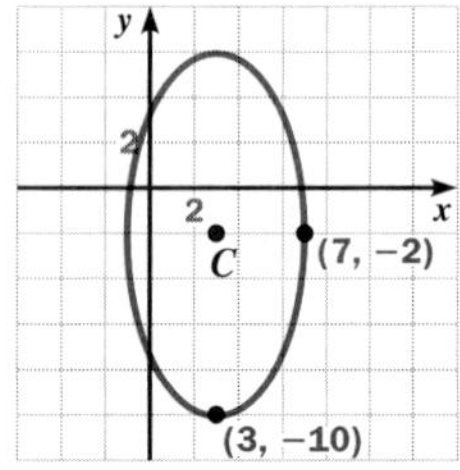

15. center: (–3, 5); radius: 4

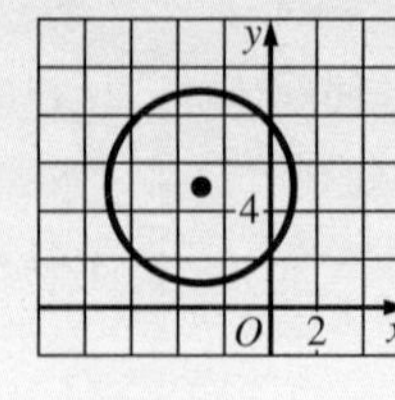

16. center: (0, 4); radius: 7

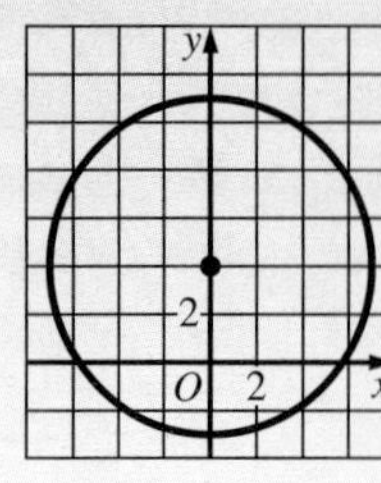

17. center: (3, 0); radius: $\sqrt{10}$

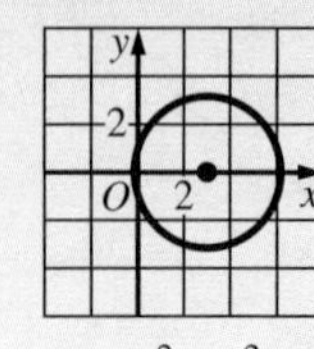

18. $(x - 2)^2 + y^2 = 9$
19. $(x + 1)^2 + (y - 8)^2 = 7$
20. $(x - 3)^2 + (y + 4)^2 = 12$
21. $(x + \sqrt{5})^2 + (y + 1)^2 = 36$
22. $x^2 + (y - k)^2 = 16$
23. $(x - p)^2 + (y - q)^2 = p^2$
24. center: (0, 0); vertices: $(0, \pm 9)$; foci: $(0, \pm 3\sqrt{5})$; vertical

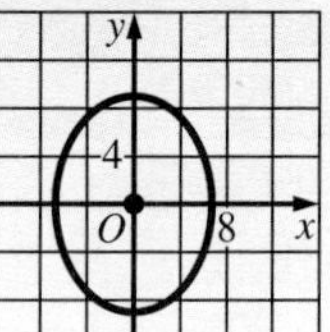

Identify the center, the vertices, and the foci of each hyperbola with the given equation. Tell whether the hyperbola is *horizontal* or *vertical*. Find equations of the asymptotes. Then graph the equation. *Section 11.5*

30. $\frac{y^2}{4} - \frac{x^2}{25} = 1$

31. $\frac{x^2}{49} - \frac{y^2}{36} = 1$

32. $\frac{(x-4)^2}{36} - \frac{(y-1)^2}{81} = 1$

33. $\frac{(y-2)^2}{9} - \frac{(x+3)^2}{49} = 1$

34. $16(y+4)^2 - (x+1)^2 = 64$

35. $25(x+2)^2 - 9(y-5)^2 = 225$

Write an equation for each hyperbola with the given characteristics. Then sketch its graph. *Section 11.5*

36. asymptotes: $y = \pm\frac{3}{4}x$; foci (0, 10) and (0, −10)

37. center (1, 3); vertex (7, 3); focus (10, 3)

38. center (2, −4); vertex (2, 1); focus (2, 9)

For each equation of a conic, rewrite the equation in standard form, identify the conic, and state its important characteristics. *Section 11.6*

39. $4x^2 - 8x - y^2 - 6y - 9 = 0$

40. $x^2 - 8x + 7 + 9y^2 = 0$

41. $x^2 + 4y - 6x + 17 = 0$

42. $x^2 + 4y^2 - 24y - 72 = 0$

Identify and graph each equation of a degenerate conic. *Section 11.6*

43. $x^2 + 9y^2 = 0$

44. $x^2 - 4x + 4 - y^2 = 0$

45. $x^2 - 25y^2 = 0$

46. $y^2 + 2y + 1 + x^2 = 0$

CHAPTER 12

For each graph:

a. Find the degree of each vertex.

b. Draw an equivalent graph.

c. Color your graph so that vertices that share an edge are different colors. Try to use as few colors as possible. *Section 12.1*

1.

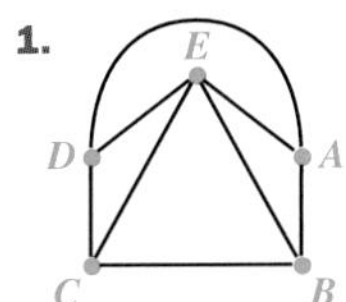

2.

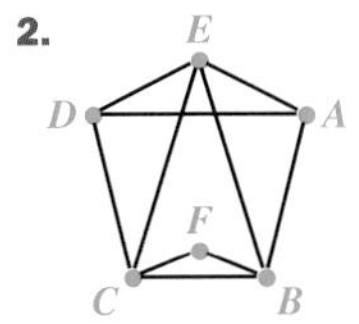

3.

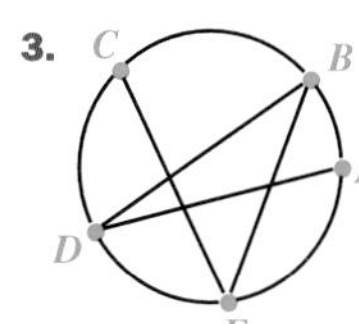

A map of the Rocky Mountain states is shown at the right. Use this map in Exercises 4 and 5. *Section 12.1*

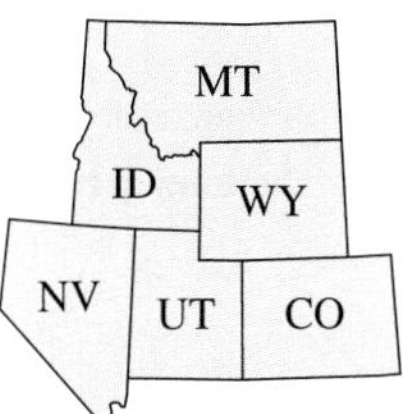

4. Draw a graph with each vertex corresponding to one of the 6 states in the map.

5. Label the vertices of the graph to show how you could color the map with as few colors as possible.

25. center: (0, 0);
vertices: (±7, 0);
foci: $(\pm 2\sqrt{6}, 0)$;
horizontal

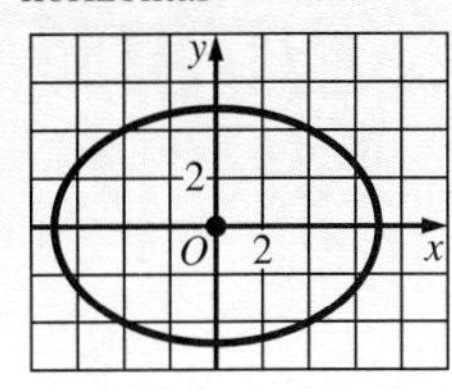

26. center: (−2, 5);
vertices: (−2, 15), (−2, −5);
foci: (−2, 13), (−2, −3);
vertical

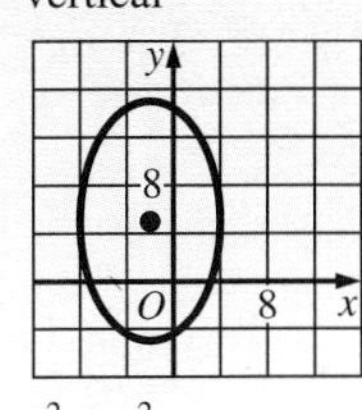

27. $\frac{x^2}{36} + \frac{y^2}{49} = 1$

28. $\frac{(x+1)^2}{16} + \frac{(y-4)^2}{4} = 1$

29. $\frac{(x-3)^2}{16} + \frac{(y+2)^2}{64} = 1$

30. center: (0, 0);
vertices: (0, ±2);
foci: $(0, \pm\sqrt{29})$;
vertical; $y = \pm\frac{2}{5}x$

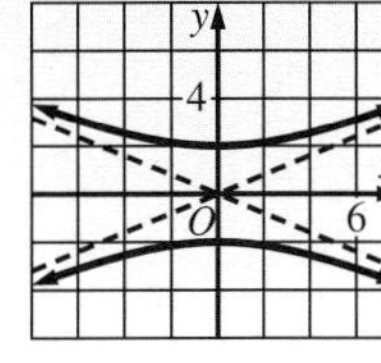

31. center: (0, 0);
vertices: (±7, 0);
foci: $(\pm\sqrt{85}, 0)$;
horizontal; $y = \pm\frac{6}{7}x$

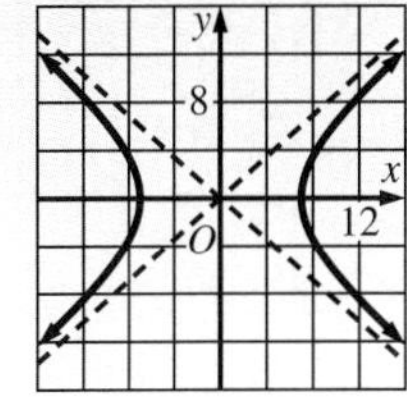

32. center: (4, 1);
vertices: (10, 1), (−2, 1);
foci: $(4 \pm \sqrt{117}, 1)$;
horizontal;
$y = \pm\frac{3}{2}(x-4) + 1$

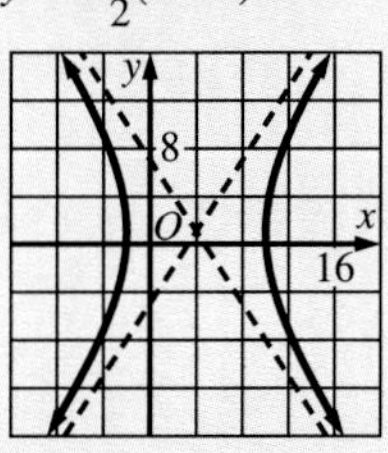

33. center: (−3, 2);
vertices: (−3, 5), (−3, −1);
foci: $(-3, 2 \pm \sqrt{58})$; vertical;
$y = \pm\frac{3}{7}(x+3) + 2$

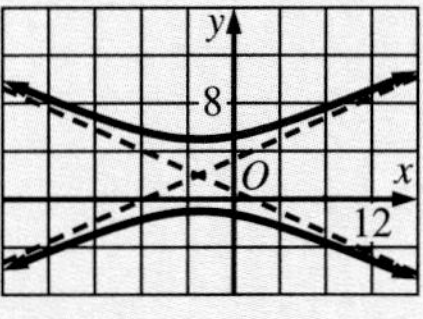

34. center: (−1, −4);
vertices: (−1, −2), (−1, −6);
foci: $(-1, -4 \pm 2\sqrt{17})$;
vertical; $y = \pm\frac{1}{4}(x+1) - 4$

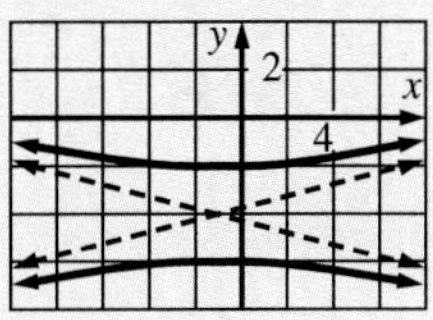

35. center: (−2, 5);
vertices: (1, 5), (−5, 5);
foci: $(-2 \pm \sqrt{34}, 5)$;
horizontal; $y = \pm\frac{5}{3}(x+2) + 5$

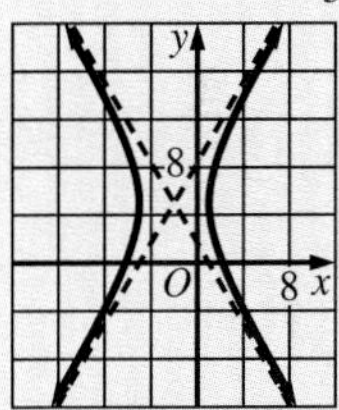

36. $\frac{y^2}{36} - \frac{x^2}{64} = 1$

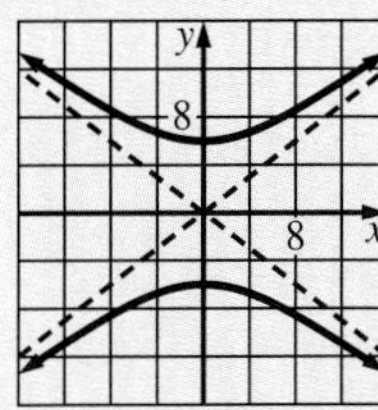

37. $\frac{(x-1)^2}{36} + \frac{(y-3)^2}{45} = 1$

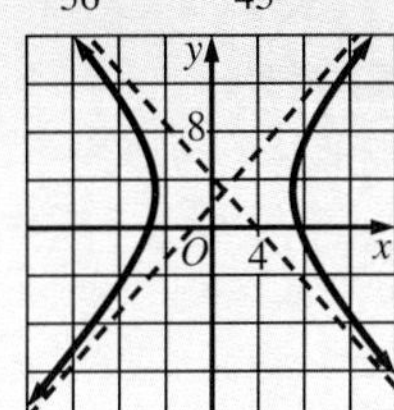

38–46. See answers in back of book.

Chapter 12

1–5. See answers in back of book.

Write a matrix representing each directed graph. Let an element be 1 if the row vertex has an edge directed to the column vertex. Otherwise let the element be 0.
Section 12.2

6.

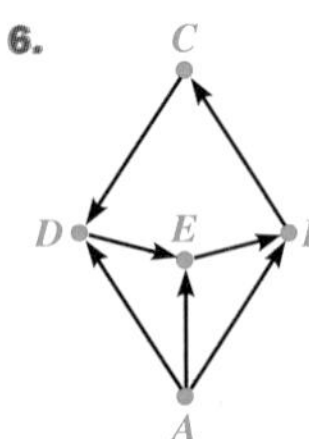

7.

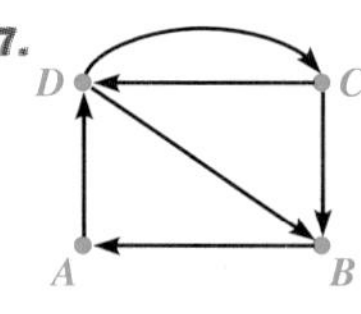

8.

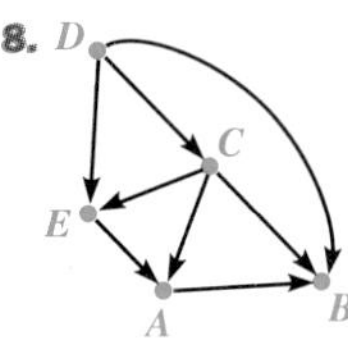

For each matrix, draw a directed graph. Assume that a 1 means that the row vertex has an edge directed to the column vertex, and a 0 means no such edge exists.
Section 12.2

9.
$$\begin{array}{c} \\ A \\ B \\ C \\ D \end{array}\begin{array}{c} \begin{array}{cccc} A & B & C & D \end{array} \\ \begin{bmatrix} 0 & 0 & 1 & 0 \\ 1 & 0 & 0 & 1 \\ 1 & 1 & 0 & 0 \\ 1 & 0 & 1 & 0 \end{bmatrix} \end{array}$$

10.
$$\begin{array}{c} \\ A \\ B \\ C \\ D \\ E \end{array}\begin{array}{c} \begin{array}{ccccc} A & B & C & D & E \end{array} \\ \begin{bmatrix} 0 & 0 & 0 & 0 & 1 \\ 1 & 0 & 0 & 1 & 0 \\ 1 & 0 & 0 & 0 & 1 \\ 0 & 0 & 1 & 0 & 1 \\ 0 & 1 & 0 & 0 & 0 \end{bmatrix} \end{array}$$

11.
$$\begin{array}{c} \\ A \\ B \\ C \\ D \end{array}\begin{array}{c} \begin{array}{cccc} A & B & C & D \end{array} \\ \begin{bmatrix} 0 & 1 & 0 & 1 \\ 0 & 0 & 1 & 0 \\ 1 & 0 & 0 & 0 \\ 0 & 1 & 1 & 0 \end{bmatrix} \end{array}$$

Calculate each expression. *Section 12.3*

12. $\frac{7!}{5!}$ **13.** $(3!)(6!)$ **14.** $\frac{8!}{2!\,2!\,4!}$ **15.** $\frac{n!}{(n-2)!}$

Simplify each expression. *Section 12.3*

16. ${}_5P_2$ **17.** ${}_7P_3$ **18.** ${}_6P_6$ **19.** ${}_8P_2$

Give an expression for the number of distinguishable permutations of the letters of each word. Then evaluate the expression. *Section 12.3*

20. ALMANAC **21.** MINIMUM **22.** NONSENSE

Simplify each expression. *Section 12.4*

23. ${}_8C_5$ **24.** ${}_9C_3$ **25.** ${}_{15}C_2$ **26.** ${}_{12}C_8$

A frozen yogurt shop offers 11 different toppings that can be added to a frozen yogurt sundae. How many possibilities are there if you order the given number of toppings combined on one sundae? *Section 12.4*

27. one topping **28.** two toppings **29.** three toppings

For each expression:

a. Find *n* such that $(a + b)^n$ has a term containing the given powers of *a* and *b*.

b. Find the coefficient of the term in the binomial expansion $(a + b)^n$. *Section 12.5*

30. a^5b^3 **31.** ab^5 **32.** a^3b^6 **33.** a^4b^4

6.
$$\begin{array}{c} \\ A \\ B \\ C \\ D \\ E \end{array}\begin{array}{c} \begin{array}{ccccc} A & B & C & D & E \end{array} \\ \begin{bmatrix} 0 & 1 & 0 & 1 & 1 \\ 0 & 0 & 1 & 0 & 0 \\ 0 & 0 & 0 & 1 & 0 \\ 0 & 0 & 0 & 0 & 1 \\ 0 & 1 & 0 & 0 & 0 \end{bmatrix} \end{array}$$

7.
$$\begin{array}{c} \\ A \\ B \\ C \\ D \end{array}\begin{array}{c} \begin{array}{cccc} A & B & C & D \end{array} \\ \begin{bmatrix} 0 & 0 & 0 & 1 \\ 1 & 0 & 0 & 0 \\ 0 & 1 & 0 & 1 \\ 0 & 1 & 1 & 0 \end{bmatrix} \end{array}$$

8.
$$\begin{array}{c} \\ A \\ B \\ C \\ D \\ E \end{array}\begin{array}{c} \begin{array}{ccccc} A & B & C & D & E \end{array} \\ \begin{bmatrix} 0 & 1 & 0 & 0 & 0 \\ 0 & 0 & 0 & 0 & 0 \\ 1 & 1 & 0 & 0 & 1 \\ 0 & 1 & 1 & 0 & 1 \\ 1 & 0 & 0 & 0 & 0 \end{bmatrix} \end{array}$$

9–11. Answers may vary. Examples are given.

9.

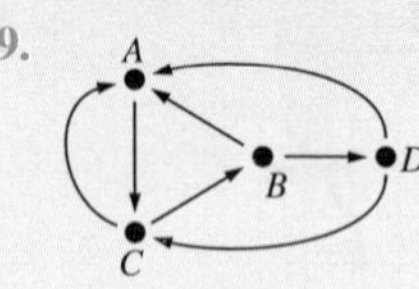

10.

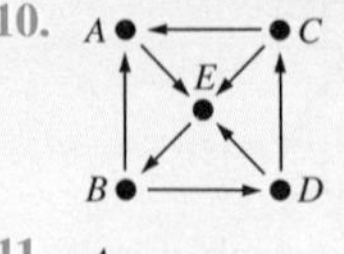

11.

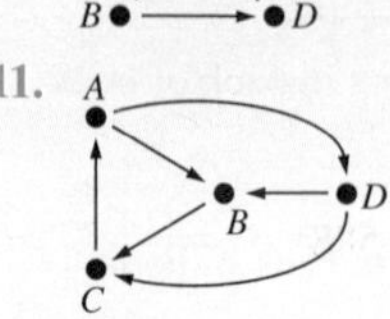

12. 42
13. 4320
14. 420
15. $n(n-1)$, or $n^2 - n$
16. 20
17. 210
18. 720
19. 56
20. $\frac{7!}{3!} = 840$
21. $\frac{7!}{3!2!} = 420$
22. $\frac{8!}{3!2!2!} = 1680$
23. 56
24. 84
25. 105
26. 495
27. 11
28. 55
29. 165
30. a. 8
  b. 56
31. a. 6
  b. 6
32. a. 9
  b. 84
33. a. 8
  b. 70

Use the binomial theorem to expand each power of a binomial. *Section 12.5*

34. $(x + y)^7$ **35.** $(x - y)^8$ **36.** $(x + 3y)^5$

37. $(2x - 3y)^6$ **38.** $(x^2 + y)^4$ **39.** $(x^3 - y^2)^7$

CHAPTER 13

The targets shown consist of a large square divided into smaller, congruent squares. Suppose a randomly thrown dart hits each target. Find the probability that the dart hits the target's shaded region. *Section 13.1*

1.

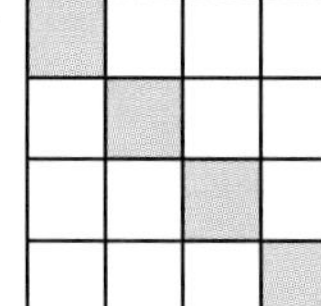

2.

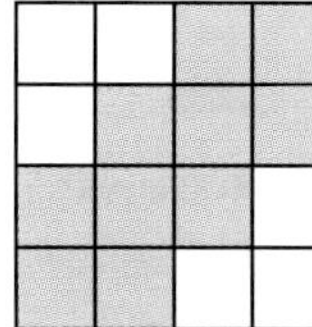

3.

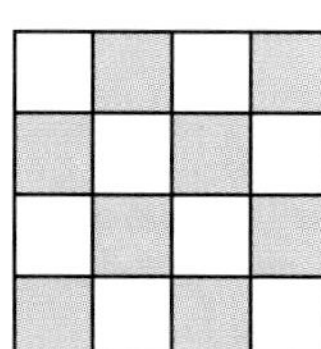

Find the probability of choosing each type of card at random from a standard deck of 52 playing cards. (See page 615 for a description of such a deck.) *Section 13.1*

4. a diamond **5.** an ace **6.** a black card

7. not a heart **8.** not the 5 of clubs **9.** a red six

10. not a face card **11.** a jack or queen **12.** a card neither black nor red

Suppose a six-sided die is rolled and a fair coin is flipped. Find the probability of each event. *Section 13.2*

13. the die shows 6, and the coin comes up heads

14. the die shows 2, 3, or 5, and the coin comes up tails

15. the die does not show 4, and the coin comes up heads or tails

A baseball player's batting average is the probability that the player will get a hit in his or her next official at-bat. Suppose a player batting .250 is followed by a player batting .320, and both have official at-bats. Find the probability of each event. *Section 13.2*

16. both players get hits **17.** exactly one of the two players gets a hit

18. at least one player gets a hit **19.** neither player gets a hit

Suppose two cards are drawn at random from a standard deck of 52 playing cards. (See page 615 for a description of such a deck.) The first card is *not* put back in the deck before the second card is drawn. Find the probability of each event. *Section 13.3*

20. the first card is a face card, and the second card is not a face card

21. the second card is a heart, given that the first card is a spade

22. the second card is a 10, given that the first card is a queen

23. at least one of the cards is a diamond

24. neither card is a face card

34. $x^7 + 7x^6y + 21x^5y^2 + 35x^4y^3 + 35x^3y^4 + 21x^2y^5 + 7xy^6 + y^7$

35. $x^8 - 8x^7y + 28x^6y^2 - 56x^5y^3 + 70x^4y^4 - 56x^3y^5 + 28x^2y^6 - 8xy^7 + y^8$

36. $x^5 + 15x^4y + 90x^3y^2 + 270x^2y^3 + 405xy^4 + 243y^5$

37. $64x^6 - 576x^5y + 2160x^4y^2 - 4320x^3y^3 + 4860x^2y^4 - 2916xy^5 + 729y^6$

38. $x^8 + 4x^6y + 6x^4y^2 + 4x^2y^3 + y^4$

39. $x^{21} - 7x^{18}y^2 + 21x^{15}y^4 - 35x^{12}y^6 + 35x^9y^8 - 21x^6y^{10} + 7x^3y^{12} - y^{14}$

Chapter 13

1. $\frac{1}{4}$, or 0.25
2. $\frac{5}{8}$, or 0.625
3. $\frac{1}{2}$, or 0.5
4. $\frac{1}{4}$, or 0.25
5. $\frac{1}{13}$, or about 0.0769
6. $\frac{1}{2}$, or 0.5
7. $\frac{3}{4}$, or 0.75
8. $\frac{51}{52}$, or about 0.981
9. $\frac{1}{26}$, or about 0.0385
10. $\frac{10}{13}$, or about 0.769
11. $\frac{2}{13}$, or about 0.154
12. 0
13. $\frac{1}{12}$, or about 0.0833
14. $\frac{1}{4}$, or 0.25
15. $\frac{5}{6}$, or about 0.833
16. 0.08
17. 0.41
18. 0.49
19. 0.51
20. $\frac{40}{221}$, or about 0.181
21. $\frac{13}{51}$, or about 0.255
22. $\frac{4}{51}$, or about 0.0784
23. $\frac{15}{34}$, or about 0.441
24. $\frac{10}{17}$, or about 0.588

Write a formula to describe the probability distribution for a binomial experiment with *n* trials, each with probability *p* of success. Then find the probability that the experiment will have exactly *k* successful trials. *Section 13.4*

25. $n = 6, p = 0.25; k = 2$ **26.** $n = 4, p = 40\%; k = 3$ **27.** $n = 10, p = \frac{4}{5}; k = 5$

Suppose 10% of the population are left-footed kickers. What is the probability that, on a soccer team of 11 players, exactly *k* players are left-footed kickers? *Section 13.4*

28. $k = 2$ **29.** $k = 5$ **30.** $k = 0$ **31.** $k = 11$

The following values are from a normal distribution with mean 24 and standard deviation 5. Standardize each value. *Section 13.5*

32. 29 **33.** 20 **34.** 16 **35.** 35

A group of test scores is normally distributed with mean 500 and standard deviation 100. Find the percent of scores in the group that satisfy each of the given conditions. *Section 13.5*

36. below 650 **37.** above 700 **38.** between 390 and 540

CHAPTER 14

For each trigonometric ratio, find θ. Express the answer in two ways:

a. in decimal degrees, to the nearest tenth of a degree

b. in degrees, minutes, and seconds, to the nearest second *Section 14.1*

1. $\sin \theta = 0.5623$ **2.** $\cos \theta = 0.8920$ **3.** $\sin \theta = 0.4337$

4. $\cos \theta = 0.0682$ **5.** $\sin \theta = 0.1706$ **6.** $\cos \theta = 0.3625$

Find the measure of $\angle A$ in decimal degrees, to the nearest tenth of a degree. *Section 14.1*

7.

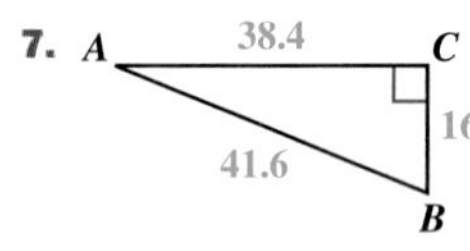

8.

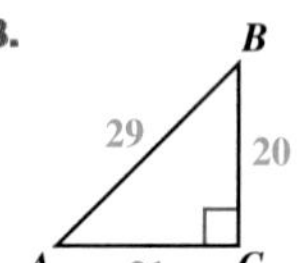

9.

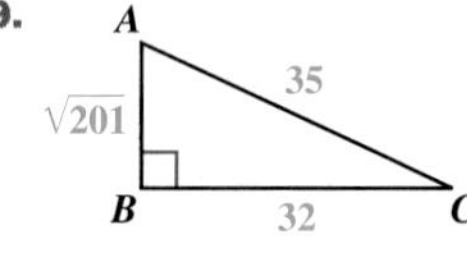

Find each value. *Section 14.2*

10. tan 62° **11.** tan 47° 33′ **12.** tan 25.6° **13.** tan 60°

Solve $\triangle ABC$ using the given measures. *Section 14.2*

14. $a = 14; b = 23$ **15.** $c = 17; \angle A = 38°$

16. $\tan B = \frac{4}{3}; c = 42$ **17.** $\sin A = \frac{7}{25}; b = 18$

An angle θ is in standard position with the given point *P* on its terminal side. Find $\sin \theta$, $\cos \theta$, and $\tan \theta$. *Section 14.3*

18. $P(6, 2.5)$ **19.** $P(7, -24)$ **20.** $P(-4.5, 6)$ **21.** $P(-6\sqrt{2}, -7)$

25. ${}_6C_k(0.25)^k(0.75)^{6-k}$; 0.297
26. ${}_4C_k(0.4)^k(0.6)^{4-k}$; 0.154
27. ${}_{10}C_k\left(\frac{4}{5}\right)^k\left(\frac{1}{5}\right)^{10-k}$; 0.0264
28. 0.213
29. 0.00246
30. 0.314
31. 1×10^{-11}
32. 1
33. −0.8
34. −1.6
35. 2.2
36. 93.32%
37. 2.28%
38. 52.0%

Chapter 14

1. a. 34.2°
   b. 34°12′54″
2. a. 26.9°
   b. 26°52′28″
3. a. 25.7°
   b. 25°42′9″
4. a. 86.1°
   b. 86°5′22″
5. a. 9.8°
   b. 9°49′22″
6. a. 68.7°
   b. 68°44′46″
7. 22.6°
8. 43.6°
9. 66.1°
10. 1.8807
11. 1.0932
12. 0.4791
13. 1.7321
14. $c \approx 26.9$; $\angle A \approx 31.3°$; $\angle B \approx 58.7°$
15. $\angle B = 52°$; $a \approx 10.5$; $b \approx 13.4$
16. $\angle B \approx 53.1°$; $\angle A \approx 36.9°$; $a = 25.2$; $b = 33.6$
17. $\angle A \approx 16.3°$; $\angle B \approx 73.7°$; $a = 5.25$; $c = 18.75$

18–27. Answers are given in the order $\sin \theta$; $\cos \theta$; $\tan \theta$.

18. $\frac{5}{13} \approx 0.3846$; $\frac{12}{13} \approx 0.9231$; $\frac{5}{12} \approx 0.4167$
19. $-\frac{24}{25} = -0.96$; $\frac{7}{25} = 0.28$; $-\frac{24}{7} \approx -3.4286$
20. $\frac{4}{5} = 0.8$; $-\frac{3}{5} = -0.6$; $-\frac{4}{3} \approx -1.3333$
21. $-\frac{7}{11} \approx -0.6364$; $-\frac{6\sqrt{2}}{11} \approx -0.7714$; $\frac{7}{6\sqrt{2}} \approx 0.8250$

For each angle θ, use a calculator to find $\sin \theta$, $\cos \theta$, and $\tan \theta$. *Section 14.3*

22. $\theta = 220°$ **23.** $\theta = 114°$ **24.** $\theta = 310°$

25. $\theta = 67°$ **26.** $\theta = 248°$ **27.** $\theta = 153°$

For Exercises 28–33, find the area of $\triangle ABC$. If there is not enough information, explain why. *Section 14.4*

28. $a = 14.7, b = 8.5, \angle C = 42°$ **29.** $b = 24.6, c = 17.3, \angle A = 37°$ **30.** $b = 13, c = 45.6, \angle B = 17°$

31. $a = 9, c = 11, \angle B = 56°$ **32.** $b = 25, c = 18, \angle A = 22°$ **33.** $a = 9.5, b = 16, \angle A = 29°$

Find the area of each sector having central angle θ and radius r. *Section 14.4*

34. $\theta = 90°; r = \frac{7}{4}$ **35.** $\theta = 240°; r = \frac{15}{8}$ **36.** $\theta = 108°; r = 4.5$

37. $\theta = 265°; r = 14$ **38.** $\theta = 7.5°; r = 8$ **39.** $\theta = 72°; r = 3.6$

Solve each $\triangle DEF$. *Section 14.5*

40. $d = 15, \angle D = 33°, \angle E = 65°$ **41.** $e = 26, \angle D = 21°, \angle F = 49°$

42. $f = 27, \angle D = 18°, \angle E = 115°$ **43.** $d = 8.5, \angle E = 29°, \angle F = 73°$

44. $e = 4\frac{2}{5}, \angle E = 127°, \angle F = 15°$ **45.** $f = 11, \angle D = 34°, \angle F = 67°$

Solve each $\triangle ABC$. If there are two solutions, give both. If there is no solution, explain why. *Section 14.5*

46. $b = 26, c = 17, \angle C = 33°$ **47.** $a = 19, b = 28, \angle A = 105°$ **48.** $a = 8, b = 34, \angle A = 40°$

49. $b = 12, c = 9.5, \angle C = 26°$ **50.** $a = 29, b = 14, \angle B = 30°$ **51.** $a = 23, c = 12, \angle A = 53°$

Find the missing side length of each triangle. *Section 14.6*

52.

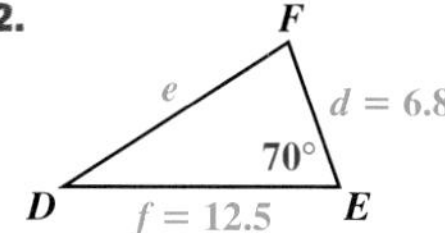

53.

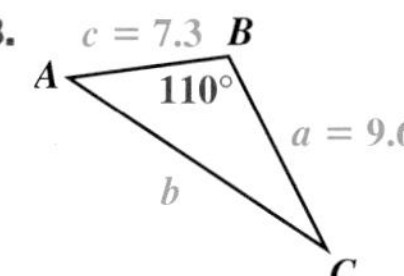

54.

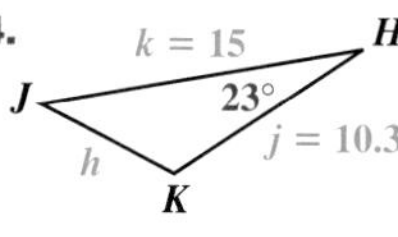

Find the angle measures of each $\triangle XYZ$. *Section 14.6*

55. $x = 17, y = 10, z = 14$ **56.** $x = 38, y = 24, z = 31$

57. $x = 4.6, y = 5.2, z = 8.3$ **58.** $x = \frac{13}{5}, y = \frac{7}{2}, z = \frac{25}{8}$

CHAPTER 15

Find all angles θ for $0° \le \theta \le 360°$ that satisfy each equation. Give answers to the nearest degree. *Section 15.1*

1. $\cos \theta = 1$ **2.** $\sin \theta = 0.4$ **3.** $\sin \theta = -0.8660$ **4.** $\cos \theta = 0.5$

5. $\sin \theta = 0.8660$ **6.** $\cos \theta = 0.7675$ **7.** $\cos \theta = -0.6250$ **8.** $\sin \theta = -0.1128$

9. $\cos \theta = 0.2145$ **10.** $\sin \theta = 0.8164$ **11.** $\sin \theta = -0.5722$ **12.** $\cos \theta = -0.1700$

22. –0.6428; –0.7660; 0.8391
23. 0.9135; –0.4067; –2.2460
24. –0.7660; 0.6428; –1.1918
25. 0.9205; 0.3907; 2.3559
26. –0.9272; –0.3746; 2.4751
27. 0.4540; –0.8910; –0.5095
28. about 41.8 square units
29. about 128.1 square units
30. not enough information; Given b and c, you need to know the measure of the included angle, $\angle A$.
31. about 41.0 square units
32. about 84.3 square units
33. not enough information; Given a and b, you need to know the measure of the included angle, $\angle C$.
34. about 2.41 square units
35. about 7.36 square units
36. about 19.1 square units
37. about 453 square units
38. about 4.19 square units
39. about 8.14 square units
40. $\angle F = 82°; e \approx 25.0; f \approx 27.3$
41. $\angle E = 110°; d \approx 9.92; f \approx 20.9$
42. $\angle F = 47°; d \approx 11.4; e \approx 33.5$
43. $\angle D = 78°; e \approx 4.21; f \approx 8.31$
44. $\angle D = 38°; d \approx 3.39; f \approx 1.43$
45. $\angle E = 79°; d \approx 6.68; e \approx 11.7$
46. $\angle A \approx 90.6°; \angle B \approx 56.4°; a \approx 31.2$; or $\angle A \approx 23.4°; \angle B \approx 123.6°; a \approx 12.4$
47. no solution; $\angle A$ is obtuse and $a < b$.
48. no solution; $a < b \sin A$
49. $\angle A \approx 120.4°; \angle B \approx 33.6°; a \approx 18.7$; or $\angle A \approx 7.6°; \angle B \approx 146.4°; a \approx 2.9$
50. no solution; $b < a \sin B$
51. $\angle B \approx 102.4°; \angle C \approx 24.6°; b \approx 28.1$
52. about 12.0
53. about 13.9
54. about 6.8
55. $\angle X \approx 88.6°; \angle Y \approx 36.0°; \angle Z \approx 55.4°$
56. $\angle X \approx 86.4°; \angle Y \approx 39.1°; \angle Z \approx 54.5°$
57. $\angle X \approx 30.0°; \angle Y \approx 34.4°; \angle Z \approx 115.6°$
58. $\angle X \approx 45.8°; \angle Y \approx 74.7°; \angle Z \approx 59.5°$

Chapter 15

1. 0°, 360°
2. 23.6°, 156.4°
3. 240°, 300°
4. 60°, 300°
5. 60°, 120°
6. 40°, 320°
7. 129°, 231°
8. 187°, 354°
9. 78°, 282°
10. 55°, 125°
11. 215°, 325°
12. 100°, 260°

Evaluate each expression. *Section 15.2*

13. $\cos 170°$ **14.** $\sin 380°$ **15.** $\sin(-110°)$ **16.** $\cos(-55°)$

17. $\sin 156°$ **18.** $\cos(-450°)$ **19.** $\cos 230°$ **20.** $\sin(-25°)$

Convert each angle measure from radians to degrees (to the nearest degree). *Section 15.2*

21. $\frac{3\pi}{2}$ **22.** $\frac{5\pi}{6}$ **23.** 4.2 **24.** $-\frac{\pi}{3}$

25. -3.2 **26.** 4π **27.** $\frac{3\pi}{8}$ **28.** -5.6

Convert each angle measure from degrees to radians. Express each answer as a decimal to the nearest hundredth of a radian, and in terms of π. *Section 15.2*

29. 60° **30.** $-45°$ **31.** 150° **32.** 450°

33. 195° **34.** 226° **35.** $-130°$ **36.** 144°

Graph each function. State the amplitude and the period of each graph. *Section 15.3*

37. $y = 4\cos 2x$ **38.** $y = -2\sin\frac{2}{3}x$ **39.** $y = \frac{3}{4}\cos 3x$ **40.** $y = -\frac{4}{3}\sin\left(-\frac{2\pi}{3}x\right)$

Write a sine function with the given maximum *M*, minimum *m*, and period. *Section 15.3*

41. $M = 4, m = -4$, period $= 2$ **42.** $M = 5, m = -1$, period $= \frac{2\pi}{3}$

43. $M = 3, m = -7$, period $= 5\pi$ **44.** $M = -2, m = -5$, period $= \frac{1}{4}$

Graph each function. Show at least one period. *Section 15.4*

45. $y = 2\sin\left(x - \frac{\pi}{3}\right)$ **46.** $y = -3\cos\frac{\pi}{2}(x - 5)$ **47.** $y = \frac{5}{4}\sin 2\left(x + \frac{2\pi}{3}\right)$ **48.** $y = 1.5\cos\frac{\pi}{4}(x + 1)$

Write a sine function for each graph. *Section 15.4*

49.

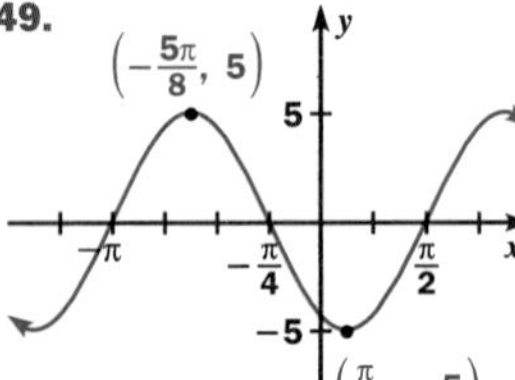

50.

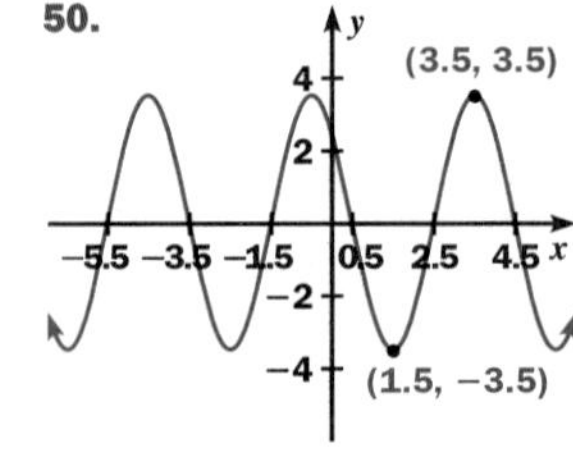

51.

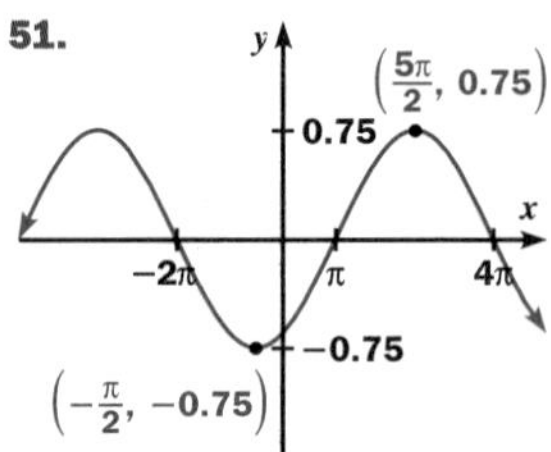

Evaluate each expression. *Section 15.5*

52. $\tan\frac{\pi}{4}$ **53.** $\tan 360°$ **54.** $\tan\frac{\pi}{3}$ **55.** $\tan 135°$

56. $\tan 38°$ **57.** $\tan\left(-\frac{7\pi}{6}\right)$ **58.** $\tan 0.62$ **59.** $\tan\pi$

Graph each function over two periods. *Section 15.5*

60. $y = \frac{1}{2}\tan\frac{x}{3}$ **61.** $y = 2\tan\frac{x}{2} + 3$ **62.** $y = -\frac{1}{3}\tan\frac{\pi}{4}(x - 2)$

13. –0.9848
14. 0.3420
15. –0.9397
16. 0.5736
17. 0.4067
18. 0
19. –0.6428
20. –0.4226
21. 270°
22. 150°
23. 241°
24. –60°
25. –183°
26. 720°
27. 68°
28. –321°
29. 1.05; $\frac{\pi}{3}$
30. –0.79; $-\frac{\pi}{4}$
31. 2.62; $\frac{5\pi}{6}$
32. 7.85; $\frac{5\pi}{2}$
33. 3.40; $\frac{13\pi}{12}$
34. 3.94; $\frac{113\pi}{90}$
35. –2.27; $-\frac{13\pi}{18}$
36. 2.51; $\frac{4\pi}{5}$
37–40. See answers in back of book.
41. $y = 4\sin\pi x$
42. $y = 3\sin 3x + 2$
43. $y = 5\sin\frac{2}{5}x - 2$
44. $y = \frac{3}{2}\sin 8\pi x - \frac{7}{2}$
45–48. See answers in back of book.
49–52. Answers may vary. Examples are given.
49. $y = 5\sin\frac{4}{3}\left(x - \frac{\pi}{2}\right)$
50. $y = 3.5\sin\frac{\pi}{2}(x - 2.5)$
51. $y = 0.75\sin\frac{1}{3}(x - \pi)$
52. 1
53. 0
54. $\sqrt{3} \approx 1.7321$
55. –1
56. 0.7813
57. $-\frac{\sqrt{3}}{3} \approx -0.5774$
58. 0.7139
59. 0

60.

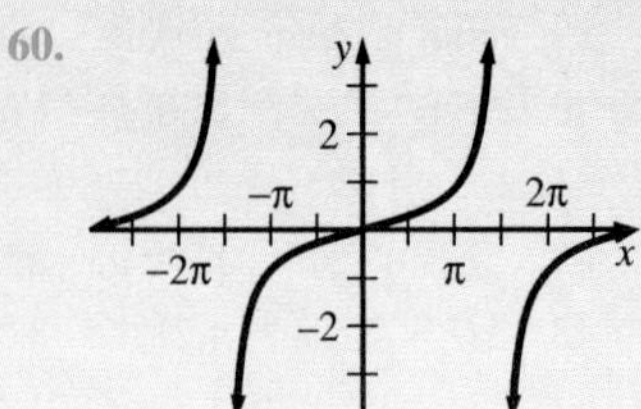

61.

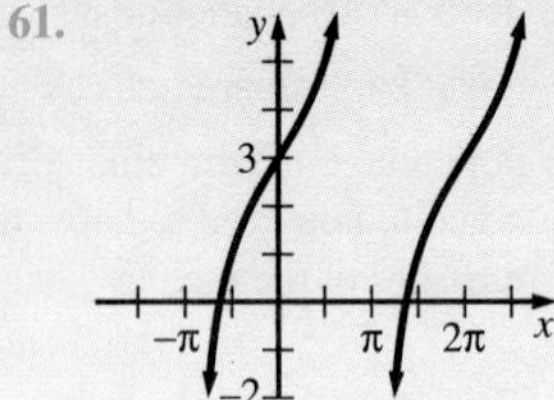

62.

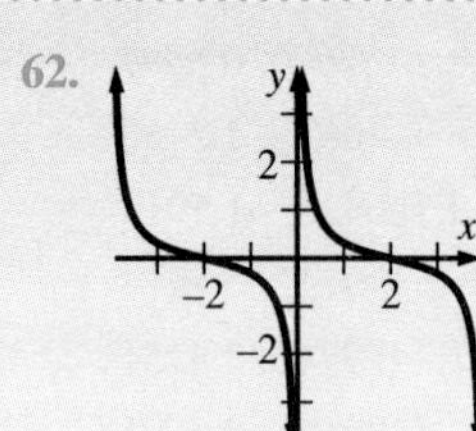

Toolbox

NUMBER OPERATIONS AND PROPERTIES

Numbers and Number Lines

The *real numbers* consist of positive numbers, negative numbers, and zero. You can graph real numbers on a *number line*.

The *coordinate* of point A is 3.

Arrows show that the line and the numbers continue without end in both directions.

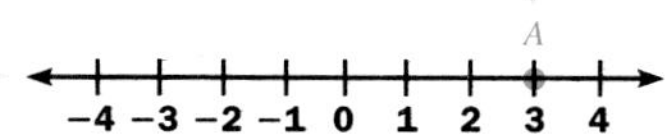

The *graph* of $x = 3$ is point A.

The set of real numbers includes each of the following sets of numbers:

whole numbers: $0, 1, 2, 3, \ldots$

integers: $\ldots, -3, -2, -1, 0, 1, 2, 3, \ldots$

positive integers: $1, 2, 3, \ldots$

negative integers: $-1, -2, -3, \ldots$

The three dots mean that the list continues.

EXAMPLE

Graph the numbers $-2\frac{1}{4}$, -2.5, and -2 on a number line. Then list the numbers in order from least to greatest.

SOLUTION

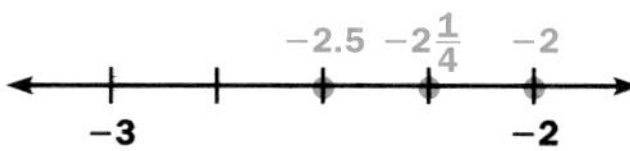

Numbers on a number line increase from left to right.

From least to greatest, the numbers are -2.5, $-2\frac{1}{4}$, and -2.

PRACTICE

Refer to the number line below. Name the point with each coordinate.

1. -0.5

2. 1

3. $-\frac{4}{3}$

4. $\frac{9}{4}$

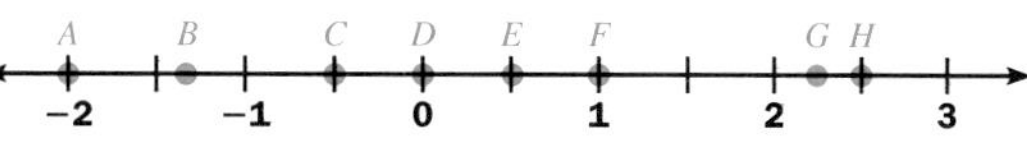

Refer to the number line above. Name the coordinate of each point.

5. D

6. A

7. H

8. E

ANSWERS Toolbox

Numbers and Number Lines

1. C
2. F
3. B
4. G
5. 0
6. -2
7. $2\frac{1}{2}$
8. $\frac{1}{2}$

9. Refer to the number line for Exercises 1–8. Name each point whose coordinate is:

a. positive **b.** negative **c.** a whole number **d.** a real number

Graph each set of numbers on a separate number line. Then list the numbers in order from least to greatest.

10. $\frac{3}{2}, -2, -0.25$ **11.** $2\frac{1}{3}, 3\frac{1}{2}, 1\frac{2}{3}$ **12.** $-\frac{3}{4}, -1.45, -1\frac{1}{4}$

Properties of Real Numbers

The following properties are useful when you are calculating with numbers or simplifying algebraic expressions.

Property	Example	Summary
Commutative $a + b = b + a$ $ab = ba$	$-2 + 3 = 3 + (-2)$ $-2(3) = 3(-2)$	You can add or multiply numbers in any order without changing the result.
Associative $(a + b) + c = a + (b + c)$ $(ab)c = a(bc)$	$(-8 + 6) + 5 = -8 + (6 + 5)$ $(-8 \cdot 6) \cdot 5 = -8 \cdot (6 \cdot 5)$	When you add or multiply three or more numbers, you can regroup the numbers without changing the result.
Distributive $a(b + c) = ab + ac$	$7(60 + 2) = 7(60) + 7(2)$	When a sum is multiplied by a number, you can distribute the multiplication to each of the numbers being added.
Absolute Value $\lvert x \rvert = x$ if $x \geq 0$ $\lvert x \rvert = -x$ (the opposite of x) if $x < 0$	$\lvert 8 \rvert = 8$ $\lvert -8 \rvert = -(-8) = 8$	The absolute value of a number is its distance from 0 on a number line. This distance is always a nonnegative number.

EXAMPLE 1

Tell what property is used in the following: $2(4a) = (2 \cdot 4)a = 8a$.

SOLUTION

The factors have been regrouped. The associative property is used.

PRACTICE

Tell what property is used in each lettered step.

1. a. $7(\lvert -3 \rvert + \lvert 5 \rvert) = 7(3 + 5)$
b. $= 7(3) + 7(5)$
$= 21 + 35$

2. a. $(4a)(7b) = (4a \cdot 7)b$
b. $= (7 \cdot 4a)b$
c. $= (7 \cdot 4)ab$
$= 28ab$

3. a. $7x + (8 + 3x) = 7x + (3x + 8)$
b. $= (7x + 3x) + 8$
c. $= (7 + 3)x + 8$
$= 10x + 8$

9. a. *E, F, G, H*
b. *A, B, C*
c. *D, F*
d. *A, B, C, D, E, F, G, H*

10.
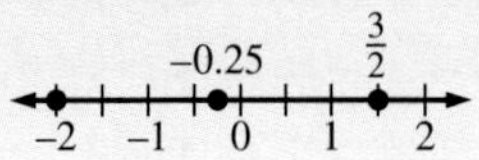
$-2, -0.25, \frac{3}{2}$

11.
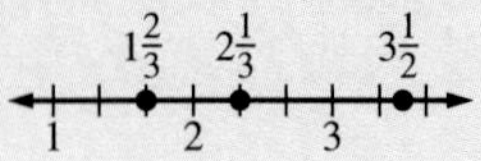
$1\frac{2}{3}, 2\frac{1}{3}, 3\frac{1}{2}$

12.
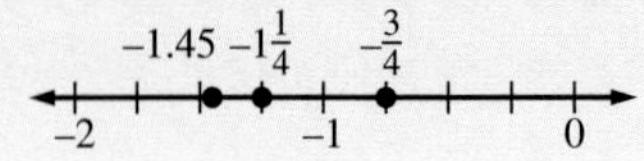
$-1.45, -1\frac{1}{4}, -\frac{3}{4}$

Properties of Real Numbers

1. a. absolute value property
b. distributive property
2. a. associative property
b. commutative property
c. associative property
3. a. commutative property
b. associative property
c. distributive property

You can show that a statement is false in general if you can find a specific example, called a *counterexample*, for which the statement is false.

EXAMPLE 2

Tell whether each property is true for all real numbers *a* and *b*. Give an example to support your answer.

a. $0 \cdot a = 0$

b. $a \div b = b \div a$

SOLUTION

a. True; the product of any number and 0 is 0; for example, $0 \cdot 7 = 0$.

b. False; for example, $2 \div 10 = 0.2$ and $10 \div 2 = 5$.

PRACTICE

Tell whether each property is true for all nonzero real numbers *a*, *b*, *c*, and *d*. Give an example to support your answer.

4. $(a + b)(a - b) = a^2 - b^2$

5. $a^2 + b^2 = (a + b)^2$

6. $a - b = a + (-b)$

7. $a(b - c) = ab - ac$

8. $1 \cdot a = 1$

9. $a \div b = a \cdot \frac{1}{b}$

10. $\frac{a}{b} + \frac{c}{d} = \frac{a + c}{b + d}$

11. $\frac{a}{b} \cdot \frac{c}{d} = \frac{ac}{bd}$

12. $|a| - |b| = |a - b|$

13. $0 + a = 0$

Percent Change

Recall that $n\%$ means $\frac{n}{100}$. For example, $54\% = \frac{54}{100} = 0.54$. To calculate a percent change, use the following formula:

$$\text{percent change} = \frac{\text{amount of change}}{\text{original amount}} \cdot 100$$

EXAMPLE

a. A stock that once sold for \$10 per share now sells for \$12 per share. Find the percent change in the price.

b. A bracelet, originally priced at \$40, is on sale for \$30. Find the percent change in the price.

SOLUTION

a.
$$\text{percent change} = \frac{\text{new price} - \text{old price}}{\text{old price}} \cdot 100$$
$$= \frac{12 - 10}{10} \cdot 100$$
$$= 20$$

There is a 20% increase in the price.

b.
$$\text{percent change} = \frac{\text{new price} - \text{old price}}{\text{old price}} \cdot 100$$
$$= \frac{30 - 40}{40} \cdot 100$$
$$= -25$$

There is a 25% decrease in the price.

4–13. Examples given may vary.

4. True; $(2 + 3)(2 - 3) = 5(-1) = -5$; $2^2 - 3^2 = 4 - 9 = -5$.

5. False; $1^2 + 2^2 = 5$, but $(1 + 2)^2 = 9$.

6. True; $3 - 5 = -2$; $3 + (-5) = -2$.

7. True; $4(1 - 3) = 4(-2) = -8$ and $4(1) - 4(3) = 4 - 12 = -8$.

8. False; $1 \cdot 13 = 13$.

9. True; $12 \div 3 = 4$ and $12 \cdot \frac{1}{3} = 4$.

10. False; $\frac{1}{2} + \frac{3}{4} = \frac{5}{4}$ and $\frac{1+3}{2+4} = \frac{2}{3}$.

11. True; $\frac{1}{2} \cdot \frac{3}{4} = \frac{3}{8}$ and $\frac{1 \cdot 3}{2 \cdot 4} = \frac{3}{8}$.

12. False; $|3| - |7| = 3 - 7 = -4$ and $|3 - 7| = |-4| = 4$.

13. False; $0 + (-27) = -27$.

PRACTICE

Find the percent change. Round answers to the nearest tenth.

1. The Goldsteins bought a house in 1990 for \$124,000. They sold the house in 1995 for \$128,500.
2. A car purchased for \$12,400 is worth only \$9810 one year later.
3. The price of lettuce was \$.69/lb during the summer and \$1.39/lb during the following winter.

Exponents and Powers

The expression b^n is called a *power*. The number b is called the *base*, and the number n is called the *exponent*. The expression means you use b as a factor n times.

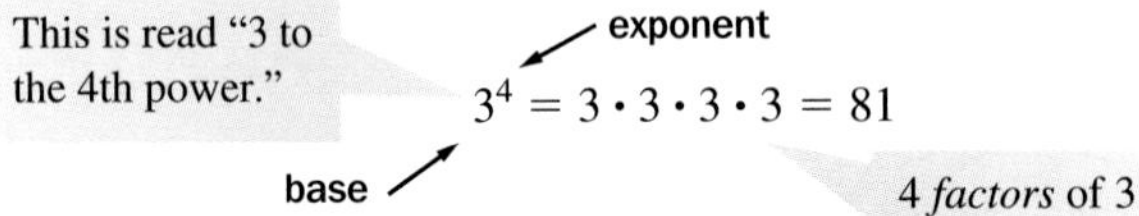

Here are some properties of exponents for $b > 0$:

Property	Summary
Product property: $b^m \cdot b^n = b^{m+n}$	When you multiply powers with the same base, add the exponents.
Quotient property: $\frac{b^m}{b^n} = b^{m-n}$	When you divide powers with the same base, subtract the exponents.
Power property: $(b^m)^n = b^{mn}$	When you raise a power to a power, multiply the exponents.
Zero exponent: $b^0 = 1$	When you raise any nonzero base to the 0 power, you get 1.
Negative exponent: $b^{-n} = \frac{1}{b^n}$	A nonzero base raised to a negative power is the same as the reciprocal of the base raised to the opposite power.

EXAMPLE

Simplify using properties of exponents.

a. $(-5)^3$

b. $\frac{2^8}{2^5}$

c. $(x^4)^2$

SOLUTION

a. $(-5)^3 = (-5)(-5)(-5)$
$= -125$

b. $\frac{2^8}{2^5} = 2^{8-5}$
$= 2^3$
$= 8$

c. $(x^4)^2 = x^{4 \cdot 2}$
$= x^8$

Percent Change

1. 3.6% increase
2. 20.9% decrease
3. 101.4% increase

PRACTICE

Simplify using properties of exponents.

1. 4^3	**2.** $(-2)^5$	**3.** 12^0	**4.** $(-3)^{-4}$
5. $3^2 \cdot 3^3$	**6.** $5^0 \cdot 5^4$	**7.** $2^7 \cdot 2^{-7}$	**8.** $5^{-3} \cdot 5^2$
9. $a^7 \cdot a^3$	**10.** $\frac{b^8}{b^3}$	**11.** $c^{-3} \cdot c^{-2}$	**12.** $\frac{d^2}{d^5}$
13. $(a^2)^3$	**14.** $(b^{-3})^2$	**15.** $(c^5)^0$	**16.** $(d^{-1})^{-1}$

Square Roots

A number x is a *square root* of a number y if $x^2 = y$. For example, 5 and -5 are the square roots of 25 because $5^2 = 25$ and $(-5)^2 = 25$. Every positive number has two square roots.

25 has two square roots. ⟶ $\sqrt{25} = 5$ and $-\sqrt{25} = -5$ $\quad$ $\sqrt{0} = -\sqrt{0} = 0$ ⟵ 0 has only one square root.

EXAMPLE 1

Simplify.

a. $-\sqrt{81}$

b. $\sqrt{\frac{4}{9}}$

SOLUTION

a. $81 = 9^2$, so $-\sqrt{81} = -9$

b. $\frac{4}{9} = \left(\frac{2}{3}\right)^2$, so $\sqrt{\frac{4}{9}} = \frac{2}{3}$

Here are two properties of square roots for a and b not both negative and $b \neq 0$:

Property	Summary
Product property: $\sqrt{ab} = \sqrt{a} \cdot \sqrt{b}$	The square root of the product of a and b equals the product of the square roots of a and b.
Quotient property: $\sqrt{\frac{a}{b}} = \frac{\sqrt{a}}{\sqrt{b}}$	The square root of the quotient of a and b equals the quotient of the square roots of a and b.

EXAMPLE 2

Simplify.

a. $\sqrt{48}$

b. $\sqrt{3} \cdot \sqrt{6}$

c. $\frac{\sqrt{3}}{\sqrt{27}}$

SOLUTION

a. $\sqrt{48} = \sqrt{16 \cdot 3} = \sqrt{16} \cdot \sqrt{3} = 4\sqrt{3}$

b. $\sqrt{3} \cdot \sqrt{6} = \sqrt{3 \cdot 6} = \sqrt{3 \cdot 3 \cdot 2} = 3\sqrt{2}$

c. $\frac{\sqrt{3}}{\sqrt{27}} = \sqrt{\frac{3}{27}} = \sqrt{\frac{1}{9}} = \frac{1}{3}$

Exponents and Powers

1. 64
2. –32
3. 1
4. $\frac{1}{81}$
5. 243
6. 625
7. 1
8. $\frac{1}{5}$
9. a^{10}
10. b^5
11. $\frac{1}{c^5}$
12. $\frac{1}{d^3}$
13. a^6
14. $\frac{1}{b^6}$
15. 1
16. d

PRACTICE

Find the square roots of each number. If the number has no real square roots, write *none*.

1. 49 **2.** -100 **3.** $\frac{1}{64}$ **4.** 0.16

Simplify, if possible.

5. $\sqrt{400}$ **6.** $-\sqrt{1.69}$ **7.** $\sqrt{-1}$ **8.** $\sqrt{\frac{100}{121}}$

9. $\sqrt{125}$ **10.** $-\frac{1}{2}\sqrt{72}$ **11.** $0.7\sqrt{300}$ **12.** $\sqrt{10} \cdot \sqrt{6}$

13. $3\sqrt{2} \cdot 4\sqrt{3}$ **14.** $\frac{\sqrt{12}}{\sqrt{3}}$ **15.** $\frac{\sqrt{6}}{\sqrt{8}}$ **16.** $\frac{\sqrt{8}}{\sqrt{18}}$

17. Is the statement $\sqrt{a^2} = a$ true for all values of a? Explain your answer.

Order of Operations

When you simplify a numerical expression, you must use the *order of operations*, a set of rules that guarantees that an expression has just one value.

Order of Operations

1. First simplify expressions inside parentheses or other grouping symbols.
2. Then evaluate powers.
3. Next, do multiplications and divisions in order from left to right.
4. Last, do additions and subtractions in order from left to right.

EXAMPLE

Simplify each expression.

a. $3(4 - 6) + 2$ **b.** $\frac{1 - 7^2}{2(5 - 1)}$ **c.** $6^3 \div 4 \cdot 5 + 3 - 4 \div 2$ **d.** $|-2 \div 5^2|$

SOLUTION

a. $3(4 - 6) + 2 = 3(-2) + 2$

$= -6 + 2$

$= -4$

b. Simplify the numerator and the denominator before dividing.

$$\frac{1 - 7^2}{2(5 - 1)} = \frac{1 - 49}{2(4)}$$

$$= \frac{-48}{8}$$

$$= -6$$

The fraction bar acts like parentheses: $(1 - 7^2) \div (2(5 - 1))$

Square Roots

1. ± 7
2. none
3. $\pm\frac{1}{8}$
4. ± 0.4
5. 20
6. -1.3
7. not possible
8. $\frac{10}{11}$
9. $5\sqrt{5}$
10. $-3\sqrt{2}$
11. $7\sqrt{3}$
12. $2\sqrt{15}$
13. $12\sqrt{6}$
14. 2
15. $\frac{\sqrt{3}}{2}$
16. $\frac{2}{3}$
17. No; the statement is true only if $a \geq 0$. If $a < 0$, $\sqrt{a^2} = -a$.

c. $6^3 \div 4 \cdot 5 + 3 - 4 \div 2 = 216 \div 4 \cdot 5 + 3 - 4 \div 2$ — Evaluate powers.

$= 54 \cdot 5 + 3 - 2$

$= 270 + 3 - 2$ — Multiply and divide in order from left to right.

$= 273 - 2$

$= 271$ — Add and subtract in order from left to right.

d. Simplify the expression inside the absolute value bars first.

$|-2 \div 5^2| = |-2 \div 25|$

$= |-0.08|$

$= 0.08$ — The absolute value of a negative number is positive.

PRACTICE

Simplify each expression.

1. $9 - 5 - 6 + 1$ **2.** $9 - (5 - 6) + 1$ **3.** $2 \cdot 5 + 7 \cdot 6$ **4.** $2(5 + 7 \cdot 6)$

5. $2(5 + 7) \cdot 6$ **6.** $|2 - 5| + |5 - 2|$ **7.** $9 - 6^2 \div 2 + 4$ **8.** $9 - (6^2 \div 2 + 4)$

9. $(9 - 6^2) \div (2 + 4)$ **10.** $(9 - 6^2) \div 2 + 4$ **11.** $\frac{1 - 5^3}{(1 + 3)^2}$ **12.** $\frac{7 + 3 \cdot 11}{1 + 8 \div 2}$

Rational Numbers and Irrational Numbers

The numbers that you graph on a number line are *real numbers*. Every real number is either *rational* or *irrational*.

A *rational number* is a number that can be written as a ratio of two integers or as a decimal that repeats or ends.

An *irrational number* cannot be written as a ratio of two integers. When written as a decimal, an irrational number neither repeats nor ends.

EXAMPLE

Tell whether each number belongs to the *whole numbers*, the *integers*, the *rational numbers*, the *irrational numbers*, and the *real numbers*.

a. $-\sqrt{4}$ **b.** 2.777. . . **c.** $\sqrt{5}$

SOLUTION

a. $-\sqrt{4} = -2$, so $-\sqrt{4}$ is an integer, a rational number $\left(-2 = -\frac{2}{1}\right)$, and a real number.

b. 2.777. . . is a decimal that repeats, so it is a rational number and a real number.

c. $\sqrt{5} = 2.236067977\ldots$, which is a decimal that neither repeats nor ends, so it is an irrational number and a real number.

Order of Operations

1. –1
2. 11
3. 52
4. 94
5. 144
6. 6
7. –5
8. –13
9. –4.5
10. –9.5
11. –7.75
12. 8

PRACTICE

Tell whether each number belongs to the *whole numbers*, the *integers*, the *rational numbers*, the *irrational numbers*, and the *real numbers*. **(See page 773.)**

1. 3.2 **2.** $\sqrt{9}$ **3.** 1.234567891011... **4.** 0

5. π **6.** $-\frac{1}{4}$ **7.** 8.535353... **8.** $\sqrt{2}$

Name a real number that fits each description. If there is no such number, write *not possible*.

9. an integer that is not a whole number

10. a real number that is not rational

11. a real number that is not positive or negative

12. a number that is rational and irrational

OPERATIONS WITH VARIABLE EXPRESSIONS

Evaluating Variable Expressions

A *variable expression* is an expression formed using variables, numbers, and operation symbols. To *evaluate* a variable expression, substitute a value for each variable and simplify the resulting expression using the *order of operations*.

EXAMPLE

Evaluate each expression when $a = -5$ and $b = 2$.

a. $\frac{a+b}{2}$ **b.** $a^3 - 2a + 4$ **c.** $7(a+b)^2 - a$

SOLUTION

a. $\frac{a+b}{2} = \frac{-5+2}{2}$

$= \frac{-3}{2}$ Simplify the numerator first.

$= -1.5$

b. $a^3 - 2a + 4 = (-5)^3 - 2(-5) + 4$ Evaluate powers.

$= -125 - 2(-5) + 4$ Multiply and divide.

$= -125 + 10 + 4$

$= -111$ Add and subtract.

c. $7(a+b)^2 - a = 7(-5+2)^2 - (-5)$

$= 7(-3)^2 - (-5)$ Simplify inside parentheses first.

$= 7(9) - (-5)$

$= 63 - (-5)$

$= 63 + 5$ Remember: Subtracting is the same as adding the opposite.

$= 68$

Rational Numbers and Irrational Numbers

1. the rational numbers, the real numbers
2. the whole numbers, the integers, the rational numbers, the real numbers
3. the irrational numbers, the real numbers
4. the whole numbers, the integers, the rational numbers, the real numbers
5. the irrational numbers, the real numbers
6. the rational numbers, the real numbers
7. the rational numbers, the real numbers
8. the irrational numbers, the real numbers

9, 10. Examples may vary.

9. any negative integer
10. π
11. 0
12. not possible

PRACTICE

Evaluate each expression when $x = 4$.

1. $\frac{1}{2}x - 4$
2. $x^2 - 5x - 2$
3. $3(7 - x) + 5$
4. $x^3 - 5x^2$
5. $3(x - 6)^2 + 3x$
6. $47 + 7.25x$
7. $\frac{6}{x - 1}$
8. $x + \frac{2x + 7}{3}$

Evaluate each expression when $r = -7$ and $s = 5$.

9. $-4(r + s) - 5$
10. $9 - 3r + s^2$
11. $9 - 3(r + s^2)$
12. $9 - (3r + s^2)$
13. $(r + 2s)^3$
14. $-6rs + rs^2$
15. $\frac{3r - 4}{2s}$
16. $\frac{r^2 + rs}{r + s}$

Simplifying Variable Expressions

For a variable expression like $3x - 2 + x$, the parts that are joined by plus signs or minus signs are called *terms*. Terms with the same variable parts, such as $3x$ and x, are called *like terms.*

To write an expression in *simplest form*, find an *equivalent expression* that has no parentheses and has all like terms combined. In simplest form, the expression $3x - 2 + x$ becomes $4x - 2$.

EXAMPLE

Simplify.

a. $-(1 - 5x + x^2)$

b. $(t^2 - 5t) + (-t^2 + 6t + 3)$

SOLUTION

a. $-(1 - 5x + x^2) = -1(1 - 5x + x^2)$

$= -1(1) - (-1)(5x) + (-1)(x^2)$ — Use the distributive property.

$= -1 - (-5x) + (-x^2)$

$= -1 + 5x - x^2$

b. $(t^2 - 5t) + (-t^2 + 6t + 3) = (t^2 - t^2) + (-5t + 6t) + 3$ — Group like terms.

$= 0t^2 + (-5 + 6)t + 3$ — Use the distributive property.

$= t + 3$ — Remember: $1 \cdot t = t$.

PRACTICE

Simplify.

1. $2(7t - 4) - 8t$
2. $-5(k + 1) + 4$
3. $-(a - 3b + 2)$
4. $-(x^2 + 7) + 7$
5. $3(-4 - 2j) + 3j$
6. $12 - 4(-a + 4)$
7. $5x + 3y - 4x + y$
8. $-a + 2ab - 3b + 7a$
9. $m^2 + 2m + 5 - 3m$
10. $(a - 5b) + (4b + 7)$
11. $(a - 5b) - (4b + 7)$
12. $a - (5b - 4b + 7)$
13. $2(3 - 5x) - (x - 1)$
14. $-1.5(4 + 2k) + (3k - 10)$
15. $(3y^2 - 4y + 1) + (y^2 + y - 5)$
16. $-(3d^2 - 7d) + 5(1 + d)$
17. $(6 - n + n^2) - (n^3 - 7n - 8)$
18. $3(c^2 + 5cd + 4d^2) - (c^2 - d^2)$

Evaluating Variable Expressions

1. –2
2. –6
3. 14
4. –16
5. 24
6. 76
7. 2
8. 9
9. 3
10. 55
11. –45
12. 5
13. 27
14. 35
15. –2.5
16. –7

Simplifying Variable Expressions

1. $6t - 8$
2. $-5k - 1$
3. $-a + 3b - 2$
4. $-x^2$
5. $-12 - 3j$
6. $-4 + 4a$
7. $x + 4y$
8. $6a + 2ab - 3b$
9. $m^2 - m + 5$
10. $a - b + 7$
11. $a - 9b - 7$
12. $a - b - 7$
13. $7 - 11x$
14. -16
15. $4y^2 - 3y - 4$
16. $-3d^2 + 12d + 5$
17. $-n^3 + n^2 + 6n + 14$
18. $2c^2 + 15cd + 13d^2$

Multiplying Variable Expressions

To find the product of two variable expressions, use the *distributive property*. (See page 774.)

EXAMPLE

Multiply.

a. $-t(2t - 3)$

b. $(x - 3)(4x + 1)$

SOLUTION

a. $-t(2t - 3) = -t(2t) - (-t)(3)$
$= (-1 \cdot 2)(t \cdot t) - (-3t)$
$= -2t^2 + 3t$

b. $(x - 3)(4x + 1) = x(4x + 1) - 3(4x + 1)$ ← Use the distributive property.
$= [x(4x) + x(1)] - [3(4x) + 3(1)]$ ← Use it again.
$= (4x^2 + x) - (12x + 3)$
$= 4x^2 + x - 12x - 3$
$= 4x^2 - 11x - 3$

PRACTICE

Multiply.

1. $b(2b - 7)$
2. $2m(m + 4)$
3. $-h(5 - h)$
4. $(r + 2)(r - 1)$
5. $(p + 4)(-p + 7)$
6. $(s - 4)(s + 4)$
7. $(2c + 3)(c + 3)$
8. $(5z - 1)(z - 2)$
9. $(8 - 3w)(4 + w)$
10. $(3j - 1)(2j - 3)$
11. $(9 - 4y)(1 + 3y)$
12. $(2k + 5)(5k + 2)$
13. $(a - 5)^2$
14. $(m + 3)^2$
15. $(2s + 7)^2$
16. $(5j - 1)^2$

Factoring Variable Expressions

The distributive property can be read in two ways:

$a(b + c) = ab + ac$ — Distribute *a* to *b* and *c*.

$ab + ac = a(b + c)$ — Factor *a* from *ab* and *ac*.

Distributing changes a product into a sum. Factoring changes a sum into a product by finding common factors. Each process is the reverse of the other.

To factor a sum, first find the factors of each addend. Then look for the *greatest common factor* (GCF).

Multiplying Variable Expressions

1. $2b^2 - 7b$
2. $2m^2 + 8m$
3. $-5h + h^2$
4. $r^2 + r - 2$
5. $-p^2 + 3p + 28$
6. $s^2 - 16$
7. $2c^2 + 9c + 9$
8. $5z^2 - 11z + 2$
9. $32 - 4w - 3w^2$
10. $6j^2 - 11j + 3$
11. $9 + 23y - 12y^2$
12. $10k^2 + 29k + 10$
13. $a^2 - 10a + 25$
14. $m^2 + 6m + 9$
15. $4s^2 + 28s + 49$
16. $25j^2 - 10j + 1$

EXAMPLE

Rewrite each expression in factored form.

a. $12a + 16b$

b. $3z^2 - 5z$

SOLUTION

a. $12a + 16b = (2 \cdot 2 \cdot 3 \cdot a) + (2 \cdot 2 \cdot 2 \cdot 2 \cdot b)$ — Factor each term.

$= (4 \cdot 3a) + (4 \cdot 4b)$

$= 4(3a + 4b)$ — The GCF of the terms is $2 \cdot 2$, or 4.

b. $3z^2 - 5z = (3 \cdot z \cdot z) - (5 \cdot z)$

$= z(3z - 5)$ — The GCF is z.

PRACTICE

Rewrite each expression in factored form.

1. $2t + 8$ **2.** $-6m + 9$ **3.** $r^2 - 5r$ **4.** $2v^2 + 7v$

5. $18j + 6k$ **6.** $-9d - 15f$ **7.** $-4g + 15g^2$ **8.** $bc + bd$

9. $21a^2 + 49$ **10.** $5c + 25c^2$ **11.** $-2pt + 3st$ **12.** $20x^2 - 50y^2$

Adding and Subtracting Rational Expressions

A *rational expression* is an expression that can be written as a fraction. If two rational expressions have the same denominator, you can add or subtract them by adding or subtracting the numerators. If the expressions have different denominators, rewrite them using the *least common denominator* (LCD).

EXAMPLE

Simplify.

a. $\frac{4}{x-2} - \frac{2x}{x-2}$

b. $\frac{4}{x-2} + 2$

SOLUTION

a. $\frac{4}{x-2} - \frac{2x}{x-2} = \frac{4-2x}{x-2}$ — Subtract the numerators.

$= \frac{-2(x-2)}{x-2}$ — Factor the numerator.

$= -2$

b. $\frac{4}{x-2} + 2 = \frac{4}{x-2} + \frac{2(x-2)}{x-2}$ — Rewrite using the LCD $x - 2$.

$= \frac{4}{x-2} + \frac{2x-4}{x-2}$ — Use the distributive property.

$= \frac{4+2x-4}{x-2}$ — Add the numerators.

$= \frac{2x}{x-2}$

Factoring Variable Expressions

1. $2(t + 4)$
2. $3(-2m + 3)$ or $-3(2m - 3)$
3. $r(r - 5)$
4. $v(2v + 7)$
5. $6(3j + k)$
6. $-3(3d + 5f)$
7. $g(-4 + 15g)$ or $-g(4 - 15g)$
8. $b(c + d)$
9. $7(3a^2 + 7)$
10. $5c(1 + 5c)$
11. $t(-2p + 3s)$ or $-t(2p - 3s)$
12. $10(2x^2 - 5y^2)$

PRACTICE

Simplify.

1. $\frac{1}{t} + \frac{5}{t}$ **2.** $\frac{1}{t} - \frac{5}{t}$ **3.** $\frac{2n}{n+1} + \frac{3}{n+1}$ **4.** $\frac{2n}{n+1} - \frac{3}{n+1}$

5. $\frac{3y}{y-1} - \frac{3}{y-1}$ **6.** $\frac{1}{4k} + \frac{3}{4k}$ **7.** $\frac{5}{x} + \frac{3}{2x}$ **8.** $\frac{5}{x} - \frac{3}{2x}$

9. $\frac{1}{3} + \frac{1}{2x-1}$ **10.** $\frac{4}{5z} + 6$ **11.** $1 - \frac{1}{r+2}$ **12.** $2 + \frac{1}{2x-1}$

13. $\frac{c}{c+4} + 1$ **14.** $\frac{c}{c+4} - 1$ **15.** $\frac{1}{y} + \frac{1}{y+1}$ **16.** $\frac{2}{m+2} - \frac{1}{m-2}$

LINEAR EQUATIONS AND INEQUALITIES

Solving Linear Equations

An *equation* is a mathematical sentence that says that two expressions are equal. A linear equation with one variable can be written in the form $ax + b = c$, where a, b, and c are real numbers and $a \neq 0$.

You *solve* an equation by finding each value of the variable that makes the equation true. To do this, undo operations to get the variable alone on one side. It's always a good idea to check your solutions in the original equation.

EXAMPLE

Solve each linear equation.

a. $\frac{x-4}{3} = -2$ **b.** $7(x+2) = 4x + 20$

SOLUTION

a. $\frac{x-4}{3} = -2$

$x - 4 = -6$ ← Multiply both sides by 3.

$x = -6 + 4$ ← Add 4 to both sides.

$x = -2$

Check

$\frac{-2-4}{3} \stackrel{?}{=} -2$

$\frac{-6}{3} \stackrel{?}{=} -2$

$-2 = -2$ ✓

b. $7(x+2) = 4x + 20$

$7x + 14 = 4x + 20$ ← Use the distributive property.

$3x + 14 = 20$ ← Subtract 4x from both sides.

$3x = 6$ ← Subtract 14 from both sides.

$x = 2$ ← Divide both sides by 3.

Check

$7(2+2) \stackrel{?}{=} 4(2) + 20$

$7(4) \stackrel{?}{=} 8 + 20$

$28 = 28$ ✓

Adding and Subtracting Rational Expressions

1. $\frac{6}{t}$
2. $-\frac{4}{t}$
3. $\frac{2n+3}{n+1}$
4. $\frac{2n-3}{n+1}$
5. 3
6. $\frac{1}{k}$
7. $\frac{13}{2x}$
8. $\frac{7}{2x}$
9. $\frac{2x+2}{3(2x-1)}$
10. $\frac{4+30z}{5z}$
11. $\frac{r+1}{r+2}$
12. $\frac{4x-1}{2x-1}$
13. $\frac{2c+4}{c+4}$
14. $-\frac{4}{c+4}$
15. $\frac{2y+1}{y(y+1)}$
16. $\frac{m-6}{(m+2)(m-2)}$

PRACTICE

Solve each linear equation.

1. $\frac{1}{8} + k = \frac{7}{8}$ **2.** $4z = -8$ **3.** $-12 = -16r$ **4.** $n - 5 = 2.4$

5. $7a + 3 = -18$ **6.** $9 - z = z + 5$ **7.** $\frac{x}{4} - 3 = 1$ **8.** $2 - \frac{y}{8} = 0$

9. $\frac{v - 2}{9} = 2$ **10.** $\frac{8 - s}{3} = s$ **11.** $4(h + 3) = h$ **12.** $-2(3x + 1) = 4x$

13. $\frac{r}{5} + 7 = 3r$ **14.** $\frac{3b + 5}{4} = b$ **15.** $2(y + 5) = 6 - 3y$ **16.** $\frac{4j}{5} + 7.2 = 5.6$

Ratios and Proportions

When one number is divided by another, the quotient is called a *ratio*. Ratios can be expressed in several equivalent ways:

$a \div b$ $\frac{a}{b}$ a to b $a : b$

When two ratios are set equal, you get an equation that is called a *proportion*. To solve a proportion, use the *means-extremes property*:

If $\frac{a}{b} = \frac{c}{d}$, then $ad = bc$.

EXAMPLE

Solve each proportion.

a. $\frac{3}{2x} = \frac{15}{8}$ **b.** $\frac{x - 1}{x + 4} = 6$

SOLUTION

a.
$$\frac{3}{2x} = \frac{15}{8}$$
$$3 \cdot 8 = 2x \cdot 15$$
$$24 = 30x$$
$$\frac{24}{30} = x$$
$$0.8 = x$$

b.
$$\frac{x - 1}{x + 4} = \frac{6}{1}$$
$$1(x - 1) = 6(x + 4)$$
$$x - 1 = 6x + 24$$
$$-1 = 5x + 24$$
$$-25 = 5x$$
$$-5 = x$$

PRACTICE

Solve each proportion.

1. $\frac{x}{4} = \frac{5}{12}$ **2.** $\frac{9}{x} = \frac{-2}{7}$ **3.** $\frac{10}{9} = \frac{3x}{5}$ **4.** $\frac{7}{15} = \frac{21}{5x}$

5. $\frac{4t - 1}{5} = \frac{3}{4}$ **6.** $\frac{5}{6} = \frac{3k + 7}{8}$ **7.** $\frac{14}{3} = \frac{x - 6}{2x + 5}$ **8.** $\frac{12}{3 - r} = \frac{16}{2r + 1}$

Solving Linear Equations

1. $\frac{3}{4}$
2. -2
3. $\frac{3}{4}$
4. 7.4
5. -3
6. 2
7. 16
8. 16
9. 20
10. 2
11. -4
12. $-\frac{1}{5}$
13. $2\frac{1}{2}$
14. 5
15. $-\frac{4}{5}$
16. -2

Ratios and Proportions

1. $1\frac{2}{3}$
2. $-31\frac{1}{2}$
3. $1\frac{23}{27}$
4. 9
5. $1\frac{3}{16}$
6. $-\frac{1}{9}$
7. $-3\frac{13}{25}$
8. $\frac{9}{10}$

Use a proportion to solve each problem.

9. In 1980, Janice Brown flew the first solar-powered aircraft 6 mi in 22 min near Marana, Arizona. During that flight, about how far did she travel in the first 2 min?

10. Carlos Rodriguez bought a 3 lb roast beef on sale for \$5.67. If he had bought a $3\frac{1}{2}$ lb roast instead, how much would the roast have cost?

11. On a map of Pennsylvania, 1 in. represents 10 mi. The distance between Philadelphia, Pennsylvania, and Allentown, Pennsylvania, is 49 mi. How far apart are these cities on the map?

Identities and False Statements

Sometimes when you solve an equation you may get a puzzling result: a statement that is always true or a statement that is never true.

EXAMPLE

Solve each linear equation.

a. $7x - 9 = -\frac{1}{2}(18 - 14x)$

b. $7x - 9 = x + 6x$

SOLUTION

a. $7x - 9 = -\frac{1}{2}(18 - 14x)$

$7x - 9 = -9 + 7x$

$-9 = -9$

The equation $-9 = -9$ is an *identity*, because it is always true. When solving an equation leads to an identity, the equation has all real numbers as solutions.

b. $7x - 9 = x + 6x$

$7x - 9 = 7x$

$-9 = 0$

The equation $-9 = 0$ is a *false statement* since $-9 \neq 0$. When solving an equation leads to a false statement, the equation has no solution.

PRACTICE

Solve each equation. If all real numbers are solutions, write *all real numbers.* If there is no solution, write *no solution.*

1. $x - 8 = -(8 - x)$

2. $x - 8 = x + 3$

3. $x - 8 = 4 - x$

4. $2t - 5 = 5 - 2t$

5. $2t - 5 = 5 + 2t$

6. $2t - 5 = 2(t - 2.5)$

7. $3(2a + 3) = 9 + 6a$

8. $3(2a + 3) = 6(a + 1)$

9. $3(2a + 3) = 2(a - 3)$

10. $20 - 8b = 4(2b - 1)$

11. $\frac{3}{4}h + 8 = 2$

12. $\frac{5 + 2r}{2} = r + 6$

13. $23z - 46 = -46$

14. $-18 + 9x = -9(2 - x)$

15. $0.4k + 1 = 0.02(50 + 20k)$

9. $\frac{6}{11}$ mi $\approx$ 0.55 mi
10. \$6.62
11. 4.9 in.

Identities and False Statements

1. all real numbers
2. no solution
3. 6
4. $2\frac{1}{2}$
5. no solution
6. all real numbers
7. all real numbers
8. no solution
9. $-3\frac{3}{4}$
10. $1\frac{1}{2}$
11. –8
12. no solution
13. 0
14. all real numbers
15. all real numbers

Solving Linear Inequalities

You solve a linear inequality such as $\frac{2}{3}n - 5 > 1$ in much the same way that you solve a linear equation. One important difference is this: When you multiply or divide both sides of an inequality by a negative number, you must reverse the direction of the inequality sign.

EXAMPLE

Solve each inequality. Graph the solution on a number line.

a. $\frac{2}{3}n - 5 > 1$ **b.** $-3(z + 2) \leq 7 + z$

SOLUTION

a.

$$\frac{2}{3}n - 5 > 1$$

$$\frac{2}{3}n - 5 + 5 > 1 + 5$$ Add 5 to both sides.

$$\frac{2}{3}n > 6$$

$$\frac{3}{2} \cdot \frac{2}{3}n > \frac{3}{2} \cdot 6$$ Multiply both sides by the reciprocal of $\frac{2}{3}$.

$$n > 9$$

The solution is all real numbers greater than 9.

The *open* circle shows that 9 is not a solution.

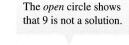

6 7 8 9 10 11

b.

$$-3(z + 2) \leq 7 + z$$ Use the distributive property.

$$-3z - 6 \leq 7 + z$$ Add 6 to both sides.

$$-3z \leq 13 + z$$ Subtract z from both sides.

$$-4z \leq 13$$

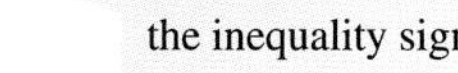

$$\frac{-4z}{-4} \geq \frac{13}{-4}$$ Divide both sides by -4. *Reverse* the direction of the inequality sign.

$$z \geq -3.25$$

The solution is all real numbers greater than or equal to -3.25.

The *solid* circle shows that -3.25 is a solution.

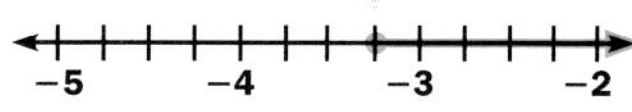

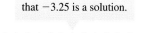

PRACTICE

Solve each inequality. Graph the solution on a number line.

1. $2x < 18$ **2.** $y - 4 \geq -7$ **3.** $-4k < -8$ **4.** $\frac{1}{3}z \leq -1$

5. $2 - p > -3$ **6.** $r + 7 \leq 5$ **7.** $11 + 2c > 35$ **8.** $3 - 7x \leq 17$

9. $1.2d + 1 < 4$ **10.** $5h - 6 \geq 2h$ **11.** $-\frac{1}{3}t + 2 < t$ **12.** $-(s - 3) \leq 3$

13. $4 - 2g > 3g - 6$ **14.** $4(x + 3) \geq -12$ **15.** $-6(w - 4) < 2w$ **16.** $5(1 - m) \leq -10m$

Solving Linear Inequalities

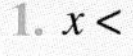

1. $x < 9$

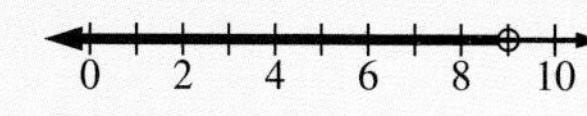

2. $y \geq -3$

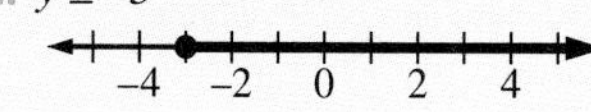

3. $k > 2$

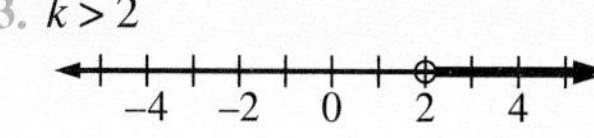

4. $z \leq -3$

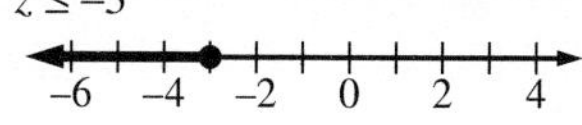

5. $p < 5$

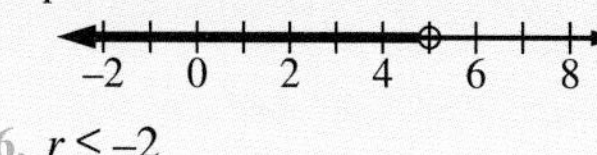

6. $r \leq -2$

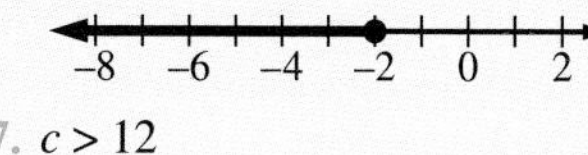

7. $c > 12$

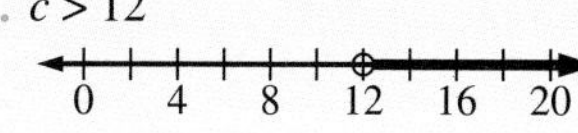

8. $x \geq -2$

9. $d < 2.5$

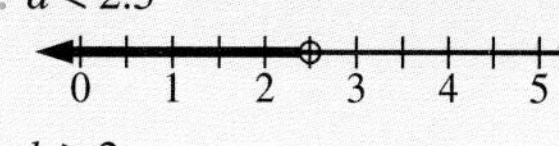

10. $h \geq 2$

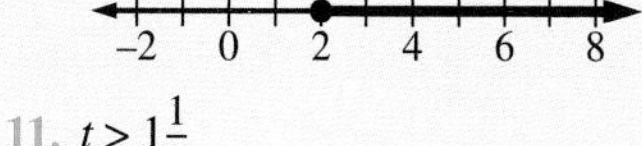

11. $t > 1\frac{1}{2}$

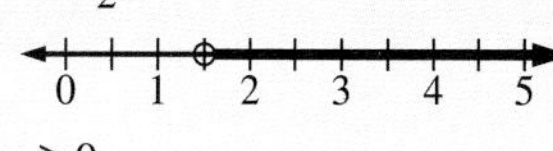

12. $s \geq 0$

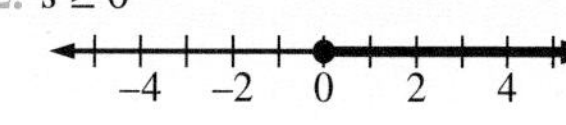

13. $g < 2$

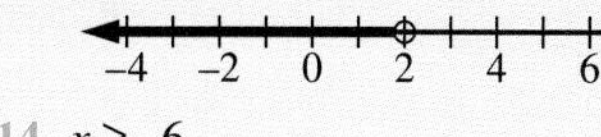

14. $x \geq -6$

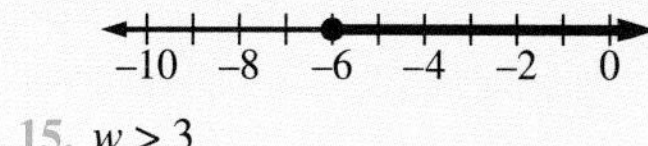

15. $w > 3$

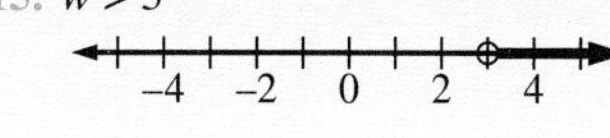

16. $m \leq -1$

Rewriting Equations and Formulas

Sometimes you need to solve an equation or a formula for a particular variable. For example, if you want to use a graphing calculator to graph the equation $2x + 3y = -6$, you need to solve the equation for y first.

EXAMPLE

Solve $2x + 3y = -6$ for y.

SOLUTION

$$2x + 3y = -6$$

$$3y = -6 - 2x$$ Subtract $2x$ from both sides.

$$\frac{1}{3}(3y) = \frac{1}{3}(-6 - 2x)$$ Multiply both sides by the reciprocal of 3.

$$y = -2 - \frac{2}{3}x$$

PRACTICE

Solve each equation or formula for the indicated variable.

1. $5x - y = 15$ for y

2. $A = \frac{1}{2}bh$ for b

3. $A = \pi r^2$ for r

4. $A = \frac{1}{2}h(b_1 + b_2)$ for b_1

5. $x = \frac{2}{3}y + 8$ for y

6. $P = 2l + 2w$ for l

7. $ax + by = c$ for y

8. $x = \frac{180(n - 2)}{n}$ for n

PROBABILITY AND STATISTICS

Experimental Probability

Probability is a ratio between 0 and 1 that measures how likely it is that an *event* will occur. If an event has a probability close to 0, it is unlikely to occur. If an event has a probability close to 1, it is almost certain to occur.

EXAMPLE 1

Express each probability as a decimal between 0 and 1, inclusive.

a. a 30% chance

b. a 1 in 5 chance

SOLUTION

a. Write the percent as a decimal.

$$30\% = 0.3$$

b. Write the probability as a fraction and convert it to a decimal.

$$\frac{1}{5} = 1 \div 5 = 0.2$$

You can calculate the *experimental probability* of an event by finding the ratio of the number of times an event occurs to the number of times an experiment is performed.

Rewriting Equations and Formulas

1. $y = 5x - 15$
2. $b = \frac{2A}{h}$
3. $r = \sqrt{\frac{A}{\pi}}$
4. $b_1 = \frac{2A}{h} - b_2$ or $b_1 = \frac{2A - b_2h}{h}$
5. $y = \frac{3}{2}x - 12$
6. $l = \frac{P - 2w}{2}$ or $l = \frac{P}{2} - w$
7. $y = \frac{c - ax}{b}$
8. $n = \frac{360}{180 - x}$

EXAMPLE 2

In the 1994 National Football League season, quarterback Troy Aikman of the Dallas Cowboys completed 233 of 361 passes. Based on this performance, find the probability that Aikman completed his first pass in the 1995 season.

SOLUTION

$$\text{probability} = \frac{\text{number of completed passes}}{\text{number of attempted passes}} = \frac{233}{361} \approx 0.65$$

PRACTICE

Express each probability as a decimal between 0 and 1, inclusive.

1. a 75% chance

2. a 1 in 3 chance

3. a 3 in 25 chance

4. a 53% chance

5. a 4 out of 4 chance

6. a 7 in 8 chance

7. It rains in San Diego, California, an average of about 42 days a year. Find the probability that it will rain in San Diego on any given day.

8. A quality control engineer at the Best Manufacturing Company selected and tested a sample of 500 batteries and found that 6 were defective.

a. Find the probability that a randomly chosen battery is defective.

b. Find the probability that a randomly chosen battery is not defective.

9. Registered voters were asked in a survey about the proposed construction of a new highway. The survey found that 132 voters are in favor of the construction, 102 are opposed, and 16 have no opinion.

a. Find the probability that a voter is in favor of the construction.

b. Suppose that 60% of voters must vote in favor of the construction in order for it to be approved. Is it likely that the construction will be approved? Explain.

Significant Digits

Because measuring tools, such as rulers and thermometers, are limited in their ability to measure precisely, measurements must be recorded with a certain number of *significant digits*. You can determine the number of significant digits using the following guidelines:

- All nonzero digits are significant. For example, 467, 46.7, and 4.67 all have three significant digits.
- For a decimal, any zeros that appear after the last nonzero digit or between two nonzero digits are significant, but any zeros that appear before the first nonzero digit are not significant. For example, 0.04067, 4067, and 46.70 all have four significant digits.
- For a whole number, any zeros that appear between two nonzero digits are significant. Unfortunately, you cannot tell whether any zeros after the last nonzero digit are significant, so you should assume that they are not significant (unless you know otherwise). For example, 47, 470, and 4700 all have two significant digits.

Experimental Probability

1. 0.75
2. 0.333 (to three decimal places)
3. 0.12
4. 0.53
5. 1
6. 0.875
7. 0.12
8. a. 0.012
   b. 0.988
9. a. 0.528
   b. No; even if those who now have no opinion decide to support the construction, only 59.2% of the voters would vote in favor of construction.

When you perform calculations using measurements having various numbers of significant digits, a general rule to follow is to give the result with no more significant digits than the measurement with the *fewest* number of significant digits. (*Note:* Many science books require that you follow more specific rules than the one given here.)

Also, when a calculation involves multiple operations, carry *all* digits through the calculation and then round the result to the appropriate number of significant digits.

EXAMPLE

Simplify. Write your answers with the appropriate number of significant digits.

a. $4.37 - 2.6$

b. 2.05×30

SOLUTION

a. $4.37 - 2.6 = 1.77$

≈ 1.8

Write the answer with two significant digits because 2.6 has two significant digits.

b. $2.05 \times 30 = 61.5$

≈ 60

Write the answer with one significant digit because 30 has one significant digit.

PRACTICE

Simplify. Write your answers with the appropriate number of significant digits.

1. $3.7 + 1.6$ **2.** $3.75 - 1.6$ **3.** 3.75×1.125 **4.** $3.7 \div 1.125$

5. $30.78 + 1.5$ **6.** $30.0 - 1.5$ **7.** 370×1.875 **8.** $3.78 \div 2.0$

9. $2.057 + 0.38$ **10.** $2.05 - 0.0375$ **11.** 0.057×0.03 **12.** $20.057 \div 4.03$

Mean, Median, and Mode

The mean, the median, and the mode are types of averages for a set of data. To find the *mean*, add the data values and divide by the number of data values. To find the *median*, write the data values in order from least to greatest and find the middle number. If there is an even number of data values, then the median is the mean of the two middle values. The *mode* is the value that appears most often. There may be no mode or more than one mode.

EXAMPLE

Find the mean, the median, and the mode of Brandon's test scores.

Brandon's Test Scores
78, 83, 88, 90, 90, 93

SOLUTION

$$\text{mean} = \frac{78 + 83 + 88 + 90 + 90 + 93}{6} = \frac{522}{6} = 87$$

$$\text{median} = \frac{88 + 90}{2} = 89$$

Since there is an even number of scores, the median is the mean of the middle two scores.

$$\text{mode} = 90$$

The score of 90 appears most often in the data set.

Significant Digits

1. 5.3
2. 2.2
3. 4.22
4. 3.3
5. 32
6. 29
7. 690
8. 1.9
9. 2.4
10. 2.01
11. 0.002
12. 4.98

PRACTICE

The tables below show the final standings for the teams in each division of the National Football Conference for the 1995 season. Use the tables for Exercises 1 and 2.

Eastern Division

Team	W	L
Dallas	**12**	**4**
Philadelphia	**10**	**6**
Washington	**6**	**10**
New York Giants	**5**	**11**
Arizona	**4**	**12**

Central Division

Team	W	L
Green Bay	**11**	**5**
Detroit	**10**	**6**
Chicago	**9**	**7**
Minnesota	**8**	**8**
Tampa Bay	**7**	**9**

Western Division

Team	W	L
San Francisco	**11**	**5**
Atlanta	**9**	**7**
New Orleans	**7**	**9**
St. Louis	**7**	**9**
Carolina	**7**	**9**

1. Find the mean, the median, and the mode for the number of wins (W) for all the teams in each division.

a. Eastern Division **b.** Central Division **c.** Western Division

2. Find the mean, the median, and the mode for the number of wins for all the teams in the National Football Conference. How do these values compare with the values you found for the divisions?

3. The television ratings for the first 20 football Super Bowl games are listed below. Find the mean, the median, and the mode for the data.

41.1, 36.8, 36.0, 39.4, 39.9, 44.2, 42.7, 41.6, 42.4, 42.3,
44.4, 47.2, 47.1, 46.3, 44.4, 49.1, 48.6, 46.4, 46.4, 48.3

Data Displays

A *bar graph* displays data that fall into distinct categories.

EXAMPLE 1

The bar graph shows the total number of people who voted in the 1994 national election in the United States.

a. About how many people voted in the 1994 national election?

b. About 20% of the United States population live in the Northeast. Does the number of votes cast in the Northeast accurately represent the population?

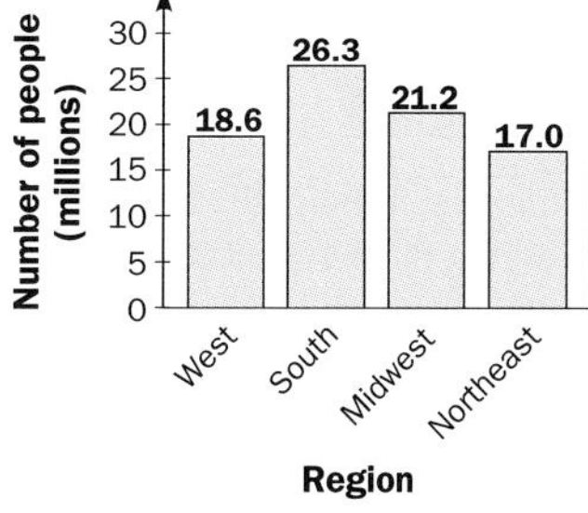

SOLUTION

a. $18.6 + 26.3 + 21.2 + 17.0 = 83.1$
About 83.1 million people voted in the 1994 national election.

b. Find the percent of all votes that came from the Northeast: $\frac{17.0}{83.1} \approx 20.5\%$.
About 20% of the people who voted lived in the Northeast. The region was accurately represented.

Mean, Median, and Mode

1. a. 7.4; 6; no mode
   b. 9; 9; no mode
   c. 8.2; 7; 7
2. 8.2; 8; 7; Answers may vary. An example is given. Only the Western Division values have a mode; it is the same as that for the entire conference. The mean for the conference data is higher than that for the Eastern Division, lower than that for the Central Division, and the same as that for the Western Division. The median for the conference data is higher than those for the Eastern and Western Divisions, and lower than that for the Central Division. Overall, the values for the conference data are most similar to those for the Western Division.
3. 43.73; 44.3; 44.4, 46.4

A *circle graph* shows how data relate to a whole and to each other.

EXAMPLE 2

The manager of a car dealership decides to record the colors of the cars that customers buy to see which colors are most popular. The data collected are shown in the table. Draw a circle graph to display the data.

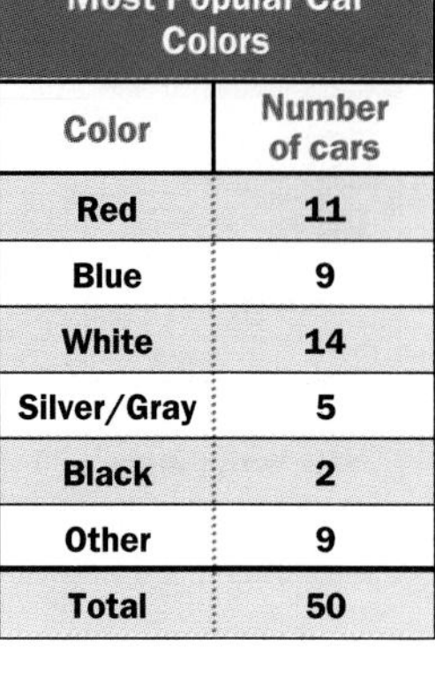

Most Popular Car Colors	
Color	**Number of cars**
Red	11
Blue	9
White	14
Silver/Gray	5
Black	2
Other	9
Total	50

SOLUTION

Step 1 Write the result for each category as a fraction of all the data collected. Then write each fraction as a decimal.

Step 2 Multiply each decimal from Step 1 by 360° to find the degree measure for each *sector*, or wedge, of the circle graph.

	Step 1	Step 2
Red:	$\frac{11}{50} = 0.22$	$0.22 \times 360° \approx 79°$
Blue:	$\frac{9}{50} = 0.18$	$0.18 \times 360° \approx 65°$
White:	$\frac{14}{50} = 0.28$	$0.28 \times 360° \approx 101°$
Silver/gray:	$\frac{5}{50} = 0.1$	$0.1 \times 360° = 36°$
Black:	$\frac{2}{50} = 0.04$	$0.04 \times 360° \approx 14°$
Other:	$\frac{9}{50} = 0.18$	$0.18 \times 360° \approx 65°$

Step 3 Draw a circle. Use a protractor to divide the circle into sectors of the appropriate size. Convert each decimal from Step 1 to a percent and label each sector appropriately.

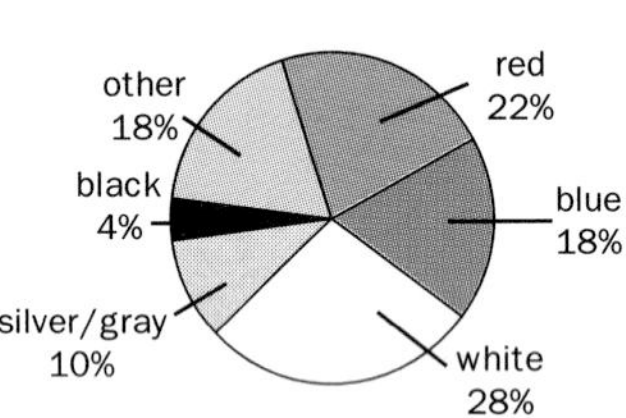

A *line graph* shows an amount and a direction of change in data over a period of time.

EXAMPLE 3

The line graph shows the number of teens, aged 16–19, who had jobs during the period 1990–1994.

a. About how many teens had jobs in 1992?

b. How does the graph show that the number of teens with jobs increased more from 1993 to 1994 than from 1992 to 1993?

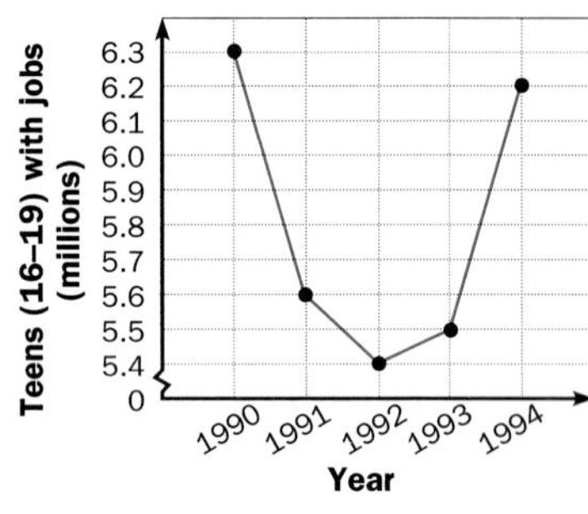

SOLUTION

a. About 5.4 million teens had jobs in 1992.

b. The line segment between 1993 and 1994 has a greater slope than the line segment between 1992 and 1993.

792

PRACTICE

Use the bar graph shown to answer Exercises 1–3.

Average time in circulation (years)

Denomination	$1	$5	$10	$20	$50	$100
Average time in circulation (years)	1.5	2	3	4	9	9

1. Which denomination is in circulation an average of 2 years? 4 years?
2. What is the difference in the average circulation time of a $10 bill and a $50 bill?
3. Which denominations are in circulation the longest? Why do you think this is so?
4. The table shows the approximate 1994 population for various cities in the United States. Draw a bar graph to represent the data.

City	New York	Los Angeles	Chicago	Houston	Philadelphia
Population	7,300,000	3,500,000	2,800,000	1,700,000	1,600,000

5. In 1993, 32.8% of the cars sold in the United States were compact or subcompact cars, 43.3% were midsize, 11.1% were large, and the rest were luxury cars. Draw a circle graph to represent the data.
6. A survey conducted in Chicago, Illinois, in 1990 showed that 1,700,857 workers commuted by car, 826,767 used public transportation, and 256,103 used another method. Draw a circle graph to represent the data.
7. **a.** The table at the right shows the number of students in high schools in the United States from 1960 to 1990. Draw a line graph to represent the data.

   b. State a conclusion you can make from the graph.

Year	Enrollment (millions)
1960	13.0
1970	19.7
1980	18.0
1990	15.3

Dimensional Analysis

When you want to convert from one unit of measure to another, you multiply by an appropriate *conversion factor* and cancel the units of measurement as if they were numbers. This method is called *dimensional analysis*.

EXAMPLE

Convert 55 mi/h to feet per minute.

SOLUTION

Set up the conversion factors so that miles and hours will cancel, leaving feet in the numerator and minutes in the denominator.

$$\frac{55 \text{ mi}}{1 \text{ h}} \times \frac{? \text{ ft}}{1 \text{ mi}} \times \frac{1 \text{ h}}{? \text{ min}} = \frac{? \text{ ft}}{1 \text{ min}}$$

$$\frac{55 \text{ mi}}{1 \text{ h}} \times \frac{\mathbf{5280 \text{ ft}}}{\mathbf{1 \text{ mi}}} \times \frac{\mathbf{1 \text{ h}}}{\mathbf{60 \text{ min}}} = \frac{? \text{ ft}}{1 \text{ min}}$$

1 h = 60 min
1 mi = 5280 ft

$$\frac{55 \cancel{\text{mi}}}{1 \cancel{\text{h}}} \times \frac{5280 \text{ ft}}{1 \cancel{\text{mi}}} \times \frac{1 \cancel{\text{h}}}{60 \text{ min}} = 4840 \text{ ft/min}$$

Data Displays

1. $5 bill; $20 bill
2. 6 years
3. $50 bills and $100 bills; Answers may vary. An example is given. I think the bills are probably handled much less and so wear out less quickly than more commonly used bills.

4. 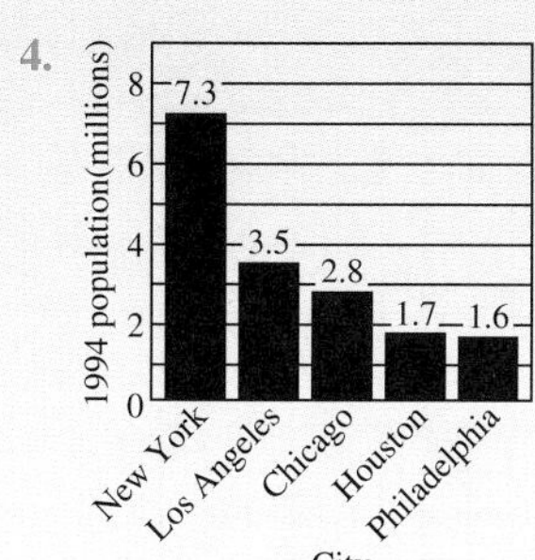

5. 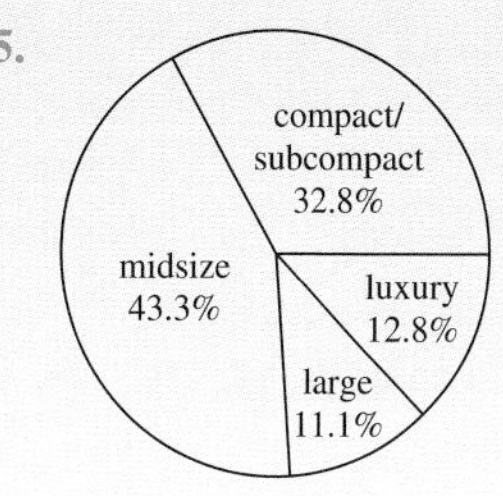

6. 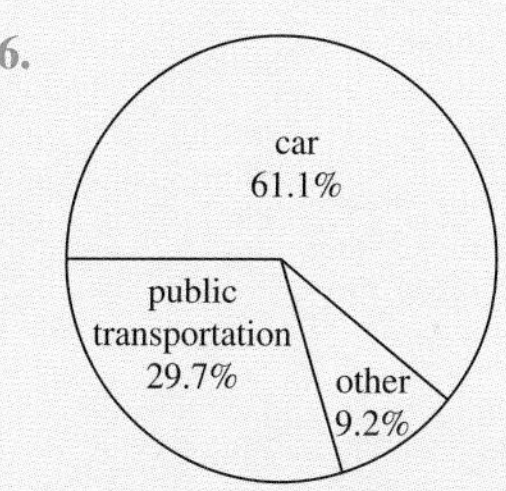

7. a. 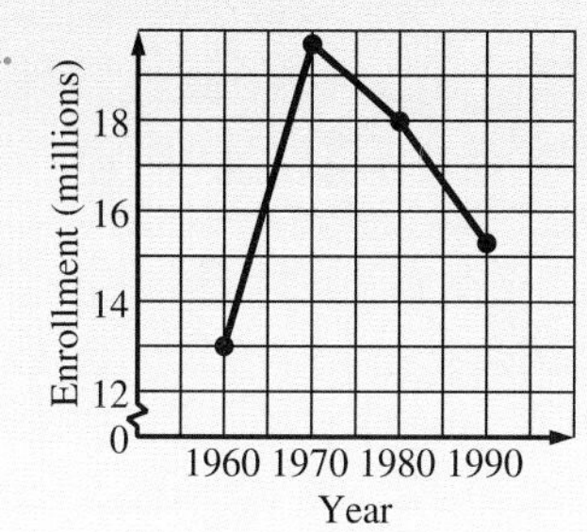

   b. Answers may vary. An example is given. It appears that enrollment peaked in 1970 and has been declining since.

PRACTICE

Use dimensional analysis for each conversion.

1. A car is traveling 80 km/h. Find this speed in meters per second.
2. Adam is taking a vacation in Mexico and wants to exchange \$1250 for Mexican pesos. On the day he goes to the bank, one United States dollar is worth 3.10 pesos. How many pesos did Adam get?
3. Sound travels at a speed of about 3650 m/s through brick. Find this speed in feet per second. (*Hint:* 1 ft ≈ 0.3048 m)
4. The average fuel efficiency rate of Mona's car is about 24 mi/gal. The car's gas tank holds 15 gal of gas. If Mona pays \$1.30 per gallon for gas, find the approximate cost of the gas in dollars per mile.

GRAPHS OF POINTS AND EQUATIONS

Graphing Points in a Coordinate Plane

A *coordinate plane* consists of a horizontal *x-axis* and a vertical *y-axis* that intersect at a point called the ***origin***, labeled ***O***. The axes divide the coordinate plane into four *quadrants* as shown.

Each point in a coordinate plane is associated with an *ordered pair* (a, b) of real numbers. The first number, a, is the *x-coordinate*. The second number, b, is the *y-coordinate*.

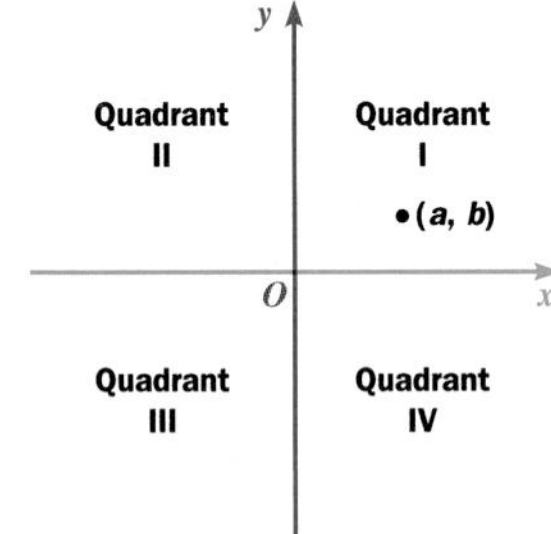

EXAMPLE

Graph the points $A(-3, 2)$ and $B(2, 0)$ in a coordinate plane. Name the quadrant, if any, in which each point lies.

SOLUTION

To graph ***A*(−3, 2)**, start at the origin and move **left 3 units** and **up 2 units**. The point is in Quadrant II.

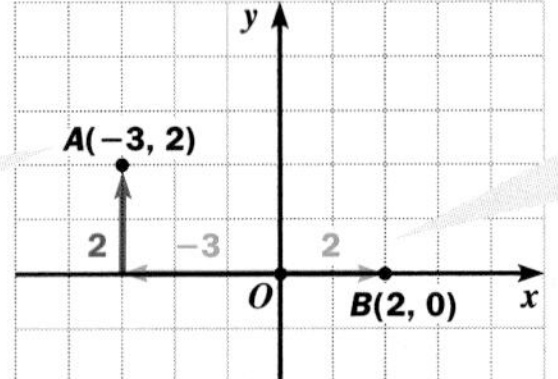

To graph ***B*(2, 0)**, start at the origin and move **right 2 units** and **vertically 0 units**. The point is not in any quadrant.

PRACTICE

Graph the following points in the same coordinate plane. Name the quadrant, if any, in which each point lies.

1. $C(1, 6)$
2. $D(-4, -5)$
3. $E(0, 3)$
4. $F(2, -4)$
5. $G(-5, 0)$
6. $H(-3, 2.5)$
7. 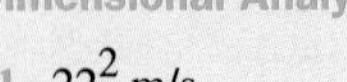 $I\left(\frac{14}{3}, \frac{14}{3}\right)$
8. $J(0, 0)$

Dimensional Analysis

1. $22\frac{2}{9}$ m/s
2. 3875 pesos
3. about 11,975 ft/s
4. about \$0.05 per mile

Graphing Points on a Coordinate Plane

1–8.

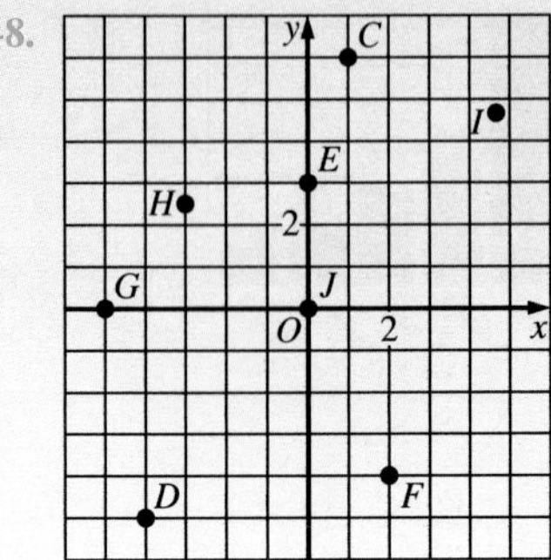

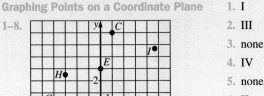

1. I
2. III
3. none
4. IV
5. none
6. II
7. I
8. none

Graphing an Equation

You can use a coordinate plane to graph an equation.

EXAMPLE

Graph $y = |x|$.

SOLUTION

Step 1 Make a table of values. Be sure to include both positive and negative values of x.

$y = \|x\|$	
x	y
−3	$\|-3\| = 3$
−2	$\|-2\| = 2$
−1	$\|-1\| = 1$
0	$\|0\| = 0$
1	$\|1\| = 1$
2	$\|2\| = 2$
3	$\|3\| = 3$

Step 2 Draw a coordinate grid. Plot the ordered pairs (x, y) from the table. Connect the plotted points.

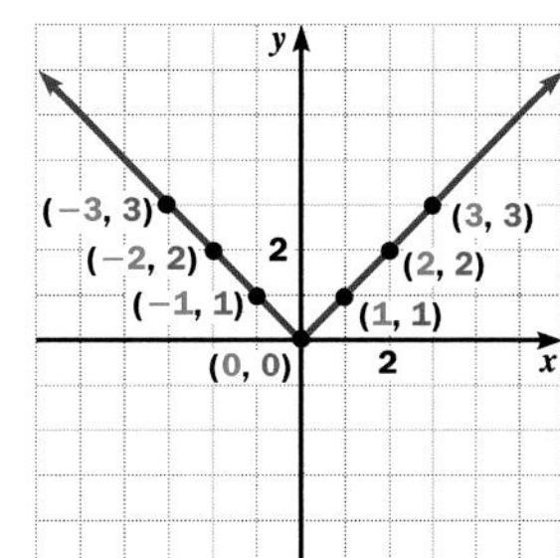

PRACTICE

Graph each equation.

1. $y = x$ **2.** $y = -4x - 5$ **3.** $y = |2x|$ **4.** $y = 3|x|$

5. $y = x^2$ **6.** $y = -\frac{1}{2}x^2$ **7.** $y = 2^x$ **8.** $y = \left(\frac{1}{2}\right)^x$

Making a Scatter Plot

A *scatter plot* is a graph that shows the relationship between two sets of data.

EXAMPLE

The table gives the number of VCRs owned by people in the United States for each year from 1985 to 1990. Make a scatter plot of the data.

Years since 1985	Number of VCRs (in millions)
0	18
1	31
2	43
3	51
4	58
5	63

SOLUTION

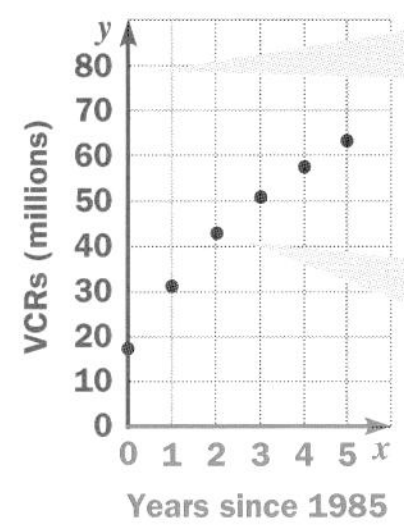

Step 1 Draw a first-quadrant coordinate grid. Use the **horizontal axis** for **years since 1985** and the **vertical axis** for **number of VCRs**.

Step 2 Plot each data pair from the table.

Graphing an Equation

1.

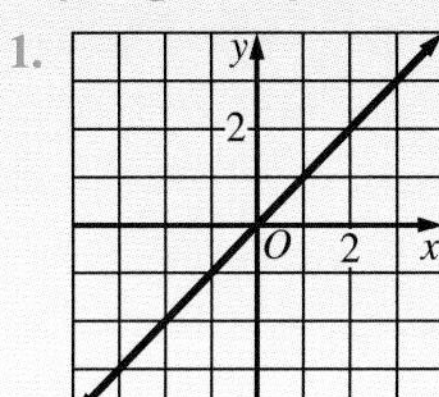

2.

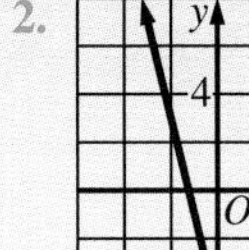

3.

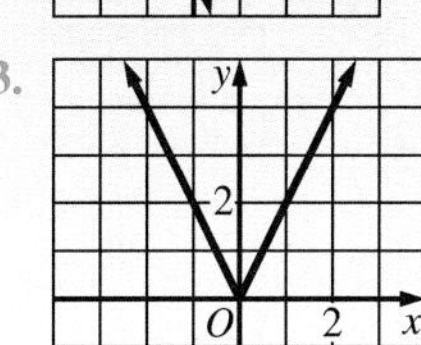

4.

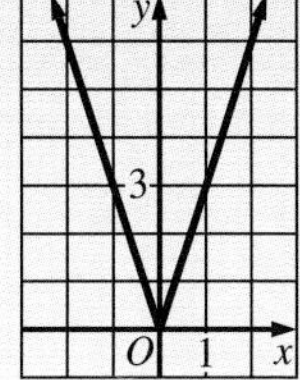

5.

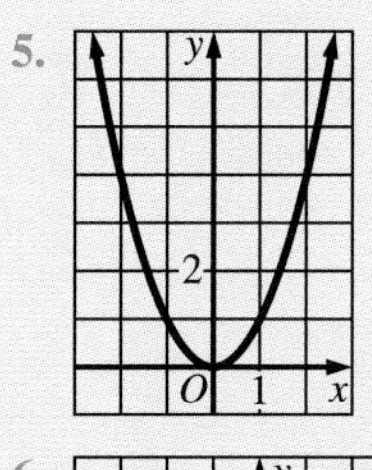

6.

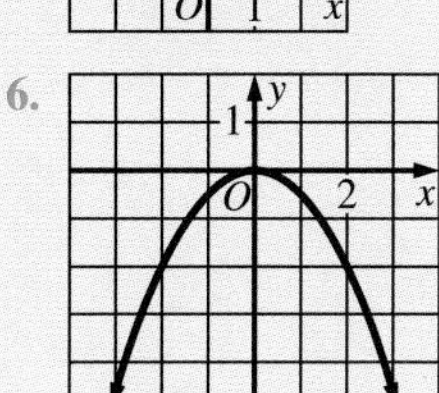

7.

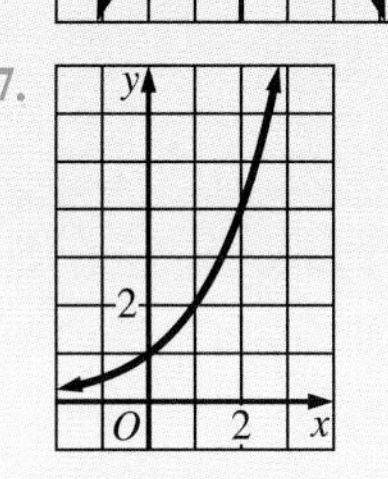

8.

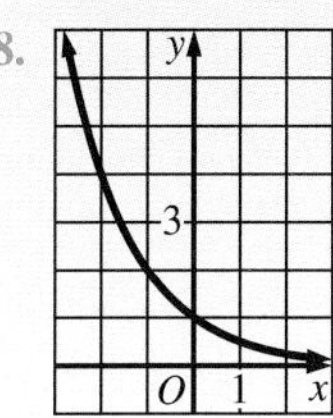

PRACTICE

Make a scatter plot of the data in each table.

1.

Energy Used While In-Line Skating					
Time skating (min)	3	7	12	18	25
Energy used (Cal)	29	67	115	172	238

2.

Plain Cheese Pizzas at Doug's Pizza Palace					
Diameter (in.)	7	10	12	14	16
Price ($)	2.49	6.49	8.49	11.49	14.49

TRANSFORMATIONS IN THE COORDINATE PLANE

A *transformation* of a geometric figure is a change in the position or size of the figure. The result of a transformation is called an *image*. Some common transformations are *translations*, *reflections*, and *dilations*.

Translating a Figure

A *translation* is a transformation that moves each point of a figure the same distance in the same direction. A translation neither flips nor turns a figure, and it does not change the figure's size or shape.

EXAMPLE

a. Translate $\triangle ABC$ (shown below) right 9 units and down 5 units. Give the coordinates of the image's vertices.

b. Write a formula representing the translation in part (a).

SOLUTION

a.

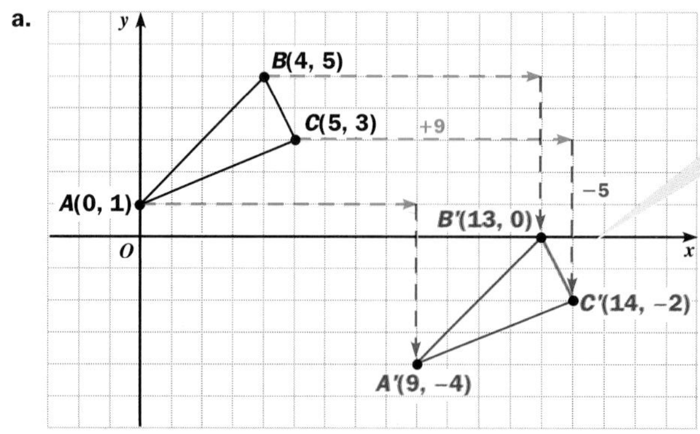

Move each vertex of $\triangle ABC$ **right 9 units** and **down 5 units** to get new vertices ***A'***, ***B'***, and ***C'***. Connect these vertices to form $\triangle$***A'B'C'***, the image of $\triangle ABC$.

To find the coordinates of ***C'***, **add 9** to the **x-coordinate** of *C* and **subtract 5** from the **y-coordinate** of *C*: **(5 + 9, 3 − 5)**, or **(14, −2)**

b. Under the translation "right 9 units and down 5 units," each point $P(x, y)$ is moved to the point $P'(x + 9, y - 5)$. You can write $P(x, y) \rightarrow P'(x + 9, y - 5)$.

Making a Scatter Plot

1.

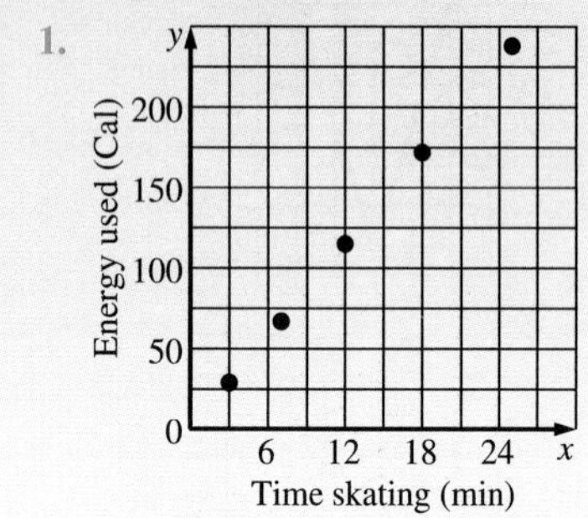

2.

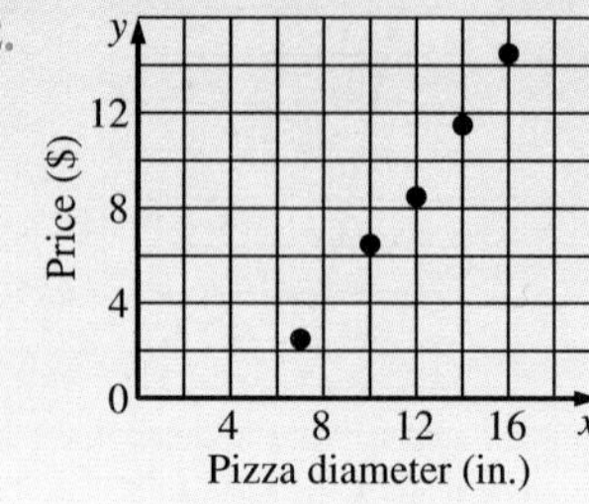

PRACTICE

Let △*DEF* be the triangle with vertices *D*(−3, 2), *E*(4, 4), and *F*(2, 0). For each of Exercises 1–6, draw △*DEF* and its image under the given translation. Write a formula representing the translation.

1. right 4 units
2. down 5 units
3. left 5 units and up 2 units
4. right 1 unit and up 1 unit
5. right 7.5 units and down 6 units
6. left 4 units and down 9 units

Reflecting a Figure

A *reflection* is a transformation that flips a figure over a line, called the *line of reflection.* A reflection does not change the size or shape of a figure.

EXAMPLE 1

Reflect △*ABC* shown on the previous page over each line.

a. the *y*-axis

b. the *x*-axis

SOLUTION

First reflect each vertex of △*ABC* over the given line to get new vertices A', B', and C'. Then connect these vertices to form $\triangle A'B'C'$, the image of △*ABC* under the given reflection.

a.

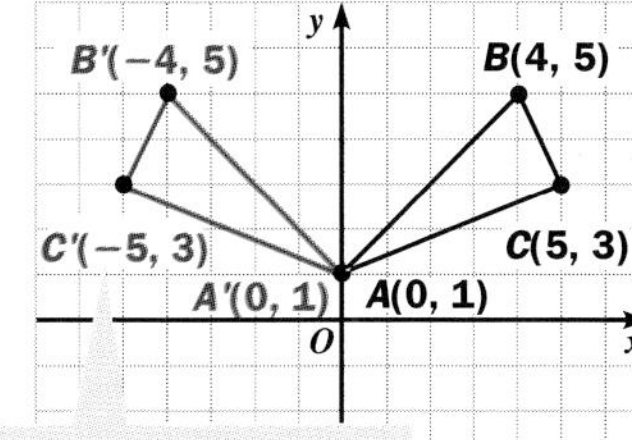

C is 5 units to the *right* of the *y*-axis, so ***C′*** is 5 units to the *left* of the *y*-axis.

b.

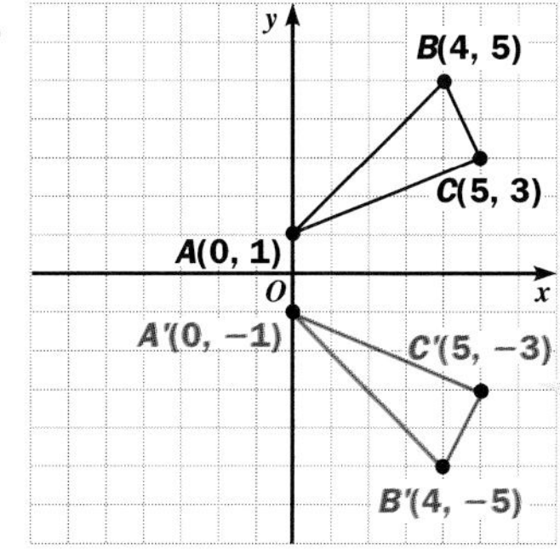

C is 3 units *above* the *x*-axis, so ***C′*** is 3 units *below* the *x*-axis.

If a figure is reflected over a line and the image coincides with the original figure, then the line is called a *line of symmetry* for the figure.

EXAMPLE 2

Find the number of lines of symmetry of an equilateral triangle.

SOLUTION

The diagram shows that an equilateral triangle has 3 lines of symmetry. Each line of symmetry passes through a vertex and is perpendicular to the side opposite the vertex.

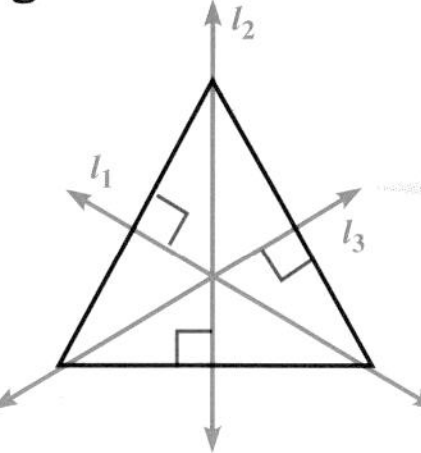

Note that if you fold a figure along a line of symmetry, the figure's edges coincide.

Translating a Figure

1.

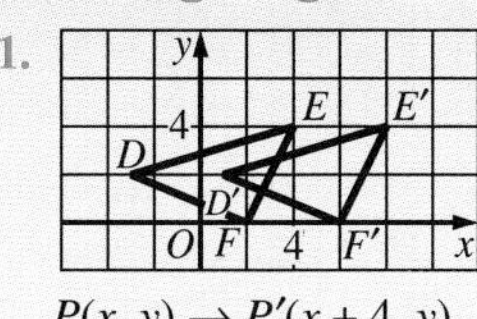

$P(x, y) \to P'(x + 4, y)$

2.

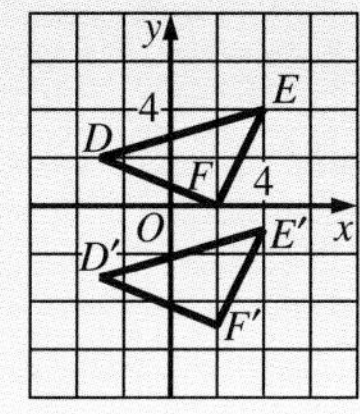

$P(x, y) \to P'(x, y - 5)$

3.

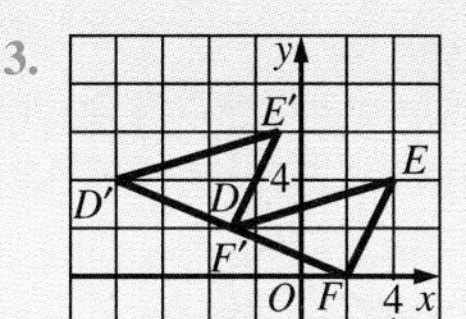

$P(x, y) \to P'(x - 5, y + 2)$

4.

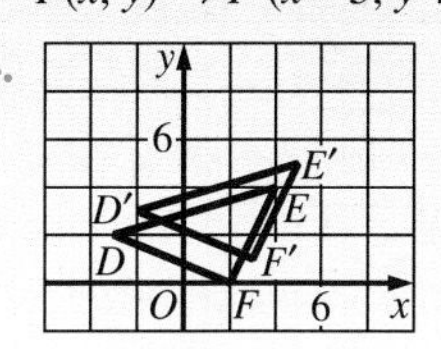

$P(x, y) \to P'(x + 1, y + 1)$

5.

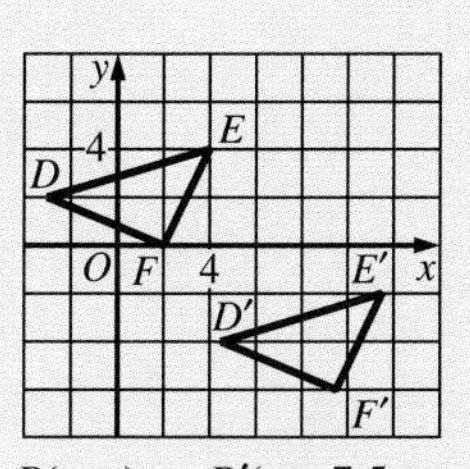

$P(x, y) \to P'(x + 7.5, y - 6)$

6.

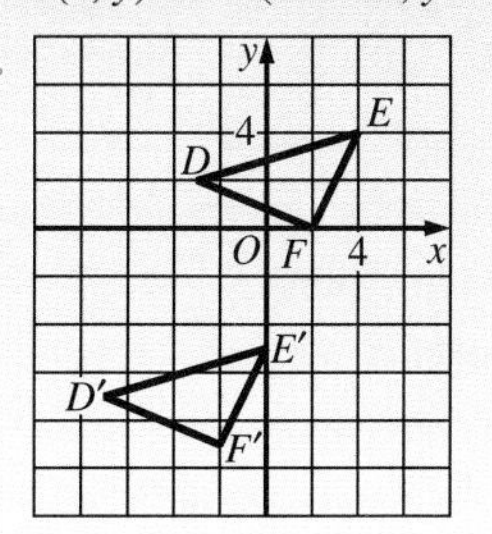

$P(x, y) \to P'(x - 4, y - 9)$

Practice

Let $\triangle GHI$ be the triangle with vertices $G(0, 1)$, $H(3, 3)$, and $I(5, 0)$. For each of Exercises 1–4, draw $\triangle GHI$ and its image when $\triangle GHI$ is reflected over the given line.

1. the x-axis **2.** the y-axis **3.** $x = 6$ **4.** $y = -2$

Give the minimum number of lines of symmetry for each type of figure.

5. a rectangle **6.** a square **7.** a parallelogram

8. an isosceles triangle **9.** a parabola **10.** a circle

Dilating a Figure

A *dilation* of a figure is a transformation whose image is similar to the original figure. The ratio of any length in the image to the corresponding length in the original figure is called the *scale factor* of the dilation.

Example

Dilate $\triangle ABC$ shown on page 796. Use $P(8, 1)$ as the *center* of the dilation and a scale factor of 2.

Solution

Step 1 Draw rays from P through the vertices of $\triangle ABC$.

Step 2 Locate point A' on $\overrightarrow{PA}$ so that $PA' = 2(PA)$, since the scale factor is 2. Locate points B' and C' similarly.

Step 3 Draw $\triangle A'B'C'$, the image of $\triangle ABC$.

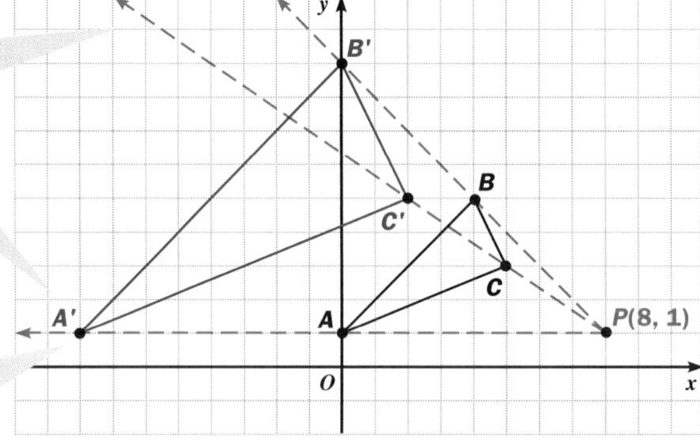

In general, when the scale factor of a dilation is greater than 1 (as in the example above), the image is larger than the original figure. When the scale factor is between 0 and 1, the image is smaller than the original figure.

Practice

Let $\triangle JKL$ be the triangle with vertices $J(-6, 0)$, $K(-6, 4)$, and $L(0, 2)$. For each of Exercises 1–4, draw $\triangle JKL$ and its image under the dilation with the given center and scale factor.

1. center: $P(0, 0)$; scale factor: 2

2. center: $P(0, 0)$; scale factor: $\frac{1}{2}$

3. center: $P(-2, -4)$; scale factor: $\frac{3}{4}$

4. center: $P(-8, 6)$; scale factor: 3

Reflecting a Figure

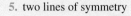

1.–4.

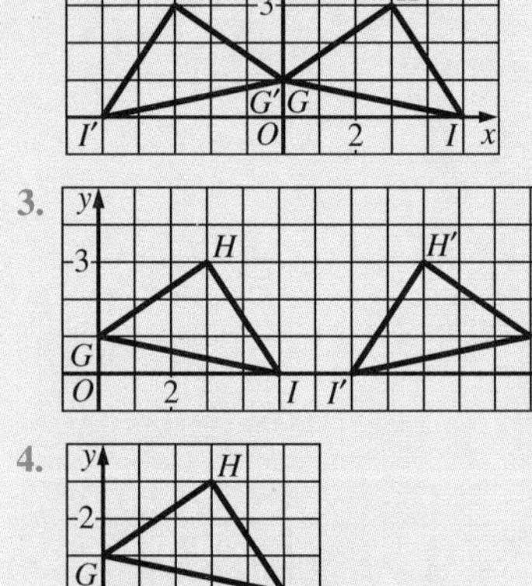

5. two lines of symmetry
6. four lines of symmetry
7. no lines of symmetry
8. one line of symmetry
9. one line of symmetry
10. infinitely many lines of symmetry

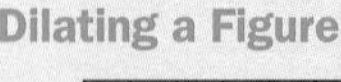

Dilating a Figure

1.

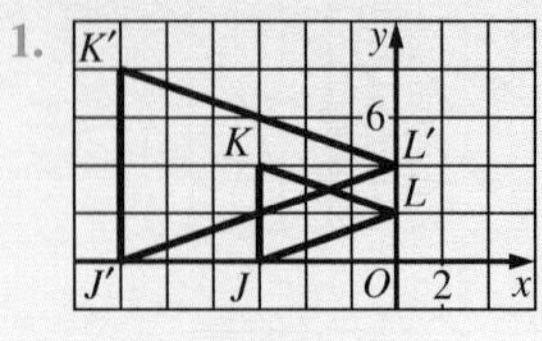

2.

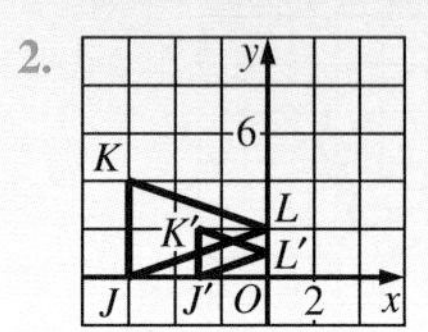

3.

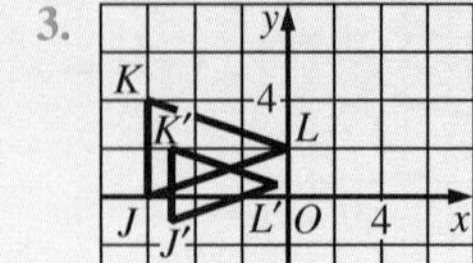

4.

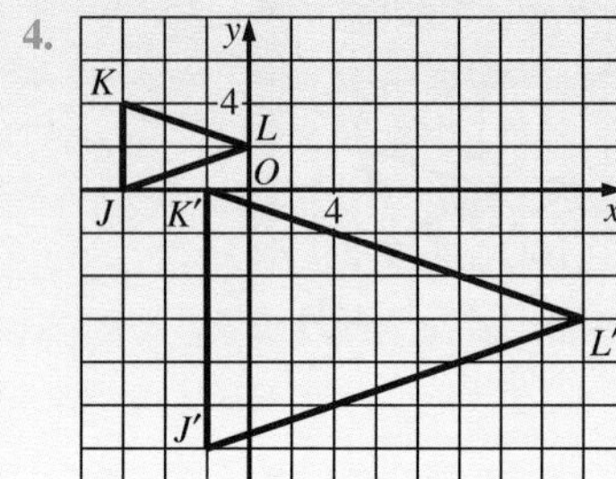

LOGIC AND GEOMETRIC RELATIONSHIPS

Logical Statements

A *conditional statement* can be written in the form "if p, then q." The phrase represented by p is the *hypothesis* and the phrase represented by q is the *conclusion.*

You write the *converse* of a conditional statement by interchanging the hypothesis and the conclusion. The converse of "if p, then q" is "if q, then p."

If-then statements are either true or false. If you find an instance where the hypothesis is true and the conclusion is false, the statement in general is false.

EXAMPLE 1

For parts (a) and (b), use the statement "If $b = \sqrt{a}$, then $b^2 = a$."

a. Identify the hypothesis and the conclusion. Then tell whether the statement is *true* or *false*.

b. Write the converse. Then tell whether the converse is *true* or *false*.

SOLUTION

a. The hypothesis is "$b = \sqrt{a}$." The conclusion is "$b^2 = a$." Since $(\sqrt{a})^2 = a$, the statement is true.

b. The converse is "If $b^2 = a$, then $b = \sqrt{a}$." If $b^2 = a$, then b can equal either $\sqrt{a}$ or $-\sqrt{a}$, so the statement is not necessarily true. The converse is false.

A statement and its converse can be combined into one statement, called a *biconditional*, using the phrase "*if and only if*." The truth of one statement in a biconditional implies the truth of the other statement.

For example, the biconditional "$b^2 = a$ if and only if $b = \pm\sqrt{a}$" is a true statement, but the biconditional "$b^2 = a$ if and only if $b = \sqrt{a}$" is false.

EXAMPLE 2

Tell whether the statement is *true* or *false*: "A number is divisible by 4 if and only if the number is divisible by 2."

SOLUTION

One conditional statement is "If a number is divisible by 2, then the number is divisible by 4." Since 10 is divisible by 2 but not divisible by 4, this conditional statement is false.

The other conditional statement, which is the converse of the first, is "If a number is divisible by 4, then the number is divisible by 2." Since every number that is divisible by 4 is an even number, the converse is true.

Since divisibility by 2 does not imply divisibility by 4, the biconditional is false.

PRACTICE

For Exercises 1–6:

a. Identify the hypothesis and the conclusion of each conditional statement. Then tell whether the statement is *true* or *false*.

b. Write the converse of each statement. Then tell whether the converse is *true* or *false*. Explain.

c. Write a biconditional using each statement and its converse. Then tell whether the biconditional is *true* or *false*. Explain.

1. If a figure is a rectangle, then it is a quadrilateral.

2. If $a = 0$ or $b = 0$, then $ab = 0$.

3. If a triangle is equilateral, then it is isosceles.

4. If $a > b$, then $a + c > b + c$.

5. If two lines intersect, then they are not parallel.

6. If $x^2 = 9$, then $x = 3$.

Angle Relationships

A *right angle* has a measure of 90°, and a *straight angle* has a measure of 180°. An angle with a measure between 0° and 90° is an *acute angle*. An angle with a measure between 90° and 180° is an *obtuse angle*.

Two angles whose measures have a sum of 90° are *complementary angles*. Two angles whose measures have a sum of 180° are *supplementary angles*. Two angles whose sides form two pairs of opposite rays are *vertical angles*. Vertical angles have the same measure.

EXAMPLE

For the diagram shown, find the values of *x* and *y*.

SOLUTION

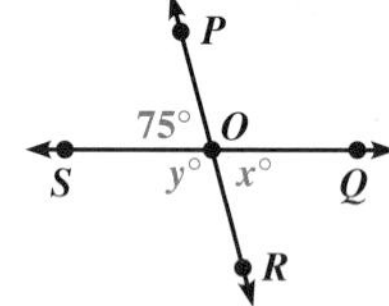

Since $\angle POS$ and $\angle QOR$ are vertical angles, their measures are equal. Therefore, $x = 75$.

Since $\angle POS$ and $\angle SOR$ are supplementary angles, $75 + y = 180$. Therefore, $y = 105$.

PRACTICE

Find the values of *x* and *y* in each diagram.

1. $(x+70)°$, $y°$, $(2x-10)°$

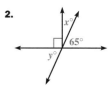

2. $x°$, $65°$, $y°$

3.

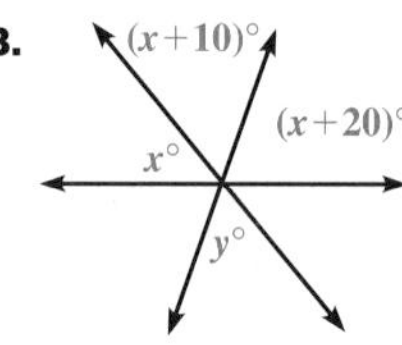

Tell if each statement is *always true, sometimes true,* or *never true.*

4. Two angles that are supplementary are acute angles.

5. Two angles that are right angles are supplementary.

6. Two angles that are vertical angles are complementary.

Logical Statements

1. a. a figure is a rectangle; it is a quadrilateral; True.
   b. If a figure is a quadrilateral, then it is a rectangle. False; a non-square rhombus is a quadrilateral that is not a rectangle.
   c. A figure is a rectangle if and only if it is a quadrilateral. False; one of the conditionals is false.
2. a. $a = 0$ or $b = 0$; $ab = 0$; True.
   b. If $ab = 0$, then $a = 0$ or $b = 0$. True; if the product of two numbers is zero, at least one of the numbers is 0.
   c. $ab = 0$ if and only if $a = 0$ or $b = 0$. True; both of the conditionals are true.
3. a. a triangle is equilateral; it is isosceles; True.
   b. If a triangle is isosceles, then it is equilateral. False; for example, an isosceles right triangle is not equilateral.
   c. A triangle is equilateral if and only if it is isosceles. False; one of the conditionals is false.
4. a. $a > b$; $a + c > b + c$; True.
   b. If $a + c > b + c$, then $a > b$; True.
   c. $a + c < b + c$ if and only if $a < b$. True; both of the conditionals are true.
5. a. Two lines intersect; they are not parallel. True
   b. If two lines are not parallel, then they intersect. False; Parallel lines are lines that lie in the same plane and do not intersect.
   c. Two lines intersect if and only if they are not parallel. False; one of the conditionals is false.
6. a. $x^2 = 9$; $x = 3$; False.
   b. If $x = 3$, then $x^2 = 9$. True; $3^2 = 9$.
   c. $x^2 = 9$ if and only if $x = 3$. False; one of the conditionals is false.

Angle Relationships

1. 80; 30
2. 25; 65
3. 50; 60
4. never true
5. always true
6. sometimes true

Triangle Relationships

The sum of the measures of the angles of a triangle is 180°. Two triangles are *congruent* if they have the same size and shape. If two triangles have the same shape but are not necessarily the same size, the two triangles are *similar*. If two pairs of corresponding angles are congruent, the triangles are similar.

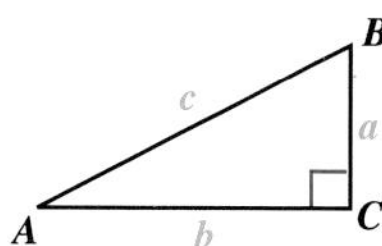

The Pythagorean theorem and its converse tell you that $\triangle ABC$ is a right triangle if and only if $a^2 + b^2 = c^2$.

EXAMPLE

Use the diagram at the right.

a. Find the value of x.

b. Find the value of y.

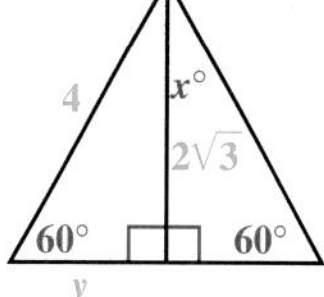

SOLUTION

a. The sum of the measures of the angles of a triangle is 180°.

$$90 + 60 + x = 180$$
$$x = 30$$

b. Use the Pythagorean theorem.

$$y^2 + (2\sqrt{3})^2 = 4^2$$
$$y^2 + 12 = 16$$
$$y^2 = 4$$
$$y = 2$$

Use the positive square root because y is a length.

PRACTICE

Find the value of *x* in each diagram.

1.

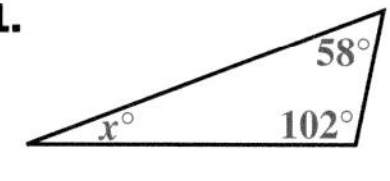

2.

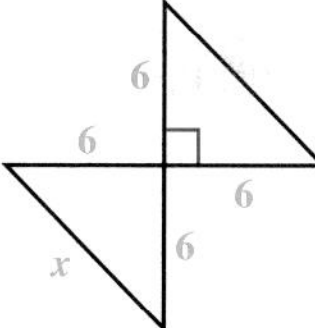

3.

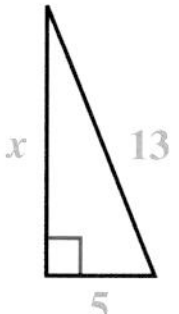

4.

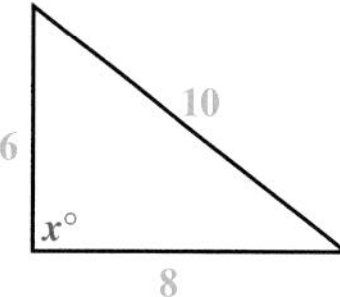

Polygons

A *polygon* is the union of several line segments that are joined end to end so as to completely enclose an area. A polygon with all angles of equal measure is *equiangular*. A polygon with all sides of equal length is *equilateral*. A polygon that is both equiangular and equilateral is a *regular polygon*.

EXAMPLE

Complete each statement with one of the words *always*, *sometimes*, or *never*. Explain your reasoning.

a. An isosceles triangle is _?_ equilateral.

b. A rectangle is _?_ a parallelogram.

SOLUTION

a. All isosceles triangles have two congruent sides, but only some have three congruent sides. Therefore, an isosceles triangle is *sometimes* equilateral.

b. All rectangles are quadrilaterals that have parallel opposite sides, so a rectangle is *always* a parallelogram.

Triangle Relationships

1. 20
2. $6\sqrt{2}$
3. 12
4. 90

PRACTICE

Complete each statement with one of the words *always*, *sometimes*, or *never*. Explain your reasoning.

1. A hexagon is _?_ a regular polygon.
2. A quadrilateral _?_ has five sides.
3. An isosceles triangle is _?_ a right triangle.
4. An obtuse triangle is _?_ equiangular.
5. A square is _?_ a regular polygon.
6. A rectangle is _?_ an equiangular quadrilateral.

Perimeter, Circumference, Area, and Volume

Refer to the *Tables, Properties, and Formulas* section on pages 817–825 for a list of specific formulas for perimeter, circumference, area, surface area, and volume.

EXAMPLE

Use the rectangular prism shown at the right.

a. Find the surface area. **b.** Find the volume.

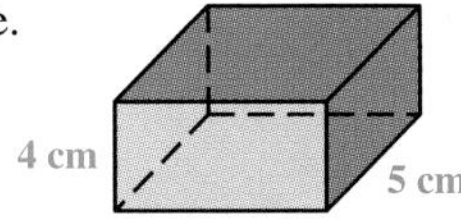

SOLUTION

a. surface area of the prism = twice the areas of the 3 faces shown

$= 2(8 \cdot 5 + 4 \cdot 8 + 4 \cdot 5)$

$= 184$

The surface area of the prism is 184 cm^2.

b. volume of the prism = **area of a base × height of the prism**

$= 8 \cdot 5 \cdot 4$

$= 160$

The volume of the prism is 160 cm^3.

PRACTICE

Find the perimeter and area of each figure.

1. a trapezoid with height 4 cm, bases with lengths 3 cm and 9 cm, and both legs with length 5 cm
2. a right triangle with legs of lengths 18 ft and 15 ft
3. a rectangle with length 9 m and width 3.5 m

Find the circumference and area of each circle. Round to the nearest hundredth.

4. a circle with radius 3
5. a circle with diameter 12
6. a circle with radius $\sqrt{5}$

Find the surface area and volume of each figure. Round to the nearest hundredth.

7. 

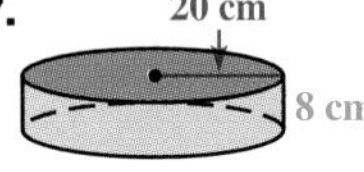

8. 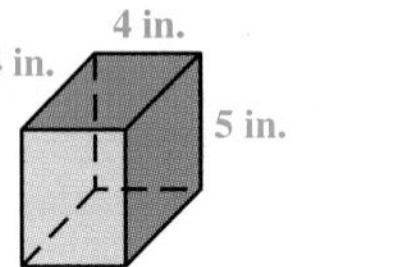

9. 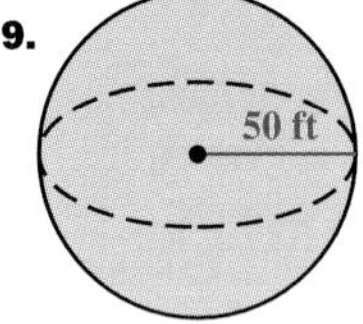

10. 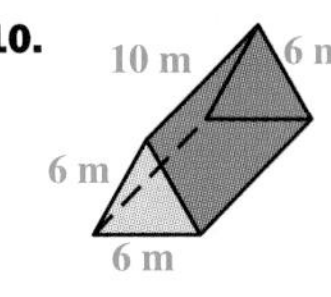

Polygons

1. sometimes; A hexagon may or may not be equiangular and equilateral.
2. never; A quadrilateral has four sides.
3. sometimes; A right triangle with two 45° angles is an isosceles triangle.
4. never; If a triangle had three obtuse angles, the sum of the measures of the angles would be greater than 180°.
5. always; Every square is both equiangular and equilateral.
6. always; Every rectangle is a quadrilateral with four 90° angles.

Perimeter, Circumference, Area, and Volume

1. 22 cm; 24 cm^2
2. $33 + 3\sqrt{61}$ ft ≈ 56.43 ft; 135 ft^2
3. 25 m; 31.5 m^2
4. 18.85; 28.27
5. 37.70; 113.10
6. 14.05; 15.71
7. 3518.58 cm^2; 10,053.10 cm^3
8. 112 $in.^2$; 80 $in.^3$
9. 31,415.93 ft^2; 523,598.78 ft^3
10. 211.18 m^2; 155.88 m^3

Technology Handbook

This handbook introduces you to the basic features of most graphing calculators. Check your calculator's instruction manual for specific keystrokes and any details not provided here.

Performing Calculations

The Keyboard

Look closely at your calculator's keyboard. Notice that most keys serve more than one purpose. Each key is labeled with its primary purpose, and labels for any secondary purposes appear somewhere near the key. You may need to press 2nd, SHIFT, or ALPHA to use a key for a secondary purpose.

On the TI-82, for example, the x^2 key can be used as follows:

- Press x^2 to square a number.
- Press 2nd and then x^2 to take a square root.
- Press ALPHA and then x^2 to type the letter I.

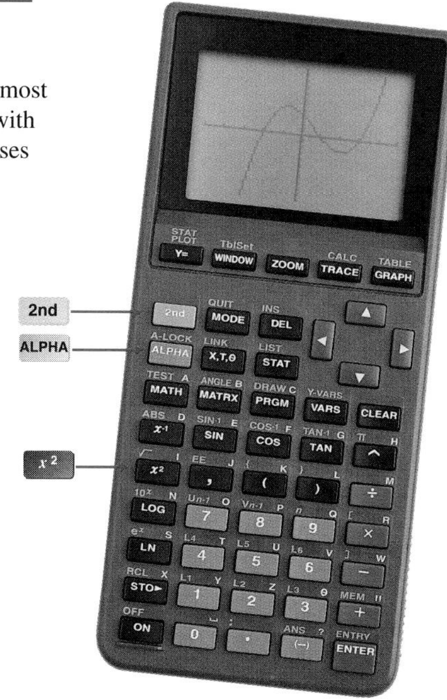

The Home Screen

Your calculator has a *home screen* where you can do calculations. You can usually enter a calculation on a graphing calculator just as you would write it on a piece of paper.

Below are some things to remember as you enter calculations on your graphing calculator.

The calculator may recognize implied multiplication.

Don't confuse the subtraction key, −, and the negation key, (−).

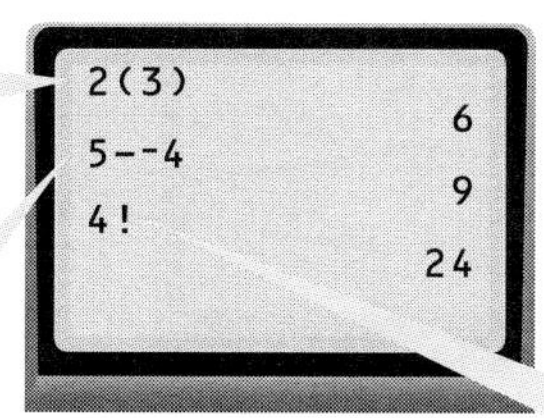

You may need to press MATH to access some operations, such as evaluating a factorial.

CALCULATOR PRACTICE

Use your calculator to find the value of each expression.

1. $-3 - 9$ **2.** $\sqrt{12.25}$ **3.** $\sqrt[3]{2.744}$ **4.** $6!$

5. 8^5 **6.** ${}_5C_2$ **7.** $\tan 75°$ **8.** $|e^{-1} - 2\pi|$

DISPLAYING GRAPHS

Entering and Graphing a Function

To graph a function, enter its equation in the form $y = f(x)$. If variables other than x and y are used in the equation, replace the independent variable with x and the dependent variable with y.

For example, to enter the equation $-3r + 5s = 10$, first rewrite it as $-3x + 5y = 10$ (assuming r is the independent variable and s is the dependent variable). Then solve for y to get $y = \frac{3}{5}x + 2$. The graph of this equation is shown.

Use parentheses. If you enter $y = 3/5x + 2$, the calculator may interpret the equation as $y = \frac{3}{5x} + 2$.

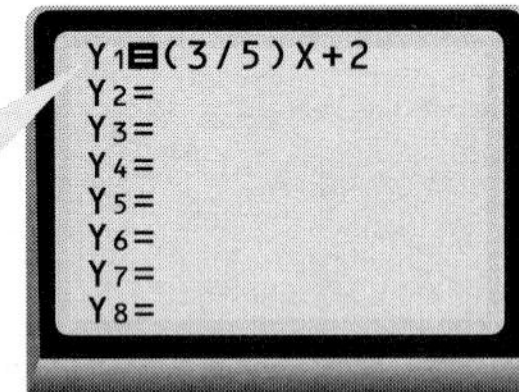

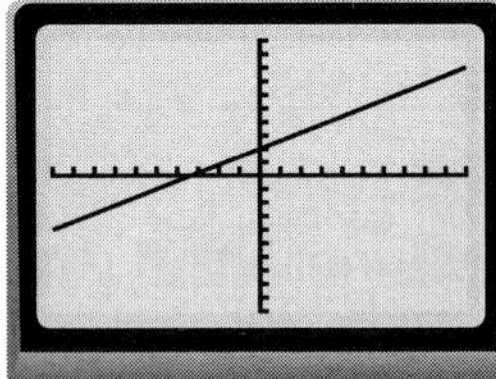

Be sure to use parentheses when groupings are implied by radical signs or fraction bars. For example, enter $y = \sqrt{x + 1} - 2$ as $y = \sqrt{\ }(x + 1) - 2$, and enter $y = \frac{1}{x + 1}$ as $y = 1/(x + 1)$.

The Viewing Window

When graphing, think of the calculator screen as a *viewing window* that lets you look at a portion of the coordinate plane.

On many calculators, the *standard window* shows values from -10 to 10 on both the x- and y-axes. You can adjust the viewing window by pressing WINDOW or RANGE and entering new values for the window variables.

The interval $-10 \le x \le 10$ will be shown on the x-axis.

The interval $-10 \le y \le 10$ will be shown on the y-axis.

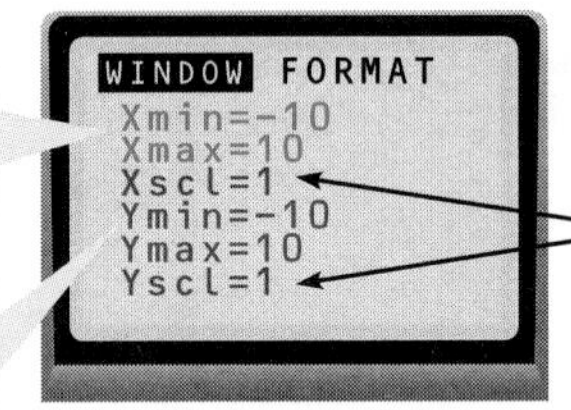

With the scale variables set equal to 1, tick marks will be 1 unit apart on both the x- and y-axes.

ANSWERS Technology Handbook

Calculator Practice

1. –12
2. 3.5
3. 1.4
4. 720
5. 32,768
6. 10
7. 3.732
8. 5.92

Squaring the Viewing Window

In a *square* viewing window, the distance between consecutive integers is the same on both the x- and y-axes. This causes graphs to appear undistorted. For example, the circle with equation $x^2 + y^2 = 49$ is shown below in both the standard window and a square window.

Standard Viewing Window

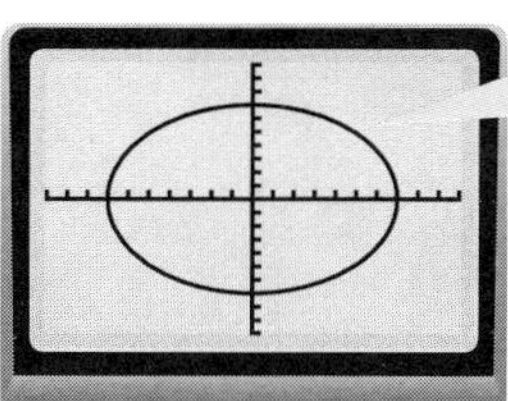

A circle looks like an ellipse in the standard window.

Square Viewing Window

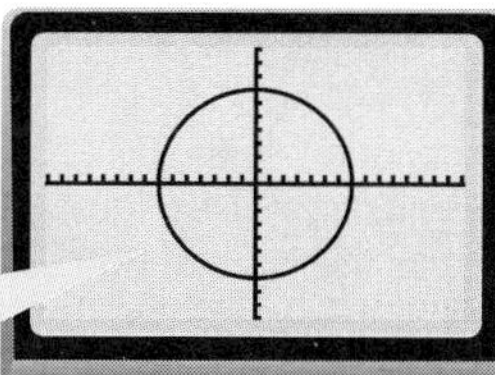

A circle appears undistorted in a square window.

Your calculator may have a feature that automatically produces a square viewing window. If not, you may be able to create one using the fact that the ratio of the screen's height to its width is often about 2 to 3. Just choose values for the window variables that make the "length" of the y-axis about two thirds the "length" of the x-axis:

$$\text{Ymax} - \text{Ymin} \approx \frac{2}{3}(\text{Xmax} - \text{Xmin})$$

The *Zoom* Feature

To magnify part of a graph, you can use your calculator's *zoom* feature. One common way of zooming involves putting a *zoom box* around the portion of the graph you want to magnify.

Fix one corner of the zoom box. Then fix the opposite corner.

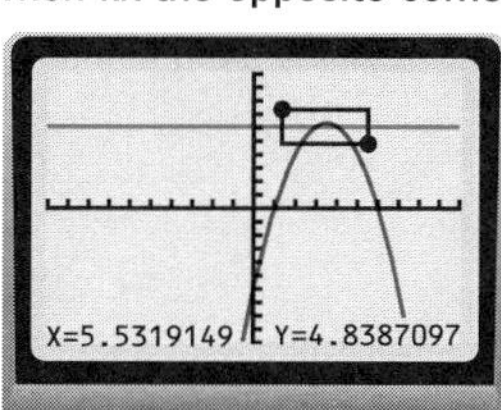

The calculator draws what is inside the zoom box at full-screen size.

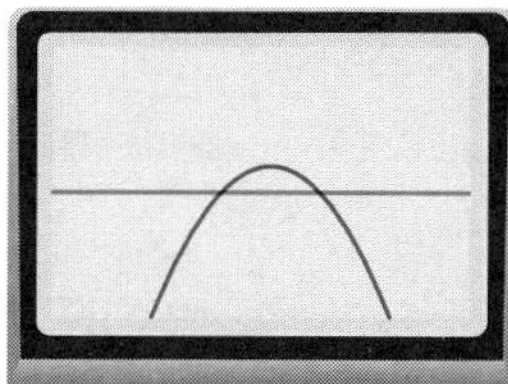

CALCULATOR PRACTICE

Enter and graph each function. Use the standard viewing window.

9. $-6x + 2y = 2$ **10.** $y = -x^2 - x + 6$ **11.** $y = e^x$ **12.** $y = |x - 2|$

13. Graph the perpendicular lines $y = x - 1$ and $y = -x + 7$ in the same coordinate plane. Adjust the viewing window so that the lines look perpendicular. Then zoom in on the point where the lines intersect.

Calculator Practice

9.

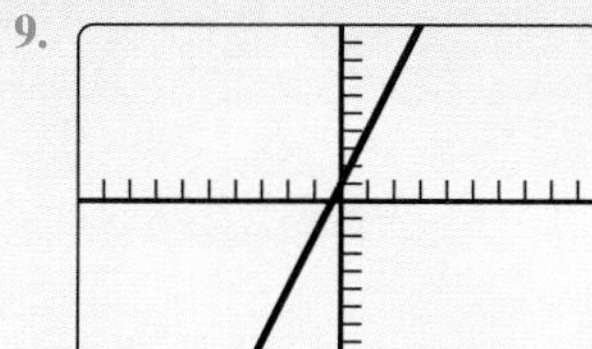

10.

11.

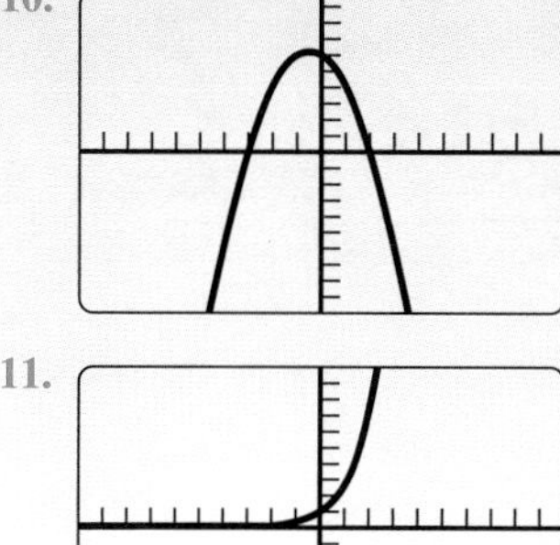

12.

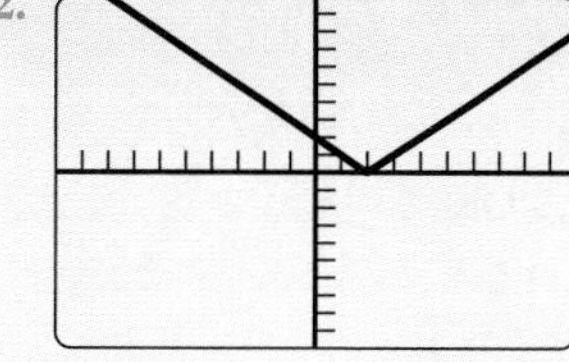

13. To make the lines $y = x - 1$ and $y = -x + 7$ look perpendicular, use a square viewing window like the one shown below. Zoom windows may vary.

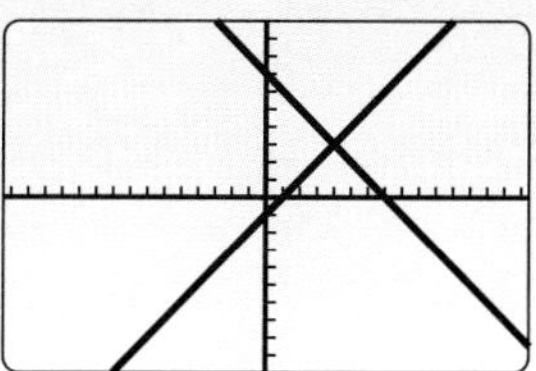

EXAMINING GRAPHS

The *Trace* Feature

After a graph is displayed, you can use your calculator's *trace* feature. When you press TRACE, a flashing cursor appears on the graph. The *x*- and *y*-coordinates of the cursor's location are shown at the bottom of the screen. You can move the cursor along the graph by pressing ◀ and ▶.

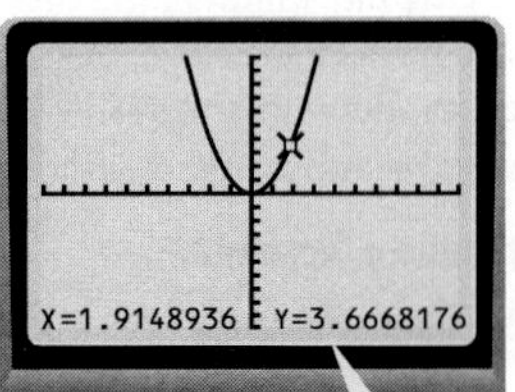

The trace cursor is at the point (1.9148936, 3.6668176) on the graph of $y = x^2$.

For example, consider the equation $y = 64x + 2100$, which gives the total pressure y (in pounds per square foot) on a diver working at a depth of x ft. You can use trace to find the depth at which the pressure reaches 4000 lb/ft^2.

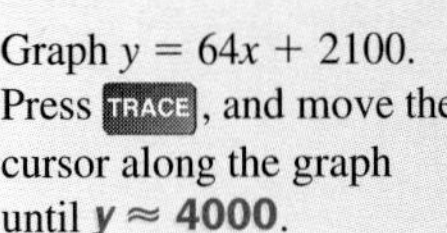

Graph $y = 64x + 2100$. Press TRACE, and move the cursor along the graph until $y \approx$ **4000**.

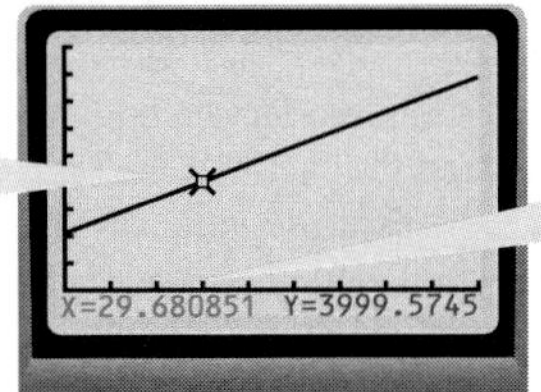

The corresponding water depth is about **30 ft**.

Friendly Windows

As you press ▶ while tracing along a graph, the *x*-coordinate may increase in "unfriendly" increments. For example, if you use the standard window on the TI-82, the *x*-increment, ΔX, is 0.21276596.

Your calculator may allow you to control ΔX directly. If not, you can control ΔX indirectly by choosing an appropriate Xmax for a given Xmin. On the TI-82, for example, choose Xmax so that:

$$\text{Xmax} = \text{Xmin} + 94(\Delta\text{X})$$

This number depends on the calculator you are using.

Suppose you want Xmin to equal **−5** and ΔX to equal **0.1**. Then:

$$\text{Xmax} = -5 + 94(0.1) = 4.4$$

Setting Xmax equal to **4.4** gives a "friendly window" in which the trace cursor's *x*-coordinate increases by **0.1** each time you press ▶.

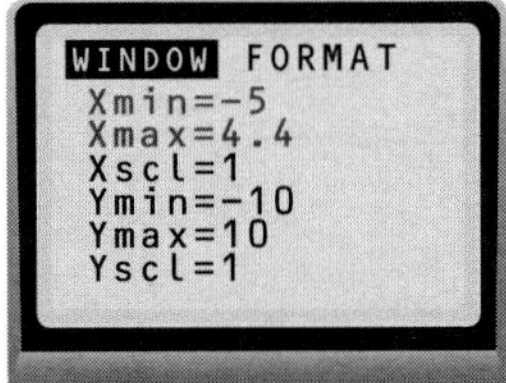

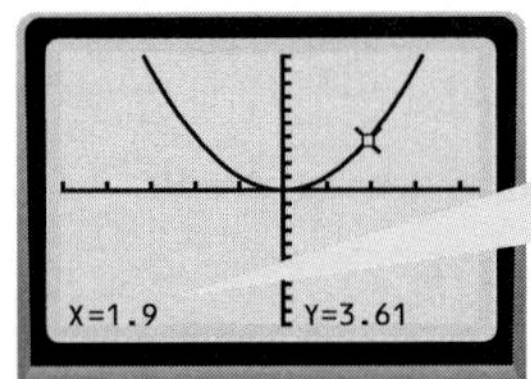

Using trace with the window settings shown gives "friendly" *x*-coordinates like 1.9.

Calculating Coordinates of Interest

Some graphing calculators have a *calculate* feature that you can use to find coordinates of interest on a graph. On the TI-82, you access the calculate menu, shown below, by pressing 2nd and then TRACE after you have graphed a function.

To evaluate the function for a given value of x, select *value* from the calculate menu.

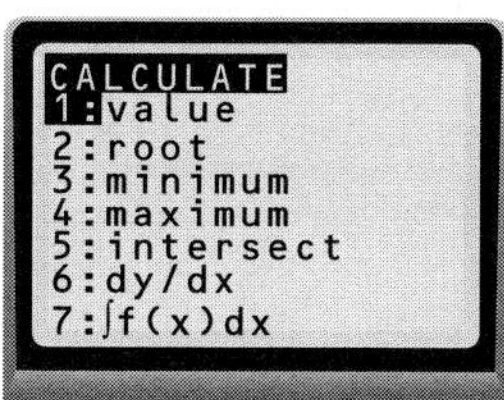

Enter the *x*-value.

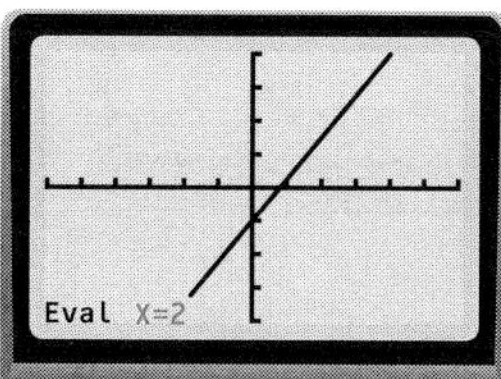

The calculator displays the corresponding *y*-value.

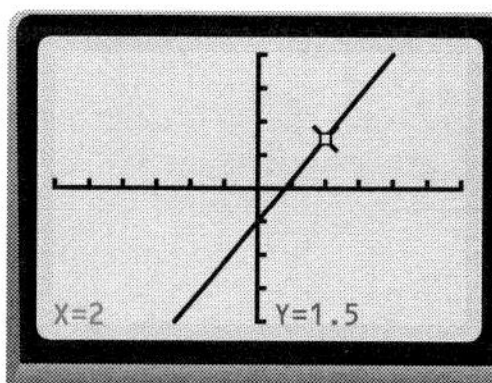

To find a zero of a function, select *root* from the calculate menu.

Use the cursor to specify lower and upper bounds for the zero.

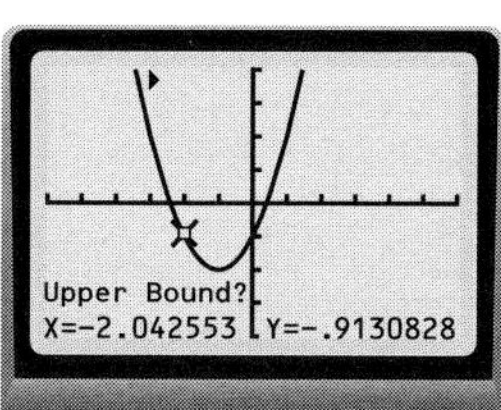

Then guess the location of the zero to help the calculator find it more quickly.

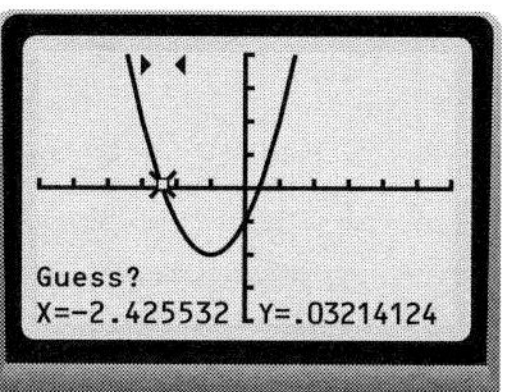

The zero is the *x*-coordinate displayed by the calculator.

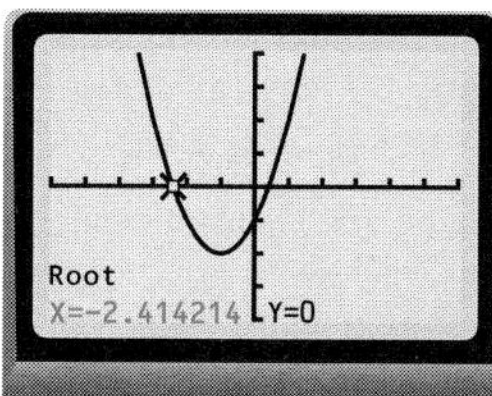

To find a local minimum or local maximum of a function, select *minimum* or *maximum* from the calculate menu.

Use the cursor to specify lower and upper bounds for the minimum or maximum.

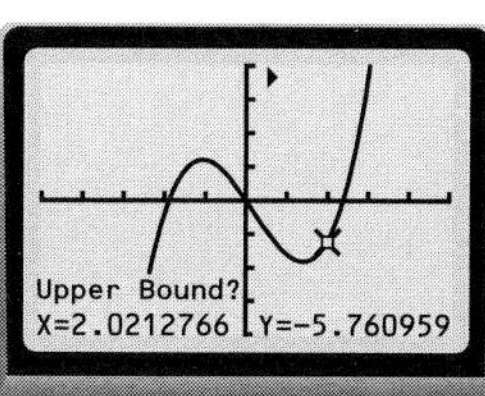

Then guess the location of the minimum or maximum.

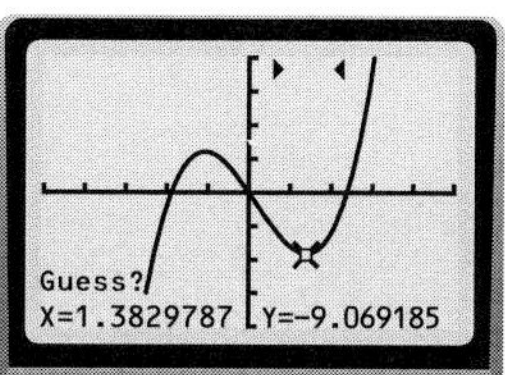

The minimum or maximum is the *y*-coordinate displayed by the calculator.

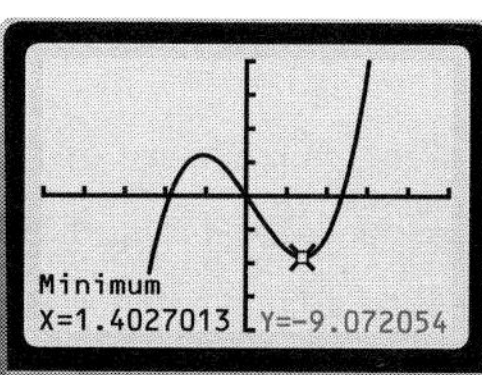

To find a point where two graphs intersect, select *intersect* from the calculate menu.

Use ▲ and ▼ to choose the two graphs.

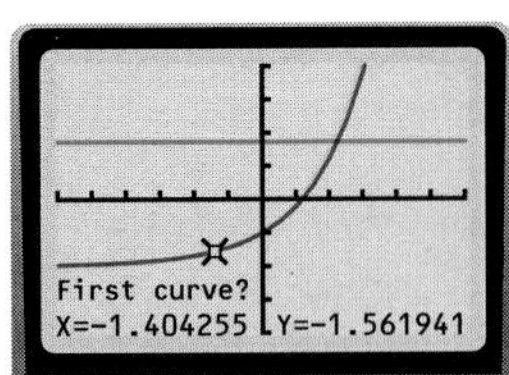

Use ◄ and ► to move the cursor near a point of intersection.

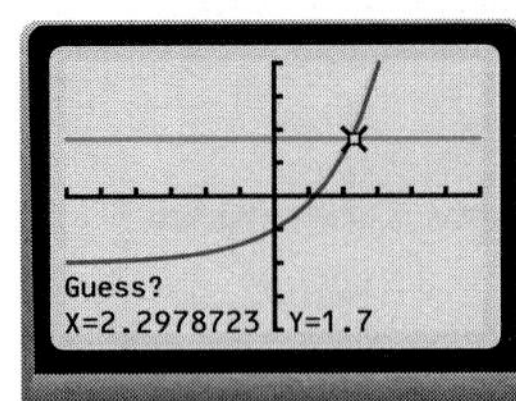

The calculator displays the coordinates of the point of intersection.

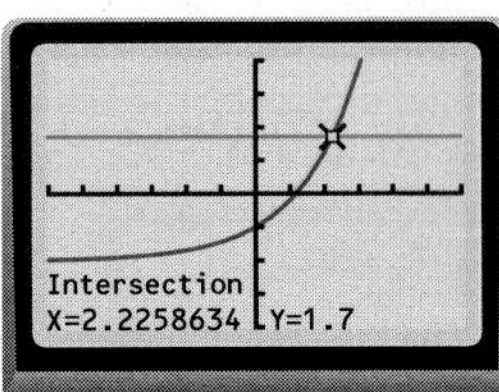

The *Table* Feature

Instead of using the trace and calculate features to find the coordinates of points on a function's graph, you may want to create a table of function values. Some calculators have a *table* feature that allows you to do this. For example, the table below shows values of the function $f(x) = x^3 + 2x^2 - 5x - 10$.

The minimum x-value is 0, and the x-increment is 1. Many calculators let you change these settings.

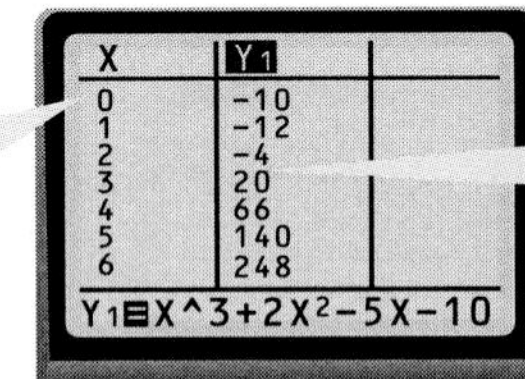

X	Y1
0	-10
1	-12
2	-4
3	20
4	66
5	140
6	248

Y1=X^3+2X²-5X-10

Since the y-values change sign between $x = 2$ and $x = 3$, a zero of f lies between these x-values.

CALCULATOR PRACTICE

14. Graph $y = \frac{6}{x^2 + 1}$. Create a friendly window such that Xmin $= -4$ and ΔX $= 0.1$. Use trace to find the value of y when $x = -0.8$.

15. Let $f(x) = x^4 - 7x^2 + x + 5$. Use the calculate feature to complete parts (a)–(d).

a. Find $f(0.5)$.

b. Find all real zeros of f.

c. Find all local minimums and local maximums of f.

d. Find all real solutions of the equation $f(x) = 9$.

16. Let $g(x) = x - \sqrt{x + 8}$. Make a table of values for g. The minimum x-value in the table should be 0, and the x-increment should be 1.

a. Use the table to find $g(6)$.

b. By repeatedly adjusting the table's minimum x-value and x-increment, find a zero of g to the nearest hundredth.

Calculator Practice

14. 3.66

15. a. 3.81

b. 2.39, 1, –0.813, –2.58

c. local minimums: –5.40, –9.14; local maximum: 5.04

d. 2.68, –2.80

16. a. 2.26

b. 3.37

COMPARING GRAPHS

Using a List to Graph a Family of Functions

Your graphing calculator may allow you to enter a *list* as part of an equation. The graph of such an equation is a family of curves, rather than a single curve. The calculator draws one curve for each number in the list.

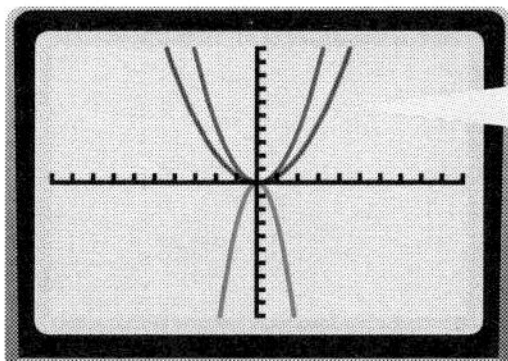

Y_1 = {1, −3, 0.5}X^2 gives the family of functions $y = 1x^2$, $y = -3x^2$, and $y = 0.5x^2$.

Y_2 = X^2 + {5, 1, −6} gives the family of functions $y = x^2 + 5$, $y = x^2 + 1$, and $y = x^2 - 6$.

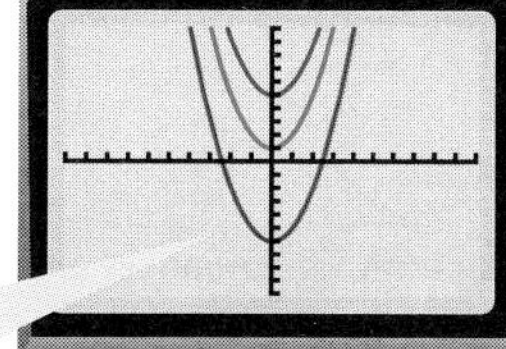

Graphing a Function and Its Inverse

Some graphing calculators have a *draw inverse* feature that lets you graph the inverse of a function without entering the inverse's equation. For example, suppose you want to graph $y = 2x + 4$ and its inverse.

Enter the equation $y = 2x + 4$.

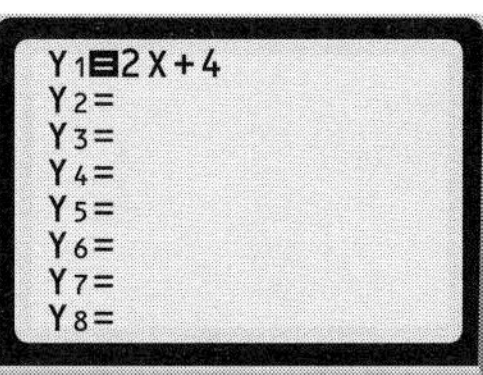

Using draw inverse, tell the calculator to graph the inverse of $y = 2x + 4$.

The calculator graphs $y = 2x + 4$ and its inverse in the same coordinate plane.

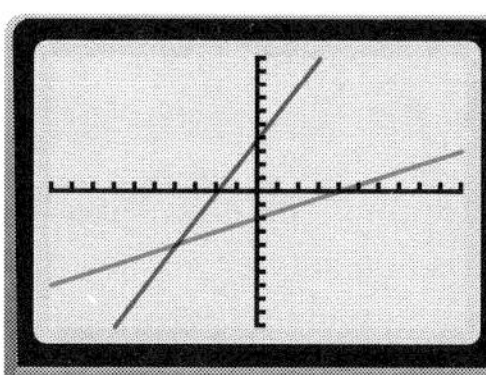

CALCULATOR PRACTICE

17. Describe how to use a list to graph the family of functions $y = x$, $y = -x$, $y = \frac{1}{2}x$, and $y = 2x$.

18. Graph each family of functions. (Set your calculator to radian mode. Adjust the viewing window so that the intervals $-2\pi \le x \le 2\pi$ and $-4 \le y \le 4$ are shown on the axes. Use Xscl $= \frac{\pi}{2}$ and Yscl $= 1$.)

a. $y = \{1, -3, 0.5\}\sin x$ **b.** $y = \cos x + \{0, 2, -2\}$

19. Use the draw inverse feature to graph each function and its inverse.

a. $y = -3x + 2$ **b.** $y = 2^x$ **c.** $y = -\sqrt{x}$

Calculator Practice

17. Enter the equation Y_1={1,–1,0.5,2}X.

18. a.

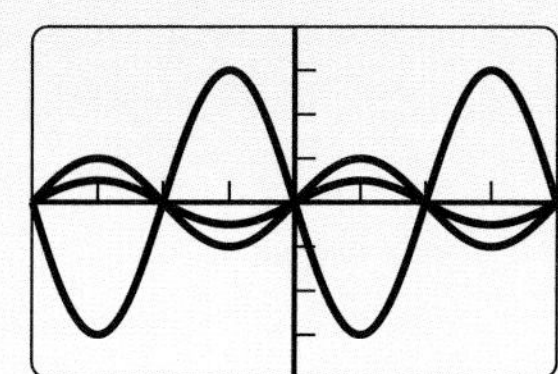

b.

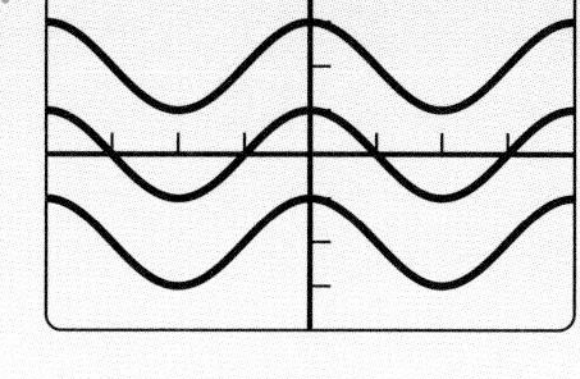

19. a.

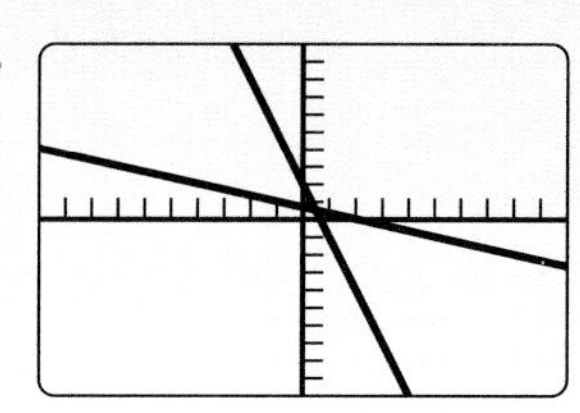

b.

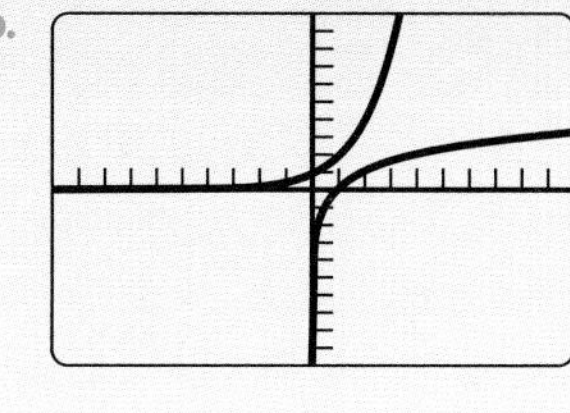

c.

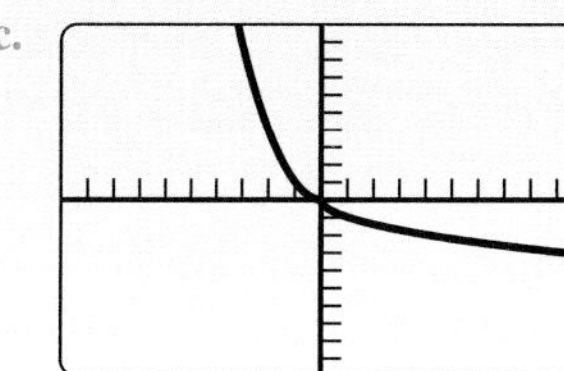

GRAPHING IN DIFFERENT MODES

Graphing Parametric Equations

Suppose an airplane flying at an altitude of **15,000 ft** is preparing to land. Once the plane begins its descent to the runway, its altitude decreases at a rate of 20 ft/s. The plane's horizontal speed is **250 ft/s**.

You can use parametric equations to describe the plane's horizontal position x and vertical position y after t seconds:

$$x = 250t$$

$$y = 15{,}000 - 20t$$

To graph these equations on your graphing calculator, switch to *parametric mode* (denoted by "par" on some calculators). Then complete these steps.

Set appropriate intervals for t, x, and y.

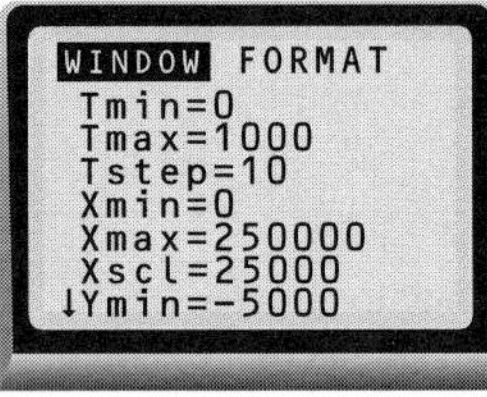

Enter the parametric equations.

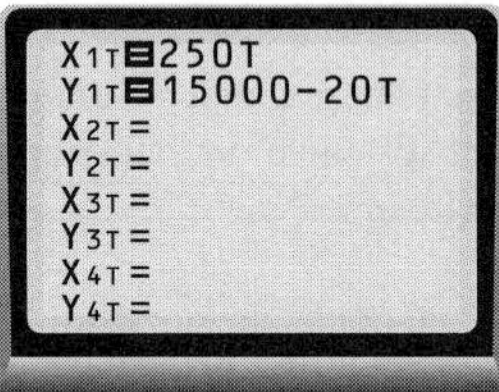

Graph the equations. You can use trace to find the plane's position as a function of time.

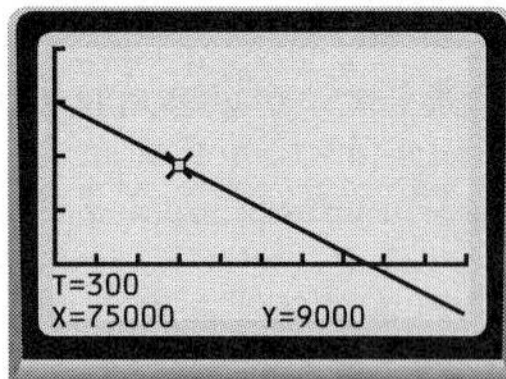

Graphing Sequences

Many graphing calculators have a *sequence mode* (often denoted by "seq") that lets you graph sequences defined either explicitly or recursively. For example, suppose you want to graph the first 10 terms of this sequence:

$$t_1 = 2$$

$$t_n = 0.6(t_{n-1}) + 3$$

On the TI-82, use the following steps. Note that you must replace t_n with u_n or v_n when entering the sequence.

Enter the starting value t_1 and the minimum and maximum values of n.

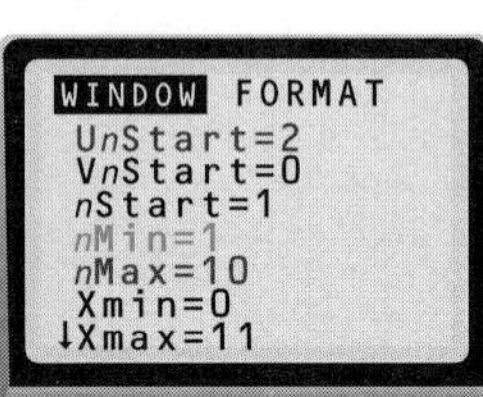

Enter the recursion equation.

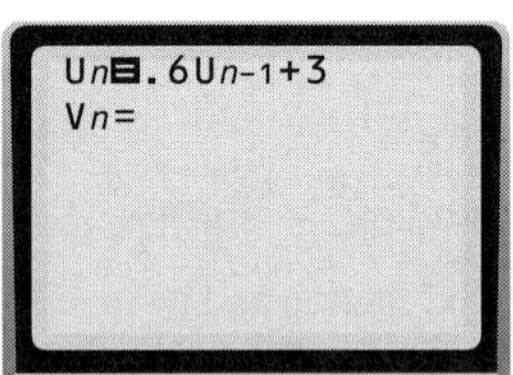

Graph the sequence in dot mode. Use trace to find values of t_n for different values of n.

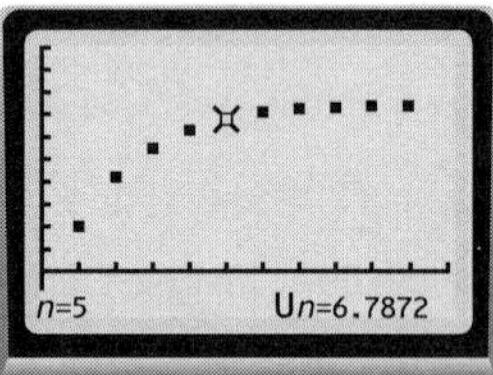

CALCULATOR PRACTICE

20. Graph the parametric equations that describe the position of the airplane as it descends for landing. (See the top of the previous page.) How long does it take for the plane to reach the ground?

21. Graph the first 10 terms of each sequence.

a. $t_1 = 10$, $t_n = t_{n-1} - 1$ **b.** $t_1 = 1$, $t_n = 1.25(t_{n-1})$ **c.** $t_1 = 6$, $t_n = 0.7(t_{n-1}) + 0.2$

USING MATRICES

Adding and Subtracting Matrices

You can use a graphing calculator to add or subtract matrices that have the same dimensions. Consider the tables below, which show the numbers of left- and right-handed males and females in two algebra classes.

Class 1	Left	Right	Both
Males	2	8	1
Females	1	13	0

Class 2	Left	Right	Both
Males	3	16	0
Females	2	9	2

To get a single table containing data for *both* classes, enter each table as a matrix and find the sum of the two matrices.

Use matrix [A] for class 1. The dimensions of [A] are 2 × 3.

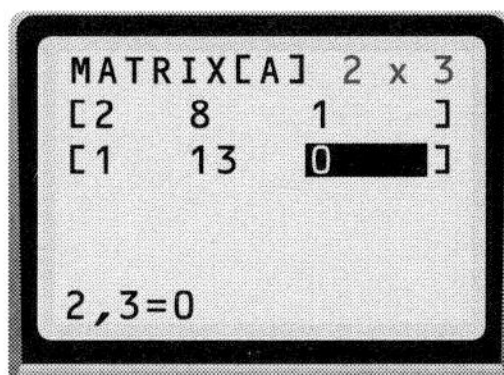

Use matrix [B] for class 2. This matrix is also 2 × 3.

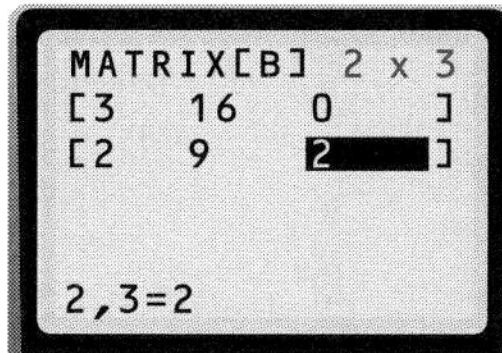

From the home screen, add the two matrices.

```
[A]+[B]
      [[5 24 1]
       [3 22 2]]
```

Multiplying a Matrix by a Scalar

Look at matrix *A* above showing the numbers of males and females in class 1 who are left- and right-handed. To find the *percent* of students who fall into each category, first divide each matrix entry by the total number of students in the class (25). Then multiply the resulting entries by 100. You can perform these operations quickly using scalar multiplication on a graphing calculator, as shown at the right.

```
(1/25)[A]
 [[.08 .32 .04]
  [.04 .52 0  ]]
Ans*100
      [[8 32 4]
       [4 52 0]]
```


Calculator Practice

20. 750 s, or 12.5 min

21. a.

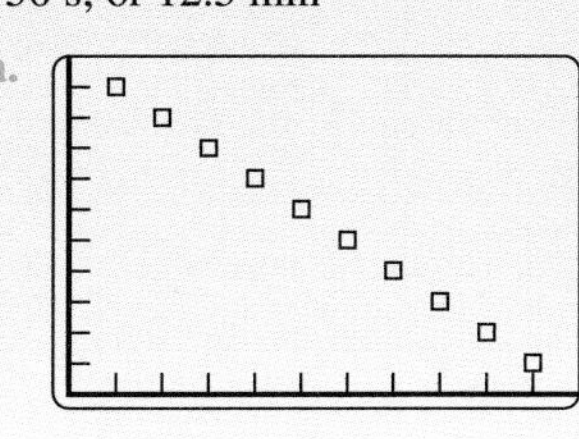

b.

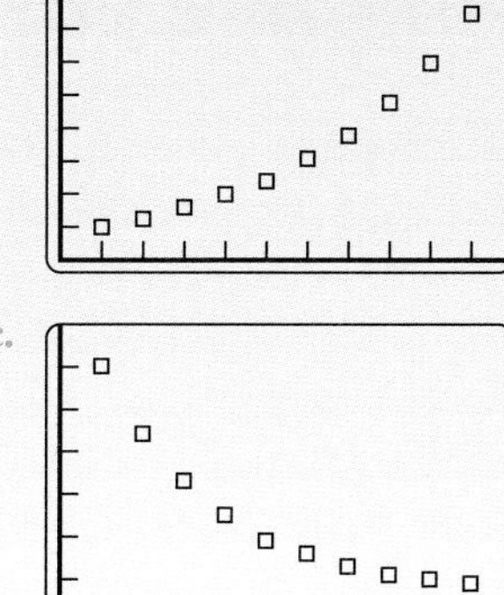

c.

Finding the Product of Two Matrices

You can multiply two matrices when the number of columns in the first matrix is the same as the number of rows in the second matrix.

For example, suppose you want to find this product: $\begin{bmatrix} 9 & 4 \\ 3 & 1 \\ 2 & 8 \\ 1 & 5 \end{bmatrix} \begin{bmatrix} 4 & 2 & 0 \\ 3 & 0 & 2 \end{bmatrix}$

Enter the matrices into a graphing calculator as A and B. Then find the product AB.

A has **2** columns and B has **2** rows, so the product AB is defined.

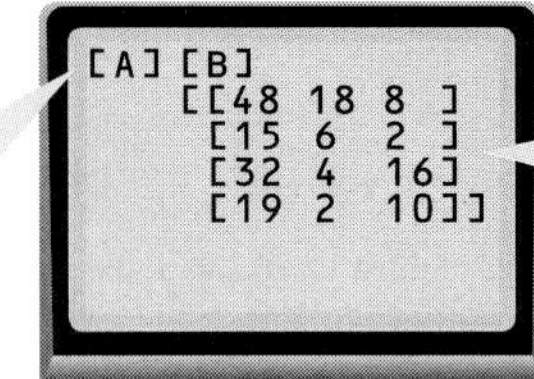

Since A is a 4 × **2** matrix and B is a **2** × 3 matrix, AB is a 4 × 3 matrix.

Finding the Inverse of a Matrix

If a square matrix A has an inverse, you can use a graphing calculator to find it. This feature is useful when you want to solve a system of linear equations. For example, suppose you want to solve this system:

$$\begin{aligned} 2x - 5y + 3z &= 35 \\ 7x - 2y - 4z &= -11 \\ -x \qquad + z &= 5 \end{aligned} \quad \longrightarrow \quad \underset{A}{\begin{bmatrix} 2 & -5 & 3 \\ 7 & -2 & -4 \\ -1 & 0 & 1 \end{bmatrix}} \underset{X}{\begin{bmatrix} x \\ y \\ z \end{bmatrix}} = \underset{B}{\begin{bmatrix} 35 \\ -11 \\ 5 \end{bmatrix}}$$

This equation, $AX = B$, has solution $X = A^{-1}B$.

Enter the matrices A and B. Then calculate $A^{-1}B$.

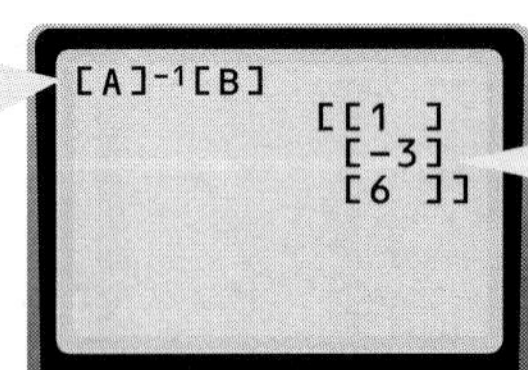

The solution of the system of equations is $x = 1$, $y = -3$, and $z = 6$.

CALCULATOR PRACTICE

Use matrices *A*, *B*, *C*, and *D* to evaluate each matrix expression. If an operation cannot be performed, write *undefined*.

$A = \begin{bmatrix} 3 & 4 \\ 6 & 7 \end{bmatrix}$ $\quad B = \begin{bmatrix} 5 & 11 \\ -3 & -7 \end{bmatrix}$ $\quad C = \begin{bmatrix} 9 & 3 \\ 12 & 4 \end{bmatrix}$ $\quad D = \begin{bmatrix} 2 & 0 & -8 & 1 \\ 5 & 7 & 3 & 9 \end{bmatrix}$

22. $A + B$ **23.** $B - C$ **24.** $2A$ **25.** AB

26. $C + D$ **27.** CD **28.** A^{-1} **29.** D^{-1}

Calculator Practice

22. $\begin{bmatrix} 8 & 15 \\ 3 & 0 \end{bmatrix}$

23. $\begin{bmatrix} -4 & 8 \\ -15 & -11 \end{bmatrix}$

24. $\begin{bmatrix} 6 & 8 \\ 12 & 14 \end{bmatrix}$

25. $\begin{bmatrix} 3 & 5 \\ 9 & 17 \end{bmatrix}$

26. undefined

27. $\begin{bmatrix} 33 & 21 & -63 & 36 \\ 44 & 28 & -84 & 48 \end{bmatrix}$

28. $\begin{bmatrix} -2.33 & 1.33 \\ 2 & -1 \end{bmatrix}$

29. undefined

STATISTICS AND PROBABILITY

Histograms, Box Plots, and Line Graphs

Many graphing calculators can display histograms, box plots, and line graphs of data that you enter. For example, suppose you want to make a histogram of the following employee commuting times from page 260:

15, 25, 30, 40, 60, 40, 25, 15, 15, 20, 30, 75, 20, 15, 65, 40, 30, 20, 15, 10, 20, 35, 45, 35, 30, 20, 10, 55, 25, 15, 5, 45, 20, 10, 5, 15, 30, 30, 50, 15, 5, 10, 45, 30, 20, 20, 10, 30, 20, 30

The steps used to enter data and perform statistical functions vary greatly from calculator to calculator. On the TI-82, follow these steps:

Press STAT, and select the "Edit..." option. Enter the data values into a list, such as L1.

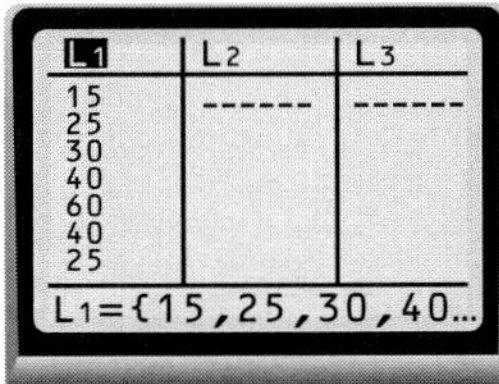

Tell the calculator to display a **histogram** using list L1. Set the frequency option to 1.

Set up a good viewing window. Then press GRAPH to display the histogram.

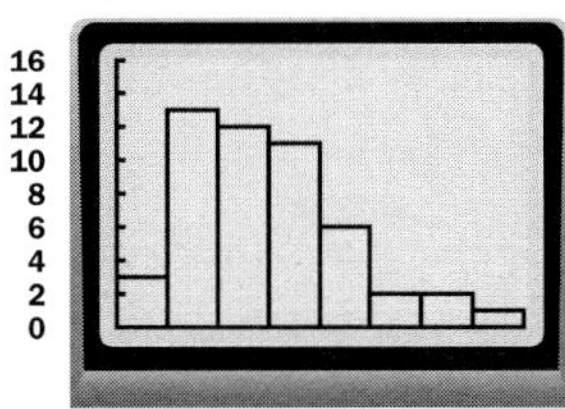

Mean and Standard Deviation

You can use a scientific or graphing calculator to find the mean and standard deviation of a set of data values. Some calculators give two types of standard deviation—the *sample standard deviation S* and the *population standard deviation* σ. In this book, the population standard deviation is always used.

For example, you can calculate the mean and standard deviation of the employee commuting times shown above.

Select the 1-variable statistics option from the menu of statistics functions.

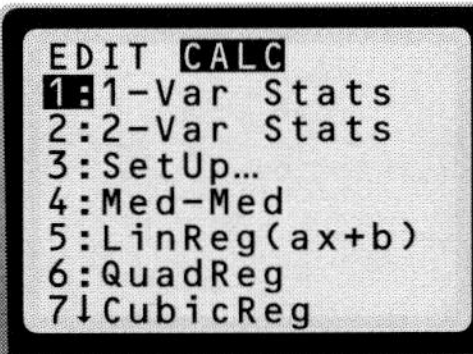

Specify the list containing the data (in this case, L1).

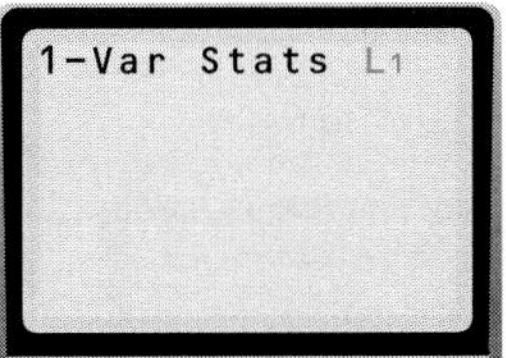

The calculator displays the **mean** and **standard deviation** of the data.

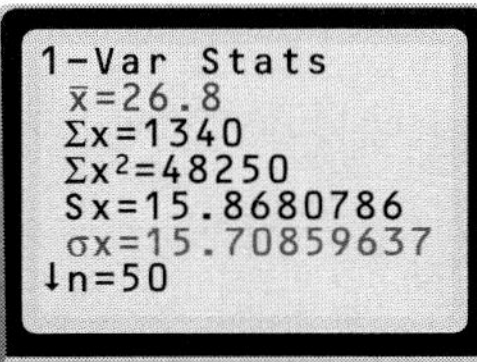

Scatter Plots and Curve Fitting

Your graphing calculator may have a curve-fitting feature that lets you fit a line, parabola, or other type of curve to a scatter plot. For example, suppose you want to make a scatter plot and find a line of fit for the following bicycle data from page 66:

Years since 1965	0	5	10	15	20	25
Number of bicycles (in millions)	21	36	43	62	79	95

On the TI-82, follow these steps:

Enter the years since 1965 in list L1. Enter the numbers of bicycles in list L2.

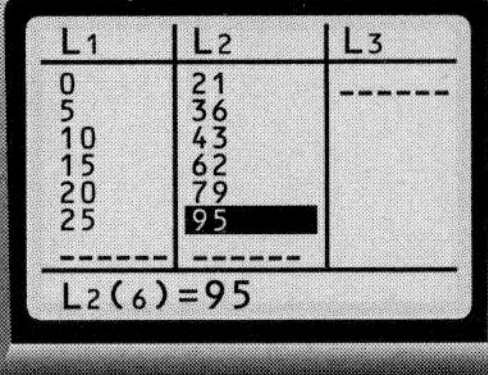

Use the linear regression feature to find an equation of the least-squares line.

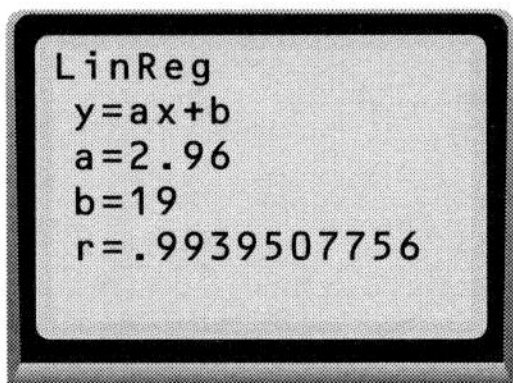

Graph the data pairs and the regression equation in the same coordinate plane.

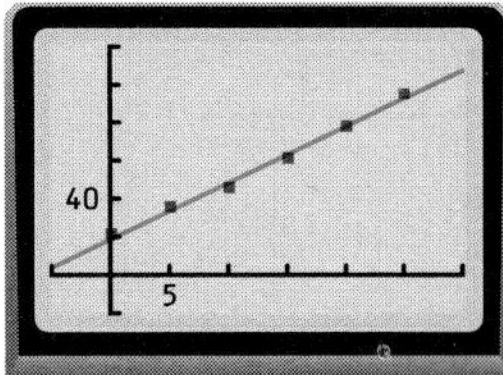

Generating Random Numbers

Most scientific and graphing calculators will generate random numbers between 0 and 1. On the TI-82, this is accomplished using the *rand* feature, as shown below.

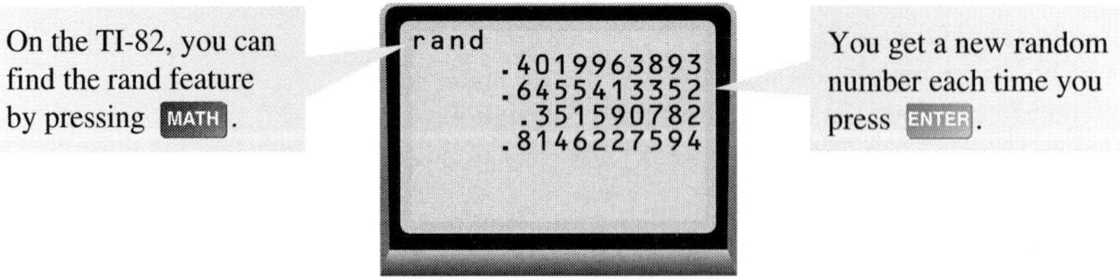

You can transform a random number x in the interval $0 < x < 1$ into a random number in any interval. For example, suppose you want to simulate rolling a six-sided die by generating random integers from 1 to 6.

CALCULATOR PRACTICE

30. Students in a history class obtained these scores on an exam:

75, 88, 63, 79, 90, 52, 81, 68, 84, 77,
92, 75, 64, 70, 57, 89, 66, 96, 75, 83

a. Make a histogram of the scores. Adjust the viewing window so that the intervals $50 \le x \le 100$ and $0 \le y \le 8$ are shown. Use Xscl = 10 and Yscl = 1.

b. Find the mean and standard deviation of the scores.

31. Use the data on swimming given in the table.

a. Find an equation of the least-squares line for the data.

b. Graph the data pairs (x, y) and the equation of the least-squares line in the same coordinate plane.

32. Explain how you can use a calculator to generate random integers from 10 to 25.

Men's Winning Times in Olympic 400 m Freestyle Swimming

x = years since 1960	y = time (seconds)
0	258.3
4	252.2
8	249.0
12	240.27
16	231.93
20	231.31
24	231.23
28	226.95
32	225.00

PROGRAMMING YOUR CALCULATOR

You may be able to program your graphing calculator to carry out mathematical procedures that are not built in. The following program for the TI-82 determines the number of real or imaginary solutions of a quadratic equation $ax^2 + bx + c = 0$. It also finds these solutions.

```
:Input "ENTER A:",A:Input "ENTER B:",B:
 Input "ENTER C:",C
:B²–4AC→D:–B/(2A)→R:(√abs D)/(2A)→S
:ClrHome
:If D>0:Then:Disp "TWO REAL:",R+S,R–S
:Else
:If D=0:Then:Disp "ONE REAL:",R
:Else
:Disp "TWO IMAGINARY:","R+SI AND R–SI","WHERE"
:Output(4,1,"R="):Output(4,3,R):Output(5,1,"S="):
 Output(5,3,S)
:End:End
```

You can use this program to solve $x^2 + 4x + 13 = 0$, for example.

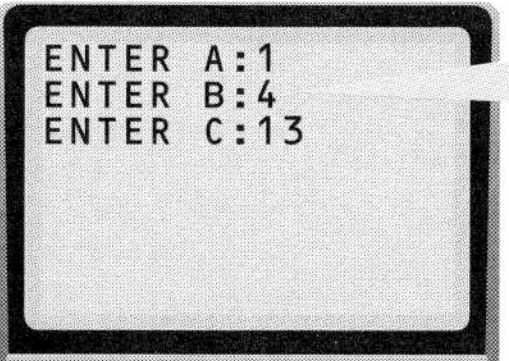

Enter the equation's coefficients.

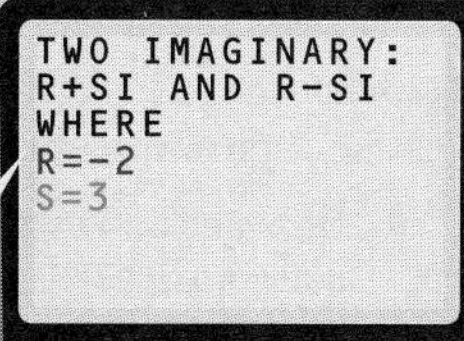

The solutions of $x^2 + 4x + 13 = 0$ are $-2 \pm 3i$.

Calculator Practice

30. a.

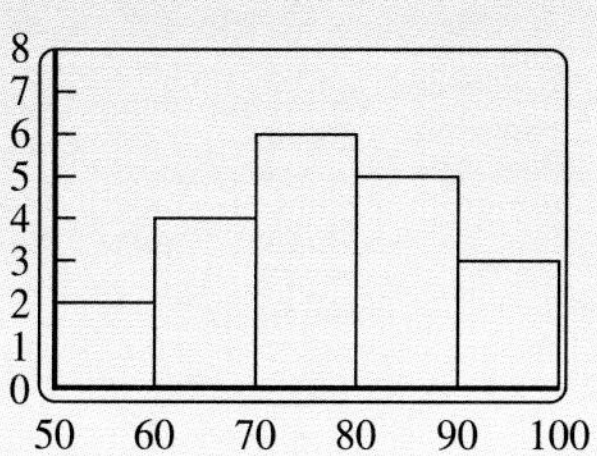

b. mean: 76.2; standard deviation: 11.8

31. a. $y = -1.06x + 255$

b.

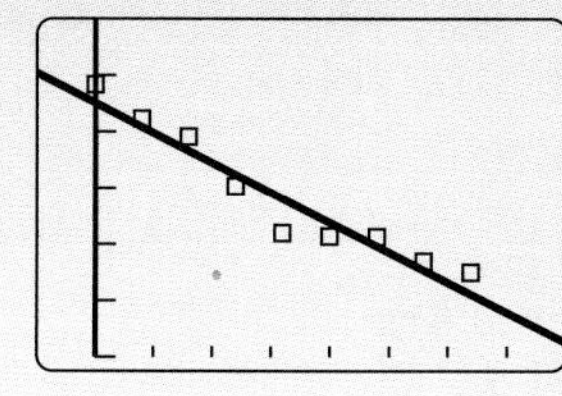

32. Answers will depend on the type of calculator used. On the TI-82, enter the expression "int(16*rand)+10."

CALCULATOR PRACTICE

33. Enter the program on the previous page into your calculator. (If you don't have a TI-82, you will have to modify some of the commands.) Use the program to solve each quadratic equation.

a. $x^2 + 3x - 18 = 0$ **b.** $25x^2 - 20x + 4 = 0$ **c.** $16x^2 = 8x - 33$

34. Write a program for your calculator that solves linear equations of the form $ax + b = 0$.

USING A SPREADSHEET

Suppose a high school volunteer club sells T-shirts to raise money. The club raises \$3000 in 1995 and projects that this amount will increase by 25% per year. In what year will the amount raised exceed \$7000?

You can solve this problem using a computer spreadsheet. A spreadsheet is made up of cells named by a column letter and a row number, such as A3 or B4. You can enter a label, a number, or a formula into a cell.

row numbers

column letters

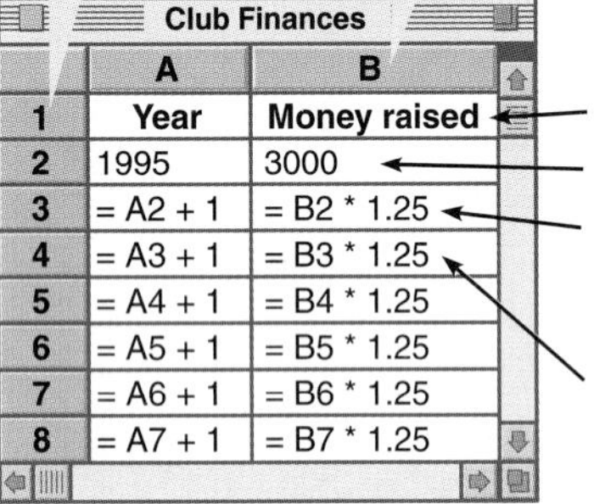

Club Finances

	A	B
1	Year	Money raised
2	1995	3000
3	= A2 + 1	= B2 * 1.25
4	= A3 + 1	= B3 * 1.25
5	= A4 + 1	= B4 * 1.25
6	= A5 + 1	= B5 * 1.25
7	= A6 + 1	= B6 * 1.25
8	= A7 + 1	= B7 * 1.25

Cell B1 contains the label "Money raised."

Cell B2 contains the number 3000.

Cell B3 contains the formula "= B2 * 1.25." This formula tells the computer to take the number in cell B2, multiply it by 1.25, and store the result in cell B3.

Instead of typing a formula into each cell individually, you can use the spreadsheet's *copy* or *fill down* command.

The computer replaces all formulas in the spreadsheet with calculated values, as shown at the right. You can see that the amount of money raised by the volunteer club will **exceed \$7000** in 1999.

Club Finances

	A	B
1	Year	Money raised
2	1995	3000.00
3	1996	3750.00
4	1997	4687.50
5	1998	5859.38
6	1999	7324.22
7	2000	9155.27
8	2001	11444.09

SPREADSHEET PRACTICE

35. Suppose you have \$600 in a savings account that pays 5% annual interest. Use a spreadsheet to answer parts (a) and (b).

a. How much money will you have in your account after 10 years if you make no deposits or withdrawals?

b. How many full years must you wait before the value of your account exceeds \$800?

Calculator Practice

33. a. 3, –6

b. 0.4

c. $0.25 \pm 1.41i$

34. The specific commands in the program will depend on the calculator used. The following program is for the TI-82:

```
:Input "ENTER A:",A
:Input "ENTER B:",B
:Disp "SOLUTION:",–B/A
```

Spreadsheet Practice

35. a. \$977.34

b. 6 years

Table of Measures

Time

60 seconds (s) = 1 minute (min)
60 minutes = 1 hour (h)
24 hours = 1 day
7 days = 1 week
4 weeks (approx.) = 1 month

365 days, 52 weeks (approx.), 12 months = 1 year
10 years = 1 decade
100 years = 1 century

Metric

Length

10 millimeters (mm) = 1 centimeter (cm)
100 cm, 1000 mm = 1 meter (m)
1000 m = 1 kilometer (km)

Area

100 square millimeters (mm^2) = 1 square centimeter (cm^2)
10,000 cm^2 = 1 square meter (m^2)
10,000 m^2 = 1 hectare (ha)

Volume

1000 cubic millimeters (mm^3) = 1 cubic centimeter (cm^3)
1,000,000 cm^3 = 1 cubic meter (m^3)

Liquid Capacity

1000 milliliters (mL) = 1 liter (L)
1000 L = 1 kiloliter (kL)

Mass

1000 milligrams (mg) = 1 gram (g)
1000 g = 1 kilogram (kg)
1000 kg = 1 metric ton (t)

Temperature — **Degrees Celsius (°C)**

0°C = freezing point of water
37°C = normal body temperature
100°C = boiling point of water

United States Customary

Length

12 inches (in.) = 1 foot (ft)
36 in., 3 ft = 1 yard (yd)
5280 ft, 1760 yd = 1 mile (mi)

Area

144 square inches ($in.^2$) = 1 square foot (ft^2)
9 ft^2 = 1 square yard (yd^2)
43,560 ft^2, 4840 yd^2 = 1 acre (A)

Volume

1728 cubic inches ($in.^3$) = 1 cubic foot (ft^3)
27 ft^3 = 1 cubic yard (yd^3)

Liquid Capacity

8 fluid ounces (fl oz) = 1 cup (c)
2 c = 1 pint (pt)
2 pt = 1 quart (qt)
4 qt = 1 gallon (gal)

Weight

16 ounces (oz) = 1 pound (lb)
2000 lb = 1 ton (t)

Temperature — **Degrees Fahrenheit (°F)**

32°F = freezing point of water
98.6°F = normal body temperature
212°F = boiling point of water

Table of Symbols

Symbol		Page
$\begin{bmatrix} 1 & 0 \\ 0 & 1 \end{bmatrix}$	matrix	16
$a_{2,1}$	element in first column and second row of a matrix A	16
x_1	x sub 1	46
$f(x)$	f of x, or the value of f at x	60
b^{-n}	$\frac{1}{b^n}, b \neq 0$	101
e	irrational number, about 2.718	116
f^{-1}	inverse of function f	142
$\log_b x$	base-b logarithm of x	149
$\log x$	base-10 logarithm of x	150
$\ln x$	base-e logarithm of x	150
$\pm$	plus-or-minus	187
$\sqrt{a}$	the nonnegative square root of a	187
$\bar{x}$	x-bar; the mean of $x_1, x_2, \ldots, x_n$	268
σ	sigma; the standard deviation of a data set	268
A^{-1}	inverse of matrix A	304
$\sqrt[n]{a}$	nth root of a	345
π	pi; irrational number, about 3.14	358
i	$\sqrt{-1}$	359
$\lvert a + bi \rvert$	the magnitude of a complex number $a + bi$	366
$-a$	opposite of a	373
$\frac{1}{a}$	reciprocal of a, $a \neq 0$	374
∞	infinity	405
Σ	summation	490
!	factorial	579
${}_nP_r$	number of permutations of r objects taken from n objects	580
${}_nC_r$	number of combinations of r objects taken from n objects	585
$P(A)$	probability of event A	611
$P(B \mid A)$	probability of event B given that event A has occurred	626
sin	sine	656
cos	cosine	656
° ′ ″	degrees, minutes, and seconds	657
$\sin^{-1}$	inverse sine	658
$\cos^{-1}$	inverse cosine	658
θ	theta; name of an angle, or measure of an angle	658
tan	tangent	663
$\tan^{-1}$	inverse tangent	664
$\cdot$	multiplication, times ($\times$)	774
$\lvert a \rvert$	absolute value of a	774
a^n	nth power of a	776
$\neq$	is not equal to	786
$>$	is greater than	787
$<$	is less than	787
$\geq$	is greater than or equal to	787
$\leq$	is less than or equal to	787
$\approx$	is approximately equal to	790
(x, y)	ordered pair	794
$\angle A$	angle A, or measure of $\angle A$	800
$\triangle ABC$	triangle ABC	801

Properties of Algebra

Properties of Real Numbers

Let a, b, and c be real numbers.

Group Properties of Real Numbers under Addition, p. 373

- **Closure:** $a + b$ is a real number.
- **Identity:** $a + 0 = 0 + a = a$. 0 is the identity for addition.
- **Inverse:** There is a real number $-a$, called the *opposite*, or *additive inverse*, of a, such that $a + (-a) = 0$.
- **Associative:** $(a + b) + c = a + (b + c)$
- **Commutative:** $a + b = b + a$

Group Properties of Real Numbers under Multiplication, p. 374

- **Closure:** ab is a real number.
- **Identity:** $a \cdot 1 = 1 \cdot a = a$. 1 is the identity for multiplication.
- **Inverse:** If a is a nonzero real number, then there is a real number $\frac{1}{a}$, called the *reciprocal*, or *multiplicative inverse*, of a, such that $a \cdot \frac{1}{a} = 1$.
- **Associative:** $(ab)c = a(bc)$
- **Commutative:** $ab = ba$

Field Properties of Real Numbers under Addition and Multiplication, p. 377: The field properties include all of the properties listed above as well as the *distributive property*:

- **Distributive:** $a(b + c) = ab + ac$

Properties of Exponents

Let b, m, and n be positive real numbers.

Zero Exponent, p. 776: $b^0 = 1$

Negative Exponent, p. 101: $b^{-n} = \frac{1}{b^n}$

Product of Powers, p. 94: $b^m \cdot b^n = b^{m+n}$

Quotient of Powers, p. 94: $\frac{b^m}{b^n} = b^{m-n}$

Power of a Power, p. 94: $(b^m)^n = b^{mn}$

Properties of Exponents and Radicals

Let a, b, and c be positive real numbers, and let m and n be positive integers.

Relationship between Exponents and Radicals, pp. 102, 346: $b^{1/n} = \sqrt[n]{b} = c$ if and only if $b = c^n$

Power or Root of a Product, p. 346

$(ab)^m = a^m \cdot b^m$ and $\sqrt[m]{ab} = \sqrt[m]{a} \cdot \sqrt[m]{b}$

Power or Root of a Quotient, p. 346

$\left(\frac{a}{b}\right)^m = \frac{a^m}{b^m}$ and $\sqrt[m]{\frac{a}{b}} = \frac{\sqrt[m]{a}}{\sqrt[m]{b}}$

Power or Root of a Power, p. 346

$\left(b^m\right)^n = b^{mn} = b^{nm} = \left(b^n\right)^m$ and

$b^{n/m} = \sqrt[m]{b^n} = \left(\sqrt[m]{b}\right)^n$

Properties of Logarithms

Let M, N, b, and c be positive real numbers with $b \neq 1$ and $c \neq 1$.

Relationship between Exponents and Logarithms, p. 149: $\log_b M = N$ if and only if $b^N = M$

Logarithm of a Product, p. 156

$\log_b MN = \log_b M + \log_b N$

Logarithm of a Quotient, p. 156

$\log_b \frac{M}{N} = \log_b M - \log_b N$

Logarithm of a Power, p. 156

$\log_b M^k = k \log_b M$ for any real k

Change of Base, p. 163: $\log_b M = \frac{\log_c M}{\log_c b}$

Visual Glossary of Graphs

Constant

$f(x) = 2$

Direct Variation

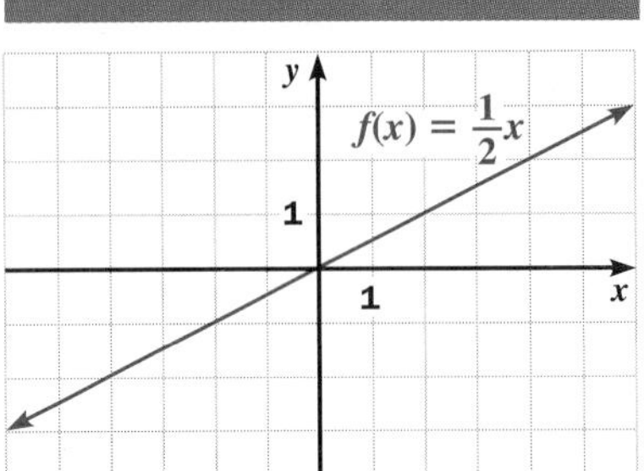

Linear

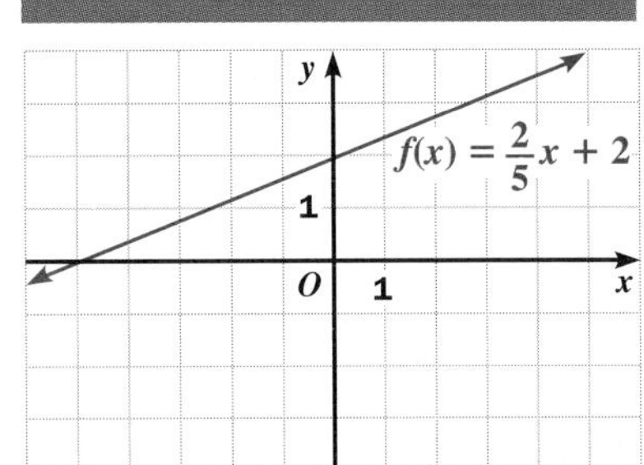

Exponential Growth

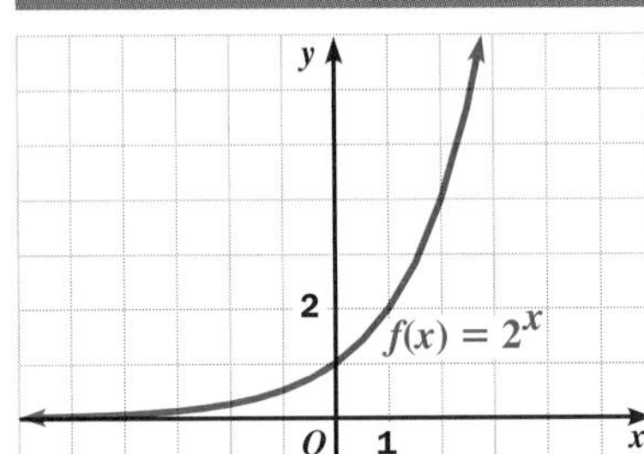

Exponential Decay

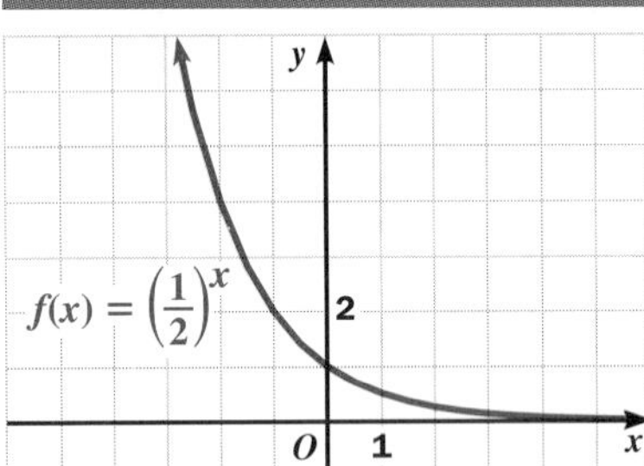

Exponential/Logarithmic (inverse functions)

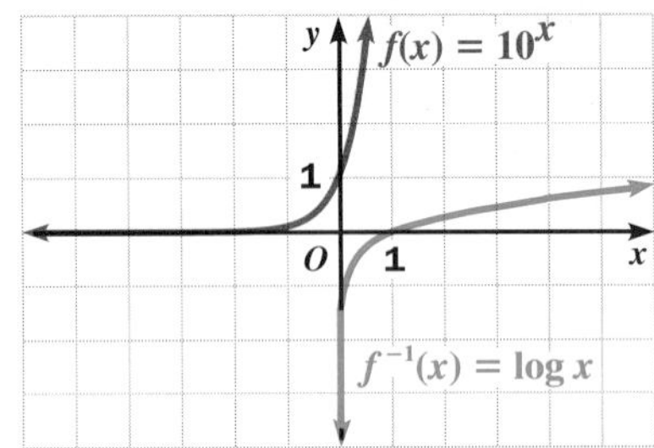

Quadratic

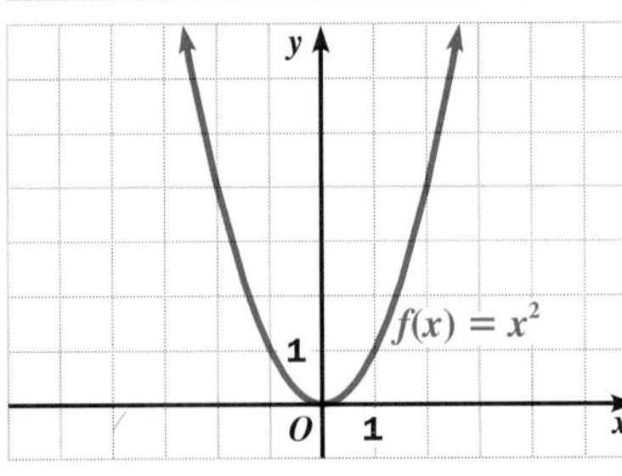

Quadratic/Square Root (inverse functions)

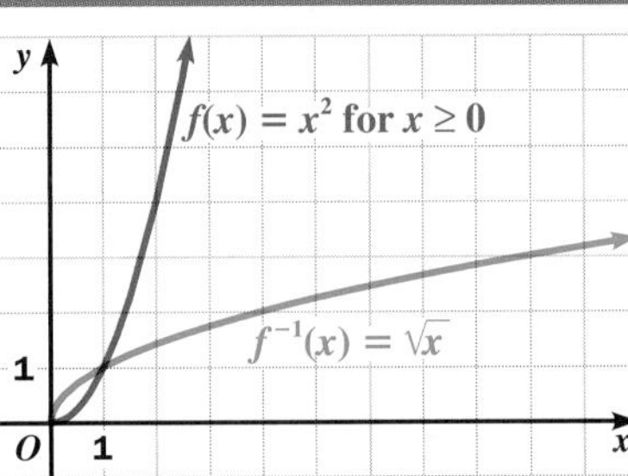

Cubic/Cube Root (inverse functions)

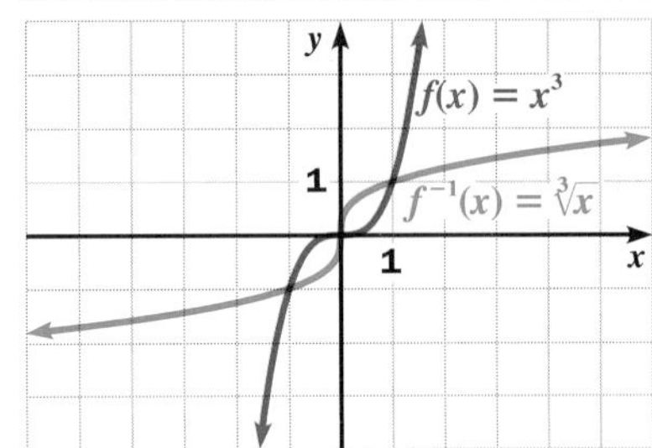

Polynomial (even degree)

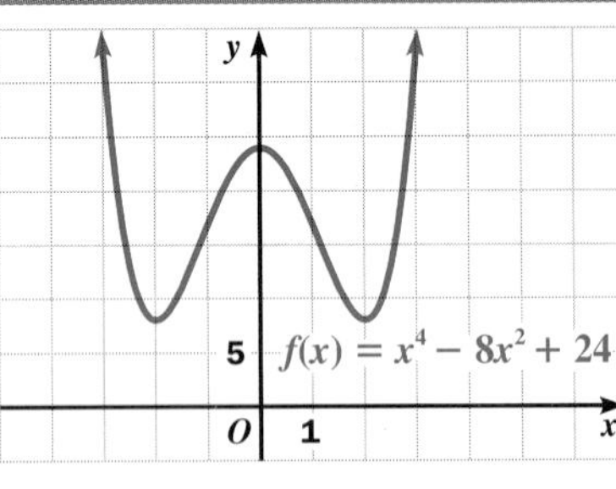

Polynomial (odd degree)

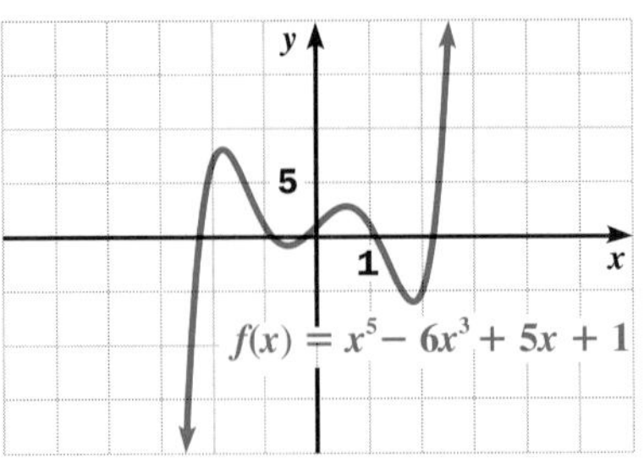

Rational

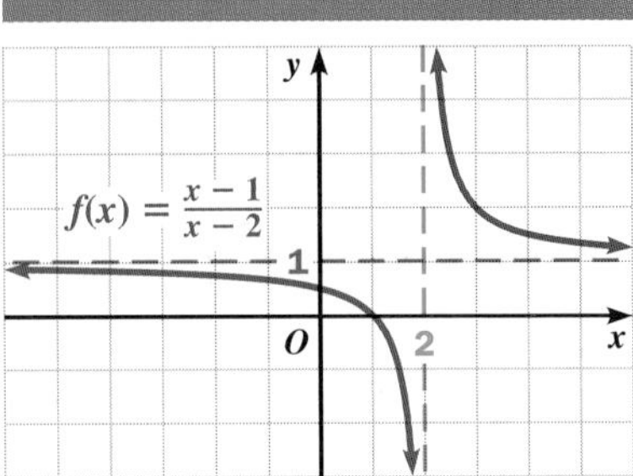

820

Sine

Cosine

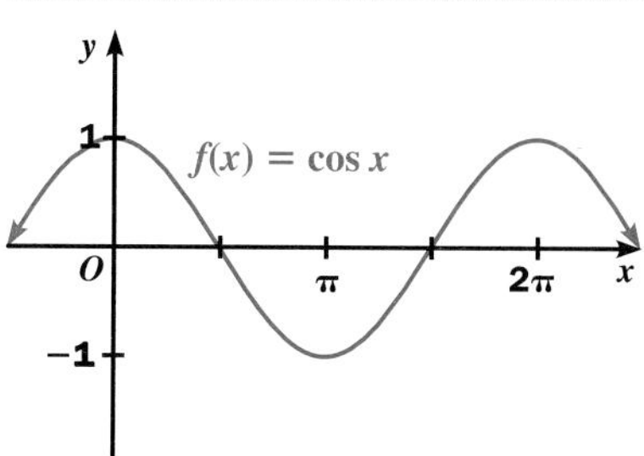

Tangent

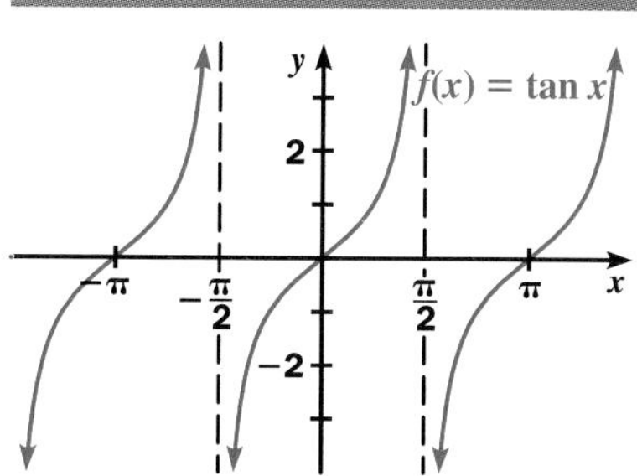

Transformations

Translation

Vertical Stretch

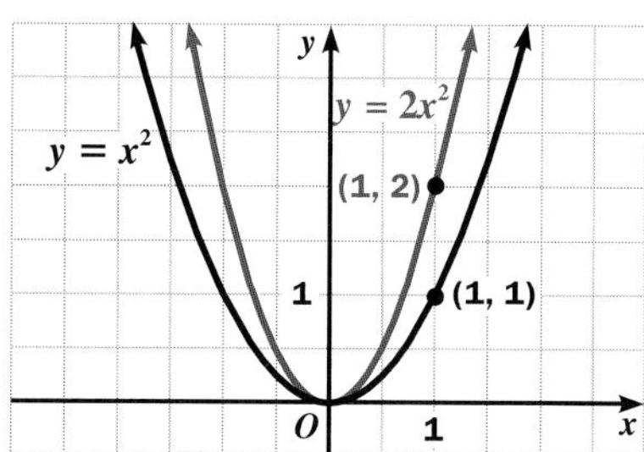

Reflection over *x*-axis

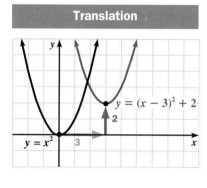

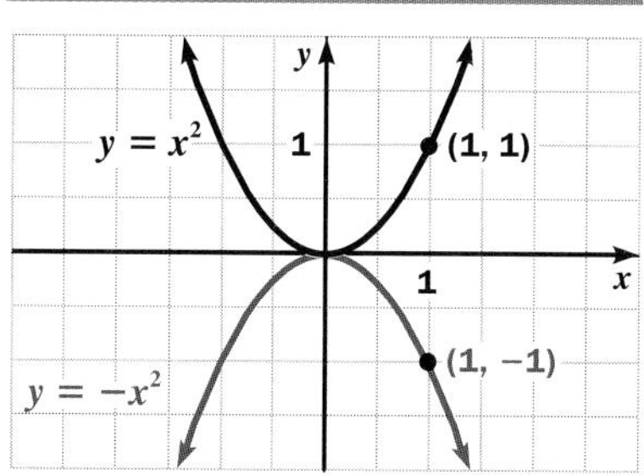

Conic Sections

Circle

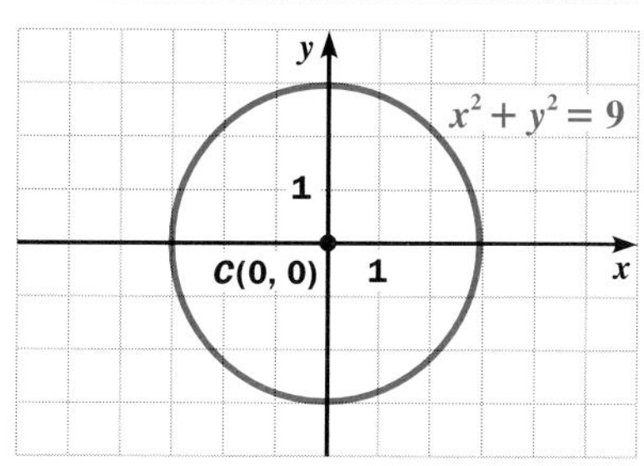

Horizontal Ellipse

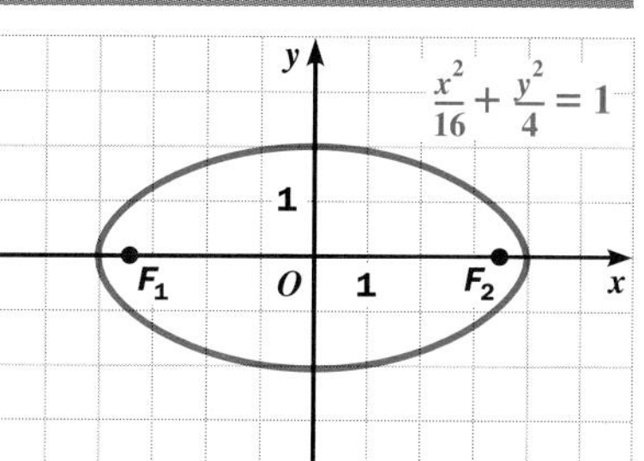

Vertical Ellipse

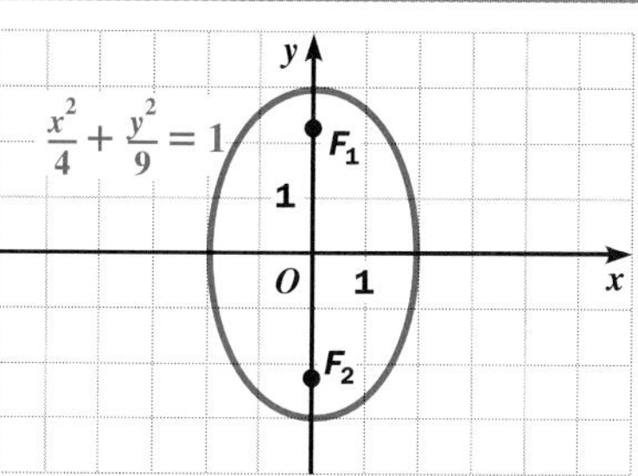

Horizontal Parabola

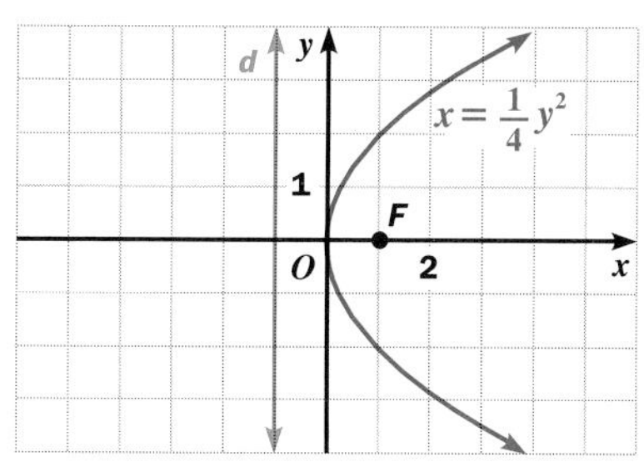

Horizontal Hyperbola

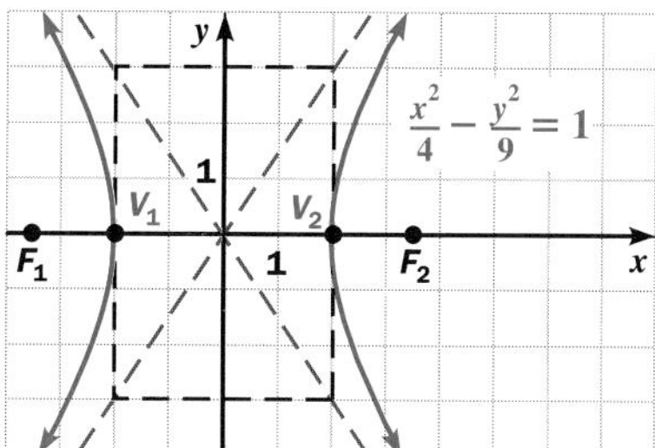

Vertical Hyperbola

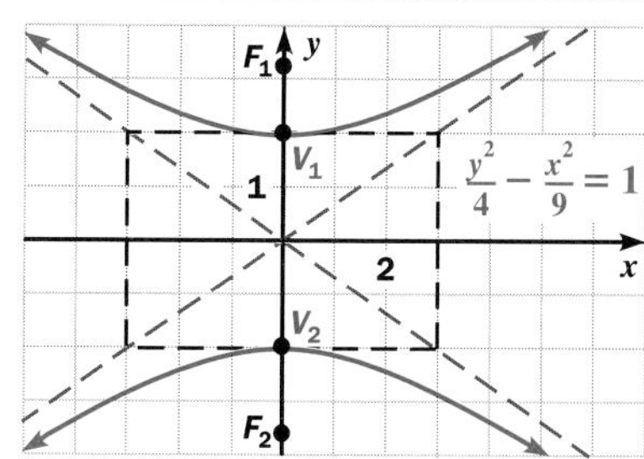

Formulas

Formulas and Theorems about Equations

The Quadratic Formula, p. 216

The solutions of any quadratic equation in the form $ax^2 + bx + c = 0$ are:

$$x = \frac{-b + \sqrt{b^2 - 4ac}}{2a} \text{ and } x = \frac{-b - \sqrt{b^2 - 4ac}}{2a}$$

Fundamental Theorem of Algebra, p. 421

A polynomial function of degree n has exactly n complex zeros, provided each double zero is counted as 2 zeros, each triple zero is counted as 3 zeros, and so on.

Rational Zeros Theorem, p. 421

Let f be a polynomial function with integral coefficients. Then the only possible rational zeros of f are $\frac{p}{q}$ where p is a divisor of the constant term of $f(x)$ and q is a divisor of the leading coefficient.

Formulas from Coordinate Geometry

The Slope of a Line, pp. 47, 53

The slope of a line through the points (x_1, y_1) and (x_2, y_2) is $\frac{\text{vertical change}}{\text{horizontal change}} = \frac{y_2 - y_1}{x_2 - x_1}$.

The Distance Formula, p. 514

The distance d between two points (x_1, y_1) and (x_2, y_2) is:

$$d = \sqrt{(x_2 - x_1)^2 + (y_2 - y_1)^2}$$

The Midpoint Formula, p. 514

The midpoint of the line segment joining the points (x_1, y_1) and (x_2, y_2) has coordinates:

$$\left(\frac{x_1 + x_2}{2}, \frac{y_1 + y_2}{2}\right)$$

Parallel and Perpendicular Lines, pp. 515, 516

If line l_1 has slope a_1 and line l_2 has slope a_2, then:

$l_1 \parallel l_2$ if and only if $a_1 = a_2$

$l_1 \perp l_2$ if and only if $(a_1)(a_2) = -1$

Formulas from Trigonometry

Area of a Triangle, p. 678

The area K of a $\triangle ABC$ is given by any of the following formulas:

$$K = \frac{1}{2}bc \sin A \quad K = \frac{1}{2}ac \sin B \quad K = \frac{1}{2}ac \sin C$$

Area of a Sector, p. 680

For any sector determined by a central angle θ in a circle of radius r:

$$\text{area of the sector} = \frac{\theta}{360} \cdot \pi r^2$$

Law of Sines, p. 684

For any $\triangle ABC$:

$$\frac{\sin A}{a} = \frac{\sin B}{b} = \frac{\sin C}{c}$$

Law of Cosines, p. 691

For any $\triangle ABC$:

$$a^2 = b^2 + c^2 - 2bc \cos A$$

$$b^2 = a^2 + c^2 - 2ac \cos B$$

$$c^2 = a^2 + b^2 - 2ab \cos C$$

Converting Degree Measure to Radian Measure, p. 714

Since there are 360° and 2π radians in a full circle, use the fact that $1° = \frac{\pi}{180}$ radians to convert degrees to radians, and use the fact that 1 radian $= \frac{180°}{\pi}$ to convert radians to degrees.

Formulas from Statistics

The Mean of a Data Set, p. 790

For the data $x_1, x_2, x_3, \ldots, x_n$, the mean of the n data values, $\bar{x}$, is given by this formula:

$$\bar{x} = \frac{x_1 + x_2 + \cdots + x_n}{n}$$

The Standard Deviation of a Data Set, p. 268

Standard deviation measures the variability of data. For the data $x_1, x_2, x_3, \ldots, x_n$, where $\bar{x}$ is the mean of the data values and n is the number of data values, the standard deviation σ is given by this formula:

$$\sigma = \sqrt{\frac{(x_1 - \bar{x})^2 + (x_2 - \bar{x})^2 + \cdots + (x_n - \bar{x})^2}{n}}$$

The Margin of Error for a Sample Proportion, p. 274

When a random sample of size n is taken from a large population, the sample proportion has a margin of error approximated by this formula:

$$\text{margin of error} \approx \pm\frac{1}{\sqrt{n}}$$

Formulas from Probability

Experimental Probability, p. 612

Suppose an experiment is performed that consists of a certain number of trials. The experimental probability of an event A is:

$$P(A) = \frac{\text{number of trials where } A \text{ happens}}{\text{total number of trials}}$$

Theoretical Probability, p. 613

Suppose a procedure can result in n equally likely outcomes. If an event A corresponds to m of these outcomes, then the theoretical probability of A is:

$$P(A) = \frac{m}{n}$$

Multiple Events, p. 619

- **Independent Events:** The occurrence of event A does not affect whether event B happens.

$$P(A \text{ and } B) = P(A) \cdot P(B)$$

- **Mutually Exclusive Events:** Events A and B cannot both happen.

$$P(A \text{ or } B) = P(A) + P(B)$$

- **Complementary Events:** Events A and B are mutually exclusive and one of the events must happen.

$$P(B) = 1 - P(A)$$

Conditional Probability, p. 627

For any events A and B:

$$P(A \text{ and } B) = P(A) \cdot P(B \mid A)$$

$$P(B \mid A) = \frac{P(A \text{ and } B)}{P(A)}$$

Binomial Distribution, p. 634

In a binomial experiment, there are two mutually exclusive outcomes, often called *success* and *failure*. For each trial, $P(\text{success}) = p$ and $P(\text{failure}) = 1 - p$. The probability of k successes in n independent trials is given by this formula:

$$P\left(\begin{matrix} k \text{ successes} \\ \text{in } n \text{ trials} \end{matrix}\right) = {}_nC_k \cdot p^k \cdot (1 - p)^{n-k}$$

The probability values for $k = 0, 1, 2, \ldots, n$ form a binomial distribution.

Normal Distribution, p. 640

In a normal distribution:

- 50% of the data lies below the mean and 50% of the data lies above the mean.
- about 68% of the distribution lies within one standard deviation of the mean.
- about 95% of the distribution lies within two standard deviations of the mean.
- about 99% of the distribution lies within three standard deviations of the mean.

***z*-score, p. 641**

The x-values in a normal distribution with mean $\bar{x}$ and standard deviation σ can be converted to a z-score by this formula:

$$z = \frac{x - \bar{x}}{\sigma}$$

A z-score gives the number of standard deviations that x lies from the mean. Obtaining a z-score for a data value is called *standardizing* the value.

Formulas from Discrete Mathematics

Explicit Formulas for Arithmetic and Geometric Sequences, p. 473

- The *n*th term of an arithmetic sequence with common difference d is given by this explicit formula:

$$t_n = t_1 + d(n - 1)$$

- The *n*th term of a geometric sequence with common ratio r is given by this explicit formula:

$$t_n = t_1 \cdot r^{n-1}$$

Arithmetic and Geometric Means, p. 475

- The term x between any two terms a and b of an arithmetic sequence is given by:

$$x = \sqrt{ab}$$

- The term x between any two terms a and b of a geometric sequence is given by:

$$x = \frac{a+b}{2}$$

Recursive Formulas for Arithmetic and Geometric Sequences, p. 481

- The *n*th term of an arithmetic sequence with common difference d is given by this recursive formula:

$$t_1 = \text{starting value}$$
$$t_n = t_{n-1} + d$$

- The *n*th term of a geometric sequence with common ratio r is given by this recursive formula:

$$t_1 = \text{starting value}$$
$$t_n = r(t_{n-1})$$

Sum of a Finite Arithmetic Series, p. 489

The sum S_n of the n terms of a finite arithmetic series $t_1 + t_2 + t_3 + t_4 + \cdots + t_n$ is:

$$S_n = \frac{n}{2}(t_1 + t_n)$$

Sum of a Finite Geometric Series, p. 496

The sum S_n of the finite geometric series $t_1 + t_1 \cdot r + t_1 \cdot r^2 + \cdots + t_1 \cdot r^{n-1}$ where $r \neq 1$ is:

$$S_n = \frac{t_1(1 - r^n)}{1 - r}$$

Sum of an Infinite Geometric Series, p. 498

The sum S of the infinite geometric series $t_1 + t_1 \cdot r + t_1 \cdot r^2 + t_1 \cdot r^3 + t_1 \cdot r^4 + \cdots$ where $|r| < 1$ is:

$$S = \frac{t_1}{1 - r}$$

The Multiplication Counting Principle, p. 578

If a choice is to be made among m things, and for each of those choices another choice is to be made among n things, then there are a total of $m \times n$ possibilities to choose from.

Permutations, p. 580

The number of permutations of r objects taken from a set of n objects is given by ${}_nP_r$ where ${}_nP_r = \frac{n!}{(n-r)!}$.

Permutations with Repeated Elements, p. 580

If a set of n elements has one element repeated q_1 times, another element repeated q_2 times, . . . , and the last element repeated q_k times, then the number of distinguishable permutations of the n elements is given by $\frac{n!}{q_1!q_2! \cdots q_k!}$.

Combinations, p. 586

The number of combinations of r objects taken from a set of n objects is given by ${}_nC_r$ where ${}_nC_r = \frac{{}_nP_r}{{}_rP_r} = \frac{n!}{(n-r)!r!}$.

The Binomial Theorem, p. 594

For any positive integer n,

$$(a + b)^n = {}_nC_0a^n + {}_nC_1a^{n-1}b + {}_nC_2a^{n-2}b^2 + \cdots + {}_nC_{n-1}ab^{n-1} + {}_nC_nb^n$$

Formulas from Geometry

To find the perimeter P of any plane figure made up of line segments, find the sum of the lengths of each side of the figure.

Formulas for the areas of several plane figures as well as the volumes and surface areas of several space figures are given on the next page.

For the formulas below, let A = area, C = circumference, V = volume, B = the area of the base of a space figure, S.A. = the (total) surface area of a space figure, and L.S.A. = the lateral surface area of a space figure. Note that $\pi \approx 3.14$.

Parallelogram

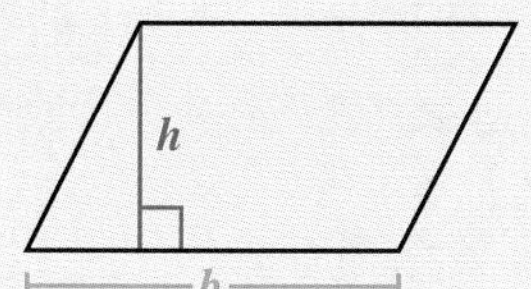

A = base × height

$A = bh$

Triangle

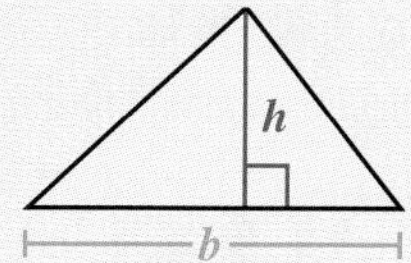

$A = \frac{1}{2} \times$ base × height

$A = \frac{1}{2}bh$

Trapezoid

$A = \frac{1}{2} \times$ sum of bases × height

$A = \frac{1}{2}(b_1 + b_2)h$

Circle

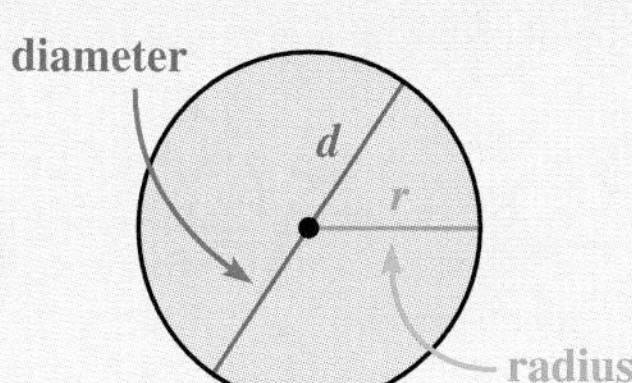

$C = 2\pi r$ or $C = \pi d$

$A = \pi r^2$

Right Rectangular Prism

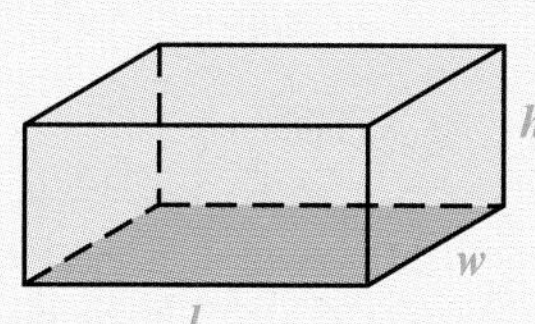

V = area of base × height

$V = Bh = lwh$

S.A. $= 2(lw + wh + lh)$

Right Cylinder

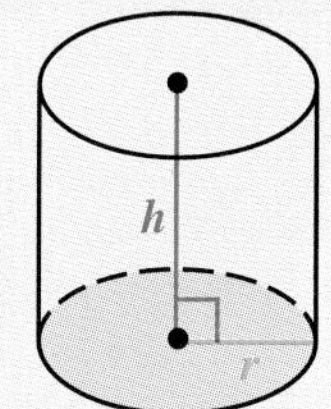

V = area of base × height

$V = Bh = \pi r^2 h$

L.S.A. $= 2\pi rh$

S.A. $= 2\pi r^2 + 2\pi rh$

Right Regular Pyramid

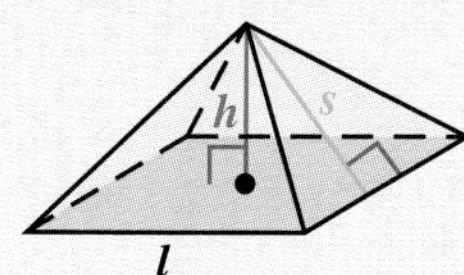

$V = \frac{1}{3} \times$ area of base × height

$V = \frac{1}{3}Bh$

L.S.A. $= \frac{1}{2} \times$ perimeter of base × slant height

For a base with n sides:

L.S.A. $= \frac{1}{2}nls$

Right Circular Cone

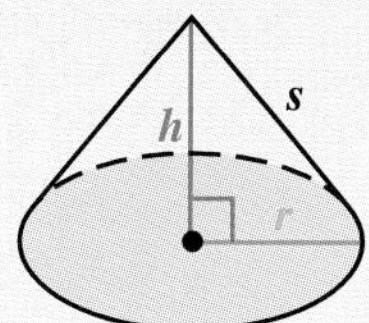

$V = \frac{1}{3} \times$ area of base × height

$V = \frac{1}{3}Bh = \frac{1}{3}\pi r^2 h$

L.S.A. =

$\frac{1}{2} \times$ circumference of base × slant height

L.S.A. $= \frac{1}{2}(2\pi r)s = \pi rs$

S.A. $= \pi r^2 + \pi rs$

Sphere

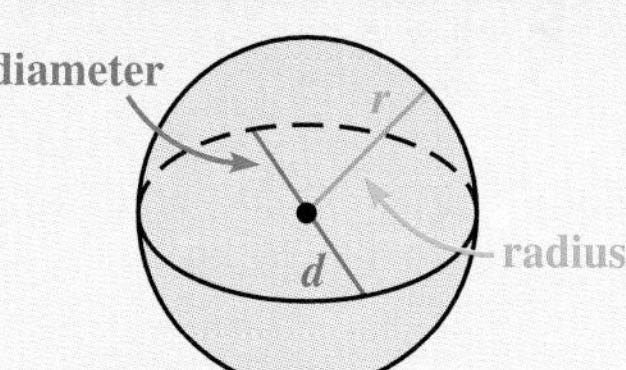

$V = \frac{4}{3}\pi r^3$ or $V = \frac{1}{6}\pi d^3$

S.A. $= 4\pi r^2$ or S.A. $= \pi d^2$

Appendix 1

Objectives

- Solve inequalities involving different types of functions.
- Write and solve inequalities arising from real-world situations, and interpret the solutions.

Appendix Notes

Example 1, *Think and Communicate* Question 1, and Exercises 1–5 and 10 can be done after completing Chapter 3.

Exercises 6–9 and 11 can be done after completing Chapter 4.

Example 2, *Think and Communicate* Question 2, and Exercises 12–18 can be done after completing Chapter 5.

Example 3, *Think and Communicate* Question 3, and Exercises 19–25 can be done after completing Chapter 8.

Example 4, *Think and Communicate* Question 4, and Exercises 26–33 can be done after completing Chapter 9.

Additional Example 1

Paul Barton puts $7500 in a mutual fund investing in bonds. He also has $8000 in cash reserves.

a. Write equations giving the predicted values $b(t)$ and $c(t)$ of Paul's bond investment and cash reserves, respectively, after t years.

For the bond investment, the initial amount is $7500 and the average annual return is 5.2%, so an equation for $b(t)$ is $b(t) = 7500(1 + 0.052)^t$, or $b(t) = 7500(1.052)^t$. For the cash reserves, the initial amount is $8000 and the average annual return is 3.7%, so an equation for $c(t)$ is $c(t) = 8000(1 + 0.037)^t$, or $c(t) = 8000(1.037)^t$.

b. Predict when Paul's bond investment will be worth more than his cash reserves.

Use a spreadsheet and the equations from part (a) to calculate $b(t)$ and $c(t)$ for different values of t. Find the t-values for which $b(t) > c(t)$.

Paul's Investments

	A	B	C
1	t	b(t)	c(t)
2	0	7500.00	8000.00
3	1	7890.00	8296.00
4	2	8300.28	8602.95
5	3	8731.89	8921.26
6	4	9185.95	9251.35
7	5	9663.62	9593.65
8	6	10166.13	9948.61
9	7	10694.77	10316.71

The inequality $b(t) > c(t)$ is true when $t \geq 5$. Paul's bond investment will be worth more than his cash reserves after 5 years.

Appendix

1 Solving Inequalities

Learn how to...

- solve inequalities involving different types of functions

So you can...

- solve problems about investing, for example

Financial planners recommend that you invest money you won't need for many years in stocks. The table at the right shows why. Over the long run, stocks have increased in value more rapidly than other types of investments.

Investment Data (1926–1995)	
Type of investment	Average annual return
Stocks	10.5%
Bonds	5.2%
Cash reserves	3.7%

EXAMPLE 1 Application: Investing

Alma Ramos puts $5000 in a mutual fund investing in stocks and $6000 in a mutual fund investing in bonds.

a. Write equations giving the predicted values $s(t)$ and $b(t)$ of Alma's stock and bond investments, respectively, after t years.

b. Predict when Alma's stock investment will be worth more than her bond investment.

SOLUTION

a. Use the **initial amounts invested** and the **average annual returns** in the table to write $s(t)$ and $b(t)$ as exponential functions of t.

$$s(t) = 5000(1 + 0.105)^t \quad \rightarrow \quad s(t) = 5000(1.105)^t$$
$$b(t) = 6000(1 + 0.052)^t \quad \rightarrow \quad b(t) = 6000(1.052)^t$$

The growth factor for each function is 1 plus the average annual return expressed as a decimal.

b. Use a spreadsheet and the equations from part (a) to calculate $s(t)$ and $b(t)$ for different values of t. Find the t-values for which $s(t) > b(t)$.

Alma's Investments

	A	B	C
1	t	s(t)	b(t)
2	0	5000.00	6000.00
3	1	5525.00	6312.00
4	2	6105.13	6640.22
5	3	6746.16	6985.52
6	4	7454.51	7348.76
7	5	8237.23	7730.90
8	6	9102.14	8132.90

The inequality $s(t) > b(t)$ is true when $t \geq 4$.

Alma's stock investment will be worth more than her bond investment after 4 years.

BY THE WAY

Although stocks make good long-term investments, they can be risky in the short run. For example, in October of 1987, stocks lost more than 20% of their value in a single day.

THINK AND COMMUNICATE

1. a. **Technology** Use a spreadsheet to find the values of $s(t)$ and $b(t)$ in Example 1 for $t = 3.0, 3.1, 3.2, \dots, 4.0$. Write a solution of $s(t) > b(t)$ that is more precise than the solution given in Example 1.
   b. Describe how you can use a spreadsheet to make the solution of $s(t) > b(t)$ as precise as desired.

In Example 1, you used a spreadsheet to solve an inequality. You can also solve inequalities using graphs.

EXAMPLE 2

Solve $x^2 - 6x + 11 < 5$.

SOLUTION

The solution is the set of x-values for which the graph of $y = x^2 - 6x + 11$ lies below the graph of $y = 5$. First find where these graphs intersect, as shown.

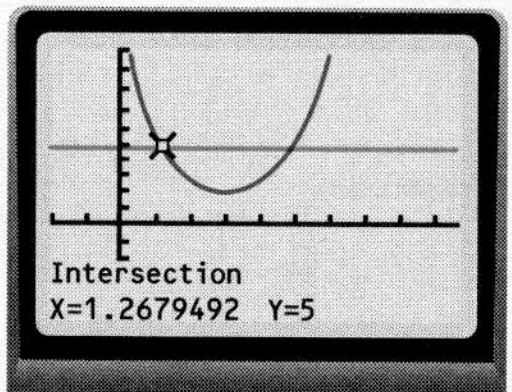

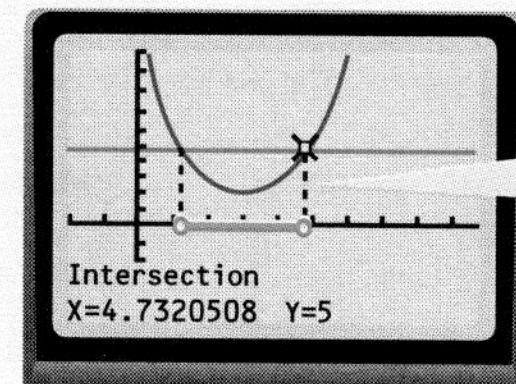

The graph of $y = x^2 - 6x + 11$ lies below the graph of $y = 5$ when $1.27 < x < 4.73$.

The solution is $1.27 < x < 4.73$.

EXAMPLE 3

Solve $\sqrt{2x-5} \geq \sqrt{x+3} - 1$.

SOLUTION

The inequality has the form $y_1 \geq y_2$ where $y_1 = \sqrt{2x-5}$ and $y_2 = \sqrt{x+3} - 1$. This inequality is equivalent to $y_1 - y_2 \geq 0$, which you can solve by graphing the function $y_3 = y_1 - y_2$ and finding the x-values for which the graph lies on or above the x-axis.

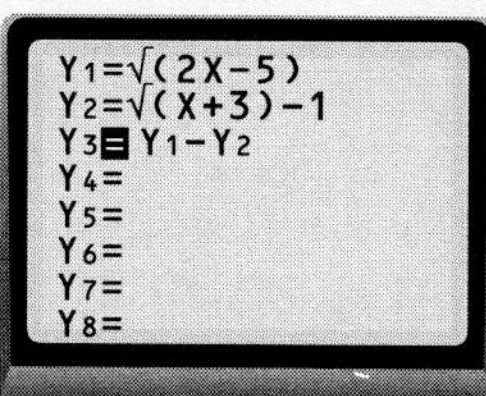

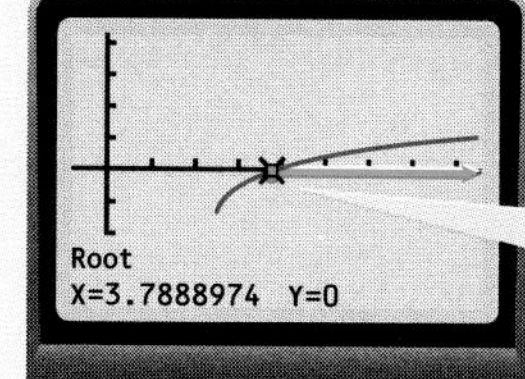

The x-intercept of the graph of y_3 is about 3.79. The graph lies on or above the x-axis when $x \geq 3.79$.

The solution is $x \geq 3.79$.

Additional Example 2

Solve $x^2 + 3x - 4 \geq 2x + 1$.

The solution is the set of x-values for which the graph of $y = x^2 + 3x - 4$ lies on or above the graph of $y = 2x + 1$. First find where these graphs intersect, as shown.

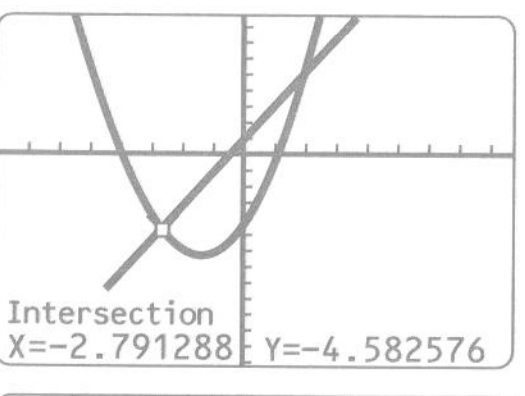

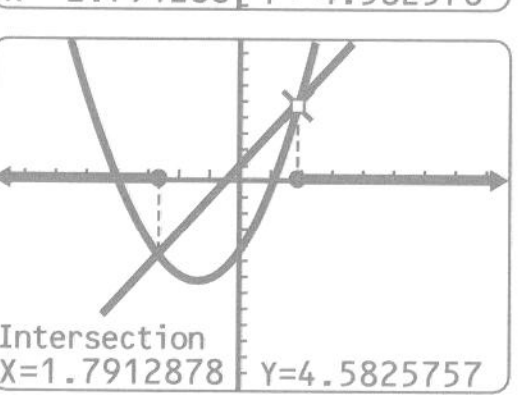

The graph of $y = x^2 + 3x - 4$ lies on or above the graph of $y = 2x + 1$ when $x \leq -2.79$ or $x \geq 1.79$. So the solution of $x^2 + 3x - 4 \geq 2x + 1$ is $x \leq -2.79$ or $x \geq 1.79$.

Additional Example 3

Solve $x < \sqrt{x+4}$.

The inequality has the form $y_1 < y_2$ where $y_1 = x$ and $y_2 = \sqrt{x+4}$. This inequality is equivalent to $y_1 - y_2 < 0$, which you can solve by graphing the function $y_3 = y_1 - y_2$ and finding the x-values for which the graph lies below the x-axis.

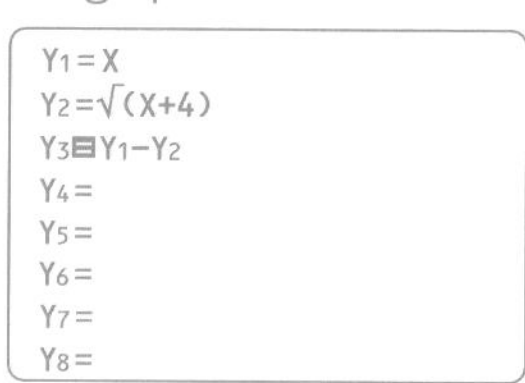

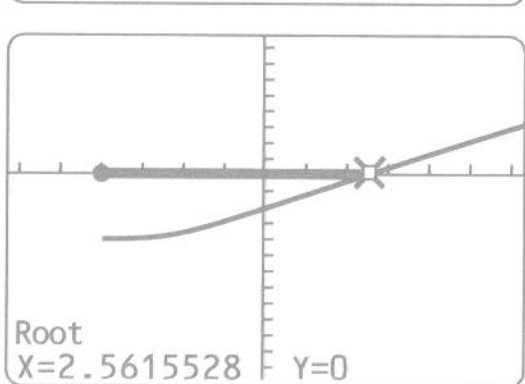

The x-intercept of the graph of y_3 is about 2.56, and the graph lies below the x-axis when $k \leq x < 2.56$, where k is the least number in the domain of y_3. To find k, note that the domain of y_3 is the set of x-values for which $\sqrt{x+4}$ is defined, that is, for which $x + 4 \geq 0$. Therefore, the domain of y_3 is $x \geq -4$, so that $k = -4$. This means that the solution of $x < \sqrt{x+4}$ is $-4 \leq x < 2.56$.

ANSWERS Appendix 1

Think and Communicate

1. a. $t \geq 3.8$
   b. Find an interval in which $s(t)$ becomes greater than $b(t)$, such as $3.7 \leq t \leq 3.8$ (so that an approximate solution of $s(t) > b(t)$ is $t \geq 3.8$). Evaluate $s(t)$ and $b(t)$ for different t-values in this interval, such as 3.70, 3.71, 3.72, . . . , 3.79, 3.80. Use the results to find a smaller interval in which $s(t)$ becomes greater than $b(t)$, such as $3.70 \leq t \leq 3.71$ (so that a better approximation of the solution of $s(t) > b(t)$ is $t \geq 3.71$). Repeat this procedure to find progressively smaller intervals in which $s(t)$ becomes greater than $b(t)$, such as $3.709 \leq t \leq 3.710$, $3.7093 \leq t \leq 3.7094$, and so on. Stop when a corresponding solution having the desired precision is obtained (for example, $t \geq 3.7094$).

Appendix Note

Using Technology When using a TI-81 or TI-82 graphing calculator to graph a truth function, set the calculator to *Dot* mode rather than *Connected* mode.

About Example 4

Using Technology On a graphing calculator or graphing software, it is usually impossible to tell from the graph of an inequality's truth function whether the function's value at a discontinuity is 0 or 1. (In Example 4, the truth function has discontinuities at $x = -2$, 2, and 4.) You can use a TI-82 graphing calculator to evaluate a truth function at a discontinuity by selecting *value* from the CALCULATE menu and entering the *x*-value at which the function is discontinuous. You can also substitute this value for *x* in both sides of the inequality by hand and determine whether the resulting inequality is true or false.

Additional Example 4

Solve $\frac{x}{6} > \frac{2}{x-1}$.

Graph the inequality's truth function on a graphing calculator or graphing software by defining the functions $y_1 = \frac{x}{6}$ and $y_2 = \frac{2}{x-1}$ and then entering the truth function as $y_3 = (y_1 > y_2)$.

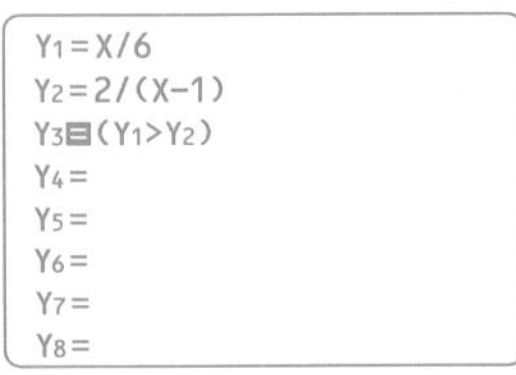

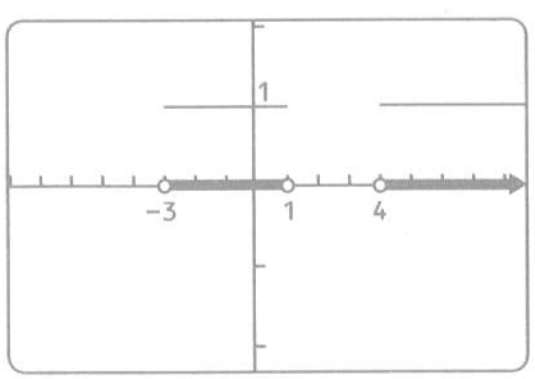

The solution of the inequality is the set of *x*-values for which the truth function equals 1: $-3 < x < 1$ or $x > 4$.

THINK AND COMMUNICATE

2. Use the graphs in Example 2 to solve $x^2 - 6x + 11 \geq 5$.

3. Consider the inequality $y_1 < y_2$ where y_1 and y_2 are as defined in Example 3. Explain why the solution of this inequality is *not* $x < 3.79$. Then solve the inequality.

Another way to solve an inequality is to graph the associated *truth function* using a graphing calculator or graphing software. The value of an inequality's truth function is 1 for *x*-values that satisfy the inequality and 0 for *x*-values that do not satisfy the inequality.

EXAMPLE 4

Solve $\frac{3}{x+2} \leq \frac{1}{x-2}$.

SOLUTION

Graph the inequality's truth function. On most graphing calculators and graphing software, you can define the functions $y_1 = \frac{3}{x+2}$ and $y_2 = \frac{1}{x-2}$ and then enter the truth function as $y_3 = (y_1 \leq y_2)$.

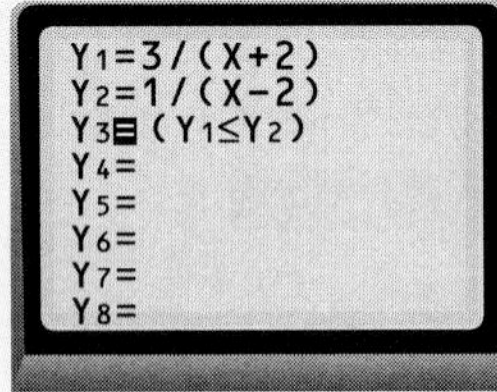

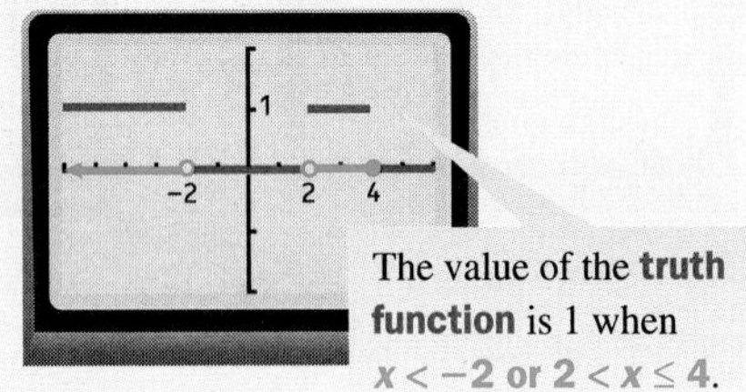

The solution is $x < -2$ or $2 < x \leq 4$.

THINK AND COMMUNICATE

4. In Example 4, how do you know that 4 is a solution of the given inequality, but -2 and 2 are not solutions?

Exercises and Applications

Technology Solve each inequality. Round all numbers in your answers to the nearest hundredth.

1. $2^x \geq 8$
2. $(0.5)^{x+1} \leq 16$
3. $6e^{0.17x} < 34$
4. $100(1.09)^x > 250(1.03)^x$
5. $(0.77)^x > 0$
6. $\ln x \geq 2$
7. $3 - \ln x \geq \sqrt{2}$
8. $\log(|x| + 1) < 0$
9. $\log_3 x \leq \log_2 x$

Think and Communicate

2. $x \leq 1.27$ or $x \geq 4.73$

3. The solution of $y_1 < y_2$ consists of only those *x*-values in the interval $x < 3.79$ for which both $y_1 = \sqrt{2x-5}$ and $y_2 = \sqrt{x+3} - 1$ are defined. The function y_1 is defined when $2x - 5 \geq 0$, or when $x \geq 2.5$. The function y_2 is defined when $x + 3 \geq 0$, or when $x \geq -3$. So y_1 and y_2 are both defined when $x \geq 2.5$. Therefore, the solution of $y_1 < y_2$ is $2.5 \leq x < 3.79$.

4. When $x = 4$, the inequality $\frac{3}{x+2} \leq \frac{1}{x-2}$ becomes $\frac{3}{4+2} \leq \frac{1}{4-2}$, or $\frac{1}{2} \leq \frac{1}{2}$, which is true. Therefore, 4 is a solution of the inequality. However, when $x = -2$, the left side of the inequality, $\frac{3}{x+2}$, is undefined, and when $x = 2$, the right side of the inequality, $\frac{1}{x-2}$, is undefined. Therefore, -2 and 2 are not solutions.

Exercises and Applications

1. $x \geq 3$
2. $x \geq -5$
3. $x < 10.20$
4. $x > 16.18$
5. all real numbers
6. $x \geq 7.39$
7. $0 < x \leq 4.88$
8. no solution
9. $x \geq 1$

10. INVESTING Look back at Example 1 and the table of investment data on page 826. Suppose Alma Ramos has \$3000 in cash reserves in addition to her stock and bond investments. Predict when Alma's stock investment will be worth more than her bond investment and cash reserves combined.

11. a. Sketch the graph of $y = \ln x$. Use the graph to explain why $\ln b > \ln a$ if and only if $b > a$, where a and b are any positive numbers.

b. Use the result from part (a) to solve the inequality $e^{2x} > 8$ algebraically. (*Hint:* Take the natural logarithm of both sides.)

c. Challenge Solve the inequality $s(t) > b(t)$ in Example 1 algebraically.

Technology Solve each inequality. Round all numbers in your answers to the nearest hundredth.

12. $x^2 - 7x + 10 \le 0$ **13.** $2x^2 - 3x - 9 < -4$ **14.** $3x^2 - 20x + 35 > 5$

15. $-0.7x^2 + 2.3x + 3.6 \le 1.8$ **16.** $-x^2 + 2x + 4 \ge 3x + 8$ **17.** $-2x^2 - 7x > x^2 + 8x + 6$

18. ENGINEERING The fuel efficiency E of an average car, measured in miles per gallon, can be modeled by the equation

$$E = -0.0177s^2 + 1.48s + 3.39$$

where s is the speed of the car in miles per hour. For what speeds is an average car's fuel efficiency at least 25 mi/gal?

Technology Solve each inequality. Round all numbers in your answers to the nearest hundredth.

19. $\sqrt{x-1} \ge 2$ **20.** $\sqrt{3x-10} \ge \sqrt{x+4}$ **21.** $6.5 - \sqrt{x} < x$

22. $\sqrt{x+1} \le \sqrt{x} + 1$ **23.** $\sqrt{2x} > \sqrt{5x-4} - 2$ **24.** $\sqrt[3]{x} + 3 < 2\sqrt{x}$

25. OCEANOGRAPHY The speed s of a wave depends on the depth d of the water through which it travels according to the equation

$$s = \sqrt{35.28d}$$

where d is in kilometers and s is in kilometers per minute. For what depths is wave speed between 7 km/min and 10 km/min?

Technology Solve each inequality. Round all numbers in your answers to the nearest hundredth.

26. $\frac{1}{x} \le 3$ **27.** $\frac{5}{x+2} > 1$ **28.** $\frac{x+4}{x-2} \ge 0$

29. $\frac{x^2+1}{x-1} < 0$ **30.** $\frac{2}{x} < x + 1$ **31.** $\frac{1}{x-4} \ge \frac{x}{x-6}$

32. DENTISTRY The pH level in your mouth t minutes after eating sugary foods can be modeled by this equation:

$$\text{pH} = \frac{65t^2 - 204t + 2340}{10t^2 + 360}$$

During what time interval after eating is the pH level in your mouth below 5?

33. Open-ended Problem Write an inequality. Solve the inequality using at least two of the methods presented in this appendix.

Closure Question

Describe four methods that you can use to solve an inequality of the form $f(x) > g(x)$.

You can solve the inequality $f(x) > g(x)$ by (1) using a spreadsheet to evaluate $f(x)$ and $g(x)$ for different values of x and identifying the x-values for which $f(x) > g(x)$; (2) graphing $y = f(x)$ and $y = g(x)$ in the same coordinate plane and finding the x values for which the graph of $y = f(x)$ lies above the graph of $y = g(x)$; (3) graphing the function $y = f(x) - g(x)$ and finding the x-values for which the graph lies above the x-axis; and (4) graphing the truth function $y = (f(x) > g(x))$ and finding the x-values for which the truth function equals 1.

10. after about 10.9 years

11. a.

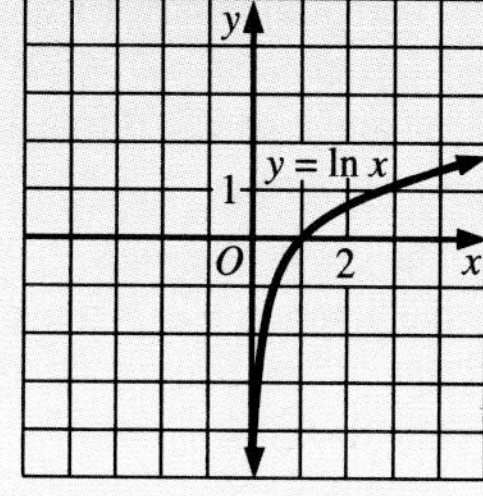

Suppose $b > a$. Then the point $(b, \ln b)$ lies to the right of the point $(a, \ln a)$. Since the graph of $y = \ln x$ rises from left to right, the point $(b, \ln b)$ must also lie above the point $(a, \ln a)$. So $\ln b > \ln a$. Conversely, suppose $\ln b > \ln a$. Then the point $(b, \ln b)$ lies above the point $(a, \ln a)$. Since the graph of $y = \ln x$ rises from left to right, the point $(b, \ln b)$ must also lie to the right of the point $(a, \ln a)$. So $b > a$. It follows that $\ln b > \ln a$ if and only if $b > a$.

b. $x > 1.04$

c. $t > 3.71$

12. $2 \le x \le 5$

13. $-1 < x < 2.5$

14. $x < 2.28$ or $x > 4.39$

15. $x \le -0.65$ or $x \ge 3.94$

16. no solution

17. $-4.56 < x < -0.44$

18. $18.9 \le s \le 64.8$

19. $x \ge 5$

20. $x \ge 7$

21. $x > 4.40$

22. $x \ge 0$

23. $0.8 \le x < 8$

24. $x > 5.74$

25. $1.39 < d < 2.83$

26. $x < 0$ or $x \ge 0.33$

27. $-2 < x < 3$

28. $x \le -4$ or $x > 2$

29. $x < 1$

30. $-2 < x < 0$ or $x > 1$

31. $2 \le x \le 3$ or $4 < x < 6$

32. $3.6 < t < 10$

33. Answers may vary. An example is given. Consider the inequality $\sqrt{4x+1} > 3$. One way to solve this inequality is to graph $y = \sqrt{4x+1}$ and $y = 3$ in the same coordinate plane and find the x-values for which the graph of $y = \sqrt{4x+1}$ lies above the graph of $y = 3$. A second way to solve the inequality is to graph its truth function, $y = (\sqrt{4x+1} > 3)$, and find the x-values for which the truth function equals 1. The solution of the inequality is $x > 2$.

Appendix 2

Objectives

- Recognize how the parameters h and k affect the domains, ranges, and graphs of $y = b^{x-h} + k$ and $y = \log_b (x - h) + k$.
- Graph functions of the form $y = b^{x-h} + k$ and $y = \log_b (x - h) + k$.

Appendix Notes

Think and Communicate Questions 1–5, Example 1, and Exercises 1–6 can be done after completing Chapter 3.

Example 2 and Exercises 7–12 can be done after completing Chapter 4.

Appendix

2 Translating Graphs of Exponential and Logarithmic Functions

Learn how to...

- **recognize how the parameters h and k affect the domains, ranges, and graphs of $y = b^{x-h} + k$ and $y = \log_b (x - h) + k$**

So you can...

- **translate graphs of exponential and logarithmic functions**

Look at the graph of the exponential function $y = 2^x$.

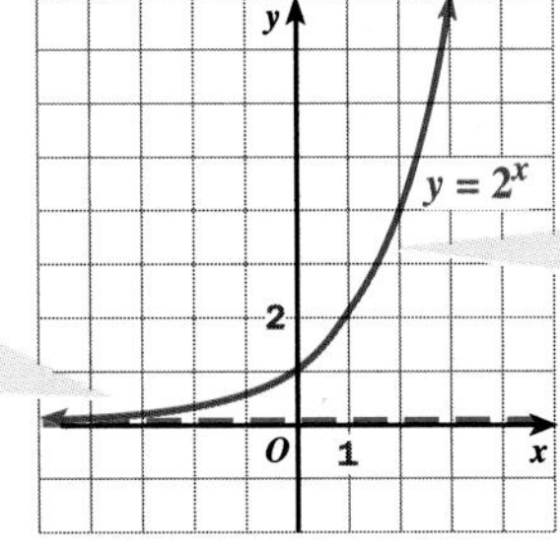

The **x-axis** is a horizontal *asymptote* of the graph. An asymptote is a line that a graph approaches more and more closely.

The domain of $y = 2^x$ is all real numbers. The range is $y > 0$.

In each of the coordinate planes below, the graph of $y = 2^x$ is shown along with graphs of some related functions.

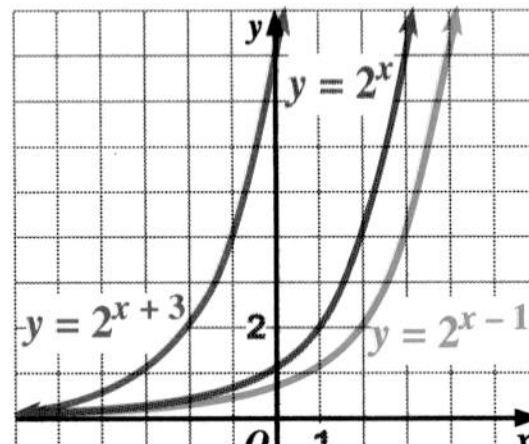

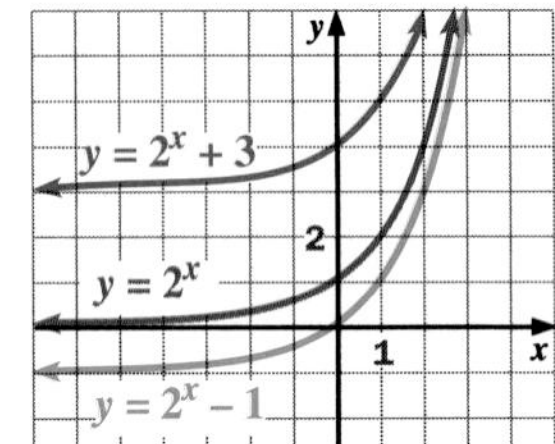

Think and Communicate

Use the graphs shown above to help you answer Questions 1 and 2.

1. a. Give the domains, ranges, and asymptotes for $y = 2^{x-1}$ and $y = 2^{x+3}$.

b. How are the graphs of $y = 2^{x-1}$ and $y = 2^{x+3}$ geometrically related to the graph of $y = 2^x$?

2. a. Give the domains, ranges, and asymptotes for $y = 2^x - 1$ and $y = 2^x + 3$.

b. How are the graphs of $y = 2^x - 1$ and $y = 2^x + 3$ geometrically related to the graph of $y = 2^x$?

3. What do you think are the domain, range, and asymptote(s) for $y = 2^{x-1} + 3$? How do you think the graph of $y = 2^{x-1} + 3$ is geometrically related to the graph of $y = 2^x$? Explain your reasoning.

ANSWERS Appendix 2

Think and Communicate

1. a. For both $y = 2^{x-1}$ and $y = 2^{x+3}$, the domain is all real numbers, the range is $y > 0$, and a horizontal asymptote is the x-axis.

b. The graph of $y = 2^{x-1}$ is the graph of $y = 2^x$ translated right 1 unit. The graph of $y = 2^{x+3}$ is the graph of $y = 2^x$ translated left 3 units.

2. a. For $y = 2^x - 1$, the domain is all real numbers, the range is $y > -1$, and a horizontal asymptote is $y = -1$. For $y = 2^x + 3$, the domain is all real numbers, the range is $y > 3$, and a horizontal asymptote is $y = 3$.

b. The graph of $y = 2^x - 1$ is the graph of $y = 2^x$ translated down 1 unit. The graph of $y = 2^x + 3$ is the graph of $y = 2^x$ translated up 3 units.

3. The graph of $y = 2^{x-1} + 3$ is the graph of $y = 2^x$ translated right 1 unit and up 3 units. The domain of $y = 2^{x-1} + 3$ is all real numbers, the range is $y > 3$, and a horizontal asymptote is $y = 3$.

You can generalize the results you obtained in the *Think and Communicate* questions as follows.

Exponential Functions of the Form $f(x) = b^{x-h} + k$

An exponential function $f(x) = b^{x-h} + k$ has these properties:

- The domain of f is all real numbers, and the range is $y > k$.
- The graph of f is the graph of $y = b^x$ translated horizontally h units and vertically k units. The function $y = b^x$ is called the *parent function* of f.
- The line $y = k$ is a horizontal asymptote of the graph of f.

EXAMPLE 1

a. Give the domain, range, and parent function of $f(x) = \left(\frac{1}{2}\right)^{x+1} - 3$.

b. Sketch the graphs of f and its parent function in the same coordinate plane. Identify any asymptotes of the graph of f.

SOLUTION

a. First write the equation for f in the form $f(x) = b^{x-h} + k$:

$$f(x) = \left(\frac{1}{2}\right)^{x-(-1)} + (-3)$$

For this function, $h = -1$ and $k = -3$.

The domain of f is all real numbers, and the range is $y > -3$.

The parent function of f is $y = \left(\frac{1}{2}\right)^x$.

b.

The graph of f is the graph of $y = \left(\frac{1}{2}\right)^x$ translated **left 1 unit** and **down 3 units**.

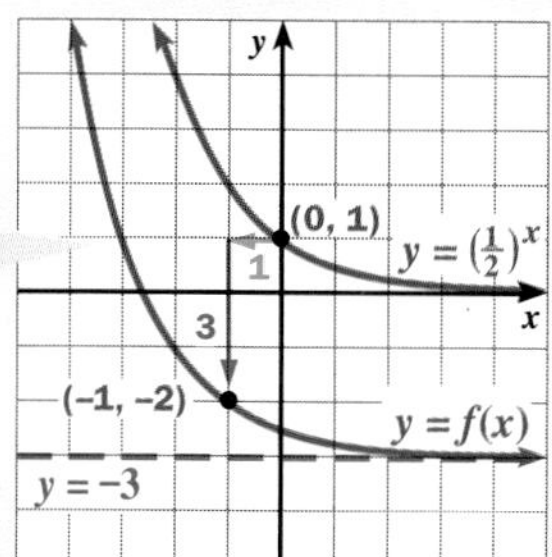

The line $y = -3$ is a horizontal asymptote of the graph of f.

THINK AND COMMUNICATE

4. In Example 1, explain how you know that the direction of the translation is left and down (rather than right and up, for example).

5. Tell how the graph of $y = b^{x-h} + k$ is geometrically related to the graph of $y = b^x$ if:

a. $h > 0$ and $k > 0$ **b.** $h > 0$ and $k < 0$

c. $h < 0$ and $k > 0$ **d.** $h < 0$ and $k < 0$

Additional Example 1

a. Give the domain, range, and parent function of $f(x) = 3^{x-2} - 4$.

Write the equation for f in the form $f(x) = b^{x-h} + k$ to get $f(x) = 3^{x-2} + (-4)$. So $h = 2$ and $k = -4$. The domain of f is all real numbers, and the range is $y > k$, or $y > -4$. The parent function of f is $y = 3^x$.

b. Sketch the graphs of f and its parent function in the same coordinate plane. Identify any asymptotes of the graph of f.

The graph of f is the graph of $y = 3^x$ translated right 2 units and down 4 units.

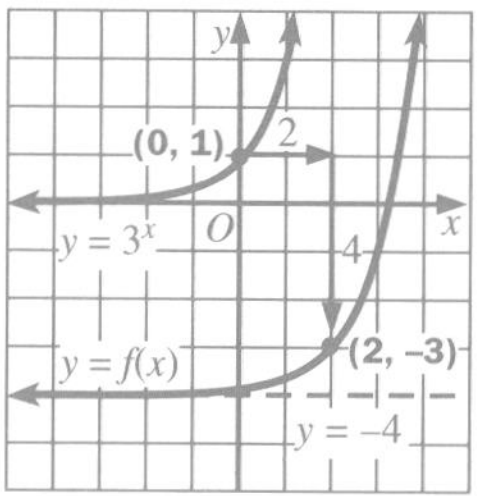

The line $y = -4$ is a horizontal asymptote of the graph of f.

Think and Communicate

4. Because h and k are negative, the translation is in the negative x- and y-directions. This means that the translation is left and down.

5. a. The graph of $y = b^{x-h} + k$ is the graph of $y = b^x$ translated right h units and up k units.

b. The graph of $y = b^{x-h} + k$ is the graph of $y = b^x$ translated right h units and down $|k|$ units.

c. The graph of $y = b^{x-h} + k$ is the graph of $y = b^x$ translated left $|h|$ units and up k units.

d. The graph of $y = b^{x-h} + k$ is the graph of $y = b^x$ translated left $|h|$ units and down $|k|$ units.

Additional Example 2

a. Give the domain, range, and parent function of $f(x) = \log_4 (x + 2) + 3$.

Write the equation for f in the form $f(x) = \log_b (x - h) + k$ to get $f(x) = \log_4 (x - (-2)) + 3$. So $h = -2$ and $k = 3$. The domain of f is $x > h$, or $x > -2$, and the range is all real numbers. The parent function of f is $y = \log_4 x$.

b. Sketch the graphs of f and its parent function in the same coordinate plane. Identify any asymptotes of the graph of f.

The graph of f is the graph of $y = \log_4 x$ translated up 3 units and left 2 units.

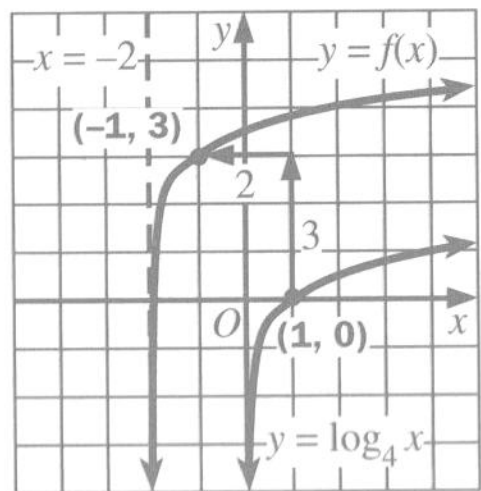

The line $x = -2$ is a vertical asymptote of the graph of f.

Closure Question

Describe the domains, ranges, and graphs of $y = b^{x-h} + k$ and $y = \log_b (x - h) + k$.

For $y = b^{x-h} + k$, the domain is all real numbers, and the range is $y > k$. The graph of $y = b^{x-h} + k$ is the graph of $y = b^x$ translated horizontally h units and vertically k units. A horizontal asymptote of the graph of $y = b^{x-h} + k$ is $y = k$. For $y = \log_b (x - h) + k$, the domain is $x > h$, and the range is all real numbers. The graph of $y = \log_b (x - h) + k$ is the graph of $y = \log_b x$ translated horizontally h units and vertically k units. A vertical asymptote of the graph of $y = \log_b (x - h) + k$ is $x = h$.

You can also translate graphs of logarithmic functions. The equations of the translated graphs have the form given below.

Logarithmic Functions of the Form $f(x) = \log_b (x - h) + k$

A logarithmic function $f(x) = \log_b (x - h) + k$ has these properties:

- The domain of f is $x > h$, and the range is all real numbers.
- The graph of f is the graph of $y = \log_b x$ translated horizontally h units and vertically k units. The function $y = \log_b x$ is the parent function of f.
- The line $x = h$ is a vertical asymptote of the graph of f.

EXAMPLE 2

a. Give the domain, range, and parent function of $f(x) = \log_2 (x - 3) + 4$.

b. Sketch the graphs of f and its parent function in the same coordinate plane. Identify any asymptotes of the graph of f.

SOLUTION

a. Note that the function $f(x) = \log_2 (x - 3) + 4$ is already in the form $f(x) = \log_b (x - h) + k$ with $h = 3$ and $k = 4$. The domain of f is $x > 3$, and the range is all real numbers. The parent function of f is $y = \log_2 x$.

b.

The graph of f is the graph of $y = \log_2 x$ translated **up 4 units** and **right 3 units.**

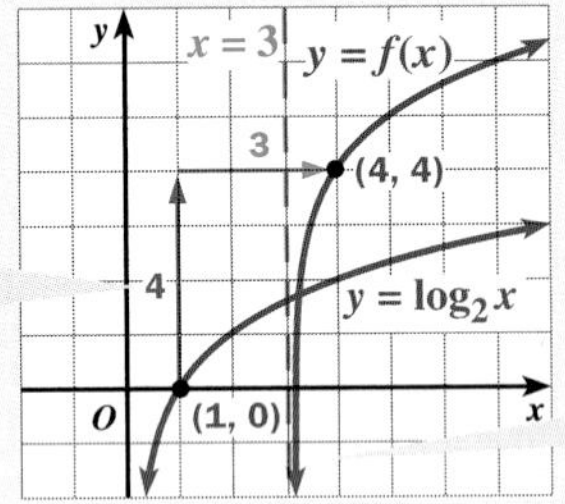

The line $x = 3$ is a vertical asymptote of the graph of f.

Exercises and Applications

For each function f:

a. Give the domain, range, and parent function of f.

b. Sketch the graphs of f and its parent function in the same coordinate plane. Identify any asymptotes of the graph of f.

1. $f(x) = 2^{x-3}$

2. $f(x) = 3^x + 1$

3. $f(x) = 4^{x-2} + 3$

4. $f(x) = e^{x+3} - 5$

5. $f(x) = \left(\frac{1}{2}\right)^{x+4} + 2$

6. $f(x) = \left(\frac{1}{3}\right)^{x-1} - 4$

7. $f(x) = \log_2 (x + 4)$

8. $f(x) = \log_2 x - 3$

9. $f(x) = \log_3 (x - 1) + 6$

10. $f(x) = \log_5 (x + 2) + 1$

11. $f(x) = \log (x - 3) - 1$

12. $f(x) = \ln (x + 5) - 2$

Exercises and Applications

For answers to odd-numbered exercises, see *Selected Answers for Appendices* on page 833.

2. a. domain: all real numbers; range: $y > 1$; parent function: $y = 3^x$

b.

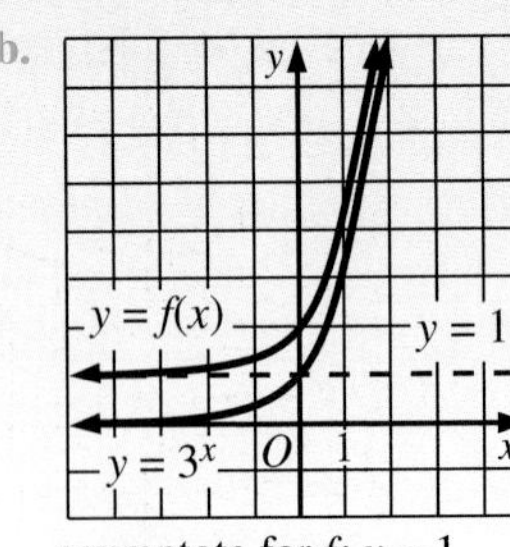

asymptote for f: $y = 1$

4. a. domain: all real numbers; range: $y > -5$; parent function: $y = e^x$

b.

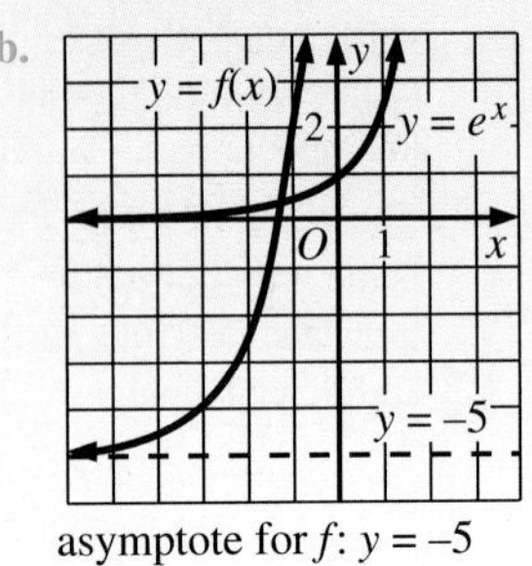

asymptote for f: $y = -5$

6. a. domain: all real numbers; range: $y > -4$; parent function: $y = \left(\frac{1}{3}\right)^x$

b.

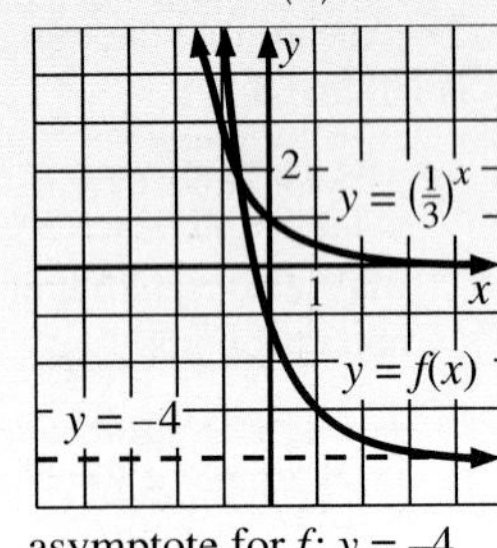

asymptote for f: $y = -4$

Selected Answers for Appendices

APPENDIX 1

Pages 828–829 Exercises and Applications

1. $x \ge 3$ **3.** $x < 10.20$ **5.** all real numbers **7.** $0 < x \le 4.88$ **9.** $x \ge 1$ **13.** $-1 < x < 2.5$ **15.** $x \le -0.65$ or $x \ge 3.94$ **17.** $-4.56 < x < -0.44$ **19.** $x \ge 5$ **21.** $x > 4.40$ **23.** $0.8 \le x < 8$ **25.** $1.39 < d < 2.83$ **27.** $-2 < x < 3$ **29.** $x < 1$ **31.** $2 \le x \le 3$ or $4 < x < 6$ **33.** Answers may vary. An example is given. Consider the inequality $\sqrt{4x+1} > 3$. One way to solve this inequality is to graph $y = \sqrt{4x+1}$ and $y = 3$ in the same coordinate plane and find the x-values for which the graph of $y = \sqrt{4x+1}$ lies above the graph of $y = 3$. A second way to solve the inequality is to graph its truth function, $y = (\sqrt{4x+1} > 3)$, and find the x-values for which the truth function equals 1. The solution of the inequality is $x > 2$.

APPENDIX 2

Page 832 Exercises and Applications

1. a. domain: all real numbers; range: $y > 0$; parent function: $y = 2^x$

b.

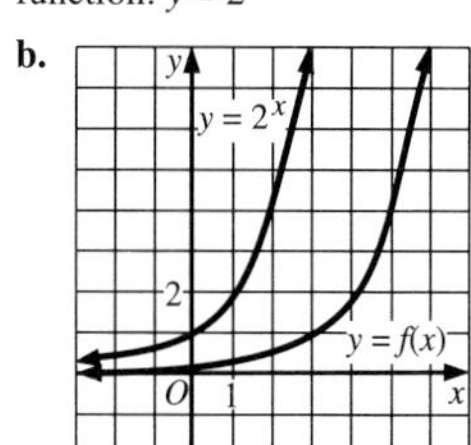

A horizontal asymptote of the graph of f is the x-axis.

3. a. domain: all real numbers; range: $y > 3$; parent function: $y = 4^x$

b.

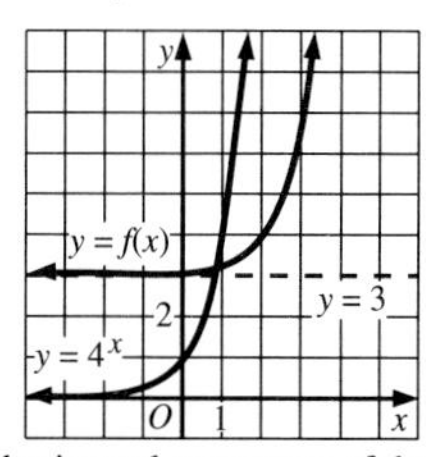

A horizontal asymptote of the graph of f is $y = 3$.

5. a. domain: all real numbers; range: $y > 2$; parent function: $y = \left(\frac{1}{2}\right)^x$

b.

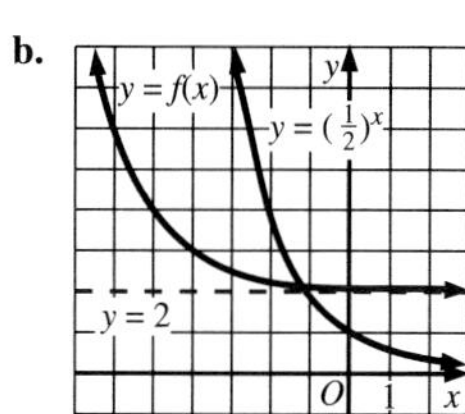

A horizontal asymptote of the graph of f is $y = 2$.

7. a. domain: $x > -4$; range: all real numbers; parent function: $y = \log_2 x$

b.

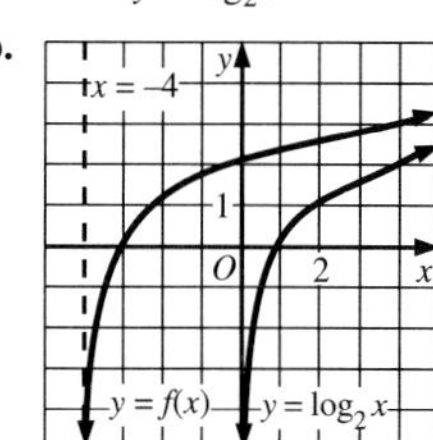

A vertical asymptote of the graph of f is $x = -4$.

9. a. domain: $x > 1$; range: all real numbers; parent function: $y = \log_3 x$

b.

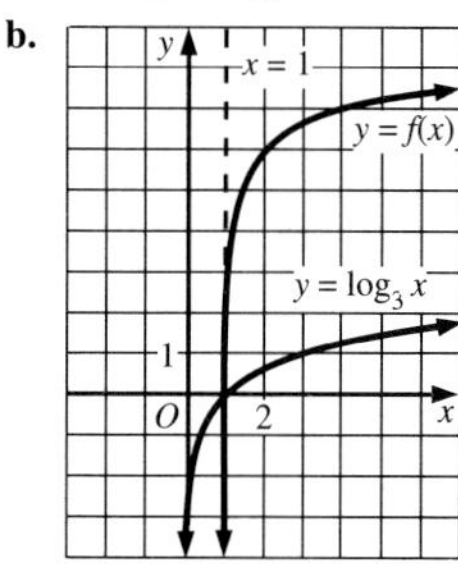

A vertical asymptote of the graph of f is $x = 1$.

11. a. domain: $x > 3$; range: all real numbers; parent function: $y = \log x$

b.

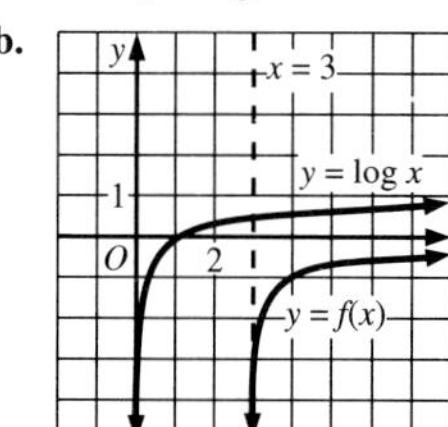

A vertical asymptote of the graph of f is $x = 3$.

8. a. domain: $x > 0$; range: all real numbers; parent function: $y = \log_2 x$

b.

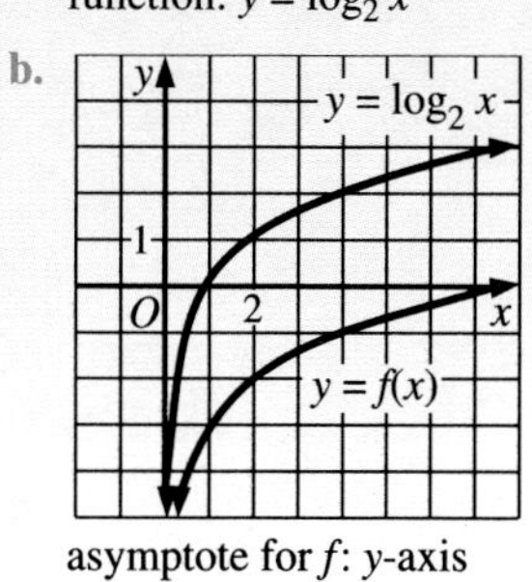

asymptote for f: y-axis

10. a. domain: $x > -2$; range: all real numbers; parent function: $y = \log_5 x$

b.

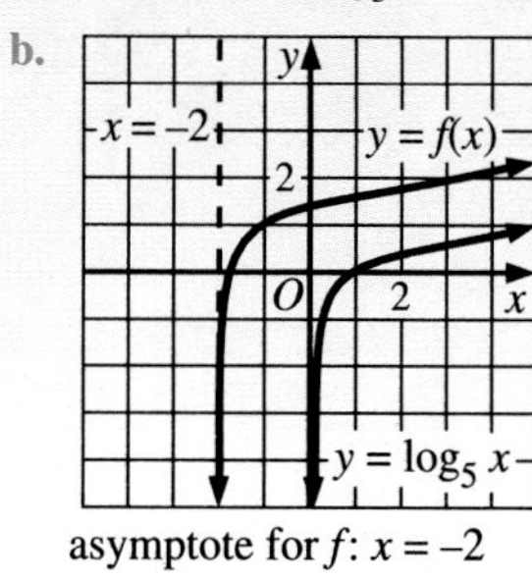

asymptote for f: $x = -2$

12. a. domain: $x > -5$; range: all real numbers; parent function: $y = \ln x$

b.

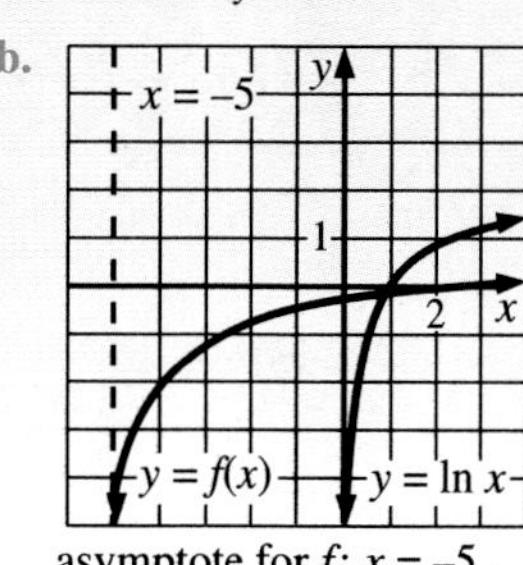

asymptote for f: $x = -5$

Glossary

absolute frequency (p. 260) The vertical axis of a histogram shows absolute frequencies when you can read the actual count of data within each interval on the horizontal axis.

amplitude (p. 720) One half the difference between the maximum and minimum values of a sine or cosine function. The functions $y = a \sin bx$ and $y = a \cos bx$ have an amplitude of $|a|$.

angle of depression (p. 665) The angle formed by a horizontal line at eye level and your line of sight to an object lower than you are.

angle of elevation (p. 665) The angle formed by a horizontal line at eye level and your line of sight to an object higher than you are.

arithmetic mean (p. 475) The term between any two terms of an arithmetic sequence. If a, x, b form an arithmetic sequence, then $x = \frac{a + b}{2}$.

arithmetic sequence (p. 473) A sequence where the difference between any term and the term before it is constant.

arithmetic series (p. 488) The indicated sum of an arithmetic sequence.

associative property (pp. 373, 374) For real numbers a, b, and c, $(a + b) + c = a + (b + c)$ and $(ab)c = a(bc)$.

asymptote (p. 427) A line that a graph approaches more and more closely.

base of a logarithmic function (p. 149) The number b, which must be positive and not equal to 1, in the logarithmic function $y = \log_b x$.

base-*b* logarithm of *x* (p. 149) The power to which the base b must be raised to equal x, denoted $\log_b x$.

biased question (p. 245) A question that produces responses that do not accurately reflect the opinions or actions of the respondents.

biased sample (p. 251) A sample that overrepresents or underrepesents part of the population.

binomial distribution (p. 634) In a binomial experiment with n independent trials, each having probability p of success, the values of ${}_nC_k \cdot p^k \cdot (1 - p)^{n - k}$ for $k = 0, 1, 2, \ldots, n$ form a binomial distribution of probabilities for the possible outcomes of the experiment.

box plot (p. 259) A display that shows the median, the quartiles, and the extremes of a numerical data set.

categorical data (p. 238) Data that are names or labels.

center (pp. 528, 536, 543) A point used to define certain conic sections. See also *circle*, *ellipse*, and *hyperbola*.

circle (p. 528) The set of all points in a plane that are the same distance from a point called the *center* of the circle.

closure property (pp. 373, 374) For real numbers a and b, the numbers $a + b$ and ab are also real.

cluster sample (p. 250) A sample chosen as a group.

combination (p. 585) A selection of elements of a set. Position is not important in a combination. The number of combinations of r objects taken from a set of n objects is given by ${}_nC_r$ where ${}_nC_r = \frac{n!}{(n - r)!r!}$.

common difference (p. 473) In an arithmetic sequence, the constant difference between any term and the term before it.

common logarithm (p. 150) The base-10 logarithm of N $(N > 0)$, denoted $\log N$.

common ratio (p. 473) In a geometric sequence, the constant ratio of any term to the term before it.

commutative group (p. 373) A group for which the commutative property holds.

commutative property (pp. 373, 374) For real numbers a and b, $a + b = b + a$ and $ab = ba$.

complementary events (p. 619) Two mutually exclusive events, one of which must happen. For complementary events A and B, $P(B) = 1 - P(A)$.

complex conjugates (p. 361) Two complex numbers of the form $a + bi$ and $a - bi$.

complex numbers (p. 359) The set of numbers that consists of all real and all imaginary numbers.

complex plane (p. 365) A coordinate plane where each point (a, b) represents the complex number $a + bi$. The complex plane has a horizontal real axis and a vertical imaginary axis.

conditional probability (p. 626) The probability of event B, given that event A has happened, denoted $P(B \mid A)$. For any events A and B, $P(B \mid A) = \frac{P(A \text{ and } B)}{P(A)}$.

conic section (p. 550) A parabola, circle, ellipse, or hyperbola produced by slicing a double cone with a plane.

constant of variation (pp. 46, 427) The nonzero constant a in the direct variation function $y = ax$ or in the inverse variation function $y = \frac{a}{x}$.

constant term (p. 390) The term in a polynomial that does not have a variable or that has a variable with 0 as the exponent.

convenience sample (p. 250) A sample chosen because it is easy to obtain.

corner-point principle (p. 323) For a polygonal feasible region, the solution of a linear programming problem will be at a corner point of the region.

correlation coefficient (p. 72) A number, denoted r, between -1 and $+1$ that measures how well data points line up.

cosine (pp. 656, 708) In a right $\triangle ABC$, the cosine of $\angle A$, which is written "cos A," is given by the equation $\cos A = \frac{\text{length of leg adjacent to } \angle A}{\text{length of hypotenuse}}$. For a unit circle, $\cos \theta$ is the x-coordinate of the point where the terminal side of an angle θ in standard position intersects the circle.

cubic function (p. 413) A polynomial function of degree 3.

cycle (p. 720) For a periodic graph, the smallest portion that repeats.

degenerate conic (p. 551) A point, a line, or a pair of intersecting lines produced when a plane intersects a double cone at its vertex.

degree of a polynomial (p. 390) The greatest exponent of the polynomial.

degree of a vertex (p. 563) In discrete mathematics, the number of other vertices connected to that vertex by edges.

dependent system (p. 299) A system of equations that has infinitely many solutions.

dependent variable (p. 10) A variable whose value is determined by the value of some other variable.

dimensions of a matrix (p. 16) The number of rows r and the number of columns c in a matrix, written $r \times c$.

direct variation (p. 46) The relationship between two variables in which the ratio of the variables is constant. If $\frac{y}{x} = a$, or $y = ax$, where a is a nonzero constant, then y varies directly with x.

directed graph (p. 569) In discrete mathematics, a graph in which the edges are arrows.

directrix (p. 521) See *parabola.*

discriminant (p. 217) The value of $b^2 - 4ac$ in the quadratic formula. It tells you how many solutions the quadratic equation $ax^2 + bx + c = 0$ has.

domain of a function (p. 60) The set of values of the independent variable for which a function is defined.

double zero of a polynomial function (p. 420) A number k is a double zero of a polynomial function f if $(x - k)^2$ is a factor of $f(x)$.

doubling time (p. 108) The time it takes for the initial value of an exponential growth function to double.

***e* (p. 116)** The irrational number e, approximately equal to 2.718, is often used as the base of exponential functions. As the value of n increases, the value of $\left(1 + \frac{1}{n}\right)^n$ approaches e.

edge of a graph (p. 563) In discrete mathematics, a connecting link between two vertices of a graph. Edges may have arrows.

element of a matrix (p. 16) A number in a matrix.

ellipse (p. 536) The set of all points in a plane such that the sum of the distances between each point and two fixed points (called the *foci*) is constant. The midpoint of the segment whose endpoints are the foci is called the *center* of the ellipse.

end behavior (p. 405) The behavior of a function $f(x)$ as x takes on large positive and large negative values.

experimental probability (p. 612) For an event A, the ratio of the number of trials where A happens to the total number of trials performed in an experiment.

explicit formula (p. 480) A formula that expresses the nth term, t_n, of a sequence as a function of its position number n.

exponential decay function (p. 96) A function of the form $y = ab^x$ where $a > 0$ and $0 < b < 1$. Its graph *falls* from left to right.

exponential growth function (p. 96) A function of the form $y = ab^x$ where $a > 0$ and $b > 1$. Its graph *rises* from left to right.

extraneous solution (p. 165) A solution that does not satisfy the original equation.

factoring (p. 222) For a quadratic expression of the form $ax^2 + bx + c$ where a, b, and c are integers, factoring means finding (if possible) integers k, l, m, and n such that $ax^2 + bx + c = (kx + m)(lx + n)$.

feasible region (p. 323) A graph of the solution of the system of inequalities representing all the given constraints in a linear programming problem.

finite sequence (p. 467) A sequence that has a last term.

focus, foci (pp. 521, 536, 543) A fixed point (focus) or two fixed points (foci) used to define certain conic sections. See also *parabola*, *ellipse*, and *hyperbola*.

function (p. 10) A pairing in which there is exactly one output value for each input value.

function notation (p. 60) The notation $y = f(x)$, which is read "y equals f of x" and which means that y is a function of x.

geometric mean (p. 475) The term between any two terms of a geometric sequence. If a, x, b form a geometric sequence and a and b are both positive, then $x = \sqrt{ab}$.

geometric probability (p. 614) A probability based on length, area, or volume.

geometric sequence (p. 473) A sequence where the ratio of any term to the term before it is constant.

geometric series (p. 495) The indicated sum of a geometric sequence.

graph (p. 563) In discrete mathematics, a collection of points, called *vertices*, connected by links, called *edges*.

group (p. 373) A number set and an operation form a group if the set is closed under the operation and the identity, inverse, and associative properties hold.

half-life (p. 109) The time it takes for the initial value of an exponential decay function to be halved.

half-planes (p. 310) A line divides the coordinate plane into two regions called half-planes.

histogram (p. 260) A graph that displays the overall shape of the distribution of numerical data. You create a histogram by dividing the data into intervals of equal width and drawing a bar over each interval so that the height indicates the frequency of the data within that interval.

hyperbola (pp. 427, 543) The set of all points in a plane such that the difference of the distances between each point and two fixed points (called *foci*) is constant. The midpoint of the segment whose endpoints are the foci is called the *center* of the hyperbola. The graph of an inverse variation equation is a hyperbola.

identity matrix (p. 304) A square matrix that has 1's on the diagonal running from the upper left to the lower right and 0's everywhere else. For any square matrix A and identity matrix I with the same dimensions, $AI = IA = A$.

identity property (pp. 373, 374) For a real number a, $a + 0 = 0 + a = a$ and $a \cdot 1 = 1 \cdot a = a$. The identity for addition is 0, and the identity for multiplication is 1.

imaginary axis (p. 365) The vertical axis of the complex plane.

imaginary numbers (p. 359) All numbers of the form $a + bi$, where a and b are real numbers and $b \neq 0$.

inconsistent system (p. 299) A system of equations that has no solution.

independent events (p. 619) Two events such that the occurrence of one event does not affect whether the other event happens. For independent events A and B, $P(A \text{ and } B) = P(A) \cdot P(B)$.

independent variable (p. 10) A variable whose value determines the value of some other variable.

infinite sequence (p. 467) A sequence that continues without end.

initial side of an angle (**p. 671**) See *standard position of an angle*.

intercept form (**p. 200**) The form $y = a(x - p)(x - q)$ for a quadratic function, where p and q are the x-intercepts of the function's graph.

interquartile range (**IQR**) (**p. 266**) The difference between the upper quartile and the lower quartile for a set of numerical data.

inverse matrix (**p. 304**) For a square matrix A, the inverse of A, denoted A^{-1}, is the matrix that, when multiplied by A, gives the identity matrix I: $A^{-1}A = A(A^{-1}) = I$. Not all square matrices have inverses.

inverse of a function *f* (**p. 142**) A function, denoted f^{-1}, that satisfies this property: $f^{-1}(b) = a$ if and only if $f(a) = b$. Sometimes the domain of f must be restricted in order for f^{-1} to exist.

inverse property (**pp. 373, 374**) For every real number a, there is a real number $-a$ such that $a + (-a) = 0$. For every nonzero real number a, there is a real number $\frac{1}{a}$ such that $a\left(\frac{1}{a}\right) = 1$.

inverse variation (**p. 427**) The relationship between two variables in which the product of the variables is constant. If $xy = a$, or $y = \frac{a}{x}$, where a is a nonzero constant, then y varies inversely with x.

leading coefficient (**p. 406**) The coefficient of the highest power of the variable in a polynomial.

least-squares line (**p. 67**) A line of fit that minimizes the sum of the squares of the vertical distances of a given set of data points from the line.

like terms (**p. 392**) Terms of polynomials that have the same power of the variable.

line of fit (**p. 65**) A line that lies as close as possible to all the points in a scatter plot. It does not necessarily have to pass through any of the points.

line of symmetry (**p. 186**) For a graph, a line that divides the graph so that the part of the graph on one side of the line is a reflection of the part on the other side.

linear programming (**p. 323**) A method for finding the best solution to a problem that involves maximizing or minimizing the value of some linear function of two variables when the values of the variables are constrained to some feasible region.

local maximum (**p. 407**) The y-coordinate of a turning point that is higher than all nearby points.

local minimum (**p. 407**) The y-coordinate of a turning point that is lower than all nearby points.

logarithmic function (**p. 149**) A function that is the inverse of the exponential function $f(x) = b^x$, denoted $f^{-1}(x) = \log_b x$.

logistic growth (**p. 118**) A type of limited growth that is basically exponential at first and then levels off.

lower extreme (**p. 259**) In a set of numerical data, the least value.

lower quartile (**p. 259**) In a set of numerical data, the median of the lower half of the data.

magnitude of a complex number (**p. 366**) In a complex plane, the distance between (0, 0) and (a, b), denoted $|a + bi|$.

major axis (**pp. 536, 543**) For an ellipse or a hyperbola, the line segment whose endpoints are the vertices.

margin of error (**p. 274**) An indication of the expected variability in a sample proportion when a random sample is taken from a large population. For a sample of size n, it is approximated by this formula:

$$\text{margin of error} \approx \pm \frac{1}{\sqrt{n}}$$

mathematical model (**p. 4**) A table, a graph, an equation, a function, or an inequality that describes a real-world situation.

matrix (**p. 16**) A group of numbers arranged in rows and columns.

matrix multiplication (**p. 23**) The product of an $m \times n$ matrix A and an $n \times p$ matrix B is the $m \times p$ matrix C such that each element $c_{i,j}$ of C is obtained by adding the products of corresponding elements as you go across the ith row of A and down the jth column of B.

maximum value of a quadratic function (**p. 195**) The value k of the function $y = a(x - h)^2 + k$ when $a < 0$.

median (**p. 259**) The middle number when you put data in order from least to greatest. When the number of data items is even, the median is the mean of the two middle numbers.

minimum value of a quadratic function (**p. 195**) The value k of the function $y = a(x - h)^2 + k$ when $a > 0$.

minor axis (pp. 536, 543) For an ellipse or a hyperbola, a line segment that has its midpoint at the center of the conic and is perpendicular to the major axis.

mutually exclusive events (p. 619) Two events that cannot both happen. For mutually exclusive events A and B, $P(A \text{ or } B) = P(A) + P(B)$.

natural logarithm (p. 150) The base-e logarithm of N ($N > 0$), denoted $\ln N$.

negative correlation (p. 72) For two variables x and y, the correlation coefficient r is negative ($-1 \leq r < 0$) when y tends to decrease as x increases.

normal distribution (p. 639) A distribution of data whose graph is a bell-shaped curve that reaches its maximum height at the mean.

numerical data (p. 238) Data that are counts or measurements.

outlier (p. 266) A data value whose distance from the nearer quartile is more than 1.5 times the interquartile range.

parabola (pp. 186, 521) The set of all points in the plane such that each point is the same distance from a point F, called the *focus*, and a line d, called the *directrix*. The graph of a quadratic function is a parabola.

parameter (p. 79) A variable, usually denoted t, upon which two other variables depend.

parametric equations (p. 79) Equations that express two variables in terms of a third variable, called the *parameter*.

Pascal's triangle (p. 593) A triangular display of numbers that gives the coefficients when you expand $(a + b)^n$ for $n = 0, 1, 2, \ldots$.

period (p. 720) For a function $y = f(x)$ whose graph repeats, the length of the interval on the x-axis covered by the smallest part of the graph that repeats. The functions $y = a \sin bx$ and $y = a \cos bx$ have a period of $\frac{2\pi}{|b|}$. The function $y = a \tan bx$ has a period of $\frac{\pi}{|b|}$.

permutation (p. 579) An arrangement of the elements of a set. Position is important in a permutation. The number of permutations of r objects taken from a set of n objects is given by ${}_nP_r$ where ${}_nP_r = \frac{n!}{(n-r)!}$.

phase shift (p. 728) A horizontal translation of a periodic function.

point-slope equation (p. 59) An equation of the form $y = y_1 + a(x - x_1)$ where (x_1, y_1) is a point on a line and a is the slope of the line.

polynomial (p. 390) A term or a sum of terms, where each term is the product of a real-number coefficient and a variable with a whole-number exponent.

population (p. 239) A complete group from which a sample is taken.

positive correlation (p. 72) For two variables x and y, the correlation coefficient r is positive ($0 < r \leq 1$) when y tends to increase as x increases.

power function (p. 170) A function of the form $y = ax^b$ where a and b are constants.

probability (p. 611) The fraction of the time that an event A is expected to happen, denoted $P(A)$.

pure imaginary number (p. 358) The square root of any negative real number. A pure imaginary number can be written in the form bi, where $b \neq 0$.

quadratic formula (p. 216) The formula $x = \frac{-b \pm \sqrt{b^2 - 4ac}}{2a}$, which gives the solutions of any quadratic equation in the form $ax^2 + bx + c = 0$.

quadratic function (p. 186) A function that involves squaring the independent variable. Quadratic functions can be written in standard form, $y = ax^2 + bx + c$; intercept form, $y = a(x - p)(x - q)$; or vertex form, $y = a(x - h)^2 + k$.

radian (p. 714) On a unit circle, the radian measure of an angle in standard position is the length of the arc that is intercepted by the angle. Radians are related to degrees by the fact that there are 2π radians or 360° in a full circle, so 1 radian $= \frac{180°}{\pi}$.

radical function (p. 344) A function that involves taking the nth root of the independent variable.

radius (p. 528) For a circle, the distance between the center of the circle and any point on the circle.

random sample (p. 250) A sample chosen so that each person or object has an equally likely chance of being selected.

range of a data set (p. 266) The difference between the maximum and minimum values in a set of numerical data.

range of a function *f* (p. 60) The set of all values of $f(x)$, where x is in the domain of f.

rational equation (p. 450) An equation that contains only polynomials or quotients of polynomials.

rational function (p. 435) A function of the form $f(x) = \frac{p(x)}{q(x)}$ where $p(x)$ and $q(x)$ are polynomials.

real axis (p. 365) The horizontal axis of the complex plane.

real numbers (p. 358) The set of numbers that includes all rational and all irrational numbers.

recursion (p. 481) The process of repeatedly applying the same operation(s) to a starting value and to each successive answer.

recursive formula (p. 481) A formula for a sequence that includes two parts: a starting value and a recursion equation for the nth term, t_n, as a function of t_{n-1}.

relative frequency (p. 260) The vertical axis of a histogram shows relative frequencies when the count of the data within each interval on the horizontal axis is expressed as a percent of all the data.

S

sample (p. 239) A subset of a population.

sample proportion (p. 273) The fraction of categorical data that fall into a particular category.

sample size (p. 274) The number of items in a sample.

sampling distribution (p. 274) A histogram that shows the distribution of sample proportions when many samples of the same size are taken from the same population of categorical data.

scalar (p. 18) A number by which a matrix is multiplied.

scalar multiplication (p. 18) The process of multiplying each element of a matrix by a scalar.

self-selected sample (p. 250) A sample of people who have volunteered to be surveyed.

sequence (p. 465) An ordered list of numbers or figures.

series (p. 488) The indicated sum when the terms of a sequence are added.

simulation (p. 29) An experiment that is used to model a situation and make predictions.

sine (pp. 656, 708) In a right $\triangle ABC$, the sine of $\angle A$, which is written "sin A," is given by the equation $\sin A = \frac{\text{length of leg opposite } \angle A}{\text{length of hypotenuse}}$. For a unit circle, $\sin \theta$ is the y-coordinate of the point where the terminal side of an angle θ in standard position intersects the circle.

skewed distribution (p. 260) A distribution that is not symmetric.

slope (p. 47) The ratio $\frac{\text{vertical change}}{\text{horizontal change}}$ for two points on a line. If (x_1, y_1) and (x_2, y_2) are two points on a line, then the slope a of the line is given by:

$$a = \frac{y_2 - y_1}{x_2 - x_1}$$

slope-intercept form (p. 53) The form $y = ax + b$ for a linear function, where a is the slope and b is the vertical intercept of the function's graph.

solving a triangle (p. 665) Finding the lengths of all sides and the measures of all angles of a triangle.

square root (p. 187) A number x is a square root of a number y if it satisfies the equation $x^2 = y$. A positive number y has two square roots, written $\sqrt{y}$ and $-\sqrt{y}$.

standard deviation (p. 268) A measure of the variability in a set of numerical data, denoted σ. For the data $x_1, x_2, x_3, \ldots, x_n$ with mean $\bar{x}$, the standard deviation is given by:

$$\sigma = \sqrt{\frac{(x_1 - \bar{x})^2 + (x_2 - \bar{x})^2 + \cdots + (x_n - \bar{x})^2}{n}}$$

standard form of a quadratic function (p. 206) The form $y = ax^2 + bx + c$.

standard form of a polynomial (p. 390) The form of a polynomial where exponents decrease from left to right.

standard normal distribution (p. 641) The normal distribution with mean 0 and standard deviation 1.

standard position of an angle (p. 671) An angle in standard position is formed by rotating a ray, called the *terminal side*, in a counterclockwise direction from the positive x-axis, called the *initial side*.

stratified random sample (p. 250) A sample chosen by dividing the population into groups and randomly selecting people or objects from each group.

symmetric distribution (p. 260) A distribution whose histogram is approximately symmetrical about a vertical line passing through the interval with the greatest frequency.

synthetic substitution (p. 391) A method of evaluating a polynomial for any value of the variable.

system of equations (p. 292) Two or more equations involving the same variables.

system of inequalities (p. 315) Two or more inequalities involving the same variables.

systematic sample (p. 250) A sample chosen using a pattern, such as every fourth person in line.

tangent (pp. 663, 736) In a right $\triangle ABC$, the tangent of $\angle A$, which is written "tan A," is given by the equation $\tan A = \frac{\text{length of leg opposite } \angle A}{\text{length of leg adjacent to } \angle A}$. For a unit circle, $\tan \theta$ is the ratio of the y-coordinate to the x-coordinate of the point where the terminal side of an angle θ in standard position intersects the circle.

term of a sequence (p. 465) An element of a sequence, denoted t_n where n is the position number.

terminal side of an angle (p. 671) See *standard position of an angle.*

theoretical probability (p. 613) For a procedure that can result in n equally likely outcomes and for an event A that corresponds to m of these outcomes, the theoretical probability of A is given by $P(A) = \frac{m}{n}$.

triple zero of a polynomial function (p. 420) A number k is a triple zero of a polynomial function f if $(x - k)^3$ is a factor of $f(x)$.

turning point (p. 407) A point that is higher or lower than all nearby points on the graph of a function.

upper extreme (p. 259) In a set of numerical data, the greatest value.

upper quartile (p. 259) In a set of numerical data, the median of the upper half of the data.

vertex form (p. 193) The form $y = a(x - h)^2 + k$ for a quadratic function, where (h, k) is the vertex of the function's graph.

vertex of a graph (p. 563) In discrete mathematics, a point used to represent some object, such as a country on a map or an animal in a food web.

vertex of a parabola (pp. 186, 521) The point where the line of symmetry crosses the parabola.

vertices of a hyperbola (p. 543) The points of intersection of the hyperbola and the line containing the foci.

vertices of an ellipse (p. 536) The points of intersection of the ellipse and the line containing the foci.

vertical intercept (p. 53) The value of the dependent variable when the value of the independent variable is 0. On a graph of a function, the vertical intercept indicates where the graph crosses the vertical axis.

***x*-intercept (p. 199)** The x-coordinate of any point where a graph crosses the x-axis.

***z*-score (p. 641)** The number of standard deviations that a given data value lies from the mean. The z-score for a data value x from a normal distribution with mean $\bar{x}$ and standard deviation σ is given by $z = \frac{x - \bar{x}}{\sigma}$.

zero of a polynomial function (p. 413) For a polynomial function f, each solution of the equation $f(x) = 0$.

Index

Credits

Cover Images

Front cover Earth Imaging/Tony Stone Images/Chicago, Inc.(tl); Dana Berry, Wright Center Tufts University(m); Seth Shostak, Science Photo Library/Photo Researchers, Inc. (m background); Roger Ressmeyer—©1989 CORBIS(r background); NASA (l background) **Back cover** Roger Ressmeyer—©1989 CORBIS; NASA (background)

Chapter Opener Writers

Linda Borcover interviewed Johnpaul Jones (pp. 652–654); Melissa Burdick Harmon interviewed Vera Rubin (pp. 386–388); Yleanna Martinez interviewed Norbert Wu (pp. 42–44), Ednaly Ortiz (pp. 138–140), Jhane Barnes (pp. 462–464); Steve Nadis interviewed Twyla Lang (pp. xxxiv–2), Finn Strong (pp. 90–92), Mark Thomas (pp. 182–184), Donna Cox (pp. 234–236), Gina Oliva (pp. 288–290), Ronald Toomer (pp. 334–336), Kija Kim (pp. 510–512), Maria Rodriguez (pp. 560–562), Robert Ward (pp. 608–610).

Acknowledgements

371 Excerpt from *Smilla's Sense of Snow* by Peter Høeg, translation by Tina Nunnally. Copyright © 1993 by Farrar, Straus & Giroux, Inc. Reprinted by permission of Farrar, Straus & Giroux, Inc. **484** Excerpts and illustrations from *Bringing the Rain to Kapiti Plain*, retold by Verna Aardema. Illustrations by Beatriz Vidal. Copyright © 1981 by Verna Aardema, text, copyright © 1981 by Beatriz Vidal, illustrations. Used by permission of Dial Books, a division of Penguin Books USA.

Stock Photography

i EarthImaging/Tony Stone Images/Chicago, Inc.(tl); Dana Berry, Wright Center Tufts University(m); Seth Shostak, Science Photo Library/Photo Researchers, Inc.(m background); Roger Ressmeyer—©1989 CORBIS(r background); **iv** Ted Thai/Time Inc.(t, m); **v** ©1993 Charles Lindsay/Mo Yung Productions(b); **vi** James King-Holmes, Science Photo Library/Photo Researchers, Inc.(t); photo by Alessandro Sanvito. Artifacts from the collection of Museo Archeologico di Napoli, Italy.(m); **viii** ©The Stock Market(t); courtesy Mark Thomas/General Motors(b); **ix** ©John Eastcott/ Yvamomativk/Photo Researchers, Inc.(t); **x** courtesy Gina Oliva; **xi** Daniel Forster Photography(tl, tr); ©1995 Chad Slattery(b); **xii** Katherine Lambert; **xiii** based on an illustration from *Navajo Architecture: Forms, History, Distributions*, by Jett and Spencer ©1981 University of Arizona Press(t); Jerry Jacka Photography(m); courtesy Jhane Barnes(b); David Austen/Stock Boston(t); Tom Herde/Boston Globe(b); **xv** courtesy Consolidated Edison Company, New York(b); **xvi** ©Ken Eward, Science Source/Photo Researchers, Inc.(t); **xvii** Comstock Photofile Limited, Toronto(t); **xviii** ©Dan McCoy/Rainbow(t); Robert W. Parvin(b); **xx** courtesy Jhane Barnes(bl); Craig Fujii/Detroit Free Press, Inc.(br); **xxi** ©Norbert Wu(tl, m); ©1993 Bob Cranston/Mo Yung Productions(tr, bl); ©1993 Charles Lindsay/Mo Yung Productions(br); **xxii** Michael Rosenfeld/ Tony Stone Images/Chicago, Inc.; **xxiv** Benn Mitchell/The Image Bank(t); **xxv** Royal Tyrell Museum/Alberta Community Development(tl); Julie Sheffield/Hannibal Courier Post(bl); **xxvii** ©Allsport/Chris Cole; **xxviii** Rhoda Sidney/Leo De Wys(b); **xxix** ©Michael Marten, Science Photo Library/Photo Researchers, Inc.(tl); Thomas Braise/Tony Stone Images/Chicago, Inc.(tr); © 1995 Chris Arend/Alaska Stock Images(bl); **xxxi** (m)Bill Gallery/Stock Boston(tl, bl); ©Mark Marten, Science Source, Photo Researchers, Inc.(tr); Galen Rowell/Mountain Light(br); **xxxii** Bruce Kliewe/Picture Cube; **xxxiii** ©P. Gontier, Photo Researchers, Inc.(t); courtesy Jhane Barnes(br); **1** Bonwire/Owen Franken/Stock Boston(tr, far r); **9** Superstock(t); ©The Stock Market/Chris Sorensen(l); ©Leonard Lessin/Peter Arnold, Inc.(b); **12** David Young Wolff/Tony Stone Images/Chicago; **13** Reuters/The Bettmann Archive(t); Ted Thai/Time Inc.(m, b); **16** Philip & Karen Smith/Tony Stone Images/Chicago, Inc.; **19** Jeff Hetler/Stock Boston(far l); Kevin R. Morris/Tony Stone Images/Chicago, Inc.(ml); ©Norm Thomas, Photo Researchers, Inc.(mr); Tony Stone Images/Chicago, Inc.(far r); **21** Michael Melford/The Image Bank(t); CMCD, Inc.(b); **22** Lambert/ Archive Photos **24** Rhoda Sidney/Leo de Wys(tl); **27** Hall Puckett/Houston Astros Baseball Club; **32** The Topps Company, Inc.; **33** Trans. No. 4971 (4) (Photo by Don Eiler) Courtesy Department of Library Services, American Museum of Natural History(l); Nevada Historical Society(r); **42** ©Norbert Wu; **43** ©1993 Bob Cranston/Mo Yung Productions(tr); ©Norbert Wu(bl); **44** ©1993 Charles Lindsay/Mo Yung Productions(tr); **44** ©1993 Bob Cranston/Mo Yung Productions(bl); **48** ©Allsport/Shaun Botterill; **49** ©1993 Charles Lindsay/ Mo Yung Productions(tr); **50** Tony Freeman/PhotoEdit; **56** Renee Lynn/Tony Stone Images/Chicago, Inc.(l); Daryl Balfour/ Tony Stone Images/Chicago, Inc. (tr); ©Animals Animals/Richard Packwood(br); **58** Superstock; **62** Greg Vaugh/Tony Stone Images/Chicago, Inc.(br); ©NYNEX Information Resources Company. Printed by permission of NYNEX Information Resources Company.(overlay); **66** Ed Pritchard/Tony Stone Images/Chicago, Inc.; **69** Julie Sheffield/Hannibal Courier Post;—courtesy Hannibal Courier Post(b) **70** David Burnett/Contact Press Images(l); Tony Freeman/PhotoEdit(r); **72** Bachmann/Uniphoto(tl); Fotoworld/ The Image Bank(tm); Harald Sund/The Image Bank(tr); David Ball/Picture Cube(bl); Colin Molyneux/The Image Bank(bm); ©Arvind Garg, Photo Researchers, Inc.(br); **75** courtesy World Data Center A for Glaciology, National Snow and Ice Data Center(br); **76** Darrell Gulin/Tony Stone Images/Chicago, Inc.(tl); Paul H. Humann(tr); ©Animals Animals/Raymond A. Mendez(bl); ©Animals Animals/Zig Leszczynski(br); **81** Superstock; **82** ©The Stock Market/Frank Rossotto; **90–91** courtesy Finn Strong, River Tank Systems(m); **90** ©E.R. Degginger, The National Audubon Society Collection/Photo Researchers, Inc.(bl); courtesy Finn Strong, River Tank Systems(bm); **91** ©Jeff Lepore, The National Audubon Society Collection/Photo Researchers, Inc.(t); ©Rod Planck, The National Audubon Society Collection/Photo Researchers, Inc.(m); **98** K. Tarusov/ITAR-TASS/Sovfoto/Eastfoto(br); **100** ©Bob Daemmrich; **103** Archive Photos; **105** John Coletti/ Picture Cube(t); Michael Dwyer/Stock Boston(m); Superstock(b); **106** Richard Pasley/Stock Boston(t); UPI/Bettmann Archive(b); **113** Roger Ressmeyer—©1995 CORBIS(l); ©James King-Holmes, Science Photo Library/Photo Researchers, Inc.(tr); photo by Alessandro Sanvito. Artifacts from the collection of Museo Archeologico di Napoli, Italy.(br); **114** Rob Crandall;

116 Archive Photos; **117** The Bettmann Archive; **118** Glen Allison/Tony Stone Images/Chicago, Inc.; **120** ©Animals Animals/George F. Godfrey(b); **121** Thomas Braise/Tony Stone Images/Chicago, Inc.(t); ©Michael Marten, Science Photo Library/Photo Researchers, Inc.(b); **122** ©Animals Animals; **123** Jonathan Nourok/PhotoEdit(b); **124** Chronis Jons/Tony Stone Images/Chicago, Inc.; **125** ©Tom McHugh, The National Audubon Society Collection/Photo Researchers, Inc.; **126** Bruce Kliewe/Picture Cube; **127** Nathan Bilow/Tony Stone Images/Chicago, Inc.(bl); Lori Adamski Peek/Tony Stone Images/Chicago, Inc.(br); **128** courtesy Finn Strong, River Tank Systems(bl); ©Animals Animals/Max Gibbs(br); **141** Eric Roth/Picture Cube(t); The Bettmann Archive(b); **144** Baron Wolman/Tony Stone Images/Chicago, Inc.; **145** Louis Bencze/Tony Stone Images; **146** Cary S. Wolinski/Stock Boston(r) **151** courtesy Revue Thommen AG(tl); AP/Wide World Photos(tr); Superstock(br); **153** ©John Sanforn, Science Source/Photo Researchers, Inc.(l); copyright © 1992 by Houghton Mifflin Company. Reproduced by permission from THE AMERICAN HERITAGE DICTIONARY OF THE ENGLISH LANGUAGE, THIRD EDITION.(r); **154** Superstock; **157** USGS; **160** ©P. Gontier, Photo Researchers, Inc.(t); **166** Tony Freeman/PhotoEdit; **168** The Bridgeman Art Library; **169** Felicia Martinez/PhotoEdit; **170** ©Jerome Wexler, Science Source/Photo Researchers, Inc.; **174** Royal Tyrell Museum/Alberta Community Development; **175** The Bettmann Archive; **176** Robert Caputo/Aurora(tl); Superstock(tr); **182–183** courtesy Mark Thomas, General Motors; **184** Michael Rosenfeld/Tony Stone Images/Chicago, Inc.(t); courtesy Mark Thomas, General Motors(m); Craig Fujii/Detroit Free Press, Inc.(b); **185** Benn Mitchell/The Image Bank; **187** A. Susterman/The Bettmann Archive(bl); Hideo Kurihara/Tony Stone Images/Chicago, Inc.(bm); James Sugar/Black Star(r); **189** C. Michael Lewis(t, mr); Tony Freeman/PhotoEdit(m far l); John Kelly/The Image Bank(ml); Superstock(m far r); ©Jeffrey Sylvester/FPG International Corp.(b); **190** Ron Kimball(l); The Image Bank(m); Superstock(r); **194** Don Smetzer/Tony Stone Images/Chicago, Inc.(l); ©Don Hebert/FPG International Corp.(r); **196** ©The Stock Market/Washnik Studio 1989; **201** Superstock; **204** Curt Fischer/Tony Stone Images/Chicago, Inc.(l); ©The Stock Market(r); **210** ©Animals Animals/Stephen Dalton; **211** ©Allsport USA/Vandystat/Michel Hans(tl); ©FPG International Corp.(tr); ©Animals Animals/Robert Maier(b); **212** Duomo/The Image Bank(r); **213** ©Travelpix 1991/FPG International Corp.; **218** Lars Ternblad/The Image Bank; **219** Bruce Ayres/Tony Stone Images/Chicago, Inc.(b); **220** courtesy Ford Motor Company(br); **226** courtesy Professor Borra, Dept. Physique, University Laval(l); reproduced from the *Dorling Kindersley Science Encyclopedia*. Published by Dorling Kindersley Pub. Inc.(br); **228** Shaun van Steÿn/Uniphoto(t); **234** Linda McCandless/New York State Agricultural Experiment Station(t); NCSA/University of Illinois at Urbana-Champaign(background); **235** Linda McCandless/New York State Agricultural Experiment Station(t); Michael Jeffords/A Prairie Gallery(tr); ©Scott Camazine, Photo Researchers, Inc.(bl); **236** Computer graphics of Cox, Onstad and Kornkven(t); **239** American Red Cross; **241** ©1995 Susan Allen Sigmon/NCAA Photos; **242** AP/Wide World Photos; **247** Steven Needham/Envision; **248** Archive Photos(tl, tr); ©Chromosohm/Sohm/Photo Researchers, Inc.(ml, mr); **251** ©Bastin & Evrard — Bruxelles; **253** Woodfin Camp and Associates; **254** ©John Eastcott/Yvamomativk/Photo Researchers, Inc.(tr);

255 Computer graphics of Cox, Onstad and Kornkven(ml, mr); **256** Lawrence Migdale/Tony Stone Images/Chicago, Inc.(t); Bob Daemmrich/Tony Stone Images/Chicago, Inc.(m); M. Richards/PhotoEdit(b); **263** © 1995 Chris Arend Photography(t); E. Spieghalter/Woodfin Camp & Associates(b); **264** Archive Photos(far l); Painting by John Trumbull/Archive Photos(ml); UPI/Bettman Archive(mr); John Elk/Stock Boston(far r); **266** ©Mark Marten, Science Source/Photo Researchers, Inc.(t); Galen Rowell/Mountain Light(br); **272** David Guttenfelder/AP/Wide World Photos; **277** Boston Public Library Print Department; **288** sign language letters reprinted by permission of the publisher, from H. Bornstein, K. Saulnier, and L. Hamilton, eds., *The Comprehensive Signed English Dictionary* (1983): C-1. Washington, DC: Gallaudet University Press. ©1983 by Gallaudet University(t); Photography by Andrew Eccles, courtesy Lady Foot Locker and *Self* Magazine(bl); **289** Photography by Andrew Eccles, courtesy Lady Foot Locker and *Self* Magazine(tl, tr, mr, bl); courtesy Gina Oliva(ml, bl); **290** Michal Heron/Woodfin Camp & Associates(t); Photography by Andrew Eccles, courtesy Lady Foot Locker and *Self* Magazine(b); **291** Greg Pease/Tony Stone Images/Chicago, Inc.; **294** Jon Riley/Tony Stone Images/Chicago, Inc.; **295** Rod Furgason/Unicorn Stock Photo(l); Jose Carrillo/PhotoEdit(m); David Young-Wolff/PhotoEdit(r); **298** ©Ted Kerasote, Photo Researchers, Inc.; **299** Photography by Andrew Eccles, courtesy Lady Foot Locker and *Self* Magazine(t); **303** Andy Sacks/Tony Stone Images/Chicago, Inc.; **307** Grant Heilman/Grant Heilman Photography(t, b); Dwight R. Kuhn(br); **309** Susan Werner(tl, tr, bm); ©Mark Gilbert/Light Sources Stock(bl); Robert W. Ginn/Unicorn Stock Photo(br); **314** ©Animals Animals/E.R. Degginger(t); Kindra Clineff/Picture Cube(b); **316** Tony Freeman/PhotoEdit; **318** Thomas Del Brase/Tony Stone Images/Chicago, Inc.; **319** Photography by Andrew Eccles, courtesy Lady Foot Locker and *Self* Magazine(t); William Sallaz/Duomo/The Image Bank(bl); Tom McCarthy/PhotoEdit(br); **320** Terry Vine/Tony Stone Images/Chicago, Inc.; **321** ©FPG International Corp.; **322** Peter Vandermark/Stock Boston; **325** ©Jonathan A. Meyers/FPG International Corp.; **328** ©Allsport/Chris Cole; **329** ©The Stock Market/Lewis Portnoy 1987(t); ©FPG International Corp.(m); Duomo/The Image Bank(b); **334** ©1995 Chad Slattery(tl); **334–335** ©Paul L. Ruben(b); **335** ©1995 Chad Slattery(t); ©Paul L. Ruben(br); **336** ©Paul L. Ruben(b); **339** ©Allsport/Ken Levine; **341** courtesy The Smithsonian Institution, from *Krakatau 1883: The Volcanic Eruption and Its Effects,* by Tom Simkin and Richard S. Fiske, Smithsonian Institution Press 1983(tr); ©Krafft-Explorer, Photo Researchers, Inc.(bl); **342** ©1995 Chad Slattery(t); Dan Feicht/Cedar Point(b); **345** ©Animals Animals/Richard Davies/OSF; **348** Dan Feicht/Cedar Point(bl); **349** ©Bob Daemmrich; **350** NASA(t); **354** Lynn Sladky/Palm Beach Post; **355** Hans Rienhard/Tony Stone Images/Chicago, Inc.(t); **356** Daniel Forster Photography; **357** ©1994 David Madison/Duomo; **358** CALVIN AND HOBBES ©1988 Watterson. Dist. by UNIVERSAL PRESS SYNDICATE. Reprinted with permission. All rights reserved.; **363** John Yurka/Picture Cube; **371** Bryn Campbell/Tony Stone Images/Chicago, Inc.; **377** Todd Powell/ProFiles West; **378** William Weems/Woodfin Camp and Associates(l); **386–387** ©Tony Hallas, Science Photo Library/Photo Researchers, Inc.(t); **386** Katherine Lambert(b); **387** ©Francoise Gohier, Photo Researchers, Inc.(b); **388** NASA(m); photo by Mark Godfrey, courtesy Vera Rubin(b);

389 MS. Arch. Selden. A. 1, fol.46r, The Bodleian Library, Oxford(l); Margarete Büsing/Ägyptisches Museum Staatliche Museen, Preußischer Kulturbesitz, Berlin(tr); **391** Roland & Sabrina Michaud/Woodfin Camp and Associates; **396** ©The Stock Market/David Sailors; **398** Fred Zahradnik/Bicycling Magazine; **401** reprinted from *Magic Squares* by *John Lee Fults* by permission of Open Court Publishing Company, La Salle, Illinois. Copyright © 1974 by Open Court(l); photo: Wan-go H.C. Weng. Collection of National Palace Museum, Taipei(r); **402** Tony Freeman/PhotoEdit(r); Envision(tl); The Bettmann Archive(bl); **407** Stacy Pick/Stock Boston; **409** ©Paul J. Sutton/Duomo; **410** courtesy of Plimoth Plantation; Plymouth, MA USA **412** NASA; **417** Brian Parker/Tom Stack & Associates; **426** © 1995 Chris Arend/Alaska Stock Images; **427** Canadian Museum of Civilization, S93-7213; **428** ©Richard R. Hansen, Photo Researchers, Inc.(tl); Bill Marchel Outdoor Observations(ml); ©Animals Animals/Charles Palek(mr); A. Forbes-Watson/VIREO(m far r); **429** Tony Freeman/PhotoEdit; **430** Katherine Lambert; **431** UPI/Bettmann Archive; **438** ©Norbert Wu; **439** Uniphoto; **446** Katherine Lambert(t); ©Julian Baum, Science Photo Library/Photo Researchers, Inc.(m); **449** John Lei/Stock Boston(l); © 1989 Robert Mathena/Fundamental Photographs/New York(r); **452** Tom Pantages(tl, tm, tr); ©1992 Richard Megna/Fundamental Photographs, New York(b); **453** Focus on Sports; **454** Alan Klehr/Tony Stone Images/Chicago, Inc.(tl); **462** courtesy Jhane Barnes(b); **463** photo by Paul Jasmin, courtesy Jhane Barnes(tl, tr) **463** courtesy Jhane Barnes(b); **464** courtesy Jhane Barnes; **465** ©Bob Daemmrich(t); ©David Sanger Photography(b); **467** ©ArSciMed/Science Photo Library/Photo Researchers, Inc.; **469** Keylight Image Library Ltd.; **470** The Bettmann Archive(tr); Uniphoto(bl); **471** courtesy Jhane Barnes; **475** Edwards Alto Trombone — division of the Getzen Company, Elkhorn, Wisconsin; **476** ©Animals Animals/E.R. Degginger; **477** Cesar Lucas/The Image Bank(tl); ©Animals Animals/M.A. Chappell(tr); ©The Stock Market/Gabe Palmer(br); **478** Jerry Jacka Photography(tl); based on an illustration from *Navajo Architecture: Forms, History, Distributions* by Jett and Spencer ©1981 University of Arizona Press(tr); ©Duomo(bl); **479** courtesy Jhane Barnes; **485** Marc Chamberlain/Tony Stone Images/Chicago, Inc.(t); ©Thomas R. Taylor, Photo Researchers, Inc.(b); **489** Uniphoto; **490** from *A Handbook of Korea* ©1993 by permission of the Korean Consulate General, NY; **492** Shelburne Museum, Shelburne, Vermont. Photograph by Ken Burris.; **493** courtesy Wu-Yi Hsiang; **495** courtesy Jhane Barnes(b); **496** ©Allsport/Clive Brunskill; **501** Cheryl Graham/Tony Stone Images/Chicago, Inc.(tr); **502** courtesy Jhane Barnes; **505** David Joel/Tony Stone Images/Chicago, Inc.(bl); ©Richard Laird/FPG International Corp.(br); **510** ©The Stock Market/Randy Duchaine(t); **511** Lou Jones/The Image Bank(b); **514** Uniphoto; **517** ©Jack Dermid, Photo Researchers, Inc.; **518** courtesy the State of California(t); **523** W.E. Carter/National Oceanic and Atmospheric Administration/Stocktrek; **525** David Austen/Stock Boston; **526** Steven Kuzma/The Ann Arbor Hands-On Museum(bl, br); **528** Kindra Clineff/Picture Cube; **532** Tom Herde/Boston Globe; **533** ©The Stock Market/John Madere; **534** courtesy U.S. Cavalry; **539** Llewellyn/Uniphoto; **540** "Black Hole" by William Wainwright, courtesy DeCordova Museum and Sculpture Park; **541** ©SIU/Photo Researchers, Inc.(t); **543** ©Hank Morgan, Photo Researchers, Inc.; **548** ©The Stock Market/Jim Brown; **552** ©FPG International Corp.; **554** Lewis Portnoy/Uniphoto; **560** D.S. Henderson/The Image Bank(l); Joseph Szkodzinski/The Image Bank(b); **561–562** courtesy Consolidated Edison Company, New York; **567** ©Tom Carroll/FPG International Corp.; **569** Uniphoto; **570** ©Animals Animals/Robert Maier(tl); **574** courtesy Consolidated Edison Company, New York; **575** ©Bob Daemmrich; **579** Dennis O'Clair/Tony Stone Images/Chicago, Inc.; **582** courtesy Consolidated Edison Company, New York(t); Jeffrey Coolidge/The Image Bank(b); **590** Robert Fried/D. Donne Bryant Stock(tl); **593** Archive Photos(b); **595** Superstock; **597** from *Precious Mirror of the Four Elements,* adapted courtesy of the Needham Research Institute, Cambridge, England; **598** Dave Jacobs/Tony Stone Images/Chicago, Inc.(t); **608–609** ©Irving Geis, Science Source/Photo Researchers, Inc.(t); **610** ©Ken Eward, Science Source/Photo Researchers, Inc.(t); Permission is granted by Birkhäuser Boston to use the picture from the cover page of the book *BIOTECHNOLOGY: PROTEINS TO PCR a Course in Strategies and Lab Techniques* by *David W. Burden* and *Donald B. Whitney*, copyright 1995(m); **611** John Swart/AP/Wide World Photos; **612** courtesy Drake University; **614** ©John Chumack, Science Source/Photo Researchers, Inc.; **615** ©Photo Researchers, Inc.; **616** Catalogue No. MNH#300, artifact catalog #101753, Department of Anthropology, Smithsonian Institution; **618** Brown Brothers, Sterling, PA; **620** MONOPOLY®, the distinctive design of the game board, the four corner squares, as well as each of the distinctive elements of the board and the playing pieces are trademarks of Hasbro, Inc. for its real estate trading game and game equipment. ©1935, 1997, Parker Brothers. All rights reserved. Photo ©Leonard Lessin/Peter Arnold, Inc.; **621** ©The Stock Market/A & J Verkaik; **622** ©The Stock Market/Kunio Owaki; **628** G. Rancinan/Sygma(t); Dale Wittner/PHOTOSeattle(b); **629** ©Keith Kent/Peter Arnold, Inc.; **630** (m)Bill Gallery/Stock Boston; **635** Frank Siteman/Stock Boston; **637** Jim Pickerell/Tony Stone Images/Chicago, Inc.; **638** ©Volker Steger/Peter Arnold, Inc.; **639** ©The Stock Market/Roy Morsch; **641** I. Devore/Anthro Photo File; **643** Mary Evans Picture Library(t); ©Leonard Lessin/Peter Arnold, Inc.(b); **644** David Joel/Tony Stone Images/Chicago, Inc.; **645** Eric Carle/Stock Boston; **652** Jim Brandenburg/Minden Pictures(l); **652–653** Jones & Jones(b); **653** ©Nigel Dennis, The National Audubon Society Collection/Photo Researchers, Inc.(br); **654** ©Animals Animals/John Chellman(t); **656** Gavin Hellier/Tony Stone Images/Chicago, Inc.; **657** C M Dixon; **664** Tom Prettyman/PhotoEdit; **666** Steve Leonard/Tony Stone Images/Chicago, Inc.; **669** Oldrich Karasek/Tony Stone Images/Chicago, Inc.; **675** Cliff Hollenbeck/Tony Stone Images/Chicago, Inc.; **684** Robert Frerck, Odyssey Productions/Chicago; **693** Neg. No. 338463 (Photo by R.T. Bird) Courtesy Department of Library Services, American Museum of Natural History(l); ©Francois Gohier, Photo Researchers, Inc.(r); **695** Osentoski & Zoda/Envision(b); **696** Comstock Photofile Limited, Toronto; **704–705** Robert W. Parvin; **706** Will Van Overbeek(t); Robert **707** ©Ray Coleman, Photo Researchers, Inc.(t); **710** ©The Stock Market/Dan McCoy(t); **711** ©Renee Lynn, Photo Researchers, Inc.(l); **715** Robert W. Parvin(l); ©Dan McCoy/Rainbow(r); **716** ©The Stock Market/Tom Martin(bl); ©Tim Davis, Photo Researchers, Inc.(br); **717** Robert Frerck, Odyssey Productions/Chicago(t); ©The Stock Market/Thomas Del Braise(b); **718** Robert W. Parvin; **719** Robert E. Daemmrich/Tony Stone Images/Chicago, Inc.; **720** Michael Newman/PhotoEdit; **722** Navy Pier Ferris Wheel, Chicago; **724** Robert W. Parvin; **725** Eric R. Berndt/Unicorn Stock Photo(l); courtesy of Leica Inc.(r); **732** P.H. Cornut/Tony Stone

Images/Chicago, Inc.(t); Superstock(b); **733** courtesy of Emil Fries Piano Hospital and Training Center; **735** Superstock **738** John Kelly/The Image Bank; **739** ©The Stock Market/Bob Krist.

Assignment Photography

Kindra Clineff **553**
Jeffrey Dunn **vi**(b), **90**(br), **91**(b), **92**(t, b), **112**(l, r)
Charlotte Fiorito **iv**(b), **xxiv** (bl, br), **xxxiv**, **1**(tl, bl), **2**(t, b), **6, 26**
Ken Karp **495**(r)
John Marshall **vii**, **xx**(tl), **x, 138–139**, **140**, **146**(l), **159**
RMIP/Richard Haynes **v**(t), **xv**(t), **xvi**(b), **xxiii**, **xxvi**, **xxvii**(br, bl), **xxviii**(tr), **xxix**(r), **3**, **12**(tl, tr), **20**, **29**, **31**, **36**, **37**, **46**, **51**, **52**, **53**, **65**(tr), **69**(b), **84**, **85**, **93**(bl, br), **94**, **95**, **98**(tl, tr, m), **99**, **107**, **111**, **115**(t, b), **120**(t), **123**(t), **127**(t), **128**(t, ml, m, mr), **130**(l, r), **131**(r), **148**, **155**, **158**, **160**(b), **162**, **165**, **167**, **176**(b), **177**, **183**(tr), **191**(r), **192**, **195**, **199**, **203**, **206**, **209**, **214**, **216**, **219**, **220**(tr), **221**, **225**, **228**, **229**, **237**, **244**, **249**, **250**, **258**, **273**, **278**, **280**, **297**, **299**(b), **324**(tl), **337**, **343**, **344**, **350**(m), **355**(b), **377**, **378**(br), **380-381**, **404**, **419**, **433**(l, m), **436**, **441**, **454**(bl), **455**(bl), **462-464**(background), **472**, **480**, **486**, **497**, **501**(l, br), **504**, **505**(t), **513**, **515**, **520**, **526**(tl, m, tr), **535**(t, bl, br), **541**(b), **547**, **555**, **560**(t), **563**, **566**, **570**(r), **573**, **577**, **585**, **586**, **590**(tr), **593**(t), **596**, **598**(b), **600**, **601**, **608**(b), **610**(b), **625**, **632**, **637**(t), **646**, **647**, **655**, **663**, **673**, **677**, **681**, **685**, **691**, **695**(t), **698**, **707**(b), **713**, **726**(tl), **740**, **741**
Jeff Reinking **xvii**(b), **653**(t), **654**(b), **661**, **668**
Tony Scarpetta **455**(tr)
Robert Schoen **xxxiii**(l), **510**(b), **511**(t), **512**
Bill Wiegand **ix**(b), **234**(b), **236**(b), **255**(tr), **271**(t)

Illustrations

Arnold Bombay **341**, **563**, **564**, **567**(t), **576**, **584**
Hannah Bonner **198**, **204**(b), **296**(t), **303**, **307**, **468**
Steve Cowden **xxvii**, **xxx**, **45**, **54**, **78**, **79**, **81**, **208**, **212**, **298**, **301**(t), **315**(t), **326**, **328**, **336**, **409**, **424**, **432**, **569**, **589**, **659**, **661**(b), **662**, **668**(b), **670**, **699**
Dartmouth Publishing **vii**, **2**, **4**, **6**(t), **12**, **24**, **25**, **45**, **62**, **84**, **93**, **97**, **99**(tl, mr), **100**, **131**, **169**, **170**, **173**, **191**, **197**, **204**(t, b), **239**, **242**, **247**, **248**, **253**, **256**(b), **264**(t), **271**, **275**, **276**(t), **351**(background), **369**(t), **371**, **377**, **381**, **413**, **428**, **454**(b), **490**, **493**, **500**, **505**, **518**, **590**, **598**
DLF Group **vii**, **xxxii**, **1**, **8**, **22**, **56**, **69**, **114**, **118**(tl), **154**, **168**(l), **172**, **534**
Eureka Cartography **176**(tm), **313**, **315**(b), **320**, **501**, **602**
Harvard Design & Mapping **512**, **517**, **532**
Phillip Jones **727**
Piotr Kaczmarek **xxv**, **14**, **49**, **296**(b), **313**, **354**, **431**, **446**(b), **657**, **661**(t)
Joe Klim **426**
Andrea Maginnis **351**
Precision Graphics **xii**, **xxi**, **xxxii**, **5**, **6**(b), **7**, **8**(b), **9**, **13**, **15**, **29**, **30**, **33**(b), **35**, **44**, **61**, **77**, **101**, **125**(r), **147**, **226**, **227**, **237**, **239**, **243**, **246**, **252**, **254**, **256**(t), **263**, **264**(t), **270**, **273**, **276**(b), **277**, **280**, **281**, **290**, **291**, **310**, **349**, **370**, **388**, **417**, **430**, **439**(m), **514**, **528**, **533**, **539**, **541**, **548**, **550**, **552**, **567**(b), **568**, **582**, **596**, **626**, **656**, **664**, **665**, **666**, **668**(t), **683**, **696**, **717**, **732**, **738**(b)
Ann Raymond **347**, **359**, **364**, **372**, **376**(t), **402**, **410**, **447**, **448**, **562**, **565**, **570**, **575**, **577**, **578**, **581**, **583**, **586**, **587**, **588**, **592**(t), **593**, **710**(tl), **711**, **740**, **741**
Patrice Rossi **159**
David Shepherd Design **48**, **55**, **69**, **74**, **82**, **119**, **146**, **166**, **168**(l)
Doug Stevens **92**, **224**, **292**, **301**(b), **302**, **669**, **671**, **680**, **689**

Selected Answers

CHAPTER 1

Page 5 Checking Key Concepts

1. about 1,178,000 people **3.** Yes; the number of degrees (in thousands) predicted by the constant growth model for the second and third years are 921 and 1049 and by the proportional growth model are about 912 and 1049. Both sets of estimates are reasonable, since they are close to the actual values for those years.

Pages 5–8 Exercises and Applications

1. Prediction estimates may vary due to rounding. Increases are rounded to the nearest dollar or the nearest percent.

a.

Average Daily Hospital Cost			
Year	**Daily charge ($)**	**$ Increase**	**% Increase**
1975	134	—	—
1980	245	111	83%
1985	460	215	88%
1990	687	227	49%

b. about $870 **c.** about $1200; It is much higher. **3.** Prediction estimates may vary due to rounding. Answers may vary. An example is given. I think a reasonable estimate is about 2300 scarves. This is based on the constant growth model, which I think is a reasonable model. **5.** Prediction estimates may vary due to rounding. **a.** constant growth model: about $6.10; proportional growth model: about $8.86 **b.** the estimate obtained from the constant growth model; Both estimates will be off because of the fact that the minimum wage did not increase at all for 9 years. **c.** 1979–1981: The estimate with the constant growth model would be slightly higher; the estimate with the proportional growth model would be slightly lower. 1974–1976: The estimate with the constant growth model would be significantly lower; the estimate with the proportional growth model would be about the same. **7, 9.** Estimates may vary due to rounding. Answers may vary. Two examples are given for each item. The first example is based on a constant growth model, the second on a proportional growth model. **7.** about $13,126; about $14,100 **9.** about $3028; about $3250 **11.** total expense; tuition **13. a.** about 43% **b.** about $42,954; about $61,424 **c.** No; the actual statistics are significantly lower than the estimates. Answers may vary. An example is given. The difference between a master's degree and a doctorate may not have as significant an effect on income as the difference between lower education levels.

21.

x	y
0	6
1	3
2	0
3	−3

23. 3.4

Page 12 Checking Key Concepts

1. number of workers; years since 1987 **3.** $y = 7147(1.374)^x$

Pages 12–15 Exercises and Applications

1. Marya; Jon-Paul **3.** 1985; 1980; Jon-Paul's data begins in 1985, while Marya's begins in 1980. **5.** 3000; Jon-Paul's (the linear model) **7.** about 39,000 teams **11. a.** Answers may vary slightly due to rounding. Let x = the number of four-year periods since 1960 and y = time in seconds.
(a) $y = 258.3 - 4.16x$ (b) $y = 258.3(0.983)^x$
(c)

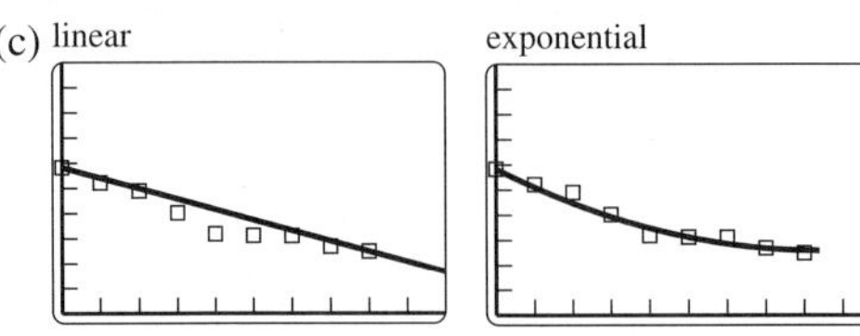

Answers may vary. An example is given. Yes; the data points are reasonably near both graphs. (d) linear: about 220.86 s, about 216.70 s; exponential: about 221.36 s, about 217.60 s **17.** Answers may vary due to rounding. Examples are given. **a.** about $22.2 billion **b.** about $22.4 billion **19.** −1 **21.** 13.5; 13; 13

Page 15 Assess Your Progress

2. Answers may vary due to rounding. Examples are given. In parts (a) and (b), let x = the number of decades since 1951 and y = the cost in dollars of a movie ticket. **a.** $y = 0.50 + 1.1x$ **b.** $y = 0.50(1.78)^x$ **c.**

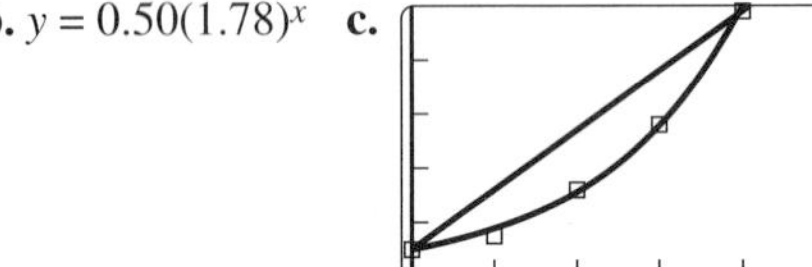

The exponential model fits the data almost exactly.

d. about $6 (linear model); about $8.93 (exponential model), which would probably be rounded to $9.00

Page 18 Checking Key Concepts

1. $\begin{bmatrix} 5 & 4 \\ 6 & 5 \end{bmatrix}$ **3.** $\begin{bmatrix} -3 & 2 \\ 6 & -1 \end{bmatrix}$ **5.** $\begin{bmatrix} 2 & 4 & 4 \\ 18 & -26 & 18 \\ 3 & -20 & 2 \end{bmatrix}$

Pages 19–21 Exercises and Applications

1. 2×4 **3.** No; the matrices do not have the same dimensions, so they do not have corresponding elements.

5. $\begin{bmatrix} 14 & 6 \\ 19 & 18 \\ 11 & -17 \end{bmatrix}$ **7.** $\begin{bmatrix} -2 & 10 \\ -19 & 6 \\ 9 & 9 \end{bmatrix}$ **9.** $\begin{bmatrix} -26 & -44 \\ 68 & -23 \end{bmatrix}$ **11.** undefined

13. $A + B$; $\begin{bmatrix} 2{,}497{,}800 & 136{,}000 & 696{,}150 & 18{,}160 \\ 2{,}989{,}800 & 5{,}075 & 674{,}225 & 62{,}050 \\ 394{,}250 & 835{,}000 & 90{,}500 & 12{,}430 \end{bmatrix}$

15. The dimensions would be 5×4; the dimensions would be 3×5 (or 5×5 if Nebraska and Missouri were added as well).

17. a. $\begin{bmatrix} -1 & 1 & 5 & 3 \\ -3 & 2 & 2 & -3 \end{bmatrix}$

b.

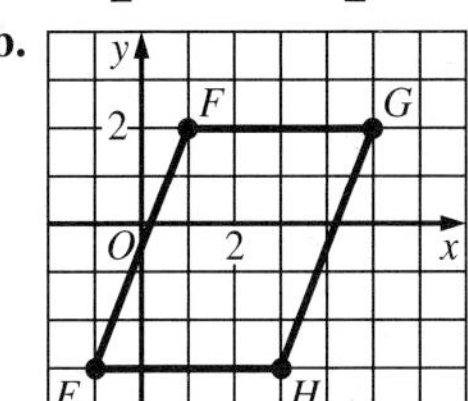

EFGH is *ABCD* moved 1 unit in the negative *x*-direction and 3 units in the negative *y*-direction.

19. enlarged by a factor of 3

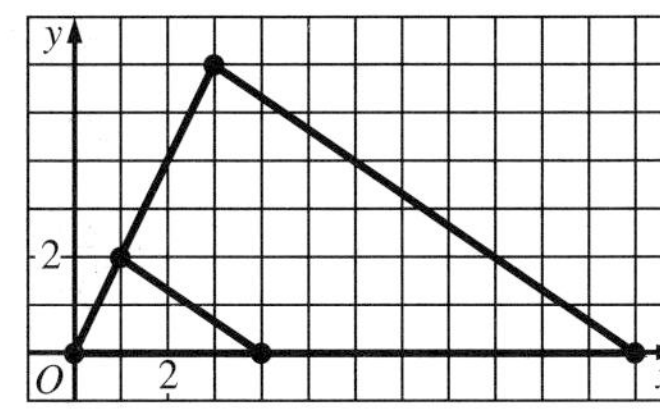

21. position moved 1 unit in the negative *x*-direction and 3 units in the positive *y*-direction

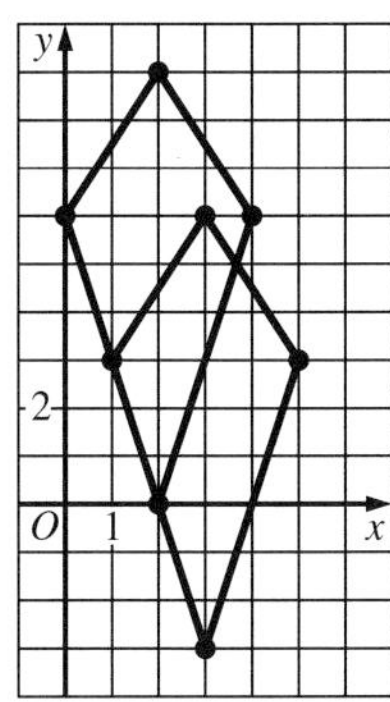

23. The element in row *m*, column *n* of *A* + *B* is $a_{m,n} + b_{m,n}$. The element in row *m*, column *n* of *B* + *A* is $b_{m,n} + a_{m,n}$, which is equal to $a_{m,n} + b_{m,n}$.

$$A + B = \begin{bmatrix} 3+(-9) & 1+10 & -2+(-3) \\ -1+0 & 5+6 & 6+1 \\ 4+14 & 13+7 & 0+(-8) \end{bmatrix} = \begin{bmatrix} -6 & 11 & -5 \\ -1 & 11 & 7 \\ 18 & 20 & -8 \end{bmatrix};$$

$$B + A = \begin{bmatrix} -9+3 & 10+1 & -3+(-2) \\ 0+(-1) & 6+5 & 1+6 \\ 14+4 & 7+13 & -8+0 \end{bmatrix} = \begin{bmatrix} -6 & 11 & -5 \\ -1 & 11 & 7 \\ 18 & 20 & -8 \end{bmatrix}$$

25. The element in row *m*, column *n* of $r(A + B)$ is $r(a_{m,n} + b_{m,n})$. The element in row *m*, column *n* of $rA + rB$ is $ra_{m,n} + rb_{m,n}$, which is equal to $r(a_{m,n} + b_{m,n})$.

$$r(A + B) = r\begin{bmatrix} -6 & 11 & -5 \\ -1 & 11 & 7 \\ 18 & 20 & -8 \end{bmatrix} = \begin{bmatrix} -6r & 11r & -5r \\ -r & 11r & 7r \\ 18r & 20r & -8r \end{bmatrix};$$

$$rA + rB = \begin{bmatrix} 3r & r & -2r \\ -r & 5r & 6r \\ 4r & 13r & 0 \end{bmatrix} + \begin{bmatrix} -9r & 10r & -3r \\ 0 & 6r & r \\ 14r & 7r & -8r \end{bmatrix} = \begin{bmatrix} -6r & 11r & -5r \\ -r & 11r & 7r \\ 18r & 20r & -8r \end{bmatrix}$$

31. Answers may vary due to rounding. Let x = the number of decades since 1960 and y = the number (in millions) of cable users. $y = 0.65(4.53)^x$ **33.** 1 **35.** 48 **37.** False; for example, $5 - 1 = 4$, but $1 - 5 = -4$. **39.** True for every real number *a* except 0; for example, $7 \cdot \frac{1}{7} = 1$. ($\frac{1}{0}$ is not defined.)

41. False; for example, $|-5| + |5| = 5 + 5 = 10$, while $|-5 + 5| = |0| = 0$.

Page 25 Checking Key Concepts

1. a. $\begin{bmatrix} 8 & 6 & -2 \\ 18 & 20 & -4 \end{bmatrix}$ **b.** undefined

3. a. $\begin{bmatrix} 16 & -23 & 3 \\ 47 & -61 & -3 \\ 45 & -57 & -1 \end{bmatrix}$ **b.** $\begin{bmatrix} -32 & -14 & 22 \\ -1 & -8 & 11 \\ 27 & 4 & -6 \end{bmatrix}$ **5.** *RS* and *SR*; No.

Pages 25–28 Exercises and Applications

1. *A*: 3×5; *B*: 5×2; *AB*: 3×2 **3.** *LM*: Yes; 8×6. *ML*: No, the number of columns in *M* does not equal the number of rows in *L*. **5.** *LM*: No; the number of columns in *L* does not equal the number of rows in *M*. *ML*: Yes; 3×5.

7. $\begin{bmatrix} 8 & -3 \\ 5 & -10 \end{bmatrix}$ **9.** $\begin{bmatrix} -24 \\ -10 \end{bmatrix}$ **11.** $\begin{bmatrix} -4 & -7 & -4 \\ 6 & 30 & -7 \end{bmatrix}$ **13.** undefined

15. $S = $

	Class	Plain	2-color sorority	Sorority	Earrings	Shirts
Day 1	40	45	12	28	25	65
Day 2	32	52	24	16	18	45

;

$T = $

	\$
Class	25
Plain	20
2-color sorority	25
Sorority	30
Earrings	15
Shirts	12

; *S* has dimensions 2×6 and *T* has dimensions 6×1.

b.

	\$
Day 1	4195
Day 2	3730

; total sales for the two days

23. a. $\begin{bmatrix} -510 & 90 & 60 & 696 \\ -1071 & 653 & -18 & 444 \\ -810 & 450 & 0 & 432 \end{bmatrix}$

b. An error statement results. *TS* cannot be computed because the number of columns of *T* and the number of rows of *S* are not equal. **25.** sometimes true **27.** always true

31. $\begin{bmatrix} -\frac{3}{2} & 0 \\ -3 & -6 \end{bmatrix}$ **33.** $\begin{bmatrix} 1 & 2 \\ -1 & -1 \end{bmatrix}$ **35.** 0.25 **37.** 0.01

39. $x \leq -9$ **41.** $x \geq -\frac{3}{5}$

Page 32 Checking Key Concepts

Answers may vary. Examples are given. In each case, repeat the simulation a number of times, say 10, and compute the average. **1.** Roll a die and let rolling a 6 represent getting the particular sticker. Count the number of trials needed to roll a 6. **3.** Use the random number feature on a calculator. Let random numbers greater than or equal to 0.2 represent individual stickers and numbers less than 0.2 represent team stickers. Count the number of trials needed to generate a number less than 0.2.

Pages 32–35 Exercises and Applications

1, 3. Answers may vary. Examples are given. In each case, repeat the simulation a number of times, say 10, and compute the average.

1. Use the random number feature on a calculator. Let the first digit of each random number represent the number of correct answers guessed. Count the number of trials needed to generate a number greater than or equal to 7. **3.** Roll a die. Let rolling a 1 or a 2 represent a hit. Count the number of trials needed to roll a 1 or a 2. **5.** B **7.** Answers may vary. Sample trial results are given. **a.** 7, 15, 18, 10, 9, 9, 13, 13, 16, 15 **b.** for the preceding sample, 18 packs; 7 packs; about 13 packs **9. a.** Simulation method may vary. For example, toss a coin. Let heads represent painted side up and tails represent painted side down. To simulate each player's turn, toss the coin 12 times. The player gets 1 point if heads come up exactly five times. **b.** Answers may vary. **c.** Answers may vary. An example is given. On each turn, each player has an equal chance of getting 1 point. However, I think the player who throws first has a slight advantage. Consider, for example, if both players score 1 point on each turn, Player 1 will win.

17. $\begin{bmatrix} 14 \\ 3 \end{bmatrix}$ **19.** $\begin{bmatrix} 7 & -5 & 10 \\ -3 & 4 & -7 \\ -4 & -12 & 16 \end{bmatrix}$

21.

x	y
–3	3
–2	2
–1	1
0	0
1	–1
2	–2
3	–3

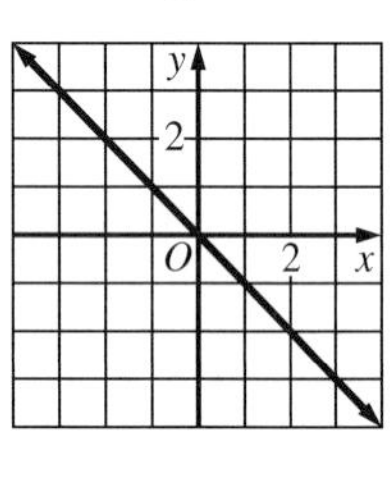

23.

x	y
–3	11
–2	9
–1	7
0	5
1	3
2	1
3	–1

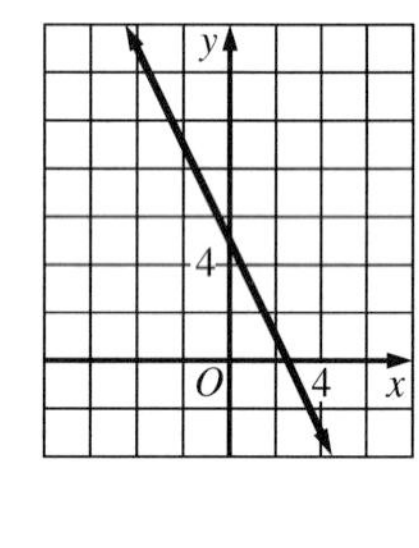

Page 35 Assess Your Progress

1. $\begin{bmatrix} 1 & 6 & 1 \\ -3 & 4 & 1 \end{bmatrix}$ **2.** $\begin{bmatrix} 9 & -6 & 6 \\ 15 & -3 & -9 \\ 0 & 12 & 3 \end{bmatrix}$ **3.** undefined **4.** undefined

5. $\begin{bmatrix} -3 & -6 & 0 \\ 6 & -5 & -5 \end{bmatrix}$ **6.** $\begin{bmatrix} 6 & -8 & 3 \\ -4 & 21 & -5 \end{bmatrix}$ **7.** $\begin{bmatrix} 29 & -14 \\ 3 & -3 \end{bmatrix}$

8. undefined **9.** $\begin{bmatrix} 1 & 2 \\ -8 & 5 \\ 19 & -9 \end{bmatrix}$ **10.** undefined

11. a. Simulation methods may vary. An example is given. For each employee, use int(6*rand)+1 to generate random numbers. Let 1 represent the employee's leaving the company. Generate random numbers until you get a 1. The number of tries is the number of years the employee stays at the company. Sample trial results are given.

Employee	1	2	3	4	5	6	7	8	9	10
No. of years employee stays	6	12	4	2	11	5	3	5	12	5
Employee	11	12	13	14	15	16	17	18	19	20
No. of years employee stays	7	6	14	5	2	1	3	6	11	5

b. for the preceding sample, about 6 years; 9 employees; 5 employees

Pages 40–41 Chapter 1 Assessment

1. function; independent variable; dependent variable **2.** matrix; element **3. a.** 4.9 million **b.** 22% **c, d.** Let x = the number of decades since 1970. **c.** $y = 20.0 + 4.9x$; about 34.7 million **d.** $y = 20.0(1.22)^x$; about 36.3 million **4. a.** $y = 24.7 + 2.3x$ **b.** $y = 24.7(1.09)^x$ **c.** Amounts are in millions of dollars.

Year	Actual Amount Spent	Amount predicted by linear function	Amount predicted by exponential function
1986	24.7	24.7	24.7
1987	27.1	27.0	26.9
1988	29.4	29.3	29.3
1989	31.6	31.6	32.0

d. The values predicted by the linear model are slightly closer to the actual values than are the values predicted by the exponential model. However, both models are reasonable.

5. $\begin{bmatrix} 8 & -2 \\ -2 & 0 \\ 7 & 31 \end{bmatrix}$ **6.** undefined **7.** $\begin{bmatrix} 7 & -18 \\ -8 & 5 \\ -7 & 69 \end{bmatrix}$ **8.** $\begin{bmatrix} 53 \\ 6 \\ 25 \end{bmatrix}$

9. $\begin{bmatrix} 3 & 0 \\ -6 & 3 \end{bmatrix}$ **10.** undefined **11.** 2×4 **12.** –2 **13.** 3×4

14. a. Let three numbers, say 1, 2, and 3, represent the letter A. Let two numbers, say 4 and 5, represent the letter B, and let the remaining number represent the letter C. Roll the die until a number representing each of the three letters comes up. Record the number of rolls it took. Repeat a number of times, say 20, and calculate the average number of rolls. **b.** 7 or 8 bottle caps

CHAPTER 2

Page 47 Checking Key Concepts

1. a. Yes; the ratio of y and x is constant; $\frac{y}{x} = 3.2$. **b.** Yes; the ratio of y and x is constant; $\frac{y}{x} = -3.2$. **c.** No; the ratio of y and x is not constant. The graph of $y = 3.2x + 1$ does not pass through the origin. **3. a.** Yes; $y = \frac{1}{2}x$. **b.** No.

Pages 48–51 Exercises and Applications

1. Yes; the common ratio is about 62.5; $y = 62.5x$. **3.** No. **5.** No. **7.** No. **9.** No. **11.** No.

13. Let l = the number of laps and d = the distance of the race in meters; $l = 0.0025d$; 7.5 laps. **19.** Yes; $y = \frac{2}{3}x$. **21.** Yes; $y = -3x$. **23.** Answers may vary. about 50 Cal **25. a.** Let x = the number of hours spent exercising and y = the number of calories used; $y = 900x$; 900 Cal/h. **b.** Yes; the constant would be less for a person who weighs less than 150 lb, because it requires less energy to move a smaller mass.
31. the basic monthly cost of the account; the cost per check

Page 54 Checking Key Concepts

1. 2 **3.** 0 **5.** $-\frac{3}{7}; \frac{1}{4}$ **7.** $y = -\frac{4}{3}x + 4$

Pages 55–57 Exercises and Applications

1. 3; 7 **3.** 1; 0 **5.** $y = -2$ **7.** $y = \frac{1}{2}x - 1.5$

9. Let x = the number of gal of water in the pitcher and y = the total weight in pounds of the pitcher and water; $y = 8.3x + 1.2$.

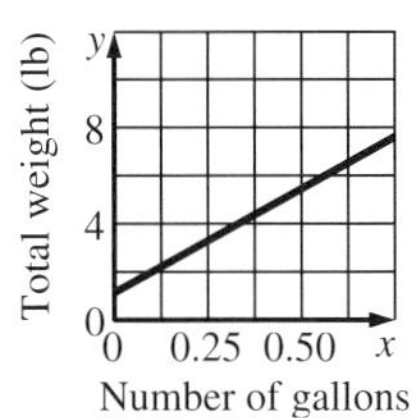

11. Let x = the number of wrong answers and y = the test score; $y = -5x + 100$.

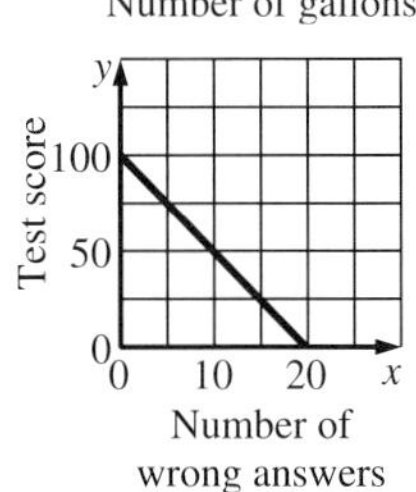

13. Let x = the number of miles driven and y = the number of gallons of gas in the tank; $y = -\frac{1}{15}x + 16$.

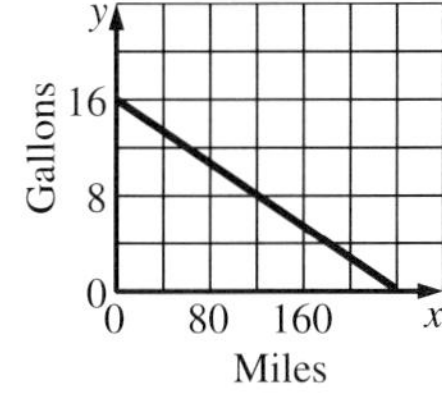

15.

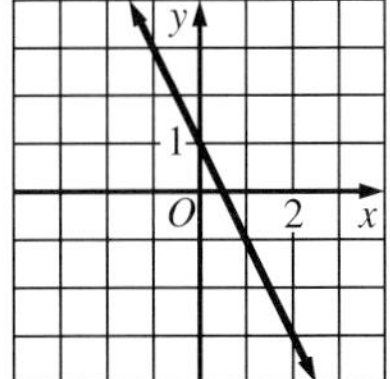

17.

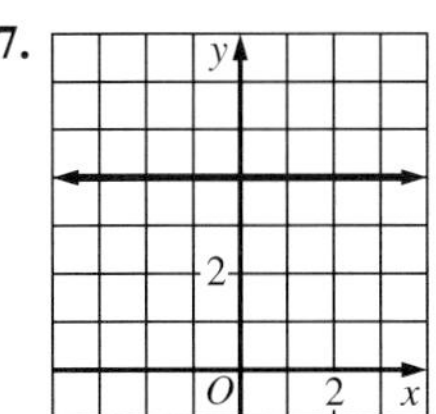

19.

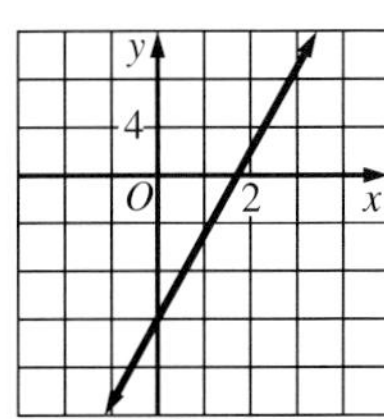

27. B **31.** A **33.** $-\frac{3}{2}$

Page 61 Checking Key Concepts

1. $y = 3 + (x - 2)$ **3.** $y = 3 - \frac{3}{2}x$ **5.** $y = 3 + (x + 2)$ or $y = 2 + (x + 3)$ **7. a.** −1 **b.** $x \geq -2$; $f(x) \geq -5$

Pages 61–64 Exercises and Applications

1. a. $y = 2 + (x - 3)$

b.

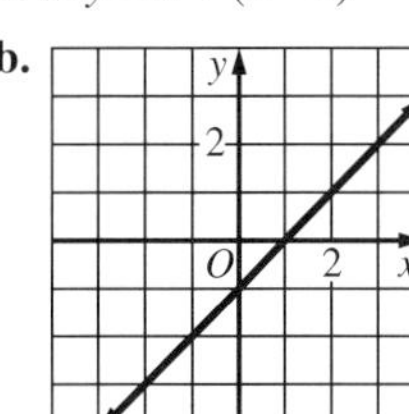

c. $y = x - 1$

3. a. $y = 1 - \frac{2}{3}(x - 3)$

b.

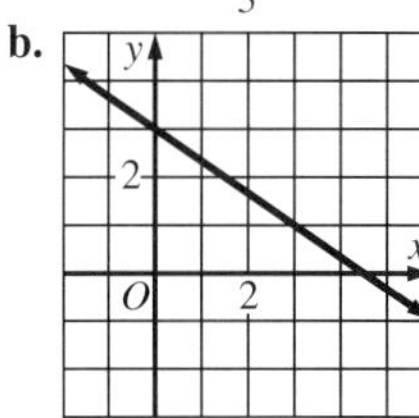

c. $y = -\frac{2}{3}x + 3$

5. a. $s = 1484 + 0.017(d - 1000)$ **b.** about 1501 m/s; about 4803 m **c.** The domain is $1000 \leq d \leq 10{,}924$ since the function describes the relation at depths below approximately 1000 m and the deepest point is 10,924 m. Using the domain, you can find that the range is $1484 \leq s \leq 1565$. (The speed at 10,924 m is 1652.708 ≈ 1653 m/s.)
11. all real numbers; all real numbers **13.** all real numbers; 2
15. a. $t = 390 - 0.5(w - 8)$ **b.** about 394 s or 6 min 34 s
17. $y = 4 + 2(x - 3)$ or $y = -4 + 2(x + 1)$
19. $y = 1 + \frac{1}{2}(x - 1)$ or $y = \frac{1}{2}(x + 1)$ **21.** $x > 0$; $f(x) > -2$
23. $x \geq 14$; $f(x) \geq 420.8$ **27.** $y = -\frac{2}{3}x - 3$ **29.** Let x = the number of years since 1993 and y = the number of families with preschoolers; $y = 11 - x$. **31.** 4 **33.** $-\frac{8}{5}$

Page 64 Assess Your Progress

1. a. For every data pair, the ratio $\frac{\text{distance}}{\text{time}} \approx 55$. Since the ratio is constant, distance varies directly with time.
b. Estimates may vary; about 9 h.

2.

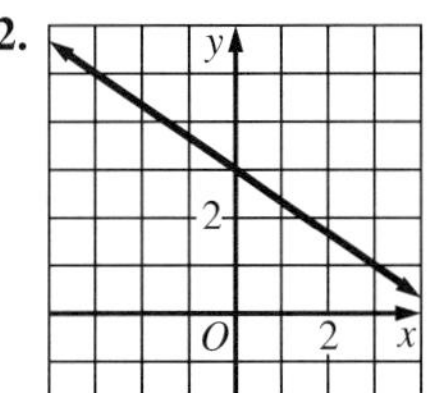

3. $f(x) \geq -1$

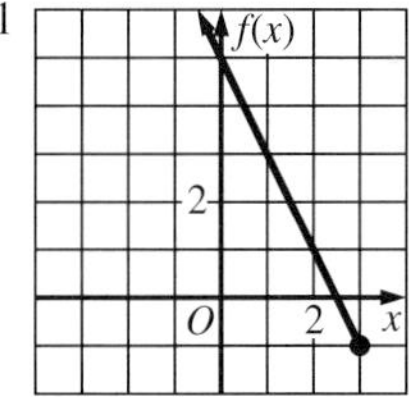

Page 68 Checking Key Concepts

1. Answers may vary. Examples are given.

a.

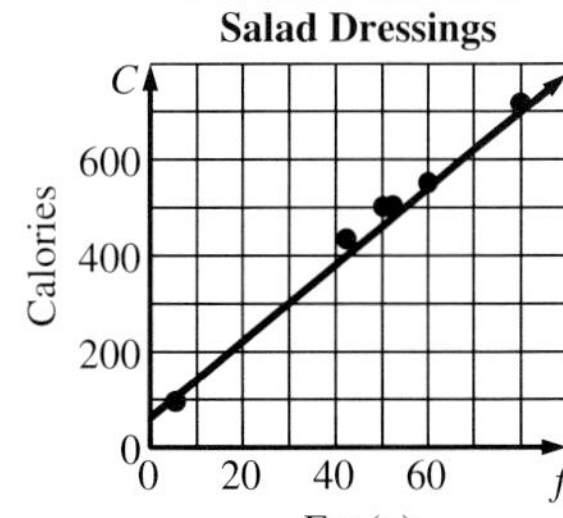

b. $C = 8f + 62$
c. about 302 g

Pages 68–71 **Exercises and Applications**

1. Answers may vary. Examples are given.

a.

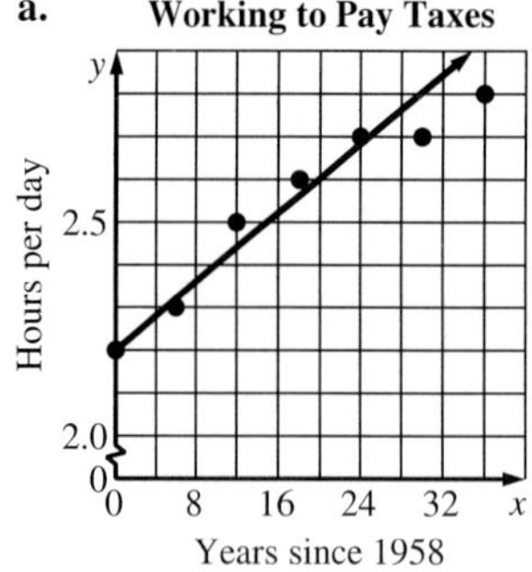

b. Let x = the number of years since 1958 and y = the number of hours worked per day to pay taxes; $y = 0.02x + 2.2$; the increase per year.
c. about $3\frac{1}{4}$ h

3. Answers may vary. Examples are given.
a.
b. Let x = the date in April, 1993 and y = the height of the river above its banks; $y = 0.45x + 0.2$.

5. a.

Year	District	m	a
1982	2	26,427	346
1982	4	15,904	282
1982	8	42,448	223
1986	2	15,671	293
1986	4	36,276	360
1986	8	36,710	306
1990	2	700	151
1990	4	11,529	–349
1990	8	26,047	160

b. Line of fit may vary.

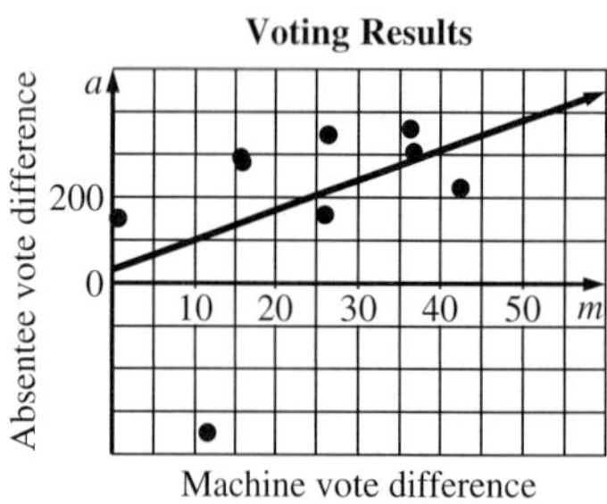

9. Answers may vary. Examples are given.
a.

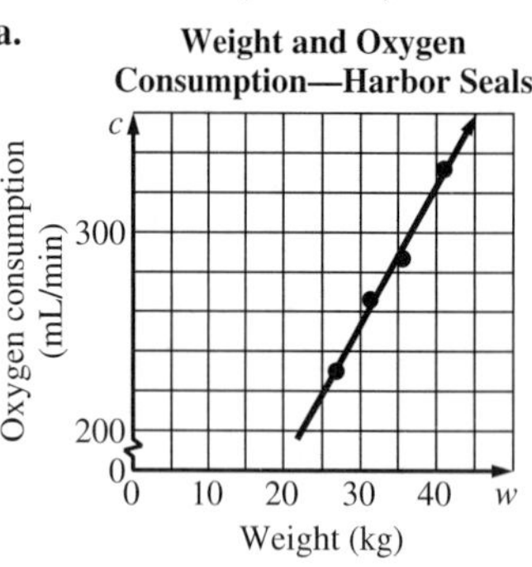

b. Let w = weight in kg and c = oxygen consumption in mL/min; $c = 7w + 43$.
c. about 218 mL/min

13. $y = 3 - 4.2(x - 7)$ **15.** $y = -3 + 0.5(x + 2)$ **17.** Yes. **19.** Yes.

Page 74 **Checking Key Concepts**

1. –0.9 **3.** –0.3

Pages 75–77 **Exercises and Applications**

1, 3. Estimates may vary. **1.** about –0.9 **3.** about 1 **5.** positive **7.** about zero

13. a.

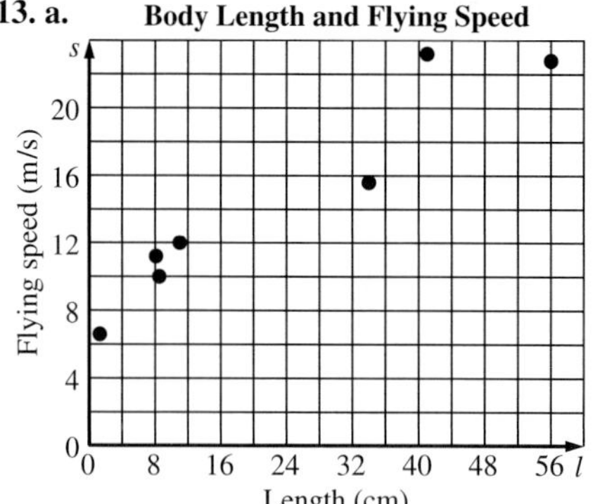

Estimates may vary; about 0.9. **b.** 0.95 **c.** 0.95; They are the same. No.

17. a.

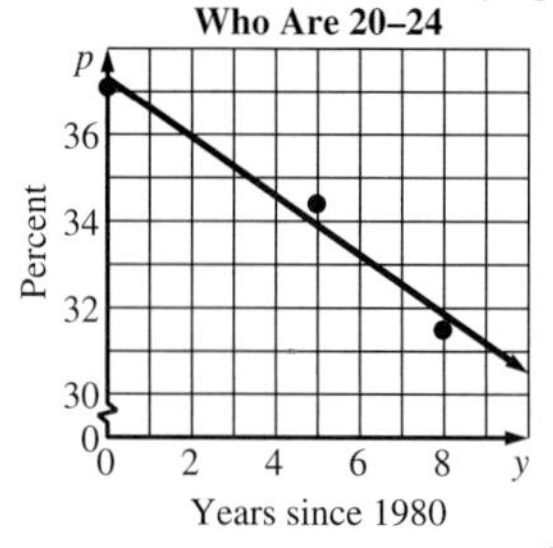

Percent of Women Marrying Who Are 25–29

Percent of Women Marrying Who Are 65 and Older

b. Answers may vary. Examples are given. Let y = the number of years since 1980 and p = the percent of women marrying who are the given age. 20–24: $p = -0.68y + 37.30$; 25–29: $p = 0.68y + 18.71$; 65 and older: $p = 1$ **c.** 20–24: about 30.5% in 1990, about 23.7% in 2000; 25–29: about 25.5% in 1990, about 32.3% in 2000; 65 and older: 1% in 1990, 1% in 2000 **19.** $y = -2$ **21.** 1 **23.** 4

Page 80 **Checking Key Concepts**

1. 3; –12 **3.** The equations are not defined for $t < 5$. **5.** $y = 2x + 1$; $0 \le x \le 6$

Pages 81–83 **Exercises and Applications**

1.

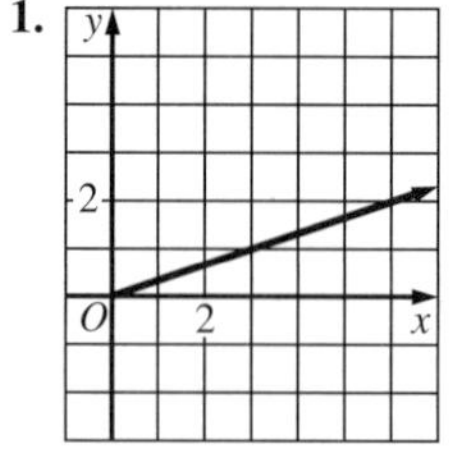

3.

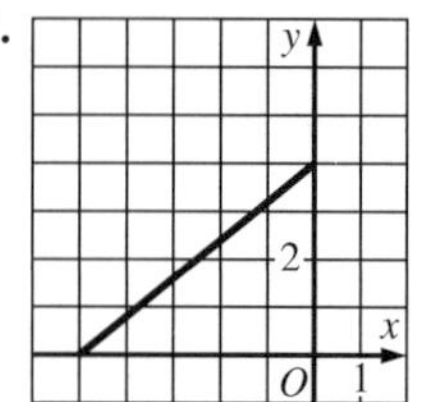

5. $y = \frac{1}{3}x$; $x \ge 0$ **7.** $y = \frac{4}{5}x + 4$; $-5 \le x \le 0$ **11. a.** 280 s after deploying the parachute **b.** $x = 12t$, $y = 3000 - 10t$, $20 \le t \le 300$ **17. a.** Let x = the number of years since 1990 and y = the number of computer books sold; $y = 560(1.15)^x$.

b. about 979 books **19.** $y = 7 - \frac{12}{5}(x + 3)$ or $y = -5 - \frac{12}{5}(x - 2)$

Page 83 Assess Your Progress

1. Lines of fit may vary.

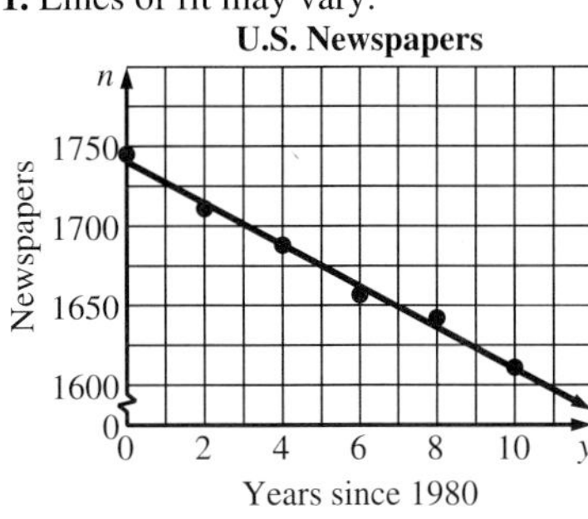

2. –0.996; strong; negative

3. a.

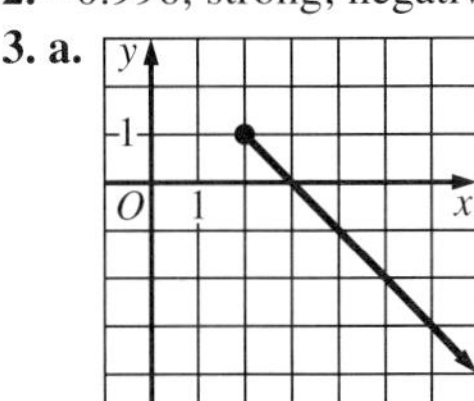

b. $y = -x + 3; x \geq 2$

Pages 88–89 Chapter 2 Assessment

1. direct variation **2.** slope-intercept; slope; y-intercept **3.** least-squares line; correlation coefficient **4.** No.

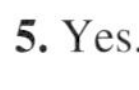

5. Yes.

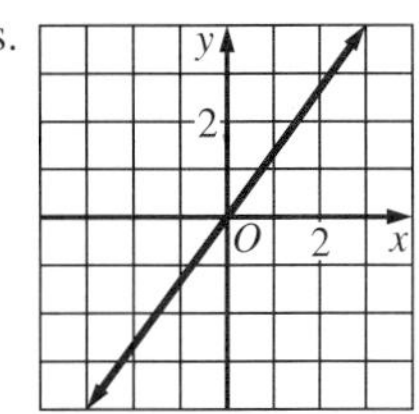

6. No. **7. a.** $p = \frac{5}{12}s$ **b.** $11\frac{1}{4}$ **8.** 6; 0 **9.** –5; 2 **10.** 0; –3

11.

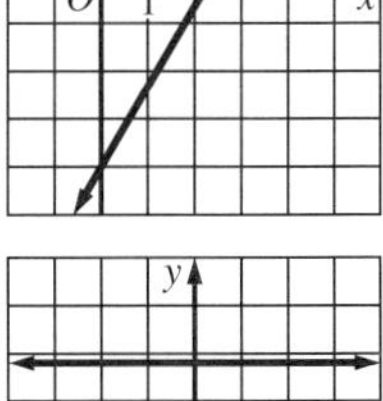

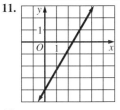

12.

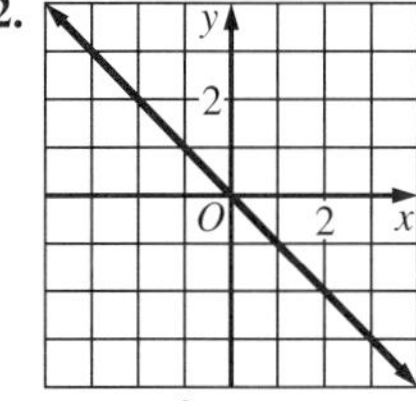

13.

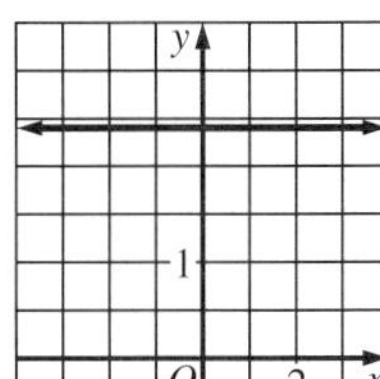

14. $y = 3 - \frac{2}{3}(x - 2)$ or $y = 1 - \frac{2}{3}(x - 5)$

15. $y = -3 + 2(x - 1)$ or $y = 1 + 2(x - 3)$

16. $y = 1 - \frac{1}{4}(x - 1)$ or $y = 2 - \frac{1}{4}(x + 3)$

17. a. $x \leq 3; y \geq -1$ **b.**

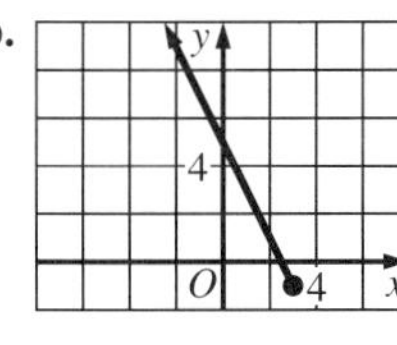

18. a.

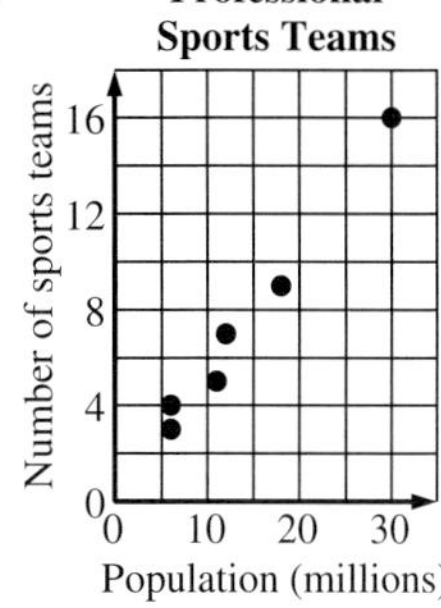

b. positive; strong **c.** 0.99 **d.** Answers may vary. The least-squares line is $y = 0.519x + 0.151$. **e.** Answers may vary. The estimate based on the least-squares line is about 4 teams; that is higher than the actual figure.

20. a.

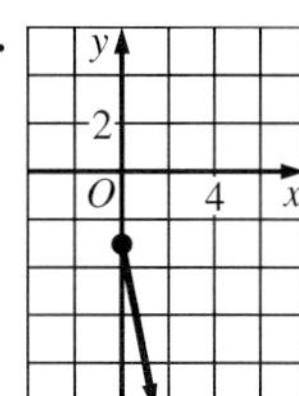

b. $y = -5x - 3; x \geq 0$

21. a.

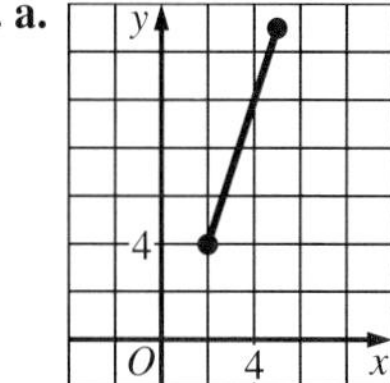

b. $y = 3x - 2; 2 \leq x \leq 5$

22. a.

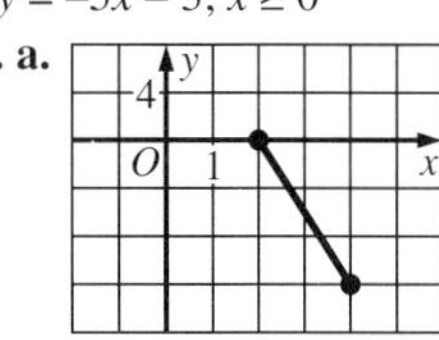

b. $y = -6x + 12; 2 \leq x \leq 4$

CHAPTER 3

Page 96 Checking Key Concepts

1. 128 leaves **3.** 512 leaves

5. a.

Number of folds	Page width (cm)	Page height (cm)	Page area (cm^2)
0	38	51	1938
1	25.5	38	969
2	19	25.5	484.5
3	12.75	19	242.25
4	9.5	12.75	121.125
5	6.375	9.5	60.5625
6	4.75	6.375	30.28125

b. Let n = the number of folds and A = the page area in square centimeters; $A = 1938\left(\frac{1}{2}\right)^n$.

Pages 97–99 Exercises and Applications

1. 4 folds **3.** 3 folds **5.** 2^4 **7.** 2^6 **9.** 160 **11.** 200 **13.** linear **15.** neither **29.** $y = x + 1$ **31.** Yes. **33.** No.

Page 104 Checking Key Concepts

1, 3. Answers are rounded to two decimal places. **1.** 5.31; the predicted population in millions in 1800 **3.** about 9.53; the predicted population in millions in 1820 **5.** The growth rate is an estimate. Also, the proportional growth rates in Column D of the spreadsheet are not all equal, only approximately equal. This indicates that the exponential equation models the data only approximately. **7.** 4 **9.** 16 **11.** about 4.64

Pages 104–106 Exercises and Applications

1. The growth is exponential, not linear. **3, 5, 7.** Answers are rounded to two decimal places. **3.** 22.94 million **5.** 7.57 million **7.** 447.38 million **11.** $\frac{1}{16}$ **13.** $\frac{1}{7}$ **15.** 36 **17.** 1 **19.** 16 **21.** 2 **23.** 5 **25.** 1 **27.** Let d = the number of decades after 1959 and v = the number of vehicles per day in thousands; $v = 75(1.36)^d$. **29.** Answer may vary due to rounding. about 119,000 vehicles per day **35. a.** Answers may vary depending on rounding and how the independent variable is defined. An example is given. Let d = the number of degrees above zero and a = the air flow per gallon of water in ft^3/min; $a = 3.0(1.06)^d$. **b.** Answers may vary due to rounding; about 4.0 ft^3/min; about 7.2 ft^3/min; about 12.9 ft^3/min **43.** 0.2 **45.** The graphs have the same slope but different y-intercepts. The graph of $y = 6 + 2x$ is the graph of $y = 3 + 2x$ shifted 3 units in the positive y-direction. **47.** The equation is simply a point-slope equation of the same line. The graphs are identical.

Page 110 Checking Key Concepts

1. exponential growth **3.** exponential growth **5.** 3 **7.** –1 **9.** –2.5 **11.** 4 (half-life)

Pages 110–114 Exercises and Applications

1. a.

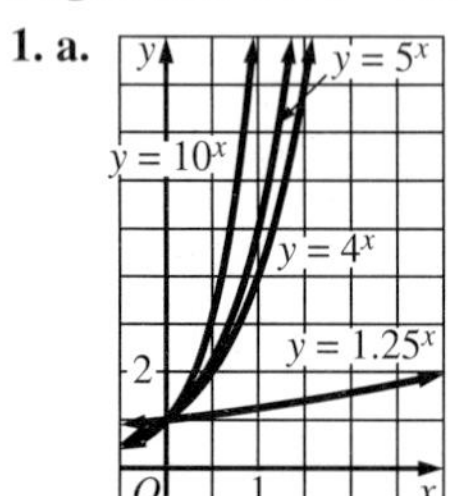

b. $y = 10^x$; $y = (1.25)^x$
c. The y-intercept of each of the graphs is 1.

3. a.

b. $y = -10^x$; $y = -5^x$; $y = -4^x$; $y = -(1.25)^x$

5. a. 1
b. exponential growth
c.

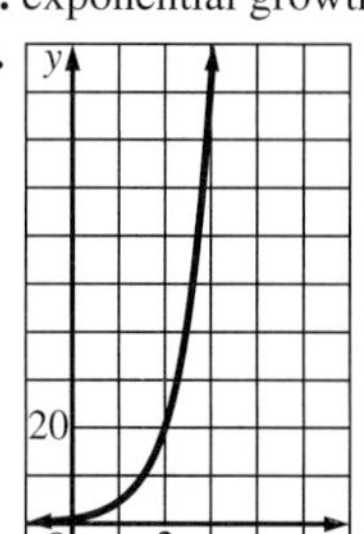

7. a. 2.5
b. exponential decay
c.

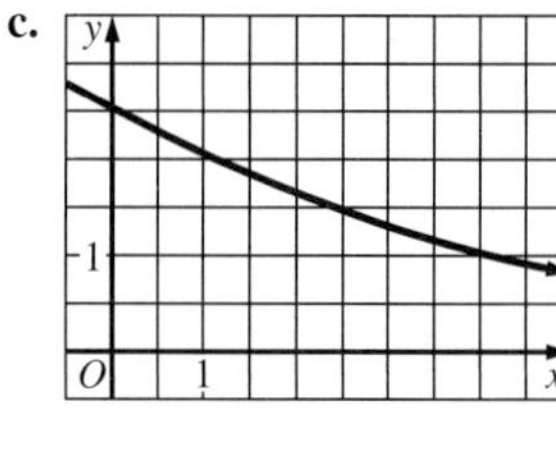

9. a. For $b = 1$, the graph of $y = ab$ is a horizontal line through $(0, a)$. **b.** For $a = 0$, the graph of $y = ab^x$ is the x-axis. **11.** C **13.** A **17.** 12 years **19.** 8 years **21.** 4 years **25, 27.** Answers are given to two decimal places. **25.** 0.74 **27.** 0.30 **29, 31.** Estimates may vary. **29.** about 3700 years ago **31.** about 10.4% **35.** about 10.6 million **37.** Estimates may vary. An example is given. about 35 years **39.** $\frac{1}{9}$

41. a. Lines of fit may vary.
b. Answers may vary. An example is given. Let P = percent of water and C = the number of calories; $C = -4.4P + 435$.
c. Estimates may vary; about 31 calories.

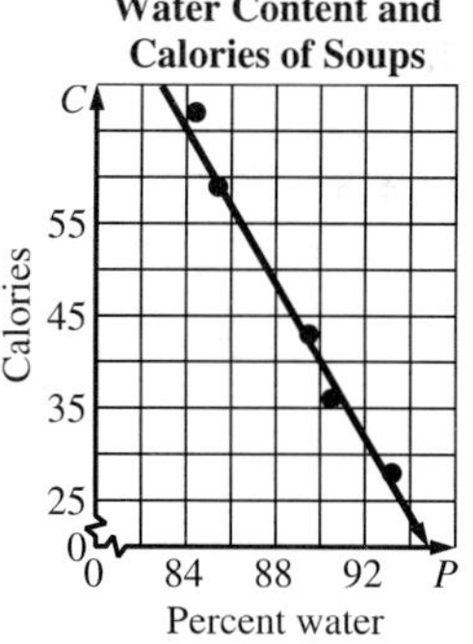

Page 114 Assess Your Progress

1. 16 **2.** $\frac{1}{16} = 0.0625$ **3.** 24 **4.** Let T = the number of teams; $T = 64\left(\frac{1}{2}\right)^n$. **5.** $2^{3/4} \approx 1.68$ **6.** 7 **7.** 2 **8.** about 267 accidents **9.** about 241 accidents **10.** about 327 accidents **11.** Estimates may vary. An example is given. in about $13\frac{1}{2}$ years

Page 119 Checking Key Concepts

1. \$1032.52; the annual interest rate that would yield the same amount after one year **3. a.** about 0.2501 g **b.** about 0.1251 g **c.** about 0.0625 g **5. a.** 100 **b.** When t is very large, $e^{-0.1t} \approx 0$, so $L(t) \approx 2000$. **c.** Estimates may vary. An example is given. about 29.4 **d.** none

Pages 119–122 Exercises and Applications

1, 3. Answers are given to two decimal places. **1.** 2.53% **3.** 7.79% **5.** \$1454.99 **7.** \$2585.71 **9.** 2 **11.** 2.704814 **13.** 2.718146 **15.** 2.718280 **17. a.** Graphs may vary.
b. Answers may vary. See table in part (c).
c. Estimates may vary.

Doubling Your Money in a Year

Compounding	n	Rate
annually	1	100%
semiannually	2	82.8%
quarterly	4	75.7%
monthly	12	71.4%
daily	365	69.4%

As the frequency of compounding increases, the interest rate needed to double your money drops. However, the rate is very high even for daily compounding. It would be extremely unusual to find an investment that offered an annual interest rate of 69.4%. **e.** Estimates may vary; about 69.3%. **19.** Estimates and equations may vary depending on rounding. An example is given. about 5.8 years; $E(t) = 179.2\left(\frac{1}{2}\right)^{t/5.8}$ **27.** $A(t) = 100(0.995)^t$ **33.** $a > 0$ and $b > 1$ **35.** Let y = the number of years since 1985 and P = public funds in billions of dollars; $P = 0.12y + 1.32$

37. $\begin{bmatrix} 39 & -8 & -10 \\ 96 & 48 & -40 \end{bmatrix}$

Page 125 **Checking Key Concepts**

1. $y = 5\left(\frac{3}{5}\right)^x$ **3.** $y = 3 \cdot 4^x$

Pages 126–129 **Exercises and Applications**

1. $y = 5\left(\frac{8}{5}\right)^x$ **3.** $y = 10\left(\frac{4}{5}\right)^x$ **5.** $y = 1.804(1.185)^x$

7. $y = 12.642(0.944)^x$

17. Let x = the number of degrees Celsius above zero and y = the oxygen absorbed in mL/L; $y = 10.015(0.98)^x$.

21. a. Let t be the number of hours after 1:00 P.M. and $P(t)$ = the number of bacteria; $P(t) = 30e^{0.116t}$. **b.** about 48 **c.** about 27 **d.** Substitute 2.75 for t in the equation.

23. a.

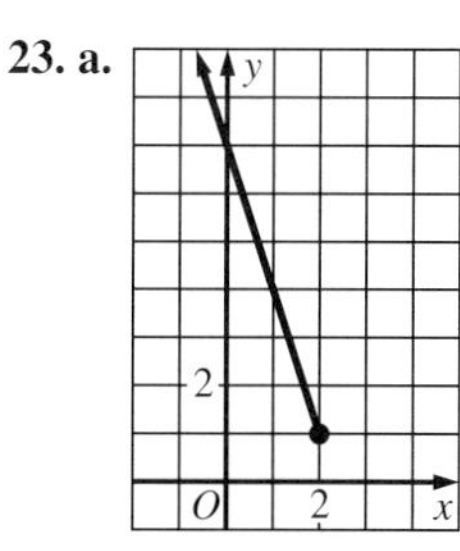

b. $y = -3x + 7;\ x \le 2$

Page 129 **Assess Your Progress**

1. \$2385.04 **2.** \$2394.25 **3.** \$2394.43

4.

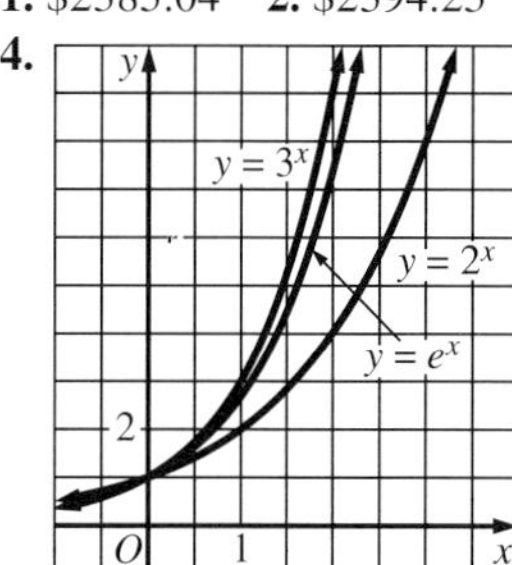

Summaries may vary. All three graphs have the same y-intercept, 1. All three are exponential, with equations of the form $y = b^x$. The greater the value of b, the steeper the graph.

5. a. Answers may vary. **b.** $y = 4.975(1.046)^x$ **c.** about 181.7 cm

Pages 134–135 **Chapter 3 Assessment**

1. exponential growth; exponential decay **2.** e **3.** 2 **4.** $\frac{1}{6\sqrt{6}}$ or $\frac{\sqrt{6}}{36}$ **5.** $\frac{1}{49}$ **6.** 1 **7. a.** $E = 179.2(0.89)^{w/52}$ **b.** about 142 eggs per year **8. a.** $s = 30{,}000(1.05)^n$ **b.** 14.2 years **9. a.** $y = 10\left(\frac{1}{2}\right)^{x/11.2}$ **b.** about 8.84 g **10. a.** \$3136.62 **b.** \$4738.15 **c.** \$11,225.04 **11. a.** $E(t) = 50(0.970)^t$ **b.** about 27 errors **13.** $y = 4.43(1.17)^x$ **14.** $y = 7.40(1.04)^x$ **15.** $y = 0.943(2.12)^x$ **16. a.** $y = 0.766(2.17)^x$ **b.** \$36.9 trillion **17. a.** Let x = the number of years since 1990 and y = the CPI; $y = 130.7(1.04)^x$. **b.** about 120.8; about 141.4

CHAPTERS 1–3

Pages 136–137 **Cumulative Assessment**

1. Let y = the winning time in seconds; $y = 44.6 - 0.367x$; $y = 44.6(0.992)^x$. **5.** undefined **7.** $\begin{bmatrix} 11 & -7 & 0 \\ 0 & -6 & -2 \end{bmatrix}$ **9.** No.

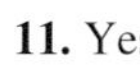

11. Yes.

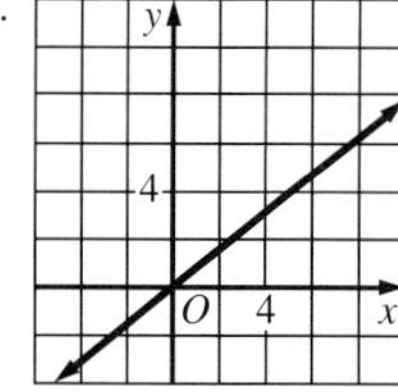

13. $y = \frac{4}{3}x + \frac{5}{3}$

15. $y = \frac{1}{2}x + \frac{11}{2}$

17.

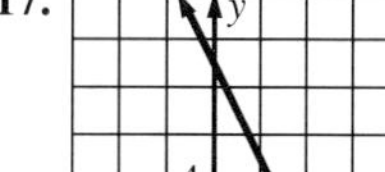

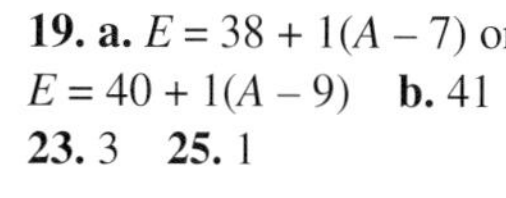

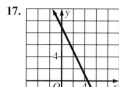

19. a. $E = 38 + 1(A - 7)$ or $E = 40 + 1(A - 9)$ **b.** 41

23. 3 **25.** 1

27. a.

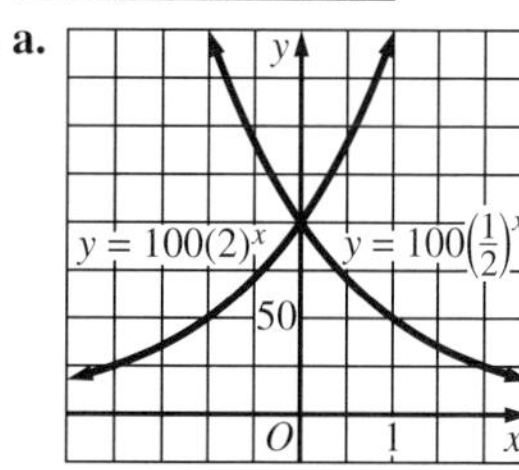

29. a. about 17.7 g

b. about 20.0 s; $A(t) = 50\left(\frac{1}{2}\right)^{t/20.0}$

CHAPTER 4

Page 145 **Checking Key Concepts**

1. 2 **3.** No.

Pages 145–147 **Exercises and Applications**

1. a.

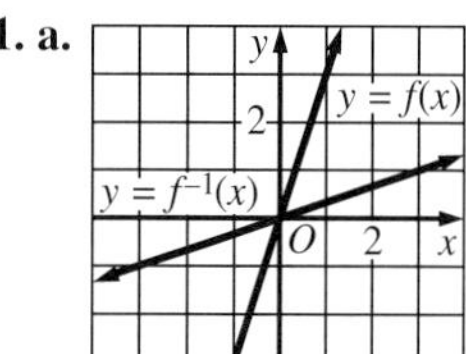

b. $f^{-1}(x) = \frac{1}{3}x$

3. a.

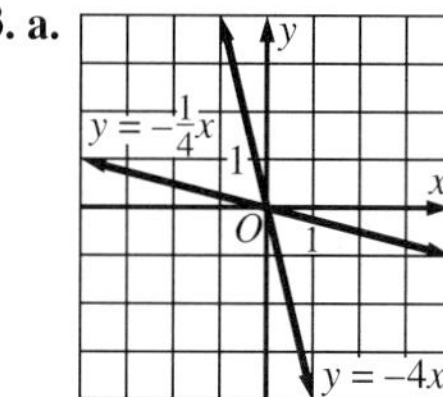

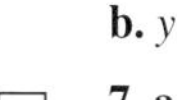

b. $y = -4x$

5. a.

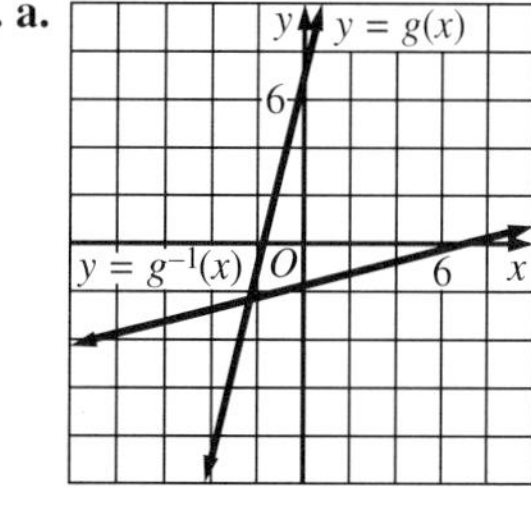

b. $g^{-1}(x) = \frac{1}{4}x - \frac{7}{4}$

7. a.

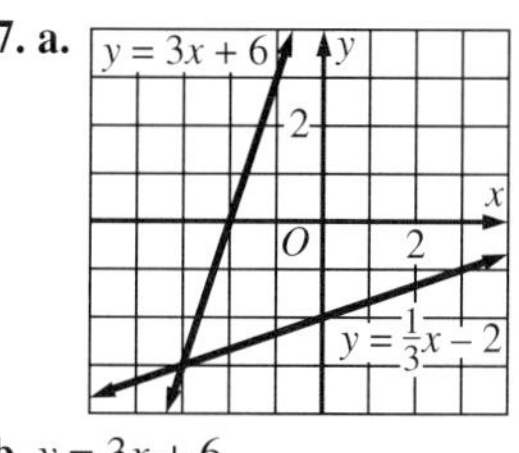

b. $y = 3x + 6$

9. E **11.** For C, the domain is the nonnegative real numbers and the range is the real numbers greater than or equal to 84,000. For S, the domain is the real numbers greater than or equal to 84,000 and the range is the nonnegative real numbers. The domain of C is the range of S and the range of C is the domain of S. **17. a.** $y = 1.849x - 28.60$; average fuel economy; number of years since 1980 **b.** 2007 **19. a.** $y = 50{,}100(1.26)^x$ **b.** about 318,000 m^3 **c.** about 25,000 m^3 **21.** $y = 4 - 2(x + 3)$ or $y = -2 - 2x$ **23.** \$563.75

Page 152 **Checking Key Concepts**

1. $f^{-1}(x) = \log_2 x$ **3.** $y = \ln x$ **5.** $\log_8 1 = 0$ **7.** $6^3 = 216$ **9.** $27^{1/3} = 3$ **11.** $\frac{2}{3}$ **13.** 2.16 **15.** −0.49

Pages 152–154 Exercises and Applications

1. a. $x > 1$ **b.** $0 < x < 1$ **c.** 1 **d.** $x \le 0$ **3.** $\log_2 8 = 3$ **5.** $\ln 1 = 0$ **7.** $\log_{64} \frac{1}{4} = -\frac{1}{3}$ **9.** $\log_{4/5} \frac{64}{125} = 3$ **11.** $2^4 = 16$ **13.** $2^{-5} = \frac{1}{32}$ **15.** $\left(\frac{1}{6}\right)^2 = \frac{1}{36}$ **17.** $64^{2/3} = 16$ **19.** 2 **21.** –3 **23.** 0 **25.** $\frac{3}{4}$ **31.** 1.08 **33.** 1.87

35. a.

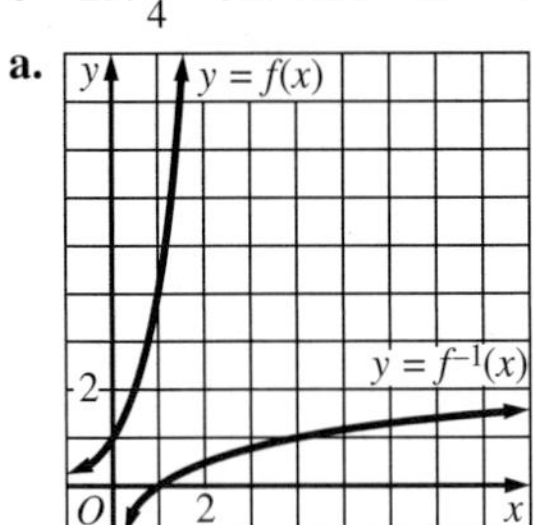

b. $f^{-1}(x) = \log_4 x$

37. a.

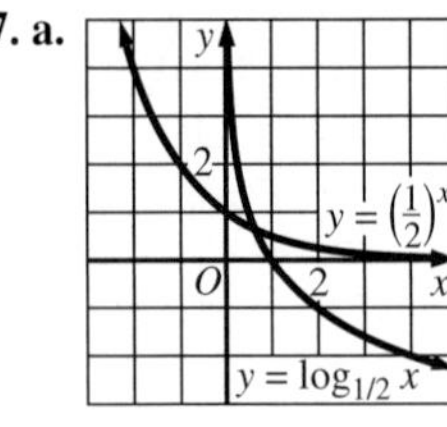

b. $y = \log_{1/2} x$

39. a. $t = 40 \ln \frac{P}{19.1}$; the number of years after 1990 that the population reaches a given population P **b.** 2008 **c.** 1980

45. a.

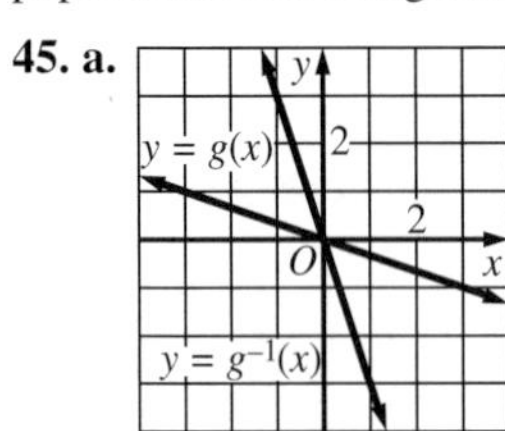

b. $g^{-1}(x) = -3x$

47. a.

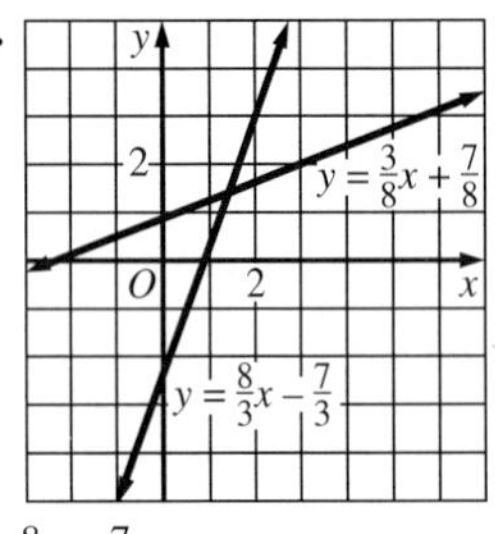

b. $y = \frac{8}{3}x - \frac{7}{3}$

49. neither **51.** exponential

Page 158 Checking Key Concepts

1. $3 \log_2 p$ **3.** $2 \log_2 p - 4 \log_2 r$ **5.** $\log_a 32$ **7.** $\log_a \frac{x^9}{y^2}$

Pages 158–161 Exercises and Applications

1. $4 \log_7 p$ **3.** $2 \log_7 q + 3 \log_7 r$ **5.** $-\log_7 q$ **7.** $9 \log_7 p + 11 \log_7 q - 7 \log_7 r$ **11.** $\log_a \frac{1}{7}$ **13.** $\log_5 \frac{u^6}{v^{10}}$ **15.** $\log x^{32}$ **17.** $\log_b 9$ **23.** $y - x$ **25.** $2x + y$ **27. a.** about 26% **29.** A **31.** substitution; the product property of exponents; definition of logarithm; substitution **35.** $\log_4 16 = 2$ **37.** 9.16

Page 161 Assess Your Progress

1. a.

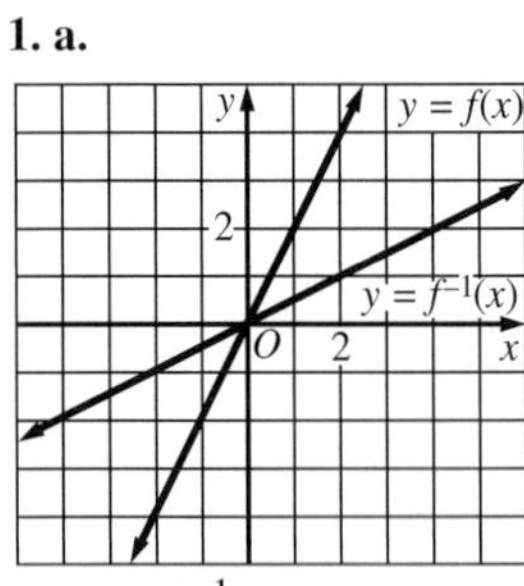

b. $f^{-1}(x) = \frac{1}{2}x$

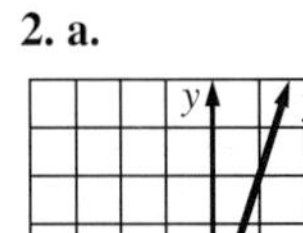

2. a.

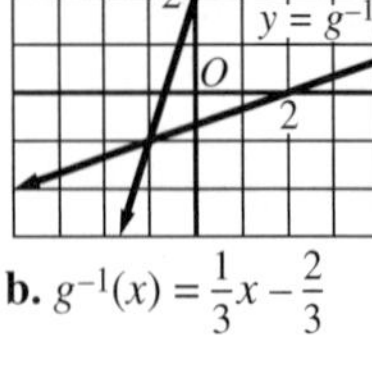

b. $g^{-1}(x) = \frac{1}{3}x - \frac{2}{3}$

3. a.

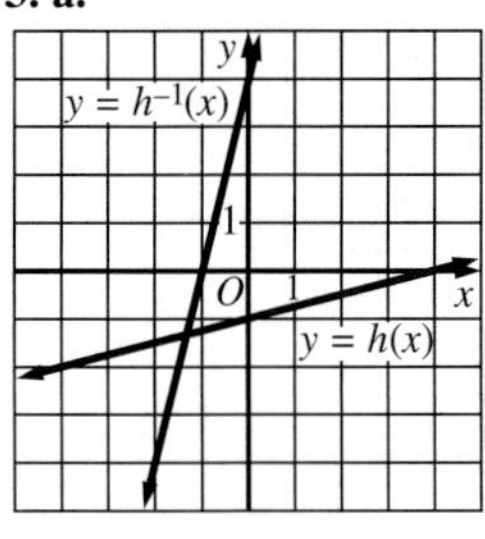

b. $h^{-1}(x) = 4x + 4$

4. a.

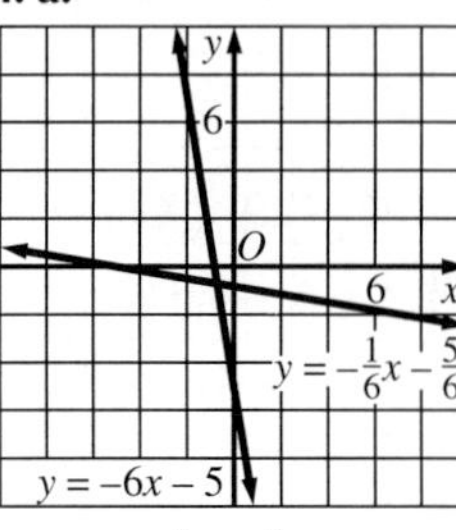

b. $y = -\frac{1}{6}x - \frac{5}{6}$

5. 2 **6.** $\frac{1}{2}$ **7.** –4 **8.** $\frac{5}{11}$ **9. a.** $t = 91.7 \ln \frac{P}{33.9}$ **b.** 1997 **10.** $3 \log p + 4 \log q - 5 \log r$ **11.** $\log_b 8$

Page 165 Checking Key Concepts

1. 1.10 **3.** 0.86 **5.** 2.32 **7.** 1.28 **9.** 32 **11.** 8

Pages 165–167 Exercises and Applications

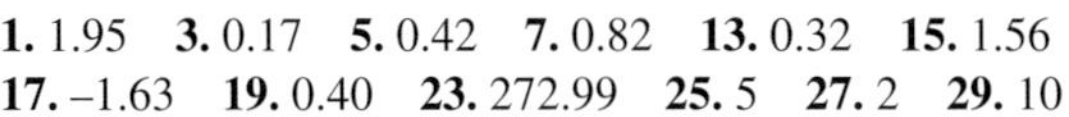

1. 1.95 **3.** 0.17 **5.** 0.42 **7.** 0.82 **13.** 0.32 **15.** 1.56 **17.** –1.63 **19.** 0.40 **23.** 272.99 **25.** 5 **27.** 2 **29.** 10

31. a.

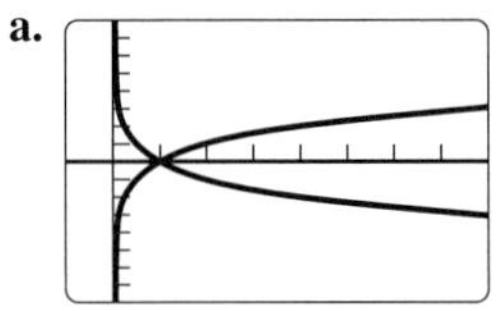

b.

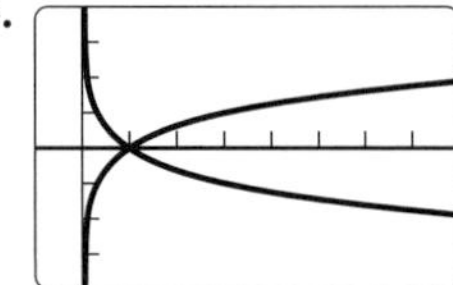

c.

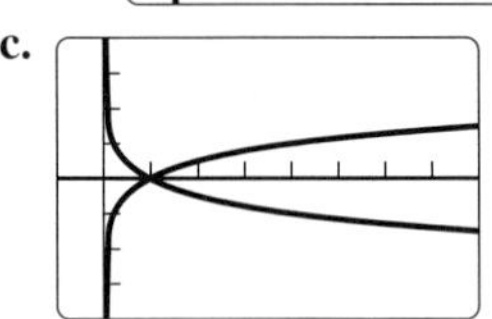

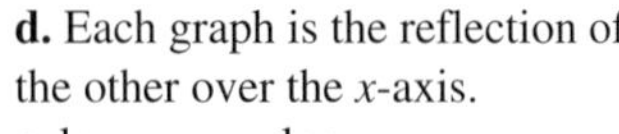

d. Each graph is the reflection of the other over the x-axis. **e.** $\log_{1/b} x = -\log_b x$

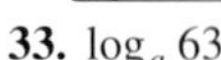

33. $\log_a 63$

35.

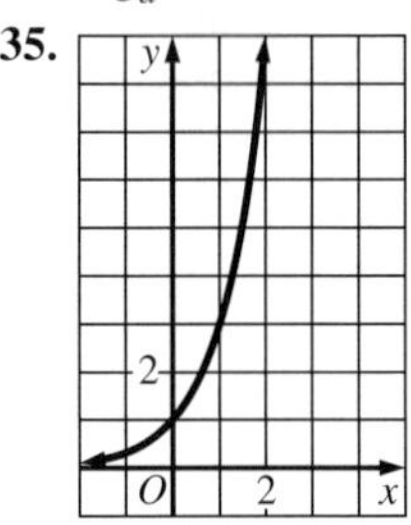

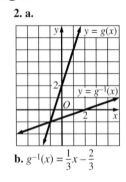

37.

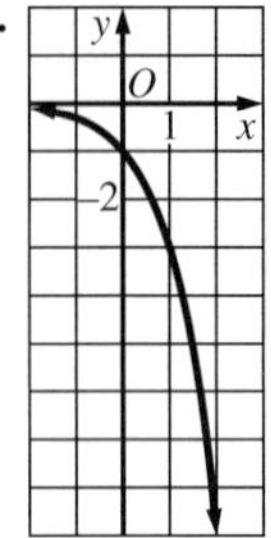

Page 171 Checking Key Concepts

1. a. **b.**

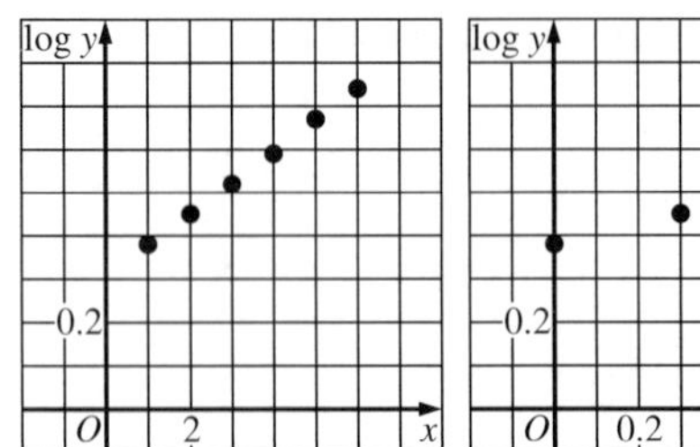

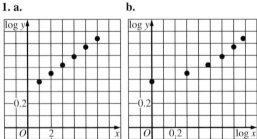

c. exponential function; $y = 2.01(1.18)^x$

3. a.

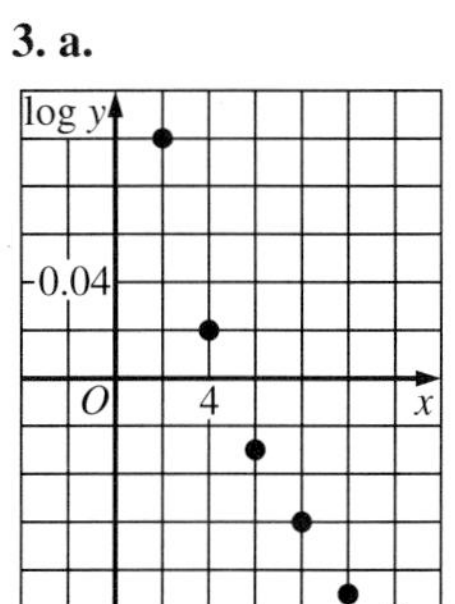

b.

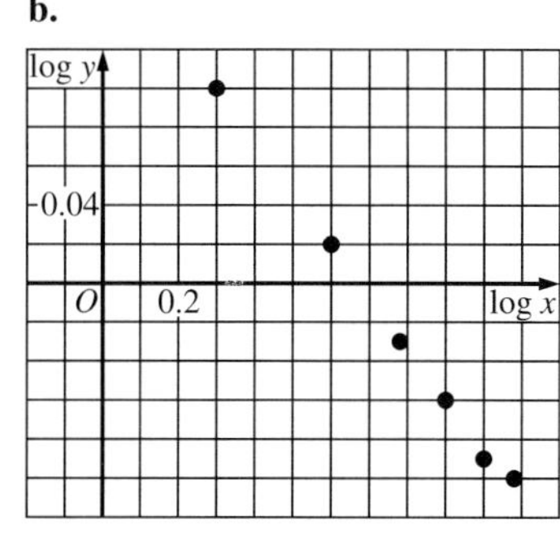

c. power function; $y = 1.50x^{-0.260}$

Pages 171–175 Exercises and Applications

3. $y = 6310(1.58)^x$ **5.** $y = 0.811(0.912)^x$ **7.** $y = 2.29(0.0380)^x$ **11.** $y = 219x^{0.720}$ **13.** $y = 0.914x^{-0.61}$ **15.** $y = 4.39x^{-1}$

17. a.

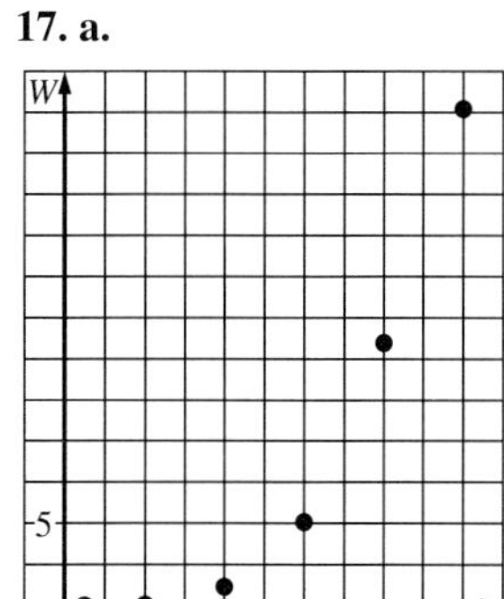

b.

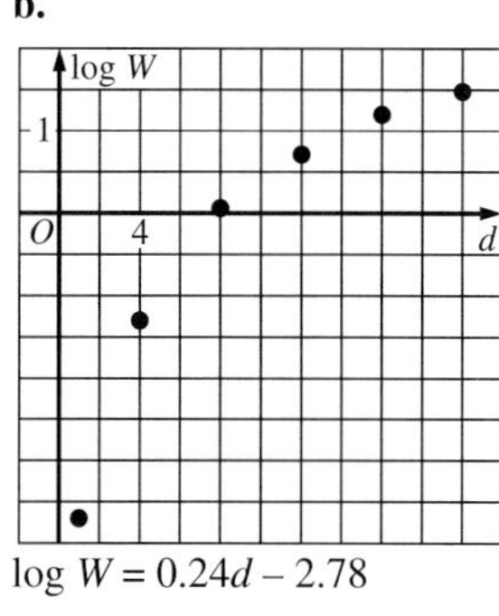

$\log W = 0.24d - 2.78$

No; power functions and exponential functions have similarly shaped graphs. **21.** D **23.** Parksosaurus: about 176 kg; Struthiomimus: about 391 kg; Allosaurus: about 5740 kg; Tyrannosaurus: about 19,500 kg **27.** 0.94 **29.** –0.85 **31.** 9; 4 **33.** –36; –16

Page 175 Assess Your Progress

1. 1.79 **2.** 2.26 **3.** 8.60 **4.** 64 **5.** no solution **6. a.** $W = 4.83(1.14)^t$ **b.** $W = 5.19t^{0.365}$ **c.** The exponential function is the better model since the correlation coefficient for the exponential function, 0.9963581796, is closer to 1 than the correlation coefficient for the power function, 0.9550944564.

Pages 180–181 Chapter 4 Assessment

1. inverse; reflection **2.** common; natural; base-5; change-of-base

3. a.

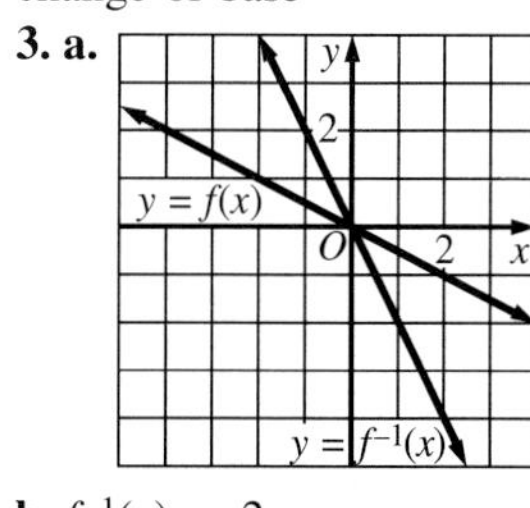

b. $f^{-1}(x) = -2x$

4. a.

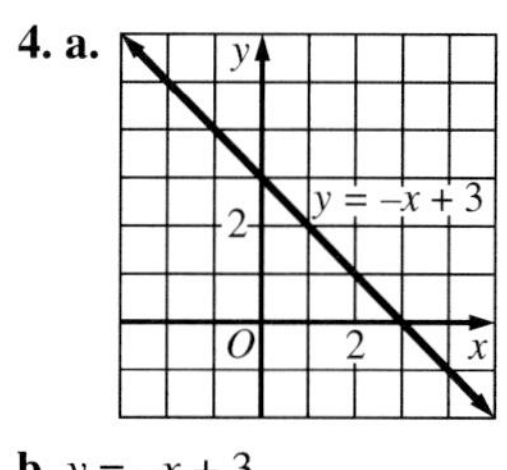

b. $y = -x + 3$

5. a.

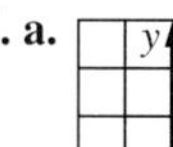

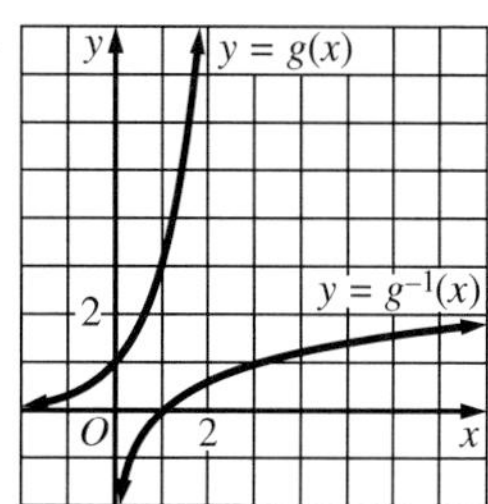

b. $g^{-1}(x) = 2 \log x$

6. a.

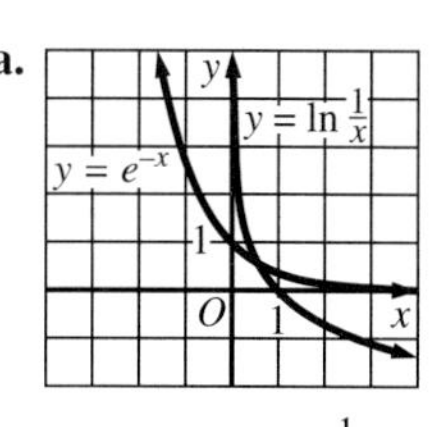

b. $y = -\ln x$ or $y = \ln \frac{1}{x}$

7. a. The equation for the function and the equation for its inverse are the same. **8. a.** Let t = time in hours after the candle is lit and h = the height of the candle in inches; $h = 10 - 0.25t$ **b.** $t = 40 - 4h$ **c.** 16 h after the candle is lit **9.** $\log_2 64 = 6$ **10.** $\ln \frac{1}{e} = -1$ **11.** $\log_{0.5} 8 = -3$ **12.** $4^{0.5} = 2$ **13.** $e^0 = 1$ **14.** $\left(\frac{1}{9}\right)^{-3/2} = 27$ **15.** $\frac{1}{3}$ **16.** –2 **17.** 4 **18.** –2 **19.** If $0 < x < 1$, then $\log x < 0$. Raising 10 to any positive power produces a number greater than 1. **20.** $3^2 = 9$ and $3^3 = 27$; Since $9 < 15 < 27$, $2 < \log_3 15 < 3$. **21.** $\log_5 p + 2 \log_5 q$ **22.** $1 - \log_5 r$ **23.** $\frac{1}{4} \log_5 r - 3 \log_5 p - \log_5 q$ **24.** ln 2 **25.** $\log_a u^3v^2$ **26.** $\log_b 25$ **28.** 4.52 **29.** 0.59 **30.** –0.22 **31.** 61 **32.** 1.70 **33.** –0.61 **34.** –0.32 **35.** Answers may vary. Examples are given. In the examples, x = years since 1980 and y = price in thousands.

a.

Type of function	Equation	Correlation coefficient
exponential	$y = 59.8(1.05)^x$	0.9965490038
power	$y = 42.8x^{0.353}$	0.9986221616

b. The power function is the better model since the correlation coefficient for the power function, 0.9986221616, is closer to 1 than the correlation coefficient for the exponential function, 0.9965490038. **c.** about \$103,000; The prediction is very close to the actual median price.

CHAPTER 5

Page 188 Checking Key Concepts

1.

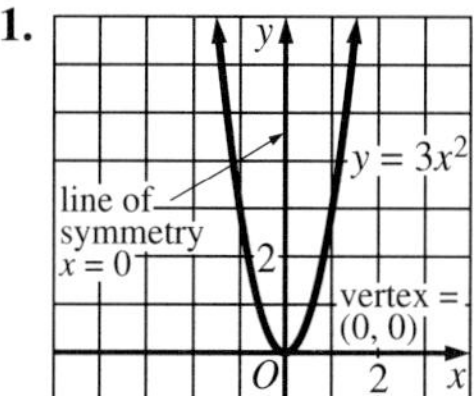

3. ±5 **5.** ±3.5 **7.** ±15.8

Pages 188–191 Exercises and Applications

1. B **3.** C **7.** ±3.5 **9.** ±0.5 **11.** ±9 **13.** ±2.6 **15.** 5 **17. a.** $r^2 = \frac{V}{\pi h} = \frac{75}{\pi}$ **b.** about 4.9 in. **25. a.** 9.25×10^{-10} **b.** about 13,000,000 mi **29.** $y = 20.0x^2$ **31.** They all have the same slope, 3. Each graph has a different vertical intercept.

Page 196 **Checking Key Concepts**

1. $y = -2(x - 1)^2 - 2$ **3.** down; maximum value: -2
5. $-3.7, -0.3$

Pages 196–198 **Exercises and Applications**

1. B **3.** A **9.** 1 (maximum) **11.** -4 (minimum)
13. -2 (minimum)

15. The vertex is $(-3, -3)$. The line of symmetry is $x = -3$. The parabola opens down.

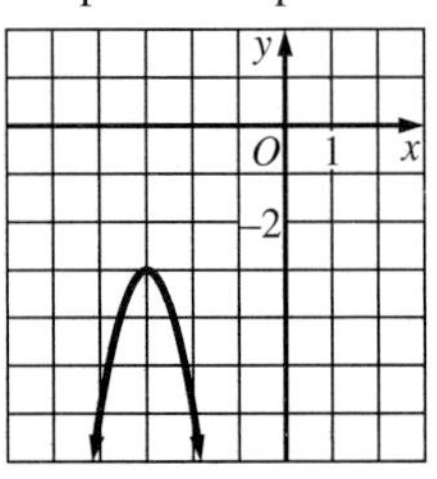

17. The vertex is $(-2, 0)$. The line of symmetry is $x = -2$. The parabola opens down.

19. The vertex is $\left(-4, \frac{1}{2}\right)$. The line of symmetry is $x = -4$. The parabola opens down.

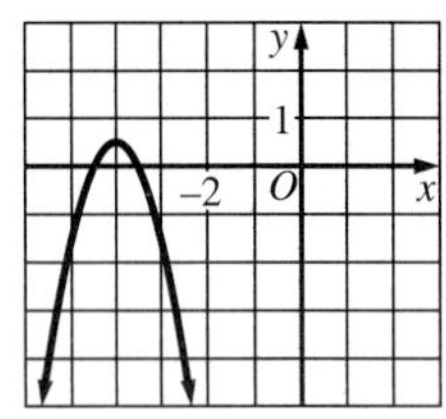

21. $y = 2(x + 2)^2 + 4$ **23.** $y = -3(x - 5)^2 + 2$ **29.** ± 3.2
31. $3(x + 2)$ **33.** $x(5x + 2)$ **35.** $g^{-1}(x) = \frac{1}{3}x - 5$

Page 202 **Checking Key Concepts**

1. a. -7; 6 **b.** $\left(-\frac{1}{2}, -253.5\right)$ **c.**

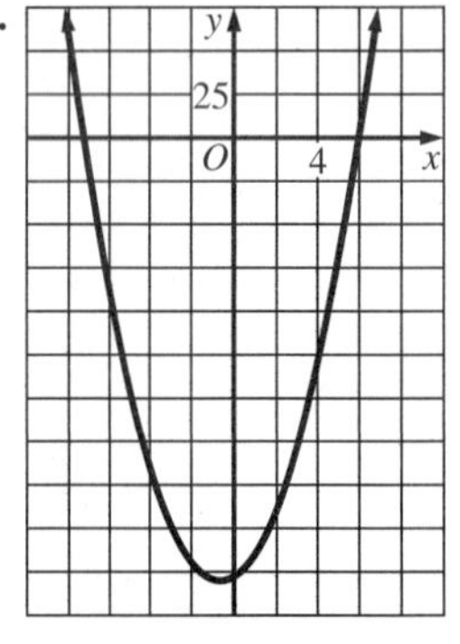

3. 72

Pages 203–205 **Exercises and Applications**

1. a. 1; -3
b. $(-1, -16)$
c.

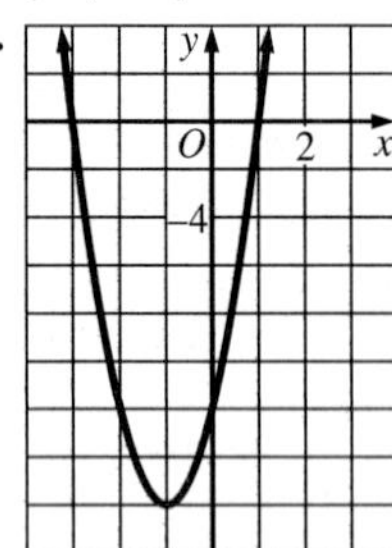

3. a. 4; -4
b. $(0, 16)$
c.

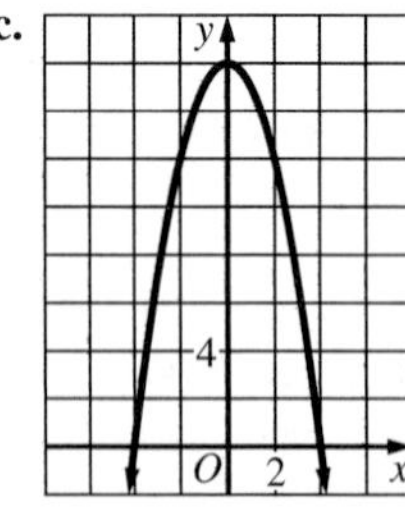

5. a. 6; -4 **b.** $(1, 125)$
c.

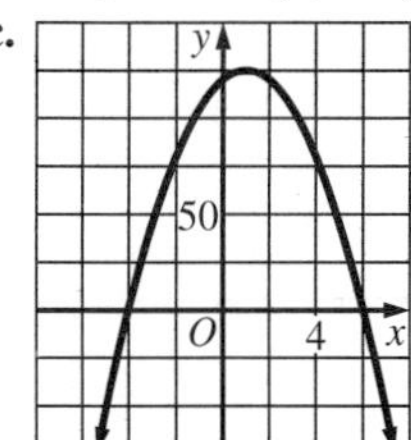

9. $y = -6(x - 4)(x - 6)$
17. D **23.** ± 5
25. -1.9; 0.9 **27.** $\log_2 x^6$

Page 205 **Assess Your Progress**

1–3. a.

1. b. ± 3.5
2. b. ± 1.7
3. b. ± 6.9
4. $y = -3(x - 3)^2 + 7$
5. $y = \frac{1}{2}(x + 1)^2 + 4$
6. $y = 2(x - 5)^2 - 2$
7. $y = \frac{1}{8}(x + 6)^2 - 1$

8. $y = 3(x + 9)(x + 4)$

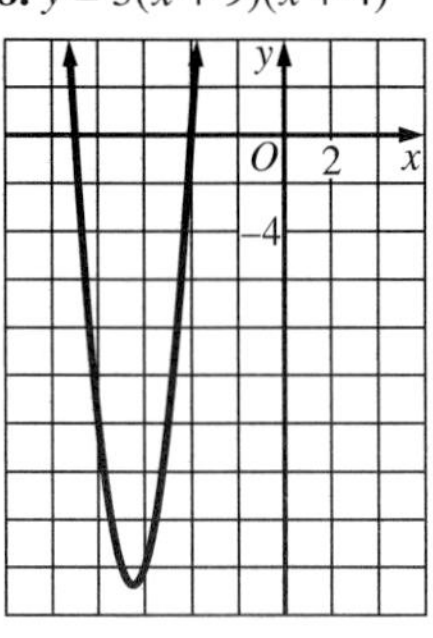

9. $y = -6(x - 3)(x - 4)$

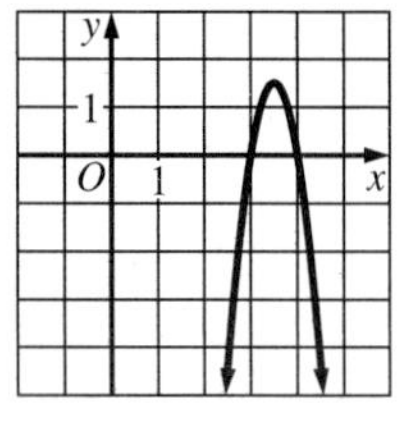

10. $y = 4(x - 3)(x + 3)$

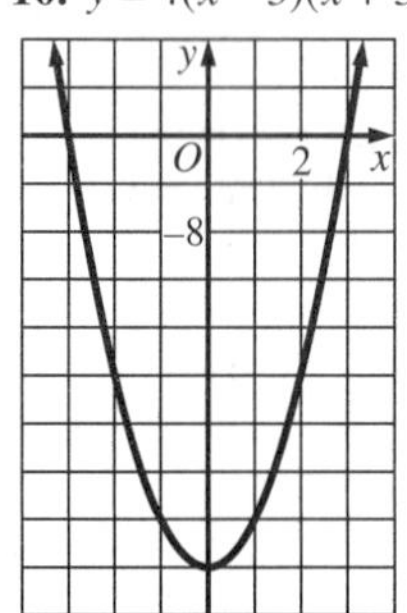

11. $y = -10(x - 2)(x + 2)$

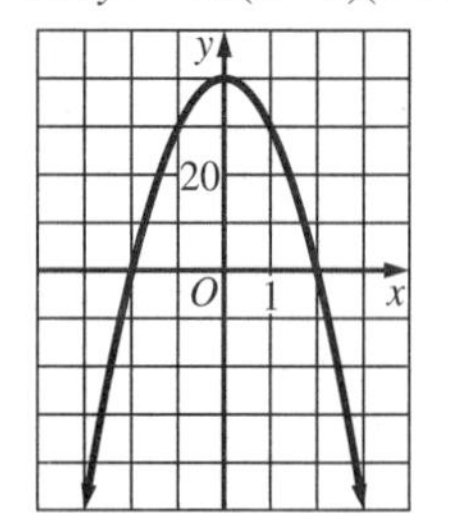

Page 209 **Checking Key Concepts**

1. a. 49 **b.** $\frac{25}{4}$ **c.** 0.01 **3.** 11

Pages 210–213 **Exercises and Applications**

1. $y = (x + 2)^2 - 1$ **3.** $y = \left(x - \frac{7}{2}\right)^2 - \frac{97}{4}$ **5.** $y = (x + 3)^2 - 9$

7. $y = 2\left(x - \frac{5}{4}\right)^2 - \frac{49}{8}$ **9. a.** $h = -16t^2 + 24t + 3$ **b.** 12 ft

c. 0.75 s after it is thrown **d.** 1.6 s after it is thrown
11, 13. Explanations may vary. Examples are given.
11. A; This graph is the only one in which the parabola opens down. **13.** B; The vertex form of this equation is $y = 2(x + 1)^2$, so the vertex of the graph is $(-1, 0)$.
17. a. $y = 0.06(x - 33.33)^2 + 20.33$

b. 20.33; The minimum oxygen consumption of which the parrot is capable is about 20.33 mL/g • h. This is achieved by flying at a speed of about 33.33 km/h. **25.** maximum; 20 **27.** maximum; 12 **29.** maximum; 11 **31.** minimum; $\frac{5}{24}$ **33.** maximum; 21 **37.** $y = -\frac{3}{8}x(x - 8)$ **39.** $y = \frac{3}{4}(x - 3)(x - 7)$ **41.** ± 2 **43.** ± 1.5 **45.** $f^{-1}(x) = 0.4x + 3.12$ **47.** $h^{-1}(x) = -\frac{1}{2}x + 2$

Page 218 Checking Key Concepts

1. a. 1.25 s **b.** 0.5 s **c.** 9 ft

Pages 218–220 Exercises and Applications

Answers may vary slightly due to rounding. **1.** –5.2; 1.2 **3.** –0.6; 1 **5.** –5; 3 **7.** –3.4; 0.9 **9.** –1.1; 2.4 **11. a.** 32.9° **b.** 324.6 ft **13.** one solution **15.** no solution **17.** two solutions **19.** no x-intercepts **21.** 1 **23.** –0.4; 2.4 **31.** 2001 **37.** (–1, –3) **39.** ± 5 **41.** ± 12 **43.** 2.31

Page 224 Checking Key Concepts

1. $y = (x - 9)(x - 9)$ **3.** $y = 2(3x - 1)(x + 2)$ **5.** 18 mi/h

Pages 225–227 Exercises and Applications

1. $y = (x - 4)(x - 1)$ **3.** $y = (x + 1)(x - 4)$ **5.** $y = (3x + 2)(3x + 2)$ **7.** $y = (4x - 5)(x + 2)$ **9.** $y = 4(3x - 5)(3x + 5)$ **13, 15, 17, 19.** Answers are rounded to the nearest tenth when necessary. **13.** –6 **15.** 1; 3 **17.** $-4\frac{1}{2}$; $1\frac{2}{3}$ **19.** –5; 0.6

21. a. graph:

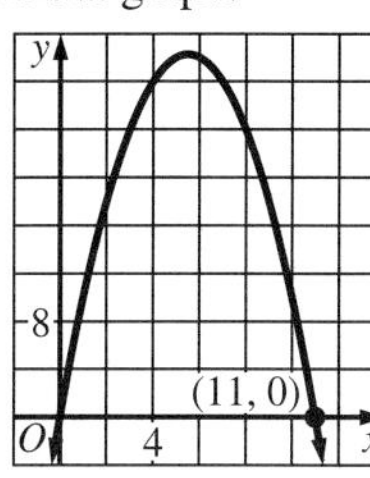

algebra: $0 = 11x - x^2$; $0 = -x(x - 11)$; $x = 0$ or $x = 11$
arithmetic: Add terms until the sum is zero: 10 + 8 + 6 + 4 + 2 + 0 + (–2) + (–4) + (–6) + (–8) + (–10) = 0; there are 11 terms. **b.** The answers are the same; preferences may vary. An example is given. I think the algebraic method was the easiest since the function was so easy to write as a product of factors. **c.** Conjecture: The sum of the first x terms of the numbers $12 + 10 + 8 + 6 + \cdots$ is given by the function $S(x) = 13x - x^2$. It takes 13 terms for the function to get the sum of zero. **27.** 0; 6 **29.** $\frac{7}{6}$ **31.** 0.4; 7.6 **33.** –0.7; 1.1 **35.** about $780

37.

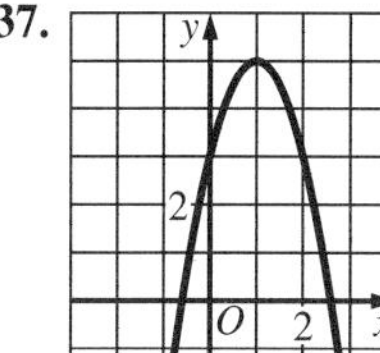

39.

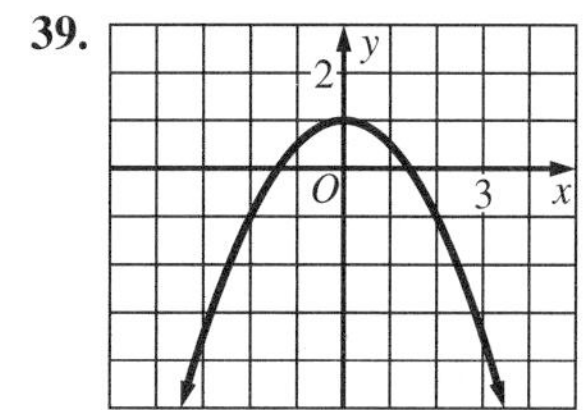

Page 227 Assess Your Progress

1. $y = 15(x + 0.3)^2 - 1.35$ **2.** $y = (x + 6)^2 - 29$ **3.** $y = 4(x - 2.5)^2 - 28$ **4.** –1.5; –0.5 **5.** no x-intercepts **6.** ± 3.2 **7.** $y = (x + 4)(x + 3)$ **8.** $y = (3x + 1)(x - 2)$ **9.** $y = (2x - 1)(x + 7)$

Pages 232–233 Chapter 5 Assessment

1. quadratic; vertex; minimum; parabola; vertex; line of symmetry **2.** standard; discriminant; x-intercepts

3.

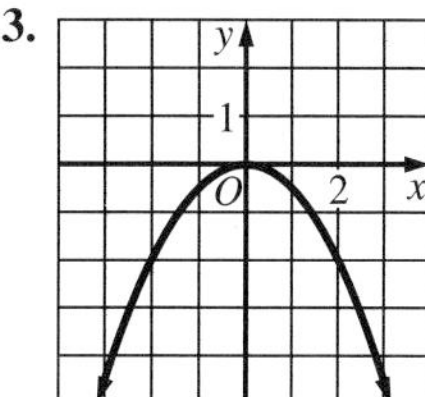

4. 3 **5.** 19.4 ft **8. a.** $y = 2(x + 1)^2 - 7$ **b.** –2.87; 0.87

9. $y = -2(x - 2)(x - (-1))$

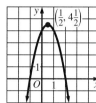

10. a. $y = \frac{1}{3}(x + 1)(x - 5)$ **b.** $y = -x(x + 4)$ **11. a.** 12.25 ft **b.** 1.5 s; Answers may vary. An example is given. I used the quadratic formula to solve the equation $-16t^2 + 20t + 6 = 0$.

12. $y = 2\left(x + \frac{9}{4}\right)^2 - \frac{57}{8}$ **13.** $y = -(x - 3)^2 + 12$ **14.** $y = \frac{1}{4}(x - 4)^2 + 7$ **15. a.** $P = -16(I - 3.75)^2 + 225$ **b.** (3.75, 225); The maximum power, 225 watts corresponds to a current of 3.75 amperes. **16.** 2.5 **17.** no x-intercepts **18.** $-\frac{2}{3}$; 6 **19.** 35 mi/h **20.** $y = (x - 4)(x - 2)$ **21.** $y = (3x - 5)(2x + 1)$ **22.** $y = (4x + 3)(x + 6)$

CHAPTER 6

Page 240 Checking Key Concepts

1. The graph on the left is a bar graph displaying the percents of class members who own a dog, a cat, a bird, another pet, or no pet. The data are categorical. The graph on the right is a histogram displaying the percentages of those students who are cat owners and have 1, 2, or 3 cats. The data are numerical. **3.** Estimates may vary; about 14,000 cat owners. **5.** Estimates may vary; about 22,000 cats.

Pages 240–243 Exercises and Applications

Opinions may vary as to whether some data are numerical or categorical. Classifications other than those given in the following answers should be accepted if they can be reasonably justified. **1.** numerical; whole numbers between, say, 100 and 300 **3.** categorical; names of colors **5.** No; the data are categorical, so a bar graph would be more appropriate. **7.** Yes; the data are categorical. **9. a.** histogram **b.** Estimates may vary. about 72 in. (6 ft) **c.** heights in inches: 72, 72, 65, 72, 74, 72, 74, 72, 74, 73, 68, 72, 70, 74, and 71; $71\frac{2}{3}$ in. (5 ft $11\frac{2}{3}$ in.); The estimate given in part (b) is fairly close. **13. a.** all cars of a given model; 100 of the cars **b.** numerical **c.** histogram **15. a.** all radio listeners in a city between the ages of 12 and 25; those selected for the survey **b.** categorical **c.** bar graph or circle graph **17. a.** all eligible voters; the voters who are surveyed **b.** categorical **c.** bar graph or circle graph **25.** –3; 1 **27.** –9; –4

29. 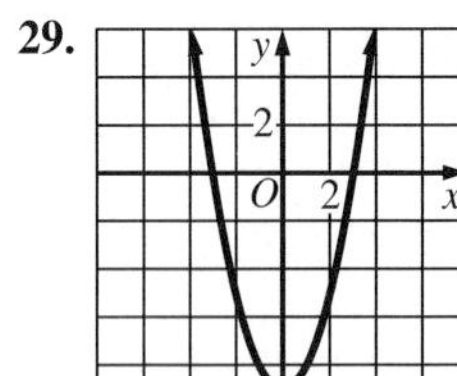

31. 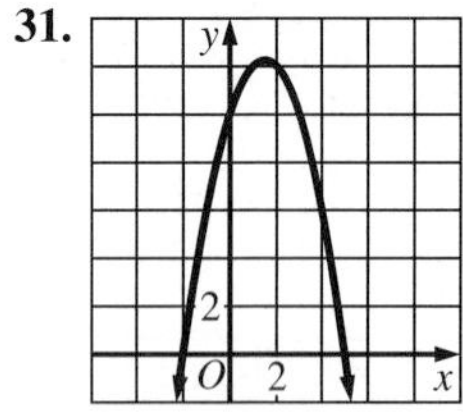

Page 247 **Checking Key Concepts**

1, 3. Answers may vary. Examples are given. **1.** The question assumes that the respondent knows all the candidates' platforms and supports one. You could provide a list of the key issues, each with all the candidates' positions given, and ask respondents to check which positions they support.
3. The first statement encourages the respondent to reply positively. Omit the first sentence.

Pages 247–249 **Exercises and Applications**

1, 3, 5, 7. Answers may vary. Examples are given. **1.** Weight is a sensitive issue and people may not wish to answer or may not answer truthfully. The information should be collected anonymously. **3.** Unless all the people surveyed are familiar with Japanese food, they may have no information about the dishes. A reasonable question might describe each of the dishes and ask respondents to rank them in order of preference.
5. The question assumes that the respondent is familiar with the facts of the case. Any presentation of the facts (as interpreted by the pollster) may be biased as well. It might be best then to ask the question as given only to those who reply affirmatively to the question, "Are you familiar with the facts of the Carter case?" **7.** I think questions A, C, and D do not need to be rewritten if question B is rephrased, "Do you think a student can maintain good grades while holding down an after-school job?" **11.** Answers may vary. Examples are given. (1) categorical; It will help determine whether there is an equal amount of interest among males and females. (2) categorical; It will help determine whether a banquet should be held at all. (3) categorical; It will help determine what type of banquet should be held. (4) numerical; It will help determine what type of banquet students can afford. **17, 19.** Survey questions may vary. Examples are given. **17.** Have you ever used the train service? If so, how often? How often do you anticipate using the service in the future? Which hours do you usually use the service? If service were expanded to include more early morning and late evening trains, do you think you would use the service during the new hours? If so, at what times and what routes? How often do you think you would use the new service? People who live or work in the area served by the trains should be surveyed. **19.** Have you watched any shows on this station on Thursday night during the past month? Give the frequency with which you have watched each of the Thursday shows listed below during the past month. Give the frequency with which you anticipate watching each of the Thursday shows listed below during the next month. People within the service area of the station should be surveyed. **21.** numerical; The data might be expressed in miles per gallon, which can vary widely.

23. either categorical (data that classify members as family members, individual members, patrons, students, and so on) or numerical (data that indicate membership totals for given museums) **25.** 0.25

Page 253 **Checking Key Concepts**

1. systematic sample; all customers of the store; The sample is biased unless it is done over a time period covering all the store's hours. Doing the survey at a particular time of day or on a particular day of the week will underrepresent customers who shop at other times. **3.** stratified random sample; all students of the school; The sample may be somewhat biased if the number of students in each grade varies too widely. It would be better to find the percent of students in each grade and randomly choose students from the grade to match this percentage.

Pages 253–256 **Exercises and Applications**

1. a. Ms. Rose: stratified random sample; Mr. Champine: cluster sample; Mr. Santanella: random sample; Mrs. Kim: convenience sample **b.** Answers may vary. An example is given. I would use Mr. Santanella's method. It is random and gives all students an equal chance of participating. **3.** convenience or cluster sample; people in the viewing area; Yes; sports fans are overrepresented. **5.** systematic sample; all customers of the taxicab company; probably not; Choosing every tenth rider in each taxicab should produce a fairly representative sample.
7. systematic sampling **9.** cluster sampling **11. a.** cluster sampling **b.** Answers may vary. An example is given. I think no one sampling method will always produce the best results; under different circumstances, different methods will produce the most representative samples. **13.** Estimates may vary; field on the left: about 70%; field on the right: about 50%.
15. Answers may vary. An example is given. Plants are numbered from 1 to 800, starting at the top of the first column and continuing from top to bottom in each column. every plant whose number is a multiple of 80 **21, 23.** Answers may vary. Examples are given. **21. a.** The first sentence encourages an exaggerated response because the respondents may want to look good to the pollster. **b.** numerical; histogram **23. a.** The wording of the question implies the respondent eats ice cream and has a favorite flavor. However, the question is not biased if possible responses include "none" and "I don't eat ice cream." **b.** categorical; bar graph **25. a.** Answers may vary. An example is given. Let x = the number of 30-year periods since period 1839–1868 and y = the number of new stamps issued. If the first and last data points are used, then the function is $y = 88(1.95)^x$; the equation fits the data reasonably well. **b.** $y = 95.5(1.86)^x$; Answers may vary. The fit is similar.

Page 257 **Assess Your Progress**

1. numerical; nonnegative decimal numbers **2.** categorical; the names of the month **3.** categorical; examples: small, medium, and large **4.** numerical; usually numbers between 0.0 and 6.0 **5.** Yes. **6.** No; the data are categorical and should be displayed in a bar graph or a circle graph.
7, 8. Answers may vary. Examples are given. **7.** The first sentence encourages a positive response. The first sentence should be omitted.

8. The question is too vague. A better question would be, "How many times a week do you floss between your teeth?" **9.** stratified random sample; residents of the county; Yes; people without phones or with unlisted phone numbers are underrepresented. **10.** self-selected sample; customers of the store; Yes; only customers (and perhaps people who are not even regular customers) who want to win a gift certificate are represented.

Page 262 Checking Key Concepts

1. Width of intervals may vary. The graph on the left is an absolute frequency histogram. The graph on the right is a relative frequency histogram. The distribution is skewed.

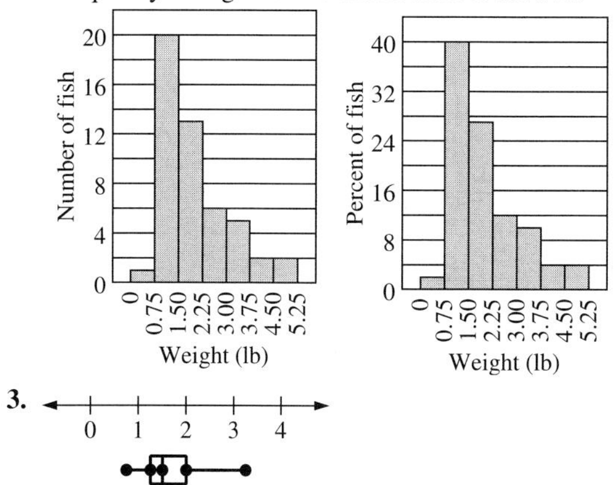

3.

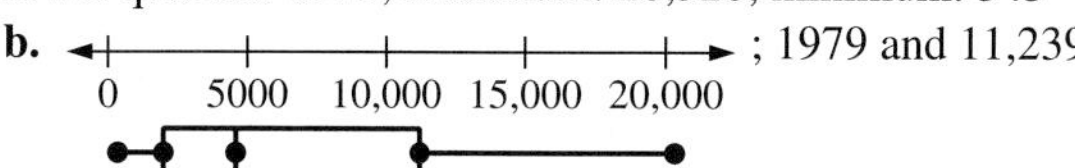

Both box plots have the same lower extreme and lower quartile. The median, upper quartile, and upper extreme of the first data set are greater than those for the second. The first harvest was better. In that harvest, half of the fish were over 1.75 lb, with 25% over 2.75 lb, and the largest fish weighed 5 lb. In the second harvest, 75% of the fish were under 2 lb and the largest fish weighed only 3.25 lb.

Pages 262–265 Exercises and Applications

3. skewed **7. a.** median: 4588.5; upper quartile: 11,239; lower quartile: 1979; maximum: 20,320; minimum: 345
b. ; 1979 and 11,239

c. The median changes to 4393, the upper quartile is 9994, the lower quartile is 1965, and the maximum is 14,494.

9. a.

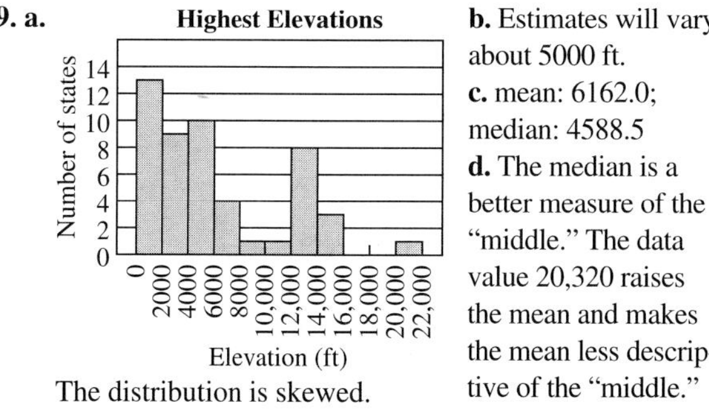

The distribution is skewed.

b. Estimates will vary; about 5000 ft. **c.** mean: 6162.0; median: 4588.5 **d.** The median is a better measure of the "middle." The data value 20,320 raises the mean and makes the mean less descriptive of the "middle."

11. The intervals on the horizontal axis are not of equal width. The data should be divided into intervals of equal width. **15.** Since at least 11 territories had become states by 1788, it indicates that the lower quartile is about 1788. Since the lower extreme is 1787, the lower half of the box plot will be very condensed.

17.

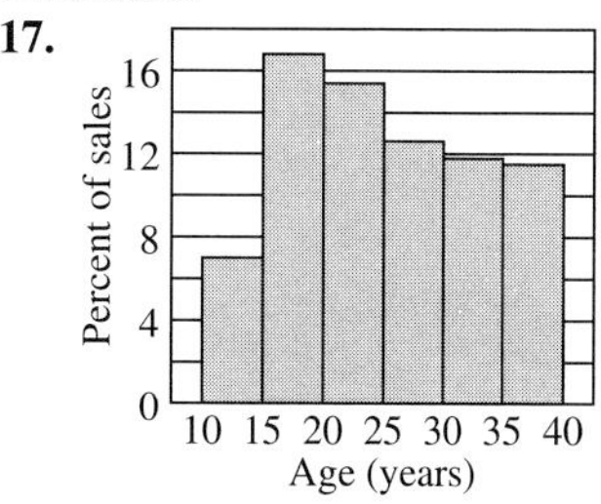

skewed distribution; The shape of the histogram is not symmetric. **21.** systematic; people at the spring play; Yes; people who are not at the spring play are not represented. **23.** $3\sqrt{6}$ **25.** 10

Page 269 Checking Key Concepts

1. No.

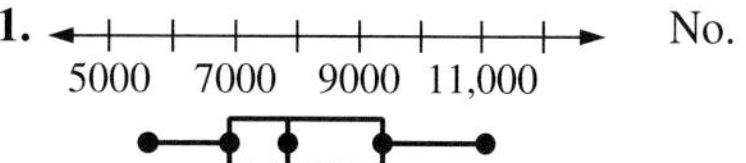

3. values between 6570.4 and 9640.8; 63.0%

Pages 270–272 Exercises and Applications

1. 12 **3.** There are none. **5.** 15; 20.4; The estimate is about 0.7σ. **7.** 6; 9.3; The estimate is about 0.6σ.
9.

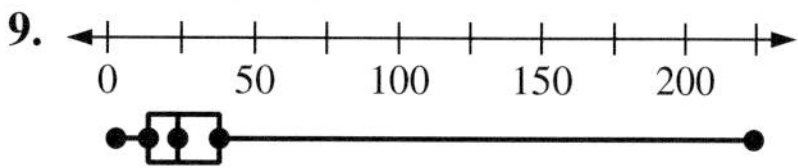

range: 217; IQR: 24; outliers: 79, 89, 97, 161, 224
11. 90%, 96% **13.** about 3.3 standard deviations **17.** The mean for each data set is 16. **19.** 0.7; 1.2; 2.5; The standard deviations vary greatly, with that of the third data set much higher than the others. **25.** $x \leq 8$ **27.** $x < 2$

Page 276 Checking Key Concepts

1. Mirdik: 56%; Wong: 44% **3.** between 52% and 60%
5. It would be ±3%.

Pages 276–279 Exercises and Applications

1. between 51.3% and 55.3% **3.** No; you need to know the sample size. **7.** The sample is self-selected and, therefore, biased. **9.** No. Answers may vary. Examples are given. Even if the intervals did not overlap (and they do), they refer to percents of two different populations. You would have to calculate actual numbers based on population figures. Also, the survey was not scientific.

11. a.

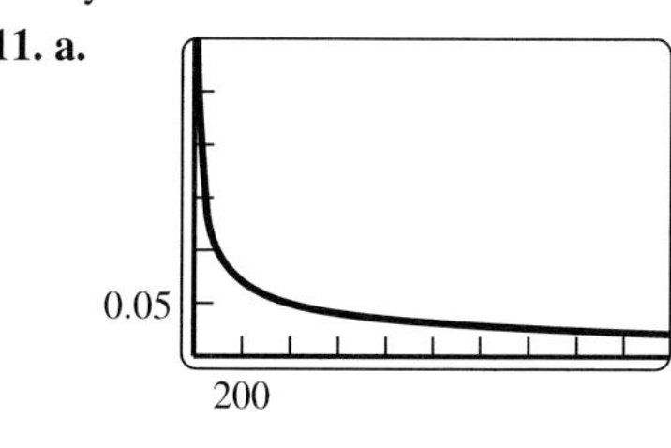

Sample size	Margin of error
100	0.1
200	0.0707
400	0.05
800	0.0354
1600	0.025
2400	0.0204

b. The larger the sample, the smaller the margin of error.
c. Using the margin of error formula, you could not take a large enough sample to have zero error, since there is no

number n for which $\frac{1}{\sqrt{n}} = 0$. In actuality, however, the margin of error would be zero if the sample consisted of the entire population. **13.** ± 1.1 (to two significant digits) **17.** 8.46; ≈ 2.90 **19.** $y = 4 + \frac{3}{5}(x - 1)$ or $y = 1 + \frac{3}{5}(x + 4)$

Page 279 Assess Your Progress

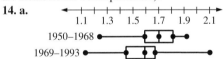

1. a. 0 2 4 6 8 10 **b.** 6 years old; 8 years old

3. a. 7; 2; There are no outliers. **b.** 1.8 **4. a.** between 48% and 56% **b.** No.

Pages 284–285 Chapter 6 Assessment

1. histogram; box plot; categorical **2.** systematic; random **3.** absolute frequencies; relative frequencies **4.** categorical; names of the species **5.** may be numerical (usually numbers between 0 and 100 or between 0 and 4.0) or categorical (letter grades A through F) **6.** numerical; numbers of kilowatt-hours **7.** Yes; the data are categorical. **8.** No; the data are categorical. The data should be displayed in a bar graph or circle graph. **9–10.** Answers may vary. Examples are given.
9. Unless the question is asked anonymously, it may not elicit an honest response since it involves sensitive personal information. The information should be collected anonymously.
10. The first sentence encourages a negative response, since it appeals to a person's respect for individual freedoms. The first sentence should be eliminated. **11.** random; all those in the area covered by the types of numbers generated (for example, with the appropriate area codes); Yes; those without phones are underrepresented **12.** systematic; all students in the class; No (unless seating was not assigned randomly). **13.** cluster; all students in the school; Yes; students who are not members of the student council are underrepresented (while students who are members are overrepresented).

14. a. 1.1 1.3 1.5 1.7 1.9 2.1
1950–1968
1969–1993

b. 1.23 and 1.7 or 1.57 and 1.81 or 1.7 and 1.91
c. 1.15 and 1.56 or 1.425 and 1.67 or 1.56 and 2.12
15. Intervals may vary. Examples are given.

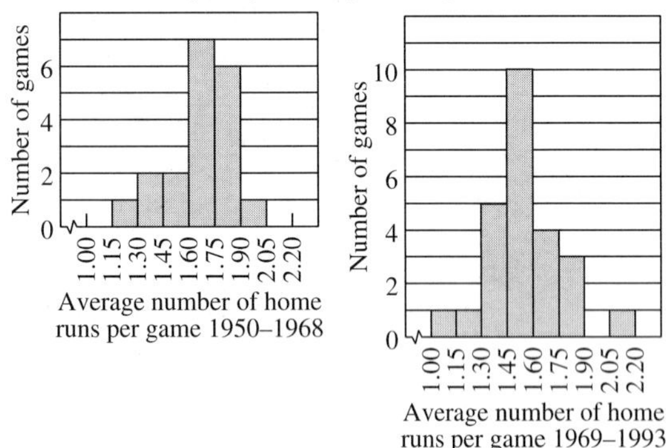

16 a. 1950-1968: 0.68; 0.24; no outliers; 1969-1993: 0.97; 0.25; 2.12 **b.** 1950-1968: 0.18; 1969-1993: 0.19

The distribution for the 1950–1968 data is skewed; the distribution for the 1969–1993 data is roughly symmetric.
17. Answers may vary. An example is given. Since the upper quartile for the later data is about the same as the median for the earlier data, it appears the average number of home runs per game decreased after 1969. I think it is reasonable to assume that the changes described were at least partially responsible.

18 a. 75% **b.** about $\pm 11\%$; between 64% and 86%

CHAPTERS 4–6

Pages 286–287 Cumulative Assessment

1. a.

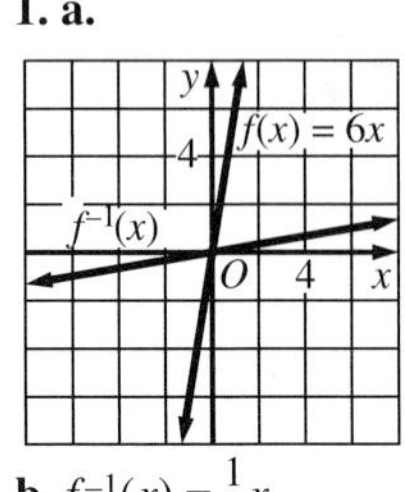

b. $f^{-1}(x) = \frac{1}{6}x$

3. a.

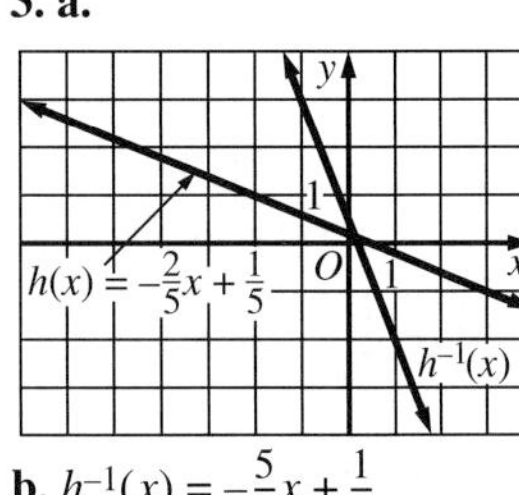

b. $h^{-1}(x) = -\frac{5}{2}x + \frac{1}{2}$

5. $\log_{81} \frac{1}{9} = -\frac{1}{2}$ **7.** −2 **9.** −2.59 **11.** $\log_a \frac{1}{9}$ **13.** 0.02
15. $y = 10(6.31)^x$ **17.** Answers may vary slightly due to differences in technology or to rounding. **a.** $y = 31.5(1.03)^x$ **b.** $y = 14.5x^{0.489}$

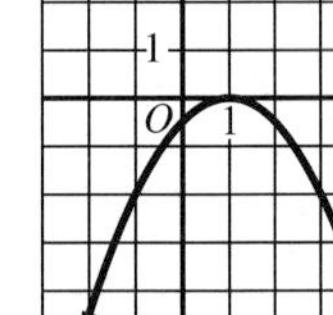

19. Summaries may vary. The graph is a parabola that opens down. Its vertex is at (1, 0) and it is wider than the graph of $y = x^2$. **21.** ± 2.2 **23.** 1.9; 5.6 **25.** $y = 2(x + 4)^2 + 1$ **27.** $y = -\frac{1}{8}(x + 5)(x - 3)$ **29.** −2 **31.** $y = (x + 8)(x - 8)$ **33.** $y = (3x - 4)^2$
35. Answers may vary. Examples are given. **a.** contributors to the station **b.** self-selected; Yes; it underrepresents listeners who do not contribute. **c.** numerical (for example, how many listeners listen to a particular type of show); categorical (types of programs that are most popular) **d.** histogram; bar graph; circle graph

39. a.

Conference	Minimum	Lower quartile	Median	Upper quartile	Maximum
Eastern	9	18	22	27	30
Western	16	16.5	18.5	24	33

b.

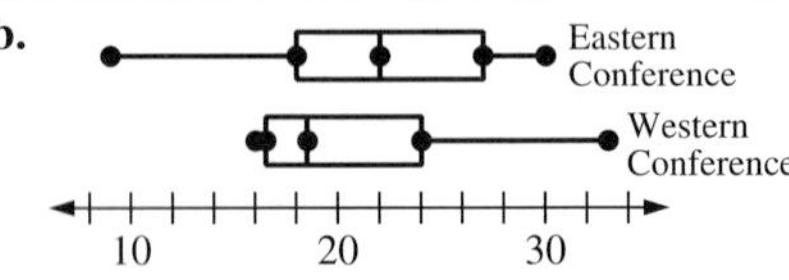

c. Eastern Conference: 21; 9; no outliers; Western Conference: 17; 7.5; no outliers. **d.** Eastern Conference: 5.62; Western Conference: 5.25; The standard deviation is greater for the Eastern Conference data.

CHAPTER 7

Page 293 Checking Key Concepts

1. To solve a system of equations means to find the coordinates of the point(s) where the graphs of the equations intersect (that is, the values of the variables that make all the equations of the system true). **3.** (1.5, 14.5) **5.** (3, 39) **7.** (2, 4.5)

Pages 293–296 Exercises and Applications

1. (2, 4) **3.** (3, 4) **5.** (2, 11) **7.** $\left(\frac{60}{41}, \frac{156}{41}\right)$ **9. a.** (5, 19); The y_1-value and the y_2-value that correspond to the x-value 5 are the same. **11.** 75 times **13.** (2.02, 120), (13.7, 295) **15.** (–0.303, –0.908), (3.303, 9.908) **17.** (–1, –3), (9, 97) **19.** (–3.32, –5), (3.32, –5) **21.** (2.98, –53) **23.** (–1.79, 0.417), (2.79, 9.58) **35.** no solution **37.** no solution

Page 300 Checking Key Concepts

1, 3, 5. Answers may vary. All the systems may be solved by graphing. Examples of algebraic solution methods are given. **1.** Add. **3.** Use substitution. **5.** Multiply both sides of the first equation by 3 and both sides of the second equation by –2, then add. **7, 9.** Answers may vary. Examples are given. **7.** $x - y = 7, x + y = 9$ **9.** $2x - 3y = -7, 3x + 2y = -4$

Pages 300–302 Exercises and Applications

1. (2, –1) **3.** $\left(-1, 5\frac{1}{2}\right)$ **5.** (–40, –27) **7.** (2, 0) **9.** Values that satisfy one equation always satisfy the other; infinitely many. **11.** no solution **13.** $\left(-40\frac{5}{6}, 26\frac{52}{63}\right)$; one solution **15.** no solution **17. a.** $10d_1 = 2d_2$ **b.** $d_1 + d_2 = 14$ **c.** $2\frac{1}{3}$ in.; $11\frac{2}{3}$ in. **23.** (3, 21) **25.** (3, 4)

27. $\begin{bmatrix} 8 & 11 \\ -11 & -2\frac{1}{2} \end{bmatrix}$ **29.** $\begin{bmatrix} -2 & 10 \\ -19 & 6 \\ 9 & 9 \end{bmatrix}$

Page 305 Checking Key Concepts

1. $\begin{bmatrix} 14 & -13 \\ -5 & -1 \end{bmatrix}\begin{bmatrix} x \\ y \end{bmatrix} = \begin{bmatrix} -6 \\ 2 \end{bmatrix}$; (0.253, 0.734)

3. $\begin{bmatrix} 81.2 & 17.4 \\ 4.6 & -12.8 \end{bmatrix}\begin{bmatrix} x \\ y \end{bmatrix} = \begin{bmatrix} 0.8 \\ 12.8 \end{bmatrix}$; (0.208, –0.925)

5. (4.63, –1.03, –0.567)

Pages 306–308 Exercises and Applications

3. $\begin{bmatrix} 5 & -12 \\ 14 & -7 \end{bmatrix}\begin{bmatrix} x \\ y \end{bmatrix} = \begin{bmatrix} 13 \\ 34 \end{bmatrix}$; (2.38, –0.90) **5. a.** 3 batches of rolls, 2 batches of muffins **b.** There is no way Charlie can use up exactly 8 cups of buttermilk and 6 eggs using the given recipes. **7.** (0.29, 6.12) **9.** (–0.70, 8.35) **11.** (3, –1) **13.** (–0.26, 0.12) **17.** (1.94, –6.43, 1.62) **19, 21.** Coefficients may vary due to rounding. Examples are given. **19.** $y = 1.42x^2 + 2.37x - 1.2$ **21.** $y = 0.19x^2 - 1.5x + 3.31$ **29.** $\left(\frac{15}{4}, -\frac{5}{4}\right)$; one solution **31.** $x \geq -\frac{3}{2}$ **33.** $x > \frac{1}{2}$ **35.** $y = (2x + 1)(x - 5)$

Page 308 Assess Your Progress

1. (–0.35, 0.25), (2.85, 16.3) **2.** $\left(-\frac{9}{5}, 3\right)$ **3.** $\left(0, \frac{3}{5}\right)$ **4.** (0.96, 1.27); one solution **5.** no solution **6.** (3, –4); one solution **7.** $y = -0.35x^2 - 3.65x - 4.8$

Page 312 Checking Key Concepts

1.

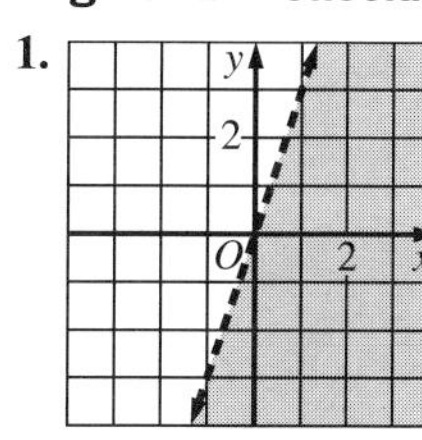

3.

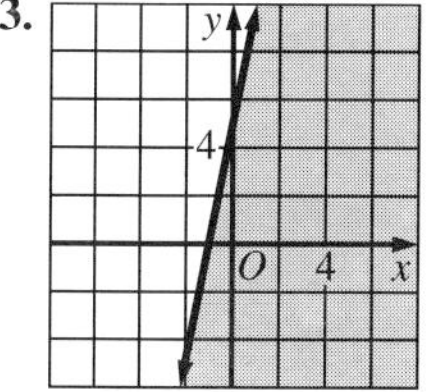

5.

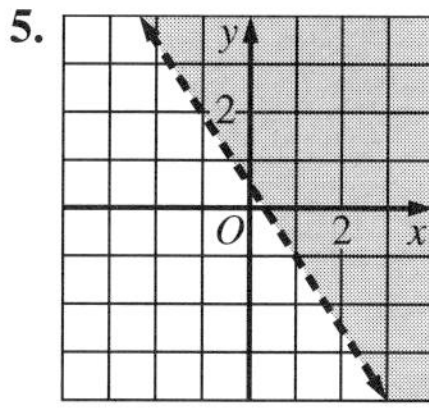

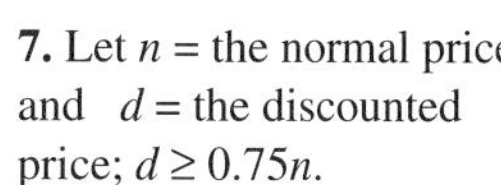

7. Let n = the normal price and d = the discounted price; $d \geq 0.75n$.

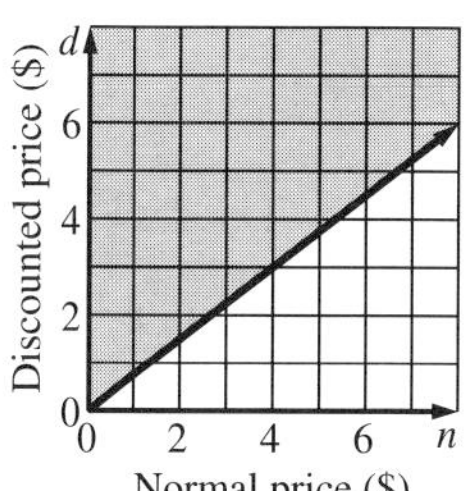

Pages 312–314 Exercises and Applications

1. Let i = the number of infants and s = the number of supervisors; $s \geq \frac{1}{3}i$.

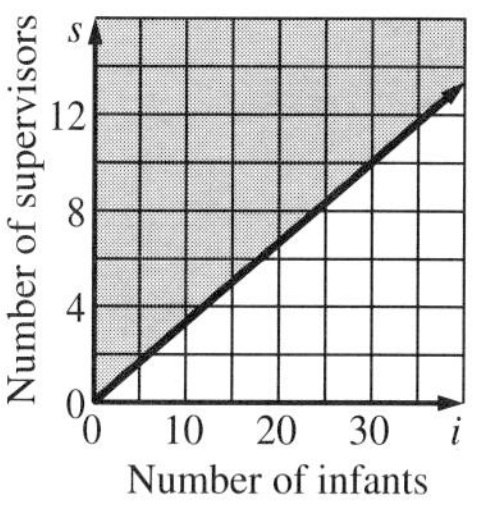

3.

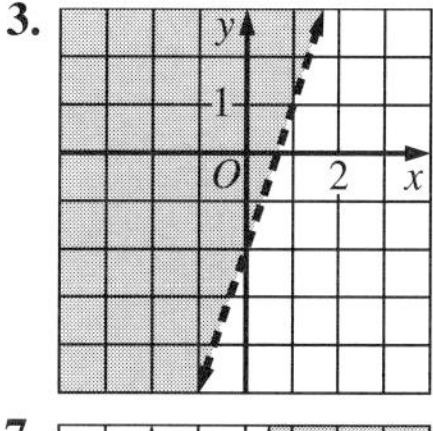

5.

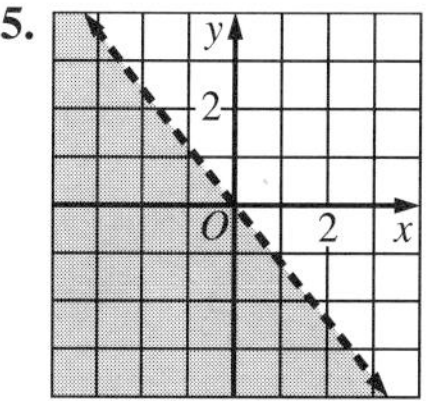

7.

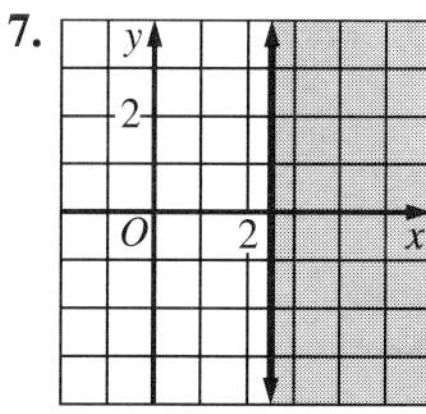

9.

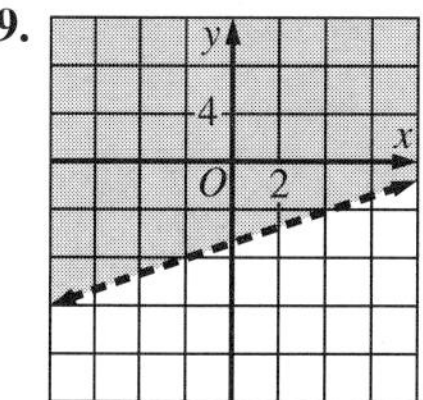

11.

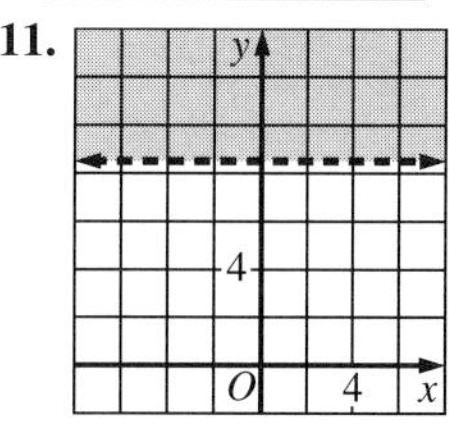

13.

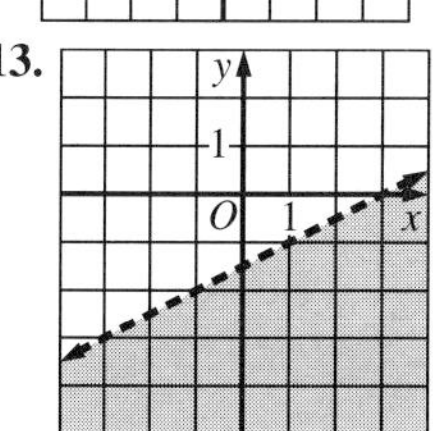

15. $x \geq -4$ **17.** $y \leq -\frac{1}{3}x + \frac{11}{3}$

23. Let f = the number of female ostriches and e = the number of eggs; $e \leq 70f$.

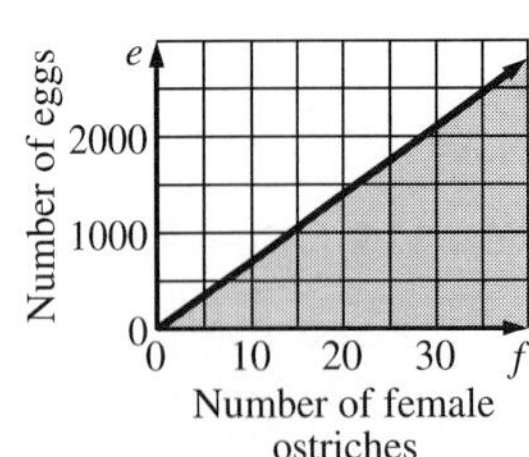

25. Let x = the blood pressure in the arm and y = the blood pressure in the ankle; $y \geq 0.9x$. (Both x and y are measured in milligrams of mercury.)

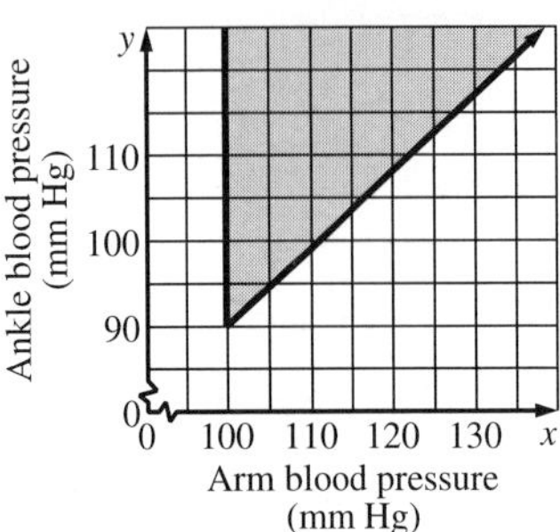

29. (15.97, 5.26)

31. a.

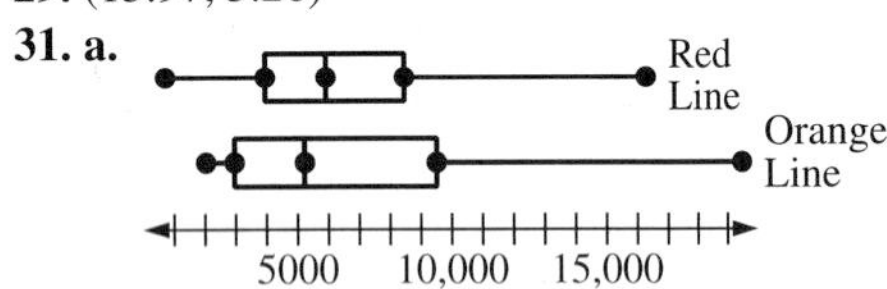

Page 317 Checking Key Concepts

1.

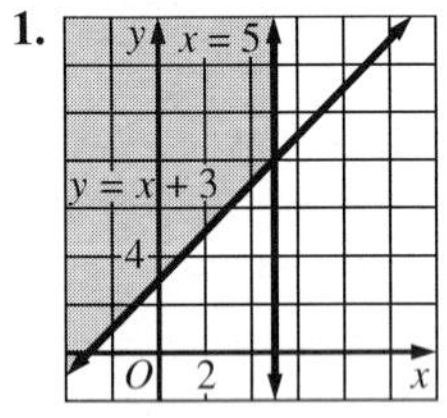

3.

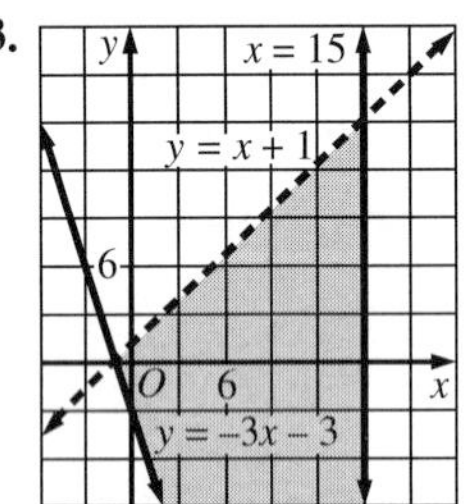

Pages 317–321 Exercises and Applications

1.

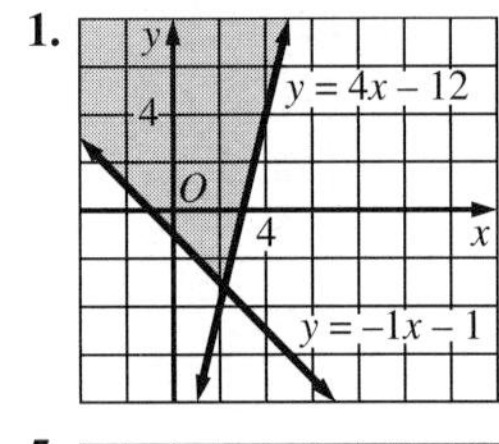

3.

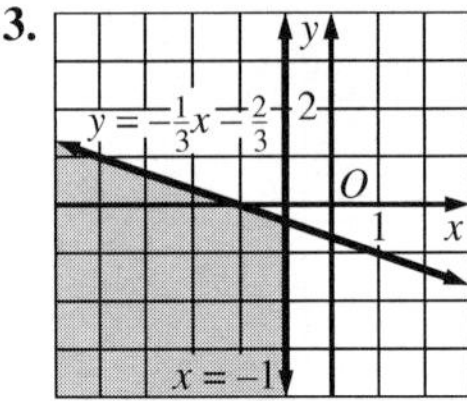

5.

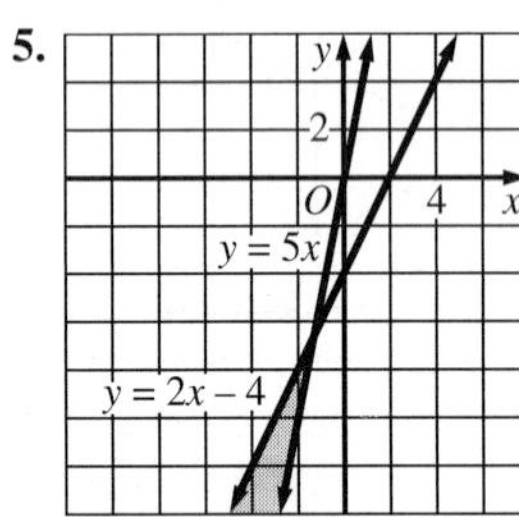

7.

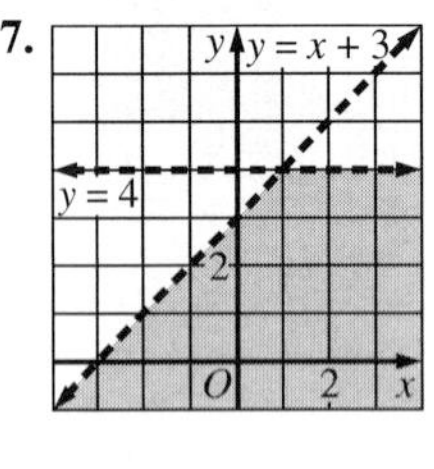

11. $x \leq 0, y \geq 3, y \leq \frac{1}{2}x + 4$ **13.** $x \geq 0, y \geq 0, y \leq -x + 4$

15. Lines of fit may vary. Least-squares lines are shown.

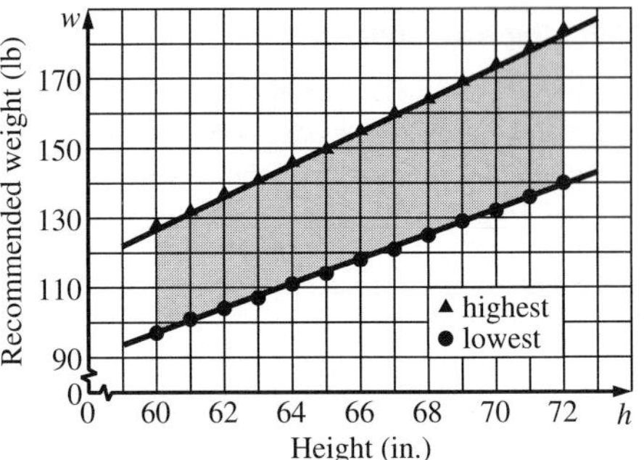

23.

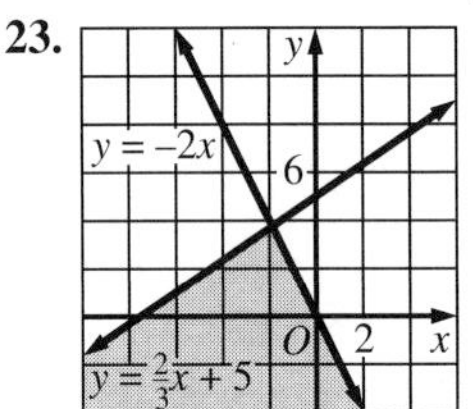

25.

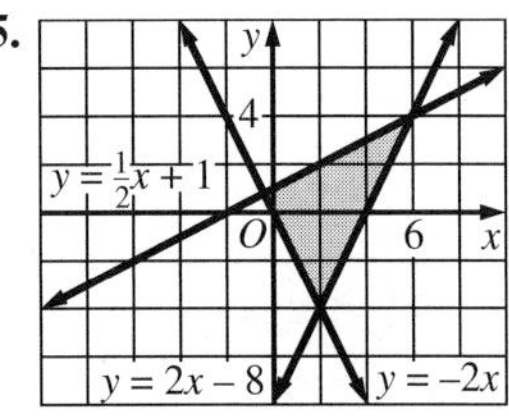

27.

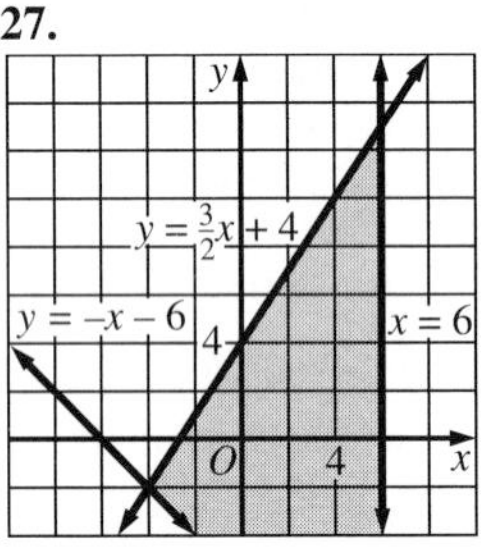

29.

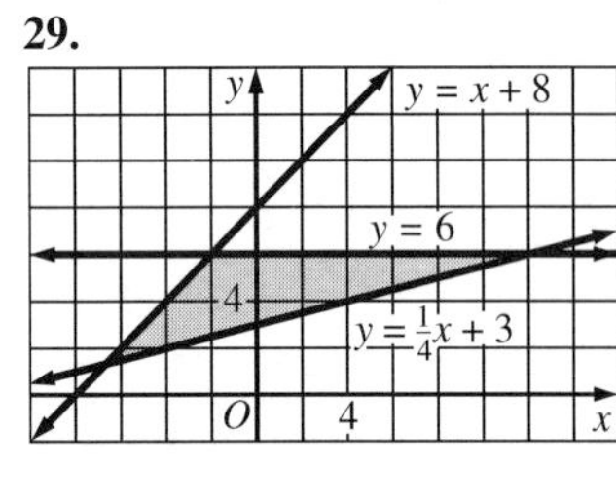

31. $x \leq 5, y \leq 3, y \geq -\frac{1}{3}x - \frac{4}{3}, y \leq 2x + 1$

33. $x \geq 0, y \geq 0, y \leq 20, y \leq x + 8, y \leq -\frac{5}{2}x + 80$

37.

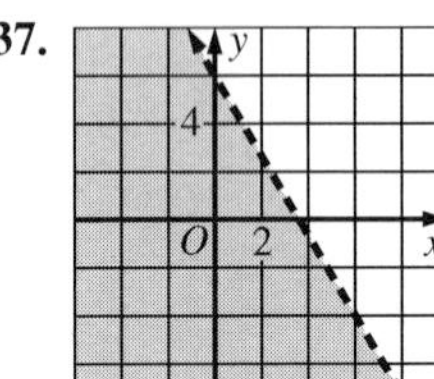

39.

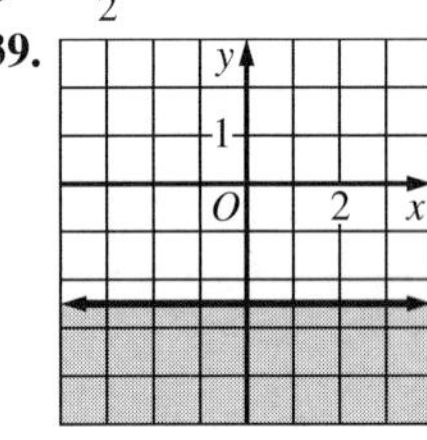

41. (5, 5) **43.** (−15, 4) **45.** 1, −4 **47.** ±1

Page 325 Checking Key Concepts

1. Let x = the number of oatmeal cookies and y = the number of chocolate chip cookies; $x \geq 24, y \geq 24, y \geq -x + 120, y \leq -x + 144$.

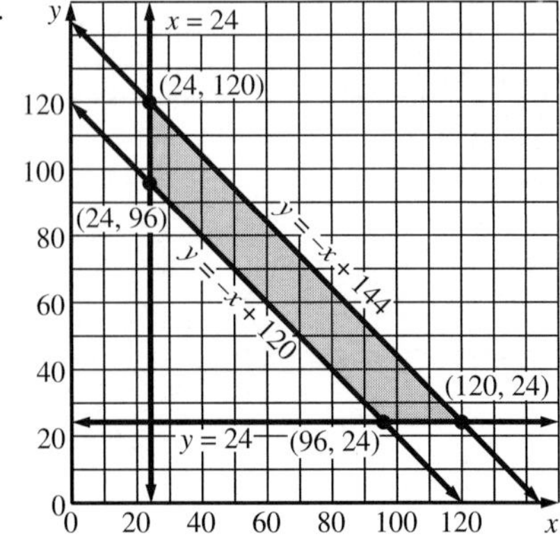

3. 96 oatmeal cookies, 24 chocolate chip cookies

Pages 325–327 Exercises and Applications

1.

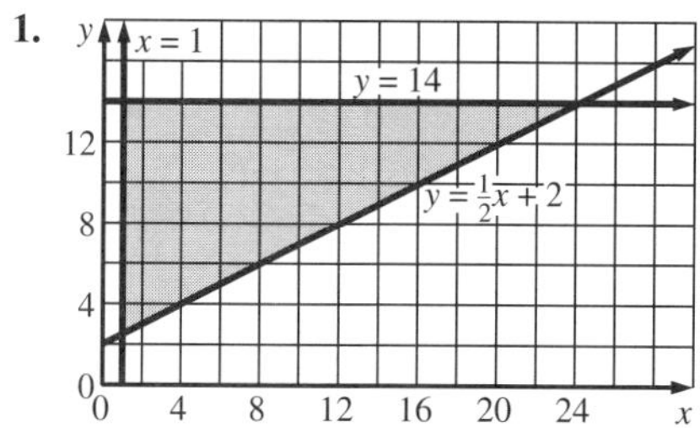

3.

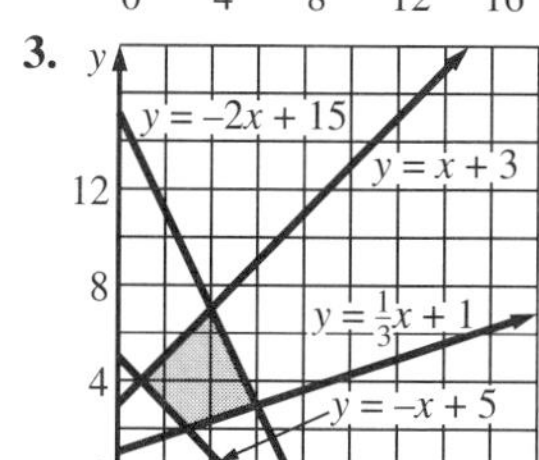
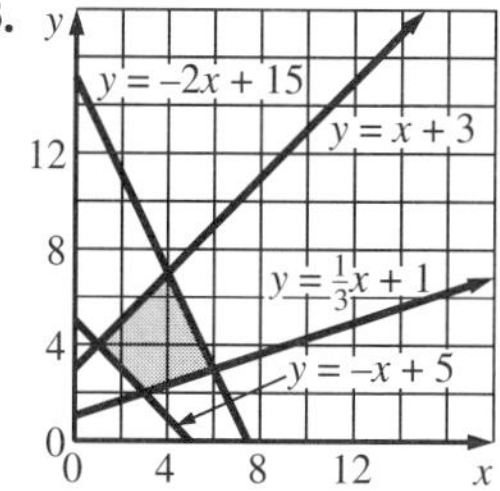

5. \$176 **7.** \$48 **9.** \$10.50 **11.** \$15

13. a. $x \geq 4$, $y \geq 0$, $y \leq -\frac{5}{4}x + 10$

b.

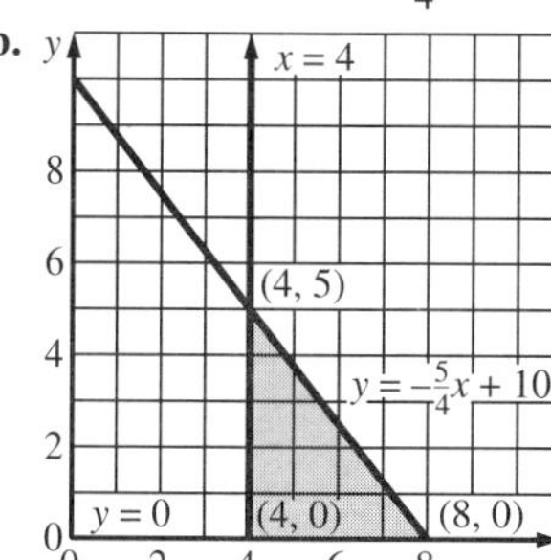

c. 4 bowls, 5 plates

19. \$102 **21.** \$32 **23.** $f^{-1}(x) = \frac{1}{2}x - \frac{7}{2}$ **25.** $h^{-1}(x) = \frac{1}{15}x + \frac{1}{5}$

27. Answers may vary. An example is given. The first part of the question might encourage a negative answer. I would omit that part of the question as well as the word "another."

Page 327 Assess Your Progress

1.

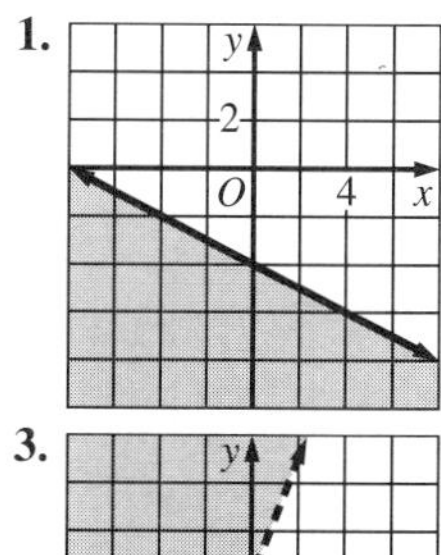

2.

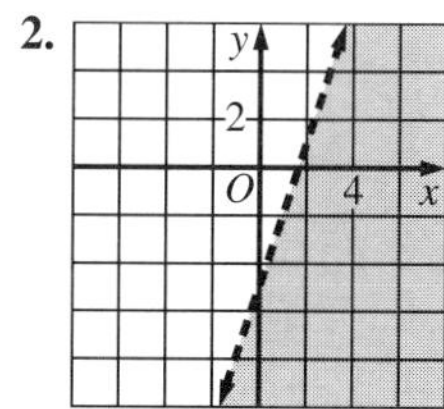

3.

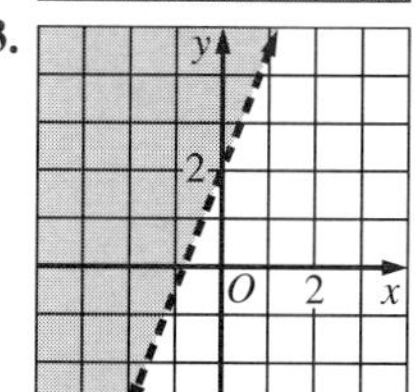

4. Let g = gross income in dollars and m = mortgage amount in dollars; $m \leq 0.28g$.

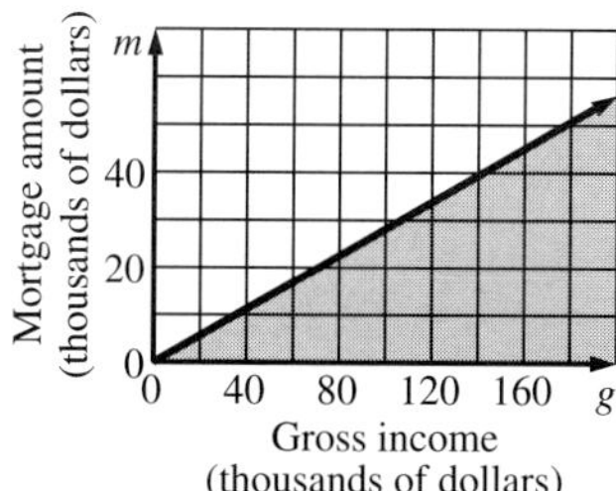

5.

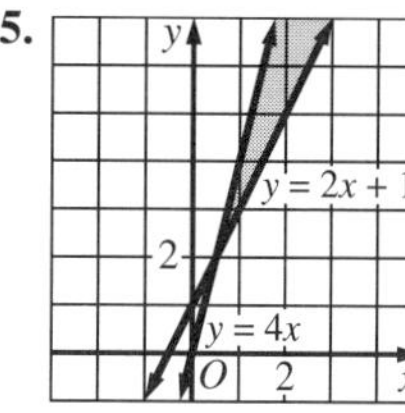

6.

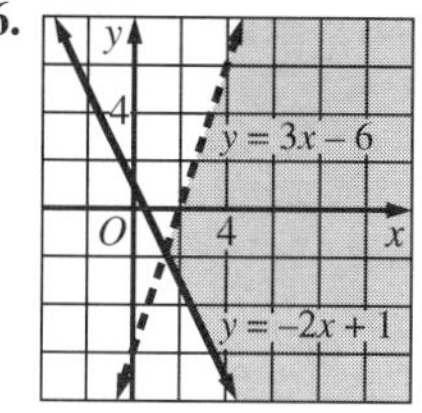

7.

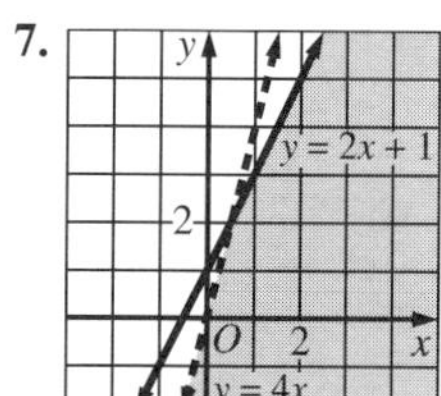

8.

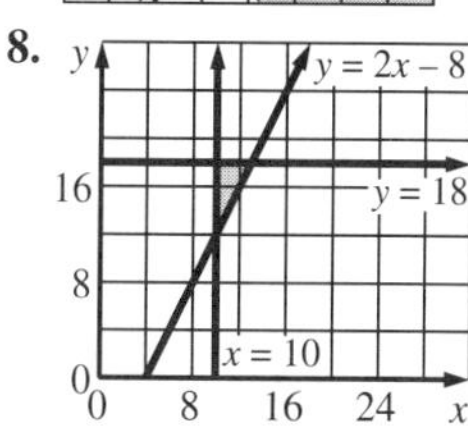

\$426

9.

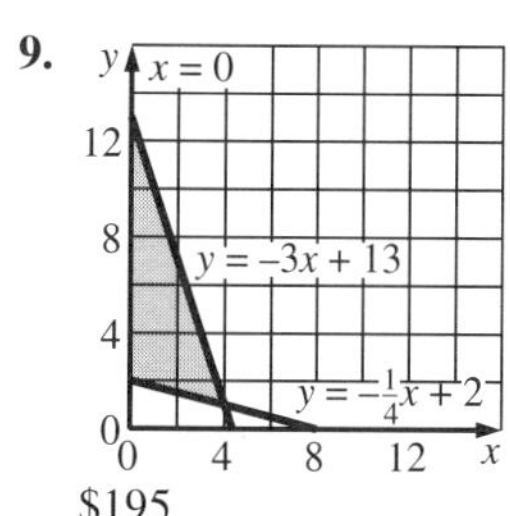

\$195

10.

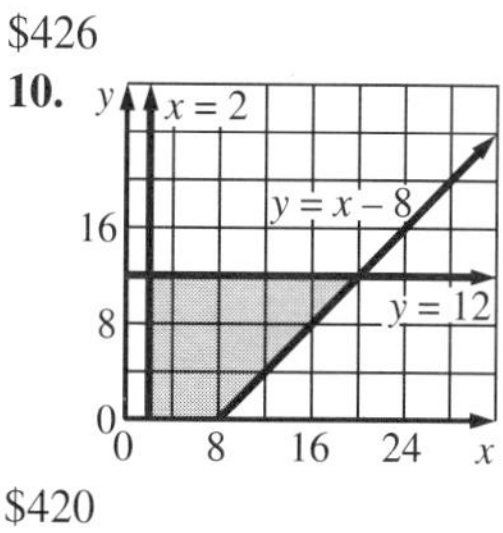

\$420

Pages 332–333 Chapter 7 Assessment

1. inconsistent; dependent **2.** inverse; identity matrix **3.** (8.07, 44.9) **4.** (0.118, 1.94); one solution **5.** no solution **6.** (1.02, 0.408); one solution **7.** no solution **8.** $\left(\frac{3}{2}, -\frac{1}{2}\right)$; one solution **9.** $\left(-2, -\frac{2}{3}\right)$; one solution

11. a. Let h = the number of hours of use and C = the cost in dollars; $C = 0.0018h + 12$ (fluorescent) and $C = 0.0072h + 0.50$ (incandescent). **b.** (2130, 15.83); If bulbs are used for 2130 h, the cost is the same for fluorescent and incandescent bulbs. **12. a.** $h = -15t^2 + 52.5t + 212.5$ **b.** about 258 ft

13.

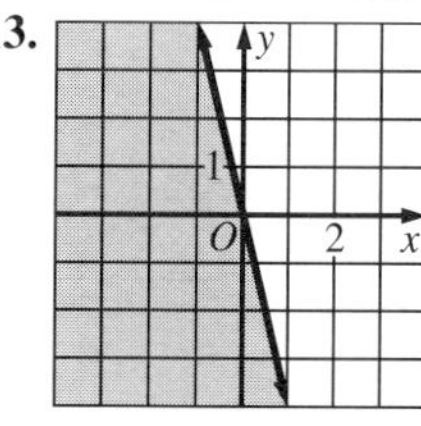

14.

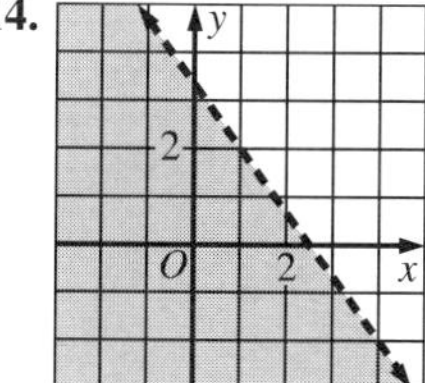

15.

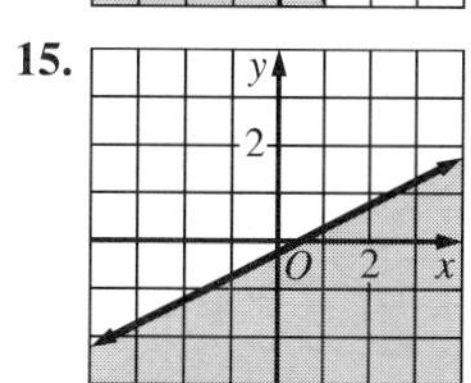

16. $x \geq 0$; $y \geq 0$; $y \leq \frac{1}{2}x + 2$; $y \leq -2x + 12$

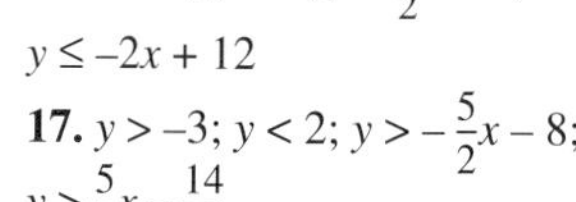

17. $y > -3$; $y < 2$; $y > -\frac{5}{2}x - 8$; $y > \frac{5}{3}x - \frac{14}{3}$

18. a.

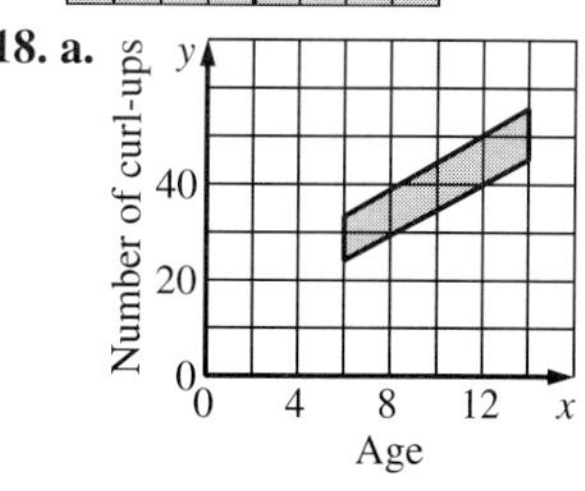

b. With the exception of the upper boundary, the region represents all scores for boys between 6 and 14 that achieve or exceed the limits for the National Award and fall short of the limits for the Presidential

Award; every point on the upper boundary represents an age and number of curl-ups that qualify for both awards.

19. a. Let b = the amount in dollars invested in the balanced fund and g = the amount in dollars invested in the growth fund; $b + g \geq 12{,}000$; $b + g \leq 20{,}000$; $b \geq \frac{2}{3}g$; $b \leq 10{,}000$.

b.

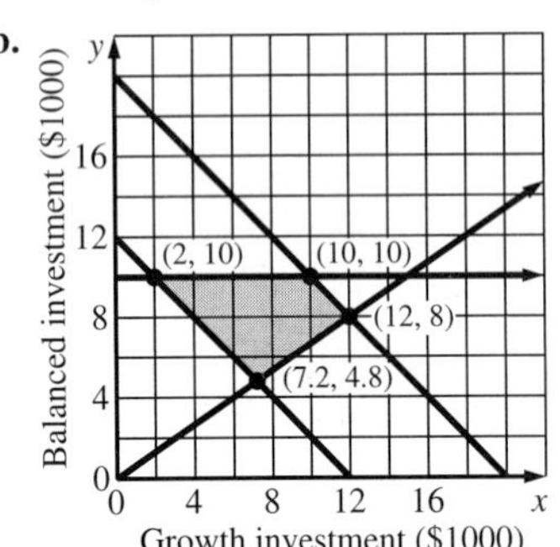

c. $12,000 in the growth fund and $8,000 in the balanced fund

CHAPTER 8

Page 340 Checking Key Concepts

1. a. $v = \sqrt{19.6h}$

b.

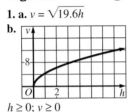

$h \geq 0$; $v \geq 0$

c. 8.85 m/s

3.

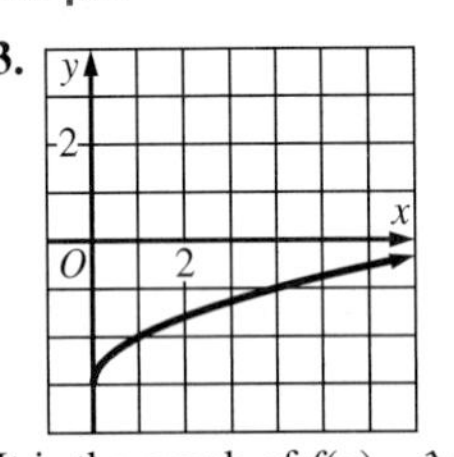

It is the graph of $f(x) = \sqrt{x}$ translated down 3 units.

Pages 340–343 Exercises and Applications

1. a, b.

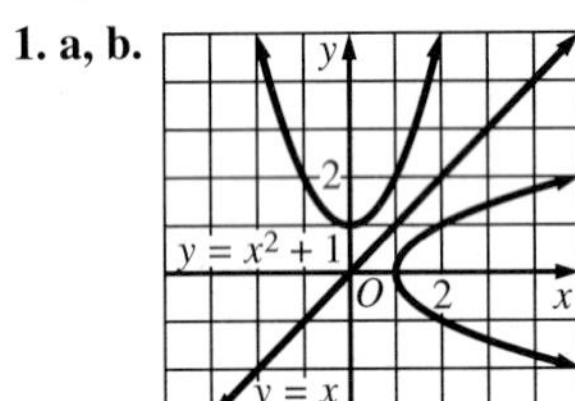

c. No; there are two outputs for each input except 1.

d. $y = \sqrt{x - 1}$ and $y = -\sqrt{x - 1}$; $x \geq 1$

3. $2\sqrt{2x}$ **5.** $5\sqrt{x}$ **7.** $10\sqrt{x}$ **11.** C **13.** F **15.** A

21.

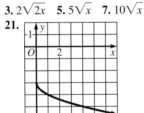

nonnegative numbers; $y \leq -3$

23.

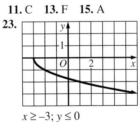

$x \geq -3$; $y \leq 0$

25.

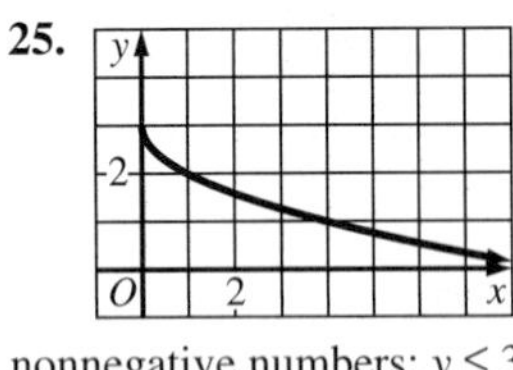

nonnegative numbers; $y \leq 3$

27.

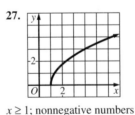

$x \geq 1$; nonnegative numbers

29.

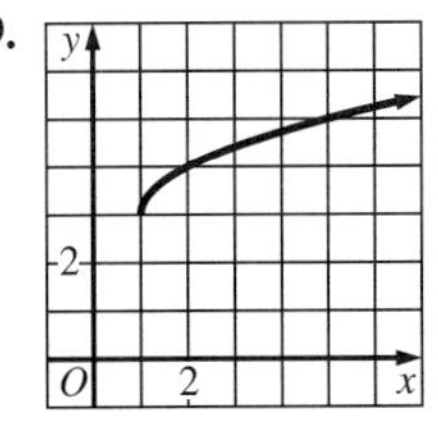

$x \geq 1$; $y \geq 3$

31.

$x \geq -3$; $y \geq -1$

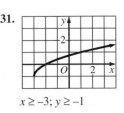

33. never **35.** sometimes **37.** sometimes **39.** always **43.** 10 **45.** 5

47.

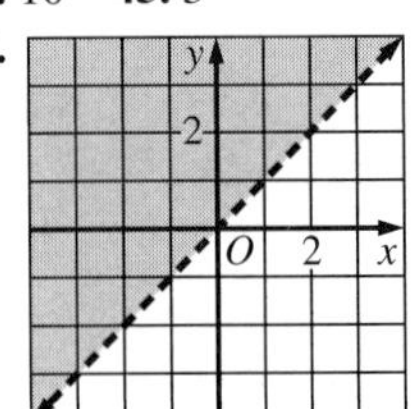

49.

Page 347 Checking Key Concepts

1. all numbers; all numbers **3.** all numbers; all numbers **5.** undefined **7.** –3 **9.** –4 **11.** $3\sqrt[3]{10}$ **13. a.** $r = \sqrt[3]{\frac{3V}{\pi}}$ **b.** 9.85 cm

Pages 347–350 Exercises and Applications

1. 15 in. **3.** 24 in. **5.** undefined **7.** 10 **9.** $\frac{1}{2}$ **11.** 5 **13.** 10 **15.** 2 **17. a.** $d = \sqrt[3]{\frac{6V}{\pi}}$ **b.** 12.4 cm **19.** all numbers; all numbers **21.** nonnegative numbers; nonnegative numbers **23. a.** $\sqrt{z}$ **b.** $\sqrt[3]{p^2}$ **c.** $\sqrt{q^3}$ **d.** $\sqrt[10]{y^7}$ **25.** $\sqrt[6]{x} \cdot \sqrt[6]{x} = x^{1/6} \cdot x^{1/6} = x^{1/6 + 1/6} = x^{1/3} = \sqrt[3]{x}$ **27.** $x^{m/n} = x^{m(1/n)} = (x^m)^{1/n} = \sqrt[n]{x^m}$

41.

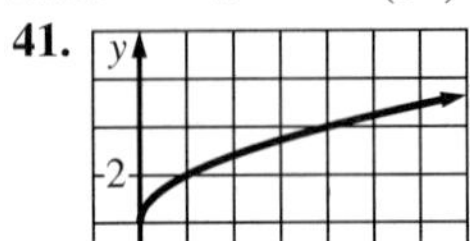

nonnegative numbers; $y \geq 1$

43.

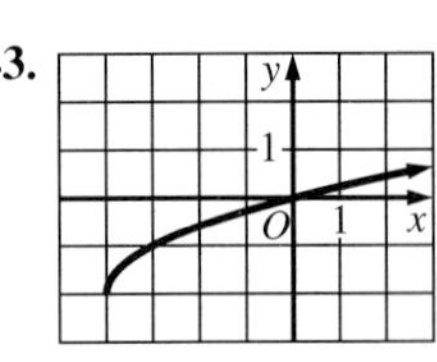

$x \geq -4$; $y \geq -2$

45. 10

Page 353 Checking Key Concepts

1. 12 **3.** –3 **5.** 20 **7.** –4 **9.** 3

Pages 353–357 Exercises and Applications

1. 49 **3.** 61 **5.** 18 **7.** 15 **9.** 1 **11.** no solution **13.** 173 **15.** –35 **19.** 27.5 m/s; 33.9 m/s; 45.0 m/s; 59.8 m/s **21.** Wind velocity increases. **23.** 7 **25.** –2; –5 **27.** 3 **29.** 9 **31.** –2 **33.** –5 **37.** 4 **39.** 2.22 **45.** 0; 2 **47.** 0; 1 **49.** 0.791 **51.** 1.55 **55.** $5\sqrt{2}$ **57.** $2\sqrt[4]{25} = 2\sqrt{5}$ **59.** two solutions **61.** no solution **63.** $\left(-\frac{4}{7}, -\frac{2}{7}\right)$ **65.** (1, 3); (–1, 5)

Page 357 Assess Your Progress

1. a.

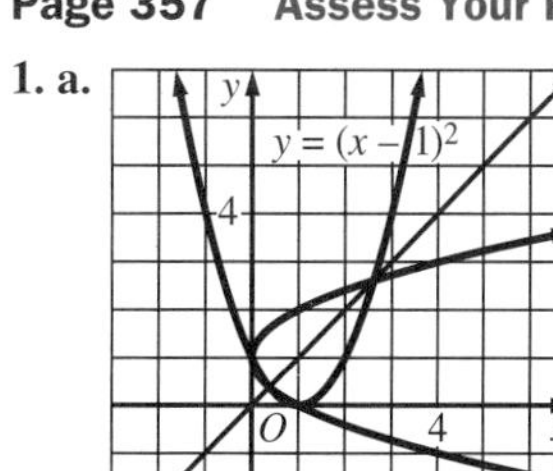

b. $y = 1 + \sqrt{x}$; $y = 1 - \sqrt{x}$, $x \geq 0$
c. The reflection of the graph is not the graph of a function since each input value has two output values. The reflection is the combined graph of two functions.

2. $T = 47.3w^{2/3}$ **3.** 659; 692.5; 725; 784; 841; 896.5; 950; 1019; 1086; 1182.5 **4.** no solution **5.** $\frac{3 + \sqrt{13}}{2} \approx 3.30$ **6.** $\pm 2\sqrt{6} \approx \pm 4.90$

Page 362 Checking Key Concepts

1. $i\sqrt{7}$ **3.** $10i\sqrt{3}$ **5.** $14 + 5i$ **7.** $5 + 6i$ **9.** $12 + 8i$ **11.** $11 + 23i$ **13.** $\frac{11}{25} + \frac{2}{25}i$ **15.** $7 \pm 2i$

Pages 362–364 Exercises and Applications

1. $6i$ **3.** $-2i\sqrt{7}$ **5, 7.** Answers may vary. Examples are given. **5.** $x^2 + 16 = 0$ **7.** $x^2 + 2 = 0$ **9.** $12 + 5i$ **11.** $18 + 3i$ **13.** $-2 + 4i$ **15.** $10 + 4i$ **17.** $-11 + 4i$ **19.** 0 **21.** $20i$ **23.** $10 + 15i$ **25.** $-45 + 18i$ **27.** $19 + 13i$ **29.** $10 + 5i$ **31.** 25 **33.** $\frac{21}{65} + \frac{38}{65}i$ **35.** $-\frac{9}{25} + \frac{13}{25}i$ **43.** $7 \pm i$ **45.** $-3 \pm i\sqrt{3}$ **47.** $-\frac{1}{2} \pm \frac{\sqrt{3}}{2}i$ **51.** 66 **53.** $\frac{21}{2}$ **55.** $y = 3.28(1.25)^x$ **57.** $\sqrt{74} \approx 8.60$ **59.** $4\sqrt{2} \approx 5.66$

Page 367 Checking Key Concepts

1.

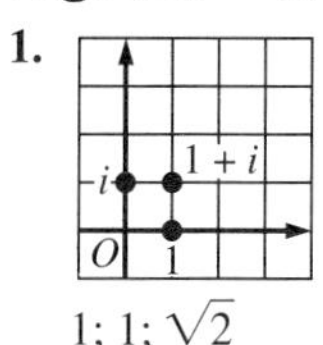

1; 1; $\sqrt{2}$

3.

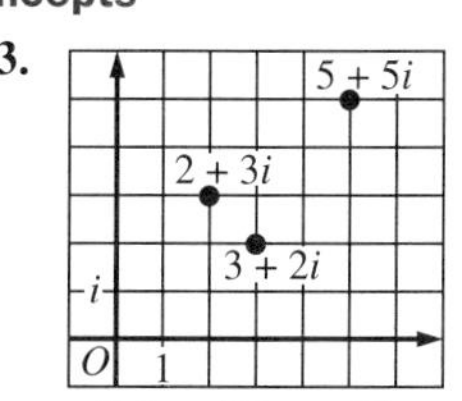

$\sqrt{13}$; $\sqrt{13}$; $5\sqrt{2}$

5.

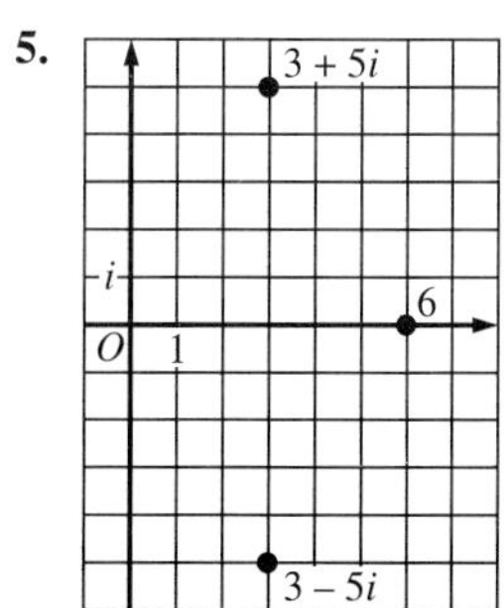

$\sqrt{34}$; $\sqrt{34}$; 6

7.

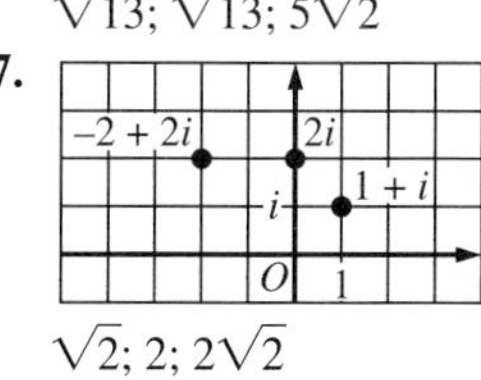

$\sqrt{2}$; 2; $2\sqrt{2}$

9.

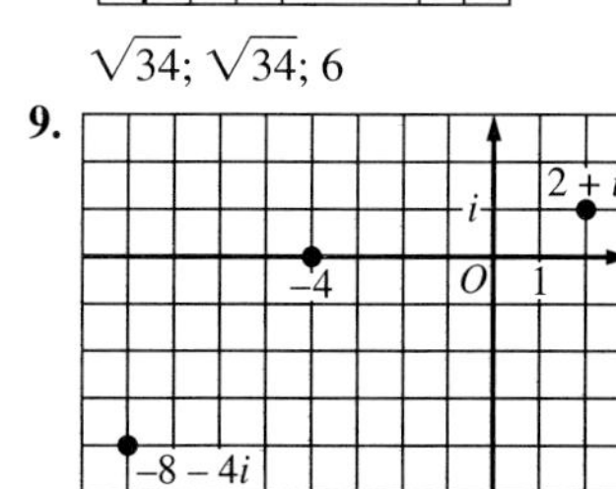

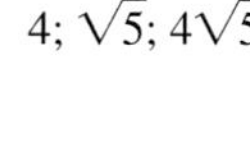

4; $\sqrt{5}$; $4\sqrt{5}$

11.

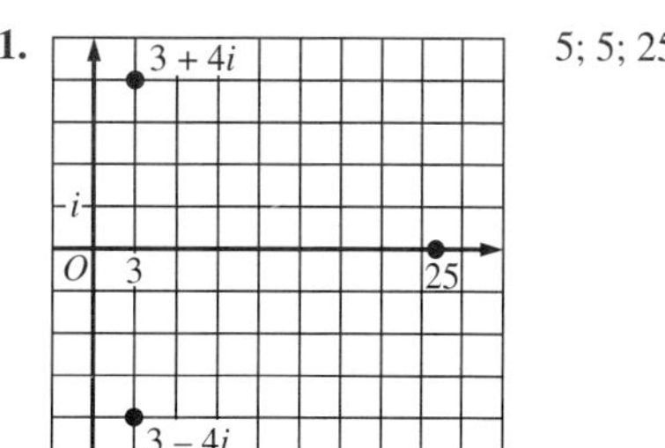

5; 5; 25

13.

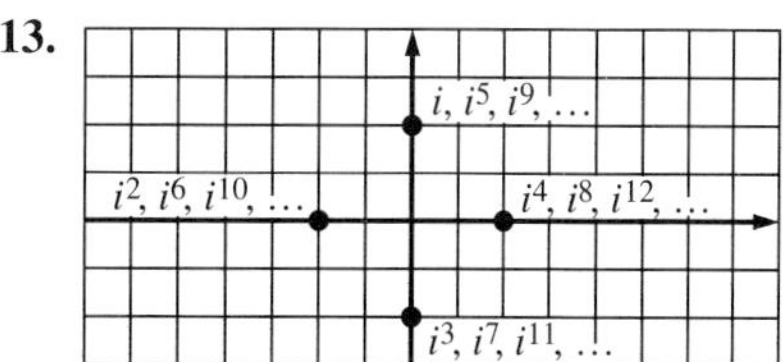

Answers may vary. The sequence with $t_n = i^n$ is the repeating sequence $i, -1, -i, 1, i, -1, -i, 1, \ldots$.

Pages 368–370 Exercises and Applications

1, 3. Choices of numbers with the same magnitude may vary. Examples are given. **1.** $1 + 2i$ and $1 - 2i$; $\sqrt{5}$; $-1 + 2i$ and $-1 - 2i$ **3.** $-2 - 3i$ and $2 - 3i$; $\sqrt{13}$; $2 + 3i$ and $-2 + 3i$

5.

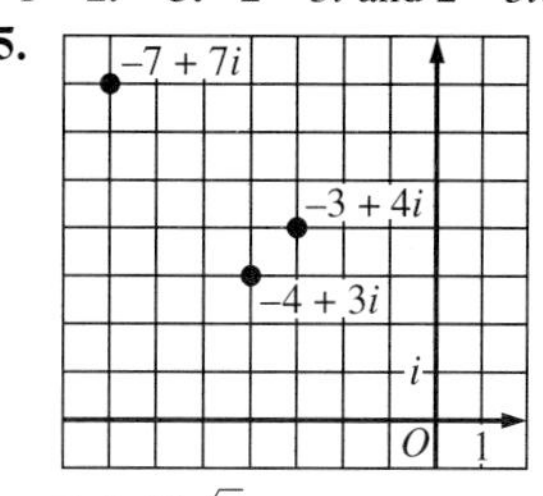

5; 5; $7\sqrt{2}$

7.

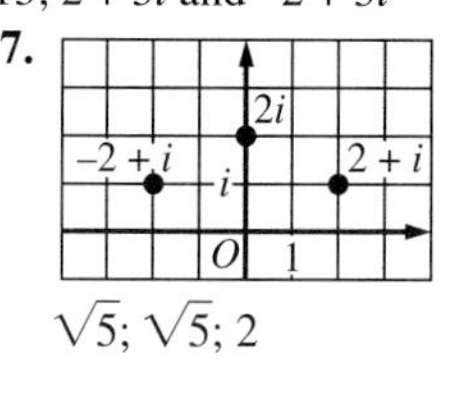

$\sqrt{5}$; $\sqrt{5}$; 2

9.

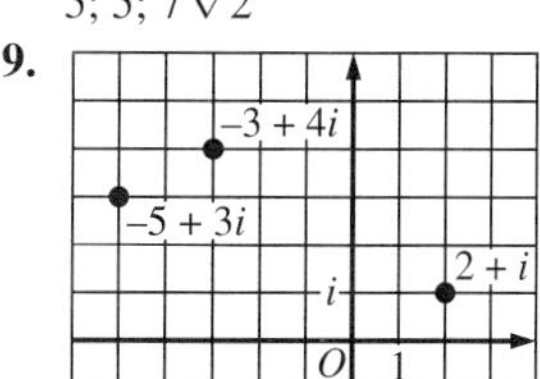

$\sqrt{5}$; $\sqrt{34}$; 5

11.

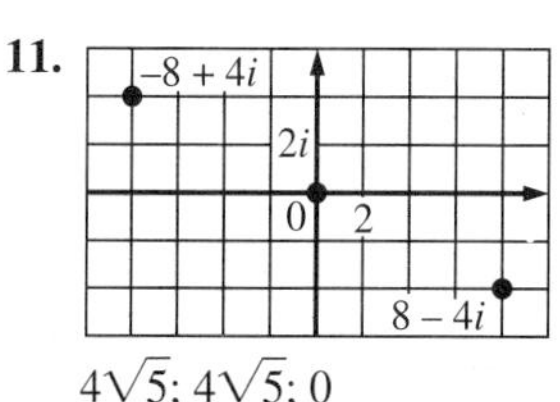

$4\sqrt{5}$; $4\sqrt{5}$; 0

13.

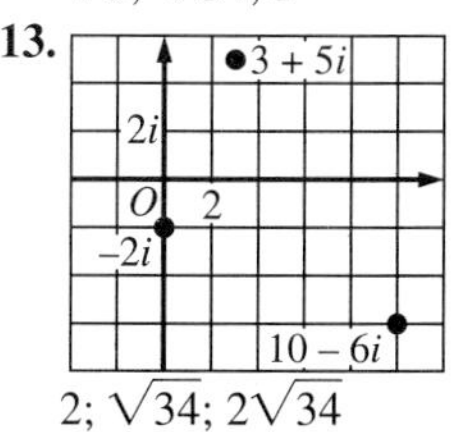

2; $\sqrt{34}$; $2\sqrt{34}$

15.

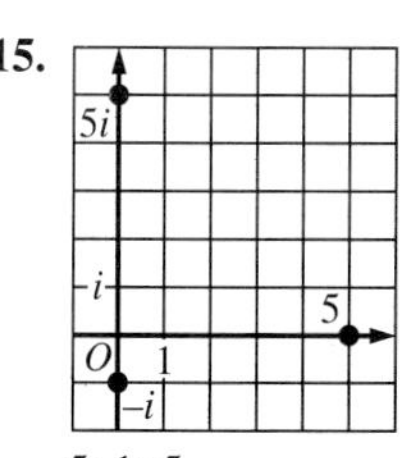

5; 1; 5

17.

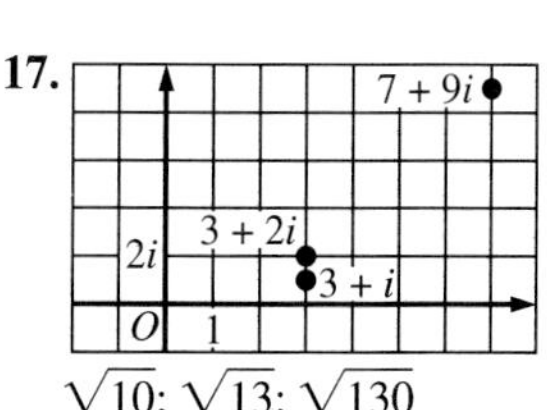

$\sqrt{10}$; $\sqrt{13}$; $\sqrt{130}$

19.

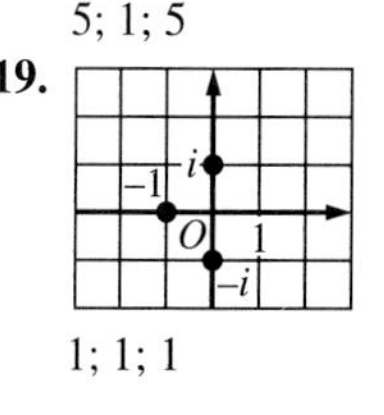

1; 1; 1

21. $5 - 2i$ **23.** $-5 - 6i$ **25.** $4 - 7i$ **27.** $-a - bi$

29. 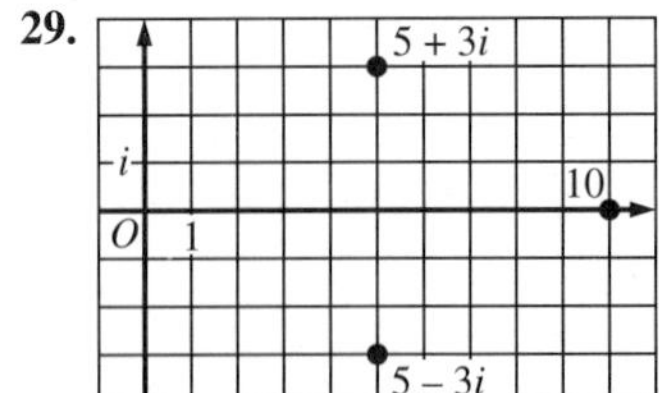

31. 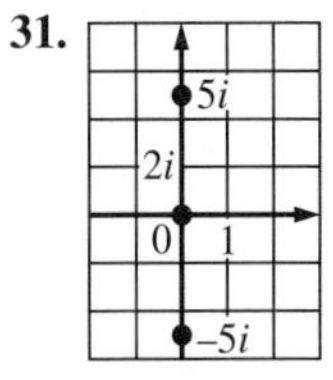

33. 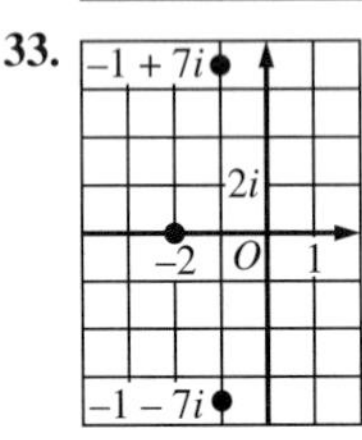

35. Answers may vary. An example is given. A complex number and its complex conjugate are images of each other reflected over the real axis.
53. 16.8; 7.1 **55.** 51.8; 25.5
57. Answers may vary. An example is given. $x^2 + 9 = 0$

59. $\begin{bmatrix} 7 & 3 \\ 13 & 5 \end{bmatrix}$ **61.** $\begin{bmatrix} 11 & 3 \\ 56 & 27 \end{bmatrix}$

Page 375 Checking Key Concepts

1. whole numbers, integers, rational numbers, real numbers, complex numbers **3.** imaginary numbers, complex numbers **5.** not possible **7.** Answers may vary. An example is given. π **9.** Yes; do not form a group; there is no identity and the inverse property does not hold. **11.** No; do not form a group; the set is not closed under addition.

Pages 375–379 Exercises and Applications

1. imaginary numbers, complex numbers **3.** irrational numbers, real numbers, complex numbers **5.** not possible **7.** Answers may vary. An example is given. 7 **9.** No; do not form a group; the set is not closed under multiplication. **11.** Yes; form a commutative group; the set is closed under multiplication and all the group properties as well as the commutative property hold. **13.** No; do not form a group; the set is not closed under multiplication. **15.** Yes; form a commutative group; the set is closed under multiplication and all the group properties as well as the commutative property hold. **17.** Yes; do not form a group; there is no identity and the inverse property does not hold. **19.** No; do not form a group; the set is not closed under multiplication. **21. a.** closure, identity, inverse, associative, commutative **b.** closure, identity, associative **25.** Counterexamples may vary. No; $\alpha \dagger \delta = \alpha$ but $\gamma \dagger \alpha = \gamma$, so there is not a unique identity.

27.

×	A	B	C	D
A	D	C	B	A
B	C	D	A	B
C	B	A	D	C
D	A	B	C	D

29. 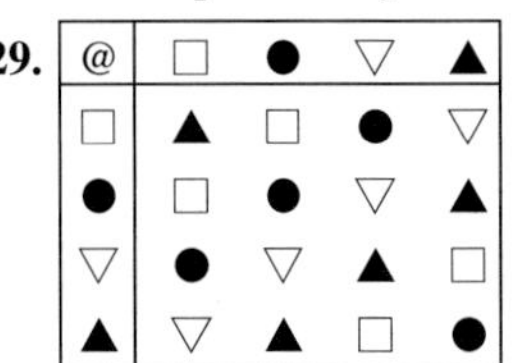

@	□	●	▽	▲
□	▲	□	●	▽
●	□	●	▽	▲
▽	●	▽	▲	□
▲	▽	▲	□	●

37. a. No. **b.** Yes. **c.** Yes. **39.** RF **41.** TRF **43.** I
45. RF

57. 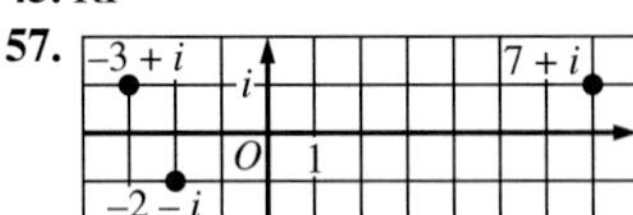

59. $y = (2x - 5)(x + 3)$
61. $y = (2x - 7)(2x + 7)$
63. $5x - 15$

Page 379 Assess Your Progress

1. $-1 + 5i$ **2.** 58 **3.** $39 - 80i$

4. Answers may vary. An example is given. **5.** $\sqrt{153}$

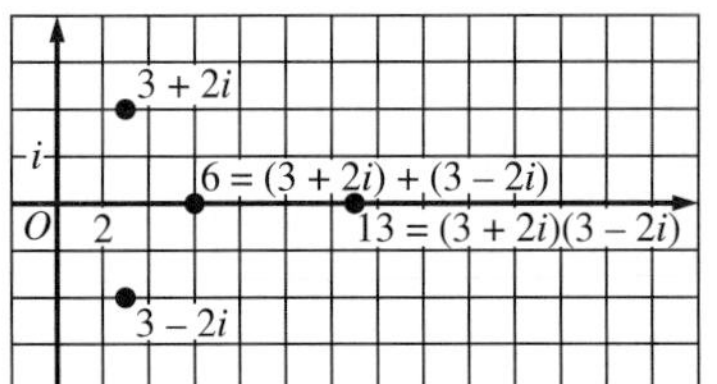

6. Yes; do not form a group; there is no identity and the inverse property does not hold.
7. Yes; do not form a group; the inverse property does not hold.

Pages 384–385 Chapter 8 Assessment

1. radical function **2.** complex plane; magnitude **3.** closure **4.** nonnegative numbers; $y \geq -4$ **5.** nonnegative numbers; $y \geq 1$ **6.** $x \geq 2$; $y \geq 3$ **7.** all real numbers; all real numbers **8.** $x \geq -5$; $y \geq 1$ **9.** nonnegative numbers; $y \geq 2$ **10.** $z^{1/9}$ **11.** $y^{3/5}$ **12.** $y^{7/4}$ **13.** $\sqrt[3]{t^4}$ **14.** $\sqrt[8]{b^3}$ **15.** $\sqrt[7]{p^5}$ **16.** 4 **17.** 13 **18.** 2 **20. a.** 33.6 m/s **b.** 32.8 m/s **c.** 14.1 m/s **21.** $3 - 2i$ **22.** $9 - 3i$ **23.** 15 **24.** $12 - 9i$ **25.** $-2 + 14i$ **26.** $\frac{38}{29} - \frac{21}{29}i$ **27–32.** Examples of numbers with the same magnitude may vary.

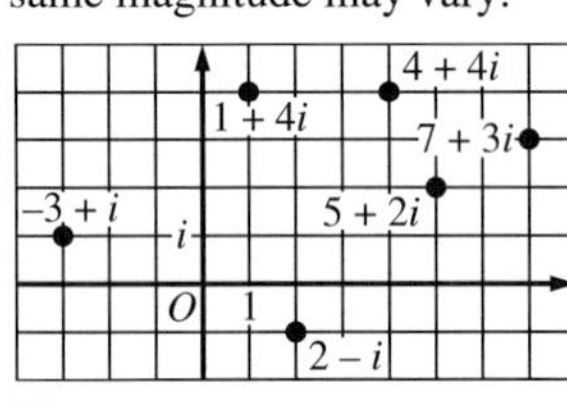

27. $\sqrt{17}$; $1 - 4i$ and $-4 + i$
28. $\sqrt{58}$; $7 - 3i$ and $-7 + 3i$
29. $4\sqrt{2}$; $4 - 4i$ and $-4 - 4i$
30. $\sqrt{5}$; $2 + i$ and $-2 + i$
31. $\sqrt{29}$; $5 - 2i$ and $-5 - 2i$
32. $\sqrt{10}$; $3 - i$ and $-3 - i$

33. 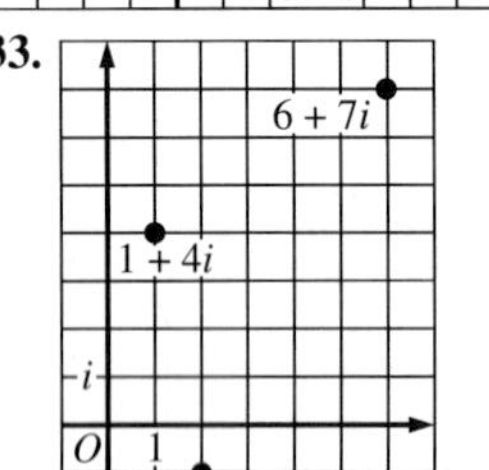

34. 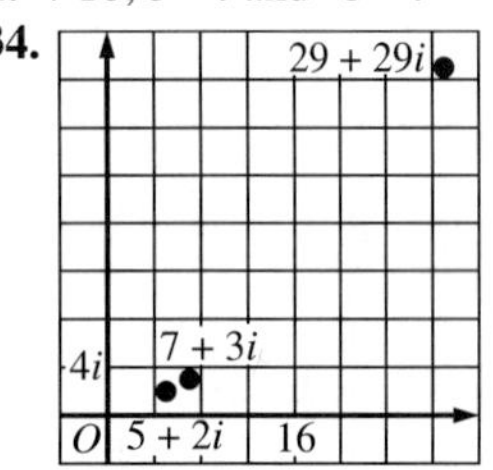

35. 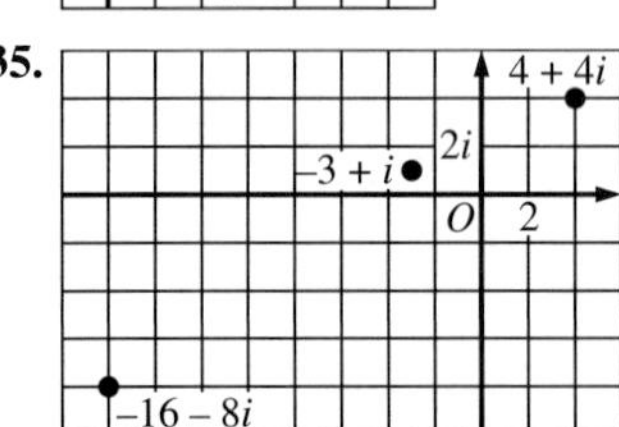

36. Yes; form a commutative group; the set is closed under addition and all the group properties as well as the commutative property hold. **37.** No. **38.** No.
39. Yes. **40.** No; for example A Δ C = C, but C Δ A = D.
41. No; for example, 3 × 7 = 1, which is not in the set.

CHAPTER 9

Page 393 Checking Key Concepts

1. No; the exponents are not whole numbers. **3.** Yes; $-0.94y^5 - 3.8y^4 - 1.6y^2 + 0.22y$; 5 **5.** 7
7. $8x^3 + 8x^2 + 4x + 9$

Pages 393–396 Exercises and Applications

1. No; the exponent 2.9 is not a whole number. **3.** Yes; $t^2 + 6t - 1$; 2 **5.** Yes; $-0.6u^{10} - \sqrt{2}u^8 + 5u^7 - \frac{3}{2}u^4 + \pi$; 10
7. No; 2^r is not the product of a real number and a whole-number power of r. **9.** 143 **11.** –2 **13.** 221 **15.** –1
21. $16x^3 + 9x^2 + 9x + 13$

23. $9.1y^5 - 2.2y^4 - 4.8y^3 - 10.8y^2 + 8.5y + 4$ **25.** $-4\sqrt{2}m^3 - 7\sqrt{5}m^2 + 23m$ **27.** $-x^3 - 10x^2 + 4$ **29.** $-v^3 + \frac{1}{14}v^2 + \frac{27}{20}v - \frac{59}{36}$ **31.** $0.29t^5 - 0.78t^4 + 2.59t^2 - 2.1t + 3.52$ **35.** A **41.** Yes; do not form a group; the inverse property does not hold. **43.** Yes; do not form a group; the inverse property does not hold. **45.** 4 **47.** $\frac{3}{2}$ **49.** $x^2 + x - 12$ **51.** $36n^2 + 12n + 1$

Page 400 Checking Key Concepts

1. $3x^2 + x - 10$ **3.** $u^3 - 4u^2 - 6u - 1$ **5.** $x + 4 + \frac{2}{x+2}$ **7.** $y^2 + 7y + 1 + \frac{4}{y-3}$

Pages 401–403 Exercises and Applications

1. $8x^2 + 26x - 7$ **3.** $-6t^5 + 12t^4 + 2t^3$ **5.** $u^3 + 2u^2 + 2u + 1$ **7.** $6m^3 - 31m^2 - 40m + 16$ **9.** $40n^3 + 66n^2 - 31n - 42$ **11.** $w^4 - 12w^3 + 54w^2 - 108w + 81$ **19.** $x + 2 + \frac{4}{x+1}$ **21.** $3t + 13 + \frac{26}{t-2}$ **23.** $y^2 + 4y - 2 + \frac{5}{y+4}$

41.

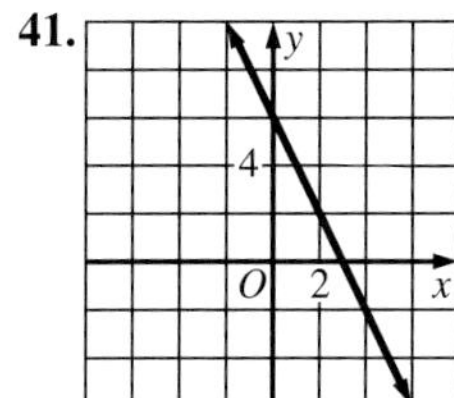

25. $6x^2 + 9x + 4$ **27.** $2x^3 - 5x^2 + 11x - 12 + \frac{18}{x+2}$ **29.** $4n - 1 + \frac{12n+4}{2n^2+7n-5}$ **33.** 47 **35.** 47 **37.** (3, 1) **39.** (–2, 1, –3)

43.

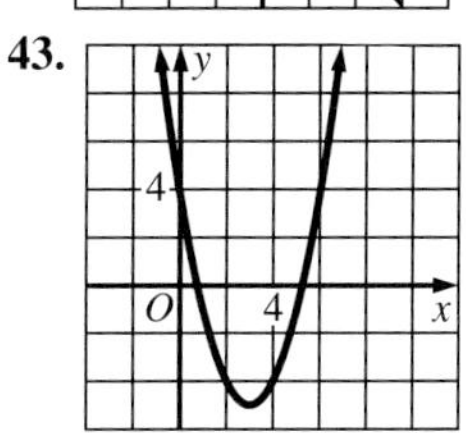

45.

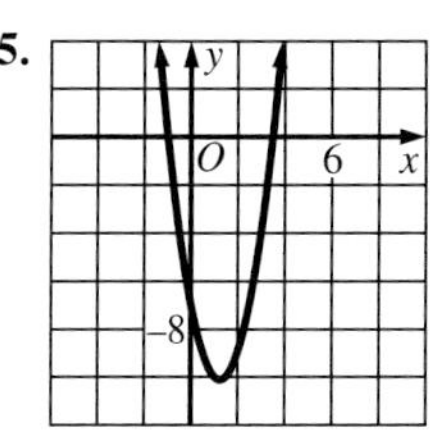

Page 408 Checking Key Concepts

1. A **3.** B

5.

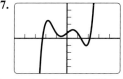

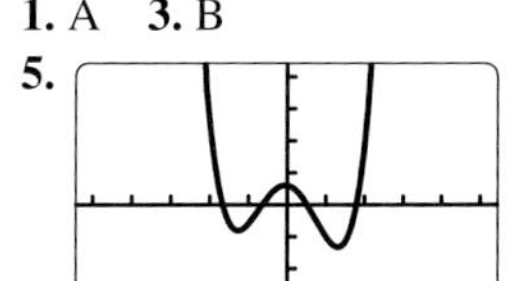

local minimums: -4.06 and -6.64
local maximum: 3.03

7.

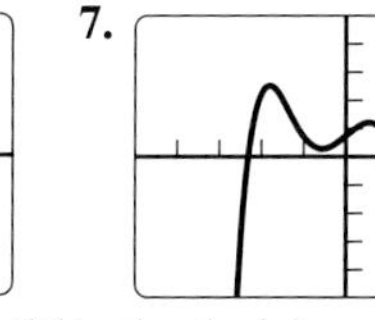

local minimums: 1.20 and –4.10;
local maximums: 10.1 and 4.80

Pages 408–411 Exercises and Applications

3.

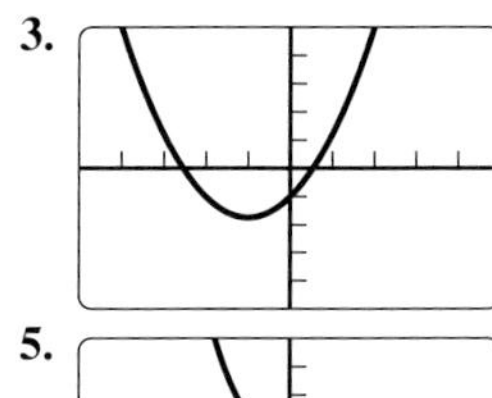

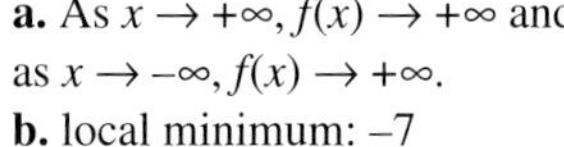

a. As $x \to +\infty, f(x) \to +\infty$ and as $x \to -\infty, f(x) \to +\infty$.
b. local minimum: –7

5.

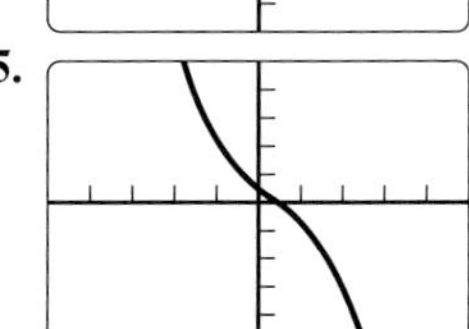

a. As $x \to +\infty, f(x) \to -\infty$ and as $x \to -\infty, f(x) \to +\infty$.
b. no local maximum, no local minimum

7.

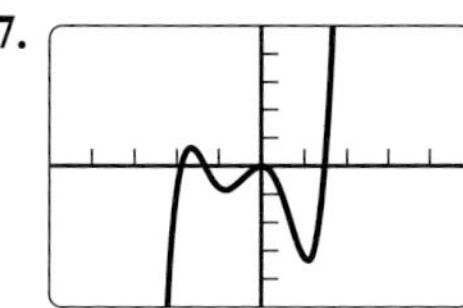

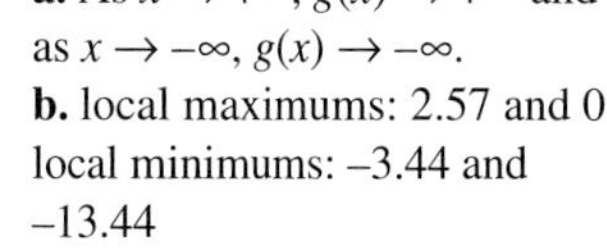

a. As $x \to +\infty, g(x) \to +\infty$ and as $x \to -\infty, g(x) \to -\infty$.
b. local maximums: 2.57 and 0; local minimums: –3.44 and –13.44

9.

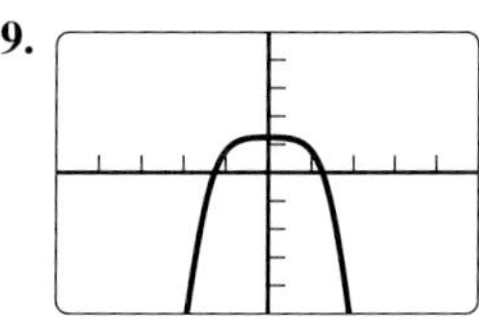

a. As $x \to +\infty, h(x) \to -\infty$ and as $x \to -\infty, h(x) \to -\infty$.
b. local maximum: 5

11.

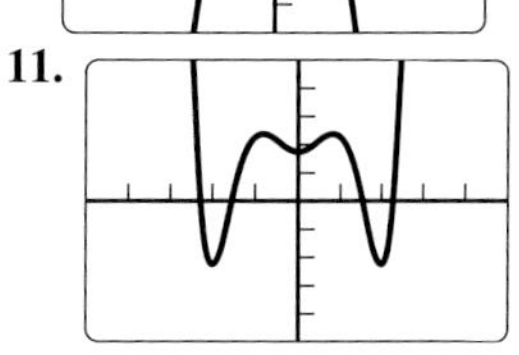

a. As $x \to +\infty, f(x) \to +\infty$ and as $x \to -\infty, f(x) \to +\infty$.
b. local minimums: –9 and 7; local maximum: 9.52

21. $y = \frac{1}{3}(x + 5) - 3$ or $y = \frac{1}{3}(x - 4)$ **23.** 3; 4 **25.** $-\frac{3}{4} \pm \frac{\sqrt{17}}{4}$

Page 411 Assess Your Progress

1. $-2x^5 + 8x^4 + x^3 - 13x + 9$; 5 **2.** 41 **3.** $3x^3 - 3x^2 - 7$ **4.** $4y^4 + 9y^3 - 8y^2 - y - 9$ **5.** $8u^3 + 10u^2 + 13u + 35$ **6.** $4x^2 - 9x + 18 - \frac{28}{x+2}$ **7.** As $x \to +\infty, f(x) \to -\infty$ and as $x \to -\infty, f(x) \to -\infty$.

8. a. $V = x(24 - 2x)(18 - 2x)$

b.

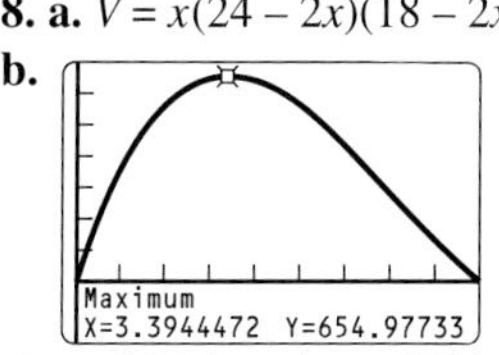

about 3.39 in.; about 655 in.^{3}

Page 416 Checking Key Concepts

1. 0.66 **3.** $f(x) = 2(x - 1)(x - 2)(x - 3)$ **5.** 2; 4; 5 **7.** –1; 2; 6

Pages 416–418 Exercises and Applications

1. 3 **3.** –2.62; –0.38; 1.50 **5.** –1.44; 1.27; 4.38 **7.** $f(x) = (x - 1)(x - 3)(x - 6)$ **9.** $f(x) = 5x(x + 2)(x - 2)$ **13.** –10; 3; 7 **15.** –1; 1; 4 **17.** –9; $\frac{1-\sqrt{5}}{2}$; $\frac{1+\sqrt{5}}{2}$ **19.** 6; $\frac{2-\sqrt{14}}{2}$; $\frac{2+\sqrt{14}}{2}$ **21.** 1989 **23.** $g(x) = 24\left(x + \frac{3}{2}\right)\left(x + \frac{4}{3}\right)\left(x + \frac{5}{4}\right)$ **25.** $f(x) = (x + 7)(x - 1 - \sqrt{5})(x - 1 + \sqrt{5})$ **31.** As $x \to +\infty$, $g(x) \to -\infty$ and as $x \to -\infty, g(x) \to -\infty$. **33.** $\frac{1}{25}$ **35.** 3 **37.** $\pm 4i$ **39.** $2 \pm 3i$

Page 423 Checking Key Concepts

1. 2.00 **3.** –2.98; –1.87; –0.70; 0.63; 1.42 **5. a.** ± 1; ± 3; $\pm\frac{1}{2}$; $\pm\frac{3}{2}$ **b.** –1; $\frac{1}{2}$; 3 **7. a.** ± 1; ± 2; ± 4; ± 8 **b.** 2 (double zero); $1 \pm i$

Pages 423–425 Exercises and Applications

1. –3.67; 0.35; 2.32 **3.** –2.24; 2.76 **5.** –2 (double zero); –0.41; 0.41; 1 **7.** –2.21; –1.67; –0.59; 0.64; 1.57; 2.26 **9. a.** ± 1; ± 13 **b.** 1, $2 \pm 3i$ **11. a.** ± 1; ± 2; ± 4; ± 8; $\pm\frac{1}{3}$; $\pm\frac{2}{3}$; $\pm\frac{4}{3}$; $\pm\frac{8}{3}$ **b.** $\frac{2}{3}$; –2 (double zero) **13. a.** ± 1; ± 2; ± 4; ± 5; ± 8; ± 10; ± 20; ± 40 **b.** –4; –1; 2; 5 **15. a.** ± 1; $\pm\frac{1}{2}$; $\pm\frac{1}{4}$; $\pm\frac{1}{8}$; $\pm\frac{1}{16}$ **b.** $-\frac{1}{4}$ (double zero); $\pm i$ **17. a.** ± 1; ± 2

b. –1 (triple zero); $\frac{5 \pm \sqrt{17}}{2}$ **21.** D
25. $f(x) = -2x(x + 7)(x + 5)$ **27.** No. **29.** No.

Page 425 Assess Your Progress

1. about 9.98 m/s **2.** 1; –2; 3 **3.** –4; $2 \pm \sqrt{2}$
4. $\frac{5}{3}$; $\frac{-3 \pm i\sqrt{7}}{4}$ **5.** $-\frac{1}{2}$; 3; $\pm 2i$

Page 429 Checking Key Concepts

1. Yes; 5. **3.** No. **5.** No. **7.** No.

Pages 430–432 Exercises and Applications

1. Yes; 2. **3.** No. **5.** No. **7.** Yes; 5.

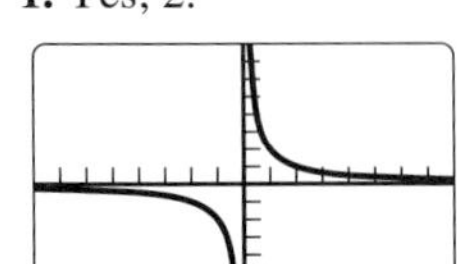

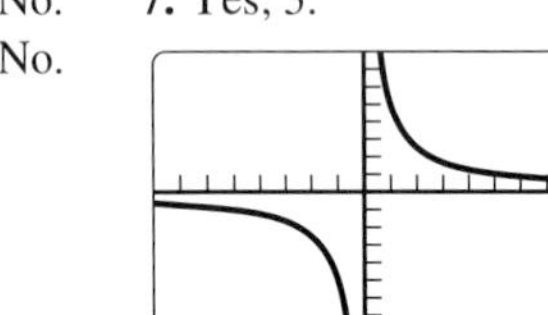

9. Yes; let l = length and w = width; $lw = 72$, so $l = \frac{72}{w}$ or $w = \frac{72}{l}$. **11.** No; let n = the number of gallons of gasoline used, d = the distance driven in miles, and let M represent the car's gasoline usage in mi/gal (a constant); $n = \frac{d}{M}$; this is a direct variation. **23.** $\frac{1}{3}$; $-\frac{1}{2}$; 4 **25.** $\frac{1}{2}$ (double zero); $-2 \pm i$
27. translated 1 unit right **29.** translated 1 unit right and 3 units up

Page 437 Checking Key Concepts

1. D **3.** B

Pages 437–440 Exercises and Applications

1. Yes; $2x + 3$ and $5x - 1$ are both polynomials. **3.** No; $2^t + 1$ is not a polynomial. **7.** $f(x) = \frac{4}{x} + 2$ **9.** $f(x) = \frac{6}{x+2} + 4$
11. $f(x) = \frac{-12}{x+8} - 10$ **17. a.** $x = \frac{3}{2}$; $y = 4$ **b.** It is the graph of $y = \frac{13}{x}$ translated $\frac{3}{2}$ units right and 4 units up. **19. a.** $x = -2$; $y = 3$ **b.** It is the graph of $y = \frac{5}{x}$ translated 2 units left and 3 units up. **21. a.** $x = -\frac{1}{4}$; $y = -5$ **b.** It is the graph of $y = \frac{2}{x}$ translated $\frac{1}{4}$ unit left and 5 units down. **23. a.** $x = \frac{4}{5}$; $y = -6$
b. It is the graph of $y = -\frac{5}{x}$ translated $\frac{4}{5}$ unit right and 6 units down. **27.** Yes; 1.5. **29.** No. **31.** $x + 7 + \frac{3}{x+1}$
33. $4x^2 - x - 1 - \frac{5}{2x-1}$

Page 440 Assess Your Progress

1. Yes; $y = \frac{-12}{x}$. **2.** 20 min **3.** $f(x) = \frac{2}{x-1} - 3$ **4.** $x = -\frac{2}{3}$; $y = 2$; It is the graph of $y = \frac{3}{x}$ translated $\frac{2}{3}$ unit left and 2 units up.

Page 445 Checking Key Concepts

1. a. $x = -7$; $x = 4$ **b.** $f(x) \to 0$ as $x \to \pm\infty$. **3. a.** $x = \frac{17}{4}$
b. $g(x) \to -\infty$ as $x \to -\infty$ and $g(x) \to +\infty$ as $x \to +\infty$.

Pages 445–448 Exercises and Applications

1. C **3.** B **7. a.** $x = -5$; $x = 2$ **b.** $f(x) \to 0$ as $x \to \pm\infty$.
9. a. $x = \frac{9}{5}$ **b.** $h(x) \to +\infty$ as $x \to -\infty$ and $h(x) \to -\infty$ as $x \to +\infty$. **11. a.** $x = -4$; $x = 2$; $x = 3$ **b.** $g(x) \to -1$ as $x \to \pm\infty$. **19.** It is the graph of $y = \frac{5}{x}$ translated 6 units left and 3 units down. **21.** It is the graph of $y = \frac{-2}{x}$ translated $\frac{1}{2}$ unit right and 7 units down. **23.** 6 **25.** 4

Page 451 Checking Key Concepts

1. 4 **3.** 1; 15 **5.** $16\frac{2}{3}$ oz

Pages 451–453 Exercises and Applications

1. 2 **3.** no solution **5.** $\frac{7}{3}$; 3 **7.** 6 **9.** no solution **11.** $-\frac{1}{4}$
17. $13 - 4i$ **19.** $8 + 6i$ **21.** $y = x^2 - x$

Page 453 Assess Your Progress

1. a. $x = 1$; $x = 5$ **b.** $f(x) \to 0$ as $x \to \pm\infty$. **2. a.** $x = -3$
b. $g(x) \to +\infty$ as $x \to -\infty$ and $g(x) \to -\infty$ as $x \to +\infty$.
3. a. $x = -5$; $x = 2$ **b.** $h(x) \to +\infty$ as $x \to \pm\infty$. **4. a.** $x = 0$; $x = -3$; $x = 3$ **b.** $f(x) \to 3$ as $x \to \pm\infty$. **5.** –1 **6.** 2
7. 5 consecutive hits

Pages 458–459 Chapter 9 Assessment

1. degree; leading coefficient; constant term **2.** inverse variation; constant of variation **3.** –32 **4.** $-2x^3 + 13x^2 - 11x + 1$
5. $4u^4 + 3u^3 - 7u^2 + 11u - 21$ **6.** $12y^3 - 10y^2 + 23y - 7$
7. $5x^2 - 6x + 9 - \frac{32}{2x+3}$ **9.** local minimums: 3.92 and –11; local maximum: 9.58 **11.** 1956 **12.** $f(x) = 2x(x + 2)(x - 3)$
13. 4; $3 + 2\sqrt{2} \approx 5.83$; $3 - 2\sqrt{2} \approx 0.17$
14. $-\frac{3}{2}$ (double zero); $1 \pm i\sqrt{3}$
15. a. No; the products dl are not constant.

b.

d (in.)	l (in.)	A (in.^{2})
$\frac{1}{8}$	1440	$\frac{\pi}{256}$
$\frac{1}{4}$	360	$\frac{\pi}{64}$
$\frac{3}{8}$	160	$\frac{9\pi}{256}$
$\frac{1}{2}$	90	$\frac{\pi}{16}$

Yes; for all corresponding values of l and A, $lA = \frac{45\pi}{8}$.

c. $l = \frac{45\pi}{8A}$ **16.** horizontal asymptote: $y = 5$; vertical asymptote: $x = -\frac{1}{6}$; The graph is the graph of $y = \frac{2}{x}$ translated $\frac{1}{6}$ unit left and 5 units up. **17.** $x = -3$ and $x = \frac{1}{4}$; As $x \to \pm\infty$, $h(x) \to +\infty$. **18.** at least 10 three-point shots **19.** –1

CHAPTERS 7–9

Pages 460–461 Cumulative Assessment

1. (2, –3); one solution **3.** (10, 4); one solution

5. $\begin{bmatrix} 0.3 & -2.1 & 4.5 \\ 5.2 & 0.2 & -3.7 \\ -1.8 & -3.0 & -1.6 \end{bmatrix}\begin{bmatrix} x \\ y \\ z \end{bmatrix} = \begin{bmatrix} 12.4 \\ 4.0 \\ 9.2 \end{bmatrix}$; (1.41, –4.27. 0.669)

7.

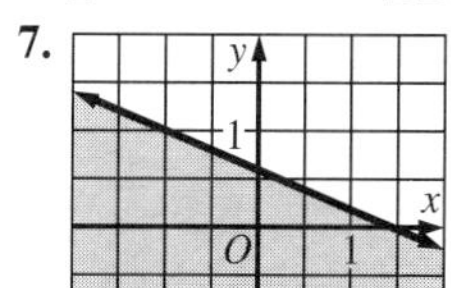

9.

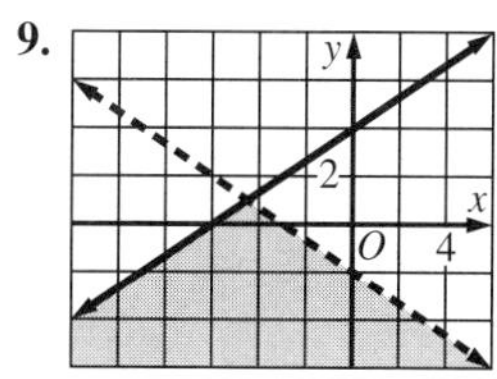

11. 4 tapes; 4 CDs

13. a.

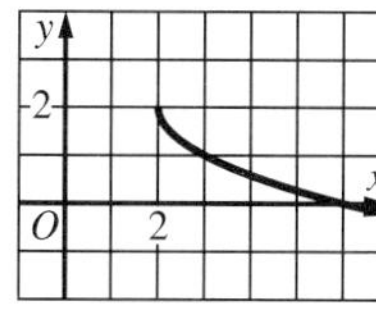

$x \geq 2$; $y \leq 2$

15. 5

17. $\pm 4i\sqrt{2}$

19.

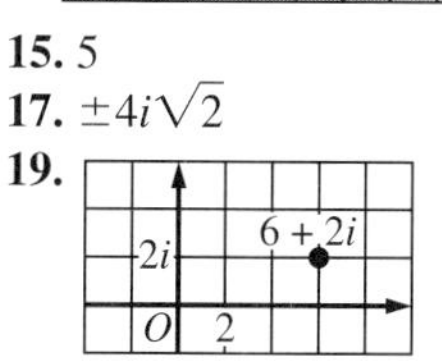

$6 + 2i$; $2\sqrt{10}$

21. $\frac{24}{25} + \frac{7}{25}i$ **25.** $1.2x^3 - x^2 + 10x - 13$

29. $y = -\frac{1}{2}(x + 2)(x - 1)(x - 4)$ **31.** possible rational zeros: ± 1; ± 2; ± 5; ± 10; ± 25; ± 50; zeros: –1; 5; $-1 \pm 3i$ **33.** Yes; 40. **35.** $x = -3$; $x = 2$; As $x \to \pm\infty$, $g(x) \to 0$. **37.** 0; 3

CHAPTER 10

Page 468 Checking Key Concepts

1. a. $\frac{1}{25}, \frac{1}{36}, \frac{1}{49}, \frac{1}{64}$ **b.** $t_n = \frac{1}{n^2}$ **3. a.** $\sqrt{5}, \sqrt{6}, \sqrt{7}, 2\sqrt{2}$ **b.** $t_n = \sqrt{n}$ **5.** 65,536

Pages 468–471 Exercises and Applications

1. 1024, 4096, 16,384, 65,536 **3.** 7, 72, 9, 144 **5.** 55,555, 555,555, 5,555,555, 55,555,555 **7.** 183, 243, 313, 393 **9.** $\sqrt{243}$, 27, $\sqrt{2187}$, 81 **11.** Yes; U.S. presidential elections are held every four years. **13.** Yes; on most streets, house lots are numbered by consecutive even or consecutive odd integers. (There are exceptions to this rule.) **17.** 8; 125; 1728 **19.** $\frac{5}{2}$; $\frac{50}{17}$; $\frac{60}{19}$ **21.** 16; 169; 1156 **23.** 16, 100, 576 **25.** –1; 1; –1 **27.** Logs are rounded to two decimal places. log 16 ≈ 1.20; log 40 ≈ 1.60; log 96 ≈ 1.98 **29.** $t_n = 6n + 37$ **31.** $t_n = \frac{3}{5^n}$ **33.** $t_n = 109 - 6n$ **35.** $t_n = \frac{1}{2n-1}(-1)^{n+1}$ **43.** finite; The circumference of the vase is finite, so the pattern can only contain a finite number of terms. **45.** infinite; The sequence 1, 3, 5, 7, 9, ... continues without end. **47.** finite; $\frac{1}{128} = 0.0078125$ **57.** –4; 7 **59.** –2 **61.** 729 **63.** $2\sqrt[5]{3}$

65.

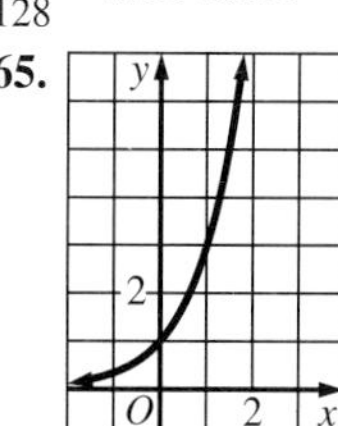

67.

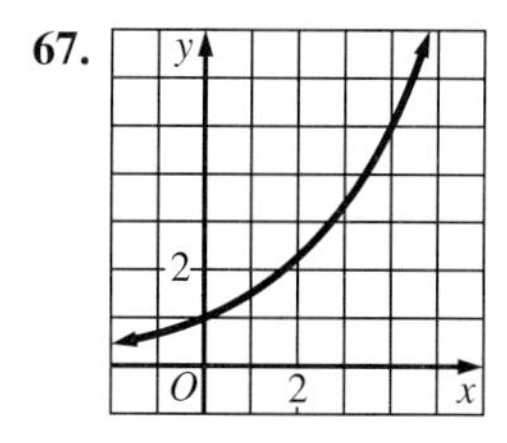

Page 476 Checking Key Concepts

1. a. arithmetic **b.** 51 **3. a.** arithmetic **b.** –7 **5. a.** neither **b.** 121 **7. a.** 14.5 **b.** 10 **9. a.** $2\sqrt{2}$ **b.** $\sqrt{6}$

Pages 476–479 Exercises and Applications

1. arithmetic **3.** neither **5.** geometric **7.** neither **9.** geometric **11.** arithmetic **13. a.** (1) 8; (4) 1; (8) –41; (11) –0.03 **b.** (1) 112; (4) 45; (8) –404; (11) –1 **17. a.** 36.5 **b.** 24 **19. a.** 10.5 **b.** $4\sqrt{5}$ **21. a.** 3 **b.** $\sqrt{8.51}$ **23. a.** 2.02 **b.** 0.4 **25. a.** neither; The sequence is 1, 6, 12, 18, 24, ... **b.** $t_1 = 1$ and for $n \geq 2$, $t_n = 6(n - 1)$. **35. a.** 1, 3, 9, 27 **b.** geometric; Each term is 3 times the preceding term. **c.** $t_n = 3^{n-1}$ **37. a.** $\frac{\sqrt{3}}{4}, \frac{3\sqrt{3}}{16}, \frac{9\sqrt{3}}{64}, \frac{27\sqrt{3}}{256}$ **b.** geometric; Each term is $\frac{3}{4}$ times the precedingterm. 39. $15\sqrt{3}$ 41. 435 43. $\pm\sqrt{2}$; 5

Page 483 Checking Key Concepts

1. 0, 6, 12, 18 **3.** –2, –7, –12, –17 **5.** 6.2, 3.1, 1.55, 0.775 **7.** $t_1 = 4$; $t_n = 7(t_{n-1})$ **9.** $t_1 = 19$; $t_n = t_{n-1} - 6$ **11.** $t_1 = \frac{1}{3}$; $t_n = (t_{n-1})^2$

Pages 483–487 Exercises and Applications

1. a. 4, –2, 1, $-\frac{1}{2}$ **b.** $t_n = 4\left(-\frac{1}{2}\right)^{n-1}$ **3. a.** 4.5, 27, 162, 972 **b.** $t_n = 4.5 \cdot 6^{n-1}$ **5. a.** 9, 9, 9, 9 **b.** $t_n = 9$ **7. a.** 0, 6, 12, 18 **b.** $t_n = 6(n - 1)$ or $t_n = 6n - 6$ **11.** $t_1 = 1.1$; $t_n = t_{n-1} + 1.1$ **13.** $t_1 = 50$; $t_n = \frac{1}{2}(t_{n-1})$ **15.** $t_1 = 3$; $t_n = (t_{n-1})^2$ **17.** $t_1 = 3$; $t_n = \sqrt{5}(t_{n-1})$ **19.** $t_1 = 2$; $t_n = t_{n-1} + 3$ **21.** $t_1 = 1$; $t_n = t_{n-1} + 1$ **27.** $t_1 = 6$; $t_2 = 7$; $t_n = 2(t_{n-1} + t_{n-2})$ **33.** $12\sqrt{3}$ **35.** $2\sqrt{6\sqrt{10}}$ **37.** Answers may vary. Justifications are given for both. Numerical data may involve measurement such as with shirts sized by neck measurements. Categorical data may involve clothing such as sweaters grouped into sizes such as small, medium, and large. **39.** 75 **41.** $-\frac{55}{14} \approx -3.93$

Page 487 Assess Your Progress

1. –19, –25, –32; neither **2.** 128, –256, 512; geometric **3.** 40, 49, 58; arithmetic **4.** 15; $4\sqrt{11}$ **5.** 31.7; 7.5 **6.** $3\sqrt{3}$; $\sqrt{15}$ **7.** explicit: $t_n = -7 + (-6)(n - 1)$ or $t_n = -6n - 1$; recursive: $t_1 = -7$; $t_n = t_{n-1} - 6$ **8.** explicit: $t_n = 3 \cdot \left(\frac{2}{3}\right)^{n-1}$; recursive: $t_1 = 3$; $t_n = \frac{2}{3}(t_{n-1})$ **9.** explicit: $t_n = 19 + 3(n - 1)$ or $t_n = 3n + 16$; recursive: $t_1 = 19$; $t_n = t_{n-1} + 3$

Page 491 Checking Key Concepts

1. a. 21 terms **b.** 1260 **3. a.** 13 terms **b.** –39 **5.** 4 + 32 + 108 + 256 + 500 + 864 **7.** 1 + 4 + 7 + 10 + 13

9. $\sum_{n=1}^{4} 3 \cdot 4^{n-1}$ **11.** $\sum_{n=1}^{9} (28 - 2n)$

Pages 491–494 Exercises and Applications

1. 3240 **3.** 15,150 **5.** 19,943 **9.** 50 **11.** 68

13. $-11 + (-10) + (-9) + (-8) + (-7) + (-6)$
15. $3 + (-11) + (-49) + (-123) + (-245) + (-427) + (-681)$
17. $\sum_{n=1}^{41} (2.3n - 2.3)$; 1886 **19.** $\sum_{n=1}^{18} \left(\frac{7}{2}n + \frac{9}{2}\right)$; $679\frac{1}{2}$
21. $\sum_{n=1}^{33} (7.3 - 0.2n)$; 128.7 **25. a.** 91 oranges **b.** $\sum_{n=1}^{12} n^2$
27. B **31.** $t_1 = 14$; $t_n = t_{n-1} - 8$ **33.** $\begin{bmatrix} -42 & -27 & 19 \\ -33 & -18 & 11 \\ -21 & -27 & 26 \end{bmatrix}$

Page 499 Checking Key Concepts

1. 728 **3.** –255 **5.** The sum does not exist because the series is infinite with $r = 4$ and $|4|$ is not less than 1. **7.** 1.5
9. $\sum_{n=1}^{6} 2 \cdot 3^{n-1}$ **11.** $\sum_{n=1}^{8} 3(-2)^{n-1}$

Pages 500–503 Exercises and Applications

1. 508 **3.** 2,222,222,222,222 **5.** 7.9921875 **7.** 61.4
9. 2.33×10^{20} **11.** 265,720 employees **13.** 18 **15.** $117\frac{1}{3}$
17. The sum does not exist because the series is infinite with $r = 1\frac{1}{3}$ and $\left|1\frac{1}{3}\right|$ is not less than 1. **19.** 6.25 **21.** The sum does not exist because the series is infinite with $r = -1$ and $|-1|$ is not less than 1. **25.** $\frac{8}{9}$ **27.** $\frac{4}{99}$ **29.** $\frac{368}{99}$ **31.** $\frac{1058}{999}$
33. $-\frac{151}{3330}$ **35.** $-\frac{359}{33}$ **45.** $\sum_{n=1}^{42} (-14n + 197) = -4368$
47. $\left(-14, 6\frac{3}{4}\right)$ **49.** $\left(\frac{26}{31}, \frac{94}{31}\right)$ **51.** $-\frac{1}{5}$ **53.** 5

Page 503 Assess Your Progress

1. a. geometric **b.** 8 terms **c.** 1275 **2. a.** arithmetic **b.** 8 terms **c.** $\frac{1}{2}$ **3. a.** geometric **b.** 7 terms **c.** 117,186
4. does not exist **5.** 125
6. $\sum_{n=1}^{8} (-17n + 131)$; 436 **7.** $\sum_{n=1}^{\infty} 3\left(\frac{2}{3}\right)^{n-1}$; 9
8. $6 + 11 + 16 + 21 + 26 + 31 + 36 + 41 + 46$; 234
9. $0.3 + 2.1 + 14.7 + 102.9 + 720.3 + 5042.1 + 35{,}294.7$; 41,177.1 **10.** $0.8 + 0.64 + 0.512 + 0.4096 + \ldots$; 4

Pages 508–509 Chapter 10 Assessment

1. geometric; common ratio; arithmetic; arithmetic mean
2. finite; series **3.** 125, 216, 343, 512; neither **4.** 1, –1, 1, –1; geometric **5.** –13, –18, –23, –28; arithmetic **6.** $t_n = n^3$
7. $t_n = (-1)^{n-1}$ **8.** $t_n = 7 - 5(n - 1)$ or $t_n = -5n + 12$ **9.** $\frac{12}{13}$
10. $\frac{1}{531{,}441}$ **11.** 58 **12. a.** 20.5 **b.** 20 **13. a.** 9 **b.** $\sqrt{77}$
14. a. 0.75 **b.** 0.6 **15. a.** 100, –10, 1, –0.1
b. $t_n = 100(-0.1)^{n-1}$ **16. a.** 100, 99.9, 99.8, 99.7
b. $t_n = 100 - 0.1(n - 1)$ or $t_n = -0.1n + 100.1$ **17. a.** 3, $3\sqrt{2}$, 6, $6\sqrt{2}$ **b.** $t_n = 3(\sqrt{2})^{n-1}$ **18.** $t_1 = 7$; $t_n = t_{n-1} - 5$

19. $t_1 = \frac{1}{3}$; $t_n = \frac{1}{3}(t_{n-1})$ **20.** $t_1 = 3$; $t_n = t_{n-1} + 5$ **21.** $t_n = 2^n$; $t_1 = 2$, $t_n = 2(t_{n-1})$; geometric; Each term is twice the preceding term. **23. a.** arithmetic **b.** 16 terms
c. $\sum_{n=1}^{16} (-8n + 28)$ **d.** –640 **24. a.** geometric **b.** 8 terms
c. $\sum_{n=1}^{8} 2 \cdot 5^{n-1}$ **d.** 195,312 **25. a.** geometric **b.** 6 terms
c. $\sum_{n=1}^{6} 8\left(-\frac{1}{2}\right)^{n-1}$ **d.** $5\frac{1}{4}$ **26. a.** arithmetic **b.** 17 terms
c. $\sum_{n=1}^{17} (0.5n + 6.5)$ **d.** 187 **27.** 20 rows; 59 seats
28. a. $\frac{1}{2} + 4 + \frac{27}{2} + 32 + \frac{125}{2}$; $112\frac{1}{2}$
b. $1000 - 250 + 62.5 - 15.625 + \ldots$; 800 **30.** $\frac{35}{1998}$

CHAPTER 11

Page 516 Checking Key Concepts

1. a. $3\sqrt{2} \approx 4.24$ **b.** (2.5, 6.5) **3. a.** $\sqrt{13} \approx 3.61$
b. (–4.5, –1)

Pages 516–519 Exercises and Applications

1. a. (1.5, –4) **b.** $\frac{4}{5}$ **c.** $-\frac{5}{4}$ **d.** $y + 4 = -\frac{5}{4}(x - 1.5)$ or $y = -\frac{5}{4}x - \frac{17}{8}$ **3. a.** 2.5 **b.** $y = \frac{3}{4}x - \frac{21}{16}$ **5. a.** $\sqrt{197} \approx 14.04$
b. $y = -14x - 60.5$ **7. a.** $\sqrt{106} \approx 10.30$ **b.** $y = -\frac{5}{9}x + \frac{2}{9}$
9. a. $\sqrt{183.38} \approx 13.54$ **b.** $y = -\frac{107}{83}x - \frac{2118}{415}$ **11. a.** $\frac{\sqrt{1445}}{8} \approx 4.75$ **b.** $y = -38x - \frac{979}{16}$ **13. a.** $\sqrt{460.85} \approx 21.47$ **b.** $y = \frac{17}{214}x + \frac{8177}{4280}$ **15.** 7 or 15 **17.** 14 or –34 **19.** $-3 \pm 4\sqrt{2}$; about 2.66 or –8.66 **21.** Answers may vary. An example is given. $QR = RS = ST = TQ = 5\sqrt{2}$. The slope of $\overline{QR}$ and $\overline{ST}$ is 1, and the slope of $\overline{RS}$ and $\overline{TQ}$ is –1. The adjacent sides of $QRST$ are perpendicular. Therefore, $QRST$ is a rhombus with four right angles, or a square. **25. a.** $10 + 2\sqrt{34} \approx 21.66$
b. Proofs may vary. An example is given. $AB = CD = \sqrt{34}$, and $BC = AD = 5$. Since both pairs of opposite sides are congruent, $ABCD$ is a parallelogram. **29. a.** (1.5, 2.5), (2.5, 6.5), (6.5, 5.5), (5.5, 1.5) **b.** (2, 4.5), (4.5, 6), (6, 3.5), (3.5, 2)
c. $4\sqrt{34}$; $4\sqrt{17}$; $4\sqrt{8.5}$; The perimeter of the next square should be $4\sqrt{4.25}$. To check the prediction, find two vertices of the next square, for example, (3.25, 5.25) and (5.25, 4.75). These give a side length of $\sqrt{4.25}$, and a perimeter of $4\sqrt{4.25}$, as predicted. **35.** $\sum_{n=1}^{7} 6(3)^{n-1} = 6558$

37. a.

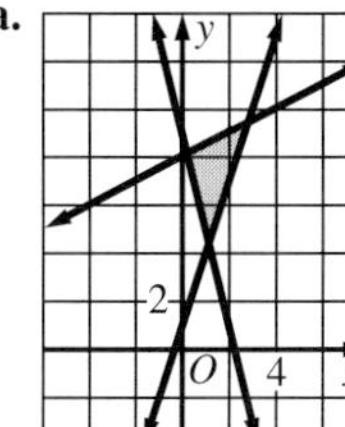

b. 55.40 **c.** $\frac{97}{7} \approx 13.86$

39. parabola that opens up and has vertex $(3, -4)$

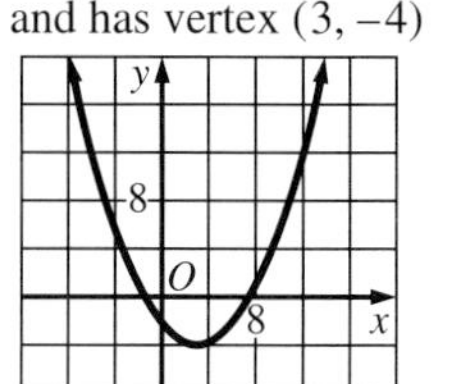

Page 524 Checking Key Concepts

1. The farther apart the directrix and focus are, the wider the parabola becomes. **3.** $y - \frac{3}{2} = \frac{1}{6}(x - 1)^2$

5. $x - 1 = -\frac{1}{12}(y + 1)^2$

Pages 525–527 Exercises and Applications

3. A

5. a. $y = -\frac{1}{12}x^2$

b.

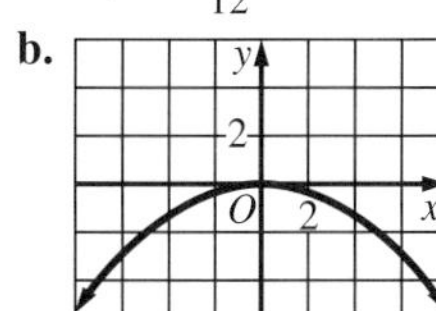

7. a. $y = \frac{1}{4}x^2$

b.

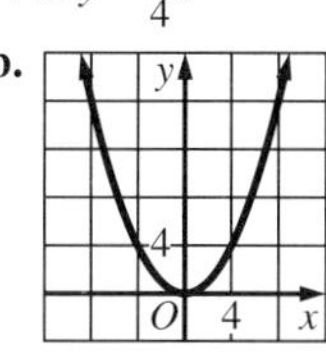

9. a. $x - 2 = -\frac{1}{24}(y - 6)^2$

b.

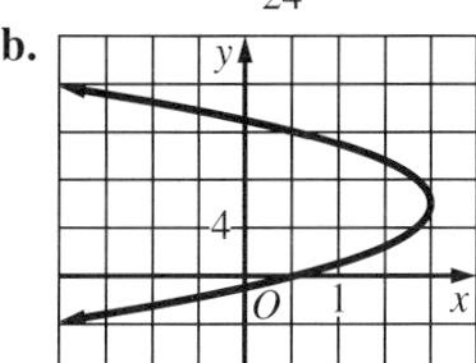

11. about 1.8 m

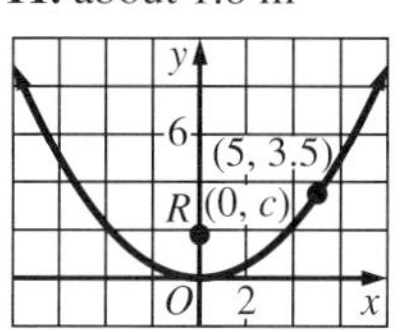

13. about 27.8 m

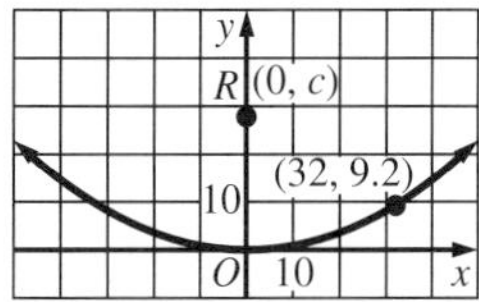

15. vertex: $(0, 0)$; focus: $\left(0, \frac{1}{12}\right)$; directrix: $y = -\frac{1}{12}$

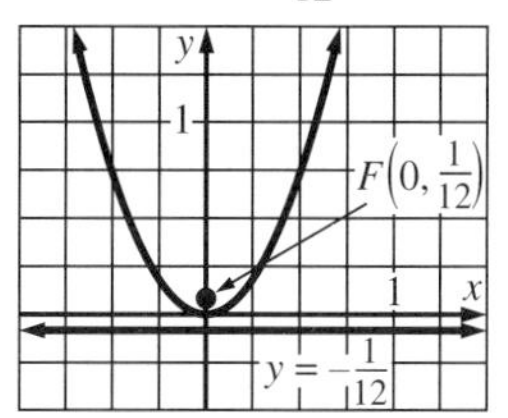

17. vertex: $(-1, 4)$; focus: $(-1, 6)$; directrix: $y = 2$

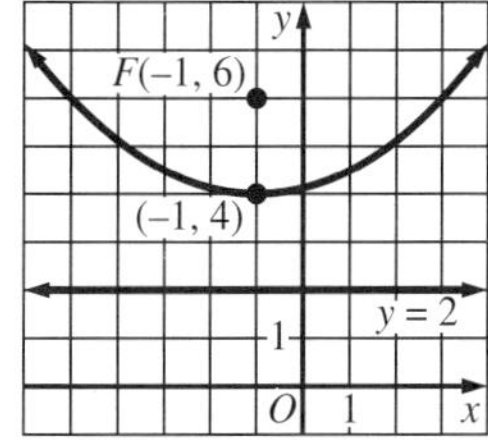

19. vertex: $(2, 1)$; focus: $\left(2\frac{1}{16}, 1\right)$; directrix: $x = 1\frac{15}{16}$

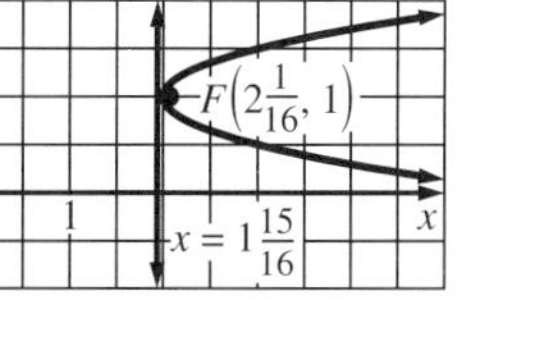

21. vertex: $(0, 2)$; focus: $\left(0, 2\frac{1}{4}\right)$; directrix: $y = 1\frac{3}{4}$

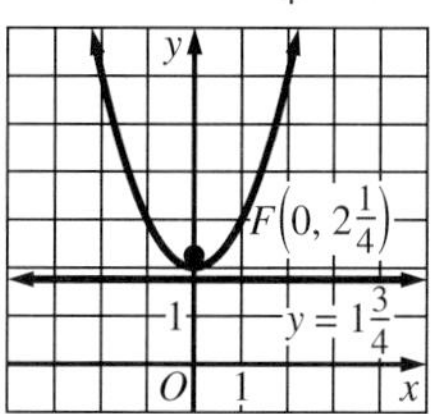

23. vertex: $\left(-\frac{3}{4}, 0\right)$; focus: $\left(\frac{1}{4}, 0\right)$; directrix: $x = -1\frac{3}{4}$

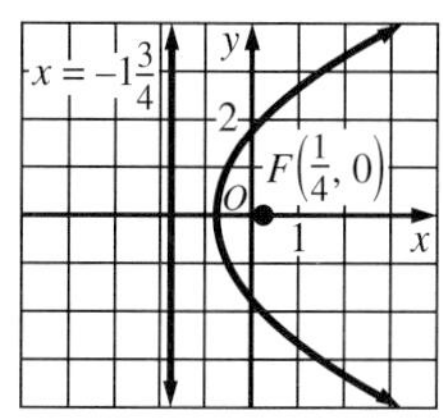

31. a. $8\sqrt{13} \approx 28.84$ **b.** $(4, -3)$ **33.** $\left(\frac{15}{4}, \frac{5}{4}\right)$

Page 527 Assess Your Progress

1. a. $\sqrt{145} \approx 12.04$ **b.** $(1, 3.5)$ **2. a.** $\sqrt{185} \approx 13.60$ **b.** $(0.5, -1)$ **3. a.** $y = -\frac{4}{5}x + 1$ **b.** $y = \frac{5}{4}x - \frac{37}{4}$

4. vertex: $(-3, 1)$; focus: $\left(-2\frac{3}{4}, 1\right)$; directrix: $x = -3\frac{1}{4}$

5. vertex: $(-5, -6)$; focus: $(-5, -1)$; directrix: $y = -11$

6. $y - 1 = \frac{1}{12}(x - 3)^2$ **7.** $x + 2\frac{1}{2} = -\frac{1}{14}y^2$

Page 531 Checking Key Concepts

1. $C(6, 4)$; $r = 12$
3. $C(-5, 0)$; $r = 6$
5. $(x + 2)^2 + (y - 3)^2 = 16$
9. $(0, -2)$

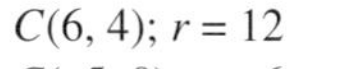

7.

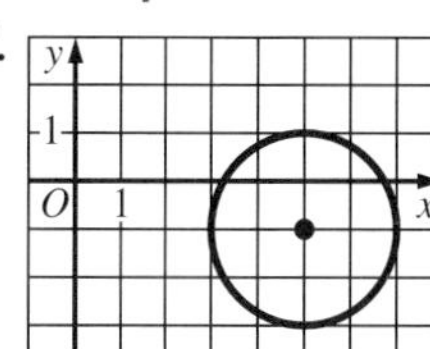

Pages 532–534 Exercises and Applications

1. $C(4, 2)$; $r = 5$

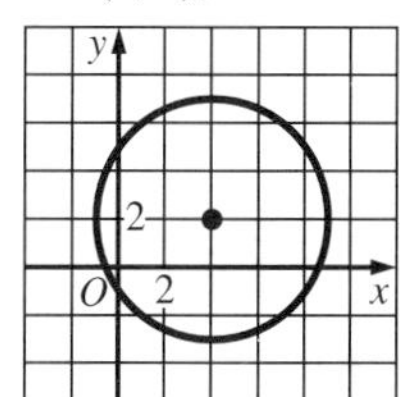

3. $C(0, 1)$; $r = 2$

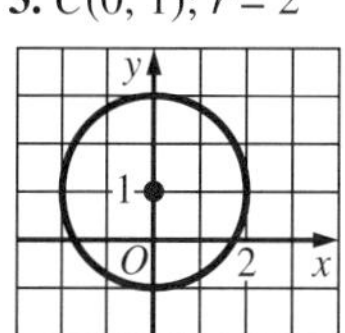

5. $C(-1, -1)$; $r = \sqrt{10} \approx 3.16$

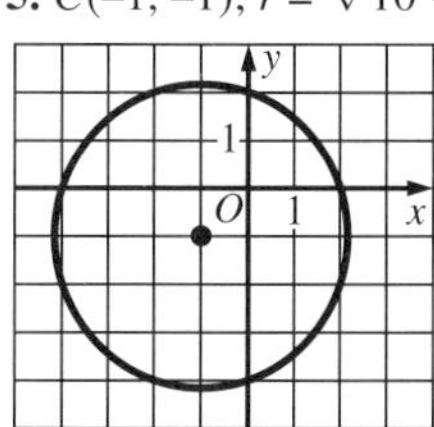

7. $C(0, -3)$; $r = \sqrt{2} \approx 1.41$

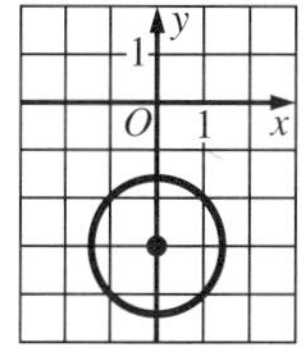

9. $C(0, 1)$; $r = 2\sqrt{2} \approx 2.83$

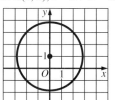

11. $x^2 + y^2 = 1$; 1 mi
13. Yes; $(0.9)^2 + (1.2)^2 < (1.6)^2$
15. $x^2 + y^2 = 49$
17. $(x + 1)^2 + (y - 6)^2 = 9$
19. $x^2 + (y - \sqrt{3})^2 = 121$
21. $(x - a)^2 + (y - b)^2 = 121$
27. Yes. **29.** Yes.
33. $(x - 3)^2 + (y - 1)^2 = 25$

35. about (3.46, 3) and (–3.46, 3)

41. $y = \frac{1}{4}(x - 3)^2$

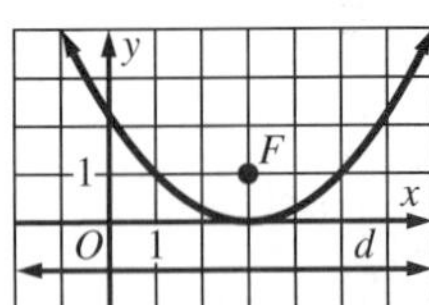

43. $y + 1 = -\frac{1}{8}(x - 4)^2$

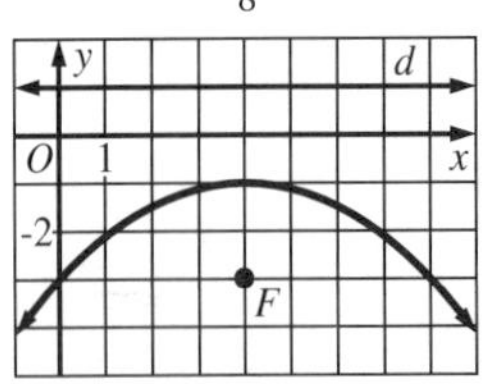

45.

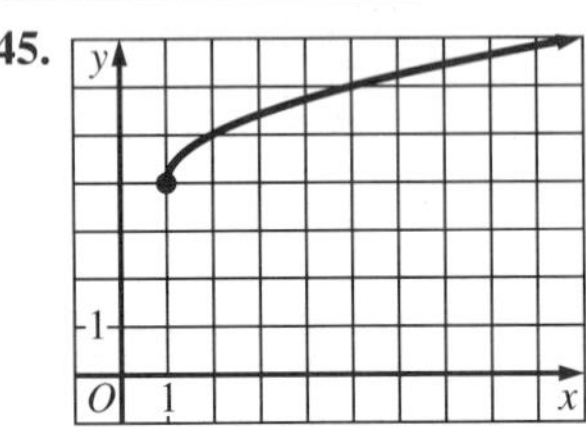

$x \geq 1;\ y \geq 4$

Page 539 Checking Key Concepts

1. (6, –4); (0, –4) and (12, –4) **3.** $\frac{(x-6)^2}{100} + \frac{(y+4)^2}{64} = 1$

5. center: (–1, 3); vertices: (–9, 3) and (7, 3); foci: $(-1 - 2\sqrt{15}, 3)$ and $(-1 + 2\sqrt{15}, 3)$; horizontal

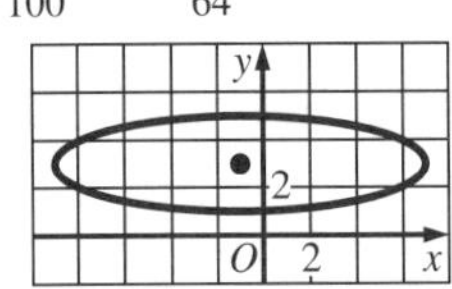

Pages 540–542 Exercises and Applications

1. center: (0, 0); vertices: (8, 0) and (–8, 0); foci: $(\sqrt{39}, 0)$ and $(-\sqrt{39}, 0) \approx (\pm 6.24, 0)$; horizontal

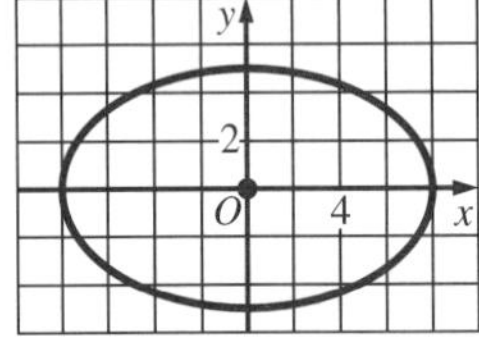

3. center: (0, 1); vertices: $(2\sqrt{2}, 1)$ and $(-2\sqrt{2}, 1) \approx (\pm 2.83, 1)$; foci: $(\sqrt{6}, 1)$ and $(-\sqrt{6}, 1) \approx (\pm 2.45, 1)$; horizontal

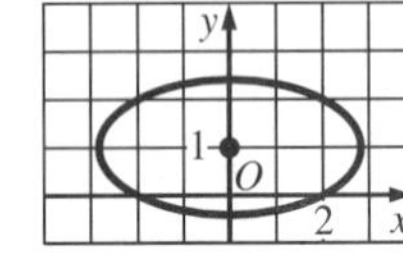

5. center: (–8, –1); vertices: (–8, 9) and (–8, –11); foci: $(-8, -1 \pm 5\sqrt{3}) \approx$ (–8, 7.66) and (–8, –9.66); vertical

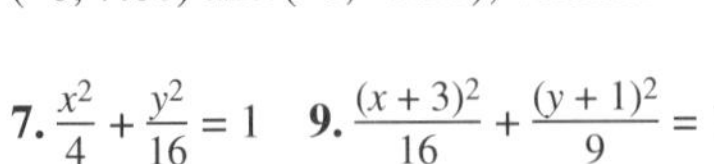

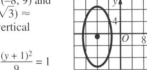

7. $\frac{x^2}{4} + \frac{y^2}{16} = 1$ **9.** $\frac{(x+3)^2}{16} + \frac{(y+1)^2}{9} = 1$

21. $\frac{(x-5)^2}{4} + \frac{(y+3)^2}{9} = 1$ **23.** $\frac{(x+2)^2}{49} + \frac{(y-4)^2}{9} = 1$

27. $(x + 2)^2 + (y + 1)^2 = 36$

29. $y = 2x$

31. $y = -\frac{b}{a}x$

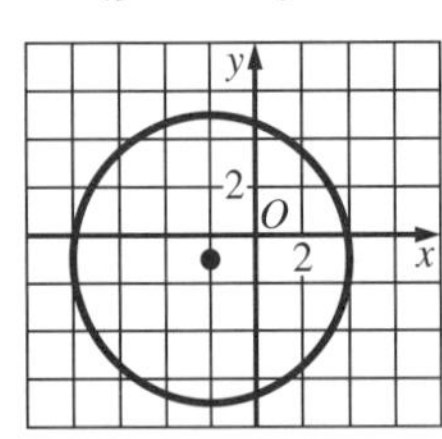

Page 542 Assess Your Progress

1. center: (–1, 5); radius: 4 **2.** center: (3, 0); radius: 5

3. (3, 5) and $\left(-\frac{77}{41}, \frac{45}{41}\right) \approx (-1.88, 1.10)$ **4.** center: (0, 5); vertices: (–8, 5) and (8, 5); foci: $(-2\sqrt{7}, 5)$ and $(2\sqrt{7}, 5)$; horizontal

5. center: (–3, 5); vertices: (–3, 19) and (–3, –9); foci: $(-3, 5 + 4\sqrt{6})$ and $(-3, 5 - 4\sqrt{6})$; vertical

6. $x^2 + (y + 5)^2 = 16$

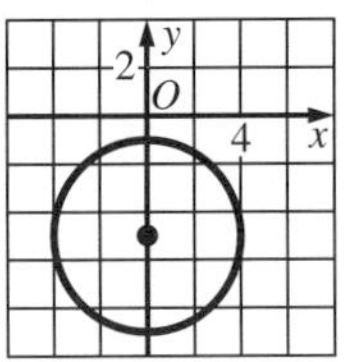

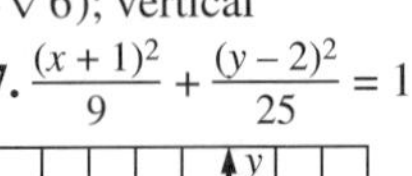

7. $\frac{(x+1)^2}{9} + \frac{(y-2)^2}{25} = 1$

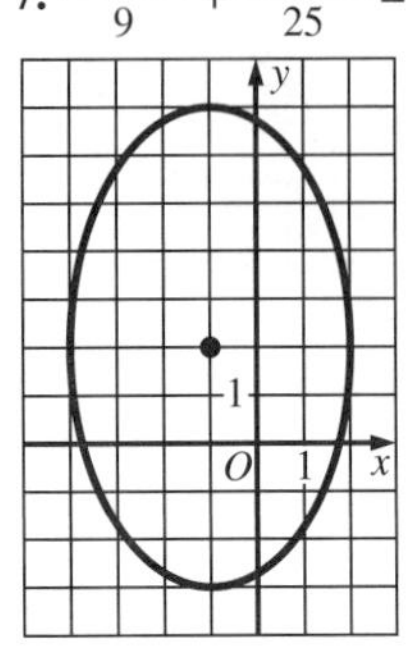

Page 546 Checking Key Concepts

1. (–2, 5); (–2, 10) and (–2, 0) **3.** $\frac{(y-5)^2}{16} - \frac{(x+2)^2}{9} = 1$

5. center: (6, 4); vertices: (2, 4) and (10, 4); foci: $(6 \pm \sqrt{65}, 4) \approx (14.06, 4)$ and (–2.06, 4); horizontal; asymptotes: $y = \pm\frac{7}{4}(x - 6) + 4$

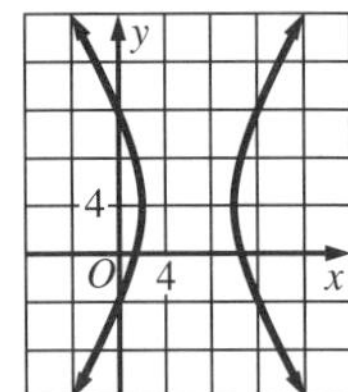

Pages 547–549 Exercises and Applications

1. center: (0, 0); vertices: (±6, 0); foci: $(\pm\sqrt{61}, 0) \approx (\pm 7.81, 0)$; horizontal; asymptotes: $y = \pm\frac{5}{6}x$

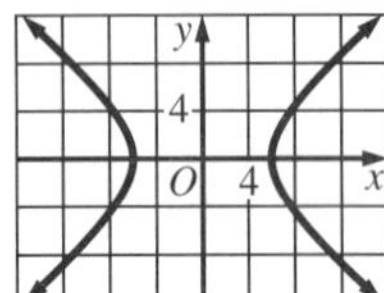

3. center: (0, 3); vertices: (±9, 3); foci: $(\pm 3\sqrt{10}, 3) \approx (\pm 9.49, 0)$; horizontal; asymptotes: $y = \pm\frac{1}{3}x + 3$

5. center: (5, –7); vertices: (5, –2) and (5, –12); foci: $(5, -7 \pm 5\sqrt{5}) \approx$ (5, 4.18) and (5, –18.18); vertical; asymptotes: $y = \pm\frac{1}{2}(x - 5) - 7$

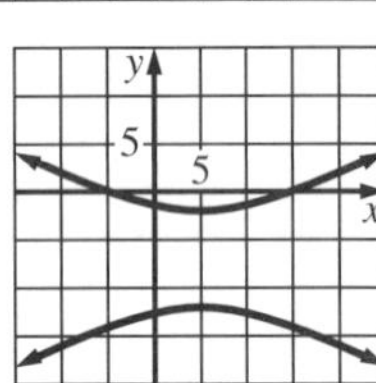

7. center: (–4, 0); vertices: (–4, ±3); foci: $(-4, \pm\sqrt{58}) \approx (-4, \pm 7.62)$; vertical; asymptotes: $y = \pm\frac{3}{7}(x + 4)$

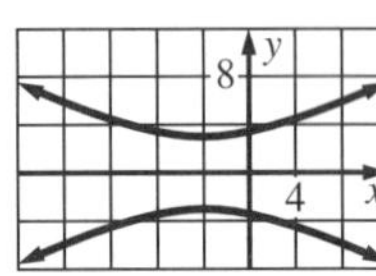

9. center: (0, 0); vertices: $(0, \pm 6\sqrt{3}) \approx (0, 10.39)$; foci: $(0, \pm\sqrt{183}) \approx (0, \pm 13.53)$; vertical; asymptotes: $y = \pm 1.2x$

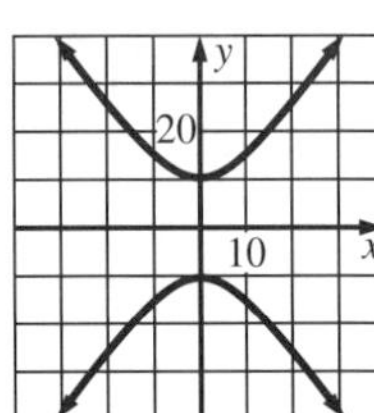

11. C **13.** A

19. a.

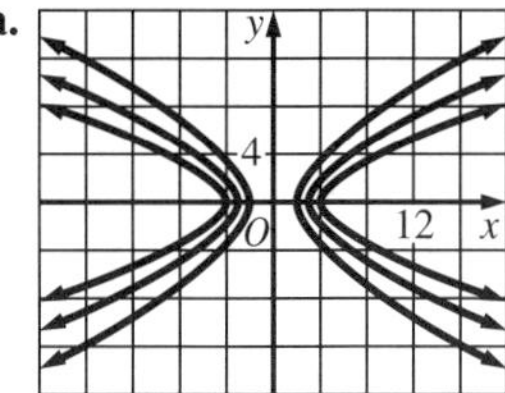

b. The vertices move out farther from the center (the origin), and the hyperbola becomes narrower.

c. Answers may vary. An example is given, using the equations $\frac{x^2}{4} - \frac{y^2}{4} = 1$, $\frac{x^2}{4} - \frac{y^2}{9} = 1$, and $\frac{x^2}{4} - \frac{y^2}{16} = 1$.

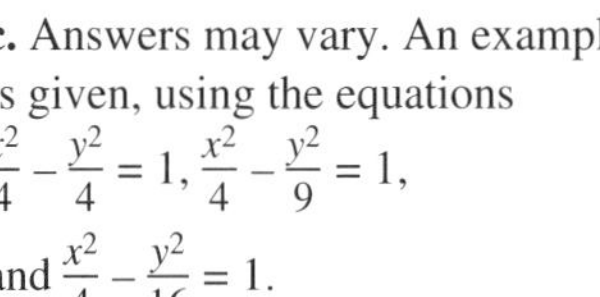

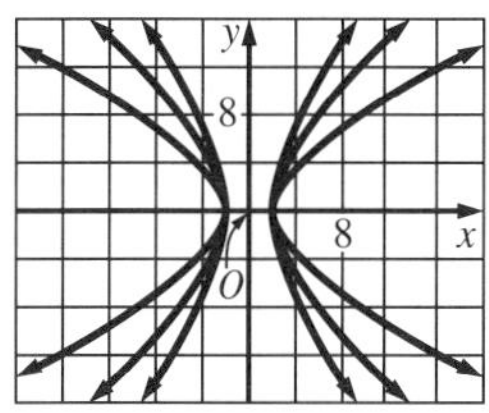

As the value of b increases, the hyperbola widens, while the vertices remain fixed.

21. $\frac{x^2}{9} - \frac{y^2}{16} = 1$

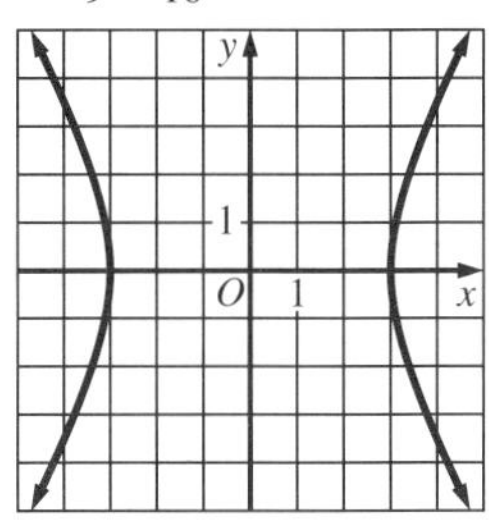

23. $\frac{x^2}{36} - \frac{y^2}{64} = 1$

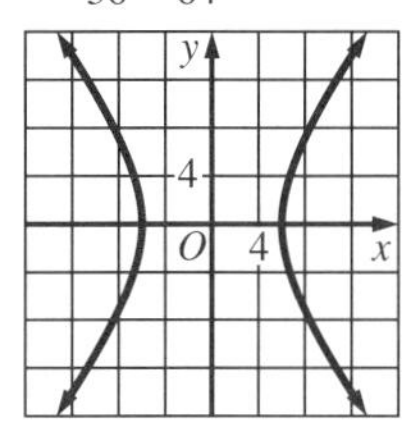

25. $\frac{(x-1)^2}{36} - \frac{(y+4)^2}{64} = 1$

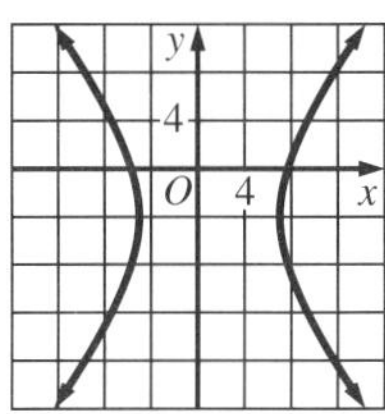

29. $\frac{(x+6)^2}{16} + \frac{(y+1)^2}{1} = 1$

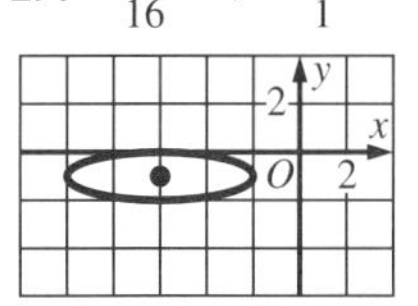

31. $\frac{(x-2)^2}{169} + \frac{(y+3)^2}{25} = 1$

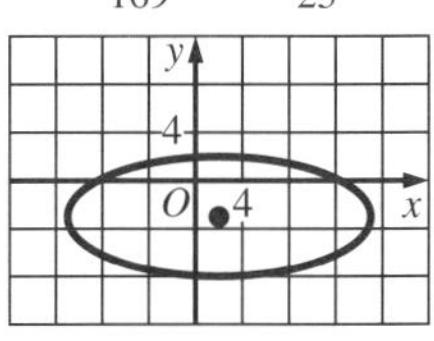

33. $y = (x-6)^2 - 4$; $y = (x-8)(x-4)$

Page 552 Checking Key Concepts

1. $(x-3)^2 + \left(y + \frac{1}{2}\right)^2 = \frac{25}{4}$; a circle with center $\left(3, -\frac{1}{2}\right)$ and radius $\frac{5}{2}$ **3.** a point, (0, 0), representing a degenerate ellipse; The graph is the point (0, 0).

Pages 552–553 Exercises and Applications

1. $\frac{(x+5)^2}{4} - \frac{(y+2)^2}{9} = 1$; a horizontal hyperbola with center (–5, –2), vertices (–3, –2) and (–7, –2), and asymptotes $y = \pm\frac{3}{2}(x+5) - 2$ **3.** $y = 6(x+1)^2$; a parabola, opening upward, with vertex (–1, 0) and axis $x = -1$

5. $x^2 + \left(y + \frac{3}{2}\right)^2 = \frac{25}{4}$; a circle with center $\left(0, -\frac{3}{2}\right)$ and radius $\frac{5}{2}$

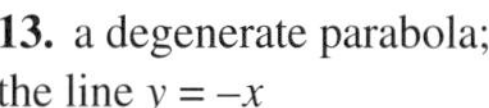

13. a degenerate parabola; the line $y = -x$

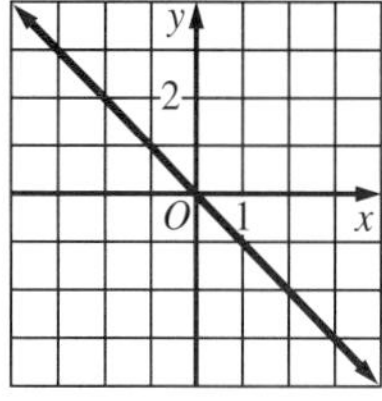

15. a degenerate ellipse; the point (2, 0)

17. a degenerate hyperbola; the intersecting lines $y = \frac{1}{2}x + \frac{3}{2}$ and $y = -\frac{1}{2}x + \frac{1}{2}$

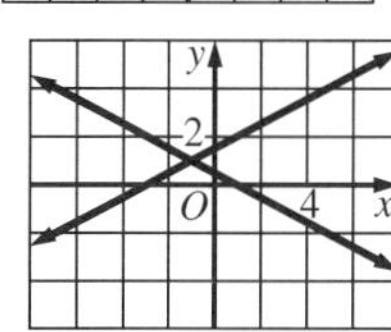

19. center: (–1, 3); vertices: (6, 3) and (–8, 3); foci: $(-1 \pm \sqrt{74}, 3) \approx$ (7.60, 3) and (–9.60, 3); horizontal; asymptotes: $y = \pm\frac{5}{7}(x+1) + 3$

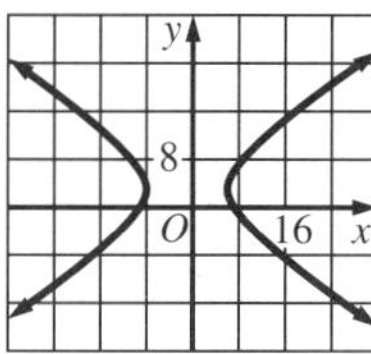

21.

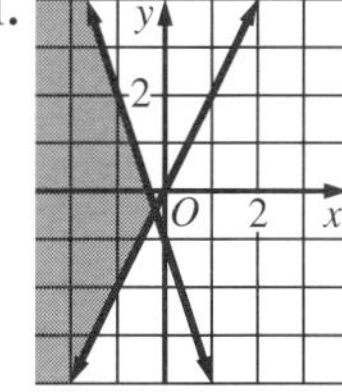

23.

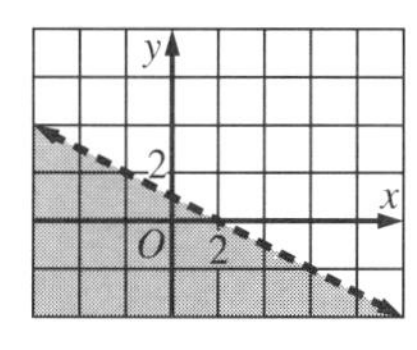

Page 553 Assess Your Progress

1. center: (0, 5); vertices: (±9, 5); foci: $(\pm 3\sqrt{13}, 5) \approx$ (10.82, 5); horizontal; asymptotes: $y = \pm\frac{2}{3}x + 5$

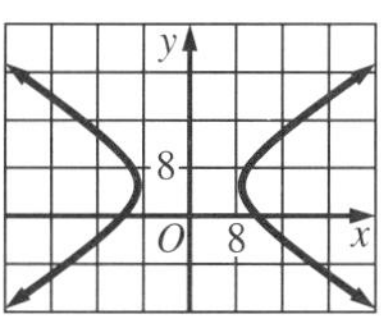

2. center: (–2, 2), vertices: (–2, 7) and (–2, –3); foci: $(-2, 2 \pm \sqrt{29}) \approx$ (–2, 7.39) and (–2, –3.39); vertical; asymptotes: $y = \pm\frac{5}{2}(x+2) + 2$

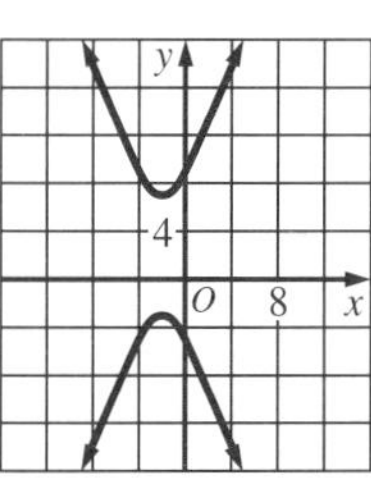

3. $(x+7)^2 + (y-4)^2 = 64$; a circle with center (–7, 4) and radius 8 **4.** $\frac{(x+2)^2}{9} - \frac{(y-5)^2}{9} = 1$; a horizontal hyperbola with center (–2, 5), vertices (1, 5) and (–5, 5), and asymptotes $y = \pm(x+2) + 5$

5. a degenerate circle; the point (0, 0)

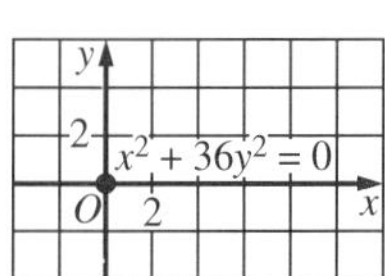

6. a degenerate hyperbola; the intersecting lines $y = \frac{1}{6}x$ and $y = -\frac{1}{6}x$

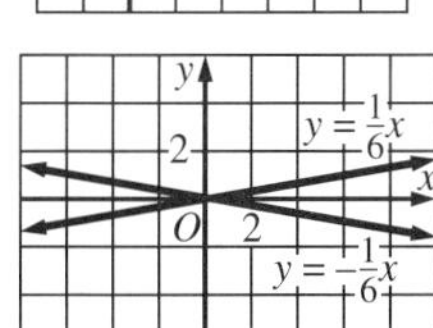

Pages 558–559 Chapter 11 Assessment

1. parallel; perpendicular **2.** parabola; focus; directrix **3.** major axis; vertices **4. a.** $\sqrt{2} \approx 1.41$ **b.** (1.5, 4.5) **5. a.** 5 **b.** (5, 3.5) **6. a.** $\sqrt{17} \approx 4.12$ **b.** (–5, 0.5)

7. a. $y = \frac{3}{5}x + \frac{36}{5}$ **b.** $y = -\frac{5}{3}x + \frac{8}{3}$ **8. a.** $\left(-\frac{1}{2}, 1\right)$

b. $\sqrt{61} \approx 7.81$; $\frac{1}{2}\sqrt{61} \approx 3.91$ **c.** $y = \frac{5}{6}x + \frac{13}{2}$

9. $x - \frac{1}{2} = \frac{1}{14}(y + 2)^2$

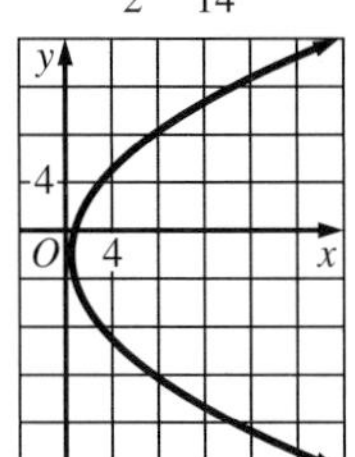

10. a circle with center (–2, 5) and radius 4

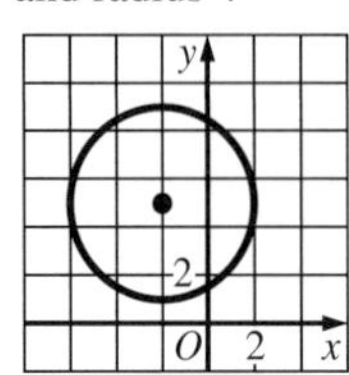

11. a horizontal hyperbola with center (1, 6), vertices (4, 6) and (–2, 6), foci (6, 6) and (–4, 6), and asymptotes $y = \pm\frac{4}{3}(x - 1) + 6$

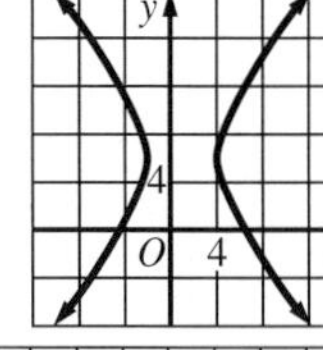

12. a vertical ellipse with center (–2, 1), vertices (–2, –7) and (–2, 9), and minor axis with endpoints (–8, 1) and (4, 1)

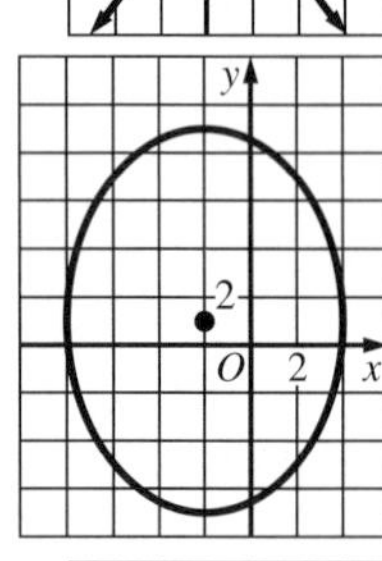

13. a horizontal ellipse with center (–1, –7), vertices (–6, –7) and (4, –7), and minor axis with endpoints (–1, –9) and (–1, –5)

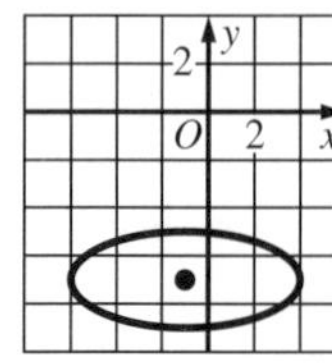

14. a. (17.2, 0) or (–17.2, 0)

b. $\frac{x^2}{316.84} + \frac{y^2}{21} = 1$

16. $y - 4 = -\frac{1}{8}(x - 3)^2$

17. $x^2 + (y - 8)^2 = 4$

18. $\frac{(x + 4)^2}{4} + \frac{(y + 2)^2}{9} = 1$

19. $\frac{(x - 6)^2}{16} - \frac{(y - 6)^2}{9} = 1$

20. $x + 1 = -\frac{1}{16}(y - 3)^2$; a parabola that opens left, has vertex (–1, 3), and axis $y = 3$ **21.** $\frac{(x - 1)^2}{4} - \frac{(y - 2)^2}{1} = 1$; a horizontal hyperbola with center (1, 2), vertices (3, 2) and (–1, 2), and asymptotes $y = \pm\frac{1}{2}(x - 1) + 2$ **22.** $x^2 + (y - 6)^2 = 11$; a circle with center (0, 6) and radius $\sqrt{11} \approx 3.32$ **23.** $\frac{(x - 3)^2}{25} + \frac{(y - 2)^2}{1} = 1$; a horizontal ellipse with center (3, 2), vertices (8, 2) and (–2, 2), and minor axis with endpoints (3, 3) and (3, 1)

24. a degenerate hyperbola; intersecting lines: $y = 5x$ and $y = -5x$

25. a degenerate circle; the point (–1, –1)

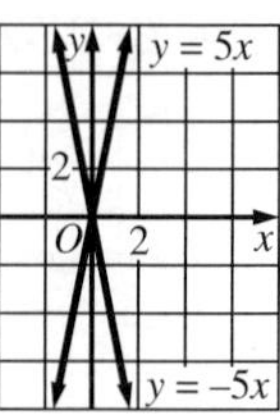

CHAPTER 12

Page 566 Checking Key Concepts

1. Graphs A and B are equivalent. Answers may vary. An example is given. Each graph consists of edges connecting *A* and *B*, *B* and *C*, *C* and *D*, and *D* and *A*. Graphs C and D are equivalent. **3.** 2 **5.** Answers may vary. An example is given. Vermont would have degree 3.

Pages 566–568 Exercises and Applications

1. A and B; In both graphs, each vertex is connected in the same way to the other vertices. **3, 5.** Answers to parts (b) and (c) may vary. Examples are given.

3. a. *A*, 2; *B*, 3; *C*, 4; *D*, 2; *E*, 3; *F*, 4; *G*, 2

b, c. Three colors are needed.

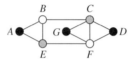

5. a. *A*, 4; *B*, 4; *C*, 4; *D*, 4; *E*, 4

b, c. Five colors are needed.

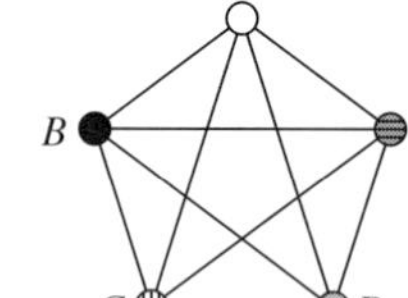

9, 11. Answers may vary. Examples are given.

9. a.

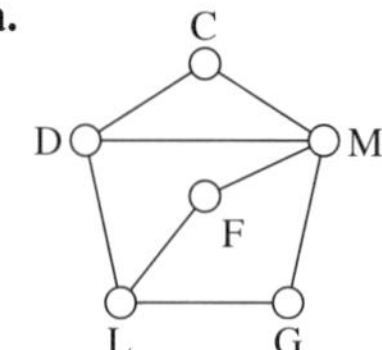

b. Yes, as shown in the graph in part (a).

11. Answers may vary. Examples are given. (1) damselfish and filefish, (2) cardinalfish and goby, and (3) moray and lionfish; and (1) damselfish, filefish, and goby, (2) moray, and (3) cardinalfish and lionfish **19.** $\left(x + \frac{3}{2}\right)^2 + y^2 = \frac{25}{4}$; a circle with center $\left(-\frac{3}{2}, 0\right)$ and radius $\frac{5}{2}$ **21.** $x - \frac{1}{2} = -\frac{1}{2}(y + 5)^2$; a parabola with vertex $\left(\frac{1}{2}, -5\right)$, focus (0, –5), and directrix $x = 1$

23. 39 **25.** undefined **27.** $\begin{bmatrix} 3 & 2 \\ -2 & -1 \end{bmatrix}$

Page 572 Checking Key Concepts

1. B **3.** A → D → F, B → D → F, and B → E → F

5. $\begin{array}{c|cccccc} & A & B & C & D & E & F \\ \hline A & 0 & 0 & 0 & 0 & 2 & 1 \\ B & 0 & 0 & 0 & 0 & 1 & 2 \\ C & 0 & 0 & 0 & 0 & 0 & 1 \\ D & 0 & 0 & 0 & 0 & 0 & 1 \\ E & 0 & 0 & 0 & 0 & 0 & 0 \\ F & 0 & 0 & 0 & 0 & 0 & 0 \end{array}$; the number of ways a course can be taken with two prerequisites

Pages 573–576 Exercises and Applications

1. a.

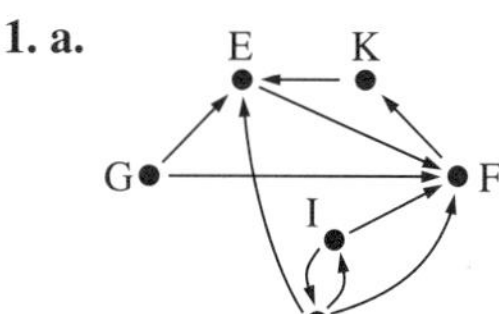

b. Barbara → Fatima → Katrina, Barbara → Ezra → Fatima → Katrina, Barbara → Ilana → Fatima → Katrina

c. No. Explanations may vary. An example is given. Ilana could have gotten the disease only from Barbara, and Barbara could have gotten it only from Ilana. Since Geoff could not have gotten the disease from either Barbara or Ilana, it is not possible that the disease began with one of these six students.

3.

	A	B	C	D
A	0	1	0	1
B	0	0	1	1
C	1	0	0	1
D	0	0	0	0

5.

	A	B	C	D	E	F
A	0	0	0	0	0	0
B	1	0	0	0	0	0
C	1	0	0	0	0	0
D	1	0	0	0	0	0
E	1	0	0	0	0	0
F	0	1	1	1	1	0

7.

	D	R	A	I	E	S	P	Sn	M	Rc	F
D	0	0	0	0	0	0	0	0	0	0	0
R	0	0	0	0	0	0	0	0	0	0	0
A	0	1	0	0	0	0	0	0	0	0	0
I	0	1	1	1	0	1	0	1	1	0	0
E	0	0	0	0	0	1	0	0	0	0	0
S	0	1	0	0	0	0	0	0	0	0	0
P	1	0	0	1	1	1	0	0	1	1	1
Sn	0	1	0	0	0	0	0	1	0	0	1
M	0	0	0	0	0	0	0	1	0	1	1
Rc	0	0	0	0	0	0	0	0	0	0	1
F	0	0	0	0	0	0	0	0	0	0	0

9.

	D	R	A	I	E	S	P	Sn	M	Rc	F
D	0	0	0	0	0	0	0	0	0	0	0
R	0	0	0	0	0	0	0	0	0	0	0
A	0	1	0	0	0	0	0	0	0	0	0
I	0	5	2	2	0	2	0	4	2	1	2
E	0	1	0	0	0	1	0	0	0	0	0
S	0	1	0	0	0	0	0	0	0	0	0
P	1	2	1	2	1	3	0	2	2	2	3
Sn	0	2	0	0	0	0	0	2	0	0	2
M	0	1	0	0	0	0	0	2	0	1	3
Rc	0	0	0	0	0	0	0	0	0	0	1
F	0	0	0	0	0	0	0	0	0	0	0

21. a. $M =$

	A	B	C	D	E	F	G
A	0	0	1	1	0	1	1
B	1	0	0	0	1	0	1
C	0	1	0	1	1	0	0
D	0	1	0	0	0	0	0
E	1	0	0	1	0	1	1
F	0	1	1	1	0	0	0
G	0	0	1	1	0	1	0

b. The row sums represent the total number of games won by the row team.

23. a. $M + M^2 =$

	A	B	C	D	E	F	G	Total
A	0	3	3	4	1	2	1	14
B	2	0	2	3	1	3	3	14
C	2	2	0	2	2	1	2	11
D	1	1	0	0	1	0	1	4
E	1	2	3	4	0	3	2	15
F	1	3	1	2	2	0	1	10
G	0	3	2	3	1	1	0	10

The row sums represent the total number of points awarded to the row team. **b.** See the last column in the matrix in part (a). The Elks win the championship. **27.** 9840

29. $\frac{625}{3} \approx 208.33$

Page 576 Assess Your Progress

1. Answers may vary. An example is given.

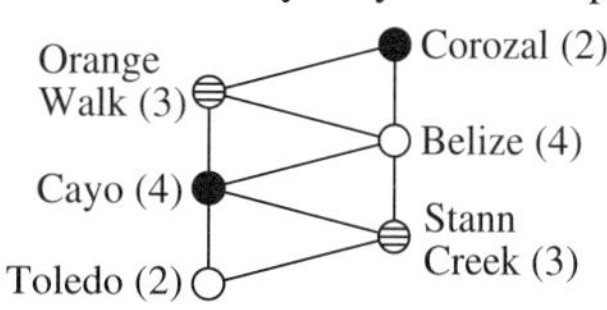

Three colors are needed. Color groupings: Corozal and Cayo, Belize and Toledo, Orange Walk and Stann Creek.

2. Two meeting times are needed.

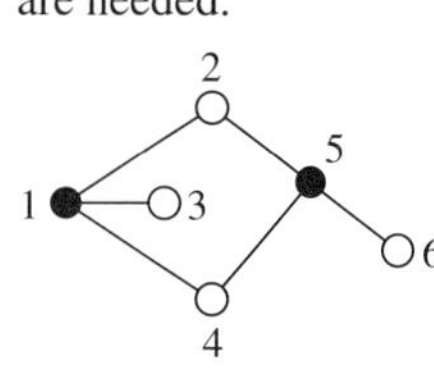

3. a.

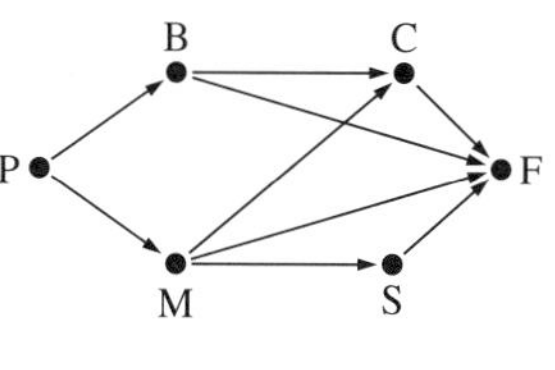

b. $M =$

	Predator					
	F	C	S	B	M	P
F	0	0	0	0	0	0
C	1	0	0	0	0	0
S	1	0	0	0	0	0
B	1	1	0	0	0	0
M	1	1	1	0	0	0
P	0	0	0	1	1	0

c. $M^2 =$

	F	C	S	B	M	P
F	0	0	0	0	0	0
C	0	0	0	0	0	0
S	0	0	0	0	0	0
B	1	0	0	0	0	0
M	2	0	0	0	0	0
P	2	2	1	0	0	0

M^2 gives the number of ways that each prey is a food source for the column predator through one intermediate source. Answers may vary. An example is given. Barnacles are an intermediate food source for fish through crabs.

$M + M^2 =$

	F	C	S	B	M	P
F	0	0	0	0	0	0
C	1	0	0	0	0	0
S	1	0	0	0	0	0
B	2	1	0	0	0	0
M	3	1	1	0	0	0
P	2	2	1	1	1	0

$M + M^2$ gives the total number of ways that each prey is a food source for the column predator either directly or through one intermediate source. Answers may vary. An example is given. Barnacles are a food source for fish in two ways, both directly and indirectly through crabs.

Page 581 Checking Key Concepts

1. a. spaghetti/corn/pie, spaghetti/corn/pudding, spaghetti/beans/pie, spaghetti/beans/pudding, spaghetti/squash/pie, spaghetti/squash/pudding, chicken/corn/pie, chicken/corn/pudding, chicken/beans/pie, chicken/beans/pudding, chicken/squash/pie, chicken/squash/pudding **b.** There are 2 choices of entrees, 3 choices of vegetables, and 2 choices of desserts, so there are $2 \times 3 \times 2 = 12$ choices of meals. **3.** 6 **5.** 6720

7. D **9.** $\frac{6!}{2!\,2!\,1!\,1!}$

Pages 581–584 Exercises and Applications

1. a. 720 **b.** 12 **3. a.** 24 **b.** 24 **7.** 7 **9.** 210 **11.** 96 ways **21.** 8 **23.** 360 **25.** 259 other cars **27.** 362,880 batting orders **31.** C

33. $\frac{5!}{2!\,2!\,1!} = 30$ **35.** $\frac{6!}{3!\,1!\,1!\,1!} = 120$ **43.** $x = 1$ and $y = 3$; $y = \frac{2}{x}$ is translated 1 unit right and 3 units up. **45.** $x = 4$ and $y = 2$; $y = \frac{9}{x}$ is translated 4 units right and 2 units up.

Page 588 Checking Key Concepts

1. combinations; ${}_9C_3$ **3.** permutations; ${}_6P_4$ **5.** ${}_{10}C_4 \cdot {}_{13}C_4$ **7. a.** 21 **b.** 35 **c.** 35 **d.** 4950

Pages 588–592 Exercises and Applications

1. 792 **3.** 1820 **5.** 1 **7.** 8 **9.** 20 **11.** 441 **13.** ${}_{500}C_{10} \approx 2.458 \times 10^{20}$ sets **15.** 56 triangles **17.** 6 ways **19.** 20 possible purchases **23.** 35,640 ways **25. a.** 38,760 ways **b.** ${}_{20}P_9 \approx 6.095 \times 10^{10}$ ways **c.** 465,120 ways **29.** 635,013,559,600 different hands **31.** 103,776 hands **33.** 156 different kinds of full houses **43.** 120 **45.** 40,320 **47.** 239,500,800 **49.** $4t^2 - 4t + 1$ **51.** $n^3 - 6n^2 + 12n - 8$ **53.** $y^4 + 4y^3 + 6y^2 + 4y + 1$ **55.** $f(x) = -x(x + 5)(x + 2)$

Page 595 Checking Key Concepts

1. 1, 8, 28, 56, 70, 56, 28, 8, 1 **3.** 252 sequences **5.** $a = 4x$; $b = y$; $n = 12$

Pages 595–599 Exercises and Applications

1. 1, 9, 36, 84, 126, 126, 84, 36, 9, 1 **5.** 2 **7.** 8 **9.** 165 **11.** $x^5 - 5x^4y + 10x^3y^2 - 10x^2y^3 + 5xy^4 - y^5$ **13.** $32y^5 - 240y^4z + 720y^3z^2 - 1080y^2z^3 + 810yz^4 - 243z^5$ **15.** $m^{14} - 7m^{12}n + 21m^{10}n^2 - 35m^8n^3 + 35m^6n^4 - 21m^4n^5 + 7m^2n^6 - n^7$ **17.** $x^{12} + 6x^{10}y^3 + 15x^8y^6 + 20x^6y^9 + 15x^4y^{12} + 6x^2y^{15} + y^{18}$ **19. a.** 12 **b.** ${}_{12}C_2$ **c.** $220a^9b^3$ **37.** D **39.** 1 **41.** 7 **43.** $576\pi \approx 1810\ \text{cm}^2$ **45.** $y + 2 = 6(x - 3)$

Page 599 Assess Your Progress

1. 9 **2.** 9 **3.** 210 **4.** 5040 **5.** 35 **6.** 8 **7.** ${}_{40}P_5 = 78{,}960{,}960$ **8.** ${}_8C_3 = 56$ **9.** ${}_{20}P_5 = 1{,}860{,}480$ **10.** 5040 possible rainbows **11.** 1080 possible combinations of classes **12.** 165 **13.** $j^6 + 6j^5k + 15j^4k^2 + 20j^3k^3 + 15j^2k^4 + 6jk^5 + k^6$ **14.** $32p^5 - 80p^4q + 80p^3q^2 - 40p^2q^3 + 10pq^4 - q^5$ **15.** $p^3 + 1.5p^2q + 0.75pq^2 + 0.125q^3$

Pages 604–605 Chapter 12 Assessment

1. degree **2.** directed graph **3.** permutation; combination **4–6. b, c.** Answers may vary. Examples are given.

4. a. A, 3; B, 3; C, 2; D, 3; E, 3; F, 2
b, c.

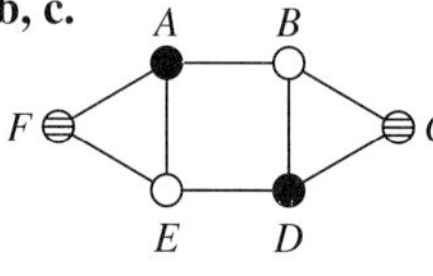

5. a. A, 4; B, 3; C, 4; D, 4; E, 3; F, 3; G, 3
b, c.

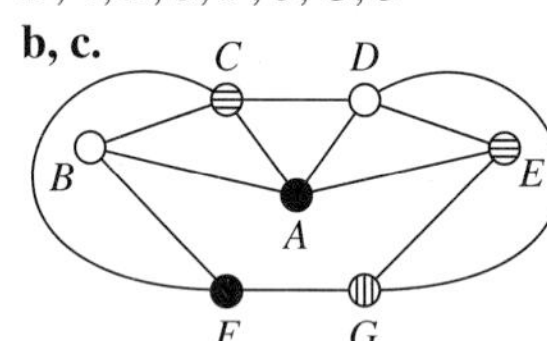

6. a. A, 2; B, 2; C, 4; D, 2; E, 2 **b, c.**

7–10. Answers may vary. Examples are given.
7. In the diagram, H = healer, I = interpreter, M = merchant, S = scribe, A = advisor, K = knight, and P = prince. The arrows are drawn from a person who requires another to a person who is required, although the arrows can be drawn in the opposite direction.

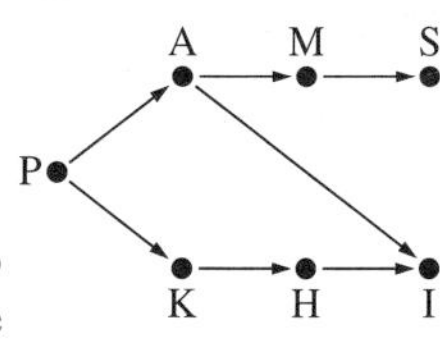

8. The arrows indicate which characters are required by others.

9.

Rows: *Character requiring another*; columns: *Character required*

	P	A	M	S	K	H	I
P	0	1	0	0	1	0	0
A	0	0	1	0	0	0	1
M	0	0	0	1	0	0	0
S	0	0	0	0	0	0	0
K	0	0	0	0	0	1	0
H	0	0	0	0	0	0	1
I	0	0	0	0	0	0	0

or

Rows: *Character required*; columns: *Character requiring another*

	P	A	M	S	K	H	I
P	0	0	0	0	0	0	0
A	1	0	0	0	0	0	0
M	0	1	0	0	0	0	0
S	0	0	1	0	0	0	0
K	1	0	0	0	0	0	0
H	0	0	0	0	1	0	0
I	0	1	0	0	0	1	0

10.

Rows: *Character requiring another*; columns: *Character required*

	P	A	M	S	K	H	I
P	0	0	1	0	0	1	1
A	0	0	0	1	0	0	0
M	0	0	0	0	0	0	0
S	0	0	0	0	0	0	0
K	0	0	0	0	0	0	1
H	0	0	0	0	0	0	0
I	0	0	0	0	0	0	0

or

Rows: *Character required*; columns: *Character requiring another*

	P	A	M	S	K	H	I
P	0	0	0	0	0	0	0
A	0	0	0	0	0	0	0
M	1	0	0	0	0	0	0
S	0	1	0	0	0	0	0
K	0	0	0	0	0	0	0
H	1	0	0	0	0	0	0
I	1	0	0	0	1	0	0

The elements of the squared matrix tell you which characters are required indirectly through one intermediate character.
11. 840 **12.** 7 **13.** 30 **14.** 360 **15.** 35 **16.** 7 **17.** 15 **18.** 210 **19.** $10^9 - 10^6 - 10^3 + 1 = 998{,}999{,}001$ **21.** ${}_{12}C_6 = 924$ **22.** 28 **23.** $x^3 - 3x^2y + 3xy^2 - y^3$ **24.** $16a^4 + 64a^3b + 96a^2b^2 + 64ab^3 + 16b^4$ **25.** $32y^5 + 80y^4z^2 + 80y^3z^4 + 40y^2z^6 + 10yz^8 + z^{10}$ **26.** ${}_{13}C_3a^{10}b^3 = 286a^{10}b^3$

CHAPTERS 10–12

Pages 606–607 Cumulative Assessment

1. neither; $t_n = \frac{n}{n + 1}$; $\frac{10}{11}$ **3.** arithmetic; $t_n = (6 - n)\sqrt{2}$; $-4\sqrt{2}$ **5.** (2) $t_1 = 2$; $t_n = \frac{1}{2}t_{n-1}$; (3) $t_1 = 5\sqrt{2}$; $t_n = t_{n-1} - \sqrt{2}$ **7.** 2, 5, 11, 23 **9.** 9842 **11.** no sum **15.** $\overleftrightarrow{PQ}$ and $\overleftrightarrow{SR}$ both have slope $\frac{2}{3}$; $\overleftrightarrow{PS}$ and $\overleftrightarrow{QR}$ both have slope $-\frac{3}{2}$. Then $\overleftrightarrow{PQ} \perp \overleftrightarrow{QR}$, $\overleftrightarrow{PQ} \perp \overleftrightarrow{PS}$, $\overleftrightarrow{SR} \perp \overleftrightarrow{PS}$, and $\overleftrightarrow{SR} \perp \overleftrightarrow{QR}$, so $\angle P$, $\angle Q$, $\angle R$, and $\angle S$ are right angles and $PQRS$ is a rectangle.

17. $(x + 4)^2 + y^2 = 8$ **19.** $\frac{x^2}{9} - \frac{y^2}{16} = 1$

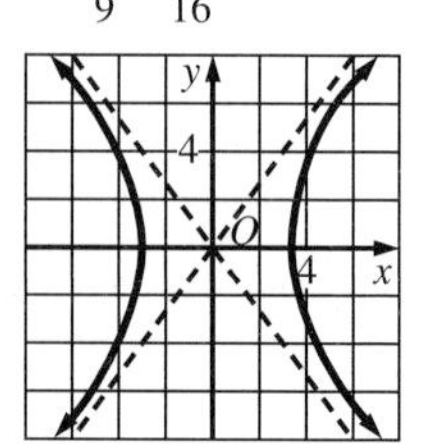

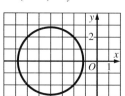

21. (4, 0); 2 **29.** 3024 **31.** 35 **33.** 311,875,200 ways

CHAPTER 13

Page 615 Checking Key Concepts

1. $\frac{8}{250}$, or 0.032 **3.** $\frac{3{,}618{,}770}{4\pi(3963)^2}$, or about 0.0183

Pages 615–617 Exercises and Applications

1. $\frac{1}{52}$, or about 0.0192 **3.** $\frac{1}{4}$, or 0.25 **5.** $\frac{1}{2}$, or 0.5

7. $\frac{3}{4}$, or 0.75 **9.** 0 **11.** $\frac{5}{9}$, or about 0.556

19. $a^3 + 3a^2b + 3ab^2 + b^3$ **21.** $16u^4 + 96u^3v + 216u^2v^2 + 216uv^3 + 81v^4$ **23.** 4, $\frac{-1-\sqrt{13}}{2}$, $\frac{-1+\sqrt{13}}{2}$

Page 621 Checking Key Concepts

1. $\frac{1}{16}$, or 0.0625 **3.** $\frac{168}{169}$, or about 0.994 **5.** $\frac{7}{16}$, or 0.4375

Pages 621–624 Exercises and Applications

3. 0.12 **5.** 0.42 **7.** 0.46 **11. a.** 0.64 **b.** 0.96 **c.** 0.32 **d.** 0.04 **13.** B **15.** about 0.000002192 **19.** 801

Page 629 Checking Key Concepts

1. $\frac{4}{9}$, or about 0.444 **3.** $\frac{4}{15}$, or about 0.267

5. $\frac{2}{15}$, or about 0.133

Pages 629–631 Exercises and Applications

1. $\frac{24}{103}$, or about 0.233 **3.** $\frac{4}{15}$, or about 0.267

5. $\frac{2}{9}$, or about 0.222 **7.** $\frac{15}{49}$, or about 0.306

13. $\frac{1}{17}$, or about 0.0588 **15.** $\frac{25}{102}$, or about 0.245

17. $\frac{1}{13}$, or about 0.0769 **19.** about 0.620

21. a. $\frac{1}{3}$, or about 0.333 **b.** $\frac{1}{12}$, or about 0.0833

c. $\frac{11}{12}$, or about 0.917 **23.** $x^3 - 3x^2y + 3xy^2 - y^3$

25. $s^6 + 3s^4t^4 + 3s^2t^8 + t^{12}$

Page 631 Assess Your Progress

1. $\frac{2}{\pi}$, or about 0.637 **2.** $\frac{625}{1296}$, or about 0.482

3. about 0.00786

Page 635 Checking Key Concepts

1. true/false test; On a true/false test, the probability of getting any question right is $\frac{1}{2}$, while on a multiple-choice test, the probability of guessing any question right from among 3 answers is $\frac{1}{3}$, from among 4 answers is $\frac{1}{4}$, and so on.

3. $\frac{15}{1024}$, or about 0.0146 **5.** For each trial, there are two possible mutually exclusive results: correct or incorrect. Each trial is also independent, so the distribution is binomial; ${}_6C_k\left(\frac{1}{2}\right)^6$, where k is the number correct.

Pages 635–638 Exercises and Applications

3. ${}_{10}C_k(0.5)^{10}$; about 0.117 **5.** ${}_7C_k(0.1)^k(0.9)^{7-k}$; 6.3×10^{-6} **7.** ${}_1C_k(0.7)^k(0.3)^{1-k}$; 0.7 **9. a.** about 0.0424 **b.** about 0.141 **c.** about 0.228 **d.** about 0.589 **15.** $P(k$ successes$)$ and $P(j$ successes$)$ both equal 0, since k and j are both positive, and there is a 0% chance of success (certain failure) on every trial of either experiment. **17.** $P(k$ successes$)$ and $P(j$ successes$)$ are the same (about 0.410). This is because the first experiment asks for the probability of 1 success in 5 trials with a 20% chance of success. You can restate this as the probability of 4 failures in 5 trials with an 80% chance of failure. This is what the second experiment describes, redefining "failure" as "success." **29.** 0.4 **31.** mean: 0.31; standard deviation: about 0.146 **33.**

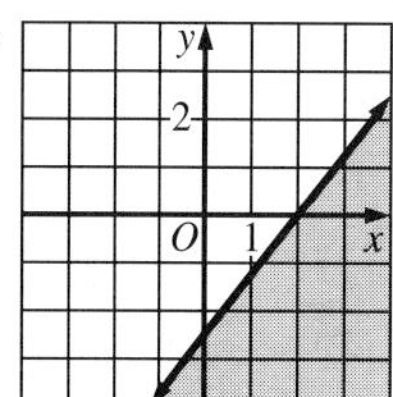

Page 642 Checking Key Concepts

1. a.

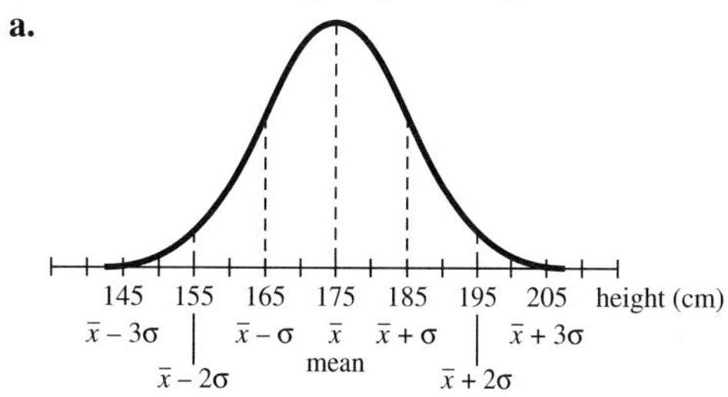

b. 95%; 84% **3.** about 21.2%; about 2.87%

Pages 642–645 Exercises and Applications

1. Answers may vary. Examples are given. **a.** about 0.475; Because about 95% of the distribution lies within two standard deviations of the mean, half of that lies between the mean and two standard deviations below the mean. **b.** about 0.495; Because about 99% of the distribution lies within three standard deviations of the mean, half of that lies between the mean and three standard deviations above the mean. **c.** about 0.976; The probability can be found by subtracting the probability for a z-score of up to –2 from the probability for a z-score of up to 3: 0.9987 – 0.0228 = 0.9759 **9.** 99.99% **13.** Answers may vary depending upon when rounding is done, and if interpolation is used. Examples are given. **a.** 34% **b.** 54% **c.** 73% **17.** about 0.322 **19.** 35 **21.** 171

Page 645 Assess Your Progress

1. about 0.0536; about 0.263 **2.** about 0.766 **3. a.** 15.9% **b.** 2.28% **4.** about 0.0446

Pages 650–651 Chapter 13 Assessment

1. independent **2.** standard normal distribution **3.** probability **4.** $\frac{1}{52}$, or about 0.0192 **5.** $\frac{1}{13}$, or about 0.0769 **6.** $\frac{1}{4}$, or 0.25 **7.** $\frac{1}{2}$, or 0.5 **8.** $\frac{3}{26}$, or about 0.115 **9.** 0 **10. a.** $\frac{1}{20}$, or 0.05

b. $\frac{25}{648}$, or about 0.0386 **11.** $\frac{1}{2}$, or 0.5 **12.** $\frac{24}{25\pi}$, or about 0.306 **13.** $0.5 - \frac{24}{25\pi}$, or about 0.194 **14. a.** 0.01 **b.** 0.81 **c.** 0.19 **d.** 0.18 **15.** about 6.57×10^{-5}

16. about 0.610 **17. a.** about 0.000244 **b.** about 0.178 **c.** about 0.132 **d.** about 0.169 **19. a.** about 15.9% **b.** about 6.68% **c.** about 22.6%

CHAPTER 14

Page 659 Checking Key Concepts

1. $\sin B = \frac{b}{c}$; $\cos B = \frac{a}{c}$ **3.** $\sin \theta = 0.5346$; $\cos \theta = 0.8451$

Pages 659–662 Exercises and Applications

1. $\sin A = 0.8000$; $\cos A = 0.6000$ **3.** $\sin A = 0.3846$; $\cos A = 0.9231$ **5.** $\sin B = 0.9756$; $\cos B = 0.2195$ **7.** 14 in. **9. a.** 43.5° **b.** 43°32′52″ **11. a.** 73.4° **b.** 73°25′2″ **13. a.** 26.4° **b.** 26°23′6″ **15. a.** 59.1° **b.** 59°5′34″ **17.** < **21.** 50.2° **31.** 10.1 **33.** 0.43 **35.** $x = \pm 6\sqrt{2}$

Page 666 Checking Key Concepts

1. a. $\tan 64.8° = \frac{a}{18.4}$ **b.** $\angle B = 25.2°$; $a \approx 39.1$; $c \approx 43.2$

3. They are alternate interior angles and are therefore congruent.

Pages 667–670 Exercises and Applications

1. a. $\frac{3}{4}$ **b.** $\frac{4}{3}$ **c.** They are reciprocals. **3.** A

5, 7, 9. Side lengths are rounded to the nearest tenth. **5.** $\angle B = 71°$; $a \approx 1.2$; $c \approx 3.7$ **7.** $\angle A \approx 52.0°$; $\angle B \approx 38.0°$; $a \approx 15.4$ **9.** $\angle A \approx 18.4°$; $\angle B \approx 71.6°$; $b = 21$; $c \approx 22.1$ **11.** about 69 ft **13. a.** $12\sqrt{2} \approx 17$ **b, c.** Methods may vary. Examples are given. **b.** $\angle DAC \approx 26.6°$; $\angle CAB \approx 63.4°$; $\triangle DAC$ and $\triangle CAB$ are right triangles with right angles at D and B. So, $\angle DAC = \tan^{-1}\left(\frac{\sqrt{8}}{\sqrt{32}}\right) \approx 26.6°$ and $\angle CAB = \tan^{-1}\left(\frac{\sqrt{32}}{\sqrt{8}}\right) \approx 63.4°$. **c.** 53.2°; By continuing the process in part (b), you can find that two of the angles of each acute triangle have measures of 63.4°. The sum of the angles of a triangle is 180°, so the measure of the third angle is about 53.2°. **15.** $\angle M$; $\angle L$; No; the tangent is the ratio of the legs of a right triangle, but $\overline{LM}$ is the hypotenuse. **23. a.** arithmetic; common difference = 90 **b.** geometric; common ratio = –1

Page 670 Assess Your Progress

1. $\frac{12}{13}$ **2. a.** about 188 m **b.** about 282 m

Page 674 Checking Key Concepts

1. a. II **b.** III **c.** I **3.** about 66°, about 246°

Pages 675–677 Exercises and Applications

1.

3.

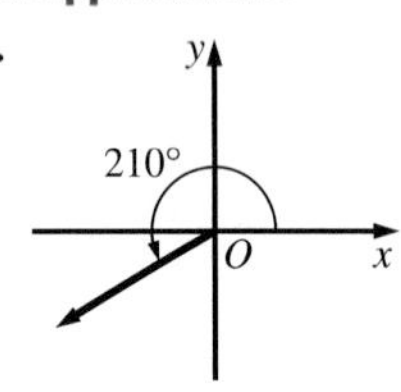

5. 205° **7. a.** 90° **b.** 270° **9.** 0.8824; 0.4706; 1.875 **11.** –0.6; –0.8; 0.75 **13.** –0.2079; –0.9781; 0.2126 **15.** 0.9298; –0.3681; –2.526 **19. a.** 60°; 120° **b.** 40°; 320° **c.** 68°; 248° **21. a.** 260° **b.** $\cos 260° = \frac{x}{98}$; about –17; $\sin 260° = \frac{y}{98}$; about –97 **c.** about 17 mi; about 97 mi **29. a.** 204 ft^2 **b.** $3\sqrt{3}$ cm^2, or about 5.2 cm^2

Page 681 Checking Key Concepts

1. about 1.8 square units **3.** about 170 square units **5.** about 11.2 square units **7.** about 7.5 square units

Pages 681–683 Exercises and Applications

1. about 11.7 square units **3.** not enough information; need the measure of included angle **5.** about 580 square units **7.** about 32.1 square units **9.** not enough information; need the measure of included angle **13.** about 25.8 square units **15.** about 28.3 square units **17.** about 5 square units **19.** about 15.1 square units **27.** 55°, 125° **29.** 26°, 334° **31.** Yes; 6. **33.** Yes; 12.

Page 683 Assess Your Progress

1. 0.6428; 0.7660; 0.8391

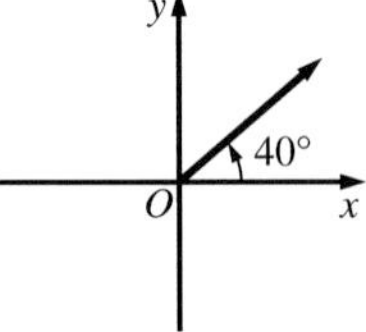

2. –0.7071; –0.7071; 1

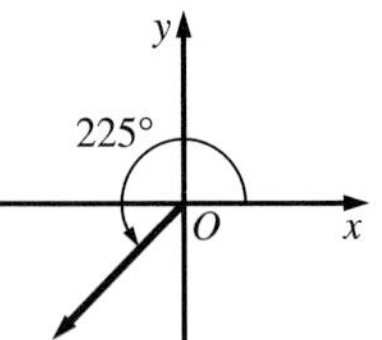

3. –0.7880; 0.6157; –1.280

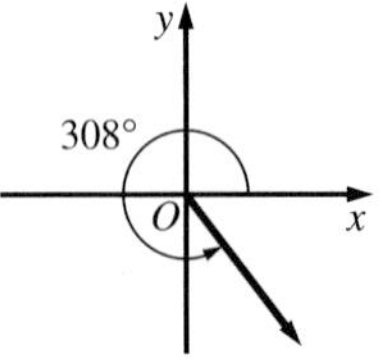

4. 0.9962; –0.0872; –11.43

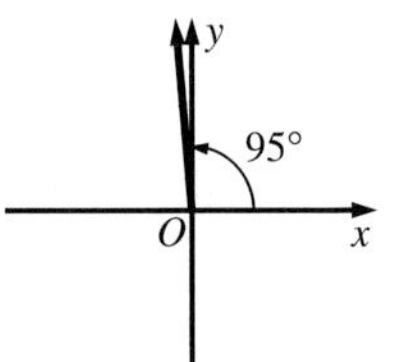

5. 0.3162; 0.9487; 0.3333 **6.** –0.2425; 0.9701; –0.25 **7.** 0.7894; –0.6139; –1.286 **8.** –0.8480; –0.5300; 1.6 **9.** about 115 square units **10.** about 0.12 square units **11.** about 10,367 square units **12.** about 5.2 square units **13.** about 106 square units

Page 688 Checking Key Concepts

1. $\angle A \approx 52.3°$; $\angle C \approx 97.7°$; $c \approx 47.6$ **3.** $\angle B = 60°$; $a \approx 10.6$; $c \approx 13.0$ **5.** $\angle B = 48°$; $a \approx 18.6$; $c \approx 10.0$ **7.** $\angle B \approx 77.1°$, $\angle C \approx 43.9°$; $c \approx 17.3$; $\angle B \approx 102.9°$; $\angle C \approx 18.1°$; $c \approx 7.8$

Pages 688–690 Exercises and Applications

1. $\angle N = 31°$; $m \approx 0.94$; $n \approx 0.54$ **3.** $\angle Q = 29°$; $r \approx 19$; $p \approx 10$ **5.** $\angle Y = 57°$; $x \approx 21$; $z \approx 14$ **7.** $\angle A = 76°$; $b \approx 2.2$; $c = 4.6$ **9.** $\angle B \approx 97.3°$; $\angle C \approx 27.7°$; $b \approx 18.2$ **11.** $\angle A \approx 19.6°$; $\angle C \approx 37.4°$; $c \approx 1.4$ **13.** $\angle B \approx 29.8°$; $\angle C \approx 94.2°$; $c \approx 24$ **15.** no solution; The side opposite the obtuse angle is not the longest. **17.** $\angle A \approx 54.7°$; $\angle B \approx 33.3°$; $b \approx 11.7$ **27.** about 9.2 square units **29.** about 25.6 square units **31.** $\log_a 64$ **33.** $\log_b \frac{1}{25}$ **35.** $\ln 1 = 0$

Page 694 Checking Key Concepts

1. about 10 **3.** about 17 **5.** $\angle J \approx 101.5°$; $\angle K \approx 34.0°$; $\angle L \approx 44.4°$

Pages 695–697 Exercises and Applications

1. about 16 **3.** about 4.05 **5.** about 14 **7.** about 10.8 **9.** about 15.4 **15.** $\angle R \approx 59.4°$; $\angle S \approx 52.6°$; $\angle T \approx 68.0°$ **17.** $\angle X \approx 117.4°$; $\angle Y \approx 57.5°$; $\angle Z \approx 5.1°$ **19.** $\angle P \approx 7.3°$; $\angle Q \approx 166.6°$; $\angle R \approx 6.1°$ **21.** $\angle J \approx 52.0°$; $\angle K \approx 95.6°$; $\angle L \approx 32.4°$ **25.** 38.9° **29.** $\angle C = 84°$; $a \approx 5.0$; $c \approx 8.5$ **31.** $\angle B = 90°$; $\angle C = 60°$; $c \approx 7.36$ **33.** $\frac{\sqrt{2}}{2} \approx 0.7071$; $-\frac{\sqrt{2}}{2} \approx -0.7071$ **35.** 1; 0 **37.** −0.5; $\frac{\sqrt{3}}{2} \approx 0.8660$ **39.** $-\frac{\sqrt{2}}{2} \approx -0.7071$; $-\frac{\sqrt{2}}{2} \approx -0.7071$

Page 697 Assess Your Progress

1. $\angle X = 39°$; $x \approx 9.1$; $z \approx 13.9$ **2.** $\angle Y \approx 55.3°$; $\angle Z \approx 39.7°$; $x \approx 2.73$ **3.** $\angle Y \approx 8.0°$; $\angle Z \approx 22.0°$; $y \approx 2$ **4.** $\angle X \approx 51.9°$, $\angle Z \approx 71.1°$, $x \approx 9.15$; $\angle X \approx 14.1°$, $\angle Z \approx 108.9°$, $x \approx 2.84$ **5.** $\angle X \approx 39.3°$, $\angle Y \approx 27.3°$; $\angle Z \approx 113.4°$ **6.** no solution; The side opposite the obtuse angle is not the longest side. **7.** $\angle X \approx 109°$; $x \approx 19.0$; $z \approx 9.4$ **8.** $\angle X = 144.2°$, $\angle Y = 25.2°$; $\angle Z = 10.5°$ **9.** no solution; $y < x \sin y$, so no triangle can be formed. **10.** $\angle X \approx 16.8°$, $\angle Z \approx 138.2°$, $z \approx 30$ **11.** $\angle Y \approx 40.8°$; $\angle Z \approx 15.2°$; $x \approx 15.8$ **12.** $\angle Y = 60°$; $\angle Z = 90°$; $y = \frac{3\sqrt{3}}{8} \approx 0.65$ **13.** $\angle Y \approx 29.3°$, $\angle Z \approx 78.7°$, $y \approx 3.3$; $\angle Y \approx 6.7°$, $\angle Z \approx 101.3°$, $y \approx 0.79$

Pages 702–703 Chapter 14 Assessment

1. tangent **2.** solve a triangle **3.** $\frac{8}{17}$ **4.** $\frac{15}{17}$ **5.** $\frac{15}{8}$ **6.** 30° **7.** about 262 ft **8. a.** 200° **b.** $\cos 200° = \frac{x}{100}$; $x \approx -94$; $\sin 200° = \frac{y}{100}$; $y \approx -34$ **c.** about 94 mi west; about 34 mi south **9.** −0.7809; −0.6247; 1.25 **10.** 76.4°, 283.6° **11.** 10 square units **12.** about 69 square units **13.** about 597 square units **14.** about 742 in.^{2} **15.** about 218 square units **16.** about 117 square units **17.** about 201 square units **18.** $\angle B \approx 46°18'$; $\angle C \approx 18°22'$; $c \approx 14$ **19.** no solution; The side opposite the obtuse angle is not the longest side. **20.** $\angle A = 45°$; $\angle B = 90°$; $c = 6$ **21.** $\angle A \approx 133.8°$, $\angle C \approx 31.2°$, $a \approx 70$; $\angle A \approx 16.2°$, $\angle C \approx 148.8°$, $a \approx 27$ **22.** $\angle A \approx 24.7°$; $\angle B \approx 35.3°$; $b \approx 75$ **23.** $\angle A \approx 60.4°$; $\angle B \approx 46.5°$; $\angle C \approx 73.1°$

CHAPTER 15

Page 710 Checking Key Concepts

1. 56°, 124° **3.** 77°, 283° **5.** 225°, 315° **7.** about 43 ft **9.** 24 ft

Pages 711–712 Exercises and Applications

1. 270° **3.** 0°, 180°, 360° **5.** 30°, 150° **7.** 27°, 153° **9.** 210°, 330° **11.** 109°, 251° **13.** 0.5 **15.** 0.2588 **17.** −1 **19.** 0.1736 **27.** 3.5 **29.** 3.5 **31.** 2 **33.** 2 **35.**

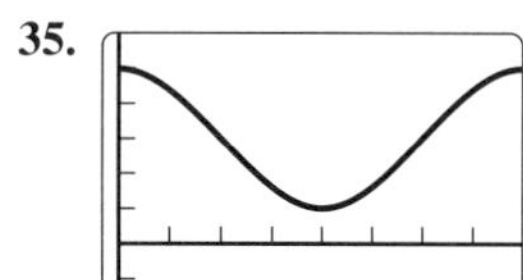

a. 5 **b.** 1 **c.** maximum: 0°, 360°; minimum: 180° **39.** m $\angle A = 50°$; $AB \approx 3$; $AC \approx 9$ **41. a.** 2π in. **b.** $\frac{3\pi}{2}$ in. **43.** 228 in.^{2}

Page 716 Checking Key Concepts

1. 450° **3.** 221° **5.** −3.49 **7.** 0.52 **9.** −1 **11.** −0.5

Pages 716–718 Exercises and Applications

1. 57° **3.** 225° **5.** 540° **7.** 1260° **9.** 337° **11.** −90° **13. a.** The first pair of equations graphs the unit circle counterclockwise from (1, 0). As each point is plotted, the second set of equations plots the sine value of the corresponding angle. **b.** The first pair of equations again plots the unit circle counterclockwise from (1, 0). As each point is plotted on the unit circle, the second pair of equations plots the cosine function of the corresponding angle. **15.** −3.14; $-\pi$ **17.** 12.57; 4π **19.** −1.13; $-\frac{13\pi}{36}$ **21.** 6.81; $\frac{13\pi}{6}$ **23.** 1.52; $\frac{29\pi}{60}$ **25.** 7.07; $\frac{9\pi}{4}$ **27.** −1 **29.** 0 **31.** 0 **33.** 0.6822 **35.** 0.5872 **37.** 0.9093 **47.** 31°; 149° **49.** 24°; 156° **51.** $\frac{1}{10}$

Page 723 Checking Key Concepts

1. a.

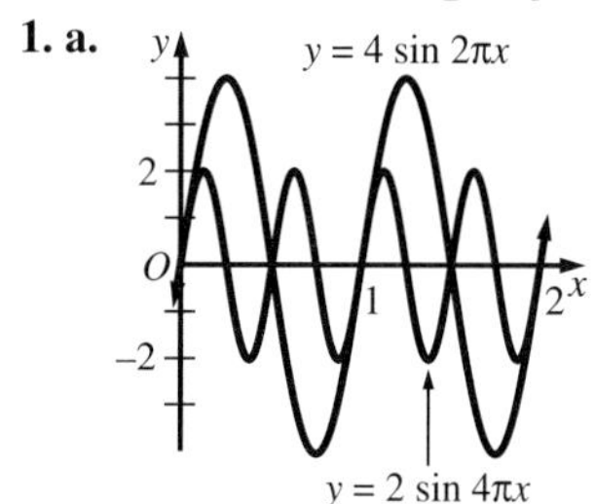

b. $y = 2 \sin 4\pi x$ has amplitude 2 and period 0.5; $y = 4 \sin 2\pi x$ has amplitude 4 and period 1.

Pages 723–726 Exercises and Applications

1. amplitude: 2; period: 1

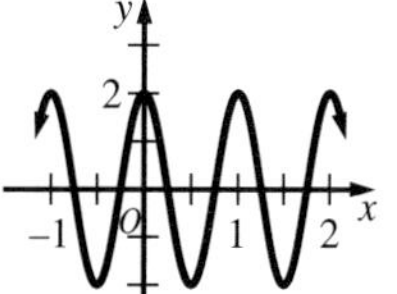

3. amplitude: $\frac{1}{2}$; period: $\frac{2\pi}{3}$

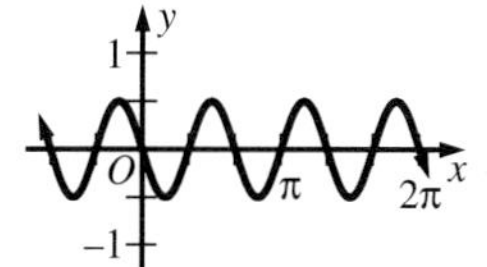

5. amplitude: 1; period: 4

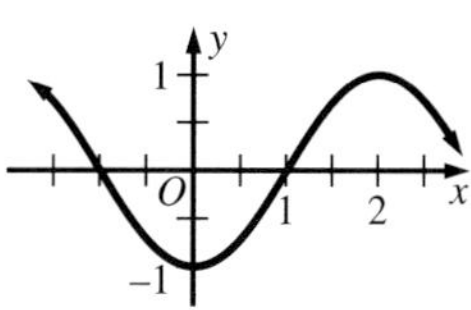

7. amplitude: 4; period: $\frac{\pi}{2}$

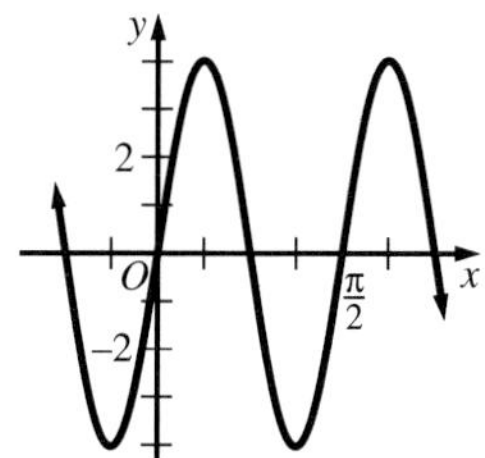

9. amplitude: $\frac{1}{3}$; period: 8π

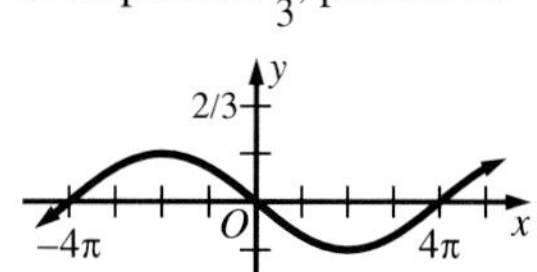

11. amplitude: 2; period: $\frac{4\pi}{5}$

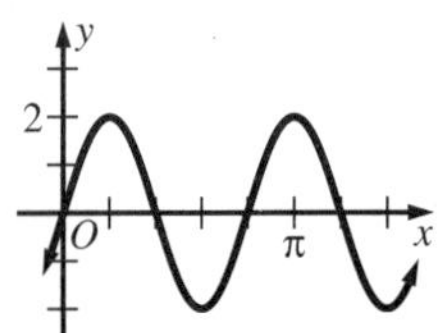

13. $y = 3 \sin 4x$

15. $y = -10 \cos \frac{1}{2}x$

17. $y = 2 \sin \frac{2}{3}x + 5$

19. $y = 3 \sin 4x - 7$ **21.** A

23. a. 1 cycle; 1.5 cycles; 2 cycles; 2.5 cycles **b.** 1.25 cycles; 1.75 cycles; 2.25 cycles **c.** $y = \sin bx$: $|b|$ cycles; $y = \cos bx$: $|b|$ cycles

33. B

39. $-540°$

41. $5 + 14i$

43. 2

45.

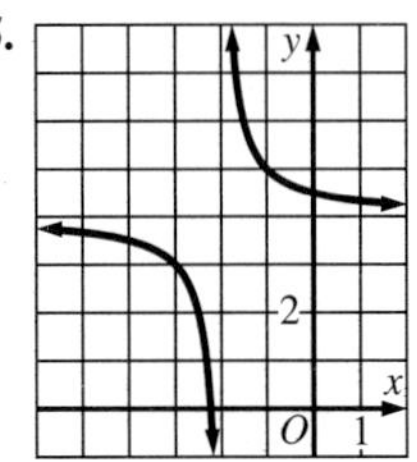

Page 726 Assess Your Progress

1. 49°, 131° **2.** 35°, 325° **3.** 328°, 212° **4.** -0.5748
5. 0.0707 **6.** 0.7071

7.

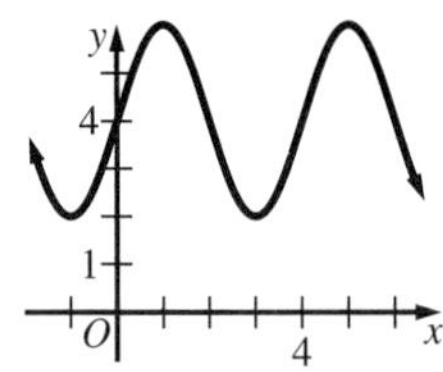

amplitude: 2; period: 4

8.

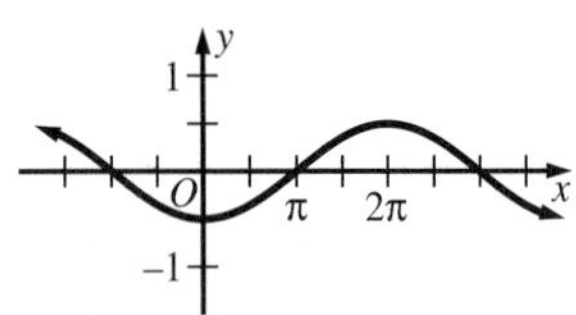

amplitude: $\frac{1}{2}$; period: 4π

Page 730 Checking Key Concepts

1.

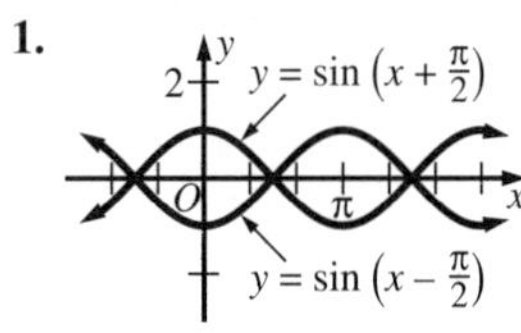

The graph of $y = \sin\left(x - \frac{\pi}{2}\right)$ is the graph of $y = \sin x$ shifted $\frac{\pi}{2}$ to the right, while the graph of $y = \sin\left(x + \frac{\pi}{2}\right)$ is the graph of $y = \sin x$ shifted $\frac{\pi}{2}$ to the left.

3.

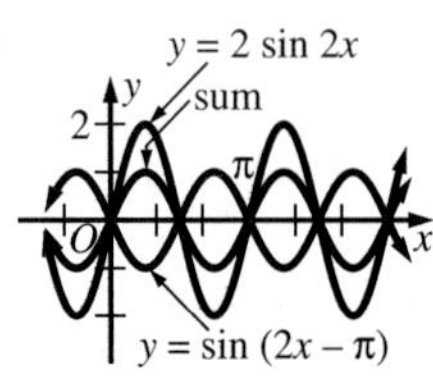

Function	Period	Range
$y = 2 \sin 2x$	π	$-2 \le y \le 2$
$y = \sin(2x - \pi)$	π	$-1 \le y \le 1$
sum	π	$-1 \le y \le 1$

Pages 731–734 Exercises and Applications

1.

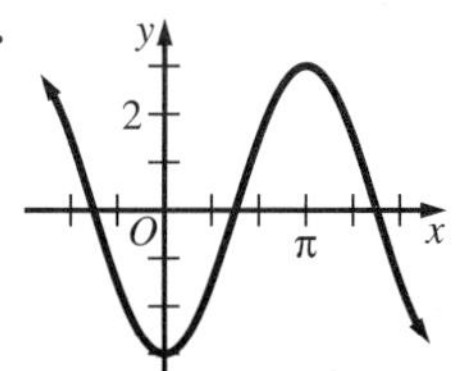

3.

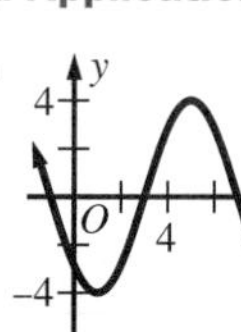

5.

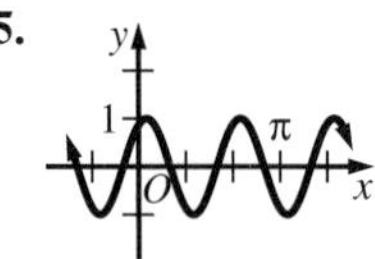

7.

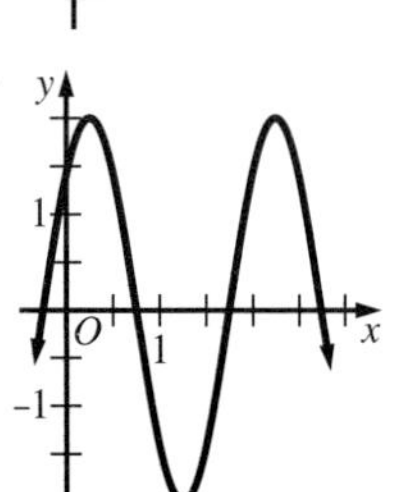

9, 11, 13, 15. Answers may vary. Examples are given.

9. $y = \sin \pi(x + 0.3)$ **11.** $y = 3 \sin \frac{\pi}{4}(x - 3)$

13. $y = 10 \cos \frac{\pi}{12}(x - 5)$ **15.** $y = 5 \cos 2\left(x + \frac{\pi}{3}\right)$

17. a. 6000 rev/min = 100 rev/s, and since the period is the reciprocal of the frequency, the period is $\frac{1}{100}$ s.

b. $y = 3.75 \cos 200\pi x$ **c.** 2: $y = 3.75 \cos 200\pi\left(x - \frac{1}{300}\right)$; 3: $y = 3.75 \cos 200\pi\left(x - \frac{1}{150}\right)$ **23.** Answers may vary. An example is given. $y = \cos\left(x - \frac{\pi}{2}\right)$; $y = \cos\left(x + \frac{3\pi}{2}\right)$; $y = \cos\left(x - \frac{5\pi}{2}\right)$

25.

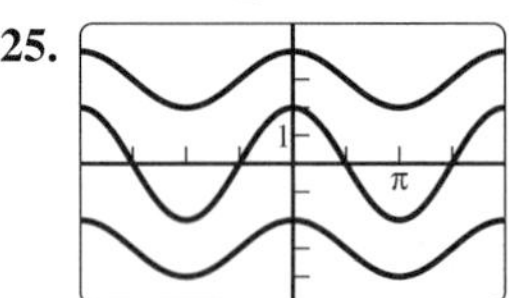

Function	Period	Range
$y = 3 + \cos x$	2π	$2 \le y \le 4$
$y = -3 + \cos x$	2π	$-4 \le y \le -2$
sum	2π	$-2 \le y \le 2$

27.

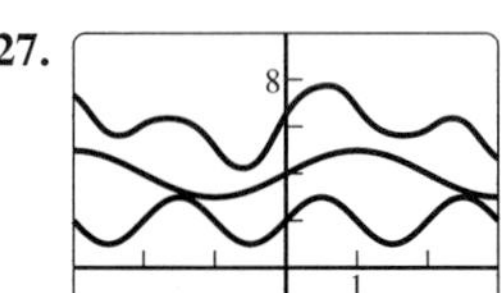

Function	Period	Range
$y = 2 + \sin \pi x$	2	$1 \le y \le 3$
$y = 4 + \sin \frac{\pi}{2}x$	4	$3 \le y \le 5$
sum	4	$4.24 \le y \le 7.76$

29.

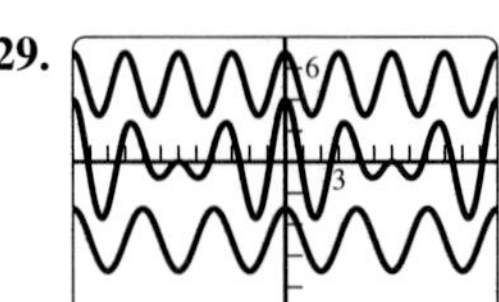

Function	Period	Range
$y = 5 + 2\cos\frac{2\pi}{3}x$	3	$3 \le y \le 7$
$y = -5 + 2\cos\frac{\pi}{2}x$	4	$-7 \le y \le -3$
sum	12	$-3.61 \le y \le 4$

33. 5 **35.** 13 **37.** 5.39 **41.** Answers may vary. An example is given. $y = 3\cos\frac{\pi}{3}x + 5$ **43.** $x = 2$; $x = 5$ **45.** $x = -6$; $x = 6$ **47.** 1.4

Page 737 Checking Key Concepts

1. 0.7813 **3.** −1

5.

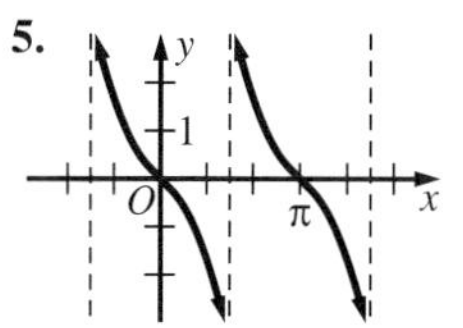

7.

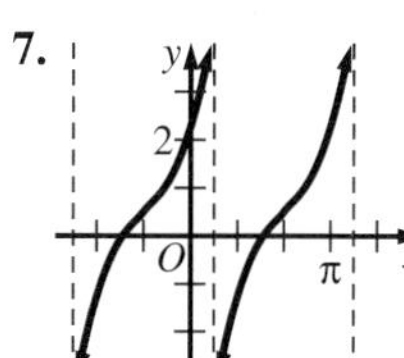

Pages 738–739 Exercises and Applications

1. 0
3. −0.5774
5. 2.4142
7. −0.3640
9. 0.1820
11. 0

19.

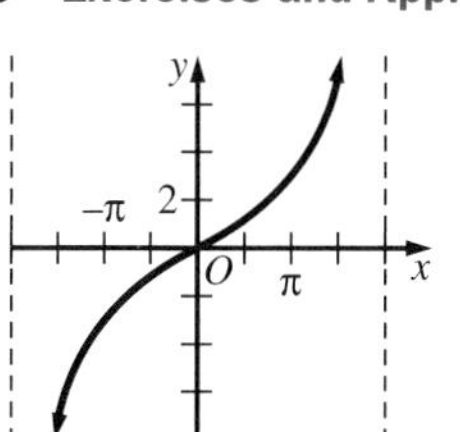

21.

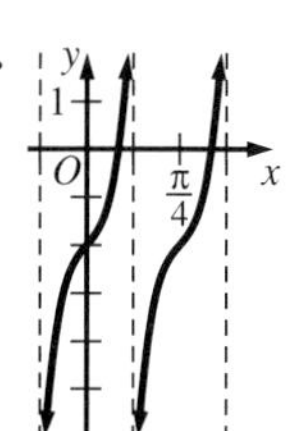

23. Answers may vary. An example is given. $y = 2\cos\left(x - \frac{\pi}{2}\right)$ **25.** 0 **27.** −1

Page 739 Assess Your Progress

1.

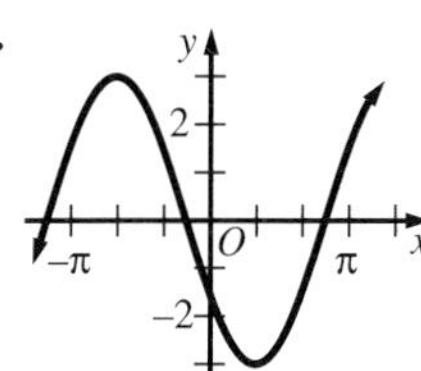

2.

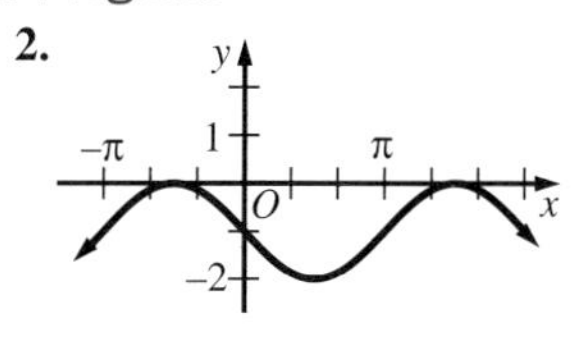

3. a. $s = c\tan\theta$ **b.** $s = 29{,}800\ \frac{\text{m}}{\text{s}}$

4.

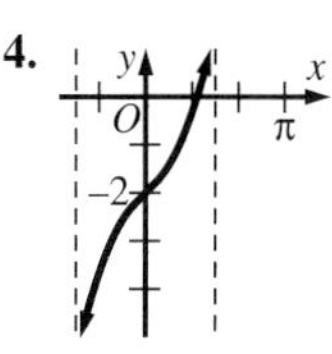

5.

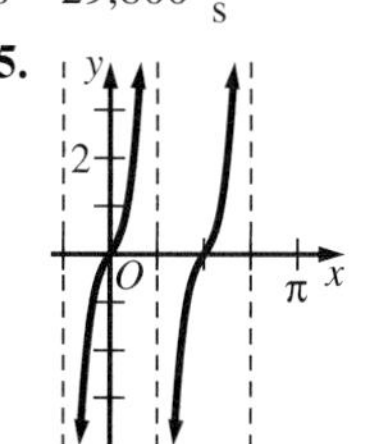

6.

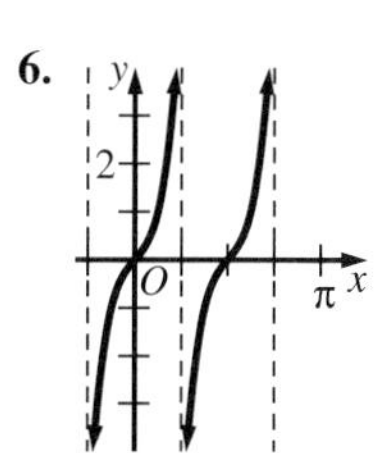

Pages 744–745 Chapter 15 Assessment

1. radians **2.** amplitude, period, phase shift **3.** sine, cosine **4. a.** 39 m **b.** 55 m **c.** 30 m **5.** 47°, 133° **6.** 137°, 223° **7.** 34°, 326° **8.** 0 **9.** 0 **10.** −1 **11.** −1 **12.** 0 **13.** 1 **15. a.** amplitude: 316,200; period: $\frac{1}{22{,}000}$ **16, 17.** Answers may vary. Examples are given. **16.** $y = 7\cos 2x$ **17.** $y = 5\cos\pi x - 1$

20.

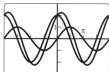

Function	Period	Range
$y = \sin x$	2π	$-1 \le y \le 1$
$y = 2\cos x$	2π	$-2 \le y \le 2$
sum	2π	$-2.24 \le y \le 2.24$

21.

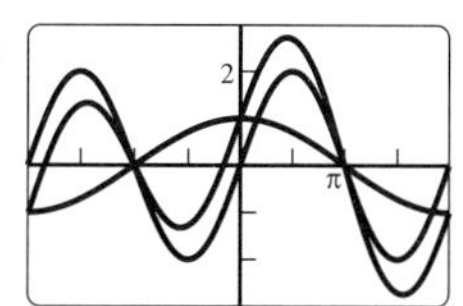

Function	Period	Range
$y = \cos\frac{1}{2}x$	4π	$-1 \le y \le 1$
$y = 2\sin x$	2π	$-2 \le y \le 2$
sum	4π	$-2.74 \le y \le 2.74$

22.

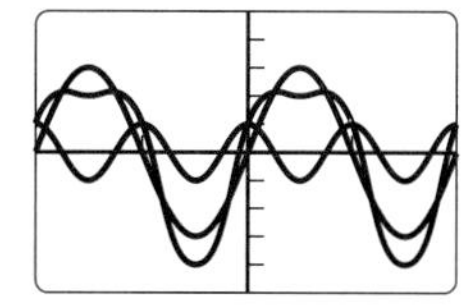

Function	Period	Range
$y = 3\sin\pi x$	2	$-3 \le y \le 3$
$y = \cos 2\pi x$	1	$-1 \le y \le 1$
sum	2	$-4 \le y \le 2.13$

23.

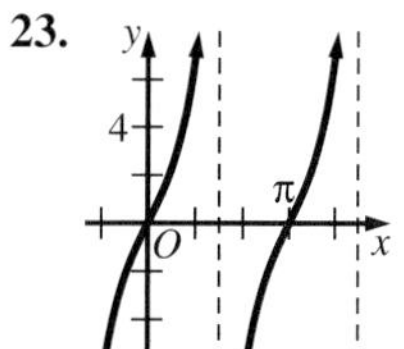

24.

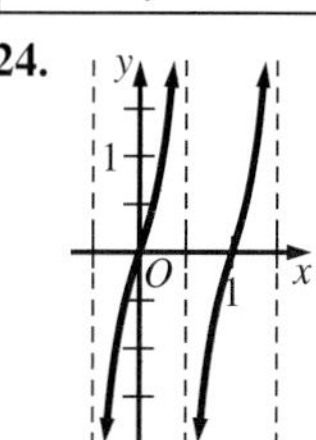

25.

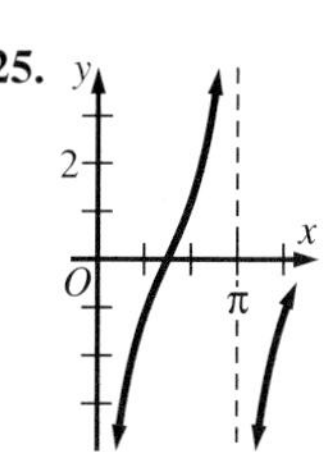

27. a. By the diagram, $\tan\theta = \frac{l}{h}$, and thus $l = h\tan\theta$. **b.** 0, 10.4 in., 31.2 in., and (when $\theta = 90°$) undefined

CHAPTERS 13–15

Pages 746–747 Cumulative Assessment

3. about 0.0588 **5.** 0.735 **7.** about 0.145 **9. a.** about 0.1587 **b.** about 0.995 **c.** about 0.4592 **11.** about 291 ft **13.** −0.7071; −0.7071; 1 **15. a.** 170° **b.** 49.2 km; 8.68 km **17.** $\angle C = 83°$; $b \approx 7.18$; $c \approx 10.1$ **19.** 1 **21.** about 0.8660 **23.** 3; 1; maximum: $\theta = 135°$ and $\theta = 315°$; minimum: $\theta = 45°$ and $\theta = 225°$ **25.** about −1.47 or $-\frac{7\pi}{15}$

27.

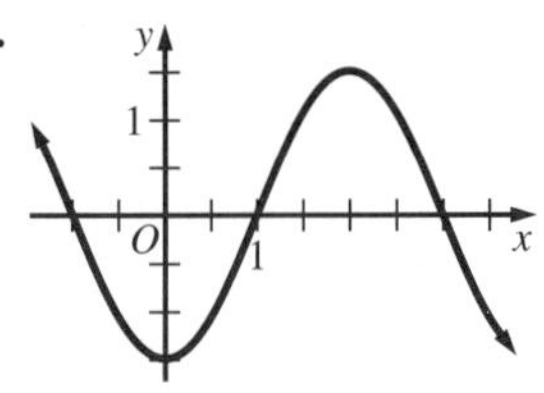

amplitude: $\frac{3}{2}$; period: 4

31.

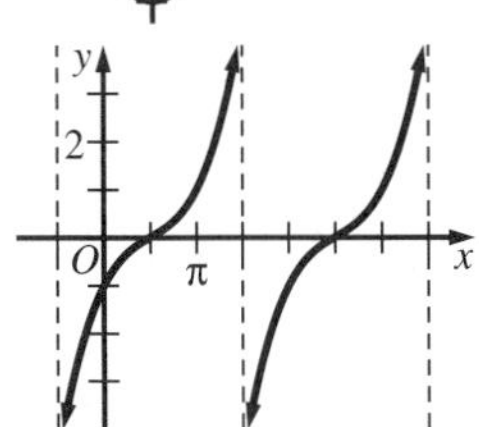

EXTRA PRACTICE

Pages 749–750 Chapter 1

1, 3, 5. Answers may vary. Examples are given.
1. 3.35 million **3.** 66.0 million; 69.3 million
5. The constant growth model provides a more accurate estimate, because the lower estimates provided by this model seem to reflect the fact that sales did not increase as much from 1985–1990 as from 1980–1985.

7.

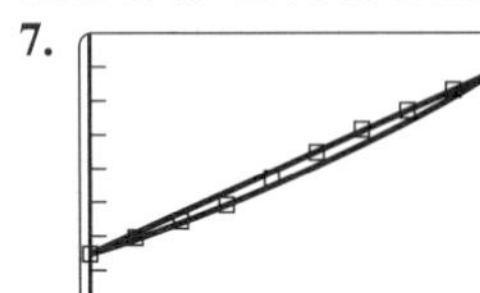

Answers may vary. An example is given. The linear model appears to fit the data better.

9. $\begin{bmatrix} 6 & -6 & 9 \\ 2 & 5 & -3 \end{bmatrix}$ **11.** $\begin{bmatrix} 12 & 0 & -3 \\ 6 & -9 & 3 \end{bmatrix}$ **13.** $\begin{bmatrix} 0 & 12 & -21 \\ 2 & -19 & 9 \end{bmatrix}$

15. $\begin{bmatrix} -10 & -18 & 34 \\ -8 & 36 & -16 \end{bmatrix}$ **17.** $\begin{bmatrix} 11 & 42 & -4 \\ -1 & -12 & 0 \\ 17 & 24 & -8 \end{bmatrix}$ **19.** $\begin{bmatrix} -1 & 0 & -8 \\ -57 & 1 & 46 \end{bmatrix}$

21. $\begin{bmatrix} 9 & 10 \\ -1 & -4 \\ 13 & 0 \end{bmatrix}$ **23.** $\begin{bmatrix} 11 & -3 & -6 \\ 8 & 51 & -2 \end{bmatrix}$ **25, 27.** Answers may vary. Examples are given. **25.** Draw cards at random from a standard deck of 52 cards, letting each suit represent one of the four answer choices, and choosing a particular suit to represent a right answer. **27.** Generate a random number from 1–5 on a calculator by using the formula "int(5*rand)+1." Let "1" and "2" represent getting on base. **29.** int(6*rand)+1
31. int(5*rand)+2

Pages 750–752 Chapter 2

1. Yes. **3.** No. **5.** Yes. **7.** No.

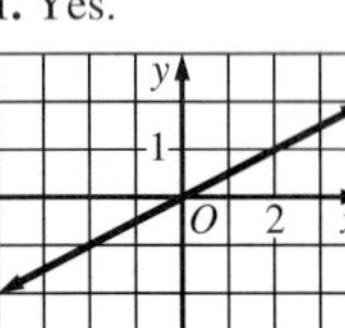

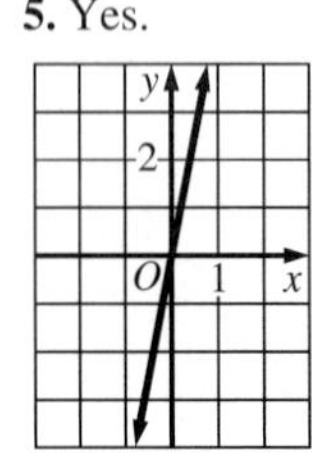

9. $y = 1.5x$; 13.5 **11.** $y = 10.3x$; 329.6

13. $y = 3x + 8$ **15.** $y = -\frac{3}{2}x - \frac{5}{2}$ **17.** $y = \frac{2}{3}x - \frac{1}{3}$

19.

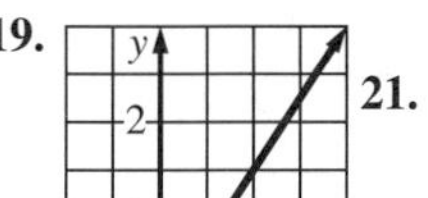

21.

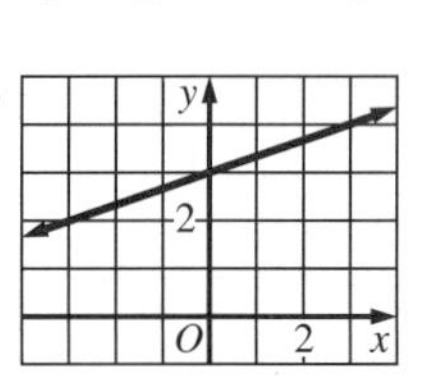

23.

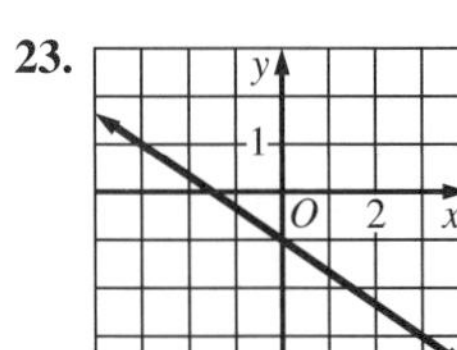

25. a. $y = 2 + \frac{1}{3}(x + 3)$

b.

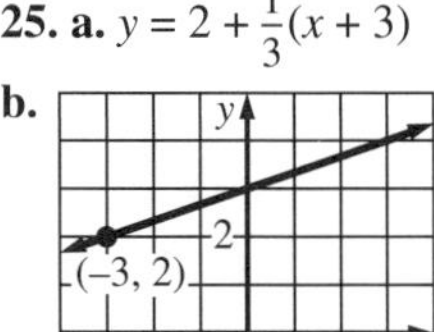

c. $y = \frac{1}{3}x + 3$

27. a. $y = -2 + 2(x - 6)$

b.

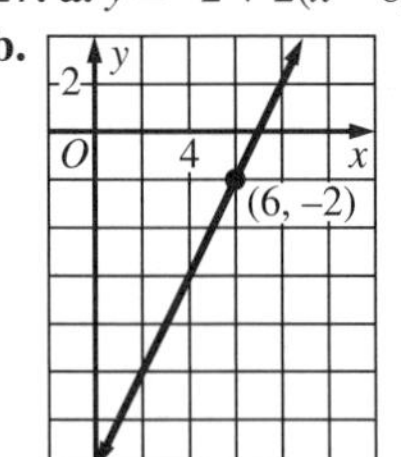

c. $y = 2x - 14$

29. a. $y = 2 + \frac{2}{3}(x - 9)$

b.

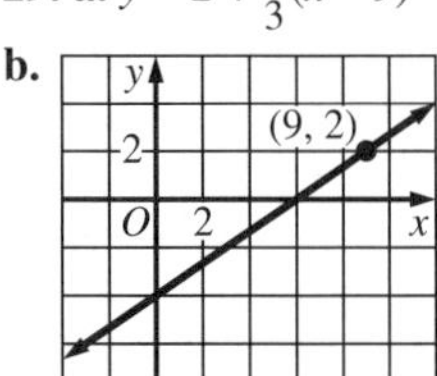

c. $y = \frac{2}{3}x - 4$

31. $y = 5 - \frac{2}{5}(x + 1)$ or $y = 3 - \frac{2}{5}(x - 4)$ **33.** $y = 6 + 2(x - 1)$ or $y = -2 + 2(x + 3)$ **35.** $y = -4 - \frac{9}{5}(x + 1)$ or $y = 5 - \frac{9}{5}(x + 6)$
37, 39. Answers may vary. Examples are given.
37. $y = 2.7x + 31$ **39.** about 80 million **41.** about zero

43.

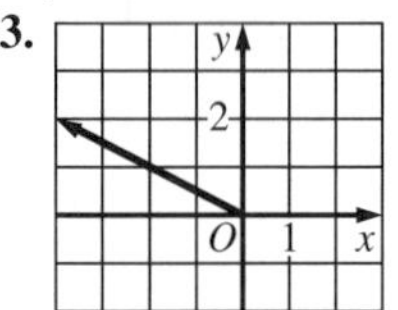

45.

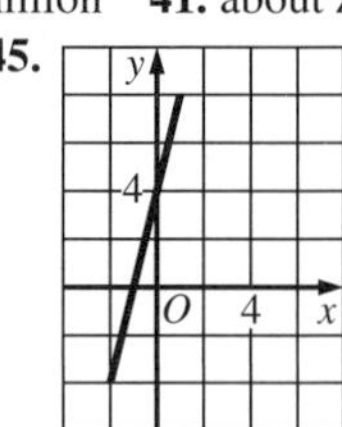

Pages 752–753 Chapter 3

1. 2^5 **3.** 2^8 **5.** 640 **7.** 8 **9.** exponential **11.** linear
13. $\frac{1}{64}$ **15.** $\frac{1}{243}$ **17.** 5 **19.** 64 **21.** $\frac{1}{5}$ **23.** 7

25. a. 7
b. exponential growth
c.

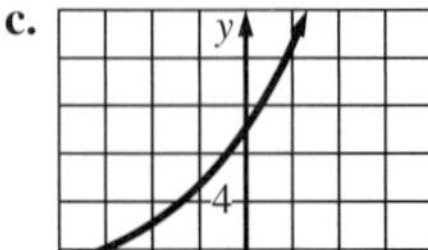

27. a. 6
b. exponential decay
c.

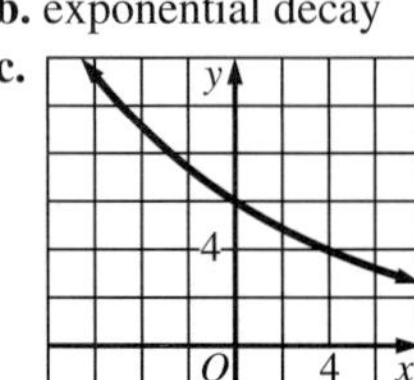

29. a. 2
b. exponential decay
c.

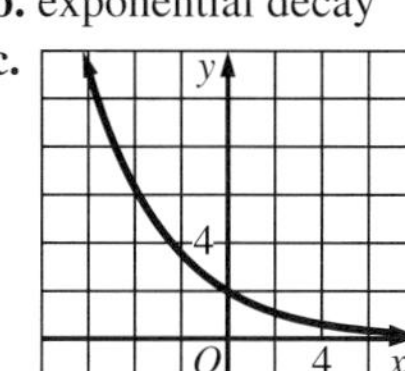

31.

33. \$1648.72 **35.** \$1529.59

37. a.

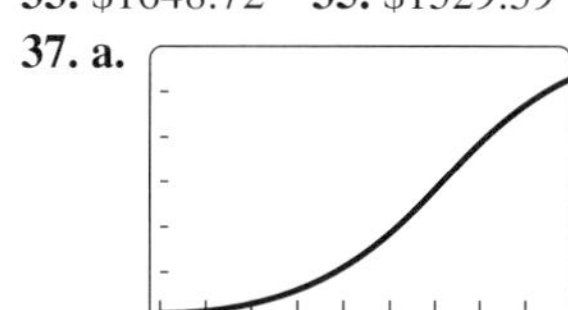

b. about 193,000 mi
c. in 1893, shortly before the turn of 1894

39. $y = 4\left(\frac{3}{2}\right)^x$

41. $y = 1.6\left(\frac{5}{2}\right)^x$ **43.** $y = 186{,}810(1.03101)^x$

45. Predictions should range from 633,746 to 636,540.

Pages 753–754 Chapter 4

1. a.

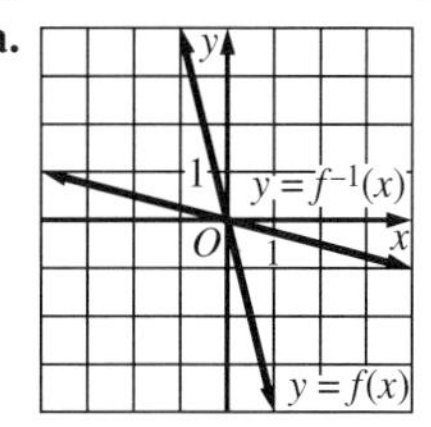

b. $f^{-1}(x) = -\frac{1}{4}x$

3. a.

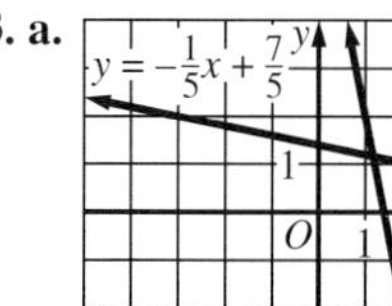

b. $y = -\frac{1}{5}x + \frac{7}{5}$

5. a.

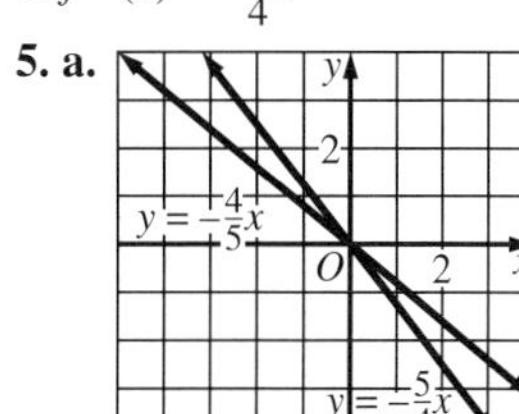

b. $y = -\frac{4}{5}x$

7. a.

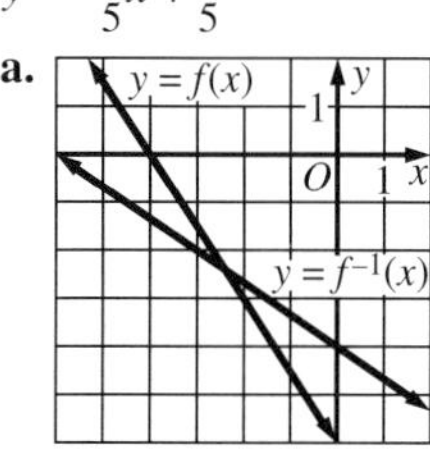

b. $f^{-1}(x) = -\frac{2}{3}x - 4$

9. $\log_5 125 = 3$ **11.** $\log_{2/3} \frac{4}{9} = 2$ **13.** $\left(\frac{1}{2}\right)^{-3} = 8$
15. $(0.36)^{1/2} = 0.6$ **17.** 7 **19.** 0 **21.** $3 \log_5 x - 2 \log_5 y$
23. $\frac{1}{2} \log_5 x + \log_5 y$ **25.** $\log_2 \frac{x}{y}$ **27.** $\log_b 45$ **29.** $\ln \frac{1}{2}$
31. 0.36 **33.** 0.86 **35.** 1.64 **37.** 0.28 **39.** 1.89 **41.** 2.07
43. –0.16 **45.** –0.79 **47.** $y = 251(63.1)^x$
49. $y = 40.4(0.549)^x$ **51.** $y = 0.00398x^{0.76}$ **53.** $y = 2.5x^{5.8}$

Pages 754–756 Chapter 5

1. C **3.** A **5.** ± 3 **7.** $\pm 5\sqrt{2} \approx \pm 7.1$ **9.** $\pm\sqrt{15} \approx \pm 3.9$
11. $\pm\sqrt{30} \approx \pm 5.5$ **13.** –4.5 **15.** minimum value; –2
17. minimum value; –5 **19.** minimum value; –10

21. a. $x = -3$ **b.** (–3, 0)
c. opens down
d.

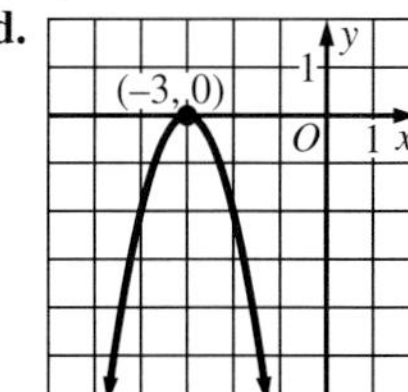

23. a. $x = -1$ **b.** (–1, 3)
c. opens down
d.

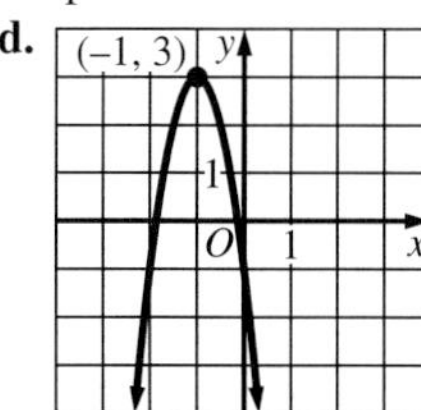

25. a. $x = -4$ **b.** (–4, –2)
c. opens down
d.

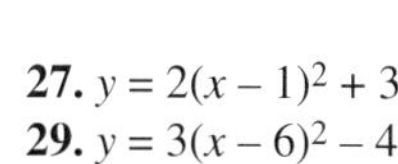

27. $y = 2(x - 1)^2 + 3$
29. $y = 3(x - 6)^2 - 4$

31. a. 2, 5 **b.** (3.5, 4.5)
c.

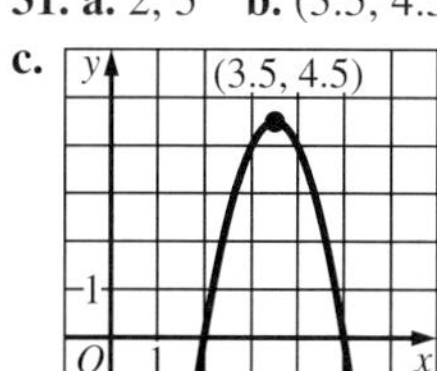

33. a. –3, –7 **b.** (–5, 6)
c.

35. a. 2, 4 **b.** (3, 3)
c.

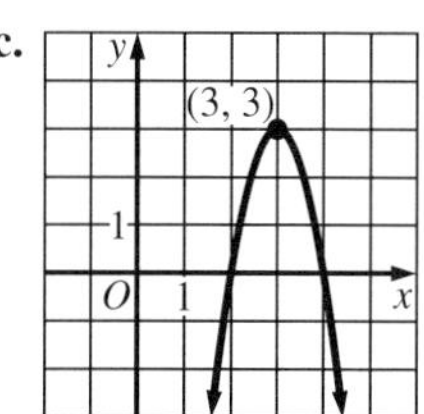

37. $y = -2(x + 5)(x - 3)$
39. $y = (x - 3)^2 - 1$
41. $y = (x - 4)^2 - 16$
43. $y = 2(x - 2)^2 - 3$
45. minimum value; 2.5
47. maximum value; –1
49. minimum value; –12

51. –2, 6 **53.** $\frac{9 \pm \sqrt{41}}{10} \approx 0.26, 1.54$
55. $\frac{-1 \pm \sqrt{41}}{4} \approx -1.85, 1.35$ **57.** $2 \pm \sqrt{3} \approx 0.27, 3.73$
59. $-6 \pm 2\sqrt{7} \approx -11.29, -0.71$ **61.** two solutions **63.** one solution **65.** one solution **67.** $y = (2x - 5)(x - 2)$
69. $y = (5x - 7)(5x + 7)$ **71.** $y = (5x - 3)(x + 4)$ **73.** $y = (3x - 2)(x + 6)$ **75.** 5, 7 **77.** $1, \frac{3}{2}$ **79.** $0, \frac{4}{5}$ **81.** 5 **83.** 2

Pages 757–758 Chapter 6

1, 3. Descriptions may vary. Examples are given. **1.** numerical; A batting average can vary from 0.0 to 1.0, with values around 0.2 to 0.25 being common. **3.** categorical; Some of the modes are train, car, bus, bicycle, and taxi. **5.** Answers may vary. An example is given. the circle graph; The circle graph allows you to see more easily how each choice is related to the results as a whole. **7, 9.** Answers may vary. Examples are given. **7.** This is a leading question, since most people would like lower taxes and public recreation facilities. The question could be ended after "Do you think this city should re-elect Mayor Fong?" **9.** Respondents might inflate their answers to look good in the eyes of the questioner. The question would be best if it could be answered anonymously.
11, 13. Answers may vary. Examples are given.
11. convenience sample; biased; The students in a few specific homerooms might not represent students as a whole.
13. self-selected sample; biased; Students who volunteer might spend more time doing homework than others.
15. median: \$29,500

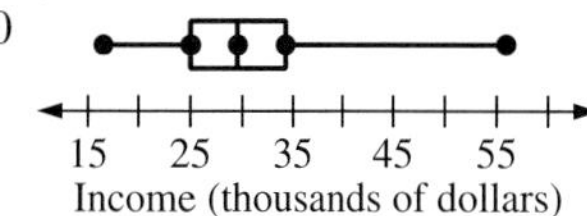

17. Answers may vary. An example is given. The median is a better measure of the center because the two highest incomes skew the mean toward the right, which makes it less representative of most of the incomes. **19.** 1st set: about 1.3; 2nd set: about 1.3; 3rd set: about 2.0; The standard deviation of the 3rd set is much greater. **21. a.** ±2% **b.** 16–20%
23. about 46.3%

Pages 758–760 Chapter 7

1. (3, 7) **3.** (2, –5) **5.** about (–3.54, 7.54) and (2.54, 1.46)
7. about (1.65, –2.31) and (–6.65, 14.31)
9. about (–1.58, 1.67) and (1.48, 13.90)
11. (–1, 4); one solution **13.** infinitely many solutions
15. (6, 5); one solution

17. $\begin{bmatrix} 11 & -8 \\ 15 & -14 \end{bmatrix}\begin{bmatrix} x \\ y \end{bmatrix} = \begin{bmatrix} 48 \\ 33 \end{bmatrix}$; (12, 10.5)

19. $\begin{bmatrix} 5 & 13 \\ 9 & -4 \end{bmatrix}\begin{bmatrix} x \\ y \end{bmatrix} = \begin{bmatrix} 40.8 \\ 2.2 \end{bmatrix}$; (1.4, 2.6)

21. $\begin{bmatrix} 9 & 1 \\ 1.6 & 1 \end{bmatrix}\begin{bmatrix} x \\ y \end{bmatrix} = \begin{bmatrix} -13.5 \\ 0 \end{bmatrix}$; about (–1.82, 2.92)

23. about (0.30, 2.49, 3.01) **25.** about (1.20, 0.58, –0.82)

27.
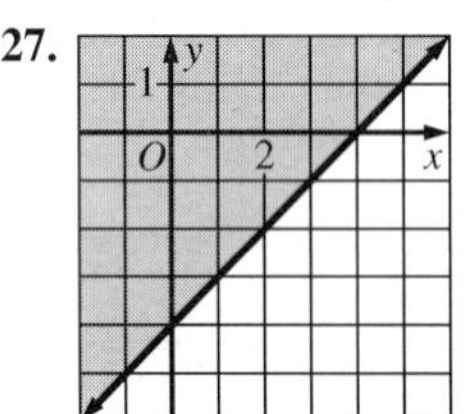

29.
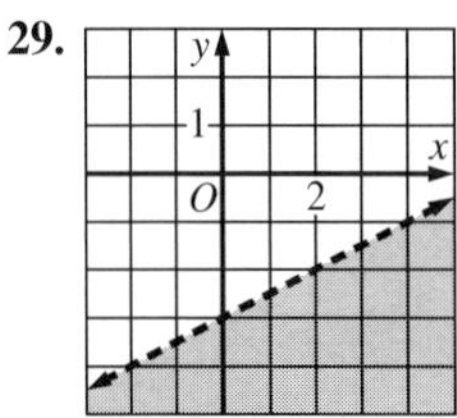

31.
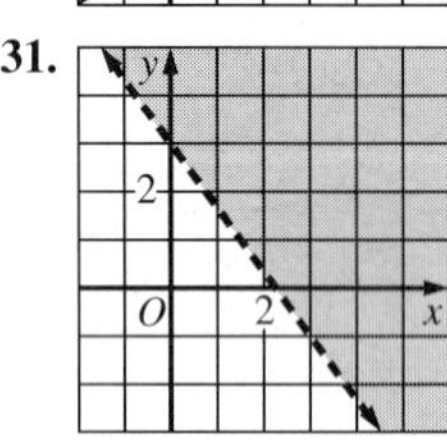

33.
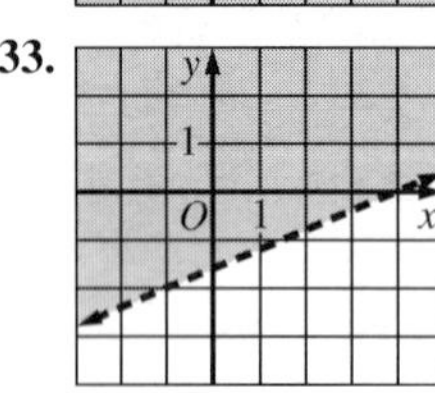

35. $y \geq 2x - 1$

37.
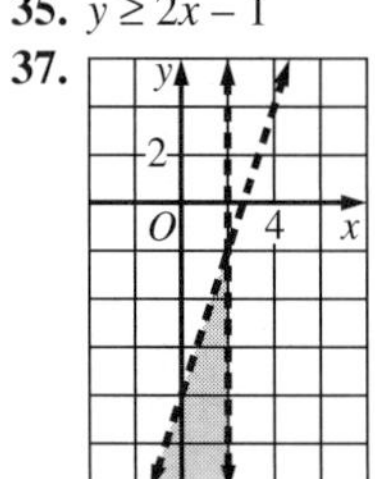

39.
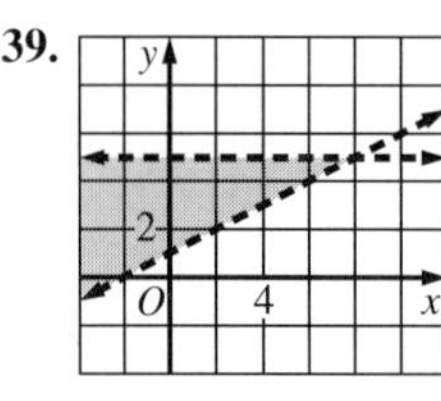

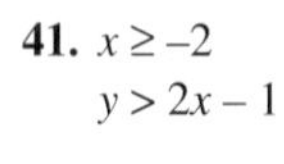

41. $x \geq -2$
$y > 2x - 1$

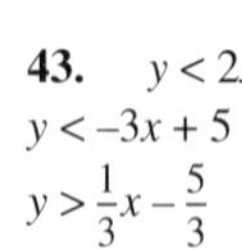

43. $y < 2x$
$y < -3x + 5$
$y > \frac{1}{3}x - \frac{5}{3}$

45.
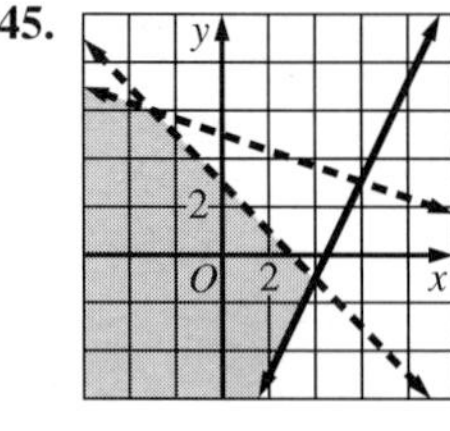

47.
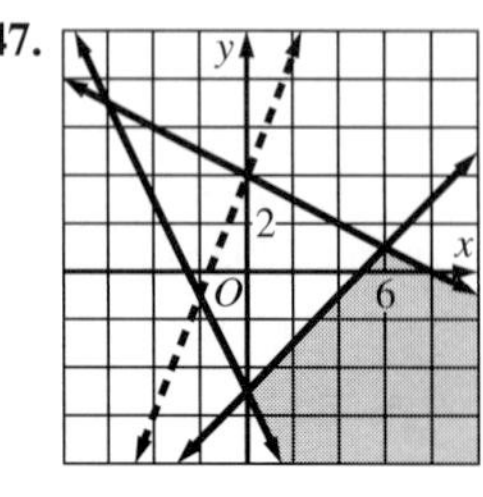

Pages 760–762 Chapter 8

1. domain: $x \geq -2$; range: $y \geq 0$

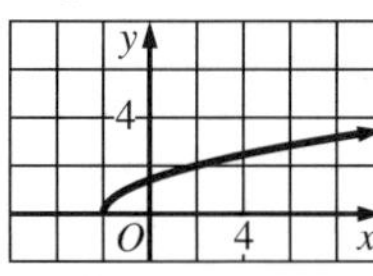

3. domain: $x \geq 0$; range: $y \geq -2$

5. domain: $x \geq 0$; range: $y \geq -3$

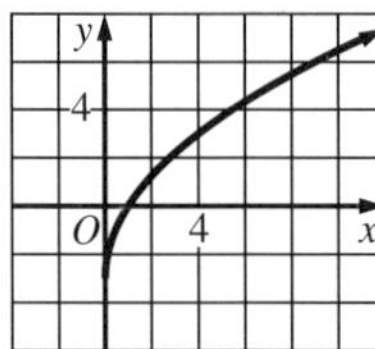

7. sometimes true

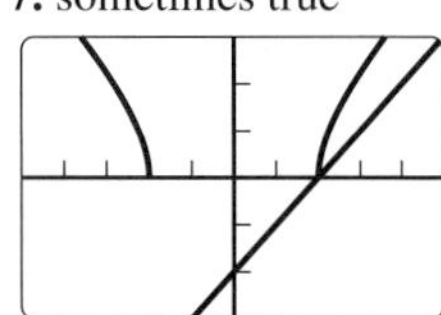

9. never true

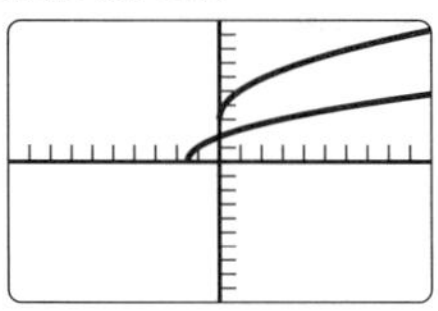

11. sometimes true

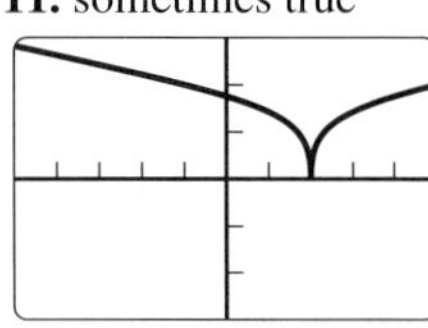

13. –1 **15.** undefined **17.** –2 **19.** $\frac{3}{2}$ **21.** domain: $x \geq 5$; range: $y \geq 0$ **23.** domain: $x \geq 0$; range: $y \geq -3$ **25.** $3\sqrt{2}$

27. $2\sqrt[4]{3}$ **29.** $x^4\sqrt{5}$ **31.** $3x^8$ **33.** 81 **35.** –21 **37.** 160
39. 103 **41.** $-\frac{4}{3}$ **43.** 72 **45.** 7 **47.** 8 **49.** –2, 3

51. $5 - 2i$ **53.** $-14 - 7i$ **55.** $45 - 28i$ **57.** $\frac{13}{10} - \frac{11}{10}i$

59. $\frac{9}{13} + \frac{32}{13}i$ **61.** $2 \pm i$ **63.** $1 \pm 3i$

65. $|3 + 2i| = \sqrt{13}$; $|3 - 2i| = \sqrt{13}$; $|6| = 6$

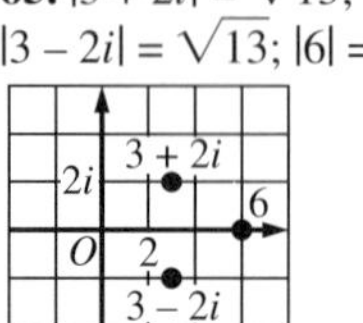

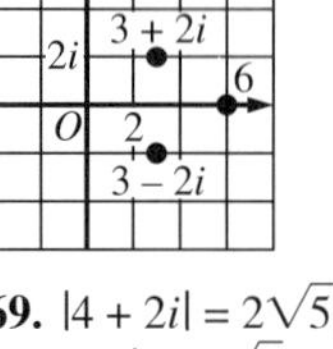

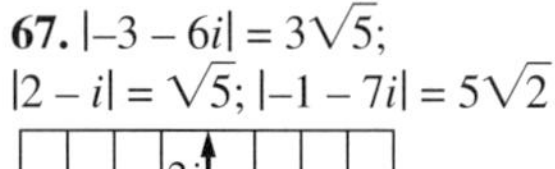

67. $|-3 - 6i| = 3\sqrt{5}$; $|2 - i| = \sqrt{5}$; $|-1 - 7i| = 5\sqrt{2}$

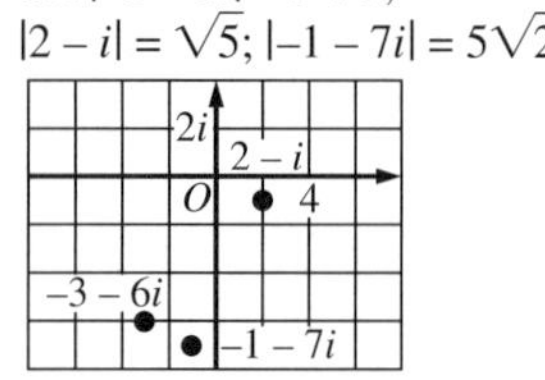

69. $|4 + 2i| = 2\sqrt{5}$; $|-3 + 3i| = 3\sqrt{2}$; $|1 + 5i| = \sqrt{26}$

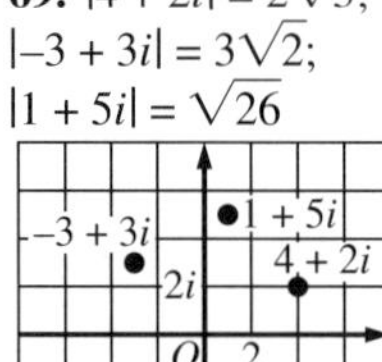

71. $|-2 + 3i| = \sqrt{13}$; $|1 + i| = \sqrt{2}$; $|-5 + i| = \sqrt{26}$

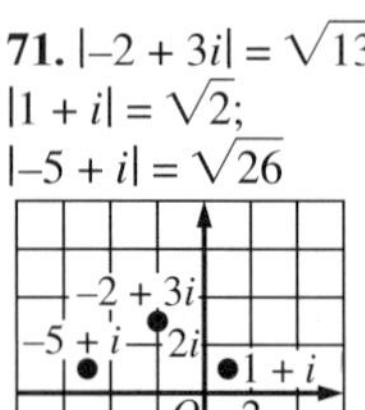

73. $|5 - 3i| = \sqrt{34}$; $|2 - i| = \sqrt{5}$; $|7 - 11i| = \sqrt{170}$

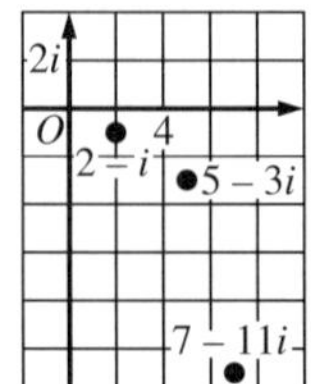

75. $-3 - 2i$ **77.** $3 + 2i$ **79.** No; for example, $3 \times 2 = 6$, but 6 is not in the set {1, 2, 3}. **81.** any integer
83. any rational number

Pages 762–764 Chapter 9

1. 55 **3.** 12 **5.** $3x^3 - 3x + 7$ **7.** $7x^4 - 5x^3 + 2x^2 - 3$ **9.** $10x^2 - 29x - 21$ **11.** $5y^3 - 11y^2 + 6y - 8$ **13.** $a^3 + 2a^2 - a - 2$ **15.** $2x - 1 - \frac{5}{x-2}$ **17.** $4c^2 + 9c + 28 + \frac{83}{c-3}$ **19.** $b^2 - 2b - 5 - \frac{21}{2b-5}$

21.

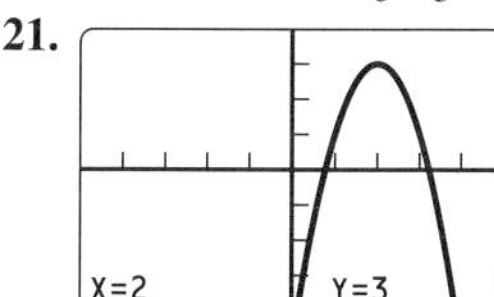

a. As $x \to -\infty, f(x) \to -\infty$. As $x \to +\infty, f(x) \to -\infty$. **b.** local (and global) maximum: (2, 3)

23. −2, 1, 5 **25.** −1, 2, 3 **27.** $y = 2x(x + 1)(x - 3)$ **29.** $y = \frac{3}{2}(x + 3)(x + 1)(x - 3)$ **31.** −0.71, 2.14, 4.57 **33. a.** $\pm 1, \pm 3, \pm\frac{1}{2}, \pm\frac{3}{2}, \pm\frac{1}{5}, \pm\frac{3}{5}, \pm\frac{1}{10}, \pm\frac{3}{10}$ **b.** $-\frac{1}{5}, \frac{3}{2}$, 1 (double root)

35. Yes; 8. **37.** No.

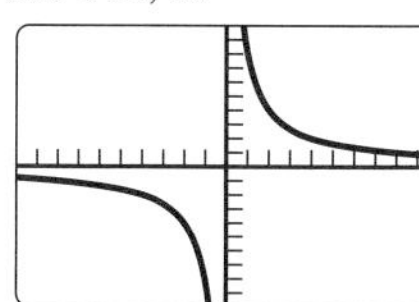

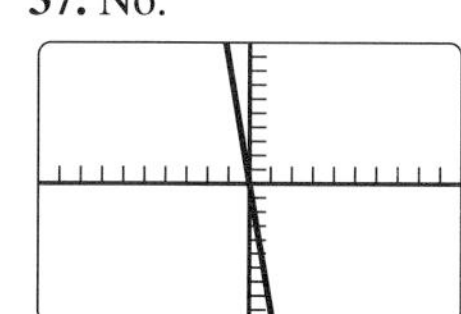

39. No; let n = the number of pounds in the bag, c = the total cost of the bag, and k = the fixed price per pound. Then $nk = c$, or $\frac{c}{n} = k$, which is a direct variation equation.

41. $y = \frac{6}{x+3} - 5$ **43. a.** $x = 3$ and $y = 2$ **b.** The graph is the graph of $y = \frac{10}{x}$ translated 3 units to the right and 2 units up.

45. a. $x = 5$ and $y = 3$ **b.** The graph is the graph of $y = \frac{9}{x}$ translated 5 units to the right and 3 units up. **47.** degree $p(x) >$ degree $q(x)$; degree $q(x) \geq 3$; roots of $q(x)$: −2, 2, 4 **49. a.** $x = 4$ and $x = -1$ **b.** As $x \to \pm\infty, f(x) \to 0$. **51. a.** $x = 0$ and $x = 1$ **b.** As $x \to \pm\infty, g(x) \to 0$. **53. a.** $x = \frac{3}{2}$ **b.** As $x \to \pm\infty, h(x) \to 0$. **55.** $\frac{3}{2}$ **57.** 4, 5 **59.** −2

Pages 764–765 Chapter 10

1. 32, −64, 128, −256 **3.** 100,001, 1,000,001, 10,000,001, 100,000,001 **5.** 0.005, 0.0005, 0.00005, 0.000005 **7.** $\frac{1}{25}, \frac{1}{36}, \frac{1}{49}, \frac{1}{64}$ **9.** $9999\frac{1}{9}, 99{,}999\frac{1}{9}, 999{,}999\frac{1}{9}, 9{,}999{,}999\frac{1}{9}$

11, 13, 15. Answers may vary. Examples are given. **11.** $t_n = -4n + 21$ **13.** $t_n = \frac{1}{2n}$ **15.** $t_n = \frac{1}{5}(-2)^{n+1}$ **17.** geometric **19.** arithmetic **21.** neither **23.** geometric **25. a.** 62.5 **b.** 22 **27. a.** 42.5 **b.** 20 **29. a.** 0.25 **b.** 0.15 **31. a.** $21\frac{1}{4}$ **b.** 9 **33.** $t_1 = 3$; $t_n = 5(t_{n-1})$ **35.** $t_1 = \frac{5}{2}$; $t_n = \frac{1}{2}(t_{n-1})$ **37.** $t_1 = \frac{4}{3}$; $t_n = \frac{1}{2}(t_{n-1})$ **39. a.** 0.2, −1, 5, −25 **b.** $t_n = 0.2(-5)^{n-1}$ **41. a.** 32, −16, 8, −4 **b.** $t_n = 32\left(-\frac{1}{2}\right)^{n-1}$ **43.** 704 **45.** 172.5 **47.** 4 + 5 + 6 + 7 + 8 + 9 + 10 **49.** −5 − 75 − 265 − 635 − 1245 − 2155

51, 53. Answers may vary. Examples are given.

51. $\sum_{n=1}^{9}(25n - 112) = 117$ **53.** $\sum_{n=1}^{16}\left(2n - \frac{5}{3}\right) = 245\frac{1}{3}$

55. $127\frac{3}{4}$ **57.** 208.336 **59.** −4.5 **61.** 12 **63.** −3

Pages 766–767 Chapter 11

1. a. $6\sqrt{2}$ **b.** $y = -x + 11$ **3. a.** $6\sqrt{10}$ **b.** $y = \frac{1}{3}x + \frac{8}{3}$ **5.** 2 or −8 **7.** 1.5 or 0.1

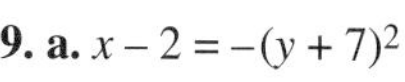

9. a. $x - 2 = -(y + 7)^2$ **b.**

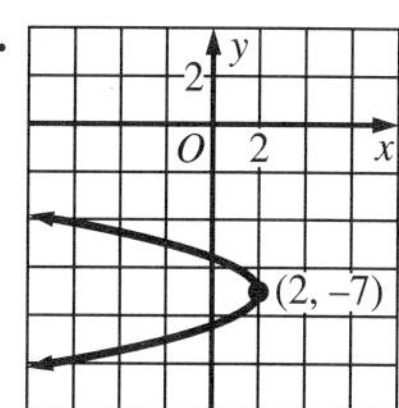

11. vertex: (0, 0); focus: $\left(0, -\frac{1}{8}\right)$; directrix: $y = \frac{1}{8}$

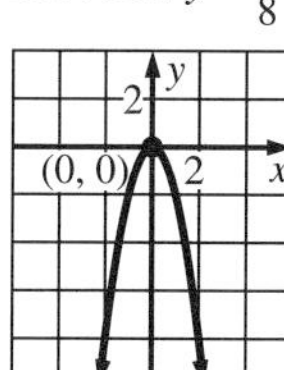

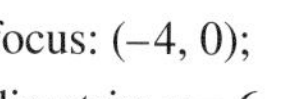

13. vertex: (−4, 3); focus: (−4, 0); directrix: $y = 6$

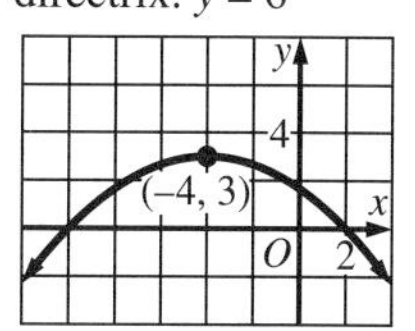

15. center: (−3, 5); radius: 4

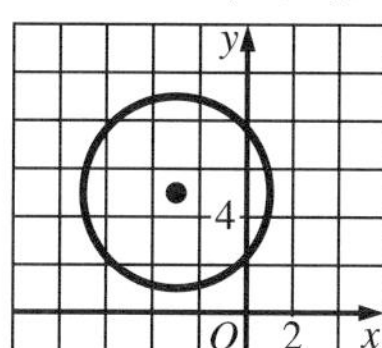

17. center: (3, 0); radius: $\sqrt{10}$

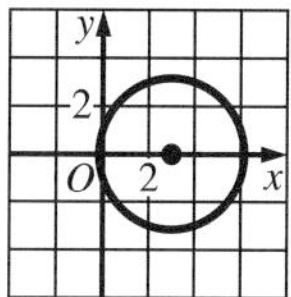

19. $(x + 1)^2 + (y - 8)^2 = 7$ **21.** $(x + \sqrt{5})^2 + (y + 1)^2 = 36$ **23.** $(x - p)^2 + (y - q)^2 = p^2$

25. center: (0, 0); vertices: (±7, 0); foci: $(\pm 2\sqrt{6}, 0)$; horizontal

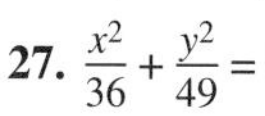

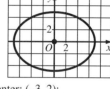

27. $\frac{x^2}{36} + \frac{y^2}{49} = 1$ **29.** $\frac{(x-3)^2}{16} + \frac{(y+2)^2}{64} = 1$

31. center: (0, 0); vertices: (±7, 0); foci: $(\pm\sqrt{85}, 0)$; horizontal; $y = \pm\frac{6}{7}x$

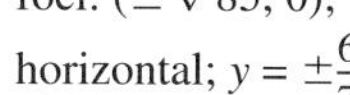

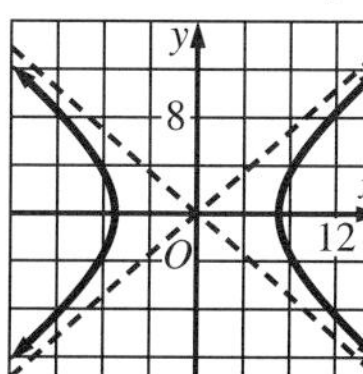

33. center: (−3, 2); vertices: (−3, 5), (−3, −1); foci: $(-3, 2 \pm \sqrt{58})$; vertical; $y = \pm\frac{3}{7}(x + 3) + 2$

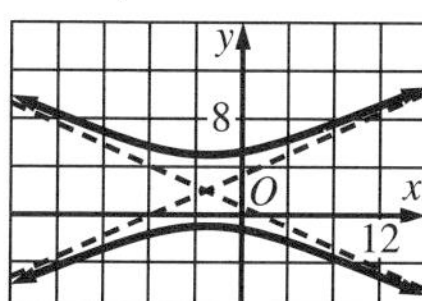

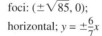

35. center: (–2, 5); vertices: (1, 5), (–5, 5); foci: $(-2 \pm \sqrt{34}, 5)$; horizontal; $y = \pm\frac{5}{3}(x + 2) + 5$

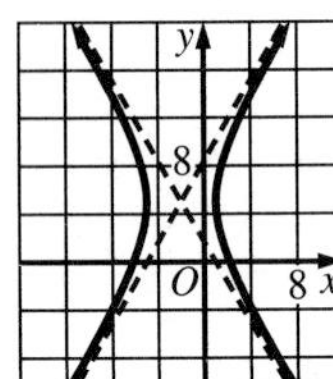

37. $\frac{(x-1)^2}{36} + \frac{(y-3)^2}{45} = 1$

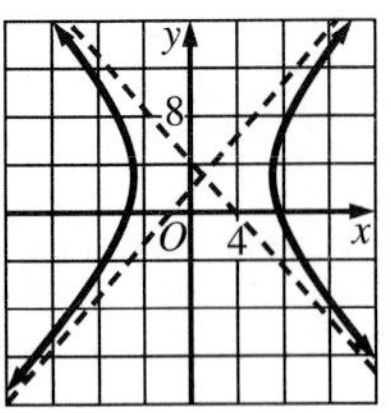

39. $\frac{(x-1)^2}{1} - \frac{(y+3)^2}{4} = 1$; horizontal hyperbola with center (1, –3), vertices (2, –3) and (0, –3), foci $(1 \pm \sqrt{17}, -3)$, and asymptotes $y = \pm 2(x - 1) - 3$ **41.** $y + 2 = -\frac{1}{4}(x - 3)^2$; parabola that opens down with vertex (3, –2), focus (3, –3), and directrix $y = -1$ **43.** degenerate ellipse; The graph is the point (0, 0). **45.** degenerate hyperbola; The graph is the pair of lines $y = \frac{1}{5}x$ and $y = -\frac{1}{5}x$.

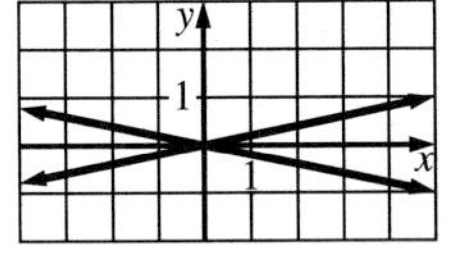

Pages 767–769 Chapter 12

1–3. b, c. Answers may vary. Examples are given.
1. a. *A*: 3; *B*: 3; *C*: 3; *D*: 3; *E*: 4 **b.**

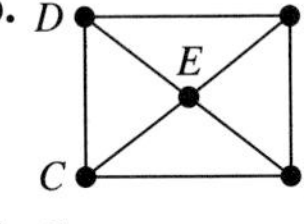

c. 3 colors: *E* red, *A* and *C* green, *B* and *D* blue

3. a. *A*: 3; *B*: 4; *C*: 3; *D*: 4; *E*: 4 **b.**

c. 4 colors: *E* red, *D* blue, *B* green, *A* and *C* yellow

5. Answers may vary. An example is given. 3 colors: UT and MT red, ID and CO green, NV and WY blue

7.
$$\begin{array}{c} \\ A \\ B \\ C \\ D \end{array} \begin{array}{c} \begin{array}{cccc} A & B & C & D \end{array} \\ \begin{bmatrix} 0 & 0 & 0 & 1 \\ 1 & 0 & 0 & 0 \\ 0 & 1 & 0 & 1 \\ 0 & 1 & 1 & 0 \end{bmatrix} \end{array}$$

9, 11. Answers may vary. Examples are given.

9.

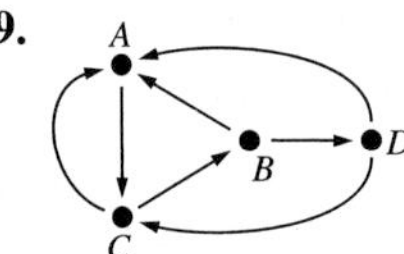

11.

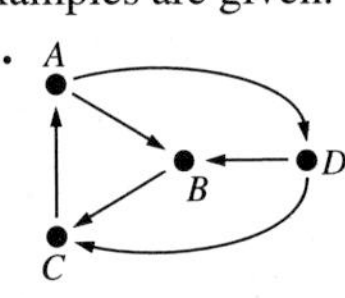

13. 4320 **15.** $n(n - 1)$, or $n^2 - n$ **17.** 210 **19.** 56 **21.** $\frac{7!}{3!2!} = 420$ **23.** 56 **25.** 105 **27.** 11 **29.** 165 **31. a.** 6 **b.** 6 **33. a.** 8 **b.** 70 **35.** $x^8 - 8x^7y + 28x^6y^2 - 56x^5y^3 + 70x^4y^4 - 56x^3y^5 + 28x^2y^6 - 8xy^7 + y^8$ **37.** $64x^6 - 576x^5y + 2160x^4y^2 - 4320x^3y^3 + 4860x^2y^4 - 2916xy^5 + 729y^6$ **39.** $x^{21} - 7x^{18}y^2 + 21x^{15}y^4 - 35x^{12}y^6 + 35x^9y^8 - 21x^6y^{10} + 7x^3y^{12} - y^{14}$

Pages 769–770 Chapter 13

1. $\frac{1}{4}$, or 0.25 **3.** $\frac{1}{2}$, or 0.5 **5.** $\frac{1}{13}$, or about 0.0769 **7.** $\frac{3}{4}$, or 0.75 **9.** $\frac{1}{26}$, or about 0.0385 **11.** $\frac{2}{13}$, or about 0.154 **13.** $\frac{1}{12}$, or about 0.0833 **15.** $\frac{5}{6}$, or about 0.833 **17.** 0.41 **19.** 0.51 **21.** $\frac{13}{51}$, or about 0.255 **23.** $\frac{15}{34}$, or about 0.441 **25.** ${}_6C_k(0.25)^k(0.75)^{6-k}$; 0.297 **27.** ${}_{10}C_k\left(\frac{4}{5}\right)^k\left(\frac{1}{5}\right)^{10-k}$; 0.0264 **29.** 0.00246 **31.** 1×10^{-11} **33.** –0.8 **35.** 2.2 **37.** 2.28%

Pages 770–771 Chapter 14

1. a. 34° **b.** 34°12′54″ **3. a.** 26° **b.** 25°42′9″ **5. a.** 10° **b.** 9°49′22″ **7.** 22.6° **9.** 66.1° **11.** 1.0932 **13.** 1.7321 **15.** $\angle B = 52°$; $a \approx 10.5$; $b \approx 13.4$ **17.** $\angle A \approx 16.3°$; $\angle B \approx 73.7°$; $a = 5.25$; $c = 18.75$ **19–27.** Answers are given in the order sin θ; cos θ; tan θ. **19.** $-\frac{24}{25} = -0.96$; $\frac{7}{25} = 0.28$; $-\frac{24}{7} \approx -3.4286$ **21.** $-\frac{7}{11} \approx -0.6364$; $-\frac{6\sqrt{2}}{11} \approx -0.7714$; $\frac{7}{6\sqrt{2}} \approx 0.8250$ **23.** 0.9135; –0.4067; –2.2460 **25.** 0.9205; 0.3907; 2.3559 **27.** 0.4540; –0.8910; –0.5095 **29.** about 128.1 square units **31.** about 41.0 square units **33.** not enough information; Given *a* and *b*, you need to know the measure of the included angle, $\angle C$. **35.** about 7.36 square units **37.** about 453 square units **39.** about 8.14 square units **41.** $\angle E = 110°$; $d \approx 9.92$; $f \approx 20.9$ **43.** $\angle D = 78°$; $e \approx 4.21$; $f \approx 8.31$ **45.** $\angle E = 79°$; $d \approx 6.68$; $e \approx 11.7$ **47.** no solution; $\angle A$ is obtuse and $a < b$. **49.** $\angle A \approx 120.4°$; $\angle B \approx 33.6°$; $a \approx 18.7$; or $\angle A \approx 7.6°$; $\angle B \approx 146.4°$; $a \approx 2.9$ **51.** $\angle B \approx 102.4°$; $\angle C \approx 24.6°$; $b \approx 15.9$ **53.** about 28.1 **55.** $\angle X \approx 88.6°$; $\angle Y \approx 36.0°$; $\angle Z \approx 55.4°$ **57.** $\angle X \approx 30.0°$; $\angle Y \approx 34.4°$; $\angle Z \approx 115.6°$

Pages 771–772 Chapter 15

1. 0°, 360° **3.** 240°, 300° **5.** 60°, 120° **7.** 129°, 231° **9.** 78°, 282° **11.** 215°, 325° **13.** –0.9848 **15.** –0.9397 **17.** 0.4067 **19.** –0.6428 **21.** 270° **23.** 241° **25.** –183° **27.** 68° **29.** 1.05; $\frac{\pi}{3}$ **31.** 2.62; $\frac{5\pi}{6}$ **33.** 3.40; $\frac{13\pi}{12}$ **35.** –2.27; $-\frac{13\pi}{18}$

37. amplitude: 4; period: π

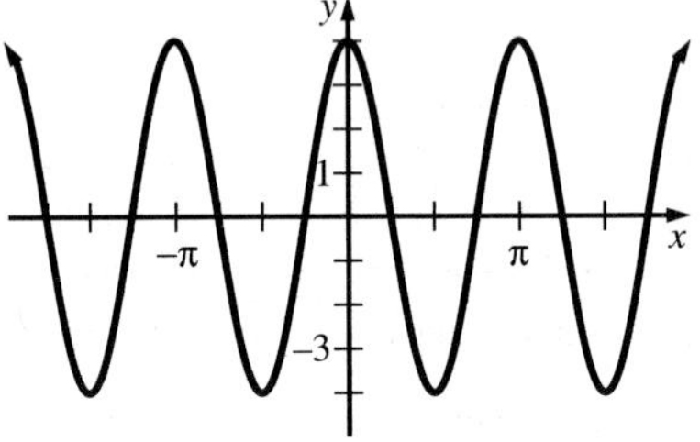

39. amplitude: $\frac{3}{4}$; period: $\frac{2\pi}{3}$

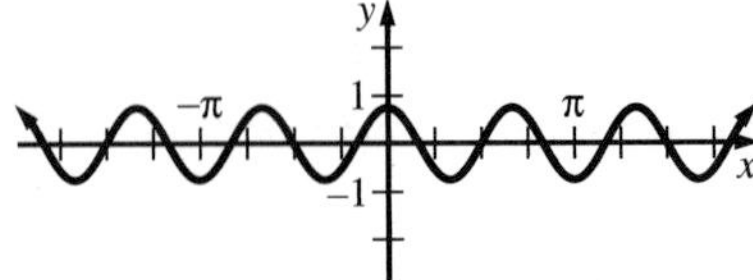

41. $y = 4 \sin \pi x$ **43.** $y = 5 \sin \frac{2}{5}x - 2$

45.

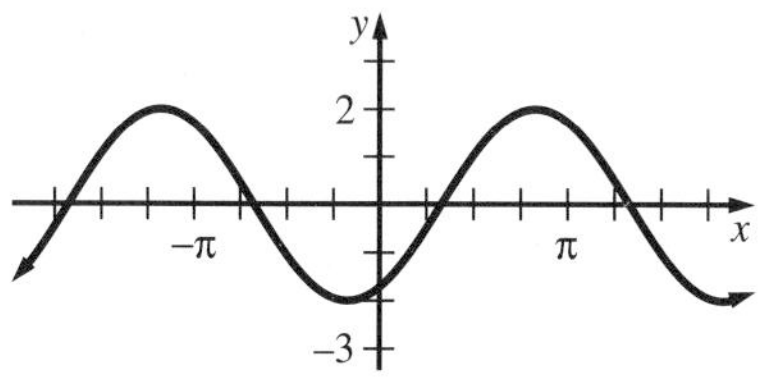

47.

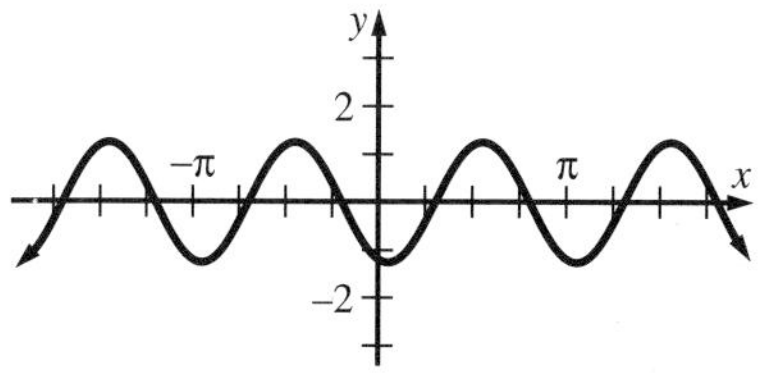

49, 51. Answers may vary. Examples are given.

49. $y = 5 \sin \frac{4}{3}\left(x - \frac{\pi}{2}\right)$

51. $y = 0.75 \sin \frac{1}{3}(x - \pi)$ **53.** 0

55. –1 **57.** $-\frac{\sqrt{3}}{3} \approx -0.5774$ **59.** 0

61.

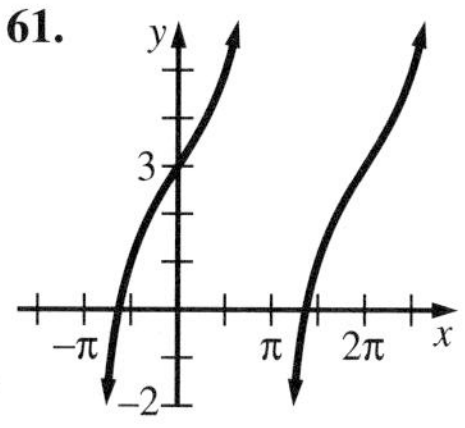

TOOLBOX

Pages 773–774 Numbers and Number Lines

1. *C* **2.** *F* **3.** *B* **4.** *G* **5.** 0 **6.** –2 **7.** $2\frac{1}{2}$ **8.** $\frac{1}{2}$

9. a. *E, F, G, H* **b.** *A, B, C* **c.** *D, F* **d.** *A, B, C, D, E, F, G, H*

10.

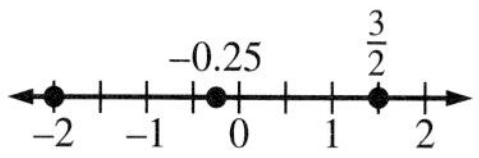

$-2, -0.25, \frac{3}{2}$

11.

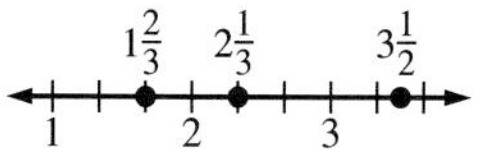

$1\frac{2}{3}, 2\frac{1}{3}, 3\frac{1}{2}$

12.

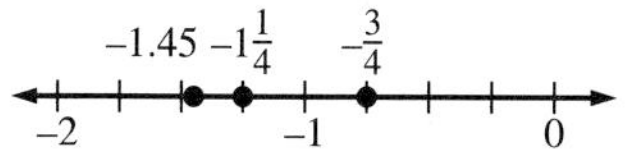

$-1.45, -1\frac{1}{4}, -\frac{3}{4}$

Pages 774–775 Properties of Real Numbers

1. a. absolute value property **b.** distributive property **2. a.** associative property **b.** commutative property **c.** associative property **3. a.** commutative property **b.** associative property **c.** distributive property **4–13.** Examples given may vary. **4.** True; $(2 + 3)(2 - 3) = 5(-1) = -5$; $2^2 - 3^2 = 4 - 9 = -5$. **5.** False; $1^2 + 2^2 = 5$, but $(1 + 2)^2 = 9$. **6.** True; $3 - 5 = -2$; $3 + (-5) = -2$. **7.** True; $4(1 - 3) = 4(-2) = -8$ and $4(1) - 4(3) = 4 - 12 = -8$. **8.** False; $1 \cdot 13 = 13$. **9.** True; $12 \div 3 = 4$ and $12 \cdot \frac{1}{3} = 4$. **10.** False; $\frac{1}{2} + \frac{3}{4} = \frac{5}{4}$ and $\frac{1+3}{2+4} = \frac{2}{3}$. **11.** True; $\frac{1}{2} \cdot \frac{3}{4} = \frac{3}{8}$ and $\frac{1 \cdot 3}{2 \cdot 4} = \frac{3}{8}$. **12.** False; $|3| - |7| = 3 - 7 = -4$ and $|3 - 7| = |-4| = 4$. **13.** False; $0 + (-27) = -27$.

Page 776 Percent Change

1. 3.6% increase **2.** 20.9% decrease **3.** 101.4% increase

Page 777 Exponents and Powers

1. 64 **2.** –32 **3.** 1 **4.** $\frac{1}{81}$ **5.** 243 **6.** 625 **7.** 1 **8.** $\frac{1}{5}$ **9.** a^{10} **10.** b^5 **11.** $\frac{1}{c^5}$ **12.** $\frac{1}{d^3}$ **13.** a^6 **14.** $\frac{1}{b^6}$ **15.** 1 **16.** *d*

Page 778 Square Roots

1. ±7 **2.** none **3.** $\pm\frac{1}{8}$ **4.** ±0.4 **5.** 20 **6.** –1.3 **7.** not possible **8.** $\frac{10}{11}$ **9.** $5\sqrt{5}$ **10.** $-3\sqrt{2}$ **11.** $7\sqrt{3}$ **12.** $2\sqrt{15}$ **13.** $12\sqrt{6}$ **14.** 2 **15.** $\frac{\sqrt{3}}{2}$ **16.** $\frac{2}{3}$ **17.** No; the statement is true only if $a \geq 0$. If $a < 0$, $\sqrt{a^2} = -a$.

Page 779 Order of Operations

1. –1 **2.** 11 **3.** 52 **4.** 94 **5.** 144 **6.** 6 **7.** –5 **8.** –13 **9.** –4.5 **10.** –9.5 **11.** –7.75 **12.** 8

Page 780 Rational Numbers and Irrational Numbers

1. the rational numbers, the real numbers **2.** the whole numbers, the integers, the rational numbers, the real numbers **3.** the irrational numbers, the real numbers **4.** the whole numbers, the integers, the rational numbers, the real numbers **5.** the irrational numbers, the real numbers **6.** the rational numbers, the real numbers **7.** the rational numbers, the real numbers **8.** the irrational numbers, the real numbers **9, 10.** Examples may vary. **9.** any negative integer **10.** π **11.** 0 **12.** not possible

Page 781 Evaluating Variable Expressions

1. –2 **2.** –6 **3.** 14 **4.** –16 **5.** 24 **6.** 76 **7.** 2 **8.** 9 **9.** 3 **10.** 55 **11.** –45 **12.** 5 **13.** 27 **14.** 35 **15.** –2.5 **16.** –7

Page 781 Simplifying Variable Expressions

1. $6t - 8$ **2.** $-5k - 1$ **3.** $-a + 3b - 2$ **4.** $-x^2$ **5.** $-12 - 3j$ **6.** $-4 + 4a$ **7.** $x + 4y$ **8.** $6a + 2ab - 3b$ **9.** $m^2 - m + 5$ **10.** $a - b + 7$ **11.** $a - 9b - 7$ **12.** $a - b - 7$ **13.** $7 - 11x$ **14.** –16 **15.** $4y^2 - 3y - 4$ **16.** $-3d^2 + 12d + 5$ **17.** $-n^3 + n^2 + 6n + 14$ **18.** $2c^2 + 15cd + 13d^2$

Page 782 Multiplying Variable Expressions

1. $2b^2 - 7b$ **2.** $2m^2 + 8m$ **3.** $-5h + h^2$ **4.** $r^2 + r - 2$ **5.** $-p^2 + 3p + 28$ **6.** $s^2 - 16$ **7.** $2c^2 + 9c + 9$ **8.** $5z^2 - 11z + 2$ **9.** $32 - 4w - 3w^2$ **10.** $6j^2 - 11j + 3$ **11.** $9 + 23y - 12y^2$ **12.** $10k^2 + 29k + 10$ **13.** $a^2 - 10a + 25$ **14.** $m^2 + 6m + 9$ **15.** $4s^2 + 28s + 49$ **16.** $25j^2 - 10j + 1$

Page 783 Factoring Variable Expressions

1. $2(t + 4)$ **2.** $3(-2m + 3)$ or $-3(2m - 3)$ **3.** $r(r - 5)$ **4.** $v(2v + 7)$ **5.** $6(3j + k)$ **6.** $-3(3d + 5f)$ **7.** $g(-4 + 15g)$ or $-g(4 - 15g)$ **8.** $b(c + d)$ **9.** $7(3a^2 + 7)$ **10.** $5c(1 + 5c)$ **11.** $t(-2p + 3s)$ or $-t(2p - 3s)$ **12.** $10(2x^2 - 5y^2)$

Page 784 Adding and Subtracting Rational Expressions

1. $\frac{6}{t}$ **2.** $-\frac{4}{t}$ **3.** $\frac{2n+3}{n+1}$ **4.** $\frac{2n-3}{n+1}$ **5.** 3 **6.** $\frac{1}{k}$ **7.** $\frac{13}{2x}$ **8.** $\frac{7}{2x}$ **9.** $\frac{2x+2}{3(2x-1)}$ **10.** $\frac{4+30z}{5z}$ **11.** $\frac{r+1}{r+2}$ **12.** $\frac{4x-1}{2x-1}$ **13.** $\frac{2c+4}{c+4}$ **14.** $-\frac{4}{c+4}$ **15.** $\frac{2y+1}{y(y+1)}$ **16.** $\frac{m-6}{(m+2)(m-2)}$

Pages 785 Solving Linear Equations

1. $\frac{3}{4}$ **2.** –2 **3.** $\frac{3}{4}$ **4.** 7.4 **5.** –3 **6.** 2 **7.** 16 **8.** 16 **9.** 20 **10.** 2 **11.** –4 **12.** $-\frac{1}{5}$ **13.** $2\frac{1}{2}$ **14.** 5 **15.** $-\frac{4}{5}$ **16.** –2

Pages 785–786 Ratios and Proportions

1. $1\frac{2}{3}$ **2.** $-31\frac{1}{2}$ **3.** $1\frac{23}{27}$ **4.** 9 **5.** $1\frac{3}{16}$ **6.** $-\frac{1}{9}$ **7.** $-3\frac{13}{25}$ **8.** $\frac{9}{10}$ **9.** $\frac{6}{11}$ mi $\approx$ 0.55 mi **10.** \$6.62 **11.** 4.9 in.

Page 786 Identities and False Statements

1. all real numbers **2.** no solution **3.** 6 **4.** $2\frac{1}{2}$ **5.** no solution **6.** all real numbers **7.** all real numbers **8.** no solution **9.** $-3\frac{3}{4}$ **10.** $1\frac{1}{2}$ **11.** −8 **12.** no solution **13.** 0 **14.** all real numbers **15.** all real numbers

Page 787 Solving Linear Inequalities

1. $x < 9$ **2.** $y \geq -3$ **3.** $k > 2$ **4.** $z \leq -3$ **5.** $p < 5$ **6.** $r \leq -2$ **7.** $c > 12$ **8.** $x \geq -2$ **9.** $d < 2.5$ **10.** $h \geq 2$ **11.** $t > 1\frac{1}{2}$ **12.** $s \geq 0$ **13.** $g < 2$ **14.** $x \geq -6$ **15.** $w > 3$ **16.** $m \leq -1$

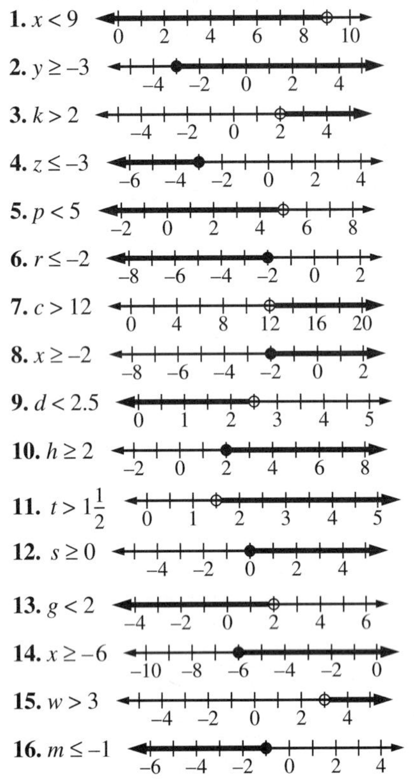

Page 788 Rewriting Equations and Formulas

1. $y = 5x - 15$ **2.** $b = \frac{2A}{h}$ **3.** $r = \sqrt{\frac{A}{\pi}}$ **4.** $b_1 = \frac{2A}{h} - b_2$ or $b_1 = \frac{2A - b_2h}{h}$ **5.** $y = \frac{3}{2}x - 12$ **6.** $l = \frac{P - 2w}{2}$ or $l = \frac{P}{2} - w$ **7.** $y = \frac{c - ax}{b}$ **8.** $n = \frac{360}{180 - x}$

Page 789 Experimental Probability

1. 0.75 **2.** 0.333 (to three decimal places) **3.** 0.12 **4.** 0.53 **5.** 1 **6.** 0.875 **7.** 0.12 **8. a.** 0.012 **b.** 0.988 **9. a.** 0.528 **b.** No; even if those who now have no opinion decide to support the construction, only 59.2% of the voters would vote in favor of construction.

Page 790 Significant Digits

1. 5.3 **2.** 2.2 **3.** 4.22 **4.** 3.3 **5.** 32 **6.** 29 **7.** 690 **8.** 1.9 **9.** 2.4 **10.** 2.01 **11.** 0.002 **12.** 4.98

Page 791 Mean, Median, and Mode

1. a. 7.4; 6; no mode **b.** 9; 9; no mode **c.** 8.2; 7; 7 **2.** 8.2; 8; 7; Answers may vary. An example is given. Only the Western Division values have a mode; it is the same as that for the entire conference. The mean for the conference data is higher than that for the Eastern Division, lower than that for the Central Division, and the same as that for the Western Division. The median for the conference data is higher than those for the Eastern and Western Divisions, and lower than that for the Central Division. Overall, the values for the conference data are most similar to those for the Western Division. **3.** 43.73; 44.3; 44.4, 46.4

Page 793 Data Displays

1. \$5 bill; \$20 bill **2.** 6 years **3.** \$50 bills and \$100 bills; Answers may vary. An example is given. I think the bills are probably handled much less and so wear out less quickly than more commonly used bills.

4.

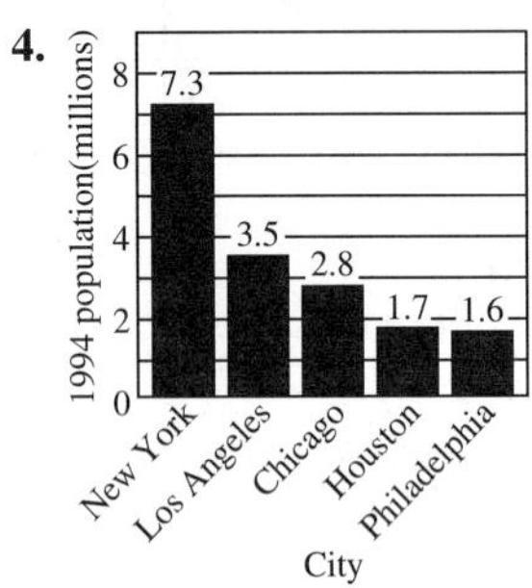

5.

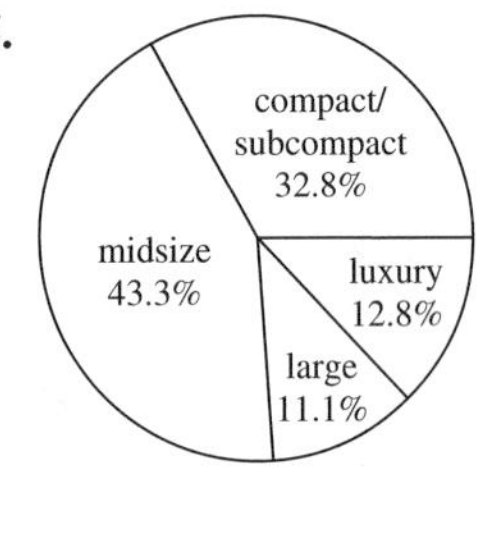

6.

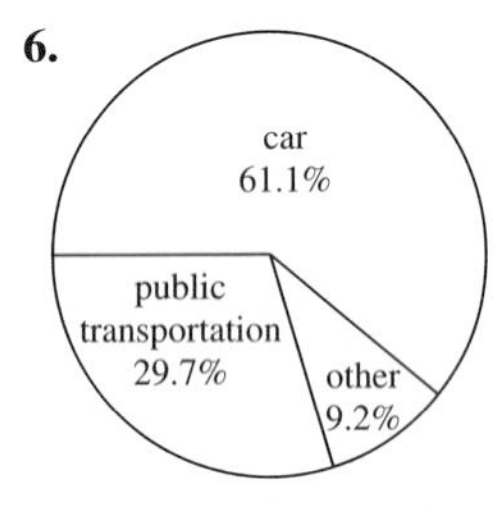

7. a.

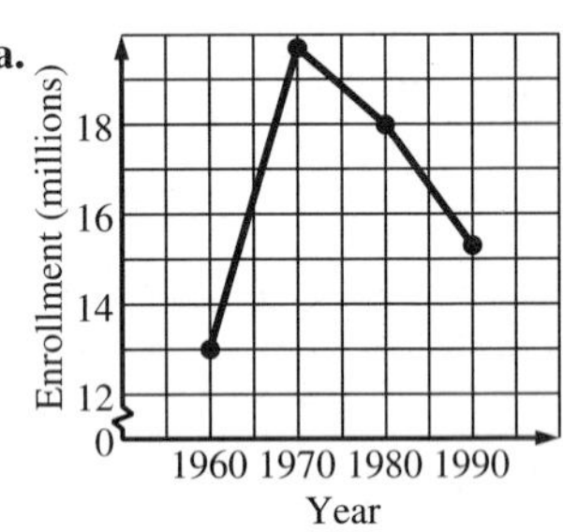

b. Answers may vary. An example is given. It appears that enrollment peaked in 1970 and has been declining since.

Page 794 Dimensional Analysis

1. $22\frac{2}{9}$ m/s **2.** 3875 pesos **3.** about 11,975 ft/s **4.** about \$0.05 per mile

Page 794 Graphing Points on a Coordinate Plane

1–8.

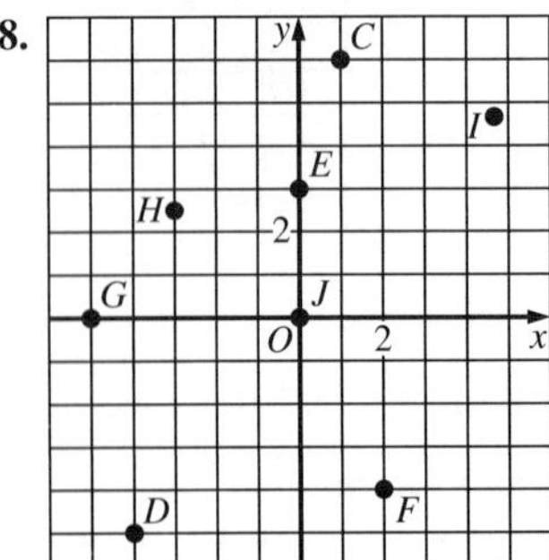

1. I **2.** III **3.** none **4.** IV **5.** none **6.** II **7.** I **8.** none

Page 795 **Graphing an Equation**

1. 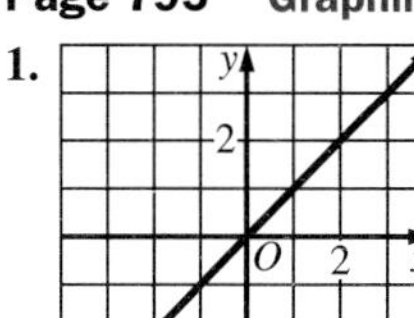

2. 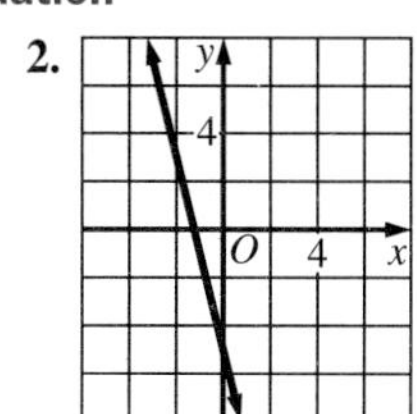

3. 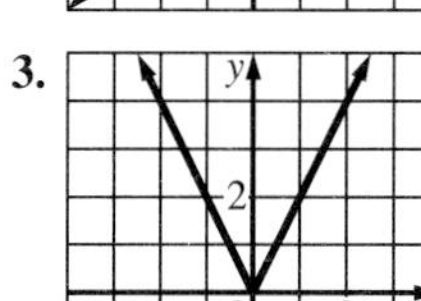

4. 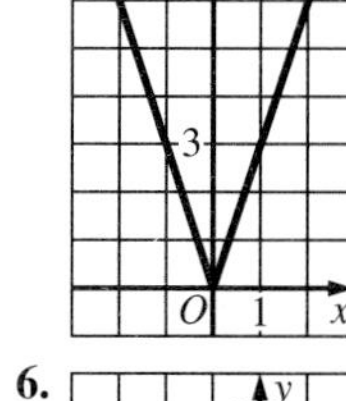

5.

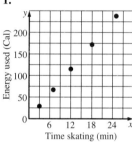

6. 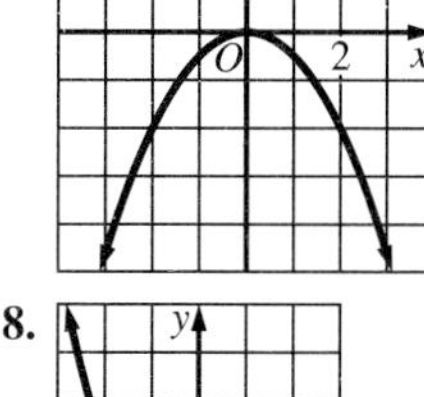

7.

8.

Page 796 **Making a Scatter Plot**

1. 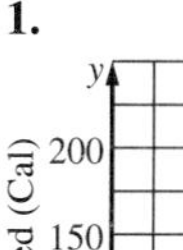

2. 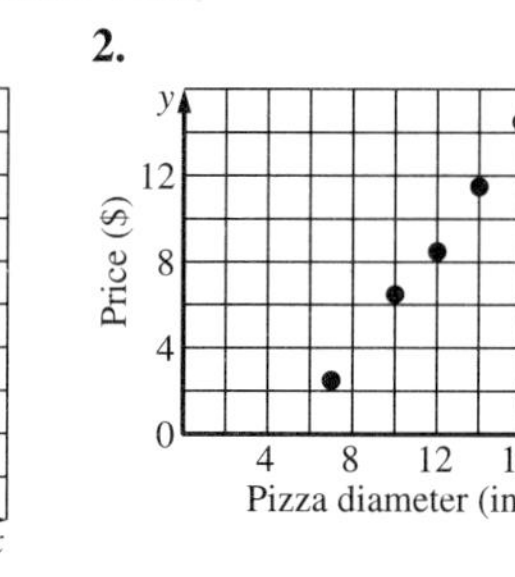

Page 797 **Translating a Figure**

1. 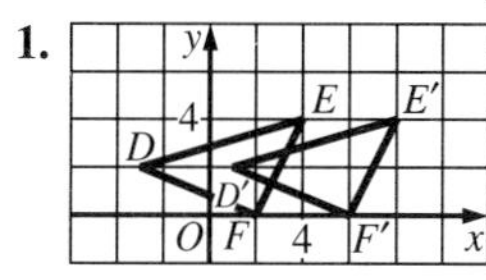

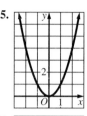

$P(x, y) \rightarrow P'(x + 4, y)$

2. 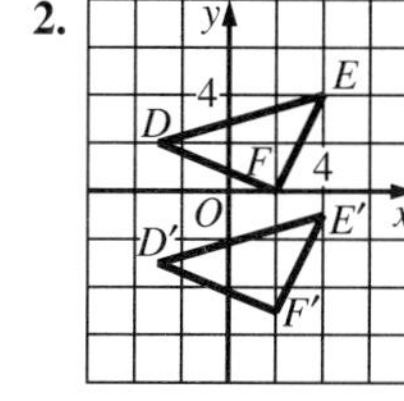

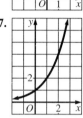

$P(x, y) \rightarrow P'(x, y - 5)$

3. 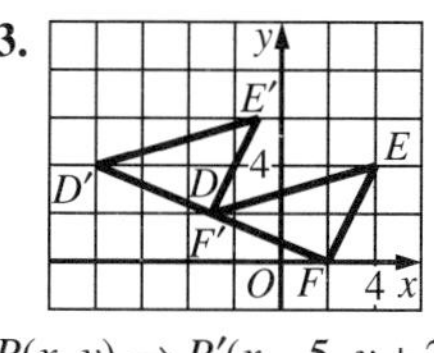

$P(x, y) \rightarrow P'(x - 5, y + 2)$

4. 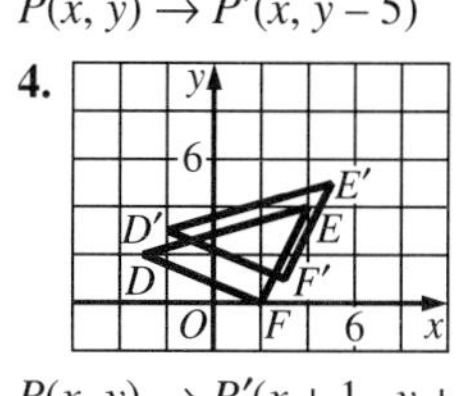

$P(x, y) \rightarrow P'(x + 1, y + 1)$

5. 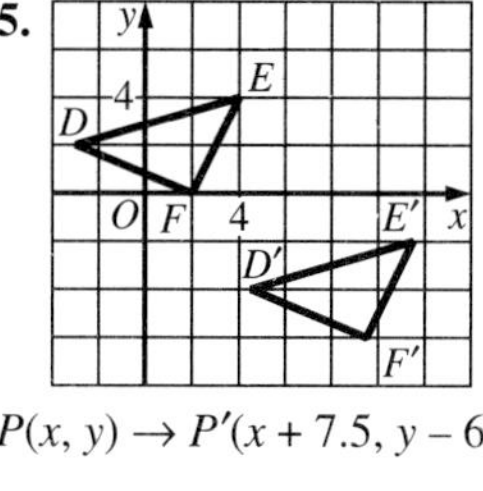

$P(x, y) \rightarrow P'(x + 7.5, y - 6)$

6. 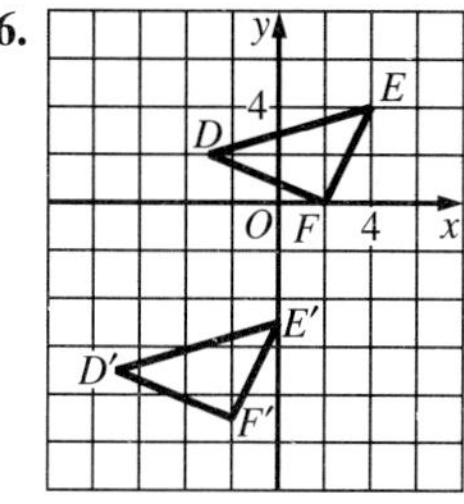

$P(x, y) \rightarrow P'(x - 4, y - 9)$

Page 798 **Reflecting a Figure**

1. 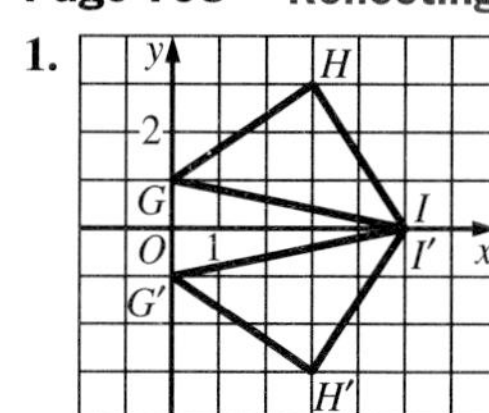

2. 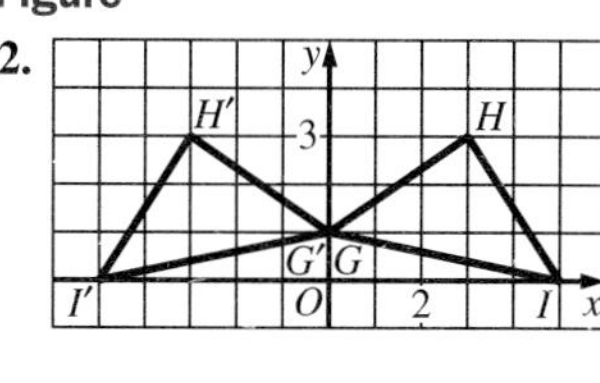

3. 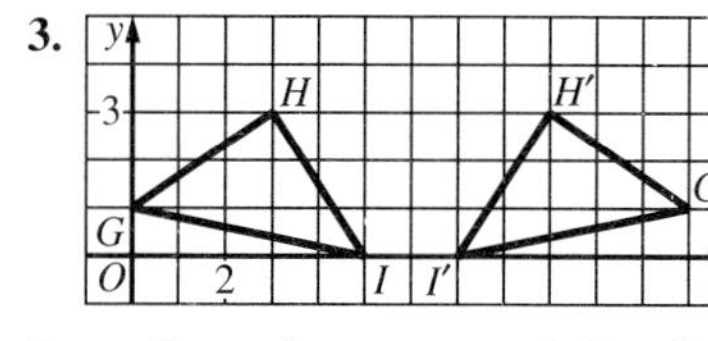

4. 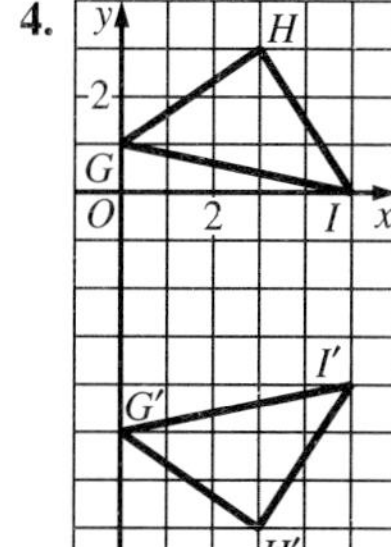

5. two lines of symmetry **6.** four lines of symmetry **7.** no lines of symmetry **8.** one line of symmetry **9.** one line of symmetry **10.** infinitely many lines of symmetry

Page 798 **Dilating a Figure**

1. 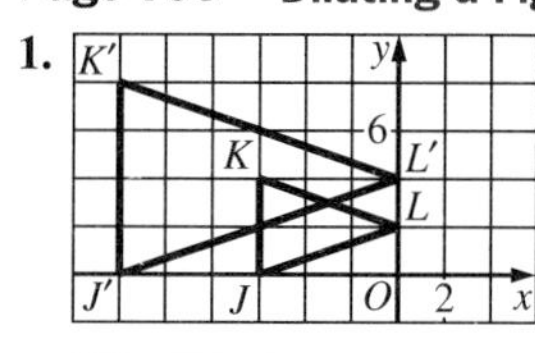

2. 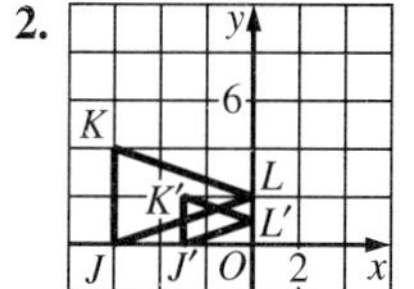

3. 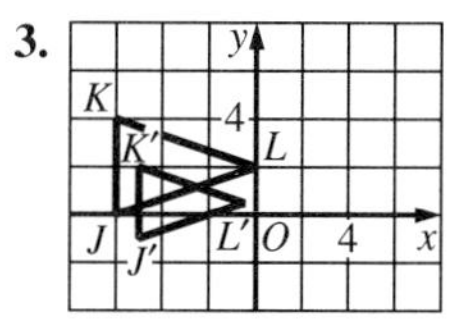

4. 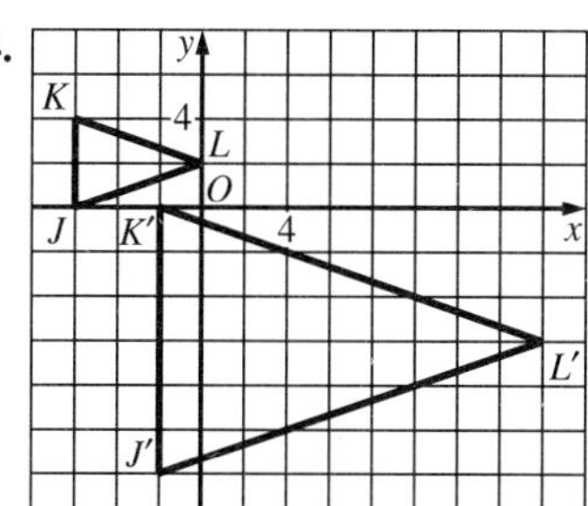

Page 800 **Logical Statements**

1. a. a figure is a rectangle; it is a quadrilateral; True. **b.** If a figure is a quadrilateral, then it is a rectangle. False; a non-square rhombus is a quadrilateral that is not a rectangle. **c.** A figure is a rectangle if and only if it is a quadrilateral. False; one of the conditionals is false. **2. a.** $a = 0$ or $b = 0$; $ab = 0$; True. **b.** If $ab = 0$, then $a = 0$ or $b = 0$. True; if the product of two numbers is zero, at least one of the numbers is 0. **c.** $ab = 0$ if and only if $a = 0$ or $b = 0$. True; both of the conditionals are true.

3. a. a triangle is equilateral; it is isosceles; True. **b.** If a triangle is isosceles, then it is equilateral. False; for example, an isosceles right triangle is not equilateral. **c.** A triangle is equilateral if and only if it is isosceles. False; one of the conditionals is false. **4. a.** $a > b$; $a + c > b + c$; True. **b.** If $a + c < b + c$, then $a < b$; True. **c.** $a + c < b + c$ if and only if $a < b$. True; both of the conditionals are true. **5. a.** Two lines intersect; they are not parallel. True. **b.** If two lines are not parallel, then they intersect. False; Parallel lines are lines that lie in the same plane and do not intersect. **c.** Two lines intersect if and only if they are not parallel. False; One of the conditionals is false. **6. a.** $x^2 = 9$; $x = 3$; False. **b.** If $x = 3$, then $x^2 = 9$. True; $3^2 = 9$. **c.** $x^2 = 9$ if and only if $x = 3$. False; one of the conditionals is false.

Page 800 Angle Relationships

1. 80; 30 **2.** 25; 65 **3.** 50; 60 **4.** never true **5.** always true **6.** sometimes true

Page 801 Triangle Relationships

1. 20 **2.** $6\sqrt{2}$ **3.** 12 **4.** 90

Page 802 Polygons

1. sometimes; A hexagon may or may not be equiangular and equilateral. **2.** never; A quadrilateral has four sides. **3.** sometimes; A right triangle with two 45° angles is an isosceles triangle. **4.** never; If a triangle had three obtuse angles, the sum of the measures of the angles would be greater than 180°. **5.** always; Every square is both equiangular and equilateral. **6.** always; Every rectangle is a quadrilateral with four 90° angles.

Page 802 Perimeter, Circumference, Area, and Volume

1. 22 cm; 24 cm^2 **2.** $33 + 3\sqrt{61}$ ft ≈ 56.43 ft; 135 ft^2 **3.** 25 m; 31.5 m^2 **4.** 18.85; 28.27 **5.** 37.70; 113.10 **6.** 14.05; 15.71 **7.** 3518.58 cm^2; 10,053.10 cm^3 **8.** 112 in.^{2}; 80 in.^{3} **9.** 31,415.93 ft^2; 523,598.78 ft^3 **10.** 211.18 m^2; 155.88 m^3

TECHNOLOGY HANDBOOK

Pages 804–815 Calculator Practice

1. –12 **2.** 3.5 **3.** 1.4 **4.** 720 **5.** 32,768 **6.** 10 **7.** 3.732 **8.** 5.92

9.

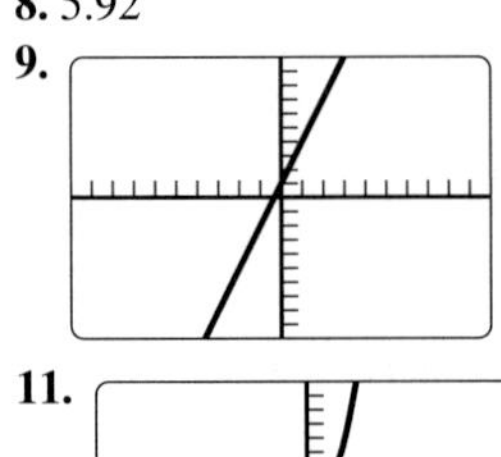

10. **11.** **12.**

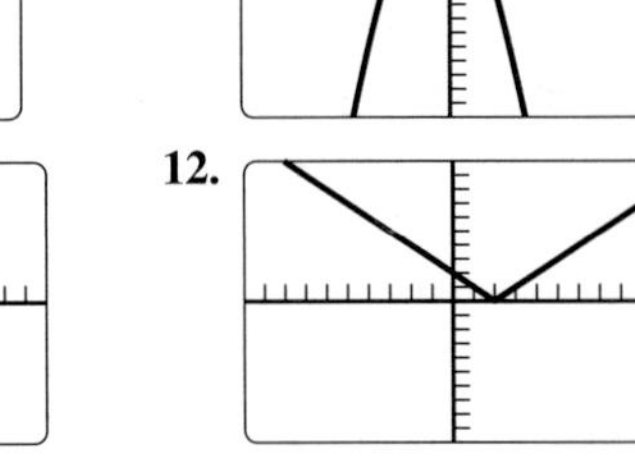

13. To make the lines $y = x - 1$ and $y = -x + 7$ look perpendicular, use a square viewing window like the one shown below. Zoom windows may vary.

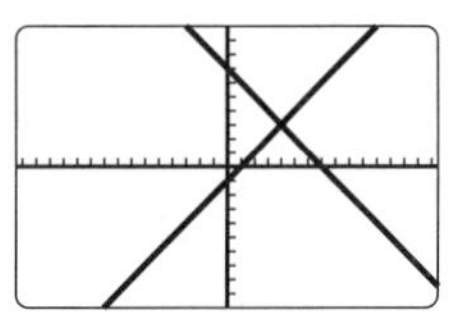

14. 3.66 **15. a.** 3.81 **b.** 2.39, 1, –0.813, –2.58 **c.** local minimums: –5.40, –9.14; local maximum: 5.04 **d.** 2.68, –2.80 **16. a.** 2.26 **b.** 3.37 **17.** Enter the equation Y_1={1,–1,0.5,2}X.

18. a.

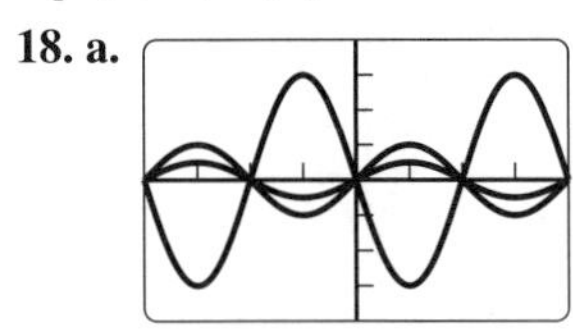

b.

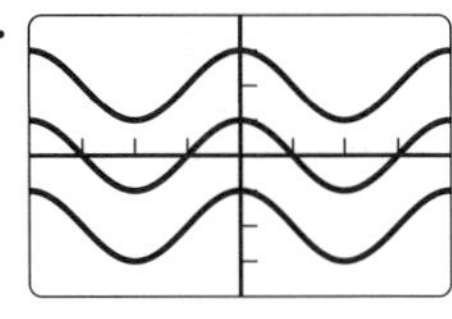

19. a.

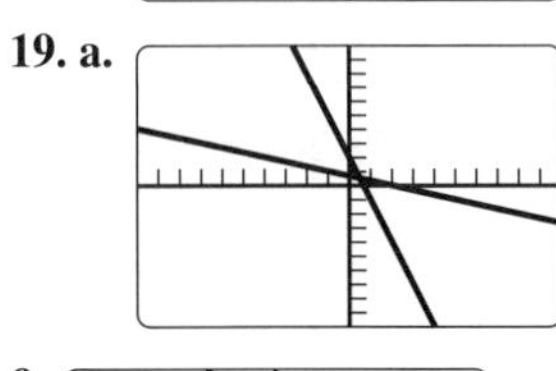

b.

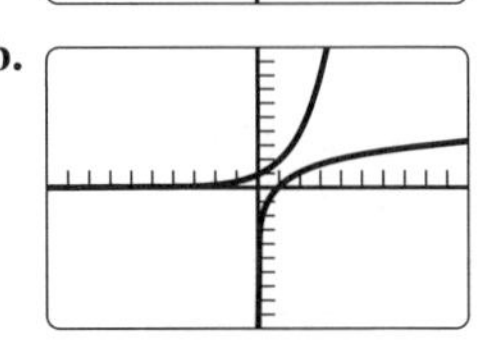

c.

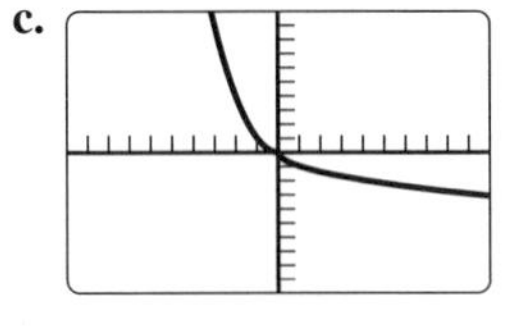

20. 750 s, or 12.5 min

21. a.

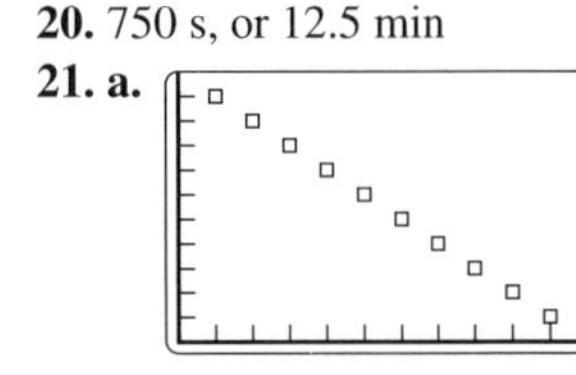

b.

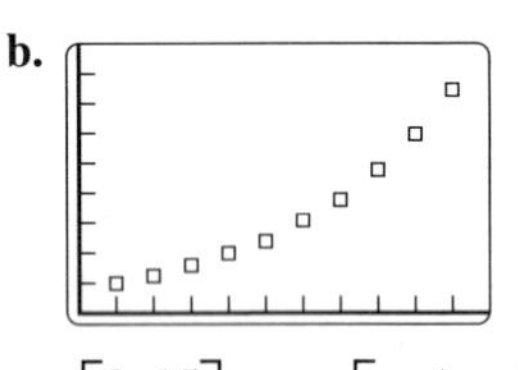

c.

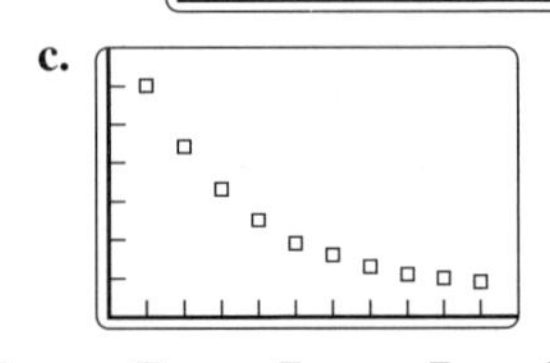

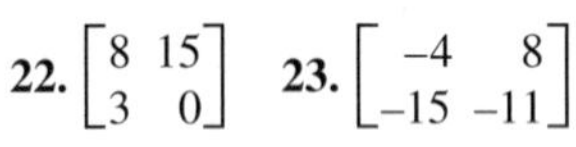

22. $\begin{bmatrix} 8 & 15 \\ 3 & 0 \end{bmatrix}$ **23.** $\begin{bmatrix} -4 & 8 \\ -15 & -11 \end{bmatrix}$

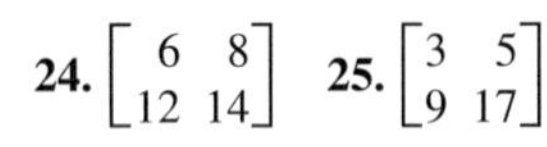

24. $\begin{bmatrix} 6 & 8 \\ 12 & 14 \end{bmatrix}$ **25.** $\begin{bmatrix} 3 & 5 \\ 9 & 17 \end{bmatrix}$

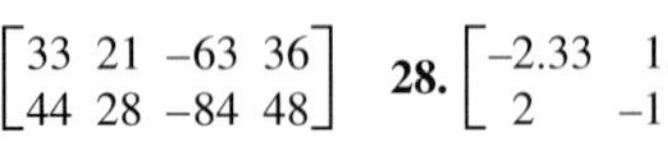

26. undefined **27.** $\begin{bmatrix} 33 & 21 & -63 & 36 \\ 44 & 28 & -84 & 48 \end{bmatrix}$ **28.** $\begin{bmatrix} -2.33 & 1.33 \\ 2 & -1 \end{bmatrix}$

29. undefined

30. a.

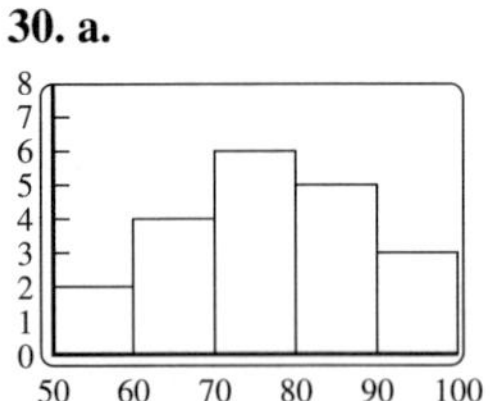

31. a. $y = -1.06x + 255$

b.

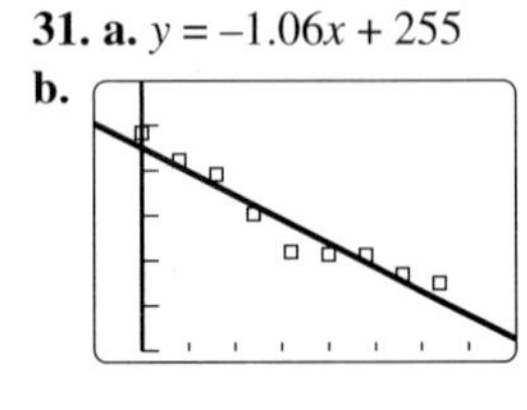

b. mean: 76.2; standard deviation: 11.8

32. Answers will depend on the type of calculator used. On the TI-82, enter the expression "int(16*rand)+10."

33. a. 3, –6 **b.** 0.4 **c.** $0.25 \pm 1.41i$ **34.** The specific commands in the program will depend on the calculator used. The following program is for the TI-82:

```
:Input "ENTER A:",A:Input "ENTER B:",B
:Disp "SOLUTION:",–B/A
```

Page 816 Spreadsheet Practice

35. a. $977.34 **b.** 6 years

Additional Answers

CHAPTER 1

Page 11 Think and Communicate

5. about 117 billion pounds **6.** the exponential function; Predictions based on the exponential function will be much larger than predictions based on the linear function. **7.** Answers may vary. An example is given. I think the linear function predicts the growth better, because I think limitations on recycling may slow the growth of plastics production.

Page 12 Checking Key Concepts

1. number of workers; years since 1987 **2.** $y = 7147 + 5440x$
3. $y = 7147(1.374)^x$ **4.** Answers may vary. A low estimate, based on the linear model, is about 77,867 workers. A high estimate, based on the exponential model, is about 444,581 workers. A mid-range estimate, the average of the low and high estimates, is about 261,224 workers.

Page 15 Exercises and Applications

17. Answers may vary due to rounding. Examples are given.
a. about \$22.2 billion **b.** about \$22.4 billion **18.** –1 **19.** –1 **20.** 27
21. 13.5; 13; 13

Page 15 Assess Your Progress

1. To model a situation using constant growth, you add (or subtract) the constant change for each period. To model a situation using proportional growth, you multiply by the growth factor. **2.** Answers may vary due to rounding. Examples are given. In parts (a) and (b), let x = the number of decades since 1951 and y = the cost in dollars of a movie ticket. **a.** $y = 0.50 + 1.1x$
b. $y = 0.50(1.78)^x$ **c.**

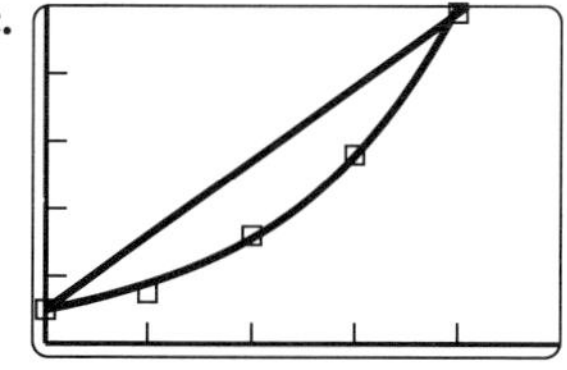

; The exponential model fits the data almost exactly.

d. about \$6 (linear model); about \$8.93 (exponential model), which would probably be rounded to \$9.00 **3.** Answers may vary. Examples are given. I use a table if I want to be able to make a growth model. I use an equation if I can determine one and also when I need to find specific values that I do not have. I use a graph when I want to visualize how the data are related. I prefer different methods depending on the situation.

Page 20 Exercises and Applications

18. a. $\begin{bmatrix} 0 & 4 & 12 & 8 \\ 0 & 10 & 10 & 0 \end{bmatrix}$

b.

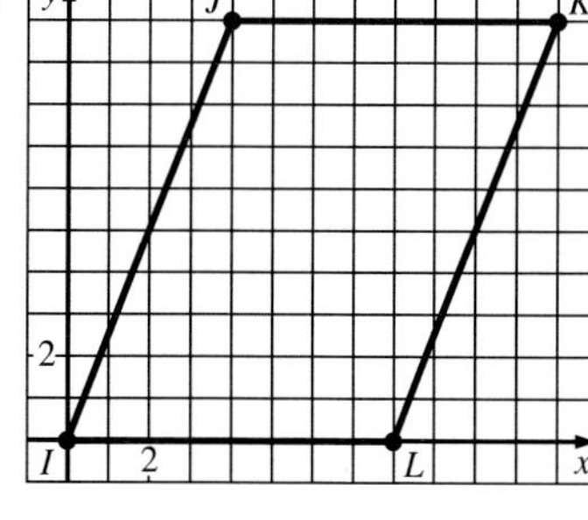

IJKL is *ABCD* enlarged by a factor of 2.

21. position moved 1 unit in the negative x-direction and 3 units in the positive y-direction

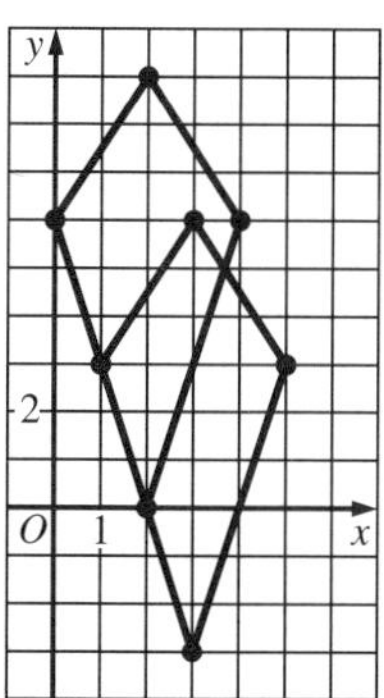

22. a. Every element is 0. **b.** They are the same matrix. **23.** The element in row m, column n of $A + B$ is $a_{m,n} + b_{m,n}$. The element in row m, column n of $B + A$ is $b_{m,n} + a_{m,n}$, which is equal to $a_{m,n} + b_{m,n}$.

$$A + B = \begin{bmatrix} 3+(-9) & 1+10 & -2+(-3) \\ -1+0 & 5+6 & 6+1 \\ 4+14 & 13+7 & 0+(-8) \end{bmatrix} = \begin{bmatrix} -6 & 11 & -5 \\ -1 & 11 & 7 \\ 18 & 20 & -8 \end{bmatrix};$$

$$B + A = \begin{bmatrix} -9+3 & 10+1 & -3+(-2) \\ 0+(-1) & 6+5 & 1+6 \\ 14+4 & 7+13 & -8+0 \end{bmatrix} = \begin{bmatrix} -6 & 11 & -5 \\ -1 & 11 & 7 \\ 18 & 20 & -8 \end{bmatrix}$$

24. The element in row m, column n of $A + (B + C)$ is $a_{m,n} + (b_{m,n} + c_{m,n})$. The element in row m, column n of $(A + B) + C$ is $(a_{m,n} + b_{m,n}) + c_{m,n}$, which is equal to $a_{m,n} + (b_{m,n} + c_{m,n})$.

$$B + C = \begin{bmatrix} -9+7 & 10+(-2) & -3+0 \\ 0+15 & 6+1 & 1+19 \\ 14+(-4) & 7+12 & -8+2 \end{bmatrix} = \begin{bmatrix} -2 & 8 & -3 \\ 15 & 7 & 20 \\ 10 & 19 & -6 \end{bmatrix};$$

$A + B$ was found in Ex. 23.

$$A + (B + C) = \begin{bmatrix} 3 & 1 & -2 \\ -1 & 5 & 6 \\ 4 & 13 & 0 \end{bmatrix} + \begin{bmatrix} -2 & 8 & -3 \\ 15 & 7 & 20 \\ 10 & 19 & -6 \end{bmatrix} = \begin{bmatrix} 1 & 9 & -5 \\ 14 & 12 & 26 \\ 14 & 32 & -6 \end{bmatrix};$$

$$(A + B) + C = \begin{bmatrix} -6 & 11 & -5 \\ -1 & 11 & 7 \\ 18 & 20 & -8 \end{bmatrix} + \begin{bmatrix} 7 & -2 & 0 \\ 15 & 1 & 19 \\ -4 & 12 & 2 \end{bmatrix} = \begin{bmatrix} 1 & 9 & -5 \\ 14 & 12 & 26 \\ 14 & 32 & -6 \end{bmatrix}$$

Page 25 Exercises and Applications

8. undefined **9.** $\begin{bmatrix} -24 \\ -10 \end{bmatrix}$ **10.** undefined **11.** $\begin{bmatrix} -4 & -7 & -4 \\ 6 & 30 & -7 \end{bmatrix}$

12. $\begin{bmatrix} -9 & -20 & -8 \\ 2 & -3 & 18 \end{bmatrix}$ **13.** undefined

Page 26 Exercises and Applications

14. a. $\begin{matrix} \text{Class} \\ \text{Plain} \\ \text{Sorority} \\ \text{2-color sorority} \end{matrix} \overset{\$}{\begin{bmatrix} 25 \\ 20 \\ 30 \\ 25 \end{bmatrix}}$ **b.** $\begin{matrix} 1992 \\ 1993 \\ 1994 \end{matrix} \overset{\$}{\begin{bmatrix} 3{,}000 \\ 9{,}200 \\ 16{,}775 \end{bmatrix}}$; total sales for the three years

15. $S = \begin{matrix} \text{Day 1} \\ \text{Day 2} \end{matrix} \overset{\begin{matrix} \text{Class} & \text{Plain} & \text{2-color sorority} & \text{Sorority} & \text{Earrings} & \text{Shirts} \end{matrix}}{\begin{bmatrix} 40 & 45 & 12 & 28 & 25 & 65 \\ 32 & 52 & 24 & 16 & 18 & 45 \end{bmatrix}}$; $T = \begin{matrix} \text{Class} \\ \text{Plain} \\ \text{2-color sorority} \\ \text{Sorority} \\ \text{Earrings} \\ \text{Shirts} \end{matrix} \overset{\$}{\begin{bmatrix} 25 \\ 20 \\ 25 \\ 30 \\ 15 \\ 12 \end{bmatrix}}$;

S has dimensions 2×6 and T has dimensions 6×1.

b. $\begin{matrix} & \$ \\ \text{Day 1} \\ \text{Day 2} \end{matrix}\begin{bmatrix} 4195 \\ 3730 \end{bmatrix}$; total sales for the two days

Page 33 Exercises and Applications

11. Answers may vary. Examples are given. **a.** To simulate the number of people arriving, generate random numbers. Let random numbers with first digit 0 or 1 represent 0 people, with first digit 2–5 represent 1 person, with first digit 6–8 represent 2 people, and with first digit 9 represent 3 people. To simulate the arrival of the bus, roll a die. Let rolling a 1 represent the arrival of a bus.

b. Sample results are given.

Min of simulation	1	2	3	4	5	6	7	8	9	10
Number of people	1	2	1	2	1	1	1	1	1	1
Bus arrives	0	1	0	0	0	0	0	0	1	0
Min of simulation	11	12	13	14	15	16	17	18	19	20
Number of people	0	0	2	2	2	0	1	0	0	1
Bus arrives	0	1	0	0	0	0	0	1	0	1
Min of simulation	21	22	23	24	25	26	27	28	29	30
Number of people	2	2	3	0	0	0	2	1	1	2
Bus arrives	0	0	0	0	0	0	0	0	0	0
Min of simulation	31	32	33	34	35	36	37	38	39	40
Number of people	2	1	2	2	1	2	1	1	0	1
Bus arrives	0	0	0	0	0	0	0	0	0	0
Min of simulation	41	42	43	44	45	46	47	48	49	50
Number of people	1	2	2	0	3	1	1	1	1	0
Bus arrives	0	0	1	0	1	0	0	1	0	1
Min of simulation	51	52	53	54	55	56	57	58	59	60
Number of people	2	2	3	2	0	1	0	1	3	2
Bus arrives	0	0	0	0	0	0	0	0	1	1

c. 74 people; 11 buses; about 7 people **d.** for the preceding sample, about 7 min; 23 min; 1 min

Page 35 Assess Your Progress

9. $\begin{bmatrix} 1 & 2 \\ -8 & 5 \\ 19 & -9 \end{bmatrix}$ **10.** undefined

11. a. Simulation methods may vary. An example is given. For each employee, use int(6*rand)+1 to generate random numbers. Let 1 represent the employee's leaving the company. Generate random numbers until you get a 1. The number of tries is the number of years the employee stays at the company. Sample trial results are given.

Employee	1	2	3	4	5	6	7	8	9	10
No. of years employee stays	6	12	4	2	11	5	3	5	12	5
Employee	11	12	13	14	15	16	17	18	19	20
No. of years employee stays	7	6	14	5	2	1	3	6	11	5

b. for the preceding sample, about 6 years; 9 employees; 5 employees

CHAPTER 2

Page 46 Think and Communicate

1. They are equal. If $\frac{y_1}{x_1}$ and $\frac{y_2}{x_2}$ are data pairs from a direct variation with constant of variation a, then $\frac{y_1}{x_1} = a$ and $\frac{y_2}{x_2} = a$. **2. a.** Moira used two data pairs from the direct variation and set them equal. **b.** The constant of variation is not exactly equal to the ratio Moira used, 0.8. **3.** The person at the center of rotation has speed 0 ft/s. The skater has no forward motion and is simply spinning in place.

Page 55 Exercises and Applications

12. Let x = the number hours and y = the height of the candle in inches; $y = -1.5x + 8$.

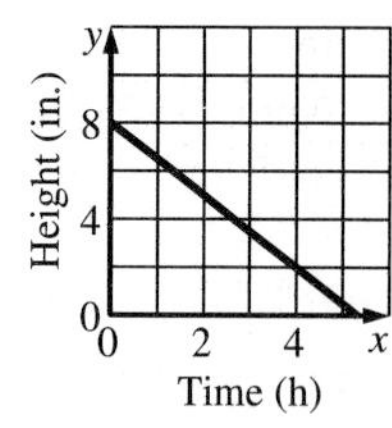

13. Let x = the number of miles driven and y = the number of gallons of gas in the tank; $y = -\frac{1}{15}x + 16$.

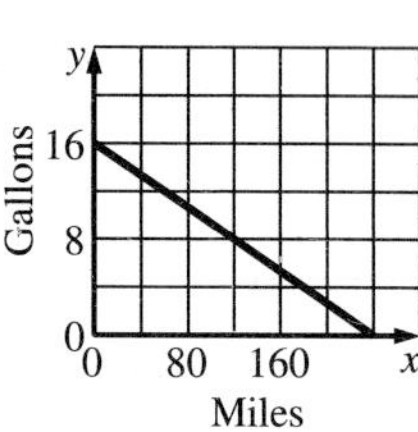

14. Answers may vary. An example is given. At the Pizza Pan, a small plain pizza costs $4.25. Additional toppings cost $.75 each.

15.

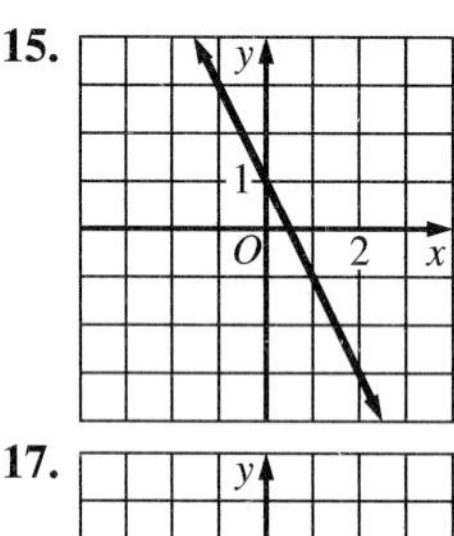

16.

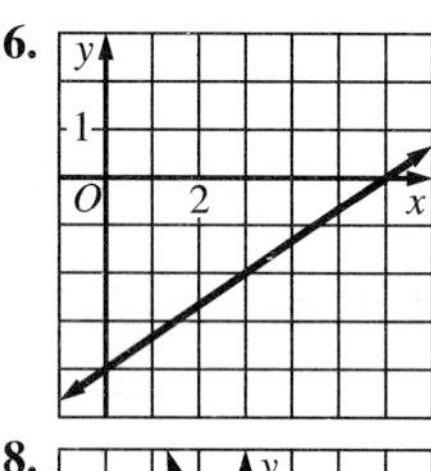

17.

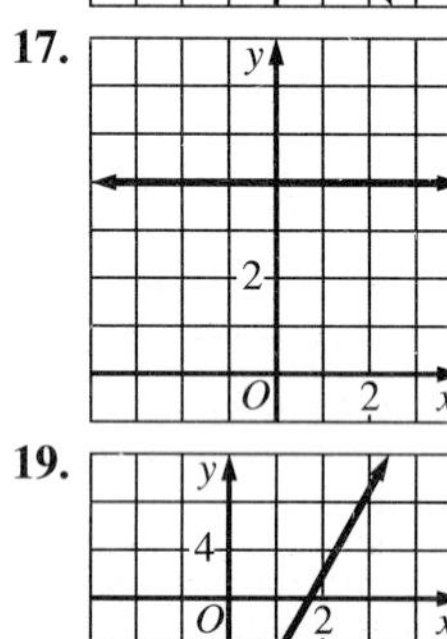

18.

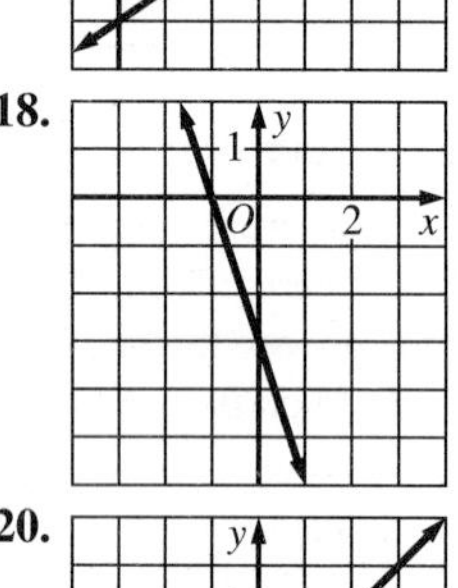

19. **20.**

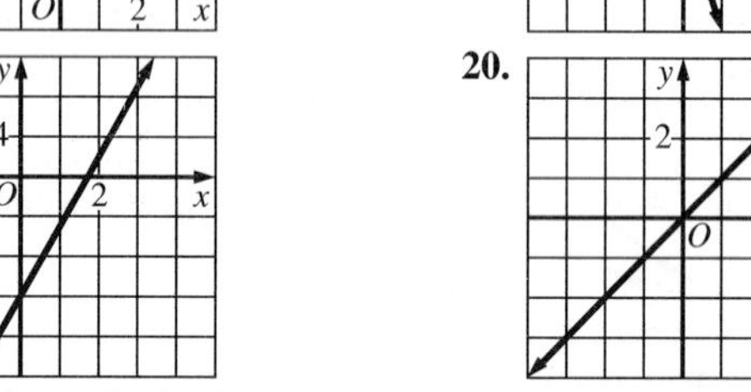

21. If the slope of a line is positive, the line rises from left to right. If the slope is negative, the line falls from left to right.

22. a.

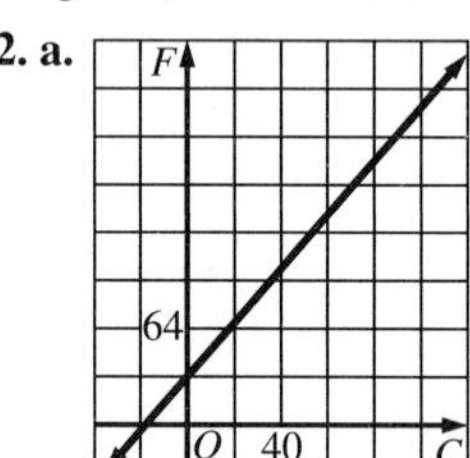

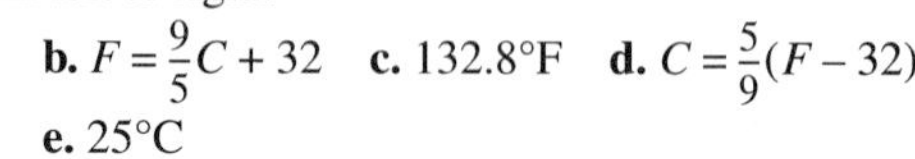

b. $F = \frac{9}{5}C + 32$ **c.** 132.8°F **d.** $C = \frac{5}{9}(F - 32)$ **e.** 25°C

Page 89 Chapter 2 Assessment

21. a.

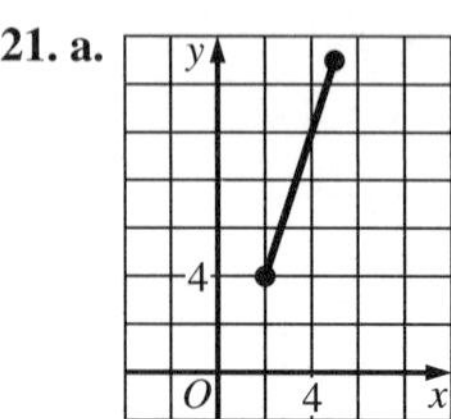

b. $y = 3x - 2$; $2 \le x \le 5$

22. a.

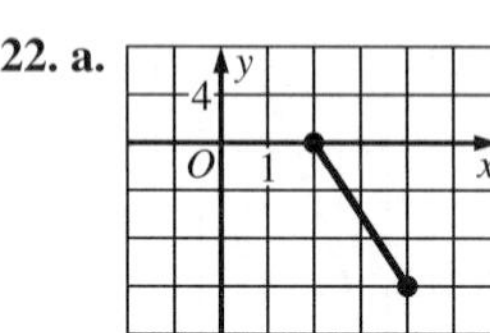

b. $y = -6x + 12$; $2 \le x \le 4$

23. Answers may vary. Check students' work.

CHAPTER 3

Page 96 Checking Key Concepts

5. a.

Number of folds	Page width (cm)	Page height (cm)	Page area (cm^2)
0	38	51	1938
1	25.5	38	969
2	19	25.5	484.5
3	12.75	19	242.25
4	9.5	12.75	121.125
5	6.375	9.5	60.5625
6	4.75	6.375	30.28125

b. Let n = the number of folds and A = the page area in square centimeters; $A = 1938\left(\frac{1}{2}\right)^n$.

Page 97 Exercises and Applications

20. a.

Bookmaking Page Sizes

C9		=0.5*D8			
	A	B	C	D	E
1	Name of page size	Number of folds	Page width (cm)	Page length (cm)	Page area (cm^2)
2	POTT	0	31	39	1209
3	folio	1	19.5	31	604.5
4	quarto	2	15.5	19.5	302.25
5	octavo	3	9.75	15.5	151.125
6	16mo	4	7.75	9.75	75.5625
7	32mo	5	4.875	7.75	37.78125
8	64mo	6	3.875	4.875	18.890625
9	128mo	7	2.4375	3.875	9.4453125

b. Let n = the number of folds and A = the page area in square centimeters; $A = 1209\left(\frac{1}{2}\right)^n$.

c.

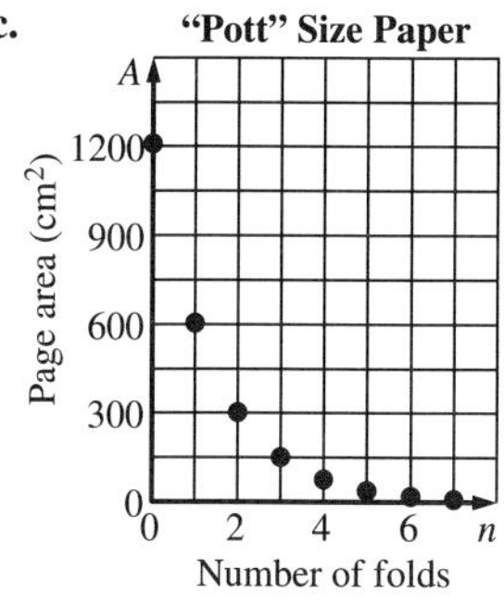

Page 99 Exercises and Applications

27. a.

Triangle number	Edge length	Perimeter	Area
1	1	3	$\frac{\sqrt{3}}{4}$
2	2	6	$\sqrt{3}$
3	4	12	$4\sqrt{3}$
4	8	24	$16\sqrt{3}$
5	16	48	$64\sqrt{3}$

b. Let n = the triangle number, l = edge length, P = perimeter, and A = area; $l = 2^{n-1}$; $P = 3(2^{n-1})$; $A = \frac{\sqrt{3}}{16}(4^n)$.

c.

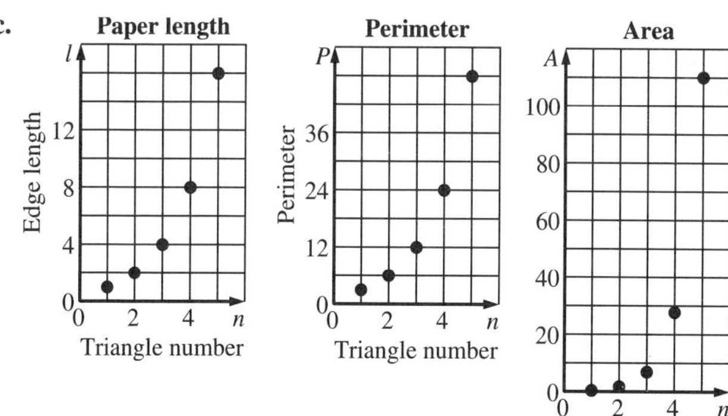

Page 106 Exercises and Applications

37.

Note	A	A#	B	C	C#	D	D#
x	0	1	2	3	4	5	6
F	440	466.2	493.9	523.3	554.4	587.3	622.3

Note	E	F	F#	G	G#	A
x	7	8	9	10	11	12
F	659.3	698.5	740.0	784.0	830.6	880

38. concert A and E, A# and F, B and F#, C and G, C# and G#, D and A (the note one octave higher than concert A)

39.

Name of scale	Notes in scale	Frequency equation
half-tone	12	$F = 440 \cdot 2^{x/12}$
quarter-tone	24	$F = 440 \cdot 2^{x/24}$
sixth-tone	36	$F = 440 \cdot 2^{x/36}$
eighth-tone	48	$F = 440 \cdot 2^{x/48}$
sixteenth-tone	96	$F = 440 \cdot 2^{x/96}$

40. Frequencies are rounded to the nearest tenth. 440, 443.2, 446.4, 449.6, 452.9, 456.2, 459.5, 462.8, 466.2, 469.5

Page 108 Think and Communicate

1. a.

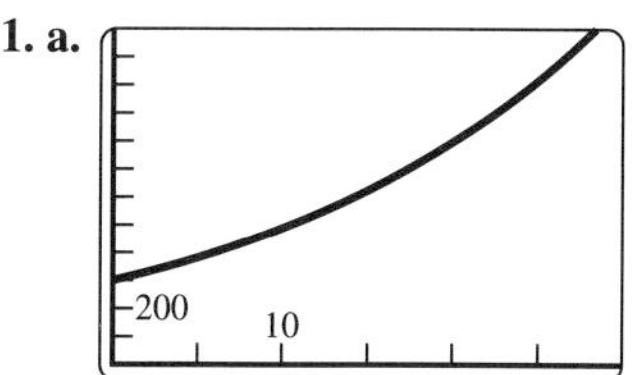

b. Using a calculator, $2^{1/14.2} \approx 1.050024169$, or about 1.05.

Page 111 Exercises and Applications

10. For $a > 0$ and $b > 1$, the graph of $y = ab^{-x}$ lies above the x-axis and represents exponential decay. For $a > 0$ and $0 < b < 1$, the graph of $y = ab^{-x}$ lies above the x-axis and represents exponential growth. For $a < 0$ and $b > 1$, the graph of $y = ab^{-x}$ lies below the x-axis and represents neither exponential growth nor decay. For $a < 0$ and $0 < b < 1$, the graph of $y = ab^{-x}$ lies below the x-axis and represents neither exponential growth nor decay. Since ab^{-x} is equal to $a\left(\frac{1}{b}\right)^x$, the graphs of $y = ab^{-x}$ and $y = a\left(\frac{1}{b}\right)^x$ are the same. The graph of $y = ab^{-x}$ is the graph of $y = ab^x$ reflected over the y-axis. **11.** C **12.** B **13.** A **14.** D **15.** The graph of $y = 8 \cdot 2^x$ is the graph of $y = 2^x$ translated 3 units left if for every point (x, y) on the graph of $y = 8 \cdot 2^x$, the point $(x + 3, y)$ is on the graph of $y = 2^x$. Suppose (x, y) is on the graph of $y = 8 \cdot 2^x$. Then $y = 2^3 \cdot 2^x = 2^{x+3}$, and hence, $(x + 3, y)$ is on the graph of $y = 2^x$.

Page 115 Exploration

1.

Compounding	n	Formula	$r = 0.05$	$r = 0.10$	$r = 0.50$	$r = 1.00$
annually	1	$\left(1 + \frac{r}{1}\right)^1$	1.05	1.10	1.50	2.00
semiannually	2	$\left(1 + \frac{r}{2}\right)^2$	1.0506	1.1025	1.5625	2.25
quarterly	4	$\left(1 + \frac{r}{4}\right)^4$	1.0509	1.1038	1.6018	2.4414
monthly	12	$\left(1 + \frac{r}{12}\right)^{12}$	1.0512	1.1047	1.6321	2.6130
daily	365	$\left(1 + \frac{r}{365}\right)^{365}$	1.0513	1.1052	1.6482	2.7146
hourly	8760	$\left(1 + \frac{r}{8760}\right)^{8760}$	1.0513	1.1052	1.6487	2.7181
every minute	525,600	$\left(1 + \frac{r}{525{,}600}\right)^{525{,}600}$	1.0513	1.1052	1.6487	2.7183
every second	31,536,000	$\left(1 + \frac{r}{31{,}536{,}000}\right)^{31{,}536{,}000}$	1.0513	1.1052	1.6487	2.7183

2. Increasing the frequency of compounding initially increases the earnings, but the effect levels off eventually. For $r = 0.05$, for example, the yield is no greater for compounding every second than for compounding daily. **3.** No; the effect of more frequent compounding eventually becomes negligible. This happens very quickly for values of r that correspond to real-world interest rates.

Page 122 Exercises and Applications

31. a. Graphs may vary depending on the choice of value for A_0. An example is given that uses $A_0 = 1$.

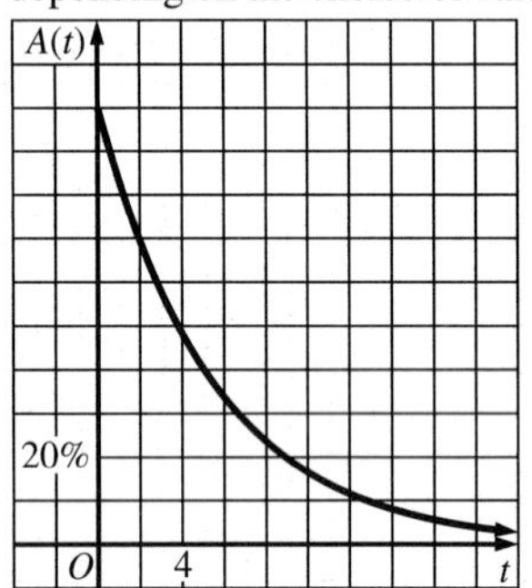

b. Estimates may vary. Examples are given.

Percent of Initial Amount Remaining	Number of Days
90	0.6
80	1.2
70	2.0
60	2.8
50	3.9
40	5.1
30	6.7
20	8.9
10	12.8

Page 126 Exercises and Applications

9. a. Let x = the apparent magnitude and y = the brightness value; $y = 251.189(0.398)^x$.

b.

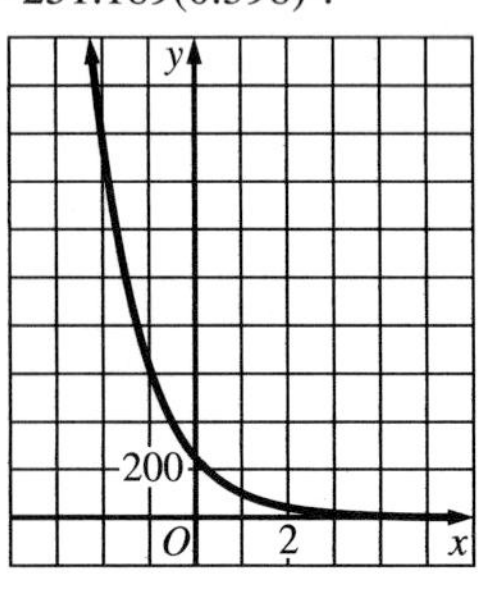

c. Answers may vary due to rounding.

Sky object	Apparent magnitude	Brightness value
Uranus	6	1
Aldebaran	1	100
Vega	0	251.2
Sirius	–1.5	1000.4
Full moon	–12.5	2.5×10^7
Sun	–26.7	1.2×10^{13}

d. Answers may vary. Examples are given.

Sky object	Apparent magnitude	Brightness value
Canopus	–0.7	478.7
Arcturus	–0.1	275.4
Alpha Centauri	0	251.2
Rigel	0.1	229.1
Betelgeuse	0.8	120.2
Deneb	1.25	79.4

Page 129 Assess Your Progress

1. \$2385.04 **2.** \$2394.25 **3.** \$2394.43

4.

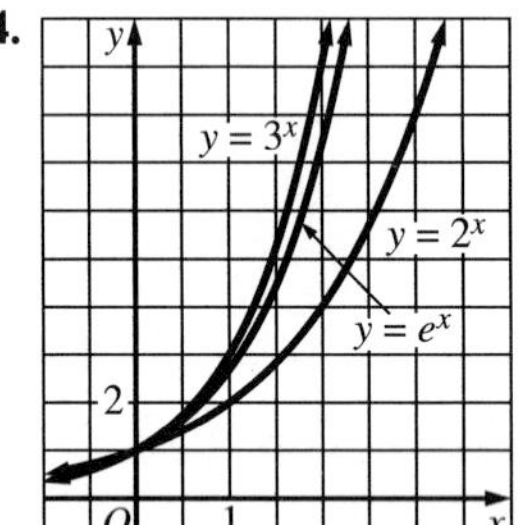

Summaries may vary. All three graphs have the same y-intercept, 1. All three are exponential, with equations of the form $y = b^x$. The greater the value of b, the steeper the graph.

5. a. Answers may vary. Check students' work. **b.** $y = 4.975(1.046)^x$ **c.** about 181.7 cm **6.** Answers may vary. An example is given. Exponential growth is growth by a constant rate. It can be described by a function of the type $y = ab^x$, where $a > 0$ and $b > 1$. Logistic growth is limited; it grows exponentially at first, then levels off. Logistic functions are often written using e. In real-world situations, exponential functions are often used to model such situations as earned interest or the growth of bacteria. One situation that can be modeled by logistic growth is the increase in an animal population, which will eventually be limited by availability of living space, food supply, and perhaps disease.

CHAPTER 4

Page 145 Exercises and Applications

6. a.

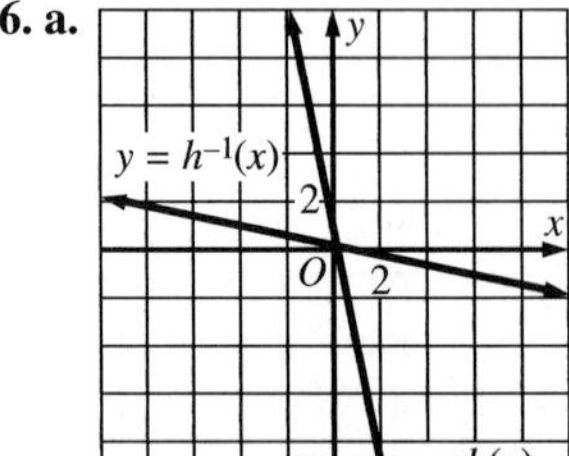

b. $h^{-1}(x) = -\frac{1}{5}x + \frac{1}{5}$

7. a.

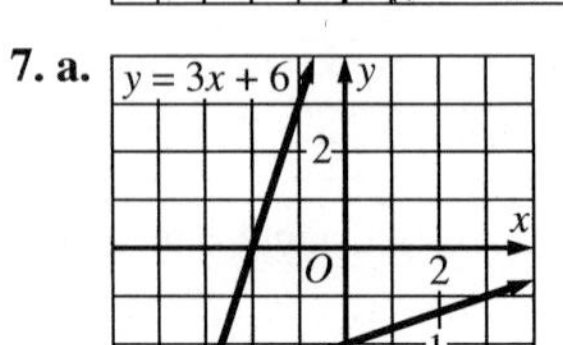

b. $y = 3x + 6$

8. a. 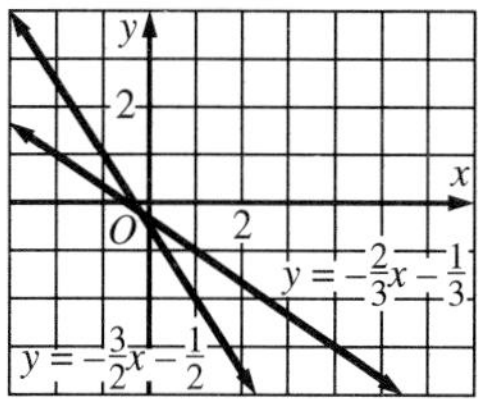 (graph) **b.** $y = -\frac{2}{3}x - \frac{1}{3}$

9. E **10. a.** Answers may vary. An example is given. You could use a measuring tape to measure around the tree, being careful to keep the tape at about the same height above the ground all the way around the tree. Since you cannot measure through the center of the tree, the only way to measure the diameter directly would be with a caliper-like device. **b.** $C = \pi d$ **c.** $d = \frac{C}{\pi}$ **d.** about 23.87 in.

Page 153 Exercises and Applications

36. a.

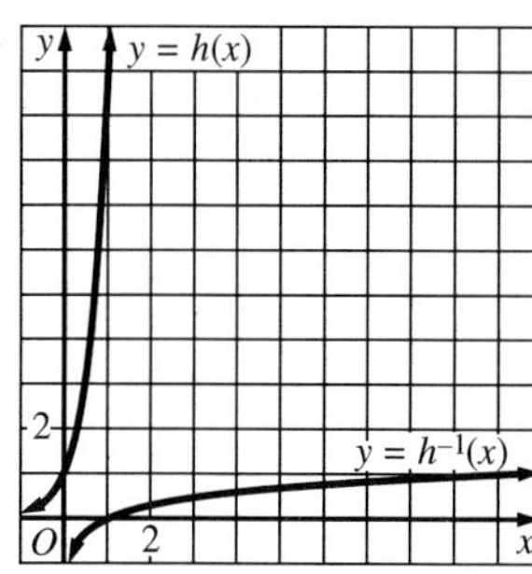

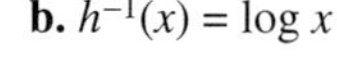

b. $h^{-1}(x) = \log x$

37. a.

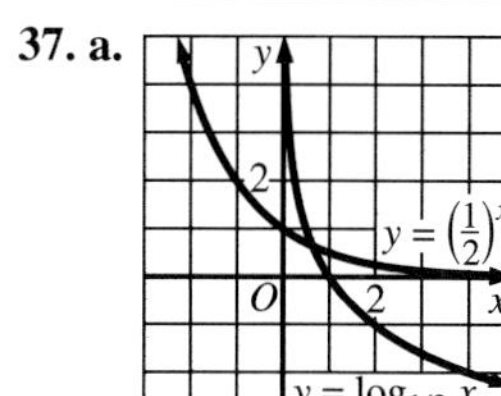

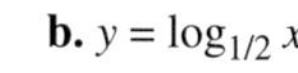

b. $y = \log_{1/2} x$

38. a.

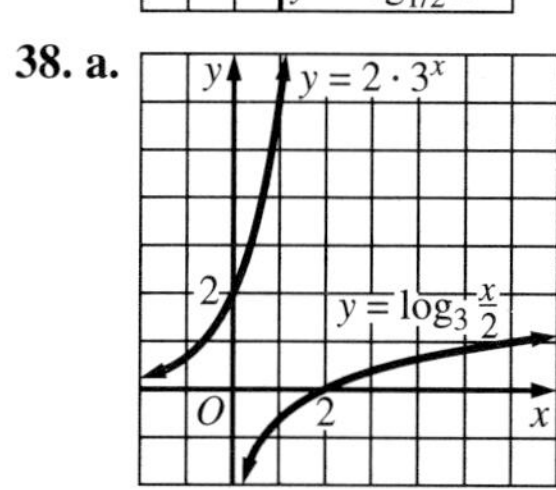

b. $y = \log_3 \frac{x}{2}$

Page 154 Exercises and Applications

39. a. $t = 40 \ln \frac{P}{19.1}$; the number of years after 1990 that the population reaches a given population P **b.** 2008 **c.** 1980

40.

N	7	10	52	100	613	1,000	4,849	10,000	91,770
Number of digits in N	1	2	2	3	3	4	4	5	5
$\log N$	0.85	1.00	1.72	2.00	2.79	3.00	3.69	4.00	4.96

The number of digits in N is one more than the first digit of the common logarithm of N. **41.** 478 digits **42.** Answers may vary. An example is given. The inverse of the exponential function in Example 2 on page 117 indicates the age in days t of a 100 microgram (μg) sample of Polonium-210 that has decayed to an amount A μg. The function is given by $t = -200 \ln \frac{A}{100}$. For example, if the amount remaining is 60 μg, the sample is about 102 days old. **43.** Answers may vary. An example is given. The domain of every logarithmic function is the positive real numbers and the range is the real numbers. The base of every logarithmic function is a positive real number not equal to 1. The logarithmic function $y = \log_b x$ is the inverse of the exponential function $y = b^x$. Some values of the function $y = \log_5 x$ are given. $\log_5 25 = 2$; $\log_5 \frac{1}{5} = -1$; $\log_5 1 = 0$; $\log_5 125 = 3$

Page 161 Assess Your Progress

5. 2 **6.** $\frac{1}{2}$ **7.** –4 **8.** $\frac{5}{11}$ **9. a.** $t = 91.7 \ln \frac{P}{33.9}$ **b.** 1997 **10.** $3 \log p + 4 \log q - 5 \log r$ **11.** $\log_b 8$ **12.** A logarithmic scale is one in which adding 1 to the given scale corresponds to multiplying a related quantity by some number. Real-world examples include the Richter scale, the pH scale, and the decibel scale.

Page 171 Checking Key Concepts

1. a.

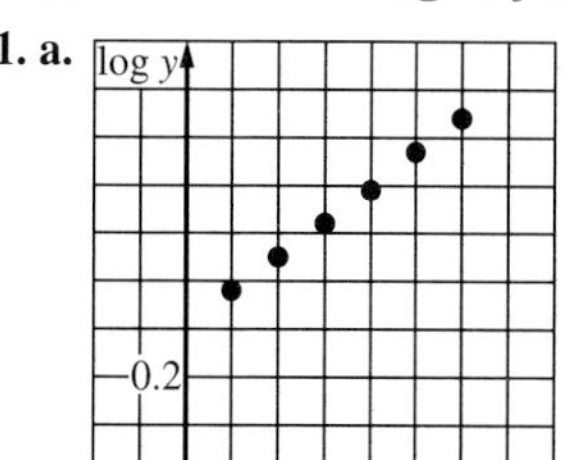

b.

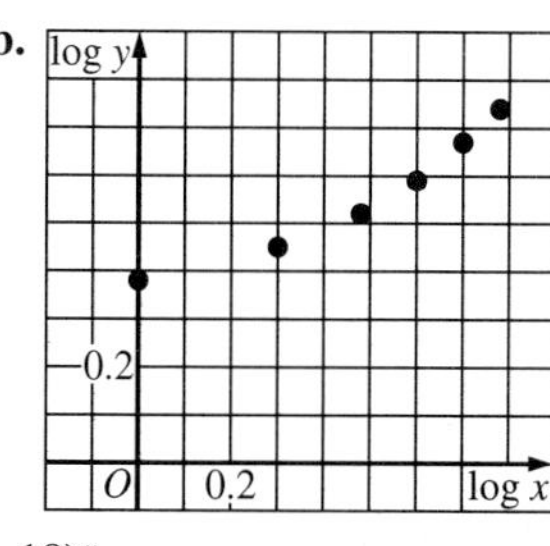

c. exponential function; $y = 2.01(1.18)^x$

2. a.

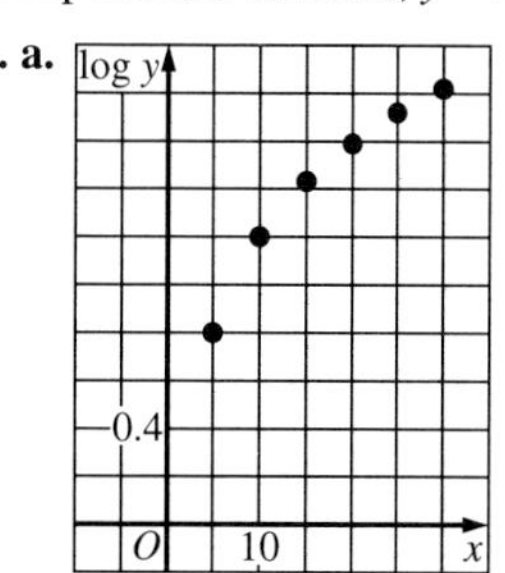

b.

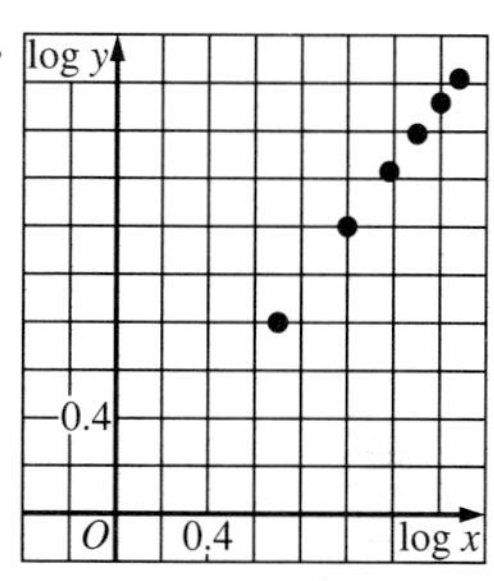

c. power function; $y = 0.771x^{1.31}$

3. a.

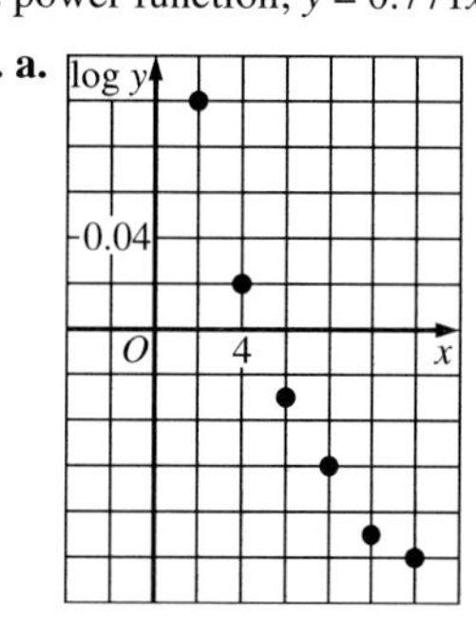

b.

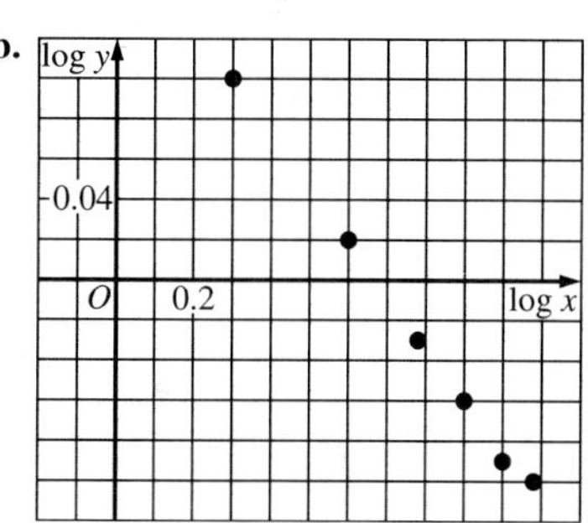

c. power function; $y = 1.50x^{-0.263}$

Page 173 Exercises and Applications

17. a.

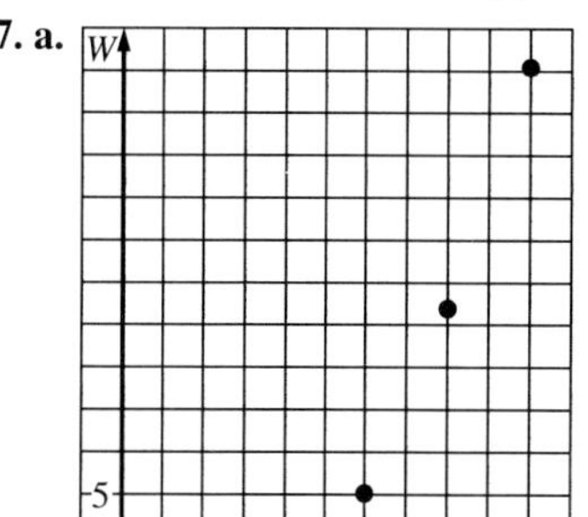

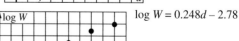

No; power functions and exponential functions have similarly shaped graphs.

b.

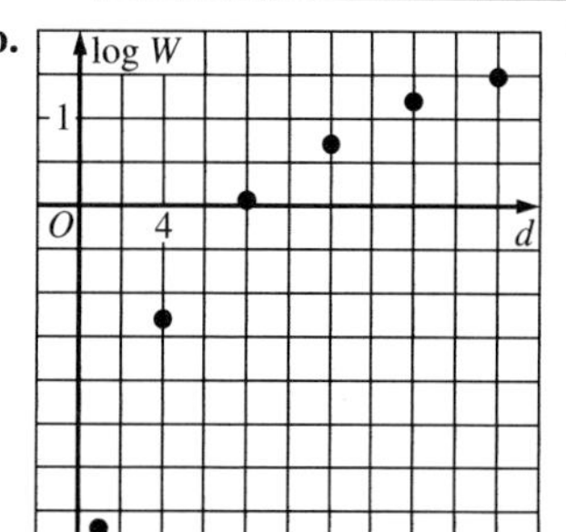

$\log W = 0.248d - 2.78$

c. Clearly the data points in Example 2 lie very near the line, so there is a linear relationship between log d and log W, which indicates the data can be modeled by a power function. The data points in the scatter plot above do not lie near the line, they appear to lie nearer a curve. This indicates there is not a linear relationship between d and log W.

18. a.

x	y = 2*x	y = 2^x	y = log(x)/log(2)	y = x^2
1	2	2	0	1
2	4	4	1	4
3	6	8	1.584962501	9
4	8	16	2	16
5	10	32	2.321928095	25
6	12	64	2.584962501	36
7	14	128	2.807354922	49
8	16	256	3	64
9	18	512	3.169925001	81
10	20	1024	3.321928095	100
11	22	2048	3.459431619	121
12	24	4096	3.584962501	144

$y = 2^x$; $y = x^2$; $y = 2x$; $y = \log_2 x$

b. Functions may vary; the functions in order from fastest to slowest rate will be $y = b^x$, $y = x^b$, $y = bx$, and $y = \log_b x$. **c.** Answers may vary. An example is given. In general, for positive values of x and b, exponential functions increase fastest, followed by power functions, linear functions, and logarithmic functions.

Page 175 Exercises and Applications

25. Answers may vary. Examples are given. In the examples, x = the number of years since 1980 and y = the number of African-American elected officials.

Type of function	Equation	Correlation coefficient
linear	$y = 249x + 4720$	0.988500544
exponential	$y = 4860(1.04)^x$	0.9814998067
power	$y = 4280x^{0.225}$	0.9906997347

According to the correlation coefficients, all three functions model the data fairly well, the best model being the power function. According to that function, I predict that in the year 2000, the number of African-American elected officials will be about 8400. (The estimates according to the linear and exponential models are 9700 and about 10,600.)

Page 181 Chapter 4 Assessment

35. Answers may vary. Examples are given. In the examples, x = years since 1980 and y = price in thousands.

a.

Type of function	Equation	Correlation coefficient
exponential	$y = 59.8(1.05)^x$	0.9965490037
power	$y = 42.8x^{0.353}$	0.9986221615

b. The power function is the better model since the correlation coefficient for the power function, 0.9986221615, is closer to 1 than the correlation coefficient for the exponential function, 0.9965490037. **c.** about \$103,000; The prediction is very close to the actual median price.

CHAPTER 5

Page 192 Exploration

2. a. The graph of $y = ax^2 + k$ is the graph of $y = ax^2$ translated $|k|$ units up if k is positive and down if k is negative. **b.** The graph of $y = a(x - h)^2$ is the graph of $y = ax^2$ translated $|h|$ units left if h is positive and right if h is negative.

Page 197 Exercises and Applications

7. a. Except for the values for $u = -3$ and $u = 2$, the ratios are about equal. **b.** $-435.07417 \approx -435$ **c.** $y = -435(x - 50)^2 + 300{,}000$

d.

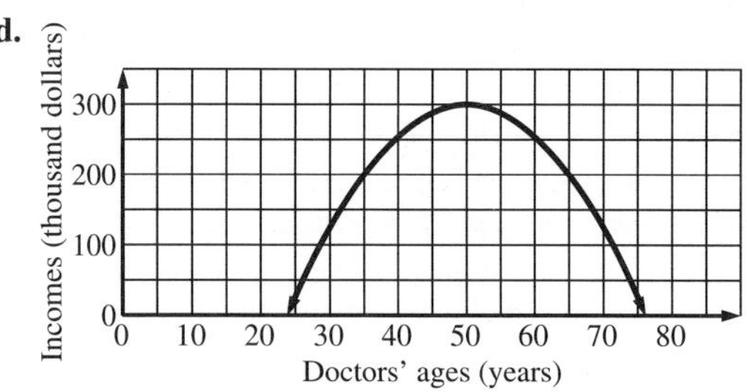

The equation is a good model for the data. If the scatter plot of the data and the equation are graphed in the same coordinate plane, the graph of the equation is nearly superimposed on the data points.

Page 198 Exercises and Applications

19. The vertex is $\left(-4, \frac{1}{2}\right)$. The line of symmetry is $x = -4$. The parabola opens down.

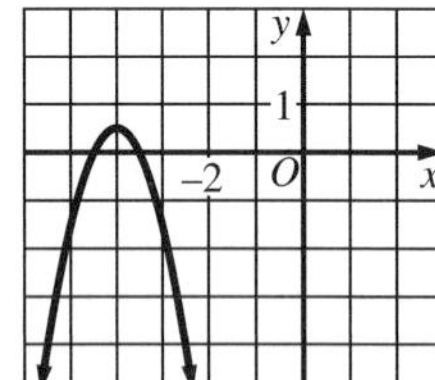

20. The vertex is (3, −4). The line of symmetry is $x = 3$. The parabola opens down.

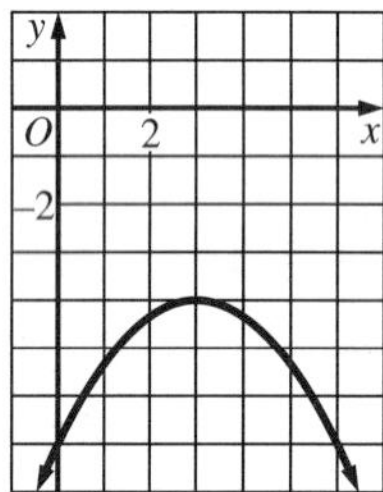

21. $y = 2(x + 2)^2 + 4$ **22.** $y = -\frac{1}{2}(x - 1)^2 - 3$ **23.** $y = -3(x - 5)^2 + 2$

24. a. about 0.1 s, about 0.9 s **b.** about 1.1 s **c.** about 1.2 s **25. a.** Let t = the time in seconds after the whelk is dropped and h = the height above the ground; $h = -16t^2 + 17$. **b.** about 1.0 s **26.** Let b be any number. When $x = h + b$, $y = a(h + b - h)^2 + k = ab^2 + k$. When $x = h - b$, $y = a(h - b - h)^2 + k = a(-b)^2 + k = ab^2 + k$. Then for every number b, the y-values are the same for $x = h + b$ and $x = h - b$, so $x = h$ is the line of symmetry for the parabola.

27.

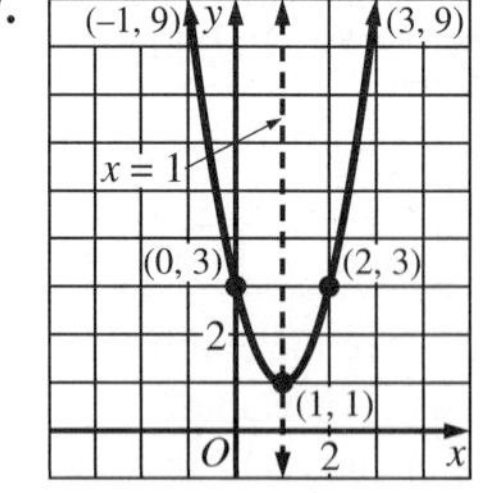

Answers may vary. An example is given. The equation $y = a(x - h)^2 + k$ is in vertex form. To graph the equation, first draw the vertex, which is at (h, k), and the line of symmetry $x = h$. Plot points that are 1 unit to the left or right of the line of symmetry, 2 units to the left or right of the line of symmetry, and so on, and draw a curve through the points. The sketch shows the graph of $y = 2(x - 1)^2 + 1$.

28. ± 0.5 **29.** ± 3.2 **30.** ± 3 **31.** $3(x + 2)$ **32.** $7(5x - 3y)$ **33.** $x(5x + 2)$ **34.** $f^{-1}(x) = e^x$ **35.** $g^{-1}(x) = \frac{1}{3}x - 5$ **36.** $h^{-1}(x) = \log_2 x$

Page 203 Exercises and Applications

7. Yes; $(2x - 1)(3x + 4) = 2\left(x - \frac{1}{2}\right)\left(3\left(x + \frac{4}{3}\right)\right) = 2(3)\left(x - \frac{1}{2}\right)\left(x + \frac{4}{3}\right) = 6\left(x - \frac{1}{2}\right)\left(x + \frac{4}{3}\right)$. **8.** $y = -\frac{4}{9}(x + 3)(x - 3)$ **9.** $y = -6(x - 4)(x - 6)$ **10.** $y = \frac{4}{9}(x + 5)(x - 1)$

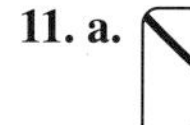

11. a. 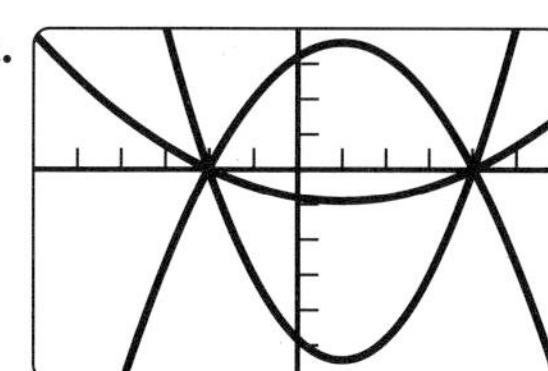

The graphs all have the same x-intercepts and line of symmetry. The width of the parabolas varies, as does the vertex. The first two parabolas open up, the third opens down.

b. $y = 3(x + 2)(x - 4)$; $y = 0.5(x + 2)(x - 4)$; $y = -2(x + 2)(x - 4)$; The values of p and q are the same for all three equations. The values of a differ.
12. a. $R = (16 + 4x)(20{,}000 - 1000x)$ **b.** \$48 **c.** \$576,000

Page 204 Exercises and Applications

15. Answers may vary. Examples are given.

a. 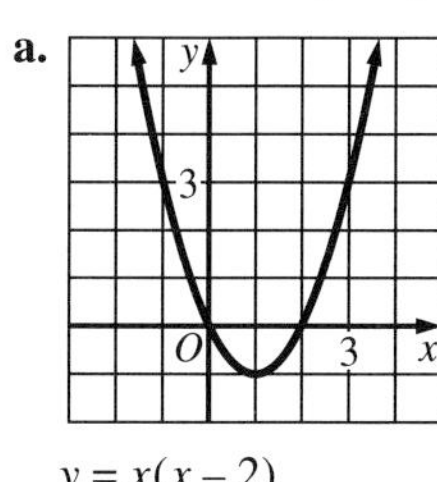

$y = x(x - 2)$

b. 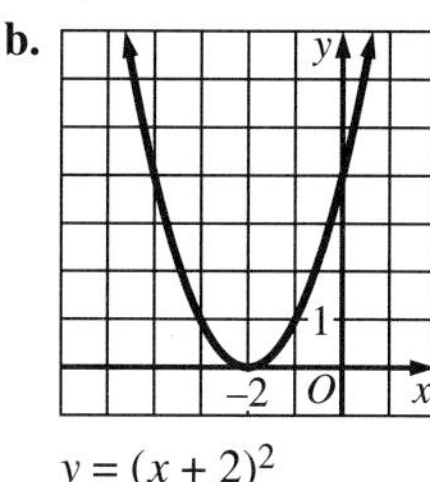

$y = (x + 2)^2$

c. 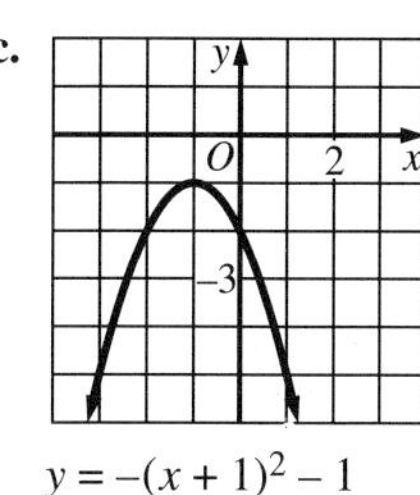

$y = -(x + 1)^2 - 1$

Page 205 Assess Your Progress

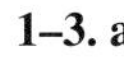

1–3. a. 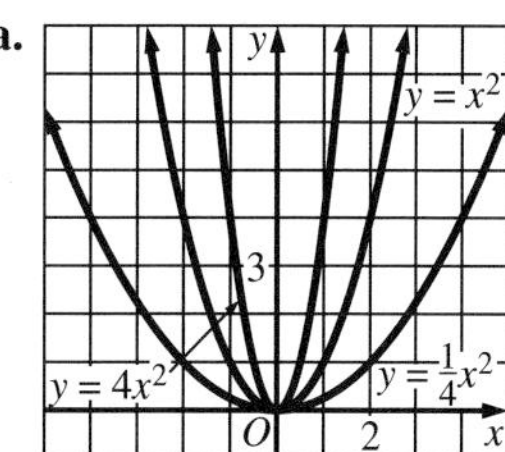

1. b. ± 3.5
2. b. ± 1.7
3. b. ± 6.9

4. $y = -3(x - 3)^2 + 7$ **5.** $y = \frac{1}{2}(x + 1)^2 + 4$ **6.** $y = 2(x - 5)^2 - 2$

7. $y = \frac{1}{8}(x + 6)^2 - 1$

8. $y = 3(x + 9)(x + 4)$

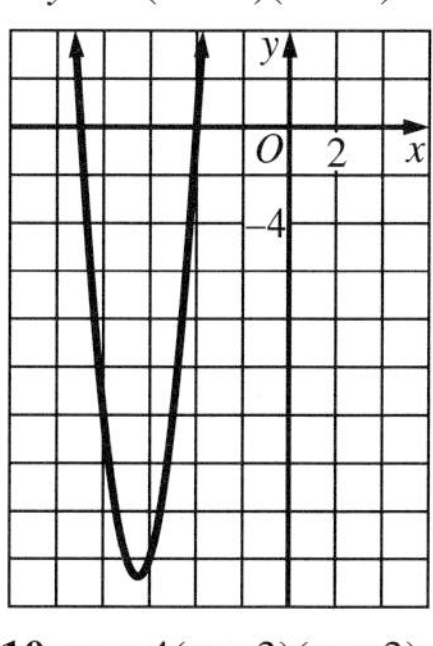

9. $y = -6(x - 3)(x - 4)$

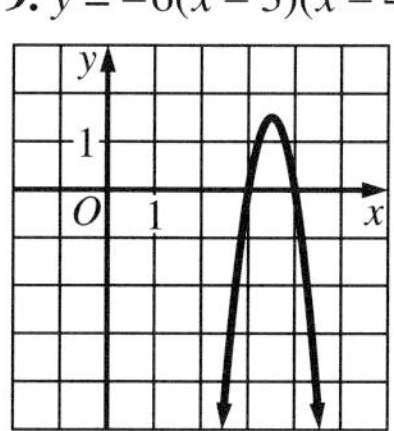

10. $y = 4(x - 3)(x + 3)$

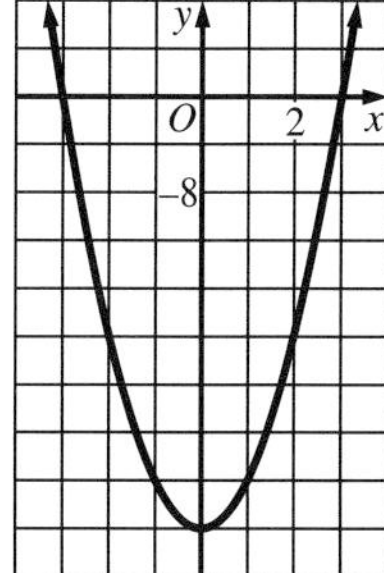

11. $y = -10(x - 2)(x + 2)$

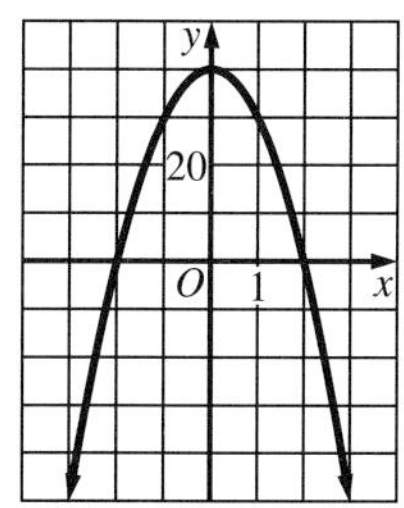

12. The x-intercepts of the graph of the equation $y = a(x - p)(x - q)$, which is in intercept form, are p and q. The x-coordinate of the vertex is $\frac{p + q}{2}$. To find the y-coordinate of the vertex, evaluate the equation for $x = \frac{p + q}{2}$. The vertex of the graph of the equation $y = a(x - h)^2 + k$, which is in vertex form, is (h, k). To find the x-intercepts, solve the equation $0 = a(x - h)^2 + k$.

Page 225 Exercises and Applications

22. a. The functions $y = 2x^2 + 3x - 1$ and $y = 3x^2 - 4x + 4$ cannot be written as products of factors; $y = 2x^2 - x - 1 = (2x + 1)(x - 1)$ and $y = 8x^2 + 2x - 3 = (4x + 3)(2x - 1)$. **b.** 17; 9; 100; –32; Three of the numbers are positive and one is negative. Two are perfect squares. **c.** A quadratic function $y = ax^2 + bx + c$ is factorable if the discriminant of $ax^2 + bx + c = 0$ is a perfect square. Examples may vary.

Page 226 Exercises and Applications

24. a. $y = \frac{\pi^2}{19.6}x^2 - \frac{\pi^2}{19.6}$ **b.** $y = \frac{\pi^2}{19.6}(x + 1)(x - 1)$

c. –1; 1; $\left(0, -\frac{\pi^2}{19.6}\right) \approx (0, -0.5)$

c, d. 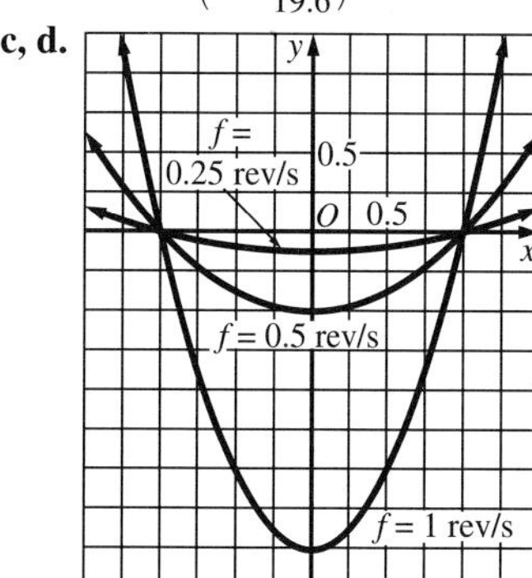

d. Answers may vary. Examples are given. (See graph.) All the graphs have the same x-intercepts. Also, as the liquid is spun faster, the parabola becomes narrower and steeper.

Page 227 Assess Your Progress

10. Answers may vary. An example is given. When an equation is written in standard form, it is simple to use the quadratic formula to solve the equation and, so, to find the x-intercepts (if any) of the related quadratic function, the line of symmetry of the parabola, and the coordinates of its vertex. The equation itself, however, gives no information directly about the graph of the quadratic function. The x-intercepts of the graph of a quadratic function in intercept form can be read directly from the equation. The equation of the line of symmetry of the graph and the x-coordinate of the vertex can be easily found from the equation. If a quadratic function is written in vertex form, the coordinates of the vertex and the equation of the line of symmetry can be read directly from the equation. In both intercept form and vertex form, you can tell if the graph of the equation opens up or down, but you must rewrite the equation to use the quadratic formula.

CHAPTER 6

Page 243 Exercises and Applications

22. a. Aiyana: 42.9%; Kelly: 71.4%; Marcus: 14.3% **b.** 42.9%; The estimate using all three samples should be most reliable, since it is based on more data. **c.** about 18 orchestras **23.** Answers may vary. **24.** –3; 2 **25.** –3; 1

26. $-\frac{1}{2}; \frac{1}{3}$ **27.** –9; –4

28. 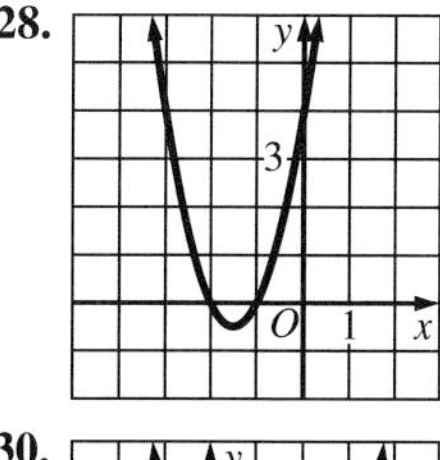

29. 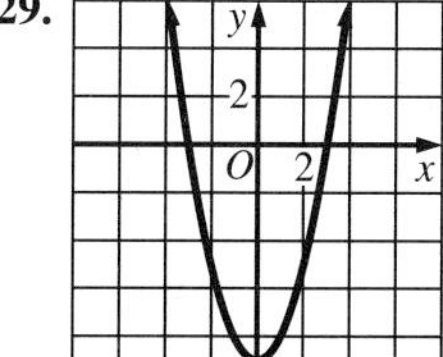

30. 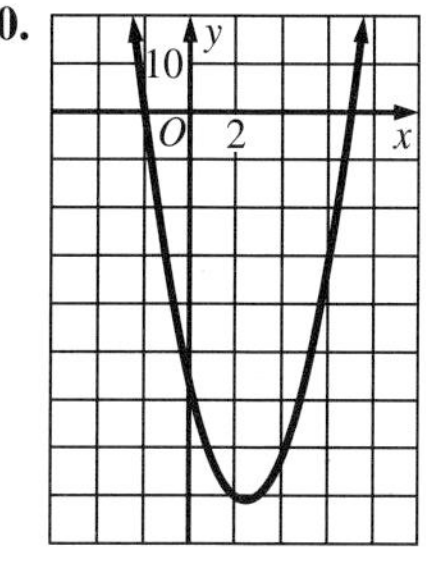

31. 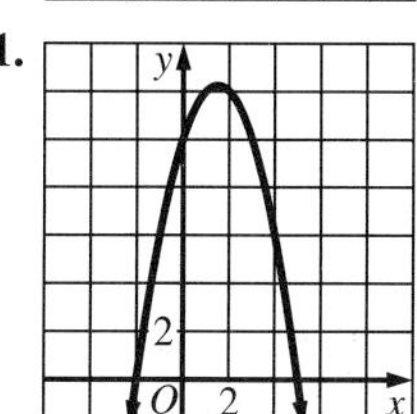

Page 244 Exploration

4. Yes; students' responses would likely be based on their actual opinions if the other questions are not read before responses are given.

Page 247 Exercises and Applications

3. Unless all the people surveyed are familiar with Japanese food, they may have no information about the dishes. A reasonable question might describe each of the dishes and ask respondents to rank them in order of preference.
4. Mentioning that the current stadium was built in 1905 could bias the question in either direction. Some people may feel the stadium is a part of tradition and should be retained. Others may feel a stadium that old must be outdated. A less biased question would be, "Are you in favor of replacing the baseball stadium with a new sports complex?" **5.** The question assumes that the respondent is familiar with the facts of the case. Any presentation of the facts (as interpreted by the pollster) may be biased as well. It might be best then to ask the question as given only to those who reply affirmatively to the question, "Are you familiar with the facts of the Carter case?" **6.** Grades are a sensitive issue and students may not wish to reveal their grade point averages. The information should be collected anonymously. **7.** I think questions A, C, and D do not need to be rewritten if question B is rephrased, "Do you think a student can maintain good grades while holding down an after-school job?" **8.** what kind of senior banquet, if any, students want to have **9–15.** Answers may vary. Examples are given. **9.** I think the first two questions are reasonable. The third question is too restrictive. The fourth question is too open-ended. I would also ask students how likely they are to attend, with response choices scaled from "not at all" to "certain." I would expand on the last question to include options such as location, refreshments (snacks, buffet, a sit-down dinner), music (a band or a DJ), extent of decorations, and the availability of mementos. **10.** Do you think a dress code is preferable for the banquet? If so, should dress be casual, semiformal (dresses, jackets and ties), or formal (gowns, tuxedos)? To rewrite the last question, I would present a variety of options with estimated prices.
11. (1) categorical; It will help determine whether there is an equal amount of interest among males and females. (2) categorical; It will help determine whether a banquet should be held at all. (3) categorical; It will help determine what type of banquet should be held. (4) numerical; It will help determine what type of banquet students can afford.

Page 249 Exercises and Applications

16–19. Survey questions may vary. Examples are given. **16.** How often do you eat at the restaurant? Have you ever had the special? If so, how often? If the special were available on the regular menu, how often do you think you would order it? People who have eaten at the restaurant should be surveyed.
17. Have you ever used the train service? If so, how often? How often do you anticipate using the service in the future? Which hours do you usually use the service? If service were expanded to include more early morning and late evening trains, do you think you would use the service during the new hours? If so, at what times and what routes? How often do you think you would use the new service? People who live or work in the area served by the trains should be surveyed. **18.** How often do you currently attend our meetings? What day of the week are you most likely to attend? For each day of the week, estimate how many of the four meetings a month you would be able to attend if the meetings were held on that day. Members of the crafts group should be surveyed.
19. Have you watched any shows on this station on Thursday night during the past month? Give the frequency with which you have watched each of the Thursday shows listed below during the past month. Give the frequency with which you anticipate watching each of the Thursday shows listed below during the next month. People within the service area of the station should be surveyed.

Page 251 Think and Communicate

6. a. A movie theater in a small town or rural area may be the only choice for entertainment, while a theater in an urban area may be one of many choices.
b. The "local paper" in a large urban area reaches a much more diverse audience than the local paper in a small town. **c.** You may be more likely to reach people at home in certain locations than in others. **7. a.** Movie attendance increases sharply around some holidays. Also, many people may be vacationing and have more time to see movies during the summer. **b.** Many people vacation during the summer and may not read the paper. **c.** During the summer or during major holiday seasons, it may be very difficult to reach people at home.

Page 253 Exercises and Applications

4. self-selected sample; all alumni; Yes; the sample is self-selected; also, those who graduated more than 20 years ago are underrepresented. **5.** systematic sample; all customers of the taxicab company; probably not; Choosing every tenth rider in each taxicab should produce a fairly representative sample.

Page 263 Exercises and Applications

2. symmetric **3.** skewed **4–6.** Answers may vary. **7. a.** median: 4588.5; upper quartile: 11,239; lower quartile: 1,979; maximum: 20,320; minimum: 345

b. 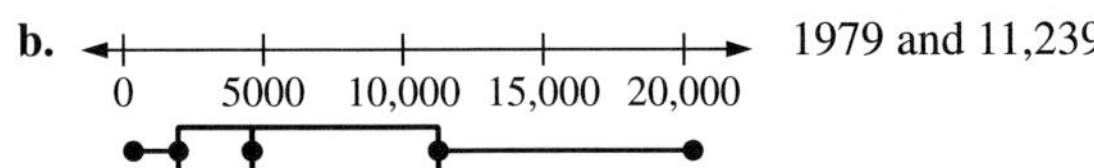

1979 and 11,239

c. The median changes to 4393, the upper quartile is 9994, the lower quartile is 1965, and the maximum is 14,494.

8. a. 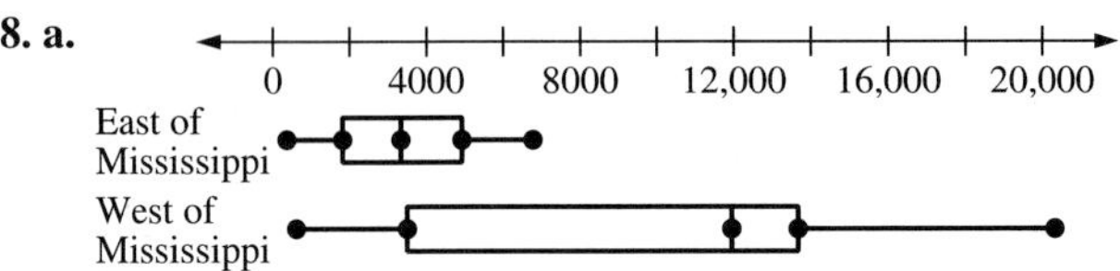

b. Rocky Mountains; the highest elevation east of the Mississippi is lower than the median of the highest elevations west of the Mississippi.

9. a. 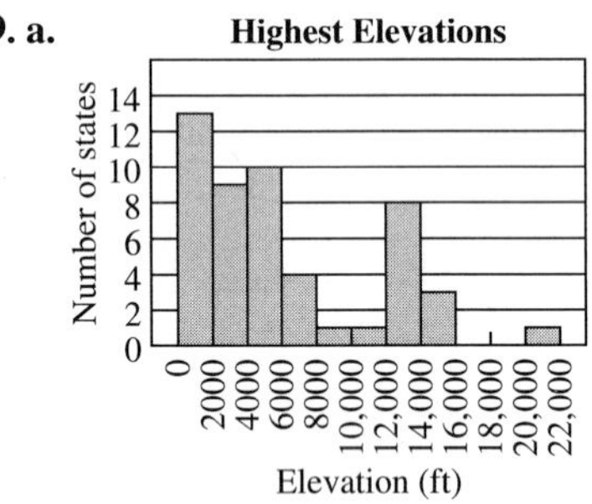

skewed
b. Estimates will vary; about 5000 ft.
c. mean: 6162.0; median: 4588.5; The estimate is fairly close to the median.
d. The median is a better measure of the "middle." The data value 20,320 raises the mean and makes the mean less descriptive of the "middle."

10. Answers may vary.

Page 277 Exercises and Applications

12. If you locate $x = 1000$ on the graph in Ex. 11, you will see that the curve has begun to level off. That is, at that point, increases in the sample size produce smaller and smaller decreases in the margin of error. The company probably feels comfortable with a margin of error of $\pm 3\%$ and feels it is not worth the additional cost and effort to use a larger sample size.

Page 285 Chapter 6 Assessment

15. Intervals may vary. Examples are given.

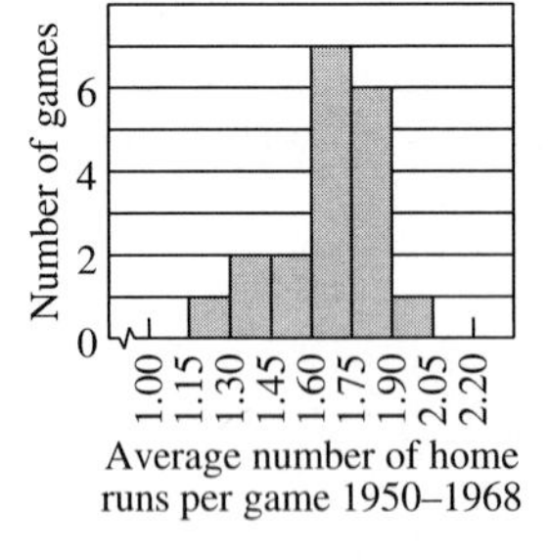

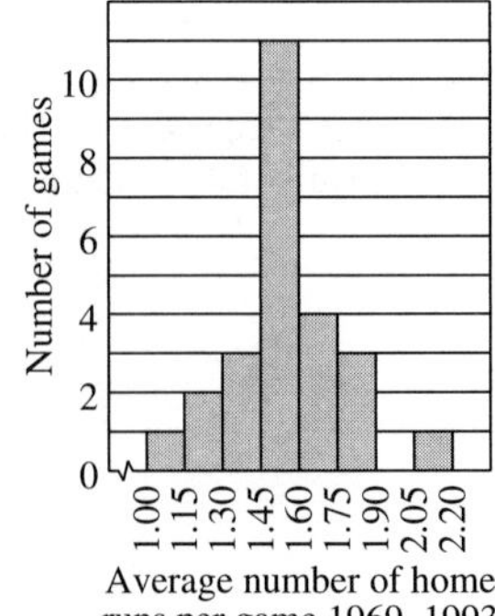

The distribution for the 1950–1968 data is skewed; the distribution for the 1969–1993 data is roughly symmetric.

CHAPTERS 4–6

Page 287 Cumulative Assessment

38. Intervals may vary.

a.

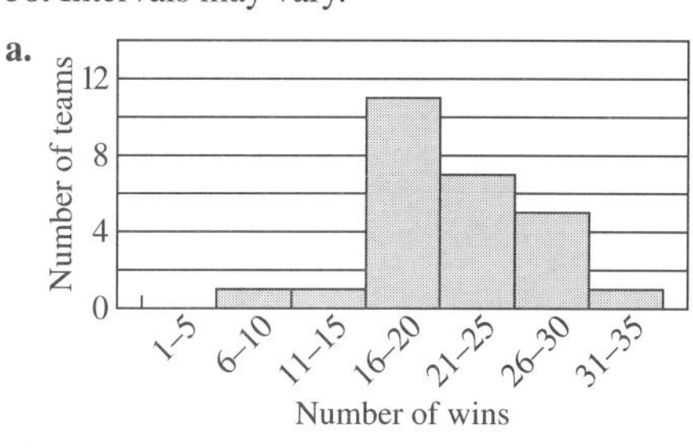
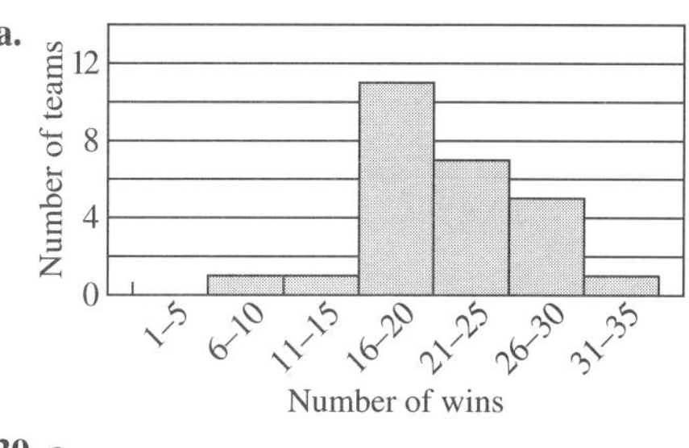

b. skewed

39. a.

Conference	Minimum	Lower quartile	Median	Upper quartile	Maximum
Eastern	9	18	22	27	30
Western	16	16.5	18.5	24	33

b.

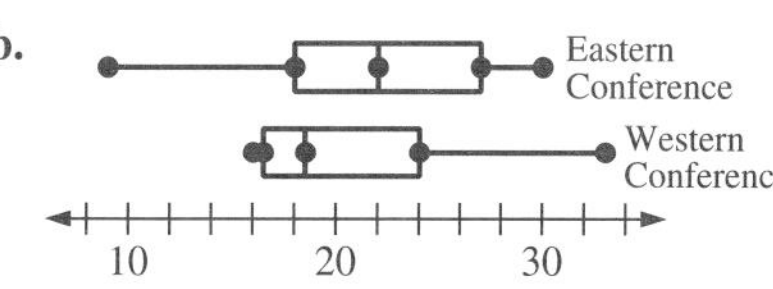

c. Eastern Conference: 21; 9; no outliers; Western Conference: 17; 7.5; no outliers. **d.** Eastern Conference: 5.62; Western Conference: 5.25; The standard deviation is greater for the Eastern Conference data.

CHAPTER 7

Page 302 Exercises and Applications

19. a.

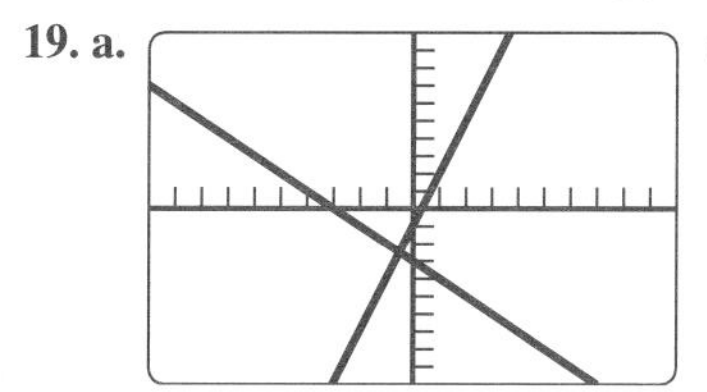

(–0.5, –2.5)

b. The graph of the new equation passes through the intersection of the graphs of the original equations.

c. If (p, q) satisfies both $Ax + By = C$ and $Dx + Ey = F$, then $Ap + Bq = C$ and $Dp + Eq = F$, so $Ap + Bq + Dp + Eq = C + F$. Then $(A + D)p + (D + E)q = C + F$ and (p, q) satisfies the sum of the equations.

Page 312 Checking Key Concepts

1.

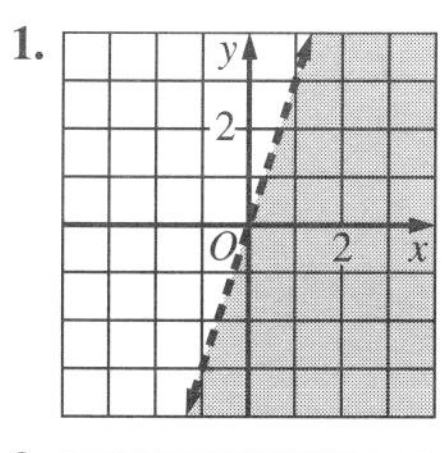

2.

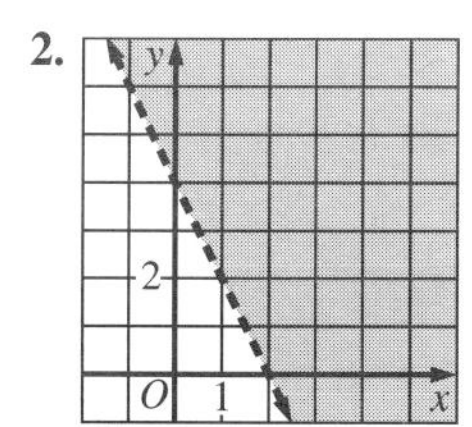

3.

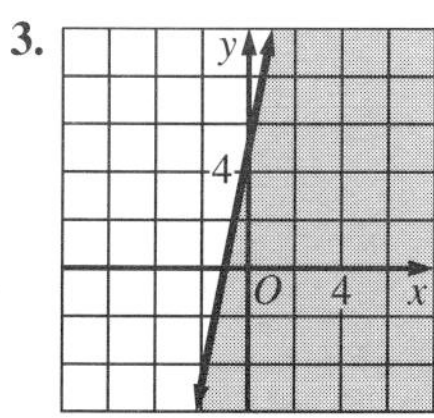

4.

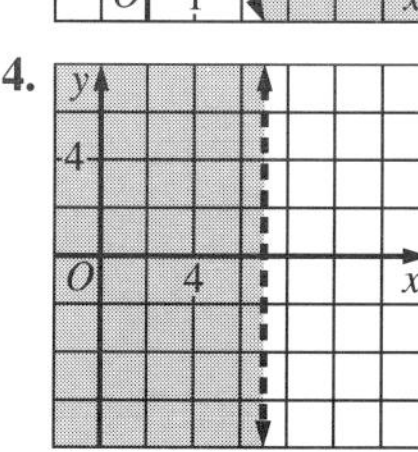

5.

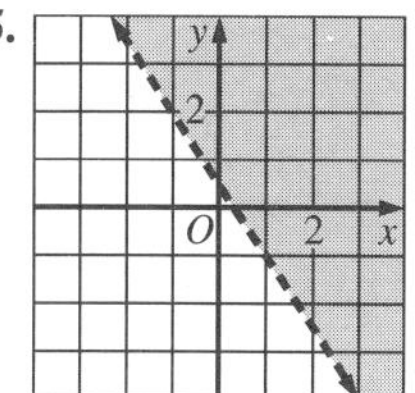

6.

7. Let n = the normal price and d = the discounted price; $d \geq 0.75n$.

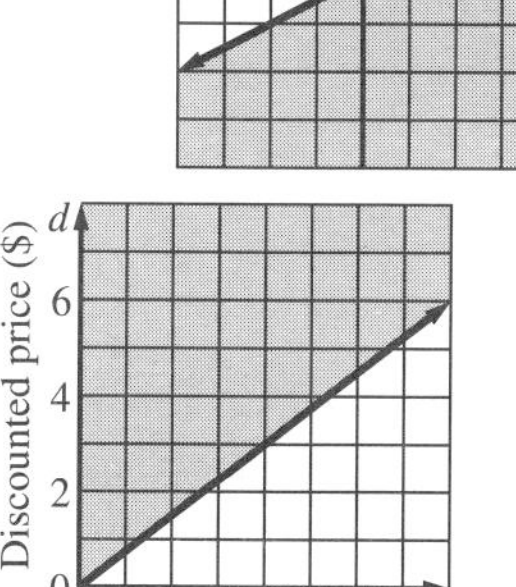

Page 312 Exercises and Applications

1. Let i = the number of infants and s = the number of supervisors; $s \geq \frac{1}{3}i$.

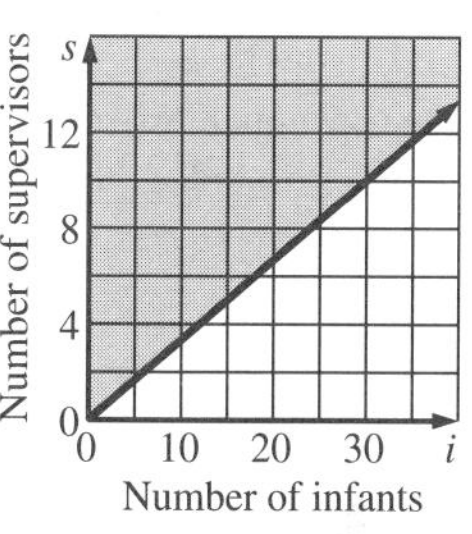

2.

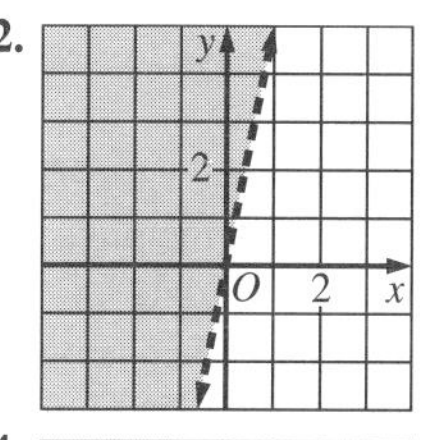

3.

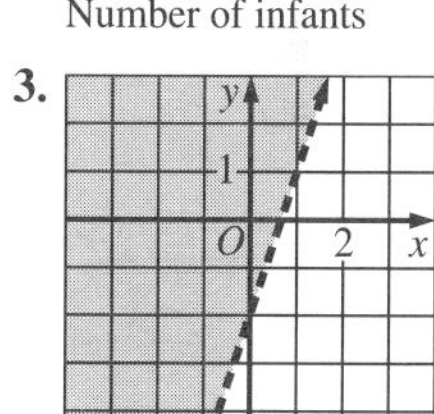

4.

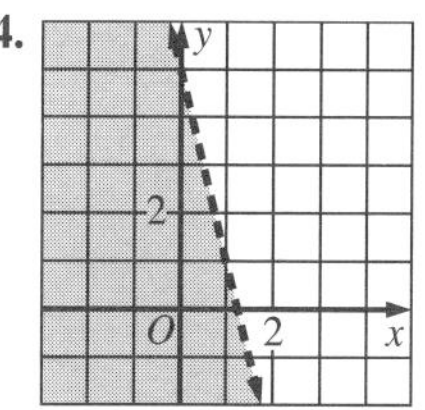

5.

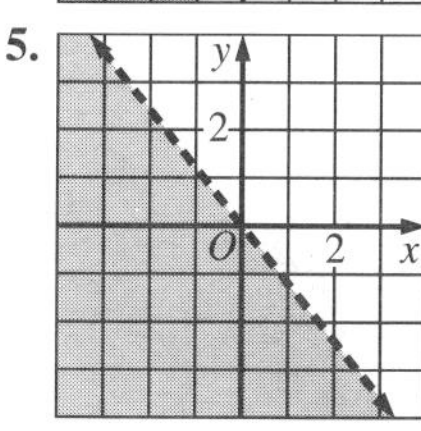

6.

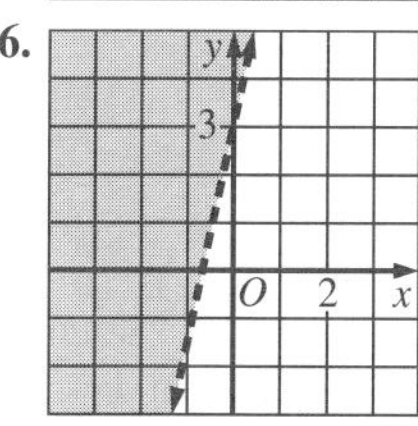

7.

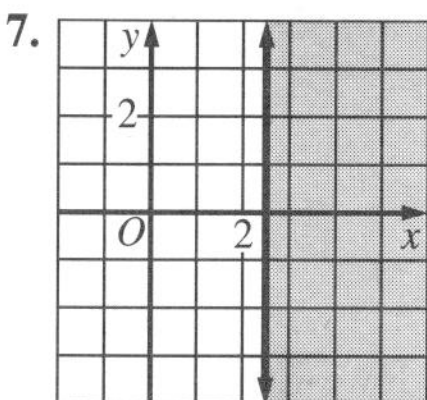

8.

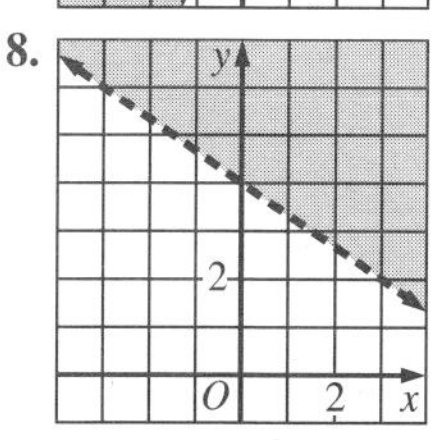

9.

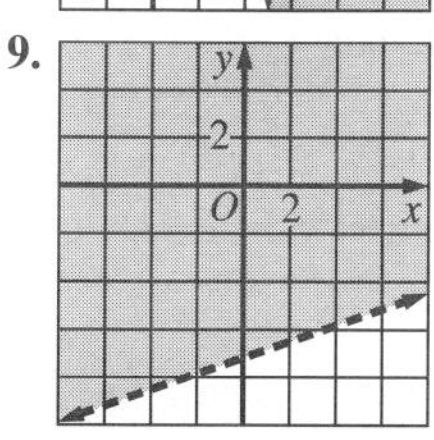

10.

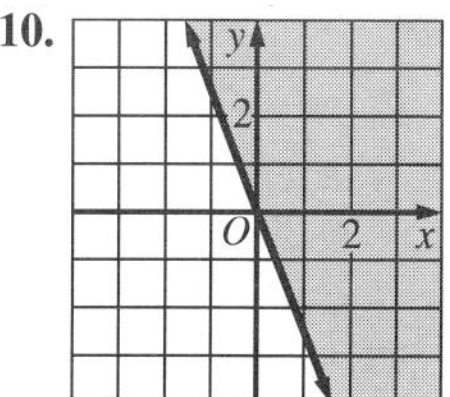

11.

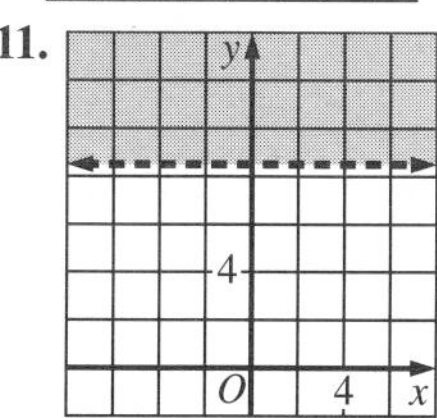

Page 313 Exercises and Applications

21. Salt water weighs more than fresh water.

22. a. Let d = the viewing distance in feet and w = the width in inches of a legible symbol; $w \geq 0.007d$.

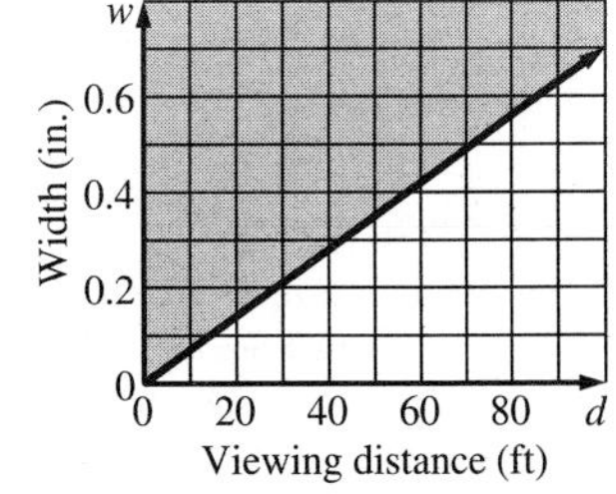

b. Let d = the viewing distance in feet and h = the height in inches of a legible symbol; $h \geq 0.03d$.

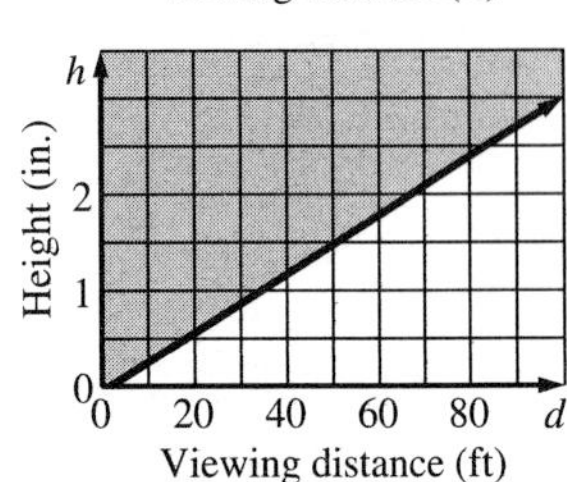

23. Let f = the number of female ostriches and e = the number of eggs; $e \leq 70f$.

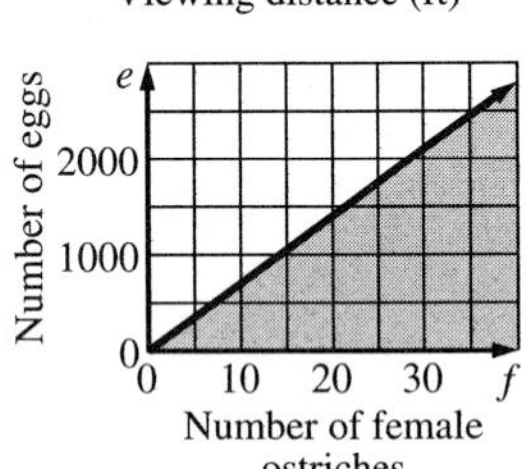

Page 314 Exercises and Applications

31. a.

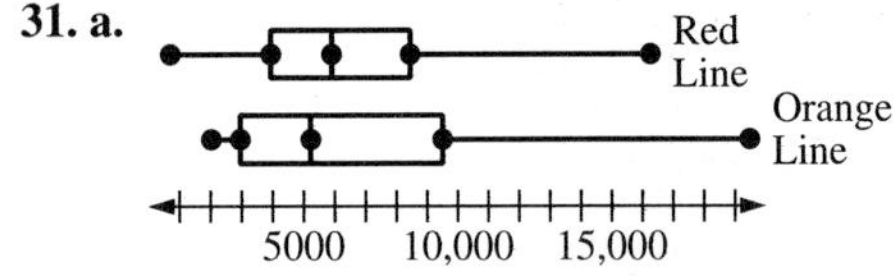

b. Summaries may vary. The medians for both the plots are nearly equal. The data are more closely grouped about the median for the Red Line than for the Orange Line.

Page 319 Exercises and Applications

14, 15. Lines of fit may vary. Least-squares lines are shown.

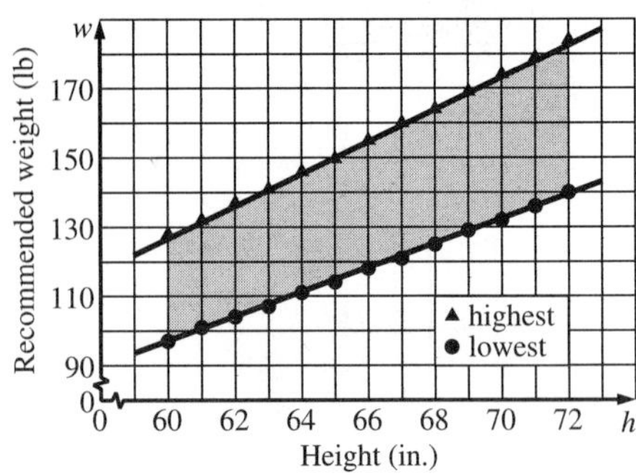

16. $w \geq 3.55h - 116$, $w \leq 4.66h - 153$, $60 \leq h \leq 72$

17. a. Lines of fit may vary. Least-squares lines are shown.

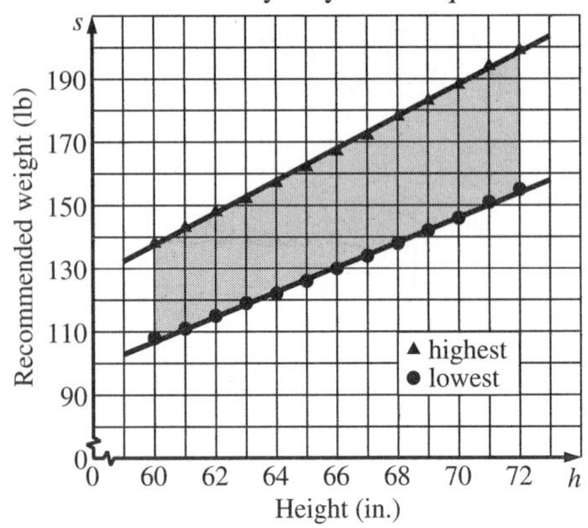

$w \geq 3.93h - 129$; $w \leq 5.09h - 168$, $60 \leq h \leq 72$

b. Summaries may vary. The regions have about the same shape, however both lowest and highest recommended weights are about 10 lb higher for the older group. **18.** Answers may vary. Check students' work.

Page 320 Exercises and Applications

19.

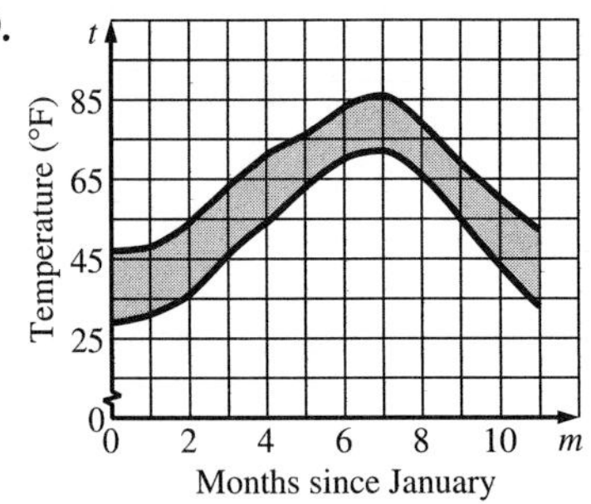

20. May through October; January through March, December

21. Answers may vary. Check students' work.

22.

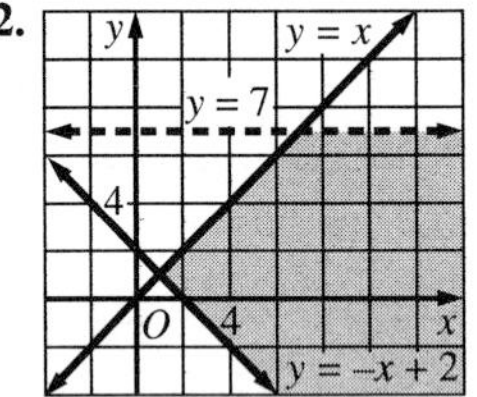

23.

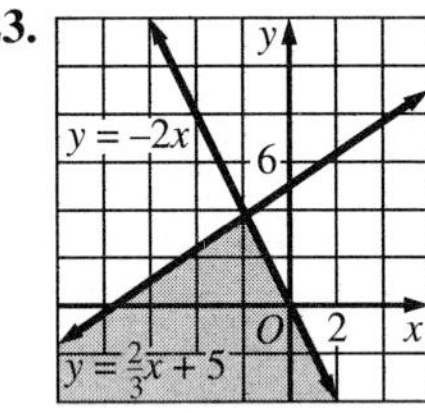

Page 321 Exercises and Applications

38.

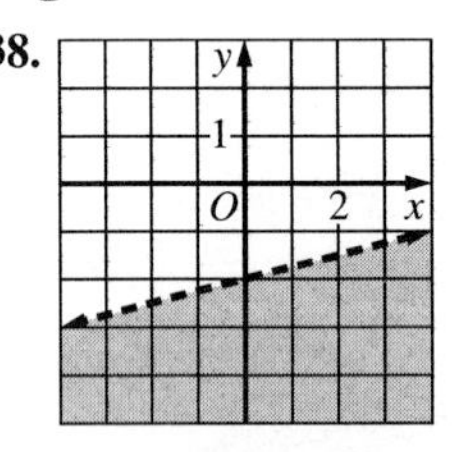

39.

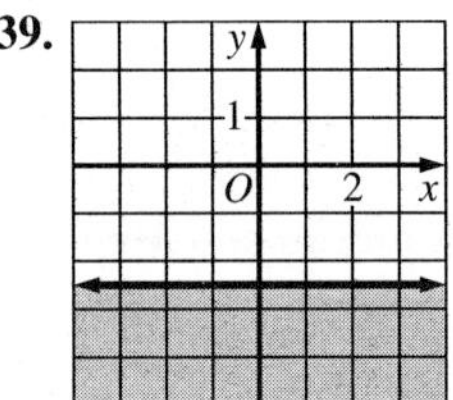

40.

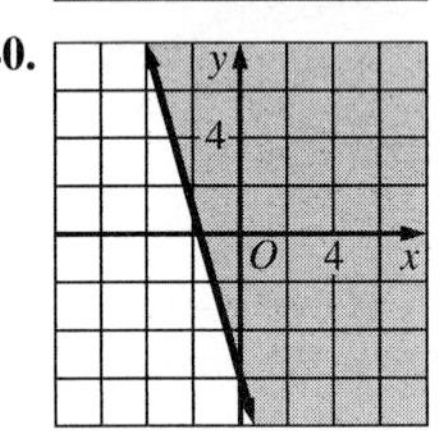

Page 325 Exercises and Applications

4.

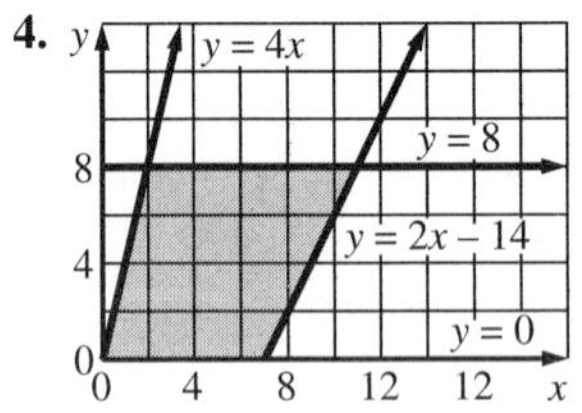

5. \$176 **6.** \$66 **7.** \$48 **8.** \$87 **9.** \$10.50 **10.** \$12 **11.** \$15

12. \$0; If x and y are both nonzero, the minimum cost is \$30.

13. a. $x \geq 4$, $y \geq 0$, $y \leq -\frac{5}{4}x + 10$

b.

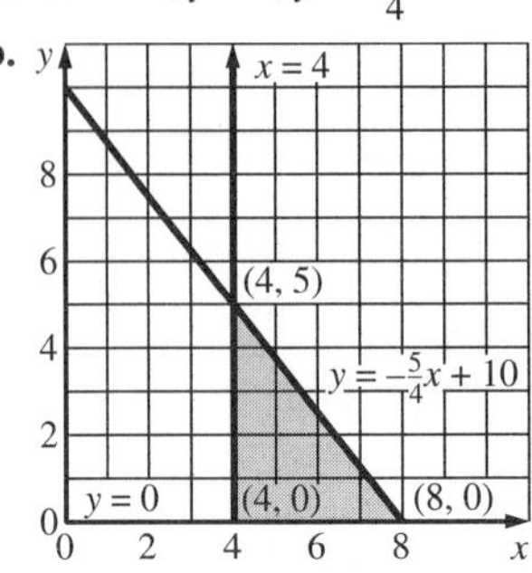

c. 4 bowls, 5 plates

14. a. Let x = the number of animals and y = the number of ovals; $y \leq 6x$, $y \geq 3x$, $y \leq -\frac{2}{3}x + 24$, $y \geq -\frac{2}{3}x + 20$

b.

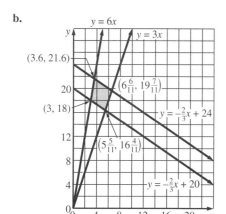

c. 3 animals, 18 ovals

Page 326 Exercises and Applications

15. a. Let a = the number of minutes spent doing aerobics and s = the number of minutes spent swimming; $a \geq 60$, $s \geq 60$, $s \geq -a + 180$, $s \leq -a + 300$, $s \geq -\frac{6}{11}a + \frac{1200}{11}$.

b.

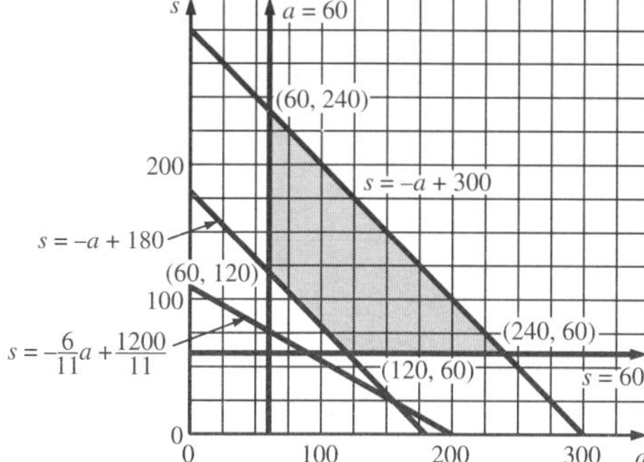

c. 60 min of aerobics, 120 min of swimming **d.** 60 min of aerobics, 240 min of swimming

16. a. Let a = the number of minutes spent doing aerobics and s = the number of minutes spent skiing; $a \geq 60$, $a \leq 180$, $s \geq 240$, $s \geq -a + 360$, $s \leq -a + 480$.

b.

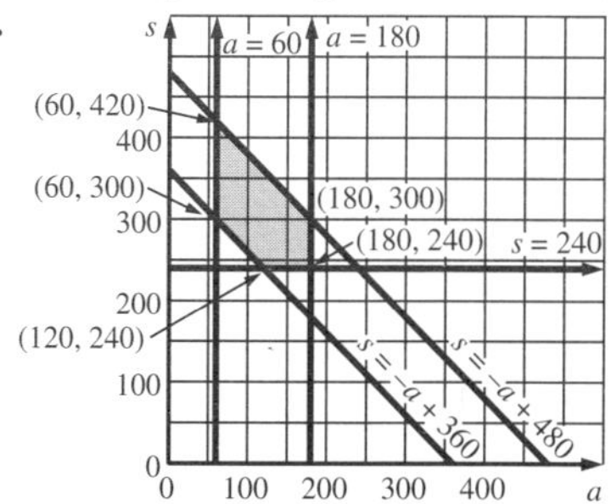

c. 60 min of aerobics, 420 min of skiing

17. Answers may vary. Check students' work. **18.** \$106 **19.** \$102 **20.** \$32 **21.** \$32 **22.** Answers may vary. An example is given. Linear programming involves using a systems of inequalities called *constraints* to describe a situation in which some quantity, such as profit or cost, must be minimized or maximized. The solution of the system is called the *feasible region* and, according to the corner-point principle, the best solution to the problem occurs at a corner-point of the feasible region.

CHAPTER 8

Page 343 Exercises and Applications

30.

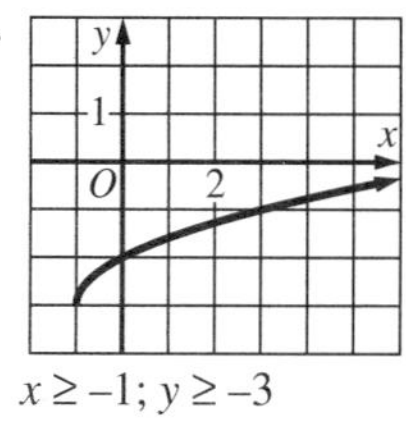

$x \geq -1$; $y \geq -3$

31.

$x \geq -3$; $y \geq -1$

32. sometimes **33.** never **34.** sometimes **35.** sometimes **36.** always **37.** sometimes **38.** always **39.** always **40.** No; $\sqrt{a+b} = \sqrt{a} + \sqrt{b}$ only if at least one of a and b is zero. **41. a.** Every graph of an equation of the form $y = a(x-h)^2 + k$ can be compared to the graph of $y = x^2$. If $0 < a < 1$, the graph of $y = x^2$ is shrunk vertically. If $a > 1$, the graph is stretched vertically. If $-1 < a < 0$, the graph is shrunk vertically and reflected over the x-axis. If $a < -1$, the graph is stretched vertically and reflected over the x-axis. The graph is translated left $|h|$ units if h is negative and right h units if h is positive. The graph is translated up k units if k if positive and down $|k|$ units if k is negative. **b.** Every graph of an equation of the form $y = a\sqrt{x-h} + k$ can be compared to the graph of $y = \sqrt{x}$. The values of a, h, and k have the effect described in part (a) above. **c.** Answers may vary. Divide the expression under the radical sign by the coefficient of x, then simplify the radical. For example, rewrite the equation $y = 3\sqrt{2x-4} + 1$ as $y = 3\sqrt{2(x-2)} + 1 = 3\sqrt{2}(\sqrt{x-2}) + 1$. **42. a.** Let s = the number of sweatshirts and t = the number of T-shirts; $t + s \geq 200$, $t + s \leq 300$, $t \geq 100$, $s \leq 150$.

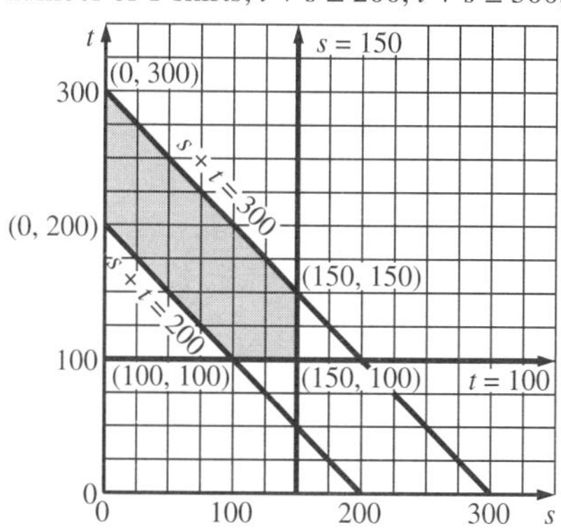

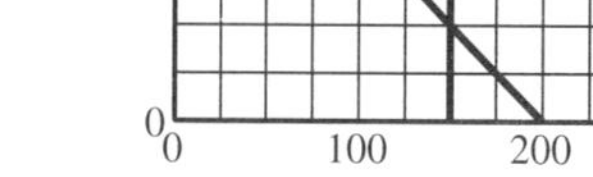

b. no sweatshirts, 200 T-shirts **c.** 150 sweatshirts, 150 T-shirts

43. 10 **44.** 2 **45.** 5 **46.** 7

47.

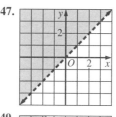

48.

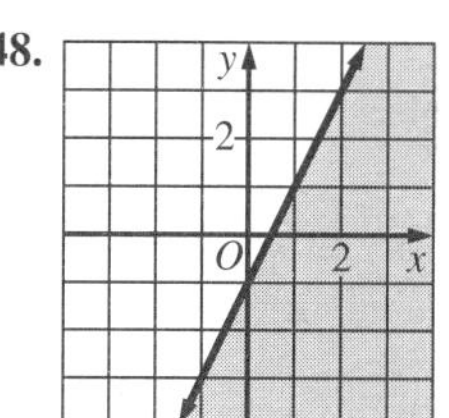

49.

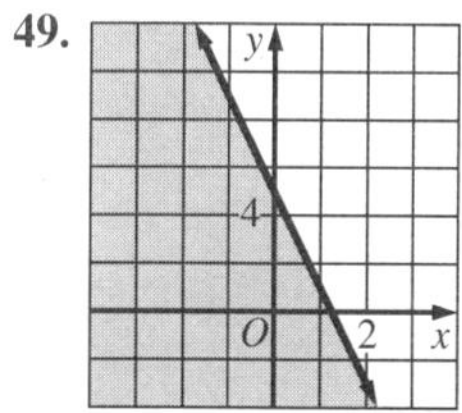

50.

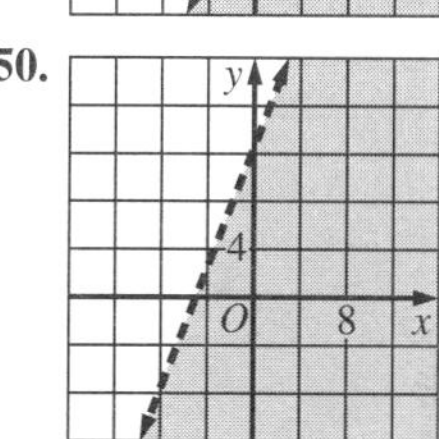

Page 345 Exploration

1.

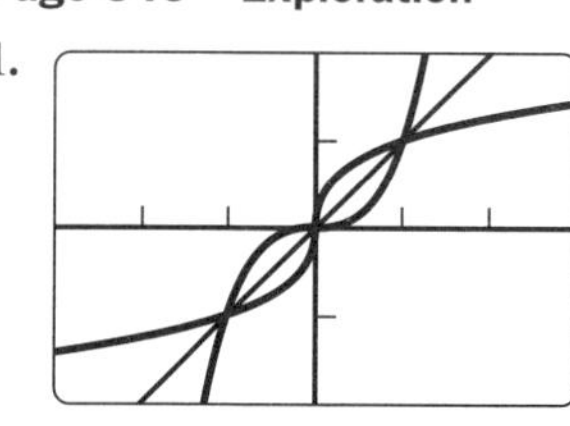

Yes.

2. $y = x^4$

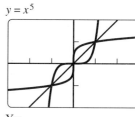

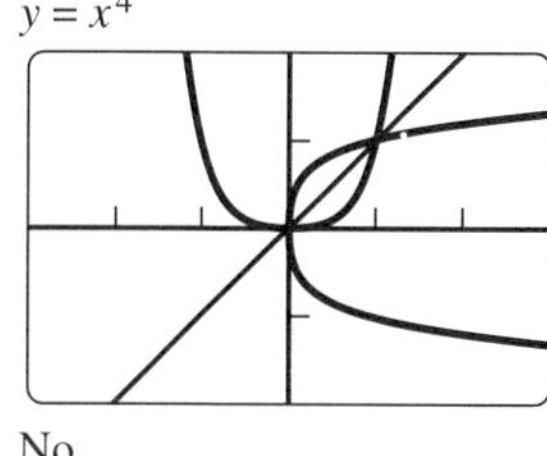

No.

$y = x^5$

Yes.

$y = x^6$

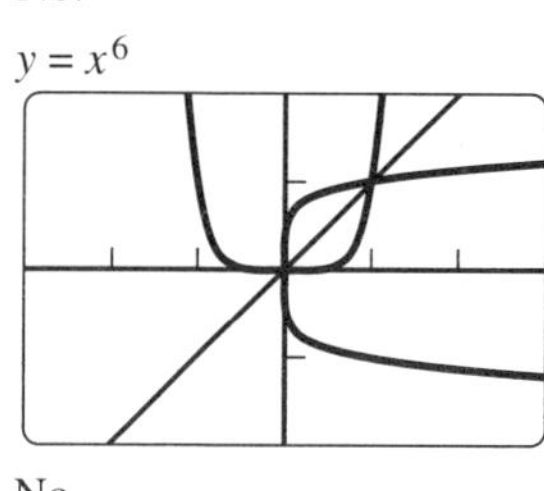

No.

Questions

1. Answers may vary. The graphs of $y = x^n$ for n odd have the shape of a sideways and backwards "S." No horizontal line intersects the graph in more than one point. The reflections of these graphs are graphs of functions. The graphs of $y = x^n$ for n even are shaped like parabolas. (For $n = 2$, the graph is a parabola.) Every horizontal line that intersects such a graph (except for the one intersecting the vertex) intersects the graph in two points. The reflections of these graphs are not the graphs of functions. **2.** n even; The domains may be restricted to $x \geq 0$ or $x \leq 0$.

Page 349 **Exercises and Applications**

38. a. $M = 70W^{3/4}$

b, d.

Animal	Weight (kg)	Metabolic rate (Cal)
Mouse	0.019	3.6
Cat	3	160
Rabbit	3.81	191
Dog	14	507
Goat	25.7	800
Chimpanzee	38	1070
Sheep	42.2	1160
Pony	253	4440
Bull	963	12,100
Elephant	3672	33,020

c. $W = \left(\frac{M}{70}\right)^{4/3}$ **d.** See table.

Page 367 **Checking Key Concepts**

5.

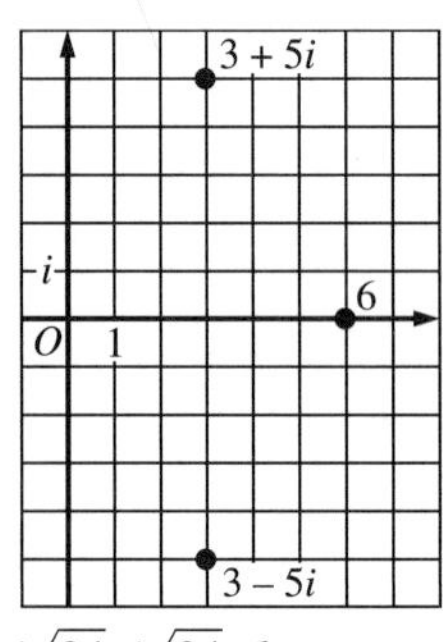

$\sqrt{34}$; $\sqrt{34}$; 6

6.

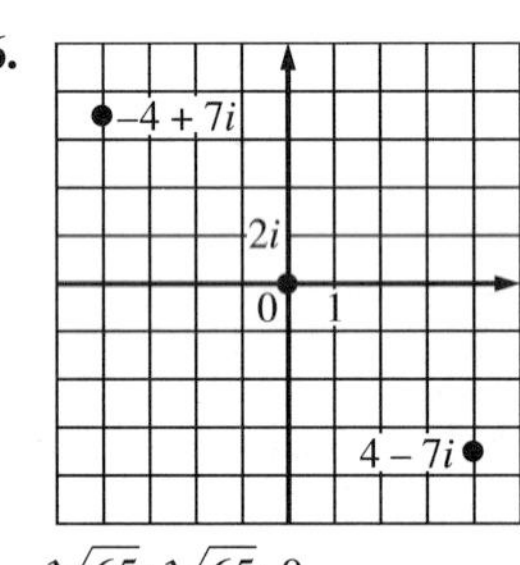

$\sqrt{65}$; $\sqrt{65}$; 0

7.

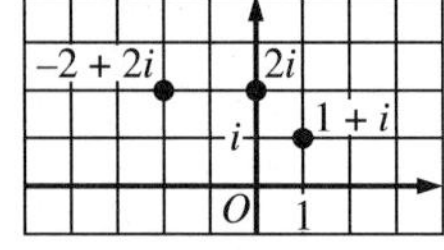

$\sqrt{2}$; 2; $2\sqrt{2}$

8.

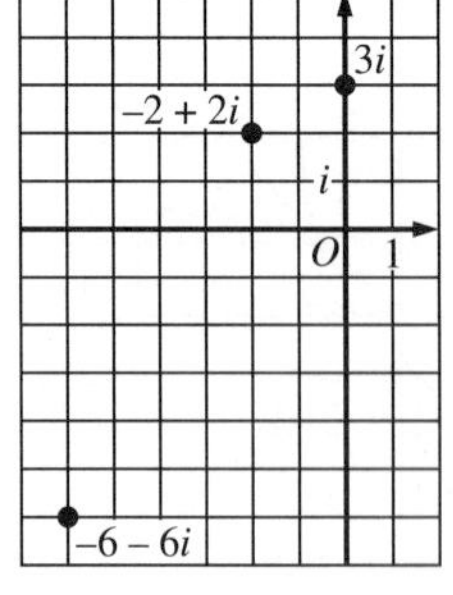

3; $2\sqrt{2}$; $6\sqrt{2}$

9.

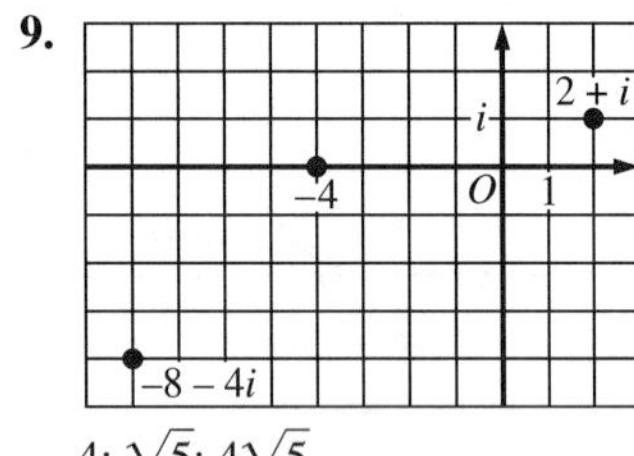

4; $\sqrt{5}$; $4\sqrt{5}$

10.

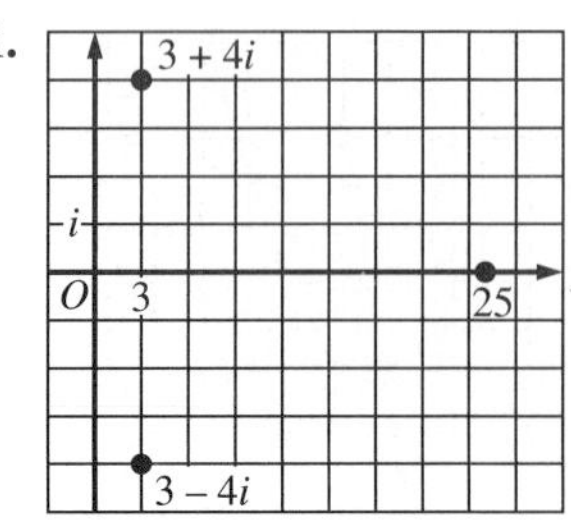

3; 2; 6

11.

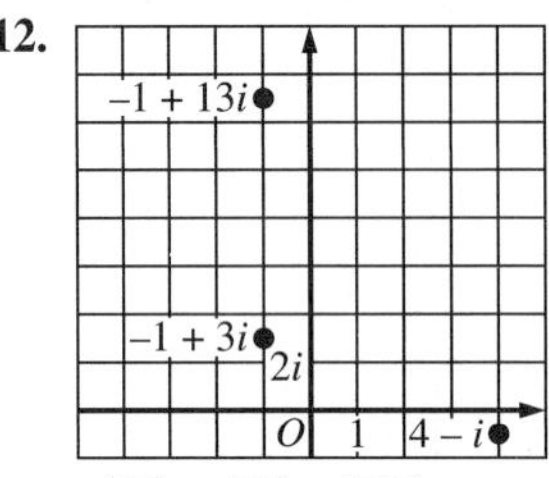

5; 5; 25

12.

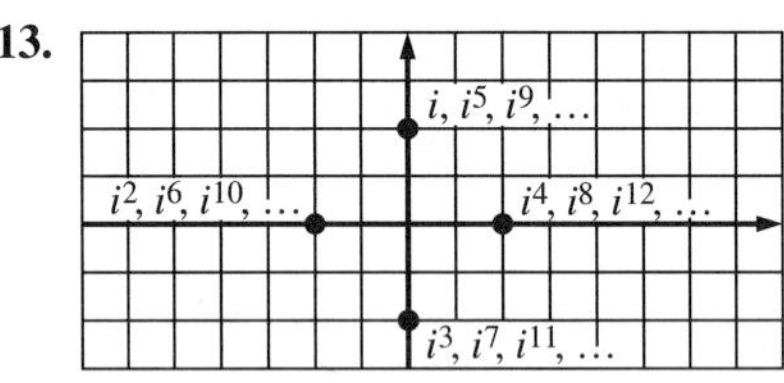

$\sqrt{10}$; $\sqrt{17}$; $\sqrt{170}$

13.

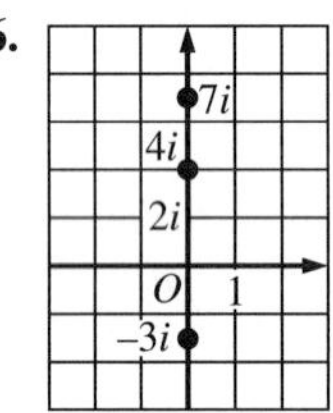

Answers may vary. The sequence with $t_n = i^n$ is the repeating sequence $i, -1, -i, 1, i, -1, -i, 1, \ldots$.

Page 368 **Exercises and Applications**

6.

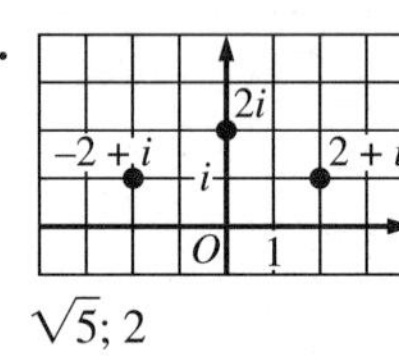

7; 3; 4

7.

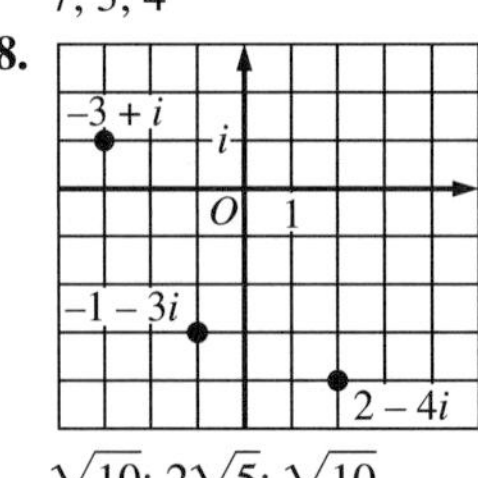

$\sqrt{5}$; 2

8.

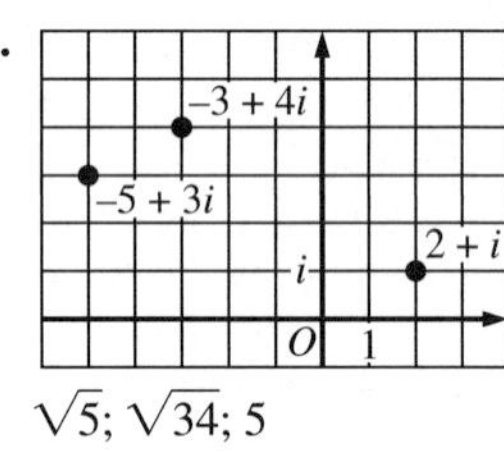

$\sqrt{10}$; $2\sqrt{5}$; $\sqrt{10}$

9.

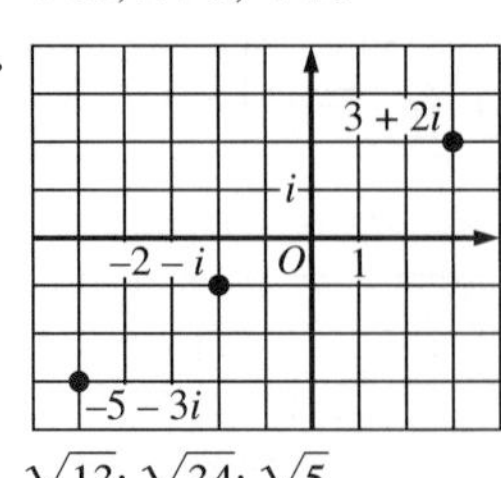

$\sqrt{5}$; $\sqrt{34}$; 5

10.

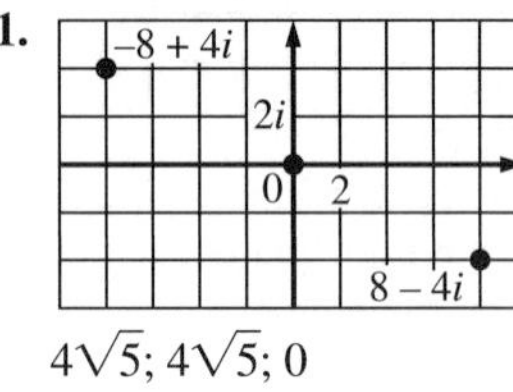

$\sqrt{13}$; $\sqrt{34}$; $\sqrt{5}$

11.

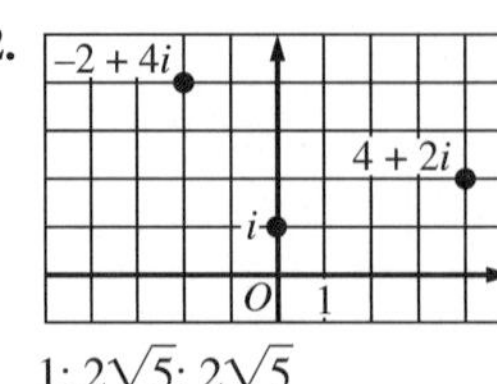

$4\sqrt{5}$; $4\sqrt{5}$; 0

12.

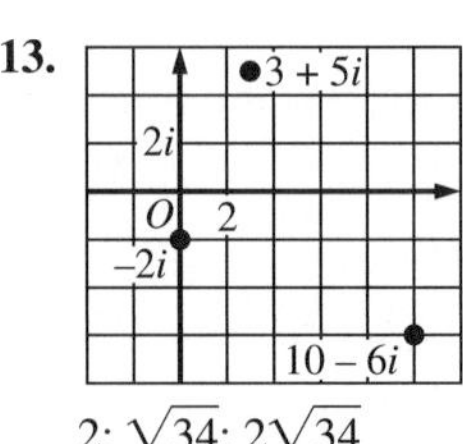

1; $2\sqrt{5}$; $2\sqrt{5}$

13.

3 + 5i, 2i, O, 2, –2i, 10 – 6i

2; $\sqrt{34}$; $2\sqrt{34}$

14.

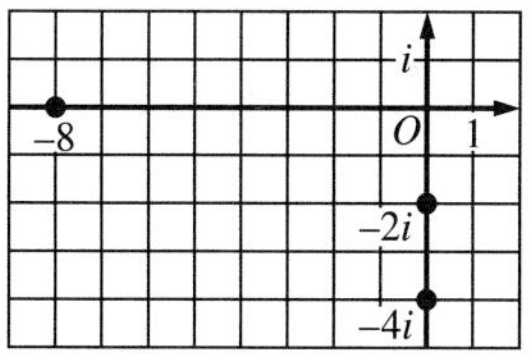

4; 2; 8

15.

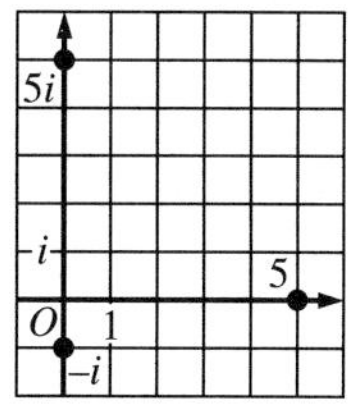

5; 1; 5

16.

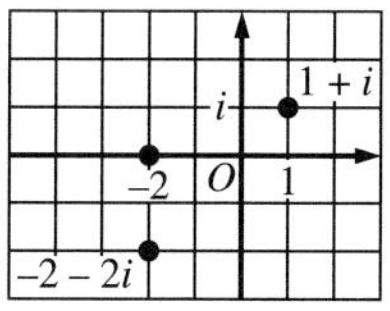

2; $\sqrt{2}$; $2\sqrt{2}$

17.

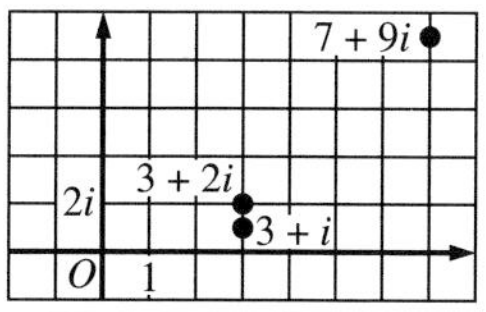

$\sqrt{10}$; $\sqrt{13}$; $\sqrt{130}$

18.

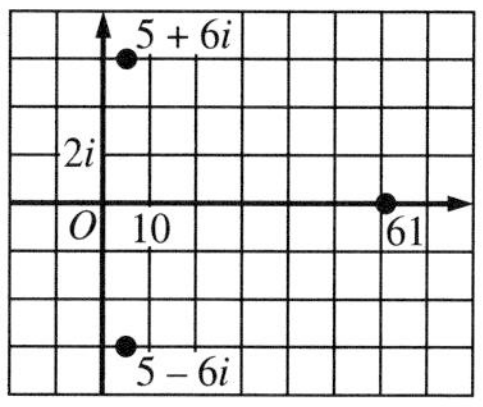

$\sqrt{61}$; $\sqrt{61}$; 61

19.

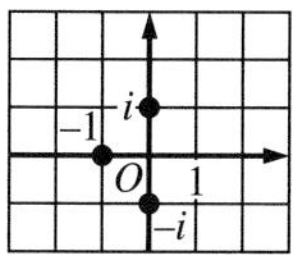

1; 1; 1

20. $-4 - i$ **21.** $5 - 2i$ **22.** $-3 + 6i$ **23.** $-5 - 6i$ **24.** $1 + i$ **25.** $4 - 7i$

26.

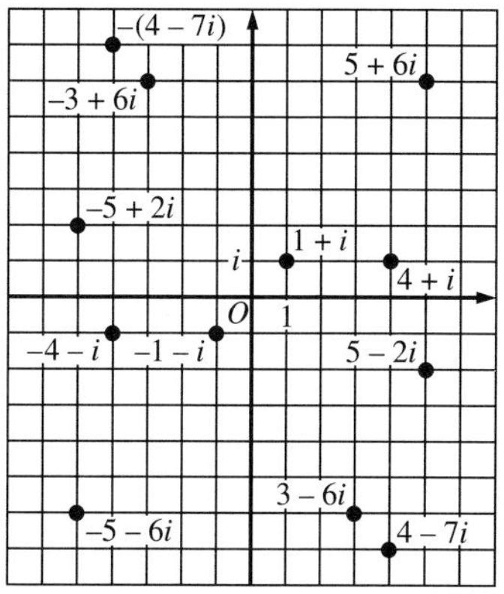

Answers may vary. An example is given. A complex number and its opposite are images of each other rotated 180° about the origin.

27. $-a - bi$ **28. a.** The square of a complex number and the square of its opposite are equal. **b.** The magnitude of the square of a complex number is the square of the magnitude of the complex number. **c.** They are opposites.

29.

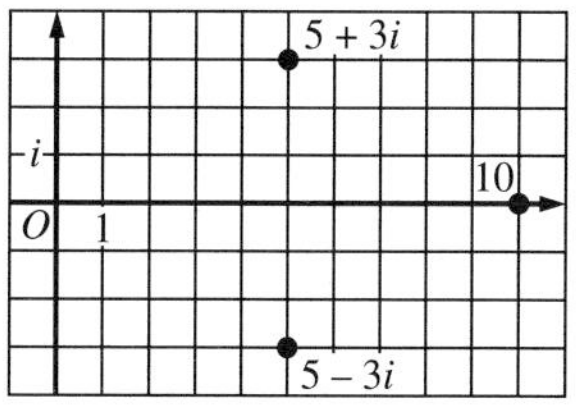

30.

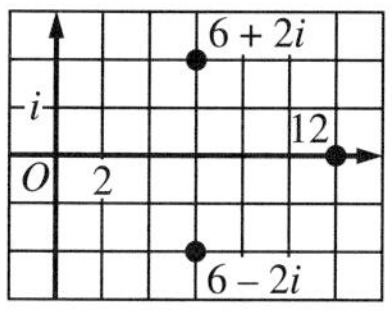

31.

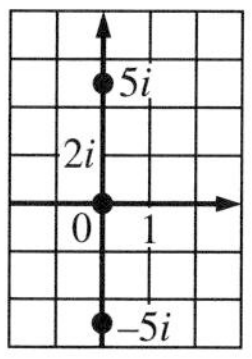

32.

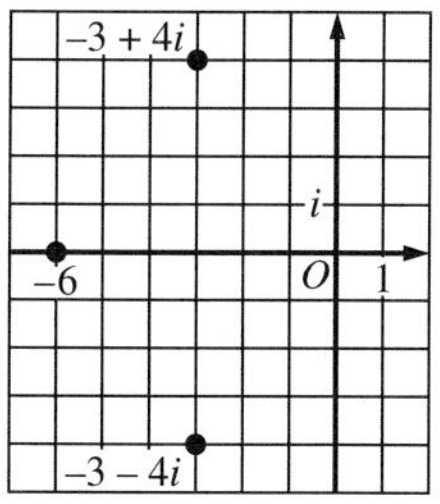

33.

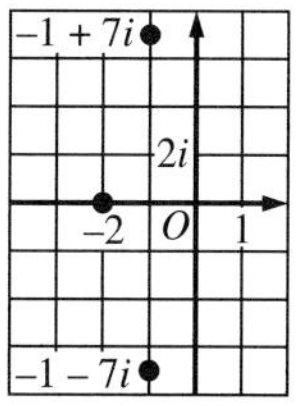

34.

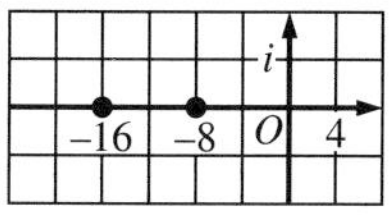

35. Answers may vary. An example is given. A complex number and its complex conjugate are images of each other reflected over the real axis.

36. $(a + bi)^2 = a^2 + 2abi + b^2i^2 = (a^2 - b^2) + 2abi$; $(a - bi)^2 = a^2 - 2abi + b^2i^2 = (a^2 - b^2) - 2abi$; The graph shows $2 + i$, its complex conjugate $2 - i$, and their squares.

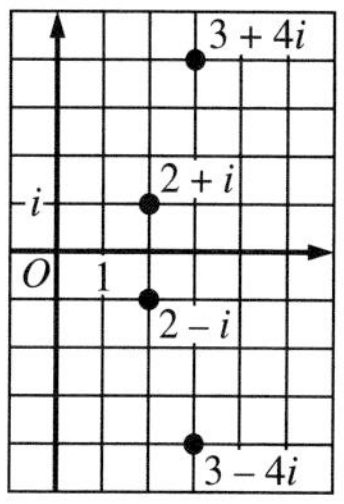

Page 375 Exercises and Applications

13. No; do not form a group; the set is not closed under multiplication. **14.** Yes; form a commutative group; the set is closed under addition and all the group properties as well as the commutative property hold. **15.** Yes; form a commutative group; the set is closed under multiplication and all the group properties as well as the commutative property hold. **16.** Yes; form a commutative group; the set is closed under addition and all the group properties as well as the commutative property hold. **17.** Yes; do not form a group; there is no identity and the inverse property does not hold. **18.** No; do not form a group; the set is not closed under addition. **19.** No; do not form a group; the set is not closed under multiplication.

Page 376 Exercises and Applications

23. a. Thursday; 4

b.

+	0	1	2	3	4	5	6
0	0	1	2	3	4	5	6
1	1	2	3	4	5	6	0
2	2	3	4	5	6	0	1
3	3	4	5	6	0	1	2
4	4	5	6	0	1	2	3
5	5	6	0	1	2	3	4
6	6	0	1	2	3	4	5

closure, identity, inverse, associative, commutative

Page 377 Exercises and Applications

34, 35.

	⊕	0	1	2	3	4	5	6	7
ABCDEFGH	0	0	1	2	3	4	5	6	7
BADCFEHG	1	1	2	3	4	5	6	7	0
GDAFCHEB	2	2	3	4	5	6	7	0	1
DGFAHCBE	3	3	4	5	6	7	0	1	2
EFGHABCD	4	4	5	6	7	0	1	2	3
FEHGBADC	5	5	6	7	0	1	2	3	4
CHEBGDAF	6	6	7	0	1	2	3	4	5
HCBEDGFA	7	7	0	1	2	3	4	5	6

Page 378 Exercises and Applications

48.

∘	I	T	R	F	TR	TF	RF	TRF
I	I	T	R	F	TR	TF	RF	TRF
T	T	I	TR	TF	R	F	TRF	RF
R	R	TR	I	RF	T	TRF	F	TF
F	F	TF	RF	I	TRF	T	R	TR
TR	TR	R	T	TRF	I	RF	TF	F
TF	TF	F	TRF	T	RF	I	TR	R
RF	RF	TRF	F	R	TF	TR	I	T
TRF	TRF	RF	TF	TR	F	R	T	I

Page 385 **Chapter 8 Assessment**

42. a, b.

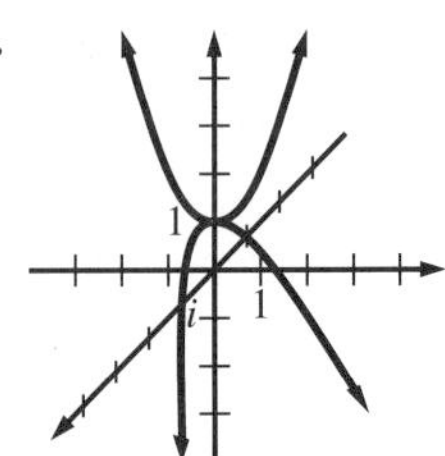

c. First determine which part of the graph to use. For equations where the right side is greater than or equal to 1, use the graph with domain the real numbers. For equations where the right side is less than 1, use the graph with domain the pure imaginary numbers.

CHAPTER 9

Page 404 **Exploration**

Steps 1 and 2: Coodinates that involve approximations may vary due to rounding. Examples are given. $y = 4x^2 - 8x - 3$: As x takes on large negative or positive values, the graph rises; one turning point, (1, –7). $y = x^3 - x^2 - 3x + 1$: As x takes on large negative values, the graph falls; as x takes on large positive values, the graph rises; two turning points, (–0.72, 2.27), (1.39, –2.42). $y = x^4 + 2x^3 - 5x^2 - 7x + 3$: As x takes on large negative or positive values, the graph rises; three turning points, (–2.26, –3.72), (–0.58, 5.10), (1.34, –7.32). $y = 2x^5 + 6x^4 - 2x^3 - 14x^2 + 5$: As x takes on large negative values, the graph falls; as x takes on large positive values, the graph rises; four turning points, (–2, –3), (–1.40, –4.66), (0, 5), (1, –3). $y = 3x^6 - 13x^4 + 15x^2 + x - 17$: As x takes on large negative or positive values, the graph rises; five turning points, (–1.46, 16.3), (–0.865, –12.66), (–0.0334, –17.02), (0.919, –10.88), (1.44, –13.60). $y = x^7 - 8x^5 + 18x^3 - 6x$: As x takes on large negative values, the graph falls; as x takes on large positive values, the graph rises. six turning points, (–1.92, –3.33), (–1.38, –8.52), (–0.349, 1.37), (0.349, –1.37), (1.38, 8.52), (1.92, 3.33).

Questions: 1. They rise; they rise. **2.** They rise; they fall. **3.** Let n be the degree of the polynomial function and t the number of turning points. **a.** $t = n - 1$ **b.** $t \leq n - 1$

Page 408 **Checking Key Concepts**

Answers that involve approximations may vary due to rounding. Examples are given. **1.** A **2.** D **3.** B **4.** C

5.

local minimums: –4.06 and –6.64; local maximum: 3.03

6.

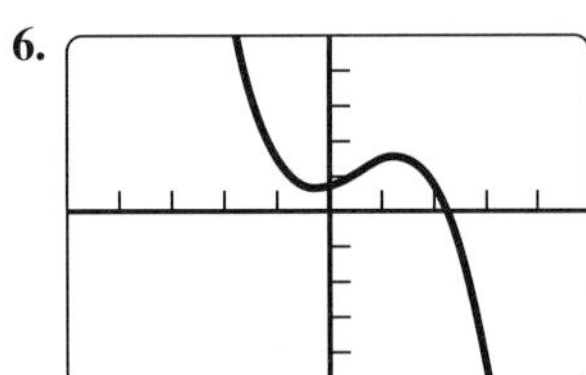

local minimum: 2.72; local maximum: 6.28

7.

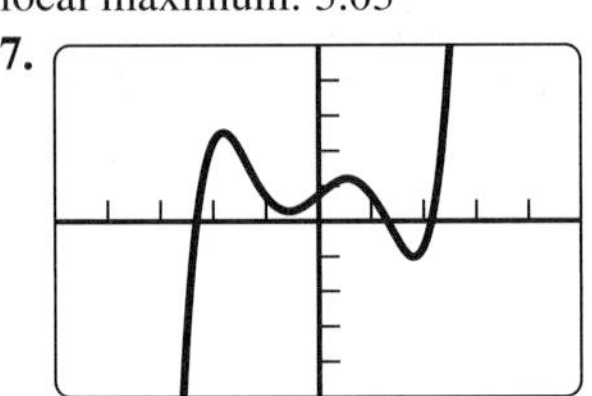

local minimums: 1.20 and –4.10; local maximums: 10.1 and 4.80

8.

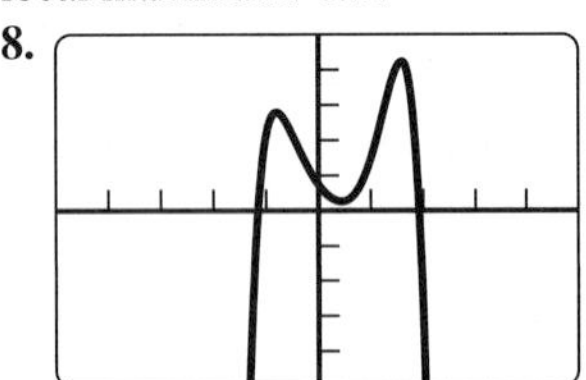

local minimum: 1.12; local maximums: 11.19 and 17.03

Page 409 **Exercises and Applications**

7.

a. As $x \to +\infty$, $g(x) \to +\infty$ and as $x \to -\infty$, $g(x) \to -\infty$.
b. local maximums: 2.57 and 0; local minimums: –3.44 and –13.44

8.

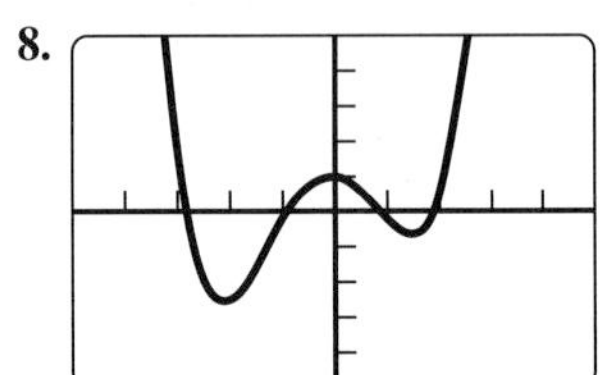

a. As $x \to +\infty$, $g(x) \to +\infty$ and as $x \to -\infty$, $g(x) \to +\infty$.
b. local minimums: –10.18 and –2.60; local maximum: 4.04

9.

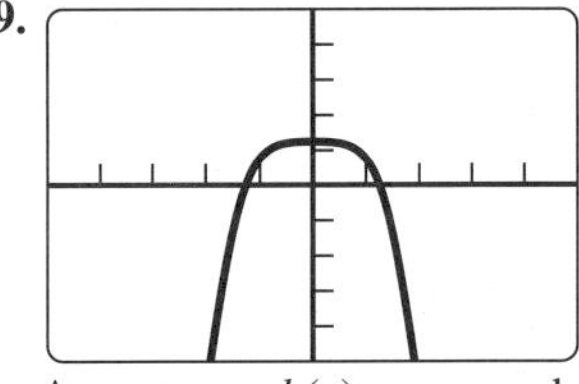

a. As $x \to +\infty$, $h(x) \to -\infty$ and as $x \to -\infty$, $h(x) \to -\infty$.
b. local maximum: 5

10.

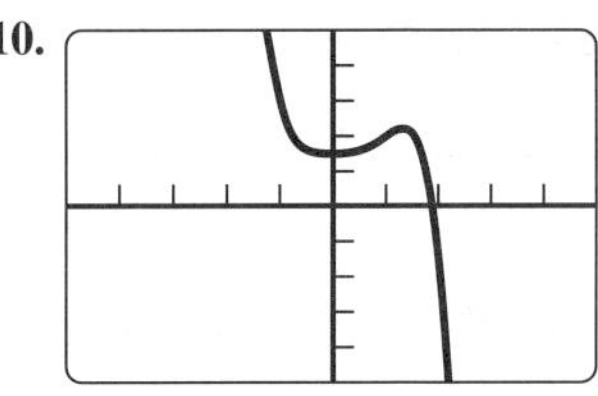

a. As $x \to +\infty$, $h(x) \to -\infty$ and as $x \to -\infty$, $h(x) \to +\infty$.
b. local minimum: 6; local maximum: 8.84

11.

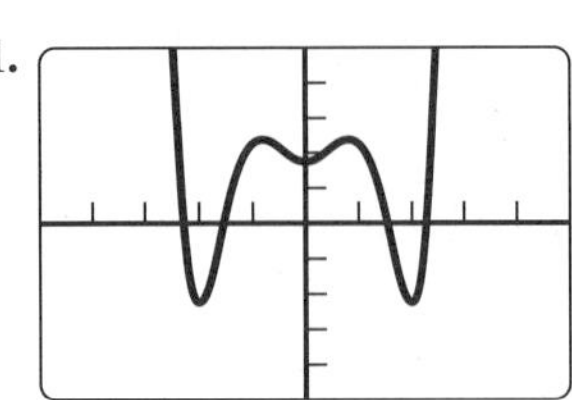

a. As $x \to +\infty$, $f(x) \to +\infty$ and as $x \to -\infty$, $f(x) \to +\infty$.
b. local minimums: –9 and 7; local maximum: 9.52

12.

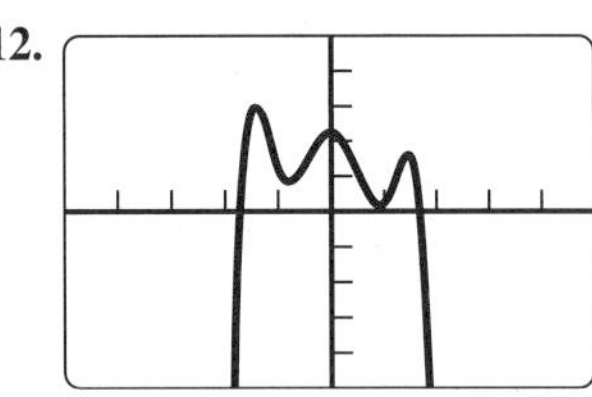

a. As $x \to +\infty$, $f(x) \to -\infty$ and as $x \to -\infty$, $f(x) \to -\infty$.
b. local minimums: 3.40 and 0.70; local maximums: 11.87, 9, and 6.40

13. a. Answers may vary. Examples are given. local minimum at (0.5, 1.8); local maximums at (0.3, 1.8) and (0.75, 2) **b.** local minimum at (0.5, 1.76); local maximums at (0.34, 1.85) and (0.76, 2); 0.76 s; 2 m/s

14.

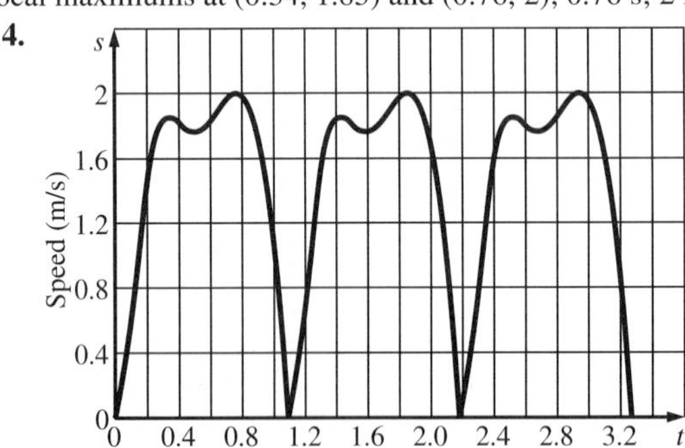

15. Answers may vary. An example is given. I think a reasonable average is about 1.3 m/s. I calculated this by determining the swimmer's speed at the middle of each 0.2 s interval and finding the average.

Page 417 **Exercises and Applications**

11.

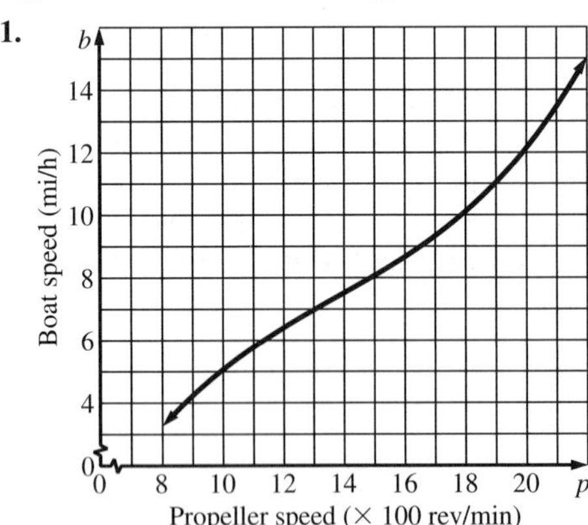

This is the original curve translated 3 units down.

12. Answers may vary. An example is given. Determine the speeds related to given propeller speeds by timing the boat over a fixed distance and dividing distance by time. Graph the boat speed versus propeller speed.

Page 419 **Exploration**

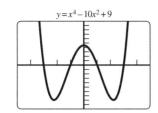

1.

$y = x^3 - x^2 - 7x - 5$

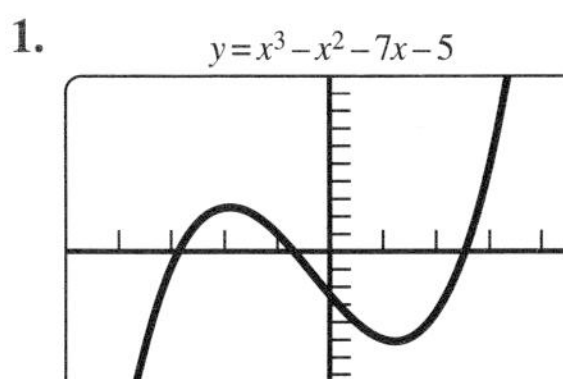

3 real zeros

$y = x^4 - 10x^2 + 9$

4 real zeros

$y = x^5 - 7x^3 - x^2 + 8x$

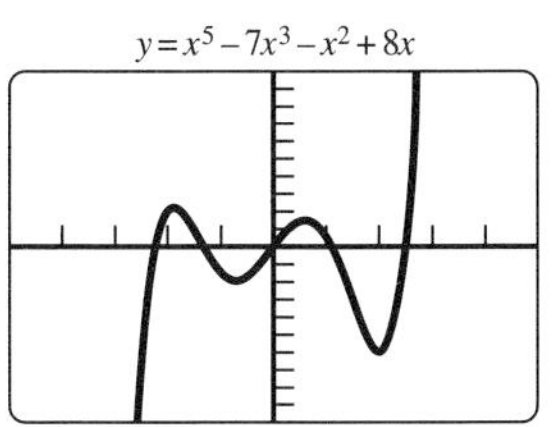

5 real zeros

$y = x^6 - 11x^4 + 29x^2 - x - 8$

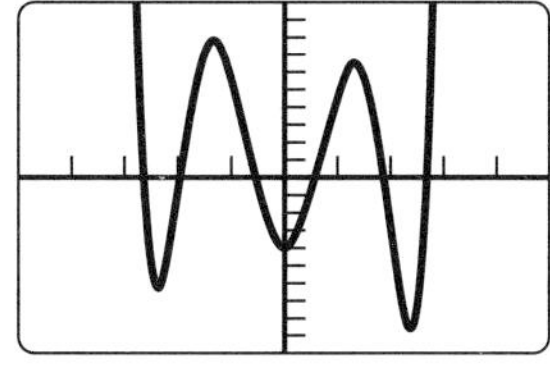

6 real zeros

2. The number of real zeros is the same as the degree.

3.

$y = x^3 + x^2 + 3x + 5$

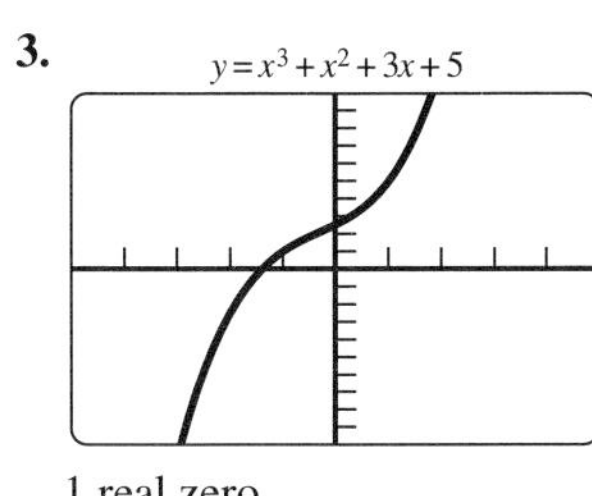

1 real zero

$y = x^4 + 2x^3 - x^2 + 6x + 7$

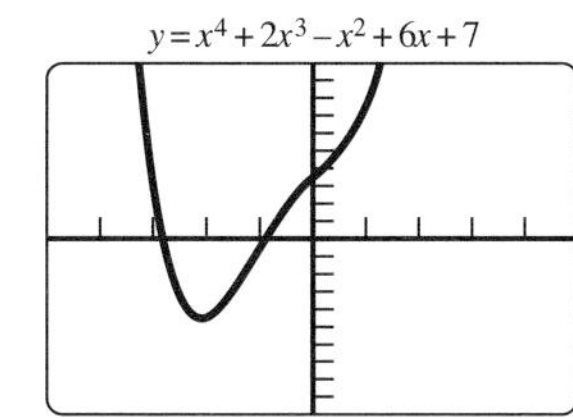

2 real zeros

$y = x^5 - x^4 - 6x^3 + 4x^2 + 8x$

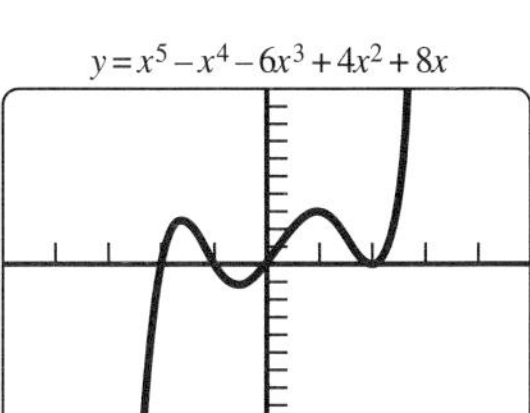

4 real zeros

$y = x^6 - 3x^4 - 7x^2 + 10$

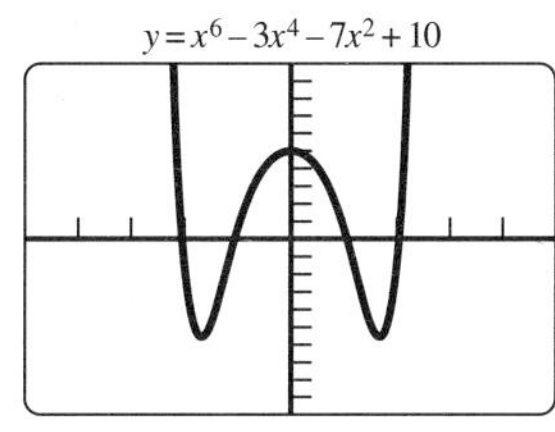

4 real zeros

No; the number of real zeros is less than or equal to the degree.

4. The number of real zeros is less than or equal to the degree.

Page 425 **Assess Your Progress**

1. about 9.98 m/s **2.** 1; –2; 3 **3.** –4; $2 \pm \sqrt{2}$ **4.** $\frac{5}{3}; \frac{-3 \pm i\sqrt{7}}{4}$ **5.** $-\frac{1}{2}$; 3; $\pm 2i$

6. $x - k$ is a factor of $f(x)$ if and only if k is a zero of f. Summaries may vary. The rational zero theorem allows you to determine all possible rational zeros. By testing the possibilities, you may identify a rational zero, k, of the function, divide the polynomial function by $x - k$, and continue testing. Once you have determined all the rational zeros, you may be able to use your knowledge of quadratic equations to determine other zeros. Real zeros may also be determined graphically.

Page 430 **Exercises and Applications**

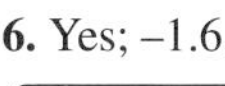

6. Yes; –1.6.

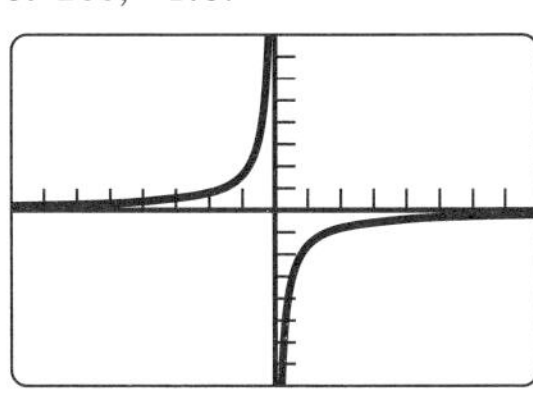

7. Yes; 5.

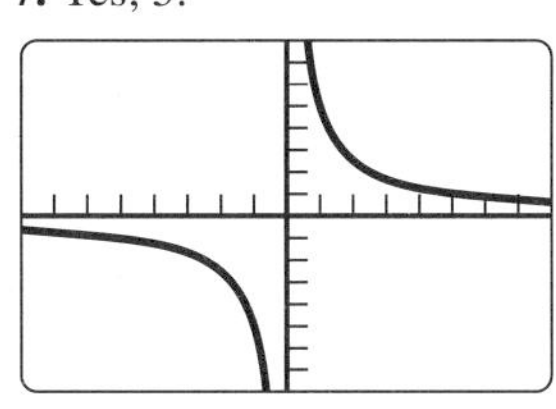

8. No. **9.** Yes; let l = length and w = width; $lw = 72$, so $l = \frac{72}{w}$ or $w = \frac{72}{l}$.

10. Yes; let n = the number of students, s = each student's share of the rent, and let R represent the total rent (a constant); $ns = R$, so $s = \frac{R}{n}$. **11.** No; let n = the number of gallons of gasoline used, d = the distance driven in miles, and let M represent the car's gasoline usage in mi/gal (a constant); $n = \frac{d}{M}$; this is a direct variation.

12. No; let r = the radius and h = the height; $V = 1 = \pi r^2 h$, so $r^2 = \frac{1}{\pi h}$.

13. a. The product of the mean distance from the sun and the apparent size of the sun is about 92.9 for all the planets in the table. **b.** $s = \frac{92.9}{d}$ **c.** 0.361

14. Check students' work. The diameter of the sun as seen from Mercury in the drawing should be about 100 times that of Pluto. For example, if the sun as seen from Pluto is depicted with a diameter of $\frac{1}{16}$ in., the image of the sun as seen from Mercury should be depicted with a diameter of about $6\frac{1}{4}$ in.

Page 433 **Exploration**

1.

$y = \frac{1}{x-1}$

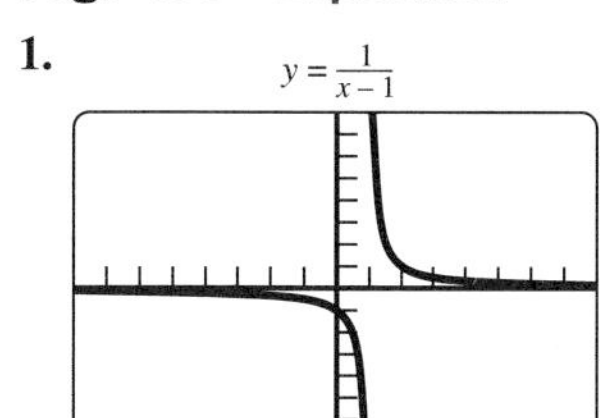

$x = 1$; $y = 0$; translated 1 unit right

$y = \frac{1}{x-3}$

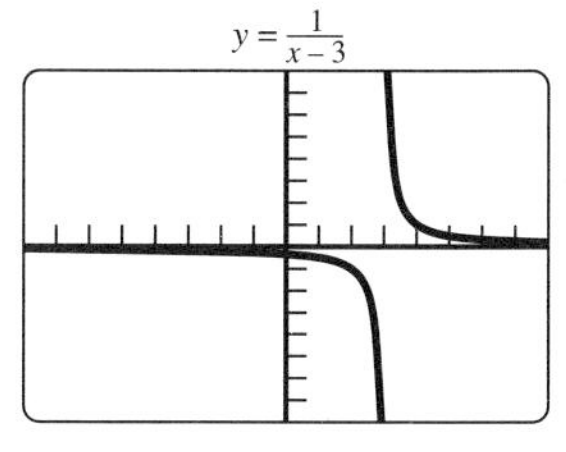

$x = 3$; $y = 0$; translated 3 units right

$y = \frac{1}{x+2}$

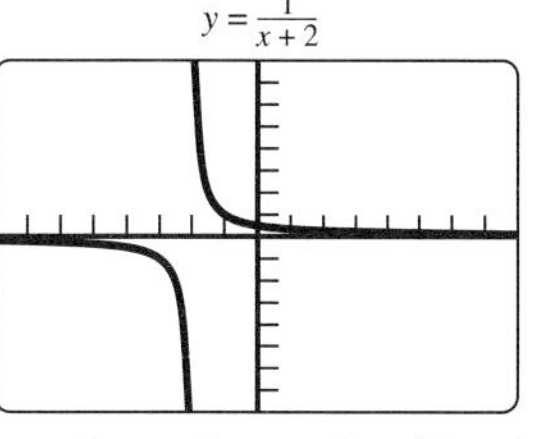

$x = -2$; $y = 0$; translated 2 units left

$y = \frac{1}{x+4}$

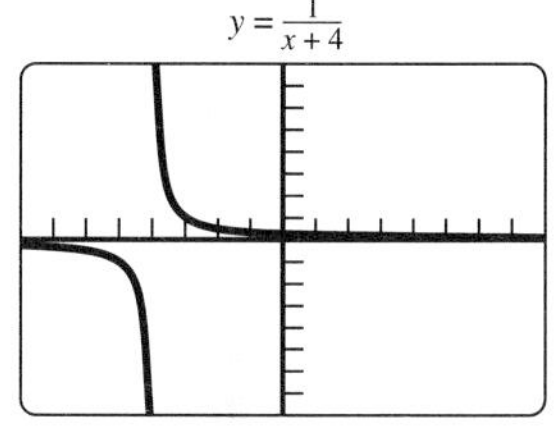

$x = -4$; $y = 0$; translated 4 units left

2.

$y = \frac{1}{x} + 1$

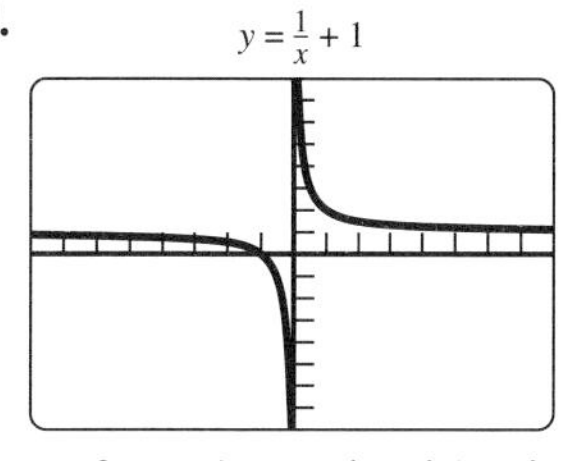

$x = 0$; $y = 1$; translated 1 unit up

$y = \frac{1}{x} + 3$

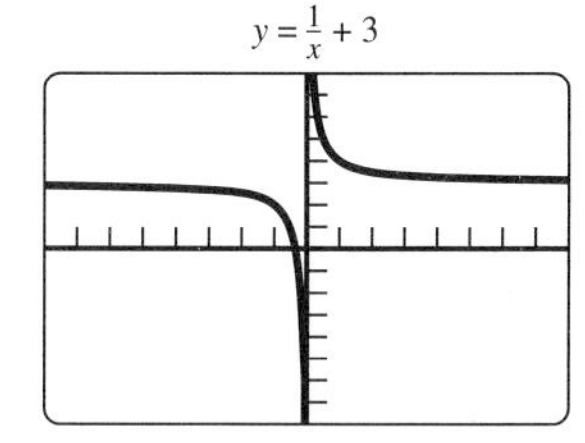

$x = 0$; $y = 3$; translated 3 units up

$y = \frac{1}{x} - 2$

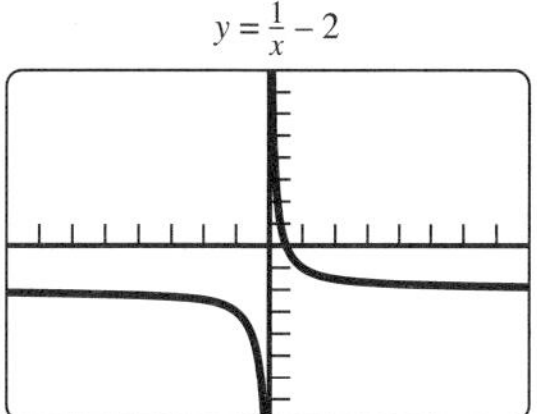

$x = 0$; $y = -2$; translated 2 units down

$y = \frac{1}{x} - 4$

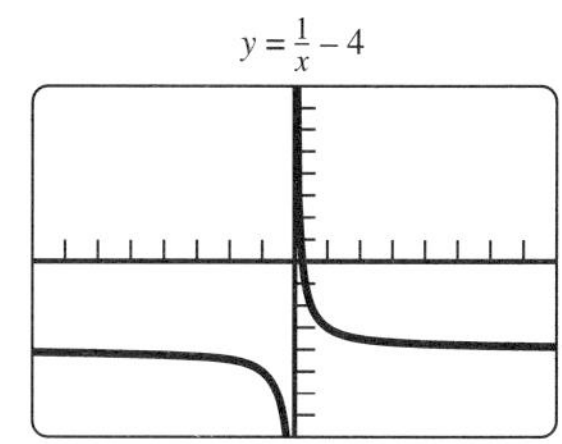

$x = 0$; $y = -4$; translated 4 units down

3. translated 4 units right; translated 1 unit left; translated 3 units up; translated 5 units down

Page 441 **Exploration**

1.

$y = \frac{1}{(x-1)(x-3)}$

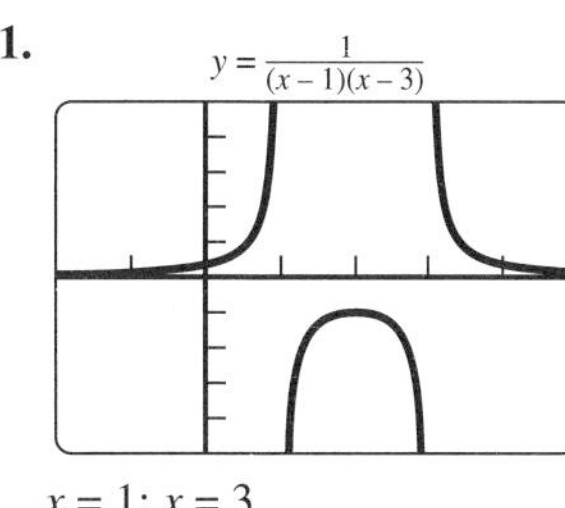

$x = 1$; $x = 3$

$y = \frac{x-1}{(x-2)(x-5)}$

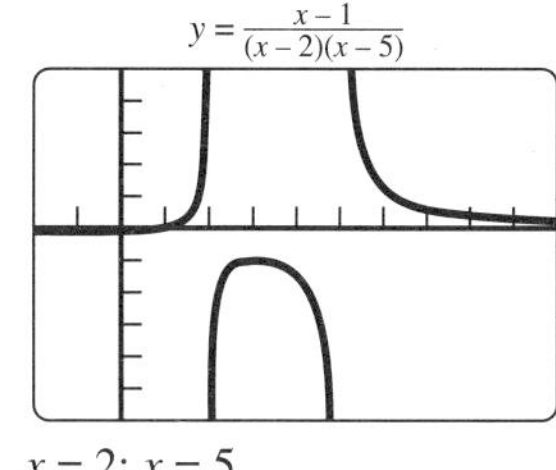

$x = 2$; $x = 5$

$y = \frac{2x^2 - 3x - 7}{(x+1)(x-4)}$

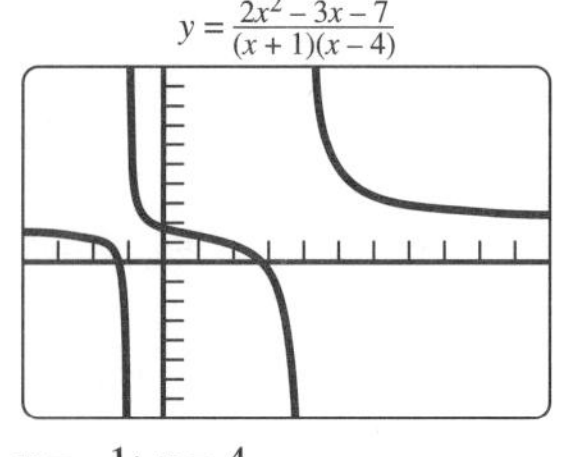

$x = -1; x = 4$

$y = \frac{4}{x(x+3)(x-3)}$

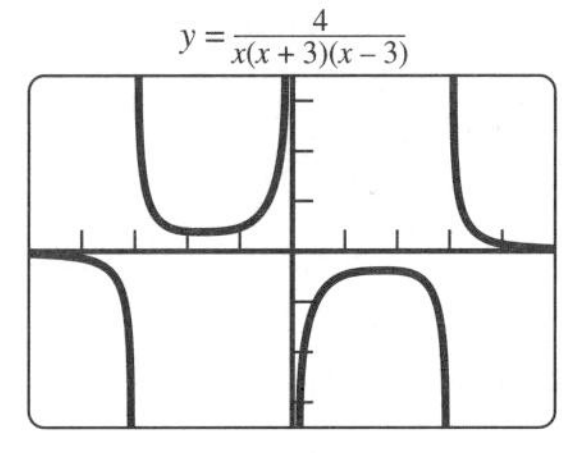

$x = -3; x = 0; x = 3$

$y = \frac{-x^2 - 6x + 2}{(x+3)(x+1)(x-4)}$

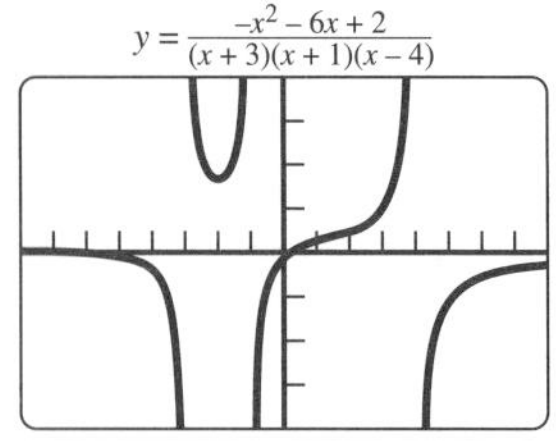

$x = -3; x = -1; x = 4$

$y = \frac{x^4 + x^3 - 5x^2}{(x+1)^2(x-2)}$

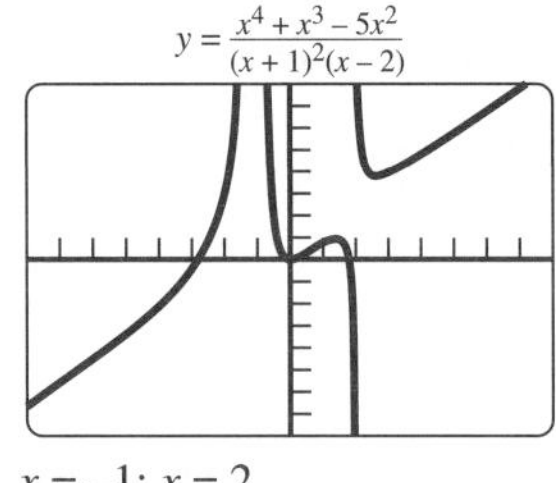

$x = -1; x = 2$

2. $y = \frac{-3x^2}{(x+3)(x-2)}$: $x = -3; x = 2$

$y = \frac{7x+2}{(x+5)(x+1)(x-3)}$: $x = -5; x = -1; x = 3$

$y = \frac{x^3}{x^2 - 16}$: $x = \pm 4$

3. The functions in Steps 1 and 2 are rational functions of the form $f(x) = \frac{p(x)}{q(x)}$. For each of the given functions, if k is a zero of $q(x)$, then $x = k$ is a vertical asymptote of the graph of f.

CHAPTER 10

Page 469 Exercises and Applications

40. $p = 5$: 0, 1, 4, 4, 1, 0, 1, 4, 4, 1; $p = 11$: 0, 1, 4, 9, 5, 3, 3, 5, 9, 4, 1, 0, 1, 4, 9, 5, 3, 3, 5, 9, 4, 1; $p = 13$: 0, 1, 4, 9, 3, 12, 10, 10, 12, 3, 9, 4, 1, 0, 1, 4, 9, 3, 12, 10, 10, 12, 3, 9, 4, 1 **41.** Answers may vary. The terms of the sequence $t_n = (n^2)_{\text{mod } p}$ repeat in blocks of p terms. In each block, if the first term (0) is omitted, the remaining $p - 1$ terms form a mirror sequence, that is the second $\frac{p-1}{2}$ terms are the first $\frac{p-1}{2}$ terms in reverse order.

42. 0, 1, 4, 2, 2, 4, 1, …

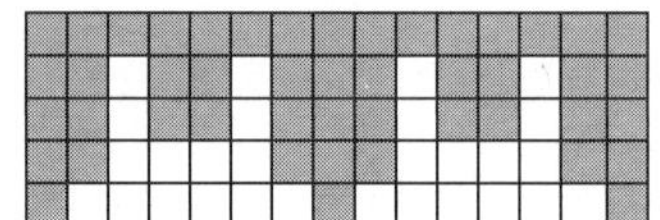

Page 472 Exploration

	No. of Pieces	
No. of Cuts	**A**	**B**
1	2	2
2	4	3
3	8	4
4	16	5
5	32	6

1. Method A: the number doubles; Method B: the number increases by 1. **2.** See table. **3.** Method A: $t_n = 2 \cdot 2^{n-1}$; Method B: $t_n = 2 + (n - 1)$

Page 473 Think and Communicate

1. The graph of the arithmetic sequence $t_n = t_1 + d(n-1) = dn + (t_1 - d)$ consists of those points on the graph of the linear function $f(x) = dx + (t_1 - d)$ for which x is a positive integer; the common difference d; the nth term t_n; the position number n. This is clear from comparing the two equations. **2.** The graph of the geometric sequence $t_n = t_1 \cdot r^{n-1} = \frac{t_1}{r} r^n$ consists of those points on the graph of the exponential function $f(x) = \frac{t_1}{r} r^x$ for which x is a positive integer; the common ratio r; the nth term t_n; the position number n. This is clear from comparing the two equations.

3. The terms of the sequence are decreasing; its graph falls from left to right. **4.** The terms of the sequence are decreasing; its graph falls from left to right.

Page 481 Think and Communicate

2. The formulas have the same form. The general explicit formula gives a value for t_n in terms of t_1 and the common ratio r. The general recursive formula indicates the value of t_1 and gives the value for t_n in terms of t_{n-1} and the common ratio r.

CHAPTER 11

Page 513 Exploration

1. Check students' work. 5 cm; $d = \sqrt{3^2 + 4^2} = 5$ cm **2.** Answers may vary. An example is given. Let the two given points be the endpoints of the hypotenuse of a right triangle. Draw a horizontal line segment and a vertical line segment to form a right triangle. Then the distance, d, between the points can be computed as follows: $d = \sqrt{(\text{change in } x\text{-coordinates})^2 + (\text{change in } y\text{-coordinates})^2}$. Given the points (−2, 1) and (3, 13), use the point (3, 1) to form a right triangle with legs of length 5 and 12. Then $d = \sqrt{5^2 + 12^2} = 13$. **3.** about $\left(2\frac{1}{2}, 4\right)$ **4.** $2\frac{1}{2}$; 4; The mean of the x-coordinates = the x-coordinate of the midpoint, and the mean of the y-coordinates = the y-coordinate of the midpoint. **5.** Answers may vary. An example is given. The x-coordinate of the midpoint is the mean of the x-coordinates of the given points. The y-coordinate of the midpoint is the mean of the y-coordinates of the given points. Given the points (3, 8) and (−1, 6), the coordinates of the midpoint of the line segment joining the points are $\left(\frac{3 + (-1)}{2}, \frac{8+6}{2}\right) = (1, 7)$.

Page 518 Exercises and Applications

26. Answers may vary. An example is given. The slope of $\overline{PQ} = -\frac{7}{3}$, and the slope of $\overline{QR} = \frac{3}{7}$, so $\overline{PQ} \perp \overline{QR}$. $\triangle PQR$ is a right triangle with hypotenuse $\overline{PR}$, and with legs $\overline{PQ}$ and $\overline{QR}$. **27.** Answers may vary. An example is given. Draw a rectangle around the extreme north, south, east, and west borders of the county. Draw a vertical line halfway between the western and eastern edges of the rectangle, and draw a horizontal line halfway between the northern and southern edges of the rectangle. The point at which the lines intersect will approximate the center of the county. It would make sense to locate the county seat at the center of the county because this location would make the county seat a reasonable distance from all parts of the county.

Page 519 Exercises and Applications

32. $PM = \sqrt{\left(\frac{x_1 + x_2}{2} - x_1\right)^2 + \left(\frac{y_1 + y_2}{2} - y_1\right)^2} = \sqrt{\left(\frac{x_2 - x_1}{2}\right)^2 + \left(\frac{y_2 - y_1}{2}\right)^2} = \frac{1}{2}\sqrt{(x_2 - x_1)^2 + (y_2 - y_1)^2}$; $QM = \sqrt{\left(\frac{x_1 + x_2}{2} - x_2\right)^2 + \left(\frac{y_1 + y_2}{2} - y_2\right)^2} = \sqrt{\left(\frac{x_1 - x_2}{2}\right)^2 + \left(\frac{y_1 - y_2}{2}\right)^2} = \frac{1}{2}\sqrt{(x_1 - x_2)^2 + (y_1 - y_2)^2}$; Since $(x_2 - x_1)^2 = (x_1 - x_2)^2$ and $(y_2 - y_1)^2 = (y_1 - y_2)^2$, $QM = \frac{1}{2}\sqrt{(x_1 - x_2)^2 + (y_1 - y_2)^2} = \frac{1}{2}\sqrt{(x_2 - x_1)^2 + (y_2 - y_1)^2}$. Thus, $PM = QM$. **33.** Answers may vary.

34. $\sum_{k=1}^{\infty} 512\left(-\frac{1}{4}\right)^{k-1} = 409.6$ **35.** $\sum_{n=1}^{7} 6(3)^{n-1} = 6558$

36. a.

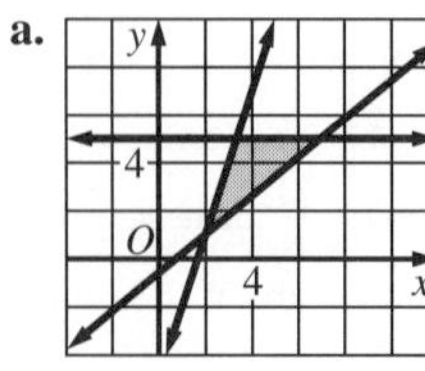

b. 46 **c.** 4

37. a. 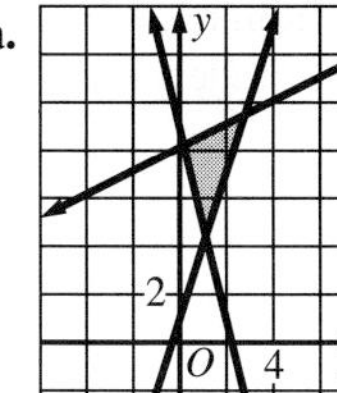 **b.** 55.40 **c.** $\frac{97}{7} \approx 13.86$

38. parabola that opens up and has vertex (5, 0)

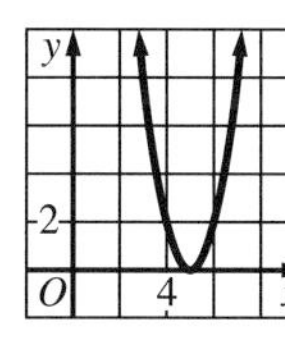

Page 525 Exercises and Applications

10. a. $x + 3 = -\frac{1}{32}(y - 3)^2$

b.

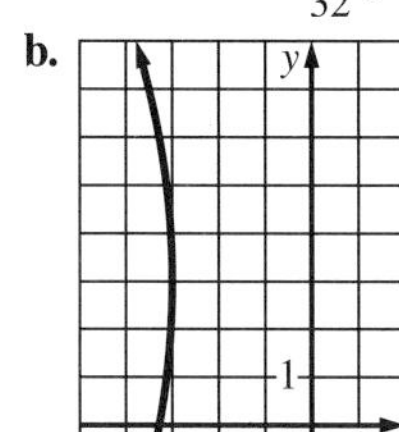

11. about 1.8 m

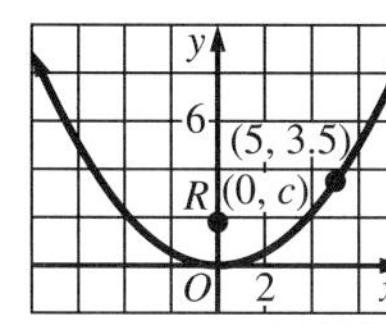

12. about 34.8 ft

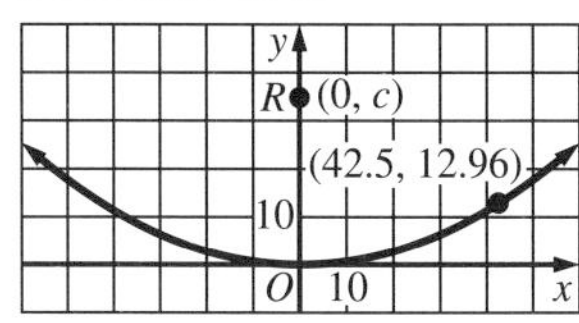

13. about 27.8 m

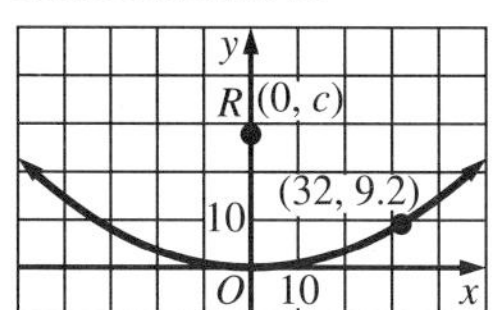

14. a. $y = -1 \pm \sqrt{\frac{x+3}{2}}$

b.

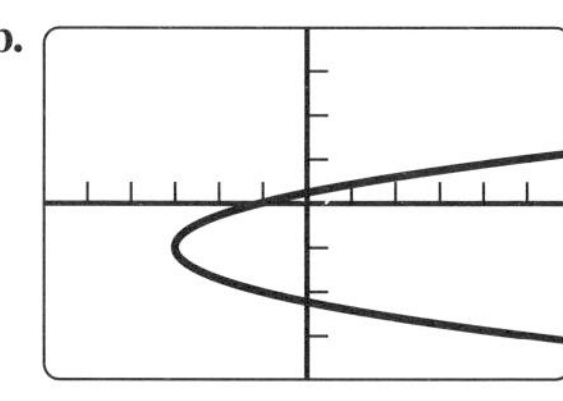

To obtain the graph of the entire parabola, you must graph $y = -1 + \sqrt{\frac{x+3}{2}}$ and $y = -1 - \sqrt{\frac{x+3}{2}}$ on the same screen.

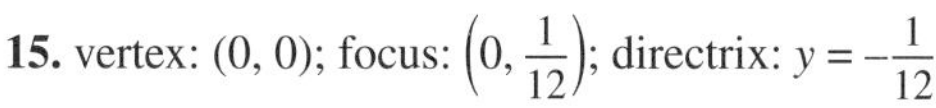

15. vertex: (0, 0); focus: $\left(0, \frac{1}{12}\right)$; directrix: $y = -\frac{1}{12}$

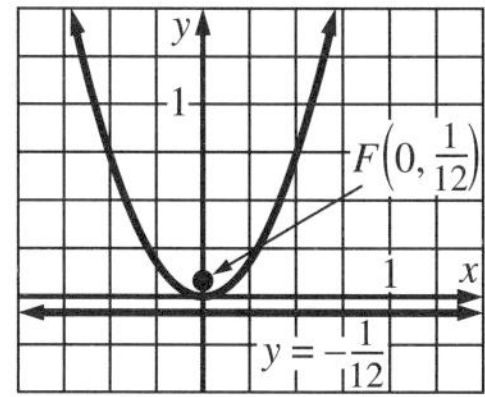

16. vertex: (0, 0); focus: $\left(-\frac{1}{2}, 0\right)$; directrix: $x = \frac{1}{2}$

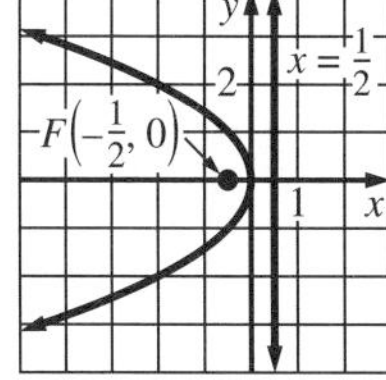

17. vertex: (–1, 4); focus: (–1, 6); directrix: $y = 2$

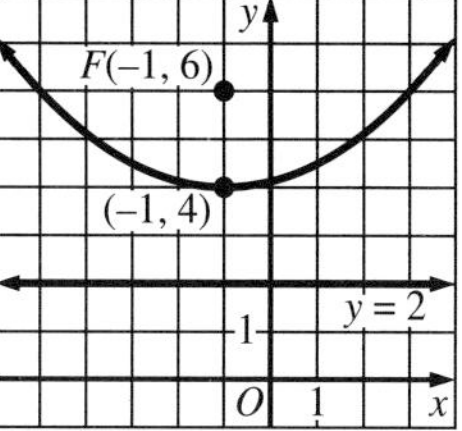

18. vertex: (3, –2); focus: (3, 2); directrix: $y = -6$

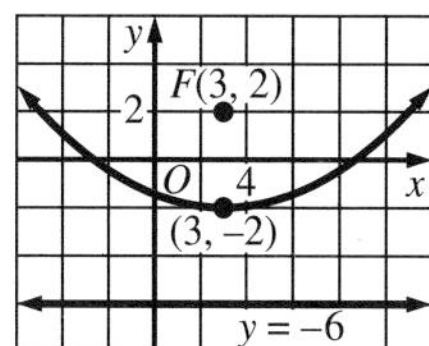

19. vertex: (2, 1); focus: $\left(2\frac{1}{16}, 1\right)$; directrix: $x = 1\frac{15}{16}$

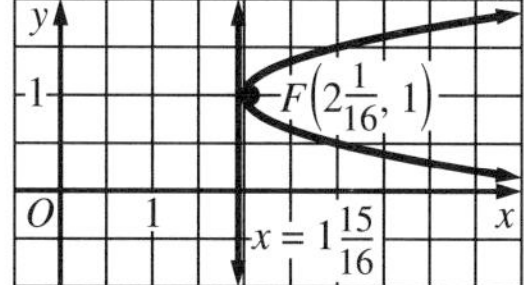

20. vertex: (0, 4); focus: (3, 4); directrix: $x = -3$

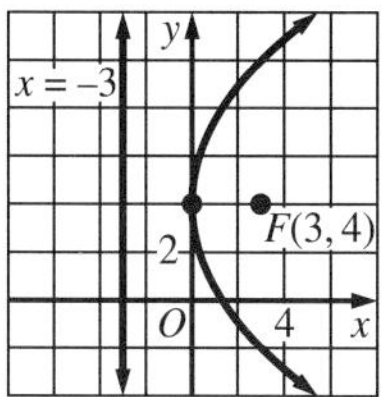

21. vertex: (0, 2); focus: $\left(0, 2\frac{1}{4}\right)$; directrix: $y = 1\frac{3}{4}$

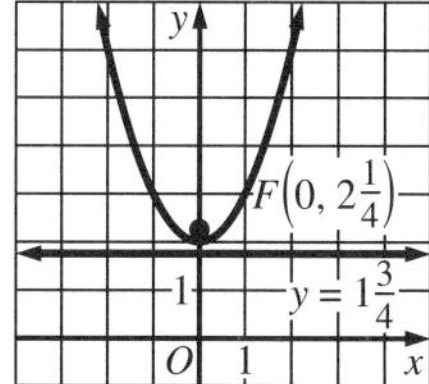

22. vertex: (5, 0); focus: $\left(5\frac{1}{8}, 0\right)$; directrix: $x = 4\frac{7}{8}$

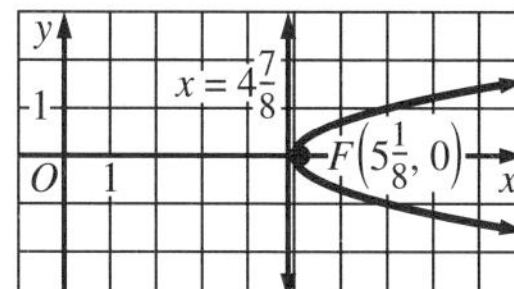

23. vertex: $\left(-\frac{3}{4}, 0\right)$; focus: $\left(\frac{1}{4}, 0\right)$; directrix: $x = -1\frac{3}{4}$

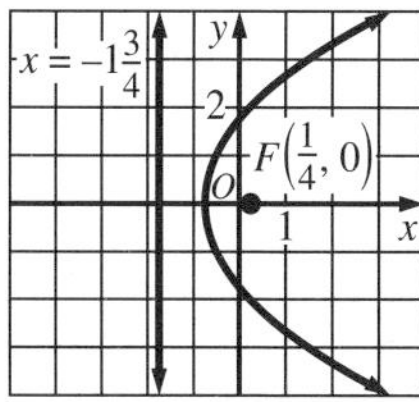

Page 526 Exercises and Applications

28.

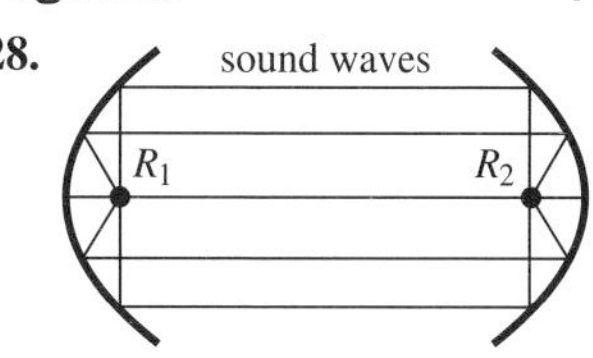

Answers may vary. An example is given. Sound waves travel from the first ring to the nearby parabolic dish, and are reflected horizontally across the room to the other dish. There, the sound waves are reflected to the second ring located at the focus of the receiving dish.

Page 527 **Assess Your Progress**

8. Answers may vary. An example is given. Before I started studying this chapter, I already knew about equations of lines and parabolas, and how to graph them. This information was helpful to me when I needed to write equations of perpendicular bisectors in Section 11.1 and when I learned about using the focus and directrix of a parabola to find its equation in Section 11.2.

Page 532 **Exercises and Applications**

5. $C(-1, -1)$; $r = \sqrt{10} \approx 3.16$

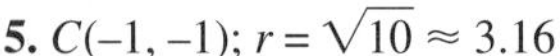

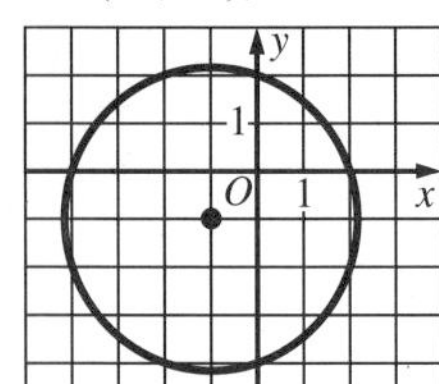

6. $C(-6, 0)$; $r = 1$

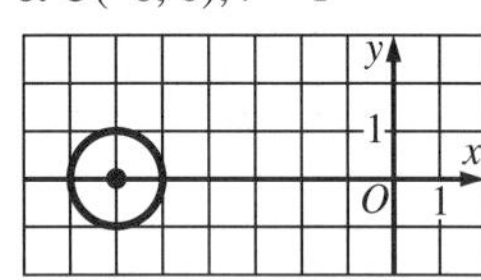

7. $C(0, -3)$; $r = \sqrt{2} \approx 1.41$

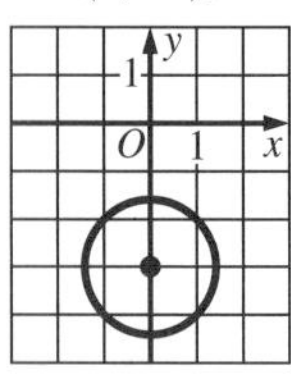

8. $C(5, 0)$; $r = 2$

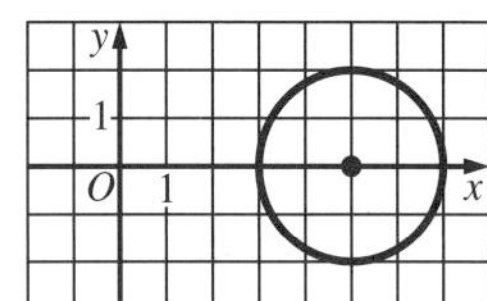

9. $C(0, 1)$; $r = 2\sqrt{2} \approx 2.83$

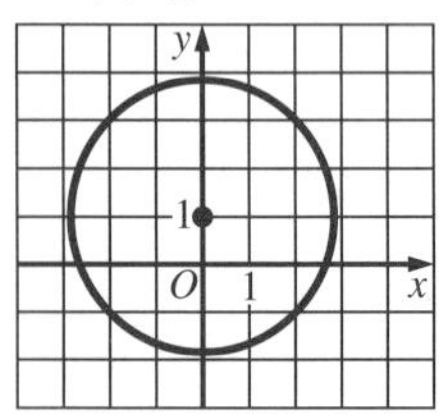

10. $x^2 + y^2 = 0.36$; 0.6 mi **11.** $x^2 + y^2 = 1$; 1 mi **12.** $x^2 + y^2 = 2.56$; 1.6 mi **13.** Yes; $(0.9)^2 + (1.2)^2 < (1.6)^2$. **14. a.** $(x - 1.6)^2 + (y - 1.6)^2 = 1$ **b.** Yes.

Page 540 **Exercises and Applications**

4. center: $(3, -3)$; vertices: $(-1, -3)$ and $(7, -3)$; foci: $(3 \pm \sqrt{7}, -3) \approx (5.65, -3)$ and $(0.35, -3)$; horizontal

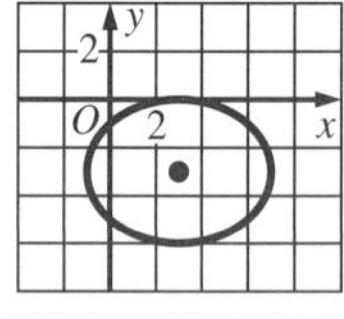

5. center: $(-8, -1)$; vertices: $(-8, 9)$ and $(-8, -11)$; foci: $(-8, -1 \pm 5\sqrt{3}) \approx (-8, 7.66)$ and $(-8, -9.66)$; vertical

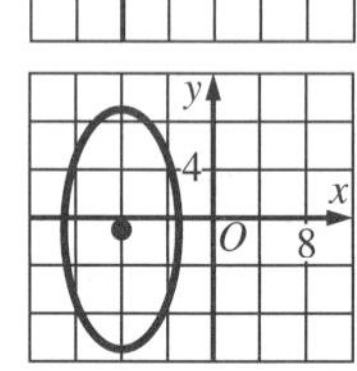

6. center: $(5, 0)$; vertices: $(14, 0)$ and $(-4, 0)$; foci: $(5 \pm 4\sqrt{5}, 0) \approx (13.94, 0)$ and $(-3.94, 0)$; horizontal

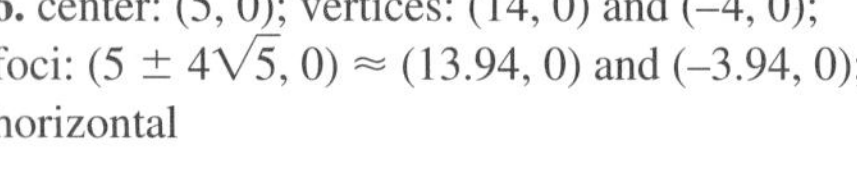

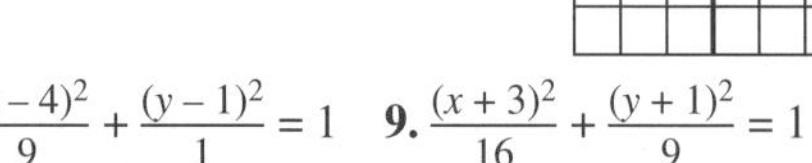

7. $\frac{x^2}{4} + \frac{y^2}{16} = 1$ **8.** $\frac{(x-4)^2}{9} + \frac{(y-1)^2}{1} = 1$ **9.** $\frac{(x+3)^2}{16} + \frac{(y+1)^2}{9} = 1$

10. a. $\frac{x^2}{36} + \frac{y^2}{16} = 1$ **b.** about 4.47 ft from the barrier **11. a.** $(4.47, 0)$ **b.** $(8.94, 0)$ **c.** $\frac{(x-4.47)^2}{36} + \frac{y^2}{16} = 1$

Page 541 **Exercises and Applications**

19. Answers may vary. Examples are given. **a.** Let V_1 be the vertex closest to F_1, and let V_2 be the vertex closest to F_2. Since the length of the string is $2a$, $V_1F_1 + V_1F_2 = 2a$. The ellipse is symmetric with respect to the axes, so $V_1F_1 = V_2F_2$. By substituting, you get $V_2F_2 + V_1F_2 = 2a$, or $V_1V_2 = 2a$. By the symmetry of the ellipse, the vertices must be $(-a, 0)$ and $(a, 0)$. **b.** Let P be the upper endpoint of the minor axis. Since the length of the string is $2a$, $PF_1 + PF_2 = 2a$. By the symmetry of the ellipse, $PF_1 = PF_2$. Therefore, $PF_1 = PF_2 = a$. Consider right triangle OPF_2, where O is the origin. By the Pythagorean theorem, $(OP)^2 = (PF_2)^2 - (OF_2)^2$; $b^2 = a^2 - c^2$. **20.** $\frac{x^2}{16} + \frac{y^2}{9} = 1$ **21.** $\frac{(x-5)^2}{4} + \frac{(y+3)^2}{9} = 1$ **22.** $\frac{(x-6)^2}{16} + \frac{(y-2)^2}{25} = 1$ **23.** $\frac{(x+2)^2}{49} + \frac{(y-4)^2}{9} = 1$

24. a. 5 **b.** 6.25 in.; 3.75 in. **c.** If the ellipse is horizontal and its center is at the origin, its equation is $\frac{x^2}{(3.125)^2} + \frac{y^2}{(1.875)^2} = 1$, or approximately $\frac{x^2}{9.77} + \frac{y^2}{3.52} = 1$.

Page 542 **Assess Your Progress**

1. center: $(-1, 5)$; radius: 4 **2.** center: $(3, 0)$; radius: 5 **3.** $(3, 5)$ and $\left(-\frac{77}{41}, \frac{45}{41}\right) \approx (-1.88, 1.10)$ **4.** center: $(0, 5)$; vertices: $(-8, 5)$ and $(8, 5)$; foci: $(-2\sqrt{7}, 5)$ and $(2\sqrt{7}, 5)$; horizontal **5.** center: $(-3, 5)$; vertices: $(-3, 19)$ and $(-3, -9)$; foci: $(-3, 5 + 4\sqrt{6})$ and $(-3, 5 - 4\sqrt{6})$; vertical

6. $x^2 + (y + 5)^2 = 16$

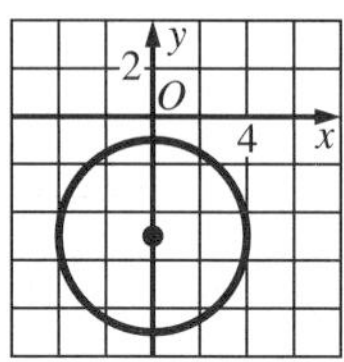

7. $\frac{(x+1)^2}{9} + \frac{(y-2)^2}{25} = 1$

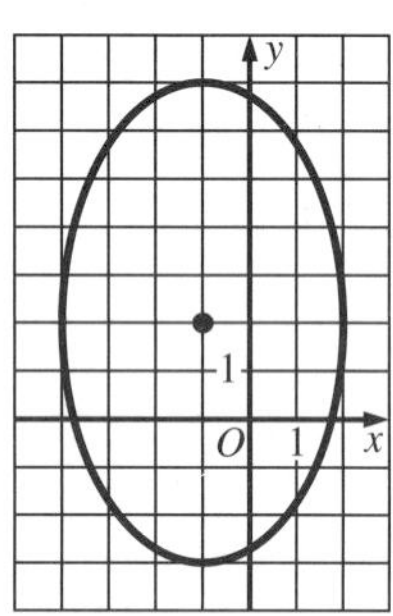

8. Answers may vary. An example is given. A circle is the set of points in a plane that are equidistant from one given point, the center. An ellipse is the set of points in a plane such that the sum of the distances from two given points, the foci, is constant. As the foci of an ellipse move closer together, the shape of the ellipse approaches a circle. The curves have similar shapes. An ellipse has two lines of symmetry, one through each axis, while a circle has an infinite number of lines of symmetry. The equation of a circle, $(x - h)^2 + (y - k)^2 = r^2$, can be written in the form of an equation of an ellipse, $\frac{(x-h)^2}{r^2} + \frac{(y-k)^2}{r^2} = 1$. Therefore, the only difference in the equations is that with an ellipse the denominators of the terms are different, whereas with a circle, the denominators are the same.

Page 547 **Exercises and Applications**

3. center: $(0, 3)$; vertices: $(\pm 9, 3)$; foci: $(\pm 3\sqrt{10}, 3) \approx (\pm 9.49, 0)$; horizontal; asymptotes: $y = \pm\frac{1}{3}x + 3$

4. center: $(1, -3)$; vertices: $(5, -3)$ and $(-3, -3)$; foci: $(1 \pm 4\sqrt{5}, -3) \approx (9.94, -3)$ and $(-7.94, -3)$; horizontal; asymptotes: $y = \pm 2(x - 1) - 3$

5. center: $(5, -7)$; vertices: $(5, -2)$ and $(5, -12)$; foci: $(5, -7 \pm 5\sqrt{5}) \approx (5, 4.18)$ and $(5, -18.18)$; vertical; asymptotes: $y = \pm\frac{1}{2}(x - 5) - 7$

6. center: $(2, 0)$; vertices: $(-7, 0)$ and $(11, 0)$; foci: $(2 \pm \sqrt{82}, 0) \approx (11.06, 0)$ and $(-7.06, 0)$; horizontal; asymptotes: $y = \pm\frac{1}{9}(x - 2)$

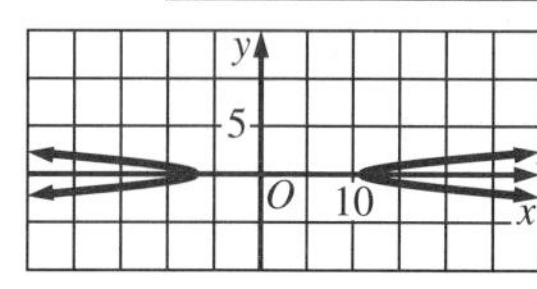

7. center: (–4, 0); vertices: (–4, ±3); foci: $(-4, \pm\sqrt{58}) \approx (-4, \pm 7.62)$; vertical; asymptotes: $y = \pm\frac{3}{7}(x + 4)$

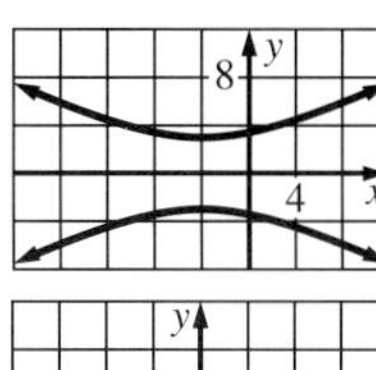

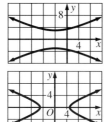

8. center: (0, 0); vertices: (±5, 0); foci: $(\pm\sqrt{29}, 0) \approx (\pm 5.39, 0)$; horizontal; asymptotes: $y = \pm\frac{2}{5}x$

9. center: (0, 0); vertices: $(0, \pm 6\sqrt{3}) \approx (0, 10.39)$; foci: $(0, \pm\sqrt{183}) \approx (0, \pm 13.53)$; vertical; asymptotes: $y = \pm 1.2x$

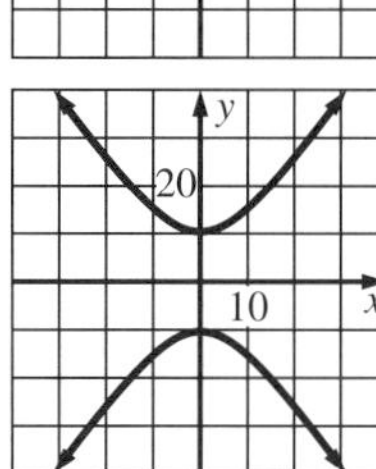

10. a, b.

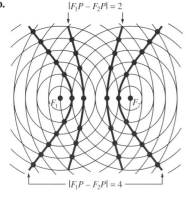

c. The vertices of the hyperbola move farther apart and the hyperbola becomes more narrow.

Page 549 Exercises and Applications

21. $\frac{x^2}{9} - \frac{y^2}{16} = 1$

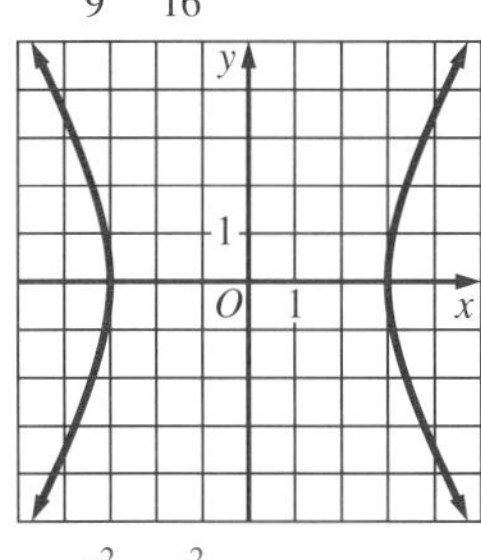

22. $\frac{y^2}{25} - \frac{x^2}{144} = 1$

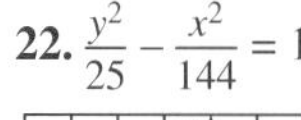

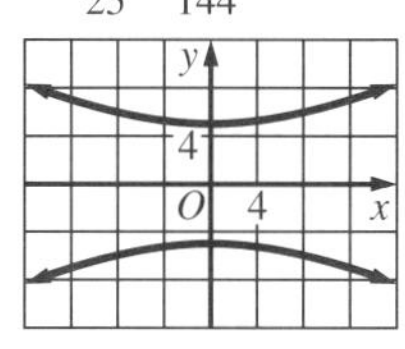

23. $\frac{x^2}{36} - \frac{y^2}{64} = 1$

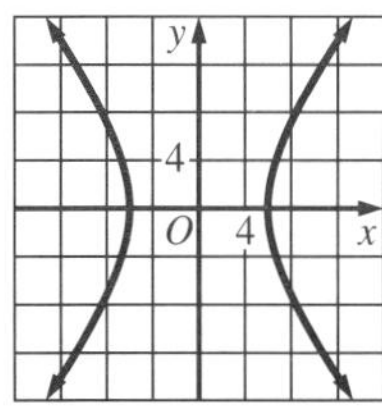

24. $\frac{(y-2)^2}{9} - \frac{(x+5)^2}{16} = 1$

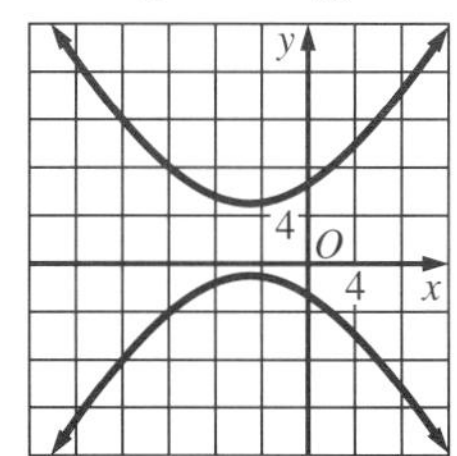

25. $\frac{(x-1)^2}{36} - \frac{(y+4)^2}{64} = 1$

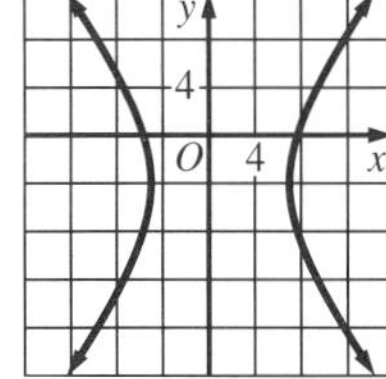

26. a. $\frac{x^2}{a^2} - \frac{y^2}{b^2} = 1$; $\frac{y^2}{b^2} = \frac{x^2}{a^2} - 1$; $y^2 = b^2\left(\frac{x^2}{a^2} - \frac{a^2}{a^2}\right) = \frac{b^2}{a^2}(x^2 - a^2)$; $y = \pm\frac{b}{a}\sqrt{x^2 - a^2}$

b. As $x \to \infty$, $\sqrt{x^2 - a^2} \to \sqrt{x^2} = x$. Thus, $\pm\frac{b^2}{a^2}\sqrt{x^2 - a^2} \to \pm\frac{b}{a}x$ as $x \to \infty$.

c. Part (b) shows that as $x \to \infty$, $y \to \pm\frac{b}{a}x$. Therefore, the graph of $\frac{x^2}{a^2} - \frac{y^2}{b^2} = 1$ approaches the graph of $y = \pm\frac{b}{a}x$ as $x \to \infty$. This explains why $y = \pm\frac{b}{a}x$ are the asymptotes of the graph of the equation $\frac{x^2}{a^2} - \frac{y^2}{b^2} = 1$. **27.** $\frac{x^2}{16} - \frac{y^2}{9} = 1$

28.

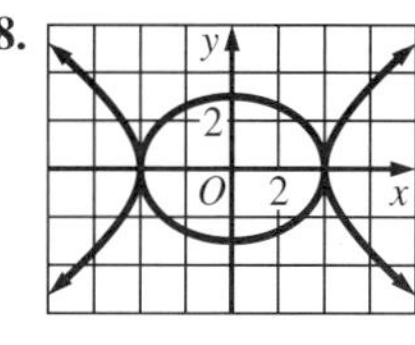

Answers may vary. An example is given. Both graphs are curves that are not functions, that are symmetric with respect to the coordinate axes, and have the same vertices and center. However, the first equation represents a hyperbola, and so has two separate branches, while the second equation represents an ellipse, and so is a single continuous curve.

29. $\frac{(x+6)^2}{16} + \frac{(y+1)^2}{1} = 1$

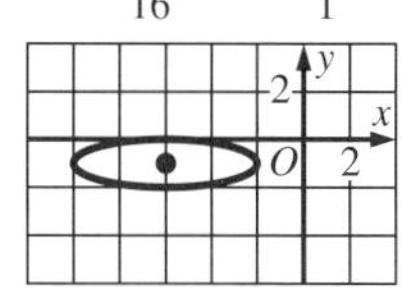

30. $\frac{(x-1)^2}{11} + \frac{(y-2)^2}{36} = 1$

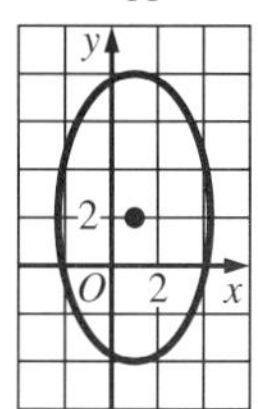

31. $\frac{(x-2)^2}{169} + \frac{(y+3)^2}{25} = 1$

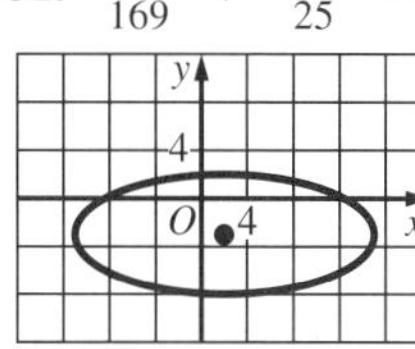

32. $y = (x-4)^2 - 36$; $y = (x-10)(x+2)$ **33.** $y = (x-6)^2 - 4$; $y = (x-8)(x-4)$

34. $y = 2\left(x - \frac{5}{4}\right)^2 - \frac{49}{8}$; $y = (2x+1)(x-3)$

Page 552 Exercises and Applications

5. $x^2 + \left(y + \frac{3}{2}\right)^2 = \frac{25}{4}$; a circle with center $\left(0, -\frac{3}{2}\right)$ and radius $\frac{5}{2}$

6. $\frac{(y-9)^2}{60} - \frac{(x+1)^2}{15} = 1$; a vertical hyperbola with center (–1, 9), vertices $(-1, 9 \pm 2\sqrt{15}) \approx (-1, 16.75)$ and (–1, 1.25), and asymptotes $y = \pm 2(x + 1) + 9$

7. a. a parabola; a hyperbola **b.** Answers may vary. An example is given. A hyperbole is an exaggerated statement, or overstatement; thus, it is appropriately linked to a hyperbola, where *t* exceeds *s*. A parable is a short story which illustrates a moral point by making a comparison; thus, it is appropriately linked to a parabola, where *t* equals, or is comparable, to *s*. An ellipsis is a symbol used to indicate an omission of one or more words. That is, the written words fall short of the intended meaning. Thus, an ellipsis is appropriately linked to an ellipse, where *t* is less than *s*. **8.** one branch of a hyperbola **9.** a parabola, an ellipse, or a circle **10.** straight up, or vertically **11.** No. Explanations may vary. An example is given. In order for the sonic boom to occur along a line, the vertex of the cone, which is at the tail of the airplane, would have to be on the ground.

Page 553 Exercises and Applications

18. Answers may vary. An example is given. The flashlight beam represents a single cone, and the wall represents the plane that cuts the cone. You get a circle when the flashlight is perpendicular to the wall. You get an ellipse when the flashlight is held at an angle relative to the wall. A single branch of a hyperbola is formed by holding the flashlight parallel to the wall. It is difficult to form a parabola, since the light must be adjusted until one edge of the beam is parallel to the wall.

19. center: (–1, 3); vertices: (6, 3) and (–8, 3); foci: $(-1 \pm \sqrt{74}, 3) \approx (7.60, 3)$ and (–9.60, 3); horizontal; asymptotes: $y = \pm\frac{5}{7}(x + 1) + 3$

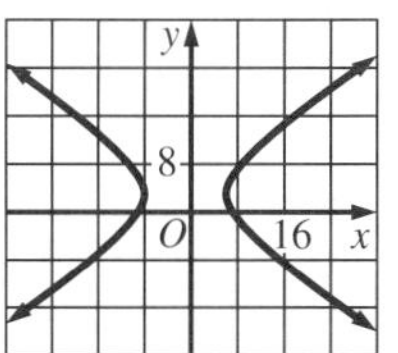

20. center: $(-2, 0)$, vertices: $(-2, \pm 4)$; foci: $(-2, \pm 2\sqrt{13}) \approx (-2, \pm 7.21)$; vertical; asymptotes: $y = \pm\frac{2}{3}(x + 2)$

21. 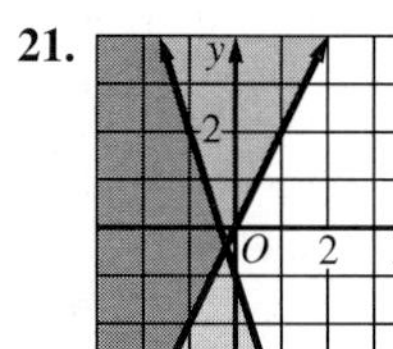 **22.** 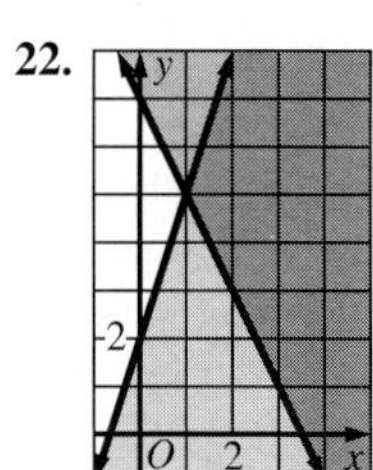 **23.** 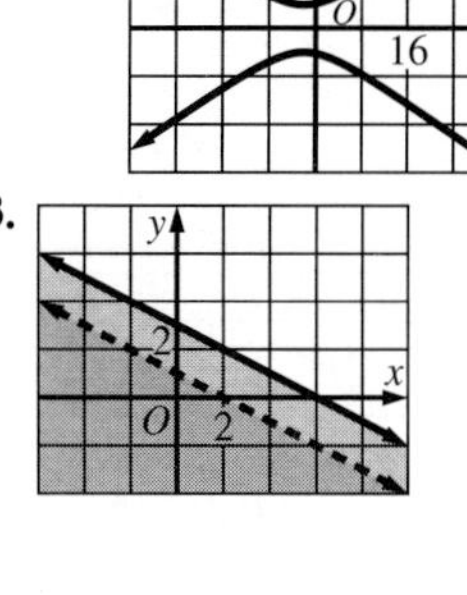

Page 553 Assess Your Progress

2. center: $(-2, 2)$, vertices: $(-2, 7)$ and $(-2, -3)$; foci: $(-2, 2 \pm \sqrt{29}) \approx (-2, 7.39)$ and $(-2, -3.39)$; vertical; asymptotes: $y = \pm\frac{5}{2}(x + 2) + 2$

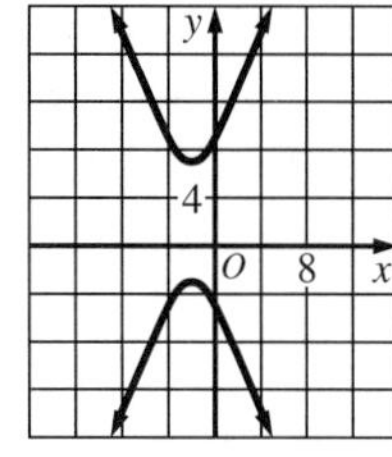

3. $(x + 7)^2 + (y - 4)^2 = 64$; a circle with center $(-7, 4)$ and radius 8

4. $\frac{(x+2)^2}{9} - \frac{(y-5)^2}{9} = 1$; a horizontal hyperbola with center $(-2, 5)$, vertices $(1, 5)$ and $(-5, 5)$, and asymptotes $y = \pm(x + 2) + 5$

5. a degenerate circle; the point $(0, 0)$

6. a degenerate hyperbola; the intersecting lines $y = \frac{1}{6}x$ and $y = -\frac{1}{6}x$

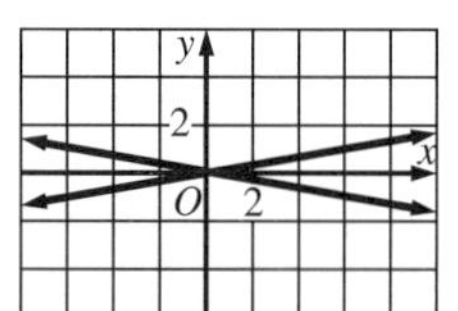

7. Answers may vary. An example is given. Ellipses and circles are closely related. You can consider a circle to be an ellipse whose foci are coincident. Both have equations of the form $\frac{(x-h)^2}{a^2} + \frac{(y-k)^2}{b^2} = 1$; in a circle, $a = b$, and in an ellipse, $a \neq b$. Their graphs look very similar. Also, hyperbolas and ellipses have equations that look similar. When a hyperbola and an ellipse have the same constants a and b and the same orientation, the two figures have the same center and vertices.

Pages 558–559 Chapter 11 Assessment

18. $\frac{(x+4)^2}{4} + \frac{(y+2)^2}{9} = 1$ **19.** $\frac{(x-6)^2}{16} - \frac{(y-6)^2}{9} = 1$ **20.** $x + 1 = -\frac{1}{16}(y - 3)^2$; a parabola that opens left, has vertex $(-1, 3)$, and axis $y = 3$

21. $\frac{(x-1)^2}{4} - \frac{(y-2)^2}{1} = 1$; a horizontal hyperbola with center $(1, 2)$, vertices $(3, 2)$ and $(-1, 2)$, and asymptotes $y = \pm\frac{1}{2}(x - 1) + 2$ **22.** $x^2 + (y - 6)^2 = 11$; a circle with center $(0, 6)$ and radius $\sqrt{11} \approx 3.32$ **23.** $\frac{(x-3)^2}{25} + \frac{(y-2)^2}{1} = 1$; a horizontal ellipse with center $(3, 2)$, vertices $(8, 2)$ and $(-2, 2)$, and minor axis with endpoints $(3, 3)$ and $(3, 1)$

24. a degenerate hyperbola; the intersecting lines $y = 5x$ and $y = -5x$

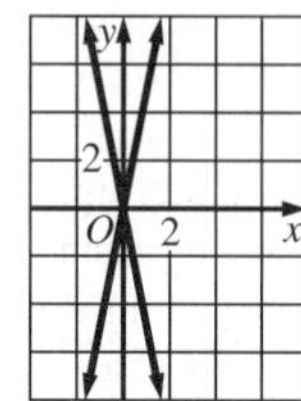

25. a degenerate circle; the point $(-1, -1)$ **26.** Answers may vary. Check students' work.

CHAPTER 12

Page 567 Exercises and Applications

6. a. **b.**

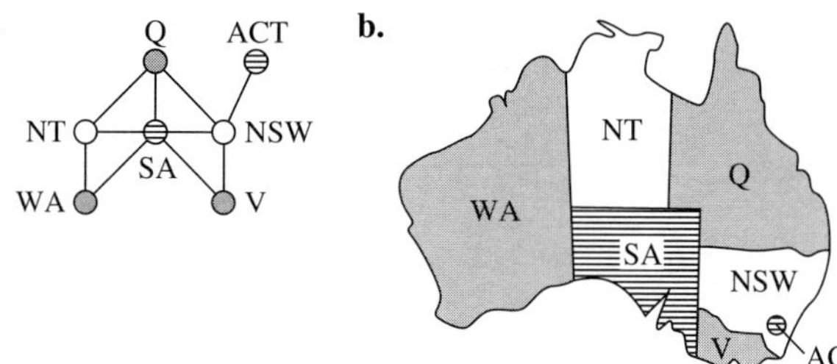

9–11. Answers may vary. Examples are given.

9. a. **b.** Yes, as shown in the graph in part (a).

10. Three tanks are needed. In the arrangement shown, the groupings are (1) moray and lionfish; (2) damselfish, filefish, and goby; and (3) cardinalfish.

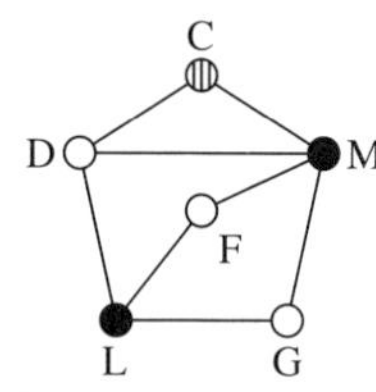

11. Answers may vary. Examples are given. (1) damselfish and filefish, (2) cardinalfish and goby, and (3) moray and lionfish; and (1) damselfish, filefish, and goby, (2) moray, and (3) cardinalfish and lionfish.

12. cardinalfish, moray, and damselfish

Page 568 Exercises and Applications

14. Answers may vary. Examples are given.

a. **b.**

c.

15. Answers may vary. An example is given.

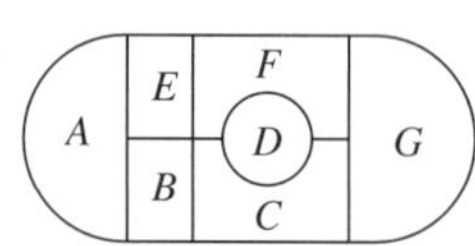

16. a. Graphs may vary. An example is given. Four storage areas are needed. **b.** 4

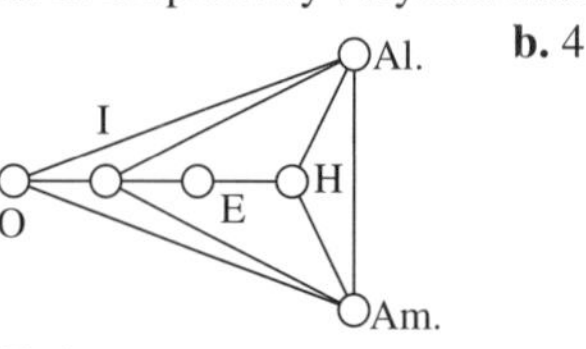

17. Answers may vary. An example is given.

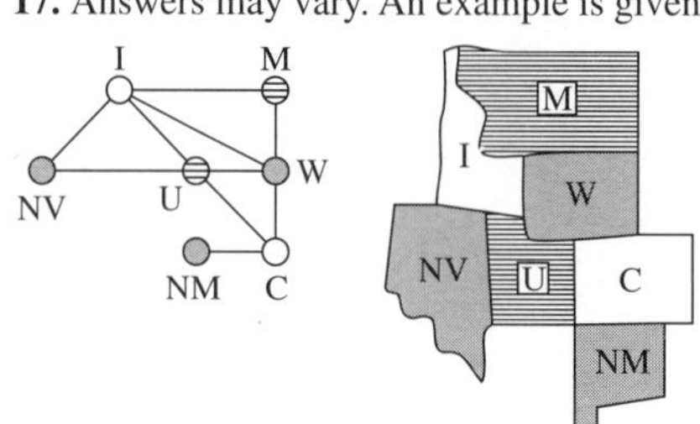

There is no other way to group the states by color so that states sharing a border have different colors.

Page 573 Exercises and Applications

4.

	A	B	C	D	E
A	0	1	1	1	1
B	0	0	1	0	0
C	0	0	0	0	0
D	0	0	0	0	0
E	0	0	0	1	0

5.

	A	B	C	D	E	F
A	0	0	0	0	0	0
B	1	0	0	0	0	0
C	1	0	0	0	0	0
D	1	0	0	0	0	0
E	1	0	0	0	0	0
F	0	1	1	1	1	0

6.

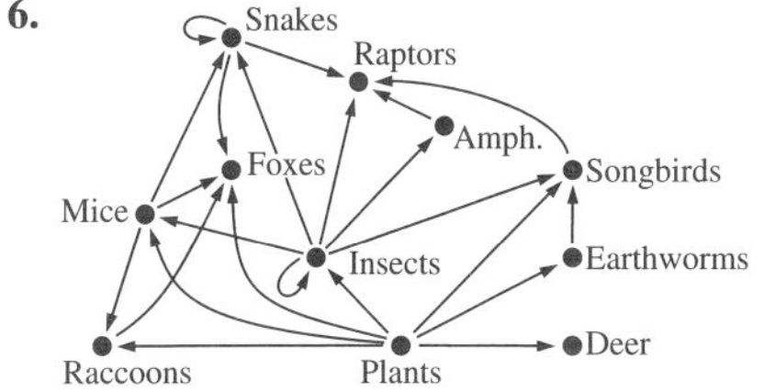

7.

	D	R	A	I	E	S	P	Sn	M	Rc	F
D	0	0	0	0	0	0	0	0	0	0	0
R	0	0	0	0	0	0	0	0	0	0	0
A	0	1	0	0	0	0	0	0	0	0	0
I	0	1	1	1	0	1	0	1	1	0	0
E	0	0	0	0	0	1	0	0	0	0	0
S	0	1	0	0	0	0	0	0	0	0	0
P	1	0	0	1	1	1	0	0	1	1	1
Sn	0	1	0	0	0	0	0	1	0	0	1
M	0	0	0	0	0	0	0	1	0	1	1
Rc	0	0	0	0	0	0	0	0	0	0	1
F	0	0	0	0	0	0	0	0	0	0	0

8.

	D	R	A	I	E	S	P	Sn	M	Rc	F
D	0	0	0	0	0	0	0	0	0	0	0
R	0	0	0	0	0	0	0	0	0	0	0
A	0	0	0	0	0	0	0	0	0	0	0
I	0	4	1	1	0	1	0	3	1	1	2
E	0	1	0	0	0	0	0	0	0	0	0
S	0	0	0	0	0	0	0	0	0	0	0
P	0	2	1	1	0	2	0	2	1	1	2
Sn	0	1	0	0	0	0	0	1	0	0	1
M	0	1	0	0	0	0	0	1	0	0	2
Rc	0	0	0	0	0	0	0	0	0	0	0
F	0	0	0	0	0	0	0	0	0	0	0

9.

	D	R	A	I	E	S	P	Sn	M	Rc	F
D	0	0	0	0	0	0	0	0	0	0	0
R	0	0	0	0	0	0	0	0	0	0	0
A	0	1	0	0	0	0	0	0	0	0	0
I	0	5	2	2	0	2	0	4	2	1	2
E	0	1	0	0	0	1	0	0	0	0	0
S	0	1	0	0	0	0	0	0	0	0	0
P	1	2	1	2	1	3	0	2	2	2	3
Sn	0	2	0	0	0	0	0	2	0	0	2
M	0	1	0	0	0	0	0	2	0	1	3
Rc	0	0	0	0	0	0	0	0	0	0	1
F	0	0	0	0	0	0	0	0	0	0	0

10. foxes; They eat mice directly and indirectly three different ways. The "3" in row M, column F shows this. Since 3 is the greatest element in row M, foxes would be most affected if all of the mice died.

Page 575 Exercises and Applications

19. In the matrices, the number and letter combination indicates the intersection of the numbered street with the lettered street. For example, 3B means the intersection of Third St. and B St.

a. $M =$

	1A	1B	1C	2A	2B	2C	3A	3B	3C
1A	0	1	0	1	0	0	0	0	0
1B	0	0	1	0	0	0	0	0	0
1C	0	0	0	0	0	1	0	0	0
2A	0	0	0	0	0	0	1	0	0
2B	0	1	0	1	0	0	0	0	0
2C	0	0	0	0	1	0	0	0	0
3A	0	0	0	0	0	0	0	1	0
3B	0	0	0	0	1	0	0	0	1
3C	0	0	0	0	0	1	0	0	0

An element of 1 means that it is possible to get from the row intersection to the column intersection by traveling one city block. An element of 0 means that this is not possible.

b. $M^2 =$

	1A	1B	1C	2A	2B	2C	3A	3B	3C
1A	0	0	1	0	0	0	1	0	0
1B	0	0	0	0	0	1	0	0	0
1C	0	0	0	0	1	0	0	0	0
2A	0	0	0	0	0	0	0	1	0
2B	0	0	1	0	0	0	1	0	0
2C	0	1	0	1	0	0	0	0	0
3A	0	0	0	0	1	0	0	0	1
3B	0	1	0	1	0	1	0	0	0
3C	0	0	0	0	1	0	0	0	0

$M^3 =$

	1A	1B	1C	2A	2B	2C	3A	3B	3C
1A	0	0	0	0	0	1	0	1	0
1B	0	0	0	0	1	0	0	0	0
1C	0	1	0	1	0	0	0	0	0
2A	0	0	0	0	1	0	0	0	1
2B	0	0	0	0	0	1	0	1	0
2C	0	0	1	0	0	0	1	0	0
3A	0	1	0	1	0	1	0	0	0
3B	0	0	1	0	1	0	1	0	0
3C	0	1	0	1	0	0	0	0	0

$M^4 =$

	1A	1B	1C	2A	2B	2C	3A	3B	3C
1A	0	0	0	0	2	0	0	0	1
1B	0	1	0	1	0	0	0	0	0
1C	0	0	1	0	0	0	1	0	0
2A	0	1	0	1	0	1	0	0	0
2B	0	0	0	0	2	0	0	0	1
2C	0	0	0	0	0	1	0	1	0
3A	0	0	1	0	1	0	1	0	0
3B	0	1	0	1	0	1	0	1	0
3C	0	0	1	0	0	0	1	0	0

M^2 represents the number of ways to get from the row intersection to the column intersection through an intermediate intersection, that is, by traveling two city blocks. M^3 represents the number of ways to get from the row intersection to the column intersection through two intermediate intersections, that is, by traveling three city blocks. M^4 represents the number of ways to get from the row intersection to the column intersection through three intermediate intersections, that is, by traveling four city blocks. **c.** No. Explanations may vary. An example is given. In matrix M, the column representing the intersection at First and A Streets, 1A, contains all zero elements, so every power of M will have a zero as the element in row 1C and column 1A.

20.

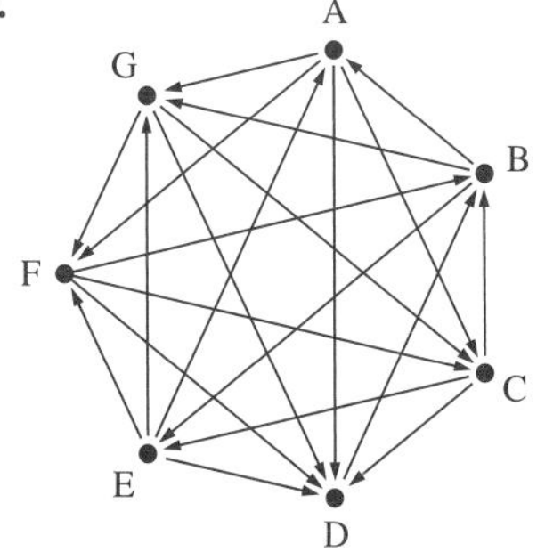

It is difficult to determine the champion using the graph. Students' choices of the winning team may vary. Possibilities include the Angels and the Elks, since each of these teams won the most games.

21. a. $M =$

	A	B	C	D	E	F	G
A	0	0	1	1	0	1	1
B	1	0	0	0	1	0	1
C	0	1	0	1	1	0	0
D	0	1	0	0	0	0	0
E	1	0	0	1	0	1	1
F	0	1	1	1	0	0	0
G	0	0	1	1	0	1	0

b. The row sums represent the total number of games won by the row team.

22. a. $M^2 =$

	A	B	C	D	E	F	G
A	0	3	2	3	1	1	0
B	1	0	2	3	0	3	2
C	2	1	0	1	1	1	2
D	1	0	0	0	1	0	1
E	0	2	3	3	0	2	1
F	1	2	0	1	2	0	1
G	0	3	1	2	1	0	0

Each element of M^2 represents the number of teams that (1) the row team defeated and (2) that defeated the column team.
b. The row sums represent the additional points each row team is awarded for defeating a team that has defeated other teams.

23. a. $M + M^2 =$

	A	B	C	D	E	F	G	Total
A	0	3	3	4	1	2	1	14
B	2	0	2	3	1	3	3	14
C	2	2	0	2	2	1	2	11
D	1	1	0	0	1	0	1	4
E	1	2	3	4	0	3	2	15
F	1	3	1	2	2	0	1	10
G	0	3	2	3	1	1	0	10

The row sums represent the total number of points awarded to the row team.
b. See the last column in the matrix in part (a). The Elks win the championship.
24. Answers may vary. An example is given. Give each team one point for each win. In the case of a tie, give each team that scored the most games one point for each team that a defeated team has beaten. You can use the row totals of a matrix M, as in Ex. 21, to determine the winner or the teams that have tied. You can use the row totals of the matrix $M + M^2$ when there is a tie to determine the winner. (In this case, the Bisons and Elks have tied with 4 wins each. The totals for row B and row E of the matrix $M + M^2$ show that, as in Ex. 23, the Elks win with 15 points.)

Page 576 Assess Your Progress

2. Two meeting times are needed. **3. a.**

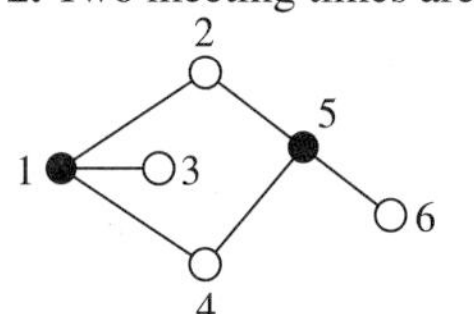

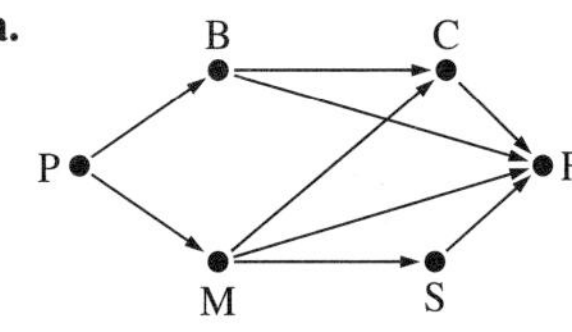

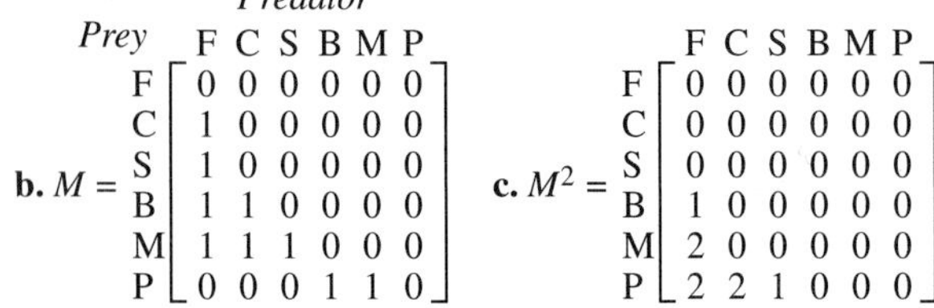

Predator

b. $M =$

Prey	F	C	S	B	M	P
F	0	0	0	0	0	0
C	1	0	0	0	0	0
S	1	0	0	0	0	0
B	1	1	0	0	0	0
M	1	1	1	0	0	0
P	0	0	0	1	1	0

c. $M^2 =$

	F	C	S	B	M	P
F	0	0	0	0	0	0
C	0	0	0	0	0	0
S	0	0	0	0	0	0
B	1	0	0	0	0	0
M	2	0	0	0	0	0
P	2	2	1	0	0	0

M^2 gives the number of ways that each prey is a food source for the column predator through one intermediate source. Answers may vary. An example is given. Barnacles are an intermediate food source for fish through crabs.

$M + M^2 =$

	F	C	S	B	M	P
F	0	0	0	0	0	0
C	1	0	0	0	0	0
S	1	0	0	0	0	0
B	2	1	0	0	0	0
M	3	1	1	0	0	0
P	2	2	1	1	1	0

$M + M^2$ gives the total number of ways that each prey is a food source for the column predator either directly or through one intermediate source. Answers may vary. An example is given. Barnacles are a food source for fish in two ways, both directly and indirectly through crabs. **4.** Answers may vary. An example is given. The graphs in discrete mathematics are different from algebraic graphs, because graphs in discrete mathematics are used to sort items into groups and to indicate relationships between groups, whereas graphs in algebra are used to indicate location and solutions. The two types of graphs are alike because both involve labeled points, numbers, and specific procedures.

Page 592 Exercises and Applications

41. Answers may vary. Examples are given. **a.** Combinations and permutations are similar because each involves making a selection from a group of items. They are different because the order of the choices is important in a permutation but is not important in a combination. For example, if you are choosing 4 plants from a group of 6 plants to give to your friend, you would find the number of possibilities by using a combination. If you are choosing 4 plants from a group of 6 plants to arrange in a window box, you would find the number of possibilities by using a permutation, since the order of the plants matters. **b.** ${}_nC_r$ represents the number of ways to choose r objects from a set of n objects when the order of the objects is not important. ${}_nP_r$ represents the number of ways to choose r objects from a set of n objects when the order of the objects is important. If you know ${}_nP_r$, you can find ${}_nC_r$ by dividing ${}_nP_r$ by $r!$. For example, using the plant selection described in part (a), ${}_6P_4 = \frac{6!}{2!} = 360$ and ${}_6C_4 = \frac{360}{4!} = 15$.

Page 597 Exercises and Applications

28. a. ${}_nC_1 = n$, which is the row number of the triangle **b.** Each number in the ${}_nC_2$ diagonal is the sum of all the numbers in the ${}_nC_1$ diagonal in the rows above that number. For example, ${}_6C_2 = {}_5C_1 + {}_4C_1 + {}_3C_1 + {}_2C_1 + {}_1C_1$.
c. The sum can be found in row $n + 1$ and in the next diagonal to the right; Yes.

Page 598 Exercises and Applications

38. Answers may vary. An example is given. **a.** A pattern of inverted equilateral triangles appears. At each row $n = 3^k$, where $k = 0, 1, 2, \ldots$, a pattern of inverted equilateral triangles with $n - 1$ numbers on a side begins.
b. A pattern of inverted equilateral triangles appears. At each row $n = 5^k$, where $k = 0, 1, 2, \ldots$, a pattern of inverted equilateral triangles with $n - 1$ numbers on a side begins. **c.** A pattern of inverted equilateral triangles appears when multiples of prime numbers are circled. At each row, $n = x^k$, where x is a prime number and $k = 0, 1, 2, \ldots$, a pattern of inverted equilateral trinagles with $n - 1$ numbers on a side begins.

Page 604 Chapter 12 Assessment

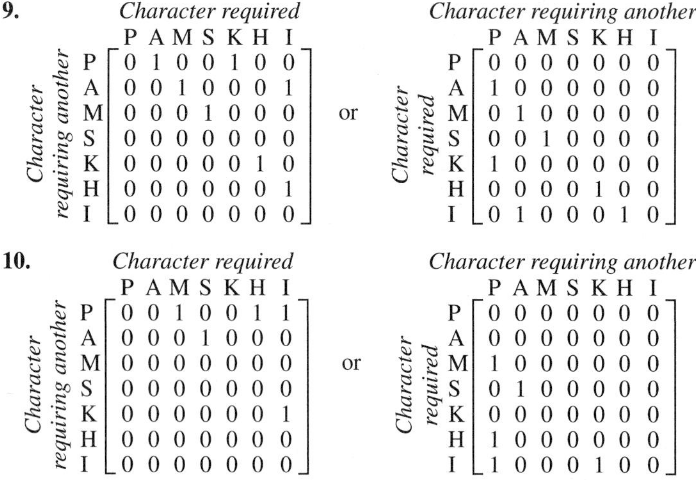

9. *Character required* (columns); *Character requiring another* (rows)

	P	A	M	S	K	H	I
P	0	1	0	0	1	0	0
A	0	0	1	0	0	0	1
M	0	0	0	1	0	0	0
S	0	0	0	0	0	0	0
K	0	0	0	0	0	1	0
H	0	0	0	0	0	0	1
I	0	0	0	0	0	0	0

or

Character requiring another (columns); *Character required* (rows)

	P	A	M	S	K	H	I
P	0	0	0	0	0	0	0
A	1	0	0	0	0	0	0
M	0	1	0	0	0	0	0
S	0	0	1	0	0	0	0
K	1	0	0	0	0	0	0
H	0	0	0	0	1	0	0
I	0	1	0	0	0	1	0

10. *Character required* (columns); *Character requiring another* (rows)

	P	A	M	S	K	H	I
P	0	0	1	0	0	1	1
A	0	0	0	1	0	0	0
M	0	0	0	0	0	0	0
S	0	0	0	0	0	0	0
K	0	0	0	0	0	0	1
H	0	0	0	0	0	0	0
I	0	0	0	0	0	0	0

or

Character requiring another (columns); *Character required* (rows)

	P	A	M	S	K	H	I
P	0	0	0	0	0	0	0
A	0	0	0	0	0	0	0
M	1	0	0	0	0	0	0
S	0	1	0	0	0	0	0
K	0	0	0	0	0	0	0
H	1	0	0	0	0	0	0
I	1	0	0	0	1	0	0

The elements of the squared matrix tell you which characters are required indirectly through one intermediate character.

CHAPTER 13

Page 616 Exercises and Applications

14. a. Answers may vary. Students' results should be close to the theoretical probability of $\frac{119}{324}$, or about 0.367. **b.** Let N_i = the number of ways of getting exactly i of a kind. Then the number of ways of rolling at least three of a kind is $N_6 + N_5 + N_4 + N_3$. There are six ways to roll six of a kind (six ones, six twos, and so on), so $N_6 = 6$. There are ${}_6C_5$ different arrangements of dice showing five of a kind. Each set can take one of six values. The remaining die can be any of five differing values, so $N_5 = {}_6C_5 \cdot 6 \cdot 5 = 180$. There are ${}_6C_4$ arrangements of four of a kind. Each can take one of six values. The remaining two dice can be any of five differing values, so $N_4 = {}_6C_4 \cdot 6 \cdot 5^2 = 2250$. There are two cases for N_3: exactly one set of three of a kind, or two different sets of three of a kind. In the first case, there are ${}_6C_3$ arrangements showing three of a kind, each taking one of six values. The remaining three dice can take any of five values, excepting the five instances where these three are the same. In the second case, there are ${}_6C_2$ pairs of two different values for the two triples. The first triple can be in any of ${}_6C_3$ arrangements of dice. Picking the first triple also fixes the second triple. So, $N_3 = ({}_6C_3 \cdot 6 \cdot (5^3 - 5)) + ({}_6C_2 \cdot {}_6C_3) = 14{,}700$. Since there are $6^6 = 46{,}656$ possible roles, the theoretical probability of rolling at least three of a kind is $\frac{6 + 180 + 2250 + 14{,}700}{46{,}656} = \frac{17{,}136}{46{,}656} = \frac{119}{324}$. Students' comparisons will vary.

Page 622 Exploration

1. Answers may vary. **2.** $\frac{1}{2}$; $\frac{1}{64}$

3. The translation of the test and answers are as follows.

1. More people live in California than in all of Australia. True.
2. The main crop of the Ukraine is wheat. True.
3. Kenya gained its independence in 1963. True.
4. Calcutta is the capital of India. False; the capital is New Delhi.
5. Shanghai is a country in Asia. False; Shanghai is a city in Asia.
6. Malawi lies on the western shore of a large lake. True.

4. Histograms may vary, but they should generally be bell-shaped. **5.** Answers may vary, though students may note that it appears to be close to 0. Students should realize that an experimental probability may vary from a theoretical probability, especially if relatively few trials are performed. **6.** Probabilities based on the histograms will vary, though the theoretical probabilities are $\frac{11}{32}$ and $\frac{21}{32}$. Theoretically, it is more likely to answer three or fewer questions correctly. Since the probabilities of being correct or incorrect are equal, you are as likely to get 0 right (miss all 6) as to get 6 right, to get 1 right (miss 5) as to get 5 right, and to get 2 right as to get 4 right. So, getting 3 or fewer right is as likely as getting 3 or more right. Because getting 4 or more right is a proper subset of getting 3 or more right, it is thus more likely to get 3 or fewer right than to get 4 or more correct.

Page 637 Exercises and Applications

22. a. about 0.991 **b.** $10p^3(1-p)^3 + 15p^4(1-p)^2 + 6p^5(1-p) + p^6$

p	0	0.1	0.2	0.3	0.4	0.5	0.6	0.7	0.8	0.9	1
conviction probability (≈)	0	0.009	0.058	0.163	0.317	0.5	0.683	0.837	0.942	0.991	1

c. $462p^6(1-p)^6 + 792p^7(1-p)^5 + 495p^8(1-p)^4 + 220p^9(1-p)^3 + 66p^{10}(1-p)^2 + 12p^{11}(1-p) + p^{12}$

p	0	0.1	0.2	0.3	0.4	0.5	0.6	0.7	0.8	0.9	1
conviction probability (≈)	0	0.000	0.012	0.078	0.247	0.5	0.754	0.922	0.988	1.00	1

d. For $p = 0$, 0.5, or 1, the juries are equally likely to convict. When $0 < p < 0.5$, the 6-member jury is more likely to convict. When $0.5 < p < 1$, the 12-member jury is more likely to convict. **23.** Suggestions will vary. The Walbert model is simplistic in its assumption that all jurors have the same probability of conviction, and in its assumption that a majority will always sway a minority and that a tie is equally likely to go either way, without regard to the differing personalities and abilities of the jurors.

Page 638 Exercises and Applications

30. mean: 180; standard deviation: about 50.9 **31.** mean: 0.31; standard deviation: about 0.146

32.

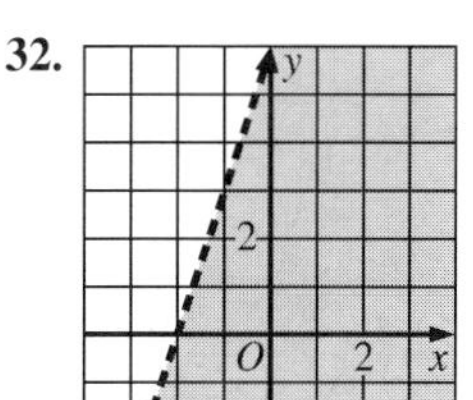

33.

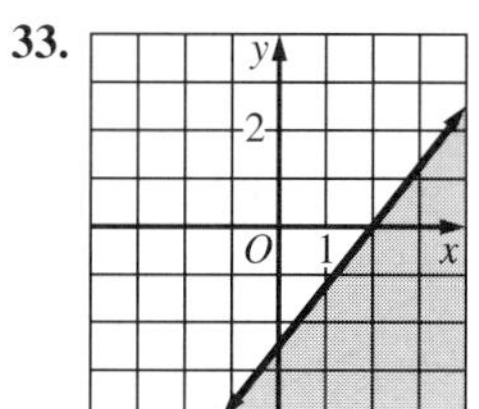

34.

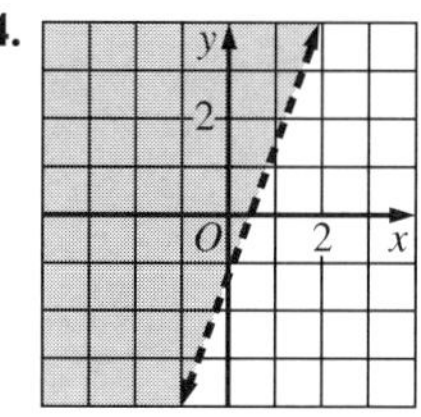

Page 642 Checking Key Concepts

1. a.

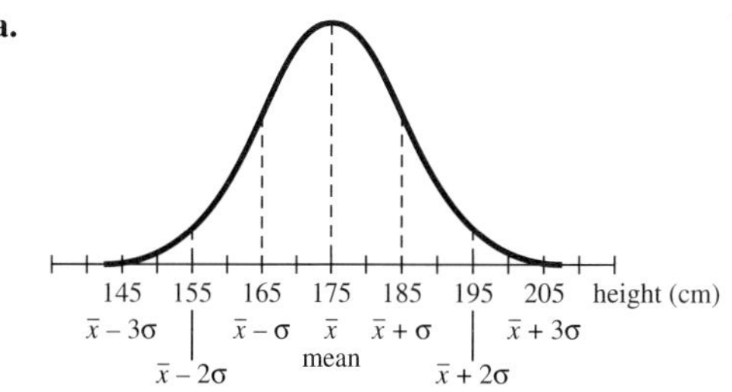

b. 95%; 84%

2. a. 0 **b.** –1 **c.** 2 **d.** $\frac{1}{3}$, or about 0.33 **e.** $\frac{15}{3}$, or about 1.67

3. about 21.2%; about 2.87%

Page 643 Exercises and Applications

6. a.

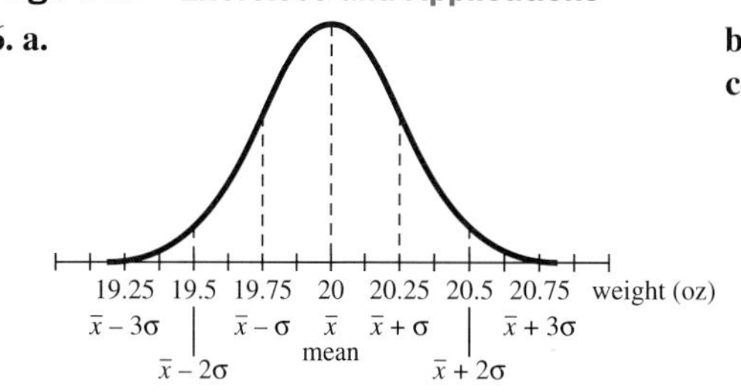

b. 50%
c. about 2.28%

Page 645 Exercises and Applications

13, 14. Answers may vary depending upon when rounding is done, and if interpolation is used. Examples are given. **13. a.** 34% **b.** 54% **c.** 73%
14. a. 2 years: about 38%; 3 years: about 34%; 4 years: about 24%; 5 years: about 27% **b.** Answers may vary. An example is given. Between the ages of 2 years and 4 years, Cathy's growth rate slowed compared to average so that her weight went from being more than 38% of girls her age to being more than just 24% of girls her age. By age 5, Cathy began to grow more rapidly so that by age 10 she was just below average weight and by age 13 she was just above average weight.

CHAPTER 14

Page 659 Exercises and Applications

7. 14 in. **8. a.** 72.4° **b.** 72°24′6″ **9. a.** 43.5° **b.** 43°32′52″ **10. a.** 13.7° **b.** 13°39′23″ **11. a.** 73.4° **b.** 73°25′2″ **12. a.** 32.5° **b.** 32°31′38″ **13. a.** 26.4° **b.** 26°23′6″ **14. a.** 38.7° **b.** 38°44′22″ **15. a.** 59.1° **b.** 59°5′34″ **16.** > **17.** <

CHAPTER 15

Page 707 Exploration

1.

Angle θ	$\sin\theta$	$\cos\theta$
0°	0	1
30°	0.5	0.8660
60°	0.8660	0.5
90°	1	0
120°	0.8660	–0.5
150°	0.5	–0.8660
180°	0	–1
210°	–0.5	–0.8660
240°	–0.8660	–0.5
270°	–1	0
300°	–0.8660	0.5
330°	–0.5	0.8660
360°	0	1

2.

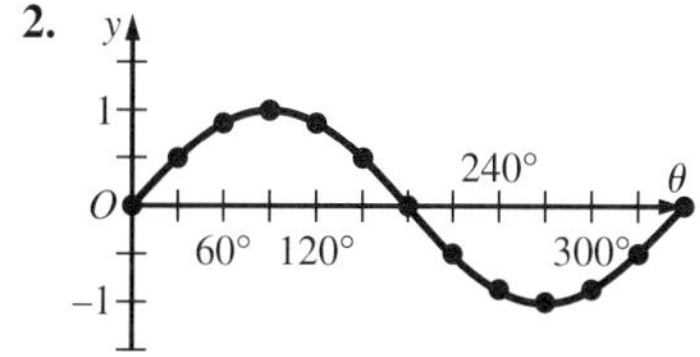

3.

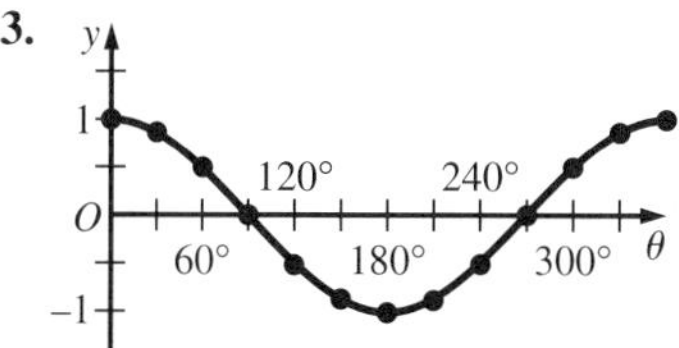

Questions

1. The graphs have the same shape, domain, and range. They have different θ-intercepts, maximum points, and minimum points. The graph of cos θ is the graph of sin θ shifted 90° to the left. **2. a.** 2 **b.** 2 **c.** 1 **d.** 2

Page 717 Exercises and Applications

39.

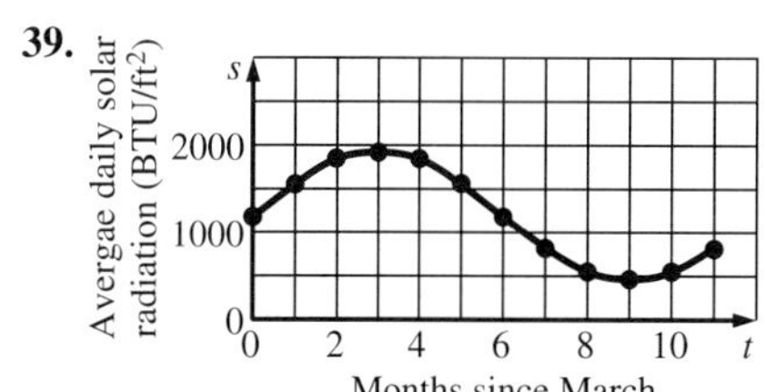

a. The graph resembles a sine function with maximum 1920 and minimum 463.

b.

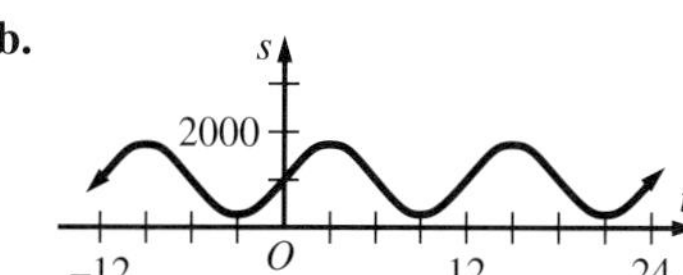

Because the amount of solar radiation in a given month is roughly the same from year to year, you can extend the graph a year forward or backward by just repeating the graph.

40. Answers may vary. An example is given. Closer to the equator, solar radiation would vary less from month to month, resulting in a flatter curve.

41.

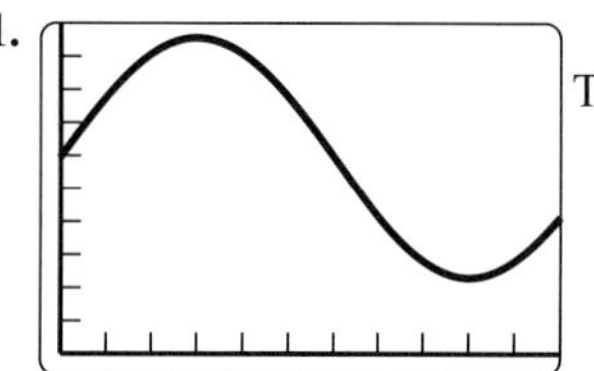

The model fits the data quite closely.

42. Answers may vary. An example is given. Solar collectors are most useful in the summer months when the sunlight is strongest. To gather more energy from the sunlight, you might tilt the solar collectors to have them more closely face the sun or have them move throughout the day to follow the path of the sun.

Page 723 Exercises and Applications

4. amplitude: 2; period: $\frac{4}{3}$

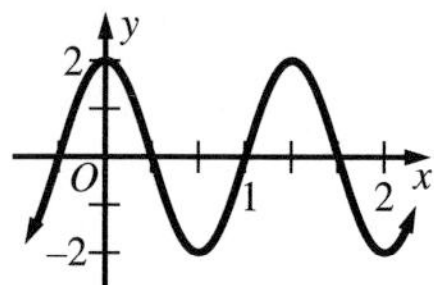

5. amplitude: 1; period: 4

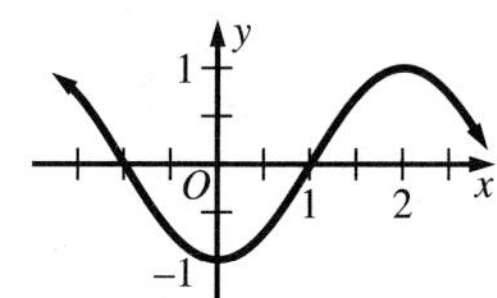

6. amplitude: 4; period: $\frac{1}{2}$

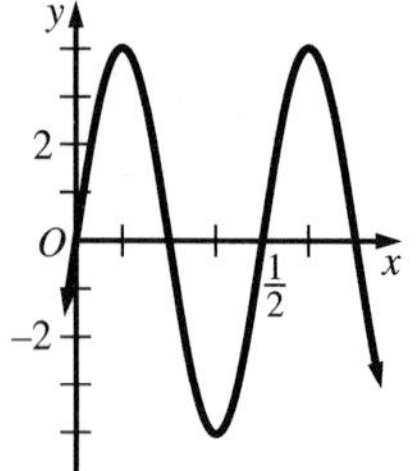

7. amplitude: 4; period: $\frac{\pi}{2}$

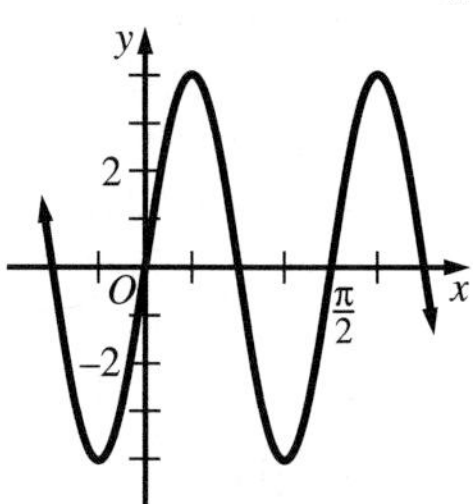

8. amplitude: $\frac{1}{3}$; period: π

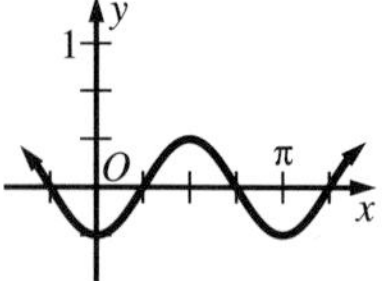

9. amplitude: $\frac{1}{3}$; period: 8π

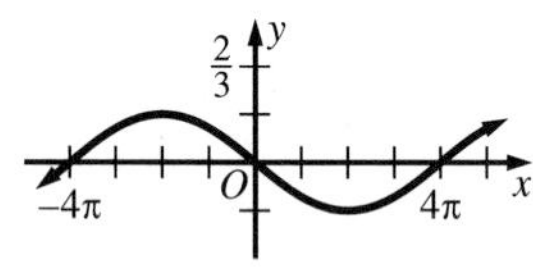

10. amplitude: 3; period: 2

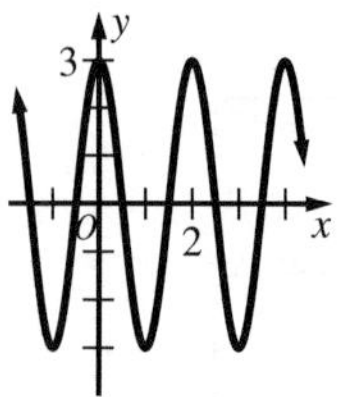

11. amplitude: 2; period: $\frac{4\pi}{5}$

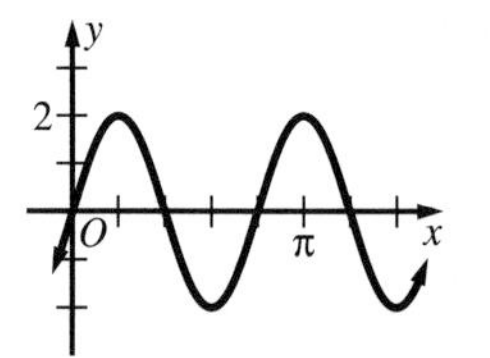

12. amplitude: 2; period: $\frac{\pi}{2}$

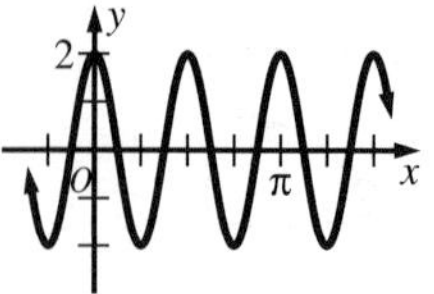

Page 724 Exercises and Applications

25.

String (frequency)	Fundamental	1st harmonic	2nd harmonic	3rd harmonic
E $\left(82\frac{1}{2}\text{ cps}\right)$	$P = \sin 165\pi t$	$P = \sin 330\pi t$	$P = \sin 495\pi t$	$P = \sin 660\pi t$
A (110 cps)	$P = \sin 220\pi t$	$P = \sin 440\pi t$	$P = \sin 660\pi t$	$P = \sin 880\pi t$
D $\left(146\frac{2}{3}\text{ cps}\right)$	$P = \sin 293\frac{1}{3}\pi t$	$P = \sin 586\frac{2}{3}\pi t$	$P = \sin 880\pi t$	$P = \sin 1173\frac{1}{3}\pi t$
G $\left(195\frac{5}{9}\text{ cps}\right)$	$P = \sin 391\frac{1}{9}\pi t$	$P = \sin 782\frac{2}{9}\pi t$	$P = \sin 1173\frac{1}{3}\pi t$	$P = \sin 1564\frac{4}{9}\pi t$
B $\left(247\frac{1}{2}\text{ cps}\right)$	$P = \sin 495\pi t$	$P = \sin 990\pi t$	$P = \sin 1485\pi t$	$P = \sin 1980\pi t$
E (330 cps)	$P = \sin 660\pi t$	$P = \sin 1320\pi t$	$P = \sin 1980\pi t$	$P = \sin 2640\pi t$

Page 726 Assess Your Progress

7.

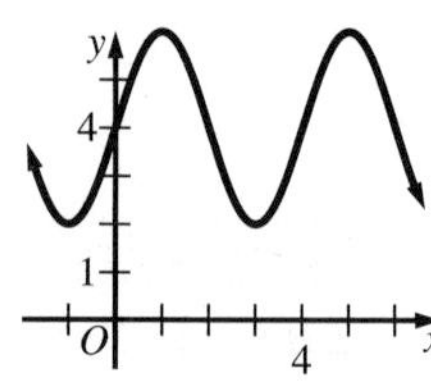

amplitude: 2; period: 4

8.

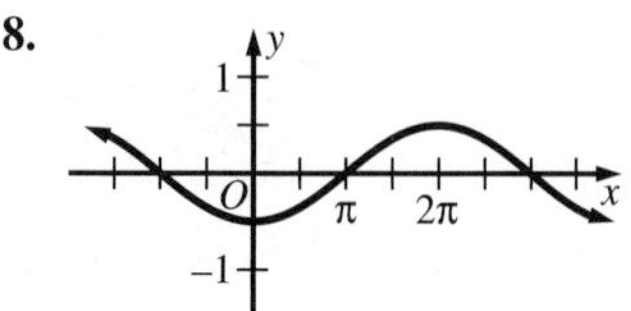

amplitude: $\frac{1}{2}$; period: 4π

9. Summaries may vary. $y = \pm(\text{amplitude}) \sin \frac{2\pi}{\text{period}}x + \frac{\text{minimum} + \text{maximum}}{2}$

Page 731 Exercises and Applications

5.

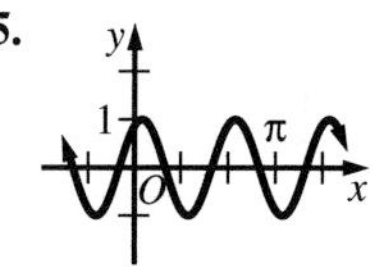

6.

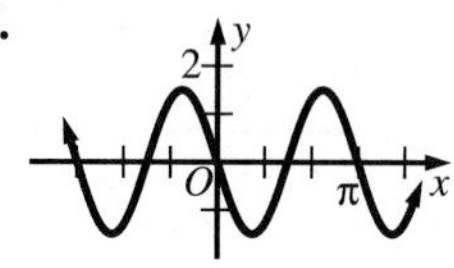

7.

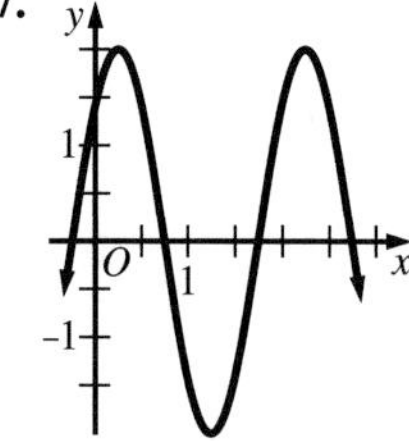

8.

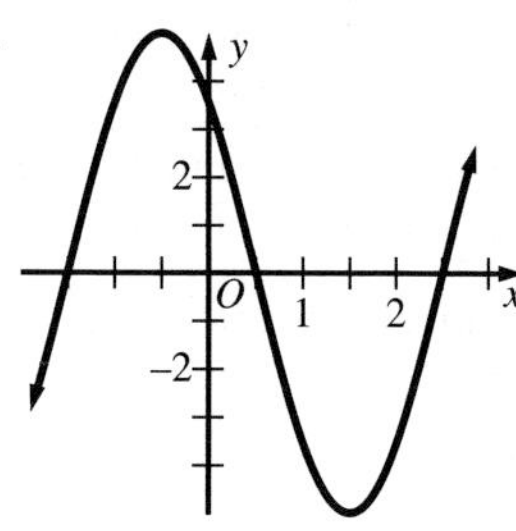

9–16. Answers may vary. Examples are given. **9.** $y = \sin \pi(x + 0.3)$ **10.** $y = 2 \sin\left(x - \frac{\pi}{6}\right)$ **11.** $y = 3 \sin \frac{\pi}{4}(x - 3)$ **12.** For Ex. 9, $y = \cos \pi(x - 0.2)$. **13.** $y = 10 \cos \frac{\pi}{12}(x - 5)$ **14.** $y = 12 \cos \frac{1}{8}(x + 5\pi)$ **15.** $y = 5 \cos 2\left(x + \frac{\pi}{3}\right)$ **16.** For Ex. 13, $y = 10 \sin \frac{\pi}{12}(x + 1)$. **17. a.** 6000 rev/min = 100 rev/s, and since the period is the reciprocal of the frequency, the period is $\frac{1}{100}$ s. **b.** $y = 3.75 \cos 200\pi x$ **c.** 2: $y = 3.75 \cos 200\pi\left(x - \frac{1}{300}\right)$; 3: $y = 3.75 \cos 200\pi\left(x - \frac{1}{150}\right)$

Page 733 Exercises and Applications

26.

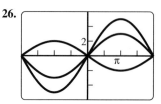

Function	Period	Range
$y = 5 \sin \frac{1}{2}x$	4π	$-5 \le y \le 5$
$y = -2 \sin \frac{1}{2}x$	4π	$-2 \le y \le 2$
sum	4π	$-3 \le y \le 3$

27.

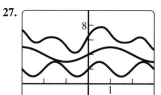

Function	Period	Range
$y = 2 + \sin \pi x$	2	$1 \le y \le 3$
$y = 4 + \sin \frac{\pi}{2}x$	4	$3 \le y \le 5$
sum	4	$4.24 \le y \le 7.76$

28.

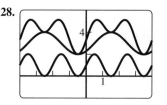

Function	Period	Range
$y = 1 + \sin \pi x$	2	$0 \le y \le 2$
$y = 3 + \sin \frac{\pi}{2}\left(x - \frac{1}{2}\right)$	4	$2 \le y \le 4$
sum	4	$2 \le y \le 5.13$

29.

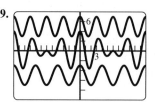

Function	Period	Range
$y = 5 + 2 \cos \frac{2\pi}{3}x$	3	$3 \le y \le 7$
$y = -5 + 2 \cos \frac{\pi}{2}x$	4	$-7 \le y \le -3$
sum	12	$-3.61 \le y \le 4$

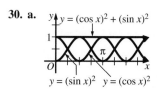

30. a. $y = (\cos x)^2 + (\sin x)^2$; $y = (\sin x)^2$; $y = (\cos x)^2$

b. The coordinates of points on the unit circle are given by $(\cos x, \sin x)$, where x is the radian measure around the circle. Thus, substituting into the standard equation for a circle, $\cos^2 x + \sin^2 x = 1$.

31. a.

Notes	Equations	Sum	Beats/s
15 cps 10 cps	$y = \sin 30\pi x + 5$ $y = \sin 20\pi x - 5$	$y = \sin 30\pi x + \sin 20\pi x$	5
14 cps 10 cps	$y = \sin 28\pi x + 5$ $y = \sin 20\pi x - 5$	$y = \sin 28\pi x + \sin 20\pi x$	4
13 cps 10 cps	$y = \sin 26\pi x + 5$ $y = \sin 20\pi x - 5$	$y = \sin 26\pi x + \sin 20\pi x$	3
12 cps 10 cps	$y = \sin 24\pi x + 5$ $y = \sin 20\pi x - 5$	$y = \sin 24\pi x + \sin 20\pi x$	2
11 cps 10 cps	$y = \sin 22\pi x + 5$ $y = \sin 20\pi x - 5$	$y = \sin 22\pi x + \sin 20\pi x$	1

b. The number of beats decreases as the notes get closer in frequency. The number of beats is the difference in frequency of the two notes.

Page 734 Exercises and Applications

40. a. Answers may vary. Check part (c).

b.

Position	Equation	Phase shift
A	$y = 60 + 50 \sin \frac{\pi}{80}(t - 140)$	140 s
B	$y = 60 + 50 \sin \frac{\pi}{80}(t - 100)$	100 s
C	$y = 60 + 50 \sin \frac{\pi}{80}(t - 60)$	60 s
D	$y = 60 + 50 \sin \frac{\pi}{80}(t - 20)$	20 s

c.

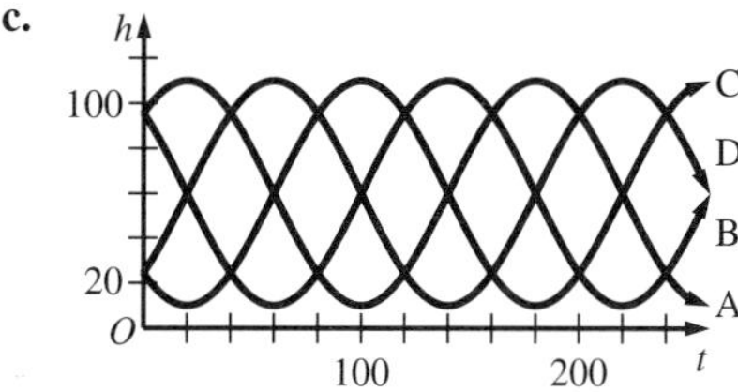

Page 739 Assess Your Progress

2.

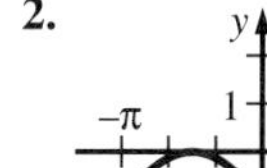

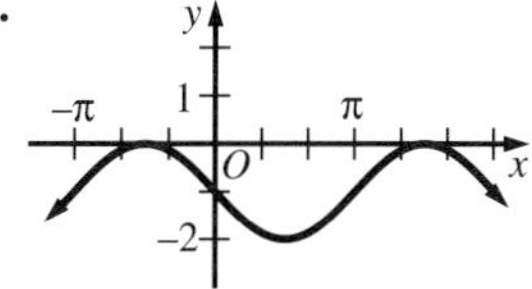

3. a. $s = c \tan \theta$ **b.** $s = 29{,}800$ m/s

4.

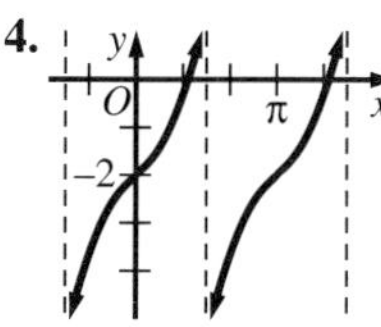

5.

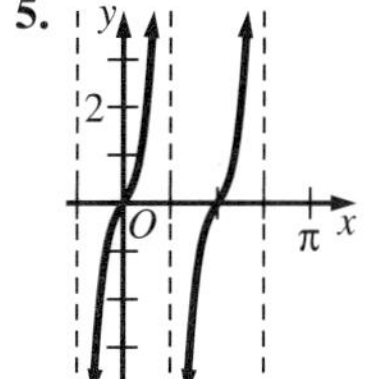

6.

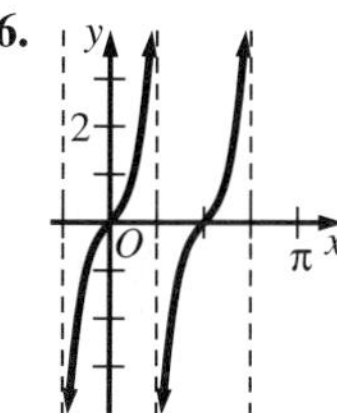

7. Answers may vary. An example is given. Sine and cosine can be treated together. $|a|$ is the amplitude of each graph. The period of each graph is $\frac{2\pi}{b}$. h is the phase shift, to the right if h is positive and to the left if h is negative. k is the vertical shift, up for positive k and downward for negative k. For the tangent function, the larger the value of a the steeper each piece of the graph. The period is $\frac{\pi}{b}$. h represents the horizontal translation, and k the vertical translation.

Page 744 Chapter 15 Assessment

14. To convert from degrees to radians, multiply the number of degrees by $\frac{\pi}{180}$. Example: $135° = 135 \times \frac{\pi}{180}$ radians $= \frac{3\pi}{4}$ radians. To convert from radians to degrees, multiply the number of radians by $\frac{180°}{\pi}$. Example: $\frac{\pi}{3}$ radians $= \frac{\pi}{3} \times \frac{180°}{\pi} = 60°$.

EXTRA PRACTICE

Page 751 Chapter 2

36.

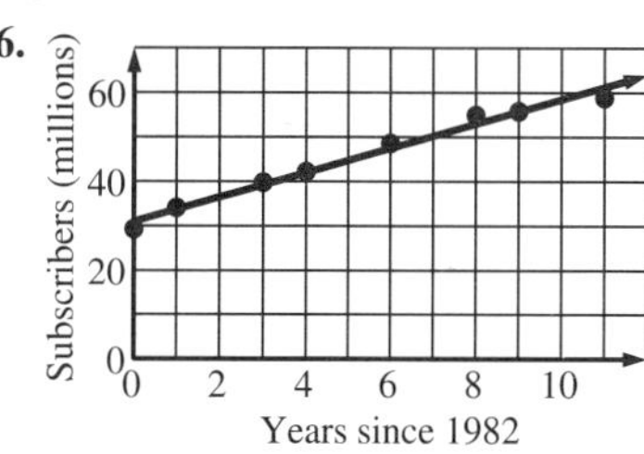

37. Answers may vary. An example is given. $y = 2.7x + 31$ **38.** the average increase per year in the number of cable television subscribers in the U.S. (in millions) **39.** Answers may vary. An example is given. about 81 million

Page 753 Chapter 4

8. a.

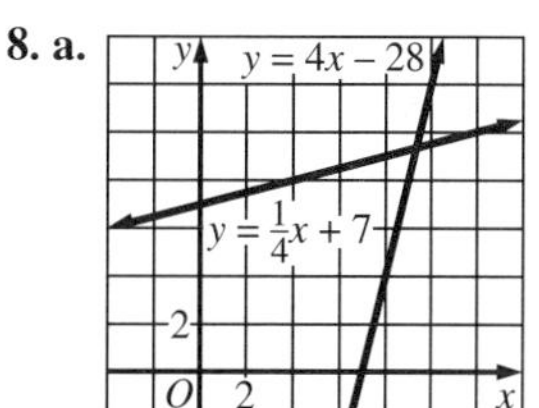

b. $y = 4x - 28$

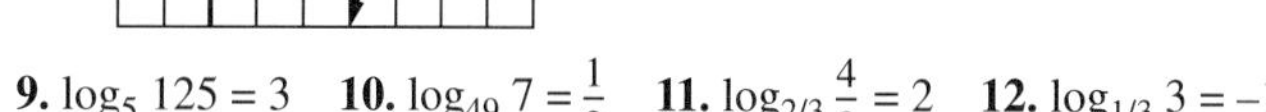

9. $\log_5 125 = 3$ **10.** $\log_{49} 7 = \frac{1}{2}$ **11.** $\log_{2/3} \frac{4}{9} = 2$ **12.** $\log_{1/3} 3 = -1$

Page 760 Chapter 8

8. always true

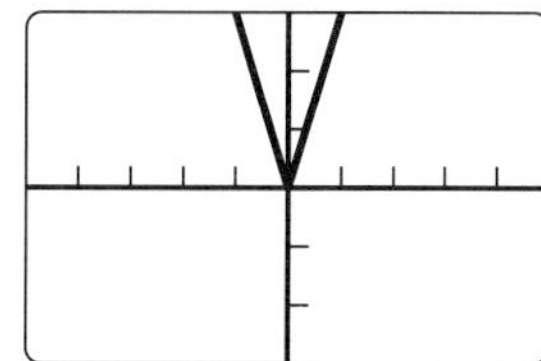

9. never true

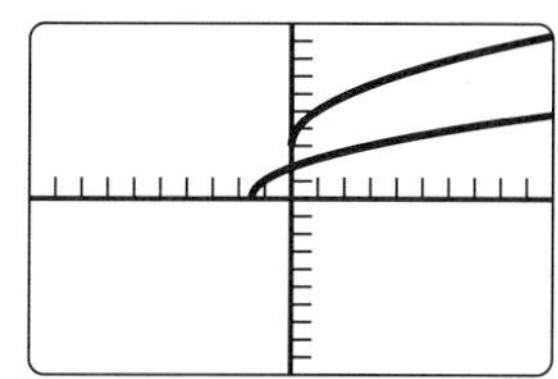

10. sometimes true

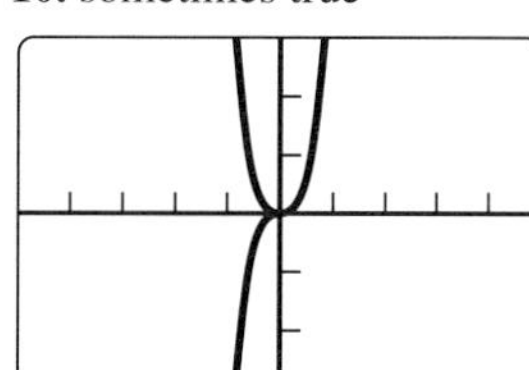

11. sometimes true

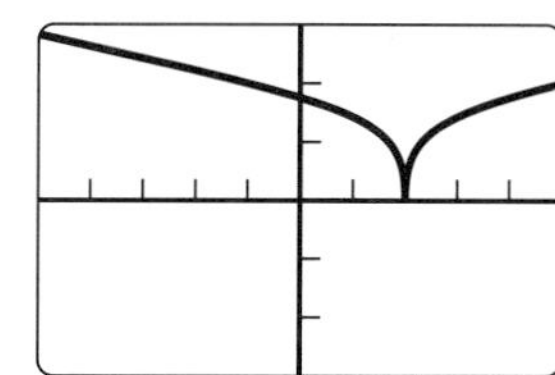

12. always true

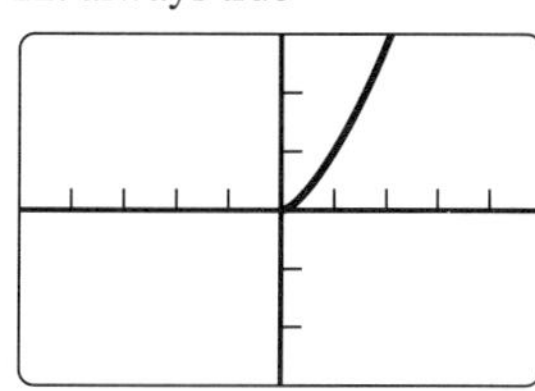

13. –1 **14.** 0.03 **15.** undefined **16.** 100 **17.** –2 **18.** undefined **19.** $\frac{3}{2}$ **20.** $\frac{1}{5}$

Page 767 Chapter 11

38. $\frac{(y+4)^2}{25} - \frac{(x-2)^2}{144} = 1$

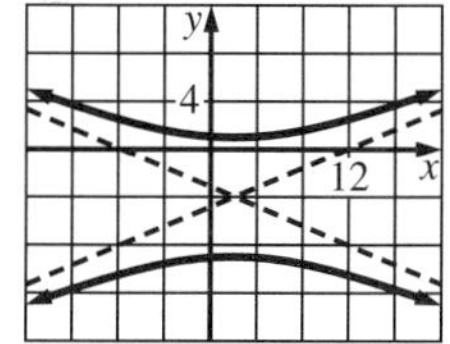

39. $\frac{(x-1)^2}{1} - \frac{(y+3)^2}{4} = 1$; horizontal hyperbola with center (1, –3), vertices (2, –3) and (0, –3), foci $(1 \pm \sqrt{5}, -3)$, and asymptotes $y = \pm 2(x - 1) - 3$

40. $\frac{(x-4)^2}{9} + \frac{y^2}{1} = 1$; horizontal ellipse with center (4, 0), vertices (1, 0) and (7, 0), and foci $(4 \pm 2\sqrt{2}, 0)$, and minor axis of length 2
41. $y + 2 = -\frac{1}{4}(x-3)^2$; parabola that opens down with vertex (3, –2), focus (3, –3), and directrix $y = -1$ **42.** $\frac{x^2}{108} + \frac{(y-3)^2}{27} = 1$; horizontal ellipse with center (0, 3), vertices $(\pm 6\sqrt{3}, 3)$, foci (±9, 3), and minor axis of length $6\sqrt{3}$
43. degenerate ellipse; The graph is the point (0, 0). **44.** degenerate hyperbola; The graph is the pair of lines $y = x - 2$ and $y = -x + 2$.

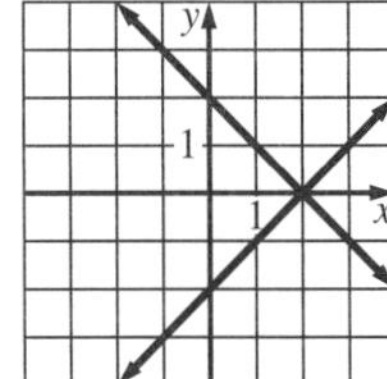

45. degenerate hyperbola; The graph is the pair of lines $y = \frac{1}{5}x$ and $y = -\frac{1}{5}x$.

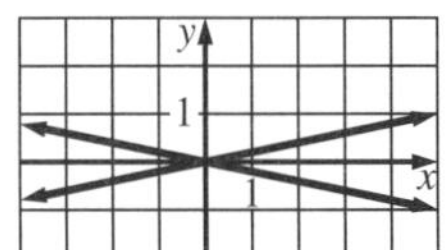

46. degenerate circle; The graph is the point (0, –1).

Page 767 Chapter 12

1–3. b, c. Answers may vary. Examples are given.

1. a. *A*: 3; *B*: 3; *C*: 3; *D*: 3; *E*: 4 **b.**

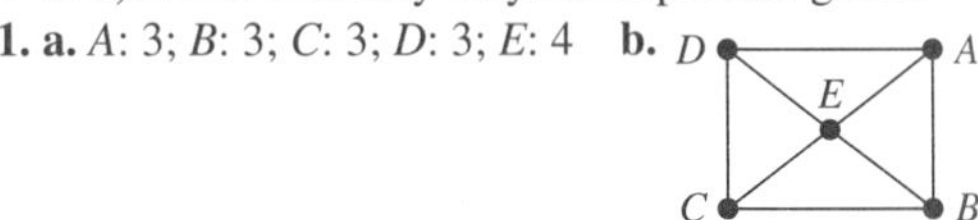

c. 3 colors: *E* red, *A* and *C* green, *B* and *D* blue

2. a. *A*: 3; *B*: 4; *C*: 4; *D*: 3; *E*: 4; *F*: 2 **b.**

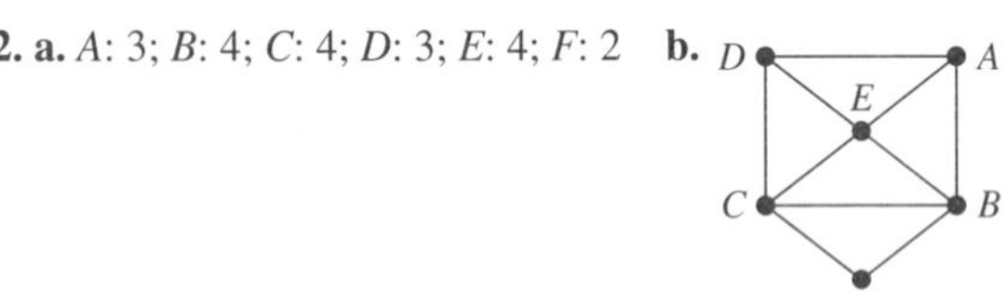

c. 3 colors: *E* and *F* red, *A* and *C* green, *B* and *D* blue

3. a. *A*: 3; *B*: 4; *C*: 3; *D*: 4; *E*: 4 **b.**

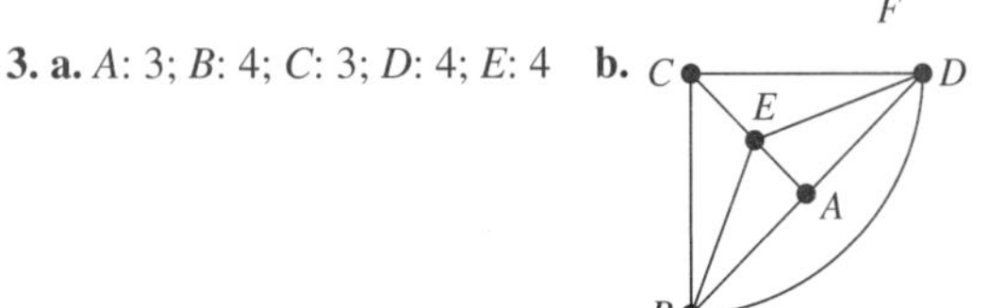

c. 4 colors: *E* red, *D* blue, *B* green, *A* and *C* yellow

4. Answers may vary. An example is given.

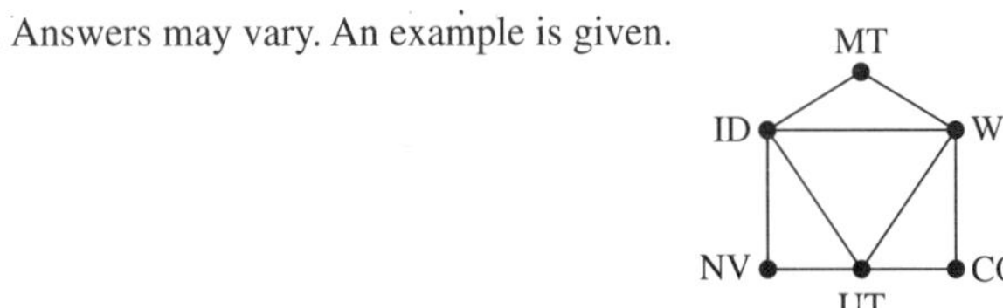

5. Answers may vary. An example is given. 3 colors: UT and MT red, ID and CO green, NV and WY blue

Page 772 Chapter 15

37. amplitude: 4; period: π

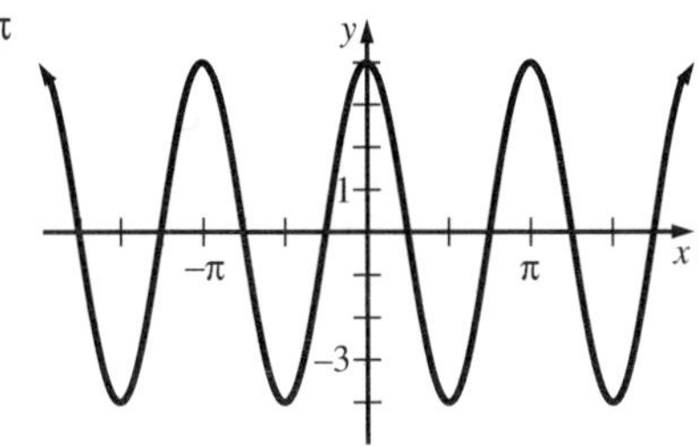

38. amplitude: 2; period: 3π

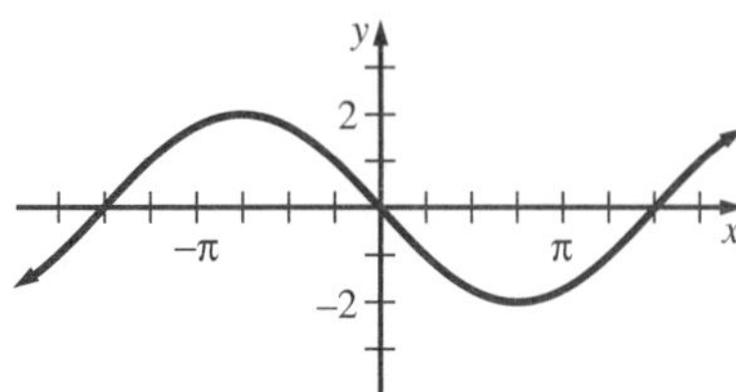

39. amplitude: $\frac{3}{4}$; period: $\frac{2\pi}{3}$

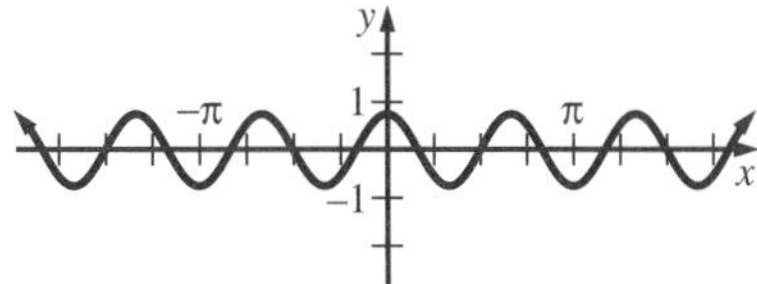

40. amplitude: $\frac{4}{3}$; period: 3

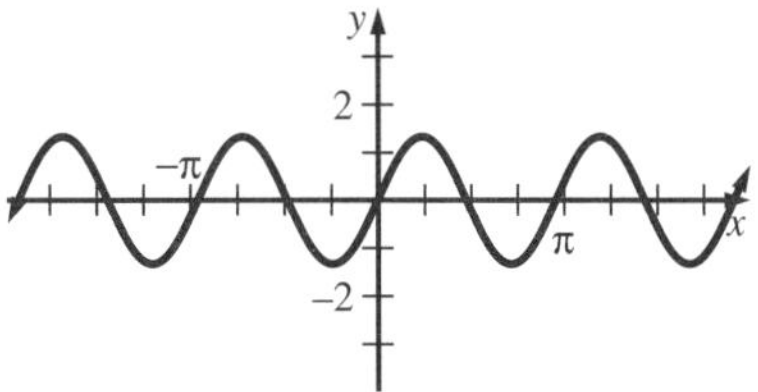

45.

46.

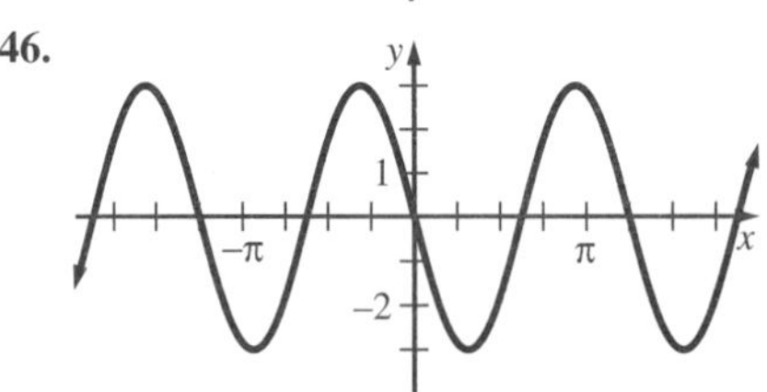

47.

48.

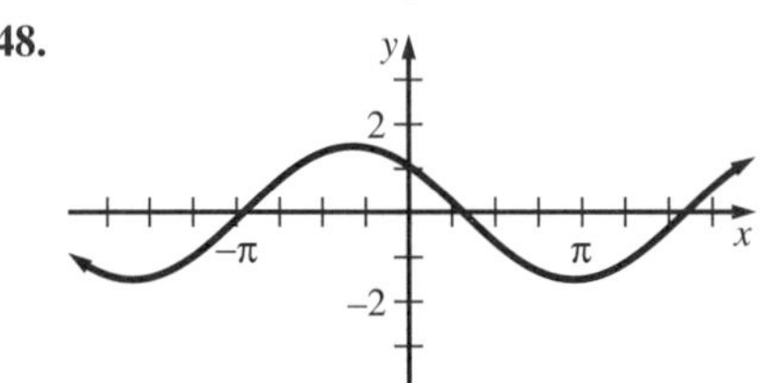